MW00714766

Boron Science

New Technologies and Applications

BORON SCIENCE

New Technologies and Applications

EDITED BY

NARAYAN S. HOSMANE

CRC Press is an imprint of the
Taylor & Francis Group, an **informa** business

Description of the Cover Figure: Lipscomb's Perception of the Element Boron in the Modern Periodic Table.

Original Artwork Credit: Jean C. Evans, Lipscomb's wife (http://www.niu.edu/boronamericas/news/index.shtml).

CRC Press
Taylor & Francis Group
6000 Broken Sound Parkway NW, Suite 300
Boca Raton, FL 33487-2742

Printed in the United States of America on acid-free paper
Version Date: 20110825

International Standard Book Number: 978-1-4398-2662-1 (Hardback)

Visit the Taylor & Francis Web site at
http://www.taylorandfrancis.com

and the CRC Press Web site at
http://www.crcpress.com

This book is dedicated to William N. Lipscomb Jr. (December 9, 1919–April 14, 2011), Emeritus Professor at Harvard University and the winner of the 1976 Nobel Prize for Chemistry. Professor Lipscomb, one of the founders of theoretical and structural polyhedral boron chemistry, had dedicated much of his career to promoting boron science, which is the subject of this book. Much of the progress in these endeavors rests on Professor Lipscomb's original work on bonding in boranes. His dedication to our global society is hereby recognized and saluted. We miss him.

Contents

Part VII Boron for Chemistry and Catalysis: Catalysis and Organic Transformations

Foreword

It has been observed many times but bears repeating: boron is a unique element. Of course, in a sense the same can be said of other elements, such as carbon and hydrogen, but the capacity of boron to confound established notions of structure and bonding—at times seeming to write its own rules—has fascinated and exasperated generations of workers. This seemingly innocuous element, tucked into the top of the periodic table between beryllium and carbon, appeared normal—even boring—for a century following its isolation in 1808 by Humphrey Davy and, independently, Gay-Lussac and Thenard. It behaved exactly as expected, forming trivalent compounds such as the boron trihalides, trialkylboranes, $B(OH)_3$, and borates, and everyone, including the eminent William Ramsey, a Nobel Laureate and discoverer of the noble gases, believed that its simplest hydride had to be BH_3 (what else could it be?). It took the great German chemist Alfred Stock to uncover the truth about the boron hydrides and, even he, working in the early twentieth century, had little notion of their incredible structures and assumed they must be similar to hydrocarbons despite their inconvenient shortage of electrons. Not until another half-century had passed did another towering scientist, William Nunn Lipscomb, finally crack the mystery of the three-center-bonded polyboranes, proving many of their structures with his ingenious apparatus that enabled the collection of x-ray diffraction data at ultracold temperatures and developing a theoretical foundation based on the concept of the three-center two-electron bond. For these achievements, Lipscomb was awarded the 1976 Nobel Prize in Chemistry.

It is highly appropriate that the editor of this volume highlighting recent advances in boron science across an amazingly broad spectrum of applications, with contributions by leading experts in their respective fields, has chosen to dedicate it to Lipscomb, a Kentuckian known as "the Colonel" to generations of his students, coworkers, and colleagues. I first met the Colonel in 1958 as a new graduate student in inorganic chemistry at the University of Minnesota, when it was suggested that I talk with him about joining his research group. I almost didn't go—after all, he was not a member of the inorganic faculty, in fact serving as physical chemistry division chair, and I had no interest in boranes (or so I thought). But I went anyway, and after an hour in his office I was hooked. The borane structures and bonding diagrams that filled several blackboards were exotic and strange to me, but there was something about this man and his mission that was irresistible, and I signed on. Only later did I discover that I was to be his first student who was neither a crystallographer nor a theoretician—it would be my job to learn how to build vacuum lines and manipulate such lovely materials as diborane and pentaborane. With the invaluable help of a summer in Riley Schaeffer's lab at Indiana University just prior to moving with the Colonel to Harvard in early 1960, I eventually managed to acquire these skills. But what struck me most forcefully about the Colonel, and still does, was his unconventional approach to research. In group discussions in which we were encouraged to think outside the box—the further outside, the better—I seldom, if ever, heard him dismiss an idea out of hand. Vacuum line explosions and other setbacks were shrugged off as the price of progress. (There were rare exceptions to his usual good humor, including an incident in which an expensive piece of x-ray diffraction equipment was dropped while being unloaded by crane from the second story of Gibbs Hall, gouging a large crater in the front lawn.)

Reflecting on the Lipscomb research philosophy after all these years, I think that a central attribute has been his uncommon willingness to risk error, as he acknowledged in an early classic paper on three-center bonding with Eberhardt and Crawford. This is not a universal attitude these days, as research funding agencies have little tolerance for false starts and dead ends, and pure exploratory synthesis is practically extinct. Moreover, it must be admitted that a free-wheeling approach is more likely to produce success when coupled with a consummate scientific imagination like the Colonel's.

To me, this book represents more than a fine up-to-date compilation of progress in the applied chemistry and physics of boron. It is, as well, a tribute to the groundbreaking studies of legions of workers over many years led by some truly inspired scientists including not only the late Lipscomb but Fred Hawthorne (who is still at it, full force, directing his new International Institute of Nano and Molecular Medicine at the University of Missouri), the late Herbert C. Brown, and their scientific protégés who together transformed boron science from a boring area to a compelling, exciting, vibrant, and growing one. One of the latter-day scientific descendants of these pioneers, Narayan Hosmane (a postdoc in my own laboratory over 30 years ago, and a major contributor in his own right) has performed a major service in bringing together this excellent fount of information in a rapidly moving field.

Russell N. Grimes
Charlottesville, Virginia

Preface

What is boron? The question itself may not seem very significant to many people. Their introduction to the word "boron" could come as the butt of a joke. Conan O'Brien made jokes about boron (http://www.wired.com/wiredscience/2009/02/conanchemistry/) by calling it either "Boring Boron" or "Boron Moron." Many people may connect boron to borax, which is one of its natural sources. Some of us with longer memories might recall a popular television show called *Death Valley Days*, aired in the 1950s, narrated by a young Ronald Reagan and sponsored by 20 Mule Team Borax. Although boron is one of the closest neighbors to carbon in the periodic table, it has neither gained the importance nor the popularity of carbon. But boron is not just about borax. How many of us know that a regular intake of boron can lessen the chance of prostate cancer? How many of us know that boron plays a direct and critical role in combating cancer through a treatment called boron neutron capture therapy (BNCT)?

If you were among those fortunate attendees of Professor M. Frederick Hawthorne's lecture at UCLA titled "From Mummies to Rockets and on to Cancer Therapy," you had the opportunity to explore the power of boron chemistry. Boron has made a significant impact in our lives through its use in fertilizers, fungicides, soaps, and detergents as well as many household glassware utensils; thus boron is silently present everywhere. Those involved in boron chemistry are beginning to realize that this silence needs to be broken. This book, titled *Boron Science: New Technologies and Applications*, attempts to do just that.

To illustrate the versatility of boron in all areas of applications, the 29 chapters of this book are divided into seven major sections, Boron for Living: Medicine (Part I); Boron for Living: Health and Nutrition (Part II); Boron for Living: Radioisotope (Part III); Boron for Living: Boron Neutron Capture Therapy (Part IV); Boron for Electronics: Optoelectronics (Part V); Boron for Energy: Energy Storage, Space, and Other Applications (Part VI); and Boron for Chemistry and Catalysis: Catalysis and Organic Transformations (Part VII). Each chapter has been rigorously reviewed by at least three reviewers. In order to maintain high quality, reviews were solicited not only from the expert authors of other chapters but also from renowned scientists who willingly shared their expertise and help in improving the quality of the chapters. The invaluable reviews from the following scientists are hereby gratefully acknowledged.

Didier Astruc, University of Bordeaux, France
S. Thomas Autrey, Pacific Northwest National Laboratory, USA
Rolf F. Barth, The Ohio State University School of Pharmacy, USA
Bhaskar C. Das, The Albert Einstein College of Medicine, USA
Mark Fox, University of Durham, UK
Hong-Jun Gao, Chinese Academy of Sciences, China
Mark M. Goodman, Emory University School of Medicine, USA
Russell N. Grimes, University of Virginia, USA
Phil Harris, Halliburton, USA
Catherine Housecroft, University of Basel, Switzerland
Frieder Jäkle, Rutgers University-Newark, USA
Stephen Kahl, University of California-San Francisco, USA
Sang Ook Kang, Korea University, Korea
George Newkome, University of Akron, USA
Forrest H. Nielsen, United States Department of Agriculture, USA
David W. Nigg, Idaho National Engineering Environmental Lab (INEEL), USA

Herbert W. Roesky, Universität Göttingen, Germany
Bakthan B. Singaram, University of California-Santa Cruz, USA
Larry G. Sneddon, University of Pennsylvania, USA
John A. Soderquist, University of Puerto Rico, PR
Morris Srebnik, The Hebrew University of Jerusalem, Israel
Matthias Tamm, Technische Universität Braunschweig, Germany
Kung K. Wang, West Virginia University, USA
Lars Wesemann, Universität Tübingen, Germany
Alexander Wlodawer, National Cancer Institute at Frederick, USA

While I would take this opportunity to thank all of the contributors for their timely submission and valuable input, the immense help from the senior project coordinator, Kari Budyk, and persuasive ability of Lance Wobus, the scientific editor of CRC Press, in making me edit this unique book will always be remembered, appreciated, and acknowledged.

Since the contributing authors are all renowned scientists in their respective disciplines, they truly deserve all of the credit for the success of this book. However, it would be a great mistake if I forgot to thank three most important individuals, Dr. Amartya Chakrabarti, his undergraduate research assistant Hiren Patel, and my wife, Sumathy Hosmane, for their immense support, patience, and help throughout the editing process, from drawing the figures to proofreading to contributing chapters.

We sincerely hope that the chapters in this book will make your reading more intriguing and fascinating than those derived from other elements, as they illuminate the multifaceted nature of the boron compounds beyond their normal place as chemical curiosities.

For the authors,
Narayan S. Hosmane

Editor

Narayan S. Hosmane was born in Gokarn near Goa, Karnatak state, Southern India, and is a BS and MS graduate of Karnatak University, India. He obtained a PhD degree in inorganic/organometallic chemistry in 1974 from the University of Edinburgh, Scotland, under the supervision of Professor Evelyn Ebsworth. After a brief postdoctoral research training in Professor Frank Glockling's laboratory at the Queen's University of Belfast, he joined the Lambeg Research Institute in Northern Ireland, and then moved to the United States to study carboranes and metallacarboranes. After postdoctoral work with Russell Grimes at the University of Virginia, in 1979 he joined the faculty at the Virginia Polytechnic Institute and State University. In 1982 he joined the faculty at the Southern Methodist University, where he became professor of chemistry in 1989. In 1998 he moved to Northern Illinois University and is currently a distinguished research professor and inaugural board of trustees professor. Dr. Hosmane is widely acknowledged to have an international reputation as "one of the world leaders in an interesting, important, and very active area of boron chemistry that is related to cancer research" and as "one of the most influential boron chemists practicing today." Hosmane has received numerous international awards that include but are not limited to the Alexander von Humboldt Foundation's Senior U.S. Scientist Award twice; the BUSA Award for Distinguished Achievements in Boron Science; the Pandit Jawaharlal Nehru Distinguished Chair of Chemistry at the University of Hyderabad, India; and the Gauss Professorship of the Göttingen Academy of Sciences in Germany. While his recent lecture at the Kishwaukee Community Hospital in DeKalb, Illinois, on "Boron and Gadolinium Neutron Capture Therapy: A New Perspective in Cancer Treatment" has resulted in the initiation of collaborative research efforts between NIU and the oncologists/surgeons at Kish Hospital, his featured lecture at the recent American Chemical Society's special symposium on nanomaterials, held in Philadelphia, brought special attention of Dr. Hosmane's work in utilizing magnetic nanomaterials for effective drug delivery in cancer research. He has published over 270 papers in leading scientific journals.

Contributors

Hitesh K. Agarwal
Division of Medicinal Chemistry
and Pharmacognosy
The Ohio State University
Columbus, Ohio

David A. Atwood
Department of Chemistry
University of Kentucky
Lexington, Kentucky

Sebastian Bauer
Institut für Anorganische Chemie
Universität Leipzig
Leipzig, Germany

Vladimir I. Bregadze
A. N. Nesmeyanov Institute of
Organoelement Compounds
Moscow, Russia

Amartya Chakrabarti
Department of Chemistry
and Biochemistry
Northern Illinois University
DeKalb, Illinois

and

Department of Chemistry
Southern Methodist University
Dallas, Texas

Petr Cígler
Institute of Organic Chemistry
and Biochemistry
Gilead Sciences and IOCB Research
Center
Prague, Czech Republic

Anthony Cirri
Department of Chemistry
and Biochemistry
Rowan University
Glassboro, New Jersey

Michael A. Corsello
Department of Chemistry and Biochemistry
Rowan University
Glassboro, New Jersey

Barada Prasanna Dash
Department of Chemistry and Biochemistry
Northern Illinois University
DeKalb, Illinois

Debra A. Feakes
Department of Chemistry and Biochemistry
Texas State University—San Marcos
San Marcos, Texas

Meika Foster
School of Molecular Bioscience
University of Sydney
New South Wales, Australia

Detlef Gabel
Department of Chemistry
University of Bremen
Bremen, Germany

Michael J. Greenhill-Hooper
Business Development
Rio Tinto Minerals
Toulouse, France

Bohumír Grüner
Institute of Inorganic Chemistry
Husinec-Rez u Prahy, Czech Republic

Sherifa Hasabelnaby
Division of Medicinal Chemistry
and Pharmacognosy
The Ohio State University
Columbus, Ohio

Brandon R. Hetzell
Department of Chemistry and Biochemistry
Rowan University
Glassboro, New Jersey

Evamarie Hey-Hawkins
Institut für Anorganische Chemie
Universität Leipzig
Leipzig, Germany

Narayan S. Hosmane
Department of Chemistry and Biochemistry
Northern Illinois University
DeKalb, Illinois

Duncan Hunter
School of Molecular Bioscience
University of Sydney
New South Wales, Australia

Paul A. Jelliss
Department of Chemistry
Saint Louis University
Saint Louis, Missouri

Subash C. Jonnalagadda
Department of Chemistry and Biochemistry
Rowan University
Glassboro, New Jersey

George W. Kabalka
Departments of Chemistry and Radiology
University of Tennessee—Knoxville
Knoxville, Tennessee

Piotr Kaszynski
Organic Materials Research Group
Chemistry Department
Vanderbilt University
Nashville, Tennessee

Jan Konvalinka
Institute of Organic Chemistry
and Biochemistry
Gilead Sciences and IOCB Research Center
and
Institute of Molecular Genetics
and
Department of Biochemistry
Charles University
Prague, Czech Republic

Kate J. Krise
Department of Chemistry and Biochemistry
Northern Illinois University
DeKalb, Illinois

J. Sravan Kumar
Department of Chemistry and Biochemistry
University of Minnesota
Duluth, Minnesota

Lauren M. Kuta
Department of Chemistry and Biochemistry
Northern Illinois University
DeKalb, Illinois

Martin Lepšík
Institute of Organic Chemistry
and Biochemistry
Gilead Sciences and IOCB Research Center
Prague, Czech Republic

Zbigniew J. Leśnikowski
Laboratory of Molecular Virology
and Biological Chemistry
Institute for Medical Biology
Lodz, Poland

Mária Lučaníková
Nuclear Research Institute Řež plc.
Řež near Prague, Czech Republic

John A. Maguire
Department of Chemistry
Southern Methodist University
Dallas, Texas

Pavel Matějíček
Department of Physical and Macromolecular
Chemistry
Charles University
Prague, Czech Republic

Venkatram R. Mereddy
Department of Chemistry and Biochemistry
University of Minnesota
Duluth, Minnesota

Amitabha Mitra
Department of Chemistry
University of Wisconsin—Madison
Madison, Wisconsin

Hiroyuki Nakamura
Department of Chemistry
Gakushuin University
Toshima-ku, Tokyo, Japan

Rosario Núñez
Institut de Ciència de Materials de Barcelona
Bellaterra, Spain

Jana Pokorná
Institute of Organic Chemistry and Biochemistry
Gilead Sciences and IOCB Research Center
and
Department of Biochemistry
Charles University
Prague, Czech Republic

Jiří Rais
Nuclear Research Institute Řež plc.
Řež near Prague, Czech Republic

Pavlína Řezáčová
Institute of Organic Chemistry and Biochemistry
Gilead Sciences and IOCB Research Center
and
Institute of Molecular Genetics
Prague, Czech Republic

Samir Samman
School of Molecular Bioscience
University of Sydney
New South Wales, Australia

Rashmirekha Satapathy
Department of Chemistry and Biochemistry
Northern Illinois University
DeKalb, Illinois

David M. Schubert
Rio Tinto Minerals
U.S. Borax Inc.
Greenwood Village, Colorado

Pavel Selucký
Nuclear Research Institute Řež plc.
Řež near Prague, Czech Republic

Hao Shen
Department of Chemistry
The Chinese University of Hong Kong
Hong Kong, People's Republic of China

Martha Sibrian-Vazquez
Department of Chemistry
Portland State University
Portland, Oregon

Igor B. Sivaev
A. N. Nesmeyanov Institute of Organoelement Compounds
Moscow, Russia

Sven Stadlbauer
Institut für Anorganische Chemie
Universität Leipzig
Leipzig, Germany

Masao Takagaki
Department of Neurosurgery
Aino College Hospital
Osaka, Japan

Francesc Teixidor
Institut de Ciència de Materials de Barcelona
Bellaterra, Spain

Rohit Tiwari
Division of Medicinal Chemistry and Pharmacognosy
The Ohio State University
Columbus, Ohio

Werner Tjarks
Division of Medicinal Chemistry and Pharmacognosy
The Ohio State University
Columbus, Ohio

Nobutaka Tomaru
Second Department of Internal Medicine
University of Tokyo
Bunkyo-Ku, Tokyo, Japan

Maria da Graça H. Vicente
Department of Chemistry
Louisiana State University
Baton Rouge, Louisiana

Clara Viñas
Institut de Ciència de Materials de Barcelona
Bellaterra, Spain

Andrea Vöge
Department of Chemistry
University of Bremen
Bremen, Germany
and
Department of Chemistry
University of Patras
Rio Patras, Greece

Zuowei Xie
Department of Chemistry
The Chinese University of Hong Kong
Hong Kong, People's Republic of China

Min-Liang Yao
Departments of Chemistry and Radiology
University of Tennessee—Knoxville
Knoxville, Tennessee

Zhu Yinghuai
Institute of Chemical and Engineering Sciences
Jurong Island, Singapore

Part I

Boron for Living: Medicine

INTRODUCTION

Boron lies next to carbon in the periodic table of elements. This causes compounds of boron to share some similarities with those of carbon but also preserves important differences. It is the combination of those similarities and differences that puts boron in a unique place for a variety of applications.

The physical and chemical properties of boron make it possible to the design boron-containing molecules with new biological characteristics and offer medicinal chemists a rare opportunity to explore and pioneer new areas of molecular design and medicinal applications. Emerging boron therapeutics show different modes of activity against a variety of biological targets. The utility of this class of compounds is likely to grow over the next decade, and boron could become widely accepted as a useful element in the development of future drugs.

The applications of boron in medicine are not limited to the development of new chemotherapeutics, but also include diagnostics and the modulation of many metabolic processes through dietary boron. The discovery of polyhedral boron compounds added new dimensions to the medicinal chemistry of boron, facilitating the quest for biologically active molecules containing boron clusters rather than only a single boron atom per molecule. Thus, boron is an essential element for life and living.

The first section consists of three chapters covering the medical uses of boron. Chapter 1 by Lesnikowski describes some of the properties of boron clusters that are important for medical applications and presents some examples of boron clusters in different types of medical applications. The following two chapters describe bioactivity of two specific types of boron compounds: carbaboranyl phosphonates (in Chapter 2 by Stadlbauer and Hey-Hawkins) and metal bis(dicarbollides) (Chapter 3 by Konvalinka et al). The former compounds are active as pesticides, bactericides and gameticides, while the later are useful as inhibitors of HIV protease. Both classes have been studied as boron delivery agents for boron neutron capture therapy in cancer treatment.

1 New Opportunities in Boron Chemistry for Medical Applications

Zbigniew J. Leśnikowski

CONTENTS

1.1 INTRODUCTION

The chemistry of boron has many facets. The pharmacological uses of boron compounds have been known for several decades. More recently it was found that some boron analogs of amino acids and their derivatives express hypolipidemic,[1] anti-inflammatory,[2] antineoplastic,[3] and antiostereoporotic[4] properties.

In animals and humans, at intracellular pH, nearly all natural boron exists as boric acid, which behaves as a Lewis acid, and forms molecular additive compounds with amino- and hydroxy acids, carbohydrates, nucleotides, and vitamins through electron donor–acceptor interactions. In most of the commercial dietary boron supplements now available, boron is chelated with amino acids or with hydroxy acids (i.e., glycine, aspartic acid, or citric acid) in combination with vitamins. However, little is known about the molecular structure of these boron chelates.[5,6]

The real stimulus for the development of bioorganic and medicinal chemistry of boron was provided by revival of interests in boron neutron capture therapy (BNCT) of cancers.[7,8] This is connected to the progress in nuclear research reactor technology and the prospective availability of suitable clinical sources of neutrons necessary for the treatment of cancers using BNCT. The discovery of polyhedral boron compounds in the 1960s[9] facilitated the quest for new boron carriers for BNCT, containing boron clusters instead of one boron atom per molecule.[10–13] The study of electron-deficient boron clusters has developed into a major area of inorganic–organometallic chemistry with considerable overlap into organic and medicinal chemistry as well. Bioorganic chemistry of polyhedral boranes is a new trend in this field. It emerged in part as a result of a quest for better boron carriers for BNCT of tumors. Among many low-molecular-weight compounds synthesized for BNCT are carborane-containing amino acids, lipids, carbohydrates, porphyrins, nucleic acid bases and nucleosides, and DNA groove binders.[10,13] A new generation of radiosensitizers for BNCT described recently includes biopolymers bearing one or more carboranyl residues. This class of boron trailers includes carboranyl peptides and proteins, carboranyl oligophosphates, and nucleic acids (DNA-oligonucleotides).[14–16] However, the interest in BNCT is characterized by ups and downs and there is still a long way ahead before it will become a useful clinical technology. Nevertheless, basic knowledge about toxicity and pharmacokinetics of particular classes of boron-containing compounds has been obtained during extensive studies on boron carriers for BNCT, and this can help in the development of the other biological applications of boron compounds.

Simultaneously new, previously unknown biological activities of boron cage molecules and their complexes with metals have been discovered, which including anti-HIV activity, anticancer activity, or in the form of conjugates with nucleosides (nucleic acid components) activity as purinergic receptor modulators.

These and other recent findings clearly show that there is a great, still unexplored potential in bioorganic chemistry of boron and medical applications of boron-containing molecules. This account is not meant to be comprehensive but rather focused on developments of new chemistries and avenues. Several aspects of biological applications of boron-containing compounds are discussed in other chapters of this book, therefore, issues such as BNCT, application in radiotherapy, or anti-HIV protease activity will be omitted. Emphasis is laid on molecules bearing or being derivatives of boron clusters. The vast field of compounds were derived from boric acid and their biological functions had to be left out due to the limited scope of the present overview; Research in this area merits focused reviews in its own right.

1.2 SOME PROPERTIES OF BORON CLUSTERS IMPORTANT FOR MEDICAL APPLICATIONS

Polyhedral heteroboranes have been the subject of intense research for over 50 years. A subset of this extensive class of compounds are dicarbadodecaboranes expressed by the general formula $C_2B_{10}H_{12}$.[9,17] Carboranes are a category of caged-boron compounds most widely explored in medicinal chemistry; however, other boron clusters, such as dodecaborate anion and boron cluster-based coordination compounds accommodating boron cages of different sizes and structural features (metallaboranes and metallacarboranes), also focus on growing attention (Figure 1.1).

Carboranes such as icosahedral dicarba-*closo*-dodecacarboranes ($C_2B_{10}H_{12}$) (**1**–**3**) are characterized by rigid geometry, high chemical and biological stability, and an exceptional hydrophobic character. These characteristics support their use in designs based on the hydrophobic nature of carboranes. If increase in water solubility of boron cluster-biomolecule conjugates is required, an ionic and hydrophilic dodecaborate unit type of (**4**) can be used. Metallacarboranes (e.g., types of **5** and **6**) offer opportunities related to the properties of metals being a part of the complex as well as unique structural features of the metallacarborane molecule as a whole.[18,19] Metallacarboranes comprise a vast family of metallocene-type complexes consisting of carborane cage ligand(s) and one or more metal atoms such as Co, Fe, Cr, Nb, Ni, Cu, Au, Pt, Ru, Re, Tc, and many others. Application of metallacarboranes as vehicles for metals for modification of biomolecules is still underexplored.

An additional advantage of boron clusters and their metal complexes is that they are abiotic and therefore chemically and biologically orthogonal to native cellular components (stable in

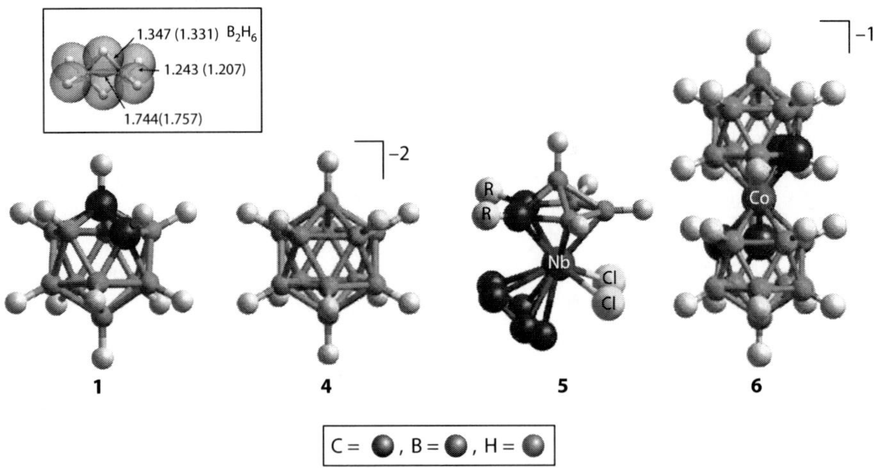

FIGURE 1.1 Example of boron clusters and boron cluster-based coordination compounds used in medicinal chemistry: 1,2-dicarba-*closo*-dodecaborane ($C_2B_{10}H_{12}$) (**1**), *closo*-dodecaborate ($B_{12}H_{12}^{2-}$) (**4**), carbon-substituted cyclopentadienyl-niobium dichloride-dicarbahexaborate [$(C_5H_5)NbCl_2(R_2C_2B_4H_4)$] (**5**), 3-cobalt-bis(1,2-dicarbollide)ate [$Co(C_2B_9H_{12})_2^-$] (**6**). Inset: An example of 3-center-2-electron (3c2e) bonds (B-H-B bridge in diborane).

biological environment) and usually resistant to catabolism, which is a desirable property for biological applications.

1.3 STRUCTURAL FEATURES

General boron cluster chemistry has been the topic of several books and review articles;[9,11,17] therefore, only basic information pertinent to that discussion in this chapter is provided. Polyhedral borane and carborane clusters are characterized by delocalized electron-deficient linkages, meaning that there are too few valence electrons for bonding to be described exclusively in terms of 2-center-2-electron (2c2e) bonds. One characteristic of electron-deficient structures is the aggregation of atoms to form 3-center-2-electron (3c2e) bonds (Figure 1.1), which typically results in the formation of trigonal faces and hypercoordination. The high connectivity of atoms in a cluster compensates for the relatively low electron density in skeletal bonds. The three-dimensional deltahedral shapes are typical of boron and carborane clusters.

The types of polyhedral boron compounds best known and most often used in medicinal chemistry are icosahedral dicarbadodecarboranes ($C_2B_{10}H_{12}$) (**1–3**) in which two BH vertices have been replaced by two CH units. They have nearly spherical geometry with 20 sides and 12 vertices, in which the carbon and boron atoms are hexacoordinated and participate in the heavily delocalized bonding. This, together with undergoing reactions typical for aromatic compounds, leads to the carborane molecule being characterized as a "pseudoaromatic" system. It is remarkable that the aromatic character of carboranes is expressed in three dimensions in contrast to two-dimensional aromaticity in planar polygonal hydrocarbons such as benzene.[20] The space occupied by dicarbadodecaborane is about 50% larger than that of the rotating phenyl group. Dicarba-*closo*-dodecaboranes (1,*X*-$C_2B_{10}H_{12}$) exist in three isomeric forms depending on the location of two carbon atoms within the cage, with $X = 2, 7, 12$, as *ortho*-, *meta*-, and *para*-carboranes, respectively (**1–3**). Numbering of boron and carbon atoms is shown in Figure 1.2.

One of the most important features of a carborane system is its ability to enter into substitution reactions at both the carbon and boron atoms without degradation of the carborane cage. The susceptibility for removal of the most electrophilic boron atom in closed-cage carborane such as 1,2-dicarba-*closo*-dodecaborane is an added advantage for its transformation into open cage form

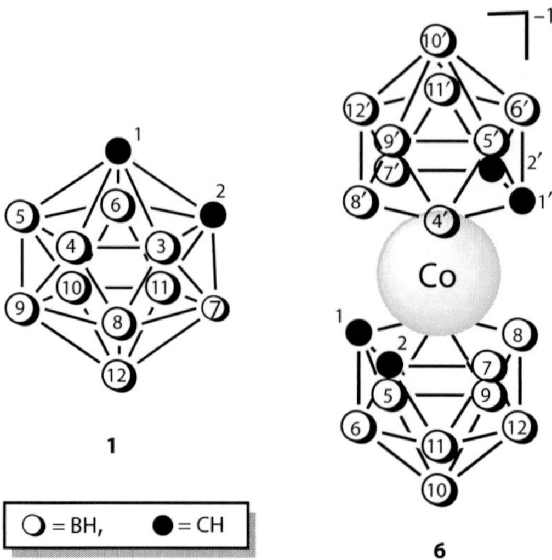

FIGURE 1.2 Numbering of boron and carbon atoms in 1,2-dicarba-*closo*-dodecaborane (*ortho*-carborane) (**1**) and 3-cobalt-bis(1,2-dicarbollide)ate, $[Co(C_2B_9H_{11})_2^-]$ (**6**).

7,8-dicarba-*nido*-undecaborate ion $(C_2B_9H_{12})_{-1}$ under basic conditions, followed by the ability to form "sandwich" type derivatives which are analogs of the transition metal cyclopentadienide derivatives.

1.4 BORON CONTENTS

The boron content per molecule is an important factor for the application of boron compounds as boron carriers for BNCT. For BNCT to be successful, a suitable number of ^{10}B atoms must be localized on or preferably within neoplastic cells, and a sufficient number of thermal neutrons must be delivered to sustain a lethal ^{10}B (n, alpha) lithium-7 reaction. It is estimated that an average of 3–7 alpha-particle traversals through the cell is required to destroy a cell which contains ca. 15 μg ^{10}B/g tumor tissues. Therefore, the development of highly boron loaded, tumor-selective drugs play an important role if BNCT is to evolve into clinically accepted treatment for cancer. At the early stages of the process of development of boron carriers for BNCT, boron compounds consisting of natural boron existing as a 19.9% ^{10}B isotope and 80.1% ^{11}B isotope mixture are used. In clinical practice, boron carriers enriched in ^{10}B isotope are applied due to a much higher neutron capture cross-section (σ) for ^{10}B than for ^{11}B, 3837 and 0.005 barns, respectively.

Boric acid $[B(OH)_3]$ and borane clusters (*closo*-$B_{12}H_{12}^{2-}$) are the only two types of boron entities used so far in the synthesis of carrier molecules for BNCT being a component of two clinically used drugs: L-4-(dihydroxyboryl)phenylalanine (BPA) and the sodium salt of thioborane anion ($Na_2B_{12}H_{11}SH$, BSH). The obvious advantage of polyhedral boron compounds over the derivatives containing one boron atom per molecule as boron donors for BNCT is their much higher boron content. More recently, a growing interest in metallacarboranes containing about 1.5 times as much boron as BSH and 18 times more boron atoms than boric acid, and double-cage boron compounds such as N-substituted bis-decaborate anion a^2-$(B_{10}H_9)(B_{10}H_8)NH_2R$ (**7**) or 1-(1,2-dicarbadode carborane)-oxabutoxy-dodecaborate $(C_2B_{10}H_{11})(CH_2)_4O(B_{12}H_{11})$ (**8**) (Figure 1.3), can be observed.

1.5 HYDROPHOBICITY

Carboranes and numerous metallacarboranes are characterized by exceptional lipophilicity or amphiphilicity. These properties make them particularly suitable for use as hydrophobic components in biologically active molecules, particularly those which interact hydrophobically with

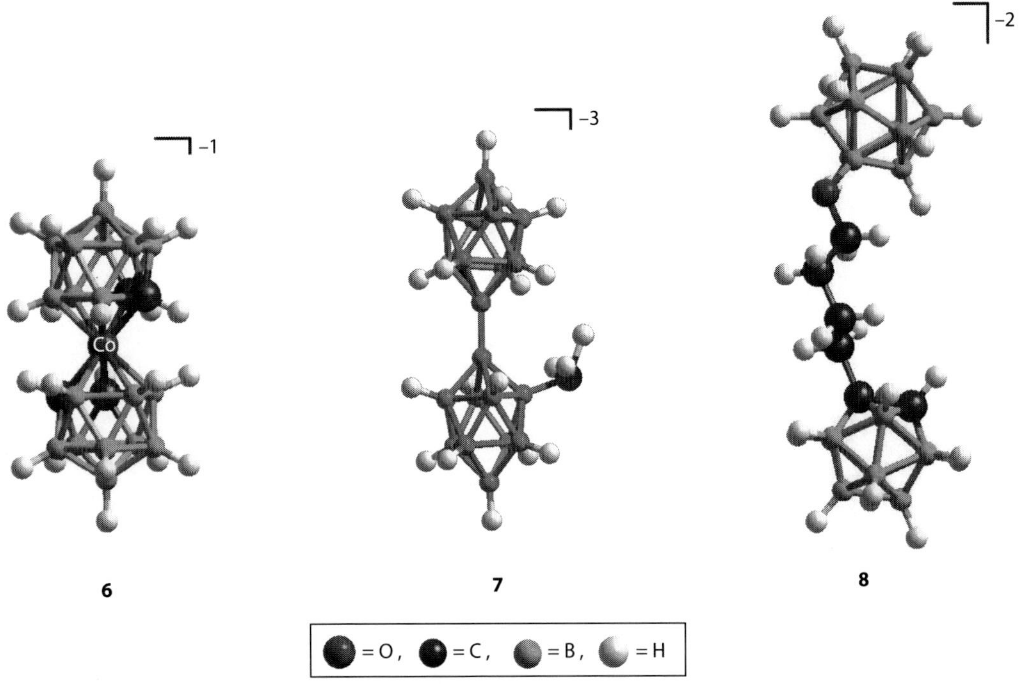

FIGURE 1.3 Examples of double-cage boron compounds for biomedical applications: 3-cobalt-bis(1,2-dicarbollide)ate, $[Co(C_2B_9H_{12})_2^-]$ (**6**); N-substituted bis-decaborate anion a^2-$[(B_{10}H_9)(B_{10}H_8)NH_2R]^{-3}$ (**7**) (a) (Adapted from E. M. Georgiev et al., *Inorg. Chem.*, **1996**, *35*, 5412–5416.) and 1-(1,2-dicarbadodecarborane)-oxabutoxy-dodecaborate $[(C_2B_{10}H_{11})(CH_2)_4O(B_{12}H_{11})]^{-2}$ (**8**) (b) (Adapted from I. B. Sivaev, S. Sjoberg, V. I. Bregadze, *J. Organomet. Chem.*, **2003**, *680*, 106–110.)

molecular targets in the cellular environment. The high hydrophobicity of many boron clusters and their derivatives can be explained by the presence of partial negative charge (density differs for different hydrogens and depends on the type of cluster), located on boron-bound hydrogen atoms in B–H groups and their "hydride-like" character. This prevents them from forming classical hydrogen bonds, which causes a lipophilic/hydrophobic character of the boron clusters (Figure 1.4).

The electronegativity of hydrogens enables boranes to form unconventional hydrogen bonds, namely dihydrogen bonds. Dihydrogen bonds, also called proton–hydride bonds, generally occur between a positively charged hydrogen atom of a proton donor *A*H (*A* = N, O, S, C, halogen) and a bond of an *M*H proton acceptor (*M* = electropositive atom, such as boron, alkali metal, or transition metal).[23] However, the dihydrogen bonds are weaker than classical hydrogen bonds; therefore, repulsive effect toward surrounding water molecules prevails, and the hydrophobic nature of many boron clusters is observed. In boranes, NH···HB, CH···HB, and SH···HB dihydrogen bonds have been found. Another type of interaction was found for CH (carborane)–Y hydrogen-bonded complexes. These complexes were, however, much less stable.[24,25]

The electronic effects of the boron atom in the carborane cage change according to the following sequence: the more remote the boron atom (electronegativity = 2.0) is from the carbon atoms (electronegativity = 2.6) in the cluster, the stronger is its electron-donating effect; consequently, the electron density in 1,2-dicarba-*closo*-dodecaborane (**1**) decreases in the order 9 (12) > 8 (10) > 4 (5,7,11) > 3 (6) > 1 (2). On the other hand, the carborane group as a whole has a strong electron-withdrawing character.

Boron clusters as modifying entities for biomolecules offer rich possibilities of tailoring the hydrophobic property due to different dipole moments and hydrogen-binding sites of the molecule though there is only limited knowledge as to how to reach this in a geometrically predefined manner.

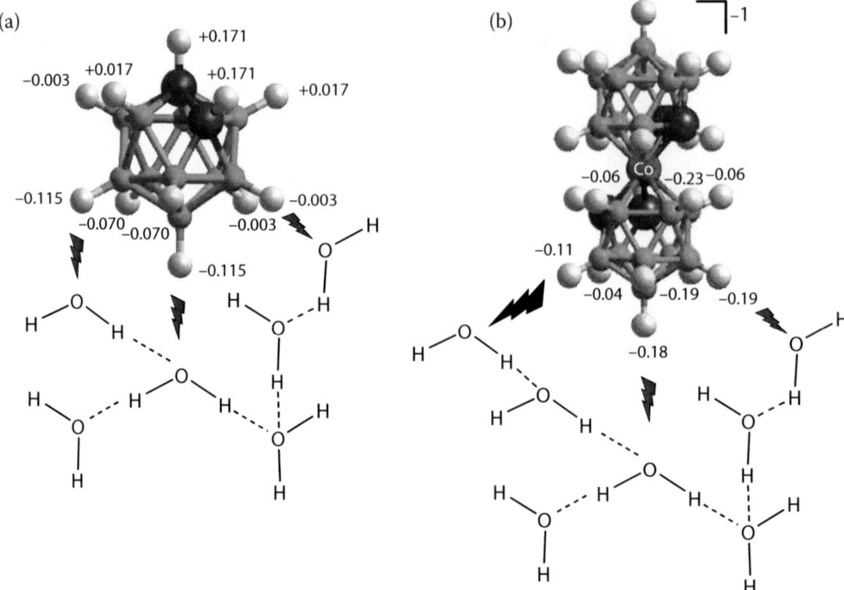

FIGURE 1.4 Hydridic character of boron cage hydrogens causes inability to contribute in hydrogen bonding with water and participates in hydrophobic character of carboranes and metallacarboranes. Charge density distribution for 1: 1 (2) = +0.172; 3 (6) = +0.017; 4 (5, 7, 11) = −0.003; 8 (10) = −0.070; 9 (12) = −0.115 determined from pK_a values of carborancarboxylic acid (a),[26] and 6: 4 (7) = −0.06; 8 = −0.23; 5 (11) = −0.04; 6 = −0.11; 10 = −0.18; 9 (12) = −0.19, calculated according to Two Atom Natural Population Analysis method (b).[27]

The hydrophobicity of dicarbadodecaboranes can be reduced by the presence of a dipole moment, which is strongly dependent on the position of the carbon atoms in the carborane cage.

Consequently, the hydrophobicity of dicarbadodecaborane ($C_2B_{10}H_{12}$) (**1–3**) isomers increases in the following order: 1,2-dicarba-*closo*-dodecaborane (*ortho*-carborane) (**1**) <1,7-dicarba-*closo*-dodecaborane (*meta*-carborane) (**2**), 1,12-dicarba-*closo*-dodecaborane (*para*-carborane) (**3**) (Figure 1.5). Removal of the most electrophilic boron atom in hydrophobic, neutral *closo*-carborane results in the formation of more hydrophilic, anionic *nido*-carboranes ($C_2B_9H_{12}^-$).

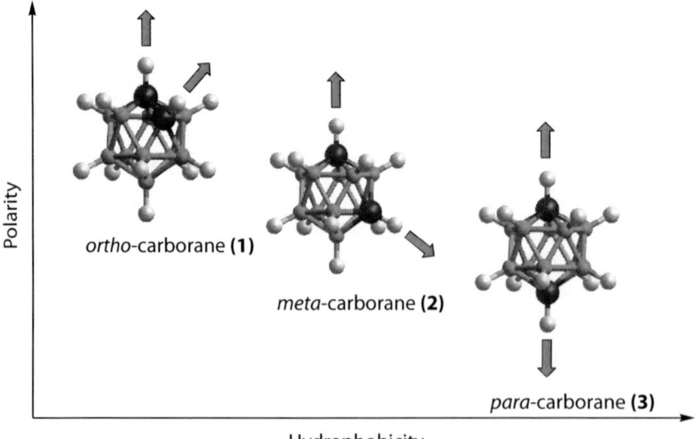

FIGURE 1.5 An effect of dipole moment on hydrophobicity of dicarbadodecaboranes: 1,2-dicarba-*closo*-dodecaborane (*ortho*-carborane) (**1**); 1,7-dicarba-*closo*-dodecaborane (*meta*-carborane) (**2**); 1,12-dicarba-*closo*-dodecaborane (*para*-carborane) (**3**).

1.6 MEDICAL APPLICATIONS OF BORON CLUSTERS

The cage-like, ball-shaped dicarbadodecaborane structure mimics well the dodecahedral volume created by rotation of the planar phenyl ring over 360°, but it is bigger and has a much more hydrophobic moiety. Its higher volume and surface area in comparison with phenyl ring may explain the high-efficacy interactions of carborane-containing biomolecules with hydrophobic domains of proteins such as receptors. These advantages were first exploited for modification of amino acids and peptides. L-Ortho-carboranylalanine and several other carborane-modified amino acids were synthesized in the late seventies. The lipophilicity of L-ortho-carboranylalanine assigned by partition coefficient measurement was much higher than that of the parent L-phenylalanine and higher than that of another lipophilic phenylalanine analog, L-adamantylalanine. Subsequently, several analogs of biologically active peptides, such as enkephalin, angiotensin, bradykinin, substance P, and insect neuropeptide pyrokinin in which the residue Phe and/or Tyr was replaced with carborane-bearing amino acid analog, have been synthesized. Often highly increased biological activity was achieved, for example, the carborane-modified pyrokinin analog exhibited a 30-fold increase in pheromonotropic activity *in vitro* and a 10-fold increase in *in vivo* studies. It was also significantly more stable toward aminopeptidase. Several homopeptides containing carborane-containing amino acid analog for antibody modification and subsequent BNCT studies have also been prepared.[15]

The recent revival of interest in the application of boron clusters in drug design was initiated by extensive studies by Endo and associates on a range of receptor modulators containing carboranes as a hydrophobic pharmacophore, such as analogs of retinoic acid, estradiol, or androgene. Some of these and other examples are briefly discussed below.

1.6.1 Retinoid Receptor Ligands Having a Dicarba-*closo*-Dodecaborane as a Hydrophobic Moiety

Retinoids and their analogs are of particular interest as chemopreventive and therapeutic agents in the field of dermatology and oncology. The biological activities of retinoids are mediated by binding to and activation of the retinoic acid receptors (RARs), with subsequent modulation of the gene transcription by the complex. High-binding activity requires a carboxylic acid moiety and an appropriate hydrophobic group which interacts with the hydrophobic cavity of the RAR ligand-binding domain.

Numerous diphenylamines bearing a 1,2-dicarba-*closo*-dodecaborane (**1**) moiety have been synthesized and their biological properties were studied. Retinoidal activity was examined in terms of the differentiation-inducing ability toward human promyelocytic leukemia HL-60 cells. 4-[4-(1,2-Dicarba-*closo*-dodecaboran-1-yl)phenylamino]benzoic acids (**12**) and 4-[3-(1,2-dicarba-*closo*-dodecaboran-1-yl)phenylamino]benzoic acids (**13**) (Figure 1.6) showed potent agonistic activity at concentrations of 10^{-8}–10^{-9} M. Among the synthesized compounds 4-[4-(2-propyl-1,2-dicarba-*closo*-dodecaboran-1-yl)phenylamino]benzoic acid (**12d**: R_1=*n*-C_3H_7, R_2=H, R_3=H) exhibited biological activity almost equal to that of the natural ligand. It was also found that the RXR antagonistic activity could be separated from retinoid agonistic activity by introduction of a methyl group on the aromatic ring or alkyl groups on the nitrogen atom. The field of carborane analogs of RAR ligands and modulators is still expanding and provides excellent candidates for a new class of therapeutic agents.[28–30]

1.6.2 Steroid Analogs Bearing Boron Cluster Modification

Steroids are a class of lipids characterized by a sterane core and additional functional groups. Hundreds of distinct steroids are found in plants, animals, and fungi. It is possible to classify steroids based on their chemical composition into cholestanes (e.g., cholesterol), cholanes (e.g., cholic acid), pregnanes (e.g., progesterone), and androstanes (e.g., testosterone and androgen). Steroids play many important biological roles, often as hormones interacting with specific receptors. The hydrophobic

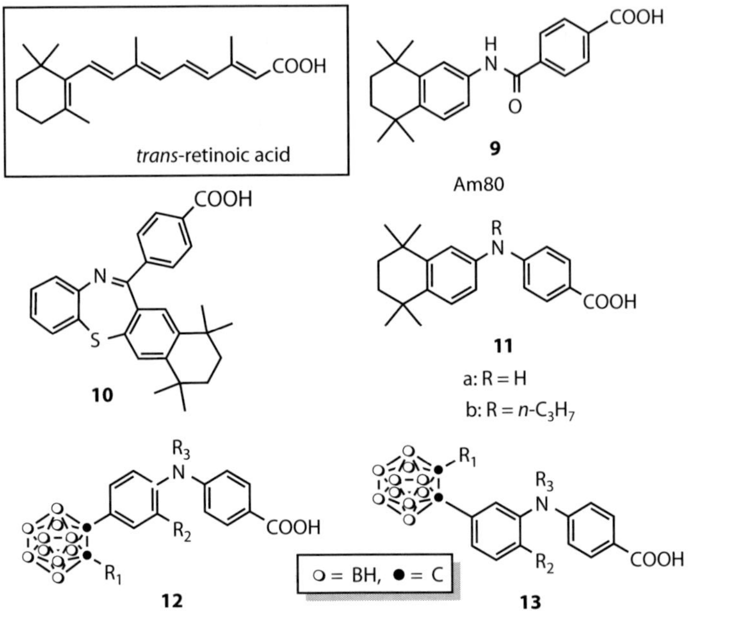

trans-retinoic acid

9

Am80

10

11

a: R = H

b: R = n-C_3H_7

12

○ = BH, ● = C

13

12	13
a: $R_1 = H, R_2 = H, R_3 = H$	a: $R_1 = H, R_2 = H, R_3 = H$
b: $R_1 = CH_3, R_2 = H, R_3 = H$	b: $R_1 = CH_3, R_2 = H, R_3 = H$
c: $R_1 = C_2H_5, R_2 = H, R_3 = H$	c: $R_1 = C_2H_5, R_2 = H, R_3 = H$
d: $R_1 = n\text{-}C_3H_7, R_2 = H, R_3 = H$	d: $R_1 = n\text{-}C_3H_7, R_2 = H, R_3 = H$
e: $R_1 = i\text{-}C_3H_7, R_2 = H, R_3 = H$	e: $R_1 = i\text{-}C_3H_7, R_2 = H, R_3 = H$
f: $R_1 = n\text{-}C_4H_9, R_2 = H, R_3 = H$	f: $R_1 = n\text{-}C_4H_9, R_2 = H, R_3 = H$
g: $R_1 = CH_3, R_2 = H, R_3 = CH_3$	g: $R_1 = CH_3, R_2 = H, R_3 = CH_3$
h: $R_1 = CH_3, R_2 = CH_3, R_3 = CH_3$	h: $R_1 = CH_3, R_2 = CH_3, R_3 = CH_3$

FIGURE 1.6 Selected examples of two types of specific nuclear receptor ligands (retinoic acid receptors: RARs and retinoid X receptors: RXRs), unmodified and modified with boron cluster pharmacophore. (Adapted from Y. Endo et al., *Bioorg. Med. Chem. Lett.*, **1999**, *9*, 3313–3318; Y. Endo et al., *Bioorg. Med. Chem. Lett.*, **1999**, *9*, 3387–3392; K. Ohta et al., *Bioorg. Med. Chem. Lett.*, **2004**, *14*, 5913–5918.)

structure of steroids is important for their biological activity, often playing the role of a scaffold, fixing the spatial positions of hydrogen-bonding functional groups. These open attractive opportunities for modification of biologically important steroids with boron cluster lipophilic pharmacophore to modulate essential hydrophobic interaction in receptor–ligand complexation.

1.6.2.1 Estrogen Analogs Having a Dicarba-*closo*-Dodecaborane as a Hydrophobic Moiety

The steroid hormone estrogen influences the growth, differentiation, and functioning of many target tissues. Estrogen plays an important role in the female and male reproductive system and also in bone maintenance, in the central nervous system, and in the cardiovascular system. The first step in the appearance of these activities is mediated by binding of hormonal ligands to the α and β estrogen receptor (ER) monomers resulting in the formation of an ER dimer; hydrophobic interactions play an important role in this process. Several novel carborane-containing estrogenic agonists have been synthesized (Figure 1.7).

The estrogenic activities of the synthesized compounds were examined by the luciferase reporter gene assay in which a rat ERK expression plasmid and a reporter plasmid, containing five copies of estrogen-response elements, were transiently transfected into COS-1 cells. 17β-Estradiol at 1×10^{-10}–1×10^{-8} M induced the expression of luciferase in a dose-dependent manner. Among the

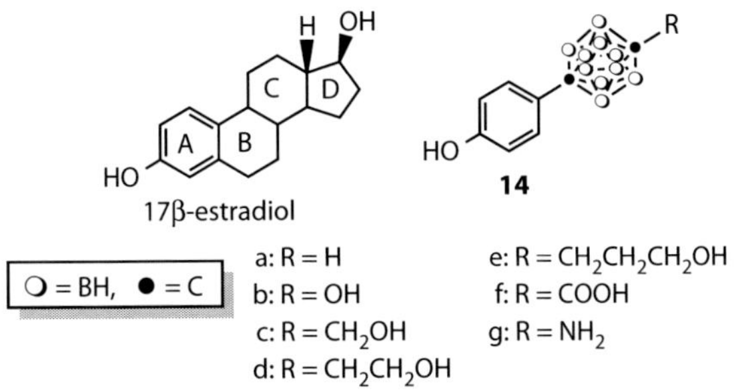

17β-estradiol

14

○ = BH, ● = C

a: R = H
b: R = OH
c: R = CH_2OH
d: R = CH_2CH_2OH
e: R = $CH_2CH_2CH_2OH$
f: R = COOH
g: R = NH_2

FIGURE 1.7 Example of highly active estrogen receptor (ER) agonists based on carborane structure. (Adapted from Y. Endo et al., *Chem. Biol.*, **2001**, *8*, 341–355.)

compounds studied, 1-hydroxymethyl-12-(4-hydroxyphenyl)-1,12-dicarba-*closo*-dodecaborane (**14c**) appeared 10 times more active than its natural counterpart, 17β-estradiol.[31]

Development of potent carborane-containing estrogenic agonists should yield novel candidates of therapeutic agents, especially selective ER modulators. Furthermore, the suitability of the spherical carborane cage for binding to the cavity of ERαLBD should provide a basis for a similar approach to develop novel ligands for other steroid receptors. In addition, the unique character of biologically active molecules containing a carborane skeleton may give rise to unusual membrane transport characteristics and metabolism compared to conventional active molecules.

1.6.2.2 Androgen Analogs Based on Boron Cluster Structure

The androgen receptor (AR) is a member of the nuclear receptor super family of ligand-regulated transcription factors and plays a key role in the development and maintenance of the male reproductive system. Its functions are initiated by the binding of the steroid hormones, testosterone, and/or 5α-dihydrotestosterone to the AR, and an intricate machinery involving translocation of AR into the nucleus, binding to specific DNA sites, formation of a transcriptional complex, and activation of the expression of specific genes occurs. AR ligands have been applied clinically for the treatment of illnesses such as aplastic anemia and prostate cancer.

Nonsteroidal androgen antagonists with a 1,12-dicarba-*closo*-dodecaborane (**3**) cage in place of the steroidal C, D rings of the testosterone or 5α-dihydrotestosterone have been developed. The second-generation, more potent AR antagonists containing cyanophenyl and nitrophenyl groups, respectively, instead of the cyclohexene ring were also proposed.[32,33]

The biological activity of the synthesized compounds was evaluated by means of a transient transactivation assay. The cotransfection assay was conducted in mouse fibroblast NIH3T3 cells, using an expression plasmid for hAR and reporter plasmids, ARE/Luci (firefly luciferase), and pRL/CMV (Renilla luciferase).

The potency of compounds with the CN and NO_2 groups in meta-position (Figure 1.8) was superior to that of hydroxyflutamide, a nonsteroidal AR antagonist used clinically for the treatment of prostate cancer. It was suggested that hydrophobic interaction of the carborane structure with the hydrophobic region of the AR ligand-binding pocket may account for the high binding affinity to AR, and owing to the bulky carborane cage, the conformation of the AR–ligand complex may not be appropriate for interaction with cellular coregulators, resulting in antagonistic activity.

1.6.2.3 Boron Cluster Bearing Cholesterol Mimics

Cholesterol is an important constituent of the mammalian cell membrane and is a frequently used component of liposomal formulations in drug-delivery technology. Numerous cholesterol derivatives containing boron cluster attached to the cholesterol skeleton as external entities have been synthesized as potential boron carriers for BNCT. In general, two types of cholesterol boron cluster

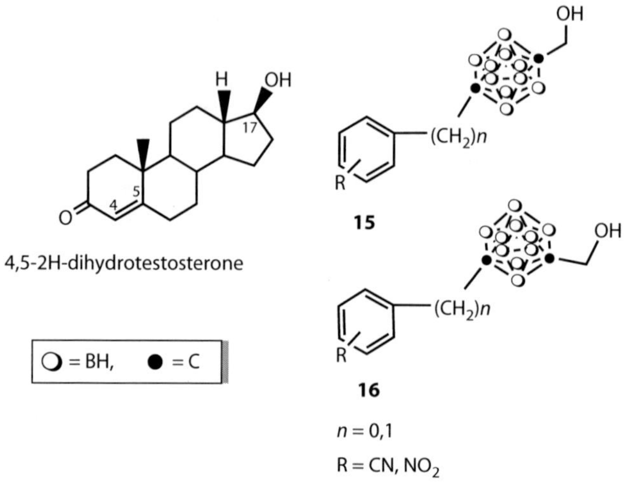

FIGURE 1.8 Selected examples of androgen receptor (AR) ligands based on carborane structure. (Adapted from T. Goto et al., *Bioorg. Med. Chem.*, **2005**, *13*, 6414–6424; S. Fujii et al., *Bioorg. Med. Chem.*, **2009**, *17*, 344–350.)

modification can be defined, the first includes cholesterols conjugated with boron clusters,[34,35] and the second includes cholesterol mimic[36] (Figure 1.9).

Recently, novel cholesterol derivatives containing boron cluster as a pharmacophor were proposed (**19–21**). A major structural feature of these boronated cholesterol mimics is replacement of the B and C rings of cholesterol with boron cluster, analogous to estradiol modification proposed earlier.[36]

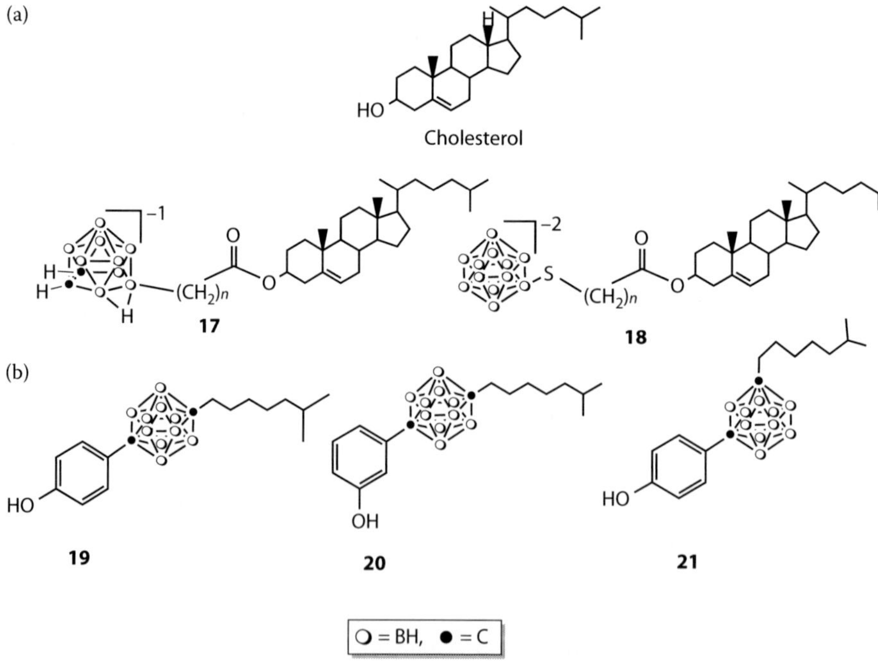

FIGURE 1.9 Boronated cholesterol conjugates (**17,18**) (a) (Adapted from D. A. Feakes, J. K. Spinler, F. R. Harris, *Tetrahedron*, **1999**, *55*, 11177–11186; H. Nakamura et al., *Tetrahedron Lett.*, **2007**, *48*, 3151–3154.), and cholesterol mimics (**19–21**) (b) (Adapted from B. T. S. Thirumamagal et al., *Bioconjugate Chem.*, **2006**, *17*, 1141–1150.)

Computational analyses indicated that all three boronated compounds have structural features and physicochemical properties that are very similar to those of cholesterol. One of the synthesized boronated cholesterol mimics (**21**) was stably incorporated into nontargeted, folate receptor (FR)-targeted, and vascular endothelial growth factor receptor-2 (VEGFR-2)-targeted liposomes. No major differences were found in appearance, size distribution, and lamellarity between conventional dipalmitoylphosphatidylcholine (DPPC)/cholesterol liposomes, nontargeted, and FR-targeted liposomal formulations of compound **21**. These results demonstrate that the novel carboranyl cholesterol mimics are excellent lipid bilayer components for the construction of nontargeted and receptor-targeted boronated liposomes for BNCT of cancer and for other applications.

1.6.3 Transthyretin Amyloidosis Inhibitors Containing Carborane Pharmacophores

Transthyretin (TTR) is a thyroxine-transport protein found in blood that has been implicated in a variety of amyloid-related diseases. Previous investigation have identified a variety of nonsteroidal anti-inflammatory drugs (NSAIADs), such as flufenamic acid or diflunisal, and structurally related derivatives that instill kinetic stabilization to TTR, thus inhibiting its dissociative fragmentation and subsequent aggregation to form putative toxic amyloid fibrils. The cyclooxygenase (COX) activity associated with these pharmaceuticals may, however, limit their potential as long-term therapeutic agents for TTR amyloid diseases. In efforts to solve this problem, several carborane analogs of NSAIADs have been synthesized and evaluated for inhibition of amyloid fibril formation.[37] Some of them also caused substantial decrease of COX activity, a highly desired property. Inhibitors were evaluated by using a 72 h stagnant fibril-formation assay. During this assay, physiological concentrations of TTR (3.6 μM) are subjected to acid-mediated partial denaturation at pH 4.4 either in the presence or absence of an inhibitor. Fibril formation is measured by optical density at 400 nm, and the results are reported as percent fibril formation (% ff), with TTR in the absence of inhibitor defined as 100% ff. The most promising of these compounds is 1-carboxylic acid-7-[3-fluorophenyl]-1,7-dicarba-*closo*-dodecborane (**28**) with ff 15 +/− 5 at 3.6 μM inhibitor concentration (Figure 1.10).

The obtained results show that hydrophobicity, steric bulk, and lack of π-interactions make the carborane functionality ideal scaffolding for use in the binding channel of TTR. Similarly, these same properties are detrimental when applied to COX inhibitors. Consequently, it has been shown that the substitution of a carborane moiety for a phenyl ring in NSAIDs with known TTR tetramer stabilizing activity retains the TTR potency of the compound while dramatically reducing the

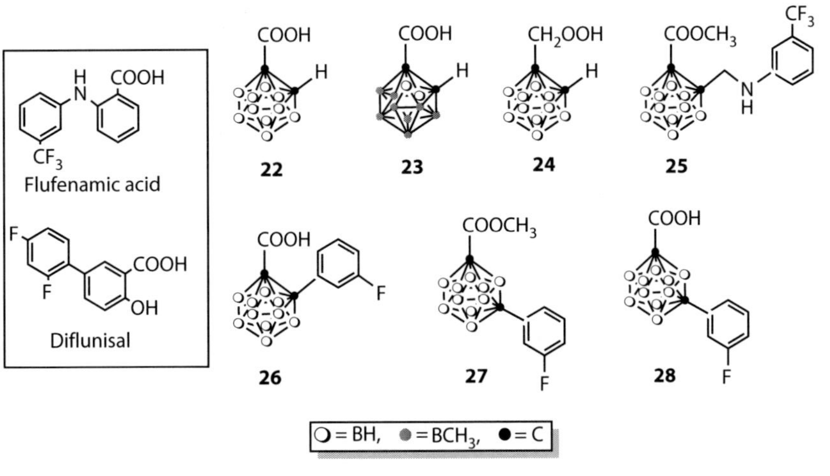

FIGURE 1.10 Transthyretin (TTR) amyloidosis inhibitors, analogs of nonsteroidal anti-inflammatory drugs flufenamic acid and diflunisal, containing carborane pharmacophors. (Adapted from R. L. Julius et al., *Proc. Nat. Acad. Sci. USA*, **2007**, *104*, 4808–4813.)

detrimental COX activity. The unique properties noted above as well as biological and chemical stability make carboranes very attractive pharmacophores in the design and synthesis of a new class of potent and selective NSAIADs and other types of pharmaceuticals.

1.6.4 α-Human Thrombin Inhibitor Containing a Carborane Pharmacophore

α-Human thrombin, a serine protease, is a potent platelet agonist involved in the blood coagulation cascade and is an attractive target for an anticoagulant agent due to its involvement in several debilitating diseases. In recent years, the development of synthetic small-molecules, thrombin and prothrombin activating factor Xa inhibitors have led to a number of potent anticoagulant compounds. These include bicyclic *trans*-lactone **29** isolated from leaves of *Lantana camara* (wild sage) and its synthetic analogs **31** and **32**, more resistant to hydrolysis in plasma. In this direction, a new architecture for size-selective serine protease inhibitors that utilize a fully methylated icosahedral *p*-carborane as a dominating hydrophobic pharmacophore was also proposed recently. Using a computational docking program, flexX, a carborane-containing inhibitor, was designed and synthesized. Computationally, this compound displayed the ability to provide ligand–protein binding interactions throughout the thrombin's main active site (S1–S3), while positioning an acylating group for facile irreversible attack at the Ser^{195} hydroxyl group (Figure 1.11).

It is worth noting that in design of the α-human thrombin inhibitor, the fully methylated *para*-carborane derivative was used instead of the dicarbadodecaborane structure most often utilized as pharmacophore.[38]

1.6.5 Adenosine Modified with Boron Cluster Pharmacophores as a New Human Blood Platelet Function Inhibitor

Adenosine is an endogenous modulator of intercellular signaling that provides homeostatic reduction in cell excitability during tissue stress and trauma. The adenine nucleoside phosphates: adenosine 5′-monophosphate (AMP), adenosine 5′-diphosphate (ADP), adenosine 5′-triphosphate (ATP), which are

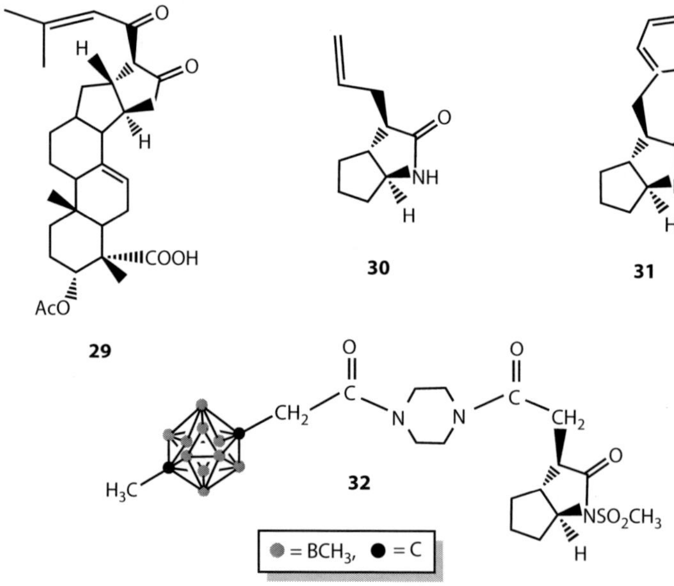

FIGURE 1.11 Naturally occurring [5,5]-*trans*-lactone thrombin inhibitor **29**, its synthetic analogs **30** and **31**, and lead thrombin inhibitor containing hydrophobic carborane pharmacophore **32**. (Adapted from M. F. Z. Page et al., *Synthesis,* **2008**, 555–563.)

maintained in equilibrium by adenylate kinase (AK), constitute the bulk of the purine nucleotide pool. In addition, ATP is the most important molecule for capturing and transferring free energy in most organisms and it is a substrate for RNA polymerases. Another adenosine phosphate cAMP (cyclic adenosine 3′,5′-monophosphate) is the second messenger for many hormones and plays an important role in the regulation of cellular metabolism. Adenosine is also known as an endogenous antiaggregating substance that inhibits platelet aggregation *in vitro* and acts synergistically with other antithrombotic drugs.

The biological activity of adenosine and its phosphates is manifested through its interaction with purinergic receptors and therefore they are considered as important targets for new drugs based on adenosine structure. Among the disadvantages of using adenosine and its derivatives as chemotherapeutics are their short half-lives in circulation and the often unsatisfactory specificity and affinity for receptors. Alterations in the structure and stereochemistry of synthetic adenosine derivatives result in the modulation of receptor-binding potency, selectivity, and biological activity; therefore, a vast array of adenosine derivatives and analogs has been synthesized and tested. Adenosine–boron cluster conjugates were, however, not available due to the lack of methods for their synthesis. Recently, several general methods for the modification pyrimidine as well as purine nucleosides with a variety of different boron cluster was proposed opening the way for biological screening and evaluation of this new class of bioorganic–inorganic conjugates.[39–41]

A small library of adenosine and 2′-deoxyadenosine conjugates with boron clusters (Figure 1.12) was compared for their effects on the responses of platelets (aggregation, protein release, and P-selectin expression) to stimulation by the agonists, ADP, and thrombin. It was found that the modification of adenosine at the 2′-C position with a *para*-carborane cluster ($C_2B_{10}H_{11}$) (**33**) results in the efficient inhibition of platelet function, including aggregation (Figure 1.13). These preliminary findings, and the new chemistry proposed, form the basis for the development of a new class of adenosine analogs that modulate human blood platelet activities.[42,43]

1.7 SUMMARY

All life is derived ultimately from the element carbon, which lies next to boron in the periodic table of elements. Boron compounds not only share some similarities with carbon but also have important differences. It is the combination of these similarities and differences that give boron its unique potential in medicine. The important similarity is that boron, like carbon, combines with hydrogen to form stable compounds that can participate in biochemical reactions and interactions. The key difference is that these compounds have distinctive geometrical shapes and electronic charge distributions with greater 3D complexity than their carbon-based equivalents. While organic carbon molecules tend to comprise rings and chains, boron hydrides are made up of clusters and cages. This 3D structure makes it possible to design molecules with specific charge distributions by varying their internal structure, and this in turn brings the potential to tune how each part of the structure relates to water molecules—if a component is hydrophobic, it is well placed to enter cells by crossing the membrane and to interact with hydrophobic domains of target proteins. If it is hydrophilic it will naturally be soluble in water.

As decades of research begin to bear fruit, the field of bioorganic and medicinal chemistry of boron is on the verge of providing a new generation of drugs and therapies. Boron clusters can find applications as enhancers of hydrophobic interactions between pharmaceuticals and their receptors and entities can be used to increase *in vivo* stability and bioavailability of compounds that are normally rapidly metabolized. They are attracting fast-growing research interest in the quest for novel drugs potentially overcoming limitations and side effects of current products. Several examples in the area of anticancer drugs, inhibitors of platelet aggregation, and modulators of important hormone receptors have been shown above.

Anti-infectious disease drugs bearing essential boron component forms another area of medicinal chemistry of boron awaiting exploration. The fact that novel boron compounds will be unfamiliar to life has potential advantages for antibiotic drugs, since pathogens will be less able to develop

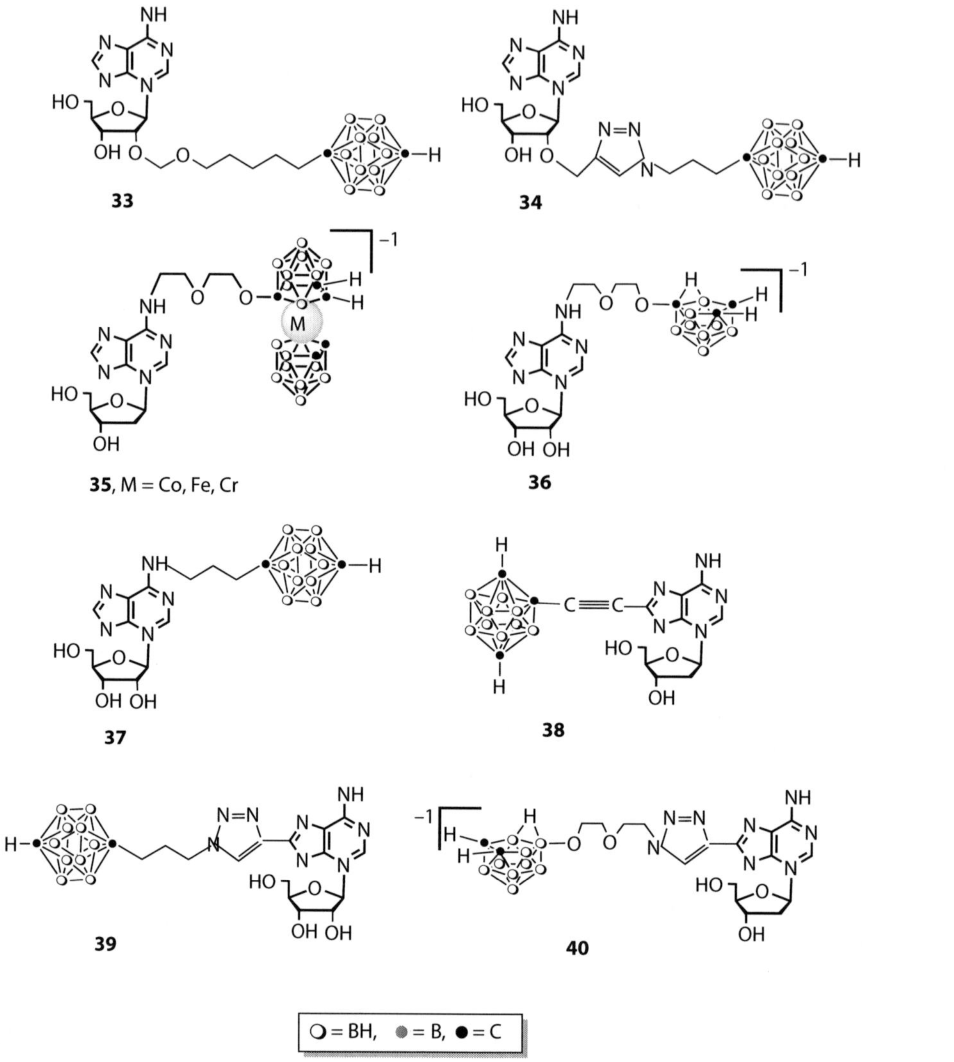

FIGURE 1.12 Selected examples of adenosine and 2′-deoxyadenosine conjugates with boron clusters: 2′-O-(*para*-carboran-1-yl)propyleneoxymethyl]-adenosine (**33**), 2′-O-{[(*para*-carboran-1-yl)-3-propan-1-yl]-1N-1,2,3-triazol-4-yl]methyl}-adenosine (**34**), 6-N-{5-[3-metal bis(1,2-dicarbollide)-8-yl]-3-oxa-pentoxy}-2′-deoxyadenosine (metal: cobalt, iron, chromium) (**35**), 6-N-{5-[7,8-*nido*-carboran-10-yl]-3-oxa-pentoxy}-adenosine (**36**), 6-N-[(7,8-*nido*-carboran-10-yl)-3-propan-1-yl]-adenosine (**37**), 8-(2-*para*-carboranethynyl)-2′-deoxyadenosine (**38**), 8-{[(*para*-carboran-1-yl)-3-propan-1-yl]-1N-1,2,3-triazol-4-yl}adenosine (**39**), 8-{5-[7,8-*nido*-carboran-10-yl]-3-oxa-pentoxy}-1N-1,2,3-triazol-4-yl}adenosine (**40**). (Adapted from A. B. Olejniczak, J. Plesek, Z. J. Leśnikowski, *Chem. Eur. J.*, **2007**, *13*, 311–318; B. A. Wojtczak et al., *Chem. Eur. J.*, **2008**, *14*, 10675–10682; B. A. Wojtczak, A. B. Olejniczak, Z. J. Leśnikowski, *Current Protocols in Nucleic Acid Chemistry*, S. L. Beaucage (Ed.), John Wiley & Sons, Inc., Chapter 4, Unit 4.37, pp. 1–26, **2009.**)

resistance against them. Also the kind of interactions would somehow be different from key–lock systems built up in living cell lines in nature for billions of years. One can thus anticipate that active substances would be less prone to development of resistance. This is an obvious advantage of boron drugs. While eventually pathogens such as bacteria and viruses are capable of evolving resistance against almost any molecule that attacks them, one can believe that it would take longer for this to happen in the case of boron-based compounds, which would therefore make it easier for humans to remain one step ahead rather than struggling to keep pace as at present.

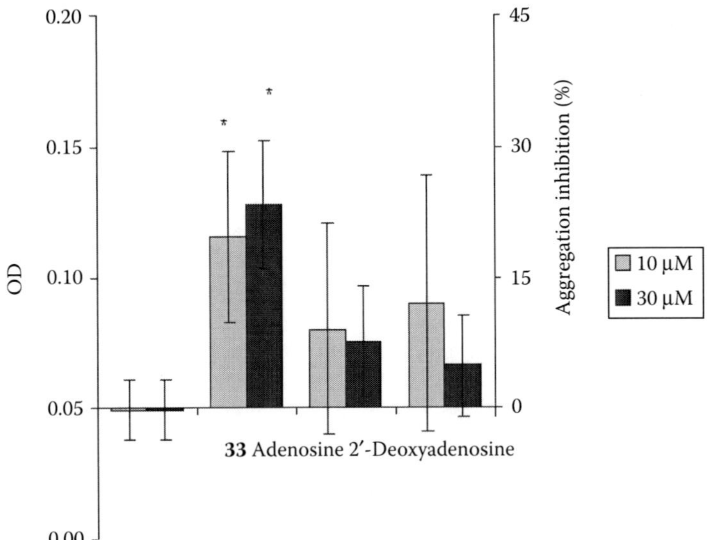

FIGURE 1.13 The effects of compounds **33**, adenosine and 2′-deoxyadenosine (10 and 30 μM) on the extent of platelet aggregation after a long-time (120 min) treatment of platelets with thrombin. Platelets were incubated with the compounds for 5 min at 37°C and then stimulated with thrombin (0.5 U/cm^3) for 120 min. The optical density (OD, wavelength 630 nm) shows the extent of platelet aggregation (left axis); the percentage of platelet aggregation inhibition (right axis). The data are expressed as mean values ± SD for three donors in triplicates; $^*p < 0.05$ (compound + thrombin versus thrombin). (Adapted from K. Bednarska et al., *ChemMedChem.*, **2010**, *5*, 749–756.)

REFERENCES

1. I. H. Hall, W. L. Williams Jr., C. J. Gilbert, A. T. McPhail, B. F. Spielvogel, Hypolipidemic activity of tetrakis-μ-(trimethylamine-boranecarboxylato)-*bis*(trimethylamine-carboxyborane)-dicopper(II) in rodents and its effect on lipid metabolism, *J. Pharmacol. Sci.*, **1984**, *73*, 973–977.
2. I. H. Hall, C. O. Starnes, A. T. McPhail, P. Wisian-Neilson, M. K. Das, F. Harchelroad Jr., B. F. Spielvogel, Anti-inflammatory activity of amine cyanoboranes, amine carboxyboranes, and related compounds, *J. Pharmacol. Sci.*, **1980**, *69*, 1025–1029.
3. I. H. Hall, C. J. Gilbert, A. T. McPhail, K. W. Morse, K. Hassett, B. F. Spielvogel, Antineoplastic activity of a series of boron analogues of α-amino acids, *J. Pharmacol. Sci.*, **1985**, *74*, 755–758.
4. K. G. Rajendran, S. Y., Chen, A. Sood, B. F. Spielvogel, I. H. Hall, The anti-osteoporotic activity of amine-carboxyboranes in rodents, *Biomed. Pharmacother.*, **1995**, *49*, 131–140.
5. C. D. Hunt, Biochemical effects of physiological amounts of dietary boron, *J. Trace Elem. Exp. Med.*, **1996**, *9*, 133–243.
6. T. A. Devirian, S. L. Volpe, The physiological effects of dietary boron, *Crit. Rev. Food Sci. Nutr.*, **2003**, *43*, 219–231.
7. R. F. Barth, J. A. Coderre, M. G. H. Vicente, T. E. Blue, Boron neutron capture therapy of cancer: Current status and future prospects, *Clin. Cancer Res.*, **2005**, *11*, 3987–4002.
8. M. F. Hawthorne, New horizons for therapy based on the boron neutron capture reaction, *Mol. Med. Today*, **1998**, *4*, 174–181.
9. J. Casanova (Ed.). *The Borane, Carborane, Carbocation Continuum*, John Wiley and Sons, Inc., New York, **1998**.
10. A. F. Armstrong, J. F. Valiant, The bioorganic and medicinal chemistry of carboranes: From new drug discovery to molecular imaging and therapy, *Dalton Trans.*, **2007**, 4240–4251.
11. V. I. Bregadze, I. B. Sivaev, S. A. Glazun, Polyhedral boron compounds as potential diagnostic and therapeutic antitumor agents, *Anti-Cancer Agents Med. Chem.*, **2006**, *6*, 75.
12. S. Sjöberg, J. Carlsson, H. Ghaneolhosseini, L. Gedda, T. Hartman, J. Malmquist, C. Naeslund, P. Olsson, W. Tjarks, Chemistry and biology of some low molecular weight boron compounds for boron neutron capture therapy, *J. Neuro-Oncol.*, **1997**, *33*, 41–52.

13. J. F. Valliant, K. J. Guenther, A. S. King, P. Morel, P. Schaffer, O. O. Sogbein, K. A. Stephenson, The medicinal chemistry of carboranes, *Coordin. Chem. Rev.*, **2002**, *232*, 173–230.
14. Z. J. Leśnikowski, DNA as a platform for new biomaterials. Metal-containing nucleic acids, *Curr. Org. Chem.*, **2007**, *11*, 355–381.
15. Z. J. Leśnikowski, Boron units as pharmacophors—New applications and opportunities of boron cluster chemistry, *Coll. Czech. Chem. Commun.*, **2007**, *72*, 1646–1658.
16. Z. J. Leśnikowski, A. B. Olejniczak, R. F. Schinazi, At the crossroads of biological and inorganic chemistry—Oligonucleotides and oligophosphates containing boron clusters, in *"Frontiers in Nucleic Acids," IHL Press, Informed Horizons,* R. F. Schinazi, D. C. Liotta (Eds), *LLC,* Tucker, GA, **2004**, pp. 577–592.
17. *Chem. Rev.* 92, **1992**, special issue on boranes and heteroboranes.
18. R. N. Grimes, Metallacarboranes in the new millennium, *Coord. Chem. Rev.*, **2000**, 200–202, 773–811.
19. N. S. Hosmane, J. Maguire, From boric acid to organometallics: Latest developments in metallacarborane chemistry, *Comments Inorg. Chem.*, **2005**, *26*, 183–215.
20. P. Rague Schleyer, K. Najafian, Are polyhedral boranes, carboranes and carbocations aromatic? in *The Borane, Carborane, Carbocation Continuum*, J. Casanova (Ed.), John Wiley and Sons, Inc., New York, **1998**, pp. 169–190.
21. E. M. Georgiev, K. Shelly, D. A. Feakes, J. Kuniyoshi, S. Romano, M. Hawthorne, Synthesis of amine derivatives of the polyhedral borane anion $[B_{20}H_{18}]^{4-}$, *Inorg. Chem.*, **1996**, *35*, 5412–5416.
22. I. B. Sivaev, S. Sjoberg, V. I. Bregadze, $[C_2B_{10}]$-$[B_{12}]$ double cage boron compounds—A new approach to the synthesis of water-soluble boron-rich compounds for BNCT, *J. Organomet. Chem.*, **2003**, *680*, 106–110.
23. R. Custelcean, J. E. Jackson, Dihydrogen bonding: Structures, energetics, and dynamics, *Chem. Rev.*, **2001**, *101*, 1963–1980.
24. J. Fanfrlík, M. Lepšík, D. Horinek, Z. Havlas, P. Hobza, Interaction of carboranes with biomolecules: Formation of dihydrogen bonds, *ChemPhysChem.*, **2006**, *7*, 1100–1005, and references therein.
25. J. Fanfrlík, D. Hnyk, M. Lepšík, P. Hobza, Interaction of heteroboranes with biomolecules, Part 2. The effect of various metal vertices and exo-substitutions, *Phys. Chem. Chem. Phys.*, **2007**, *9*, 2085–2093.
26. L. I. Zakharkin and V. A. Ol'shevskaya, Experimental estimation of electron density distribution on skeleton atoms of *o*-, *m*-, and *p*-carboranes using carboranecarbonic acids, *Zh. Obshch. Khim.*, **1987**, *57*, 368–372.
27. P. F. Costa, Functionalització de L'anió Cobalto-bis(dicarballur) Per a Nous Materials, *Ph.D. Thesis*, Barcelona, **2009**.
28. Y. Endo, T. Iijima, Y. Yamakoshi, A. Kubo, A. Itai, Structure–activity study of estrogenic agonists bearing dicarba-closo-dodecaborane. Effect of geometry and separation distance of hydroxyl groups at the ends of molecules, *Bioorg. Med. Chem. Lett.*, **1999**, *9*, 3313–3318.
29. Y. Endo, T. Yoshimi, T. Iijima, Y. Yamakoshi, Estrogenic antagonists bearing dicarba-*closo*-dodecaborane as a hydrophobic pharmacophore, *Bioorg. Med. Chem. Lett.*, **1999**, *9*, 3387–3392.
30. K. Ohta, T. Iijima, E. Kawachi, H. Kagechika, Y. Endo, Novel retinoid *X* receptor (RXR) antagonists having a dicarba-*closo*-dodecaborane as a hydrophobic moiety, *Bioorg. Med. Chem. Lett.*, **2004**, *14*, 5913–5918.
31. Y. Endo, T. Iijima, Y. Yamakoshi, H. Fukasawa, C. Miyaura, M. Inada, A. Kubo, A. Itai, Potent estrogen agonists based on carborane as a hydrophobic skeletal structure. A new medicinal application of boron clusters, *Chem. Biol.*, **2001**, *8*, 341–355.
32. T. Goto, K. Ohta, T. Suzuki, S. Ohta, Y. Endo, Design and synthesis of novel androgen receptor antagonists with sterically bulky icosahedral carboranes, *Bioorg. Med. Chem.*, **2005**, *13*, 6414–6424.
33. S. Fujii, K. Ohta, T. Goto, H. Kagechika, Y. Endo, Acidic heterocycles as novel hydrophilic pharmacophore of androgen receptor ligands with a carborane core structure, *Bioorg. Med. Chem.*, **2009**, *17*, 344–350.
34. D. A. Feakes, J. K. Spinler, F. R. Harris, Synthesis of boron-containing cholesterol derivatives for incorporation into unilamellar liposomes and evaluation as potential agents for BNCT, *Tetrahedron*, **1999**, *55*, 11177–11186.
35. H. Nakamura, M. Ueno, J-D. Lee, H. S. Ban, E. Justus, P. Fan, D. Gabel, Synthesis of dodecaborate-conjugated cholesterols for efficient boron delivery in neutron capture therapy, *Tetrahedron Lett.*, **2007**, *48*, 3151–3154.
36. B. T. S. Thirumamagal, X. B. Zhao, A. K. Bandyopadhyaya, S. Narayanasamy, J. Johnsamuel, R. Tiwari, D. W. Golightly et al., Receptor-targeted liposomal delivery of boron-containing cholesterol mimics for boron neutron capture therapy (BNCT), *Bioconjugate Chem.*, **2006**, *17*, 1141–1150.

37. R. L. Julius, O. K. Farha, J. Chiang, L. J. Perry, M. F. Hawthorne, Synthesis and evaluation of transthyretin amyloidosis inhibitors containing carborane pharmacophors, *Proc. Nat. Acad. Sci. USA*, **2007**, *104*, 4808–4813.
38. M. F. Z. Page, S. S. Jalisatgi, A. Maderna, M. F. Hawthorne, Design and synthesis of a candidate alpha-human thrombin irreversible inhibitor containing a hydrophobic carborane pharmacophore, *Synthesis,* **2008**, 555–563.
39. A. B. Olejniczak, J. Plesek, Z. J. Leśnikowski, Nucleoside-metallacarborane conjugates for base-specific metal-labeling of DNA, *Chem. Eur. J.*, **2007**, *13*, 311–318.
40. B. A. Wojtczak, A. Andrysiak, B. Grűner, Z. J. Leśnikowski, "Chemical ligation"—A versatile method for nucleoside modification with boron clusters, *Chem. Eur. J.*, **2008**, *14*, 10675–10682.
41. B. A. Wojtczak, A. B. Olejniczak, Z. J. Leśnikowski, Nucleoside modification with boron clusters and their metal complexes" in *Current Protocols in Nucleic Acid Chemistry*, S. L. Beaucage (Ed.), John Wiley & Sons, Inc., Somerset, NJ, Chapter 4, Unit 4.37, pp. 1–26, **2009**.
42. K. Bednarska, A. B. Olejniczak, B. A. Wojtczak, A. Piskała, Z. Sułowska, Z. J. Leśnikowski, Adenosine and 2′-deoxyadenosine modified with boron clusters ($C_2B_{10}H_{12}$) as new classes of human blood platelet and neutrophil function modulators, "Purines 2010," *An International Meeting on Adenine Nucleoside and Nucleotides in Biomedicine*, Tarraco, Spain, on May 30–June 2, **2010**.
43. K. Bednarska, A. B. Olejniczak, B. A. Wojtczak, Z. Sułowska, Z. J. Leśnikowski, Adenosine and 2′-deoxyadenosine modified with boron cluster pharmacophores as new classes of human blood platelet function modulators, *ChemMedChem.*, **2010**, *5*, 749–756.

2 Bioconjugates of Carbaboranyl Phosphonates

Sven Stadlbauer and Evamarie Hey-Hawkins

CONTENTS

2.1 INTRODUCTION

To date, the treatment of malignant tumors has always been accompanied by extremely negative side effects. One potentially useful approach for the selective destruction of tumor cells is boron neutron capture therapy (BNCT), a powerful form of radiotherapy involving preferential incorporation of ^{10}B-containing compounds into tumor cells, followed by irradiation of the tumor with thermal neutrons.[1–3] The high-energy fission products that are formed on absorption of a neutron allow selective destruction of the tumor cells without affecting the surrounding healthy tissue. High and selective accumulation in tumor cells is one important requirement for a BNCT agent. For successful treatment, a concentration of 20–35 μg ^{10}B per gram tumor must be achieved. The main problem to date is the availability of boron compounds that exhibit the necessary high selectivity, water solubility, and low toxicity in high concentrations.[4]

Phosphorus plays an important role in biological processes as phosphates, for example, as an essential building block in DNA and RNA, and as an energy transporter in the form of nucleoside di- and triphosphates. Additionally, many biological processes involve the phosphorylation of proteins, for example, in signal transduction. Phosphate also plays an important role in maintaining the strength and integrity of the bone skeleton. In adult humans there is an equilibrium between bone resorption caused by osteoclasts and new bone formation caused by osteoblasts.[5,6] One important step in the formation of new bone is the mineralization of calcium phosphate and calcium carbonate induced by the osteoblasts. The two processes, bone resorption and formation, are tightly regulated so that the amount of formed bone equals the amount of resorbed bone, thereby restoring the removed bone completely. Imbalances between the activities of osteoclasts and osteoblasts, especially an increased activity of the former, cause a number of various diseases, for example, osteoporosis, Paget's disease, and rheumatoid arthritis. Therapy is performed by treatment with bisphosphonates that inhibit bone resorption by interfering with the action of osteoclasts. They are strongly adsorbed on hydroxylapatite crystals and are therefore able to target bone tissue selectively. Increased formation of hydroxylapatite is found in primary bone tumors such as osteosarcoma and Ewing's sarcoma and in metastatic bone tumors originating from prostate, breast, and ovarian cancer. Due to their high affinity to bone cells, bisphosphonates are also used in the diagnosis and treatment of bone cancer.[7–9] On the basis of the above findings, it was concluded that phosphonates might be interesting boron carriers for BNCT.

To achieve high-boron concentrations, compounds bearing more than one boron atom are favored for drug development. Therefore, icosahedral polyboranes such as dodecahydro-*closo*-dodecaborate ($B_{12}H_{12}^{2-}$) and the isolobal dicarba-*closo*-dodecaborane(12)s are of great interest for incorporation in tumor-targeting entities (Scheme 2.1). Due to the icosahedral structure of the dicarba-*closo*-dodecaboranes, formation of 1,2-, 1,7-, and 1,12-isomers (generally named *ortho*-, *meta*-, and *para*-carborane) is possible. All three isomers exhibit thermal, hydrolytic, and oxidative stability. Whereas the B_{12} cluster bears two negative charges and is therefore water soluble, the neutral carboranes are highly lipophilic compounds. It was reported that the *ortho*-carbaboranyl moiety is about 324 times more hydrophobic than the phenyl ring in phenylalanine.[10] Additionally the lipophilic character of the carborane depends on the position of the carbon atoms in the cage and increases from *ortho* to *meta* to *para* isomer.[11,12]

Theoretical calculations showed that the carboranes feature an irregular electron density distribution.[13] The highest electron density was found at the positions B(9) and B(12) and the lowest electron density at the C atoms in *ortho*-carborane, and thus electron-acceptor or electron-donor effects occur depending on the substitution pattern of the cluster vertex.

The hydrogen atoms of the two CH groups are acidic, in contrast to those of the BH groups, which are hydridic. Therefore, selective functionalization of the carbon atoms is possible by electrophilic substitution or deprotonation with strong bases followed by subsequent reaction with electrophiles. However, the steric bulk of the clusters requires appropriate reactivity of the electrophile to obtain the desired product in acceptable yields. Phosphorus-containing compounds such as halophosphines are suitable for this type of functionalization of carboranes. In 1963, the first reaction with phosphines was described[14] and until today it has remained the method of choice for phosphorus functionalization of carboranes.

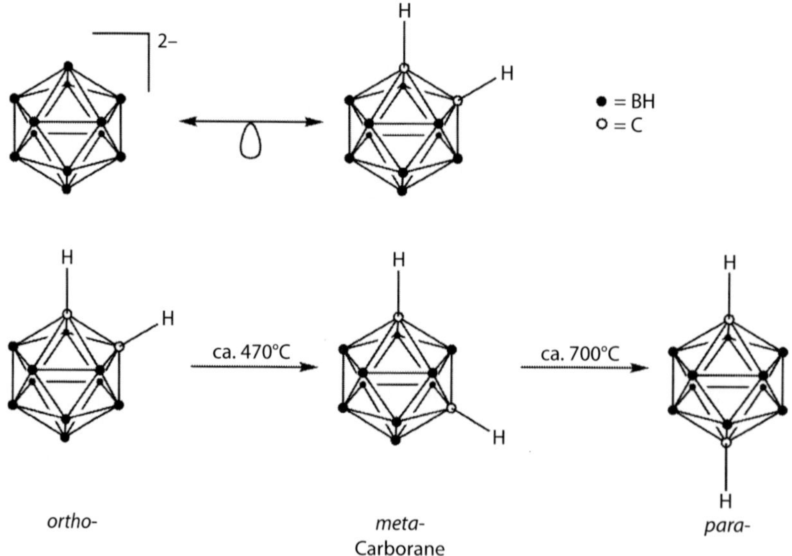

SCHEME 2.1 Dodecahydro-*closo*-dodecaborate and isolobal dicarba-*closo*-dodecaboranes.

2.2 CARBABORANYL MONOPHOSPHONATES WITH ANTICHOLINESTERASE ACTIVITY

2.2.1 SYNTHESIS OF C- AND B-SUBSTITUTED CARBABORANYL PHOSPHONATES AND PHOSPHONOTHIOATES

Esters of (thio)phosphoric and (thio)phosphonic acid are known to inhibit the enzyme acetylcholinesterase. To study the influence of the enzyme inhibition when carboranes are incorporated in (thio) phosphonates, a large number of compounds were synthesized by Russian researchers from the 1960s to 1980s bearing either a phosphonate group, which is directly connected to a carbon or boron atom of *ortho*-, *meta*-, or *para*-carborane or linked through a spacer. This area was reviewed by Godovikov et al. in 1997.[15] The synthesis of compounds with direct P–$C_{carborane}$ bonds was performed by monolithiation of *ortho*-carborane, 1-methyl-, 1-phenyl-, or 1-vinyl-*ortho*-carborane, followed by subsequent reaction with chlorophosphites (Scheme 2.2). The trivalent phosphorus compounds were converted into the corresponding phosphonates by oxidation with N_2O_4[16] or to phosphonothioates by sulfurization with elemental sulfur.[17] Synthesis of phosphonic acid thioesters was achieved by Michaelis–Arbuzov reaction,[17,18] whereas Perkow reaction with an excess of chloral leads to the corresponding dichlorovinylphosphonate derivatives.[17,19]

R = o-H, m-H, p-H, o-Me, o-Ph, o-MeC=CH2
EtO OEt
N2O4
S
R = o-H, m-H, o-Me, o-MeC=CH2
o-, m-, or p-Carborane
1. ⁿBuLi
2. ClP(OEt)2
ClSR′
R = o-Me; R′ = ⁿBu, 2,4-(NO2)2C6H3
R = o-MeC=CH2, R′ = ⁿBu
Cl3CCHO
X2 (or SO2Cl2)
R = o-Ph, o-MeC=CH2
X = Cl, Br
R = o-H, m-H, p-H, o-Me, o-Ph, o-MeC=CH2

SCHEME 2.2 Access to carbaboranyl phosphonates and phosphonothioates bearing a direct P–$C_{carborane}$ bond. (Adapted from Degtyarev, A. N. 1978. *Candidate Thesis in Chemical Sciences. Institute of Organoelement Compounds*, Academy of Sciences of the USSR, Moscow; Degtyarev, A. N. et al. 1975. *Izv. Akad. Nauk SSSR, Ser. Khim.* 2568–2573; Degtyarev, A. N. et al. 1973. *Izv. Akad. Nauk SSSR, Ser. Khim.*, 2369; Degtyarev, A. N. *Izv. Akad. Nauk SSSR, Ser. Khim.* 2099–2104.)

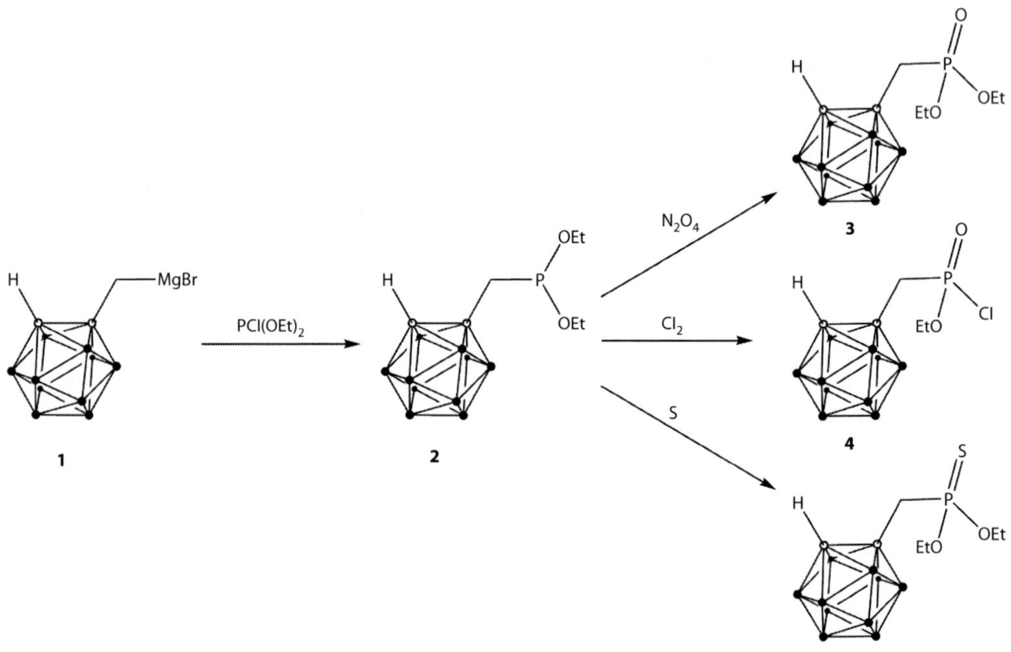

SCHEME 2.3 Preparation of methylene-bridged carbaboranyl phosphonates and phosphonothioates. (Adapted from Kazantsev, A. V.; Meiramov, M. G.; Zakharkin, L. I. 1984. *Zh. Obshch Khim.* 54: 2002–2004.)

Phosphonates and phosphonothioates which are separated from the *ortho*-carborane core by a methylene bridge were also reported.[20] In these compounds, the electronic influence of the carborane on the phosphonate group is low. Their synthesis starts from the Grignard reagent (1-bromomagnesiomethyl)-*ortho*-carborane and diethyl chlorophosphite (Scheme 2.3). Transformation into phosphonates **3** and **4** as well as into phosphonothioate **5** was achieved in the same manner as reported for the above-mentioned carborane derivatives.

A large number of compounds bearing the *o*- or *m*-carborane vertex in the ester moiety of phosphonic and thiophosphonic acids have been described. These phosphonates do not exhibit a $P–C_{carborane}$ bond, but a $P–C_{methyl}$ bond. The carbaboranyl moiety is linked to the phosphorus atom through sulfur, selenium, or a hydroxyalkyl or thioalkyl group. Carbaboranylmethyl esters of phosphonates (**6**) and phosphonothioates (**7–9**) have been synthesized by the reaction of decaborane ($B_{10}H_{14}$) and propynylphosphonic or thiophosphonic acids.[21,22] 2-Phenyl-*ortho*-carboranethiolate was esterified with *O*-ethyl chlorophosphonothioate to give **10** (Figure 2.1).[23]

In analogy to the insecticide isosystox (demeton), carbaboranylphosphonothioates **11–13** have been synthesized by reaction of lithiated *ortho*-carboranethiol or *ortho*-selenol with *O*-ethyl *S*-β-bromoethyl methylphosphonothioate (Scheme 2.4).[23,24]

B-substituted carborane-containing phosphonates with a linkage between the boron atom at the 9-position of the carborane cage and a phosphorus atom are rare; only one example each for a *meta*- and a *para*-carborane were reported.[25] These compounds were obtained by the reaction of bis(*m*- or *p*-carbaboran-9-yl)mercury (**14** or **15**) with trimethyl phosphite under UV irradiation; however, the *meta* derivative **16** was obtained only in very low yield (Scheme 2.5). An alternative three-step synthesis of **16** started with the reaction of **14** and phosphorus trichloride.[26,27] Subsequent oxidation of the dichlorophosphine with sulfuryl chloride followed by methanolysis gave target compound **16**. More recently, the palladium-catalyzed cross-coupling of 9-I-*meta*- and 2-I-*para*-carborane with dimethyl hydrogen phosphate was reported to give both compounds **16** and **17** in about 60% yield.[28]

X = Y = O, R = Me, R′ = OEt (**6**)
X = O, Y = S, R = Me, R′ = OEt (**7**)
X = S, Y = O, R = Me, R′ = OEt (**8**)
X = O, Y = S, R = Ph, R′ = OEt (**9**)

10

FIGURE 2.1 Phosphonates bearing the carborane in the ester moiety. (Adapted from Kabachnik, M. I.; Godovikov, N. N.; Rys, E. G. 1980. *Izv. Akad. Nauk SSSR, Ser. Khim.* 1455–1456; Rys, E. G.; Godovikov, N. N.; Kabachnik, M. I. 1983. *Izv. Akad. Nauk SSSR, Ser. Khim.* 2640–2644; Balema, V. P. et al. 1989. *Izv. Akad. Nauk SSSR, Ser. Khim.* 194–197.)

R = Me, X = S (**11**)
R = Ph, X = S (**12**)
R = Ph, X = Se (**13**)

SCHEME 2.4 Synthesis of *S*-β-carbaboranylthio(seleno)ethyl esters of methylphosphonothioate. (Adapted from Balema, V. P.; Rys, E. G.; Godovikov, N. N.; Kabachnik, M. I. 1989. *Izv. Akad. Nauk SSSR, Ser. Khim.* 194–197; Rys, E. G. et al. 1990. *Izv. Akad. Nauk SSSR, Ser. Khim.* 1653–1655.)

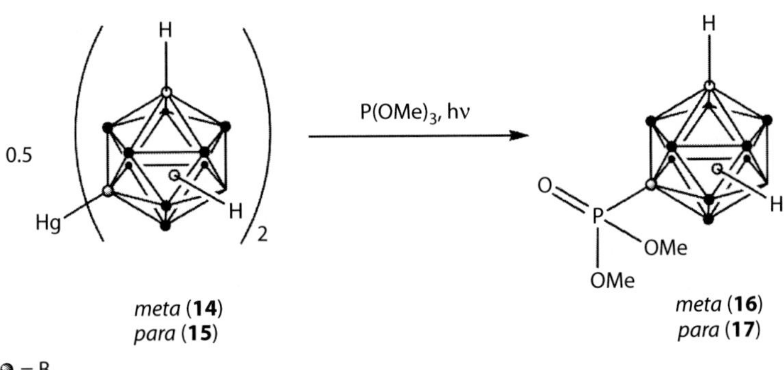

SCHEME 2.5 B9-substituted phosphonates of *meta*- and *para*-carborane. (Adapted from Bregadze, V. I. et al. 1985. *Dokl. Akad. Nauk SSSR* 285: 1127–1130.)

	R	X	Y
18	OEt	O	S
19	O-(*p*-NO_2)-Ph	O	S
20	OEt	O	Se
21	OEt	O	Se

	R	R′	Y	X
22	OEt	Me	S	I
23	OEt	Et	S	I
24	OEt	Me	Se	SO_4Me
25	9-*o*-$C_2H_2B_{10}H_9S$	Me	S	I

SCHEME 2.6 B9-substituted phosphonates bearing the boron cluster in the ester moiety. (Adapted from Rys, E. G. et al. 1990. *Izv. Akad. Nauk SSSR, Ser. Khim.* 1653–1655; Rys, E. G.; Godovikov, N. N.; Kabachnik, M. I. 1986. *Izv. Akad. Nauk SSSR, Ser. Khim.* 719–721; Balema, V. P. et al. 1990. *Izv. Akad. Nauk SSSR, Ser. Khim.* 2857–2859. Balema, V. P.; Rys, E. G.; Godovikov, N. N. 1992. *Izv. Akad. Nauk SSSR, Ser. Khim.* 459–461.)

In analogy to compounds linked to the carbon atoms of the cluster through a spacer, B-substituted phosphonates bearing the *ortho*-carborane cluster in the ester moiety were also reported. These compounds were prepared from *o*-carborane-9-thiol or -selenol. The phosphonates can be prepared either by reaction with methylchlorophosphonates[24,29] or by Michaelis–Arbuzov reaction of phosphonothioites[30,31] or -selenoites[24] (Scheme 2.6).

2.2.2 Biological Activity

As the intention was to synthesize new inhibitors of acetylcholinesterase, the anticholinesterase activity of the above-described phosphonates and phosphonothioates was studied. In 1978, Zakharova et al. examined the activity of dichlorovinyl-substituted carbaboranyl phosphonates **27–30** toward acetylcholinesterase and butyrylcholinesterase.[32] As these compounds are analogs of 2,2-dichlorovinyl dimethyl phosphate (dichlorvos or DDVP) (**26**), their activity was compared with the activity of this known insecticide. As shown in Table 2.1, compounds **27–30** exhibit remarkably lower activity toward both enzymes than the phosphate DDVP.[33] The replacement of an alkoxyl group with the bulky carborane moiety increases the steric demand of the compound and therefore hinders binding on the surface of the enzyme and the following phosphorylation of the hydroxyl group of serine. Additionally, the type of inhibition changed in such a way that the compounds exhibit reversible instead of irreversible binding. The substituents at the second carbon atom of the carborane core have only minor influence on the inhibitor activity.

In contrast to these molecules, compounds containing the carbaboranyl group in the ester moiety of the phosphonate, for example, **11** and **12**, exhibit higher antiacetylcholinesterase and antibutyrylcholinesterase activities than the isosystox analog GD-7 (**31**).[34] The high anticholinesterase activity was traced back to the electron-acceptor property of the carborane core, which increases the positive charge on the sulfur atom and therefore increases the interaction with the anionic center of the enzyme. Neither B-substituted phosphonates nor phosphonates containing carboranes linked through the B9 position to the ester moiety were evaluated; however, B-substituted phosphates were studied.[34] Due to the linkage to the electron-rich B9 atom, the carborane cage provides an electron-donating effect, which therefore lowers the positive charge on the sulfur atom. As a result the interaction with the enzyme decreases and the inhibitor activity is lower. It can be assumed that these findings for B-substituted phosphates would also hold for B-substituted phosphonates.

Besides acetylcholinesterase activity, some of the compounds show other biological activities. The *ortho*-carbaboran-9-yl thio (**32**) and seleno esters (**33**) of thiophosphonic acid were tested on

TABLE 2.1
Acetyl- and Butyrylcholinesterase Activities of Carbaboranyl Phosphonates and Thiophosphonates

Compound	Formula	k_{II} (m^{-1} min^{-1})	
		Acetylcholinesterase	Butyrylcholinesterase
26	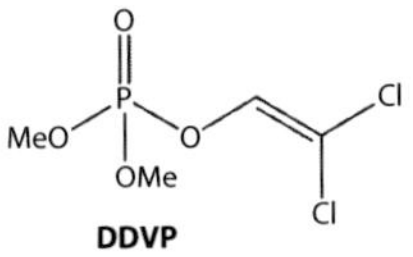 DDVP	2.30×10^4	2.50×10^5
27	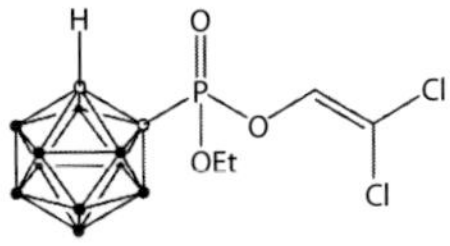	1.10×10^2	5.00×10^2
28	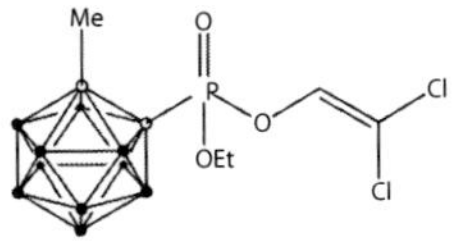	4.10×10^2	8.10×10^2
29	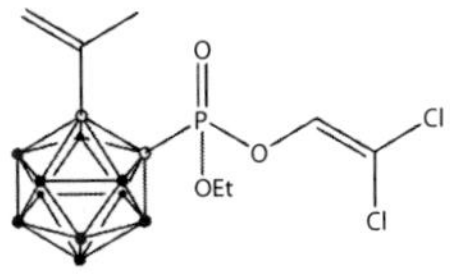	11.0×10^2	8.70×10^2
30	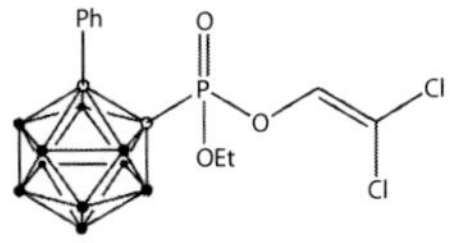	1.90×10^2	6.20×10^2
31	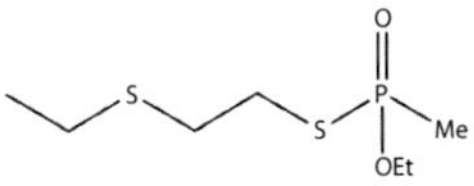 GD-7	6.00×10^4	6.80×10^3
11	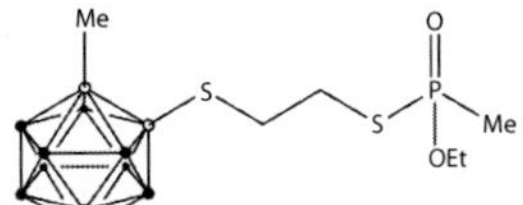	2.10×10^6	1.40×10^6
12	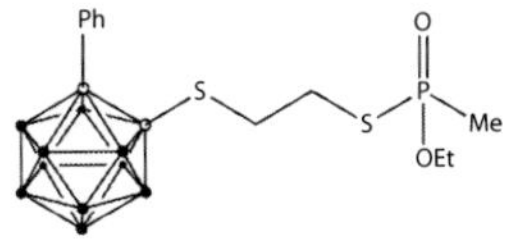	3.10×10^6	9.70×10^6

Source: Adapted from Zakharova, L. M. et al. 1978. *Izv. Akad. Nauk SSSR, Ser. Khim.* 2178–2180; Balema, V. P. et al. 1993. *Bioorg. Khim.*, 19, 1077–1080.

bacteria and were found to be active toward Gram-negative *Escherichia coli.*[35] For carbaboranyl phosphonate **34** high gametocidal activity on hybrid wheat seeds was reported, rendering them interesting in wheat production.[36] The 2,2-dichlorovinyl carbaboranyl phosphonates **27–30** have been extensively studied for their pesticide properties. However, weak insectoacaricidal and relatively low herbicidal and fungicidal activities have been found (Figure 2.2).[16]

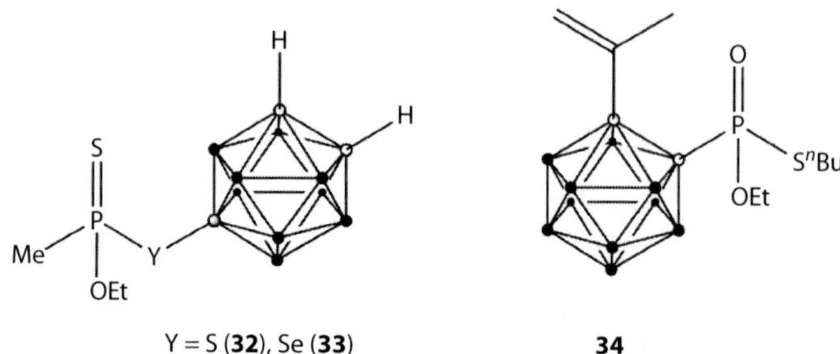

FIGURE 2.2 Bactericidal and gametocidal carbaboranyl phosphonates. (Adapted from Godovikov, N. N. et al. Abstracts of Reports at the All-Union Seminar "The Chemistry of Physiologically Active Compounds" Chernogolovka, 1989, 71; Kabachnik, M. I. et al. 1998. *O*-ethyl-*S*-butyl-(isopropenyl-*o*-carboranyl)-thiophosphonate showing gametocidal activity on rye. U.S.S.R., SU 953823 A3.)

2.3 CARBORANE- AND DODECABORATE-SUBSTITUTED MONO- AND BISPHOSPHONATES AS PHOSPHATE MIMETICS FOR USE IN BNCT

As mentioned in the introduction, clinically approved phosphonates and bisphosphonates are able to target areas of high hydroxylapatite production, which is the case for bone tumors. To use the high bone affinity of phosphonates, Lemmen and coworkers proposed carborane containing phosphonates for the therapy of bone tumors.[37,38] They prepared a diethylphosphonate linked through an ethylene spacer to the *ortho*-carborane cage (**35**). In a Michaelis–Arbuzov reaction with *ortho*-carbaboranylacetyl chloride, compound **36** was unexpectedly obtained. However, the bioactivity of both compounds was not evaluated (Figure 2.3).

For the same purpose of targeting bone tumors with boron-containing phosphonates, carbaboranylbis- and -polyphosphonic acids were described by Benedict in a patent application.[39] Two examples of the prepared compounds (**37**, **38**) are shown in Figure 2.4. The *gem*-bisphosphonate **37** was subjected to a crystal growth inhibition test to determine its potential to reduce calcium phosphate deposition. This showed that the capability of **37** to retard the formation of calcium phosphate is on the same order as the known inhibitor dichloromethylene diphosphonate. Furthermore, it was found to exhibit high skeletal affinity in an animal skeletal uptake test.

In 1999, Komagata et al. described the synthesis of bisphosphonic acids and esters linked through an alkyl spacer to *ortho*-carborane (**39**–**41**) in a patent application.[40] These boron-containing phosphonates were proposed for use as BNCT agents in the treatment of bone tumors. However, no data concerning the biological activity of these compounds were given.

Besides carboranes, dodecaborates are attractive for modification with phosphate and phosphonate entities. In 2002, Kultyshev et al. reported the synthesis of dodecaborate-containing phosphonate

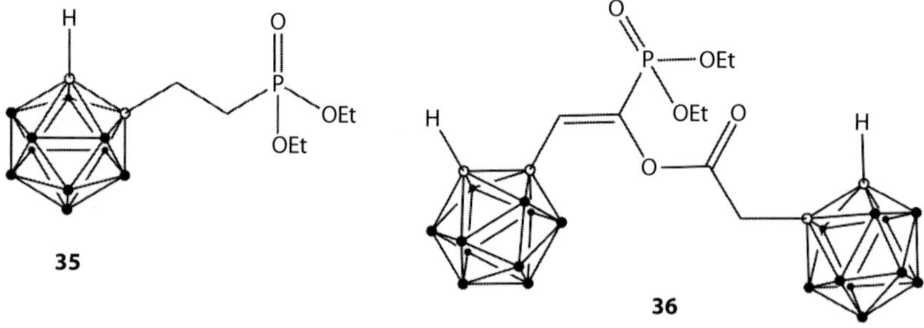

FIGURE 2.3 *ortho*-Carbaboranylalkyl- and -vinylphosphonates. (Adapted from Semioshkin, A. A. et al. 1997. *Advances in Boron Chemistry*, W. Siebert (ed.), 311. The Royal Society of Chemistry: Cambridge; Semioshkin, A. A. et al. 1998. *Russ. Chem. Bull.* 47: 1985–1988.)

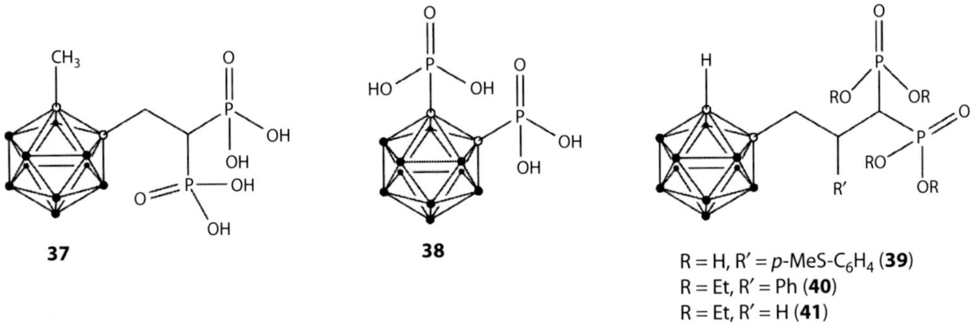

37

38

R = H, R′ = *p*-MeS-C_6H_4 (**39**)
R = Et, R′ = Ph (**40**)
R = Et, R′ = H (**41**)

FIGURE 2.4 Examples of bisphosphonic acid derivatives of *ortho*-carborane reported in the patent literature. (Adapted from Benedict, J. J. EP 0068584 A1, 1981. Boron-containing polyphosphonates for the treatment of calcific tumors; Komagata, T., Kawanabe, T., Matsushita, T. Jpn Kokai Tokyo Koho, JP 11-080177, A2, Heisei, 1999. Preparation of boron-containing bisphosphonic acids as agents for boron neutron capture therapy.)

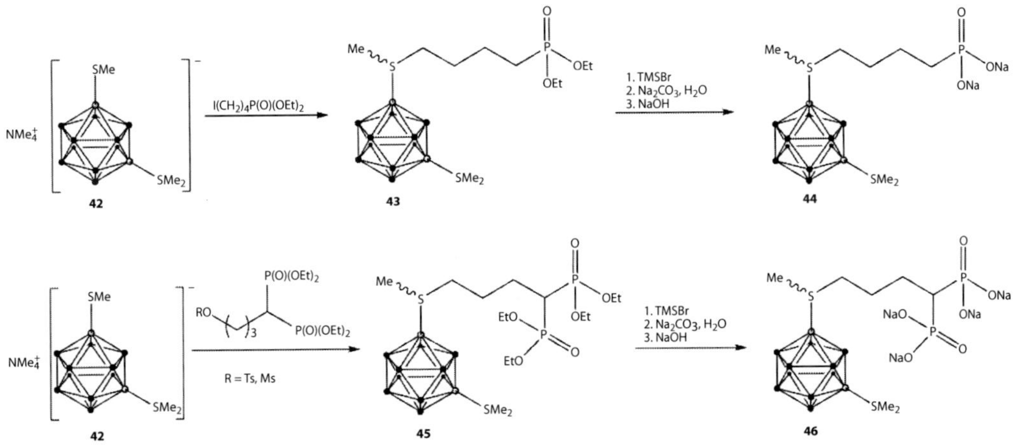

SCHEME 2.7 Synthesis of dodecaborate-containing phosphonate **44** and *gem*-bisphosphonate **46**. (Adapted from Kultyshev, R. G. et al. 2002. *J. Am. Chem. Soc.* 124: 2614–2624.)

43 and *gem*-bisphosphonate **45** by alkylation of methylthio ether **42** with either diethyl 1-iodobutylphosphonate or tosylated tetraethyl 1-hydroxy-4,4-phosphonobutane.[41] The obtained esters were converted into their highly water-soluble sodium salts **44** and **46** (Scheme 2.7). Tjarks et al. studied the biodistribution of *gem*-bisphosphonate **46** in comparison with ^{99m}Tc methylene diphosphonate (MDP) in vivo on BALB/c mice bearing K8 osteosarcomas.[42] Tumor-to-blood (T:Bl) and tumor-to-bone (T:B) ratios were 7.6 and 1.5, respectively, for **46**, whereas the T:Bl and T:B ratios for ^{99m}Tc MDP are 17.6 and 1.4, respectively. Furthermore, compound **46** was found to be nontoxic in animals in concentrations of 18 μg B/g tumor. In contrast to ^{99m}Tc MDP, high concentrations of **46** were found in spleen, kidneys, and especially in the liver. It was supposed that structural modifications may improve the suboptimal T:B ratio and yield more favorable biodistribution.

2.4 NUCLEOTIDE CONJUGATES OF CARBABORANYLMONO- AND BISPHOSPHONATES FOR USE AS AGENTS IN ANTISENSE STRATEGY AND IN BNCT

Oligonucleotides received interest as boron carriers, because it is supposed that they can selectively target tumor cells that contain overexpressed genes. Another important advantage would be that oligonucleotides accumulate not only in the cytoplasm but also in the cell nucleus. From the viewpoint of BNCT the latter fact is very important, because microdosimetric calculations showed that the effectiveness of the neutron capture reaction is 2–5 times higher if the BNCT agent is located in the cell nucleus. Carborane-modified oligonucleotides were proposed as excellent candidates to fulfill this requirement. The lipophilic carborane moiety should facilitate transport through the cell membrane. Furthermore, carborane-containing oligonucleotides were proposed as agents in antisense oligonucleotide therapy (AOT).[43] Oligonucleotides bearing the carborane in the ester moiety would be more susceptible for degradation by nucleases. Therefore, Lesnikowski and Schinazi reported in 1993 on thymidine(3′,5′)thymidine (*o*-carbaboran-1-ylmethyl)phosphonate (**50**), which is the first oligonucleotide analog modified with a carbaboranyl cage.[43] This compound bears an (*o*-carbaboran-1-ylmethyl)phosphonate (CBMP) group as 3′,5′-*O*,*O*-internucleotide linkage, which is stable toward degradation by nucleases. Starting from *O*,*O*-dimethyl-(*o*-carbaboran-1-ylmethyl) phosphonate (**47**), repeated coupling according to the standard phosphotriester method yielded the nucleotide dimer as a mixture of R_P and S_P diastereomers (Scheme 2.8).

Introduction of the CBMP group increases the lipophilicity of the oligonucleotide by neutralization of the negative charge of the phosphate group and by incorporation of the highly hydrophobic carborane core.[12] The lipophilicity of CBMP oligomers is even considerably higher than of methylphosphonate oligonucleotides. By adopting the methodology developed for the synthesis of **50**, dodecathymidylates containing the CBMP group at different locations in the 12-mer were prepared by solid-phase-automated synthesis. Interestingly, all oligonucleotides bearing the CBMP modifications were obtained as mixtures of both *closo*- and *nido*-carborane. CBMP-modified oligomers were found to be chemically stable at physiological pH and resistant toward enzymatic digestion with bovine spleen phosphodiesterase (BSPDE) and snake venom phosphodiesterase (SVPDE). When the CBMP group is located at the 5′-end of the oligonucleotide, complete protection of the oligomer against digestion by BSPDE is achieved. CBMP modifications at other locations led to digestion, but at a significantly lower rate than in unmodified oligomers. Complete protection of the oligonucleotide against digestion by SVPDE is not achieved, but increased resistance in comparison with the unmodified oligonucleotide was observed. Furthermore, enzymatic phosphorylation with T4 polynucleotide kinase was reported to be efficient for all oligonucleotides bearing the CBMP group, with the exception of oligomers bearing the CBMP group at the 5′-end.[44]

With the same purpose of delivering significant boron concentrations to the cell nucleus by mimicking naturally occurring nucleosides, Hosmane and coworkers reported in 2004 nucleoside mono- and bisphosphonate derivatives of *ortho*-carborane (Figure 2.5).[45] Adenosine and guanosine were connected through the 5′-position of the sugar moiety to the phosphorus atoms. With the aim to imitate naturally occurring adenosine diphosphate, Hosmane and coworkers designed carbaboranyl bis(adenosine diphosphate) (**57**) as a boron carrier.[46] It was supposed that this compound can be converted to the corresponding adenosine triphosphate. However, none of these compounds were subjected to biological investigations.

2.5 GLYCOSYL CONJUGATES OF CARBABORANYL BISPHOSPHONATES

Endogenous lectins are located on the surface of normal and malignant cells and serve as specific receptors for mediating endocytosis of glycoconjugates.[47–49] Transformation of normal to malignant cells is often accompanied by modification of the lectin composition and overexpression of certain lectins. For example, lactose-binding lectin (LBL) plays an important role in the metastatic growth of tumors.[50] We have therefore devised efficient syntheses for boron compounds that provide a combination of tumor-targeting systems: The use of phosphonato groups as phosphate mimics and galactosyl groups for binding to lectins at the surface of a tumor cell.[51] The 6-position of galactose is usually involved in recognition in lectins, but for proof of a general synthetic principle we synthesized galactosyl derivatives connected through the 6-position to the phosphonate.[52,53]

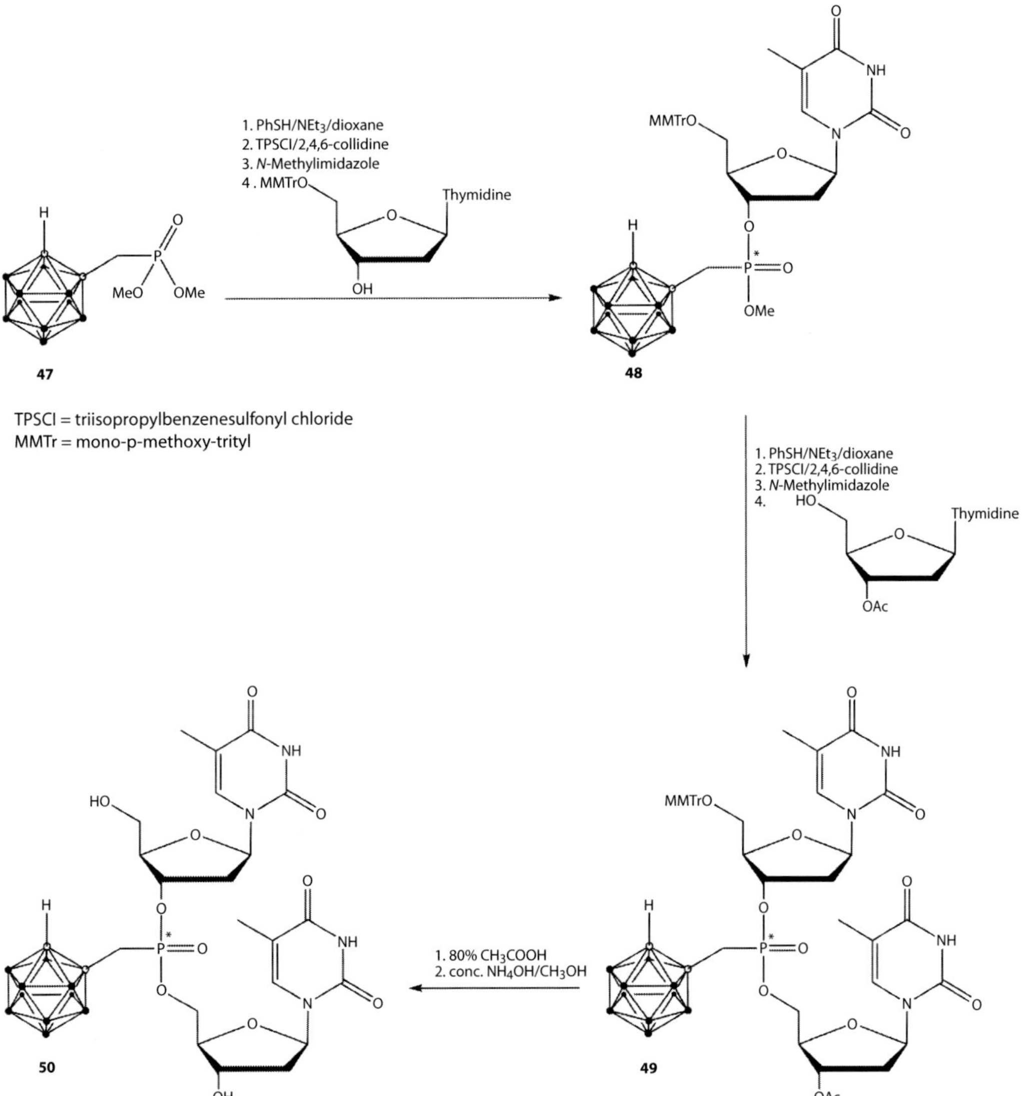

SCHEME 2.8 Synthesis of a CBMP-modified thymidine dimer **50**. (Adapted from Lesnikowski, Z. J.; Schinazi, R. F. 1993. *J. Org. Chem.* 58: 6531–6534.)

Carbaboranyl bisglycophosphonates were synthesized according to the phosphoramidite method from *meta*- or *para*-carbaboranyl bisphosphonite and isopropylidene-protected galactose (**59**) with benzimidazolium triflate (BIT) as the promoter. Due to strong P···P interactions in *ortho*-carborane derivatives, the synthesis of the respective carbaboranyl bisglycophosphonates was unsuccessful. Synthesis of disodium salts **62** and **63** starts from *meta*-carbaboranyl bisphosphonite **58**, which acts as a phosphoramidite analog. After galactosylation with **58** and oxidation with *tert*-butyl hydroperoxide (TBHP), subsequent deprotection of the phosphonate and sugar moieties led to disodium salt **62**. For improved *in vivo* stability toward phosphatases and phosphonate esterases, the corresponding bisphosphonothioate **63** was prepared accordingly by this method using 3*H*-1,2-benzodithiol-3-one-1,1-dioxide, the so-called Beaucage reagent, instead of TBHP (Scheme 2.9). Due to two chiral phosphorus atoms in **63**, a mixture of diastereomers was obtained.

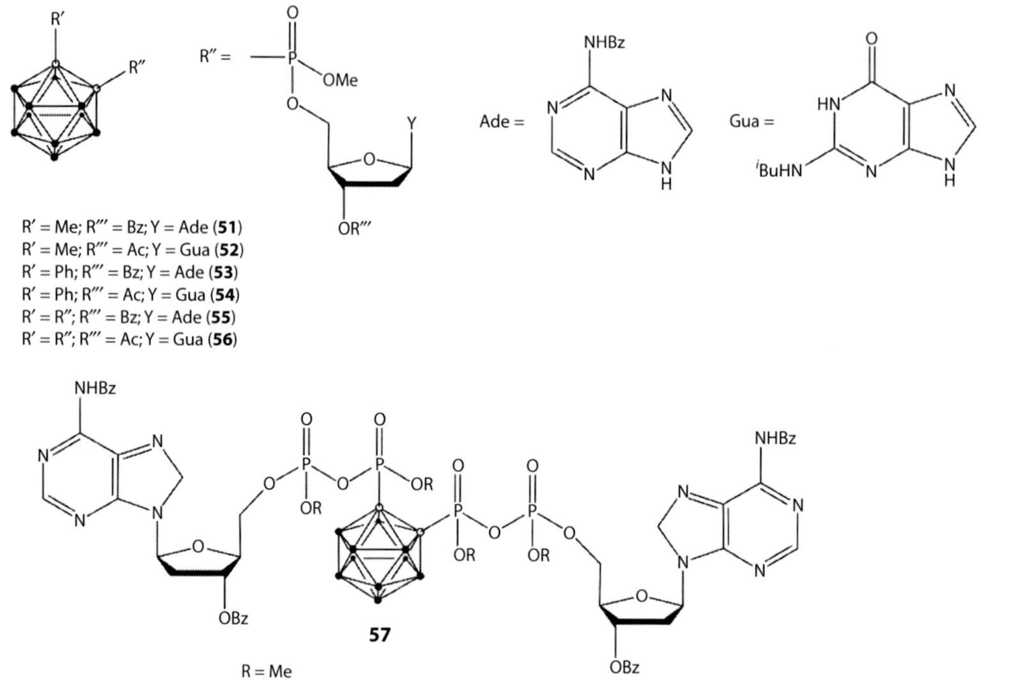
R′
R″
R″ =
O
P
OMe
O
Y
O
OR‴
Ade =
NHBz
Gua =
O
HN
^{i}BuHN
R′ = Me; R‴ = Bz; Y = Ade (51)
R′ = Me; R‴ = Ac; Y = Gua (52)
R′ = Ph; R‴ = Bz; Y = Ade (53)
R′ = Ph; R‴ = Ac; Y = Gua (54)
R′ = R″; R‴ = Bz; Y = Ade (55)
R′ = R″; R‴ = Ac; Y = Gua (56)
NHBz
OR
OBz
57
R = Me

FIGURE 2.5 Nucleosyl conjugates of carbaboranylmono- and bisphosphonates. (Adapted from Vyakaranam, K.; Hosmane, N. S. 2004. *Bioinorg. Chem. Appl.* 2: 31–42; Vyakaranam, K. et al. 2003. *Inorg. Chem. Commun.* 6: 654–657.)

To examine whether two sodium monogalactosyl phosphonate groups or two digalactosyl phosphonate moieties are better for accumulation in tumor cells, the tetra-D-galactosylated carbaboranyl bis(phosphonates) were also synthesized according to the protocol for **62** and **63**.

Starting from a carbaboranyl bisphosphonite bearing four amido groups, treatment with an excess of protected galactose (**59**) and BIT according to the procedure described above yielded bisphosphonate **64** and bisphosphonothioate **65** after oxidation and sulfurization, respectively. Cleavage of the isopropylidene moieties led to an equilibrium between α and β forms of the

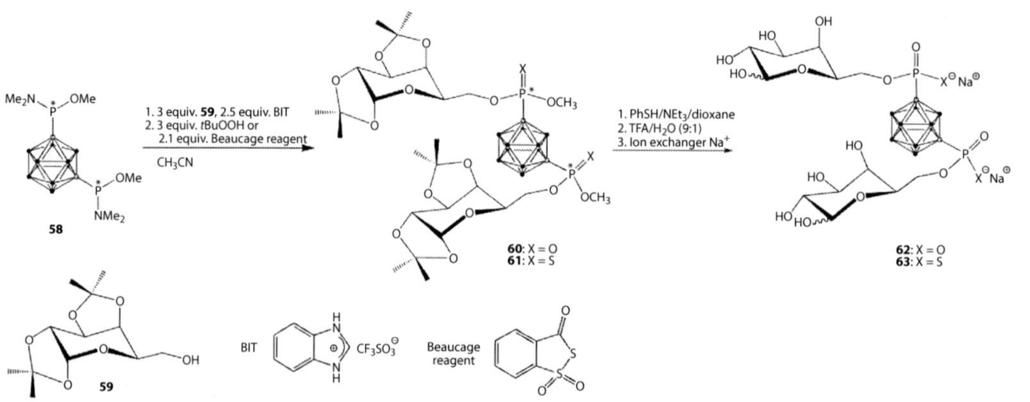
Me2N
OMe
OMe
NMe2
58
1. 3 equiv. 59, 2.5 equiv. BIT
2. 3 equiv. tBuOOH or
2.1 equiv. Beaucage reagent
CH3CN
OCH3
60: X = O
61: X = S
1. PhSH/NEt3/dioxane
2. TFA/H2O (9:1)
3. Ion exchanger Na+
62: X = O
63: X = S
OH
59
BIT
CF3SO3
Beaucage reagent

SCHEME 2.9 Synthesis of carboranediyl bis(phosphonate) **62** and bis(phosphonothioate) **63**. (Adapted from Stadlbauer, S.; Welzel, P.; Hey-Hawkins, E. 2009. *Inorg. Chem.* 48: 5005–5010; Stadlbauer, S. et al. 2009. *Eur. J. Org. Chem.* 6301–6310; Stadlbauer, S. 2008. Glycophosphonsäurekonjugate von Carbaboranen–Synthese und biologische Aktivität. Doctoral thesis, Universität Leipzig.)

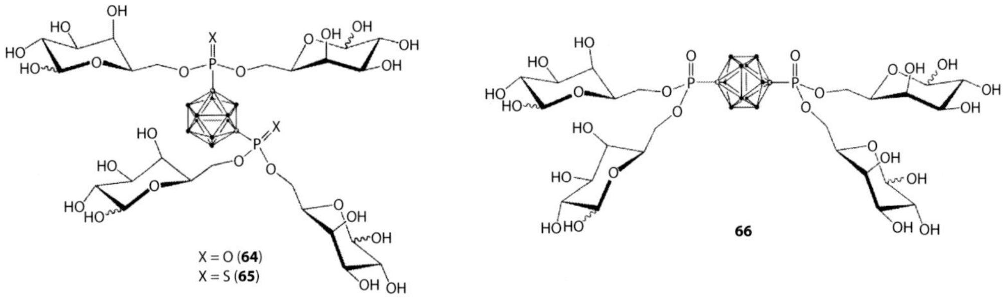

FIGURE 2.6 Tetra-D-galactosylated carbaboranyl bisphosphonates. (Adapted from Stadlbauer, S. et al. 2009. *Eur. J. Org. Chem.* 6301–6310; Stadlbauer, S. 2008. Glycophosphonsäurekonjugate von Carbaboranen–Synthese und biologische Aktivität. Doctoral thesis, Universität Leipzig.)

galactosyl groups. As the four carbohydrate units can theoretically form 16 stereoisomers, anomerization induces chirality at the phosphorus atoms, which results in the formation of several diastereomers. Due to the fast anomerization, separation of the diastereomeric mixture was impossible. For further studies on the influence of the isomeric carborane core on the biological activity, an analogous *para*-carborane derivative **66** was synthesized in the same manner (Figure 2.6).

The water solubility of compounds **62–66** was determined and found to be high in comparison with D-galactose. The values for the glycophosphonate derivatives are shown in Table 2.2. The disodium salts exhibit the highest water solubility among these compounds with 910 (**62**) and 830 g/L (**63**). These absolute values are higher than for D-galactose (650 g/L). However, the molar solubilities of about 1.35 (**62**) and about 1.18 mol/L (**63**) are lower than that of galactose (3.6 mol/L). Due to the missing ionic part of the molecules, tetragalactosyl-substituted compounds **64** and **65** are less water soluble. In general, all bis(phosphonothioates) show lower solubility than their bis(phosphonate) counterparts. The *para*-carborane unit was reported to be the most lipophilic in the series of carborane isomers;[11,12] however, no remarkable difference in solubility was observed in comparison with *meta*-carborane analog **64**. It can be concluded that in general the combination of an ionic phosphonato group and carbohydrate moieties provides high water solubility of the extremely hydrophobic carborane moieties, which is crucial for proposed application in BNCT.

The cytotoxicity of these compounds was studied on HeLa cells in a resazurin assay.[54] Bisphosphonate **62** and the diastereomeric mixture of bisphosphonothioate **63** exhibit extraordinarily low cytotoxicity. They do not show cytotoxic effects up to a concentration of 20 mM (Figure 2.7) and are thus far less toxic than the boron compounds which are currently employed in

TABLE 2.2
Water Solubility of 62–66

	Water Solubility[a]	
Compound	**g/L**	**mol/L**
62	910	1.35
63	830	1.18
64	790	0.83
65	380	0.39
66	770	0.81
D-Galactose	650	3.6

Source: Adapted from Stadlbauer, S. et al. 2009. *Eur. J. Org. Chem.* 6301–6310.

[a] The error margin is ±10 g/L.

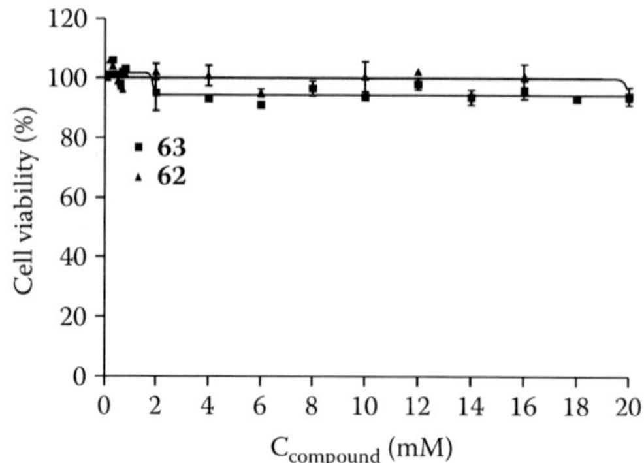

FIGURE 2.7 Cell viability of HeLa cells exposed to bisphosphonate **62** and bisphosphonothioate **63.** (Adapted from Stadlbauer, S.; Welzel, P.; Beck-Sickinger, A. B.; Gabel, D.; Hey-Hawkins, E. unpublished results.)

BNCT, that is, sodium mercaptoundecahydrododecaborane (BSH, IC_{50} 3.9 mM on V79 Chinese hamster cells),[55] and this makes them interesting candidates for BNCT. It can be assumed that the high water solubility of both compounds is responsible for their low toxicity.

In contrast, the diastereomeric mixtures of **64** and **65** show increased cytotoxicity. The EC_{50} values are about 29 mM for **64** and 14.0 ± 3.5 mM for **65**, respectively. The relatively large error margin for the EC_{50} values was traced back to osmotic effects that may occur at these high concentrations. Interestingly, there is a remarkable difference in cytotoxicity of the two compounds. The bisphosphonothioate showed increased cytotoxicity in comparison with the analogous bisphosphonate. As the only difference in the chemical structure is the P=S group instead of P=O, the sulfurized phosphorus atom must be responsible for this drastically increased cell toxicity. As **65** showed reduced water solubility, it was assumed that the P=S group increases the hydrophobicity and therefore the cytotoxicity (Figure 2.8).

Compounds **62–66** exhibited low cytotoxicity and were thus suitable for tumor selectivity studies.[54,56] Bisphosphonothioates **63** and **65** were chosen as representative examples to study in vivo toxicity in Swiss mice. Compound **65** (dosage 100 mg/kg boron) resulted in toxic side effects, while compound **63** was well tolerated and thus employed in further boron distribution studies. For these studies four female BALB/c mice with a CRL tumor were treated intraperitoneally with a dosage of 100 mg/kg boron. The mice were euthanized after certain periods of time, and then frozen thin sections were made and subsequently irradiated. The results showed high concentrations of the

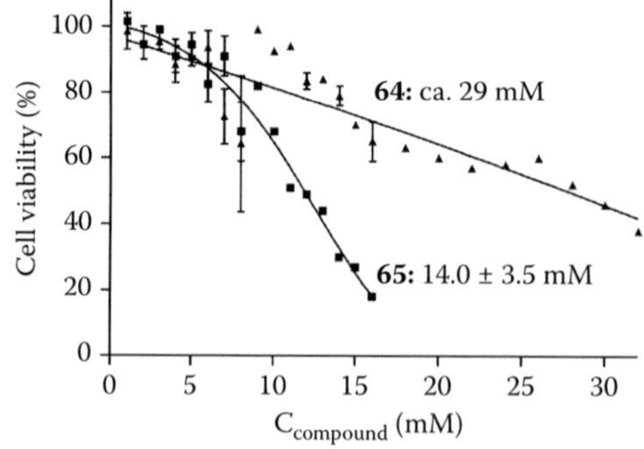

FIGURE 2.8 Cell viability of HeLa cells exposed to tetra-D-galactosylated derivatives **64** and **65**. (Adapted from Stadlbauer, S. et al. unpublished results.)

compound in the kidneys, liver, and the colon and only minor concentrations in the tumor. Within 2 h the compound was excreted through the colon. In conclusion, this compound showed a lack of selectivity for this tumor type.

2.6 GLYCOSYL CONJUGATES OF BISCARBABORANYL BISPHOSPHONATES

It is important for a successful BNCT agent to reach a very high ^{10}B concentration. Therefore, the neutron capture reaction is more effective if the molecule delivers a large number of boron atoms. To fulfill this requirement, compounds with two boron clusters bearing glycophosphonate groups were synthesized by us.[57]

Bis(*meta*-carborane) (**67**) was obtained by copper-mediated C—C coupling of monolithiated *meta*-carborane. Functionalization of the C—H groups of **67** with amidophosphites gave P-chiral (**68**) and achiral (**69**) bisphosphonite derivatives (Scheme 2.10).

Reaction of **68** with protected galactose (**59**) and benzimidazolium triflate as activator followed by oxidation or sulfurization, deprotection, and conversion into the sodium salt provided bis(phosphonate) **70** and bis(phosphonothioate) **71** (Figure 2.9) in a total yield of 5 (**70**) and 25% (**71**).

Similarly, the corresponding tetragalactosyl esters were available starting from tetraamido-substituted bisphosphonite **69**. Deprotection provided **72** and **73** in a total yield of 61 (**72**) and 24% (**73**), respectively. Both compounds partly flocculate from aqueous solution; this indicates reduced water solubility due to the biscarborane unit.

In general, the water solubility showed the same trend as for the analogous monocarborane derivatives. The disodium salts exhibited the highest water solubility among these compounds of 700 (**70**) and 636 g/L (**71**). The molar solubilities are 0.83 mol/L for **70** and about 0.75 mol/L for **71**, respectively. As expected, the neutral tetragalactosyl-substituted compounds **72** and **73** were much less water soluble. In general, all bis(phosphonothioates) exhibited lower solubility than their phosphonate counterparts (Table 2.3).

The presence of the second lipophilic carborane cluster increases the cytotoxicity drastically. Thus, the EC_{50} value is 19.5 mM for bis(phosphonate) **70** and 2.1 mM for bisphosphonothioate **71** (Figure 2.10). Compounds **72** and **73** have lower water solubility than the other derivatives, which results in lower EC_{50} values in HeLa cells (480.9 μM (**72**), 174.9 μM (**73**); Figure 2.11). The morphology of the HeLa cells after treatment with the EC_{50} concentration of **72** and **73** was studied. Besides intact HeLa cells, a large fraction of cell fragments, that is, cell nuclei, were clearly observed under the fluorescence microscope. Apparently, the extremely hydrophobic bis(*meta*-carborane) unit drills into the cell membrane and with increasing concentration leads to disruption of the membrane. Therefore, both compounds are unsuitable for application in BNCT.

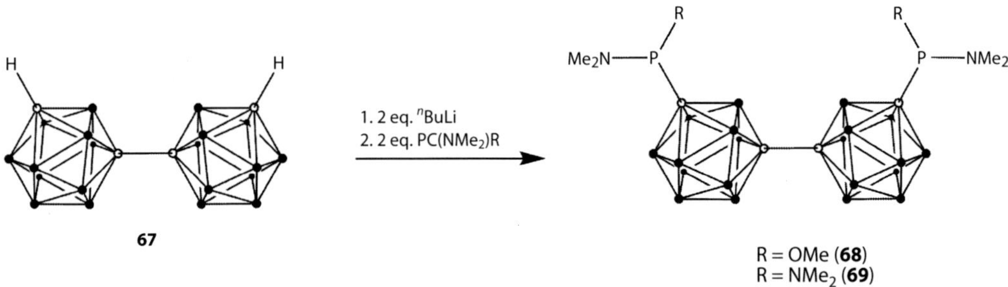

SCHEME 2.10 Synthesis of bisphosphonito-substituted bis(*meta*-carborane) derivatives **68** and **69**. (Adapted from Stadlbauer, S. et al. 2010. *Eur. J. Org. Chem.* 3129–3139.)

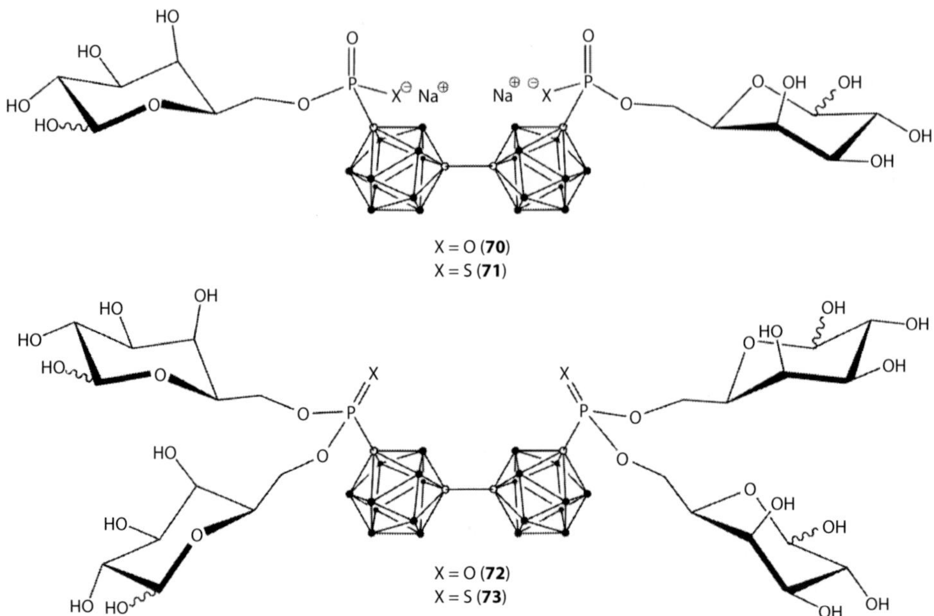

FIGURE 2.9 Bis(carboranediyl) bis(phosphonates) and bis(phosphonothioates). (Adapted from Stadlbauer, S. et al. 2010. *Eur. J. Org. Chem.* 3129–3139.)

TABLE 2.3
Water Solubility of Biscarborane Derivatives 70–73

	Water Solubility[a]	
Compound	**g/L**	**mol/L**
70	700	0.83
71	600	0.71
72	480	0.44
73	370	0.33
D-Galactose	650	3.6

Source: Adapted from Stadlbauer, S. et al. 2010. *Eur. J. Org. Chem.* 3129–3139.
[a] The error margin is ±20 g/L.

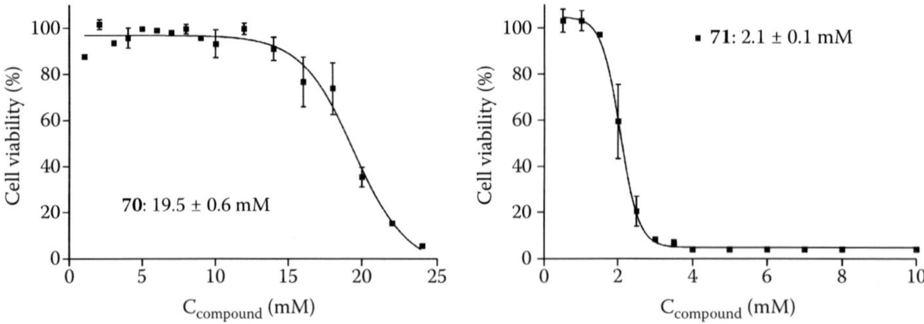

FIGURE 2.10 Cell viability of HeLa cells exposed to biscarborane derivatives **70** and **71**. (Adapted from Stadlbauer, S. et al. unpublished results.)

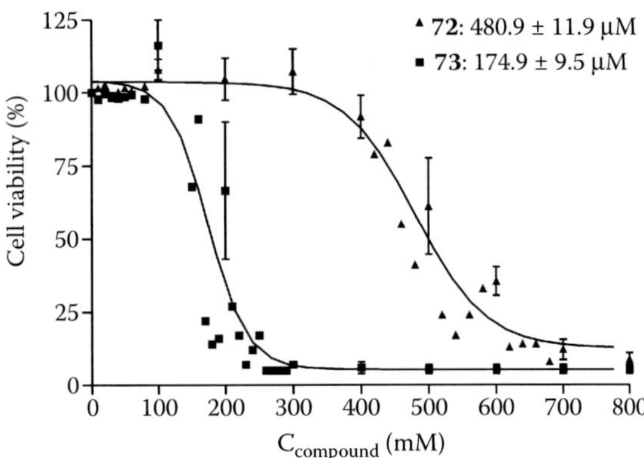

FIGURE 2.11 Cell viability of HeLa cells exposed to bisphosphonate **72** and bisphosphonothioate **73**. (Adapted from Stadlbauer, S. et al. unpublished results.)

2.7 SUMMARY

Boron neutron capture therapy (BNCT) is a promising method for the treatment of cancer. However, one critical issue is the availability of boron compounds which exhibit the necessary tumor selectivity and low cytotoxicity in high concentrations. Phosphorus-containing boron compounds are therefore interesting, especially for the treatment of bone cancer due to their capability of targeting calcium-rich tumor tissue. Besides their use in BNCT, carbaboranyl phosphonates are known as highly bioactive compounds. When the carbaboranyl group is incorporated into the ester moiety of phosphonic acid esters, these compounds show high anticholinesterase activity. Compounds in which the phosphonate is directly attached to the carborane exhibit gametocidal activity, while phosphonates connected through sulfur or selenium to the cluster cage exhibit bactericidal activity. Nucleoside-linked carbaboranyl mono- and bisphosphonates may allow selective accumulation within the cell nucleus and thus increase the effectiveness of the neutron capture reaction. Incorporated into oligonucleotide strands, they can serve as agents in antisense oligonucleotide therapy, because they show increased resistance to digestion by nucleases. A recent approach is the use of glycosyl conjugates of *meta*-carbaboranyl bis(phosphonates) and bis(phosphonothioates) providing a combined tumor-targeting system. These compounds are interesting candidates for future applications in BNCT due to their high water solubility and low cytotoxicity. Boron-rich biscarborane derivatives were synthesized to increase the boron concentration. However, they do not meet the criteria of low cytotoxicity.

In conclusion, there is still much to discover in this field, as comprehensive biological assessments of boron-containing phosphonates as potential tumor-targeting BNCT agents are still rare.

REFERENCES

1. Hawthorne, M. F. 1993. The role of chemistry in the development of boron neutron capture therapy of cancer. *Angew. Chem. Int. Ed.* 32: 950–984.
2. Soloway, A. H.; Tjarks, W.; Barnum, B. A.; Rong, F.-G.; Barth, R. F.; Codogni, I. M.; Wilson, J. G. 1998. The chemistry of neutron capture therapy. *Chem. Rev.* 98: 1515–1562.
3. Valliant, J. F.; Guenther, K. J.; King, A. S.; Morel, P.; Schaffer, P.; Sogbein, O. O.; Stephenson, K. A. 2002. The medicinal chemistry of carboranes. *Coord. Chem. Rev.* 232: 173–230.
4. Hawthorne, M. F.; Lee, M. W. 2003. A critical assessment of boron target compounds for boron neutron capture therapy. *J. Neurooncol.* 62: 33–45.
5. Henriksen, K.; Neutzsky-Wulff, A. V.; Bonewald, L. F.; Karsdal, M. A. 2009. Local communication on and within bone controls bone remodeling. *Bone* 44: 1026–1033.
6. Narducci, P.; Bareggi, R.; Nicolin V. 2009. Receptor activator for nuclear factor kappa B ligand (RANKL) as an osteoimmune key regulator in bone physiology and pathology. *Acta Histochem.* 113: 73–81.

7. Hassfjell, S. P.; Bruland, O. S.; Hoff, P. 1997. ^{212}Bi-DOTMP: An alpha particle emitting bone-seeking agent for targeted radiotherapy. *Nuc. Med. Biol.* 24: 231–237.
8. Sparidans, R. W.; Twiss, I. M.; Talbot, S. 1998. Bisphosphonates in bone diseases. *Pharm. World Sci.* 20: 206–213.
9. Stresing, V.; Daubiné, F.; Benzaid, I.; Mönkkönen, H.; Clézardin, P. 2007. Bisphosphonates in cancer therapy. *Cancer Lett.* 257: 16–35.
10. Fauchère, J. L.; Do, K. Q.; Jow, P. Y. C.; Hansch, C. 1980. Unusually strong lipophilicity of "fat" or "super" amino acids, including a new reference value for glycine. *Experientia* 36: 1203–1204.
11. Sjöberg, S.; Carlsson, J.; Ghaneolhosseini, H.; Gedda, L.; Hartmann, T.; Malmquist, J.; Naeslund, C.; Olsson, P.; Tjarks, W. 1997. Chemistry and biology of some low molecular weight boron compounds for boron neutron capture therapy. *J. Neuro-Oncol.* 33: 41–52.
12. Lesnikowski, Z. J.; Shi, J.; Schinazi, R. F. 1999. Nucleic acids and nucleosides containing carboranes. *J. Organomet. Chem.* 581: 156–169.
13. Bregadze, V. I. 1992. Dicarba-*closo*-dodecaboranes $C_2B_{10}H_{12}$ and their derivatives. *Chem. Rev.* 92: 209–223 and references cited herein.
14. Alexander, R. P.; Schröder, H.-J. 1963. Chemistry of decaborane-phosphorus compounds. IV. Monomeric, oligomeric, and cyclic phosphinocarboranes. *Inorg. Chem.* 2: 1107–1110.
15. Godovikov, N. N.; Balema, V. P.; Rys, E. G. 1997. Carborane-containing organophosphorus compounds. Synthesis and properties. *Russ. Chem. Rev.* 66: 1017–1032.
16. Degtyarev, A. N. 1978. *Candidate Thesis in Chemical Sciences. Institute of Organoelement Compounds*, Academy of Sciences of the USSR, Moscow.
17. Degtyarev, A. N.; Godovikov, N. N.; Bregadze, V. I.; Kabachnik, M. I. 1975. Synthesis and chemical properties of some carboranylphosphonous acid esters. *Izv. Akad. Nauk SSSR, Ser. Khim.* 2568–2573.
18. Degtyarev, A. N.; Godovikov, N. N.; Bregadze, V. I.; Kabachnik, M. I. 1973. Arbuzov rearrangement in organophosphorus carborane derivatives. *Izv. Akad. Nauk SSSR, Ser. Khim.* 2369.
19. Degtyarev, A. N.; Godovikov, N. N.; Bregadze, V. I.; Matrosov, E. I.; Shcherbina, T. M.; Kabachnik, M. I. 1978. Synthesis and chemical properties of some carboranylmethyl-phosphinite esters. *Izv. Akad. Nauk SSSR, Ser. Khim.* 2099–2104.
20. Kazantsev, A. V.; Meiramov, M. G.; Zakharkin, L. I. 1984. Synthesis and certain reactions of *O,O*-diethyl (1-*o*-carboranylmethyl)phosphonite. *Zh. Obshch Khim.* 54: 2002–2004.
21. Kabachnik, M. I.; Godovikov, N. N.; Rys, E. G. 1980. Synthesis of *S*-(*o*-carboran-1-ylmethyl) thiophosphonates. *Izv. Akad. Nauk SSSR, Ser. Khim.* 1455–1456.
22. Rys, E. G.; Godovikov, N. N.; Kabachnik, M. I. 1983. *o*-Carboranyl-containing esters of pentavalent phosphorus. *Izv. Akad. Nauk SSSR, Ser. Khim.* 2640–2644.
23. Balema, V. P.; Rys, E. G.; Godovikov, N. N.; Kabachnik, M. I. 1989. Synthesis of *S*-(carboran-1-yl) thiophosphates and thiophosphonates. *Izv. Akad. Nauk SSSR, Ser. Khim.* 194–197.
24. Rys, E. G.; Balema, V. P.; Godovikov, N. N.; Kabachnik, M. I. 1990. Carboranyl selenium-containing esters of pentavalent phosphorus acids. *Izv. Akad. Nauk SSSR, Ser. Khim.* 1653–1655.
25. Bregadze, V. I.; Kampel, V. Ts.; Matrosov, E. I. et al. 1985. Synthesis and structure of β-carboranylphosphonates. *Dokl. Akad. Nauk SSSR* 285: 1127–1130.
26. Kampel, V. Ts.; Bregadze, V. I.; Ermanson, L. V. et al. 1992. (Boron-carboranyl)dichlorophosphines. *Metalloorg. Khim.* 5: 1024–1027.
27. Tumanskii, B. L.; Kampel´, V. Ts.; Bregadze, V. I. et al. 1986. Interaction of boron-centered carborane radicals with phosphites and the addition of carborane-containing and other phosphorane radicals to 3,6-di-tert-butyl-*o*-benzoquinone. *Izv. Akad. Nauk SSSR, Ser. Khim.* 458–461.
28. Zakharkin, L. I.; Guseva, V. V.; Ol´shevskaya, V. A. 2001. Synthesis of dimethyl (*m*-carboran-9-yl)phosphonate and dimethyl (*p*-carboranyl)phosphonate by palladium-catalyzed cross-coupling of 9-I-*m*- and 2-I-*p*-carboranes with dimethyl hydrogen phosphate. *Russ. J. Gen. Chem.* 71: 903–904.
29. Rys, E. G.; Godovikov, N. N.; Kabachnik, M. I. 1986. Synthesis of *S*-carboran-9-yl esters of pentavalent phosphorus acids. *Izv. Akad. Nauk SSSR, Ser. Khim.* 719–721.
30. Balema, V. P.; Rys, E. G.; Godovikov, N. N.; Kabachnik, M. I. 1990. Synthesis and some transformations of *O,O*-diethyl *S*-(*o*-carboran-9-yl) thiophosphite. *Izv. Akad. Nauk SSSR, Ser. Khim.* 2857–2859.
31. Balema, V. P.; Rys, E. G.; Godovikov, N. N. 1992. *O*-ethyl *S,S*-bis(o-carboran-9-yl) dithiophosphite. *Izv. Akad. Nauk SSSR, Ser. Khim.* 459–461.
32. Zakharova, L. M.; Degtyarev, A. N.; Agabekyan, R. S.; Bregadze, V. I.; Godovikov, N. N.; Kabachnik, M. I. 1978. Anticholinesterase activity of dichlorovinyl esters of carboranylphosphonic and carbaboranylmethylphosphinic acids. *Izv. Akad. Nauk SSSR, Ser. Khim.* 2178–2180.

33. Brestkin, A. P.; Khovanskikh, A. E.; Maizel, E. B. et al. 1986. Cholinesterases of aphids. II. Anticholinesterase potency and toxicity of different organophosphorous inhibitors for spring grain aphid schizaphis gramina rond. *Insect. Biochem.* 16: 701–707.
34. Balema, V. P.; Rys, E. G.; Sochilina, E. E. et al. 1993. Anticholinesterase activity of carboranyl containing thio- and selenoesters of pentavalent phosphoric acids. *Bioorg. Khim.*, 19, 1077–1080.
35. Godovikov, N. N.; Limanov, V. E.; Balema, V. P.; Aref́eva, L. I.; Rys, E. G. Abstracts of Reports at the All-Union Seminar *"The Chemistry of Physiologically Active Compounds"* Chernogolovka, 1989, 71.
36. Kabachnik, M. I.; Godovikov, N. N.; Bregvadze, V. I. et al. 1998. *O*-ethyl-*S*-butyl-(isopropenyl-*o*-carboranyl)-thiophosphonate showing gametocidal activity on rye. U.S.S.R., SU 953823 A3.
37. Semioshkin, A.; Lemmen, P.; Inyushin, S.; Ermanson, L. 1997. Alkyl carboranyl sulfonates as available synthons for carboranyl derivatives, especially for carborane-containing phosphonic acid esters. *Advances in Boron Chemistry*, W. Siebert (ed.), 311–314. The Royal Society of Chemistry: Cambridge.
38. Semioshkin, A. A., Inyushin, S. G., Ermanson, L. V., Petrovskii, P. V., Lemmen, P., Bregadze, V. I. 1998. Interactions of carborane-containing electrophiles with triethyl phosphite. Synthesis of new carborane-containing phosphonates. *Russ. Chem. Bull.* 47: 1985–1988.
39. Benedict, J. J. EP 0068584 B1, 1985. Boron containing polyphosphonates for the treatment of calcific tumors.
40. Komagata, T., Kawanabe, T., Matsushita, T. Jpn Kokai Tokyo Koho, JP 11-080177, A2, Heisei, 1999. Preparation of boron containing bisphosphonic acids as agents for boron neutron capture therapy.
41. Kultyshev, R. G.; Liu, J.; Liu, S.; Tjarks, W.; Soloway, A. H.; Shore, S. G. 2002. S-Alkylation and S-amination of methyl thioethers – derivatives of *closo*-$[B_{12}H_{12}]^{2-}$. synthesis of a boronated phosphonate, *gem*-bisphosphonates and dodecaborane-*ortho*-carborane oligomers. *J. Am. Chem. Soc.* 124: 2614–2624.
42. Tjarks, W.; Barth, R. F.; Rotaru, J. H. et al. 2001. *In vivo* evaluation of phosphorus-containing derivatives of dodecahydro-*closo*-dodecaborate for boron neutron capture therapy of gliomas and sarcomas. *Anticancer Res.* 21: 841–846.
43. Lesnikowski, Z. J.; Schinazi, R. F. 1993. Carboranyl oligonucleotides. 1. Synthesis of thymidine(3′,5′) thymidine (*o*-carboranyl-1-ylmethyl)phosphonate. *J. Org. Chem.* 58: 6531–6534.
44. Lesnikwoski, Z. J.; Lloyd Jr, R. M.; Schinazi, R. F. 1997. Comparison of physicochemical and biological properties of (*o*-carboran-1-ylmethyl)phosphonate and methylphosphonate oligonucleotides. *Nucl. Nucl.* 16: 1503–1505.
45. Vyakaranam, K.; Hosmane, N. S. 2004. Novel carboranyl derivatives of nucleoside mono- and diphosphites and phosphonates: A synthetic investigation. *Bioinorg. Chem. Appl.* 2: 31–42.
46. Vyakaranam, K.; Rana, G.; Delaney, S.; Ledger, S.; Hosmane, N. S. 2003. The first carboranyl bis(adenosine diphosphate) (CBADP): A synthetic investigation. *Inorg. Chem. Commun.* 6: 654–657.
47. Wadhwa, M. S.; Rice, K. G. 1995. Receptor mediated glycotargeting. *J. Drug Target.* 3: 111–127.
48. Lis, H.; Sharon, N. 1998. Lectins: Carbohydrate-specific proteins that mediate cellular recognition. *Chem. Rev.* 98: 637–674.
49. Yamazaki, N.; Kojima, S.; Bovin, N. V.; André, S.; Gabius, S.; Gabius, H.-J. 2000. Endogeneous lectins as targets for drug delivery. *Adv. Drug Del. Rev.* 43: 225–244.
50. Dean, B.; Oguchi, H.; Cai, S.; Otsuji, E.; Tashiro, K.; Hakomori, S.; Toyokuni, T. 1993. Synthesis of multivalent β-lactosyl clusters as potential tumor metastasis inhibitors. *Carbohydr. Res.* 245: 175–192.
51. Stadlbauer, S.; Hey-Hawkins, E. Patent WO2009021978 (A3). Novel chemical compounds and the use thereof in medicine, especially for use in tumor therapy.
52. Stadlbauer, S.; Welzel, P.; Hey-Hawkins, E. 2009. Access to carbaboranyl glycophosphonates—An Odyssey. *Inorg. Chem.* 48: 5005–5010.
53. Stadlbauer, S.; Lönnecke, P.; Welzel, P.; Hey-Hawkins, E. 2009. Highly water-soluble carbaborane-bridged bis(glycophosphonates). *Eur. J. Org. Chem.* 6301–6310.
54. Stadlbauer, S. 2008. Glycophosphonsäurekonjugate von Carbaboranen—Synthese und biologische Aktivität. Doctoral thesis, Universität Leipzig.
55. Justus, E.; Awad, D.; Hohnholdt, M.; Schaffran, T.; Edwards, K.; Karlsson, G.; Damian, L.; Gabel, D. 2007. Synthesis, liposomal preparation, and *in vitro* toxicity of two novel dodecaborate cluster lipids for boron neutron capture therapy. *Bioconjugate Chem.* 18: 1287–1293.
56. Stadlbauer, S.; Welzel, P.; Beck-Sickinger, A. B.; Gabel, D.; Hey-Hawkins, E. unpublished results.
57. Stadlbauer, S.; Lönnecke, P.; Welzel, P.; Hey-Hawkins, E. 2010. Bis-carbaborane- bridged bis-glycophosphonates as boron-rich delivery agents for BNCT. *Eur. J. Org. Chem.* 3129–3139.

3 Medicinal Application of Carboranes

*Inhibition of HIV Protease**, †

Pavlína Řezáčová, Petr Cígler, Pavel Matějíček, Martin Lepšík, Jana Pokorná, Bohumír Grüner, and Jan Konvalinka

CONTENTS

3.1 INTRODUCTION

The aspartic protease of the human immunodeficiency virus (HIV PR) is one of the most extensively studied enzymes known to man. More than 400 x-ray structures of the protein in the presence or absence of ligands or inhibitors have been determined to date, and the symmetric depictions of this dimeric enzyme have become one of the icons of modern structural biology. The HIV protease is crucial for the production of infectious viral particles (Kohl, 1988), and PR inhibitors (PIs) are potent and specific anti-HIV drugs. The successful rational design of HIV PIs is one of the most striking examples of structure-based drug design. In this chapter, we explain why the design of novel, potent PIs is still urgently needed and that derivatives of cobalt(III) bis(dicarbollide)(1-) ion [(1,2-$C_2B_9H_{11}$)-3,3′-Co(III)](1-), a compound well known to the carborane community (Hawthorne, 1965;

* The authors wish to dedicate this work to the memory of Dr. Jaromír Plešek, an outstanding scientist, distinguished colleague, mentor, and friend.

† This work was supported by the European Commission 6th Framework #QLRT2001-02360 and in part by research projects nos. AV0Z50520514, AV0Z40320502, and AV0Z0550506 awarded by the Academy of Sciences of the Czech Republic, Grant IAAX00320901 from the Grant Agency of the Academy of Sciences of the Czech Republic, and MSM0021620857, LC512, and LC523 from the Ministry of Education of the Czech Republic.

Sivaev and Bregadze, 1999), act as unexpected class of specific HIV PIs. We shall review their activity, specificity, mechanism of action, and binding mode to HIV protease.

3.2 ROLE OF HIV PROTEASE IN THE VIRAL LIFE CYCLE

HIV protease is indispensable for the proteolytic processing of viral polyproteins which takes place in the late stage of the viral life cycle (see Figure 3.1).

The HIV life cycle begins with the attachment of the viral particle to the plasma membrane of the host cell by the interaction of viral glycoproteins with a host cell transmembrane glycoprotein known as CD4 and with a chemokine receptor (either CXCR4 or CCR5, depending on the host cell type). High-affinity binding of the trimeric envelope protein of the virus (SU) to CD4 and to the chemokine receptor brings about a structural change in the envelope protein, allowing the viral transmembrane (TM) protein to penetrate the cell membrane. This causes fusion of the cell and virus membranes, leading to the subsequent entry of the virus capsid into the cytoplasm of the host cell. Next, the viral reverse transcriptase (RT) transcribes viral RNA into complementary DNA, which is transported, along with the viral and cellular proteins, as a preintegration complex into the cell nucleus. Another viral enzyme, the viral integrase, then catalyzes the integration of proviral DNA into the host cell genome. Integrated DNA is later activated by mechanisms that are not entirely understood and transcribed into both spliced and nonspliced RNAs using the host cell transcription machinery and auxiliary viral gene products. Viral RNA is then translated into the large polyproteins Gag, GagPol, and Env. While the Env polyprotein is transported to the Golgi complex and processed by a host cell serine protease into TM and SU, the Gag and GagPol polyproteins are transported to the plasma membrane and processed by the viral protease (PR; Figure 3.2), which is encoded within the GagPol polyprotein, into individual viral enzymes and structural proteins.

The HIV protease is a relatively small enzyme comprised of two identical monomers, each containing 99 aminoacids. It belongs to the class of aspartic proteases with two catalytic aspartates

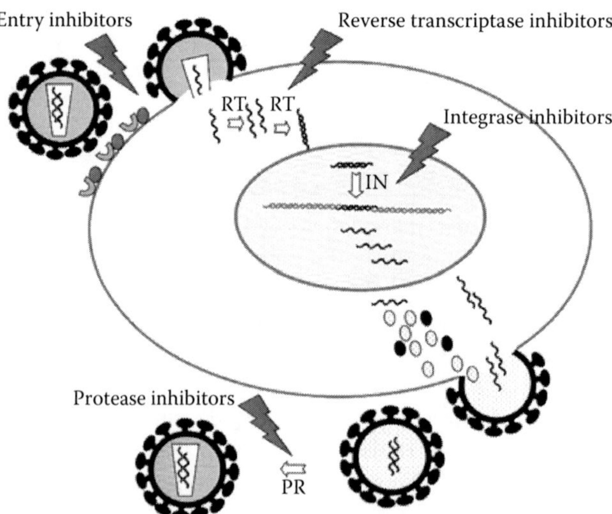

FIGURE 3.1 Replication cycle of HIV. On the left top, a virion attached to the CD4 receptor is shown. After penetration, the viral enzymes reverse transcriptase and integrase produce the double-stranded DNA copy and integrate it into the host chromosome, respectively. The viral genome is transcribed and translated by the host cell apparatus, and viral proteins are delivered to the membrane assembly site in the form of polyprotein precursors. On the bottom right, the noninfectious virion with immature morphology buds off from the cell. Condensation of the viral core and infectivity is associated with protease-mediated cleavage of the polyprotein precursors (bottom). For details, see text. Gray arrows mark targets of clinically used anti-HIV drugs.

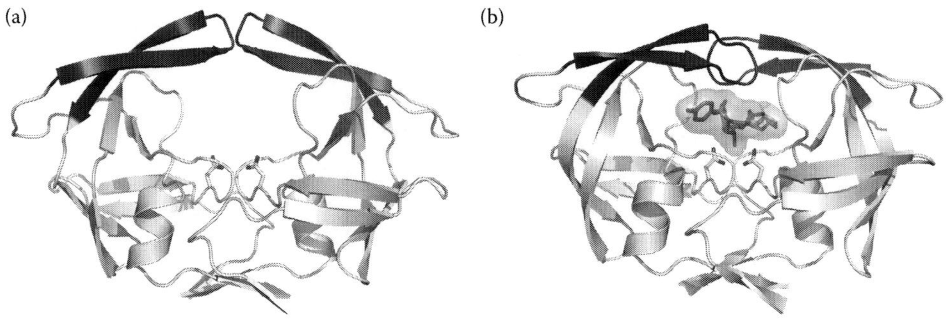

FIGURE 3.2 (See color insert.) Three-dimensional structure of HIV protease (HIV PR) in open and closed conformation. (a) Overall structure of the apo-form of HIV PR. One monomer is colored blue and the second is pink. Two catalytic aspartates are represented by sticks with carbon color corresponding to the chain color and oxygen atoms colored red. Flaps (residues 43–58) in open conformation are highlighted in darker colors. (b) Overall structure of HIV PR in complex with the FDA-approved drug darunavir bound in the active site. Active site-bound darunavir is shown as a stick model (carbon atoms in green, sulfur in yellow, oxygens and nitrogens in red and blue, respectively) with its solvent accessible surface in semitransparent green. The figure was generated with PyMol (DeLano, 2002) using the structure of free HIV-1 PR [PDB code 1HHP (Spinelli, 1991)] and the structure of a highly mutated patient-derived HIV-1 PR [PDB code 3GGU (Saskova, 2009)].

in the active site contributed by each monomer. Dimerization of the protease is mostly mediated by its N- and C- termini (residues 1–4 and 96–99) intertwining into a four-stranded antiparallel β-sheet. Other functionally important parts of the enzyme are two flexible β-hairpins, called flaps, closing over the active site cavity on the substrate or inhibitor binding (*cf.* Figure 3.2a and b).

3.3 CLINICALLY USED HIV PROTEASE INHIBITORS

For each step of the viral life cycle, specific small-molecule inhibitors could be designed that block a particular event and thus halt viral replication (*cf.* Figure 3.1). Since 1981, more than 20 different antiviral drugs have been introduced into clinical practice. More virostatics are now available for HIV than for all other viruses combined. This very successful development has been covered in a number of excellent reviews (Anderson, 2009; De Clercq, 2005, 2009; Gulnik, 2009; Mastrolorenzo, 2007; Prejdova, 2004; Wlodawer, 1998, 2002; Yin, 2006).

Inhibitors of the HIV protease belong among the most effective anti-HIV drugs. If the HIV PR is inactivated either by genetic mutation or chemical inhibition, the virus loses its infectivity, thus stopping the spread of HIV. Specific inhibitors of HIV PR are therefore effective anti-HIV agents. Indeed, the introduction of the HIV protease inhibitors in 1995 as a part of highly active anti-retroviral therapy (HAART) brought about decreased mortality and a prolonged life expectancy for HIV-positive patients. At present, there are nine HIV PIs approved by the US Food and Drug Administration as antiviral agents (De Clercq, 2009; Gulnik, 2009; Pokorná 2009). However, for several reasons there is a continuing need for the development of novel HIV PIs that would be more potent, more specific, and, perhaps most importantly, less prone to the development of viral resistance.

The emergence of HIV drug resistance against PIs is associated with accumulation of numerous mutations (mainly) in the HIV PR. The mutation rate of the HIV is very high due to the lack of proofreading activity of the viral reverse transcriptase, and the rapid viral replication in infected persons lead to a selection of viral species resistant to the virostatics that are currently available. The pattern of mutations associated with viral resistance is very complex, and a proper description of the mechanism of resistance development is outside the scope of this chapter (see Weber, 2009 and the references therein). The mutations are selected not only in the protease substrate-binding cleft, but also outside the active site of the enzyme (Figure 3.3).

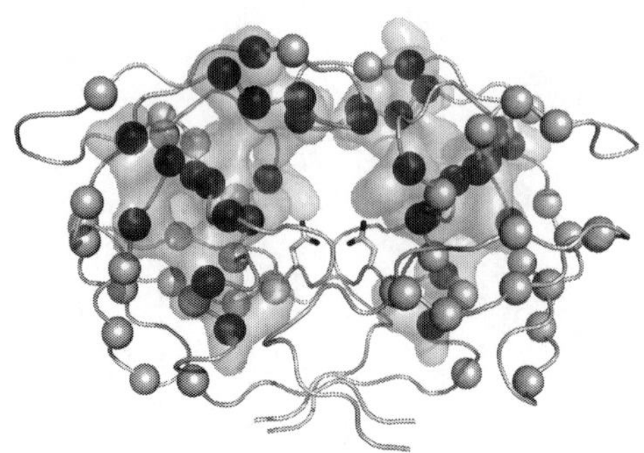

FIGURE 3.3 (See color insert.) HIV PR resistance-associated mutations. The three-dimensional crystal structure of the HIV PR dimer depicting mutations associated with resistance to clinically available protease inhibitors (Johnson, 2008). Mutated residues are represented by their Cα atoms (spheres) and colored in red and yellow for major and minor mutations, respectively. For major mutations in residues affecting substrate and/or inhibitor binding, the transparent solvent accessible surface is also shown in red. Active site aspartates are represented in stick models; the inhibitor bound in the enzyme active site is omitted from the figure for clarity. The figure was generated from the structure of highly mutated patient-derived HIV-1 PR [PDB code 3GGU (Saskova, 2009)] with the program PyMol (DeLano, 2002).

In addition to the resistance development, the use of PIs in the clinics is further affected by their tolerability, toxicity, and adverse effects. The PIs often interfere with lipid metabolism and trafficking pathways. The side effects might decrease the willingness of patients to undergo treatment and thus contribute indirectly to the evolution of resistance. Therefore, the need for development of novel PIs with broad specificity against PI-resistant HIV mutants, better pharmacokinetic properties, lower toxicity, and simple dosage is still very urgent.

Ten protease inhibitors are currently on the market (see Figure 3.4). They all are competitive inhibitors of HIV PR (targeting the enzyme active site) and all but for one are peptidomimetics of the polyprotein cleavage sites. HIV PIs approved for clinical application are often referred to as "first-generation" and "second-generation" drugs.

"First-generation compounds" (Figure 3.4a–d) were designed against the wild-type HIV PR. They typically have poor bioavailability, require coadministration of a cytochrome P450 inhibitor ("booster") to improve the half-life of the drug in the circulation, and often show severe side effects (Barbaro, 2009; Duvivier, 2009; Flint, 2009; Gulnik, 2009; Mallewa, 2008). The "second-generation drugs" (Figure 3.4e–j) were designed to inhibit the HIV PR mutants resistant to the first-generation inhibitors, to minimize the side effects, and to increase the oral availability of the drugs. Lopinavir and darunavir are currently the most widely used compounds of this class. Lopinavir (LPV, ABT-378) was developed by Abbott (marketed in a coformulation with ritonavir under the name of Kaletra) and was designed to inhibit resistant PR species that contain the common mutation V82A. The scissile bond of the parental peptide substrate is replaced with a hydroxyethylene peptidomimetic (Sham, 1998).

Darunavir is the most recently approved HIV PI (DRV, TMC-114, UIC-94017, trade name Prezista) (Koh, 2003). This compound, developed by Tibotec, retains its activity even toward highly mutated resistant PR species. Interestingly, DRV might bind not only to the active site but also to a surface pocket in the flaps of HIV PR, and it could also inhibit dimerization of the PR (Koh, 2007; Kovalevsky, 2006, 2008). A number of mutations in the HIV PR are apparently needed to render the enzyme resistant toward DRV. Saskova et al. (2008) recently characterized a virus strain isolated from an HIV-positive patient under DRV treatment that harbored 22 mutations in the PR region and was highly resistant toward the drug.

The only truly nonpeptidic compound among clinically approved HIV PIs is tipranavir (TPV, PNU-140690, approved under the name Aptivus), which was designed at Pharmacia & Upjohn and

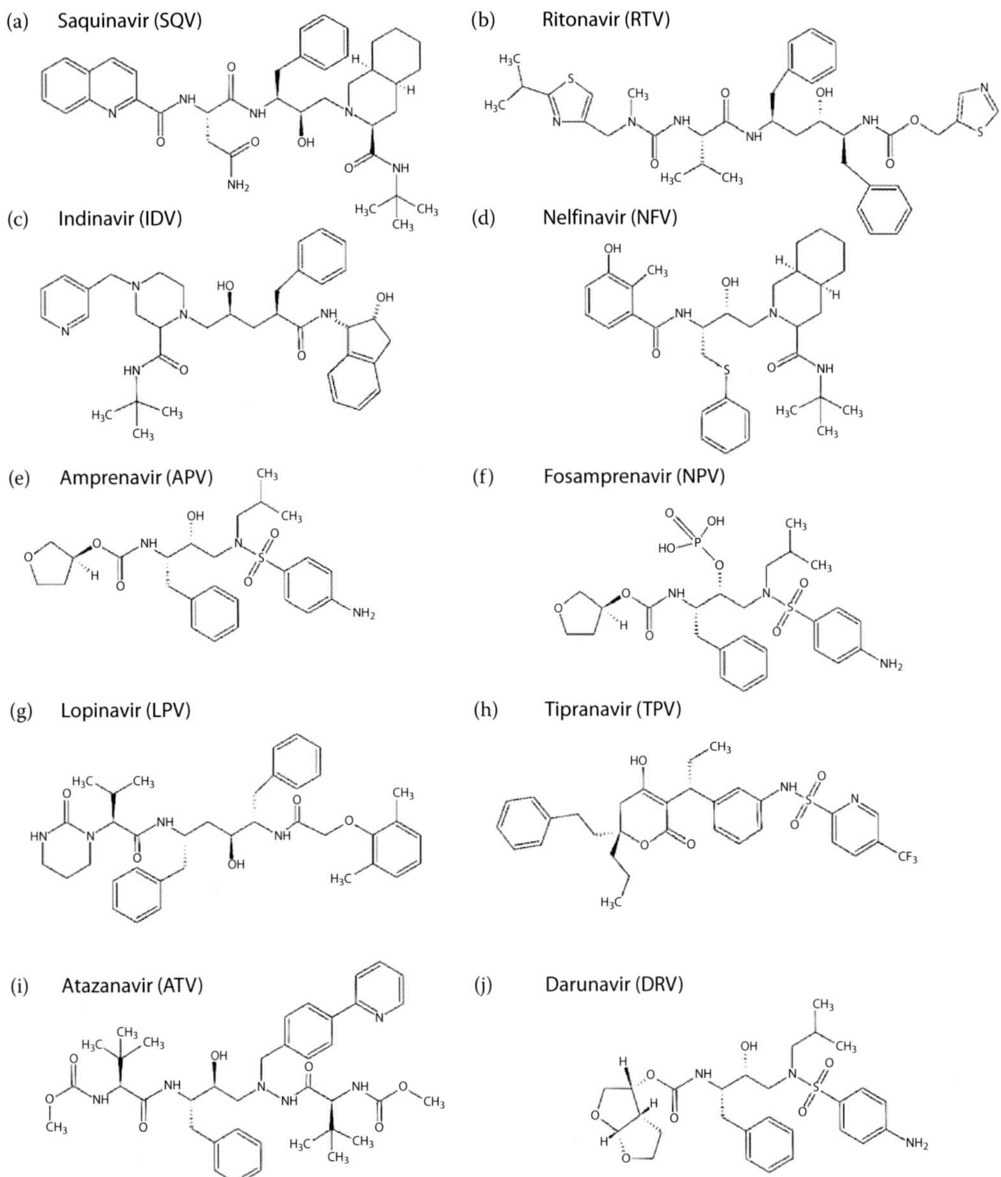

FIGURE 3.4 Chemical structures of the HIV protease inhibitors approved for clinical use. All inhibitors bind in the enzyme active site and mimic its substrate structure and interactions. (a–d) First-generation PIs, (e–j) second-generation PIs.

marketed by Boehringer-Ingelheim. Chemically, it is 4-hydroxy-5,6-dihydro-2-pyrone sulfonamide (Poppe, 1997). Interestingly, its interactions with HIV PR binding cleft resemble those of inhibitors based on peptide bond isosteres (Muzammil, 2007).

3.4 FROM PSEUDOMIMETICS TO NONPEPTIDIC HIV PROTEASE INHIBITORS

An ideal HIV PI should have low molecular weight, the ability to make multiple interactions with the backbone of the PR binding cleft, and maximal flexibility in order to fit properly into the variable binding clefts of resistant PR species. Since peptides and pseudopeptides are usually poor drugs in terms of bioavailability and stability, nonpeptide compounds with the potential to engage in a variety of noncovalent interactions with the PR-binding cleft have been and still are extensively sought after. As early as 1994, Lam et al. (1994) introduced cyclic ureas as a class of PIs. Cyclic urea provides a structural scaffold enabling the design of multiple interactions with the side chains and peptide backbone of the protease. A particular feature of these compounds is their ability to mimic the "flap water," a highly conserved element in HIV PR structures. The flap water has been shown to interact with the main chain atoms of the closed flaps and the substrate and almost all peptidic inhibitors known to date (Erickson-Viitanen, 1994; Lam, 1996).

Very unusual chemistry has sometimes been used for the design of novel HIV PIs. Even inorganic Nb-containing polyoxometalates have been reported to inhibit the HIV PR with submicromolar potency in tissue cultures (Judd, 2001). The inhibitors were reported to be noncompetitive, and modeling suggested that they bind to a pocket on the outer surface of the flaps. C_{60} fullerene was shown to be another rather unusual nonpeptidic inhibitor (Bosi, 2003; Friedman, 1993; Sijbesma, 1993).

3.5 CARBORANE CLUSTERS AS HIV PIs

We searched for other types of unconventional, versatile chemical structures that would fit into the PR-binding cleft, be biologically stable, and would enable facile chemical modification. While screening a number of structural motifs, we identified icosahedral boranes and carboranes, namely 12-vertex metal bis(dicarbollides), as promising frameworks for a novel class of nonpeptide PIs (Cígler, 2005; Kozisek, 2008; Řezáčová, 2009).

Boron-containing compounds in general and carboranes in particular have been extensively studied for their use in boron neutron capture therapy (Hawthorne, 2003; Soloway, 1998) and in radioimaging (Hawthorne, 1999). Carboranes are also used as stable hydrophobic pharmacophores, usually replacing bulky aromatic amino acid side chains (Armstrong, 2007; Lesnikowski, 2007; Sivaev, 2009; Valliant, 2002). These approaches are extensively reviewed by very competent authors in other chapters of this book. In the following sections, we shall concentrate exclusively on metal bis(dicarbollides) as HIV PIs.

Our main attention has been focused on ionic metal bis(dicarbollides) that consist of two dicarbollide subclusters sandwiching the central metal atom. In metal bis(dicarbollides), the equal 11-vertex dicarbollide subclusters are connected by a *commo* metal vertex, forming two 12-vertex metal dicarbollide subclusters. These *closo* 26-electron compounds 12-vertex geometry resembling "a peanut-shape" were described as early as 1965 by Hawthorne (Hawthorne, 1965). Among other transition metal metallacarboranes (Saxena, 1993), the cobalt bis(1,2-dicarbollide) ion (Sivaev, 1999) shows certain unique features: synthetic availability, wide possibilities of *exo*-skeletal modifications, high stability, charge delocalization, low nucleophilicity, strong acidity of conjugated acids, and high hydrophobicity (Plesek, 1992). These properties are reflected in the unique solution properties and ion-pairing behavior of this ion, which in turn led to its known applications in extraction chemistry (Rais, 2004) and in the developments of lowest coordinating anions (Reed, 1998) and compounds for radioimaging (Hawthorne, 1999). However, the metal bis(dicarbollides) have never been considered as biologically active compounds or pharmacophores.

3.6 STRUCTURE–ACTIVITY ANALYSIS OF SELECTED CARBORANES: SINGLE-CLUSTER COMPOUNDS

Table 3.1 summarizes 31 cobaltacarborane compounds tested as inhibitors, inclusive some already known simpler derivatives (Grüner, 1998; Hawthorne, 1968; Plesek, 1978, 2002; Sivaev, 2002), and new compounds, both recently published (Cígler, 2005; Farras, 2008; Grüner, 2005; Kozisek, 2008; Kubat, 2007; Řezáčová, 2009; Uchman, 2010), and unpublished. A panel of ionic species comprising parent 1,2- and 1,7- bis(dicarbollides) and cobalt bis(dicarbollides) substituted with various polar and hydrophobic groups was selected for the initial study. The series consisted mainly of known

TABLE 3.1
Structures and Inhibitory Constants of Metallacarborane Inhibitors of HIV-1 PR

	Compound Description	*In Vitro* Enzyme Assay		
Compound No.	**Structure**[a]	**IC_{50}**	**Mechanism**	**K_i (nM)**
GB-18[1,12,13]	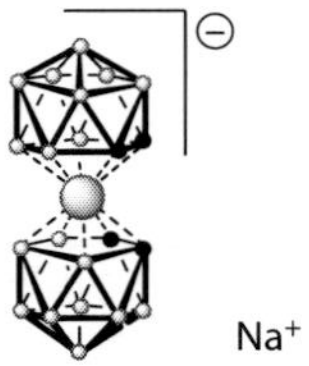 $\ominus$ Na^+	1.1 μM	Competitive	66 ± 30
GB-129[1]	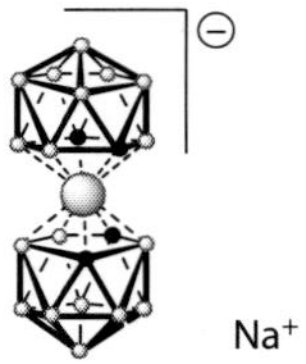 $\ominus$ Na^+	1.4 μM	Non-competitive	N.D.
GB-8[2,12]	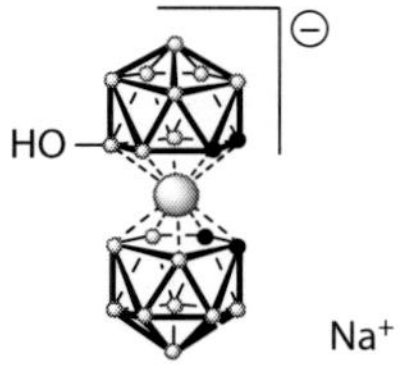 HO– $\ominus$ Na^+	13.5 μM	Competitive	6800 ± 500
GB-189[3]	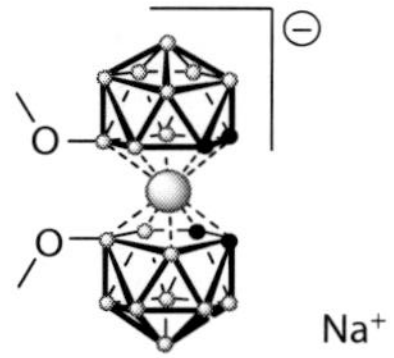 O– O– $\ominus$ Na^+	4.7 μM	Competitive	N.D.

continued

TABLE 3.1 (continued)
Structures and Inhibitory Constants of Metallacarborane Inhibitors of HIV-1 PR

	Compound Description	*In Vitro* Enzyme Assay		
Compound No.	**Structure**[a]	**IC_{50}**	**Mechanism**	**K_i (nM)**
GB-12[4]	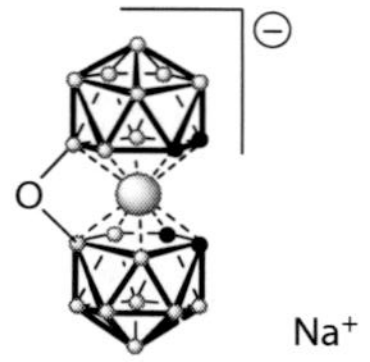 O; ⊖; Na^+	5.2 μM	Non-competitive	N.D.
GB-16[5]	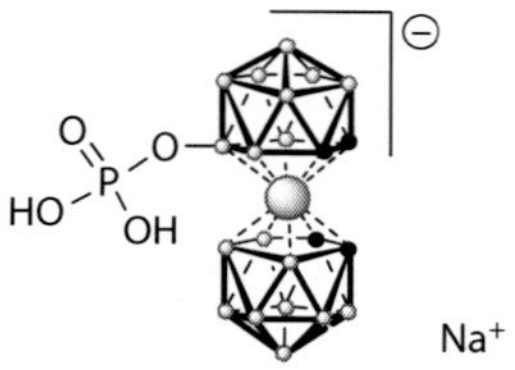 O, O, P, HO, OH; ⊖; Na^+	6.2 μM	Competitive	N.D.
GB-152[1]	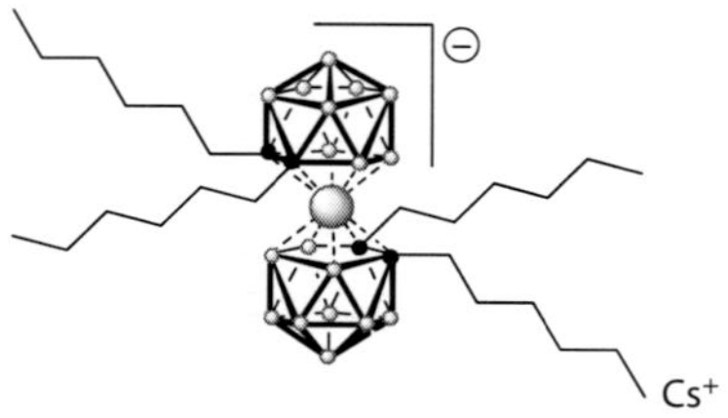 ⊖; Cs^+	160 nM	Non-competitive	N.D.
GB-28[6]	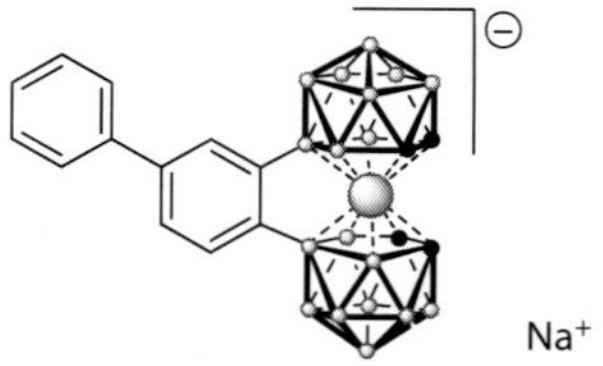 ⊖; Na^+	290 nM	Non-competitive	N.D.

GB-1[7,12]	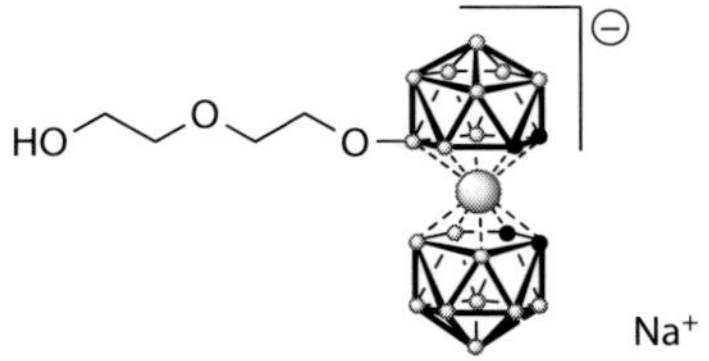 HO, O, O, Na⁺	6.1 μM	Competitive	2500 ± 400
GB-21[12,13]	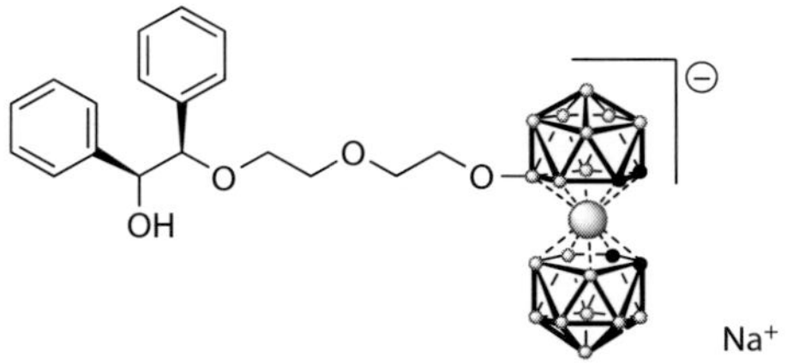 OH, O, O, O, Na⁺	130 nM	Competitive	20 ± 5
GB-46[14]	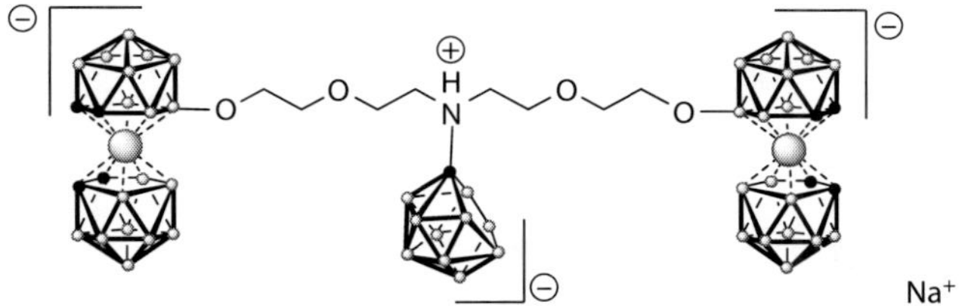 O, O, H, N, O, O, Na⁺	50 nM	Non-competitive	N.D.
GB-179[8,15]	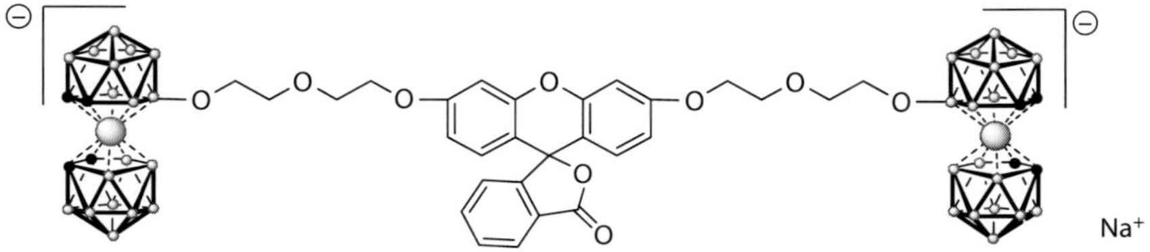 O, O, O, O, O, O, O, O, Na⁺	100 nM	Non-competitive	N.D.

continued

TABLE 3.1 (continued)
Structures and Inhibitory Constants of Metallacarborane Inhibitors of HIV-1 PR

	Compound Description	*In Vitro* Enzyme Assay		
Compound No.	**Structure**[a]	**IC_{50}**	**Mechanism**	**K_i (nM)**
GB-128[9]	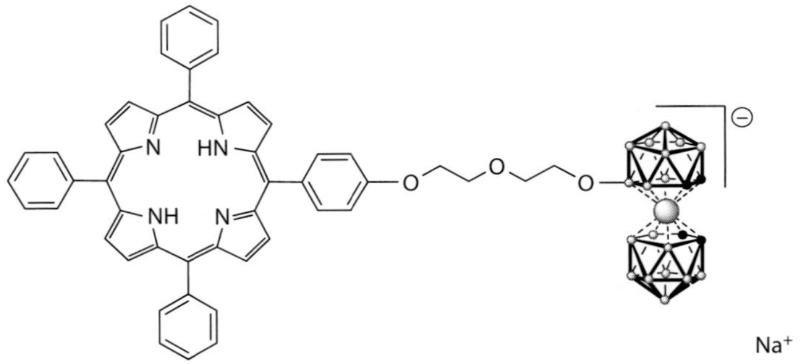	290 nM	Non-competitive	N.D.
GB-101[9,15]	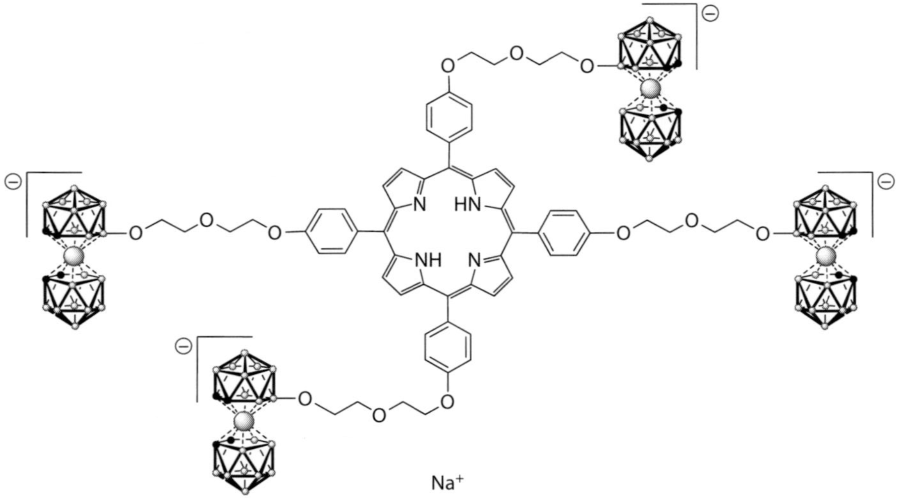	77 nM	Non-competitive	N.D.

GB-75[10]	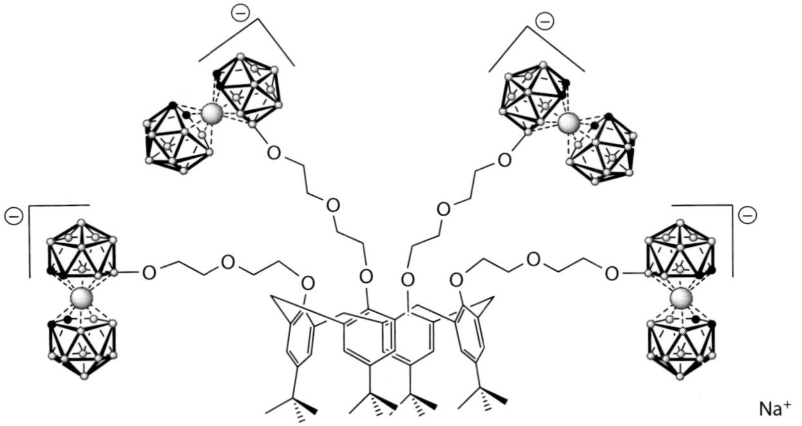	90 nM	Non-competitive	N.D.
GB-80[12,13,14]	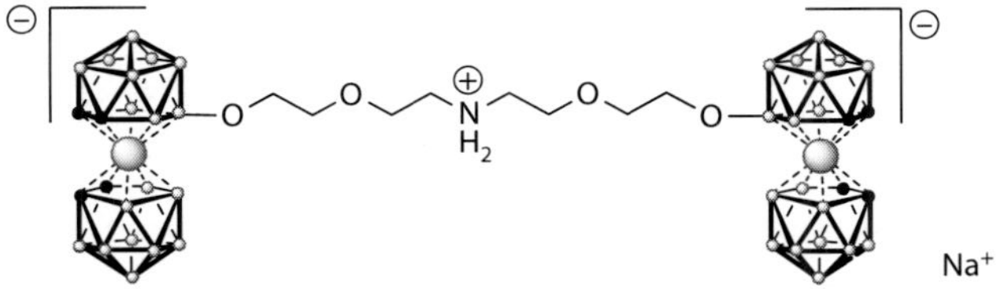	140 nM	Competitive	4.9 ± 2.1
GB-50[14]	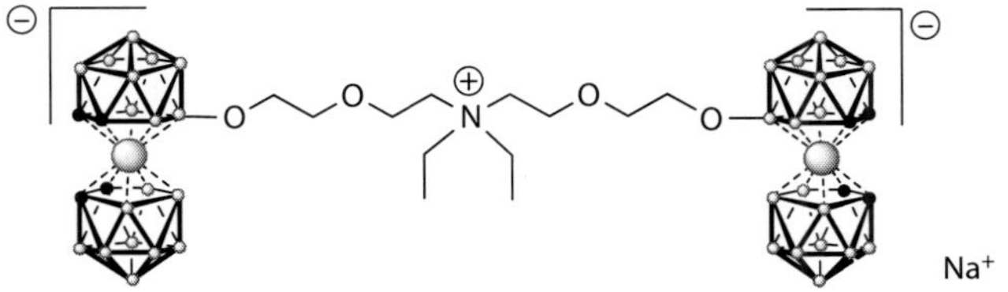	160 nM	Concentration dependent[15]	N.D.

continued

TABLE 3.1 (continued)
Structures and Inhibitory Constants of Metallacarborane Inhibitors of HIV-1 PR

	Compound Description	*In Vitro* Enzyme Assay		
Compound No.	**Structure[a]**	**IC_{50}**	**Mechanism**	**K_i (nM)**
GB-48[12,13,14]	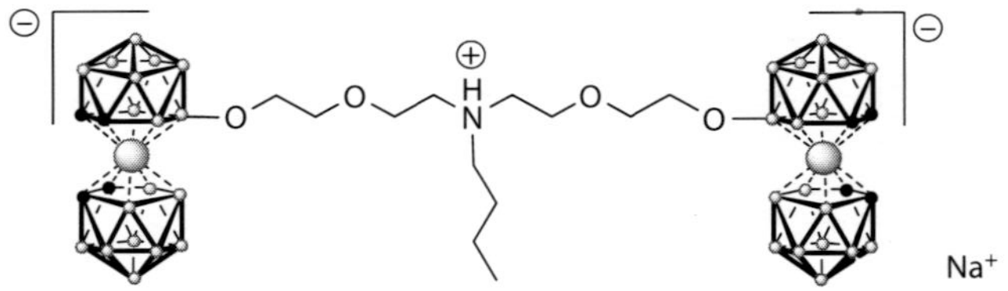 Na^+	100 nM	Competitive	2.2 ± 1.2
GB-162	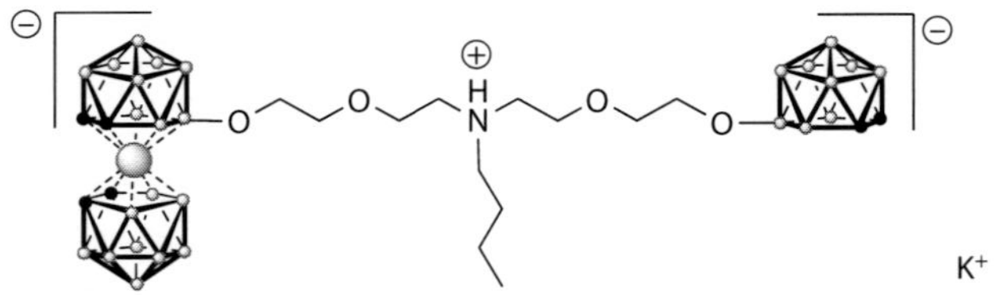 K^+	190 nM	Noncompetitive	N.D.
GB-155[14]	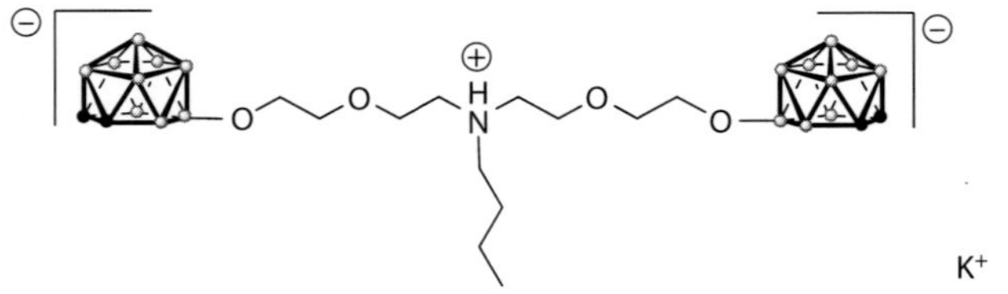 K^+	8.5 μM	N.D.	N.D.
GB-105[14]	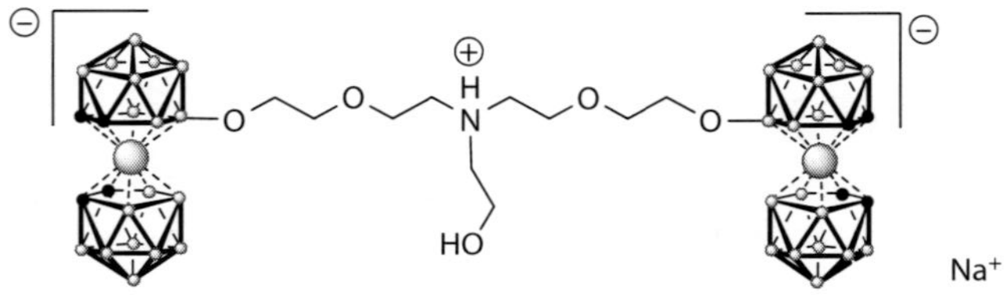 Na^+	140 nM	Competitive	4.7 ± 1.2

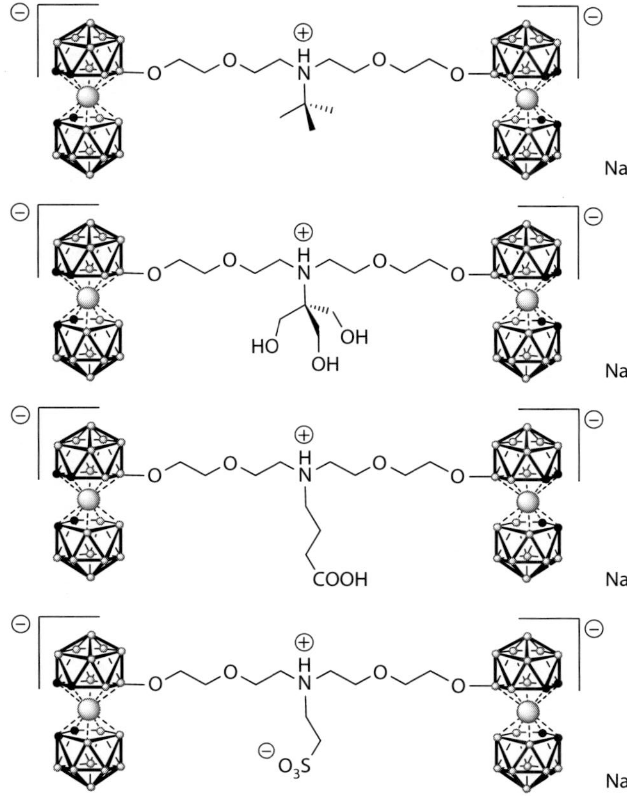

GB-106[14]		130 nM	Competitive	2.7 ± 1.1
GB-120[14]		190 nM	Noncompetitive	N.D.
GB-124[14]		110 nM	Competitive	4.2 ± 1.5
GB-134[14]		110 nM	Noncompetitive	N.D.

continued

TABLE 3.1 (continued)
Structures and Inhibitory Constants of Metallacarborane Inhibitors of HIV-1 PR

	Compound Description	*In Vitro* Enzyme Assay		
Compound No.	**Structure**[a]	**IC_{50}**	**Mechanism**	**K_i (nM)**
GB-104[14]	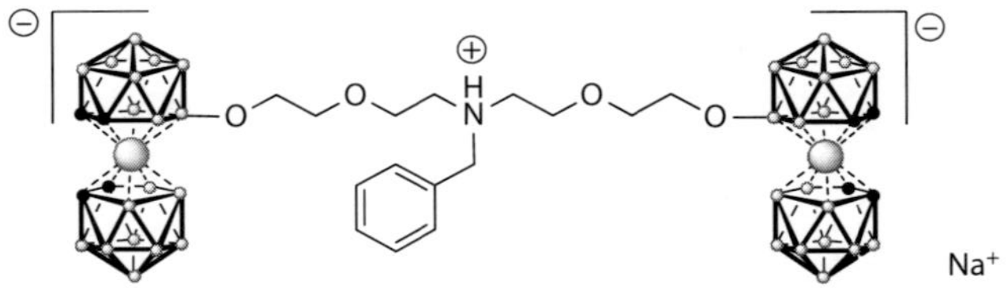	140 nM	Concentration dependent[15]	N.D.
GB-57[14]	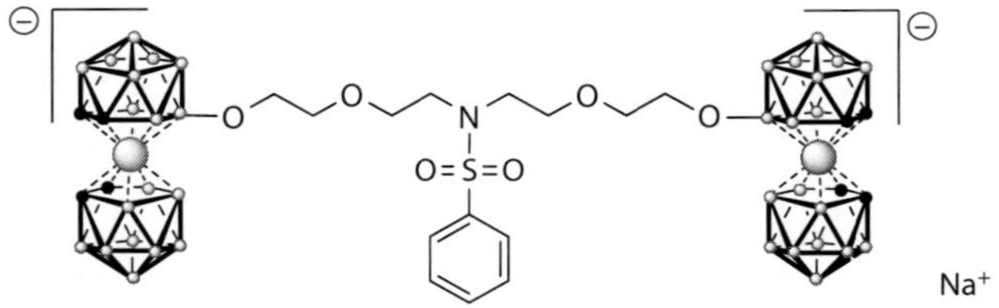	70 nM	Concentration dependent[15]	N.D.
GB-198	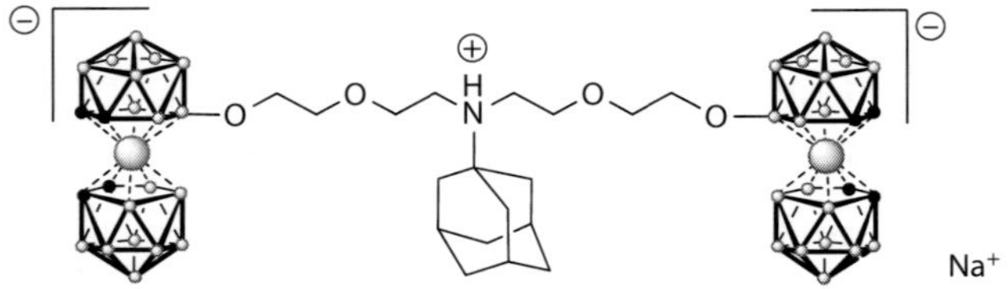	190 nM	Concentration dependent[15]	N.D.
GB-71[14]	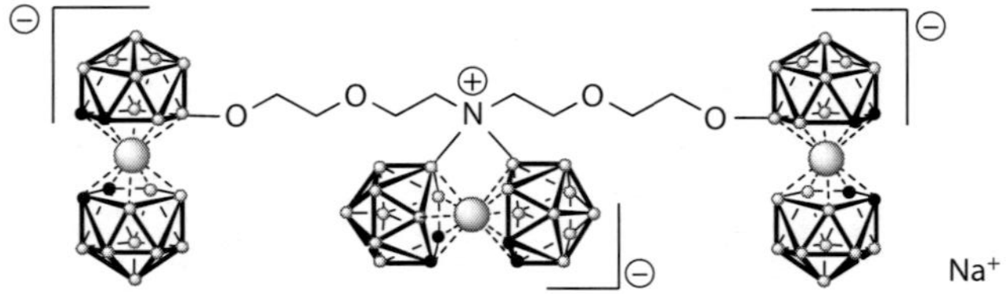	250 nM	Mixed	N.D.

GB-119[14]	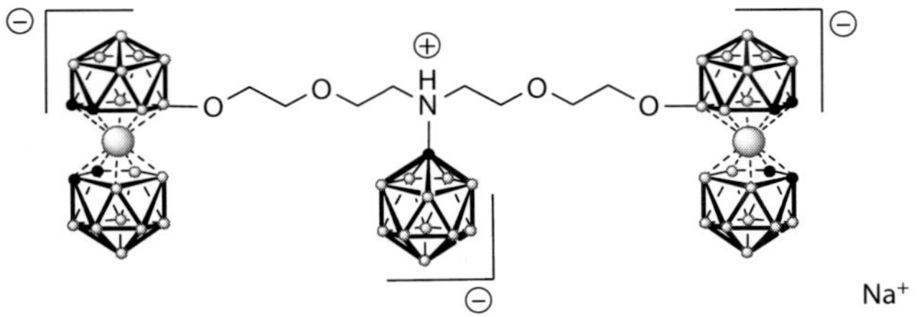 Na^+	58 nM	Concentration dependent[15]	N.D.
GB-79[11]	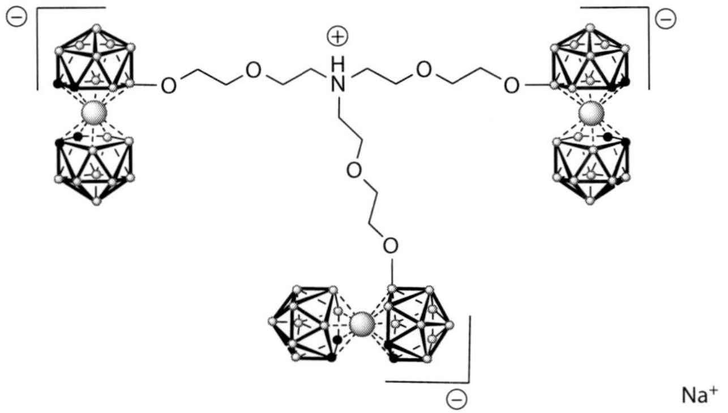 Na^+	130 nM	Concentration dependent[15]	N.D.

[a] Color coding: small gray spheres, BH groups or B (if substituted); black, CH groups or C (if substituted); large gray sphere, Co atom.

The synthesis published in: [1] (Hawthorne, 1968), [2] (Plesek, 2002), [3] (Plesek, submitted), [4] (Plesek, 1978), [5] (Plesek, 2002), [6] (Grüner, 1998), [7] (Sivaev, 2002), [8] (Uchman, 2010a), [9] (Kubat, 2007), [10] (Grüner, 2005), [11] (Farras, 2008).

Anti HIV PR activity published in: [9] (Kubat, 2007), [12] (Cígler, 2005), [13] (Kozisek, 2008), [14] (Řezáčová, 2009).

[15] The inhibition mechanism depends on used inhibitor concentration.

compounds, which were prepared according to established procedures. The aim was to elucidate the effect of the carborane-cluster organization (GB 1 vs. GB 129) and the effects of groups potentially able to form hydrogen bonds [–OH, $-(OCH_2CH_2)_2OH$, $-OP(O)(OH)_2$, and $-(OCH_2CH_2)_2-O-C(C_6H_5)-C(C_6H_5)OH$ in GB-8, GB-1, GB-16, and GB-21, respectively], hydrophobic and steric interactions (alkyl as in GB-152 and GB-156 vs. sterically voluminous aryl or arylelene substituents as in GB-21 and GB-28) on the interactions with the HIV PR-binding cleft.

The compounds were tested for their ability to inhibit cleavage of the peptide substrate KARVNle*NphEANle-NH_2 by recombinant HIV protease. The peptide substrate is modified in the vicinity of the scissile peptide bond in order to follow its hydrolysis spectrophotometrically as previously described (Weber, 2002). Inhibitory concentrations needed to decrease the HIV PR activity to one half of its maximal value (IC_{50} values) were determined from the plot of the initial velocity of the enzymatic reaction versus the inhibitor concentration. The inhibition constants (K_i values) were obtained by fitting the initial velocity data to the competitive tight binding inhibition equation according to Williams and Morrison (Williams, 1979). The mechanisms of inhibition were determined from Lineweaver–Burk plots (Lineweaver, 1934). The characteristic pattern of competitive, noncompetitive, or mixed mode of inhibition of a given inhibitor was distinguished using double reciprocal fits of initial rates versus substrate concentration. Experiments were carried out at three fixed inhibitor concentrations.

The nonmodified parental cages, GB-18 and GB-129 $[1,7-(C_2B_9H_{11})2-2,2'-Co)](1-)$, show comparable IC_{50} values. Interestingly, the *meta*-isomer GB-129 exhibits a noncompetitive mechanism of action, suggesting that it might bind outside the enzyme substrate-binding cleft. Addition of hydrophilic substituents mostly decreased the affinity of the compounds for the enzyme (GB-8, GB-189, GB-12, GB-16, and GB-1). It should be noted that neutral derivatives substituted with various ammonium groups were tested for comparison and all were found to be less efficient than the parent sandwich. The series also contained GB-152, which bears alkyl substituents on the cage carbon atoms. In addition to increasing the number of potential hydrophobic interactions, these substituents prevent the occurrence of interactions of the slightly acidic cobalt bis(dicarbollide) {CH} sites. Indeed, on this modification, the mechanism of inhibition switched from competitive to noncompetitive. However, it is difficult to distinguish whether this effect is caused by the specific modification of the acidic {CH} sites or if it is due to the increased hydrophobicity of the compound. A noncompetitive mode of inhibition might also be explained by the formation of inhibitor aggregates binding outside of the enzyme active site cavity to other functionally important regions of the enzyme. Indeed, the aggregation of cobalt bis(dicarbollides) in aqueous solutions has been described and is discussed in Section 3.11.

3.7 STRUCTURE–ACTIVITY ANALYSIS OF SELECTED METALLACARBORANES: DOUBLE (TRIPLE)-CLUSTER COMPOUNDS

The crystal structure of GB-18 complexed to HIV PR showed that two cobaltacarborane cages are needed for efficient binding into the HIV-1 PR active site cleft (Cígler, 2005). Therefore, we attempted to connect the two parental cages with a hydrophilic linker in order to increase binding affinity to the PR. We used an *N*-substituted bis(ethyleneglycol) amine linker that enables modification of the substitution pattern of the central part. Indeed, connecting the two clusters led to an approximately 14-fold improvement of the K_i value (*cf.* GB-18 and GB-80 in Table 3.1). These double-cluster compounds are easily synthetically accessible. This approach, based on consecutive reactions of different building blocks (exemplified in Scheme 3.1, see also Cígler 2005; Farras, 2008; Řezáčová, 2009), is quite simple, efficient, and gives moderate-to-high yields.

Generally, the oxonium ring in carborane-dioxane **1** is opened by nucleophilic attack of *N*-nucleophile giving rise to a zwitterionic *N*-nucleophile **2**. This intermediate can be reacted again with carborane-dioxanate **1** providing double-cluster compounds **3** (e.g., GB-80). If a boron building block with primary amine function is used as the ring opener, three-cluster molecules can be

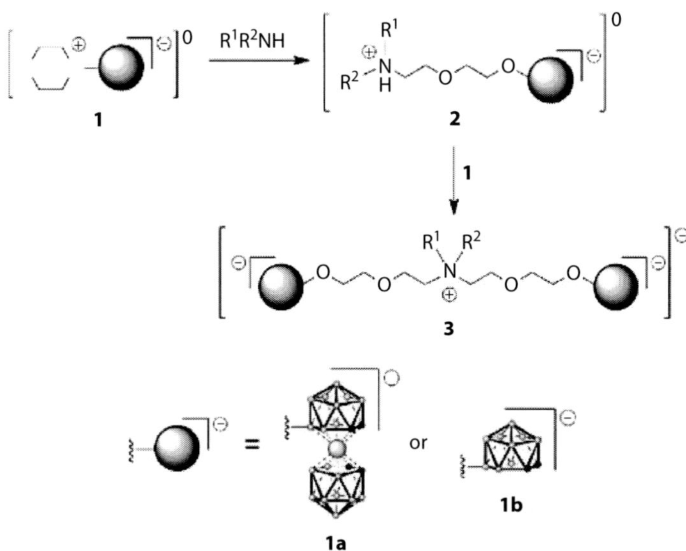

SCHEME 3.1 Synthetic pathway to double- and multicluster inhibitors of HIV protease based on *N*-substituted bis(ethyleneglycol) amine linker. The carborane-dioxanate building block **1** can be either 8-dioxane-cobalt bis(1,2-dicarbollide) **1a** or 11-vertex 10-dioxane-*nido*-7,8-dicarbaundecaborate **1b** (bridge hydrogen atom is not shown here for the clarity).

obtained (GB-46, GB-71, GB-119). Notably, when the *N*-nucleophile is an amine, the resulting compound is zwitterionic–anionic bearing total charge 1–. The protonation of basic amine moiety and formation of zwitterionic structure is caused by proximity of negatively charged carborane cluster. When sulfamide (see GB-57) or amide is used as *N*-nucleophile, the more acidic nitrogen atom cannot be protonated and double-charged anionic molecules are formed. Similarly, *O*-nucleophiles containing two or more *O*-atoms can be used for ring opening by carborane-dioxanate **1** (see GB-179, GB-101, GB-75). The total charge of resulting molecules is then given directly by actual number of negatively charged clusters. Indeed, the presence of other charged groups (like sulfonic acid in GB-134) should be included in calculation of the total charge.

In this manner, symmetric or asymmetric structural motifs are available as shown in Scheme 3.1 (*cf.* GB-48, GB-162, and GB-155). These reactions on cobalt bis(dicarbollide)(1-) cages are based on Plesek's original synthetic concept of ring opening of the B(8) dioxane- cobalt bis(dicarbollide) zwitterion or closely related compounds and have become a very versatile tool for incorporating metallaborane clusters into various molecules and materials. For a recent review of this area, see Semioshkin (2008).

All double-cluster compounds show inhibition potency with IC50 values in the submicromolar range. The only outlier is GB-155, documenting the need for a cobalt bis(dicarbollide) full sandwich for the effective binding to HIV PR. Compound GB-162 with one cobalt bis(dicarbollide) moiety replaced for an 11-vertex dicarbollide ion [7,8-C2B9H12]- exhibits a slightly weaker inhibitory activity but different, noncompetitive inhibition mode. However, a rigorous comparison of the inhibition kinetic constants is possible only for the compounds that exhibit a competitive mode of inhibition.

The inhibition constants (K_i values) were determined only for the competitive inhibitors. These values can be directly compared in order to draw structure–activity conclusions. The comparison of inhibitory activity of GB-48 with that of GB-106 suggests that further improvement of the *in vitro* inhibitory potency could be gained by the substitution of the central secondary amino group of the linker in GB-80 with small hydrophobic groups, for example, the butyl moiety (GB-48) or *tert*-butyl (1,1-dimethylethyl) moiety (GB-106). This position is relatively tolerant of substitution by small derivatives with a polar group (*cf.* e.g., hydroxyethyl in GB-105 or carboxypropyl in GB-124).

Strikingly, the substitution of the central amino group of the linker by other, sometimes rather similar substituents, often leads to a switch of the inhibition mode from competitive to noncompetitive (*cf.* compounds GB-134, GB-104, and GB-57). Some of these substituents involve a bridged cobalt(III) bis(dicarbollide) substituent in GB-71, adamantyl derivative in GB-198, sulfonamide in GB-57, benzyl in GB-104, ethylsulfate in GB-134, and tris(hydroxy)*tert*-butyl moiety in GB-120. One of the most active compounds from this series of noncompetitive inhibitors is GB-46, bearing carbadodecaborane {-7-NH-7-$CB_{10}H_{12}$} residue in the central part of the molecule *N*,*N*-attached symmetrically to the two diethyleneglycol linkers. Another group of structurally divergent noncompetitive inhibitors includes tetraphenylporphyrin derivatives GB-128 and GB-101, and calix[4]arene conjugate GB-75.

It is very difficult to discuss the structure–activity relationships for noncompetitive or mixed-type inhibitors until structural information on the mode of binding for these compounds is available. Mixed and noncompetitive types of inhibition suggest specific binding of compounds outside the enzyme active site. The binding site might involve a functionally important part of the enzyme such as the dimerization domain or the flap region. However, the noncompetitive mode of inhibition might also be the result of aggregation or oligomerization of the inhibitor and binding of such aggregates onto the surface of the enzyme. As we shall discuss below, the behavior of aggregates of cobalt(III) bis(1,2-dicarbollide) compounds in aqueous solutions is fairly complex and depends on numerous factors, including compound concentration. Indeed, for some of the compounds listed in Table 3.1 we observed a concentration-dependent inhibition mode (at high concentrations the inhibition mechanism was competitive, while at low inhibitor concentrations we observed a transition to mixed or noncompetitive inhibition mechanisms). The role of dicarbollide aggregation in the noncompetitive inhibition of HIV PR will be further analyzed. It should be noted that the overall charge of some potent but noncompetitive inhibitors GB-46, GB-57, GB-101, GB-119, GB-134, GB-179 is two minus and for others, for example, GB-75, it is even higher. The charge can undoubtedly contribute to different water solubility and other phenomena, but the possible effect on the inhibition mechanism of the compounds has not yet been fully elucidated.

3.8 ANTIVIRAL ACTIVITY AND INHIBITION OF RESISTANT HIV PR SPECIES

The antiviral potency of selected metallacarborane derivatives was investigated by analyzing their effect on HIV-1 infectivity in HIV-1_{NL43} infected MT-4 T-cell tissue culture (Cígler, 2005). Several compounds (most notably GB-48 and GB-80) show antiviral activity at submicromolar or low micromolar concentrations. However, the limited solubility of the compounds leading to aggregation on dilution in aqueous solution (e.g., tissue culture medium) results in experimental variation in these measurements, especially for the more hydrophobic compounds. Clearly, further investigation of the aggregation properties of cobalt bis(dicarbollides) in aqueous solutions and delivery systems for tissue culture experiments will be necessary to overcome problems with compound solubility and to achieve optimal comparison.

An important feature of the HIV PIs presented in this chapter is their ability to inhibit HIV PR variants originating from resistant viral strains selected under prolonged treatment of patients with numerous PIs. Several HIV PR variants carrying specific signature mutations for various FDA-approved HIV PIs were prepared by site-directed mutagenesis. These variants, as well as highly resistant HIV PR species amplified from HIV-positive patients failing antiretroviral therapy with HIV PIs, were used in the inhibition assays (Kozisek, 2008; Řezáčová, 2009). Cobaltacarboranes with a competitive mode of action (most notably compounds GB-48, GB-80, GB-105, and GB-106) show a low relative loss of activity as indicated by their low vitality values for all tested HIV PR variants and thus prove to be potent inhibitors of the mutated HIV-1 PRs (Figure 3.5). The ability of these compounds to inhibit a structurally diverse set of HIV PR mutants is clearly caused by their unusual binding mode, specifically by a certain degree of freedom in accommodating the cobaltacarborane clusters within the enzyme.

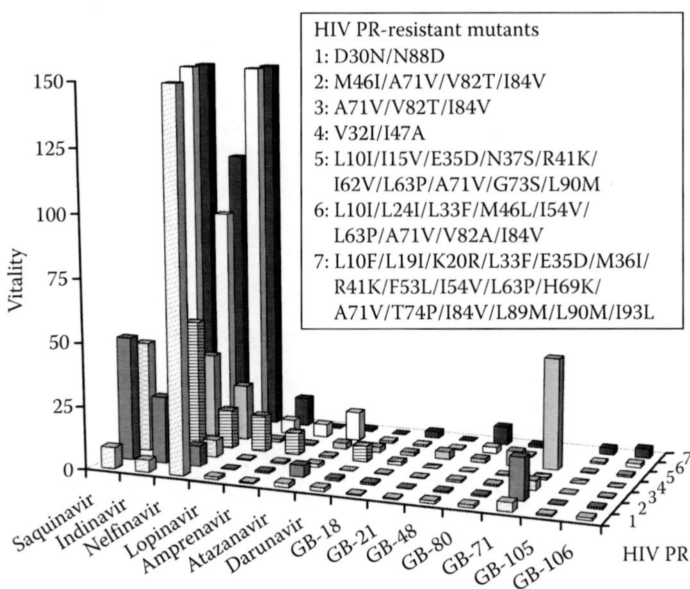

FIGURE 3.5 **(See color insert.)** Metallacarboranes inhibit resistant HIV PR variants. Vitality values of seven clinical inhibitors and seven cobaltacarborane compounds analyzed with the panel of HIV-1 PR-resistant species (Kozisek, 2008; Řezáčová, 2009). Mutations in the HIV-1 PR variants are shown in the figure inset. The vitality is a measure of the enzymatic fitness of a particular mutant in the presence of a given inhibitor and is defined as $(K_i\, k_{cat}/K_m)_{MUT}/(K_i\, k_{cat}/K_m)_{WT}$, where K_i is inhibition constant, K_{cat} and K_m are enzymatic constants and MUT and WT is mutated and wild-type enzyme variant, respectively (Gulnik, 1995).

3.9 STRUCTURAL STUDIES OF HIV PR-METALLACARBORANE BINDING

The binding mode of metallacarboranes in the HIV PR binding site and the details of their interactions were revealed by crystal structures of enzyme–inhibitor complexes. Two crystal structures of HIV PR with metallacarborane compounds are available to date (Figure 3.6a–c). The crystal structure of wild-type HIV PR in complex with the parent cobalt bis(1,2-dicarbollide) ion (GB-18) was determined at 2.15 Å resolution with an R factor of 17.6% and R_{free} of 23.6% and deposited to protein data bank (PDB) under the accession code 1ZTZ (Cígler, 2005). The crystal structure of the wild-type HIV-1 PR in complex with GB-80 ($[H_2N\text{-}(8\text{-}(C_2H_4O)_2\text{-}1,2\text{-}C_2B_9H_{10})(1',2'\text{-}C_2B_9H_{11})\text{-}3,3'\text{-}Co)_2]$ Na), which has two cobalt bis(1,2-dicarbollide) clusters connected with a linker, was determined at 1.7 Å resolution with an R factor of 17.6% and R_{free} of 21.2% and can be accessed under the PDB code 3I8W (Řezáčová, 2009).

Both compounds for which the crystal structures are available exhibit a competitive inhibition mechanism, and thus, the binding of these compounds into the enzyme active site cavity was expected. The crystal structures, however, revealed quite a unique binding mode. Two cobalt bis(1,2-dicarbollide) clusters occupy the flap-proximal part of the enzyme active site and do not interact with the catalytic aspartates (Figure 3.6a). The inhibitor binding sites are two symmetrical hydrophobic pockets formed by the side chains of HIV PR residues Pro-81, Ile-84, and Val-82 and covered by the flap residues Ile-47, Gly-48, and Ile-54. These pockets approximately correspond to the S3 and S3′ substrate-binding subsites (Figure 3.6b). On average, over 80% of contacts of the cobalt bis(1,2-dicarbollide) cluster with HIV PR are made with nonpolar atoms, and on binding, the clusters lose around 90% of their total solvent-accessible surface area.

Strikingly, with the metallacarborane inhibitors bound, the overall structure of the PR resembles the structure of the apoenzyme with open-flap conformation (PDB ID code 1HHP, Spinelli, 1991) rather than the closed conformation typical for almost all active site inhibitors (Figure 3.7). The

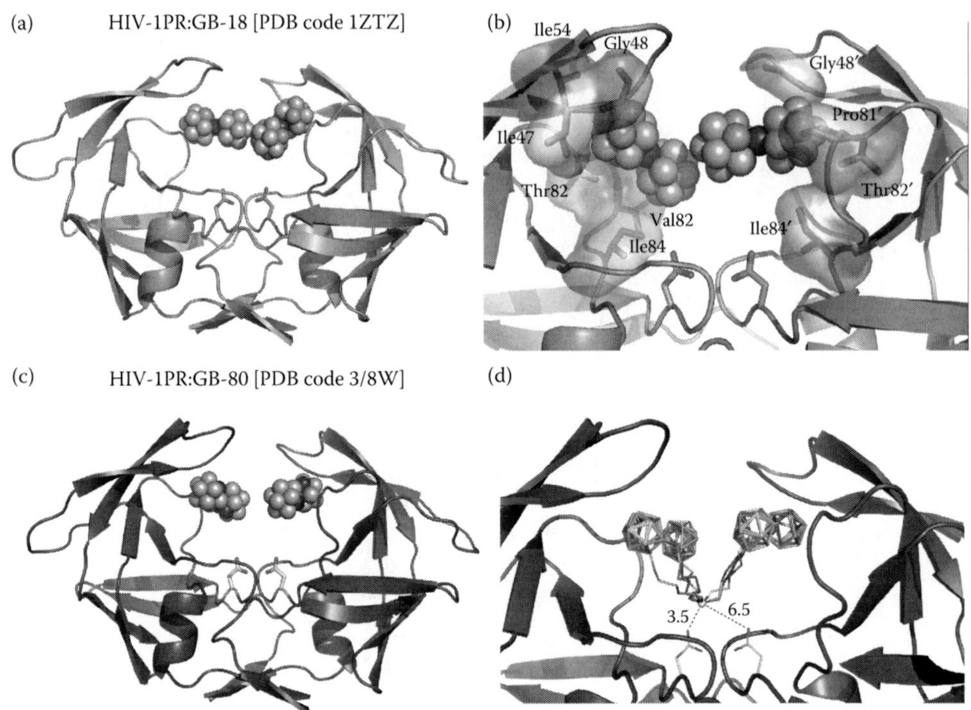

FIGURE 3.6 (See color insert.) Metallacarborane binding to HIV PR. Crystal structures of wild-type HIV-1 PR in complex with the parent cobalt bis(1,2-dicarbollide) ion (GB-18) and compound GB-80 containing two parent clusters connected with a linker. (a) Overall structure depicting the asymmetric binding of two cobalt bis(1,2-dicarbollide) clusters of GB-18 into the symmetric HIV PR dimer. The enzyme is shown by means of animation with two catalytic aspartates shown in sticks. The metallacarborane atoms are depicted as spheres. (b) Detail of the GB-18 inhibitor:enzyme binding. The cluster occupies a hydrophobic pocket formed by enzyme residues Pro81, Val82, and Ile84 and is covered by the flap residues Ile47, Gly48, and Ile54. This site corresponds approximately to the S3 and S3′ substrate-binding subsites. Interacting residues are highlighted in sticks with their van der Waals surfaces. (c) Overall structure depicting the symmetric binding of two cobalt bis(1,2-dicarbollide) clusters of GB-80. Representation corresponds to panel A. (d) Detail of the GB-80:enzyme interaction. Models of the four most probable conformers of the linker are shown. The dashed lines and numbers represent distances to catalytic aspartates in Å. The figure was generated using the program PyMol (DeLano, 2002).

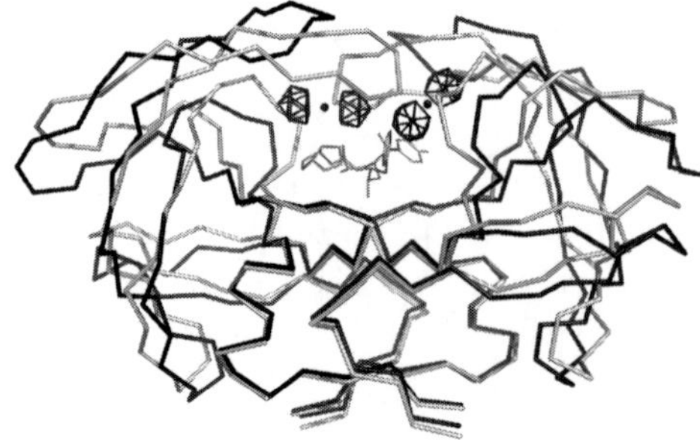

FIGURE 3.7 (See color insert.) Comparison of metallacarborane and darunavir binding to HIV PR. Superposition of HIV PR–compound GB-18 complex with HIV PR–darunavir complex structure. Protease complex with darunavir [PDB code 3GGU (Saskova, 2009)] is represented in orange with darunavir shown as a stick model, and color coding for PR–GB-18 complex is the same as in Figure 3.6. The figure was generated using the program PyMol (DeLano, 2002).

inhibitor binding prevents the flaps from closing over the substrate-binding site, and the flaps are apparently held in the semiopen conformation by the binding of the inhibitor molecules to the flap-proximal part of the substrate-binding site. As the conformation and mobility of the flap in HIV PR is functionally very important, this finding might suggest that these compounds act by locking the flap closure in addition to filling the specific bonding pockets in the active site cleft. Interestingly, the metallacarborane clusters do not make any contacts with the catalytic aspartates, and the metallacarborane binding sites lie "above" the canonical binding site for peptide substrates and/or inhibitors (Figure 3.7).

The comparison of the two available crystal structures revealed some structural differences pointing out to a certain degree of freedom in the metallacarborane cluster binding into the enzyme pockets. The inhibitor binding sites in both monomers of the HIV PR remain essentially the same for the binding of GB-18 and GB-80; nevertheless, there are some differences in metallacarborane cluster position resulting in the protein–inhibitor interactions. While the HIV PR complex structure with parent cobalt bis(dicarbollide) ion (GB-18) shows an asymmetric binding of the two cobalt bis(1,2-dicarbollide) clusters into two symmetrical binding sites of in the enzyme dimer, in the HIV PR:GB-80 complex, the two cobalt bis(1,2-dicarbollide) clusters occupy their binding sites symmetrically (Figure 3.8). The symmetry in the binding of the two clusters of GB-80 into the

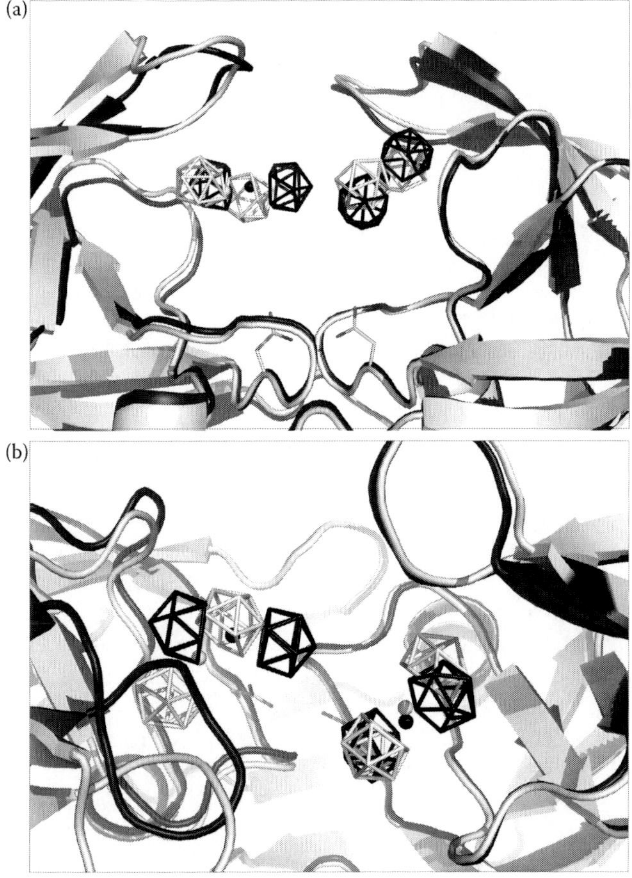

FIGURE 3.8 (See color insert.) Comparison of the GB-18 and GB-80:HIV-1 PR complex structures. (a) Close-up top view into the enzyme active site showing differences in the HIV PR:GB-18 (green) and HIV PR:GB-80 (gray) structures. Carbon and boron atoms in cobaltacarboranes are shown as sticks with the central cobalt atom represented by a sphere. (b) Top view into the enzyme active. Representation and color coding are analogous to panel A. The figure was generated using the program PyMol (DeLano, 2002).

symmetric protease dimer is also reflected in a higher symmetry of the HIV PR:GB-80 crystals (Řezáčová, 2009).

The difference in position of the clusters of GB-80 and GB-18 can be illustrated by a shift of the central cobalt atoms in GB-80 by 1.5 and 1.3 Å compared to the corresponding atoms of GB-18 clusters. Also, the GB-80 clusters are rotated by 37° and 44° with respect to the two clusters of cobalt bis(1,2-dicarbollide) ion GB-18 (Figure 3.8b). As a result of these cluster position rearrangements, the extent of interactions with individual protein residues differs. Compound GB-80 forms more interactions with flap residues (Ile 47, Gly 48, and Ile 54) than with the bottom of the binding site, while the GB-18 clusters interact more with residues Pro-81, Ile-84, and Val-82. This freedom in the position of metallacarborane clusters within the HIV PR binding site was also confirmed by molecular modeling studies and could be a benefit for the efficient binding to the enzyme variants resistant toward clinical protease inhibitors (Kozisek, 2008). Such mutant PR variants often contain mutations of some residues belonging to the GB-18 and GB-80 binding site; however, as our inhibition studies indicate, these mutations do not greatly affect the inhibition efficiency of metallacarborane inhibitors (*cf.* Figure 3.5; Kozisek, 2008; Řezáčová, 2009).

While in the GB-80 complex structure, the cobalt bis(1,2-dicarbollide) cluster could be modeled into well-defined electron density, no continuous map for the linker connecting the two cobalt bis(1,2-dicarbollide) clusters was observed and thus this part is missing in the crystal structure coordinates. The lack of a well-defined electron density map reflects the inherent flexibility of the linker and the presence of several alternative linker conformations in the crystal. As the knowledge of the linker conformation is important for the rational drug design, we employed the molecular dynamics/quenching (MD/Q) and QM/MM computational procedure to explore the conformational space of the linker (Řezáčová, 2009). Based on these calculations, four lowest-energy conformers within the range of 3 kcal mol^{-1} were suggested to be copopulated in the complex (Figure 3.6d).

3.10 NONCOVALENT INTERACTIONS OF HETEROBORANES WITH BIOMOLECULES: THEORETICAL CONSIDERATIONS

The crystal structure of the HIV PR/metallacarborane complex elucidated the binding of those unusual compounds to proteins. To comprehend the nature and properties of interactions of heteroboranes with aminoacids and proteins, methods of molecular modeling were employed.

The inorganic deltahedral boranes (boron hydrides) and the heteroboranes possess unique surface characteristics which predetermine the type of their noncovalent interactions. Due to the electropositivity of boron, the exo-skeletal boron-bound hydrogen atoms are hydridic, bearing a slightly negative charge. This feature is enhanced in the anionic species, as the surplus electrons are delocalized over the whole cage, an effect known as 3D-aromaticity (Chen, 2005; King, 2001). Further variability is brought in by less electropositive heteroatoms, such as carbon in carboranes, since the hydrogen atoms attached to them are slightly positively charged.

The presence of hydridic hydrogens renders boranes capable of forming a special type of noncovalent H···H interaction called a dihydrogen bond (DHB) (Figure 3.9a). DHBs generally occur between the positively charged hydrogen of a proton donor *A*–H (*A* = N, O, S, C, halogen) and the σ-bond of an *M*-H proton acceptor (*M* = electropositive atom, such as B, alkali, or transition metal) (reviewed in Belkova, 2005; Crabtree, 1996; Custelcean, 2001). In the realm of heteroborane interactions, spectroscopic and structural experimental evidence has been gathered for C–H···H–B (Glukhov, 2005), N–H···H–B (Klooster, 1999; Richardson, 1995), O–H···H–B (Epstein, 1998), and S–H···H–B (Planas, 2005) DHB types. The DHBs are characterized by short hydrogen–hydrogen contacts in the range of 1.7–2.2 Å, less than the sum of hydrogen van der Waals radii of 2.4 Å (Crabtree, 1996). The strength of DHBs is comparable to that of classical hydrogen bonds; the IR and NMR spectroscopy-derived formation enthalpies ($-\Delta H$) range from 2.5 to 3.3 kcal mol^{-1} for neutral boranes and 4.0–7.6 kcal mol^{-1} for anionic boron hydrides (Epstein, 1998), while the theoretically calculated stabilization energies span the values of 6.1–7.6 kcal mol^{-1} (Custelcean, 2001).

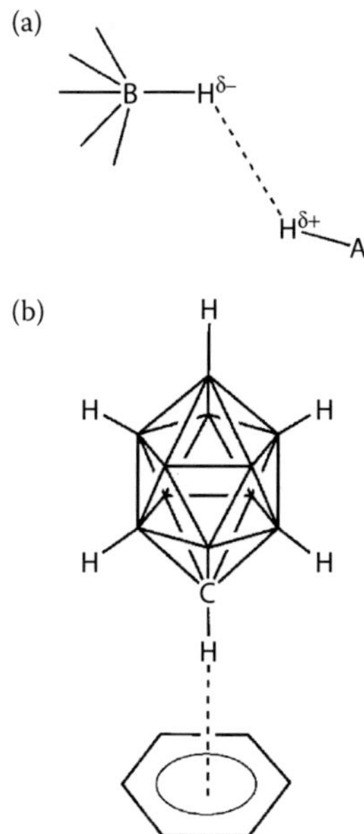

FIGURE 3.9 Types of noncovalent interactions between carboranes and biomolecules. (a) Schematic drawing of a B–H···H–A dihydrogen bond between a borane molecule and a biomolecule. (b) C–H···π type of interaction between carboranes and aromatic systems. The vertices of icosahedron represent BH groups or boron atoms, if H is attached.

A less common type of noncovalent interaction of heteroboranes occurs with aromatic systems. The structural motifs can be classified as either $B_2H\cdots\pi$ type, as for example, in the case of the bridging hydrogen of n-$B_{18}H_{22}$ (Hamilton, 2006), or C–H···π type (Figure 3.9b), which were observed in noncovalent complexes of icosahedral carboranes (Raston, 2004). Both these interaction types are dispersion driven with stabilization energy of 3.6–5.0 kcal mol^{-1} (Li, 2007; Sedlak, 2010) and should therefore be considered as weak hydrogen bonds (Li, 2007). With the growing use of substituted boron clusters in medicinal chemistry (Lesnikowski, 2007; Sivaev, 2009; Valliant, 2002), it is becoming increasingly important to study the noncovalent interactions of heteroboranes with biomolecules. HIV PR could be a useful model system for studies on the structure and energetics of heteroborane:biomolecule complexes by a combination of experimental and theoretical techniques.

We chose the monocarborane anion $CB_{11}H_{12}^-$ and amino acids (Figure 3.10) as a model system representing contacts between GB-18 and hydrophobic PR residues. Using quantum chemical (QM) methods, we determined that DHBs stabilize these complexes by 4.2–5.8 kcal mol^{-1} per interaction (Fanfrlik, 2006). In other model complexes where substituted metallacarboranes bound a peptide, we found that by increasing the total negative charge of the cage due to various metal vertices as well as by introducing *exo*-skeletal substitutions, the stabilization energies of the complexes increased; this was, however, counteracted by increased solvation energies (Fanfrlik, 2007).

The next step toward a more realistic theoretical description of the HIV PR/GB-18 complex was to take into account the whole protein surroundings of the metallacarborane inhibitor molecule. Using a hybrid quantum mechanics/molecular mechanics (QM/MM) treatment, we were able to study the structures and interactions of GB-18 molecules within the active site of HIV PR in the QM part, while including the two HIV PR dimers in the MM region (Fanfrlik, 2008). In conjunction with a detailed x-ray structure analysis, the calculations suggest that the "structural water molecules" Wat50 and Wat128 in the active site were in fact sodium cations that acted as counterions to the closely positioned anionic inhibitors (Fanfrlik, 2008). Such a complex bears resemblance to the Na^+-bridged aggregates of 2–5 molecules of GB-18 which are formed in solution, as observed by SAXS and cryo-TEM techniques (see below).

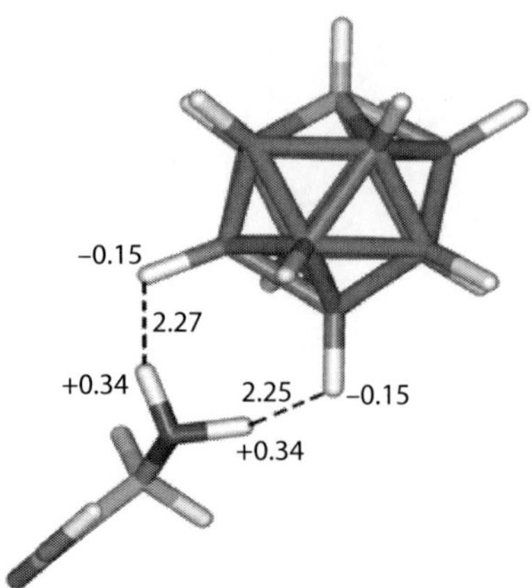

FIGURE 3.10 (See color insert.) Interaction between $CB_{11}H_{12}^-$ with glycine via two dihydrogen bonds. Partial charges on the interacting hydrogens as well as distances in Å are shown.

3.11 AGGREGATION OF METALLACARBORANES IN AQUEOUS MEDIA

For the potential biomedical use of metallacarboranes as HIV inhibitors, their properties and behavior in aqueous media is an important problem to analyze. Especially, it is crucial to understand their self-association and aggregation properties. Each anion of cobalt bis(dicarbollide) consists of two *nido* clusters, the surface of which is composed of nine hydridic hydrogen atoms that cannot form classical hydrogen bonds and two positively charged hydrogen atoms attached to carbon atoms. Since the overall negative charge is delocalized, cobalt bis(dicarbollides) exhibit superacidity and their salts are fully dissociated in water. All these attributes lead to a peculiar behavior of cobalt bis(dicarbollides) in aqueous media: the presence of delocalized negative charge and free-moving counterions manifests itself in a distinct amphiphilic character despite the lack of amphiphilic topology.

The amphiphilicity is closely related to the surface activity. An accumulation of cobalt bis(dicarbollide) anions at the "oil/water" interface was studied by Wipff et al. using a molecular dynamics approach (Chaumont, 2004; Chevrot, 2006; Chevrot et al., 2007a,b). Experimental studies on the surface tension measurements of cobalt bis(dicarbollide) were published by Borisova (Borisova, 2004; Popov, 2001) and van Mau et al. (1984). It is a well-established fact that some surface-active compounds like alkylsulfonates can form multimolecular aggregates, micelles, in aqueous solutions. The question is, whether the water-soluble salts of cobalt bis(dicarbollide) could also associate in aqueous media. Somewhat surprisingly, the detailed analysis of the self-assembly of cobaltacarboranes in aqueous solutions was only provided recently (Matějíček, 2006), 40 years after the discovery of cobalt bis(dicarbollides) by Hawthorne (1965, 1971).

In aqueous solutions of sodium cobalt bis(1,2-dicarbollide), GB-18, large aggregates with radii of approximately 100 nm were observed by light-scattering and microscopy techniques (Matějíček, 2006). The important fact is that the radius of the nanoparticles substantially varies with concentration, ionic strength, and the aging of solution. Nevertheless, an inner structure of such huge nanoparticles, the dimensions of which exceed the size of cobaltacarborane molecules by several orders of magnitude, is still unexplored. Preliminary small-angle x-ray scattering (SAXS) results suggest that compact subunits consisting of few molecules (2–5) could form these large and loose aggregates, which are probably swelled substantially by water (Matějíček, in preparation). The cobalt bis(1,2-dicarbollide) compact sub-units visualized by the cryo-TEM technique as dark stains are shown in Figure 3.11a.

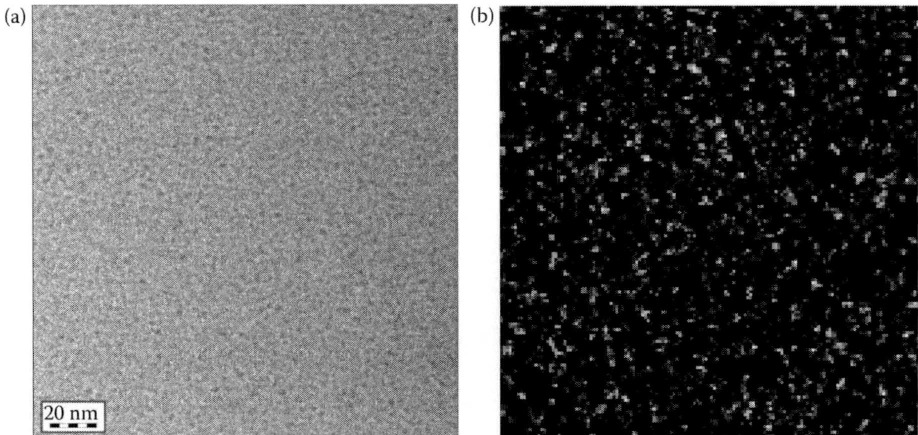

FIGURE 3.11 **(See color insert.)** Aggregation of cobalt dicarbollides in aqueous solutions. (a) Typical cryo-TEM micrograph of sodium cobalt bis(dicarbollide) GB-18 in aqueous solution. (J. Ruokolainen, Helsinki, unpublished work.)(b) Typical fluorescence lifetime imaging (FLIM) picture of fluorescein-cobalt bis(dicarbollide) conjugate (GB-179) aggregates deposited on the bottom of an aqueous solution in which the diameter of the aggregates is roughly 400 nm. (P. Jurkiewicz, Prague, unpublished work; Uchman, 2010a.)

The fact that the metallacarborane aggregates can rather be interpreted as temporal fluctuations of small subunits is partly supported by quantum mechanics calculations, which suggest that a mutual attraction of metallacarborane anions in the implicit solvent model is relatively weak and it only slightly beats a contribution of electrostatic repulsion (Matějíček, 2009).

We paid detailed attention to the self-assembly of nonsubstituted cobalt bis(dicarbollides), because it has a substantial impact on the biochemical behavior of metallacarboranes, such as inhibition of the HIV protease in various assay systems. Exoskeletal substitution of metallacarboranes leads to a complex aggregation behavior in aqueous solutions. Although the measure of hydrophobicity of boron cluster conjugates (e.g., the water–octanol partition coefficient) correlates fairly well with the association tendency in the solution (Matějíček, 2008), a capability to aggregate is often controlled by electrostatic or other specific interactions between the linker, metallacarborane anion, and its counterion. For example, a presence of highly charged pending groups like ATP attached to cobalt bis(dicarbollide) decreases its tendency to aggregate (Matějíček, 2008). On the contrary, any amino-functionalities, which can be protonated, form sparingly soluble zwitterions with metallacarboranes and the aggregation is pronounced in these cases (Hao, 2007; Kubat, 2007; Matějíček, 2008). We also extensively studied two types of fluorescein-metallacarborane conjugates, such as GB-179 listed in Table 3.1, bearing different overall charge (Uchman, 2010a), where the presence of the fluorescent probe allows the visualization of large aggregates using a fluorescence imaging as shown in Figure 3.11b.

Our findings revealed that cobalt bis(dicarbollides) can form large and ill-defined aggregates in water almost regardless of the nature of exoskeletal substituents. This behavior is undesired for the potential medicinal applications of metallacarboranes. To increase solubility of the sparingly soluble metallacarborane conjugates and prevent the aggregation, extensive studies on the interaction of cobalt bis(dicarbollides) with various biocompatible amphiphilic systems are strongly needed. In recent years, several papers dealing with the interaction of cobalt bis(dicarbollides) with cyclodextrins (Chetcuti, 1991; Rak, 2008, 2010; Uchman, 2010a), phospholipid bilayers (Amblard, 1983; Atwell, 1986; Uchman, 2010a; van Mau, 1984), surfactants (Matějíček, 2008; Rak, 2010), and hydrophilic polymers (Matějíček, 2009; Rak, 2010; Uchman, 2010b) have been published. For example, a formation of the insoluble composite of cobalt bis(dicarbollide) with poly(ethylene oxide), PEO (identical to PEG), can be exploited for the preparation of polymeric nanoparticles (Matějíček, 2009;

Matějíček, 2011), potential drug-delivery carriers of boron clusters. Moreover, recent results of Uchman et al. (2010b) suggest that similar micelles can be prepared with the compounds showing high inhibitor potency against HIV PR, such as GB-179.

In our preliminary unpublished experiments, we noticed that some proteins, for example, HIV PR, can precipitate on addition of the parent metallacarborane GB-18 at elevated concentrations. On the other hand, in the case of dansyl and fluorescein moieties attached to one cobaltacarborane cluster, we observed almost quantitative solubilization of the probes in the presence of serum albumin with no traces of secondary aggregates. It was further estimated that approximately three fluorescein-metallacarborane conjugate (single-cluster compound) molecules are bound to one human serum albumin molecule. Interestingly, no such binding was observed for the double-cluster fluorescein conjugate (GB-179, see Table 3.1), representing the model compound of potent HIV protease inhibitors.

3.12 SUMMARY

The history of the rational design and clinical application of HIV PIs is a fascinating and, indeed, an iconic example of successful drug design that helped to control a deadly human plague. A plethora of inhibitors of HIV-replicative enzymes, including protease inhibitors, is now available for physicians to control the spread of the virus in an infected individual. However, the need for the development of safer, cheaper drugs that would be active against various multiresistant HIV species is still urgent. Moreover, HIV PR is a unique model for the analysis of ligand–enzyme interactions and the understanding of enzyme microevolution under selection pressure. The discovery of metal bis(dicarbollides) as specific and potent inhibitors of HIV protease opens a new, unexpected pathway to the design of antivirals.

We prove the inhibition efficiency of substituted metallacarboranes toward wild-type HIV PR as well as toward resistant variants. Interestingly, our study revealed various modes of inhibition for various compounds in our series: competitive, mixed, and noncompetitive. The competitive mode of inhibition implies that the inhibitor competes with the substrate for binding to the enzyme active site and this was, indeed, confirmed by two protein–ligand complex structures. The other inhibitory modes, mixed and non-competitive, suggest specific binding of compounds outside the enzyme active site or might also be explained by the activity of inhibitor aggregates that bind outside of the enzyme active site cavity.

In this chapter, we have reviewed the structure–activity relationship of 31 derivatives and analyzed the published and unpublished data on their mechanism of inhibition, interaction with the protein target, and three-dimensional structures and computional modeling of their complexes with the HIV protease, as well as the behavior of these molecules in solution.

Clearly, much remains to be learned before such compounds can be considered as tools for the treatment of humans; specifically, the aggregation of cobalt bis(dicarbollides) in aqueous solution seems to be a serious obstacle for the development of these compounds as potential drugs. However, the research reviewed in this chapter opens up new possibilities for rational drug design and, at least, teaches us one important lesson: expect the unexpected.

ACKNOWLEDGMENTS

The authors would like to thank Josef Plestil, Piotr Jurkiewicz, Janne Ruokolainen, Milan Kožíšek for providing some figures and unpublished data, and Devon Maloy for critical proofreading of the manuscript.

REFERENCES

Amblard, G., Issaurat, B., Depenoux, B. et al. 1983. Zero-current bilayer-membrane potential 2. Diffusion potential of hydrophobic anions. *J. Electroanal. Chem.* 144: 373–390.

Anderson, J., Schiffer, C., Lee, S. K. et al. 2009. Viral protease inhibitors. *Handb. Exp. Pharmacol.* 189: 85–110.

Armstrong, A. F. and Valliant, J. F. 2007. The bioinorganic and medicinal chemistry of carboranes: From new drug discovery to molecular imaging and therapy. *Dalton Trans.* 2007: 4240–4251.

Atwell, R. J., Sridharan, R., and De Levie, R. 1986. The translocation of cobalt dicarbamide anions across a lipid bilayer membrane: The effect of solution resistance. *Proc. Indian Acad. Sci.* 97: 431–436.

Barbaro, G. and Iacobellis, G. 2009. Metabolic syndrome associated with HIV and highly active antiretroviral therapy. *Curr. Diab. Rep.* 9: 37–42.

Belkova, N. V., Shubina, E. S., and Epstein, L. M. 2005. Diverse world of unconventional hydrogen bonds. *Acc. Chem. Res.* 38: 624–631.

Borisova, T. 2004. Adsorption distribution of the ion associates and ion-exchange at the water/1,2-dichloroethane interface by extracting with cesium dicarbollide. *Proc. Estonian Acad. Sci. Eng.* 1: 18–22.

Bosi, S., Da Ros, T., Spalluto, G. et al. 2003. Fullerene derivatives: An attractive tool for biological applications. *Eur. J. Med. Chem.* 38: 913–923.

Chaumont, A., Galand, N., Schurhammer, R. et al. 2004. Accumulation of host–guest ion complexes with different counterions at the water-supercritical CO_2 interface: A molecular dynamics study. *Russ. Chem. Bull.* 53: 1459–1465.

Chen, Z. F., and King, R. B. 2005. Spherical aromaticity: Recent work on fullerenes, polyhedral boranes, and related structures. *Chem. Rev.* 105: 3613–3642.

Chetcuti, P. A., Moser, P., and Rihs, G. 1991. Metallacarborane complexes as guest for cyclodextrins—Molecular-structure of the inclusion complex Cs[Closo-3,3,3-(Co)3–3,1,2-Rec2b9h11.Alpha-Cd].8h2o. *Organometallics* 10: 2895–2897.

Chevrot, G., Schurhammer, R., and Wipff, G. 2006. Surfactant behavior of "ellipsoidal" dicarbollide anions: A molecular dynamics study. *J. Phys. Chem. B* 110: 9488–9498.

Chevrot, G., Schurhammer, R., and Wipff, G. 2007a. Molecular dynamics study of dicarbollide anions in nitrobenzene solution and at its aqueous interface. Synergistic effect in the Eu(III) assisted extraction. *Phys. Chem. Chem. Phys.* 9: 5928–5938.

Chevrot, G., Schurhammer, R., and Wipff, G. 2007b. Synergistic effect of dicarbollide anions in liquid–liquid extraction: A molecular dynamics study at the octanol–water interface. *Phys. Chem. Chem. Phys.* 9: 1991–2003.

Cígler, P., Kozisek, M., Řezáčová, P. et al. 2005. From nonpeptide toward noncarbon protease inhibitors: Metallacarboranes as specific and potent inhibitors of HIV protease. *Proc. Natl. Acad. Sci. USA* 102: 15394–15399.

Crabtree, R. H., Siegbahn, P. E. M., Eisenstein, O. et al. 1996. A new intermolecular interaction: Unconventional hydrogen bonds with element-hydride bonds as proton acceptor. *Acc. Chem. Res.* 29: 348–354.

Custelcean, R. and Jackson, J. E. 2001. Dihydrogen bonding: Structures, energetics, and dynamics. *Chem. Rev.* 101: 1963–1980.

De Clercq, E. 2005. New approaches toward anti-HIV chemotherapy. *J. Med. Chem.* 48: 1297–1313.

De Clercq, E. 2009. The history of antiretrovirals: Key discoveries over the past 25 years. *Rev Med Virol* 19: 287–299.

DeLano, W. L. 2002. *The PyMOL Molecular Graphics System*. DeLano Scientific, Palo Alto, CA, USA.

Duvivier, C., Kolta, S., Assoumou, L. et al. 2009. Greater decrease in bone mineral density with protease inhibitor regimens compared with nonnucleoside reverse transcriptase inhibitor regimens in HIV-1 infected naive patients. *AIDS* 27: 817–824.

Epstein, L. M., Shubina, E. S., Bakhmutova, E. V. et al. 1998. Unusual hydrogen bonds with a hydride atom in boron hydrides acting as proton acceptor. Spectroscopic and theoretical studies. *Inorg. Chem.* 37: 3013–3017.

Erickson-Viitanen, S., Klabe, R. M., Cawood, P. G. et al. 1994. Potency and selectivity of inhibition of human immunodeficiency virus protease by a small nonpeptide cyclic urea, DMP 323. *Antimicrob. Agents Chemother.* 38: 1628–1634.

Fanfrlik, J., Brynda, J., Rezac, J. et al. 2008. Interpretation of protein/ligand crystal structure using QM/MM calculations: Case of HIV-1 protease/metallacarborane complex. *J. Phys. Chem. B* 112: 15094–15102.

Fanfrlik, J., Hnyk, D., Lepšík, M. et al. 2007. Interaction of heteroboranes with biomolecules—Part 2. The effect of various metal vertices and exo-substitutions. *Chem. Phys. Chem.* 9: 2085–2093.

Fanfrlik, J., Lepšík, M., Horinek, D. et al. 2006. Interaction of carboranes with biomolecules: Formation of dihydrogen bonds. *Chem. Phys. Chem.* 7: 1100–1105.

Farras, P., Teixidor, F., Kivekas, R. et al. 2008. Metallacarboranes as building blocks for polyanionic polyarmed aryl-ether materials. *Inorg. Chem.* 47: 9497–9508.

Flint, O. P., Noor, M. A., Hruz, P. W. et al. 2009. The role of protease inhibitors in the pathogenesis of HIV-associated lipodystrophy: Cellular mechanisms and clinical implications. *Toxicol. Pathol.* 37: 65–77.

Friedman, S. H., DeCamp, D. L., Sijbesma R. P. et al. 1993. Inhibition of the HIV-1 protease by fullerene derivatives: Model building studies and experimental verification. *J. Am. Chem. Soc.* 115: 6506–6509.

Glukhov, I. V., Lyssenko, K. A., Korlyukov, A. A. et al. 2005. Nature of weak inter- and intramolecular contacts in crystals 2. Character of electron delocalization and the nature of *X*-H. . .H-*X* (*X* = C, B) contacts in the crystal of 1-phenyl-*o*-carborane. *Russ. Chem. Bull.* 54: 547–559.

Grüner, B., Heřmánek S., and Plešek J. 1998. In: Final report of the EC-INCO-Copernicus Project, Contract CIPA-CT 93–1333, "New trends in the separation of 137Cs and 90Sr and transplutonium elements from radioactive HLW by borane and heteroborane anions." F. Teixidor Ed., Luxembourg, European Commission, Nuclear Science and Technology.

Grüner, B., Mikulášek, L., Báča, J. et. al. 2005. Cobalt bis(dicarbollides)(1-) covalently attached to the calix[4] arene platform—The first combination of organic bowl shaped matrices and inorganic metallaborane cluster anions. *J. Organomet. Chem.* 2005: 2022–2039.

Gulnik, S. V., Afonina, E., and Eissenstaat, M. 2009. HIV-1 protease inhibitors as antiretroviral agents. In *Book. HIV-1 Protease Inhibitors as Antiretroviral Agents* (Lu, C., and Li, A. P., eds), John Wiley and Sons, Inc. Hoboken, NJ.

Gulnik, S. V., Suvorov, L. I., Liu, B. et al. 1995. Kinetic characterization and cross-resistance patterns of HIV-1 protease mutants selected under drug pressure. *Biochemistry* 34(29): 9282–9287.

Hamilton, E. J. M., Kultyshev, R. G., Du, B. et al. 2006. A stacking interaction between a bridging hydrogen atom and aromatic pi density in the n-B18H22–benzene system. *Chem. Eur. J.* 12: 2571–2578.

Hao, E., Sibrian-Vazquez, M., Serem, W. et al. 2007. Synthesis, aggregation and cellular investigations of porphyrin-cobaltacarborane conjugates. *Chem. Eur. J.* 13: 9035–9042.

Hawthorne, M. F. and Francis, J. N. 1971. Synthesis and properties of cobalt complexes containing the bidentate .pi.-bonding B8C2H104-ligand. *Inorg. Chem.* 10: 863–864.

Hawthorne, M. F. and Lee, M. W. 2003. A critical assessment of boron target compounds for boron neutron capture therapy. *J. Neuro-Oncol.* 62: 33–45.

Hawthorne, M. F. and Maderna, A. 1999. Applications of radiolabeled boron clusters to the diagnosis and treatment of cancer. *Chem. Rev.* 99: 3421–3434.

Hawthorne, M. F., Young D. C., Andrews T. D. et al. 1968. Pi-dicarbollyl derivatives of transition metals. Metallocene analogs. *J. Am. Chem. Soc.* 90: 879–896.

Hawthorne, M. F., Young, D. C., and Wegner, P. A. 1965. Carbametallic boron hydride derivatives I. Apparent analogs of ferrocene and ferricinium ion. *J. Am. Chem. Soc.* 87: 1818–1819.

Johnson, V. A., Brun-Vezinet, F., Clotet, B. et al. 2008. Update of the drug resistance mutations in HIV-1. *Top. HIV Med.* 16: 138–145.

Judd, D. A., Nettles, J. H., Nevins, N. et al. 2001. Polyoxometalate HIV-1 protease inhibitors. A new mode of protease inhibition. *J. Am. Chem. Soc.* 123: 886–897.

King, R. B. 2001. Three-dimensional aromaticity in polyhedral boranes and related molecules. *Chem. Rev.* 101: 1119–1152.

Klooster, W. T., Koetzle, T. F., Siegbahn, P. E. M. et al. 1999. Study of the N-H center dot center dot center dot H-B dihydrogen bond including the crystal structure of BH3NH3 by neutron diffraction. *J. Am. Chem. Soc.* 121: 6337–6343.

Koh, Y., Matsumi, S., Das, D. et al. 2007. Potent inhibition of HIV-1 replication by novel non-peptidyl small molecule inhibitors of protease dimerization. *J. Biol. Chem.* 282: 28709–28720.

Koh, Y., Nakata, H., Maeda, K. et al. 2003. Novel bis-tetrahydrofuranylurethane-containing nonpeptidic protease inhibitor (PI) UIC-94017 (TMC114) with potent activity against multi-PI-resistant human immunodeficiency virus *in vitro. Antimicrob. Agents Chemother.* 47: 3123–3129.

Kohl, N. E., Emini, E. A., Schleif, W. A. et al. 1988. Active human immunodeficiency virus protease is required for viral infectivity. *Proc. Natl. Acad. Sci. USA* 85: 4686–4690.

Kovalevsky, A. Y., Ghosh, A. K., and Weber, I. T. 2008. Solution kinetics measurements suggest HIV-1 protease has two binding sites for darunavir and amprenavir. *J. Med. Chem.* 51: 6599–6603.

Kovalevsky, A. Y., Liu, F., Leshchenko, S. et al. 2006. Ultra-high resolution crystal structure of HIV-1 protease mutant reveals two binding sites for clinical inhibitor TMC114. *J. Mol. Biol.* 363: 161–173.

Kozisek, M., Cígler, P., Lepšík, M. et al. 2008. Inorganic polyhedral metallacarborane inhibitors of HIV protease: A new approach to overcoming antiviral resistance. *J. Med. Chem.* 51: 4839–4843.

Kubat, P., Lang, K., Cígler, P. et al. 2007. Tetraphenylporphyrin-cobalt(III) bis(1,2-dicarbollide) conjugates: From the solution characteristics to inhibition of HIV protease. *J. Phys. Chem. B* 111: 4539–4546.

Lam, P. Y., Jadhav, P. K., Eyermann, C. J. et al. 1994. Rational design of potent, bioavailable, nonpeptide cyclic ureas as HIV protease inhibitors. *Science* 263: 380–384.

Lam, P. Y., Ru, Y., Jadhav, P. K. et al. 1996. Cyclic HIV protease inhibitors: Synthesis, conformational analysis, P2/P2′ structure–activity relationship, and molecular recognition of cyclic ureas. *J. Med. Chem.* 39: 3514–3525.

Lesnikowski, Z. J. 2007. Boron units as pharmacophores—New applications and opportunities of boron cluster chemistry. *Collect. Czech. Chem. Commun.* 72: 1646–1658.

Li, H. Z., Min, D. H., Shore, S. G. et al. 2007. Nature of "hydrogen bond" in the diborane-benzene complex: Covalent, electrostatic, or dispersive? *Inorg. Chem.* 46: 3956–3959.

Lineweaver, H. and Burk, D. 1934. The determination of enzyme dissociation constants. *J. Am. Chem. Soc.* 56: 658–666.

Mallewa, J. E., Wilkins, E., Vilar, J. et al. 2008. HIV-associated lipodystrophy: A review of underlying mechanisms and therapeutic options. *J. Antimicrob. Chemother.* 62: 648–660.

Mastrolorenzo, A., Rusconi, S., Scozzafava, A. et al. 2007. Inhibitors of HIV-1 protease: Current state of the art 10 years after their introduction. From antiretroviral drugs to antifungal, antibacterial and antitumor agents based on aspartic protease inhibitors. *Curr. Med. Chem.* 14: 2734–2748.

Matějíček, P., Brus, J., Jigounov, A. et al., 2011. On the structure of polymeric composite of metallacarborane with poly(ethylene oxide). *Macromolecules* doi:10.1021/ma200502t.

Matějíček, P., Cígler, P., Olejniczak, A. B. et al. 2008. Aggregation behavior of nucleoside–boron cluster conjugates in aqueous solutions. *Langmuir* 24: 2625–2630.

Matějíček, P., Cígler, P., Prochazka, K. et al. 2006. Molecular assembly of metallacarboranes in water: Light scattering and microscopy study. *Langmuir* 22: 575–581.

Matějíček, P., Zednik, J., Uselova, K. et al. 2009. Stimuli-responsive nanoparticles based on interaction of metallacarborane with poly(ethylene oxide). *Macromolecules* 42: 4829–4837.

Muzammil, S., Armstrong, A. A., Kang, L. W. et al. 2007. Unique thermodynamic response of tipranavir to human immunodeficiency virus type 1 protease drug resistance mutations. *J. Virol.* 81: 5144–5154.

Planas, J. G., Vinas, C., Teixidor, F. et al. 2005. Self-assembly of mercaptane-metallacarborane complexes by an unconventional cooperative effect: A C–H center dot center dot center dot S–H center dot center dot center dot H–B hydrogen/dihydrogen bond interaction. *J. Am. Chem. Soc.*127: 15976–15982.

Plešek, J. 1992. Potential applications of the boron cluster compounds. *Chem. Rev.* 92: 269–278.

Plešek, J., Grüner, B., Báča, J. et al. 2002. Syntheses of the B(8)-hydroxy- and B(8,8′)- dihydroxy-derivatives of the bis(1,2- dicarbollido)-3-cobalt(1-)ate ion by its reductive acetoxylation and hydroxylation. Molecular structure of $[8,8'\text{-}\mu\text{-}CH_3C(O)_2 < (1,2\text{-}C_2B_9H_{10})_2\text{-}3\text{-}Co]^0$ zwitterion determined by x-ray diffraction analysis. *J. Organomet. Chem.* 649(2): 181–190.

Plešek, J., Grüner, B., Šícha, V., Böhmer, V., Císařová, I., unpublished results.

Plešek, J., Heřmánek, S., Todd, L. J. et al. 1978. Redazin. *Collect. Czech. Chem. Commun.* 41: 3509–3523.

Pokorná, J., Machala, L., Řezáčová, P. et al. 2009. Current and novel inhibitors of HIV protease. *Viruses* 1: 1209–1239.

Popov, A. and Borisova, T. 2001. Adsorption of dicarbollylcobaltate(III) Anion {(pi-(3)-1,2-B(9)C(2)H(11))(2) Co(III)(-)} at the water/1,2-dichloroethane interface. Influence of counterions' in nature. *J. Colloid Interface Sci.* 236: 20–27.

Poppe, S. M., Slade, D. E., Chong, K. T. et al. 1997. Antiviral activity of the dihydropyrone PNU-140690, a new nonpeptidic human immunodeficiency virus protease inhibitor. *Antimicrob. Agents Chemother.* 41: 1058–1063.

Prejdova, J., Soucek, M., and Konvalinka, J. 2004. Determining and overcoming resistance to HIV protease inhibitors. *Curr. Drug Targets: Infect. Disord.* 4: 137–152.

Rais, J. and Grüner, B. 2004. Extractions with cobalt bis(dicarbollide) ions. In *Ion Exchange, Solvent Extraction* (Marcus, Y. and SenGupta, A. K., eds.) vol. 17, pp. 243–334, New York: Marcel Dekker.

Rak, J., Kaplanek, R., and Kral, V. 2010. Solubilization and deaggregation of cobalt bis(dicarbollide) derivatives in water by biocompatible excipients. *Bioorg. Med. Chem. Lett.* 20(3): 1045–1048.

Rak, J., Tkadlecovka, M., Cígler, P. et al. 2008. NMR study of complexation of metallacarboranes with cyclodextrins. *Chem. List* 102: 209–212.

Raston, C. L. and Cave, G. W. V. 2004. Nanocage encapsulation of two ortho-carborane molecules. *Chem. Eur. J.* 10: 279–282.

Reed, C. A. 1998. Carboranes: A new class of weakly coordinating anions for strong electrophiles, oxidants, and superacids. *Acc. Chem. Res.* 31: 133–139.

Řezáčová, P., Pokorná, J., Brynda, J. et al. 2009. Design of HIV protease inhibitors based on inorganic polyhedral metallacarboranes. *J. Med. Chem.* 52: 7132–7141.

Richardson, T. B., de Gala, S., Crabtree, R. H. et al.1995. Unconventional hydrogen bonds: Intermolecular B-H center dot center dot center dot H–N interactions. *J. Am. Chem. Soc.* 117: 12875–12876.

Saskova, K. G., Kozisek, M., Lepšík, M. et al. 2008. Enzymatic and structural analysis of the I47A mutation contributing to the reduced susceptibility to HIV protease inhibitor lopinavir. *Protein Sci.* 17: 1555–1564.

Saskova, K. G., Kozisek, M., Řezáčová, P. et al. 2009. Molecular characterization of clinical isolates of human immunodeficiency virus resistant to the protease inhibitor darunavir. *J. Virol.* 83: 8810–8818.

Saxena, A. K. and Hosmane, N. S. 1993. Recent advances in the chemistry of carborane metal-complexes incorporating D-Block and F-block elements. *Chem. Rev.* 93: 1081–1124.

Sedlak, R., Fanfrik, J., Hnyk, D. et al., 2010. Interactions of boranes and carboranes with aromatic systems: CCSD(T) complete basis set calculations and DFT-SAPT analysis of energy components. *J. Phys. Chem. A.* 114: 11304–11311.

Semioshkin, A. A., Sivaev, I. B., and Bregadze, V. I. 2008. Cyclic oxonium derivatives of polyhedral boron hydrides and their synthetic applications. *Dalton Trans.* 2008: 977–992.

Sham, H. L., Kempf, D. J., Molla, A. et al. 1998. ABT-378, a highly potent inhibitor of the human immunodeficiency virus protease. *Antimicrob. Agents Chemother.* 42: 3218–3224.

Sijbesma, R., Srdanov, G., Wudl, F. et al. 1993. Synthesis of a fullerene derivative for the inhibition of HIV enzymes. *J. Am. Chem. Soc.* 115: 6510–6512.

Sivaev, I. B. and Bregadze, V. I. 1999. Chemistry of cobalt bis(dicarbollides). A review. *Collect. Czech. Chem. Commun.*64: 783–805.

Sivaev, I. B. and Bregadze, V. I. 2009. Polyhedral boranes for medical applications: Current status and perspectives. *Eur. J. Inorg. Chem.* 2009: 1433–1450.

Sivaev, I. B., Starikova, Z. A., Sjoberg, S. et al. 2002. Synthesis of functional derivatives of the [3,3′-Co(1,2-$C_2B_2H_{11})_2$](-) anion. *J. Organomet. Chem.* 649: 1–8.

Soloway, A. H., Tjarks, W., Barnum, B. A. et al. 1998. The chemistry of neutron capture therapy. *Chem. Rev.* 98: 1515–1562.

Spinelli, S., Liu, Q. Z., Alzari, P. M., Hirel, P. H., and Poljak, R. J. 1991. The three-dimensional structure of the aspartyl protease from the HIV-1 isolate BRU. *Biochimie* 73: 1391–1396.

Uchman, M., Cígler, P., Grüner, B. et al. 2010b. Micelle-like nanoparticles of block copolymer poly(ethylene oxide)-*block*-poly(methacrylic acid) incorporating fluorescently substituted metallacarboranes designed as HIV protease inhibitor interaction probes. *J. Colloid Interface Sci.* 348: 129–136.

Uchman, M., Jurkiewicz, P., Cígler, P. et al. 2010a. Interaction of fluorescently substituted metallacarboranes with cyclodextrins and phospholipid bilayers: Fluorescence and light scattering study. *Langmuir* 26: 6268–6275.

Valliant, J. F., Guenther, K. J., King, A. S. et al. 2002. The medicinal chemistry of carboranes. *Coord. Chem. Rev.* 232: 173–230.

van Mau, N. D., Isaaurat, B., and Amblard, G. 1984. Adsorption of hydrophobic anions on phospholipid monolayers. *J. Colloid Interface Sci.* 101: 1–9.

Weber, I. T. and Agniswamy, J. 2009. HIV-1 protease: Structural perspectives on drug resistance. *Viruses* 1: 1110–1136.

Weber, J., Mesters, J. R., Lepšík, M. et al. 2002. Unusual binding mode of an HIV-1 protease inhibitor explains its potency against multi-drug-resistant virus strains. *J. Mol. Biol.* 324: 739–754.

Williams, J. W. and Morrison, J. F. 1979. The kinetics of reversible tight-binding inhibition. *Methods Enzymol.* 63: 437–467.

Wlodawer, A. 2002. Rational approach to AIDS drug design through structural biology. *Annu. Rev. Med.* 53: 595–614.

Wlodawer, A. and Vondrasek, J. 1998. Inhibitors of HIV-1 protease: A major success of structure-assisted drug design. *Annu. Rev. Biophys. Biomol. Struct.* 27: 249–284.

Yin, P. D., Das, D., and Mitsuya, H. 2006. Overcoming HIV drug resistance through rational drug design based on molecular, biochemical, and structural profiles of HIV resistance. *Cell. Mol. Life Sci.* 63: 1706–1724.

Part II

Boron for Living: Health and Nutrition

This section, covering nutritional aspects of boron, consists of a single chapter, Chapter 4, by Samman, Foster and Hunter. Boron has been found to be an essential nutrient for both plants and animals, including man. Current difficulties in advancing knowledge of its functions include the limited availability of boron data in tables of food composition, and challenges in the analysis of boron in food and in biopsy materials. The focus on a role for phytochemicals in disease prevention should be expanded to consider the possibility that some of their biological effects are intertwined with those of boron. In view of the distribution of boron in the food chain, mainly in foods of plant origin, the reported benefits of the consumption of vegetables, fruit, and cereals may be contributed in part by boron. Thus, boron is an integral part of health and nutrition.

Accordingly, Chapter 4 assesses the literature on dietary sources and intake of boron in humans, and evaluates the impact of boron on metabolism in health and disease. It was found that boron was capable of influencing steroid hormone metabolism; however, the specific condition that controls the nature (and direction) of the response remains to be determined.

4 The Role of Boron in Human Nutrition and Metabolism

Samir Samman, Meika Foster, and Duncan Hunter

CONTENTS

4.1 INTRODUCTION

In 1923, boron was accepted as being an essential nutrient initially for broad beans (*Vicia faba*) and subsequently for all plants. Boron is required for the development of zebrafish (*Danio rerio*) (Rowe and Eckert, 1999) and for reproduction and development in frogs (*Xenopus laevis*) (Fort et al., 1999). Experimental studies suggest that boron is a potentially essential nutrient for humans also (WHO, 1996; Nielsen, 2008). The aim of the present review is to assess the literature on dietary sources and intake of boron in humans; and to evaluate the impact of boron on metabolism in health and disease.

Boron, atomic number 5, is the first member of the metalloid or semiconductor family of elements, and as such has intermediate properties between metals and nonmetals. The boron atom is small, with only three valence electrons. It acts as a Lewis acid, accepting hydroxyl ions, and thus leaving an excess of protons. Boron conjugates with organic compounds containing hydroxyl groups and is therefore capable of interacting with substances of biological interest, including nutrients and pharmacologically active agents (Hunter, 2009).

The average concentration of boron in the earth's crust is 17 ppm, and most soils fall within the range of 3–100 ppm. Boron occurs in high concentrations in sedimentary rocks and in clay-rich marine sediment due to the relatively high concentration of boron in seawater. Deposits of boron are found in association with volcanic activity and where marshes or lakes have evaporated under arid conditions. Coastal soils contain up to 50 times as much boron as do inland soils while, generally, humid soils are lower in boron content than slightly leached soils. The concentration of boron in soil is influenced also by the presence of other minerals, and soil pH and texture (Butterwick et al., 1989; Steinnes, 2009).

4.2 BORON IN THE FOOD SUPPLY

4.2.1 Boron Content of Food

The boron content of some foods from a number of countries has been published, including foods from the United States (Gates Zook and Lehmann, 1968; Gormican, 1970; Hunt et al., 1991), Finland (Koivistoinen, 1980), Australia (Naghii et al., 1996), South Korea (Choi and Jun, 2008), Singapore (Bloodworth, 1989), the United Kingdom (Ysart et al., 1999), France (Biego et al., 1998), and Turkey (Simsek et al., 2003). Consistently, foods of plant origin are reported to be rich sources of boron. The boron content of bottled water from a number of countries ranges from 0 ppm (Austria) to 4.3 ppm (USA) (Allen et al., 1989). Examples of the boron content of foods are shown in Table 4.1.

The concentration of boron within plants varies considerably. Geological influences, in addition to variations in the analytical methods used for the determination of boron concentrations, result in wide-ranging estimates of the boron content of a particular food (Naghii and Samman, 1993). For instance, carrots from the United States, analyzed by means of Inductively Coupled Plasma Analysis, were reported to contain 0.75 ppm (Hunt et al., 1991) whereas carrots from Finland, analyzed by means of a colorimetric method, were reported to contain 3.2 ppm (Koivistoinen, 1980). Other examples can be seen elsewhere (Naghii and Samman, 1993; Naghii et al., 1996).

The preferred method for the analysis of boron is inductively coupled plasma atomic emission spectroscopy (ICP-AES). Inductively coupled plasma mass spectroscopy (ICP-MS) is the most widely used nonspectrophotometric method for analysis of boron, as it uses small volumes of sample, is fast, and applies to a wide range of materials. When ICP equipment is unavailable, colorimetric/spectrophotometric methods can be utilized. However, many of these methods are subject to interference and should be used with caution (WHO, 1998). Part of the discrepancy in the nutritional composition of boron may be related to differences in methods of analysis.

4.2.2 Intake of Boron in Humans

The daily intake of boron from the American diet is estimated to be approximately 1 mg/day (Anderson et al., 1994; Meacham and Hunt, 1998; Rainey et al., 1999), with intakes being slightly

TABLE 4.1
Concentration of Boron in Selected Australian Foods and the Amount in a Typical Serving Size

Food		mg B/100 g	mg B/Typical Serving Size
Fruit			
	Apple—red	0.32	0.32
	Apricots—dried	2.11	0.53
	Avocado	2.06	2.06
	Currants	1.74	0.26
	Dates (California)	1.08	0.38
	Grapes—red	0.50	0.50
	Orange	0.25	0.33
	Peach	0.52	0.57
	Peaches—dried	3.24	0.81
	Pear	0.32	0.48
	Prunes	1.88	0.94
	Raisins—seedless	4.51	0.67
Vegetables			
	Beans—red kidney	1.40	1.82
	Beetroot—canned	0.32	0.16
	Broccoli	0.31	0.15
	Carrot	0.30	0.21
	Celery	0.50	0.32
	Chick peas	0.71	0.92
	Lentils	0.74	0.96
	Mushroom	0.16	0.02
	Olive	0.35	0.07
	Onion	0.20	0.07
	Parsley	0.59	0.03
	Potato	0.18	0.14
Nuts			
	Almond	2.82	0.42
	Brazil nuts	1.72	0.34
	Cashew nuts	1.15	0.17
	Hazel nuts	2.77	0.68
	Peanut butter	1.92	0.38
	Pistachio nuts	1.20	0.18
	Walnuts (California)	1.63	0.24
Cereals			
	Bran (wheat)	0.32	0.02
	Bread (white + maize flour)	0.10	0.02
Animal foods			
	Cheese—slice	0.05	0.01
	Chicken	0.04	0.05
	Egg yolk	0.06	0.18
	Honey	0.50	0.14
	Lamb	0.02	0.01

Source: Data adapted from Naghii, M.R., Lyons Wall, P.M., Samman, S. *J. Am. Coll. Nutr.* 1996;15:614–619.

TABLE 4.2
Boron Intake in Population Samples of Men and Women

	Boron Intake (mg/day)		
Country	**Men**	**Women**	**Reference**
Australia	2.3	2.2	Naghii et al. (1996)
Germany	1.7	1.6	Rainey and Nyquist (1998)
Kenya	2.0	1.8	Rainey and Nyquist (1998)
Korea		0.9	Kim et al. (2008)
Mexico	2.1	1.8	Rainey and Nyquist (1998)
USA	1.2	0.9	Anderson et al. (1994)
USA	1.2	1.0	Rainey et al. (1999)
USA	0.7	0.9	Hunt and Meacham (2001)
USA	1.1	0.9	Rainey and Nyquist (1998)

higher in men and in vegetarians (Rainey et al., 1999). Coffee and milk, although low in boron content, make the highest contribution (12%) to the total intake of boron in the reported population; an observation similar to what has been previously reported (Anderson et al., 1994). Other prominent dietary contributors include wine, dried fruits, and nuts. In the Korean diet (Kim et al., 2008), the consumption of fruit, vegetables, and cereals contributes the majority of boron in the diet, with rice being the single highest contributor. Mean estimates of boron intake that are higher than those of the US population have been obtained for population samples in Australia (Naghii et al., 1996), Germany, Mexico, and Kenya (Rainey and Nyquist, 1998). In an Australian group of healthy subjects, the mean intake was reported to be 2.2 mg/day, an amount that is similar for men and women (Naghii et al., 1996). The intake of boron was significantly correlated with the intakes of dietary fiber and dietary plant protein, confirming that the main sources of dietary boron are obtained from plant foods.

Dietary exposure estimates of boron from the UK diet (Ysart et al., 1999) show a population intake of 1.5 mg/day, slightly lower than the previous survey data from the United Kingdom. The estimate is similar to that reported for the French population (1.6 mg/day) (Biego et al., 1998). The consumption of beverages, including wine, make a significant contribution to the total intake of boron in France (Biego et al., 1998), and in the Australian sample (Naghii et al., 1996), such that when wine drinkers were excluded from the latter analysis the total intake of boron was decreased by approximately 50% (Naghii et al., 1996). Thus, estimates of boron intake have been carried out in a number of countries with varied food supplies. The range of intakes appears to be approximately 1–2 mg/day (Table 4.2), and generally slightly higher in males as compared to females. A possible explanation for the gender difference in boron intake may be the higher intake of calories in men as compared to women, given the positive relationship between energy intake and boron intake (Naghii et al., 1996).

The concentration of boron in human milk is approximately 30–40 μg/L (Hunt et al., 2004, 2005), which is similar to concentrations of boron that are found in cow's milk (Anderson, 1992). The boron concentrations in human milk do not change significantly over time in mothers of term infants, but are reported to decline gradually in mothers of preterm infants (Hunt et al., 2004). The intake of boron is 0.5 mg/day in infants and toddlers (Hunt and Meacham, 2001), and 1 mg/day for preschool children (Pieczyńska et al., 2003). School lunches in the United States provide approximately 0.5 mg (Murphy et al., 1971).

4.2.3 Variations in the Intake of Boron

The intake range of 1–2 mg boron/day currently estimated for adults depends on differences in methods of food analysis (Naghii and Samman, 1993), varied agricultural methods, regional differences that may exist in the water and soil content of boron (Hunt et al., 1991), and individual food preferences—whether individuals choose to consume foods that are rich or poor dietary sources of boron (Naghii et al., 1996).

4.2.3.1 Regional Differences

Turkey is a country that has high- and low-borate-containing soils. In many instances, foods obtained from high-borate regions contained higher levels of boron than those obtained from low-boron regions, or when compared with values in the literature (Simsek et al., 2003). For instance, pistachio nuts are reported to contain fivefold higher levels of boron than shown by others (Naghii et al., 1996). Also, hazelnuts display a wide range in their content of boron (14–24 ppm), depending on the variety and region of cultivation (Simsek and Aykut, 2007). However, from limited statistical analysis of the data, Simsek et al. (2003) suggest that there is no effect of the region of cultivation on the boron content of the food in Turkey.

4.2.3.2 Methods of Food Production: Organic and Conventional Agriculture

As part of ongoing research on the influence of agricultural methods on the composition of food (Samman et al., 2008; Hunter et al., in press), we have critically evaluated data that compare the micronutrient content of plant foods produced by organic and conventional agricultural methods (Hunter et al., in press). Of the total of 33 studies that were selected systematically, boron content was reported in eight studies (Amodio et al., 2007; Fjelkner-Modig et al., 2000; Lester et al., 2007; Peck et al., 2006; Warman, 2005; Warman and Havard, 1997, 1998; Wszelaki et al., 2005). The foods that were analyzed tended to be those that contribute significantly to the dietary intake of boron, that is, vegetables and fruit. Boron was reported to be higher in organic foods more frequently than in conventional foods (Figure 4.1), with an overall difference of +8.2%. On analysis of the combined data, however, we found that this difference did not reach statistical significance due to the large variations within studies (Hunter et al., in press).

In individual reports where statistical analysis was applied by the authors, organically produced carrots and cabbages were found to contain a higher level of boron than their conventional counterparts

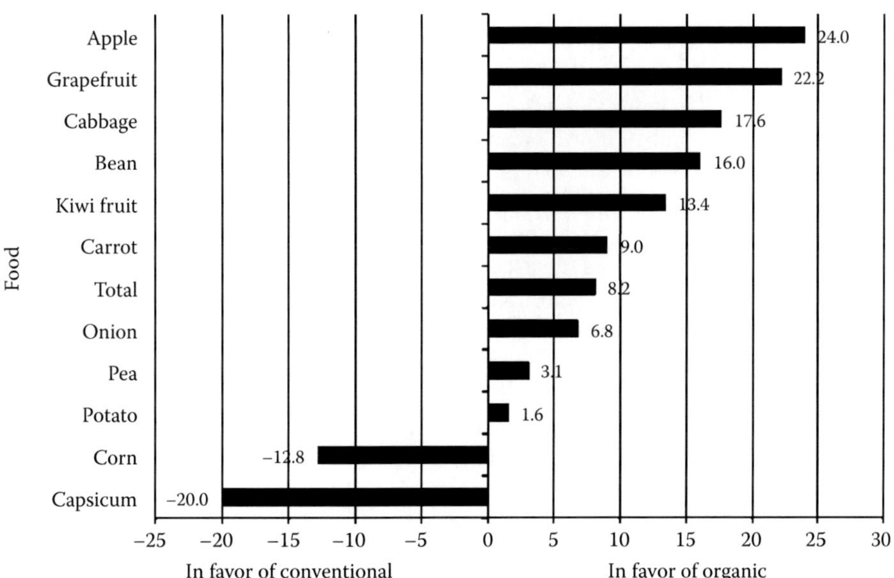

FIGURE 4.1 Percent differences in the boron content of conventional compared to organic food. Positive values indicate a higher content in organically produced foods.

(Warman and Havard, 1997). In contrast, organic onions were lower in boron than those produced by conventional methods (Gundersen et al., 2000; Warman, 2005).

4.2.4 Recommended Intake Levels of Boron

The acceptable safe range for boron intake is 1–13 mg/day, and a daily intake of up to 20 mg/day can be achieved with a diet high in nuts, dried fruits, and wine (WHO, 1996). While there is no Recommended Dietary Allowance for boron, 20 mg/day is the upper limit of intake based on extrapolation from toxicological studies in rats (Institute of Medicine, 2001). In view of the known distribution of boron in the food supply, dietary advice that encourages the consumption of a plant-based diet, including a moderate intake of wine, promotes the intake of boron, along with other potentially bioactive phytochemicals (Cook and Samman, 1996).

4.2.5 Conclusion

From evidence published so far, it appears that soil conditions and agricultural methods have limited effect on the boron composition of plant foods. This suggests that plants have effective mechanisms for maintaining homeostasis of boron, possibly mediated by specific boron transporters (Takano et al., 2002). The lack of a clear effect of agricultural methods on the plant composition of boron is consistent with the limited impact of agricultural methods on other nutrients (Dangour et al., 2009), and may well reflect the essentiality of boron for the integrity of the cell wall in plants (O'Neill et al., 1996).

4.3 ABSORPTION AND METABOLISM OF BORON

4.3.1 Metabolic Balance Studies

Anionic trace elements, such as boron, are absorbed from the gastro-intestinal tract usually with high efficiency. In a pioneering metabolic balance study in two subjects, Kent and McCance (1941) showed that the major route of excretion of boron is through the urine. Boron was provided in food form or as supplemental boric acid for 28 days. Regardless of the form of boron, whether it was obtained from supplements or from food, the element was absorbed with equal efficiency, and 93–94% of the ingested dose was recovered mainly in the urine and to a lesser extent in the feces. The bioavailability of boron was unaffected by dietary fiber, and appeared to be similar regardless of whether subjects were consuming white or brown bread. These observations were confirmed later with the use of modern methodology (Hunt et al., 1997; Sutherland et al., 1999), and urinary excretion was established as the primary mechanism for maintaining homeostasis of physiological levels of boron.

The metabolic balance study by Sutherland et al. (1999) highlighted a potential interaction between boron and zinc. The authors reported boron as being in negative balance during a period of zinc repletion, as compared with the positive boron balance observed during consumption of the baseline diet and the low zinc diet. An interaction has been observed also with calcium whereby postmenopausal women who were supplemented with boron showed a decrease in the excretion of calcium (Hunt et al., 1997), and an increase in serum ionized calcium (Nielsen, 2004). Based on these data and others derived from experimental animals, boron is postulated to interact with a range of other micronutrients, particularly under situations of metabolic stress (Nielsen, 1991, 2008); however, the mechanisms remain to be elucidated.

4.3.2 Supplementation Trials

In men maintaining their habitual diet, the mean concentrations of urinary boron in samples obtained on two different occasions were 1.87 and 1.90 mg boron/day, respectively. Following supplementation with 10 mg boron/day for 4 weeks, urinary boron excretion increased to 10.2 mg/day and accounted for 84% of the ingested dose (Naghii and Samman, 1997). Similarly, individuals living in a high-boron area in Turkey excrete higher amounts (approximately 5-fold higher) of boron than those living in communities with low-boron exposure (Korkmaz et al., 2007a,b). In female students supplemented with 3 mg boron/day over 10 months, urinary excretion increased substantially, and was more pronounced in athletic women as compared to sedentary women (Meacham et al., 1995).

4.3.3 Boron Homeostasis

Limited evidence is available regarding the molecular mechanism of boron homeostasis. Based on analysis of plasma boron concentrations in seven sibships, Barr et al. (1996) observed smaller variances within than between families and hypothesized that boron levels in humans are under genetic control. The data presented by Barr et al. (1996) need to be interpreted with caution as the blood samples were collected from individuals living in a rural region of northern Chile, and their dietary intakes were not determined. Also, the study was retrospective and utilized blood samples that had been stored for over 20 years. More recent evidence suggests the involvement of a sodium-coupled boron transporter in animal cells, expressed in the basolateral membrane, which determines the steady-state concentration of borate in the cytoplasm and hence maintains borate homeostasis (Park et al., 2004).

4.3.4 Conclusion

Studies under controlled metabolic-ward conditions and in free-living subjects show that the urinary excretion of boron is a reflection of its intake. In addition, preliminary evidence suggests that boron may interact with other micronutrients depending on the physiological or nutritional status of the host individual. It has been suggested that such interactions can be demonstrated more clearly in the presence of metabolic stress.

4.4 BORON AND STEROID HORMONES

4.4.1 Controlled Intervention Studies in Humans

In a landmark study of the effect of boron on steroid hormone metabolism (Nielsen et al., 1987), postmenopausal women lived in a metabolic ward under controlled conditions for 167 days. For the first 119 days, the participants were fed basal diets, which provided 0.25 mg boron/day. The interventions also included dietary periods of magnesium and aluminum supplementation. In two additional dietary periods, a subsample of the women was provided with a boron supplement of 3 mg/day. The study demonstrated, for the first time, that supplementation of postmenopausal women with a low dose (3 mg/day) of boron increases significantly the plasma concentrations of 17b-estradiol and testosterone. The study showed also that boron supplementation decreases significantly the urinary excretion of calcium and magnesium.

In a later study in women who were deprived of dietary magnesium, boron supplementation resulted in a decrease in the serum 17b-estradiol concentration (Nielsen, 2004), and the plasma progesterone concentration was influenced by a boron × magnesium interaction. This study confirmed the existence of an interaction between boron and other nutrients, specifically magnesium, under conditions of metabolic stress (Nielsen, 2008).

The effect of boron on plasma steroid hormones was examined further in a metabolic study by Beattie and Peace (1993). Utilizing a study design that tested the effect of boron per se, the authors found no effect on plasma estradiol or testosterone concentrations in postmenopausal women. The authors suggest that the lack of concordance between their results and the findings of Nielsen et al.

(1987) may be due to differences in the composition of the basal diet, and methodological challenges that interfered with the accurate estimation of steroid hormone concentrations in postmenopausal women. One could also speculate that the baseline boron status was higher in the study by Beattie and Peace (1993) as compared to the human volunteers in the study by Nielsen et al. (1987), and this may have contributed to differences in the outcome.

4.4.2 Boron Supplementation in Men

In male bodybuilders, supplementation with low doses (2.5 mg/day) of boron has no apparent effect on plasma steroid hormone concentrations or performance related to bodybuilding (Ferrando and Green, 1993). The results suggest that the impact of physical exertion and training associated with bodybuilding overshadow the effect, if any, of low doses of boron. In contrast, higher doses of boron (10 mg/day) given to sedentary men resulted in a significant increase in the plasma estradiol concentration and a trend for an increase in the plasma testosterone concentration (Naghii and Samman, 1997).

4.4.3 Boron and Steroid Hormones in Rats

In rats, boron intake has a marked effect on the plasma testosterone concentration. As the intake of boron increased from 2 to 25 mg, plasma testosterone concentrations tended to decrease, and the response in plasma was paralleled by a decrease in the testosterone concentration in the testes (Samman et al., 1998).

4.4.4 Conclusion

From the studies described above and those reported by others, as reviewed by Nielsen (1991, 1997, 2000, 2008), boron is capable of influencing steroid hormone metabolism; however, the specific conditions that control the nature (and direction) of the response remain to be determined.

4.5 EFFECTS OF BORON IN HEALTH AND DISEASE

Changes in the concentrations of steroid hormones in plasma have implications for a range of chronic diseases, such as osteoporosis, arthritis, and cardiovascular disease. Studies investigating these diseases have been carried out mainly in animal models, and have been reviewed elsewhere (Nielsen, 1991, 1997, 2000, 2008; Mastromatteo and Sullivan, 1994; Naghii and Samman, 1993; Samman et al., 1998; Devirian and Volpe, 2003).

4.5.1 Reproductive Health

Exposure to boron through higher concentrations of the element in drinking water or from high environmental and workplace exposure appears to affect reproductive function. Yazbeck et al. (2005) showed that the birth rate was higher in regions of France with high boron concentrations in the drinking water (>0.3 mg/L) as compared with the general population exposed to low boron levels (<0.09 mg/L). In workers exposed to boron through industrial means, reproductive outcomes are compromised, such as, delays in pregnancy, higher rates of induced abortions, and a decrease in the number of live births (Chang et al., 2006). A shift in the gender balance of offspring has been reported also, with boron workers having a significantly higher percentage of male offspring, possibly due to a decrease in Y-bearing sperm (Robbins et al., 2008). Further, the concentration of boron in semen is significantly lower in controls as compared to individuals who are exposed to industrial levels of boron (Xing et al., 2008). Potentially, the effect of boron on reproductive function could be mediated by perturbations in the metabolism of steroid hormones, such as an increase in the plasma estradiol concentrations following boron supplementation in men (Naghii and Samman, 1997).

4.5.2 Longevity

In a survey based in northern France, exposure to high levels of boron (>0.3 mg/L) in the drinking water was associated with a significantly lower mortality rate (8.8%) as compared to that of a low-boron reference area (9.8%) (Yazbeck et al., 2005). More recently, Huang et al. (2009) reported a significant correlation between total exposure to boron and the ratio of people over the age of 90 years per 100,000 inhabitants. The so-called "90-rate" was higher in areas where the bioavailability of boron from soils was greater due to the fine soil texture and low pH. Similar benefits were reported also for selenium and molybdenum, which suggests that the relationship between the soil content of boron and longevity is confounded by the presence of other nutrients in soil, and subsequently in the food supply. The mechanism of a potential effect of boron on longevity is not known; however similar observations have been shown in the drosophila fly, which has an increased life span following exposure to low concentrations of boron per se (Massie et al., 1990).

4.5.3 Brain Function

On the basis of a series of experiments in humans (Penland, 1994, 1995a), it appears that boron intake may influence brain and psychological function. The studies suggest that in healthy older adults a boron-restricted diet results in decreased brain electrical activity, as measured by changes in electroencephalogram (EEG) parameters. Further, boron deprivation resulted in impaired measures of motor speed and dexterity, and the cognitive processes of attention, perception, and short-term memory (Penland, 1998). In a study designed to evaluate the effects of boron on the somatic and psychological symptoms of menopause, boron supplementation (3 mg/day) was found to have no effect in the majority of the 46 participants (Penland, 1998). The effects of boron on brain function may depend on the status of other nutrients. Vitamin D-deficient rats fed a low-boron diet (0.158 ppm) had a diminished calcium and phosphorus balance compared to those on a higher-boron diet (2.72 ppm) (Hegsted et al., 1991), and boron-deficient (0.1 mg/kg diet) rats were less active than their boron-adequate (3.1 mg/kg diet) counterparts when fed safflower oil in contrast to fish oil (Nielsen and Penland, 2006). It has been hypothesized that the effects of boron on brain function and consequent behavior may result from changes in membranes that influence nerve-impulse transmission (Nielsen, 2000).

It is estimated from human cadavers that the boron concentration of the adult brain is approximately 0.06 ppm, being similar to the concentration in muscle tissue (0.1 ppm) but less than that of other major organs (Naghii and Samman, 1993; Devirian and Volpe, 2003). Studies in animals suggest that the boron concentration of the brain remains relatively constant. The feeding of 3-week-old rats with 20 mg/kg boron in their drinking water for 21 days resulted in a steady increase of boron levels in the brain during the first 9 days of treatment followed by a rapid return to control levels. Since the boron concentration in the blood continued to rise up to day 21 of treatment, with the concentration surpassing that of the brain by day 13, the authors suggest the presence of a homeostatic mechanism that eliminates excess boron from the brain against its own concentration gradient (Magour et al., 1982).

4.5.4 Wound Healing

The topical application of boron has long been associated with wound healing; the boron content of thermal waters is believed to contribute to their beneficial healing effects (Chebassier et al., 2004) and a study conducted in 1941 reported that a mixture of boric acid powder and potassium permanganate was effective in the treatment of contaminated and infected traffic and war wounds (Gyorffy, 1941). Application of a 3% boric acid solution has been shown to improve the healing of deep wounds, significantly reducing the time required in intensive care (Blech et al., 1990). The mechanisms of action of boron in wound healing are still largely unknown, but may include an involvement in the synthesis and release of extracellular matrix macromolecules, including proteoglycans and collagen, and cytokines (Benderdour et al., 1998). Boron has also been shown to induce the synthesis of vascular endothelial growth factor (VEGF), which plays a critical role in angiogenesis (Dzondo-Gadet et al., 2002). Further, boron has been suggested to improve remodeling of the extracellular matrix by inducing keratinocyte secretion of matrix metalloproteinases (Chebassier et al., 2004). There is a need for additional studies to further elucidate the mechanism of action of boron in wound healing and for randomized controlled trials to be conducted to explore a role for boron in the routine treatment of wounds.

4.5.5 Cancer

A number of surveys have been carried out to explore the relationship between boron and a range of different cancers. The investigations have differed in their designs, sampling procedures, and depth of dietary assessments, as well as in the types of cancers studied.

4.5.5.1 Prostate Cancer

The association between boron intake and risk of prostate cancer was investigated in an American population. The study, which followed a cross-sectional case–control design with a small number of cases ($n = 95$), demonstrated an inverse association between higher intakes of boron and the risk of prostate cancer (Cui et al., 2004). In a subsequent observational study, a similar association was observed between boron exposure from ground water (across the state of Texas) and risk of prostate cancer (Barranco et al., 2007). In contrast to these reports, a large cohort study involving 35,244 men found that boron intake (derived from diet plus multivitamin preparations) was not associated with prostate cancer risk (Gonzalez et al., 2007). The median intake of boron in the latter study was 1.4 mg/day, which was obtained almost entirely (98%) from food. Total boron intake and specific rich sources of boron, such as nuts (peanuts) and wine, were not shown to be protective against prostate cancer. The discrepancy in the results of the prostate cancer studies may relate to differences in sample size and other methodological issues involving the measurement of dietary intake.

4.5.5.2 Breast Cancer

A study of the relationship between dietary factors and breast tumor characteristics was carried out in premenopausal women (Touillaud et al., 2005). Tumors were characterized in terms of their estrogen receptor (ER) status. Tumors with ER-positive status, as compared to ER-negative, are known to respond to the proliferative effect of estrogen, and are more likely to react to endocrine therapy. In the study, a low intake of boron (0.8 mg/day) was associated with a lower risk of ER-negative tumors as compared to a higher intake (1.03 mg/day). A similar relationship was observed with phytochemicals, such as phytosterols and kaempferol. Although these data are preliminary, they suggest that the effect of boron is confounded by the presence of phytosterols that are known to interfere with the entero-hepatic circulation of steroids, steroid hormones, and cholesterol (Samman et al., 2004).

4.5.5.3 Cervical Cancer

A cervical histopathology study was undertaken in a sample of women from boron-rich and boron-poor areas in Turkey. The differences in boron status of the women was confirmed by assessment of their urinary excretion of boron, which was approximately sevenfold higher in those living in the boron-rich regions. A small but significant number of women from boron-poor areas had cytopathological indicators of cervical cancer, in contrast to none in the boron-rich areas, suggesting a protective effect of boron. In a subsample of women, micronucleus frequencies in buccal cells were determined but no differences by region were observed (Korkmaz et al., 2007a,b).

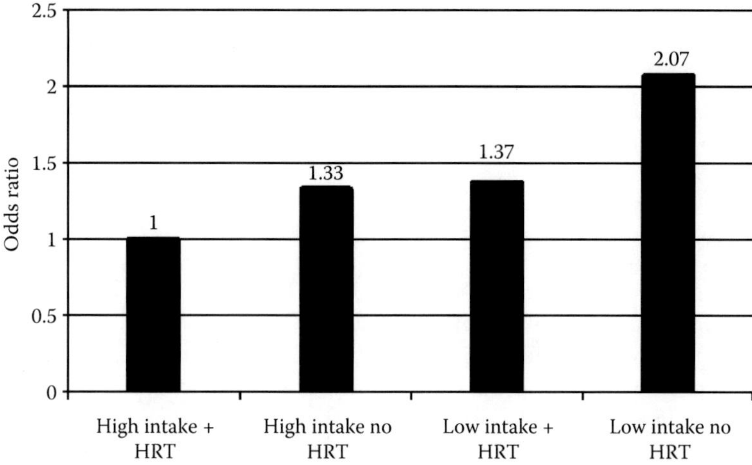

FIGURE 4.2 Effect of boron, with or without hormone replacement therapy (HRT), on the risk (odds ratio) of lung cancer in women. (Data adapted from Mahabir, S. et al., *Am. J. Epidemiol.* 2008;167:1070–1080.)

4.5.5.4 Lung Cancer

In a case–control study, boron intake was found to be inversely associated with the risk of lung cancer in women (Mahabir et al., 2008). The results suggested that women who had a boron intake >1.3 mg/day (quartile 1) were at significantly lower risk of cancer than those consuming <0.8 mg/day (quartile 4). In a further subanalysis of the data, an interaction with hormone replacement therapy (HRT) was identified, with those on low boron intake and no HRT being at greater risk of developing lung cancer, indicated as an increase in the Odds Ratio (Figure 4.2), than those undergoing HRT treatment. The main contributors of boron intake in cases and controls were coffee, wine, apples and pears, and peanuts and peanut butter. These data add support to the reported interaction between boron and steroid hormones in postmenopausal women (Nielsen et al., 1987), and suggest a synergistic effect with HRT.

4.5.5.5 *In Vitro* Studies

In cell culture studies, boron (as boric acid) elicits cellular responses that are consistent with a reduction in the risk of hormone-sensitive cancers, such as prostate cancer (Barranco and Eckhert, 2004; Gallardo-Williams et al., 2004; Henderson et al., 2009) and breast cancer (Scorei et al., 2008).

4.5.5.6 Conclusion

To date there is insufficient evidence to determine whether boron is protective against cancer. However, the findings from studies in humans and *in vitro* warrant further research to determine the conditions under which boron may act to modify cancer risk.

4.5.6 Cardiovascular Disease

In humans, boron supplements are reported to lower the platelet count and potentially reduce the risk of thrombosis (Nielsen et al., 1991), while experimental evidence has been obtained for the likely usefulness of boron-containing thrombin inhibitors in the treatment of cardiovascular disorders (Tapparelli et al., 1993). In an acute feeding study, the concentration of activated factor VII was unaffected following the consumption of a high-fat meal with or without supplemental boron (Wallace et al., 2002), and in a longer-term supplementation trial, boron had no effect on plasma lipoprotein–cholesterol concentrations or the propensity of low-density lipoproteins for oxidative modification (Naghii and Samman, 1997). Thus, evidence in humans that boron affects biomarkers of cardiovascular health is conflicting. However, recent studies in animal models suggest that boron deprivation raises the concentrations of plasma homocysteine (Nielsen, 2009), and insulin (Bakken and Hunt, 2003), both of which have been suggested as risk factors for heart disease.

4.6 SUMMARY

Boron is a potentially essential nutrient and appears to have a widespread role in nutrition and metabolism. It is found in foods of plant origin and its usual intake in humans is in the range of 1–2 mg per day. The available information from human studies suggests that boron is absorbed easily and is excreted rapidly in urine. In some studies, boron has been shown to impact on steroid hormone metabolism and related interactions, such as reproductive health and hormone-sensitive cancers. Current difficulties in advancing knowledge of its functions include the limited availability of boron data in tables of food composition, and challenges in the analysis of boron in food and in biopsy materials. The focus on a role for phytochemicals in disease prevention should be expanded to consider the possibility that some of their biological effects are intertwined with those of boron. In view of the distribution of boron in the food chain, mainly in foods of plant origin, the reported benefits of the consumption of vegetables, fruit, and cereals may be contributed in part by boron.

REFERENCES

Allen, H.E., Halley-Henderson, M.A., Hass, C.N. Chemical composition of bottled mineral water. *Arch. Environ. Health* 1989;44:102–116.

Amodio, M.L., Colelli, G., Hasey, J.K., Kader, A.A. A comparative study of composition and postharvest performance of organically and conventionally grown kiwifruits. *J. Sci. Food Agric.* 2007;87:1228–1236.

Anderson, D.L., Cunningham, W.C., Lindstrom, T.R. Concentrations and intakes of H, B, S, K, Na, Cl, and NaCl in foods. *J. Food Comp. Anal.* 1994;7:59–82.

Anderson, R.R. Comparison of trace elements in milk of four species. *J. Dairy Sci.* 1992;75:3050–3055.

Bakken, N.A., Hunt, C.D. Dietary boron decreases peak pancreatic *in situ* insulin release in chicks and plasma insulin concentrations in rats regardless of vitamin D or magnesium status. *J. Nutr.* 2003;133:3577–3583.

Barr, R.D., Barton, S.A., Schull, W.J. Boron levels in man:Preliminary evidence of genetic regulation and some implications for human biology. *Med. Hypotheses* 1996;46:286–289.

Barranco, W.T., Eckhert, C.D. Boric acid inhibits human prostate cancer cell proliferation. *Cancer Lett.* 2004;216:21–29.

Barranco, W.T., Hudak, P.F., Eckhert, C.D. Evaluation of ecological and *in vitro* effects of boron on prostate cancer risk (United States). *Cancer Causes Control.* 2007;18:71–77.

Beattie, J.H., Peace, H.S. The influence of a low-boron diet and boron supplementation on bone, major mineral and sex steroid metabolism in postmenopausal women. *Br. J. Nutr.* 1993;69:871–884.

Benderdour, M., Hess, K., Dzondo-Gadet, M., Nabet, P., Belleville, F., Dousset, B. Boron modulates extracellular matrix and TNFa synthesis in human fibroblasts. *Biochem. Biophys. Res. Commun.* 1998;246:746–751.

Biego, G.H., Joyeux, M., Hartemann, P., Debry, G. Daily intake of essential minerals and metallic micropollutants from foods in France. *Sci. Total Environ.* 1998;217:27–36.

Blech, M.F., Martin, C., Borrelly, J., Hartemann, P. Treatment of deep wounds with loss of tissue. Value of a 3 percent boric acid solution. *Presse Med.* 1990;19:1050–1052.

Bloodworth, B.C. *Report on Boric Acid in Food.* Singapore:Ministry of Health, Department of Scientific Services, 1989:1–12.

Butterwick, L., de Oude, N., Raymond, K. Safety assessment of boron in aquatic and terrestrial environments. *Ecotoxicol. Environ. Safety* 1989;17:339–371.

Chang, B.L., Robbins, W.A., Wei, F., Xun, L., Wu, G., Li, N., Elashoff, D.A. Boron workers in China: Exploring work and lifestyle factors related to boron exposure. *AAOHN J.* 2006;54:435–443.

Chebassier, N., El Houssein, O., Viegas, I., Dréno, B. *In vitro* induction of matrix metalloproteinase-2 and matrix metalloproteinase-9 expression in keratinocytes by boron and manganese. *Exp. Dermatol.* 2004;13;484–490.

Choi, M.K., Jun, Y.S. Analysis of boron content in frequently consumed foods in Korea. *Biol. Trace Elem. Res.* 2008;126:13–26.

Cook, N.C., Samman, S. Flavonoids: Chemistry, metabolism, cardioprotective effects and dietary sources. *J. Nutr. Biochem.* 1996;7:66–76.

Cui, Y., Winton, M.I., Zhang, Z.F., Rainey, C., Marshall, J., De Kernion, J.B., Eckhert, C.D. Dietary boron intake and prostate cancer risk. *Oncol. Rep.* 2004;11:887–892.

Dangour, A.D., Dodhia, S.K., Hayter, A., Allen, E., Lock, K., Uauy, R. Nutritional quality of organic foods: A systematic review. *Am. J. Clin. Nutr.* 2009;90:680–685.

Devirian, T.A., Volpe, S.L. The physiological effects of dietary boron. *Crit. Rev. Food Sci. Nutr.* 2003;43:219–231.

Dzondo-Gadet, M., Nzietchueng, R.M., Hess, K., Nabet, P., Bellevilee, F., Dousset, B. Action of boron at the molecular level: Effects on transcription and translation in an acellular system. *Biol Trace Elem Res.* 2002;85:23–33.

Ferrando, A.A., Green, N.R. The effect of boron supplementation on lean body mass, plasma testosterone levels, and strength in male bodybuilders. *Int. J. Sport Nutr.* 1993;3:140–149.

Fjelkner-Modig, S., Bengtsson, H., Stegmark, R., Nystrom, S. The influence of organic and integrated production on nutritional, sensory and agricultural aspects of vegetable raw materials for food production. *Acta Agric. Scand. Section B Soil Plant Sci.* 2000;50:102–113.

Fort, D.J., Stover, E.L., Strong, P.L., Murray, F.J., Keen, C.L. Chronic feeding of a low boron diet adversely affects reproduction and development in *Xenopus laevis. J. Nutr.* 1999;129:2055–2060.

Gates Zook, E. Lehmann, J. Mineral composition of fruits. *J. Am. Diet. Assoc.* 1968;52:225–231.

Gallardo-Williams, M.T., Chapin, R.E., King, P.E., Moser, G.J., Goldsworthy, T.L., Morrison, J.P., Maronpot, R.R. Boron supplementation inhibits the growth and local expression of IGF-1 in human prostate adenocarcinoma (LNCaP) tumors in nude mice. *Toxicol. Pathol.* 2004;32:73–78.

Gormican, A. Inorganic elements in foods used in hospital menus. *J. Am. Dietet. Assoc.* 1970;56:397–403.

Gonzalez, A., Peters, U., Lampe, J.W., White, E. Boron intake and prostate cancer risk. *Cancer Causes Control.* 2007;18:1131–1140.

Gundersen, V., Bechmann, I.E., Behrens, A., Sturup, S. Comparative investigation of concentrations of major and trace elements in organic and conventional Danish agricultural crops. 1. Onions (*Allium cepa Hysam*) and peas (*Pisum sativum Ping Pong*). *J. Agric. Food Chem.* 2000;48:6094–6102.

Gyorffy, I. Potassium permanganate–boric acid powder in wound treatment. *Chirurg.* 1941;13:45.

Hegsted, M., Keenan, M.J., Siver, F., Wozniak, P. Effect of boron on vitamin D deficient rats. *Biol. Trace Elem. Res.* 1991;28:243–255.

Henderson, K., Stella, S.L., Kobylewski, S., Eckhert, C.D. Receptor activated Ca(2+) release is inhibited by boric acid in prostate cancer cells. *PLoS One.* 2009;4:e6009.

Huang, B., Zhao, Y., Sun, W., Yang, R., Gong, Z., Zou, Z., Ding, F., Su, J. Relationships between distributions of longevous population and trace elements in the agricultural ecosystem of Rugao County, Jiangsu, China. *Environ. Geochem. Health.* 2009;31:379–390.

Hunt, C.D., Butte, N.F., Johnson, L.K. Boron concentrations in milk from mothers of exclusively breast-fed healthy full-term infants are stable during the first four months of lactation. *J. Nutr.* 2005;135:2383–2386.

Hunt, C.D., Friel, J.K., Johnson, L.K. Boron concentrations in milk from mothers of full-term and premature infants. *Am. J. Clin. Nutr.* 2004;80:1327–1333.

Hunt, C.D., Herbel, J.L., Nielsen, F.H. Metabolic responses of postmenopausal women to supplemental dietary boron and aluminum during usual and low magnesium intake: Boron, calcium, and magnesium absorption and retention and blood mineral concentrations. *Am. J. Clin. Nutr.* 1997;65:803–813.

Hunt, C.D., Meacham, S.L. Aluminum, boron, calcium, copper, iron, magnesium, manganese, molybdenum, phosphorus, potassium, sodium, and zinc: Concentrations in common western foods and estimated daily intakes by infants;toddlers;and male and female adolescents, adults, and seniors in the United States. *J. Am. Diet. Assoc.* 2001;101:1058–1060.

Hunt, C.D., Shuler, T.R., Mullen, L.M. Concentration of boron and other elements in human foods and personal-care products. *J. Am. Diet. Assoc.* 1991;91:558–568.

Hunter, D., Foster, M., McArthur, J.O., Ojha, R., Petocz, P., Samman, S. Evaluation of the micronutrient composition of plant foods produced by organic and conventional agricultural methods. *Crit. Rev. Food Sci. Nutr.* (in press).

Hunter, P. Not boring at all: Boron is the new carbon in the quest for novel drug candidates. *EMBO Reports* 2009;10:125–128.

Institute of Medicine. *Dietary Reference Intakes: Arsenic, Boron, Nickel, Silicon, and Vanadium.* National Research Council, Ed. Washington, DC: National Academy Press, 2001, pp. 502–553.

Kent, N.L., McCance, R.A. The absorption and excretion of minor elements by man: Silver, gold, lithium, boron and vanadium. *Biochem. J.* 1941;35:837–844.

Kim, M.H., Bae, Y.J., Lee, Y.S., Choi, M.K. Estimation of boron intake and its relation with bone mineral density in free-living Korean female subjects. *Biol. Trace Elem. Res.* 2008;125:213–222.

Koivistoinen, P. Mineral element composition of Finnish foods. *Acta Agric. Scand.* 1980(Suppl.);22:7–165.

Korkmaz, M., Sayli, U., Sayli, B.S., Bakirdere, S., Titretir, S., Yavuz Ataman, O., Keskin, S. Estimation of human daily boron exposure in a boron-rich area. *Br. J. Nutr.* 2007a;98:571–575.

Korkmaz, M., Uzgören, E., Bakirdere, S., Aydin, F., Ataman, O.Y. Effects of dietary boron on cervical cytopathology and on micronucleus frequency in exfoliated buccal cells. *Environ. Toxicol.* 2007b;22:17–25.

Lester, G.E., Manthey, J.A., Buslig, B.S. Organic vs conventionally grown red whole grapefruit and juice: Comparison of production inputs, market quality, consumer acceptance, and human health-bioactive compounds. *J. Agric. Food Chem.* 2007;55:4474–4480.

Magour, S., Schramel, P., Ovcar, J., Mäser, H. Uptake and distribution of boron in rats: Interaction with ethanol and hexobarbital in the brain. *Arch. Environm. Contam. Toxicol.* 1982;11:521–525.

Mahabir, S., Spitz, M.R., Barrera, S.L., Dong, Y.Q., Eastham, C., Forman, M.R. Dietary boron and hormone replacement therapy as risk factors for lung cancer in women. *Am. J. Epidemiol.* 2008;167:1070–1080.

Massie, H.R., Whitney, S.J., Aiello, V.R., Sternick, S.M. Changes in boron concentration during development and ageing of *Drosophila* and effect of dietary boron on life span. *Mech. Ageing Dev.* 1990;53:1–7.

Mastromatteo, E., Sullivan, F. Summary: International Symposium on the health effects of boron and its compounds. *Environ. Health Perspect.* 1994;102(Suppl 7):139–141.

Meacham, S.L., Hunt, C.D. Dietary boron intakes of selected populations in the United States. *Biol. Trace Elem. Res.* 1998;66:65–78.

Meacham, S.L., Taper, L.J., Volpe, S.L. Effect of boron supplementation on blood and urinary calcium, magnesium, and phosphorus, and urinary boron in athletic and sedentary women. *Am. J. Clin. Nutr.* 1995;61:341–345.

Murphy, E.W., Page, L., Watt, B.K. Trace minerals in type A school lunches. *J. Am. Diet. Assoc.* 1971;58:115–122.

Naghii, M.R., Lyons Wall, P.M., Samman, S. The boron content of selected foods and the estimation of its daily intake among free-living subjects. *J. Am. Coll. Nutr.* 1996;15:614–619.

Naghii, M.R., Samman, S. Role of boron in nutrition and metabolism. *Prog. Food Nutr. Sci.* 1993;14:331–349.

Naghii, M.R., Samman, S. The effect of boron supplementation on its urinary excretion and selected cardiovascular risk factors in healthy male subjects. *Biol. Trace Elem. Res.* 1997;56:273–286.

Nielsen, F.H. Nutritional requirements for boron, silicon, vanadium, nickel, and arsenic: Current knowledge and speculation. *FASEB J.* 1991;5:2661–2667.

Nielsen, F.H. Boron in human and animal nutrition. *Plant and Soil* 1997;193:199–208.

Nielsen, F.H. The emergence of boron as nutritionally important throughout the life cycle. *Nutrition* 2000;16:512–514.

Nielsen, F.H. Is boron nutritionally relevant? *Nutr. Rev.* 2008;66:183–191.

Nielsen, F.H., Hunt, C.D., Mullen, L.M., Hunt, J.R. Effect of dietary boron on mineral, estrogen, and testosterone metabolism in postmenopausal women. *FASEB J.* 1987;1:394–397.

Nielsen, F.H., Mullen, L.M., Nielsen, E.J. Dietary boron affects blood cell counts and hemoglobin concentrations in humans. *J. Trace Elem. Exp. Med.* 1991;4:211–223.

Nielsen, F.H., Penland, J.G. Boron deprivation alters rat behaviour and brain mineral composition differently when fish oil instead of safflower oil is the dietary fat source. *Nutr. Neurosci.* 2006;9:105–112.

Nielsen, F.H. Boron deprivation decreases liver *S*-adenosylmethionine and spermidine and increases plasma homocysteine and cysteine in rats. *J. Trace Elem. Med. Biol.* 2009;23:204–213.

Nielsen, F.H. The alteration of magnesium, calcium and phosphorous metabolism by dietary magnesium deprivation in postmenopausal women is not affected by dietary boron deprivation. *Magnesium Res.* 2004;17:197–210.

O'Neill, M.A., Warrenfeltz, D., Kates, K., Pellerin, P., Doco, T., Darvill, A.G., Albersheim, P. Rhamnogalacturonan-II, a pectic polysaccharide in the walls of growing plant cell, forms a dimer that is covalently cross-linked by a borate ester. *In vitro* conditions for the formation and hydrolysis of the dimer. *J. Biol. Chem.* 1996;271:22923–22930.

Park, M., Li, Q., Shcheynikov, N., Zeng, W., Muallem, S. NaBC1 is a ubiquitous electrogenic Na+ -coupled borate transporter essential for cellular boron homeostasis and cell growth and proliferation. *Mol. Cell.* 2004;16:331–341.

Peck, G.M., Andrews, P.K., Reganold, J.P., Fellman, J.K. Apple orchard productivity and fruit quality under organic, conventional, and integrated management. *Hortscience* 2006;41:99–107.

Penland, J.G. Dietary boron, brain function and cognitive performance. *Environ. Health Perspectives.* 1994;102(Suppl 7):65–72.

Penland, J.G. Qualitative analysis of EEG effects following experimental marginal magnesium and boron deprivation. *Magnesium Res.* 1995a;8:341–358.

Penland, J.G. The importance of boron nutrition for brain and psychological function. *Biol. Trace Elem. Res.* 1998;66:299–317.

Pieczyńska, J., Borkowska-Burnecka, J., Biernat, J., Grajeta, H., Zyrnicki, W., Zechałko-Czajkowska, A. Boron content in daily meals for preschool children and school youth. *Biol. Trace Elem. Res.* 2003;96:1–8.

Rainey, C., Nyquist, L. Multi-country estimation of dietary boron intake. *Biol. Trace Elem. Res.* 1998;66:79–86.

Rainey, C.J., Nyquist, L.A, Christensen, R.E., Strong, P.L., Culver, B.D., Coughlin, J.R. Daily boron intake from the American diet. *J. Am. Diet. Assoc.* 1999;99:335–340.

Robbins, W.A., Wei, F., Elashoff, D.A., Wu, G., Xun, L., Jia, J. Y:X sperm ratio in boron-exposed men. *J. Androl.* 2008;29:115–121.

Rowe, R.I., Eckhert, C.D. Boron is required for zebrafish embryogenesis. *J. Exp. Biol.* 1999;202:1649–1654.

Samman, S., Chow, J.W.Y., Foster, M.J., Ahmad, Z.I., Phuyal, J.L., Petocz, P. Fatty acid composition of edible oils derived from certified organic and conventional agricultural methods. *Food Chem.* 2008;109:670–674.

Samman, S., Lyons Wall, P.M., Lai, N.T., Sullivan, D.R. Intake of selected phytochemicals and prevention of coronary heart disease: Practical implications. In: *Antioxidants and Cardiovascular Disease.* (Eds) Nath, R., Khullar, M. Singal, P.K. New Delhi: Narosa Publishing House. 2004;pp. 194–202.

Samman, S., Naghii, M.R., Lyons Wall, P.M., Verus, A.P. The nutritional and metabolic effects of boron in humans and animals. *Biol. Trace Elem. Res.* 1998;66:227–235.

Scorei, R., Ciubar, R., Ciofrangeanu, C.M., Mitran, V., Cimpean, A., Iordachescu, D. Comparative effects of boric acid and calcium fructoborate on breast cancer cells. *Biol. Trace Elem. Res.* 2008;122:197–205.

Simsek, A., Aykut, O. Evaluation of the microelement profile of Turkish hazelnut (*Corylus avellana* L.) varieties for human nutrition and health. *Int. J. Food Sci. Nutr.* 2007;58:677–688.

Simsek, A., Velioglu, Y.S., Coskun, A.L., Sayli, B.S. Boron concentrations in selected foods from borate-producing regions in Turkey. *J. Sci. Food Agric.* 2003;83:586–592.

Steinnes, E. Soild and geomedicine. *Environ. Geochem. Health* 2009;31:523–535.

Sutherland, B., Woodhouse, L.R., Strong, P., King, J.C. Boron balance in humans. *J. Trace Elem. Exp. Med.* 1999;12:271–284.

Takano, J., Noguchi, K., Yasumori, M., Kobayashi, M., Gajdos, Z., Miwa, K., Hayashi, H., Yoneyama, T., Fujiwara, T. Arabidopsis boron transporter for xylem loading. *Nature.* 2002;420:337–340.

Tapparelli, C., Metternich, R., Ehrhardt, C., Zurini, M., Claeson, G., Scully, M.F., Stone, S.R. *In vitro* and *in vivo* characterization of a neutral boron-containing thrombin inhibitor. *J. Biol. Chem.* 1993;268:4734–4741.

Touillaud, M.S., Pillow, P.C., Jakovljevic, J., Bondy, M.L., Singletary, S.E., Li, D., Chang, S. Effect of dietary intake of phytoestrogens on estrogen receptor status in premenopausal women with breast cancer. *Nutr. Cancer.* 2005;51:162–169.

Wallace, J.M., Hannon-Fletcher, M.P., Robson, P.J., Gilmore, W.S., Hubbard, S.A., Strain, J.J. Boron supplementation and activated factor VII in healthy men. *Eur. J. Clin. Nutr.* 2002;56:1102–1107.

Warman, P.R. Soil fertility, yield and nutrient contents of vegetable crops after 12 years of compost or fertilizer amendments. *Biol. Agric. Hort.* 2005;23:85–96.

Warman, P.R., Havard, K.A. Yield, vitamin and mineral contents of organically and conventionally grown carrots and cabbage. *Agric. Ecosystems Env.* 1997;61:155–162.

Warman, P.R., Havard, K.A. Yield, vitamin and mineral contents of organically and conventionally grown potatoes and sweet corn. *Agric. Ecosystems Env.* 1998;68:207–216.

Wszelaki, A.L., Delwiche, J.F., Walker, S.D., Liggett, R.E., Scheerens, J.C., Kleinhenz, M.D. Sensory quality and mineral and glycoalkaloid concentrations in organically and conventionally grown redskin potatoes (*Solanum tuberosum*). *J. Sci. Food Agric.* 2005;85:720–726.

World Health Organisation. *Trace Elements in Human Nutrition and Health.* Geneva: World Health Organisation;1996;pp. 175–179.

World Health Organisation. Boron. *International Programme on Chemical Safety, Environmental Health Criteria Monograph #204.* Geneva: World Health Organisation;1998.

Xing, X., Wu, G., Wei, F., Liu, P., Wei, H., Wang, C., Xu, J. et al. Biomarkers of environmental and workplace boron exposure. *J. Occup. Environ. Hyg.* 2008;5:141–147.

Yazbeck, C., Kloppmann, W., Cottier, R., Sahuquillo, J., Debotte, G., Huel, G. Health impact evaluation of boron in drinking water: A geographical risk assessment in Northern France. *Environ. Geochem. Health.* 2005;27:419–427.

Ysart, G., Miller, P., Crews, H., Robb, P., Baxter, M., De L'Argy, C., Lofthouse, S., Sargent, C., Harrison, N. Dietary exposure estimates of 30 elements from the UK Total Diet Study. *Food Addit. Contam.* 1999;16:391–403.

Part III

Boron for Living: Radioisotope

One major use of radioisotopes is in nuclear medicine. Of the 30 million people who are hospitalized each year in the United States, one third are treated with nuclear medicine. More than 10 million nuclear-medicine procedures are performed on patients and more than 100 million nuclear-medicine tests are performed each year in the United States alone. A comparable number of such procedures are performed in the rest of the world.

What is a radioisotope? When a combination of neutrons and protons, which does not already exist in nature, is produced artificially, the atom will be unstable and is called a radioactive isotope or radioisotope. There are also a number of unstable natural isotopes arising from the decay of primordial uranium and thorium. Overall there are some 1800 radioisotopes. At present there are up to 200 radioisotopes used on a regular basis, and most must be produced artificially.

A very effective role for radioisotopes in nuclear medicine is the use of short-lived positron emitters such as 11C, 13N, 15O, or 18F in a process known as positron emission tomography. Incorporated in chemical compounds that selectively migrate to specific organs in the body, diagnosis is effected by detecting annihilation gamma rays–two gamma rays of identical energy emitted when a positron and an electron annihilate each other. These gamma rays have the very useful property that they are emitted in exactly opposite directions. When both are detected, a computer system may be used to reconstruct where the annihilation occurred. By attaching a positron emitter to a protein or a glucose molecule, and allowing the body to metabolize it, we can study the functional aspect of an organ such as the human brain. Therefore, the radioisotopes play a crucial role in nuclear medicine that extends the lifespan of living people.

Part III consists of two chapters. Chapter 5 by Kabalka discusses the use of boron in the incorporation of stable and unstable radioisotopes in molecules using organoborane reagents. The isotopes usually incorporated are those of H, C, N, O, and the halides. The growing need for structurally complex compounds labeled with such isotopes will continue to be a challenge. It is clear that organoboration will play an ever-increasing role in this area of research. On the other hand, Chapter 6 by Tjarks, et al. discusses the syntheses and uses of (radio)halogenated boron clusters in tumor imaging and therapy. The undesired *in vivo* release of radiohalogens, frequently observed in therapeutics and diagnostics radiolabeled at carbon atoms, has been a major motivation for the exploration of boron clusters as "prosthetic groups" for radiohalogens in radiopharmaceuticals. Not only are the halogen–boron bonds stronger than carbon–halogen bonds, the boron–halogen bonds are less susceptible to enzymatic and hydrolytic cleavage than the carbon–halogen bonds; no enzymatic system has yet been reported for the *in vivo* cleavage of the boron–halogen bond. The fact that novel boron compounds, especially those based on the polyhedral boron hydride design, are unfamiliar to life gives them an advantage in living systems. Since pathogens develop less or no resistance against those boron-based drugs, enzymatic systems will be less able to metabolize them. Also, the kind of interactions would be somewhat different from that of the natural key–lock systems that have been developed over billions of years.

5 Isotope Incorporation Using Organoboranes

George W. Kabalka

CONTENTS

5.1 INTRODUCTION

Compounds labeled with isotopes have played an important role in chemistry, biology, and medicine since they were first used as tracers by Hevesey.[1,2] Both stable[3–5] and radioactive[6,7] isotopes were utilized in early investigations, but the situation changed dramatically with the invention of the cyclotron by Lawrence in 1930 and the construction of the nuclear reactor by Fermi in 1942 that enabled access to radioisotopes on a regular basis. Radioisotope use in medicine was also accelerated by advances in radiation-detection techniques.[8,9] The development of single-photon emission computerized tomography (SPECT)[10,11] and positron emission tomography (PET)[12–14] revolutionized diagnostic medical imaging by facilitating noninvasive, *in vivo*, three-dimensional imaging of living systems after administration of appropriate compounds labeled with short-lived nuclides.[15,16] PET and SPECT techniques complement the traditional x-ray CAT scan and are responsible for the tremendous upsurge in the use of radiopharmaceuticals in hospitals today.

The use of stable isotopes for labeling organic molecules trailed radioisotope usage due to the lack of availability of sufficient quantities of useful stable isotopes (carbon-13, nitrogen-15, oxygen-17, etc.) as well as a paucity of appropriate analytical techniques. The situation changed dramatically when, in 1969, the Division of Biology and Medicine of U.S. Atomic Energy Commission established a program to make increased quantities of the stable isotopes of carbon, oxygen, and nitrogen available at significantly reduced costs. At about the same time, rapid advances in instrumentation were occurring. Multinuclear magnetic resonance and gas chromatography–mass spectrometry techniques were developed, which could be applied to the analysis of stable isotopic mixtures as well as identifying locations of the stable isotope in a given molecule.[17–20] On the medical front, the development of multinuclear, whole-body MRI scanners enhanced the need for compounds labeled with NMR-active, stable isotopes such as carbon-13, nitrogen-15, and oxygen-17.[21,22]

The preparation of stable and unstable isotopically labeled compounds generally begins with a few basic building blocks such as carbon dioxide, carbonate salts, nitric oxide, water, and halide salts.[23–25] These are the species that are generated in the cyclotrons and reactors or that are the most amenable to cryogenic distillation (the most common method for separating stable isotopes). The chemical stability of these building blocks has traditionally been a significant barrier to the synthesis of complex isotopically labeled agents.[26–28] The problem can be quite challenging for the syntheses of radiopharmaceuticals containing short-lived isotopes such as carbon-11 ($t_{1/2}$ = 20.4 min), nitrogen-13 ($t_{1/2}$ = 10 min), oxygen-15 ($t_{1/2}$ = 2 min), and fluorine-18 ($t_{1/2}$ = 110 min). Ironically, these are generally the isotopes of greatest interest for studies involving humans due to their decay characteristics (positron emission resulting in the emission of two 511 keV annihilation photons which can be readily detected outside the body).[29] In addition, radioisotopes are generally supplied in very high specific activities (which is a measure of the ratio of radioactive atoms versus their nonradioactive isotopes (carrier) in the sample.) Generally, no-carrier-added radiotracers contain very little mass. For example, a fluorine-18-labeled sample of commercially available $Na^{18}F$ (a common isotope in medical imaging) is delivered from the cyclotron with a specific activity approaching 1.0×10^9 Ci/mmol;[30] since a typical laboratory synthesis is carried out with 50 milliCuries or less of $Na^{18}F$, the total mass of fluorine-18 in the reaction is of the order of 50×10^{-12} mol (50 fmol). From the chemist's perspective the short half-lives and low concentrations of these radionuclides provide formidable challenges (the low concentrations often render classic (e.g., S_N2) reaction pathways ineffective).

Prior to the mid-1970s, incorporation of carbon isotopes generally involved the use of CO_2 incorporation through Grignard (and organolithium) reagents or cyanide incorporation through S_N2 chemistry. For long-lived and stable isotopes this chemistry was often acceptable because the incorporation could be carried out early in the synthesis and desired functional groups, if any, could be added after the molecular framework was completed. Although not particularly efficient, the early incorporation was often necessary because of the stringent reaction conditions required during the installation of the carbon fragment (i.e., the Grignard reagent's basicity or the acidity of the hydrolysis media). The situation was more complex for carbon-11 because its 20-min half-life limited the number of reactions that could be accomplished after installation of the radioactive element. A parallel situation existed for the synthesis of molecules containing positron emitting nitrogen, oxygen, and halogen isotopes.

The application of modern organometallic methodology (organoboron, organotin, organosilicon, etc.) to the synthesis of isotopically labeled compounds began in the 1970s. The bulk of the early work was done with the organoboranes.[31,32] This was a consequence of the fact that a wide variety of functionally substituted organoboranes could be prepared using the simple hydroboration techniques developed by Nobel Laureate Herbert C. Brown and his research associates.[33–35] The new chemistry avoided the harsh conditions employed in more traditional Grignard and solvolysis reactions (Scheme 5.1).[36]

Brown and others had also demonstrated the versatility of the organoboranes in a wide variety of organic syntheses. These initial studies demonstrated an important aspect of the organoboranes: they could be prepared from molecules containing a wide variety of functional groups.[37] This is important from a medical viewpoint since functional groups are generally responsible for physiological activity. More importantly, the early research in Brown's laboratories demonstrated that a variety of elements would replace boron under appropriate reaction conditions, and these elements possessed both stable and radioactive isotopes having importance in biology, medicine,

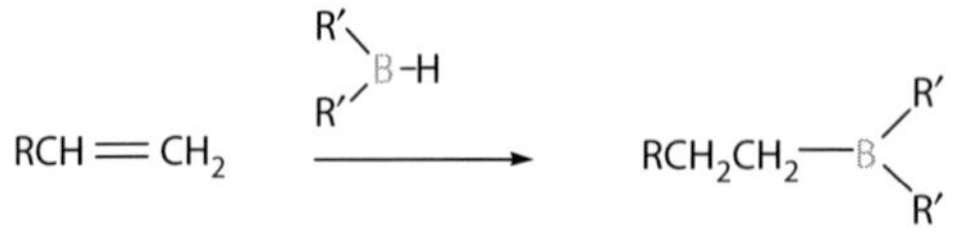

$$RCH{=}CH_2 \xrightarrow{R'_2B\text{-}H} RCH_2CH_2\text{—}BR'_2$$

Where: R′ = H, alkyl, aryl, vinyl, alkoxy, halo, etc.

R may contain: –OH, –SH, –NH2 –S–S–, –CN –$CONH_2$, –CO_2R, –CO_2H etc.

SCHEME 5.1 Overview of the synthesis of organoboranes via hydroboration.

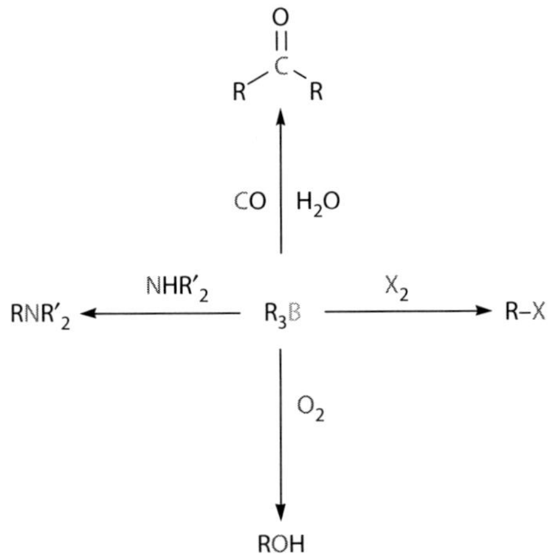

SCHEME 5.2 Early boron-substitution reactions.

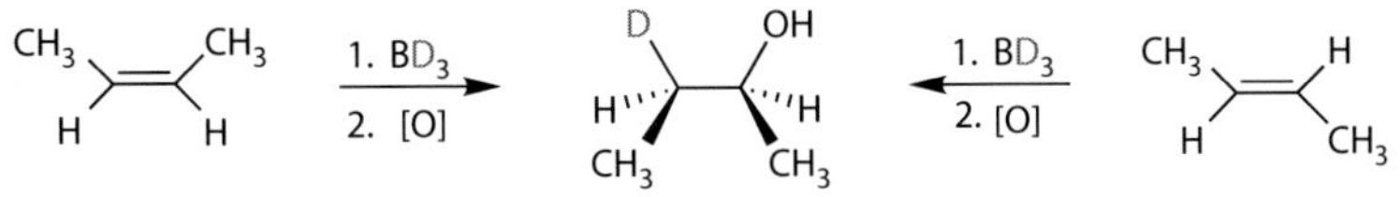

SCHEME 5.3 The use of deuterium to determine the stereochemistry of the hydroboration–oxidation sequence.

and agriculture (Scheme 5.2). The reported reaction conditions were, however, quite stringent and the reaction times were too lengthy for utilization in most radiopharmaceutical syntheses. In addition, little information was available concerning the mechanism and kinetics of many of the organoborane reactions known at that time.

By the mid-1970s, only the hydrogen isotopes, tritium and deuterium, had been incorporated into organic molecules using organoborane chemistry. Soon after the initial report describing the hydroboration reaction, Brown utilized deuteration reactions to determine the stereochemistry and regiochemistry of the hydroboration reaction and the related solvolysis reactions of organoboranes.[38,39] Tritiated alkenes were also prepared using a hydroboration–tritiation sequence involving tritiated BH_3.[40] In later studies, deuteroboration chemistry was used to confirm the stereochemistry of both the hydroboration[41] and hydroboration/oxidation[42] reactions (Scheme 5.3).

The versatility of the organoboranes as synthetic precursors provided the basis for subsequent studies centered on the incorporation of stable and radioisotopes of elements other than hydrogen.

5.2 CARBON

Carbon isotopes have played an important role in biological, chemical, medical, and agricultural research. Ironically, the first radioactive isotope of carbon to be used in research was carbon-11,[43] but because of its short half-life (20 min), it was soon displaced by carbon-14 in radiotracer studies. There were two carbon-insertion reactions that appeared to be suited for incorporating carbon isotopes using organoboranes: carbon monoxide[44] and cyanide ion (Scheme 5.4).[45]

The first use of organoboranes for incorporating carbon isotopes was reported in 1979.[46] In this case the carbonylation reaction was used to prepare a trialkylcarbinol, a ketone, and an aldehyde (Scheme 5.5).

The reaction could also be used for carbon-13 incorporation into a wide variety of reagents.[47] Additionally, it was reported that the cyanide incorporation reactions were readily applicable to incorporation of both carbon-14 and the stable carbon-13 isotopes (Scheme 5.6).[48–50]

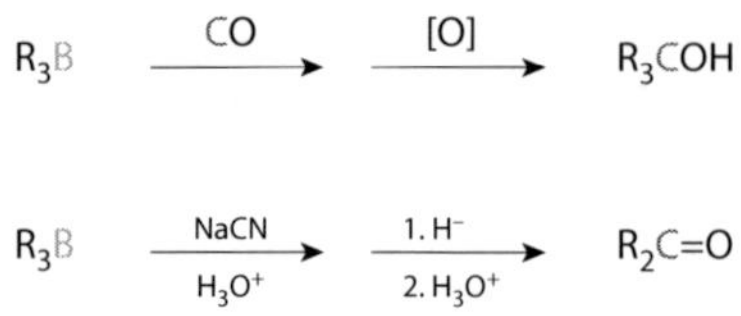

$$R_3B \xrightarrow{CO} \xrightarrow{[O]} R_3COH$$

$$R_3B \xrightarrow[H_3O^+]{NaCN} \xrightarrow[2.\ H_3O^+]{1.\ H^-} R_2C{=}O$$

SCHEME 5.4 The reaction of organoboranes with carbon monoxide and cyanide ion.

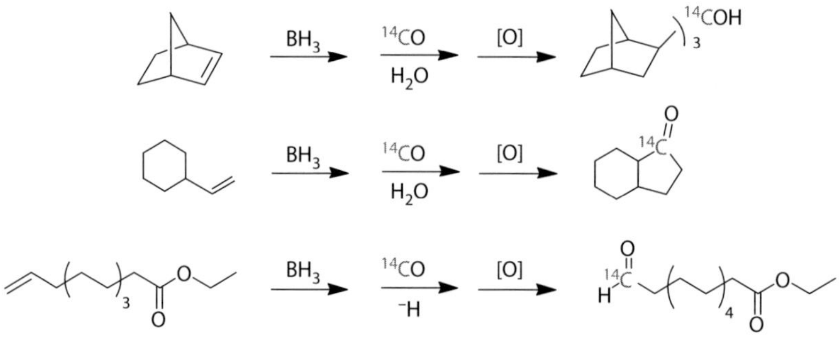

SCHEME 5.5 Introduction of carbon-14 utilizing organoborane carbonylation chemistry.

Starting in the 1980s, there was a renewed interest in reactions that could be used to incorporate carbon-11. Its short half-life (20 min) and the fact that it was a positron emitter made it especially valuable for medical imaging. As one might imagine, the short half-life presented significant challenges to the synthetic chemist.[51,52] In the 1980s, most of the carbon-11 insertion reactions involved incorporation of carbon-11-labeled carbon dioxide or cyanide ion using classical S_N2 and Grignard chemistry. More recently, microwave[53] and solid-state reactions[54] have also been utilized. The reaction of organoboranes with carbon monoxide appeared to be an ideal method for incorporating carbon isotopes. However, the reported reaction times (hours) for most carbonylation reactions were far too long for incorporation of carbon-11. Interestingly, hydride-induced reactions had been reported that appeared to be complete in less than an hour.[55] On further investigation, it was discovered that organoborane carbonylation reactions involving hydride mediation were complete in less than 15 min! These reactions involve a single migration of an organic group to the carbon monoxide and can be used to prepare alcohols,[56] aldehydes,[57] and carboxylic acids (through subsequent oxidation reactions) (Scheme 5.7). The organoborane cyanidation reaction can also be utilized to make complex carbon-11-labeled molecules in less than an hour (Scheme 5.8).[58]

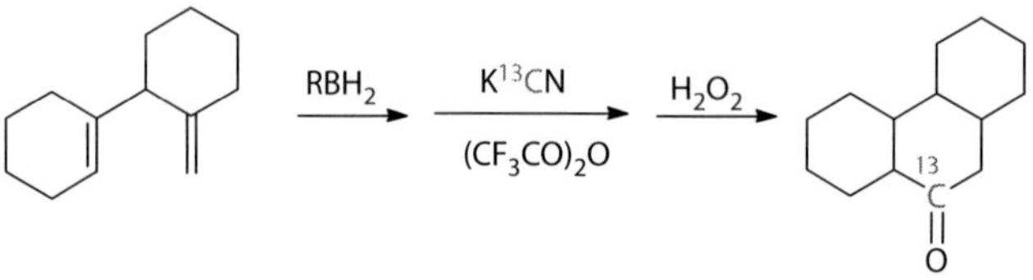

SCHEME 5.6 Introduction of carbon-13 using the cyanidation of organoboranes.

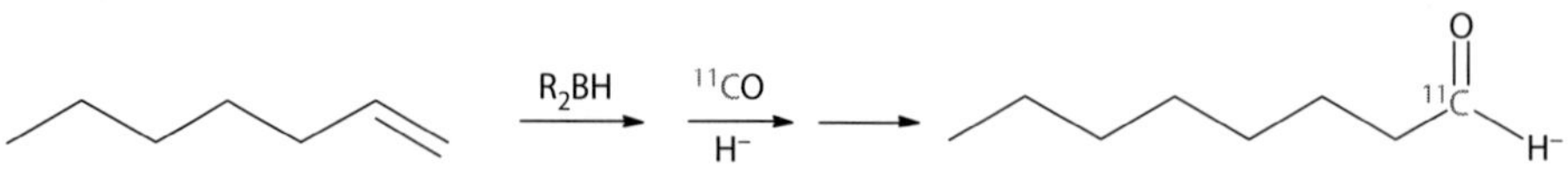

SCHEME 5.7 Introduction of carbon-11 using hydride-induced carbonylation.

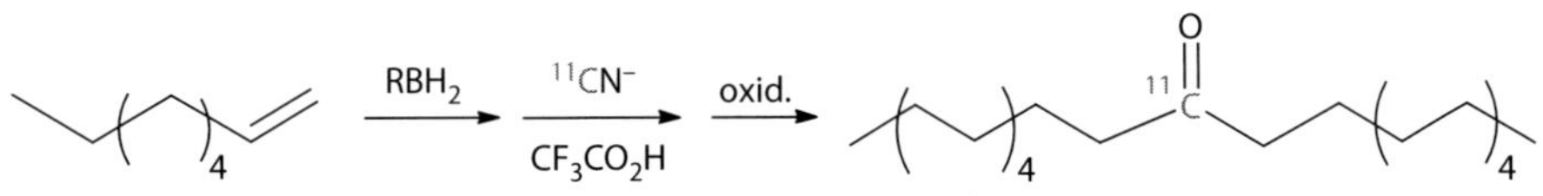

$$\xrightarrow{RBH_2} \xrightarrow[CF_3CO_2H]{^{11}CN^-} \xrightarrow{oxid.}$$

SCHEME 5.8 Introduction of carbon-11 via the reaction of radiolabeled cyanide ion.

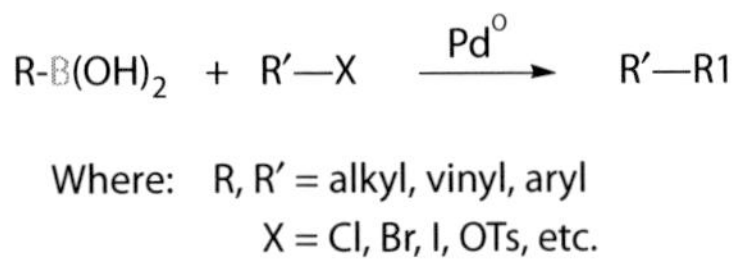

$$\text{R-B(OH)}_2 + \text{R'—X} \xrightarrow{Pd^0} \text{R'—R1}$$

Where: R, R′ = alkyl, vinyl, aryl
X = Cl, Br, I, OTs, etc.

SCHEME 5.9 The Suzuki–Miyaura coupling of organoboronic acids and organic halides.

Even though it was well established by 1990 that organoboranes could be used to incorporate carbon isotopes, including carbon-11, it took another decade before boranes were once again examined as precursors for carbon-isotope incorporation. The reason for the apparent lack of interest was that, in general, researchers found the available preparative routes to the aromatic and vinylic organoborane intermediates to be rather limited. Trialkylboranes were readily accessible through the hydroboration reaction but only if the appropriate alkenes were available. However, the preparation of aromatic borane derivatives was more problematic in that transmetallation reactions were generally required.[59] The necessity of using reactive (and basic) Grignard and organolithium reagents often precluded the presence of interesting and useful functional groups.

The situation changed dramatically when Akira Suzuki and Norio Miyaura discovered a new palladium-catalyzed carbon–carbon bond-forming reaction. The reaction involves the insertion of palladium into carbon–halogen bonds through an oxidative addition, transfer of an organic group from boron to palladium to form a palladium with two organic groups attached, and then reductive elimination of the palladium to generate a new carbon–carbon bond (Scheme 5.9).[60,61]

The coupling reactions generally occur at, or near, room temperature and require the presence of only mild organic and inorganic bases. The reactions can also be utilized to insert carbon oxides by simply adding carbon monoxide or carbon dioxide (Scheme 5.10).[62,63]

As significant as Suzuki chemistry is, the true catalyst for the increased use in the new carbon–carbon bond-forming methodology was a borylation reaction first reported by Ishiyama, which provided a straightforward route to the desired arylboronic ester (and thus the acid) starting materials (Scheme 5.11).[64]

The new boronation reaction is not confined to halide-starting materials! Aryl triflates also react with pinacol diboron and related diboron reagents.[65] These developments have changed the face of organic synthesis by making a wide variety of boronic acids and esters readily available in the

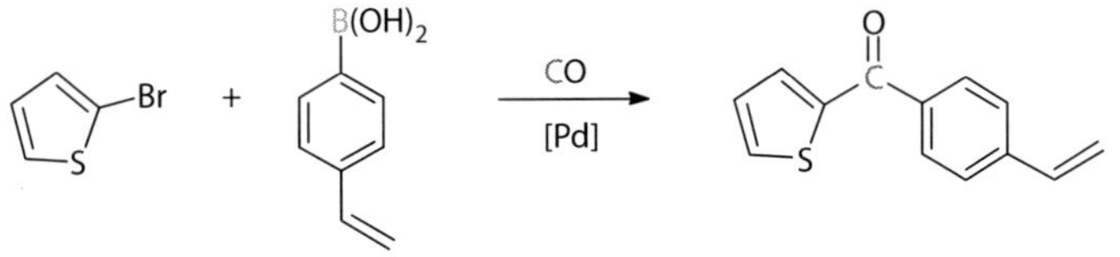

$$\xrightarrow[\text{[Pd]}]{CO}$$

SCHEME 5.10 Preparation of ketones using Suzuki–Miyaura coupling.

$$\text{Ar—X} + (RO)_2B\text{—}B(OR)_2 \xrightarrow[KOAc]{PdCl_2} \text{Ar—B}(OR)_2$$

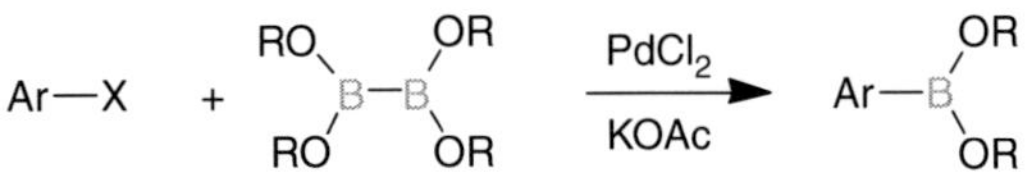

SCHEME 5.11 The preparation of organoboronic acids using diboron esters.

laboratory and from commercial sources. The recent use of chiral catalysts in the preparation of boronic-ester intermediates,[66,67] Suzuki coupling,[68–70] the readily available functionalized boronic acid (and ester) starting materials, and the nontoxic nature of boron (except for roaches and termites!) has served to make the new boron-based coupling chemistry very attractive to the synthetic and medicinal chemistry communities. The renewed interest has carried over to the short-lived isotope community where carbon-11 is a very important isotope because of regulatory concerns in the medical field. (Since the chemical and physiological properties of a chemical substance are not dependent on the carbon-isotope ratio, toxicity data for carbon-11 analogs of stable molecules are readily available.) Langstrom and his coworkers were among the first to reexamine organoboron chemistry for carbon-11 incorporation using Suzuki methodology.[71] Their initial focus was on coupling carbon-11-labeled-methyl iodide to an alkylboron reagent (Scheme 5.12).

Other syntheses using the Suzuki methodology include the preparation of carbon-11-labeled palmitic acid with carbon-11 at the terminal position[72] and a diaryl alkyne, glutamate-receptor agent, Figure 5.1.[73]

The methylation of aromatic groups has also been found to provide carbon-11-labeled products in high yields and high radiochemical purity.[74] It is important to remember that applications of the Suzuki technology are not restricted to alkylation chemistry.

As noted earlier, carbon-11 carbonylation reactions had been limited by the paucity of organoborane preparative routes. Suzuki and Ishiyama solved the problem and then reported new palladium-catalyzed carbonylation reactions that hold great promise.[75] Zeisler demonstrated that, using the new palladium-catalyzed carbonylation reactions, carbon-11-labeled benzophenone could be prepared from phenylboronic acid[76] in yields that were equivalent to those obtained earlier using trialkylboranes (Scheme 5.13).[77]

In an illustration of the of the reaction's functional group tolerance, the reaction was used to prepare carbon-11-labeled 2-(2-benzoylphenoxy)-*N*-phenylacetamide from phenylboronic acid (Scheme 5.14).[78]

$^{11}CH_3I$ + [alkyl-9-BBN] $\xrightarrow{Pd^0}$ [alkyl]$^{11}CH_3$

SCHEME 5.12 The preparation of carbon-11-labeled methyl derivatives using Suzuki–Miyaura chemistry.

CH_3 S N $^{11}CH_3$ CN

FIGURE 5.1 A carbon-11-labeled glutamate receptor agent prepared using Suzuki–Miyaura chemistry.

$B(OH)_2$ + I $\xrightarrow[Pd]{^{11}CO}$ O ^{11}C

SCHEME 5.13 The insertion of carbon-11 via the Suzuki–Miyaura chemistry.

$B(OH)_2$ + I O NH O $\xrightarrow[Pd]{^{11}CO}$ O ^{11}C O NH O

SCHEME 5.14 The use of palladium-catalyzed carbonylation chemistry in the preparation of functionally substituted molecules.

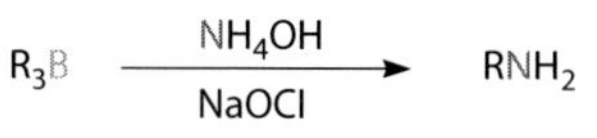

$$R_3B \xrightarrow[\text{NaOCl}]{\text{NH}_4\text{OH}} RNH_2$$

SCHEME 5.15 Replacement of boron by a nitrogen atom.

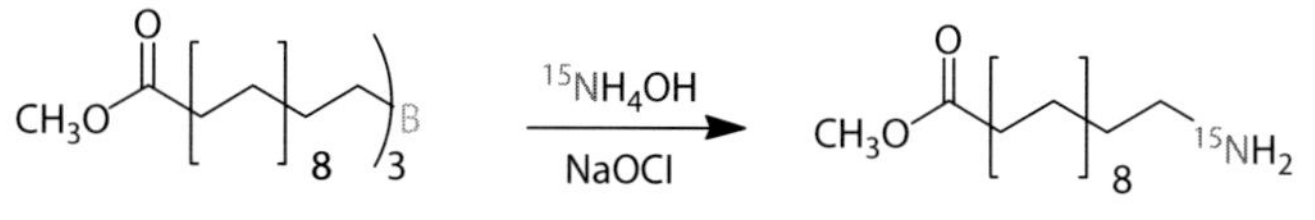

SCHEME 5.16 Incorporation of nitrogen-15 via organoborane chemistry.

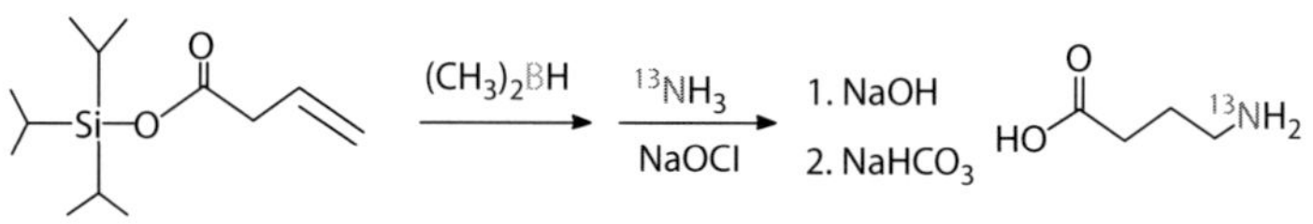

SCHEME 5.17 The preparation of nitrogen-13-labeled γ-aminobutyric acid.

More recently, Langstrom and his collaborators investigated the effect of various bases[79] and leaving groups on the carbonylation reaction and its application to the preparation of carbon-11-labeled molecules.[80] These studies provide the groundwork for the ready application of the Suzuki chemistry to routine radiopharmaceutical production.[81] There is reason to believe that advances will appear at an increasing rate as PET researchers become aware of the versatility of organoboranes in synthesis. Interestingly, borane itself has been utilized to efficiently capture carbon-11 carbon monoxide.[82]

5.3 NITROGEN

Nitrogen isotopes have a number of applications in biology and agriculture. More recently, nitrogen-13 has been studied for possible use in medicine because it is a positron emitter. H. C. Brown laid the foundation for the use of organoborane precursors in amine preparations. He and his coworkers found that chloramines and hydroxylamine-*O*-sulfonic acid would react with organoboranes to yield amines.[83,84] Unfortunately, isotopically enriched analogs of these reagents are not readily available (chloramine is not stable and cannot be conveniently stored or shipped.) The situation, of course, is especially problematic for nitrogen-13 which has a 10 min half-life.

The discovery that chloramines could be prepared *in situ* through the straightforward reaction of sodium hypochlorite and ammonium salts provided the solution to the problem. Even more fascinating was the revelation that the reaction could be carried out in the presence of organoborane precursors to provide a rapid *in situ* amination process (Scheme 5.15).[85]

The reaction can also be used to synthesize dialkylamines.[86] The new chemistry was used in the preparation of nitrogen-15-labeled reagents in exceptionally high yields (Scheme 5.16).[87]

This methodology was also employed in the preparation of a series of nitrogen-13-labeled amines[88] including nitrogen-13-labeled γ-aminobutyric acid and putrescine.[89] Polymeric precursors were developed in an effort to increase the efficiency of the amination reactions but, to date, the instability of the organoborane polymers have been problematic (Scheme 5.17).[90]

5.4 OXYGEN

Oxygen isotopes are also of value in biology, medicine, and agriculture. It is well known to scientists in the organoborane field that low-molecular-weight organoboranes are pyrophoric (they auto ignite in air), a property noted by Franklin in his initial report.[91] In fact, the bright green flame that is generated when an organoborane is ignited is analytical confirmation of the presence of boron.[92–95]

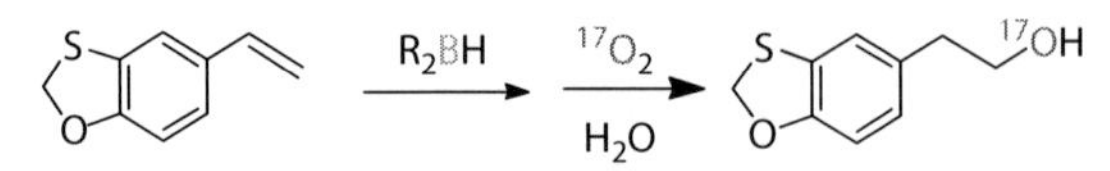

SCHEME 5.18 Incorporation of oxygen-17 via reaction of an organoborane with oxygen gas.

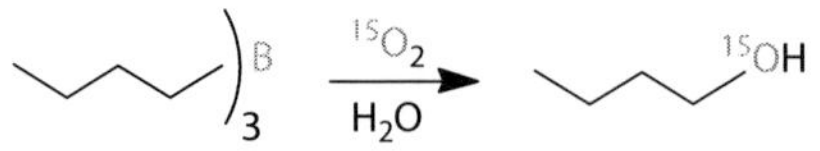

SCHEME 5.19 Preparation of oxygen-15-labeled butanol.

Numerous oxidation methods have been developed to simplify the oxidation of organoboranes over the years but none are particularly useful for incorporating specific isotopes of oxygen.[96–100] A controlled reaction of organoboranes with oxygen gas was discovered in the 1970s and it was found to be useful in the production of organic alcohols.[101,102] The reaction was later applied to the synthesis of oxygen-17-labeled alcohols (Scheme 5.18).[103]

Interest in oxygen-15-labeled alcohols increased when it was discovered that the biodistribution of carbon-11-labeled butanol was superior to oxygen-15-labeled water for use in PET blood flow studies.[104] The short half-life of oxygen-15 (2 min) made it possible to carry out a blood pool study using PET prior to injection of a longer-lived biologically active agent such as fluorine-18-labeled deoxyglucose. Such a dual study would not be feasible using the longer-lived carbon-11-labeled butanol. A simple organoborane oxidation route was soon developed to generate oxygen-15-labeled butanol (Scheme 5.19).[105]

The pyrogenicity of low-molecular-weight organoborane derivatives led others to modify the chemistry by adsorbing the tributylborane precursors on alumina.[106,107] Oxygen-15-labeled butanol has found limited use in the PET community.[108–111]

5.5 IODINE AND THE HALOGENS

Radiohalogens have played a long and important role in medicine.[112] In earlier work, Brown and his collaborators discovered that organoboranes could be used as precursors to a variety of organic halides.[113,114] Unfortunately, the reactions required the presence of a strong base that could lead to the destruction of sensitive functional groups as well as induce potential side reactions such as dehydrohalogenation and ether formation (Scheme 5.20).

Subsequent mechanistic studies revealed that the halogenation reactions proceed through an electrophilic attack on an electron-rich boron–carbon bond by an electron-deficient halogen species.[115] This was a significant discovery because it meant that strong bases could be avoided by simply oxidizing a halide ion in the presence of a Lewis base. Mild oxidants such as chloramine-T, *N*-chlorosuccinimide, and hydrogen peroxide could all be used to successfully convert halide ions to species which could halogenate organoboranes in less than a minute![116] This halogenation system proved to be ideal for radiohalogenation reactions because radiohalogens are provided in halide form, normally sodium salt, for safety reasons (halogen molecules are quite volatile) (Scheme 5.21).[117–119]

The iodination reactions proceed readily at the no-carrier-added level and tolerate a variety of functional groups. For instance, the reaction was used to prepare a series of ω-substituted fatty acids (Scheme 5.22).[120]

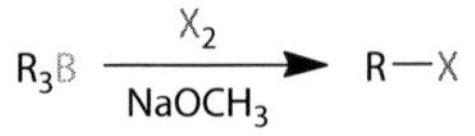

$$R_3B \xrightarrow[NaOCH_3]{X_2} R{-}X$$

SCHEME 5.20 Direct halogenation of trialkylboranes.

$$R_3B + Na^{123}I \xrightarrow{NCS} R{-}^{123}I + R_2BOH$$

SCHEME 5.21 The iodination of trialkylboranes derivatives utilizing iodide salts under oxidizing conditions (*N*-chlorosuccinimide).

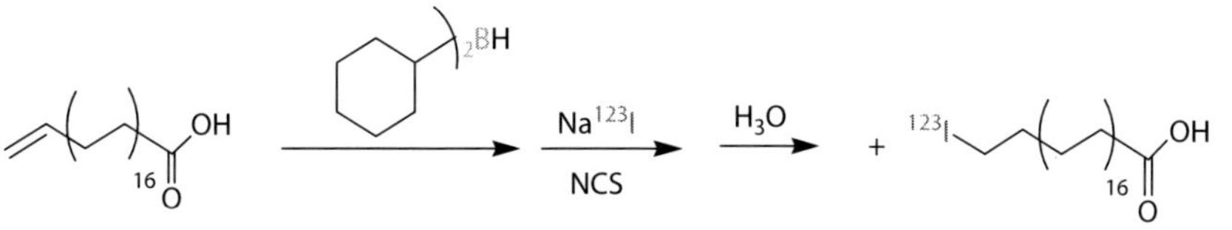

SCHEME 5.22 Preparation of a terminally substituted fatty acid.

Additionally, the oxidative iodination sequence was used to prepare 17-α-iodovinylestradiol, the first use of a terminal iodovinyl group as a method of increasing the *in vivo* retention time of a radioiodine (Figure 5.2).[121] Its use was also implemented in the preparation of a series of radioiodinated tellurium-substituted fatty acids[122] as well as other physiologically active agents.[123–125] It should be noted that the method works equally well for preparing radiobrominated reagents.[126]

Living systems are quite efficient at removing iodine from simple organic molecules. The vinyl iodides proved to be fairly stable *in vivo*, but aryl iodides are generally the most stable with respect to enzymatic dehydrohalogenation. Until recently, it has been very difficult to prepare arylboronic acids containing functional groups of interest in medicine because of the necessity of using transmetallation reactions to generate the necessary boron precursor. However, the advent of Suzuki–Miyaura chemistry, noted earlier, changed the situation dramatically. Boron–halogen exchange reactions of arylboronic esters have been found them to be ideal precursors (Scheme 5.23).[127]

Very recently, the halogenation reaction has been used for the introduction of astatine-211 (an alpha emitter) into a variety of physiologically active reagents of potential use in cancer therapy (Scheme 5.24).[128]

The recent development of crystalline, air-stable organotrifluoroborate salts[129,130] as reaction intermediates has carried over into the radiohalogen arena (Scheme 5.25).[131]

The method can be used to prepare numerous radioiodinated aryl, vinyl, and alkynyl iodides[132,133] and is suitable for preparing bromine-76 derivatives.[134]

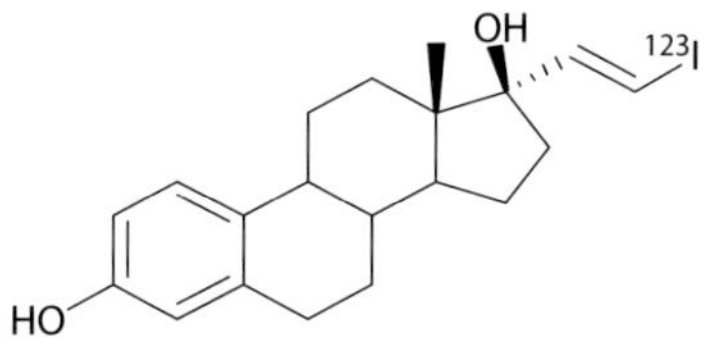

FIGURE 5.2 The first radioiodinated iodovinylestradiol.

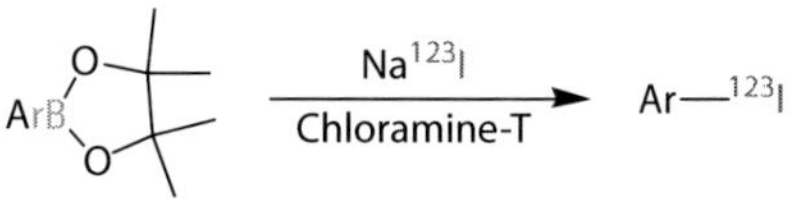

SCHEME 5.23 Radiohalogenation of arylboronic esters under oxidative conditions.

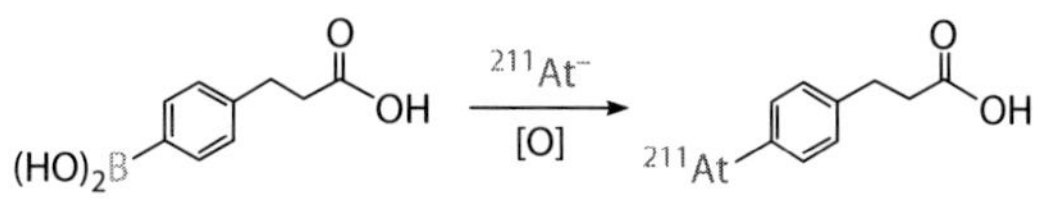

SCHEME 5.24 Incorporation of astatine-211 using the oxidative halogenation reaction.

KF_3B — $\xrightarrow[CH_3CO_3H]{Na^{123}I}$ — ^{123}I

SCHEME 5.25 The use of a trifluoroborate precursor for the synthesis of a cyclooxygenase inhibitor.

$Ar{-}BF_3K$ + $^{+}NHR_3$ OH^{-} ⟶ $Ar{-}\overset{-}{B}F_3$ $\overset{+}{N}R_3$ $\xrightarrow{Na^{123}I}$ $Ar{-}^{123}I$

Where: (sphere) represents a polystyrene lattice

SCHEME 5.26 The use of polymeric organotrifluororate reagents in radioiodination reactions.

An interesting extension of this chemistry involves the preparation of ionic polymeric organotrifluoroborate derivatives. The reagents are designed to be used as precursors in the rapid synthesis of radioiodinated, nuclear medicine imaging agents. Since the starting reagents are both solid state (insoluble) and ionic, the targeted products can be rapidly isolated using simple filtration techniques (Scheme 5.26).[135]

5.6 CONCLUSION

The growing need for structurally complex compounds labeled with stable and radioisotopes will continue to challenge organic chemists. It is clear that organoborane reagents will play an ever-increasing role in the area of research, Figure 5.3.

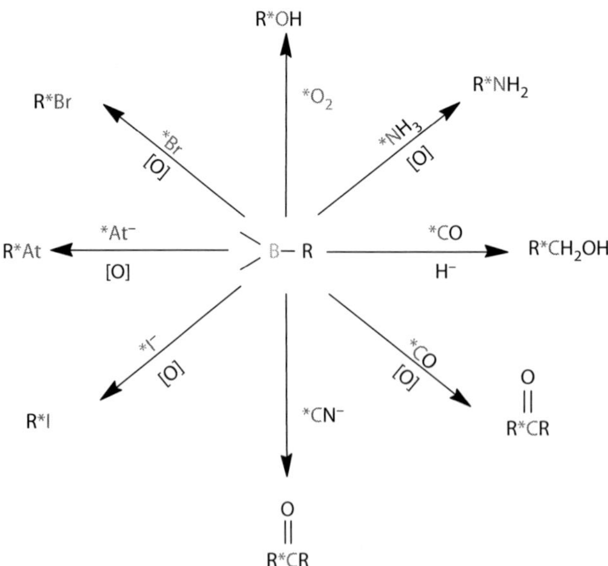

FIGURE 5.3 Incorporation of stable and radioisotopes via organoborane chemistry.

REFERENCES

1. Hevesy, G., Paneth, F. Z. 1913. The solubility of lead sulphides and lead chromates. *Zeits. Fur Anorg. Chem.* 82:323–328.
2. Hevesy, G., Hofer, E. 1934. Elimination of water from the human body. *Nature* (London) 134:879.
3. Lewis, G. N. 1933. The mobility of ions in (HHO)-H_2O_2. *J. Am. Chem. Soc.* 55:3503–3506.
4. Rittenburg, D., Fox, M., Keston, A. S., Ratner, S. 1937. The nitrogen isotope (N-15) as a tool in the study of the intermediary metabolism of nitrogenous compounds. *J. Am. Chem. Soc.* 59:1768.
5. Wood, H. G., Werkman, C. H., Hemingway, A., Nier, A. O. 1942. Assimilation of heavy carbon dioxide by heterotrophic bacteria. *J. Biol. Chem.* 143:133–145.
6. Hevesy, G., Hobbie, R. 1931. Lead content of rocks. *Nature* (London) 128:1038–1039.
7. Blumgart, H. L., Yens, O. C. 1925. The rate of blood flow as determined by a new method. *Am. J. Physiol.* 72:216–218.
8. Freedman, G. S. 1973. *Tomographic Imaging in Nuclear Medicine*. New York: Society of Nuclear Medicine.
9. Phelps, M. E. 1978. ECAT—New computerized tomographic imaging-system for positron-emitting radiopharmaceuticals. *J. Nucl. Med.* 19:635–647.
10. Budinger, T. F. 1980. Physical attributes of single-photon tomography. *J. Nucl. Med.* 21:579–592.
11. Strauss, L., Bostel, F., Clorius, J. H., Raptow, E., Wellman, H., Georgi, P. 1982. Single-photon emission computed-tomography (SPECT) for assessment of hepatic-lesions. *J. Nucl. Med.* 23:1059–1065.
12. Budinger, T. F. 1981. Revival of clinical nuclear-medicine brain imaging. *J. Nucl. Med.* 22:1094–1097.
13. Osman, R., Phelps, M. E., Huang, S.-C., Henze, E., Selin, C. E., Schelbert, H. R. 1982. Positron tomography with deoxyglucose for estimating local myocardial glucose-metabolism. *J. Nucl. Med.* 23:557–586.
14. Yonekura, Y., Benua, R. S., Brill, A. B., Som, P., Yeh, S. D. J., Kemeny, N. E., Fowler, J. S. et al., 1982. Increased accumulation of 2-deoxy-2-[F-18]fluoro-D-glucose in liver metastases from colon-carcinoma. *J. Nucl. Med.* 23:1133–1137.
15. Bergmann, H. 1995. *Radioactive Isotopes in Clinical Medicine and Research.* Basel: Birkhaeuser.
16. Murphy, P. S., Bergstrom, M. 2009. Radiopharmaceuticals for oncology drug development: A pharmaceutical industry perspective. *Curr. Pharm. Des.* 15:957–965.
17. Berezin, A. A. 2009. Stable isotopes in nanotechnology. *Nanotechnology Perceptions.* 5:27–36.
18. Stuerup, S., Hansen, H.R, Gammeigaard, B. 2008. Application of enriched stable isotopes as tracers in biological systems: A critical review. *Analyt. Bioanalyt. Chem.* 390:541–554.
19. Sutzman, E. W. 2007. Stable isotope chemistry and measurement: A primer. In *Stable Isotopes in Ecology and Environmental Science.* Michewer, R. and Lajtha, K. (eds), Oxford: Blackwell Publishing, Ltd.
20. Kohen, A. L., Limbach, H. 2006. *Isotope Effects in Chemistry and Biology*. Boca Raton, FL: CRC Press.
21. Golman, K., Petersson, J. S., Magnusson, P., Johansson, E., Aakeson, P., Chai, C., Hansson, G., Maansson, S. 2008. Cardiac metabolism measured noninvasively by hyperpolarized 13C MRI. *Mag. Reson. Med.* 59:1005–1013.
22. Sood, R. 2004. Off-resonance binomial preparatory pulse technique for high sensitivity MRI of H2O17. *Magn. Reson. Imag.* 22:181–195.
23. Baillie, T. A. 1977. *Stable Isotopes*. Baltimore, MD: University Park Press.
24. Calvin, M., Heidelberger, C., Reid, J. C., Tolbert, B. M., Yankwich, P. M. 1949. *Isotopic Carbon*. New York: Wiley.
25. Raaen, V. F. 1968. *Carbon-14*. New York: McGraw-Hill.
26. Ott, D. G. 1981. *Synthesis with Stable Isotopes*. New York: Wiley-Interscience.
27. Murray, A., Williams, D. L. 1958. *Organic Synthesis with Isotopes.* New York: Interscience.
28. Allard, M., Fouquet, E., James, D., Szlosek-Pinaud, M. 2008. State of art in ^{11}C labelled radiotracer synthesis. *Curr. Med. Chem.* 15:235–277.
29. Rocha, A. F. G., Harhbert, J. C. 1978. *Textbook of Nuclear Medicine: Basic Science.* Lea and Febiger: Philadelphia.
30. Kilbourn, M. J. 1990. *Fluorine-18 Labeling of Radiopharmaceuticals.* Washington, DC: National Academy Press.
31. Kabalka, G. W. 1978. Radionuclide incorporation via oranoboranes In *Aspects of Mechanism and Organometallic Chemistry.* Brewster, J. H. (ed.), New York: Plenum Press.
32. Kabalka, G. W. 1984. Incorporation of stable and radioactive isotopes via oganoborane chemistry. *Acc. Chem. Res.* 17:215–221.

33. Brown, H. C. 1975. *Organic Synthesis Via Boranes*. New York: Wiley.
34. Brown, H. C. 1972. *Boranes in Organic Chemistry*. Ithaca, NY: Cornell University Press.
35. Brown, H. C., Zaidlewicz, M., Negishi, E. 1982. *Comprehensive Organometallic Chemistry*, Vol. 7. Wilkinson, G., Stone, F. G. A. (eds.), Oxford: Pergamon Press.
36. Kabalka, G. W. 2007 Isotope incorporation via organoboranes. *J. Labelled Compd. Radiopharm.* 50: 888–894.
37. Brown, H. C. 1962. *Hydroboration*. Benjamin: New York; Cragg, G. M. L. 1973. *Organoboranes in Organic Synthesis*. New York: Marcel Dekker.
38. Brown, H. C., Murray, K. J. 1961. Protonolysis and deuterolysis of tri-2- norbornylborane—Evidence for retention of configuration in protonolysis of organoboranes. *J. Org. Chem.* 26:631–632.
39. Brown, H. C., Rao, B. C. S. 1957. Hydroboration of olefins. A remarkably fast room-temperature addition of diborane to olefins. *J. Org. Chem.* 22:1136–1137.
40. Nam, N. H., Russo, A. J., Nystrom, R. F. 1963. Preparation of labelled olefins by tritioboration procedures. *Chem. Ind.* (London) 47:1876–1877.
41. Kabalka, G. W., Newton, R. J., Jacobus, J. 1978. Stereochemistry of the hydroboration alkenes. *J. Org. Chem.* 43:1567–1569.
42. Kabalka, G. W., Bowman, N. S. 1973. The stereochemistry of the hydroboration reactions. *J. Org. Chem.* 38:1607–1608.
43. Ruben, S., Hassid, W. Z., Kamen, M. D. 1939. Radioactive carbon in the study of photosynthesis. *J. Am. Chem. Soc.* 61:661–663.
44. Hillman, M. E. D. 1962. Carbonylation of organoboranes .1. Carbonylation of trialkylboranes—A novel synthesis of trialkylcarbinols. *J. Am. Chem. Soc.* 84:4715–4720.
45. Pelter, A., Hutchings, M. G., Smith, K. (1972–1999) (1975). Chemistry of organoborates. III. Protonation of trialkylcyanoborates. *J. Chem. Soc. Perkin Trans. 1* 2:142–145.
46. Kabalka, G. W., Gooch, E. E., Collins, C. J., Raaen, V. F. 1979. A new method for isotopic labeling of organic compounds involving organoboranes. *J. Chem. Soc. Chem. Commun.* 14:607–608.
47. Kabalka, G. W., Delgado, M. C., Sastry, U., Sastry, K. A. R. 1982. Synthesis of carbon-13 labeled carboxylic acids via organoborane reactions. *J. Chem. Soc. Chem. Commun.* 21:1273–1274.
48. Mohammadi, M., Kabalka, G. W., Finn, R. D. 1987. A convenient synthesis of carbon-13-labeled 1-keto-7-methoxyoctahydrophenanthrene. *J. Labelled Compd. Radiopharm.* 24:317–322.
49. Kabalka, G. W. 1980. Incorporation of carbon-13 and carbon-14 via organoborane technology. *Syn. Commun.* 10:93–97.
50. Kabalka, G. W., Delgado, M. C., Kunda, U. S., Kunda, S. A. 1984. Synthesis of carbon-13-labeled aldehydes, carboxylic acids, and alcohols via organoborane chemistry *J. Org. Chem.* 49:174–176.
51. Antoni, G., Kihlberg, T., Langstrom, B. 2003. ^{11}C - Labeling chemistry and labeled compounds. *Handbook of Nuclear Chemistry*. 4:119–165.
52. Wolf, A. P., Redvanly, C. S. 1977. Carbon-11 and radiopharmaceuticals. *Int. J. Appl. Rad. Isot.* 28:29–48.
53. Elander, N., Jones, J. R., Lu, Shui-Yu, Stone-Elander, S. 2000. Microwave-enhanced radiochemistry. *J. Chem. Soc. Rev.* 29:239–249.
54. Scott, P. J. H. 2009. Methods for the incorporation of carbon-11 to generate radiopharmaceuticals for PET imaging. *Angew, Chem.: Int. Ed.* 48:6001–6004.
55. Brown, H. C., Coleman, R. A., Rathke, M. W. 1968. Reaction of carbon monoxide at atmospheric pressure with trialkylboranes in presence of lithium trimethoxyaluminohydride. A convenient procedure for conversion of olefins into aldehydes via hydroboration. *J. Am. Chem. Soc.* 90:495–501.
56. Kothari, P. J., Finn, R. D., Kabalka, G. W., Vora, M. M., Boothe, T. E., Emran, A. M. 1985. 1-^{11}C-butanol: Synthesis and development as a radiopharmaceutical for blood flow measurements. *Int. J. Appl. Rad. Isot.* 36:412–413.
57. Tang, D. Y., Lipman, A., Meyer, G.-J., Wan, C.-N., Wolf, A. P. 1979. ^{11}C-Labeled octanal and benzaldehyde. *J. Labelled Compd. Radiopharm.* 16:435–440.
58. Kothari, P. J., Finn, R. D., Kabalka, G. W., Boothe, T. E., Vora, M. M., Emran, A. 1986. Carbon-11 labeled dialkylketones: Synthesis of 9-[^{11}C]-heptadecan-9-one. *Int. J. Rad. Appl. Instrum.* [A] 37:471–473.
59. Kabalka, G. W., Sastry, U., Sastry, K. A. R., Knapp, F. F., Srivastava, P. C. 1983. Synthesis of arylboronic acids via the reaction of borane complexes with arylmagnesium halides. *J. Organometal. Chem.* 259:269–274.
60. Miyaura, N., Suzuki, A., Yanagi, T. 1981. The palladium-catalyzed cross-coupling reaction of phenylboronic acid with haloarenes in the presence of bases. *Synth. Commun.* 11:513–519.
61. Miyaura, N., Suzuki, A. 1995. Palladium-catalyzed cross-coupling reactions of organoboron compounds. *Chem. Rev.* 95:2457–2483.

62. Suzuki, A., Brown, H. C. 2003. *Organic Synthesis Via Boranes* Vol. 3 *Suzuki Coupling.* Milwaukee, WI: Aldrich Chemical Company.
63. Neumann, H., Brennfuhrer, A., Beller, M. 2008. A general synthesis of diarylketones by means of a three-component cross-coupling of aryl and heteroaryl bromides, carbon monoxide, and boronic acids. *Chem. Eur. J.* 14:3645–3652.
64. Ishiyama, T., Murata, M., Miyaura, N. 1995. Palladium(0)-catalyzed cross-coupling reaction of alkoxy-diboron with haloarenes—a direct procedure for arylboronic esters. *J. Org. Chem.* 60:7508–7510.
65. Murata, M., Oyama, T., Watanabe, S., Masuda, Y. 2000. Palladium-catalyzed borylation of aryl halides or triflates with dialkoxyborane: a novel and facile synthetic route to arylboronates. *J. Org. Chem.* 65:164–168.
66. Chen, I.-Hon, Yin, L., Itano, W., Kanai, M., Shibasaki, M. 2009. Catalytic asymmetric synthesis of chiral tertiary organoboronic esters through conjugate boration of beta-substituted cyclic enones. *J. Am. Chem. Soc.* 131:11664–11665.
67. Burks, H. E., Kliman, L.T., Morken, J. P. 2009. Asymmetric 1,4-dihydroxylation of 1,3-dienes by catalytic enantioselective diboration *J. Am. Chem. Soc.* 131:9134–9135.
68. Ma, L., Jin, R., Lu, G., Bian, Z., Ding, M., Gao, L. 2007. Synthesis and applications of 3-[6-(hydroxymethyl)pyridin-2-yl]-1,1′-bi-2-naphthols or 3,3′-bis[6-(hydroxymethyl)pyridin-2-yl]-1,1′-bi-2-naphthols. *Synthesis* 16:2461–2470.
69. Baudoin, O. 2005. The asymmetric Suzuki coupling route to axially chiral biaryls. *Eur. J. Org. Chem.* 20:4223–4229.
70. Jensen, J. F., Johannsen, M. 2003. New air-stable planar chiral ferrocenyl monophosphine ligands: Suzuki cross-coupling of aryl chlorides and bromides. *Org. Lett.* 5:3025–3028.
71. Rahman, O., Langstrom, B., Kihlberg, T., Llop, J. 2005. Methods for carbon isotope labeling synthesis of ketones and amines by Suzuki coupling reactions using carbon- isotope monoxide. *PCT Int. Appl.* 36pp. WO 2005/066100A1.
72. Hostettler, E. D., Fallis, S., McCarthy, T. J., Welch, M. J., Katzenllenbogan, J. A. 1998. Improved methods for the synthesis of [*omega*-C-11]palmitic acid *J. Org. Chem.* 63:1348–1351.
73. Hamill, T. G., Krause, S., Ryan, C., Bonnefous, C., Govek, S., Seiders, T. J., Cosford, N.D.P., et al., 2005 Synthesis, characterization, and first successful monkey imaging studies of metabotropic glutamate receptor subtype 5 (mGluR5) PET radiotracers. *Synapse* 56:206–216.
74. Hostettle, E. D., Terry, G. E., Burns, H. D., 2005. An improved synthesis of substituted [C-11]toluenes via Suzuki coupling with [C-11] methyl iodide. *J. Labelled Compd. Radiopharm.* 48:629–634.
75. Suzuki, A. 2004. *Organoboranes in Organic Synthesis.* Hokkaido, Japan: Hokkaido University Press.
76. Zeisler, S. K., Nader, M., Theobald, A., Oberdorfer F. 1997. Conversion of no-carrier- added [C-11]carbon dioxide to [C-11]carbon monoxide on molybdenum for the synthesis of C-11-labelled aromatic ketones. *Appl. Radiat. Isot.* 48:1091–1095.
77. Kothari, P. J., Finn, R. D., Kabalka, G. W., Vora, M. M., Boothe, T. E., Emran, M., Mohammadi, M. 1986. C-11 labeled dialkylketones-synthesis of 9-[C-11]heptadecan-9-one. *Appl. Radiat. Isot.* 37:471–473.
78. Nader, M. W., Oberdorfer, F. 2002. Syntheses of [carbonyl-C-11]2-(2-benzoylphenoxy)-*N*-phenylacetamide from [C-11]carbon monoxide by the Suzuki and the Stille reactions. *Appl. Radiat. Isot.* 57:681–685.
79. Rahman, O., Llop, J., Langstrom, B. 2004. Organic bases as additives to improve the bases as additives to improve the radiochemical yields of [C-11]ketones prepared by the Suzuki coupling reaction. *Appl. Radiat. Isot.* 12:2674–2678.
80. Rahman, O., Kihlberg, T., Langstrom, B. 2004. Synthesis of 11C-/13C-ketones by Suzuki coupling. *Eur. J. Org. Chem.* 3:474–478.
81. Hostetler, E. D., Terry, G. E., Burns, H. D. 2005. An improved synthesis of substituted [11C]toluenes via Suzuki coupling with [11C]methyl iodide. *J. Labelled Compd. Radiopharm.* 48:629–634.
82. Audrain, H., Martearello, L., Gee, A., Bender, D. 2004. Utilization of [C-11]-labelled boron carbonyl complexes in palladium carbonylation reaction. *Chem. Commun.* 5:558–559.
83. Brown, H. C., Heydkamp, W. R., Breuer, E., Murphy, W. S. 1964. Comparison of effect of substituents at 2-position of norbornyl system with their effect in representative secondary aliphatic and alicyclic derivative. Evidence for absence of non-classical stabilization of norbornyl cation. *J. Am. Chem. Soc.* 86:3565–3566.
84. Rathke, M. W., Inoue, N., Varma, K. R., Brown, H. C. 1966. A stereospecific synthesis of alicyclic and bicyclic amines via hydroboration. *J. Am. Chem. Soc.* 88:2870–2871.
85. Kabalka, G. W., Sastry, K. A. R., McCollum, G. W., Yoshioka, H. 1981. A convenient synthesis of alkyl amines via the reaction of organoboranes with ammonium hydroxide. *J. Org. Chem.* 46:4296–4298.

86. Kabalka, G. W., McCollum, G. W., Kunda, S. A. R. 1984. Synthesis of dialkylamines via the reaction of organoboranes with *N*-chloroalkylamines. *J. Org. Chem.* 49:1656–1658.
87. Kabalka, G. W., Sastry, K. A. R., McCollum, G. W., Lane, C. A. 1982. Synthesis of nitrogen-15-labeled primary amines via organoborane reactions. *J. Chem. Soc., Chem. Commun.* 1:62.
88. Kothari, P. J., Finn, R. D., Kabalka, G. W., Vora, M. M., Boothe, T. E., Emran, A. M. 1986. Synthesis of nitrogen-13 labeled alkylamines via amination of organoboranes. *App. Radiat. Isot.* 37:469–470.
89. Kabalka, G. W., Wang, Z., Green, J. F., Goodman, M. M. 1992. Synthesis of isomerically pure nitrogen-13 labeled *gamma*-aminobutyric acid and putrescine. *App. Radiat. Isot.* 43:389–391.
90. Kabalka, G. W., Green, J. F., McCollum, G. 1989. The Synthesis of a polymeric organoborane for the preparation of [^{15}O]-butanol. *J. Labelled Compd. Radiopharm.* 24:76.
91. Franklin, E., Duppa, B. 1859. On boric ethide. *Proc Royal Soc (London)* 10:568–570.
92. Jones, F., Taylor, R. L. 1881. On boron hydride. *J. Chem. Soc. Trans.* 3:213–219.
93. von Spindler, O. 1905. Detection of boric acid [in foods]. *Chem.-Zeit.* 29:566–567.
94. Goding, R. F., Cason, L. R. 1946. Test for boric acid as a preservative in milk. *Am. J. Clin. Path. Tech. Sect.* 10:95.
95. Greenwood, N. N. 1991. She burns green, Rosie—we're rich!. *Proc. Royal Inst. of Great Britain* 63:153–174.
96. Brown, H. C., Garg, C. P. 1961. Chromic acid oxidation of organoboranes. A convenient procedure for converting olefins into ketones via hydroboration. *J. Am. Chem. Soc.* 83:2951–2952.
97. Kabalka, G. W., Hedgecock, H. C. Jr. 1975. Mild and convenient oxidation procedure for the conversion of organoboranes to the corresponding alcohols. *J. Org. Chem.* 12:1776–1779.
98. Kabalka, G. W., Slayden, S. W. 1977. Oxidation of organoboranes with trimethylamine N-oxide dihydrate. *J. Organomet. Chem.* 125:273–280.
99. Brown, H. C., Snyder, C., Rao, B. C. S., Zweifel, G. 1986. Organoboranes for synthesis. 2. Oxidation of organoboranes with alkaline hydrogen peroxide as a convenient route for the *cis* hydration of alkenes via hydroboration. *Tetrahedron* 42:5505–5510.
100. Kabalka, G. W., Shoup, T. M., Goudgaon, N. M. 1989. Sodium perborate: A mild and convenient reagent for efficiently oxidizing trialkylboranes. *Tetrahedron Lett.* 30:1483–1486.
101. Brown, H. C., Midland, M., Kabalka, G. W. 1971, The stoichiometrically controlled reaction of organoboranes with oxygen under very mild conditions to achieve essentially quantitative conversion to alcohols. *J. Am. Chem. Soc.* 93:1024.
102. Brown, H. C., Midland, M. M., Kabalka, G. W. 1986. Organoboranes for synthesis. 5. Stoichiometrically controlled reaction of organoboranes with oxygen under mild conditions to achieve quantitative conversion to alcohols. *Tetrahedron* 42:5523–5530.
103. Kabalka, G. W., Reed, T.J., Kunda, S. A. 1983. Synthesis of oxygen-17-labeled alcohols via organoborane reactions. *Syn. Commun.* 13:737–740.
104. Dischino, D. D., Welch, M. J., Kilbourne, M. R., Raichle, M. E. 1983. Relationship between lipophilicity and brain extraction of carbon-11-labeled radiopharmaceuticals. *J. Nucl. Med.* 24:1030–1038.
105. Kabalka, G. W., Lambrecht, R. M., Sajjad, M., Fowler, J. S., Kunda, S. A., McCollum, G. W., MacGregor, R. 1985. *Int. J. Appl. Rad. Isot.* 36:853–855.
106. Berridge, M. S., Franceschini, M. P., Tewson, T. J., Gould, K. L. 1986. Preparation of oxygen-15 butanol for positron tomography. *J. Nucl. Med.* 27:834–837.
107. Berridge, M. S., Cassidy, E. H., Terris, A. H., 1990. A routine, automated synthesis of oxygen-15-labeled butanol for positron tomography. *J. Nucl. Med.* 31:1727–1731.
108. Moerlein, S. M., Gaehle, G. G., Lechner, K. R., Bera, R. K., Welch, M. J. 1993. Automated production of oxygen-15-labeled butanol for PET measurement of regional cerebral blood flow. *Appl. Radiat. Isot.* 44:1213–1218.
109. Jonsson, C., Pagani, M., Ingvar, M., Thurfjell, L., Kimiaei, S., Jacobsson, H., Larsson, S. A. 1998. Resting state rCBF mapping with single-photon emission tomography and positron emission tomography: magnitude and origin of differences. *Eur. J. Nucl. Med.* 25:157–165.
110. Herzog, H., Seitz, R. J., Gellmann, L., Kops, E. R., Juelicher, F., Schlaug, G., Kleinschmidt, A., Mueller-Gaertner, H. W. 1996. Quantitation of regional cerebral blood flow with O-15-butanol and positron emission tomography in humans. *J. Cereb. Blood Flow Metab.* 16:645–664.
111. Moerlin, S. M., Gaehle, G. G., Lechner, K. R., Bera, R. K., Welch, M. J. 1993. Automated production of O-15 labeled butanol for PET measurement of regional cerebral blood-flow. *Appl. Radiat. Isot.* 44:1213–1218.
112. Grampa, G., Marinoni, F. 1952. Applications of radioactive iodine in biology and medicine; Diagnostic and therapeutic applications. *Farm. Soc. Chim. Ital.* 7:569–599.

113. Brown, H. C., Rathke, M.W., Rogic, M. M. 1968. A fast reaction of organoboranes with iodine under the influence of base. A convenient procedure for the conversion of terminal olefins into primary iodides via hydroboration-iodination. *J. Am. Chem. Soc.* 90:5038–5040.
114. Brown, H. C., Lane, C. F. 1970. Base-induced reaction of organoboranes with bromine—A convenient procedure for anti-Markovnikov hydrobromination of terminal olefins via hydroboration-bromination. *J. Am. Chem. Soc.* 92:6660–6661.
115. Brown, H. C., DeLue, N. R., Kabalka, G. W., Hedgecock, H. C. 1976. Consistent inversion in base-induced reaction of iodine with organoboranes-convenient procedure for synthesis of optically-active iodides. *J. Am. Chem. Soc.* 98:1290–1291.
116. Kabalka, G. W., Gooch, E. E. 1980. A mild and convenient procedure for the conversion of alkenes into alkyl iodides via the reaction of iodine monochloride with organoboranes. *J. Org. Chem.* 45:3578–3580.
117. Kabalka, G. W., Gooch, E. E. 1981. Rapid and mild syntheses of radioiodine labeled radiopharmaceuticals. *J. Nucl. Med.* 22:908–912.
118. Kabalka, G. W., Sastry, K. A. R., Pagni, P. G. 1982. Rapid incorporation of radiobromine via the reaction of labeled sodium bromide with organoboranes. *J. Radioanal. Chem.* 74:315–321.
119. Goodman, M. M., Kung, M. P., Kabalka, G. W., Kung, H. F., Switzer, R. 1994. Synthesis and characterization of radioiodinated N-(3-iodopropen-2-yl)-2-*beta*-carbomethoxy-3-*beta*-(4-chlorophenyl)tropanes—Potential dopamine reuptake site imaging agents. *J. Med. Chem.* 37:1535–1542.
120. Kabalka, G. W., Gooch, E. E., Otto, C. A. 1981. Rapid synthesis of radioiodinated *omega*-iodofatty acids. *J. Radioanalyt. Chem.* 65:115–121.
121. Kabalka, G. W., Gooch, E. E., Sastry, K. A. R. 1981. Rapid and mild syntheses of radioiodine-labeled radiopharmaceuticals. *J. Nucl. Med* . 22:908–912.
122. Srivastava, P. C., Callahan, F. F., Owen, A. P., Kabalka, G. W., Sastry. K. A. R. 1985. Myocardial imaging agents-Synthesis, characterization, and evaluation of (*Z*) and (*Z,E*)-18-[Br-82]bromo-5-tellura-17-octadecenoic acids. *J. Med. Chem.* 28:408–413.
123. Kabalka, G. W., Varma, R. S., Jineraj, U. K., Huang, L., Painter, S. K. 1985. Synthesis of i-125 labeled *omega*-iodoundecyl cholesteryl ether *J. Label. Compd. Radiopharm.* 22:333–338.
124. Goodman, M. M., Kabalka, G. W., Marks, R. C., Knapp, F. F., Lee, J., Liang, Y. 1992. Synthesis and evaluation of radioiodinated 2-(2-(R/S)-aminopropyl)-5- iodothiophenes as brain imaging agents. *J. Med. Chem.* 35:280–285.
125. Srivastava, P. C., Knapp, F. F. 1984. [(E)-1-[I-123]iodo-1-penten-5-yl]triphenylphosphonium iodide—Convenient preparation of a potentially useful myocardial perfusion agent. *J. Med. Chem.* 27:978–981.
126. Kabalka, G. W., Sastry, K. A. R., Pagni, P. G. 1982. Rapid incorporation of radiobromine via the reaction of labeled sodium bromide with organoboranes. *J. Radioanal. Chem.* 74:315–321.
127. Kabalka, G. W., Akula, M. R., Zhang, J. 2003. A facile synthesis of radioiodinated (*Z*)-vinyl iodides via vinylboronates. *Nuc. Med. Biol.* 39:369–372.
128. Meyer, G.-J., Krull, D. Grote, M., Knapp, W. H., Kabalka, G. W. 2009. Initial studies using organotrifluorborates as precursors for astatine-211 labelled compounds. *Nuklearmedizin* 48:A142.
129. Molander, G. A., Ellis, N. 2007. Organotrifluoroborates: Protected boronic acids that expand the versatility of the Suzuki coupling reaction. *Acc. Chem. Res.* 40:275–286.
130. Vedejs, E., Chapman, R. W., Fields, S. C., Schrimpf, L. S. 1995. Conversion of arylboronic acids into potassium aryltrifluoroborates—Convenient precursors of arylboron difluoride Lewis-acids. *J. Org. Chem.* 60:3020–3027.
131. Kabalka, G. W., Mereddy, A. R. 2004. A facile no-carrier-added radioiodination procedure suitable for radiolabeling kits. *Nucl. Med. Biol.* 31:935–938.
132. Kabalka, G. W., Mereddy, A. R. 2005. A facile synthesis of radioiodinated alkynyl iodides using potassium alkynyltrifluoroborates, *J. Labelled Compd. Radiopharm.* 48:359–362.
133. Kabalka, G. W.,Yao, M.-L., 2009. No-carrier-added radiohalogenations utilizing organoboranes: The synthesis of iodine-123 labeled curcumin. J. *Organometal. Chem.* 694:1638–1641.
134. Kabalka, G. W., Mereddy, A. R., Green, J. F. 2006. The no-carrier-added synthesis of bromine-76 labeled alkenyl and alkynyl bromides using organotrifluoroborates. *J. Labelled Compd. Radiopharm.* 49:11–15.
135. Yong, L., Yao, M. L., Green, J. F., Hall, K., Kabalka, G. W. 2009. Syntheses and characterization of polymer-supported organotrifluoroborates: Applications in radioiodination reactions. *Chem. Commun.* 46:2623–2625.

6 Boron Cluster (Radio) Halogenation in Biomedical Research

Hitesh K. Agarwal, Sherifa Hasabelnaby, Rohit Tiwari, and Werner Tjarks

CONTENTS

6.1 INTRODUCTION

Physicochemical versatility combined with chemical reactivity and high stability under physiological conditions make boron clusters very attractive choices for a variety of applications including conventional drug design and development (Armstrong and Valliant, 2007; Lesnikowski, 2009), boron neutron capture therapy (BNCT) (Sivaev et al., 2002; Soloway et al., 1998; Tjarks, 2000; Tolmachev and Sjöberg, 2002), and, in the case of (radio)halogenated boron clusters, tumor imaging and therapy (Armstrong and Valliant, 2007; Hawthorne and Maderna, 1999). In BNCT, radiohalogenation of boron clusters has been used to explore pharmacokinetic profiles of boronated tumor-targeting agents (Chen et al., 1994; Mizusawa et al., 1985; Paxton et al., 1992; Varadarajan et al., 1991), whereas other research efforts in this area were directed to more general applications in tumor imaging and therapy (Armstrong and Valliant, 2007; Bruskin et al., 2004; Green et al., 2008; Hawthorne and Maderna, 1999; Shchukin et al., 2004; Tran et al., 2007; Wilbur et al., 2004d; Winberg et al., 2004).

Boron clusters, highly substituted with cold iodine, have been extensively studied as contrast agents in x-ray imaging (Srivastava et al., 1996; Vaca et al., 2006). Most of the conventional x-ray contrast agents are iodinated benzene derivatives with a limited number of iodine atoms (Srivastava et al., 1996). In contrast, boron clusters provide a platform for a much higher iodine load combined with a superior *in vivo* stability.

Sivaev and Bregadze discussed the major types of boron clusters used in biomedical applications (Sivaev and Bregadze, 2009). These can be classified into two groups: (1) neutral clusters, which include *ortho*-, *meta*-, and *para*-dicarbadodecaboranes (*closo*-carboranes, (**1–3**, Figure 6.1), and (2) anionic clusters, which include dicarbaundecaborates (1-) (*nido*-carborane anions, **4–6**), 1-carbadodecaborate (1-) (**7**), and the twofold negatively charged dodecahydro-*closo*-dodecaborate (2-) (**8**) and decahydro-*closo*-decaborate (2-) clusters (**9**) (Sivaev and Bregadze, 2009).

Two strategies have been employed for the preparation of radiopharmaceuticals containing radiohalogenated boron clusters: (1) initial conjugation of the boron cluster to a biomolecule (e.g., antibodies, liposomes, carbohydrates, biotin, and streptavidin) followed by radiohalogenation of the entire conjugate, or (2) initial radiohalogenation of the boron cluster followed by attachment to a biomolecule (Nestor et al., 2003; Orlova et al., 2006; Persson et al., 2007; Sjöström et al., 2003; Wilbur et al., 2004b, 2008). In many cases, bioconjugates containing radiolabeled boron clusters are

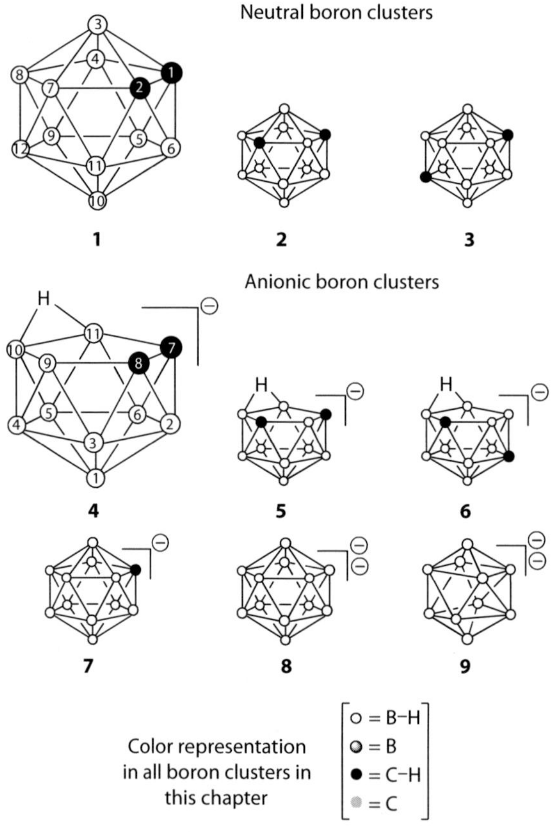

FIGURE 6.1 Neutral and anionic boron clusters, *o*-carborane **(1)**, *m*-carborane **(2)**, *p*-carborane **(3)**, *nido*-*o*-carborane **(4)**, *nido*-*m*-carborane **(5)**, *nido*-*p*-carborane **(6)**, 1-carbadodecaborate anion **(7)**, dodecahydro-*closo*-dodecaborate anion **(8),** and decahydro-*closo*-decaborate anion **(9)**.

found to be less vulnerable to *in vivo* dehalogenation than those possessing radiohalogen–carbon bonds (Nestor et al., 2003; Sjöström et al., 2003; Wilbur et al., 2007, 2009b).

Most radiohalogenations have been carried out with anionic boron clusters because their electron-rich nature allows for the convenient introduction of radioactive halogens through electrophilic halogen species generated *in situ* with oxidizing agents, such as Chloramine T or Iodogen (Bruskin et al., 2004; Green et al., 2008, 2009; Sjöström et al., 2003; Tolmachev et al., 2002; Wilbur et al., 2004b, 2008). Also, the use of negatively charged carborane clusters in bioconjugates improves their overall aqueous solubility, as reviewed by Soloway et al. and Hawthorne and coworkers (Hawthorne, 1993; Soloway et al., 1998; Wilbur et al., 2004d).

The undesired *in vivo* release of radiohalogens, frequently observed in therapeutics and diagnostics radiolabeled at carbon atoms, has been a major motivation for the exploration of boron clusters as "prosthetic groups" for radiohalogens in radiopharmaceuticals (Conti et al., 1995; Kinsella et al., 1988; Klecker et al., 1985a; Mariani et al., 1996; Mester et al., 1996; Prusoff et al., 1960). In the case of conventional carbon-based radioiodinated agents, release of radioiodine leads to the accumulation of significant quantities of iodine in the thyroid and stomach (Adam and Wilbur, 2005; Klecker et al., 1985b; Tolmachev et al., 1999). For example, 5-iodo-2′-deoxyuridine undergoes degradation within a few minutes as a result of enzymatic cleavage at the *N*-glycosidic bond by nucleoside phosphorylases followed by rapid dehalogenation of the resulting 5-iodouracil (Ghosh and Mitra, 1992; Klecker et al., 1985a). Similarly, directly astatinated proteins displayed rapid *in vivo* deastatination apparently due to a weaker bond formation with the protein leading to astatine accumulation in liver, spleen, neck, and stomach (Orlova et al., 2000b; Wilbur et al., 2004a, 2007). Radiopharmaceuticals containing halogens linked to boron atoms of cage structures, such as those shown in Figure 6.1, provide three major advantages over the radiopharmaceuticals having conventional carbon–halogen bonds. (1) The boron–halogen bond is presumably less susceptible to enzymatic and hydrolytic cleavage than the carbon–halogen bond due to the fact that no enzymatic system has been reported for the *in vivo* cleavage of the boron–halogen bond (Orlova et al., 2004). (2) Boron–halogen bonds are stronger than carbon–halogen bonds. For example, the boron–iodine bond is stronger than the carbon–iodine bond (220 kJ/mol vs. 209 kJ/mol) (Kerr and Stocker, 1993; Tolmachev et al., 1999). Similarly, the boron–bromine bond is stronger than the carbon–bromine bond (448 kJ/mol vs. 280 kJ/mol) (Kerr and Stocker, 1993). (3) Iodine attached to boron atoms of the *closo*-carborane (B-iodocarboranes) display an extraordinary chemical stability toward nucleophiles under various chemical reaction conditions (Zakharkin and Kalinin, 1971).

As discussed above, (radio)halogenated boron clusters have attracted considerable attention in recent years, in particular as cancer diagnostics and therapeutics. Therefore, in the following sections, we review both "cold" and "hot" halogenation strategies for various types of boron clusters. Chlorine is excluded as a review topic because it does not play any role in this field of research.

6.2 COLD BORON CLUSTER HALOGENATION

6.2.1 Neutral Boron Clusters

6.2.1.1 1,2-Dicarbadodecaborane (*o*-Carborane)

In *o*-carborane (**1**, Figure 6.1), boron atoms 9 and 12 are the most electron-rich atoms followed by boron atoms 8 and 10 whereas boron atoms 3 and 6, which are adjacent to the two carbon atoms, have the highest electron deficiency (Grimes, 1970). Therefore, substitution reactions of *o*-carborane with electrophilic halogen species preferentially take place at the boron atoms 8, 9, 10, and 12 (Grimes, 1970).

Fluorination of *o*-carborane (**1**) can be achieved in several ways. Depending on the reaction conditions (temperature, solvent, and the reactants ratio), electrophilic fluorination of **1** in the presence of antimony pentafluoride (SbF_5) led to regioselective fluorination producing 9-fluoro- (**10**),

9,12-difluoro- (**11**), 8,9,12-trifluoro- (**12**), and 8,9,10,12-tetrafluoro-*o*-carborane (**13**, Scheme 6.1) (Lebedev et al., 1990).

o-Carborane can also be fluorinated through nucleophilic deboronation and subsequent reaction with boron trifluoride etherate ($BF_3O[C_2H_5]_2$) (Scheme 6.1). Deboronation of *o*-carborane can be achieved by treatment with the fluoride anion, as in tetrabutylammonium fluoride (TBAF), or

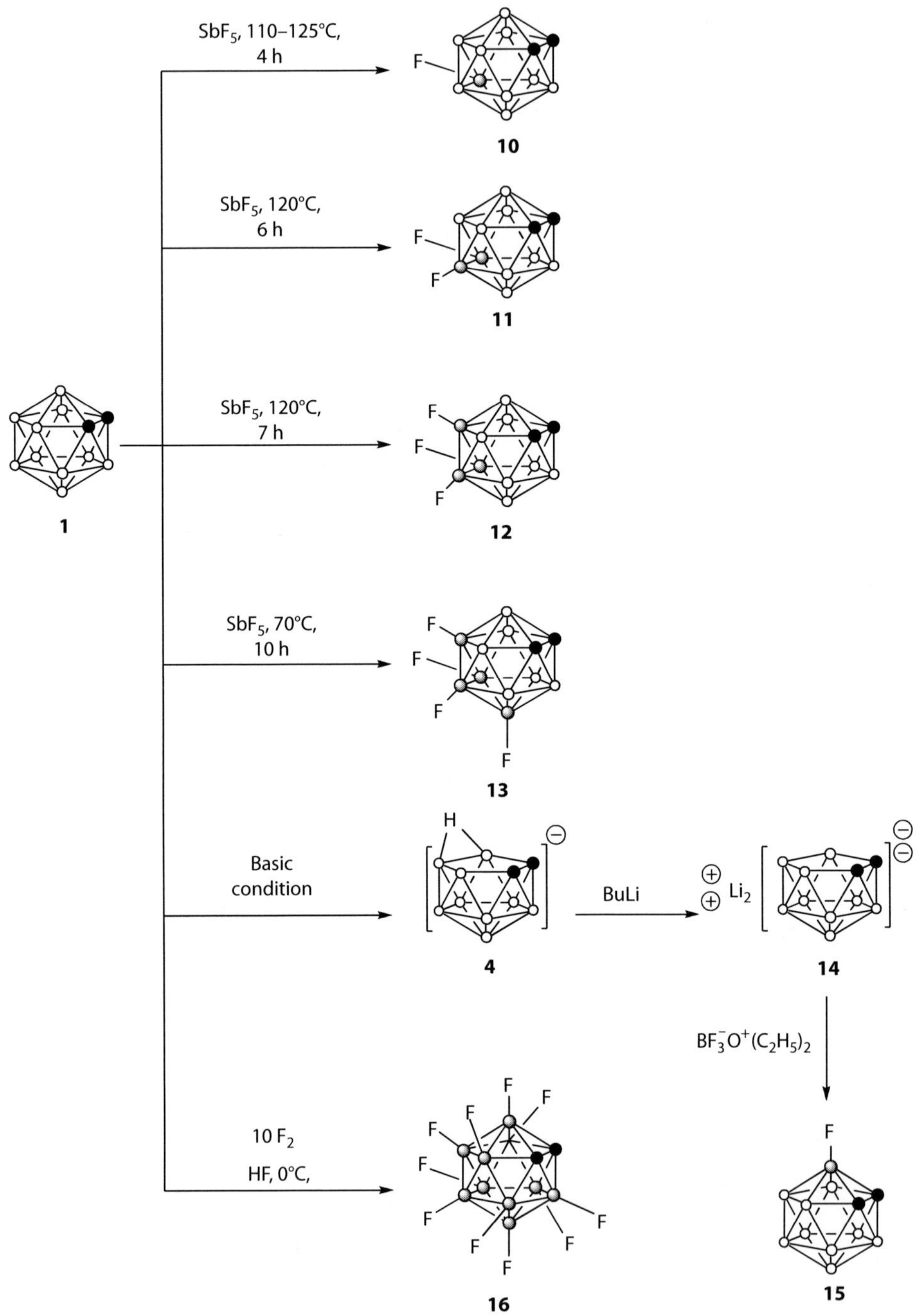

SCHEME 6.1 Electrophilic fluorination of *o*-carborane (**1**). 9-Fluoro-*o*-carborane (**10**), 9,12-difluoro-*o*-carborane (**11**), 8,9,12-trifluoro-*o*-carborane (**12**), 8,9,10,12-tetrafluoro-*o*-carborane (**13**), dianion (**14**), 3-fluoro-*o*-carborane (**15**), and decafluoro *o*-carborane (**16**).

Brønsted bases (**4**, Figure 6.1) (Barbera et al., 2002; Fox and Wade, 1999; Yamazaki et al., 2005). The resulting *nido-o*-carborane (**4**) is treated with BuLi to abstract the "bridge" hydrogen to produce **14**, the dianion form of **4** (Scheme 6.1) (Yamazaki et al., 2005). Compound **14** is then reacted with boron trifluoride etherate ($BF_3O[C_2H_5]_2$) to yield reconstructed monofluorinated 3-fluoro-*closo-o*-carborane (**15**, 70% yield, Scheme 6.1) (Roscoe et al., 1970).

The level of fluorination of *ortho*-, *meta*-, and *para*-carboranes (Figure 6.1) was difficult to control when molecular fluorine (F_2) was used as the halogenation agent (Kongpricha and Schroeder, 1969). The reaction of *o*-carborane (**1**) with excess F_2 in hydrofluoric acid (HF) furnished decafluoro-*o*-carborane (**16**, 30% yield, Scheme 6.1) as the sole product. No fluorination at the carbon atoms was observed.

Similar to fluorination, iodination of *o*-carborane (**1**) has been carried out primarily through electrophilic iodination or by nucleophilic deboronation followed by reconstruction of the carborane cage with BI_3 (Scheme 6.2) (Yamazaki et al., 2005). Depending on the stoichiometric quantity of molecular iodine (I_2), the electron-rich borons 9 and 12 of **1** can be iodinated easily in the presence of $AlCl_3$ to give 9-iodo-*o*-carborane (**17**) and/or 9,12-diiodo-*o*-carborane (**18**) (Li et al., 1991; Yamazaki et al., 2005), whereas the reaction of compound **14** with boron triiodide (BI_3) afforded 3-iodo-*o*-carborane (**19**, 74% yield) (Yamazaki et al., 2005).

When both iodination methods are combined, as shown in Scheme 6.3, several other iodocarboranes can be obtained (Yamazaki et al., 2005). For example, the electrophilic iodination of **19** gave 3,9,12-triiodo-*o*-carborane (**21**, 55% yield), whereas deboronation and reconstruction afforded 3,6-diiodo-*o*-carborane (**22**, 52% yield). Electrophilic iodination of **22** with 1 equivalent of I_2 in the presence of a catalytic amount of $AlCl_3$ produced 3,6,9-triiodo-*o*-carborane (**23**, 88% yield) as the major product, whereas electrophilic iodination with two equivalents I_2 and one equivalent $AlCl_3$ afforded 3,6,9,12-tetraiodo-*o*-carborane (**24**, 77% yield).

Iodination of **1** with iodine monochloride (ICl) in the presence of triflic acid (CF_3SO_3H) produced octaiodo-*o*-carborane (**20**, 67% yield, Scheme 6.2) (Srivastava et al., 1996). On the other

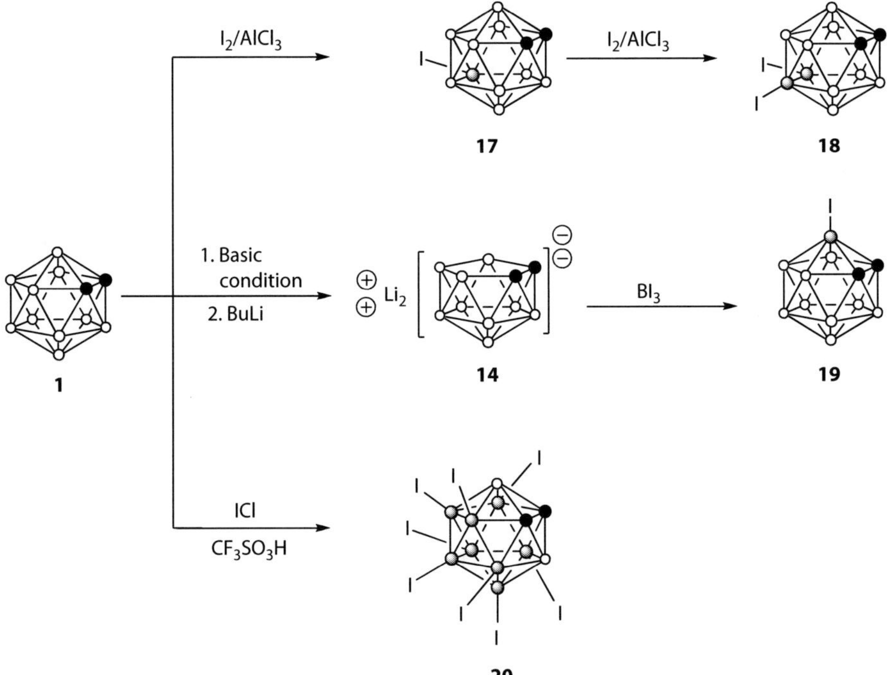

SCHEME 6.2 Iodination of *o*-carborane (**1**). Dianion (**14**), 9-iodo-*o*-carborane (**17**), 9,12-diiodo-*o*-carborane (**18**), 3-iodo-*o*-carborane (**19**), and octaiodo-*o*-carborane (**20**).

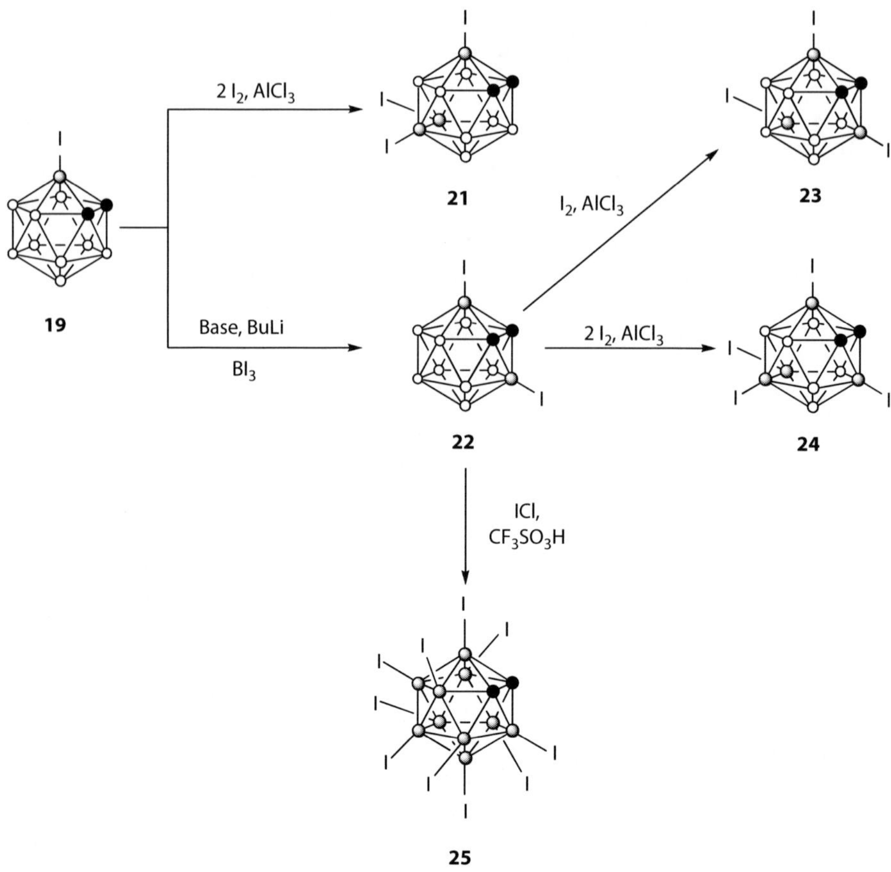

SCHEME 6.3 Iodination of 3-iodo-*o*-carborane (**19**) and 3,6-diiodo-*o*-carborane (**22**). 3,9,12-Triiodo-*o*-carborane (**21**), 3,6,9-triiodo-*o*-carborane (**23**), 3,6,9,12-tetraiodo-*o*-carborane (**24**), and decaiodo-*o*-carborane (**25**).

hand, similar reaction conditions applied to **22** afforded decaiodo-*o*-carborane (**25**, 73% yield, Scheme 6.3) (Teixidor et al., 2006).

Depending on the temperature and the reaction time, the solvent-free iodination of **1** with I_2 in a sealed tube allowed the regioselective synthesis of 9-iodo-*o*-carborane (**17**, 97% yield), 9,12-diiodo-*o*-carborane (**18**, 62% yield), and 8,9,10,12-tetraiodo-*o*-carborane (**26**, 93% yield, Scheme 6.4) (Barbera et al., 2008; Vaca et al., 2006). Shorter reaction times were required when mono/di-C-alkyl- and mono/di-C-aryl-substituted *o*-carboranes were iodinated using the same method (Barbera et al., 2008). Tetraiodination of 1,2-diphenyl-*o*-carborane (**27**) and 1,2-dimethyl-*o*-carborane (**28**) required only 3.5 h and 2.5 h, respectively, to give **29** and **30**. These results were in agreement with previous studies showing that the presence of methyl substituents at the carbon atoms increased the electron density at the boron atoms of **1** (Boer et al., 1966; Potenza and Lipscomb, 1966).

Depending on the stoichiometric quantity of Br_2 and the reaction temperature, electrophilic bromination of **1** with Br_2 in the presence of catalytic amounts of $AlCl_3$ gave 9-bromo-*o*-carborane (**31**, 95% yield), 9,12-dibromo-*o*-carborane (**32**), and 8,9,12-tribromo-*o*-carborane (**33**), as shown in Scheme 6.5 (Smith et al., 1965). Under similar reaction conditions, 1,2-dimethyl-*o*-carborane (**28**) was found to be tetrabrominated (**34**), presumably due to the electron donating effect of the methyl groups. 3-Bromo-*o*-carborane (**35**, 65% yield) was obtained by the reaction of **14** with a boron tribromide-methylsulfide complex [$BBr_3 \cdot S(CH_3)_3$] (Li and Jones, 1990).

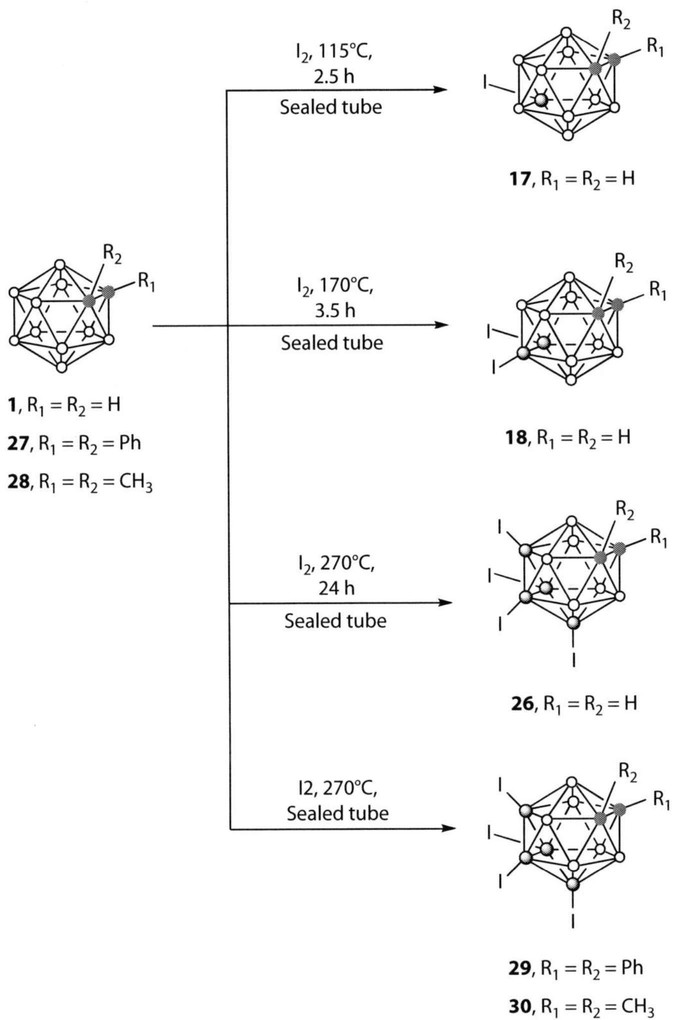

SCHEME 6.4 Solvent-free iodination. 9-Iodo-*o*-carborane (**17**), 9,12-diiodo-*o*-carborane (**18**), 8,9,10,12-tetraiodo-*o*-carborane (**26**), 1,2-diphenyl-*o*-carborane (**27**), 1,2-dimethyl-*o*-carborane (**28**), 8,9,10,12-tetraiodo-1,2-diphenyl-*o*-carborane (**29**), and 8,9,10,12-tetraiodo-1,2-dimethyl-*o*-carborane (**30**).

6.2.1.2 1,7-Dicarbadodecaborane (*m*-Carborane)

In *m*-carborane (**2**, Figure 6.1), boron atoms 9 and 10 have the highest electron density and are therefore most susceptible to electrophilic halogenation (Grimes, 1970). However, *m*-carborane (**2**) is in general far less reactive toward electrophilic halogen species than *o*-carborane (**1**). Iodination and bromination of **2** is 11 and 7 times, respectively, slower than for **1** (Grimes, 1970).

Fluorination of *m*-carborane (**2**) with SbF_5 afforded 9-fluoro-*m*-carborane (**36**) and 9,10-difluoro-*m*-carborane (**37**) (Scheme 6.6) (Lebedev et al., 1990). The reaction of **38**, which was synthesized according to the procedure discussed above for compound **14**, with a boron trifluoride etherate complex afforded 2-fluoro-*m*-carborane (**39**, 65% yield) (Roscoe et al., 1970). Compound **2** was also reacted with BuLi and subsequently perchloryl fluoride in ether at −15°C to yield 1,7-difluoro-*m*-carborane (**40**), which was converted to dodecafluoro-*m*-carborane (**41**, 30% yield) by treatment with an excess of molecular fluorine in an HF suspension (Kongpricha and Schroeder, 1969).

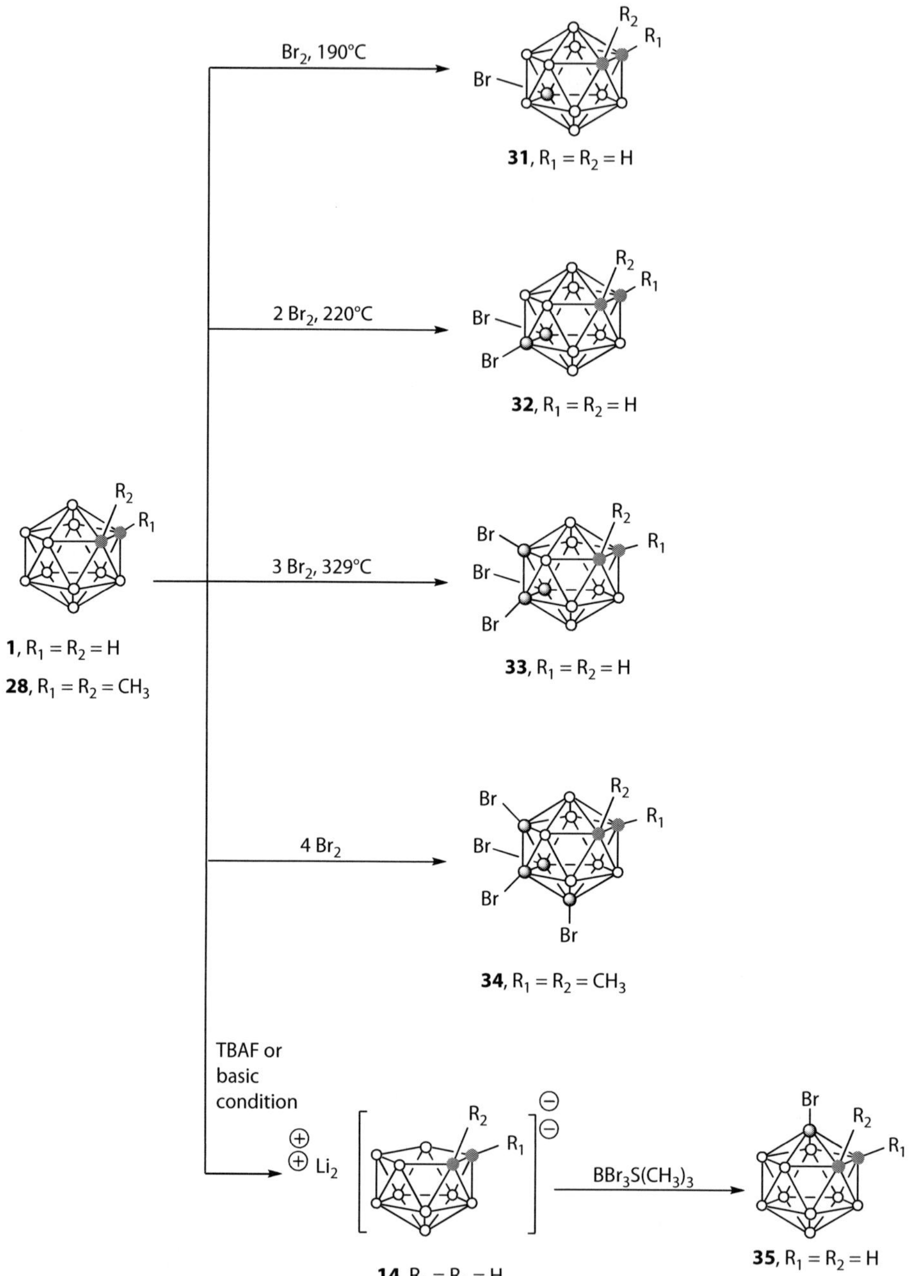

SCHEME 6.5 Electrophilic bromination of *o*-carborane (**1**). 1,2-Dimethyl-*o*-carborane (**28**), 9-bromo-*o*-carborane (**31**), 9,12-dibromo-*o*-carborane (**32**), and 8,9,12-tribromo-*o*-carborane (**33**), tetrabromo-1,2-dimethyl *o*-carborane (**34**), and 3-bromo-*o*-carborane (**35**).

Iodination of **2** with I_2 in the presence of $AlCl_3$ produced a mixture of 9-iodo-*m*-carborane (**42**) and 9,10-diiodo-*m*-carborane (**43**, 60% yield, Scheme 6.7) (Zheng et al., 1995). When **2** was reacted with $ICl/AlCl_3$, however, **43** (90% yield) was the primary product. On the other hand, reaction of **2** with ICl in the presence of triflic acid afforded octaiodo-*m*-carborane (**44**). Bromination of **2** with Br_2 using appropriate stoichiometric quantities furnished 9-bromo- (**45**) or 9,10-dibromo-*m*-carborane (**46**) (Smith et al., 1965).

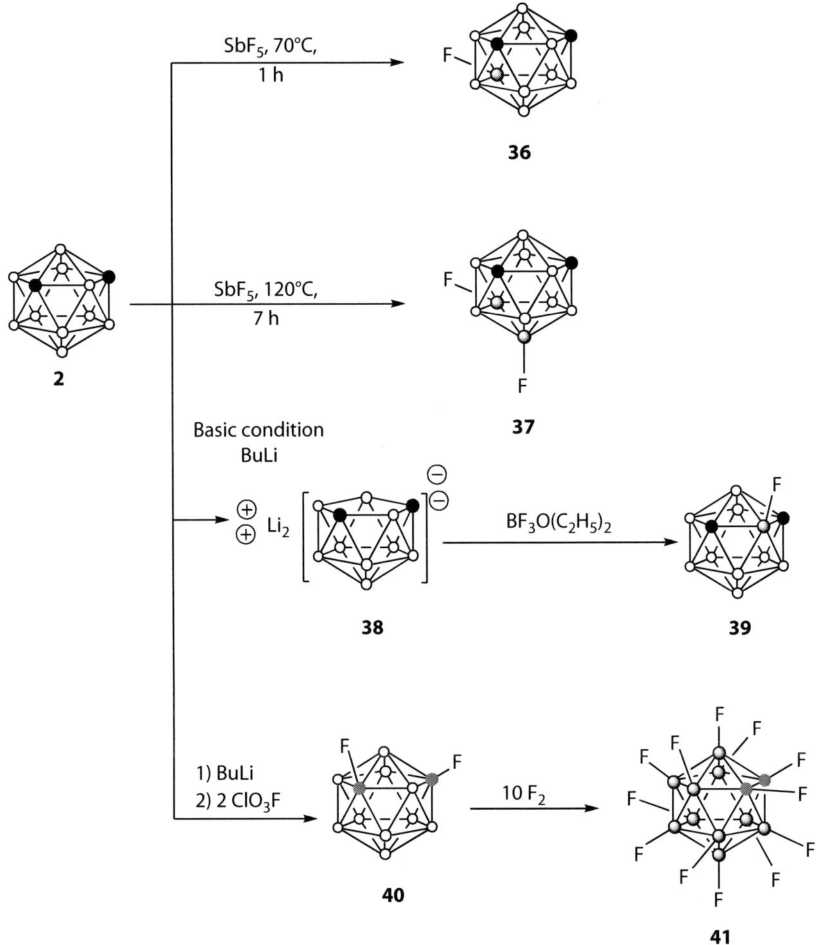

SCHEME 6.6 Fluorination of *m*-carborane (**2**). 9-Fluoro-*m*-carborane (**36**) 9,10-difluoro-*m*-carborane (**37**), dianion (**38**), 2-fluoro-*m*-carborane (**39**), 1,7-difluoro-*m*-carborane (**40**), and dodecafluoro-*m*-carborane (**41**).

6.2.1.3 1,12-Dicarbadodecaborane (*p*-Carborane)

p-Carborane (**3**) is the least reactive of the three carboranes toward electrophilic species because all of its boron atoms are fairly electron deficient as a result of direct bonding with the carbon atoms of the cluster (Jiang et al., 1995; Zakharkin et al., 1970). Because of its overall symmetry there are no preferential halogenation sites, which can lead to uncontrolled formation of isomers and levels of halogenation. Iodination of **3** with I_2 in the presence of $AlCl_3$ produced only 2-iodo-*p*-carborane (**47**, 87% yield) (Scheme 6.8) whereas treatment with $ICl/AlCl_3$ generated diiodo-*p*-carborane (**48**, 85% yield) as a mixture of all five possible isomers [2,3-diiodo- (**a**, 7% yield), 2,4-diiodo- (**b**), 2,7-diiodo- (**c**, 14% yield), 2,8-diiodo- (**d**), 2,9-diiodo-*p*-carborane (**e**, 11% yield)] (Jiang et al., 1995). Decaiodo-*p*-carborane (**49**, 85% yield) was obtained by treating **3** with ICl in the presence of triflic acid. Electrophilic bromination of **3** with Br_2 in the presence of $AlCl_3$ furnished monobromo-*p*-carborane and dibromo-*p*-carborane (Srivastava et al., 1996).

6.2.1.4 Preparation of Halogenated Carboranes by Halogen Exchange

Iodinated carboranes can be used as starting materials for the synthesis of other B-halogenated carboranes (Grushin et al., 1988). The B–I bond does not easily undergo nucleophilic substitution and requires activation. For example, the reaction of 9-iodo-*o*-carborane (**17**) with Cu_2Br_2 proceeded

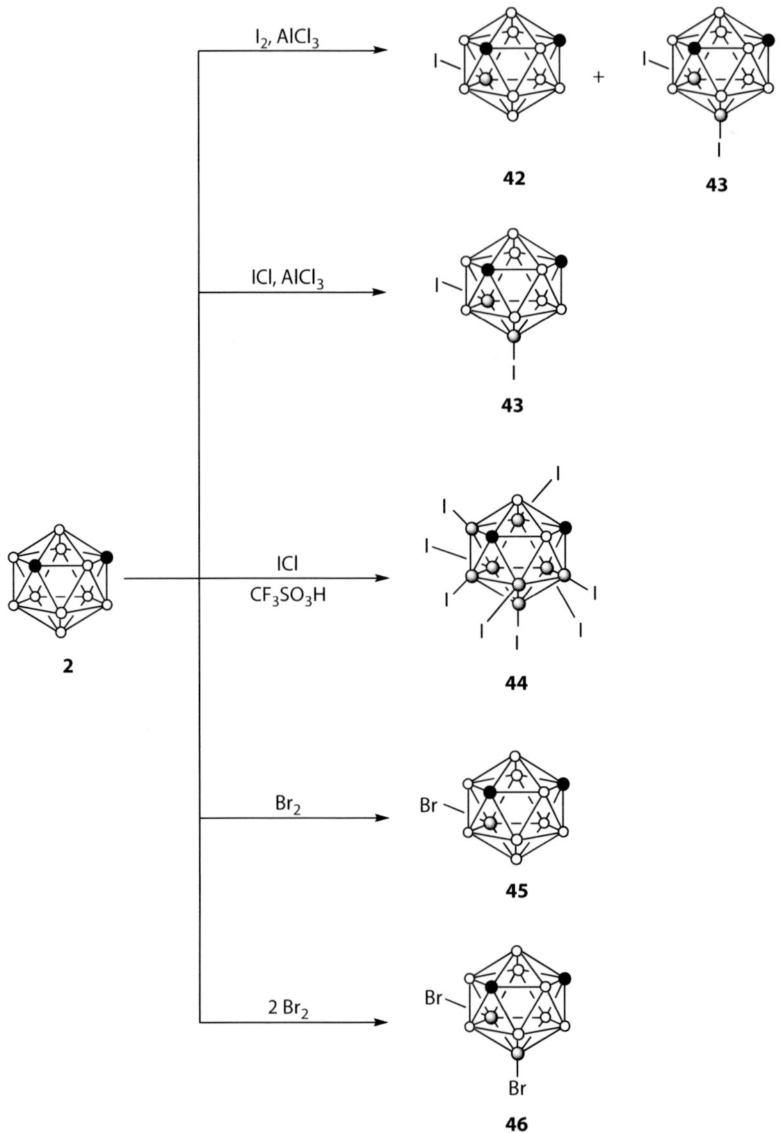

SCHEME 6.7 Iodination and bromination of *m*-carborane (**2**). 9-Iodo-*m*-carborane (**42**), 9,10-diiodo-*m*-carborane (**43**), octaiodo-*m*-carborane (**44**), 9-bromo-*m*-carborane (**45**), and 9,10-dibromo-*m*-carborane (**46**).

only in an autoclave at 270–350°C to give 9-bromo-*o*-carborane (**31**, Scheme 6.9). B–I activation of 9-iodo-*m*-carborane (**42**) was also accomplished through palladium-catalyzed nucleophilic substitution. Using tetrabutyl ammonium bromide as the halogenation source, 9-bromo-*m*-carborane (**45**) was obtained applying this method (Scheme 6.10) (Marshall et al., 2001). Another way of B–I activation involves the conversion of iodinated carboranes into the corresponding phenyl (B-carboranyl) iodonium salts (Scheme 6.11) (Marshall et al., 2001). The reactions of the phenyl (*o*-carboran-9-yl) iodonium salt (**50**) with F^- and Br^- (Grushin et al., 1985, 1991; Marshall et al., 2001) resulted in 9-fluoro- (**10**, 97% yield) and 9-bromo-*o*-carborane (**31**, 93% yield), respectively (Scheme 6.11). Using *m*- and *p*-carborane as precursors, similar reaction conditions produced 9-fluoro-*m*-carborane (**36**, 89% yield), 9-bromo-*m*-carborane (**45**, 85% yield), 2-fluoro-*p*-carborane (99.5% yield), and 2-bromo-*p*-carborane (96% yield), respectively (Grushin et al., 1985, 1991).

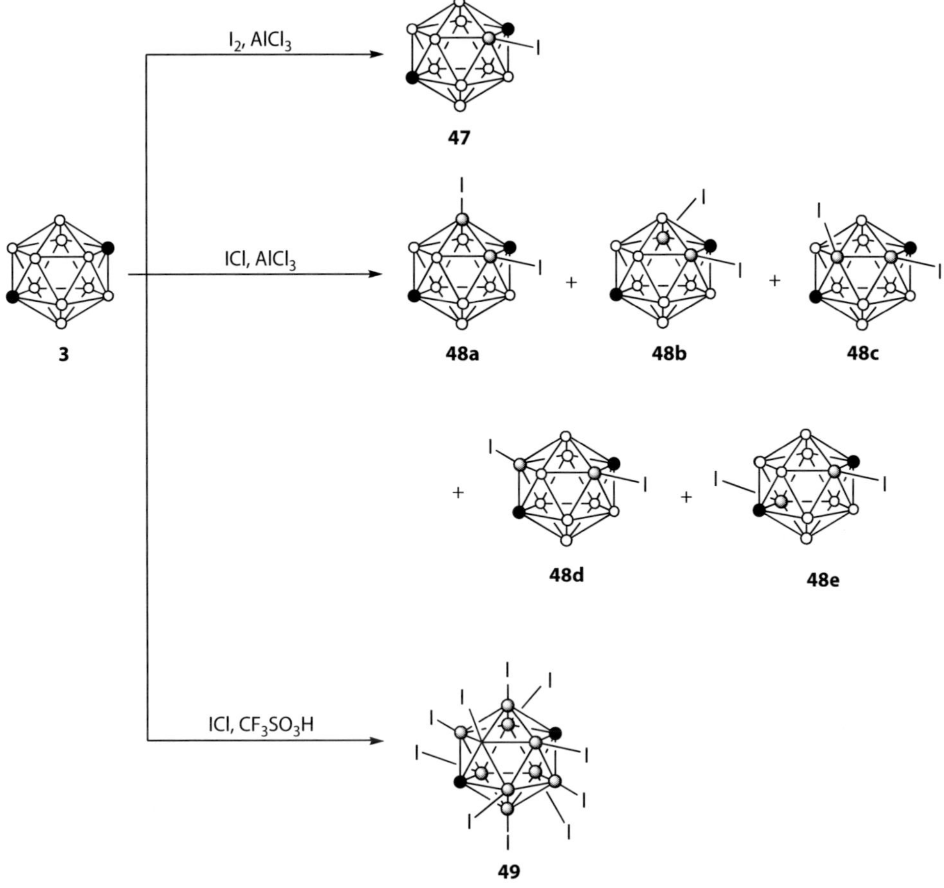

SCHEME 6.8 Iodination of *p*-carborane (**3**). 2-Iodo-*p*-carborane (**47**), diiodo-*p*-carborane (**48a–e**), and decaiodo-*p*-carborane (**49**).

+ Cu_2Br_2 — Autoclave, 270–350°C →

17 31

SCHEME 6.9 Synthesis of 9-bromo-*o*-carborane (**31**).

+ $Bu_4N^+Br^-$ — $(Ph_3P)_4Pd$ →

42 45

SCHEME 6.10 Bromination of 9-iodo-*m*-carborane (**42**). 9-Bromo-*m*-carborane (**45**).

I
+
17
$K_2S_2O_8$, H^+
50
$X^{\ominus}$
X
+
I
10, X = F
31, X = Br

SCHEME 6.11 Nucleophilic substitution of phenyl(*o*-carboran-9-yl) iodonium salt (**50**) with halogen anions (F, Br).

6.2.2 Anionic Boron Clusters

6.2.2.1 7,8-Dicarbaundecaborate (1-) (*nido-o*-Carborane Anion)

In 7,8-dicarbaundecaborate (1-) (**4**, Figure 6.1), boron atoms 9 and 11 have the highest electron density and hence are most susceptible to electrophilic halogenation (Barbera et al., 2002). Deboronation of halogenated *closo*-carboranes under basic conditions is another commonly used approach for the synthesis of the halogenated *nido*-carboranes (Barbera et al., 2002; Fox and Wade, 1999).

Deboronation of 9-fluoro-*o*-carborane (**10**) with KOH furnished 5-fluoro- (**51**) and 6-fluoro-*nido-o*-carborane (**52**) in a 2:1 ratio in 82% total yield (Scheme 6.12), whereas deboronation of 8,9,12-trifluoro-*o*-carborane (**12**) with NaOH yielded 1,5,6-trifluoro- (**53**) and 5,6,10-trifluoro-*nido-o*-carborane (**54**) in a 2:1 ratio in 73% total yield (Fox and Wade, 1999).

Reaction of 9-iodo-*o*-carborane (**17**) with TBAF yielded 5-iodo- (**55**, 94% yield), and 6-iodo-*nido-o*-carborane (**56**) (Scheme 6.13) (Fox and Wade, 1999). On the other hand, reaction of the trimethylammonium salt of **4** (**57**) with 1 equivalent of I_2 resulted in 9-iodo-*nido-o*-carborane (**58**, 88% yield), whereas reaction with 2.5 equivalents of I_2 afforded 9,11-diiodo-*nido-o*-carborane (**59**,

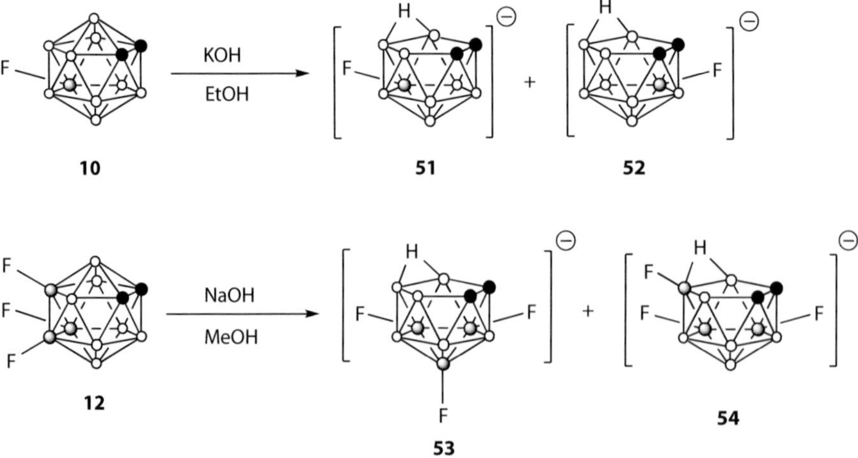

SCHEME 6.12 Synthesis of fluorinated *nido-o*-carborane derivatives (**51–54**). 9-Fluoro-*o*-carborane (**10**), 8,9,12-trifluoro-*o*-carborane (**12**), 5-fluoro-*nido-o*-carborane (**51**), 6-fluoro-*nido-o*-carborane (**52**), 1,5,6-trifluoro-*nido-o*-carborane (**53**), and 5,6,10-trifluoro-*nido-o*-carborane (**54**).

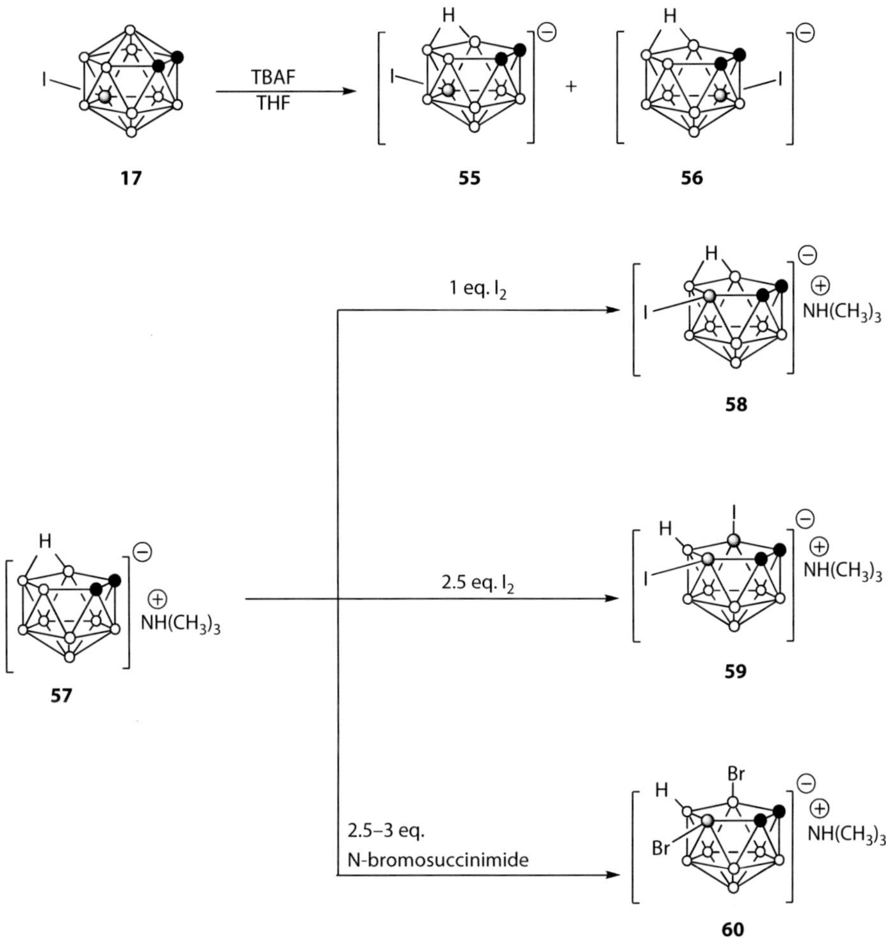

SCHEME 6.13 Synthesis of halogenated *nido-o*-carborane derivatives (**56–60**). 9-Iodo-*o*-carborane (**17**), 5-iodo-*nido-o*-carborane derivative (**55**), 6-iodo-*nido-o*-carborane (**56**), trimethylammonium salt of *nido-o*-carborane (**57**), 9-iodo-*nido-o*-carborane (**58**), 9,11-diiodo-*nido-o*-carborane (**59**), and 9,11-dibromo-*nido-o*-carborane (**60**).

78% yield) (Pak et al., 1994). The 9,11-dibromo-*nido-o*-carborane anion (**60**, 82% yield) was synthesized by the reaction of the trimethylammonium salt of **4** with *N*-bromosuccinimide (Santos et al., 2000).

6.2.2.2 1-Carbadodecaborate (1-)

In 1-carbadodecaborate (1-) (**7**), boron atom 12 is the antipodal to the carbon atom. It has the highest electron density and hence is the first site of electrophilic substitution followed by the boron 7 (Jelinek et al., 2002).

When **7** was exposed to anhydrous hydrogen fluoride at 25°C or 140°C, the 12-fluoro- (**61**, 96% yield) and the 7,12-difluoro-1-carbadodecaborate anion (**62**, 89% yield), respectively, were obtained (Scheme 6.14) (Ivanov et al., 1998a, 2002). Fluorination of **7** with F_2 in liquid HF resulted in the undecafluoro-1-carbadodecaborate anion (**63**, 74% yield) (Finze et al., 2007; Ivanov et al., 1998b).

Iodination of **7** with I_2 in acetic acid at room temperature occurs at position 12 (**64**), whereas the disubstituted 7,12-diiodo anion (**65**) was obtained at 100°C (Scheme 6.15) (Finze, 2008). Compound **65** was obtained in 68% yield by reacting **7** initially with mercuric trifluoroacetate to afford the 12-(mercuriotrifluoroacetate)-1-carbadodecaborate anion (**66**, 72% yield) followed by reaction with

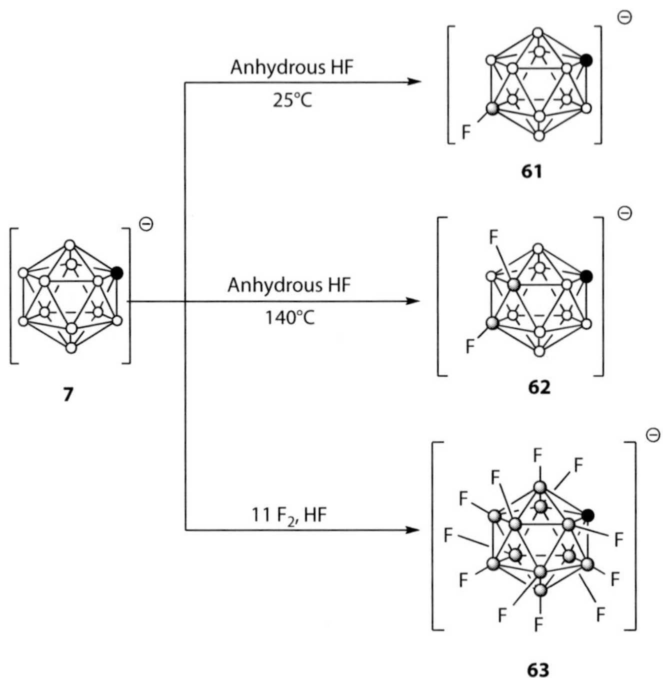

SCHEME 6.14 Fluorination of 1-carbadodecaborate anion (**7**) (**61–63**). 12-Fluoro-1-carbadodecaborate anion (**61**), 7,12-difluoro-1-carbadodecaborate anion (**62**), and undecafluoro-1-carbadodecaborate anion (**63**).

I_2 in 1,2-dimethoxyethane (DME) (Jelinek et al., 2002). On the other hand, the reaction of **7** with ICl provided the hexaiodo derivative **67** (70% yield) (Tsang et al., 2000; Xie et al., 1998, 1999). The reaction of **67** with Br_2 in the presence of triflic acid in a sealed tube resulted in a mixed iodo/bromo-1-carbadodecaborate anion (**68**, 90% yield) (Tsang et al., 2000). Reaction of both **7** and **67** with excess ICl afforded the undecaiodo-1-carbadodecaborate anion (**69**, 64% yield) (Tsang et al., 2000; Xie et al., 1998).

Bromination of **7** with a stoichiometric amount of bromine resulted in the monobromo-1-carbadodecaborate anion (**70**, 71% yield) (Scheme 6.16) (Jelinek et al., 2002). The use of excess of *N*-bromosuccinimide (Jelinek et al., 2002) or Br_2 (Xie et al., 1999) in the reaction with **7** gave the 7,12-dibromo anion (**71**, 90% yield) and the hexabromo-1-carbadodecaborate anion (**72**, 70% yield), respectively. Reaction of **72** with I_2 in the presence of triflic acid in a sealed tube resulted in a mixed bromo/iodo-1-carbadodecaborate anion (**73**, 80% yield) (Tsang et al., 2000). Reaction of **7** with Br_2 in the presence of triflic acid at 200°C in a sealed tube for 4 days generated the undecabromo-1-carbadodecaborate anion (**74**, 82% yield). When the temperature was increased to 250°C, even the carbon atoms were brominated to give the dodecabromo-1-carbadodecaborate anion (**75**, 85% yield) (Xie et al., 1998).

6.2.3 Dodecahydro-*closo*-Dodecaborate (2-) and Decahydro-*closo*-Decaborate (2-)

Fluorination of dodecahydro-*closo*-dodecaborate (2-) (**8**) with anhydrous hydrogen fluoride yielded a mixture of mono to dodecafluorinated *closo*-dodecaborate derivatives (Ivanov et al., 1998a). In contrast, the action of anhydrous hydrogen fluoride led to the decomposition of decahydro-*closo*-decaborate (2-) (**9**) even at temperature less than 0°C. The reaction of **8** with F-TEDA [1-chloromethyl-4-fluoro-1,4-diazoniabicyclo[2.2.2]-octane bis (tetrafluoroborate)] (Grushin et al., 1991) resulted in a mixture of mono- to dodecafluorinated derivatives (F_1–F_{12}) depending on the temperature

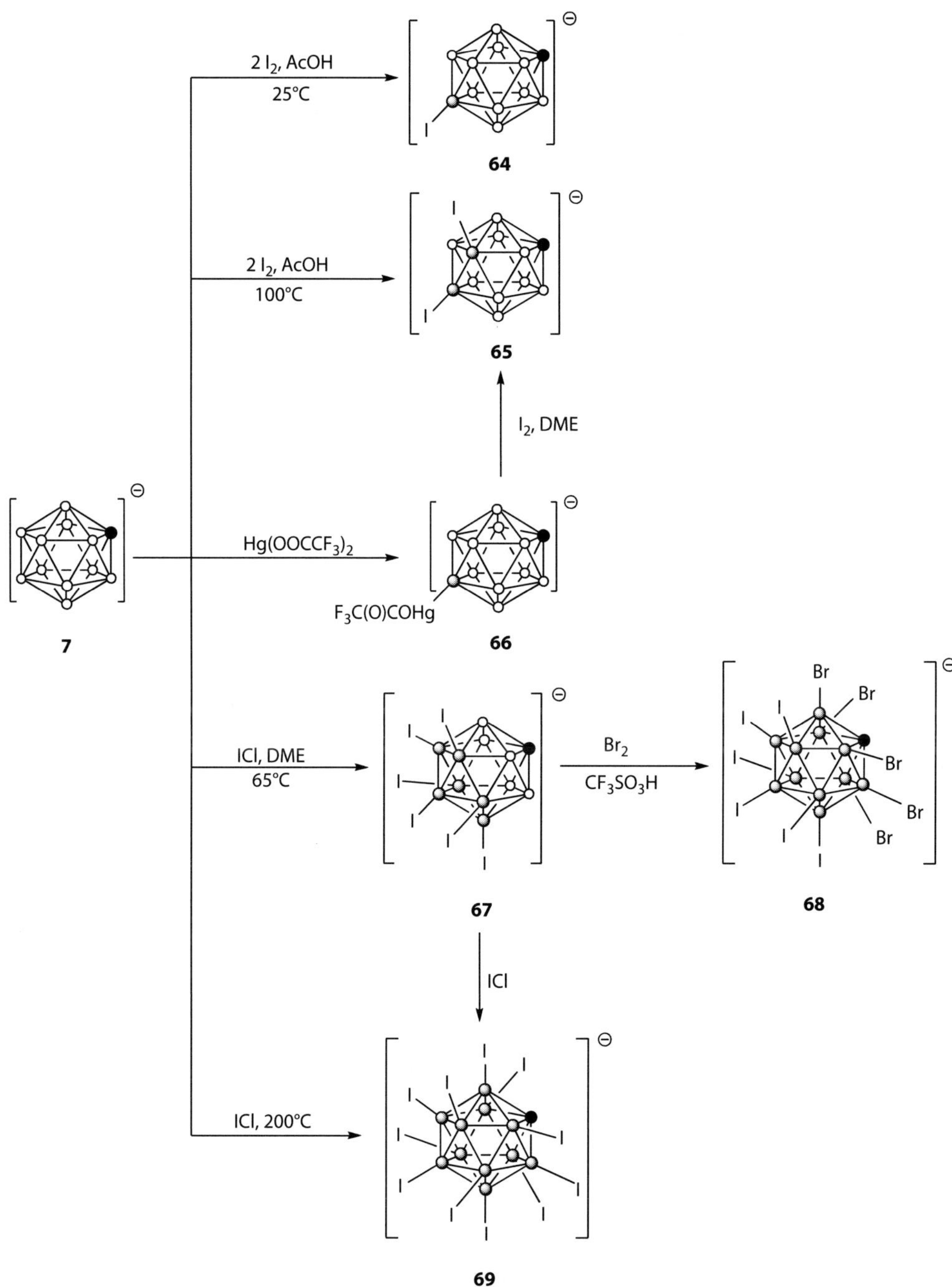

SCHEME 6.15 Iodination of 1-carbadodecaborate anion (**7**) (**64–69**). 12-Iodo-1-carbadodecaborate anion (**64**), 7,12-diiodo-1-carbadodecaborate anion (**65**), hexaiodo-1-carbadodecaborate anion (**67**), mixed halogen (**68**), and undecaiodo-1-carbadodecaborate anion (**69**).

(25–100°C), the solvent, and the stoichiometric quantity of the fluorinating agent. The reaction of **9** with 1 equivalent of F-TEDA at 25°C afforded a mixture of mono- to tetrafluorinated derivatives ($B_{10}H_{10-n}F_n^{2-}$) (n = 1–4).

Partially iodinated and brominated derivatives of **8** [$B_{12}H_6Br_6^{2-}$ (62% yield), $B_{12}H_2Br_{10}^{2-}$, and $B_{12}H_{10}I_2^{2-}$] and **9** [$B_{10}H_3Br_7^{2-}$ (94% yield), $B_{10}H_7I_3^{2-}$ (8% yield), $B_{10}H_6I_4^{2-}$, $B_{10}H_4I_6^{2-}$ (85% yield)] were

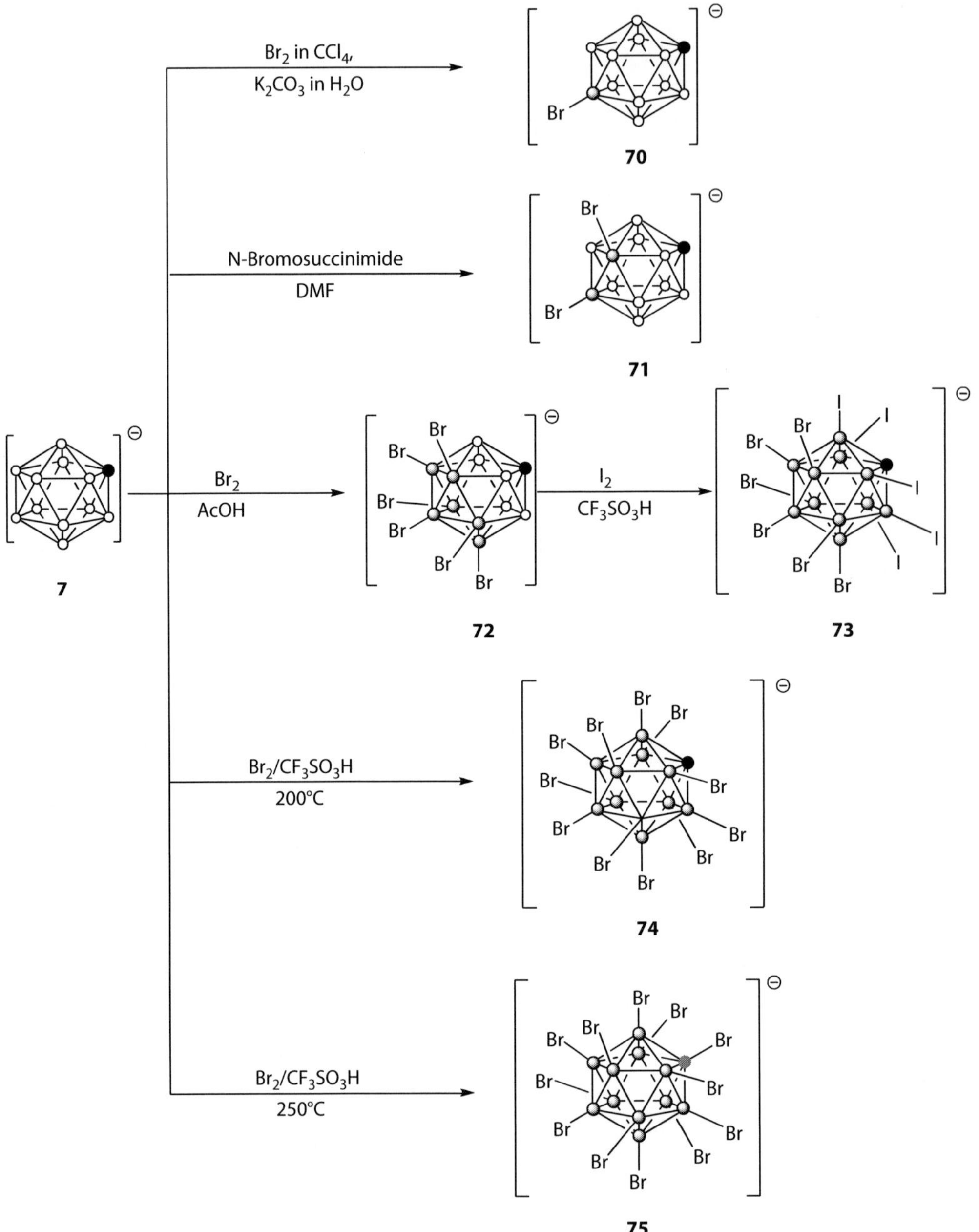

SCHEME 6.16 Bromination of 1-carbadodecaborate anion (**7**) (**70–75**). 12-Bromo-1-carbadodecaborate anion (**70**), 7,12-dibromo-1-carbadodecaborate anion (**71**), hexabromo-1-carbadodecaborate anion (**72**), mixed halogen (**73**), undecabromo-1-carbadodecaborate anion (**74**), and dodecabromo-1-carbadodecaborate anion (**75**).

synthesized by controlled addition of molecular halogens (Br_2, I_2) at temperatures ranging from 5°C to 90°C (Knoth et al., 1964). Diiodination of **8** resulted in two isomers, namely 1,7-diiodododecaborate (**76**, *meta*) and 1,12-diiodododecaborate (**77**, *para*) in a 5:1 ratio (Figure 6.2). This observation was in agreement with the results of theoretical calculations carried out with **8**, which suggested that electron-donating substituents at the cage have an *ortho*-directing effect whereas electron-withdrawing substitutents, such as iodine, have a *meta*- and *para*-directing effect (Jiang et al., 1995).

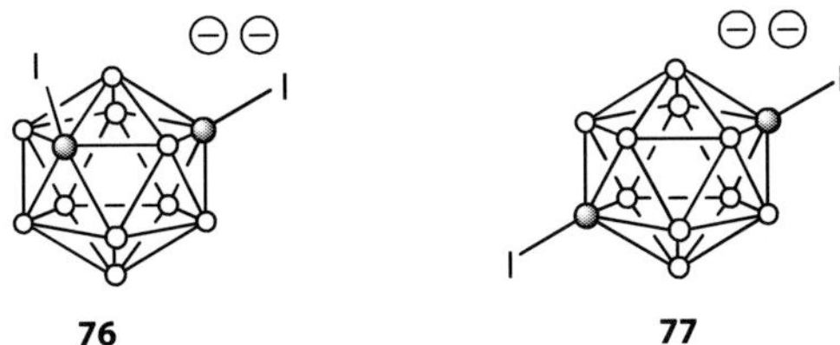

FIGURE 6.2 Meta-1,7-diiodododecaborate (**76**) and para-1,12-diiodododecaborate (**77**) isomeric forms of diiodododecaborate.

6.3 RADIOHALOGENATION OF BORON CLUSTERS

Radioactive iodine, bromine, and astatine are most commonly used for the radiolabeling of boron clusters (Bruskin et al., 2004; Shchukin et al., 2004; Sjöström et al., 2003; Tolmachev et al., 2002, 2004; Wilbur et al., 2004b, 2008). Less frequently used radiolabeling agents include rhenium, technetium (^{99m}Tc), and tritium (^{3}H) (Adam and Wilbur, 2005; Green et al., 2006; Varadarajan et al., 1991). To the best of our knowledge, radiofluorination (^{18}F) of boron clusters has not yet been explored.

Radiohalogenation of boron clusters is either carried out by oxidative radiohalogenation or catalytic halogen exchange. The former is carried with oxidizing agents, such as Chloramine-T and Iodogen, (Bruskin et al., 2004; Green et al., 2008, 2009; Sjöström et al., 2003; Tolmachev et al., 2002; Wilbur et al., 2004b, 2008), whereas the latter takes place in a nonoxidizing environment (Eriksson et al., 2003; Winberg et al., 2003, 2005).

As will be discussed in the following sections, radiohalogenation of boron clusters often takes place subsequent to their attachment to unstable biomolecules. A main objective of exploring and developing these fairly mild radiohalogenation procedures was therefore to circumvent some of the harsh halogenation conditions described in Section 6.2 of this review.

Iodine-125 (^{125}I, $t_{1/2}$ = 60 d) and ^{131}I ($t_{1/2}$ = 8 d) are the major radioactive isotopes of iodine whereas ^{76}Br ($t_{1/2}$ = 16.2 h) is the major radioactive isotope of bromine used for the labeling of boron clusters in medicinal research (Bruskin et al., 2004; Tran et al., 2007; Winberg et al., 2004, 2005). In one case, a mixture of bromine radioisotopes (^{76}Br, ^{77}Br and ^{82}Br) has been used for the labeling of boron clusters (Tolmachev et al., 2002). Astatine-211 (^{211}At, $t_{1/2}$ = 7.2 h) also has been used for boron cluster labeling (Orlova et al., 2000a). Bromine-76 and ^{77}Br are mainly used for PET and SPECT imaging, respectively, whereas ^{125}I, ^{131}I, ^{82}Br, and ^{211}At are primarily used for therapeutic applications (Adam and Wilbur, 2005; Tolmachev and Sjöberg, 2002).

6.3.1 Radiohalogenation by Oxidative Agents

6.3.1.1 Radioiodination

Wilbur et al. synthesized the radioiodinated monocarbon carborane derivatives **78**, **79**, **80**, and **81** in 76–93% labeling yields using N-chlorosuccinimide as the oxidizing agent and Na$^{125/131}$I as the radiohalogen source (Figure 6.3) (Wilbur et al., 2004d). The amino succinyl groups of **79** and **81** are intended for an ultimate conjugation to a biomolecule. The biodistribution of **79** and **81** following coinjection into mice was studied. Apparently, both derivatives did not undergo significant dehalogenation since the localization of radioiodine in the neck, which would be indicative of uptake in the thyroid, was minimal. Renal excretion of **81** from blood was more rapid than for **79**. The former compound also accumulated to a higher extent in the liver and intestines.

Orlova et al. optimized the reaction conditions for the synthesis of **83** starting from **82** using Chloramine-T as the oxidizing agent and Na^{125}I as the radiohalogen source (Figure 6.4). Variables included the pH, reaction times, and reagent concentrations (Orlova et al., 2004). Already after

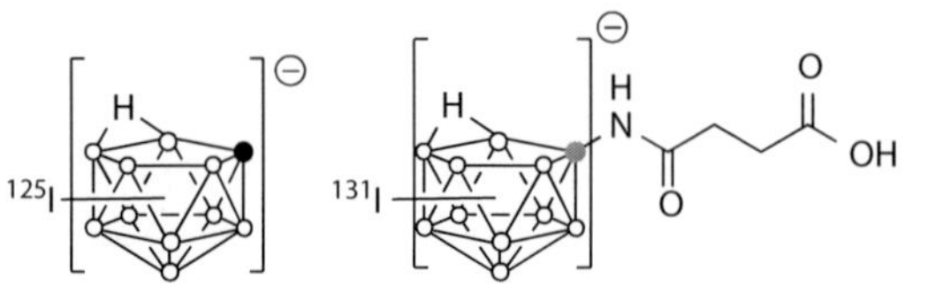

FIGURE 6.3 Radioiodinated *nido*- and *closo*-derivatives of monocarbon carboranes.

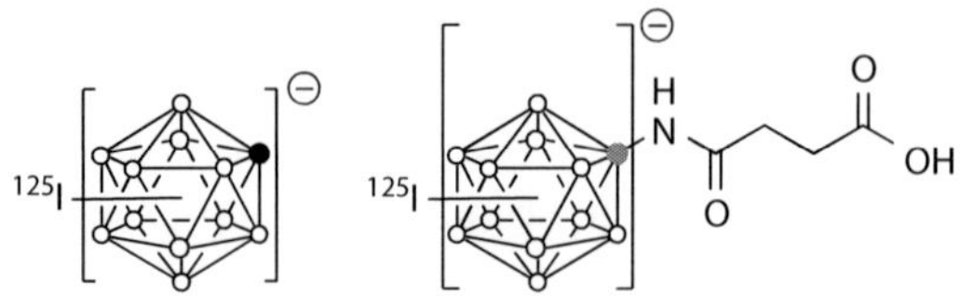

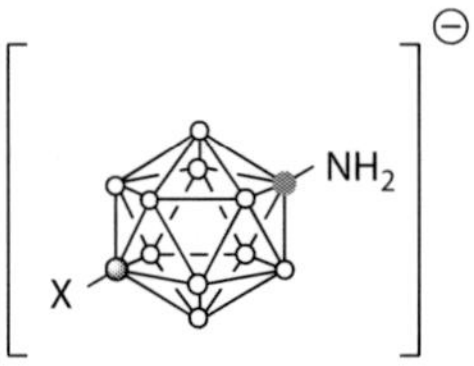

FIGURE 6.4 Amino-*closo*-monocarbon carborane (**82**) and 12-^{125}I-1-amino-*closo*-monocarbon carborane (**83**).

1 min reaction time, a yield of approximately 80% was achieved for **83**, which increased only slightly with longer reaction times. A pH range of 5–7 was optimal for radioiodination. At higher pH, the yields decreased drastically. Maximum yields for **83** were obtained with a concentration ratio of 8.8 mM for Chloramine-T and 0.315 mM for **82**.

Lebdea et al. studied the effect of ionizing radiation on the radioiodination of derivatives of dodecahydro-*closo*-dodecaborate (2-) (Figure 6.5) (Lebeda et al., 2000). The authors suggested that radiation may facilitate the formation of positively charged iodine species. Compounds **84** and **85** were radioiodinated with Na^{125}I in the presence and absence of Chloramine-T. Radioiodination of both compounds was also carried out under γ-radiation exposure from an external ^{60}Co source with doses of 11.7, 23.4, 94, and 280 Gy in the absence of Chloramine-T. Compound **84** was also labeled in the presence of the α-particle emitter ^{211}At with doses of 1.33, 21.1, and 284 Gy in the absence of Chloramine-T. In the presence of Chloramine-T, labeling yields for **86** and **87** were ~94% and ~99%, respectively. In absence of Chloramine-T, only very low radiation levels could be observed for **87** (~2%). However, yields improved proportionately with increasing external doses of both γ- and α-radiation. α-Radiation at a dose of 284 Gy improved the yields of **86** to ~16%. Similarly, γ-radiation at a dose of 280 Gy improved the yields for **86** and **87** to ~27% and 56%, respectively.

Sjöberg and coworkers synthesized compounds **90**, **91**, and **93** with the intention to study the stability of the latter in rat liver homogenates (Figure 6.6) (Ghirmai et al., 2004; Tolmachev et al., 2004). The *nido-o*-carborane derivative **90** was radioiodinated with Na^{125}I using Chloramine-T for 5 min in PBS buffer to give **^{125}I-90** (**91**) with 95% labeling yield (Ghirmai et al., 2004). In another approach, **90** was conjugated first with dextran T10 (molecular weight ~10 kDa) followed by boron cluster radioiodination with Na^{125}I solution in the presence of Iodogen® or Chloramine-T to give **93** with 69–85% labeling yield (Tolmachev et al., 2004). For a comparative biological evaluation, albumin was also radioiodinated using Chloramine-T and Na^{125}I. Conjugate **93** and **^{125}I-albumin** were subjected to

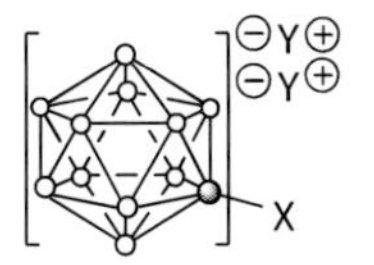

84, X = H, Y = $(C_2H_5)_3NH$
85, X = H, Y = $(C_4H_9)_4N$
86, X = ^{125}I, Y = $(C_2H_5)_3NH$
87, X = ^{125}I, Y = $(C_4H_9)_4N$
88, X = ^{211}At, Y = $(C_2H_5)_3NH$
89, X = $^{76/77/82}Br$, Y = $(C_2H_5)_3NH$

FIGURE 6.5 Derivatives of dodecahydro-*closo*-dodecaborate (2-) (**84–89**).

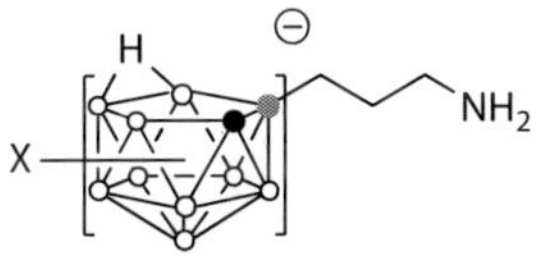

90, X = H
91, X = ^{125}I
92, X = ^{211}At

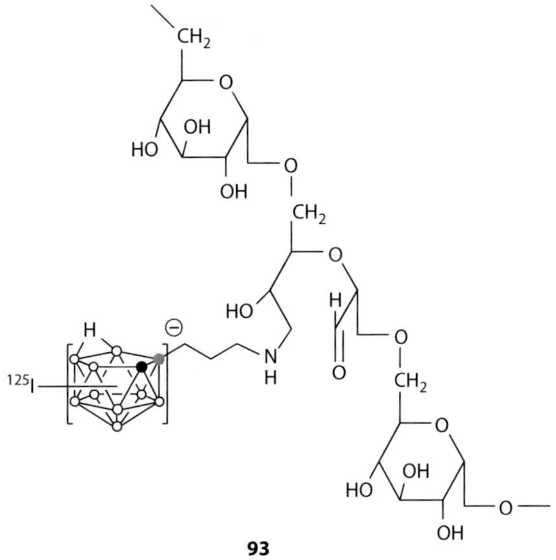

93

FIGURE 6.6 Derivatives of 7-(3′-aminopropyl)-7,8-dicarba-*nido*-undecaborate (–1) (**90–92**) and dextran-T10 conjugate of **91** (**93**).

metabolic studies in rat liver homogenate. The degradation of **^{125}I-albumin** increased rapidly after 10 h of incubation leaving only 20% of intact radioiodinated conjugate after 24 h of incubation. On the other hand, **93** proved to be fairly stable with only about 30% dehalogenation after 24 h.

In a similar study, mercapto-undecahydro-*closo*-dodecaborate(2-) [BSH] was conjugated with allyldextran (molecular weight ~76 kDa) to give dextran-propylmercapto-*closo*-dodecaborate conjugate **94** (Figure 6.7) (Tolmachev et al., 1999). Both **94** and docecahydro-*closo*-dodecaborate **8** (Figure 6.1, Cs^+ and Na^+ salts) were radioiodinated using Chloramine-T and/or Iodogen® and a ^{125}I solution to yield **95** and ^{125}I-undecahydro-*closo*-dodecaborate (**^{125}I-8**). The *in vivo* stabilities of both compounds were evaluated in rats. Conjugate **95** appeared to be very stable because the measured levels of radioactivity in the thyroid were not very high after 20 h. In addition, renal excretion of **^{125}I-8** was rapid with no apparent accumulation in the thyroid.

Valliant et al. synthesized several mono- and bifunctional cage-radioiodinated derivatives of *nido-o*-carborane (**4**, Figure 6.1) as model compounds for targeted radiopharmaceuticals and diagnostics (Figure 6.8) (Green et al., 2008, 2009). These included the monofunctional glucose

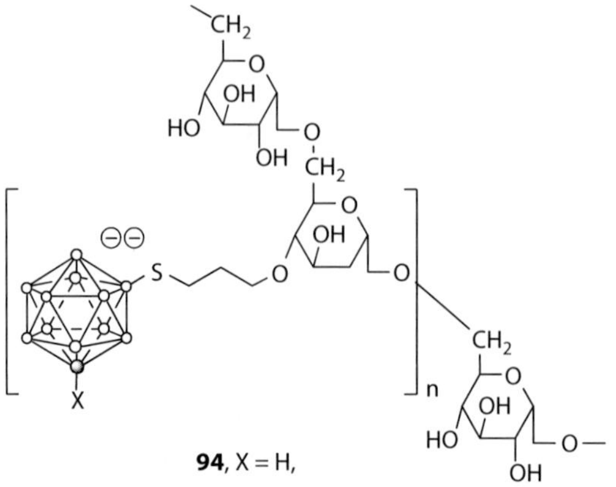

94, X = H,

95, X = ^{125}I

FIGURE 6.7 Dextran conjugate of propyl mercapto-*closo*-dodecaborate conjugate (**94**) and radioiodinated **94** (**95**).

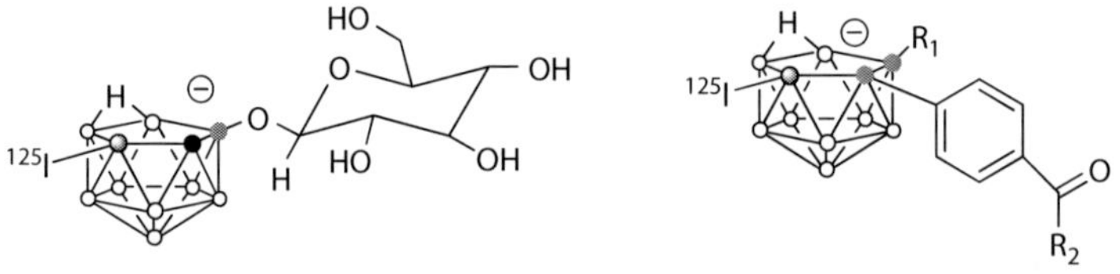

96

97, $R_1 = CH_2$-D-glucose, $R_2 = OH$

98, $R_1 = CH_2$-D-glucose, $R_2 = NH(CH_2)_2N(C_2H_5)_2$

99, $R_1 = H$, $R_2 = NH(CH_2)_2N(C_2H_5)_2$

FIGURE 6.8 Glycosylated conjugates of radioiodinated 7,8-dicarba-*nido*-undecaborate (–1) (**96**–**99**).

derivative **96**, the bifunctional glucose/benzoic acid derivative **97**, the bifunctional glucose/diethyl(aminoethyl)benzamide derivative **98**, and the monofunctional *N,N*-diethyl(aminoethyl) benzamide derivative **99**. Radioiodination of the noniodinated precursor molecules was carried out with Na[^{125}I]/Chloramine-T. The use of the stronger oxidizing agent Iodogen® not only produced higher radiolabeling yields in the case of **96** but also led to N-dealkylation of **98** and **99**. Overall radiolabeling yields between 29% and 92% were found. The *in vitro* uptake of **98** and **99** in B16F10 murine melanoma cells was evaluated. Compound **98** showed significantly lower cell-binding capacity (0.62%) following 24 h of incubation compared to **99** (30.7%). This was attributed to higher hydrophilicity of **98**, which has a calculated log *P* of 0.82 as compared to a calculated log *P* of 1.53 for **99**.

The isothiocyanate group has a long history in conjugating radiolabeled boron clusters to amino groups of bioactive proteins such as antibodies (Scheme 6.17 and Figure 6.9) (Boorsma and Streefkerk, 1976; Cheng et al., 2004; Nestor et al., 2003; Orlova et al., 2006; Persson et al., 2007; Sivaev et al., 1999; Sneath et al., 1974, 1976; Tran et al., 2007; Varadarajan et al., 1991). Early on, radioiodinated 7-(*p*-isothiocyanatophenyl)-9-iodododecahydro-7,8-dicarba-*nido*-undecaborate (1-) (**101**, Scheme 6.17) was synthesized from the corresponding precursor molecule **100** with Na^{125}I/Chloramine-T in 85–95% labeling yield (Mizusawa et al., 1985; Varadarajan et al., 1991). Techniques for the conjugation of **100** and **101** with antibodies were developed and some of these conjugates were identified as promising BNCT agents in biodistribution studies with tumor-bearing rodents.

In a similar approach, (4-isothiocyanatobenzylammonio)-iodododecahydro-*closo*-dodecaborate (1-) (**103,** Figure 6.9) was synthesized by Sivaev et al. and subsequently radioiodinated (**104**) and radiobrominated (**105**) using Chloramine-T as the oxidizing agent (Figure 6.9) (Bruskin et al.,

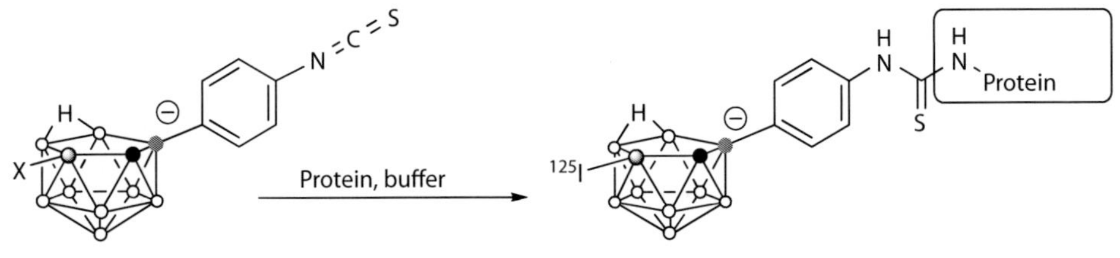

100, X = H
101, X = ^{125}I
102, X = ^{76}Br

101-Protein

Protein, buffer

SCHEME 6.17 Conjugation of radioiodinated boron clusters containing a phenylisothiocyanate group [7-(*p*-isothiocyanatophenyl)-9-iodododecahydro-7,8-dicarba-*nido*-undecaborate (1-) (**101**)].

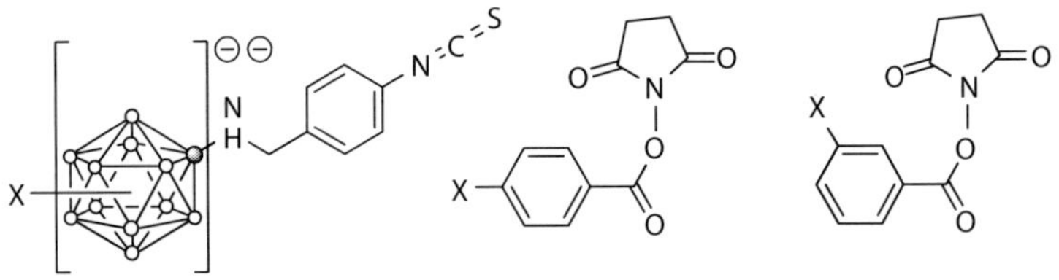

103, X = H
104, X = ^{125}I
105, X = ^{76}Br

106, X = ^{125}I
107, X = ^{211}At
108, X = ^{76}Br

109, X = ^{125}I
110, X = ^{211}At

FIGURE 6.9 Radiolabeled boron clusters (**103–105**) and radiolabeled N-succinimidylbenzoates (**106–110**).

2004; Nestor et al., 2003; Sivaev et al., 1999). Compound **104** was then conjugated in 68% labeling yield to the chimeric monoclonal antibody (cMAb) U36 (**104-U36**), which recognizes the CD44v6 antigen that is frequently overexpressed in squamous cell carcinomas of the head and neck (Nestor et al., 2003). For a comparative biological evaluation, cMAb U36 was directly labeled with Na^{125}I/Chloramine-T and also conjugated to *N*-succinimidyl-4-^{125}I-benzoate (**106**) to afford **^{125}I-U36** (90–95% yield) and **106-U36** (60% yield), respectively. The three conjugates (**104-U36**, **106-U36**, and **^{125}I-U36**) showed comparable *in vitro* affinity for the SCC25 cell line (squamous cell carcinomas of the head and neck cells). However, they had different biodistribution patterns in NMRI mice. The radioactivity level in the thyroid was about 10 times higher for **^{125}I-U36** compared to the other two conjugates, whereas **104-U36** showed a somewhat higher accumulation in the liver as compared to **106-U36** and **^{125}I-U36**.

Another biodistribution study was carried out with **104-U36** and **106-U36** in NMRI mice bearing SCC325 xenografts (Cheng et al., 2004). Overall, **104-U36** and **106-U36** displayed similar biodistribution profiles. For both conjugates, significantly higher radioactivity levels were observed in the tumor as compared to blood and normal organs. However, a somewhat higher accumulation of radioactivity in the thyroid was found for **106-U36** as compared to **104-U36** indicating increased *in vivo* stability of the latter.

Orlova et al. optimized the radioiodination conditions for **103** and the conjugation of the resulting **104** to Trastuzumab to obtain **104-Trast** (Orlova et al., 2006). Trastuzumab (Herceptin) recognizes the HER/neu receptor, which is overexpressed on a subgroup of breast cancers (Cho et al., 2003). Initially, **84** (Figure 6.5) was used as a surrogate for the **103** to optimize reagent concentrations [Chloramine-T (0.03 μmol), **84** (1 nmol), reaction time (30 s), and pH (4–7.5)] for the radioiodination procedure. Using these optimized conditions, **103** was radioiodinated with Na^{125}I in 90% labeling yield. The yields for **104-Trast** ranged from 55% to 60% using 600 μg of mAb and 1 μg of **104** as well as 1 h incubation time at both 25°C and 37°C. Conjugate **104-Trast** was stable for at least 48 h under physiological conditions and retained its immunoreactivity during this time period.

Persson et al. conducted comparative *in vitro* affinity, accumulation, and retention studies (SKBR-3 cells) with directly labeled Trastuzumab (**^{125}I-Trast**) and the conjugates of **101**, **104**, and **106** with Trastuzumab (**101-Trast**, **104-Trast**, **106-Trast**) (Persson et al., 2007). Conjugate **104-Trast** showed about 33% higher accumulation than **106-Trast** despite a lower affinity (K_d 3.2 nM vs. K_d 0.77 nM). This indicates that the accumulation in the case of **104-Trast** may be due to radio-catabolites rather than the conjugate itself. Conjugate **104-Trast** also produced a 42–55% better cellular retention of radioactivity than the other three conjugates.

Apart from antibodies, radiohalogenated boron clusters have also been conjugated with other biomolecules, such as antibody fragments, nucleosides, and proteins such as bovine serum albumin (BSA). Tran et al. explored a novel approach in tumor imaging and conjugated **104** and **106** with the anti-HER2 affibody ligand, $Z_{HER2:342}$ (Tran et al., 2007). These conjugates (**104-$Z_{HER2:342}$** and **106-$Z_{HER2:342}$**) were then subjected to *in vitro* cell binding in SKOV 3 cells and *in vivo* biodistribution studies in mice bearing xenografts of the same cell line. In contrast to the corresponding Trastuzumab conjugates (**104-Trast** and **106-Trast**), there was no difference between the *in vitro* cellular retention of **104-$Z_{HER2:342}$** and **106-$Z_{HER2:342}$**. A more favorable *in vivo* biodistribution pattern was found for **106-$Z_{HER2:342}$**. After 24 h, selective accumulation of radioactivity in the tumor was only found in the case of **106-$Z_{HER2:342}$** whereas **104-$Z_{HER2:342}$** also showed high radioactivity levels in thyroid, liver, intestine, and kidney. The uptake of radioactivity into the tumor was about 4.7 times higher for **106-$Z_{HER2:342}$** than for **104-$Z_{HER2:342}$**.

The oligomeric *nido*-carboranyl phosphate diester **111** was synthesized by Hawthorne et al. and radiolabeled with Na^{125}I/Iodogen to obtain the corresponding radiohalogenated derivative **112** (Figure 6.10) (Chen et al., 1994). The oligomer **112** was conjugated with the ^{131}I-labeled monoclonal antibody T84.66 (**^{131}I-mT84.66**) to obtain doubly labeled **112-^{131}I-mT84.66** (Chen et al., 1994). The biodistribution of this conjugate was evaluated in nude mice with LS-174T colon carcinoma xenografts in comparison with **112** and **^{131}I-mT84.66** alone. The uptake of **^{131}I-mT84.66** in the tumor was ~7 times higher than that of **112-^{131}I-mT84.66** and ~35 times higher than that of **112**.

Lin et al. synthesized a conjugate of BSA with ^{125}I-3-(*nido*-carboranyl) propionate (**115**) and studied its *in vitro* and *in vivo* stability (Scheme 6.18) (Lin et al., 2009). Tetrafluorophenyl-3-(*nido*-carboranyl) propionate (**113**) was synthesized and radioiodinated using Na^{125}I in the presence of *N*-chlorosuccinimide as the oxidizing agent to obtain **114** in 88–98% labeling yield. Compound **114** was then conjugated with BSA to obtain **115** in 99.3% purity and a labeling yield of 58–75%. When

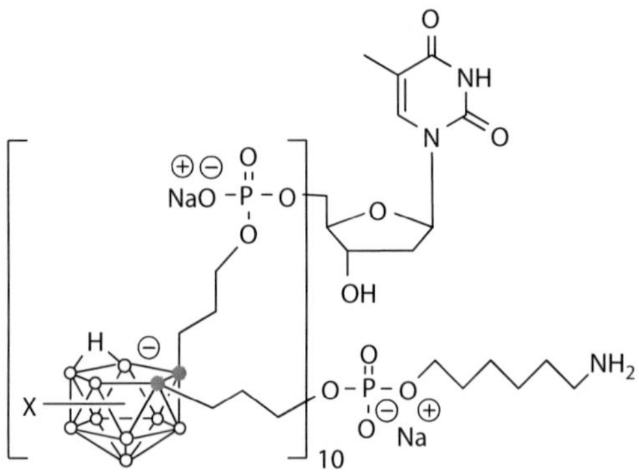

111, X = H,

112, X = ^{125}I

FIGURE 6.10 Structure of carboranyl oligophosphate C.

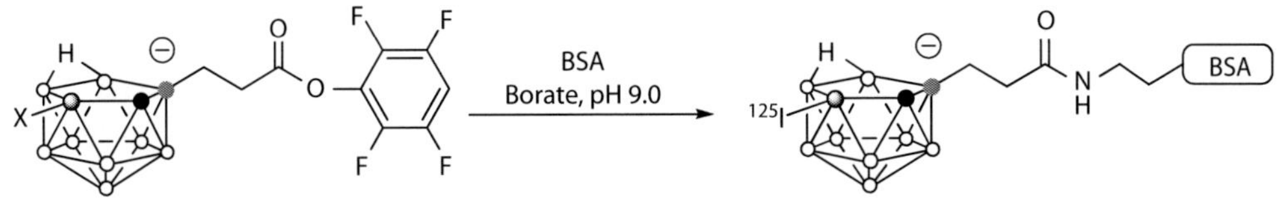

SCHEME 6.18 ^{125}I-3-(*nido*-carboranyl) propionate conjugate with BSA.

BSA alone was radioiodinated using the same procedure to afford **^{125}I-BSA**, a labeling yield of 38–45% combined with only 93.9% purity was found. Conjugate **115** was tested for its *in vivo* and *in vitro* stability and compared with **^{125}I-BSA**. Conjugate **115** was stable for at least 72 h in PBS buffer at pH 7.0 with a radiochemical purity of 98%, whereas that of **^{125}I-BSA** was only 81%. The biodistribution of **115** in mice showed preferential localization in liver and kidney, whereas radioactivity stemming from **^{125}I-BSA** was higher in blood and thyroid compared to **115** indicating higher *in vivo* dehalogenation of **^{125}I-BSA**.

6.3.1.2 Radioastatination

Tolmachev et al. reported that the first radioastatination of a boron cluster (Orlova et al., 2000a,b). Bis(triethylammonium) dodecahydro-*closo*-dodecaborate (2-) (**84**) was astatinated using varying incubation times, buffer pH, and substrate concentrations. Under optimized conditions, ^{211}At-bis(triethylammonium) dodecahydro-*closo*-dodecaborate (2-) (**88**) was synthesized with up to 75% labeling yield in PBS buffer at pH 7.4 using a solution of ^{211}At in methanol and Chloramine-T as the oxidizing agent (Figure 6.5).

Tolmachev and coworkers also astatinated conjugates of 7-(3-aminopropyl)-7,8-dicarba-*nido*-undecaborate (**90**) with the human epidermal growth factor (hEGF) in PBS using a solution of ^{211}At in methanol and Chloramine-T as the oxidizing agent (Figures 6.6 and 6.11) (Sjöström et al., 2003). Initially, hEGF was modified either by a glutaraldehyde cross-linker (hEGF-GA) or Traut's reagent (hEGF-TR), and subsequently *m*-maleimidobenzoyl-N-hydroxysulfosuccinimide ester (Sulfo-MBS), for conjugation with **90**, which was followed by astatination to give **116** and **117**. For a com-

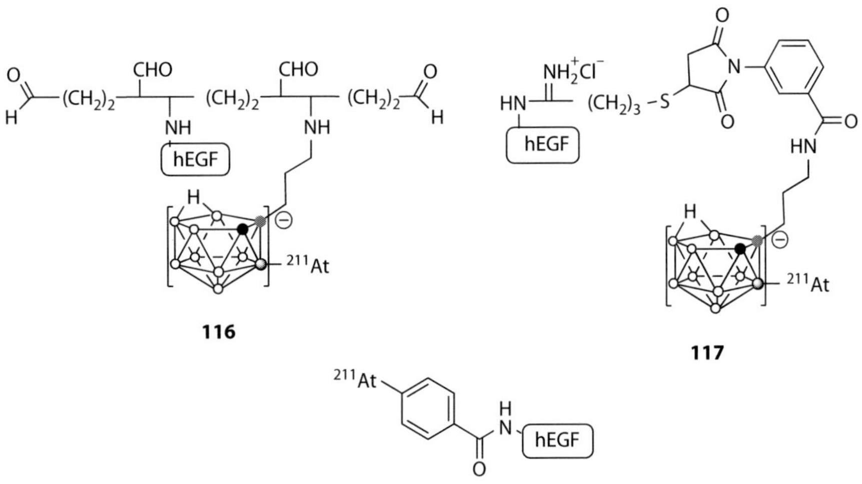

FIGURE 6.11 Human epidermal growth factor (hEGF) conjugates. **^{211}At-hEGF-GA-90** (**116**), **^{211}At-hEGF-TR-90** (**117**), and **^{211}At-AB-hEGF** (**118**).

parative biological evaluation, hEGF was also astatinated directly to give **^{211}At-hEGF** and indirectly using *N*-succinimidyl astatobenzoate **107** to produce **118**. The highest labeling yield was observed for **116** (68%) followed by **118** (44%), **117** (32%), and **^{211}At-hEGF** (5%). Conjugates **116** and **118** were equally resistant toward deastatination in PBS followed by **^{211}At-hEGF**. After 15 h of incubation, more than 80% of the astatine was still associated with hEGF in the cases of **116** and **118**. However, only 18% of astatine was still bound to directly labeled hEGF (**^{211}At-hEGF**).

Wilbur and coworkers synthesized several compounds containing astatinated boron clusters (**133**, **136**, **139**, **142**, **145**, **148**, **151**, **154**, **157**, **166**, **171**, **177**, **179**, **184**, **187**, **190**, **193**, and **197**) and the biodistribution patterns of these compounds and some of their antibody conjugates were studied (Wilbur et al., 2000, 2004a, 2007, 2009a). Some conjugates were radiolabeled with either ^{125}I or ^{211}At and were then coinjected in mice to investigate their *in vivo* stability. In the lung and spleen, astatine concentrations were higher than those of iodine, whereas the iodine concentrations were higher in the neck, which is indicative of uptake in the thyroid (Wilbur et al., 2004a).

In an earlier study, four benzamide derivatives (**119**, **122**, **125**, and **128**; Figure 6.12), seven *nido*-carboranyl derivatives (**131**, **134**, **137**, **140**, **143**, **146**, and **149**; Figure 6.13), and two compounds containing two *nido*-carboranyl cages (Venus flytrap complexes, VFC) (**152** and **155**; Figure 6.13) were synthesized and radiolabeled with ^{125}I and ^{211}At (Wilbur et al., 2004a). Radioiodination of the benzamide derivatives **119**, **122**, **125**, and **128** to obtain the corresponding iodinated derivatives **120**, **121**, **126**, and **129** was achieved in 90–95% yields whereas the yields for the corresponding astatinated compounds **121**, **124**, **127**, and **130** were only 33–55%. In contrast, astatination of the *nido*-carborane cages in **131**, **134**, **137**, **140**, **143**, **146**, **149**, **152**, and **155** improved the yields to 39–72%. Radioiodination yields for these compounds were in a similar range with 28–95%. The benzamide derivatives **121**, **124**, **127**, and **130** showed rapid *in vivo* deastatination in BALB/c mice whereas all astatinated *nido*-carborane derivatives (**133**, **136**, **139**, **142**, **145**, **148**, **151**, **154**, and **157**) appeared to be slightly more stable under these conditions. Even though the ^{211}At-VFC derivatives **154** and **157** were very stable *in vivo*, they localized extensively in the liver and were excreted through the hepatobiliary tract.

Wilbur et al. also synthesized the *nido*-carboranyl- and *closo*-decaboranyl iso(thio)cyanate derivatives **158–163** (Figure 6.14) and conjugated them with an Fab′ fragment derived from the monoclonal antibody 107-1A4, which recognized the PMSA antigen on prostate cells (Wilbur et al., 2007). These conjugates were radioiodinated or radioastatinated and their biodistributions were investigated in BALB/c mice in comparison with the directly radiohalogenated nonboronated Fab′ fragment (**^{125}I-Fab′**) and the nonboronated Fab′ fragment conjugated with radiolabeled *p*-benzoate esters (**106-Fab′** and **107-Fab′**). The labeling yields for direct radioiodination and radioastatination were 89% and <1%, respectively, whereas the yields for indirect labeling were 32% and 14%, respectively. The directly labeled Fab′ fragment (**^{125}I-Fab′**) proved to be unstable with significant levels of radioactivity found in the neck region (thyroid) of the mice suggesting dehalogena-

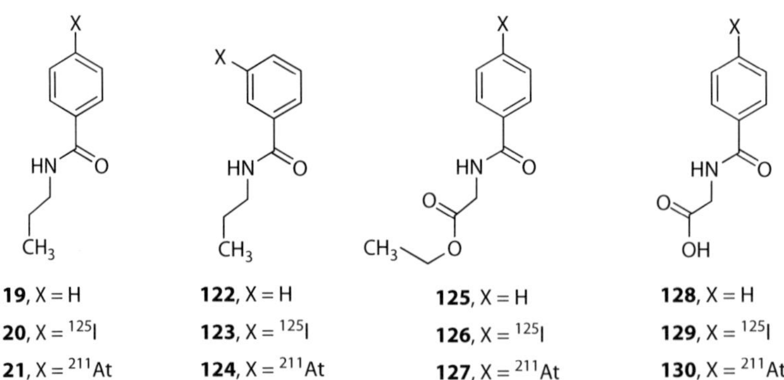

FIGURE 6.12 Benzamide derivatives **119–130**.

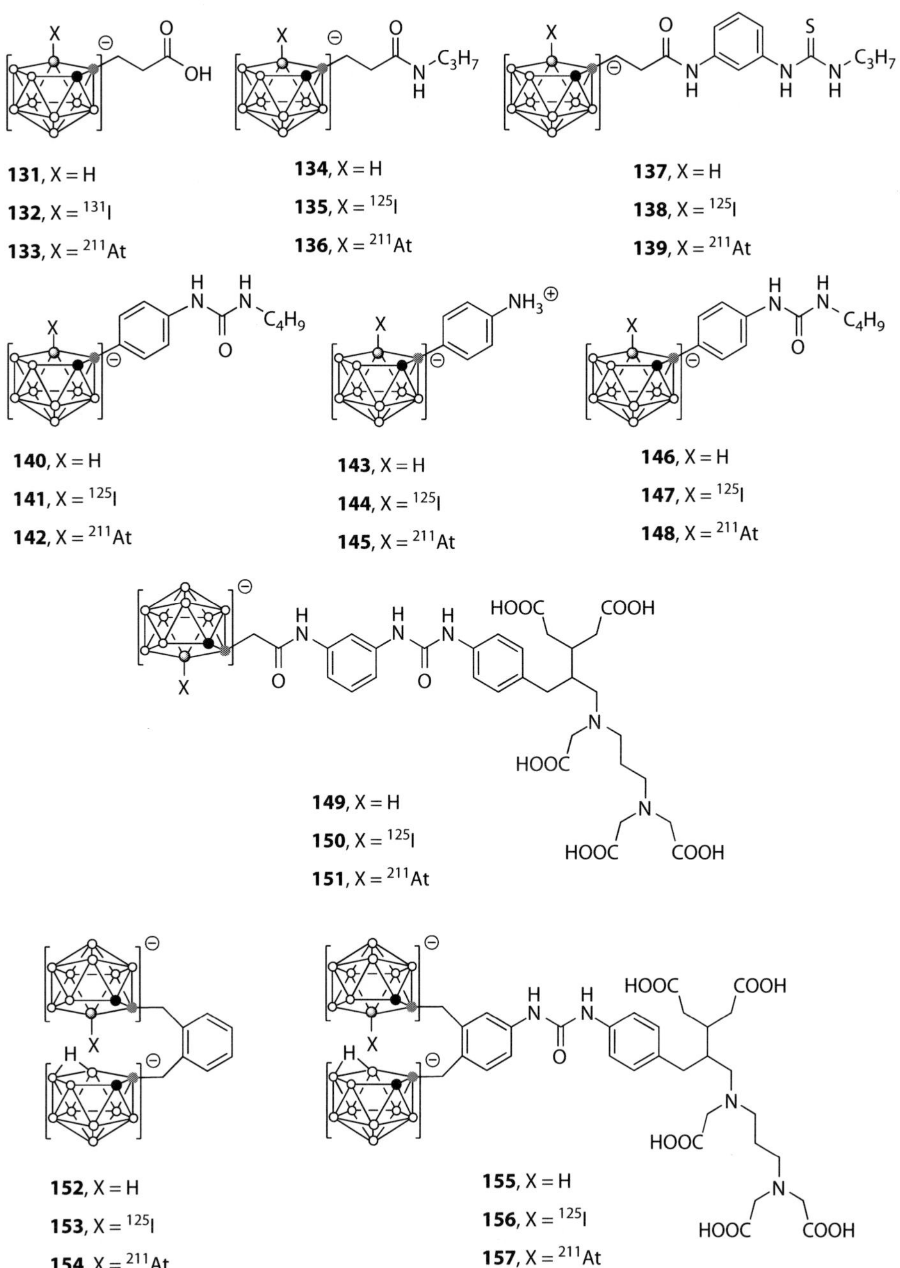

FIGURE 6.13 Biomolecules containing boron clusters (**131–157**) for astatination.

tion of the protein. In the case of the indirectly labeled Fab′ fragments **106-Fab′** and **107-Fab′**, only the astatinated form (**107-Fab′**) was unstable producing very high concentrations of radioactivity in the neck region (thyroid).

Attachment of **Fab′** with the *nido*-carboranes **158** and **159** caused aggregation of the conjugate and changed the *in vivo* behavior of the Fab′ fragment significantly. On the other hand, Fab′ conjugated with *closo*-decaborate **160** displayed a similar biodistribution pattern as the Fab′ fragment itself with high accumulation in the kidney and somewhat lower accumulation in the neck region (thyroid) (Figure 6.14) (Wilbur et al., 2007). Unfortunately, the **160-Fab′** conjugate was radiolabeled

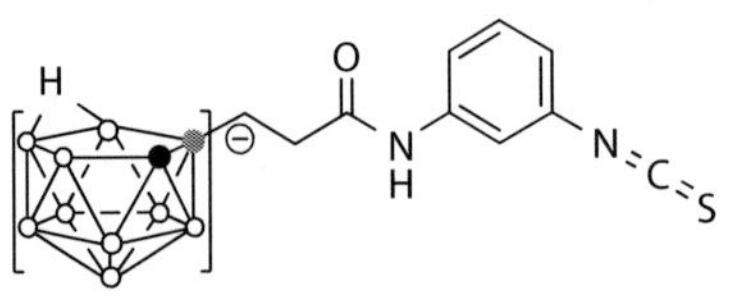

158

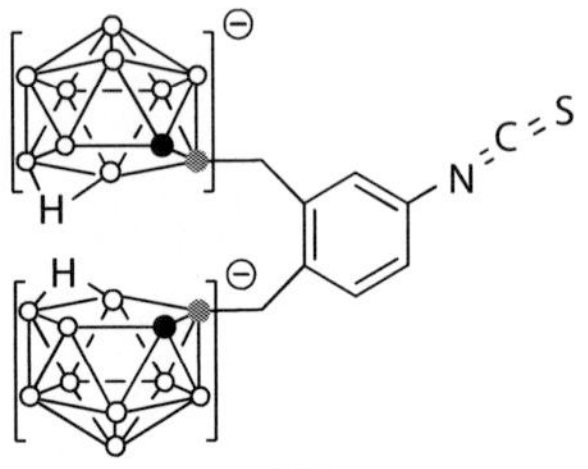

159

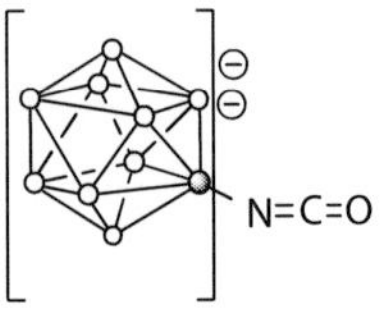

160

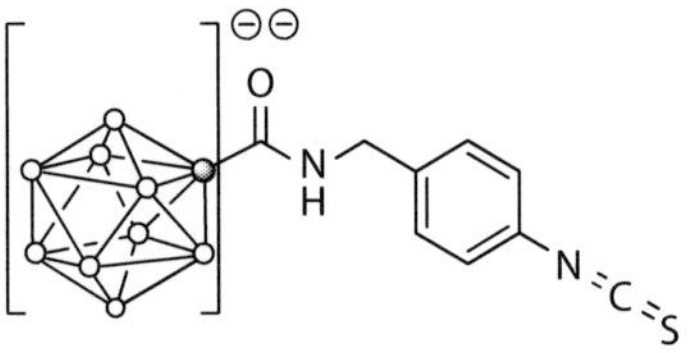

161

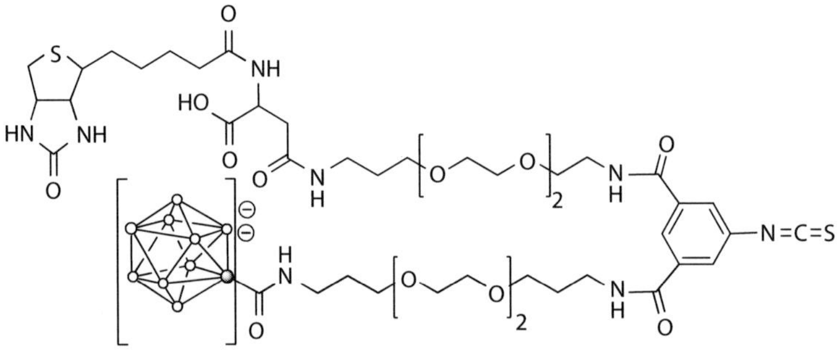

162

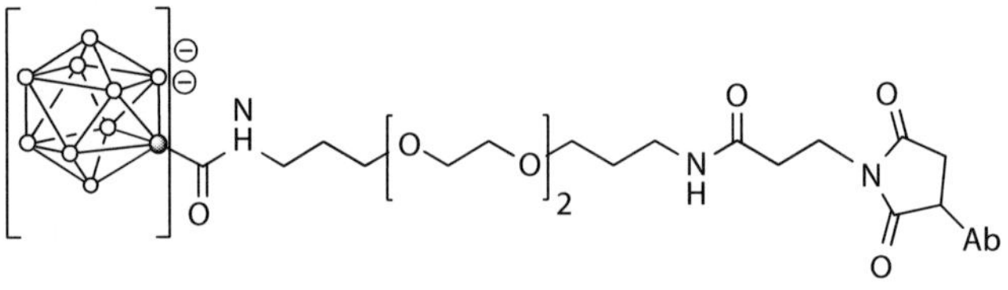

163

FIGURE 6.14 Protein-reactive boron clusters with isocyanate and isothiocyanate groups (**158–163**).

with very low yields (33% radioiodination and 36% astatination). However, increasing the distance between the dianionic cage and **Fab′**, as in the **161-Fab′** conjugate, improved the radiolabeling yield to 60% (astatination) and 72% (iodination). The radiolabeled conjugates **^{125}I/^{211}At-161-Fab′** retained the *in vivo* stability of radiolabeled **^{125}I/^{211}At-160-Fab′** and even showed significantly reduced accumulations in the kidney. Similarly, **162-Fab′** was astatinated with 58% and iodinated with 70% labeling yields. Compound **162** was attached to Fab′ through the amino groups of lysine residues to yield **162-Fab′**, whereas **163** was conjugated to the protein through sulfhydryl groups and radiolabeled to yield **163-Fab′**. The **163-Fab′** conjugate had the highest radiolabeling yields with 75% astatination and 84% iodination. The biodistribution profiles of **^{125}I/^{211}At-162-Fab′** and **^{125}I/^{211}At-163-Fab′** were comparable to that of **^{125}I/^{211}At-160-Fab′**.

Compound **163** was also conjugated with the CD45 specific mAb 30F11 and the nonspecific antibody CA12.10C12 (Wilbur et al., 2009b). Conjugates **163–30F11** and **163-CA12.10C12** were labeled with ^{125}I and ^{211}At and their biodistribution in BALB/c mice was compared with the corresponding antibody conjugates labeled with the m-benzoates **109** and **110** (Figure 6.9) [**109/110–**

30F11 and **109/110-CA12.10C12**]. In accordance with the results of the studies with **Fab′** fragments discussed above (Wilbur et al., 2007), the astatinated conjugates of **163–30F11** and **163-CA12.10C12** showed lower deastatination than **110–30F11** and **110–CA12.10C12**. Furthermore, very high concentrations of **110–30F11** and **163–30F11** in the CD45-rich cells of the spleen indicated that these conjugates did not lose their antigen-binding properties.

Propyl amide and biotin conjugates of *closo*-decaborate (2-) (**164**, **165**, **166**, and **172**) and *closo*-dodecaborate (2-) (**169**, **174**) were synthesized to compare their radiolabeling properties and *in vivo* biodistribution patterns (Figure 6.15) (Wilbur et al., 2009a). The *closo*-decaborates showed several advantages over *closo*-dodecaborates including higher reactivity, higher labeling yields, rapid clearance from tissues, and lower kidney concentrations. The tetrabutylammonium salt **164** and the triethylammonium salt **169** were radiohalogenated with ^{211}At and ^{125}I, whereas the triethylammonium salt **167** was only labeled with ^{125}I. The triethylammonium salts of the biotin derivatives **172** and **174** were radiolabeled with ^{131}I and ^{125}I, respectively. The radioiodination yields for the *closo*-decaborates **165**, **168**, **173** (82–96%) and those of the *closo*-dodecaborate **170** and **171** (79–86%) were comparable. However, the astatination yield for *closo*-decaborate **166** was significantly higher (84%) than that for *closo*-dodecaborate **171** (53%). Presumably, as a result of their different counter ions, dodecaborates **170** and **171** (with triethylammonium cations) were excreted through the kidneys whereas decaborates **165** and **166** (with tetrabutylammonium cations) were excreted through the hepatobiliary system. This was confirmed by the finding that *closo*-decaborate **168** (with triethylammonium cations) was also excreted through the kidney. In order to avoid any counter ion effect on their biodistribution profiles, *closo*-decaborate derivative **172** and *closo*-dodecaborate derivative

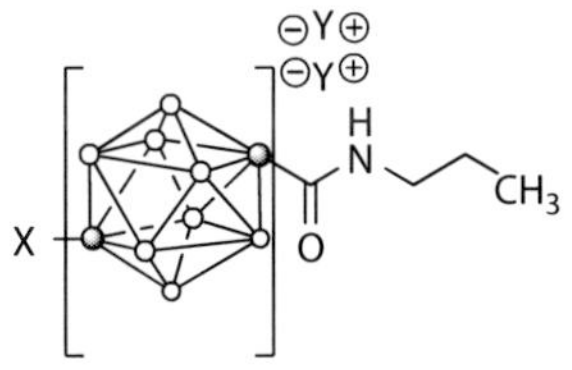

164, X = H, Y = Bu_4N
165, X = ^{125}I, Y = Bu_4N
166, X = ^{211}At, Y = Bu_4N
167, X = H, Y = Et_3NH
168, X = ^{125}I, Y = Et_3NH

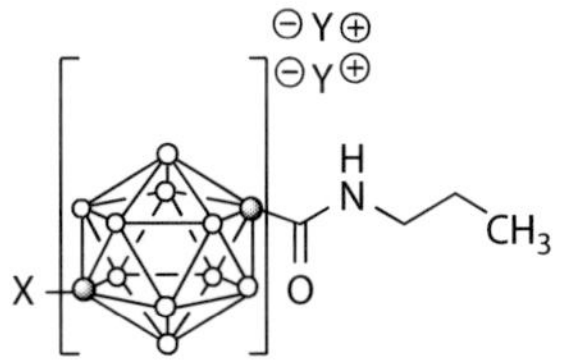

169, X = H, Y = Et_3NH
170, X = ^{125}I, Y = Et_3NH
171, X = ^{211}At, Y = Et_3NH

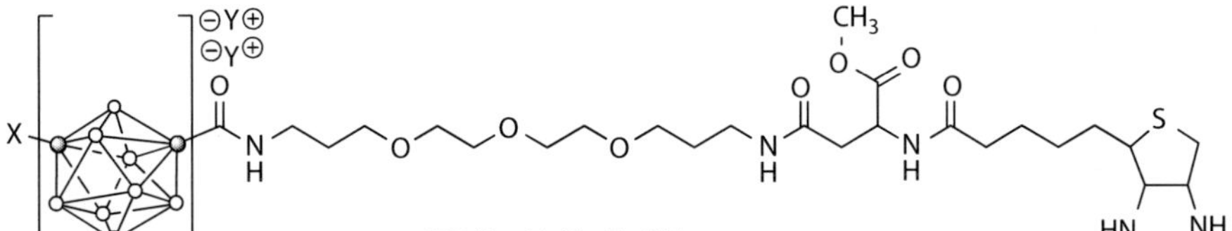

172, X = H, Y = Et_3NH
173, X = ^{131}I, Y = Et_3NH

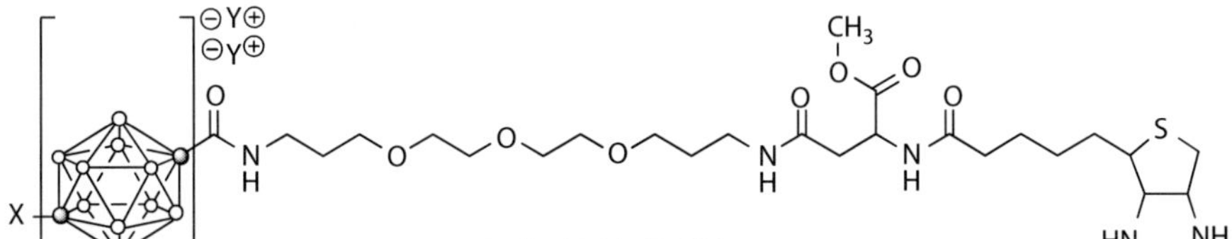

174, X = H, Y = Et_3NH
175, X = ^{125}I, Y = Et_3NH

FIGURE 6.15 Conjugation reagents for radiolabeling proteins.

174 were radiolabeled with different iodine isotopes to obtain ^{125}I-**173** and ^{131}I-**175**, which were then coinjected into mice. In this study, the radioactivity levels for both isotopes were higher in the kidney than in the liver.

Wilbur et al. explored the biotin–streptavidin approach in the context of radiohalogenated boron cluster conjugation for cancer therapy extensively (Wilbur et al., 1997a,b, 1998a,b,c, 2000, 2001, 2002, 2004b,c, 2008). Boron clusters were conjugated with either biotin (Figure 6.16) or recombinant streptavidin (Figure 6.17) and radiohalogenated (^{125}I and ^{211}At) to study the *in vivo* biodistribution patterns. Various radioastatinated and/or radioiodinated *nido*-carboranyl- (**181–187**, Figure 6.16) and benzoate-containing (**176–179**, Figure 6.16) biotin conjugates were synthesized (Wilbur et al., 2004b). At these conjugates, *N*-methyl- (**176, 177**) or carboxylate groups (**178–187**) were introduced at or adjacent to the biotinamide bonds to decrease biotinidase activity. In the case of the

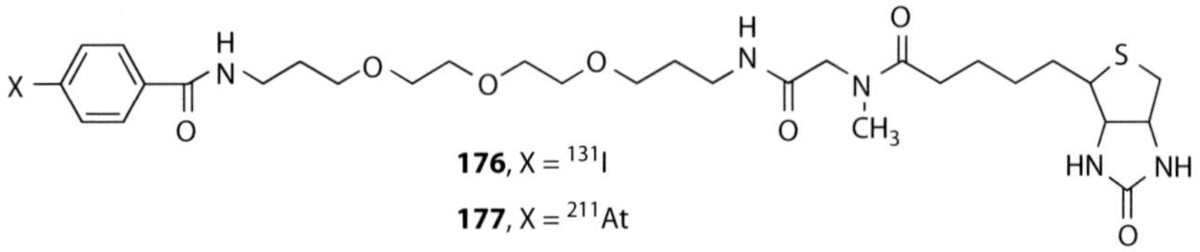

176, X = ^{131}I
177, X = ^{211}At

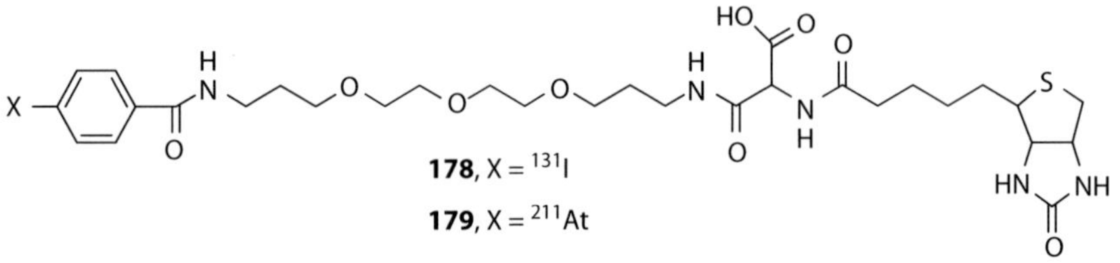

178, X = ^{131}I
179, X = ^{211}At

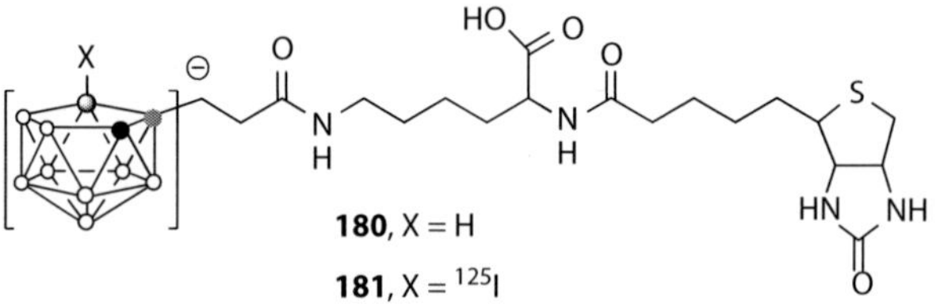

180, X = H
181, X = ^{125}I

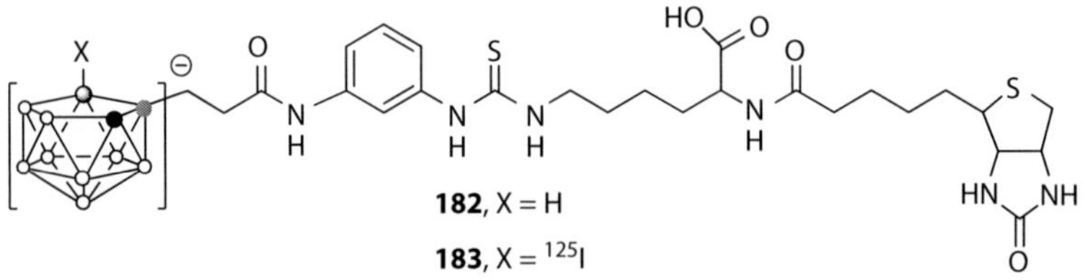

182, X = H
183, X = ^{125}I
184, X = ^{211}At

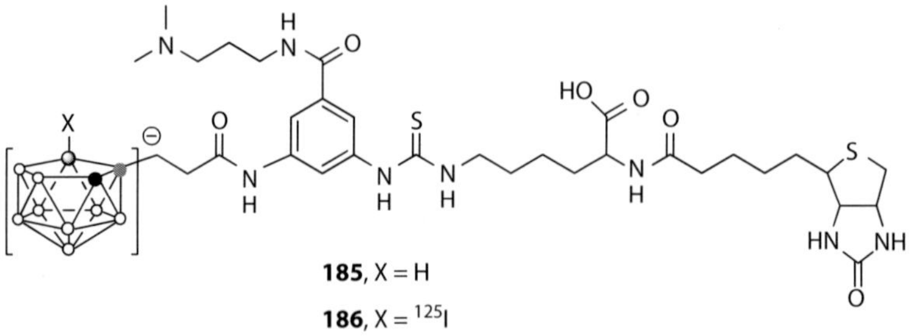

185, X = H
186, X = ^{125}I
187, X = ^{211}At

FIGURE 6.16 Biotin conjugates of radiolabeled aryl groups and *nido*-carborane clusters.

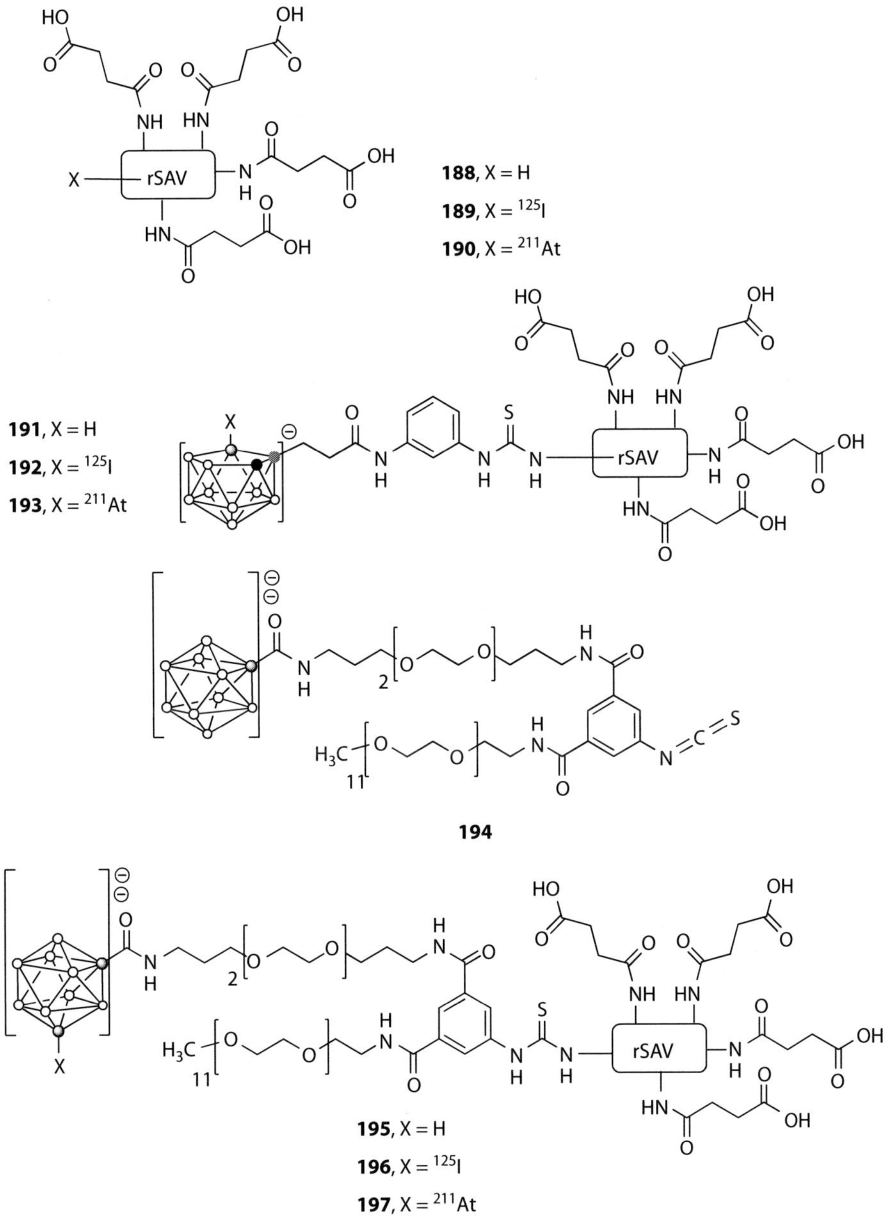

FIGURE 6.17 Radiolabeled recombinant streptavidin.

N-methylated derivatives **176** and **177**, the alteration led to a decreased binding between avidin and streptavidin. The carboranyl compounds **183** and **186** showed somewhat better *in vivo* stability than **176**, as was evident by a higher level of radioactivity in the neck region (thyroid) caused by the latter. The radioactivity in the neck region (thyroid) as a result of deastatination of **179** was significantly higher than the radioactivity from deiodination of **178** indicating a week C-^{211}At bond in the former. Similarly, the B-^{211}At bonds in **184** and **187** appeared to be less stable than the B-^{125}I bonds in **183** and **186** but differences were not nearly as pronounced as between the C-^{211}At and the C-^{125}I bonds in **178** and **179**, respectively.

Recombinant streptavidin (rSAV) was conjugated with the *nido*-carborane derivative **158** and the *closo*-decaborane derivative **194** to obtain conjugates **191** and **195** (Figure 6.17) (Wilbur et al.,

2008). Both conjugates as well as nonboronated rSAV (**188**) were radiohalogenated with ^{125}I and ^{211}At to obtain conjugates **189**, **190**, **192**, **193**, **196**, and **197**. Succinic acid groups were also attached to these rSAVs to reduce their renal uptake. Radioastatination of **188** resulted in low radiolabeling yields for **190** (18%) in comparison with the boronated rSAVs **193** (68%) and **197** (50%). Compounds **190** and **193** were dehalogenated in BALB/c mice approximately to the same extent whereas **197** appeared to be 2–3 times more stable than the former two derivatives.

6.3.1.3 Radiobromination

Compared with radioiodination and astatination, only very few radiobromination studies with boron clusters have been conducted. Tolmachev et al. standardized the aqueous conditions for the radiobromination of the triethylammonium salt of *closo*-dodecaborate **84** to obtain **89** as a mixture containing the bromine isotopes ^{76}Br, ^{77}Br, and ^{82}Br (Figure 6.5) (Tolmachev et al., 2002). The radiobrominated triethylammonium salt of **89** was synthesized using a radiobromide solution and Chloramine-T as the oxidizing agent. The radiobromination process was optimized by varying the concentrations of the substrate and Chloramine-T, the temperature, the pH, and the reaction times. Labeling was very fast and only 3 min of incubation at room temperature resulted in more than 90% yields. The labeling yields increased by increasing both substrate and Chloramine-T concentrations and reached a maximum at a concentration of 105 nM of *closo*-dodecaborate with as little as 3.2 mM of Chloramine-T. Radiolabeling yields were consistent in the pH range from 2 to 7 but decreased drastically at pH 8.

The triethylammonium salt of *closo*-dodecaborate **84** was also radiobrominated in the presence of tyramine to study the selective labeling of *closo*-dodecaborate in competition with phenolic groups (Figure 6.5) (Tolmachev et al., 2002). Labeling was performed in two ways. In the first approach, equimolar ratios of both substrates were radiolabeled at varying pH. Radiobromination of **84** decreased by increasing the pH from 4 to 7.4. Compound **84** was more effectively brominated between pH 4 and 7 than tyramine. On the other hand, tyramine bromination increased rapidly in the pH range of 7–7.4 with more than 90% labeling at pH 7.4. In the second approach, ratios of tyramine to *closo*-dodecaborate were varied from 1:1 to 30:1. In a mixture containing tyramine and *closo*-dodecaborate in a ratio of 10:1, equal amounts of both substrates were radiolabeled at pH 6.0. However, when pH was reduced to 5.5, radioactivity was distributed in a 4:1 ratio. When the ratio of tyramine to **84** was increased to 30:1 at pH 5.5, equal amounts of both substrates were radiolabeled.

Using optimized conditions (Tolmachev et al., 2002), the isothiocyanate derivative **103** (Figure 6.9) was radiobrominated with $Na^{76}Br$ and Chloramine-T as the oxidizing agent to afford **105** in 96–98% labeling yield (Figure 6.9) (Bruskin et al., 2004). Compound **105** was then conjugated with Trastuzumab to obtain **105-Trast** in 81% labeling yield under optimized reaction conditions. **105-Trast** was stable for more than 72 h in both PBS and serum. On the other hand, significantly lower coupling yields (49%) were obtained by using *N*-succinimidyl *p*-[^{76}Br]bromobenzoate (**108**) for labeling. In addition, the modified mAb retained its antigen-binding properties as it was easily replaced by nonradioactive mAb in a competitive antigen-binding assay.

Isothiocyanate derivative **100** (Scheme 6.17) was also radiobrominated to furnish **102** (93–95% labeling yield), which was subsequently conjugated with Trastuzumab (56% labeling yield) (Winberg et al., 2004). As observed for **105-Trast**, **102-Trast** was stable in both PBS and serum for more than 72 h and retained its antigen-binding abilities.

6.3.2 Radiohalogenation by Catalytic Halogen Exchange

Stanko and Iroshnikova reported the first radiolabeling of iodo-*o*-*closo*-carborane (Scheme 6.2) via isotopic exchange reaction using $Na^{125}I$ and iron (II) sulfate as a catalyst (Stanko and Iroshnikova, 1970). Adapting a method by Marshall et al. for Pd-catalyzed "cold" halogen exchange, as described in detail in Section 6.2.1.4., Sjöberg and coworkers developed a strategy for the radiohalogenation of *closo*-carboranes based on Pd-catalyzed isotopic exchange with commercially available $Na^{125}I$ (Marshall et al., 2001; Winberg et al., 2003, 2004, 2005).

6.3.2.1 Radiohalogenation of Simple Carboranyl Structures

Sjöberg et al. synthesized several radiohalogenated carboranes using Pd-catalyzed isotope exchange reactions (Eriksson et al., 2003; Winberg et al., 2003, 2005). Using tris(dibenzylideneacetone)-dipalladium as a catalyst, 2-^{125}iodo-*para*-carborane (**198**) was synthesized starting from $Na^{125}I$ and 2-iodo-*para*-carborane (**47**) (Scheme 6.19) (Eriksson et al., 2003). Since $Na^{125}I$ was obtained as an aqueous solution containing stabilizing NaOH, *tetra*-butylammonium hydrogen sulfate ($QHSO_4$) was added to the reaction mixture to quench the hydroxide anion and prevent its nucleophilic attack on *p*-carborane. A maximum labeling yield of 83% for **198** was achieved by using 0.04 equivalents of Pd catalyst and 0.4 equivalents of $QHSO_4$ in relationship to **47** at 100°C for 40 min in acetonitrile. Higher quantities of $QHSO_4$ decreased the yield.

Since this reaction depended on the presence of $QHSO_4$, Winberg et al. explored a different catalyst, *trans*-di-μ-acetato-bis[2-(di-*o*-tolylphosphino)benzyl]dipalladium (II) (Herrmann's catalyst; **208**), which produced similar labeling yields without $QHSO_4$ (Figure 6.18) (Winberg et al., 2003). Radioiodinated 2-^{125}I-*p*-carborane (**198**), 9-^{125}I-*o*-carborane (**200**), 3-^{125}I-*o*-carborane (**202**), 9-^{125}I-*m*-carborane (**204**), 1-phenyl-3-^{125}I-*o*-carborane (**206**), and 1,2-diphenyl-3-^{125}I-*o*-carborane (**207**) were synthesized using Herrmann's catalyst (**208**). A quantity of only 0.1 mol% of catalyst was sufficient to achieve 90% radiolabeling yield after 5 min reaction time at 100°C in acetonitrile. Increasing the amount of catalyst to 5 mol% further increased the labeling yields to 93%, 95%, 98%, and 95% for **198**, **200**, **202**, and **204**, respectively. On the other hand, the labeling yield for sterically hindered

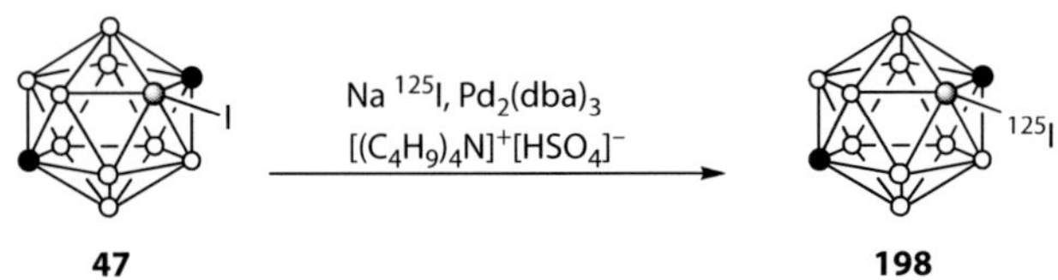

SCHEME 6.19 Synthesis of 2-^{125}I-*para*-carborane (**198**) using $Pd_2(dba)_3$ catalyst.

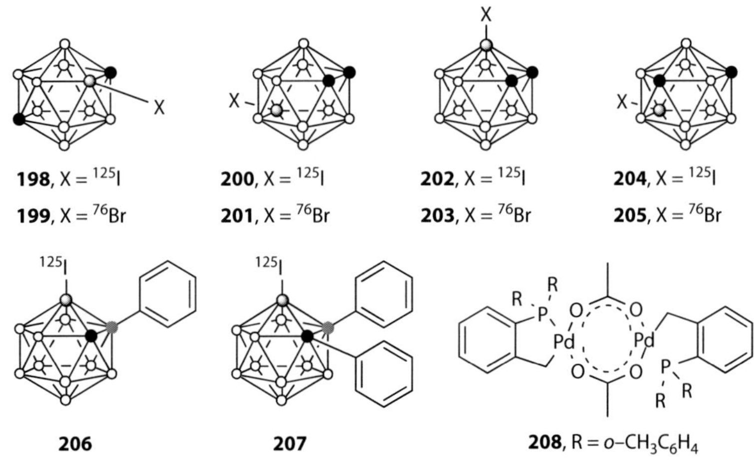

FIGURE 6.18 Synthesis of radiolabeled 2-*p*-carborane (**198**, **199**), 9-*o*-carborane (**200**, **201**), 3-*o*-carborane (**202**, **203**), 9-*m*-carborane (**204**, **205**), 1-phenyl-3-^{125}I-*o*-carborane (**206**), and 1,2-diphenyl-3-^{125}I-*o*-carborane (**207**) using Herrmann's catalyst (**208**).

carboranes were lower with 83% and 65% for **206** and **207**, respectively, after 5 min and 58% and 86%, respectively, after 20 min reaction time.

Tolmachev et al. used nonoxidative conditions to synthesize radiohalogenated dodecaborates using copper sulfate as the halogen exchange catalyst (Korsakov et al., 2003; Shchukin et al., 2004). Korsakov et al. synthesized radioiodinated bis(triethylammonium) undecahydro-^{125}I-*closo*-dodecaborate (**86**) from the corresponding cold undecahydro-^{127}I-*closo*-dodecaborate precursor through halogen exchange reaction with Na^{125}I in the presence of acetamide and copper sulfate in acetonitrile (Figure 6.5) (Korsakov et al., 2003). Maximum labeling yields of 78% were obtained when 0.115 μmol of catalyst and 2.2 μmol of substrate were reacted with Na^{125}I at 170°C for 30 min. Increased reaction times did not change the yield drastically whereas substrate concentrations below 0.5 μmol had a negative impact on labeling.

Shchukin et al. developed a method of radioiodination under aqueous conditions (Shchukin et al., 2004). Bis(triethylammonium) undecahydro-^{125}I-*closo*-dodecaborate (**86**) was synthesized with 90–95% labeling yield by reacting corresponding cold undecahydro-^{127}I-*closo*-dodecaborate precursor with Na^{125}I in the presence of a copper catalyst. A radiolabeling yield of higher 90% was achieved when 1 μmol of copper sulfate was reacted for 3 h at 100°C. However, as little as 0.01 μmol of copper sulfate was necessary to produce a radiolabeling yield of ~80%.

Using the procedure described above for radioiodination with Herrmann's catalyst (**208**), Sjöberg et al. also synthesized radiobrominated **199**, **201**, **203**, and **205** (Figure 6.18) from the corresponding cold iodine precursors (Winberg et al., 2003, 2005). Compound **205** was obtained in 64% whereas **199**, **201**, and **203** were radiolabeled with more than 90% yields. Maximum yields were obtained at 110°C temperature and 40 min reaction time.

6.3.2.2 Radioiodination of 3-Carboranyl Thymidine Analogs (3CTAs) through Isotope Exchange

Thymidine analogs substituted with a *closo*-carborane cluster at the N3 position (3-carboranyl thymidine analogs or 3CTAs), such as **N5** (**209**, Figure 6.19) (Lunato et al., 1999), have been synthesized and evaluated as boron delivery agents for BNCT because of their tumor cell selective accumulation, which is facilitated by phosphorylation through human thymidine kinase 1 (hTK1) and subsequent intracellular entrapment (Al-Madhoun et al., 2002, 2004; Barth et al., 2004, 2008; Byun et al., 2005, 2006; Johnsamuel et al., 2004; Lunato et al., 1999; Narayanasamy et al., 2006; Thirumamagal et al., 2006; Tjarks, 2000; Tjarks et al., 2007; Yan et al., 2002). For the same reason,

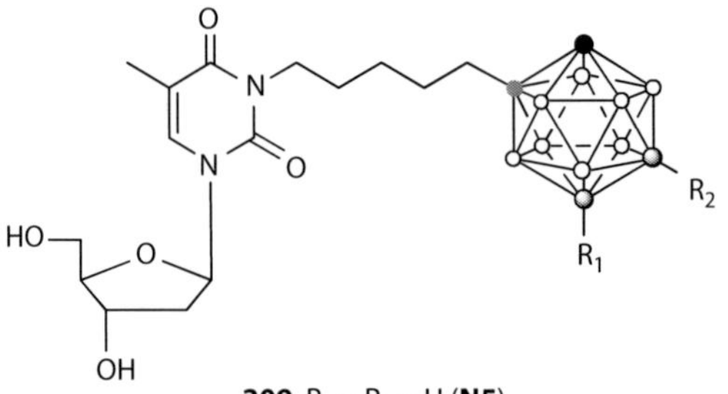

209, $R_1 = R_2 = H$ (**N5**)

210, $R_1 = I, R_2 = H$

211, $R_1 = H, R_2 = I$

212, $R_1 = Br, R_2 = H$

213, $R_1 = H, R_2 = Br$

214, $R_1 = {}^{125}I, R_2 = H$

215, $R_1 = H, R_2 = {}^{125}I$

FIGURE 6.19 Various carborane cage halogenated and radioiodinated derivatives of **N5**.

3CTAs that are carborane cluster radioiodinated may be attractive candidates for radiotherapeutic and diagnostic applications. Such molecules may be less susceptible to dehalogenation, and thus, superior to conventional carbon radioiodinated nucleosides, such as the various radioactive variants of 5-iodo-2′-deoxyuridine (see Section 6.1 for discussion).

Two methods were developed for the synthesis of compounds **210/211** (Tiwari et al., 2009a; Tiwari, R. et al., 2009b, unpublished work). The first method employed the direct cold iodination of **209** with 1 equivalent of ICl in the presence of 10 equivalents of $AlCl_3$ to yield a mixture of the geometric isomers **210** and **211** along with significant quantities of diiodinated product, "and decomposed nucleoside." In the second method, compound **17** (Scheme 6.2) was attached to the N3 position of thymidine through a pentylene spacer by adapting conventional procedures described previously for the synthesis of noniodinated 3CTAs using *o*-carborane (**1**) as a starting material (Al-Madhoun et al., 2002, 2004; Barth et al., 2004; Johnsamuel, 2004; Byun et al., 2005, 2006; Lunato et al., 1999; Narayanasamy et al., 2006; Thirumamagal et al., 2006; Tjarks et al., 2007; Yan et al., 2002. This method produced a mixture of **210** and **211** without decomposition).

Initially, halogen exchange conditions were optimized by subjecting **210/211** to 10 equivalents of NaBr and in the presence of Herrmann's catalyst (**208**) (25 mol%) in DMF for 1 h, which resulted in 100% exchange to afford **212/213** in 52% yield. Finally, the isotope exchange reaction of **210/211** was carried out with $Na^{125}I$ in the presence of 10 mol% of Herrmann's catalyst in DMF at 110°C for 1 h to yield **214/215**.

Carborane cage radioiodinated 3CTAs, such as **214/215**, constitute a novel type of radiopharmaceuticals. They should be excellent tools to study the general suitability of low-molecular-weight compounds as carriers for radioiodinated boron clusters in the diagnosis and therapy of cancer.

ACKNOWLEDGMENT

The preparation of this chapter was financially supported by The Ohio State University College of Pharmacy, NIH Grant R01 CA127935, and the Scholar Exchange Program of the Egyptian Education Ministry (salary to S. H.).

REFERENCES

Adam, M. J. and Wilbur, D. S. 2005. Radiohalogens for imaging and therapy. *Chem. Soc. Rev.*, 34, 153–63.

Al-Madhoun, A. S., Johnsamuel, J., Barth, R. F., Tjarks, W., and Eriksson, S. 2004. Evaluation of human thymidine kinase 1 substrates as new candidates for boron neutron capture therapy. *Cancer Res.*, 64, 6280–86.

Al-Madhoun, A. S., Johnsamuel, J., Yan, J. et al. 2002. Synthesis of a small library of 3-(carboranylalkyl)thymidines and their biological evaluation as substrates for human thymidine kinases 1 and 2. *J. Med. Chem.*, 45, 4018–28.

Armstrong, A. F. and Valliant, J. F. 2007. The bioinorganic and medicinal chemistry of carboranes: From new drug discovery to molecular imaging and therapy. *Dalton Trans.*, 38, 4240–51.

Barbera, G., Vaca, A., Teixidor, F. et al. 2008. Designed synthesis of new *ortho*-carborane derivatives: From mono- to polysubstituted frameworks. *Inorg. Chem.*, 47, 7309–16.

Barbera, G., Vinas, C., Teixidor, F., Welch, A. J., and Rosair, G. M. 2002. Retention of the B(3)-X (X = Br, I) bond in *closo-o*-carborane derivatives after nucleophilic attack. The first synthesis of [3-X-7-R-7,8-*nido*-$C_2B_9H_{10}$]- (X = Br, I). Crystal structure of [$HNMe_3$][3-I-7,8-*nido*-$C_2B_9H_{11}$]. *J. Organomet. Chem.*, 657, 217–23.

Barth, R. F., Yang, W., Al-Madhoun, A. S. et al. 2004. Boron-containing nucleosides as potential delivery agents for neutron capture therapy of brain tumors. *Cancer Res.*, 64, 6287–95.

Barth, R. F., Yang, W., Wu, G. et al. 2008. Thymidine kinase 1 as a molecular target for boron neutron capture therapy of brain tumors. *Proc. Natl. Acad. Sci.*, 105, 17493–97.

Boer, F. P., Potenza, J. A. and Lipscomb, W. N. 1966. Inductive rule in carboranes. Charge distributions in $B_8C_2H_{10}$ isomers. *Inorg. Chem.*, 5, 1301–2.

Boorsma, D. M. and Streefkerk, J. G. 1976. Peroxidase-conjugate chromatography isolation of conjugates prepared with glutaraldehyde or periodate using polyacrylamide-agarose gel. *J. Histochem Cytochem.*, 24, 481–6.

Bruskin, A., Sivaev, I., Persson, M. et al. 2004. Radiobromination of monoclonal antibody using potassium [^{76}Br] (4-isothiocyanatobenzyl-ammonio)-bromo-decahydro-*closo*-dodecaborate (bromo-DABI). *Nucl. Med. Biol.*, 31, 205–11.

Byun, Y., Narayanasamy, S., Johnsamuel, J. et al. 2006. 3-Carboranyl thymidine analogues (3CTAs) and other boronated nucleosides for boron neutron capture therapy. *Anticancer Agents Med. Chem.*, 6, 127–44.

Byun, Y., Yan, J., Al-Madhoun, A. S. et al. 2005. Synthesis and biological evaluation of neutral and zwitterionic 3-carboranyl thymidine analogues for boron neutron capture therapy. *J. Med. Chem.*, 48, 1188–98.

Chen, C.-J., Kane, R. R., Primus, F. J. et al. 1994. Synthesis and characterization of oligomeric *nido*-carboranyl phosphate diester conjugates to antibody and antibody fragments for potential use in boron neutron capture therapy of solid tumors. *Bioconjugate Chem.*, 5, 557–64.

Cheng, J., Persson, M., Tolmachev, V. et al. 2004. Targeting of a head and neck squamous cell carcinoma xenograft model using the chimeric monoclonal antibody U36 radioiodinated with a *closo*-dodecaborate-containing linker. *Acta Oto-Laryngol.*, 124, 1078–85.

Cho, H.-S., Mason, K., Ramyar, K. X. et al. 2003. Structure of the extracellular region of HER2 alone and in complex with the Herceptin Fab. *Nature*, 421, 756–60.

Conti, P. S., Alauddin, M. M., Fissekis, J. R., Schmall, B., and Watanabe, K. A. 1995. Synthesis of 2′-fluoro-5-[^{11}C]-methyl-1-beta-D-arabinofuranosyluracil ([^{11}C]-FMAU): A potential nucleoside analog for *in vivo* study of cellular proliferation with PET. *Nucl. Med. Biol.*, 22, 783–9.

Eriksson, L., Tolmachev, V. and Sjöberg, S. 2003. Feasibility of palladium-catalyzed isotopic exchange between sodium [^{125}I]I and 2-iodo-para-carborane. *J. Labelled Comp. Radiopharm.*, 46, 623–31.

Finze, M. 2008. Carba-*closo*-dodecaborates with one or two alkynyl substituents bonded to boron. *Inorg. Chem.*, 47, 11857–67.

Finze, M., Reiss, G. J. and Zahres, M. 2007. $[1\text{-}H_2N\text{-}CB_{11}F_{11}]^-$—Synthesis and reactions of a functionalized fluorinated carbadodecaborate anion. *Inorg. Chem.*, 46, 9873–83.

Fox, M. A. and Wade, K. 1999. Deboronation of 9-substituted- *ortho*- and -*meta*-carboranes. *J. Organomet. Chem.*, 573, 279–91.

Ghirmai, S., Malmquist, J., Lundquist, H., Tolmachev, V., and Sjöberg, S. 2004. Synthesis and radioiodination of 7-(3′-ammoniopropyl)-7,8-dicarba-*nido*-undecaborate(-1), (ANC). *J. Labelled Comp. Radiopharm.*, 47, 557–69.

Ghosh, M. K. and Mitra, A. K. 1992. Brain parenchymal metabolism of 5-iodo-2′-deoxyuridine and 5′-ester prodrugs. *Pharm. Res.*, 9, 1048–52.

Green, A. E., Causey, P. W., Louie, A. S. et al. 2006. Microwave-assisted synthesis of 3,1,2- and 2,1,8-Re(I) and ^{99m}Tc(I)-metallocarborane complexes. *Inorg. Chem.*, 45, 5727–9.

Green, A. E. C., Harrington, L. E., and Valliant, J. F. 2008. Carborane-carbohydrate derivatives—Versatile platforms for developing targeted radiopharmaceuticals. *Can. J. Chem.*, 86, 1063–9.

Green, A. E. C., Parker, S. K., and Valliant, J. F. 2009. Synthesis and screening of bifunctional radiolabelled carborane-carbohydrate derivatives. *J. Organomet. Chem.*, 694, 1736–46.

Grimes, R. N. 1970. *Carboranes*. New York: Academic Press Inc.

Grushin, V. V., Bregadze, V. I., and Kalinin, V. N. 1988. Synthesis and properties of carboranes(12) containing boron-element bonds. *J. Organomet. Chem. Libr.*, 20, 1–68.

Grushin, V. V., Demkina, I. I., and Tolstaya, T. P. 1991. Reactions of aryl(B-carboranyl)iodonium cations with the fluoride anion. Synthesis of icosahedral *o*-carboran-9-yl, *m*-carboran-9-yl, and *p*-carboran-2-yl fluorides. *Inorg. Chem.*, 30, 4860–3.

Grushin, V. V., Shcherbina, T. M., and Tolstaya, T. P. 1985. The reactions of phenyl(B-carboranyl)iodonium salts with nucleophiles. *J. Organomet. Chem.*, 292, 105–17.

Hawthorne, M. F. 1993. The role of chemistry in the development of cancer therapy by the boron-neutron capture reaction. *Angew. Chem.*, 105, 997–1033.

Hawthorne, M. F. & Maderna, A. 1999. Applications of radiolabeled boron clusters to the diagnosis and treatment of cancer. *Chem. Rev.*, 99, 3421–34.

Ivanov, S. V., Lupinetti, A. J., Miller, S. M. et al. 2002. Regioselective fluorination of $CB_{11}H_{12}$-. New weakly coordinating anions. *Inorg. Chem.*, 34, 6419–20.

Ivanov, S. V., Lupinetti, A. J., Solntsev, K. A., and Strauss, S. H. 1998a. Fluorination of deltahedral *closo*-borane and -carborane anions with N-fluoro reagents. *J. Fluorine Chem.*, 89, 65–72.

Ivanov, S. V., Rockwell, J. J., Polyakov, O. G. et al. 1998b. Highly fluorinated weakly coordinating monocarborane anions. 1-H-$CB_{11}F_{11}^-$, 1-CH_3-$CB_{11}F_{11}^-$, and the structure of $[N(n\text{-}Bu)_4]_2[CuCl(CB_{11}F_{11})]$. *J. Am. Chem. Soc.*, 120, 4224–5.

Jelinek, T., Baldwin, P., Scheidt, W. R., and Reed, C. A. 2002. New weakly coordinating anions. 2. Derivatization of the carborane anion $CB_{11}H_{12}$. *Inorg. Chem.*, 32, 1982–90.

Jiang, W., Knobler, C. B., Curtis, C. E., Mortimer, M. D., and Hawthorne, M. F. 1995. Iodination reactions of icosahedral *para*-carborane and the synthesis of carborane derivatives with boron–carbon bonds. *Inorg. Chem.*, 34, 3491–8.

Johnsamuel, J., Lakhi, N., Al-Madhoun, A. S. et al. 2004. Synthesis of ethyleneoxide modified 3-carboranyl thymidine analogues and evaluation of their biochemical, physicochemical, and structural properties. *Bioorg. Med. Chem.*, 12, 4769–81.

Kerr, J. A. and Stocker, D. W. 1993. Strengths of chemical bonds. D. R. Lide, (Ed.), *CRC Handbook of Chemistry and Physics.* Boca Raton, FL: CRC Press.

Kinsella, T. J., Collins, J., Rowland, J. et al. 1988. Pharmacology and phase I/II study of continuous intravenous infusions of iododeoxyuridine and hyperfractionated radiotherapy in patients with glioblastoma multiforme. *J. Clin. Oncol.*, 6, 871–9.

Klecker, R. W., Jr., Jenkins, J. F., Kinsella, T. J. et al. 1985a. Clinical pharmacology of 5-iodo-2′-deoxyuridine and 5-iodouracil and endogenous pyrimidine modulation. *Clin. Pharmacol. Ther.*, 38, 45–51.

Klecker, R. W. J., Jenkins, J. F., Kinsella, T. J. et al. 1985b. Clinical pharmacology of 5-iodo-2′-deoxyuridine and 5-iodouracil and endogenous modulation. *Clin. Pharamcol. Ther.*, 38, 45–51.

Knoth, W. H., Miller, H. C., Sauer, J. C. et al. 1964. Chemistry of boranes. IX. Halogenation of $B_{10}H_{10}^{-2}$ and $B_{12}H_{12}^{-2}$. *Inorg. Chem.*, 3, 159–67.

Kongpricha, S. and Schroeder, H. 1969. Icosahedral carboranes. XII. Direct fluorination of *o*-, *m*-, and *p*-carborane. *Inorg. Chem.*, 8, 2449–52.

Korsakov, M. V., Shchukin, E. V., Korsakova, L. N. et al. 2003. Feasibility of isotopic exchange in the system [^{125}I] iodide—undecahydro-iodo-*closo*-dodecaborate(2-) anion. *J. Radioanal. Nucl. Chem.*, 256, 67–71.

Lebeda, O., Orlova, A., Tolmachev, V. et al. 2000. Effect of ionizing radiation on the labeling of *closo*-dodecaborate(2-) anion with ^{125}I. *Spec. Publ. - R. Soc. Chem.*, 253, 148–51.

Lebedev, V. N., Balagurova, E. V., Polyakov, A. V. et al. 1990. Selective fluorination of *o*- and *m*-carboranes. Synthesis of 9-monofluoro-, 9,12-difluoro-1,8,9,12-trifluoro-, and 8,9,10,12-tetrafluoro-*o*-carboranes and 9-monofluoro-, and 9,10-difluoro-*m*-carboranes. Molecular structure of 8,9,10,12-tetrafluoro-*o*-carborane. *J. Organomet. Chem.*, 385, 307–18.

Lesnikowski, Z. J. 2009. Nucleoside-boron cluster conjugates—beyond pyrimidine nucleosides and carboranes. *J. Organomet. Chem.*, 694, 1771–75.

Li, J. and Jones, M., Jr. 1990. A simple synthesis of 3-bromo-*o*-carborane. *Inorg. Chem.*, 29, 4162–3.

Li, J., Logan, C. F., and Jones, M., Jr. 1991. Simple syntheses and alkylation reactions of 3-iodo-*o*-carborane and 9,12-diiodo-*o*-carborane. *Inorg. Chem.*, 30, 4866–8.

Lin, R., Liu, N., Yang, Y. et al. 2009. Radioiodination of protein using 2,3,5,6-tetrafluorophenyl 3-(*nido*-carboranyl) propionate (TCP) as a potential bi-functional linker: Synthesis and biodistribution in mice. *Appl. Radiat. Isot.*, 67, 83–7.

Lunato, A. J., Wang, J., Woollard, J. E. et al. 1999. Synthesis of 5-(carboranylalkylmercapto)-2′-deoxyuridines and 3-(carboranylalkyl)thymidines and their evaluation as substrates for human thymidine kinases 1 and 2. *J. Med. Chem.*, 42, 3378–89.

Mariani, G., Di Sacco, S., Volterrani, D. et al. 1996. Tumor targeting by intra-arterial infusion of 5-[^{123}I]iodo-2′-deoxyuridine in patients with liver metastases from colorectal cancer. *J. Nucl. Med.*, 37, 22S–25S.

Marshall, W. J., Young, R. J., Jr., and Grushin, V. V. 2001. Mechanistic features of boron-iodine bond activation of B-iodocarboranes. *Organometallics*, 20, 523–33.

Mester, J., Degoeij, K., and Sluyser, M. 1996. Modulation of [5-^{125}I]iododeoxyuridine incorporation into tumour and normal tissue DNA by methotrexate and thymidylate synthase inhibitors. *Eur. J. Cancer*, 32A, 1603–8.

Mizusawa, E. A., Thompson, M. R., and Hawthorne, M. F. 1985. Synthesis and antibody-labeling studies with the p-isothiocyanatobenzene derivatives of 1,2-dicarba-*closo*-dodecarborane(12) and the dodecahydro-7,8-dicarba-*nido*-undecaborate(-1) ion for neutron-capture therapy of human cancer. Crystal and molecular structure of Cs^+[*nido*-7-(*p*-C_6H_4NCS)-9-I-7,8-$C_2B_9H_{11}$]. *Inorg. Chem.*, 24, 1911–16.

Narayanasamy, S., Thirumamagal, B. T., Johnsamuel, J. et al. 2006. Hydrophilically enhanced 3-carboranyl thymidine analogues (3CTAs) for boron neutron capture therapy (BNCT) of cancer. *Bioorg. Med. Chem.*, 14, 6886–99.

Nestor, M., Persson, M., Cheng, J. et al. 2003. Biodistribution of the chimeric monoclonal antibody U36 radioiodinated with a *closo*-dodecaborate-containing linker. Comparison with other radioiodination methods. *Bioconjugate Chem.*, 14, 805–10.

Orlova, A., Bruskin, A., Sivaev, I. et al. 2006. Radio-iodination of monoclonal antibody using potassium [^{125}I]-(4-isothiocyanatobenzylammonio)-iodo-decahydro-*closo*-dodecaborate(iodo-DABI). *Anticancer Res.*, 26, 1217–23.

Orlova, A., Lebeda, O., Tolmachev, V. et al. 2000a. *Closo*-dodecaborate (2-) anion as a prosthetic group for labelling proteins with astatine. *Spec. Publ. - R. Soc. Chem.*, 253, 144–7.

Orlova, A., Lebeda, O., Tolmachev, V. et al. 2000b. *Closo*-dodecaborate(2-) anion as a potential prosthetic group for attachment of astatine to proteins. Aspects of the labelling chemistry with chloramine-T. *J. Labelled Comp. Radiopharm.*, 43, 251–60.

Orlova, A., Sivaev, I., Sjöberg, S., Lundqvist, H., and Tolmachev, V. 2004. Radioiodination of ammonio-*closo*-monocarborane, 1-H_3N-1-$CB_{11}H_{11}$. Aspects of labelling chemistry in aqueous solution using chloramine-T. *Radiochim. Acta*, 92, 311–5.

Pak, R. H., Kane, R. R., Knobler, C. B., and Hawthorne, M. F. 1994. Synthesis and structural characterization of [Me_3NH][*nido*-9,11-I_2-7,8-$C_2B_9H_{10}$] and [Me_3NH][*nido*-9-I-7,8-$C_2B_9H_{11}$]. *Inorg. Chem.*, 33, 5355–7.

Paxton, R. J., Beatty, B. G., Varadarajan, A., and Hawthorne, M. F. 1992. Carboranyl peptide-antibody conjugates for neutron-capture therapy: Preparation, characterization, and *in vivo* evaluation. *Bioconjugate Chem.*, 3, 241–7.

Persson, M., Sivaev, I., Winberg, K. J. et al. 2007. *In vitro* evaluation of two polyhedral boron anion derivatives as linkers for attachment of radioiodine to the anti-HER2 monoclonal antibody trastuzumab. *Cancer Biother. Radiopharm.*, 22, 585–96.

Potenza, J. A. and Lipscomb, W. N. 1966. Molecular structure of carboranes. Molecular and crystal structure of *o*-$B_{10}Br_4H_6C_2(CH_3)_2$. *Inorg. Chem.*, 5, 1483–8.

Prusoff, W. H., Jaffe, J. J., and Gunther, H. L. 1960. Pharmacology of 5-iododeoxyuridine, an analog of thymidine, in the mouse. *Biochem. Pharmacol.*, 3, 110–21.

Roscoe, J. S., Kongpricha, S., and Papetti, S. 1970. Icosahedral carboranes. XIV. Preparation of boron-substituted carboranes by boron-insertion reactions. *Inorg. Chem.*, 9, 1561–3.

Santos, E. C., Pinkerton, A. B., Kinkead, S. A. et al. 2000. Syntheses of *nido*-9,11-X-2–7,8-$C_2B_9H_{10}$- anions (C = Cl, Br or I) and the synthesis and structural characterization of $N(C_2H_5)_4$[commo-3,3′-Co(4,7-Br-2–3,1,2-$CoC_2B_9H_9)_2$]. *Polyhedron*, 19, 1777–81.

Shchukin, E., Orlova, A., Korsakov, M., Sjöberg, S., and Tolmachev, V. 2004. Copper-mediated isotopic exchange between [^{125}I]iodide and bis(triethylammonium) undecahydro-12-iodo-*closo*-dodecaborate in aqueous media. *J. Radioanal. Nucl. Chem.*, 260, 295–9.

Sivaev, I. B., Bregadze, V. I., and Kuznetsov, N. T. 2002. Derivatives of the *closo*-dodecaborate anion and their application in medicine. *Russ. Chem. Bull.*, 51, 1362–74.

Sivaev, I. B. and Bregadze, V. V. 2009. Polyhedral boranes for medical applications: Current status and perspectives. *Eur. J. Inorg. Chem.*, 2009, 1433–50.

Sivaev, I. B., Bruskin, A. B., Nesterov, V. V. et al. 1999. Synthesis of schiff bases derived from the ammoniaundecahydro-*closo*-dodecaborate(1-) anion, [$B_{12}H_{11}$NH = CHR]$^-$, and their reduction into monosubstituted amines [$B_{12}H_{11}NH_2CH_2R$]$^-$: A new route to water soluble agents for BNCT. *Inorg. Chem.*, 38, 5887–93.

Sjöström, A., Tolmachev, V., Lebeda, O. et al. 2003. Direct astatination of a tumour-binding protein, human epidermal growth factor, using *nido*-carborane as a prosthetic group. *J. Radioanal. Nucl. Chem.*, 256, 191–7.

Smith, H. D., Knowles, T. A., and Schroeder, H. 1965. Chemistry of decaborane-phosphorus compounds. V. Bromo carboranes and their phosphination. *Inorg. Chem.*, 4, 107–11.

Sneath, R. L., Jr., Soloway, A. H., and Dey, A. S. 1974. Protein-binding polyhedral boranes. 1. *J. Med. Chem.*, 17, 796–9.

Sneath, R. L., Jr., Wright, J. E., Soloway, A. H. et al. 1976. Protein-binding polyhedral boranes. 3. *J. Med. Chem.*, 19, 1290–4.

Soloway, A. H., Tjarks, W., Barnum, B. A. et al. 1998. The chemistry of neutron capture therapy. *Chem. Rev.*, 98, 1515–62.

Srivastava, R. R., Hamlin, D. K., and Wilbur, D. S. 1996. Synthesis of highly iodinated icosahedral mono- and dicarbon carboranes. *J. Org. Chem.*, 61, 9041–4.

Stanko, V. I. and Iroshnikova, N. G. 1970. Isotopic exchange of *o*-, *m*-, and *p*-B-iodocarboranes. *Zh. Obshch. Khim.*, 40, 311–5.

Teixidor, F., Barbera, G., Vinas, C., Sillanpaeae, R., and Kivekaes, R. 2006. Synthesis of boron-iodinated *o*-carborane derivatives. Water stability of the periodinated monoprotic salt. *Inorg. Chem.*, 45, 3496–8.

Thirumamagal, B. T., Johnsamuel, J., Cosquer, G. Y. et al. 2006. Boronated thymidine analogues for boron neutron capture therapy. *Nucleosides Nucleotides Nucleic Acids*, 25, 861–6.

Tiwari, R., Toppino, A., Byun, Y. et al. 2009a. Cage-iodination of *o*-carboranyl nucleoside analogs for the diagnosis and treatment of cancer. *Abstracts of Papers, 237th ACS National Meeting*, Salt Lake City, UT, United States, March 22–26, 2009.

Tjarks, W. 2000. The use of boron clusters in the rational design of boronated nucleosides for neutron capture therapy of cancer. *J. Organomet. Chem.*, 614–615, 37–47.

Tjarks, W., Tiwari, R., Byun, Y., Narayanasamy, S., and Barth, R. F. 2007. Carboranyl thymidine analogues for neutron capture therapy. *Chem. Commun. (Camb)*, 39, 4978–91.

Tolmachev, V., Bruskin, A., Sivaev, I., Lundqvist, H., and Sjöberg, S. 2002. Radiobromination of *closo*-dodecaborate anion. Aspects of labelling chemistry in aqueous solution using chloramine-T. *Radiochim. Acta*, 90, 229–35.

Tolmachev, V., Bruskin, A., Sjöberg, S., Carlsson, J., and Lundqvist, H. 2004. Preparation, radioiodination, and *in vitro* evaluation of a *nido*-carborane-dextran conjugate, a potential residualizing label for tumor targeting proteins and peptides. *J. Radioanal. Nucl. Chem.*, 261, 107–12.

Tolmachev, V., Koziorowski, J., Sivaev, I. et al. 1999. *Closo*-dodecaborate(2-) as a linker for iodination of macromolecules. Aspects on conjugation chemistry and biodistribution. *Bioconjugate Chem.*, 10, 338–45.

Tolmachev, V. and Sjöberg, S. 2002. Polyhedral boron compounds as potential linkers for attachment of radiohalogens to targeting proteins and peptides. *Collect. Czech. Chem. Commun.*, 67, 913–35.

Tran, T., Orlova, A., Sivaev, I., Sandstroem, M. and Tolmachev, V. 2007. Comparison of benzoate- and dodecaborate-based linkers for attachment of radioiodine to HER2-targeting affibody ligand. *Int. J. Mol. Med.*, 19, 485–93.

Tsang, C.-W., Yang, Q., Sze, E. T.-P. et al. 2000. Weakly coordinating nature of a carborane cage bearing different halogen atoms. Synthesis and structural characterization of icosahedral mixed halocarborane anions, 1-H-$CB_{11}Y_5X_6$- (X, Y = Cl, Br, I). *Inorg. Chem.*, 39, 5851–8.

Vaca, A., Teixidor, F., Kivekaes, R., Sillanpaeae, R., and Vinas, C. 2006. A solvent-free regioselective iodination route of *ortho*-carboranes. *Dalton Trans.*, 41, 4884–5.

Varadarajan, A., Sharkey, R. M., Goldenberg, D. M., and Hawthorne, M. F. 1991. Conjugation of phenyl isothiocyanate derivatives of carborane to antitumor antibody and *in vivo* localization of conjugates in nude mice. *Bioconjugate Chem.*, 2, 102–10.

Wilbur, D. S., Chyan, M.-K., Hamlin, D. K. et al. 2004a. Reagents for atatination of biomolecules: Comparison of the *in vivo* distribution and stability of some radioiodinated/astatinated benzamidyl and *nido*-carboranyl compounds. *Bioconjugate Chem.*, 15, 203–23.

Wilbur, D. S., Chyan, M.-K., Hamlin, D. K., and Perry, M. A. 2009a. Reagents for astatination of biomolecules. 3. Comparison of *closo*-decaborate(2-) and *closo*-dodecaborate(2-) moieties as reactive groups for labeling with astatine-211. *Bioconjugate Chem.*, 20, 591–602.

Wilbur, D. S., Chyan, M.-K., Hamlin, D. K. et al. 2007. Reagents for astatination of biomolecules. 2. Conjugation of anionic boron cage pendant groups to a protein provides a method for direct labeling that is stable to *in vivo* deastatination. *Bioconjugate Chem.*, 18, 1226–40.

Wilbur, D. S., Chyan, M.-K., Pathare, P. M. et al. 2000. Biotin reagents for antibody pretargeting. 4. Selection of biotin conjugates for *in vivo* application based on their dissociation rate from avidin and streptavidin. *Bioconjugate Chem.*, 11, 569–83.

Wilbur, D. S., Hamlin, D. K., Buhler, K. R. et al. 1998a. Streptavidin in antibody pretargeting. 2. Evaluation of methods for decreasing localization of streptavidin to kidney while retaining its tumor binding capacity. *Bioconjugate Chem.*, 9, 322–30.

Wilbur, D. S., Hamlin, D. K., Chyan, M.-K., and Brechbiel, M. W. 2008. Streptavidin in antibody pretargeting. 5. Chemical modification of recombinant streptavidin for labeling with the alpha-particle-emitting radionuclides 213Bi and 211At. *Bioconjugate Chem.*, 19, 158–70.

Wilbur, D. S., Hamlin, D. K., Chyan, M.-K., Kegley, B. B., and Pathare, P. M. 2001. Biotin reagents for antibody pretargeting. 5. Additional studies of biotin conjugate design to provide biotinidase stability. *Bioconjugate Chem.*, 12, 616–23.

Wilbur, D. S., Hamlin, D. K., Chyan, M.-K. et al. 2004b. Biotin reagents in antibody pretargeting. 6. Synthesis and *in vivo* evaluation of astatinated and radioiodinated aryl- and *nido*-carboranyl-biotin derivatives. *Bioconjugate Chem.*, 15, 601–16.

Wilbur, D. S., Hamlin, D. K., Meyer, D. L. et al. 2002. Streptavidin in antibody pretargeting. 3. Comparison of biotin binding and tissue localization of 1,2-cyclohexanedione and succinic anhydride modified recombinant streptavidin. *Bioconjugate Chem.*, 13, 611–20.

Wilbur, D. S., Hamlin, D. K., Pathare, P. M., and Weerawarna, S. A. 1997a. Biotin reagents for antibody pretargeting. Synthesis, radioiodination and *in vitro* evaluation of water soluble, biotinidase resistant biotin derivatives. *Bioconjugate Chem.*, 8, 572–84.

Wilbur, D. S., Hamlin, D. K., Sanderson, J., and Lin, Y. 2004c. Streptavidin in antibody pretargeting. 4. Site-directed mutation provides evidence that both arginine and lysine residues are involved in kidney localization. *Bioconjugate Chem.*, 15, 1454–63.

Wilbur, D. S., Hamlin, D. K., Srivastava, R. R., and Chyan, M. K. 2004d. Synthesis, radioiodination, and biodistribution of some *nido*- and *closo*-monocarbon carborane derivatives. *Nucl. Med. Biol.*, 31, 523–30.

Wilbur, D. S., Pathare, P. M., Hamlin, D. K., Buhler, K. R., and Vessella, R. L. 1998b. Biotin reagents for antibody pretargeting. 3. Synthesis, radioiodination, and evaluation of biotinylated starburst dendrimers. *Bioconjugate Chem.*, 9, 813–25.

Wilbur, D. S., Pathare, P. M., Hamlin, D. K., and Weerawarna, S. A. 1997b. Biotin reagents for antibody pretargeting. 2. Synthesis and *in vitro* evaluation of biotin dimers and trimers for cross-linking of streptavidin. *Bioconjugate Chem.*, 8, 819–32.

Wilbur, D. S., Stayton, P. S., To, R. et al. 1998c. Streptavidin in antibody pretargeting. Comparison of a recombinant streptavidin with two streptavidin mutant proteins and two commercially available streptavidin proteins. *Bioconjugate Chem.*, 9, 100–7.

Wilbur, D. S., Thakar, M. S., Hamlin, D. K. et al. 2009b. Reagents for astatination of biomolecules. 4. Comparison of maleimido-*closo*-decaborate(2-) and *meta*-[211At]astatobenzoate conjugates for labeling anti-CD45 antibodies with [211At]astatine. *Bioconjugate Chem.*, 20, 1983–91.

Winberg, K. J., Barbera, G., Eriksson, L. et al. 2003. High yield [^{125}I]iodide-labeling of iodinated carboranes by palladium-catalyzed isotopic exchange. *J. Organomet. Chem.*, 680, 188–92.

Winberg, K. J., Mume, E., Tolmachev, V., and Sjöberg, S. 2005. Radiobromination of *closo*-carboranes using palladium-catalyzed halogen exchange. *J. Labelled Comp. Radiopharm.*, 48, 195–202.

Winberg, K. J., Persson, M., Malmstrom, P.-U., Sjöberg, S., and Tolmachev, V. 2004. Radiobromination of anti-HER2/NEU/ERBB-2 monoclonal antibody using the *p*-isothiocyanatobenzene derivative of the [^{76}Br] undecahydro-bromo-7,8-dicarba-*nido*-undecaborate(1-) ion. *Nucl. Med. Biol.*, 31, 425–33.

Xie, Z., Tsang, C.-W., Sze, E. T.-P. et al. 1998. Highly chlorinated, brominated, and iodinated icosahedral carborane anions: 1-H-$CB_{11}X_{11}$-, 1-CH_3-$CB_{11}X_{11}$- (X = Cl, Br, I); 1-Br-$CB_{11}Br_{11}$. *Inorg. Chem.*, 37, 6444–51.

Xie, Z., Tsang, C.-W., Xue, F., and Mak, T. C. W. 1999. Synthesis and structural characterization of new weakly coordinating anions. Crystal structure of Ag(1-CH_3-$CB_{11}H_5X_6$) (X = H, Cl, Br, I). *J. Organomet. Chem.*, 577, 197–204.

Yamazaki, H., Ohta, K., and Endo, Y. 2005. Regioselective synthesis of triiodo-*o*-carboranes and tetraiodo-*o*-carborane. *Tetrahedron Lett.*, 46, 3119–22.

Yan, J., Naeslund, C., Al-Madhoun, A. S. et al. 2002. Synthesis and biological evaluation of 3′-carboranyl thymidine analogues. *Bioorg. Med. Chem. Lett.*, 12, 2209–12.

Zakharkin, L. I. and Kalinin, V. N. 1971. Action of nucleophilic reagents on B-halocarboranes. *Izv. Akad. Nauk SSSR, Ser. Khim.*, 10, 2310–2.

Zakharkin, L. I., Kalinin, V. N., and Podvisotskaya, L. S. 1970. Comparative reactivity of *o*-, *m*-, and *p*-carboranes. *Izv. Akad. Nauk SSSR, Ser. Khim.*, 6, 1297–302.

Zheng, Z., Jiang, W., Zinn, A. A., Knobler, C. B., and Hawthorne, M. F. 1995. Facile electrophilic iodination of icosahedral carboranes. Synthesis of carborane derivatives with boron–carbon bonds via the palladium-catalyzed reaction of diiodocarboranes with Grignard reagents. *Inorg. Chem.*, 34, 2095–100.

Part IV

Boron for Living: Boron Neutron Capture Therapy

Part IV discusses boron neutron capture therapy (BNCT) that is divided into six chapters. BNCT is a binary form of cancer therapy that is based on the preferential localization of B-10-containing compounds in cancer cells and then irradiating the cells with thermal neutrons. Boron-10 has an extremely high tendency of absorb neutron, giving an excited B-11 that immediately undergoes a fission reaction producing high-energy alpha particles $^{4}He_{22+}$ and recoil $^{7}Li_{3}$ ions. These particles deposit their energies at distances of essentially a cell diameter in a biological medium. Thus, cell death will occur within the cell containing the B-10, sparing those cells not containing boron. The discovery of polyhedral boron compounds added new dimensions to the medicinal chemistry of boron, facilitating the quest for biologically active molecules containing boron clusters rather than only a single boron atom per molecule. The major difficulties to be overcome in BNCT are the design and syntheses of tumor-seeking boron drugs and the development of efficient methods for drug delivery; the chapters in this section address these points.

Chapter 7 (by Zhu, Maguire & Hosmane) and Chapter 10 (by Sibrian-Vazquez and Vicente) discuss the use of macromolecules in boron delivery. While Chapter 7 presents a general review of the use of liposomes, dendrimers, polymers, magnetic nanoparticles and carbon, boron nitride and boron nanotubes as boron delivery agents, Chapter 10 extends the discussion to include porphyrins, nucleosides, antibodies, and liposomes. Chapter 8 (by Nakamura) and Chapter 12 (by Feakes) concentrate the use of liposomes in boron delivery. Liposomal delivery is a method of supplying large amounts of boron to cells and may well prove to be the most efficient drug delivery method. Chapter 9 (by Bregadze and Sivaev) describes the syntheses of various polyhedral boron compounds that show promise as boron delivery drugs. Chapter 11 discusses a companion therapy that uses gadolinium neutron capture as a cancer therapy (GdNCT). Gadolinium-157 has a thermal-neutron-capture cross section of 255,000 barn, which is 65 times greater than that of B-10. The neutron-capture reaction, 157Gd(n,γ)158Gd, releases Auger electrons, internal conversion electrons, γ-rays, and x-rays. The reaction releases a total kinetic energy of 7.7 MeV, almost double that released by 10B(n,α)7Li. The disadvantage of GdNCT is that, to be cytotoxic, the metal must be next to the DNA in the cell nucleus. Gadolinium is extensively used as a magnetic resonance image enhancement agent, and the combination of both elements in compounds could results in potent diagnostic and therapeutic compounds.

7 Recent Developments in Boron Neutron Capture Therapy Driven by Nanotechnology*

Zhu Yinghuai, John A. Maguire, and Narayan S. Hosmane

CONTENTS

7.1 INTRODUCTION: BNCT BACKGROUND

It has been widely recognized that boron neutron capture therapy (BNCT) is potentially a promising and powerful binary anticancer therapy in which compounds containing the ^{10}B isotope are selectively introduced into tumor cells and then irradiated with thermal neutrons. The ^{10}B nucleus adsorbs a neutron forming an excited ^{11}B nucleus that undergoes a rapid fission reaction, producing a high-energy α-particle (1.47 MeV) and Li-7 ion (0.84 MeV), in addition to a low-energy gamma γ-ray (478 keV). These particles may cross a cell nucleus and thus destroy a tumor cell, see Figure 7.1 for a proposed cell-killing mechanism. The linear energy transfer (LET) of these heavily charged particles have a range of about one cell diameter [1,2], which confines radiation damage to the cell from which they arise, hence minimizing cytotoxic effects on the surrounding tissue. Therefore, if boron can be selectively accumulated, the target region can be dosed with neutrons at a sizeable flux, but have a minimal effect on the boron-free regions in the beam path. Compared to conventional radiotherapy, boron-10 has the advantages of being nonradioactive without neutron irradiation and easily incorporated with various materials. In principle, the required boron concentration is generally

* Supports by grants from Institute of Chemical and Engineering Sciences, A*STAR, Singapore, the National Science Foundation (CHE-0906179), the Robert A. Welch Foundation (N-1322), and the awards from Alexander von Humboldt Foundation and NIU Inaugural Board of Trustees Professorship are hereby acknowledged.

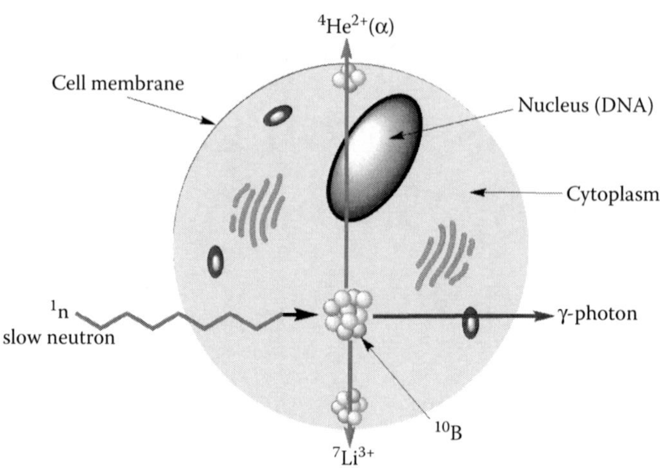

FIGURE 7.1 BNCT-killing mechanism.

estimated at 10^9 ^{10}B atoms (natural abundance 19.9%) per cell, which translates to approximately 35 μg ^{10}B per gram of tissue [3]. To prevent damage to healthy tissue in the path of the neutron beam, the surrounding tissue should contain not more than 5 μg of $^{10}B/g$ of tissue.

BNCT has been developed primarily for the treatment of malignant brain tumors, such as glioblastoma multiforme (GBM) [4]. GBM is one of the most malignant forms of cancer that infiltrates the brain so aggressively that surgery is rarely able to remove all of the cancerous tissue; it is virtually untreatable and is inevitably fatal [5,6]. It was postulated that the reduction of the blood–brain barrier (BBB) in the vicinity of tumors could be exploited to selectively increase the concentration of boron in brain tumor over normal brain. However, the invasive nature of GBM means that neoplastic cells could invade regions of the brain that is protected by the BBB; it must be considered a whole brain disease. In addition to treating difficult brain and melanoma tumors, BNCT has been successfully employed in other cancer therapies in last decades. For example, in Finland, patients with recurrent head and neck tumors have been treated with BNCT and showed an excellent response [7]. Other potential candidates for BNCT such as lung and liver tumors have also been investigated [8,9]. Selective localization of boron in the tumor cells is necessary to produce substantial therapy ratios of tumor-to-blood and tumor-to-brain during BNCT. Consequently, improvements in boron delivery vehicles will dramatically improve BNCT effectiveness. However, in common with all other forms of radio- and chemotherapy, the "bottle-neck" obstacle to be overcome in BNCT is the relatively nonspecific *in vivo* distribution for most of the boron-containing compounds. When BNCT compounds are intravenously injected, they will generally distribute in both tumor and healthy cells, and methods must be developed to maximize the tumor/healthy tissue ratio. In the last decades, a variety of carrier molecules such as boronated polyamines, liposomes, porphyrins, nucleosides, and other molecules have been investigated as possible boron delivery agents to tumor cells [1,10–17]. Despite these efforts, only two low-molecular-weight compounds, (Figure 7.2), $Na_2B_{12}H_{11}SH$ (BSH) and 4-borono-L-phenylalanine (L-BPA), are currently used in clinical trials. The results on these compounds are not universally promising, owing to their low tumor-to-blood and tumor-to-brain tissue ^{10}B ratios [10–17]. While there are methods to overcoming the BBB, an increased toxicity associated with BBB disruption has also been found [18]. Hence, it is important to develop potential precursor molecules and more sophisticated delivery systems for BNCT in the preclinical and clinical evaluations in future. A promising trend has been in the development of nanomaterials, such as liposomes, nanotubes, and magnetic nanoparticles, as both boron host molecules and delivery agents. These developments are detailed in this chapter.

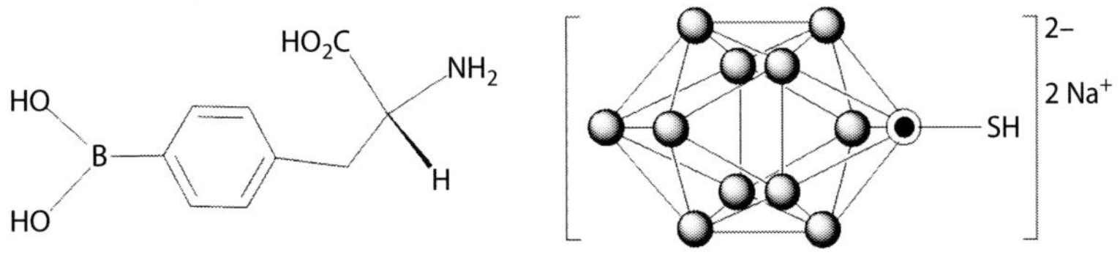

FIGURE 7.2 Structures of BPA and BSH.

7.2 NANOSCALE BNCT AGENTS

7.2.1 Nanoscaled Material-Based Drug Delivery

Nanomaterials composed of assorted sizes and shapes with various chemical and surface properties have been constructed. These classes of material have been used in biotechnology, especially in nanomedicine. Nanomaterials including organic and inorganic components demonstrate unique properties that make them highly promising candidates for both drug delivery and imaging applications [19–21]. The morphological diversity of scaffold structures in nanomaterials facilitates the presentation and encapsulation of drugs, biomolecules, and imaging agents. Nanomaterial-based drug delivery systems provide structural and dynamic properties that are complementary to the conventional lipid-based delivery vehicles, such as liposomes. Magnetic nanomaterials can be used in both MRI imaging and therapeutic applications. Nanomaterial-based drug systems provide the advantages of being able to penetrate cell membranes through minuscule capillaries in the cell walls of the rapidly dividing tumor cells [22], while at the same time having low cytotoxicity in normal cells. However, there are clearly hurdles to be overcome in implementing these delivery systems, such as long-term stability and reproducible *in vivo* behavior.

In theory, BNCT nanomaterials could incorporate a large number of boron atoms, thereby lowering the dose requirement for delivering critical amounts of boron to tumor cells. For this reason, the syntheses and evaluations of boron-carrying nanomaterials as possible BNCT therapeutic drug delivery agents have received a great deal of interest [23]. Nanomaterials have been found to have favorable interaction with the brain blood vessel endothelial cells of mice [24], and thus have the possibility of being transported to other brain tissues, making them potential BNCT agents. Their role as boron carriers can be further enhanced if the nanocarriers are capable of avoiding detection and clearance by the immune system, specifically the reticuloendothelial system (RES). The size of the nanocarriers, though offering advantages, also presents problems. For example, it has been shown that liposomes having diameters >50 nm have difficulty penetrating the BBB [25], and thus the optimum dimension of the nanoparticles as drug carriers has become an important area of research. Currently, several different types of nanomaterials as BNCT drug-targeting agents are under investigation in terms of their effectiveness and applications [23].

7.2.2 Liposomes-Based BNCT Agents

Liposomes are lipid-based natural nanomaterials and conventional drug delivery vectors because they are highly biocompatible, biodegradable, and nontoxic. Liposomes can also be targeted by embedding suitable targeting proteins in their lipid bilayers. Once their delivery function has been achieved, liposomes may break down inside cells. Liposomes are composed of cholesterol and phospholipids, which can be altered to engineer different chemical and physical characteristics. Because of their enormous diversity and flexibility, liposomes have been recognized as potentially effective carriers for the selective delivery of active drugs to tumors. Previous studies have demonstrated that the uptake of drugs encapsulated in liposomes is higher in tumor cells when compared with the surrounding normal cells. This can be partly attributed to the relative ease of modification of the surfaces of liposomes to achieve selective localization in tumor cells [26]. In order to make them more effective, ligands, such as antibodies and their fragments, folate, transferrin, and some peptides [27], are linked to the liposomes. With such modifications, care must be taken that the attached ligands do not alter the properties of the encapsulated compounds. BNCT agents can be encapsulated inside liposomes that circulate throughout the blood stream where the drug is then released via diffusion through the liposome or by liposomal degradation. Liposomes as potential BNCT delivery agents have been well investigated [28]. Hawthorne and coworkers have made major contributions in developing liposomes as boron delivery agents to tumors [29]. Several of the compounds developed for this purpose are shown in Figure 7.3 [29–32]. For the compound $Na_3[\alpha^2\text{-}B_{20}H_{17}NH_2CH_2CH_2NH_3]$ (Figure 7.3a) as a BNCT agent, preliminary *in vivo* tests with mice demonstrated improved tumor uptake and retention [30]. Dual-chain *nido*-carborane compounds, such as shown in (Figure 7.3c), have been used to prepare stable, high-boron-containing liposomes [31,32]. Cationic liposomes, which, because of their ability to target the cell nuclei, are currently used in gene therapy and also have been used as carriers for sugar-derived carborane compounds (Figure 7.3d) in BNCT. These new delivery systems may reduce the amount of boron required for successful BNCT treatment [33]. It has also been demonstrated that the administration of ^{10}B-polyethylene-glycol (PEG)-binding liposomes, with mean diameters of about 116 nm, can suppress the growth of AsPC-1 tumors in nude mice implanted with human pancreatic carcinoma xenografts [34]. The conjugate was able to avoid the phagocytosis by the reticuloendothelial system (RES) and, thus, could circulate in the body for longer periods of time. Additionally, the development of a new generation of magnetic liposomes is in progress. In this technology, liposomes are loaded with both a therapeutic drug and a ferromagnetic material that give them magnetic properties. These magnetic liposomes could then act as drug carriers under the influence of an external magnetic field. Magnetic liposomes containing doxorubicin have been shown to be rather effective drug carriers to osteosarcoma-bearing hamsters in that they increased the concentration of doxorubicin in the tumor [35]. However, the concentration of doxorubicin in the tumor appears to have dropped several weeks after administration (see Figure 7.4). In addition, problems such as the propensity to fuse together in aqueous solutions and thus release their payload prematurely have to be addressed before engineering liposomes' nanomaterial with drugs.

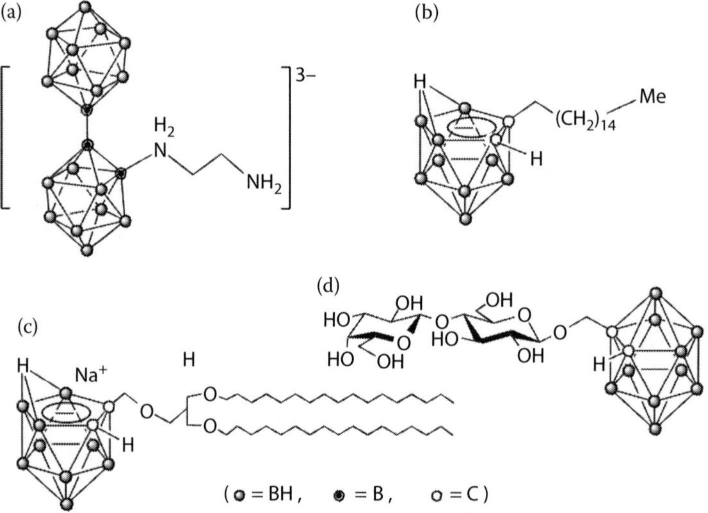

FIGURE 7.3 Structures of amino polyhedron (a) boron anion, (b) carboranyl anion, (c) dual-chain *nido*-carborane, and (d) sugar-derived carborane for liposome encapsulation.

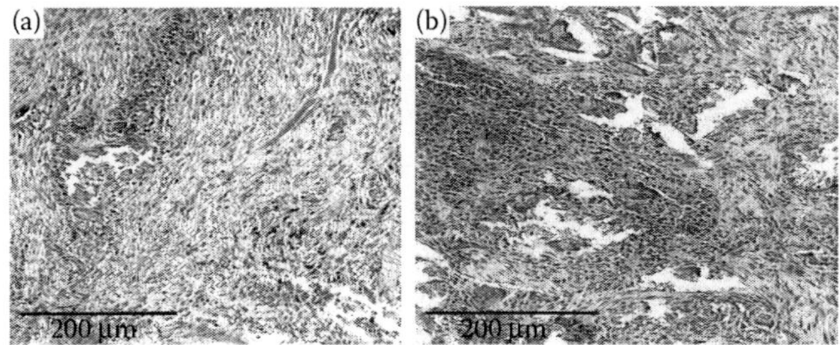

FIGURE 7.4 Light micrographs of Berlin blue-stained osteosarcoma excised 1 h (a) and 2 weeks (b) after IV administration of magnetic doxorubicin liposomes under magnetic force of 0.4 T. Blue spots represent magnetite particles, and they are concentrated and deposited in the endothelium of proliferated tumor vessels and the tumor interstitium in (a) but not (b). (Adapted from Renner, M. W. et al. *Anti-Cancer Agents Med. Chem.* **2006**, *6*, 145–157.)

7.2.3 Dentritic Polymer-Based BNCT Agents

Polymer-derived drug delivery systems have attracted much attention since the idea of using polymers to transport proteins was first raised in the 1970s [36,37]. Modified polymer carriers can improve water solubility, prolong vascular circulation time, and therefore selectively increase the accumulation in tumor tissue through the so-called enhanced permeation and retention (EPR) mechanism [38,39].

Dendrimers are uniform-sized and water-soluble macromolecules that have distinctive characteristics such as high degrees of branching, multivalency, unique molecular weight, and modifiable surface functionality [40]. In general, structurally there are two types of dendrimers, one is a globular structure with a central core and radiating branches, and the other is composed of highly branched polymer chains, with no core in the center. Globular dendrimers are built from a series of branches around an inner core and have the potential to carry therapeutic agents either in their internal cavities or on their periphery by means of covalent bonding [41]. Furthermore, they can be engineered to deliver either hydrophobic or hydrophilic molecules and thus offer a broad range of possibilities as therapeutic delivery agents [42]. A frequently used dendrimer scaffold is the polyamidoamine (PAMAM) dendrimer that is designed to have a water-soluble backbone and is widely used in drug delivery [43]. The PAMAM dendrimers are biocompatible, nonimmunogenic, water soluble, and possess modifiable surface amine groups for binding to various molecules. Anticancer drugs loaded onto PAMAM dendrimers have been tested on mice with promising results [44]. Barth et al. have reported tumor boron concentrations of ~20 μg/g tumor in rats bearing the implants of the F98 glioma when a heavily boranated PAMAM dendrimer was targeted at the epidermal growth factor receptor (EGFR), which is often overexpressed in brain tumors [45]. In a different *in vivo* study, a G-5 PAMAM dendrimer carrying 1100 boron atoms and linked to an EGFR-recognizing cetuximab (Figure 7.5a) resulted in a substantial increase of ^{10}B accumulation in brain tumors [42].

Covalent attachment of drug molecules to the dendrimer periphery allows regulation of drug loading and release, and several dendrimer delivery systems synthesized in this way have produced positive results [42]. However, problems arise with dendrimers that encapsulate drugs in their core. Drugs tagged to the dendrimers in this way often encounter difficulty during the drug-release process, which is not easy to control. A recent effort by Gillies and Frechet has led to a new release methodology that makes use of polyethylene oxide (PEO) hybrids and dendrimers tagged with pH-sensitive hydrophobic acetal groups on the periphery [42]. On hydrolysis at slightly acidic pHs, the hydrophobic groups were lost, which prompted the release of the load. The introduction of PEO chains to boron-containing PAMAM dendrimer conjugates has been explored as a means of reducing their uptake by the liver; at present, the results of this work are inconclusive [42]. A better knowledge of the interaction of PAMAM dendrimers with the RES is required to optimize their boron delivery abilities to tumor cells, while at the same time ensuring that they do not remain in

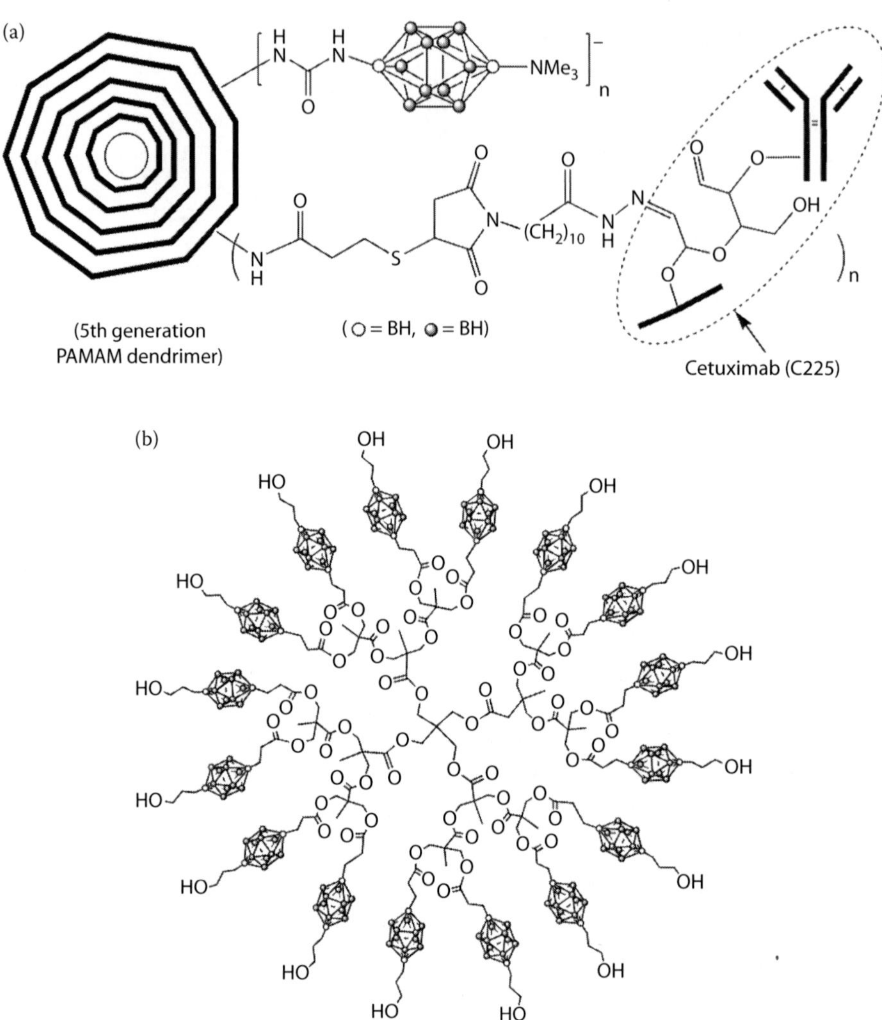

FIGURE 7.5 (a) Conjugation scheme for linkage of a boron-containing dendrimer to the monoclonal antibody, C225 (cetuximab), and (b) aliphatic polyester dendrimer containing 16 carborane cages.

the system so long as to cause harm. At present, research is continuing on the biological applications of different dendrimers [46]. Despite their extensive study, steps are required to modify the amine groups on the dendrimer surfaces to minimize their toxicity and liver accumulation. To investigate the use of nontoxic, water-soluble, biocompatible dendrimers as BNCT agents, Parrot et al. have synthesized aliphatic polyester dendrimers containing as many as 16 carborane cages (see Figure 7.5b). The 16-cage compound is currently being subjected to bio-investigations to assess its potentiality as a BNCT agent [47].

The development of dendrimers as efficient boron agents for BNCT will require advances in the release mechanism of the boron compounds, better tumor-targeting capabilities of the dendrimers, and more reliable and efficient methods of manufacturing.

7.2.4 Magnetic Nanoparticle-Based BNCT Agents

One of the biggest obstacles in chemotherapy and drug discovery is that most of the compounds are relatively nonspecific. When the therapeutic drugs are infused intravenously, the drug will generally

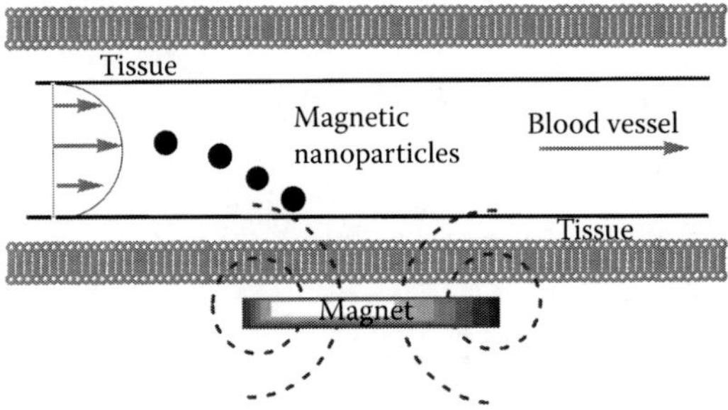

FIGURE 7.6 Magnetic drug delivery mechanism.

distribute in and attack both tumorous and healthy cells, which results in deleterious side effects. In magnetically targeted therapy (see Figure 7.6), a drug is attached to a biocompatible magnetic nanoparticle carrier, usually in the form of a ferrofluid, and is injected into the patient through the circulatory system. In general, these fluids are superparamagnetic; they can be attracted to a magnetic field but do not retain residual magnetism after removing the field. When these particles enter the blood stream, external, high-gradient magnetic fields can be used to concentrate the complex at a specific target site within the body. They also have been used as MRI contrast agents. Once the drug/carrier is correctly concentrated, the drug can be released, either through enzymatic activity or changes in physiological conditions, and be taken up by the tumor cells [48]. Advantage of the methodology exists in the decrease of the required amount of the cytotoxic drugs, thus reducing the associated side effects. This technique has been examined as a means to target cytotoxic drugs to brain tumors. Studies have demonstrated that particles as large as 1–2 μm could be concentrated at the site of intracerebral rat glioma-2 (RG-2) tumors; a later study demonstrated that 10–20 nm magnetic particles were even more effective in targeting these tumors in rats [49,50]. Studies of magnetic targeting in humans demonstrated that the infusion of ferrofluids was well tolerated in most of the patients, and the ferrofluid could be successfully directed to advanced sarcomas without associated organ toxicity. In addition, the preclinical and experimental results indicated that it is possible to overcome the limitations associated with magnetically targeted drug delivery, such as the possibility of embolization of the blood vessels in the target region due to accumulation of the magnetic carriers, and so on [51]. Studies on magnetic targeting are one of the most active areas of drug delivery research [52–54]. Recently, FeRx Inc. was granted fast-track status to proceed with multicenter Phase I and II clinical trials of their magnetic-targeting system for liver tumors. Application of this technique therefore can be considered appropriate vectors for the use of BNCT treatment.

A persistent, technical challenge has been the controlled synthesis of uniform and stable core–shell magnetic nanoparticles. Recently, we have successfully immobilized *ortho*-carborane cages on modified magnetic nanoparticles by the catalytic azide–alkyne cycloaddition between 1-R-2-butyl-*ortho*-$C_2B_{10}H_{10}$ (R = Me, Ph) and propargyl group-enriched magnetic nanoparticles (Figure 7.7) [55].

Magnetic core (1) — NaOH/DMF → Magnetic core (2) — CuBr/DMF → Magnetic core (3, 4)

(R = Me (3), Ph (4); ● = BH; ○ = C)

FIGURE 7.7 Synthesis of encapsulated magnetic nanocomposites.

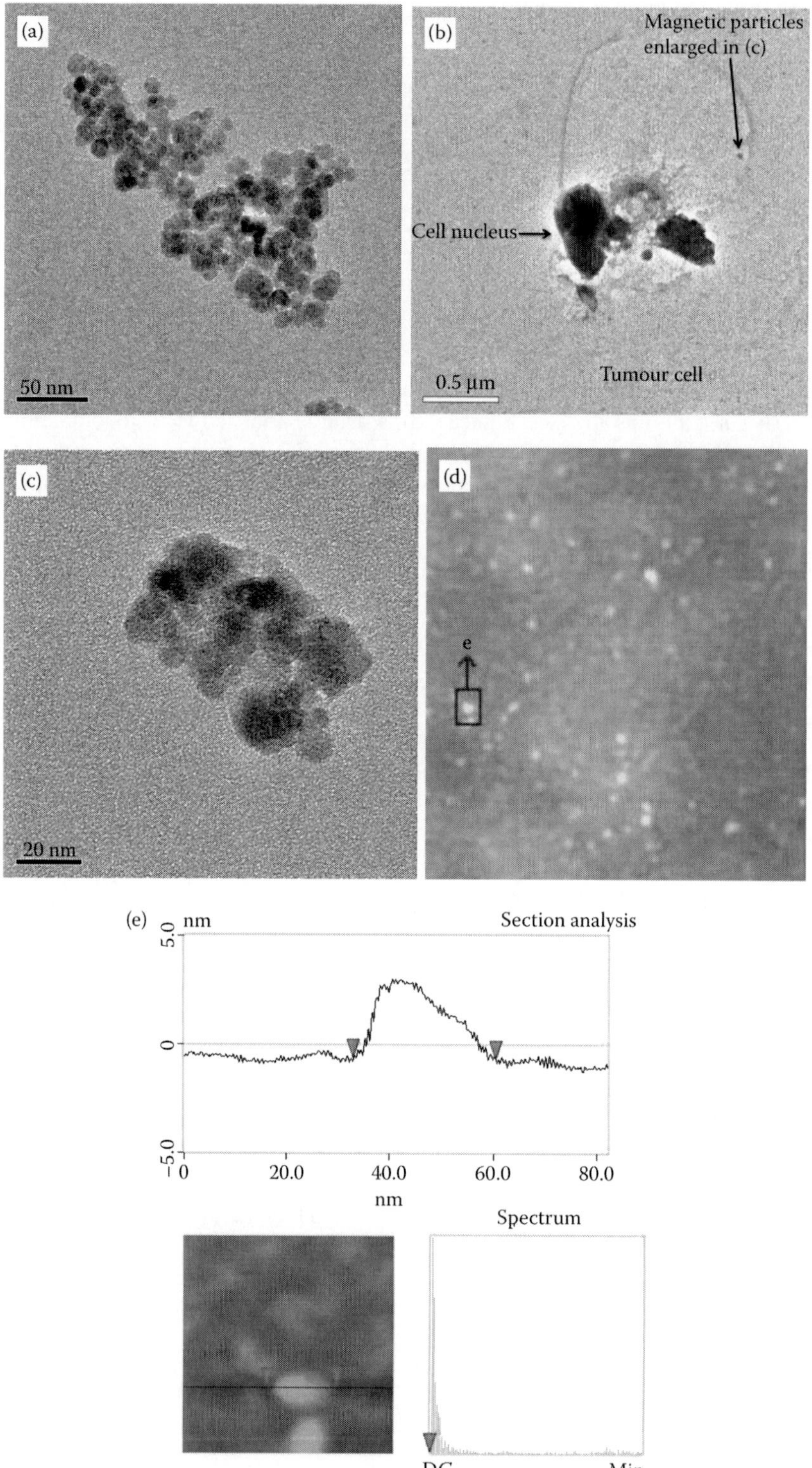

FIGURE 7.8 TEM images of the free (a) and inside tumor cells (b,c) of and AFM (d,e) of compound **3**.

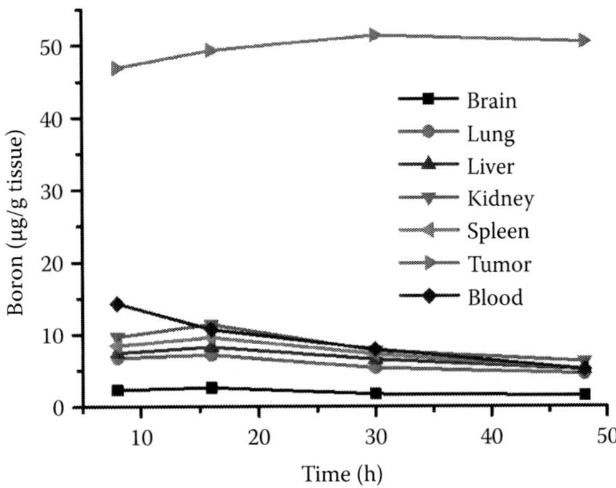

FIGURE 7.9 Boron tissue distribution of encapsulated magnetic nanocomposites **3** with external magnet. (Adapted from Zhu, Y. H., *Boron enriched magnetic nanocomposites: Effective boron drug delivery vehicles in cancer treatment*, the US patent application number: 61/105,492, filed on 15, Oct, **2008**; Zhu, Y. et al., *Encapsulated Magnetic Nanocomposites: High Effective Boron Drug Delivery Vehicle in Cancer Treatment.* P387, 11th Pacific Polymer Conference 2009, 6–10, Dec. **2009**, Cairns, Australia.)

A loading amount of 9.83 mmol boron atom per gram starch-matrixed magnetic nanoparticles has been reached. After modification, aggregation of the modified magnetic nanoparticles has been observed in the TEM image (Figure 7.8a), which may be caused by solvent evaporation of TEM samples. This phenomenon has been generally reported [56–58]. In contrast, most of the functionalized magnetic nanoparticles of compound 3 in Figure 7.7 have been found separated in an atomic force microscope (AFM) image (see Figure 7.8d and e). The resulting nanocomposites have been found to be highly tumor-targeted vehicles under the influence of external magnetic field (1.14 T), giving a high boron concentration of 51.4 μg/g tumor and high ratios of around 10:1 tumor to normal tissues (Figure 7.9). These results have significance in the treatment of not only brain tumors, but also of other kinds of cancers through BNCT. This research is just in its initial stages and issues related to magnetic nanoparticles, such as particle size control, surface functionalizations, and their environmental compatibility need to be addressed.

7.2.5 Nanotube-Based BNCT Agents

7.2.5.1 Carbon Nanotubes

The carbon nanotube (CNT) is an allotrope of carbon that takes the form of a cylindrical arrangement of carbon atoms. CNTs are either single-walled (SWCNT) or multiwalled (MWCNT). Since their discovery in 1991 [59], CNTs have attracted a great deal of attention, with numerous applications proposed for their use [60]. CNTs have many applications in material science due to their strength and unique electrical properties. They have also been used in biomedicine as carriers of drugs and other biomolecules. CNTs possess the unique property of being able to enter various cells without showing apparent toxic effects. For instance, it has been reported that peptide-functionalized SWCNTs were able to cross cell membranes and concentrate in the cytoplasm of 3T6, 3T3 fibroblasts, and phagocytic cells, without causing their death or inflicting other damages [61]. Similar results were obtained in HL60 cells, where it was found that functionalized SWCNTs could help transport large attached groups into cells without themselves exhibiting cell toxicity [62]. *In vitro* studies of folic acid and fluorescent tag-conjugated SWCNTs have shown them to be effective in targeting HeLa cells, which are rich in folic acid receptors and are thought to stem from cervical cancer [63]. The stability and flexibility of CNTs are likely to prolong the circulation time and bioavailability of these macromolecules, thus enabling highly effective gene and drug therapies. The

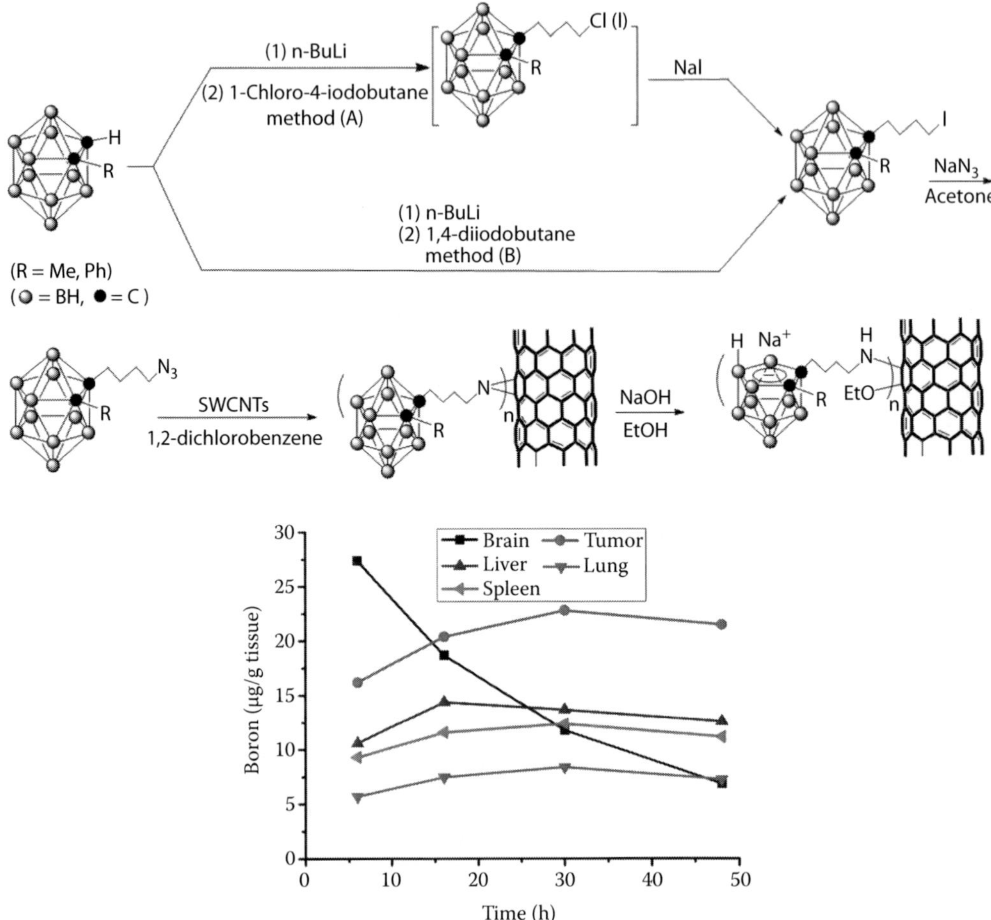

FIGURE 7.10 Synthesis and distributions of boron by SWCNTs supported *nido*-carboranes in saline in different tissues. (Adapted from Zhu, Y. H. et al., *J. Am. Chem. Soc.* **2005**, *127*, 9875–9880.)

reports of enhanced water solubility of CNTs through side-wall derivation with biologically important moieties [64–81] caused us to explore the feasibility of using suitably derived SWCNTs as boron delivery agents for use in BNCT [82]. We have successfully attached *nido*-carborane units to the side-walls of SWCNTs to produce high-boron-containing, water-soluble SWCNTs (Figure 7.10) [82]. These were then used to treat mice bearing EMT6 tumor cells, a mammary carcinoma. A favorable tumor-to-blood ratio of 3.12 and a boron concentration of 21.5 μg/g tumor were obtained 48 h after administration. In addition, it was observed that retention in tumor tissue was higher than in the blood and other tissues including lung, liver, and spleen. Although the initial results were good enough for a possibly successful BNCT trial, current research is involved in modifications to improve both the selectivity and boron concentration.

In addition to side-wall attachments, it is possible to encapsulate boron carriers in the empty cavities of CNTs [83]. In this regard, it has been reported that *ortho*-carboranes were successfully encapsulated inside SWCNTs (see Figure 7.11) [84]. If high-yield methods could be developed for producing this material, it could develop into an effective carrier for BNCT treatment. Overall, the use of CNTs in future generations of BNCT drug delivery systems may improve therapeutic effectiveness, and decrease undesirable side effects. Nevertheless, the long-term toxicity of CNTs must be completely assessed. Further research is also needed to fully reveal the mechanisms of transport across cell membranes. Other issues to be addressed include devising cheap, large-scale

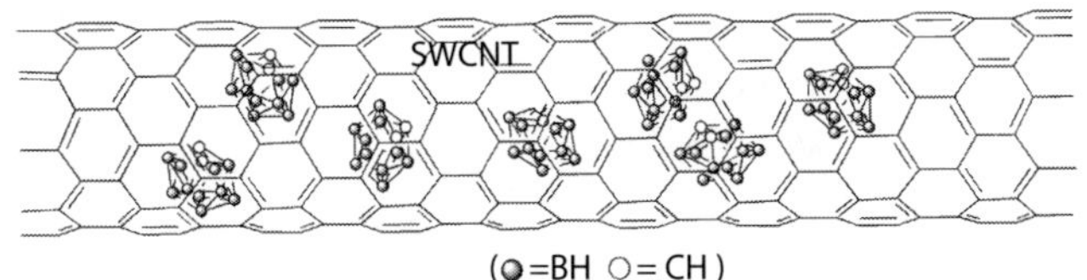

FIGURE 7.11 Schematic representation of *o*-carborane molecules within an SWCNT. (Adapted from Morgan, D. A.; Sloan, J.; Green, M. L. H., *Chem. Commun.* **2002**, *20*, 2442–2443.)

methods and techniques to embed small drug molecules, proteins, peptides, and genes inside and onto CNTs.

7.2.5.2 Boron Nanotubes

Potentially the most effective nanomaterial for use as a BNCT agent would be the boron nanotube (BNT) itself. BNTs are one of the latest materials resulting from advancements in nanotechnology. Interest in their fabrication is largely due to the variable conductivity limitations present in the well-documented CNTs. The BNTs can lead to the creation of a series of boron nanostructures, such as boron nanoribbons [85] and nanowires [86] (see Figure 7.12).

The first successful synthesis of a single-walled BNT was reported quite recently by Ciuparu et al. [87]. They managed to produce nanotubes with diameters of approximately 3 nm and lengths of 16 nm. Unfortunately, these novel BNTs were found to be highly sensitive to high-energy electron beams and hence detailed structural characteristics could not be obtained. Nevertheless, this encouraging breakthrough has increased the interest in BNTs. For instance, a recent study by Kuntsmann and Quandt found new forms of radially constricted, single-walled, zigzag BNTs that may serve as intermediates in the formation of an ideal nanotubular system [88].

Although BNTs have the potential to be ideal BNCT agents, there remain many issues that need to be resolved before these materials can be applied to drug delivery. The major challenges are in the large-scale fabrication of BNTs and in developing strategies to make BNTs water soluble. At present, most of the research work is focused on determining the characteristics of BNTs and developing new applications for them [89]. However, it might not be long before BNTs can be engaged in drug delivery.

7.2.5.3 Boron Nitride Nanotubes

Another potential delivery agent for BNCT is the boron nitride nanotube (BNNT) which is ~44% boron by mass. BNNTs have been investigated as carriers of boron atoms to enhance the ablative efficacy and selective targeting of BNCT for tumor treatments [90–92]. Ciofani et al. have

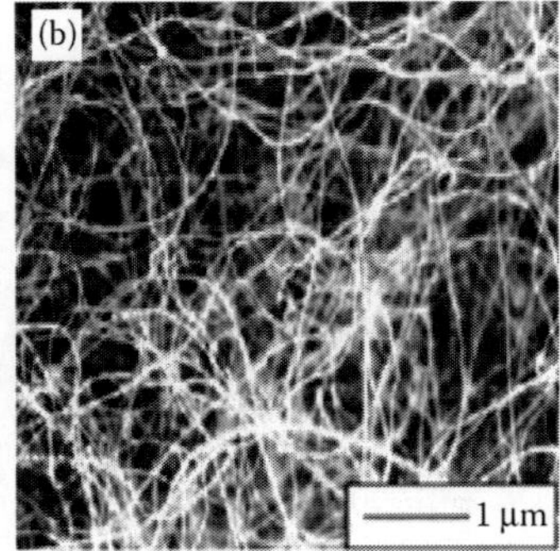

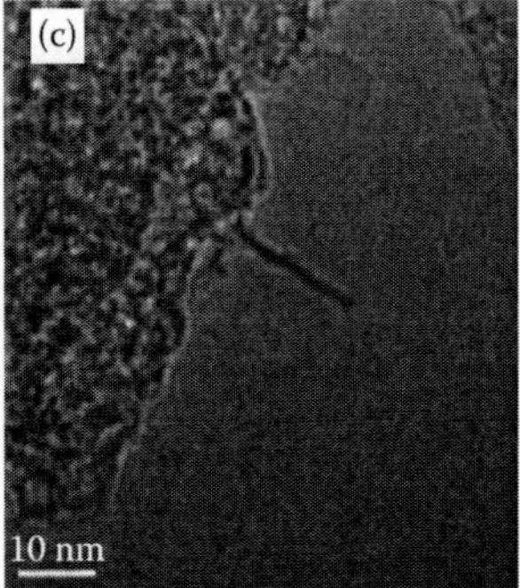

FIGURE 7.12 Micrographs of novel boron nanomaterials: (a) SEM image of boron nanoribbons, (b) SEM image of boron nanowires, and (c) TEM image of BNTs. (Adapted from Quandt, A.; Boustani, I., *Chem. Phy. Chem.* **2005**, *6*, 2001–2008.)

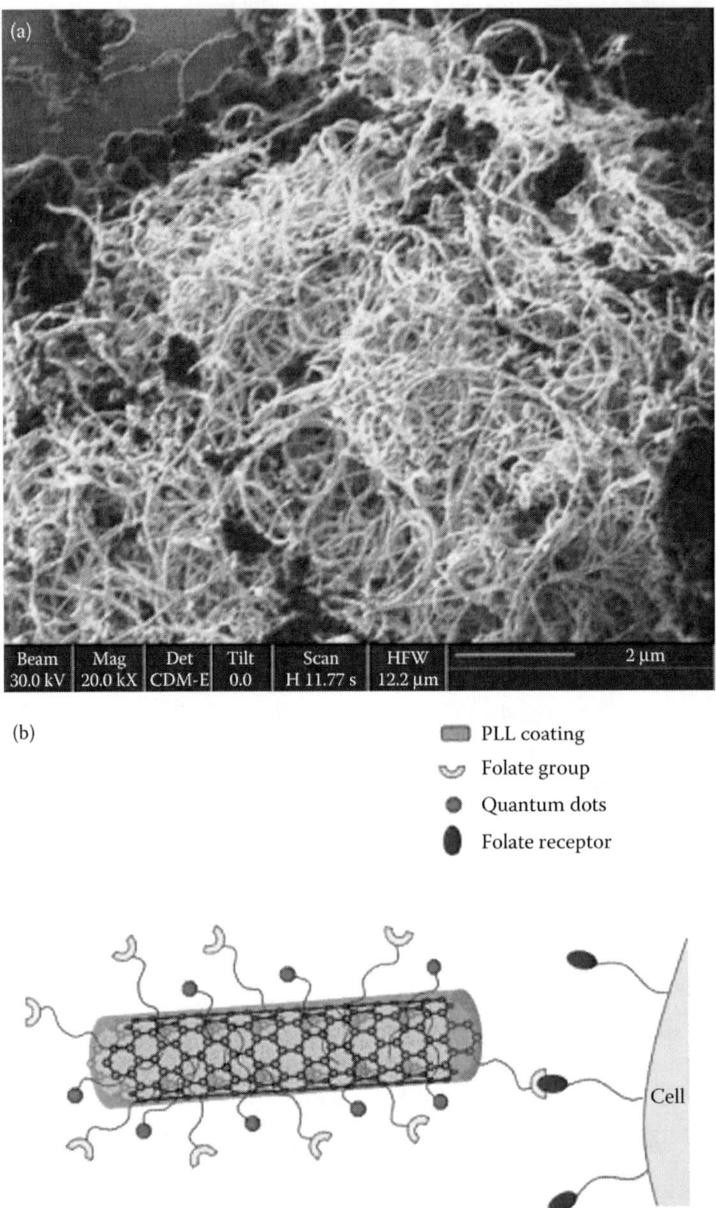

FIGURE 7.13 (a) FIB image of well-dispersed functionalized BNNTs and (b) diagram of the proposed nanovector. (Adapted from Ciofani, G. et al. *Nanoscale Res. Lett.* **2009**, *4*, 113–121.)

functionalized the boron nitride nanotubes with folic acid as selective tumor-targeting ligand and with a fluorescent probe (quantum dots) to enable their tracking [90] (Figure 7.13). Initial *in vitro* studies have confirmed substantive and selective uptake of these nanovectors by glioblastoma multiforme cells, and thus confirms their potential clinical application for BNCT therapy.

BNNTs containing Fe catalysts (see Figure 7.14) have been synthesized and their magnetic properties have been studied [91]. *In vitro* studies have demonstrated the feasibility of influencing the uptake of the BNNTs by human neuroblastoma SH-SY5Y cells through the use of external magnetic fields [91]. Therefore, the magnetic BNNTs have the potential for use as nanovectors for magnetic drug targeting including usage in BNCT.

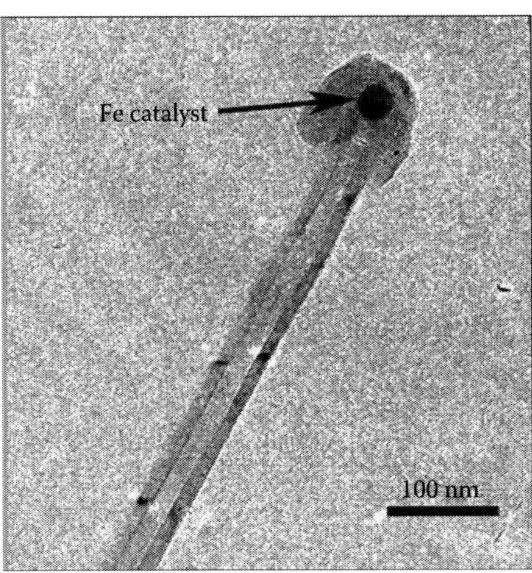

FIGURE 7.14 TEM image of a multiwalled BNNT containing a Fe particle at one tip. (Adapted from Ciofani, G. et al. *Curr. Nanosci.* **2009**, *5*, 33–38.)

An alternate method of magnetizing BNNTs was described by Huang Y. et al., in which an ethanol-thermal process was used for *in situ* formation of dense and uniformly distributed Fe_3O_4 nanoparticles on the surfaces of multiwalled BNNTs [92]. By virtue of the magnetic Fe_3O_4 coatings, the BNNTs could be physically manipulated in a relatively low magnetic field. These BNNT-based magnetic nanocomposites should be applicable to the targeted drug delivery and BNCT [92].

7.3 SUMMARY

BNCT is a potentially promising binary radiotherapy treatment. Highly selective and effective BNCT agents are urgently required to promote BNCT clinical trials from Phase I and Phase II to Phase III. In recent years, much effort has been directed toward designing and synthesizing new boron compounds as well as developing nanomaterial-based BNCT agents; to date, a majority of the studies have proved reasonably promising. Conversely, further *in vivo* studies and clinical trails are needed to establish them as appropriate boron carriers; this is especially so with the relatively novel BNTs and magnetic nanoparticles. In addition, more application of BNCT treatment to various tumors should also be investigated. More advanced forms of BNTs can be anticipated as the interest in their synthesis and future applications increases. To be suitable BNCT agents, the problem of their low water solubility needs to be resolved by chemical modification. In the case of magnetic nanoparticles, strategies are required to counter their tendency of embolization and their unclear cytotoxicity must be resolved. In summary, the recent progress in the studies of nanostructured materials as BNCT agents have been reviewed with the hope of generating interest in nanomaterial-based delivery agents.

REFERENCES

1. Soloway, A. H.; Tjarks, W.; Bauman, B. A.; Rong, F. G.; Barth, R. F.; Codogni, I. M.; Wilson, J. G., The Chemistry of Neutron Capture Therapy. *Chem. Rev.* **1998**, *98*, 1515–1562 and references therein.
2. Hawthorne, M. F., The role of chemistry in the development of boron neutron capture therapy of cancer. *Angew. Chem. Int. Ed. Engl.* **1993**, *32*, 950–984.
3. Fairchild, R. G.; Bond, V. P., Current status of 10-B neutron capture therapy: enhancement of tumour dose via beam filtration and dose rate, and the effects of these paramenters on minimum boron content: a theoretical evaluation. *Int. J. Radiat. Oncol. Biol. Phys.* **1985**, *11*, 831–840.
4. Laws, E. R.; Shafferey, M. E., The inherent invasiveness of cerebral gliomas: Implications of clinical management. *Int. J Dev. Neurosci.* **1999**, *17*, 413–420.

5. Barth, R. F.; Coderre, J. A.; Vicente, M. G. H.; Blue, T. E., Boron neutron capture therapy of cancer: Current status and future prospects. *Clin. Cancer. Res.* **2005**, *11*, 3987–4002.
6. Perry, A.; Schmidt, R. E., Cancer therapy-associated CNS neuropathology: An update and review of the literature. *Acta Neuropathol.* **2006**, *111*, 197–212.
7. Kouri, M.; Kankaanranta, L.; Seppälä, T.; Tervo, L.; Rasilainen, M.; Minn, H.; Eskola, O.; Vähätalo, J.; Paetau, A.; Savolainen, S.; Auterinen, I.; Jääskelainen, J.; Joensuu, H., Undifferentiated sinonasal carcinoma may respond to single-fraction boron neutron capture therapy. *Radiother. Oncol.* **2004**, *72*, 83–85.
8. Wittig, A.; Malago, M.; Collette, L., Uptake of two ^{10}B-compounds in liver metastases of colorectal adenocarcinoma for extracorporeal irradiation with BNCT (EORTC trial 11001). *Int. J. Cancer* **2008**, *122*, 1164–1171.
9. Suzuki, M.; Sakurai, Y.; Hagiwara, S.; Masunaga, S.; Kinashi, Y.; Nagata, K.; Maruhashi, A.; Kudo, M.; Ono, K., First attempt of boron neutron capture therapy (BNCT) for hepatocellular carcinoma. *Jpn. J. Clin. Oncol.* **2007**, *37*, 376–381.
10. Wu, G.; Barth, R. F.; Yang, W.; Lee, R. J.; Tjarks, W.; Backer, M. V.; Backer, J. M., Boron containing macromolecules and nanovehicles as delivery agents for neutron capture therapy. *Anti-Cancer Agents Med. Chem.* **2006**, *6*, 167–184.
11. Esfand, R.; Tomalia, D. A., Polyamidoamine (PAMAM) dendrimers: from biomimicry to drug delivery and biomedical applications. *Drug. Discov. Today* **2001**, *6*, 427–436.
12. Rij, C.; Wilhelm, A.; Sauerwein, W.; Loenen, A., Boron neutron capture therapy for glioblastoma multiforme. *Pharm. World Sci.* **2005**, *27*, 92–95.
13. Zhu, Y.; Koh, C. Y.; Maguire, J. A.; Hosmane, N. S., Recent developments in the Boron Neutron Capture therapy (BNCT) driven by nanotechnology. *Cur. Chem. Biol.* **2007**, *1*, 141–149 and references therein.
14. Lesnikowski, Z. J.; Paradowska, E.; Olejniczak, A. B.; Studzinska, M.; Seekamp, P.; Schuβler, U.; Gabel, D.; Schinazi, R. F.; Plesek, J., Towards new boron carriers for boron neutron capture therapy: metallacarboranes and their nucleoside conjugates. *Bioorg. Med. Chem.* **2006**, *13*, 4168–4175.
15. Renner, M. W.; Miura, M.; Easson, M. W.; Vicente, M. G. H., Recent progress in the syntheses and biological evaluation of boronated porphyrins for Boron Neutron-Capture Therapy. *Anti-Cancer Agents Med. Chem.* **2006**, *6*, 145–157.
16. Azab, A. K.; Abu, A. H.; Srebnik, M., Chapter 5: Boron neutron capture therapy. In *Contemporary Aspects of Boron: Chemistry and Biological Applications*, Eds. Ali, H. A., Dembitsky, V., Srebnik, M. Elsevier Science, Amsterdam, The Netherlands, **2005**, pp. 338–366.
17. Barth, R. F.; Coderre, J. A.; Vicente, M. G. H.; Blue, T. E.; Miyatake, S-I., Chapter 26, Boron neutron capture therapy of brain tumours: Current status and future prospects. In *High-Grade Gliomas: Diagnosis and Treatment*, Ed. Barnett, G., Humana Press, Totowa, NJ, **2007**, pp. 431–461.
18. Shapiro, W. R.; Voorhies, R. M.; Hiesiger, E. M.; Sher, P. B.; Basler, G. A.; Lipschutz, L. E., Pharmacokinetics of Tumor Cell Exposure to [^{14}C] Methotrexate after Intracarotid Administration without and with Hyperosmotic Opening of the Blood-Brain and Blood-Tumor Barriers in Rat Brain Tumors: A Quantitative Autoradiographic Study. *Cancer Res.*, **1988**, *48*, 694–701.
19. Jain, R. K., Normalizing Tumor Vasculature with Anti-Angiogenic Therapy: A New Paradigm for Combination Therapy. *Nat. Med.* **2001**, *7*, 987–989.
20. Santini, J.; Cima, M.; Langer, R., Langer, R., A Controlled-release Microchip. *Nature*, **1999**, *397*, 335–338.
21. Sartor, O.; Dineen, M. K.; Perez-Marreno, R.; Chu, F. M.; Carron, G. J.; Tyler, R. C., An eight-month clinical study of LA-2575 30.0 mg: a new 4-month, subcutaneous delivery system for leuprolide acetate in the treatment of prostate cancer. *Urology*, **2003**, *62*, 319–323.
22. Hobbs, K.; Monsky, W.; Yuan F. et al., Regulation of transport pathways in tumor vessels: role of tumor type and microenvironment. *Proc. Natl. Acad. Sci. USA*, **1998**, *95*, 4607–4612.
23. Yih, T. C.; Al-Fandi, M., Engineered Nanoparticles as Precise Drug Delivery Systems. *J. Cell Biochem.* **2006**, *97*, 1184–1190.
24. Kreuter, J.; Alyyautidin, R. N.; Kharkevich, D. A.; Ivanov, A. A., Passage of peptides through the blood-brain barrier with colloidal polymer particles (nanoparticles). *Brain. Res.* **1995**, *674*, 171–174.
25. Lu, D. R.; Mehta, S. C.; Chen, W., Selective boron drug delivery to brain tumors for boron neutron capture therapy. *Adv. Drug Deliv. Rev.* **1997**, *26*, 231–247.
26. Gabizon, A.; Price, D. C.; Huberty, J.; Bresalier, R. S.; Papahadjopoulos, D., Effect of liposome composition and other factors on the targeting of liposomes to experimental tumors: biodistribution and imaging studies. *Cancer Res.* **1990**, *50*, 6371–6378.
27. Torchilin, V. P., Recent advances with liposomes as pharmaceutical carriers. *Nat. Rev. Drug Discov.* **2005**, *4*, 145–160.

28. Wu, G.; Barth, R. F.; Yang, W. et al., Boron containing macromolecules and nanovehicles as delivery agents for neutron capture therapy. *Anti-Cancer Agents Med. Chem.* **2006**, *6*, 167–184.
29. Hawthorne, M. F.; Feakes, D. A.; Shelly, K., Chapter 2, Recent results with liposomes as boron delivery vehicles from boron neutron capture therapy. In *Cancer Neutron Capture Therapy*, Ed. Mishima, Y., Plenum Press: New York, **1996**, p. 27–36.
30. Hawthorne, M. F.; Shelly, K., Liposomes as drug delivery vehicles for boron agents. *J. Neuro-Oncol.* **1997**, *33*, 53–58.
31. Li, T.; Hamdi, J.; Hawthorne, M. F., Unilamellar liposomes with enhanced boron content. *Bioconjug. Chem.* **2006**, *17*, 15–20.
32. Nakamura, H.; Miyajima, Y.; Takei, T.; Kasaoka, S.; Maruyama, K., Synthesis and vesicle formation of a *nido*-carborane cluster lipid for boron neutron capture therapy. *Chem. Commun.* **2004**, 1910–1911.
33. Ristori, S.; Oberdisse, J.; Grillo, I.; Donati, A.; Spalla, O., Structural characterization of cationic liposomes loaded with sugar-based carboranes. *Biophys. J.* **2005**, *88*, 535–547.
34. Yanagie, H.; Matuyama, K.; Takizawa, T. et al., Application of boron-entrapped stealth liposomes to inhibition of growth of tumour cells in the *in vivo* boron neutron-capture therapy model. *Biomed. & Pharmacother.* **2006**, *60*, 43–50.
35. Nobuto, H.; Sugita, T.; Kubo, T. et al., Evaluation of systemic chemotherapy with magnetic liposomal doxorubicin and a dipole external electromagnet. *Int. J. Cancer* **2004**, *109*, 627–635.
36. Ringsdorf, H., Structure and properties of pharmacologically active polymers. *J. Polym. Sci. Polym. Symp.* **1975**, *51*, 135–153.
37. Kopecek, J., Soluble biomedical polymers. *Polym. Med. (Wroclaw)*, **1977**, *7*, 191–221.
38. Gillies, E. R.; Frechet, J. M. J., Designing Macromolecules for Therapeutic Applications: Polyester Dendrimer-Poly(ethylene oxide) "Bow-Tie" Hybrids with Tunable Molecular Weight and Architecture. *J. Am. Chem. Soc.* **2002**, *124*, 14137–14146.
39. Seymour, L. W., Passive tumor targeting of soluble macromolecules and drug conjugates. *Crit. Rev. Ther. Drug Carrier Syst.* **1992**, *9*, 135–187.
40. Newkome, G. R.; Moorefield, C. N.; Vogtle, F., *Dendritic Macromolecules: Concepts, Synthesis, Perspectives.* Germany, VCH: Weinheim, **1996**.
41. Niederhafner, P.; Sebestik, J.; Jezek, J., Peptide dendrimers. *J. Peptide. Sci.* **2005**, *11*, 757–788.
42. Gillies, E. R.; Frechet, J., Dendrimers and dendritic polymers in drug delivery. *Drug Discov. Today*, **2005**, *10*, 35–43.
43. Esfand, R.; Tomalia, D. A., Poly(amidoamine) (PAMAM) dendrimers: from biomimicry to drug delivery and biomedical applications. *Drug Discov. Today* **2001**, *6*, 427–436.
44. Latallo, J.; Candido, K.; Cao, Z. et al., Nanoparticle targeting of anticancer drug improves therapeutic response in animal model of human epithelial cancer. *Cancer Res.* **2005**, *65*, 5317–5324.
45. Barth, R. F.; Yang, W.; Adams, D. M. et al., Molecular Targeting of the Epidermal Growth Factor Receptor for Neutron Capture Therapy of Gliomas. *Cancer Res.* **2002**, *62*, 3159–3166.
46. Shukla, S.; Wu, G.; Chatterjee, M. et al., Synthesis and Biological Evaluation of Folate Receptor Targeted Boronated PAMAM Dendrimers as Potential Agents for Neutron Capture Therapy. *Bioconjug. Chem.* **2003**, *14*, 158–167.
47. Parrott, M. C.; Marchington, E. B.; Valliant, J. F.; Adronov, A., Synthesis and Properties of Carborane-Functionalized Aliphatic Polyester Dendrimers. *J. Am. Chem. Soc.* **2005**, *127*, 12081–12087.
48. Alexiou, C.; Arnold, W.; Klein, R. J. et al., Locoregional Cancer Treatment with Magnetic Drug Targeting. *Cancer Res.* **2000**, *60*, 6641–6648.
49. Pulfer, S. K.; Ciccotto, S. L.; Gallo, J. M., Distribution of small magnetic particles in brain tumor-bearing rats. *J. Neuro-Oncol.* **1999**, *41*, 99–105.
50. Sincai, M.; Ganga, D.; Ganga, M.; Argherie, D.; Bica, D., Antitumor effect of magnetite nanoparticles in cat mammary adenocarcinoma. *J. Magn. Magn. Mater.* **2005**, *293*, 438–441.
51. Gallo, J. M.; Hafeli, U., Preclinical Experiences with Magnetic Drug Targeting: Tolerance and Efficacy. Cancer Res., 56: 4694–4701, 1996; and Clinical Experiences with Magnetic Drug Targeting: A Phase I Study with 4′-Epidoxorubicin in 14 Patients with Advanced Solid Tumors. *Cancer Res.* **1997**, *57*, 3063–3064.
52. Veyret, R.; Delair, T.; Elaissari, A., Polyelectrolyte functionalized magnetic emulsion for specific isolation of nucleic acids. *J. Magn. Magn. Mater.* **2005**, *293*, 171–176.
53. Sang, J. S.; Reichel, J.; Bo, H.; Schuchman, M.; Sang, B. L., Magnetic Nanotubes for Magnetic-Field-Assisted Bioseparation, Biointeraction, and Drug Delivery. *J. Am. Chem. Soc.* **2005**, *127*, 7316–7317.
54. Zebli, B.; Susha, A. S.; Sukhorukov, G. B.; Rogach, A. L.; Parak, W. J., Magnetic Targeting and Cellular Uptake of Polymer Microcapsules Simultaneously Functionalized with Magnetic and Luminescent Nanocrystals. *Langmuir*, **2005**, *21*, 4262–4265.

55. (a) Zhu, Y. H., *Boron enriched magnetic nanocomposites: Effective boron drug delivery vehicles in cancer treatment*, the US patent application number: 61/105,492, filed on 15, Oct, **2008**.
(b) Zhu, Y.; Lin, Y.; Shen, S. et al., *Encapsulated Magnetic Nanocomposites: High Effective Boron Drug Delivery Vehicle in Cancer Treatment.* P387, 11th Pacific Polymer Conference 2009, 6–10, Dec, **2009**, Cairns, Australia.
56. Zhou, J.; Leuschner, C.; Kumar, C.; Hormes, C. K.; Soboyejo, W. O., Sub-Cellular Accumulation of Magnetic Nanoparticles in Breast Tumor and Metasteses. *Biomaterials*, **2006**, *27*, 2001–2008.
57. Beaune, G.; Dubertret, B.; Clément, O.; Vayssettes, C.; Cabuil, V.; Ménager, C., Giant Vesicles Containing Magnetic Nanoparticles and Quantum Dots: Feasibility and Tracking by Fiber Confocal Fluorescence Microscopy. *Angew Chem. Int. Ed.* **2007**, *46*, 5421–5424.
58. Bergemann, C.; Muller-Schulte, D.; Oster, J.; Brassard, L.; Lübbe, A. S., Magnetic ion-exchange nano- and macroparticles for medical, biochemical and molecular biological applications. *J. Magn. Magn. Mater.* **1999**, *194*, 45–52.
59. Iijma, S., Helical microtubules of graphitic carbon. *Nature*, **1991**, *354*, 56–58.
60. Cuenca, A.; Jiang, H.; Hochwald, S.; Delano, M.; Cance, W.; Grobmyer, S., Emerging implications of nanotechnology on cancer diagnostics and therapeutics. *Cancer*, **2006**, *107*, 459–466.
61. Pantarotto, D.; Briand, J. P.; Prato, M.; Bianco, A., Translocation of Bioactive Peptides Across Cell Membranes by Carbon Nanotubes. *Chem. Commun.* **2004**, 16–17.
62. Cherukuri, P.; Bachilo, S. M.; Litovsky, S. H.; Weisman, R. B., Near-Infrared Fluorescence Microscopy of Single-Walled Carbon Nanotubes in Phagocytic Cells. *J. Am. Chem. Soc.*, **2004**, *126*, 15638–15639.
63. Kam, N. W.; O'Connell, M.; Wisdom, J. A.; Dai, H., Carbon nanotubes as multifunctional biological transporters and near-infrared agents for selective cancer cell destruction. *Proc. Natl. Acad. Sci. USA*, **2005**, *102*, 11600–11605.
64. Pantarotto, D.; Partidos, C. D.; Graff, R. et al., Synthesis, Structural Characterization, and Immunological Properties of Carbon Nanotubes Functionalized with Peptides. *J. Am. Chem. Soc.* **2003**, *125*, 6160–6164.
65. Kam, N. W. S.; Jessop, T. C.; Wender, P. A.; Dai, H., Nanotube Molecular Transporters: Internalization of Carbon Nanotube-Protein Conjugates into Mammalian Cells. *J. Am. Chem. Soc.* **2004**, *126*, 6850–6851.
66. Furtado, C. A.; Kim, U. J.; Gutierrez, H. R.; Pan, L.; Dickey, E. C.; Eklund, P. C., Debundling and Dissolution of Single-Walled Carbon Nanotubes in Amide Solvents. *J. Am. Chem. Soc.* **2004**, *126*, 6095–6105.
67. Hu, H.; Ni, Y.; Montana, V.; Haddon, R. C.; Parpura, V., Chemically Functionalized Carbon Nanotubes as Substrates for Neuronal Growth. *Nano. Lett*. **2004**, *4*, 507–511.
68. Georgakilas, V.; Tagmatarchis, N.; Pantarotto, D.; Bianco, A.; Briand, J. P., Amino acid functionalisation of water soluble carbon nanotubes. *Chem. Commun.* **2002**, 3050–3051.
69. Fu, K.; Li, H.; Zhou, B.; Kitaygorodskiy, A.; Allard, L. F.; Sun, Y. P., Deuterium Attachment to Carbon Nanotubes in Deuterated Water. *J. Am. Chem. Soc.* **2004**, *126*, 4669–4675.
70. Georgakilas, V.; Kordatos, K.; Prato, M.; Guldi, D. M.; Holzinger, M.; Hirsch, A., Organic Functionalization of Carbon Nanotubes. *J. Am. Chem. Soc.* **2002**, *124*, 760–761.
71. Guldi, D. M.; Marcaccio, M.; Paolucci, D. et al., Single-Wall Carbon Nanotube–Ferrocene Nanohybrids: Observing Intramolecular Electron Transfer in Functionalized SWNTs. *Angew. Chem. Int. Ed.* **2003**, *42*, 4206–4209.
72. Holzinger, M.; Abraham, J.; Whelan, P. et al., Functionalization of single-walled carbon nanotubes with (R-)oxycarbonyl nitrenes. *J. Am. Chem. Soc.* **2003**, *125*, 8566–8580.
73. Sun, Y. P.; Huang, W.; Lin, Y. et al., Soluble dendron-functionalized carbon nanotubes: Preparation, characterization, and properties. *Chem. Mater.* **2001**, *13*, 2864–2869.
74. Peng, H.; Alemany, L. B.; Margrave, J. L.; Khabashesku, V. N., Sidewall carboxylic acid functionalization of single-walled carbon nanotubes. *J. Am. Chem. Soc.* **2003**, *125*, 15174–15182.
75. Peng, H.; Reverdy, P.; Khabashesku, V. N.; Margrave, J. L., Sidewall Functionalization of Single-Walled Carbon Nanotubes with Organic Peroxides. *Chem. Commun.* **2003**, 362–363.
76. Huang, W.; Fernando, S.; Lin, Y.; Zhou, B.; Allard, L. F.; Sun, Y. P., Preferential solubilization of smaller single-walled carbon nanotubes in sequential functionalization reactions. *Langmuir*, **2003**, *19*, 7084–7088.
77. Dyke, C. A.; Tour, J. M., Unbundled and Highly Functionalized Carbon Nanotubes from Aqueous Reactions. *Nano. Lett.* **2003**, *3*, 1215–1218.
78. Zhu, W.; Minami, N.; Kazaoui, S.; Kim, Y., Fluorescent chromophore functionalized single-wall carbon nanotubes with minimal alteration to their characteristic one-dimensional electronic states. *J. Mater. Chem.* **2003**, *13*, 2196–2201.

79. Umek, P.; Seo, J. W.; Hernadi, K. et al., Addition of Carbon Radicals Generated from Organic Peroxides to Single Wall Carbon Nanotubes. *Chem. Mater.* **2003**, *15*, 4751–4755.
80. Hu, H.; Zhao, B.; Hamon, M. A.; Kamaras, K.; Itkis, M. E.; Haddon, R. C., Sidewall Functionalization of Single-Walled Carbon Nanotubes by Addition of Dichlorocarbene. *J. Am. Chem. Soc.* **2003**, *125*, 14893–14900.
81. Chen, R. J.; Choi, H. C.; Bangsaruntip, S. et al., An investigation of the mechanisms of electronic sensing of protein adsorption on carbon nanotube devices. *J. Am. Chem. Soc.* **2004**, *126*, 1563–1568.
82. Zhu, Y. H.; Ang, T. P.; Carpenter, K.; Maguire, J. A.; Hosmane, N. S.; Takagaki, M., Substituted Carborane-Appended Water-Soluble Single-Wall Carbon Nanotubes: New Approach to Boron Neutron Capture Therapy Drug Delivery. *J. Am. Chem. Soc.* **2005**, *127*, 9875–9880.
83. Dai, H., Carbon Nanotubes: Synthesis, Integration, and Properties. *Acc. Chem. Res.* **2002**, *35*, 1035–1044.
84. Morgan, D. A.; Sloan, J.; Green, M. L. H., Direct imaging of o-carborane molecules within single walled carbon nanotubes. *Chem. Commun.* **2002**, *20*, 2442–2443.
85. Xu, T.; Zheng, J. G.; Wu, N.; Nicholls, A. et al., Crystalline Boron Nanoribbons: Synthesis and Characterization. *Nano. Lett.* **2004**, *4*, 963–968.
86. Cao, L. M.; Tian, H.; Zhang, Z.; Zhang, X. Y.; Gao, C. X.; Wang, W. K., Nucleation and growth of feather-like boron nanowire nanojunctions. *Nanotechnology*, **2004**, *15*, 139–142.
87. Ciuparu, D.; Klie, R.; Zhu, Y.; Pfefferle, L., Synthesis of Pure Boron Single-Wall Nanotubes. *J. Phys. Chem.* **2004**, *108*, 3967–3969.
88. Kuntsmann, J.; Quandt, A., Constricted boron nanotubes. *Chem. Phys. Lett.* **2005**, *402*, 21–26.
89. Quandt, A.; Boustani, I., Boron Nanotubes. *Chem. Phy. Chem.* **2005**, *6*, 2001–2008.
90. Ciofani, G.; Raffa, V.; Meniassi, A.; Cuschieri, A., Folate Functionalized Boron Nitride Nanotubes and their Selective Uptake by Glioblastoma Multiforme Cells: Implications for their Use as Boron Carriers in Clinical Boron Neutron Capture Therapy. *Nanoscale Res. Lett.* **2009**, *4*, 113–121.
91. Ciofani, G.; Raffa, V.; Yu, J.; Chen, Y.; Obata, Y.; Takeoka, S.; Meniassi, A.; Cuschieri, A., Boron Nitride nanotubes: A novel vector for targeted magnetic drug delivery, *Curr. Nanosci.* **2009**, *5*, 33–38.
92. Huang, Y.; Lin, J.; Bando, Y.; Tang, C.; Zhi, C.; Shi, Y.; Takayama-Muromachi, E.; Golberg, D., BN nanotubes coated with uniformly-distributed Fe_3O_4 nanoparticles: novel magneto-operable nanocomposites. *J. Mater. Chem.* **2010**, *20*, 1007–1011.

8 Liposomal Boron Delivery System for Neutron Capture Therapy of Cancer

Hiroyuki Nakamura

CONTENTS

8.1 INTRODUCTION

The high accumulation and selective delivery of ^{10}B into the tumor tissue are the most important requirements to achieve efficient boron neutron capture therapy for cancer (BNCT), because the cell-killing effect of BNCT depends on the nuclear reaction of two essentially nontoxic species, boron-10 (^{10}B) and thermal neutrons, whose destructive effect is well documented in boron-loaded tissues (Barth, 2003, 2009; Hawthorne, 1993; Soloway et al., 1998). Two boron compounds, sodium mercaptoundecahydrododecaborate ($Na_2^{10}B_{12}H_{11}SH$; ^{10}BSH) (Soloway et al., 1967) and L-*p*-boronophenylalanine (L-^{10}BPA) (Snyder et al., 1958), have been clinically utilized for the treatment of patients with malignant brain tumors (Nakagawa and Hatanaka, 1997) and malignant melanoma (Mishima et al., 1989). According to the theoretical estimations as well as clinical data, three important parameters should be considered in the development of boron carriers for fatally damaging tumor cells with BNCT: (1) boron concentrations in the tumor should be in the range of 20–35 μg ^{10}B/g; (2) the tumor/normal tissue ratio should be greater than 3; and (3) the toxicity should be sufficiently low (Barth, R. F. et al., 2005). Recently, BNCT has been applied to various cancers, including head and neck cancer (Aihara et al., 2006; Kato et al., 2004), lung cancer, hepatoma (Suzuki et al., 2007), chest wall cancer, and mesothelioma (Ono et al., unpublished). Therefore, the development of new boron carriers is one of the most important issues that should be resolved to extend the application of BNCT to various cancers.

In the last decade, boron carrier development has taken two directions: small boron molecules and boron-conjugated biological vehicles. Unlike approaches using pharmaceuticals, boron carriers require high tumor selectivity and should be essentially nontoxic. Therefore, the approach to use biological vehicles has become one of the recent trends to accumulate a large amount of ^{10}B in tumor tissues.

Liposomes, whose size typically ranges in mean diameters from 50 to 200 nm, display some unique pharmacokinetic characteristics. Liposomes exhibit preferential extravasation and accumulation at the site of solid tumors due to increased endothelial permeability and reduced lymphatic drainage in these tissues, which has been defined as enhanced permeability and retention effect (Matsumura and Maeda, 1986; Maeda et al., 2000). Therefore, liposomes are efficient drug delivery vehicles, because encapsulated drugs can be delivered to tumors selectively. Liposomal boron delivery system (BDS), in this context, is also considered to be promising for BNCT due to the possibility of carrying a large amount of ^{10}B compound. Two approaches have been investigated for the use of liposomes as boron delivery vehicles: (1) encapsulation of boron compounds into liposomes, and (2) incorporation of boron-conjugated lipids into the liposomal bilayer. In this chapter, new technologies for liposomal BDS using both boron-encapsulation and boron–lipid liposome approaches are described.

8.2 BORON COMPOUND–ENCAPSULATED LIPOSOME APPROACH

8.2.1 Carcinoembryonic Antigen–Targeted Liposomes

Yanagie and coworkers first investigated a BSH-encapsulated liposome that was conjugated with a monoclonal antibody specific for carcinoembryonic antigen (CEA) (Yanagie et al., 1989, 1991). They prepared a new murine monoclonal antibody (2C-8) by injecting mice intraperitoneally (IP) with CEA producing human pancreatic cancer cell line, AsPC-1. This anti-CEA monoclonal antibody was conjugated with large multilamellar liposomes incorporated $Cs_2{}^{10}BSH$. The liposome was prepared from egg yolk phosphatidylcholine, cholesterol, and dipalmitoylphosphatidylethanolamine (1/1/0.05), and $Cs_2{}^{10}BSH$ was encapsulated. The liposomes were treated with dithiothreitol and suspended in the *N*-hydroxysuccinimidyl-3-(2-pyridyldithio)propionate-treated antibody solution for conjugation. This immunoliposome was shown to bind selectively to human pancreatic carcinoma cells (AsPC-1) bearing CEA on their surface. The therapeutic effects of locally injected BSH-encapsulated immunoliposome on AsPC-1 xenografts in nude mice were evaluated. After intratumoral injection of the immunoliposomes, boron concentrations in tumor tissue and blood were 49.6 ± 6.6 and 0.30 ± 0.08 ppm, respectively. Thermal neutron irradiation (2×10^{12} n/cm^2) suppressed tumor growth in mice with intratumoral injection of BSH-encapsulated immunoliposomes and hyalinization and necrosis were found in the immunoliposome-treated tumors (Yanagie et al., 1997).

8.2.2 Various Boron Compound–Encapsulated PEG Liposomes

Hawthorne and coworkers reported the preparation of boron-encapsulated liposomes from distearoylphosphatidylcholine (DSPC) and cholesterol. They encapsulated the hydrolytically stable borane anions $B_{10}H_{10}^{2-}$, $B_{12}H_{11}SH^{2-}$, $B_{20}H_{17}OH^{4-}$, $B_{20}H_{19}^{3-}$, and the normal form and photoisomer of $B_{20}H_{18}{}^{2-}$ in liposomes as their water-soluble sodium salts. Selective boron accumulation in tumor was observed in the use of these liposomes, although the boron compounds used do not normally exhibit affinity for tumors and are normally rapidly cleared from the body. The highest tumor concentrations achieved the therapeutic range (>15 μg of boron per gram of tumor), but more favorable results were obtained with the two isomers of $B_{20}H_{18}^{2-}$. These boron compounds are capable of reacting with intracellular components after they have been deposited within tumor cells by the liposome, thereby preventing the borane ion from being released into blood (Shelly et al., 1992). The PEG-conjugated liposome was prepared with 5% PEG-200-distearoyl phosphatidylethanolamine and an apical-equatorial (*ae*) isomer of the $B_{20}H_{17}NH_3^{3-}$ ion, $[1\text{-}(2'\text{-}B_{10}H_9)\text{-}2\text{-}NH_3B_{10}H_8]^{3-}$ was encapsulated into the liposome. This liposome exhibited a long circulation lifetime due to escape from the

4– 2– 3–

2 OH⁻ / –H$_2$O   NH$_2^-$

OH   NH$_3$

o BH
● B

$[B_{20}H_{17}OH]^{4-}$ $[n\text{-}B_{20}H_{18}]^{2-}$ $[1\text{-}(2'\text{-}B_{10}H_9)\text{-}2\text{-}NH_3B_{10}H_8]^{3-}$

FIGURE 8.1 Hydrolysis and amination of n-$B_{20}H_{18}^{2-}$.

reticuloendothelial system (RES), resulting in the continued accumulation of boron in the tumor over the entire 48 h experiment and reaching a maximum of 47 μg of boron per gram of tumor (Feakes et al., 1994). Preparation of $B_{20}H_{17}OH^{4-}$ and $[1\text{-}(2'\text{-}B_{10}H_9)\text{-}2\text{-}NH_3B_{10}H_8]^{3-}$ from n-$B_{20}H_{18}^{2-}$ is shown in Figure 8.1.

8.2.3 Folate Receptor–Targeted Liposomes

Expression of folate receptor (FR) is frequently amplified among human tumors. Lee and coworkers developed boron-containing folate receptor-targeted liposomes (Pan, X. Q. et al., 2002). Two negatively charged boron compounds, $Na_2[B_{12}H_{11}SH]$ and Na_3 $(B_{20}H_{17}NH_3)$, as well as five weakly basic boronated polyamines, SPD-5, SPM-5, ASPD-5, ASPM-5, and SPM-5,10, as shown in Figure 8.2, were incorporated into liposomes by a pH-gradient-driven remote-loading method with varying loading efficiencies. Greater loading efficiencies were obtained with lower-molecular-weight boron derivatives, using ammonium sulfate as the trapping agent, compared to those obtained with sodium citrate. The *in vitro* boron uptake of folate-conjugated liposomes was investigated using human KB squamous epithelial cancer cells. Higher cellular boron uptake (up to 1584 μg/10^9 cells) was observed in the case of FR-targeted liposomes than in the case of nontargeted control liposomes (up to 154 μg/10^9 cells).

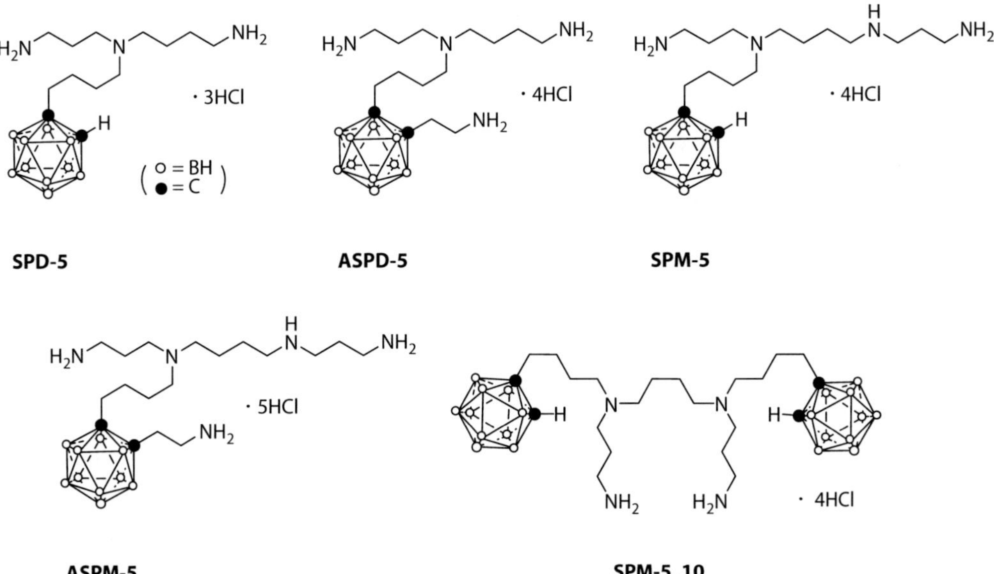

FIGURE 8.2 Structures of boronated polyamine derivatives.

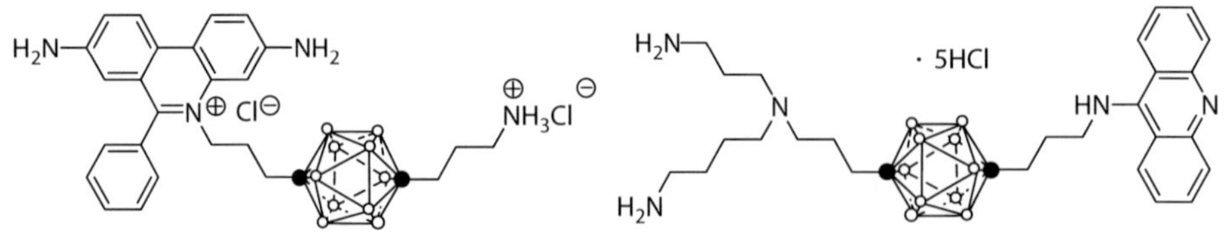

FIGURE 8.3 Structures of WSP1 and WSA1.

8.2.4 Epidermal Growth Factor Receptor–Targeted Liposomes

Epidermal growth factor receptor (EGFR) tyrosine kinase plays a fundamental role in signal transduction pathways and the uncontrolled activation of this EGFR-mediated signaling may be due to overexpression of the receptors in numerous tumors. Kullberg and coworkers investigated EGF-conjugated PEGylated liposome delivery vehicle, containing water-soluble boronated phenanthridine, WSP1, or water-soluble boronated acridine, WSA1, for EGFR targeting. In the case of WSA1 a ligand-dependent uptake was obtained and the boron uptake was as good as if free WSA1 was given. No ligand-dependent boron uptake was seen for WSP1-containing liposomes. *In vitro* boron uptake by glioma cells (6.29 ± 1.07 μg/g cells) was observed with WSA1-encapsulated EGF-conjugated PEGylated liposomes (Kullberg et al., 2003) (Figure 8.3).

Cetuximab, a recombinant chimeric monoclonal antibody, binds to the extracellular domain of the EGFR, thereby preventing the activation and subsequent dimerization of the receptor. Lee and coworkers developed cetuximab-immunoliposomes as an alternative immunoliposome for targeting of EGFR(+) glioma cells through a cholesterol-based membrane anchor, maleimido-PEG-cholesterol (Mal-PEG-Chol), to conjugated cetuximab to liposomes. BSH-encapsulated cetuximab-immunoliposomes were evaluated for targeted delivery to human EGFR gene- transfected $F98_{EGFR}$ glioma cells. Much greater (~8-fold) cellular uptake of boron was obtained using cetuximab-immunoliposomes in EGFR(+) $F98_{EGFR}$ compared with nontargeted human IgG-immunoliposomes (Pan et al., 2007).

8.2.5 Transferrin Receptor–Targeted Liposomes

Transferrin (TF) receptor–mediated endocytosis is a normal physiological process by which TF delivers iron to the cells and higher concentration of TF receptor has been observed on most tumor cells in comparison with normal cells. Maruyama and coworkers developed TF-conjugated PEG liposome. This liposome showed a prolonged residence time in the circulation and low RES uptake in tumor-bearing mice, resulting in enhanced extravasation of the liposomes into the solid tumor tissue, where the liposomes were internalized into tumor cells by receptor-mediated endocytosis (Ishida et al., 2001). The TF-conjugated PEG liposomes and PEG liposomes encapsulating ^{10}BSH were prepared and their tissue distributions in Colon 26 tumor-bearing mice after IV injection were compared with those of bare liposomes and free ^{10}BSH. When TF-PEG liposomes were injected at a dose of 35 mg ^{10}B/kg, a prolonged residence time in the circulation and low uptake by the reticuloendothelial system (RES) were observed in Colon 26 tumor-bearing mice, resulting in enhanced accumulation of ^{10}B into the solid tumor tissue. TF-PEG liposomes maintained a high ^{10}B level in the tumor, with concentrations over 30 μg of boron per gram of tumor for at least 72 h after injection. On the other hand, the plasma level of ^{10}B decreased, resulting in a tumor/plasma ratio of 6.0 at 72 h after injection. Administration of ^{10}BSH encapsulated in TF-PEG liposomes at a dose of 5 or 20 mg ^{10}B/kg and irradiation with 2×10^{12} neutrons/cm^2 for 37 min produced tumor growth suppression and improved long-term survival compared with boron-loaded PEG liposomes and bare liposomes, and free ^{10}BSH. Masunaga and coworkers evaluated biodistribution of ^{10}BSH and $Na_2{}^{10}B_{10}H_{10}$-encapsulated TF-PEG liposomes in SCC VII tumor-bearing mice (Masunaga et al.,

2006). The kinetics in the ^{10}B concentration in tumors loaded with both liposomes were similar except that ^{10}B concentrations were greater 24 h after the administration of $Na_2{}^{10}B_{10}H_{10}$ than ^{10}BSH in TF-PEG liposomes and ^{10}B concentration in tumors was 35.6 μg of boron per gram of tumor with injection of $Na_2{}^{10}B_{10}H_{10}$-encapsulated TF-PEG liposomes via the tail vein at a dose of 35 mg ^{10}B/kg.

8.3 BORON LIPID–LIPOSOME APPROACH

Since a demonstration of liposomes as models for the biomembrane mimics were reported by Bangham (Bangham et al., 1965) and a first totally synthetic bilayer vesicle of didodecacyldimethylammonium bromide by Kunitake (Kunitake, 1992; Kunitake and Okahata, 1977), various self-organization bilayer membranes have been synthesized (Menger and Gabrielson, 1995). Generally, amphiphiles of liposomal membranes consist of a long hydrocarbon chain, which is called a tail, and a hydrophilic part (Allen, 1998; Maruyama, 2000); (Torchilin and Weissig, 2002); (Betageri et al., 1993). In the meanwhile, development of lipophilic boron compounds embedded within the liposome bilayer, provides an attractive method to increase the overall efficiency of incorporation of boron-containing species, as well as raise the gross boron content of the liposomes in the formulation. Various boron lipids have been developed recently and those are classified into two groups: *nido*-carborane conjugates (**1–3**) and *closo*-dodecaborane conjugates (**4–11**) as shown in Figures 8.4 and 8.5, respectively. Less toxicity has been observed in the boron lipids belonging to the latter group.

8.3.1 *NIDO*-CARBORANE AMPHIPHILE

Hawthorne and coworkers first introduced *nido*-carborane as a hydrophilic moiety into the amphiphile and this single-tailed *nido*-carborane amphiphile was utilized for liposomal boron delivery using tumor-bearing mice (Feakes et al., 1995; Watson-Clark et al., 1998). They synthesized the *nido*-carborane amphiphile **1** (Figure 8.4) from the reaction of decaborane and 1-octadecyne followed by degradation of the resulting carborane cage under basic conditions. Boronated liposomes were prepared from DSPC, cholesterol, and **1**. After the injection of liposomal suspensions in BALB/c mice bearing EMT6 mammary adenocarcinomas, the time-course biodistribution of boron was examined. At the low injected doses normally used (5–10 mg ^{10}B/kg), peak tumor boron concentrations of 35 μg of boron per gram of tumor and tumor/blood boron ratios of approximately

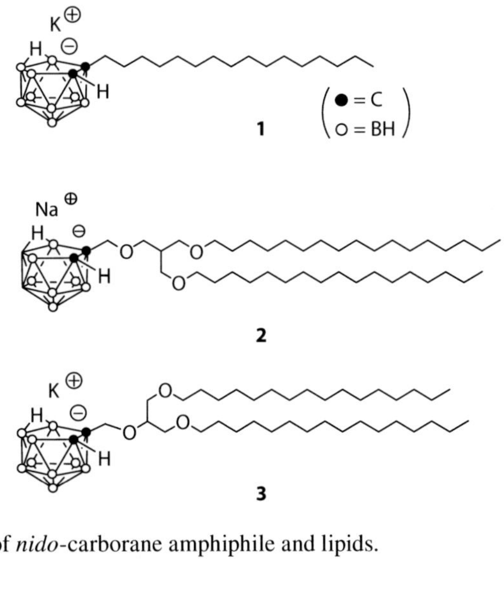

FIGURE 8.4 Structures of *nido*-carborane amphiphile and lipids.

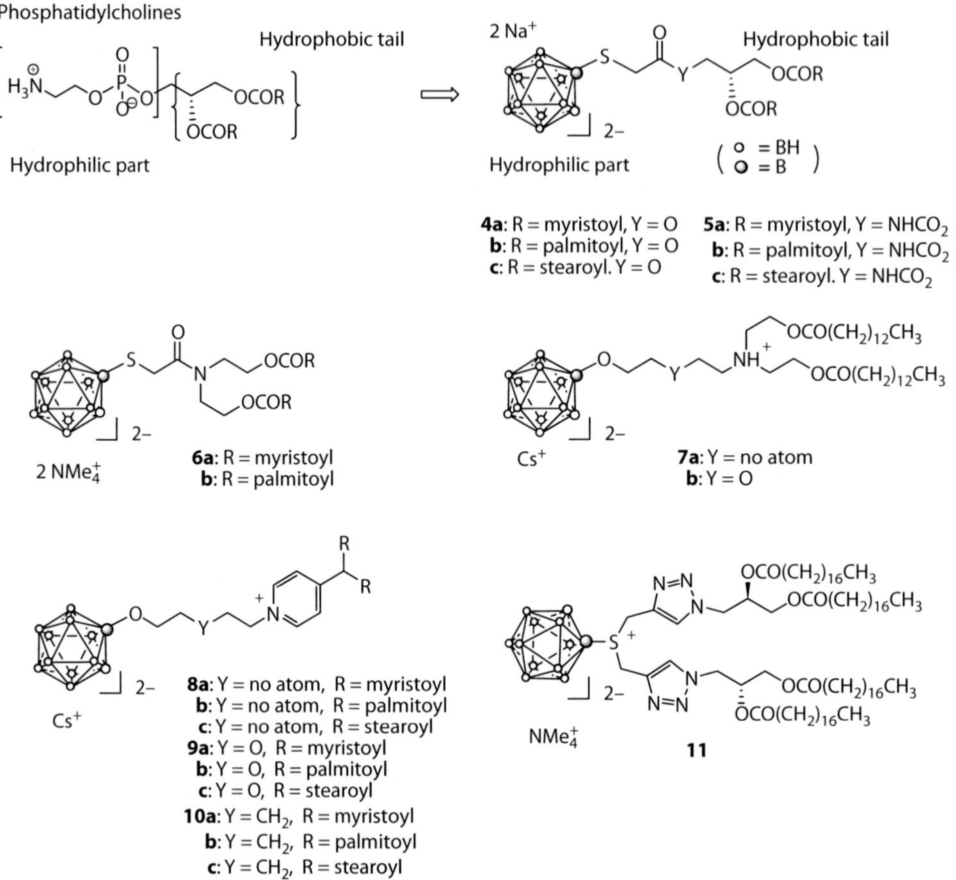

FIGURE 8.5 Structures of phosphatidylcholine and *closo*-dodecaborane lipids.

8 were achieved. These values are sufficiently high for the successful application of BNCT. The incorporation of both **1** and the hydrophilic species, $Na_3[1\text{-}(2'\text{-}B_{10}H_9)\text{-}2\text{-}NH_3B_{10}H_8]$, within the same liposome demonstrated significantly enhanced biodistribution characteristics, exemplified by maximum tumor boron concentration of 50 μg of boron per gram of tumor and tumor/blood boron ratio of 6.

8.3.2 *NIDO*-CARBORANE LIPIDS

Nakamura and coworkers developed *nido*-carborane lipid **2**, which consists of the *nido*-carborane moiety as the hydrophilic functionality conjugated with two long alkyl chains as the lipophilic functionality. Chemical synthesis of *nido*-carborane lipid **2** is shown in Scheme 8.1. Reaction of two equivalents of heptadecanol with 3-chloro-2-chloromethyl-1-propene using NaH as base gave the diether **12** in 93% yield and the hydroboration of **12** gave the corresponding alcohol **13** in 71% yield. The alcohol **13** was converted into the propargyl ether **14** in 48% yield by the treatment with propargyl bromide and the decaborane coupling of **6** was carried out in the presence of acetonitrile in toluene under reflux condition to give the corresponding *ortho*-carborane **15** in 80% yield. The degradation of the carborane cage by the treatment with sodium methoxide in methanol afforded the *nido*-carborane lipid **2** in 57% yield.

Analysis under a transmission electron microscope by negative staining with uranyl acetate showed stable vesicle formation of *nido*-carborane lipid **2**. The *nido*-carborane lipid **2** (CL) was incorporated into DSPC liposomes in a concentration-dependent manner (Nakamura et al., 2004).

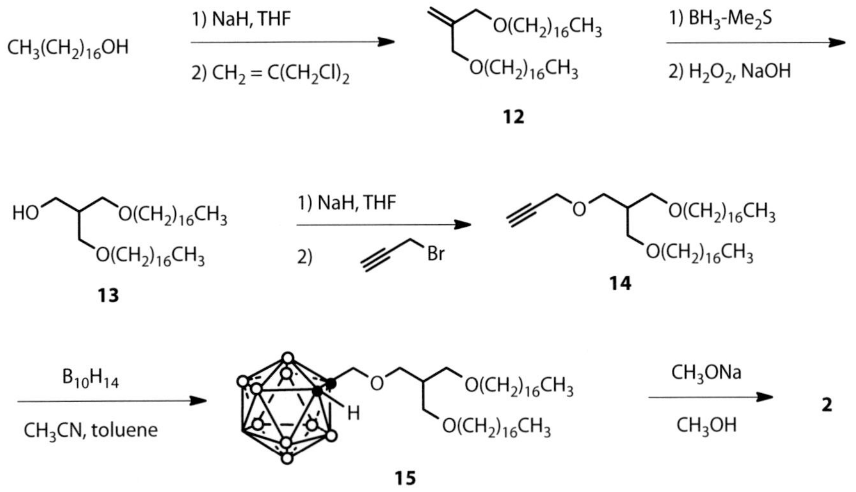

SCHEME 8.1 Synthesis of *nido*-carborane lipid **2** (CL).

Furthermore, TF could be introduced to the surface of *nido*-carborane lipid liposomes (Tf(+)-PEG-CL liposomes) by coupling TF to the PEG-CO_2H moieties of Tf(−)-PEG-CL liposomes. The biodistribution of Tf(+)-PEG-CL liposomes injected intravenously into colon 26 tumor-bearing BALB/c mice revealed that Tf(+)-PEG-CL liposomes accumulated in tumor tissues and stayed there for a sufficiently long time to increase tumor/blood boron ratio, although Tf(-)-PEG-CL liposomes were gradually released from tumor tissues with time. A boron concentration of 22 μg of boron per gram of tumor was achieved by injecting Tf(+)-PEG-CL liposomes (7.2 mg ^{10}B/kg) into tumor-bearing mice. As noted earlier, BSH-encapsulated Tf(+)-PEG liposomes accumulated in tumors at 35.5 μg of boron per gram of tissue 72 h after administration of 35 mg ^{10}B/kg. Therefore, ^{10}B delivery to tumor tissues by Tf(+)-PEG-CL liposomes would be more efficient than that by BSH-encapsulated Tf(+)-PEG liposomes based on dose-dependent drug delivery efficacy. However, significant acute toxicity was observed in 50% of the mice when Tf(+)-PEG-CL liposomes were injected at a dose of 14 mg ^{10}B/kg. Injection of Tf(+)-PEG-CL liposomes at a dose of 7.2 mg ^{10}B/kg and irradiation with 2×10^{12} neutrons/cm^2 for 37 min at the KUR atomic reactor suppressed tumor growth and the average survival rate of mice not treated with Tf(+)-PEG-CL liposomes was 21 days, whereas that of treated mice was 31 days (Miyajima et al., 2006).

Hawthorne and coworkers also synthesized *nido*-carborane lipid **3** as shown in Scheme 8.2. The reaction of two equivalents of 1-hexadecanol with epichlorohydrin using sodium hydride gave the corresponding alcohol **16**. Propargylation of **16** followed by the decaborane coupling afforded the carborane **18**. Finally, degradation of the carborane cage of **18** proceeded in the presence of KOH in ethanol to give the *nido*-carborane lipid **3**.

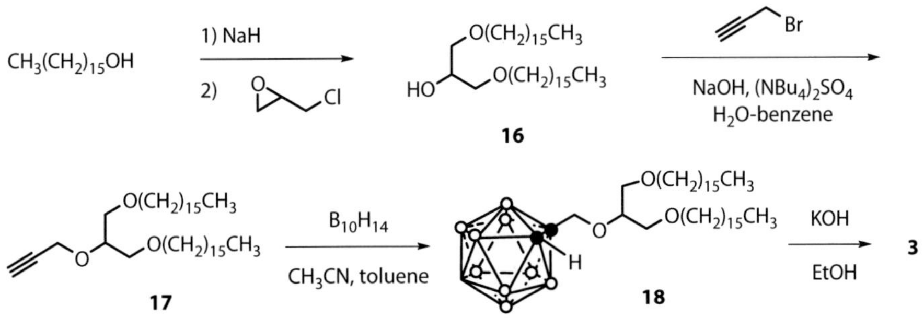

SCHEME 8.2 Synthesis of *nido*-carborane lipid **3**.

DSPC-free liposomes prepared from **3** and cholesterol exhibited a size distribution pattern of 40–60 nm, which was in the range normally associated with selective tumor uptake. Animal studies of the liposomes, containing **3**, DSPC, and cholesterol in varied proportions, were performed using male BALB/c mice (about 10 g body weight) bearing small EMT-6 tumors. Typically, 200 μL of each liposome suspension was injected into the tail vein, and the behavior of the mice was followed for up to 48 h. Unfortunately, in each case, the liposomes were found to be very toxic: no mouse survived longer than 48 h following injection of doses ranging from 6 to 30 mg of boron per kg of body weight (Li et al., 2006).Due to the significant toxicity in both cases, *nido*-carborane framework is not suitable for use in BNCT.

8.3.3 *closo*-Dodecaborate Lipids

In order to solve the problem of the significant toxicity of liposomes prepared from *nido*-carborane lipids **2** and **3**, *closo*-dodecaborate has been focused on as an alternative hydrophilic function of boron lipids. BSH is known as a water-soluble divalent "*closo*-type" anion cluster and significantly lowered toxicity (Haritz et al., 1994), and thus has been utilized for clinical treatment of BNCT. Nakamura and coworkers succeeded in the synthesis of double-tailed *closo*-dodecaborate lipids **4a–c** and **5a–c**, which have a $B_{12}H_{11}S$-moiety as a hydrophilic function with chirality similar to natural phospholipids, such as distearoylphosphatidylcholine (DSPC), in their lipophilic tails (Lee et al., 2007; Nakamura et al., 2007a).

Synthesis of the hydrophobic tail functions of **22** is shown in Scheme 8.3. Reaction of the chiral alcohol **19** with 1.2 equivalents of bromoacetyl bromide gave the ester **20**, quantitatively, and the deprotection of **20** was carried out using catalytic amounts of *p*-TsOH in MeOH to give the corresponding diol **21**. Ester formation from diol **21** using various carboxylic acids was promoted by dicyclohexylcarbodiimide in the presence of catalytic amounts of *N,N*-dimethylaminopyridine in CH_2Cl_2 to afford the precursors **22a–c** in 61–75% yields. Synthesis of the hydrophobic tail functions of **27** is shown in Scheme 8.4. The chiral alcohol **19** was first protected with benzyl bromide using NaH and the resulting dioxolane **23** was converted into the diol **24** using aqueous AcOH in 83% yield. Ester formation of **24** using various carboxylic acids was carried out in a manner similar to give **25a–c**, quantitatively. Deprotection of the benzyl group of **25a–c** by hydrogenation gave the corresponding alcohols **26a–c** (89– >99% yields), which then reacted with chloroacetyl isocyanate in CH_2Cl_2 to give **27a–c** in 74–98% yields.

Introduction of BSH into the hydrophobic tail functions **22** and **27** was examined using the "protected BSH (**28**)," which was prepared according to the Gabel's protocol (Gabel et al., 1993), as shown in Scheme 8.5. S-Alkylation of **28** with **22a–c** proceeded in acetonitrile at 70°C for 12–24 h, giving the corresponding *S*-dialkylated products **29**, which were immediately treated

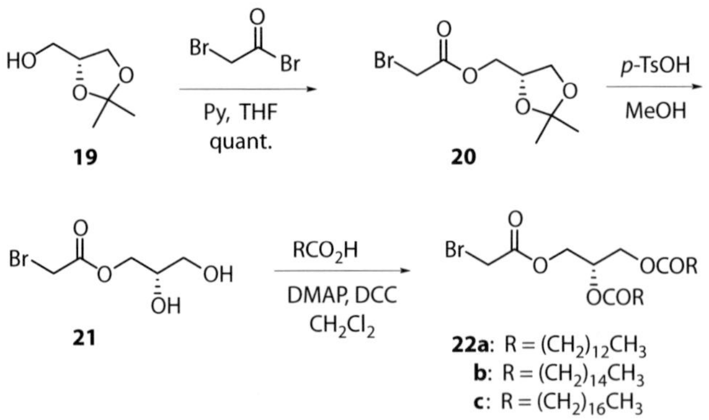

SCHEME 8.3 Synthesis of hydrophobic tail functions **22a–c**.

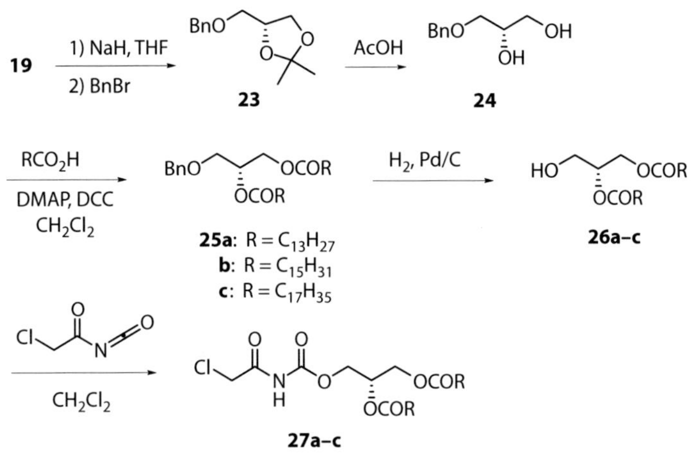

SCHEME 8.4 Synthesis of hydrophobic tail functions **27a–c**.

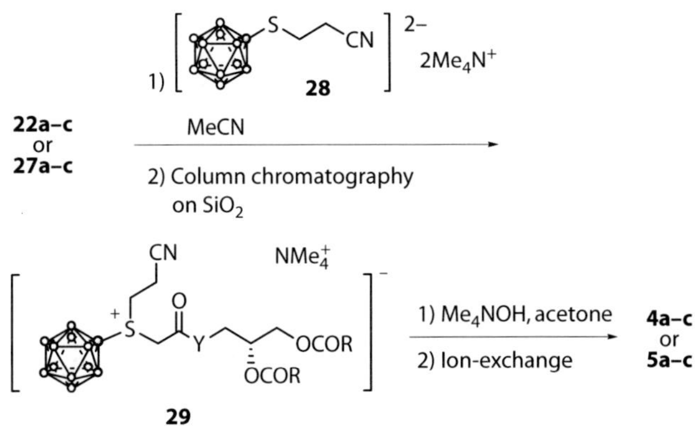

SCHEME 8.5 Synthesis of *closo*-dodecaborate lipids **4** and **5**.

with tetramethylammonium hydroxide (1 equiv.) in acetone to give **4a–c** in 76–91% yields, as tetramethylammonium salts. In a similar manner, the **5a–c** were obtained from **27a–c** in 54–83% yields.

Calcein-encapsulation experiments revealed that the liposomes, prepared from boron cluster lipids **4**, DMPC, PEG-DSPE, and cholesterol, are stable at 37°C in FBS solution for 24 h.

The time-dependent biodistribution experiment of boronated liposomes prepared from closo-dodecaborate lipid **4c** and injected intravenously into colon 26 tumor-bearing BALB/c mice (20 mg ^{10}B/kg) showed high ^{10}B accumulation in the tumor tissue (23 μg of boron per gram of tumor) 24 h after injection (Nakamura et al., 2009). In addition to determining ^{10}B concentration in various organs, neutron irradiation of the mice was carried out 24 h after administration of the boron liposomes in the JAEA atomic reactor (JRR-4). Tumor growth rate in mice administered with boron liposomes was significantly suppressed, although the administration of saline did not reduce tumor growth after neutron irradiation (Ueno et al., 2010).

Gabel and coworkers synthesized *closo*-dodecaborate lipids **6a** and **6b** as shown in Scheme 8.6. The introduction of the boron cluster was achieved by alkylation with "protected BSH" (**28**) and subsequent alkaline removal of the cyanoethyl protecting group (Gabel et al., 1993). In method A, 2 equivalents of the chloroanhydride of the fatty acids was allowed to react with diethanolamine. The resulting products *N,N*-(2-dimyristoyloxyethyl)- and *N,N*-(2-dipalmitoyloxyethyl)-amine (**30a** and **30b**, respectively) were reacted with chloroacetylchloride in the presence of triethylamine to obtain the chloroacetamides **31a** and **31b**. The reaction of **31a,b** and the tetramethylammonium salt of **28**

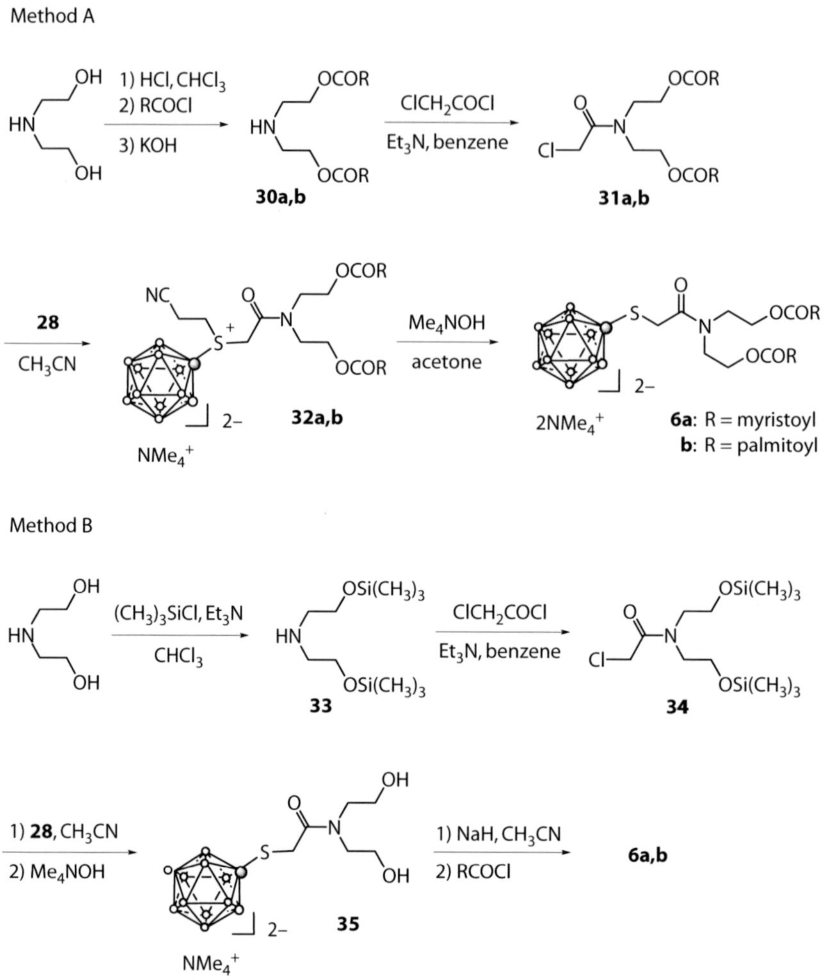

SCHEME 8.6 Synthesis of *closo*-dodecaborate lipids **6**.

produced sulfonium salts **32a,b**. The products **6a** and **6b** were obtained from the reaction of sulfonium salts **32a,b** with tetramethylammonium hydroxide in acetone. The yields of lipids **6a** and **6b** are 48–55% (overall yield from diethanolamine 25–28%).

The lipids were also able to be obtained by means of method B (Scheme 8.6). Two equivalents of chlorotrimethylsilane were reacted with diethanolamine in the presence of triethylamine. The resulting trimethylsilyloxy derivative **33** was reacted with chloroacetylchloride in the presence of triethylamine to give **34** in 84% yield. The alkylation of **34** with **28,** followed by deprotection with tetramethylammonium hydroxide in acetone, and subsequent esterification with the alkanoylchloride gave the products **6a** and **6b**. The overall yield of **6a** from diethanolamine was 46%.

Differential scanning calorimetry showed that **6a** and **6b** alone exhibit a main phase transition at 18.8 and 37.9°C, respectively. These temperatures were quite comparable to the transition temperatures of DMPC and DPPC, (24.3 and 41°C, respectively). Liposomes prepared from boron lipids, DSPC, and cholesterol (1:1:1 mole ratio) were successfully prepared by thin-film hydration and extrusion. The mean diameters of the liposomes containing **6a** and **6b** in combination with DSPC and cholesterol were found to be 135 and 123 nm, respectively. The liposomes had ξ-potentials of −67 and −63 mV, respectively, reflecting the double-negative charge of the head group. Liposomes prepared from **6a** were slightly less toxic to V79 Chinese hamster cells ($IC_{50} = 5.6$ mM) than

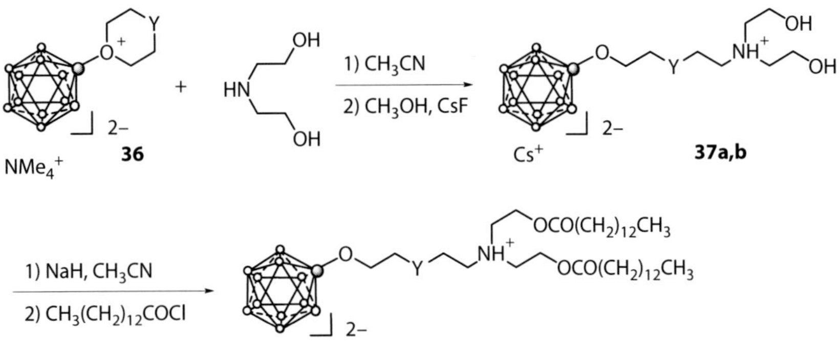

36 NMe$_4^+$ + HN(CH$_2$CH$_2$OH)$_2$ → 1) CH_3CN 2) CH_3OH, CsF → 37a,b Cs$^+$

1) NaH, CH_3CN 2) $CH_3(CH_2)_{12}COCl$ → Cs$^+$

7a: Y = no atom
b: Y = O

SCHEME 8.7 Synthesis of *closo*-dodecaborate lipids **7**.

unformulated BSH (IC_{50} = 3.9 mM), while liposomes prepared from **6b** were not toxic even at 30 mM (Justus et al., 2007).

Schaffran and coworkers synthesized new boron-containing lipids, which consist of a diethanolamine frame with two myristoyl chains bonded as esters (Schaffran et al., 2009b). Butylene or ethyleneoxyethylene units provide a link between the doubly negatively charged dodecaborate cluster and the amino function of the frame, obtained by nucleophilic attack of diethanolamine on the tetrahydrofuran and dioxane derivatives, respectively, of *closo*-dodecaborate (Scheme 8.7). The thermotropic behavior was found to be different for the two lipids, with the butylene lipid **7a** showing sharp melting transitions at surprisingly high temperatures. Toxicity *in vitro* and *in vivo* varied greatly, with the butylene derivative **7a** being more toxic than the ethyleneoxyethylene derivative **7b**.

Furthermore, Schaffran and coworkers developed pyridinium lipids with the dodecaborate cluster. The lipids consist of a pyridinium core with C12, C14, and C16 chains as lipid backbone, connected through the nitrogen atom through a butylene, pentylene, or ethyleneoxyethylene linker to the oxygen atom on the dodecaborate cluster as headgroup (Schaffran et al., 2009a). Synthesis of pyridinium lipids with the dodecaborate cluster is shown in Scheme 8.8. The lipids were obtained by nucleophilic attack of 4-(bisalkylmethyl)pyridine on the tetrahydrofurane **36**, the dioxane **37**, and a newly prepared tetrahydropyrane derivative **38**, respectively, of *closo*-dodecaborate. All these boron lipids form closed vesicles in addition to some bilayers in the pure state and in the presence of helper lipids. The thermotropic behavior was found to be increasingly complex and polymorphic with increasing alkyl chain length. Except for two lipids (**9a** and **9b**), all lipids showed low *in vitro* toxicity, and longer alkyl chains led to a significant decrease in toxicity. The choice of the linker

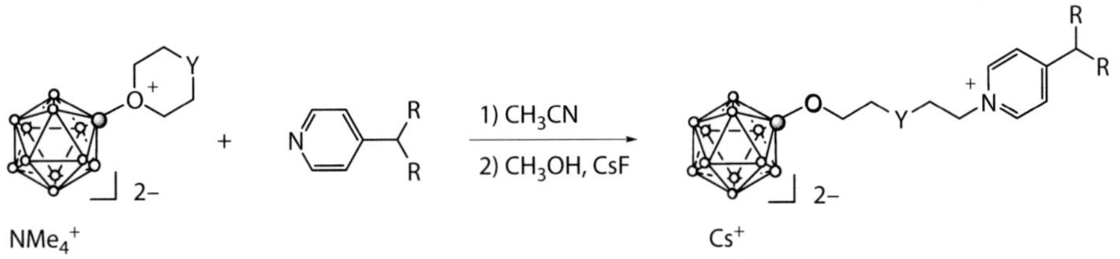

NMe$_4^+$ + 4-(CHR$_2$)pyridine → 1) CH_3CN 2) CH_3OH, CsF → Cs$^+$

36: Y = no atom
37: Y = O
38: Y = CH_2

8a: Y = no atom, R = $(CH_2)_{11}CH_3$
b: Y = no atom, R = $(CH_2)_{13}CH_3$
c: Y = no atom, R = $(CH_2)_{15}CH_3$
9a: Y = O, R = $(CH_2)_{11}CH_3$
b: Y = O, R = $(CH_2)_{13}CH_3$
c: Y = O, R = $(CH_2)_{15}CH_3$
10a: Y = CH_2, R = $(CH_2)_{11}CH_3$
b: Y = CH_2, R = $(CH_2)_{13}CH_3$
c: Y = CH_2, R = $(CH_2)_{15}CH_3$

SCHEME 8.8 Synthesis of tetrahydropyran derivatives **8–10** of the *closo*-dodecaborate cluster.

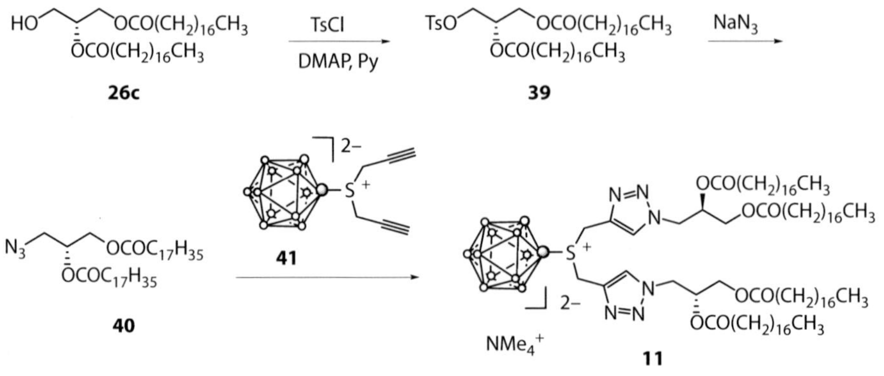

SCHEME 8.9 Synthesis of the *closo*-dodecaborate lipid **11** by the click cycloaddition reaction.

played no major role with respect to their ability to form liposomes and their thermotropic properties, but the toxicity was influenced by the linkers in the case of short alkyl chains.

El-Zaria and Nakamura developed a new method that utilizes the click cycloaddition reaction to functionalize BSH with organic molecules (El-Zaria and Nakamura, 2009; El-Zaria et al., 2010). *S,S*-bis(propynyl)sulfonioundecahydro-*closo*-dodecaborate (1-) tetramethylammonium salt (*S,S*-dipropargyl-$SB_{12}H_{11}^-$: **41**) was prepared from BSH with propargyl bromide. Compound **41** acts as a powerful building block for the synthesis of a broad spectrum of 1,4-disubstituted 1,2,3-triazole products in high yields based on the click cycloaddition reaction mediated by Cu(II) ascorbate. The reactions require only benign reaction conditions and simple workup and purification procedures; an unsymmetric bis-triazole BSH derivative could also be synthesized by the stepwise click reaction. Synthesis of the *closo*-dodecaborate lipid with four-tailed moieties was achieved by the click cycloaddition reaction of **41** with 3-*O*-azidoacetyl-1,2-*O*-distearoyl-*sn*-3-glycerol **40**, which was readily prepared from the corresponding alcohol **26c** in two steps as shown in Scheme 8.9.

8.4 BORON CHOLESTEROL–LIPOSOME APPROACH

Cholesterol is indispensable for the formation of stable liposomes, especially in blood circulation. Therefore, the development of boronated cholesterol derivatives is considered to be an alternative approach to boron embedment in the liposome bilayer. The first boronated cholesterol derivatives were reported by Feakes et al. (1999) They introduced a *nido*-carborane into the cholesterol framework through ether or ester bonds. Although they synthesized *nido*-carborane-conjugated cholesterols **42a–b** (Figure 8.6), the evaluation of their liposomes has not been reported yet.

Tjarks and coworkers developed ortho-carboranyl phenol **43** as a cholesterol mimic, which was utilized as a lipid bilayer component for the construction of nontargeted and receptor-targeted boronated liposomes. The major structural feature of the boronated cholesterol mimic is the physicochemical similarity between cholesterol and carborane frameworks (Endo et al., 1999). Cholesterol analog **43** was stably incorporated into non-, FR-, and vascular endothelial growth factor receptor-2 (VEGFR-2)-targeted liposomes. No major differences in appearance, size distribution, and lamellarity were found among conventional DPPC/cholesterol liposomes, nontargeted, and FR-targeted liposomal formulations of this carboranyl cholesterol derivative. FR-targeted boronated liposomes were taken up extensively by FR-overexpressing KB cells *in vitro*, and the uptake was effectively blocked in the presence of free folate. There was no apparent *in vitro* cytotoxicity in FR-overexpressing KB cells and VEGFR-2-overexpressing 293/KDR cells when these were incubated with boronated FR- and (VEGFR-2)-targeted liposomes, respectively, although the former accumulated extensively in KB cells and the latter effectively interacted with VEGFR-2 by causing autophosphorylation and protecting 293/KDR cells from SLT (Shiga-like toxin)-VEGF cytotoxicity (Thirumamagal et al., 2006).

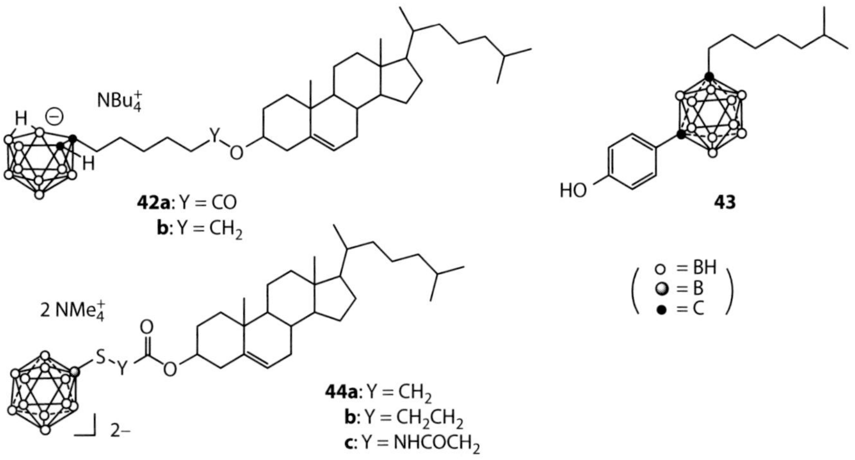

42a: Y = CO
b: Y = CH_2

43

44a: Y = CH_2
b: Y = CH_2CH_2
c: Y = $NHCOCH_2$

FIGURE 8.6 Structures of cholesterol-boron cluster conjugates (**42** and **44**) and ortho-carboranyl phenol **43**.

Nakamura, Gabel, and coworkers developed *closo*-dodecaborate-conjugated cholesterols **44a–c**. The closo-dodecaborate-conjugated cholesterol **44a** liposome, which was prepared from dimyristoylphosphatidylcholine, cholesterol, **44a**, and PEG-conjugated distearoylphosphatidylethanolamine (1:0.5:0.5:0.1), exhibited higher cytotoxicity than BSH at the same boron concentration and the IC_{50} values of **44a** liposome and BSH toward colon 26 cells were estimated to be 25 and 78 ppm of boron concentration, respectively (Nakamura et al., 2007b).

8.5 SUMMARY

Recent developments of liposomal BDS are summarized in this chapter. Two approaches to encapsulation of boron compounds into liposomes and incorporation of boron-conjugated lipids into the liposomal bilayer have been investigated. Since the leakage on storage has been observed in boron-encapsulated liposomes, the combination of both approaches would be more potent to carry a large amount of ^{10}B compounds into tumor. In general drug delivery systems, liposomes are used for selective delivery of drugs to tumors in an effort to avoid the drugs from undesirably accumulating in other organs. However, a large amount of liposomes that are administrated accumulate in the liver and may cause severe side effects for patients. BDS, in this context, is a safer system because boron compounds delivered with liposomes are nontoxic unless neutron capture reaction of boron takes place. Therefore, the boron compounds accumulated in other organs will not cause side effects for the patient under condition that the boron compounds are nontoxic. Liposomal drug delivery system is variable depending on tumoral blood vessel formations and diffusion processes for the liposomes to reach deeper portion in tumors, and thus it is important to choose the target cancers that are considered to be suitable for treatment with BDS. In this regard, liposomal BDS targeted to brain tumor is not a suitable strategy although brain tumors are still the major target for BNCT. Since successful BNCT is highly dependent on the selective and significant accumulation of boron-10 in tumor cells, liposomal BDS would be one of the efficient approaches for the treatment of a variety of cancers by means of BNCT.

REFERENCES

Aihara, T., Hiratsuka, J., Morita, N. et al. 2006. First clinical case of boron neutron capture therapy for head and neck malignancies using 18f-bpa pet. *Head Neck* 28: 850–5.

Allen, T. M. 1998. Liposomal drug formulations: Rationale for development and what we can expect for the future. *Drugs* 56: 747–56.

Bangham, A. D., Standish, M. M., and Watkins, J. C. 1965. Diffusion of univalent ions across the lamellae of swollen phospholipids. *J. Mol. Biol.* 13: 238–52, IN26–7.

Barth, R. 2003. A critical assessment of boron neutron capture therapy: An overview. *J. Neuro-Oncol.* 62: 1–5.

Barth, R. F. 2009. Boron neutron capture therapy at the crossroads: Challenges and opportunities. *Appl. Radiat. Isot.* 67: S3–6.

Barth, R. F., Coderre, J. A., Vicente, M. G. et al. 2005. Boron neutron capture therapy of cancer: Current status and future prospects. *Clin. Canc. Res.* 11: 3987–4002.

Betageri, G. V., Jenkins, S. A., and Parsons, D. L. 1993. *Liposome Drug Delivery Systems*. Basel: Technomic.

El-Zaria, M. E., Genady, A. R., and Nakamura, H. 2010. Synthesis of triazolyl methyl-substituted amino- and oxy-undecahydrododecaborates for potential application in boron neutron capture therapy. *New J. Chem.* 34: 1612–22.

El-Zaria, M. E. and Nakamura, H. 2009. New strategy for synthesis of mercaptoundecahydrododecaborate derivatives via click chemistry: Possible boron carriers and visualization in cells for neutron capture therapy. *Inorg. Chem.* 48: 11896–902.

Endo, Y., Iijima, T., Yamakoshi, Y. et al. 1999. Potent estrogenic agonists bearing dicarba-closo-dodecaborane as a hydrophobic pharmacophore. *J. Med. Chem.* 42: 1501–04.

Feakes, D. A., Shelly, K., and Hawthorne, M. F. 1995. Selective boron delivery to murine tumors by lipophilic species incorporated in the membranes of unilamellar liposomes. *Proc. Natl. Acad. Sci.* 92: 1367–70.

Feakes, D. A., Shelly, K., Knobler, C. B. et al. 1994. Na3[b20h17nh3]: Synthesis and liposomal delivery to murine tumors. *Proc. Natl. Acad. Sci.* 91: 3029–33.

Feakes, D. A., Spinler, J. K., and Harris, F. R. 1999. Synthesis of boron-containing cholesterol derivatives for incorporation into unilamellar liposomes and evaluation as potential agents for BNCT. *Tetrahedron* 55: 11177–86.

Gabel, D., Moller, D., Harfst, S. et al. 1993. Synthesis of *s*-alkyl and *s*-acyl derivatives of mercaptoundecahydrododecaborate, a possible boron carrier for neutron capture therapy. *Inorg. Chem.* 32: 2276–78.

Haritz, D., Gabel, D., and Huiskamp, R. 1994. Clinical phase-I study of $Na_2B_{12}H_{11}SH$ (BSH) in patients with malignant glioma as precondition for boron neutron capture therapy (BNCT). *Int. J. Radiation Oncology Biol. Phys.* 28: 1175–81.

Hawthorne, M. F. 1993. The role of chemistry in the development of boron neutron capture therapy of cancer. *Angew. Chem. Int. Ed. Engl.* 32: 950–84.

Ishida, O., Maruyama, K., Tanahashi, H. et al. 2001. Liposomes bearing polyethyleneglycol-coupled transferrin with intracellular targeting property to the solid tumors *in vivo*. *Pharm. Res.* 18: 1042–8.

Justus, E., Awad, D., Hohnholt, M. et al. 2007. Synthesis, liposomal preparation, and *in vitro* toxicity of two novel dodecaborate cluster lipids for boron neutron capture therapy. *Bioconjugate Chem.* 18: 1287–93.

Kato, I., Ono, K., Sakurai, Y. et al. 2004. Effectiveness of bnct for recurrent head and neck malignancies. *Appl. Radiat. Isot.* 61: 1069–73.

Kullberg, E. B., Carlsson, J., Edwards, K. et al. 2003. Introductory experiments on ligand liposomes as delivery agents for boron neutron capture therapy. *Int. J. Oncol.* 23: 461–7.

Kunitake, T. 1992. Synthetic bilayer membranes: Molecular design, self-organization, and application. *Angew. Chem. Int. Ed. Engl.* 31: 709–26.

Kunitake, T. and Okahata, Y. 1977. A totally synthetic bilayer membrane. *J. Am. Chem. Soc.* 99: 3860–61.

Lee, J.-D., Ueno, M., Miyajima, Y. et al. 2007. Synthesis of boron cluster lipids: Closo-dodecaborate as an alternative hydrophilic function of boronated liposomes for neutron capture therapy. *Org. Lett.* 9: 323–26.

Li, T., Hamdi, J. and Hawthorne, M. F. 2006. Unilamellar liposomes with enhanced boron content. *Bioconjugate Chem.* 17: 15–20.

Maeda, H., Wu, J., Sawa, T. et al. 2000. Tumor vascular permeability and the EPR effect in macromolecular therapeutics: A review. *J. Controlled Release* 65: 271–84.

Maruyama, K. 2000. *In vivo* targeting by liposomes. *Biol. Pharm. Bull.* 23: 791–99.

Masunaga, S.-i., Kasaoka, S., Maruyama, K. et al. 2006. The potential of transferrin-pendant-type polyethyleneglycol liposomes encapsulating decahydrodecaborate-^{10}B (GB-10) as ^{10}B-carriers for boron neutron capture therapy. *Int. J. Radiation Oncology Biol. Phys.* 66: 1515–22.

Matsumura, Y. and Maeda, H. 1986. A new concept for macromolecular therapeutics in cancer chemotherapy: Mechanism of tumoritropic accumulation of proteins and the antitumor agent SMANCS. *Cancer Res.* 46: 6387–92.

Menger, F. M. and Gabrielson, K. D. 1995. Cytomimetic organic chemistry: Early developments. *Angew. Chem. Int. Ed. Engl.* 34: 2091–106.

Mishima, Y., Ichihashi, M., Hatta, S. et al. 1989. New thermal neutron capture therapy for malignant melanoma: Melanogenesis-seeking 10b molecule-melanoma cell interaction from *in vitro* to first clinical trial. *Pigment Cell Res.* 2: 226–34.

Miyajima, Y., Nakamura, H., Kuwata, Y. et al. 2006. Transferrin-loaded nido-carborane liposomes: Tumor-targeting boron delivery system for neutron capture therapy. *Bioconjugate Chem.* 17: 1314–20.

Nakagawa, Y. and Hatanaka, H. 1997. Boron neutron capture therapy: Clinical brain tumor studies. *J. Neuro-Oncol.* 33: 105–15.

Nakamura, H., Lee, J.-D., Ueno, M. et al. 2007a. Synthesis of closo-dodecaboryl lipids and their liposomal formation for boron neutron capture therapy. *NanoBioTechnology* 3: 135–45.

Nakamura, H., Miyajima, Y., Takei, T. et al. 2004. Synthesis and vesicle formation of a nido-carborane cluster lipid for boron neutron capture therapy. *Chem. Commun.*: 1910–11.

Nakamura, H., Ueno, M., Ban, H. S. et al. 2009. Development of boron nanocapsules for neutron capture therapy. *Appl. Radiat. Isot.* 67: S84–S87.

Nakamura, H., Ueno, M., Lee, J.-D. et al. 2007b. Synthesis of dodecaborate-conjugated cholesterols for efficient boron delivery in neutron capture therapy. *Tetrahedron Lett.* 48: 3151–54.

Pan, X., Wu, G., Yang, W. et al. 2007. Synthesis of cetuximab-immunoliposomes via a cholesterol-based membrane anchor for targeting of EGFR. *Bioconjugate Chem.* 18: 101–08.

Pan, X. Q., Wang, H., Shukla, S. et al. 2002. Boron-containing folate receptor-targeted liposomes as potential delivery agents for neutron capture therapy. *Bioconjugate Chem.* 13: 435–42.

Schaffran, T., Burghardt, A., Barnert, S. et al. 2009a. Pyridinium lipids with the dodecaborate cluster as polar headgroup: Synthesis, characterization of the physical–chemical behavior, and toxicity in cell culture. *Bioconjugate Chem.* 20: 2190–98.

Schaffran, T., Lissel, F., Samatanga, B. et al. 2009b. Dodecaborate cluster lipids with variable headgroups for boron neutron capture therapy: Synthesis, physical–chemical properties and toxicity. *J. Organomet. Chem.* 694: 1708–12.

Shelly, K., Feakes, D. A., Hawthorne, M. F. et al. 1992. Model studies directed toward the boron neutron-capture therapy of cancer: Boron delivery to murine tumors with liposomes. *Proc. Natl. Acad. Sci.* 89: 9039–43.

Snyder, H. R., Reedy, A. J., and Lennarz, W. J. 1958. Synthesis of aromatic boronic acids. Aldehydo boronic acids and a boronic acid analog of tyrosine1. *J. Am. Chem. Soc.* 80: 835–38.

Soloway, A. H., Hatanaka, H., and Davis, M. A. 1967. Penetration of brain and brain tumor. VII. Tumor-binding sulfhydryl boron compounds remove. *J. Med. Chem.* 10: 714–17.

Soloway, A. H., Tjarks, W., Barnum, B. A. et al. 1998. The chemistry of neutron capture therapy. *Chem. Rev.* 98: 1515–62.

Suzuki, M., Sakurai, Y., Hagiwara, S. et al. 2007. First attempt of boron neutron capture therapy (BNCT) for hepatocellular carcinoma. *Japanese Journal of Clinical Oncology* 37: 376–81.

Thirumamagal, B. T., Zhao, X. B., Bandyopadhyaya, A. K. et al. 2006. Receptor-targeted liposomal delivery of boron-containing cholesterol mimics for boron neutron capture therapy (BNCT). *Bioconjugate Chem.* 17: 1141–50.

Torchilin, V. P. and Weissig, V. 2002. *Liposomes*. New York: Oxford.

Ueno, M., Ban, H. S., Nakai, K. et al. 2010. Dodecaborate lipid liposomes as new vehicles for boron delivery system of neutron capture therapy. *Bioorg. Med. Chem.* 18: 3059–65.

Watson-Clark, R. A., Banquerigo, M. L., Shelly, K. et al. 1998. Model studies directed toward the application of boron neutron capture therapy to rheumatoid arthritis: Boron delivery by liposomes in rat collagen-induced arthritis. *Proc. Natl. Acad. Sci.* 95: 2531–4.

Yanagie, H., Fujii, Y., Takahashi, T. et al. 1989. Boron neutron capture therapy using ^{10}B entrapped anti-CEA immunoliposome. *Human Cell* 2: 290–6.

Yanagie, H., Tomita, T., Kobayashi, H. et al. 1997. Inhibition of human pancreatic cancer growth in nude mice by boron neutron capture therapy. *Br. J. Cancer* 75: 660–5.

Yanagie, H., Tomita, T., Kobayashi, H. et al. 1991. Application of boronated anti-CEA immunoliposome to tumour cell growth inhibition in *in vitro* boron neutron capture therapy model. *Br. J. Cancer* 63: 522–6.

9 Polyhedral Boron Compounds for BNCT

Vladimir I. Bregadze and Igor B. Sivaev

CONTENTS

9.1 INTRODUCTION

One of the most important and exciting events in the chemistry of the twentieth century was a discovery of polyhedral boron hydrides at the end of 1950s. It was shown that boron atoms in boron hydrides are linked by unusual three-centered two-electron bonds or multicentered bonds. The establishment of three-centered two-electron bonds made a true revolution in the theory of chemical bonding and William Lipscomb was awarded the Nobel Prize in 1976 "for his studies on the structure of boranes illuminating problems of chemical bonding" [1].

Most known polyhedral boron hydrides are decahydro-*closo*-decaborate $[B_{10}H_{10}]^{2-}$, dodecahydro-*closo*-dodecaborate $[B_{12}H_{12}]^{2-}$ anions, and dicarba-*closo*-dodecaboranes $[C_2B_{12}H_{12}]$ (see Figure 9.1). The first derivatives of $[B_{10}H_{10}]^{2-}$ and $[B_{12}H_{12}]^{2-}$ anions were prepared by M. F. Hawthorne, and A. R. Pitochelli [2], and the synthesis and properties of carboranes were first reported at the end of 1963 by chemists from the the Soviet Union and the United States [3,4].

In contrast to the previously known boron hydrides, the polyhedral boron hydrides were shown to be exceptionally stable. One of the most striking features of the carboranes is the capability of the two carbon atoms and ten boron atoms to adopt the icosahedral geometry in which the carbon and boron atoms are hexacoordinated. This feature of carboranes gives rise to the unusual properties of these molecules and their carbon and boron derivatives. Due to these features of icosahedral carboranes, their chemistry has developed extensively, and the results obtained are summarized in R. Grimes' monograph and review [5]. Investigation of the properties of polyhedral boron hydrides resulted in the conclusion that these compounds have aromatic properties. It was the first example of nonplanar three-dimensional aromatic compounds and resulted in the development of the concept of three-dimensional aromaticity that is generally accepted at the present time [6,7].

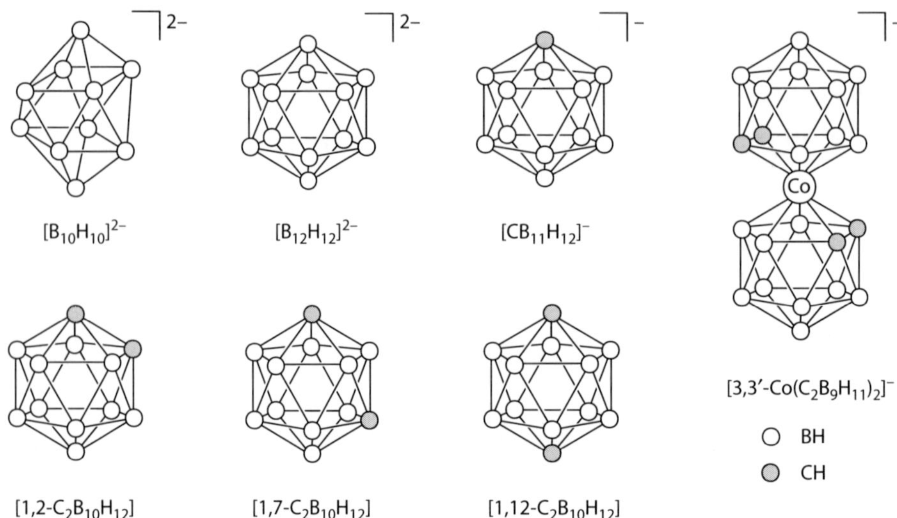

FIGURE 9.1 Main types of polyhedral boron hydrides.

Besides the interest in the basic aspects of the chemistry of polyhedral boron compounds, considerable effort has been put in the development of high-energy fuels for aircraft and rocket propulsion [6]. Other potential applications of the boron cluster compounds are presented in the review by J. Plešek [8].

At the same time, an interest in practical application of polyhedral boron hydrides provoked active development of their chemistry. Moreover, synthesis of boron compounds for medical applications has been one of the main driving forces of development of boron hydride chemistry for the past 25 years. This review is devoted to perspectives of application of polyhedral boron hydrides for boron neutron capture therapy of cancer (BNCT) [9].

9.2 MAIN TYPES OF POLYHEDRAL BORON HYDRIDES

All polyhedral boron hydrides proposed for BNCT (Figure 9.1) can be divided into two main groups: neutral and anionic ones. The first group includes *ortho*-, *meta*-, and *para*-isomers of icosahedral dicarboranes $[C_2B_{10}H_{12}]$. All these compounds are highly hydrophobic due to hydridic character of the BH groups preventing the formation of classical hydrogen bonds with water. On the other hand, the CH groups in carboranes are weakly acidic that make them available for "normal" organic chemistry [5]. Owing to this feature, the carborane cages can be easily functionalized and incorporated into organic structures to give a wide range of various carborane-containing analogs of biomolecules [10].

Removal of one of the boron atoms adjacent to both the carbon atoms of *ortho*-carborane by action of base results in the formation of anionic 7,8-dicarba-*nido*-undecaborate (*nido*-carborane) $[7,8\text{-}C_2B_9H_{12}]^-$. This approach is often used to increase the water solubility of carborane-containing biomolecules [10]. Recently, an efficient method of direct functionalization of the parent *nido*-carborane has been described [11].

Dodecahydro-*closo*-dodecaborate anion $[B_{12}H_{12}]^{2-}$, the isoelectronic and isostructural analog of carboranes, is a typical representative of anionic polyhedral boron hydrides. Sodium salts of the parent *closo*-dodecaborate and their derivatives have a good solubility in water, which is an important precondition of many medical applications. The sodium salt $Na_2[B_{12}H_{12}]$ was found to be practically nontoxic with the approximate lethal dose for rats >7.5 g/kg of body weight, which is roughly comparable with that of sodium chloride [12]. However, functionalization of this purely inorganic system till recently was problematic due to the absence of a clear reaction center. There are two general approaches to the synthesis of functional derivatives of *closo*-dodecaborate [13]. The first one includes an introduction of primary substituent (–OH, –SH, $–NH_2$) followed by modification of this substituent using standard methods of organic chemistry [14–17]. Another approach is based on the preparation of cyclic oxonium derivatives of *closo*-dodecaborate followed by the cycle opening with various nucleophilic reagents [18]. This method is very efficient for the synthesis of derivatives with pendant functional groups connected to the boron cage through flexible spacer of 5–6 atoms and could be successfully applied to other types of polyhedral boron hydrides [19].

The decahydro-*closo*-decaborate anion $[B_{10}H_{10}]^{2-}$ is another member of the $[B_nH_n]^{2-}$ family [2]. However, the chemistry of *closo*-decaborate is much less studied than the *closo*-dodecaborate chemistry. The carba-*closo*-dodecaborate anion $[CB_{11}H_{12}]^-$ combines advantages of both its isostructural and isoelectronic analogs, $[C_2B_{10}H_{12}]$ and $[B_{12}H_{12}]^{2-}$: the good solubility in water as the sodium salt and the presence of carbon atom available for functionalization using standard methods of organic chemistry [20]. It should be noted that the practical importance of the carba-*closo*-dodecaborate anion was very limited up to recently when convenient methods for the synthesis of the parent cluster and its *C*-phenyl derivatives were developed [21]. A few examples of incorporation of $[CB_{11}H_{12}]^-$ into various porphyrins have been reported quite recently [22].

Cobalt bis(dicarbollide) anion $[3,3'\text{-}Co(1,2\text{-}C_2B_9H_{11})_2]^-$ is a boron moiety, which was also proposed for use in medicinal chemistry relatively recently. Cobalt atom in this stable complex is held between two η^5-bonding $[C_2B_9H_{11}]^{2-}$ ligands derived from *nido*-carborane [23a]. The sodium salt of this metallacarborane cluster demonstrates good solubility in water; however, the anion in itself is rather lipophilic and thus could be advantageous in medical applications. The problem of synthesis of monosubstituted functional derivatives of cobalt bis(dicarbollide) was solved a few years ago when nucleophilic opening of 1,4-dioxane oxonium derivative was applied [23b]. At present, this approach is widely used for the synthesis of various functional derivatives of cobalt bis(dicarbollide) [18]. The other functionalization methods include Pd-catalyzed cross-coupling of 8-iodo derivative [24] or modification of 8-amino derivative [25]; however, both these methods did not find wide synthetic application.

The polyhedral boranes described above are available from commercial sources; however, they are all very expensive due to the absence of large-scale industrial production and could find practical applications only in some exclusive areas where no alternative exists. The use of polyhedral boranes in some of the fields described below obviously has no alternative, in other fields the price of modern pharmaceuticals is comparable with price of boron hydride derivatives, and in some cases the boron hydride price is negligible in comparison with the price of drug as a whole.

9.3 MAIN PRINCIPLES AND REQUIREMENTS FOR BORON NEUTRON CAPTURE THERAPY

BNCT is the most known of medical applications of boron hydrides. BNCT is a binary method for the treatment of cancer, which is based on the nuclear reaction of two essentially nontoxic species, nonradioactive ^{10}B and low-energy thermal neutrons. The neutron-capture reaction by ^{10}B produces α-particle, $^4He^{2+}$, and $^7Li^{3+}$ ion together with 2.4 MeV of kinetic energy and a 480 keV photon. These high-linear-energy transfer ions dissipate their kinetic energy before traveling one cell diameter (5–9 μm) in biological tissues, ensuring their potential for precise cell-killing. High accumulation and selective delivery of boron into the tumor tissue are the most important requirements to achieve efficient neutron-capture therapy of cancer. The most important requirements for the development of boron compounds are (1) achieving minimal tumor concentrations in the range of 20–35 μg of ^{10}B per gram of tumor tissue, (2) selective delivery of boronated compounds to tumor cells, while at the same time, the boron concentration in the cells of surrounding normal tissue should be kept low to minimize the damage to normal tissue, and (3) sufficiently low toxicity. Ideally, only tumor cells will be destroyed without damage to healthy tissues in the irradiated bulk. The absence of an adverse effect on the surrounding healthy tissues is attributed to the fact that the

TABLE 9.1
Thermal Neutron-Capture Cross Sections of Isotopes with the Largest Capture Cross Sections and Cross Sections of Isotopes of Some Physiologically Important Elements

Isotope	Capture Cross Section (Barn)	Isotope	Capture Cross Section (Barn)	Average Content in Tissues (%)
^{10}B	3.8×10^3	^{1}H	0.33	(10.0)
^{113}Cd	2.0×10^4	^{12}C	3.4×10^{-3}	(18.0)
^{149}Sm	4.2×10^4	^{14}N	1.8	(3.0)
^{151}Eu	5.8×10^3	^{16}O	1.8×10^{-4}	(65.0)
^{155}Gd	6.1×10^4	^{31}P	0.18	(1.16)
^{157}Gd	2.6×10^5	^{32}S	0.53	(0.20)

Source: Adapted from J. W. Kennedy, F. S. Macias, J. M. Miller, *Nuclear and Radiochemistry* (3rd edn.), John Wiley, New York, **1981**.

thermal neutron-capture cross sections of elements involved in tissues are 4–7 orders of magnitude smaller than that of the ^{10}B isotope (Table 9.1).

Despite the fact that the thermal neutron-capture cross section of the ^{10}B isotope is smaller (see Table 9.1) than those of some other elements (^{113}Cd, ^{149}Sm, ^{151}Eu, ^{155}Gd, or ^{157}Gd), at present the ^{10}B isotope is a virtually alternativeless nucleus for neutron-capture therapy for cancer because it readily forms stable covalent compounds.

At the present time, the two boron compounds that have been extensively used in clinical BNCT trials are L-*p*-dihydroxy-borylphenylalanine (BPA) and disodium mercaptoundecahydro-*closo*-dodecaborate (BSH) (Figure 9.2). In some cases, effectiveness of the treatment can be improved by combination of BPA and BSH. It should be noted, that current BNCT agents were developed more than 30 years ago [13b,27,28] and are far from ideal. These low-molecular-weight compounds are cleared easily from the cancer cells and blood. Therefore, high accumulation and selective delivery of boron compounds into tumor tissues are most important to achieve effective BNCT and to avoid damage of adjacent healthy cells (Figure 9.2).

Although one of clinically used BNCT agents, L-*p*-dihydroxy-borylphenylalanine, contains only one boron atom, there are at least three reasons for using rather expensive polyhedral boron hydrides for synthesis of BNCT agents: (1) amount of boron atoms in molecule (the use of boron clusters containing 10 or more boron atoms, in principle, allows delivery of at least 10-fold more boron

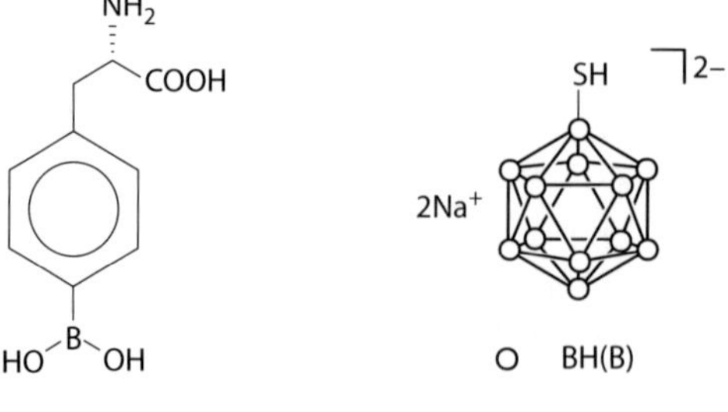

FIGURE 9.2 Design of BNCT compounds.

atoms using the same tumor-targeting vector), (2) extremely high stability, and (3) rather low toxicity of boron clusters.

The main subject of BNCT is brain tumors—high-grade gliomas, and specifically glioblastoma multiform (>300 patients treated), which are extremely resistant to all current forms of therapy, including surgery, chemotherapy, radiotherapy, immunotherapy, and gene therapy [29]. Metastatic melanomas, which cannot be treated by either surgical excision or stereotactic radiosurgery, are other candidates for the BNCT treatment [30]. The use of BNCT for treatment of the head and neck recurrent tumors (squamous cell carcinomas, sarcomas, parotid tumor) [31] and adenocarcinoma of the colon that had metastasized to the liver [32] have been reported recently.

Therapeutical treatment of brain cancers such as gliomas is a very serious problem due to necessity to transport therapeutic agents across the blood–brain barrier (BBB). It is estimated that more than 98% of all low-molecular-weight drugs and practically 100% of high-molecular-weight drugs developed for central nervous system disorders do not across the BBB. A very restricted number of lipophilic small molecules (MW < 400 Da) cross the BBB by free diffusion. All the other molecules do not cross the BBB at all, or they are transported across the BBB through catalyzed transport, owing to specific interaction between the therapeutic and certain BBB transport systems [33]. It was shown that BSH does not penetrate either the BBB or the cellular membrane, and therefore passively accumulates in the intracellular space of the bulk tumor, that is, where the BBB is disrupted, whereas BPA is actively transported across the BBB and the cellular membranes and into the tumor cells [34].

The BBB transporters can be divided into three groups: carrier-mediated transporters, active efflux transporters, and receptor-mediated transporters. The carrier-mediated and active efflux transporters generally transport small molecules, whereas endogenous large molecules enter the brain from blood through receptor-mediated transport. The carrier-mediated transporters include the Glut1 glucose carrier, the MCT1 monocarboxylic acid carrier, the LAT1 or CAT1 amino acid carriers, and the CNT2 purine nucleoside carrier. Generally, these systems are responsible for the transport of nutrients in the blood-to-brain direction [35]. Since the vascular density in the brain is very high, once molecules have penetrated the BBB, they distribute rapidly to the whole brain tissue. Nutrients provide both building blocks and energy supply to drive the cell metabolism and synthesize the necessary cellular components. In contrast to normal cells, tumor cells have an elevated requirement for certain constituents necessary for cell replication (amino acids, nucleic acid precursors, carbohydrates, and so on). Thus, the limiting factor in the treatment of brain cancers is the delivery of therapeutic agents to the brain across the BBB. That is why the most widespread chemical effort has been related to the design and synthesis of boron-containing cellular building blocks.

9.4 MAIN TYPES OF BORON CARRIERS

Many different types of boron-containing compounds have been designed and synthesized for testing as BNCT drugs over the past 30 years. In order to fulfill the requirements for the BNCT agents these new compounds should consist of a boron-containing part connected through a hydrolytically stable linkage to a tumor-targeting part responsible for delivering of boron fragment to the tumor cell and its retention there. Thus, the structure of these BNCT agents can be presented in the following way (Figure 9.3).

In this chapter, we present data on conjugates of polyhedral boron compounds with different tumor-targeting carriers that have been synthesized and evaluated as potential agents for BNCT.

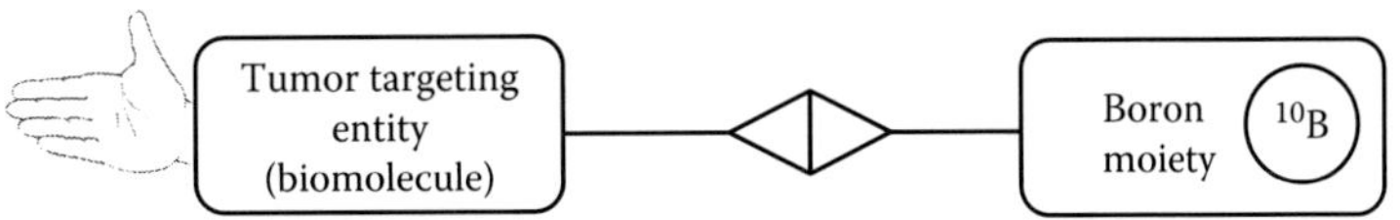

FIGURE 9.3 Design of BNCT compounds.

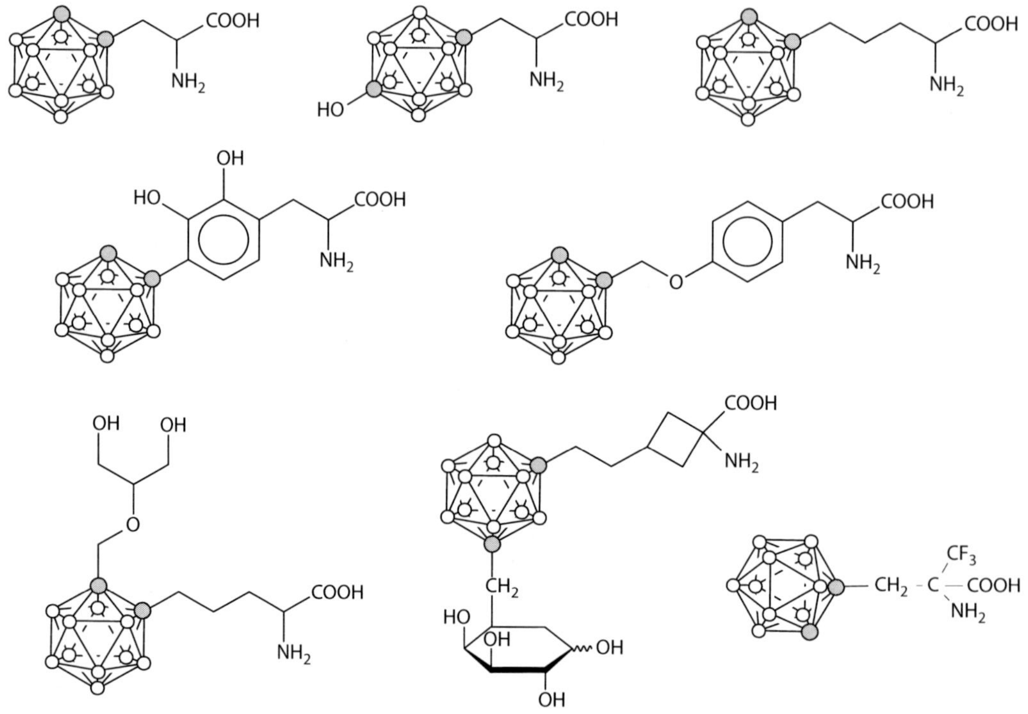

FIGURE 9.4 Some carborane-containing amino acids [36–47].

9.4.1 Amino Acids

The interest in the development of carborane-containing amino acids arose soon after the first carboranes were synthesized and *o*-carboranylalanine was one of the earliest-described carborane-based analogs of biomolecules. It was first synthesized as the racemic mixture independently by Brattsev [36] and Zakharkin [37]. Other methods of its synthesis in high yield [38–40] as well as the stereoselective synthesis of the L-isomer [41] have been proposed, and then several more useful stereoselective syntheses of the L- and D-isomers have been developed [42–44] (Figure 9.4).

At the present time, a number of various carborane-containing amino acids, including the *p*-carborane analog of tyrosine [45], carborane derivatives of 1-amino-cyclobutanecarboxylic acid [46], and α-trifluoromethyl-*o*-carboranylalanine [47] have been synthesized (Figure 9.4). Synthesis and the use of carborane-containing amino acids have been reviewed recently [48].

Owing to high hydrophobicity of the carborane cluster, most carborane-containing amino acids are insoluble in water. An introduction of different types of hydrophilic groups as well as transformation of the *closo*-carborane cage to the *nido*-carborane were proposed to increase the water solubility of carborane-based amino acids. Amino acids based on anionic boron hydrides were prepared through the nucleophilic ring-opening of their cyclic oxonium derivatives [14a,23a] (Figure 9.5). More recently, this approach has been applied to the synthesis of boron-containing amino acids derived from lysine and tyrosine [49].

9.4.2 Nucleotides and Nucleosides

Boron-containing nucleosides are considered as promising candidates for BNCT because of their potential to be retained in rapidly dividing tumor cells after 5′-monophosphorylation by phosphorylating enzymes, thymidine kinase 1 and deoxycytidine kinase. Thymidine kinase 1 is pyrimidine

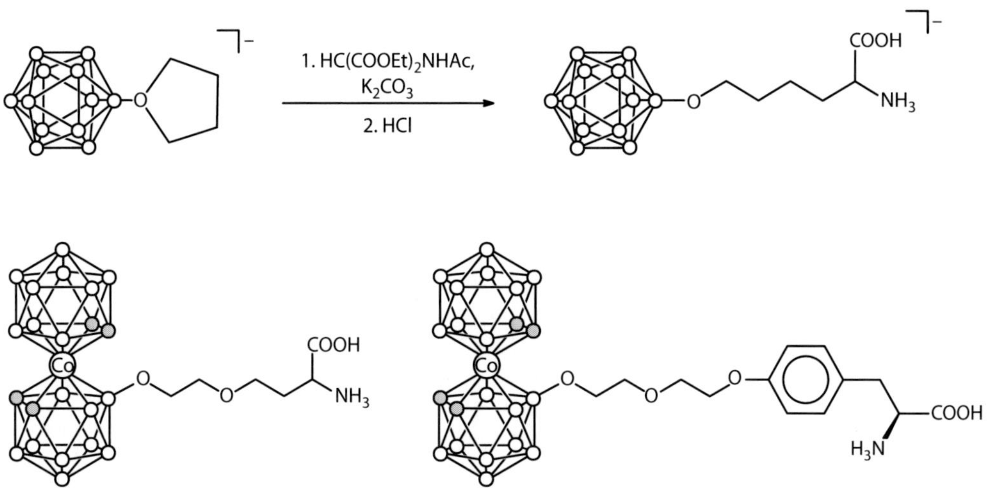

FIGURE 9.5 Anionic boron-containing amino acids. (Adapted from M. Ya. Berzina et al. manuscript in preparation.)

specific, phosphorylating thymidine and 2′-deoxyuridine, whereas deoxycytidine kinase has a broad substrate specificity, phosphorylating 2′-deoxycytidine, 2′-deoxyadenosine, and 2′-deoxyguanosine. Cellular efflux of such 5′-monophosphates would be retarded due to the negatively charged phosphate moiety. Design strategy for BNCT nucleoside prodrugs should focus on structures that enter tumor cells either by passive diffusion or through nucleoside membrane transporter and are selectively trapped intracellularly as anabolically and catabolically stable nontoxic 5′-monophosphates and/or 5′-diphosphates [50].

Syntheses and results of biological studies of carborane-containing nucleosides have been reviewed several times for the last 10 years [50–56]. Based on the results of phosphoryl transfer assays of a few series of carborane-containing nucleosides, three-dimensional quantitative structure–activity relationship has been developed [57]. Structures of some carborane-containing nucleosides synthesized are illustrated in Figure 9.6.

Synthesis of a number of boron-containing nucleosides prepared by reaction of the 1,4-dioxane derivative of cobalt bis(dicarbollide) with the canonical nucleosides thymidine, 2′-*O*-deoxycytidine, 2′-*O*-deoxyadenosine, and 2′-*O*-deoxyguanosine has been described. Direct reaction of the oxonium cycle opening results in a mixture of *O*- and *N*-boronated nucleosides that were separated by chromatography [58,59]. Syntheses of ferra bis(dicarbollide)-based [59] and *closo*-dodecaborate-based [60] nucleosides were reported as well. Structures of some of these nucleosides are shown in Figure 9.7.

9.4.3 Boron-Containing Peptides and Antibodies

In the 1960s, the use of antibodies to deliver boron to tumor cells was suggested. In the 1970s, nearly simultaneously in several laboratories the investigations of boron-conjugation chemistry with model protein substrates such as bovine serum albumin and human γ-globulin were initiated. The studies focused on the type of boron compounds that were to be used, the linkages that would be required, and the effect of such covalent attachment and its boron moiety on the physiochemical properties of the conjugate formed. The results of these early studies were surveyed in Hawthorne's review [9a]. However, some fundamental strategic problems arose soon after the first boron-containing conjugates were obtained. The minimum number of boron atoms that must be attached to each receptor-targeted biomolecule to achieve necessary therapeutic dose 10^9 boron atoms per

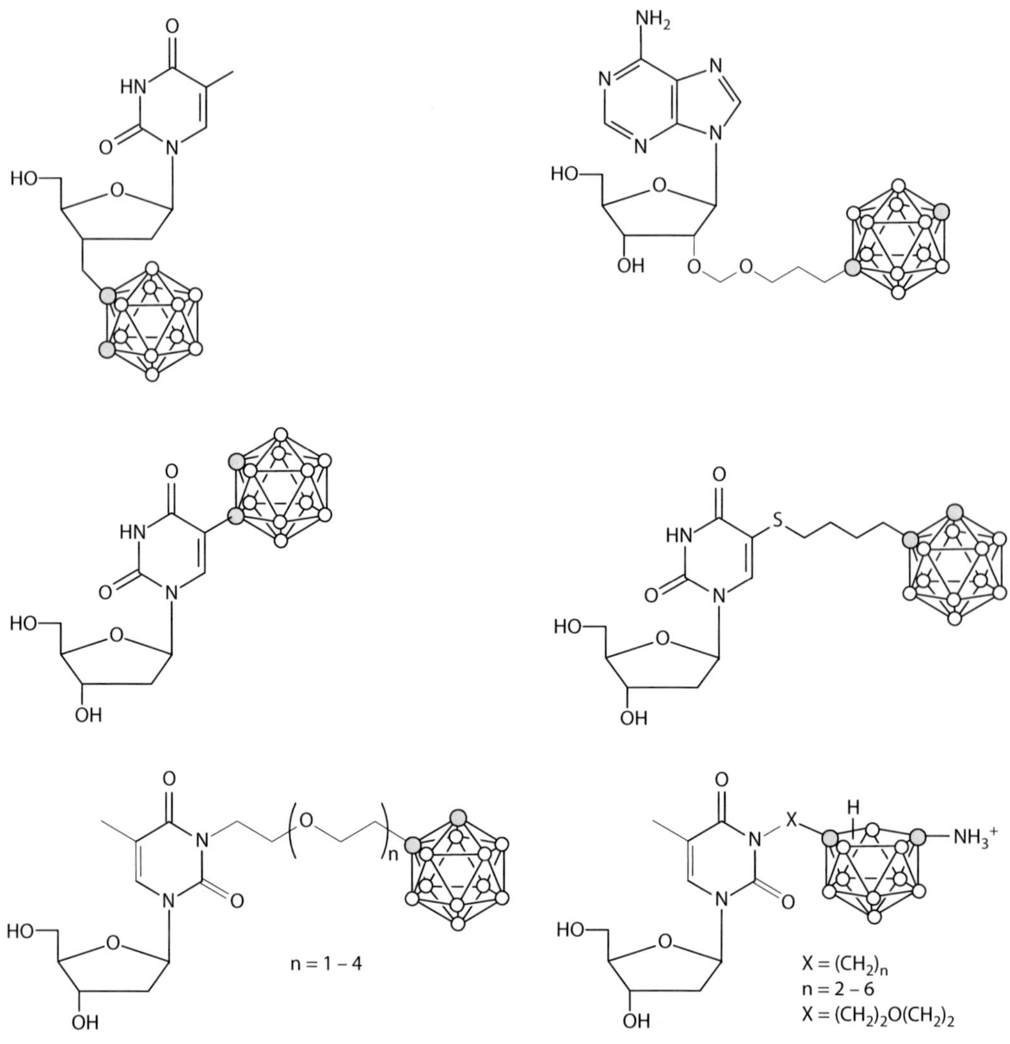

FIGURE 9.6 Some carborane-containing nucleosides [50–57].

cell is $10^9/R$ where R is the average effective number of receptor sites available on each targeted cell. It was estimated that the boronated antibody molecule must contain 10^3 ^{10}B atoms (approx. 100 boron cages) to provide the necessary therapeutic boron concentration in tumor [61]. However, it was demonstrated that the introduction of approximately 1300 ^{10}B atoms per molecule of 17–1A Mab monoclonal antibody using *N*-succinimidyl 3-(2-undecahydro-*closo*-dodecaboranyldithio) propionate $[B_{12}H_{11}SS(CH_2)_2COO(N(CO)_2(CH_2)_2)]^{2-}$ resulted in 90% loss of its immunoreactivity [62]. This result indicates that attaching 100 functional groups on the antibody molecule apparently results in the loss of its targeting specificity. It means that the minimum modification of the antibody molecule is needed for retention of its immunoreactivity. Therefore, the synthesis of a boronated macromolecule containing up to 100 boron cages, and the development of a linkage technology to attach such structures to antibodies are required for the preparation of boron-containing antibodies for BNCT.

The initial approach included the use of a preformed macromolecule containing a large number of functional groups to which the boron polyhedron could be covalently attached. The first macromolecule that has been used as a platform for delivery of boron compounds was polylysine, a polymer having multiple reactive amino groups. The protein-binding polyhedral boron derivative,

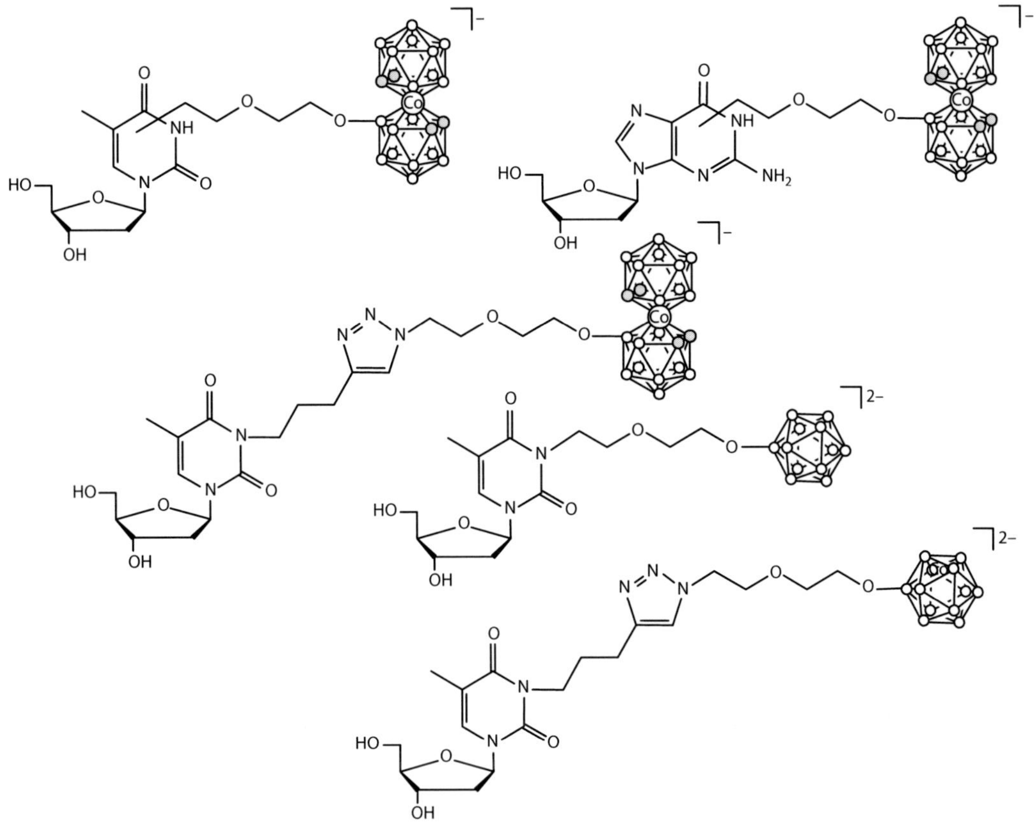

FIGURE 9.7 Cobalt bis(dicarbollide)- and *closo*-dodecaborate-based nucleosides [58,59].

isocyanato-*closo*-dodecaborate $[B_{12}H_{11}NCO]^{2-}$, was linked to polylysine and subsequently to the anti-B16 melanoma mAb IB16–6 [63]. The bioconjugate had an average of 2700 boron atoms per molecule and retained 58% of the native antibody immunoreactivity. Other bioconjugates prepared by this method had more than 1000 boron atoms per molecule of antibody and retained 40–90% of the immunoreactivity. Using site-specific linkage of boronated polylysine to the carbohydrate moieties of anti-TSH antibody resulted in a bioconjugate that had approximately 6000 boron atoms with retention of its immunoreactivity [64]. One of the limitations of this approach is that the polymer itself is not a discrete and homogeneous entity and that heterogeneity is markedly increased following boronation since the number of boron groups attached to each polymeric molecule could vary.

Dendrimers having a well-defined structure and a large number of reactive terminal groups became one of the most attractive polymers that have been used as boron carriers. Initially, second- and fourth-generation polyamidoamino (PAMAM) dendrimers, which have 12 and 48 reactive terminal amino groups, respectively, were reacted with the isocyanato derivative of *closo*-decaborate anion $[Me_3NB_{10}H_8NCO]^-$. The boronated dendrimer was then linked to mAb IB16–6 directed against the murine B16 melanoma [65]. More recently, the fifth-generation PAMAM dendrimers were boronated with the same polyhedral borane anion and linked to the anti-EGFR mAb cetuximab or the EGFRvIII specific mAb L8A4. The resulting bioconjugate contained ~1100 boron atoms per molecule and was found to retain the native antibody immunoreactivity [66].

Boronated PAMAM dendrimers can be targeted to tumor using epidermal growth factors [67,68], vascular endothelial growth factor [69], and folic acid, a vitamin that is transported into cells through folate receptor-mediated endocytosis [70].

Other boronated dendrimers [71,72] and dendrimer-like structures [73,74] that potentially could be used as boron delivery agents for BNCT were synthesized. Other type of polymers that could be used as boron carriers are dextranes, glucose polymers consisting mainly of a linear α-1,6-glucosidic linkage with some degree of branching via a 1,3-linkage. Several examples of $[B_{12}H_{11}SH]^{2-}$ coupling to modified-dextran derivatives were described [75–77].

Two different approaches were proposed for tumor targeting of boronated polymers—streptavidin–biotin strategy [78] and bispecific antibodies [65b,79]. Both methods were considered in details in the earlier review [9b].

9.4.4 Lipoproteins and Liposomes

The use of nanocontainers, such as low-density lipoproteins (LDL) or liposomes is the important approach directed to selective delivery of therapeutics into tumors. One of the observed differences between tumor cells and their normal counterparts is the rate of metabolism of low-density lipoproteins (LDLs). The LDL vesicle comprises a phospholipids/cholesterol shell with a diameter of approximately 15–20 nm, filled with cholesteryl and glyceryl esters of long-chain alkyl carboxylic acids. This difference is based on the increased need that tumor cells possess for cholesterol to facilitate new membrane formation. The overexpression of the LDL receptors on the tumor cell membrane is responsible for its LDL accretion. This provides a basis for cellular differentiation and the targeting of tumor cells with boron if cholesteryl esters of the LDL core are replaced with a boron species that would simulate cholesterol in its physiochemical properties. This concept was proposed by Kahl in the early 1990s. The initial compounds synthesized were esters of carborane carboxylic acid with various fatty acid alcohols [80]. Later, some other derivatives of cholesterol were synthesized [81,82] and LDLs were proposed as tumor delivery agents for carborane-containing porphyrins [83].

While LDLs are natural lipoproteins with a proclivity for those tumor cells in which the receptors for these vesicles are overexpressed, liposomes can be considered as related synthetic vesicles. The liposomes consist of a phospholipids bilayer that forms a spherical shell surrounding an aqueous core. Modification of the liposomal surface by PEGylation or attachment of antibodies or receptor ligands will increase their circulation time and improve their selective targeting. Design strategies for boron-containing liposomes for BNCT focus on both nontargeted and tumor-targeted formulations. The latter includes liposomes conjugated to transferrin [84], EGF [85], antibodies [86], vascular endothelial growth factor [87], α(v)-integrin-specific arginine–glycine–aspartate peptides [88], and folic acid [89].

The construction of boron-containing liposomes is accomplished by encapsulation of boron compounds in aqueous core of the liposome, or by incorporation of boron-containing lipids in the liposome bilayer. Most of the liposomes designed for BNCT were based on the encapsulation of boron compounds such as BSH and BPA, anionic polyhedral boron hydrides with or without simple substitution patterns, various carborane derivatives (amines, polyamines, acridines, carbohydrates, and so on.) in aqueous core of liposome [90–94]. Problems arising as a result of the encapsulation of these compounds into liposomes are low encapsulation efficiency, changes in the physical–chemical behavior of liposomes, and boron leakage on storage and in contact with serum. Many of the problems encountered in the encapsulation approach can be avoided by using the second approach. Phospholipids are common lipid bilayer components of liposomes and they have been proven to be effective anchors for boron compounds in the form of single- [89a,95] or dual-chain *nido*-carborane [84c,96,97] or *closo*-dodecaborate [98,99] phospholipid mimetics. Cholesterol is another major component of the mammalian cell membrane and most liposomal formulations. Therefore, the development of carborane- [80,81] and *closo*-dodecaborate- [100] containing derivatives of cholesterol is potentially an effective approach for delivery of boron to cancer cells through both liposomes and LDL. More recently, a novel strategy for the design and synthesis of carboranyl cholesterol mimics has been proposed. In this mimic, both the B and the C rings of cholesterol were replaced with a

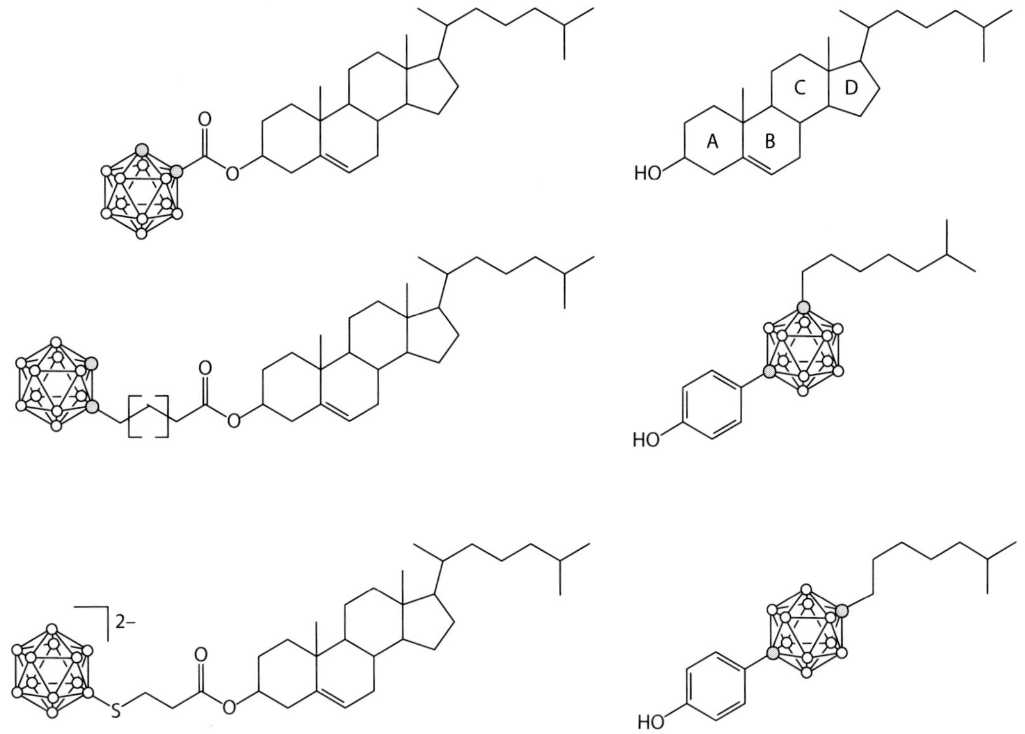

FIGURE 9.8 Some boron-containing analogues of cholesterol. (Adapted from B. T. S. Thirumamagal et al. *Bioconjugate Chem.* **2006**, *17*, 1141–1150.)

carborane cluster (Figure 9.8). The novel carboranyl cholesterol mimics are excellent lipid bilayer components for the construction of nontargeted and receptor-targeted boronated liposomes for BNCT of cancer [87].

Recently, carborane derivatives loaded into liposomes as an efficient delivery system for BNCT have been studied [101].

The synthesis of BSH-derived lipid with four-tailed moieties was achieved by the "click" cycloaddition reaction of [*S*,*S*-dipropargyl-$SB_{12}H_{11}]^-$ with 3-*O*-azidoacetyl-1,2-*O*-disteraroyl-*sn*-3-glycerol, which may be useful in the liposomal boron delivery system for BNCT. [*S*,*S*-dipropargyl-$SB_{12}H_{11}]^-$ reacts with 3-azidopropyl-*o*-carborane to give a high-boron-content compound having two different boron clusters (one *closo*-dodecaborate and two *o*-carboranes) (Figure 9.9) [102].

Synthesis of *closo*-dodecaborate derivatives of glycerol is important because these compounds are standard starting materials for the preparation of lipids, precursors of liposomes that can be used as drug carriers. The easy and convenient methods of preparation of dodecaborate derivatives of glycerol with a variety of the total charges of cluster products were presented. Since the total charge of the *closo*-dodecaborate-based lipososomes is important, the preparation of some *closo*-dodecaborate clusters with glycerol fragment with (–2), (–1), and (0) charges of the molecule, respectively, has been studied. The reaction of the oxonium derivative of dodecaborate with solketal followed by removal of protection group results in dianionic dodecaborate derivative with glycerol fragment. Similar reaction with 3-dimethylamino-propane 1,2-diole gives the monoanionic glycerol derivative. Neutral glycerol derivative was obtained by the reaction of the oxonium derivative of dodecaborate with protected piperazine derivative of glycerol followed by acidic hydrolysis of the protecting group resulting in a double quaternization of the piperazine nitrogen atoms (Figure 9.10) [103].

More recently, carborane and dodecaborate derivatives loaded into liposomes were used for their delivery to tumor cells for BNCT [104–106].

FIGURE 9.9 Synthesis of new BSH derivatives via "click" chemistry. (Adapted from M. Ueno et al. *Bioorg. Med. Chem.*, **2010**, DOI:10.1016/j.bmc.2010.03.050.)

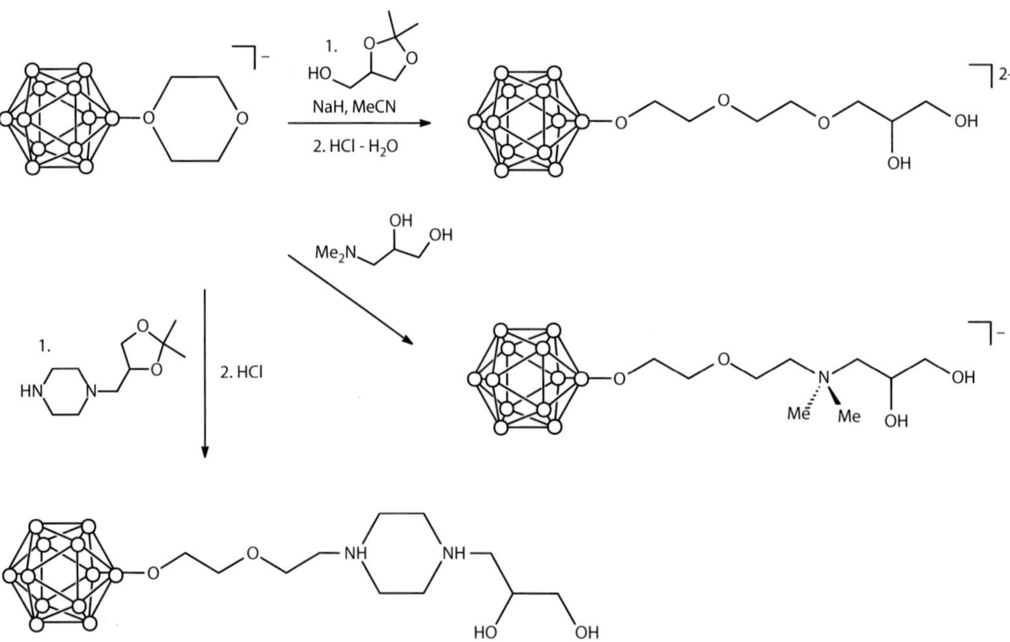

FIGURE 9.10 Synthesis of *closo*-dodecaborate derivatives of glycerol. (Adapted from A. Semioshkin et al. *J. Organomet. Chem.*, **2010**, *695*, 370–374.)

Details of investigation of boron-containing lipids and liposomes will not be presented here since liposomal boron delivery for NCT was reviewed recently [107] and a special chapter in this book is devoted to this subject.

9.4.5 Porphyrins and Phthalocyanines

The interest in synthesis of boronated porphyrins was caused by their possible application as a BNCT agent and as dual sensitizers for BNCT and photodynamic therapy (PDT). PDT has become a clinically established bimodal cancer therapy whereby superficial tumors loaded with a photosensitizing porphyrin or related macrocycle are irradiated with red laser light to form an excited triplet state that reacts with molecular oxygen and other substrates to generate highly cytotoxic species (e.g., single oxygen, superoxide anion, hydroxyl radicals) that cause irreversible destruction of tumor cells [108]. It was reported that the LDL receptor-mediated pathway plays a key role in the delivery of porphyrins to tumor cells [109].

Two approaches have been used for the synthesis of these compounds—condensation of boron-containing building blocks, on the one hand, and attachment of boron cages to natural or synthetic porphyrins and phthalocyanines, on the other hand. The first carborane-containing porphyrins were reported by Haushalter and Rudolph in 1978 [110], and this field is under extensive development [111–114]. Most of the prepared carborane-containing porphyrins are based on the meso-tetraphenylporphyrin skeleton and contain from one to eight closo- or nido-carborane cages. Besides carborane-based porphyrins, syntheses of meso-tetraarylporphyrins containing carba-closo-dodecaborate [22a], closo-dodecaborate [115], and cobalt bis(dicarbollide) [116] moieties were reported (Figure 9.11).

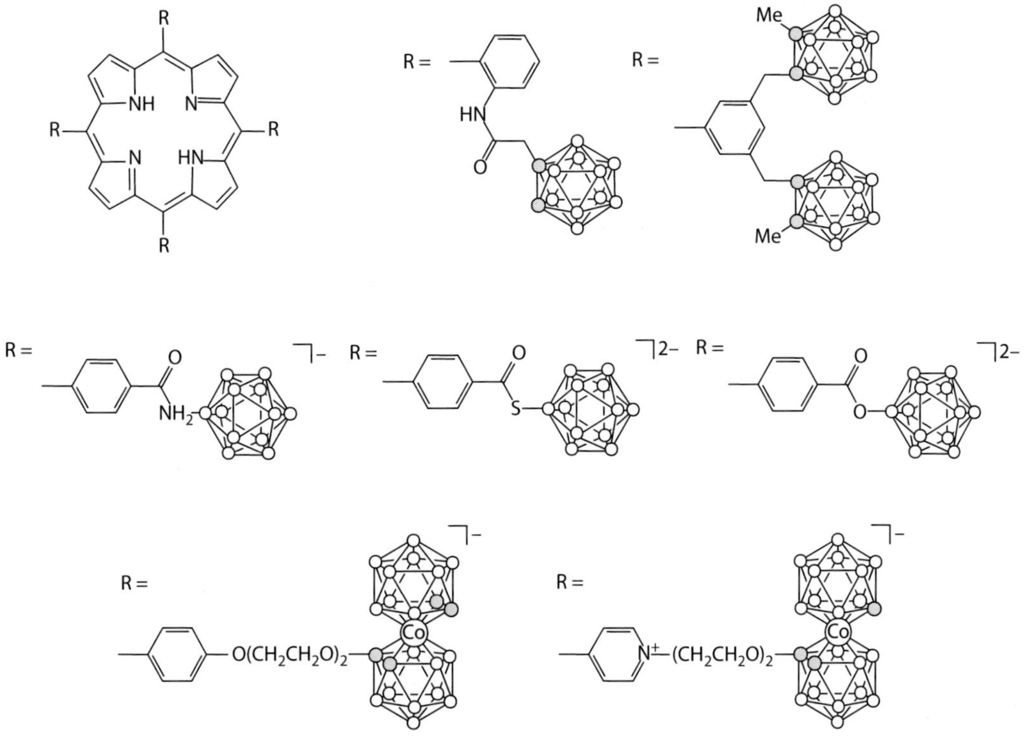

FIGURE 9.11 Some boron-containing porphyrins. (Adapted from V. I. Bregadze et al. *J. Porphyrins Phthalocyanines* **2001**, *5*, 767–781.)

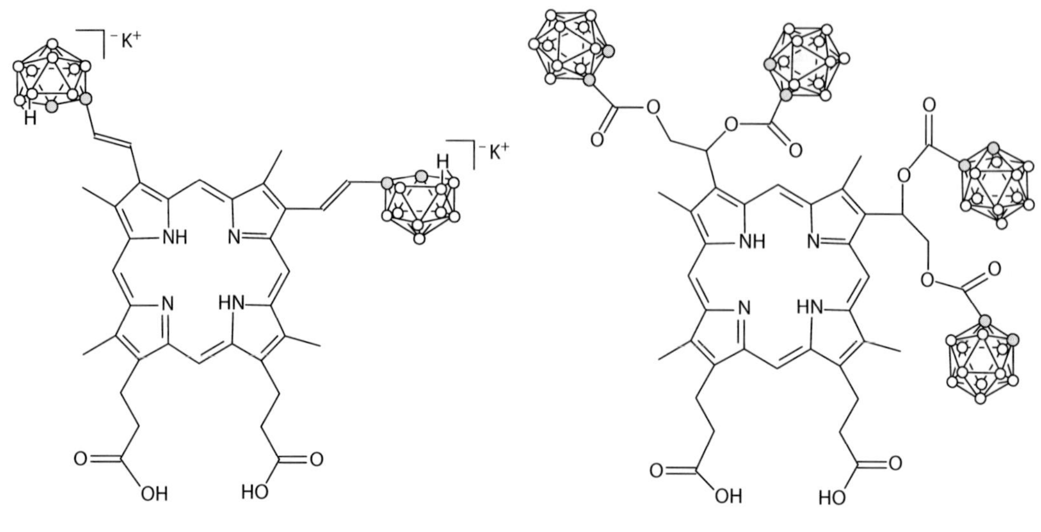

VCDP BOPP

FIGURE 9.12 VCDP and BOPP carborane-containing porphyrins [119–122].

A number of carboranyl porphyrins have been obtained on the basis of the natural porphyrin derivatives, such as deuteroporphyrin IX and hematoporphyrin IX, and two of them, VCDP and BOPP (Figure 9.12), have been extensively studied in animals. BOPP was reported to have a tumor–normal brain ratio from 13:1 to 400:1 for different glioma models [117,118]. High boron levels in tumor (>60 μg ^{10}B/g tumor) were achieved in these animal studies. However, data obtained from a human Phase I clinical trial showed that under intravenous injection BOPP does not deliver therapeutic concentrations of boron to the tumors of glioblastoma patients, and dose escalation is prevented by the toxicity of this compound. Nevertheless, BOPP has shown some promise as an effective PDT photosensitizer [119–121]. More recently, it was demonstrated that convection-enhanced delivery of BOPP significantly enhances the boron concentration in tumors and produces very favorable tumor–brain and tumor–blood ratios [122].

More recently, syntheses of anionic boron hydride derivatives of naturally occurring porphyrin systems pyropheophorbide *a* [123], chlorine e_6 [22b,22c,124,125], and bacteriochlorin *p* [126] have been reported (Figures 9.13 through 9.15).

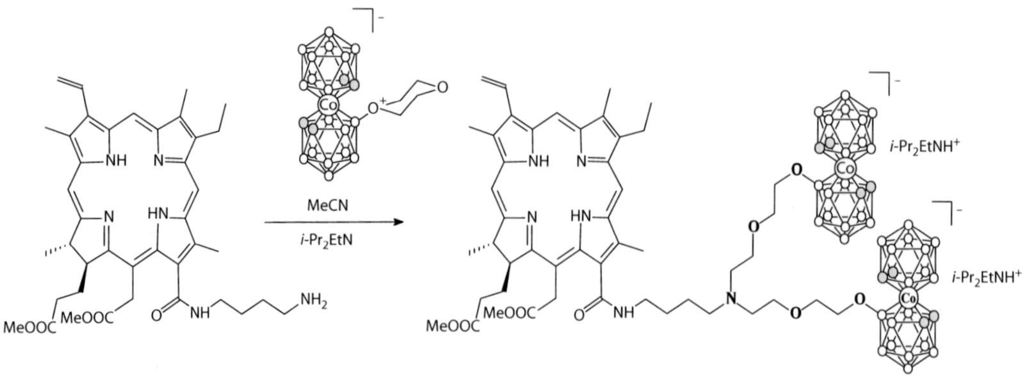

MeCN
i-Pr$_2$EtN

FIGURE 9.13 Conjugate of chlorine e_6 with two cobalt bis(dicarbollide) anions. (Adapted from V. I. Bregadze et al. *Appl. Radiat. Isotop.*, **2009**, *67*, N.7–8, Supl.1, P.S101-S104.)

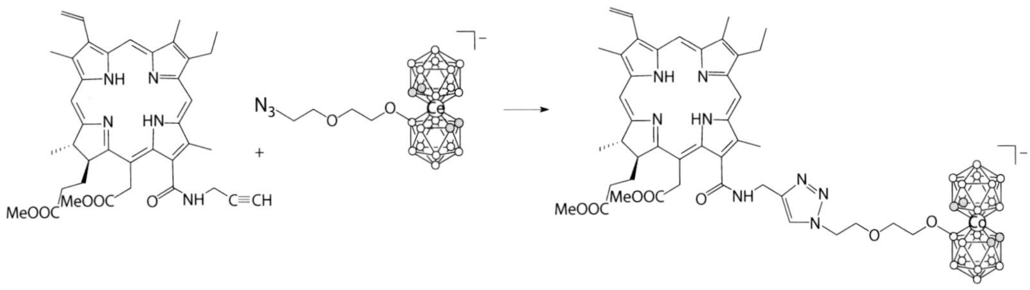

FIGURE 9.14 Novel types of boronated chlorine e_6 conjugate through "click chemistry." (Adapted from V. I. Bregadze et al., *Appl. Organomet. Chem.*, **2009**, *23*, 370–374.)

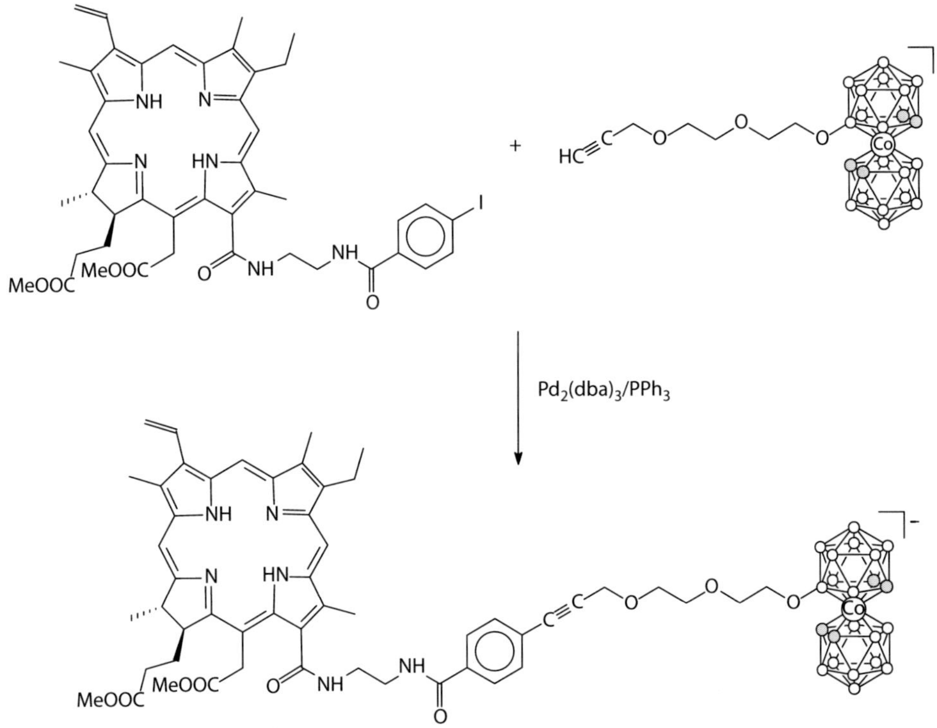

FIGURE 9.15 Novel types of boronated chlorine e_6 conjugate through Sonogashira reaction. (Adapted from M. A. Grin et al., *Russ. Chem. Bull., Int. Ed.*, **2010**, *59*, 219–224.)

In vitro study of the prepared conjugates revealed their effective accumulation in A549 human lung adenocarcinoma cells. Each molecule transports 36 boron atoms to the cell, and these compounds can be considered as potential candidates for being used in BNCT.

It was established by laser scanning confocal microscopy that both conjugates penetrate and accumulate in the human lung adenocarcinoma A549 cells cytoplasm.

Synthesis of boron hydride-based phthalocyanines has received much less development. After the first paper reviewed this field [111], only several reports have been published [127–131] (Figure 9.16).

9.4.6 Carbohydrates

Synthesis of the first carborane-containing carbohydrates was described more than 25 years ago as a way of compensating for the hydrophobicity of the carborane cage and to enhance the water

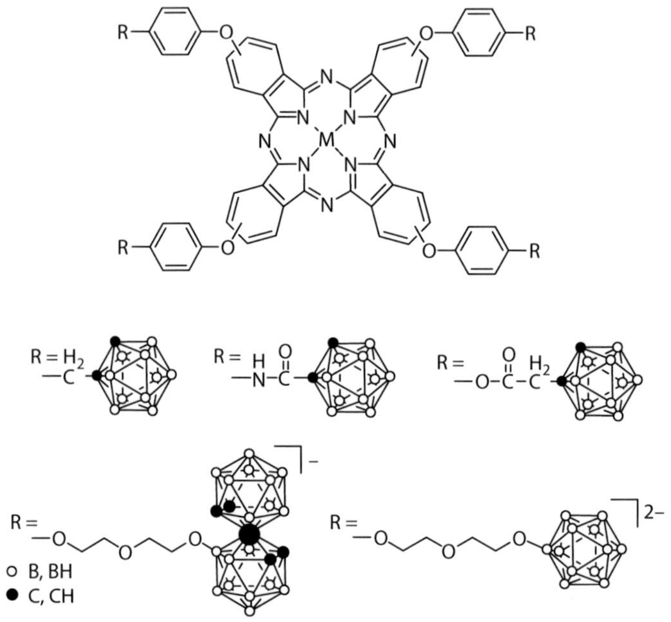

FIGURE 9.16 Some boron-containing phthalocyanines.

solubility of carborane biomolecules [132–134]. The concept still persists; however, nowadays much more attention is paid to using carbohydrates as tumor-targeting agents. This form of biomolecular recognition involves binding of a carbohydrate to a lectin receptor. Endogenous lectins are found on surfaces of many normal and malignant cells and involved in various biological functions, acting as specific receptors and/or mediating endocytosis of specific glycoconjugates. This feature has stimulated some interest in carbohydrate-mediated delivery (glycotargeting) of drugs to cells expressing the corresponding lectins [135]. Transformation of a normal cell to a tumor cell often results in the change of lectin composition of the cell surface and is usually accompanied by overexpression of certain lectins. Attachment of boron moiety to an oligosaccharide ligand of the lectin will lead to the preparation of boron-containing neoglycoconjugates, which can be used for targeted delivery of boron to the tumor tissues. Syntheses of various boron-containing conjugates have been reported by several research groups [136–140] and have been reviewed very recently [141]. Some examples of lactose conjugates with various polyhedral boron hydrides are illustrated in Figure 9.17.

9.4.7 Other Boron Carriers

Recently, some other boron carriers were used to deliver polyhedral boranes into tumor cells. One of the most interesting and new direction is developments in the BNCT driven by nanotechnology [142–147]. This subject will not be discussed here since a special chapter is planned in this book. Carboranylquinazolines [148], carboranyl aminoalcohols [149], carboranyl-α-acyloxy-amides [150], carboranyl glycophosphonates [151], conjugates of icosahedral dicarboranes with cobalt bis(dicarbollide) [152], and platinum complexes containing carborane [153] have been obtained as potential candidates for BNCT.

ACKNOWLEDGMENTS

The authors thank the Russian Foundation for Basic Research (Grants 08–03–00463, 10–03–00698, and 10–03–91331) for financial support.

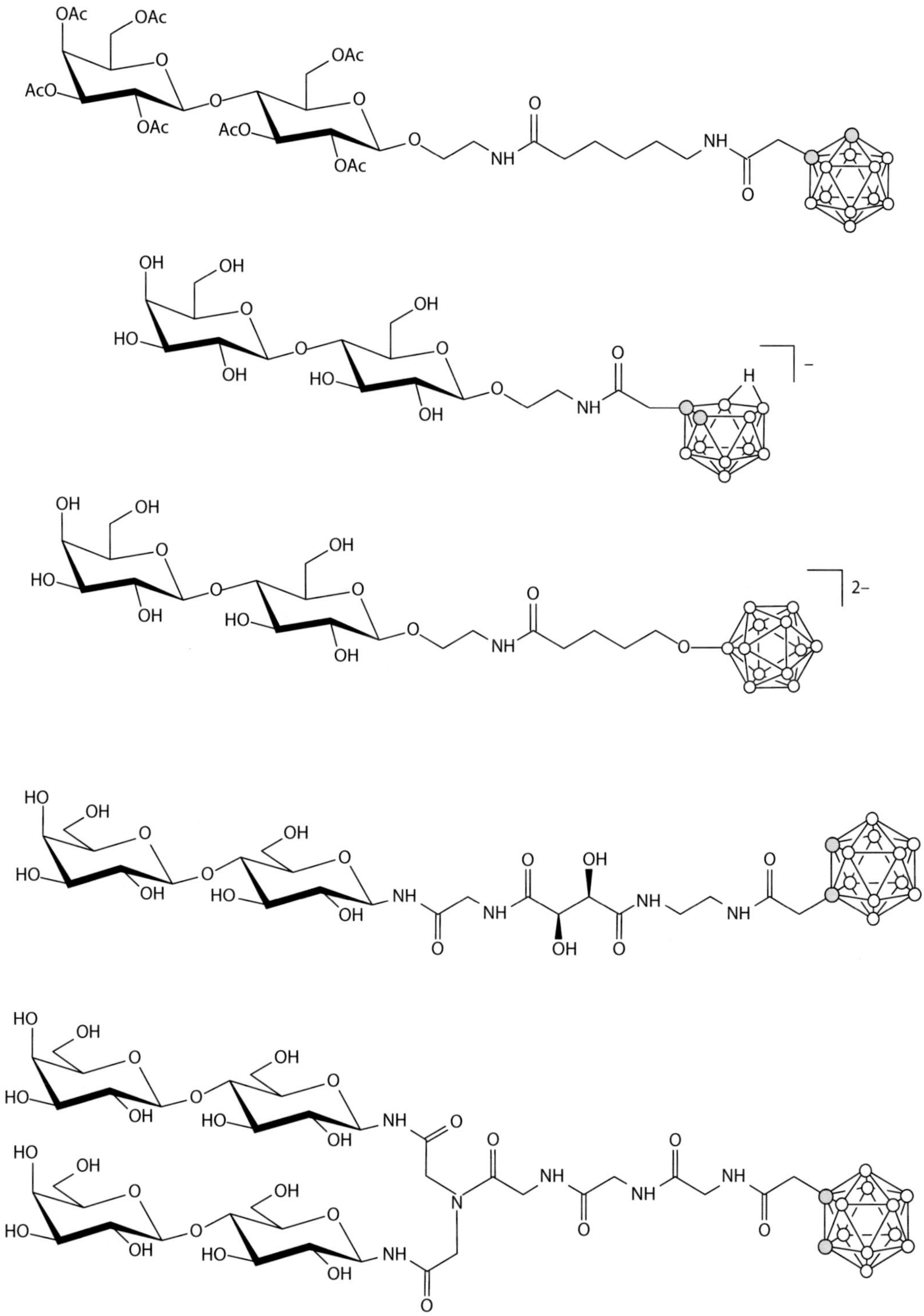

FIGURE 9.17 Boron-containing lactoses [137].

REFERENCES

1. (a) W. N. Lipscomb. *Boron Hydrides*, W. A. Benjamin Inc., New York, **1963**; (b) W. H. Lipscomb. In: *Nobel Lectures, Chemistry 1971–1980* (Ed. S. Forsren), World Scientific Publishing, Singapore, **1993**, 224.
2. (a) M. F. Hawthorne, A. R. Pitochelli. The reactions of bis-acetonitrile decaborane with amines. *J. Am. Chem. Soc.*, **1959**, *81*, 5519; (b) M. F. Hawthorne, A. R. Pitochelli. The isolation of the icosahedral $B_{12}H_{12}^{-2}$ ion. *J. Am. Chem. Soc.*, **1960**, *82*, 3228–3220.
3. (a) T. L. Heying, J. W. Ager, S. L. Clark, D. J. Mangold, H. L. Goldstein, M. Hillman, R. J. Polak, J. W. Szymanski. **A n**ew series of organoboranes. I. Carboranes from the reaction of decaborane with acetylenic Compounds. *Inorg. Chem.*, **1963**, *2*, 1089–1092; (b) M. M. Fein, J. Bobinski, N. Mayes, N. Schwartz, M. S. Cohen. Carboranes. I. The preparation and chemistry of 1-isopropenylcarborane and its derivatives (a new family of stable clovoboranes). *Inorg. Chem.*, **1963**, *2*, 1111–1115.
4. (a) L. I. Zakharkin, V. I. Stanko, V. A. Brattsev, Yu. A. Chapovsky, Yu. T. Struchkov. The structure of $B_{10}C_2H_{12}$ (barene) and its derivatives. *Bull. Acad. Sci. USSR, Div. Chem. Sci.*, **1963**, *12*, 1911; (b) L. I. Zakharkin, V. I. Stanko, V. A. Brattsev, Yu. A. Chapovsky, O. Yu. Okhlobystin. Synthesis of new class of organoboron compounds. $B_{10}C_2H_{12}$ (barene) and its derivatives. *Bull. Acad. Sci. USSR, Div. Chem. Sci.*, **1963**, *12*, 2074.
5. (a) R. N. Grimes, *Carboranes*, Academic Press, New York, **1970**; (b) V. I. Bregadze. Dicarba-*closo*-dodecaborane $C_2B_{10}H_{12}$ and their derivatives. *Chem. Rev.*, **1992**, *92*, 209–223.
6. Hawthorne, M.F. Boron hydrides. In: *The Chemistry of Boron and Its Compounds* (Ed. E. L. Muetterties), John Wiley & Sons, Inc., New York, **1967**, pp. 223–323.
7. R. B. King. Three-dimensional aromaticity in polyhedral boranes and related molecules, *Chem. Rev.*, **2001**, *101*, 1119–1152.
8. (a) J. Plešek. Potential applications of the boron cluster compounds. *Chem. Rev.* **1992**, *92*, 269–278; (b) I. B. Sivaev, V. I. Bregadze. Polyhedral boron hydrides in use: Current status and perspectives. In: *Organometallic Chemistry Research Perspectives* (Ed. R. P. Irwin), Nova Publishers, New York, **2007**, 1–59.
9. (a) M. F. Hawthorne. The role of chemistry in the development of boron neutron capture therapy of cancer. *Angew. Chem., Int. Ed. Engl.* **1993**, *32*, 950–984; (b) A. H. Soloway, W. Tjarks, B. A. Barnum, F.-G. Rong, R. F. Barth, I. M. Codogni, J. G. Wilson. The chemistry of neutron capture therapy, *Chem. Rev.*, **1998**, *98*, 1515–1562; (c) R. F. Barth, J. A. Coderre, M. G. H. Vicente, T. E. Blue. Boron neutron capture therapy of cancer: Current status and future prospects. *Clin. Cancer Res.*, **2005**, *11*, 3987–4002; (d) C. Salt, A.J. Lennox, M. Takagaki, J.A. Maguire, N.S. Hosmane. Boron and gadolinium neutron capture therapy. *Russ. Chem. Bull., Int. Ed.*, **2004**, *53*, 1871–1888; (e) G. Rana, K. Vyakaranam, J. A. Maguire, N. S. Hosmane. Boron compounds as therapeutic drugs. In: *Metallotherapeutic Drugs and Metal-Based Diagnostic Agents: The Use of Metals in Medicine* (Eds. M. Gielen, E. R. T. Tiekink), Wiley-VCH: New York, **2005**, 19–50.
10. (a) C. Morin. The chemistry of boron analogues of biomolecules. *Tetrahedron*, **1994**, *50*, 12521–12569; (b) J. F. Valliant, K. J. Guenther, A. S. King, P. Morel, P. Schaffer, O. O. Sogbein, K. A. Stephenson. The medicinal chemistry of carboranes. *Coord. Chem. Rev.*, **2002**, *232*, 173–230; (c) A. F. Armstrong, J. F. Valliant. The bioinorganic and medicinal chemistry of carboranes: from new drug discovery to molecular imaging and therapy. *Dalton Trans.*, **2007**, 4240–4251.
11. M. Yu. Stogniy, E. N. Abramova, I. A. Lobanova, I. B. Sivaev, V. I. Bragin, P. V. Petrovskii, V. N. Tsupreva, O. V. Sorokina, V. I. Bregadze. Synthesis of functional derivatives of 7,8-dicarba-*nido*-undecaborate anion by ring-opening of its cyclic oxonium derivatives. *Collect. Czech. Chem. Commun.*, **2007**, *72*, 1676–1688.
12. W. H. Sweet, A. H. Soloway, R. L. Wright. Evaluation of boron compounds for use in neutron capture therapy of brain tumors. II. Studies in man. *J. Pharmacol. Exptl. Therap.*, **1962**, *137*, 263–266.
13. (a) I. B. Sivaev, V. I. Bregadze, S. Sjöberg. Chemistry of *closo*-dodecaborate anion $[B_{12}H_{12}]^{2-}$: A review. *Collect. Czech. Chem. Commun.*, **2002**, *67*, 679–727; (b) I. B. Sivaev, V. I. Bregadze, N. T. Kuznetsov. Derivatives of the *closo*-dodecaborate anion and their application in medicine. *Russ. Chem. Bull.*, **2002**, *51*, 1362–1374.
14. D. Gabel, D. Moller, S. Harfst, J. Rösler, H. Ketz. Synthesis of S-alkyl and S-acyl derivatives of mercaptoundecahydrododecaborate, a possible boron carrier for neutron capture therapy. *Inorg. Chem.*, **1993**, *32*, 2276–2278.
15. (a) T. Peymann, E. Lork, D. Gabel. Hydroxoundecahydro-*closo*-dodecaborate(2–) as a Nucleophile. Preparation and structural characterization of *O*-alkyl and *O*-acyl derivatives of hydroxoundecahydro-*closo*-dodecaborate(2–). *Inorg. Chem.*, **1996**, *35*, 1355–1360; (b) A. A. Semioshkin, P. V. Petrovski,

I. B. Sivaev, E. G. Balandina, V. I. Bregadze. Synthesis and NMR spectra of the hydroxyundecahydro-*closo*-dodecaborate $[B_{12}H_{11}OH]^{2-}$ and its acylated derivatives. *Russ. Chem. Bull,.* **1996**, *45*, 683–686; (c) I. B. Sivaev, S. Sjöberg, V. I. Bregadze, D. Gabel. Synthesis of alkoxy derivatives of dodecahydro-*closo*-dodecaborate anion $[B_{12}H_{12}]^{2-}$. *Tetrahedron Lett.*, **1999**, *40*, 3451–3454.

16. I. B. Sivaev, A. B. Bruskin, V. V. Nesterov, M. Yu. Antipin, V. I. Bregadze, S. Sjöberg. Synthesis of Schiff bases derived from the ammoniaundecahydro-*closo*-dodecaborate(1-) anion, $[B_{12}H_{11}NH=CHR]^-$, and their reduction into monosubstituted amines $[B_{12}H_{11}NH_2CH_2R]$-: A new route to water soluble agents for BNCT, *Inorg. Chem,.* **1999**, *38*, 5887–5893.
17. S. Hoffmann, E. Justus, M. Ratajski, E. Lork, D. Gabel. $B_{12}H_{11}$-containing guanidinium derivatives by reaction of carbodiimides with $H_3N–B_{12}H_{11}$(1-). A new method for connecting boron clusters to organic compounds. *J. Organomet. Chem.*, **2005**, *690*, 2757–2760.
18. (a) I. B. Sivaev, A. A. Semioshkin, B. Brellochs, S. Sjöberg, V. I. Bregadze. Synthesis of oxonium derivatives of the dodecahydro-*closo*-dodecaborate anion $[B_{12}H_{12}]^{2-}$. Tetramethylene oxonium derivative of $[B_{12}H_{12}]^{2-}$ as a convenient precursor for the synthesis of functional compounds for boron neutron capture therapy. *Polyhedron*, **2000**, *19*, 627–632; (b) I. B. Sivaev, N. Yu. Kulikova, E. A. Nizhnik, M. V. Vichuzhanin, Z. A. Starikova, A. A. Semioshkin, V. I. Bregadze. Practical synthesis of 1,4-dioxane derivative of the *closo*-dodecaborate anion and its ring opening with acetylenic alkoxides. *J. Organomet. Chem.*, **2008**, *693*, 519–525.
19. A. A. Semioshkin, I. B. Sivaev, V. I. Bregadze. Cyclic oxonium derivatives of polyhedral boron hydrides and their synthetic applications. *Dalton Trans.*, **2008**, 977–992.
20. S. Körbe, P. J. Schreiber, J. Michl. Chemistry of the carba-*closo*-dodecaborate(-) Anion, $CB_{11}H_{12}^-$. *Chem. Rev.*, **2006**, *106*, 5208–5249.
21. (a) A. Franken, B. T. King, J. Rudolph, P. Rao, B. C. Noll, J. Michl. Preparation of [*closo*-$CB_{11}H_{12}]^-$ by dichlorocarbene insertion into [*nido*-$B_{11}H_{14}]^-$. *Collect. Czech. Chem. Commun.*, **2001**, *66*, 1238–1249; (b) S. Körbe, D. B. Sowers, A. Franken, J. Michl. Preparation of 1-*p*-halophenyl and 1-*p*-biphenylyl substituted monocarbadodecaborate anions [*closo*-1-Ar – $CB_{11}H_{11}]^-$ by insertion of arylhalocarbenes into [*nido*-$B_{11}H_{14}]^-$., *Inorg. Chem.*, **2004**, *43*, 8158–8161.
22. (a) V. A. Ol'shevskaya, A. V. Zaitsev, V. N. Luzgina, T. T. Kondratieva, O. G. Ivanov, E. G. Kononova, P. V. Petrovskii et al. Novel boronated derivatives of 5,10,15,20-tetraphenylporphyrin: Synthesis and toxicity for drug-resistant tumor cells. *Bioorg. Med. Chem.*, **2006**, *14*, 109–120; (b) V. A. Ol'shevskaya, A. N. Savchenko, A. V. Zaitsev, E. G. Kononova, P. V. Petrovskii, A. A. Ramonova, V. V. Tatarskiy et al. Novel metal complexes of boronated chlorin e_6 for photodynamic therapy. *J. Organomet. Chem.*, **2009**, *694*, 1632–1637; (c) S. B. Kahl, Z. Yao, M. S. Koo. Synthesis, toxicology and biodistribution of the first porphyrin bearing the *clos*o-monocarbaborane anion $[HCB_{11}H_{11}]$. In: *A New Option Against Cancer - Proc. 13th Int. Congress Neutron Capture Therapy* (Eds. A. Zonta, S. Altieri, L. Roveda, R. Bath), ENEA, Florence, **2008**, 177.
23. (a) I. B. Sivaev, V. I. Bregadze. Chemistry of cobalt bis(dicarbollides). A review. *Collect. Czech. Chem. Commun.*, **1999**, *64*, 783–805.; (b) I. B. Sivaev, Z. A. Starikova, S. Sjöberg, V. I. Bregadze. Synthesis of functional derivatives of the $[3,3'\text{-}Co(1,2\text{-}C_2B_9H_{11})_2]^-$ anion, *J. Organomet. Chem.*, **2002**, *649*, 1–8.
24. (a) I. P. Beletskaya, V. I. Bregadze, V. A. Ivushkin, P. V. Petrovskii, I. B. Sivaev, S. Sjöberg, G. G. Zhigareva. New B-substituted derivatives of *m*-carborane, *p*-carborane, and cobalt bis(1,2-dicarbollide) anion. *J. Organomet. Chem.*, **2004**, *689*, 2920–2929; (b) I. Rojo, F. Teixidor, C. Viñas, R. Kivekäs, R. Sillanpää. Relevance of the electronegativity of boron in η^5-coordinating ligands: Regioselective monoalkylation and monoarylation in cobaltabisdicarbollide $[3,3'\text{-}Co(1,2\text{-}C_2B_9H_{11})_2]^-$ clusters. *Chem. Eur. J.*, **2003**, *9*, 4311–4323.
25. V. Šícha, J. Plešek, M. Kvíčalová, I. Císařová, B. Grüner. Boron(8) substituted nitrilium and ammonium derivatives, versatile cobalt bis(1,2-dicarbollide) building blocks for synthetic purposes. *Dalton Trans.*, **2009**, 851–860.
26. J. W. Kennedy, F. S. Macias, J. M. Miller, *Nuclear and Radiochemistry* (3rd edn.), John Wiley, New York, **1981**.
27. V. A. Brattsev, J. H. Morris, G. N. Danilova. Sulfur introduction into $B_{12}H_{12}^{2-}$ by $[(NH_2)_2CS]_2Cl_2$ – a new halogen-like electrophilic agent. In: *Boron Chemistry at the Beginning of the 21st Century* (Ed. Yu. N. Bubnov), URSS Editorial, Moscow, **2003**, 321.
28. (a) I. B. Sivaev, V. I. Bregadze. *L*-4-Boronophenylalanine (all around the one molecule). *ARKIVOC*, **2008**, iv, 47–61, and references therein; (b) Y. Hattori, T. Asano, M. Kirihata, Y. Yamaguchi, T. Wakamiya. Development of the first and practical method for enantioselective synthesis of ^{10}B-enriched *p*-borono-l-phenylalanine. *Tetrahedron Lett.*, **2008**, *49*, 4977–4980.

29. (a) R. F. Barth, J. A. Coderre, M. G. H. Vicente, T. E. Blue, S.-I. Miyatake. Boron neutron capture therapy of brain tumors: Current status and future prospects. In: *High-Grade Gliomas: Diagnosis and Treatment* (Ed. G. H. Barnett). Humana Press, Totowa, New Jersey, **2008**, 431; (b) B. H-Stenstam, L. Pellettieri, K. Sköld, A. Rezaei, A. Brun. Neuropathological postmortem evaluation of BNCT for GBM. *Acta Neurol. Scand.*, **2007**, *116*, 169–176; (c) T. Yamamoto, K. Nakai, A. Matsumura. Boron neutron capture therapy for glioblastoma. *Cancer Lett.*, **2008**, *262*, 143–152; (d) S.-I. Miyatake, S. Kawabata, K. Yokoyama, T. Kuroiwa, H. Michiue, Y. Sakurai, H. Kumada et al. Survival benefit of boron neutron capture therapy for recurrent malignant gliomas. *J. Neurooncol.*, **2009**, *91*, 199–206.
30. R. F. Barth, W. Yang, R. T. Bartus, J. H. Rotaru, A. K. Ferketich, M. L. Moeschberger, M. M. Nawrocky, J. A. Coderre, E. K. Rofstad. Neutron capture therapy of intracerebral melanoma: enhanced survival and cure after blood-brain barrier opening to improve delivery of boronophenylalanine. *Int. J. Radiat. Oncol. Biol. Phys.*, **2002**, *52*, 858–868.
31. (a) I. Kato, K. Ono, Y. Sakurai, M. Ohmae, A. Maruhashi, Y. Imahori, M. Kirihata, M. Nakazawa, Y. Yura. Effectiveness of BNCT for recurrent head and neck malignancies. *Appl. Radiat. Isot.* **2004**, *61*, 1069–1073; (b) T. Aihara, J. Hiratsuka, N. Morita, M. Uno, Y. Sakurai, A. Maruhashi, K. Ono, T. Harada. First clinical case of boron neutron capture therapy for head and neck malignancies using ^{18}F-BPA PET. *Head & Neck*, **2006**, 850–855; (c) L. Kankaanranta, T. Seppälä, H. Koivunoro, K. Saarilahti, T. Atula, J. Collan, E. Salli et al. Boron neutron capture therapy in the treatment of locally recurred head and neck cancer. *Int. J. Radiat. Oncol. Biol. Phys.*, **2007**, *69*, 475–482; (d) N. Fuwa, M. Suzuki, Y. Sakurai, K. Nagata, Y. Kinashi, S. Masunaga, A. Maruhashi et al. Treatment results of boron neutron capture therapy using intra-arterial administration of boron compounds for recurrent head and neck cancer. *Br. J. Radiol.*, **2008**, *81*, 749–752.
32. T. Pinelli, A. Zonta, S. Altieri, S. Barni, A. Braghieri, P. Pedroni, P. Bruschi et al. TAOrMINA: From the first idea to the application to the human liver. In: *Research and Development in Neutron Capture Therapy* (Eds. M. W. Sauerwein, R. Moss, A. Wittig), Monduzzi Editore, Bologna, **2002**, 1065–1072.
33. W. M. Pardridge. The blood-brain barrier and neurotherapeutics. *NeuroRx*, **2005**, *2*, 1–2.
34. J. A. Coderre, J. C. Turcotte, K. J. Riley, P. J. Binns, O. K. Harling, W. S. Kiger III. Boron neutron capture therapy: Cellular targeting of high linear energy transfer radiation. *Technol. Cancer Res. Treat.*, **2003**, *2*, 355–375.
35. L. Juillerat-Jeanneret. The targeted delivery of cancer drugs across the blood–brain barrier: chemical modifications of drugs or drug-nanoparticles? *Drug Discov. Today*, **2008**, *13*, 1099–1106.
36. V. A. Brattsev, V. I. Stanko. β-(*o*-Barenyl) alanine. *Zh. Obshch. Khim.*, **1969**, *39*, 1175–1176.
37. L. I. Zakharkin, A. V. Grebennikov, A. I. L'vov. Synthesis of some nitrogenous barene derivatives. *Russ. Chem. Bull.*, **1970**, *19*, 97–102.
38. O. Leukart, M. Caviezel, A. Eberle, E. Escher, A. Tun-Kyi, R. Schwyzer. *L-o*-Carboranylalanine, a boron analogue of phenylalanine. *Helv. Chim. Acta.*, **1976**, *59*, 2184–2187.
39. I. M. Wyzlic, A. H. Soloway. A general, convenient way to carborane-containing amino acids for boron neutron capture therapy. *Tetrahedron Lett.*, **1992**, *33*, 7489–7490.
40. I. M. Wyzlic, W. Tjarks, A. H. Soloway, D. J. Perkins, M. Burgos, K. P. O'Reilly. Synthesis of carboranyl amino acids, hydantoins, and barbiturates. *Inorg. Chem.*, **1996**, *35*, 4541–4547.
41. J. L. Fauchere, O. Leukart, A. Eberle, R. Schwyzer. The synthesis of [4-carboranylalanine, 5-leucine]-enkephalin (including an improved preparation of *t*-butoxycarbonyl-*L-o*-carboranylalnine, new derivatives of *L*-propargylglycine, and a note on melanotropic and opiate receptor binding characteristics). *Helv. Chim. Acta.*, **1979**, *62*, 1385–1395.
42. W. Karnbrock, H.-J. Musiol, L. Moroder. Enantioselective synthesis of *S-o*-carboranylalanine via methylated bislactim ethers of 2,5-diketopiperazines. *Tetrahedron*, **1995**, *51*, 1187–1196.
43. P. A. Radel, S. B. Kahl. Enantioselective synthesis of *L*- and *D*-carboranylalanine. *J. Org. Chem.*, **1996**, *61*, 4582–4588.
44. P. Lindström, C. Naeslund, S. Sjöberg. Enantioselective synthesis and absolute configurations of the enantiomers of *o*-carboranylalanine. *Tetrahedron Lett.*, **2000**, *41*, 751–754.
45. I. Ujvary, R. Nachman. Synthesis of 3-(12-hydroxy-*p*-carboranyl)propionic acid, a hydrophobic, *N*-terminal tyrosine-mimetic for peptides. *Peptides*, **2001**, *22*, 287–290.
46. (a) R. R. Srivastava, R. R. Singhaus, G. W. Kabalka. Synthesis of 1-amino-3-[2-(1,7-dicarba-*closo*-dodecaboran(12)-1-yl)ethyl]cyclobutanecarboxylic acid: A potential BNCT agent. *J. Org. Chem.*, **1997**, *62*, 4476–4478; (b) R. R. Srivastava, G. W. Kabalka. Syntheses of 1-amino-3-[2-(7-(2-hydroxyethyl)-1,7-dicarba-*closo*-dodecaboran(12)-1-yl)ethyl]cyclo-butanecarboxylic acid and its *nido*-analogue: Potential BNCT agents. *J. Org. Chem.*, **1997**, *62*, 8730–8734; (c) B. C. Das, G. W. Kabalka, R. R. Srivastava, W. Bao, S. Das, G. Li. Synthesis of a water soluble boron neutron capture therapy agent:

1-amino-3-[2-(7-{3-[2-(2-hydroxymethyl-ethoxy)-1-(2-hydroxy-1-hydroxymethyl-ethoxymethyl)ethoxy] propyl}-1,7-di-carba-*closo*-dodecaboran-1-yl)ethyl]cyclobutanecarboxylic acid. *J. Organomet. Chem.*, **2000**, *614–615*, 255–261; (d) G. M. Kabalka, B. C. Das, S. Das, G. Li, R. Srivastava, N. Natarajan, M. K. Khan. Synthesis of 1-amino-3-{2-[7-(6-deoxy-α/β-*D*-galactopyranos-6-yl)-1,7-dicarba-*closo*-dodecaboran(12)-1-yl]ethyl}cyclobutane carboxylic acid hydrochloride. *Collect. Czech. Chem. Commun.*, **2002**, *67*, 836–842.

47. (a) I. P. Beletskaya, V. I. Bregadze, S. N. Osipov, P. V. Petrovskii, Z. A. Starikova, S. V. Timofeev. New nonnatural α-amino acid derivatives with carboranyl fragments in α- and β-positions. *Synlett*, **2004**, 1247–1248; (b) S. V. Timofeev, V. I. Bregadze, S. N. Osipov, I. D. Titanyuk, P. V. Petrovskii, Z. A. Starikova, I. V. Glukhov, I. P. Beletskaya. New carborane-containing amino acids and their derivatives. Crystal structures of *n*-protected carboranylalaninates. *Russ. Chem. Bull.*, **2007**, *56*, 791–797.
48. G. W. Kabalka, M.-L. Yao. The synthesis and use of boronated amino acids for boron neutron capture therapy. *Anti-Cancer Agents Med. Chem.*, **2006**, *6*, 111–125.
49. I. A. Lobanova, M. Ya. Berzina, I. B. Sivaev, P. V. Petrovskii, V. I. Bregadze. New approach to synthesis of amino acids based on cobalt bis(dicarbollide). *Russ. Chem. Bull.*, **2010**, *59*, 2302–2308.
50. Y. Byun, S. Narayanasamy, J. Johnsamuel, A. K. Bandyopadhyaya, R. Tiwari, A. S. Al-Madhoun, R. F. Barth, S. Eriksson, W. Tjarks. 3-Carboranyl thymidine analogues (3CTAs) and other boronated nucleosides for boron neutron capture therapy. *Anti-Cancer Agents Med. Chem.*, **2006**, *6*, 127–144.
51. A. H. Soloway, J.-C. Zhuo, F. G. Rong, A. J. Lunato, D. H. Ives, R. F. Barth, A. K. M. Anisuzzaman, C. D. Barth, B. A. Barnum. Identification, development, synthesis and evaluation of boron-containing nucleosides for neutron capture therapy. *J. Organomet. Chem.*,**1999**, *581*, 150–155.
52. Z. J. Lesnikowski, J. Shi, R. F. Schinazi. Nucleic acids and nucleosides containing carboranes. *J. Organomet. Chem.*, **1999**, *581*, 156–169.
53. W. Tjarks. The use of boron clusters in the rational design of boronated nucleosides for neutron capture therapy of cancer. *J. Organomet. Chem.*, **2000**, *614–615*, 37–47.
54. Z. J. Lesnikowski. Boron clusters—A new entity for DNA-oligonucleotide modification. *Eur. J. Org. Chem.*, **2003**, 4489–4500.
55. W. Tjarks, R. Tiwari, Y. Byun, S. Narayanasamy, R. F. Barth. Carboranyl thymidine analogues for neutron capture therapy. *Chem. Commun.*, **2007**, 4978–4991.
56. Z. J. Lesnikowski. Nucleoside–boron cluster conjugates—Beyond pyrimidine nucleosides and carboranes. *J. Organomet. Chem.*, **2009**, *694*, 1771–1775.
57. A. K. Bandyopadhyaya, R. Tiwari, W. Tjarks. Comparative molecular field analysis and comparative molecular similarity indices analysis of boron-containing human thymidine kinase 1 substrates. *Bioorg. Med. Chem.*, **2006**, *14*, 6924–6932.
58. (a) A. B. Olejniczak, J. Plešek, O. Kříž, Z. J. Lesnikowski. A nucleoside conjugate containing a metallacarborane group and its incorporation into a DNA oligonucleotide. *Angew. Chem. Int. Ed.*, **2003**, *42*, 5740–5743; (b) Z. J. Lesnikowski, E. Paradowska, A. B. Olejniczak, M. Studzinska, P. Seekamp, U. Schüßler, D. Gabel, R. F. Schinazi, J. Plešek. Towards new boron carriers for boron neutron capture therapy: metallacarboranes and their nucleoside conjugates. *Bioorg. Med. Chem.*, **2005**, *13*, 4168–4175; (c) A. B. Olejniczak, J. Plešek, Z. J. Lesnikowski. Nucleoside–metallacarborane conjugates for base-specific metal labeling of DNA. *Chem. Eur. J.*, **2007**, *13*, 311–318.
59. (a) A. B. Olejniczak, P. Mucha, B. Grüner, Z. J. Lesnikowski. DNA-dinucleotides bearing a 3',3'-cobalt- or 3',3'-iron-1,2,1',2'-dicarbollide complex. *Organometallics*, **2007**, *26*, 3272–3274; (b) B. A. Wojtczak, A. Andrysiak, B. Grüner, Z. J. Lesnikowski. "Chemical Ligation": A versatile method for nucleoside modification with boron clusters, *Chem. Eur. J.*, **2008**, *14*, 10675–10682.
60. A. Semioshkin, J. Laskova, B. Wojtczak, A. Andrysiak, I. Godovikov, V. Bregadze, Z. J. Lesnikowski. Synthesis of *closo*-dodecaborate based nucleoside conjugates. *J. Organomet. Chem.*, **2009**, 694, 1375–1379.
61. F. Alam, A. H. Soloway, R. F. Barth. Boron containing immunoconjugates for neutron capture therapy of cancer and for immunocytochemistry. *Antibody Immunoconjugates Radiopharm.*, **1989**, *2*, 145–163.
62. F. Alam, A. H. Soloway, J. E. McGuire, R. F. Barth, W. E. Carey, D. M. Adams. Dicesium *N*-succinimidyl 3-(undecahydro-*closo*-dodecaboranyldithio)propionate, a novel heterobifunctional boronating agent. *J. Med. Chem.*, **1985**, *28*, 522–525.
63. F. Alam, A. H. Soloway, R. F. Barth, N. Mafune, D. M. Adams, W. H. Knoth. Boron neutron capture therapy: linkage of a boronated macromolecule to monoclonal antibodies directed against tumor-associated antigens. *J. Med. Chem.*, **1989**, *32*, 2326–2330.
64. S. Novick, M. R. Quastel, S. Marcus, D. Chipman, G. Shani, R. F. Barth, A. H. Soloway. Linkage of boronated polylysine to glycoside moieties of polyclonal antibody; boronated antibodies as potential delivery agents for neutron capture therapy. *Nucl. Med. Biol.*, **2002**, *29*, 159–167.

65. (a) R. F. Barth, D. M. Adams, A. H. Soloway, F. Alam, M. V. Darby. Boronated starburst dendrimer-monoclonal antibody immunoconjugates: Evaluation as a potential delivery system for neutron capture therapy. *Bioconjugate Chem.*, **1994**, *5*, 58–66; (b) L. Liu, R. F. Barth, D. M. Adams, A. H. Soloway, R. Reisfeld. Critical evaluation of bispecific antibodies as targeting agents for boron neutron capture therapy of brain tumors. *Anticancer Res.*, **1996**, *16*, 2581–2588.
66. (a) G. Wu, R. F. Barth, W. Yang, M. Chatterjee, W. Tjarks, M. J. Ciesielski, R. A. Fenstermaker. Site-specific conjugation of boron-containing dendrimers to anti-EGF receptor monoclonal antibody Cetuximab (IMC-C225) and its evaluation as a potential delivery agent for neutron capture therapy. *Bioconjugate Chem.*, **2004**, *15*, 185–194; (b) R. F. Barth, G. Wu, W. Yang, P. J. Binns, K. J. Riley, H. Patel, J. A. Coderre, W. Tjarks, A. K. Bandyopadhyaya, B. T. Thirumamagal, M. J. Ciesielski, R. A. Fenstermaker. Neutron capture therapy of epidermal growth factor (+) gliomas using boronated cetuximab (IMC-C225) as a delivery agent. *Appl. Radiat. Isot.*, **2004**, *61*, 899–903.
67. J. Capala, R. F. Barth, M. Bendayan, M. Lauzon, D. M. Adams, A. H. Soloway, R. A. Fenstermaker, J. Carlsson. Boronated epidermal growth factor as a potential targeting agent for boron neutron capture therapy of brain tumors. *Bioconjugate Chem.*, **1996**, *7*, 7–15.
68. W. Yang, R. F. Barth, G. Wu, A. K. Bandyopadhyaya, B. T. Thirumamagal, W. Tjarks, P. J. Binns et al. Boronated epidermal growth factor as a delivery agent for neutron capture therapy of EGF receptor positive gliomas. *Appl. Radiat. Isot.*, **2004**, *61*, 981–985.
69. M. V. Backer, T. I. Gayunutdinov, V. Patel, A. K. Bandyopadhyaya, B. T. Thirumamagal, W. Tjarks, R. F. Barth, K. Claffey, J. M. Backer. Vascular endothelial growth factor selectively targets boronated dendrimers to tumor vasculature. *Mol. Cancer Ther.*, **2005**, *4*, 1423–1429.
70. S. Shukla, G. Wu, M. Chatterjee, W. Yang, M. Sekido, L. A. Diop, R. Muller et al. Synthesis and biological evaluation of folate receptor-targeted boronated PAMAM dendrimers as potential agents for neutron capture therapy. *Bioconjugate Chem.*, **2003**, *14*, 158–167.
71. M. C. Parrott, E. B. Marchington, J. F. Valliant, A. Adronov. Synthesis and properties of carborane-functionalized aliphatic polyester dendrimers. *J. Am. Chem. Soc.*, **2005**, *127*, 12081–12089.
72. (a) R. Núñez, A. González, C. Viñas, F. Teixidor, R. Sillanpää, R. Kivekäs. Approaches to the preparation of carborane-containing carbosilane compounds. *Org. Lett.*, **2005**, *7*, 231–233; (b) R. Núñez, A. González-Campo, C. Viñas, F. Teixidor, R. Sillanpää, R. Kivekäs. Boron-functionalized carbosilanes: Insertion of carborane clusters into peripheral silicon atoms of carbosilane compounds. *Organometallics*, **2005**, *24*, 6351–6357; (c) A. González-Campo, C. Viñas, F. Teixidor, R. Núñez, R. Sillanpää, R. Kivekäs. Modular construction of neutral and anionic carboranyl-containing carbosilane-based dendrimers. *Macromolecules*, **2007**, *40*, 5644–5652.
73. (a) J. Thomas, M. F. Hawthorne. Dodeca(carboranyl)-substituted closomers: toward unimolecular nanoparticles as delivery vehicles for BNCT. *Chem. Commun.*, **2001**, 1884–1885; (b) L. Ma, J. Hamdi, F. Wong, M. F. Hawthorne. Closomers of high boron content: Synthesis, characterization, and potential application as unimolecular nanoparticle delivery vehicles for boron neutron capture therapy. *Inorg. Chem.*, **2006**, *45*, 278–285.
74. Ya. Z. Voloshin, O. A. Varsatskii, Yu. N. Bubnov. Cage complexes of transition metals in biochemistry and medicine. *Russ. Chem. Bull.*, **2007**, *56*, 577–605.
75. A. Holmberg, L. Meurling. Preparation of sulfhydrylborane-dextran conjugates for boron neutron capture therapy. *Bioconjugate Chem.*, **1993**, *4*, 570–573.
76. (a) J. Carlsson, L. Gedda, C. Gronvik, T. Hartman, A. Lindström, P. Lindström, H. Lundqvist et al. Strategy for boron neutron capture therapy against tumor cells with over-expression of the epidermal growth factor-receptor. *Int. J.Radiat. Oncol. Biol. Phys.*, **1994**, *30*, 105–115; (b) L. Gedda, P. Ollson, J. Ponten, J. Carlsson. Development and *in vitro* studies of epidermal growth factor – dextran conjugates for boron neutron capture therapy. *Bioconjugate Chem.*, **1996**, *7*, 574–591; (c) P. Olsson, L. Gedda, H. Goike, L. Liu, V. P. Collins, J. Ponten, J. Carlsson. Uptake of a boronated epidermal growth factor-dextran conjugate in CHO xenografts with and without human EGF-receptor expression. *Anticancer Drug Des.*, **1998**, *13*, 279–289.
77. S. C. Mehta, D. R. Lu. Targeted drug delivery for boron neutron capture therapy. *Pharm. Res.*, **1996**, *13*, 344–51.
78. T. Sano. Boron-enriched streptavidin potentially useful as a component of boron carriers for neutron capture therapy of cancer. *Bioconjugate Chem.*, **1999**, *10*, 905–911.
79. F. J. Primus, R. H. Pak, K. J. Rickard-Dickson, G. Szalai, J. L. Bolen, R. R. Kane, M. F. Hawthorne. Bispecific antibody mediated targeting of *nido*-carboranes to human colon carcinoma cells. *Bioconjugate Chem.*, **1996**, *7*, 532–535.

80. S. B. Kahl. Comparison of three methods for the synthesis of carborane carboxylic acid esters. *Tetrahedron Lett.*, **1990**, *31*, 1517–1520.
81. D. A. Feakes, J. K. Spinler, F. R. Harris. Synthesis of boron-containing cholesterol derivatives for incorporation into unilamellar liposomes and evaluation as potential agents for BNCT. *Tetrahedron*, **1999**, *55*, 11177–11186.
82. (a) B. Ji, G. Peacock, D. R. Lu. Synthesis of cholesterol–carborane conjugate for targeted drug delivery. *Bioorg. Med. Chem. Lett.*, **2002**, *12*, 2455–2458; (b) F. Alanazi, H. Li, D. S. Halpern, S. Ǿie, D. R. Lu. Synthesis, preformulation and liposomal formulation of cholesteryl carborane esters with various fatty chains. *Int. J. Pharm.*, **2003**, *255*, 189–197; (c) G. F. Peacock, B. Ji, C. K. Wang, D. R. Lu. Cell culture studies of a carborane cholesteryl ester with conventional and PEG liposomes. *Drug. Deliv.*, **2003**, *10*, 29–34; (d) G. Peacock, R. Sidwell, G. Pan, S. Ǿie, D. R. Lu. *In vitro* uptake of a new cholesteryl carborane ester compound by human glioma cell lines. *J. Pharm. Sci.*, **2004**, *93*, 13–19; (e) G. Pan, S. Ǿie, D. R. Lu. Uptake of the carborane derivative of cholesteryl ester by glioma cancer cells is mediated through LDL receptors. *Pharm. Res.*, **2005**, *21*, 1257–1262.
83. P. Dozzo, M.-S. Koo, S. Berger, T. M. Forte, S. B. Kahl. Synthesis, characterization, and plasma lipoprotein association of a nucleus-targeted boronated porphyrin. *J. Med. Chem.*, **2005**, *48*, 357–359.
84. (a) K. Maruyama, O. Ishida, S. Kasaoka, T. Takizawa, N. Utoguchi, A. Shinohara, M. Chiba, H. Kobayashi, M. Eriguchi, H. Yanagie. Intracellular targeting of sodium mercaptoundecahydrododecaborate (BSH) to solid tumors by transferrin-PEG liposomes for boron neutron-capture therapy (BNCT). *J. Control Rel.* **2004**, *98*, 195–207; (b) Y. Miyajima, H. Nakamura, Y. Kuwata, J.-D. Lee, S. Masunaga, K. Ono, K. Maruyama. Transferrin-loaded *nido*-carborane liposomes:Tumor-targeting boron delivery system for neutron capture therapy. *Bioconjugate Chem.*, **2006**, *17*, 1314–1320.
85. (a) E. Bohl Kullberg, N. Bergstrand, J. Carlsson, K. Edwards, M. Johnsson, S. Sjöberg, L. Gedda. Development of EGF-conjugated liposomes for targeted delivery of boronated DNA-binding agents. *Bioconjugate Chem.*, **2002**, *13*, 737–743; (b) E. Bohl Kullberg, M. Nestor, L. Gedda. Tumor-cell targeted epidermal growth factor liposomes loaded with boronated acridine: Uptake and processing. *Pharm. Res.*, **2003**, *20*, 229–236; (c) E. Bohl Kullberg, J. Carlsson, K. Edwards, J. Capala, S. Sjöberg, L. Gedda. Introductory experiments on ligand liposomes as delivery agents for boron neutron capture therapy. *Int. J. Oncol.*, **2003**, *23*, 461–467; (d) E. Bohl Kullberg, Q. Wei, J. Capala, V. Giusti, P.-U. Malmström, L. Gedda. EGF-receptor targeted liposomes with boronated acridine: Growth inhibition of cultured glioma cells after neutron irradiation. *Int. J. Radiat. Biol.*, **2005**, *81*, 621–629.
86. Q. Wei, E. Bohl Kullberg, L. Gedda. Trastuzumab-conjugated boron-containing liposomes for tumor-cell targeting; development and cellular studies. *Int. J. Oncol.*, **2003**, *23*, 1159–1165.
87. B. T. S. Thirumamagal, X. B. Zhao, A. K. Bandyopadhyaya, S. Narayanasamy, J. Johnsamuel, R. Tiwari, D. W. Golightly et al. Receptor-targeted liposomal delivery of boron-containing cholesterol mimics for boron neutron capture therapy (BNCT). *Bioconjugate Chem.*, **2006**, *17*, 1141–1150.
88. G. C. Krijger, M. M. Fretz, U. D. Woroniecka, O. M. Steinebach, W. Jiskoot, G. Storm, G. A. Koning. Tumor cell and tumor vasculature targeted liposomes for neutron capture therapy. *Radiochim. Acta*, **2005**, *93*, 589–593.
89. (a) J. J. Sudimack, D. Adams, J. Rotaru, S. Shukla, J. Yan, M. Sekido, R. F. Barth, W. Tjarks, R. J. Lee. Folate receptor-mediated liposomal delivery of a lipophilic boron agent to tumor cells *in vitro* for neutron capture therapy. *Pharm. Res.*, **2002**, *19*, 1502–1508; (b) X. Q. Pan, H. Wang, S. Shukla, M. Sekido, D. M. Adams, W. Tjarks, R. F. Barth, R. J. Lee. Boron-containing folate receptor-targeted liposomes as potential delivery agents for neutron capture therapy. *Bioconjugate Chem.*, **2002**, *13*, 435–442; (c) X. Q. Pan, H. Wang, R. J. Lee. Boron delivery to a murine lung carcinoma using folate receptor-targeted liposomes. *Anticancer Res.*, **2002**, *22*, 1629–1633; (d) S. M. Stephenson, W. Yang, P. J. Stevens, W. Tjarks, R. F. Barth, R. J. Lee. Folate receptor-targeted liposomes as possible delivery vehicles for boron neutron capture therapy. *Anticancer Res.*, **2003**, *23*, 3341–3345.
90. (a) K. Shelly, D. A. Feakes, M. F. Hawthorne, P. G. Schmidt, T. A. Krisch, W. F. Bauer. Model studies directed toward the boron neutron-capture therapy of cancer: boron delivery to murine tumors with liposomes. *Proc. Natl. Acad.Sci. USA*, **1992**, *89*, 9039–9043; (b) D. A. Feakes, K. Shelly, C. B. Knobler, M. F. Hawthorne. $Na_3[B_{20}H_{17}NH3]$: Synthesis and liposomal delivery to murine tumors. *Proc. Natl. Acad. Sci. USA*, **1994**, *91*, 3029–3033.
91. S. C. Mehta, J. C. K. Lai, D. R. Lu. Liposomal formulations containing sodium mercaptoundecahydrododecaborate (BSH) for boron neutron capture therapy. *J. Microencapsul.*, **1996**, *13*, 269–279.
92. J. Carlsson, E. Bohl Kullberg, J. Capala, S. Sjöberg, K. Edwards, L. Gedda. Ligand liposomes and boron neutron capture therapy. *J. Neurooncol.*, **2003**, *62*, 47–59.

93. S. Ristori, J. Oberdisse, I. Grillo, A. Donati, O. Spalla. Structural characterization of cationic liposomes loaded with sugar-based carboranes. *Biophys. J.*, **2005**, *88*, 535–547.
94. S. Rossi; R. F. Schinazi, G. Martini. ESR as a valuable tool for the investigation of the dynamics of EPC and EPC/cholesterol liposomes containing a carboranyl-nucleoside intended for BNCT. *Biochim. Biophys. Acta*, **2005**, *1712*, 81–91.
95. D. A. Feakes, K. Shelly, M. F. Hawthorne. Selective boron delivery to murine tumors by lipophilic species incorporated in the membranes of unilamellar liposomes. *Proc. Natl. Acad. Sci. USA*, **1995**, *92*, 1367–1370.
96. H. Nakamura, Y. Miyajima, T. Takei, S. Kasaoka, K. Maruyama. Synthesis and vesicle formation of a *nido*-carborane cluster lipid for boron neutron capture therapy. *Chem. Commun.*, **2004**, 1910–1911.
97. T. Li, J. Hamdi, M. F. Hawthorne. Unilamellar liposomes with enhanced boron content. *Bioconjugate Chem.*, **2006**, *17*, 15–20.
98. (a) J.-D. Lee, M. Ueno, Y. Miyajima, H. Nakamura. Synthesis of boron cluster lipids: *closo*-dodecaborate as an alternative hydrophilic function of boronated liposomes for neutron capture therapy. *Org. Lett.*, **2007**, *9*, 323–326; (b) H. Nakamura, J.-D. Lee, M. Ueno, Y. Miyajima, H. S. Ban. Synthesis of *closo*-dodecaboryl lipids and their liposomal formation for boron neutron capture therapy. *Nanobiotechnol.*, **2007**, *3*, 135–145.
99. (a) E. Justus, D. Awad, M. Hohnholt, T. Schaffran, K. Edwards, G. Karlsson, L. Damian, D. Gabel, *Bioconjugate Chem.*, **2007**, *18*, 1287–1293; (b) T. Schaffran, F. Lissel, B. Samatanga, G. Karlsson, A. Burghardt, K. Edwards, M. Winterhalter, R. Peschka-Süss, R. Schubert, D. Gabel, *J. Organomet. Chem.*, **2009**, *694*, 1708–1712; (c) T. Schaffran, A. Burghardt, S. Barnert, R. Peschka-Süss, R. Schubert, M. Winterhalter, D. Gabel. Pyridinium lipids with the dodecaborate cluster as polar headgroup: Synthesis, characterization of the physical-chemical behavior, and toxicity in cell culture. *Bioconjugate Chem.*, **2009**, *20*, 2190.
100. H. Nakamura, M. Ueno, J.-D. Lee, H. S. Ban, E. Justus, P. Fan, D. Gabel. Synthesis of dodecaborate-conjugated cholesterols for efficient boron delivery in neutron capture therapy. *Tetrahedron Lett.*, **2007**, *48*, 3151–3154.
101. S. Altieri, M. Balzi, S. Bortolussi, P. Bruschi, L. Ciani, A. M. Clerici, P. Faraoni et al. Carborane derivatives loaded into liposomes as efficient delivery systems for boron neutron capture therapy. *J. Med. Chem.*, **2009**, *52*, 7829–7835.
102. M. E. El-Zaria, H. Nakamura. New strategy for synthesis of mercaptoundecahydrododecaborate derivatives via click chemistry: Possible boron carriers and visualization in cells for neutron capture therapy. *Inorg. Chem.*, **2009**, *48*, 11896–11902.
103. A. Semioshkin, J. Laskova, O. Zhidkova, I. Godovikov, Z. Starikova, V. Bregadze, D. Gabel. Synthesis and structure of novel *closo*-dodecaborate-based glycerols. *J. Organomet. Chem.*, **2010**, *695*, 370–374.
104. S. Altieri, M. Balzi, S. Bortolussi, P. Bruschi, L. Ciani, A. M. Glerici, P. Faraoni et al. Carborane derivatives loaded into liposomes as efficient delivery systems for boron neutron capture therapy. *J. Med. Chem.*, **2009**, *52*, 7829–7835.
105. B. Feng, K. Tomizawa, H. Michiue, S.-i. Miyatake, X.-J. Han, A. Fujimura, M. Seno, M. Kirihata, H. Matsui. Delivery of sodium borocaptate to glioma cells using immunoliposome conjugated with anti-EGFR antibodies by ZZ-His. *Biomaterials*, **2009**, *30*, 1746–1755.
106. M. Ueno, H. S. Ban, K. Nakai, R. Inomata, Y. Kaneda, A. Matsumura, H. Nakamura. Dodecaborate lipid liposomes as new vehicles for boron delivery system of neutron capture therapy. *Bioorg. Med. Chem.*, **2010**, *18*, 3059–3065.
107. H. Nakamura. Liposomal boron delivery for neutron capture therapy. In: *Methods in Enzymology* (Eds. M. I. Simon; B. R. Crane; A. Crane), Chapter 10, **2009**, Vol. 465, pp. 179–208.
108. M. G. H. Vicente. Porphyrin-based sensitizers in the detection and treatment of cancer: Recent progress. *Curr. Med. Chem. - Anti-Cancer Agents*, **2001**, *1*, 175–194.
109. (a) S. Bonneau, C. Vever-Bizet, P. Morliere, J.-C. Maziere, D. Brault. Equilibrium and kinetic studies of the interactions of a porphyrin with low-density lipoproteins. *Biophys. J.*, **2002**, *83*, 3470–3481; (b) S. Novick, B. Laster, M. R. Quastel. Positive cooperativity in the cellular uptake of a boronated porphyrin. *Int. J. Biochem. Cell Biol.*, **2006**, *38*, 1374–1381.
110. (a) R. C. Haushalter, R. W. Rudolph. *meso*-Tetracarboranylporphyrins. *J. Am. Chem. Soc.*, **1978**, *100*, 4628–4629; (b) R. C. Haushalter, W. M. Butler, R. W. Rudolph. The preparation and characterization of several meso-tetracarboranylporphyrins. *J. Am. Chem. Soc.*, **1981**, *103*, 2620–2627.
111. V. I. Bregadze, I. B. Sivaev, D. Gabel, D. Wöhrle. Polyhedral boron derivatives of porphyrins and phthalocyanines. *J. Porphyrins Phthalocyanines*, **2001**, *5*, 767–781.

112. R. P. Evstigneeva, A. V. Zaitsev, V. N. Luzgina, V. A. Ol'shevskaya, A. A. Shtil. Carboranylporphyrins for boron neutron capture therapy of cancer. *Curr. Med. Chem. - Anti-Cancer Agents.* **2003**, *3*, 383–392.
113. M. W. Renner, M. Miura, M. W. Easson, M. G. H. Vicente. Recent progress in the syntheses and biological evaluation of boronated porphyrins for boron neutron capture therapy. *Anti-Cancer Agents Med. Chem.*, **2006**, *6*, 145–157.
114. V. A. Ol'shevskaya, A. V. Zaytsev, A. N. Savchenko, A. A. Shtil, C. S. Cheong, V. N. Kalinin. Boronated porphyrins and chlorins as potential anticancer drugs. *Bull. Korean Chem. Soc.*, **2007**, *28*, 1910–1916.
115. M.-S. Koo, T. Ozawa, R. A. Santos, K. R. Lamborn, A. W. Bollen, D. F. Deen, S. B. Kahl. Synthesis and comparative toxicology of a series of polyhedral borane anion-substituted tetraphenyl porphyrins. *J. Med. Chem.*, **2007**, *50*, 820–827.
116. (a) E. Hao, M. G. H. Vivente, *Chem. Commun.* **2005**, 1306–1308; (b) E. Hao, T. J. Jensen, B. H. Courtney, M. G. H. Vicente. Synthesis and cellular studies of porphyrin–cobaltacarborane conjugates. *Bioconjugate Chem.*, **2005**, *16*, 1495–1502; (c) M. Sibrian-Vazquez, E. Hao, T. J. Jensen, M. G. H. Vicente. Enhanced cellular uptake with a cobaltacarborane—porphyrin—HIV-1 Tat 48 – 60 conjugate. *Bioconjugate Chem.*, **2006**, *17*, 928–934; (d) E. Hao, M. Sibrian-Vazquez, W. Serem, J. C. Garno, F. R. Fronczek, M. G. H. Vicente. Synthesis, aggregation and cellular investigations of porphyrin–cobaltacarborane conjugates. *Chem. Eur. J.*, **2007**, *13*, 9035–9047; (e) E. Hao, M. Zhang, W. E, K. M. Kadish, F. R. Fronczek, B. H. Courtney, M. G. H. Vicente. Synthesis and spectroelectrochemistry of *N*-cobaltacarborane porphyrin conjugates. *Bioconjugate Chem.*, **2008**, *19*, 2171–2181.
117. J. S. Hill, S. B. Kahl, A. H. Kaye, S. S. Stylli, M. S. Koo, M. F. Gonzales, N. J. Verdaxis, C. I. Johnson. Selective tumor uptake of a boronated porphyrin in an animal model of cerebral glioma. *Proc. Natl. Acad. Sci. USA*, **1992**, *89*, 1785–1789.
118. C. P. Ceberg, A. Brun, S. B. Kahl, M. S. Koo, B. R. R. Persson, L. G. Salford. A comparative study on the pharmacokinetics and biodistribution of boronated porphyrin (BOPP) and sulfhydryl boron hydride (BSH) in the RG2 rat glioma model. *J. Neurosurg.*, **1995**, *83*, 86–92.
119. J. S. Hill, S. B. Kahl, S. S. Stylli, Y. Nakamura, M.-S. Koo, A. H. Kaye. Selective tumor kill of cerebral glioma by photodynamic therapy using a boronated porphyrin photosensitizer. *Proc. Natl. Acad. Sci. USA*, **1995**, *92*, 12126–12130.
120. J. Tibbitts, J. R. Fike, K. R. Lamborn, A. W. Bollen, S. B. Kahl. Toxicology of a boronated porphyrin in dogs. *Photochem., Photobiol.* **1999**, *69*, 587–594.
121. J. Tibbitts, N. C. Sambol, J. R. Fike, W. F. Bauer, S. B. Kahl. Plasma pharmacokinetics and tissue biodistribution of boron following administration of a boronated porphyrin in dogs. *J. Pharm. Sci.*, **2000**, *89*, 469–477.
122. T. Ozawa, J. Afzal, K. R. Lamborn, A. W. Bollen, W. F. Bauer, M.-S. Koo, S. B. Kahl, D. F. Deen. Toxicity, biodistribution, and convection-enhanced delivery of the boronated porphyrin BOPP in the 9L intracerebral rat glioma model. *Int. J.Rad. Oncol. Biol. Phys.*, **2005**, *63*, 247–252.
123. M. Ratajski, J. Osterloh, D. Gabel. Boron-containing chlorins and tetraazaporphyrins: Synthesis and cell uptake of boronated pyropheophorbide A derivatives. *Anti-Cancer Agents Med. Chem.*, **2006**, *6*, 159–166.
124. (a) V. I. Bregadze, I. B. Sivaev, I. A. Lobanova, R. A. Titeev, D. I. Brittal, M. A. Grin, A. F. Mironov. Conjugates of boron clusters with derivatives of natural chlorin and bacteriochlorin. *Appl. Radiat. Isotop.*, **2009**, *67*, S101–S104; (b) V. I. Bregadze, A. A. Semioshkin, J. N. Las'kova, M. Ya. Berzina, I. A. Lobanova, I. B. Sivaev, M. A. Grin et al. Novel types of boronated chlorine *e*6 conjugates via "Click Chemistry". *Appl. Organomet. Chem.*, **2009**, *23*, 370–374.; (c) M. A. Grin, R. A. Titeev, D. I. Brittal, A. V. Chestnova, A. V. Feofanov, A. F. Mironov, I. A. Lobanova, I. B. Sivaev, V. I. Bregadze. Synthesis of cobalt bis(dicarbollide) conjugates with natural chlorins by the Sonogashira reaction. *Russ. Chem. Bull.*, **2010**, *59*, 219–224.
125. V. A. Ol'shevskaya, R. G. Nikitina, A. N. Savchenko, M. V. Malshakova, A. M. Vinogradov, G. V. Golovina, D. V. Belykh et al. Novel boronated chlorin e_6-based photosensitizers: Synthesis, binding to albumin and antitumour efficacy. *Bioorg. Med. Chem.*, **2009**, *17*, 1297–1306.
126. (a) M. A. Grin, A. A. Semioshkin, R. A. Titeev, E. A. Nizhnik, J. N. Grebenyuk, A. F. Mironov, V. I. Bregadze. Synthesis of a cycloimide bacteriochlorin *p* conjugate with the *closo*-dodecaborate anion. *Mendeleev Commun.*, **2007**, *17*, 14–15; (b) M. A. Grin, R. A. Titeev, O. M. Bakieva, D. I. Brittal, I. A. Lobanova, I. B. Sivaev, V. I. Bregadze, A. F. Mironov. New boron_containing bacteriochlorin *p* cycloimide conjugate. *Russ. Chem. Bull.*, **2008**, *57*, 2230–2232.
127. F. Giuntini, Y. Raoul, D. Dei, M. Municchi, G. Chiti, C. Fabris, P. Colautti, G. Jori, G. Roncucci. Synthesis of tetrasubstituted Zn(II)-phthalocyanines carrying four carboranyl-units as potential BNCT and PDT agents. *Tetrahedron Lett.*, **2005**, *46*, 2979–2982.

128. O. Tsaryova, A. Semioshkin, D. Wöhrle, V. I. Bregadze. Synthesis of new carborane-based phthalocyanines and study of their activities in the photo-oxidation of citronellol. *J. Porphyrins Phthalocyanines*, **2005**, *9*, 268–274.
129. A. Semioshkin, O. Tsaryova, O. Zhidkova, V. Bregadze, D. Wöhrle. Reactions of oxoniumderivatives of $[B_{12}H_{12}]^{2-}$ with phenols, and synthesis and photochemical properties of a phthalocyanine containing four $[B_{12}H_{12}]^{2-}$ groups. *J. Porphyrins Phthalocyanines*, **2006**, *10*, 1293–1300.
130. H. Li, F. R. Fronczek, M. G. H. Vicente. Synthesis and properties of cobaltacarborane-functionalized Zn(II)-phthalocyanines. *Tetrahedron Lett.*, **2008**, *49*, 4838–4830.
131. H. Li, F. R. Fronczek, M. G. H. Vicente. Cobaltacarborane–phthalocyanine conjugates: Syntheses and photophysical properties. *J. Organomet. Chem.*, **2009**, *694*, 1607–1611.
132. (a) J. L. Maurer, A. J. Serino, M. F. Hawthorne. Hydrophilically augmented glycosyl carborane derivatives for incorporation in antibody conjugation reagents. *Organometallics*, **1988**, *7*, 2519–2524; (b) J. L. Maurer, F. Berchier, A. J. Serino, C. B. Knobler, M. F. Hawthorne. Glycosylcarborane derivatives and the determination of the absolute configuration of a diastereomeric triol from X-ray diffraction. *J. Org. Chem.*, **1990**, *57*, 838–843.
133. W. Tjarks, A. K. M. Anisuzzaman, L. Liu, A. H. Soloway, R. F. Barth, D. J. Perkins, D. M. Adam. Synthesis and *in vitro* evaluation of boronated uridine and glucose derivatives for boron neutron capture therapy. *J. Med. Chem.*, **1992**, *35*, 1628–1633.
134. W. V. Dahlhoff, J. Bruckmann, K. Angermund, C. Krüger. 1-[1,7-dicarba-*closo*-dodecaboran(12)-1-yl] aldoses: Novel boron neutron capture candidates. *Liebigs Ann. Chem.*, **1993**, 831–835.
135. N. Yamazaki, S. Kojima, N. V. Bovin, S. Andre, S. Gabius, H.-J. Gabius. Endogenous lectins as targets for drug delivery. *Adv. Drug Deliv. Rev.*, **2000**, *43*, 225–244.
136. (a) L. F. Tietze, U. Bothe. *Ortho*-Carboranyl glycosides of glucose, mannose, maltose and lactose for cancer treatment by boron neutron-capture therapy. *Chem. Eur. J.*, **1998**, *4*, 1179–1183; (b) L. F. Tietze, U. Bothe, I. Schuberth. Preparation of a new carboranyl lactoside for the treatment of cancer by boron neutron capture therapy: Synthesis and toxicity of fluoro carboranyl glycosides for *in vivo* ^{19}F-NMR spectroscopy. *Chem. Eur. J.*, **2000**, *6*, 836–842; (c) L. F. Tietze, U. Bothe, U. Griesbach, M. Nakaichi, T. Hasegawa, H. Nakamura, Y. Yamamoto, *Bioorg. Med. Chem.* **2001**, *9*, 1747–1752; (c) L. F. Tietze, U. Bothe, U. Griesbach, M. Nakaichi, T. Hasegawa, H. Nakamura, Y. Yamamoto. Carboranyl bisglycosides for the treatment of cancer by boron neutron capture therapy. *ChemBioChem*, **2001**, *2*, 326–334; (d) L. F. Tietze, U. Griebach, I. Schuberth, U. Bothe, A. Marra, A. Dondoni. Novel carboranyl *C*-glycosides for the treatment of cancer by boron neutron capture therapy. *Chem. Eur. J.*, **2003**, *9*, 1296–1302.
137. (a) A. V. Orlova, A. I. Zinin, N. N. Malysheva, L. O. Kononov, I. B. Sivaev, V. I. Bregadze. Conjugates of polyhedral boron compounds with carbohydrates. 1. New approach to the design of selective agents for boron neutron capture therapy of cancer. *Russ. Chem. Bull.*, **2003**, *52*, 2766–2768; (b) L. O. Kononov, A. V. Orlova, A. I. Zinin, B. G. Kimel, I. B. Sivaev, V. I. Bregadze. Conjugates of polyhedral boron compounds with carbohydrates. 2. Unexpected easy *closo*- to *nido*-transformation of a carborane–carbohydrate conjugate in neutral aqueous solution. *J. Organomet. Chem.*, **2005**, *690*, 2769–2774; (c) A. V. Orlova, N. N. Kondakov, A. I. Zinin, B. G. Kimel, L. O. Kononov, I. B. Sivaev, V. I. Bregadze. Conjugates of polyhedral boron compounds with carbohydrates. 3. The first synthesis of a conjugate of the dodecaborate anion with a disaccharide lactose as a potential agent for boron neutron capture therapy of cancer. *Russ. Chem. Bull.*, **2005**, *54*, 1352–1353; (d) A. V. Orlova, L. O. Kononov, B. G. Kimel, I. B. Sivaev, V. I. Bregadze. Conjugates of polyhedral boron compounds with carbohydrates. 4. Hydrolytic stability of carborane–lactose conjugates depends on the structure of a spacer between the carborane cage and sugar moiety. *Appl. Organomet. Chem.*, **2006**, *20*, 416–420; (e) A. V. Orlova, N. N. Kondakov, A. I. Zinin, B. G. Kimel, L. O. Kononov, I. B. Sivaev, V. I. Bregadze. A universal approach to the synthesis of carbohydrate conjugates of polyhedral boron compounds as potential agents for boron neutron capture therapy. *Russ. J. Bioorg. Chem.*, **2006**, *32*, 568–577; (f) L. M. Likhosherstov, O. S. Novikova, L. O. Kononov, A. V. Orlova, I. B. Sivaev, V. I. Bregadze. Conjugates of polyhedral boron compounds with carbohydrates. 5. Synthesis of glycoconjugates of *closo-ortho*-carborane and *N*-acyl-β-lactosylamines with various spacers. *Russ. Chem. Bull.*, **2007**, *56*, 2105–2108; (g) L. M. Likhosherstov, O. S. Novikova, L. O. Kononov, I. B. Sivaev, V. I. Bregadze. Conjugates of polyhedral boron compounds with carbohydrates. 6. Synthesis of glycoconjugates of *closo-ortho*-carborane with β-lactosylamine and β-D-galactopyranosylamine derivatives as galectin bi- and trivalent ligands. *Russ. Chem. Bull.*, **2009**, *58*, 446–449.
138. (a) G. B. Giovenzana, L. Lay, D. Monty, G. Palmisano, L. Panza. Synthesis of carboranyl derivatives of alkynyl glycosides as potential BNCT agents. *Tetrahedron*, **1999**, *55*, 14123–14136; (b) S. Ronchi, D. Prosperi, C. Thimon, C. Morin, L. Panza. Synthesis of mono- and bisglucuronylated carboranes.

Tetrahedron Asym., **2005**, *16*, 39–44; (c) C. Di Meo, L. Panza, D. Capitani, L. Mannina, A. Banzato, M. Rondina, D. Renier, A. Rosato, V. Crescenzi. Hyaluronan as carrier of carboranes for tumor targeting in boron neutron capture therapy. *Biomacromolecules*, **2007**, *8*, 552–559; (d) C. Di Meo, L. Panza, F. Campo, D. Capitani, L. Mannina, A. Banzato, M. Rondina, A. Rosato, V. Crescenzi. Novel types of carborane-carrier hyaluronan derivatives *via* "click chemistry". *Macromol. Biosci.*, **2008**, *8*, 670–681.

139. P. Basak, T. L. Lowary. Synthesis of conjugates of *L*-fucose and *ortho*-carborane as potential agents for boron neutron capture therapy. *Can. J. Chem.*, **2002**, *80*, 943–948.
140. B. Leichtenberg, D. Gabel. Synthesis of a $(B_{12}H_{11}S)^{2-}$ containing glucuronoside as potential prodrug for BNCT. *J. Organomet. Chem.*, **2005**, *690*, 2780–2782.
141. A. V. Orlova, L. O. Kononov. Synthesis of conjugates of polyhedral boron compounds with carbohydrates. *Russ. Chem. Rev.*, **2009**, *78*, 629–642.
142. Zhu Yinghuai, Koh Cheng Yana, J. A. Maguire, N. S. Hosmane. Recent developments in boron neutron capture therapy (BNCT) driven by nanotechnology. *Curr. Chem. Biol.*, **2007**, *1*, 141–149.
143. N. S. Hosmane, Zhu Yinghuai, J. A. Maguire, W. Kaim, M. Takagaki. Nano and dendritic structured carboranes and metallacarboranes: From materials to cancer therapy. *J. Organomet. Chem.*, **2009**, *694*, 1690–1697.
144. W. Kaim, N. S. Hosmane. Multidimensional potential of boron-containing molecules in functional materials. *J. Chem. Sci.*, **2010**, *122*, 7–18.
145. K. Jiang, A. Loni, L. T. Canham, J. L. Coffer. Incorporation and characterization of boron neutron capture therapy agents into mesoporous silicon and silicon nanowires. *Phys. Status Solidi A*, **2009**, *206*, 1361–1364.
146. K. Jiang, J. L. Coffer, J. G. Gillen, T. M. Brewer. Incorporation of cesium borocaptate onto silicon nanowires as a delivery vehicle for boron neutron capture therapy. *Chem. Mater.*, **2010**, *22*, 279–281.
147. D. C. Kennedy, D. R. Duguay, Li-Lin Tay, D. S. Richeson, J. P. Pezacki. SERS detection and boron delivery to cancer cells using carborane labelled nanoparticles. *Chem. Commun.*, **2009**, 6750–6752.
148. A. R. Genady. Promising carboranylquinazolines for boron neutron capture therapy: Synthesis, characterization, and *in vitro* toxicity evaluation. *Eur. J. Med. Chem.*, **2009**, *44*, 409–416.
149. C.-H. Lee, G. F. Jin, J. G. Joung, J.-D. Lee, H. S. Ban, H. Nakamura, J.-K. Cho, O. Kang. New types of potential BNCT agents, *o*-carboranyl aminoalcohols. *Tetrahedron Lett.*, **2009**, *50*, 2960–2963.
150. S. C. Jonnalagadda, J. S. Cruz, R. J. Connell, P. M. Scott, V. R. Mereddy. Synthesis of α-carboranyl-α-acyloxy-amides as potential BNCT agents. *Tetrahedron Lett.*, **2009**, *50*, 4314–4317.
151. S. Staudlbauer, P. Welzel, E. Hey-Hawkins. Access to carbaboranyl glycophosphonates: An Odyssey. *Inorg. Chem.*, **2009**, *48*, 5005–5010.
152. V. Šícha, P. Farràs, B. Štíbr, F. Teixidor, B. Grüner, C. Viñas. Syntheses of C-substituted icosahedral dicarbaboranes bearing the 8-dioxane-cobalt bisdicarbollide moiety. *J. Organomet. Chem.*, **2009**, *694*, 1599–1601.
153. J. Yoo, Y. Do. Synthesis of stable platinum complexes containing carborane in a carrier group for potential BNCT agents. *Dalton Trans.*, **2009**, 4978–4986.

10 Boron Tumor Delivery for BNCT

Recent Developments and Perspectives

Martha Sibrian-Vazquez and Maria da Graça H. Vicente

CONTENTS

10.1 INTRODUCTION

10.1.1 The Neutron-Capture Reaction

BNCT is based on the very high cross section of boron-10 nuclei for neutron capture, which is followed by the spontaneous fission of the resulting excited boron-11 nuclei to produce high-linear energy transfer (high-LET) alpha-particles and recoiling lithium-7 nuclei according to the equation:[1–4]

$$^{10}B + {}^{1}n \rightarrow {}^{7}\mathrm{Li}^{3+} + {}^{4}\mathrm{He}^{2+} + \gamma + 2.4\ \mathrm{MeV}$$

The high-LET particles have only limited path lengths in tissue (less than 10 μm) and therefore BNCT has the potential to be highly localized to ^{10}B-containing tumor cells in the presence of normal boron-free cells. However, one of the major challenges in BNCT development has been the discovery of tumor-selective boron agents with the ability to deliver therapeutic boron concentrations (15–30 μg/g tumor) to targeted tumors with minimal normal tissue toxicity.[5] The two boron compounds currently used clinically for BNCT of malignant brain tumors, melanomas, and squamous cell carcinomas are the sodium mercaptoundecahydro-*closo*-dodecaborate (**1**) designated

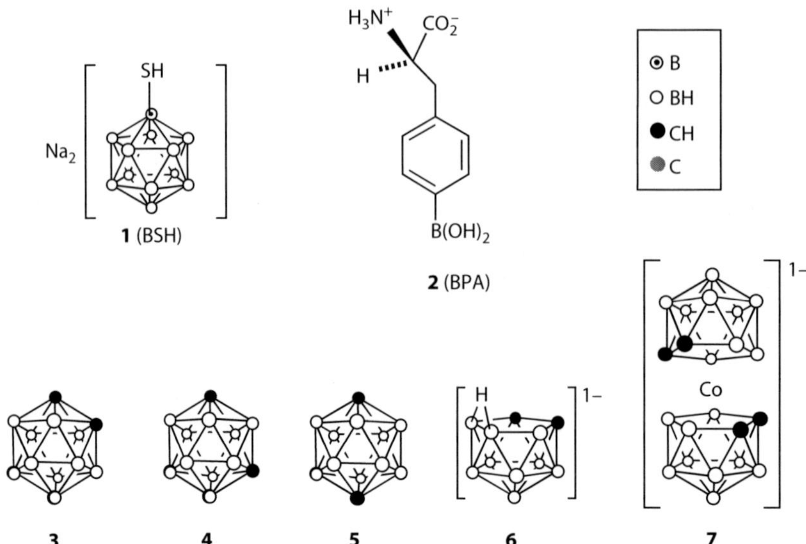

FIGURE 10.1 Structures of BSH, BPA (clinically approved for use in BNCT) and common boron clusters currently used in BNCT drug development. The cluster atom representations are used throughout this chapter.

BSH, and the amino acid (*L*)-4-dihydroxy-borylphenylalanine (**2**), known as BPA (Figure 10.1).[5–11] These were selected in the 1960s among a large number of boronated small molecules tested, based on their low toxicity and recorded effectiveness as boron delivery agents for BNCT, although both have only moderate selectivity and low retention times in tumors.[5] Over the past 25 years, other classes of boron-containing compounds, both low- and high-molecular weight, including boronated amino acids, peptides, nucleosides, porphyrin derivatives, lipids, and monoclonal antibodies (MAb) have emerged. Among these, boronated nucleosides, porphyrin derivatives, and MAb-containing molecules are particularly promising boron carriers that could increase tumor specificity and boron retention times in diseased tissues. In addition, the use of fluorescent molecules can facilitate tumor detection as well as quantification of tumor-localized boron and treatment planning. Although hundreds of potential boron carriers for BNCT have been designed and synthesized in the last two decades, only a few have actually been assessed in *in vivo* biological investigations, and in part as a consequence of this no additional boron carriers are yet approved for use in the clinic. Detailed reviews of compound development for BNCT have been recently published;[12–15] this chapter summarizes the main classes of boron carriers and discusses strategies toward the development of efficient boron delivery agents for BNCT.

10.1.2 General Requirements and Strategies

An efficacious boron delivery agent should display the following characteristics: (1) low normal tissue uptake and toxicity, (2) high tumor uptake and high tumor-to-normal tissue and tumor-to-blood concentration ratios, (3) deliver boron tumor concentrations in the range or above 15–30 µg ^{10}B/g tumor, (4) rapid clearance from blood and normal tissues, and (5) persistence in tumor during the irradiation treatment. In addition, BNCT agents should be easy to quantify within tumors to facilitate treatment planning, and to be amphiphilic, to favor diffusion across biological barriers and retention within tumors. The overall molecular weight of a boronated agent is also of importance since it influences its biological rate of diffusion and tumor boron microdistribution.

The major challenges in compound development for BNCT have been the requirements for selective tumor targeting and the delivery of therapeutic boron concentrations to tumors with minimal normal tissue toxicity. The necessary tumor boron concentration for effective BNCT has been estimated to be between 15 and 30 μg/g tumor, depending on the microdistribution of the boron-10 atoms;[16,17] less boron is necessary if it localizes within tumor cells rather than extracellularly, and in particular near or within cell nuclei. In order to overcome these challenges, most recently developed boronated agents consist of stable boron cluster(s) conjugated to a tumor-targeting moiety, such as peptide or MAb. For example, the use of a cell-penetrating peptide,[18] and the targeting of the epidermal growth factor receptor (EGFR)[19] which is highly expressed in gliomas and squamous cell carcinomas of the head and neck,[20] are very promising approaches. In addition, the investigation of alternative routes and procedures for the *in vivo* administration of boronated agents to improve tumor uptake is also of crucial importance.[21] Drug penetration of the blood–brain barrier (BBB) constitutes a great challenge in glioma therapy, in particular when high amount of drug is needed within the tumor for efficient treatment, as in the case of BNCT. Convection-enhanced delivery (CED),[22] developed in the 1990s, appears to be a promising methodology for the delivery of high amounts of boronated agents to intracranial animal tumors, with very high tumor-to-brain and tumor-to-blood boron concentration ratios and very low systemic toxicity.[23–29] CED involves drug infusion under high pressure using intracranial catheters, and has been used to efficiently deliver a variety of drugs and toxins to brain tumors and surrounding parenchyma, in both preclinical and clinical trials.[22] Although CED has been shown to significantly enhance overall tumor boron uptake while reducing systemic toxicity, this technique does not necessarily increase boron tumor cell uptake that mainly depends on the chemical properties of the boronated agents and their mechanisms of tumor cell uptake. Besides intratumoral injection and CED, the intra-arterial (carotid) administration of BPA and/or BSH has been shown by Barth and coworkers to approximately double the amount of boron uptake into tumors (F98 rat glioma) compared with that obtained by IV administration;[30–32] this amount further increases when the BBB is disrupted, using either mannitol or cereport. Furthermore, the preloading of specific amino acids, for example, L-DOPA, has been shown to enhance (2.7-fold) the uptake of BPA in rat C6 glioma tumors, while no significant differences were observed in normal brain tissue.[33] The above studies demonstrate the importance of the optimization of the delivery protocol for boronated drugs.

As discussed below, the isomeric carboranes *ortho*-, *meta*-, and *para*-$C_2B_{10}H_{12}$ (**3–5**, respectively), and the open cage *nido*-$C_2B_9H_{12}^-$ (**6**), have been the clusters of choice for attachment to tumor-targeting molecules (mostly *via* their carbon atoms) because of their high boron content, known chemistry, amphiphilic properties, and their high photochemical, kinetic, and hydrolytic stabilities (Figure 10.1).[34,35] In addition, stable metallacarboranes, such as the Co(III) complex of bis(dicarbollide) $[3,3'\text{-Co}(1,2\text{-}C_2B_9H_{11})_2]^-$ designated cobaltacarborane (**7**),[36,37] have been recently explored as the source of boron in various classes of boronated agents. Furthermore, computational methods can assist in boronated-drug design for use in BNCT.[38]

10.2 BORON DELIVERY AGENTS

10.2.1 Derivatives of BSH and Other Boron Clusters

BSH contains a thiol group and the dodecahydro-*closo*-dodecaborate anion $[B_{12}H_{12}]^{2-}$ as the boron cluster; this cluster is easily prepared ^{10}B-enriched for use in BNCT, and it is readily mono-functionalized.[39] However, further functionalization of the boron cluster in BSH is challenging due to lack of chemo-regioselectivity. Therefore, most derivatives of BSH prepared to date involve either derivatization *via* the thiol group, or the replacement of the boron cage by a more hydrophobic carborane cluster or the polyhedral $[B_{22}H_{22}]^{2-}$ anion which has nearly double the boron content. Carboranes (see Figure 10.1) are easily functionalized at the C–H bonds using metallation reactions in the presence of an organometallic reagent, followed by reaction with various electrophiles. Alternatively, the *ortho*-carborane cluster can be prepared by reaction of alkynes with neutral decaborane ($B_{10}H_{14}$) in the presence of a weak Lewis base, such as CH_3CN, R_2S, or R_3N (R = alkyl group), and its *meta*- and *para*-isomers are obtained via thermal isomerization of *ortho*-carboranes.[34,35]

Porphyrin derivatives of BSH (Figure 10.2) have been prepared via the reaction of $[B_{12}H_{11}SH]^{2-}$ with porphyrin acyl chlorides,[40] or the reaction of amino-porphyrins with $[B_{12}H_{11}SCH_2CH_2COCl]^{2-}$.[41] Tetra anionic porphyrin **8** and a related BSH-derivative of protoporphyrin IX dimethyl ester were prepared using the first methodology, as well as an octa-anionic 5,10,15,20-tetraphenylporphyrin thioester derivative. Mn(III)-Porphyrin **9**, designated STA-BX909, was shown to deliver a higher

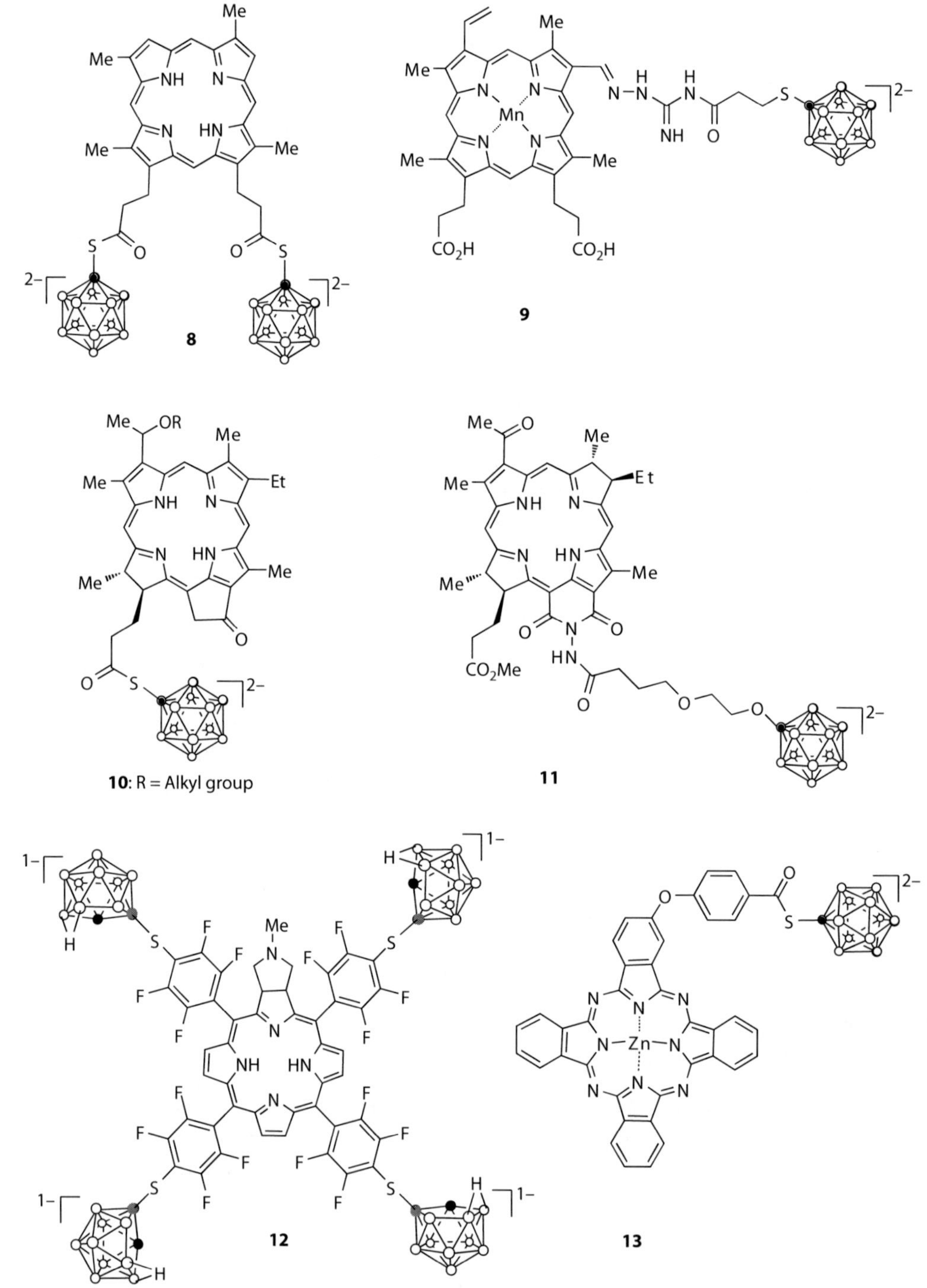

FIGURE 10.2 Porphyrin derivatives of BSH and related clusters.

amount of boron to 9L rat brain tumors than BSH, and with higher tumor-to-blood boron concentration ratio.[41]

Chlorins **10** (Figure 10.2) were prepared using a similar route, from methyl pyropheophorbide-a (a chlorophyll-a derivative) by functionalization of the vinyl group using various alcohols ROH (*R* = methyl, propyl, pentyl, heptyl, nonyl), followed by cleavage of the ester group, activation with oxalyl chloride, and reaction with the tetramethylammonium salt of $[B_{12}H_{11}SH]^{2-}$.[42,43] Chlorins **10** readily accumulated within V79 Chinese hamster cells, with the exception of the nonyl derivative, localized mainly in the mitochondria, and showed moderate cytotoxicity.[43] The synthesis of bacteriochlorin derivative **11** containing a *closo*-dodecaborate dianion has recently been reported, via the nucleophilic opening of the oxonium derivative of this cluster.[44] Another chlorin derivative **12** (Figure 10.2) was prepared from the nucleophilic substitution of the *para*-phenyl fluoride of a tetra(pentafluorophenyl)chlorin using 1-mercapto-*ortho*-carborane, rather than the more hydrophilic $[B_{12}H_{11}SH]^{2-}$ anion, followed by deboronation of the *ortho*-carborane cages with potassium fluoride.[45] This tetraanionic chlorin was found to be nontoxic in the dark, but showed extensive photosensitizing ability both *in vitro* and *in vivo*, toward melanotic melanomas.

Zn(II)-phthalocyanine **13** bearing one dodecaborate anion was prepared via the reaction of the sodium salt of $[B_{12}H_{11}SH]^{2-}$ with a carboxyl-activated phthalocyanine, in 71% overall yield.[46] Phthalocyanine **13** was found to be an efficient photosensitizer and to have a high quantum yield for singlet oxygen generation (~0.5 in DMF) on light activation. Another Zn(II)-phthalocyanine containing four $[B_{12}H_{12}]^{2-}$ cages has also been synthesized as a mixture of regioisomers, via nucleophilic opening of an oxonium derivative of dodecahydro-*closo*-dodecaborate, as in the case of **9**.[47]

The BPA currently used in BNCT clinical trials is complexed with fructose to increase its water solubility. Carbohydrate derivatives of BSH and carboranes, including glucose, mannose, ribose, gulose, fucose, galactose, maltose, and lactose derivatives have also been reported, and a few of these (**14**–**18**) are shown in Figure 10.3.[48–60] Carbohydrate-functionalized molecules usually have increased water solubility, and can show enhanced interaction with tumor cell receptors via carbohydrate-mediated cell recognition processes, and therefore increased tumor cell uptake. The *C*-glycosylated carboranes are potentially more robust toward enzymatic cleavage by glycohydrolases. Biological evaluations of these compounds have revealed low toxicity and often poor uptake by tumor cells, probably due to their high hydrophilicity and consequently poor ability for crossing cellular membranes. However, it has been reported that some of these compounds might selectively accumulate within the glycerophospholipid membrane bilayer, as well as in other areas of the tumor, including the vasculature. Some of these compounds were also prepared containing fluoride to allow for their detection in tumor cells by ^{19}F-NMR spectroscopy or ^{18}F-PET. Boronated polysaccharides, such as hyaluronan, have also been reported and shown to accumulate within tumor cells overexpressing certain polysacharide receptors, for example, CD44.[61]

Nitroimidazole derivatives of BSH and other clusters have also been reported, since the nitroimidazole unit has been shown to be selectively taken up and retained in poorly vascularized tumor areas.[62,63] Examples of such molecules are shown in Figure 10.4 (**19**, **20**). While compound **19** is dianionic and highly water soluble, **20** is only partially soluble in water and it was shown to readily accumulate within tumor cells.

BSH has also been coupled to a cyclic amino acid, a Tyr^3-octreotate derivative, which was shown to be a selective ligand for somatostatin receptors (see Section 10.2.2).[64,65]

Boronated lipids **21** (Figure 10.5) containing a double tail as the lipophilic part and a carborane head group carrying two negative charges have been synthesized.[66] These boronated lipids form stable liposomes with equimolar amounts of distearoyl phosphatidylcholine (DSPC) and cholesterol. While liposomes prepared from **21** (R=$C_{13}H_{27}$) were slightly less toxic toward V79 Chinese hamster cells (IC_{50} = 5.6 μM) than free $Na_2B_{12}H_{11}SH$ (IC_{50} = 3.9 μM), liposomes prepared from **21** (R=$C_{15}H_{31}$) were not toxic even at 30 μM concentration.

In addition, BSH has been incorporated into lipids **22**, which have a $B_{12}H_{11}S$-moiety as the hydrophilic function, and were designed to mimic phosphatidylcholines.[67–69] These boronated lipids were

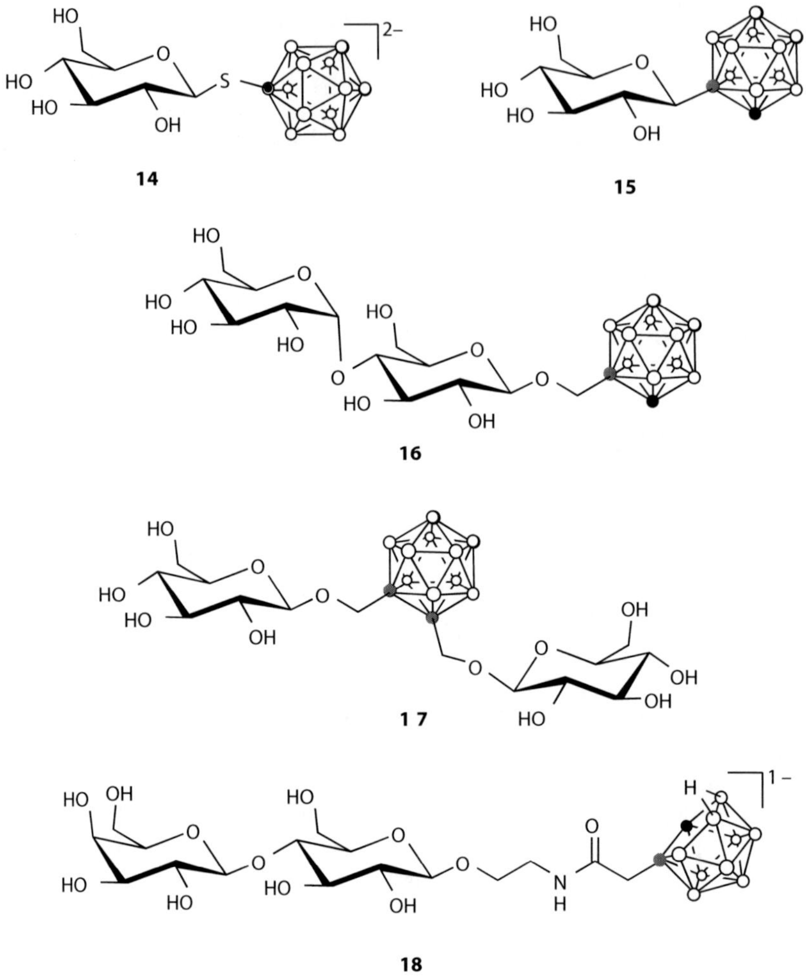

FIGURE 10.3 Carbohydrate derivatives of BSH and other clusters.

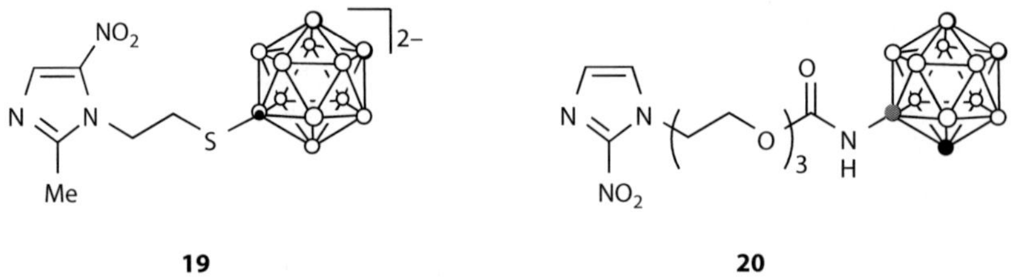

FIGURE 10.4 Nitroimidazole derivatives of BSH and other clusters.

obtained by S-alkylation of $B_{12}H_{11}SH$ using bromoacetyl and chloroacetocarbamate derivatives of the corresponding diacylglycerols. Recently, liposomes were prepared from lipid **22** (*R*=stearoyl), cholesterol, distearoylphosphatidylcholine (DSPC) and pegylated distearoylphosphatidylethanolamine (PEG-DSPE) via a reverse-phase evaporation methodology.[69] These liposomes were shown to deliver 22 ppm of boron to colon tumors in BALB/c mice, 24 h after tail vein injection of a 20 mg/kg boron dose.

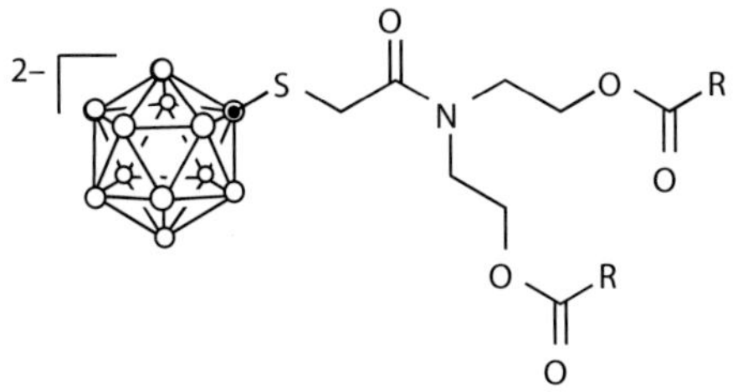

FIGURE 10.5 Lipid derivatives of BSH.

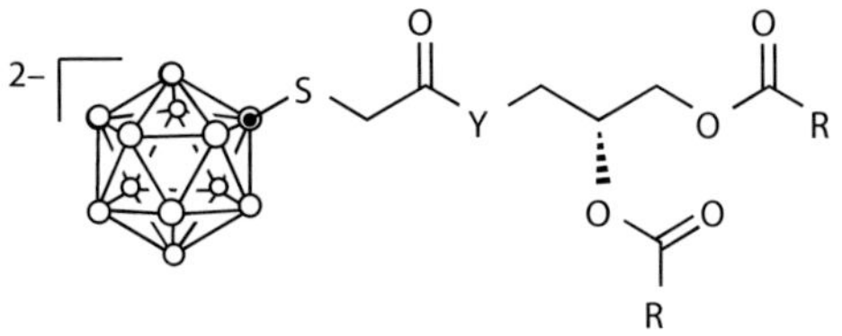

10.2.2 Amino Acids and Peptides

Derivatives of BPA and other boron-containing amino acids such as glycine, alanine, aspartic acid, tyrosine, cysteine, methionine, serine, as well as nonnaturally occurring amino acids, have been synthesized and some of these are shown in Figure 10.6 (**23**–**30**).[70–84] Amino acids have a tendency for selective accumulation within tumor cells, due to increased amino acid transport across plasma membranes of rapidly growing malignant cells, where they could be used for protein synthesis. Additionally, tumor cells might overexpress specific peptide receptors, such as the somatostatin receptor. Recently reported boronated amino acids contain one or more boron clusters, mainly carboranes due to their easy synthesis from the corresponding alkynyl derivatives, and concomitantly larger amounts of boron by weight compared with BPA. The advantages of such compounds are that they can potentially deliver higher concentrations of boron to target tumors without increased toxicity. However, due to the hydrophobicity of the dicarba-*closo*-dodecaboranes, some of the resulting amino acids were not water soluble and consequently hydrophilic substituents (such as hydroxyl groups) were introduced on these molecules; alternatively, the hydrophobic *closo*-clusters were converted into the amphiphilic and anionic open-face *nido*-carboranes. However, most carborane derivatives of BPA evaluated to date have shown lower ability for tumor cell targeting than BPA. On the other hand some unnatural boronated cyclic amino acids, such as boronated derivatives of 1-aminocyclobutane-1-carboxylic acid (e.g., **27** and **28**) and the 1-amino-3-boronocyclopentanecarboxylic acid **29** designated ABCPC, have shown promise for use in BNCT.[78–83] In particular, **29** was recently shown to have superior tumor selectivity both *in vitro* and *in vivo* compared with BPA.[83] In addition, the biological evaluation of the racemic mixtures of the *cis* and *trans* isomers of **29** in BALB/c mice bearing EMT-6 mammary carcinomas revealed a remarkable difference in the tumor-targeting ability of the two mixtures, indicating that further isomeric separation of ABCPC into the *D* and *L* forms might lead to enhanced tumor-targeting and BNCT efficacy. Furthermore, this type of unnatural amino acid shows significantly higher metabolic stability.

Since an efficient synthetic route to zwitterionic [3,3′-Co(8-$C_4H_8O_2$-1,2-$C_2B_9H_{10}$)(1′,2′-$C_2B_9H_{11}$)] was reported,[36,37] a variety of nucleophiles was shown to open the dioxane ring of this compound, and this methodology has allowed the preparation of a variety of cobaltacarborane-substituted organic molecules. Boronated amino acid **30** (Figure 10.6) was prepared using this methodology and shown to increase the killing effect of neutron radiation on melanoma B-16 tumor cells.[84]

Boronated peptides containing 2 or 3 amino acid residues, such as **31** (Figure 10.7) have shown low toxicity and good tumor-localizing properties, although little advantage over the use of BPA and other boronated amino acids for BNCT has been documented.[85–88] Additionally, boronated cyclic peptides **32** and its BSH analog, have recently been synthesized for application in BNCT, although no biological evaluation has yet been reported.[64,65] Such cyclic peptides are ligands for somatostatin receptors, some subtypes of which are overexpressed on tumor cells. Other peptide ligands have been designed and synthesized for the selective targeting of an overexpressed tumor cell receptor, such as the EGFR, or to increase tumor cell uptake, for example, a cell-penetrating peptide. Such strategy seems to be particularly promising for the selective delivery of high amounts of boron into

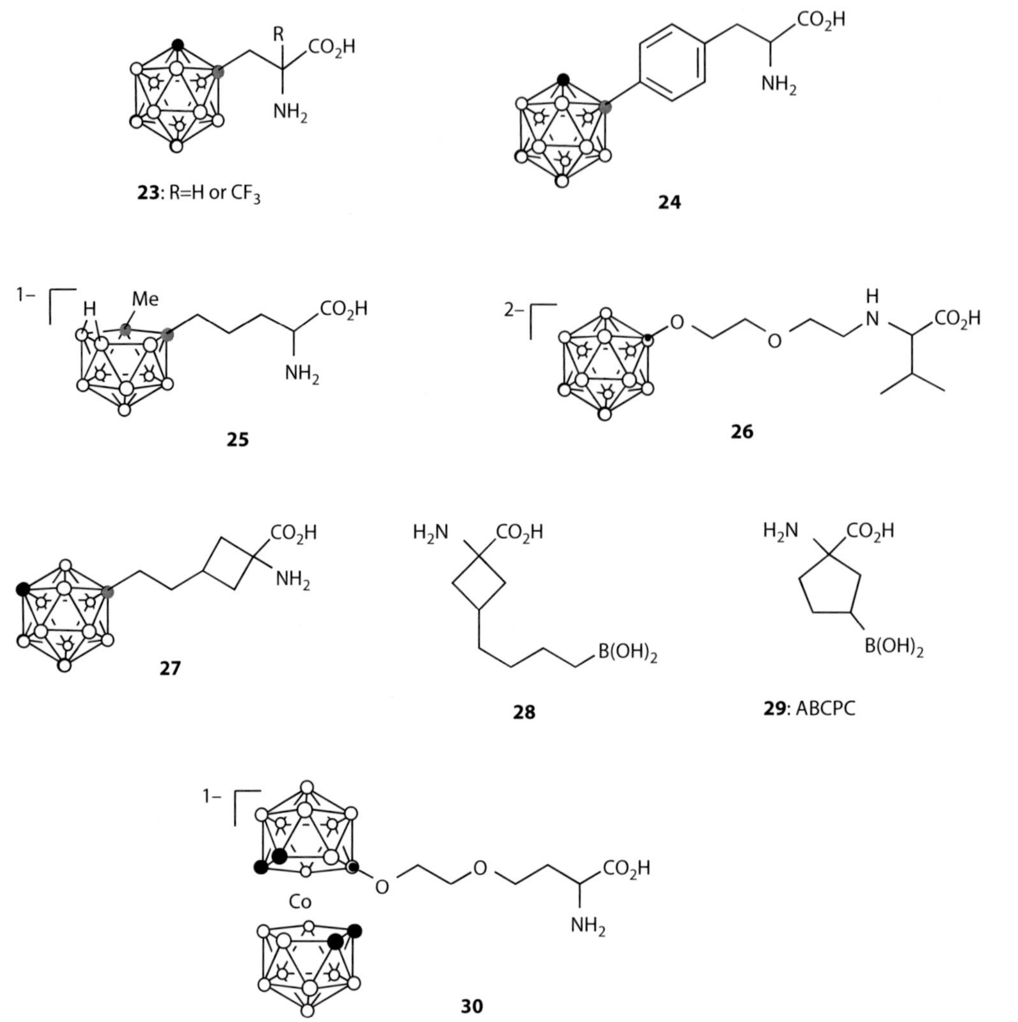

23: R=H or CF_3

24

25

26

27

28

29: ABCPC

30

FIGURE 10.6 Boronated amino acids.

tumor cells. One such molecule, porphyrin **33** (Figure 10.7) containing three cobaltacarborane moieties and a cell-penetrating peptide sequence from the human immunodeficiency virus I transcriptional activator (HIV-1 Tat)[89] with the sequence GRKKRRQRRRPPQ, was recently reported by us;[18] this compound was found to accumulate within human HEp2 carcinoma cells to a much greater extent than boronated porphyrins containing no peptide sequence, and to have low cytotoxicity. On the other hand, the BPA-containing liposome **34** containing the HIV-1 Tat peptide and two fatty acids as the hydrophobic counterpart, has been recently designed.[90] This BPA liposome shows high performance in terms of film stability and ability for encapsulating BSH, but its *in vitro* and *in vivo* evaluations have not yet been reported.

10.2.3 Nucleosides

Boron-containing analogs of the biochemical precursors of nucleic acids, including purines, pyrimidines, thymines, nucleosides, and nucleotides, have been synthesized and several were evaluated in cellular and animal studies.[91–102] Figure 10.8 shows several representative molecules of this class (**35–42**) containing carborane cages linked via various spacers to either the base or the carbohydrate

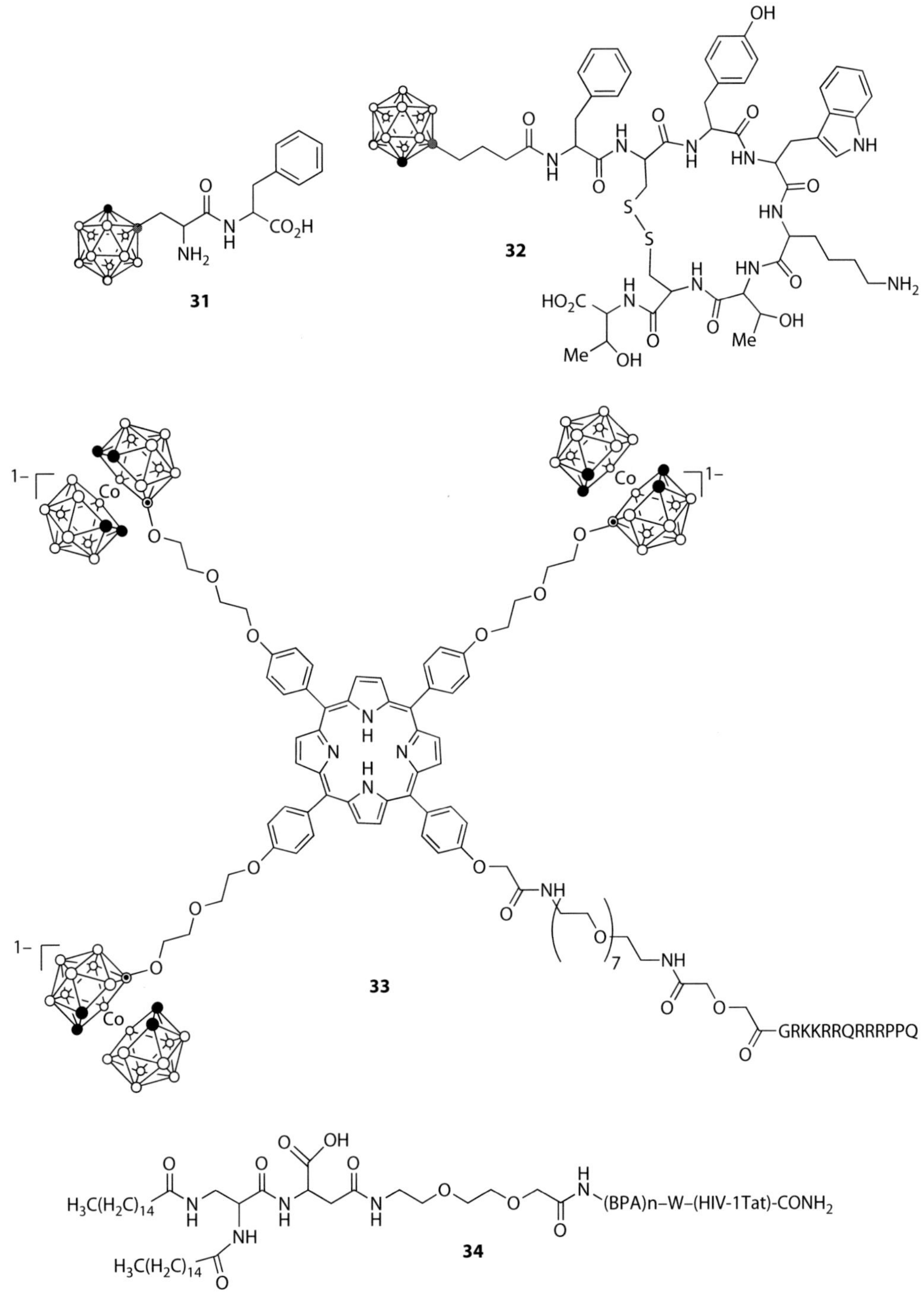

FIGURE 10.7 Boron-containing peptides.

group. Boronated nucleosides could selectively accumulate within tumor cells due to their conversion to the corresponding monophosphates by cytosolic kinases, which are upregulated in actively proliferating tumor cells. Furthermore, their subsequent incorporation into the DNA of tumor cells could place boron in close proximity to DNA, the most critical molecular target in BNCT. Since nucleoside membrane transport depends on the degree of cell proliferation and the activity levels of

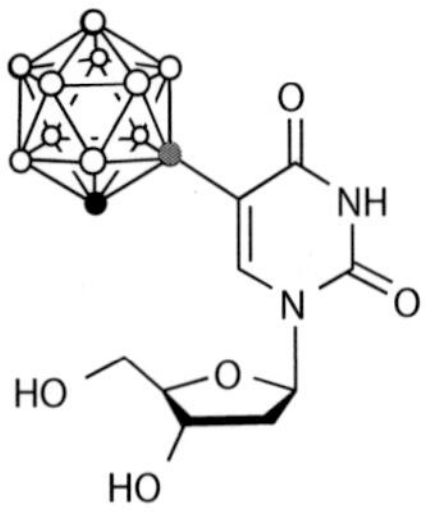

35: CDU

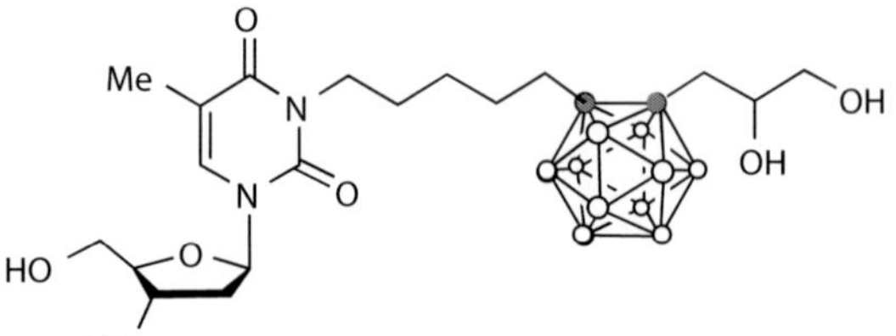

36: N5-2OH

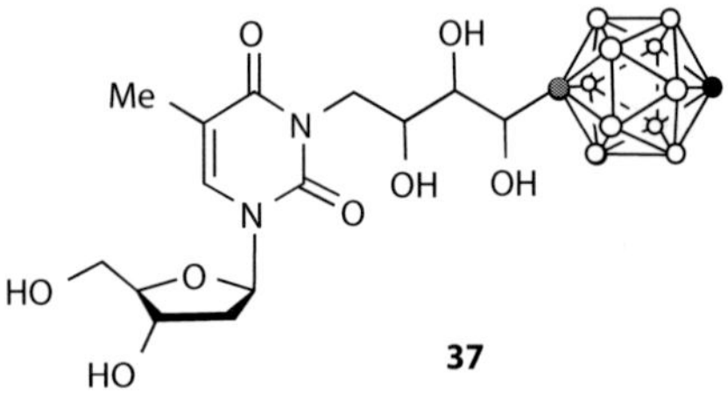

37

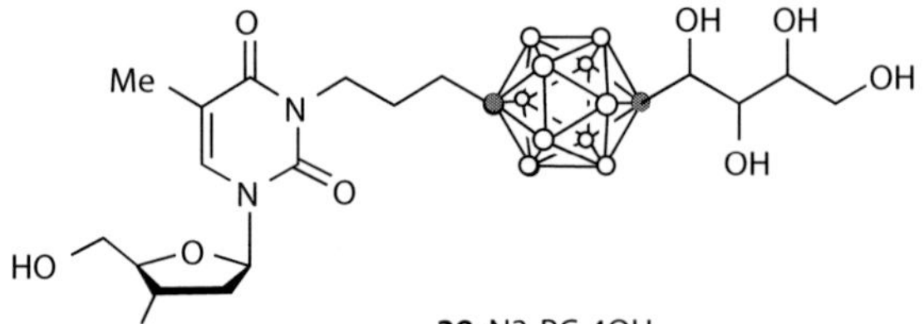

38: N3-PC-4OH

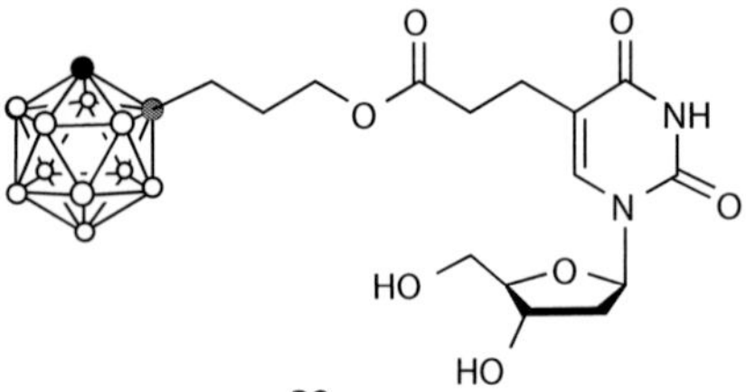

39

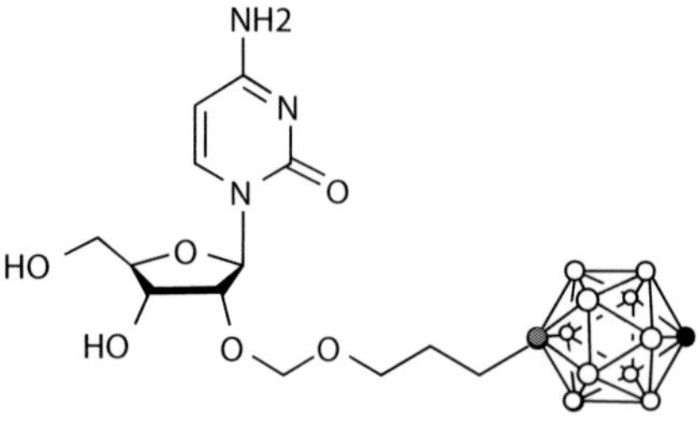

40

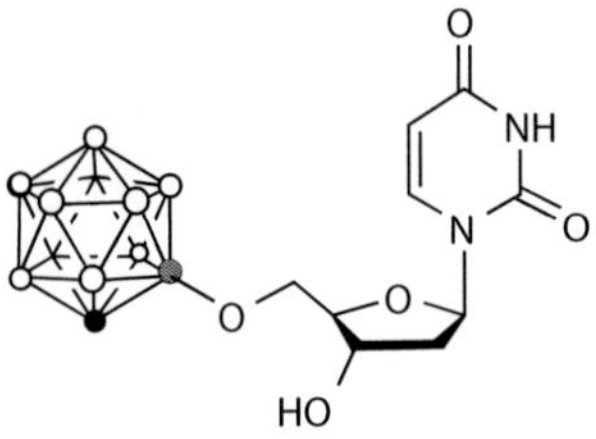

41

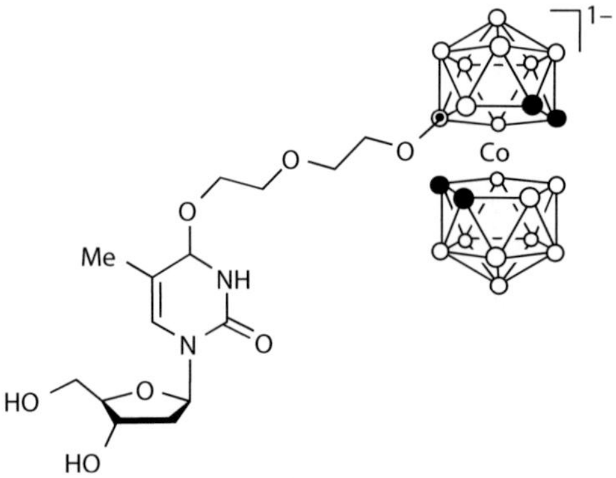

42

FIGURE 10.8 Boronates nucleosides.

N5-2OH nucleoside kinases, preferential intracellular accumulation and retention of boronated nucleosides are expected.[102]

Among this class of compounds the development of 3-carboranyl thymidine analogs (3CTAs) has received special attention in recent years. In particular, the 3-(dihydroxypropyl-carboranyl-pentyl)thymidine derivative designated N5-2OH (**36**) was shown to be very promising, due to its low toxicity, selective tumor cell uptake and high rate of phosphorylation into the corresponding nucleotide. The two stereochemically pure forms of N5-2OH as well as their geometrical isomers bearing either a *meta*- or *para*-carborane cluster were prepared and evaluated as boron delivery vehicles for BNCT.[103] The (*R*)-N5-2OH isomer was administrated to rats bearing intracerebral F98 gliomas; subsequent neutron irradiation resulted in prolonged survival compared to controls. On the other hand, results from determination of human thymidine kinase 1 (TK1) phosphorylation rates and docking studies using (*R*)-N5-2OH and its *meta*- and *para*-carboranyl derivatives, showed that the di-hydroxypropyl group attached to the carborane cluster may only function as a hydrophilicity-enhancing structural element but does not increase the binding affinity to the active site of human TK1. It has been reported that CED of the borononucleoside N5-2OH was effective in selectively delivering potentially therapeutic amounts of boron to rats bearing intracerebral implants of the F98 glioma.[104] Hydrophilic analogs of N5-2OH (e.g., **37**, **38**) bearing a *p*-carboranyl cluster and multiple hydroxyl groups have been recently reported.[105–107] Compounds **37** and **38** showed higher relative phosphorylation rates than N5-2OH as well as lower toxicity than N5-2OH.

Boronated nucleosides bearing a cobaltacarborane group (e.g., **42**) have also been synthesized.[108,109] Compound **42** was found to have low cytotoxicity and to readily accumulate within V79 hamster cells.

10.2.4 Porphyrin Derivatives

The syntheses of boron-containing porphyrin derivatives, including chlorin, bacteriochlorin, tetrabenzoporphyrin, corrole, and phthalocyanine macrocycles, have recently been reviewed.[43,110–112] This class of compounds shows several advantages for use as BNCT delivery vehicles, including: (a) their demonstrated tumor selectivity and long retention times in tumors; (b) low dark toxicities; (c) their ability for sensitizing the production of cytotoxic oxygen species on light activation, which constitutes the basis for their use in the photodynamic therapy (PDT)[113,114] of tumors; (d) their ability for DNA and RNA binding and for inducing cell apoptosis upon brief exposure to light; (e) their fluorescent properties, which facilitate the detection of tissue-localized boron and treatment planning; (f) their ability to form stable *in vivo* complexes with a variety of metal ions while retaining their biodistribution and pharmacokinetic properties, and (g) the FDA has approved two porphyrin-based drugs for the PDT treatment of melanoma, early and advanced stage cancer of the lung, digestive tract, genitourinary tract, and the wet form of age-related macular degeneration, and several others are currently undergoing clinical trials for the treatment of a large range of diseases.[114–116] The observed preferential accumulation and retention of porphyrin derivatives in tumors has been attributed to several factors, including their *in vivo* association to plasma proteins (particularly to low-density lipoproteins), the high content of macrophages in tumor cells (which can account for 20–50% of the total tumor cell population) and their lower intracellular pH (due to the higher level of lactic acid), as well as to passive diffusion and to the high vascular permeability and inefficient lymphatic clearance of neoplastic tissues.[117]

The first boronated porphyrins were reported in 1978 by Haushalter, Rudolph, and coworkers;[118,119] however, the investigation of porphyrin derivatives as boron delivery agents for BNCT did not start until the late 1980s.[120,121] Perhaps the most investigated boronated porphyrins in biological studies are BOPP (**43**)[25,122–127] and CuTCPH (**44**),[128–132] shown in Figure 10.9. The syntheses of these porphyrins were reported in 1990, either by functionalization of a preformed porphyrin macrocycle (in the case of BOPP)[133] or by total synthesis from the condensation of a carboranyl-substituted benzaldehyde with pyrrole (in the case of CuTCPH).[134] Both these boronated porphyrins **43** and **44** have been shown in multiple investigations to deliver very high amounts of boron to tumor-bearing mice, with very high tumor-to-blood and tumor-to-brain boron concentration ratios and retention times in tumors. However, BOPP was found to be toxic to rats[25] and dogs[127] following IV administration of 35–100 mg/kg BOPP, and its human maximum tolerated dose is reported to be 4 mg/kg;[135] at this low dose suboptimal amounts of boron are delivered to glioma tumors for effective BNCT, although PDT using BOPP was observed to enhance the patient survival times. The toxicity of BOPP might

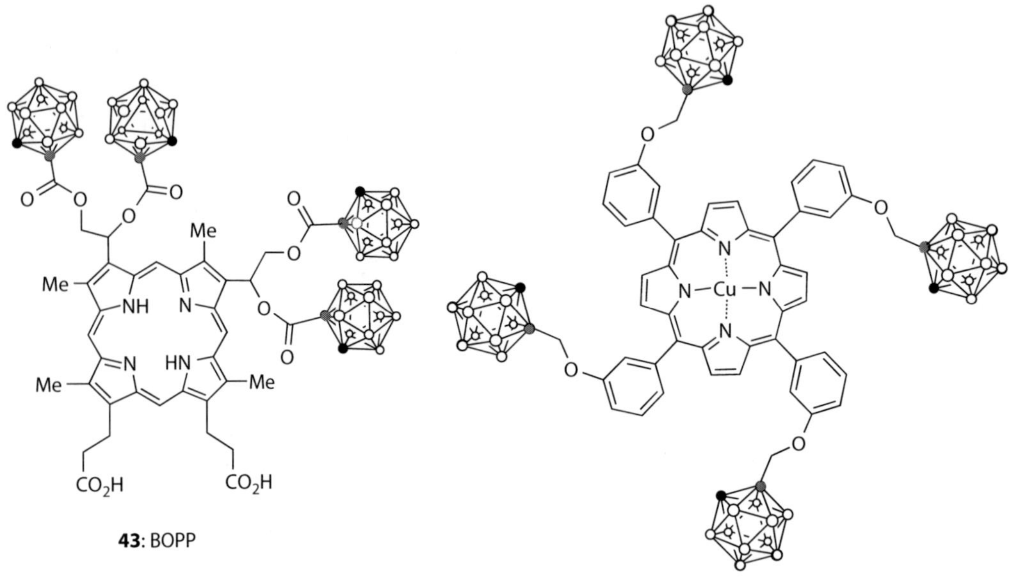

FIGURE 10.9 Boronated porphyrins BOPP and CuTCPH.

be related to its possible *in vivo* hydrolysis and subsequent release of 1-carboxyl-*ortho*-carborane. More recently, and in order to improve the performance of BOPP (**43**) and CuTCPH (**44**) which contain potentially hydrolysable ester or ether bonds between the porphyrin and the carborane clusters, respectively, boronated porphyrins containing chemically stable carbon–carbon linkages and/or a larger amount of boron by weight have been proposed as delivery vehicles for BNCT. Several representatives of such molecules are shown in Figures 10.10 through 10.12.

Boronated porphyrin derivatives are typically prepared either by direct functionalization of a preformed macrocycle (protoporphyrin-type or *meso*-tetraphenylporphyrin-type) or via the tetramerization of boron-substituted precursors, that is, aldehydes and/or pyrroles or dipyrromethanes.[112] The free-base and various metalated porphyrin derivatives, mainly the Zn(II) and Cu(II) complexes, have been proposed for use as BNCT agents. While the Zn(II) complexes generally enhance the fluorescent and photosensitizing properties of these macrocycles, the Cu(II) complexes usually display short triplet half-lives, therefore usually reducing normal tissue photosensitization, and could allow tumor detection via PET (using ^{64}Cu) or SPECT (using ^{67}Cu). In addition, the synthesis of boronated texaphyrins containing Gd(III) or Lu(III) has also been reported; in particular the Gd(III) complex could find application in combined gadolinium and boron NCT, as well as being imaged and quantified using MRI.[136]

Carborane-substituted porphyrins bearing icosahedral carboranes, such as *ortho*, *meta*- or *para*-carboranes, generally have poor water solubility and require solubilization agents (such as Cremophor EL, DMSO, or liposomes) for biological applications; on the other hand their *nido*-carborane analogs are highly water soluble in the form of potassium or sodium salts. Therefore, porphyrins **45–48** (Figure 10.10) all showed significant water solubility, although their hydrophobic character, as estimated from their distribution between 1-octanol and PBS at pH 7.4, is higher (partition coefficient, $P > 41$) than porphyrins bearing four carboxylate or sulfonate groups [e.g., $TPP(CO_2H)_4$ and $TPP(SO_3H)_4$][137] due to the amphiphilic nature of the *nido*-carboranyl groups. Furthermore, the zinc(II) complexes of *nido*-carboranylporphyrins have been shown to have lower hydrophobic character than the corresponding free bases, lower tendency for aggregation (in part due to the protection of the inner nitrogen atoms from protonation)[138] while usually more cytotoxic, possibly as a result of a more efficient generation of singlet oxygen,[139] and to accumulate within cells

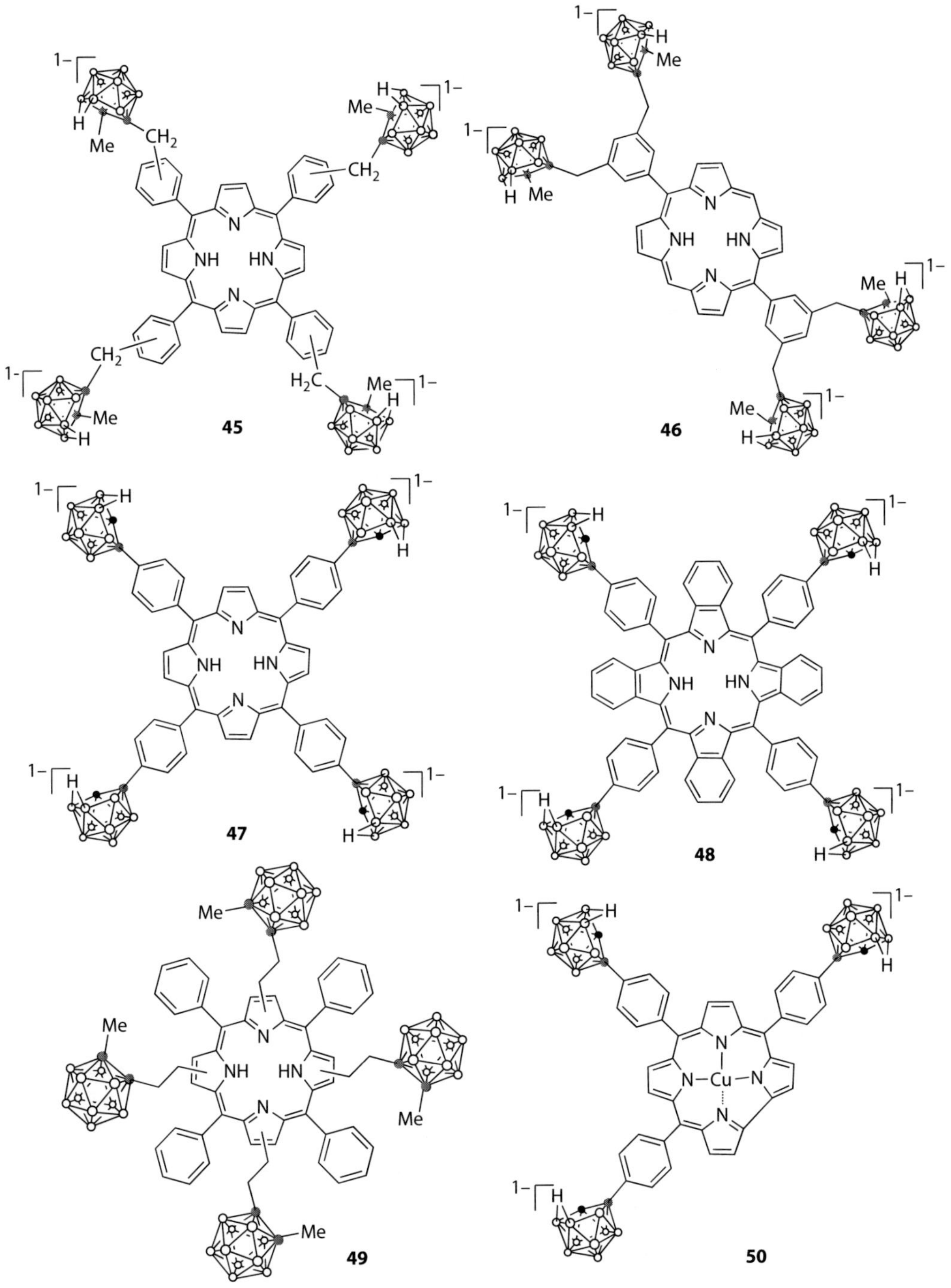

FIGURE 10.10 Boronated porphyrin derivatives with carbon–carbon linkages.

to a lesser extent than the corresponding metal-free porphyrins.[137] Since during the base- or fluoride-induced deboronation reaction of the *ortho*-carborane clusters of tetracarboranylporphyrins into the corresponding *nido*-carboranes, the most electropositive boron atom from each carborane cluster is regioselectively removed, mixtures of stereoisomeric *nido*-carboranylporphyrins are obtained, as has been shown by ^{1}H-NMR and *ab initio* calculations;[140] such a cocktail of at least ten isomers could possibly contribute to the observed biological efficacy of this type of compound.

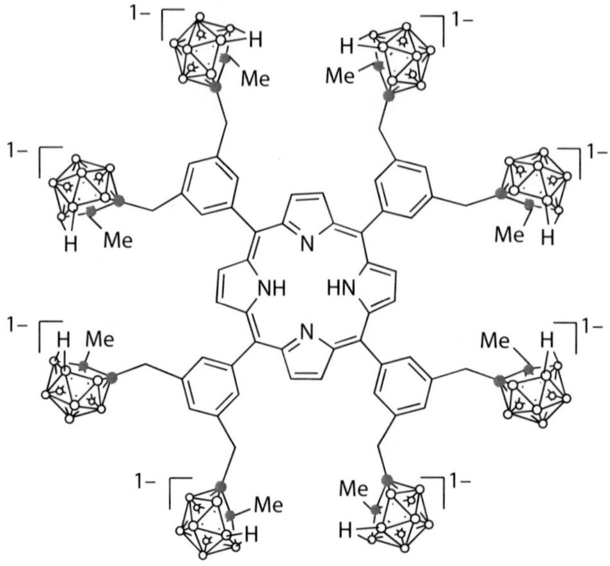

51

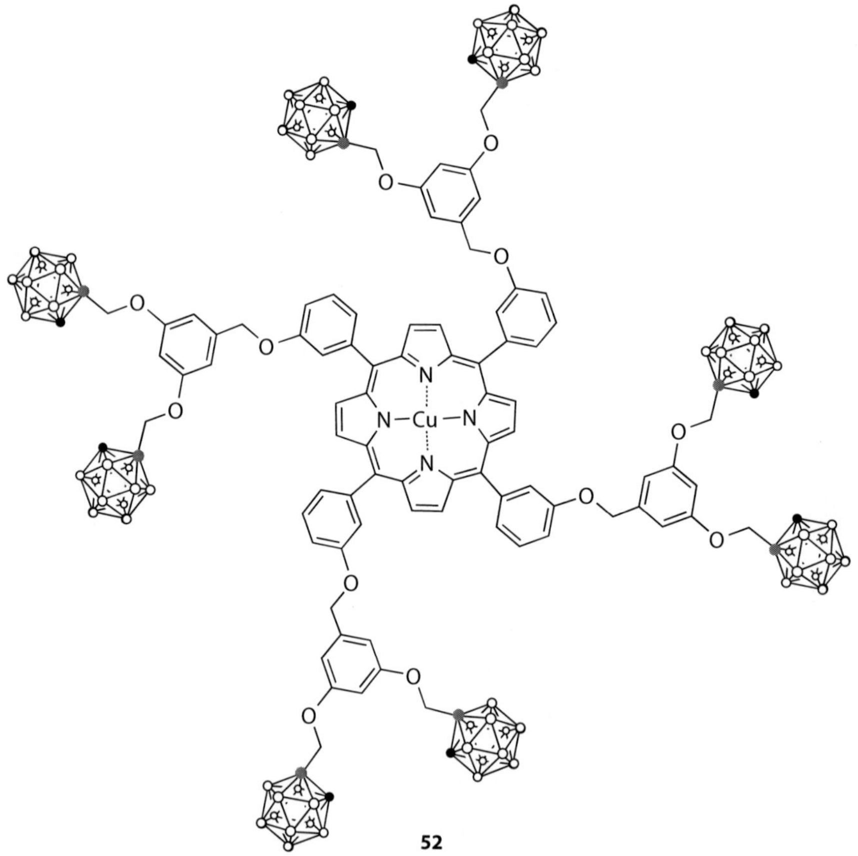

52

FIGURE 10.11 Boronated porphyrins of high boron content.

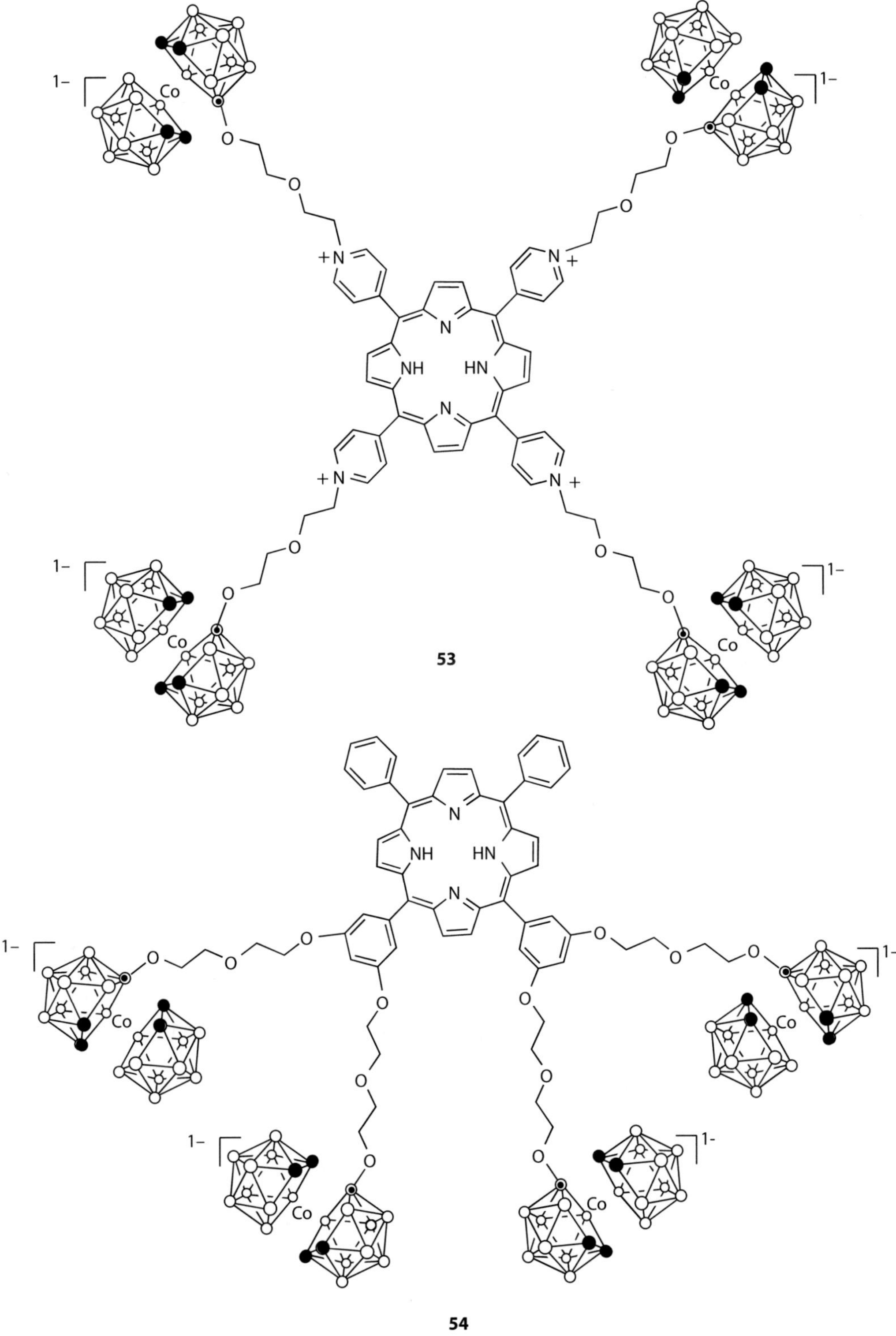

FIGURE 10.12 Cobaltacarborane-functionalized porphyrins.

Tetra-carboranylporphyrins **45** (Figure 10.10), both *meta*- and *para*-phenyl substituted, as well as their corresponding Zn(II) complexes, were evaluated in rat 9L gliosarcoma, mouse B16 melanoma, and human U-373MG glioblastoma cells, and shown to have low dark cytotoxicities ($IC_{50} > 150$ μM in all cell lines), to readily accumulate within cells and to localize preferentially in the cell lysosomes;[137] while no apparent differences were observed between the *in vitro* properties of the *para*- and the *meta*-substituted porphyrins, the Zn(II) complexes were found to be more cytotoxic than the corresponding free-bases and to accumulate within cells to a significant lower extent.

Porphyrin **46**, designated H_2DCP, also containing four carborane cages but on only two sides of the porphyrin ring, and its zinc(II) derivative were reported to selectively deliver therapeutic concentrations of boron to BALB/c mice bearing EMT-6 mammary carcinomas, with very high (>100:1) tumor-to-brain boron concentration ratios and low toxicity.[141] The Zn(II) complex delivered 1.2–1.7 times higher amounts of boron to tumors than the metal-free one, although with lower tumor-to-blood boron concentration ratio (4.7:1 vs. 9.8:1, 48 h after IP injections). Porphyrin **47**, designated H_2TCP, was shown to have low cytotoxicity in the dark toward hamster lung V79 fibroblasts ($CS_{50} > 300$ μM), but to cause effective DNA photodamage.[142] H_2TCP and its tetrabenzo derivative **47**, designated H_2TBP, were recently synthesized in both ^{10}B-enriched and nonenriched forms, and shown to have very low toxicity even at a high dose of 160 mg/kg administered intraperitoneally into BALB/c mice.[143] In addition, H_2TCP (**47**) was found to selectively deliver significant levels of boron to melanotic melanomas subcutaneously transplanted in C57/BL6 mice, via either intratumoral (~60 ppm ^{10}B) or intravenous (~6 ppm ^{10}B) administration; in spite of about 10-fold difference in the tumor boron concentrations, similar delay in tumor growth (5–6 days) was induced by neutron irradiation in comparison with control groups of mice.[144–146] This finding might reflect an inhomogeneous distribution of the boronated porphyrin among the various compartments of the tumor, particularly in the case of intratumoral (IT) administration. In agreement with these results, the delivery of porphyrins **46**, **47,** and **48** by CED into rats bearing intracranial F98 gliomas, showed that >150 μg/g boron was delivered from a very low dose of 1.0 mg/kg porphyrin;[26–28] however, efficacy BNCT studies conducted at the MITR showed that the delay in tumor growth of F98 rat gliomas following neutron irradiation was similar to those obtained with BPA, which delivered much lower tumor boron concentrations compared with the boronated porphyrins. Histopathological examination of brains from tumor-bearing rats revealed that most porphyrin accumulated within the tumor-associated macrophages rather than within tumor cells, which might in part explain the relatively low-treatment efficacy.[26–28] CED of BOPP[25] and the highly water-soluble octa-anionic boronated porphyrin TABP-1[24] was shown to deliver extremely high amounts of boron to rat brain tumors, with high tumor-to-normal brain and tumor-to-blood boron ratios and minimal toxicity; for example, CED administration of 1.0 mg of TABP-1 resulted in 201 μg/g boron concentration in U-87 MG intracerebral human glioblastoma xenografts, whereas 1.0 mg of BOPP produced 422 μg/g boron concentration in 9L intracerebral rat gliomas. However, no therapy studies were conducted to evaluate the animals' mean survival times following boron delivery. Furthermore, large standard deviations were found in the above studies for the tumor boron concentrations, which suggest a nonuniform delivery of boron to tumors.

Ortho-carboranylporphyrins, such as **49**[147] and others bearing water-solubilizing groups, including phosphonates and quaternary ammonium salts,[148] have been synthesized but their biological evaluation has not yet been reported. A series of water-soluble boronated corroles, including **50** have also been synthesized, and shown to have low dark cytotoxicity and to accumulate within human glioma T98G cells, localizing mainly in the cell lysosomes.[149]

Boronated porphyrins of high boron content have the potential to deliver high amounts of boron to targeted tumors with low toxicity. The octa-carboranylporphyrin **48** (Figure 10.11), designated H_2OCP, was shown to have low dark toxicity toward V79 lung fibroblasts ($CS_{50} \geq 250$ μM), to readily accumulate within human glioblastoma T98G cells and to localize mainly in the cell lysosomes.[150] H_2OCP (**48**) was also found to be taken-up by the T98G cells to a significant higher extent than its zinc(II) complex analog, although its tumor cell uptake *in vitro* was lower than that of the related tetracarboranylporphyrin **42** (*para*-isomer), maybe as a result of its higher hydrophilic character. On the other hand, the *ortho*-carboranylporphyrin **49** was found to deliver a high amount of boron (46 μg/g) to BALB/c mice bearing EMT6 mammary carcinomas, 48 h after a series of IP injections with very high (>90:1) tumor-to-brain and tumor-to-blood boron concentration ratios, although a related copper(II) tetra(*ortho*-carboranyl)porphyrin was found to deliver even higher amounts of boron (169 μg/g) to tumors at nearly twice the boron dose, and with higher tumor-to-brain and tumor-to-blood ratios (>400:1).[151]

Another methodology for the synthesis of boronated porphyrins of high boron content is via the one-step high-yielding reaction of hydroxyphenyl- and pyridyl-substituted porphyrins with zwitterionic $[3,3'\text{-Co}(8\text{-C}_4H_8O_2\text{-}1,2\text{-C}_2B_9H_{10})(1',2'\text{-C}_2B_9H_{11})]$[36,37] to produce cobaltacarborane-substituted porphyrins, such as those shown in Figure 10.12.[18,152–154] The synthesis of a series of five zwitterionic cobaltacarboranyl-pyridyl porphyrins containing one to four cobaltabisdicarbollide anions, including **53** (Figure 10.12), has been reported.[152] The uptake of this series of compounds by human carcinoma HEp2 cells was shown to increased with the number of cobaltacarborane moieties linked to the porphyrin macrocycle; porphyrin **53** bearing four cobaltacarboranes, accumulated the most within cells of this series of compounds, approximately 16 times more than a porphyrin bearing only one cobaltacarborane group. In another study, a series of seven anionic cobaltacarboranylporphyrins containing two to eight cobaltacarborane moieties, including porphyrin **54**, was synthesized as the corresponding cesium or potassium salts, and in 85–97% yields.[153,154] Of this series of anionic porphyrins, tetra(cobaltacarboranyl)porphyrin **54** and the hexa-cobaltacarboranylporphyrin analog accumulated the most within human HEp2 carcinoma cells, while having very low toxicity.[154] These results suggest that up to six cobaltacarborane groups can be attached to a porphyrin macrocycle to observe high tumor cell uptake. Another efficient way to increase boronated porphyrin uptake into tumor cells is via its conjugation to a cell-penetrating peptide. Such sequences, for example, the HIV-1 Tat (48–60) peptide, are normally found in the protein transduction domains responsible for rapid and efficient cellular internalization. Tri(cobaltacarboranyl)porphyrin **33** (Figure 10.7) bearing the HIV-1 Tat peptide with the sequence GRKKRRQRRRPPQ was shown to accumulate within cells ~10-fold compared with related cobaltacarboranylporphyrins with the same number of cobaltacarboranes but without the HIV-1 Tat peptide.[18] BOPP (**43**) has also been conjugated to a nuclear localizing sequence peptide[155], but the biological evaluation of this compound has not yet been reported.

Among the porphyrin derivatives, boronated chlorins, bacteriochlorins, and phthalocyanines are particularly promising for application as dual BNCT and PDT sensitizers because of their strong absorptions of light wavelengths above 640 nm, which penetrate deeper into most human tissues.[43,45,112] Such a combination of selective therapies could lead to the enhancement of localized tumor treatment via the targeting of different mechanisms of tumor cell destruction. Some of the most recently reported boronated chlorins are shown in Figures 10.2 and 10.13 (**10**–**12**, **55**–**58**).[42–45,156–162] Among these, TPFC (**10**) (Figure 10.2) was shown to readily accumulate within melanoma B16F1 cells, and to have low dark cytotoxicity but efficient phototoxicity.[45] TPFC also exhibited selectivity for targeting B16 melanomas in C57B1/6 mice following IV injection, with high tumor-to-skin and tumor-to-blood boron concentration ratios, as well as low toxicity and effective photodynamic activity. Boronated chlorin **55** was also found to have very low dark cytotoxicity (IC_{50} > 500 μM) toward human glioma T98G cells while being toxic in the presence of red light (IC_{50} = 80 μM at 0.55 J/cm^2 light dose), and to accumulate within T98G cells to a significantly higher extent than chlorin e_6, a chlorophyll degradation product.[156,157] Chlorins **56**–**58** were prepared by direct functionalization of pheophorbide-a, a chlorophyll-a derivative. Chlorins **58** and bacteriochlorin analog containing cobaltacarborane groups were recently synthesized via aminolysis of pheophorbide-a using various diaminoalkanes, followed by reaction with zwitterionic $[3,3'\text{-Co}(8\text{-C}_4H_8O_2\text{-}1,2\text{-C}_2B_9H_{10})(1',2'\text{-C}_2B_9H_{11})]$.[163,164] The preliminary biological evaluation of chlorins **58** showed readily accumulation of these compounds within human A549 lung adenocarcinoma cells.[164]

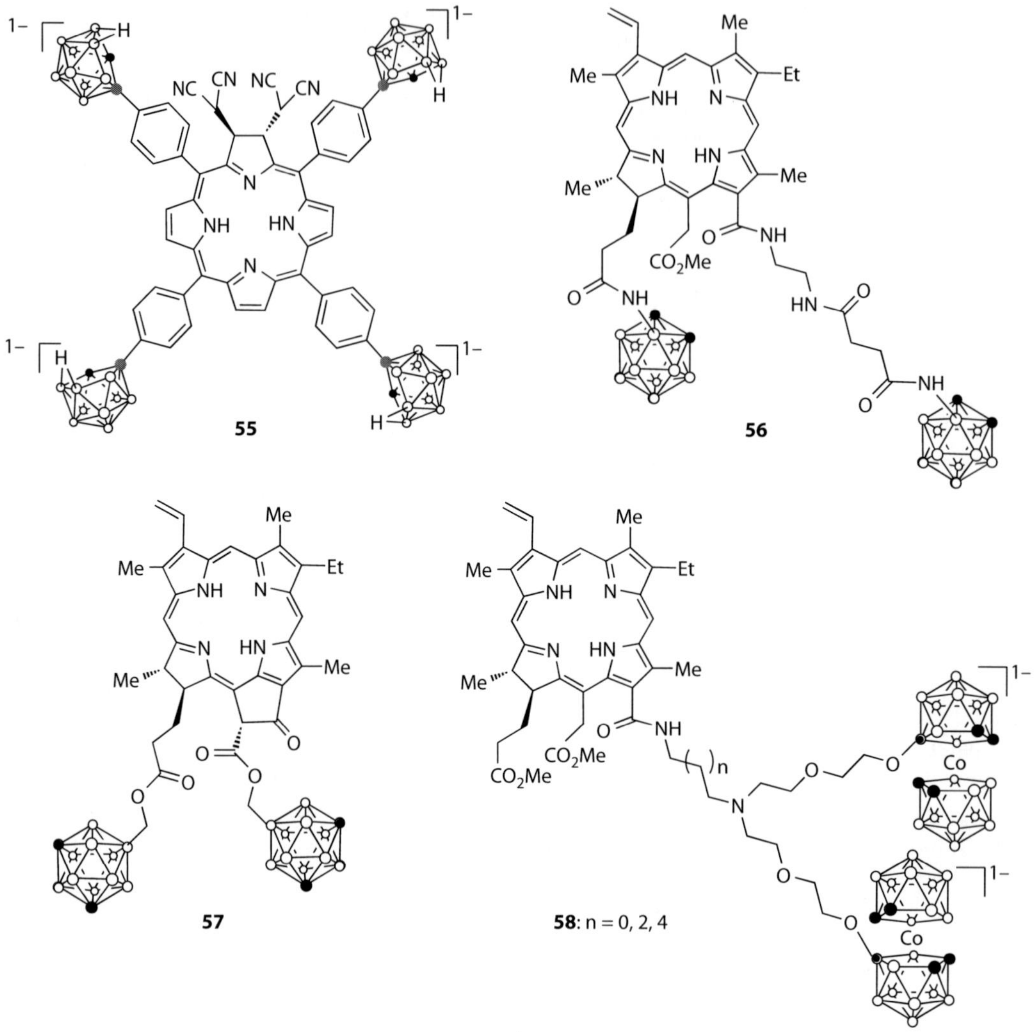

FIGURE 10.13 Boronated chlorines.

On the other hand, recently reported boronated phthalocyanines are shown in Figure 10.14 (**59**, **60**).[165–169] The a-substituted Zn(II)-phthalocyanines **59**, as a mixture of regioisomers, were synthesized ^{10}B-enriched and found to efficiently accumulate within B16F1 melanotic melanoma cells and to induce extensive photodamage.[165,166] Furthermore, following IV injection to C57BL mice bearing melanoma tumors and activation by 600–700 nm light or thermal neutrons, **59** was observed to cause significant delay in tumor growth. Pure Zn(II)-pthtalocyanines of the A_3B type containing one or two cobaltacarborane moieties, such as **60**, were recently synthesized via cyclotetramerization of two phthalonitriles, but their biological properties have not yet been reported.[168,169]

10.2.5 Other DNA Binders

Besides porphyrin derivatives,[138] several other boronated DNA-binding molecules, such as alkylating agents, intercalators, minor-groove binders, polyamines, and metal complexes have been reported.[170] Some examples of such molecules are shown in Figure 10.15, and include derivatives of aziridines (**61**), acridines (**62**), phenanthridines (**63**), Pt(II) complexes (**64**, **65**), trimethoxyindoles (**66**), carboranylpolyamines, di- and tri-benzimidazoles, and dequalinium derivatives (**67**).[171–178]

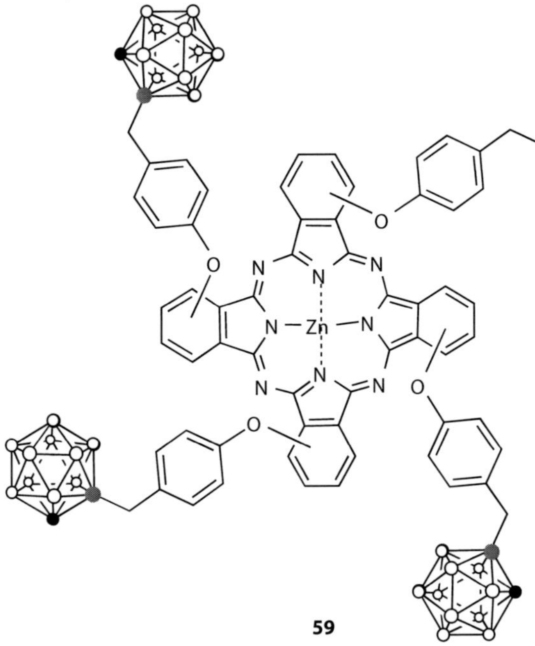

59

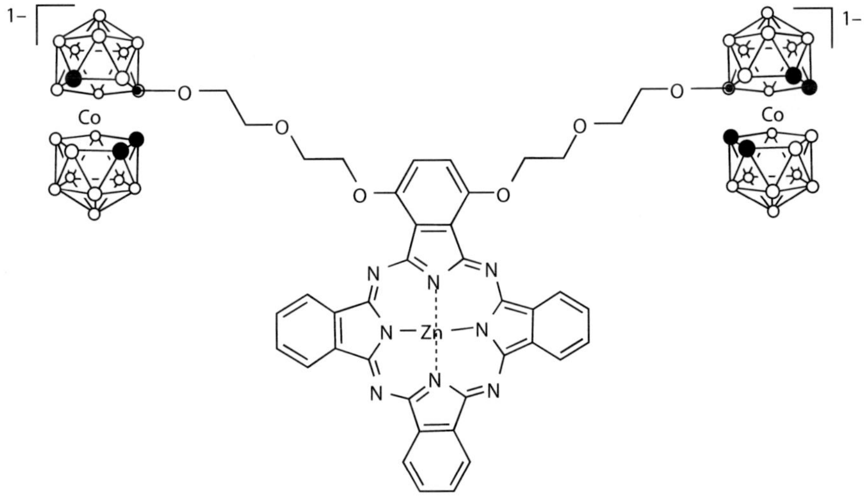

60

FIGURE 10.14 Boronated phthalocyanines.

From the reported biological evaluations of these classes of compounds, low tumor selectivity and significant toxicity are often observed for these polycationic molecules, although some promising derivatives with low cytotoxicity have been synthesized, but no additional biological investigations are yet reported. In any case, several of these molecules have been shown to indeed intercalate or bind to DNA, a critical target in BNCT. In addition, the cationic nature of some molecules of this class might facilitate their binding to tumor cell plasma membranes and accumulation within cells. On the other hand, boronated Pt(II) anticancer agents have the potential for DNA intercalation and subsequent irreversible tumor cell damage. The synthesis of mono-, di-, and tri-nuclear Pt(II) complexes bearing *closo*-carborane clusters (e.g., **64** and **65**) have been reported and these compounds were shown to bind plasmid DNA and to have *in vitro* anticancer activity against L1210 leukemia

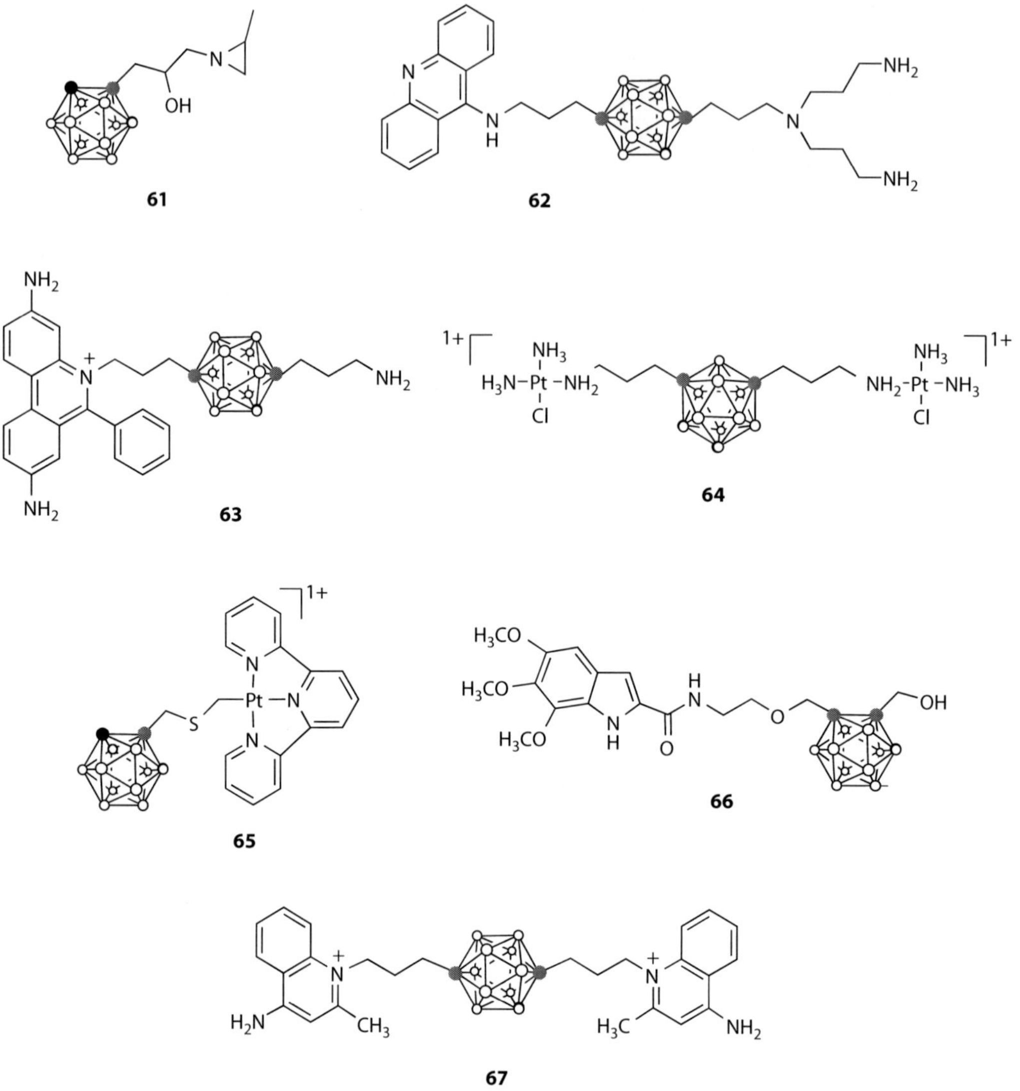

FIGURE 10.15 Boronated DNA binding molecules.

cells. Some of the complexes, in particular those containing one *para*-carborane cluster or multiple carborane cages, displayed low water solubility. Among the boronated Pt(II) complexes evaluated, **64** was shown to have a similar cytotoxic effect to cisplatin in the absence of neutrons.[175]

10.2.6 Antibodies

Monoclonal antibodies (MAb) represent a promising class of targeted tumor cell agents due to their very high specificity for the target; however, noninternalizable boronated MAb might only deliver boron extracellularly. One such target is the epidermal growth factor receptor (EGFR) and its mutant isoform, EGFRvIII, which are overexpressed in gliomas and squamous cell carcinomas of the head and neck, while low in normal brain.[19] Barth and coworkers have used the anti-EGFR MAb cetuximab, which specifically binds to the extracellular domain of human EGFR, and MAb L8A4 which binds EGFRvIII and not wild-type EGFR. Recent studies performed by Barth and coworkers[29,179] show that intracerebral CED of a boronated dendrimer-epidermal growth factor (BD-EGF) bioconjugate on F98 glioma-bearing rats was therapeutically more effective than its intratumoral injection. The boronated dendrimer-EGF bioconjugate was obtained by reaction of a fourth-generation polyamidoamine (PAMAM) dendrimer with the ^{10}B-enriched methylisocyanato borane anion $Na(CH_3)_3NB_{10}H_8NCO$, followed by conjugation to EGF derivatized with the heterobifunctional reagent m-maleimidobenzoyl-*N*-hydroxysulfosuccinimide ester (mMBS). Biodistribution studies indicated that after 24 h CED of the BD-EGF bioconjugate, EGFR overexpressing tumors showed 43% more radioactivity, a twofold increase in tumor boron concentration, as well as a 7.3 increase in the volume of distribution within the infused cerebral hemisphere and a twofold increase in tumor uptake of BD-EGF as compared to IT injection. A similar boron delivery system designated BD-C225 was prepared by conjugation of PAMAM to MAb cetuximab via the heterobifunctional reagents *N*-succinimidyl 3-(2-pyridyldithio)-propionate and *N*-(*k*-maleimido undecanoic acid)-hydrazide.[180] This conjugate was specifically taken up by F98 EGFR glioma cells *in vitro* as compared to receptor negative F98 wild-type cells (41.6 vs 9.1 μg/g). The amount of boron in F98 EGFR gliomas after 24 h following CED or IT injection was 77.2 and 50.8 μg/g, respectively. The mean survival times were 54.5 and 70.9 days for BNCT treatment after 24 h CED injection of BD-C225, either alone or in combination with IV administration of BPA. Further studies[181,182] were reported to determine the efficacy of BD-C225 and boronated BD-L8A4 on rats bearing composite tumors, which were produced by implanting equal numbers of F98 EGFR and F98npEGFRvIII cells. Results from these studies show that a significant increase in survival time is only observed when both BD-C225 and L8A4 are used in combination. The corresponding percentages of increased life spans were 97% for the combination of BD-C225 and BD-L8A4 vs. 36% for CED of BD-C225 alone and 29% for CED of L8A4. A similar approach has been used aimed at targeting VEGF receptors.[183] The synthesis of a VEGF-driven boronated dendrimer using human recombinant $VEGF_{121}$ has been reported from a fifth generation of PAMAM dendrimer containing 128 primary amino groups, linked to 105–110 decaborate molecules. The dendrimers were labeled with the fluorescent dye Cy5, for biodistribution and uptake studies in tumor-bearing mice. *In vivo* near-IR fluorescent imaging studies showed that this bioconjugate accumulated in tumor vasculature and that its uptake was mediated by VEGFR-2-positive cells. Using the 4T1 mouse tumor model, it was found that VEGF-BD/Cy5 selectively accumulated in the peripheral areas of growing tumors, where tumor neovascularization was most active. VEGF-BD/Cy5 was internalized via receptor-mediated endocytosis and localized mostly in the perinuclear region with a distinct punctate pattern.

Boron-targeted delivery in BNCT has also been achieved via the use of immunoliposomes, which are usually prepared using the following methodologies: (a) by conjugation of tumor-targeting or receptor ligands to a functional group on the liposome surface, or (b) by postinsertion of tumor-targeting or receptor ligands into a preformed liposome via the micelles of an antibody–lipid conjugate. An immunoliposome has been recently reported for delivery of BSH into brain tumors.[184] In this system, BSH was encapsulated into a nickel–liposome and recombinant ZZ-His; the IgG Fc-binding motif was used as an adaptor to conjugate anti-EGFR MAb to the nickel–liposome. The biological efficiency of this immunoliposome was evaluated both *in vitro* and *in vivo* using parental U87 glioma cells (PAU87), human wild-type EGFR, and EGFRvIII-transfected U87 glioma cells (U87 WT and U87 DEGFR). Western blotting and lipid analysis showed that the binding of the nickel–liposome to ZZ and of ZZ to the MAb was specific as compared to controls that did not contain the nickel lipid. The delivery of BSH within glioma cells was evaluated using the anti-BSH MAb; these studies showed that the uptake of the immunoliposomes was temperature dependent and that their internalization within cells followed an endocytic mechanism with BSH localizing in the cell nuclei. *In vivo* biodistribution studies indicated that in immunoliposome-treated mice, the amount of ^{10}B in the tumor reached 28.36 ± 7.63 μg/g 24 h after the injection and remained high at 48 h (21.38 ± 5.31 μg/g). On the other hand in liposome-only treated mice, the ^{10}B content of tumors at 24 and 48 h was 3.45 and 2.97 μg/g, respectively.

The use of maleimido-PEG-cholesterol **68** (Figure 10.16) to incorporate MAb cetuximab (C225) into liposomes, via either surface conjugation or postinsertion, was reported.[185] The resulting

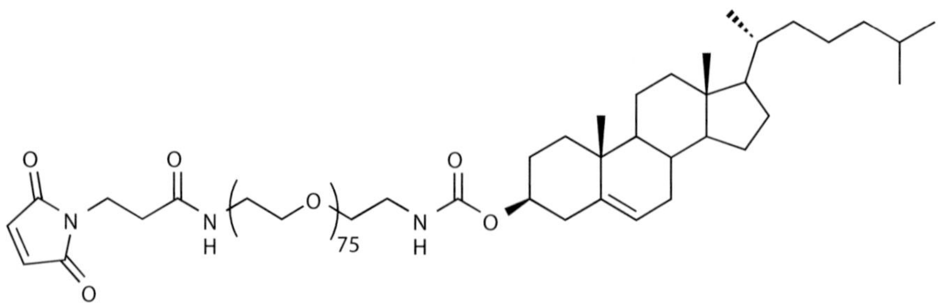

68

FIGURE 10.16 Maleimido-PEG-cholesterol.

immunoliposomes (C225-ILs) were used to encapsulate dodecahydro-*closo*-dodecaborate $[B_{12}H_{12}]^{2-}$ and their biological efficiency was evaluated *in vitro* using F98 EGFR cells. Fluorescence microscopy and flow cytometry studies showed that C225-ILs internalized in EGFR overexpressing F98 cells *in vitro*, presumably by receptor-mediated endocytosis, and that they distributed throughout the cell, including the plasma membrane, intracellular vesicles, and the cytosol. The immunoliposomes C225-ILs were found to deliver 509.75 ± 148 μg $^{10}B/10^9$ F98 EGFR cells, which corresponds to an eightfold increase in cellular uptake of boron as compared with that of nontargeted human IgG-liposomes.

10.2.7 Liposomes

The use of liposomes as boron delivery vehicles for BNCT has been extensively studied, since liposomes can transport large amounts of boronated molecules, both hydrophobic and hydrophilic in nature, selectively into tumor cells. Three methods for incorporation of boron into liposomes have been reported: incorporation of hydrophobic molecules into lipid bilayers, entrapment of hydrophilic molecules into the liposomes' aqueous core, and the use of boronated lipids. Recently, PEG-based liposomes (DPPC/cholesterol/DSPC-PEG2000) have been used as effective carriers of boronated compounds within tumor tissues. Intravenously injected ^{10}B-PEG-liposomes on human pancreatic carcinoma xenografts in nude mice inhibited tumor growth after thermal neutron irradiation.[186] The cationic liposome formulation (DOTAP/DOPC/DOPE) was used to deliver *ortho*-carboranyl β-lactoside (LCOB) and 1-methyl-*ortho*-carboranyl-2-hexylthioporphyrazine (H_2Pz-COB) within DHD/K12/TRb (DHD) rat colon carcinoma and B16-F10 murine melanoma cell lines.[187] Results from these studies show that a thirty-fold increase on the cellular uptake of ^{10}B was obtained for the boronated compounds in the liposome formulation as compared with that of the BPA–fructose complex.

Selective tumor targeting by liposome-based systems is generally achieved by covalent conjugation of targeting molecules to amine or carboxylate functional groups, available on the liposomes' surface. Targeting of receptors that are overexpressed in tumor cells is a strategy that has been exploited for the selective delivery of high amounts of boron to tumor cells. Transferrin, folate, EGFR, and VEGFR have been the receptors of choice for the selective delivery of ^{10}B to tumor tissues. Transferrin PEG-liposomes have been used to encapsulate BSH.[188] Delivery of ^{10}B for this system was studied in oral squamous cell carcinoma (SCC)-bearing mice. Results form this study show that the amount of boron delivered is higher compared with controls to which BSH and PEG-liposomes were administered.

Hawthorne and coworkers reported the synthesis and biological properties of *nido*-carborane lipid **69** (Figure 10.17).[189] DSPC/cholesterol liposomes containing **69** were shown to deliver up to 71 mg/g boron to EMT6 tumors in BALB/c mice. More recently, "double-tail" boronated lipids **70** and **71** were synthesized and evaluated.[190,191] A transferrin-loaded *nido*-carborane-PEG (Tf(+)-PEG-CL liposomes) was prepared, containing *nido*-carborane lipid **71** (Figure 10.17) as a

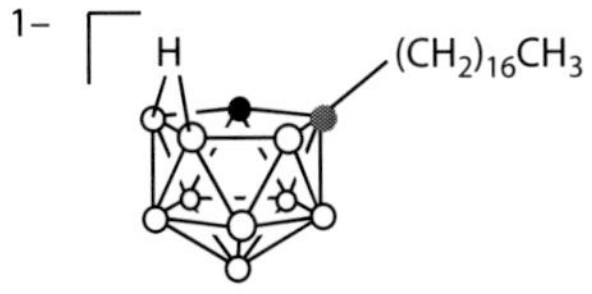

FIGURE 10.17 Boronated lipids.

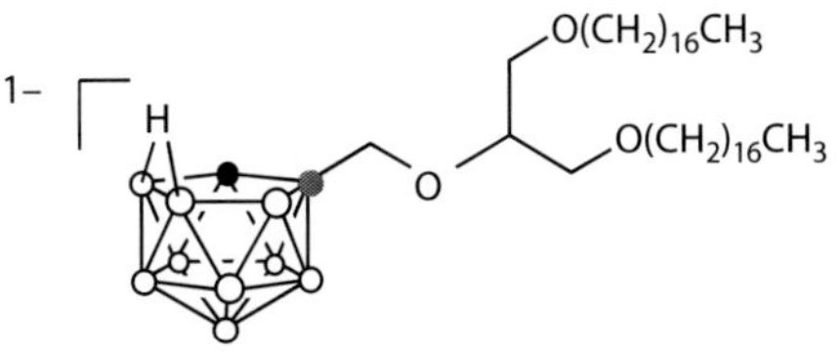

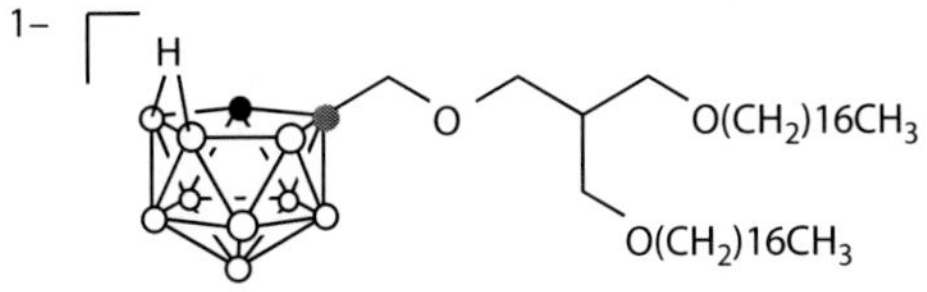

double-tailed boron lipid at 25% molar ratio toward DSPC with cholesterol. Conjugation of the PEG-CO_2H moieties to transferrin was achieved by using EDC and S-NHS as the coupling reagents in MES buffer.[191] Biodistribution studies showed that (Tf(+)-PEG-CL liposomes) in which ^{125}I-tyraminyl inulins were encapsulated accumulated preferentially in tissues with a boron concentration of 22 ppm after injection of the Tf(+)-PEG-CL liposomes at 7.2 mg/kg b.w. in tumor-bearing mice. After neutron irradiation, the average survival rate of mice treated with Tf(+)-PEG-CL liposomes was 31 days as compared to nontreated controls. However, the injection of higher boron dose resulted in mortality, which was also observed with liposomes containing **70**. Another transferrin-conjugated PEG-containing liposome was used to encapsulate BSH.[192] The delivery of ^{10}B by these (TF-PEG-BSH) liposomes to human U87D glioma cells was found to be efficient (25.48 ± 8.62 μg/g) compared with BSH and BSH-PEG controls and to also provide a higher survival rate, 21.8 ± 1.3 days versus 16.2 ± 2.3 days after neutron irradiation.

Design and synthesis of novel boronated lipids is also an area of interest, since the use of boron-containing lipids avoids problems associated with possible leakage of encapsulated boronated material. Boronated cholesterol mimics **72–74** (Figure 10.18) have been synthesized.[193] The major structural feature of these boronated cholesterol mimics is the replacement of the B and the C ring of cholesterol with a carborane cluster. These novel carboranyl derivatives were synthesized as lipid bilayer components of nontargeted and receptor-targeted boronated liposomes. Folate (FR) and vascular endothelial growth factor (VEGF) targeted liposomes incorporating cholesterol mimics **72–74** were prepared and their efficiency for ^{10}B delivery was evaluated in FR overexpressing KB cells and VEGFR-2 overexpressing 293/KDR cells. Targeted liposomes containing these boronated

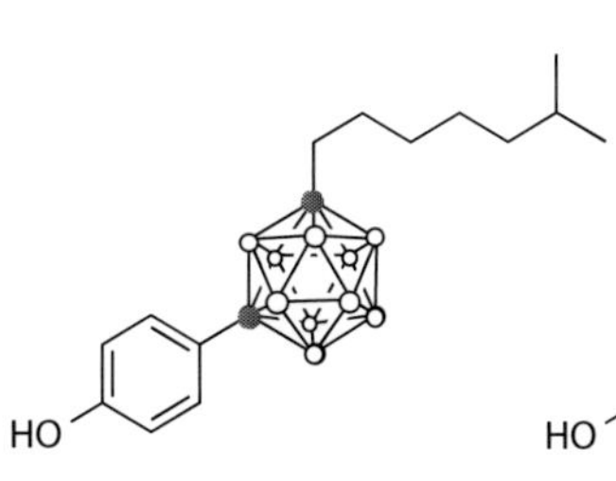

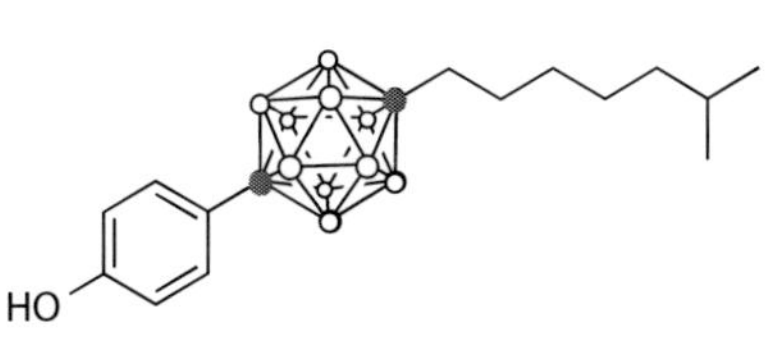

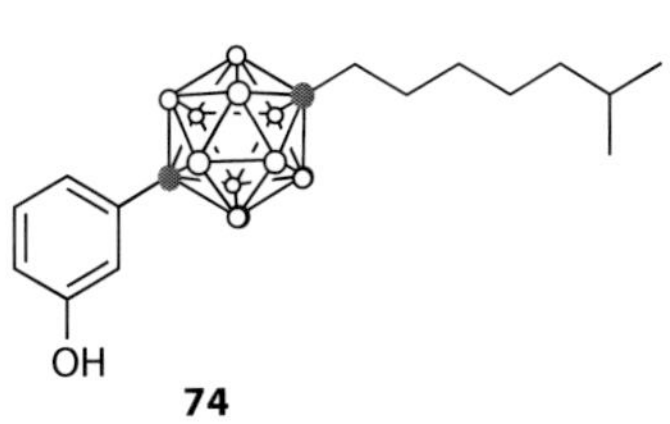

FIGURE 10.18 Boronated cholesterol mimics.

cholesterol mimics showed no cytotoxicity *in vitro* and the boron release from the targeted liposomes was higher than the nontargeted liposome system.

Other nanomaterials containing boron have been synthesized and proposed as delivery agents for BNCT.[194,195]

10.3 SUMMARY

With the advances in boron cluster chemistry, a large number of boronated compounds have been synthesized and proposed as selective BNCT delivery agents. Several of these are currently undergoing biological evaluation and further improvement for use in BNCT. In particular, a number of agents containing high percentage of boron by weight and low toxicity show promise for application in BNCT, although very little preclinical data are currently available. Simultaneously, alternative methods for the selective delivery of boronated agents are also being investigated and among these CED appears to be promising. CED of small boronated molecules (nucleosides and porphyrins), and of high-molecular-weight boronated MAb and EGF-containing particles has been shown to be effective in selectively delivering high amounts of boron into intracerebral animal tumors. Nevertheless, while IT administration of boronated compounds in general achieves very high tumor-to-normal tissues and tumor-to-blood boron concentration ratios, BNCT agents with the ability to penetrate tumor cell membranes and to localize intracellularly in close proximity or within cell nuclei are the most desired. Furthermore, the visualization and quantification of tumor-localized boron greatly assists treatment planning and improvement in drug design and delivery. Particularly promising classes of BNCT agents include boronated amino acids (such as **29**), nucleosides (such as **36**), porphyrin derivatives of high boron content (such as **33**), antibodies, as well as a combination of these. Indeed, a combination of various boronated agents with different tumor cell-targeting profiles might be desired for the optimization of BNCT, as has been previously demonstrated, for example, on studies using both BPA and BSH. On another hand, of the hundreds of boronated compounds that have been synthesized and proposed as delivery vehicles for BNCT, only a small number have been evaluated in biological studies, and most of these were investigated in animal tumor models that do not necessarily translate into their potential use for BNCT of human gliomas. More normalized and even centralized BNCT studies on the various classes of BNCT agents and their possible combinations and/or delivery optimizations might lead to further insight into their applicability and overall acceptance of BNCT. However, such clinical biodistribution and therapy studies are still difficult to conduct, particularly in the US and Europe, for a variety of reasons including availability of adequate neutron sources and cost. Nevertheless, the outlook for BNCT is clearly bright and it is anticipated that new boronated agents will soon be approved for clinical use in BNCT.

ACKNOWLEDGMENTS

MGHV is thankful for the continuing support of our work on the development of boronated porphyrin derivatives, from the National Institutes of Health, current grant number R01 CA 098902.

REFERENCES

1. Hawthorne, M. F. The role of chemistry in the development of boron neutron-capture therapy of cancer. *Angew. Chem. Int. Ed. Eng.* **1993**, *32*, 950–984.
2. Soloway, A. H.; Tjarks, W.; Barnum, B. A.; Rong, F. G.; Barth, R. F.; Codogni, I. M.; Wilson, J. G. The chemistry of neutron capture therapy. *Chem. Rev.* **1998**, *98*, 1515–1562.
3. Barth, R. F.; Soloway, A. H.; Goodman, J. H.; Gahbauer, R. A.; Gupta, N.; Blue, T. E.; Yang, W. L.; Tjarks, W. Boron neutron capture therapy of brain tumors: An emerging therapeutic modality. *Neurosurg.* **1999**, *44*, 433–451.

4. Barth, R. F.; Coderre, J. A.; Vicente, M. G. H.; Blue, T. E. Boron neutron capture therapy of cancer: Current status and future prospects. *Clin. Cancer Res.* **2005**, *11*, 3987–4002.
5. Barth, R. F.; Joensuu, H. Boron neutron capture therapy for the treatment of gliobtastomas and extracranial tumours: As effective, more effective or less effective than photon irradiation? *Rad. Oncol.* **2007**, *82*, 119–122.
6. Diaz, A. Z. Assessment of the results from the phase I/II boron neutron capture therapy trials at the Brookhaven National Laboratory from a clinician's point of view. *J. Neuro-Oncol.* **2003**, *62*, 101–109.
7. Busse, P. M.; Harling, O. K.; Palmer, M. R.; Kiger III, W. S.; Kaplan, J.; Kaplan, I.; Chuang, C. F. et al., A critical examination of the results from the Harvard-MIT NCT program phase I clinical trial of neutron capture therapy for intracranial disease. *J. Neuro-Oncol.* **2003**, *62*, 111–121.
8. Nakagawa, Y.; Pooh, K.; Kobayashi, T.; Kageji, T.; Uyama, S.; Matsumura, A.; Kumada, H. Clinical review of the Japanese experience with boron neutron capture therapy and a proposed strategy using epithermal neutron beams. *J. Neuro-Oncol.* **2003**, *62*, 87–99.
9. Hideghety, K.; Sauerwein, W.; Wittig, A.; Gotz, C.; Paquis, P.; Grochulla, F.; Haselsberger, K. et al., Tissue uptake of BSH in patients with glioblastoma in the EORTC 11961 phase I BNCT trial. *J. Neuro-Oncol.* **2003**, *62*, 145–156.
10. Joensuu, H.; Kankaanranta, L.; Seppala, T.; Auterinen, I.; Kallio, M.; Kulvik, M.; Laakso, J. et al., Boron neutron capture therapy of brain tumors: clinical trials at the Finnish facility using boronophenylalanine. *J. Neuro-oncol.* **2003**, *62*, 123–134.
11. Capala, J.; Stenstam, B. H.; Skold, K.; af Rosenschold, P. M.; Giusti, V.; Persson, C.; Wallin, E. et al., Boron neutron capture therapy for glioblastoma multiforme: clinical studies in Sweden. *J. Neuro-Oncol.* **2003**, *62*, 135–144.
12. Vitale, A. A.; Hoffman, G.; Pomilio, A. B. Boron-containing bioactive molecules: An approach to boron neutron capture therapy. *Mol. Med. Chem.* **2005**, *8*, 1–49.
13. Yamamoto, T.; Nakai, K.; Matsumura, A. Boron neutron capture therapy for glioblastoma. *Cancer Lett.* **2008**, *2*, 143–152.
14. Bregadze, V. I.; Sivaev, I. B.; Glazun, S. A. Polyhedral Boron Compounds as Potential Diagnostic and Therapeutic Antitumor Agents. *Curr. Med. Chem., Anti-Cancer Agents* **2006**, *6*, 75–110.
15. Sivaev, I. B.; Bregadze, V. I. Polyhedral Boranes for Medical Applications: Current Status and Perspectives. *Eur. J. Inorg. Chem.* **2009**, *11*, 1433–1450.
16. Fairchild, R. G.; Bond, V. P. Current status of b-10-neutron capture therapy - enhancement of tumor dose via beam filtration and dose-rate, and the effects of these parameters on minimum boron content - a theoretical evaluation. *Int. J. Radiat. Oncol. Biol. Phys.* **1985**, *11*, 831–840.
17. Gabel, D.; Foster, S.; Fairchild, R. G. The monte-carlo simulation of the biological effect of the b-10(n,alpha)li-7 reaction in cells and tissue and its implication for boron neutron-capture therapy. *Radiat. Res.* **1987**, *111*, 14–25.
18. Sibrian-Vazquez, M.; Hao, E.; Jensen, T. J.; Vicente, M. G. H. Enhanced cellular uptake with a cobaltacarborane-porphyrin-HIV-1 Tat 48–60 conjugate. *Bioconj. Chem.* **2006**, *17*, 928–934.
19. Barth, R. F.; Wu, G.; Yang, W.; Binns, P. J.; Riley, K. J.; Patel, H.; Coderre, J. A. et al., Neutron capture therapy of epidermal growth factor (plus) gliomas using boronated cetuximab (IMC-C225) as a delivery agent. *Appl. Rad. Isotopes* **2004**, *61*, 899–903.
20. Burgess, A. W. EGFR family: Structure physiology signalling and therapeutic targets. *Growth Factors* **2008**, *26*, 263–274.
21. Barth, R. F. Boron neutron capture therapy at the crossroads: Challenges and opportunities. *Appl. Rad. Isotopes* **2009**, *67*, S3–S6.
22. Ferguson, S.; Lesniak, M. S. Convection enhanced drug delivery of novel therapeutic agents to malignant brain tumors. *Curr. Drug Deliv.* **2007**, *4*, 169–180.
23. Barth, R. F.; Yang, W. L.; Al-Madhoun, A. S.; Johnsamuel, J.; Byun, Y.; Chandra, S.; Smith, D. R.; Tjarks, W.; Eriksson, S. Evaluation of human thymidine kinase 1 substrates as new candidates for boron neutron capture therapy. *Cancer Res.* **2004**, *64*, 6287–6295.
24. Ozawa, T.; Santos, R. A.; Lamborn, K. R.; Bauer, W. F.; Koo, M.-S.; Kahl, S. B.; Deen, D. F. *In vivo* evaluation of the boronated porphyrin TABP-1 in U-87 MG intracerebral human glioblastoma xenografts. *Mol. Pharm.***2004**, *1*, 368–374.
25. Ozawa, T.; Afzal, J.; Lamborn, K. R.; Bollen, A. W.; Bauer, W. F.; Koo, M. S.; Kahl, S. B.; Deen, D. F. Toxicity, biodistribution, and convection-enhanced delivery of the boronated porphyrin BOPP in the 9L intracerebral rat glioma model. *Int. J. Rad. Oncol. Biol. Phys.* **2005**, *63*, 247–252.
26. Kawabata, S.; Barth, R. F.; Yang, W. L.; Wu, G.; Gottumukkala, V.; Vicente, M. G. H. *Proc. 13th World Congress of Neurological Surgery* **2005**, *1*, 975–979.

27. Kawabata, S.; Barth, R. F.; Yang, W.; Wu, G.; Binns, P. J.; Riley, K. J.; Ongayi, O.; Gottumukkala, V.; Vicente, M. G. H. *Proc. 12th International Congress on Neutron Capture Therapy for Cancer* **2006**, *1*, 123–126.
28. Kawabata, S.; Yang, W.; Barth, R. F.; Wu, G.; Huo, T.; Binns, P. J.; Riley, K. J.; Ongayi, O.; Gottumukkala, V.; Vicente, M. G. H., Convection enhanced delivery of carboranylporphyrins for neutron capture therapy of brain tumors. *J Neurooncol.* **2010**, DOI 10.1007/s11060-010-0376-5.
29. Yang, W.; Barth, R. F.; Wu, G.; Huo, T.; Tjarks, W.; Ciesielski, M.; Fenstermaker, R. A. et al. Convection enhanced delivery of boronated EGF as a molecular targeting agent for neutron capture therapy of brain tumors. *J. Neuro-oncol.* **2009**, *95*, 355–365.
30. Barth, R. F.; Yang, W.; Rotaru, J. H.; Moeschberger, M. L.; Joel, D. D.; Nawrocky, M. M.; Goodman, J. H.; Soloway, A. H. Boron neutron capture therapy of brain tumors: Enhanced survival following intracarotid injection of either sodium borocaptate or boronophenylalanine with or without blood-brain barrier disruption. *Cancer Res.* **1997**, *57*, 1129–1136.
31. Barth, R. F.; Yang, W.; Rotaru, J. H.; Moeschberger, M. L.; Boesel, C. P.; Soloway, A. H.; Joel, D. D.; Nawrocky, M. M.; Ono, K.; Goodman, J. H. Boron neutron capture therapy of brain tumors: Enhanced survival and cure following blood-brain barrier disruption and intracarotid injection of sodium borocaptate and boronophenylalanine. *Intl. J. Radiat. Oncol. Biol. Phys.* **2000**, *47*, 209–218.
32. Barth, R. F.; Yang, W.; Bartus, R. T.; Rotaru, J. H.; Ferketich, A. K.; Moeschberger, M. L.; Nawrocky, M. M.; Coderre, J. A.; Rofstad, E. K. Neutron capture therapy of intracerebral melanoma: Enhanced survival and cure after blood-brain barrier opening to improve delivery of boronophenylalanine. *Intl. J. Radiat. Oncol. Biol. Phys* **2002**, *52*, 858–868.
33. Capuani, S.; Gili, T.; Bozzali, M.; Russo, S.; Porcari, P.; Cametti, C.; Muolo, M. et al. Boronophenylalanine uptake in C6 glioma model is dramatically increased by L-DOPA preloading. *Appl. Rad. Isotopes* **2009**, *67*, S34–S36.
34. Grimes, R. N. *Carboranes*; New York: Academic Press, **1970**.
35. Bregadze, V. I. Dicarba-*closo*-dodecaboranes $C_2B_{10}H_{12}$ and their derivatives. *Chem. Rev.* **1992**, *92*, 209–223.
36. Plesek, J.; Hermanek, S.; Franken, A.; Cisarova, I.; Nachtigal, C. Dimethyl sulfate induced nucleophilic substitution of the [bis(1,2-dicarbollido)-3-cobalt(1-)]ate ion. Syntheses, properties and structures of its 8,8'-mu-sulfato, 8-phenyl and 8-dioxane derivatives. *Collect. Czech. Chem. Commun.* **1997**, *62*, 47–56.
37. Teixidor, F.; Pedrajas, J.; Rojo, I.; Vinas, C.; Kivekas, R.; Sillanpaa, R.; Sivaev, I.; Bregadze, V.; Sjoberg, S. Methylation and demethylation in cobaltabis(dicarbollide) derivatives. *Organometallics* **2003**, *22*, 3414–3423.
38. Tiwari, R.; Mahasenan, K.; Pavlovicz, R.; Li, C.; Tjarks, W. Carborane clusters in computational drug design: a comparative docking evaluation using autodock, flexx, glide, and surflex. *J. Chem. Inf. Model.* **2009**, *49*, 1581–1589.
39. Sivaev, I. B.; Bregadze, V. I.; Kuznetsov, N. T. Derivatives of the *closo*-dodecaborate anion and their application in medicine. *Russ. Chem. Bull. Int. Ed.* **2002**, *51*, 1362–1365.
40. Gabel, D.; Harfst, S.; Moller, D.; Ketz, H.; Peymann, T.; Rosler, J. In *Current Topics in the Chemistry of Boron*, Kabalka, G. W. Ed., Cambridge: The Royal Society of Chemistry, 1994; pp. 161–164.
41. Matsumura, A.; Shibata, Y.; Yamamoto, T.; Yoshida, F.; Isobe, T.; Nakai, K.; Hayakawa, Y. et al. A new boronated porphyrin (STA-BX909) for neutron capture therapy: an *in vitro* survival assay and *in vivo* tissue uptake study. *CancerLett.* **1999**, *141*, 203–209.
42. Harfst, S.; Moller, D.; Ketz, H.; Rosler, J.; Gabel, D. Reversed-phase separation of ionic organoborate clusters by high-performance liquid-chromatography. *J. Chromatogr. A* **1994**, *678*, 41–48.
43. Ratajski, M.; Osterloh, J.; Gabel, D. Boron-Containing Chlorins and Tetraazaporphyrins: Synthesis and Cell Uptake of Boronated Pyropheophorbide A Derivatives. *Curr. Med. Chem., Anti-Cancer Agents* **2006**, *6*, 159–166.
44. Grin, M. A.; Semioshkin, A. A.; Titeev, R. A.; Nizhnik, E. A.; Grebenyuk, J. N.; Mironov, A. F.; Bregadze, V. I. Synthesis of a cycloimide bacteriochlorin p conjugate with the *closo*-dodecaborate anion. *Mendeleev Comm.* **2007**, *17*, 14–25.
45. Hao, E.; Friso, E.; Miotto, G.; Jori, G.; Soncin, M.; Fabris, C.; Sibrian-Vazquez, M.; Vicente, M. G. H. Synthesis and biological investigations of tetrakis(p-carboranylthio-tetrafluorophenyl)chlorin (TPFC). *Org. Biomol. Chem.* **2008**, *6*, 3732–3740.
46. Fabris, C.; Jori, G.; Giuntini, F.; Roncucci, G. J. Photosensitizing properties of a boronated phthalocyanine: studies at the molecular and cellular level. *Photochem. Photobiol. B: Biol.* **2001**, *64*, 1–7.
47. Semioshkin, A.; Tsaryova, O.; Zhidkova, O.; Bregadze, V.; Wohrle, D. Reactions of oxonium derivatives of $(B_{12}H_{12})_2$- with phenols, and synthesis and photochemical properties of a phthaloocyanine containing four $(B_{12}H_{12})_2$- groups. *J. Porphyrins Phthalocyanines* **2006**, *10*, 1293–1300.

48. Giovenzana, G. B.; Lay, L.; Monty, D.; Palmisano, G.; Panza, L. Synthesis of carboranyl derivatives of alkynyl glycosides as potential BNCT agents. *Tetrahedron* **1999**, *55*, 14123–14136.
49. Ronchi, S.; Prosperi, D.; Thimon, C.; Morinc, C.; Panza, L. Synthesis of mono- and bisglucuronylated carboranes. *Tetrahedron: Asymmetry* **2005**, *16*, 39–44.
50. Tietze, L. F.; Bothe, U.; Griesbach, U.; Nakaichi, M.; Hasegawa, T.; Nakamura, H.; Yamamoto, Y. ortho-Carboranyl glycosides for the treatment of cancer by boron neutron capture therapy. *Bioorg. Med. Chem.* **2001**, *9*, 1747–1752.
51. Orlova, A. V.; Zinin, A. I.; Malysheva, N. N.; Kononov, L. O.; Sivaev, I. B.; Bregadze, V. I. Conjugates of polyhedral boron compounds with carbohydrates. 1. New approach to the design of selective agents for boron neutron capture therapy of cancer. *Russ. Chem.Bull., Int. Ed.* **2003**, *52*, 2766–2768.
52. Orlova, A. V.; Kononov, L. O.; Kimel, B. G.; Sivaev, I. B.; Bregadze, V. I. Conjugates of polyhedral boron compounds with carbohydrates. 4. Hydrolytic stability of carborane-lactose conjugates depends on the structure of a spacer between the carborane cage and sugar moiety. *Appl. Organometal. Chem.* **2006**, *20*, 416–420.
53. Tietze, L. F.; Bothe, U. Ortho-carboranyl glycosides of glucose, mannose, maltose and lactose for cancer treatment by boron neutron-capture therapy. *Chem. Eur. J.* **1998**, *4*, 1179–1183.
54. Raddatz, S.; Marcello, M.; Kliem, H.-C.; Troster, H.; Trendelenburg, M. F.; Oeser, T.; Granzow, C.; Wiessler, M. Synthesis of new boron-rich building blocks for boron neutron capture therapy or energy-filtering transmission electron microscopy. *Chem. Biol. Chem.* **2004**, *5*, 474–482.
55. Kononov, L. O.; Orlova, A. V.; Zinin, A. I.; Kimel, B. G.; Sivaev, I. B.; Bregadze, V. I. Conjugates of polyhedral boron compounds with carbohydrates. 2. Unexpected easy *closo-* to *nido*-transformation of a carborane-carbohydrate conjugate in neutral aqueous solution. *J. Organomet. Chem.* **2005**, *690*, 2769–2774.
56. Tietze, L. F.; Bothe, U.; Schuberth, I. Preparation of a new carboranyl lactoside for the treatment of cancer by boron neutron capture therapy: Synthesis and toxicity of fluoro carboranyl glycosides for *in vivo* F-19-NMR spectroscopy. *Chem. Eur. J.* **2000**, *6*, 836–842.
57. Tietze, L. F.; Griebach, U.; Schuberth, I.; Bothe, U.; Marra, A.; Dondoni, A. Novel carboranyl C-glycosides for the treatment of cancer by boron neutron capture therapy. *Chem. Eur. J.* **2003**, *9*,1296–1302.
58. Basak, P.; Lowary, T. L. Synthesis of conjugates of L-fucose and ortho-carborane as potential agents for boron neutron capture therapy. *Can. J. Chem.* **2002**, *80*, 943–948.
59. Swenson, D. H.; Laster, B. H.; Metzger, R. L. Synthesis and evaluation of a boronated nitroimidazole for boron neutron capture therapy. *J. Med. Chem.* **1996**, *39*, 1540–1544.
60. Lechtenberg, B.; Gabel, D. Synthesis of a (B12H11S)(2-) containing glucuronoside as potential prodrug for BNCT. *J. Organometal. Chem.* **2005**, *690*, 2780–2782.
61. Meo, C. D.; Panza, L.; Capitani, D.; Mannina, L.; Banzato, A.; Rondina, M.; Renier, D.; Rosato, A.; Crescenzi, V. Hyaluronan as carrier of carboranes for tumor targeting in boron neutron capture therapy. *Biomacromolecules* **2007**, *8*, 552–559.
62. Scobie, M.; Threaddgill, M. D. Tumor-Targeted Boranes. 4. Synthesis Of Nitroimidazole-Carboranes With Polyether-Isoxazole Links. *J. Org. Chem.* **1994**, *59*, 7008–7013.
63. Wood, P. J.; Scobie, M.; Threaddgill, M. D. Uptake and retention of nitroimidazole-carboranes designed for boron neutron capture therapy in experimental murine tumours: Detection by B-11 magnetic resonance spectroscopy. *Int. J. Radiat. Biol.* **1996**, *70*, 587–592.
64. Schirrmacher, E.; Schirrmacher R.; Beck, C.; Mier, W.; Trautman, N.; Rosh, F. Synthesis of a Tyr(3)-octreotate conjugated *closo*-carborane [HC2B10H10]: a potential compound for boron neutron capture therapy. *Tetrahedron Lett.* **2003**, *44*, 9143–9145.
65. Mier, W.; Gabel, D.; Haberkorn, U.; Eisenhut, M. Z. Conjugation of the *closo*-borane mereaptoundecahydrododecaborate (BSH) to a tumour selective peptide. *Anorg. Allg. Chem.* **2004**, *630*, 1258–1262.
66. Justus, E.; Awad, D.; Hohnholt, M.; Schaffran, T.; Edwards, K.; Karlsson, G.; Damian, L.; Gabel, D. Synthesis, liposomal preparation, and *in vitro* toxicity of two novel dodecaborate cluster lipids for boron neutron capture therapy. *Bioconjugate Chem.* **2007**, *18*, 1287–1293.
67. Nakamura, H.; Lee J-D.; Ueno, M.; Miyajima, Y.; Ban, H-S. *Nanobiotechnol.* **2007**, *3*, 135–145.
68. Lee, J.-D.; Ueno, M.; Miyajima, Y.; Nakamura, H. Synthesis of boron cluster lipids: *closo*-dodecaborate as an alternative hydrophilic function of boronated liposomes for neutron capture therapy. *Org. Lett.* **2007**, *9*, 323–326.
69. Nakamura, H.; Ueno, M.; Ban, H.-S.; Nakai, K.; Tsuruta, K.; Kaneda, Y.; Matsumura, A. Development of boron nanocapsules for neutron capture therapy. *Appl. Rad. Isotopes* **2009**, *67*, S84–S87.
70. Diaz, S.; Gonzalez, A.; De Riancho, S. G.; Rodriguez, A. Boron complexes of S-trityl-L-cysteine and S-tritylglutathione. *J. Organomet. Chem.* **2000**, *610*, 25–30.

71. Lindstrom, P.; Naeslund, C.; Sjoberg S. Enantioselective synthesis and absolute configurations of the enantiomers of o-carboranylalanine. *Tetrahedron Lett.* **2000**, *41*, 751–754.
72. Masunaga, S.-I.; Ono, K.; Kirihata, M.; Takagaki, M.; Sakurai, Y.; Kinashi, Y.; Kobayashi, T. et al., Potential of alpha-amino alcohol p-boronophenylalaninol as a boron carrier in boron neutron capture therapy, regarding its enantiomers. *J. Cancer Res. Clin. Oncol.* **2003**, *129*, 21–28.
73. Ujvary, I.; Nachman, R. J. Synthesis of 3-(12-hydroxy-p-carboranyl)propionic acid, a hydrophobic, N-terminal tyrosine-mimetic for peptides. *Peptides* **2001**, *22*, 287–290.
74. Beletskaya, I. P.; Bregadze, V. I.; Osipov, S. N.; Petrovskii, P. V.; Starikova, Z. A.; Timofeev, S. V. New nonnatural alpha-amino acid derivatives with carboranyl fragments in alpha- and beta-positions. *Synlett* **2004**, 1247–1248.
75. Morin, C.; Thimon, C. Synthesis and evaluation of boronated lysine and bis(carboranylated) gamma-amino acids as monomers for peptide assembly. *Eur. J. Org. Chem.* **2004**, 3828–3832.
76. Timofeev, S. V.; Bregadze, V. I.; Osipov, S. N.; Titanyuk, I. D.; Petrovskii, V.; Starikova, Z. A.; Glukhov, I. V.; Beletskayab, I. P. New carborane-containing amino acids and their derivatives. Crystal structures of n-protected carboranylalaninates. *Russ. Chem. Bull. Int. Ed.* **2007**, *56*, 791–797.
77. Semioshkin, A.; Nizhnik, E.; Godovikov, I.; Starikova, Z.; Bregadze, V. Reactions of oxonium derivatives of [B-12 H-12](2-) with amines: Synthesis and structure of novel B-12-based ammonium salts and amino acids. *J. Organometallic Chem.* **2007**, *692*, 4020–4028.
78. Kabalka, G. W.; Yao, M.-L. The synthesis and use of boronated amino acids for boron neutron capture therapy. *Curr. Med. Chem. Anti-Cancer Agents* **2006**, 111–126.
79. Kabalka, G. W.; Yao, M.-L. Synthesis of 1-amino-3-[(dihydroxyboryl)methyl]-cyclobutanecarboxylic acid as a potential therapy agent. *J. Org. Chem.* **2004**, *69*, 8280–8286.
80. Kabalka, G. W.; Yao, M.-L. Synthesis of a novel boronated 1-aminocyclobutanecarboxylic acid as a potential boron neutron capture therapy agent. *Appl. Organomet Chem* **2003**, *17*, 398–402.
81. Kabalka, G. W.; Wu, Z. Z.; Yao, M.-L.; Natarajan, N. The syntheses and *in vivo* biodistribution of novel boronated unnatural amino acids. *App. Rad. Isotopes* **2004**, *61*, 1111–1115.
82. Kabalka, G. W.; Wu, Z.; Yao, M.-L. Synthesis of a series of boronated unnatural cyclic amino acids as potential boron neutron capture therapy agents. *Appl. Organometal. Chem.* **2008**, *22*, 516–522.
83. Kabalka, G. W.; Yao, M.-L.; Marepally, S. R.; Chandra, S. Biological evaluation of boronated unnatural amino acids as new boron carriers. *App. Rad. Isotopes* **2009**, *67*, S374–S379.
84. Sivaev, I. B.; Starikova, Z. A.; Sjoberg, S.; Bregadze, V. I. Synthesis of functional derivatives of the [3,3′-Co(1,2-C2B2H11)(2)](-) anion. *J. Organomet. Chem.* **2002**, *649*, 1–8.
85. Takagaki, M.; Powell, W.; Sood, A.; Spielvogel, B. F.; Hosmane, N. S.; Kirihata, M.; Ono, K. et al. Boronated dipeptide borotrimethylglycylphenylalanine as a potential boron carrier in boron neutron capture therapy for malignant brain tumors. *Radiat Res.* **2001**, *156*, 118–122.
86. Wakamiya, T.; Yamashita, T.; Fujii, T.; Yamaguchi, Y.; Nakano, T.; Kirihata, M. Syntheiss of 4-boronphenylalanine-containig peptides for boron neutron capture therapy of cancer cells. *Peptide Sci.* **1999**, *36*, 209–212.
87. Spielvogel, B. F.; Rana, G.; Vyakaranam, K.; Grelck, K.; Dicke, K. E.; Dolash, B. D.; Li, S.-J. et al., A novel approach to the syntheses of functionalized, water-soluble icosahedral carboranyl anions. Crystal structure of methyl N-[(trimethylamineboryl)carbonyl]-L- tyrosinate: A synthon for novel carboranyl-peptides. *Collect Czech Chem. Commun.* **2002**, *67*, 1095–1108.
88. Ivanov, D.; Bachovchin, W. W.; Redfield, A. G. Boron-11 pure quadrupole resonance investigation of peptide boronic acid inhibitors bound to alpha-lytic protease. *Biochemistry* **2002**, *41*, 1587–1590.
89. Vives, E. Cellular uptake of the Tat peptide: an endocytosis mechanism following ionic interactions. *J. Mol. Recog.* **2003**, *16*, 265–271.
90. Shirakawa, M.; Yamamto, T.; Nakai, K,; Aburai, K.; Kawatobi, S.; Tsurubuchi, T.; Yamamoto, Y.; Yokoyama, Y.; Okuno, H.; Matsumura, A. Synthesis and evaluation of a novel liposome containing BPA-peptide conjugate for BNCT. *Appl. Rad. Isotopes* **2009**, *67*, S88–S90.
91. Tjarks, W. The use of boron clusters in the rational design of boronated nucleosides for neutron capture therapy of cancer *J. Organomet. Chem.* **2000**, *614–615*, 37–47.
92. Schinazi, R. F.; Hurwitz, S. J.; Liberman, I.; Juodawlkis, A. S.; Tharnish, P.; Shi, J.; Liotta, D. C.; Coderre, J. A.; Olson, J. Treatment of isografted 9L rat brain tumors with beta-5-o-carboranyl-2′-deoxyuridine neutron capture therapy. *Clin. Cancer Res.* **2000**, *6*, 725–730.
93. Al-Madhoun, A. S.; Johnsamuel, J.; Yan, J.; Ji, W.; Wang, J.; Zhuo, J.-C.; Lunato, A. J. et al., Synthesis of a small library of 3-(Carboranylalkyl)thymidines and their biological evaluation as substrates for human thymidine kinases 1 and 2. *J. Med.Chem.* **2002**, *45*, 4018–4028.

94. Yan, J.; Naeslund, C.; Al-Madhoun, A. S.; Wang, J.; Ji, W.; Cosquer, G. Y.; Johnsamuel, J.; Sjoberg, S.; Eriksson, S.; Tjarks, W. Synthesis and biological evaluation of 3′-Carboranyl thymidine analogues. *Bioorg. Med. Chem. Lett.* **2002**, *12*, 2209–2212.
95. Vyakaranam, K.; Rana, G.; Delaney, S.; Ledger, S.; Hosmane, N. S. The first carboranyl bis(adenosine diphosphate) (CBADP): a synthetic investigation. *Inorg. Chem. Commun.* **2003**, *6*, 654–657.
96. Olejniczak, A. B.; Semenuk, A.; Kwiatkowski, M.; Lesnikowski Z. J. Synthesis of adenosine containing carborane modification. *J. Organomet. Chem.* **2003**, *680*, 124–126.
97. Schinazi, R. F.; Hurwitz, S. J.; Liberman, I.; Glazkova, Y.; Mourier, N. S.; Olson, J.; Keane, T. Tissue disposition of 5-*o*-carboranyluracil - A novel agent for the boron neutron capture therapy of prostate cancer. *Nucleosides, Nucleotides and Nucleic Acids* **2004**, *23*, 291–306.
98. Johnsamuel, J.; Lakhi, N.; Al-Madhoun A. S.; Byun, Y.; Yan, J.; Eriksson, S.; Tjarks, W. Synthesis of ethyleneoxide modified 3-carboranyl thymidine analogues and evaluation of their biochemical, physicochemical, and structural properties. *Bioorg. Med. Chem.* **2004**, *12*, 4769–4781.
99. Al-Madhoun, A. S.; Johnsamuel, J.; Barth, R. F.; Tjarks, W.; Eriksson, S. Evaluation of human thymidine kinase 1 substrates as new candidates for boron neutron capture therapy. *Cancer Res.* **2004**, *64*, 6280–6286.
100. Wojtczak, B.; Semenyuk, A.; Olejniczak, A. B.; Kwiatkowski, M.; Lesnikowski, Z. J. General method for the synthesis of 2′-*O*-carboranyl-nucleosides. *Tetrahedron Lett.* **2005**, *46*, 3969–3972.
101. Byun, Y.; Yan, J.; Al-Madhoun A. S.; Johnsamuel, J.; Yang, W.; Barth, R. F.; Eriksson, S.; Tjarks, W. Synthesis and biological evaluation of neutral and zwitterionic 3-carboranyl thymidine analogues for boron neutron capture therapy. *J. Med. Chem.* **2005**, *48*, 1188–1198.
102. Byun, Y.; Narayanasamy, S.; Johnsamuel, J.; Bandyopadhyaya, A. K.; Tiwari, R.; Al-Madhoun, A. S.; Barth, R. F.; Eriksson, S.; Tjarks, W. *Curr.* Carboranyl Thymidine Analogues (3CTAs) and Other Boronated Nucleosides for Boron Neutron Capture Therapy *Med. Chem. Anti-Cancer Agents* **2006**, *6*, 127–144.
103. Byun, Y.; Thirumamagal, B. T. S.; Yang, W. L.; Eriksson, S.; Barth, R. F.; Tjarks, W. Preparation and biological evaluation of B-10-enriched 3-[5-{2-(2,3-dihydroxyprop-1-yl)-o-carboran-1-yl}pentan-1-yl] thymidine (N5-2OH), a new boron delivery agent for boron neutron capture therapy of brain tumors. *J. Med. Chem.* **2006**, *49*, 5513–5523.
104. Barth, R. F.; Yang, W.; Wu, G.; Swindall, M.; Byun, Y.; Narayanasamy, S.; Tjarks, W. et al., Thymidine kinase 1 as a molecular target for boron neutron capture therapy of brain tumors. *Proc. Nat. Acad. Sci. USA* **2008**, *105*, 17493–17497.
105. Thirumamagal, B. T. S.; Johnsamuel, J.; Cosquer, G. Y.; Byun, Y.; Yan, J.; Narayanasamy, S.; Tjarks, W.; Barth, R. F.; Al-Madhoun, A. S.; Eriksson, S. Boronated thymidine analogues for boron neutron capture therapy. *Nucleosides Nucleot. Nucleic Acids* **2006**, *25*, 861–866.
106. Narayanasamy, S.; Thirumamagal, B. T. S.; Johnsamuel, J.; Byun, Y.; Al-Madhoun, A. S.; Usova, E.; Cosquer, G. Y. et al., Hydrophilically enhanced 3-carboranyl thymidine analogues (3CTAs) for boron neutron capture therapy (BNCT) of cancer. *Bioorg. Med. Chem.* **2006**, *14*, 6886–6899.
107. Tjarks, W.; Tiwari, R.; Byun, Y.; Narayanasamy, S.; Barth, R. F. Carboranyl Thymidine Analogues for Neutron Capture Therapy. *Chem. Comm.* **2007**, *47*, 4978–4991.
108. Olejniczak, A. B.; Plešek, J.; Křiž O.; Lesnikowski Z. J. A nucleoside conjugate containing a metallacarborane group and its incorporation into a DNA oligonucleotide. *Angew. Chem. Int. Ed.* **2003**, *43*, 5740–5743.
109. Lesnikowski, Z. J.; Paradowska, E.; Olejniczak, A. B.; Studzinska, M.; Seekamp, P.; Schusler, U.; Gabel, D.; Schinazi, R. F.; Plešek. Towards new boron carriers for boron neutron capture therapy: metallacarboranes and their nucleoside conjugates. *Bioorg. Med. Chem.* **2005**, *13*, 4168–4175.
110. Renner, M. W.; Miura, M.; Easson, M. W.; Vicente, M. G. H. Recent Progress in the Syntheses and Biological Evaluation of Boronated Porphyrins for Boron Neutron Capture Therapy. *Curr. Med. Chem., Anti-Cancer Agents* **2006**, *6*, 145–157.
111. Ol'shevskaya, V. A.; Zaytsev, A. V.; Savchenko, A. N.; Shtil, A. A.; Cheong, C. S.; Kalinin, V. N. Boronated porphyrins and chlorins as potential anticancer drugs. *Bull. Korean Chem. Soc.* **2007**, *28*, 1910–1916.
112. Vicente, M. G. H.; Sibrian-Vazquez, M. In *The Handbook of Porphyrin Science*, Kadish, K. M.; Smith, K. M.; Guilard, R. Eds., Singapore: World Scientific Publishers, 2010, Vol. 4, Chapter 18, pp. 191–248.
113. Dougherty, T. J.; Gomer, C. J.; Henderson, B. W.; Jori, G.; Kessel, D.; Korbelik, M.; Moan, J.; Peng, Q. Photodynamic therapy. *J. Natl. Cancer Inst.* **1998**, *90*, 889–905.
114. Pandey, R. K.; Zhang, G. In *The Porphyrin Handbook*, Kadish, K. M.; Smith, K. M.; Guilard, R. Eds., Boston: Academic Press, 2002; Vol. 6, pp. 157–230.

115. Brown, S.; Brown, E. A.; Walker, I. The present and future role of photodynamic therapy in cancer treatment. *Lancet Oncol.* **2004**, *5*, 497–508.
116. Huang, Z. A review of progress in clinical photodynamic therapy. *Techn. Cancer Res. Treat.* **2005**, *4*, 283–293.
117. Osterloh, J.; Vicente, M. G. H. Mechanisms of porphyrinoid localization in tumors. *J. Porphyrins and Phthalocyanines* **2002**, *6*, 305–324.
118. Haushalter, R. C.; Rudolph, R. W. Meso-Tetracarboranylporphyrins. *J. Am. Chem. Soc.* **1978**, *100*, 4628–4629.
119. Haushalter, R. C.; Butler, W. M.; Rudolph, R. W. The Preparation And Characterization Of Several Meso-Tetracarboranylporphyrins. *J. Am. Chem. Soc.* **1981**, *103*, 2620–2627.
120. Kahl, S. B.; Koo, M.-S.; Laster, B. H.; Fairchild, R. G. Boronated porphyrins in NCT: Results with a new potent tumor localizer. *Strahlenther. Onkol.* **1989**, *165*, 131–134.
121. Miura, M.; Gabel, D.; Fairchild, R. G.; Laster, B. H.; Warkentien, L. S. Synthesis and *in vivo* studies of a carboranyl porphyrin. *Strahlenther. Onkol.* **1989**, *165*, 134–137.
122. Kahl, S. B.; Joel, D. D.; Nawrocky, M. M.; Micca, P. L.; Tran, K. P.; Finkel, G. C.; Slatkin, D. N. Uptake of a *nido*-carboranylporphyrin by human glioma xenografts in athymic nude-mice and by syngeneic ovarian carcinomas in immunocompetent mice. *Proc. Nat. Acad. Sci. USA* **1990**, *87*, 7265–7269.
123. Fairchild, R. G.; Kahl, S. B.; Laster, B. H.; Kalef-Ezra, J.; Popenoe, E. A invitro determination of uptake, retention, distribution, biological efficacy, and toxicity of boronated compounds for neutron-capture therapy–a comparison of porphyrins with sulfhydryl boron hydrides. *Cancer Res.* **1990**, *50*, 4860–4865.
124. Hill, J. S.; Kahl, S. B.; Kaye, A. H.; Stylli, S. S.; Koo, M. S.; Gonzales, M. F.; Vardaxis, N. J.; Johnson, C. I. Selective tumor uptake of a boronated porphyrin in an animal-model of cerebral glioma. *Proc. Nat. Acad. Sci. USA* **1992**, *89*, 1785–1789.
125. Hill, J. S.; Kahl, S. B.; Stylli, S. S.; Nakamura, Y.; Koo, M.-S.; Kaye, A. H. Selective tumor kill of cerebral glioma by photodynamic therapy using a boronated porphyrin photosensitizer. *Proc. Natl. Acad. Sci. USA* **1995**, *92*, 12126–12130.
126. Ceberg, C. P.; Brun, A.; Kahl, S. B.; Koo, M. S.; Persson, B. R. R.; Salford, L. G. A comparative-study on the pharmacokinetics and biodistribution of boronated porphyrin (bopp) and sulfhydryl boron hydride (bsh) in the rg2 rat glioma model. *J. Neurosurg.* **1995**, *83*,86–92.
127. Tibbitts, J.; Fike, J. R.; Lamborn, K. R.; Bollen, A. W.; Kahl, S. B. Toxicology of a boronated porphyrin in dogs. *Photochem. Photobiol.* **1999**, *69*, 587–594.
128. Miura, M.; Micca, P.; Heinrichs, J.; Gabel, D.; Fairchild, R.; Slatkin, D. Biodistribution And Toxicity Of 2,4-divinyl-*nido*-ortho-carboranyldeuteroporphyrin IX in mice. *Biochem. Pharmacol.* **1992**, *43*, 467–476.
129. Miura, M.; Micca, P. L.; Fisher, C. D.; Heinrichs, J. C.; Donaldson, J. A.; Finkel, G. C.; Slatkin, D. N. Synthesis of a nickel tetracarboranylphenylporphyrin for boron neutron-capture therapy: Biodistribution and toxicity in tumor-bearing mice. *Int. J. Cancer* **1996**, *68*, 114–119.
130. Miura, M.; Micca, P.; Fisher, C.; Gordon, C.; Heinrichs, J.; Slatkin, D.N. Evaluation of carborane-containing porphyrins as tumour targeting agents for boron neutron capture therapy. *Br. J. Radiol.* **1998,** *71*, 773–781.
131. Kreimann, E. L.; Miura, M.; Itoiz, M. E.; Heber, E.; Garavaglia, R. N.; Batistoni, D.; Rebagliati, R. J. et al., Biodistribution of a carborane-containing porphyrin as a targeting agent for Boron Neutron Capture Therapy of oral cancer in the hamster cheek pouch. *Arch. Oral Biol.* **2003**, *48*, 223–232.
132. Miura, M.; Morris, G. M.; Micca, P. L.; Nawrocky, M. M.; Makar, M. S.; Cook, S. P.; Slatkin, D. N. Synthesis of copper octabromotetracarboranylphenylporphyrin for boron neutron capture therapy and its toxicity and biodistribution in tumour-bearing mice. *Br. J. Radiol.* **2004,** *77*, 573–580.
133. Kahl, S. B.; Koo, M. S. Synthesis of tetrakis-carborane-carboxylate esters of 2,4-bis-(α,β-dihydroxyethyl)-deuteroporphyrin IX. *J. Chem. Soc. Chem. Commun.* **1990**, *24*, 1769–1771.
134. Miura, M.; Gabel, D.; Oenbrink, G.; Fairchild, R. G. Preparation of carboranyl porphyrins for boron neutron-capture therapy. *Tetrahedron Lett.* **1990**, *31*, 2247–2250.
135. Rosenthal, M. A.; Kavar, B.; Uren, S.; Kaye, A. H. Promising survival in patients with high-grade gliomas following therapy with a novel boronated porphyrin. *J. Clin. Neurosci.* **2003**, *10*, 425–427.
136. Bandyopadhyaya, A. K.; Narayanasamy, S.; Barth, R. F.; Tjarks, W. Synthesis of novel texaphyrins containing lanthanides and boron. *Tetrahedron Lett.* **2007**, *48*, 4467–4469.
137. Vicente, M. G. H.; Edwards, B. F.; Shetty, S. J.; Hou, Y.; Boggan, J. E. Syntheses and preliminary biological studies of four meso-tetra[(*nido*-carboranylmethyl)phenyl]porphyrins. *Bioorg. Med. Chem.* **2002**, *10*, 481–492.

138. Lauceri, R.; Purrello, R.; Shetty, S. J.; Vicente, M. G. H. Interactions of anionic carboranylated porphyrins with DNA. *J. Am. Chem. Soc.* **2001**, *123*, 5835–5836.
139. Vicente, M. G. H.; Gottumukkala, V.; Wickramasinghe, A.; Anikovsky, M.; Rodgers, M. A. Singlet Oxygen Generation and Dark Toxicity of a *nido*- and a *closo*-Carboranylporphyrin. *J. Proc. SPIE* **2004**, *5315*, 33–40.
140. Bobadova-Parvanova, P.; Oku, Y.; Wickramasinghe, A.; Hall, R. W.; Vicente, M. G. H. Ab initio and 1H NMR study of the Zn(II) complexes of a *nido*- and a *closo*-carboranylporphyrin. *J. Porphyrins Phthalocyanines* **2004**, *8*, 996–1006.
141. Vicente, M. G. H.; Wickramasinghe, A.; Nurco, D. J.; Wang, H. J. H.; Nawrocky, M. M.; Makar, M. S.; Miura, M. Synthesis, toxicity and biodistribution of two 5,15-di[3,5-(*nido*-carboranylmethyl)phenyl] porphyrins in EMT-6 tumor bearing mice. *Bioorg. Med. Chem.* **2003**, *11*, 3101–3108.
142. Vicente, M. G. H.; Nurco, D. J.; Shetty, S. J.; Osterloh, J.; Ventre, E.; Hegde, V.; Deustch, W. A. Synthesis, dark toxicity and induction of *in vitro* DNA photodamage by a tetra(4-*nido*-carboranylphenyl)porphyrin. *J. Photochem. Photobiol. B: Biol.* **2002**, *68*, 123–132.
143. Gottumukkala, V.; Ongayi, O.; Baker, D. G.; Lomax, L. G.; Vicente, M. G. H. Synthesis, cellular uptake and animal toxicity of a tetra(carboranylphenyl)-tetrabenzoporphyrin. *Bioorg. Med. Chem.* **2006**, *14*, 1871–1879.
144. Fabris, C.; Vicente, M. G. H.; Hao, E.; Friso, E.; Borsetto, L.; Jori, G.; Miotto, G. et al., Tumour-localizing and -photosensitising properties of *meso*-tetra(4-*nido*-carboranylphenyl)porphyrin (H_2TCP). *J. Photochem. Photobiol. B: Biol.* **2007**, *89*, 131–138.
145. Soncin, M.; Friso, E.; Jori, G.; Hao, E.; Vicente, M. G. H.; Miotto, G.; Colautti, P. et al. Tumor-localizing and radiosensitizing properties of meso-tetra(4-*nido*-carboranylphenyl)porphyrin (H2TCP). *J. Porphyrins Phthalocyanines* **2008**, *12*, 866–873.
146. Jori, G.; Soncin, M.; Friso, E.; Vicente, M. G. H.; Hao, E.; Miotto, G.; Colautti, P. et al., A novel boronated-porphyrin as a radio-sensitizing agent for boron neutron capture therapy of tumours: *In vitro* and *in vivo* studies. *Appl. Rad. Isotopes* **2009**, *67*, S321–S324.
147. Clark, J. C.; Fronczek, F. R.; Vicente, M. G. H. Novel carboranylporphyrins for application in boron neutron capture therapy (BNCT) of tumors. *Tetrahedron Lett.* **2005**, *46*, 2365–2368.
148. Easson, M. W.; Fronczek, F. R.; Jensen, T. J.; Vicente, M. G. H. Synthesis and *in vitro* properties of trimethylamine- and phosphonate-substituted carboranylporphyrins for application in BNCT. *Bioorg. Med. Chem.* **2008**, *16*, 3191–3208.
149. Luguya, R.J.; Fronczek, F. R.; Smith, K. M.; Vicente, M. G. H. Carboranylcorroles. *Tetrahedron Lett.* **2005**, *46*, 5365–5368.
150. Gottumukkala, V.; Luguya, R.; Fronczek F. R.; Vicente, M. G. H. Synthesis and cellular studies of an octa-anionic 5,10,15,20-tetra[3,5-(*nido*-carboranylmethyl)phenyl]porphyrin (H2OCP) for application in BNCT. *Bioorg. Med. Chem.* **2005**, *13*, 1633–1640.
151. Wu, H. T.; Micca, P. L.; Makar, M. S.; Miura, M. Total syntheses of three copper (II) tetracarboranylphenylporphyrins containing 40 or 80 boron atoms and their biological properties in EMT-6 tumor-bearing mice. *Bioorg. Med. Chem.* **2006**, *14*, 5083–5092.
152. Hao, E.; Jensen, T. J.; Courtney, B. H.; Vicente, M. G. H. Synthesis and cellular studies of porphyrin-cobaltacarborane conjugates. *Bioconj. Chem.* **2005**, *16*, 1495–1502.
153. Hao, E.; Vicente, M. G. H. Expeditious synthesis of porphyrin-cobaltacarborane conjugates. *Chem. Commun.* **2005**, 1306–1308.
154. Hao, E.; Sibrian-Vazquez, M.; Serem, W.; Garno, J. C.; Fronczek, F. R.; Vicente, M. G. H. Synthesis, aggregation and cellular investigations of porphyrin-cobaltacarborane conjugates. *Chem. Eur. J.* **2007**, *13*, 9035–9042.
155. Dozzo, P.; Koo, M. S.; Berger, S.; Forte, T. M.; Kahl, S. B. Synthesis, characterization, and plasma lipoprotein association of a nucleus-targeted boronated porphyrin. *J. Med. Chem.* **2005,** *48*, 357–359.
156. Luguya, R.; Fronczek, F. R.; Smith, K. M.; Vicente, M. G. H. Synthesis of novel carboranylchlorins with dual application in boron neutron capture therapy (BNCT) and photodynamic therapy (PDT). *App. Rad. Isotopes* **2004**, *61*, 1117–1123.
157. Luguya, R.; Jensen, T. J.; Smith, K. M.; Vicente, M. G. H. Synthesis and cellular studies of a carboranylchlorin for the PDT and BNCT of tumors. *Bioorg. Med. Chem.* **2006**, *14*, 5890–5897.
158. Kuchin, R. A. V.; Mal'shakova, M. V.; Belykh, D. V.; Ol'shevskaya, V. A.; Kalinin, V. N. Synthesis of boronated derivatives of chlorin e(6) with amide bond. *Dokl. Chem.* **2009**, *425*, 80–83.
159. Ol'shevskaya, V. A.; Nikitina, R. G.; Savchenko, A. N.; Malshakova, M. V.; Vinogradov, A. M.; Golovina, G. V.; Belykh, D. V. et al., Novel boronated chlorin e(6)-based photosensitizers: Synthesis, binding to albumin and antitumour efficacy. *Bioorg. Med. Chem.* **2009**, *17*, 1297–1306.

160. Ol'shevskaya, V. A.; Savchenko, A. N.; Zaitsev, A. V.; Kononova, E. G.; Petrovskii, P. V.; Ramonova, A. A.; Tatarskiy, V. V. et al., Novel metal complexes of boronated chlorin e(6) for photodynamic therapy. *J. Organomet. Chem.* **2009**, *694*, 1632–1637.
161. Ol'shevskaya, V.; Zaitsev, A.; Savchenko, A.; Kononova, E.; Petrovskii, P.; Kalinin, V. Synthesis of boronated derivatives of pheophorbide a. *Dokl. Chem.* **2008**, *423*, 294–298.
162. Luzgina, V. N.; Ol'shevskaya, V. A.; Sekridova, A. V.; Mironov, A. F.; Kalinin, V. N.; Pashchenko, V. Z.; Gorokhov, V. V.; Tusov, V. B.; Shtil, A. A. Synthesis of boron-containing derivatives of pyropheophorbide (a) under-bar and investigation of their photophysical and biological properties. *Russ. J. Org. Chem.* **2007**, *43*, 1243–1251.
163. Bregadze, V. I.; Sivaev, I. B.; Lobanova, I. A.; Titeev, R. A.; Brittal, D. I.; Grin, M. A.; Mironov, A. F. Conjugates of boron clusters with derivatives of natural chlorin and bacteriochlorin. *Appl. Rad. Isotopes* **2009**, *67*, S101–S104.
164. Mironov, A. F.; Grin, M. A. Synthesis of chlorin and bacteriochlorin conjugates for photodynamic and boron neutron capture therapy. *J. Porphyrins Phthalocyanines* **2008**, *12*, 1163–1172.
165. Giuntini, F.; Raoul, Y.; Dei, D.; Municchi, M.; Chiti, G.; Fabris, C.; Colautti, P.; Jori, G.; Roncucci, G. Synthesis of tetrasubstituted Zn(II)-phthalocyanines carrying four carboranyl-units as potential BNCT and PDT agents. *Tetrahedron Lett.* **2005**, *46*, 2979–2982.
166. Friso, E.; Roncucci, G.; Dei, D.; Soncin, M.; Fabris, C.; Chiti, G.; Colautti, P. et al., A novel B-10-enriched carboranyl-containing phthalocyanine as a radio- and photo-sensitising agent for boron neutron capture therapy and photodynamic therapy of tumours: *in vitro* and *in vivo* studies. *Photochem. Photobiol. Sci.* **2006**, *5*, 39–50.
167. Tsaryova, O.; Semioshkin, A.; Wohrle, D.; Bregadze, V. I. Synthesis of new carboran-based phthalocyanimes and study of their activities in the photoxidation of citronellol. *J. Porphyrins Phthalocyanines* **2005**, *9*, 268–274.
168. Li, H.; Fronczek, F. R.; Vicente, M. G. H. Synthesis and properties of cobaltacarborane-functionalized Zn(II)-phthalocyanines. *Tetrahedron Lett.* **2008**, *49*, 4828–4830.
169. Li, H. R.; Fronczek, F. R.; Vicente, M.G. H. Cobaltacarborane-phthalocyanine conjugates: Syntheses and photophysical properties. *J. Organomet. Chem.* **2009**, *694*, 1607–1611.
170. Crossley, E. L.; Ziolkowski, E. J.; Coderre, J. A.; Rendina, L. M. Boronated DNA-binding compounds as potential agents for boron neutron capture therapy. *Mini-Rev. Med. Chem.* **2007**, *7*, 303–313.
171. Ghaneolhosseini, H.; Tjarks, W.; Sjoberg, S. Synthesis of novel boronated acridines- and spermidines as possible agents for BNCT. *Tetrahedron* **1998**, *54*, 3877–3884.
172. Bateman S. A.; Kelly, D. P.; Martin, R. F,; White, J. M. DNA binding compounds. VII. Synthesis, characterization and DNA binding capacity of 1,2-dicarba-*closo*-dodecarborane bibenzimidazoles related to the DNA minor groove biner Hoechst 33258. *Aust. J. Chem.* **1999**, *52*, 291–301.
173. Gedda, L.; Ghaneolhosseini, H.; Nilsson, P.; Nyholm, K.; Pettersson, J.; Sjoberg, S.; Carlsson, J. DNA binding compounds. VII - Synthesis, characterization and DNA binding capacity of 1,2-dicarba-*closo*-dodecaborane bibenzimidazoles related to the DNA minor groove binder Hoechst 33258. *Anti-Cancer Drug Design* **2000**, *15*, 277–286.
174. Woodhouse, S. L.; Ziolkowski, E. J.; Rendina, L. M. Synthesis and anti-cancer activity of dinuclear platinum(II) complexes containing bis(thioalkyl)dicarba-*closo*-dodecaborane(12) ligands. *J. Chem. Soc., Dalton Trans.* **2005**, 2827–2829.
175. Woodhouse, S. L.; Rendina, L. M. Multinuclear platinum(II)-amine complexes containing bis(aminopropyl)dicarba-*closo*-dodecaborane(12) ligands. *J. Chem. Soc., Dalton Trans.* **2004**, *21*, 3669–3677.
176. Adams, D. M.; Ji, W.; Barth, R. F.; Tjarks, W. Comparative *in vitro* evaluation of dequalinium B, a new boron carrier for neutron capture therapy (NCT). *Anticancer Res.* **2000**, *20*, 3395–3402.
177. Zakharkin, L. I.; Ol'shevskaya, V. A.; Spryshkova, R. A.; Grigor'eva, E. Y.; Ryabkova, V. I.; Borisov, G.I. Synthesis of bis(dialkylaminomethyl)-*o*- and *m*-carboranes and study of these compounds as potential preparations for boron neutron capture therapy. *Pharm. Chem. J.* **2000**, *34*, 301–304.
178. El-Zaria, M. E.; Doerfler, U.; Gabel, D. J. Synthesis of [(aminoalkylamine)-N-amino-alkyl]azanonaborane(11) derivatives for boron neutron capture therapy. *J. Med. Chem.* **2002**, *45*, 5817–5819.
179. Yang, W.; Barth, R. F.; Wu, G.; Tjarks, W.; Binns, P.; Riley, K. Boron neutron capture therapy of EGFR or EGFRvIII positive gliomas using either boronated monoclonal antibodies or epidermal growth factor as molecular targeting agents. *Appl. Rad. Isotopes* **2009**, *67*, S328–S331.
180. Wu, G.; Yang, W.; Barth, R. F.; Kawabata, S.; Swindall, M.; Bandyopadhyaya, A. K.; Tjarks, W. et al., Molecular targeting and treatment of an epidermal growth factor receptor-positive glioma using boronated cetuximab. *Clin. Cancer Res.* **2007**, *13*, 1260–1268.

181. Yang, W.; Wu, G.; Barth, R. F.; Swindall, M. R.; Bandyopadhyaya, A. K.; Tjarks, W.; Tordoff, K. et al., Molecular targeting and treatment of composite EGFR and EGFRvIII-positive gliomas using boronated monoclonal antibodies. *Clin. Cancer Res.* **2008**, *14*, 838–891.
182. Backer, M. V.; Gaynutdinov, T. I.; Patel, V. P.; Bandyopadhyaya, A. K.; Thirumamagal, B. T. S.; Tjarks, W.; Barth, R. F.; Claffey, K.; Backe, J. M. Vascular endothelial growth factor selectively targets boronated dendrimers to tumor vasculature. *Mol. Cancer Therap.* **2005**, *4*, 1423–1429.
183. Yang, W.; Barth, R. F.; Wu, G.; Kawabata, S.; Sferra, T. J.; Bandyopadhyaya, A. K.; Tjarks, W. et al., Vascular endothelial growth factor selectively targets boronated dendrimers to tumor vasculature. *Clin. Cancer Res.* **2006**, *12*, 3792–3802.
184. Feng, B.; Tomizawa, K.; Michiue, H.; Miyatake, S-I.; Han, X-J.; Fujimura, A.; Seno, M.; Kirihata, M.; Matsui, H. Delivery of sodium borocaptate to glioma cells using immunoliposome conjugated with anti-EGFR antibodies by ZZ-His. *Biomaterials* **2009**, *30*, 1746–1755.
185. Pan, X.; Gong, W.; Yang, W.; Barth, R. F.; Tjarks, W.; Lee, R. J. Synthesis of cetuximab-immunoliposomes via a cholesterol-based membrane anchor for targeting of EGFR. *Bioconjugate Chem.* **2007**, *18*, 101–108.
186. Yanagie, H.; Maruyama, K.; T. Takizawa, T.; Ishida, O.; Ogura, K.; Matsumoto, T.; Sakurai, Y. et al., Application of boron-entrapped stealth liposomes to inhibition of growth of tumour cells in the *in vivo* boron neutron-capture therapy model. *Biomed. Pharmacotherapy* **2006**, *60*, 43–50.
187. Altieri, S.; Balzi, M.; Bortolussi, S.; Bruschi, P.; Ciani, L.; Clerici, A. M.; Faraoni, P. et al., Carborane derivatives loaded into liposomes as efficient delivery systems for boron neutron capture therapy. *J. Med. Chem.* **2009**, *52*, 7829–7835.
188. Ito, Y.; Kimura, Y.; Shimahara, T.; Ariyoshi, Y.; Shimahara, M.; Miyatake, S.; Kawabata, S.; Kasaoka, S; Ono, K. Disposition of TF-PEG-Liposome-BSH in tumor-bearing mice. *Appl. Rad. Isotopes* **2009**, *67*, S109–S110.
189. Watson-Clark, R. A.; Banquerigo, M. L.; Shelly, K.; Hawthorne, M. F.; Brahn, E. Model studies directed toward the application of boron neutron capture therapy to rheumatoid arthritis: Boron delivery by liposomes in rat collagen-induced arthritis. *Proc. Natl. Acad. Sci. USA* **1998**, *95*, 2531–2534.
190. Li, T.; Hamdi, J.; Hawthorne, M. F. Unilamellar liposomes with enhanced boron content. *Bioconjugate Chem.* **2006**, *17*, 15–20.
191. Miyajima, Y.; Nakamura, H.; Kuwata, Y.; Lee, J-D.; Masunaga, S; Ono, K.; Maruyama, K. Transferrin-loaded *nido*-carborane liposomes: Tumor-targeting boron delivery system for neutron capture therapy. *Bioconjugate Chem.* **2006**, *17*, 1314–1320.
192. Doi, A.; Kawabata, S.; Iida, K.; Yokoyama, K.; Kajimoto, Y.; Kuroiwa, T.; Shirakawa T. et al. Tumor-specific targeting of sodium borocaptate (BSH) to malignant glioma by transferrin-PEG liposomes: a modality for boron neutron capture therapy. *J. Neurooncol.* **2008**, *87*, 287–294.
193. Thirumamagal, B. T. S.; Zhao, X. B.; Bandyopadhyaya, A. K.; Narayanasamy, S.; Johnsamuel, J.; Tiwari, R.; Golightly, D. W. et al., Receptor-targeted liposomal delivery of boron-containing cholesterol mimics for boron neutron capture therapy (BNCT). *Bioconjugate Chem.* **2006**, *17*, 1141–1150.
194. Zhu, Y.; Yan, K. C.; Maguire, J. A.; Hosmane, N. S. Recent Developments in the Boron Neutron Capture Therapy (BNCT) Driven by Nanotechnology. *Curr. Chem. Biol.* **2007**, *1*, 141–149.
195. Wu, G.; Barth, R. F.; Yang, W.; Lee, R. J.; Tjarks, W.; Backer, M. V.; Backer, J. M. In *Neutron capture therapy of cancer: Nanoparticles and high molecular weight boron delivery agents in Nanotechonolgy for cancer therapy*. Mansoor M. Amiji (Ed.), CRC Press, **2006**, Chapter 6, pp. 77–103.

11 Future Applications of Boron and Gadolinium Neutron Capture Therapy

Masao Takagaki, Nobutaka Tomaru, John A. Maguire, and Narayan S. Hosmane

CONTENTS

11.1 BORON NEUTRON CAPTURE THERAPY AND ITS LIMITATIONS

This chapter is meant primarily for researchers in nonmedical fields. We will discuss some biological findings generated in studies involving boron neutron capture therapy (BNCT) and gadolinium neutron capture therapy (GdNCT). For more complete understanding of how these therapies are used in cancer treatment, we also include a number of illustrations and figures from clinical studies. Because of the authors' interests, this chapter concentrates mainly on the use of neutron capture therapy (NCT) for brain tumors. There are other aspects of NCT, such as its application to head and neck tumors that show great promise, but, due to space limitations, will not be specifically addressed.

BNCT is based on the fission of ^{10}B after it absorbs a thermal neutron and produces an α particle and a recoil lithium ion. The high linear energy transfer (LET) α particles and their recoil ^{7}Li

particles release a large amount of kinetic energy (3.3 MeV). This energy is transferred only within the total combined trajectories of the resultant particles, a distance of less than 14 μm (Figure 11.1). If compounds with large amounts of ^{10}B can be preferentially localized in cancer cells, the energy transfer will be confined to the tumor cells, avoiding serious collateral radiation injury to surrounding healthy tissue [1]. However, the distribution of the radiation-absorbed dose is mostly dependent on the microdistribution of ^{10}B in the tumor cells. Consequently, owing to the heterogeneity of tumor tissues, effective treatment remains a challenge because it is difficult to achieve a sufficiently uniform boron distribution in tumors [2]. This uneven distribution, and the resulting dose distribution, is one of the major reasons for tumor recurrence after BNCT. Since 80–90% of recurrence occurs in peritumoral parenchyma, a key potential benefit of BNCT is selective destruction of tumor cells invading into the periphery, which are seen in magnetic resonance imaging (MRI) as abnormal T2 high-density areas around tumors [3].

For more complete tumor destruction, a promising alternative to ^{10}B is the ^{157}gadolinium (^{157}Gd) atom, which has a large thermal neutron capture cross section of 255,000 barn, 65 times greater than ^{10}B. In a single thermal neutron capture reaction, ^{157}Gd releases Auger electrons, IC electrons, γ-rays, and x-rays. The $^{157}Gd(n,\gamma)^{158}Gd$ reaction releases total kinetic energy of 7.7 MeV, almost double that released by $^{10}B(n,\alpha)^{7}Li$ [4]. Since Gd-based NCT (GdNCT) dose distribution is more uniform than in BNCT, it is likely to be more effective for destroying both pathologically heterogeneous malignant tumors and peripherally invading tumor cells. In this chapter, we present findings from *in vitro* and *in vivo* studies regarding the tumoricidal effect of GdNCT on experimental brain tumors, and discuss issues in the treatment of glioblastoma.

In a thermal neutron capture reaction, the stable isotope of ^{10}B divides into two heavy ionizing particles, ^{4}He and ^{7}Li. Shared by each particle, the reaction releases a total kinetic energy of 3.3 MeV. This is distributed within combined trajectories totaling 14 μm, which is almost equal to the typical diameter of a tumor cell. A single hit of these heavy ionizing particles can kill tumor cells without serious radiation injury to the surrounding healthy parenchyma. The two compounds currently approved for clinical trials are *p*-borophenylalanine (BPA) and borocaptate sodium (BSH). Because of the heterogeneous micro-distribution of the absorbed dose in the tumor by BPA or BSH, this highly localized energy transfer may not be evenly distributed throughout the tumor.

As shown in Figure 11.2, in BNCT, as it is currently undertaken, there are three components that are both biologically selective and nonselective. While the two components resulting from the $^{10}B(n,\alpha)^{7}Li$ and $^{14}N(n,p)^{14}C$ reactions are therapeutically useful, γ-ray contamination, from core γ and secondary γ-rays originating in the reactor, may be a source of radiation damage to normal brain tissues. Consequently, to minimize risk, minimal neutron doses are used. Moreover, therapeutic depth is inevitably determined by the maximum tolerable dose that can be received at the brain surface (Figure 11.3).

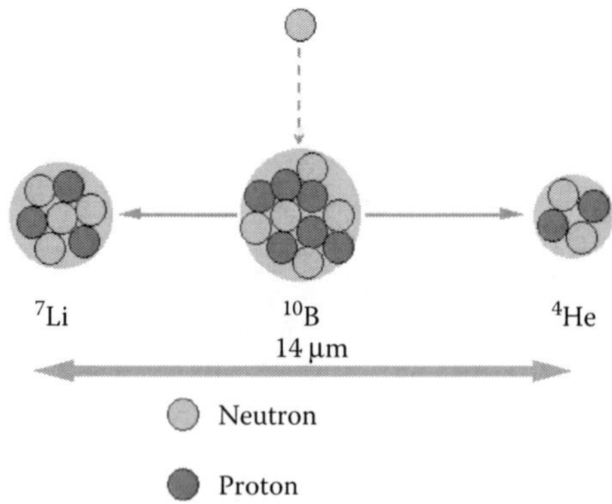

FIGURE 11.1 $^{10}B(n,\alpha)^{7}Li$ neutron capture reaction.

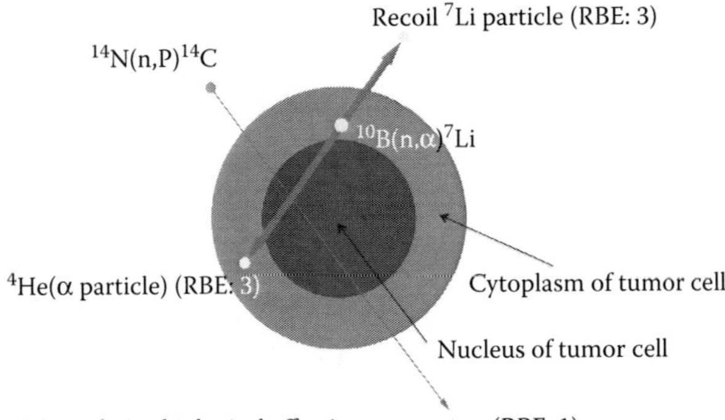

FIGURE 11.2 Physical circumstances of BNCT. The three physical components include particles from the therapeutically useful $^{10}B(n,\alpha)^{7}Li$ reaction, particles emanating from the reactor, and detrimental particles resulting from bombardment from $^{14}N(n,p)^{14}C$, which produces both primary and secondary γ-rays.

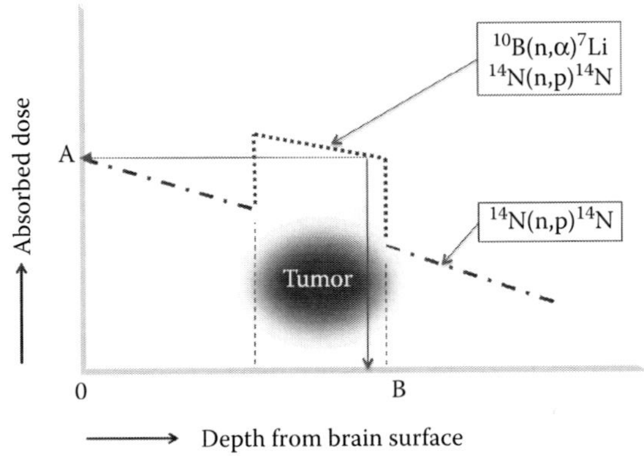

FIGURE 11.3 The maximum therapeutic depth is inevitably determined by the maximum tolerable dose that can be received at the brain surface. (A) Maximum tolerable dose at brain surface; (B) maximum therapeutic depth.

Successful BNCT depends on three factors: shallow seating of the tumor in the brain; limited tumor infiltration; and location of tumor outside of the eloquent cortex (those areas of the brain which, if removed, will result in loss of sensory processing or linguistic ability, or paralysis). Therapeutic depth is limited, because thermal neutrons undergo a Maxwellian distribution in the brain, fluence decreases three dimensionally along the incident axis (Figure 11.4). Consequently, related to the incidental axis, both vertically and horizontally, the neutron dose exponentially decreases. If tumor cells are present beyond the therapeutic isodose contour, the absorbed dose will not be sufficiently lethal, and peri-lesional recurrence is likely.

Successful neutron therapy has to compromise between two conflicting requirements: maximum 3D uptake of the deep tumor dose and minimum accumulation in normal surface tissues. The heterogeneous distribution of boron affects the radiation sensitivity of tumor tissues, which are characterized by a cell proliferation cycle that unevenly distributes quiescent cells and resting cells. Each type of cell has a different uptake of BPA and BSH [5]. The effectiveness of BNCT is greatly affected by this biological and neutron dose heterogeneity.

As illustrated in Figure 11.5, delivering boron to peripherally invading tumor cells (satellite lesions) is crucially important for preventing tumor recurrence. Unfortunately, fully effective delivery is yet to be achieved.

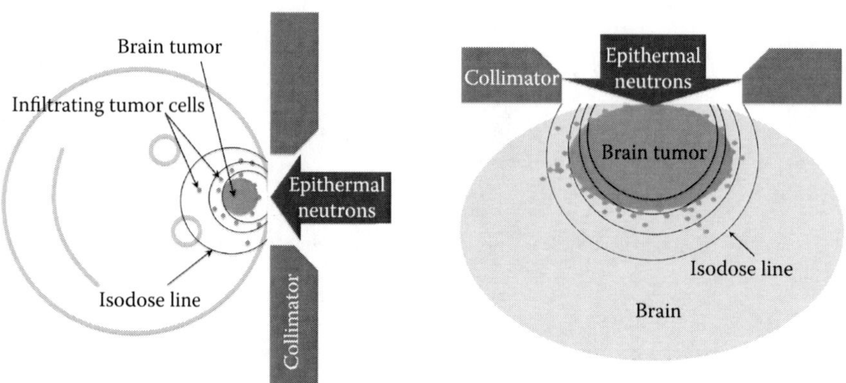

FIGURE 11.4 Schematic drawing of iso-dose distribution. All tumor cells must be present within the therapeutic isodose line.

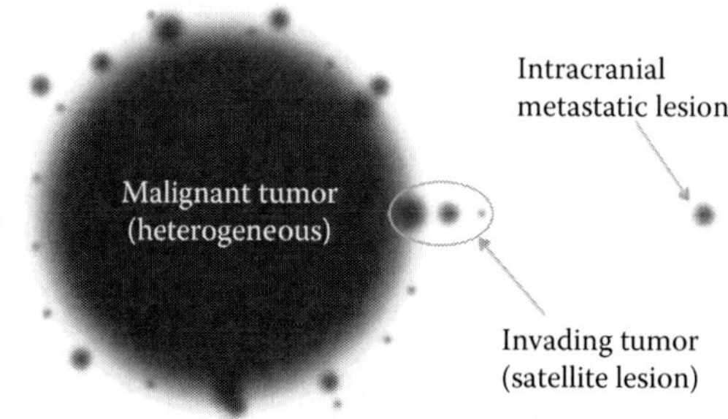

FIGURE 11.5 Schematic diagram of a highly malignant tumor with highly invasive tumor cells: invasive cells form satellite lesions or intracranial metastatic lesions or both, often far from the original tumor.

To deliver the requisite thermal neutrons into deep lesions, epithermal neutrons are currently the standard radiation source for BNCT. Epithermal neutrons are thermalized in the scalp or at the brain surface, the thermal neutron dose reaching its maximum intensity at a depth of 2 cm, almost equal to the total thickness of scalp and skull (Figure 11.6) [6]. Thus, thermal neutron doses are acquired in the highest concentration at the brain surface when the epithermal neutrons are irradiated onto the scalp surface [7]. This modality allows BNCT without craniotomy. To enable thermal neutrons to penetrate as deeply as possible, however, deeply seated tumors can be treated by BNCT after a craniotomy to expose the brain.

Figure 11.7 shows a picture of a clinical BNCT treatment in a study at the Research Reactor Institute of Kyoto University [7]. Here, a sedated patient is being treated while sitting on the BNCT platform, so as to receive neutrons to the head, which was fixed using a heat-flexible holder. After setting the platform in position, the neutron shutter was opened. Then the platform was moved to the irradiation point. Depending on boron concentration and neutron flux, the total irradiation period typically ranges from 40 to 60 min.

To obtain the maximum neutron dose in the main tumor mass, the patient's head is kept fixed and orientated to the collimator. Gold wires for thermal neutron monitoring and thermoluminescent dosimeters (TLDs) for γ-rays are applied to the scalp surface (Figure 11.8). A wire is withdrawn via remote maneuver to determine the thermal neutron flux on the scalp surface during BNCT. Patients are tranquilized during BNCT. The radius of the collimator opening usually varies from patient to patient; in the procedure shown in Figure 11.7, the radius diameter was 12 cm.

Figure 11.9 shows a series of MRI images taken before and after epithermal neutron-based BNCT without a craniotomy. The onset for this 30-year-old female patient was marked by epileptic

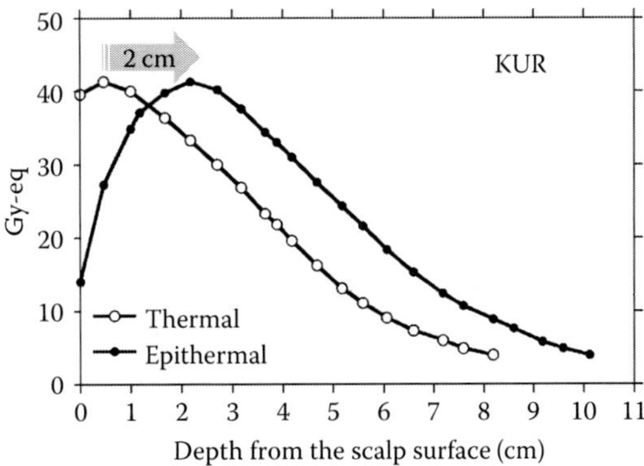

FIGURE 11.6 Depth distribution of neutrons in BNCT. Owing to loss of the kinetic energy by epithermal neutrons in elastic collisions with the hydrogen atoms of water in tissues, thermal neutrons are available in their maximum amount at a depth of 2 cm from the incident surface.

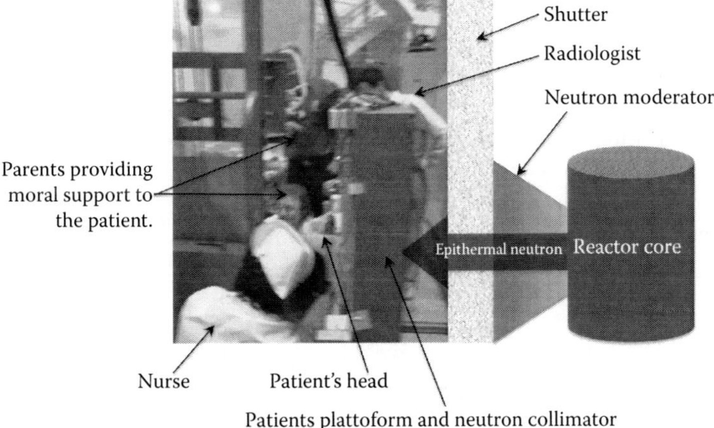

FIGURE 11.7 BNCT at the treatment platform of Research Reactor Institute of Kyoto University.

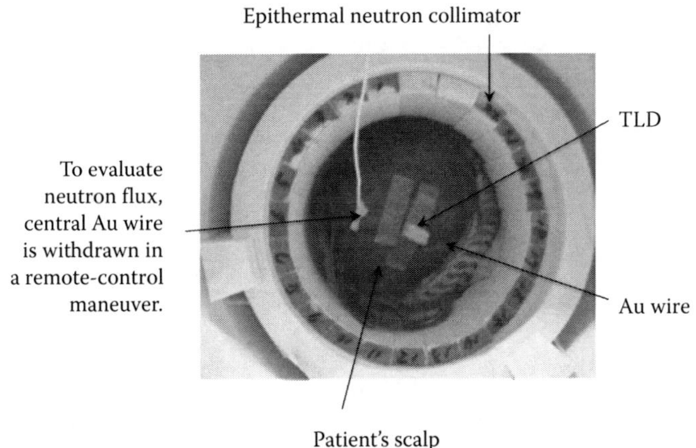

FIGURE 11.8 View of a patient's scalp through an irradiation hole. Radiation monitors are applied to the scalp surface.

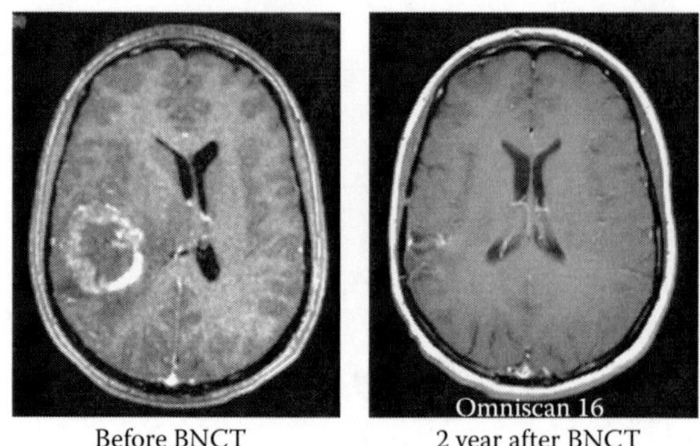

FIGURE 11.9 GdDTPA-enhanced T1-weighted MRI images taken before and after epithermal neutron-based BNCT without craniotomy. An irregular, round, thick rim with peripheral edema and mid-line shift can be seen (left). Two years later, after successive surgical removal and BNCT, MRI revealed no tumor recurrence (right), and no neurological deficits were apparent. Unfortunately, although there was no evidence of recurrence of the original tumor, the patient died 3 years after BNCT when, far from the original location, the tumor recurred in the right caudate-head.

events in the left foot. Head imaging revealed a ring-like enhanced tumor in the right parietal cortico-medullary junction. Using BPA + BSH (fractionated BNCT), BNCT was successfully carried out twice, the second time after a 2-week interval. The patient died 3 years after onset, when the tumor recurred in the ipsilateral deep basal ganglia. When recurrence occurs after BNCT of grade IV glioma, it usually appears in infiltrating lesions far from the main tumor mass. Tumor recurrence after BNCT typically manifests as peri-lesions or satellite lesions that might have received only sublethal damage during BNCT.

The main challenge for NCT has always been targeting. Figure 11.10 (upper) shows tumors visualized via T1-weighted gadolinium diethylene triamine pentaacetic acid (GdDTPA)-enhanced MR imaging. Gross tumor volume (GTV) shows up reasonably well. However, T2-weighted MR images reveal peri-focal edema surrounding the tumor, a high-density area containing infiltrating tumor cells [8]. These high-density areas appropriately delineate the primary tumor volume (PTV). This PTV is the minimal target area for B/GdNCT directed at malignant brain tumors.

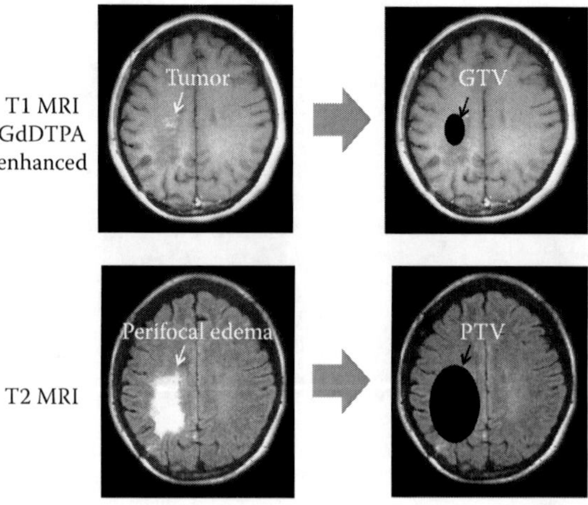

FIGURE 11.10 Upper, T1-weighted MRI. GdDTPA enhancement reveals GTV region. Lower, T2-weighted MRI.

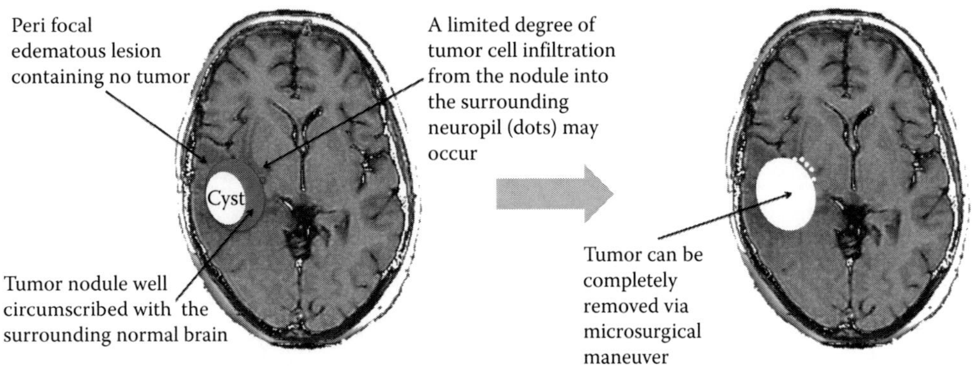

FIGURE 11.11 Principles of treatment for low-grade glioma.

Peripheral edema are visualized with T2-weighted MRI; this reveals PTV, the appropriate targeting area for B/GdNCT.

Low-grade glioma (e.g., juvenile pilocytic astrocytoma) is typically grossly circumscribed, often comprising a cyst (bright area) with a mural nodule (dark area) (Figure 11.11). A limited amount of tumor cell infiltration from the nodule into the surrounding neuropil (dots) may occur. Most of the remaining cyst wall consists of gliotic neuropil and the true extent of neoplastic cell distribution is difficult to determine. In such cases, complete surgical resection is necessary to reduce the risk of tumor recurrence. Low-grade glioma can be completely removed by microsurgical maneuvers without any adjunctive treatment. For grade I, and for many grade II gliomas, no other radiation therapy or chemotherapy are indicated.

A typical high-grade glioma (malignant brain tumor) is schematically represented in Figure 11.12. Three morphological zones are usually exhibited in high-grade glioma: the glioblastoma, a central area of necrosis (white area) surrounded by a zone of densely cellular tumor cells (dark area), which is again surrounded by an irregular region of infiltration (dots). The extent of the outer zone of infiltration is quite variable, and the most distal penetration by individual tumor cells cannot be determined using any conventional histological method. The highly variable extent and density of infiltration may lead to significant biopsy sampling errors, particularly since stereotactic procedures provide such a limited amount of tissue. Figure 11.12 schematically shows intracranial metastasis, infiltration to corpus callosum, and dissemination to the CSF system through the ependyma.

In high-grade gliomas, targeting is not limited to the main tumor mass: the target must be widened to include infiltrating tumor cells residing far from the edge of the main tumor mass.

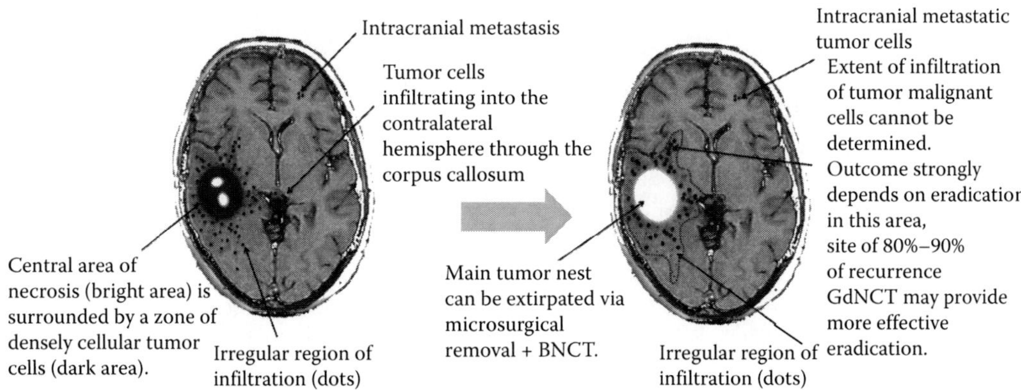

FIGURE 11.12 Principles of treatment for malignant brain tumors.

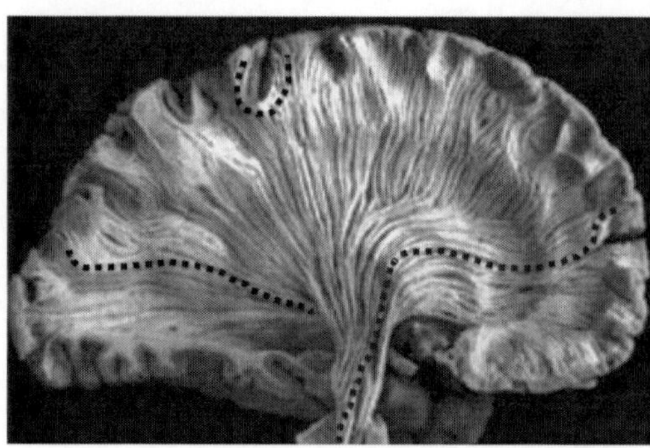

FIGURE 11.13 Typical infiltration of malignant tumor cells along a neuronal tract (outlined by dotted lines). (Adapted from F. A. Mettler, *J. Neuropathol. Exp. Neurol.* **1955**, *14*, 115–141.)

Treatment for malignant glioma has become more integrated over the last 10 years, and standard treatment now generally includes surgical debulking (maximum extirpation), adjuvant chemotherapy, and various forms of radiation therapy. Even so, these recent refinements have hardly improved patient survival time from initial treatment. Long-term survival is still the exception. Physicians have to wonder why treatment of high-grade gliomas remains ineffective despite the array medical skills and resources being deployed; they remain frustrated by the biological character of tumor cells, which are highly infiltrative and radiation resistant. In particular, microscopic tumor cells often spread more than 4 cm beyond the edge of surgical resection [9]. As there is no reliable method for properly specifying the target for radiation therapy, clinicians have to use chemotherapy to fight the spreading infiltration of tumor cells throughout the brain parenchyma.

The effectiveness of GdNCT also depends on timing treatment to evenly transfer the radiation-absorbed dose into the spreading tumor cells. Early GdNCT followed by chemotherapy might be the most successful method as long as the infiltrating tumor cells are quite small in number. Thus, early GdNCT, soon after onset, may lead to longer survival.

Malignant brain tumor cells have a tendency to infiltrate throughout the brain along neuronal tracts (Figure 11.13). If tumor cells invade the other hemisphere through neuronal tracts that traverse the corpus callosum, chemotherapy is the only available treatment. Other modalities, including BNCT, are definitely contraindicated.

11.2 USE OF GdNCT FOR BRAIN TUMOR THERAPY

11.2.1 Principles and Concepts of GdNCT

11.2.1.1 The $^{157}Gd(n,\gamma)^{158}Gd$ Reaction

Figure 11.14 schematically presents the $^{157}Gd(n,\gamma)^{158}Gd$ reaction. The ^{157}Gd atom, which occurs in 15.6% natural abundance, releases Auger electrons, IC electrons, γ-rays, and x-rays after thermal neutron capture across its wide capture cross section of 255,000 barn, which is 65 times greater than the thermal neutron capture cross section of ^{10}B. Although Auger electrons travel miniscule distances of only several 100 Å, they are able to kill tumor cells if the ^{157}Gd is located close to a critical target, such as DNA [10,11]. Concomitant x-rays are negligibly lethal owing to their relatively low kinetic energy. Thus, γ-rays, IC electrons, and Auger electrons are mainly responsible for the tumoricidal effect of GdNCT.

To successfully employ this modality of NCT, it is essential to have a source of well-thermalized neutrons. Suitable thermal neutron energy spectra can be obtained at the NCT facilities of the Kyoto University Research Reactor Institute (KUR), at the Japan Atomic Energy Research Institute (JRR4) and at reactors in a number of other countries. A Maxwellian distribution, with good elimination of concomitant γ-rays from the irradiation field, can be achieved by scattering with Bi atoms (bisthmus

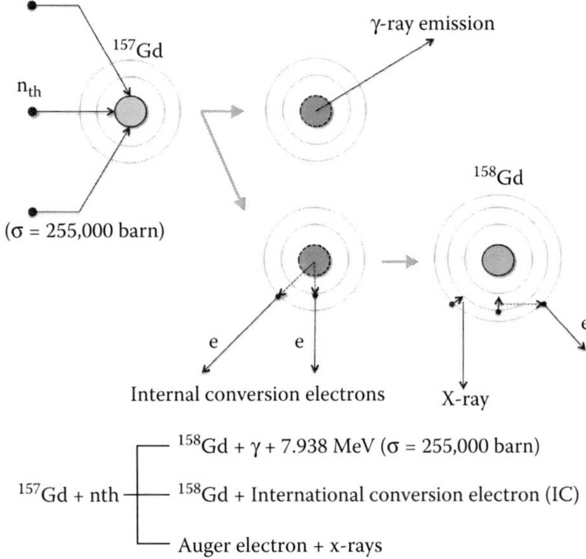

FIGURE 11.14 ^{157}Gd(n,γ)^{158}GdNCR.

scatter). The Cd ratio, that is, the proportional fluence of thermal neutrons (<10 eV) and nonthermal neutrons (>10 eV), is very large (150), while the γ-ray contamination in the irradiation field is small. These facilities also yield epithermal neutron fields, depending on how the D_2O filter is set. A Cd ratio adjustable from 150 to 9.4, for example, can be achieved by filtering the beam with a boral plate containing cadmium. Generally, thermal neutron modes are applied to superficial tumors, such as cutaneous melanoma, while epithermal modes are more effective for brain tumors.

Excitation occurs when a neutron is captured by a ^{157}Gd atom, energy is released by various types of radiation. Cascading energy released from the excited ^{157}Gd nucleus mainly comprises γ-rays and IC electrons. Excited intrinsic orbital energy releases Auger electrons. The total kinetic energy of the various particles is 7.9 MeV, almost double the energy released by ^{10}B(n,α)^{7}Li.

As the data in Table 11.1 illustrate, the most promising nuclei for NCT are ^{10}B and ^{157}Gd atoms, which are stable isotopes with large thermal neutron capture cross sections. Gadolinium is toxic, while boron exhibits low toxicity. Gadolinium compounds tend to be chemically unstable, and must be stabilized by chemical procedures such as chelation.

TABLE 11.1
Thermal Neutron Capture Cross Section (σ) of Candidate Nuclides for NCT

Nuclide	Cross Section (σ)	Nuclide	Cross Section (σ)
^{3}He	5500	^{155}Gd	58,000
^{6}Li	953	^{157}Gd	240,000
^{10}B	3837	^{174}Hf	400
^{113}Cd	20,000	^{199}Hg	2000
^{135}Xe*	2,720,000	^{235}U	678
^{149}Sm	41,500	^{241}Pu*	1375
^{151}Eu	5900	^{242}Am*	8000

Note: Asterisk (*) indicates that the nuclides are radioactive.
Capture cross section (σ) are given in barns, where 1 barn = 10^{-24} cm^2.

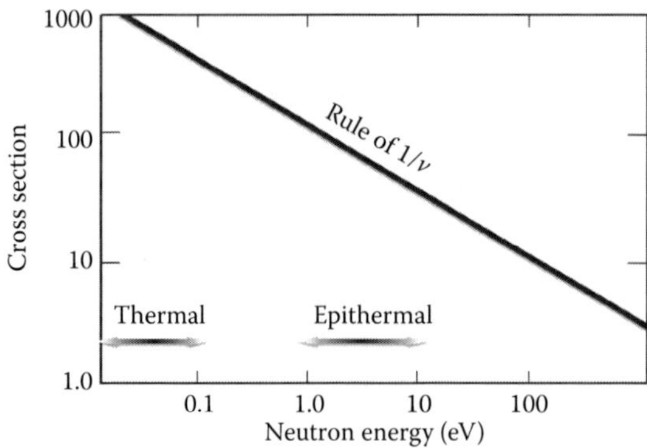

FIGURE 11.15 Rule of 1/*v*.

The neutron capture cross section increases exponentially as the neutron energy decreases (Rule of 1/*v*) (Figure 11.15). The cross section is extraordinarily wide for very low-energy neutrons (thermal neutrons). On the other hand, by their nature, thermal neutrons cannot reach deep tissue; as they pass through tissues, they are rapidly used in $^{14}N(n,p)^{14}C$ and other reactions.

The ^{157}Gd atom captures neutrons across a very large cross section of 255,000 barns. The ^{157}Gd atom is thus comparable to the Thousand-armed Goddess of Mercy (千手観音), or Kannon, a manifestation of the Buddha that accepts prayers and is believed to have the power to grant anyone the blessing of complete relief (Figure 11.16). If we scale the diameter of the ^{157}Gd nucleus to 1 cm, the nominal 2-km width of the outstretched hands would comfortably capture any passing thermal neutrons. (Figure 11.17).

If we were to scale the nuclear diameter of the ^{157}Gd atom to around 1 cm, its thermal neutron cross section of 255,000 barn would scale up to 2 km.

The concept of LET is essential for understanding the difference between GdNCT and BNCT (Figure 11.18). LET is a measure of the ionizing power of a particle as it travels through the atmosphere. Typically, this measurement is used for quantifying the effects of ionizing radiation on biological specimens. LET is closely related to the energy loss per unit distance, d*E*/d*x*, which is

The Goddess of Mercy (Kannon) having 1000 hands not to miss the capture of mercy for peoples (Kyoto Sanju-San Gen Doh)

FIGURE 11.16 The thousand-armed Goddess of Mercy (Kannon) at Kyoto Sanju-San Gen Doh. ^{157}Gd in tissue is the target and thermal neutrons are the ammunition launched at the target. Like the Kannon's outstretched arms, ^{157}Gd has a wide embrace and unfailingly captures thermal neutrons.

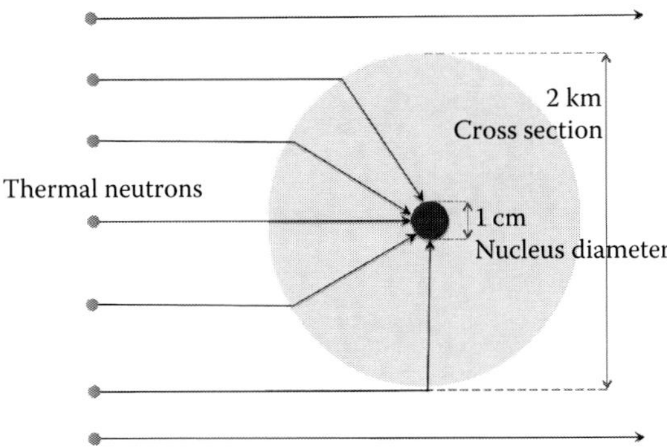

FIGURE 11.17 Capture capability of GdNCT.

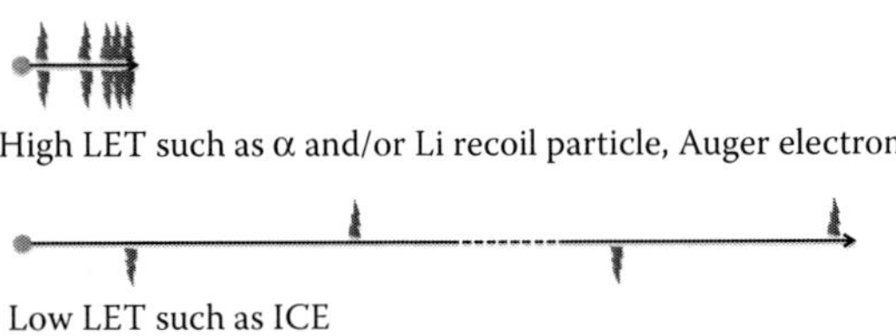

FIGURE 11.18 Schematic illustration of LET.

transferred to the materials surrounding the particle track in secondary ionizing radiations. In short, except for Auger electron reactions, GdNCT is a low-LET, and BNCT a high-LET, radiation therapy.

LET is a measure of energy transferred to the atmosphere by an ionizing particle traveling through it. Typically, this measurement is used for quantifying the effects of ionizing radiation on biological tissues.

High-LET Auger electrons cause lethal damage to cells by breaking double-stranded DNA (Figure 11.19). As the trajectory of Auger electrons is so limited (a few hundred Å), ^{157}Gd atoms must be in close proximity to the DNA. While rupture of both strands is lethal to tumor cells, a single-stranded breakage is reparable via DNA polymerase and is usually sublethal. Thus, the more Gd atoms that can be positioned close to tumor DNA by ligand binding, the more potent will be the killing effect from double-strand DNA breakage [10].

High-LET Auger electrons lethally damage tumor cells by breaking the double-stranded DNA. As the trajectory of Auger electrons is so small (a few hundred Å), to be highly tumoricidal, the ^{157}Gd atoms must bind to the DNA. High lethality cannot be expected when Gd atoms are mainly located in interstitial spaces or cytoplasm or both.

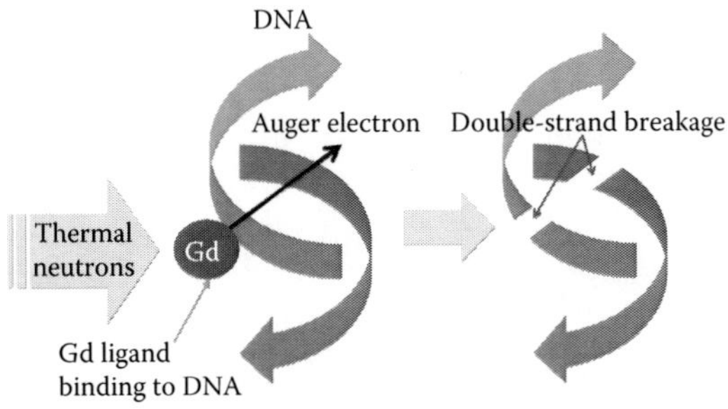

FIGURE 11.19 Lethal damage to double-strand DNA.

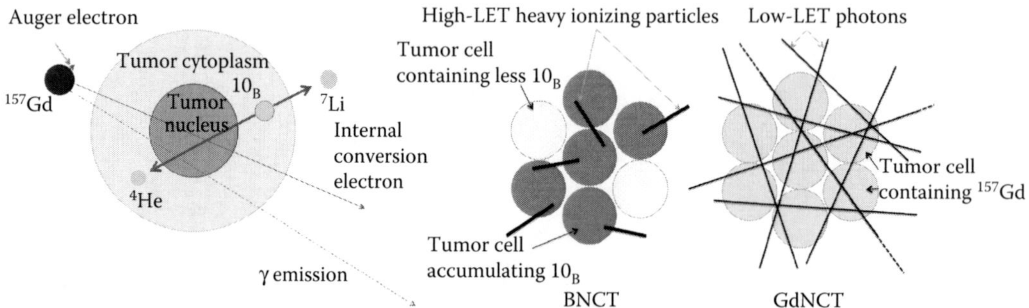

FIGURE 11.20 Schematic illustration of radio-biological differences between GdNCT and BNCT.

The radio-biological mechanisms of GdNCT and BNCT are quite different. In BNCT, heavy particles transfer their total kinetic energy to tumor cells and their neighboring cells, and the micro-distribution of kinetic energy in tissues is uneven. In contrast, the low-energy ionizing particles used in GdNCT travel farther and their kinetic energy is more uniformly distributed throughout tumor tissues (Figure 11.20).

The different physical characteristics of GdNCT and BNCT are shown in Table 11.2. In BNCT, while the high-LET heavy particles that are released are capable of killing tumor cells, the micro-distribution of the radiation absorbed dose is uneven due to the heterogeneous boron distribution in tumor tissues, and highly localized energy transfer from the particles can occur within ranges that do not exceed the dimensions of tumor cells. In contrast, while low-LET ionizing particles are emitted in GdNCT, the absorbed dose distribution is more uniform. Thus, GdNCT can theoretically be used to resolve the shortcoming of heterogeneous dose distribution in BNCT, assuming a uniform distribution of the gadolinium reagent near the DNA.

The vascular endothelium is highly vulnerable to radiation, which causes vascular necrosis. While necrosis of tumor tissues at the brain surface is therapeutically desirable, doses received by vascular tissues should, in principle, be kept within tolerable limits. In GdNCT, the vascular dose is negligible, because low LET causes little radiation damage (Figure 11.21). There is also very little possibility of endothelial damage from Auger electrons generated from intravascular Gd neutron capture reactions (GdNCR). Indeed, a preliminary study revealed no vascular necrosis from GdNCT after continuous infusion of high concentrations of GdDTPA (see Section 11.3.2).

TABLE 11.2
Difference in Physical Characteristics of GdNCT and BNCT

	^{157}Gd(n,γ)^{158}Gd	^{10}B(n,α)^{7}Li
Cross section	255,000 barn	3985 barn
Total kinetic energy	7.98 MeV	3.33 MeV
Heavy ion	—	High LET α and Li particle (RBE = 2–3,14 (μ)
Auger electron	High LET (RBE = unknown, several 100 Å)	— —
High-energy electron	Internal convergent electron (RBE = 1, several 100 μ)	—
High-energy photon	Prompt γ emission (RBE = 1, several 100 cm)	—

Note: RBE, relative biological effectiveness compared with standard (250 keV x-ray) radiation. Numerical values in parentheses indicate RBE values and trajectories of radiation particles.

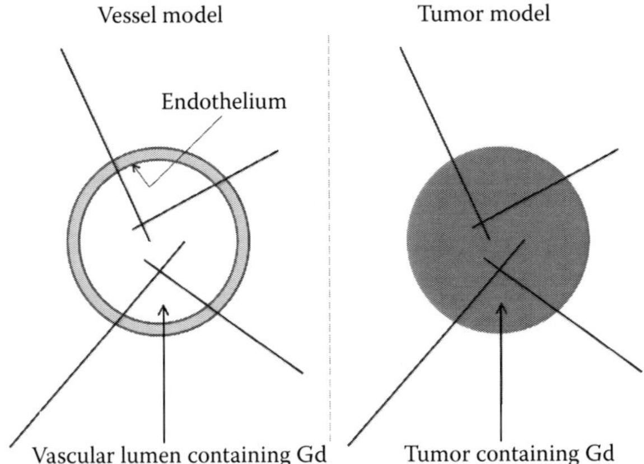

FIGURE 11.21 Different radiation dose distribution in blood vessels and tumors.

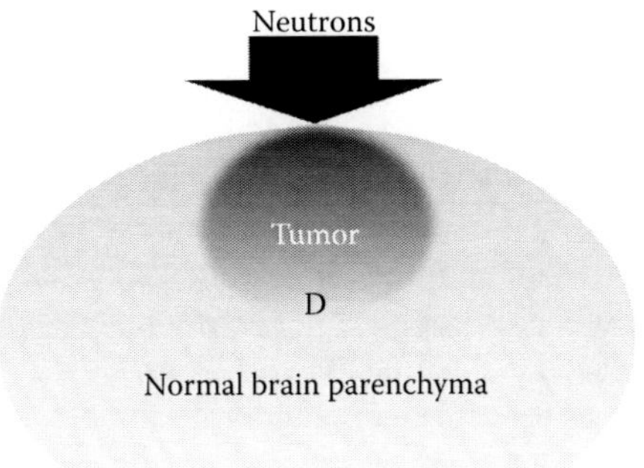

FIGURE 11.22 Self-shielding by ^{157}Gd.

In GdNCT, owing to the generally longer trajectories of the radiation (i.e., γ-rays, electrons, and x-rays) released by GdNCT, the larger the tumor, the greater is the general density of the absorbed dose in the tumor. Thus, since the radiation dose is directly dependent on tumor size, the tumoricidal effect of GdNCT is directly proportional to the size of tumor (bulk effect).

In GdNCT, the vascular dose is negligibly small because low-LET radiation causes little damage to vascular cells.

Gd has an extremely large nuclear cross section for thermal neutron capture. Tissues containing Gd avidly absorb thermal neutrons and its presence in the upper levels of tissue tend to shield the lower levels from neutron bombardment. Consequently, the density of neutron capture decreases depending on tissue-transit depth. As a result, the number of thermal neutrons able to reach deep tumors (D) is inevitably insufficient for a therapeutic dose (Figure 11.22). To minimize this shielding effect, we estimate that the optimum ^{157}Gd concentration in tumors should be around 200 ppm ^{157}Gd (<1000 ppm ^{n}Gd).

As ^{157}Gd has a large thermal neutron capture cross section, few thermal neutrons can make it past the superficial tumor and deeper tissues (D) do not receive efficacious doses.

11.2.1.2 Peri-Tumoral Radiation Effect of GdNCT

In GdNCT, the dose distribution extends slightly beyond the tumor mass, and radiation injury may also occur even in the peri-region of tumor infiltration (C) (Figure 11.23). However, this radiation

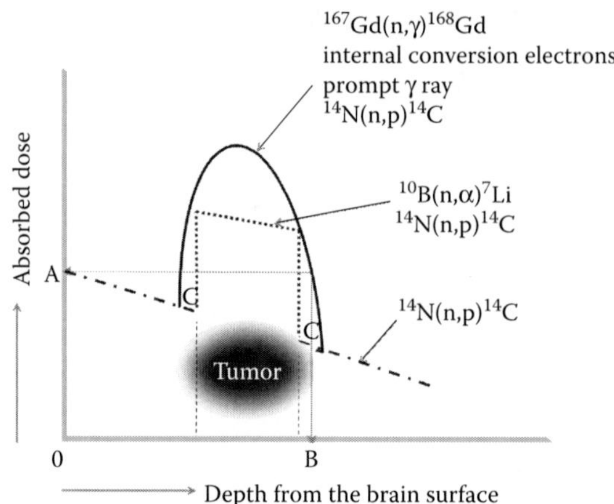

FIGURE 11.23 Limits of therapeutic dosage. (A) Maximum tolerable dose of the brain surface; (B) maximum therapeutic depth; (C) the peri-region of tumor infiltration.

halo is not strongly tumoricidal beyond 100 μm from the edge of the main tumor [12]. The thickness of the halo (100 μm) is equivalent to a thickness of only about of 7 tumor cells. It is difficult to deliver boron compounds to this marginal area. Of course, neurons and other normal brain parenchyma are present in this region, but the clinical implications are probably minimal.

Currently, the strategy for treating malignant brain tumors that neurosurgeons commonly consent to is maximum surgical extirpation of the tumor followed by adjuvant therapies such as radiation and chemotherapy. In line with this, for postoperative adjunctive and fractionated GdNCT, after tumor extirpation, Gd can be placed in the tumor cavity. Thus, GdNCT may provide a highly acceptable, noninvasive alternative treatment for malignant brain tumors. In future, if tumor malignancy can be determined, without any surgical exploration, by radio-pathological methods, fractionated GdNCT may become a highly reliable and acceptable treatment option, that readily gains consent as a treatment for malignant brain tumors.

In GdNCT, tumor cells infiltrating close to the main tumor mass are susceptible to peripheral radiation from the main tumor (Figure 11.24). Tumor cells that have infiltrated far from the main tumor, however, cannot be treated even by Gd/BNCT. In particular, tumor cells infiltrating to the contralateral hemisphere are beyond the scope of Gd/BNCT. Effective elimination of distantly infiltrating tumor cells, the prime cause of recurrence, remains a major challenge.

Red spots indicate infiltrating tumors.

11.2.2 *In Vitro* GdNCT

11.2.2.1 Tumor Cell Killing Effect

For *in vitro* GdNCT, colony formation assay is the standard method for evaluating the effects of biological radiation (Figure 11.25). The procedures are outlined below. A suspension of 5×10^3/mL C6 gliosarcoma cells in a logarithmic growth phase is irradiated in 1 mL Eagle's minimum essential medium, supplemented with 10% heat-inactivated fetal bovine serum (MEM(FCS+)) containing GdDTPA (Magnevist®, Shelling AG, Berlin, Germany) at concentrations of 0, 500, and 2500 ppm ^{n}Gd (respectively equivalent to 0, 78, and 390 ppm ^{157}Gd). For each sample, 1.5 mL of a sample containing 5×10^3 suspended cells/mL is transferred onto a Teflon tube (1 cm diameter, 3 cm high). Such tubes do not generate any secondary radiation during thermal neutron bombardment. Thermal neutron fluence is evaluated by averaging the activity of two Au foils symmetrically attached on the Teflon tube surface, along the thermal neutron incidence as shown in Figure 11.25. The thermal neutron fluence ranges from 0 to 1×10^{13} nvt. The γ-ray dose is monitored by TLD attached to the

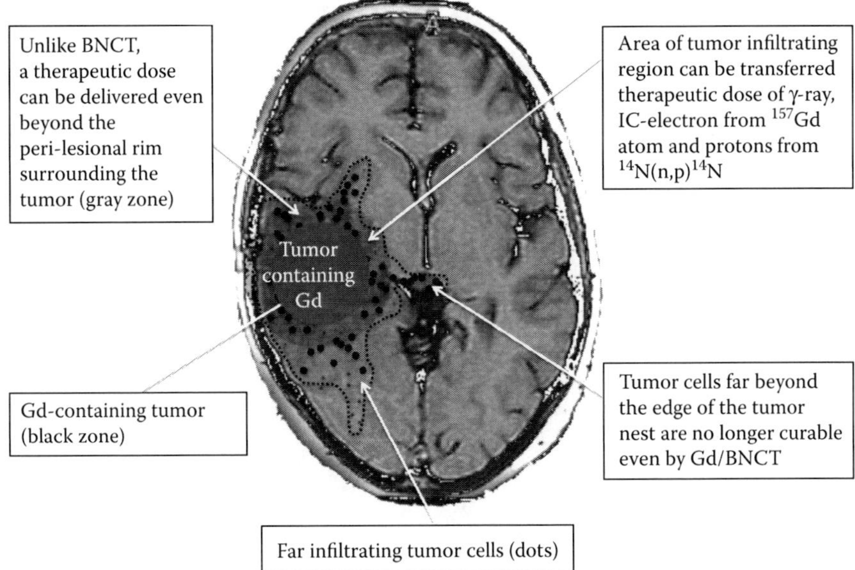

FIGURE 11.24 Illustration of a typical malignant brain tumor.

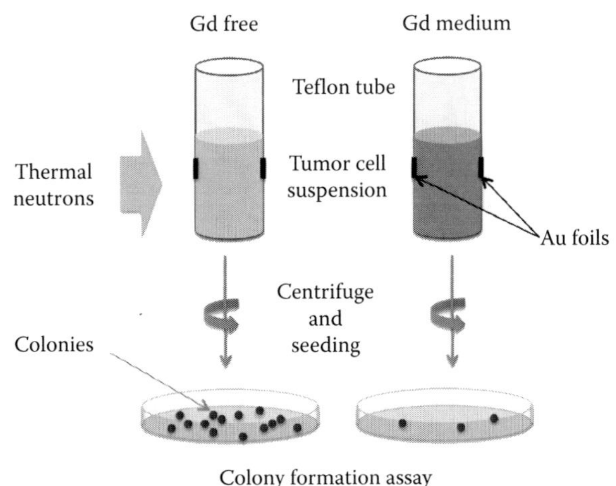

FIGURE 11.25 Colony formation assay procedures for *in vitro* GdNCT.

outer surface of the Teflon tube on the exposed side. The γ-ray dose ranges from 0 to 5 Sv. Immediately after irradiation, 60 and/or 180 μL of the samples (equivalent to 300 and/or 900 irradiated cells) are pipetted into 6 cm Petri dishes containing 6 mL MEM(FCS+) and incubated for 10 days to form colonies in a humidified 5% carbon dioxide atmosphere at 37°C. The colonies are fixed and stained with formaldehyde 1% toluidine blue solution and then counted macroscopically.

11.2.2.2 *In Vitro* Survival

Figure 11.26 shows surviving fractions after *in vitro* GdNCT. Without a sigmoidal shoulder on the plot, survival decreases as a function of the thermal neutron dose: with 500 ppm Gd(+) medium, a 37% survival dose (D_{37}) is obtained with 3.55×10^{12} n/cm^2; with 2500 ppm Gd(+) medium, D_{37} was 1.40×10^{12} n/cm^2; and with Gd-free medium, D_{37} was 6.80×10^{12} n/cm^2. Compared with control

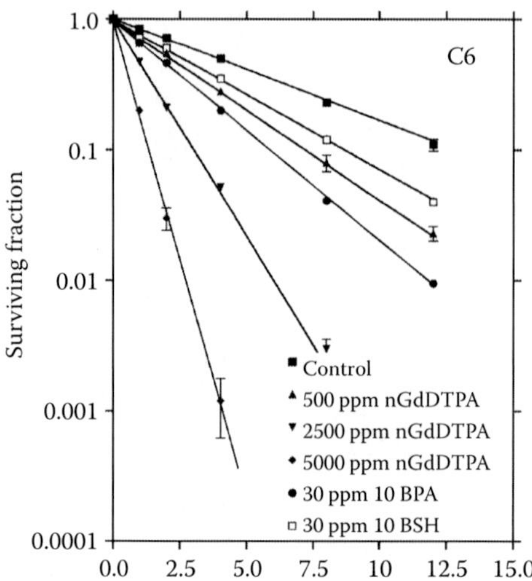

FIGURE 11.26 Surviving fractions after GdNCT using various Gd concentrations, and comparative results for BNCT.

survival, radiation in GdNCT was 1.92 times more lethal with 500 ppm ^{n}Gd(+) and 4.86 times more lethal with 2500 ppm ^{n}Gd(+). Lethality with BPA at 30 ppm ^{10}B is similar to that of 500 ppm ^{n}Gd.

11.2.3 *In Vivo* GdNCT

The relevant institutional ethical committee must approve the design of each experiment involving animals that are to be conducted. A minimum number of animals are used and they were cared for according to International Standard Guide for the Care and Use of Laboratory Animals.

11.2.3.1 Brain Tumor

11.2.3.1.1 GdNCT in Brain Tumor Models

Fisher 344 rats ($N = 20$) were anesthetized by 2 mg/kg intraperitoneal injection of barbiturate. The head of each rat was placed on a stereotactic frame and a burr hole was drilled 2 mm lateral and caudal to the mid-point of the coronal suture on the right parietal region. To induce tumor formation, using a 27G Hamilton syringe, in aseptic conditions, 10^6/10 μL C6 cell suspension was slowly injected into the right caudate putamen at the depth of 3 mm from the brain surface. Immediately after removal of the syringe, the burr hole was closed with bone wax.

11.2.3.1.2 Pharmacokinetics of Gd in Tumors

The absolute ^{157}Gd concentration can be evaluated by comparing the emission of prompt γ-rays from the nuclide with those from hydrogen, which is constantly present in live organic matter (Figure 11.27) [13]. To avoid overestimation of ^{157}Gd, it is important to prevent dehydration of specimens before evaluation.

In situ quantification of ^{157}Gd is achieved after irradiation with a cold neutron beam through a burr hole. Prompt γ-rays are then monitored using a germanium scintillation counter. In the future, this method may be used with computed tomography to visualize the quantitative distribution of ^{157}Gd and ^{10}B [14].

Figure 11.28 shows the changing Gd concentration in rat blood after injection: it rapidly decreases with a two-component decay. The half-life of the initial decay is approximately 19 min. After bolus injection of 0.4 mL GdDTPA into a tail vein, ^{157}Gd concentration was evaluated by prompt γ-ray assay. A peak concentration of 80 ppm was observed just after injection. Then, after similar

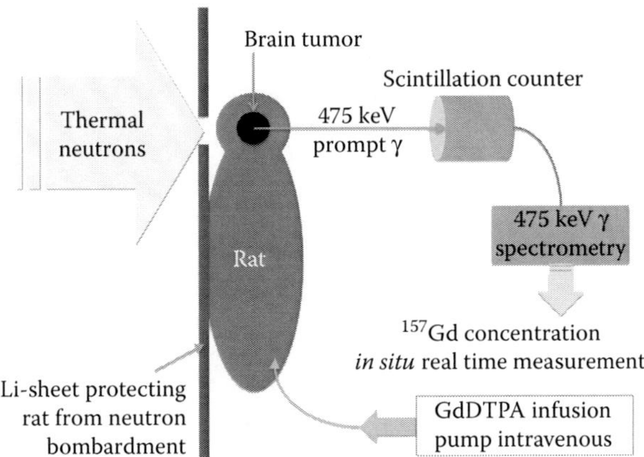

FIGURE 11.27 *In situ* quantification of ^{157}Gd in a rat model.

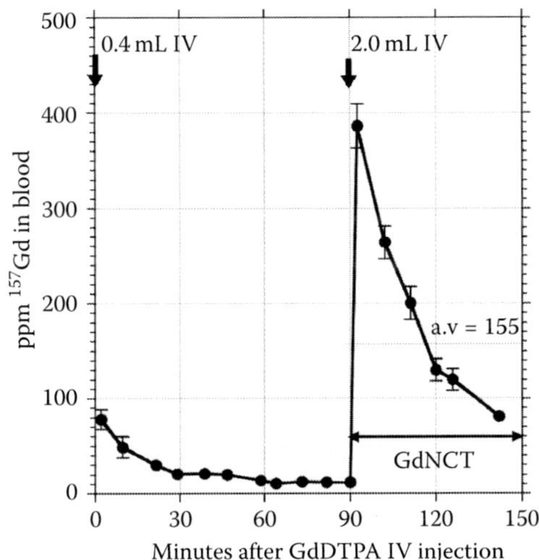

FIGURE 11.28 Time course of Gd concentration in blood in a rat model of C6 brain tumor.

administration of 2 mL GdDTPA, the Gd concentration proportionally increased to 400 ppm. Gadolinium concentration in the tumor was assumed to be half the blood concentration (*T/B* ratio = 0.5). Measuring ^{157}Gd concentration in animals with no brain tumor via GdDTPA administration in the same manner described above, the Gd concentration in total normal brain tissue was found to be less than about 10% of the concentration found in blood. Since blood vessels usually account for about 10% of the volume of the brain, it can be concluded that GdDTPA does not pass through the normal blood–brain barrier and is not distributed in normal parenchyma.

Immediately after a 2.0 mL GdDTPA bolus injection, thermal neutrons were irradiated onto the brain surface for 60 min, and during this time the average Gd concentration in brain tumor tissue was estimated to be 80 ppm ^{157}Gd. Peak concentration of Gd in the blood was almost proportional to the total amount of the intravenous injection. In this model, the average concentration of Gd in the blood was found to be 155 ppm ^{157}Gd (= 993 ppm ^{n}Gd). Assuming that the *T/B* ratio, 0.54, is the same for humans and rats, the average concentration of Gd in tumors during human GdNCT can be estimated to be approximately 80 ppm ^{157}Gd (i.e., 512 ppm ^{n}Gd).

A Kaplan–Meier plot of GdNCT for rat brain tumor is shown in Figure 11.29. C6 rats that did not receive any treatment, surviving for 16.4 ± 0.6 days, all died at about the same time. We found no

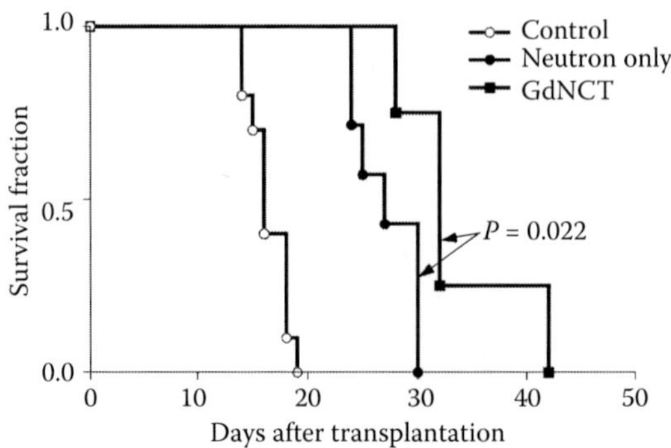

FIGURE 11.29 Kaplan–Meier plot of rat brain tumor GdNCT.

statistically significant difference ($p = 0.134$) in survival for rats receiving only neutron bombardment (27.0 ± 0.9 days) and those receiving neutrons and 0.4 mL ^{n}GdDTPA (27.1 ± 1.0 days). Statistically significant longer survival (33.5 ± 3.0 days) is apparent after treatment with neutrons and 2 + 1 mL ^{n}GdDTPA.

There are two major drawbacks in *in vivo* GdNCT experiments.

11.2.3.1.2.1 Self-Shielding Effect of Neutrons Owing to the large thermal neutron capture cross section of ^{157}Gd, if the Gd concentration is too high, efficient capture in the upper levels drastically decreases the neutron dose received by deeper tissues. To ensure that tumor tissues receive the proper neutron dose, the most effective concentration of ^{157}Gd might be less than 200 ppm (1000 ppm ^{n}Gd), which is a concentration sufficient to kill the tumors [15].

11.2.3.1.2.2 Tumor Volume Effect The tumor dose is dependent on the size of the tumor itself. Theoretically, the γ-ray dose increases proportionately with tumor volume (dose $\propto R^3$, where R is tumor radius). Therefore, GdNCT is less effective with in low-volume tumor models such as rat brain tumor and completely extirpated tumors.

Even so, GdNCT is a promising noninvasive therapy for malignant tumors. Malignant brain tumors, such as glioblastoma, should be treated nonsurgically.

Experiments using tumor models in larger animals are needed to investigate the tumor volume effect. Such an investigation has been carried out using a cat model essentially similar to the rat brain tumor model. A suspension of 10^6 C6 gliosarcoma cells in 10 μL 10% agar-MEM (CSF free) was implanted into the right parietal region at a depth of 5 mm from the brain surface through a burr hole made 5 mm caudal to the bregma. Although this brain tumor model was heterozygous, the pathological specimens revealed tumor invasion in peri-tumoral parenchyma and central necrotic lesion in the tumor caused by inner high pressure and hypo vascularity due to rapid tumor growth (Figure 11.30). This type of malignancy seems to provide an effective brain tumor model.

After intraperitoneal injection of 0.2 mL/kg pentobarbiturate, under general anesthesia, along with continuous intravenous administration of GdDTPA 2 mL/kg/h into an exposed femoral vein, cat brain tumors were irradiated through the scalp with thermal neutrons for 50 min. The γ-ray dose was monitored using a TLD on the scalp surface. By shrouding with thermal neutron absorbing ^{6}LiF flexible sheets, special care was taken to protect the whole body of the animals, especially the eyes and buccal mucosa, from thermal neutron bombardment. Using prompt γ-ray spectrometry (PGS), Gd concentration in blood was measured every 10 min after the start of continuous GdDTPA infusion. Brain tumor Gd concentrations were estimated by peak ground accelerations, as described above for rat brain tumors (Figure 11.31). The average ratio of Gd concentration in cat brain tumor compared with blood (*T*/*B* ratio) was similarly estimated to be 0.5.

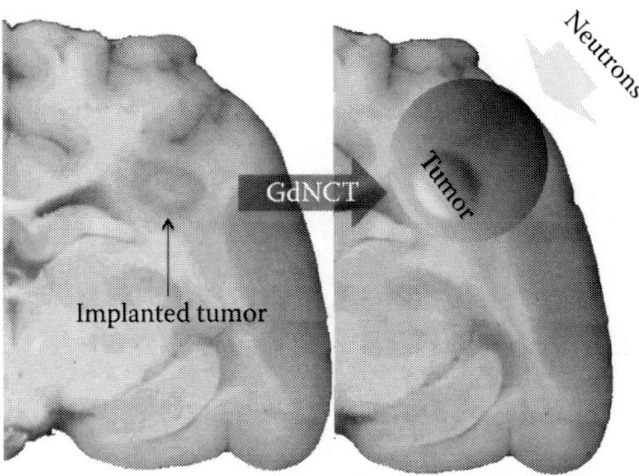

FIGURE 11.30 GdNCT in cat brain tumor model.

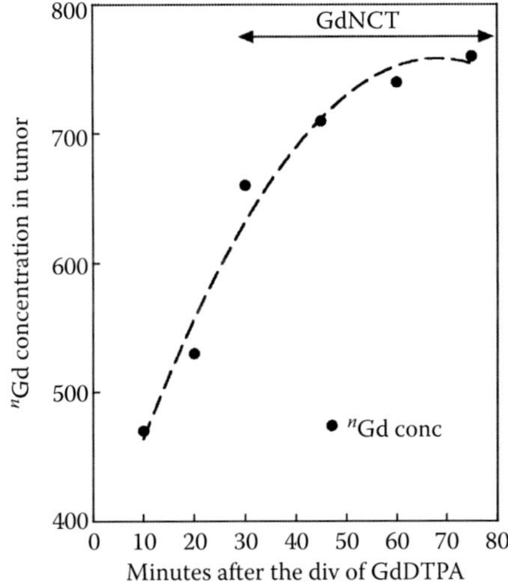

FIGURE 11.31 Gd concentration in tumor during continuous venous infusion of GdDTPA.

After GdNCT, the cats survived without any observable neurological abnormalities. One week after GdNCT, the cats were sacrificed and brain tumor pathology was investigated. As Figure 11.32 shows, after GdNCT, the tumor cells were necrotic and almost completely destroyed, and a large number of lymphocyte aggregates were present in samples; however, vial tumor cuffing remained unaffected. These vial tumor cells might be highly radiation resistant or able to survive with sublethal damage or both. No cellular damage was observed in normal brain samples exposed to radiation during GdNCT, not even in samples of parietal parenchyma surrounded by a superior sagittal sinus.

Radiation bombardment of the whole body of the cats was monitored by TLD (Table 11.3). Whole-body exposure was generally high, showing the difficulty of adequately protecting small bodies weighing around 4 kg. The most important finding was that, for cats 3 and 4, the scalp surface dose was statistically significantly higher on the brain tumor. Meanwhile, no other particularly major differences in whole body bombardment were found for different cats irrespective of the presence of brain tumor.

In the cat model, average Gd concentration in brain tumor tissue was confirmed to be around 700 ppm ^{n}Gd (= 110 ppm^{157}Gd). A similar concentration in human brain tumors can be achieved by using ^{157}Gd-enriched DTPA. The cat model study shows that GdNCT, along with high-dose

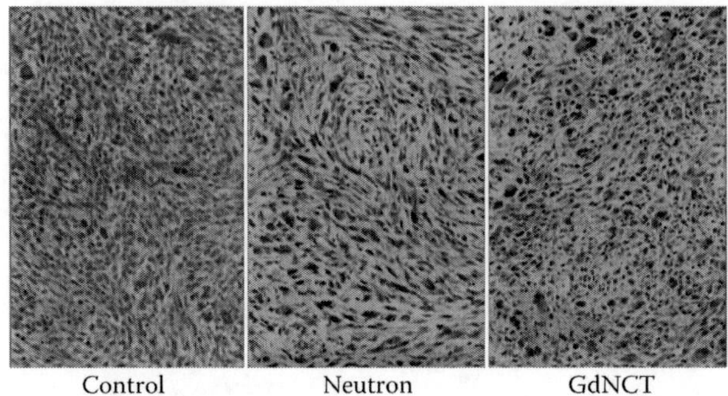

FIGURE 11.32 Pathology samples taken 1 week after GdNCT.

TABLE 11.3
Whole-Body Radiation Exposure TLD Results for GdNCT Cats with and without Brain Tumors

Site	Cat 1 Normal Brain	Cat 2 Normal Brain	Cat 3 Brain Tumor	Cat 4 Brain Tumor
Scalp surface	260	230	1580	500
Thyroid	360	310	340	440
Ant. chest	320	280	220	230
Shoulder	240	210	250	180
Abdomen	300	190	215	180
Genital	130	140	120	100
NVT ($\times 10^{13}$)	1.9	2.0	1.5	1.7

administration of GdDTPA at levels almost 10 times greater than the clinical dose used for MRI enhancement, had a tumoricidal effect on C6 brain tumors without serious injury to normal brain and blood vessels, even after a high dose of thermal neutron exposure of 2.0×10^{13} n/cm^2. Furthermore, for effective GdNCT, it is essential to increase the period of retention of Gd in the tumor.

Although the results of *in vitro* studies of GdNCT have been promising [16–18], *in vivo* studies of GdNCT have been hampered by the challenge of maintaining the Gd concentration in tumors during GdNCT. Akine has reported excellent *in vivo* results with murine ascites tumor after intraperitoneal administration of Gd [19]. Meanwhile, to maintain Gd concentration in tumor tissues during GdNCT, Ichikawa experimented with Gd-containing microcapsules [20,21]. In the studies described above, Gd concentrations were maintained during GdNCT by continuous intravenous administration of GdDTPA. No collateral damage was detected in pathological examinations after GdNCT and, despite the high concentration of Gd in blood vessels, about 1400 ppm ^{n}Gd, good results were obtained and no evidence of vascular damage was observed, especially in the susceptible peri-sinus region.

Other potential modes for administering GdDTPA, high-dose intraventricular or intrathecal injection or both, may have a supplemental killing effect when there is subependymal or intraventricular tumor dissemination.

In GdNCT, macroscopic dose distribution in tumor tissues is almost uniform. Theoretically, in a spherical tumor with radius R, the dose intensity, $\phi(r)$, uniformly caused by the ^{157}Gd(n,γ/e)^{158}Gd nuclear reaction in tumor can be expressed as

$$d^2\phi(r)/d^2r + 2d\phi(r)/rdr + B^2\phi(r) = 0$$

$$\phi(r) = (\pi N\sigma\Phi/r)\sin(\pi r/R)$$

where r is the distance from the tumor center, N is the ^{157}Gd concentration, σ is 255,000 barn, Φ is the thermal neutron flux, and B is a constant whose value depends on the size of the spheroid. Now, the total dose distribution is estimated as follows:

$$\gamma^* + {}^{157}\mathrm{Gd}(n,\gamma/e)^{158}\mathrm{Gd} + {}^{14}N(n,p)^{14}\mathrm{C}$$
$$= \gamma^* + [N({}^{157}\mathrm{Gd})\sigma({}^{157}\mathrm{Gd})\mathrm{RBE}(\gamma/e) + N({}^{14}N)\sigma({}^{14}N)\mathrm{RBE}(p)]\phi(r)$$

where γ^* is contamination caused by the reactor core and structural materials, RBE is the relative biological effectiveness, and N the concentration in tissue of ^{157}Gd or ^{14}N or both.

The maximum contribution of γ-rays to the tumor-absorbed dose was less than 50% [22]. In a spherical 3 cm-diameter tumor model, approximately 25% of the absorbed dose was found to be attributable to γ-rays, and more than 50% could be accounted for by IC electrons. Selective high dosing in tumor tissues is attributable to IC electrons, which have a trajectory in tissues of approximately 100 μm [23]. Unlike as in BNCT, the peri-tumoral dose distribution does not decrease sharply. The absorbed dose of γ-rays has been estimated to decrease to 10% at 2–3 cm from the tumor margin [24]. This peri-tumoral dose distribution might have lethal effects on tumor cells invading the peri-tumoral lesion, which manifest as a high-density region in T_2-weighted MRI. The dose distribution of high-energy γ-rays and electrons in tissues immediately adjacent to the main tumor mass might be tumoricidal for infiltrating tumor cells that are close to the main tumor.

The deleterious effects of unavoidable exposure to spatial γ* and protons, caused by ^{14}N, precludes extended periods of irradiation. This frustrating limitation can be somewhat alleviated by combining GdNCT with BNCT, especially for deep-seated tumors. The optimal ^{n}Gd concentration in tumors has been reckoned to be around several 1000 ppm. This optimal concentration coincides well with theoretical simulation results obtained using a two-dimensional neutron-coupled γ-ray transport code (DOT 3.5) for GdNCT [25]. GdNCT is largely dependent on the balance between Gd concentration in tumors and thermal neutron fluence. Downward adjustment of the thermal neutron fluence to compensate for the large capture cross section of the ^{157}Gd atom is the most difficult aspect of GdNCT technique. The optimum ^{157}Gd concentration in tumors is estimated to be around 200 ppm ^{157}Gd (<1000 ppm ^{n}Gd). If the ^{n}Gd concentration in tumor tissues were to exceed 1000 ppm, fewer thermal neutrons would reach deeply seated tumor tissues. This early neutron elimination makes the therapy ineffective for deep brain tumors. Possible solutions to this problem are discussed in Section 11.3.

11.2.3.2 Cutaneous Melanomas

Subcutaneous melanomas were investigated in larger tumor models. In 10 days after subcutaneous implantation of 1 mm cubes of Green's melanotic melanoma into the thighs of Syrian golden hamster model (N = 30), the melanomas had grown to the size of a thumb tip (approx. $12 \times 12 \times 12 \div 3$ mm^3 = 576 mm^3). For continuous infusion of GdDTPA, the femoral vein on the contralateral side was cannulated. During continuous 2 mL/kg/h infusion of GdDTPA, thermal neutrons were irradiated on the thigh. During the protocols, the whole body of each animal was protected from neutron bombardment by a Li sheet (Figure 11.33). After the start of infusion, Gd concentration in blood was determined every 10 min by PGS. After GdNCT, tumor size was measured at intervals and tumor volume was estimated by dividing the result of multiplying the 3 measured dimensions by 3.

As the data in Figure 11.34 demonstrate, thermal neutron irradiation without Gd had little effect on tumor growth, which recommenced after a single week of inhibition. In contrast, GdNCT was able to effectively destroy melanoma tissues. All tumor growth was strongly inhibited: tumor size dramatically decreased and was well restrained for more than 2 weeks after treatment. Complete destruction was also observed in 2 of 10 hamsters. Thus, remission can be achieved by GdNCT with continuous infusion of GdDTPA.

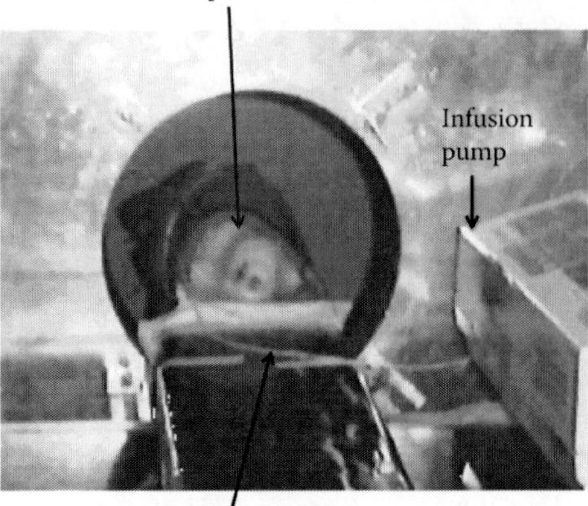

FIGURE 11.33 Hamster with cutaneous melanoma in thigh placed at the thermal neutron irradiation port of the Research Reactor Institute of Kyoto University.

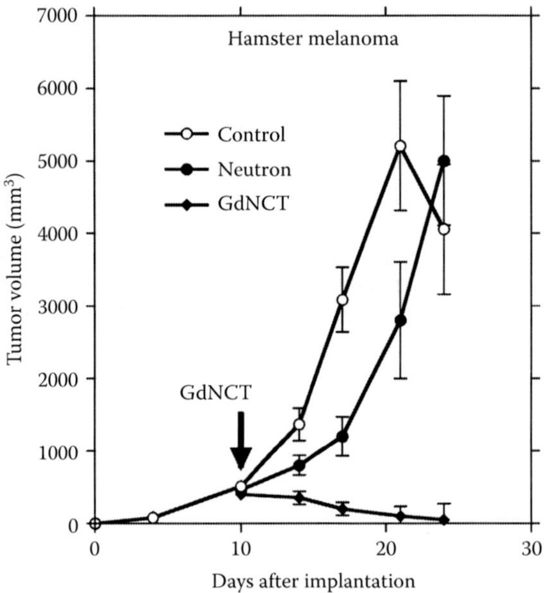

FIGURE 11.34 Effect of GdNCT on growth of subcutaneous melanoma in hamsters.

11.2.3.3 Pharmacokinetic Study of GdDTPA for Human Brain Tumors

In this section, the measurements of Gd concentration in several brain tumor samples, obtained from fresh surgical specimens, are compared to the results for blood samples from the same patients [26]. The issue of whether the Gd concentration for this compound would be sufficiently high to treat brain tumors using NCT is also discussed.

During surgical operations, 97 tissue samples from 31 patients were obtained. Until analysis, the samples were stored at −20°C (Tables 11.4 through 11.7). Tissue specimens were collected from those parts of the tumors where Gd enhancement was most apparent in preoperative MRI studies. In preliminary studies, both malignant and benign intracranial tumor samples were collected.

Gd concentration was measured using inductively coupled plasma atom emission spectrophotometry. Introduction of error due to blood contamination in tumor tissue was excluded by measuring the cyan-methemoglobin content of the tissue.

TABLE 11.4
Gd Concentration in Glioma: High-Grade, Relatively Malignant

After Inj. (min)	Case (Initials)	Tumor ppm Gd	Blood ppm Gd	*T/B* Ratio
15	C. H.	3.12	38.0S	0.082
20	Y. T.	7.40	42.52	0.174
40	C. H.	10.24	24.56	0.417
60	K. M.	0.55	24.14	0.023
>60	Y. T.	2.07	28.69	0.072

TABLE 11.5
Gd Concentration in Glioma: Low-Grade, Relatively Benign

After Inj. (min)	Case (Initials)	Tumor ppm Gd	Blood ppm Gd	*T/B* Ratio
0	F. O.	2.60	—	—
10	K. S.	0.93	54.43	0.017
15	Y. N.	1.30	41.34	0.031
30	F. O.	9.68	45.88	0.211
80	M. Y.	UD	17.72	—

Three of the seven gliomas were found to be malignant (Table 11.4). The *T/B* ratio for Gd peaked at 30–40 min after bolus injection of GdDTPA. At 60 min after injection of GdDTPA, the ratio dropped to less than 10%. No major differences between malignant and low-grade gliomas were observed, even though the latter included oligodendroglioma and ependymoma (Table 11.5). For meningiomas, the *T/B* ratio was around 20–30%, and showed a tendency to gradually increase in the later phase, even up to more than 60 min after Gd administration (Table 11.6). In one case of recurrent meningioma, the *T/B* ratio was more than 100% and reached a peak of 644% after 85 min (Table 11.7). This makes meningioma a good candidate for GdNCT, especially if the tumor is situated in a surgically inaccessible site. For each different type of metastatic tumor, the ratio was different due to the varied histologies of the primary lesions. Chondroma, pituitary adenoma, craniopharyngioma, and other types of tumor mostly had low *T/B* ratios (<10%).

TABLE 11.6
Gd Concentration in Meningiomas

After Inj. (min)	Case (Initials)	Tumor ppm Gd	Blood ppm Gd	*T/B* Ratio
15	M. K.	1.33	61.85	0.022
	T. F.	0.72	48.18	0.015
25	H. Y.	10.20	43.52	0.234
30	M. K.	2.67	47.04	0.057
40	M. K.	5.13	29.98	0.171
	N. S.	18.68	30.52	0.612
55	Y. F.	8.50	26.09	0.326
60	K. S.	0.60	31.89	0.019
	M. K.	2.80	29.08	0.096
80	N. S.	5.76	17.92	0.321
135	K. S.	5.35	12.89	0.415

TABLE 11.7
Gd Concentration in Recurrent Meningiomas

After Inj. (min)	Case (Initials)	Tumor ppm Gd	Blood ppm Gd	*T/B* Ratio
40	T. T.	100.48	26.68	3.766
85	T. T.	77.12	11.96	6.448

In MRI studies, it has been a general assumption that even 45 min after the injection of GdDTPA, the enhancement effect is not clearly apparent. Yoshida [27] has reported, however, as measured by T1-relaxation time in MRI images, that there is a stronger uptake of Gd in tumor tissues than in blood. Moreover, in preliminary experiments, it was found that most of the brain tumors tested absorbed and held GdDTPA for 120–135 min, although normal brain tissue does not behave similarly (data not shown). In one case of recurrent meningioma, the *T/B* ratio did not much exceed 100% after GdDTPA administration. The oil/water distribution coefficient of GdDTPA is <0.0001, causing rapid excretion of gadolinium from tissues. Consequently, rather than being present intracellularly, almost all Gd-bearing compounds are thought to be located in the extracellular spaces of tumors.

As noted above, in GdNCT using GdDTPA, IC electrons and γ-rays, not Auger electrons, are the main killers of tumor cells. In cerebral tissues, blood vessels occupy only 4.2% of the whole area of any section in any plane, and the thickness of the endothelium of intraparenchymal blood vessels with a diameter of 50 μm is about 3.5 μm. Therefore, under conditions where the ***T/B*** ratio of GdDTPA is higher than 10%, damage to cerebral blood vessels by centralized γ-rays is negligible compared to be damage on solid tumor masses that have diameters larger than 1 cm (also see Figure 11.21).

Hara (1987, unpublished data) has reported, in a rabbit model, that a subarachnoidal injection of 0.25 mL/kg Gd-DTPA of 0.125 mmol/mL did not have any effect on EEG readings. Therefore, in the treatment of disseminated malignant brain tumors by neuron capture therapy, intrathecal administration of GdDTPA might be effective.

In conclusion, GdNCT might be effective for the treatment of many types of brain tumors. Enriched ^{157}Gd-bearing GdDTPA is a promising agent for GdNCT. In the application of GdDTPA for GdNCT, it is necessary to keep the concentration of GdDTPA in tumor tissues stable during long periods of tumor exposure to epithermal neutrons. Intrathecal administration of GdDTPA and NCT might become a useful modality for treating disseminated malignant brain tumors. The development of Gd compounds that can be incorporated into tumor cells or attached to DNA is also required to increase the lethality to tumor cells of high-LET Auger electrons.

The *T/B* ratios varied according to type of tumor and specimen. Despite careful surgical excision to harvest the samples, the pieces of removed tumor tend to be slightly coagulated. Generally, except in recurrent malignant meningioma, *T/B* ratios were less than 0.3–0.5.

11.3 FROM PROGRESS TO ADVANCEMENT

11.3.1 Brain Tumors

Table 11.8 contains a summary of the bio-distribution of GdDTPA, BPA, and BSH. BSH is passively distributed in tumor cells and heterogeneously in tumor tissues. Through metabolic activity, BPA is actively accumulated inside tumor cells and heterogeneously distributed in tumor tissues. GdDTPA is distributed in interstitial spaces and is rapidly washed out. BSH and GdDTPA are unable to pass through the blood–brain barrier.

Given these characteristics, the distribution of GdDTPA, BSH, and BPA in brain tumor tissues is schematically shown in Figure 11.35. The basic layout is shown in the upper left, where the main

TABLE 11.8
Bio-Distribution and Pharmacokinetic Parameters of GdDTPA, BSH, and BPA

(Containing in)	GdDTPA	BSH	BPA
Tumor vessel	+	+	+
Brain vessel	–	–	+
Concentration Ratio	**GdDTPA**	**BSH**	**BPA**
T/B	0.5	1	3
T/N	>10	>10	3

Note: GdDTPA and BSH are not distributed in normal parenchyma except in vascular component which accounts for 4–6% of brain parenchyma. As an amino acid analogue, BPA easily passes through the blood-brain barrier. GdDTPA, BSH, and BPA permeates into tumor tissue through the tumor vessel.
T/B, tumor-to-blood concentration.
T/N, tumor-to-normal brain concentration.

tumor is surrounded by a broken circle that encompasses tumor cells and a blood vessel (thick broken circle). Normal glial cells (pentagon) and normal brain blood vessels are susceptible to malignant brain tumors such as glioblastoma multiforme and other tumors. The distribution of GdDTPA is shown in gray (upper right). GdDTPA does not enter normal brain parenchyma or tumor cells. It is only distributed in the interstitial spaces of tumor tissues and blood vessels. While

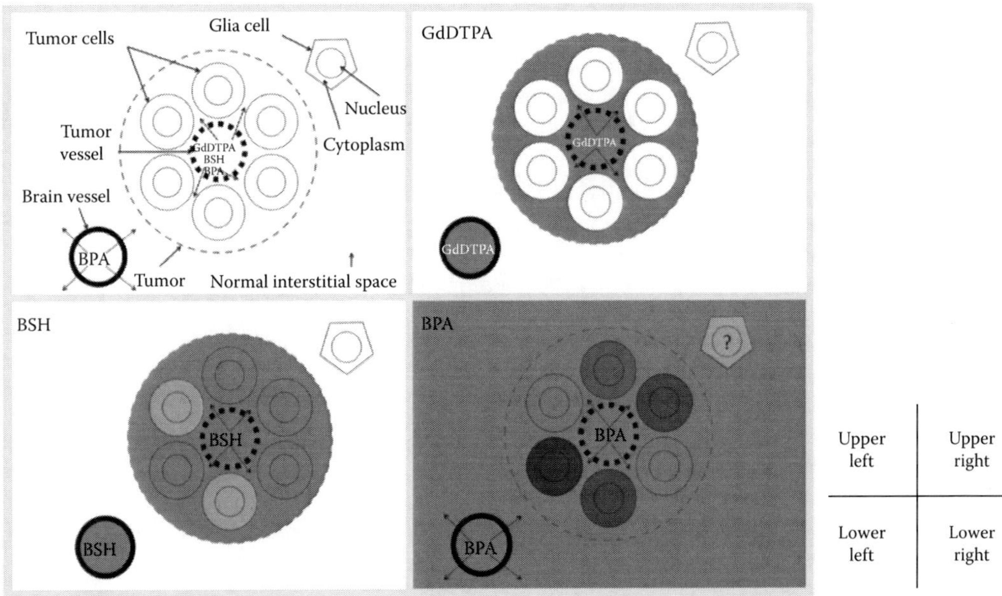

FIGURE 11.35 Bio-distribution of GdDTPA, BPA, and BSH. Upper left: Tumor model showing tumor cells arranged concentrically around a tumor blood vessel. Beyond the tumor is normal brain parenchyma containing a normal brain blood vessel and a glial cell. Upper right: GdDTPA is distributed only in interstitial spaces in blood vessels and tumor tissues. Lower left: BSH is passively and heterogeneously distributed in tumor interstitial spaces through tumor–blood barrier, in blood vessels, and in tumor cells. Unable to pass through the blood–brain barrier, BSH does not reach the normal brain except in blood vessels. Lower right: BPA is distributed throughout the brain and is actively and heterogeneously accumulated in tumor cells. Radiation can cause serious brain damage if boron is disseminated in normal brain tissues. It may also be accumulated in normal glial cells.

GdDTPA has been distributed in tumor cells in *in vitro* [28], there is no evidence that it enters tumor cells *in vivo*. BSH distribution is shown in gray area (lower left). Because it cannot pass through the blood–brain barrier, BSH does not reach normal brain parenchyma. Thus, BSH is distributed only in tumors and in blood vessels, and uptake in tumor cells is passive. Microdistribution in tumor cells is mostly heterogeneous. BPA distribution is shown in the gray area (lower right). BPA is uniformly located throughout the brain but is actively and heterogeneously accumulated in tumor cells.

As it passes through the blood–brain barrier, the most problematic characteristic of BPA is the way it disseminates in normal parenchyma. Exposure to radiation when BPA is generally distributed in normal parenchyma can cause serious brain injury, although tumor cells can be selectively killed if a high *T/B* ratio of approximately 3:1 can be achieved. While BSH actively accumulates in tumor cells (*T/B* ratio < 1.0) [29], if there is a very high concentration of BSH in normal blood vessels, high-LET particles may cause serious vascular damage. In contrast, GdDTPA-based NCT causes hardly any serious injury to either vascular or normal brain parenchyma.

More selective uptake and longer retention in tumor tissues or tumor cells or both are desirable for future Gd compounds. In preliminary studies, Gd-HP-DO3A (gadoteridol) was slightly more effective than GdDTPA for GdNCT. Commercially available Gd compounds are listed in Table 11.9 and Figure 11.36 shows the structures of the Gd sequestering agents. Evaluation of the IC_{50} of GdDTPA and Gd-HP-DO3A was carried out using Celltiter 96 Aqueous One Solution Cell Proliferation Assay and the results are presented in Figure 11.37. Gd-HP-DO3A is less toxic and more suitable for ongoing *in vivo* GdNCT studies.

11.3.1.1 Limitations of BNCT

As described earlier, infiltrating tumors (satellite lesions) and intracranial metastatic lesions far from the main tumor mass are beyond the scope of current modes of B/GdNCT. However, to prevent the iatrogenic spread of tumor cells, it is preferable to treat malignant brain tumors noninvasively. Here, GdNCT promises to provide more effective noninvasive treatment options for malignant brain tumors than BNCT. Successful treatment of larger tumors is also more likely with GdNCT. Even so, unless the entire brain can be uniformly irradiated with thermal neutrons, NCT is likely to be ineffective against distantly infiltrating tumor cells. Thus, after NCT, adjuvant therapies, such as temozolomide administration, must be used against satellite or intracranial metastatic lesions (Figure 11.38).

The main tumor mass treatable by GdNCT (dark gray area) is surrounded by numerous infiltrating tumor cells and/or satellite tumor masses (black area). Tumor cells infiltrating far from the

TABLE 11.9
Commercially Available Gd Compounds

Gd-Ligand	Chemical Name (Registered)	Chemical Formula	MW	Tonicity	Chem. Struct.	Since
Gd-DTPA	Meglumine gadopentetate (Magnevist)	$C_{14}H_{20}GdN_3O_{10}C_7H_{17}NO_5$	742.79	Ionic	Linear	1988
Gd-HP-DO3A	Gadoteridol (ProHance)	$C_{17}H_{29}GdN_4O_7$	558.69	Non-ionic	Macrocyclic	1994
Gd-DTPA-BMA	Gadodiamide (Ominiscan)	$C_{16}H_{28}GdN_5O_9$	645.72	Non-ionic	Linear	1996
Gd-DOTA	Meglumine gadotetate (Magnescope)	$C_{16}H_{25}GdN_4O_8C_7H_{17}NO_5$	753.86	Ionic	Macrocyclic	2001

Gd-HP-DO3A

Gd-DTPA

Gd-DOTA

Gd-DTPA-BMA

FIGURE 11.36 Commercially available Gd sequestering agents.

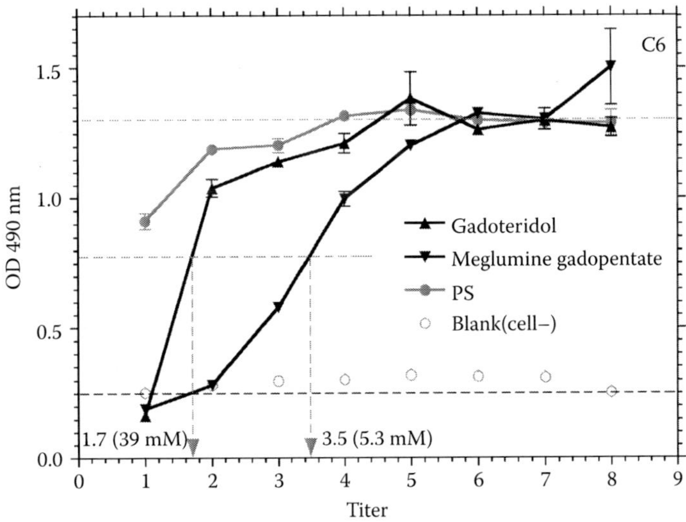

FIGURE 11.37 IC_{50} results for assays of GdDTPA (▼) and Gd-HP-DO3A (▲) by Celltiter 96 Aqueous One Solution Cell Proliferation. PS (•), physiological saline.

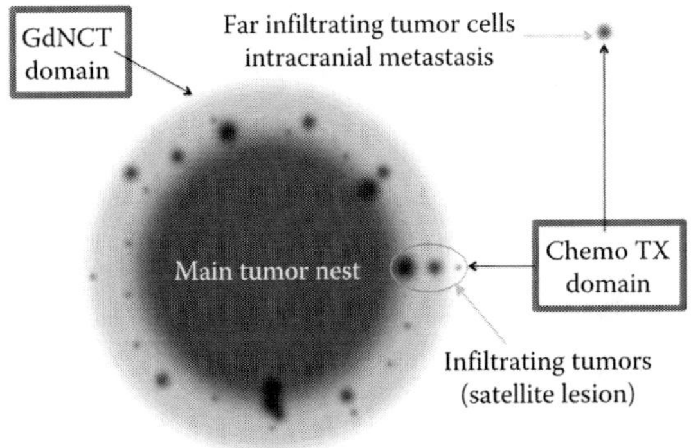

FIGURE 11.38 Schematic representation of a malignant brain tumor.

main tumor, and intracranial metastatic lesions are beyond the scope of GdNCT and require chemotherapy.

11.3.1.2 Complex Gd–B Compounds Required for Future NCT Applications

Complexes containing Gd and B are of great interest both for therapy and for evaluating boron distribution with MRI during NCT. If MRI shows high-Gd accumulation in a tumor, such compounds may also help with diagnosis and decision making for Gd/BNCT. Preliminary studies showed that, if both ^{n}Gd and ^{10}B atoms could be used for therapeutic effect, the optimal ratio of ^{n}Gd to ^{10}B in tumors should be about 2.7. If the ratio ^{n}Gd/^{10}B is smaller than 3 or the ratio of ^{10}B/^{157}Gd is larger than 3, as a rule, the Gd atom will not be effective in therapy; however, it may still be useful as a contrast agent in MRI (Figure 11.39).

If therapeutic concentrations of ^{10}B and ^{157}Gd in tumors are, respectively, 30 and 200 ppm, going by the atomic number, ratios of the complexes of ^{n}Gd–^{10}B can be simply calculated as >2.4 for ^{n}Gd and >0.43 for ^{157}Gd. The lower ratio for ^{n}Gd would make such compounds less effective for therapy, but still useful as a diagnostic contrast medium.

11.3.1.3 Hybrid NCT: GdNCT-Fast Neutron Therapy

The use of accelerators to provide the neutron source is important for establishing NCT as a practical cancer therapy. Cyclotron-based neutron sources have been reported to be better than reactors [30]. Moreover, accelerators require less space, are safer than reactors, and the neutron energy spectrum is easier to modify. If 100–800 eV neutrons can be used, gradual thermalization in tissues will result in a more ideal thermal neutron distribution. Thus, exhaustion of thermal neutrons before they reach deep tissues, owing to the previously mentioned shielding effect, can be avoided. While thermal neutrons inevitably produce 0.63 MeV protons via $^{14}N(n,p)^{14}C$ reactions, neutrons with more than 5 eV kinetic energy produce recoil protons with various amounts of kinetic energy via elastic collisions with water in tissues. This results in the release of [$^{\bullet}$OH] radicals, which are nonselective but simultaneously therapeutic (Figure 11.40) (T. Kobayashi, personal communication). The biological effect of any recoil proton depends on its energy, which, in turn, depends on the energy of the incident fast neutron that releases it. A mixed range of neutrons and low-energy neutrons spanning several hundred electrovolts may be a better choice for hybrid NCT therapy.

Neutrons with energy of more than 5 eV produce recoil protons in elastic collisions with hydrogen atoms in tissue water, and [$^{\bullet}$OH] radicals are concomitantly released. These are nonselective but simultaneously therapeutic.

The feasibility of applying a form of hybrid NCT using Gd/BNCT and fast neutrons with moderately high-energy and epithermal neutrons from high-energy accelerators is currently being investigated [31,32].

It is clear that the development of accelerators is crucial both for establishing a safe and practical neutron source and for enabling the potential of hybrid NCT. If accelerator-based neutron sources

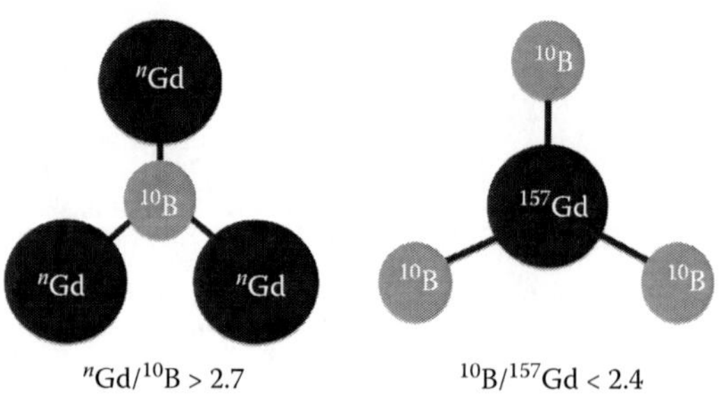

FIGURE 11.39 Diagrams of B–Gd compounds.

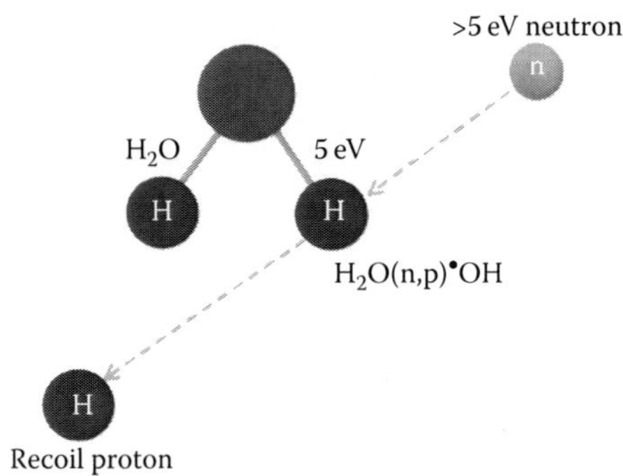

FIGURE 11.40 Useful release of •OH radicals.

become available, integrated modalities using GdNCT and chemotherapy are likely to become the best treatment option for malignant brain tumors. Moreover, because of the low dosage from accelerator-based sources can be efficiently captured by the large neutron-capture cross section of the ^{157}Gd atom, accelerator-based NCT is more suitable for GdNCT than BNCT. In addition, fractionated GdNCT promises to be more effective in dealing with the biological heterogeneity of tumor tissues. Currently, a criterion for approving treatment is the possibility of 2-year survival of the patient after onset. In the future, eradication of malignant brain tumors may be achieved through vaccination; in the interim, GdNCT will be able to achieve 2-year survival.

Malignant grade IV glioma is a prime candidate for noninvasive treatment and here Gd/BNCT might be the best option for avoiding surgical invasion and the risk of iatrogenic tumor dispersion. Although surgical biopsy is usually necessary for definitive diagnosis, Gd/BNCT should be applied without delay after definitive diagnosis of malignant brain tumor. In the future, if radio-pathological or noninvasive methods were able to provide definitive specification, malignant brain tumors could be completely treated nonsurgically.

Main tumor masses revealed by GdDTPA-enhanced T1 images are treatable using GdNCT, which can uniformly distribute a killing dose to the tissues in the tumor mass and to the region of infiltration surrounding the tumor rim. In fact, to prevent recurrence, it is also crucially important to destroy infiltrating tumor cells. Gd/BNCT would be capable of doing this if Gd drug delivery systems able to seek tumor cells become available.

Once main tumor masses can be successfully eliminated, infiltrating tumor cells will become the next target, and total eradication can be anticipated.

11.3.2 Vascular Lesions

11.3.2.1 GdNCT to Prevent Restenosis after Carotid/Coronary Stenosis

To prevent restenosis after percutaneous transluminal coronary angioplasty (PTCA), a variety of drugs have been investigated [33–35]. No drug has proved effective as yet. In animal studies, gene therapy has been successful in preventing intimal hyperplasia or cell proliferation, but its clinical effectiveness remains unknown. Exciting results have recently been reported using β-emitters to inhibit restenosis after PTCA [36].

While the study off the effectiveness of NCT for treating malignant brain tumors has been a primary focus, this technology has tremendous potential for inhibiting the cell proliferation that leads to restenosis. A study was undertaken to test the effectiveness of GdNCT in inhibiting intimal hyperplasia or thrombosis of vascular wall and whether or not it caused collateral tissue damage.

At their origins, the internal carotid and coronary arteries are susceptible to atheromatous plaque formation, which is liable to lead to clinically significant stenosis and possible lethal cerebral or cardiac thrombosis (Figure 11.41). While such constrictions can be opened, for example, by insertion of a stent into stenotic carotid artery [carotid artery stent (CAS)], there is a high risk of restenosis

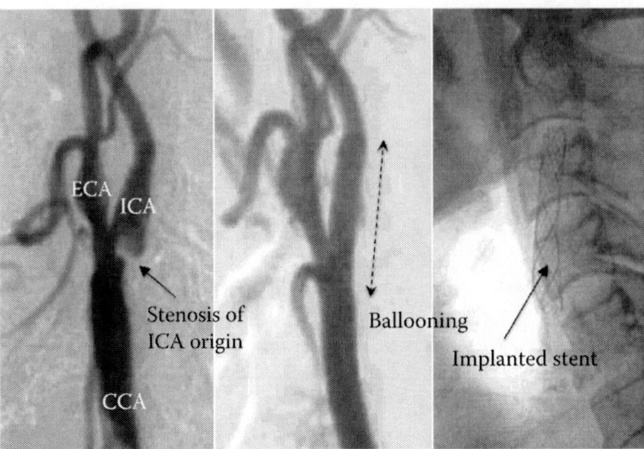

FIGURE 11.41 Left, Carotid bifurcation with severe stenosis at the origin of the internal carotid artery (arrow). Middle, Balloon catheter inserted in stenotic artery to enlarge the constriction. Right, After enlargement, a stent is implanted to prevent early restenosis. CCA, common carotid artery; ICA, internal carotid artery; ECA, external carotid artery.

at the ends of implanted stents. Research is being undertaken to prevent restenosis after CAS [37]. In this section, we present the results of tests using GdNCT to prevent vascular stenosis.

Anesthetized Wistar rats ($n = 21$), weighing about 400 g were used for this study. After bolus injection of heparin (50 IU/kg), using the cut-down method, a distal femoral artery was exposed and a guide wire (0.025 in.) was inserted into the origin of the common femoral artery. Using the wire, approximately 2 cm lengths along the common femoral artery were injured by rubbing with the wire. The blood flow of the common femoral artery was reconfirmed after injury. The rats were randomly assigned to either of the two following groups. The GdNCT group ($n = 11$) received 45 min of continuous thermal neutron irradiation (Cd ratio = 160) of the medial femoral region while also receiving continuous IV. infusion of GdDTPA 4 mL/kg via the contralateral femoral vein. Control ($n = 10$) received only thermal neutron irradiation for 45 min without GdDTPA loading immediately after vascular injury. The remaining areas of the bodies of the rats were shrouded from thermal neutron bombardment with flexible Li-sheets. All GdNCT rats survived for one month, and were subsequently examined. Thermal neutron fluence and γ-ray dose were monitored by Au foils and TLDs on the medial femoral skin corresponding with the common femoral artery. Absorbed radiation dose on the common femoral artery wall was estimated to be about 12 Gy, of which more

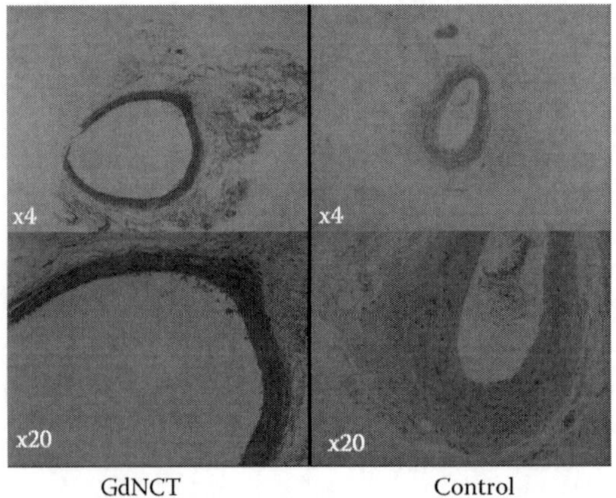

FIGURE 11.42 Pathological findings for arterial walls of the control group and the GdNCT group.

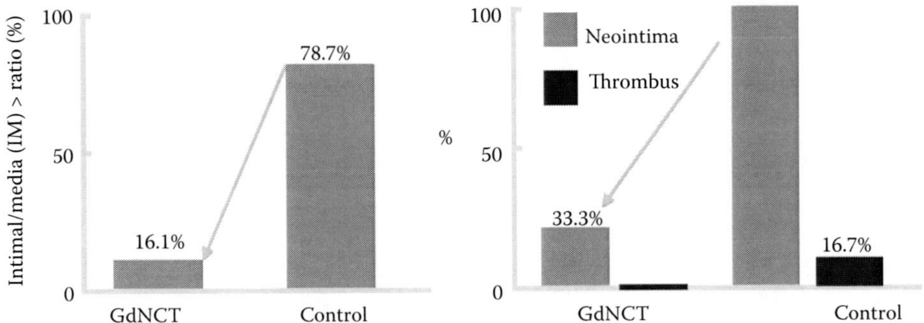

FIGURE 11.43 Inhibition by GdNCT of intimal hyperplasia (left) and thrombus formation (right) after rat-model vascular injury.

than 50% was contributed by IC electrons. During the neutron exposure ^{n}Gd concentration in blood was around 1400 ppm. Further one month later, the rats were killed, after which the femoral artery and abdominal aorta were perfusion-fixed by glutaraldehyde at a pressure of 100 mmHg, and the arteries were excised for microscopic examination.

On microscopic examination, while neither intimal hyperplasia nor thrombus formation were apparent in the GdNCT group, slight intimal hyperplasia of the injured arteries was observed in all control samples (Figure 11.42). In GdNCT group samples, elastic lamina was almost completely preserved. This finding is important because proteoglycan sulfate, which protects against thrombosis, is present in this internal elastic layer. No venous thrombosis was observed in either group.

The intimal/media ratio (IM ratio) was $16.1 \pm 17.8\%$ in the GdNCT group samples, and $78.7 \pm 23.9\%$ in control (mean ± SD, $P < 0.001$). In general, it was found that three to eight extra cell layers had been formed on the intimal surface of injured arteries in control. This study revealed that GdNCT is able to completely inhibit intimal hyperplasia and thrombus formation after vascular injury in rats (Figure 11.43), and that GdNCT does not lead to any serious vascular damage. The results are exciting because catheter-based β- or γ-radiation after PTCA still presents the risk of acute thrombotic obstruction or late restenosis. GdNCT seems to have great potential for preventing restenosis and vascular events after PTCA and angioplasty or vasospasm in angina patients [38]. Further immunohistochemical studies involving antibodies against growth factors, including PGFG, FGF, and VEGF, are being carried out to see if GdNCT inhibits any of these factors.

ACKNOWLEDGMENTS

Supports by grants from the National Science Foundation (CHE-0906179), the Robert A. Welch Foundation (N-1322), and the awards from Alexander von Humboldt Foundation and NIU Inaugural Board of Trustees Professorship are hereby acknowledged.

REFERENCES

1. G. L. Locher, Biological Effects and Therapeutic Possibilities of Neutrons. *Am. J. Roentgenol.*, **1936**, *36*, 1.
2. M. Matsuda, T. Yamamoto, H. Kumada, K. Nakai, M. Shirakawa, T. Tsurubuchi, A. Matsumura, Dose distribution and clinical response of glioblastoma treated with boron neutron capture therapy. *Appl. Radiat. Isot.*, **2009**, *67*(7–8 Suppl), S19–S21.
3. A. Giese, M. Westphal, Glioma invasion in the central nervous system. *Neurosurg.*, **1996,** 39, 235.
4. C. G. Greenwood, C. W. Reich, H. A. Baader, H. R. Koch, D. S. Bretig, Collective and two-quasiparticle states in Gd-158 observed through study of radioactive neutron-capture in Gd-157. *Nucl. Phys.*, **1978**, *A304*, 327.

5. K. Ono, S. Masunaga, Y. Kinashi, M. Takagaki, M. Akaboshi, T. Kobayashi, K. Akuta, Radiobiological evidence suggesting heterogeneous microdistribution of boron compounds in tumors: its relation to quiescent cell population and tumor cure in neutron capture therapy. *Int. J. Radiat. Oncol. Biol. Phys.*, **1996**, *34*, 1081.
6. Y. Onizuka, S. Endo, M. Ishikawa, M. Hoshi, M. Takada, T. Kobayashi, Y. Sakurai et al., Microdosimetry of epithermal neutron field at the Kyoto University reactor. *Radiat. Prot. Dosim.*, 2002, 99, 383.
7. T. Kobayashi, K. Kanda, Y. Ujeno, M. R. Ishida, Biomedical irradiation system for boron neutron capture therapy at the Kyoto University Reactor. *Basic Life Sci.*, **1990**, *54*, 321.
8. K. Okamoto, J. Ito, N. Takahashi, K. Ishikawa, T. Furusawa, S. Tokiguchi, K. Sakai, MRI of high-grade astrocytic tumors: early appearance and evolution. *Neuroradiology*, **2002**, *44*, 395.
9. D. Schiffer, P. Cavalla, A. Dutto, L. Borsotti, Cell proliferation and invasion in malignant gliomas. *Anticancer Res.*, **1997**, *17*, 61.
10. R. F. Martin, G. D'Cunha, M. Pardee, B. J. Allen, Induction of DNA double-strand breaks by 157Gd neutron capture. *Pigment Cell Res.*, **1989**, *2*, 330.
11. T. Goorley, R. Zamenhof, H. Nikjoo, Calculated DNA damage from gadolinium Auger electrons and relation to dose distributions in a head phantom. *Int. J. Radiat. Biol.*, **2004**, *80*, 933.
12. V. M. Mitin, V. N. Kulakov, V. F. Khokhlov, I. N. Sheino, A. M. Arnopolskaya, N. G. Kozlovskaya, K. N. Zaitsev, A. A. Portnov, Comparison of BNCT and GdNCT efficacy in treatment of canine cancer. *Appl. Radiat. Isot.*, **2009**, *67*, S299–S301.
13. T. Kobayashi, K. Kanda, Microanalysis system of ppm order concentrations in tissue for neutron capture therapy by prompt gamma-ray spectrometry. *Nucl. Instrum. Methods*, **1983**, *204*, 525.
14. T. Kobayashi, Y. Sakurai, M. Ishikawa, A non-invasive dose estimation system for boron neutron capture therapy under a clinical irradiation by PG-SPECT – conceptual study and fundamental experiments using HPGe and CdTe Semiconductor detectors. *Med. Phys.*, **2000**, *27*, 2124.
15. M. Takagaki, N. S. Hosmane, Gadolinium neutron capture therapy for malignant brain tumors. *Aino J.*, **2007**, *6*, 39.
16. G. D. Stasio, P. Casalbore, R. Pallini, B. Gilbert, F. Sanita, M. T. Ciotti, G. Rosi et al., Gadolinium in human glioblastoma cells for gadolinium neutron capture therapy. *Cancer Res.*, **2001**, *61*, 4272.
17. M. Takagaki, Y. Oda, M. Matsumoto, H. Kikuchi, T. Kobayashi, K. Kanda, Y. Ujeno, Gadolinium neutron capture therapy for brain tumors – Biological Aspects. *Biological Aspects of Brain Tumors*, K. Tabuchi (Ed.), Springer-Verlag, **1991**, p. 494.
18. Y. Akine, N. Tokita, K. Tokuuye, M. Satoh, Y. Fukumori, H. Tokumitsu, R. Kanamori, T. Kobayashi, K. Kanda, Neutron-capture therapy of murine ascites tumor with gadolinium-containing microcapsules. *J. Cancer Res. Clin. Oncol.*, **1992**, *119*, 71.
19. Y. Akine, N. Tokita, K. Tokuuye, M. Satoh, H. Churei, C. Le Pechoux, T. Kobayashi, K. Kanda, Suppression of rabbit VX-2 subcutaneous tumor growth by gadolinium neutron capture therapy. *Jpn. J. Cancer Res.*, **1993**, *84*, 841.
20. U. M. Le, Z. Cui, Long-circulating gadolinium-encapsulated liposomes for potential application in tumor neutron capture therapy. *Int. J. Pharm.*, **2006**, *7*, 105.
21. H. Ichikawa, T. Watanabe, H. Tokumitsu, Y. Fukumori, Formulation considerations of gadolinium lipid nanoemulsion for intravenous delivery to tumors in neutron-capture therapy. *Curr. Drug Deliv.*, **2007**, *4*, 131.
22. K. Wangerin, C. N. Culbertson, T. Jevremovic, A comparison of the COG and MCNP codes in computational neutron capture therapy modeling, Part II: gadolinium neutron capture therapy models and therapeutic effects. *Health Phys.* **2005**, *89*, 135.
23. T. Goorley, H. Nikjoo, Electron and photon spectra for three gadolinium-based cancer therapy approaches. *Radiat Res.*, **2000**, *154*, 556.
24. J. T. Masiakowski, J. L. Horton, L. J. Peter, Gadolinium neutron capture therapy for brain tumors: a computer study. *Med. Phys.*, **1992**, *19*, 1277.
25. T. Matsumoto, Transport calculations of depth-dose distributions for gadolinium neutron capture therapy. *Phys. Med. Biol.*, **1992**, *37*, 155.
26. Y. Oda, M. Takagaki, S. Miyatake, H. Kikuchi, Gadolinium atom on neutron capture therapy. *Intraoperative Radiation Therapy*, M. Abe and M. Takahashi (Ed.), Pergamon Press, **1988**, p. 156.
27. K. Yoshida, M. Furuse, Y. Kaneoke, K. Saso, S. Inao, Y. Motegi, K. Ichihara, A. Izawa, Assessment of T1 time course changes and tissue-blood ratios after Gd-DTPA administration in brain tumors. *Magn. Reson. Imaging*, **1989**, *7*, 9.
28. G. DeStasio, P. Casalbore, R. Pallini, B. Gilbert, F. Sannita, M. T. Ciotti, G. Rosi et al., Gadolinium in human glioblastoma cells for gadolinium neutron capture therapy. *Cancer Res.*, **2001**, *15*, 4272.

29. M. Takagaki, Y. Oda, S. Miyatake, T. Kobayashi, Y. Sakurai, M. Osawa, K. Mori, K. Ono, Boron neutron capture therapy: preliminary study of BNCT with sodium borocaptate ($Na_2B_{12}H_{11}SH$) on glioblastoma. *J. Neuro. Oncol.*, **1997**, *35*, 177.
30. H. Tanaka, Y. Sakurai, M. Suzuki, T. Tanaka, S. Masunaga, Y. Kinashi, G. Kashino et al., Improvement of dose distribution in phantom by using epithermal neutron source based on the Be(p,n) reaction using a 30 MeV proton cyclotron accelerator. *Appl. Radiat. Isot.*, **2009**, 67(7–8 Suppl), S258–S261.
31. K. Langen, A. J. Lennox, T. K. Kroc, P. M. DeLuca, Feasibility of the utilization of BNCT in the fast neutron therapy beam at Fermilab. **2000**, *FERMILAB-TM-2118*. http://www.osti.gov/accomplishments/documents/fullText/ACC0226.pdf.
32. J. Burmeister, M. Yudelev, C. Kota, R. Maughan, A. Wittig, R. Hentschel, W. Sauerwein, Boron Neutron Capture Enhancement of Fast Neutron Therapy at the Wayne State University / Karmanos Cancer Center Neutron Therapy Facility and at CIRCE in Essen. *Neutron for Therapy*, Satellite Symposium to NEUDOS-11, at iThemba LABS, Cape Town, October 12, abstract, **2009**, 22.
33. M. Butt, D. Connolly, G. Y. Lip, Drug-eluting stents: a comprehensive appraisal. *Future Cardiol.*, **2009**, *5*, 141.
34. T. J. Kiernan, B. P. Yan, I. Cruz-Gonzalez, R. J. Cubeddu, A. Caldera, G. D. Kiernan, V. Gupta, Pharmacological and cellular therapies to prevent restenosis after percutaneous transluminal angioplasty and stenting. *Cardiovasc. Hematol. Agents Med. Chem.*, **2008**, *6*, 116.
35. R. Zargham, Preventing restenosis after angioplasty: a multistage approach. *Clin. Sci. (Lond)*, **2008**, *114*, 257–64.
36. S. B. King III, D. O. Williams, P. Chougule, J. L. Klein, R. Waksman, R. Hilstead, J. Macdonald, K. Anderberg, J. R. Crocker, Endovascular beta-radiation to reduce restenosis after coronary balloon angioplasty: results of the beta energy restenosis trial (BERT). *Circulation*, **1998**, *97*, 2025.
37. M. Sadek, N. S. Cayne, H. J. Shinn, I. C. Turnbull, M. L. Martin, P. J. Faries, Safety and efficacy of carotid angioplasty and stenting for radiation-associated carotid artery steno. *J. Vasc. Surg.*, **2009**, *50*, 1308.
38. M. B. Leon, P. S. Teirstein, J. W. Moses, P. Tripuraneni, A. J. Lansky, S. Jani, Localized intracoronary gamma-radiation therapy to inhibit the recurrence of restenosis after stenting. *N. Engl. J. Med.*, **2001**, *344*, 2506.
39. F. A. Mettler, The experimental anatomophysiologic approach to study of diseases of the basal ganglia. *J. Neuropathol. Exp. Neurol.* **1955**, *14*, 115–141.

12 Design and Development of Polyhedral Borane Anions for Liposomal Delivery

Debra A. Feakes

CONTENTS

12.1 INTRODUCTION

The successful application of boron neutron capture therapy (BNCT) is dependent on the identification and preparation of boron-containing compounds that can be delivered to and retained by the tumor in significant amounts (>15 μg boron/g tumor) (Fairchild and Bond, 1985). Compounds targeted for application in BNCT must either accumulate in the tumor through natural mechanisms or, in the case of compounds that have no natural propensity for the tumor, be delivered to the tumor using a tumor-specific delivery modality. In addition to the required tumor boron concentration, the selected compounds should yield *in vivo* biodistributions with tumor to normal tissue (including blood) ratios of three or higher and be sufficiently nontoxic (Hawthorne, 1993; Soloway et al., 1998).

Researchers have developed a wide variety of derivatives of species known to accumulate in the tumor or which have the potential for *in vivo* metabolization. Included in this category are boron derivatives of amino acids (Radel and Kahl, 1996; Lindström et al., 2000; Das et al., 2001; Kabalka and Yao, 2003), nucleosides, and nucleotides (Schinazi and Prusoff, 1978; Sood et al., 1989; Yamamoto et al., 1989; Li et al., 1996; Barth et al., 2004), monoclonal antibodies (Hawthorne, 1991), antisense agents (Lesnikowski and Schinazi, 1995; Zhuo et al., 2000), polyamines (Cai and Soloway, 1996; Cai et al., 1997; Ghaneolhosseini et al., 1998), and porphyrins (Kahl and Koo, 1990; Hill et al., 1992; Genady and Gabel, 2002). For compounds that have no inherent tumor specificity, researchers have investigated the utility of tumor-specific delivery vehicles. Although low-density lipoproteins incorporating lipophilic boron-containing compounds have been investigated (Laster et al., 1991), the primary system studied in this category is the utilization of unilamellar liposomes as delivery vehicles (Shelly et al., 1992; Feakes et al., 1994, 1995, 1999; Alanazi et al., 2003; Peacock et al., 2004; Li et al., 2006).

Although liposomes in general do not concentrate in tumors (Mayhew and Papahadjopoulos, 1983), liposomes of a specific size and composition have demonstrated an ability to selectively deposit within the tumor, most likely as a result of the immature vasculature characteristic of rapidly growing tumor masses (Straubinger et al., 1983). Once delivered to the tumor mass, the liposomes bind to the cell wall through an undefined mechanism and are endocytosed in coated pits. After endocytosis, the contents of the liposomes are distributed among the cellular organelles and cytoplasm (Straubinger et al., 1990). Liposomes are an appealing approach to the delivery of boron-containing compounds to tumors for application in BNCT. The vesicles extend the circulation lifetime of the compound, relative to an unincorporated compound, can incorporate a wide variety of boron-containing compounds with no inherent tumor specificity, and reduce any potential toxic effects of the compound. Liposomes investigated for application in BNCT are composed of a combination of pure synthetic phospholipid and cholesterol. The combination of cholesterol and pure synthetic phospholipid is necessary to encapsulate high concentrations of solution within the aqueous core of the liposomes without deleterious effects to the serum stability of the liposomes and the alteration of the surface presented to the cell membrane which could result in a loss of specificity (Wallingford and Williams, 1985).

Liposomes are capable of delivering both hydrophilic and lipophilic boron-containing compounds. Hydrophilic boron-containing compounds are encapsulated within the aqueous core while lipophilic boron-containing compounds are embedded within the lipid bilayer. Although liposomes have the potential to deliver lipophilic boron-containing compounds, both independently and in conjunction with hydrophilic boron-containing compounds, the development of lipophilic boron-containing compounds for liposomal delivery is reviewed in another area of this book. Within this chapter, the development of hydrophilic boron-containing compounds for encapsulation within the aqueous core of unilamellar liposomes is reviewed.

12.2 POLYHEDRAL BORANE ANION BUILDING BLOCK, THE $[B_{10}H_{10}]^{2-}$ ION

Compounds suitable for encapsulation within the aqueous core of liposomes should be soluble in water at relatively high concentrations, hydrolytically stable, and have a large boron content with a low osmotic stress. The high concentrations of boron required to achieve therapeutic levels *in vivo* necessitates the encapsulation of hyperosmotic solutions of the polyhedral borane salts with up to three times the osmotic pressure of the external solution.

The $[B_{10}H_{10}]^{2-}$ ion, also known as GB10, is prepared in high yield from the reaction of decaborane and triethylamine (Figure 12.1) (Hawthorne and Pilling, 1967). The sodium salt is very soluble in water, relatively inert, and nontoxic. Additionally, for every unit, five boron atoms are delivered per unit charge, thereby providing a high boron content with low osmotic stress. The sodium salt of the $[B_{10}H_{10}]^{2-}$ ion was the first compound reported that was encapsulated in unilamellar liposomes and investigated for application in BNCT (Shelly et al., 1992). The injected dose of boron for the murine biodistribution experiment was 110 µg B (~6 mg B/kg body weight). Notably, the injected

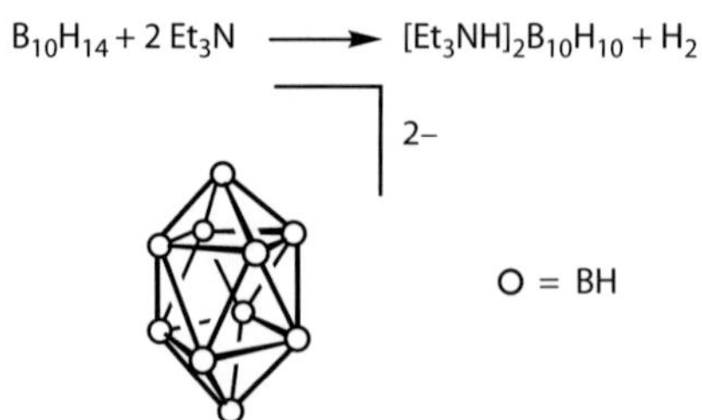

FIGURE 12.1 Synthesis and structure of the $[B_{10}H_{10}]^{2-}$ ion.

doses for liposome experiments tend to be much lower than doses used for other methodologies (Sweet et al., 1962; Soloway et al., 1967; Slatkin et al., 1986; Gabel et al., 1987; Coderre et al., 1988; Clendonen et al., 1990). Although the boron was delivered to the tumor by the unilamellar liposomes, the boron concentration diminished rapidly from all tissues within the 48 h experiment. Therapeutically useful boron concentrations were not achieved; however, the utility of the unilamellar liposome delivery system was established. More recently, researchers have increased the boron concentration delivered to the tumor by encapsulating the $[B_{10}H_{10}]^{2-}$ ion in liposomes with pendant PEG groups (Masunaga et al., 2006).

12.3 THE $[B_{20}H_{18}]^{2-}$ ION

The $[B_{10}H_{10}]^{2-}$ ion can be oxidatively coupled to form the normal isomer of the $[B_{20}H_{18}]^{2-}$ anion (Figure 12.2) (Chamberland and Muetterties, 1964; Kaczmarczyk et al., 1962). The $[B_{20}H_{18}]^{2-}$ anion has a wealth of characteristics that makes it an attractive candidate for encapsulation in unilamellar liposomes. Compared to the $[B_{10}H_{10}]^{2-}$ anion, the $[B_{20}H_{18}]^{2-}$ anion has twice the boron content per unit charge, thereby increasing the potential dose delivered at the same osmotic stress. The $[B_{20}H_{18}]^{2-}$ ion contains two electron-deficient three center-two electrons bonds which are susceptible to nucleophilic attack (Pilling et al., 1964), as shown in Figure 12.2 using the hydroxide ion (Hawthorne et al., 1963). Additionally, the normal isomer can be photolyzed to form the photoisomer, $[i\text{-}B_{20}H_{18}]^{2-}$, which also contains two electron-deficient three center-two electron bonds (Hawthorne and Pilling, 1966). The photoisomer is also susceptible to nucleophilic attack, but at an even higher rate of reaction. The $[B_{20}H_{18}]^{2-}$ ion can also be reversibly reduced to form the $[B_{20}H_{18}]^{4-}$ anion, an ion with the same boron content, but at a higher charge and, as a result, a higher osmotic stress (Hawthorne et al., 1963, 1965b).

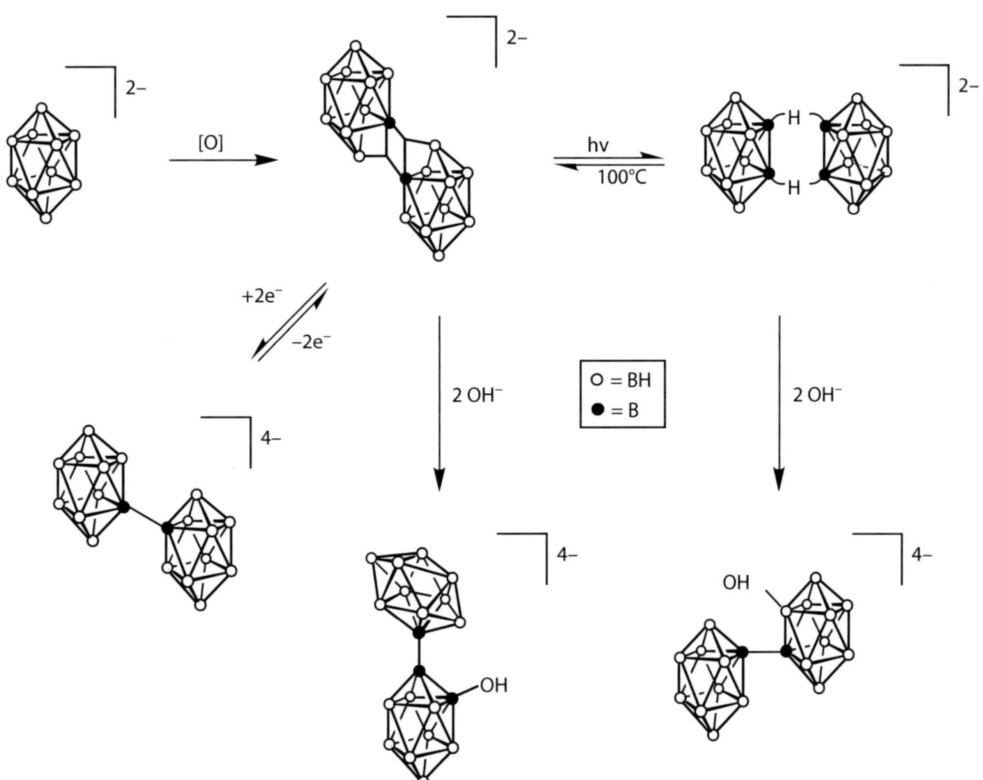

FIGURE 12.2 Synthesis and reactivity of the $[B_{20}H_{18}]^{2-}$ anion.

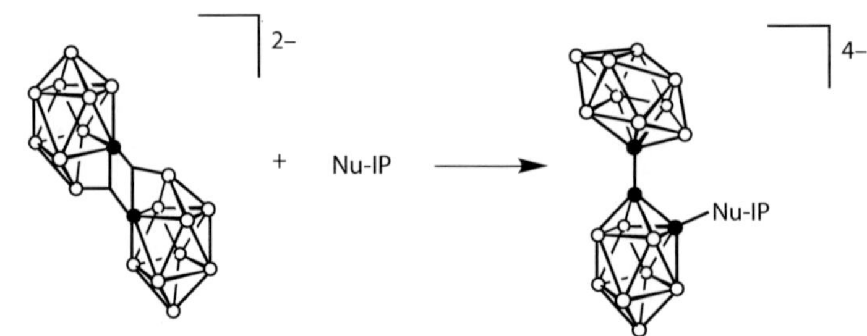

FIGURE 12.3 Hypothesized reaction between the $[B_{20}H_{18}]^{2-}$ anion and a nucleophilic site on an intracellular protein (Nu-IP).

The susceptibility of the $[B_{20}H_{18}]^{2-}$ anion to nucleophilic attack provided a potential means for the retention of the boron within the tumor cell once delivered by the unilamellar liposomes. Shelly and co-workers proposed that the $[B_{20}H_{18}]^{2-}$ anion had the potential to react with intracellular proteins (Figure 12.3), thereby covalently bonding the boron-containing species within the tumor cell (Shelly et al., 1992) and increasing the retention time within the tumor cell.

The sodium salt of the normal isomer of the $[B_{20}H_{18}]^{2-}$ ion was encapsulated in unilamellar liposomes and investigated in a murine biodistribution experiment (Shelly et al., 1992). The injected dose of boron was 273 µg B (~15 mg/kg body weight). The higher injected dose, as compared to the dose used in the $[B_{10}H_{10}]^{2-}$ ion biodistribution experiment, was indicative of the higher boron content within the $[B_{20}H_{18}]^{2-}$ ion. While the boron in other tissues, including blood and liver, cleared rapidly, the tumor boron concentration was retained significantly more than observed in the biodistribution of the encapsulated $[B_{10}H_{10}]^{2-}$ ion. Between 6 and 24 h, the tumor boron concentration decreased by only 11% as compared to the $[B_{10}H_{10}]^{2-}$ ion which decreased by 46% in the same time period. As anticipated, based on the higher reactivity, the biodistribution of the encapsulated photoisomer exhibited higher tumor retention with promising overall tumor boron concentrations (13.9 µg B/g tumor and a tumor/blood ratio of 12 at 48 h). The biodistribution of unencapsulated photoisomer exhibited rapid clearance from all tissues and demonstrated the necessity of liposomal encapsulation.

12.4 DERIVATIVES OF THE $[B_{20}H_{18}]^{2-}$ ION

Hawthorne and coworkers developed a working hypothesis based on the biodistributions of the $[B_{10}H_{10}]^{2-}$ anion and the isomers of the $[B_{20}H_{18}]^{2-}$ ion: *polyhedral borane anions with the potential to form covalent bonds to intracellular protein moieties would be retained within the tumor cells after liposomal delivery, while polyhedral borane anions which are not capable of reacting with intracellular protein moieties would be cleared from all tissues* (Shelly et al., 1992). This hypothesis was supported by the biodistribution of liposomally encapsulated $[B_{20}H_{19}]^{3-}$ and, in a separate experiment, the $[B_{20}H_{17}OH]^{4-}$ ion.

12.4.1 THE $[B_{20}H_{19}]^{3-}$ ION

The $[B_{20}H_{18}]^{4-}$ anion is prepared by the reversible reduction of the $[B_{20}H_{18}]^{2-}$ ion in a solution of sodium metal in liquid ammonia (Figure 12.4) (Hawthorne et al., 1963, 1965b). In aqueous solutions, the ion is in equilibrium with the protonated species, $[B_{20}H_{19}]^{3-}$. Although the $[B_{20}H_{19}]^{3-}$ ion has been recognized since 1963 (Hawthorne et al., 1963), it was not structurally characterized in the solid state until 1996 (Watson-Clark et al., 1996). Reduction of the $[B_{20}H_{18}]^{2-}$ ion results in the loss of the electron-deficient bonding region. As a result, the ion is unreactive to intracellular protein moieties and would not be expected to form a covalent bond with the intracellular protein moieties

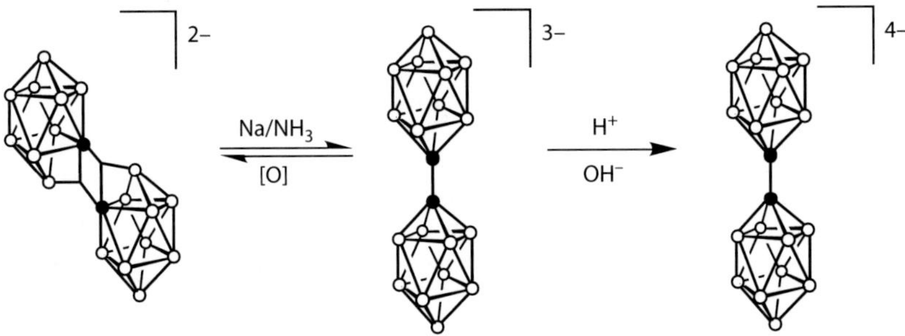

FIGURE 12.4 Reversible reduction of the $[B_{20}H_{18}]^{2-}$ ion. [O] is an oxidizing agent.

once delivered by the unilamellar liposomes. Biodistribution experiments of encapsulated $[B_{20}H_{19}]^{3-}$ confirmed the hypothesis (Shelly et al., 1992). The potassium salt of the $[B_{20}H_{19}]^{3-}$ ion was encapsulated in unilamellar liposomes and investigated in a murine biodistributions experiment. The injected dose of boron was 134 μg B (~8 mg/kg body weight). The liposomes displayed a very high tumor boron uptake (27.7 μg B/g tumor at 6 h), particularly considering the low injected dose; however, the tumor boron concentration decreased throughout the time course experiment (11.2 μg B/g tumor at 48 h), consistent with the inability to form covalent bonds with intracellular protein moieties.

12.4.2 The $[B_{20}H_{17}OH]^{4-}$ Ion

Isomers of the $[B_{20}H_{17}OH]^{4-}$ anion can be prepared from either the normal or the photoisomer of the $[B_{20}H_{18}]^{2-}$ anion by reaction with the hydroxide ion in aqueous solution (Hawthorne et al., 1963, 1965b). When the normal isomer of the $[B_{20}H_{18}]^{2-}$ ion is allowed to react with the hydroxide ion, the kinetic isomer of $[B_{20}H_{17}OH]^{4-}$ is formed (Figure 12.5). The kinetic isomer is characterized by having an apical–equatorial boron atom connection between the two cages and the hydroxy substituent located on the equatorial belt adjacent to the intercage connection. The kinetic isomer is designated $[ae\text{-}B_{20}H_{17}OH]^{4-}$. An acid-catalyzed rearrangement of the kinetic isomer yields the thermodynamic isomer, characterized by having an apical–apical boron atom connection between the two cages and the hydroxyl substituent located on the equatorial belt adjacent to the intercage connection. The thermodynamic isomer is designated as $[a^2\text{-}B_{20}H_{17}OH]^{4-}$.

When the photoisomer of the $[B_{20}H_{18}]^{2-}$ ion is allowed to react with the hydroxide ion, the kinetic isomer of $[B_{20}H_{17}OH]^{4-}$ is formed (Figure 12.6). The kinetic isomer is characterized by having an equatorial–equatorial boron atom connection between the two cages and the hydroxy substituent located on the equatorial belt adjacent to the terminal boron apex. The kinetic isomer is designated $[e^2\text{-}B_{20}H_{17}OH]^{4-}$. An acid-catalyzed rearrangement of the kinetic isomer yields the thermodynamic isomer, characterized by having an apical–apical boron atom connection between the two cages and the hydroxy substituent located on the equatorial belt adjacent to the terminal boron apex.

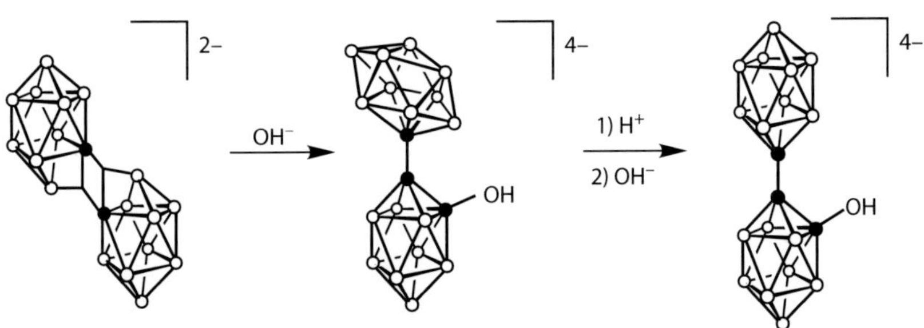

FIGURE 12.5 Reaction of the normal isomer of the $[B_{20}H_{18}]^{2-}$ ion with the hydroxide ion.

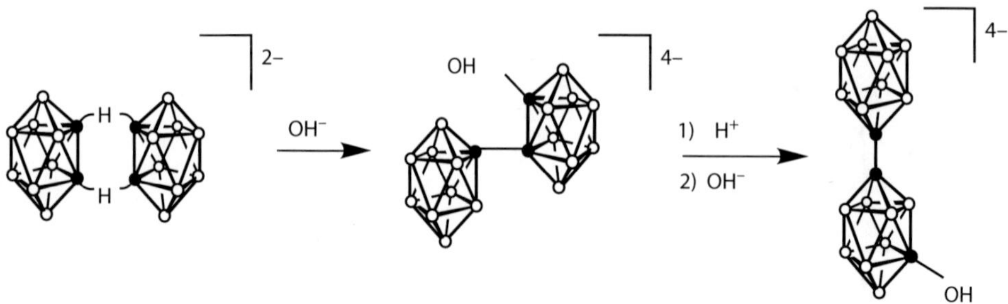

FIGURE 12.6 Reaction of the photoisomer of the $[B_{20}H_{18}]^{2-}$ ion with the hydroxide ion.

The hydroxy isomers lack the reactivity of the three-center, two-electron bond characteristic of the $[B_{20}H_{18}]^{2-}$ isomers. As a result, the anion would be expected to be unreactive to intracellular protein moieties and clear from the tissues in biodistribution experiments. The biodistribution of liposomally encapsulated $[e^2\text{-}B_{20}H_{17}OH]^{4-}$, with a boron dose of 200 μg B (~11 mg/kg body weight), exhibited rapid clearance of the boron concentration from all tissues, consistent with the working hypothesis (Shelly et al., 1992). At 48 h, the tumor boron concentration was only 7.3 μg B/g tumor.

12.4.3 The $[B_{20}H_{17}NH_3]^{3-}$ Ion

Isomers of the $[B_{20}H_{17}NH_3]^{3-}$ ion were initially prepared by allowing the normal isomer of the $[B_{20}H_{18}]^{2-}$ anion to react with sodium acetylide in liquid ammonia (Figure 12.7) (Feakes et al., 1994). In aqueous solutions, the protonated form of the amine substituent predominates. Analogous to the hydroxy-substituted series, the kinetic isomer of the product, $[ae\text{-}B_{20}H_{17}NH_3]^{3-}$, was formed first. Acid-catalyzed rearrangement yields the thermodynamic isomer, $[a^2\text{-}B_{20}H_{17}NH_3]^{3-}$. Since the original synthesis, a simpler method for the preparation of the $[B_{20}H_{17}NH_3]^{3-}$ ion has been determined and reported (Georgiev et al., 1996).

The $[B_{20}H_{17}NH_3]^{3-}$ isomers lack the electron-deficient region of the $[B_{20}H_{18}]^{2-}$ ion and, as a result, should not, based on the working hypothesis of the time, be retained in the tumor cells once delivered by the unilamellar liposomes. The biodistribution of liposomally encapsulated $[ae\text{-}B_{20}H_{17}NH_3]^{3-}$ ion, with a boron dose of 198 μg B (~11 mg/kg body weight), yielded highly unanticipated results (Feakes et al., 1994). Over the first 30 h of the biodistribution experiment, the tumor boron concentration increased, reaching an observed maximum of 32.3 μg of B/g of tumor. After 30 h, the tumor boron concentration decreased; however, the 48-h tumor boron concentration of 25.4 μg of B/g of tumor was well within the therapeutic range. Even higher tumor boron concentrations were achieved by encapsulating the polyhedral borane anion in liposomes containing 5 mol% polyethylene glycol (PEG)-distear-oylphosphatidylethanolamine. The inclusion of the PEG increases the circulation lifetime by preventing opsonins from labeling the liposomes as a foreign particle and thus preventing the clearance of the liposomes in the liver and the spleen (Lasic, 1992). The $[ae\text{-}B_{20}H_{17}NH_3]^{3-}$ ion

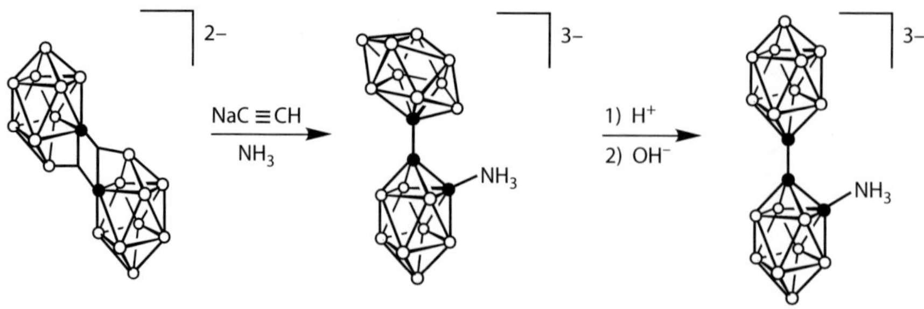

FIGURE 12.7 Formation of $[a^2\text{-}B_{20}H_{17}NH_3]^{3-}$ from the normal isomer of the $[B_{20}H_{18}]^{2-}$ ion.

was the first polyhedral borane anion to exhibit tumor boron accretion. Based on the unexpected biodistribution data, results that were inconsistent with the working hypothesis at the time, several additional biodistribution experiments were completed (Feakes et al., 1994).

Unencapsulated [*ae*-$B_{20}H_{17}NH_3]^{3-}$, with a dose of 175 μg B (~10 mg/kg body weight), was investigated in a biodistribution experiment. Although the unencapsulated compound exhibited a higher propensity for tumor retention than other unencapsulated polyhedral borane anions, the high tumor boron concentrations observed with liposomal encapsulation are clearly a result of the specificity of the liposomal delivery.

The thermodynamic isomer, [a^2-$B_{20}H_{17}NH_3]^{3-}$, was also investigated in a murine biodistribution experiment. The dose was 288 μg B (~16 mg/kg body weight). Analogous to the kinetic isomer, the biodistribution experiment exhibited an accretion of tumor boron; however, the tumor boron concentration, once the maximum of 40.0 μg B/g tumor was achieved, remained essentially constant over the entire time-course experiment. For both isomers, therapeutic boron concentrations (>15 μg boron/g tumor) were present at all measured time points.

The relevance of the ammonia substituent in the unexpected biodistribution results was investigated by obtaining the biodistribution results of liposomally encapsulated $[B_{10}H_9NH_3]^-$ion. Analogous to the $[B_{10}H_{10}]^{2-}$ ion, the $[B_{10}H_9NH_3]^-$ ion exhibited no significant accumulation or retention of tumor boron concentration.

Comparison of the biodistributions of the isomers of the $[B_{20}H_{17}NH_3]^{3-}$ ions to the biodistributions of the $[B_{20}H_{19}]^{3-}$ ion, the $[B_{20}H_{17}OH]^{4-}$ ion, and the $[B_{10}H_9NH_3]^-$ion indicate that the unusual tumor boron accretion and retention of the $[B_{20}H_{17}NH_3]^{3-}$ isomers was not a direct result of the reduced 20 boron atom structure, the location of the substituent on the reduced ion, or the presence of the ammonia substituent. Investigation of the oxidation potentials of these compounds using cyclic voltammetry indicated that the $[B_{20}H_{17}NH_3]^{3-}$ ions oxidized significantly easier than any of the other 20 boron atom species and led researchers to a modified hypothesis for the retention of polyhedral borane anions in the tumor cells: *anions with the potential to oxidize to more reactive species in vivo would be retained within the tumor cells after liposomal delivery.* The product of the proposed oxidation of the $[B_{20}H_{17}NH_3]^{3-}$ ion would be the $[B_{20}H_{17}NH_3]^-$ion (Figure 12.8), an anion that again was characterized by the electron-deficient bonding region, analogous to the $[B_{20}H_{18}]^{2-}$ ion. The oxidized anion would then be susceptible to reaction by nucleophilic moieties on intracellular protein moieties, thereby covalently binding the ion within the tumor cell (Figure 12.9).

12.4.4 The $[B_{20}H_{17}SH]^{4-}$ Ion

Sulfhydryl derivatives of polyhedral borane anions have been of interest to BNCT researchers for many years because of the mercaptoundecahydro-*closo*-dodecaborate dianion ($[B_{12}H_{11}SH]^{2-}$, abbreviated borocaptate sodium (BSH) that has been used in a variety of preclinical and clinical BNCT trials (Hawthorne, 1993; Soloway et al., 1998; Hawthorne and Lee, 2003). Although only the free sodium salt of the $[B_{12}H_{11}SH]^{2-}$ has been used in the trials; researchers have investigated the biodistribution of BSH encapsulated in liposomes (Shelly et al., 1992; Maruyama et al., 2004; Doi et al.,

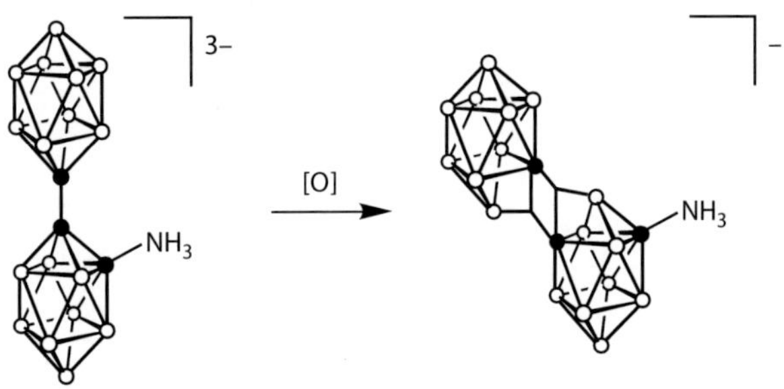

FIGURE 12.8 Proposed oxidation of the [a^2-$B_{20}H_{17}NH_3]^{3-}$ ion. [O] is an oxidizing agent.

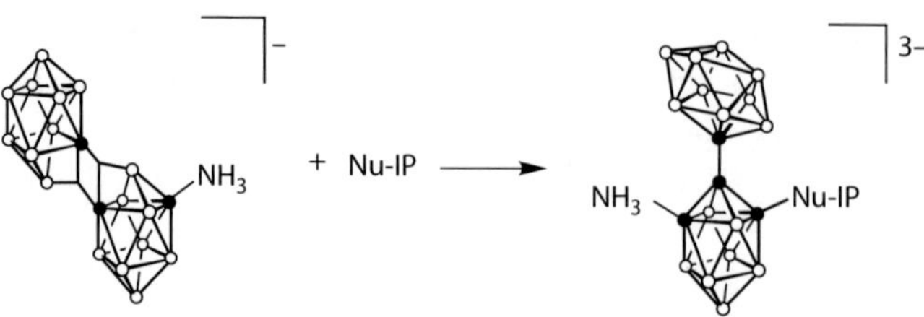

FIGURE 12.9 Hypothesized reaction between the $[B_{20}H_{17}NH_3]^-$ ion and a nucleophilic site on an intracellular protein (Nu-IP).

2008; Feng et al., 2009). The murine biodistribution of BSH encapsulated in liposomes with a bilayer composed only of distearylphosphatidylcholine (DSPC) and cholesterol, at an injected dose of 126 μg B (~7 mg/kg body weight), exhibited more favorable results than the biodistribution of the unsubstituted $[B_{10}H_{10}]^{2-}$ ion (Shelly et al., 1992). Blood boron concentration was cleared rapidly; however, the boron species was not cleared from the tumor as rapidly. Between 6 and 30 h, the tumor boron concentration decreased by only 25% as compared to the $[B_{10}H_{10}]^{2-}$ ion which decreased by 46% between 6 and 24 h. Therefore, the BSH anion does possess some degree of inherent tumor specificity. Even with the increased retention, the low doses used for the biodistribution experiment did not yield therapeutically useful values. Early observations of the retention of BSH within the tumor mass were attributed to the possible formation of disulfide bonds with the cysteine residues of proteins (Soloway et al., 1967). Although the reaction of BSH with serum albumins has been investigated, no evidence currently exists that BSH forms a disulfide bond with these species (Tang et al., 1995). Any existing interactions are likely to be electrostatic in nature. Regardless of the type of interaction that exists within the body, BSH has demonstrated some degree of tumor specificity and retention. As a result, the synthesis and evaluation of the $[B_{20}H_{17}SH]^{4-}$ ion was proposed.

The $[B_{20}H_{17}SH]^{4-}$ ion has the potential to react with intracellular protein substituents through a variety of mechanisms. The ion has the potential to form disulfide bonds in a manner similar to that proposed, but not observed, for the BSH ion (Figure 12.10). Formation of the disulfide bonds would covalently bond the polyhedral borane anion within the tumor cells.

The $[B_{20}H_{17}SH]^{4-}$ ion also has the potential to oxidize to a more reactive species, the $[B_{20}H_{17}SH]^{2-}$ ion, analogous to that proposed for the $[B_{20}H_{17}NH_3]^{3-}$ ion (Figure 12.11a). Formation of the $[B_{20}H_{17}SH]^{2-}$ ion would yield an ion that has the potential to react with nucleophilic sites on intracellular proteins, analogous to that proposed for the isomers of the $[B_{20}H_{18}]^{2-}$ ion (Figure 12.11b). And, finally, the $[B_{20}H_{17}SH]^{2-}$ ion also has the potential to react with the cysteine residues of intracellular proteins (Figure 12.11c).

The synthesis of the $[B_{20}H_{17}SH]^{4-}$ ion was achieved by allowing the normal isomer of the $[B_{20}H_{18}]^{2-}$ ion to react with the protected thiol nucleophile, $KSC(O)OC(CH_3)_3$, followed by removal of the protecting group by the addition of aqueous acid (Figure 12.12) (Feakes et al., 1999). No evidence of the anticipated product, $[ae\text{-}B_{20}H_{17}SH]^{4-}$, based on the reactions yielding the $[B_{20}H_{17}OH]^{4-}$ anion (Figure 12.5) and the $[B_{20}H_{17}NH_3]$3- anion (Figure 12.7), was observed. Instead, the

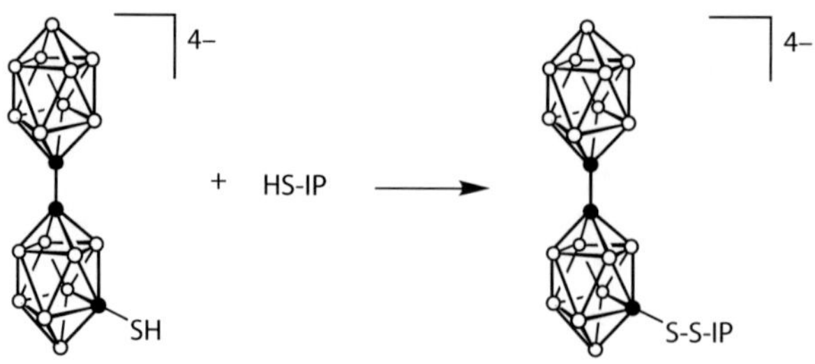

FIGURE 12.10 Hypothesized reaction between the $[a^2\text{-}B_{20}H_{17}SH]^{4-}$ ion and a cysteine residue of an intracellular protein (HS-IP).

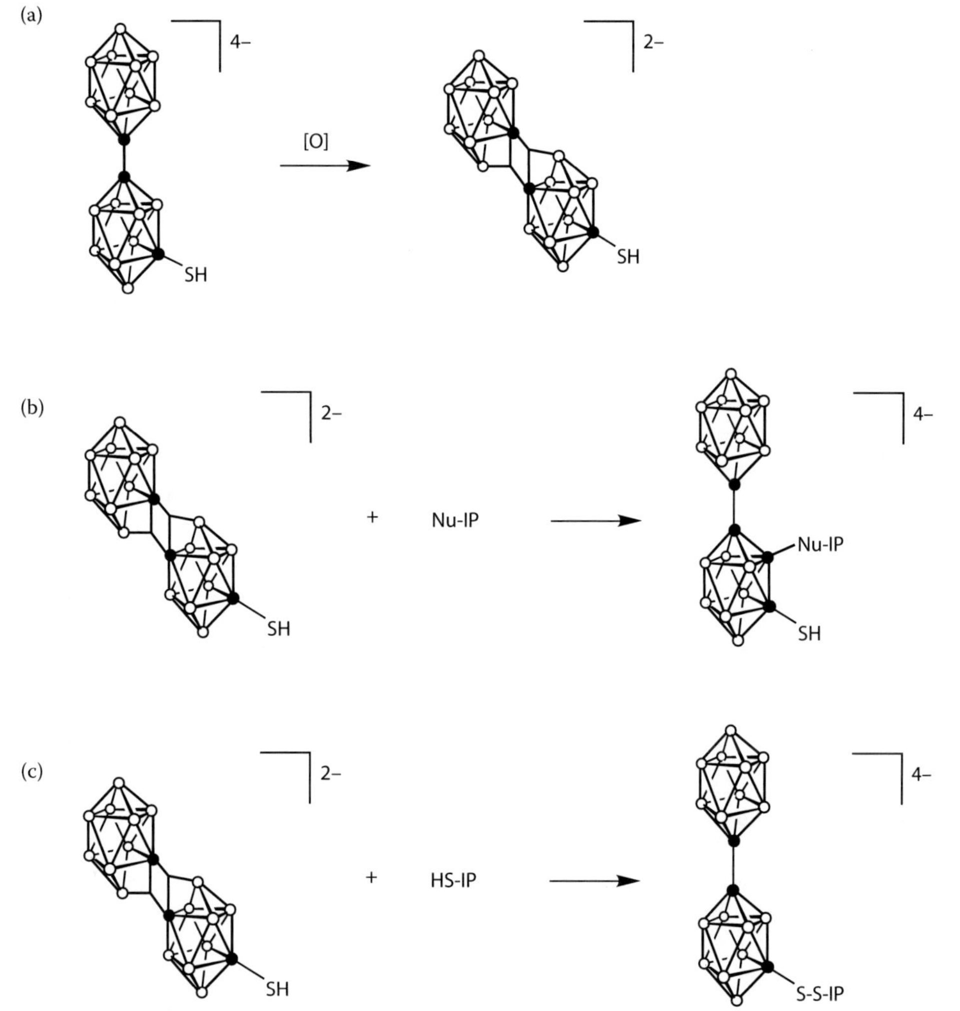

FIGURE 12.11 (a) Proposed oxidation of the $[a^2\text{-}B_{20}H_{17}SH]^{4-}$ ion, [O] is an oxidizing agent; (b) Hypothesized reaction between the $[B_{20}H_{17}SH]^{2-}$ ion and a Nu-IP; (c) Hypothesized reaction between the $[B_{20}H_{17}SH]^{2-}$ ion and an HS-IP.

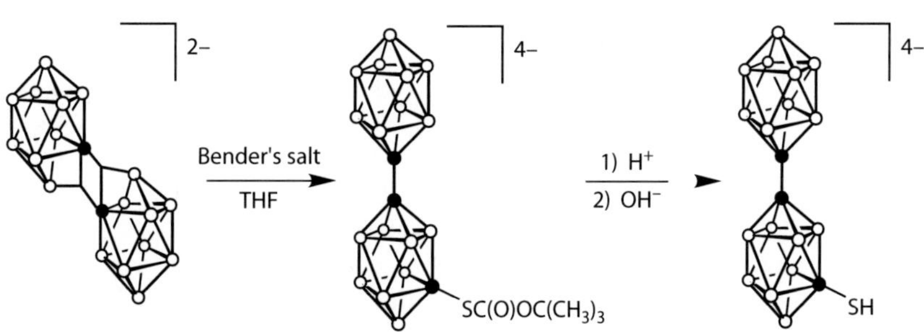

FIGURE 12.12 Preparation of the $[a^2\text{-}B_{20}H_{17}SH]^{4-}$ ion.

[a^2-$B_{20}H_{17}SH]^{4-}$ ion was formed directly in relatively high yield. The substitution pattern observed for the [a^2-$B_{20}H_{17}SH]^{4-}$ ion was also unexpected. The sulfhydryl substituent was located on the equatorial belt adjacent to the terminal apex, analogous to that obtained for the reaction of the photoisomer of the $[B_{20}H_{18}]^{2-}$ ion (Figure 12.6). The reaction mechanism proposed for the reaction of the normal isomer of the $[B_{20}H_{18}]^{2-}$ ion with nucleophiles (Hawthorne et al., 2002, 1965a) does not explain the formation of the unusual isomer. Current investigations within our laboratory are seeking to elucidate whether the formation of the unusual isomer is a result of the large steric requirement of the protected thiol group or the electronic properties of the sulfur atom (D. A. Feakes, unpublished results).

Liposomally encapsulated sodium salt of $[B_{20}H_{17}SH]^{4-}$, with an injected dose of 210 μg B (~11 mg/kg body weight), yielded a biodistribution that exhibited tumor accretion throughout the entire time-course experiment (Feakes et al., 1999). The tumor boron concentration at 48 h was 46.7 μg B/g tumor and would be expected to increase if the time-course experiment had been extended. Therapeutic levels of boron (>15 μg of B per g of tumor) are present in the tumor at all measured points in the time-course experiment.

The reactivity of the $[B_{20}H_{17}SH]^{4-}$ ion is under investigation (McVey et al., 2008). Matrix-assisted laser desorption ionization mass spectrometry (MALDI-MS) was utilized to evaluate the reaction mixtures of three representative polyhedral borane anions, $[B_{20}H_{18}]^{2-}$, $[B_{20}H_{17}OH]^{4-}$, and $[B_{20}H_{17}SH]^{4-}$, with serum albumins. In each case, the mass of the serum albumin increased; however, MALDI-MS cannot differentiate increases in mass as a result of covalent bonding from increases in mass as a result of strong electrostatic interactions. Native polyacrylamide gel electrophoresis, in the absence of sodium dodecyl sulfate and reducing agents, was used as an initial evaluation of the presence of a covalent bond. A change in migration pattern was only observed in the reaction of the $[B_{20}H_{17}SH]^{4-}$ ion with the serum albumin, indicating that a covalent bond was formed with this anion, but not with the other two anions being investigated. The reactivity of the sulfhydryl group was investigated using 5,5′-dithiobis(2-nitrobenzoic acid), commonly referred to as Ellman's reagent (Ellman, 1959). Ellman's reagent is commonly used to ascertain the number of free sulfhydryl groups present in a protein. If a covalent bond is formed between the $[B_{20}H_{17}SH]^{4-}$ and the serum albumin, no reaction with the Ellman's reagent would be observed whereas if a covalent bond is not formed between the $[B_{20}H_{17}SH]^{4-}$ and the serum albumin, a reaction with the Ellman's reagent would be observed. The results of the Ellman's reagent investigation clearly indicated that the $[B_{20}H_{17}SH]^{4-}$ ion formed a disulfide bond with the serum albumin. Ongoing investigations are evaluating the potential of the $[B_{20}H_{17}SH]^{4-}$ to oxidize *in vivo* as well as the reactivity of the resulting $[B_{20}H_{17}SH]^{2-}$ ion (D. A. Feakes, unpublished results).

12.5 EVALUATION OF THE POTENTIAL OF LIPOSOMAL ENCAPSULATION FOR APPLICATION IN BNCT

Liposomal encapsulation of hydrophilic boron-containing compounds has several appealing aspects to their usefulness in BNCT. The unilamellar liposomes are created from biologically inert components, cholesterol and DSPC, and have been widely accepted as a potential drug delivery vehicle (Juliano and Stamp, 1978; Forssen and Tokes, 1983; Rahman et al., 1986). The hydrophilic compounds that are encapsulated within the aqueous core are not required to have any inherent tumor specificity because the unilamellar liposomes provide the tumor selectivity. As a result, a wide variety of compounds can be encapsulated within the aqueous core and delivered to the interior of tumor cells. The injected dose of boron that is used with unilamellar liposomes is significantly smaller than those utilized in other investigations (Sweet et al., 1962; Soloway et al., 1967; Slatkin et al., 1986; Gabel et al., 1987; Coderre et al., 1988; Clendonen et al., 1990). The combination of smaller boron doses and the isolation of the compound during the liposomal delivery aid in continuing concerns regarding the toxicity of some boron-containing compounds.

While several positive aspects exist for the potential application of liposomal delivery for BNCT, there are also limitations. Liposomes do not cross the blood–brain barrier, as evidenced by the low brain boron concentrations in prior biodistribution experiments. Therefore, one of the targets of BNCT, *glioblastoma multiforme* (GBM), a particularly lethal brain tumor, cannot be treated by intravenous injection with compounds delivered by unilamellar liposomes. The unilamellar liposomes used in the reported studies contain hypertonic solutions of polyhedral borane salts to maximize the boron content of the formulations and, consequently, the injected dose. Larger concentrations of the polyhedral borane salts cannot be utilized without deleterious effects to the stability of the lipid bilayer. As a result, the advantage of low injected doses through the utilization of unilamellar liposomes is also impeded by an inability to raise the injected dose. The preparation of small unilamellar liposomes is characterized by low encapsulation efficiencies (~3%) (Deamer and Uster, 1983). The low encapsulation efficiencies are increasingly important when boron-10 enriched compounds are prepared at a much higher expense. Researchers continue to design liposomes with modified bilayers in order to improve the tumor-targeting abilities as well as the longevity of the liposome formulations within the circulatory system (Kullberg et al., 2003; Koning et al., 2004; Martini et al., 2004; Maruyama et al., 2004; Kullberg et al., 2005; Masunaga et al., 2006; Yanagie et al., 2006; Doi et al., 2008; Feng et al., 2009). Modifications of the liposome bilayer have been reviewed in another section of the current chapter.

Unilamellar liposomes are capable of delivering both hydrophilic boron-containing compounds, encapsulated in the aqueous core of the liposomes, and lipophilic boron-containing compounds, embedded in the lipid bilayer. The use of lipophilic boron-containing compounds as well as modifications to the lipid bilayer is discussed thoroughly elsewhere in this book and will not be duplicated in this chapter. It is noteworthy, however, that liposome formulations containing hydrophilic boron-containing compounds, such as the $[B_{20}H_{17}NH_3]^{3-}$ ion, and lipophilic boron-containing compounds, such as K[*nido*-7-$CH_3(CH_2)_{15}$-7,8-$C_2B_9H_{11}$], concurrently have been created and investigated (Feakes et al., 1995). The results of the biodistribution experiments clearly indicate that the injected dose of boron can be increased in this manner without deleterious effects to the stability of the liposomes or the tumor boron concentration. An added benefit to the incorporation of compounds, such as K[*nido*-7-$CH_3(CH_2)_{15}$-7,8-$C_2B_9H_{11}$], in the bilayer is that the anion imparts a negative charge to the lipid bilayer, a characteristic which is believed to enhance tumor specificity (Fraley et al., 1981; Straubinger et al., 1983).

12.6 CONCLUSIONS

The water-soluble salts of polyhedral borane anions, encapsulated in tumor-selective unilamellar liposomes, have demonstrated therapeutically useful boron concentrations for application in BNCT. Although the number of compounds that have been investigated remain relatively small compared to other classes of compounds currently being investigated, hypothesis have been developed which enable the *a priori* prediction of which compounds are likely to be retained within the tumor cells and which compounds are likely to be cleared from all tissues, including the tumor. Anions which have the potential to react with intracellular protein moieties are retained by the tumor cells once delivered by the liposomes. In this group of anions, the $[B_{20}H_{18}]^{2-}$ anion and its derivatives, such as $[B_{20}H_{17}NH_3]^{-}$ and $[B_{20}H_{17}SH]^{2-}$, are representative. Anions which have the potential to oxidize to a more reactive species, capable of covalently bonding with intracellular protein moieties, are retained by the tumor cells once delivered by the liposomes. The reduced derivatives of the $[B_{20}H_{18}]^{2-}$ anion, such as the $[B_{20}H_{17}NH_3]^{3-}$ anion and the $[B_{20}H_{17}SH]^{4-}$ anion, are representative. Once delivered by the liposomes, anions which have the potential to form disulfide bonds are retained by the tumor cells. Although the $[B_{12}H_{11}SH]^{2-}$ anion has not exhibited this characteristic, the $[B_{20}H_{17}SH]^{4-}$ certainly has done so. Anions which lack any of these reactivities are rapidly cleared from all tissues, including the tumor.

Although significant progress has been made in this research area, there is a great deal of research still remaining. A wide variety of compounds can be prepared based on the current understanding of modes of tumor retention. The substituents that have been investigated to date target predominantly the reactivity of the sulfhydryl group of cysteine residues, but other amino acid substituents could be targeted with appropriate derivatives. While initial experiments have been completed regarding the reactivity of the polyhedral borane anions with protein substituents, a clear understanding of the potential reaction sites remains to be gained. And, finally, few experiments have been completed which explore the best modes of injection. For example, an early experiment successfully investigated the possibility of increasing tumor boron concentration using a double-injection protocol (Shelly et al., 1992), but none have been completed since that time. Blood–brain barrier disruption (Barth et al., 2000, 2002) or convection-enhanced delivery (Groothuis, 2000; Yang et al., 2009) could increase the boron concentration within the tumor and enhance the application of unilamellar liposomes in the treatment of *GBM*. And, combinations of different boron-containing compounds may enable BNCT to finally achieve a place in the treatment of cancer.

ACKNOWLEDGMENTS

The author would like to thank the Robert Welch Foundation (AI-0045) for support and would like to dedicate this chapter to M. Frederick Hawthorne for the knowledge and passion for boron chemistry that he generates in each of his students.

REFERENCES

Alanazi, F., Hengguang, L., Halpern, D. S., Svein, Ø., and Lu, D. R. 2003. Synthesis, preformulation and liposomal formulation of cholesteryl carborane esters with various fatty chains. *Int. J. Pharm.* 255:189–197.

Barth, R. F., Yang, W., Al-Madhoun, A. S., Johnsamuel, J., Byun, Y., Chandra, S., Smith, D. R., Tjarks, W., and Eriksson, S. 2004. Boron-containing nucleosides as potential delivery agents for neutron capture therapy of brain tumors. *Cancer Res.* 64(17):6287–6295.

Barth, R. F., Yang, W., Bartus, R. T., Rotaru, J. H., Ferketich, A. K., Moeschberger, M. L., Nawrocky, M. M., Coderre, J. A., and Rofstad, E. K. 2002. Neutron capture therapy of intracerebral melanoma: Enhanced survival and cure after blood–brain barrier opening to improve delivery of boronophenylalanine. *Int. J. Rad. Onc. Bio. Phys.* 52(3):858–868.

Barth, R. F., Yang, W., Rotaru, J. H., Moeschberger, M. L., Boesel, C. P., Soloway, A. H., Joel, D. D., Nawrocky, M. M., Ono, K., and Goodman, J. H. 2000. Boron neutron capture therapy of brain tumors: Enhanced survival and cure following blood–brain barrier disruption with intracarotid injection of sodium borocaptate and boronophenylalanine. *Int. J. Rad. Onc. Bio. Phys.* 47(1):209–218.

Cai, J. and Soloway, A. H. 1996. Synthesis of carboranyl polyamines for DNA targeting. *Tet. Lett.* 37(52):9283–9286.

Cai, J., Soloway, A. H., Barth, R. F., Adams, D. M., Hariharan, J. R., Wyzlic, I. W., and Radcliffe, K. 1997. Boron-containing polyamines as DNA targeting agents for neutron capture therapy of brain tumors: Synthesis and biological evaluation. *J. Med. Chem.* 40:3887–3896.

Chamberland, B. L. and Muetterties, E. L. 1964. Chemistry of boranes. XVIII. Oxidation of $B_{10}H_{10}^{2-}$ and its derivatives. *Inorg. Chem.* 3:1450–1456.

Clendonen, N. R., Barth, R. F., Gordon, W. A., Goodman, J. H., Alam, F., Staubus, A. E., Boesel, C. P. et al. 1990. Boron neutron capture therapy of a rat glioma. *Neurosurg.* 26:47–55.

Coderre, J. A., Kalef-Ezra, J. A., Fairchild, R. G., Micca, P. L., Reinstein, L. E., and Glass, J. D. 1988. Boron neutron capture therapy of a murine melanoma. *Cancer Res.* 48:6313–6316.

Das, B. C., Das, S., Li, G., Bao, W., and Kabalka, G. W. 2001. Synthesis of a water soluble carborane containing amino acid as a potential therapeutic agent. *Synlett* 9:1419–1420.

Deamer, D. W. and Uster, P. S. 1983. Liposome preparation: Methods and mechanism. In *Liposomes*, ed. M. J. Ostro, 31–35. New York: Dekker.

Doi, A., Kawabata, S., Iida, K., Yokoyama, K., Kajimoto, Y., Kuroiwa, T., Shirakawa, T. et al. 2008. Tumor-specific targeting of sodium borocaptate (BSH) to malignant glioma by transferrin-PEG liposomes: a modality forboron neutron capture therapy. *J. Neurooncol.* 87(3):287–294.

Ellman, G. L. 1959. Tissue sulfhydryl groups. *Arch. Biochem. Biophys.* 82:70–77.

Fairchild, R. G. and Bond, V. P. 1985. Current status of boron-10-neutron capture therapy: Enhancement of tumor dose via beam filtration and dose rate, and the effects of these parameters on minimum boron content: A theoretical evaluation. *J. Radiat. Oncol. Biol. Phys.* 11:831–840.

Feakes, D. A., Shelly, K., and Hawthorne, M. F. 1995. Selective boron delivery to murine tumors by lipophilic species incorporated in the membranes of unilamellar liposomes. *Proc. Natl. Acad. Sci. USA* 92:1367–1370.

Feakes, D. A., Shelly, K., Knobler, C. B., and Hawthorne, M. F. 1994. $Na_3[B_{20}H_{17}NH_3]$: Synthesis and liposomal delivery to murine tumors. *Proc. Natl. Acad. Sci. USA* 91:3029–3033.

Feakes, D. A., Waller, R. C., Hathaway, D. K., and Morton, V. S. 1999. Synthesis and *in vivo* murine evaluation of $Na_4[1\text{-}(1'\text{-}B_{10}H_9)\text{-}6\text{-}SHB_{10}H_8]$ as a potential agent for boron neutron capture therapy. *Proc. Natl. Acad. Sci. USA* 96:6406–6410.

Feng, B., Tomizawa, K., Michiue, H., Miyatake, S., Han, X. J., Fujimura, A., Seno, M., Kirihata, M., and Matsui, H. 2009. Delivery of sodium borocaptate to glioma cells using immunoliposome conjugated with anti-EGFR antibodies by ZZ-His. *Biomaterials* 30(9):1746–1755.

Forssen, E. A. and Tokes, Z. A. 1983. Improved therapeutic benefits of doxorubicin by entrapment in anionic liposomes. *Cancer Res.* 43:546–550.

Fraley, R., Straubinger, R., Rule, G., Springer, L., and Papahadjopoulos, D. 1981. Liposome-mediated delivery of deoxyribonucleic acid to cells: Enhanced efficiency of delivery related to lipid composition and incubation conditions. *Biochemistry* 20:6978–6987.

Gabel, D., Holstein, H., Larsson, B., Gille, L., Ericson, G., Sacker, D., Som, P., and Fairchild, R. G. 1987. Quantitative neutron capture radiography for studying the biodistribution of tumor-seeking boron-containing compounds. *Cancer Res.* 47:5451–5454.

Genady, A. R. and Gabel, D. 2002. Synthesis and optical properties of novel covalent and non-covalent porphyrin dimers. *J. Por. Phth.* 6(6):382–388.

Georgiev, E. M., Shelly, K., Feakes, D. A., Kuniyoshi, J., Romano, S., and Hawthorne, M. F. 1996. Synthesis of amine derivatives of the polyhedral borane anion $[B_{20}H_{18}]^{4-}$. *Inorg. Chem.* 35:5412–5416.

Ghaneolhosseini, H., Tjarks, W., and Sjöberg, S. 1998. Synthesis of novel boronated acridines and spermidines as possible agents for BNCT. *Tetrahedron* 54(15):3877–3884.

Groothuis, D. R. 2000. The blood-brain and blood–tumor barriers: A review of strategies for increasing drug delivery. *Neuro Oncol.* 2(1):45–59.

Hawthorne, M. F. 1991. Biochemical applications of boron cluster chemistry. *Pure Appl. Chem.* 63:327–334.

Hawthorne, M. F. 1993. The role of chemistry in the development of boron neutron capture therapy of cancer. *Angew. Chem. Int. Ed. Engl.* 32: 950–984.

Hawthorne, M. F. and Lee. M. W. 2003. A critical assessment of boron target compounds for boron neutron capture therapy. *J. Neuro-Onc.* 62:33–45.

Hawthorne, M. F. and Pilling, R. L. 1966. Photoisomerization of the $B_{20}H_{18}^{-2}$ ion. *J. Am. Chem. Soc.* 88(16):3873–3874.

Hawthorne, M. F. and Pilling, R. L. 1967. The preparation of the $B_{10}H_{10}^{-2}$ ion. *Inorg. Synth.* 9:16–19.

Hawthorne, M. F., Pilling, R. L., and Garrett, P. M. 1965a. A study of the reaction of hydroxide ion with $B_{20}H_{18}^{-2}$. *J. Am. Chem. Soc.* 87(21):4740–4746.

Hawthorne, M. F., Pilling, R. L., and Stokely, P. F. 1965b. The preparation and rearrangement of the three isomeric $B_{20}H_{18}^{-4}$ ions. *J. Am. Chem. Soc.* 87(9):1893–1899.

Hawthorne, M. F., Pilling, R. L., Stokely, P. F., and Garrett, P. M. 1963. The reaction of the $B_{20}H_{18}^{-2}$ ion with hydroxide ion. *J. Am. Chem. Soc.* 85(22):3704–3705.

Hawthorne, M. F., Shelly, K., and Li, F., 2002. The versatile chemistry of the $[B_{20}H_{18}]^{2-}$ ions: Novel reactions and structural motifs. *Chem. Comm.* 6:547–554.

Hill, J. S., Kahl, S. B., Kaye, A. H., Stylli, S. S., Koo, M.-S., Gonzales, M. F., Vardaxis, N. J., and Johnson, C. I. 1992. Selective tumor uptake of a boronated porphyrin in an animal model of cerebral glioma. *Proc. Nat. Acad. Sci. USA* 89:1785–1789.

Juliano, R. L. and Stamp, D. 1978. Pharmacokinetics of liposome-encapsulated antitumor drugs. Studies with vinblastine, actinomycin D, cytosine arabinoside, and daunomycin. *Biochem. Pharmacol.* 27:21–27.

Kabalka, G. W. and Yao, M-L. 2003. Synthesis of a novel boronated 1-aminocyclobutanecarboxylic acid as a potential boron neutron capture therapy agent. *Appl. Organomet. Chem.* 17:398–402.

Kaczmarczyk, A., Dobrott, R. D., and Lipscomb, W. N. 1962. Reactions of $B_{10}H_{10}^{-2}$ ion. *Proc. Natl. Acad. Sci. USA* 48:729–733.

Kahl, S. B. and Koo, M. S. 1990. Synthesis of tetrakiscarboranecarboxylate esters of 2,4-bis-(α, β-dihydroxyethyl) deuteroporphyrin IX. *J. Chem. Soc. Chem. Commun.* 24:1769–1771.

Koning, G. A., Fretz, M. M., Woroniecka, U., Storm, G., and Krijger, G. C. 2004. Targeting liposomes to tumor endothelial cells for neutron capture therapy. *Appl. Radiat. Isot.* 61(5):963–967.

Kullberg, E. B., Nestor, M., and Gedda, L. 2003. Tumor-cell targeted epidermal growth factor liposomes loaded with boronated acridine: Uptake and processing. *Pharm. Res.* 20(2):229–236.

Kullberg, E. B., Wei, Q., Capala, J., Giusti, V., Malmström, P. U., and Gedda, L. 2005. EGF-targeted liposomes with boronated acridine: Growth inhibition of cultured glioma cells after neutron irradiation. *Int. J. Radiat. Biol.* 81(8):621–629.

Lasic, D. 1992. Liposomes. (Synthetic lipid microspheres serve as multipurpose vehicles for the delivery of drugs, genetic material and cosmetics). *Am. Sci.* 80:20–31.

Laster, B. H., Kahl, S. B., Popenoe, E. A., Pate, D. A., and Fairchild, R. G. 1991. Biological efficacy of boronated low-density lipoprotein for boron neutron capture therapy as measured in cell culture. *Cancer Res.* 51:4588–4593.

Lesnikowski, Z. J. and Schinazi, R. F. 1995. Boron neutron capture therapy of cancers: Nucleic bases, nucleosides, and oligonucleotides as potential boron carriers. *Polish J. Chem.* 69(6):827–840.

Li, H., Hardin, C., and Shaw, B. R. 1996. Hydrolysis of thymidine boranomonophosphate and stepwise deuterium substitution of the borane hydrogens. ^{31}P and ^{11}B NMR studies. *J. Am. Chem. Soc.* 118:6606–6614.

Li, T., Hamdi, J., and Hawthorne, M. F. 2006. Unilamellar liposomes with enhanced boron content. *Bioconj. Chem.* 17(1):15–20.

Lindström, P., Naeslund, C., and Sjöberg, S. 2000. Enantioselective synthesis and absolute configurations of the enantiomers of o-carboranylalanine. *Tet. Lett.* 41:751–754.

Martini, S., Ristori, S., Pucci, A., Bonechi, C., Becciolini, A., Martini, G., and Rossi, C. 2004. Boronophenylalanine insertion in cationic liposomes for boron neutron capture therapy. *Biophys. Chem.* 111(1):27–34.

Maruyama, K., Ishida, O., Kasaoka, S., Takizawa, T., Utoguchi, N., Shinohara, A., Chiba, M., Kobayashi, H., Eriguchi, M., and Yanagie, H. 2004. Intracellular targeting of sodium mercaptoundecahydrododecaborate (BSH) to solid tumors by transferrin-PEG liposomes for boron neutron-capture therapy. *J. Control Release* 98(2):195–207.

Masunaga, S., Kasaoka, S., Maruyama, K., Nigg, D., Sakurai, Y., Nagata, K., Suzuki, M., Kinashi, Y., Maruhashi, A., and Ono, K. 2006. The potential of transferrin-pendant-type polyethyleneglycol liposomes encapsulating decahydrodecaborate-(10)B (GB-10) as (10)B-carriers for boron neutron capture therapy. *Int. J. Radiat. Oncol. Biol. Phys.* 66(5):1515–1522.

Mayhew, E. and Papahadjopoulos, D. 1983. Therapeutic applications of liposomes. In *Liposomes*, ed. M. J. Ostro, 109–156. New York: Dekker.

McVey, W. J., Matthews, B., Motley, D. M., Linse, K. D., Blass, D. P., Booth, R. E., and Feakes, D. A. 2008. Investigation of the interactions of polyhedral borane anions with serum albumins. *J. Inorg. Biochem.* 102:943–951.

Peacock, G., Sidwell, R., Guangliang, P., Svein, Ø., and Lu, D. R. 2004. *In vitro* uptake of a new cholesteryl carborane ester compound by human glioma cell lines. *J. Pharm. Sci.* 93(1):13–19.

Pilling, R. L., Hawthorne, M. F., and Pier, E. A. 1964. The boron-11 nuclear magnetic resonance spectrum of $B_{20}H_{18}^{-2}$ at 60 Mc./sec. *J. Am. Chem. Soc.* 86:3568–3569.

Radel, P. A. and Kahl, S. B. 1996. Enantioselective synthesis of L- and D-carboranylalanine. *J. Org. Chem.* 61:4582–4588.

Rahman, A., Fumagalli, A., Barbieri, B., Schein, P., and Casazza, A. M. 1986. Antitumor and toxicity evaluation of free doxorubicin and doxorubicin entrapped in cardiolipin liposomes. *Cancer Chemother. Pharmacol.* 16:22–27.

Schinazi, R. F. and Prusoff, W. H. 1978. Synthesis and properties of boron and silicon substituted uracil or 2′-deoxyuridine. *Tet. Lett.* 50:4981–4984.

Shelly, K., Feakes, D. A., Hawthorne, M. F., Schmidt, P. G., Krisch, T. A., and Bauer, W. F. 1992. Model studies directed toward the boron neutron-capture therapy of cancer: Boron delivery to murine tumors with liposomes. *Proc. Natl. Acad. Sci. USA* 89:9039–9043.

Slatkin, D., Micca, P., Forman, A., Gabel, D., Wielopolski, L., and Fairchild, R. G. 1986. Boron uptake in melanoma, cerebrum and blood from sodium sulfhydryl borane ($Na_2B_{12}H_{11}SH$) and sodium disulfide borane ($Na_4B_{24}H_{22}S_2$) administered to mice. *Biochem. Pharmacol.* 35:1771–1776.

Soloway, A. H., Hatanaka, H., and Davis, M. A. 1967. Penetration of brain and brain tumor. VII. Tumor-binding sulfhydryl boron compounds. *J. Med. Chem.* 10:714–717.

Soloway, A. H., Tjarks, W., Barnum, B. A., Rong, F.-G., Barth, R. F., Codogni, I. M., and Wilson, J. G. 1998. The chemistry of neutron capture therapy. *Chem. Rev.* 98:1515–1562.

Sood, A., Shaw, B. R., and Spielvogel, B. F. 1989. Boron-containing nucleic acids. Synthesis of cyanoborane adducts of 2′-deoxynucleosides. *J. Am. Chem. Soc.* 111:9234–9235.

Straubinger, R. M., Hong, K., Friend, D. S., and Papahadjopoulos, D. 1983. Endocytosis of liposomes and intracellular fate of encapsulated molecules: Encounter with a low pH compartment after internalization in coated vesicles. *Cell* 32:1069–1079.

Straubinger, R. M., Papahadjopoulos, D., and Hong, K. 1990. Endocytosis and intracellular fate of liposomes using pyranine as a probe. *Biochemistry* 29:4929–4939.

Sweet, W. H., Soloway, A. H., and Wright, R. L. 1962. Evaluation of boron compounds for use in neutron capture therapy of brain tumors. II. Studies in man. *J. Pharmacol. Exp. Ther.* 137:263–266.

Tang, P. P., Schweizer, P., Bradshaw, K. M., and Bauer, W. F. 1995. ^{11}B nuclear magnetic resonance studies of the interaction of borocaptate sodium with serum albumin. *Biochem. Pharm.* 49(5):625–632.

Wallingford, R. H. and Williams, L. E. 1985. Is stability a key parameter in the accumulation of phospholipid vesicles in tumors? *J. Nucl. Med.* 26:1180–1185.

Watson-Clark, R., Knobler, C. B., and Hawthorne, M. F. 1996. Synthesis and structure of the elusive $[a^2\text{-}B_{20}H_{18}]^{3-}$ Anion. *Inorg. Chem.* 35:2963.

Yamamoto, Y., Seko, T., and Nemoto, H. 1989. New method for the synthesis of boron-10 containing nucleoside derivatives for neutron-capture therapy via palladium-catalyzed reaction. *J. Org. Chem.* 54:4734–4736.

Yanagie, H., Maruyama, K., Takizawa, T., Ishida, O., Ogura, K., Matsumoto, T., Sakurai, Y. et al. 2006. Application of boron-entrapped stealth liposomes to inhibition of growth of tumour cells in the *in vivo* boron neutron-capture therapy model. *Biomed. Pharmacother.* 60(1):43–50.

Yang, W., Barth, R. F., Wu, G., Huo, T., Tjarks, W., Ciesielski, M., Fenstermaker, R. A. et al. 2009. Convection enhanced delivery of boronated EGF as a molecular targeting agent for neutron capture therapy of brain tumors. *J. Neurooncol.* 95(3):355–365.

Zhuo, J.-C., Tjarks, W., Soloway, A. H., Cai, J., Wang, J., Barth, R. F., Adams, D. M., Eriksson, S., Lunato, A. J., and Ji, W. 2000. Synthesis and evaluation of DNA tumor targeting agents for neutron capture therapy. *In Special Publication of the Royal Society of Chemistry 253 (Contemporary Boron Chemistry)* 127–130.

Part V

Boron for Electronics: Optoelectronics

INTRODUCTION

Boron for electronics is the highlight of the chapters in this section, with contributions from Andrea Vöge, Detlef Gabel, Piotr Kaszynski, and Paul A. Jelliss. Boron compounds in the field of nonlinear optics, liquid crystals, and photoluminescence are the major part of discussions in this section. The "structure–property relationship" is a commonly used phrase because the structural features of certain compounds dictate their properties. Evidently, the structure–property relationship of boron-based materials plays a vital role in the field of electronics, as described in the following chapters.

Nonlinear optics is one of the rapidly growing fields of science that describes a branch of optics where matter responds in a nonlinear fashion to incident light beams. Nonlinear optical materials have gained tremendous importance lately due to their possible applications in information storage and optical communications. Although organic molecules, with their diverse structural variations, are effective as nonlinear optical materials, more studies are now being performed on polyhedral boron cluster compounds as nonlinear optical materials. Carboranes have long been known for their nonlinear optical behavior, while numerous substituted carboranes are now emerging as candidates for this kind of behavior. In addition, photoluminescence properties of polyhedral boron cluster compounds are equally important in the field of electronics. Boranes, carboranes, and metallacarboranes, by virtue of their structural features, are steadily making ground in the field of optoelectronics.

Liquid crystals are also an important class of compounds; they have become an inseparable part of our day-to-day lives through their presence in almost all display devices. They are mainly composed of an anisotropic rigid core and a terminal alkyl substituent. Several derivatives of *closo*-boranes, along with a metallacarbollide, form liquid crystals, and many more clusters are under investigation. Ongoing research in this field should provide us with a better understanding of the structure–property relationships of similar type of materials.

13 Boron Derivatives for Application in Nonlinear Optics

Andrea Vöge and Detlef Gabel

CONTENTS

13.1 OVERVIEW: BORON IN NLO

This chapter gives an overview of research activities in the field of boron in nonlinear optics (NLO) in recent years. Different types of boron derivatives are of interest for applications in NLO: some act as electron donors, for example, the dodecaborate and tetracoordinate boron, while some act as acceptors, such as *closo*-carboranes and tricoordinate boron.

Over the past decades dodecahydro-*closo*-dodecaborate(2-) derivatives have been investigated experimentally and theoretically; studies have shown that the cluster has favorable properties for application as nonlinear optical materials, compounds in which the dodecaborate acts as an electron donor and is connected to an electron-withdrawing substituent. In contrast to dodecaborate, neutral icosahedral *closo*-carboranes behave as strong electron acceptors; several investigations and theoretical studies confirm this effect.

In the past, a large number of *trivalent boron-containing compounds*—in this case the boron acts as an electron acceptor because of its vacant *p*-orbital—with nonlinear optical properties have been prepared and investigated as well. Also zwitterionic molecules with *tetracoordinate negatively charged borate* as an electron donor seem to be interesting materials for NLO.

13.2 INTRODUCTION

13.2.1 What is NLO?

Photonics is the key technology of our information era and incorporates optical technologies for storage, transfer, and processing of information. In this age, the increasing transfer of a great deal of data has always demanded new possibilities to transport plenty of information within a short span of time. In telecommunications, optical signal processing allows the parallel conversion of information in briefer time than serial electronic data processing. The advantage of photons, having neither mass nor charge, over electrons with respect to data transfer is their lower attenuation, and so they do not interfere with each other during information transmission.

NLO deals with the investigation of the interaction between intensive light and materials. A nonlinear optical material which is passed by a high electric field will be polarized, and in turn this polarization affects the properties of the incident light beam.

In May 1960, Maiman made a seminal development toward the progression of NLO: he generated the first ruby light amplification by stimulated emission of radiation (*laser*). From then on, changes of optical properties of material by strong irradiation fields have been demonstrated by many experiments.

Franken et al. performed the first nonlinear optical experiment in 1961 by using this ruby laser: they generated the second-harmonic generation (SHG), that is, frequency-doubled light, by passing the incident monochromatic light through crystalline quartz (Figure 13.1). Thus, the availability of coherent high-intensity laser light permits one to show that a material can differ in its linear response behavior by interaction with light.

Frequency doubling is one of the most important nonlinear optical effects and has great significance in its application to photonic devices. Information and image processing technologies, for example, optical information storage on optical discs, can be improved by using light of shorter wavelengths (Prasad and Williams, 1991).

Since the discovery of SHG, NLO has attracted increasing interest, and scientists all over the world have carried out research in this important field. Further nonlinear optical effects were discovered, such as sum-frequency generation (SFG), third-harmonic generation (THG), two photon absorption (TPA), Kerr effect, Pockels effect, and so on.

13.2.2 Materials for NLO

In recent years, a lot of research has been carried out toward the design, preparation, and investigation of new nonlinear optical *organic materials*, because of their potential applications in optical

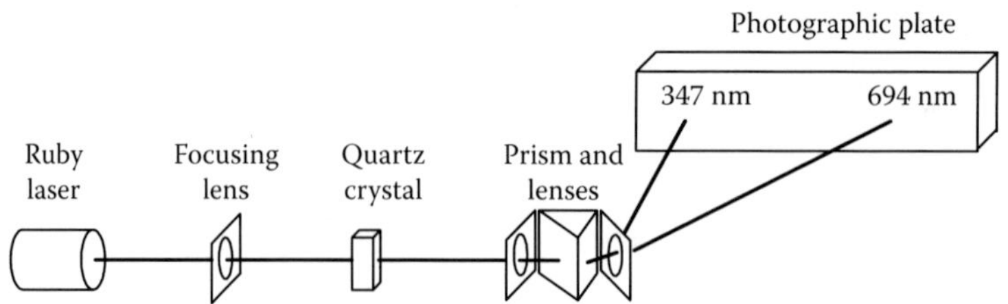

FIGURE 13.1 Experiment for frequency doubling (SHG). (Adapted from Franken, P. A. et al., *Phys. Rev. Lett.* **1961**, 7, 4, 118.)

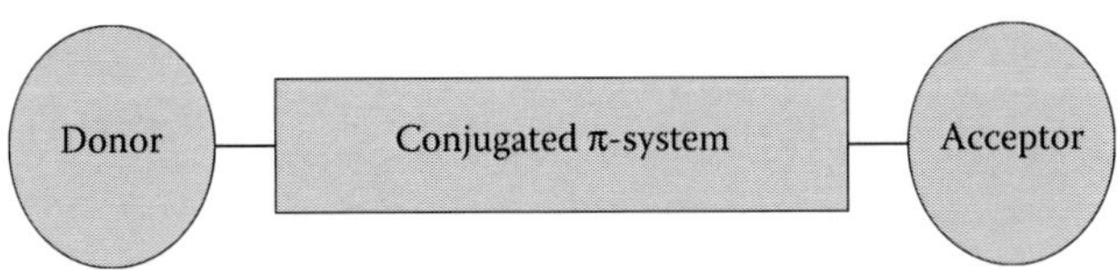

FIGURE 13.2 Scheme of a push–pull- or CT-chromophore.

communications and information storage. Typically, organic materials for NLO are composed of electron donors and acceptors that are connected via π-conjugated electron systems to enable interaction with radiation (push–pull or charge-transfer (CT) molecules) (Figure 13.2). Correlations between substituent patterns, spacer length between electron donor and acceptor, and the nonlinear optical properties of these chromophores could be elucidated.

Because of their comparatively simple preparation, the numerous possibilities of variation in structure, and their oftentimes better nonlinear optical properties, organic compounds are more suitable than traditional *inorganic crystals*, for example, lithium niobate ($LiNbO_3$), potassium dihydrogen phosphate (KH_2PO_4, KDP), potassium titanyl phosphate ($KTiOPO_4$, KTP), and β-barium borate (BaB_2O_4, BBO). The polarization of organic molecules induced by external fields occurs much faster and more intensively than in inorganic compounds. In the latter case the electrons are bonded by the atom core and only lattice vibrations are possible (Miyata et al., 1994).

The interaction of NLO-active organic molecules with intensive light induces a dipole moment, which results in strong hyperpolarization (see Section 13.3.1). In the simplest case of neutral compounds with uncharged donor and acceptor moieties, the CT and the zwitterionic resonance structures are shown below (Scheme 13.1).

A condition for second-order nonlinear organic materials to double the frequency of an incident laser light is a *noncentrosymmetric* configuration on the molecular and macroscopic level. The term macroscopic implies that the compounds must have a noncentrosymmetric crystal structure, because only materials without a center of symmetry have anisotropic polarizability.

In centrosymmetric molecules all tensor elements β_{ijk} are equal to zero, because the potential energy is the same in one and in the opposite direction, and thus no nonlinear response appears (Prasad and Williams, 1991).

4-Nitroaniline is the standard of comparison for nonlinear optical compounds. The molecule is noncentrosymmetric on the molecular level and has a high hyperpolarizability ($\beta = 34.5 \times 10^{-30}$ esu, Oudar and Chemla, 1977), but it crystallizes centrosymmetrically; therefore, it does not show second-order nonlinear optical effects in the crystalline phase (Miyata et al., 1994). In contrast, 3-nitroaniline crystallizes in a noncentrosymmetric space group, but the β value is much lower ($\beta = 6.0 \times 10^{-30}$ esu, Oudar and Chemla, 1977) than the value of 4-nitroaniline, because the charge transfer in the latter case (*para*-isomer) is much stronger.

It is still a big challenge to design compounds which crystallize in noncentrosymmetric space groups because roughly 75% of all organic molecules crystallize centrosymmetrically (Marder et al., 1989). Dipole–dipole interactions are the driving force for centrosymmetric crystallization. There are several approaches to avoid this undesirable effect: Marder et al. have described stilbazolium salts, with different counteranions (Figure 13.3), some of which exhibit strong SHG signals in comparison with the standard urea, which is used as reference for the determination of nonlinear optical properties in the Kurtz powder method (see Chapter 2.5).

Furthermore, these salts crystallize with noncentrosymmetric packing. An explanation for this required crystallization could be that Coulomb interactions disrupt the dipole–dipole interactions.

SCHEME 13.1 CT in a donor-π-conjugated system–acceptor chromophore.

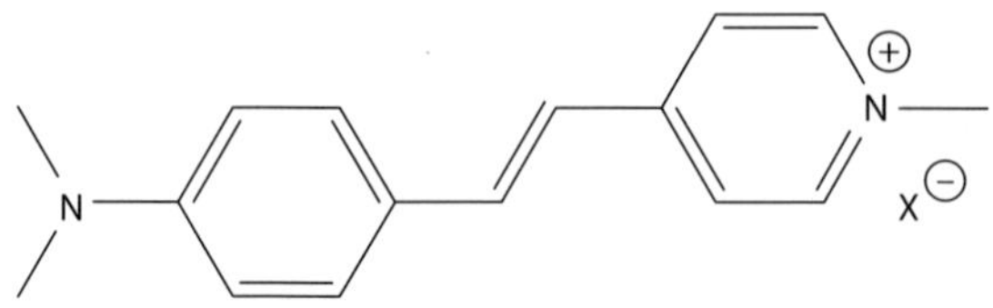

FIGURE 13.3 *N,N*-Dimethylamino-*N′*-methylstilbazolium salts.

Another way of avoiding centrosymmetric crystallization is to use chiral molecules. One example is dinitrophenylmethylalaninate described by Oudar and Chemla (1977), where a chiral methylalaninate group was inserted at the amino group of 2,4-dinitroaniline to have an impact on the crystallization. The methylalaninate moiety has the same effect on the hyperpolarizability β as the amino group.

For nonlinear optical molecules which do not possess the desired centrosymmetric crystallization, there are very potent methods to incorporate these chromophores into polymer hosts ("electric field poling") (Singer et al., 1986, 1987; Hubbard et al., 1989; Bjorklund et al., 1991), self-assembly of molecular layers (Li et al., 1990; Katz et al., 1991), and Langmuir–Blodgett assembly of films (Roberts et al., 1990; Kajikawa et al., 1992; Marder et al., 1994).

13.3 PRINCIPLES OF NLO

The nonlinear optical effect is a phenomenon where a substance is polarized by the field of a very intensive light. This induced polarization generates an optical harmonic by which the light is changed by passing through the medium.

13.3.1 Nonlinear Optical Effects on Molecular Level

The interaction of a molecule with an electric field results in a shift of the electron density. This polarization induces a molecular dipole μ_i whose absolute value is in the case of low light intensities, proportional to the applied electric field E:

$$\mu_i = \alpha \cdot E$$

where α is the molecular polarizability, a proportionality constant.

In the case of using intensive laser light the formula must be extended by using a Taylor series, and the induced dipole moment for a single molecule is

$$\mu_i = \alpha \cdot E + \beta \cdot E^2 + \gamma \cdot E^3 + \cdots$$

The first term, the linear part containing the polarizability α, will be extended by the first nonlinear optical term including the first hyperpolarizability β (second-order polarization), the second nonlinear optical term containing the second hyperpolarizability γ (third-order polarization), and so on. The most important nonlinear optical effects are based on the first hyperpolarizability β because with increasing order of the hyperpolarizabilities the effects are getting weaker ($\beta \sim 1/E^2 > \gamma \sim 1/E^3$ etc.).

13.3.2 Nonlinear Optical Effects in Macroscopic Materials

The oscillating electric field of the light wave induces electric dipoles by the passage through a dielectric medium, which leads to a macroscopic electric polarization of the medium. In the case of less intensive light the dipole moments are linear and the resulting macroscopic polarization P is

$$P = \varepsilon_0 \cdot \chi \cdot E$$

where P is proportional to the electric field E of the light wave. ε_0 is the dielectric constant of the vacuum, and χ the electric susceptibility independent of the electric field. By applying high electric field strength, for example, by the use of laser, this proportionality is not followed, because the electric dipoles of the material can no longer respond linearly to these high electric fields. In addition to the linear part $\chi^{(1)} \cdot E$ we must also consider the nonlinear susceptibility coefficients $\chi^{(2)}$, $\chi^{(3)}$, and so on:

$$P = \varepsilon_0(\chi^{(1)} \cdot E + \chi^{(2)} \cdot E^2 + \chi^{(3)} \cdot E^3 + \cdots)$$

The coefficients $\chi^{(n)}$ are applied for the description of the macroscopic susceptibilities instead of the molecular coefficients α, β, and γ. The first-order susceptibility $\chi^{(1)}$ describes linear processes such as refraction and absorption. Nonlinearities are characterized by the susceptibilities of higher order ($\chi^{(2)}$, $\chi^{(3)}$...). The even- and odd-order terms lead to different nonlinear responses. $\chi^{(2)}$ is contributed to the polarization only in noncentrosymmetric media, and $\chi^{(3)}$ contributing in any media irrespective of the symmetry (Prasad and Williams, 1991).

13.3.3 Second-Order Nonlinear Optical Effects

As a wide variety of second-order nonlinear optical effects exist, only a few examples will be mentioned here.

Pockels effect, named after the physicist F. C. A. Pockels, is a second-order nonlinear optical process where the light is doubly refracted by a crystal when applying an external electric field. This birefringence is proportional to the electric field applied.

The occurrence of the *SFG* is also a second-order nonlinear optical effect: light of two different frequencies, when shining on a nonlinear optical medium, will generate light of a new frequency, which is the sum of the two original frequencies of the incident light (see Section 13.3.3.1, Figure 13.5).

The most important effect based on the first hyperpolarizability β is the *frequency doubling* (SHG), which is explained in more detail in the following Section 13.3.3.1. The coefficient β provides information about the efficiency of a nonlinear optical material in generating such a second harmonic.

13.3.3.1 Second-Harmonic Generation

In 1961, Franken et al. discovered for the first time the generation of the SHG by passing a beam of intensive monochromatic red light (wavelength 694 nm) of a ruby laser through a quartz crystal. Not only the wavelength of the irradiate red light beam, but also the light with double the frequency, could be detected, that is, half-wavelength (ultraviolet, 347 nm). Figure 13.4 shows a schematic representation of the SHG.

Figure 13.5 explains the origin of the second harmonic: the nonlinear optical material interacts simultaneously with two photons with the same frequency ν, which are "combined" to a new photon, resulting in the emission of light with twice the energy and a frequency of 2ν. The Planck relation between the energy of a photon E and the frequency ν is $E = h \cdot \nu$, where h is the Planck

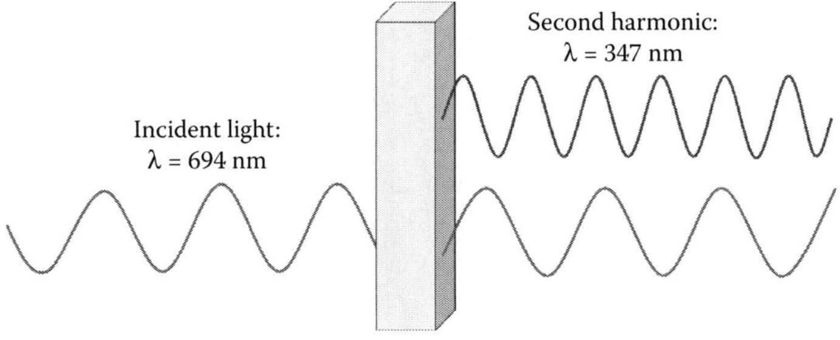

FIGURE 13.4 SHG on a nonlinear optical material.

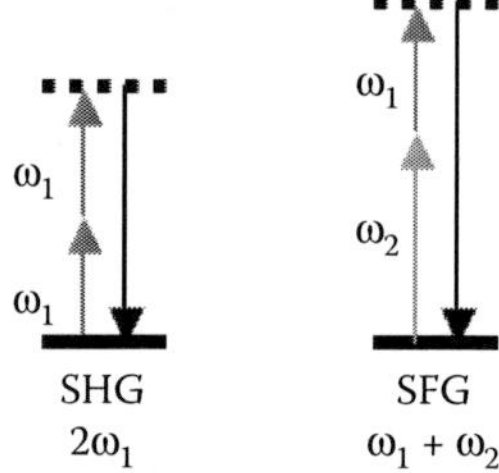

FIGURE 13.5 Second harmonic generation (SHG) and sum-frequency generation (SFG).

constant. The correlation between the wavelength λ and the frequency ν is $\nu = c/\lambda$; c is the speed of light. Expressing the frequency by the angular frequency $\omega = 2\pi\nu$, the reduced Planck constant is used ($\hbar = h/2\pi$). Insertion of the two terms into the Planck equation gives the relation between the energy and the angular frequency: $E = \hbar \cdot \omega$, and hence when the nonlinear optical material is excited by two photons of the same frequency the emitted light has double the energy.

The SHG finds application in the development of new optical information storage systems where near-infrared laser light can be converted into deep-blue light. The capacity for storage information on optical discs rises strongly by frequency doubling (Prasad and Williams, 1991).

13.3.3.2 Improvement of the Hyperpolarizability

The hyperpolarizability β of organic chromophores can be increased by the improvement of molecular parameters, for example, the length and nature of the π-conjugated system and the electron donor and acceptor strengths (Marder et al., 1991).

A variety of investigations of *donor and acceptor properties* resulted in the following series (Nalwa and Miyata, 1997):

$$\text{Donor strength: } OH < Br < OC_6H_5 < OCH_3 < SCH_3 < N_2H_3 < NH_2 < N(CH_3)_2$$
$$\text{Acceptor strength: } SO_2CH_3 < CN < CHO < COCF_3 < NO < NO_2 < CH(CN)_2$$

By means of electric-field-induced SHG (EFISH)-experiments (for a description of this method see Section 13.3.5), the influence of substituents in the *para*-configuration on the hyperpolarizability has been investigated, for example, by Dulcic and Sauteret (1978): for example, bis-alkylation of the amino group of 4-nitroaniline increases the β value.

Not only does the choice of donor and acceptor substituents, respectively, influence the hyperpolarizability, but the *π-bridge* can also be diversified in length and nature. Studies related to the π-conjugation length were carried out by Huijts and Hesselink (1989) based on EFISH experiments with CT-chromophores bearing a methoxy group as an electron donor and a nitro substituent as an acceptor. The results of these investigations demonstrated that the hyperpolarizabilities rise with the number of π-bonds between electron-donating and electron-withdrawing moieties. It is not possible, however, to argue that the extension of conjugation length should be maximal because with growing size of the molecule there will be a loss of transparency, as the molecules will absorb light of increasing wavelength. An *efficiency-transparency-trade-off* should be made to reach a compromise between high hyperpolarizability on one hand and sufficient transparency in the laser operating wavelength region and the region of frequency doubling on the other hand (Burland et al., 1992; Papagni et al., 2002).

A higher information density in optical data-storage systems can be obtained by shortening the wavelength of the applied laser. Thus, nonlinear optical organic materials that are able to operate with near-infrared lasers (820–880 nm GaAs, AlGaAs diode lasers, or Ti:sapphire lasers) are of substantial interest. The goal is to design organic chromophores that have high nonlinearities and full transparency at 820–880 nm and above 410 nm (Papagni et al., 2002).

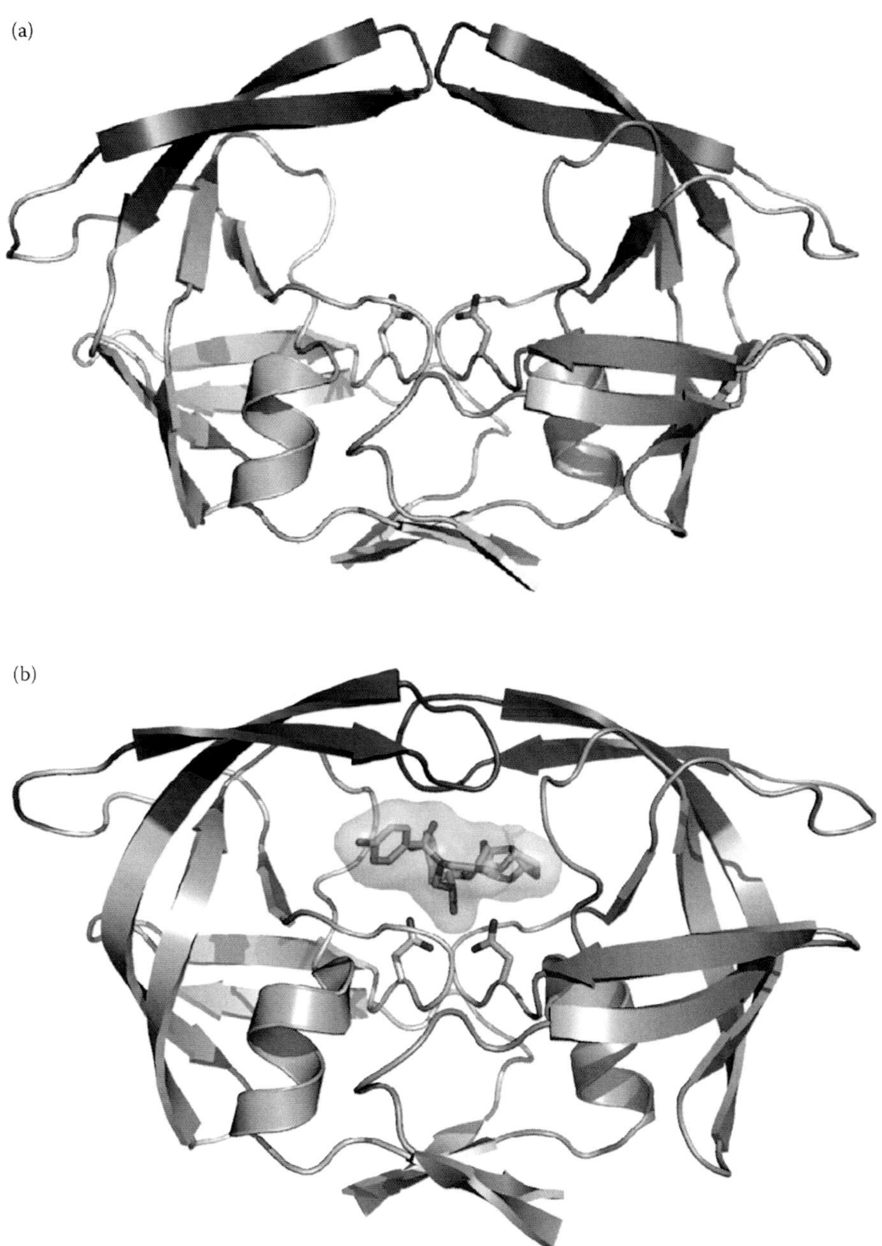

FIGURE 3.2 Three-dimensional structure of HIV protease (HIV PR) in open and closed conformation. (a) Overall structure of the apo-form of HIV PR. One monomer is colored blue and the second in pink. Two catalytic aspartates are represented by sticks with carbon color corresponding to the chain color and oxygen atoms colored red. Flaps (residues 43–58) in open conformation are highlighted in darker colors. (b) Overall structure of HIV PR in complex with the FDA-approved drug darunavir bound in the active site. Active site-bound darunavir is shown as a stick model (carbon atoms in green, sulfur in yellow, oxygens and nitrogens in red and blue, respectively) with its solvent accessible surface in semitransparent green. The figure was generated with PyMol (DeLano, 2002) using the structure of free HIV-1 PR [PDB code 1HHP (Spinelli, 1991)] and the structure of a highly mutated patient-derived HIV-1 PR [PDB code 3GGU (Saskova, 2009)].

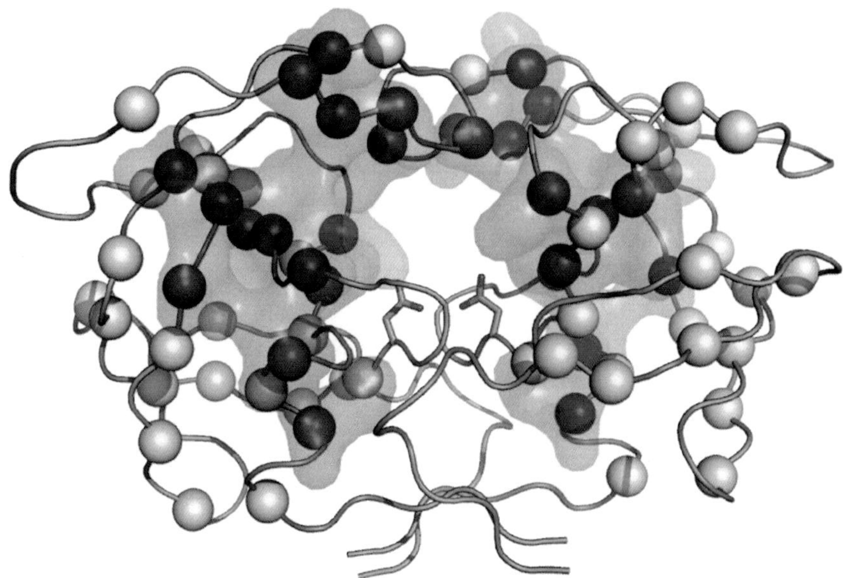

FIGURE 3.3 HIV PR resistance-associated mutations. The three-dimensional crystal structure of the HIV PR dimer depicting mutations associated with resistance to clinically available protease inhibitors (Johnson, 2008). Mutated residues are represented by their Cα atoms (spheres) and colored in red and yellow for major and minor mutations, respectively. For major mutations in residues affecting substrate and/or inhibitor binding, the transparent solvent accessible surface is also shown in red. Active site aspartates are represented in stick models; the inhibitor bound in the enzyme active site is omitted from the figure for clarity. The figure was generated from the structure of highly mutated patient-derived HIV-1 PR [PDB code 3GGU (Saskova, 2009)] with the program PyMol (DeLano, 2002).

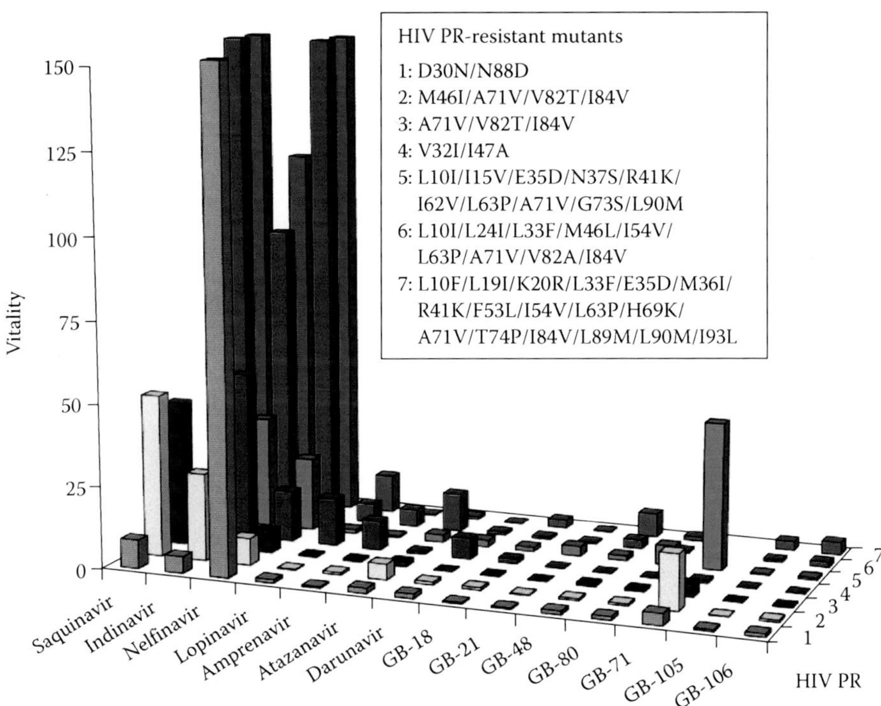

FIGURE 3.5 Metallacarboranes inhibit resistant HIV PR variants. Vitality values of seven clinical inhibitors and seven cobaltacarborane compounds analyzed with the panel of HIV-1 PR-resistant species (Kozisek, 2008; Řezáčová, 2009). Mutations in the HIV-1 PR variants are shown in the figure inset. The vitality is a measure of the enzymatic fitness of a particular mutant in the presence of a given inhibitor and is defined as $(K_i\ k_{cat}/K_m)_{MUT}/(K_i\ k_{cat}/K_m)_{WT}$, where K_i is inhibition constant, K_{cat} and K_m are enzymatic constants and MUT and WT is mutated and wild-type enzyme variant, respectively (Gulnik, 1995).

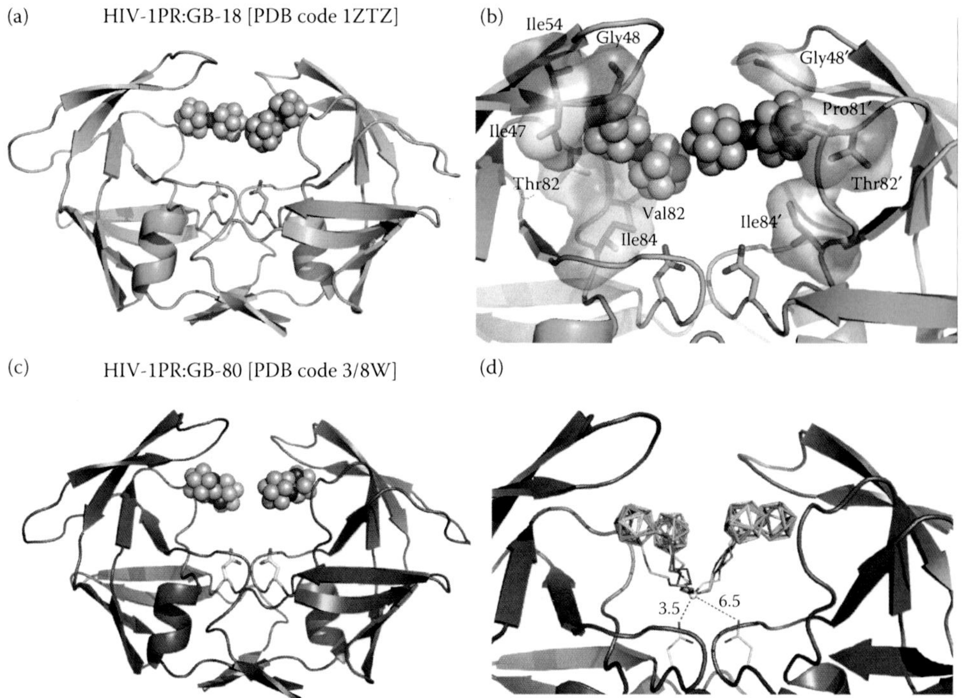

FIGURE 3.6 Metallacarborane binding to HIV PR. Crystal structures of wild-type HIV-1 PR in complex with the parent cobalt bis(1,2-dicarbollide) ion (GB-18) and compound GB-80 containing two parent clusters connected with a linker. (a) Overall structure depicting the asymmetric binding of two cobalt bis(1,2-dicarbollide) clusters of GB-18 into the symmetric HIV PR dimer. The enzyme is shown by means of animation with two catalytic aspartates shown in sticks. The metallacarborane atoms are depicted as spheres. (b) Detail of the GB-18 inhibitor:enzyme binding. The cluster occupies a hydrophobic pocket formed by enzyme residues Pro81, Val82, and Ile84 and is covered by the flap residues Ile47, Gly48, and Ile54. This site corresponds approximately to the S3 and S3′ substrate-binding subsites. Interacting residues are highlighted in sticks with their van der Waals surfaces. (c) Overall structure depicting the symmetric binding of two cobalt bis(1,2-dicarbollide) clusters of GB-80. Representation corresponds to panel A. (d) Detail of the GB-80:enzyme interaction. Models of the four most probable conformers of the linker are shown. The dashed lines and numbers represent distances to catalytic aspartates in Å. The figure was generated using the program PyMol (DeLano, 2002).

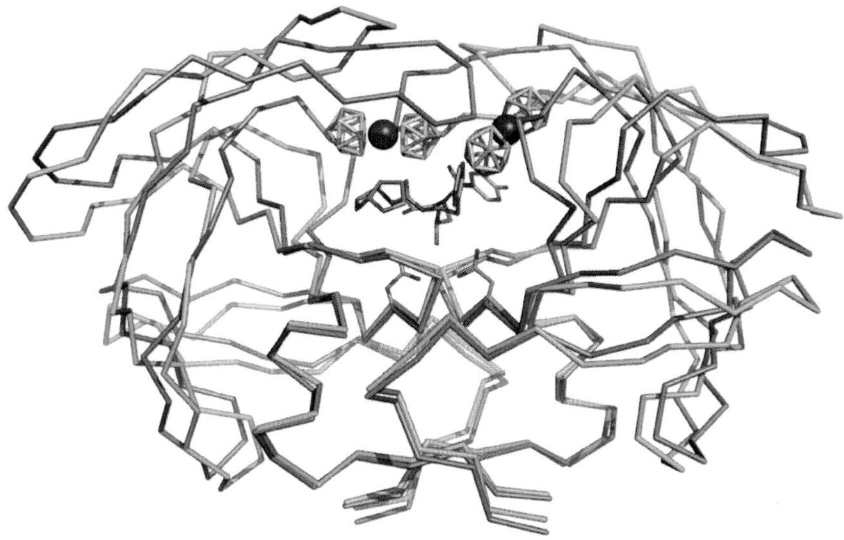

FIGURE 3.7 Comparison of metallacarborane and darunavir binding to HIV PR. Superposition of HIV PR–compound GB-18 complex with HIV PR–darunavir complex structure. Protease complex with darunavir [PDB code 3GGU (Saskova, 2009)] is represented in orange with darunavir shown as a stick model, and color coding for PR–GB-18 complex is the same as in Figure 3.6. The figure was generated using the program PyMol (DeLano, 2002).

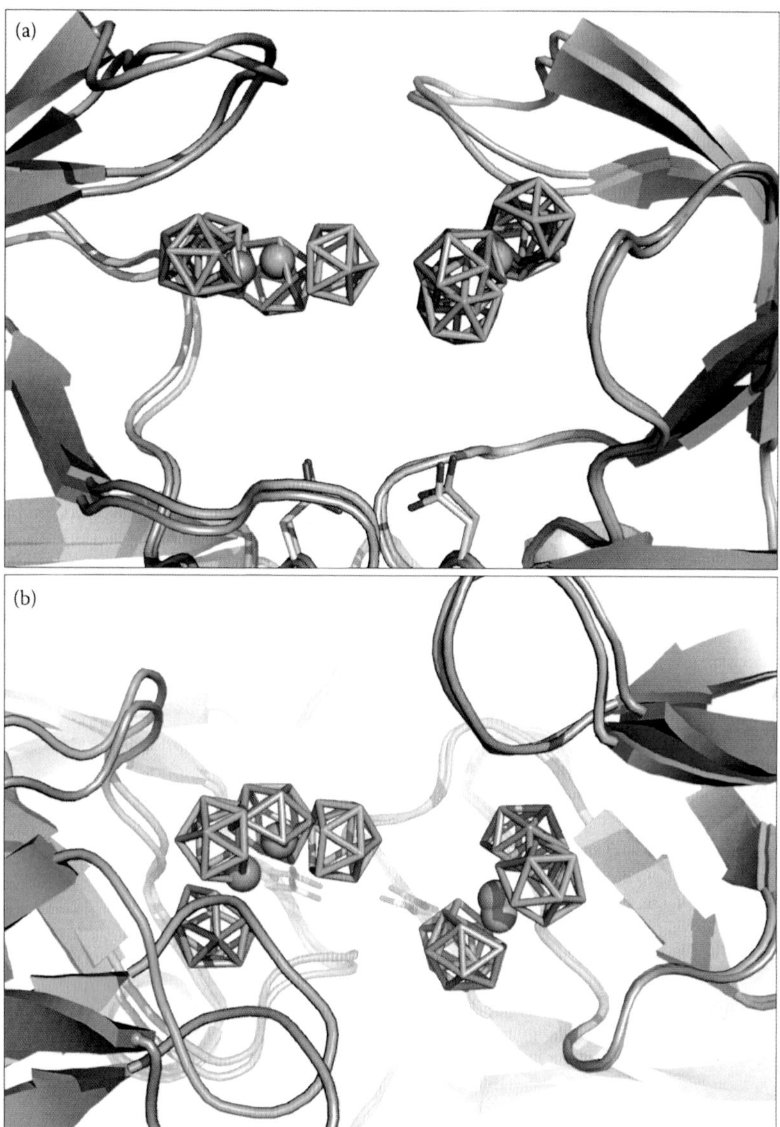

FIGURE 3.8 Comparison of the GB-18 and GB-80:HIV-1 PR complex structures. (a) Close-up top view into the enzyme active site showing differences in the HIV PR:GB-18 (green) and HIV PR:GB-80 (gray) structures. Carbon and boron atoms in cobaltacarboranes are shown as sticks with the central cobalt atom represented by a sphere. (b) Top view into the enzyme active. Representation and color coding are analogous to panel A. The figure was generated using the program PyMol (DeLano, 2002).

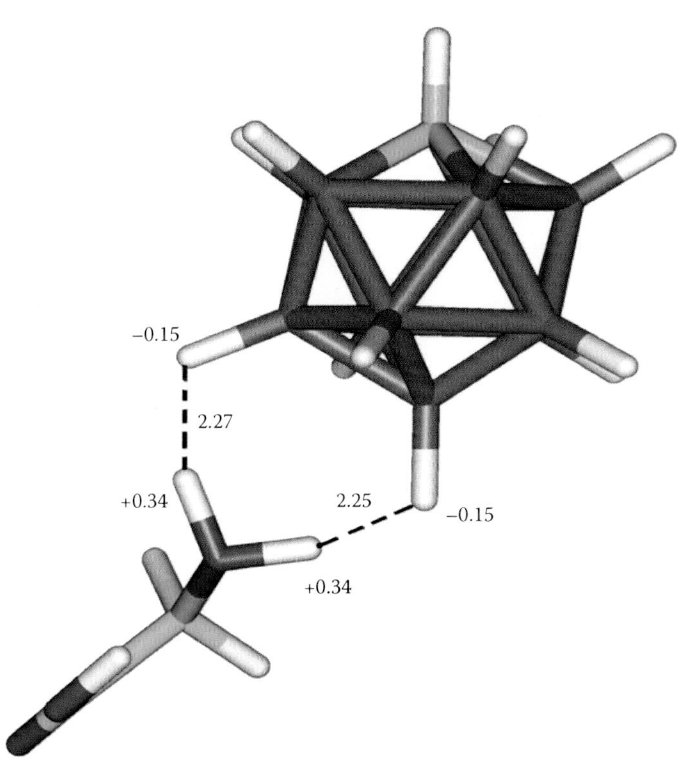

FIGURE 3.10 Interaction between $CB_{11}H_{12}^-$ with glycine via two dihydrogen bonds. Partial charges on the interacting hydrogens as well as distances in Å are shown. Color coding: boron–purple, carbon–green, nitrogen–blue, oxygen–red, hydrogen–white.

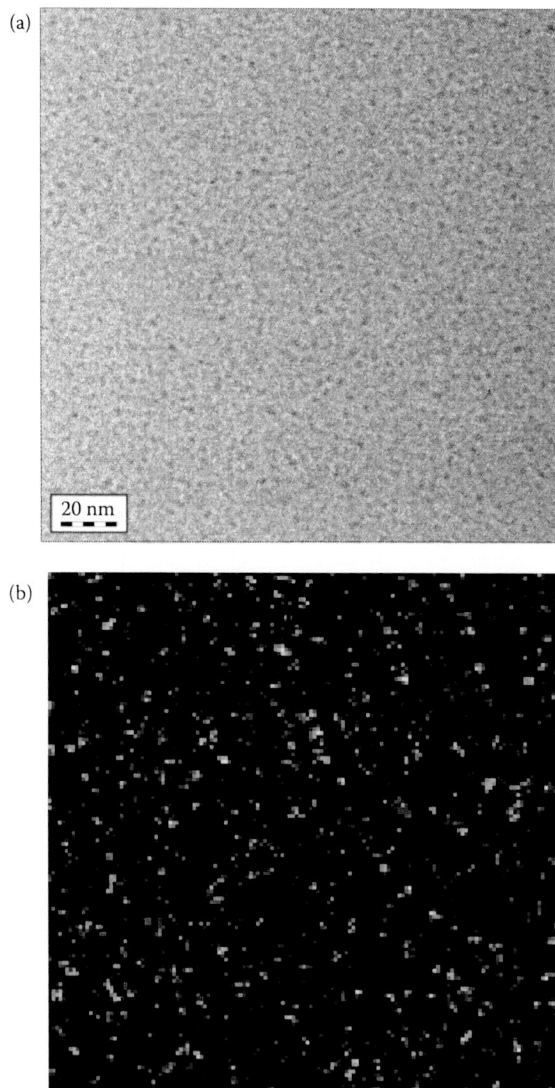

FIGURE 3.11 Aggregation of cobalt dicarbollides in aqueous solutions. (a) Typical cryo-TEM micrograph of sodium cobalt bis(dicarbollide) GB-18 in aqueous solution. (J. Ruokolainen, Helsinki, unpublished work.) (b) Typical fluorescence lifetime imaging (FLIM) picture of fluorescein-cobalt bis(dicarbollide) conjugate (GB-179) aggregates deposited on the bottom of an aqueous solution in which the diameter of the aggregates is roughly 400 nm. (P. Jurkiewicz, Prague, unpublished work; Uchman, 2010a.)

13.3.4 Third-Order Nonlinear Optical Effects

In the case of third-order nonlinear phenomena the materials can be centrosymmetric in its molecular structure. One example is the optical *Kerr effect* reported for the first time by J. Kerr in 1877 and 1878: exposing a material to an electric field the refraction index of an optical medium changes, proportional to the square of the applied field. A double refraction can be generated with the difference between Kerr and Pockels effect being that in the latter case the double refraction is linearly proportional to the electric field.

A further third-order nonlinear optical effect is the *TPA*, where a molecule is able to absorb two photons simultaneously to reach an excited state. The energy difference between the excited and ground state relate exactly to the energy of the photons.

The frequency tripling (*THG*) is a third-order effect as well and was observed in crystals, glasses, and liquids in 1965 by Maker and Terhune and in gases for the first time 1967 by New and Ward. The tripled frequency results in light of one-third of the original wavelength.

13.3.5 Methods for Measuring Nonlinear Optical Properties

The *Kurtz powder technique* developed by S. W. Kurtz and T. T. Perry in 1968 offered the first possibility for the determination of nonlinear optical properties of crystalline organic compounds. In comparison with a standard such as quartz or urea, the capability of materials to generate the second harmonic is detected. The pulverized sample is irradiated by a laser and the intensity of frequency-doubled light is measured. As this method is influenced by many factors, such as particle size, it is only semi-quantitative. The results give no precise values for the hyperpolarizability β. A further disadvantage is that the method is limited to molecules that crystallize in noncentrosymmetric space groups.

The *EFISH method* (Singer and Garito, 1981) permitted for the first time the establishment of a correlation between molecular structure of organic chromophores and the first hyperpolarizability β. In this method an electric field is applied to a solution of the nonlinear optical materials, resulting in an alignment of the dipoles. A direct determination of β with the EFISH method is not possible; the third-order polarization γ is measured, the dipole moment μ must be known, and with these values the hyperpolarizability β can be calculated. The EFISH technique is not readily applied to salts as the solutions conduct electricity.

Clays and Persoons presented a new technique in 1992 for the determination of the hyperpolarizabilities of nonlinear optical compounds in solution using the *hyper-Rayleigh scattering* (*HRS*). The values for first hyperpolarizabilities can be obtained from second-order scattered light intensity. This method has several advantages over the EFISH technique: it needs no electric field, and thus the experiment is simplified, the hyperpolarizability β can be directly measured, and the β values for salts can be determined as well.

13.4 BORON DERIVATIVES FOR NLO

13.4.1 Polyhedral Boron Cluster Derivatives for NLO

Various calculations and investigations of icosahedral borane derivatives, such as dodecahydro-*closo*-dodecaborates(2-), which have electron-donating properties, and *closo*-carboranes, as electron acceptors, have demonstrated very interesting optical behaviors. Both clusters are promising candidates for nonlinear optical materials because they have some exceptional advantages: in addition to a remarkable thermal and chemical stability, they have a high degree of electron delocalization within the polyhedral cages and a large number of polarizable electrons. In 1954, Lipscomb et al. developed the concept of three-center-two-electron (3c-2e) bonding for boranes. Neutral *closo*-carboranes $C_2B_{10}H_{12}$ and *closo*-dodecaborates $B_{12}H_{12}^{2-}$ are isoelectronic and they exhibit unusual high stability, and thus the concept of aromaticity can be extended to these boron deltahedra (Eberhardt et al., 1954; Aihara, 1978; Schleyer and Najafian, 1998; Jursic, 1999; King, 2001a,b).

In this chapter, the polyhedral boron clusters are represented such that every corner carries one boron which is connected to one hydrogen atom, unless a substituent is drawn explicitly. In the case of carboranes the boron is partially substituted by a carbon.

13.4.1.1 Results of Quantum Mechanical Calculations for Boron Clusters

The most commonly used model for understanding the relationship between the hyperpolarizability β and molecular structure is the two-state model (Oudar and Chemla, 1977). This model provides only a rough description but it allows a qualitative understanding of the nonlinear optical properties of molecules. For push–pull compounds with electron donors and acceptors, the β value depends mainly on the intramolecular polarization, the oscillator strength, and the excited state of the material (Allis and Spencer, 2001). There are three different possibilities for a refinement of the two-state model: semi-empirical, *ab initio* and density functional theory methods (Li et al., 1992; Kanis et al., 1994; Kurtz and Dudis, 1998).

Several theoretical studies were carried out with regard to second-order nonlinear effects for carboranes and boranes and some examples are noted below.

Abe et al. (1998) have performed *ab initio* calculations for an electron-rich dodecaborate directly connected to electron-deficient 1,2-, 1,7-, and 1,12-dicarba-*closo*-dodecaboranes, respectively, which are also known as *ortho*-, *meta*-, and *para*-carboranes, and they determined high hyperpolarizabilities between 3.4 and 3.6×10^{-30} cm^5 esu^{-1} (Figure 13.6).

The problem with most chromophores with large second-order nonlinear responses is that an extension of the π-conjugation length between the electron donor and the acceptor increases the hyperpolarizability β, but at the same time causes a bathochromic shift of the CT absorption band, and thus the prerequisite of high transparency to visible light is no longer fulfilled. The inorganic dodecaborate–carborane conjugate provided a solution to this problem because the CT is possible by through-space interaction and thus they should be highly transparent (Abe et al., 1998), similar to both parent species.

Murphy et al. (1993b, Part 2) determined the β values for the hyperpolarizability of neutral diphenylsubstituted *o*-carboranes by semi-empirical quantum chemical ZINDO-SOS computational calculations. The results are mentioned in Section 13.4.1.2 in conjunction with further investigations of the prepared *ortho*-carborane derivatives.

The structures and nonlinear optical properties for tropylium-*nido*-, *conjuncto*-, and *closo*-borane derivatives, as well as for the *ortho*-, *meta*-, and *para*-carboranes, have been calculated by semi-empirical methods (Littger et al., 2000; Allis and Spencer, 2000, 2001; Taylor et al., 2001).

In 2000, Allis and Spencer carried out theoretical studies of new classes of molecules that are based on charged aromatic electron donors and acceptors linked through polyhedral cluster bridges. These compounds displayed large calculated nonlinear optical responses; a *closo*-carborane derivative which exhibits an extraordinary hyperpolarizability is shown in Figure 13.7 (left side). The β value calculated with semi-empirical methods for this tropylium- and cyclopentadienyl-substituted *p*-carborane is 998×10^{-30} cm^5 esu^{-1}.

A precursor of a polyhedral *para*-carborane with tropylium- and cyclopentadienyl-moieties was described by Taylor et al. (2001) (see Section 13.4.1.2).

Calculations have been carried out for various polyhedral clusters as a function of spacer between donor and acceptor substituents: *ortho*-, *meta*-, and *para*-carboranes and monocarborate. The

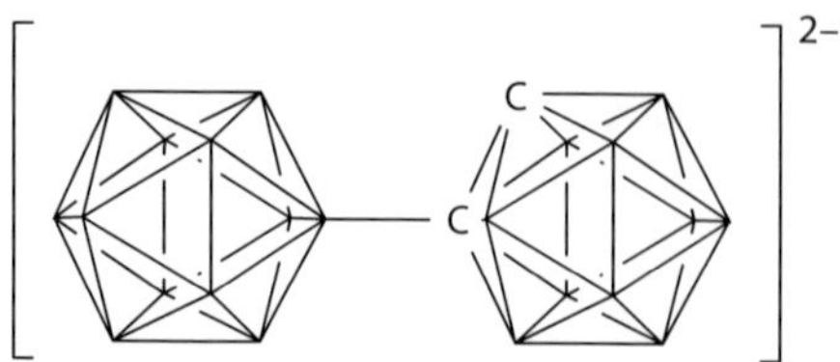

FIGURE 13.6 Dodecaborate-*o*-carborane dyad. (Adapted from Abe, J. et al., *Inorg. Chem.* **1998**, 37, 172.)

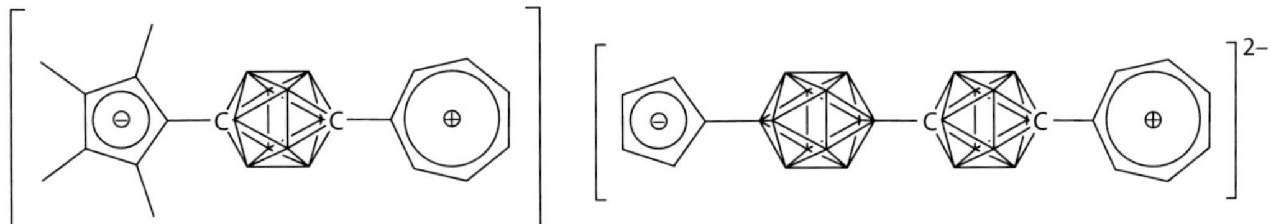

FIGURE 13.7 Calculated neutral *para*-[1-$(C_7H_6)C_2B_{10}H_{10}$-12-(C_5Me_4)] and anionic *para*-[1-$(C_7H_6)C_2B_{10}H_{10}$-12-$B_{12}H_{10}(C_5H_4)]^{2-}$. (Adapted from Allis, D. G., J. T. Spencer, *J. Organomet. Chem.* **2000**, 614, 309.)

hyperpolarizabilities were shown to range from 8.6 to 1226.0×10^{-30} cm^5 esu^{-1}; the highest value was exhibited by the anionic dodecaborate-carborane conjugate with sigma-bonded tropylium- and cyclopentadienyl-substituents in *para*-position (Figure 13.7, right side).

For a dodecaborate-containing analogue of the neutral 1,2-dicarba-*closo*-dodecaborane (Figure 13.7, left side) semiempirical calculations have been performed for the dodecaborate as polyhedral cluster bridge by Allis and Spencer in 2001 as well. For the *para*-substituted dodecaborate a hyperpolarizability of 1208.7×10^{-30} cm^5 esu^{-1} was determined (Figure 13.8).

In addition to *ortho*-, *meta*-, *para*-carboranes, and dodecaborates, further molecules with polyhedral borane bridges were calculated and they all displayed large nonlinear optical responses. An explanation for the large hyperpolarizabilities of these polyhedral boron-based compounds is that the highest occupied molecular orbitals are localized on the electron-donating aromatic rings, for example, the cyclopentadienyl substituent, while the lowest unoccupied molecular orbitals are located on the aromatic acceptor rings, for example, the tropylium residue; thus, they have two relatively independent, highly polarized regions (Allis and Spencer, 2001).

The preparative availability of the theoretically investigated disubstituted *para*-dodecaborates, which are shown in Figures 13.7 and 13.8 will pose problems, because electrophilic substitutions take place not only in the desired *para*-position (1,12-isomer); rather, the 1,7-isomer will be the main product, as predicted by calculations of Hoffmann and Lipscomb (1962): an inductively electron-withdrawing substituent will direct the second substitution to boron 7 and 12 in a ratio 5:1. In the case of an electron-donating substituent, the electrophilic substitution occurs in *meta*-position 2. Examples for this behavior are the halogenation (Knoth et al., 1964) and amination (Hertler and Raasch, 1964) of the B_{12}-species.

13.4.1.2 Prepared Boron Cluster Compounds for NLO

At this point not many nonlinear optical compounds based on boron clusters as electron donors and acceptors, respectively, have been described in the literature. A few examples limited to carborane and dodecaborate derivatives are now introduced.

13.4.1.2.1 Carboranes in Nonlinear Optical Compounds

Preliminary investigations of nonlinear optical behavior of icosahedral carboranes were published by Murphy et al. (1993a, Part 1). *ortho*-Carborane has a strong dipole moment of 4.45 D in the ground state, and thus the insertion of electron-donating and electron-accepting groups possibly

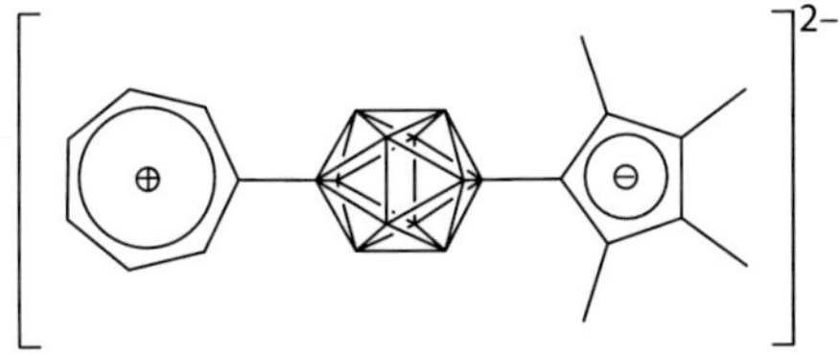

FIGURE 13.8 Calculated dianion $[1-(C_7H_6)B_{12}H_{10}-12-(C_5Me_4)]^{2-}$. (Adapted from Allis, D. G., J. T. Spencer, *Inorg. Chem.* **2001**, 40, 3373.)

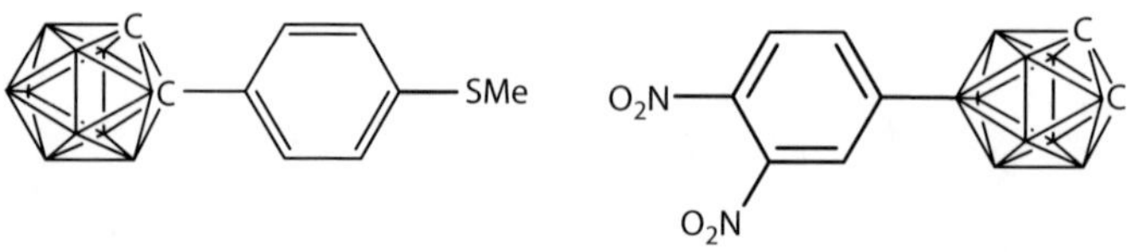

FIGURE 13.9 1-(4-Methylthiophenyl)carborane and 9-(3,4-dinitrophenyl)carborane. (Adapted from Murphy, D. M., D. M. P. Mingos, J. M. Forward, *J. Mater. Chem.* **1993a**, 3, 67 (Part 1).)

enhance this dipole moment. EFISH-measurements of phenylsubstituted *o*-carboranes, where the substitution of the carborane takes place at one carbon atom, one boron atom or at both atoms, have been carried out. Two examples of such phenylsubstituted *o*-carboranes are shown in Figure 13.9.

Results of EFISH-experiments confirmed that carboranes have hyperpolarizabilities, but not as large as for donor- and acceptor-bearing aromatic organic compounds; 1-(4-methylthiophenyl) carborane exhibits a β value for the hyperpolarizability of 5×10^{-30} esu and the 9-(3,4-dinitrophenyl)carborane of 3.5×10^{-30} esu.

Further investigations of the nonlinear optical properties of neutral phenylsubstituted *o*-carboranes, where the carboranes are substituted on both carbon atoms have been described by Murphy et al. as well (1993b, Part 2). One example of a diphenylsubstituted carborane is shown in Figure 13.10.

The β value for the hyperpolarizability of this compound obtained from semiempirical quantum chemical ZINDO-SOS computational calculations is 13.2×10^{-30} cm^5 esu^{-1}, but unfortunately the 1,2-diphenylcarboranes tend to crystallize centrosymmetrically, and hence the determination of the SHG efficiency in comparison with urea is usually not possible because of symmetry reasons. 1,2-Bis(4-nitrophenyl)carborane shows an SHG efficiency 0.14 times of that of urea.

The linkage of a 1-carba-*closo*-dodecaborate(1-) with tropylium at the boron in position 12 was described by Grüner et al. (1999, 2000) (Figure 13.11). For this overall neutral colorless monocarborate derivative a high second-order polarizability of 7.2×10^{-30} esu was determined and this zwitterion is transparent above a wavelength of 400 nm.

Interestingly, compounds containing two acceptors (*ortho*-, *meta*-, *para*-carboranes, and fullerene)—which nevertheless differ in the strength of electron-accepting properties—connected via a phenylethinyl-π-system display unexpectedly high hyperpolarizabilities (Lamrani et al., 2000). The comparison of *o*-, *m*-, and *p*-carboranes shows that the *p*-carborane derivative exhibits the highest β value (1189×10^{-30} esu) (Figure 13.12).

Cyclic voltammetry measurements elucidated that in this kind of molecule the carborane behaves as a donor because of its smaller acceptor strength compared with the fullerene.

Subsequently, Tsuboya et al. (2002) investigated ethenyl- and iminylphenyl-π-bridged *ortho, meta*, and *para*-carborane-ferrocene dyads in which a real donor–acceptor system is present. Figure 13.13 shows two examples of such ferrocene–carborane dyads.

Tsuboya et al. investigated the nonlinear response of these compounds by the HRS technique and compared them with dyads bearing a nitro group instead of a carborane. The *trans*-ethenylphenyl-

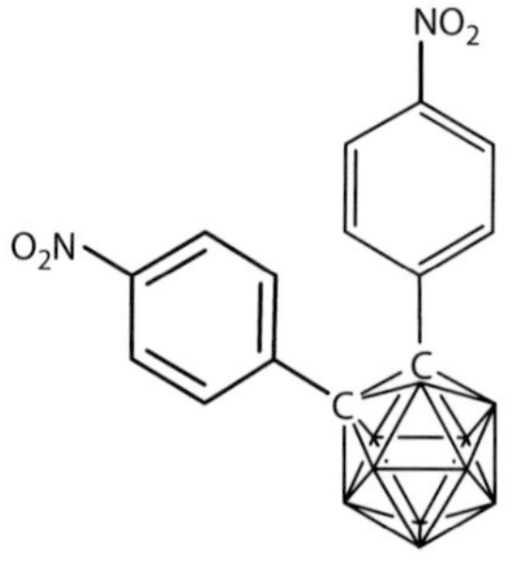

FIGURE 13.10 1,2-Bis(4-nitrophenyl)carborane. (Adapted from Murphy, D. M. et al., *J. Mater. Chem.* **1993b**, 3, 139 (Part 2).)

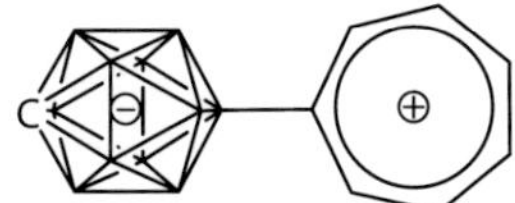

FIGURE 13.11 Tropylium-substituted 1-carba-*closo*-dodecaborate(1-). (Adapted from Grüner, B. et al., *J. Am. Chem. Soc.* **1999**, 121, 3122.)

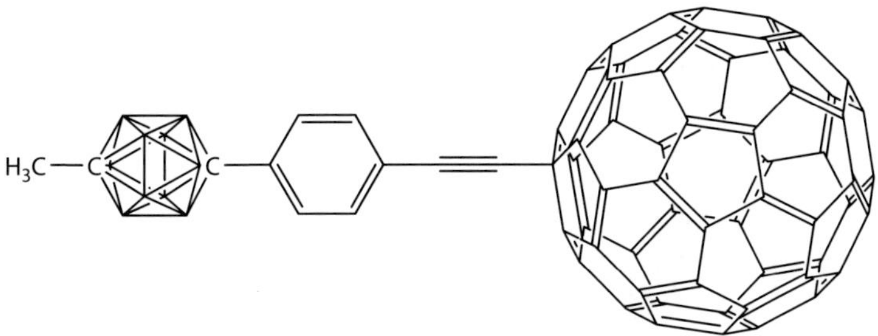

FIGURE 13.12 Fullerene-*p*-carborane dyad. (Adapted from Lamrani, M. et al., *Chem. Commun.* **2000**, 1595.)

bridged carborane derivatives show large hyperpolarizabilities in the range of 110–130 × 10^{-30} esu, slightly smaller than that of the nitro-substituted molecule ($\beta = 152 \times 10^{-30}$ esu). When changing the olefin to an imine spacer the hyperpolarizabilities become smaller: the highest β value is shown for the *para*-carborane with 107 × 10^{-30} esu, whereas the values for the *ortho*- and *meta*-carborane derivatives range between 53 and 60 × 10^{-30} esu. In all cases the electron-withdrawing ability of the *para*-carborane is better than that of *ortho*- and *meta*-carborane.

Taylor et al. (2001) dealt with the preparation of *closo*-icosahedral carborane-containing compounds for nonlinear optical applications as well. The values for the nonlinear answer of different tropylium $[C_7H_7]^+$ and cyclopentadienide $[C_5H_5]^-$ polyhedral boron derivatives calculated by Allis and Spencer in 2000 proved to be very high (see Section 13.4.1.1); thus, it is of interest as to whether these compounds could also be preparatively accessible. The very stable precursor of a new class of nonlinear optical materials could be prepared (Figure 13.14, right side) and experiments for deprotonation and hydride abstraction to get the zwitterionic molecule were quoted to be in progress, but are yet to be reported.

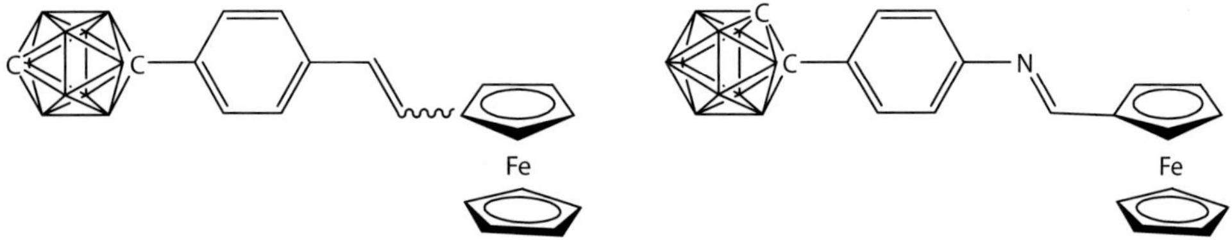

FIGURE 13.13 Carborane–ferrocene-π-conjugated dyads (left side: ethenylphenyl-, right side: iminylphenyl-bridged). (Adapted from Tsuboya, N. et al., *J. Mater. Chem.* **2002**, 12, 2701.)

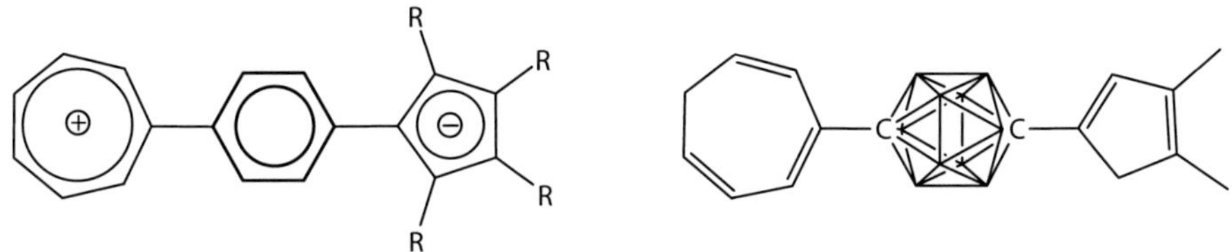

FIGURE 13.14 Organic [5.6.7]quinarene derivative and the precursor for a tropylium-cyclopentadienide derivative of *p*-carborane. (Adapted from Taylor, J. et al., *Inorg. Chem.* **2001**, 40, 3381.)

After removing a hydride of the cycloheptatriene and a proton of cyclopentadiene the bis-substituted *p*-carborane can be compared with the analogous organic [5.6.7]quinarene derivative shown in Figure 13.14 (left side), but there is one major difference between them: the bridging benzene unit allows electronic communication between the tropylium and the cyclopentadienide unit and thus a fully electron-delocalized system exists, while the boron-based polyhedral bridge restricts the electron transport, which leads to a highly polarized molecule (Taylor et al., 2001).

13.4.1.2.2 Dodecaborates in Nonlinear Optical Compounds

Molecules with dodecaborate as an electron-donating substituent in nonlinear optical systems have been investigated much less intensively. Bernard et al. (2005) described noncentrosymmetric π-conjugated systems containing the dodecahydro-*closo*-dodecaborate as a donor and a cyano group as an acceptor substituent (Figure 13.15). The connection of the cluster with the π-system takes place via a Schiff base similar to the procedure described by Sivaev et al. in 1999. The donor effect of the dodecaborate could be verified by comparing the UV–Vis absorption of these cluster-containing Schiff bases with the boron-free molecule having a methyl group instead of the cluster. In Figure 13.15, the first dodecaborate-containing compounds with TPA properties are shown.

In 2006, Bernard et al. also examined centrosymmetric boron cluster-containing molecules. Although these molecules did not show second-order nonlinear optical effects because of their centrosymmetry, they are interesting candidates for third-order TPA nonlinear processes. Centrosymmetric materials are more suitable for TPA applications in the visible range than noncentrosymmetric ones, because of their lower coloration (Wang et al., 2001). The presence of donor substituents at each arm of tripodal systems increases the TPA properties (Brunel et al., 2001). Bernard et al. described, for the first time, branched octupolar systems with three dodecaborate units (Figure 13.16).

A UV–Vis absorption spectrum of this branched molecule showed that it is transparent in the visible range, an important criterion for application as a TPA material. Unfortunately, however, the Schiff bases of centrosymmetric π-systems—in contrast to the imines of noncentrosymmetric ones—are hydrolytically quite instable, and thus they are not suitable for applications as nonlinear optical materials. The difference in chemical stability might lie in the possibility of electron transfer and hence stabilization by conjugation of π-electrons in noncentrosymmetric dodecaborate derivatives (Bernard et al., 2006).

A higher stability than that of the imines is desirable; a direct connection of boron to carbon of a π-conjugated system is much more stable than Schiff bases. Peymann et al. (1998) were the first to describe palladium-catalyzed couplings with monoiodinated dodecaborate and methyl, phenyl, and *n*-octadecyl Grignard reagents.

Recently, based on this opportunity, dodecaborate-containing donor–acceptor compounds with nitro-, perfluorophenyl-, and *o*-carborane as acceptors (Figure 13.17) could be achieved by palladium-catalyzed coupling reactions (Vöge, dissertation 2009). Papagni et al. (2002) prepared pentafluoro stilbenes with dimethyl- and diphenylamino substituents as electron donors. They demonstrated that in these push–pull compounds the incorporated perfluorinated aromatic system serves as an inductive electron-withdrawing acceptor and hence this perfluoro stilbene seemed to be a possibility of a connection with the dodecaborate (Figure 13.17).

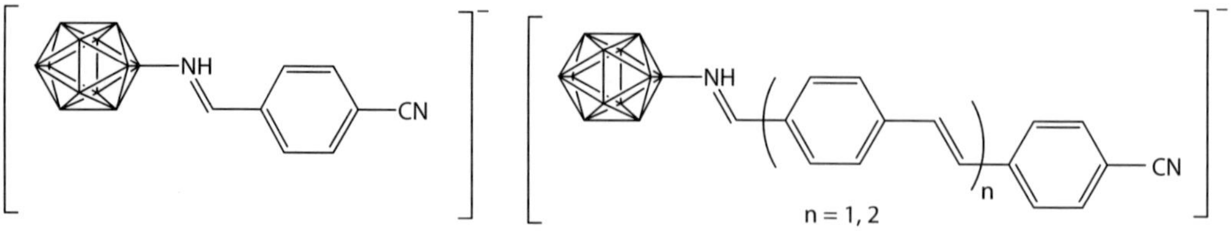

NH CN NH CN n n = 1, 2

FIGURE 13.15 Noncentrosymmetric π-conjugated dodecaborate-containing Schiff bases. (Adapted from Bernard, R. et al., *Dalton Trans.* **2005**, 3065.)

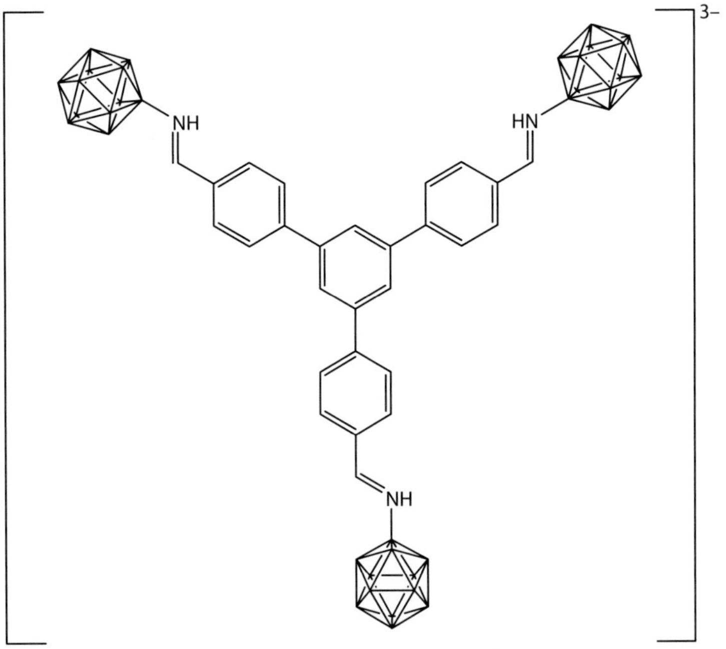

FIGURE 13.16 Centrosymmetric π-conjugated dodecaborate-containing molecule. (Adapted from Bernard, R. et al., *Inorg. Chem.* **2006**, 45, 8743.)

A molecule could be prepared for the first time in which both clusters, dodecaborate as donor and *o*-carborane as acceptor substituent, were connected via a stable π-conjugated system. The advantage of this compound is a high transparency in the wavelength region above 400 nm, yet the hyperpolarizabilities β of these compounds remain to be determined.

The hyperpolarizability β of 2,4-dinitroaniline is 21×10^{-30} esu (Oudar and Chemla, 1977). The insertion of an additional electron-donating dodecaborate at the amino group of this dinitroaniline

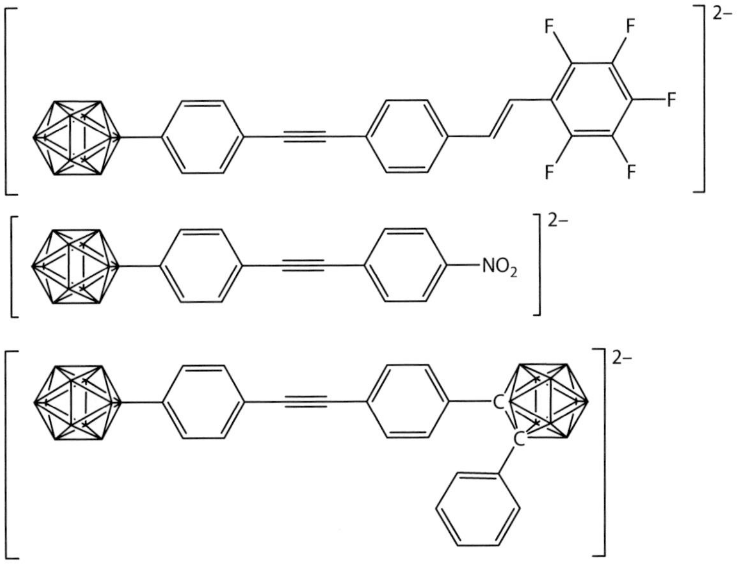

FIGURE 13.17 Dodecaborate-containing π-conjugated donor–acceptor systems. (Adapted from Vöge, A. et al., *J. Organomet. Chem.* **2009**, 694, 1698.)

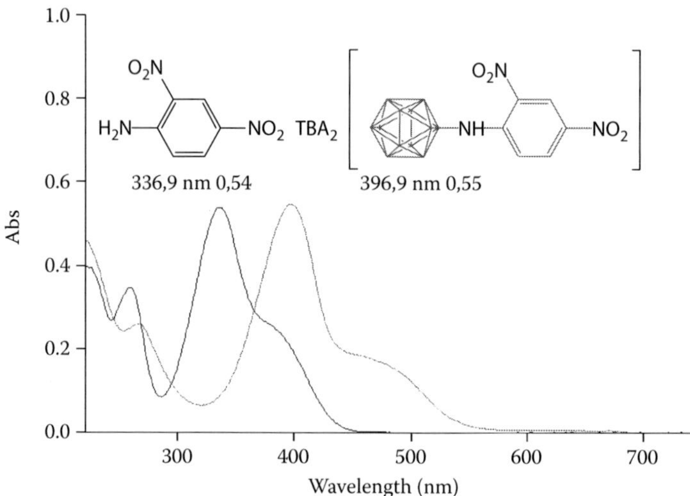

FIGURE 13.18 Comparison of the UV–Vis absorption spectra of 2,4-dinitroaniline and 2,4-dinitrophenyl-aminoundecahydro-*closo*-dodecaborate in acetonitrile (25 μM) (TBA = tetrabutylammonium). (Adapted from Vöge, A. et al., *J. Organomet. Chem.* **2009**, 694, 1698.)

should increase the hyperpolarizability and potentially avoid centrosymmetric crystallization, because the insertion of the negatively charged cluster leads to a salt. As described before with this kind of salts (see Section 13.2.2, Marder et al., 1994), coulombic forces are able to disturb dipole–dipole interactions, which are mainly responsible for centrosymmetric crystallization. The cluster-containing dinitroaniline could be obtained by nucleophilic aromatic substitution of 1-chloro-2,4-dinitrobenzene with deprotonated 1-ammonio-undecahydro-*closo*-dodecaborate (Vöge et al., 2009). The UV–Vis absorption spectra of the dodecaborate-substituted 2,4-dinitroaniline showed a bathochromic shift of 60 nm compared with the unsubstituted 2,4-dinitroaniline (Figure 13.18). This result confirms that the dodecaborate exerts an electron-donating effect in addition to that of the amino group (Vöge, 2009).

The drawback of this stronger CT character is that this red shift leads to less transparency in the region above 400 nm. Usually, Schiff bases and aminated dodecaborates are protonated because of their high pK_a values and hence the conjugate base shows high basicity, which can be attributed to the strong electron-donating ability of the cluster (Hertler et al., 1964). Analysis using ^{1}H NMR and ESI mass spectrometry showed that in the case of the cluster containing dinitroaniline the nitrogen is no longer protonated, because of the strongly electron-withdrawing nitro groups. An electronic interaction between the cluster and the aromatic ring is probable, as indicated by their intense color.

Unfortunately, x-ray analysis elucidated that this compound crystallizes in a centrosymmetric $P2_1/c$ space group, despite the fact that it is a salt.

13.4.2 Tri- and Tetracoordinate Boron in NLO

13.4.2.1 Tricoordinate Boron

Over recent decades, a large number of electron-withdrawing trivalent boron-containing materials with large second-order nonlinearities have been prepared and investigated. The first theoretical calculations of organoboranes were made in 1991 by Kanis et al. who described the application of the semiempirical ZINDO method for calculating the quadratic hyperpolarizabilities of unusual inorganic boron-containing substituents in stilbene chromophores as shown in Figure 13.19.

The calculations show that a nitro group and the -BR_2 group are similar in their electron-accepting properties, as the hyperpolarizabilities are related: for chromophores bearing boron dichloride the β value of 45×10^{-30} cm^5 esu^{-1} is nearly as great as for a nitro group ($\beta = 55.5 \times 10^{-30}$ cm^5 esu^{-1}).

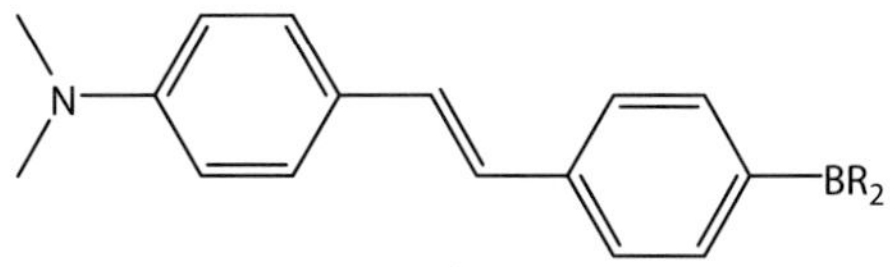

R = H, F, Cl

FIGURE 13.19 Calculated tricoordinate boron as an acceptor substituent in stilbene chromophores. (Adapted from D. R. Kanis, M. A Ratner, T. J. Marks, *Chem. Mater.* **1991**, 3, 19.)

For achieving moisture- and air-stable organoboranes, the use of 2,4,6-trimethylphenyl (= mesityl) groups has been established; the two bulky electron-rich moieties prevent hydrolysis by kinetic hindrance of the *p*-orbital. For this reason the organoboron derivatives described below all bearing two mesityl groups.

Push–pull *E*-alkenes with nonlinear optical properties and with dimesitylborane as π-electron acceptor were described by Yuan in 1990 (Figure 13.20).

Measurements of second-order hyperpolarizabilities gave high β values between 2.6 and 18×10^{-30} esu for these alkenes and they all showed transparency at wavelengths above 400 nm. The vinyl borane with the tricoordinate phosphorus as an electron-donating group seems to be a particularly interesting candidate for application as a nonlinear optical material because of its crystallization in the desired noncentrosymmetric space group $P2_12_12_1$.

Lequan et al. (1991) carried out various investigations with dimesitylborane derivatives as well. They described the preparation and characterization of two trivalent boron-containing dimesitylboranes, [4-(dimethylamino)biphenyl-4′-yl]dimesitylborane (BNB) and [4-(dimethylamino)-phenylazophenyl-4′-yl]dimesitylborane (BNA) (Figure 13.21).

In 1992, they published the results of EFISH measurements of these compounds. The measured hyperpolarizability coefficients for BNB and BNA are quite high: β is 42×10^{-30} and 72×10^{-30} esu, respectively (Lequan et al., 1992a). The experimental studies confirmed the calculations made by Kanis et al. in 1991: the dimesitylborane group has an electron-accepting strength comparable to a nitro group.

In the same year, Lequan et al. presented [4-(dimethylamino)stilben-*E*-4′-yl]dimesitylborane (BNS), dimesitylboron-containing *E*- and *Z*-stilbenes (Figure 13.22), and the measurement of their second-order optical nonlinearities (Lequan et al., 1992b).

The *E*-stilbene derivative exhibited a high hyperpolarizability β of 60×10^{-30} esu, the *Z*-isomer had a lower value. Replacement of the azo-unit (N = N) in BNA by an olefin (C = C) in BNS resulted in a slightly reduced hyperpolarizability.

In 1997, the same working group (Branger et al., 1997) published results with an azo-dye containing dimesitylboryl units used in the side chains in a polyurethane matrix (Scheme 13.2). The main

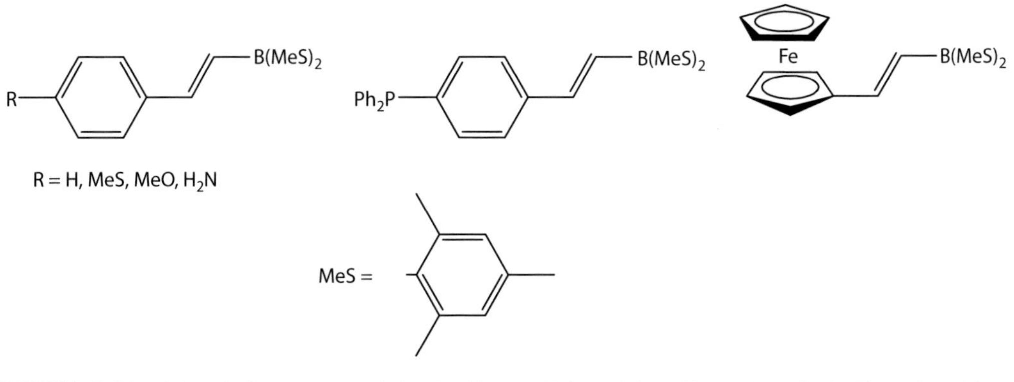

R = H, MeS, MeO, H_2N

MeS =

FIGURE 13.20 Dimesitylborane-containing *E*-alkenes. (Adapted from Yuan, Z. et al., *J. Chem. Soc. Chem. Commun.* **1990**, 1489.)

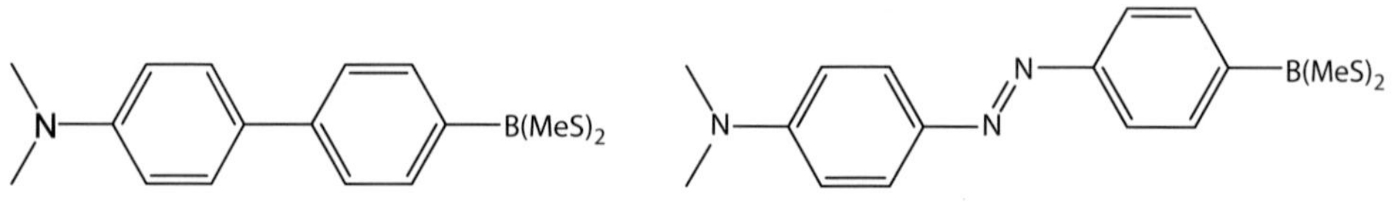

FIGURE 13.21 [4-(Dimethylamino)biphenyl-4′-yl]dimesitylborane and [4-(dimethylamino)-phenylazophenyl-4′-yl]dimesitylborane. (Adapted from Lequan, M. Lequan, R. M. Lequan, and K. Chane-Ching, *J. Mater. Chem.* **1991**, 1, 6, 997.)

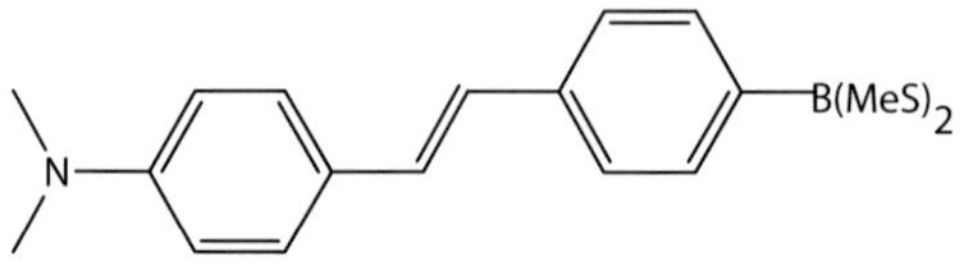

FIGURE 13.22 [4-(Dimethylamino)stilben-E-4′-yl]dimesitylborane. (Adapted from Lequan, M. et al., *Adv. Mater. Opt. Elect.* **1992b**, 1, 243.)

aim of this procedure was to achieve nonlinear optical materials in which the optical properties are maintained for a long time and at high temperatures.

These new polymers show high nonlinear responses and high glass transition temperatures. In comparison with the DR1 azo-dye with a nitro group as electron donor, the optical properties of the boron-containing polyurethanes are slightly better. These polymers seem to be interesting new materials with potential application in NLO.

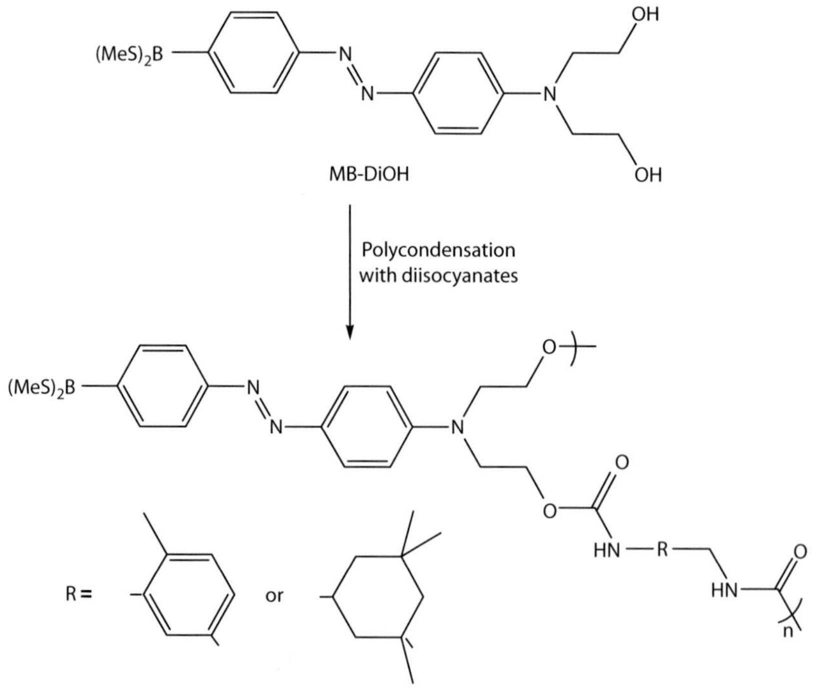

SCHEME 13.2 Polyurethane matrix containing an azo-dye with dimesitylboron as side chains. (Adapted from Branger, C. et al., *Chem. Phys. Lett.* **1997**, 272, 265.)

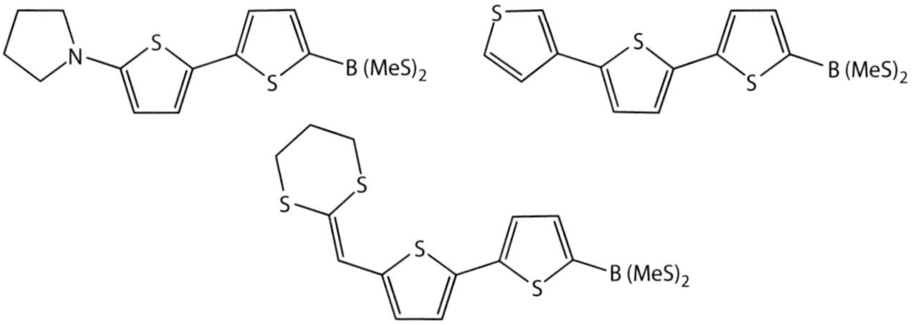

FIGURE 13.23 Dimesitylborane connected via a bithiophene bridge to donor moieties. (Adapted from Branger, C. et al., *J. Mater. Chem.* **1996**, 6, 4, 555.)

The same research group (Branger et al., 1996) described the synthesis and nonlinear optical properties of molecules with dimesitylboron as an electron acceptor, bithiophene as an unsaturated chain, and different substituents as donors (pyrrolidin-1-yl, 3-thienyl and dithianylidene) (Figure 13.23).

EFISH measurements revealed large molecular hyperpolarizabilities of 31 and 37×10^{-30} esu for the pyrrolidine and dithionylidene derivatives, respectively. These investigations again confirmed the good electron-accepting character of the dimesitylboryl group.

13.4.2.2 Tetracoordinate Boron

Tetracoordinate boron has a negative charge and in comparison with tricoordinate boron no longer has a vacant *p*-orbital. As described before, tricoordinate boron acts as a strong electron acceptor; in contrast, the tetracoordinate moiety acts as an electron-donating substituent. Two variations of tetracoordinate boron in materials with interesting nonlinear optical properties are mentioned below; zwitterionic molecules with tetracoordinate negatively charged borate as the electron donor and pyridine/borane Lewis base adducts.

13.4.2.2.1 Why Zwitterions for NLO?

Most of the typical organic molecules with large second-order polarizabilities investigated in the past have the composition described above: a neutral electron donor is linked to an uncharged electron acceptor moiety through a π-conjugated system. In the excited state both electron donor and acceptor are charged (Scheme 13.3).

The lower part of Scheme 13.3 shows the case of a charged electron acceptor, such as phosphonium and ammonium, combined with a neutral donor.

The polarization caused by excitation leads to an increase in the dipole moment ($\Delta\mu = \mu_e - \mu_g$; e = excited state, g = ground state). The hyperpolarizability depends on the change of the dipole moment, and hence materials for nonlinear optical applications should have an electronic transition with a large value for $\Delta\mu$ (the absolute value is relevant) (Chane-Ching et al., 1991).

The correlation between the static first-order hyperpolarizability β_0 and the change of the dipole moment ($\mu_e - \mu_g$) is shown as a sum over states (SOS) approximation (Lambert et al., 1996):

$$\beta_0 \propto \mu_{eg}^2(\mu_e - \mu_g)/\tilde{\nu}_{eg}^2$$

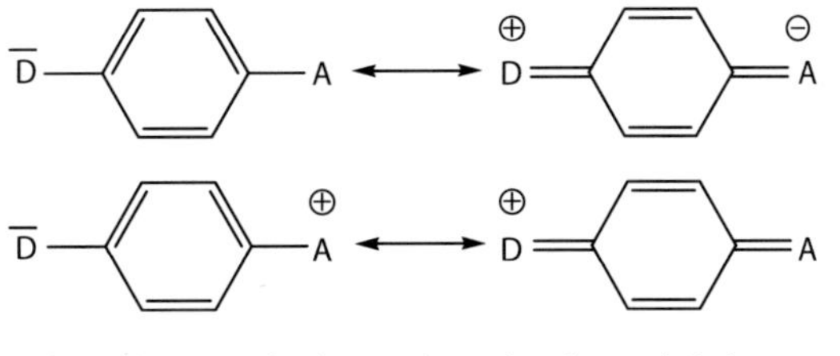

SCHEME 13.3 Examples of nonlinear optical organic molecules and their resonance structures.

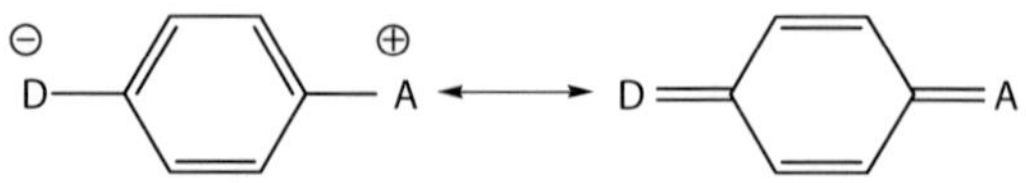

SCHEME 13.4 Resonance structure for a zwitterion.

where μ_{eg} (eg = excited state/ground state) is a transition dipole moment.

To achieve a large hyperpolarizability one possibility is to lower the CT transition energy $\tilde{\nu}_{eg}$, for example, by extending the π-system. The drawback is that by prolonging the conjugation length the transparency gets lost, and therefore it seems to be more appropriate to maximize $\Delta\mu$ by designing transparent zwitterionic molecules (Scheme 13.4).

Zwitterions exhibit a high ground state dipole moment and presumably a small dipole moment in the first excited singlet state (Lambert et al., 1996).

13.4.2.2.2 Borate-Containing Zwitterions Investigated

In 1991, Chane-Ching et al. described a zwitterionic compound with tetracoordinate negatively charged borate as an electron donor and a phosphonium cation with a positive charge as an acceptor (Figure 13.24).

The most common nonlinear optical molecules, with a donor—π-system—acceptor structure are, as a rule, neutral in the ground state and become charged in the excited state by electron transfer, in contrast with {4′-[Methyl(diphenyl)phosphonio]biphenyl-4-yl}triphenylborate (PBB), which has a high dipole moment in the ground state and a lower one in the excited state. The hyperpolarizability β of PBB is comparable to that of *p*-nitroaniline and fortunately PBB shows transparency above 400 nm.

Additional borate zwitterions with ammonium as acceptor have been prepared by Lambert et al. in 1996 (Figure 13.25).

These electron donors and acceptors generate a polarization by an inductive and/or field effect rather than by resonance effects, and hence the internal CT transitions are at shorter wavelengths than those of the analogous donor–acceptor push–pull systems. Therefore, the zwitterions are more transparent. The hyperpolarizabilities of these zwitterions are in the same range as those of the typical donor—π-system—acceptor compounds.

13.4.2.2.3 Tetracoordinate Boron in Pyridine/Borane Lewis Base Adducts

Marder et al. (1994) described stilbazolium salts, for example, *N,N*-dimethylamino-*N′*-methylstilbazolium *p*-toluenesulfonate (DAST) (Figure 13.26a), which exhibit some of the highest reported values for SHG (1000 times more efficient than the standard urea) and the desirable noncentrosymmetric crystallization. Due to their ionic nature, the stilbazolium compounds, are not well suited for poling into polymer hosts (see Section 13.2.2) (Lesley et al., 1998). The idea was to prepare similar compounds that carry no net charge, in order to optimize the poling abilities. Therefore, neutral Lewis acidic borane (BF_3 and $B(C_6F_5)_3$) adducts with pyridine derivatives and dimethylamino and methoxy groups as donors were prepared and investigated by Lesley et al. in 1998 (Figure 13.26b).

In comparison with the *N*-unsubstituted pyridine the hyperpolarizabilities by insertion of the Lewis acidic borane increase by a factor of approximately two. The β values determined by EFISH experiments ranged between 30 and 72.5×10^{-30} esu for the pyridine/BF_3 and $B(C_6F_5)_3$ Lewis adducts.

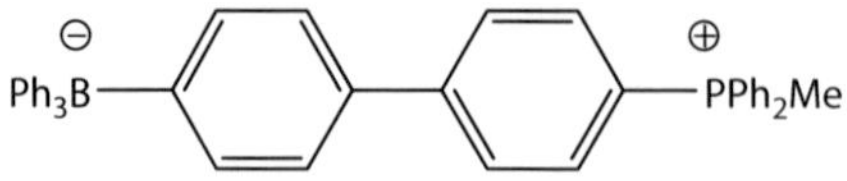

FIGURE 13.24 {4′-[Methyl(diphenyl)phosphonio]biphenyl-4-yl}triphenylborate. (Adapted from Chane-Ching, K. et al., *J. Chem. Soc. Faraday Trans.* **1991**, 87, 14, 2225.)

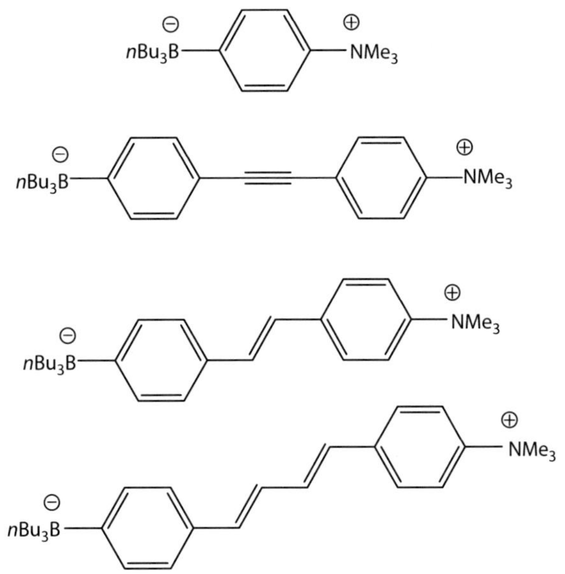

FIGURE 13.25 Ammonium/borate zwitterions. (Adapted from Lambert, C. et al., *Angew. Chem. Int. Ed. Engl.* **1996**, 35, 6, 644.)

In 2001, Su et al. confirmed this tendency by calculation of pyridine, styryl pyridine, and phenylethynyl pyridine/borane adducts using the quantum chemical AM1/Finite Field method. These theoretical data are in good agreement with the experimental values determined by Lesley et al., 1998.

13.5 SUMMARY

Research of recent years has shown that compounds containing boron in various forms are highly promising materials with oftentimes very interesting optical properties.

Icosahedral borane derivatives have been proposed with high calculated values of hyperpolarizability β. These theoretical investigations demonstrated the electron-donating properties of the

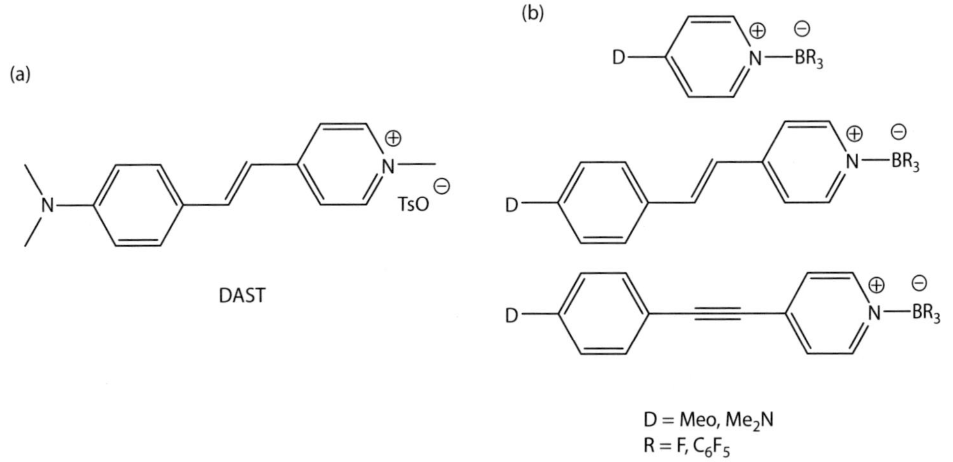

FIGURE 13.26 DAST (a) (Adapted from Marder, S. R., J. W. Perry, C. P. Yakymyshyn, *Chem. Mater.* **1994**, 6, 1137.) and pyridine/borane Lewis adducts (b) (Adapted from Lesley, M. J. G. et al., *Chem. Mater.* **1998**, 10, 1355.).

electron-rich dodecahydro-*closo*-dodecaborate(2-), in contrast to the electron deficiency of *closo*-carboranes.

Compounds containing boron clusters with conjugated bonds have been prepared and their nonlinear optical properties have been investigated. The electron-donating character of the dodecahydro-*closo*-dodecaborate, which was predicted from calculations, could be confirmed experimentally. Investigations of various prepared substituted carboranes showed their electron-withdrawing character. In materials containing two acceptors, fullerene and carborane, the latter exhibited smaller acceptor strength.

In addition to tricoordinate boron which acts as a strong electron acceptor because of its vacant *p*-orbital, negatively charged electron-donating tetracoordinate boron in two variations, zwitterionic molecules and as pyridine/borane Lewis base adducts, shows interesting nonlinear optical properties.

Calculations and experimental studies of tricoordinate boron in organoboranes elucidated that borane moieties have hyperpolarizabilities comparable to the nitro group. Tetracoordinate boron no longer has a vacant *p*-orbital and hence the negatively charged borate acts as an electron-donating substituent. The hyperpolarizabilities of zwitterions such as compounds with positively charged phosphonium as electron acceptors and borate as electron donors are comparable to that of *p*-nitroaniline and they are more transparent compared with the analogous donor–acceptor push–pull systems.

Theoretical and experimental investigations of Lewis base adducts of boranes with pyridine derivatives with an overall neutral net charge show that they possess large hyperpolarizability values.

As of now there are no commercial boron-containing nonlinear optical materials, but the prospects are bright for future use of such materials.

REFERENCES

J. Abe, N. Nemoto, Y. Nagase, Y. Shirai, T. Iyoda, A new class of carborane compounds for second-order nonlinear optics: Ab Initio molecular orbital study of hyperpolarizabilities for 1-(1',X'-Dicarba-*closo*-dodecaborane-1'-yl)-*closo*-dodecaborate Dianion (X = 2, 7, 12). *Inorg. Chem.* **1998**, 37, 172.

J.-I. Aihara, Three-dimensional aromaticity of polyhedral boranes. *J. Am. Chem. Soc.* **1978**, 100, 3339.

D. G. Allis, J. T. Spencer, Polyhedral-based nonlinear optical materials. Part 1. Theoretical investigation of some new high nonlinear optical response compounds involving carboranes and charged aromatic donors and acceptors. *J. Organomet. Chem.* **2000**, 614, 309.

D. G. Allis, J. T. Spencer, Polyhedral-based nonlinear optical materials. 2.1 theoretical investigation of some new high nonlinear optical response compounds involving polyhedral bridges with charged aromatic donors and acceptors. *Inorg. Chem.* **2001**, 40, 3373.

R. Bernard, D. Cornu, P. L. Baldeck, J. Čáslavský, J.-M. Létoffé, J.-P. Scharff, P. Miele, Synthesis, characterization and optical properties of π-conjugated systems incorporating *closo*-dodecaborate clusters: new potential candidates for two-photon absorption processes. *Dalton Trans.* **2005**, 3065.

R. Bernard, D. Cornu, J.-P. Scharff, R. Chiriac, P. Miele, Synthesis, characterization, and UV-vis linear absorption of centrosymmetric π-systems incorporating *closo*-dodecaborate clusters. *Inorg. Chem.* **2006**, 45, 8743.

G. C. Bjorklund, S. Ducharme, W. Fleming, D. Jungbauer, W. E. Moerner, J. D. Swalen, R. J. Twieg, C. G. Willson, D. Y. Yoon, In *Materials for Nonlinear Optics: Chemical Perspectives* **1991**; S. R. Marder, J. E. Sohn, G. D. Stucky (Eds.), ACS Symposium Series Vol. 455; American Chemical Society: Washington, DC, 216–225.

C. Branger, M. Lequan, R. M. Lequan, M. Barzoukas, A. Fort, Boron derivatives containing a bithiophene bridge as new materials for non-linear optics. *J. Mater. Chem.* **1996**, 6, 4, 555.

C. Branger, M. Lequan, R. M. Lequan, M. Large, F. Kajzar, Polyurethanes containing boron chromophores as sidechains for nonlinear optics. *Chem. Phys. Lett.* **1997**, 272, 265.

J. Brunel, A. Jutand, I. Ledoux, J. Zyss, M. Blanchard-desce, boomerang-shaped octupolar molecules derived from triphenylbenzene. *Synth. Met.* **2001**, 124, 195.

D. M. Burland, R. D. Miller, O. Reiser, R. J. Twieg, C. A. Walsh,The design, synthesis, and evaluation of chromophores for second-harmonic generation in a polymer waveguide. *J. Appl. Phys.* **1992**, 71, 410.

K. Chane-Ching, M. Lequan, R. M. Lequan, A. Grisard, D. Markovitsi, A new zwitterionic salt for non-linear optics: {4'- [Methyl(diphenyl)phosphonio] biphenyl-4-y1)triphenylborate. *J. Chem. Soc. Faraday Trans.* **1991**, 87, 14, 2225.

K. Clays, A. Persoons, Hyper-rayleigh scattering in solution. *Rev. Sci. Instrum.* **1992**, 63, 3, 3285.

A. Dulcic, C. Sauteret, The regularities observed in the second order hyperpolarizabilities of variously disubstituted benzenes. *J. Chem. Phys.* **1978**, 69, 8, 3453.

W. H. Eberhardt, B. Crawford, W. N. Lipscomb, The valence structure of the boron hydrides. *J. Chem. Phys.* **1954**, 22, 989.

P. A. Franken, A. E. Hill, C. W. Peters, G. Weinreich, Generation of optical harmonics. *Phys. Rev. Lett.* **1961**, 7, 4, 118.

B. Grüner, Z. Janoušek, B. T. King, J. N. Woodford, C. H. Wang, V. Všetečka, J. Michl, Synthesis of 12-Substituted 1-Carba-*closo*-dodecaborate anions and first hyperpolarizability of the 12-$C_7H_6^+$-$CB_{11}H_{11}$-Ylide. *J. Am. Chem. Soc.* **1999**, 121, 3122.

B. Grüner, Z. Janoušek, B. T. King, J. N. Woodford, C. H. Wang, V. Všetečka, J. Michl, Synthesis of 12-Substituted 1-Carba-*closo*-dodecaborate anions and first hyperpolarizability of the 12-$C_7H_6^+$-$CB_{11}H_{11}$ Ylide. *J. Am. Chem. Soc.* **2000**, 122, 11274.

W. R. Hertler, M. S. Raasch, Chemistry of boranes. XIV. Amination of $B_{10}H_{10}^{-2}$ and $B_{12}H_{12}^{2-}$ with Hydroxylamine-O-sulfonic Acid. *J. Am. Chem. Soc.* **1964**, 86, 3661.

R. Hoffmann, W. N. Lipscomb, Sequential substitution reactions on $B_{10}H_{10}^{-2}$ and $B_{12}H_{12}^{-2}$. *J. Chem. Phys.* **1962**, 37, 520.

M. A. Hubbard, T. J. Marks, J. Yang, G. K. Wong, Poled polymeric nonlinear optical materials. enhanced second harmonic generation stability of crosslinkable matrix/chromophore essembles. *Chem. Mater.* **1989**, 1, 167.

R. A. Huijts, G. L. Hesselink, Length dependence of the second-order polarizability in conjugated organic molecules. *Chem. Phys. Let.* **1989**,156, 209.

B. S. Jursic, Complete basis set Ab Initio computational study of three-dimensional aromaticity in highly symmetric hydrogen clusters. *J. Mol. Struct. (Theochem)* **1999**, 490, 81.

K. Kajikawa, T. Anzai, H. Takezoe, A. Fukuda, S. Okada, H. Matsuda, H. Nakanishi, T. Abe, H. Ito, Second-harmonic generation in langmuir and langmuir-blodgett films of a polymer with pendant chromophore. *Chem. Phys. Lett.* **1992**,192, 113.

D. R. Kanis, M. A. Ratner, T. J. Marks, Nonlinear optical characteristics of novel inorganic chromophores using the zindo formalism. *Chem. Mater.* **1991**, 3, 19.

D. R. Kanis, M. A. Ratner, T. J. Marks, Design and construction of molecular assemblies with large second-order nonlinearities. quantum chemical aspects. *J. Chem. Rev.* **1994**, 94, 195.

H. E. Katz, G. Scheller, T. M. Putvinski, M. L. Schilling, W. L. Wilson, C. E. D. Chidsey, Polar orientation of dyes in robust multilayers by zirconium phosphate-phosphonate interlayers. *Science* **1991**, 254, 1485.

J. Kerr, On rotation of the plane of polarization by reflection from the pole of a magnet. *Phil. Mag.* **1877**, 3, 321.

J. Kerr, Reflection of polarized light from the equatorial surface of a magnet. *Phil. Mag.* **1878**, 5, 161.

R. B. King, Three-dimensional aromaticity in polyhedral boranes and related molecules. *Chem. Rev.* **2001**, 101, 1119.

R. B. King, Skyrmion models for three-dimensional aromaticity in deltahedral boranes. *Chem. Phys. Lett.* **2001**, 338, 237.

W. H. Knoth, H. C. Miller, J. C. Sauer, J. H. Balthis, Y. T. Chia, E. L. Muetterties, Chemistry of boranes. IX. Halogenation of $B_{10}H_{10}^{-2}$ and $B_{12}H_{12}^{-2}$. *Inorg. Chem.* **1964**, 3, 2, 159.

S. K. Kurtz, T. T. Perry, A powder technique for the evaluation of nonlinear optical materials. *J. Appl. Phys.* **1968**, 39, 8, 3798.

H. A. Kurtz, D. S. Dudis, *Rev. in Computational Chemistry*, K. B. Lipkowitz, D. B. Boyd (Eds.), Wiley-VCH, New York, **1998**, 12, 241–279.

C. Lambert, S. Stadler, G. Bourhill, C. Bräuchle, Polarized n-electron systems in a chemically generated electric field: Second-order nonlinear optical properties of ammonium/borate zwitterion. *Angew. Chem. Int. Ed. Engl.* **1996**, 35, 6, 644.

M. Lamrani, R. Hamasaki, M. Mitsuishi, T. Miyashita, Y. Yamamoto, Carborane-fullerene hybrids as a seemingly attractive–attractive dyad with high hyperpolarizability. *Chem. Commun.* **2000**, 1595.

M. Lequan, R. M. Lequan, K. Chane-Ching, Trivalent boron as acceptor chromophore in asymmetrically substituted 4,4'-Biphenyl and azobenzene for non-linear optics. *J. Mater. Chem.* **1991**, 1, 6, 997.

M. Lequan, R. M. Lequan, K. Chane-Ching, M. Barzoukas, A. Fort, H. Lahoucine, G. Bravic, D. Chasseau, J. Gaultier, Crystal structures and non-linear optical properties of borane derivatives. *J. Mater. Chem.* **1992**, 2, 7, 719.

M. Lequan, R. M. Lequan, K. Chane-Ching, A.-C. Callier, M. Barzoukas, A. Fort, Second-order optical non-linearities of {4-(dimethylamino) stilben Z & E 4'-yl}Dimesityl borane. *Adv. Mater. Opt. Elect.* **1992**, 1, 243.

M. J. G. Lesley, A. Woodward, N. J. Taylor, T. B. Marder, I. Cazenobe, I. Ledoux, J. Zyss, A. Thornton, D. W. Bruce, A. K. Kakkar, Lewis acidic borane adducts of pyridines and stilbazoles for nonlinear optics. *Chem. Mater.* **1998**, 10, 1355.

D. Li, M. A. Ratner, T. J. Marks, C. Zhang, J. Yang, G. K. Wong, Chromophoric self-assembled multilayers. Organic superlattice approaches to thin-film nonlinear optical materials. *J. Am. Chem. Soc.* **1990**, 112, 7389.

D. Li, T. J. Marks, M. A. Ratner, Nonlinear optical phenomena in conjugated organic chromophores. theoretical investigations via a π-electron formalism. *J. Phys. Chem.* **1992**, 96, 4325.

R. Littger, J. Taylor, G. Rudd, A. Newlon, D. G. Allis, S. Kotiah, J. T. Spencer, *Contemp. Boron Chem., Spec. Publ. No. 253*, M. G. Davidson, A. K. Hughes, T. B. Marder, K. Wade (Eds.), The Royal Soc. of Chem., Cambridge, U. K., **2000**, 67.

T. H. Maiman, Stimulated optical radiation in ruby. *Nature* **1960**, 187, 4736, 493.

P. D. Maker, R. W. Terhune, Study of optical effects to an induced polarization third order in the electric field strength. *Phys. Rev.* **1965**, 137, 3A, A801.

S. R. Marder, J. W. Perry, W. P. Schaefer, Synthesis of organic salts with large second-order nonlinearities. *Science* **1989**, 245, 626.

S. R. Marder, D. N. Beratan, L.-T. Cheng, Approaches for optimizing the first electronic hyperpolarizability of conjugated organic molecules. *Science* **1991**, 252, 103.

S. R. Marder, J. W. Perry, C. P. Yakymyshyn, Organic salts with large second-order optical nonlinearities. *Chem. Mater.* **1994**, 6, 1137.

S. Miyata, T. Hosomi, T. Suzuki, T. Watanabe, H. Yamamoto, A. Hayashi, DE68910224T2, Material für die nichtlineare optik. *Patent* **1994**.

D. M. Murphy, D. M. P. Mingos, J. M. Forward, Synthesis of Icosahedral carboranes for second-harmonic generation. Part 1. *J. Mater. Chem.* **1993**, 3, 67.

D. M. Murphy, D. M. P. Mingos, J. L. Haggitt, H. R. Powell, S. A. Westcott, T. B. Marder, N. J. Taylor, D. R. Kanis, Synthesis of Icosahedral carboranes for second-harmonic generation. Part 2. *J. Mater. Chem.* **1993**, 3, 139.

H. S. Nalwa, S. Miyata (Hrsg.), *Nonlinear Optics of Organic Molecules and Polymers*, 1. Auflage, CRC Press, Inc., Boca Raton, **1997**.

G. H. C. New, J. F. Ward, Optical third-harmonic generation in gases. *Phys. Rev. Lett.* **1967**, 19, 10, 556.

J. L. Oudar, D. S. Chemla, Hyperpolarizabilities of the nitroanilines and their relations to the excited state dipole moment. *J. Chem. Phys.* **1977**, 66, 2664.

A. Papagni, S. Maiorana, P. del Buttero, D. Perdicchia, F. Cariati, E. Cariati, W. Marcolli, Synthesis and spectroscopic and NLO-properties of "Push-Pull" Structures incorporating the inductive electron-withdrawing pentafluorophenyl group. *Eur. J. Org. Chem.* **2002**, 1380.

T. Peymann, C. B. Knobler, M. F. Hawthorne, Synthesis of alkyl and aryl derivatives of *closo*-$B_{12}H_{12}^{2-}$ by the Palladium-Catalyzed coupling of *closo*-$B_{12}H_{11}I^{2-}$ with grignard reagents. *Inorg. Chem.* **1998**, 37, 7, 1544.

P. N. Prasad, D. J. Williams, *Introduction to Nonlinear Optical Effects in Molecules and Polymers*, John Wiley & Sons, Inc., New York, **1991.**

G. G. Roberts, In *Langmuir-Blodgett Films* **1990**; G. G. Roberts (Ed.), Plenum Press: New York, 351–368.

P. v. R. Schleyer, K. Najafian, Stability and three-dimensional aromaticity of *closo*-monocarbaborane anions, CBn-1Hn$^-$, and *closo*-dicarboranes, C2Bn-2Hn. *Inorg. Chem.* **1998**, 37, 3454.

K. D. Singer, A. F. Garito, Measurements of molecular second order optical susceptibilities using dc induced second harmonic generation. *J. Chem. Phys.* **1981**, 75, 7, 3572.

K. D. Singer, J. E. Sohn, S. J. Lalama, Second harmonic generation in poled polymer films. *Appl. Phys. Lett.* **1986**, 49, 248.

K. D. Singer, S. J. Lalama, J. E. Sohn, R. D. Small, In *Nonlinear Optical Properties of Organic Molecules and Crystals* **1987**; D. S. Chemla, J. Zyss (Eds.); Academic Press: San Diego, Vol. 1, 460.

I. B. Sivaev, A. B. Bruskin, V. V. Nesterov, M. Y. Antipin, V. I. Bregadze, S. Sjöberg, Synthesis of schiff bases derived from the ammoniaundecahydro-*closo*-dodecaborate(1-) anion, $[B_{12}H_{11}NHdCHR]^-$, and their reduction into monosubstituted amines $[B_{12}H_{11}NH_2CH_2R]^-$: A new route to water soluble agents for BNCT. *Inorg. Chem.* **1999,** 38, 25, 5887.

Z. M. Su, X. J. Wang, Z. H. Huang, R. S. Wang, J. K. Feng, J. Z. Sun, Nonlinear optical properties for pyridines and stilbazoles adducted by borane. *Synth. Met.* **2001**, 119, 583.

J. Taylor, J. Caruso, A. Newlon, U. Englisch, K. Ruhlandt-Senge, J. T. Spencer, Polyhedral-based nonlinear optical materials. 3. synthetic studies of cyclopentadiene- and cycloheptatriene-substituted polyhedral compounds: synthesis of 1,12-[(CH)CBH(CHMe)] and related species. *Inorg. Chem.* **2001**, 40, 3381.

N. Tsuboya, M. Lamrani, R. Hamasaki, M. Ito, M. Mitsuishi, T. Miyashita, Y. Yamamoto, Nonlinear optical properties of novel carborane–ferrocene conjugated dyads. electron-withdrawing characteristics of carboranes. *J. Mater. Chem.* **2002**, 12, 2701.

A. Vöge, E. Lork, B. S. Sesalan, D. Gabel, N-Arylammonio- and N-Pyridinium-substituted derivatives of dodecahydro-*closo*-dodecaborate(2-). *J. Organomet. Chem.* **2009**, 694, 1698.

A. Vöge, *Dissertation: Dodecahydro-closo-dodecaboratderivate für die Anwendung in der nichtlinearen Optik* **2009** (http://elib.suub.uni-bremen.de/diss/docs/00011457.pdf).

C.-K. Wang, P. Macak, Y. Luo, H. Ågren, Effects of π centers and symmetry on two-photon absorption cross sections of organic chromophores. *J. Chem. Phys.* **2001**, 114, 22, 9813.

Z. Yuan, N. J. Taylor, T. B. Marder, I. D. Williams, S. K. Kurtz, L.-T. Cheng, Three coordinate phosphorus and boron as π-donor and π-acceptor moieties respectively, in conjugated organic molecules for nonlinear optics: Crystal and molecular structures of E-Ph-CH = CH-B(mes)2, E-4-MeO-C_6H_4-CH = CH-B(mes)2 and E-Ph2P-CH = CH-B(mes) [mes = 2,4,6-$Me_3C_6H_2$]. *J. Chem. Soc. Chem. Commun.* **1990**, 1489.

14 *closo*-Boranes as Structural Elements for Liquid Crystals

Piotr Kaszynski

CONTENTS

14.1 INTRODUCTION

Since the discovery of unusual thermal behavior of cholesteryl benzoate over 120 years ago,[1] the field of liquid crystals (LCs) has grown to an impressive size, and its significance in contemporary technologies is apparent. More than a century of research has resulted in nearly 100,000 organic, organometallic, ionic, and polymeric compounds characterized as thermotropic LCs.[2] With respect to their molecular shapes, these compounds are divided into several classes, which include calamitic (or rod-like, Figure 14.1a), bent-core (Figure 14.1b), and discotic, and they form nematic, smectic (lamellar), banana, and columnar types of supramolecular structures (Figure 14.2).[3–5]

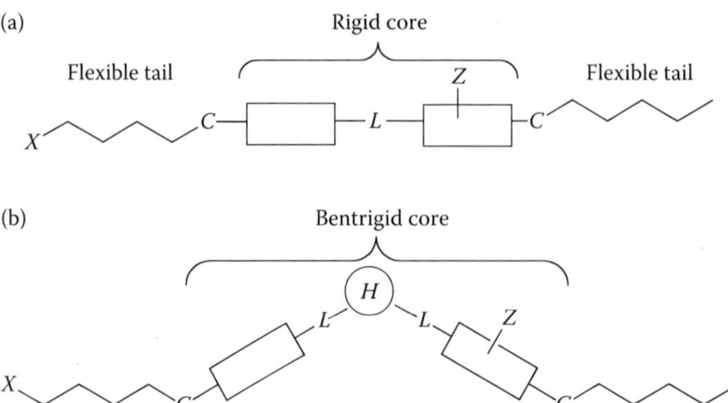

FIGURE 14.1 A general schematic representation of (a) rod-like and (b) bent-core liquid crystal molecules with two terminal alkyl chains. *L* is the linking group, *C* is the terminal chain connector, *Z* is a lateral substituent, *X* is a chain substituent, and *H* is the head ring.

From the point of view of physics, LCs are partially oriented fluids that exhibit anisotropic optical, dielectric, magnetic, and mechanical properties. The most important property of LCs is the reorganization of their supramolecular structures on external stimuli such as electric and magnetic fields,[6] temperature, and mechanical stress, which lead to changes in their optical properties. In particular, electric filed-induced control of optical properties of LCs (electro-optical effects[7] based on the Frėedericksz transition[8]) is at the heart of the multi-billion dollar liquid crystal display (LCD) industry.[9] Most current LCD technologies rely on nematic[10,11] and to a lesser extent on ferroelectric[12] LCs, while the recently discovered bent-core[13–15] and orthoconic[16,17] LCs still require significant investment into fundamental research and development. These and other applications and technologies continue to drive the search for new liquid crystal materials, and provide impetus to continue fundamental studies on new, often exotic,[18] classes of compounds.

Liquid crystal molecules typically comprise an anisometric rigid core, such as rod-like, bent, or disk-like, and at least one terminal alkyl substituent (Figure 14.1).[19] In calamitic LCs the elongated rigid core contains two or more carbocyclic or heterocyclic rings connected directly or through a short linking group *L* (Figure 14.1a). Bent-core mesogens can be described as two calamitic mesogenic units connected through an angular head group *H*, such as *m*-phenylene (Figure 14.1b).

In LCs the intermolecular core–core interactions are ordering, while the alkyl chains increase the entropy. Consequently, the liquid crystalline state originates from a delicate balance between these ordering and disordering factors, and the choice of the rigid core element is critical for materials properties. The rings in the rigid core are typically planar aromatics, such as benzene (**A** in Figure 14.3) and its heteroaromatic analogs, and alicyclic (e.g., cyclohexane, **B**, and bicyclo[2.2.2] octane, **C**), although a large number of other rings have been explored as structural elements of

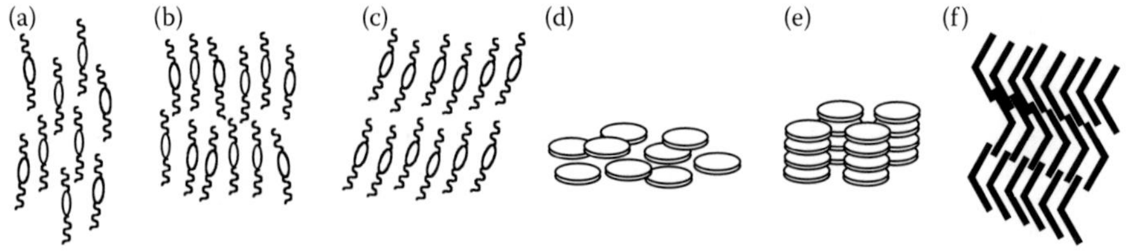

FIGURE 14.2 Molecular arrangements in selected liquid crystalline phases: (a) nematic (N), (b) SmA, (c) SmC, (d) nematic discotic (N_D), (e) columnar discotic (Col), and (f) banana phase. The ovals represent the rigid cores and the lines are the flexible tails (Figure 14.1a). Discs consist of a rigid core and radially distributed flexible tails. The chevron represents a bent rigid core with two flexible tails propagating along the arms (Figure 14.1b).

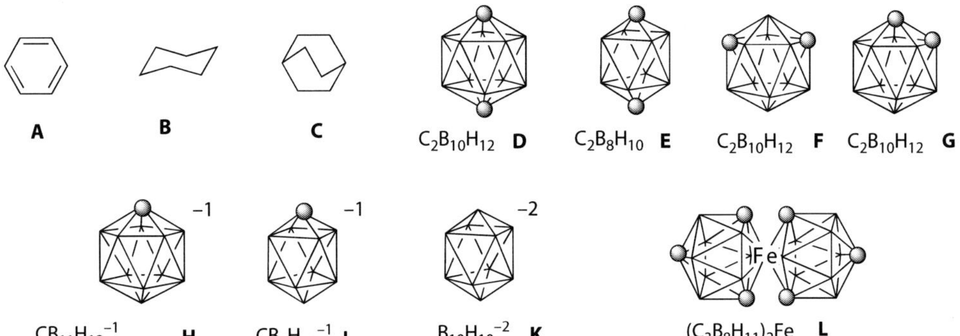

FIGURE 14.3 Skeletal representations of carbocycles **A–C** and boron clusters **D–L**. In structures **D–L** each vertex corresponds to a B–H fragment and the CH groups are marked by filled circles.

LCs.[19] Among them are carboranes **D** and **E** and other 12- and 10-vertex *closo*-boranes, which uniquely combine some aspects of the electronic structure of conventional aromatics with steric features of bicyclo[2.2.2]octane.[20] Therefore, they are attractive for designing new materials with unusual combination of properties and for addressing questions in the area of fundamental and applied liquid crystal research and technology.

14.2 *closo*-BORANES

The 12- and 10-vertex *closo*-boranes **D–K** and tricarbolide **L** belong to an extensive family of inorganic boron hydrides and are characterized by molecular and electronic features uncommon for organic rings that are used as structural components of mesogenic materials.[19] Reminiscent of bicyclo[2.2.2]octane (**C**), *closo*-boranes such as *p*-carboranes **D** and **E** (Figure 14.3) are spherocylindrical, and their substitution in the antipodal positions with appropriate groups gives rise to extended molecular shapes that are conducive to the formation of a mesogenic state.

A detailed comparison demonstrated that 12-vertex and 10-vertex *closo*-boranes, represented by carboranes **D** and **E** respectively, are larger than organic carbocycles, such as bicyclo[2.2.2]octane (**C**), and have higher order rotational axes[20,21] (Figure 14.4). Calculations supported by experimental data show that the van der Waals diameter curved by the *p*-carboranes **D** and **E** rotating around the C···C axis is 7.4 Å and 7.2 Å respectively, which compares to 6.7 Å for bicyclo[2.2.2]octane (**C**) and benzene (**A**).[22]

closo-Boranes exhibit three-dimensional σ-aromaticity and a large HOMO–LUMO gap.[23] Consequently, the clusters have only marginal electronic absorption above 200 nm.[23–25] In addition, the charge in the *closo*-borates, such as **H–K**, is highly delocalized placing them among the least nucleophilic anions.[26] The extent of the electronic interaction between the σ-aromatic electron

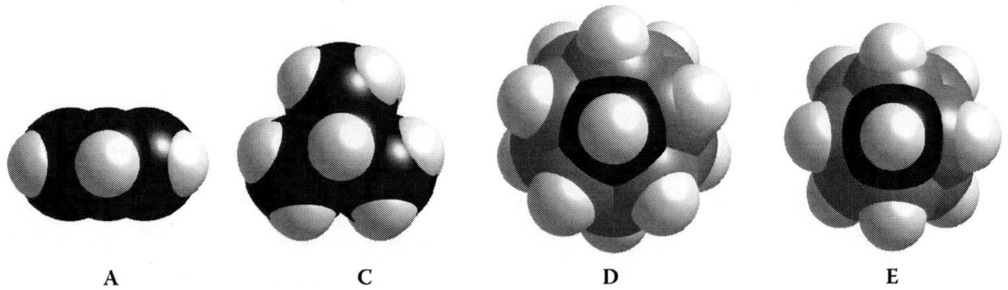

FIGURE 14.4 Space-filling models of benzene (**A**), bicyclo[2.2.2]octane (**C**), and carboranes **D** and **E** viewed along the C(1)···C(4), C(1)···C(12), and C(1)···C(10) axes, respectively.

manifold and π-substituents vary between 10-vertex and 12-vertex clusters. Photophysical,[27–31] structural,[28,29,32] and theoretical[23,28,29,32] investigations have shown that the 10-vertex *closo*-boranes interact more strongly with π substituents than the 12-vertex analogs. This is related to the symmetry and energy of molecular orbitals, and also their amplitude at the substitution site.[23] In general, these factors are more favorable for 10-vertex clusters for which electronic interactions with substituents at the apical positions are closer to those observed in benzene analogs. In contrast, electronic interactions in 12-vertex derivatives are uniformly weak and reminiscent of those in bicyclo[2.2.2]octane derivatives. Under certain circumstances, however, the presence of the 12-vertex *p*-carborane **D** in a molecule manifests itself by the appearance of new strong absorption bands in the UV region.[30] Low absorption in the UV region is desired for LCs used in conventional display applications.

The delocalization of the skeletal σ electrons results in high electronic polarizability of the boron clusters,[33,34] but, due to molecular geometry, they exhibit low polarizability anisotropy.[29,35] This results in relatively high isotropic refractive index n and low birefringence (Δn) of carborane-containing LC.[27,36,37]

The symmetry of the clusters and the strength of interactions with substituents affect the distribution and depth of conformational minima in their derivatives. This, in turn, has an effect on the molecule's ability to maintain rigidity and high aspect ratio, which consequently impact on the mesophase stability. In general, high-order rotational axes, fivefold in 12-vertex and fourfold in 10-vertex clusters, result in relatively shallow minima on the conformational potential energy surface when compared with their carbocyclic analogs.[29,38]

Analysis of results in Table 14.1 demonstrates that for most carborane derivatives the calculated $\Delta E^{\ddagger}$ for the rotation about the ring—substituent bond is close to zero and the $\Delta G^{\ddagger}_{298}$ of about 1.5 kcal/mol is solely due to entropy of activation. The $\Delta H^{\ddagger}$ values are calculated negative for several derivatives due to underestimation of thermal contribution to the activation enthalpy. An exception is 1-phenyl 10-vertex *p*-carborane, which exhibits strong cage–Ph electronic interactions and a reasonably high value of rotational TS energy. The calculated value of activation energy, $\Delta E^{\ddagger} = 2.3$ kcal/mol, is markedly higher than that for 1-phenylbicyclo[2.2.2]octane, but lower than that for biphenyl (Table 14.1). Another example of a reasonably deep conformational minimum and hence high rotational rigidity is the 1-isobutyl derivative of 10-vertex *p*-carborane (**E**).

TABLE 14.1
Calculated Barriers to Internal Rotation at $T = 298$ K[a]

R \ A	D			E			C			A		
	$\Delta E^{\ddagger}$	$\Delta H^{\ddagger}$	$\Delta G_{298}^{\ddagger}$	$\Delta E^{\ddagger}$	$\Delta H^{\ddagger}$	$\Delta G_{298}^{\ddagger}$	$\Delta E^{\ddagger}$	$\Delta H^{\ddagger}$	$\Delta G_{298}^{\ddagger}$	$\Delta E^{\ddagger}$	$\Delta H^{\ddagger}$	$\Delta G_{298}^{\ddagger}$
Et	0.3	–0.3	1.3	0.2	–0.4	1.6	5.4	4.9	5.8	1.4	0.8	2.5
i-Bu	2.1	1.6	3.0	1.9	1.4	2.8	4.3	3.8	5.1	4.5	3.9	5.5
$CH_2=CH$	0.6	0.0	1.5	0.6	0.1	1.9	1.5	1.1	2.4	2.2	1.8	3.2
CHO	0.4	–0.2	1.5	0.2	–0.4	1.5	1.1	0.6	1.9	7.9	7.6	8.1
COOH	0.0	–0.6	1.2	0.7	0.2	1.8	0.9	0.3	2.2	6.2	5.8	6.8
Ph	0.0	–0.6	1.9	2.3	1.8	3.4	0.9	0.4	2.1	3.9	3.4	4.8

[a] Obtained at the MP2/6–31G(d) level of theory using appropriate symmetry constraints and B3LYP/6-31G(d) thermodynamic corrections. Transition state structures were located using the TS or QST3 keyword.

TABLE 14.2
Substituent Orientation in and Point Group Symmetry of the Conformational GS for Selected Derivatives[a]

	D	E	C	A
n-Alkyl	E(C_s)[b]	S(C_s)[b]	S(C_1)	S(C_s)
CH=CH_2	E(C_s)	S(C_s)	E(C_1)	E(C_s)
CHO	E(C_s)	S(C_s)	E(C_1)	E(C_s)
COOH	E(C_s)[c]	S(C_s)[c]	E(C_1)	E(C_s)
Ph	E(C_s)[d]	S(C_{2v})[d]	E(C_1)	S(D_2)

[a] E, eclipsed; S, staggered. MP2/6–31G(d) level of theory.
[b] Related experimental structure: ref. 29
[c] Related experimental structure: ref. 40
[d] Related experimental structure: ref. 30

Conformational analysis shows that substituents in the 10-vertex *p*-carborane (**E**) derivatives generally prefer the staggered (S) orientation in the rotational GS (Table 14.2). In contrast, most substituents in 12-vertex *p*-carborane (**D**) and also in bicyclo[2.2.2]octane (**C**) are found in the eclipsed orientation relative to the ring in the conformational GS, which is related to the odd-order rotational axes. In the carboxyl and formyl derivatives of bicyclo[2.2.2]octane (**C**) the carbonyl group eclipses the C–H bond of the ring. In contrast, the orientation of the carbonyl group in 12-vertex *p*-carborane-1-carboxylic acid has practically no thermodynamic preference, while in the formyl derivative, the C=O group assumes the staggered orientation relative to the cage.

Bicyclo[2.2.2]octane (**C**) is unique among the four structural elements. It exhibits some flexibility and adopts a twisted conformation in the GS (D_3 point group symmetry) with the twist angle calculated at about 17° (MP2/6–31G(d)). Therefore, most of its derivatives lack symmetry and consequently all conformers are chiral.

Carbonyl, vinyl, and alkoxy groups capable of conjugation with the aromatic π system are coplanar with the benzene ring, while the alkyl group is perpendicular to the ring. In the conformational GS the phenyl group is staggered (45° dihedral angle) as a result of a compromise between steric interactions of the *ortho* hydrogen atoms and π–π overlap. These conformational preferences are consistent with those found in a number of experimental solid-state structures of mesogenic compounds.[39]

Results of theoretical conformational analysis for *closo*-boranes are consistent with the orientation of the substituents observed in experimental structures of selected liquid crystalline compounds, which include alkyl,[29] alkynyl,[29] carboxyl,[40] and phenyl[30] derivatives of carboranes **D** and **E**. The distribution of conformational minima calculated for derivatives of *p*-carboranes in Table 14.2 appears to be valid also for their isosteric charged analogs: monocarbarborates, [*closo*-1-$CB_{11}H_{12}]^-$ (**H**) and [*closo*-1-$CB_9H_{10}]^-$ (**J**), and decaborate [*closo*-$B_{10}H_{10}]^{2-}$ (**K**). This is evident from the solid-state structures of alkyl and carboxy derivatives of the [*closo*-1-$CB_9H_{10}]^-$ (**J**).[41,42]

Conformational properties of zwitterionic derivatives of the [*closo*-1-$CB_{11}H_{12}]^-$ (**H**), [*closo*-1-$CB_9H_{10}]^-$ (**J**), and decaborate [*closo*-$B_{10}H_{10}]^{2-}$ (**K**) were also studied computationally and experimentally. The pyridinium substituent adopts a staggered orientation relative to the 10-vertex *closo*-borate cage,[31,42] which is the same as the conformational minimum of the nonpolar 1-phenyl-*p*-carborane analog. The ground state (GS) orientations of the pentamethylenesulfonium and quinuclidinium rings relative to the *closo*-borate cages are shown in Figure 14.5. In the 10-vertex derivatives, the lone pair of the sulfonium substituent eclipses the C–B in 1-substituted

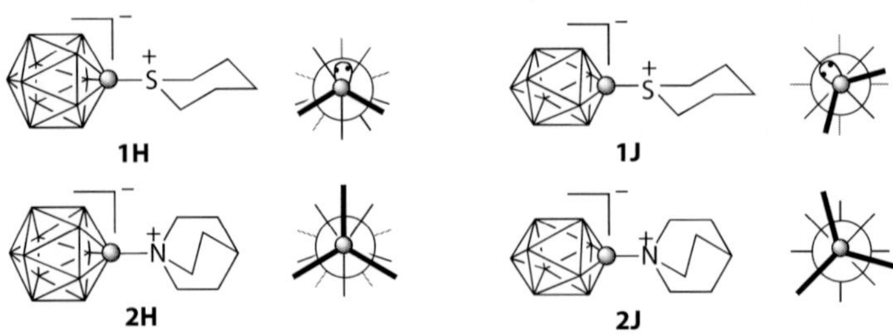

FIGURE 14.5 Extended Newman projection along the long molecular axes of **1** and **2** showing main conformations. The bars represent the substituent and the circle is the nitrogen or sulfur atom.

[*closo*-1-CB_9H_{10}]$^-$ (**1J**) or B–B bonds in the 10-substituted [*closo*-1-CB_9H_{10}]$^-$ and 1-substituted [*closo*-$B_{10}H_{10}$]$^{2-}$ derivatives.[31,42] In contrast, in the 12-vertex monocarbaborate derivative **1H** the lone pair of sulfonium is found in the staggered orientation relative to the cage either at the C(1) or B(12) position. The same opposite preference for the substituent orientation is found for the quinuclidinium, which favors an eclipsed conformer for the 10-vertex[31,41] (**2J**) and staggered in the 12-vertex cluster[43] derivatives (**2H**).

The distribution of conformational minima of the substituents and the symmetry of the clusters define the dynamic molecular aspect ratio of the disubstituted derivatives and impact on their mesogenic properties. As a consequence of molecular symmetry, the two antipodal substituents in the 12-vertex boron clusters can adopt antiperiplanar conformational minimum (Figure 14.6),[29,30] which gives rise to extended molecular shape and is favorable for the mesogenic state. In contrast, in the 10-vertex analogs all conformers are chiral and the substituents are offset from the coplanarity by 45° or 135°,[23,44] which destabilizes the mesophase.

This general property of 10- and 12-vertex clusters is illustrated in Figure 14.6 with a series of carboxylic acids substituted with a propyl group. The conformations in Figure 14.6 are consistent with experimental molecular structures of mesogenic derivatives[29–31] and relative stability of the mesophase in isostructural derivatives.

The orientation of the pyridinium ring in zwitterionic derivatives of the [*closo*-1-CB_9H_{10}]$^-$ is the same as the phenyl in derivatives of *p*-carborane **E**, and in consequence all conformers of acid **3J**

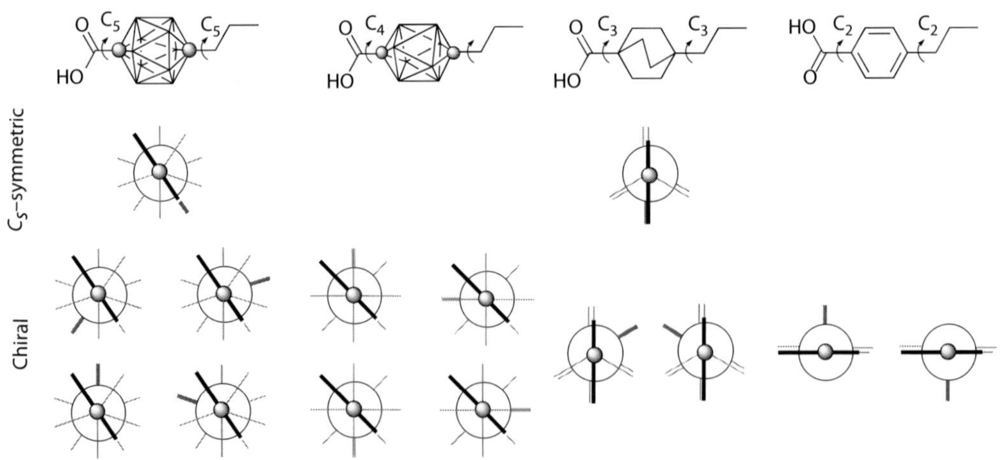

FIGURE 14.6 Possible relative orientations of the carboxyl and propyl groups in the conformational GSs. The carboranes, bicyclo[2.2.2]octane and benzene are shown in extended Newman projections along the long molecular axis. The propyl and carboxyl groups shown as bars. The short end of the bar represents the C=O group and the long end the C–OH group. For simplicity the BCO ring is presented in its D_{3h} symmetry.

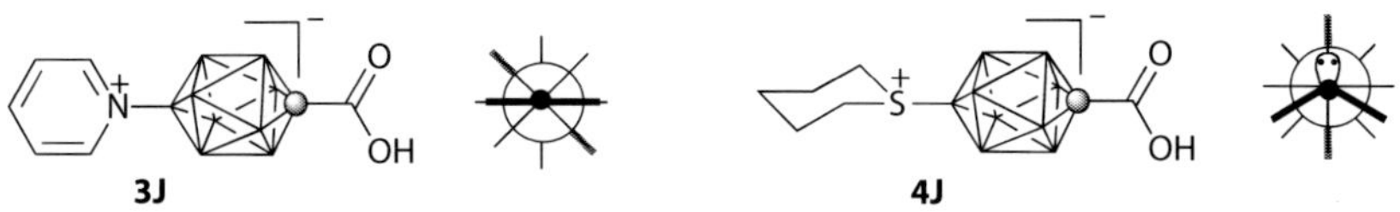

FIGURE 14.7 Extended Newman projections for GS conformation of carboxylic acids **3J** and **4J**.

are chiral (Figure 14.7).[42] Conformational properties of the sulfonium group result, however, in a C_s-symmetric acid **4J** in its conformational GS.[42]

The pentamethylenesulfonium group attached to the monocarborane cage undergoes epimerization at the *S* center, which affects the molecular aspect ratio, and, consequently, bulk properties of the mesogenic compounds. Results of DFT and MP2 level calculations shown in Table 14.3 demonstrate that the epimerization process is easier for both derivatives of the monocarborates **H** and **J** than that for a trialkylsulfonium, $(Alk)_3S^+$, and particularly facile for the *B*-isomers. For the latter the enthalpy of activation $\Delta H^{\ddagger}$ is calculated at about 24 kcal/mol, which suggests that the epimerization process is relatively fast even at ambient temperature. In contrast, epimerization of the *C*-isomers **1H** and **1J** requires elevated temperatures, presumably above 100°C. Overall, the data show that the 10-vertex derivatives have higher enthalpy of activation $\Delta H^{\ddagger}$ by about 1 kcal/mol, shorter cage–S bonds, and higher dipole moments than the 12-vertex analogs. The length of the cage–S bond affects the stability of the axial epimer and, consequently, the position of the axial/equatorial equilibrium. Thus, for the *B*-isomers with longer cage–S bonds the ΔG values are small and the concentration of

TABLE 14.3
Activation Energy for Epimerization at the Sulfur Center in Selected Derivatives[a]

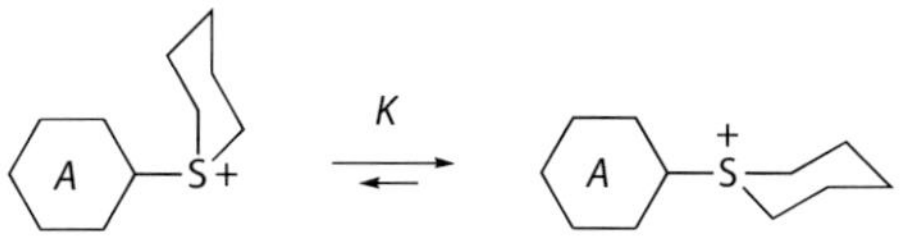

Compound	$\Delta H^{\ddagger}$ (kcal/mol)	ΔG_{298} (kcal/mol)	K_{298}	$d_{S\text{-}X}$[b] (Å)	μ[c] (D)
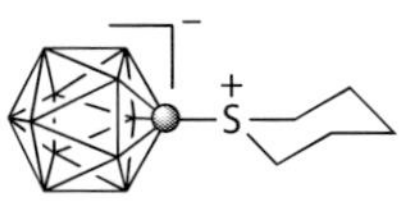 1H	30.0	2.3	51	1.799	14.1
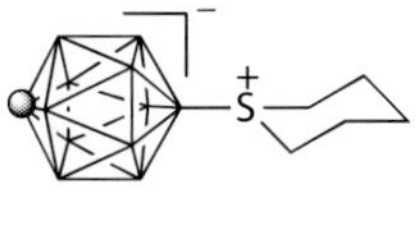	23.3	0.82	4	1.894	8.5
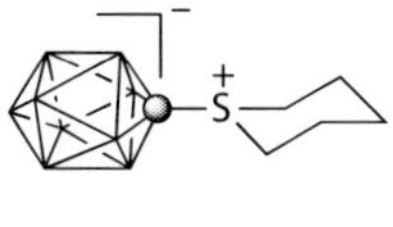 1J	31.2	2.2	42	1.750	14.4
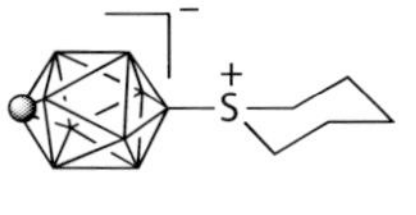	24.2	0.4	2	1.854	8.75

[a] MP2/6–31G(d,p) level calculations with B3LYP/6–31G(d,p) thermodynamic corrections.
[b] Interatomic distance between the S and the cage's C or B atoms in the equatorial epimer.
[c] Dipole moment of the equatorial epimer.

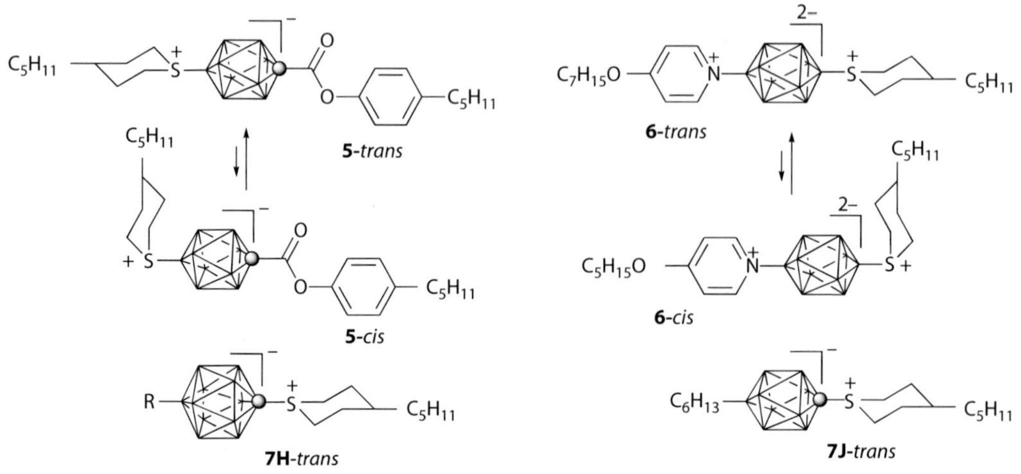

FIGURE 14.8 The *trans* and *cis* isomers of selected compounds.

the axial conformer is high. In contrast, the steric factors for the C(1)-isomers are high and about 2.2 kcal/mol, which corresponds to small concentration (~1%) of the axial form at the equilibrium.

The activation parameters and the equilibrium constants are expected to be sensitive to solvent polarity. The TS is generally more polar by about 1 D and the axial epimer is less polar by about 0.5 D that the equatorial form. Thus, the more polar solvent is expected to decrease activation energy for epimerization, and increase the preference for the equatorial form.

Overall, the *B*-isomers are expected to have high concentration of the axial form (about 25%) with a fast rate of epimerization at ambient temperature. In contrast, the C(1) isomers are predicted to have <2% of the axial epimer in the equilibrium and significantly lower rates of epimerization.

Theoretical results are consistent with experimental data for compounds **5–7** (Figure 14.8). While relatively high concentrations of the *cis* isomer of 10-sulfonium derivatives of the $[closo\text{-}1\text{-}CB_9H_{10}]^-$ and $[closo\text{-}B_{10}H_{10}]^{-2}$ clusters have been observed by nuclear magnetic resonance (NMR), no epimerization has been detected for C(1) sulfonium derivatives of $[closo\text{-}1\text{-}CB_9H_{10}]^-$ or $[closo\text{-}1\text{-}CB_{11}H_{12}]^-$ at ambient temperature. Temperature-dependent NMR measurements for **5**[45] and **6**[31] established that the equilibrium mixture contains 22% (MeCN, $\Delta G_{298} = 0.76 \pm 0.05$ kcal/mol) and 20% (toluene, $\Delta G_{298} = 0.81 \pm 0.05$ kcal/mol), respectively, of the *cis* isomer at ambient temperature. Kinetic measurements established activation parameters, $\Delta H^{\ddagger} = 22.8 \pm 0.5$ kcal/mol and $\Delta S^{\ddagger} = -3 \pm 1$ cal/mol · K, and a rate of epimerization $k = 7.8 \pm 0.1 \times 10^{-4}$ s^{-1} at ambient temperature for **6**.[31]

In summary, data presented in Table 14.3 demonstrate that the *B*-sulfonium derivatives of 10- and 12-vertex monocarbaborates **H** and **J** have lower activation energies to epimerization at the sulfur center and smaller energy difference between the axial and equatorial epimers than for the C(1) isomers. The latter is related to the significant difference in the cage–S bond length. The concentration of the axial epimer is calculated at 25–30% for the *B*-isomers, and an order of magnitude smaller for the C(1) isomers.

All these factors, the number of conformational equilibria, rotation around the cage-substituent bond, and epimerization (axial/equatorial equilibrium) impact on the dynamic aspect ratio of the mesogens and affect the phase stability.

14.3 LIQUID CRYSTALLINE DERIVATIVES OF *CLOSO*-BORANES

The largest group of LCs containing boron clusters comprises derivatives of carboranes. To date, over 200 low-molecular-weight calamitic mesogens containing *p*-carboranes **D** and **E** (Table 14.4), and several bent-core mesogens based on *m*-carborane (**F**, Figure 14.3) have been prepared and

TABLE 14.4
Liquid Crystalline Derivatives of Type *I*

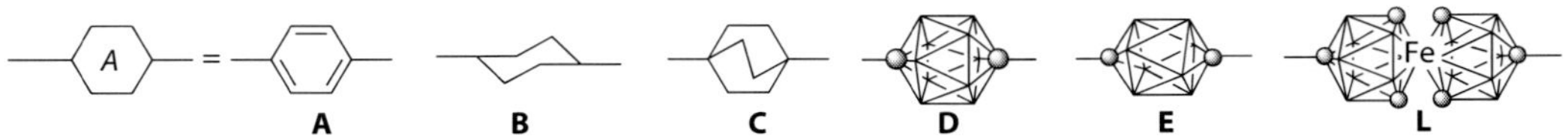

A = A, B, C, D, E, L

Structure	*A*	Substituents	References
 C_5H_{11}–(A)–C_5H_{11}	A, C, D, E	**8**	35,44
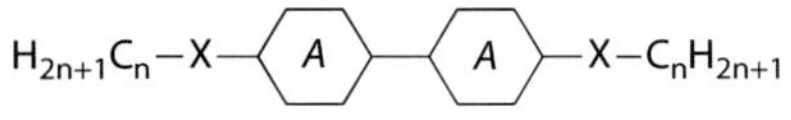 $H_{2n+1}C_n$–X–(A)–(A)–X–C_nH_{2n+1}	A, C, D, E	**9**, X=CH_2CH_2, n=1, 3, 5	29,35,50,51
	A, D, E	**10**, X=CC, n=5	29,51
 C_5H_{11}–(A)–C_6H_4–OR	C, D	**11**, R=CH_3	52
	A, C, D	**12**, R=$C_{12}H_{25}$	53
	A, C, D	**13**, R=$(CH_2)_6C_6F_{13}$	53
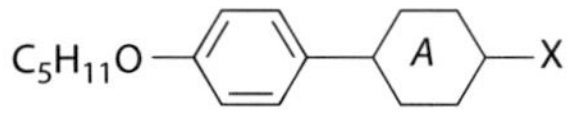 $C_5H_{11}O$–C_6H_4–(A)–X	A, D	**14**, X=COOH	54
	A, D	**15**, X=$COOC_5H_{11}$	54
	A, D	**16**, X=CH=CHCOOH	54
	A, D	**17**, X=CH=CHCOOEt	54
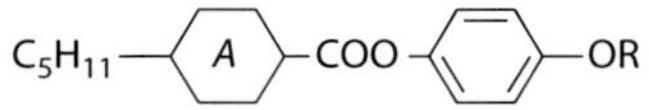 C_5H_{11}–(A)–COO–C_6H_4–OR	A, C, D	**18**, R=C_nH_{2n+1}, n=5, 10, 12	27,53,55
	A, C, D	**19**, R=$(CH_2)_6C_6F_{13}$	53
	A, B, C, D	**20**, R=$(CH_2)_3C_2F_5$	55
	A, B, C, D	**21**, R=CH_2Ph	55
	A, B, C, D	**22**, R=H	55
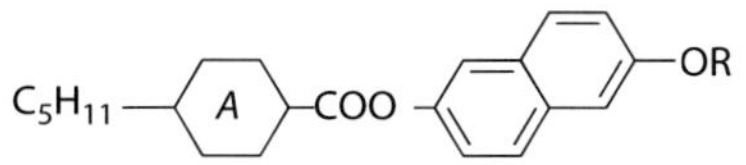 C_5H_{11}–(A)–COO–$C_{10}H_6$–OR	A, C, D	**23**, R=$(CH_2)_6C_6H_{13}$	53
	A, C, D	**24**, R=$(CH_2)_6C_6F_{13}$	53
	A, C, D	**25**, R=CH_2Ph	53
	A, C, D	**26**, R=H	53
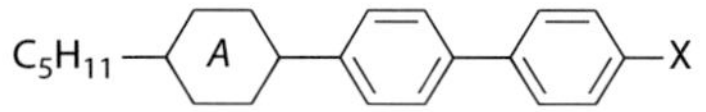 C_5H_{11}–(A)–C_6H_4–C_6H_4–X	A, B, C, D, E	**27**, X=CN	30,50,56
	A, B, C, D, E	**28**, X=OC_8H_{17}	30,56,57
	A, C, D, E	**29**, X=OC_6H_{13}	22
	A, C, D, E	**30**, X=C_7H_{15}	22

continued

TABLE 14.4 (continued)
Liquid Crystalline Derivatives of Type *I*

Structure	*A*	Substituents	References
C_5H_{11}–A–Ph–Ph(F)(X)(X)	A, B, C, D, E	**31**, X=H	58
	A, B, C, D, E	**32**, X=F	37,58
C_5H_{11}–A–Ph–Ph(F,F)–X–C_nH_{2n+1}	A, C, D, E	**33**, X=CH_2, $n = 6$	22
	A, B, C, D, E	**34**, X=O, $n = 2, 6$	22,37
$H_{2n+1}C_n$–X–Ph–A–Ph–X–C_nH_{2n+1}	A, C, D, E	**35**, X=O, $n = 1, 7$	30,59
	A, C, D, E	**36**, X=COO, $n = 6$	30,59
	A, C, D, E	**37**, X=OOC, $n = 5, 6, 7$	59
$C_5H_{11}O$–Ph–carborane–dioxane–X		**38**, X=C_nH_{2n+1}	36
		39, X=$C_6H_4C_nH_{2n+1}$	36
		40, X=$C_6H_4OC_nH_{2n+1}$	36
C_5H_{11}–carborane–CH_2CH_2–cyclohexyl–Ph–F		**41**	60,61
C_5H_{11}–A–COO–Ph–Ph–X	A, B, C, D	**42**, X=CN	56
	A, C, D, E	**43**, X=OC_8H_{17}	56,62,63
	A, C, D, E	**44**, X=C_8H_{17}	63
	A, D	**45**, X=$O(CH_2)_6C_6F_{13}$	53
	A, D	**46**, X=$OC_{12}H_{25}$	53
C_2H_5–CH(CH_3)–CH_2–A–COO–Ph–Ph–C_5H_{11}	A, C, D	**47**	38

Structure			
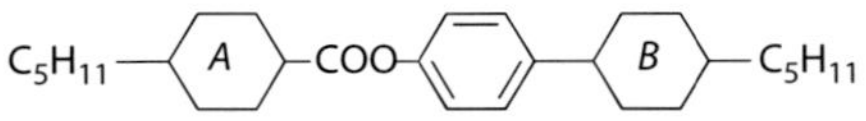 C_5H_{11}–A–COO–⬡–B–C_5H_{11}	A, B, C, D	**48**, B = A, B, C, D	52
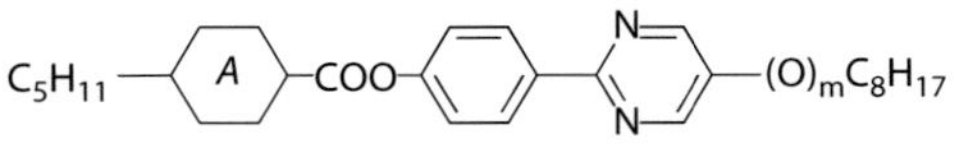 C_5H_{11}–A–COO–⬡–(N, N)–$(O)_mC_8H_{17}$	A, C, D, E	**49**, $m = 0, 1$	63
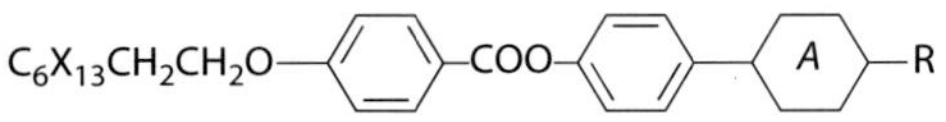 $C_6X_{13}CH_2CH_2O$–⬡–COO–⬡–A–R	A, C, D	X=H, F	64
	A, D	**50**, $R=C_5H_{11}$	65
	A, D	**51**, $R=CO_2CH(CH_3)C_6H_{13}(S)$	
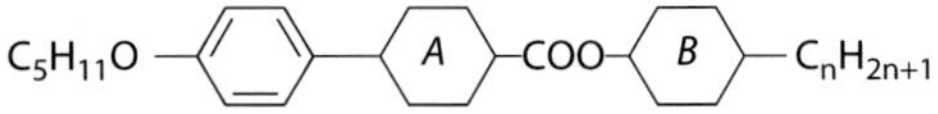 $C_5H_{11}O$–⬡–A–COO–B–C_nH_{2n+1}	A, D	**52**, *B*=A, B, C; $n = 5$	52
	D	**53**, *B*=A; $n = 7$	52
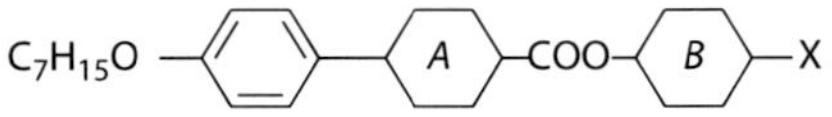 $C_7H_{15}O$–⬡–A–COO–B–X	D, E	**54**, *B*=A; $X=C_5H_{11}$, OC_4H_9	52,66
	E	**55**, *B*=B, C; $X=C_5H_{11}$	66
	E	**56**, *B*=B, X=CN	66
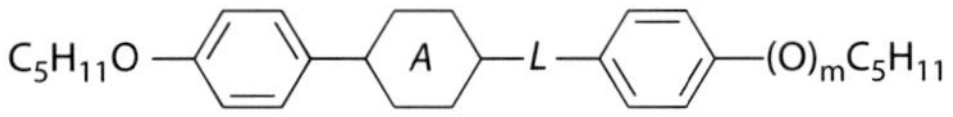 $C_5H_{11}O$–⬡–A–*L*–⬡–$(O)_mC_5H_{11}$	A, D	**57**, $L=CH_2CH_2$; COO; CH=CHCOO; CH=N; CH=CH; CONH; CH_2CH_2OOC; $m = 0, 1$	54
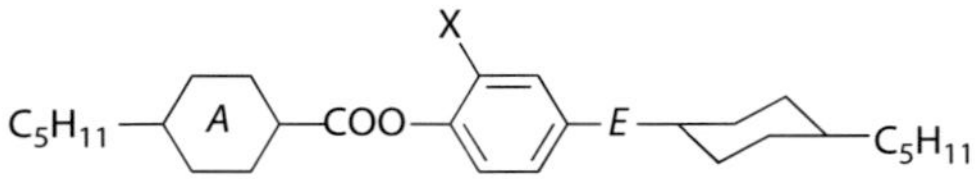 C_5H_{11}–A–COO–⬡(X)–*E*–C_5H_{11}	A, B, C, D	**58**, X=H, F $E=1,4\text{-}C_6H_4$, $\text{-}CH_2CH_2\text{-}$	67
 $H_{2n+1}C_n$–X–⬡–OOC–A–COO–⬡–X–C_nH_{2n+1}	A, B, C, D, E	**59**, X=O, $n = 1–22$	40
	A, B, C, D, E	**60**, $X=CH_2$, $n = 4$	21,59
	A, C, D, E	**61**, X=COO, $n = 3, 10$	56,59
	A, C, D, E	**62**, X=OOC, $n = 3$	59
 C_5H_{11}–OOC–A–COO–C_5H_{11}	A, B, D	**63**	56
 $C_nH_{2n+1}(O)_m$–⬡–NCH–A–CHN–⬡–$(O)_mC_nH_{2n+1}$	A, D	**64**, $m = 1$; $n = 1–10$	68
	A, D	**65**, $m = 0$; $n = 5, 6, 7$	68
	L	**66**, $m = 1$; $n = 4–18$	69

investigated. The 12-vertex *p*-carborane **D** has also been used in the preparation of several liquid crystalline polymers.[46] Mesogenic derivatives of monocarbaborates **H** and **J** are less numerous, and only relatively recent developments[42,47,48] in the chemistry of the [*closo*-1-CB_9H_{10}]$^-$ cluster (**J**) opened access to highly polar and ionic derivatives shown in Table 14.5. There are only a handful of derivatives of [*closo*-$B_{10}H_{10}$]$^{2-}$ dianion (**K**) and iron tricarbollide **L**, which are shown in Tables 14.4 and 14.5, respectively.

TABLE 14.5
Liquid Crystalline Derivatives of *closo*-Borates

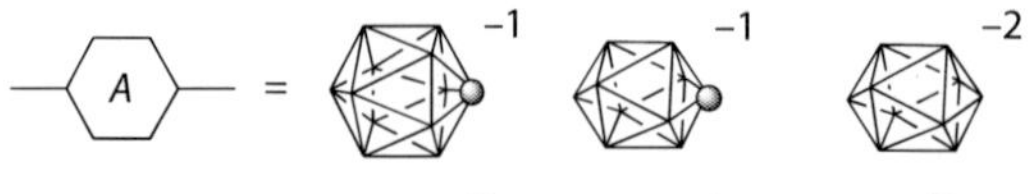

Structure	A	Substituents	References
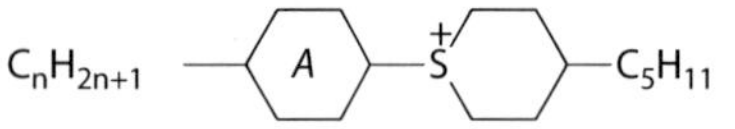	H, J	**7**, n = 5, 6, 10	41,76
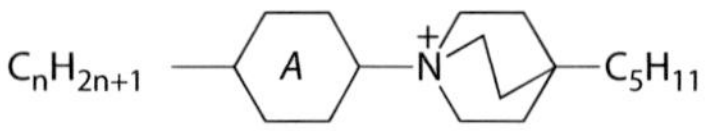	H, J	**68**, n = 5, 6, 10	41,76
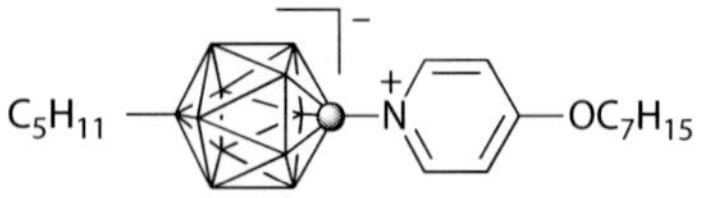		**69**	76
		70, *B*=A; X=OC_4H_9	66
		71, *B*=A, B, C; X=C_5H_{11}	45,66
		72, *B*=A; X=CN	45,66
		5, *B*=A; X=C_5H_{11}	45
		73, *B*=A; X=OC_4H_9	45
	H, J	**74**, R=$O(CH_2)_8CH_3$	77,78
	H, J	**75**, R=$OOC(CH_2)_7CH_3$	77
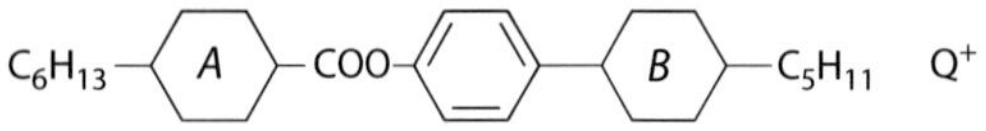	H, J	**76**, *B*=B, D	77,78
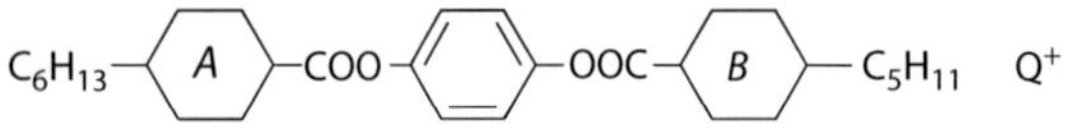	H, J	**77**, *B*=B, D	77,79
		78, R=$O(CH_2)_8CH_3$	77
		79, R=$OOC(CH_2)_7CH_3$	77
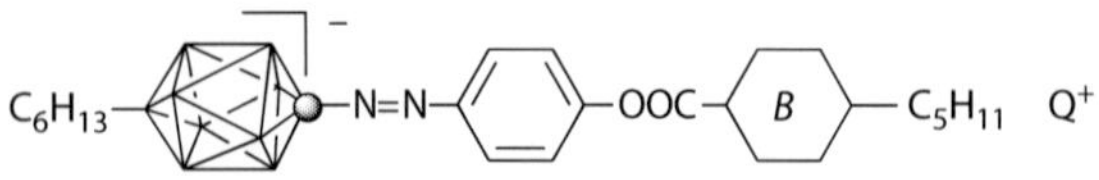		**80**, *B*=B, D	77
		6, *B*=1-$^+SC_5H_9$, R=C_5H_{11}	31,79
		81, *B*=1-Pyridine$^+$, R=OC_7H_{15}	21
		82, *B*=1-$^+NC_7H_{12}$, R=C_5H_{11}	31
		**83**	31

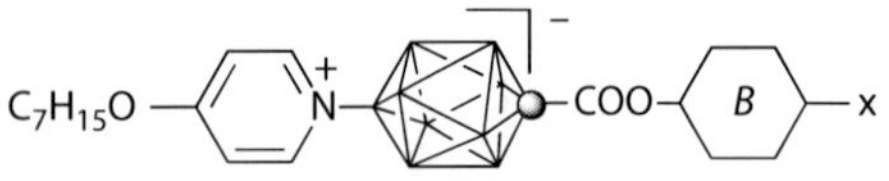

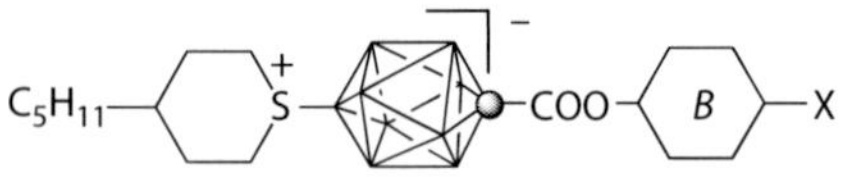

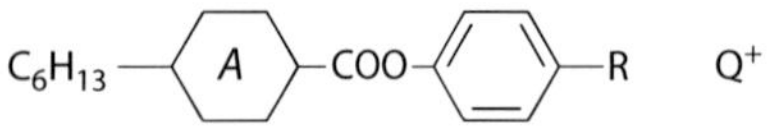

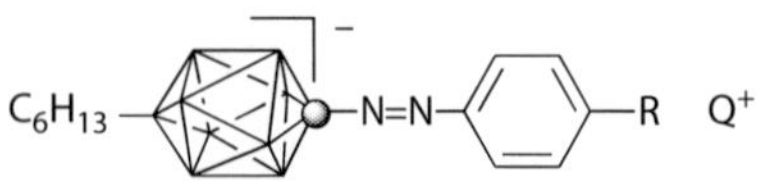

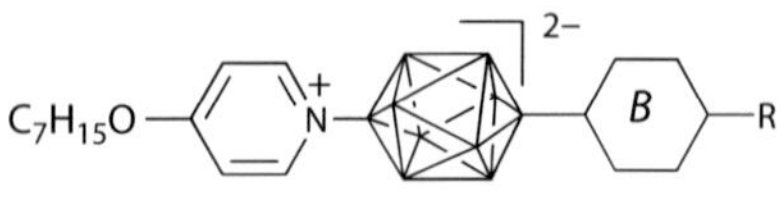

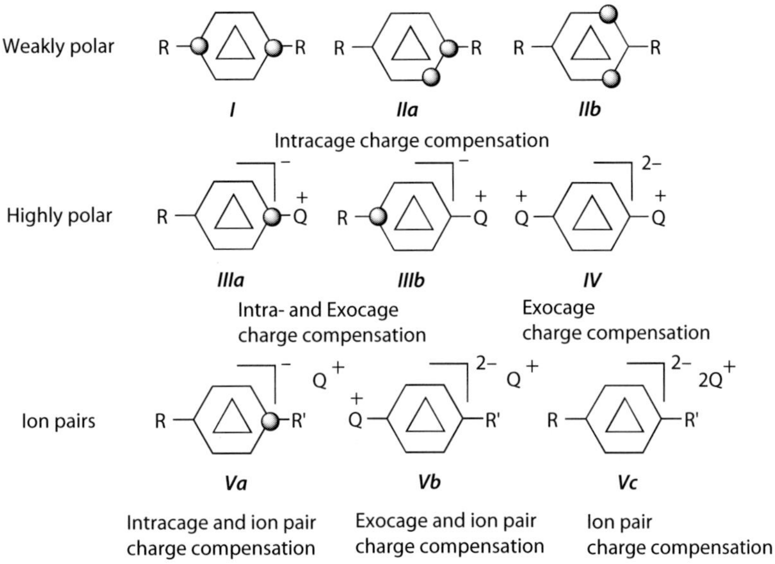

FIGURE 14.9 A general representation of five classes (**I–V**) of derivatives of *closo*-boranes (**D–K**). The full circles in **I–V** represent the C atoms, and Q is an onium fragment such as ammonium, sulfonium, or pyridinium.

All liquid crystalline derivatives containing *closo*-boranes can be classified as weakly and moderately polar, zwitterionic (highly polar and quadrupolar), and ion pairs (Figure 14.9) depending on how the double negative charge of the parent *closo*-borate is compensated.[20] The electrically neutral carboranes in which the charge is compensated internally by the carbon atoms give rise to the weakly and moderately polar mesogens of type *I* and *II* ($\mu = 0$ D for **D** and **E**, $\mu = 2.85$ D for **F**, $\mu = 4.53$ D for **G**).[49] The singly charged monocarborates can form highly polar zwitterionic mesogens of type *III* in which the dipole moment is in the range of 8–16 D. Substitution of *closo*-borate dianions with onium fragments (exocage charge compensation) leads to highly quadrupolar bis-zwitterionc derivatives of type *IV*. The negative charge of the monocarbaborates can also be compensated by a counterion in ion pairs of type *Va*. Similar ion pairs are envisioned for derivatives of [*closo*-$B_{10}H_{10}$]$^{2-}$ in which the cluster is substituted with one onium fragment and has one counterion (*Vb*) or the charge is compensated with two counterions (*Vc*).

Most LCs containing *closo*-boranes are derived from *p*-carboranes **D** and **E** (type *I*). Mesogens of type *II* are still unknown, and the major obstacle in their preparation lies in the availability of isomerically pure disubstituted carboranes. This problem was solved recently[47] for the [*closo*-1-CB_9H_{10}]$^-$ cluster (**J**) and highly polar (type *III*) and ionic mesogens (type *Va*) containing either [*closo*-1-CB_9H_{10}]$^-$ or [*closo*-1-$CB_{11}H_{12}$]$^-$ cluster have become accessible. The differential reactivity of the apical and equatorial positions of the [*closo*-$B_{10}H_{10}$]$^{2-}$ cluster permits the isolation of some 1,10-disubstituted isomers in the pure state, and several such highly quadrupolar liquid crystalline derivatives have been investigated.[31] There are no mesogenic derivatives containing the 12-vertex analog [*closo*-$B_{12}H_{12}$]$^{2-}$ mainly due to difficulties with isolation of the pure 1,12-isomer and cage functionalization. Ionic liquid crystals (ILCs) of type *Vb* and *Vc* can be designed using the [*closo*-$B_{10}H_{10}$]$^{2-}$ cluster (**K**), but such materials have not been studied to date.

14.4 STRUCTURAL EFFECTS ON THERMAL PROPERTIES

Investigation of liquid crystalline derivatives of boron clusters follows three general paths: (1) comparative studies of the ring effect on properties in series of isostructural mesogens, (2) behavior of homologous series, and (3) studies of mesogens with unique structures. The first two types of analysis are mainly used for derivatives of carboranes **D** and **E** shown in Table 14.4 for which isostructural carbocyclic analogs are easily accessible. Derivatives of monocarbaborates, **H** and **J**, and also decaborate **K** do not have carbocyclic analogs and comparative analysis is limited.

14.4.1 Derivatives of *p*-Carboranes

To date only single-, two-, three- and four-ring low-molecular-weight mesogens containing *p*-carborane in the rigid core have been investigated. Mesophase thermal stability of such *p*-carborane derivatives typically is lower than that of the bicyclo[2.2.2]octane and benzene analogs and follows the order **C** > **A** > **D** > **E**. The only exceptions are compounds with two carborane units such as **9** and **48**, which exhibit higher clearing temperatures* than those of benzene analogs.[51,52]

The clearing temperatures show dependence on the position of the carborane unit in the rigid core; a centrally located carborane typically is less favorable than that in a terminal position. For instance, a greater destabilization of the mesophase is observed on substitution of *p*-carborane **D** for benzene (**A**) in rigid three-ring mesogen **35** (for $n = 7$, $\Delta T_{NI} = -137$ K)[30] than in **28** ($\Delta T_{NI} = -72$ K),[57] **29** ($\Delta T_{NI} = -70$ K),[22] or **30** ($\Delta T_{NI} = -86$ K, Table 14.4).[22] In more flexible three-ring monoesters, the impact of the substitution of **D** for **A** is smaller, for example, in **52** (B = **A**) $\Delta T_{NI} = -86$ K, while in **48** (B = **A**) $\Delta T_{NI} = -16$ K.[52] In diester **59** ($n = 4$) the effect of **D** for **A** substitution is diminished further ($\Delta T_{NI} = -60$ K).[40]

The majority of *p*-carborane derivatives with nonfluorinated alkyl tails display exclusively nematic phases, and only a few compounds exhibit smectic behavior. For example, a monotropic smectic A (SmA) phase was found in the three-ring mesogen **28D**,[57] and **43D** exhibits a monotropic smectic C (SmC) phase in addition to the nematic phase.[62] In contrast, the carbocyclic analogs **28** and **43** and many other three-ring carbocyclic mesogens exhibit smectic polymorphism. Substitution of carborane for a carbocyclic ring in these derivatives results in dramatic destabilization of the smectic behavior. For instance, SmC is destabilized in **43D** by about 100 K relative to the bicyclo[2.2.2]octane analog **43C**.[62] In **28D** the SmA phase is less stable by about 175 K than that in **28C** or **28A**.[57] Overall, the smectic phase stability appears to follow the order **A** > **C** >> **D** ~ **E**. An example of this highly nematogenic character of *p*-carborane is the pair **59A[12]/59D[12]** in which the terephthalate has a SmC phase, while the *p*-carborane analog exhibits exclusively a nematic phase that persists up to at least $n = 22$.[40]

The smectic phase destabilization and lower mesophase stability observed for carborane derivatives has been ascribed to ring size disparity and unfavorable broadening of the molecule by the bulky carborane cage (Figure 14.4); the broadening reduces molecule's ability to pack efficiently in the lamellar arrangements that characterize smectic phases. The observed structural effects on mesogenic behavior have also been discussed in the context of the conformational properties of ring *A*.[29,35,40,44,51]

Several specific structure–property relationship issues are described below. Previous reviews have covered the chemistry of boron clusters in the context of preparation of mesogenic derivatives,[70] their electronic properties,[23] and some general aspects of structure–property relationships.[20]

14.4.1.1 Homologous Series

To date, four homologous series of carborane and one series of tricarbollide mesogens have been investigated. Series **38, 40, 64** were investigated up to $n = 10$, series **66L** to $n = 18$, and diesters **59D** were prepared to $n = 22$. In all series of carborane derivatives no smectic phases were detected. For instance, diester **59D[22]** exhibits only a nematic phase, while in the hydrocarbon analogs the smectic phases onset at $n = 5$ and completely replace the nematic phase at $n = 12$.

Analysis of homologous series demonstrates that the mesophase in all carborane-containing compounds is destabilized with increasing chain length faster than in carbocyclic analogs. The ring

* Temperature at which liquid crystal becomes an ordinary isotropic liquid.

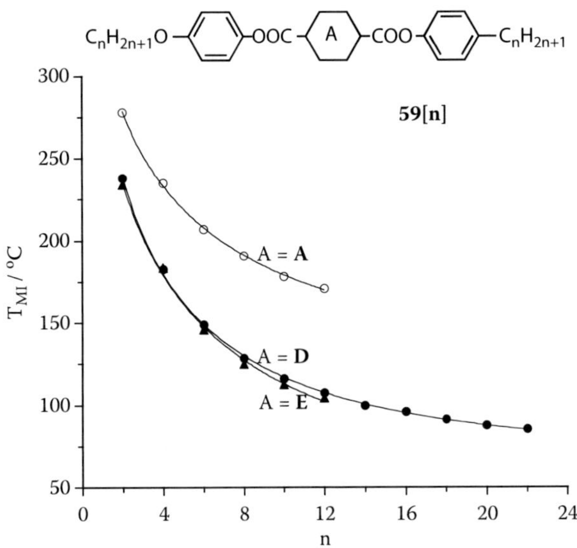

FIGURE 14.10 N–I or SmA-I transition temperatures for diesters **59[n]** and the best fit lines: **59A[n]** (open circles), $T_{MI} = 123 + \exp(5.87 - 0.58 \times \sqrt{n})$. $R^2 \geq 0.997$; **59D[n]** (full circles) $T_{MI} = 70 + \exp(6.16 - 0.73 \times \sqrt{n})$; **59E[n]** (triangles), $T_{MI} = 49 + \exp(6.09 - 0.61 \times \sqrt{n})$.

effect on the mesogenic properties was quantified by fitting the T_{MI} values for either odd or even members of each series to an empirical 4-parameter exponential function equation 14.1.[36,37,40,68,69] For example, numerical analysis of the T_{MI} values for each series (n = even) using a three-parameter function (equation 14.1, $c = 0.5$) gave limiting values $T_{MI}(\infty)$ of about 125°C for the carbocyclic derivaties **59A[n]–59C[n]**, while the limit for the diesters containing carboranes **D** and **E** is 70°C and 49°C, respectively (Figure 14.10).[40] This leads to the order of effectiveness of the structural units: **A** ~ **B** ~ **C** > **D** > **E**.

$$T_{MI}(n) = T_{MI}(\infty) + \exp(a \times n^c + b) \tag{14.1}$$

The observed significant destabilization of the mesophase in the carborane series has been accounted for by the high conformational flexibility of the carborane derivatives as compared to the carbocyclic analogs (Figure 14.6). This, in turn, results in faster decrease of the dynamic aspect ratio for the carborane derivatives than for the carbocyclic analogs and hence faster mesophase destabilization with increasing chain length (Figure 14.11).

14.4.1.2 Linking Group *L*

Comparative analysis of the effect of the linking group L (Figure 14.1a) on mesogenic properties of three-ring compounds **57** demonstrates different trends for carborane **D** and benzene **A** series.[54] As shown in Figure 14.12, variation of the linking group has little effect on the mesophase stability of the

FIGURE 14.11 Least restricted rotation around the C–C(=O) bond results in a cone curved by the alkoxyphenyl groups in **59D[4]**.

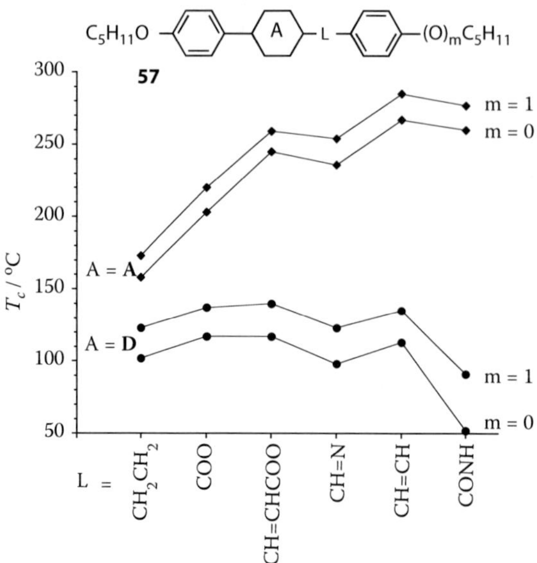

FIGURE 14.12 A plot of clearing temperatures T_C for two series of biphenyl mesogens **57A[m]** (diamonds) and *p*-carborane mesogens **57D[m]** (circles). The lines are guide for the eye.

carborane derivatives **57D**. In contrast, the clearing temperature of the analogous benzene series **57A** generally increases with the increasing π conjugation between the linking group and the adjacent benzene ring. A particularly striking difference in phase stability is observed for the carboxamide linking group (L = CONH). The amide group typically increases the clearing temperature of the mesogens when replacing an ester linkage due to the formation of intermolecular H-bonds. In carborane derivatives an effective intermolecular H-bond cannot be formed due to the shielding effect of the large carborane cage. This conclusion was supported by *ab initio* calculations, which demonstrated a much smaller exotherm for the H-bond formation in a dimeric model of carborane derivatives than for the benzamide analog.[54]

Overall, *p*-carborane interacts with the linking group L in a way similar to a saturated system such as cyclohexane. As a consequence, the choice of the linking group L has a relatively small impact on mesogenic properties of the compound. However, the bulk of the *p*-carborane diminishes the effectiveness of the carboxamide group in stabilization of the mesophases by obstructing the formation of the intermolecular H-bonds.

14.4.1.3 Terminal Connector *C*

The nature of the connector C joining the benzene ring with the terminal alkyl chain (Figure 14.1a) also has a distinctly different impact on the phase stability in derivatives of *p*-carborane (**D**) than in their isostructural carbocycles.[59] Results for several series demonstrated that mesophase stability in mesogens containing **A**, **C**, **D**, and **E** is in agreement with general trends[19,71] and follows the order (Alk)CH_2CH_2– < (Alk)OOC– < (Alk)CH_2O– < (Alk)COO–. However, as shown in Figures 14.13 and 14.14, the connecting groups (Alk)CH_2CH_2– and (Alk)OOC– destabilize the mesophase significantly stronger for carboranes (**D** and **E**) than for carbocyclic derivatives (**A** and **C**). The effect is also well illustrated in Figure 14.12 for series **57** in which the introduction of the connecting oxygen atom results in a larger increase of mesophase stability by an average of 6 ± 2 K for the 12-vertex *p*-carborane derivatives than for their benzene analogs.[54] Other examples of a substantial effect of the –CH_2O– → –CH_2CH_2– substitution is observed in pairs **54/55** ($\Delta T_{NI} = -32$ K, Table 14.4) and **70/71** ($\Delta T_{NI} = -42$ K, Table 14.5).

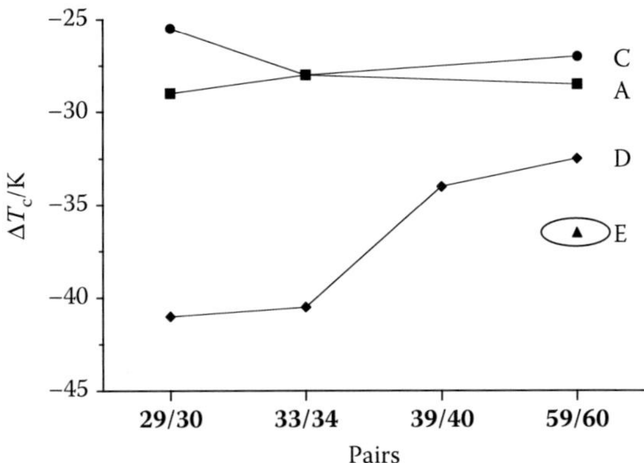

FIGURE 14.13 The change in the clearing temperature ΔT_c upon substitution $-OCH_2- \rightarrow -CH_2CH_2-$ in selected pairs of compounds. The lines are guides for the eye.

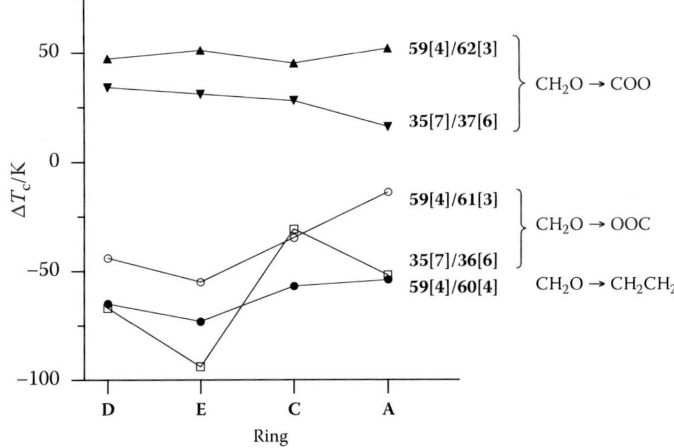

FIGURE 14.14 The change in the clearing temperature ΔT_c upon replacing of the $-OCH_2-$ connecting group with another in selected pairs of compounds. The lines are guides for the eye.

The origin of this phenomenon is unclear, but it appears to be general and is observed for compounds in which the *p*-carborane is connected directly to the substituted benzene ring or through a spacer.

In the pair of two-ring derivatives **9/10** a replacement $-CH_2CH_2- \rightarrow -C{\equiv}C-$ results in destabilization of the nematic phase by over 150 K for carborane derivatives, while in the biphenyl analog the Sm-I transition is 20 K higher.[29] This dramatic difference in behavior of the two sets of compounds has been attributed to the less correlated motion of the terminal alkyl chains in **10D** than in **10A** due to different topology of the rigid core elements: cylindrical versus planar.

14.4.1.4 Lateral Substituent *Z*

The 12-vertex carborane unit provides an effective shielding for lateral substitution of the core with fluorine atoms ($Z = F$ in Figure 14.1a). Typically, a lateral substituent decreases the aspect ratio of the mesogenic molecule, which reduces the stability of the mesophase. It has been demonstrated that the T_{NI} in such derivatives approximately linearly depends on the size of the lateral substituent.[71,72]

Results demonstrated that lateral fluorination in series **33**, **34**, and **58** (X=F) decreases the T_{NI} as compared to the respective unsubstituted compounds **30**, **29**, and **58** (X=H), and the effect of phase

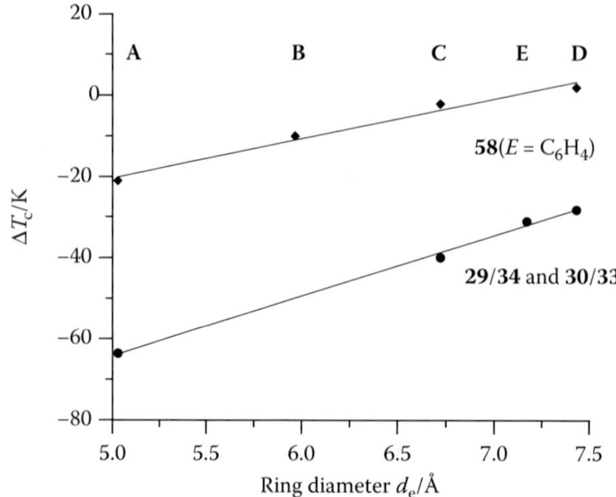

FIGURE 14.15 Average change in clearing temperatures (ΔT_{MI}) on fluorination in series **29/34** and **30/33** and also **58** plotted as a function of effective VDW diameter of ring *A*.

destabilization appears to be proportional to size[22] of ring A (Figure 14.15). The inclusion of the carborane in the studies complements previous findings for shielding effectiveness of the lateral fluorination in the series consisting of benzene, cyclohexane, and bicyclo[2.2.2]octane derivatives,[71] and permits quantitative analysis of the effect as a function of the ring size.

14.4.1.5 Terminal Chain Fluorination $X = R_F$

Partial fluorination of the terminal alkyl chain ($X = R_F$ in Figure 14.1a) induces smectic behavior by microsegregation of fluorous and nonfluorous molecular components.[73] This method is also effective for generation of smectic phases in carborane derivatives, which otherwise exhibit exclusively nematic behavior.[53,55,64,65] The incorporation of a fluorinated two-carbon[55] fragment C_2F_5 into the molecular structure of **18** induces the smectic behavior only in carbocycles **20**, and in general lowers the clearing point by as much as 25 K in **20D** relatively to **18D** (Figure 14.16). In contrast, longer fluorinated alkyl chain C_6F_{13} increases the clearing temperatures and induces smectic behavior even

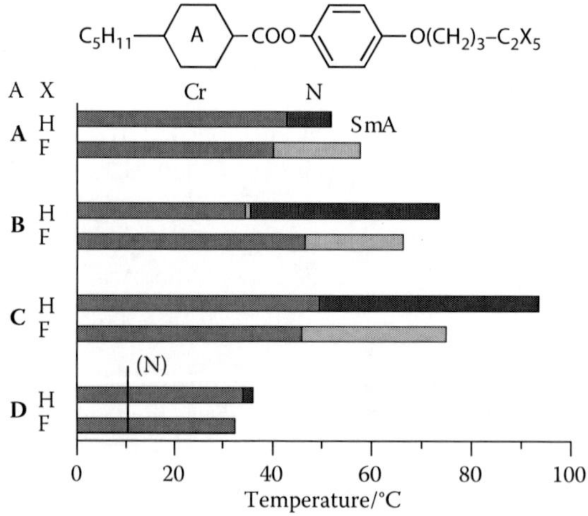

FIGURE 14.16 Transition temperatures for series **18** and **20**.

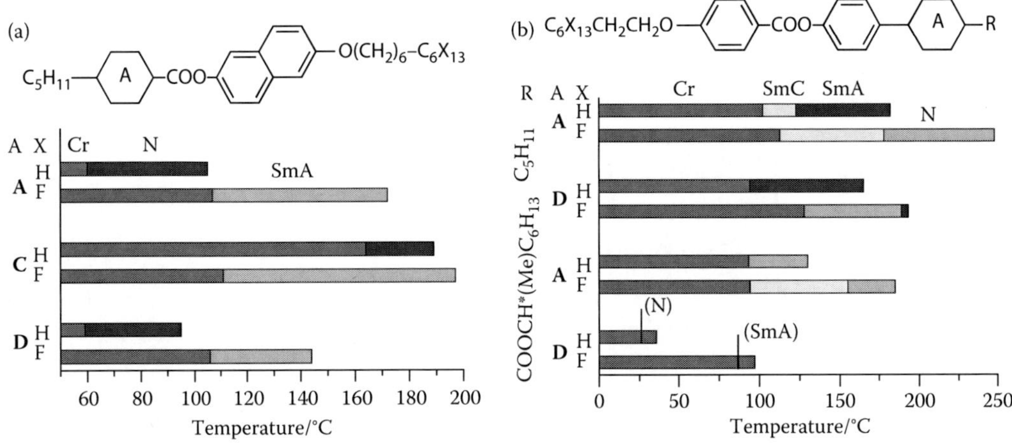

FIGURE 14.17 Transition temperatures for series **23** and **24** (a), and **50** and **51** (b).

in carborane derivatives **19D**, **24D** (Figure 14.17a), **45D**, and **50D**, and **51D** (Figure 14.17b) as compared to their hydrocarbon analogs.[53,64,65] Such carborane derivatives are candidates for materials exhibiting the so-called de Vries smectic phases[74] and applications in FLC displays.

14.4.2 Derivatives of *m*-Carboranes

Unlike *p*-carboranes, which gives rise to linear molecules when substituted on the carbon atoms, *m*-carborane (**F**) provides an angular substitution pattern (~120°) characteristic for the head group *H* of bent-core mesogens[14] (Figure 14.1b). However, in contrast to typical planar aromatic head groups, *m*-carborane is voluminous and sterically demanding, which is not optimal for the formation of lamellar phases.

The performance of *m*-carborane as the head group *H* was tested using several members of a homologous series of symmetric compounds **67[n]** (Figure 14.18), which were investigated by optical, calorimetric, x-ray diffraction, and electro-optical methods.[75] Results show that only the two shortest members of the series, **67[9]** and **67[10]**, exhibit an intercalated lamellar B6 phase (SmA_c) with the layer spacing of about 25 Å measured for the former. For comparison, analogous series of esters were prepared with adamantane as the head group *H* (Figure 14.18). The compounds exhibit

$C_nH_{2n+1}O$ … OC_nH_{2n+1}

67[n]

FIGURE 14.18 Bent-core mesogens containing *m*-carborane or adamantane head group.

a monotropic weakly birefringent soft crystalline or crystalline phase with homochiral domains, which do not undergo polarization switching in an electric field.[75]

The difference in behavior of the two series of compounds, mesogenic derivatives of *m*-carborane versus nonmesogenic adamantane analogs, was rationalized by higher conformational mobility of the *m*-carborane–Ph junction (fivefold symmetry rotational potential) than the adamantane–Ph (threefold symmetry rotational potential), which leads to less ordered phases in the former.

14.4.3 Derivatives of *closo*-Monocarbaborates

There are nearly 30 derivatives of monocarbaborates [*closo*-1-$CB_{11}H_{12}]^-$ (**H**) and [*closo*-1-$CB_9H_{10}]^-$ (**J**), which have been specifically designed and investigated as polar or ILCs (Table 14.5).

14.4.3.1 Polar Derivatives

First zwitterionic derivatives of clusters **H** and **J** were two-segment compounds **7**, **68**, and **69** of the general structure *IIIa* (Figure 14.9) in which the onium fragment at the C(1) position is a substituted quinuclidinium, sulfonium, or pyridinium.[20,41,76] All zwitterions are high-melting solids. The quinuclidinium derivatives **68** melt with decomposition above 350°C, while the sulfonium **7** and pyridinium **69** melt about 200°C. Although most of these compounds exhibit rich solid state and possibly soft crystalline polymorphism, they do not form liquid crystalline phases. Nevertheless, solution studies in a nematic host revealed reasonably high virtual N–I transition temperatures above 100 °C (*vide infra*). The observed difference in thermal behavior of the compounds has been attributed to the less favorable packing of the sulfonium and pyridinium in the crystal lattice than the quinuclidinium.

The development of synthetic access[42] to the amino acid [*closo*-1-CB_9H_8-10-NH_3-1-COOH] opened access to derivatives of type *IIIb* (Figure 14.9). The substitution of the onium fragment to the 10- position of the {*closo*-1-CB_9} cluster resulted in series **5**, **70–73**, which are characterized by a lower net molecular dipole moment than in compounds of type *IIIa* (~10 D vs. 15 D), greater structural versatility, and higher solubility in organic mesogens.[45,66] Surprisingly, most of these derivatives exhibit nematic phases with the clearing temperatures in the range of 100–220°C.

Mesogenic properties of pyridinium derivatives, **70–72**, have been compared to those of the non-zwitterionic analogs **54–56** in which the polar N^+–B^- fragment is replaced with the isosteric nonpolar fragment C–C.[66] The two series have practically the same geometrical and conformational properties and provide a unique means to assess the effect of molecular dipole moment on phase stability. Analysis of the two series demonstrated that the polar compounds **71** with alicyclic fragments, cyclohexane (*B* = **B**) and bicyclo[2.2.2]octane (*B* = **C**), have higher N–I transition temperatures than the non-zwitterionic carborane analogs **55** by about 55 K (Figure 14.19). Surprisingly, the nematic phase in benzene derivatives **70** and **71** (*B* = **A**) is markedly less stabilized relative to the corresponding carborane derivatives, and in the cyano ester **72** is even less stable by 4 K than in **56**. The observed trend in the ΔT_{NI} for the three phenol esters parallels that in the Hammett substituent parameter σ_p and the more electron-donating the greater phase stabilization.[66]

The sulfonium derivatives **5** and **73** form less stable nematic phases than the pyridinium analogs.[45] The butoxy derivative **73** has lower T_{NI} than the pyridinium **70** by 59 K and the pentyl derivative **5** crystallizes at about 60°C, 54 K below T_{NI} for pyridinium **71**, without showing a mesophase. Two other derivatives **84a**[42] and **84b**[45] permitted the assessment of the substituent effect on melting temperature. Thus, extending the alkoxy chain from methoxy (**84a**) to butoxy (**84b**) lowers the mp by 17 K, and appending a pentyl chain in **73** lowers the melting point by additional 38 K.

84a, X = OCH_3
84b, X = OC_4H_9

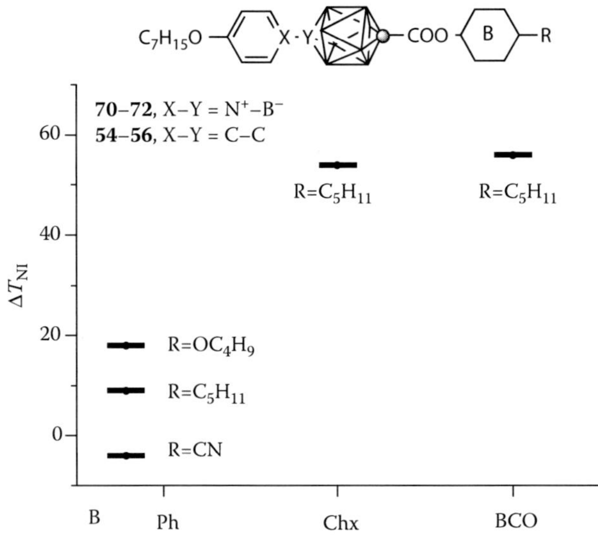

FIGURE 14.19 A plot of the ΔT_{NI} between polar (**70–72**) and nonpolar analogs (**54–56**).

14.4.3.2 Ionic Liquid Crystals

An appropriate substitution of the [*closo*-1-$CB_9H_{10}]^-$ and [*closo*-1-$CB_{11}H_{12}]^-$ clusters leads to highly anisometric anions that are suitable for ILCs (ILC, type *Va* in Figure 14.9). A typical ILC comprises a large organic cation that drives the mesogenicity, and a small counterion, and such LCs are of interest for applications in dye-sensitized solar cells.[80] ILCs in which the anisometric anion drives the mesogenic behavior offer a unique access to solid-state battery electrolytes with anisotropic Li^+ conduction. In this context, the high delocalization of the negative charge in the clusters **H** and **J** and their low nucleophilicity are favorable for obtaining high ion conductivity.[81]

To date, over a dozen salts of the general structure *Va* (Figure 14.9) have been prepared and investigated for their liquid crystalline properties.[77,78] Results demonstrated that derivatives of **H** and **J** with the NMe_4^+ counterion are nonmesogenic, while those with substituted pyridinium cation form liquid crystalline phases.[77,78] Among the two-ring derivatives, only diazene **78** exhibits a mesophase, while all three-ring mesogens display a SmA phase. A comparison of the series of esters **74J**, **75J**, and **77J** with the azo analogs **78**, **79**, and **80** demonstrates that the esters have higher clearing points and richer polymorphism than the diazenes (Table 14.5). Derivatives of the 12-vertex cluster **H** show higher degrees of order that those of the 10-vertex analogs. For instance, substitution of **H** for the 10-vertex anion **J** increases the T_C by 16 K and eliminates the SmA phase in favor of a soft crystalline phase in the pair **76J/76H** (*B* = **B**).[78]

Spectroscopic analysis of diazene **80** (*B* = **D**) found $\pi \rightarrow \pi^*$ (λ_{max} = 305 nm, log ε = 4.20) and $n \rightarrow \pi^*$ (λ_{max} = 410 nm, log ε = 2.7) transitions, which are blue shifted relative to typical absorption bands observed in azobenzene derivatives.[77]

14.4.4 Derivatives of *closo*-Decaborate

Several three-ring, highly quadrupolar derivatives of the [*closo*-$B_{10}H_{10}]^{2-}$ cluster **K** (type *IV*, Figure 14.9) containing quinuclidinium, pyridinium, or sulfonium were prepared taking advantage of the methods developed by Knoth[82] and Hawthorne.[83] The 1,10-bispyridinium derivative **81** exhibits a narrow range nematic phase with the clearing temperature of 225°C.[21] Replacement of one of the pyridinium rings with sulfonium in **6** lowers the transition temperatures and expands the range of the nematic phase.[31] Similar substitution of pyridinium with quinuclidinium in **82** replaces the nematic behavior with an ordered smectic phase, presumably B or L, and increases the clearing

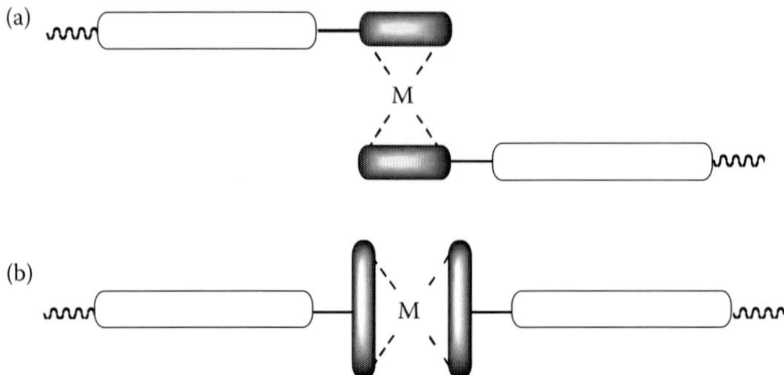

FIGURE 14.20 Topology of a metallocene-based mesogen (a) and tricarbollide derivative **66L** (b). The gray element represents the plane of the π system that is coordinated to the metal, and the white element is a mesogenic substituent.

temperature by 60 K. The second substitution of pyridinium for quinuclidinium in **83** completely eliminates mesogenic behavior and increases the melting point to about 360°C, at which temperature the material decomposes.[31]

Compound **81** and its nonpolar analog **35E** allow for an assessment of quadrupole moment impact on nematic phase stability. Thus, the double replacement of the nonpolar C–C fragment in **35E** with the polar N^+–B^- fragment in **81** increases the T_{NI} by 120 K. Both compounds exhibit a strong absorption band in the UV, 319 nm (MeCN) for **81** and 278 nm (cyclohexane) for **35E**, which has been ascribed to the cage-to-ring (former) or ring-to-cage (latter) π, π* excitation. Both compounds are also fluorescent with a large Stoke shift of about 1.4 eV,[21,30] which appears to be a general characteristics of aryl derivatives of 10-vertex *closo*-boranes. Some photophysical properties of mesogenic *closo*-boranes were reviewed before.[20,23]

14.4.5 Derivatives of Bis(tricarbollide)Fe(II)

The availability[84] of the 9,9′-diaminobis(tricarbollide)Fe(II) made it possible to investigate a series of metallomesogens **66L** that are characterized by a new topology of the π ligand in which the π face is perpendicular to the long molecular axis (Figure 14.20).[69] The orange Schiff bases display nematic, SmA, and for longer homologues also SmC phases. The clearing temperatures in the series are in the range of >370°C for $n = 4$ and 209°C for $n = 18$, and are in between those for phenylene-1,4-diamine and 4,4′-diaminobiphenyl Schiff base analogs. The compounds exhibit a strong absorption band at 307 nm (log $\varepsilon = 4.9$) ascribed to the π–π* excitation from the arene to the bis(tricarbollide)Fe(II) unit that has a large MO density near the Fe center.

14.5 DIELECTRIC AND OPTICAL PROPERTIES OF PURE MATERIALS

Dielectric properties were investigated for pure compounds **6**,[31] **32B**,[58] **32C**,[58] **32E**,[58] and **38[4]**.[36] As expected from the orientation of the net molecular dipole moment along the long molecular axis, all five compounds have positive dielectric anisotropy $\Delta\varepsilon$ ranging from +0.4 (**38[4]**)[36] to +6.3 (**32E**)[58] away from the N–I transition. Temperature dependence of the dielectric parameters obtained for the first four mesogens exhibits normal behavior as shown for **38[4]** in Figure 14.21.[36]

Analysis of dielectric data using the Maier–Meier theory,[85] which relates molecular and bulk properties of the mesogen, allowed for partial conformational analysis of the dioxane derivative **38[4]**.[36]

Birefringence (or optical anisotropy, $\Delta n = n_e - n_o$) was investigated for two compounds, ester **18**[27] and dioxane **38[4]**.[36] Since the *p*-carborane cage **D** has large polarizability and nearly spherical shape, its derivatives are expected to exhibit high isotropic refractive index and low birefringence.[20,29,35] Indeed, in series **18** the carborane derivative **18D** has the largest isotropic refractive

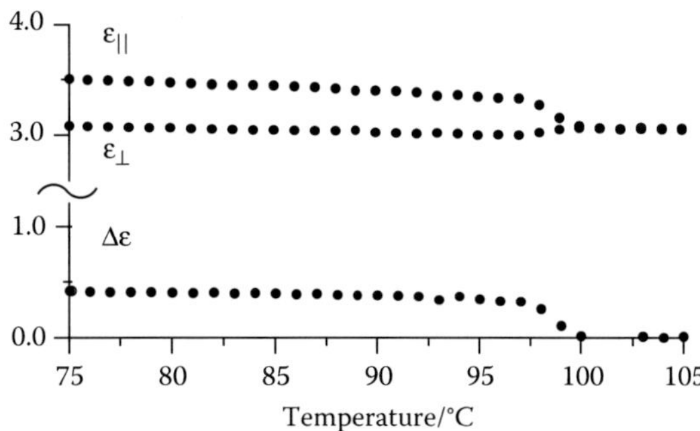

FIGURE 14.21 Dielectric parameters for **38[4]** as a function of temperature. Estimated standard deviation for $\varepsilon_{\parallel}$ and $\varepsilon_{\perp}$ is 0.1, and for $\Delta\varepsilon$ is 0.01.

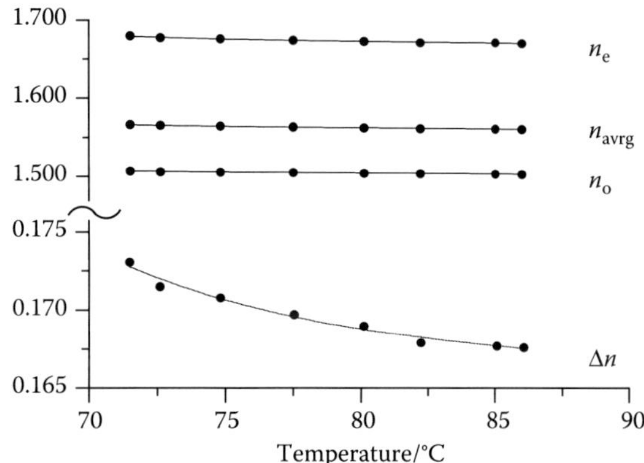

FIGURE 14.22 Refractive indices and birefringence measured for **38[4]** at 589 nm as a function of temperature.

index, $n_{avrg} = 1.533$, and lowest anisotropy, $\Delta n = 0.057$ ($T = T_{NI}$–13°C, $\lambda = 589$ nm).[27] For dioxane **38[4]** the optical parameters are $n_{avrg} = 1.560$ and $\Delta n = 0.1675$ at $T = T_{NI}$–15°C ($\lambda = 589$ nm) and their temperature dependence is shown in Figure 14.22.[36]

The birefringence of **6** in a planar cell was estimated at $\Delta n = 0.31$ at $\lambda = 527$ nm using the interference fringe method and fitting with a Sellmeier-type equation.[79]

Bis-zwitterions derived from the [*closo*-$B_{10}H_{10}]^{2-}$ cluster (**K**) having one π and one σ onium substitents undergo a photo-induced directional cage-to-π ring intramolecular charge transfer which results in an increase of the net molecular dipole moment. Such compounds are expected to exhibit nonlinear properties.[23,50] Analysis of absorption spectra in different solvents demonstrated a moderate negative solvatochromic effect for **6**, **82**, and several of their precursors.[20,23,31] Solution measurements for **6** gave the value of $\beta = 45 \pm 10 \times 10^{27}$ m^{-3} obtained by the hyper-Rayleigh method at 1064 nm in CH_2Cl_2.[31] Investigation of a planarly aligned sample of **6** revealed a low level of optical second-harmonic generation in glassy and polycrystalline phases, but not in a fluid nematic phase.[79] The main components of the second-order nonlinear susceptibility were calculated at χ^2_{xxx} = 2.3 ± 0.3 pmV^{-1} and $\chi^2_{xyy} = 0.4 \pm 0.1$ pmV^{-1}. This indicates a low degree of spontaneous polar order in the sample confined in a planar cell.

14.6 MESOGENIC MIXTURES CONTAINING *closo*-BORANE ADDITIVES

Many mesogens containing boron clusters have been investigated in binary mixtures[27,57,62,64] and also as low concentration additives to liquid crystalline hosts.[22,36–38,41,44,45,51,58,63,65] Complete phase

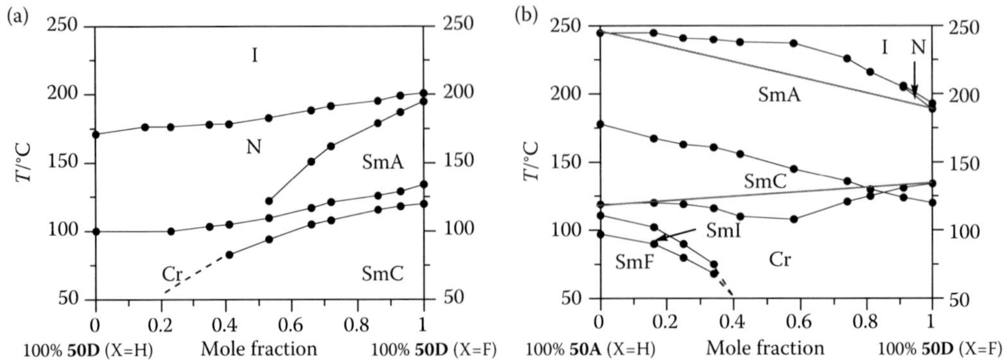

FIGURE 14.23 Isobaric binary phase diagram for (a) **50D** (X=H) and **50D** (X=F), and (b) **50A** (X=F) and **50D** (X=F).

diagrams provide information about miscibility of these partially inorganic compounds with all-organic LCs. Low concentration studies permit one to extrapolate the N–I transition temperature (virtual N–I transition) and electro-optical parameters, which in turn give insight into solute–solvent interactions. Such solution studies are an important first step toward mixture formulation and tuning of material properties for practical applications.

14.6.1 Phase Diagrams of Mesogenic Mixtures

Analysis of a number of phase diagrams demonstrated that the N–I transition typically has linear dependence on the concentration[27,57,62,64] and that the smectic behavior of the host is quickly suppressed[27] as shown[64] for the pair **50D** (X=H) and **50D** (X=F) in Figure 14.23a. In rare cases an expansion of the smectic phase is observed[57,62,64] as shown for **50A** (X=F) and **50D** (X=F) in Figure 14.23b.

The nearly linear dependence of T_{NI} on concentration has been used to extrapolate $[T_{NI}]$ for compounds that do not form liquid crystalline phases and also for several mesogens.[41,44,45,51,66] An example of such an analysis for **71** (B = **A**) is shown in Figure 14.24.[45] It has been observed that some 10-v p-carborane derivatives have nonlinear behavior at low concentrations <10 mol%.[44,51] This has been ascribed to the symmetry of the cluster and conformational properties of its derivatives (see Figure 14.6), and, consequently lower compatibility with the host. A particularly striking example is **8E** for

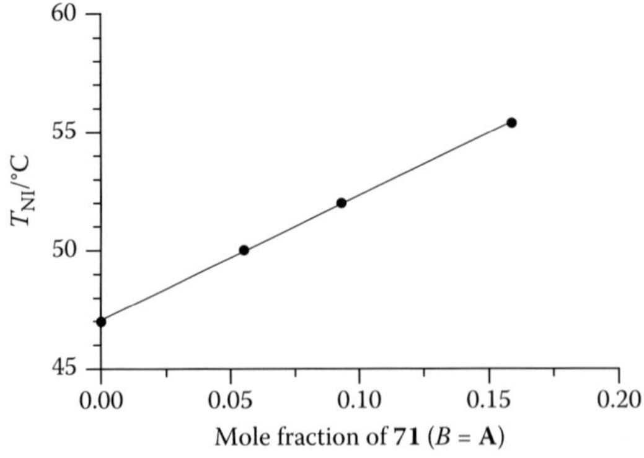

FIGURE 14.24 Nematic–isotropic transition temperature (T_{NI}) as a function of the mole fraction of **71** (B = **A**) in ClEster host (Chart 14.1).

which the virtual [T_{NI}] is significantly lower than that for the 12-vertex analog **8D** and the value dramatically depends on the inclusion of the datapoint for the host (Table 14.6).

The virtual [T_{NI}] values for mesogens are often lower than those for pure compounds and are host dependent as seen for **5** in Table 14.6.

TABLE 14.6
Virtual N–I Transition Temperatures [T_{NI}] for Selected Mesogens

Compound	Host[a]	T_{NI} (°C)[b]	References
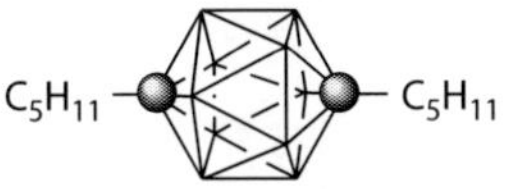 C_5H_{11} — C_5H_{11} **8D**	MeBCO	[−18]	44
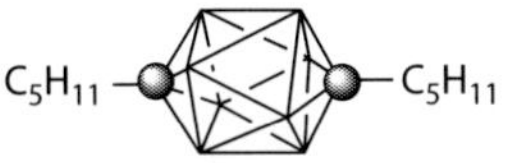 C_5H_{11} — C_5H_{11} **8E**	MeBCO	[−109][c] [−65]	44 35
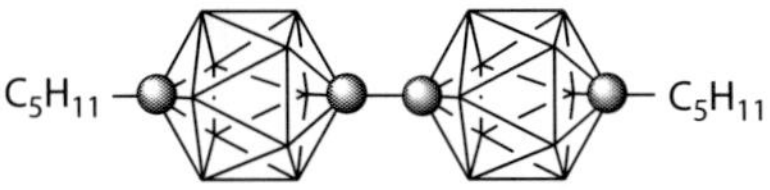 C_5H_{11} — C_5H_{11} **9D [3]**	MeBCO	[156], 172.5	35
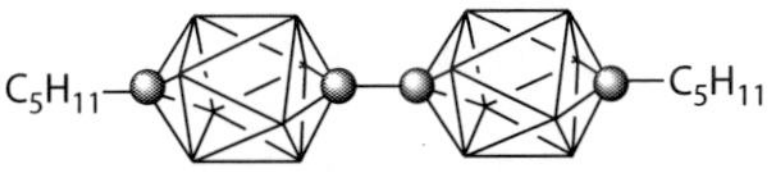 C_5H_{11} — C_5H_{11} **9E [3]**	MeBCO	[75], 98	35
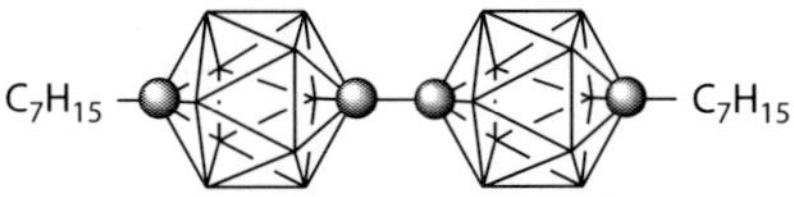 C_7H_{15} — C_7H_{15} **9D [5]**	MeBCO	[142], 148	51
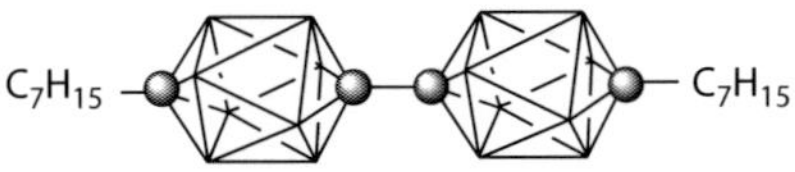 C_7H_{15} — C_7H_{15} **9E [5]**	MeBCO	[70], (82.5)	51
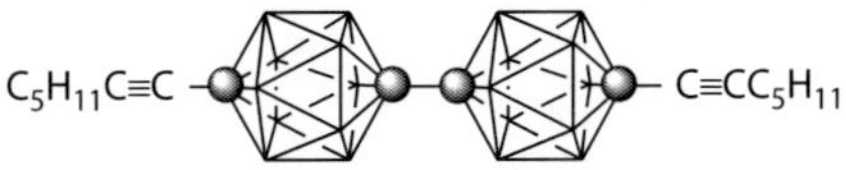 $C_5H_{11}C{\equiv}C$ — $C{\equiv}CC_5H_{11}$ **10D**	MeBCO	[−31]	51
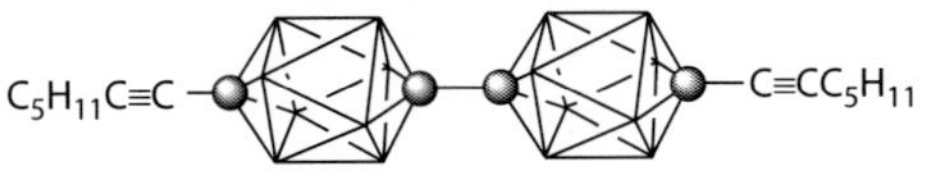 $C_5H_{11}C{\equiv}C$ — $C{\equiv}CC_5H_{11}$ **10E**	MeBCO	[−73]	51
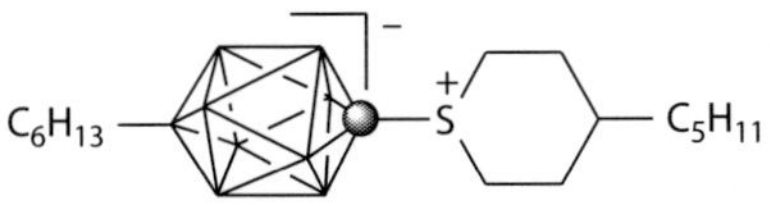 C_6H_{13} — S^+ — C_5H_{11} **7J[6]**	ClEster	[92][d]	41
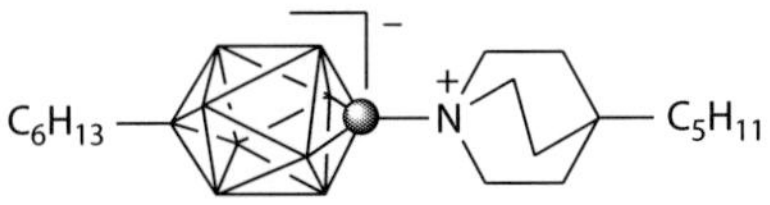 C_6H_{13} — N^+ — C_5H_{11} **68J[6]**	ClEster	[139]	41
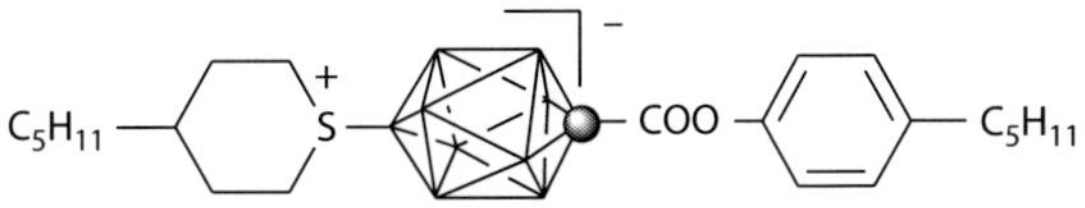 C_5H_{11} — S^+ — COO — C_5H_{11} **5**	6-CHBT ClEster	[112] [89]	45
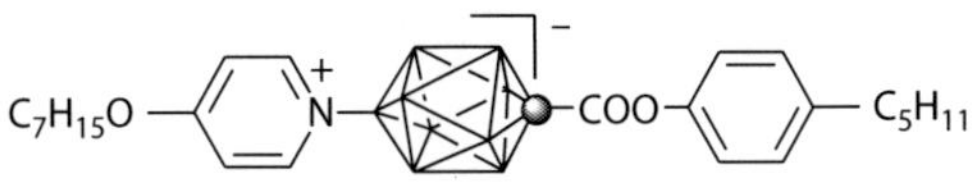 $C_7H_{15}O$ — N^+ — COO — C_5H_{11} **71** (*B* = A)	ClEster	[103], (114)	45

continued

TABLE 14.6 (continued)
Virtual N–I Transition Temperatures [T_{NI}] for Selected Mesogens

Compound	Host[a]	T_{NI} (°C)[b]	References
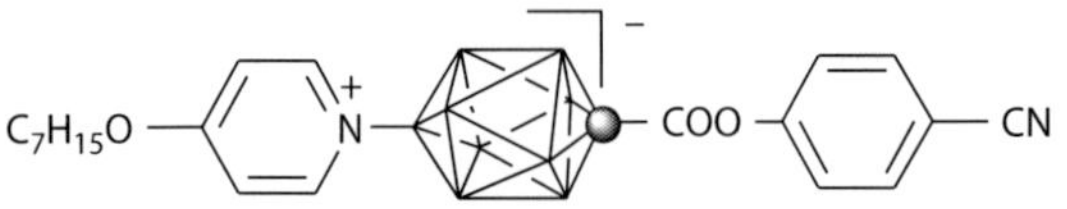 $C_7H_{15}O$–pyridinium–carborane–COO–C$_6$H$_4$–CN **72**	ClEster	[96], (129)	45
 C_5H_{11}–thianium–carborane–COO–C$_6$H$_4$–OC_4H_9 **73**	6-CHBT	[133], (98)	45

[a] Host's structure shown in Chart 14.1.
[b] Virtual temperatures in brackets and monotropic transitions in parentheses. For all extrapolated values the datapoint for the pure host is included unless specified otherwise. Typical precision of the extrapolated values ±2 K.
[c] Datapoint for pure host not included.
[d] Highly nonlinear behavior of the $T_{NI}(c)$ curve. Calculated from the 9.5 mol% point.

14.6.2 Dielectric Studies of Nematic Solutions

Dielectric parameters for compounds with negative dielectric anisotropy, series **34[2]** and **34[6]**,[22,37] were extrapolated from low concentration solutions in 6-CHBT ($\Delta\varepsilon = +7.1$, Chart 14.1) or commercial ZLI-2857 mixture ($\Delta\varepsilon = -1.4$). Compounds with positive dielectric anisotropy, **5** (B = **A**),[45] **7J[6]**,[41] **31**,[58] **32**,[37,58] **38[4]**,[36] **56**,[45] **68J[6]**,[41] **71**,[45] **72**,[45] and **73**,[45] were investigated in solution of ClEster with weakly negative dielectric anisotropy ($\Delta\varepsilon = -0.59$), 6-CHBT and ZLI-4792 with moderately moderate positive $\Delta\varepsilon$ (+7.1 and +5.3, respectively), and also in ZLI-1132 with high $\Delta\varepsilon$ (+11.5). Results showed that compounds containing carboranes **D** and **E** exhibit moderate dielectric parameters typical for this class of compounds (Table 14.7). In contrast, large values of dielectric anisotropy were extrapolated for all zwitterion additives.[41,45] Compounds **7J[6]** and **68J[6]** (type *IIIa*, Figure 14.9) have $\Delta\varepsilon$ reaching the value of +70.[41] Unfortunately, they have limited solubility and exhibit nonlinear behavior of ε with increasing concentration indicating strong aggregation in the solution. The derivatives of type *IIIb*, compounds **5**, **71–73**, are more soluble in ClEster host, and show linear dependence of ε on concentration. The extrapolated values $\Delta\varepsilon$ the cyano derivative **72** is +113(!), which is the highest ever recorded for a nematic compound and is related to an impressive longitudinal component of the dipole moment $\mu_{||} = 20.2$ D.[45] The sulfonium derivatives **5** and **73** are also soluble in 6-CHBT and ZLI-1132 hosts and their extrapolated $\Delta\varepsilon$ values are lower and about +30.

Analysis of the extrapolated dielectric values using the Maier–Meier relationship[85] allowed for a better understanding of the solute–solvent interactions and aggregation in the solution. Results

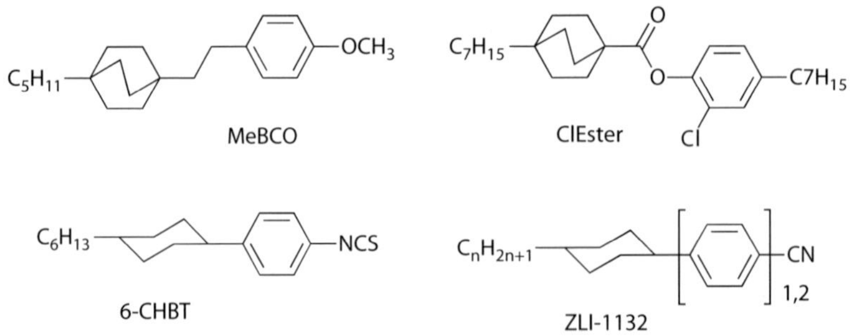

MeBCO

ClEster

6-CHBT

ZLI-1132

CHART 14.1 Nematic hosts.

TABLE 14.7
Extrapolated Dielectric Parameters for Selected Mesogens

Compound	Host[a]	$\Delta\varepsilon$[b]	$\varepsilon_{\|\|}$[b]	References
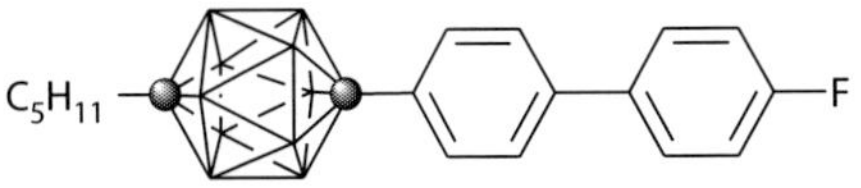 C_5H_{11} … F **31D**	6-CHBT	4.5	7.2	58
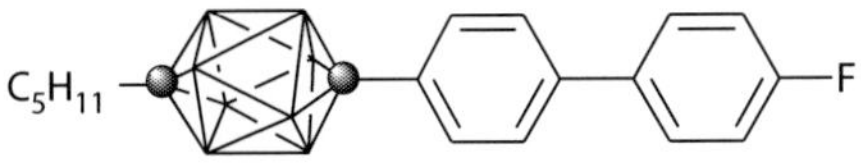 C_5H_{11} … F **31E**	6-CHBT	4.1	6.8	58
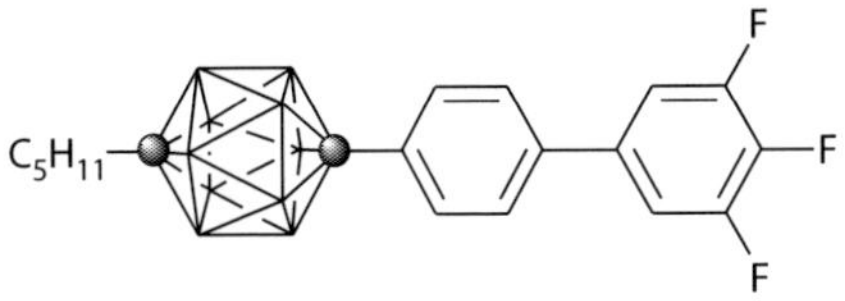 C_5H_{11} … F, F, F **32D**	6-CHBT	6.1	10.4	58
	ZLI-4792	8.7	12.7	37
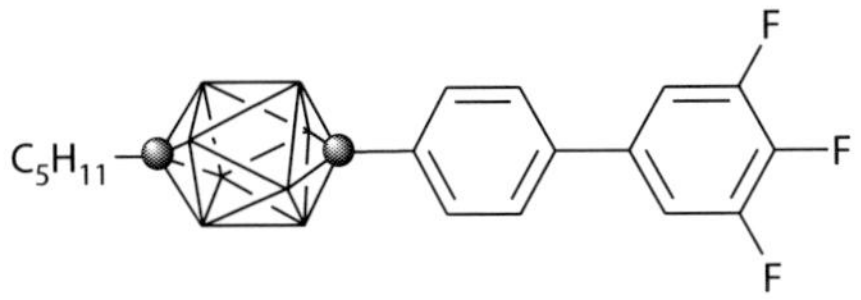 C_5H_{11} … F, F, F **32E**	6-CHBT	9.2	13.5	58
	ZLI-4792	9.5	13.6	37
		6.3[c]	10.7[c]	
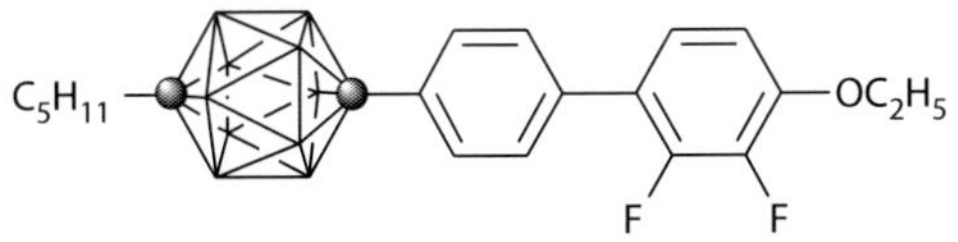 C_5H_{11} … OC_2H_5, F, F **34D [2]**	ZLI-2857	–4.6	4.1	37
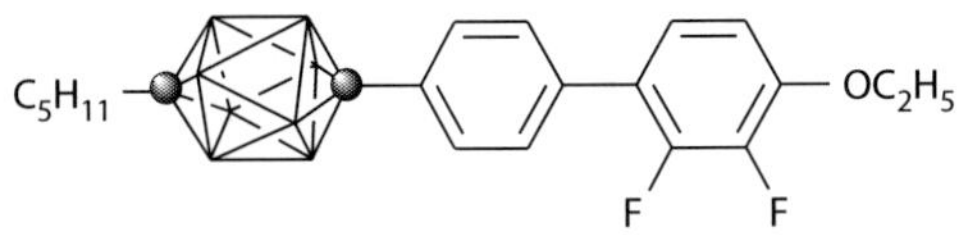 C_5H_{11} … OC_2H_5, F, F **34E [2]**	ZLI-2857	–4.7	4.2	37
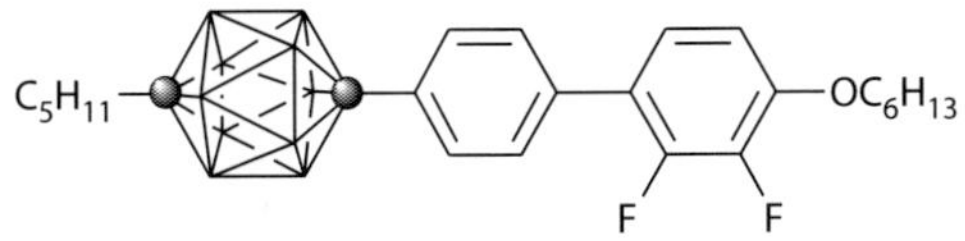 C_5H_{11} … OC_6H_{13}, F, F **34E [2]**	6-CHBT	–5.3	1.1	22
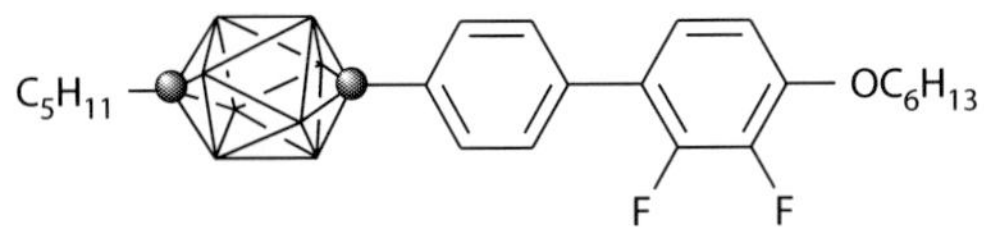 C_5H_{11} … OC_6H_{13}, F, F **34E [6]**	6-CHBT	–4.3	2.5	22
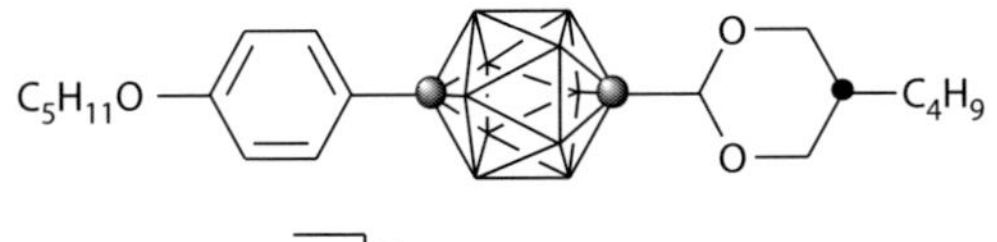 $C_5H_{11}O$ … O, O, C_4H_9 **38 [4]**	6-CHBT	0.4	4.5	36
		0.4[c]	3.45[c]	
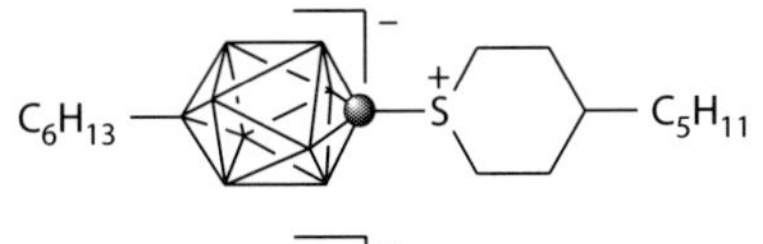 C_6H_{13} … S^+ … C_5H_{11} **7J [6]**	ClEster	61[e]	84[e]	41
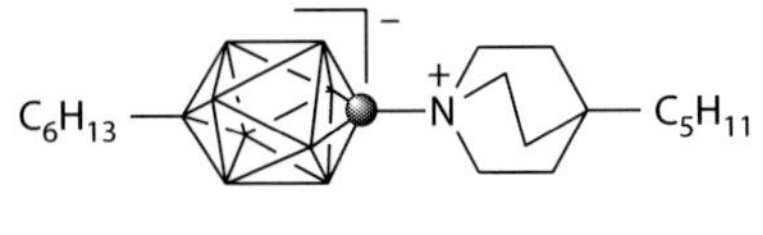 C_6H_{13} … N^+ … C_5H_{11} **68J[6]**	ClEster	70[e]	88[e]	41

continued

TABLE 14.7 (continued)
Extrapolated Dielectric Parameters for Selected Mesogens

Compound	Host[a]	$\Delta\varepsilon$[b]	$\varepsilon_{\|\|}$[b]	References
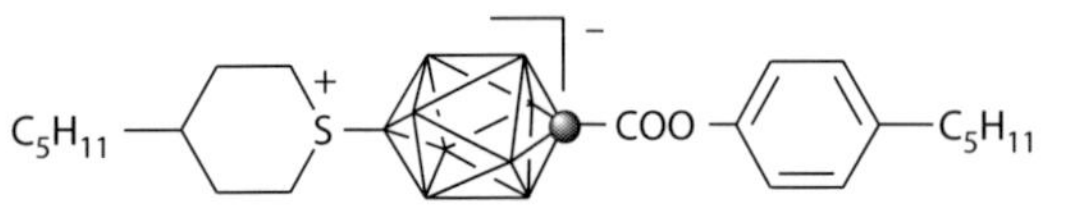 5	6-CHBT	29.5	41.3	45
	ClEster	25.3	35	
	ZLI-1132	22[d]	31[d]	
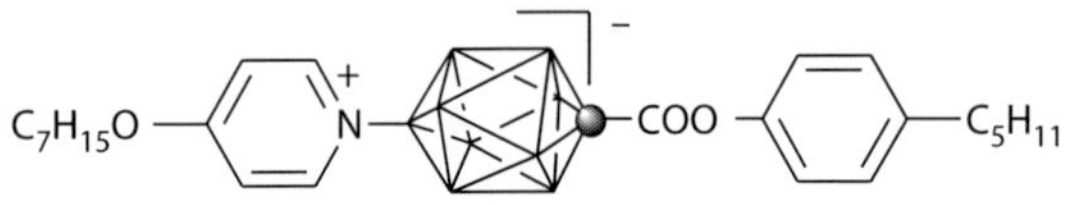 71 (B = **A**)	ClEster	42	55	45
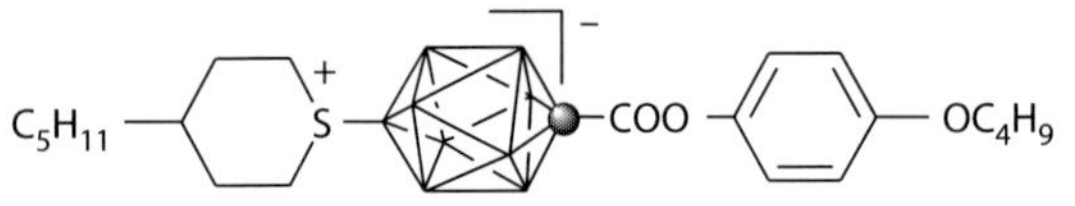 73	6-CHBT	36	48	45
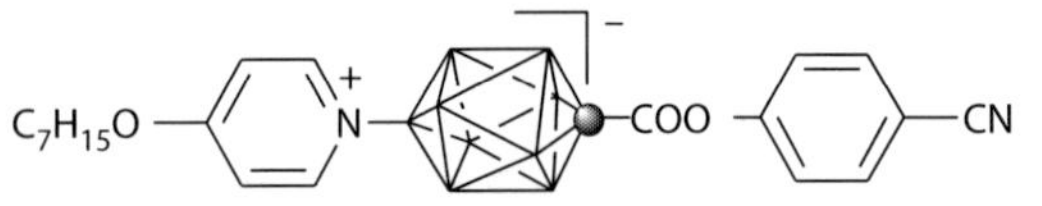 72	ClEster	113	136	45
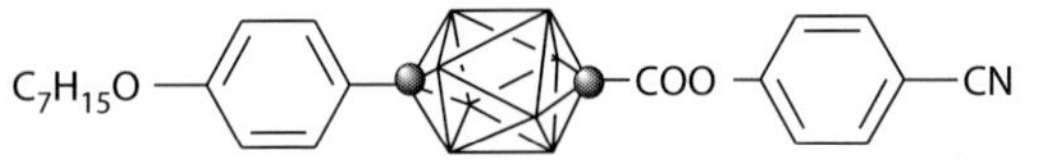 56	ClEster	15	21	45

[a] Host's structure shown in Chart 14.1.
[b] For all extrapolated values the datapoint for the pure host is included.
[c] Measured for pure compound.
[d] Single concentration extrapolation.
[e] Nonlinear behavior. Extrapolated to infinite dilution.

demonstrate that 12-vertex caboranes have relatively low local order parameter in the host.[22,58] This has been explained by the large size of the carborane cages and high conformation flexibility. Analysis of dielectric data for zwitterions **7J[6]** and **68J[6]** obtained in ClEster permitted the estimation of the association constant *K* assuming a simple model of dimerization to nonpolar pairs (Figure 14.25).[41] The resulting values of *K* are consistent with the solubility data: lower *K*, nearly twice higher solubility in the host, and lower mp are observed for sulfonium **7J[6]** than for quinuclidinium **68J[6]**.[41]

14.6.3 Optical Studies of Nematic Solutions

Refractive indices for **32** and **34[2]** that were extrapolated from ZLI-4792 solutions demonstrate again that in comparison with carbocyclic analogs, carborane derivatives have higher isotropic refractive index.[37] Derivatives containing the more spherical 12-vertex carborane **D** have lower birefringence Δn than the carbocyclic analogs, while the derivatives of the cylindrical 10-vertex carborane **E** have highest optical anisotropy. These trends are consistent with calculated values of electronic polarizability α and its anisotropy $\Delta\alpha$ for the mesogens.[20,29,35] Overall, the trends in the two series of compounds follow: **D** > **E** > **B** for n and α, and **E** > **B** > **D** for Δn and $\Delta\alpha$.

14.6.4 Effect on Viscosity of Carborane Additives

Solution studies for **34[2]** and **32** in ZLI-4792 host demonstrated that carborane derivatives increase rotational viscosity γ_1 of the nematic phase significantly stronger than the carbocyclic analogs.[37]

$2M \xrightleftharpoons{K} M_2$

$\mu = 0$ D

C_6H_{13} — C_5H_{11} C_6H_{13} — C_5H_{11}

$K = 110 \pm 2.5$ Lmol^{-1} $K = 58 \pm 1$ Lmol^{-1}

FIGURE 14.25 Idealized dimerization of two polar molecules M to a nonpolar dimer M_2 (upper) and results for **68J[6]** and **7J[6]** in ClEster host calculated for $\varepsilon = 3.03$ (lower).

The extrapolated viscosity follows the order: **D** > **E** > **B** in both series of mesogens. Similar analysis of pyridinium zwitterions **71** and **72** in ClEster host showed that they both increase viscosity γ_1 of the phase, and the increase is larger than for the sulfonium **5**.[45] The effect is host dependent and in 6-CHBT the sulfonium zwitterions **5** and **73** have significantly lower impact on viscosity.

14.6.5 Additives to an Orthoconic Antiferroelectric Host

Carborane derivatives that exhibit smectic phases are potential modifiers of orthoconic smectics. One such chiral derivative, compound **51D** (X=F, Figure 14.26) that displays Cr 97 (SmA* 93) I phase sequence, was investigated as a low concentration additive to its isostructural benzene analog **51A** (X=F).[65] Dielectric and optical analysis demonstrated that addition of 10 mol% of **51D** (X=F) to **51A** (X=F) maintains orthoconic properties (45° tilt angle), while 20 mol% of **51D** (X=F) changes the tilt angle to 40°. Contrary to expectations, in both mixtures the helical pitch p_o was significantly shorter than that in the host (Figure 14.27), which results in lower dielectric permittivity. It was noted that the effect is not linear and lower concentration of **51D** (X=F) (10 mol%) gives tighter helical pitch than the higher concentration of the additive. Spontaneous polarization P_s remains practically unchanged on addition of **51D** (X=F), while the viscosity γ_ϕ is systematically lowered (Figure 14.28) and, consequently, switching time is shortened by up to 25%. This finding is important from the standpoint of practical applications. Comprehensive analysis of the host and the two solutions revealed that the carborane dopant affects primarily the interlayer intermolecular interactions, which results in tighter pitch and consequently in higher relaxation frequency f_r and elastic constant K_φ.

14.6.6 Additives to a Ferroelectric Mixture

Four series of compounds, **43**, **44**, and **49** ($m = 0$, 1) were evaluated as 10% w/w additives to a ferroelectric liquid crystal mixture.[63] The carborane derivatives of biphenyl **43** and **44** (A = **D**, **E**) slightly increased the tilt angle Θ of the SmC material and decreased rotational viscosity γ_s, while

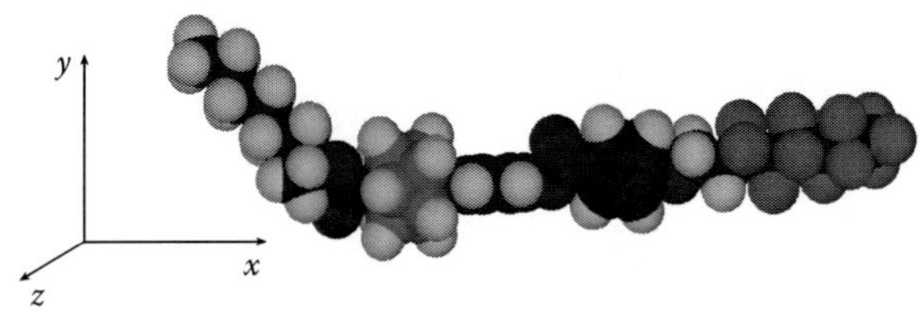

FIGURE 14.26 Space-filling model for **51D** (X=F) obtained by full geometry optimization at the HF/6–31(d) level of theory and shown in the Gaussian standard orientation.

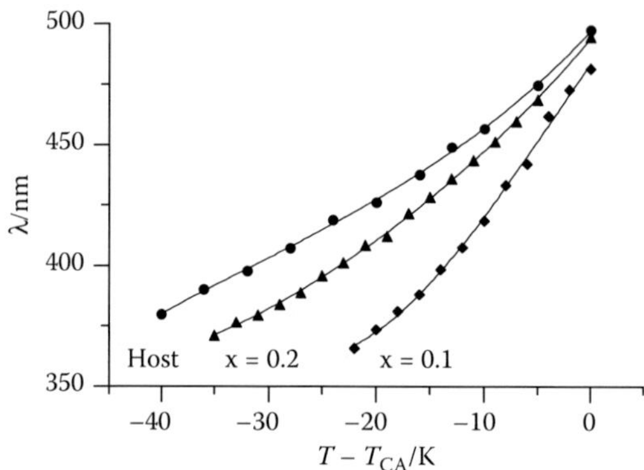

FIGURE 14.27 The wavelength of selective reflection λ as a function of relative temperature for pure host **51A** (X=F) (circles) and 0.1 (diamonds) and 0.2 mole fraction x (triangles) of **51D** (X=F) obtained on cooling.

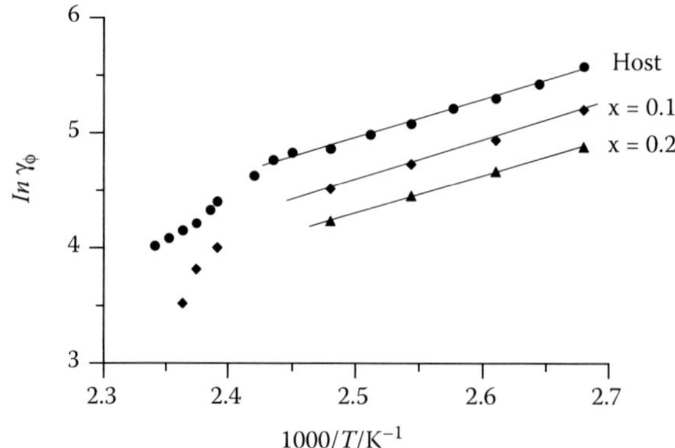

FIGURE 14.28 The Arrhenius plot of rotational viscosity γ_ϕ for pure host **51A** (X=F) (circles) and 0.1 (diamonds) and 0.2 mole fraction x (triangles) of **51D** (X=F).

analogous phenylpyrimidines **49** (A = **D**, **E**; $m = 0, 1$) reduced the tilt angle Θ by up to 34%, increased γ_s and also destabilized the SmC* phase by up to 14 K. In contrast, the carbocyclic analogs increased the Θ value up to 35%, reduced γ_s, and increased stability of the SmC* phase. Among the investigated compounds, carborane **49D** ($m = 1$) showed an exceptional ability to increase spontaneous polarization P_s^o and rotational viscosity γ_s (both normalized for Θ).

It was postulated that the markedly different effect of carborane derivatives on host's properties results from the poor packing and the larger size of the carborane units in the more densely packed tilted phase than in the orthogonal one (tilt of the smectic phase results in more efficient packing of molecules). The anomalous behavior of **49D** ($m = 1$) in the FLC mixture, was ascribed to relatively strong intermolecular interactions between the dopant and the host, which would suppress molecular motions and result in the formation of a more ordered phase. A similar stabilization of the E phase in a 3:2 mixture of **28A** and **28D** was explained by interactions between the carborane and electron-rich benzene rings.[57]

14.6.7 Chirality Transfer

A series of three isostructural esters **47** (Table 14.4) containing a chiral alkyl substituent were used to probe factors that affect transfer of molecular chirality to the bulk liquid crystalline phase.[38] The helical pitch p_o measured as a function of ester's concentration, c, in the host (Figure 14.29) gave the

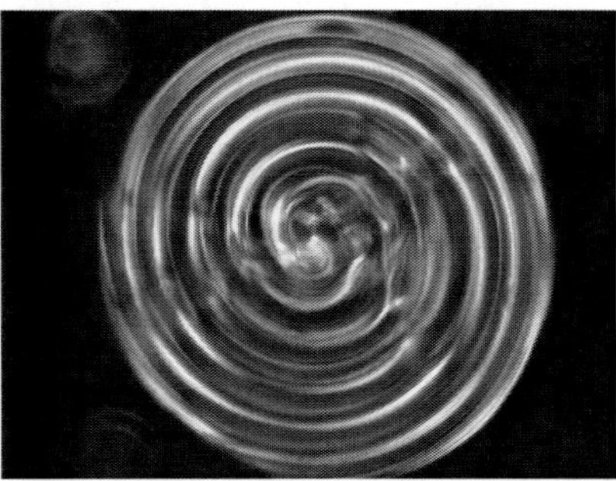

FIGURE 14.29 A microdroplet of **47D** (0.55 mol%) in a benzoate nematic host suspended in glycerol at 25°C. The droplet diameter is 0.23 mm. The helical pitch p_o is twice the distance between the spiral disclination lines.

coefficient β_M (helical twisting power or HTP, $\beta_M = (p_o c)^{-1}$), which quantifies the effectiveness of chirality transfer from the additive to the host.

Results showed that the twisting power β_M in series **47** follows the order **A** ≥ **C** > **D** in three nematic hosts. For instance, values of β_M 7.7 μm^{-1} for **47A**, 5.7 μm^{-1} for **47C**, and 1.6 μm^{-1} for **47D** were measured in 6-CHBT.[38] The observed trend correlates with the size of ring A: the larger shielding of the chirality center, the less interactions with the host molecules, and the less chirality transfer to the bulk. The observed trend also correlates with the order of rotation axis: the higher the order, the more conformationally flexible the molecule, and the less chirality transfer to the bulk. Overall, the results suggest that lower conformational freedom and higher structural rigidity of both the chiral additive and the host lead to higher β_M values and more efficient chirality transfer.

14.7 SUMMARY

Boron clusters are effective and unusual structural elements for designing calamitic and bent-core mesogens. Their structural and electronic diversity combined with judicious choice of substituents makes it possible to address fundamental aspects of liquid crystal research, to design LCs with unusual properties, and to prepare technologically important materials. Properties of these formally inorganic–organic hybrid mesogens, such as photophysical behavior, polarity, and phase structure, can be tuned by judicious choice of the cluster and substituents.

To date, seven *closo*-boranes and one metallocarbollide have been used as structural elements of LCs. Until recently, the emphases were placed on *p*-carboranes **D** and **E**, and extensive structure–property relationship studies of their derivatives resulted in a good level of understanding of impact of these structural elements on mesogenic properties. The work with clusters **D** and **E** is currently focused on mesogens exhibiting de Vries phases. The new frontier of research on boron cluster-containing LCs involves highly polar nematics and ILCs that are derived from the anions [*closo*-1-$CB_{11}H_{12}]^-$ (**H**) and [*closo*-1-$CB_9H_{10}]^-$ (**J**) and which have potential technological applications. Both types of materials require further fundamental research and development for a better understanding of structure–property relationships in these mesogens.

Research in the area of boron cluster-containing LCs has demonstrated a synergy between materials science and inorganic chemistry that drives the development of synthetic and structural chemistry of boron clusters. An example of such a synergy is the chemistry of the 1,10-disubstituted [*closo*-1-$CB_9H_{10}]^-$ anion (**J**), which was developed specifically for applications in LCs. Synthetic and structural chemistry of *m*-carborane (**F**) and *o*-carborane (**G**) and also [*closo*-$B_{12}H_{12}$] requires further development for application in LCs. Other clusters that are attractive for the design of LCs are the much less studied 6-vertex hexaborate [*closo*-$B_6H_6]^{2-}$ and 1,4-dicarbahexaborane [*closo*-1,4-$C_2B_4H_6$].

ACKNOWLEDGMENT

Support for this project has been provided by the National Science Foundation Grants (DMR-9703002, DMR-00111657, DMR-0606317, DMR-0907542).

REFERENCES

1. Reinitzer, F. Beiträge zur kenntniss des cholesterins, *Monatsh. Chem.* **1888**, *9*, 421–441.
2. Vill, V. *LiqCryst 4.8* database (LCI Publisher GmbH, Hamburg, Germany, www.lci-publisher.com) and references therein.
3. Demus, D. Types and classification of liquid crystals, in *Liquid Crystals, Applications and Uses*, B. Bahadur, Ed.; World Scientific, New Jersey, 1990, Vol. 1, pp. 1–36, and references therein.
4. Goodby, J. W.; Gray, G. W. Guide to the nomenclature and classification of liquid crystals, in *Handbook of Liquid Crystals*; Demus, D., Goodby, J. W., Gray, G. W., Spiess, H.-W., Vill, V., Eds.; Wiley-VCH: New York, 1998; Vol. 1, p. 17–23.
5. Leadbetter, A. J. Structural classification of liquid crystal, in *Thermotropic Liquid Crystals*; Gray, G. W., Ed.; Wiley: New York, 1987, p. 1–27.
6. Blinov, L. M. Behavior of liquid crystals in electric and magnetic fields, in *Handbook of Liquid Crystals*; Demus, D., Goodby, J. W., Gray, G. W., Spiess, H.-W., Vill, V., Eds.; Wiley-VCH: New York, 1998; Vol. 1, p. 477–534, and references therein.
7. Blinov, L. M.; Chigrinov, V. G. *Electrooptic Effects in Liquid Crystal Materials*; Springer-Verlag: New York, 1994, and references therein.
8. Fréedericksz, V.; Zolina, V. Forces causing the orientation of an anisotropic liquid, *Trans. Faraday. Soc.* **1933**, *29*, 919–930.
9. Tremblay, J.-F. Making fat profits in flat screens, in *Chem. Eng. News*, ACS, 2004; Vol. 82, June 28, pp. 25–26.
10. Pauluth, D.; Tarumi, K. Advanced liquid crystals for television, *J. Mater. Chem.* **2004**, *14*, 1219–1227.
11. Kirsch, P.; Bremer, M. Nematic liquid crystals for active matrix displays: Molecular design and synthesis, *Angew. Chem., Int. Ed.* **2000**, *39*, 4216–4235, and references therein.
12. S. T. Lagerwall, *Ferroelectric and Antiferroelectric Liquid Crystals*, New York, 1999, and references therein.
13. Walba, D. M. Ferroelectric liquid crystal conglomerates, in *Topics in Stereochemistry*; Green, M. M., Nolte, R. J. M., Meijer, E. W., Eds.; Wiley&Sons: Hoboken NJ, 2003; Vol. 24, p. 475–518, and references therein.
14. Reddy, R. A.; Tschierske, C. Bent-core liquid crystals: polar order, superstructural chirality and spontaneous desymmetrisation in soft matter systems, *J. Mater. Chem.* **2006**, *16*, 907–961, and references therein.
15. Pelzl, G.; Weissflog, W. Mesophase behaviour at the borderline between calamitic and banana-shaped mesogens, in *Thermotropic Liquid Crystals: Recent Advances*, A. Ramamoorthy, Ed.; Springer, Dordrecht, 2007, pp. 1–58. Jakli, A; Bailey, C.; Harden, J. Physical properties of banana liquid crystals, in *Thermotropic Liquid Crystals: Recent Advances*, A. Ramamoorthy, Ed.; Springer, Dordrecht, 2007, pp. 59–83, and references therein.
16. D'havé, K.; Dahlgren, A.; Rudquist, P.; Lagerwall, J. P. F.; Andersson, G.; Matuszczyk, M.; Lagerwall, S. T.; Dabrowski, R.; Drzewinski, W. Antiferroelectric LC with 45 deg tilt–a new class of promising electro-optic materials, *Ferroelectrics*, **2000**, *244*, 115–128.
17. Dabrowski, R.; Gasowska, J.; Otón, J.; Piecek, W.; Przedmojski, J.; Tykarska, M. High tilted antiferroelectric liquid crystalline materials, *Displays* **2004**, *25*, 9–19.
18. Tschierske, C. Non-conventional liquid crystals–the importance of micro-segregation for self-organisation, *J. Mater. Chem.* **1998**, *8*, 1485–1508, and references therein.
19. Demus, D. Chemical structure and mesogenic properties, in *Handbook of Liquid Crystals*; Demus, D., Goodby, J. W., Gray, G. W., Spiess, H.-W., Vill, V., Eds.; Wiley-VCH: New York, 1998; Vol. 1, p. 133–187, and references therein.
20. Kaszynski, P.; Douglass, A. G. Organic derivatives of *closo*-boranes: a new class of liquid crystal materials, *J. Organomet. Chem.* **1999**, *581*, 28–38, and references therein.
21. Kaszynski, P.; Huang, J.; Jenkins, G. S.; Bairamov, K. A.; Lipiak, D. Boron clusters in liquid crystals, *Mol. Cryst. Liq. Cryst.* **1995**, *260*, 315–332.

22. Januszko, A.; Glab, K. L.; Kaszynski, P.; Patel, K.; Lewis, R. A.; Mehl, G. H.; Wand, M. D. The Effect of carborane, bicyclo[2.2.2]octane and benzene on mesogenic and dielectric properties of laterally fluorinated three-ring mesogens, *J. Mater. Chem.* **2006**, *16*, 3183–3192.
23. Kaszynski, P. *closo*-boranes as p structural elements for advanced anisotropic materials, in *Anisotropic Organic Materials-Approaches to Polar Order*; Glaser, R., Kaszynski, P., Eds.; ACS Symposium Series: Washington, D.C., 2001; Vol. 798, p. 68–82, and references therein.
24. Wright, J. R.; Klingen, T. J. Investigation of gamma-ray induced polymer formation in carboranes.1. Separation and structural identification of polymer formed from 1-vinyl-*ortho*-carborane, *J. Inorg. Nucl. Chem.* **1970**, *32*, 2853–2861.
25. Thibault, R. M.; Hepburn, D. R., Jr.; Klingen, T. J. Charge scavenging in gamma-radiolysis of cyclohexane solutions of carboranes, *J. Phys. Chem.* **1974**, *78*, 788–792.
26. Reed, C. A. Carboranes: A new class of weakly coordinating anions for strong electrophiles, oxidants, and superacids, *Acc. Chem. Res.* **1998**, *31*, 133–139, and references therein.
27. Douglass, A. G.; Czuprynski, K.; Mierzwa, M.; Kaszynski, P. An assessment of carborane-containing liquid crystals for potential device application, *J. Mater. Chem.* **1998**, *8*, 2391–2398.
28. Pakhomov, S.; Kaszynski, P.; Young, V. G., Jr. 10-Vertex *closo*-boranes as potential π linkers for electronic materials, *Inorg. Chem.* **2000**, *39*, 2243–2245.
29. Kaszynski, P.; Pakhomov, S.; Tesh, K. F.; Young, V. G., Jr. Carborane-containing liquid crystals: Synthesis and structural, conformational, thermal, and spectroscopic characterization of diheptyl and diheptynyl derivatives of *p*-carboranes, *Inorg. Chem.* **2001**, *40*, 6622–6631.
30. Kaszynski, P.; Kulikiewicz, K. K.; Januszko, A.; Douglass, A. G.; Tilford, R. W.; Pakhomov, S.; Patel, M. K.; Ke, Y.; Radziszewski, G. J.; Young, V. G., Jr. unpublished results.
31. Balinski, A.; Harvey, J. E.; Kaszynski, P.; Brady, E.; Young, V. G., Jr. unpublished results.
32. Kaszynski, P.; Pakhomov, S.; Young, V. G., Jr. Investigations of electronic interactions between *closo*-boranes and triple-bonded substituents, *Collect. Czech. Chem. Commun.* **2002**, *67*, 1061–1083.
33. Kaczmarczyk, A.; Kolski, G. B. Polarizability of the closed-cage boron hydride $B_{10}H_{10}^{2-}$, *J. Phys. Chem.* **1964**, *68*, 1227–1229.
34. Kaczmarczyk, A.; Kolski, G. B. The polarizabilities and diamagnetic susceptibilities of polyhedral boranes and haloboranes, *Inorg. Chem.* **1965**, *4*, 665–671.
35. Piecek, W.; Kaufman, J. M.; Kaszynski, P. A Comparison of mesogenic properties of one- and two-ring dipentyl derivatives of *p*-carboranes, bicyclo[2.2.2]octane, and benzene, *Liq. Cryst.* **2003**, *30*, 39–48.
36. Nagamine, T.; Januszko, A.; Kaszynski, P.; Ohta, K.; Endo, Y. Mesogenic, optical, and dielectric properties of 5-substituted 2-[12-(4-pentyloxyphenyl)-*p*-carboran-1-yl] [1,3]dioxanes, *J. Mater. Chem.* **2006**, *16*, 3836–3843.
37. Jasinski, M.; Jankowiak, A.; Januszko, A.; Bremer, M.; Pauluth, D.; Kaszynski, P. Evaluation of carborane-containing nematic liquid crystals for electro-optical applications, *Liq. Cryst.* **2008**, *35*, 343–350.
38. Januszko, A.; Kaszynski, P.; Drzewinski, W. Ring effect on helical twisting power of optically active mesogenic esters derived from benzene, bicyclo[2.2.2]octane, and *p*-carborane carboxylic acids, *J. Mater. Chem.* **2006**, *16*, 452–461.
39. Haase, W.; Athanassopoulou, M. A. Crystal structures of LC mesogens, in *Structure and Bonding*; Mingos, D. M. P., Ed.; Springer: Berlin, 1999; Vol. 94, p. 139–197, and references therein.
40. Kaszynski, P.; Januszko, A.; Ohta, K.; Nagamine, T.; Potaczek, P.; Young, V. G., Jr.; Endo, Y. Conformational effects on mesophase stability: numerical comparison of carborane diester homologous series with their bicyclo[2.2.2]octane, cyclohexane and benzene analogues, *Liq. Cryst.* **2008**, *35*, 1169–1190.
41. Ringstrand, B.; Kaszynski, P.; Januszko, A.; Young, V. G., Jr. Polar derivatives of the [*closo*-1-CB_9H_{10}]$^-$ cluster as positive de additives to nematic hosts, *J. Mater. Chem.* **2009**, *19*, 9204–9212.
42. Ringstrand, B.; Kaszynski, P.; Young, V. G., Jr.; Janousek, Z. The anionic amino acid [*closo*-1-CB_9H_8-1-COO-10-NH_3]$^-$ and dinitrogen acid [*closo*-1-CB_9H_8-1-COOH-10-N_2] as key precursors to advanced materials: synthesis and reactivity, *Inorg. Chem.* **2010**, *49*, 1166–1179.
43. Douglass, A. G.; Janousek, Z.; Kaszynski, P.; Young, V. G., Jr. Synthesis and molecular structure of 12-iodo-1-(4-pentylquinuclidin-1-yl)-1-carba-*closo*-dodecaborane, *Inorg. Chem.* **1998**, *37*, 6361–6365.
44. Douglass, A. G.; Both, B.; Kaszynski, P. Mesogenic properties of single ring compounds: dipentyl derivatives of *p*-carboranes and bicyclo[2.2.2]octane, *J. Mater. Chem.* **1999**, *9*, 683–686.
45. Ringstrand, B.; Kaszynski, P. High $\Delta\varepsilon$ nematic liquid crystals: fluxional zwitterions of the [*closo*-1-CB_9H_{10}]$^-$ cluster, *J. Mater. Chem.* **2011**, *21*, 90–95.

46. Antipov, E. M.; Vasnev, V. A.; Stamm, M.; Fischer, E. W.; Platé, N. A. First observation of a columnar mesophase in a carborane-containing main-chain semiflexible copolymer, *Macromol. Rapid Commun.* **1999**, *20*, 185–189. Antipov, E. M.; Polikarpov, V. M.; Voitekunas, V. Y; Vasnev, V. A.; Stamm, M.; Platé, N. A. A comparative analysis of the structure of thermotropic liquid-crystalline boron-containing copolyethers and their organic analogs, *Vysokomolekul. Soedin. A&B*, **1999**, *41*, 432–441. Voitekunas, V. Y.; Vasnev, V. A.; Markova, G. D.; Dubovik, I. I.; Vinogradova, S. V.; Papkov, V. S.; Abdullin, B. M. Carborane-containing liquid-crystalline polyarylates, *Vysokomolekul. Soedin. A&B*, **1997**, *39*, 933–940. Kricheldorf, H. R.; Bruhn, C.; Russanov, A.; Komarova, L. LC-polyimides, 9. Poly(ester-imide)s of N-4′-carboxyphenyl)trimellitimide containing carborane 1,7-dicarboxylic acid, or isophthalic acid, *J. Polym. Sci. A*, **1993**, *31*, 279–282. Kats, G. A.; Komarova, L. G.; Rusanov, A. L. Synthesis of new para-carborane and metal-carborane cluster thermotropic liquid-crystalline polymers, *Vysokomolekul. Soedin. B*, **1992**, *34*, 62–66.
47. Ringstrand, B.; Balinski, A.; Franken, A.; Kaszynski, P. A practical synthesis of isomerically pure 1,10-difunctionalized derivatives of the [*closo*-1-CB_9H_{10}] anion, *Inorg. Chem.* **2005**, *44*, 9561–9566.
48. Ringstrand, B.; Kaszynski, P.; Franken, A. Synthesis and reactivity of [*closo*-1-CB_9H_9-1-N_2]: functional group interconversion at the carbon vertex of the {*closo*-1-CB_9} cluster, *Inorg. Chem.* **2009**, *48*, 7313–7329.
49. Laubengayer, A. W.; Rysz, W. R. The dipole moments of the isomers of dicarbadecaborane, $B_{10}H_{10}C_2H_2$, *Inorg. Chem.* **1965**, *4*, 1513–1514.
50. Kaszynski, P.; Lipiak, D. A new role for boron in advanced materials, in *Materials for Optical Limiting*; Crane, R., Lewis, K., Stryland, E. V., Khoshnevisan, M., Eds.; MRS: Boston, 1995; Vol. 374, p. 341–347.
51. Czuprynski, K.; Kaszynski, P. Homostructural two-ring mesogens: a comparison of *p*-carboranes, bicyclco[2.2.2]octane and benzene as structural elements, *Liq. Cryst.* **1999**, *26*, 775–778.
52. Ohta, K.; Januszko, A.; Kaszynski, P.; Nagamine, T.; Sasnouski, G.; Endo, Y. Structural effects in three-ring mesogenic derivatives of *p*-carborane and their hydrocarbon, Analogs *Liq. Cryst.* **2004**, *31*, 671–682.
53. Jankowiak, A.; Jatczak, M.; Kaszynski, P. unpublished results.
54. Nagamine, T.; Januszko, A.; Ohta, K.; Kaszynski, P.; Endo, Y. The effect of the linking group on mesogenic properties of three-ring derivatives of *p*-carborane and biphenyl, *Liq. Cryst.* **2008**, *35*, 865–884.
55. Januszko, A.; Kaszynski, P. A Comparison of smectic phase induction in a series of isostructural two-ring esters by tail fluorination and tail elongation, *Liq. Cryst.* **2008**, *35*, 705–710.
56. Douglass, A. G.; Mierzwa, M.; Kaszynski, P. Liquid crystals containing *p*-carborane, *SPIE* **1997**, *3319*, 59–62.
57. Czuprynski, K.; Douglass, A. G.; Kaszynski, P.; Drzewinski, W. Carborane-containing liquid crystals: A comparison of 4-octyloxy-4′-(12-pentyl-1,12-dicarbadodecaboran-1-yl with its hydrocarbon analogs, *Liq. Cryst.* **1999**, *26*, 261–269.
58. Glab, K. L.; Januszko, A.; Kaszynski, P., unpublished results.
59. Jankowiak, A.; Kaszynski, P.; Tilford, W. R.; Ohta, K.; Januszko, A.; Nagamine, T.; Endo, Y. Ring-alkyl connecting group effect on mesogenic properties of *p*-carborane derivatives and their hydrocarbon analogues, *Beils. J. Org. Chem.* **2009**, *5*, 83.
60. Douglass, A. G.; Pakhomov, S.; Reeves, B.; Janousek, Z.; Kaszynski, P. Triphenylsilyl as a protecting group in the synthesis of 1,12-heterodisubstituted *p*-carboranes, *J. Org. Chem.* **2000**, *65*, 1434–1441.
61. Differential Scanning Calorimeter analysis revealed melting to a nematic phase at 95°C and transition to an isotropic liquid at 149°C (Cr 95 N 149 I).
62. Douglass, A. G.; Czuprynski, K.; Mierzwa, M.; Kaszynski, P. Effects of carborane-containing liquid crystals on the stability of smectic phases, *Chem. Mater.* **1998**, *10*, 2399–2402.
63. Januszko, A.; Kaszynski, P.; Wand, M. D.; More, K. M.; Pakhomov, S.; O'Neill, M. Three-ring mesogens containing *p*-carboranes: Characterization and comparison with the hydrocarbon analogs in the pure state and as additivies to a ferroelectric mixture, *J. Mater. Chem.* **2004**, *14*, 1544–1553.
64. Januszko, A.; Glab, K. L.; Kaszynski, P. Induction of smectic behavior in a carborane-containing mesogen. Tail-fluorination of a three-ring nematogen and its miscibility with benzene analogues, *Liq. Cryst.* **2008**, *35*, 549–553.
65. Piecek, W.; Glab, K. L.; Januszko, A.; Perkowski, P.; Kaszynski, P. Modification of electro-optical properties of an orthoconic chiral biphenyl smectogen with its isostructural carborane analogue, *J. Mater. Chem.* **2009**, *19*, 1173–1182.

66. Ringstrand, B.; Kaszynski, P. How much can an electric dipole stabilize a nematic phase? Polar and non-polar isosteric derivatives of [*closo*-1-CB_9H_{10}]$^-$ and [*closo*-1,10-$C_2B_8H_{10}$], *J. Mater. Chem.* **2010**, *20*, 9613–9615.
67. Ringstrand, B.; Vroman, J.; Jensen, D.; Januszko, A.; Kaszynski, P.; Dziaduszek, J.; Drzewinski, W. Comparative studies of three- and four-ring mesogenic esters containing *p*-carborane, bicyclo[2.2.2] octane, cyclohexane, and benzene, *Liq. Cryst.* **2005**, *32*, 1061–1070.
68. Nagamine, T.; Januszko, A.; Ohta, K.; Kaszynski, P.; Endo, Y. A comaprison of mesogenic properties of *p*-carborane-1,12-dicarbaledyde Schiff's bases with their therephthaldehyde analogues, *Liq. Cryst.* **2005**, *32*, 985–995.
69. Januszko, A.; Kaszynski, P.; Grüner, B. Liquid crystalline derivatives of Bis(tricarbollide)Fe(II), *Inorg. Chem.* **2007**, *46*, 6078–6082.
70. Kaszynski, P. Four decades of organic chemistry of boron clusters: A synthetic toolbox for constructing liquid crystal materials. A review, *Collect. Czech. Chem. Commun.* **1999**, *64*, 895–926.
71. Toyne, K. J. Liquid crystal behavior in relation to molecular structure, in *Thermotropic Liquid Crystals*; Gray, G. W., Ed.; Wiley: New York, 1987, p. 47–57, and references therein.
72. Kelly, S. M.; Schad, H. Some novel nematic liquid crystals of negative dielectric anisotropy, *Mol. Cryst. Liq. Cryst.* **1984**, *110*, 239–261; Gray, G. W.; Kelly, S. M. Laterally substituted 4-n-alkylphenyl 4-alkylbicyclo[2.2.2]octane-1-carboxylates, *Mol. Cryst. Liq. Cryst.* **1981**, *75*, 109–119. Gray, G. W. Influence of composition and structure on the liquid crystals formedby non-amphiphilic systems, in *Liquid Crystals and Plastic Crystals*, G. W. Gray and P. A. Winsor, Eds.; Wiley& Sons, New York, 1974, Vol. 1 p. 125–135.
73. Guittard, F.; Géribaldi, S. Molecular design of highly fluorinated liquid crystals, in *Anisotropic Organic Materials-Approaches to Polar Order*; Glaser, R., Kaszynski, P., Eds.; ACS Symposium Series: Washington, D.C., 2001; Vol. 798, p. 180–194, and references therein.
74. Roberts, J. C.; Kapernaum, N.; Song, Q.; Nonnenmacher, D.; Ayub, K.; Giesselmann, F.; Lemieux, R. P. Design of liquid crystals with de vries-like properties: frustration between sma- and smc-promoting elements, *J. Am. Chem. Soc.* **2010**, *132*, 364–370, and references therein.
75. Pociecha, D.; Ohta, K.; Januszko, A.; Kaszynski, P.; Endo, Y. Symmetric bent-core mesogens with *m*-carborane and adamantane as the central units, *J. Mater. Chem.* **2008**, *18*, 2978–2982.
76. Ringstrand, B.; Pakhomov, S.; Kaszynski, P.; Douglass, A. G., unpublished results.
77. Ringstrand, B.; Monobe, H.; Kaszynski, P. Anion-driven mesogenicity: Ionic liquid crystals based on the [*closo*-1-CB_9H_{10}]$^-$ cluster, *J. Mater. Chem.* **2009**, *19*, 4805–4812.
78. Johnson, L.; Ringstrand, B.; Kaszynski, P. unpublished results.
79. Miniewicz, A.; Samoc, A.; Samoc, M.; Kaszynski, P. Observation of second harmonic generation in an oriented glassy nematic phase of a [$B_{10}H_{10}$]$^-$ derivative, *J. Appl. Phys.* **2007**, *102*, 033108.
80. Kumar, S.; Pal, K. S. Synthesis and characterization of novel imidazolium-based ionic discotic liquid crystals with a triphenylene moiety, *Tetrahedron Lett.* **2005**, *46*, 2607–2610. Yazaki, S.; Kamikawa, Y.; Yoshio, M.; Hamasaki, A.; Mukai, T.; Ohno, H.; Kato, T. Ionic liquid crystals: Self-assembly of imidazolium salts containing an L-glutamic acid moiety, *Chem. Lett.* **2008**, *37*, 538–539.
81. Bonhote, P.; Dias, A.-P.; Papageorgiou, N.; Kalyanasundaram, K.; Grätzel, M. Hydrophobic, highly conductive ambient-temperature molten salts, *Inorg. Chem.* **1996**, *35*, 1168–1178. Ohno, H. Functional design of ionic liquids, *Bull. Chem. Soc. Jpn.* **2006**, *79*, 1665–1680. Susan, M. A. B. H.; Kaneko, T.; Noda, A.; Watanabe, M. Ion gels prepared by in situ radical polymerization of vinyl monomers in an ionic liquid and their characterization as polymer electrolytes, *J. Am. Chem. Soc.* **2005**, *127*, 4976–4983.
82. Knoth, W. H. Chemistry of boranes. XXVI.1 inner diazonium salts 1,10-$B_{10}H_8(N_2)_2$, -$B_{10}Cl_8(N_2)_2$, and -$B_{10}I_8(N_2)_2$, *J. Am. Chem. Soc.* **1966**, *88*, 935–939.
83. Leyden, R. N.; Hawthorne, M. F. Synthesis of diazonium derivatives of decahydrodecaborate(2-) from arylazo intermediates, *Inorg. Chem.* **1975**, *14*, 2444–2446.
84. Grüner, B.; Backovsky, J.; Sillanpää, R.; Kivekäs, R.; Cisarová, I.; Teixidor, F.; Viñas, C.; Stíbr, B. Amino-substituted ferra-bis(tricarbollides)—Metallatricarbaboranes designed for linear molecular constructions, *Eur. J. Inorg. Chem.* **2004**, 1402–1410.
85. S. Urban, Static dielectric properties of nematics, in *Physical Properties of Liquid Crystals: Nematics*, D. A. Dunmur, A. Fukuda, and G. R. Luckhurst Eds. IEE, London, **2001**, pp. 267–276, and references therein.

15 Photoluminescence from Boron-Based Polyhedral Clusters

Paul A. Jelliss

CONTENTS

15.1 INTRODUCTION: WHY STUDY PHOTOLUMINESCENCE IN BORANES, CARBORANES, AND METALLACARBORANES?

Boranes and carboranes are comprised of σ-aromatic cage-like structures with boron and carbon vertices ranging from small tetrahedral to supra-icosahedral clusters. Despite more than four decades of work with this class of inorganic compound, there have been few reports of their photophysical properties, with a distinct paucity of information regarding luminescence. Carboranes have nevertheless been incorporated into photophysically active metal complexes, usually as ancillary components.[1–3] But what of boranes, carboranes, and metallacarboranes (where a transition metal atom can either be an integral vertex in the polyhedral skeletal framework or an exterior component thereof) as chromophores themselves and their resultant optoelectronic behavior? In this arena, there are very few contributors.

So why study molecular luminescence and particularly with such exotic species as boranes, carboranes, and metallacarboranes? In the long term one can capitalize on interesting optoelectronic activity in a number of ways. Luminescent species have been prized as probes, sensors, and sensitizers as well as photocatalysts.[4–17] Our own work is beginning to focus on a certain group of rhenacarborane complexes as photostable dye sensitizers in the so-called dye-sensitized solar cells (DSSCs) and given the universal interest in the exploration for alternative energy sources, this alone constitutes a good reason to investigate and adapt such molecules. More broadly speaking, however, boranes, carboranes, and metallacarborane complexes, which display properties such as luminescence, could not only be useful as sensitizers, mediators, and sensors but potentially do so in physiological environments where they will be metabolically unrecognizable.[18] The advantage of this is that because of their unique three-dimensional structures, there will be little chance of stable boranes, caboranes, or metallacarboranes with nonlabile ligands directly interfering with enzymatic processes, something we have demonstrated in other work that we are carrying out to adapt rhenacarborane complexes as drug-delivery vehicles for crossing the blood–brain barrier.[19]

There is a deficit of fundamental information regarding the optical and electronic responses from boranes, carboranes, and metallacarboranes and their dependence on any structural variation. Indeed there are multiple distinct methods of tuning optoelectronic properties of these molecules by substituent placement on the carborane cage carbon or boron vertices, and in metallacarborane complexes, changing the metal identity and substitution of other ligands on the metal center. The works described below show that we have just begun to scratch the surface of this phenomenon in boron-based polyhedral clusters. They have been divided into discussions of boranes and carboranes, with metallacarboranes subdivided into species comprising an exopolyhedral metal–ligand group and those comprising endopolyhedral metal–ligand moieties.

15.2 BORANES

Volkov et al. have reported basic photophysical investigations of decaborane(14) derivatives and adducts.[20] This work was motivated by the desire to find new lumophores for measuring and visualizing synchrotron radiation, requiring certain physicochemical requirements including high-quantum yield emission relative to synchrotron radiation with spectral features that match the spectral sensitivity of alkali–metallic photocathodes of photomultipliers and photoresistors of the visible range as well as with the human eye. The investigators were particularly interested in decaborane derivatives because of their chemical stability under synchrotron operating conditions and their potential to demonstrate the shortest possible "afterglow." Molecules of the type $B_{10}H_{12}[Py(X)]_2$ (**1**), where Py(X) = pyridine or an alkyl-substituted variant thereof (Figure 15.1), displayed intense fluorescence in the range λ_{em} = 525–579 nm, which intensified on cooling to 78 K.[21]

An example is shown in Figure 15.2, where fluorescence from two crystalline modifications of the basic derivative $B_{10}H_{12}Py_2$ is observed, presumably involving distinct crystal structures.

Although the emission intensities were notably less than that of the industry standard sodium *o*-oxybenzoic salicylate with λ_{em} = 450 nm (used at that time at the Synchotron Radiation Center at the Institute of Nuclear Physics of the Siberian Branch of the Russian Academy of Sciences), the emission wavelength was a much better match for the maximum spectral sensitivity of the eye (550–560 nm).

Interestingly, fluorescence following UV-excitation was also reported by the same group for *closo*-borane systems based on pyridinium salts of $[B_{10}H_{10}]^{2-}$ (**2**) and $[B_{12}H_{12}]^{2-}$ (**3**) (Figure 15.1) and these are listed in Table 15.1.[22]

Despite these interesting observations of luminescence, no further probing has been carried out on these particular systems to examine the underlying cause of the phenomenon. Therefore, what role the borane cage framework plays in photophysical acitivity relative to that of the pyridine moiety is not certain.

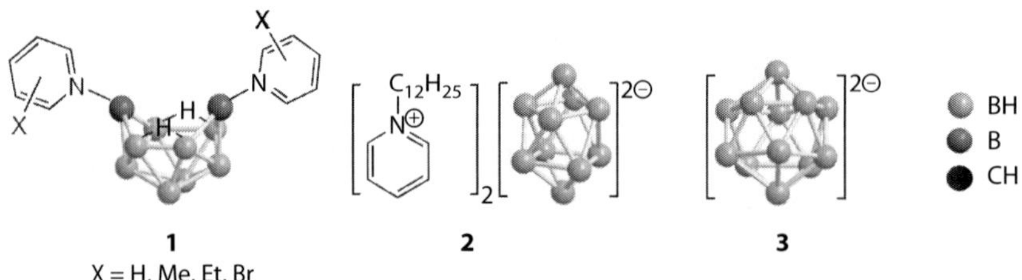

FIGURE 15.1 Structures of the borane derivatives in which fluorescence is detected in Volkov's work: Derivatives of decaborane(14) $B_{10}H_{12}[Py(X)]_2$ (**1**); salt $[PyC_{12}H_{25}]^+_2$ $[B_{10}H_{10}]^{2-}$ (**2**); (c) anion $B_{12}H_{12}{}^{2-}$ (**3**). The key for BH, B, and CH spheres in cage structures applies in all Figures and Schemes.

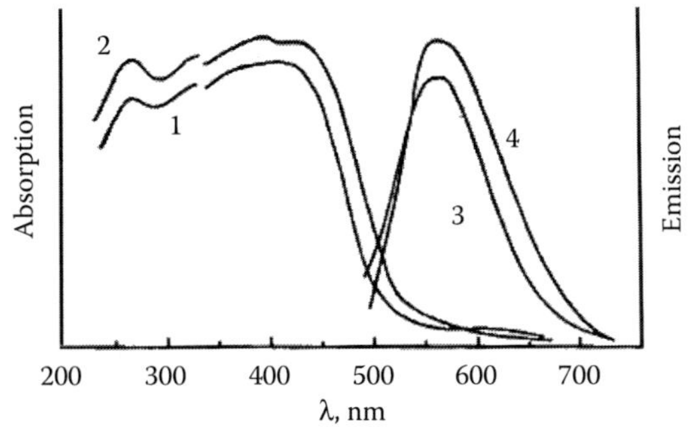

FIGURE 15.2 Diffuse reflectance (lines 1, 2) and fluorescence (lines 3, 4) spectra of two crystalline modifications of $B_{10}H_{12}Py_2$ (**1**). (Volkov, V. V.; Il'inchik, E. A.; Yur'eva, O. P.; Volkov, O. V., Luminescence of the derivatives of boranes and adducts of decaborane(14) of the $B_{10}H_{12}[Py(X)]_2$ type. *J. Appl. Spectroscopy* **2000**, *67*, 864–870. Copyright Wiley-VCH Verlag GmbH & Co. KGaA. Reproduced with permission.)

TABLE 15.1
Photoluminescence Data for *closo*-Boranes 2 and 3

Borane		λ_{em} (nm)
$[PyH]_2^+ [B_{10}H_{10}]^{2-}$	(**2a**)	518
$[PyH]_2^+ [B_{10}H_{10}]^{2-}$	(**2b**)	535
$[PyC_{12}H_{25}]_2^+ [B_{10}H_{10}]^{2-}$	(**2c**)	536
$[PyC_{12}H_{25}]_2^+ [B_{10}Cl_{10}]^{2-}$	(**2d**)	486
$[PyC_{12}H_{25}]_2^+ [B_{10}Br_{10}]^{2-}$	(**2e**)	520
$[PyC_{12}H_{25}]_2^+ [B_{12}H_{12}]^{2-}$	(**3**)	467

Source: Adapted from Virovets, A. V. et al., *Zh. Strukt. Khim.* **1994**, *35*, 72–77.

Shortly after this discovery, however, Kaszynski et al. reported fluorescence from a related *para*-1,10-disubstituted decaborate(2–) system, 1,10-$(RC_5H_4N)_2$-*closo*-$B_{10}H_8$ (R = C_8H_{17} (**4a**), R = $C_7H_{15}O$ (**4b**)) (Scheme 15.1), where pyridyl moieties are directly bound to opposing boron vertices on the 10-vertex cluster framework to afford a dizwitterionic moiety.[23]

This work cast valuable light on these systems, as UV-vis spectroscopic analysis clearly identified a charge-transfer band centered at λ_{max} = 319 nm for **4b**, for example, and excitation at this wavelength led to emission at λ_{em} = 480 nm (Figure 15.3). Semiempirical (MNDO) calculations established that the highest occupied molecular orbital (HOMO) was primarily localized on the boron cage cluster, while the lowest occupied molecular orbital (LUMO) was mostly concentrated

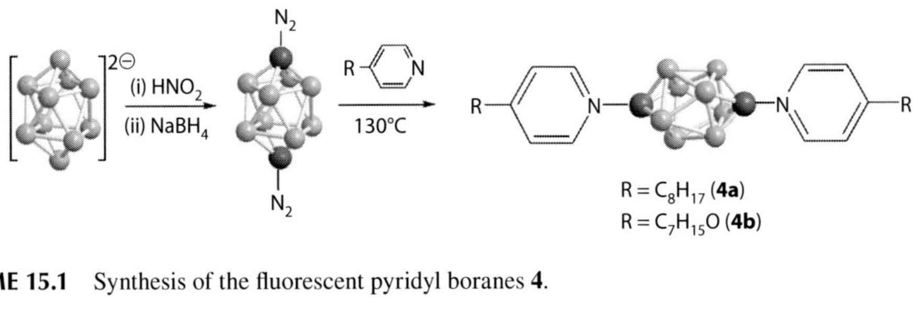

SCHEME 15.1 Synthesis of the fluorescent pyridyl boranes **4**.

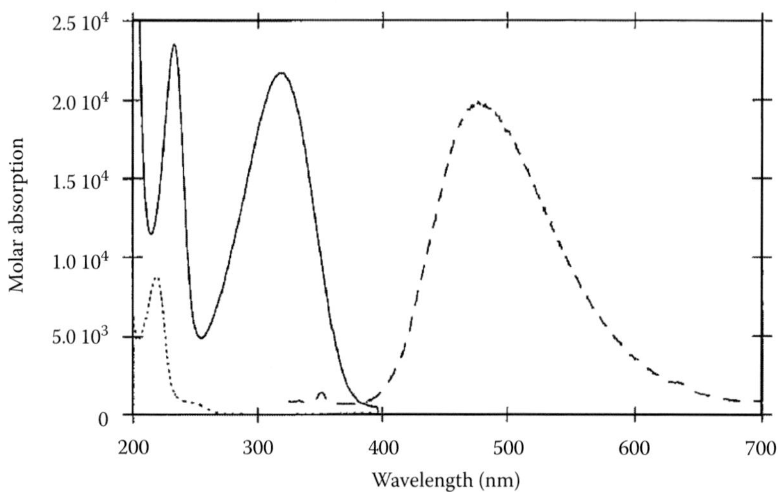

FIGURE 15.3 Electronic spectra of compound **4b**: absorption (solid line); emission (dashed line). Also shown is the absorption spectrum of the precursor pyridine $NC_5H_4OC_7H_{15}$ (dotted line). (From Kaszynski, P. et al., *Mol. Cryst. Liq. Cryst.* **1995**, *260*, 315–332. Reproduced with permission of Taylor & Francis.)

on the pyridine fragments, supporting the notion that the cage-to-ring charge transfer absorption was responsible for photoemission of the blue–green light.

Further results from this group suggest that this absorption is a characteristic of all 10-vertex *closo* cluster derivatives including the carboranes CB_9H_{10} and 1,10-$C_2B_8H_{10}$ substituted at the apical positions and additional fluorescence results based on these clusters are forthcoming.[24]

15.3 CARBORANES

The above-mentioned work of Volkov et al. was the first to report any kind of photophysical activity in carboranes, which are probably more ubiquitous now than simple boranes given their generally greater thermodynamic stability.[20] Once again pyridine derivatives of 9-Py-*nido*-7,8-$C_2B_9H_{11}$ (**5**) were studied (Figure 15.4) and only limited steady-state photoluminescence data were reported (Table 15.2) with no discussion of the fundamental orbital contributions to any emissive states or indeed whether these systems were strictly fluorescent (short lifetime emission from a singlet photoexcited state) or phosphorescent (generally longer lifetime emission from a triplet photoexcited state).

o-Carborane, *closo*-1,2-$C_2B_{10}H_{12}$ (**6**), itself has been studied by Kunkely and Vogler.[25] Although nonemissive in solution at ambient temperature, solid samples displayed weak luminescence (λ_{em} = 395 nm), which was attributed to a singlet photoexcited state. On the other hand, an EtOH glass sample at 77 K was shown to be strongly blue phosphorescent (λ_{em} = 441 nm, λ_{ex} = 260 nm) with considerable (unassigned) vibrational structure and a very long decay lifetime (τ = 4 s) (Figure 15.5a). In an effort to assuage criticism that this strong phosphorescence was due to an

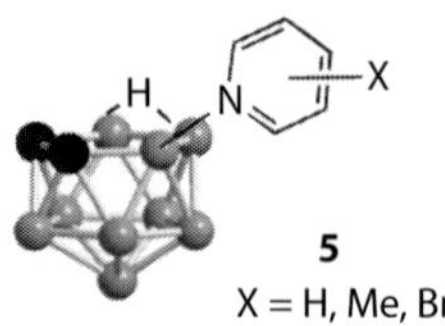

FIGURE 15.4 Structure of *nido*-carborane derivatives, $C_2B_9H_{11}Py(X)$ (**5**). See Table 15.2 for specific pyridyl groups.

TABLE 15.2
Photoluminescence Data from 9-Pyridine Derivatives 5 of *nido*-Carborane

Carborane		λ_{em} (nm)
$C_2B_9H_{11}Py$	(**5a**)	465
$C_2B_9H_{11}$•(3-MePy)	(**5b**)	486
$C_2B_9H_{11}$•(3-BrPy)	(**5c**)	493
$C_2B_9H_{11}$•(3,4-Me_2Py)	(**5d**)	461
$C_2B_9H_{10}$Br•Py	(**5e**)	476
$C_2B_9H_{10}$Br•(3-MePy)	(**5f**)	466
$C_2B_9H_{10}$Br•(3-BrPy)	(**5g**)	495

Source: Adapted from Volkov, V. V. et al., *J. Appl. Spectrosc.* **2000**, *67*, 864–870.

organic aromatic impurity, Kunkely and Vogler demonstrated similar vibrationally structured photoluminescent behavior ($\lambda_{em} = 462$ nm) with the hexabromo anion [7,8,9,10,11,12-Br_6-*closo*-1-$CB_{11}H_6]^-$ (**7**), albeit with a considerably shorter lifetime, $\tau < 10^{-3}$ s (Figure 15.5b).[26] This decrease in decay lifetime along with the observation of electronic transitions at lower energy were attributed to the presence of the bromine atoms and their associated spin-orbit coupling effect.

Although the precise nature of the photoexcited species was not described, Kunkely and Vogler did provide an account of the relationship between electronic absorption spectra of *o*-carborane versus that of [*closo*-$B_{12}H_{12}]^{2-}$. Substitution of two BH^- groups by two neutral CH groups leads to a lowering of symmetry (from I_h to C_{2v}) and an associated drop in the HOMO–LUMO gap. This is accompanied by the relaxation of symmetry restrictions, permitting the observed spin-forbidden transition, at least at 77 K. Hence, *o*-carborane's lowest energy absorption lies in the UV (at $\lambda_{ex} = 260$ nm) rather than the VUV region.

Owing to these reports, the basic steady-state measurements have also been made with *m*- and *p*-carborane (**8** and **9**, respectively). Both are weakly luminescent in the solid state at room temperature (λ_{em}(**8**) = 380 nm, λ_{em}(**9**) = 345 nm) and at 77 K (λ_{em}(**8**) = 390 nm, λ_{em}(**9**) = 320 nm) though spectra are not available.[27]

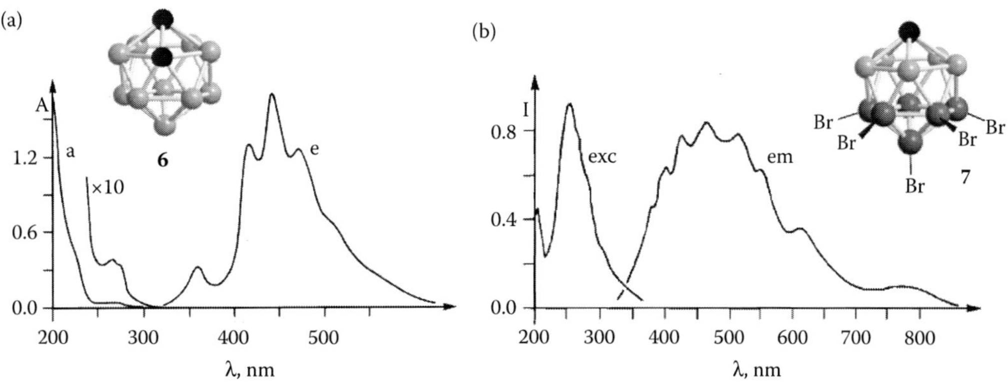

FIGURE 15.5 (a) Electronic absorption and emission spectra of *o*-carborane (**6**) under Ar: absorption by a 1.48×10^{-2} M solution in MeCN at room temperature, 1 cm cell; emission by the solid at 77 K, $\lambda_{ex} = 260$ nm. (b) Electronic excitation ($\lambda_{em} = 462$ nm) and emission ($\lambda_{ex} = 250$ nm) spectrum of a 1.05×10^{-2} M solution of $[Ag]^+$[7,8,9,10,11,12-Br_6-*closo*-1-$CB_{11}H_6$] (**7**) in EtOH under Ar at 77 K. (Reprinted from *Inorg. Chim. Acta*, 357, Kunkely, H.; Vogler, A., Is o-carborane photoluminescent? 4607–4609, Copyright 2004, with permission from Elsevier; Reprinted from *Inorg. Chem. Commun.*, 8, Kunkely, H.; Vogler, A., Luminescence of silver 7,8,9,10,11,-12-hexabromo-closo-1-carbododecaborate, 992–993, Copyright 2005, with permission from Elsevier.)

o- and *m*-Carboranes have also served as influential structural components in luminescent hyperconjugated polymers that demonstrate aggregation-induced emission (AIE).[28,29] These AIE-active alternating polymers with *o*- or *m*-carborane and *p*-phenylene–ethynylene sequences (**10** and **11**, respectively) were assembled by Sonogashira–Hagihara palladium coupling polycondensation reactions and are shown in Scheme 15.2. Considerably higher M_n polymer molecular weights (26,600–36,400 g/mol) were obtained for the *m*-carborane systems **11** than for the *o*-carborane ones **10** (2800–4100 g/mol), a result attributed to the less bent nature of the former monomer.

Although apparently not the primary chromophores in these luminescent species, Chujo has pointed out notable red-shifts in electronic absorptions and attributed these to some hyperconjugative interaction of the π-aromatic systems with the carborane σ-aromatic framework. DFT computational studies at the B3LYP/6–31G(d)//B3LYP/6–31(d) level of theory on model systems of the *m*-carborane system **11** confirm this, more so for the ground-state LUMO than for HOMO. Fluorescence spectra of *o*-carborane polymers **10a–c** with electron-donating π-conjugated linkers were nonemissive in THF solution at ambient temperatures, while **10d** with its electron-withdrawing π-conjugated linker demonstrated blue emission ($\lambda_{em} = 405$ nm, $\lambda_{ex} = 345$ nm). AIE emission was shown for polymers **10a–c** on using a mixed THF/H_2O solvent combination, with strong orange luminescence ($\lambda_{em} = 550–583$ nm) for a THF:H_2O ratio of 1:99 (Figure 15.6a). Estimated quantum yields for **10a** and **10b** indicated AIE photoemission onset at around 50% H_2O content in THF/H_2O

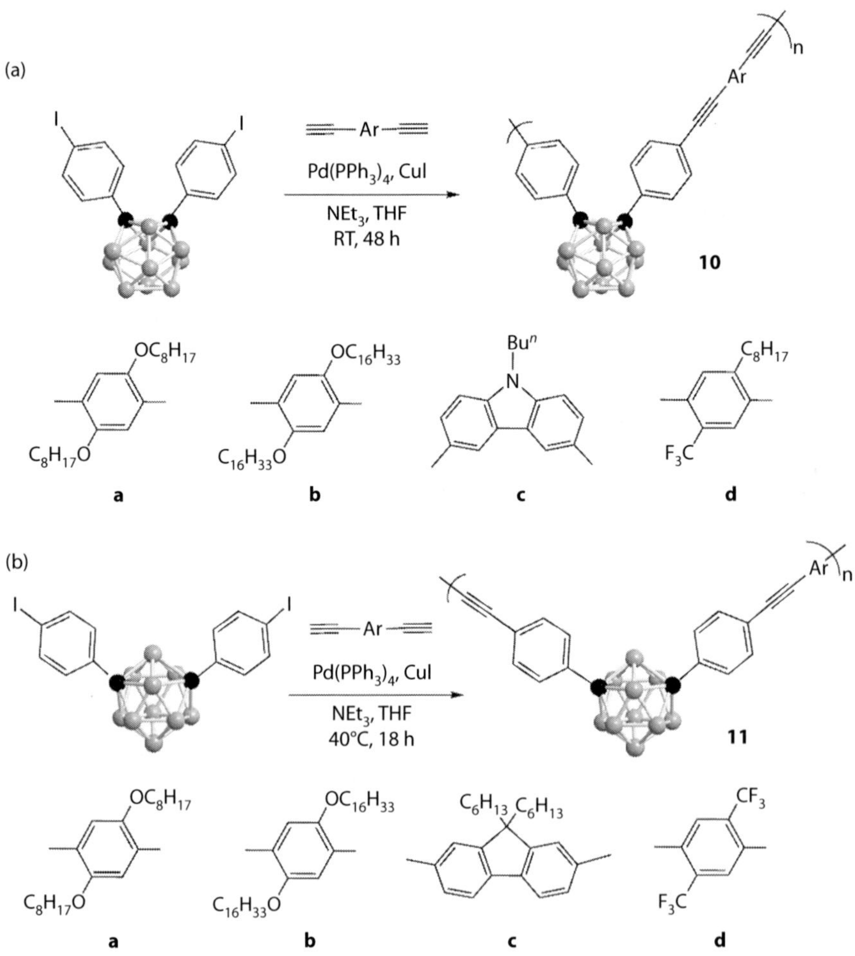

(a)

Ar; Pd(PPh$_3$)$_4$, CuI; NEt$_3$, THF; RT, 48 h

10

a: OC_8H_{17}, $C_8H_{17}O$; b: $OC_{16}H_{33}$, $C_{16}H_{33}O$; c: Bu^n, N; d: C_8H_{17}, F_3C

(b)

Ar; Pd(PPh$_3$)$_4$, CuI; NEt$_3$, THF; 40°C, 18 h

11

a: OC_8H_{17}, $C_8H_{17}O$; b: $OC_{16}H_{33}$, $C_{16}H_{33}O$; c: C_6H_{13} C_6H_{13}; d: CF_3, F_3C

SCHEME 15.2 Synthesis of carborane-*p*-phenylene-ethynylene polymers with (a) *o*-carborane (**10**) and (b) *m*-carborane (**11**).

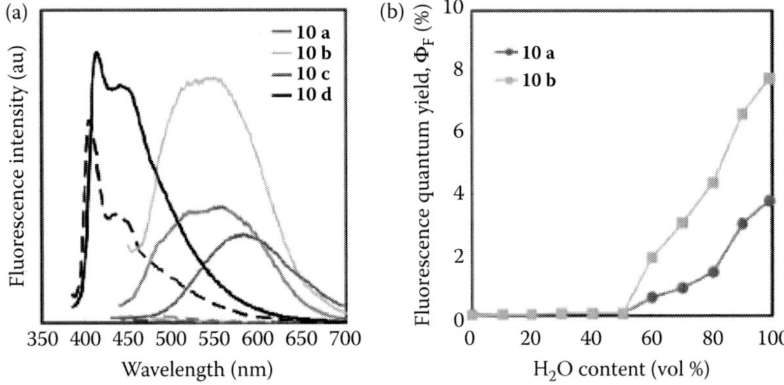

FIGURE 15.6 (a) Fluorescence spectra of **10a–d** in THF (1×10^{-5} mol/L, dashed lines) and mixed solvent of THF/H_2O 1/99 (v/v) (1×10^{-5} mol/L, solid lines). (b) Dependence of quantum yields of **10a** and **10b** on solvent compositions of the THF/H_2O mixture. (Reprinted with permission from Kokado, K.; Chujo, Y., Emission via aggregation of alternating polymers with *o*-carborane and *p*-phenylene-ethynylene sequences. *Macromolecules*, *42*, 1418–1420. Copyright 2009 American Chemical Society.)

mixtures, rising to an efficiency of 3.8% for **10a** and *ca*. 7.5% for **10b** (Figure 15.6b), and producing aggregates of 38.0 and 131.2 nm in size, respectively, as determined by dynamic light-scattering measurements.

In contrast, all the variants of the *m*-carborane-based polymer systems **11** displayed bright-blue photoluminescence in $CHCl_3$ solution (Figure 15.7) with quantum yields ranging from 11% to 25%. Chujo concluded that this was due to a lack of a variable C–C bond in the *m*-carborane systems, the presence of which in the *o*-carborane polymers was effecting fluorescence quenching. Hence, the carborane was playing a vital role in the photophysical responses and the impact of the carborane C–C bond on optoelectronic properties has important implications for the metallacarborane photoluminescence discovered in our laboratory at Saint Louis University and which is described further below.

In related work by Carter and Coughlin, polyfluorene polymers with *p*- and *o*-carborane incorporated into the backbone have been synthesized and extensively studied.[30,31] The motivation behind this effort stems from an idea to improve the stability of blue-emitting polyfluorenes in optoelectronic devices based on the favorable physicochemical properties of the carborane moiety. Polymerizations

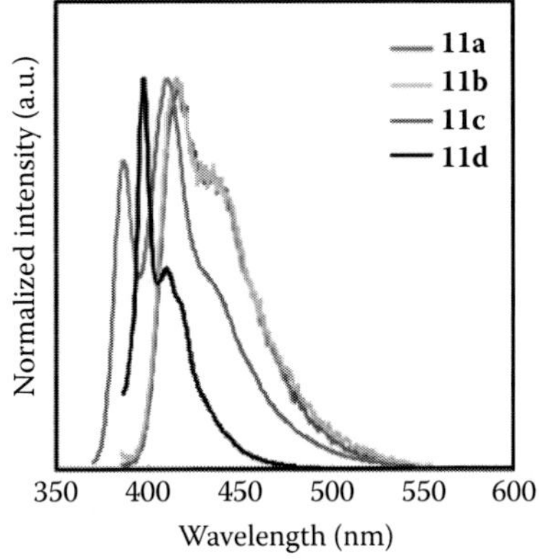

FIGURE 15.7 Photoluminescence spectra of polymers **11a–d** in $CHCl_3$ (1.0×10^{-7} M). (Reproduced with permission from Kokado, K.; Tokoro, Y.; Chujo, Y. *Macromolecules* **2009**, *42*, 2925–2930. American Chemical Society.)

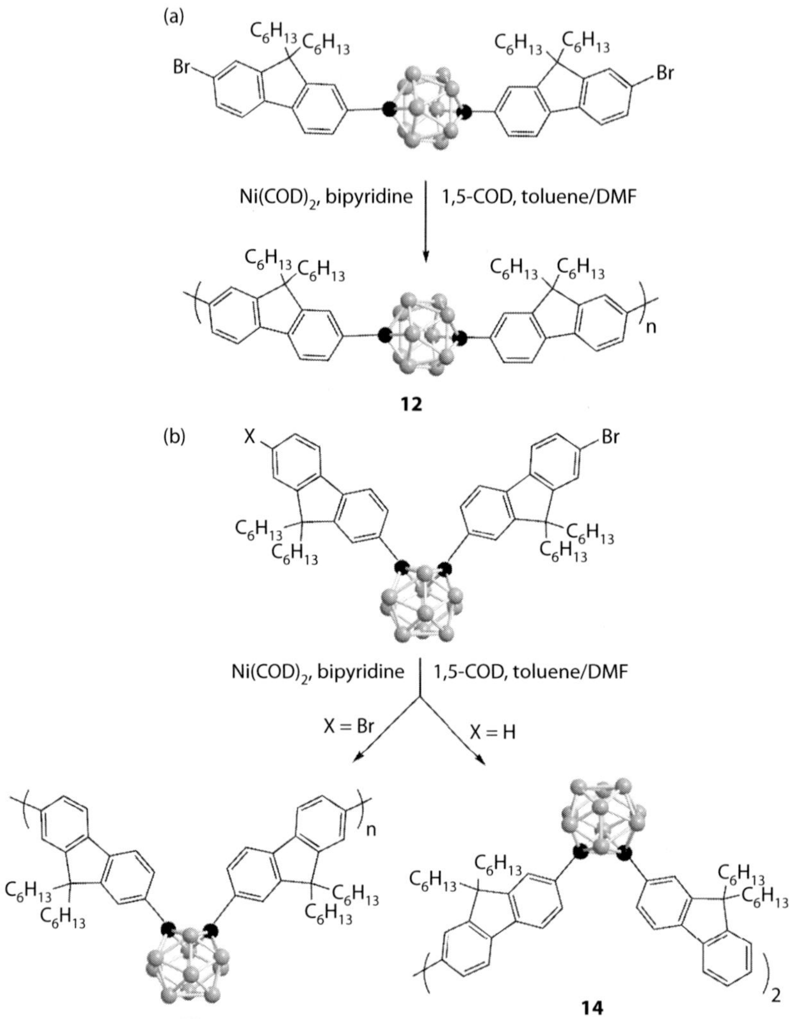

SCHEME 15.3 Synthesis of carborane-containing polyfluorenes **12** and **13** with (a) *p*-carborane, and (b) *o*-carborane. Also shown is the synthesis of the *o*-carborane fluorene dimer **14**.

of fluorene–carborane–fluorene monomer units were achieved with Yamamoto-type Ni(0) dehalogenative reactions affording a *p*-carborane polymer **12a** and *o*-carborane polymer **12b** as shown in Scheme 15.3, with reported molecular weights M_n = 50 kg/mol and 2.4–10 kg/mol, respectively.

Polymer **12** demonstrated a bright-blue emission at λ_{em} = 400 nm, which was bathochromically shifted some 15 nm from the fluorene dimer shown in Figure 15.8. UV–vis measurements on **12** also revealed a notable 20 nm red-shift of the polymer absorptions relative to the same fluorene dimer, suggesting that the carborane extends the effective conjugation length of the polymer and is thus an integral component of the chromophore. Furthermore, annealing studies of the polymer up to temperatures of 150°C revealed that the carborane cage was effectively suppressing the green emission, which often plagues blue-emitting polyfluorenes as a result of aggregation-based excimer formation. This was an encouraging discovery because the availability of solution-processable color-stable blue emitters is critically important in the polymer LED industry.

The lower average molecular weight of polymer **13** was attributed to intramolecular effects resulting from the kinked, hindered conformation of the *o*-carborane monomer. Incorporation of

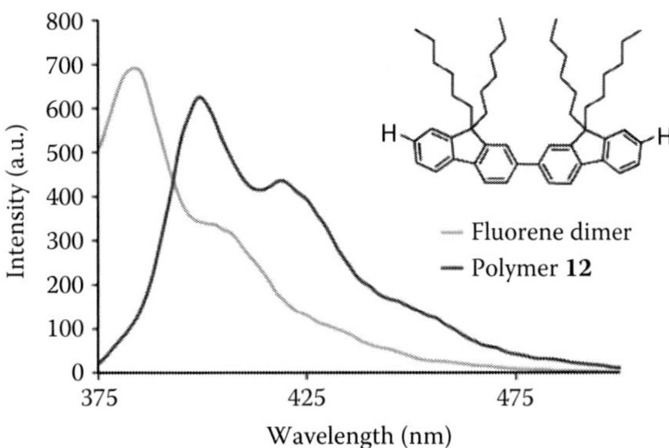

FIGURE 15.8 Fluorescence spectra of *p*-carborane-polyfluorene **12** and a dihexylfluorene dimer for comparison. (Peterson, J. J. et al., Polyfluorene with *p*-carborane in the backbone. *J. Chem. Soc., Chem. Commun.* **2009**, 4950–4952. Reproduced by permission of The Royal Society of Chemistry.)

the *o*-carborane moiety again caused a bathochromic shift in the fluorene-based fluorescence emission maximum (λ_{em} = 395 nm) in addition to a 14 nm red shift in UV–vis absorptions compared with the dihexylfluorene dimer, the repeat structure between the carborane units, which would again seem to confirm the carborane's active role in the chromophore. In addition to this blue emission, a new emission at λ_{em} = 565 nm was observed for polymer **13**, as well as for a dimeric species **14**, synthesized as a smaller model for the aforementioned polymer (Scheme 15.3). Combined with the fluorene-based emission, solid-state samples of the polymer thus appeared orange under UV irradiation (Figure 15.9).

This complex solution fluorescence was reduced to single bright green solid-state emission from thin films of **13** at λ_{max} = 520 nm (Figure 15.10), which was shown to be thermally stable following annealing at 180°C for 2 h. Based on previous observations of fluorene copolymers with electron-withdrawing groups, this has been attributed to an increase in the efficiency of energy transfer to the carborane ligand in combination with enhancement of AIE, the latter of which was also observed for the Chujo polymer systems **10** discussed above. The 565 and 415 nm emissions have distinct excitations at λ_{ex} = 380 and 390 nm, respectively, with unavoidable overlap between the two. Now because the monomer precursor to **13** showed no lower-energy fluorescence, the implication made was that the *o*-carborane segments were probably being stimulated through some energy-transfer process, rather than direct excitation by the fluorenyl emission at *ca*. 415 nm.

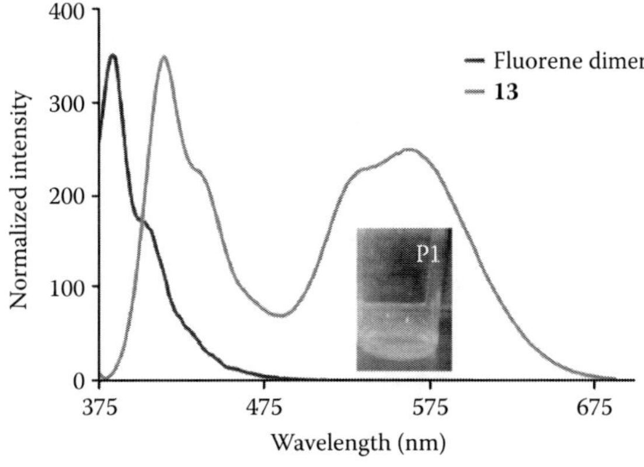

FIGURE 15.9 Fluorescence spectra of a $CHCl_3$ solution of *o*-carborane-polyfluorene **13** and a dihexylfluorene dimer for comparison. (Reprinted with permission from Peterson, J. J. et al., Carborane-containing polyfluorene: *o*-carborane in the main chain. *Macromolecules*, *42*, 8594–8598. Copyright 2009 American Chemical Society.)

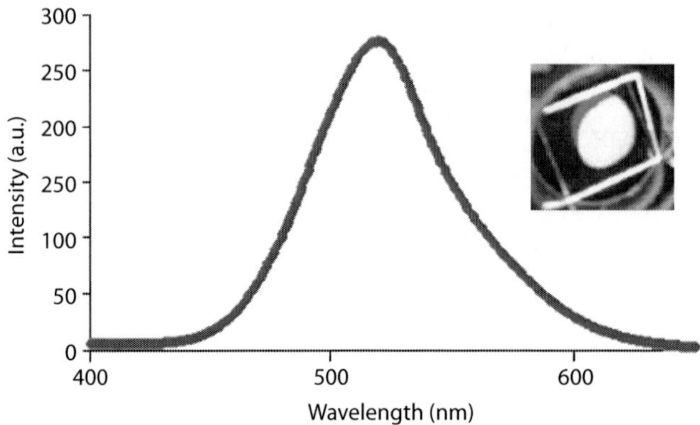

FIGURE 15.10 Film fluorescence from *o*-carborane-polyfluorene **13**. That of **14** has a similar appearance. (Reprinted with permission from Peterson, J. J. et al., Carborane-containing polyfluorene: *o*-carborane in the main chain. *Macromolecules*, *42*, 8594–8598. Copyright 2009 American Chemical Society.)

Base degradation of the *o*-carborane cage to a [*nido*-7,8-$C_2B_9H_{10}$]$^-$ system in **15** (Figure 15.11) had a significant impact on luminescent properties. Indeed, while the peak at 565 nm diminished, a new, highly intense blue emission at λ_{em} = 430 nm grew in with increasing base concentration in solutions of **14** (Figure 15.12). These changes were matched with the progress of base degradation by NMR spectroscopy. In order to affirm that these shifts in photophysical response were due to a structural change in the carborane, and not the introduction of a charge center, the material was protonated to afford a neutral *nido*-7,8-$C_2B_9H_{11}$ system. As expected, the fluorescence spectra did

FIGURE 15.11 The *nido*-carborane-polyfluorene **15** formed by KOH/EtOH degradation of **14**.

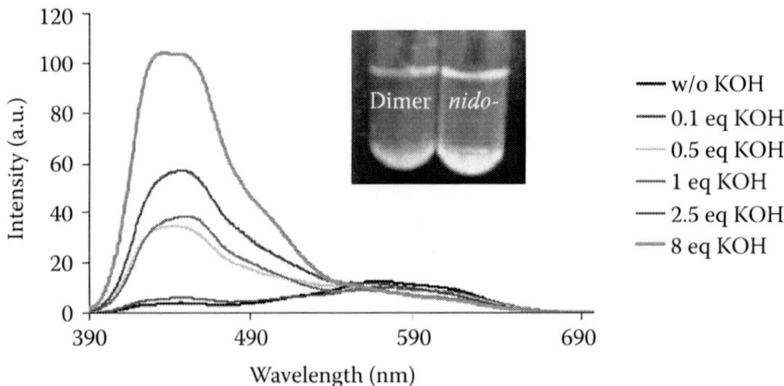

FIGURE 15.12 Solution fluorescence in EtOH/THF of *o*-carborane dimer **14** treated with increasing amounts of KOH. (Reprinted with permission from Peterson, J. J. et al., Carborane-containing polyfluorene: *o*-carborane in the main chain. *Macromolecules*, *42*, 8594–8598. Copyright 2009 American Chemical Society.)

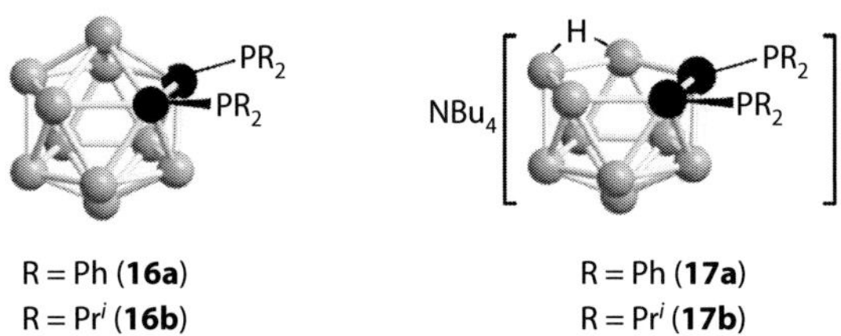

FIGURE 15.13 *Closo* and *nido* diphosphine carborane species **16** and **17**, respectively.

not change. The authors have attributed this to further enhanced conjugative interaction between the carborane and fluorenyl units in the polymer.

The change in photophysical response on carborane cage degradation from *closo* to *nido* is an interesting phenomenon highlighted in the next work. Laguna et al. have described blue photoluminescence from the *nido*-carborane diphosphines [7,8-$(PR_2)_2$-*nido*-7,8-$C_2B_9H_{10}]^-$ (R = Ph (**17a**), Pr^i (**17b**)) (Figure 15.13) as a carborane-based event that interestingly does not occur with the parent *closo* species [1,2-$(PR_2)_2$-*closo*-1,2-$C_2B_{10}H_{10}$] (**16**).[32]

Data reported for compound **17a** revealed that excitation at $\lambda_{ex} = 332$ nm afforded emissions in the solid state at $\lambda_{em} = 482$ nm at room temperature and $\lambda_{em} = 465$ nm at 77 K, while an acetone solution produced $\lambda_{em} = 429$ following excitation at $\lambda_{ex} = 379$ nm. As the data for compound **17b** showed little difference in photophysical behavior, the authors concluded that this emission could not result from a phosphine-located excitation. At this time no other rationale was offered for the lack of luminescence from the parent *closo* species **16** other than a significant change in electronic structure. Nevertheless, the ligating ability of the phosphine arms in the *nido* compounds **17** was exploited to produce luminescent gold complexes, which is discussed in the next section.

15.4 METALLACARBORANES

15.4.1 Metallacarboranes with Exopolyhedral Metal–Ligand Fragments

The *nido*-carborane-diphosphine ligands **17** have also demonstrated metal-perturbed intraligand emissions when bound within trigonal–planar Au complexes. The gold complexes [Au($(PR_2)_2C_2B_9H_{10}$)L] (R = Ph, L = PPh_3 (**18a**), L = PPh_2Me (**18b**), L = P(4-Me-$C_6H_4)_3$ (**18c**); R = Pr^i, L = PPh_3 (**18d**), L = PPh_2Me (**18e**), L = P(4-Me-$C_6H_4)_3$ (**18f**)) were actually reported by Laguna et al. to show dual emissions depending on the nature of the carborane–phosphine PR_2 groups.[32] As with the *closo* systems described above, analogous gold complexes derived from compounds **16** were nonemissive. Data for solid-state and solution measurements at room temperature and at 77 K are presented in Table 15.3.

The luminescence spectra of the diphenyl-*nido*-diphosphine derivatives **18a–c** displayed high- and low-energy emissions in the solid state, both at room temperature and at 77 K. Generally, excitations are all at about $\lambda_{ex} = 330$ nm with high-energy emissions at $\lambda_{em} = 520$ nm and the lower-energy emissions around $\lambda_{em} = 610–670$ nm. An example of photoluminescence spectra for both emissions is given for complex **18a** in Figure 15.14. The additional lower-energy orange emissions observed for **18a–c** have been attributed to metal-to-ligand charge transfer (MLCT) transitions in the gold–phosphine regime rather than the carborane. The diisopropyl-*nido*-diphosphine derivatives **18d–f**, on the other hand, show only the high-energy emissions, typically around $\lambda_{em} = 510$ nm.

The high-energy emissions common to all the complexes **18** were described as primarily carborane cage intraligand in character, rather much as was observed with the free ligands **17**. The transitions were clearly perturbed by metal-center contributions and possibly with some influence from the phosphorus substituents, as verified by computational analysis. The same influence was not observed in the free ligands **17**, however, and thus it is specifically the substituents on the gold monophosphine that are affecting the photoluminescence properties. That said, the influence appears to be minimal given the subtle variation in data in Table 15.3, at least for the diisopropyl-*nido*-diphosphine deriva-

TABLE 15.3
Excitation and Emission Data for the Gold Complexes of *nido*-Diphosphine-Carboranes, $[Au((PR_2)_2C_2B_9H_{10})L]$ (18)

	λ_{ex} (nm)	λ_{em} (nm) (solid, RT)	λ_{em}, nm (solid, 77 K)	λ_{em}, nm (acetone solution)
R = Ph, L = PPh_3 (**18a**)	332	540	529	529
	460	670	676	603
R = Ph, L = PPh_2Me (**18b**)	335	521	513	530
	412	640	653	576
R = Ph, L = $P(4\text{-Me-}C_6H_4)_3$ (**18c**)	337	513	506	534
	459	614	633	(broad)
R = Pr^i, L = PPh_3 (**18d**)	337	508	507	514
R = Pr^i, L = PPh_2Me (**18e**)	327	512	505	512
R = Pr^i, L = $P(4\text{-Me-}C_6H_4)_3$ (**18f**)	322	508	503	512

Source: Adapted from Crespo, O. et al., *Inorg. Chem.* **2003**, *42*, 2061–2068.

tives **18d–f**. DFT and TDDFT computational analysis (Figure 15.15) on a model complex $[Au((PR_2)_2C_2B_9H_{10})(PH_3)]$ (**18g**) suggested that spin-allowed transitions originate from HOMO and HOMO-1, which are carborane-based orbitals comprising *ca.* 56 and 45% carborane, respectively, with contributions of 33% and 41% from the PPh_2 groups, and 9 and 12% from the gold(I) center, respectively. The electronic transition targets appeared to be the diphenylphosphino-based orbitals LUMO + 1 through LUMO + 6, which have some carborane character and little or no contribution from the gold. The orbital analysis would therefore seem to concur with the author's proposed assignment of an admixture of intraligand and metal-to-ligand charge transfer in the photoexcited state.

Additionally, the emissions were all blue-shifted at 77 K as compared with those at room temperature in the solid state (Table 15.3) and this was identified as a case of luminescence rigidochromism, with any rigidity likely to be associated with the carborane cages.

Square planar adducts of the ligands **17** with MCl(PPh_3) groups (M = Ni (**19a**), M = Pd (**19b**), M = Pt (**19c**)) were reported by Dou et al. (Scheme 15.4).[33] The syntheses proceeded by concomitant

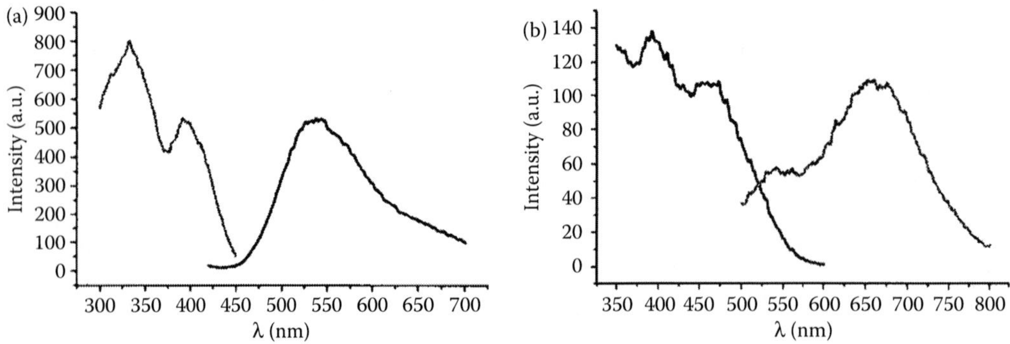

FIGURE 15.14 Emission and excitation spectra for the (a) high-energy emission and (b) low-energy emission in compound **18a** in the solid state, at room temperature. (Reprinted with permission from Crespo, O. et al., Luminescent *nido*-carborane-diphosphine anions $[(PR_2)_2C_2B_9H_{10}]^-$(R = Ph, iPr). Modification of their luminescence properties upon formation of three-coordinate gold(I) complexes. *Inorg. Chem*, *42*, 2061–2068. Copyright 2003 American Chemical Society.)

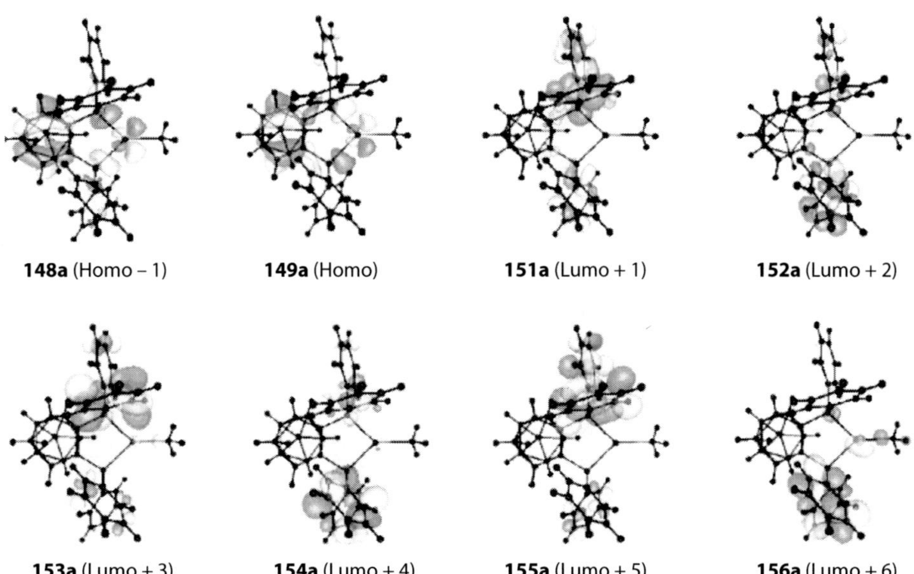

FIGURE 15.15 DFT analysis of the model complex **18g**. (Reprinted with permission from Crespo, O. et al., Luminescent *nido*-carborane-diphosphine anions $[(PR_2)_2C_2B_9H_{10}]$–(R = Ph, iPr). Modification of their luminescence properties upon formation of three-coordinate gold(I) complexes. *Inorg. Chem, 42*, 2061–2068. Copyright 2003 American Chemical Society.)

ethanol-induced degradation of the diphenyl-*closo*-diphosphines **16** with coordination of the diphosphine to the metal center. Interestingly, although the palladium complex **19b** was found to be emissive (λ_{em} = 485 nm), the carborane was not at all implicated as the chromophore, with assignment instead of an MLCT event based on the monophosphine ligand.

Using a similar synthetic methodology, the same authors also reported a tetranuclear gold complex $[Au_4(7,8\text{-}(PPh_2)_2\text{-}7,8\text{-}C_2B_9H_{10})_2(PPh_3)_2]$ (**20**) using the same ligand, which is bound in a bidentate fashion to two of the gold atoms in the cluster (Scheme 15.5).[34] This complex was also luminescent in CH_2Cl_2 solution at room temperature with λ_{em} = 468 nm, λ_{ex} = 392 nm and the origin of this photophysical activity was curiously purported to be exclusively metal-centered in origin.

It is interesting to note, given the absence of photophysical activity noted for the *closo*-carborane species **16**, that when incorporated into the W–Ag–S heterobimetallic cluster species **21** (Figure 15.16), luminescence was reported with λ_{em} = 436 nm, λ_{ex} = 355 nm.[35] In this particular example, the photoexcitation responsible for this emissive behavior has been attributed to a ligand-to-ligand charge transfer based on a L_{red}–M–L_{ox} architecture (with reducing and oxidizing ligands connected through the W–Ag–S cluster. Although the authors suggested that the *closo*-carborane-diphosphine was a potential donor in this situation, they recognized Laguna's previous observations of lack of emission in the free ligand, and ultimately implicated the sulfide ligand as the active donor, relegating the carborane as an ancillary component.

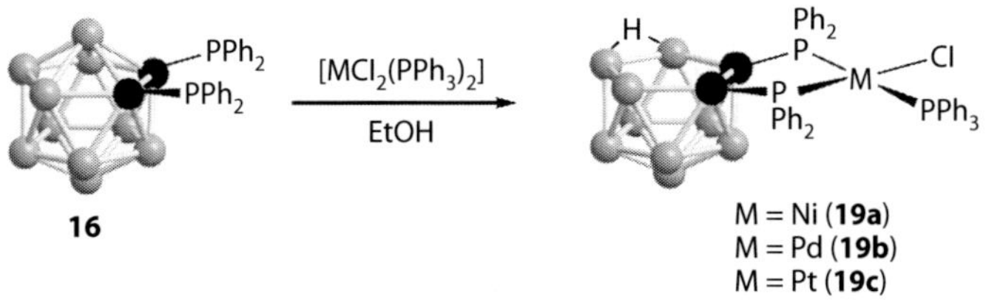

SCHEME 15.4 Formation of the complexes $[MCl(PPh_3)(7,8\text{-}(PPh_2)_2\text{-}7,8\text{-}C_2B_9H_{10})]$ (**19**).

SCHEME 15.5 Formation of the complex $[Au_4\{7,8\text{-}(PPh_2)_2\text{-}7,8\text{-}C_2B_9H_{10}\}_2(PPh_3)_2]$ (**20**).

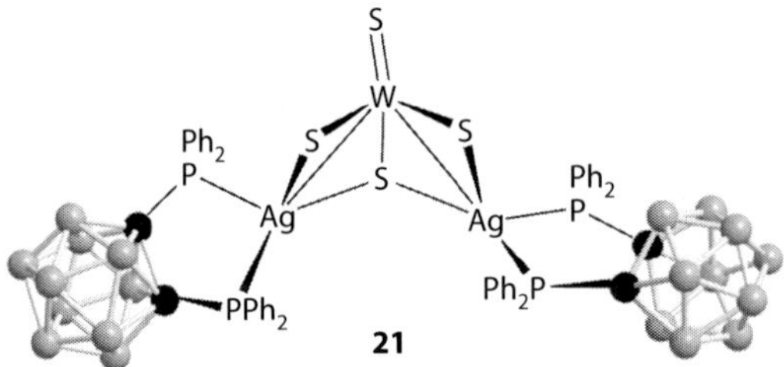

FIGURE 15.16 Structure of the luminescent W–Ag–S heterobimetallic cluster species **21**.

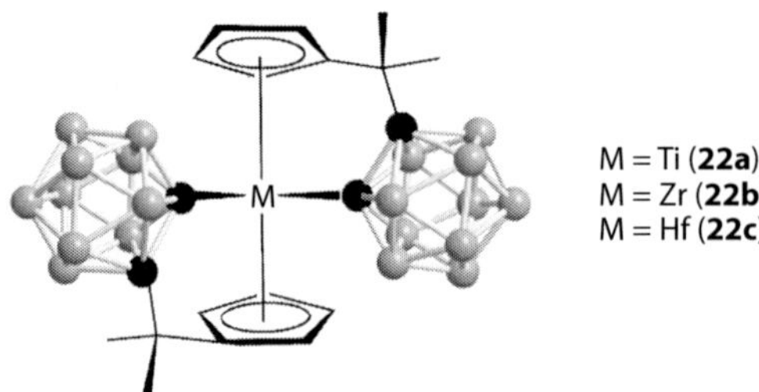

FIGURE 15.17 Structures of the Group 4 metallacarboranes $[M(\eta^5:\eta^1\text{-}C_5H_4CMe_2C_2B_{10}H_{10})_2]$ (**22**).

Although exo-metal-*closo*-caborane systems would appear to be poor candidates for photoluminescent materials, Do et al. have discovered the Group 4 metal systems $[M(\eta^5:\eta^1\text{-}C_5H_4CMe_2C_2B_{10}H_{10})_2]$ (M = Ti (**22a**), M = Zr (**22b**), M = Hf (**22c**)) (Figure 15.17) that possess some very interesting optoelectronic properties.[36,37]

Mechanoluminescence was observed for the Zr and Hf complexes **22b** and **22c** only with emission wavelengths of λ_{em} = 435 nm and 459 nm, respectively (Figure 15.18), which resembled their solid-state photoluminescence spectra in form with maxima at λ_{em} = 446 and 420 nm, and which in turn were very close to their THF solution emission profiles (Table 15.4).

That complexes **22b** and **22c** are mechanoluminescent at all occurs by virtue of their chiral structure and crystallization in noncentrosymmetric crystal systems, a prerequisite for mechanoluminescence. Because crystals of suitable size could not be procured this could not be further specifically assigned to (elastic) piezo- or (nonelastic) triboluminescence. The red-shifting of the mechanoemissive peaks by some 15 nm relative to the photoluminescence outputs was attributed to generation of distinct excited states by fracture-induced symmetry change. It was suggested that the absence of this phenomenon in the titanium complex **22a** would seem to indicate facile quenching or possibly a forbidden transition because there is significant difference between the band structures of the ground and excited states, as suggested by the large Stokes shift observed in its photoluminescence spectrum in THF solution, in turn affording a much lower-energy emission at λ_{em} = 602 nm. The quantum yield for this solution emission of **22a** (Φ = 0.04%) was commensurate with excited state instability of titanocene complexes, while the generally long lifetimes (τ = 1–3 μs) for all three

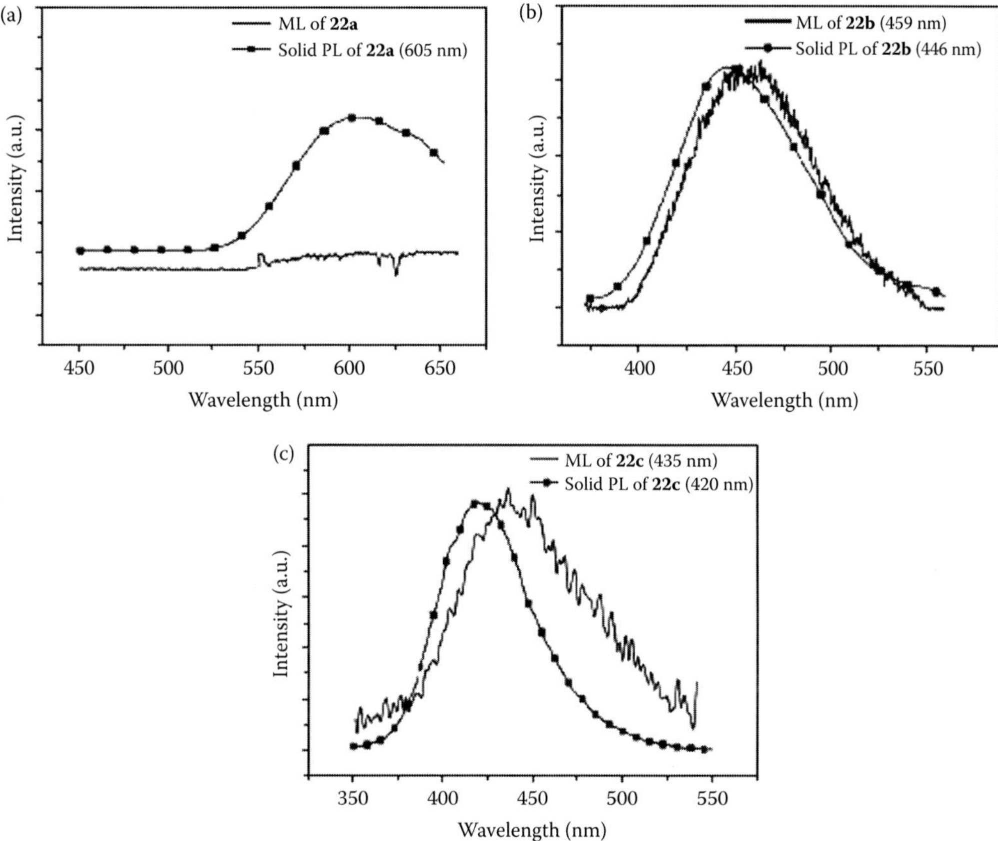

FIGURE 15.18 Solid photoluminescence (PL) and mechanoluminescence (ML) spectra for complexes **22**. (Reprinted from *J. Organomet. Chem.*, 694, Shin, C. H.; Han, Y.; Lee, M. H.; Do, Y., Group 4 *ansa*-metallocenes derived from *o*-carborane and their luminescent properties.1623–1631, Copyright 2009, with permission from Elsevier.)

complexes was suggestive of ligand-to-metal charge transfers (LMCT), which had been previously observed in $[MCl_2(\eta^5\text{-}C_5H_5)_2]$ (M = Ti, Zr, Hf) complexes.

Complexes **22b** and **22c** also turned out to provide the first examples of electroluminescence from a metallacarborane. Measurements were made by incorporating the complexes into a hole-transporting polymer of poly(*N*-vinylcarbazole) (PVK) with blending concentrations varying from 0.5% to 5.0% and spin-coating to produce a single-layer device with the structure glass/indium-tin-oxide/**22** + PVK/Al:Li(95:5). Spectra showed that green electroluminescence from these devices

TABLE 15.4
Photophysical Data for Complexes 22 in 10^{-4} M THF Solution at Room Temperature

Compound	λ_{em} (nm)	Φ (%)	τ (μs)
22a	602	0.04	2.6
22b	441	0.3	1.4
22c	419	1.5	1.2

Source: Adapted from Hong, E. et al., *Adv. Mater.* **2001**, *13*, 1094–1096. and Shin, C. H. et al., *J. Organomet. Chem.* **2009**, *694*, 1623–1631.

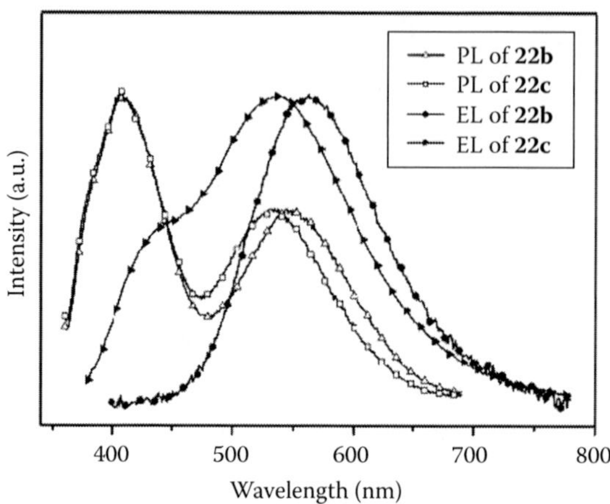

FIGURE 15.19 Electroluminescence (EL) and photoluminescence (PL) spectra of **22b**- and **22c**-doped PVK films. (Reprinted from *J. Organomet. Chem.*, 694, Shin, C. H.; Han, Y.; Lee, M. H.; Do, Y., Group 4 ansa-metallocenes derived from *o*-carborane and their luminescent properties.1623–1631, Copyright 2009, with permission from Elsevier.)

nicely matched their observed photoluminescences (λ_{em} = 562 nm for **22b** and λ_{em} = 537 nm for **22c**), while they were substantially red-shifted from their solution emissions (Figure 15.19). The reason for this was accredited to exciplex formation between the metallacarborane complex and the PVK host, with the former acting as a weak electron acceptor (as it bears an electron-deficient metal center) and the latter as a weak electron donor. This was verified by photoluminescence studies using a variety of other polymeric hosts, such as polyfluorene, polystyrene, and poly(methylmethacrylate), for example. These polymers do not bear an amino donor, and it was only once a small molecular amino system such as 9-ethylcarbazole or triphenylamine was introduced that the characteristic green emission was switched on, hence confirming exciplex formation.

The same paper reported that a silyl-indenyl derivative [$Me_2Si(\eta^5\text{-}C_9H_7)(\eta^5{:}\eta^1\text{-}C_5H_4\text{-}3\text{-}CMe_2C_2B_{10}H_{10})ZrCl$] (**23**) (Figure 15.20) was employed in a functioning organic light-emitting diode (OLED), where it was incorporated into an electron-transporting layer.

Complex **23** was also photoluminescent in both the solid state and in solution (Figure 15.21), although because it crystallized in a centrosymmetric space group, it did not demonstrate the mechanoluminescence observed with complexes **22**. The energies of emission and excitation (λ_{em} = 549 nm, λ_{ex} = 441 nm in chloroform solution) were perceptibly lower than for the Zr congener **22b** and were described to be indicative of reduced LMCT transition energy courtesy of a stabilization of the Zr d^0-centered LUMO by the poor Cl donor ligand.

Another very recent report by Laguna et al. described the emissive properties of phosphine–gold derivatives of *o*-, *m*-, and *p*-carborane.[27] The resulting gold complexes [$(\mu\text{-}1,n\text{-}C_2B_{10}H_{10})\{Au(PR_3)\}_2$]

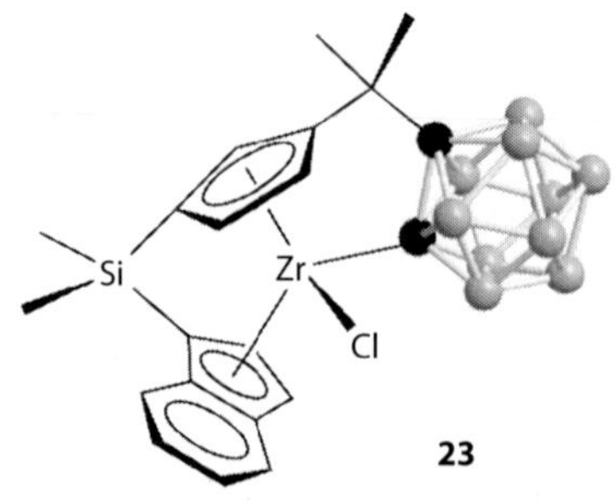

FIGURE 15.20 Structure of [$Me_2Si(\eta^5\text{-}C_9H_7)(\eta^5{:}\eta^1\text{-}C_5H_4\text{-}3\text{-}CMe_2C_2B_{10}H_{10})ZrCl$] (**23**).

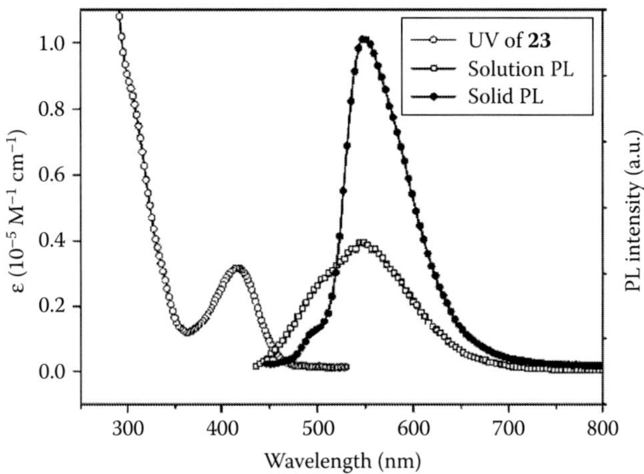

FIGURE 15.21 UV–vis and photoluminescence (PL) spectra (solid and solution) of complex **23**. (Reprinted from *J. Organomet. Chem.*, 694, Shin, C. H.; Han, Y.; Lee, M. H.; Do, Y., Group 4 ansa-metallocenes derived from *o*-carborane and their luminescent properties.1623–1631, Copyright 2009, with permission from Elsevier.)

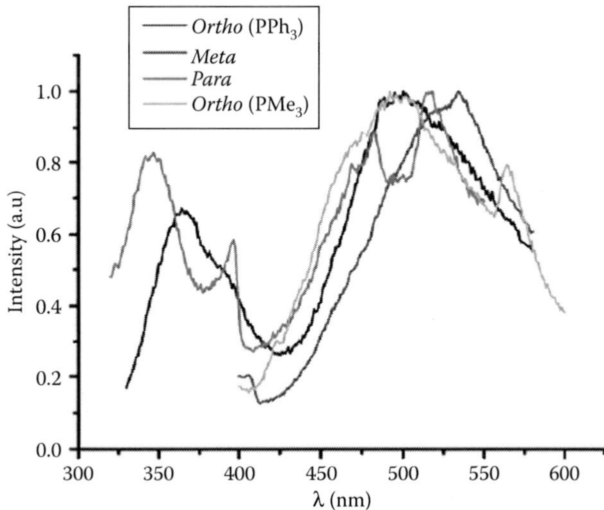

FIGURE 15.22 Emission (λ_{ex} = 310 nm) spectra of the organogold complexes [(μ-1,*n*-$C_2B_{10}H_{10}$){Au(PR_3)}$_2$] (*n* = 2, R = Ph (**24a**), Me (**24b**); *n* = 7, R = Ph (**24c**); *n* = 12, R = Ph (**24d**)). (Crespo, O.; et al. *J. Chem. Soc., Dalton Trans.* 2009, 3807–3813. Reproduced by permission of The Royal Society of Chemistry.)

(*n* = 2, R = Ph (**24a**), Me (**24b**); *n* = 7, R = Ph (**24c**); *n* = 12, R = Ph (**24d**)) all displayed two emission bands, whose maxima appeared at about 380 and 520 nm (Figure 15.22). The higher-energy emissions were attributed to carborane-centered intraligand excited states, while the lower-energy bands were assigned as MLCT in origin. Although no time-resolved data have been measured on these complexes, DFT calculations revealed HOMOs with substantial metal contributions and ligand-based (carborane + phosphine) LUMOs, which certainly supported the MLCT assignment.

15.4.2 Metallacarboranes with Endopolyhedral Metal–Ligand Fragments

Perhaps quite fittingly, one of the original metallacarborane discoveries was the subject of the first reported luminescence from this class of metallacarborane—the Ni^{IV} sandwich complex

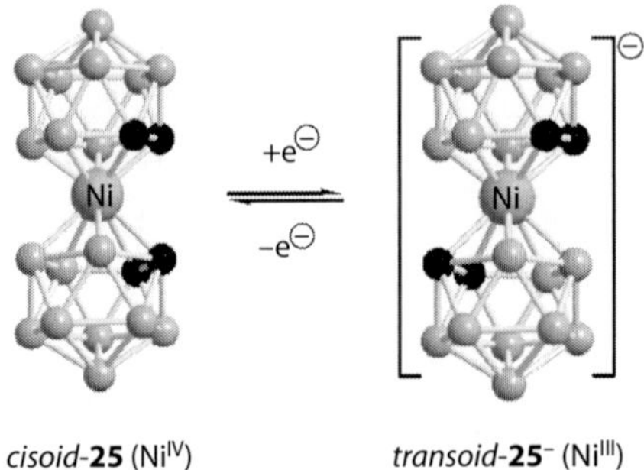

SCHEME 15.6 Electrochemical exchange between *cisoid* and *transoid* rotor configurations of complexes **25** and **25**$^-$, respectively.

[*commo*-3,3′-Ni(1,2-$C_2B_9H_{11})_2$] (**25**).[38] It was found that a corkscrew rotary motion of the carborane cage ligands relative to one another could be induced either electro- or photochemically. Of the former process, a simple one-electron reduction to Ni^{III} was sufficient to afford such rotation of the cage carbon substituents from a *cisoid* disposition to a *transoid* one (Scheme 15.6).

The same result was achieved by photoexcitation of the complex at $\lambda_{ex} \approx 450$ nm. Radiative relaxation occurred from the *transoid* rotor configuration, which DFT calculations showed to be more stable in the excited state, producing a very weak emission band at $\lambda_{em} \approx 900$ nm (and of course returning to the *cisoid* configuration in the ground state) (Figure 15.23).

This photocontrolled rotary motor could be understood in terms of MO energy Walsh correlation diagrams for the Ni(IV) species calculated by DFT (Figure 15.24). These showed that the HOMO has a minimum at the *cisoid* configuration and a maximum near the *transoid* configuration. In the minimum energy configuration, this MO is bonding between the two carborane cages, while the

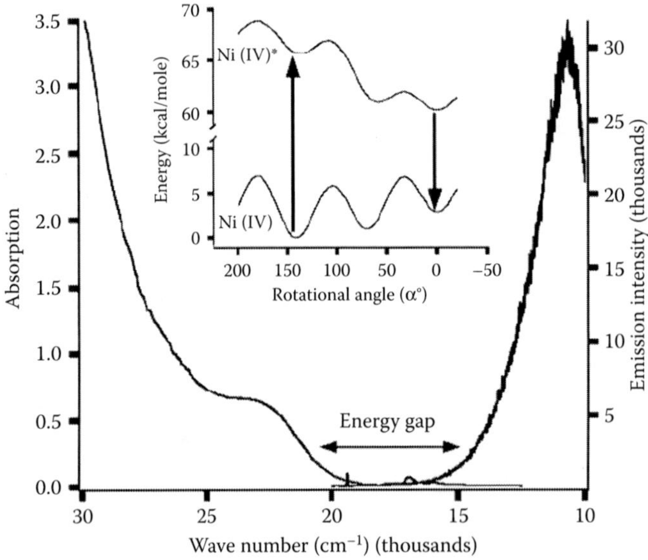

FIGURE 15.23 Absorption (left) and emission (right) spectra of complex **25**. The inset shows the calculated energies of the ground and excited electronic states as a function of rotation angle. (From Hawthorne, M. F. et al., *Science* **2004**, *303*, 1849–1851. Reproduced with permission of The American Association of the Advancement of Science.)

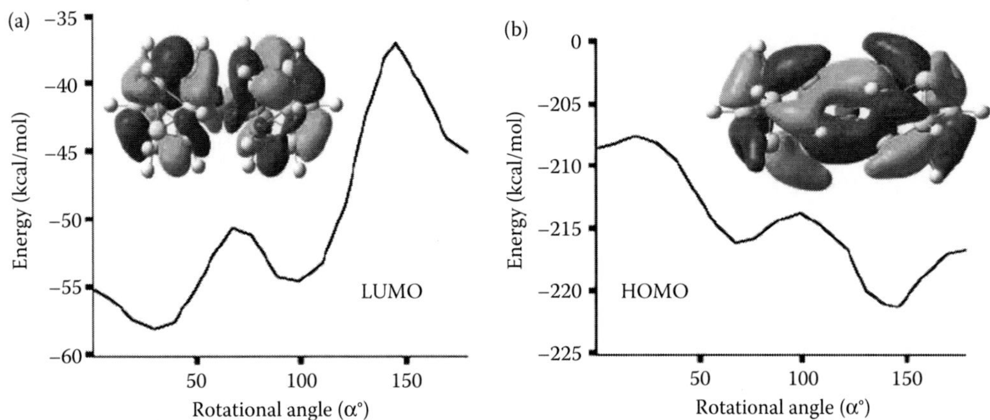

FIGURE 15.24 Energies of the (a) LUMO and (b) HOMO of complex **25** as a function of cage rotation angle. (From Hawthorne, M. F. et al., *Science* **2004**, *303*, 1849–1851. Reproduced with permission of The American Association of the Advancement of Science.)

strongly antibonding LUMO possesses a nodal plane through the Ni atom, bisecting the metal axle. The photoexcitation process displaces an electron from HOMO to LUMO and provides part of the rotational driving force with an additional increment of driving force resulting from the loss of the electron from the HOMO with its minimum at the *cisoid* configuration. Delivery of the electron to the antibonding LUMO should also cause an increase in carborane ligand–ligand distance as was evidenced by strong enhancement of the symmetric stretching mode in the resonance Raman spectrum and corroborated by TDDFT calculations.

The only other reports of metallacarborane photoluminescence come from our own laboratory at Saint Louis University, and as with the photophysical activity observed with complex **25**, we found that dynamic effects in the carborane ligand also played a critical role in our discoveries. The following results have been recently collected together and discussed.[39]

Iodination of the rhenacarborane compound Cs[3,3,3-$(CO)_3$-*closo*-3,1,2-$ReC_2B_9H_{11}$] (**26**) afforded [Z][3,3,3-$(CO)_3$-8-I-*closo*-3,1,2-$ReC_2B_9H_{10}$] ($Z^+ = NEt_4^+$ (**27a**), $Z^+ = [Fe(\eta^5\text{-}C_5H_5)_2]^+$ (**27b**)) (Figure 15.25).[40] Despite this seemingly subtle transformation of the complex anion (the net change is a simple iodination of the β-B vertex), the impact on the optoelectronic properties was quite profound. While compound **26** was found to be nonemissive (in any solvent at any temperature) and was irreversibly oxidized in MeCN or CH_2Cl_2 solution (as measured by cyclic voltammetry (CV) and differential pulse voltammetry (DPV)), complex **27** displayed a quasi-reversible 2-electron oxidation in MeCN or two sequential, fully reversible oxidations in CH_2Cl_2 (Figure 15.25a).

Although weakly emissive in solution at ambient temperature, in MeTHF glass at 77 K, bright phosphorescence was observed at $\lambda_{em} = 455$ nm resulting from excitation at $\lambda_{ex} = 275$ nm (Figure 15.25b). Time-resolved measurements revealed a single-exponential lifetime decay of $\tau = 1.65$ ms, supporting the notion of a triplet state as the emissive species. This might have been entirely expected given the presence of the iodine, which afforded an additional internal heavy atom effect leading to enhanced spin–orbit coupling. It was difficult to locate the precise nature of the photoemissive behavior, but the broadness of the emission peak was suggestive of a ^3dd state. However, the energy and strength of the absorption and the relatively high energy of the emission suggested significant carborane perturbation was likely. Thus, mixing of the highest lying Re *d* orbitals with ground-state carborane cage and carbonyl orbitals was likely to impact on the energy of photoexcitation and emission. Although our initial explanation of this phenomenon focused on involvement of the B–I σ* antibonding orbital, on reflection, we felt that this mechanism may have been an oversimplification

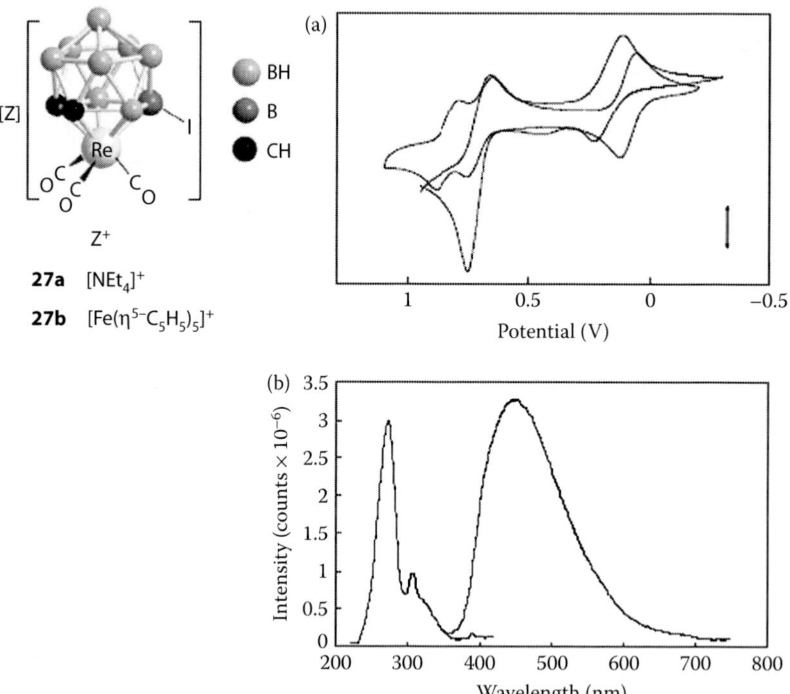

FIGURE 15.25 (a) CV of 652 μM solutions of compound **27b** in MeCN (—) and CH_2Cl_2 (—) at a scan rate of 0.1 V s^{-1}. (b) Photoluminescence spectra of a 50 μM solution of compound **27a** measured in MeTHF glass at 77 K: excitation (λ_{em} = 455 nm); emission (λ_{ex} = 275 nm). (From Jelliss, P. A. *Comment. Inorg. Chem.* **2008**, *29*, 1–25. Reproduced with permission of Taylor & Francis.)

and that the nature of the orbital was a somewhat more complicated, diffuse MO on the carborane cage. Our subsequent work with ruthenacarboranes supported this notion as described subsequently.

Starting with the ruthenacarborane complex [3,3,3-$(CO)_3$-*closo*-3,1,2-$RuC_2B_9H_{11}$] (**28**) we synthesized the 2,2′-bipyridyl complexes [3-CO-3,3-(κ^2-4,4′-R_2-2,2′-$(NC_5H_3)_2$)-*closo*-3,1,2-$RuC_2B_9H_{11}$] (R = H (**29a**), R = $(CH_2)_8Me$ (**29b**), R = Bu^t (**29c**)) as well as the TMEDA complex [3-CO-3,3-(κ^2-$Me_2N(CH_2)_2NMe_2$)-*closo*-3,1,2-$RuC_2B_9H_{11}$] (**30**) (Figure 15.26).[41] Our initial hypothesis was that the 2,2′-bipyridyl complexes **29** would demonstrate the most interesting photophysical properties, given the rich optoelectronic chemistry of 2,2′-bipyridyl complexes of Ru^{II}, such as the archetypal [Ru(κ^2-2,2′-$(NC_5H_4)_2)_3$]$^{2+}$ complexion.

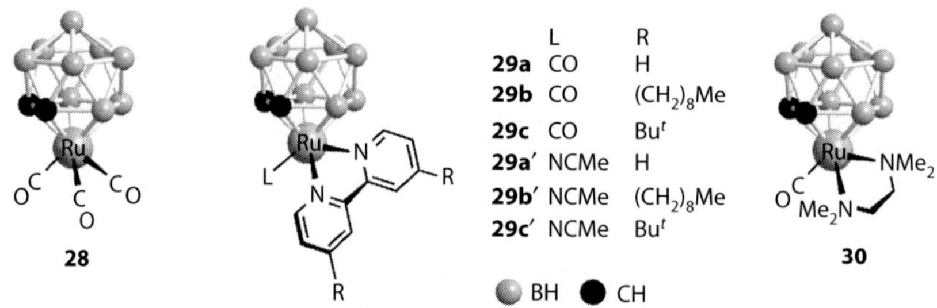

FIGURE 15.26 The ruthenacarborane complexes **28**, **29**, and **30**. (From Jelliss, P. A. *Comment. Inorg. Chem.* **2008**, *29*, 1–25. Reproduced with permission of Taylor & Francis.)

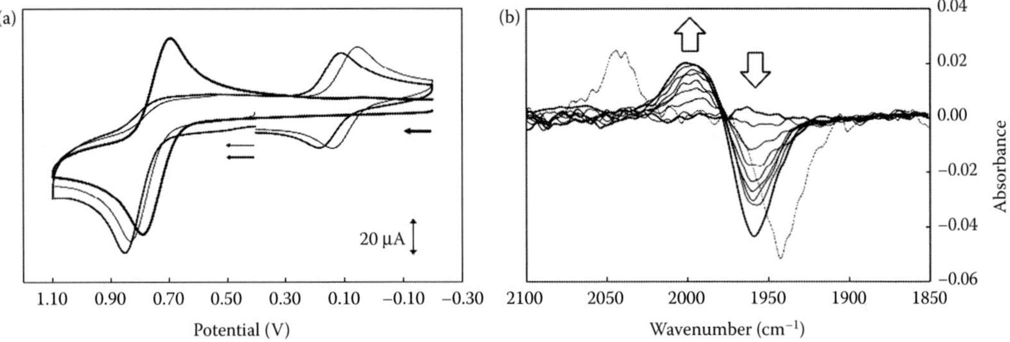

FIGURE 15.27 (a) Oxidative-anodic CVs (scan rate 1.00 V s^{-1}) of MeCN (1.0 mM) solutions of complexes **29a** (—), **29c** (—), and **30** (—). (From Jelliss, P. A. *Comment. Inorg. Chem.* **2008**, *29*, 1–25. Reproduced with permission of Taylor & Francis.)

CV measurements revealed that complex **30** underwent an expected reversible $Ru^{III/II}$ oxidation in MeCN, but we were somewhat surprised to note that the complexes **29** were completely irreversibly oxidized at around the same potential in the same solvent (Figure 15.27a).

Spectroelectrochemical analysis carried out by Professor Mike Shaw at Southern Illinois University Edwardsville (SIUE), again coupled with EPR analysis of the oxidized product **30**$^+$ strongly suggested a metal-centered oxidation had occurred. For the 2,2′-bipyridyl complexes **29**, widening of the CV scan range revealed a new reversible redox process some 0.7 V lower in potential, pursuant to the irreversible oxidation. This time spectroelectrochemical analysis indicated a considerably smaller positive shift in ν_{max}(CO) of only 37 cm^{-1} (Figure 15.27b). Furthermore, the rate of growth of the CO absorption for **29**$^+$ was notably slower than that for the consumption of the neutral precursor **30**. Hence, it was very likely that the CO was being labilized in **29**$^+$ and dissociating to be replaced by the solvent. If CVs were run in MeCN, then this coordinating solvent was affording observable products **29′**$^+$, which then displayed (following some diffusion away from the electrode as confirmed by simulation) the reversible $Ru^{III/II}$ couple at 0.7 V lower potential. Very importantly, we believed that the change in ν_{max}(CO) for complexes **28** and **29** on one-electron oxidation could only be accounted for by a structural deformation that occurred for complexes **29** but not **30**, namely that of a *closo* → *isocloso* polyhedral rearrangement (Scheme 15.7). Stone et al. have previously observed this kind of structural reorganization for *closo*-3,1,2-MC_2B_9 systems, where a boron from the upper B_5 pentagonal belt migrates to within bonding distance of the metal, at the expense of the cage C–C connectivity.[42,43] We rationalized this process for complexes **29** and **30** in terms of simple Wade–Williams rules[44,45] arguments: an induced deficiency of polyhedral skeletal

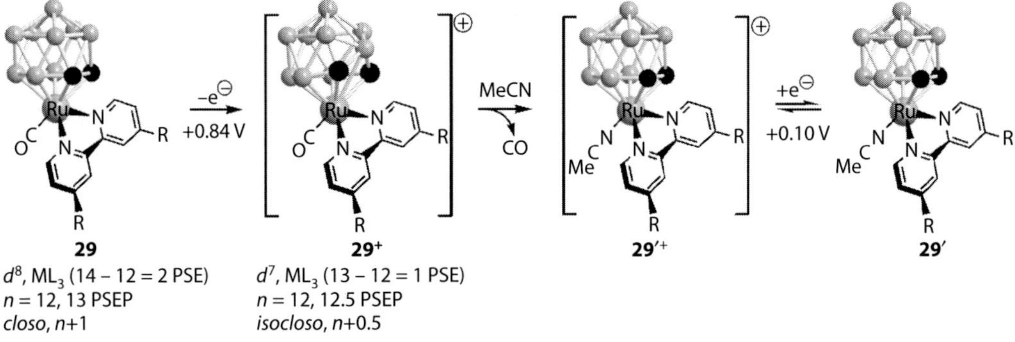

SCHEME 15.7 Substitution lability on electrochemical oxidation of the complexes **29** in MeCN. (From Jelliss, P. A. *Comment. Inorg. Chem.* **2008**, *29*, 1–25. Reproduced with permission of Taylor & Francis.)

electrons (PSE) should stimulate the rearrangement depending on the donor ability of the κ^2-*N*-donor ligands. For 2,2′-bipyridyl, the poorer overall donor, the *closo* architecture cannot be preserved, the *closo* → *isocloso* modification occurs and CO loss ensues in the oxidized species. For TMEDA, a stronger if somewhat bulkier κ^2-donor, the electron density at the metal in **29**$^+$ is still elevated enough to maintain the *closo* structure and $d\pi^*$-backbonding is sufficient to curtail CO loss, at least on the CV timescale.

Spectroelectrochemical and EPR probing has helped us interpret some important facets of the molecular orbital structures of complexes **29** and **30** and subsequently to understand their observed photophysical responses. Far from displaying interesting photophysical properties, the 2,2′-bipyridyl complexes **29** were shown to be completely nonluminescent at any temperature. Complex **30**, on the other hand, while only weakly emissive at room temperature, was strongly phosphorescent in MeTHF glass at 77 K (λ_{em} = 450 nm, λ_{ex} = 290 nm, τ = 0.77 ms) with features that were reminiscent of the photoemission from rhenium complex **27**. The optically benign nature of complexes **29** was straightforwardly attributed to the presence of *dd* states lower in energy than the MLCT transitions normally responsible for Ru-polypyridyl phosphorescence. These *dd* states are of course significantly perturbed by backbonding with both the carborane and carbonyl ligands and are primarily responsible for efficient nonradiative dispersion of energy. MLCT states are necessarily absent for complex **30** and the observed emission is very characteristic of a *dd* triplet state. The question then arose as to why the *dd* states for complex **30** were photoemissive, while those of complexes **29** were not. Again we believed the answer may be rooted in simple Wade–Williams electron-counting rules. We realized that oxidation of complexes **29** resulted in an $n + 1 \rightarrow n + 0.5$ adjustment in PSE, sufficient to cause CO loss by virtue of a depletion of electron density at the metal. Photoexcitation of the complexes **29** promoted an electron from a mostly metal *d* orbital to a carborane-delocalized antibonding MO (known from the reductive spectroelectrochemical and EPR measurements), which represented a formal $n + 1 \rightarrow n$ (*closo* → *isocloso*) change. One might therefore continue to expect photodissociation to dominate, even at 77 K. We also knew that complex **30** tolerated one-electron oxidation much better than the complexes **29**, we believed without any associated distortion of the metallacarborane cage framework. Thus, the same $n + 1 \rightarrow n$ photoexcitation of complex **30**, which was arguably more disruptive to the polyhedral skeletal bonding than the one-electron oxidation, may have resulted in a more photostable *isocloso* species leading to either photodissociation (**30′**) or the observed emissive triplet manifold (**30″**) (Scheme 15.8). In this respect, complex **30** does indeed share similar distorted d^6 metallacarborane photophysical characteristics with complex **27**,

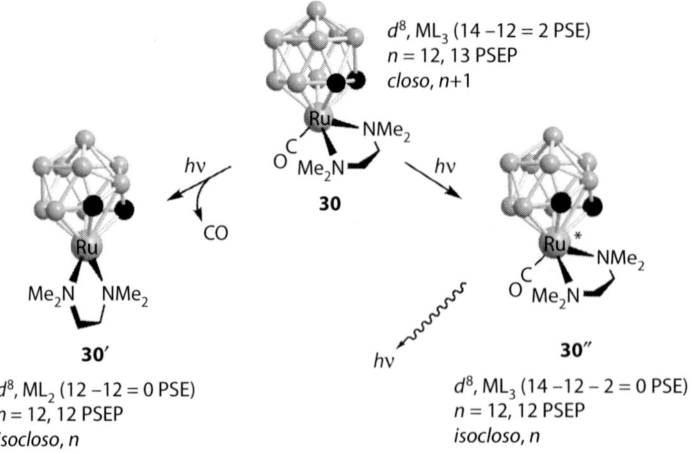

SCHEME 15.8 Photochemical and physical excitation of complex **30**. (From Jelliss, P. A. *Comment. Inorg. Chem.* **2008**, *29*, 1–25. Reproduced with permission of Taylor & Francis.)

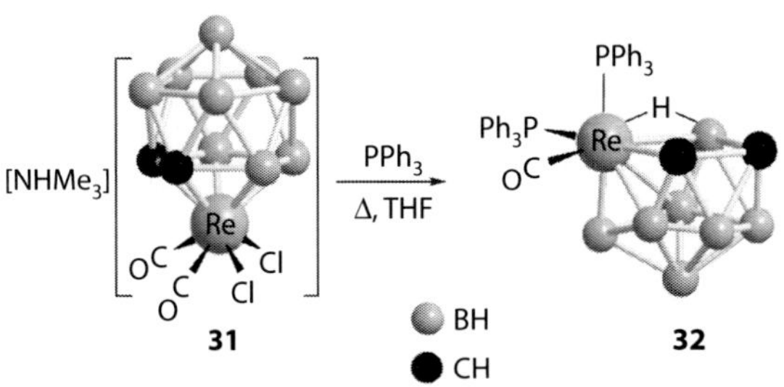

SCHEME 15.9 Synthesis of complex **32**. (From Jelliss, P. A. *Comment. Inorg. Chem.* **2008**, *29*, 1–25. Reproduced with permission of Taylor & Francis.)

although we note here that complex **30** phosphoresces without the benefit of an additional heavy atom effect, namely that of iodine.

Till now, our only observations of photoluminescence had been in frozen solvent glasses at 77 K. We still wondered whether or not one might observe ambient temperature solution luminescence from a metallacarborane, and quite by accident, we succeeded in doing so. We found that refluxing $[NHMe_3][3,3\text{-}Cl_2\text{-}3,3\text{-}(CO)_2\text{-}closo\text{-}3,1,2\text{-}ReC_2B_9H_{11}]$ (**31**) with PPh_3 in THF afforded a remarkable new metallacarborane compound, $[7,10\text{-}\mu\text{-}H\text{-}7\text{-}CO\text{-}7,7\text{-}(PPh_3)_2\text{-}isonido\text{-}7,8,9\text{-}ReC_2B_7H_9]$ (**32**) (Scheme 15.9).[46]

The most striking feature structurally was the degradation of the carborane cage by two boron vertices to give an *isonido*-7,8,9-ReC_2B_7 architecture. The open $\overparen{ReCCB}$ quadrilateral face was revealed by a single-crystal x-ray diffraction study and the bridging Re–H–B endopolyhedral hydrogen atom in that face was spectroscopically observed in the ^1H NMR spectrum (δ–7.61).

Electrochemical measurements on CH_2Cl_2 solutions of complex **32** displayed what appeared to be a fully reversible one-electron oxidation, but distortions of the CV trace at high scan rates indicated that the process was not a simple electron transfer (Figure 15.28). Of course it was surprising to us that a formally Re^{III} metallacarborane complex could even undergo such oxidation to a Re^{IV} cationic species without some indication of CO lability. Recall that the ruthenium complexes **29** with their comparatively more powerful donor 2,2′-bipyridyl ligands were distinctly labile in their oxidized Ru^{III} cationic state. Simulation studies along with, once again, an educated interpretation of the Wade–Williams counting rules, allowed us to devise a simple $\overrightarrow{E}C_{rev}\overleftarrow{E}$ square

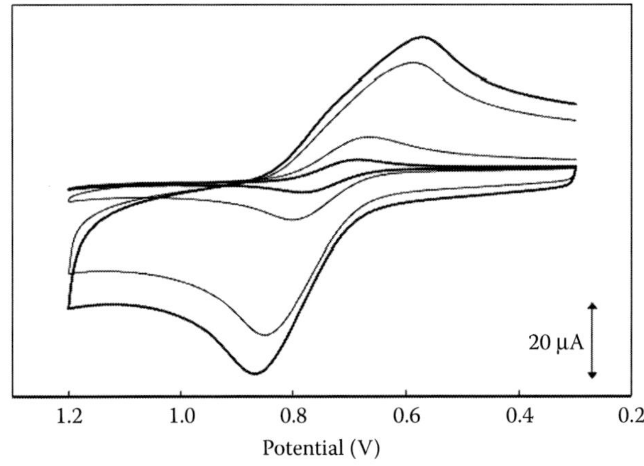

FIGURE 15.28 CVs for a CH_2Cl_2 solution (0.68 mM) of complex **32** at scan rates 0.02, 0.10, 1.00, and 1.50 V s^{-1} relative to $Ag/AgNO_3$ (MeCN, 10 mM). (From Jelliss, P. A. *Comment. Inorg. Chem.* **2008**, *29*, 1–25. Reproduced with permission of Taylor & Francis.)

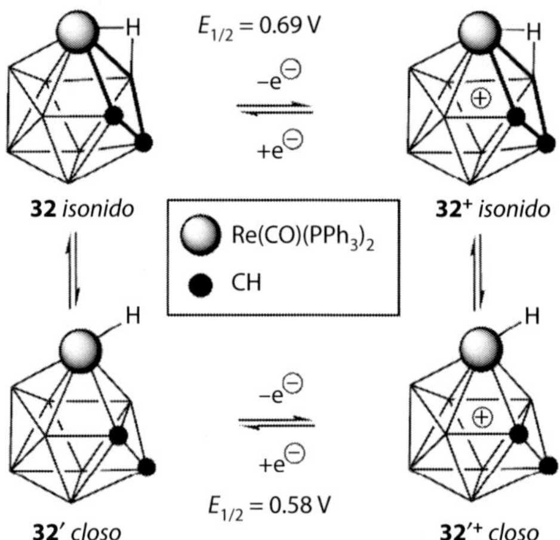

SCHEME 15.10 Square scheme ($\vec{E}C_{rev}\overleftarrow{E}$) mechanism in the oxidation of complex **32**. Potentials quoted relative to ferrocene. (From Jelliss, P. A. *Comment. Inorg. Chem.* **2008**, *29*, 1–25. Reproduced with permission of Taylor & Francis.)

Scheme for the oxidation of complex **32** (Scheme 15.10). Our rationale was that complex **32** ought to comprise a *closo* (fully deltahedral) framework by those counting rules. The loss of an electron to give **32**$^+$ was thus followed by a subtle polyhedral rearrangement to connect the *trans*-facial boron and carbon atoms. The new *closo* geometry **32′**$^+$ displayed a slightly more cathodic response on the return CV scan, the consequence of which was to reopen the cage back to the *isonido* geometry.

More surprisingly, complex **32** displayed ambient temperature luminescence in CH_2Cl_2 solution (λ_{em} = 442 nm) with an impressive quantum yield (Φ = 0.012, following self-absorbance correction) for an organometallic complex (Figure 15.29). Solutions of complex **32** in this and other solvents all

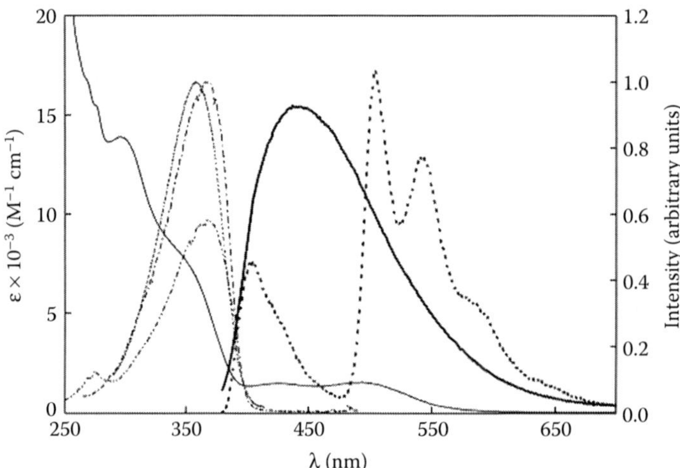

FIGURE 15.29 UV-vis spectrum of complex **32** in CH_2Cl_2 (50 μM) (—) (left ordinate). Photoluminescence spectra in CH_2Cl_2 (50 μM, 298 K): excitation (λ_{em} = 442 nm) (····); emission (λ_{ex} = 360 nm) (**—**). Photoluminescence spectra in MeTHF (50 μM, 77 K): excitation (λ_{em} = 405 nm) (-···-); excitation (λ_{em} = 504 nm) (-··-); emission (λ_{ex} = 368 nm) (---). Photoluminescence intensity (right ordinate) normalized relative to maximum excitation intensities. (From Jelliss, P. A. *Comment. Inorg. Chem.* **2008**, *29*, 1–25. Reproduced with permission of Taylor & Francis.)

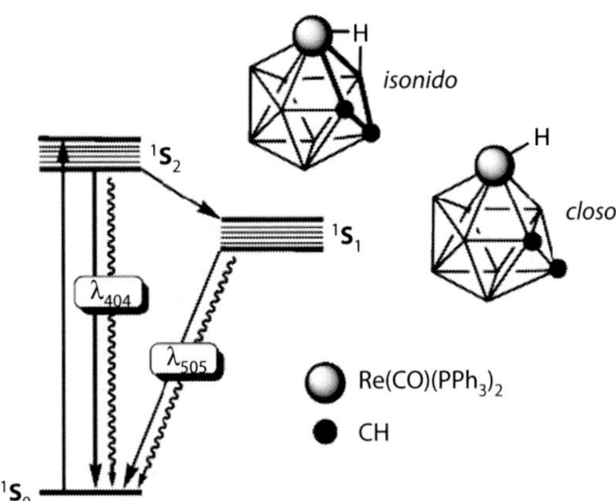

SCHEME 15.11 Energy-level diagrams for dual fluorescence of complex **32** at 77 K. (From Jelliss, P. A. *Comment. Inorg. Chem.* **2008**, *29*, 1–25. Reproduced with permission of Taylor & Francis.)

demonstrated bright-turquoise-blue luminescence when placed over a hand-held long-wavelength UV lamp.

Although the photoemission profile was broad, the UV–vis absorption and photoexcitation spectra (λ_{ex} = 360 nm) suggested that this was not due to ^3dd phosphorescence. This was corroborated by pulsed laser time-resolved measurements, which revealed a variable, biexponential fluorescent decay with essentially a nanosecond time regime—for example, at the emission peak (442 nm), τ_1 = 0.44 ns (74%) and τ_2 = 1.1 ns (26%). Measurements in MeTHF glass at 77 K exposed dual-fluorescent singlet manifolds, with a higher-energy violet emission at λ_{em} = 405 nm (τ_1 = 0.18 ns (92%) and τ_2 = 0.74 ns (8%)) and an even more intense lower-energy green emission at λ_{em} = 505 nm (τ_1 = 0.40 ns (88%) and τ_2 = 4.7 ns (12%)). Both emissions displayed vibronic structure, with the lower-energy emission clearly resolved enough to quantify this at $\Delta\nu \approx 1400$ cm^{-1}. This fortunately provided a strong clue as to the nature of these excited singlet states, with absorptions in the ground-state IR spectrum of **32** at 1430 and 1434 cm^{-1} assigned to aryl C–C *n* stretching modes in the phosphines. After ruling out a number of options for identifying the chromophore, a $\sigma_{ReP} \rightarrow \sigma_{PC*}$ transition was assigned as the primary photoexcitation leading to the first (higher-energy) singlet manifold $^1\mathbf{S}_2$ (Scheme 15.11). Thus, an electron was being excited from a metal-perturbed phosphine donor pair to one or more P–C σ^* orbitals, which we believe accounted for the observed vibronic structure. This implied that the metal-bound phosphine was the active chromophoric center rather that the carborane itself, which was nevertheless rather unique behavior for a Re^{III}–PR$_3$ complex. But to account for the dual fluorescence, there must have also been a subtle distortion of the complex to afford the second emissive singlet manifold, $^1\mathbf{S}_1$. We can draw on our accompanying electrochemical measurements and our experiences with complexes **27** and **30** described above to suggest that this distortion again emanates from an $n + 1$ *isonido* → *n closo* (or *isocloso*) transformation. This was logical from the perspective of a photoexcitation that displaces an electron from an orbital

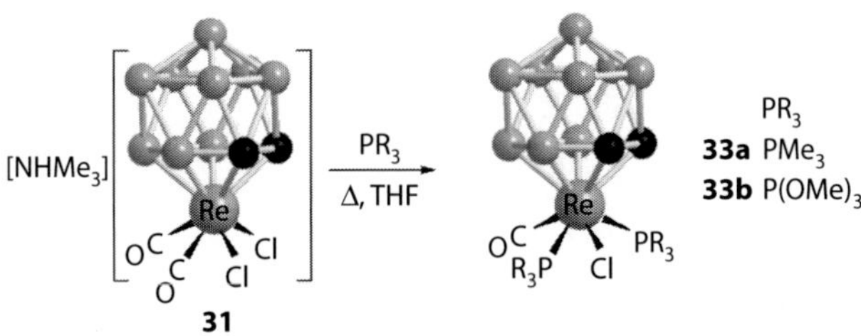

SCHEME 15.12 Synthesis of the complexes **33** without carborane cage degradation.

associated with polyhedral skeletal bonding, which the P $\rightarrow$ Re donor bond would be, to one that is completely exopolyhedral.

We have more recently synthesized and characterized the complexes [3-Cl-3-CO-3,3-L_2-*closo*-3,1,2-$ReC_2B_9H_{11}$] (L = PMe_3 (**33a**) or $P(OMe)_3$ (**33b**)) (Scheme 15.12) using a similar synthetic methodology.[47] Curiously, x-ray crystallographic analysis revealed that both complexes retain the overall *closo*-3,1,2-$ReC_2B_9H_{11}$ 12-vertex architecture, yet display surprisingly similar optoelectronic attributes to complex **32**, including bright-blue luminescence in solution at ambient temperature and completely reversible one electron $Re^{IV/III}$ redox couples. We hope to complete time-resolved fluorescence measurements and provide a more complete analysis of these unusual metallacarborane chromophores in the near future.

15.5 SUMMARY

In conclusion, we have seen that boranes, carboranes, and metallacarboranes are set to evolve into a novel group of chromophores for photoluminescence studies and optoelectronic applications. The role played by the polyhedral cage frameworks in optoelectronic functionality appears to range from ancillary in nature to one where the cage orbital manifold plays a dynamic role in both ground and photoemissive states. As things stand, most research groups have carried out only steady-state luminescence analysis, with limited time-resolved probing of these systems. Hopefully, more time-resolved techniques such as resonance Raman and time-resolved infra-red spectroscopy will be employed to take an even closer look at the photoexcited states involved in photoluminescence. Coupled with DFT and TDDFT computational techniques, only then will we acquire a more comprehensive understanding of these observed behaviors.

REFERENCES

1. Base, K.; Grinstaff, M. W., Generation of an unprecedented excited state oxidant in a coordinately unsaturated platinum complex. *Inorg. Chem.* **1998,** *37*, 1432–1433.
2. Calhorda, M. J.; Crespo, O.; Gimeno, M. C.; Jones, P. G.; Laguna, A.; López-de-Luzuriaga, J. M.; Pérez, J. L.; Ramón, M. A.; Veiros, L. F., Synthesis, structure, luminescence, and theoretical studies of tetranuclear gold clusters with phosphinocarborane ligands. *Inorg. Chem.* **2000,** *39*, 4280–4285.
3. Weinstein, J. A.; Tierney, M. T.; Davies, E. S.; Base, K.; Robeiro, A. A.; Grinstaff, M. W., Probing the electronic structure of platinum(II) chromophores: Crystal structures, NMR structures, and photophysical properties of six new bis- and di-phenolate/thiolate Pt(II) diimine chromophores. *Inorg. Chem.* **2006,** *45*, 4544–4555.
4. Barbieri, A.; Accorsi, G.; Armaroli, N., Luminescent complexes beyond the platinum group: the d^{10} avenue. *J. Chem. Soc., Chem. Commun.* **2008**, 2185–2193.
5. Lo, K. K.-W., Luminescent transition metal complexes as biological labels and probes. *Struct. Bonding* **2007,** *123*, 205–245.
6. De Cola, L.; Belser, P.; Von Zelewsky, A., Design, synthesis and photophysics of ruthenium and osmium complexes through 20 years of collaboration. *Inorg. Chim. Acta* **2007,** *360*, 775–784.
7. Robertson, N., Optimizing dyes for dye-sensitized solar cells. *Angew. Chem., Int. Ed. Engl.* **2006,** *45*, 2338–2345.
8. Piszczek, G., Luminescent metal-ligand complexes as probes of macromolecular interactions and biopolymer dynamics. *Archives Biochem. Biophys.* **2006,** *453*, 54–62.
9. Lo, K. K.-W.; Hui, W.-K.; Chung, C.-K.; Tsang, K. H.-K.; Lee, T. K.-M.; Li, C.-K.; Lau, J. S.-Y.; Ng, D. C.-M., Luminescent transition metal complex biotin conjugates. *Coord. Chem. Rev.* **2006,** *250*, 1724–1736.
10. Yam, V. W.-W.; Wong, K. M.-C., Luminescent molecular rods–transition-metal alkynyl complexes. *Topics Curr. Chem.* **2005,** *257*, 1–32.
11. Lo, K. K.-W.; Hui, W.-K.; Chung, C.-K.; Tsang, K. H.-K.; Ng, D. C.-M.; Zhu, N.; Cheung, K.-K., Biological labeling reagents and probes derived from luminescent transition metal polypyridine complexes. *Coord. Chem. Rev.* **2005,** *249*, 1434–1450.

12. Polo, A. S.; Itokazu, M. K.; Iha, N. Y. M.; Yukie, N., Metal complex sensitizers in dye-sensitized solar cells. *Coord. Chem. Rev.* **2004,** *248*, 1343–1361.
13. Lees, A. J., Luminescent metal complexes as spectroscopic probes of monomer/polymer environments. *Mol. Supramol. Photochem.* **2001,** *7*, 209–255.
14. Yam, V. W.-W., Molecular design of luminescent-based materials. *Pure Appl. Chem.* **2001,** *73*, 543–548.
15. Vogler, A.; Kunkely, H., Luminescent metal complexes: diversity of excited states. *Topics Curr. Chem.* **2001,** *213*, 143–182.
16. Demas, J. N.; DeGraff, B. A., Applications of luminescent transition platinum group metal complexes to sensor technology and molecular probes. *Coord. Chem. Rev.* **2001,** *211*, 317–351.
17. de Silva, A. P.; Fox, D. B.; Huxley, A. J. M.; McClenaghan, N. D.; Roiron, J., Metal complexes as components of luminescent signalling systems. *Coord. Chem. Rev.* **1999,** *185–186*, 297–306.
18. Unique properties of boron harnessed to develop new drugs and diagnostics. *Science Daily (European Science Foundation)* 23 October 2008.
19. Hawkins, P.; Nonaka, N.; Banks, W. A.; Jelliss, P. A.; Shi, X., Permeability of the blood-brain barrier to a rhenacarborane. *J. Pharmacol. Exp. Ther.* **2009,** *329*, 608–614.
20. Volkov, V. V.; Il'inchik, E. A.; Yur'eva, O. P.; Volkov, O. V., Luminescence of the derivatives of boranes and adducts of decaborane(14) of the $B_{10}H_{12}[Py(X)]_2$ type. *J. Appl. Spectroscopy* **2000,** *67*, 864–870.
21. Il'inchik, E. A.; Volkov, V. V.; Dunaev, S. T., *Zh. Strukt. Khim.* **1996,** *37*, 59–67.
22. Virovets, A. V.; Vakulenko, N. N.; Volkov, V. V.; Podberezskaya, N. V., *Zh. Strukt. Khim.* **1994,** *35*, 72–77.
23. Kaszynski, P.; Huang, J.; Jenkins, G. S.; Bairamov, K. A.; Lipiak, D., Boron clusters in liquid crystals. *Mol. Cryst. Liq. Cryst.* **1995,** *260*, 315–332.
24. Kaszynski, P.; Douglass, A. G., Organic derivatives of *closo*-boranes: a new class of liquid crystal materials. *J. Organomet. Chem.* **1999,** *581*, 28–38.
25. Kunkely, H.; Vogler, A., Is *o*-carborane photoluminescent? *Inorg. Chim. Acta* **2004,** *357*, 4607–4609.
26. Kunkely, H.; Vogler, A., Luminescence of silver 7,8,9,10,11,12-hexabromo-*closo*-1-carbododecaborate. *Inorg. Chem. Commun.* **2005,** *8*, 992–993.
27. Crespo, O.; Gimeno, M. C.; Laguna, A.; Ospino, I.; Aullón, G.; Oliva, J. M., Organometallic gold complexes of carborane. Theoretical comparative analysis of *ortho*, *meta*, and *para* derivatives and luminescence studies. *J. Chem. Soc., Dalton Trans.* **2009**, 3807–3813.
28. Kokado, K.; Chujo, Y., Emission via aggregation of alternating polymers with *o*-carborane and *p*-phenylene-ethynylene sequences. *Macromolecules* **2009,** *42*, 1418–1420.
29. Kokado, K.; Tokoro, Y.; Chujo, Y., Luminescent *m*-carborane-based π-conjugated polymer. *Macromolecules* **2009,** *42*, 2925–2930.
30. Peterson, J. J.; Simon, Y. C.; Coughlin, E. B.; Carter, K. R., Polyfluorene with *p*-carborane in the backbone. *J. Chem. Soc., Chem. Commun.* **2009**, 4950–4952.
31. Peterson, J. J.; Werre, M.; Simon, Y. C.; Coughlin, E. B.; Carter, K. R., Carborane-containing polyfluorene: *o*-carborane in the main chain. *Macromolecules* **2009,** *42*, 8594–8598.
32. Crespo, O.; Gimeno, C.; Jones, P. G.; Laguna, A.; López-de-Luzuriaga, J. M.; Monge, M.; Pérez, J. L.; Ramón, M. A., Luminescent *nido*-carborane-diphosphine anions $[(PR_2)_2C_2B_9H_{10}]^-$ (R = Ph, iPr). Modification of their luminescence properties upon formation of three-coordinate gold(I) complexes. *Inorg. Chem.* **2003,** *42*, 2061–2068.
33. Zhang, D.; Dou, J.; Li., D.; Wang, D., Synthesis and characterization of three group 10 complexes containing *nido*-carborane diphosphine: $[MCl(PPh_3)\{7,8\text{-}(PPh_2)_2\text{-}7,8\text{-}C_2B_9H_{10}\}]$ (M = Ni, Pd, Pt). *Inorg. Chim. Acta* **2006,** *359*, 4243–4249.
34. Zhang, D.; Dou, J.; Li, D.; Wang, D., Synthesis, characterization and luminescence of the tetranuclear gold cluster: $[Au_4\{7,8\text{-}(PPh_2)_2\text{-}7,8\text{-}C_2B_9H_{10}\}_2(PPh_3)_2]$. *J. Coord. Chem.* **2007,** *60*, 825–831.
35. Dou, J.; Zhang, D.; Zhu, Y.; Li, D.; Wang, D., Synthesis, characterization and luminescence property of heterobimetallic trinuclear Mo(W)–Ag–S clusters containing 1,2-bis(diphenylphosphino)-1,2-dicarba-*closo*-dodecaborane. *Inorg. Chim. Acta* **2007,** *360*, 3387–3393.
36. Hong, E.; Jang, H.; Kim, Y.; Jeoung, S. C.; Do, Y., Mechano- and electroluminescence of a dissymmetric hafnium carborane complex. *Advanced Mater.* **2001,** *13*, 1094–1096.
37. Shin, C. H.; Han, Y.; Lee, M. H.; Do, Y., Group 4 *ansa*-metallocenes derived from *o*-carborane and their luminescent properties. *J. Organomet. Chem.* **2009,** *694*, 1623–1631.
38. Hawthorne, M. F.; Zink, J. I.; Skelton, J. M.; Bayer, M. J.; Liu, C.; Livshits, E.; Baer, R.; Neuhauser, D., Electrical or photocontrol of the rotary motion of a metallacarborane. *Science* **2004,** *303*, 1849–1851.
39. Jelliss, P. A., Luminescent Metallacarboranes. *Comment. Inorg. Chem.* **2008,** *29*, 1–25.

40. Fischer, M. J.; Jelliss, P. A.; Phifer, L. M.; Rath, N. P., Halogenated rhenacarboranes. Optoelectronic behavior of the iodinated rhenacarborane complex anion $[3,3,3\text{-(CO)}_3\text{-8-I-}closo\text{-3,1,2-ReC}_2B_9H_{10}]^-$. *Inorg. Chim. Acta* **2005,** *358*, 1531–1544.
41. Jelliss, P. A.; Mason, J.; Nazzoli, J. M.; Orlando, J. H.; Rath, N. P.; Shaw, M. J.; Vinson, A., Synthesis, photophysical and electrochemical characterization of ruthenacarborane complexes: Unexpected luminescence from the complex $[3\text{-CO-3,3-}\kappa^2\text{-Me}_2N(CH_2)_2NMe_2\text{-}closo\text{-3,1,2-RuC}_2B_9H_{11}]$. *Inorg. Chem.* **2006,** *45*, 370–385.
42. Attfield, M. J.; Howard, J. A. K.; Jelfs, A. N. d. M.; Nunn, C. M.; Stone, F. G. A., Chemistry of polynuclear metal complexes with bridging carbene or carbyne ligands. Part 66. Carbaboranetungsten–platinum complexes. Polyhedral rearrangements of a 12-vertex cage system; crystal structures of $[PtW(CO)_2(PEt_3)_2\{\eta^6\text{-}C_2B_9H_8(CH_2C_6H_4Me\text{-4})Me_2\}]\cdot CH_2Cl_2$, $[PtW(\mu\text{-H})\{\mu\text{-}\sigma\text{:}\eta^5\text{-} C_2B_9H_7(CH_2C_6H_4\text{-Me-4})Me_2\}(CO)_2(PMe_3)(PEt_3)_2]$, and related compounds. *J. Chem. Soc., Dalton Trans.* **1987**, 2219–2233.
43. Carr, N.; Mullica, D. F.; Sappenfield, E. L.; Stone, F. G. A., Alkylidyne (carbaborane) complexes of the Group 6 metals. 9. The *closo* to *hyper-closo* transformations in a tungstacarborane cage system: crystal structure of $[NEt_4][W_2(\mu\text{-}CC_6H_4Me\text{-4})(CO)_2(\eta^5\text{-7,8-Me}_2\text{-7,8-}C_2B_9H_9)[\eta^6\text{-7,8-Me}_2\text{-7,8-}C_2B_9H_8\text{-10-}(CH_2C_6H_4Me\text{-4})]]$. *Organometallics* **1992,** *11*, 3697–3704.
44. Wade, K., *Adv. Inorg. Chem. Radiochem.* **1976,** *18*, 1.
45. Williams, R. E., *Adv. Inorg. Chem. Radiochem.* **1976,** *18*, 67.
46. Buckner, S. W.; Fischer, M. J.; Jelliss, P. A.; Luo, R.; Minteer, S. D.; Rath, N. P.; Siemiarczuk, A., Dual fluorescence from an *isonido* Re^{III} rhenacarborane phosphine complex, $[7,10\text{-}\mu\text{-H-7-CO-7,7-(PPh}_3)_2\text{-}isonido\text{-7,8,9-ReC}_2B_7H_9]$. *Inorg. Chem.* **2006,** *45*, 7339–7347.
47. Jelliss, P. A.; Rath, N. P.; Siemiarczuk, A.; Xu, S.; unpublished results, 2008.

Part VI

Boron for Energy: Energy Storage, Space, and Other Applications

INTRODUCTION

From electronics we will now enter the realm of energy. This part includes a thorough review on the usefulness of boron compounds in the field of energy. A wide variety of materials composed of boron that are particularly useful for energy-related applications will be discussed in terms of their properties and syntheses in the chapters of this part. The contributing authors are David M. Schubert, Michael J. Greenhill-Hooper, Amitabha Mitra, David A. Atwood, Bohumir Grűner, Jiří Rais, Pavel Selucký, Mária Lučaníková, Amartya Chakrabarti, Lauren M. Kuta, Kate J. Krise, John A. Maguire, and Narayan S. Hosmane.

Hydrogen is considered to be one of the most valuable and "green" energy sources in the near future. The major problem associated with its use as an energy source is the hazards related to its storage and transportation. Extensive research efforts have been initiated worldwide to get a feasible solution; they are being supported by a number of governmental and nongovernmental funding agencies. Among many hydrides and other hydrogen-containing materials being tested as possible sources of hydrogen, the metal borohydrides, by virtue of their lightweight and stabilities, have been a unique choice. Even, some boranes, for example, ammonia borane, are also currently being investigated as hydrogen storage materials. Thus, boron compounds are going to dominate the field of energy production in the near future.

The electron-deficient nature of boron contributes to several interesting applications of boron-based materials; one of them is in the oilfield technology. Borates and perborates are important classes of boron compounds that play a crucial role as additives in several oilfield operations. While borate acts as a polymer crosslinker or a set retarder for the prevention of early setting of cement on the borehole walls, perborates help to degrade and remove the polymer residues. Another useful application of boron compounds is as additives in the aluminum industries. Boron additives act as preventive agents to avoid the oxidation of aluminum metals, thus saving substantial energy loss caused by the oxidation of aluminum and magnesium metals. Boron compounds, besides being energy producers, have proven to be a great savior of energy as well.

Cobalt bis(dicarbollide), commonly known as "CoSan", is an example of a boron compound which serves the role of a scavenger for the extremely biotoxic Cs-137 ion produced from the nuclear wastes. Although it has long been synthesized, the importance of this particular compound has developed only recently. Cobalt bis(dicarbollide) anion and its chloroprotected form have now being used as extractants for high-level liquid nuclear wastes.

The properties of materials in their nanodimensions vary from those in their bulk form. These variations open up several new applications. A variety of boron and boron-based nanomaterials designed and prepared to date have been found to be useful in energy, space, and other applications. Boron nitride nanotubes (BNNTs) are useful for hydrogen storage and space applications, while boron nanorods (BNRs) are considered to be useful as rocket fuels. In the following chapters, detailed discussions are made on these versatile aspects of boron compounds in different fields of applications.

16 Boron Chemistry for Hydrogen Storage

David M. Schubert

CONTENTS

16.1 INTRODUCTION

As a clean and renewable energy carrier, hydrogen is an important component of most alternative energy portfolios. Hydrogen has a very high energy content by weight and combines with oxygen to produce only water with release of energy. For these reasons, hydrogen is an excellent fuel that can be used to power fuel cells, combustion engines, and turbines, without production of carbon dioxide greenhouse gas at the point of use [1,2]. However, a major drawback for the practical use of hydrogen is its low volumetric density, making pure hydrogen impractical to store and transport [3]. Thus, implementation of hydrogen technologies requires materials that can safely store and release large amounts of hydrogen under practical conditions [4].

In recent years, governmental agencies around the world, including the U.S. Department of Energy (DOE), have sponsored extensive research on hydrogen storage materials, including those based on boron [5,6]. These efforts have resulted in significant advances in hydrogen storage technologies. Among the many options investigated, boron materials stand out as leading candidates for advanced hydrogen storage. Combining excellent hydrogen densities, release rates, and safety characteristics, some boron materials can hold substantially more hydrogen by weight and volume than even pure liquid hydrogen. Reviews covering various aspects of boron-containing hydrogen storage materials include those by Wang and Kang, Umegaki et al., Hamilton et al., and Orimo et al. [7–10].

Nonmilitary government-funded research on hydrogen storage has focused largely on automotive applications, where metrics were established relating to practical system requirements for fuel cell vehicles (FCVs) [11]. These include volumetric and gravimetric parameters needed to deliver sufficient amounts of hydrogen at appropriate rates to power a vehicle 300 miles without unusual intrusion on the interior space of the vehicle. To meet these requirements, the DOE set challenging targets for onboard hydrogen storage relating to the weight, volume, and cost of hydrogen storage systems [11]. Changes in DOE priorities have led to a suspension of most government-sponsored automotive hydrogen storage programs in the United States in 2010. Nevertheless, considerable interest remains for development of nonautomotive applications of hydrogen storage, including portable and backup power supplies for civilian and military applications, where boron-based hydrogen storage systems can provide considerably more power than conventional high-capacity batteries at equivalent weight and volume. A number of these applications are now either commercialized or in development. Civilian uses include portable electronics, underground mining vehicles, forklifts, airport tows, camping and recreation, hand tools, remote sensors, and emergency and backup power supplies. Military uses include soldier power packs, unmanned aerial vehicles, and backup power supplies [12].

16.2 WHY HYDROGEN STORAGE?

Although hydrogen has excellent attributes as an energy carrier, its practical use in transportation and other applications presents significant challenges. Although hydrogen contains considerably more energy by weight than other conventional fuels, such as natural gas, gasoline, and ethanol, it has low volumetric density. The high gravimetric energy density of hydrogen is illustrated in Table 16.1, showing the energy content of one kilogram of hydrogen (120 MJ/kg on lower heating value [LHV] basis) compared with the same weight of other fuel options: gasoline, natural gas, and ethanol [13].

Unfortunately, hydrogen has very low volumetric energy density. A volume of 1 L of hydrogen gas has an energy content of only 0.011 MJ at normal temperature and pressure, compared with about 32 MJ for a liter of gasoline (on LHV basis). Even when compressed under high pressure or liquified at –253°C, hydrogen still has relatively low energy content compared with the same volume of other fuels. The energy content of hydrogen under different conditions compared with the same volume of gasoline is illustrated in Table 16.2 [13]. When high-pressure or cryogenic containment systems are included, the poor volumetric energy density of hydrogen becomes even more pronounced.

The low volumetric density of hydrogen presents a particular problem for transportation use. For example, consider a vehicle having a 53-L (14 gallon) fuel tank. This tank will hold about 39.2 kg of

TABLE 16.1
Energy Content per kilogram for Hydrogen Compared with Other Fuels (on LHV Basis)

Fuel	MJ/kg	Proportion
Hydrogen	120	
Natural gas	47.2	
Gasoline	43.7	
Ethanol	26.9	

TABLE 16.2
Energy Content of 1.0 L of Hydrogen at 20°C and Various Pressures Compared with the Same Volume of Gasoline (on LHV Basis) [13]

Fuel	MJ/L	Proportion
H_2 at 20°C, 1 atm (14.7 psi)	0.01	
H_2, 3000 psi	1.82	
H_2, 10,000 psi	4.71	
H_2, liquid (−253°C)	8.14	
Gasoline at 20°C	32.36	

gasoline with an energy content of about 1713 MJ (LHV basis). The volume of hydrogen compressed at a pressure of 3000 psi needed to supply the same amount of energy is about 941 L, requiring a fuel tank nearly 18 times larger. Yet, the weight of hydrogen in this tank is only about 14.3 kg. If the hydrogen is compressed to a very high pressure of 10,000 psi, the required volume decreases to about 364 L, but this is still almost seven times larger than the gasoline tank. Even if the hydrogen is liquified at −253°C, the equivalent energy volume of liquid hydrogen is 210 L, or about four times larger, not counting the substantial volume of the cryogenic containment system that would be needed. In addition, the amount of energy required to liquefy hydrogen or compress it is substantial [1].

16.3 CHEMICAL HYDRIDES

Although hydrogen has very low volumetric density, many chemical compounds of hydrogen, referred to as chemical hydrides, can pack substantially more hydrogen into a given volume than is found even in pure liquid hydrogen. In order to achieve high hydrogen densities by weight, the elements to which hydrogen are bonded must be of low atomic mass, limiting the range of elements from which to choose. For this reason, hydride compound-based light elements have drawn attention. In this category, boron is unique among the light elements in its ability to form a wide range of stable chemical hydrides, a number of which have suitable properties for use in hydrogen storage applications, as described below.

The use of boron hydride compounds, and particularly sodium borohydride, to generate hydrogen dates back to the 1940s, when early developments of alkali metal borohydrides were carried out in the research group of Hermann I. Schlesinger at the University of Chicago, USA [14,15]. Sodium borohydride, and related boron hydride compounds, later become important reagents for organic synthesis, largely through the extensive work of Schlesinger's PhD student and 1979 Nobel Prize winner Herbert C. Brown and his coworkers. However, the use of boron hydrides for hydrogen storage received less attention until recent times.

In 2003, the DOE issued a "grand challenge" to the scientific community for hydrogen storage research and development. This challenge called for establishment of three hydrogen storage Centers of Excellence, focusing on (1) chemical hydrogen storage, (2) metal hydrides, and (3) carbon-based materials and sorbents. Selected in 2004, these Centers of Excellence, together with existing DOE hydrogen storage efforts and independently funded projects, made up the framework of the U.S. National Hydrogen Storage Project. In order to meet the requirement to deliver sufficient amounts of hydrogen at appropriate rates to power a vehicle 300 miles without unacceptable intrusion on the interior space of the vehicle, the DOE set specific targets for onboard hydrogen storage systems. These include both volumetric and gravimetric parameters for hydrogen storage systems, as well as targets for costs and delivery rates [11]. These targets were revised during the course of this project, with final project objectives calling for development by 2015 of onboard hydrogen storage systems achieving a gravimetric target of 9 wt% H_2 (equivalent to 90 g/kg or 3 kWh/kg) and a volumetric target of 2.7 kWh/L, with a cost of $2/kWh. An interim target of 6 wt% H_2 (60 g/kg, 2 kWh/kg) with volume and cost goals of 1.5 kWh/L and $4/kWh was set for 2010. It should be noted that the system weight includes not only the hydrogen storage material, but also the weight of all other containment and hydrogen conditioning equipment that might be needed leading up to the fuel cell.

Driven by the challenging automotive storage targets set by the DOE, and consequent need to utilize light elements to achieve them, both the Metal Hydride and Chemical Hydrogen Storage Centers focused significant attention on boron-based materials, including metal borohydride and boron–nitrogen (B–N) systems. In fact, most of the activities within the Chemical Hydrogen Storage Center involved boron compounds. Changes in priorities at the DOE resulted in the discontinuation of the Hydrogen Storage Centers of Excellence in 2010, with the focus of programs shifting away from automotive applications to hydrogen storage for stationary and nonautomotive portable power systems. The Hydrogen Storage Centers, along with other members of the international research community, accomplished major advances in the development of boron-based materials for hydrogen storage and produced numerous scientific publications during this period. The residual momentum of these programs continues with ongoing developments in both the academic and private sectors. A number of companies and academic groups continue to actively pursue the development of these applications, and a number of boron hydride systems are currently in commercial use.

16.4 METAL TETRAHYDROBORATES

The BH_4^- anion is formally called tetrahydroborate, and thus its metal salts and complexes are correctly called metal tetrahydroborates. However, these compounds are often referred to for convenience by the more informal name metal borohydrides, following the nomenclature first given to them by earlier workers in the field [16]. These compounds comprise an extensive class having the general composition $M(BH_4)_n$, where M may be Li, Na, K, Mg, Ca, Sc, Ti, V, Cr, Mn, Zn, Zr, Al, U, and so on, and n is 2–4. Metal borohydrides can contain substantial amounts of hydrolytically or thermally accessible hydrogen on both a weight and volume basis, depending on the atomic number and valence requirements of the metal [17]. The first examples of metal borohydrides were prepared in the research group of A. Stock in Germany [18] and in the groups of H. I. Schlesinger in the United States [19,20] and E. Wiberg in Germany [21]. There is now a vast technical literature associated with these compounds. Some aspects of the more promising metal borohydrides for hydrogen storage applications are discussed briefly here.

Crystal structures of the light metal borohydrides have been reviewed with emphasis on their hydrogen storage properties [22]. These compounds exhibit a wide range of properties, with variations relating to structural differences stemming from the unusual characteristics of the B–H bond and its interactions with metals of different sizes and electronegativities. The BH_4^- anion can form bridging M–H–B bonds with metals, similar to the 3-center/2-electron B–H–B bonds found in diborane. The BH_4^- anion most often acts as a bidentate ligand to metals, but is also known to coordinate in mono and tridentate modes. Metal borohydrides can occur as essentially ionic salts with

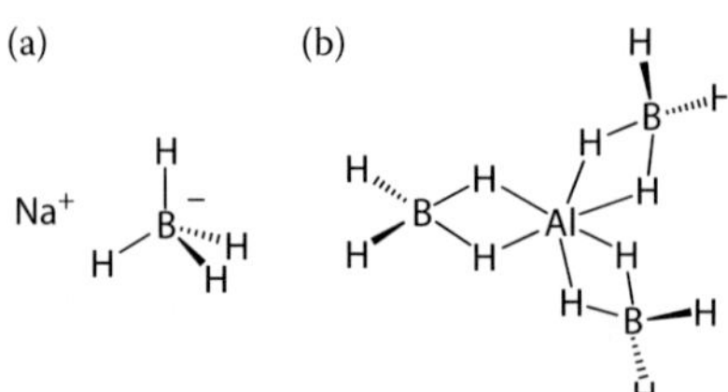

FIGURE 16.1 Sodium borohydride, $Na^+BH_4^-$ (a) is essentially an ionic salt, whereas aluminum borohydride, $Al(BH_2)_3$ (b), is a covalent compound.

TABLE 16.3
Comparison of Liquid Hydrogen with Selected Metal Borohydrides in Terms of Gravimetric and Volumetric Hydrogen Content

Compound	wt% H_2	Density[a] g/cm^3	g H_2/L	MJ H_2/L
Liq. H_2 (−253°C)	100	0.7099	71	10.1
$LiBH_4$	18.5	0.666	123	17.5
$NaBH_4$	10.7	1.074	115	16.3
$Mg(BH_4)_2$	14.9	0.989	147	20.9
$Ca(BH_2)_2$	11.6	1.12	130	18.4

[a] Multiple crystalline phases having different densities exist for most metal borohydrides.

some metal cations, such as Na^+ (Figure 16.1a), and primarily as covalent complexes with other metal cations, such as Al^{3+} (Figure 16.1b). It can be noted that $NaBH_4$ is an air-stable solid, whereas $Al(BH_4)_3$ is a pyrophoric liquid. However, most of the metal borohydrides are solids at room temperature and exhibit salt-like properties. Table 16.3 lists the more extensively studied lighter metal borohydrides and provides a comparison of their theoretical hydrogen contents with that of pure liquid hydrogen. As discussed below, not all of the theoretical hydrogen content may be accessible for some metal borohydrides under practical release condition, but available hydrogen on a volumetric basis generally exceeds that of liquid hydrogen by a significant amount.

16.4.1 Sodium Borohydride

Sodium borohydride, $NaBH_4$, contains 10.7 wt% H_2. It is currently the most commercially important of the boron hydride compounds. Several thousand tons of this material are manufactured annually for use as a reducing agent in various industries applications, including the manufacture of specialty chemicals, pharmaceuticals, paper, and in waste water treatment. Its use in hydrogen storage is also the oldest and most extensively developed among the metal borohydrides.

Sodium borohydride, or correctly sodium tetrahydroborate, is a white crystalline solid that is stable in dry air to ~300°C. It is highly soluble in water and slowly decomposes until the solution becomes alkaline. Decomposition is more rapid in acidic solution, but $NaBH_4$ is kinetically stable in alkaline solutions when the pH is above about 12.9. For this reason, 2%–3% sodium hydroxide is typically added to $NaBH_4$ solutions to provide stability. Although kinetically stable, aqueous solutions of $NaBH_4$ are thermodynamically unstable, and decomposition can be catalyzed by a number of transition metals.

The reaction of sodium borohydride with water, given by Equation 16.1, results in the production of four moles of hydrogen per mole of $NaBH_4$ with the formation of sodium metaborate. This reaction

$$NaBH_4 + 4H_2O \xrightarrow{\text{cat.}} NaB(OH)_4 + 4H_2 \quad (16.1)$$

is frequently presented in the literature in the form of Equation 16.2, indicating that the boron-containing product is anhydrous sodium metaborate, $NaBO_2$. However, this common portrayal is

$$NaBH_4 + 2H_2O \xrightarrow{\text{cat.}} NaBO_2 + 4H_2 \quad (16.2)$$

chemically unrealistic, since $NaBO_2$ does not exist in aqueous solution. Instead, it exists as the hydroxy-hydrated borate salt $Na[B(OH)_4]$ that contains the tetrahydroxyborate anion, $B(OH)_4^-$.

This borate salt crystallizes without interstitial water above 53.6°C and as the hydrate $Na[B(OH)_4]{\cdot}2H_2O$ below this temperature. Both of these forms are well-established commercial products which, as articles of commerce, are referred to by their equivalent formulas, $NaBO_2{\cdot}2H_2O$ or $Na_2O{\cdot}B_2O_3{\cdot}4H_2O$ and $NaBO_2{\cdot}4H_2O$ or $Na_2O{\cdot}B_2O_3{\cdot}8H_2O$. Based on earlier literature, these compounds in commercial use are commonly called sodium metaborate "4 Mol" and "8 Mol," referring to their oxide formulas (23–25). The full dehydration of these sodium metaborates to $NaBO_2$ requires heating in dry air above 200°C, as given by Equation 16.3.

$$Na\left[B(OH)_4\right]{\cdot}xH_2O \xrightarrow{>200^\circ C} NaBO_2\ (\text{amorphous}) + (2+x)H_2O \tag{16.3}$$

Anhydrous sodium metaborate reacts exothermically with water, rehydrating to $Na[B(OH)_4]{\cdot}xH_2O$, where x is 0 or 2, depending on temperature. Since sodium borohydride–water systems for practical hydrogen generation do not involve conditions where anhydrous sodium metaborate can form, the hydrogen release reaction is more accurately represented by Equation 16.1.

Examination of Equation 16.1 provides important insights into this system. Most important is the recognition that the sodium metaborate product ("spent fuel") contains the same number of hydrogen atoms per mole as the sodium borohydride reactant ("fuel"). This observation emphasizes the fact that borohydride does not act as a source of hydrogen so much as a reducing agent for water. The reduction of water to H_2 is a two-electron process, given by Equation 16.4. Since each B–H bond in the borohydride anion, BH_4^-, effectively supplies two electrons (Equations 16.5 and 16.6), each molar equivalent of borohydride can reduce four equivalents of water, as expressed by Equation 16.7. Thus, $NaBH_4$ can be considered a source of electrons, or reducing power, rather

$$H_2O + 2e^- \longrightarrow O^{2-} + H_2 \tag{16.4}$$

$$BH_4^- + 4H_2O \longrightarrow B(OH)_4^- + 8H^+ + 8e^- \tag{16.5}$$

$$8H^+ + 8e^- \longrightarrow 4H_2 \tag{16.6}$$

$$BH_4^- + 4H_2O \longrightarrow B(OH)_4^- + 4H_2 \tag{16.7}$$

than hydrogen. Furthermore, using incorrect Equation 16.2 to describe the $NaBH_4$–water system gives the impression that it contains >11 wt% available hydrogen on a stoichiometric basis, which is nearly equivalent to gasoline in energy content by weight. However, correct Equation 16.1 shows that the actual stoichiometric hydrogen content is about 7.3 wt%, or about 88% that of gasoline. Nevertheless, this is still a substantial energy content compared to many other proposed hydrogen storage materials.

Sodium borohydride has been employed as a source of hydrogen gas almost since the compound was first prepared in the 1940s. Indeed, in addition to their classic 1953 paper on the subject [15], the 1950 patent by Schlesinger and Brown titled, "Methods of Preparing Alkali Metal Borohydrides," describes the use of $NaBH_4$ in aqueous solution for hydrogen generation, as well as the use of catalytic accelerators, such as cobalt chloride [14]. Earlier government reports from the Schlesinger group also proposed this application to the military sponsors of this research. Numerous additional patents and publications describing methods to use $NaBH_4$ in hydrogen generation appeared with increasing frequency from the 1950s onwards. Patents specifically describing the use of $NaBH_4$ to provide a regulated flow of hydrogen to proton exchange membrane (PEM) fuel cells were published in the 1960s. For example, a 1968 patent assigned to Mine Safety Appliances Corp. describes mixtures of $NaBH_4$ with hydrated compounds that release hydrogen on heating [26]. Also, a 1969 patent assigned to Union Carbide Corp describes a portable, self-contained, hydrogen-generating apparatus for fuel cell applications involving release of hydrogen from alkaline aqueous $NaBH_4$ solutions using a Raney nickel catalyst [27]. This patent also describes a method of regulating hydrogen pressure from the device and use of $NaBH_4$ pellets as fuel concentrate. Starting in the late 1990s, the former company Millennium Cell Inc further developed practical systems for hydrogen delivery for automobiles, portable devices, and stationary power systems [28–30]. Devices of this kind are often referred to as HYDROGEN ON DEMAND®, a registered trademark of Millennium Cell Inc.

In general, the concept of HYDROGEN ON DEMAND (HOD), illustrated in Figure 16.2, entails a system allowing an aqueous solution of $NaBH_4$ to be pumped into a catalyst chamber containing selected transition metals (e.g., Co, or Ru) fixed to a porous support material. The evolved hydrogen gas is separated from the resulting sodium metaborate solution, which is returned to a spent fuel tank. The moist hydrogen gas stream, which may be hot due to heat of reaction, is passed through an optional heat exchanger/coolant loop to adjust humidity, and is then supplied to a fuel cell or other hydrogen-using device. The pressure of generated hydrogen gas can be used to control the fuel addition rate to the catalyst chamber, allowing for a constant supply of hydrogen at practical pressures for fuel cell and other applications. Many variations on this general concept have been described.

HOD technology has been commercialized for use in a variety of applications, including military and civilian fuel cell power generators. An advantage of $NaBH_4$ HOD is that it utilizes a safe, energy-dense, water-based fuel. Aqueous solutions of sodium borohydride are not flammable, even when directly exposed to an open flame. Yet, a 30 wt% $NaBH_4$ solution contains 6.7 wt% releasable H_2. Another advantage of $NaBH_4$ is its relatively low heat of reaction with water compared with other metal hydrides and borohydrides, resulting in less waste heat. Furthermore, the reaction of $NaBH_4$ with water does not involve side reactions and formation of volatile by-products other than water. Thus, it can be used to deliver a controllable stream of hydrogen free of unwanted contaminates. The delivered hydrogen contains only water vapor, partly resulting from the heat generated by the hydrolysis reaction.

The lower solubility of sodium metaborate compared with sodium borohydride places a practical limitation on HOD systems. Considering Equation 16.1, it is clear that the need for a liquid discharge from the catalyst chamber of the HOD system requires maintenance of the sodium metaborate

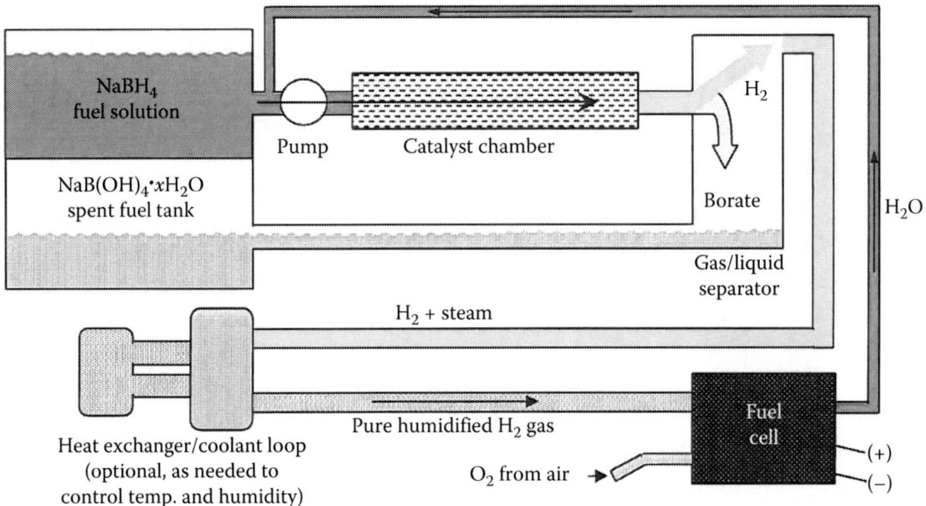

FIGURE 16.2 Schematic of HYDROGEN-ON-DEMAND® system, based on diagrams published by the former company Millennium Cell Inc.

"spent fuel" concentration below its solubility limit. Unfortunately, the lower solubility of sodium metaborate creates an upper limit on the "fuel" concentration, and thus on the total energy density of the system. Although the room-temperature water solubility of $NaBH_4$ is about 35 wt%, the substantially lower solubility of sodium metaborate may require lower practical working concentrations to prevent the metaborate from precipitating in the catalyst chamber. The heat of reaction can assist in maintaining sodium metaborate solubility, but the room-temperature stable phase, $Na[B(OH)_4]{\cdot}2H_2O$, will subsequently crystallize in the spent fuel tank, presenting a handling problem. This problem is somewhat alleviated by engineering designs allowing for product water generated by the PEM fuel cell to be recycled back to the catalyst chamber feed, as shown in Figure 16.2. Other proposed solutions involve mechanical handling of sodium borohydride and metaborate as solids, slurries, or gels [27,31–33].

Direct borohydride fuel cells (DBFCs) have also been developed in which $NaBH_4$ is decomposed directly at the fuel cell membrane, avoiding a separate hydrogen generation step [34]. DBFCs are potentially cheaper to manufacture than tradition fuel cells because they do not require platinum catalysts. Instead, nonprecious metals, such as Ni, can be used. They also have low operating temperatures and high power densities, reaching up to 9.3 Wh/g $NaBH_4$. One problem associated with DBFCs is hydrogen evolution at the anode, which leads to lower efficiencies and safety issues. Improvements in electrocatalysts to solve this problem are an active area of research. Hydrogen evolution during the oxidation of borohydride on the fuel cell anode can be prevented by shifting the anode potential to values more positive than the reversible hydrogen electrode (RHE) potential; however, this reduces the advantages of DBFC. Use of new anode catalyst materials has been claimed to allow the complete eight electron oxidation of the BH_4^- anion without hydrogen evolution in an electrode potential region negative to the RHE [35]. Further developments in this area are anticipated.

Sodium borohydride is primarily manufactured by the Schlesinger process, which involves reaction of trimethylborate with sodium hydride at 250–270°C, as given by Equation 16.8 [36]. Trimethylborate is made from boric acid and methanol (Equation 16.9) and sodium hydride is made from sodium metal and hydrogen (Equation 16.10). Thus, sodium borohydride is made from metallic sodium, boric acid, and hydrogen, as expressed by net Equation 16.12.

$$B(OMe)_3 + 4NaH \rightarrow NaBH_4 + 3NaOMe \tag{16.8}$$

$$B(OH)_3 + 3MeOH \rightarrow B(OMe)_3 + 3H_2O \tag{16.9}$$

$$4Na + 2H_2 \rightarrow 4NaH \tag{16.10}$$

$$3NaOMe_3 + 3H_2O \rightarrow 3MeOH + 3NaOH \tag{16.11}$$

$$B(OH)_3 + 4Na^\circ + 2H_2 \rightarrow NaBH_4 + 3NaOH \tag{16.12}$$

Production of 1 mol of $NaBH_4$ requires the consumption of 4 mol of metallic sodium. Although $NaBH_4$ has substantial energy content, the use of reactive metal in its manufacture results in less than optimal efficiency. Schlesinger et al. described in their original reports on $NaBH_4$ an alternate synthesis method involving grinding sodium hydride together with boric oxide, as given by Equation 16.13, where $NaBH_4$ product can be separated by solvent extraction

$$4NaH + B_2O_3 \rightarrow NaBH_4 + 3NaBO_2 \tag{16.13}$$

[11,36]. This method does not eliminate the use of reactive metal, but does avoid the need for distillation of trimethylborate ester, which is another energy-intensive step. It can be noted that the synthesis of metal borohydrides via chemical–mechanical milling methods has gained increasing attention in recent years.

Another process that has been carried out on a commercial scale was developed by Bayer Co. and used for production of $NaBH_4$ for captive use [37]. This process is sometimes called the "Bayer

Process," but should not be confused with more well-known aluminum process of the same name. This $NaBH_4$ process involves reaction of anhydrous borax, sodium metal, and silica at 700°C in a one-pot batch process, as given by Equation 16.14. This process avoids the need to

$$Na_2B_4O_7 + 16Na + 8H_2 + 7SiO_2 \xrightarrow{700°C} 4NaBH_4 + 7Na_2SiO_3 \tag{16.14}$$

rigorously distill trimethylborate, but still requires consumption of 4 mol sodium metal per mole of $NaBH_4$ produced.

Other advanced methods for the manufacture of $NaBH_4$ have been investigated by Toyota Motor Co and academic researchers in Japan. Most notable is the use of magnesium hydride for direct reduction of boron oxide feedstocks, including anhydrous borax, for initial synthesis of $NaBH_4$ (Equation 16.15) and reduction of anhydrous sodium metaborate for regeneration of spent borohydride (Equation 16.16) [38]. Aluminum metal has also been used instead of magnesium and catalysts employed to improve kinetics.

$$Na_2B_4O_7 + Na_2CO_3 + 8MgH_2 \rightarrow 4NaBH_4 + 8MgO + CO_2 \tag{16.15}$$

$$NaBO_2 + 2MgH_2 \rightarrow NaBH_4 + 2MgO \tag{16.16}$$

Wu et al. reviewed and analyzed potentially more efficient manufacturing processes for $NaBH_4$, including methods for the recycle of sodium metaborate back to borohydride [37]. To date, these efforts have not been sufficiently successful to make $NaBH_4$ economically viable for widespread use as an automotive fuel. After extensive analysis concluding that it cannot meet established performance targets, the DOE made a no–go decision in 2007 on $NaBH_4$ as an onboard automotive hydrogen storage system. However, it was acknowledged that $NaBH_4$ may serve as an intermediate in the manufacture of other hydrogen storage materials for initial fill requirements of an automotive fleet, but would not be feasible as an onboard hydrogen carrier. Nevertheless, $NaBH_4$ remains both practical and convenient for many smaller-scale fuel cell applications that may not require the same level of regeneration efficiency necessary for large automotive fleets.

16.4.2 Lithium Borohydride

Lithium borohydride, $LiBH_4$, has attracted attention because of its high hydrogen content (18.5 wt%) compared with other metal borohydrides [39,40]. It is a hygroscopic white crystalline solid that is decomposed by water. Thus, it is generally handled under inert atmosphere. Lithium borohydride occurs as two phases under ambient pressure. The higher temperature phase forms above ~107°C [41]. Lithium borohydride is currently manufactured in far smaller volumes than sodium borohydride.

Schlesinger and Brown first prepared $LiBH_4$ in the 1930s by reaction of ethyl lithium with diborane under pressure, as given by Equation 16.17 [20]. Schlesinger and coworkers later described the synthesis of $LiBH_4$ by reaction of lithium hydride with diborane in diethyl ether to give the solvate $LiBH_4{\cdot}OEt_2$, from which the ether was subsequently removed by heating at 70–100°C under vacuum [42], as in Equation 16.18. These reactions are inconvenient because of the need to handle diborane. In addition, they reported that $LiBH_4$ can be obtained by reaction of lithium hydride with trimethylborate in a manner similar to the process now used for manufacture of $NaBH_4$ [36]. They further described a method using the metathesis reaction of $NaBH_4$ with LiCl in isopropylamine, as given by Equation 16.19 [43]. However, the amine solvent was difficult to remove from the product. Brown and co-workers later described exchange reactions in aprotic solvents in which $LiBH_4$ is soluble, typically ethers, but found that these reactions are slow, requiring energetic mixing and long reaction times, and also the solvents are often difficult to remove from the product [44]. However, they showed that unsolvated $LiBH_4$ can be obtained in high yield by reaction of lithium hydride with borane-dimethyl sulfide in ethyl ether at 25°C, followed by removal of solvent and dimethyl sulfide by distillation, as given by Equation 16.20. Alternatively, reaction of KBH_4 with LiCl is reported to proceed well in tetrahydrofuran (THF) to give the THF solvate [45]. Another method involves reaction of boron trifluoride with a stoichiometric excess of lithium hydride in an ether solvent, such as THF, as in Equation 16.21 [46], where the BF_3 may be supplied as a more easily handled solvent adduct, such as BF_3·THF.

$$\mathrm{LiEt + 2B_2H_6 \longrightarrow LiBH_4 + BEt_3} \tag{16.17}$$

$$\mathrm{LiH + ½\,B_2H_6 \xrightarrow{Et_2O} LiBH_4{\cdot}OEt_2 \xrightarrow[-\,Et_2O]{70\text{–}100^\circ C,\ vac.} LiBH_4} \tag{16.18}$$

$$M\mathrm{BH_4} + \mathrm{Li}X \rightarrow \mathrm{LiBH_4} + MX\ (M = \mathrm{Na, K};\ X = \mathrm{Cl, Br}) \tag{16.19}$$

$$\mathrm{LiH + BH_3{\cdot}SEt_2 \xrightarrow[-SEt_2]{OEt_2} LiBH_4} \tag{16.20}$$

$$\mathrm{4LiH + BF_3 \xrightarrow{THF} LiBH_4 + 3LiF} \tag{16.21}$$

Lithium borohydride liberates very little hydrogen below 380°C. Although $LiBH_4$ has high thermal stability, it is rapidly hydrolyzed by water. This presents a disadvantage compared with $NaBH_4$, which is stable in alkaline solution. As a result, $LiBH_4$, and most other light metal borohydrides, cannot be used in HOD systems of the kind described above, involving catalytic release of hydrogen from aqueous fuel solutions. Hydrogen storage systems have been described involving mechanical mixing of solid lithium borohydride with water [31], but most proposed applications of $LiBH_4$ involve thermal hydrogen release.

On heating, $LiBH_4$ exhibits three thermal events [47]. There is an initial reversible polymorphic phase transition at ~107°C [41]. The compound then fuses at 268–286°C, a process accompanied by loss of ~2 wt% H_2. Major hydrogen gas evolution then commences at ~380°C, with release of ~80% of the H_2 in $LiBH_4$ taking place in the 380–680°C temperature range. The highest rate of hydrogen loss is reported to occur between 483°C and 492°C. Thermal decomposition of $LiBH_4$ occurs with release of three of its four hydrogen atoms in a reaction approximately described by Equation 16.22 [39,48]. This decomposition is complex and involves a

$$\mathrm{2LiBH_4 \xrightarrow{\Delta} LiH + 2B + 3H_2} \qquad 13.9\ \text{wt\%}\ \mathrm{H_2} \tag{16.22}$$

number of intermediates, including $Li_2B_{12}H_{12}$, which contains the kinetically stable [*closo*-$B_{12}H_{12}]^{2-}$ anion (48–51). Although $LiBH_4$ contains a total of 18.5 wt% hydrogen, Equation 16.22 shows that not all of this hydrogen is thermally accessible at practical temperatures. The experimentally observed thermally releasable hydrogen content of $LiBH_4$ is ~13.5 wt% [51], which is still considerable compared with many other proposed hydrogen storage materials. Thermal decomposition of $LiBH_4$ is exothermic, with an estimated enthalpy change of 67 kJ/mol of H_2 evolved, indicating that the reverse rehydrogenation reaction is thermodynamically unfavorable [39]. Studies of this rehydrogenation reaction report that it requires extreme conditions such as 600°C under 35 MPa H_2 pressure [52].

The high thermal stability of $LiBH_4$ and need for extreme regeneration conditions reduces its practicality as a hydrogen carrier. Consequently, efforts have been made to find ways to destabilize $LiBH_4$ to enhance hydrogen release rates and facilitate rehydrogenation at lower temperatures. It has been shown that mixing $LiBH_4$ with various other materials can significantly reduce hydrogen release temperatures. These materials include metals (e.g., Mg, Al), metal hydrides (e.g., CaH_2,

MgH_2), metal borohydrides ($NaBH_4$), amides ($LiNH_2$), oxides (e.g., SiO_2, SnO_2, TiO_2), and halides ($TiCl_3$, TiF_3), as well as activated carbon and carbon nanotubes (53–58). Some of these systems rely on the formation of more thermodynamically stable products, and thus do not improve reversibility. However, enhanced reversibility has been claimed in a number of cases. The addition of SiO_2 to $LiBH_4$ is reported to decrease the hydrogen loss temperature to ~200°C [39]. Mechanically milling $LiBH_4$ with 0.5 mol MgH_2 and a small amount of $TiCl_3$ catalyst is reported to form a system that can reversibly store 8–10 wt% H_2 [55]. The modified thermal dehydrogenation reaction for this system can be expressed by Equation 16.23, where formation of MgB_2 improves the thermodynamic pathway for

$$2LiBH_4 + MgH_2 \underset{H_2}{\overset{\Delta}{\rightleftarrows}} 2LiH + MgB_2 + 4H_2 \quad (16.23)$$

dehydrogenation and regeneration of $LiBH_4$, decreasing in enthalpy change for this process by 25 kJ/mol of H_2 released compared with pure $LiBH_4$.

Ball milling $LiBH_4$ in a 3:1 mole ratio with TiF_3 is reported to result in a reduction in the dehydrogenation onset temperature to ~100°C with a hydrogen capacity reaching 5.0 wt% at 250°C [56]. Hydrogenation/dehydrogenation kinetics were reported to be greatly improved when $LiBH_4$ was incorporated into activated carbon [57]. The $LiBH_4$-activated carbon composite exhibited an onset of hydrogen release at 220°C and a rate of dehydrogenation one order of magnitude faster than that of pure $LiBH_4$ with reduced temperature and hydrogen pressure conditions required for restoring the hydride. It was also reported that milling $LiBH_4$ together with single-walled carbon nanotubes (SWNTs) resulted in a composite material exhibiting dehydrogenation and rehydrogenation reactions at significantly reduced temperature and pressure conditions [58]. A composite of $LiBH_4$ with 30 wt% SWNTs discharged 11.4 wt% H_2 within 50 min at 450°C and more than 6.0 wt% H_2 could be recharged at 400°C under an initial hydrogen pressure of 10 MPa. A high-pressure polymorph of lithium borohydride also has been proposed as a thermally destabilized alternative [59].

The lithium borohydride–lithium amide system has also attracted attention. Reactant mixtures of composition $(LiNH_2)_x(LiBH_4)_{1-x}$ can be formed by mixing $LiBH_4$ with $LiNH_2$ [60]. This system is dominated by a compound having an ideal stoichiometry of $Li_4BN_3H_{10}$. Within this system, a specific phase of composition $LiB_{0.33}N_{0.67}H_{2.67}$ was reported, having theoretical hydrogen content of 11.9 wt%. This compound releases >10 wt% H_2 when heated. The evolved gas is reported to contain 2–3 mol% NH_3. It was also found that a maximum amount of hydrogen and a minimum amount of ammonia are evolved for compositions $(LiNH_2)_x(LiBH_4)_{1-x}$, where $x = 0.667$ [60]. A crystalline phase of composition $Li_3BN_2H_8$ was reported to form when $LiNH_2$ and $LiBH_4$ are heated or ball milled together in a 2:1 mole ratio [61]. Other phases were also identified in this system.

16.4.3 Magnesium Borohydride

Magnesium borohydride, $Mg(BH_4)_2$, contains 14.9 wt% hydrogen. It is a white crystalline compound that occurs as two phases under ambient pressure. The normal room-temperature phase converts to a high-temperature phase above 184°C. This high-temperature phase persists after cooling back to room temperature [41]. The theoretical hydrogen release capacity of $Mg(BH_4)_2$ is 9.6 wt%. Although this is lower than that of $LiBH_4$ (13.5 wt%), it is still substantial. $Mg(BH_4)_2$ also has a lower decomposition temperature and more favorable enthalpy for hydrogen release than $LiBH_4$. The combination of these properties have prompted considerable interest in $Mg(BH_4)_2$ for hydrogen storage applications [62,63].

First reported by Wiberg and Bauer in 1950, a variety of synthesis methods have since been described for $Mg(BH_4)_2$ [21]. Synthesis routes to unsolvated $Mg(BH_4)_2$ have been reviewed [64]. The diethyl ether solvate of $Mg(BH_4)_2$ can be prepared by reaction of MgH_2 with B_2H_6 in diethyl ether under pressure, and the ether solvent can be subsequently removed by heating under vacuum at 150–180°C [65]. Exchange reactions of $NaBH_4$ with $MgCl_2$ in various solvents, such as ethanol, dimethylformamide, and THF, also provide $Mg(BH_4)_2$ as the corresponding solvates, $Mg(BH_4)_2{\cdot}n$L (n = 1–6; L = solvent), in which the solvent remains even after drying under vacuum [66,67]. Attempts to remove solvent from these compounds typically results in decomposition. However, reaction of $NaBH_4$ with $MgCl_2$ in refluxing diethyl ether gives the Et_2O solvate, which can be unsolvated under conditions that provide pure $Mg(BH_4)_2$ [64]. Another reported synthesis method for $Mg(BH_4)_2$ involves ball milling mixtures of $MgCl_2$ and $NaBH_4$ in diethyl ether [64]. Additional methods for the synthesis of unsolvated $Mg(BH_4)_2$ include reaction of MgH_2 with the borane-base adduct $Et_3N{\cdot}BH_3$, followed by removal of Et_3N [68, Equation 16.24], and reaction of dibutyl magnesium with borane-base adducts such as $H_3B{\cdot}SMe_2$, as shown in Equation 16.25 [69].

$$MgH_2 + 2H_3B{\cdot}NEt_3 \rightarrow Mg(BH_4)_2{\cdot}2NEt_3 \quad (16.24)$$

$$3MgBu_2 + 8H_3B{\cdot}SMe_2 \rightarrow 3Mg(BH_4)_2{\cdot}2SMe_2 + 2BBu_3{\cdot}SMe_2 \quad (16.25)$$

The enthalpy change for decomposition of $Mg(BH_4)_2$ is estimated to be ~40 kJ/mol of H_2 released [70,71]. Hydrogen loss occurs in several steps starting at ~290°C. Thermal decomposition of $Mg(BH_4)_2$ can be approximated by Equations 16.26 and 16.27 [71]. The tetragonal

$$Mg(BH_4)_2 \rightarrow MgH_2 + 2B + 3H_2 \quad (16.26)$$

$$MgH_2 \rightarrow Mg + H_2 \quad (16.27)$$

α-phase of $Mg(BH_4)_2$ converts into a cubic β-phase at ~184°C. The compound releases most of its theoretically available hydrogen when heated to ~450°C under vacuum or low hydrogen pressure, and essentially all of it when heated to 500–550°C. The two largest steps have maximum H_2 release rates occurring at 295°C (~4.7 wt%) and 325°C (~9.6 wt%). A third smaller step with maximum release rate at 410°C yields an additional 3.3–3.7 wt% H_2. A fourth minor step occurs slowly above 500°C under vacuum to provide a final 0.8–1.0 wt% H_2. The temperatures associated with these steps vary when dehydrogenation is conducted under hydrogen pressure, but exhibit the same general trend [70,71]. Above 450°C MgB_2 is formed from the solid products of these reactions. As with the other metal borohydrides, the actual decomposition pathway is more complex than Equations 16.26 and 16.27 would suggest, and involves formation of kinetically stable polyhedral boranes, including the *closo*-$B_{12}H_{12}^{2-}$ anion [72,48].

The amine complex of magnesium borohydride, $Mg(BH_4)_2{\cdot}2NH_3$, was prepared by thermal decomposition of $Mg(BH_4)_2{\cdot}6NH_3$. The compound was structurally characterized and exhibits N–H···H–B dihydrogen bonds. $Mg(BH_4)_2{\cdot}2NH_3$, which contains 16.0 wt% H_2, decomposes endothermically starting at 150°C, and exhibits a maximum hydrogen release rate at 205°C, making it an attractive candidate for hydrogen storage [73].

16.4.4 Calcium Borohydride

Calcium borohydride, $Ca(BH_4)_2$, contains 11.6 wt% hydrogen. Although this compound contains less hydrogen by weight than lithium and magnesium borohydride, its accessible hydrogen content is still sufficiently high to make it attractive as a hydrogen storage candidate. Also, $Ca(BH_4)_2$ systems appear to have more favorable thermodynamics for reversible hydrogen storage. For these reasons, $Ca(BH_4)_2$ has been the object of considerable research interest [74].

In 1954, Kollonitsch and coworkers described syntheses of both $Mg(BH_4)_2$ and $Ca(BH_4)_2$ by reactions of $NaBH_4$ with the corresponding metal chlorides in alcohol solutions at −50°C to −20°C [75]. A 1965 patent assigned to Callery Chemical Co. describes the synthesis of $Ca(BH_4)_2$ by the reaction of calcium hydride with diborane in THF, followed by removal of solvent from the initially formed solvate by heating at 155°C under vacuum [76]. Notably, this patent also suggests the use of $Ca(BH_4)_2$ as a source of hydrogen. Titov also described in 1964 the synthesis of the THF solvate of $Ca(BH_4)_2$ by reaction of $NaBH_4$ with $CaCl_2$ in THF [77]. Current commercial synthesis of calcium borohydride is primary conducted by modifications of this general method (Equation 16.28), but other solvents and proprietary methods may be used. Calcium borohydride is most commonly available commercially as the THF adduct, $Ca(BH_2)_4{\cdot}2THF$, which is typically unsolvated by heating under vacuum at 200°C.

$$CaCl_2 + 2NaBH_4 \underset{-2NaCl}{\overset{THF}{\rightleftarrows}} CaBH_4{\cdot}2THF \underset{-2\ THF}{\overset{150\text{–}200\ °C,\ vac.}{\rightleftarrows}} CaBH_4 \qquad (16.28)$$

On heating, $Ca(BH_4)_2$ is reported to undergo a polymorphic transformation at 140–167°C and then decompose in two steps between 347°C and 497°C [74,78], as given by Equation 16.29. Calcium hydride and other intermediates form after the first step, but CaH_2 is the only crystalline phase observed after the second step with a total weight loss of ~9.6 wt% H_2.

$$3Ca(BH_4)_2 \rightarrow CaB_6 + 2CaH_2 + 10H_2 \quad 9.6\%\ H_2 \qquad (16.29)$$

The enthalpy change for this reaction is estimated to be ca. −53 kJ/mol of H_2 released, which is significantly less exothermic than the corresponding reaction of $LiBH_4$. This should allow the reverse hydrogenation reaction to occur with a maximum theoretical reversible hydrogen storage capacity of ~9.6%. It is reported that ball milling a mixture of CaB_6 and CaH_2 at 400–440°C under 70 MPa H_2 pressure gives $Ca(BH_4)_2$. The kinetics of this reaction are very slow. However, doping with catalysts is reported to enhance the kinetics of this reaction and facilitate regeneration under milder conditions. For example, adding $TiCl_3$ enables rehydrogenation at 350°C under 9 MPa H_2 pressure in a reaction described by Equation 16.30 [79]. The resulting rehydrogenated material was shown to contain ~3.8 wt% H_2, or more than half the amount in the starting sample.

$$7Ca(BH_4)_2 + 4TiCl_3 \rightarrow 6CaCl_2 + 4TiB_2 + CaB_6 + 28H_2 \qquad (16.30)$$

Thus, it was demonstrated that calcium borohydride can be prepared from its proposed decomposition products, suggesting that it may have potential for use as a reversible hydrogen storage material.

16.4.5 Mixed Metal Borohydrides

The high thermal stability of the metal borohydrides has prompted an interest in developing new metal borohydride materials having more favorable thermodynamic properties. One approach is the use of mixed cation borohydrides, providing an opportunity to tailor their thermodynamic properties [80,81]. In some cases, this approach is similar to, if not mechanistically the same as, using metals or metal-based additives to facilitate dehydrogenation/rehydrogenation reactions of monometallic borohydrides. However, the synthesis of discrete crystalline mixed metal borohydrides has allowed the design and study of interesting hydrogen storage materials.

Examples can be cited for the mixed alkali metal and alkaline-earth metal borohydrides. It was reported that a Li–Mg borohydride could be formed by mechanical milling and subsequent heating of a mixture of $LiBH_4$ and $Mg(BH_4)_2$ in 1:1 mole ratio. This dual-cation borohydride exhibits a lower dehydrogenation onset temperature [82]. Also, composite materials made by combining $LiBH_4$ and $Ca(BH_4)_2$ are reported to show hydrogen release temperatures lower than for either component alone [83]. Composites of composition $(LiBH_4)_x(1 - x)[Ca(BH_4)_2]$, where H_2 capacity varies with x, were prepared. At intermediate compositions, such as $x = 4$, dehydrogenation was complete below 400°C with release of 10 wt% H_2. Partial reversibility of this system was also reported.

16.5 BORON–NITROGEN SYSTEMS

The desire for materials having high gravimetric hydrogen densities has prompted increasing interest in materials containing only light elements. Meeting this criterion, hydrogen-rich B–N compounds have emerged as particularly promising hydrogen storage materials. These compounds have been the subject of intense study in recent years. The chemistry of B–N compounds in chemical hydrogen storage has been reviewed [8,9].

16.5.1 Ammonium Borohydride

Ammonium borohydride, $[NH_4]^+[BH_4]^-$, is a white crystalline solid having a remarkably high hydrogen content of 24.5 wt% [84]. This corresponds to 245 g H_2/kg or 155 g H_2/L. Described as having the highest thermodynamically and kinetically accessible hydrogen content of any solid material, ammonium borohydride has attracted considerable attention for its potential use in hydrogen storage [85]. This material releases ~20 wt% H_2 in a complex three-step process when heated from 50°C up to ~160°C, as given by Equation 16.31 [86].

$$\left[NH_4\right]^+\left[BH_4\right]^- \xrightarrow{50-160°C} BNH_x + \left(\frac{8-x}{2}\right)H_2, \quad x \approx 3.3 \tag{16.31}$$

Thermolysis of ammonium borohydride produces nearly pure hydrogen, with only a trace of borazine evolved at higher temperatures. The primary disadvantage of ammonium borohydride for many practical applications is its poor thermal stability. Ammonium borohydride slowly decomposes even at room temperature. Thus, methods to stabilize ammonium borohydride are needed to make this material practical for use in hydrogen storage applications. Ammonium borohydride also serves as a synthetic precursor for ammonia borane (AB), as discussed below.

16.5.2 Ammonia Borane

Ammonia borane, H_3NBH_3, also called borazane, is one of the most interesting and extensively studied materials for hydrogen storage. Called AB for short, ammonia borane is a white crystalline solid containing 19.6 wt% hydrogen. It is stable in air and water and has appropriate thermal stability for practical use in a range of hydrogen storage applications [9,87,88].

The amount of hydrogen contained in AB in terms of both weight and volume is impressive, although not quite as high as ammonium borohydride. This volumetric hydrogen density of AB is illustrated in Table 16.4, showing that a 1.0-L volume of AB contains more than twice the weight of hydrogen as found even in liquid hydrogen. These data do not include the volume of any required containment vessel, which can be significant for liquid H_2.

TABLE 16.4
Comparison of the Weights of Hydrogen contained in 1.0 L of Highly Compressed Hydrogen, Liquified Hydrogen, AB, and Ammonium Borohydride

Form of Hydrogen	g H_2/L	Weight Proportion of H_2
Compressed H_2 at 10,000 psi	40	
Liquid H_2 (−253°C)	71	
H_3NBH_3 (AB)	146	
Ammonium borohydride, NH_4BH_4	154	

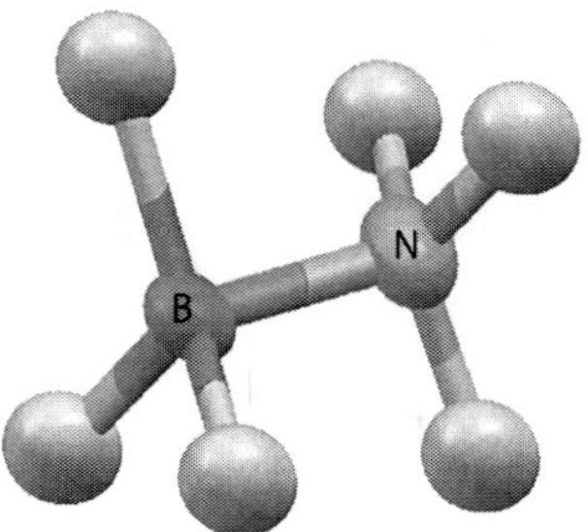

FIGURE 16.3 The structure of H_3NBH_3 (AB).

16.5.2.1 Structure

AB is the Lewis acid–base adduct between NH_3 and BH_3. It is isoelectronic with ethane and has an analogous structure, as shown in Figure 16.3. The fact that it is a solid and not a gas is attributed to intermolecular hydrogen–hydrogen bonding and dipole–dipole interactions, as supported by structural data [89,90].

Single-crystal structures of AB have been published, derived from both x-ray and neutron diffraction data collected at various temperatures [89,90]. The H_3NBH_3 molecule in the solid state has a staggered geometry with a B–N dative bond distance of 1.59 Å. The extended structure exhibits nearly linear N–H···H and bent B–H···H hydrogen–hydrogen bonds, as illustrated in Figure 16.4 [89]. The structure of AB clearly suggests an intermolecular pathway for H_2 formation.

16.5.2.2 Synthesis

AB is the Lewis acid–base adduct that might be anticipated to form in the reaction of ammonia with diborane, B_2H_6. However, early developments in the synthesis of AB were complicated by formation of a higher-molecular-weight compound in this reaction. This compound was identified as a salt referred to as the diammoniate of diborane, $[H_2B(NH_2)_2]^+[BH_4]^-$ (DADB) [91–95]. This product is the result of asymmetric cleavage of diborane, whereas H_3NBH_3 is the product of symmetric cleavage. Details of this chemistry were worked out in careful studies done in the early 1950s in the research group of Robert Parry at the University of Michigan, which are summarized in a more recent review article [96]. The first definitive synthesis of AB was published in 1955, based on the PhD thesis of Sheldon Shore, then a student in Parry's laboratory [97]. This publication described the synthesis of AB from reactions of lithium borohydride with ammonium chloride or sulfate in

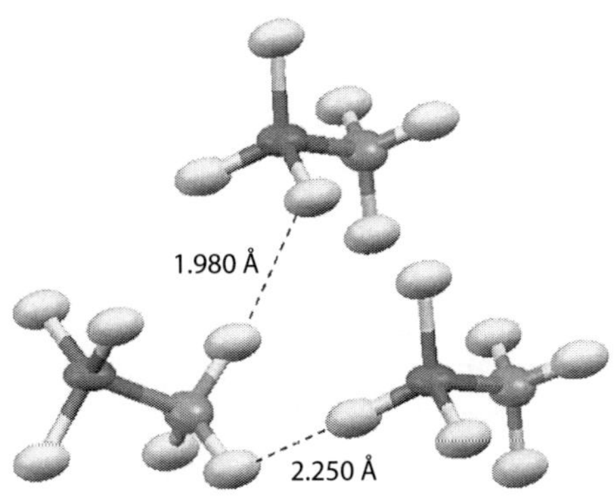

FIGURE 16.4 Hydrogen–hydrogen close approaches in H_3NBH_3 (AB).

diethyl ether, as given by Equations 16.32 and 16.33. This method was reported to give AB in ~45% yield with formation

$$LiBH_4 + NH_4Cl \xrightarrow{Et_2O} LiCl + H_3NBH_3 + H_2 \tag{16.32}$$

$$2LiBH_4 + (NH_4)_2SO_4 \xrightarrow{Et_2O} 2H_3NBH_3 + Li_2SO_4 + 2H_2 \tag{16.33}$$

of some DADB coproduct. Since AB is soluble in ether and DADB is not, these products are easily separated. The same publication describes the synthesis of AB by the metathesis reaction of DADB with ammonium chloride, as given by

$$[H_2B(NH_3)_2]^+[BH_4]^- + NH_4Cl \xrightarrow[\text{trace } NH_3]{Et_2O} [H_2B(NH_3)_2]^+Cl^- + H_3NBH_3 + H_2 \tag{16.34}$$

This latter reaction was part of the elegant experimental work to elucidate the identity of DADB and demonstrate that it was not the alternative possible isomeric salt $[H_2N(BH_3)_2]^-[NH_4]^+$ [92,93] Shore and coworkers later published improved synthetic methods for both DADB and AB [98]. Despite having the same empirical formulas, AB does not react readily with water, whereas DADB is immediately hydrolyzed. Upon standing, DADB slowly converts into AB. Thus, AB has more suitable properties for use in hydrogen storage applications.

Another early synthesis of AB described by Schaeffer and Basile in 1954 involves the reaction of diborane with lithium amide [99]. It was found that diborane gas does not react with lithium amide solid. However, when diethyl ether is added to produce a slurry, reaction between these reagents occurs readily according to Equation 16.35, even at −64°C. It was reported that care is required to regulate the reaction to avoid release of large amounts of hydrogen. Polymeric by-products, resulting from dehydrogenation of AB, are formed in reaction.

$$B_2H_6 + LiNH_2 \xrightarrow{Et_2O} LiBH_4 + H_3NBH_3 \tag{16.35}$$

More recent work carried out at the Pacific Northwest National Laboratory (PNNL), USA, showed that ammonium borohydride, formed by metathesis reactions of NH_4X ($X = Cl$, F) and MBH_4 ($M = Na$, Li) in liquid NH_3, decomposes in organic ether solvents, such as THF, to give AB in close to quantitative yields, as shown in Equation 16.36 [100]. AB prepared by this one-pot synthetic method

$$NH_4X + MBH_4 \xrightarrow{NH_3\text{ (liq.)}} [NH_4][BH_4] \xrightarrow{R_2O} H_3NBH_3 + H_2 \qquad X = Cl,\ F;\quad M = Na,\ Li \tag{16.36}$$

was shown to have sufficient purity and thermal stability to meet the requirements for onboard automotive hydrogen storage. The kinetics of the reaction of sodium borohydride and ammonium chloride and sulfate in liquid ammonia were described in the 1970s by William Jolly and coworkers at the University of California, Berkeley [101]. These same reactions were described earlier, in then classified government reports from Aerojet Engineering Corp and the Schlesinger research group at the University of Chicago, dating from the 1940s [102,103]. However, pure AB was neither isolated nor definitively characterized in these earlier studies.

Other methods for the synthesis of AB have been described by Shore and coworkers at Ohio State University involving reactions of ammonia with borane adducts of appropriate amines [104].

As discussed above, the direct reaction of diborane with ammonia results in asymmetric cleave of diborane to give DADB. However, if diborane is cleaved prior to reaction with ammonia by a Lewis base, such as an alkyl or aryl amine or dialkylsulfide, then AB can be obtained in a suitable solvent, as given by Equation 16.37 [104]. The ratio of AB to DADB

$$L:BH_3 + NH_3 \xrightarrow{Et_2O} H_3NBH_3 + L \quad (L = \text{selected amine}) \tag{16.37}$$

produced in these reactions depends on both the choice of solvent and Lewis base. Shore showed that by using a suitable base, such as dimethylaniline, high-quality AB can be obtained in high yield [104].

Syntheses of AB via reactions of borate esters with lithium and aluminum hydride reagents have been described. One example of such a method is illustrated by Equation 16.38 [105]. A number of

$$B(OMe)_3 + LiAlH_4 + NH_4Cl \xrightarrow[30\ h]{THF} H_3NBH_3 + Al(OMe)_3 + LiCl + H_2 \tag{16.38}$$

other synthetic methods for AB have been reported involving variations on the above general reactions [106].

16.5.2.3 Hydrogen Release

The U.S. Army studied B–N compounds, including AB, in the 1980s, as materials to supply hydrogen for gas laser systems [107,108]. On heating, AB releases more than two molar equivalents of hydrogen, as shown in Equation 16.39 where $x = 1 - 2$ [109–114].

$$H_3NBH_3 \xrightarrow{110-150°C} NBH_x + \left(\frac{6-x}{2}\right)H_2 \quad 16.3\ \text{wt\%}\ H_2 \quad \text{for } x = 1 \tag{16.39}$$

There is some confusion in the literature regarding the dehydrogenation onset temperature of AB [115]. A number of reports describe slow release of hydrogen starting at 70°C. Others describe either dehydrogenation onset or significant H_2 release rates commencing at 80°C. Hydrogen release by thermolysis of AB is a complex process that proceeds in steps. Studies have provided a detailed understanding of this thermal hydrogen release process [9,115]. Once AB is heated above 110°C rapid release of one equivalent of hydrogen occurs. The initial step of hydrogen release, occurring primarily between 110°C and 120°C, involves an intermolecular process resulting in formation of polyaminoborane, $(NH_2BH_2)_n$, as given by Equations 16.40. Thus, AB can release up to one equivalent of hydrogen below 100°C.

$$n\ H_3NBH_3 \xrightarrow{70-120°C} (NH_2BH_2)_n + nH_2 \quad 6.5\%\ H_2 \tag{16.40}$$

The polyaminoborane product of this first step is a linear polymer that still contains ~14 wt% hydrogen. On further heating this material undergoes more hydrogen loss through both intermolecular and intramolecular processes, leading to formation of polyiminoborane, $(NHBH)_n$, as given by Equation 16.41.

$$(NH_2BH_2)_n \xrightarrow{120-150°C} (NHBH)_n + nH_2 \quad 7.0\%\ H_2 \tag{16.41}$$

The intermolecular process involves cross-linking between polymer chains. The intramolecular processes involve B–N double bond formation and cyclization reactions within polymer chains. As a result, the polyiminoborane product may have a complex network structure, the exact nature of which depends on how the thermolysis reaction is carried out. The combination of these steps results in release of ~13.5 wt% hydrogen from AB in the 70–150°C range. At somewhat higher temperatures further cross-linking of polyiminoborane results in release of additional amounts of hydrogen. Heating to much higher temperatures ultimately leads to the formation of boron nitride (BN), as given by Equation 16.42.

$$(\mathrm{NHBH})_n \xrightarrow{>500^\circ \mathrm{C}} \mathrm{BN} + \mathrm{H_2} \qquad 7.5\%\ \mathrm{H_2} \tag{16.42}$$

The practical use of AB for hydrogen storage does not involve high-enough temperatures to completely convert AB into BN. In addition, if the spent material is to be recycled, BN may be too stable to allow for practical regeneration of AB. Instead, thermal decomposition of AB is continued to the extent that the spent fuel has a composition of BNH_x, where x is 1 – 2, as in Equation 16.39. In this way, AB can be used to supply more than 2 mol of hydrogen, or ~15 wt% hydrogen at modest temperatures.

The thermolysis of AB is an overall exothermic process. Once hydrogen release is initiated, sufficient heat may be generated to make the process self-sustaining, depending on reactor design. Initiating the process requires the input of induction energy to start decomposition. The induction energy required to release 2 mol of hydrogen gas is ~16.2 kJ/mol of H_3NBH_3, or 8.1 kJ/mol of H_2. This corresponds to only 3.35% of the chemical energy content of the hydrogen generated [112].

The first step of AB thermal dehydrogenation (Equation 16.39) yields almost entirely H_2 gas. Unfortunately, the subsequent higher temperature steps of AB dehydrogenation produce substantial amounts of volatile by-products, including lower-molecular-weight boranes, ammonia, and borazine, $B_6N_6H_6$, the B–N isoelectronic analogue of benzene. These volatile by-products are particularly undesirable because they can poison PEM fuel cells. The amounts of volatile products generated from AB are observed to be greater when higher temperatures and higher heating rates are used for decomposition. Although nonhydrogen volatiles can be removed using sorbents in a subsequent purification step, this presents an added complication. Thus, methods to facilitate AB decomposition more selectively to generate H_2 of higher purity are needed.

Catalytic hydrogen release from AB is of interest for several reasons. The self-sustaining nature of its thermal decomposition makes AB difficult to use in the regulated generation of hydrogen. Also, the uncontrolled thermolysis of AB produces a number of volatile impurities that poison PEM fuel cells. In addition, it is desirable for fuel cell applications to have relatively fast hydrogen release rates at temperatures below 80°C. In order to address these issues, studies have been made on catalyst systems having potential to direct decomposition pathways and facilitate release of higher-purity hydrogen from AB at or near room temperature [9].

Studies have been done to understand the fundamental aspects of AB catalysis identified a number of homogeneous catalysts for dehydrogenation that operate at lower temperatures in nonaqueous solvents. Aside from enhancing H_2 release rates, this work also showed that the type of catalyst has a strong effect on both total amount of H_2 released and the nature of the dehydrogenated B–N product. For example, the iridium "pincer" complex [(POCOP)Ir(H)$_2$, where POCOP = η^3-1,3-(OP-*tert*-Bu)$_2$C$_6$H$_3$], shown in Figure 16.5a, catalyzes dehydrogenation of AB with excellent rates [116]. However, this catalyst is reported to be highly selective for the formation of a proposed cyclic pentamer product $[H_2NBH_2]_5$ (Figure 16.5a), leading to release of only 1 mol of hydrogen per mole of AB, as given by Equation 16.43,

$$\mathrm{H_3NBH_3} \xrightarrow[\mathrm{THF,\ RT}]{\mathrm{Ir\ cat.}} 1/5\left[\mathrm{H_2NBH_2}\right]_5 + \mathrm{H_2} \tag{16.43}$$

although this is likely an oversimplification of the actual reaction. Alternatively, the nickel carbene complex Ni(NHC)$_2$ (NHC = N-heterocyclic carbene), Figure 16.5b, yields more than 2 mol of H_2 at

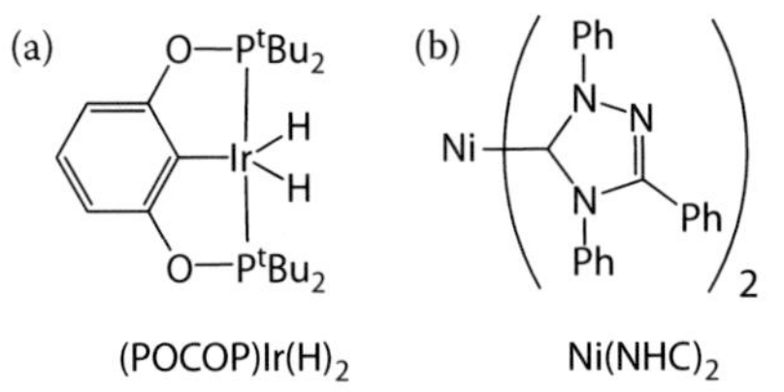

(a) (POCOP)Ir(H)$_2$ (b) Ni(NHC)$_2$

FIGURE 16.5 Experimental homogeneous catalysts for AB dehydrogenation, (a) (POCOP)Ir(H)$_2$ and (b) Ni(NHC)$_2$, where NHC is a nitrogen heterocyclic carbene.

60°C, but the rate of release is slow. Instead of a cyclic pentamer, this system produces a polymeric material proposed to have the structure, shown in Figure 16.6b, composed of linked borazine moieties [117].

Studies show that AB can be hydrolyzed in water by the action of various chemical reagents or catalysts. Metal salts, including those of Co^{2+}, Ni^{2+}, and Cu^{2+}, can be employed to catalyze hydrolysis [118–121]. Although the practical use of AB for hydrogen storage typically involves solid-state thermolysis, there may be potential for hydrolytic hydrogen generation from AB for some applications. In addition, mechanical mixing of AB with certain metal complexes in the absence of solvent can significantly reduce hydrogen release temperatures. For example, solid-state mechanical mixing of cobalt, nickel, or copper chlorides was reported to result in significant hydrogen evolution even at 60°C. The copper salt was found to be more efficient in this reaction than the other metal salts, releasing 2 mol of hydrogen over 4 h at 60°C with a Cu/H_3NBH_3 mole ratio of 0.15. A significant amount of H_2 is released from this system even at room temperature, albeit slowly [118].

Thermal decomposition of AB in polar, aprotic solvents is slow and results in mixtures of oligomeric products, including cyclic amino- and imino-boranes [122,123]. Dehydrogenation of AB in ionic liquids, such as 1-butyl-3-methylimidazolium chloride (bmimCl) has been investigated [124]. Compared to solid-state dehydrogenation, thermal decomposition of AB in bmimCl shows no induction period, with hydrogen evolution commencing immediately on subjecting the sample to heat. Furthermore, AB was found to release considerably more hydrogen when heated in ionic liquid media at 85–95°C compared with the solid-state reaction, with up to 1.5 equiv. of H_2 released at 95°C over a 22-h period. This system evolved 5.4 wt% H_2 at 95°C, including the weight of the bmimCl solvent. Previous studies of neat AB at 88°C showed that formation of DADB is a critical decomposition step. Consistent with this conclusion, the addition of DADB to AB significantly decreases the induction time for hydrogen release. Studies by NMR spectroscopy of the ionic liquid system suggest that DADB may also be an intermediate in AB dehydrogenation in this media. Since the ionic salt DADB is reported to convert into polyiminoborane with hydrogen release on heating, the activating effect of the ionic liquid may be related to the ability of these solvents to stabilize polar intermediates and transition states.

Researchers at PNNL showed that nanocomposites formed by dispersing AB in mesoporous silica or high-surface-area carbon cryogels exhibit an increase in both the rate and extent of hydrogen release from AB at reduced temperatures [125–127]. For example, composite materials

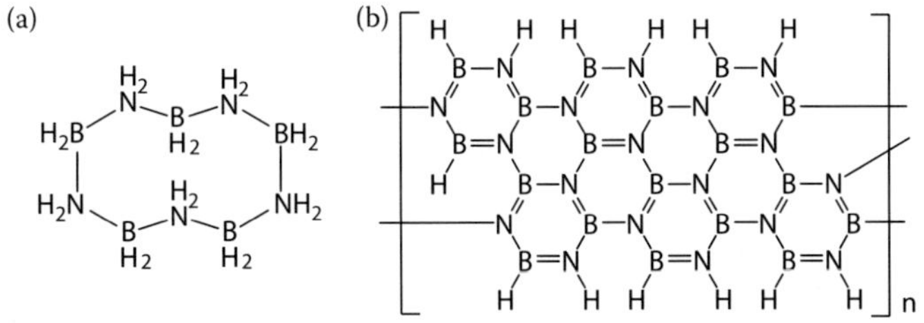

(a) (b)

FIGURE 16.6 Cyclic pentamer, $[H_2NBH_2]_5$ (a) and linked borazine structure (b).

composed of intimate mixtures of AB and SBA-15 mesoporous silica in a 1:1 weight ratio showed H_2 release at substantially lower temperatures than observed for neat AB. The H_2 generated from these systems also was higher in purity than that obtained from dehydrogenation of neat AB under the same conditions, being substantially free of the unwanted borazine by-product. Related studies showed that AB encapsulated in a mesoporous carbon framework (CMK-3) exhibited substantially reduced dehydrogenation temperatures with the majority of the H_2 release occurring in the 75–95°C temperature range, compared with 110–145°C for neat AB [128]. This system exhibited a one-step dehydrogenation process, in contrast to the two-step process of neat AB. However, the H_2 generated was found to contain some ammonia impurity. It was further reported that doping the CMK-3 framework with 5 wt% Li^+ completely suppressed the release of volatile by-products, including ammonia and borazine, when heated to 150°C. Lithium doping of the framework further reduced the dehydrogenation onset temperature to ~55°C. The 1:1 weight ratio AB/Li–CMK-3 system was reported to release ~7 wt% H_2 at 60°C, or ~3.5 wt% hydrogen for the entire system.

Dehydrogenation of AB can also be accomplished through acid catalysis using either Lewis or Brønsted acids. Addition of strong acids can initiate hydrogen release from AB even at 25°C, according to Equation 16.44 [129].

$$H_3NBH_3 + H^+ + 3H_2O \longrightarrow NH_4^+ + B(OH)_3 + 3H_2 \tag{16.44}$$

Such acid-catalyzed H_2 release can also be accomplished using immobilized acids, in the form of zeolites or ion exchange resins, providing interesting possibilities for engineered hydrogen delivery devices [130]. The ultimate boron-containing product resulting from AB hydrolysis is typically an ammonium borate. Depending on the reaction stoichiometry, this can be either ammonium pentaborate, $NH_4[B_5O_6(OH)_4]\cdot 2H_2O$, forming at lower pH values, or ammonium tetraborate, $NH_4[B_4O_5(OH)_4]\cdot{\sim}2H_2O$, forming at higher pH values. Both of these ammonium borates are important commercial products that have been manufactured on multiton industrial scale for many decades. Both compounds have well-established crystal structures [131,132]. It can be noted that $NH_4[B_4O_5(OH)_4]\cdot 2H_2O$ is often called ammonium biborate as an article of commerce and has a variable amount of interstitial water.

Addition of the strong Lewis acids, such as $B(C_6F_5)_3$, also initiates hydrogen release via the formation of boronium intermediates. These intermediates then further react with AB to form oligomeric or polymer aminoboranes with release of H_2. The relative concentration of Lewis acid needs to be kept low (≤0.5 mol%) to avoid chain termination resulting in aminodiborane, $B_2H_5(\mu\text{-}NH_2)$, and concentrated solutions can also produce large amounts of borazine.

16.5.2.4 Regeneration

A number of strategies have been proposed for regeneration of AB spent fuel [9]. One approach involves regeneration via solvolysis reactions of AB. A transition metal-catalyzed solvolysis system was reported that converts AB into $[NH_4][B(OMe)_4]$ with liberation of hydrogen, as given by Equation 16.45 [133]. The latter could be converted back into AB by treatment with NH_4Cl and $LiAlH_4$ (Equation 16.46). This approach would require reduction of $Al(OMe)_3$ back to $LiAlH_4$. A disadvantage of this method is a requirement for the consumption of a strong reducing agent.

$$H_3NBH_3 + 4MeOH \rightarrow NH_4B(OMe)_4 + 3H_2 \tag{16.45}$$

$$NH_4B(OMe)_4 + LiAlH_4 + NH_4Cl \rightarrow H_3NBH_3 + Al(OMe_3)_3 + MeOH + H_2 + LiCl + NH_3 \tag{16.46}$$

Another approach to AB regeneration involves digestion of spent material followed by reduction. The dehydrogenation of AB produces a complex mixture of products of general composition BNHx, all of which contain B–N linkages that must be broken. Acidic compounds (HX) can be used to cleave these bonds to release the nitrogen component with the formation of B–X bond-containing products. These B–X bonds in these digestion products can then be reduced using a reducing agent to form B–H bonds. In order to have an energetically efficient recycle system for AB regeneration, it is necessary to use reactions that are not overly endothermic or exothermic. Various methods of digestion and reduction have been explored in efforts to find practical and thermodynamically favorable regeneration systems.

Efforts made at Los Alamos National Laboratory toward regeneration of BNH_x-spent fuel involved digestion with 1,2-benzenedithiol to form ammonia adducts of dithioboron compounds [134]. The B–S bonds in these compounds were then reduced using a tin hydride, which could be in turn regenerated by efficient methods. Another method proposed by PNNL involves digestion of BNH_x-spent fuel in a suitable alcohol to form a borate ester that can be reduced by a metal hydride catalyst, which can be regenerated by reaction with hydrogen. Although B–O bonds are generally regarded as too strong to allow energy-efficient reduction, calculations indicate that judicious choice of alcohols can provide borate esters having sufficient hydride affinity to undergo facile reduction by a metal hydride. Borate esters produced by digestion with suitably substituted phenols appear to be capable of acting as efficient regeneration intermediates. In addition to hydrides of precious metals, base metal catalysts have also shown promise in for hydride transfer to borate esters in this regeneration cycle.

16.5.3 Metal Amidoboranes

Another approach to altering the thermal stability of AB involves chemical modification. The replacement of one N–H hydrogen atom in AB with a metal results in metal amidotrihydroborates, which are commonly called metal amidoboranes (MABs). MABs have the general composition $M(NH_2BH_3)_n$, where M can be a wide variety of elements, including both main group and transition metals [135–137]. Numerous MAB compounds have been prepared [135]. Although the replacement of one hydrogen atom in AB by a metal results in a penalty in terms of gravimetric hydrogen content, these compounds still carry substantial amounts of hydrogen. With H_2 contents in the 10.1–13.7 wt% range, MABs derived from the light alkali and alkaline earth metals, including $LiNH_2BH_3$, $NaNH_2BH_3$, $Mg(NH_2BH_3)_2$, and $Ca(NH_2BH_3)_2$, have been the most intensively studied. However, transition MABs can also contain substantial amounts of hydrogen. For example, $Ti(NH_2BH_3)_4$ contains 12.0 wt% H_2. Mixed MABs have also been prepared, including $LiAl(NH_2BH_3)_4$ (13.1 wt% H_2) and $Li_2[Zn(NH_2BH_3)_4]$ (10.1 wt% H_2) [135].

MABs, such as LiAB and NaAB, can be prepared by reactions of AB with metal hydrides in solution, as given by Equation 16.47. The synthesis of other MABs has been accomplished by exchange reactions with alkali metal MABs, as in the example given by Equation 16.48 [135]. Syntheses of MABs by high-energy ball milling of AB with alkali metal hydrides have also been reported [137].

$$NH_3BH_3 + MH \rightarrow M(NH_2BH_3) + H_2\,(g) \quad M = Li, Na \tag{16.47}$$

$$MgCl_2 + 2\,NaNH_2BH_3 \rightarrow Mg(NH_2BH_3)_2 + 2\,NaCl \tag{16.48}$$

MABs offer certain advantages over AB. A number of crystal structures have been determined for MABs [136,137]. A feature of the alkali-MABs is the presence of both positively charged ($-NH_2$) and negatively charged ($-BH_3$) hydrogen atoms. This provides a relatively low-energy pathway for local combination of $H^{\delta+}$ and $H^{\delta-}$ atoms to more readily produce H_2. Kinetic studies of hydrogen release from MABs have been carried out providing evidence that the metal atom also assists in hydride transfer to the amine group [138–140]. The thermal decomposition of AB generates hydrogen-containing impurities, including low-molecular-weight boron hydrides, borazine, and aminoboranes. It has been demonstrated for several MABs that their thermolysis generates large amounts of hydrogen without these unwanted impurities [135–137], which is a significant advantage over AB. Ammonia generation during thermolysis of $LiNH_2BH_3$ and $NaNH_2BH_3$ has been reported. For example, lithium amidoborane, $LiNH_2BH_3$, and sodium amidoborane, $NaNH_2BH_3$, are reported to release 7.5 and 10.9 wt% H_2, respectively, when heated to 91°C. Another problem associated with AB is its tendency to foam during thermolysis. At least some of the MABs do not foam during thermolysis. For example, $Ca(NH_2BH_3)_2$ (10.1 wt% H_2) releases hydrogen over a temperature range of 100°C to 170°C without foaming [137].

Other complex MAB derivatives have been reported. For example, reaction of LiH with two equivalents of AB, or the reaction of $LiNH_2BH_3$ with a second equivalent of AB, results in the crystalline complex $LiNH_2BH_3{\cdot}H_3NBH_3$. Thermolysis of $Li(NH_2BH_3){\cdot}H_3NBH_3$ is reported to produce 14% H_2 in a stepwise manner with maximum release rates occurring at ~80°C and ~140°C, without emission of detectable amounts of borazine or aminoborane [141]. An ammoniate of composition $Ca(NH_2BH_3)_2{\cdot}2NH_3$ has also been described, which releases >8 wt% H_2 when heated to 150°C without emission of borazine [142].

16.5.4 Guanidinium Borohydrides

Guanidinium borohydride (GBH) is a compound related to both the metal and ammonium borohydrides discussed above [143]. An air-stable white solid having the composition $[C(NH_2)_3]^+[BH_4]^-$, GBH contains 13.5 wt% H_2, of which 10.8 wt% is thermally accessible. This corresponds to four molar equivalents of H_2. With a calculated density of 0.905 g cm^{-3}, GBH can supply 97.7 g H_2/L. Although not as high as AB or most of the metal borohydrides of interest for hydrogen storage, GBH still has a thermally releasable hydrogen density that is nearly one-third greater than pure liquid hydrogen.

GBH is reported to be thermally stable at 55°C and to very slowly lose hydrogen at 60°C. Rapid hydrogen loss occurs when GBH is heated to 110°C, at which point dehydrogenation is exothermic and self-sustaining. Hydrogen yields above 10 wt% are reported for self-sustaining GBH thermolysis, which can reach temperatures as high as 450°C. The hydrogen obtained from GBH thermolysis was reported to be 95–97% pure, but contained 3–5% ammonia. It was further found that addition of ethylenediamine bis(borane), $H_2N(CH_2)_2NH{\cdot}2BH_3$, completely suppressed the formation of the ammonia impurity.

The synthesis of GBH in 65%–75% yield was reported from the room-temperature metathesis of guanidinium carbonate or sulfate with sodium borohydride in anhydrous isopropanol solvent. The single-crystal x-ray structure of GBH shows stacks and layers of one-dimensional ribbons of $C(NH_2)_3^+$ and BH_4^- units. Within these ribbons, four of the six protic N–H guanidinium hydrogen atoms form close approaches to the four hydridic B–H hydrogen atoms of adjacent borohydride groups. These dihydrogen interactions provide a low-energy pathway for thermal elimination of four of the five equivalents of H_2 theoretically present in GBH [143].

Other GBHs have been described, mostly resulting from rocket propellant research. Guanidinium hydrotriborate, $[C(NH_2)_3][B_3H_8]$, was prepared by the reaction of guanidinium sulfate with sodium hydrotriborate [144]. This compound decomposes slowly when stored at room temperature and violently when heated to ~100°C with release of 12.3 wt% H_2. The related triaminoguanidinium hydrotriborate, $[C(NHNH_2)_3][B_3H_8]$ was prepared by the reaction of pentaborane with triaminoguanidine, which has a potential hydrogen storage capacity of 11.0 wt%, assuming a loss of eight equivalents of H_2 [145].

16.5.5 B–N Sorbent Materials

Boron nitride nanostructures have drawn attention for their storage potential. Hydrogen storage in various nanoporous sorbent materials has been a subject of considerable interest. A primary advantage of these materials is their ease of reversibility [146]. This also creates a disadvantage because hydrogen is only weakly held within these materials, requiring low temperatures to achieve acceptable levels of hydrogen uptake. Sorbent materials typically retain hydrogen around the temperature of liquid nitrogen (–196°C). Such materials studied for hydrogen storage applications include zeolitic structures, microporous polymers, metal–organic frameworks, and various forms of carbon, such as activated carbon and carbon nanofibers and nanotubes.

Another class of materials proposed for hydrogen storage is BN nanotubes. These interesting materials, which are isoelectronic and isostructural with carbon nanotubes, were first described by Sheldon Shore and coworkers in 1993 [147]. The use of nanostructured materials having high surface areas for hydrogen adsorption and desorption under practical temperature and pressure conditions depends on the binding energy of the material. Therefore, a key objective in the design of new sorbent materials is the achievement of higher binding energies for hydrogen. BN nanotubes have been estimated to have 40% greater binding energies for hydrogen than analogous carbon nanotubes, in the appropriate range for room-temperature hydrogen storage [148].

16.5.6 Organoboron Compounds

Organoboron boron compounds show promise for hydrogen storage. Many hydrogen storage systems suffer from thermodynamic imbalance between their hydrogen-containing and hydrogen-depleted states. In order to have an efficient and recyclable hydrogen carrier, it is desirable that the free energy for hydrogen release be close to zero. One approach to designing such thermoneutral systems is through the use of B–N bond-containing heterocycles that can undergo dehydrogenation with $\Delta G \approx 0$ kcal mol^{-1} [149].

The dehydrogenation of C–C bonds is thermodynamically unfavorable. For example, the dehydrogenation of ethane to ethylene ($H_3C\text{–}CH_3 \rightarrow H_2C\text{=}CH_2 + H_2$) is substantially endergonic, with $\Delta G = +23.9$ kcal mol^{-1}. On the other hand, the analogous dehydrogenation of AB, which is isoelectronic with ethane, ($H_3NBH_3 \rightarrow H_2NBH_2 + H_2$) is exergonic, with $\Delta G = -13.6$ kcal mol^{-1} [149]. On the basis of computational results, the replacement of two methylene groups in cyclohexane with an H_2NBH_2 group results in a material that has 7.1 wt% accessible hydrogen for the loss of three equivalents of H_2 and $\Delta G = 1.9$ kcal mol^{-1}, as shown in Equation 16.49 for R=H [150]. The near thermoneutrality of this 1,2-azaboracyclohexane system for

R
N–H
BH
⇌
R
N
BH
+ 3 H_2 (16.49)

dehydrogenation/rehydrogenation makes it highly attractive for reversible hydrogen storage. Compounds of this type have been prepared. As an example, calculations indicate that Equation 16.49, where R = *t*-butyl, has $\Delta G = -0.4$ kcal mol^{-1}, with aspects of this systems demonstrated experimentally [151].

Another strategy for hydrogen storage is through the use of "frustrated Lewis pairs." For example, a phosphonium borate compound has been described that reversibly stores hydrogen in a practical temperature range [152]. The phosphino-borane compound shown on the left-hand side of Equation 16.50 was shown to accept 1 equiv. of hydrogen, at 1 atm and 25°C,

$Mes_2P\text{–}C_6F_4\text{–}B(C_6H_5)_2 \underset{-H_2\ 100°C\ \text{toluene}}{\overset{+H_2\ 25°C}{\rightleftharpoons}} Mes_2HP^+\text{–}C_6F_4\text{–}B^-H(C_6H_5)_2$ (16.50)

Mes = 2,4,6 -trimethylphenyl

forming the phosphonium borate zwitterion on the right. The latter compound releases hydrogen again on heating to 100°C. Although the hydrogen storage capacity of this compound is only 0.25 wt%, well below the level required for practical use, this system suggests new possibilities for the use of boron compounds.

Other examples involve combinations of sterically hindered *N*-heterocyclic Lewis bases and $B(C_6F_5)_3$ that also form "frustrated" Lewis pair systems capable of activating hydrogen [153,154]. This was demonstrated, for example, for the lutidine-tri(pentafluorophenyl)borane system shown in Scheme 16.1.

The equilibrium between the adduct **1** and free Lewis acid and base allows this system to be exploited for frustrated Lewis pair reactivity. Addition of H_2 to this mixture at a pressure of 1 atm was shown to result in the formation of salt **2** through cleavage of hydrogen [153].

16.6 BORON RESOURCES

The proposed use of boron materials for hydrogen storage in large automotive fleets involving hundreds of millions of vehicles has prompted questions regarding the adequacy of boron resources to meet these enormous demands. For this reason, U.S. Borax Inc. assessed borate reserves in both the United States and the world and compared this assessment with potential initial fill requirements of a recyclable boron-based fuel for a large fleet of FCVs.

Assessment of world boron resources in terms of known borate minerals reserves involves analysis and reconciliation of publicly available information on borate reserves in various parts of the world. Analysis of key sources was used to estimate total borate resources projected out to 2050 adjusted for estimated boron demands from other competing industrial applications. This demand-adjusted assessment of boron resources was then compared with estimated requirements for boron materials needed for hydrogen storage over the period relevant to FCV deployment based on DOE scenarios [155].

The assessment of borate resources concludes that global borate ore reserves are sufficient for foreseen automotive hydrogen storage applications, and that sufficient borate reserves exist in the United States to satisfy the initial fill requirements of an AB fueled U.S. FVC fleet. In this analysis it is assumed that AB will be recycled to a large extent following initial fill of an automotive fleet. These conclusions are based on published data indicating that more than 1 billion B_2O_3 tonnes are

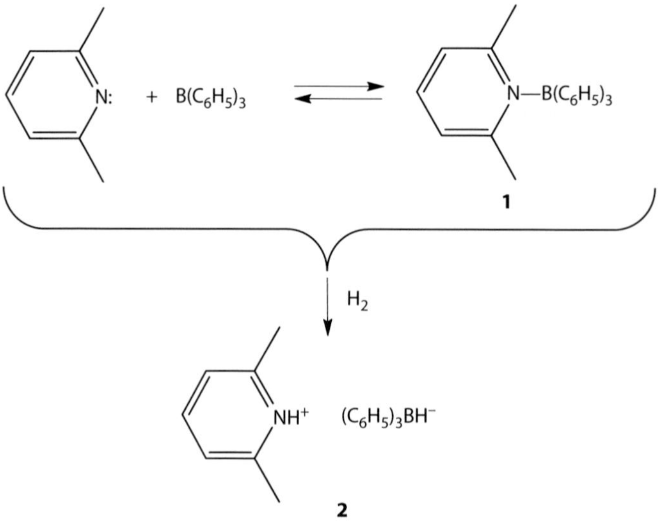

N: + $B(C_6H_5)_3$ ⇌ N—$B(C_6H_5)_3$

1

H_2

NH^+ $(C_6H_5)_3BH^-$

2

SCHEME 16.1 Hydrogen activation by frustrated Lewis pair.

available worldwide from known borate reserves, of which over 100 million B_2O_3 ton are located in the United States [156,157]. The cumulative boron requirement is estimated at 60 million B_2O_3 ton for a U.S. fleet of 360 million FCVs and 360 million B_2O_3 ton for a global fleet of 2 billion vehicles.

16.7 SUMMARY

Boron-based materials stand out among the many materials studied for hydrogen storage applications. Combining excellent hydrogen densities, release rates, and safety characteristics, a number of boron materials can hold substantially more releasable hydrogen by weight and volume than pure liquid hydrogen.

Hydrogen-on-demand technologies developed for commercially available sodium borohydride offer convenient and compact power for portable devices, underground mining vehicles, and backup power systems. Developing technologies based on other metal borohydrides, AB, and MABs have the potential to provide even higher power densities than the more developed sodium borohydride systems.

REFERENCES

1. Züttel, A. Borgschulte, A. Schlapbach, L. eds, *Hydrogen as a Future Energy Carrier*, Wiley-VCH Verlag GmbH & Co, KGaA, Weinheim, Germany, 2008.
2. Satyapal, S.; Petrovic, J.; Thomas, G., Gassing up with hydrogen, *Sci. Am.* **2007**, *296*, 80–87.
3. Committee on Alternatives and Strategies for Future Hydrogen Production and Use, *The Hydrogen Economy: Opportunities, Costs, Barriers, and R&D Needs*, National Research Council, National Academies Press, Washington, DC, 2004.
4. Schlapbach, L.; Züttel, A., Hydrogen-storage materials for mobile applications, *Nature* **2001**, *414*, 353.
5. Milliken, J., *The Advanced Energy Initiative—Challenges, Progress, and Opportunities*, Online U.S. DOE Office of Energy Efficiency and Renewable Energy, 2007.
6. Aardahl, C. L.; Rassat, S. D., Overview of systems considerations for on-board chemical hydrogen storage, *Int. J. Hydrogen Energy* **2009**, *34*, 6676–6683.
7. Wang, P.; Kang, X.-D., Hydrogen-rich boron-containing materials for hydrogen storage, *Dalton Trans.* **2008**, 5400–5413.
8. Umegaki, T.; Yan, J.-M.; Xhang, X.-B.; Shioyama, H.; Kuriyama, N.; Xu, Q., Boron- and nitrogen-based hydrogen storage materials, *Int. J. Hydrogen Energy* **2009**, *34*, 2303–2311.
9. Hamilton, C. W.; Baker, R. T.; Staubitz, A.; Manners, I., B–N compounds for chemical hydrogen storage, *Chem. Soc. Rev.* **2009**, *38*, 279–293.
10. Orimo, S.; Nakamori, Y.; Eliseo, J. R.; Zuttel, A.; Jensen, C. M., Complex hydrides for hydrogen storage, *Chem. Rev.* **2007,** *107*, 4111–4132.
11. Details of U.S. DOE hydrogen storage program and targets, available at http://www1.eere.energy.gov/hydrogenandfuelcells/mypp/pdfs/storage.pdf
12. Motyka, T., Hydrogen storage solutions in support of DoD warfighter portable power applications, *WSTIAC Quarterly*, 9, 83–87, available online at http://ammtiac.alionscience.com/pdf/AQV4N1_ART07.pdf
13. National Institute of Standards and Testing reference data, available online at http://webbook.nist.gov
14. Schlesinger, H. I.; Brown, H. C., Method of preparing alkali metal borohydrides, U.S. Patent 2,534,533.
15. Schlesinger, H. I.; Brown, H. C.; Finholt, A. E.; Gilbreath, J. R.; Hoekstra, H. R.; Hyde, E. K., Sodium borohydride, its hydrolysis and its use as a reducing agent and in the generation of hydrogen, *J. Am. Chem. Soc.* **1953**, *75*, 215–219.
16. Schlesinger, H. I.; Brown, H. C.; Abraham, B.; Bond, A. C.; Davidson, N.; Finholt, A. E.; Gilbreath, J. R. et al., New developments in the chemistry of diborane and the borohydrides. General summary, *J. Am. Chem. Soc.* **1953**, *75*, 186–190.
17. Schubert, D. M., Boron compounds—Boron hydrides, heteroboranes and their metalla derivatives, in *Kirk-Othmer Encyclopedia of Chemical Technology*, 5th edn, John Wiley & Sons, Inc., New York, 2002. Published online DOI: 10.1002/04712389-61.021518151903082.

18. Stock, A.; Sfitterlin, W.; Kurzen, F., Boron hydrides. XX. Potassium diborane $K_2(B_2H_6)$, *Z. Anorg. Allgem. Chem.* **1935**, *225*, 225–42.
19. Schlesinger, H. I.; Sanderson, R. T.; Burg, A. B., Metallo borohydrides. I. Aluminum borohydride, *J. Am. Chem. Soc.* **1940**, *62*, 3421–3425.
20. Schlesinger, H. I.; Brown H. C., Metallo borohydrides III. Lithium borohydride, *J. Am. Chem. Soc.* **1940**, *62*, 3429–3435.
21. Wiberg, E.; Bauer, R., Magnesium borohydride, $Mg(BH_4)_2$, *Zeit. Naturforsch.* **1950**, *5b*, 397–398.
22. Filinchuk, Y.; Chernyshov, D.; Dmetriev, V., Light metal borohydrides: Crystal structures and beyond, *Z. Kristallogr.* **2008**, *223*, 649–659.
23. Schubert, D. M.; Brotherton, R. J., Borates inorganic chemistry, in R. B. King, ed. *Encyclopedia of Inorganic Chemistry*, 2nd edn, Wiley, New York, 2005, p. 499.
24. Schubert, D. M., Borates in industrial use, in eds, *Group 13 Chemistry III, Industrial Applications*, H. W. Roesky and D. A. Atwood, *Structure and Bonding Series*, Vol. 105, Chapter 1, Springer-Verlag, Berlin, 2003, p. 1.
25. Smith R. A.; McBroom R. B., Boron oxides, boric acid, and borates, in *Kirk–Othmer Encyclopedia of Chemical Technology*, 4th edn, John Wiley & Sons, Inc., New York, 1992, pp. 365–413.
26. Hiltz, R. H., Gas generation, U.S. Patent 3,405,068, assigned to Mine Safety Appliances Co, 1968.
27. Litz, L. M.; Rothfleisch, J. E., Hydrogen generation, U.S. Patent 3,459,510, assigned to Union Carbide Corp, 1969.
28. Amendola, S. C.; Binder, M.; Kelly, M. T.; Petillo, P. J.; Sharp–Goldman, S. L., Advances in hydrogen energy, in P. and L. Kluwer, eds, Academic/Plenum Publishers, 2000, pp. 69–86.
29. Amendola, S. C.; Sharp-Goldman, S. L.; Janjua, M. S.; Kelly, M. T.; Spencer, M. T.; Petillo, P. J.; Binder, M., A safe, portable hydrogen gas generator using aqueous borohydride solution and Ru catalyst, *Int. J. Hydrogen Energy* **2000**, *25*, 969–975.
30. Wu, Y.; Mohring, R. M., Sodium Borohydride for Hydrogen Storage, *Prepr. Pap. - Am. Chem. Soc., Div. Fuel Chem.* **2003**, *48*, 940.
31. Jorgensen, S. W.; Perry, B. K., Hydrogen generation system using stabilized borohydrides for hydrogen storage, U.S. Patent 6,811,764 B2, assigned to General Motors Corp, 2004.
32. Goldstein, J.; Givon, M., A system for hydrogen storage and generation, Int. Pat. Appl., WO 2007/096857 A1, Hydrogen Ltd.
33. Stanic, V.; Carrington, D., Hydrogen production from borohydrides and glycerol, Int. Patent WO 2008/144038 A1, Enerfuel Inc., 2008.
34. Ma, J.; Choudhury, N. A.; Sahai, Y., A comprehensive review of direct borohydride fuel cells, *Renewable and Sustainable Energy Reviews*, **2010**, *14*, 183–199.
35. Wang, K.; Jiang, K.; Lu, J.; Zhuang, L.; Cha, C.; Hu, X.; Chen, G. Z., Eight-electron oxidation of borohydride at potentials negative to reversible hydrogen electrode, *J. Power Sources* **2008**, *185*, 892–894.
36. Schlesinger, H. I.; Brown, H. C.; Finholt, A. E., The preparation of sodium borohydride by the high temperature reaction of sodium hydride with borate esters, *J. Am. Chem. Soc.* **1953**, *75*, 205–209.
37. Wu, Y.; Kelly, M. T.; Ortega, J. V., Review of chemical processes for the synthesis of sodium borohydride, Report for DOE Cooperative Agreement DE-FC36-04GO14008, 2004.
38. Kojima, Y.; Haga, T., Recycling process of sodium metaborate to sodium borohydride, *Int. J. Hydrogen Energy* **2003**, *28*, 989–993.
39. Züttel, A.; Wenger, P.; Rensch. S.; Sudan, P.; Mauron, P.; Emmenegger, C., $LiBH_4$, A new hydrogen storage material, *J. Power Sources* **2003**, *118*, 1–7.
40. Züttel, A.; Rentsch, S.; Fischer, P.; Wenger, P.; Sudan, P.; Mauron, P.; Emmenegger, C., Hydrogen storage properties of $LiBH_4$, *J. Alloys Compd.* **2003,** *356–357*, 515–520.
41. Filinchuk, Y.; Chernyshov, D.; Cerny, R., The lightest borohydride probed by synchrotron diffraction: Experiment calls for a new theoretical revision, *J. Phys. Chem. C.* **2008**, *112*, 10579–10584.
42. Schlesinger, H. I.; Brown, H. C.; Hoekstra, H. R.; Rapp, L. A., Reactions of diborane with alkali metal hydrides and their addition compounds. New synthesis of borohydrides, *J. Am. Chem. Soc.* **1953**, *75*, 199–204.
43. Schlesinger, H. I.; Brown, H. C.; Hoekstra, H. R.; Hyde, E. K., The preparation of other borohydrides by metathesis reactions utilizing the alkali metal borohydrides, *J. Am. Chem. Soc.* **1953**, *75*, 209–213.
44. Herbert C.; Brown, H. C.; Choi, Y. M.; Narasimhan, S., Addition compounds of alkali metal hydrides. 22. convenient procedures for the preparation of lithium borohydride from sodium borohydride and borane-dimethyl sulfide in simple ether solvents, *Inorg. Chem.* **1982**, *21*, 3657–3661.
45. Paul, R; Joseph, N., Process for the production of lithium borohydrides, U.S. Patent 2,726,926, assigned to Rhone Poulenc, 1955.

46. Hauk, D; Wietelmann, U., Process for the preparation of lithium borohydride, U.S. Patent 7,288,236, assigned to Chemetal GmbH, 2007.
47. Fedneva, E. M.; Alpatova, V. L.; Mikheeva, V.I., Thermal Stability of Lithium Borohydride, *Russ. J. Inorg. Chem.* **1964**, *9*, 826–827.
48. Muetterties, E. L.; Merrifield, R. E.; Miller, H. C.; Knoth, W. H. Jr.; Downing, J. R., Chemistry of boranes. III., The infrared and raman spectra of $B_{12}H_{12}$ - and related anions *J. Am. Chem. Soc.* **1962**, *84*, 2506.
49. Orimo, S.; Nakamori, Y.; Ohba, N.; Miwa, K.; Aoki, M.; Towata, S.; Züttel, A., Experimental studies of intermediate compound of $LiBH_4$, *Appl. Phys. Lett.* **2006**, *89*, 021920.
50. Miwa, K.; Ohba, N.; Towata, S.; Nakamori, Y.; Orimo, S., First-principles study on lithium borohydride $LiBH_4$, *Phys. Rev. B*, **2004**, *69*, 245120.
51. Orimo, S.; Nakamori, Y.; Kitahara, G.; Miwa, K.; Ohba, N.; Towata, S.; Züttel, A., Dehydriding and rehydriding reactions of $LiBH_4$, *J. Alloys Compd.* **2005**, *404–406*, 427–430.
52. Kang, X.-D.; Wang, P.; Ma, L.-P.; Cheng, H.-M., Reversible hydrogen storage in $LiBH_4$ destabilized by milling with Al, *Appl. Phys. A* **2007**, *89*, 963–966.
53. Vajo J. J.; Skeith, S. L.; Mertens, F., Reversible storage of hydrogen in destabilized $LiBH_4$, *J. Phys. Chem. B* **2005**, 109, 3719.
54. Au, M.; Jurgensen, A.; Ziegler, K., Modified lithium borohydrides for reversible hydrogen storage, *J. Phys. Chem. B* **2006**, *110*(51), 26482–26487.
55. Au, M.; Jurgensen, A., Modified lithium borohydrides for reversible hydrogen storage, *J. Phys. Chem. B* **2006**, *110*, 7062–7067.
56. Guo, Y. H.; Yu, X. B.; Gao, L.; Xia, G. L.; Guo, Z. P.; Liu, H. K., Significantly improved dehydrogenation of $LiBH_4$ destabilized by TiF_3, *Energy Environ. Sci.* **2010**, *3*, 465–470.
57. Fang, Z. Z.; Wang, P.; Rufford, T. E.; Kang, X. D.; Lu, G. Q.; Cheng, H. M. Kinetic- and thermodynamic-based improvements of lithium borohydride incorporated into activated carbon, *Acta Materialia* **2008**, *56*, 6257–6263.
58. Fang, Z. Z.; Wang, P.; Rufford, T. E.; Kang, X. D.; Lu, G. Q.; Cheng, H.M., Improved reversible dehydrogenation of lithium borohydride by milling with as-prepared single-walled carbon nanotubes, *J. Phys. Chem. C* **2008**, *112*, 17023–17029.
59. Filinchuk, Y.; Chernyshov, D.; Nevidomskyy, A.; Dmitriev, V. *Angew. Chem. Intl. Ed.* **2008**, *47*, 529–532.
60. Meisner, G. P.; Scullin, M. L.; Balogh, M. P.; Pinkerton, F. E.; Meyer, M. S., Hydrogen release from mixtures of lithium borohydride and lithium amide: A phase diagram study, *J. Phys. Chem. B* **2006**, *110*, 4186–4192.
61. Chater, P. A.; David, W. I. F.; Johnson, S. R.; Edwards, P. P.; Anderson, P. A., Synthesis and crystal structure of $Li_4BH_4(NH_2)_3$, *Chem. Commun.* **2006**, 2439–2441.
62. Matsunaga, T.; Buchter, F.; Miwa, K.; Towata, S.; Orimo, S.; Züttel, A., Magnesium borohydride: A new hydrogen storage material, *Renew. Energy* **2008**, *33*, 193–196.
63. Matsunaga, T.; Buchter, F.; Mauron, P.; Bielman, M.; Nakamori, Y.; Orimo, S.; Ohba, N.; Miwa, K.; Towata, S.; Züttel, A., Hydrogen storage properties of $Mg[BH_4]_2$, *J. Alloys Compd.* **2008**, *459*, 583–588.
64. Soloveichik, G. L.; Andrus, M.; Gao, Y.; Zhao, J.-C.; Kniajanski, S., Magnesium borohydride as a hydrogen storage material: Synthesis of unsolvated $Mg(BH_4)_2$, *Int. J. Hydrogen Energy* **2009**, *34*, 2144–2152.
65. Plesek, J.; Hermanek, S., Chemistry of boranes. IV. Preparation, properties and behavior of magnesium borohydride towards lewis bases, *Collect. Czech. Chem. Commun.* **1966**, *31*, 3845–3858.
66. Mikheeva, V. I.; Konoplev, V. N., Reaction of sodium borohydride with anhydrous magnesium chloride in an *N,N*-dimethylformamide solution, *Russ. J. Inorg. Chem.* **1965**, *10*, 2108–2114.
67. Konoplev, V. N.; Silina, T. A., Phase diagram of tetrahydrofuran and magnesium tetrahydroborate, *Russ. J. Inorg. Chem.* **1974**, *19*, 2534–2537.
68. Chlopek, K.; Frommen, C.; Leon, A.; Zabara, O.; Fichtner, M., Synthesis and properties of magnesium tetrahydroborate, $Mg(BH_4)_2$, *J. Mater. Chem.* **2007**, *33*, 3496–3503.
69. Zanella, P.; Crociani, L.; Masciocchi, N.; Giunchi, G., Facile high yield synthesis of pure, crystalline $Mg(BH_4)_2$, *Inorg. Chem.* **2007**, *46*, 9039–9041.
70. Soloveichik, G. L.; Gao, Y.; Rijssenbeek, J.; Andrus, M.; Kniajanski, S.; Bowman, R. C., Jr.; Hwang, S.-J.; Zhao, J.-C., Magnesium borohydride as a hydrogen storage material: Properties and dehydrogenation pathway of unsolvated $Mg(BH_4)_2$, *Int. J. Hydrogen Energy* **2009**, *34*, 916–928.
71. Li, H.-W.; Kikuchi, K.; Sato, T.; Nakamori, Y.; Ohba, N.; Aoki, M.; Miwa, K.; Towata, S. I.; Orimo, S. I., Synthesis and hydrogen storage properties of a single-phase magnesium borohydride $Mg(BH_4)_2$, *Mater. Trans.* **2008**, *49*, 2224–2228.

72. Hwang, S.-J.; Bowman, R. C., Jr.; Reiter, J. W.; Rijssenbeek, J.; Soloveichik, G. L.; Zhao, J. C.; Kabbour, H.; Ahn, C. C., NMR Confirmation for formation of $[B_{12}H_{12}]^{2-}$ complexes during hydrogen desorption from metal borohydrides, *J. Phys. Chem. C* **2003**, *112*, 3164–3169.
73. Soloveichik, G.; Her, J.-H.; Stephens, P. W.; Gao, Y.; Rijssenbeek, J.; Andrus, M.; Zhao, J.-C., Ammine magnesium borohydride complex as a new material for hydrogen storage: Structure and properties of $Mg(BH_4)_2{\cdot}2NH_3$, *Inorg. Chem.* **2008**, *47*, 4290–4298.
74. Rönnebro, E.; Majzoub, E. H., Calcium borohydride for hydrogen storage: Catalysis and reversibility, *J. Phys. Chem. B* **2007**, *111*, 12045–12047.
75. Kollonitsch, J.; Fuchs, O.; Gabor, V., New and known complex borohydrides and some of their applications in organic syntheses, *Nature* **1954**, *173*, 125–126.
76. Pearson, R. K., Calcium Borohydride Production, U.S. Patent 3,224,832, assigned to Callery Chemical Co, 1965.
77. Titov, L. V., Synthesis of calcium borohydride, *IZD-VO Akad. Nauk. SSSR.* **1964**, *154*, 654–656.
78. Kim, J.-H.; Jin, S.-A; Shim, J.-H.; Cho, Y. W., Thermal decomposition behavior of calcium borohydride $Ca(BH_4)_2$, *J. Alloys Compd.* **2008**, *461*, L20–L22.
79. Aoki, M.; Miwa, K.; Noritake, T.; Ohba, N.; Matsumoto, M.; Li, H.-W.; Nakamori, Y.; Towata, S.; Orimo, S., Structural and dehydriding properties of $Ca(BH_4)_2$, *Appl. Phys. A* **2008**, *92*, 601–605.
80. Nakamori, Y.; Miwa, K.; Li, H. W.; Ohba, N.; Towata, S.-I.; Orimo, S.-I., Tailoring of Metal borohydrides for hydrogen storage applications, *Mater. Res. Soc. Proceedings*, Fall 2006, Vol. 971E.
81. Li, H.-W.; Orimo, S.; Nakamori, Y.; Miwa, K.; Ohba, N.; Towata, S.; Züttel, A., Materials designing of metal borohydrides: Viewpoints from thermodynamical stabilities, *J. Alloys Compd.* **2007**, *446–447*, 315–318.
82. Fang, Z. Z.; Kang, X.-D.; Wang, P.; Li, H.-W.; Orimo, S.-I., Unexpected dehydrogenation behavior of $LiBH_4/Mg(BH_4)_2$ mixture associated with the *in situ* formation of Dual-Cation borohydride, *J. Alloys Compd.* **2010**, *491*, L1–L4.
83. Lee, J. Y.; Ravnsbæk, D.; Lee, Y. S.; Kim, Y.; Cerenius, Y.; Shim, J.-H.; Jensen, T.-R.; Hur, N. H.; Cho, Y. W., Decomposition reactions and reversibility of the $LiBH_4 - Ca(BH_4)_2$ composite, *J. Phys. Chem. C* **2009**, *113*, 15080–15086.
84. Parry, R. W.; Schultz, D. R.; Girandot, P. R., The preparation and properties of hexamminecobalt(III) borohydride, hexamminechromium(III) borohydride and ammonium borohydride, *J. Am. Chem. Soc.* **1958**, *80*, 1–3.
85. Karkamkar, A.; Kathmann, S. M.; Schenter, G. K.; Hildebrant, D. J.; Hess, N.; Gutowski, M.; Autrey, T., Thermodynamic and structural investigations of ammonium borohydride, a solid with the highest content of thermodynamically and kinetically accessible hydrogen, *Chem. Mater.* **2009**, *21*, 4356–4358.
86. Pyykkö, P.; Wang, C., Theoretical study of H_2 splitting and storage by Boron–Nitrogen-based systems: A bimolecular case and some qualitative aspects, *Phys. Chem. Chem. Phys.* **2010**, *12*, 149–155.
87. Stephens, F. H.; Pons, V.; Baker R. T., Ammonia-Borane: The hydrogen source par excellence?, *Dalton Trans.* **2007**, *25*, 2613–2626.
88. Marder, T. B., Will we soon be filling our cars with ammonia borane?, *Angew. Chem.* **2007**, *46*, 8116–8118.
89. Hess, N. J.; Schenter, G. K.; Hartman, M. R.; Daemen, L. L.; Proffen, T.; Kathmann, S. M.; Mundy, C. J.; Hartl, M.; Heldebrant, D. J.; Stowe, A. C.; Autrey, T., Neutron powder diffraction and molecular simulation study of the structural evolution of ammonia borane from 15 to 340 K, *J. Phys. Chem. A* **2009**, *113*, 5723–5735.
90. Yang, J. B.; Lamsal, J.; Cai, Q.; James, W. J.; Yelon, W. B., Structural evolution of ammonia borane for hydrogen storage, *Appl. Phys. Lett.* **2008**, *92*, 091916.
91. Schlesinger, H. I.; Burg, A. B., Hydrides of boron. VIII. The structure of the diammoniate of diborane and its relation to the structure of diborane, *J. Am. Chem. Soc.* **1938**, *60*, 290–299.
92. Schultz, D. R.; Parry, R. W., Chemical evidence for the structure of the "Diammoniate of Diborane." I. Evidence for the borohydride ion and for the dihydro-diammineboron(III) cation, *J. Am. Chem. Soc.* **1958**, *80*, 4–8.
93. Shore, S. G.; Parry, R.W., Chemical evidence for the structure of the "diammoniate of diborane." II. The preparation of ammonia-borane, *J. Am. Chem. Soc.* **1958**, *80*, 8–12.
94. Parry, R. W.; Shore, S. G., Chemical evidence for the structure of the "diammoniate of diborane." III. The reactions of borohydride salts with lithium halides and aluminum chloride, *J. Am. Chem. Soc.* **1958**, *80*, 15–20.
95. Shore, S. G.; Girardot, P. R.; Parry, R. W., A Tracer study of the reaction between sodium and the "diammoniate of diborane", *J. Am. Chem. Soc.* **1958**, *80*, 20–24.

96. Parry, R. W., Symmetric and asymmetric cleavage of the lighter boron hydrides and of metal salts—The role of the dielectric constant, *J. Organometal. Chem.* **2000**, *614–615*, 5–9.
97. Shore, S. G.; Parry, R. W., The crystalline compound ammonia-borane, H_3NBH_3, *J. Am. Chem. Soc.* **1955**, *77*, 6084–6085.
98. Shore, S. G.; Boeddeker, K. W., Large scale synthesis of $H_2B(NH_3)_2^+BH_4^-$ and H_3NBH_3, *Inorg. Chem.* **1964**, *3*, 914–915.
99. Schaeffer, G. W.; Basile, L. J., The reaction of lithium amide with diborane, *J. Am. Chem. Soc.* **1955**, *77*, 331–332.
100. Heldebrant, D.J.; Karkamkar, A.; Linehan, J.C.; Autrey, T., Synthesis of ammonia borane for hydrogen storage applications, *Energy Environ. Sci.* **2008**, *1*, 156–160.
101. Briggs, T. S.; Jolly, W. L., Kinetics of the reaction of ammonium ion with hydroborate ion in liquid ammonia, *Inorg. Chem.* **1975**, *14*, 2267–2268.
102. Armstrong, D. L., *Investigation of Liquid Rocket Propellants*, Report No. 367, Project NR220 023, Aerojet Engineering Corp., Azusa, CA, 1949.
103. Schlesinger, H. I., Hydrides and Borohydrides of Light Elements Report No. XXIX, Contract No. N6ori-20, 1947.
104. Shore, S. G.; Chen, X., Methods for synthesizing ammonia borane, U.S. Patent Appl. 20090104102, 2009.
105. Ramachandran, P. V.; Gagare, P. D., Preparation of ammonia borane in high yield and purity, methanolysis, and regeneration, *Inorg. Chem.* **2007**, *46*, 7810–7817.
106. Ramachandran, P. V.; Gagare, P. D.; Bhimapaka, C. R., U.S. Patent Appl. 0243122 A1, 2007.
107. Chew, W. M.; Murfree, J. A.; Martignoni, P., Nappier, H. A., Ayers, O. E., Amine boranes as hydrogen generating propellants, U.S. Patent 4,157,927, assigned to The United States of America as represented by the Secretary of the Army, 1978.
108. Artz, G. D.; Grant, L. R., Solid propellant hydrogen generator, U.S. Patent 4,468,263, assigned to The United States of America as represented by the Secretary of the Army, 1984.
109. Hu, M. G.; Geanangel, R. A.; Wendlandt, W. W., The thermal decomposition of ammonia borane, *Thermochim. Acta* **1978**, *23*, 249–255.
110. Geanangel, R. A.; Wendlandt, W. W., A TG–DSC study of the thermal dissociation of $(NH_2BH_2)_x$, *Thermochim. Acta* **1985**, *86*, 375.
111. Sit, V.; Geanangel, R. A.; Wendlandt, W. W., The thermal dissociation of NH_3BH_3, *Thermochim. Acta* **1987**, *113*, 379–382.
112. Wolf, G.; Baumann, J.; Baitalow, F.; Hoffmann, F. P., Calorimetric process monitoring of thermal decomposition of B–N–H compounds, *Thermochim. Acta* **2000**, *343*, 19–25.
113. Baitalow, F.; Baumann, J.; Wolf, G.; Jaenicke-Rößler, K.; Leitner, G., Thermal decomposition of BNH compounds Investigated by using Combined Thermo-Analytical methods. *Thermochim. Acta* **2002**, *391*, 159–168.
114. Baumann, J.; Baitalow, F.; Wolf, G., Thermal decomposition of polymeric aminoborane $(H_2BNH_2)_x$ under hydrogen release, *Thermochim. Acta* **2005**, *430*, 9–14.
115. Rassat, S. D.; Aardahl, C. L.; Autrey, T.; Smith, R. S., The thermal stability of ammonia borane: A case study for exothermic hydrogen storage materials, *Energy Fuels* **2010**, *24*, 2596–2606.
116. Denney, M. C.; Pons, V.; Hebden, T. J.; Heinekey, D. M.; Goldberg, K. I., Efficient catalysis of ammonia borane dehydrogenation, *J. Am. Chem. Soc.* **2006**, *128*, 12048–12049.
117. Keaton, R. J.; Blacquiere, J. M.; Baker, R. T., Base metal catalyzed dehydrogenation of ammonia–borane for chemical hydrogen storage, *J. Am. Chem. Soc.* **2007**, *129*, 1844–1845.
118. Kalidindi, S. B.; Sanyal, U.; Jagirdar, B. R., Nanostructured Cu and Cu@Cu_2O core shell catalysts for hydrogen generation from ammonia–borane, *Phys. Chem. Chem. Phys.* **2008**, *10*, 5870–5874.
119. Kalidindi, S. B.; Indirani, M.; Jagirdar, B. R., First row transition metal ion-assisted ammonia-borane hydrolysis for hydrogen generation, *Inorg. Chem.* **2008**, *47*, 7424–7429.
120. He, T.; Xiong, Z.; Chu, G. H.; Zhang, T.; Chen, P., Nanosized Co- and Ni-catalyzed ammonia borane for hydrogen storage, *Chem. Mater.* **2009**, *21*, 2315–2318.
121. Mohajeri, N.; Tabatabaie-Raissi, A.; Bokerman, G., Catalytic dehydrogenation of amine borane complexes, U.S. Patent 7578992, 2009.
122. Wang, J. S.; Geanangel, R. A., ^{11}B NMR studies of the thermal decomposition of ammonia-borane in solution, *Inorg. Chim. Acta* **1988**, *148*, 185–190.
123. Shaw, W. J.; Linehan, J. C.; Szymczak, N. K.; Heldenbrandt, D. J.; Yonker, C.; Camaioni, D. M.; Baker, R. T.; Autrey, T., *In situ* multinuclear NMR spectroscopic studies of the thermal decomposition of ammonia borane in solution, *Angew. Chem., Int. Ed.* **2008**, *47*, 7493–7496.

124. Bluhm, M. E.; Bradley, M. G.; Butterick, R.; Kusari, U.; Sneddon, L. G., Amineborane-based chemical hydrogen storage: Enhanced ammonia borane dehydrogenation in ionic liquids, *J. Am. Chem. Soc.* **2006**, *128*, 7748–7749.
125. Autrey, T. S.; Gutowska, A.; Shin, Y.; Li, L., Materials for storage and release of hydrogen and methods for preparing and using same, U.S. Patent 7,316,788, 2008.
126. Gutowska, A.; Li, L.; Shin, Y.;Wang, C. M.; Li, X. S.; Linehan, J. C.; Smith, R. S. et al., Nanoscaffold mediates hydrogen release and the reactivity of ammonia borane, *Angew. Chem. Int. Ed.* **2005**, *44*, 3578–3582.
127. Sepehri, S.; Feaver, A.; Shaw, W. J.; Howard, C. J.; Zhang, Q.; Autrey, T.; Cao, G., Spectroscopic studies of dehydrogenation of ammonia borane in carbon cryogel, *J. Phys. Chem. B* **2007,** *111*, 14285–14289.
128. Li, L.; Yao, X.; Sun, C.; Du, A.; Cheng, L.; Zhu, Z.; Yu, C. et al., Lithium-catalyzed dehydrogenation of ammonia borane within mesoporous carbon framework for chemical hydrogen storage, *Adv. Funct. Mater.* **2009**, *19*, 265–271.
129. Stephens, F. H.; Baker, R. T.; Matus, M. H.; Grant, D. J.; Dixon, D. A., Acid initiation of ammonia-borane dehydrogenation for hydrogen storage, *Angew. Chem. Int. Ed.* **2006**, *46*, 746–749.
130. Chandra, M.; Xu, Q., Dissociation and hydrolysis of ammonia-borane with solid acids and carbon dioxide: An efficient hydrogen generation system, *J. Power Sources* **2006**, *159*, 855–860.
131. Becker, P.; Held, P.; Bohaty, L., Crystal growth and optical properties of the polar hydrated pentaborates $Rb[B_5O_6(OH)_4]\cdot 2H_2O$ and $NH_4[B_5O_6(OH)_4]\cdot 2H_2O$ and structure redetermination of the ammonium compound, *Cryst. Res. Technol.* **2000**, *35*, 1251–1262.
132. Janda, R.; Heller, G., Die Kristallstruktur von Synthetischem Ammoniumtetraboratdihydrate, *Zeit. Kristallogr.* **1981**, *154*, 1–9.
133. Basu, S.; Brockman, A.; Gagare, P.; Zheng, Y.; Ramachandran, P. V.; Delgass, W. N.; Gore, J. P., Chemical kinetics of Ru-catalyzed ammonia borane hydrolysis, *J. Power Sources* **2009**, *188*, 238–243.
134. Davis, B. L.; Dixon, D. A.; Garner, E. B.; Gordon, J. C.; Matus, M. H.; Scott, B.; Stephens, F. H., Efficient regeneration of partially spent ammonia borane fuel, *Angew. Chem. Int. Ed.* **2009**, *48*, 6812–6816.
135. Burrell, A. K.; Davis, B. J.; Thorn, D. L.; Gordon, J. C.; Baker, R. T.; Semelsberger, T. A.; Tumas, W.; Diyabalanage, H. V. K.; Diyabalanage, H. V. K.; Shrestha, R. P., Metal amidoboranes, *Int. Pat. Appl.* WO 2008/143780 A1.
136. Xiong, Z.; Yong, C. K.; Wu, G.; Chen, P.; Shaw, W.; Karkamkar, A.; Autrey, T. et al., High-capacity hydrogen storage in lithium and sodium amidoboranes, *Nat. Mater.* **2008**, *7*, 138–141.
137. Diyabalanage, H. V. K.; Shrestha, R. P.; Semelsberger, T. A.; Scott, B. L.; Bowden, M. E.; Davis, B. L.; Burrell, A. K., Calcium amidotrihydroborate: A hydrogen storage material, *Angew. Chem., Int. Ed.* **2007**, *46*, 8995–8997.
138. Luedtke, A. T.; Autrey, T., Hydrogen release studies of alkali metal amidoborane, *Inorg. chem.* **2010**, *49*, 3905–3910.
139. Lee, T. B.; McKee, M. L., Mechanistic study of $LiNH_2BH_3$ formation from $(LiH)_4$ + NH_3BH_3 and subsequent dehydrogenation, *Inorg. Chem.* **2009**, *48*, 7564–7575.
140. Kim, D. Y.; Singh, N. J.; Lee, H. M.; Kim, K. S., Hydrogen-release mechanisms in lithium amidoboranes, *Chem. Eur. J.* **2009**, *15*, 5598–5604.
141. Wu, C.; Wu, G.; Xiong, Z.; Han, X.; Chu, H.; He, T.; Chen, P., $LiNH_2BH_3\cdot NH_3BH_3$: Structure and hydrogen storage properties, *Chem. Mater.* **2010**, *22*, 3–5.
142. Chua, Y. S.; Wu, G.; Xiong, Z.; He, T.; Chen, P., Calcium amidoborane ammoniate - synthesis, structure, and hydrogen storage properties, *Chem. Mater.* **2009**, *21*, 4899–4904.
143. Goshens, T. J.; Hollins, R. A., New chemical hydrogen storage materials exploiting the self-sustaining thermal decomposition of guanidinium borohydride, *Chem., Commun.* **2009**, 3089–3091.
144. Titov, L. V.; Levicheva, M. D.; Dubikhina, G. N., Guanidinium octahydrotriborate and cleavage of $B_3H_8^-$ anion by anhydrous hydrazine, *Izv. Akad. Nauk SSSR, Ser. Khim.* **1976**, *8*, 1856.
145. Carvalho, D. A. L.; Shust, N. W., Triborohydride-8 salt preparation, U.S. Patent 3,564,561, assigned to American Cyanamid Co., 1971.
146. van den Berg A. W. C.; Areán, C. O., Materials for hydrogen storage: Current research trends and perspectives, *Chem. Commun.* **2008**, 668–681.
147. Hamilton, E. J. M.; Dolan, S. E.; Mann, C. M.; Colijn, H. O.; McDonald, C. A.; Shore, S. G., Preparation of amorphous boron nitride and its conversion to a turbostratic, tubular form, *Science* **1993**, *260*, 659–661.
148. Jhi, S.-H.; Kwon, Y.-K., Hydrogen adsorption on boron nitride nanotubes: A path to room-temperature hydrogen storage, *Phys. Rev. B* **2004**, *69*, 245407.

149. Campbell, P. G.; Zakharov, L. N.; Grant, D. J.; Dixon, D. A.; Liu, S.-H., Hydrogen Storage by Boron-Nitrogen Heterocycles: A simple Route for Spent Fuel Regeneration, *J. Am. Chem. Soc.* **2010**, *132*, 3289–3291.
150. Dixon, D. A.; Gutowski, M., Thermodynamic properties of molecular borane amines and the $[BH_4^-][NH_4^+]$ salt for chemical hydrogen storage systems from *ab initio* electronic structure theory, *J. Phys. Chem. A* **2005**, *109*, 5129–5135.
151. Matus, M. H.; Liu, S.-Y.; Dixon, D. A., Dehydrogenation Reactions of cyclic $C_2B_2N_2H_{12}$ and C_4BNH_{12} Isomers, *J. Phys. Chem.* **2010**, *114*, 2644–2654.
152. Welch, G. C; Ronan R. San Juan, R. R.; Masuda, J. D.; Stephan, D. W., Reversible, metal-free hydrogen activation, *Science* **2006,** *314*, 1124–1126.
153. Geier, S. J.; Stephan, D. W. *J. Am. Chem. Soc.* **2009**, *131*, 3476–3477.
154. Sumerin, V.; Schulz, F.; Nieger, M.; Atsumi, M.; Wang, K.; Leskelä, M.; Pyykkö, P.; Repo, T.; Rieger, B., Experimental and theoretical treatment of hydrogen splitting and storage in boron–nitrogen systems, *J. Organometal. Chem.* **2009**, *694*, 2654–2660.
155. DOE publication, Analysis of the transition to hydrogen fuel cell vehicles & potential hydrogen energy infrastructure requirements, available online.
156. Garret, D. E., Borates: A Handbook of Deposits, Processing, Properties, and Use, Academic Press, San Diego, CA, 1998, p. 442.
157. Anonymous *The Economics of Boron* 11th Edition, Roskill Information Services: London, 2006, Chapters 1 and 2.

17 Oilfield Technology and Applications Involving Borates

Michael J. Greenhill-Hooper

CONTENTS

17.1 INTRODUCTION

A wide range of chemicals, from commodity inorganic salts and simple organic compounds to complex custom-synthesized polymers, are employed in various oilfield chemical applications [1,2]. Although at first sight borates may be considered simply as water-soluble inorganic salts, they differ from commonly used salts like, for example, chlorides and formates, in that they are not employed to control ionic strength, water activity, or fluid density. For the major borate applications, they instead deliver other very specific properties as a result of a unique combination of chemical effects.

In order to understand the context behind the use of borates in oilfield chemicals, and before describing their applications in detail, it is helpful to provide a brief description of the functional chemicals and additives that are used in various oilfield operations. This section also provides a short introduction to borate compounds, chemistry, and aqueous solution behavior relevant to their use or potential use in oilfield applications.

17.1.1 Background on Oilfield Operations

It is convenient to group oilfield operations into the following categories:

- Drilling and well-construction
- Well and reservoir logging
- Oil and gas production
- Reservoir stimulation
- Enhanced oil recovery (EOR)

Figure 17.1 illustrates an overview of the principal activities conducted in these operations, plus a list of the types of functional additives that are used either alone or incorporated into specially formulated fluids with other additives and materials [3–5].

17.1.1.1 Drilling and Well-Construction Activities

Drilling and well-construction activities include operations that involve drilling and stabilizing the well bore, preparing it for service, and conducting ongoing well maintenance (the so-called completion and workover operations). Well drilling is always conducted with specially formulated drilling fluids, often referred to as muds. These can be either water or oil based (mineral or synthetic). They contain various additives to give the fluids specific properties. The principal requirements are a fluid with:

- The correct rheological and solid suspending capability (allowing transport of drill cuttings to the surface)
- The ability to form a good filter cake on the formation wall and minimize fluid loss to the formation
- The correct density (hydrostatic pressure to balance the pressure of formation fluids)
- Good lubricating characteristics (for cooling the drill bit)
- Good compatibility with sensitive rock types, notably shale

Completion operations, also conducted in specially formulated (completion) fluids, involve the stabilization of the drilled well bore by placing and cementing of steel casing in the hole. Workover operations include remedial well treatments and require the use of similar fluids, with the proviso that they are often simpler and do not require all the functionalities of drilling fluids (often they are clear brines with controlled specific gravity). The phenomenon of "lost circulation" refers to the loss (or potential loss) of large volumes of valuable drilling fluids into the formation rock when rock fractures or vulgar features are encountered during drilling operations. Lost circulation treatments range from the crude plugging of large fractures with ground nut husks to the use of polymer gels. It is technically challenging to pump cement into place in a hot reservoir and oil well cements used to stabilize the well-bore wall and keep the casing in place require various additives including accelerators, retarders, and dispersants.

17.1.1.2 Well- and Reservoir-Logging

Logging activities, designed to map out oil- and water-bearing zones in the reservoir, detect leaks of fluids into the formation, and provide estimates of formation rock porosity. Logging is employed

Drilling & well-construction

Activities:
- Drilling
- Completion/Cementing
- Workover

Formulated fluids/muds containing:
- Emulsifiers (oil-based muds)
- Base oil (oil-based muds)
- Weighting/density modifying agents
- Lubricants
- Dispersants
- Fluid loss agents (filter cake forming)
- Viscosity modifiers
- Gelled polymers, natural fibres & husks (lost circulation prevention)
- KCl/$CaCl_2$/other soluble salts (water activity, density control)
- Corrosion inhibitors
- Biocides
- Cement additives–set retarders, dispersants, polymers, and rubber
- Chemical tracers (for logging)

Well & reservoir logging

Oil & gas production

Activities:
- Water treatment and injection
- Gas/oil/water separation
- Asset protection
- Antifouling/scaling

Functional additives:
- Scale/corrosion/hydrate inhibitors
- Biocides/H_2S scavengers
- Demulsifiers/defoamers
- Water clarifiers
- Asphaltene dispersants

Reservoir stimulation

Activities:
- Acid stimulation
- Hydraulic fracturing

Formulated fluids/chemicals/functional additives:
- HCl/HF/formic and acetic acid
- Corrosion inhibitors
- Gelled polymers, cross-linkers, fluid loss agents, proppants
- Oxidizing and enzymatic agents (fluid breakers)

Enhanced oil recovery

Activities:
- Polymer-augmented water flooding
- Miscible flooding (CO_2 solvent)
- Immiscible flooding (chemical EOR)
- Steam and thermal recovery

Chemicals/functional additives:
- CO_2, alkalis, surfactants, polymers, solvents

FIGURE 17.1 Oilfield activities and use of chemical and material additives.

periodically throughout the life of the reservoir and not solely confined to the drilling and exploration phases. Logging techniques often require the use of chemical tracers.

17.1.1.3 Oil and Gas Production

Once the reservoir is in production, a variety of functional chemicals are required to prevent unwanted chemical and biological reactions and discourage precipitation or evolution of undesirable solids or gases in the oil and water streams. They are also used to protect assets from corrosion and fouling and help in oil, gas, and water separation and cleanup processes. These chemicals often work in tandem with mechanical processes (as in the case in oil/water separation, where separators and cyclones are employed).

17.1.1.4 Reservoir Stimulation

Reservoir stimulation refers to processes that are employed to enhance the flow and release of oil and gas from low-permeability reservoirs. Most reservoirs are subject to stimulation treatments at some point of time. Two types of processes are employed—acid stimulation and hydraulic fracturing. Acid stimulation enhances permeability and fluid flow by partially dissolving the formation rock using mineral or organic acids. Obviously, corrosion inhibitors are essential for protecting metalwork from acid attack during this operation. Hydraulic fracturing increases permeability by mechanically fracturing the rock with the injection of specially formulated fluids under high pressure. These contain particles ("proppants") often suspended in a gelled polymer or a surfactant solution; the proppants enter into the cracks in the rock created by this process and maintain them propped open when the pressure is released.

17.1.1.5 Enhanced Oil Recovery

A number of techniques have been developed over the years to release oil trapped in reservoirs after primary recovery (oil drive produced as a result of original reservoir pressure) and secondary recovery (e.g., injected-water flooding) processes. After primary and secondary recovery as much as 65% of the ultimately recoverable oil may still be trapped in the reservoir. Most EOR processes involve the injection of gases or fluids to drive residual oil out of the reservoir. Some of these are categorized as miscible or immiscible flooding processes. Miscible processes using carbon dioxide or hydrocarbon solvents facilitate oil mobilization by reducing the oil/water interface and capillary forces in the rock pores that otherwise inhibit oil release. Immiscible processes can also involve the use of carbon dioxide and inert gases, but often refer to the injection of water-based fluids sometimes containing quite complex mixtures of polymer, alkali–polymer, polymer–surfactant, and alkali–surfactant–polymer (ASP) that help to mobilize the oil by lowering the oil/water interfacial tension and improving the water wettability of the reservoir rock. These are often collectively referred to as "chemical EOR" processes. In its simplest form, only polymer is present in the injected water. Polymer-augmented or -enhanced water floods could be viewed as an extension of secondary recovery methods. The role of the polymer, the only water flood additive employed in this process, is to increase the viscosity, and control the flow profile of the water flood and ensure that it is not diverted past the oil-bearing zones into higher-permeability regions. Achieving the correct water viscosity in relation to that of the oil also helps to optimize its displacement and mobilization from the rock pores. Surfactants are included to lower oil/water interfacial tension and alkalis support the role of the surfactant by saponifying carboxylic components that are naturally present in some crude oils. In addition to these methods, steam flooding and *in situ* combustion techniques are also employed to enhance oil recovery.

CORE CONCEPTS

Oilfield operations can be classified into several categories: ***drilling and exploration, oil and gas production, reservoir stimulation, and enhanced oil recovery.*** All of these require the use of complex fluids containing a range of functional chemical additives.

17.1.2 Background on Borate Compounds, Chemistry, and Solution Behavior

17.1.2.1 Boron and Boron–Oxygen Chemistry

Boron is the fifth element in the periodic table [6]. It has electron structure $1s^22s^22p$ and resides in group IIIA of the periodic table, in the same group as aluminum. However, because of its high-ionization energies, unlike a metal, boron does not form compounds in which it is present as a univalent cation. Boron has an atomic weight of 10.81 and exists in the form of two stable isotopes—^{10}B (20%) and ^{11}B (80%). The ^{10}B isotope has an unusually high neutron capture cross section and this property is directly exploited in one particular oilfield application. Boron is electron deficient. It tends to form three-coordinate trigonal covalent compounds with other elements. The resulting boron compounds have sp^2 orbital hybridization. This leaves a lone p orbital perpendicular to the molecular plane that is responsible for these compounds being Lewis acids and for their tendency to bond with electron donor compounds. Boron forms very strong bonds with oxygen and consequently boron oxides and borates have a very rich chemistry [7,8]. Both in the solid state and in aqueous solution, borates exist in which the boron atom is either 3- or 4-coordinate with respect to its bonding with oxygen atoms. A three-coordinate boron center has the charge-neutral planar trigonal structure previously described, while with four coordination, arising when it bonds with an electron donor (e.g., OH^-), the vacant p orbital is filled, it adopts tetrahedral geometry, and acquires a negative charge that requires compensation with an associated cation.

17.1.2.2 Boric Oxide and Boric Acid

Boric oxide, B_2O_3, the simplest and only commercially important oxide of boron, exists primarily as an amorphous glass that has a 3-D arrangement of three-coordinate boron centers (BO_3) that share oxygen atoms [9]. It finds application mainly in glass and ceramics. It is water soluble, although it dissolves relatively slowly in an exothermic reaction with water. It is often referred to as anhydrous boric acid, as it converts to and behaves like boric acid on contact with water.

Boric acid (or more precisely orthoboric acid), formula H_3BO_3, is a neutral three-coordinate boron compound. It has a solid-state structure similar to that of graphite, with the molecules arranged in planes [10]. Within the planes the molecules are linked by hydrogen bonding but the separate planes are held together by weaker Van der Waals forces. Boric acid is water soluble, releasing into solution the trigonal species H_3BO_3 (see Figure 17.2).

17.1.2.3 Alkali Metal and Ammonium Borates

A wide range of alkali metal or ammonium borates exist and are commercially available. These also contain mixtures of trigonally and tetrahedrally coordinated boron atoms [11]. All can be represented in terms of the basic boric oxide building block and compensating amounts of alkali metal oxide or the ammonium ion:

$$m\mathrm{M_2O} \cdot n\mathrm{B_2O_3} \cdot x\mathrm{H_2O}$$

Many are available in hydrated or anhydrous ($x = 0$) form. Generally, the hydrated alkali metal borates are crystalline solids that dissolve quite readily in water, while the anhydrous products are glassy and dissolve much more slowly. Some examples of these compounds are illustrated in Table 17.1, which include many of the commercially important borate compounds, but should not be regarded as an exhaustive list [12].

Alkali metal and ammonium borates are considered to be reasonably soluble in water. Table 17.2 provides solubility data for several alkali metal and other metal borates at different temperatures [8].

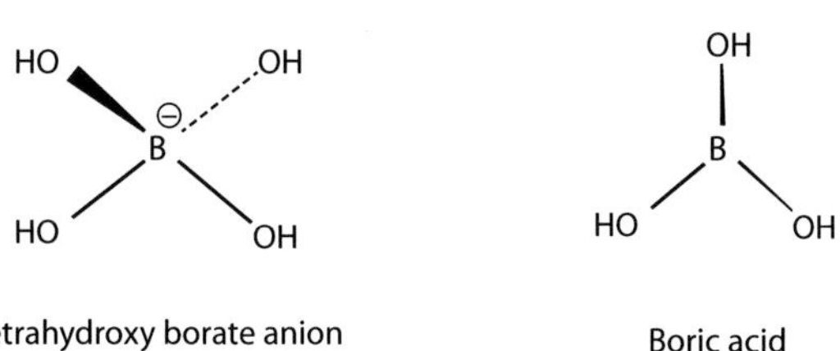

FIGURE 17.2 Borate monomers existing in aqueous borate solutions.

TABLE 17.1
Examples of Commercially Important Borates

Chemical Formula	Name(s)	Primary Applications
B_2O_3	Boric oxide (anhydrous boric acid)	Glass and ceramics, metallurgy (flux)
H_3BO_3	Boric acid	Glass, ceramics, frits, enamels, fire retardants, metallurgy, corrosion inhibition, adhesives, personal care products
$Na_2O \cdot 2B_2O_3 \cdot 5H_2O$	Disodium tetraborate pentahydrate (borax pentahydrate)	Glasses, glass fibers, glazes, enamels, fire retardants, detergents, adhesives, corrosion inhibition, agriculture
$Na_2O \cdot 2B_2O_3 \cdot 10H_2O$	Disodium tetraborate decahydrate (borax decahydrate)	Soaps, detergents, personal care products, wire drawing, corrosion inhibition
$Na_2O \cdot 2B_2O_3$	Disodium tetraborate (anhydrous borax)	Glasses, frits, glazes, enamels, metallurgy
$Na_2O \cdot 4B_2O_3 \cdot 4H_2O$	Disodium octaborate tetrahydrate	Fire retardant, corrosion inhibitor, clearing products, agriculture
$Na_2O \cdot B_2O_3 \cdot 8H_2O$ ($NaBO_2 \cdot 4H_2O$)	Sodium metaborate tetrahydrate	Corrosion inhibitor, clearing products, photographic developer, adhesives, textile processing
$Na_2O \cdot B_2O_3 \cdot 4H_2O$ ($NaBO_2 \cdot 2H_2O$)	Sodium metaborate dihydrate	
$K_2O \cdot 2B_2O_3 \cdot 4H_2O$	Dipotassium tetraborate tetrahydrate	Lubricating oil additive, welding, soldering, and brazing fluxes, buffering agent
$(NH_4)_2O \cdot 5B_2O_3 \cdot 8H_2O$ ($NH_4B_5O_8 \cdot 4H_2O$)	Ammonium pentaborate tetrahydrate	Electrolytic capacitors, corrosion inhibition, fire retardant, welding, soldering, and brazing fluxes
$NaBO_3 \cdot H_2O$	Sodium perborate monohydrate	Oxidative beach (detergents)
$NaBO_3 \cdot 4H_2O$	Sodium perborate tetrahydrate	
$2ZnO \cdot 3B_2O_3 \cdot 3.5H_2O$	Zinc borate	Fire retardant, wood preservative
$Na_2O \cdot 2CaO \cdot 5B_2O_3 \cdot 16H_2O$	Ulexite (mineral borate)	Glass and glass fibers
$2CaO \cdot 3B_2O_3 \cdot 5H_2O$	Colemanite (mineral borate)	

TABLE 17.2
Water Solubility of Borates (wt%)

	20°C	40°C	60°C	80°C	100°C
H_3BO_3	4.72	8.08	12.97	19.10	27.53
$Na_2O \cdot 2B_2O_3 \cdot 5H_2O$/ $Na_2O \cdot 2B_2O_3 \cdot 10H_2O$[a]	2.58	6.00	16.40	23.38[b]	34.63[b]
$Na_2O \cdot B_2O_3 \cdot 8H_2O$[a]	20.0	27.9	38.3[c]	43.7[c]	52.4[c]
$Na_2O \cdot 4B_2O_3 \cdot 4H_2O$	9.5	27.8	35.0	—	—
$NaBO_3 \cdot H_2O$	1.5	—	—	—	—
$NaBO_3 \cdot 4H_2O$	2.3	—	—	—	—
$Na_2O \cdot 2CaO \cdot 5B_2O_3 \cdot 16H_2O$	0.5	—	—	—	—
$2ZnO \cdot 3B_2O_3 \cdot 3.5H_2O$	0.28	—	—	—	—

[a] Wt% anhydrous salt.
[b] Transition point from borax deca- to pentahydrate.
[c] Transition point from tetra- to dihydrate.

The marked temperature dependence of borate solubility may be useful in certain oilfield chemical applications. It is also notable that water solubility varies with the ratio of *m*/*n* (the compound's molar content of M_2O/B_2O_3); with solubility maxima corresponding approximately with the compounds disodium octaborate ($m/n = 0.25$) and sodium metaborate ($m/n = 1.0$) and with a minimum that corresponds with disodium tetraborate (borax) ($m/n = 0.5$) [13,14].

17.1.2.4 Borate Solution Chemistry and Behavior

When borates are dissolved in water at concentrations up to 0.1 mol dm^{-3} (with respect to boron),* then a mixture of the boric acid molecule and the monomeric tetrahydroxy borate ion exist in equilibrium (see Figure 17.2).

At higher concentrations polyborate ions can form from the condensation of these monomeric building blocks [15]. These can influence several solution phenomena, but for most known oilfield chemical applications these higher-concentration conditions are irrelevant. In the case of disodium tetraborate (borax) pentahydrate, dissolving in water results in the release of equimolar amounts of these two species.

$$Na_2O \cdot 2B_2O_3 \cdot 5H_2O \;\rightarrow\; 2Na^+ + 2H_3BO_3 + 2B(OH)_4^-$$

Borax forms mildly alkaline solutions and is an excellent pH buffer. This is due to the weak Lewis acidity of boric acid and its underlying acid dissociation reaction:

$$H_3BO_3 + H_2O \;\Leftrightarrow\; B(OH)_4^- + H^+$$

$$K_a = \frac{[B(OH)_4^-]\,[H^+]}{[H_3BO_3]}; \quad pK_a = 9.24 \text{ at } 20°C \text{ [16]}$$

Since a solution of borax can be considered the same as that of boric acid that has been half-neutralized with a strong base, then under these conditions the pH of the solution equals the pK_a of the parent acid, that is, approximately 9.2. Moreover, the 1:2 mole ratio of Na_2O to B_2O_3 for borax is close to that of the "isohydric point," a consequence of which is that the pH of borax solutions varies very little with concentration [17].

If a more alkaline borate is required (e.g., pH 10–11), then sodium metaborate is an obvious choice. In this case boric acid can be considered to be fully neutralized with an equivalent amount of strong base (NaOH) and when dissolved in water the boron exists almost entirely in the form of the tetrahydroxy borate anion:

$$Na_2O \cdot B_2O_3 \cdot 8H_2O + H_2O \;\rightarrow\rightarrow\; 2Na^+ + 2B(OH)_4^-$$

In fact, both borax, but more particularly sodium metaborate, have recently been proposed for use in chemical EOR as alkaline agents and the robust pH buffering property of borax is possibly responsible for its apparently unexpectedly good performance in laboratory trials (see Section 17.3.2).

A useful property of the tetrahydroxyborate ion in solution is its ability to form soluble ion-pair complexes with cations, notably the "hardness" ions, calcium and magnesium (M^{2+}) [18]:

$$M^{2+} + B(OH)_4^- \;\Leftrightarrow\; MB(OH)_4^+$$

* Equivalent to 10 g/L borax decahydrate.

It has been suggested that this property may obviate the need to use softened water in chemical EOR processes based on ASP flooding in high-water-hardness carbonate reservoirs [19]. The use of conventional strong alkalis can result in precipitation with hardness ions, leading to loss of alkali and scaling problems, unless soft injection water is employed [20].

17.1.2.5 Borate Ester Formation

Boric acid and the tetrahydroxy borate anion are capable of forming organic esters, for example, with alcohols and carboxylic acids [21]. These contain B–O–C bonds, as illustrated below for the reaction between boric acid and primary alcohols (ROH):

$$H_3BO_3 + 3R(OH) \Leftrightarrow B(OR)_3 + 3H_2O$$

This reaction involves the elimination of water. The reverse hydrolysis reaction occurs quite readily, meaning in effect that in aqueous solution the ester exists in equilibrium with free borate (the tetrahydroxy borate anion, the active ester-forming species in aqueous solution). In water this susceptibility to hydrolysis means that with monohydric alcohols the equilibrium is much to the left and virtually no ester is formed [22]. However, with diols, polyols, and polyhydroxycarboxylic acids, significant amounts of the esters can exist in aqueous solution in equilibrium with un-complexed borate. Here the borate ion can form more stable bi-dentate monoesters and even tetra-dentate diesters with these compounds, containing five- or six-member rings, depending on whether it reacts with a 1,2 or 1,3 diol (Figure 17.3).

This reaction is heavily exploited in several oilfield chemical applications (Sections 17.2.1.1, 17.2.2, and 17.3.3). Of particular importance is the formation of tetra-dentate borate diesters that can be used to cross-link neighboring polyol-rich polymers in aqueous solution and promote an enhancement in solution viscosity and gelation.

17.1.2.6 Perborate

The ability of borate to form solid-stable compounds with the oxidizing agent hydrogen peroxide formed the basis for bleach technology employed in domestic laundry detergents for most of the twentieth century. Such compounds, referred to as perborates (the sodium perborates are listed in Table 17.1), rely for their excellent stability on the existence of peroxy di-borate anions (Figure 17.4) [23].

Sodium perborate dissolves in water liberating hydrogen peroxide and the tetrahydroxyborate anion:

$$2NaBO_3 \cdot H_2O \left(\text{or } Na_2B_2O_4(OH)_4\right) + 4H_2O \rightarrow 2Na^+ + 2B(OH)_4^- + 2H_2O_2$$

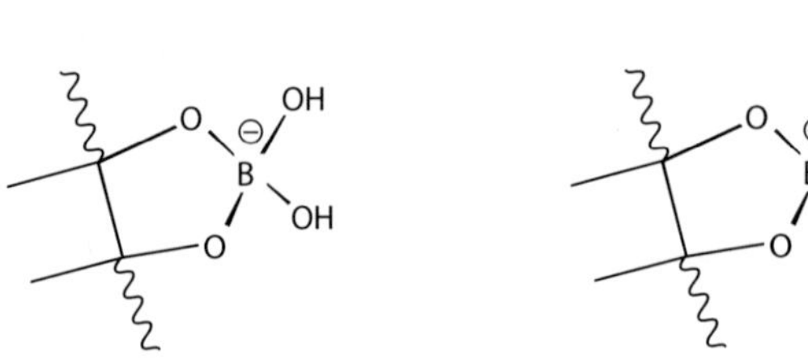

FIGURE 17.3 Borate esters formed with diols in aqueous solution.

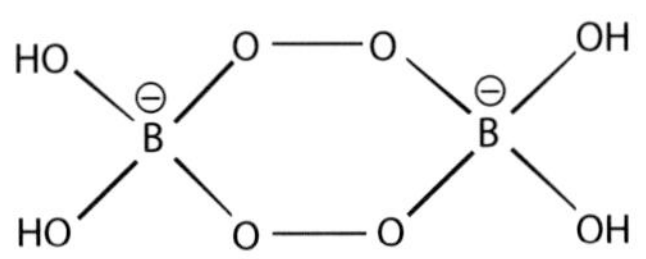

FIGURE 17.4 The peroxy di-borate anion present in the sodium perborate crystal.

Its solutions are naturally alkaline (pH 10–10.5) and these conditions promote oxidation reactions with hydrogen peroxide, especially those involving the nucleophilic attack by the perhydroxyl anion, HOO^- [24]. With the use of an appropriate metal or nonmetal catalyst, free radical oxidation and other redox oxidation pathways are possible with this oxidant. Perborates find application in at least one oilfield chemical activity, namely as a so-called oxidative polymer breakers for spent fracturing fluids (see 17.2.1.2).

17.1.2.7 Sparingly Soluble Borates

Finally, in this brief introduction to borate chemistry, it is necessary to return to Tables 17.1 and 17.2 and recognize the existence of sparingly soluble borates, notably zinc, calcium, magnesium, and sodium–calcium borates. Several of these exist in the form of upgraded minerals, rather than refined (recrystallized) compounds. Certain oilfield chemical applications exist, which demand borates that dissolve slowly and mineral borates such as colemanite and ulexite have been employed for this reason. Related to this topic is the *in situ* formation of sparingly soluble borate salts. Although it was earlier stated that borates have a good tolerance toward calcium and magnesium ions, there exist certain conditions (pH, ion concentration) where borates precipitate out of solution and are believed to form protective films on crystal or mineral surfaces in contact with the solution. These mostly amorphous films are believed to inhibit crystal or mineral hydration reactions (possibly acting as semipermeable membranes). Oil well cement set retardation is one existing oilfield chemical application for borates that is probably already exploiting this effect and there is scope to extend this to other uses.

CORE CONCEPTS

The element boron forms strong bonds with oxygen. The resultant ***borates*** include ***boric acid***, ***boric oxide***, alkali, ammonium, and alkaline earth metal borates. All except for the alkaline earth borates are reasonably water soluble. When dissolved borates release ***boric acid***, a Lewis acid, and its conjugate base, the ***tetrahydroxyborate anion. Borax,*** a form of ***di-sodium tetraborate***, releases equimolar amounts of these species and consequently is an excellent mildly alkaline pH buffer. Borates exhibit several important behaviors in solution, including the sequestration of water hardness ions, the ability to form borate esters with polyols, and to deliver the oxidant hydrogen peroxide into solution from the dissolution of ***sodium perborate***.

17.2 MAJOR EXISTING OILFIELD CHEMICAL APPLICATIONS FOR BORATES

The main existing uses of borates by the oil and gas exploration and production industry are summarized in Table 17.3.

Each of these will be described in more detail, along with the underlying borate chemical (or nuclear) property exploited in the application.

TABLE 17.3
Major Existing Oilfield Chemical Applications for Borates

Activity	Function	Borate Property
Hydraulic fracturing	Fluid gelation	Borate ester formation (polymer cross-linking)
	Polymer breaker	Oxidizing function of perborate
Lost circulation prevention	Fluid gelation	Borate ester formation (polymer cross-linking)
Completion and workover	"Nondamaging" fluid additive	Sparingly soluble borates
Oil well cementing	Cement set retardant	Sparingly soluble borates
Reservoir logging	Chemical tracer	High neutron capture cross section

17.2.1 Hydraulic Fracturing Fluids

Hydraulic fracturing is one of the two so-called reservoir stimulation techniques commonly used to improve the permeability of reservoirs and increase fluid flow and in turn oil and gas production rates; the other method being acid stimulation or acidizing. It is the fastest-growing sector within oilfield chemicals and demand is dominated by the US market. It also represents the largest single use of borates in the upstream oil industry. Growth in recent years has been fueled by the desire to treat gas as well as oil fields, by high recent and forecast oil prices and by the fact that there are many aging fields in the United States that are considered as good candidates for such stimulation. Both stimulation and EOR treatments are favored in the current climate, where it is considered easier to focus on improving yields and recovery rates from existing fields rather than explore for and develop new fields as these are mostly now only to be found in geographically and politically challenging places. Unlike EOR flooding projects that can take several years to complete and represent a significant investment hydraulic fracturing is quickly implemented and conducted on a much smaller scale. Hydraulic fracturing is also much more widely practiced with an estimated 30,000 operations conducted annually across the world.

In hydraulic fracturing operations water-based fluids containing suspended fine particles are injected into the reservoir and pressurized to fracture and create cracks and fissures in the reservoir rock and improve its permeability toward fluids. The particles, usually sand, bauxite, or ceramic based, and referred to as "proppants," are transported into the fractures and prop them open when the pressure is released. A typical operation involves the use of up to 1,000,000 gallons of fluid and very powerful pumps capable of achieving fluid pumping rates in the range 1000–5000 gallons per minute. Borates are used in two phases of the hydraulic fracturing process, the first of these representing the much larger application:

- Fluid gelation—addition of borate as a polymer solution cross-linker and gel strengthening agent in the fracturing fluid.
- Fluid clean-up (fluid breaker)—use of perborate as an oxidizing agent to chemically break down the polymer fluid residues.

17.2.1.1 Fluid Gelation

Fracture fluids fall into several distinct categories and their selection depends on the operating conditions in the reservoir:

- Low-viscosity fluids—high-rate water fracs (HRWFs)—for low-permeability ("tight") reservoirs frequently associated with gas production.
- Linear gels—polymer viscosified (but not exceptionally shear-thinning) water-based formulations.

- Cross-linked gels—having higher low-shear viscosities and improved proppant suspension compared with linear gels, often at lower polymer loadings; these high-performance fluids are used in deeper and hotter reservoirs.
- "Hybrids"—many shale treatments are HRWFs followed by a linear or cross-linked gel.
- Surfactant-stabilized foams.
- Viscoelastic surfactant solutions.

In practice, around 30% of fracturing operations use cross-linked gels and 50% of these in turn use borates and boron compounds to cross-link the polymer and provide the necessary rheological properties.

17.2.1.1.1 Cross-Linking with Borates: Conventional Approaches

The tetrahydroxy monoborate ion $B(OH)_4^-$ is able to form intra- or interchain cross-links between molecules of certain water-soluble polymers in aqueous solution; resulting in increased viscosity (particularly at low shear), gel strength and an improved particle suspending property. For this to happen, the polymers are required to have adjacent (vicinal 1,2 or 1,3) and correctly orientated (*cis*) hydroxyl groups to interact with the borate ion. As discussed previously (Section 17.1.2), the interaction takes the form of a borate ester and the relevant chemical equilibria are illustrated in Figure 17.5.

The polymer is represented as a diol, but in reality it would be a polyol with many hydroxyl groups potentially available on the polymer chain to complex with the borate ion.

The first step—the liberation of the borate ion—is governed by the dissolution and dissociation of boric acid, a weak Lewis acid (p*K*a in the range 9.0–9.2, depending on temperature), as previously discussed. Thus, the amount of monoborate ion available for involvement in the subsequent complexation reactions with the polyol polymer is governed, by, among other factors, solution pH; the more alkaline the solution, the more the borate is available. The second step is the formation of a bi-dentate monoester (1:1) complex (BL^-) between the borate ion and the polyol polymer, and the final step, giving rise to the cross-linking reaction, is the formation of a 2:1 polyol:borate ion tetra-dentate diester complex (BL_2^-). Interchain cross-links of this type enhance gel structure and the gel's

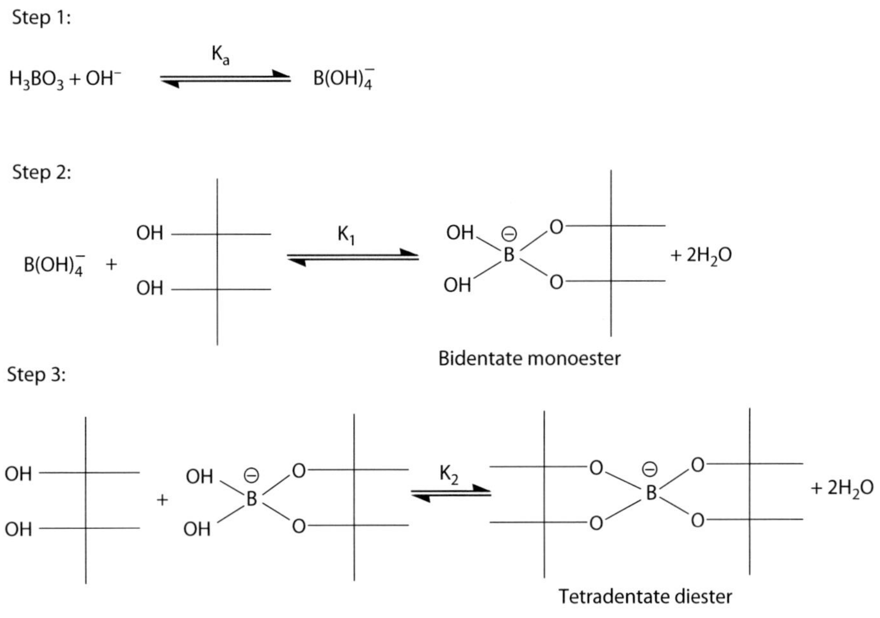

Step 1:

$$H_3BO_3 + OH^- \overset{K_a}{\rightleftharpoons} B(OH)_4^-$$

Step 2:

Step 3:

FIGURE 17.5 Chemical equilibria giving rise to the borate cross-linking reaction with polyols.

elastic modulus; a dynamic rheological property measured using a rheometer in oscillatory mode is closely related to the interchain cross-link and entanglement density [25].

A variety of synthetic, semisynthetic, and natural polyol polymers are capable of being cross-linked with borate. Among the synthetics, PVA is a good example that interacts very strongly with borates and even boric acid. For polysaccharide-based polymers (derivatives of cellulose and various natural gums), the strength of interaction depends much on the conformation of the available monosaccharide hydroxyl groups in the polymer backbone or side chains. The monosaccharide units in these polymers exist in the cyclic hemiacetal (pyranose) form with their hydroxyl groups present in either an axial or equatorial configuration. For borate to complex with adjacent hydroxyl groups they must have axial–equatorial orientation with respect to each other; ring strain in the cyclic ester is otherwise too great. The structure of glucose, mannose, and galactose, the monosaccharide building blocks for most of these polymers, is illustrated in Table 17.4, along with the pairs of sites favorable for interaction with borate. It is noteworthy that as well as bi-dentate monoester complexes with 1,2 diols that give rise to five-member rings, borate does complex with 1,3 diols to produce six-member rings as is the case with the C4,6 positions in these monosaccharides.

Natural guar gum, and its carboxymethyl hydroxypropyl derivative (CMHPG), is the workhorse polymer used in linear or cross-linked gels for hydraulic fracturing. Derived from the guar bean it is classed as a galactomannan, consisting of a backbone of mannose units with branching galactose units. The guar structure is shown in Figure 17.6a. Consistent with the complexing behavior of borates with mannose and galactose monosaccharides in Table 17.4, the sites in the guar polymer that favor interaction with borate are the C2,3 positions in the mannose backbone units (some of these may be blocked by hydroxypropyl groups in the case of the guar derivative) and the C3,4 and C4,6 positions in the galactose branch units (Figure 17.6) [6]. Due to the C1,4 mannose backbone linkage, borate cannot form an ester with the C4,6 diol pair of carbons in the polymer-bound mannose units, even though it can with the mannose monomer. For similar reasons, C1,2 diol ester formation with polymer-bound galactose side chains is prevented in the polymer, although it is possible with monomeric galactose.

Figure 17.6b shows the structure of a bi-dentate monoester formed between a borate ion and the galactose side chain of the guar gum molecule (C3,4). If the boron center in this ester then complexes

TABLE 17.4
Monosaccharides and Favored Sites for Complexation with Borate

Glycoside/Monosaccharide	Preferred Sites of Borate Esterification
α-D-glucopyranose	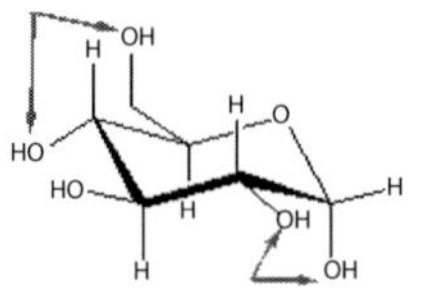 C-1/C-2 C-4/C-6
α-D-mannopyranose	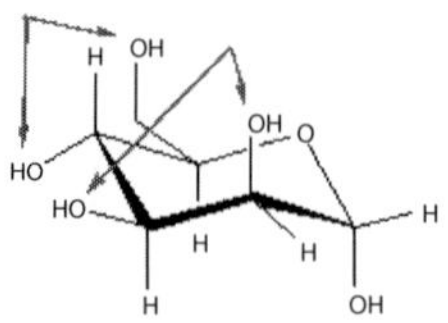 C-2/C-3 C-4/C-6
α-D-galactopyranose	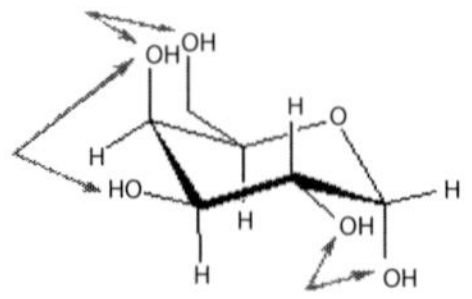 C-1/C-2 C-3/C-4 C-4/C-6

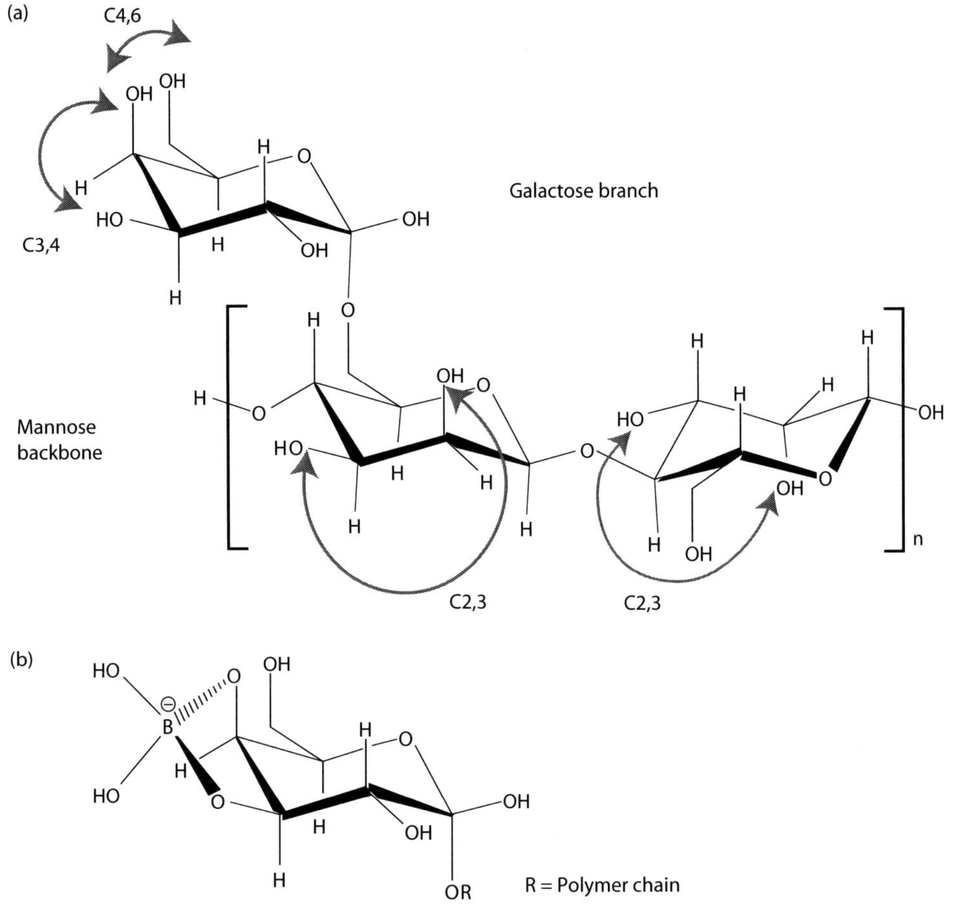

FIGURE 17.6 (a) Chemical structure of guar gum and preferred sites for complexing with borate and (b) bi-dentate monoester formed between galactose side chain (C3,4) on a guar gum molecule and borate ion.

in the same way with a galactose (or mannose group) in a neighboring polymer molecule then a tetra-dentate diester is formed and acts to cross-link the two molecules.

^{11}B NMR is a useful technique for detecting the formation of these borate ester complexes [26]. The boric acid/mono-borate exchange process (step 1, Figure 17.5) is more rapid than the acquisition rate for the technique and free or uncomplexed borate shows up as a single peak in the NMR spectrum; its position dependent on the ratio of 3- and 4-coordinate boron and therefore pH. However, the exchange with the borate ester complexes is sufficiently slow to enable the boron centers in these environments to appear as separate signals, whose position is invariant with pH. Even with this technique, however, it has proved quite difficult to detect the tetradentate diester borate ester complex thought to be responsible for cross-linking and for this reason some researchers have proposed that cross-linking might arise from an ionic association between the monoester BL^- sites on one chain and metal ions (e.g., calcium, sodium) adsorbed on the neighboring chain [6]. It is certainly true that there are plenty of sites—an estimated 12,000—on each guar molecule (molecular weight of 2 million) for cross-linking [27].

When first employed in this application in the 1960s, well-fracturing operations were conducted typically at relatively modest temperatures (up to 150°F/65°C) [28]. However, as reservoir conditions became harsher (deeper wells), it was discovered that the traditional application of borates was limited because at higher temperatures it was observed that borates had reduced ability to gel the polymer and maintain the necessary viscosity for proppant suspension. This effect can be explained

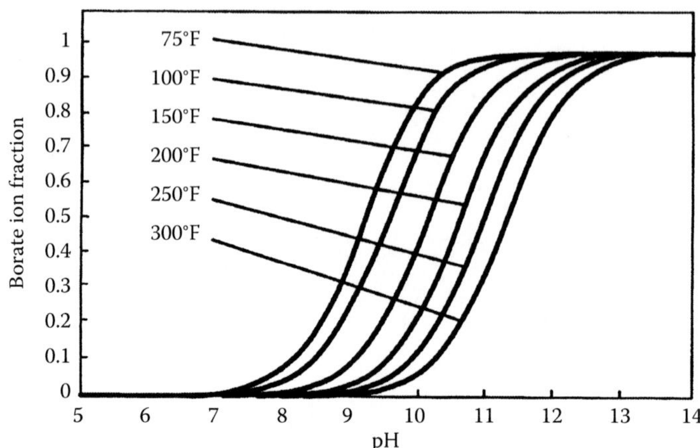

FIGURE 17.7 Fraction of boron present as mono-borate ion in solution as a function of pH and temperature (7.2 lb/Mgal boric acid in 2% KCl). (Adapted from Harris, P C; *Journal of Petroleum Technology*, pp. 264–269, March 1993 (SPE 24339). Copyright 1993 by SPE.)

primarily by the influence of temperature on the equilibrium between boric acid and the mono-borate ion, step 1 in Figure 17.5, according to Mesmer et al. [29]. In fact, the acid dissociation constant for boric acid increases with temperature, resulting in a reduction of the amount of boron in the form of the mono-borate ion for a fixed total boron solution concentration. Using Mesmer's data the influence of pH and temperature on the fraction of boron in the form of the mono-borate ion in solution has been calculated by Harris and is illustrated in Figure 17.7 [30]. This in turn influences the other equilibria in Figure 17.5, and consequently leads to a reduction in the amount of the mono- and diester complexes that form as temperature increases and a diminution in the gelling power of borates.

In order to counter this effect and to achieve the necessary gel rheology for proppant suspension above 200°F (93°C), and even as high as 300°F (149°C), the pH range for borate gels has been extended from the original levels (8.5–9.5) to much higher ones (up to pH 12). Harris has also shown that as well as displacing upward the temperature at which the maximum viscosity is obtained from the borate cross-linking effect, an increase in pH actually serves to delay the onset of cross-linking [30]. This is illustrated in Figure 17.8.

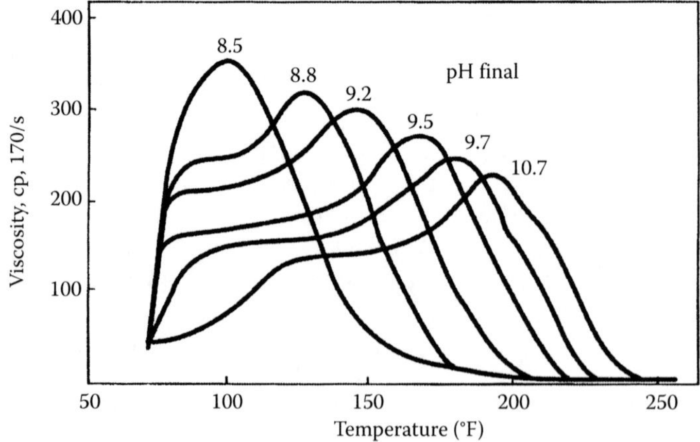

FIGURE 17.8 Effect of pH and temperature on the viscosity of borate cross-linked guar gum gels (30 lb/Mgal). (Adapted from Harris, P C; *Journal of Petroleum Technology*, pp. 264–269, March 1993 (SPE 24339). Copyright 1993 by SPE.)

Although increasing the pH can delay the onset of peak viscosity, it is nevertheless the case that a gel that is cross-linked to the maximum extent at the target reservoir temperature will be over cross-linked at surface temperatures. A fluid that is cross-linked to the correct level at reservoir temperatures could easily experience a fivefold increase in the concentration of the mono-borate ion at surface temperatures and this can result in the so-called gel syneresis, the formation of a densely cross-linked gel phase in contact with excess water. Although this process is reversible due to the labile nature of the borate ester bonds, an over-cross-linked gel can still cause high-friction pressures during initial pumping operations.

Another important consideration when targeting a higher solution pH for high-temperature cross-linking is the source of water used to prepare the cross-linked fluids and its composition. Seawater and brackish inland water contain high levels of ions, among which is magnesium, and this effectively resists attempts to increase the pH of the solution by addition of a strong base (NaOH). In order to adjust solution pH to a desired value (e.g., 10.7 for application at or even above 174°F/79°C), the magnesium ion must be precipitated out of solution as $Mg(OH)_2$ by the use of sacrificial amounts of NaOH. Until this has been done, then the pH of the solution is effectively buffered at a lower pH [31]. However, this does not prevent the successful formulation of borate cross-linked gels in seawater and brackish water [32,33].

In the 1990s, considerable effort was expended in corporate laboratories to improve on the performance of borate cross-linkers in this application, notably to overcome their deficiency at high temperatures. The dual objective was to develop systems that operated at a higher pH but which were not consequently over cross-linked at surface temperatures. The technical solutions that emerged were based on delayed chemical release mechanisms, either for the borate source, the separately added alkali, or both.

17.2.1.1.2 Cross-Linking with Borates: Advances Involving Delayed Borate Release

One of the earliest approaches to slowing the release of the borate was simply to use conventional borates but with a coarser particle size [34]. More recently, alternative approaches have proved more successful.

17.2.1.1.3 Use of Sparingly Soluble or Slowly Dissolving Borates

Certain calcium or mixed calcium sodium mineral borates, for example, colemanite, $2CaO{\cdot}3B_2O_3{\cdot}5H_2O$, and ulexite, $Na_2O{\cdot}2CaO{\cdot}5B_2O_3{\cdot}16H_2O$ have been proposed for use as they are only sparingly soluble in water, at least at surface temperatures. They can be used in the form of aqueous or nonaqueous (e.g., diesel, glycol) suspensions [35–38]. Another option has proved to be the use of slowly dissolving refined anhydrous borates such as anhydrous borax, $Na_2O \cdot 2B_2O_3$ and boric oxide, B_2O_3 [39–42].

17.2.1.1.4 Borate Complexation

This idea involves the use of low-molecular-weight polyols that are capable of forming ester complexes with the borate ion [43–48]. Examples include glyoxal, gluconate, glucoheptonate, sorbitol, and mannitol. In effect, the complexing mechanism is the same as for the guar gum and the two compete for the borate. The low-molecular-weight polyols are often referred to as cross-linker chelants and are claimed not only to control the release of borate, when precombined with it, but also to act as stabilizers regulating fluid gelation and viscosity [43]. In addition, they help to solubilize the borate and enable the development of concentrated liquid borate products [49]. In one example of the use of the polyol–borate complex "organo-complexed-borate cross-linker" colloidal particles are formed by reaction between borate and the chelant [50]. The resulting particles contain surface borate ions that are complexed with the chelant molecules. These can be displaced by the gaur chains that form multiple attachments on the particle surface which act as cross-link points. The cross-link delay time can be increased by raising the pH as this strengthens the complex between the chelant and borate. Related to these chelants are other types of borate ester, all of which are susceptible to hydrolysis and which will deliver borate into a solution at varying rates. They can be supplied as nonaqueous solutions, presumably further slowing borate release [51]. Finally, a preformed complex between the borate and guar gum itself, or starch (also capable of complexing with borate) can apparently be used as a means of delaying borate release and preventing premature hydration of the gum [52–54].

17.2.1.1.5 Cross-Linking with Borates: Advances Involving Delayed Alkali Release

Sodium hydroxide was the traditional alkali system to be employed to raise the pH of fracture fluids and promote cross-linking with borate. However, it is not easy to delay the release of this highly water-soluble alkali and slower release systems were targeted instead. Chief among these was magnesium oxide. By altering the degree of calcination, it is possible to control the rate of dissolution/alkali release with this material [42]. The dissolution reaction is simply

$$MgO + H_2O \Leftrightarrow Mg^{2+} + 2(OH)^-$$

This is effective up to 150°F (65°C), but at higher temperatures, the magnesium ion (whatever the source, from the alkali or make-up water) precipitates out the hydroxide ion and prevents the attainment of the desired pH:

$$Mg^{2+} + 2(OH)^- \rightarrow Mg(OH)_2 \downarrow$$

In order to reach higher temperatures with this alkali, it is necessary either to sequester the magnesium, which is not normally effective or economically viable, or to precipitate it out, this being the preferred option. This can be achieved by adding hydroxide, as was previously mentioned [31,32], or instead by adding another anion whose magnesium salts are even less soluble than magnesium hydroxide, for example, fluoride in the form of its potassium salt [42,55]. In this way, and when combined with the use of the appropriate borate source, cross-linked fresh water gels can be successfully formulated to work at temperatures up to 275°F (135°C). A common form of delivering the magnesium oxide is as slurry in diesel. Gel delay times are shown to correlate well with the amount of this alkali slurry employed [56]. Performance data on other alkali- or pH-buffered cross-linked activators used with borate cross-linkers have been published. They may or may not be based on magnesium hydroxide, but the objective is always to reduce the pH decline encountered with the NaOH–borate system with increasing temperature [57]. In practice, gum hydration should not occur under alkaline conditions, whereas subsequent cross-linking does require a high alkalinity. Therefore, the use of a slowly dissolving base and in some cases an additional acid buffer can provide the necessary pH profile [58].

Normally, the borate and alkaline agent (pH buffer or cross-link activator) are added as separate products during the preparation of the cross-linked gel. However, it has been proposed to combine these into a single product [59,60]. This product benefits from the use of a base (monoethanolamine) that not only provides the required pH buffering, but that also complexes with boric acid and dramatically increases its water solubility. In this way, the product is a highly concentrated source of borate in liquid form. Similar formulations have been used for many years now in agricultural applications involving the use of borates as trace micronutrients. Borates are highly soluble in glycols and unsurprisingly these also feature in the formulation of these products. Other examples of products where the borate and alkaline cross-link activator are coformulated exist [37]. Even multicomponent packages containing all the necessary ingredients, in dry form and suitably encapsulated for timed release (gum, borate, alkali plus magnesium ion precipitating agent), have been proposed [61].

In the mid-1990s, there was a move to develop cross-linked guar gels that contain lower polymer concentrations. Polymer residues from the fluid remain even after cleanup processes have been employed. The problem is more noticeable with higher polymer concentration gels and this reservoir damage can lead to a loss of permeability, somewhat negating the gains from the fracturing process [57,60,62]. Previously, cross-linked gels contained between 40 and 60 lb/Mgal of guar gum (4.8–7.2 g/L), but levels have now dropped as low as 20 lb/Mgal, only just above the polymer overlap concentration of 18 lb/Mgal (2.2 g/L). The use of borates as cross-linkers and improvements in the borate systems available have helped achieve this reduction in polymer loadings. Borate levels are substantially lower than that of the gum, and today are in the range 0.5–4.2 lb/Mgal (0.06–0.5 g/L). Achieving further reduction in polymer loading for borate cross-linked gels requires a slightly different approach; for example, the creation of nitrogen foams from these systems or in rare cases the inclusion of emulsified oil with the cross-linked gel remaining as the external phase [63].

17.2.1.1.6 Competing Cross-Linking Systems

What distinguishes borate cross-linking from other cross-linking chemistries is the labile and reversible nature of the cross-linking bonds. The gels can reheal after breakdown under high-shear conditions experienced during pumping. This is not the case with rival zirconate- or titanate-based systems [26,42,64]. This is an extremely useful property and offsets the poorer performance of borate systems at a high temperature. It is believed that the lifetime of the borate–polymer ester bond is around 1 ms; that the enthalpy for the cross-linking reaction is –20 kJ/mol; and the activation energy for bond breaking is around 24 kJ/mol [30,65]. The Arhhenius activation energy for cross-link bond formation (based on gelation times) has been separately estimated as 53 kJ/mol [66]. These values support the idea of labile bonds. Oscillatory rheological measurements have also demonstrated the existence of a single relaxation process (Maxwell behavior) associated with the borate cross-links [67].

Borate cross-linked fracture fluids are also believed to cause less damage to the reservoir and less likely to impair permeability than rival cross-linkers [26,50,68]. This is partly due to the fact that borate cross-links can be broken down after fracturing simply by reducing pH. That is not to say that chemical (oxidative) or enzymatic means for effecting cleanup of the reservoir are not required to break down the polymer chains and flush away the fluid residues, but this process is more effective with borates because of the reversible nature of the cross-link bond. Some metal ion cross-linked gels have poor cleanup properties and soluble precipitates can be formed when they react with certain chemical "breakers."

In addition to these advantages, borates are considered benign to the environment and safe to handle.

Despite the recent improvements in their high-temperature performance and the aforementioned advantages, and occasional use at bottom-hole static temperatures above 300°F (150°C), borates are usually employed alone below 200°F (93°C). Instead transition metal cross-linkers are used for higher-temperature wells. Some combined (borate plus transition metal ion) cross-link systems are claimed to be useful where it is necessary to provide a boost to the viscosity of borate cross-linked fluids at the high-temperature end where the viscosity otherwise tends to decline [69–71]. Boron–zirconium chelates are examples of this [72]. Recent studies have also demonstrated that the viscosity of borate cross-linked gels is adversely affected by high pressure as well as high temperature, which is apparently not the case with gels cross-linked with zirconium (IV) or titanium (IV) ions [73].

In summary, borates have recently undergone a renaissance in terms of their use in cross-linked gels for hydraulic fracturing operations. They remain the cross-linker of choice in the industry. Many laboratories continue to work on the further refinement of borate-based products and technology for use in this application, recognizing the fundamental advantages of borates.

17.2.1.2 Fluid Cleanup (Fluid Breaker)

At the end of the hydraulic fracturing process, and when gels are employed, these must be destroyed to allow fluids to flow freely through the freshly fractured zones and benefit from the permeability improvements achieved by fracturing. Chemical or enzymatic gel "breakers" are employed for this purpose. Traditionally, enzymatic breakers operate under mildly acidic conditions at low temperatures, while oxidizing chemical breakers operate at moderate-to-high temperatures, either alone, or with catalysts, if the temperature is below 50°C. Breakers are required to break down polymer chains and reduce the viscosity of polymer solutions not only in fracturing operations but also in other reservoir processes that employ such fluids, notably gravel packing, reservoir permeability control, and occasionally drilling. Guar gum and HPG are the polymers mostly employed in fracturing operations but in other applications cellulosics and polyacrylamides can be employed.

Sodium perborate has been used as an oxidizing chemical breaker, although in this category its use is much less prevalent than that of, for example, persulfates, hypochlorites, and alkaline earth metal peroxides [74]. When deployed it dissolves in solution to release hydrogen peroxide, the active oxidant. A perceived drawback of sodium perborate, at least when looking for an instantaneous rather than slow release effect, is its low solubility in water (2.3% at 20°C). One proposed solution to this problem is its use with low-molecular-weight polyols that are known to enhance perborate solubility by virtue of the same borate ester complexing reaction that is exploited in some borate cross-link gel delay systems [75]. Isotropic aqueous liquid detergent formulations containing high levels of dissolved sodium perborate as the bleaching agent have been formulated using the same principle. They contain polyhydroxycarboxylate salts that enhance perborate solubility and benefit also from the hardness ion sequestering power of the complex that forms [76].

Sometimes rather than having an instantly available breaker it is desirable to develop one that is encapsulated or coated with a membrane material that delays its release into solution. Another way of slowing breaker release is to employ a granule. A means of achieving this with perborate is to make a pellet or prill from granules of a sparingly soluble perborate salt, for example, calcium perborate [77].

As was stated earlier, borate cross-linked gels are generally found to be easier to break down and cause less damage to the formation compared with transition metal cross-linked gels, although there have been isolated reports of damage [78,79]. Certainly, the ability to dissociate borate ester cross-links by simply reducing pH can help reduce polymer solution viscosity and facilitate clean-up following fracturing and gravel packing. Controlled acid release systems based on hydrolyzable esters are a means of achieving this [80]. Some literature specifically proposes the use of hydrolyzable esters with perborate breakers [81].

CORE CONCEPTS

The ability of borates to form borate ester complexes with polyol-based polymers, for example, guar gum and its derivatives, leads to their use as ***cross-linkers*** for gels used in ***hydraulic fracturing***, an operation used to ***stimulate*** improved permeability, fluid flow, and production rates in reservoirs. The gels are required to support ***proppant*** particles that prop open cracks in the rock created by pressurising the gelled fluid. Borates are the most widely used cross-linker and technology advances involving delayed borate and alkali release systems have extended their temperature range in recent years. They have several advantages over rival cross-linkers including the rehealing property of their gels after shear and the fact that they leave fewer residues in the formation after treatment and cleanup. Perborate can be used as an ***oxidative breaker*** to degrade and remove the polymer residues from these gelled fluids.

17.2.2 Lost Circulation Treatments

Polymer gel plugs or pill treatments are often employed to form temporary seals on holes encountered in the well bore that would otherwise permit the loss to the formation of large quantities of costly drilling fluids. They can also be used to divert injected fluids into the required zones in the formation and isolate other regions. They frequently employ the same components used in cross-linked fracture fluid gels, the main difference being that the gels are stronger and contain higher concentrations of the water-soluble gums and borate cross-linker [34,53]. Compared with the quantities of drilling fluids employed in a typical well, plug volumes are considerably smaller; perhaps several hundreds of barrels versus several thousands of drilling fluid.

Fluid loss pills can either be solid-free or can contain dispersed solids that assist in forming a good seal and preventing fluid loss. Solid-free formulations are based on the use of soluble sodium borates along with guar-based polymers [34]. For a suspension product, it is necessary to use a sparingly soluble borate, such as ulexite or colemanite. This type of product, referred to as PBS (Polymer/Borate/Salt) Plug has been reported to perform well in the field [82]. The polymer serves to suspend the particles and the particles in turn dissolve slightly to release borate ions that are capable of cross-linking the gel. The polymers employed are a mixture of xanthan gum, cellulose, and starch; the borate interacting with the terminal mannose sugar cis hydroxyl groups in the repeating trisaccharide side chains of the gum to cause cross-linking. The use of ulexite and anhydrous borax have also been proposed as a means of delaying the development of gelation in these lost circulation material (LCM) fluids until they are in place [83].

17.2.3 Completion and Workover Fluids

It is important to conduct completion operations (cementing, casing perforation, and gravel packing) in fluids that are nondamaging to the formation and that will not, for example, leave filter cake residues that impair permeability. Low- or no-solids fluids (clear brines) are often employed for this reason and any solids chosen as bridging agents coupled with fluid loss control additives should be readily soluble in acid solutions or formation brines. Calcium carbonate has been used for this reason, as have been saturated, sized (milled) salt solutions. Commercial systems based on the use of ulexite suspensions were employed until recently as nondamaging low specific gravity completion and workover fluids [84,85]. The fluids, with specific gravities in the range 1.0–1.2 g/cm^3, were used in operations in older reservoirs where fluid pressures are generally lower. A typical fluid would contain 29 g/L ulexite along with 10 g of a viscosifying and suspending polymer (HEC or CMC) plus fluid loss control additives (pregelatinized starch and calcium lignosulfonate). The base fluid is salt water brine, whose specific gravity can be adjusted by varying the brine concentration. The sparingly soluble borate acts as a bridging agent and can be readily dissolved away afterwards by the connate formation water, injected water or nonsaturated brines (e.g., 3% KCl). The particle size of the sparingly soluble borate is important for their action as bridging agents. Although there are similarities between the composition of these fluids and those of the lost circulation polymer gel plugs discussed in the previous section, the completion fluids are much less viscous. For these products a viscosity-enhancing polymer is required to assist in suspending the ulexite particles, but it is usually selected from the cellulose ethers as these are not gelled by borate ions released from the ulexite. The field performance of ulexite-based completion fluids is reported to be good, specifically in relation to subsequent cleanup, formation damage, and impact on reservoir productivity [86].

17.2.4 Oil Well–Cementing

Borates are used as set retardants in oil well cementing operations. Cements are important materials widely used in oil wells to fix the steel casing to the drilled borehole wall (primary cementing), to plug, seal, or repair holes in the formation and casing or to isolate certain zones (remedial cementing). The technical challenge is to develop cement slurries that remain fluid until pumped in place and then set to form a hardened mass with sufficient compressive strength. Primary cementing operations involve pumping cement slurry down the casing and then upward through the annular space between the casing and the wellbore. The cement slurry sets, providing support for the casing and establishing hydraulic isolation within the wellbore. The cement must remain pumpable throughout its journey as it experiences changing temperatures. In deep wells, bottom-hole circulating temperatures (BHCTs) can reach 600°F (315°C) and under these conditions cement will set very rapidly unless set retarders are included in the formulations. Setting times can be delayed for more than 10 h by the use of retarders. Borates are one of a number of different chemical classes that are used as retarders. Also employed are lignosulfonates, natural gums, and other carbohydrate derivatives, phosphonates (e.g., ethylenediamine-tetramethylene phosphonic acid) and various synthetic ter- and copolymers. Borates employed include boric acid, borax (penta- and decahydrate), sodium and potassium pentaborate, and di-sodium octaborate tetrahydrate. They are always used in concert with other retarders, mainly to extend their useful temperature ranges. The mass efficiency of borates may be related to their B_2O_3 content, but it seems also that the more acidic borate salts (lower ratio of metal oxide to B_2O_3) perform better, meaning that the penta- and octaborates are particularly popular in this application [87]. Typical dose rates for borates are in the range 0.5–4%, based on the weight of cement and depending on the desired setting time, well depth, and temperature. The mechanism by which borates operate to retard setting may be different from that of other retarders, notably the organics. It is believed that, in the case of Ordinary Portland Cements (OPC), where setting is based on the formation of calcium silicate hydrates, borates interact with the calcium present in the hydrating phases to form an amorphous calcium borate layer on the cement grains that retards their hydration [88].

From the earliest reports of borate use in this application it has been commonplace to use borates in combination with other retarders [89]. As advances were made to meet the needs of deeper and higher-temperature wells, synergistic combinations of borates with various organic coretarders were discovered, including their use with lignosulfonates [90–92], 2-acrylamido-2-methylpropane sulfonic acid (AMPS) [93], acrylamide/acrylic acid/sodium vinyl sulfonate terpolymers [94], calcium salt of phosphonic ethylenediamine-N,N,N′,N′-tetrakis (methylene) acid (Dequest®) [95], and CMC and dextrin [96]. A good example of synergy is when borax is used with a sodium calcium lignosulfonate, where the individual additives retard the setting time of a class H OPC cement by around 1 h, but this increased to 12 h when they are used in combination [92].

The majority of oil well cements are based on OPC, with various classes (A–H and J) defined by the American Petroleum Institute (API). Each of the nine classes has slightly different compositions and is suitable for use in different well environments. Borates appear to function as retarders for all these classes. In addition, there are reports that they can also be used to retard the setting of special cement types. These include magnesium oxychloride (Sorel) cements, prepared by mixing magnesium oxide with magnesium chloride solutions. These cements are capable of being dissolved by conventional acids and are used to provide temporary seals and plugs in the wellbore. Borates are used with sugars to retard the setting of this type of cement [97]. Magnesium phosphate-based oil well cements can also employ borate–sugar combination retarders [98]. These strong, quick-setting cements rely on chemistry rather different to that of OPCs. The setting reaction:

$$MgO + KH_2PO_4 + H_2O \rightarrow MgKPO_4 \cdot 6H_2O$$

is apparently retarded by the addition of the borate-sugar retarder. Calcium aluminate cements also employ borates as set regulators (setting times for these cements are very fast) and they are claimed to bring other advantages, specifically reduced carbonation of the cement compared with the action of organic (carboxylic) retarders [99].

An interesting and separate application proposed for borate is as a component of a cement fluid loss additive. Cements require a minimum amount of water for proper hydration and setting and if water from the cement filters into the formation, the setting process can be compromised. Cements therefore contain fluid loss additives. A dried, preformed borate cross-linked polyvinyl alcohol (PVA)–sorbitol complex has been proposed for use as a fluid cement fluid loss additive [100]. This may have wider application in other oil well fluids.

CORE CONCEPTS

Oil well cements are required to fix steel casing to the drilled borehole wall, for down-hole repairs and to isolate zones in the formation. Set retarders are required to prevent premature setting while the cement is being pumped into place through high-temperature zones. In this application, borates are used as set retarders or assists for other retarders, such as lignosulfonates, cellulosic derivatives, and synthetic polymers. They are effective retarders for OPC and other cements.

17.2.5 Reservoir–Logging

The use of borate in this application relies not on a borate chemical effect, but instead on a property of the boron nucleus. It is employed in just one of a number of techniques used to map out, or "log" important producing characteristics of oil and gas reservoirs, namely porosity, water saturation, thickness of the oil and gas-bearing zones, and lithography. Termed *Pulsed Neutron Logging*, this reservoir mapping tool relies on measurable differences existing between the neutron capturing power of the atomic nuclei distributed in the various reservoir materials (notably the fluids–oil, water, and the reservoir rock). Other physical parameters of the formation, such as electrical resistance and naturally occurring radioactivity, are exploited by other logging techniques to obtain similar information.

Pulsed neutron logging is a technique that is generally confined to use in "cased holes" as opposed to "open holes," that is, after the casing is set in the bore hole. It is particularly useful for providing information on porosity and fluid composition and is used to monitor

- The progress of fluids in the reservoir over time during water flooding and tertiary recovery projects
- Unwanted leakage of fluids or gas through channels in producer or injector cement sheaths surrounding well casings
- Fracture porosity in tight formations

Borates are used as tracers in this application primarily because of their exceptional ability to capture thermal neutrons. The relevant nuclear reaction is as follows:

$$^{10}\mathrm{B} + n_{\mathrm{thermal}} \rightarrow \left[{}^{11}\mathrm{B}\right] \rightarrow {}^{4}\ \mathrm{He}\ (\alpha) + {}^{7}\mathrm{Li} + 2.3\mathrm{MeV}$$

The ability of an isotope to absorb thermal neutrons is quantified by its neutron capture cross section, Σ. The ^{10}B isotope has a cross section of 3836 (capture units), considerably larger than most other nuclei, notably those likely to be present naturally in the reservoir (H, Cl, K, Na, C, O)—see Table 17.5, below [101].

In the application of this technique a tool or sonde (probe) is lowered into the cased well. It is capable of generating a burst of high-energy (14 keV) neutrons that pass through the borehole fluid (freshwater), the casing and cement and into the formation. This cloud of neutrons gradually lose energy by collisions with atoms in the fluids and rock until they are in the "thermal" energy range. At this point, they are susceptible to being captured by atomic nuclei, frequently resulting in the emission of gamma rays. A scintillometer in the tool is able to detect these gamma rays and from this can track the decay of the neutron cloud, which obeys the following exponential relationship:

$$N = N_{\mathrm{o}}\mathrm{e}^{(-t/\tau)}$$

TABLE 17.5
Neutron Capture Cross Section of Atomic Nuclei

Atomic Number	Isotope	Natural Abundance	Σ–Capture Units
1	Hydrogen	99.985	0.0032
5	Boron 10	19.78	3836
6	Carbon 12	98.89	0.0034
11	Sodium 23	100	0.400
12	Magnesium 24	78.7	0.052
17	Chlorine 35	75.53	44
19	Potassium 40	0.0018	70
20	Calcium 40	96.95	0.430

where

N = number of gamma rays observed in time t,
N_o = number of gamma rays observed in time $t = 0$,
t = elapsed time (μs),
τ = time constant of the decay process (μs).

The capture cross section, Σ, is then calculated using the following equation:

$$\Sigma = 4550/\tau$$

The value of Σ measured by the logging tool (Σ_{log}) is a function of the individual values of the water (Σ_w) and oil (Σ_o) phases and the matrix rock (Σ_{matrix}), the relative volume fractions of these in the formation and the porosity. If two measurements of Σ_{log} are made, one base measurement, taken after the injection of a dilute brine phase and another after the introduction of a second brine phase with a boron tracer added to increase the value of Σ_w sufficiently to cause a detectable change in Σ_{log}, then it is possible to calculate, for example, the porosity of the formation. Addition of borax at a level of 12 g/L is sufficient to raise the value of Σ_w by 100 units. Achieving this increase with sodium chloride would require the use of very concentrated brines (260,000 ppm). This type of measurement has been used to estimate the pore volume of tight formations where this is mostly attributed to the presence of fractures and that is difficult to quantify by conventional open-hole logging techniques [102].

Apart from having the advantage of an anomalously high neutron capture cross section, boron is available in readily water soluble form as borax. This generally shows good compatibility with dilute formation brines and seawater and consequently the risks of formation damage occurring are low. Ease of handling, its good environmental profile, low cost, and availability are additional benefits.

CORE CONCEPTS

The ^{10}B nucleus has an unusually high affinity for thermal neutrons. This has led to its use as a tracer in fluids injected into the formation with the purpose of logging or mapping the reservoir; determining rock porosity, fluid leakages, monitoring the progress of fluids during water flooding, and EOR processes. Only low-concentration borate solutions are required in ***Pulsed neutron logging*** operations that involves the use of a tool that emits high-energy neutrons and is able to detect gamma rays that result from their reaction with the boron nuclei.

TABLE 17.6
Potential New Oilfield Chemical Applications for Borates

Activity	Function	Borate Property
Drilling fluids	Various–fluid loss, viscosity modifier, shale inhibitor, pH control, corrosion inhibitor, lubricant, antimicrobial agent (including H_2S scavenger)	Various–borate ester formation, polymer cross-linker, alkaline pH buffer, corrosion inhibition, sparingly soluble salt formation, biostatic effect, oxidizing property
Enhanced oil recovery	Alkaline agent	Alkaline pH buffering, water hardness tolerance
Water flooding–profile control	Fluid gelation	Borate ester formation (polymer cross-linking)

17.3 POTENTIAL NEW OILFIELD CHEMICAL APPLICATIONS FOR BORATES

Table 17.6 lists oilfield activities that currently do not employ borates in significant quantities, but where there has been considerable research activity, promising results, and the possibility of application potential.

17.3.1 Drilling Fluids

Borates are not used to any significant extent today in conventional water- or oil-based drilling fluids/muds. Table 17.7 summarizes the limited published literature in this area.

TABLE 17.7
Literature Citing Potential Uses of Borates in Drilling Fluids

References	Description of Borate Use and Property Exploited
103	Borates added to silicate-based muds; benefits include lower mud viscosity, improve well-bore stability.
104	Borax used as a cross-linker for starch-based drilling fluids.
105	Use of water-soluble boron compounds in combination with alkali metal carboxylates to produce water-based muds with improved metal corrosion characteristics.
106	Optional use of borate as a corrosion inhibitor in drilling fluids.
107	Use of borate-cross-linked starch in "clayless" drilling fluids for drilling horizontal, deviated holes in sandstone, and carbonate reservoirs; benefits include good filtration, rheology, lubrication, and biological resistance.
108	Use of borate-cross-linked HPG as viscosifier in water-based muds for use in drilling deviated horizontal wells; benefits include the ability to delay development of gelling and drill cuttings suspension power until the fluid is in place and the use of pH reduction to lower fluid viscosity at the surface and facilitate cuttings recovery.
109	Use of boric acid and glycerol in modular drilling fluid system; benefits include improved lubrication, reduced mud viscosity, reduced bit "balling," reduced cohesion of drilled rock (easier drilling)—attributed to the adsorption on edges of clay particles of a surface active borate ester formed between boric acid and glycerol.
110	Use of borate with an alkali, alkaline earth, or ammonium salt and optionally a mono- or oligosaccharide in water-based muds; benefits include improved clay/shale stabilization, and alkaline pH buffering.
111	Use of sodium perborate in drilling fluids; benefits relate to oxidation of hydrogen sulfide.

From this table it is concluded that no single property of borate is the dominant reason for its use or proposed use in drilling fluids, but instead several seem to have been identified:

- Corrosion inhibition
- Antimicrobial effect
- Improved lubrication
- Solids dispersion and reduced mud viscosity
- Shale stabilization
- Polymer cross-linking where strong gel property is required

Borates are not used in mainstream drilling fluids to cross-link with the viscosity modifying polymer. Xanthan gum is today the most widely used polymer in water-based muds and this is not cross-linked. Indeed cross-linking would lead to an unacceptably high–low shear viscosity for the drilling fluid. The few cited examples where borate cross-linked muds have been proposed are special cases where the extreme down-hole conditions require a mud with an exceptionally high cuttings suspending power. This effect could, however, be usefully employed in developing a fluid loss additive. Preformed borate cross-linked polyvinyl alcohol, guar, or starch particles could find application in drilling fluids in much the same way as has been proposed for use in fracturing fluids [100].

An interesting and surprising recent discovery concerns the shale stabilizing property of borates [110]. Water-sensitive shale and clay strata are often encountered in drilling operations and water-based muds have to contain shale inhibitors or stabilizers to prevent them from hydrating and swelling these clays. If this process is unchecked it will lead to erosion of the walls of the bore hole and cause weaknesses to develop in the formation. Oil-based muds do not cause this type of problem and are often preferred for this and other reasons. For water-based muds the existing chemical methods for stabilizing clays range from the use of potassium chloride—an inexpensive solution, but with limited effectiveness, through to the use of synthetic polymers specifically developed for this application. In the mid-range in terms of price and performance are the KCl–glycol and silicate-inhibited muds. In laboratory trials, water containing a source of borate and potassium ions and optionally glucose syrup was found to be equivalent to, if not better than a commercial KCl–glycol system, in terms of clay stabilization. Results of one set of trials using the standard industry "bottle rolling test" are illustrated in Figure 17.9.

The test involves roller mixing aqueous solutions of candidate inhibitors with a fixed quantity of 2–4 mm clay particles for 16 h at 20°C. At the end of the test the solutions are passed through a

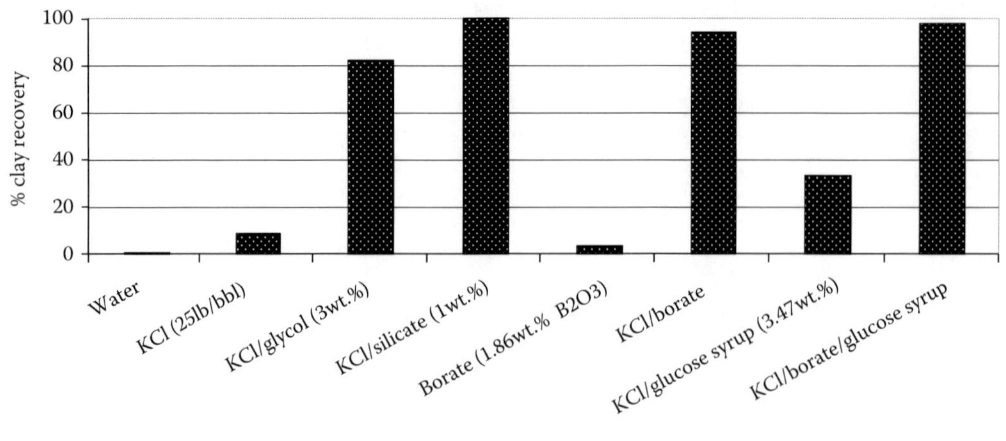

FIGURE 17.9 Clay stabilizing performance of borates versus conventional inhibitors. (Bottle rolling test; London clay (2–4 mm); 16 h at 20°C.)

500 μm sieve and the collected undispersed clay particles rinsed, dried, and weighed. A high clay recovery (on the sieve) equates to a good inhibition performance. From Figure 17.9, it is clear that borates seem to work synergistically with KCl, although other cations, notably cesium and ammonium, also performed well with borates. Likewise, glucose syrup with KCl did not work particularly well, but when combined with KCL/borate, a very good result was obtained.

Further tests demonstrated that inhibition performance was maintained in fully formulated muds. A further advantage of the use of KCl and optionally glucose syrup with borate is that the relatively low water solubility of borate is enhanced by the presence of both of these additives. In addition, borate-based muds are naturally mildly alkaline and buffer well against changes in pH.

CORE CONCEPTS

Borates are not used in significant quantities in drilling fluids at present. Where they are used it is for corrosion inhibition, antimicrobial effects, solids dispersion, and lubrication. Normally, the polymer cross-linking property of borates would result in gels that are too strong for normal drilling fluids, but in demanding circumstances this effect may be useful. Recent studies have demonstrated the effectiveness of borates as ***shale inhibitors*** for water-based muds, in synergy with other mud additives.

17.3.2 Enhanced Oil Recovery

In the late 1970s and early 1980s, much attention was focused on "chemical" EOR processes. These involved the injection of previously water-flooded reservoirs with micellar surfactant/polymer solutions to stimulate residual oil recovery. This oil, frequently accounting for 65% of the original oil in place in the reservoir, is often trapped in the reservoir rock as isolated ganglia. The oil or water wettability of the reservoir rock plus the oil/water interfacial tension are important parameters that determine how easily the trapped oil can be released. Although highly simplified, the Laplace equation provides an estimate of the pressure required to displace oil trapped in an "idealized" cylindrical pore (depicted in Figure 17.10):

$$P_c = 2\gamma \cos \theta / r$$

where:

P_c = the capillary pressure, or pressure difference across the curved oil/water interface;
γ = oil/water interfacial tension (i.f.t.);
r = pore diameter (strictly radius of curvature of the interface);
θ = contact angle between oil droplet and the rock surface, measured through the water phase (in the figure it has a value less than 90° and the rock is therefore water wet).

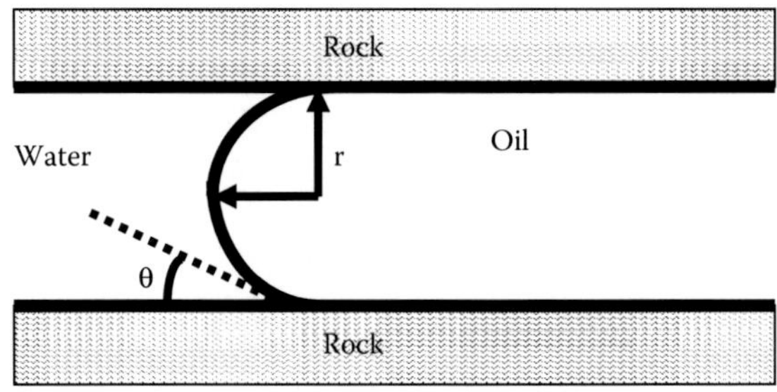

FIGURE 17.10 Simplified depiction of oil filament trapped in cylindrical rock capillary.

In reality the situation is more complex, but the Laplace equation recognizes the existence of a static pressure difference across the curved oil/water interface. For flow to occur and for the oil filament to be displaced from the pore it is necessary to exceed this pressure difference. With rock pores in the range of a few tens of microns and with an oil/water i.f.t. of 20–30 mN/m, very high pressure gradients are required to displace the trapped oil. These forces will not be overcome under normal water flooding conditions. This is where chemical EOR becomes useful. Carefully chosen surfactants or blends are capable of lowering the oil/water i.f.t. from 30 down to 10^{-3} mN/m and so can dramatically lower the capillary forces trapping the oil.

Several micellar surfactant/polymer systems were field trialed in the late 1970s and early 1980s. However, they were not judged to be a great success and turned out to be expensive in terms of the chemicals consumed; injected slugs containing up to 10% of surfactant [112]. Shortly afterward (mid-1980s), the price of crude oil plunged and such processes were deemed uneconomic. More recently, there has been a revival in interest in chemical EOR, driven by the return of higher oil prices, but also because of the economic imperative to extract as much oil as possible from the large number of depleted (by water flooding) reservoirs, particularly in the United States, S. America, China, and Indonesia. In its current reincarnation, chemical EOR processes contain much lower levels of surfactant (0.1%) and rely on the inclusion of cheaper alkaline agents (and polymers) to achieve the necessary ultra-low i.f.t.. The alkali stimulates the "saponification" of indigenous surfactants (petroleum acids) present in the crude oil and these complement the effect of the added synthetic surfactant in these ASP (alkali–surfactant–polymer) floods:

$$HA + OH^- \Leftrightarrow A^- + H_2O$$

where HA and A^- are the petroleum acid and its dissociated surface active anion.

It also reduces adsorption of the (mostly anionic) surfactant on the reservoir rock, essentially by enhancing the negative surface charge. In some cases (alkaline-polymer floods), where there are high levels of saponifiable crude oil acids present in the crude oil (high acid number) added surfactant is not even required. The polymer is present to assist in mobility control and to ensure that the injected chemical slug remains intact and promotes the formation of an oil bank ahead of it.

Over 30 ASP flooding projects are either in progress or have been completed since the mid-1990s [113–118]. Almost all have employed strong alkalis such as sodium carbonate or sodium hydroxide, although some also mention the use of sodium silicate and sodium tripolyphosphate. Although the alkaline agent is always cheaper than the surfactant in the formulation and it is added to prevent surfactant loss, it is prone to being consumed by a number of generally undesirable chemical reactions, in addition to the desired saponification of petroleum acids:

- Through ion-exchange reactions with the rock surface:

$$\text{Rock-H} + Na^+ + OH^- \Leftrightarrow \text{Rock-Na} + H_2O$$

- Through rock dissolution reactions—the dissolved minerals are then prone to consuming more alkali or precipitating elsewhere and causing reservoir plugging.
- Through reactions (precipitation) with divalent hardness ions (Ca^{2+}, Mg^{2+}) present in the reservoir or make-up brines—in the reservoir this consumes alkali and can also cause plugging, and to avoid its occurrence in the ASP injection slugs may require the use of artificially softened water.

In order to overcome some of the problems caused by the use of strong alkalis, sodium bicarbonate has been suggested as an alternative, especially for use in limestone or dolomite reservoirs [20]. Bicarbonate is known to be less reactive toward these rocks and therefore subsequent precipitation of dissolved rock minerals should be reduced. Despite it being less alkaline it is still apparently able to yield adequately low o/w i.f.t. However, it is still not an ideal choice and reservoirs with even relatively low contents of gypsum in the rock (or sulfate ion in the reservoir brine), CO_2 in the gas/oil phase or calcium in the reservoir brine can completely consume this alkali.

A recent publication by Flaaten et al. has claimed that borates, notably sodium metaborate, unlike conventional alkalis, exhibit a remarkable tolerance toward hardness ions, in this application [19]. This discovery will allow (notably carbonate) reservoirs with hard saline brines to benefit from the use of ASP flooding technology, where previously this was not possible. The metaborate ion is known to form ion-pair complexes with hardness ions, for example, the reaction with calcium ion is as follows:

$$Ca^{2+} + B(OH)_4^- \Leftrightarrow CaB(OH)_4^+$$

This reaction has an equilibrium constant of 18 (pK_{Ca} value of approximately 1.25) at 40°C. This means in effect that a 1 wt% (0.025 mol/L) solution of borax, capable of releasing approximately 0.1 mol/L of metaborate ions into solution (with pH adjusted to 10) is able to sequester approximately half of the calcium ion content of a solution. Borates are known for their use as water softeners in detergent and household cleaning products. Although they are not classed as strong sequestrants, they do exhibit extensive zones of miscibility in mixed borate/hardness ion solutions, a fact that makes them attractive for ASP flooding.

In Flaaten's work it was possible, with the use of salt and hardness tolerant surfactants to develop an ASP formulation containing 0.75 wt.% sodium metaborate prepared in a highly saline brine containing as much as 6000 ppm of hardness ions. This was not possible without precipitation in the case of the alkali sodium carbonate. Metaborate has a pH similar to sodium carbonate (close to 11, the exact value depending on concentration) and the metaborate-based ASP formulation demonstrated good recovery of light oil (API 45°) from Berea sandstone rock in laboratory core floods. Apart from its remarkable hardness tolerance, from the currently available data it appears that borate is at least able to match, in other respects, the performance of the conventional alkalis, sodium hydroxide and sodium carbonate. In radial core floods conducted using light oil (API 39°) and a sandstone core, it proved possible to develop ASP formulations based on sodium metaborate that gave results similar to those containing the existing alkalis.* Exact comparisons between the alkalis are not possible, however, as the formulations also differ slightly in terms of the amounts (range 0.1–0.2 wt%) and types of surfactant employed. Key data from these core flood experiments are illustrated in Table 17.8.

The borate formulation was not previously optimized using phase behavior or oil/water i.f.t. screening tests and yet still performed reasonably well in terms of tertiary oil recovery, chemical (strong alkali, surfactant, and polymer) loss, and in addition no injectivity problems were encountered.

A separate investigation of the ability of solutions of sodium borate combined with surfactant to produce the necessary low i.f.t. against a range of different crude oils was conducted. Once again the approach was to take a previously optimized conventional alkali–surfactant solution, and substitute the borate (a mix of sodium metaborate and tetraborate) directly in place of the conventional alkali without further optimization. The borate was employed at a concentration necessary to maintain the ionic strength equivalent to that of the conventional mixture. The results are illustrated in Figure 17.11.

Considering that no attempt was made to optimize the borate-surfactant systems, the results are encouraging and suggest that borate, as expected, behaves like the conventional alkalis in terms of

* Coreflood temperature 68°F, crude oil viscosity 7.3 cP, produced water containing 3.3% total dissolved solids (of which 2700 ppm Ca^{2+} + Mg^{2+}); surfactant concentration range in ASP slug 0.1–0.2 wt.%; polyacrylamide concentration in ASP and polymer slugs–550 mg/L; injection sequence – 3 pore volumes (Vp) produced water, 0.3 Vp ASP slug, 0.3 Vp polymer slug, 3 Vp injection water.

TABLE 17.8
Comparison of Alkali Performance in Laboratory ASP Radial Core Floods

Alkali (wt%)	Waterflood Recovery V_p	Tertiary Oil Recovery V_p	Total Oil Recovery %OOIP	Alkalinity Loss mg/100 g Rock	Surfactant Loss mg/100 g Rock	Polymer Loss mg/100 g Rock
$NaBO_2$ (0.82[a])	0.365	0.078	65.1	10.73	1.25	1.84
NaOH (0.50)	0.390	0.091	69.7	5.89	1.45	1.36
NaOH (0.75)	0.347	0.084	62.1	6.22	2.88	1.63
Na_2CO_3 (1.25)	0.375	0.105	68.8	26.45	4.52	0.95
Na_2CO_3 (1.50)	0.367	0.071	65.4	24.32	1.57	1.72

[a] Plus 0.42% disodium tetraborate (borax) pentahydrate.

promoting i.f.t. reduction with EOR surfactants. The apparent trend observed in the data, namely the poorer performance of borate with the heavier (lower API gravity) oils, cannot be considered particularly significant. If it the data were plotted against crude oil acid number, a correlation might be easier to explain since the amount of saponifiable crude oil components might well influence how it responds to alkali and to what extent the alkali can promote i.f.t. reduction. The measured pH for the sodium borate/surfactant solutions were in the range 9.6–10.4, significantly lower than those of the same formulations with carbonate (~pH 11) and sodium hydroxide (pH 13).

On the basis of a recently published article, however, it may be premature to conclude that the only potential performance benefit connected with the use of borate in this application is its hardness ion tolerance [119]. This article hints at another phenomenon as yet not fully understood. In this work crude oil-laden thin limestone plates or short cores were contacted with aqueous solutions of sodium carbonate, sodium hydroxide, sodium metaborate, di-sodium tetraborate (borax), and sodium chloride. The spontaneous uptake of these very simple salt solutions by the limestone and the consequent expulsion/recovery of oil was then observed visually and analyzed over a period of several days. This phenomenon is known as capillarity-driven natural imbibition (CDNI), and is the spontaneous spreading of the water phase into the rock. In mixed-wet or oil-wet carbonate reservoirs, where high-dissolved salts cause incompatibilities with and prevent the use of many injected chemicals (ionic surfactants and common alkalis) CDNI is often the only mechanism available for oil recovery. What was surprising was that with solutions of sodium metaborate, greater than 50%

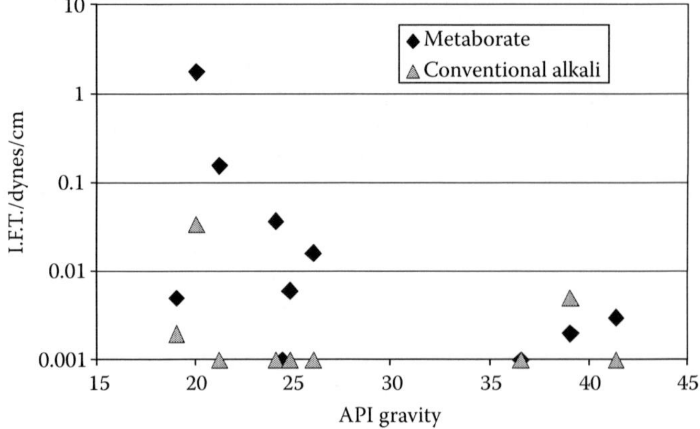

FIGURE 17.11 Oil/water interfacial tensions of EOR surfactant solutions with sodium metaborate and conventional alkalis: Effect of oil API gravity.

of the oil was recovered at the end of 3 days. Sodium metaborate outperformed sodium carbonate (40% recovery) and both were substantially better than sodium hydroxide (<4% of the original oil in place was recovered). Even borax with its much lower pH (9.5) compared with the other alkalis was only slightly inferior to sodium carbonate. Borax's good performance was attributed to its ability to maintain pH constant over a wide range of concentrations, that is, despite probable losses due to adsorption (ion-exchange reactions with the rock) or consumption in the saponification process. The good imbibition performance of borates was particularly unexpected since in separate dynamic contact angle measurements on calcite and quartz plates, it was carbonate, rather than borate, that appeared most able to change the plates from their initial oil-wet state to being more water wet.

It remains to be seen if this phenomenon is in any way connected with that observed in separate studies conducted by Zhang and Austad [120,121]. In this work, a synergistic effect of the potential determining ions, Ca^{2+}, Mg^{2+}, and SO_4^{2-}, leads to improved spontaneous imbibition of seawater containing these ions into, and a correspondingly higher recovery of crude oils (acid numbers 2.07 and 0.55 mg KOH/g) from chalk cores. The role of the sulfate anion is to promote the desorption of negatively charged carboxylic (saponified) crude oil components that otherwise promote oil wetting of the rock surface. Through adsorption, the sulfate ion achieves this in two ways, by

- Increasing the negative surface charge (negative value of the measured zeta potential) of the chalk and encouraging the desorption (due to increased charge repulsion) of these negatively charged components.
- Promoting, at a high temperature, the coadsorption of magnesium ions (with which it forms ion-pair complexes), which then displace bridging calcium ions bonding the adsorbed carboxylic components to the chalk surface. Having displaced the calcium ions, the divalent metal ion/carboxylic component complex then tends to be desorbed.

It is possible, but not yet proven, that the borate ion acts in a way similar to the sulfate ion. Like the sulfate anion, it does form soluble ion-pair complexes with calcium and magnesium ions and this may promote coadsorption on the chalk surface. However, in the absence of reliable data on borate ion adsorption and zeta potential data for chalk, it is difficult to confirm the mechanism behind the observed imbibition performance of borate.

In summary, the proposed use of borates in chemical EOR, notably ASP flooding, is a very recent development, with significant promise. It is true that no field trials have yet been conducted and there is no experience in the field of preparing and handling borate-based ASP fluids. Nevertheless, borates are considered safe for humans and the environment, are noncorrosive, and are readily available and affordable. Their hard-water tolerance makes them particularly suited for use in high-hardness and high-salinity carbonate reservoirs containing acidic crudes, where conventional alkalis have failed or cannot be used. This is especially relevant in cases where environmental restrictions prohibit the use of locally available soft water, or where it is uneconomic to artificially soften the injected water.

CORE CONCEPTS

Recent laboratory studies have demonstrated the potential utility of borates as alkaline agents in chemical enhanced oil recovery. Compared with existing alkalis, ***sodium metaborate*** has an unusually high tolerance toward the hardness ions, Ca^{2+} and Mg^{2+}, paving the way for the implementation of alkali–surfactant–polymer floods for the large number of high-hardness saline carbonate reservoirs. In the absence of surfactants, borate solutions exhibit a strong tendency for spontaneous ***imbibition***, or uptake into oil-wet or mixed-wet carbonate cores, with consequently improved recovery of oil compared with solutions of other salts and alkalis.

17.3.3 Water Flooding: Profile Control

In secondary recovery processes involving the displacement of oil in the reservoir by injected water, success relies on obtaining good volumetric sweep efficiencies. In other words, the injected water should displace the oil in such a way that it is driven ahead as an even front. When predicting volumetric sweep efficiencies, reservoir engineers refer to the mobility ratio, *M*, that is a function of the viscosities (μ_o, μ_w) of the oil and displacing water phases, the permeability of the rock toward these phases (k_{ro}, k_{rw}), and the average saturation of the oil and water phases (s_o, s_w) immediately ahead and behind the displacing front:

$$M = (k_{rw}/\mu_w)\, s_w\, (\mu_o/k_{ro})\, s_o$$

Good volumetric sweep efficiency generally requires values of *M* less than or equal to 1.0. Increasing the water phase viscosity to more closely match that of the oil phase is one way of achieving this (resulting in a reduction in *M*, according to the above equation). This can be achieved by adding viscosifying polymers to the water flood. Polymer augmented water floods usually rely on the use of either partially hydrolyzed polyacrylamide polymers or biopolymers such as xanthan gum.

However, even when the mobility ratio has been adjusted by the addition of polymers, reservoir heterogeneities, particularly in the vertical direction mean that high permeability "thief" zones often exist and these present a path of lower resistance to the invading water. In order to prevent the oil from being bypassed under these circumstances and also to reduce the excessive production of water, *in situ* permeability modification is often attempted. This process requires the use of more viscous gel-like polymer solutions that are effectively cross-linked. The same polymers are used, but cross-linked with Cr(III) (not so popular now as dichromate (Cr(VI), a known toxin, is often added and then reduced to Cr(III)) or Al(III). As was the case with cross-linked gels for fracture fluids, it is necessary to delay the gelation until the polymer solution has been pumped into the reservoir. There are only a few reports of the use of borates as the cross-linker in this type of application. In one example, boric acid was used with a biopolymer (KUSP 1) and sodium hydroxide to create a time-delayed gel [122,123]. In another example, borate esters with rates of hydrolysis decreasing in the order: triethoxyborane > triethanolamine borate > diethanolpropanolamine borate > tripropanolamine borate, were employed to delay the cross-linking of guar polymers and polyvinyl alcohol from less than 1 min to up to 21.5 h in this application [124,125].

A totally different approach to blocking high-permeability thief zones during water flooding has been proposed that also involves borates [126–129]. The technique, thus far restricted to laboratory trials, is referred to as thermal precipitation. In essence, it relies on the precipitation of a salt from its saturated solution caused by a reduction in temperature. The salt precipitating in the rock pores tends to reduce both reservoir porosity and permeability. Only salts that exhibit a marked reduction in solubility going from high (80–100°C) to low (24°C) temperatures are suitable and sodium borate is an excellent example of this (solubility falling from over 100 g to less than 5 g per 100 g of water over this temperature interval). The approach then taken is to inject a nearly saturated solution of sodium borate at a high temperature into a heated sandstone core and then allow the core to cool. Cooling by heat conduction from the initially hot solution to the cooler reservoir is a realistic scenario and should produce a uniform distribution of the precipitate. On cooling, precipitation occurred in the core samples and the measured permeability was found to have been reduced by 63%. Potassium carbonate was also found to be effective in these studies. Although the end result was similar in the case of both salts, sodium borate tended to form larger crystals that were attached to the pore surfaces at random locations. In contrast, potassium carbonate formed finer crystals that were more uniformly distributed. The solubility of sodium borate, however, varied much more strongly with temperature than potassium carbonate.

CORE CONCEPTS

Borates have been proposed for use in different ways to reduce the permeability of zones in the reservoir in order to control the path or ***profile*** taken by injected fluids in ***water flooding***. In one approach, borates are used to form cross-linked polymer gels that reduce permeability and block off fluid access to ***thief zones***. Another concept exploits the sharp reduction in borate salt solubility at low temperatures to cause a hot nearly saturated solution to form precipitates and block the rock pores, after being injected and on gradual cooling.

17.4 SUMMARY

Today's oil and gas exploration and production industry demands the use of high-performance chemical additives for use in the formulation of a range of complex functional fluids. There are several important examples of where borates are well established as the industry's preferred chemical additive, enhancing specific fluid properties, notably rheology, but also acting as nondamaging temporary solids that control fluid filtration into the reservoir, as oxidants to assist in reservoir cleanup following fluid treatments and as set retarders of oil well cements. Additionally, new applications are emerging as the industry discovers more about the rich and unique chemistry of borates. A promising example of this is their potential for use as alkaline agents in chemical EOR processes, where they may open the way for the extraction of residual oil trapped in highly saline carbonate reservoirs; something that is not easily achievable with existing alkalis. Borates are affordable, readily available, chemically compatible, safe and environmentally acceptable, all of which augers well for their continued and growing use in this industry.

REFERENCES

1. Fink, J. K.; *Oil Field Chemicals*, Gulf Professional Publishing (Elsevier Science), Burlington, Massachusetts, 2003.
2. Kelland, M. A.; *Production Chemicals for the Oil and Gas Industry*, CRC Press Taylor & Francis Group, Boca Raton, FL, 2009.
3. Hyne, N. J.; *Nontechnical Guide to Petroleum Geology, Exploration, Drilling and Production*, 2nd ed., Penn Well Corporation, Tulsa, Ok, 2001.
4. Lyons, W.; *Standard Handbook of Petroleum & Natural Gas Engineering*, Vol. 1, Gulf Publishing Company, Houston, TX, 1996.
5. Economides, M. J., Hill, A. D., and Ehlig-Economides, C.; *Petroleum Production Systems*", Prentice-Hall Engineering Series, Englewood Cliffs, NJ, 1994.
6. Jansen, L. H.; *Kirk-Othmer Encyclopedia of Chemical Technology*, 5th ed., Volume 4, Wiley-Interscience, A John Wiley & Sons, Inc., Hoboken, New Jersey, Boron Elemental, pp. 132–138, 2004.
7. Thompson, R. and Welch, A. J. E.; *Mellor's Comprehensive Treatise on Inorganic and Theoretical Chemistry*, Vol. V: *Boron, Part A: Boron–Oxygen Compounds*; Longmann, London, 1980.
8. Briggs, M.; *Kirk-Othmer Encyclopedia of Chemical Technology*, 5th ed.,Volume 4, Wiley-Interscience, A John Wiley & Sons, Inc., Hoboken, NJ, Boron Oxides, Boric acid and Borates, pp. 241–294, 2004.
9. Mozzi, R. L. and Warren, B. E.; Structure of vitreous boron oxide, *J. Appl. Crstallogr.*, 3, 251 1970.
10. Zachariasen, W. H.; Precise structure of ortho boric acid, *Acta Cryst.* 7, 305–310, 1954.
11. Giese, R. F.; A refinement of the crystal structure of borax, *Can. Mineral.*, 9(part 4), 573, 1968.
12. Heller, G.; in *Gmelin Handbook of Inorganic Chemistry*, ed. K. Niedenzu and K. C. Buschbeck, Band 28, Tiel 7, Springer-Verlag, Berlin, pp. 2–4, 1975.
13. Nies, N. P. and Hulbert, R. W.; Solubility isotherms in the system sodium oxide-boric oxide-water. Revised solubility-temperature curves of boric acid, borax, sodium pentaborate, and sodium metaborate, *J. Chem. Eng. Data*, 12(3), 303, 1967.
14. Nies, N. P. and Hulbert, R. W.; Solubility isotherms in the system sodium oxide-boric oxide-water: Revised solubility-temperature curves of boric acid, borax, sodium pentaborate and sodium metaborate, *J. Chem. Eng. Data*, 13(1), 131, 1968.

15. Ingri, N.; *Equilibrium Studies of Polyanions Containing BIII, SiIV, GeIV and VV, Swedish Chemical Journal*, 75, 199, 1963.
16. Kolthoff, I. M.; Change of the dissociation constant of boric acid with the concentration of this acid, *Rec. Trav. Chim.*, 45, 501–507, 1926.
17. Kister, R. and Helvaci, C.; Boron and borates, in *Industrial Minerals and Rocks*, 6th ed., ed. D. Carr, Littleton, Colorado. pp. 85–89, 1994.
18. Reardon, E. J.; Dissociation constants for alkali earth and sodium borate ion pairs from 10 to 50°C, *Chem Geol.*, 18, 309–325, 1976.
19. Flaaten, A. K., Nguyen, Q. P., Zhang, J., Mohammadi, H., and Pope, G. A.; *ASP Chemical Flooding Without the Need for Soft Water*, 2008 SPE Annual Technical Conference and Exhibition, Denver, Colorado, September, 21–24, 2008.
20. Lorenz, P. B. and Peru, D. A.; Guidelines help select reservoirs for $NaHCO_3$. *Oil & Gas Journal, Technology*, 87(37), 53–57, 1989.
21. Steinberg, H.; *Organoboron Chemistry, Volume 1: Boron-oxygen and boron-sulfur compounds*, Interscience Publishers, John Wiley & Sons Inc., London, 1964.
22. Gerrard, W.; *The Organic Chemistry of Boron*, Academic Press, London & New York, pp. 11–13, 1961.
23. Carrondo, C. T. and Skapski, A. C.; Refinement of the x-ray crystal structure of the industrial bleaching agent di sodium tet, *Acta Crystallogr.*, B34, 3551, 1978.
24. Jakobi, G. and Lőhr, A., *Detergents and Textile Washing: Principles and Practice, Wiley-VCH, Weinheim, Cambridge,* pp. 78–80, 1987.
25. Mark, J. E.; *Physical Properties of Polymers Handbook*, 2nd ed., Springer Science + Business Media, New York, 2007.
26. Dawson, J. C.; *A Thermodynamic Study of Borate Complexation with Guar and Guar Derivatives*; 66th Annual Technical Conference and Exhibition of SPE, Dallas, TX, October 6–9, 1991 (SPE 22837).
27. Pezron, E., Ricard, A., and Leibler, L.; Rheology of galactomannan-borax gels; *J. Polymer Science: Part B–Polymer Physics*, John Wiley & Sons, NY, 28, 2445–2461, 1990.
28. US 3,058,909, assigned to Atlantic Refining Company.
29. Mesmer, R. E., Baes, C. F., and Sweeton, F. H.; Acidity measurements at elevated temperatures, VI. Boric acid equilibria. *Inorganic Chemistry*, 6, 537–543, 1972.
30. Harris, P C; Chemistry and rheology of borate-cross-linked fluids at temperatures to 300°F, *Journal of Petroleum Technology*, 45(3), 264–269, 993 (SPE 24339).
31. de Kruijf, A. S., Roodhart, L. P., and Davies, D. R.; and flow mechanics of borate-crosslinked fracturing fluids. *SPE Production and Facilities*, 8(3), 165–170, 1993.
32. Harris, P. C. and van Batenburg, D.; *A Comparison of Freshwater- and Seawater-Based Borate-Crosslinked Fracturing Fluids;* International Symposium on Oilfield Chemistry, Houston, TX, February 16–19, 1999 (SPE 50777).
33. EP 0594363; assigned to Halliburton.
34. US 3,227,212; assigned to Halliburton.
35. US 4,619,776; assigned to Texas United Chemical Company.
36. US 2004-0067854; application by Texas United Chemical Company.
37. US 6,024,170; assigned to Halliburton.
38. US 2003-0092584; application by Baker Hughes.
39. US 5,565,513; assigned to Benchmark Research & Technology.
40. US 5,565,513, 5,488,083, 6,225,264, 6,251,838; assigned to Benchmark Research & Technology.
41. US 2003-0220203; application by Benchmark Research & Technology.
42. Ainely, B. R., Nimerick, K. H., and Card, R. J.; *High Temperature, Borate-Crosslinked Fracturing Fluids: A Comparison of Delay Methodology*, Production Operations Symposium, Oklahoma City, March 21–23, 1993 (SPE 25463).
43. US 5,082,579, 5,145,590, 5,160,643, CAN 2,065,575; assigned to B J Services.
44. US 5,877,127, 5,445,223, CAN 2,073,806; assigned to Schlumberger.
45. WO 2006-095291; application by Schlumberger.
46. US 3,215,634; assigned to Jersey Production Research Company.
47. WO 2003-025340; application by Baker Hughes.
48. US 3,096,284; assigned to Halliburton.
49. US 5,160,445, 5,252,236, 5,266,224, 5,310,489; assigned to The Zirconium Technology Corporation.
50. Brannon, H. D., Ault, M. G.; *New Delayed Borate-Crosslinked Fluid Provides Improve Fracture Conductivity in High Temperature Applications*; 66th Annual Technical Conference and Exhibition of SPE, Dallas, Texas, October 6–9, 1991 (SPE 22838).

51. US 2003-0234105; application by B J Services.
52. US 3,763,964, 3,766,984; assigned to Dow Chemical Company.
53. US 5,372,732; assigned to Halliburton.
54. Canadian Patent 2,166,596; assigned to Grain Processing Corporation.
55. US 5,259,455; assigned to Dowell Sclumberger.
56. Cawlezel, K. E. and Elbel, J. L.; A new system for controlling the cross-linking rate of borate fracturing fluids, *SPE Production Engineering*, 7(3), 275–279, 1992 (SPE 20077).
57. Nimerick, K. H., Temple, H. L., and Card, R. J.; *New pH-Buffered Low Polymer Borate Crosslinked Fluids for Hydraulic Fracturing*, Gas Technology Conference, Calgary, Alberta, Canada April 28–May 1, 1996 (SPE 35638).
58. US 3,974,077; assigned to Dow Chemical Company.
59. US 5,827,804; assigned to Halliburton.
60. Powell, R. J., McCabe, M. A., Slabaugh, B. F., Terracina, J. M., Yaritz, J. G., and Ferrer, D.; Applications of a new, efficient hydraulic fracturing fluid system. *SPE Prod. & Facilities* 14(2), 139–143, May 1999 (SPE 56204).
61. US 2004-0235675; application by Schlumberger.
62. Harris, P. C. and Heath, J. J.; *Rheological Properties of Low Gel Loading Borate Fracture Gels*, 1997 SPE Annual Technical Conference and Exhibition, San Antonio, TX, October 5–8 (SPE 52399).
63. Kakadijian, S., Rauseo, O., Marquez, R., Gabay, R., Tirado, Y., and Blanco, J.; *Crosslinked Emulsion to be used as Fracturing Fluids*, SPE International Conference on Oilfield Chemistry, Conference proceedings, pp. 497–506, Houston, Texas, USA, February 13–16, 2001.
64. Kramer, J., Prud'homme, R. K., Wiltzius, P., Mirau, P., and Knoll, S.; Comparison of galactomannan crosslinking with organotitanates and borates. *Colloid & Polymer Sci.* 266, 145–155, 1988.
65. Prud'homme, R. K., Constien, V., and Knoll, S.; *Polymers in Aqueous Media*, ed. J. E. Glass, *Advances in Chemistry Series No 223*, American Chemical Soc., Washington, DC, vol 6, p. 94, 1989.
66. Al-Anazi, H. A. and Nasr-El-Din, H. A.; *Optimisation of Borate-Based gels used for Wellbore Diversion during Well Stimulation: A case Study*, SPE/DOE Improved Oil Recovery Symposium, Tulsa, OK, 19–22 April 1998 (SPE 39699).
67. Kesevan, S. and Prud'homme, R. K.; Rheology of Guar and HPG Cross-linked by Borate. *Macromolecules* 25, 2026–2032, 1992.
68. US 6,242,390; assigned to Schlumberger Technology Corporation.
69. US 6,214,773; assigned to Halliburton.
70. US 5,165,479; assigned to Halliburton.
71. US 4,514,309; assigned to Hughes Tool Company.
72. US 5,217,632; assigned to Zirconium Technology Corporation.
73. Parris, M. D., MacKay, B. A., Rathke, J. W., Klingler, R. J., and Gerard, R. E., II; Influence of Pressure on Boron Cross-Linked Polymer Gels. *Macromolecules* 41, 8181–8186, 2008.
74. Hossaini, M. and Gabrysch, A.; *The Effect of Formation damage on HEC Gel Return Permeability and Techniques to Overcome Potential Damage*, Formation Damage Control Symposium, Lafayette, Louisiana, February 14–15, 1996 (SPE 31107).
75. US 6,918,445; assigned to Halliburton Energy Services, Inc.
76. Woods, W. G.: *Solubilization and Stabilization of Aqueous Sodium Perborate*, 3rd CESIO International Surfactants Congress and Exhibition, London, June 1–5, 1992, (proceedings, section D, 'Applications', 73–88).
77. US 5,624,886; assigned to B J Services Company.
78. Nasr-El-Din, H. A., Al-Mohammed, A. M., Al-Aamri, A. D., and Al-Fuwaires, O. A.; *A Study of Gel Degradation, Surface tension, and Their Impact on the Productivity of Hydraulically Fractured Gas Wells*, SPE Annual Technical Conference and Exhibition, November 11–14, 2007, Anaheim, California (SPE 109690).
79. Siddiqui, M. A., Nasr-El-Din, H. A., Al-Anazi, M. S., and Bartko, K. M.; *Formation Damage in Gas Sandstone Formations by High-temperature Borate Gels Due to Long Term Shut-in Periods*, SPE/DOE Symposium on Improved Oil Recovery, April 17–21, 2004, Tulsa, Oklahoma (SPE 89476).
80. WO/2005/083029; patent application by Halliburton Energy Services, Inc.
81. WO/2000/057022; patent application by Cleansorb Limited.
82. Powell, J. W., Stagg, T. O., and Reiley, R. H.; *Thixotropic, Crosslinking Polymer/Borate/Salt Plug: Development and Application*, International Arctic Technology Conference, Anchorage, AK, May 29–31, 1991 (SPE 22068).
83. US 2003-0008778; inventor R. A. Donaldson.

84. US 4,620,596; assigned to Texas United Chemical Corporation.
85. Mondshine, T; Completion fluids weight range expanded. *Oil & Gas Journal (Technology)*, 83(46), 45–152, Nov 18, 1985.
86. Dyke, C. G. and Crockett, D. A.; *Prudhoe Bay Rig Workovers: Best Practices for Minimizing Productivity Impairment and Formation Damage*, SPE Western Regional Meeting, Anchorage, AK, May 26–28, 1993 (SPE 26042).
87. Bensted, J., Callaghan, I. C., and Lepre, A.; Comparative study of the efficiency of various borate compounds as set retarders of class G oil well cement. *Cement and Concrete Research*, 21(4), 663–668, 1991.
88. Csetenyi, L. J. and Glasser, F. P.; Phase equilibrium Study in the CaO-Na2O-B2O3-H2O system at 25°C and 55°C. *Advances in Cement Research* 7, 3–19, 1995.
89. US 2,006,426; assigned to J E Weiler.
90. US 3,748,159; assigned to The Halliburton Company.
91. US 3,821,985; assigned to The Halliburton Company.
92. US 3,662,830, assigned to Dow Chemical.
93. US 4,941,536; assigned to Halliburton.
94. US 4,500,357; assigned to Halliburton.
95. EP 0614 859; assigned to Sofitech N.V.
96. US 4,137,093; assigned to Instituto Mexicano del Petroleo.
97. US 5,220,960; assigned to Halliburton.
98. US 2005263285; application by Halliburton.
99. Sugama, T., Carciello, N. R., and Gray, G.; Alkali carbonation of calcium aluminate cements: influence of set-retarding admixtures under hydrothermal conditions. *J Materials Sci.* 27, 4909–4916, 1992.
100. US 20060189487; patent application by Halliburton.
101. Sommer, F. S. and Jenkins, D. P.; *Channel Detection Using Pulsed Neutron Logging in a Borax Solution*, SPE Asia Pacific Oil & gas Conference & Exhibition, Singapore, February 8–10, 1993 (SPE 25383).
102. Bigelow, E. L.; *A Log-Inject-Log Application To Resolve Porosity in Tight, Fractured Formations*, SPE Permian Basin Oil and Gas Recovery Conference, Midland, Texas, March 16–18, 1994 (SPE 27644).
103. SU 1699991; Russian patent, assigned to Moscow Institute of the Petroleum and Gas Industry.
104. HU 33210; Hungarian patent assigned to Magyar Szenhidrogenipari Kutato-Fejleszto Intezet.
105. DE 198 40 632; German patent assigned to Clariant.
106. US 6,405,809; assigned to Halliburton.
107. RU 2186820; Russian patent assigned to Obshchestvo s Orgranichennoi Otvetstvennost'yu "PermNIPIneft", Russia.
108. US 20030236171; application by Halliburton.
109. US 6,105,691; assigned to Spectral, Inc.
110. WO/2003/052023; application by BP Exploration Operating Company Limited and Borax Europe Limited.
111. US 4,548,720; assigned to Diamond Shamrock Chemicals Company.
112. D. W. Green and G. P. Willhite, eds, *Enhanced Oil Recovery*, Society of Petroleum Engineers (SPE) Textbook Series, Richardson, Texas, Vol. 6, pp. 239–300, 1998.
113. Shutang, G., Huabin, L., Zhenyu, Y., Pitts, M. J., Surkalo, H., and Wyatt, K.; *Alkaline/Surfactant/Polymer Pilot Performance of the West Central Saertu, Daqing Oil Field. SPE Reservoir Engineering* 11(3), 181–188, 1996.
114. Demin, W., Jiecheng, C., Junzheng, W., Zhenyu, Y., Yuming, Y., and Hongfu, L.; *Summary of ASP Pilots in Daqing Oil Field*, 1999 SPE Asia Pacific Improved Oil Recovery Conference, Kuala Lumpur, Malaysia, 25–26 October 1999 (SPE 57288).
115. Zhang, J., Wang, K., and He, F.; *Ultimate Evaluation of the Alkali/Polymer Combination Flooding Pilot Test in XingLongTai Oil Field*, 1999 SPE Asia Pacific Improved Oil Recovery Conference, Kuala Lumpur, Malaysia, 25–26 October 1999 (SPE 57291).
116. Vargo, J., Turner, J., Vergnani, B., Pitts, M., Wyatt, K., Surkalo, H., and Patterson, D., Alkaline-surfactant-polymer flooding of the Cambridge minnelusa field. *SPE Reservoir Eval. & Eng.* 3(6), 552–558, 2000.
117. Manrique, E., De Carvajal, G., Anselmi, L., Romero, C., and Chaćcon, L., *Alkali/Surfactant/Polymer at VLA 6/9/21 Field in Maracaibo Lake: Experimental Results and Pilot Design*, 2000 SPE/DOE Improved Oil Recovery Symposium, Tulsa, Oklahoma, April 3–5, 2000 (SPE 59363).
118. Delshad, M., Han, W., Pope, G. A., and Sepehrnoori, K., *Alkaline/Surfactant/Polymer Flood Predictions for the Karamay Oil Field*, 1998 SPE/DOE Improved Oil Recovery Symposium, Tulsa, OK, April 19–22, 1998.

119. Zhang, J., Nguyen, Q. P., Flaaten, A. K., and Pope, G. A.; *Mechanisms of Enhanced Natural Imbibition with Novel Chemicals*, 2008 SPE/DOE Improved Oil Recovery Symposium, Tulsa, OK, April 19–23, 2008 (SPE 113453).
120. Zhang, P. and Austad, T.; Wettability and oil recovery from carbonates: Effects of temperature and potential determining ions. *Colloids and Surfaces A: Physicochem. Eng. Aspects*, 279, 179–187, 2006.
121. Zhang, P., Tweheyo, M. T., and Austad, T.; Wettability alteration and improved oil recovery by spontaneous imbibition of seawater into chalk: Impact of the potential determining ions Ca^{2+}, Mg^{2+} and SO_4^{2-}, *Colloids and Surfaces A: Physicochem. Eng. Aspects*, 301, 199–208, 2007.
122. McCool, C. S., Green, D. W., Willhite, G. P., Shaw, A. K., Bhattacharya, S., and Singh, A.; *Permeability Reduction by Treatment with KUSP1 Biopolymer Systems*; 1997 SPE International Symposium on Oilfield Chemistry, Houston, TX, February 18–21, 1997 (SPE 37298).
123. Asghari, K., Performance and properties of KUSP 1 boric acid gel system for permeability modification purposes. *Petroleum Science and Technology*, 20, (9–10), 1141–1150, 2002.
124. EP 0343294; application by Conoco Inc.
125. Canadian patent CA 1303841, assigned to Conoco Inc.
126. Acock, A. M. and Reis, J. C.; *Oil Recovery Improvement through Profile Modification by Thermal Precipitation*, SPE/DOE 9th Symposium on Improved Oil Recovery, Tulsa, OK, April 17–20, 1994 (SPE/DOE 27831).
127. Ameri, S., Aminian, K., Wasson, J. A., and Durham, D. L.; *Improved CO2 enhanced oil recovery: Mobility control by in-situ chemical precipitation*, US DOE Fossil Energy Report DOE/MC/22044-15 (DE91002243), June 1991.
128. Reis, J. C.; *Oil recovery improvement through profile modification by thermal precipitation. Final report*, DOE Fossil Energy Report DOE/BC/14660-11 (DE94000122), April 1994.
129. Reis, J. C; *Oil recovery improvement through profile modification by thermal precipitation. Annual report, October 1, 1990–September 30, 1991*, DOE Fossil Energy Report DOE/BC/14660-8 (DE93008776), 1991.

18 Inhibition of Molten Aluminum Oxidation with Boron

Amitabha Mitra and David A. Atwood

CONTENTS

18.1 INTRODUCTION

Aluminum is one of the most abundant elements in the earth's crust and is a resource that will be virtually impossible to exhaust (Altenpohl, 1998). However, aluminum forms exceedingly strong bonds to oxygen and the lattice energy of Al_2O_3 is ~15,000 kJ/mol, one of the highest known. By comparison, the lattice energy of MgO is ~3800 kJ/mol. Thus, the primary consideration in the manufacture and recycling of aluminum is the energy required to reduce the element and free it from such a stablebonding environment. This requires approximately 13 kWh/kg in the primary refining (Altenpohl, 1998). It is ideal, then, to recycle aluminum products to take advantage of the energy that has already been expended in obtaining the element. For example, primary aluminum smelting requires ~118,000 MJ of energy while recycling requires only ~7000 MJ. Recycling-used aluminum requires only about 5% of the energy required to make fresh aluminum from bauxite ore.

The oxidation of aluminum is difficult to prevent. Indeed, all aluminum products have a surface coating of aluminum oxide, unless specifically coated to prevent this oxidation. Fortunately, aluminum oxide provides a protective coating on aluminum. During recycling the oxidation of aluminum as "dross" is very costly (Hryn et al., 2002; De et al., 2004; Hryn and Ningileri, 2006). An average of 4% of the processed aluminum, ~400,000 metric tons a year with a value of ~$700,000,000, is lost as dross (Moore et al., 2002). Moreover, incorporation of oxide dross in the aluminum compromises homogeneity, and mechanical properties such as fatigue performance, yield strength, and ductility in the finished product. Preventing, managing, and removing dross lead to a significant increase in the energy required by the industry. In the United States the energy loss due to oxidative melt loss is about 70 trillion BTU per year.

The majority of "aluminum" products are actually alloys, primarily with magnesium. Magnesium actually facilitates the formation of dross and will eventually cause "breakaway oxidation" where substantially more than 4% of the aluminum is oxidized, in some cases nearing 50%.

The minimization of dross, and the unnecessary use of energy, during aluminum alloy recycling is a critical issue in the aluminum industry. The United States Department of Energy has supported research to develop new technologies to address this problem. The goal is to reduce oxidative melt loss by 50% through the use of techniques such as limiting exposure of the melt to the furnace atmosphere and addition of additives to the melt. This would result in an estimated annual energy saving in excess of 58 trillion BTU. Currently, however, there are no industrial-scale technologies capable of preventing the oxidative melt loss of aluminum.

One potential solution to this problem is the use of low levels of boron additives, which are known to dramatically inhibit the oxidation of aluminum alloys. This publication will review what is currently known about this "boron additive effect." In particular, it will explore the chemistry of boron with magnesium since the conversion of periclase (MgO) to spinel ($MgAl_2O_4$) is the key step that leads to "breakaway" oxidation and significant dross formation.

18.2 OXIDATION ACCELERATION WITH ADDITIVES

The oxidation of liquid metals is accelerated by additives such as magnesium and silicon that can be incorporated as alloying elements or "seeded" into the melt as oxides. This process has been commercialized as the Directed Metal Oxidation, DIMOX™ or DMO process for creating metal–metal oxide composites (Newkirk et al., 1986; Kennedy, 1991; Jayaram, 1992; Fareed, 1995; Koh et al., 1995; Park and Kim, 2001). The addition of dopants, internally or externally, distinguishes DMO from reaction bonding techniques. For example, aluminum ordinarily oxidizes slowly at the surface of the melt but when periclase (MgO) or spinel ($MgAl_2O_4$) is added, the oxidation occurs rapidly to provide a uniform distribution of metal and oxide throughout the resulting ceramic. When MgO is added to molten aluminum there is a short latency period where the aluminum in the alloy continues to oxidize slowly before rapid, "breakaway," oxidation occurs throughout the melt. Breakaway oxidation occurs *after* the formation of crystalline spinel ($MgAl_2O_4$). In the latency period the MgO is converted into $MgAl_2O_4$ to form columns and three-dimensional networks that introduce fresh molten aluminum and MgO to the surface. When optimized, the DMO process can cause 50% of the aluminum to be oxidized and results in useful aluminum–alumina composites. Oxidation can also occur adventitiously during aluminum alloy recycling, leading to significant loss of recoverable aluminum.

The DMO process consists of a brief period of rapid oxidation, incubation with little oxidation, then substantial oxidation and formation of the bulk metal–metal oxide composite.

The DMO of Al–Mg–Si alloys to form alumina–aluminum composites provides a representative example of this process and reveals the importance of spinel (Figure 18.1) (Salas et al., 1991; Antolin et al., 1992). After heating to ~877°C the molten alloy surface rapidly oxidizes to form a MgO-covered $MgAl_2O_4$ layer. Further heating to ~1127°C causes the duplex MgO–$MgAl_2O_4$ layer to slowly thicken. Finally, there is an onset of significant oxide growth marked by the spread of a metal-rich zone over the surface oxides through microscopic channels in the thickened duplex layer. The spinel channels remain in contact with the molten alloy throughout the process and serve as

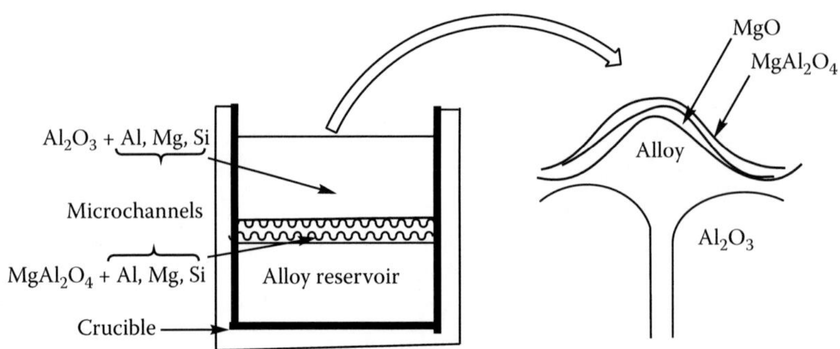

FIGURE 18.1 Schematic drawing of the bulk oxide growth process (left) and details of the near-surface-layered structure (right). (Redrawn from Vlach, K. C., O. Salas et al. 1991. *J. Mater. Res.* **6**(9): 1982–1995.)

a conduit for bringing magnesium and aluminum to the surface and oxygen down into the melt. During the final rapid growth stage small composite nodules grow and coalesce to form a planar growth front which persists until growth is complete. The external surface of the composite is covered by a metastable MgO and molten alloy layer, which prevents the formation of a protective Al_2O_3 layer that would limit the reaction (Kimura, 2001). Aluminum oxidation occurs substantially above the spinel–metal layer and increases with an increase in the amount of spinel that forms.

Thus, the formation of MgO followed by appearance of a surface layer of Mg–$MgAl_2O_4$ precedes bulk oxidation. The conversion of MgO and Al_2O_3 only requires the migration of aluminum cations into the periclase lattice and dissolution of MgO into the molten metal (Salas et al., 1991; Vlach et al., 1991). In the context of aluminum alloy recycling to *reduce* aluminum oxidation and increase recoverable elemental aluminum, the conversion of MgO to $Mg_2Al_2O_4$ must be prevented or inhibited.

18.3 OXIDATION INHIBITION BY BERYLLIUM AND BORON

It has been known since the 1960s that boron and beryllium inhibit the oxidation of magnesium and aluminum (Thiele, 1962a, b; Emley, 1966; Belitskus, 1974; Cochran et al., 1977). For example, when beryllium is added in amounts as low as 0.001% (by weight; or in the range of 5–100 ppm) the oxidation of Al-3.5 Mg alloy was inhibited for at least 93 h at 800°C (0.01% weight gain) whereas without beryllium breakaway oxidation started shortly after melting (Wikle, 1978). Dusting of elemental boron decreased the oxidation of Al–4.5 Mg alloy for 73 h at 750°C (0.06% weight gain). Boron doping suppressed spinel formation during liquid-phase sintering of alumina coated with magnesia aluminosilicate (MgO–Al_2O_3–SiO_2) glass in the temperature range 1400–1460°C (Nakajima and Messing, 1998). However, the addition of crystalline spinel to the molten metal overcomes the inhibitory effect of boron and beryllium and leads to rapid oxidation (Cochran et al., 1977).

Oxidation inhibition of molten elemental magnesium was achieved with boron and beryllium reagents when added to the molding sand prior to casting (Emley, 1966). The effect of beryllium as an alloying element (as BeO) is to reduce magnesium diffusion in the surface MgO layer. Beryllium and boron appear to have some interaction with magnesium in the absence of aluminum. Beryllium concentrates in the MgO surface layer and has the effect of "tightening" the lattice and reducing magnesium ion diffusion. The Pilling–Bedworth ratio, R, defined as the ratio of the volumes of the metal oxide to the metal volume, is 0.81 for magnesium (Pilling and Bedworth, 1923). A value of $R < 1$ implies that the magnesium oxide layer is porous and nonprotective, and thus unable to prevent propagation of fissures and cracks that would facilitate further oxidation. Beryllium may be protective for MgO by having a slightly higher affinity than magnesium towards oxygen. This would lead to a uniform thick protective oxide layer instead of thick nodular oxide features (Czerwinski, 2004). However, the Lux–Flood acidity parameter of BeO (−2.2) indicates that it is basic and only slightly less so than MgO (−4.5) (Huheey et al., 1993). In contrast, B_2O_3 is a Lux–Flood acid with an acidity parameter of 1.5, making it more suitable than BeO to combine with surface MgO oxygen atoms.

18.4 SOLID ALLOY OXIDATION INHIBITION WITH BORON

The effect of coating 2024 (92.5% Al, 1.52% Mg) and 5052 (96.2% Al, 2.60% Mg) alloys with H_3BO_3 and B_2O_3 was recently explored (B. C. Yearwood and A. Mitra, unpublished; Atwood, 2004). Squares of 5052 and 2024 alloy were coated by immersion in a saturated solution of H_3BO_3. Each square was heated in an alumina crucible at 800°C from 3.5 to 9.5 h. The uncoated alloys exhibited a black surface coating on the partially disintegrated alloy, whereas the coated alloy squares were an ash gray and intact (Figure 18.2). This indicates that the coated material was protected from oxidation. In another experiment, samples of 2024 and 5052 alloy disks coated with a dusting of

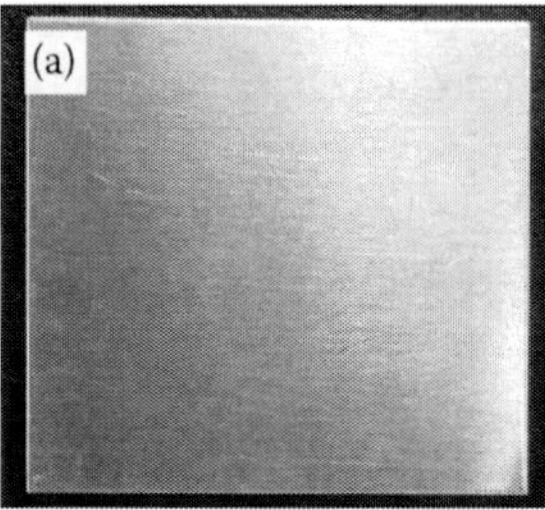

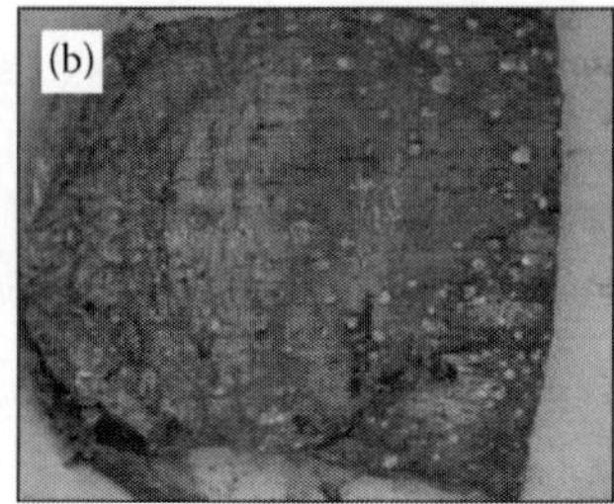

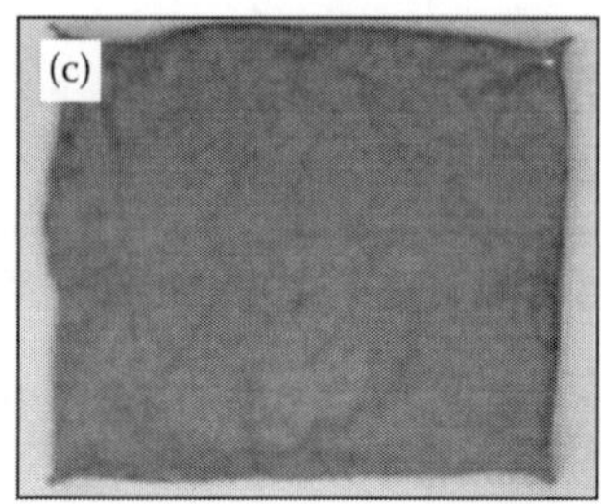

FIGURE 18.2 (a) Pristine 5052 alloy; (b) uncoated alloy heated at 800°C for 4 h; (c) alloy coated with H_3BO_3 and heated at 800°C for 4 h.

TABLE 18.1
Mass Change on Heating Uncoated and Coated Alloys

Alloy	Coating	Heating Time	% Change
5052	None	30 min	0.022
5052	B_2O_3	30 min	−1.065
5052	None	1 h	+0.030
5052	B_2O_3	1 h	−0.951
2024	None	1 h	+0.028
2024	B_2O_3	1 h	−0.18
2024	B_2O_3	6 h	+0.047
2024	B_2O_3	6 h	−0.318

B_2O_3 were heated at 800°C. The alloys were coated by immersion in a B_2O_3 solution to provide a thin surface coating. After 6 h of heating, uncoated 2024 alloy exhibited a black surface coating with pockets or craters on the bottom of the disk. Coated 2024 alloy had a gray surface coating without discernible pockets, showing the protective effect of the coating composition against oxidation-induced spinel formation. The effect of coating was further evident by the change in weight of the alloy disks (Table 18.1). The increase in weight for the uncoated alloy samples (e.g., 0.03% for 5052 alloy after 1 h of heating and 0.047% for 2024 alloy after 6 h of heating) can be ascribed to the oxide and spinel formation. No such weight gain for the alloy disks coated with B_2O_3 indicates the protective effect of boron compound against oxidation; the slight loss in weight in these cases being probably due to the volatilization of the boron compound.

18.5 COMPOUND FORMATION BETWEEN BORON AND MgO

Various forms of magnesium borates formed when MgO and B_2O_3 were heated to 850°C (Kitamura et al., 1988). The interaction of the boron compounds with MgO was investigated by powder XRD, solid-state NMR, and IR studies (Mitra et al., 2010). Mixtures of magnesium oxide and boron oxide or boric acid powder in 1:1, 10:1, and 100:1 weight ratios were heated in an electrical furnace at 400°C, 600°C, 800°C, and 1000°C. Crystalline and noncrystalline species were observed in the resulting product mixtures. This included three- and four-coordinate boron species (Table 18.2). Heating temperature and, more importantly, the nature of the boron starting material determined the nature of the compounds formed. Powder XRD revealed that mixtures of boron oxide and magnesium oxide produced magnesium borate hydrate ($MgB_6O_{10}{\cdot}7H_2O$) and magnesium pyroborate (suanite, $Mg_2B_2O_5$), whereas mixtures of boric acid and magnesium oxide led to the formation of kotoite ($Mg_3(BO_3)_2$) and suanite. This implies that, at low levels, the boron reagents react differently

TABLE 18.2
Products Identified by XRD: $MgO–B_2O_3$ and $MgO–B(OH)_3$ Combinations[a,b]

B_2O_3:MgO	600°C	800°C	1000°C
1:1	M	M + S	M + S + C + MB
1:10	M	M	M
1:100	M	M	M
$B(OH)_3$:MgO			
1:1	M	M + U	M + S + K
1:10	M	M	M + U
1:100	M	M	M

[a] M = MgO, S = suanite ($Mg_2B_2O_5$), C = magnesite ($MgCO_3$), MB = magnesium borate hydrate ($MgB_6O_{10}·7H_2O$), K = kotoite ($Mg_3(BO_3)_2$), U = unidentified.

[b] Adapted from Mitra, A., B. C. Yearwood et al. 2010. *Main Gr. Chem.* **9**(1–2): 193–201.

with magnesium oxide before dehydration converts boric acid into boron oxide, although the products in each reaction match those in the $MgO–B_2O_3$ phase diagram reasonably well given the nonequilibrium conditions (Mutluer and Timucin, 1975). An unidentified compound was also detected along with MgO for the boric acid–magnesium oxide combinations of 1:10 at 1000°C and 1:1 at 800°C. Notably, the addition of B_2O_3 or H_3BO_3 did not alter the crystallinity of MgO. However, the solid-state NMR showed that the change in mixing temperature caused a structural change in the boron species, producing more three-coordinate boron (isotropic chemical shift in the range 8.3–16.5 ppm) relative to four-coordinate boron (isotropic chemical shift in the range −0.80–1.60 ppm). This is illustrated in Figure 18.3 where the three-coordinate boron peak intensity increases on going

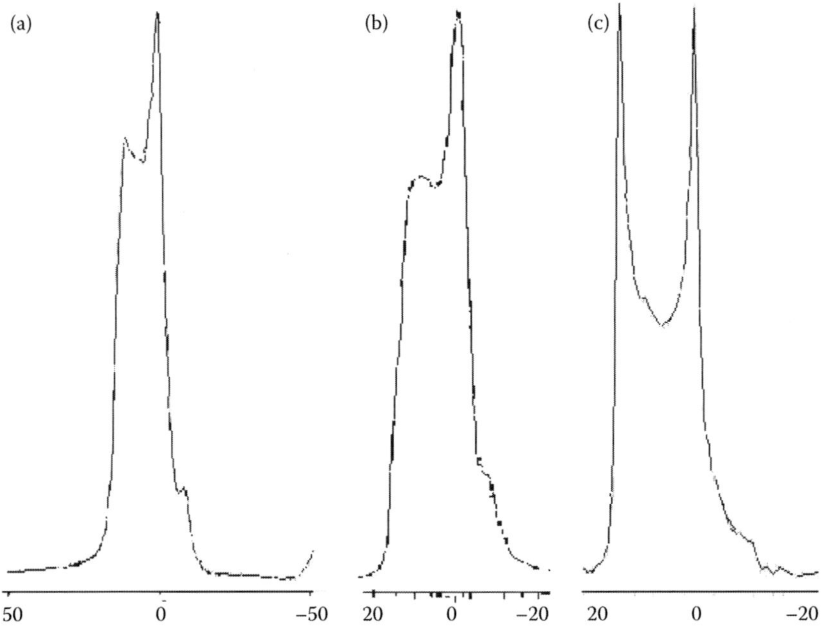

FIGURE 18.3 Solid-State ^{11}B NMR of: (a) B_2O_3 at 25°C, (b) 1:10 B_2O_3:MgO mixture after heating at 600°C, and (c) 1:10 B_2O_3:MgO mixtures after heating at 1000°C. (Redrawn from Mitra, A., B. C. Yearwood et al. 2010. *Main Gr. Chem.* **9**(1–2): 193–201.)

from 600°C to 1000°C mixing temperature. This probably indicates that the mixing temperature has an effect on network formation involving bridging oxygen in the B–O–Mg system.

It can be concluded that, as expected, combinations of boron reagents with MgO leaves the oxide unchanged so that periclase was the most significant compound, but that suanite is formed at higher temperatures. Continuous heating of the $MgO–B_2O_3$ mixtures over long periods of time is likely to result in increasing amounts of suanite.

18.6 BORON MIGRATION AND MgO LATTICE

According to the "reactive element effect," the reactive element ion, such as beryllium, diffuses in to the native oxide grain boundaries and prevents the outward diffusion of substrate metal cations (Czerwinski and Smeltzer, 1993; Czerwinski and Szpunar, 1998; Czerwinski, 2000, 2004). The inhibitory effect of boron on aluminum alloy oxidation is clearly a surface phenomenon given the effectiveness of very low levels of boron. This could occur through a combination of boron migration into the MgO lattice and/or boron bonding to the defect-rich MgO surface (Choudhary and Pandit, 1991).

When the mixed oxide, $MgO–B_2O_3$ is formed by coprecipitation of $Mg(OH)_2$ and H_3BO_3 (0–10% by weight) the resulting solids are characterized as periclase (MgO). Boron does not alter the structure of MgO. It does, however, change the surface properties by increasing the acidic sites and decreasing the basic sites (Aramendia et al., 1996, 1999). There is also a significant reduction in MgO surface area and pore volume. In one representative example the surface area was 18 m^2/g with 10 weight% boron compared to 63 m^2/g with no boron (Aramendia et al., 1999). Characterization of the boron environment by diffuse reflectance IR spectroscopy (DRIFT) and ^{11}H B Magic Angle Spinning (MAS) NMR indicates a trigonal environment. This is observed in the structures of suanite (Figure 18.4) and kotoite (Figure 18.5). Trigonal boron environments are found for surface boron atoms in borated aluminum catalysts when dehydrated and tetrahedral boron with hydration and with boria amount greater than 2%. The observation of an increase in surface acidity and trigonal boron environments is common for combinations of metal oxides with sources of boron oxide. This also occurs, for example, in $Al_2O_3–B_2O_3$ (Mazza et al., 1992; Xu et al., 1992; Colorio et al., 1996; Bautista et al., 1998) and $TiO_2–B_2O_3$ (Xu et al., 1992) systems. When combined at the % level the boron is relatively evenly dispersed throughout the MgO lattice (and Al_2O_3 and TiO_2 lattices) and not localized at the surface. Migration of boron into the Al_2O_3 lattice could also strengthen the "protective" coating of alumina that exists on the surface of the alloy substrates where the degradation of the heated alloy would begin. This would follow the Pilling and Bedworth rules for metals and metal oxides. In the boria–alumina mixed oxides prepared by a sol–gel method with boron–aluminum atomic ratios varying from 0.013 to 1.643 changes in the local microstructure of alumina

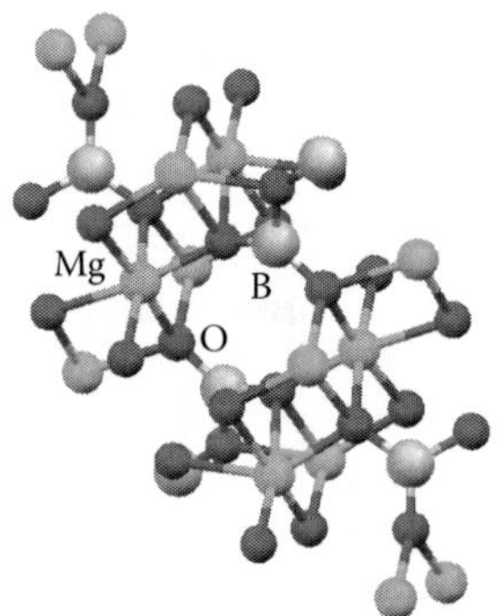

FIGURE 18.4 Crystal structure of suanite, $Mg_2B_2O_5$. (Structure redrawn from Guo, G.-C., W.-D. Cheng et al. 1995. *Acta Cryst.* **C51**(12): 2469–2471.)

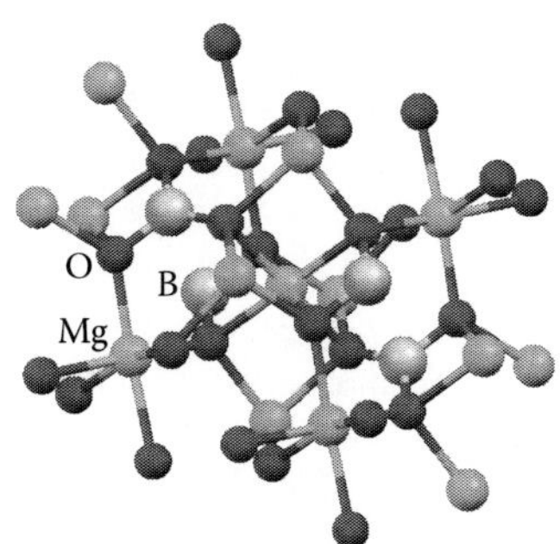

FIGURE 18.5 Crystal structure of kotoite, $Mg_2B_2O_6$. (Structure redrawn from Effenberger, H. and F. Pertlik 1984. *Zeit. Kristallogr.* **166**(1–2): 129–140.)

were observed. Introduction of boron alters the octahedral arrangement around aluminum with creation of some tetrahedral aluminum species (Dumeignil et al., 2005). More importantly, however, the incorporation of boria into alumina prevents the thermal conversion of γ-Al_2O_3 into α-Al_2O_3 (Bautista et al., 1998). This would be similar to boria preventing the conversion of MgO to $MgAl_2O_4$ since the two materials have similar hexagonal closest packed oxide structures.

18.7 CONCLUSIONS

The combination of MgO and either boric acid or boron oxide produces not only, as expected, predominantly unchanged MgO, but also suanite (Figure 18.4). This could indicate that suanite is the thermodynamically preferred Mg-B oxide at 1000°C. At this temperature kotoite was observed with boric acid but not boron oxide, indicating that differences in the initial bonding of the boron reagents on the MgO lattice can influence the resulting product.

Boron additives appear to inhibit the formation of $MgAl_2O_4$ by providing a covalent surface coating on the MgO crystallites. The charge density of B(III) is exceptionally high, 1663 C/m^3, making covalent bond formation more likely than migration of boron into the MgO bulk. By comparison, the charge densities of Mg(II) (120 C/m^3) and Al(III) (770 C/m^3) are much lower, allowing the two metals to readily migrate as cations to occupy octahedral sites in the closest packed oxide lattice of spinel. The structures of suanite (Figure 18.4) and kotoite (Figure 18.5) reveal that boron oxide bonding disrupts the MgO lattice, in keeping with the likelihood of surface bonding. With surface bonding only boron oxide would be able to limit the transformation of MgO to $MgAl_2O_4$ at very low levels, as observed.

REFERENCES

Reduction of Oxidative Melt Loss. Retrieved February 3, 2010, from http://www1.eere.energy.gov/industry/aluminum/pdfs/meltlossreduction.pdf.

Altenpohl, D. G. 1998. *Aluminum: Technology, Applications, and Environment. A Profile of a Modern Metal.* Washington, DC and Warrendale, PA: The Aluminum Association and The Minerals, Metals & Materials Society.

Antolin, S., A. S. Nagelberg et al. 1992. Formation of Al_2O_3/Metal composites by the directed oxidation of molten aluminum-magnesium–silicon alloys: Part 1, microstructural development. *J. Am. Ceram. Soc.* **75**: 447–454.

Aramendia, M. A., V. Borau et al. 1996. Synthesis and characterization of various MgO and related systems. *J. Mater. Chem.* **6**(12): 1943–1949.

Aramendia, M. A., V. Borau et al. 1999. Synthesis and characterization of MgO–B_2O_3 mixed oxides prepared by coprecipitation; selective dehydrogenation of propan-2-ol. *J. Mater. Chem.* **9**(3): 819–825.

Atwood, D. A. 2004. *Compositions and Methods for Reducing Oxidation of Metal Alloys During Heating.* U. S. Utility Patent Application.

Bautista, F. M., J. M. Campelo et al. 1998. Structural and textural characterization of $AlPO_4$–B_2O_3 and Al_2O_3–B_2O_3 (5–30 wt% B_2O_3) systems obtained by boric acid impregnation. *J. Catal.* **173**(2): 333–344.

Belitskus, D. L. 1974. Effect of H_3BO_3, BCl_3, and BF_3 pretreatments on oxidation of molten Al–Mg alloys in air. *Oxidation Metals* **8**: 303–306.

Choudhary, V. R. and M. Y. Pandi. 1991. Surface properties of magnesium oxide obtained from magnesium hydroxide: Influence on preparation and calcination conditions of magnesium hydroxide. *Appl. Catal.* **71**(2): 265–274.

Cochran, C. N., D. L. Belitskus et al. 1977. Oxidation of aluminum–magnesium melts in air, oxygen, flue gas, and carbon dioxide. *Metall. Trans.* **8B**: 323–332.

Colorio, G., J. C. Vedrine et al. 1996. Partial oxidation of ethane over alumina-boria catalysts. *Appl. Catal.* **137**(1): 55–68.

Czerwinski, F. 2000. On the use of the micromarker technique for studying the growth mechanism of thin oxide films. *Acta Mater.* **48**(3): 721–723.

Czerwinski, F. 2004. The early stage oxidation and evaporation of Mg–9%Al–1%Zn alloy. *Corrosion Sci.* **46**(2): 377–386.

Czerwinski, F. and W. W. Smeltzer, 1993. The early-stage oxidation kinetics of CeO[sub 2] sol-coated nickel. *J. Electrochem. Soc.* **140**(9): 2606–2615.

Czerwinski, F. and J. A. Szpunar, 1998. The influence of crystallographic orientation of nickel surface on oxidation inhibition by ceria coatings. *Acta Mater.* **46**(4): 1403.

De, A. K., A. Mukhopadhyay et al. 2004. Numerical simulation of early stages of oxide formation in molten aluminium–magnesium alloys in a reverberatory furnace. *Modell. Simul. Mater. Sci. Eng.* **12**(3): 389–405.

Dumeignil, F., M. Rigole et al. 2005. Characterization of boria–alumina mixed oxides prepared by a sol–gel method. 1. NMR characterization of the xerogels. *Chem. Mater.* **17**(9): 2361–2368.

Effenberger, H. and F. Pertlik, 1984. Refinement of the crystal structures of the isotypic compounds $M_3(BO_3)_2$ (M = Mg, Co, Ni) (structural type: Kotoite). *Zeit. Kristallogr.* **166**(1–2): 129–140.

Emley, E. F. 1966. *Principles of Magnesium Technology*. New York: Pergamon Press.

Fareed, A. S. 1995. Ceramic matrix composite fabrication and processing: Directed metal oxidation. In: *Handbook on Continuous Fiber-Reinforced Ceramic Matrix Composites* R. L. Lehman, S. K. El-Rahaiby and J. B. Wachtman, Jr. (eds), West Lafayette, IN: Ceramics Information Analysis Center, pp. 301–324.

Guo, G.-C., W.-D. Cheng et al. 1995. Monoclinic $Mg_2B_2O_5$. *Acta Cryst.* **C51**(12): 2469–2471.

Hryn, J. and S. Ningileri, 2006. Reduction of oxidative melt loss of aluminum and its alloys. Final Report, pp. 1–50.

Hryn, J. N., M. J. Pellin et al. 2002. *Method to Decrease Loss of Aluminum and Magnesium Melts*. Chicago, IL: The University of Chicago.

Huheey, J. A., E. A. Keiter et al. 1993. *Inorganic Chemistry: Principles of Structure and Reactivity*. New York: HarperCollins College, 321pp.

Jayaram, V. 1992. Directed metal oxidation and infiltration. *Met. Mater. Proc.* **4**(1): 51–58.

Kennedy, C. R. 1991. Directed metal oxidation and pressureless metal infiltration: New technologies for the fabrication of reinforced ceramics and metals. *Mater. Sci. Monogr.* **68**: 691–700.

Kimura, A. 2001. Breakaway behavior of surface oxide film on aluminium–silicon–magnesium alloy powder particles at high temperature in a vacuum. *Mater. Trans.* **42**(7): 1373–1379.

Kitamura, T., K. Sakane et al. 1988. Formation of needle crystals of magnesium pyroborate. *J. Mater. Sci. Lett.* **7**: 467–469.

Koh, S. C., D. K. Kim et al. 1995. Fabrication of Al_2O_3/Al composites by directed metal oxidation of Al–Mg alloy. *J. Mater. Syn. Proc.* **3**(2): 105–110.

Mazza, D., M. Vallino et al. 1992. Mullite-type structures in the systems alumina-Me_2O (Me = Na, K) and alumina-boron oxide. *J. Am. Ceram. Soc.* **75**(7): 1929–1934.

Mitra, A., B. C. Yearwood et al. 2010. Oxide compound formation between boron and excess magnesium. *Main Gr. Chem.* **9**(1–2): 193–201.

Moore, J. F., K. Attenkofer et al. 2002. X-ray diffraction studies of liquid aluminum alloy oxidation. Retrieved February 6, 2010, from http://www.aps.anl.gov/apsar2002/MOORE1.PDF.

Mutluer, T. and M. Timucin, 1975. Phase equilibriums in the magnesium oxide–boron oxide system. *J. Am. Ceram. Soc.* **58**(5–6): 196–197.

Nakajima, A. and G. L. Messing, 1998. Liquid-phase sintering of alumina coated with magnesium aluminosilicate glass. *J. Am. Ceram. Soc.* **81**(5): 1163–1172.

Newkirk, M. S., A. W. Urquhart et al. 1986. Formation of lanxide™ ceramic composite materials. *J. Mater. Res.* **1**(1): 81–89.

Park, H. S. and D. K. Kim, 2001. Effect of silica surface dopants on the formation of alumina/aluminum composites by the directed metal oxidation of an aluminum alloy. *J. Am. Ceram. Soc.* **84**(11): 2526–2530.

Pilling, N. B. and R. E. Bedworth, 1923. The oxidation of metals at high temperatures. *J. Inst. Met.* **29**: 529–591.

Salas, O., H. Ni et al. 1991. Nucleation and growth of Al_2O_3/metal composites by oxidation of aluminum alloys. *J. Mater. Res.* **6**(9): 1964–1981.

Thiele, V. W. (1962a). Die oxydation von aluminium- und aluminiumlegierungs-schmelzen. *Aluminum* **12**: 780.

Thiele, V. W. (1962b). Die oxydation von Aluminium- und aluminiumlegierungs-schmelzen. *Aluminum* **11**: 707.

Vlach, K. C., O. Salas et al. 1991. A thermogravimetric study of the oxidative growth of Al_2O_3/Al alloy composites. *J. Mater. Res.* **6**(9): 1982–1995.

Wikle, K. G. 1978. Improving aluminum castings with beryllium. *AFS Trans.* **86**: 513–518.

Xu, B.-Q., T.-X. Cai et al. 1992. The selective synthesis of but-2-ene from ethanol over alumino-borate B–C1 catalyst. *J. Chem. Soc., Chem. Commun.* 1228–1229.

19 Recent Progress in Extraction Agents Based on Cobalt Bis(Dicarbollides) for Partitioning of Radionuclides from High-Level Nuclear Waste*

Bohumír Grüner, Jiří Rais, Pavel Selucký, and Mária Lučaníková

CONTENTS

19.1 INTRODUCTION

The cobalt bis(dicarbollide)(1-) ion $[(1,2\text{-}C_2B_9H_{12})_2\text{-}3,3'\text{-}Co](1\text{-})$ (CD$^-$) was synthesized and described by Hawthorne (1965); however, its use as a perfect hydrophobic anion was awaited until 1976 when at first its extraction properties were published. The cobalt bis(dicarbollide) belongs to a class of the

* Dedicated to Professor Bohumil Štíbr, on the occasion of his 70th birthday.

low-nucleophilic, low-coordinating anions (Strauss, 1993; Reed, 1998). A characteristic feature of the bis-icosahedral CD^- ion is good solubility of its free conjugated acids and most of their salts in medium-polarity solvents such as ethers, nitrosolvents, halogenated solvents, and so on, to which they can be extracted from the water phase together with alkali metal cations. The unique importance of CD^- in the design of extraction agents lies in the extraordinary chemical and thermal stability due to "pseudoaromaticity" and to the completely filled electronic shell of the Co(III) cation that is furthermore sterically shielded by two bulky dicarbollide ligands. The central cobalt atom thus behaves as chemically inert and stable toward a nucleophilic attack. A comprehensive review of the general chemistry of the cobalt bis(dicarbollide) anion was published by Sivaev and Bregadze (1999).

The discovery that the cobalt bis(dicarbollide)(1-) anion can be used as an efficient radionuclide extractant was partly fortuitous, and also partly due to the fact that both the Institute of Inorganic Chemistry (IIC) and the Nuclear Research Institute (NRI) are located in the same area—Řež, Czech Republic. For many years at the NRI we have been interested in developing hydrophobic univalent anions suitable for the extraction of $^{137}Cs^+$, and at IIC the cobalt bis(dicarbollide)(1-) anion has been a compound of interest since the 1970s. Even from the beginning, preliminary tests involving this anion gave promising results and revealed extremely favorable properties. We found that during the two-phase (water/organic) extraction process the CD^- anion distributed well into the organic phase—nitrobenzene—due to its high hydrophobicity and, still more importantly, due to its exceptional stability in three molar nitric acid environment without the formation of undissociated cobalt bis(dicarbollide)(1-) acid. This property enabled the extraction of biotoxic ^{137}Cs directly from an original PUREX process raffinate without any preliminary correction of acidity. The separation of valuable and biotoxic elements from spent nuclear fuel is important for both economic and ecological reasons and, as a rule, solvent extraction processes are used routinely worldwide.

Thus, several patents soon appeared after the disclosure of the excellent properties of cobalt bis(dicarbollide) (1-) and its chloro-protected variant CCD^-, which is even more stable against decomposition by nitric acid (Rais, 1974, 1976; Kyrš, 1976, 1980). The full story of the early and later developments until 2004, leading to full-scale plant testing in Russia already in 1985, has been presented elsewhere (Rais and Grüner, 2004).

The basic chemistry of CD^- extraction of $^{137}Cs^+$ involves the transfer of the dissociated ion pair consisting of Cs^+ cation with cobalt bis(dicarbollide) anion from the water phase to the nitrobenzene phase. No chemical reaction or complexation occurs and only hydration and solvation forces apply. At first sight, the high extractability of the Cs^+ ion into nitrobenzene could be connected with a special affinity of this organic solvent toward Cs^+. However, this is not the case: Indeed, Cs^+ is much less solvated by nitrobenzene than by water, and the extraction is in fact facilitated by the extremely high affinity of the cobalt bis(dicarbollide) anion toward the organic solvent that makes the overall affinity of the ion pair toward nitrobenzene positive. From the acidic media, Cs^+ is extracted due to the rather high extraction-exchange constant $K(Cs^+/ H^+)$ because H^+ is still much less solvated than Cs^+. Thus, hydrophobicity of the cationic particle is the decisive factor in extractions with cobalt bis(dicarbollide) anion. Some less hydrophobic cations such as Sr^{2+} or Eu^{3+} (Am^{3+}) can be made more hydrophobic by complexing them with various neutral ligands.

This chapter summarizes recent progress (after approx. 2003) in two parallel lines of development in the field of CD^--based extractants for the treatment of high-level liquid waste (HLLW). The first part deals with the progress in use of chloroprotected ion CCD^- in synergic mixtures with organic complexing species for the partitioning of radionuclides. Since this field of Cs^+ and Sr^{2+} extractions is sufficiently developed into technological scale, the new entries relate mainly to technological aspects originating from the process design. This is followed by exploratory research in the field of extractions of trivalent nuclides (lanthanides and actinides Ln/An) using synergic mixtures of CD^- with various organic ionophores. The second part deals with development of ionic species merging CD^- anion and covalently bonded selective groups for the extraction of polyvalent nuclides. The latter area has been opened quite recently and the results have character of basic research focused on viable synthetic ways to new effective compounds and on studies leading to basic understanding of their solution and extraction properties. Therefore, the two parts also differ in their style and character of narration.

Our previous review on extraction with cobalt bis(dicarbollide) anions covering the progress from the beginning to around 2003 was published in 2004 (Rais and Grüner, 2004). Then onward, a number of studies on classical systems with chloroprotected cobalt bis(dicarbollide)(1-), CCD^-, as well as with sophisticated boron compounds containing metal-selective groups have appeared. The use of CCD^- for Cs^+ and Sr^{2+} extraction and early stages of development in the area of more selective extraction agents has also been outlined in two reviews by Grimes devoted to striking properties and applications of boron cluster compounds (Grimes, 2000, 2004).

Research in the area of CCD^- with new synergists continues with no signs of decreasing activity. The area of extraction with CCD^- and a synergist was recently reviewed by Babain from Khlopin Radium Institute, KRI (St. Petersburg, Russia). The author describes the practical application of CCD^- in worldwide nuclear technologies (Babain, 2010). Several other reviews as well dealing with or at least commenting on CCD^- technology for retreatment of PUREX waste have appeared since 2004. Paiva and Malik (2004) described extractants for the treatment of radioactive wastes up to 2004. A short review of possible practical extraction processes, while also mentioning CCD^-, was written by Choppin (2005), whereas a comprehensive review of multicoordinate ligands for actinide/lanthanide separations was published by Dam and Werboon (2007). Tachimori and Morita wrote an overview of extraction chemistry of reprocessing and also treated CCD^- processes quite recently (Tachimori, 2010). Analogously, the technological aspects of extractions with chloroprotected cobalt bis(dicarbollide) were reviewed by Hill in the same monograph (Hill, 2010). The Russian technological extraction process called CCD^- UNEX is mentioned in most contemporary reviews. Rather voluminous literature on diverse UNEX variants by scientists from KRI, Russia and INEL, USA mainly concerns the technological aspects of the proposed separation lines, and hence it is only laterally connected with the chemistry of cobalt bis(dicarbollides) themselves.

CD^- anions are extremely hydrophobic and behave as very strong superacids that exist in dissociate form in moderately polar and polar organic solvents even when in contact with strong aqueous acid (Plešek, 1992). This has allowed opening of quite new vistas for the study of the synergies of CCD^- with metal complexes. Such studies are not practicable with any other organic acid anion since none of them behave as a superacid, and being in contact with strong mineral acid they associate. Thus, a new branch of synergetic extractions from acidic aqueous media with CD^- has started. Mostly, two aspects have been under study recently: (1) testing of newly proposed synergists which, by themselves, do not extract well, but in conjunction with cobalt bis(dicarbollide) anion the extraction improves even by several orders of magnitude, and (2) the study of selective reagents capable of separating trivalent lanthanides from trivalent actinides (minor actinides, MA), that is, processes often designed as Ln/MA separations.

The new area of selective ionic ligands based on cobalt bis(dicarbollide)(1-) was opened about 10 years ago. These species contain both the hydrophobic boron cluster and a selective group/s attached by covalent bond in a single molecule. The entries into this field stem mainly from the synthetic efforts carried out at the Institute of Inorganic Chemistry, Řež, Czech Republic and evaluation of the extraction properties at the Nuclear Research Institute Řež plc., Czech Republic. Such compounds can conceivably even surpass the synergetic mixtures in possible countercurrent line because of the ease of control of the extractant composition (in a synergetic mixture the two components will generally decompose chemically and by radiation to differing extents, but not if only one extractant is used). Nevertheless, the desired compounds are still in the stages of laboratory development. Generally, if the specific group of neutral complexant is attached to negatively charged anions, the properties of the selective group might be greatly modified in comparison with the neutral synergic extractant. Thus, again the unprecedented possibilities exist for studies on neutral synergist bound to negative moiety. The cost of these relatively complicated compounds might seem, at first sight, prohibitive, but, if a reagent of exceptional properties is to be found, its price might be a secondary factor.

In this chapter, we review the achievements in the area of extractions with boron extractants accomplished in the last five years. The literature on extraction with cobalt bis(dicarbollide) and synergist, the UNEX process, and new boron-selective extractants is ample, and other connected science, such as ion-selective electrodes with boron anions, boron anion room-temperature ionic liquids, modeling of the mechanism of the extraction, IR studies on extracts, the state of CD^- in water solutions, or a description of some of the new ideas on solvation, resolvation theory, and so on, could not be covered here.

19.2 SYNERGISTS AND CCD^-

19.2.1 General

As argued previously, in the system with ion pairs (either dissociated in the organic phase or not) each ion present and extracted in the system influences the distribution of the other ions (Rais, 2004). In the systems under question, at least the following ions (as free or in the form of ion pairs) can be discerned in the organic phase: Cs^+, HL^+, SrL_n^{2+}, Ln $(Am)L_m^{3+}$, NO_3^-, CCD^- where presumably $n = 1, 2$, $m = 1, 2, 3$ and each ion can compete with all other ions in the ionic or ion pair forms. Thus, for example, increasing the concentration of Ln^{3+} synergist M or Sr^{2+} synergist N will generally lead also to a decrease in D_{Cs} since Cs^+ competes with the HM^+, HN^+ formed, and so on.

Whereas previously attention was paid mainly to "hard donors" permitting only extraction of trivalent lanthanides and trivalent actinides (minor actinides, MA) as a group (Rais and Grüner, 2004), in the last five years mainly systems with "soft donors" (molecules containing S or N atoms) have been studied.

19.2.2 Systems Involving Separations of Cs^+, Sr^{2+} (and Other Nuclides), Modifications of the UNEX Process

The original variant of extraction with dicarbollides was aimed at the extraction of ^{137}Cs and ^{90}Sr from acidic HLLW as proposed by Czechoslovak scientists and developed to pilot plant tests in collaboration with KRI, Russia (Rais and Grüner, 2004). This process is often referred to as CCD – PEG, since a two-component extractant is used, namely CCD^- for Cs^+ and polyethylene glycol (PEG) for Sr^{2+} extractions. PEG wraps as a helix around Sr^{2+} with –O– atoms oriented toward an ion and $–CH_2CH_2–$ groupings outward making the complexed Sr^{2+}, PEG hydrophobic and extractable. The CCD – PEG segment line forms a part of the complete retreatment cycle UREX+, and, as such it was tested; for example, see the report by Vandegrift (2004). One technical note is appropriate here: in several articles one PEG-specific compound, namely Slovafol 909 (mean composition: *p*-nonylphenol nonaethylene glycol of Slovakian provenience), is mentioned. Its production was apparently stopped some years ago and even if some quantities of Slovafol 909 are still scattered through different laboratories, currently available compounds, for example, Dow Chemical Triton X-100 and Triton X-114 can be alternatives.

In contrast to the original CCD-PEG process, the UNEX process in its present form focuses on the extraction of quite a number of components from PUREX waste: Cs^+ ($+Rb^+$), Sr^{2+} ($+Ba^{2+}$), trivalent lanthanides, and Am^{3+}, Cm^{3+}. This approach simplifies the extraction line, and would be allegedly economically profitable; however, it can cause other difficulties. The control of the solvent composition in the long run might be difficult. Individual synergists might decompose differently and pass differently into the aqueous phase. In addition, the required stripping agent is of a complicated nature and usually involves a mixture of compounds, some of them in quite high concentrations, and these will form secondary waste. Although these problems can be partially solved as shown below, the separate extractions in two serial countercurrent lines should perhaps be considered too. For example, the sum of all Ln + MA could be extracted as a first step at 3 M HNO_3 by some hard donor and stripped by 7–8 M HNO_3 as proposed earlier (making use of steep ca. –3 slope log–log dependence of $D_{M(3+)}$ on $c(H^+)_{aq}$) (Rais, 1994). Further, Cs^+ and Sr^{2+} could be extracted in a tandem line by the classical dicarbollide process, maybe with the use of a new solvent (phenyltrifluoromethyl sulfone, FS-13 proposed in Russia). It seems that no such scheme has been checked as yet.

Russian and U.S. scientists have made several important improvements in the area of the UNEX process and a number of technological studies concerning the process in detail have been carried out (Herbst, 2002a,b).

A comprehensive, instructive, and interesting history of development until 2003 is given in a paper by Herbst (2003). Sometimes, contradictory demands had to be solved. Thus, concomitant extraction of Zr, Mo, and Fe, mainly from dissolved calcine solutions, led to the exhaustion of the organic phase by an accumulation of these metal ions. HF had to be added, which in its turn caused excessive corrosion of the equipment. This was again overcome by adding aluminum nitrate, but later a simple scrub by 0.2 M NH_4NO_3 proved to be sufficient. The last UNEX solvent composition (0.08 M CCD^-, 0.02 M CMPO, and 0.5 vol.% PEG-400 in FS-13, UNEX A) is arguably the ideal one. This composition, according to the authors, withstands 14 M HNO_3 at temperatures up to 120°C and the mixture is fire and explosion safer than the PUREX process. The radiation damage and losses into the outgoing aqueous phase are small and the solvent is considered as not toxic (we have found no literature data on toxicity tests).

In a countercurrent mode on centrifugal extractors, the process was tested both with acidic dissolved calcine INEEL waste (Herbst, 2002a,b) and with original acidic waste (Herbst, 2003). In the latter paper (Herbst, 2003), a comprehensive table of the details of six tests performed with UNEX solvent until 2003 have been published. The four latest tests were performed using phenyl trifluoromethyl sulfone FS-13. In all tests, the concentration of H^+CCD^- was 0.08 M and the concentration of PEG 400 varied between 0.35 and 0.6 vol.% with a value 0.4 vol.% in the most recent test. Am^{3+} and M^{3+} synergist, diphenyl-*N*,*N*-di-*n*-butylcarbamoylmethyl phosphine oxide, CMPO (denoted also as Ph_2Bu_2CMPO in some reports), was used in nearly all cases, in the most recent test at 0.01 M concentration. This maximally attainable concentration of CMPO of about 0.01 M seems to be insufficient for efficient extraction of all lanthanides and MA present in the PUREX HLLW, and this was probably a reason why the feed had to be diluted and the flow of the organic phase increased. The feed was adjusted in all cases; the most recent variant was an adjustment consisting of a dilution of feed by 30 vol.% of 0.5 M HF. A strip of all nuclides of interest by one stripping solution was studied with several different variants; in the most recent test, 0.56 M guanidinium carbonate plus 0.03 M diethylene triamine pentaacetic acid (DTPA) was used.

A further important long-run test was performed at the Mining and Chemical Combine, Zheleznogorsk, Russia, as described by Alekseenko (2005). Tested with actual waste (however, not containing Mo and Zr, and with low content of Eu^{3+} and Ce^{3+} 53 and 9 mg/L, respectively, and with added F^- at 1.7 g/L), the flowsheet was as simple as possible, consisting only in extraction, scrub, and strip sections. The solvent (UNEX A, but with 0.015 M CMPO and 0.4 vol.% of PEG-400) was recycled 51 times. An additional 150 mg/L of PEG-400 was added to the strip agent of the usual composition (i.e., 180 g/L of guanidine carbonate; 30 g/L of DTPA) to replenish possible losses of this reagent.

A new regenerable strip reagent for the UNEX A process solvent was developed in 2005. Following a thorough search (Law, 2005), various nitrogen bases were tested as possible strip agents. Finally, methylammonium carbonate, MAC, was selected as the best. As noted by the authors of the cited report, the dimethylammonium cation would still be a better choice, but this is not easily washed from the organic phase (Law, 2005). In reality, the strip solution was prepared by passing CO_2 through an aqueous methylamine solution completed by final heating of the solution to 60°C to drive all reactions to termination. Hence, the exact composition of MAC solution was not known. A solution of 2.5 M MAC + 20 g/L of DTPA was proposed for use as a strip solution and tested in the countercurrent mode with centrifugal extractors. The strip solution was regenerated by distillation of MAC. Still later, the solution of 2.2 M MAC + 10 g/L of NTA (nitrilotriacetic acid) was studied by the Russian authors (Egorov, 2005). As a possible variant, we could also imagine a process in which Cs^+ would be stripped by a solution of [dimethylammonium]$^+$ with a much higher tendency to strip Cs^+ and thus enabling, possibly important, Cs^+ concentration in the strip. The dimethylammonium cation could then be washed out in a tandem line by [methylammonium]$^+$ with further processing as already described.

Several auxiliary technological problems were addressed in the following articles and reports. These concerned radiation stability of the CCD + PEG extractant (Mincher, 2007) and of strip solution, and final treatment of UNEX solvent FS-13 into the form of solid infusible resin formed by condensation with phenol and paraform (Rzhekhina, 2007). Interestingly, highly irradiated 0.12 M H^+CCD^- in flurosulfone FS-13 (γ dose 432 kGy) yielded some amount of yellow compound in the aqueous phase of undetermined composition but the effect on the extraction ability of CCD^- was not observed, testing again its extraordinary radiation stability (Mincher, 2009).

UNEX A solvent is therefore a rather efficient variant, but due to the low loading capacity for lanthanides, it is not so useful for original PUREX solutions. The authors themselves precisely expressed this opinion in their recent paper (Peterman, 2007).

Further development was in fact aimed at a different goal, namely the extraction of Cs^+, Sr^{2+}, and MA (Am^{3+} and Cm^{3+}) without the extraction of macroamounts of trivalent lanthanides. This would largely relieve the demands on reagent concentration in the organic phase and maybe it is more in agreement with the contemporary trends in the treatment of PUREX waste with the goal to increase the economy of radioactive waste repository by preliminary separation of MAs. The new processes of this UNEX B variant were proposed by Russian scientists in laboratory experiments—see the following text—and are yet to be tested in their countercurrent mode.

Another group at KRI, Russia developed a different process of extraction of elements. The process documented in several communications and tentatively, for brevity, denoted here as UNEX HDBP, uses in one of its latest variants 1.1 M dibutyl hydrogen phosphate (HDBP), 0.23 M H^+CCD^-, 0.065 M Slovafol 909, all in meta-nitrobenzo trifluoride (mNBTF) solvent (Rodionov, 2008). The Eu^{3+} content in the organic phase under such conditions could exceed 0.163 M. Moreover, almost no precipitates and suspensions were observed in the two-phase extraction system, even at a Eu^{3+} concentration in the initial solution of up to 100 g/L. Thus, it is deemed by the authors that the process could be used for the treatment of PUREX waste from high-burn-up fuel. Still another variant of the process, tentatively denoted here as UNEX ZS HDBP, uses a zirconium salt of dibutyl hydrogen phosphate, ZS HDBP, as an extractant for trivalent lanthanides and actinides. This variant, in which the following organic solution was used: 0.05 M H^+CCD^-, 0.07 M ZS HDBP (at Zr:HDBP ratio 1:9) and 0.5% PEG, all in mNBTF solvent, was tested on a countercurrent line (Zilberman, 2009). The chemistry in the presence of zirconium salt of HDBP, which is rather convoluted, is summarized in the above recent report (Zilberman, 2009). It is supposed that the complex extractant consists of a core (Zr atom valence-bonded with two CCD^- anions and two DBP anions) and a shell (outer sphere) consisting of two HDBP molecules and two CCD^- molecules hydrogen bonded to the core (Shishkin, 2009).

19.2.3 Molecular Dynamics Study of Dicarbollides

In several studies, Chevrot and Wipff et al. investigated the interfacial behavior of dicarbollide in aqueous–organic systems considering several different solvents (Chevrot, 2006, 2007). Molecular dynamics simulations permit one to see freeze-frame dissected kinetics of the transfer and the processes adjacent to the interface layer. The authors consider dicarbollide anion as an "ellipsoid" (more exactly it would be an "ovoid") and they connect its surface activity to its adsorption at the interface that would also be responsible for enhanced extraction of cations (Chevrot, 2006). Adsorption of CCD^- at the water/organic interface has also been previously studied and proved to occur (Popov, 2001). In an analogous MD study, Rose and Benjamin recently investigated the hydration of univalent ions in the organic phase (Rose, 2009). They found conspicuous analogies of clusters of ions with water molecules in organic and gaseous phases. This is also well supported by our own study (Rais, 2008).

19.2.4 Systems Focused on Extraction Separation of Ln/MA at pH ≥ 1

In the basic screening studies, the soft-donor synergists previously studied for Ln/MA separations in their pure form were now tried in the mixtures with CCD^- (sometimes bromoprotected cobalt bis(dicarbollide)(1-), $BrCD^-$ was used, probably with the same effect for extraction in the systems examined as for CCD^-). The exact nature of the added $BrCD^-$ salt used is sometimes not specified. This is an omission, since the cation in different salts, for example, Cs^+BrCD^-, Na^+BrCD^-, and H^+BrCD^- may influence the results. Very often, the addition of CCD^- ($BrCD^-$) leads to an increase in D_M by several orders of magnitude, although often at the cost of some loss of selectivity and the need to adjust the pH of the aqueous phase as described below.

Krejzler et al. studied the extraction of Eu^{3+} and Am^{3+} by two so-called hemi-BTP compounds (diethylhemi-BTP, 6-(5,6-diethyl-1,2,4-triazin-3-yl)-2,2′-bipyridine, R= ethyl and di(benzyloxyphenyl) hemi-BTP, 6-[5,6-di(benzyloxyphenyl)-1,2,4-triazin-3-yl]-2,2′-bipyridine, R = benzyloxyphenyl) in the presence of CCD^- (Krejzler, 2006), see Structure 19.1. Mutual separation was feasible only from the aqueous solution of pH = 4 ($SF_{Am/Eu} = 40$) and for excess of L over CCD^-, whereas an excess of CCD^- led to an antagonistic effect. Since BTP compounds alone enable extractions up to 1 M HNO_3, the addition of CCD^- is, in this case, counterproductive. The proposed composition of the complex in the organic phase was $(ML_2NO_3)^{2+}\cdot 2CCD^-$ (Krejzler, 2006).

Similarly, antagonism was found recently by Bhattacharyya et al. for an excess of CCD^- when the neutral ligand was ethyl-bis-triazinylpyridine (Et-BTP, R = ethyl) (Bhattacharyya, 2009), see Structure 19.2. For 0.02 M Et-BTP and 0.005 M CCD^- the best and technologically satisfactory distribution ratios were obtained for 0.1 M aqueous HNO_3 and nitrobenzene as a solvent ($D_{Am} = 28.7$, $D_{Eu} = 0.3$, SF = 96). If nitrophenyl octyl ether (NPOE) was used as a solvent, still better separation was obtained ($D_{Am} = 114$, $D_{Eu} = 0.23$, SF = 495). Am^{3+} could be quantitatively stripped by 0.01 M EDTA at pH = 3.5.

The antagonistic effects had also been observed previously for systems with tris-2-pyridyl 1,3,5-triazine (TPTA) and *n*-propyl-bis-triazinylpyridine (BTP), if the concentration of CCD^- was comparable to that of synergist (Rais and Grüner, 2004). The mechanism of observed antagonistic effects is not fully elucidated. One explanation could be that protonized form of ligand in conjunction with the highly hydrophobic CCD^- anion competes for extraction of the ML_3^{3+} cation. Hence, excess L over CCD^- is needed for efficient extraction.

CMP(O)- and *N*-acyl(thio)urea-tetrafunctionalized cavitands on the addition of $BrCD^-$ were studied as extractants for Eu^{3+} and Am^{3+} by Reinoso-García et al. (2005). CMP(O)- cavitands are

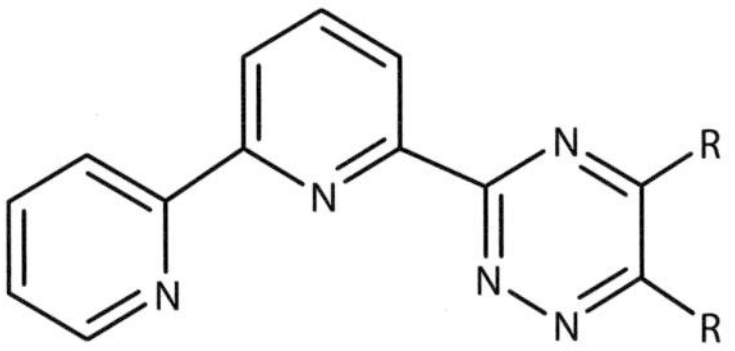

STRUCTURE 19.1 Hemi-BTP compounds (diethylhemi-BTP, 6-(5,6-diethyl-1,2,4-triazin-3-yl)-2,2′-bipyridine, for R = ethyl).

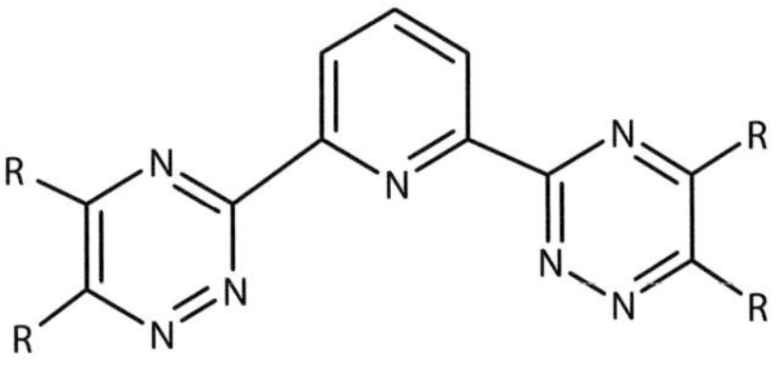

STRUCTURE 19.2 Alkyl-bis-triazynylpyridines (ethyl-bis-triazinylpyridine, Et-BTP, for R = ethyl).

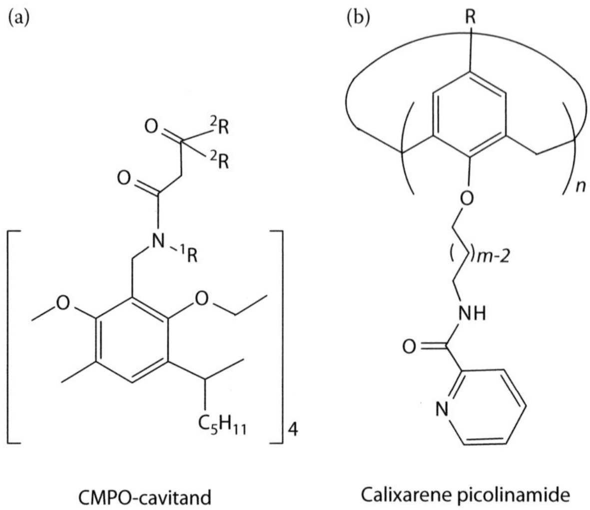

(a) (b)

CMPO-cavitand Calixarene picolinamide

STRUCTURE 19.3 (a) Tetrafunctionalized cavitands (tetrakis(diphenyl-(N-methylcarbamoyl) methyl phosphine oxide) cavitand for 1R = H, 2R = Ph), (b) calixarene-based picolinamides (5,11,17,23,29,35, 41,47-octakis(phenylmethoxy)-49,50,51,52,53,54,55,56-octakis{3-[(pyridine-2-carboxy)amino]propoxy} calix[8]arene for $n = 8$, $m = 3$, $R = $ OBn).

more effective and, from these, the tetrakis(diphenyl-(*N*-methylcarbamoyl) methyl phosphine oxide) cavitand, $R^1 = $ H, $R^2 = $ Ph, Structure 19.3, is particularly effective. With this compound at a concentration 0.001 M and 0.003 M $BrCD^-$ in nitrophenyl hexylether (NPHE), the extraction was feasible up to 3 M HNO_3 ($D_M = 2–4$) with a slight preference for Am^{3+} over Eu^{3+}.

From the various calixarene-based picolinamides studied by Casnati et al. (2005), possibly even the best compound, namely (5,11,17,23,29,35,41,47-octakis(phenylmethoxy)-49,50,51,52,53,54,55,56-octakis{3-[(pyridine-2-carboxy)amino]propoxy}calix[8]arene, $n = 8$, $m = 3$, R = OBn), see Structure 19.3, exhibited some extraction only at 0.01 M HNO_3 at a concentration of 0.0002 M and 0.003 M $BrCD^-$ in NPHE with a rather low separation factor ($D_{Am} = 3.3$, $D_{Eu} = 1.3$, $SF_{Am/Eu} = 2.54$). In addition, another ligand studied more in detail ($n = 6$, $m = 3$, $R = H$, PAR4) again operated best only at dilute acid, 0.005–0.01 M, and thus showed no extraordinary properties (Galletta, 2006). Another calixarene–picolinamide, originating from the same working group, namely hexakis-[3-*N*-(6-carboxymethylpicolinamide)propyloxy]calix[6]arene, could extract Ln/ MA even from 1 M HNO_3 (at its concentration 0.001, 0.0015 M $BrCD^-$, NPHE solvent) with about 3–4 times preference for Am^{3+} over Eu^{3+}. However, as far as this compound is concerned, we have retrieved only one article dealing mainly with its radiation stability (Mariani, 2007), whereas the original communication appeared in a less known conference proceeding (Macerata, 2006). In continuation, the upper-rim CMPO-substituted calix[6]arene and calix[8]arene extractants were prepared (Sansone, 2008), but these appeared to be less efficient than the lower-rim compounds prepared previously, even if used in conjunction with $BrCD^-$.

19.2.5 Systems Enabling Ln/An Extractions at High Nitric Acid Concentrations

For any practical applications in treatment technologies of PUREX wastes, the acidic HNO_3 solutions must be considered. Several synergists are effective here either for Ln/MA or Cs/Sr/Ln separations.

Some CMP(O) tripodands in synergy with $BrCD^-$ can extract Eu^{3+} and Am^{3+} even from 3 M HNO_3 as shown by Reinoso-García et al. (2006). From the compounds, possibly the best are those substituted with R = Ph and R = OEt as depicted in Structure 19.4.

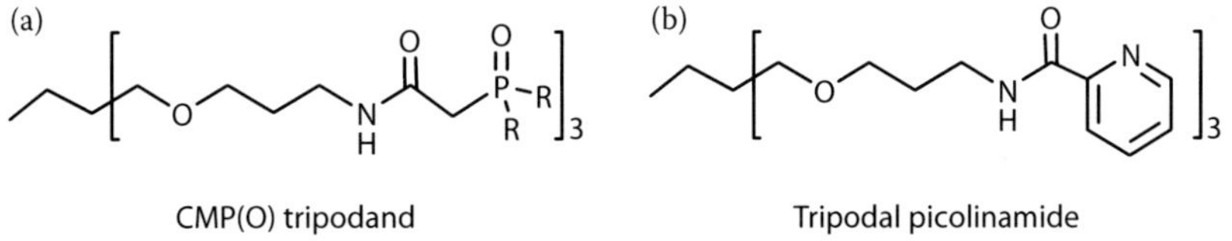

STRUCTURE 19.4 (a) CMP(O) tripodand and (b) tripodal picolinamide.

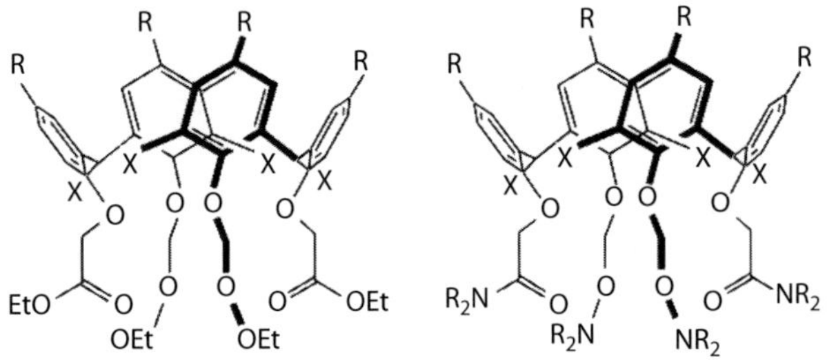

STRUCTURE 19.5 Adamantylated thiacalix[4]arenes.

The above authors measured $D_{Am} = 357$ and $D_{Eu} = 222$ from 3 M HNO_3 aqueous phase and 0.001 M CMPO tripodand with Ph substituents plus 0.001 M $HBrCD^-$ in nitrobenzene, whereas for ethoxy substituents and 1 M HNO_3 the respective values were 16 and 14.3. Hence, some minor separation of Am/Eu occurred. It is interesting that when the tripodand was covalently bound to the cobalt bis(dicarbollide) anion, the extraction efficiency dropped about three orders of magnitude. Reinoso-García, Verboom et al. also studied various tripodal *N*-acyl(thio)urea and picolin(thio) amides (Reinoso-García, 2005a,b). From 13 synthesized and tested compounds, apparently only the picolinamide derivative (see Struture 19.4) was able to extract Eu^{3+} and Am^{3+} in the presence of $BrCD^-$ from 3 M HNO_3 (0.001 M ligand and 0.003 M $BrCD^-$ in NPHE, aqueous 2 M HNO_3, $D_{Eu} = 2.99$ and $D_{Am} = 5.34$).

Recently, Kovalev et al. studied extraction in the presence of some adamantylated thiacalix[4] arenes (see Structure 19.5) in synergy with CCD^- (Kovalev, 2008). These compounds display various degrees of selectivity. The compound on the left ($X = S$, R = 1-Ad) was selective for Eu^{3+} ($D_{Eu} > 203$, $D_{Cs} = 7.9$, $D_{Sr} = 5.5$ for 0.011 M ligand plus 0.022 M CCD^- in dichloroethane, 0.5 M HNO_3) whereas two compounds of the structure on the right side (first with $X = S$, R = 1-Ad, $NR_2 = NEt_2$, and second with $X = S$, R = 1-Ad, $NR_2 = N(CH_2CH_2)_2O$, respectively) were highly selective for Sr^{2+}. The kinetics of the extraction were slow and data listed in the paper by Kovalev (2008) refer to 24 h of shaking and consequently these systems are of little interest for practical use in radiochemical technology.

19.2.6 Prospective Synergetic Systems for Ln/MA Separations at High Nitric Acid Concentrations

The prospective systems were recently reviewed and summarized by Alyapyshev et al. (2008). These, as noted above, might be considered as conceptual technological systems for the extraction of trivalent actinides without extracting weight amounts of lanthanides, UNEX B processes. No such process was as yet apparently tried in the countercurrent line.

The major work in the area was done by scientists from KRI, Russia. In several papers, the extraction of elements present in PUREX waste solutions by diamides of dipicolinic acid (DPA, see Structure 19.6) was evaluated (Alyapyshev, 2004, 2009; Romanovskiy, 2006). Other similar reagents with good extraction properties are diamides of 2,2′-dipyridyl-6,6′-dicarboxylic acid (Dyp, see Structure 19.6).

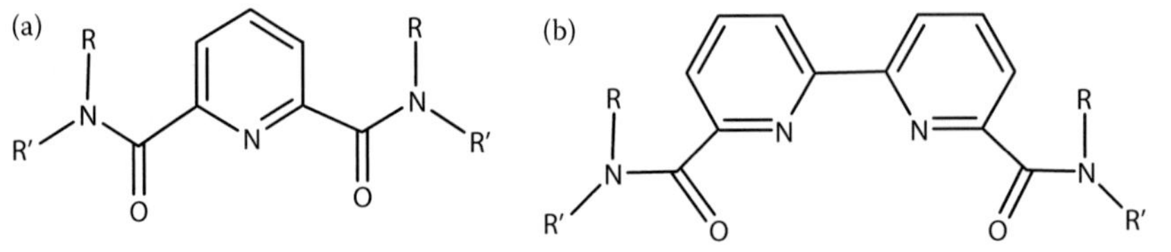

STRUCTURE 19.6 (a) Diamides of dipicolinic acid, DPA and (b) diamides of 2,2′-dipyridyl-6,6′-dicarboxylic acid, Dyp.

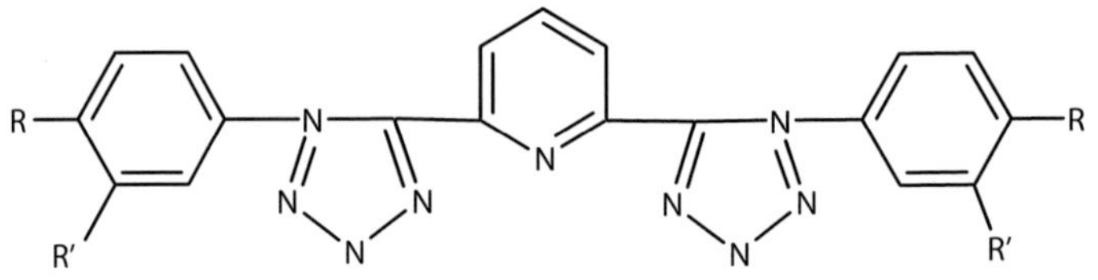

STRUCTURE 19.7 2,6-bis(1-aryl-1H-tetrazol-5-yl)pyridines (ATP) compounds.

In a detailed paper by Romanovskiy et al. (2006) various substituted DPAs were studied: tetra-ethyl DPA (TEDPA, R = R′ = Et), tetra-*n*-butyl DPA (TBDPA, R = R′ = n-Bu), *N,N*′-dibenzyl DPA (DBzDPA, R = H, R′ = $-CH_2Ph$), *N,N*′-diphenyl-*N,N*′-dimethyl DPA (PhMDPA, R = Me, R′ = Ph), *N,N*′-diheptyl DPA (DHpDPA, R = H, R′ = C_7H_{15}), and tetra-i-butyl DPA (TiBDPA, R = R′ = i-Bu).

From those, TBDPA appeared to be particularly interesting and was proposed for further testing as a variant reagent to more expensive CMPO (Paulenova, 2008; Peterman, 2007; Romanovskiy, 2006). However, efficient separation of Ln/MA is not supposed to be feasible with these compounds. For this purpose, Dyp extractants were used.

In a paper by Alyapyshev (2008) a good separation and extraction was obtained for reagent Dypl (Dypl, R = Et, R′ = Ph in the Structure 19.6) at a concentration of 0.03 M plus 0.01 M H^+CCD^- in mNBTF from 0.5 M HNO_3 ($D_{Am} = 8.5$, $D_{Eu} = 0.46$, $SF_{Am/Eu} = 18$). A similar good extraction and separation was found for Dyp2 (R = R′ = Bu in the Structure 19.6) (Alyapyshev, 2009; Alyapyshev et al., 2009a,b,c). Finally, a further DPA derivative was found that enables Am/Ln separation, namely Et(pFPh)DPA, *N,N*′-diethyl-*N,N*′-di(para)-fluorophenyl-dipicolinamide (compound in Structure 19.6 with R = Et and R′ = p-FPh). Relatively high-separation factors ($SF_{Am/Eu} \sim 6$) were found for extraction from acidic solutions (1–4 M) by using Et(pFPh)DPA in FS-13 (Paulenova, 2009).

Quite recently, 2,6-bis(1-aryl-1H-tetrazol-5-yl)pyridines (ATP) compounds (see Structure 19.7) were used. From various derivatives synthesized and studied by Smirnov et al. (Smirnov, 2007, 2009), NATP (R = R′ = H) in a mixture with CCD^- was probably the most effective and provided large $SF_{Am/Ln}$ factors up to 100 for extractions from relatively concentrated HNO_3 solutions. Moreover, NATP according to the authors is very stable, as long as for 20 days in 6 M HNO_3 at 99°C without detectable decomposition. This is in variance to many other N donor compounds and even to the CMPO extractant.

19.3 COBALT BIS(DICARBOLLIDE) EXTRACTANTS WITH COVALENTLY ATTACHED SELECTIVE GROUPS

19.3.1 General

The CD^- ion can act as a potent extractant for Cs^+. However, as has been mentioned in the preceding review (and in the previous part, without the presence of a synergist, this ion alone is unable to extract target polyvalent nuclides even Sr^{2+}, and in particular trivalent lanthanides and actinides,

Ln/An. Alternative approach consists in bonding a selective group to the cage of CD^- anion by a covalent bond. In this part, we summarize recent progress in the development of extractants based on cobalt bis(dicarbollide)(1-) ion modified by *exo*-skeletal substituents capable of binding target cations. The first families of such ionic ligands for Cs^+ extraction consisted of 4,8′,8,4′-bis(phenylene) cobalt bis(dicarbollides) tested as extraction agents, which proved appreciably increased selectivity in comparison with that of CCD^- and enabled Cs^+ extraction from highly acidic and basic solutions (Selucký, 1997). Also C(1,1′)-CH_3, C(2,2′)-alkylether (Vinas, 1998a), or C(1,2) diphenyl and C(1,7, 1′,7′) tetraphenyl (Vinas et al., 1998b)-substituted *ortho* and *meta* cobalt bis(dicarbollides) were tested as Cs^+ extractans and carriers in supporting liquid membranes, but their efficiency was limited only to a low-acidity range. The first, but still the last class of selective extractants suitable for Sr^{2+} partitioning, were based on B(8)-oxymethylene crown CD^- derivatives (Grüner et al, 2002a), but their efficiency did not exceed significantly that of synergic mixtures containing the respective crown ethers and CCD^- present in the same concentrations.

Early attempts at the design of compounds for trivalent Ln/An extraction comprised carbon substitution at CD^- ion with lipophilic alkylyether chains (Vinas et al., 1998c) or at B(8) site by diethyleneglycol chains with terminal phenoxy or phosponic acid ester groups (Plešek, 2002a). More potent at high nitric acid concentrations proved that compounds substituted by phosphoric acid residues attached either to B(8) cage position or to B(8,8′) boron atoms in a bridging manner (Plešek et al., 2002b). Most of these compounds appeared in two reviews by Grimes devoted to emerging trends in applications of boron compounds (Grimes, 2000, 2004). A detailed survey on these compounds was reported in our preceding review article (Rais and Grüner, 2004). Some of these and new compounds, discussed in more detail below, were also reported in a review by Dam and Werboon (Dam, 2007). In this chapter, we focus on recently reported compounds, particularly those suitable for trivalent Ln/An group separation. In most cases, these clearly proved to be of improved efficiency when compared with organic ionophores or their synergic mixtures with the chloroprotected CCD^- and the above, previously published series.

The design of new ionic extractants stems from recent availability of synthetic methods providing broader variety of B-substituted ammonium derivatives of CD^- anion used as structural building blocks in the extractant design and synthesis. Among the newly developed families of derivatives, a series of compounds that contain *N,N*-dialkyl carbamoyl methyl dialkylphenyl or diphenyl phosphine oxide groups (CMPO) covalently bound *via* amidic nitrogen to the cage of CD^- anion seem highly efficient. About six different structural types of extractants containing ligating CMPO functions have been prepared and tested. Extraordinary extraction efficacy has been observed for ionic ligands, in which CD^- ions and CMPO functions are preorganized on narrow or wide rims of the calix[4]arene platform. Another recent class of compounds effective for the extraction of trivalent Ln/An is represented by ionic analogs of organic tetraoctyl diglycolyl diamide (TODGA).

19.3.2 CMPO Derivatives

19.3.2.1 Synthesis and Structures

The CMPO derivatives have been designed as efficient extractants for the trivalent lanthanide/actinide group separation in the TRUEX process studied in the United States and Russia, see, for example, the review by Paiva and Malik (2004). Demonstration tests on the use of synergic mixtures of CMPO (*N,N*-di-isopropyl carbamoyl methyl alkyl phenyl phosphine oxide) and chloroderivative CCD^- for lanthanide/actinide extraction have been carried out in the development of the UNEX process, see Section 19.2.2 of this review.

We have been interested in the synthesis of compounds combining CD^- ion and CMPO groups within a single molecule over the past years. If we consider a rather high price of cobalt bis(dicarbollide) and difficulties associated with the isolation of ionic species from product mixtures, emphasis has been laid on the use of synthetic methods providing high yields, clean reaction pathways, and good purity of the products after relatively simple isolation steps. In this respect, acylation reactions of deprotonated ammonium derivatives of cobalt bis(dicarbollide) used here as building blocks, with nitrophenyl esters of alkyl phenyl phosphoryl acetic acid (active esters, ae1–3 see Scheme 19.1 and Figure 19.1), proved as very reliable, high yielding, and clean to substitute the cobalt bis(dicarbollide) cage. In contrast, several other initially tested methods (i.e., alkylations reacting CD-dioxane with *N*-monoalkyl CMPO species or acylation reaction of ammonium derivatives followed the Arbuzov reaction) led to rich mixtures of poorly separable products. It should be noted that the reaction with active esters was adapted from the method originally designed for modifications by Böhmer et al. (Arnaud-Neu et al 1996).

The first successful venture into this field has been represented by covalent bonding of a "classical" CMPO moiety to the cage of CD^- via a long diethylene glycol chain to give compounds of general formulation $[(8\text{-}^2R,Ph\text{-}P(O)(CH_2)_nC(O)N(^1R)(CH_2CH_2O)_2\text{-}1,2\text{-}C_2B_9H_{10})(1',2'\text{-}C_2B_9H_{11})\text{-}3,3'\text{-}Co(III)]^-$ (where $^1R = H$, C_4H_9, C_6H_5, $CH_2C_6H_5$, $t\text{-}C_8H_{17}$, $C_{12}H_{25}$, $^2R = C_6H_5$ or C_8H_{17}, $n = 1,2$) (Type I, see Figure 19.1) (Grüner, 2002b; Selucký, 2008). The synthesis has been based on a series of ammonium derivatives that are easily accessible by the ring opening of the Plešek's CD-dioxane compound, $[(8\text{-}(O(CH_2CH_2)_2O)\text{-}1,2\text{-}C_2B_9H_{10})\text{-}(1',2'\text{-}C_2B_9H_{11})\text{-}3,3'\text{-}Co]^0$ (Plešek, 1997) (for the scope of such reactions see the recent review) (Semioshkin, 2008). The knowledge on the synthesis and effects of structure factors on extraction properties is currently the most elaborated among other families of such species. Compounds differing in the substitution at the phosphine oxide and amide ends, and those containing methylene and ethylene unit (carbamoyl methyl and ethyl diphenyl phosphine oxide, CEPO) between both ends, were prepared and compared (Grüner, 2002b; Selucký, 2008). The list of members of this structural family consists of ca. eight compounds tested (Ia–Ih, Figure 19.1). The best extraction properties were reported for a compound with two phenyl rings at the phosphine oxide end, and with the amide end substituent $^1R = H$, Ia, but finally a compound bearing the $(Ph_2\text{-}P(O)CH_2C(O)N(t\text{-}C_8H_{17})\text{-})$ function, Ic, was subsequently used for detailed extraction tests. This compound can be easily available on a larger scale, as verified by the synthesis of several 10-g batches prepared for macroscale Eu^{3+}/Am^{3+} extraction tests (Selucký, 2008). The limiting step of the whole synthesis is the preparation of the 1-dioxane derivative with best yields ranging between 60–70%. Other two steps, the ring cleavage by *t*-octyl amine and the reaction with the active ester (see Scheme 19.1), are essentially quantitative. Published were x-ray diffraction studies on the two derivatives of this series, the sodium salt $Na_2(Ic)_2$ (Selucký, 2008) and on the neutral

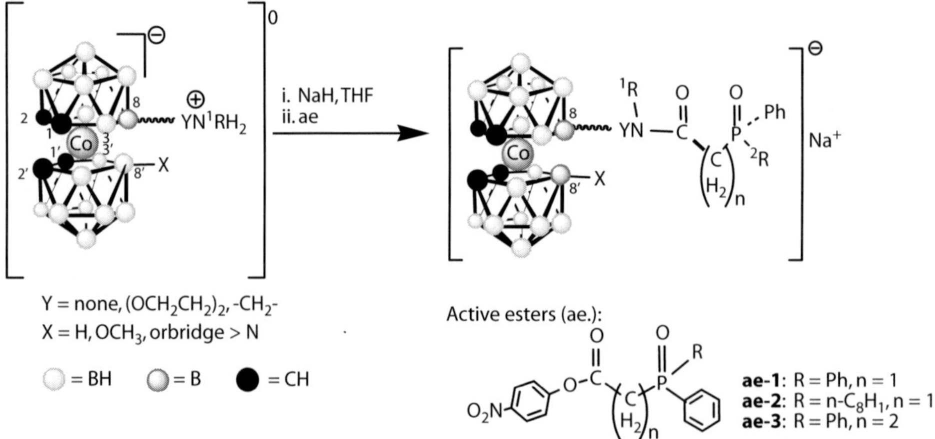

SCHEME 19.1 General method used for conversion of ammonium derivatives of the cobalt bis(dicarbollide) into CMPO derivatives by acylation using active esters **ae1-3**.

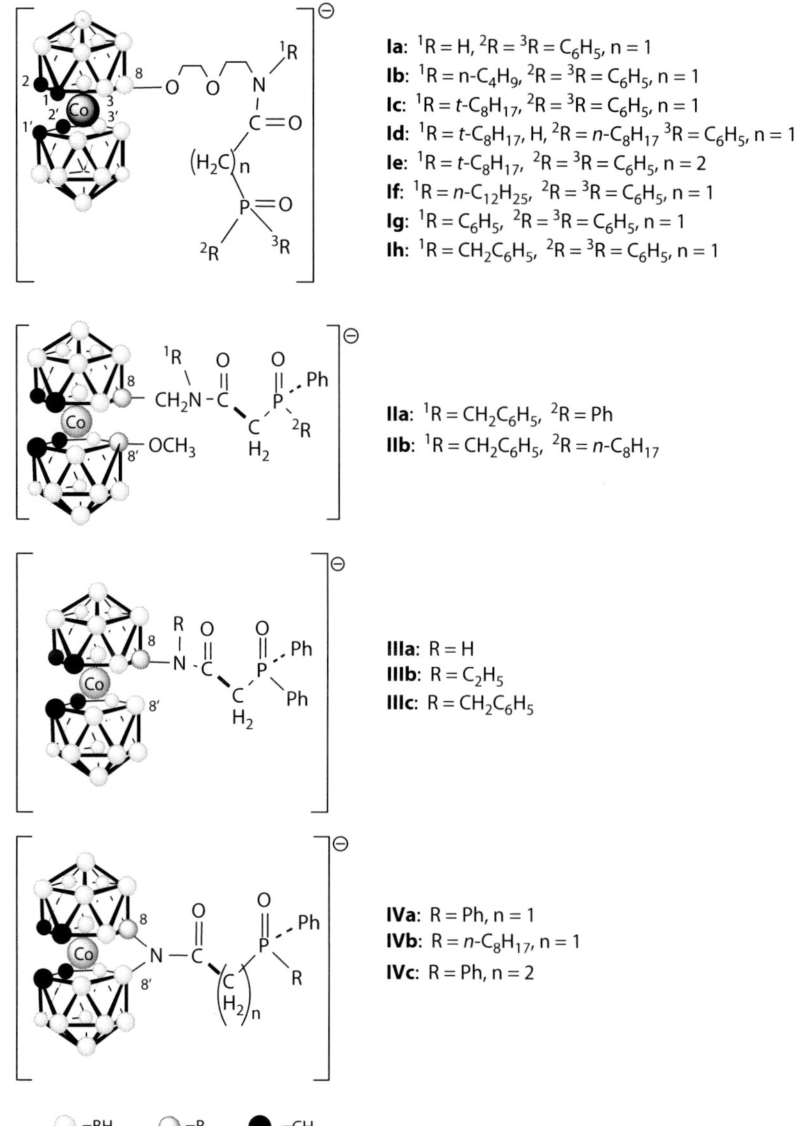

FIGURE 19.1 Schematic structures of CMPO derivatives of Types **I–IV**.

lanthanum complex with another *N*-benzyl-substituted ligand, $Ln(Ih)_3.3H_2O$ (Grüner, 2002b). The structure suggests a 1:3 La^{3+}-to-ionic ligand ratio and that both oxygen atoms from phosphine oxide and amidic group fully participate in bidentate bonding in the solid state (mean interatomic distances around the La^{3+} ion are (La–O = C) 2.52 Å; (La–O = P) 2.49 Å). It should be noted, however, that in extractions with organic CMPO derivatives, the ligands are assumed to act in a monodentate manner; only the phosphorus end group is to be considered to complex tightly the Ln^{3+}/An^{3+} cation, whereas the amidic end participates only by weak bonding, catching protons from the acid solution, and acting as a buffer.

This series of compounds was followed by several different families of anionic CMPO derivatives with the amidic nitrogen attached to the cage at a shorter distance (Grüner, 2009). These are reported in this chapter in the order roughly corresponding to a decreasing trend in their extraction efficiency (see Scheme 19.1 and Figure 19.1).

Type (II) is based on the use of a derivative of CD^- anion containing a benzyl-ammonium group N attached to the cage B(8) position via methylene group and, in addition to this substitution, the second carbollide ligand is substituted with the CH_3O- moiety at the boron B(8′) position. The synthesis and x-ray structure of this precursor was published in a recent paper by Grüner, (2009). The reaction of this ammonium derivative with two different reactive esters (see Scheme 19.1) generated compounds of the $[(8\text{-}(^2RPhP(O)CH_2(CO)N^1RCH_2\text{-}1,2\text{-}C_2B_9H_{10})(8'\text{-}CH_3O\text{-}1',2'\text{-}C_2B_9H_{10})\text{-}3,3'\text{-}Co]^-$ ($^1R = CH_2C_6H_5$, $^2R = C_6H_5$ or C_8H_{17}, IIa,b) formulation (Grüner, 2009). The molecular structure of the Ca^{2+} complex with three ionic ligands IIa was determined by crystallography. The structure revealed an octahedral arrangement of three ligands around the Ca^{2+} central atom. Each ligand is bonded to the calcium cation by both C(O) and P(O) functions (the mean distances around the calcium ion are (Ca–O = C) 2.30 Å and (Ca–O = P) 2.31 Å). The methoxy group at the B(8′) site does not participate in the calcium ligation, pointing out of the complex and showing hydrogen bond interaction with solvating CH_3OH molecules (Grüner, 2009).

The third series of compounds (III), which is structurally similar to the preceding series, but in which the amidic nitrogen is bonded directly to boron B(8) and the MeO-group is absent, was prepared using the recently reported 8-ammonium derivatives of cobalt bis(dicarbollide), $[(8\text{-}RH_2N\text{-}1,2\text{-}C_2B_9H_{10})(1',2'\text{-}C_2B_9H_{11})\text{-}3,3'\text{-}Co]^-$ (Šícha, 2009) in reactions with the active ester ae-1 (see Scheme 19.1). The series of three compounds of the $[(8\text{-}(Ph_2P(O)CH_2(CO)N^1R\text{-}1,2\text{-}C_2B_9H_{10})(1',2'\text{-}C_2B_9H_{11})\text{-}3,3'\text{-}Co]^-$ (R = H, C_2H_5 and $CH_2C_6H_5$, IIIa, IIIb, and IIIc) (see Figure 19.1) formula was prepared and tested (Grüner, 2009).

The synthesis of the last reported type (IV) of simpler compounds was based on the use of a long-known bridged zwitterionic ammonium derivative $[(8,8'\text{-}\mu\text{-}(H_2N) < (1,2\text{-}C_2B_9H_{10})_2\text{-}3,3'\text{-}Co]^o$ (Plešek, 1976) and its reactions with active esters ae-1–3 (see Scheme 19.1) to yield the expected products $[(8,8'\text{-}\mu\text{-}R(C_6H_5)P(O)(CH_2)_nC(O)N < (1,2\text{-}C_2B_9H_{10})_2\text{-}3,3'\text{-}Co]^-$ (R = C_6H_5, C_8H_{17}, n = 1,2). Compounds differing in the substitution at the phosphine oxide end and those with methylene and ethylene (CEPO) units interconnecting the phosphine oxide and amide moieties were prepared for comparison (Figure 19.1). The boron cage forms an inherent part of the CMPO (or CEPO) functional group in the structure of these wine-red compounds. Of this family of compounds, the cesium complex of the ionic ligand CsIVa (R = C_6H_5, n = 1) was characterized structurally. The structure shows the formation of almost linear chains of cesium atoms surrounded from two sides by the anionic cages positioned almost perpendicularly to this chain and parallel to each other, thus forming walls of a channel structure (Grüner, 2009). The Cs^+ cations are coordinated with the P(O) and C(O) groups with mean Cs–O = C and Cs–O = P interatomic distances of 3.03 and 3.17 Å, respectively. If we consider extraction chemistry, the structure of the Eu^{3+} complex of IVa should be more convincing. However, this could not be fully refined anisotropically due to asymmetry of the ligand and the presence of both dextrorotary and laevorotary arrangement of terminal phenyl phosphine oxide groups. Nevertheless, the clearly resolved arrangement around the central Eu^{3+} points again to a 3:1 ligand-to-metal complex with mean Eu–O = C and Eu–O = P interatomic distances of 2.46 Å and 2.37 Å, respectively, corresponding closely to bidentate bonding (Grüner, 2009).

A different type of CMP and CMPO compounds (V) has the cobalt bis(dicarbollide) anion bonded via a long pendant chain to a tripodal ligand bearing three CMP(O) groups (see Figure 19.2) (Reinoso-García, 2006). The CD^- ion was attached either via an ammonium group (Va,b) or by a carboxamide bond (Vc,d). Only the second type of compound Vc,d containing carboxamide link represents a true anion under neutral conditions, but even these compounds tend to be easily protonated. Their salts exhibited quite a poor solubility in less polar solvents which, along with the excess of complexing groups over the charge compensating ions and the easy protonation, were probably the reasons for rather poor extraction efficiency. Best D_{Eu} values were observed for species Vc, but only in the low-acidity range (D_{Eu}= 18.0, D_{Am}= 25.8 at 1.27×10^{-3} M ligand concentration in 0.1 M HNO_3).

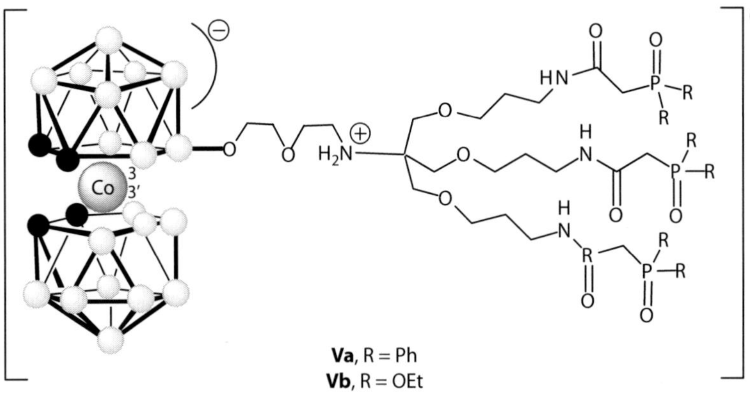

FIGURE 19.2 CMP(O) tripodants of the Type **V**.

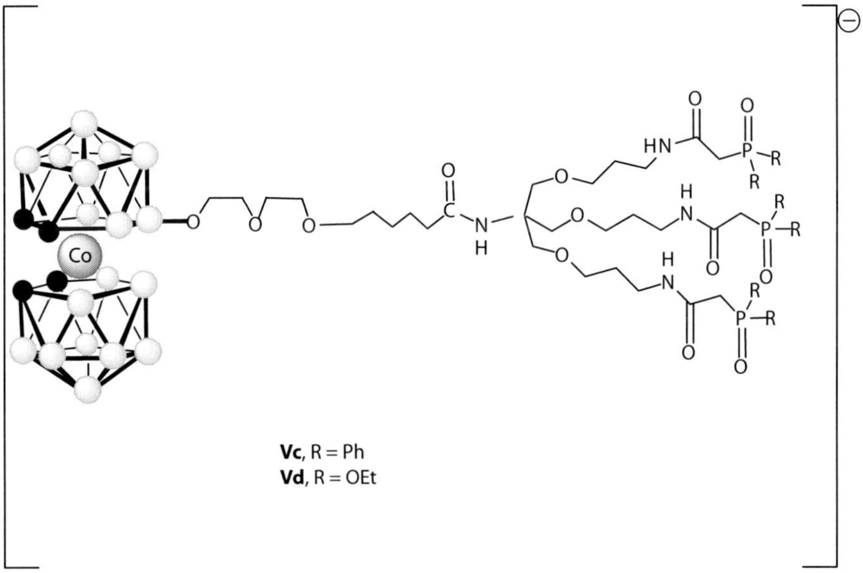

19.3.2.2 Extraction Properties of CMPO Derivatives, Series I–IV

Results plotted in Graphs 19.1 and 19.2 show trends in extraction efficiency observed within a series of sodium salts of structurally different types of CMPO derivatives· The extractants are compared at low 1×10^{-3} M concentrations in two solvents to clearly distinguish them at 1 M nitric acid concentration. Extraction ability at higher nitric acid concentrations (>1 M) decreases approximately in the following order: **I** (**II**) > **III** > **IV** (Grüner, 2002b, 2009; Selucký, 2008).

It should be noted that for extraction from the real PUREX feed, at least 0.2–0.5 M concentrations of the extractant are necessary due to a high Ln^{3+}/An^{3+} load. At 1×10^{-2} M concentrations, a large part of compounds from the **I–III** (**IV**) series are effective for Eu^{3+} extraction up to 3 M HNO_3 concentration (see data for selected compounds in Table 19.1).

The highest extraction efficiency was observed for the series **I** with a long pendant group between the **CD**$^-$ cage and the CMPO group (Selucký, 2008). The high-extraction efficiency probably results from high flexibility of the diethyleneglycol connector, thus enabling the CMPO moiety to adopt favorable geometric orientation around the Eu^{3+}/Am^{3+} cation. Another factor contributing to the

TABLE 19.1
Eu^{3+} Extraction with Selected Members of CMPO Compounds from Series I–IV

Compound	D_{Eu} 1 M HNO_3	D_{Eu} 3 M HNO_3	Reference
Na**Ib**[a]	>100	34.2	Selucký (2005)
Na**Ib**[b]	>100	21.2	Selucký (2005)
Na**Ic**[c]	90.5	21.8	Selucký (2005)
Na**If**[d]	>100	>100	Selucký (2005)
Na**If**[a]	>100	54.4	Selucký (2005)
Na**IIa**[c]	>100	79.4	Grüner (2009)
Na**IIIa**[c]	>100	2.58	Grüner (2009)
Na**IIIb**[c]	>100	2.80	Grüner (2009)
Na**IIIc**[c]	>100	7.58	Grüner (2009)
Na**IVa**[c]	77.2	1.31	Grüner (2009)

1×10^{-2} M extractants in:
[a] HMK/D mixture (1:1)
[b] Toluene
[c] Nitrobenzene
[d] 1,2-Dichloroethane; $^{152,154}Eu$ tracers in respective nitric acid solutions

efficiency is that the amidic group is not directly connected to the cage and thus the effect of the strongly acidic cobalt bis(dicarbollide) on protonation is partly suppressed. These compounds showed enhanced efficiency in aromatic solvents and are also slightly less effective (see Graphs 19.1 and 19.2), in other low-polar diluents, that is, mixtures of lauryonitrile/dodecane (D) or hexyl methyl ketone (HMK)/D (or hydrogenated tetrapropylene, that is, branched dodecane, TPH). It was reported that the efficiency of Eu^{3+}/Am^{3+} extraction for **Ic** ($Ph_2C(O)CH_2C(O)N(t\text{-}C_8H_{17})$-function) is approximately four orders of magnitude better than that of organic CMPO compounds and still approximately two orders of magnitude better than that of its synergic mixtures with ion 1-Cl_6^-. The

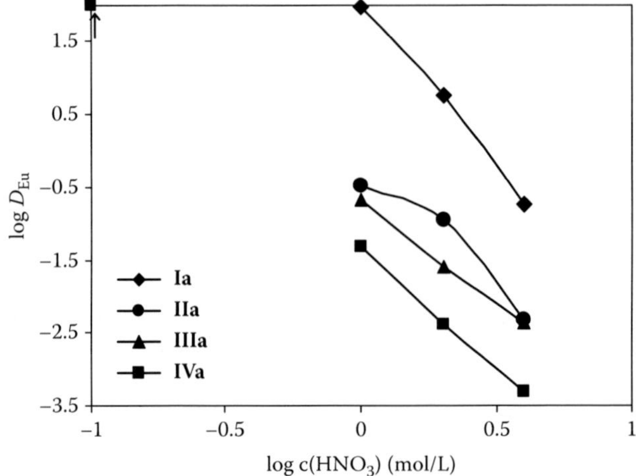

GRAPH 19.1 Plot of dependence of log D_{Eu} on logarithm of nitric acid concentration in toluene as the solvent. The representatives of four types of extractants **Ia**, **IIa**, **IIIa**, and **IVa** (for their structures see Figure 19.1) are compared at 1×10^{-3} M concentrations. (↑, the log D_{Eu} value were determined as higher than 2.)

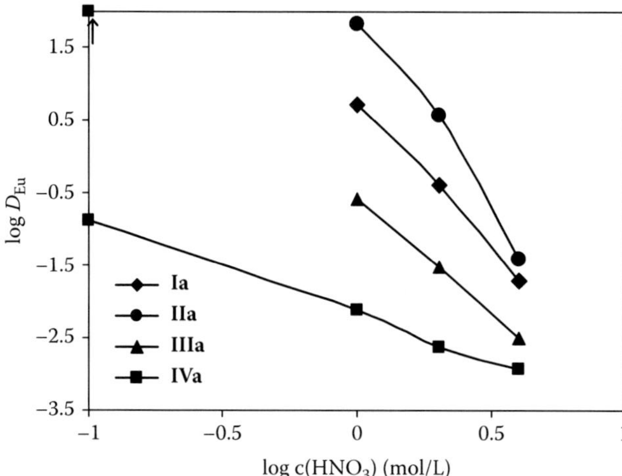

GRAPH 19.2 Plot of dependence of log D_{Eu} on logarithm of nitric acid concentration in HMK/TPH (1:1) as the solvent. The representatives of four types of extractants **Ia**, **IIa**, **IIIa**, and **IVa** (for their structures see Figure 19.1) are compared at 1×10^{-3} M concentrations. (↑, the log D_{Eu} value were determined as higher than 2.)

composition of extracted species was studied for **Ic** and **Id** that contain the $(C_8H_{17})PhC(O)CH_2C(O)N(t\text{-}C_8H_{17})$-function. The log–log dependences of D_{Eu} on extractant concentrations yielded straight lines with slopes of the curves close to 3. The slopes were 3.06 and 2.76 or 2.62 and 2.89 for 1 M and 3 M and **Ic** or **Id** in the HMK/dodecane mixture, respectively. The results indicate a 1:3 composition of the extracted species. This corresponds to a complete Eu^{3+} charge compensation by three singly charged ligating anions (L^-). The assumed extraction mechanism was proposed to correspond to the following equation:

$$Eu^{3+}_{aq} + 3HL_{or} = (Eu^{3+}, 3L^-)_{or} + 3H^+_{aq}.$$

where HL is the acidic form of the extractant, and X_{aq} and X_{or} correspond to the X species in aqueous and organic phases, respectively. The increase in distribution ratios on extractant concentration with approximately third power confirms that suitable distribution ratios could be achieved even at high acidity. Recent results on Ln^{3+}/An^{3+} separations from other fission products present in the PUREX raffinate, using ligand **Ic**, are presented in Table 19.2. Tests were carried out with two different solvent systems (corresponding to both the light (HMK/TPH) or heavy (nitrobenzene/bromoform) organic phase). A good separation of lanthanides is achieved for the majority of fission products with separation factors ranging from 50 to 1000.

The compounds of the series **II** at 1×10^{-3} M concentrations exhibit high distribution coefficients D_{Eu}/D_{Am} from 1 M HNO_3 comparable (or even higher) than in the series **I**. This is apparently due to separation of the function group from the strongly acidic cobalt bis(dicarbollide) moiety by the methylene unit resulting in sufficient flexibility of function group and lower tendency for protonation. Interesting is the clear preference of this family of extractans for HMK/D mixtures over aromatic solvents (see Graphs 19.1 and 19.2). This behavior is unique and differs from that observed for all other series **I**, **III**, and **IV**.

The distribution ratios for **I** and **II** are about one order of magnitude higher than those of series **III** with direct attachment of the amidic nitrogen to the B(8) atom of CD^- anion and even approximately two orders of magnitude higher than those of compounds with bridged CMPO groups (**IV**). D_{Eu} of the last type in nitrobenzene is approximately comparable to that of the synergic mixtures.

TABLE 19.2
Extraction and Stripping of the Fission Products from Model PUREX Feed with CMPO Derivative Ic

		D_{Me}^{extr}		D_{Me}^{wash}		D_{Me}^{strip}	
Metal [M]		A	B	A	B	A	B
Ag	7.39×10^{-5}	2.13	4.00	44.6	>100	>100	>100
Ba	1.04×10^{-3}	0.021	0.018	>100	3.58	>100	2.07
Cd	8.96×10^{-5}	<0.001	0.013	NM	1.70	NM	20.0
Ce	2.15×10^{-3}	1.50	2.03	>100	>100	<0.001	0.059
Cr	9.53×10^{-4}	<0.001	0.005	<0.001	0.325	<0.001	NM
Cs	2.25×10^{-3}	<0.001	<0.001	NM	NM	NM	NM
Cu	1.66×10^{-4}	<0.001	<0.001	NM	NM	NM	NM
Eu	1.23×10^{-4}	2.00	3.21	>100	>100	0.057	0.066
Fe	1.82×10^{-2}	0.069	1.64	47.2	16.3	7.93	0.437
Gd	1.35×10^{-4}	1.67	2.42	>100	>100	0.214	0.125
La	9.34×10^{-4}	1.16	1.34	11.9	>100	0.057	0.066
Mo	3.63×10^{-4}	0.100	0.024	0.920	0.199	3.21	4.15
Nd	2.71×10^{-3}	1.54	2.19	>100	>100	0.171	0.069
Ni	4.49×10^{-4}	<0.001	0.070	<0.001	12.2	NM	<0.001
Pd	1.46×10^{-3}	>100	0.110	>100	55.9	9.18	0.003
Pr	8.61×10^{-4}	1.67	2.26	>100	>100	0.045	0.067
Rb	4.03×10^{-4}	0.025	0.010	33.1	<0.001	>100	NM
Rh	4.13×10^{-4}	0.079	0.035	$>10^2$	24.4	>100	NM
Ru	2.00×10^{-3}	0.037	—	10.4	—	38.5	—
Sm	5.52×10^{-4}	2.18	3.23	>100	>100	0.345	0.078
Sr	1.04×10^{-3}	<0.001	<0.001	NM	NM	NM	NM
Y	5.64×10^{-4}	0.683	1.66	42.3	99.7	0.089	0.059
Zr	5.01×10^{-3}	<0.001	<0.001	NM	NM	NM	NM

Note: Extraction conditions (System A)—0.04 M **Ic** in HMK/TPH (1:1); model fission product mixture in 3 M HNO_3 (1:1), 0.1 M $H_2C_2O_4$; extraction conditions (System B)—0.06 M Ic in nitrobenzene/bromoform (9:1); model fission product mixture in 3 M HNO_3 (1:1), 0.1 M $H_2C_2O_4$, 0.05 M HEDTA. Washing (System A)—1 M HNO_3, 0.1 M $H_2C_2O_4$; Washing (System B)—1 M HNO_3, 0.2 M $H_2C_2O_4$. Stripping—0.25 M $(NH_4)_2H_3$DTPA + 1 M ammonium citrate, pH 7.5.

On the other hand, the advantage of ligands of type **IV** is their solubility in low-polar solvents (HMK/ D or aromatics) that is not attainable for the synergic system.

Reduced complexing ability of compounds **III** with direct attachment can be attributed to a steric crowding that limits the ability of the CMPO group to complex the Ln^{3+}/An^{3+} cation tightly. The influence of structural effects is even more pronounced for the rigid arrangement around the N-bridge in compounds **IVa,b**. The latter two types of ionic ligands are also most prone to protonation due to the inherent strong-acid effect of the **1**$^-$ ion. Nevertheless, as seen from the published results (Grüner, 2009), extraction efficiency for the two different forms of the N-bridged compound (enol **IVa** and keto **IVa**) was almost the same. It can be assumed that the free ligand occurs under strongly acidic conditions in the protonated form and reverts to the anionic form on M^{3+} complexation.

If we consider the role of modifications of the CMPO moiety on the extraction behavior, it can be said that extraction properties can be substantially affected by the substitutions on both the phosphorus and nitrogen ends. The replacement of the phenyl for the octyl group at the phosphine oxide end leads expectedly to a slightly better solubility, but often at the expense of slightly reduced efficiency. Considering a more demanding synthesis of the active esters (Selucký, 2008), the advantage of this modification seems dubious. Hence, more important is the substitution at the amidic end, which leads to increased efficiency and to improvements in solubility as well. The most efficient extractants of **I** and **II** families were those with $^{1}R = H$. But from the viewpoint of usability, absolute values of distribution coefficients are not necessarily the decisive factor and finally, other compounds have been selected for larger-scale tests due to better solubility in low-polar solvents.

The elongation of the alkane link between C(O) and P(O) groups from methylene (CMPO) to ethylene (CEPO) led to a sharp drop in efficiency. Expected improvements in the An^{3+}/Ln^{3+} selectivity were not confirmed and the separation factor SF remained close to 1 similarly as for all other ionic ligands across the **I–IV** series (Grüner, 2009; Selucký, 2008).

The chemical stability of selected extractants across the **I–IV** series tested in 3 M HNO_3 was good without observing any decomposition over a month.

19.3.3 Calix[4]arene with Covalently Bonded CD⁻ Anions and CMPO Groups

Six calix[4]arene derivatives bearing two cobalt bis(dicarbollide)(1-) ions at their narrow rim and two CMPO groups were reported, differing in the length of alkane spacers between the platform and CMPO groups (see Figure 19.3) (Mikulášek, 2006, 2007). The synthesis started from *t*Bu-calix[4]arene diether derivatives with appropriate precursors of amino groups (usually nitriles or eventually phthalimido and Boc derivatives). These were O-alkylated by a **1**-dioxane derivative, followed by reduction of the nitrile derivatives formed by $BH_3{\cdot}SMe_2$ (or deprotection in the case of use of the respective phtalimido- or Boc-protected amino groups), which resulted in a series of diamines. These were finally converted into the CMPO derivatives **VIa–f** (see Figure 19.3) by acylation with the active ester **ae-1** (for structure see Scheme 19.1). Pure *cone* conformers were isolated, those of derivatives **VIa** and **VIc** on a 10-g scale for larger-scale tests. These ionic ligands showed exceptionally high extraction abilities for trivalent actinides and lanthanides, even when compared with derivatives of the preceding **I** and **II** series, being able to extract target ions from 4 M HNO_3 even at 1×10^{-5} M concentration in a HMK/TPH solvent (see Graph 19.3). Only for species with the shortest and longest spacer between the CMPO function and the spacer (**VIa** and **VIf**), a visible decrease in efficiency was observed.

Also a 1:1 synergic mixture of two calix[4]arenes, the first substituted with four CD⁻ anions and the second with four CMPO functions, showed similarly high extraction efficiencies as the most potent **VIb–e** compounds. But due to low solubility of the components in low-polar solvents, the

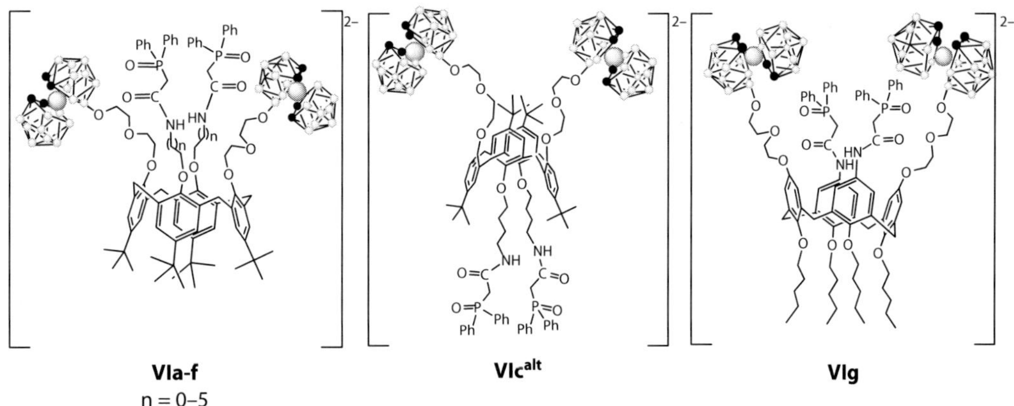

FIGURE 19.3 Schematic structures of calix[4]arene derivatives of Type **VI**.

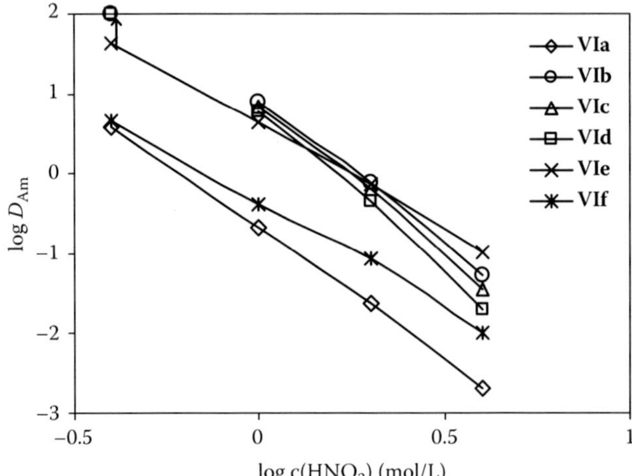

GRAPH 19.3 Comparison of the extraction properties (dependence of log D_{Am} on logarithm of nitric acid concentration) of low-rim calix[4]arenes bearing two ions and two CMPO groups in opposite positions of the platform. The compounds differ in length of the alkyl spacer (for **VIa** = C_2H_4, **VIb** = C_3H_6, **VIc** = C_4H_8, **VId** = C_5H_{10}, **VIe** = C_6H_{12}, **VIf** = C_7H_{14}) between the platform and CMPO function. Extractants are at very low concentrations 1×10^{-5} M in HMK/TPH (1:1). (↑, the log D_{Eu} value were determined as higher than 2.)

comparable properties of this synergic mixture could be observed only in nitrobenzene (Mikulášek, 2007). On the other hand, the mixtures with one tetrasubstituted calix[4]arene and the second component being free, tested in a 1:4 ratio, showed a rather low extraction ability (Mikulášek, 2006). This indicated that the preorganization of both groups on the calix[4]arene platform is essential to attain high extraction efficiency and suggested that two modified calix[4]arenes might be responsible for the extraction of a single M^{3+} cation to form a tight complex with an encapsulated metal ion (see schematic Figure 19.4). Indeed, the dependence of log D_{Eu} versus log*c* measured for compound **VIc**, where the spacer is butyl, corresponded to a straight line with a slope close to 2. This suggests an extraction mechanism, in which Eu^{3+} is trapped between two molecules of the extractant. Independent complexation studies with **VIc** for La^{3+}, Eu^{3+}, and Yb^{3+} in methanol carried

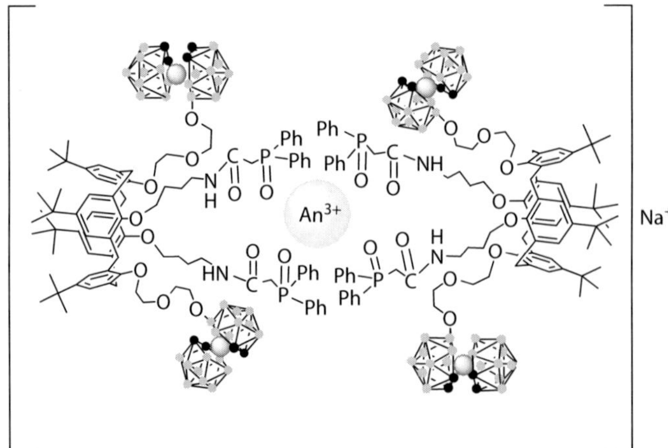

FIGURE 19.4 Schematic drawing of the assumed Eu/Am complex with ligand **VIc**. Two lower-rim-substituted calix[4]arenes participate on bonding of one Ln/An cation.

TABLE 19.3
Cycle Experiments of Eu^{3+} Extraction and Stripping Using the Calix[4]arene VIa (Ethylene Spacer between Platform and the CMPO Functions)

	D_{Eu}		
	1st Step	**2nd Step**	**3rd Step**
Extraction	>100	>100	>100
Stripping	<0.001	<0.001	<0.001

Note: Extraction conditions—4×10^{-2} M extractant in HMK/ TPH (1:1); 1×10^{-2} M $Eu(NO_3)_3$ in 3 M HNO_3; Washing—1 M HNO_3, 0.2 M oxalic acid; Stripping—0.25 M $(NH_4)_2H_3DTPA$ + 1 M ammonium citrate, pH ca. 7; Regeneration—3 M HNO_3.

out by UV-spectrophotometry and microcalorimetry revealed the formation of strong 1:1 and 1:2 metal-to-ligand complexes with log β ranging from 9.6 to 11.7 (Mikulášek, 2007).

The 1:1 ratio between CD^- and the CMPO group and their mutual distance seems essential to achieve the high efficiency observed. This follows from observations that compounds similar to **VIc**, but with 1:3 to 3:1 ratios of the two functional groups, showed a marked decrease in their extraction ability (Grüner, 2010). Also important is the preorganization of both groups at the same site of the platform in the *cone* conformation. A significant drop in $D_{Eu/Am}$ extraction was observed for the *1,3-alternate* **VIc**^{alt} conformer when compared with **VIc** (Mikulášek, 2006).

A calix[4]arene substituted at the upper-rim **VIg** was synthesized from calix[4]arene-O-tetra pentyloxy derivative conformationally fixed at the low rim by alkyl chains and substituted with two nitro groups and two phenolic hydroxyls at the wide rim in the opposite positions at the platform. After O-alkylation with two equivalents of the CD-dioxane derivative and on reduction of the nitro functions to the ammonium group, two CMPO functions were introduced by reaction with the active ester ae-1 (see Figure 19.3, **VIg**) (Grüner et al., unpublished results). However, in contrast to organic upper-rim CMPO calix[4]arenes (Schmidt, 2003), a lower separation efficiency was observed for this compound with respect to the low-rim series. This can be possibly explained by a different complexation mode. No An^{3+}/Ln^{3+} selectivity was observed, similarly as for the low-rim series.

Due to extraordinary high extraction efficiency, low-rim calix[4]arenes are primarily suitable for preconcentration and removal of traces of target radionuclides from low-concentrated waste solutions (Mikulášek, 2007). On the other hand, at least some of them might be of use for large-scale Ln^{3+}/An^{3+} partitioning. An effective method for reextraction could be found, provided that the calix[4]arene **VIa** with the short spacer and a bit-reduced efficiency was used (see Table 19.3). Cycle experiments have shown excellent extraction and stripping properties of **VIa**. Furthermore, the tests carried out with 0.01 M **VIa** in HMK/TPH and europium macroquantities (0.01 M $Eu(NO_3)_3$ in 3 M HNO_3) revealed very fast kinetics for extraction and striping steps; the equilibrium for extraction and reextraction was attained within 5 and 15 min., respectively.

19.3.4 Ionic Analogs of the Organic TODGA Extractant

The last reported series of selective ionic extractants consists of a *N,N′*-dialkyl diglycolyl complexing group as a platform to which two cobalt bis(dicarbolide)(1-) anions are covalently attached by amidic bonds. The extractants differing in the substitution of the amide nitrogen (by hydrogen, butyl-, hexyl-, octyl-, *t*-octyl-, dodecyl-, toluidine-, and benzyl-groups, **VIIa–h**) (see Figure 19.5) were prepared using acylation reactions of deprotonated ammonium derivatives of the **1**$^-$ ion with diglycolic acid dichloride (Lučaníková, 2009, 2011; Grüner, 2010). It has been found that compounds with long diethyleneglycol connectors between the cobalt bis(dicarbollide) cage and the diglycolylamide group extract trivalent lanthanides and actinides very effectively from acidic nitric acid solutions (to pH 1). The dependence of D_{Eu} on the acidity for particular derivatives is depicted in Graph 19.4.

Compound

VIIa R = H
VIIb R = C_4H_9
VIIc R = n-C_6H_{13}
VIId R = n-C_8H_{17}
VIIe R = n-$C_{12}H_{25}$
VIIf R = t-C_8H_{17}
VIIg R = C_6H_4-4-CH_3
VIIh R = $CH_2C_6H_5$

FIGURE 19.5 Schematic structures of ionic diglycolylamide-based extractants of Type **VII**.

As demonstrated, this family of extractants allows for a good extraction of trivalent lanthanides and actinides from most fission products present in the simulated PUREX feed. Most compounds showed slightly higher preference for lanthanides over actinides with SF = 3–7. Trivalent radionuclides can be effectively stripped using complexing agents. The extraction efficiency was significantly higher (2–3 orders of magnitude) than that of comparable organic TODGA molecules without CD^- anion and synergic mixtures of these two compounds with the same concentrations of

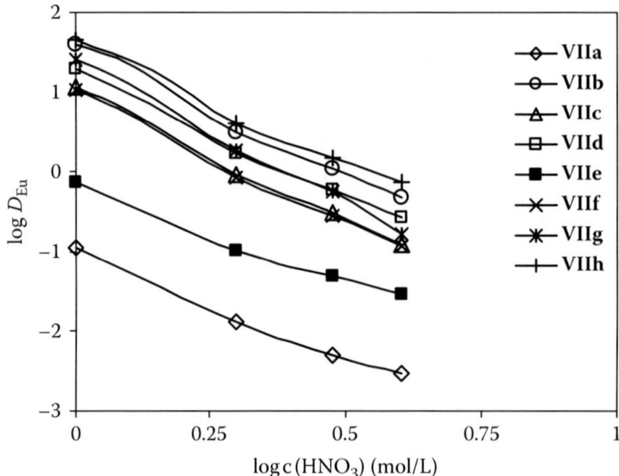

GRAPH 19.4 Plot of dependence of log D_{Eu} on logarithm of nitric acid concentration in HMK/*D* (1:1) as the solvent. The extractants differing in substituent on the amidic nitrogen atoms **VIIa** = H, **VIIb** = C_4H_9, **VIIc** = C_6H_{13}, **VIId** = C_8H_{17}, **VIIe** = t-C_8H_{17}, **VIIf** = $C_{12}H_{25}$, **VIIg** = C_6H_4-4-CH_3, **VIIh** = CH_2-C_6H_5 (for the structures see Figure 19.5) are compared at 1×10^{-3} M concentrations. *Note:* The log D_{Eu} value for all extractants at 0.1 M HNO_3 were determined as higher than 2.

functional groups. Noteworthy is also the markedly faster extraction kinetics for the covalently bonded compounds. The most perspective compound for larger-scale tests was **VIId** and was tested in more detail. Table 19.4 shows the extraction ability of this compound in different solvents. With the exception of HMK, all solvents are effective for the 1×10^{-3} M ligand concentration up to 1 M HNO_3, 1,2-dichlorethane up to 3 M HNO_3, both aromatic solvents even up to 4 M HNO_3. Pure HMK exhibited the lowest distribution ratios due to competition of the protonated solvent. Dependences of Eu^{3+} extraction on the concentration the extractant **VIId** (at 1 M and 3 M HNO_3) gave straight lines with slopes 2.2 and 2.4. These values indicate that the dominant species extracted is $[Eu(L)_2]^-$; the Eu^{3+} cation is probably trapped between two molecules of the ligand and resulting negative charge is compensated by one proton.

19.3.5 Conclusions and Outlook on Covalently Bonded Compounds

The development of covalently bonded species has been initiated approximately 10 years ago within EU projects on radionuclide partitioning. From this time, a variety of compounds able to act as

TABLE 19.4
Eu^{3+} and Am^{3+} Extraction with Cobalt Bis(dicarbollide) Analog of TODGA-Like Derivative (VIId, N-substituted with *n*-octyl) into Different Solvents

Solvent	Efficiency Selectivity	c(HNO_3) (M)					
		0.01	**0.1**	**1**	**2**	**3**	**4**
Nitrobenzene[a]	D_{Eu}	>100	>100	3.27	0.248	0.062	0.033
	D_{Am}	>100	>100	0.983	0.073	0.018	0.008
	$SF_{Eu/Am}$	—	—	3.33	3.40	3.44	4.13
DClE	D_{Eu}	>100	>100	>100	14.1	3.36	0.934
	D_{Am}	>100	>100	38.6	2.80	0.636	0.218
	$SF_{Eu/Am}$				5.04	5.28	4.28
HMK	D_{Eu}	>100	47.0	0.046	0.010	0.008	0.006
	D_{Am}	>100	16.8	0.023	0.004	0.002	0.002
	$SF_{Eu/Am}$		2.80	2.00	2.50	4.00	3.00
HMK/D (1:1)	D_{Eu}	>100	>100	19.5	1.71	0.607	0.268
	D_{Am}	>100	>100	5.91	0.364	0.108	0.041
	$SF_{Eu/Am}$			3.30	4.70	5.62	6.54
Cumene	D_{Eu}	>100	>100	>100	>100	67.0	24.0
	D_{Am}	>100	>100	>100	53.9	12.8	3.87
	$SF_{Eu/Am}$	—	—			5.23	6.20
DOS	D_{Eu}	>100	>100	>100	22.4	3.18	1.80
	D_{Am}	>100	>100	44.8	3.23	0.775	0.352
	$SF_{Eu/Am}$				6.93	4.10	5.11
Toluene	D_{Eu}	>100	>100	>100	>100	31.5	11.7
	D_{Am}	>100	>100	>100	24.6	5.74	1.95
	$SF_{Eu/Am}$					5.49	6.00

Source: From Lučaníková, M. et al. 2010. *Radiochim. Acta*: Accepted.
Note: 1×10^{-3} M **VIId** in respective solvents; $^{152,154}Eu$ and ^{241}Am tracers, variable nitric acid concentration.
[a] Turbidity of the aqueous phase was observed
D, dodecane; DClE, 1,2-dichlorethane; DOS, di-n-octyl sebacate; HMK, hexyl methyl ketone

effective extractants has gradually emerged along with knowledge about extraction properties, complexation ability, and solution behavior of particular structural archetypes. On the other hand, the positions with respect to the use of the cobalt bis(didarbollide) in technology are still firmly governed by synergic mixtures. As we have shown above, we hope that the effect of covalent bonds for properly selected systems is not only synergic, but also cooperative, otherwise leading to inaccessible high increase in extraction properties. This is a consequence of simultaneous action of several factors, such as tight metal complexation, charge compensation at short distance, and the hydrophobic behavior of CD^- anion. Some of the compounds discussed above are easily available in high yields, are more efficient and soluble in low-polar and better acceptable solvents. Thus, we believe that new extractants with covalently bonded selective groups can at least be a good alternative for considerations about future viable solutions for trivalent actinides/lanthanides partitioning.

19.4 EPILOGUE

Cobalt bis(dicarbollide) process for extracting cesium from acidic HLLW, the first variants of synergist of the PEG type for Sr^{2+}, and of hard donor type for Eu^{3+} extractions originated from a small country of Czechoslovakia nearly 40 years ago, see review by Kyrš (1994). The two large nations, Russia and the United States, started to develop the process in the 1920s and 1930s. Now, the nascent stage of CCD^- is over, but CCD^- chemistry and applications are still or possibly even more active. Even if in future the CCD^- extractions in nuclear technology may be surpassed by an alternative process, it has been with great pleasure to be involved in this fascinating area of research and we shall never stop to believe that these and other applications of CCD^- are worth studying.

ACKNOWLEDGMENTS

Professor Jaromír Plešek, from the IIC, who passed away in April 2010, was one of the most active researchers in the field and concordantly his contributions deserve to be addressed here.

This work has been supported by Grant Agency of the Czech Republic, Grant No. 104/09/866 and 104/08/0006 (to J.R. and M.L.). Partial support from the E. U. I.P. Europart (6th F.P., F16W-CT-2003–508854) and the Ministry of Education of the Czech Republic (Project LC523) is appreciated.

REFERENCES

Alekseenko, S., Babain, V. A., Bondin, V. et al. 2005. Testing of UNEX-process on Centrifugal Contactor Mocup of Mining and Chemical Combine. In *Global 2005* Paper No. 347, Tsukuba, Japan, 9–13 October, 2005.

Alyapyshev, M. Y., Babain, V. A., Borisova, N. E., Kiseleva, R. N., Safronov, D. V., and Reshetova, M. D. 2008. New systems based on 2,2′-dipyridyl-6,6′-dicarboxylic acid diamides for Am–Eu separation. *Mendeleev Commun.* 18: 336–337.

Alyapyshev, M. Y., Babain, V. A., Eliseev, I. I. J. et al. 2009b. Am-lanthanides separation using new solvents based on 2,2-dipyridyl-6,6-dicarboxylic acid diamides. In *Global 2009* p. 1113, Paris, France, September 6–11, 2009.

Alyapyshev, M. Y., Babain, V. A., Eliseev, I. I. J., Kiseleva, R., and Romanovskiy, V. 2009a. Actinides and lanthanides separation with extraction systems on the base of chlorinated cobalt dicarbollide. In *Global 2009* p. 1094, Paris, France, September 6–11, 2009.

Alyapyshev, M. Y., Babain, V. A., Eliseev, I. J., Kiseleva, R., and Romanovskiy, V. 2009c. Actinides and lanthanides separation with extraction systems on the base of chlorinated cobalt dicarbollide. In *Global 2009* p. Paper 9113, Paris, France, September 6–11, 2009.

Alyapyshev, M. Y., Babain, V. A., and Smirnov, I. V. 2004. Extractive properties of synergistic mixtures of dipicolinic acid diamides and chlorinated cobalt dicarbollide. *Radiochemistry* 46: 270–276.

ArnaudNeu, F., Bohmer, V., Dozol, J. F. et al. 1996. Calixarenes with diphenylphosphoryl acetamide functions at the upper rim. A new class of highly efficient extractants for lanthanides and actinides. *J. Chem. Soc.-Perkin Trans. 2*: 1175–1182.

Babain, V. A. 2010. Simultaneous removal of radionuclides by extractant mixtures in ion exchange and solvent extraction. In *Simultaneous Removal of Radionuclides by Extractant Mixtures in Ion Exchange and Solvent Extraction* (Moyer, B. A., ed) vol. 19, pp. 359–380, Boca Raton, USA: Taylor & Francis.

Bhattacharyya, A., Mohapatra, P. K., Roy, A., Gadly, T., Ghosh, V. K., and Manchanda, V. K. 2009. Ethyl-bis-triazinylpyridine (Et-BTP) for the separation of Americium(III) from trivalent lanthanides using solvent extraction and supported liquid membrane methods. *Hydrometallurgy* 99: 18–24.

Casnati, A., Della Ca', N., Fontanella, M. et al. 2005. Calixarene-based picolinamide extractants for selective An/Ln separation from radioactive waste, *Eur. J. Org. Chem.* 2338–2348.

Chevrot, G., Schurhammer, R., and Wipff, G. 2006. Surfactant behavior of ªEllipsoidal° dicarbollide anions: A molecular dynamics study. *J. Phys. Chem.* B 110: 9488–9498.

Chevrot, G., Schurhammer, R., and Wipff, G. 2007a. Synergistic effect of dicarbollide anions in liquid–liquid extraction: A molecular dynamics study at the Octanol–water interface. *Phys. Chem.-Chem. Phys.* 9: 1991–2003.

Chevrot, G., Schurhammer, R., and Wipff, G. 2007b. Molecular dynamics study of dicarbollide anions in nitrobenzene solution and at its aqueous interface. Synergistic effect in the Eu(III) assisted extraction. *Phys. Chem.- Chem. Phys.* 9: 5928–5938.

Choppin, G. R. 2005. Solvent extraction processes in the nuclear fuel cycle. *Solvent Extraction Research and Development, Japan* 12: 1–10.

Dam, H. H., Reinhoudt, D. N., and Verboom, W. 2007. Multicoordinate ligands for actinide/lanthanide separations. *Chem. Soc. Rev.* 36: 367–377.

Egorov, G. F., Tkhorgnitsky, G. P., Romanovskiy, V. N. et al. 2005. Radiation stability of regenerated stripping solutions for high-level waste processing. *J. Radioanal. Nucl. Chem.* 266: 349–353.

Galletta, M., Macerata, E., Mariani, M. et al. 2006. A study on synergistic effects and protonation of a selected calixarene based picolinamide ligand used in the An/Ln separation. *Czech. J. Phys.* 56: D453–D458.

Grimes, R. N. 2000. Metallacarboranes in the new millenium. *Coord. Chem. Rew.* 200–202: 773–811.

Grimes, R. N. 2004. Boron clusters come to age. *J. Chem. Ed.* 81: 658–672.

Grüner, B., Plešek, J., Báča, J. et al. 2002a. Crown ether substituted cobalto bis(dicarbollide) ion as selective extraction agents for removal of Cs^+ and Sr^{2+} from nuclear waste. *New J. Chem.* 26: 867–875.

Grüner, B., Plešek, J., Báča, J. et al. 2002b. Cobalt bis(dicarbollide) ions with covalently bonded CMPO groups as selective extraction agents for lanthanide and actinide cations from highly acidic nuclear waste solutions. *New J. Chem.* 26: 1519–1527.

Grüner, B., Kvíčalová, M., J., Plešek, J. et al. 2009. Cobalt bis(dicarbollide) ions functionalized by CMPO-like groups attached to boron by short bonds; efficient extraction agents for separation of trivalent f-block elements from highly acidic nuclear waste. *J. Organomet. Chem.* 694: 1678–1689.

Grüner, B., Mikulášek, L., Dordea, C., Böhmer, V., Selucký, P., and Lučaníková, M. 2010. Anionic alkyl diglycoldiamides with covalently bonded cobalt bis(dicarbollide)(1-) ions for lanthanide and actinide extractions. *J. Organomet. Chem.* 695: 1261–1264.

Grüner, B., Böhmer, V., Selucký, P. unpublished results.

Hawthorne, M. F., Young D. C., Wegner P. A. 1965. Carbametallic boron hydride derivatives. I. apparent analogs of ferrocene and ferricinium ion. *J. Am. Chem. Soc.* 87: 1818–1819.

Herbst, R. S., Law, J. D., Todd, T. A. et al. 2002a. Development and testing of a cobalt dicarbollide based solvent extraction process for the separation of cesium and strontium from acidic tank waste. *Sep. Sci. Technol.* 37: 1807–1831.

Herbst, R. S., Law, J. D., Todd, T. A. et al. 2002b. Universal solvent extraction (UNEX) flowsheet testing for the removal of cesium, strontium, and actinide elements from radioactive, acidic dissolved calcine waste. *Solvent Extr. Ion Exch.* 20: 429–445.

Herbst, R. S., Law, J. D., Todd, T. A. et al. 2003. Development of the universal extraction (UNEX) process for the simultaneous recovery of Cs, Sr, and actinides from acidic radioactive wastes. *Sep. Sci. Technol.* 38: 2685–2708.

Hill, C. 2010. Overview of recent advances in An(III)/Ln(III) separation by solvent extraction. In *Book. Overview of Recent Advances in An(III)/Ln(III) Separation by Solvent Extraction* (Moyer, B. A., ed.) vol. 19, pp. 119–191, Boca Raton, USA: Taylor & Francis.

Kovalev, V. V., Khomich, E. V., Shokova, E. A., Babain, V. A., and Alyapyshev, M. Y. 2008. Synergistic extraction of cesium, strontium, and europium with adamantylated thiacalix[4]arenes in the presence of chlorinated cobalt dicarbollide. *Russ. J. Gen. Chem.* 78: 19–25.

Krejzler, J., Narbutt, J., Foreman, M. R. S. J., Hudson, M. J., Čásenský, B., and Madic, C. 2006. Solvent extraction of Am(III) and Eu(III) from nitrate solution using synergistic mixtures of N-tridentate heterocycles and chlorinated cobalt dicarbollide. *Czech. J. Phys.* 56: D459–D467.

Kyrš, M. 1994. 23 Years of dicarbollide extraction of metals research and applications. *J. Radioanal. Nucl. Chem. Lett.* 187: 185–195.

Kyrš, M., Heřmánek, S., Rais, J., Plešek, J., USSR Patent USSR 508, 476; 30 March 1976.

Kyrš, M., Heřmánek, S., Rais, J., Plešek, J., CZ. Patent CS 182, 913; 29 July 1980.

Law, J. D., Herbst, R. S., Peterman, D. R. et al. 2005. Development of a regenerable strip reagent for treatment of acidic, radioactive waste with cobalt dicarbollide-based solvent extraction processes. *Solv. Extr. Ion Exch.* 23: 59–83.

Lučaníková, M., Selucký, P., Rais, J., Grüner, B., and Kvíčalová, M. 2009. Cobalt bis(dicarbollide) ion derivates covalently bonded with diglycolyldiamide for lanthanide and actinide extractions. In Book of Abstracts, *APSORC 2009–Asia-Pacific Symposium on Radiochemistry*, pp. 201–201, Napa, California, U.S.A, November 29–December 4, 2009.

Lučaníková, M., Selucký, P., Rais, J., Grüner, B., and Kvíčalová, M. 2011. Cobalt bis(dicarbollide) ion derivates covalently bonded with diglycolyldiamide for lanthanide and actinide extractions. *Radiochim. Acta* 99: in press.

Macerata, E., Casnati, A., D'Arpa, O. et al. 2006. Study of the extracting properties of a selected calixarene based picolinamide ligand in different media for the An/Ln separation. In *Proceedings of International Youth Nuclear Congress 2006*, Stockholm, Sweden, from June 18–23, 2006, Paper No. 268.

Mariani, M., Macerata, E., Galletta, M. et al. 2007. Partitioning of minor actinides: Effects of Gamma irradiation on the extracting capabilities of a selected calixarene-based picolinamide ligand. *Rad. Phys. Chem.* 76: 1285–1289.

Mikulášek, L., Grüner, B., Danila, C., Böhmer, V., Časlavský, J., and Selucký, P. 2006. Synergistic effect of ligating and ionic functions, prearranged on a calix[4]arene. *Chem. Commun.* 4001–4003.

Mikulášek, L., Grüner, B., Dordea, C. et al. 2007. Tert-butyl-calix[4]arenes substituted at the narrow rim with cobalt bis(dicarbollide)(I-) and CMPO groups—new and efficient extractants for lanthanides and actinides. *Eur. J. Org. Chem.* 4772–4783.

Mincher, B. J., Herbst, R. S., Tillotson, R. D., and Mezyk, S. P. 2007. The radiation chemistry of CCD-PEG, a solvent-extraction process for Cs and Sr from dissolved nuclear fuel. *Solv. Extr. Ion Exch.* 23: 747–755.

Mincher, B. J., Modolo, G., and Mezyk, S. P. 2009. The effects of radiation chemistry on solvent extraction. *Solvent Extr. Ion Exch.* 27: 331–353.

Paiva, A. P., and Malik, P. 2004. Recent advances on the chemistry of solvent extraction applied to the reprocessing of spent nuclear fuels and radioactive wastes. *J. Radioanal. Nucl. Chem.* 261: 485–496.

Paulenova, A., Alyapyshev, M. Y., Babain, V. A., Herbst, R. S., Law, J. D., and Todd, T. A. 2008. Extraction of lanthanides and actinides with diamides of dipicolinic acid. In NRC7—*Seventh International Conference on Nuclear and Radiochemistry*, Budapest, Hungary, August 24–29, 2008.

Paulenova, A., Lapka, J., Alyapyshev, M. Y. et al. 2009. Lanthanide and Am extraction by a modified UNEX extraction mixture. In *Global 2009,* p. 987, Paris, France: French Nuclear Energy Society by Omnipress, September 6–11, 2009.

Peterman, D. R., Herbst, R. S., Law, J. D. et al. 2007. Diamide derivatives of dipicolinic acid as actinide and lanthanide extractants in a variation of the UNEX process. In *Global 2007,* p. 1106, Boise, Idaho: American Nuclear Society (ANS), September 9–13, 2007.

Plešek, J., Heřmánek, S., Todd, L. J., and Wright, W. F. 1976. Zwitterionic compounds of the 8,8′-X$(C_2B_9H_{10})_2$Co series with monoatomic O,S, Se, Te, N bridge between dicarbollide ligands. *Collect. Czech. Chem. Commun.* 41: 3509–3515.

Plešek, J. 1992. Potential applications of the boron cluster compounds. *Chem. Rev.* 92: 269–278.

Plešek, J., Grüner, B., Cisařová, I., Báča, J., Selucký, P., and Rais, J. 2002b. Functionalized cobalt bis(dicarbollide) ions as selective extraction reagents for removal of $M2^+$ and $M3^+$ cations from nuclear waste, crystal and molecular structures of the 8,8′-mu-CIP(O)(O)(2)(1,2-$C_2B_9H_{10}$)(2)-3,3′-Co HN(C_2H_5)(3) and 8,8′-μ-Et_2NP(O)(O)(2) (1,2-$C_2B_9H_{10}$)(2)-3,3′-Co $(HN(CH_3)_3)$. *J. Organomet. Chem.* 657: 59–70.

Plešek, J., Grüner, B., Heřmánek, S. et al. 2002a. Synthesis of functionalized cobaltacarboranes based on the *closo*-$[(1,2\text{-}C_2B_9H_{11})_2\text{-}3,3'\text{-Co}]^-$ ion bearing polydentate ligands for separation of M^{3+} cations from nuclear waste solutions. Electroctrochemical and liquid-liquid extraction study of selective transfer of M^{3+} metal cations to an organic phase. Molecular structure of the [8-(2-CH_3O-C_6H_4-O-$(CH_2CH_2O)_2$-*closo-commo*-(1,2-$C_2B_9H_{10}$)-(1′,2′-$C_2B_9H_{11}$)-3,3′-Co]Na determined by x-ray diffraction analysis. *Polyhedron* 21: 975–986.

Plešek, J., Heřmánek, S., Franken, A., Císařová, I., and Nachtigal, C. 1997. Dimethyl sulfate induced nucleophilic substitution of the bis(1,2-dicarbollido)-3-cobalt(1-) ate ion. Syntheses, properties and structures of its 8,8′-mu-sulfato, 8-phenyl and 8-dioxane derivatives. *Collect. Czech. Chem. Commun.* 62: 47–56.

Popov, A. and Borisova, T. 2001. Adsorption of dicarbollylcobaltate(III) anion [1,2-$B_9C_2H_{11}$)2Co(III)]- at the water/1,2-Dichloroethane interface. Influence of counterions in nature. *J. Coll. Interf. Sci.* 236: 20–27.

Rais, J., Kyrš, M., and Heřmánek, S. 1974. Czech Patent CS 153, 933; 15 June 1974.

Rais J., Selucký, P., and Kyrš, M. 1976. Extraction of cesium into nitrobenzene in the presence of univalent polyhedral borate anions. *J. Inorg. Nucl. Chem.* 38: 1376–1378.

Rais, J., Tachimori, S., Selucký, P., and Kadlecová, L. 1994. Synergetic extraction in systems with dicarbollide and bidentate phosphonate. *Sep. Sci. Technol.* 29: 261–274.

Rais, J. 2004. Principles of extraction of electrolytes in separation chemistry. In *Ion Exchange, Solvent Extraction* (Marcus, Y. and SenGupta, A. K., eds.) vol. 17, pp. 335–385, New York: Marcel Dekker.

Rais, J. and Grüner, B. 2004. Extractions with cobalt bis(dicarbollide) ions. In *Ion Exchange, Solvent Extraction* (Marcus, Y. and SenGupta, A. K., eds.) vol. 17, pp. 243–334, New York: Marcel Dekker.

Rais, J. and Okada, T. 2008. Quantitized hydration energies of ions and structure of hydration shell from the experimental gas-phase data. *J. Phys. Chem. B* 112: 5393–5402.

Reed, C. A. 1998. Carboranes: A new class of wealky coordination anions for strong electrophiles, oxidants and superacids. *Accounts Chem. Res.* 31: 133–139.

Reinoso-García, M. M., Dijkman, A., Verboom, W. et al. 2005b. Metal complexation by tripodal N-Acyl(thio) urea and picolin(thio)amide compounds: Synthesis/extraction and potentiometric studies. *European J. Org. Chem.* 2131–2138.

Reinoso-García, M. M., Janczewski, D., Reinhoudt, D. N. et al. 2006. CMP(O) Tripodands: Synthesis, potentiometric studies and extractions. *New J. Chem.* 30: 1480–1492.

Reinoso-García, M. M., Verboom, W., Reinhoudt, D. N., Brisach, F., Arnaud-Neu, F., and Liger, K. 2005a. Solvent extraction of actinides and lanthanides by CMP(O)- and N-Acyl(thio)urea-tetrafunctionalized cavitands: Strong synergistic effect of cobalt bis(dicarbollide) ions. *Solv. Extr. Ion Exch.* 23: 425–437.

Rodionov, S. A., Viznyi, A. N., Esimantovskii, V. M., and Zilberman, B. Y. 2008. Extraction of long-lived radionuclides from nitric acid solutions using an extractant based on dibutyl hydrogen phosphate and chlorinated cobalt dicarbollide. *Radiochemistry* 50: 626–632.

Romanovskiy, V. N., Babain, V. A., Alyapyshev, M. Y. et al. 2006. Radionuclide extraction by 2,6-pyridinedicarboxylamide derivatives and chlorinated cobalt dicarbollide. *Sep. Sci. Technol.* 41: 2111–2127.

Rose, D. and Benjamin, I. 2009. Free energy of transfer of hydrated ion clusters from water to an immiscible organic solvent. *J. Phys. Chem. B* 113: 9296–9303.

Rzhekhina, E. K., Karkozov, V. G., Alyapyshev, M. Y. et al. 2007. Reprocessing of spent solvent of the UNEX process. *Radiochemistry* 49: 493–498.

Sansone, F., Galletta, M., Macerata, E. et al. 2008. Upper rim CMPO-substituted calix[6]- and calix[8]arene extractants for the An3+/Ln3+ separation from radioactive waste. *Radiochim. Acta* 96: 235–239.

Selucký, P., Rais, J., Alexová, J., and Hiklová, S. 2005. Testing of New Extraction Agents, Rep. RAWRA, December 2005.

Selucký, P., Šistková, N. V., and Rais, J. 1997. Bisphenylene COSAN as highly selective Cs^+ extraction agent for Cs. *J. Radioanal. Nucl. Chem.* 224: 89–94.

Selucký, P., Rais, J., Lučaníková, M. et al. 2008. Lanthanide and actinide extractions with cobalt bis(dicarbollide) extractants with covalently bonded CMPO functions. *Radiochim. Acta* 96: 273–284.

Semioshkin, A. A., Sivaev, I. B., and Bregadze, V. I. 2008. Cyclic oxonium derivatives of polyhedral boron hydrides and their synthetic applications. *Dalton Trans.* 977–992.

Shishkin, D. N., Galkin, B. Y., and Fedorov, Y. S. 2009. Synergistic extraction of REE and TPE from HNO3 solutions with a mixture of CCD and HDPB ZS (1:8) in a polar diluent. *Radiochemistry* 51: 30–33.

Schmidt, C., Saadioui, M., Bőhmer, V. et al. 2003. Modification of calix 4 arenes with CMPO-functions at the wide rim. Synthesis, solution behavior, and separation of actinides from lanthanides. *Org. Biomol. Chem.* 1: 4089–4096.

Sivaev, I. B. and Bregadze, V. I. 1999. Chemistry of cobalt bis(dicarbollides). A review. *Collect. Czech. Chem. Commun.* 64: 783–805.

Smirnov, I. V., Babain, V. A., and Chirkov, A. V. 2007. New hydrolytically stable solvent for Am/Eu separation in acidic media. In *Global 2007* p. 126, Boise, Idaho, September 9–13, 2007.

Smirnov, I. V., Chirkov, A. V., Babain, V. A., Pokrovskaya, E. Y., and Artamonova, T. A. 2009. Am and eu extraction from acidic media by synergistic mixtures of substituted bis-tetrazolyl pyridines with chlorinated cobalt dicarbollide. *Radiochim. Acta* 97: 593–601.

Strauss, S. H. 1993. The search for larger and more weakly coordinating anions. *Chem. Rev.* 93: 927–942.

Šícha, V., Plešek, J., Kvíčalová, M., Císařová, I., and Grüner, B. 2009. Boron substituted 8-nitrilium and 8-ammonium derivatives, versatile cobalt bis(1,2-dicarbollide) building blocks for synthetic purposes. *Dalton Trans.* 851–860.

Tachimori, S. and Morita, Y. 2010. Overview of solvent extraction chemistry for reprocessing. In *Overview of Solvent Extraction Chemistry for Reprocessing* (Moyer, B. A., ed.) vol. 19, pp. 1–63, USA: Boca Raton.

Vandegrift, G. F., Regalbuto, M. C., Aase, S. B. et al. 2004. Lab-scale demonstration of the UREX+ Process. In Waste Management, WM 04 Conference, Tucson, AZ, USA, February 29–March 4, 2004.

Vinas, C., Gomez, S., Bertran, J., Teixidor, F., Dozol, J. F., and Rouquette, H. 1998a. New polyether substituted metallacarboranes as extractants for Cs-137 and Sr-90 from nuclear wastes. *Inorg. Chem.* 37: 3640–3643.

Vinas, C., Bertran, J., Gomez, S. et al. 1998b. Aromatic substituted metallacarboranes as extractants of Cs-137 and Sr-90 from nuclear wastes. *J. Chem. Soc., Dalton Trans.* 2849–2853.

Vinas, C., Gomez, S., Bertran, J., Teixidor, F., Dozol, J. F., and Rouquette, H. 1998c. Cobalt bis(dicarbollide) derivatives as extractans of europium from nuclear wastes. *Chem Commun.* 191–192.

Zilberman, B. Y., Fedorov, Y. S., Shmidt, O. V. et al. 2009. Dibutyl phosphoric acid and its acid zirconium salt as an extractant for the separation of transplutonium elements and rare earths and for their partitioning. *J. Radioanal. Nucl. Chem.* 279: 193–208.

20 Boron-Based Nanomaterials
Technologies and Applications

Amartya Chakrabarti, Lauren M. Kuta, Kate J. Krise, John A. Maguire, and Narayan S. Hosmane

CONTENTS

20.1 INTRODUCTION

Boron, the only nonmetal among the elements of Group 13 of the periodic table, exhibits a very unique combination of properties. A very high melting point of 2076°C, low density, a Knoop hardness of 2160–2900, and high Young's modulus make it very useful in a diverse range of industrial applications.[1] The electron-deficient nature of elemental boron and the presence of vacant p-orbitals lead to the unusual structural complexity of boron-based materials. These characteristics of boron have prompted its existence in clusters containing delocalized multicenter bonding as seen in the B_{12} icosahedrons. Moreover, it may cause an increase of ligancy even to its adjacent atoms in the boron-containing compounds. The interesting features of pure boron and boron-based materials have made boron chemistry structurally varied and exciting.

In general, the properties of bulk materials vary when the structure of that material takes shape in nanodimension. Certain physical and mechanical properties change drastically with the change of materials size. Similar phenomena have been observed in the cases of optical and electrical properties. For example, gold nanoparticles exhibit a variety of colors depending on its shape and size.[2] Similarly, boron nanomaterials possess a variety of interesting properties that are suitable for a diverse field of applications. The high surface-to-volume ratio of boron nanoparticles coupled with boron's high energy content make such nanoparticles attractive as fuel additives in conventional hydrocarbon fuels for airbreathing propulsion systems. Their robust thermal and electrical stabilities, compared to metal nanowires, make one-dimensional boron nanostructures, such as nanowires, nanorods, nanobelts, or nanotubes promising materials as nanoscale electrically conducting interconnects for nanoelectronic devices. Boron-based nanoparticles are now being extensively investigated as a boron source in boron neutron capture therapy.

This chapter is divided into three major parts. In the first part, synthesis of boron and boron-based nanomaterials will be discussed, with an emphasis on different aspects of methodologies currently being employed. The second section will describe the general properties of the different boron-based nanomaterials governed by their structural features with several important application fields, followed by the summary and a brief outlook on future research directions.

20.2 SYNTHESIS OF BORON AND BORON-BASED NANOMATERIALS

Over the past two decades boron and boron-based nanostructures have attracted substantial attention among researchers all across the world. The superiority of boron-based nanomaterials over other one-dimensional nanostructures in different contexts of properties enforces substantial research being performed in many areas. In the following section we will mostly cover reported synthesis of boron nanostructures and boron-based materials in addition to our own contributions to this field.

Depending on the dimension and typical features of the nanostructures, all boron nanomaterials can broadly be accommodated in the following categories: nanowires, nanotubes, nanobelts, nanoribbons, nanorods, and nanosheets. Apart from the pure boron compounds, there are several boron-based nanomaterials, that is, combination of boron with other elements, prepared, characterized, and reported recently. Among all these nanomaterials, we have chosen a few of the most important substances, including boron nitride, boron carbide, and nanostructures, guided by the scope and limitations of this chapter.

20.2.1 Boron Nanowires

The nanowires are one of the most important classes of compounds among all the nanostructured materials due to their potential as building blocks for functional nanoscale electronics. Higher electrical conductivity of boron-based nanomaterials in the form of tubes, layers, or fullerene-like solids has been predicted,[3] which essentially directs the quest of synthetic strategies toward the feasible routes to prepare boron nanowires in bulk quantities.

Boron nanowires, prepared using radiofrequency (rf) magnetron sputtering techniques, have been reported by Cao et al.[4] A very high-purity mixture of boron and boron oxide was subjected to an rf power of 80 W under argon gas at 800°C. A neatly well-aligned array of boron nanowires was found on a silicon substrate with a typical growth time of 6 h (Figure 20.1a). The nanowires, which can easily be stripped off the substrate, are of uniform length and diameter. The nanowires have been well characterized via parallel electron energy loss spectroscopy (PEELS), energy-dispersive x-ray spectroscopy (EDS), transmission electron microscopy (TEM), and scanning electron microscopy (SEM). While the system temperature was proven to play an important role in the growth of the nanowires, the nature of the substrate was reported to be insignificant in the formation of the same.

This general method of synthesizing low-dimension boron nanostructures from B/B_2O_3 mixtures under various conditions has been exploited by Gao and coworkers to produce single-crystalline boron nanowires,[5a] nanocones,[5b] and nanotubes.[5c] The nanowires were grown on an Fe_3O_4-coated Si(001) substrate from a 2:4:1 mixture of B_2O_3, B, graphite, respectively. The mixture was slowly heated under an Ar(g) flow to a temperature of 1000–1100°C and the nanowires grew on the substrate via a vapor–liquid–solid (VLS) process.[5a] The nanocones were grown on an Fe_3O_4-coated Si(111) substrate from a solid mixture of pure B_2O_3 and B in a 1:5 molar ratio at 1000–1100°C by the passage of a mixture of 95% Ar and 5% H_2.[5b] On the other hand, single-crystal boron nanotubes were obtained over a substrate coated with Fe_3O_4 catalyst from heating a B_2O_3 and B mixture in a 4:3 mass ratio, respectively, at 1000–1200°C under a flow of Ar gas.[5c] This demonstrates the sensitivity of the boron nanostructures to the conditions of synthesis.

Preparation of superconducting MgB_2 nanowires prompted the synthesis of boron nanowires via chemical vapor transportation process.[6] Although Wu et al. published their work with an emphasis on the preparation of MgB_2, it described a simple technique to produce boron nanowires in bulk quantities. A mixture of boron, iodine, and silicon powder was taken in a sealed quartz tube and subjected to 1100°C temperature. In the comparatively cooler part of the tubular reactor, magnesium oxide (MgO) substrate (previously coated with a thin gold film) was kept. A temperature gradient of 100°C was maintained between the boron source and the substrate, which helped the BI_3 vapor formed in the hotter region to decompose on the substrate. The fluffy product found on the substrate was characterized and found to be composed of boron nanowires of hundreds of micrometers in length, having diameters around 50–100 nm. The product was then used to synthesize MgB_2 nanowires by allowing the same to react with Mg vapor.

The first reported synthesis of crystalline boron nanowires was accomplished by the chemical vapor deposition (CVD) technique.[7] In this method, a mixture of 5% diborane (B_2H_6) and argon gas, further diluted by pure Ar, was allowed to pass over a NiB catalyst powder on an alumina substrate. The gas mixture continued to flow over a 30-min period of time while the temperature inside the furnace was kept at 1100°C. The NiB particles were found to be merged together onto the alumina substrate. The nanowires, resulting from this method, were cultivated into both straight and curly formations gauged several micrometers in length and diameters of 20–200 nm (Figure 20.1b). The electron diffraction studies revealed the crystalline structure of the nanowires and the conductivity measurements demonstrated a semiconductor nature, rather than a metallic conductor.

A laser ablation technique for the preparation of boron nanowires was published almost contemporary to the previous work.[8] Crystalline boron nanowires with tetragonal structure have been reported to form via laser treatment on boron rods, doped with 10% of a half-and-half mixture of Ni and Co metal powders. A Nd/YAG laser of 532 nm wavelength, 10 Hz frequency, and 3.5 W power was applied on the B/NiCo target. It was then kept in a furnace at 1250°C, under flowing argon gas for 30 min. The x-ray diffraction data predicted that nanowires with crystalline structure and crystals are mostly single crystal in nature. The role of the catalyst in formation of the nanostructure has also been examined and proven to be vital.

FIGURE 20.1 SEM images of boron nanowires produced via (a) radiofrequency (rf) magnetron sputtering technique, (Adapted from (a) Cao, L. et al., *Adv. Mater.*, **2001**, *13*, 1701–1704; (b) Cao, L. et al., *J. Phys. Condens. Matter.*, **2002**, *14*, 11017–11021.) (b) chemical vapor deposition (CVD) method, (Adapted from Otten, C. J. et al., *J. Am. Chem. Soc.*, **2002**, *124*, 4564–4565.) and (c) laser ablation process. (Adapted from Meng, X. M. et al., *Chem. Phys. Lett.*, **2003**, *370*, 825–828.)

In contrast to the laser ablation technique mentioned above, a similar methodology was employed to achieve amorphous boron nanowires.[9] The process involved irradiation of boron powder, kept inside a long alumina tube at 1300°C under argon atmosphere, with a laser beam (248 nm wavelength) for 5 h at a pulse frequency of 10 Hz. The product was collected on a silicon substrate; it consisted of smooth nanowires that were several tens of micrometers long and 30–60 nm in diameter (Figure 20.1c).

Cao et al. have also extended their work on the magnetron sputtering techniques toward the synthesis of feather-like boron nanowire nanojunctions.[10] Maintaining almost similar methodology, they had enhanced the argon flow rate during the process of synthesis and amazingly observed a feather-like growth of boron nanowires in the product obtained (Figure 20.2). All the structures have multiple T- and Y-junctions that are greatly desired features for the applications of the nanowires in nanoscale electronic and mechanical devices.

A typical method for preparing semiconductor nanowires, the vapor–liquid–solid (VLS) process, has been adopted by Yun et al. for synthesizing boron nanowires.[11] They deviated from the usual methodology of VLS processes by implementing an oxide-assisted route to control the growth of the product. The regular version of VLS method involves metal catalysts, which generates the solid–liquid alloy to help the growth of the nanowires. In this method, a mixture of boron and boron oxide powder was used to grow the nanowires on gold-coated silicon substrates. In a tube furnace, 800–1100°C temperature was employed and the nanowires were generated via vapor transport method and were collected on the cooler region of the tube. The product has shown evidence of the presence of branches in the structure, typically in the form of Y-junctions (Figure 20.3).

A similar work on the CVD technique was reported for boron nanowires with a different and more controlled morphology.[12] Amorphous and smooth boron nanowires were fabricated using a gold catalyst within the temperature region of 800–950°C. A mixture of nitrogen, hydrogen, and diborane gases, in a typical ratio of 10:50:1, was passed over the catalyst for a period of 5 h. The system was brought back to ambient temperature, after the completion of the deposition. The most

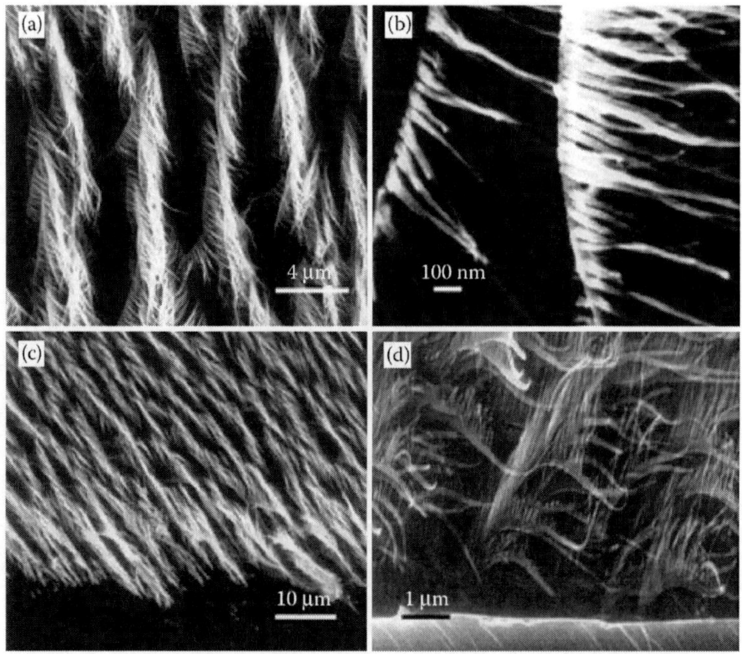

FIGURE 20.2 Multiple Y- and/or T-junction boron nanofeather arrays grown on Si substrates: (a) low-magnification top-view SEM image, (b) high-magnification SEM image, (c) side-view SEM image, and (d) cross-sectional SEM image. (Adapted from Cao, L. M. et al., *Nanotechnology*, **2004**, *15*, 139–142.)

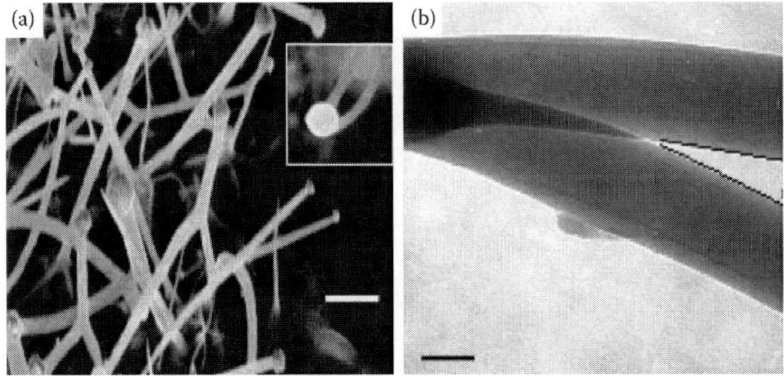

FIGURE 20.3 Boron nanowire Y junctions (a) SEM image of the Y-junctions on a silicon substrate coated with a 10-nm-thick gold film. Scale bar: 1 μm; (b) low-magnification bright-field TEM image of a Y-junction with the angle of diffusion of 20°. Scale bar: 50 nm. (Adapted from (a) Yun, S. H. et al., *Appl. Phys. Lett.*, **2005**, *87*, 113109(1–3). (b) Yun, S. H. et al., *Nano Lett.*, **2006**, *6*, 385–389.)

interesting structural feature of the product is the formation of nanochains with a periodic spacing of nanospheres between them (Figure 20.4b). This particular product was found to form at around 950°C, whereas regular nanowires were obtained in the region of 800–850°C.

20.2.2 Boron Nanotubes

Although theoretical studies, conducted during the late 1990s, predicted better stability and higher metallic conductivities of boron nanotubes (BNT) than their carbon counterparts,[3,13,14] no reports were found, until recently, on boron nanotubes. The first synthesis of boron nanotubes was published in 2004 by Ciuparu et al.[15]

The process involved a 6 mm quartz reactor containing a silica template (Mg-MCM-41), placed in a furnace undergoing continuous hydrogen flow at 870°C. Then boron trichloride (BCl_3, purity 99.9%) was passed over the template at the flow rate of 1.5 L/min for 45 min. The furnace and substances were then allowed to cool under constant helium flow. The reaction setup was then placed under a fumed hood to get rid of the excess unreacted BCl_3 that was accomplished by burning the same with a hydrogen flare. Although the high sensitivity of the sample toward electron beam prevented some crucial characterization, the conventional TEM method showed the presence of nanotubes that were of about 16 nm in length and 3 nm in diameter (Figure 20.5). They have also concluded that the Mg metal in the support material has a pronounced effect on the synthesis of the

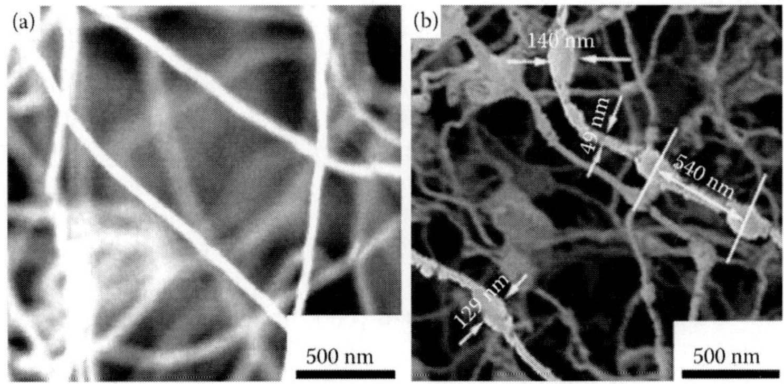

FIGURE 20.4 Field emission scanning electron microscopy (FESEM) images of boron nanowires formed at (a) 800°C and (b) 950°C. (Adapted from Yang, Q. et al., *J. Mater. Sci.*, **2006**, *41*, 3547–3552.)

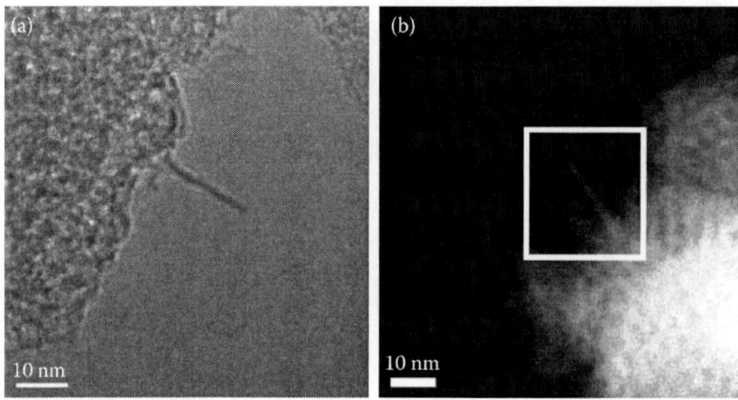

FIGURE 20.5 (a) High-resolution TEM image of pristine boron nanotube. (b) High-resolution Z-contrast image in STEM mode of the catalyst-supported formation of boron nanotube. (Adapted from Ciuparu, D. et al., *J. Phys. Chem. B*, **2004**, *108*, 3967–3969.)

boron nanotube since other substrates, for example, siliceous MCM-41, failed to produce any product under similar reaction conditions.

A more recent report by Gao and coworkers described a large-scale production of boron nanotubes.[5c] The nanotubes were grown over Fe_3O_4 nanoparticle catalysts from a 4:3 mass ratio of pure B_2O_3 and B, respectively. Studies on individual BNTs found them to possess metallic properties with an averaged conductivity of 40 Ω-L cm^{-1} and also exhibit high luminescent efficiency and stability. These results are of interest since recent theoretical results have predicted the possibility of stable boron sheets that can be rolled to give nanotubes that are metallic when their diameters are greater than 17 Å, with smaller-diameter tubes being semiconductors whose band gaps vary with size and chirality. On the other hand, tubes rolled from triangular lattices have been predicted to be always metallic.[13]

20.2.3 Boron Nanobelts

Nanobelts are a simple modification of nanowire structures with a rectangular cross-sectional shape and a low width-to-thickness ratio. There are only a few reported syntheses of the boron nanobelts available in the published literature.

Wang et al. observed a nanobelt structure while preparing boron nanowires via the laser ablation technique.[16] A typical Nd:YAG laser with a wavelength of 355 nm, pulse width varying between 5 and 7 ns, and frequency of 10 Hz was used for the ablation process. Previously, hot-pressed boron pellets were targeted by the laser while kept at 700–1000°C under the continuous flow of Ar. This simple technique did not involve any catalysts and successfully produced nanobelts of several tens of micrometer length with a width-to-thickness ratio of 5. A low-resolution TEM image is displayed in Figure 20.6a, clearly showing the belt structure.

Another recently published work in this area revealed that boron nanobelt structures could be obtained through intensive ion bombardment techniques.[17] The experimental setup is depicted schematically in Figure 20.7. During the experiment, both the substrate and the compound were evaporated by an e-beam generated from an e-gun. High-purity solid boron was used as the boron source and oxidized silica coated with Au was the substrate. The substrate was heated up to 1100°C followed by the bombardment via helicon plasma source located on top of the substrate. The sample obtained was investigated thoroughly via FESEM and TEM and showed the formation of boron nanobelts of width-to-thickness ratio of 3:2.

A CVD method via a vapor–liquid–solid mechanism has been further described to prepare boron nanobelts in both crystalline and amorphous forms by Ni and Li[18] The boron source for the process

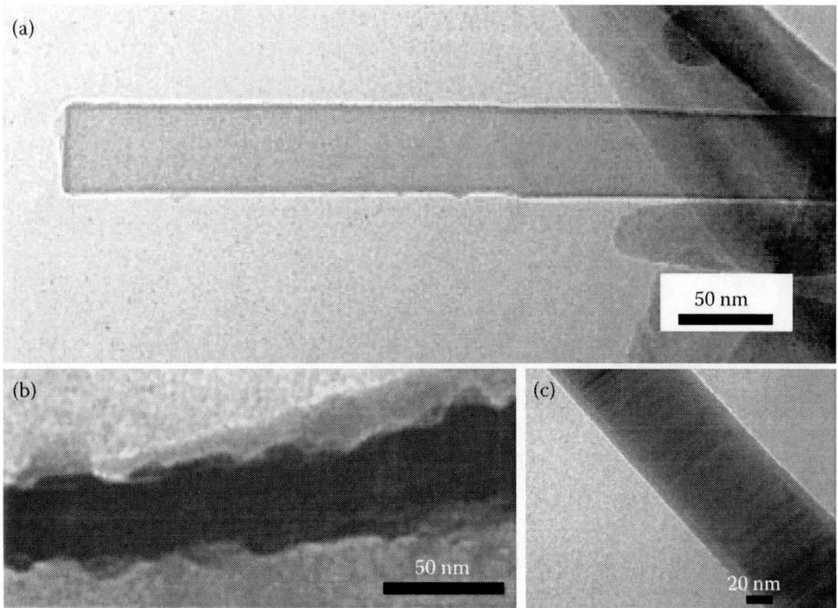

FIGURE 20.6 TEM images of boron nanobelts prepared by (a) the laser ablation technique. (Adapted from Wang, Z. et al., *Chem. Phys. Lett.*, **2003**, *368*, 663–667.) (b) An intensive ion bombardment method. (Adapted from Lia, W. T.; Boswell, R., and Gerald, J. D. F. *J. Vac. Sci. Technol. B*, **2008**, *26*, L7–L9.) (c) The chemical vapor deposition process. (Adapted from Ni, H. and Li, X. *J. Nano Res.*, **2008**, *1*, 10–22.)

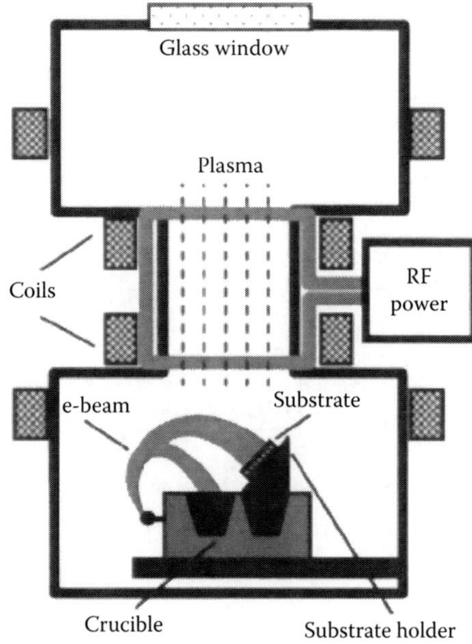

FIGURE 20.7 Schematic of the apparatus for boron nanobelt growth. (Adapted from Lia, W. T.; Boswell, R., and Gerald, J. D. F. *J. Vac. Sci. Technol. B*, **2008**, *26*, L7–L9.)

was powdered boron mixed with silicon and gold powders in a weight ratio of 40:1:1 and the mixture was kept in a quartz tube. Silicon substrate with two different coatings was used to get the amorphous and the crystalline products. For the amorphous product, 20 nm gold coating was used while a 3 nm thickness of gold was sputter-coated on the substrate for the crystalline one. An average temperature of 1150°C was used in the process with the variation of the holding time of 20 and 115 min for the amorphous and the crystalline nanobelts, respectively. Both the products exhibited belt-like structures with a width-to-height ratio of almost 2 (Figure 20.6c).

20.2.4 Boron Nanoribbons

Boron nanoribbons are also a different form of nanowires with a diverse morphology grown at relatively lower temperatures. To the best of our knowledge, there are only two available articles on the synthesis of this particular type of nanostructure.[19,20] Both of them described the structure as "grass-like" with "fork-like" slots in the middle.

The first reported synthesis of catalyst-free boron nanoribbons was in the year 2004 by Xu et al.[19] α-Tetragonal, single-crystal growth of boron nanoribbons was observed via this low-temperature (630–750°C) and low-pressure (200 mTorr) pyrolysis of diborane gas. A "puffy ball" like growth was found on the silicon substrate at the aforementioned temperature zone of the furnace (Figure 20.8a). The SEM and TEM studies disclosed the typical width of the nanoribbons to be about 800–3200 nm and thickness about 15–20 nm (Figure 20.8b).

Following the previous work, Jash and Trenary recently published their finding on boron nanoribbon structures.[20] They adopted a similar methodology to synthesize this material with the exception that a Ni-coated silicon substrate had been employed in their process. They have also discovered a tendency of growing new ribbons from the edge of the wide ribbons (Figure 20.8c, inset) while the length, width, and thickness were almost similar to that of the work by Xu.

20.2.5 Boron Nanorods

Nanorods are fundamentally the miniature version of nanowires, typically grown via similar methods used for nanowires. The structural features of these nanostructures also resemble nanowires apart from their length that falls in the range of hundreds of nanometers. If utilized properly, they can be extremely useful as seed materials for several boron-based nanostructures.

Yang et al. first synthesized boron nanorods in 2007.[21] A chemical vapor deposition technique was demonstrated on the formation of boron nanorods structures. An aluminum-coated silicon substrate was placed in a tube furnace and kept at a temperature of 700–950°C. A mixture of argon,

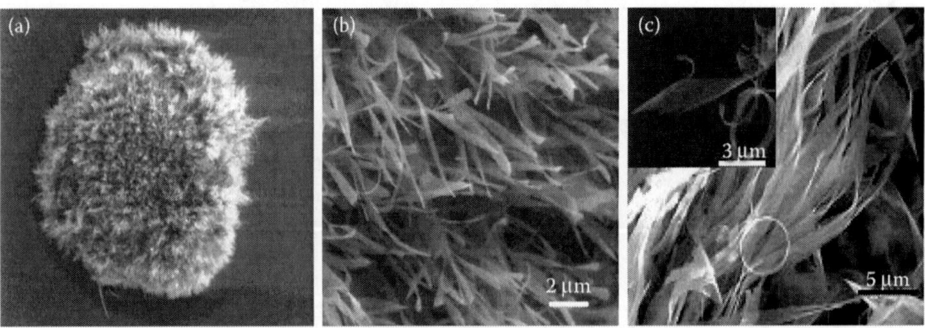

FIGURE 20.8 SEM image of boron nanoribbons formed through (a) pyrolysis of diborane gas ("puffy ball"); (b) pyrolysis of diborane gas ("grass-like") (Adapted from Xu, T. T. et al., *Nano Lett.*, **2004**, *4*, 963–968.); and (c) the chemical vapor deposition method. (Adapted from Jash, P. and Trenary, M. *Journal of Physics: Conference Series*, **2009**, *176*, 012011(1–11).)

hydrogen, and diborane, with a flow ratio of 50:20:1 was passed over for a period of 5 h. After completion of the deposition, the furnace was cooled to the ambient temperature and the product was characterized via various techniques including FESEM, TEM, and EDX and so on. The morphology of the nanorods was strictly guided by the temperature of the substrate and well-aligned array of nanorods were found in the region between 800°C and 900°C. Bearing a single-crystal lattice structure, the nanorods were 20–50 nm in diameter (Figure 20.9a).

Most of the nanomaterials have been prepared using drastic reaction conditions and required highly toxic and flammable precursor gases. In our laboratory, we have developed a "greener" and more energy-efficient synthetic strategy to prepare boron nanorods.[22] This top-down synthesis involved the reduction of nontoxic boron oxide (B_2O_3) in liquid lithium with the assistance of ultrasonication during the reaction and used a relatively lower temperature of 200–250°C. The product was thoroughly washed with alcohol, hot water, and dilute acid and base solutions. TEM images revealed dominance of nanorods in the product morphology with the diameter ranging from 20 to 40 nm and length of 80–200 nm (Figure 20.9b). The preferential formation of boron nanorods over nanoparticles can be associated with the streaming-like ultrasonic propagation,[23] which can transfer the acoustic momentum efficiently to the liquid lithium,[24] leading to directed lithiation in a B_2O_3 powder forming rod-like shapes.

20.2.6 Boron Nitride Nanotubes

Boron nitride, being isoelectronic with carbon, generated a great deal of interest among researchers as soon as Iijima reported the synthesis of carbon nanotubes.[25] Possibility of a hexagonal graphitic structure of boron nitride was predicted by Rubio et al.,[26] while the first synthesis of boron nitride nanotubes (BNNTs) was reported by Chopra et al.[27] Since then, several methods are being employed successfully for preparing boron nitride nanotubes. The most commonly used methods for the preparation of BNNTs include arc discharge,[28] chemical vapor deposition,[29] laser ablation,[30] mechanothermal,[31] pyrolysis,[32] and other chemical syntheses techniques.[33] Recently, Golberg et al. adequately documented all the synthetic methodologies currently being used in their review article.[34] In this section, we are going to discuss the current techniques reported after their publication in 2007.

A simple chemical reaction of a nitrogen and hydrogen gas mixture in the presence of previously ball-milled boron–nickel powder at 1025°C was utilized by Lim et al.[35] This method successfully generated BNNTs of diameter and lengths ranging from 20 to 40 nm and less the 250 nm, respectively. The process does not involve any harmful precursors and requires a comparatively lower temperature than most of the other reported methods.

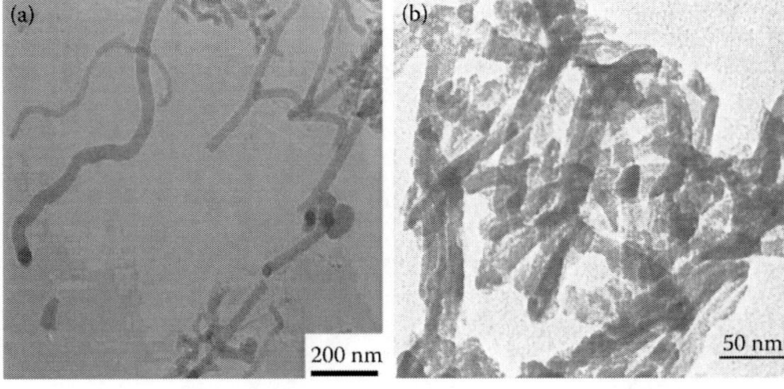

FIGURE 20.9 TEM images of boron nanorods prepared by (a) CVD technique (Adapted from Yang, Q. et al., *Eur. Phys. J. B*, **2007**, *56*, 35–39.) and (b) top-down synthesis at moderate reaction conditions. (Adapted from Chakrabarti, A. et al., *J. Nanomater.*, **2010**, 589372(1–5).)

An almost contemporary work to the previous report developed a unique yet simple method of BNNT synthesis with the assistance of nanoporous anodized aluminum oxide (AAO) template;[36] though the first template-assisted synthesis of BNNTs was reported earlier.[37] In the former method, thermolysis of a boron precursor was used to generate the nanotubular structure inside the template that had been etched out later using hydrofluoric acid and repeated washing with water, methanol, and acetone. The process of incorporation of the polymer inside the nanochannels of the template is known as liquid-phase infiltration (LPI) technique, which has been successfully improvised to make nanotubes of almost 60 μm length and 200 nm diameter (Figure 20.10).

Ball milling of boron powder followed by annealing at 1100°C under a nitrogen flow resulted in considerably longer BNNTs.[38] The formation of nanotubes was supported by an iron catalyst and found to be dependent on the time of annealing, which was optimized to be 15 h. The nanotubes were grown following a metal catalytic growth mechanism that is supported further by the special bamboo-like feature of the tubes. The longest tube formed was measured to be almost 1.05 mm in length, with an average diameter ranging from 50 to 200 nm. Conversely, a similar attempt of producing BNNTs from B_2O_3 powder resulted in nanotubes with a slightly different morphology and shorter length.[39] The report mentioned ball milling of B_2O_3 under nitrogen atmosphere for 100 hours followed by annealing the same at 1200°C for 6 h under a continuous flow of ammonia gas. The filament product had a crystalline structure with bamboo-like features, and a few of the filaments displayed a cylindrical morphology. Tubes with larger diameters (more than 80 nm) fall in the first category while the ones with smaller diameter (<50 nm) were more of cylindrical nature.

A mechanothermal process has been employed to grow BNNTs from hexagonal boron nitride (h-BN) powders.[40] This simple, though effective, method involved two steps for preparing BNNTs. At first, h-BN powders were ball-milled with liquid ammonia in acetone for a varied time period of 10–100 h under atmospheric pressure and ambient temperature. This method generated highly disordered or amorphous nanostructures that were converted into crystalline BNNTs via isothermal annealing under nitrogen at temperatures varied from 950°C to 1300°C for about 10 h. The effect of milling time and annealing temperature were thoroughly studied and it was found that longer nanotubes of about 1 μm length were obtained at 1300°C in comparison with the annealing temperature of 950°C.

A floating catalyst method, which has been used for the large-scale production of carbon nanotubes, was applied in the synthesis of double-walled BNNTs by Kim et al.[41] Borazine was chosen as the precursor for BN formation because of its chemical composition (1:1 B/N ratio) and high volatility. In this catalytic chemical vapor deposition (CCVD) technique, a mixture of ammonia and nitrogen gas (in a flow ratio of 100:3 sscm), borazine vapor along with a nickelocene catalyst flowed

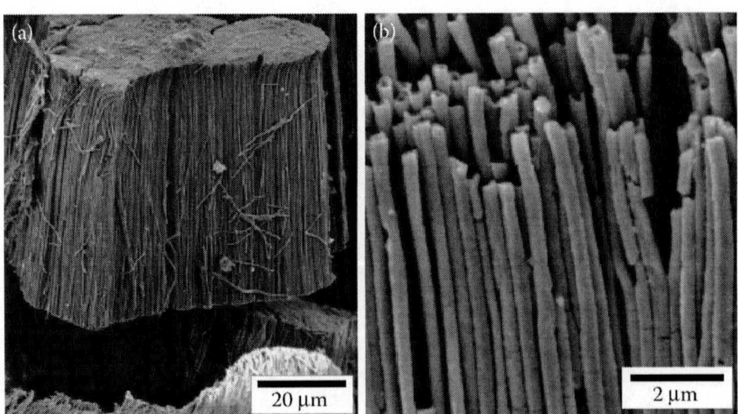

FIGURE 20.10 SEM image of as-synthesized BNNTs (a) cross-sectional view; (b) back-side view. (Adapted from Bechelany, M. et al., *J. Phys. Chem. C*, **2007**, *111*, 13378–13384.)

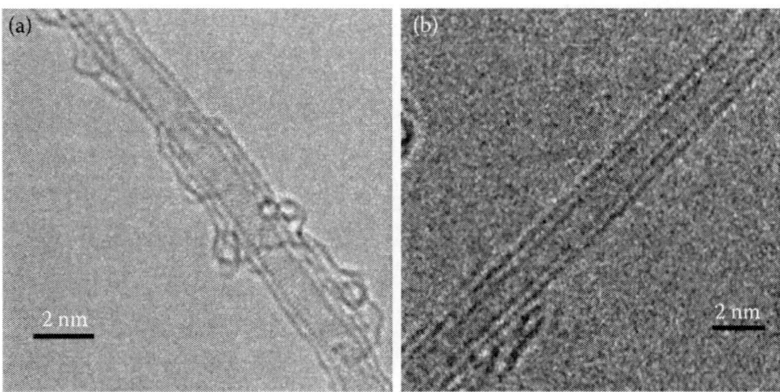

FIGURE 20.11 TEM images of BNNTs. (a) As-grown tubes with carbon contamination; (b) Tubes after the oxidation step. (Adapted from Kim, M. J. et al., *Nano Lett.*, **2008**, *10*, 3298–3302.)

through the hot zone of a tube furnace. The growth temperature was recorded to be 1200°C and the deposition was collected over a grid kept in the downstream region of the tube. Nanotubes of 100–200 nm length and 2 nm diameter were formed during the process (Figure 20.11a). There was also some carbon deposition observed on the tube surfaces, which was attributed to the decomposition of the nickelocene catalyst (Figure 20.11a). However, oxidation of the BNNTs at 600°C for 30 min in the presence of air removed most of the carbonaceous materials, while the structural integrity of the BNNTs remained unaltered (Figure 20.11b).

A simple thermal chemical vapor deposition technique was adopted by Lee et al. to form BNNTs using a conventional tube furnace and a quartz tube vacuum chamber.[42] A mixture of boron, iron oxide, and magnesium oxide was taken in the quartz tube and placed inside a horizontal tube furnace. After the quartz tube was evacuated, 200 sccm of ammonia gas was purged into the tube and the temperature of the furnace was raised up to 1200°C. Within an hour, an alumina boat in the tube was found to be coated with white powder containing the nanotubes with an average length of about 10 μm and diameters of 10–100 nm. Any substrate can also be placed over the boat to get the BNNTs deposited on it. The nanotubes were thought to be grown via a VLS mechanism, where the metal oxides play a key role as catalysts.

Smith et al. reported a pressurized vapor/condenser (PVC) technique of preparing BNNTs scalable in gram quantities.[43] This unique method implied very high temperature (over 4000°C) to produce nanotubes of small diameters and few walls. The formation of BNNTs via the reaction of boron vapor with nitrogen and its different stages has been depicted in the following schematic (Figure 20.12). This process differs from the conventional methods of synthesizing BNNTs in several ways. Firstly, it does not involve the usage of any catalysts. The product obtained is highly crystalline with extraordinary lengths compared to their diameters, producing textures similar to that of cotton fibers (Figure 20.12, inset). Liquid boron droplets formed during the course of the reaction acted as sites for nucleation growth of the nanotubes in the form of clusters.

High temperatures (above 1100°C) are required for the preparation of BNNTs and are the major contributors to the high production cost of the final product. A diverse approach to form hexagonal BNNTs via the microwave plasma CVD (MPCVD) method, reported by Guo and Singh, involved a comparatively low substrate temperature of 800°C.[44] A precursor gas mixture, containing diborane, ammonia, and hydrogen was employed to grow the BNNTs on an oxidized silicon substrate, coated with a nickel catalyst. A typical reaction chamber is depicted in Figure 20.13. In this case, formation of the nanotubes was found to be dependent on the thickness of the catalyst layer. Interestingly, no catalyst particles were noticed on the tip of the tubes that is typical for the BNNTs grown in a VLS mechanism. A different growth pattern of the nanotubes have been suggested, which was initiated through the formation of three-dimensional catalyst islands. Nanodimension of the catalyst particles

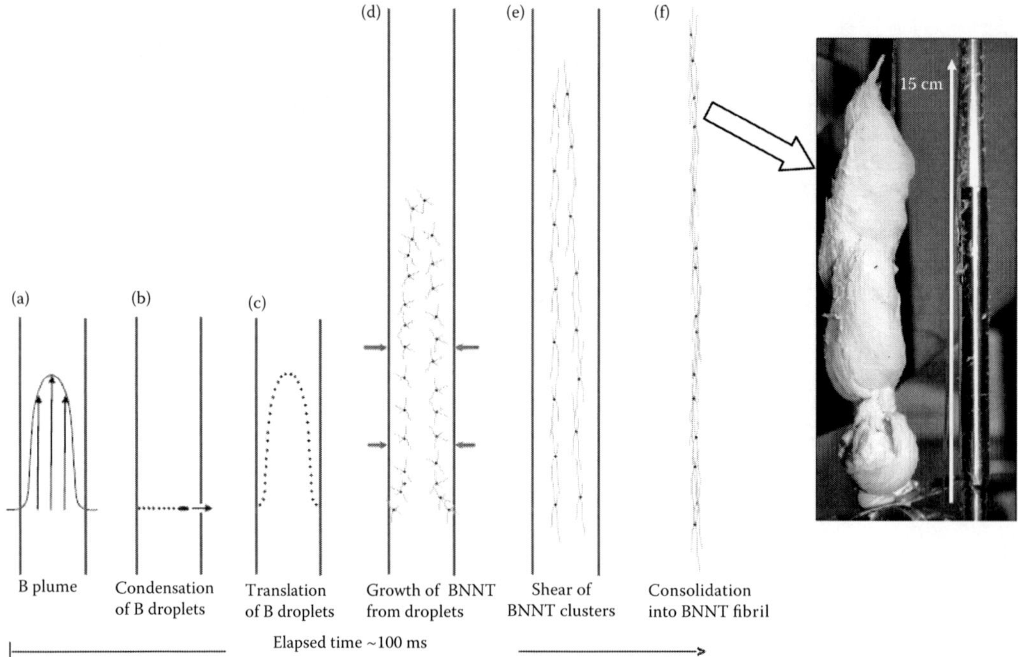

FIGURE 20.12 Schematic showing the stages of growth of BNNT fibrils via the PVC method. (a) Velocity profile of the buoyant boron vapor plume (not to scale); (b) Condenser (metal wire normal to the page, indicated by a small arrow) traverses plume, leaving a trail of boron droplets in its wake; (c) Droplets translate downstream (downstream = up in this case), in accordance with the velocity profile; (d) BNNTs nucleate and grow from droplets; (e) Growth continues as BNNT clusters shear in the flow and begin to interconnect; (f) Fibril takes its final shape as BNNT clusters consolidate into a single networked structure (Inset: Picture of BNNT fibril mass, result of a 200 mg PVC BNNT reaction). (Adapted from Smith, M. W. et al., *Nanotechnology*, **2009**, *20*, 505604 (1–6).)

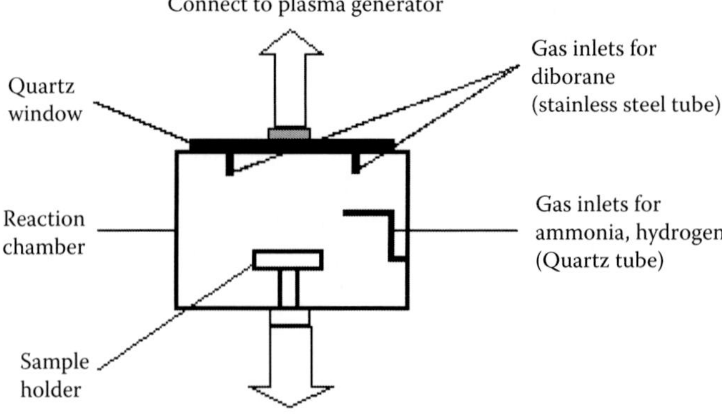

FIGURE 20.13 Schematic of the reaction chamber for the synthesis of BNNTs using the MPCVD technique. (Redrawn from Guo, L. and Singh, R. N. *Nanotechnology*, **2008**, *19*, 065601(1–6).)

reduced the melting point of the metal and heat generated by the plasma helped to melt the catalyst. In this case, the gas-phase reaction of diborane and ammonia produced BN that, while passing over the molten catalyst, gets diffused, saturated, and deposited in a one-dimensional form. If the catalyst layer is too thick, the saturation of BN leading to the precipitation does not take place. A typical layer thickness below 2 nm was found to be very effective for BNNT growths.

20.2.7 Boron Nitride Nanorods

A conventional ball milling of boron carbide (B_4C) followed by the annealing method generated conical boron nitride nanorods.[45] Pure B_4C powder was ball-milled first in the presence of nitrogen gas at 300 kPa pressure inside a container. Then the milled product was subjected to annealing at 1300°C for 8 h with ferric nitrate-coated silicon wafers. SEM micrograph revealed conical boron nitride nanorods with a typical length of 5 μm and diameter of 60 nm (Figure 20.14a). While one end of the nanorods was bearing the conical geometry, in some of the cases catalyst particle growths were seen on the other end of them (Figure 20.14b). The spherical bulb-like particles have been characterized to be an Fe–Si alloy that typically acted as the seed for the growth of the nanostructures.

Recent research work by Museur et al. showed UV-laser-assisted nucleation of boron nitride nanorods from hexagonal boron nitride (h-BN).[46] Pallets of h-BN powder were used as the target compound for the synthesis and they were subjected to a UV-laser irradiation in their ablation regime for laser pulse duration in picoseconds domain. The ambient gas pressure was kept relatively high (above 500 bar) and oxygen pressure was below 0.2 mbar. The appearance of the nanorods was noticed typically with the pulse duration of 5 ps or shorter.

20.2.8 Boron Nitride Nanocages

Boron nitride nanocages with silver nanoparticles encapsulated in them have been synthesized by Oku et al.[47] Reaction of urea and boric acid in the presence of silver nitrate produced BN matrices at 700°C. The ingredients were previously dissolved in deionized water and dried consequently to produce a homogenous mixture and annealed separately at 300°C and 700°C. The presence of silver nanoparticles and sometimes silver oxide nanoparticles were detected under TEM (Figure 20.15a) and x-ray diffraction studies. A similar methodology has been developed by Xing et al. in their preparation of zinc oxide- and titanium oxide-encapsulated BN nanocages.[48]

The presence of boron nitride nanocages was also observed by Xu et al. during their synthesis of BNNTs.[32] In the preparation of BNNTs using an autoclave pyrolysis method, many hollow boron

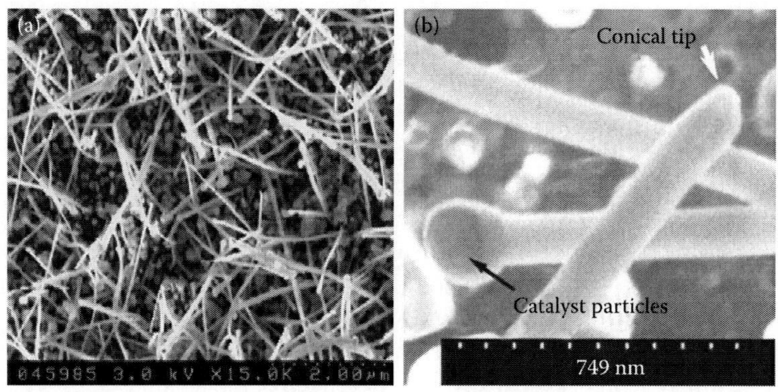

FIGURE 20.14 SEM micrographs of (a) boron nanorods (Adapted from Zhang, H. et al., *J. Am. Ceram. Soc.*, **2006**, *89*, 675–679.) and (b) with conical tip and catalyst particles. (Adapted from Zhang, H. et al., *Phys. Rev. B*. **2006**, *74*, 045407(1–9).)

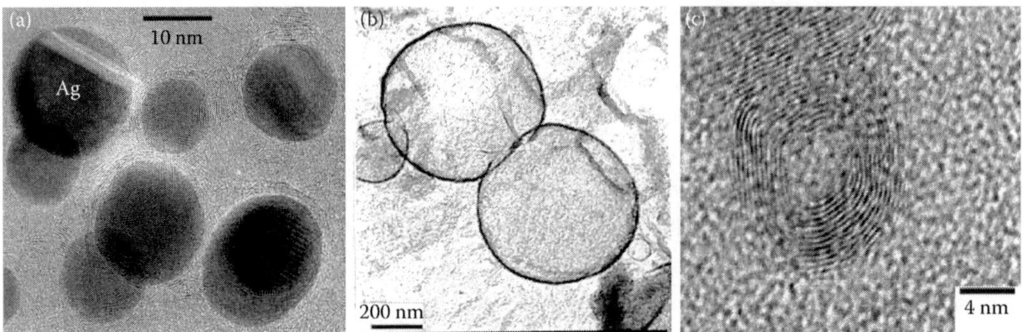

FIGURE 20.15 TEM images of boron nanocages formed by (a) annealing at 700°C (Adapted from Oku, T. et al., *J. Mater. Chem.*, **2000**, *10*, 255–257.); (b) pyrolysis method (Adapted from Xu, L. et al., *Chem. Mater.*, **2003**, *15*, 2675–2680.); and (c) nitrogenation of amorphous boron nanoparticles. (Adapted from Yeadon, M. et al., *J. Mater. Chem.*, **2003**, *13*, 2573–2576.)

nitride nanocages were found coexisting with the tubular structures (Figure 20.13b). Boron nitride nanocages were also obtained via molecular beam nitridation of FeB nanoparticles.[49] FeB nanoparticles were generated via ball-milling of ferric chloride and sodium borohydride and annealing the same at 500°C for 3 h. The nitridation of the resulting nanoparticle was carried out in the beamline of an ultrahigh vacuum transmission electron microscope in the presence of boron-doped silicon wafer substrate. The reaction temperature was maintained at 1000°C and, from experimental observations, it was concluded that boron nanocages were produced through the reaction of nitrogen gas with FeB nanoparticles where iron acted as the catalyst.

Another alternative method was reported by Zhang et al. for the synthesis of boron nitride nanocages.[50] Arc discharge in a mixture of diborane and nitrogen gas produced amorphous boron nanoparticles that were then annealed under ammonia and nitrogen gas flow at 800°C for 4 h on an aluminum oxide crucible. Nanocages of pentagonal shape accompanied by some spherical ones were observed under a microscope (Figure 20.15c). The size of the large cages was about 200 nm while the circular ones were about 12–20 nm.

20.2.9 Boron Nitride Nanosheets

The two-dimensional carbon layer structure, commonly termed as graphene, has gained much attention lately due to its unique properties attributed by the presence of high-quality crystal lattices in two-dimensional form.[51] Among the many potential applications for graphene are, sensors, electronic displays, light-emitting diodes, and thermally reinforced composites. Similar potentials for the graphene analog of boron nitride were soon proposed. Several attempts have been reported on the synthesis and applications of boron nitride nanosheets.

The most common method of preparation of boron nitride nanosheets (BNNSs) is exfoliation of boron nitride nanoparticles. Boron nitride nanocrystals dispersed in a 1,2-dichloroethane solution of poly(m-phenylenevinylene-co-2,5-dictoxy-p-phenylenevinylene) were ultrasonicated to produce a few-layered hexagonal-boron nitride structures (Figure 20.16a).[52] Dimethyl formamide was also successfully utilized by Zhi et al. as an exfoliating solvent to break down nanoparticles into nanosheets with a thickness of 1.2 nm and composed of only three atomic layers (Figure 20.16b).[53] A recent work by Lin et al. reported that soluble BNNSs using octadecylamine and amine terminated polyethyleneglycol as Lewis bases to interact with the electron-deficient boron atom of boron nitride.[54] All the exfoliation processes involve the dispersion of the nanoparticles in suitable solvents via ultrasonication, followed by centrifugation to isolated the layer-structured nanosheets; the speed of centrifugation has been found to be a determining factor of the number of layers formed during the procedure.[53]

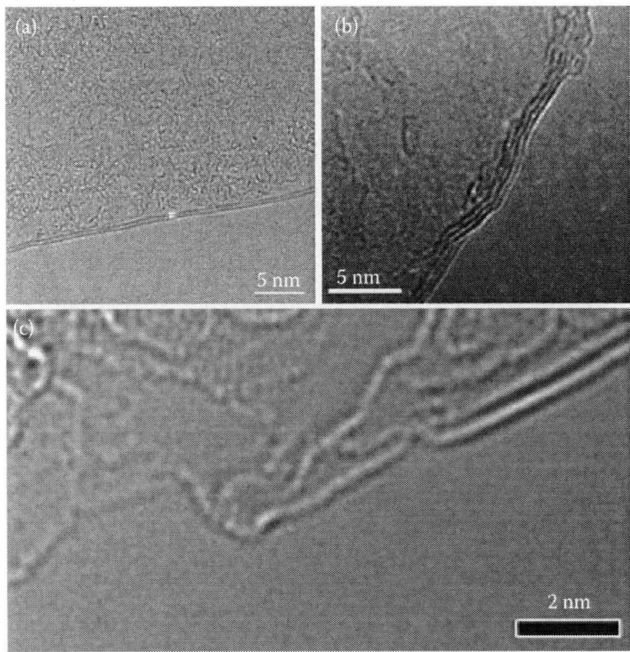

FIGURE 20.16 TEM Images of BNNSs with (a) two atomic layers (Adapted from Han, W.-Q. et al., *Appl. Phys. Lett.*, **2008**, *93*, 223103(1–3).); (b) three atomic layers (Adapted from Zhi, C. et al., *Adv. Mater.*, **2009**, *21*, 2889–2893.); and (c) a few atomic layers. (Adapted from Nag, A. et al., ACS Nano, **2009**, *4*, 1539–1544.)

A graphene analog of boron nitride has been reported by reacting boric acid and urea in different proportions at 900°C.[55] It also has been found in the same study that the number of BN layers decreased with the increment of urea in the mixture. Only a few-layered BNNs were obtained with a boric acid to urea ratio of 1:48 (Figure 20.16c). There are also a few reports on the synthesis of BNNSs by chemical vapor deposition techniques. Gao et al. adopted a method where a mixture of boron oxide and melamine was heated to 1000–1350°C, under a continuous nitrogen flowrate.[56] A narrow size distribution of the product was achieved in this method with an average thickness of 25–30 nm. A gas mixture of BF_3-N_2H_2 was used in the MPCVD technique to form BNNSs on a silicon substrate.[57] A thorough study on the dependency of the formation of nanosheets on the gas flow rates was performed in their experiment as well.

20.2.10 Boron Carbide Nanowires

Development of one-dimensional structures based on boron carbide is also contemporary to the other boron-based nanomaterials. Growth of boron carbide nanowires by plasma-enhanced chemical vapor deposition (PECVD) was reported by Zhang et al.[58] Orthocarborane (*closo*-1,2-dicarbadodecaborane, $C_2B_{10}H_{12}$) was used as the precursor material for the synthesis. The deposition temperature was kept in between 1100 and 1200°C while the plasma power was 50 W. Iron, though not added as an ingredient, was present as a contaminant and acted as an active catalyst site to grow the wires typically seen in the vapor deposition methods. A mixture of cylindrical nanowires having smooth as well as rough and faceted surface was obtained along with linear arrays of rhombohedral nanostructures.

A template-assisted synthesis of boron carbide nanowires was published by Pender and Sneddon.[59] In this method, a linked cage 6,6′-$(CH_2)_6(B_{10}H_{13})_2$ was selectively chosen as the single-source molecular precursor due to the fact that the ratio of boron to carbon in it closely matches that of boron carbide (B_4C). This precursor compound is air stable with a melting point of 96°C,[60] which helped

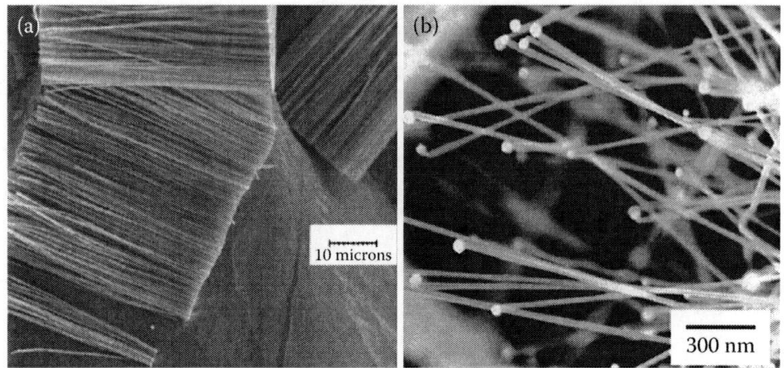

FIGURE 20.17 SEM micrograph of boron carbide nanowires prepared via (a) a template-assisted pyrolysis method. (Adapted from Pender, M. J. and Sneddon, L. G. *Chem. Mater.*, **2000**, *12*, 280–283.) (b) Vapor–liquid–solid (VLS) mechanism. (Adapted from (a) Ma, R. and Bando, Y. *Chem. Mater.*, **2002**, *14*, 4403–4407; (b) Ma, R. and Bando, Y. *Chem. Phys. Lett.*, **2002**, *364*, 314–317.)

its incorporation into the template in the liquid form. Alumina membranes with a pore size of 200 ± 50 nm were used as the template and the precursor compound was embedded inside the pores at 140°C, followed by pyrolysis at about 1025°C under the flow of argon for 3 h. The alumina template was easily etched out via washing with 48% hydrofluoric acid for 36 h. SEM images revealed long uniform arrays of crystalline nanowires with a diameter of about 250 nm and an average length of 45 μm (Figure 20.17a).

Ma and Bando reported a systemic investigation of the growth of boron carbide nanowires via a vapor–liquid–solid mechanism.[61] A template- and catalyst-free carbothermal route of nanowire production has been employed with boron powder, boron oxide, and carbon black, mixed in a 2:1:1 ratio, as the precursor. The component mixture was subjected to 1650°C temperature for 2 h under an argon flow inside a high-frequency induction furnace and resulted in boron carbide nanowires of diameters ranging from 50 to 200 nm (Figure 20.17b).

20.2.11 Boron Carbide Nanorods and Nanofibers

In a method described by Wei et al., boron carbide nanorods were prepared from carbon nanotubes.[62] Carbon nanotubes, mixed with boron powder in 1:4 atomic ratio, was heated up to 1150°C under argon gas. The as-grown structure varied from that of the parent nanotubes. They were found to be straight with occasional bents (Figure 20.18a), contrary to the tubular structure of the CNTs. The length of the nanorods corresponded to that of the nanotubes, while the diameters of them were slightly larger.

A different approach has been adopted by Welna et al. to produce boron carbide nanofibers from a polymeric precursor.[63] Poly(norbornenyldecaborane) was utilized via electrostatic spinning followed by pyrolysis to fabricate the nanofibers. The fiber diameter varied with the temperature of pyrolysis process; the higher the temperature, the narrower the diameter.

A more recent finding by Jazirehpour and Alizadeh included a carbothermal vapor deposition technique for the synthesis of boron carbide nanorods.[64] The effect of catalysts on the growth of the nanostructure was observed. The simple process involved the heating of boron oxide, activated carbon and sodium chloride powder at 1200–1400°C for 90 min under continuous flow of argon. The cobalt–boron catalyst prepared via mechanical alloying in a planetary mill was dispersed over the substrate using an ethanolic solution of the same. Surface morphology of the nanorods exhibited a dependency on the density of the catalyst dispersion over the substrate. Typically, a few micrometer-long nanorods were observed with diameters ranging from 30 to 120 nm (Figure 20.18b).

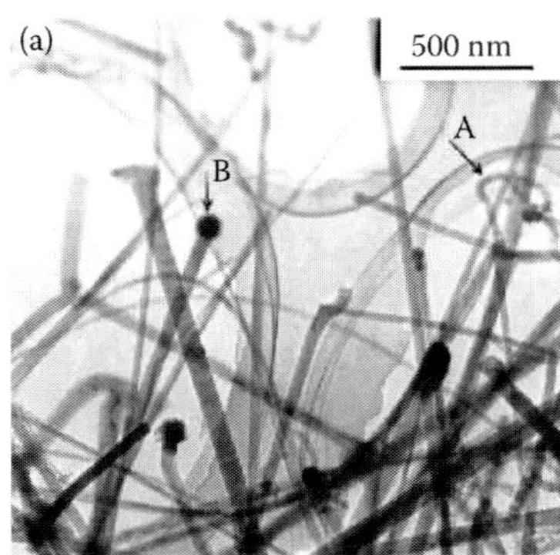

FIGURE 20.18 Boron carbide nanorods (a) TEM image (arrow A shows the tubular structures of the boron-doped CNTs, arrow B shows a large catalyst ball with a diameter of 90 nm). (Adapted from Wei, J. et al., *J. Mater. Chem.*, **2002**, *12*, 3121–3124.) (b) Higher-magnification SEM micrograph. (Adapted from Jazirehpour, M. and Alizadeh, A. *J. Phys. Chem. C*, **2009**, *113*, 1657–1661.)

20.3 PROPERTIES AND APPLICATIONS OF BORON AND BORON-BASED NANOMATERIALS

20.3.1 Boron Nanostructures

One of the major reasons for an increased interest in boron nanostructures is that boron itself has some attractive properties: it has a very low density of 2.340 g/cm^3; it has a high melting point (2076°C); and a Mohs hardness of 9.3 (diamond = 10.0). These should be reflected in boron's nanomaterials. Chemically it is of interest in that it has more valence orbitals (4) than valence electrons (3). Therefore, it tends to form so-called electron-deficient compounds having delocalized multi-centered bonds.

Nanostructured materials, bearing characteristics similar to their bulk forms, have some added advantages and improved properties by virtue of their size and dimension. The size of the materials in nanometer scale essentially increases the surface-to-volume ratio with an increment of the higher number of surface atoms, as well as surface energy of the compound.[65] The spatial confinement and reduced defects in the structure play a vital role as well. For certain nanomaterials, the high surface-to-volume ratio may increase their catalytic activities; the mechanical properties and chemical stabilities may also be improved due to the perfection imparted to the structure in their nanoforms. Gao and coworkers have carried out studies of the electric transport and field emission properties of boron nanowires,[5a] nanocones,[5b] and nanotubes.[5c] Electrical properties of the materials may vary according to their structural variations, for example, boron nanowires showed higher conductivities than undoped bulk boron.[66a] The electrical conductivity of the nanowires was found to be independent of mechanical strain, indicating that boron nanowires might be suitable building blocks for flexible nanoelectronics.[66b] On the other hand, electric conductivity studies of boron nanotubes showed that they had metal-like transport properties with conductivities 10^4–10^5 times greater than boron nanowires, and approached those of metals and carbon nanotubes (CNTs).[5c] The field emission behavior of all the nanostructures were tested. In general, it was found that those structures showing good electrical conductivity demonstrated superior field emission behavior. Boron nanotubes were found to support maximum emission current densities of ~2×10^{11} A m^{-2}, which is comparable with CNTs (10^{11}–10^{12} A m^{-2}). The maximum emission currents of individual BNTs (~40–80 μA) are 20–40 times higher than that of individual nanowires (~2 μA). Films of BNTs in luminescent tubes were found to luminescence brighter than films composed of BNTs and BNWs, approaching that of CNTs. These studies show that BNTs have properties approaching those of CNTs without suffering from problems associated with tube chirality.[67]

A very recent study done by Sun et al. showed semiconductor-metal superconductor transitions in crystalline boron nanowires under high pressure.[68] It is evident from their study that boron nanowires tend to metalize at higher pressure more than bulk β-rhombohedral boron and exhibit super conductivity at 1.5 K at 84 GPa. This tendency can solely be attributed to the size of the material that takes effect at a higher magnitude in nanostructures than in bulk.

From the novel properties of boron arise novel applications of these materials. The unique combination of several properties made boron-based nanomaterials attractive for a variety of applications, including nanoelectronic devices, hydrogen storage materials, as well as drug delivery agents for boron neutron capture therapy (BNCT). Miniaturization of electronic devices demands development of their components in the nanoscale. The potential of the nanowires as attractive building blocks in nanoscale electronics has long been realized and their utilization in similar fields has started taking effect.[69] Boron nanowires and nanorods should also start playing increasingly important roles in the field of nanoelectronics as well as high-temperature electron devices.

20.3.2 Boron Nitride Nanostructures

Bulk boron nitride exhibits excellent chemical, mechanical, and electrical properties, for example, high melting point, thermal conductivity, electrically insulating nature, chemical inertness,[70] which are expected to be inherited by their nanoforms. Nonetheless, BNNTs are semiconducting in nature with very high mechanical and thermal stabilities. Such combinations of excellent properties open up a variety of applications for BNNTs.

The ongoing demand for a better hydrogen storage medium has been partly answered by the promising outcomes of the utilization of BNNTs in this area. Ma et al. reported preliminary studies on hydrogen uptake by multiwalled and bamboo-type BNNTs at ambient temperature.[71] They have shown hydrogen adsorption of BNNTs up to 2.6% under 10 MPa. Tang et al. claimed that catalyzed collapsed BNNTs may uptake hydrogen up to 4.2%, although the effect of the presence of the platinum catalyst in the process is not completely understood.[72] Later, Chen et al. showed the possibility of electrochemical storage of hydrogen in BNNTs via cyclic voltammogram studies.[73] However, hydrogen storage via perfectly well-structured BNNTs is still under question,[74] and a perfect solution is yet to be discovered.

Unlike CNTs, BNNTs are less investigated in the biomedicinal applications. The major problem remains the insolubility of BNNTs in aqueous medium, as well as any other common organic solvents; this can be overcome via surface functionalization of the BNNTs. There are a few published reports available on noncovalent, as well as covalent, functionalizations of BNNTs. Zhi et al. reported BNNTs, wrapped in a polymer, were soluble in organic solvents, for example, chloroform, tetrahydrofuran.[75] Functionalization of BNNTs by amine-terminated oligomeric poly(ethylene glycol),[76] stearoyl chloride,[77] interaction with Lewis bases,[78] ammonia plasma irradiation[79] were also studied. Recently, Chen et al. cited the noncytotoxicity of BNNTs and emphasized the need for water-soluble BNNTs.[80]

Boron neutron capture therapy (BNCT) is one of the most promising methods for cancer treatments. This binary method involves selective introduction of B-10 containing compounds to tumor cells, followed by irradiation with thermal neutrons.[81] BNCT is focused mainly on the treatment of malignant brain tumors, for example, glioblastoma multiforme, which is virtually untreatable via surgery. Two recent publications by Ciofani et al. described BNNTs as a drug carrier for BNCT.[82,83]

A noncovalent polymer coating with poly-l-lysine (PLL) was successfully used to get the BNNTs in aqueous solution, while folic acid was used as the targeting ligand for tumor cells (Figure 20.19).[82] The amino group of PLLs can easily be linked to the carboxylic group of the folates. The preliminary studies exhibited substantial cellular uptake of the functionalized BNNTs by malignant glioblastoma cells.

The magnetic properties of BNNTs were taken into account for the first time by Ciofani et al. to utilize them as magnetic drug delivery agents.[83] The magnetic properties of BNNTs have been successfully characterized and the behavior was attributed to the presence iron impurities in them from

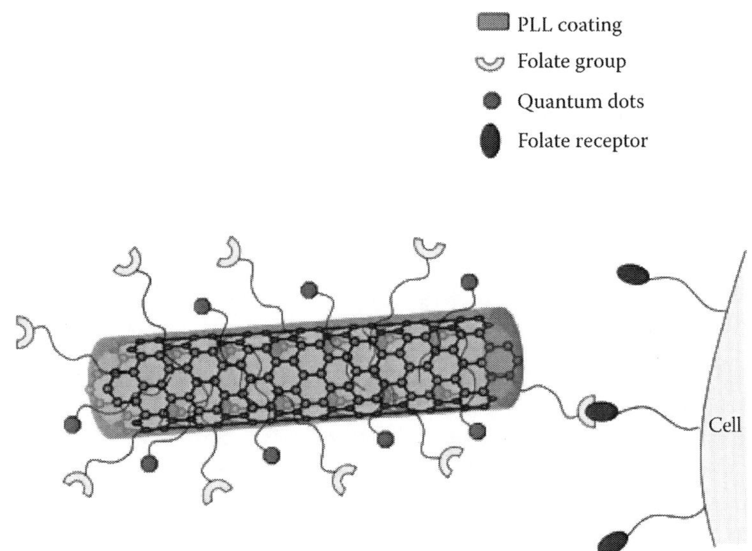

FIGURE 20.19 Schematic representation of the proposed nanovector. (Adapted from Ciofani, G. et al., *Nanoscale Res. Lett.*, **2009**, *4*, 113–121.)

the course of their synthesis. *In vitro* studies on human neuroblastoma SH-SY5Y cells showed that cellular uptake of fluorescent-labeled BNNTs can be modulated by external magnetic field.

One of the most important prospective applications of BNNTs is in space crafts. The excellent thermal and oxidation stability of BNNTs can be successfully utilized in preparations for coatings of space shuttles. Composites of boron nitride have long been prepared and studied for such purposes. Boron nitride/carbon fiber composite exhibited fascinating wear resistance properties, making them suitable for many applications including aircraft brakes.[84] Glass composites, reinforced by BNNTs were prepared for the first time in NASA Glenn Research Center.[85] Besides, BNNTs can play a major role in the prevention of space radiation exposures to the astronauts, one of the crucial problems faced during the space missions.[86] The high neutron-capture cross section of ^{10}B was taken advantage of in preparation of the ^{10}B isotopically enriched BNNTs and proposed for radiation shielding in space aircrafts.[86] In their work, Chen et al. synthesized ^{10}BNNTs using a ball-milling and annealing process starting from enriched ^{10}B powders. This large-scale synthetic procedure will certainly enable the testing and utilization of such materials in the proposed area.

Besides such usefulness of BNNTs, they are also being utilized in polymer composites,[87] field-emitting devices,[88] and electrical insulators.[89] Mass scale production of catalyst-free BNNTs is still a challenge to the scientific community and current vigorous research will surely accomplish this basic need to take it further to the next level.

Two-dimensional hexagonal boron nitride structures bear some interesting properties as well, leading to their potential applications in diverse fields. BNNSs have shown outstanding optical properties; they exhibit strong emission in the ultraviolet range.[56,57] A composite made up of BNNSs with polymethyl methacrylate (PMMA) showed improved thermal and mechanical properties without sacrificing the transparency of the material.[53] Moreover, Nag et al. observed a very high CO_2 adsorption by BNNSs.[55] The solubility of BNNSs achieved by Lin and coworkers will undoubtedly make way for this material to be used in several other directions.[54]

20.3.3 Boron Carbide Nanostructures

Boron carbide is a lightweight refractory material with high hardness and Young's modulus, and excellent thermoelectric properties.[90] The material is a water-resistant ceramic and has very high neutron capture cross section.[91] The electronic structure of boron carbide nanowires is found to be similar to that of bulk boron carbide.[92]

Recently, photoluminescence properties of boron carbide nanowires were being studied and reported.[93] Photoluminescence spectrum of a thin film made out of B_4C nanowires exhibited a broad band at 638 nm, which strongly suggests the potential application of boron nanowires in visible optical devices. Nonetheless, nanowire and nanorod structures based on boron carbide find their application in different areas, including field emission devices and thermoelectric energy converters, neutron adsorbent in nuclear industries, and especially in composite materials as reinforcing agents.[91,94]

20.4 SUMMARY AND FUTURE OUTLOOK

The development of nanostructured materials goes hand in hand with the ongoing progress of nanotechnology and characterization tools. With more sophisticated microscopic devices, investigations of nanomaterials become easier, with an eventual progression toward newer inventions. In this chapter, we detailed the technologies involved in the preparation of boron and boron-based nanostructures with major focus on recent developments in this field. The potential applications of these nanomaterials were discussed as well. While boron nanostructures are part of the fast-growing field of nanoelectronics, boron nitride nanotubes have emerged as a drug delivery agent for BNCT by virtue of its surface functionalization. Other boron nanostructures, specifically boron nanorods and nanowires could be useful for the same purposes. Functionalization of boron nanorods or nanowires is yet to be achieved and should be a prime focus at this point.

ACKNOWLEDGMENTS

Supports by grants from the National Science Foundation (CHE-0906179), the Robert A. Welch Foundation (N-1322), and the awards from Alexander von Humboldt Foundation and NIU Inaugural Board of Trustees Professorship are hereby acknowledged.

REFERENCES

1. Adams, R.M., ed. *Boron, Metallo-Boron, Compounds and Boranes*. Interscience Publishers, New York, **1964**.
2. Martin, C. R. and Kohli, P. The emerging field of nanotube biotechnology. *Nat. Rev.*, **2003**, *2*, 29–37.
3. (a) Gindulyte, A.; Lipscomb, W. N. and Massa, L. Proposed boron nanotubes. *Inorg. Chem.*, **1998**, *37*, 6544–6545. (b) Boustani, I.; Quandt, A.; Hernandez, E. and Rubio, A. New boron based nanostructured materials. *J. Chem. Phys.*, **1999**, *110*, 3176–3185.
4. (a) Cao, L.; Zhang, Z.; Sun, L.; Gao, C.; He, M.; Wang, Y.; Li. Y. et al., Well-aligned boron nanowire arrays. *Adv.Mater.*, **2001**, *13*, 1701–1704. (b) Cao, L.; Liu, J.; Gao, C.; Li, Y.; Li, X.; Wang, Y. Q.; Zhang, Z. et al., Synthesis of well-aligned boron nanowires and their structural stability under high pressure. *J. Phys. Condens. Matter.*, **2002**, *14*, 11017–11021.
5. (a) Liu. F.; Taain, J.; Bao, L.; Yang, T.; Shen, C.; Lai, X.; Xiao, Z. et al., Fabrication of vertically aligned single crystal nano wire arrays and investigation of their field-emission behavior. *Adv. Mater.* **2008**, *20*, 2609–2615. (b) Wang, X.; Tian, J.; Yang, T.; Bao, L.; Hui, C.; Liu, F.; Shen, C.; Gu, C.; Xu, N. and Gao, H. Single crystal boron nanocones: electric transport and field emission properties. *Adv. Mater.* **2007**, *19*, 4480–4485. (c) Liu, F.; Shen, C.; Su, Z.; Ding, X.; Deng, S.; Chen, J.; Xu, N. and Gao, H. Metal-like single crystalline boron nanotubes: Synthesis and *in situ* study on electric transport and field emission properties. *J. Mater. Chem.* **2010**, *20*, 2197–2205.
6. Wu, Y.; Messer, B. and Yang, P. Superconducting MgB_2 nanowires. *Adv. Mater.* **2001**, *13*, 1487–1489.
7. Otten, C. J.; Lourie, O. R.; Yu, M.-F.; Cowley, J. M.; Dyer, M. J.; Ruoff, R. S. and Buhro, W. E. Crystalline boron nanowires. *J. Am. Chem. Soc.*, **2002**, *124*, 4564–4565.
8. Zhang, Y.; Ago, H.; Yumura, M.; Komatsu, T.; Ohshima, S.; Uchida, K. and Iijima, S. Synthesis of crystalline boron nanowires by laser ablation. *Chem. Commun.*, **2002**, 2806–2807.
9. Meng, X. M.; Hu, J. Q.; Jiang, Y.; Lee, C. S. and Lee, S. T. Boron nanowires synthesized by laser ablation at high temperature. *Chem. Phys. Lett.*, **2003**, *370*, 825–828.

10. Cao, L. M.; Tian, H.; Zhang, Z.; Zhang, X. Y.; Gao, C. X. and Wang, W. K. Nucleation and growth of feather-like boron nanowire nanojunctions. *Nanotechnology*, **2004**, *15*, 139–142.
11. (a) Yun, S. H.; Wu, J. Z.; Dibos, A.; Gao, X. and Karlsson, U. O. Growth of inclined boron nanowire bundle arrays in an oxide-assisted vapor-liquid-solid process. *Appl. Phys. Lett.*, **2005**, *87*, 113109(1–3). (b) Yun, S. H.; Wu, J. Z.; Dibos, A.; Zou, X. and Karlsson, U. O. Self-assembled boron nanowire y-junctions. *Nano Lett.*, **2006**, *6*, 385–389.
12. Yang, Q.; Sha, J.; Wang, L.; Su, Z.; Ma, X.; Wang, J. and Yang, D. Morphology and diameter controllable synthesis of boron nanowires. *J. Mater. Sci.*, **2006**, *41*, 3547–3552.
13. (a) Boustani, I. and Quandt, A. Nanotubules of bare boron clusters: Ab initio and density functional study. *Europhys. Lett.*, **1997**, *39*, 527–532. (b) Boustani, I.; Quandt, A.; Hernández, E. and Rubio, A. New boron based nanostructured materials. *J. Chem. Phys.* ***1999***, 110, 3176. (c) Yang, X.; Ding, Y. and Ni, J. *Ab initio* predictions of stable boron sheets and boron nanotubes: Structure, stability, and electronic properties. *Phys. Rev. B*, **2008**, *77*, 041402(1–4).
14. (a) Sebetcia, A.; Meteb, E. and Boustani, I. Free standing double walled boron nanotubes. *J. Phys. Chem. Solids*, **2008**, *69*, 2004–2012. (b) Wang, J.; Liu, Y. and Li, Y.-C. A new class of boron nanotube. *Chem. Phys. Chem.*, **2009**, *10*, 3119–3121.
15. Ciuparu, D.; Klie, R. F.; Zhu, Y. and Pfefferle, L. Synthesis of pure boron single-wall nanotubes. *J. Phys. Chem. B*, **2004**, *108*, 3967–3969.
16. Wang, Z.; Shimizu, Y.; Sasaki, T.; Kawaguchi, K.; Kimura, K. and Koshizaki, N. Catalyst-free fabrication of single crystalline boron nanobelts by laser ablation. *Chem. Phys. Lett.*, **2003**, *368*, 663–667.
17. Lia, W. T.; Boswell, R. and Gerald, J. D. F. Boron nanobelts grown under intensive ion bombardment. *J. Vac. Sci. Technol. B*, **2008**, *26*, L7–L9.
18. Ni, H. and Li, X. Synthesis, structural and mechanical characterization of amorphous and crystalline boron nanobelts. *J. Nano Res.*, **2008**, *1*, 10–22.
19. Xu, T. T.; Zheng, J.-G.; Wu, N.; Nicholls, A. W.; Roth, J. R.; Dikin, D. A. and Ruoff, R. S. Crystalline boron nanoribbons: Synthesis and characterization. *Nano Lett.*, **2004**, *4*, 963–968.
20. Jash, P. and Trenary, M. Synthesis of crystalline boron nanoribbons and calcium hexaboride nanowires by low pressure chemical vapor deposition. *Journal of Physics: Conference Series*, **2009**, *176*, 012011(1–11).
21. Yang, Q.; Sha, J.; Wang, L.; Yuan, Z. and Yang, D. Aligned single crystal Al-catalyzed boron nanorods on Si substrates. *Eur. Phys. J. B*, **2007**, *56*, 35–39.
22. Chakrabarti, A.; Xu, T.; Paulson, L. K.; Krise, K. J.; Maguire, J. A. and Hosmane, N. S. Synthesis of Boron nanorods from non-toxic boron oxide smelted in liquid lithium. *J. Nanomater.*, **2010**, 589372(1–5).
23. (a) Marmottant, P. and Hilgenfeldt, S. Controlled vesicle deformation & lysis by single oscillating bubbles. *Nature*, **2003**, *423*, 153–156. (b) Rooney, J. A. Other nonlinear acoustic phenomena, in *Other Nonlinear Acoustic Phenomena in Ultrasound Its Chemical, Physical, and Biological Effects*, Ed. Suslick, S. VCH, New York, **1988**.
24. Scopigno, T.; Balucani, U.; Cunsolo, A.; Masciovecchio, C.; Ruocco, G.; Sette, F. and Verbeni, R. Phonon-like and single-particle dynamics in liquid lithium. *Europhys. Lett.*, **2000**, *50*, 189–195.
25. Iijima, S. Hellical microtubules of graphitic carbon. *Nature*, **1991**, *354*, 56–58.
26. (a) Rubio, A.; Corkill, J. and Cohen, M. L. Theory of graphitic boron nitride nanotubes. *Phys. Rev. B*, **1994**, *49*, 5081–5084. (b) Blase, X.; Rubio, A.; Louie, S. G. and Cohen, M. L. Stability and band gap constancy of boron nitride nanotubes. *Europhys. Lett.*, **1994**, *28*, 335–340.
27. Chopra, N. G.; Luyken, R. J.; Cherry, K.; Crespi, V. H.; Cohen, M. L.; Louie, S. G. and Zettl. A. Boron nitride nanotubes. *Science*, **1995**, *269*, 966–967.
28. (a) Loiseau, A.; Willaime, F.; Demoncy, N.; Hug, G. and Pascard, H. Boron nitride nanotubes with reduced numbers of layers synthesized by arc discharge. *Phys. Rev. Lett.*, **1996**, *76*, 4737–4740. (b) Terrones, M.; Hsu, W. K.; Terrones, H.; Zhang, J. P.; Ramos, S.; Hare, J. P.; Castillo, R. et al., Metal particle catalysed production of nanoscale BN structures. *Chem. Phys. Lett.*, **1996**, *259*, 568–573.
29. (a) Lourie, O. R.; Jones, C. R.; Bartlett, B. M.; Gibbons, P. C.; Ruoff, R. S. and Buhro, W. E. CVD growth of boron nitride nanotubes. *Chem. Mater.*, **2000**, *12*, 1808–1810. (b) Ma, R. Z.; Bando, Y.; Sato, T. and Kurashima, K. Growth, morphology, and structure of boron nitride nanotubes. *Chem. Mater.*, **2001**, *13*, 2965–2971. (c) Wang, J. S.; Kayastha, V. K.; Yap, Y. K.; Fan, Z. Y.; Lu, J. G.; Pan, Z. W.; Ivanov, I. N.; Puretzky, A. A. and Geohegan, D. B. Low temperature growth of boron nitride nanotubes on substrates. *Nano Lett.*, **2005**, *5*, 2528–2532.
30. Golberg, D.; Bando, Y.; Eremets, M.; Takemura, K.; Kurashima, K. and Yusa, H. Nanotubes in boron nitride laser heated at high pressure. *Appl. Phys. Lett.*, **1996**, *69*, 2045–2047.
31. Chen, Y.; Conway, M.; Williams, J. S. and Zou, J. Large-quantity production of high-yield boron nitride nanotubes. *J. Mater. Res.*, **2002**, *17*, 1896–1899.

32. Xu, L.; Peng, Y.; Meng, Z.; Yu, W.; Zhang, S.; Liu, X. and Qian, Y. A co-pyrolysis method to boron nitride nanotubes at relative low temperature. *Chem. Mater.*, **2003**, *15*, 2675–2680.
33. (a) Golberg, D.; Bando, Y.; Han, W.; Kurashima, K. and Sato, T. Single-walled B-doped carbon, B/N-doped carbon and BN nanotubes synthesized from single-walled carbon nanotubes through a substitution reaction. *Chem. Phys. Lett.*, **1999**, *308*, 337–342. (b) Terauchi, M.; Tanaka, M.; Suzuki, K.; Ogino, A. and Kimura, K. Production of zigzag-type BN nanotubes and BN cones by thermal annealing. *Chem. Phys. Lett.*, **2000**, *324*, 359–364.
34. Golberg, D.; Bando,Y.; Tang, C. and Zhi, C. Boron nitride nanotubes. *Adv. Mater.*, **2007**, *19*, 2413–2432.
35. Lim, S. H.; Luo, J.; Ji, W. and Lin, J. Synthesis of boron nitride nanotubes and its hydrogen uptake. *Catal. Today*, **2007**, *120*, 346–350.
36. Bechelany, M.; Bernard, S.; Brioude, A.; Cornu, D.; Stadelmann, P.; Charcosset, C.; Fiaty, K. and Miele, P. Synthesis of boron nitride nanotubes by a template-assisted polymer thermolysis process. *J. Phys. Chem. C*, **2007**, *111*, 13378–13384.
37. Shelimov, K. B. and Moskovits, M. Composite nanostructures based on template-grown boron nitride nanotubules. *Chem. Mater.*, **2000**, *12*, 250–254.
38. Chen, H.; Chen, Y.; Liu, Y.; Fu, L.; Huang, C. and Llewellyn, D. Over 1.0 mm-long boron nitride nanotubes. *Chem. Phys. Lett.*, **2008**, *463*, 130–133.
39. Li, Y.; Zhou, J.; Zhao, K.; Tung, S. and Schneider, E. Synthesis of boron nitride nanotubes from boron oxide by ball milling and annealing process. *Mater. Lett.*, **2009**, *63*, 1733–1736.
40. Singhal, S. K.; Srivastava, A. K.; Pant, R. P.; Halder, S. K.; Singh, B. P. and Gupta, A. K. Synthesis of boron nitride nanotubes employing mechanothermal process and its characterization. *J. Mater. Sci.*, **2008**, *43*, 5243–5250.
41. Kim, M. J.; Chatterjee, S.; Kim, S. M.; Stach, E. A.; Bradley, M. G.; Pender, M. J.; Sneddon, L. G. and Maruyama, B. Double-walled boron nitride nanotubes grown by floating catalyst chemical vapor deposition. *Nano Lett.*, **2008**, *10*, 3298–3302.
42. Lee, C. H.; Wang, J.; Kayatsha, V. K.; Huang, J. Y. and Yap, Y. K. Effective growth of boron nitride nanotubes by thermal chemical vapor deposition. *Nanotechnology*, **2008**, *19*, 455605(1–5).
43. Smith, M. W.; Jordan, K. C.; Park, C.; Kim, J.-W.; Lillehei, P. T.; Crooks, R. and Harrison, J. S. Very long single- and few-walled boron nitride nanotubes via the pressurized vapor/condenser method. *Nanotechnology*, **2009**, *20*, 505604 (1–6).
44. Guo, L. and Singh, R. N. Selective growth of boron nitride nanotubes by plasma-enhanced chemical vapor deposition at low substrate temperature. *Nanotechnology*, **2008**, *19*, 065601(1–6).
45. (a) Zhang, H.; Yu, J.; Chen, Y. and Gerald, J. F. Conical boron nitride nanorods synthesized via the ball-milling and annealing method. *J. Am. Ceram. Soc.*, **2006**, *89*, 675–679. (b) Zhang, H.; FitzGerald, J. D.; Chadderton, L. T.; Yu, J. and Chen, Y. Growth and structure of prismatic boron nitride nanorods. *Phys. Rev. B*. **2006**, *74*, 045407(1–9).
46. Museur, L.; Petitet, J.-P.; Michel, J.-P.; Marine, W.; Anglos, D.; Fotakis, C. and Kanaev, A. V. Picosecond laser structuration under high pressures: Observation of boron nitride nanorods. *J. Appl. Phys.*, **2008**, *104*, 093504(1–9).
47. Oku, T.; Kusunose, T.; Niihara, K. and Suganuma, K. Chemical synthesis of silver nanoparticles encapsulated in boron nitride nanocages. *J. Mater. Chem.*, **2000**, *10*, 255–257.
48. Xing, G.; Chen, G.; Song, X.; Yuan, X.; Yao, W. and Yan, H. ZnO and TiO_2 nanoparticles encapsulated in boron nitride nanocages. *Microelectron. Eng.*, **2003**, *66*, 70–76.
49. Yeadon, M.; Lin, M.; Loh, K. P.; Boothroyd, C. B.; Fud, J. and Hu, Z. Direct observation of boron nitride nanocage growth by molecular beam nitridation and liquid-like motion of Fe–B nanoparticles. *J. Mater. Chem.*, **2003**, *13*, 2573–2576.
50. Zhang, W. S.; Zheng, J. G.; Li, W. F.; Geng, D. Y. and Zhang, Z. D. Synthesize and characterization of hollow boron-nitride nanocages. *J. Nanomater.*, **2009**, 264026(1–4).
51. Allen, M. J.; Tung, V. C. and Kaner, R. B. Honeycomb carbon: A review of graphene. *Chem. Rev.*, **2010**, *110*, 132–145.
52. Han, W.-Q.; Wu, L.; Zhu, Y.; Watanabe, K. and Taniguchi, T. Structure of chemically derived mono- and few-atomic-layer boron nitride sheets. *Appl. Phys. Lett.*, **2008**, *93*, 223103(1–3).
53. Zhi, C.; Bando, Y.; Tang, C.; Kuwahara, H. and Golberg, D. Large-scale fabrication of boron nitride nanosheets and their utilization in polymeric composites with improved thermal and mechanical properties. *Adv. Mater.*, **2009**, *21*, 2889–2893.
54. Lin, Y.; Williams, T. V. and Connell, J. W. Soluble, exfoliated hexagonal boron nitride nanosheets. *J. Phys. Chem. Lett.*, **2010**, *1*, 277–283.

55. Nag, A.; Raidongia, K.; Hembram, K. P. S. S.; Datta, R.; Waghmare, U. V. and Rao, C. N. R. Graphene analogues of BN: Novel synthesis and properties. ACS Nano, **2009**, *4*, 1539–1544.
56. Gao, R.; Yin, L.; Wang, C.; Qi, Y.; Lun, N.; Zhang, L.; Liu, Y.-X.; Kang, L. and Wang, X. High-yield synthesis of boron nitride nanosheets with strong ultraviolet cathodoluminescence emission. *J. Phys. Chem. C*, **2009**, *113*, 15160–15165.
57. Yu, J.; Qin, L.; Hao, Y.; Kuang, S.; Bai, X.; Chong, Y.-M.; Zhang, W. and Wang, E. Vertically aligned boron nitride nanosheets: chemical vapor synthesis, ultraviolet light emission, and superhydrophobicity. ACS Nano, **2010**, *4*, 414–422.
58. Zhang, D.; Mcilroy, D. N.; Geng, Y. and Norton, M. G. Growth and characterization of boron carbide nanowires. *J. Mater. Sci. Lett.*, **1999**, *18*, 349–351.
59. Pender, M. J. and Sneddon, L. G. An efficient template synthesis of aligned boron carbide nanofibers using a single-source molecular precursor. *Chem. Mater.*, **2000**, *12*, 280–283.
60. Pender, M. J.; Wideman, T.; Carroll, P. J. and Sneddon, L. G. Transition metal promoted reactions of boron hydrides. 15.[11] Titanium-catalyzed decaborane-olefin hydroborations: one-step, high-yield syntheses of monoalkyldecaboranes. *J. Am. Chem. Soc.* **1998**, *120*, 9108–9109.
61. (a) Ma, R. and Bando, Y. Investigation on the growth of boron carbide nanowires. *Chem. Mater.*, **2002**, *14*, 4403–4407. (b) Ma, R. and Bando, Y. High purity single crystalline boron carbide nanowires. *Chem. Phys. Lett.*, **2002**, *364*, 314–317.
62. Wei, J.; Jiang, B.; Li, Y.; Xu, C.; Wu, D. and Wei, B. Straight boron carbide nanorods prepared from carbon nanotubes. *J. Mater. Chem.*, **2002**, *12*, 3121–3124.
63. Welna, D. T.; Bender, J. D.; Wei, X.; Sneddon, L. G. and Allcock, H. R. Preparation of boron-carbide/carbon nanofibers from a poly(norbornenyldecaborane) single-source precursor via electrostatic spinning. *Adv. Mater.*, **2005**, *17*, 859–862.
64. Jazirehpour, M. and Alizadeh, A. Synthesis of boron carbide core-shell nanorods and a qualitative model to explain formation of rough shell nanorods. *J. Phys. Chem. C*, **2009**, *113*, 1657–1661.
65. Cao, G. *Nanostructures & Nanomaterials: Synthesis, Properties & Applications*, Imperial College Press, London, **2004**.
66. (a) Wang, D.; Lu, J. G.; Otten, C. J. and Buhro, W. E. Electrical transport in boron nanowires. *Appl. Phys. Lett.*, **2003**, *83*, 5280–5282. (b) Tian, J.; Cai, J.; Hui, C.; Zhang, C.; Bao, L.; Gao, M.; Shen, C.; Gao, H. Boron nanowires for flexable electronics. *App. Phys. Lett.* **2008**, *93*, 122105.
67. Kunstmann, J. and Quandt, A. Broad boron sheets and boron nanotubes: An *ab initio* study of structural, electronic, and mechanical properties. *Phys. Rev. B.* **2006**, *74*, 035413(1–14).
68. Sun, L.; Matsuoka, T.; Tamari, Y.; Shimizu, K.; Tian, J.; Tian, Y.; Zhang, C. et al., Pressure-induced superconducting state in crystalline boron nanowires. *Phys. Rev. B*, **2009**, *79*, 140505(1–4).
69. (a) Gudiksen, M. S.; Lauhon, L. J.; Wang, J.; Smith, D. C. and Lieber, C. M. Growth of nanowire superlattice structures for nanoscale photonics and electronics. *Nature*, **2002**, *415*, 617–620. (b) Tong, L.; Gattass, R. R.; Ashcom, J. B.; He, S.; Lou, J.; Shen, M.; Maxwell, I. and Mazur, E. Subwavelength-diameter silica wires for low-loss optical wave guiding. *Nature*, **2003**, *426*, 816–819. (c) Gu, Y.; Kwak, E.-S.; Lensch, J. L.; Allen, J. E.; Odom, T. W. and Lauhon, L. J. Near-field scanning photocurrent microscopy of a nanowire photodetector. *Appl. Phys. Lett.*, **2005**, *87*, 043111(1–3).
70. Pouch, J. J. and Alterovitz, A., eds. *Synthesis and Properties of Boron Nitride, Material Science Forum.* Trans. Tech. Publ.: Switzerland, Vols. 54–55, **1990**.
71. Ma, R.; Bando, Y.; Zhu, H.; Sato, T.; Xu, C. and Wu, D. Hydrogen uptake in boron nitride nanotubes at room temperature. *J. Am. Chem. Soc.*, **2002**, *124*, 7672–7673.
72. Tang, C.; Bando, Y.; Ding, X.; Qi, S. and Golberg, D. Catalyzed collapse and enhanced hydrogen storage of bn nanotubes. *J. Am. Chem. Soc.*, **2002**, *124*, 14550–14551.
73. Chen, X.; Gao, X. P.; Zhang, H.; Zhou, Z.; Hu, W. K.; Pan, G. L.; Zhu, H. Y.; Yan, T. Y. and Song, D. Y. Preparation and electrochemical hydrogen storage of boron nitride nanotubes. *J. Phys. Chem. B*, **2005**, *109*, 11525–11529.
74. Zhou, Z.; Zhao, J.; Chen, Z.; Gao, X.; Yan, T.; Wen, B. and Schleyer, P. von R. Comparative study of hydrogen adsorption on carbon and bn nanotubes. *J. Phys. Chem. B*, **2006**, *110*, 13363–13369.
75. Zhi, C.; Bando, Y., Tang, C.; Xie, R.; Sekiguchi, T. and Golberg, D. Perfectly dissolved boron nitride nanotubes due to polymer wrapping. *J. Am. Chem. Soc.*, **2005**, *127*, 15996–15997.
76. Xie, S.-Y.; Wang, W.; Fernando, K. A. S.; Wang, X.; Lin, Y. and Sun, Y.-P. Solubilization of boron nitride nanotubes. *Chem. Commun.*, **2005**, 3670–3672.
77. Zhi, C.; Bando, Y.; Tang, C.; Honda, S.; Sato, K.; Kuwahara, H. and Golberg, D. Covalent functionalization: Towards soluble multiwalled boron nitride nanotubes. *Angew. Chem. Int. Ed.*, **2005**, *44*, 7932–7935.

78. Pal, S.; Vivekchand, S. R. C.; Govindaraj, A. and Rao, C. N. R. Functionalization and solubilization of BN nanotubes by interaction with Lewis bases. *J. Mater. Chem.*, **2007**, *17*, 450–452.
79. Ikuno, T.; Sainsbury, T.; Okawa, D.; Frechet, J. M. J. and Zettl, A. Amine-functionalized boron nitride nanotubes. *Solid State Commun.*, **2007**, *142*, 643–646.
80. Chen, X.; Wu, P.; Rousseas, M.; Okawa, D.; Gartner, Z.; Zettl, A. and Bertozzi, C. R. Boron nitride nanotubes are noncytotoxic and can be functionalized for interaction with proteins and cells. *J. Am. Chem. Soc.*, **2009**, *131*, 890–891.
81. Locher, G. L. Biological effects and therapeutic possibilities of neutrons. *Am. J. Roentgenol. Radiat. Ther.*, **1936**, *36*, 1–13.
82. Ciofani, G.; Raffa, V.; Menciassi, A. and Cuschieri, A. Folate functionalized boron nitride nanotubes and their selective uptake by glioblastoma multiforme cells: Implications for their use as boron carriers in clinical boron neutron capture therapy. *Nanoscale Res. Lett.*, **2009**, *4*, 113–121.
83. Ciofani, G.; Raffa, V.; Yu, J.; Chen, Y.; Obata, Y.; Takeoka, S.; Menciassi, A. and Cuschieri, A. Boron nitride nanotubes: A novel vector for targeted magnetic drug delivery. *Curr. Nanosci.*, **2009**, *5*, 33–38.
84. Seghi, S.; Lee, J. and Economy, J. High density carbon fiber/boron nitride matrix composites: Fabrication of composites with exceptional wear resistance. *Carbon*, **2005**, 43, 2035–2043.
85. Bansal, N. P.; Hurst, J. B. and Choi, S. R. Boron nitride nanotubes-reinforced glass composites. *J. Am. Cer. Soc.*, **2005**, *89*, 388–390.
86. Yu, J.; Chen, Y.; Elliman, R. G. and Petravic, M. Isotopically enriched ^{10}BN Nanotubes. *Adv. Mater.*, **2006**, *18*, 2157–2160.
87. (a) Zhi, C. Y.; Bando, Y.; Tang, C.; Honda, S.; Sato, K.; Kuwahara, H. and Golberg, D. Characteristics of boron nitride nanotube-polyaniline composites. *Angew. Chem. Int. Ed.*, **2005**, *44*, 7929–7932. (b) Zhi, C. Y.; Bando, Y.; Tang, C.; Honda, S.; Kuwahara, H. and Golberg, D. Boron nitride nanotubes/polystyrene composites. *J.Mater. Res.*, **2006**, *21*, 2794–2800.
88. Zhu, H.-Y.; Klein, D. J.; March, N. H. and Rubio, A. Small band-gap graphitic CBN layers. *J. Phys. Chem. Solids.*, **1998**, *59*, 1303–1319.
89. Golberg, D.; Dorozhkin, P. S.; Bando, Y.; Dong, Z.-C.; Tang, C. C.; Uemura, Y.; Grobert, N.; Reyes-Reyes, M.; Terrones, H. and Terrones, M. Structure, transport and field-emission properties of compound nanotubes: CNx vs. BNCx ($x < 0.1$). *Appl. Phys. A*, **2003**, *76*, 499–507.
90. Weimer, A. *Carbide, Nitride, and Boride Materials Synthesis and Processing*. Chapman & Hall, New York, **1997**.
91. Lazzari, R.; Vast, N.; Besson, J. M.; Baroni, S. and Dal Corso, A. Atomic structure and vibrational properties of icosahedral B_4C boron carbide. *Phys. Rev. Lett.*, **1999**, *83*, 3230–3233.
92. McIlroy, D. N.; Zhang, D.; Cohen, R. M.; Wharton, J.; Geng, Y.; Norton, M. G.; De Stasió, G. et al., Electronic and dynamic studies of boron carbide nanowires. *Phys. Rev. B*, **1999**, *60*, 4874–4879.
93. Li-Hong, B.; Chen, L.; Yuan, T.; Ji-Fa, T.; Chao, H.; Xing-Jun, W.; Cheng-Min, S. and Hong-Jun, G. Synthesis and photoluminescence property of boron carbide nanowires. *Chin. Phys. B*, **2008**, *17*, 4585–4591.
94. (a) Telle, R. in *Structure and Properties of Ceramics, Materials Science and Technology*; VCH, Weinheim, Germany, Vol. 11, **1994.** (b) Sezer, A. O. and Brand, J. I. Chemical vapor deposition of boron carbide. *Mater. Sci. Eng. B*, **2001**, *79*, 191–202.

Part VII

Boron for Chemistry and Catalysis: Catalysis and Organic Transformations

INTRODUCTION

The final section of this book focuses on another perspective of boron-based materials, that is, boron for chemistry and catalysis. It highlights the syntheses, properties, and applications of boron compounds, including titanacarborane monoamides, phosphorus-substituted carboranes, boron and organoboron halides, polyhedral boron hydrides, organoboranes, borates, carborane, and boron cluster compounds. The crucial roles of boron compounds in various organic syntheses, along with their catalytic activities, are portrayed in the following chapters by Hao Shen, Zuowei Xie, Sebastian Bauer, Evamarie Hey-Hawkins, Min-Liang Yao, George W. Kabalka, Igor B. Sivaev, Vladimir I. Bregadze, Subash Jonnalagadda, J. Sravan Kumar, Anthony Cirri, Venkatram R. Mereddy, Barada Prasanna Dash, Rashmirekha Satapathy, John A. Maguire, Narayan S. Hosmane, Rosario Núñez, Francesc Teixidor, Clara Viñas, Michael A. Corsello, Brandon R. Hetzell, Andrea Vöge, and Detlef Gabel.

Catalysts are an inevitable part of the modern world of chemistry. They are known to be useful in increasing the reaction rates, but today they are increasingly being used in asymmetric syntheses, specifically in the pharmaceutical industries, to impart chirality in the desired product. Phosphorus-substituted carborane compounds have an advantage in reactivity due to the presence of electron-deficient boron in their structure and a phosphorus atom that is either directly linked to the carborane cage or linked through a spacer moiety. They exhibit pronounced catalytic activities for hydrogenation, hydroformylation, polymerizations, and several coupling and cross-coupling reactions, including Sonogashira coupling. A different class of boron compounds, the titanacarborane monoamides with constrained geometry, has recently been synthesized and tested for catalytic activities on C–N bond formation/cleavage reactions. The positive outcome of such a compound's activity with amines and many unsaturated organic molecules should surely trigger more research endeavors in this direction.

Boron compounds play an essential part in many organic syntheses, from C–O bond breaking in a variety of ethers, esters, and acetals, to asymmetric allylation of carbonyl compounds. Boron halides are useful tools for C–O bond breaking reactions, and they also act as coupling agents for reactions of aromatic aldehydes with alkynes, allylmetals, etc. While boron halides have proven to be effective as halogen transfer reagents for many reactions, organoboron halides act as reagents to form enol borinates and many other reaction intermediates for several boron derivatives. Chiral organoboranes, on the other hand, help in asymmetric allylation of carbonyl compounds. The versatility of organoboron reagents has been successfully utilized in Suzuki–Miyaura cross-coupling reactions, which are one of the most effective and popular carbon–carbon bond forming reactions developed recently.

Polyhedral boron cluster compounds are useful in many ways. They have found applications ranging from medicinal fields to the area of materials chemistry, where they can be used as building blocks for many dendritic structured species. These dendritic macromolecules, based on polyhedral boron hydrides, besides being very promising boron delivery agents for boron neutron capture therapy (BNCT), can find their way into several other applications, including catalysis, optoelectronics, ionic liquids, and optical sensors. The fascinating chemistry of carborane cage-appended dendrimeric structures has been reported recently. Moreover, the oxonium derivatives of polyhedral boron hydrides have also been successfully synthesized and can act as a starting material for many exciting, yet useful, macromolecules.

Another interesting application of boron compounds is in syntheses involving ionic liquids, one of the fastest-growing fields in chemistry. Ionic liquids are basically salts with low melting points. These ions are weakly coordinating and thus can act as solvents for many chemical reactions. The high thermal stability and low vapor pressures of ionic liquids make them even more attractive as solvents. Boron clusters, along with tetravalent boron anions, are now being used as anions for ionic liquids. Although there are not many examples of boron-based ionic liquids, some excellent features of the cluster compounds of boron may lead many researchers to explore more in the field of ionic liquids.

21 Constrained-Geometry Titanacarborane Monoamides

From Synthesis and Reactivity to Catalytic Applications

Hao Shen and Zuowei Xie

CONTENTS

21.1 INTRODUCTION

Ligands are an essential part of organometallic compounds, which impose a dominant control over both chemical and physical properties of the resulting metal complexes.[1–12] Cyclopentadienyl (Cp), a six-electron π-ligand, has long been a ubiquitous component of metallocenes, and it is hard to imagine what organometallic chemistry would be without this ligand.[13] In recent decades, various types of ligands have been synthesized for different purposes, among which early-transition metal complexes bearing linked cyclopentadienyl-amido ligands, regularly referred to as constrained-geometry complexes (CGCs), stand out and have been found wide interests in both academia and industry since their first description in 1990.[14,15] For example, a rare earth metal CGC is recently reported to be an active catalyst for the hydroamination reaction of carbodiimides,[16] which is compatible with both primary and secondary amines via an insertion–protonation pathway. On the other hand, catalytic systems based on group 4 metal compounds give access to a large array of polymers with unique material properties and considerable commercial values.[4–12,17–29]

In this connection, replacement of a uninegative cyclopentadienyl (Cp^-) in the classical constrained-geometry ligand by an isolobal, dinegative dicarbollide ion $C_2B_9H_{11}^{2-}$, can reduce the overall charge of the resulting metallocene by one unit but leave the gross structural and metal frontier orbital properties unchanged. Such new metal/charge combinations would have an impact on the properties of resultant metal complexes. With this in mind, we synthesized constrained-geometry titanacarborane monoamides $[\sigma{:}\eta^1{:}\eta^5\text{-}(OCH_2)(R_2NCH_2)C_2B_9H_9]Ti(NR_2)$ ($R = $ Me, Et), studied their reactivity, and explored their catalytic activity in C–N bond forming/breaking reactions. The results are summarized in this chapter.

21.2 SYNTHESIS

The *nido* species $[Me_3NH][\mu\text{-}7,8\text{-}CH_2OCH_2\text{-}7,8\text{-}C_2B_9H_{10}]$ (**7.1-1**) was prepared from the *closo*-carborane $\mu\text{-}1,2\text{-}CH_2OCH_2\text{-}1,2\text{-}C_2B_{10}H_{10}$ in 95% isolated yield,[30] which was realized by using the traditional method in KOH/EtOH solution and followed by treatment with Me_3NHCl. Treatment of this dicarbollide salt with 1 equiv. of $M(NMe_2)_4$ (M = Zr, Hf) in toluene gave the expected amine elimination products $[\eta^5\text{-}(CH_2OCH_2)C_2B_9H_9]M(NMe_2)_2(NHMe_2)$ (M = Zr (**7.1-2a**), Hf (**7.1-2b**)) in almost quantitative yields (Scheme 21.1).[30] Unprecedented ring-opening products $[\sigma:\eta^1:\eta^5\text{-}(OCH_2)(R_2NCH_2)C_2B_9H_9]Ti(NR_2)$ (R = Me (**7.1–3a**), Et (**7.1-3b**)), however, were isolated in high yields from the reactions of $Ti(NR_2)_4$ (R = Me, Et) with $[Me_3NH][\mu\text{-}7,8\text{-}CH_2OCH_2\text{-}7,8\text{-}C_2B_9H_{10}]$ (**7.1-1**) in refluxing toluene (Scheme 21.2).[30] This method serves as a convenient and practical synthetic route for the preparation of constrained-geometry half-sandwich titanacarboranes with two different functional sidearms in a single step. The oxide sidearm in the complexes is strongly bonded to the Ti center by taking the advantage of its high oxophilicity whereas the other (amine sidearm) is hemilabile in nature. The molecular structures of $[\sigma:\eta^1:\eta^5\text{-}(OCH_2)(R_2NCH_2)C_2B_9H_9]Ti(NR_2)$ (**7.1-3**) (R = Me, Et) are confirmed by single-crystal x-ray analyses to adopt three-legged piano stool structures containing an η^5-dicarbollyl ligand, one amido unit, a tethered amine and an alkoxy group in the basal positions.

In order to acquire some insight of the reaction pathway, the nuclear magnetic resonance (NMR) experiments on the reaction of $Ti(NR_2)_4$ (R = Me, Et) with $[Me_3NH][\mu\text{-}7,8\text{-}CH_2OCH_2\text{-}7,8\text{-}C_2B_9H_{10}]$ (**7.1-1**) were performed. On heating the NMR solution at 80°C, this reaction generated a major species that was slowly converted to the above isolated product. Attempts to isolate the intermediate formed at 80°C failed, which could be ascribed to its instability. However, this intermediate can be trapped using an unsaturated molecule. After heating a solution of **7.1-1** with 1 equiv. of $Ti(NEt_2)_4$ at 80°C for 6 h (monitored by ^{11}B NMR spectra), addition of an excess amount of CS_2 gave an

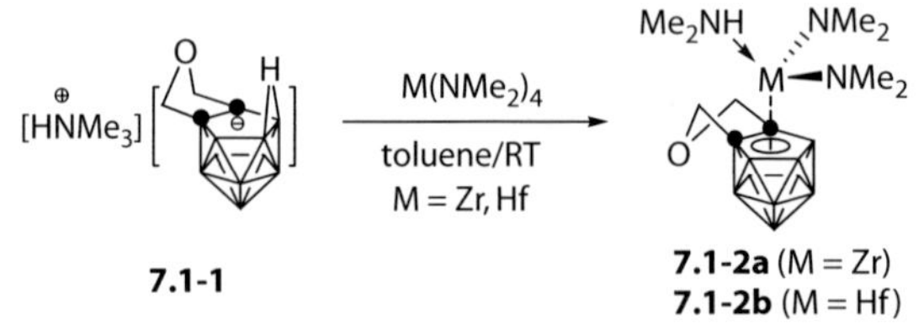

SCHEME 21.1 Synthesis of **7.1-2a,b**.

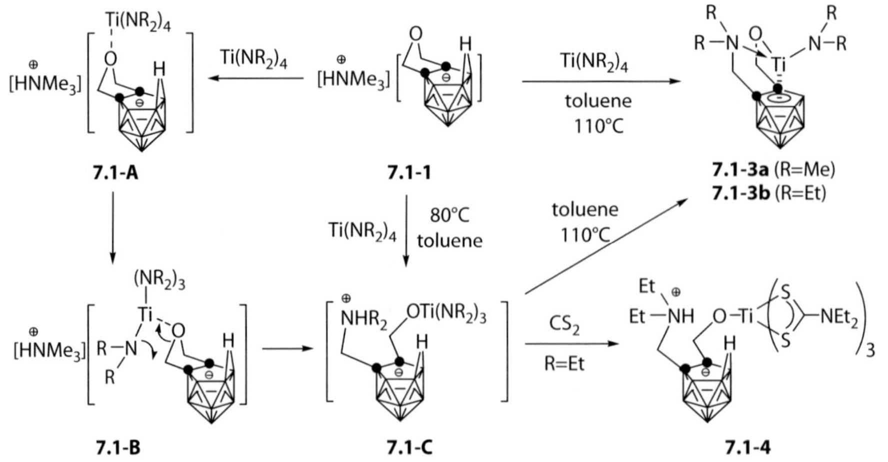

SCHEME 21.2 Reaction of **7.1-1** with titanium amides.

insertion product [σ-(OCH_2)(Et_2NHCH_2)$C_2B_9H_{10}$]Ti(η^2-S_2CNEt_2)$_3$ (**7.1-4**) that was finally isolated in 36% yield, as shown in Scheme 21.2.

Given in the above results, a possible reaction pathway for this C–O bond cleavage/C–N bond formation is proposed (Scheme 21.2). Coordination of the oxygen atom to the Ti atom and σ-bond metathesis between the C–O bond and the N–Ti bond in **7.1-B** lead to the formation of the intermediate **7.1-C**. CS_2 inserts into the Ti–N bonds in **7.1-C** to produce the product [σ-(OCH_2)(R_2NHCH_2)$C_2B_9H_{10}$]Ti(η^2-S_2CNR_2)$_3$ (**7.1-4**). On the other hand, intramolecular amine elimination of **7.1-C** affords the final product **7.1-3**, which is promoted by heat. As a result of size effect,[31,32] the higher oxophilicity of the Ti atom over the Zr and Hf atoms is anticipated. This leads to two different reaction pathways in the reaction of $M(NMe_2)_4$ (M = Ti vs. Zr, Hf) with [Me_3NH][μ-7,8-CH_2OCH_2-7,8-$C_2B_9H_{10}$] (**7.1-1**). The C–O bond cleavage is only observed in the reaction of the Ti species.

21.3 REACTIVITY

Metal amide complexes could be generally prepared by amine exchange reactions of amides or alkyl species with amines. It is found that complex [σ:η^1:η^5-(OCH_2)(Me_2NCH_2)$C_2B_9H_9$]Ti(NMe_2) (**7.1-3a**) reacts readily with 1 equiv. of 4-methoxy-aniline or 2 equiv. of 2-amino-3-picoline in toluene at room temperature to give [σ:η^1:η^5-(OCH_2)(Me_2NCH_2)$C_2B_9H_9$]Ti[NH(C_6H_4-4-OMe)] (**7.1-5**) or [σ:η^1:η^5-(OCH_2)(Me_2NCH_2)$C_2B_9H_9$]Ti[σ:η^1-(2-NH-3-CH_3-C_5H_3N)][η^1-C_5H_3N-2-NH_2-3-CH_3] (**7.1-6**) in 61% and 46% yields, respectively (Scheme 21.3). The absence of the Me_2N amido group and the presence of the corresponding aromatic proton resonances in the ^{1}H NMR spectra of both products strongly support the formation of new titanium amides.

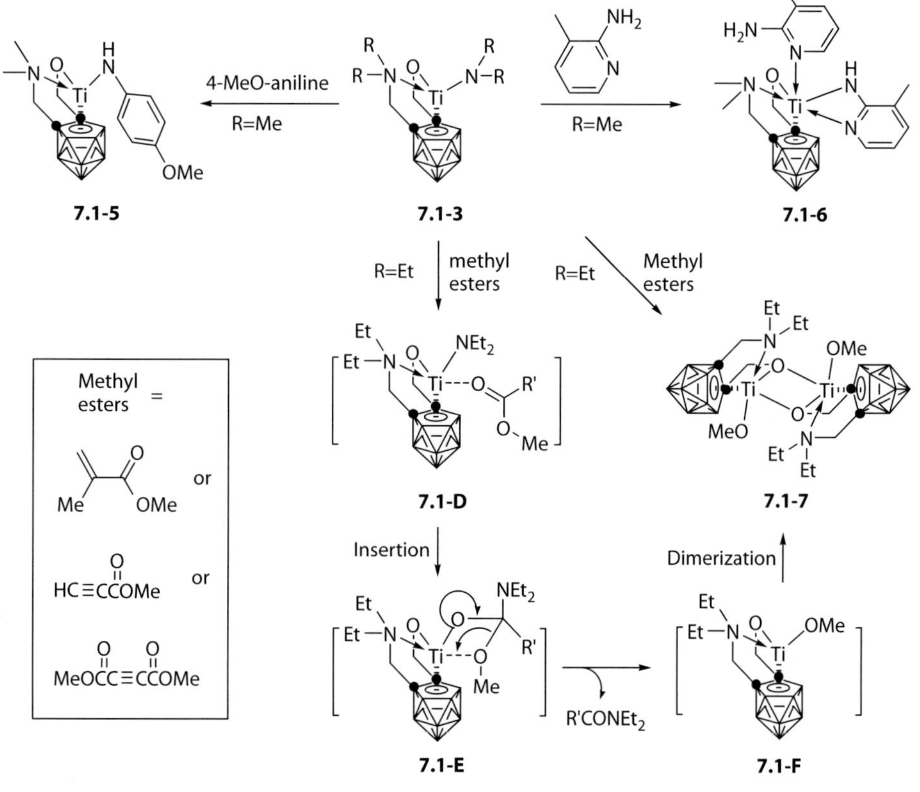

SCHEME 21.3 Reaction of **7.1-3a** with amines and esters.

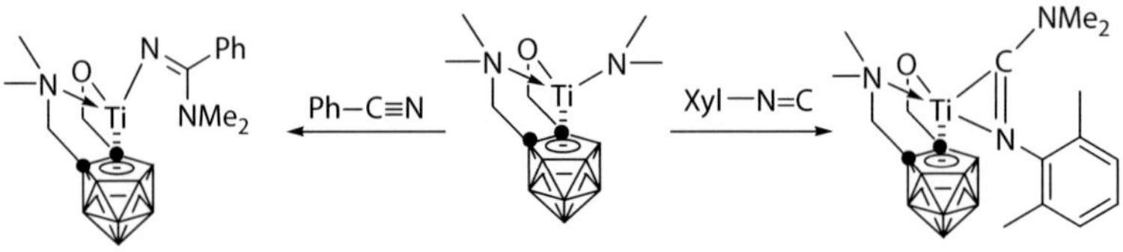

SCHEME 21.4 Reaction of **7.1-3a** with nitrile and isonitrile.

An x-ray analysis reveals that $[\sigma{:}\eta^1{:}\eta^5\text{-}(OCH_2)(Me_2NCH_2)C_2B_9H_9]Ti[\sigma{:}\eta^1\text{-}(2\text{-}NH\text{-}3\text{-}CH_3\text{-}C_5H_3N)][\eta^1\text{-}C_5H_3N\text{-}2\text{-}NH_2\text{-}3\text{-}CH_3]$ (**7.1-6**) adopts a distorted-octahedral geometry by an η^5-dicarbollyl ligand, one oxygen and one nitrogen from the sidearms, and three nitrogen atoms from the two picoline units. This structure clearly implies that the coordination number of the Ti atom can be increased by 2 units, facilitating the coordination and subsequent insertion of unsaturated molecules.

An exchange reaction can also be observed between the Ti amide complex and methyl esters, affording a Ti methoxide complex $[\{\sigma{:}\eta^1{:}\eta^5\text{-}(\mu\text{-}OCH_2)(Et_2NCH_2)C_2B_9H_9\}Ti(OMe)]_2$ (**7.1-7**).[30] Treatment of $[\sigma{:}\eta^1{:}\eta^5\text{-}(OCH_2)(Et_2NCH_2)C_2B_9H_9]Ti(NEt_2)$ (**7.1-3b**) with methyl esters such as methyl metacrylate, methyl propiolate or dimethyl acetylenedicarboxylate affords the same dimeric complex **7.1-7** in around 80% isolated yields, as shown in Scheme 21.3. On the basis of the fact that $HC{\equiv}CCONEt_2$ is detected in the reaction, a possible reaction pathway is then proposed (Scheme 21.3). Coordination of the respective ester to the Ti center gives **7.1-D**. Migratory insertion and sigma-tropic rearrangement release $HC{\equiv}CCONEt_2$ and afford **7.1-F** via **7.1-E**. Finally, dimerization produces the methoxide complex **7.1-7**.

On the other hand, reactivity of **7.1-3** toward unsaturated organic molecules is examined as well. For example, nitrile ($Ph{-}C{\equiv}N$) and isonitrile (Xyl–NC) can insert into the Ti–N bond in **7.1-3a**, as shown in Scheme 21.4, to exclusively generate the corresponding insertion products $[\sigma{:}\eta^1{:}\eta^5\text{-}(OCH_2)(Me_2NCH_2)C_2B_9H_9]Ti[N{=}C(NMe_2)Ph]$ (**7.1-8**) and $[\sigma{:}\eta^1{:}\eta^5\text{-}(OCH_2)(Me_2NCH_2)C_2B_9H_9]Ti[\eta^2\text{-}C(NMe_2){=}N(C_6H_3\text{-}2,6\text{-}Me_2)]$ (**7.1-9**) in 92% and 72% isolated yields, respectively.

Complex **7.1-3** also shows reactivity toward X=C=Y type of molecules. For example, S=C=S, ^{n}Bu–N=C=S, Cy–N=C=N–Cy, $Ph_2C{=}C{=}O$, and Ph–N=C=O are able to insert into the Ti–N σ-bond in **7.1-3a,b** to form an array of insertion products $[\sigma{:}\eta^1{:}\eta^5\text{-}(OCH_2)(Me_2NCH_2)C_2B_9H_9]Ti(\eta^2\text{-}S_2CNMe_2)$ (**7.1-10**), $[\sigma{:}\eta^1{:}\eta^5\text{-}(OCH_2)(R_2NCH_2)C_2B_9H_9]Ti[\eta^2\text{-}SC(NR_2)N\text{–}{}^nBu]$ (R = Me (**7.1-11a**), R = Et (**7.1-11b**)), $[\sigma{:}\eta^1{:}\eta^5\text{-}(OCH_2)(Me_2NCH_2)C_2B_9H_9]Ti[\eta^2\text{-}CyNC(NMe_2)NCy]$ (**7.1-12**), $[\sigma{:}\eta^1{:}\eta^5\text{-}(OCH_2)(Me_2NCH_2)C_2B_9H_9]Ti[\sigma{:}\eta^1\text{-}OC(NMe_2){=}CPh_2]$ (**7.1-13**), and $[\sigma{:}\eta^1{:}\eta^5\text{-}(OCH_2)(Me_2NCH_2)C_2B_9H_9]Ti[\eta^2\text{-}OC(NMe_2)NPh]$ (**7.1-14**) in 76–94% isolated yields, as shown in Scheme 21.5. Among these, the reactions of $Ph_2C{=}C{=}O$ and Ph–N=C=O are of particular interest.

It has been documented that the C=C or C=O bond of a ketene can insert into either a Ti–O or a M–N bond.[33–36] In our reaction, only C=O insertion product $[\sigma{:}\eta^1{:}\eta^5\text{-}(OCH_2)(Me_2NCH_2)C_2B_9H_9]Ti[\sigma{:}\eta^1\text{-}OC(NMe_2){=}CPh_2]$ (**7.1-13**) is isolated. On the other hand, it has been reported that Ph–N=C=O can insert into either a Ti–N or Ti–O bond.[37–39] In the present case, reaction of $[\sigma{:}\eta^1{:}\eta^5\text{-}(OCH_2)(Me_2NCH_2)C_2B_9H_9]Ti(NMe_2)$ (**7.1-3a**) with Ph–N=C=O affords exclusively $[\sigma{:}\eta^1{:}\eta^5\text{-}(OCH_2)(Me_2NCH_2)C_2B_9H_9]Ti[\eta^2\text{-}OC(NMe_2)N\text{–}Ph]$ (**7.1-14**), which indicates the higher reactivity of the Ti–N bond and the inertness of the Ti–O bond. It is also found that the newly formed Ti–N bond is still active, which can react with benzaldehyde to give a six-membered metallacyclic complex **7.1-15** (Scheme 21.5). The Ti–O bond remains intact in all these reactions.

21.4 CATALYTIC APPLICATIONS

Given the aforementioned amine exchange and a series of insertion reactions, reactivity of the insertion products toward amines is investigated in the hope to realize a catalytic cycle. It is found that

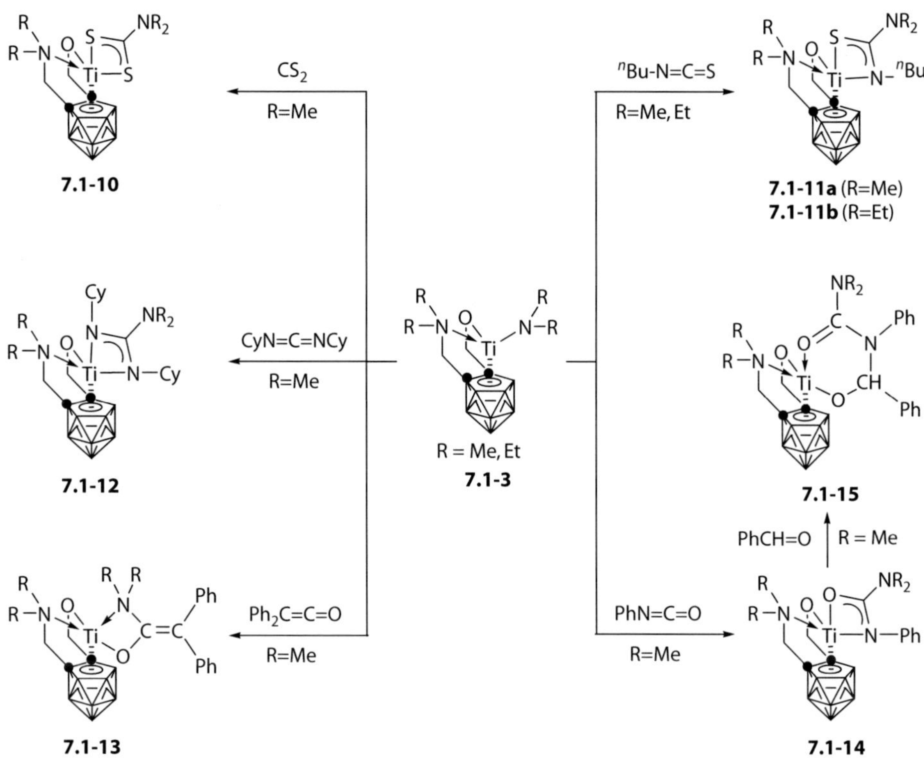

SCHEME 21.5 Reaction of **7.1-3** with unsaturated molecules.

only [σ:η^1:η^5-(OCH_2)(Me_2NCH_2)$C_2B_9H_9$]Ti[η^2-CyNC(NMe_2)NCy] (**7.1-12**) reacts with Me_2NH to regenerate [σ:η^1:η^5-(OCH_2)(Me_2NCH_2)$C_2B_9H_9$]Ti(NMe_2) (**7.1-3a**) with the formation of CyN=C(NMe_2)NHCy as indicated by NMR spectra (Scheme 21.6).[40] This result clearly implies that the catalytic hydroamination reaction of carbodiimides using [σ:η^1:η^5-(OCH_2)(Me_2NCH_2)$C_2B_9H_9$] Ti(NMe_2) as the catalyst should be feasible. Subsequently, the substrate scope is expanded and we found that this complex can effectively catalyze the hydroamination reactions. Table 21.1 summaries the results for catalytic addition of various amines into carbodiimides using complex **7.1-3a** as a catalyst.[40]

Table 21.1 clearly shows that complex [σ:η^1:η^5-(OCH_2)(Me_2NCH_2)$C_2B_9H_9$]Ti(NMe_2) (**7.1-3a**) is a very robust, effective, and elegant catalyst for the addition of primary and secondary aliphatic and aromatic amines to carbodiimides with good functional group tolerance. It also works very well for less nucleophilic pyrrole, indole, and benzotriazole.[40]

Scheme 21.7 shows a possible catalytic cycle that is proposed according to the aforementioned exchange-insertion reactions. Insertion of RN=C=NR (**7.1-16**) into the Ti–N bond in **7.1-G** gives guanidinates **7.1-H** or **7.1-H′**.[41–45] Protonation of the guanidinate complexes **7.1-H/H′** with R_1R_2NH

CyN=C=NCy

Me_2NH

CyNHC(NMe_2)=NCy

7.1-3a **7.1-12**

SCHEME 21.6 Inter-conversion between **7.1-3a** and **7.1-12**.

TABLE 21.1
Catalytic Addition of Amines to Carbodiimides

RN=C=NR + R_1R_2NH —(3 ~ 5 mol% **7.1-3a**, toluene)→ R_1R_2N–C(=NR)–NHR —(R_2=H, 1,3 - H shift)→ R_2N=C(NHR)(NHR)

Product	Isolated Yield (%)	Product	Isolated Yield (%)
N-$(CH_2)_7CH_3$; ^{i}Pr-NH, NH-^{i}Pr	92	Ph–N=C(NH-Cy)(NH-Cy)	96
pyrrolidin-1-yl; ^{i}Pr-NH, N-^{i}Pr	97	Ph–N=C(NH-^{i}Pr)(NH-^{i}Pr)	97
piperidin-1-yl; ^{i}Pr-NH, N-^{i}Pr	95	MeO–C_6H_4–N=C(NH-^{i}Pr)(NH-^{i}Pr)	97
Et-N-Et; ^{i}Pr-NH, N-^{i}Pr	93	Br–C_6H_4–N=C(NH-^{i}Pr)(NH-^{i}Pr)	95
^{n}Pr-N-^{n}Pr; ^{i}Pr-NH, N-^{i}Pr	90	O_2N–C_6H_4–N=C(NH-^{i}Pr)(NH-^{i}Pr)	92
pyrrol-1-yl; Cy-NH, N-Cy	71	O_2N (meta)–C_6H_4–N=C(NH-^{i}Pr)(NH-^{i}Pr)	87
pyrrol-1-yl; ^{i}Pr-NH, N-^{i}Pr	70	(^{i}Pr-HN)(^{i}Pr-HN)C=N–C_6H_4–N=C(NH-^{i}Pr)(NH-^{i}Pr)	92
benzotriazol-1-yl; ^{i}Pr-NH, N-^{i}Pr	71	2,6-Cl$_2$$C_6H_3$–N=C(NH-^{i}Pr)(NH-^{i}Pr)	94
indol-1-yl; ^{i}Pr-NH, N-^{i}Pr	82	1-naphthyl–N=C(NH-^{i}Pr)(NH-^{i}Pr)	90
3-Me-indol-1-yl; ^{i}Pr-NH, N-^{i}Pr	70	2-naphthyl–N=C(NH-^{i}Pr)(NH-^{i}Pr)	92
indolin-1-yl; ^{i}Pr-NH, N-^{i}Pr	88	3-Me-pyridin-2-yl–N=C(NH-^{i}Pr)(NH-^{i}Pr)	91

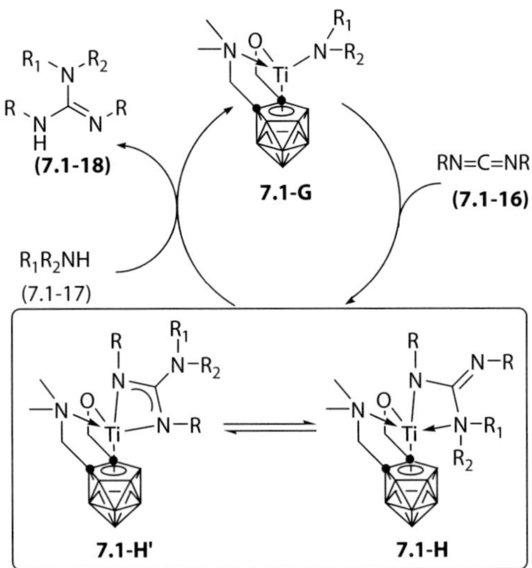

SCHEME 21.7 Proposed catalytic cycle for the hydroamination of carbodiimides.

$[M]{=}N{-}R^1$ ⇌ ([2 + 2], $R^2N{=}C{=}NR^2$ / retro - [2 + 2]) $[M]\langle N(R^1), N(R^2)\rangle C{=}N{-}R^2$ ⇌ (retro - [2 + 2] / $R^1N{=}C{=}NR^2$, [2 + 2]) $[M]{=}N{-}R^2$

7.1-M **7.1-N** **7.1-O**

SCHEME 21.8 C–N bond-forming and -breaking reaction via [2+2]cycloaddition and retro-[2+2]cycloaddition.

(**7.1-17**) releases the products **7.1-18**, and regenerates the amide species **7.1-G** to complete this catalytic cycle.

The hemilabile nature of amine sidearm may play a role in the catalytic cycle since it could reversibly coordinate to the Ti center, thus stabilizing a highly reactive intermediate. Recently, several catalyst systems are reported to be able to effectively catalyze the hydroamination of carbodiimides and offer good yields, which includes alkaline metal catalysts,[43] aluminium catalysts,[46–48] and rare earth metal catalysts.[16,41,42,44,49] The mechanisms are also proposed to undergo the similar insertion–protonation pathway. All of these mechanisms are significantly different from that is proposed for imido catalyst systems.[45,50,51] The latter involves Ti=N species (Scheme 21.8).

As shown in Scheme 21.8, group 4 metal imido complexes (**7.1-M**) are reported to react with carbodiimides via [2 + 2] cycloaddition to give guanidinate(2–) complexes (**7.1-N**). The latter equilibrate with new imido complexes **7.1-O** and new carbodiimides.[52,53] On the basis of this, hydroamination of carbodiimides and transamination of guanidines appear in literatures using group 4 metal imido complexes as catalysts.[50] However, according to the mechanism, [M]=N catalysts only work for primary amines. In sharp contrast, the titanacarborane monoamide **7.1-3a** works very well for the reactions of both primary and secondary amines with carbodiimides (Scheme 21.7). Therefore, it largely broadens the substrate scope.[40]

The first catalytic reconstruction of guanidines with amines mediated by an imido catalyst was reported in 2003,[50] which is based on C–N bond cleavage of a guanidinate(2–) ligand shown in Scheme 21.8. On the other hand, we recently disclosed another possible mechanism for the C–N bond cleavage in a zirconacarborane guanidinate(1–) complex, as shown in Scheme 21.9.[54] This process may operate in the transamination of highly substituted guanidines.

Inspired by this C–N bond cleavage reaction,[54] the catalytic activity of $[\sigma{:}\eta^1{:}\eta^5\text{-}(OCH_2)(Me_2NCH_2)C_2B_9H_9]Ti(NMe_2)$ (**7.1-3a**) for the transamination of guanidines is then examined.[55] It is

7.1-I 7.1-J 7.1-K 7.1-L

SCHEME 21.9 C–N bond-forming and -breaking reaction via sigmatropic rearrangement.

found that **7.1-3a** can also effectively catalyze the reconstruction of guanidines.[55] The results are compiled in Table 21.2, which shows a very broad substrate scope compared with those catalyzed by imido species. The yields are almost quantitative spanning a broad substrate scope of primary, secondary, heterocyclic, aliphatic, and aromatic amines, except for Et_2NH and $^nPr_2NH$ that offers good yields. No further transamination reaction of the products is observed, although excess amount of amine is used in the reactions. This could be ascribed to the higher stability of the products in this reaction system. Good functional group tolerance is also observed in these reactions.[55]

Scheme 21.10 shows a catalytic cycle for the transamination of guanidines that is proposed according to the above results. An amine exchange reaction of $[\sigma{:}\eta^1{:}\eta^5\text{-}(OCH_2)(Me_2NCH_2)C_2B_9H_9]Ti(NMe_2)$ (**7.1-3a**) with R''_2NH gives new amide species [Ti]–NR''_2 (**7.1-P**). Protonation reaction of [Ti]–NMe_2 (**7.1-3a**) or [Ti]–NR''_2 (**7.1-P**) with guanidine R–N=C(NR'_2)NH–R leads to the formation of [Ti]–[η^2-C(NR'_2)$(NR)_2$] (**7.1-Q**) to enter the catalytic cycle. Isomerization and the subsequent C–N bond cleavage, which is promoted by heat, from the amide [Ti]–NR'_2 complex (**7.1-S**) and carbodiimide. Amine exchange reaction with R''_2NH releases R'_2NH and generates [Ti]–NR''_2 species (**7.1-S′**). Insertion of carbodiimide into the Ti–N bond and isomerization produce [Ti]–[η^2-C(NR''_2)$(NR)_2$] (**7.1-Q′**). Guanidine exchange affords product R–N=C(NR''_2)NH–R and regenerates

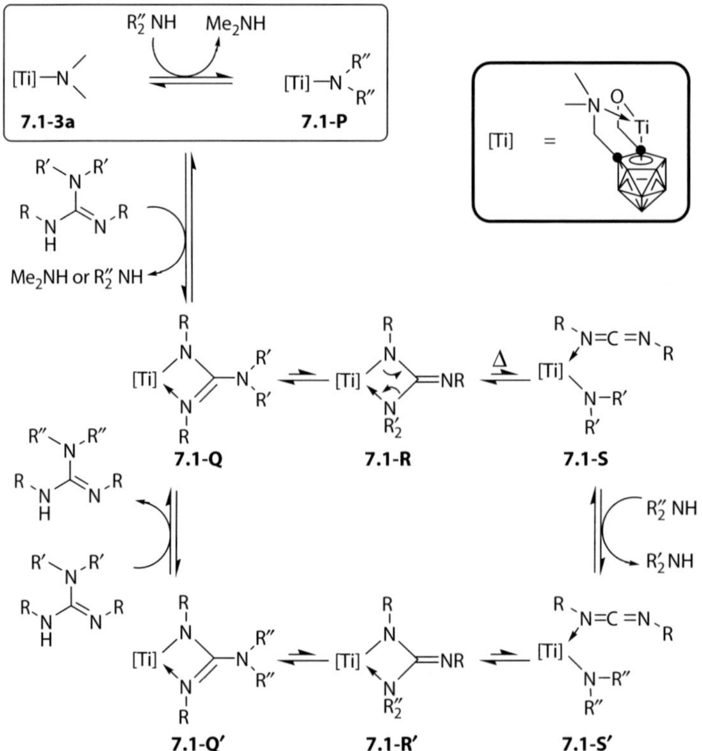

SCHEME 21.10 Proposed mechanism for the transamination of guanidines.

TABLE 21.2
Catalytic Transamination of Guanidines with Amines

$$\underset{(R={}^iPr\text{ or }Cy)}{Me_2N\text{–}C(=NR)\text{–}NHR} + \underset{(R'_2NH)}{R'NH_2} \xrightarrow[\substack{[D_6]\text{benzene} \\ \text{or toluene} \\ 110\text{–}115^\circ C}]{\substack{5\sim10\text{ mol \%} \\ \mathbf{7.1\text{-}3a}}} R'N=C(NHR)_2\ \left(R'_2N\text{–}C(=NR)\text{–}NHR\right) + Me_2NH$$

Product	Yield (%)[a]	Product	Yield (%)[a]
$^iPrN=C(NH^iPr)_2$	>95(94)	$Et_2N\text{–}C(=N^iPr)NH^iPr$	75
$CyN=C(NH^iPr)_2$	>95(92)	$^nPr_2N\text{–}C(=N^iPr)NH^iPr$	65
$CH_3(CH_2)_7N=C(NH^iPr)_2$	>95	$^iPrN=C(NHCy)_2$	>95(97)
$PhN=C(NH^iPr)_2$	>95	$CyN=C(NHCy)_2$	>95(98)
$p\text{-Br-}C_6H_4N=C(NH^iPr)_2$	>95	$PhN=C(NHCy)_2$	>95
$p\text{-OMe-}C_6H_4N=C(NH^iPr)_2$	>95	(pyrrolidin-1-yl)$C(=NCy)NHCy$	>95(97)
(pyrrolidin-1-yl)$C(=N^iPr)NH^iPr$	>95	(piperidin-1-yl)$C(=NCy)NHCy$	>95(96)
(piperidin-1-yl)$C(=N^iPr)NH^iPr$	>95	(indolin-1-yl)$C(=NCy)NHCy$	>95(83)
(indolin-1-yl)$C(=N^iPr)NH^iPr$	>95	$Et_2N\text{–}C(=NCy)NHCy$	78

[a] Yields determined by integration of ^{1}H NMR relative to internal standard of ferrocene; isolated yield within parentheses from scale-up reactions.

complex [Ti]–[η^2-C(NR'_2)(NR)$_2$] (**7.1-Q**) to complete the catalytic cycle.[55] It is noted that no Ti=N species is involved in this mechanism.

21.5 SUMMARY AND PERSPECTIVE

Constrained-geometry titanacarborane monoamides [σ:η^1:η^5-(OCH_2)(R_2NCH_2)$C_2B_9H_9$]Ti(NR_2) (**7.1-3**) exhibit a high reactivity toward amines and unsaturated organic molecules, which is successfully applied to the catalytic C–N bond forming and breaking procedures. [σ:η^1:η^5-(OCH_2)(Me_2NCH_2)$C_2B_9H_9$]Ti(NMe_2) (**7.1-3a**) is able to catalyze the hydroamination of carbodiimides and transamination of guanidines with a broad substrate scope of primary, secondary, heterocyclic, aliphatic, and aromatic amines, showing very good tolerance to many functional groups. The oxide sidearm strongly bonds the Ti atom, whereas the amine sidearm is hemilabile and probably coordinates to the metal center reversibly to stabilize the intermediates. On the other hand, the less stearic hindrance around the metal center (evidenced by the fact that the coordination number can be increased by 2 units, *vide supra*) and electronic deficiency of the titanium (with only 12 electrons) facilitate the approach of a substrate molecule and subsequent protonation or insertion reaction. In this regard, titanacarborane monoamide complexes are isolobal to constrained-geometry rare earth metal complexes. On the basis of the structural feature of **7.1-3**, amine exchange and insertion reactions, it is anticipated that this complex may also find applications in hydroboration and hydrophosphination reactions. It is noteworthy that [σ:η^1:η^5-(OCH_2)(R_2NCH_2)$C_2B_9H_9$]Ti(NR_2) (**7.1-3**) is a metal-centered chiral complex. We are working on the synthesis of a pure enantiomer from a planar chiral dicarbollyl ligand. Hopefully, the resultant complex could be applied to asymmetrical catalysis in the future.

ACKNOWLEDGMENT

This work was supported by a grant from the Research Grants Council of the Hong Kong Special Administration Region (Project No. 404609).

REFERENCES

1. W. S. Knowles, Asymmetric hydrogenations. *Angew. Chem. Int. Ed.* 41, 2002: 1998.
2. R. Noyori, Asymmetric catalysis: science and opportunities. *Angew. Chem. Int. Ed.* 41, 2002: 2008.
3. K. B. Sharpless, Searching for new reactivity. *Angew. Chem. Int. Ed.* 41, 2002: 2024.
4. J. A. Gladysz, Frontiers in metal-catalyzed polymerization. *Chem. Rev.* 100, 2000: 1167.
5. N. S. Hosmane, J. A. Maguire, in R. H. Crabtree, D. M. P. Mingos, eds., *Comprehensive Organometallic Chemistry III*, vol. 4, Chapter 3.05 (Oxford: Elsevier, 2007), p. 175.
6. S. O. Kang, J. Ko, Chemistry of *o*-carboranyl derivatives. *Adv. Organomet. Chem.* 47, 2001: 61.
7. H. Braunschweig, F. M. Breitling, Constrained geometry complexes-Synthesis and applications. *Coord. Chem. Rev.* 250, 2006: 2691.
8. X. Li, Z. Hou, Organometallic catalysts for copolymerization of cyclic olefins. *Coord. Chem. Rev.* 252, 2008: 1842.
9. Z. Xie, Advances in the chemistry of metallacarboranes of f-block elements. *Coord. Chem. Rev.* 231, 2002: 23.
10. Z. Xie, Cyclopentadienyl-Carboranyl Hybrid Compounds: A New Class of Versatile Ligands for Organometallic Chemistry. *Acc. Chem. Res.* 36, 2003: 1.
11. Z. Xie, Group 4 metallocenes incorporating constrained-geometry carboranyl ligands. *Coord. Chem. Rev.* 250, 2006: 259.
12. L. Deng, Z. Xie, Advances in the chemistry of carboranes and metallacarboranes with more than 12 vertices. *Coord. Chem. Rev.* 251, 2007: 2452.
13. A. Togni, R. L. Halterman, eds., *Metallocenes: Synthesis, Reactivity Applications*, vols. 1 and 2 (New York: Wiley-VCH, 1998), p. V.
14. P. J. Shapiro, E. E. Bunel, W. Schaefer, J. E. Bercaw, Scandium complex [{(η^5-C_5Me_4)Me_2Si(η^1-$NCMe_3$)}(PMe_3)ScH$]_2$: a unique example of a single-component α-olefin polymerization catalyst. *Organometallics* 9, 1990: 867.
15. W. E. Piers, P. J. Shapiro, E. E. Bunel, J. E. Bercaw, Coping with extreme Lewis acidity: strategies for the synthesis of stable, mononuclear organometallic derivatives of scandium. *Synlett*, 1990: 74.
16. W.-X. Zhang, M. Nishiura, Z. Hou, Catalytic addition of secondary amines to carbodiimides by a half-sandwich yttrium complex: an efficient route to N,N′,N″,N″-tetrasubstituted guanidines. *Synlett*, 2006: 1213.

17. J. C. Stevens, F. J. Timmers, D. R. Wilson, G. F. Schmidt, P. N. Nickias, R. K. Rosen, G. W. Knight, S. Lai, Constrained geometry addition polymerization catalysts, processes for their preparation, precursors therefor, methods of use, and novel polymers formed therewith. *Eur. Pat. Appl.*, 1991: EP-416815–A2.
18. J. M. Canich, G. G. Hlatky, H. W. Turner, Aluminum-free monocyclopentadienyl metallocene catalysts for olefin polymerization. *PCT Appl.* 1992: WO 92–00333.
19. P. J. Shapiro, W. D. Cotter, W. P. Schaefer, J. A. Labinger, J. E. Bercaw, Model Ziegler-Natta α-olefin polymerization catalysts derived from $[\{(\eta^5\text{-}C_5Me_4)SiMe_2(\eta^1\text{-}NCMe_3)\}(PMe_3)Sc(\mu^2\text{-}H)]_2$ and $[\{(\eta^5\text{-}C_5Me_4)SiMe_2(\eta^1\text{-}NCMe_3)\}Sc(\mu^2\text{-}CH_2CH_2CH_3)]_2$. synthesis, structures, and kinetic and equilibrium investigations of the catalytically active species in solution. *J. Am. Chem. Soc.* 116, 1994: 4623.
20. T. K. Woo, P. M. Margl, J. C. W. Lohrenz, P. E. Blöchl, T. Ziegler, Combined static and dynamic density functional study of the Ti(IV) constrained geometry catalyst $(CpSiH_2NH)TiR^+$. 1. Resting States and Chain Propagation. *J. Am. Chem. Soc.* 118, 1996: 13021.
21. G. Trouvé, D. A. Laske, A. Meetsma, J. H. Teuben, Synthesis of Group 4 metal compounds containing cyclopentadienyl ligands with a pendant alkoxide function: molecular structure of $\{[\eta^5\text{:}\eta^1\text{-}C_5H_4(CH_2)_2O]TiCl_2\}_2$ and $[\eta^5\text{:}\eta^1\text{-}C_5H_4(CH_2)_3O]TiCl_2$. *J. Organomet. Chem.* 511, 1996: 255.
22. Y.-X. Chen, T. J. Marks, "Constrained geometry" dialkyl catalysts. Efficient syntheses, C-H bond activation chemistry, monomer-dimer equilibration, and α-olefin polymerization catalysis. *Organometallics* 16, 1997: 3649.
23. Y.-X. Chen, P.-F. Fu, C. L. Stern, T. J. Marks, A novel phenolate "constrained geometry" catalyst system. Efficient synthesis, structural characterization, and α-olefin polymerization catalysis. *Organometallics* 16, 1997: 5958.
24. K. C. Hultzsch, T. P. Spaniol, J. Okuda, Half-sandwich alkyl and hydrido complexes of yttrium: convenient synthesis and polymerization catalysis of polar monomers. *Angew. Chem. Int. Ed.* 38, 1999: 227.
25. J. T. Park, S. C. Yoon, B.-J. Bae, W. S. Seo, I.-H. Suh, T. K. Han, J. R. Park, Cyclopentadienyl-hydrazido titanium complexes: synthesis, structure, reactivity, and catalytic properties. *Organometallics* 19, 2000: 1269.
26. A. L. McKnight, R. M. Waymouth, Group 4 *ansa*-cyclopentadienyl-amido catalysts for olefin polymerization. *Chem. Rev.* 98, 1998: 2587.
27. U. Siemeling, Chelate complexes of cyclopentadienyl ligands bearing pendant O-donors. *Chem. Rev.* 100, 2000: 1495.
28. H. Butenschön, Cyclopentadienylmetal complexes bearing pendant phosphorus, arsenic, and sulfur ligands. *Chem. Rev.* 100, 2000: 1527.
29. V. C. Gibson, S. K. Spitzmesser, Advances in non-metallocene olefin polymerization catalysis. *Chem. Rev.* 103, 2003: 283.
30. H. Shen, H.-S. Chan, Z. Xie, Synthesis, Structure, and Reactivity of $[\sigma\text{:}\eta^1\text{:}\eta^5\text{-}(OCH_2)(Me_2NCH_2)C_2B_9H_9]Ti(NR_2)$ (R = Me, Et). *Organometallics* 26, 2007: 2694; H. Shen, Z. Xie, *J. Organomet. Chem.* 694, 2009: 1652.
31. G. Chandra, M. F. Lappert, Amido-derivatives of metals and metalloids. Part VI. reactions of titanium(iv), zirconium(iv), and hafnium(iv) amides with protic compounds. *J. Chem. Soc. A*, 1968: 1940.
32. G. M. Diamond, R. F. Jordan, Synthesis of group 4 metal rac-(EBI)M$(NR_2)_2$ complexes by amine elimination. scope and limitations. *Organometallics* 15, 1996: 4030.
33. L. Vuitel, A. Jacot-Guillarmod, Preparation of β-hydroxyesters from carbonyl compounds and ketene in the presence of titanium alkoxides. *Synthesis* 11, 1972: 608.
34. C. Blandy, D. Gervais, Action of diphenylketene on alkyltitanates $Ti(OR)_4$. *Inorg. Chim. Acta* 47, 1981: 197.
35. F. Ando, Y. Kohmura, J. Koketsu, Insertion of ketene and diphenylketene to the pnicogen-heteroatom bonds. *Bull. Chem. Soc. Jpn.* 69, 1987: 1564.
36. R. Liu, C. Zhang, Z. Zhu, J. Luo, X. Zhou, L. Weng, Reactivity of lanthanocene amide complexes toward ketenes: unprecedented organolanthanide-induced conjugate electrophilic addition of ketenes to arenes. *Chem.Eur. J.* 12, 2006: 6940.
37. H. Wang, H.-S. Chan, J. Okuda, Z. Xie, Synthesis, structural characterization, and catalytic properties of group 4 metal complexes incorporating a phosphorus-bridged indenyl-carboranyl constrained-geometry ligand. *Organometallics* 24, 2005: 3118.
38. T. E. Patten, B. M. Novak, Organotitanium(IV) compounds as catalysts for the polymerization of isocyanates: the polymerization of isocyanates with functionalized side chains. *Macromolecules* 26, 1993: 436.
39. R. Ghosh, M. Nethaji, A. G. Samuelson, Reversible double insertion of aryl isocyanates into the Ti-O bond of titanium(IV) isopropoxide. *J. Organomet. Chem.* 690, 2005: 1282.

40. H. Shen, H.-S. Chan, Z. Xie, Guanylation of amines catalyzed by a half-sandwich titanacarborane amide complex. *Organometallics* 25, 2006: 5515.
41. Q. Li, S. Wang, S. Zhou, G. Yang, X. Zhu, Y. Liu, Highly atom efficient guanylation of both aromatic and secondary amines catalyzed by simple lanthanide amides. *J. Org. Chem.* 72, 2007: 6763.
42. S. Zhou, S. Wang, G. Yang, Q. Li, L. Zhang, Z. Yao, Z. Zhou, H. Song, Synthesis, structure, and diverse catalytic activities of [ethylenebis(indenyl)]lanthanide(III) amides on N-H and C-H addition to carbodiimides and ε-caprolactone polymerization. *Organometallics* 26, 2007: 3755.
43. T.-G. Ong, J. S. O'Brien, I. Korobkov, D. S. Richeson, Facile and atom-efficient amidolithium-catalyzed C-C and C-N formation for the construction of substituted guanidines and propiolamidines. *Organometallics* 25, 2006: 4728.
44. W.-X. Zhang, M. Nishiura, Z. Hou, Catalytic addition of amine N-H bonds to carbodiimides by half-sandwich rare-earth metal complexes: efficient synthesis of substituted guanidines through amine protonolysis of rare-earth metal guanidinates. *Chem. Eur. J.* 13, 2007: 4037.
45. F. Montilla, D. del Río, A. Pastor, A. Galindo, Use of vanadium complexes as catalysts in the synthesis of guanidines: new experimental data and DFT analysis of the carbodiimide interaction with the catalyst. *Organometallics* 25, 2006: 4996.
46. C. N. Rowley, T.-G. Ong, J. Priem, T. K. Woo, D. S. Richeson, Amidolithium and amidoaluminum catalyzed synthesis of substituted guanidines: an interplay of DFT modeling and experiment. *Inorg. Chem.* 47, 2008: 9660.
47. C. N. Rowley, T.-G. Ong, J. Priem, D. S. Richeson, T. K. Woo, Analysis of the critical step in catalytic carbodiimide transformation: proton transfer from amines, phosphines, and alkynes to guanidinates, phosphaguanidinates, and propiolamidinates with Li and Al catalysts. *Inorg. Chem.* 47, 2008: 12024.
48. W.-X. Zhang, D. Li, Z. Wang, Z. Xi, Alkyl aluminum-catalyzed addition of amines to carbodiimides: a highly efficient route to substituted guanidines. *Organometallics* 28, 2009: 882.
49. Y. Wu, S. Wang, L. Zhang, G. Yang, X. Zhu, C. Liu, C. Yin, J. Rong, Efficient guanylation of aromatic and heterocyclic amines catalyzed by cyclopentadienyl-free rare earth metal amides. *Inorg. Chim. Acta* 362, 2009: 2814.
50. T.-G. Ong, G. P. A. Yap, D. S. Richeson, Catalytic construction and reconstruction of guanidines: Ti-mediated guanylation of amines and transamination of guanidines. *J. Am. Chem. Soc.* 125, 2003: 8100.
51. F. Montilla, A. Pastor, A. Galindo, Guanylation of aromatic amines catalyzed by vanadium imido complexes. *J. Organomet. Chem.* 689, 2004: 993.
52. R. L. Zuckerman, R. G. Bergman, Mechanistic investigation of cycloreversion/cycloaddition reactions between zirconocene metallacycle complexes and unsaturated organic substrates. *Organometallics* 20, 2001: 1792.
53. T.-G. Ong, G. P. A. Yap, D. S. Richeson, Catalytic C=N bond metathesis of carbodiimides by group 4 and 5 imido complexes supported by guanidinate ligands. *Chem. Commun.* 2003: 2612.
54. H. Shen, H.-S. Chan, Z. Xie, Reaction of $[\sigma:\eta^5\text{-}(C_9H_6)C_2B_9H_{10}]Zr(NMe_2)(DME)$ with guanidines: metallacarborane-mediated C-N bond cleavage and 1,5-sigmatropic rearrangement. *J. Am. Chem. Soc.* 129, 2007: 12934.
55. H. Shen, Z. Xie, Titanacarborane amide-catalyzed transamination of guanidines. *Organometallics* 27, 2008: 2685.

22 Phosphorus-Substituted Carboranes in Catalysis

Sebastian Bauer and Evamarie Hey-Hawkins

CONTENTS

22.1 INTRODUCTION

Enantioselective catalysis is a very important technique in modern industrial chemistry, as is shown by the continuously increasing sales figures for enantiomerically pure substances. In 2003, chiral technology held a market share of 7.0 billion USD and was estimated to grow to 14.9 billion USD by the end of 2009. The estimated annual growth of conventional methods such as enantiomer separation or the use of chiral pools is 7%, whereas asymmetric synthetic methods are estimated to grow by 12% [1].

Catalytic reactions need to produce the desired products with high productivity and simultaneously be chemo-, regio-, and stereoselective, and are as such an ongoing challenge in academic and industrial research. To control the selectivity of the reaction, complexes with different steric and electronic properties have been used in technical processes [2]. The demands on catalysts are enormous. They should perform the reaction at a high rate and produce a good yield. Furthermore, they should be cost efficient, easy to handle, environmentally friendly ("green chemistry"), and have low energy consumption [3]. In the pharmaceutical industry it is of importance to use enantiomerically pure substances, as in most cases only one enantiomer interacts in the desired way with the receptor.

The use of metals in catalytic systems has been known since the early 1900s, when heterogeneous hydrogenation using a nickel catalyst was reported by Sabatier [4]. The first example of homogeneous catalysis was reported by Halpern, Harrod, and James in 1961 using $RuCl_3$ [5]. One of the most important milestones in homogeneous catalysis was the synthesis of $[RhCl\{P(C_6H_5)_3\}_3]$ by Wilkinson and co-workers [6]. The discovery of chiral catalysts by Knowles and Horner [7], independently from each other, was a breakthrough in asymmetric catalysis involving olefins. Since then, a series of ligands with different steric, electronic, and chemical properties have been synthesized, and investigations into their catalytic properties performed. The chirality was not only implemented at the phosphorus atom but also in the carbon backbone of the ligand, such as in (*S*,*S*)-CHIRAPHOS and (*S*,*S*)-(+)-DIOP (Figure 22.1). With carbon-center chirality, the development of enantiomerically pure ligands was significantly easier as it is possible to take advantage of the chiral pool. This has led to highly active and selective catalysts for several different applications, the best known of which is the BINAP ligand, first synthesized by Noyori and co-workers in 1980 [8].

Along with approaches to ligand synthesis, theoretical concepts have also been developed which explain the influence of the ligand on the catalytically active metal center. Tolman [9] developed a concept for the determination of the degree of sterical shielding of the coordinated metal center by phosphine ligands. To control the enantioselectivity, Knowles [10] developed a quadrant model, which is based on the steric blocking of coordination sites at the metal center. Catalytic cycles have been confirmed by spectroscopic investigations, quantum-mechanical calculations, and the synthesis of model substances in the investigation of their behavior. This led to further indications of the development of an ideal catalyst for a substrate by manipulation of the steric and electronic properties.

Bidentate phosphine ligands play a major role in coordination chemistry and its catalytic application. Ethylene and *ortho*-phenylene bridges are favored as C_2-symmetric backbone. This allows the formation of stable five-membered rings by coordination to a transition metal. It has been shown that using a phosphorus donor atom is the most viable option, owing to its lone pair of electrons, which promotes formation of σ bonds, and the availability of empty σ* orbitals, which facilitate π back-bonding [11]. With these orbitals the phosphorus atom can be used as the π-acceptor for complexation to late transition metals. With the use of electron-withdrawing substituents the σ-donor

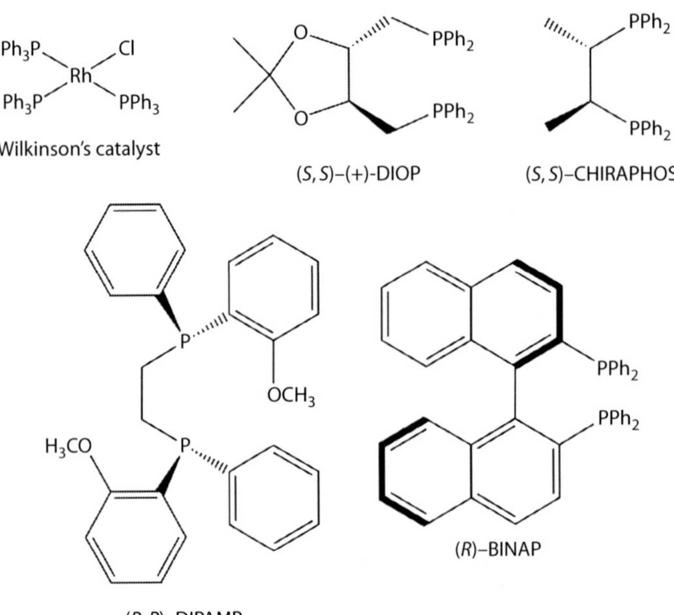

FIGURE 22.1 Wilkinson's catalyst and bisphosphine ligands used in homogeneous catalysis.

properties decrease, whereas the ability for π back-bonding increases. Therefore, the substituents at the phosphorus atoms have a direct influence on the σ-donor–π-acceptor properties of the phosphine ligands, which can be tuned by varying the substituents [12].

22.2 STRUCTURE AND PROPERTIES OF CARBORANES AND SYNTHESIS OF THE LIGANDS

Boranes, boron clusters, and in particular, carboranes are of special interest due to their unique properties that cannot be found in organic counterparts. These unique properties are based either on the element boron, due to its electron deficiency, or on the structural feature of the cluster compound. Borane clusters as a class of materials have a wide range of potential applications. This is not only due to their unique electronic and nuclear features; the fields of application, to name but a few, range from materials science through medical applications to catalysis, which will be described in more detail below [13]. Carboranes can be applied as liquid crystals in electro-optical displays [14], non-linear optics [15], and ion-selective electrodes [16] in the materials science arena. If carboranes are vaporized and fired at high temperatures they create boron films that are applied in Tokamak reactors for nuclear fusion [17]. Boranes have furthermore found application in airbag propellant systems in cars [18], as the stationary phase in gas chromatography [19] and in metal ion extraction systems, for example, for nuclear waste [20]. In medical applications, boron neutron capture therapy (BNCT), a special field of anti-cancer therapy, is noteworthy.

Of the borane clusters and heteroboranes, 1,2-dicarba-*closo*-dodecaborane(12) is of most interest due to its C_2 symmetry, whereby it can be regarded as an analogue of a bidentate *ortho*-phenylene backbone. The volume of the carborane(12) cluster is approximately 50% larger than the volume of a benzene molecule rotating along a C_2 axis [21]. There are three dicarba-*closo*-dodecaborane(12) isomers—*ortho*, *meta*, and *para*—each of which has specific electronic properties (Figure 22.2). The strongest electron-withdrawing effect is observed in the *ortho* isomer, whilst this effect is weakest in the *para* isomer. Depending on the position of substitution, the carborane moiety may behave as an electron acceptor, as in the case of the *ortho*-carboran-1-yl substituent ($\sigma_i = +0.38$), or as an electron donor, such as the *ortho*-carboran-9-yl substituent ($\sigma_i = -0.23$) [22].* In these cases the steric properties of the *ortho*-carborane moiety stay unaffected. These factors can influence the reactivity of substitution reactions at the carborane cluster and the properties of the substituent itself.

The substitution of one or more hydrogen atoms at the carborane cluster structure by heteroatoms such as carbon, nitrogen, oxygen, phosphorus, or sulphur led to a wide range of possible ligands for catalytically active metals. The hydrogen atoms at the carbon atoms behave similar to acidic protons, whereas the hydrogen atoms at boron have a more hydridic character. This behavior allows the selective removal of such protons by (most commonly) using organolithium compounds as bases.

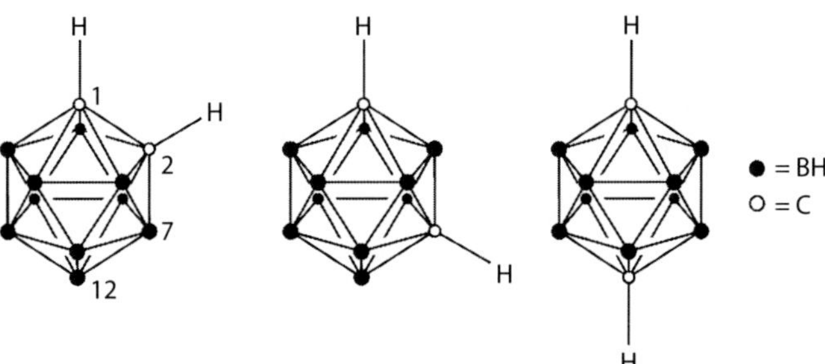

FIGURE 22.2 Three isomers of dicarba-*closo*-dodecaborane(12). From left to right: *ortho*-, *meta*-, and *para*-dicarba-*closo*-dodecaborane(12).

* Indicates in all cases the induction constant at the defined position of the carborane cluster.

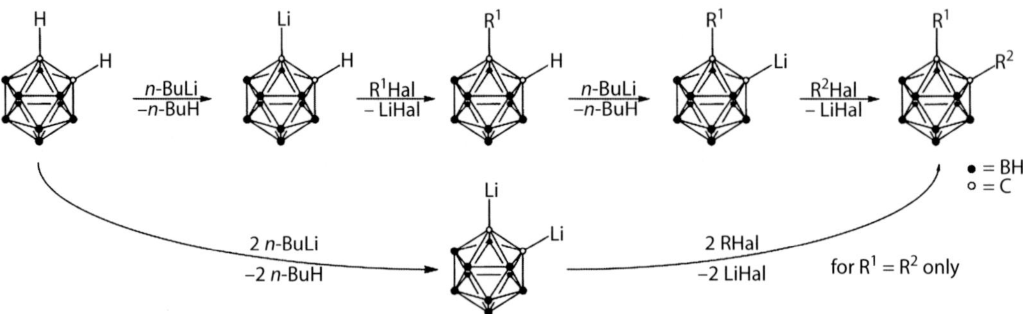

FIGURE 22.3 Synthesis of symmetrical and unsymmetrical C-substituted carboranes.

The obtained lithium salt can be widely functionalized by treatment with a whole range of electrophiles. Figure 22.3 shows the two possible pathways for substitution at both of the carbon atoms. One pathway (bottom) leads to a symmetrically substituted carborane, in which both substituents at the carbon atoms are introduced simultaneously after dilithiation of the carborane cluster. The other pathway (top) gives an unsymmetrically substituted carborane cluster. After selective mono-lithiation of one position and reaction of the lithium salt with an electrophile, the mono-substituted compound is obtained. It is then possible to use this compound as monodentate ligand or to introduce a second substituent. This can be achieved by lithiation of the hydrogen atom on the second carbon atom, followed by introduction of another electrophile. If a phosphorus substituent is introduced in the first step (R^1), *n*-BuLi can cleave the P–C and the C–H bond in the next step [23]. It is therefore possible to form ligands in which R^1 and R^2 can differ dramatically from each other in chemical nature.

The coupling of a phosphorus atom to a boron atom can be done by synthesis of the iodo-substituted carborane and subsequent palladium-catalyzed cross-coupling reaction (Figure 22.4) [24]. Unfortunately, this reaction only works with the *meta*- and *para*-carborane and not with the *ortho* isomer, because of cleavage of 9-iodo-*ortho*-carborane with triethylamine in toluene at 85–90°C. Attachment of an iodine atom to the B9 atom is done via electrophilic iodation. The mechanism is similar to the Friedel–Crafts reaction using $AlCl_3$ as the Lewis acid.

Besides direct bonding of a phosphorus atom to the carborane cage at a carbon or boron atom, introduction of a spacer between the carborane cluster and the phosphorus atom is possible. Introduction of a chalcogen spacer, excluding oxygen, between the carbon atom and the phosphorus atom is quite easy and can be carried out by mono- or dilithiation, followed by addition of the elemental chalcogen. Once the chalcogen atom is inserted into the lithium–carbon bond, the lithium salt can be further functionalized by an electrophile. To introduce an oxygen atom as a spacer, a hydroboration reaction must be performed after the lithiation step [25]. The obtained hydroxycarborane can be treated with a chlorophosphite and a base to obtain the phosphonito carborane [26].

A chalcogen spacer at a boron atom can be added by a Friedel–Crafts-like reaction. To do this, elemental sulphur or the corresponding chalcogen chloride is used, with aluminium(III)

FIGURE 22.4 Synthesis of boron-substituted carboranylphosphonates.

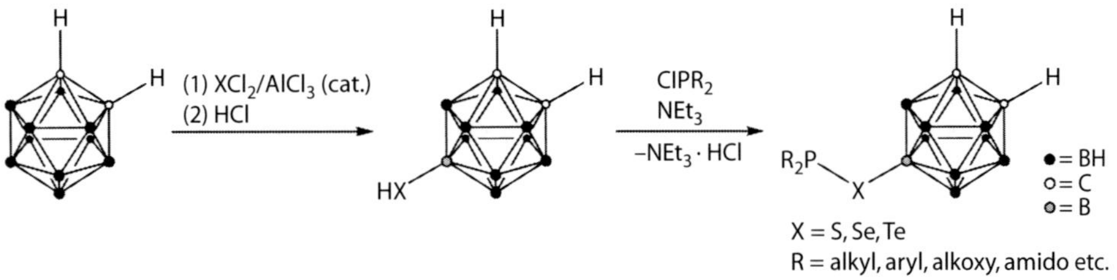

FIGURE 22.5 Synthesis of boron-substituted carboranylchalcogenophosphines.

chloride as the Lewis acid. The obtained 9-chalcogenyl-dicarba-*closo*-dodecaborane(12) can be treated with a chlorophosphine in the presence of a base to obtain the desired carboranylphosphite (Figure 22.5).

To perform asymmetric catalysis it is necessary to have at least one chiral center in the molecule, which induces the chirality in the catalytic cycle. In phosphinocarboranes, there are three distinct positions where this can occur

- In the carbon backbone of the phosphorus or carborane substituent
- At the phosphorus atom itself, by adding three different substituents
- Or in the carborane cluster itself, due to the planar chirality of decapped clusters

The first two possibilities are often employed by the use of the aforementioned methods. In most known compounds, the chirality is located in the carbon backbone, as it is often difficult to synthesize enantiomerically pure P-chiral substances. Thus, it is possible to use the chiral pool, but the chirality is in that case in the second coordination sphere. The third possibility, to introduce planar chirality into the ligand, is often associated with ferrocenes. In carboranes this can be achieved by degradation of the neutral icosahedral *closo* cluster to an anionic *nido* cluster. This deboronation, or the so-called decapping, is carried out by the treatment of the substituted *ortho* cluster with a strong Lewis base such as alkoxides, amines, and fluoride (Figure 22.6) [27]. The counter ion of the obtained negatively charged cluster can be widely varied and is often an ammonium ion such as tetraethylammonium or piperidinium. The nomenclature of planar chirality in carboranes follows the Schlögl convention for planar-chiral ferrocenes [28].

Phosphorus-substituted carborane ligands can be mono-, bi-, or multidentate when they react with a metal center. Factors affecting this behavior include the structural features of the carborane cage (i.e., whether it is the *ortho*, *meta*, or *para* isomer, or has the *closo* or *nido* structure), the substitution pattern of the carborane cluster, the number and kind of donor atoms, and the electronic and steric properties of the ligand. All these factors have an influence on the catalytic activity of the metal complex. For example, the carborane cluster can act as an electron acceptor or electron donor at the phosphorus atom [22,29]. The donor atoms adjacent to phosphorus can act as labile ligands that block the coordination site at the metal until a substrate approaches. In *nido* cluster compounds, the metal ion can bind in two ways: through the phosphorus substituent or by the decapped face of the cluster, whereby the former option is the most stable. In *nido*-carborane chemistry, B–H–Rh and

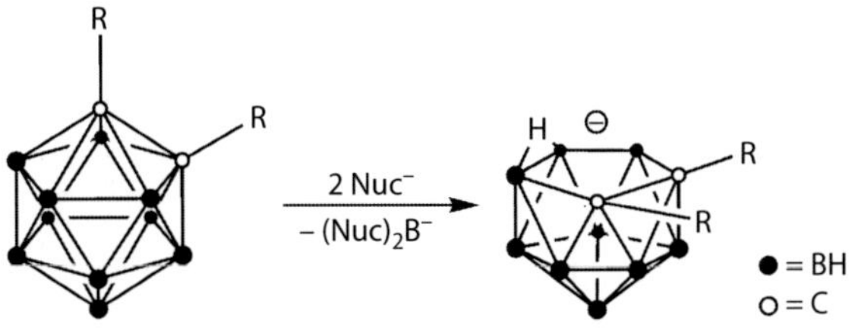

FIGURE 22.6 Deboronation of a *closo*-carborane cluster to the *nido* cluster compound.

B–H–Ru bonds are often observed between the metal and the B2 or B11 atom, in addition to the expected metal–phosphorus bond.

Tests on the catalytic activity of the phosphorus-substituted carboranes can be performed in two ways. First, it is possible to use the previously synthesized and isolated metal complexes; second, the metal complex can be generated *in situ*. In both cases metal precursors with labile leaving groups must be used which generate the coordination site for the ligand. In most cases, x-ray crystal structure analysis was performed to elucidate the structural properties of the ligand sphere around the metal atom and the absolute configuration in chiral ligands. Unfortunately, prediction of the catalytic activity or selectivity is not possible.

Besides the phosphorus-substituted carboranes presented in this review, metallacarboranes have been used in homogeneous catalysis. First attempts were presented by Hawthorne et al. [30], who used $[Rh(C_2B_9H_{11})H(PPh_3)_2]$ in hydrogenation, hydroformylation, and hydrosilylation reactions. Jordan and co-workers [31] investigated the hydrogenation of alkynes to alkenes using carborane-based isoelectronic analogues of metallocenes. Several groups have performed investigations into polymerization catalysis using similar compounds [32].

22.3 CATALYSIS

22.3.1 HYDROGENATION

Hydrogenation reactions are benchmarked to standard substances such as simple alkenes (e.g., 1-hexene), and asymmetric hydrogenations to α,β-unsaturated functionalized olefins (e.g., dimethyl itaconate) and α-dehydro amino acid esters. Thus, numerous ligands have been tested with a great diversity of structural and electronic properties.

The catalytic properties of 1,2-bis(diphenylphosphino)-*closo*-carborane(12)s (**1**) were first studied in 1985 by Hart and Owen [33] in the hydrogenation and hydroformylation of 1-hexene, 1,3- and 1,5-cyclooctadiene. The catalyst, $[RhCl(PPh_3)(\mathbf{1})]$, yielded between 70% and 100% at 110°C, under 6 MPa initial hydrogen pressure. Under milder conditions, (40°C, 0.1 MPa) no conversion was reported.

Hey-Hawkins and coworkers [34] synthesized the P-chiral 1,2-bis(chlorophosphino)-*closo*-carborane(12)s **2–4**, cyclic compound **5** and rhodium complexes thereof (Figure 22.7). Compound **3** with $[Rh(cod)_2]BF_4$ (cod = 1,5-cyclooctadiene) as metal precursor was the most active ligand in the hydrogenation of dimethyl itaconate, resulting in full conversion of the substrate in 10 min at room temperature and 2 MPa H_2 pressure. To obtain full conversion with the other ligands, higher temperatures and pressure were needed. The more electron-donating *tert*-butyl group at the phosphorus

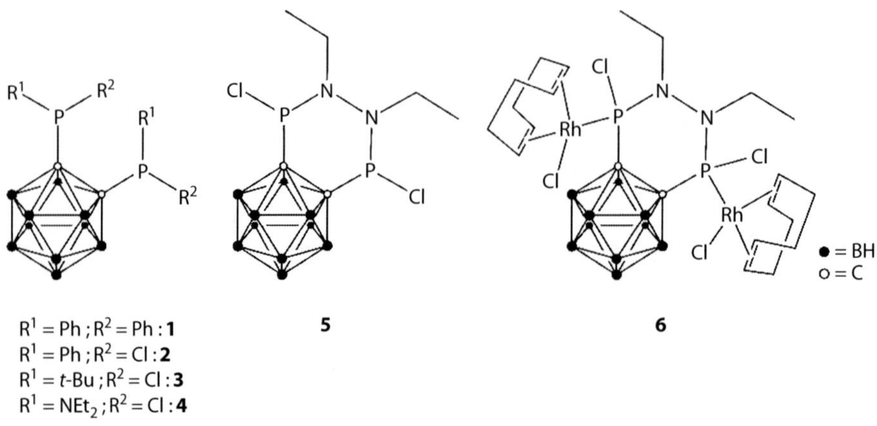

FIGURE 22.7 1,2-Bis(phosphino)-*closo*-carborane(12)s applied in hydrogenation reactions.

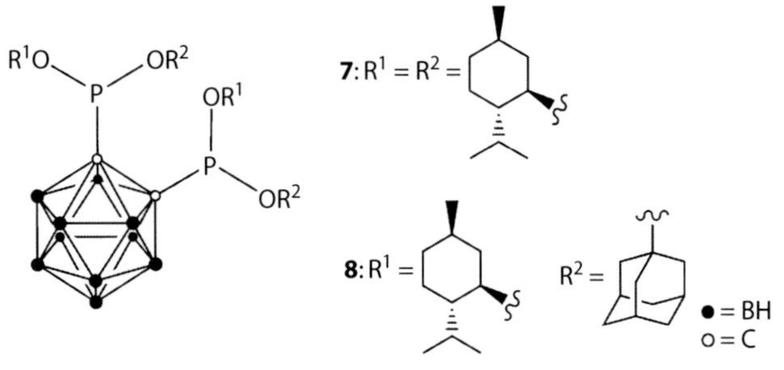

FIGURE 22.8 1,2-Bis(phosphonito)-*closo*-carborane(12)s.

atom in **3** could be responsible for the higher activity. The hydrogenation product was obtained as a racemic mixture in all cases, because the used ligands were only diastereomerically pure. Unfortunately, no conversion was observed with the isolated rhodium complexes.

Another group of ligands that were investigated with regard to their catalytic activity by Hey-Hawkins and co-workers [35] are 1,2-bis(phosphonito)-*closo*-carborane(12)s (Figure 22.8). While both ligands have chiral centers in the carbon backbone, compound **8** also has an additional chiral center at the phosphorus atom. The ligands **7** and **8** were enantiomerically pure compounds and were used with $[Rh(cod)_2]BF_4$ or $[\{Rh(\mu\text{-}Cl)(cod)\}_2]$ as metal precursors in the hydrogenation of dimethyl itaconate. Both ligands gave full conversion with $[Rh(cod)_2]BF_4$ as the precursor compound, whereas conversions of just 20–30% were observed with $[\{Rh(\mu\text{-}Cl)(cod)\}_2]$. This could be explained by the stability of the chloro bridge; even if it is opened, the remaining chlorine atom can block a coordination site at the rhodium center. Although different conversion rates were obtained, the utilization of a different metal precursor had no effect on the enantioselectivity of the hydrogenation. It could be shown that the *R* enantiomer was formed in excess regardless of the ligand, but with low enantiomeric excess (Table 22.1). As the substituents at the phosphorus atom have almost the same size, steric discrimination of the formation of one enantiomer or chiral induction through chirality at the phosphorus atom or the (–)-menthyl groups apparently does not take place.

Kang and co-workers [36] synthesized ligands **9** and **10**, which are described as phosphinooxazoline (PHOX) analogues (Figure 22.9). PHOX has been successfully applied in transition-metal catalyzed asymmetric hydrogenation, the Heck reaction and allylic alkylation. The ligands **9** and **10** were investigated in the hydrogenation of several α,β-unsaturated functionalized olefins using rhodium, and in the hydrogenation of unfunctionalized olefins with iridium as the catalytically active metal. Kang and co-workers [37] also synthesized 1-*N*,*N*-dimethylaminomethyl-2-diphenylphosphino-*closo*-carborane(12), a P,N-chelating ligand, the non-chiral analogue of PPFA, which has itself been successfully applied in catalysis [38].

The rhodium-catalyzed asymmetric hydrogenation of α,β-unsaturated functionalized olefins was performed using $[Rh(nbd)_2]BF_4$ (nbd = norbornadiene) as the metal precursor to generate the

TABLE 22.1
Rhodium-Catalyzed Hydrogenation of Dimethyl Itaconate

Ligand	Conversion (%)	ee (%)
7	100	10.3 (*R*)
(R_P,R_P)-**8**	100	3.1 (*R*)
(R_P,S_P)-**8**	100	14.5 (*R*)

Reaction conditions: 1 mol% $[Rh(cod)_2]BF_4$; 1.1 mol% ligand; 1 mmol dimethyl itaconate; dichloromethane/methanol = 1:1; p = 1 MPa; room temperature. Conversion and ee determined by HPLC.

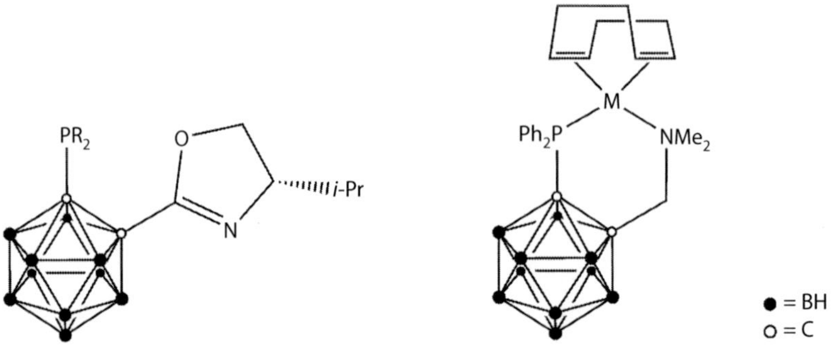
PR2
O
i-Pr
N
M
Ph2P
NMe2
● = BH
○ = C

R = Ph : **9**
R = Cy : **10**

M = Rh : **11**
M = Ir : **12**

FIGURE 22.9 Ligands **9** and **10** (left) and metal complexes **11** and **12** (right).

catalytically active species *in situ*. Yields obtained were in the range of 20–90% with enantioselectivities of 65–96% for the *S* enantiomer (Table 22.2). Comparing phosphine substituents, with phenyl-containing substrates the diphenylphosphine moiety gave a higher enantiomeric excess (ee), whereas the dicyclohexylphosphine substituent gave a higher ee with (*Z*)-2-acetamidoacrylate. This shows the strong dependence of the enantioselectivity on the ligand–substrate combination.

The asymmetric hydrogenation of unfunctionalized olefins does not give high enantioselectivities with conventional Rh- or Ru-phosphine-type catalysts. However, Pfaltz and co-workers [39] performed hydrogenations with iridium-based PHOX catalysts for a range of unfunctionalized olefins and reached enantioselectivities up to 99% and full conversion. By using **9** and **10** as ligands with iridium as catalytically active metal, enantioselectivities up to 98% for the *S* enantiomer and yields up to 97% were obtained (Table 22.3) [39]. In reactions with the most bulky substrate, *trans*-α-methylstilbene, a maximum of 40% yield and of 82% enantioselectivity was reached. Changing the substituents at the phosphorus atom, depending on the steric hindrance at the substrate, led to an increase or decrease in enantioselectivity, affected by the occupation of the semi-hindered region

TABLE 22.2
Rhodium-Catalyzed Hydrogenation of α,β-Unsaturated Functionalized Olefins

Substrate	Ligand	Yield	ee
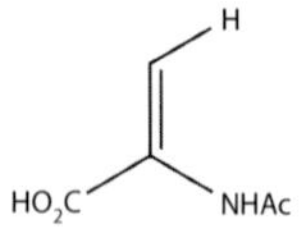 H HO2C NHAc	**9**	22	65
	10	50	84
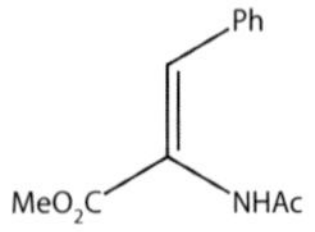 Ph MeO2C NHAc	**9**	90	96
	10	85	83
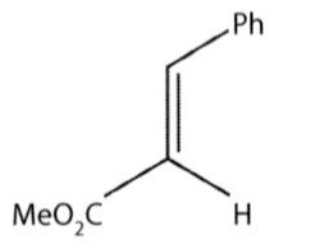 Ph MeO2C H	**9**	24	94
	10	20	87

Reaction conditions: Catalyst loading 2.0 mol%; solvent THF; $p = 1$ MPa; $t = 24$ h; room temperature.

TABLE 22.3
Iridium-Catalyzed Hydrogenation of Unfunctionalized Olefins

Substrate	Ligand	Yield	ee
Ph	**9**	40	73
	10	30	82
CO_2Et	**9**	96	93
	10	89	98
CH_2OH	**9**	90	95
	10	97	96
MeO	**9**	85	88
	10	95	90

Reaction conditions: Catalyst loading 2.0 mol%; solvent CH_2Cl_2; $p = 1$ MPa; $t = 24$ h; room temperature.

during the catalytic cycle. Thus, the steric bulk of both the substrate and the ligand have an influence on the ee and yield.

1-*N*,*N*-dimethylaminomethyl-2-diphenylphosphino-*closo*-carborane(12) was investigated for the hydrogenation of cyclohexene [37]. The precursor metal complex was synthesized by treating the ligand with $[Rh(cod)(thf)_2]BF_4$ or $[Ir(cod)(thf)_2]BF_4$. Reactions were performed in several solvents and it could be shown that the reaction is highly dependent on the solvent (Table 22.4). In all cases,

TABLE 22.4
Solvent and Metal Dependence of the Hydrogenation of Cyclohexene

Catalyst	Solvent	Conversion (%)
11	Methanol	100
12	Methanol	52.8
Wilkinson's catalyst	Methanol	59.2
11	1,2-Dichloroethane	98.8
12	1,2-Dichloroethane	90.0
Wilkinson's catalyst	1,2-Dichloroethane	87.4
11	Toluene	84.7
12	Toluene	17.4
Wilkinson's catalyst	Toluene	98.1

Reaction conditions: 0.03 mmol catalyst; 36.5 mmol cyclohexene; $T = 80°C$; $p = 2$ MPa; $t = 50$ min. Conversion determined by GC.

the rhodium complex **11** performed better than the iridium complex **12** and, with exception of toluene as the solvent, better than the benchmark catalyst, Wilkinson's catalyst. With **11**, the yield increases in polar solvents, and the reaction proceeds to 100% conversion in methanol. Complex **12** performed the best with 1,2-dichloroethane as solvent. The poorest yield for both complexes was obtained when toluene was used as the solvent.

closo-Carboranes(12) with a spacer between the carborane cluster and the phosphorus atom, for example, only one atom, such as oxygen or sulphur, or a two atom spacer consisting of a CH_2 group and a heteroatom, such as CH_2–N or CH_2–O, are also known. Lyubimov et al. [26,40] synthesized *ortho-* and *meta*-carboranes that are mono-substituted at a carbon atom or at the B9 atom. The substituents at the carborane cluster was either (*R*)-BINOL phosphonite or in case of compounds **14** and **24** (*S*)-BINOL phosphonite (Figures 22.10 and 22.11). These substances were tested in the rhodium-catalyzed asymmetric hydrogenation of dimethyl itaconate and various α-dehydro amino acid esters.

Boron-substituted compounds **13–20** were treated *in situ* with $[Rh(cod)_2]BF_4$ to generate the catalytically active complex (Scheme 22.1) and tested for activity in asymmetric hydrogenation.

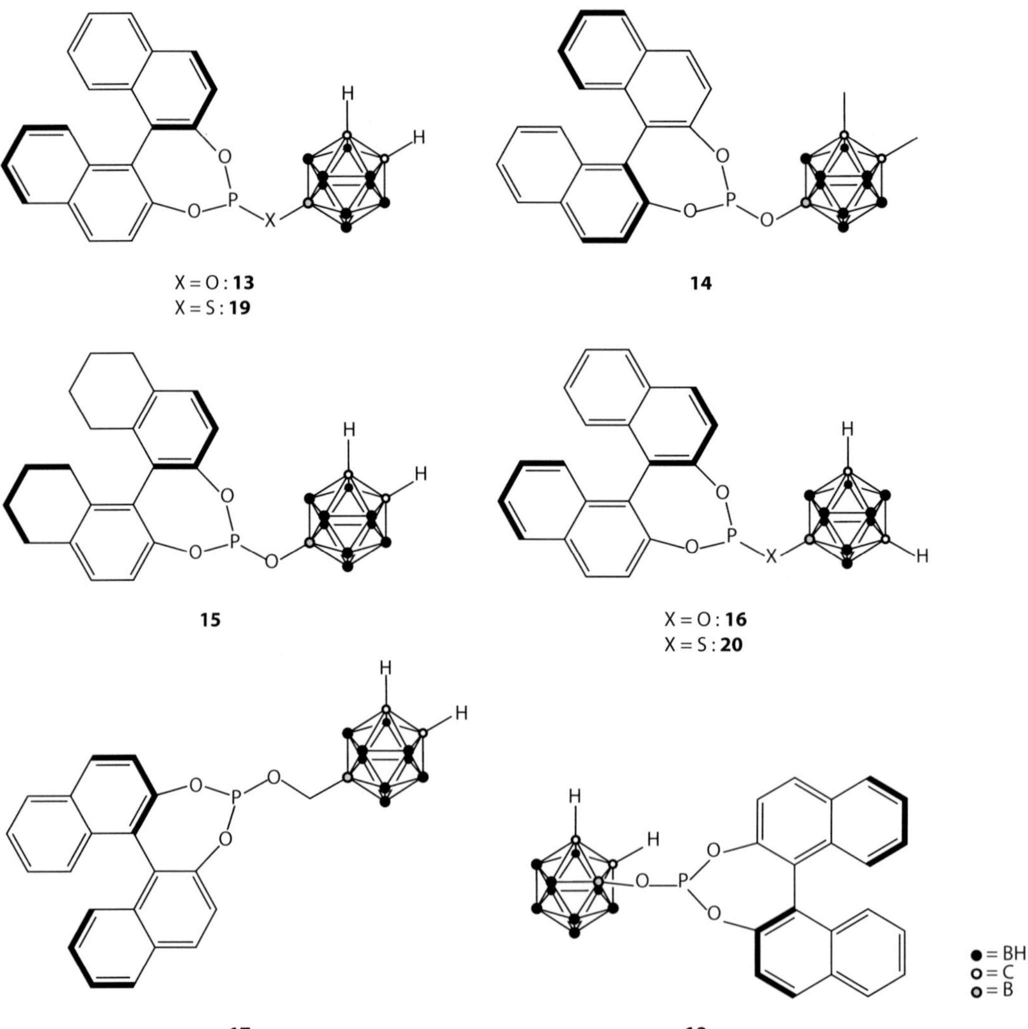

FIGURE 22.10 Boron-substituted carboranes with chalcogen spacer.

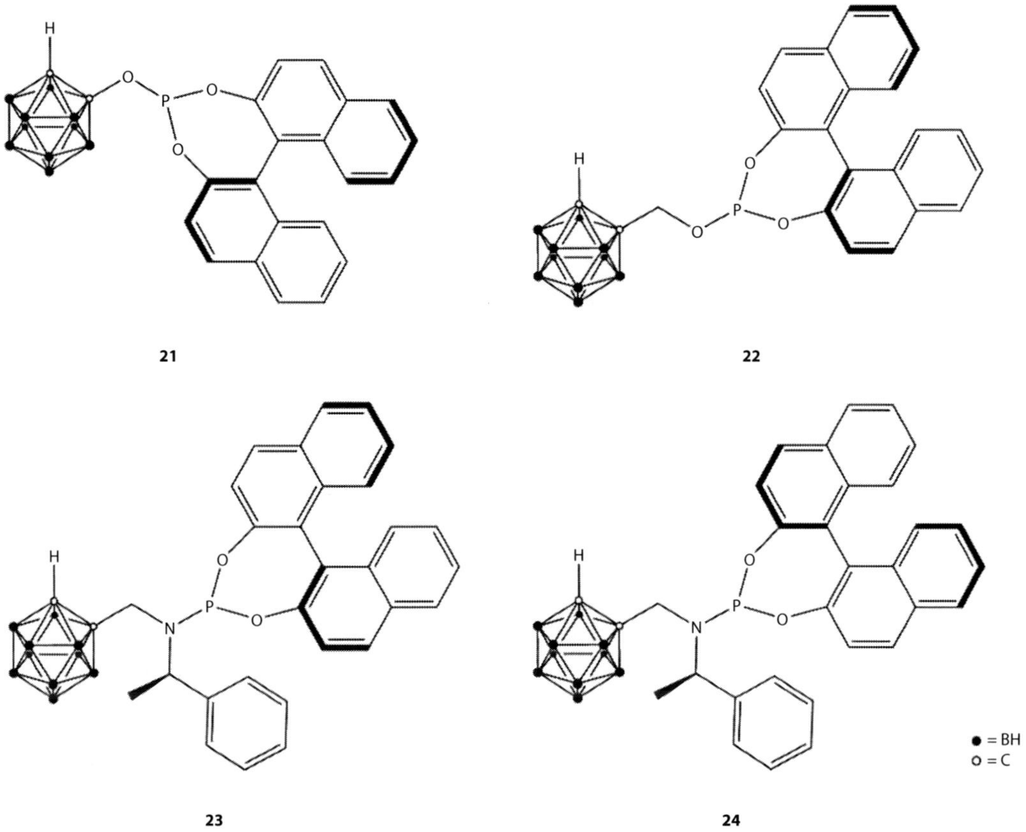

FIGURE 22.11 Carbon-substituted carboranes with spacer.

These results were compared with those for isolated rhodium complexes **16Rh**, **19Rh**, and **20Rh**, which were formed by reaction of the corresponding ligand **16**, **19**, or **20** with $[Rh(cod)_2]BF_4$. To investigate the influence of the electron-donor–electron-acceptor properties of the carborane moiety on asymmetric hydrogenation, several different boron substitution schemes were investigated.

In the asymmetric hydrogenation of dimethyl itaconate, the 9-*O*-boron-substituted compounds gave the best performance with full conversions and ee values of 99.8% (Table 22.5). Use of the *ortho* or *meta* isomer of the carborane(12) moiety had no effect on conversion or ee. Substituents at the cluster carbon atoms (compound **14**) or the use of a partly hydrogenated H_8-BINOL backbone had almost no influence on the ee. For the 9-*S*-boron-substituted compound, the *ortho* isomer **19** gave an ee of 98%, whereas the *meta* isomer **20** yielded an ee of 88%. This can be explained by the donor–acceptor properties of the carboranyl substituents. The less electron-donating 9-*meta*-carboranyl group ($\sigma_i = -0.12$) leads to a decrease in enantioselectivity in comparison with the more

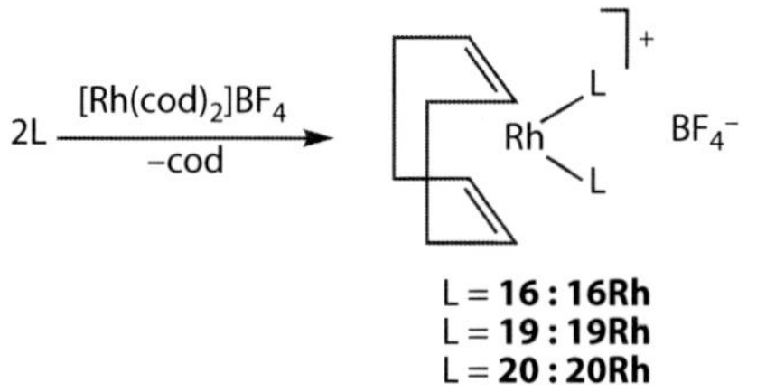

SCHEME 22.1 Synthesis of rhodium complexes with monodentate ligands.

TABLE 22.5
Asymmetric Hydrogenation of Dimethyl Itaconate

Catalyst	Time (h)	Pressure (MPa)	Conversion (%)	ee (%)
13[a]	17	10	100	99.8 (*R*)
14	16	1	100	98 (*S*)
15	16	1	100	99.5 (*R*)
16[a]	18	10	100	99.7 (*R*)
16Rh[c]	17	10	100	99.8 (*R*)
17[a]	20	0.5	100	95 (*R*)
18[a]	18	0.5	100	96 (*R*)
19[b]	18	0.5	100	98 (*R*)
19[b]	14	1	100	98 (*R*)
19Rh[c]	18	0.5	100	99 (*R*)
20[b]	14	0.5	100	88 (*R*)
20Rh[c]	16	0.5	100	93 (*R*)
21	20	0.5	100	65 (*R*)
22	18	0.5	100	88 (*R*)
23	20	0.5	80	47 (*R*)
24	20	0.5	70	60 (*S*)

[a] Reaction conditions: Ratio substrate/$[Rh(cod)_2]BF_4$/ligand = 1/0.01/0.022; solvent CH_2Cl_2; room temperature.
[b] Reaction conditions: Ratio substrate/$[Rh(cod)_2]BF_4$/ligand = 1/0.01/0.02; solvent CH_2Cl_2; room temperature.
[c] Reaction conditions: Ratio substrate/complex = 1/0.01; solvent CH_2Cl_2; room temperature.

electron-donating 9-*ortho*-carboranyl group ($\sigma_i = -0.23$). The ee of 95% obtained with the 3-*O*-boron-substituted compound indicates a weaker electron-donating effect ($\sigma_i = +0.11$). The introduction of a methylene spacer between the phosphite and the carborane cluster in compound **17** resulted in a smaller enantiodiscrimination in comparison with **13**.

The carbon-substituted carborane **21** had a low ee of 65%, as the position of substitution at the cluster is less electron accepting than in the other ligands ($\sigma_i = +0.38$). This effect can be reduced by introducing a methylene spacer between the strongly electron-withdrawing 1-*ortho*-carboranyl group and the phosphorus atom. This led to an increased ee of 88%, which is still lower than for the boron-substituted ligands. Phosphoramidite ligands **23** and **24** gave moderate enantioselectivities of 47% and 60%, respectively, and conversions of 80% and 70%, respectively. This difference can be explained by the more electron-donating nature of the nitrogen atom compared with oxygen. As both ligands only differ in the configuration of the BINOL group, the difference in the activity and selectivity can be explained by a substrate–catalyst matched–mismatched situation, where **23** with the (*S*)-BINOL is the matched situation.

To compare the activity of the active rhodium complexes formed *in situ*, the isolated cationic rhodium complexes **16Rh**, **19Rh,** and **20Rh** were tested in the asymmetric hydrogenation of dimethyl itaconate. The sulphur-substituted compounds **19Rh** and **20Rh** gave slightly lower ee but showed the same trend in terms of enantioselectivity. Complex **16Rh** gave the same result as the complex formed *in situ*. This complex and ligand **15** were further tested with supercritical CO_2 as solvent (Table 22.6). Both substances revealed a dramatically lower reaction time of just 2 h. For ligand **15** optimization of the reaction conditions was performed, whereby the enantiomeric excess increased on increasing the relative amount of H_2 in the mixture. The enantiomeric excess was slightly lower: 92% for **16Rh** and 93% for **15**. This opens the field for usage of these ligands in alternative "green" chemistry.

TABLE 22.6
Asymmetric Hydrogenation of Dimethyl Itaconate in Supercritical CO_2

Catalyst	Pressure H_2 (MPa)	Total Pressure (MPa)	Time (min)	Conversion (%)	ee (%)
15/[Rh(cod)$_2$]BF$_4$[a]	2.5	20	60	100	60 (*R*)
15/[Rh(cod)$_2$]BF$_4$[a]	2.5	12.5	60	100	80 (*R*)
15/[Rh(cod)$_2$]BF$_4$[a]	8	20	45	100	92 (*R*)
16Rh[b]	10	20	120	100	93 (*R*)

[a] Reaction conditions: Ratio substrate/ligand/[Rh(cod)$_2$]BF$_4$ = 1/0.02/0.01; T = 40°C.
[b] Reaction conditions: Ratio substrate/complex = 1/0.005; room temperature.

In the hydrogenation of α-dehydro amino acid esters, specifically of methyl 2-acetamidoacrylate, the same dependence of the enantioselectivity on the electron-donor–electron-acceptor properties of the carborane substituent and the ligand structure could be observed (Table 22.7) [26,40b,c]. Independently of the heteroatom or carborane isomer, all of the ligands substituted directly on a boron atom exhibited high enantioselectivities and full conversion. The use of sulphur gave slightly better ee values than the oxygen derivatives, whilst the more electron-withdrawing 3-*ortho*-carboranyl ligand **17** gave lower enantioselectivity. This is in agreement with previously reported results. Interestingly, the introduction of a methylene spacer between the heteroatom and the carborane cluster gave an excellent ee of 96%, which is contrary to the previous observations. The carbon-substituted ligands, which are even more electron withdrawing, gave lower conversion and/or lower enantioselectivities. Introduction of a methylene spacer in these cases increased the yield and ee. As in the hydrogenation of dimethyl itaconate a substrate–catalyst matched–mismatched situation in **23** yielded a higher conversion and ee than in **24**.

Further investigations with (*Z*)-methyl 2-acetamido-3-phenylacrylate using the oxo-bridged ligands **14** and **15** and the sulphur-bridged ligands **19** and **20** revealed that the sulphur-substituted ligands performed better. Furthermore, ligand **19** with its more strongly electron-donating 9-*ortho*-carboranyl group gave a higher ee than **20**. Using more electron-withdrawing phenyl substituents in the substrate made the enantiomeric excess lower, whereas **19** performed better than **20**. This trend continued in case of a fluorophenyl or chlorophenyl substituent in the substrate, and with the more weakly electron-donating 9-*meta*-carboranyl substituent in **20** incomplete conversion was observed at 18 h reaction time. Nevertheless, investigations on the influence of hydrogen pressure revealed that an increase has no influence on the enantioselectivity. In this series the ligands **14** and **15** were investigated in the asymmetric hydrogenation of (*Z*)-methyl 2-acetamido-3-cymantrenylacrylate. The strongly electron-withdrawing and bulky cymantrenyl group led to a remarkable increase in enantioselectivity. This shows that the electronic effects of the substituents in the enamides play a major role in the enantiocontrol.

In conclusion, the methylene spacer has an influence on the electron-donor–electron-acceptor properties, depending on the position at the carborane cluster, and therefore on the enantioselectivity of the hydrogenation reaction. For boron substitution enantioselectivity decreases with decreasing donor ability of the cluster, and vice versa for carbon substitution.

nido-Carboranes were also tested in hydrogenation reactions. Teixidor and co-workers [41] investigated the hydrogenation of 1-hexene to hexane with the (mono-phosphinocarborane)rhodium complexes **25–28**, and comparisons were made to analogous (mono-thiocarborane)rhodium complex **29** (Figure 22.12). Higher temperatures were shown to favor isomerization to 2-hexenes over hydrogenation, though the use of ambient temperature for (mono-phosphino)rhodium complexes led to a very low activity (Table 22.8). In contrast, at ambient temperature and/or low hydrogen pressure (mono-thiocarborane)rhodium complex **29** gave good yields. It was also shown that hydrogenation

TABLE 22.7
Asymmetric Hydrogenation of α-Dehydro Amino Acid Esters

Substrate	Catalyst	Pressure (MPa)	Time (h)	Conversion (%)	ee (%)
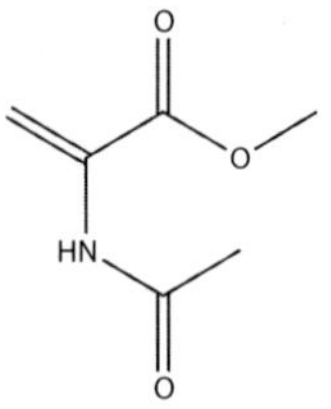	**13**[a]	0.5	17	100	86 (*S*)
	14	1	16	100	80 (*R*)
	15	1	16	100	76 (*S*)
	16[a]	0.5	17	100	85 (*S*)
	17[a]	0.5	17	100	80 (*S*)
	18[a]	0.5	17	100	96 (*S*)
	19[b]	0.5	18	100	94 (*S*)
	19[b]	1	14	100	94 (*S*)
	20[b]	0.5	16	100	95 (*S*)
	20[b]	1	12	100	95 (*S*)
	21[a]	0.5	20	60	45 (*S*)
	22[a]	0.5	18	100	70 (*S*)
	23[a]	0.5	20	90	80 (*S*)
	24[a]	0.5	20	75	68 (*R*)
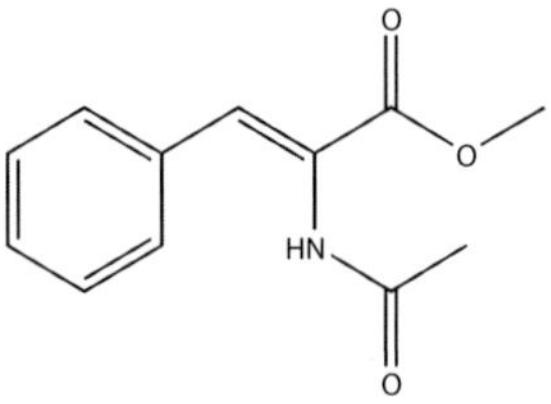	**14**	1	16	100	30 (*R*)
	15	1	16	100	46 (*S*)
	19[b]	1	16	100	85 (*S*)
	20[b]	1	18	100	70 (*S*)
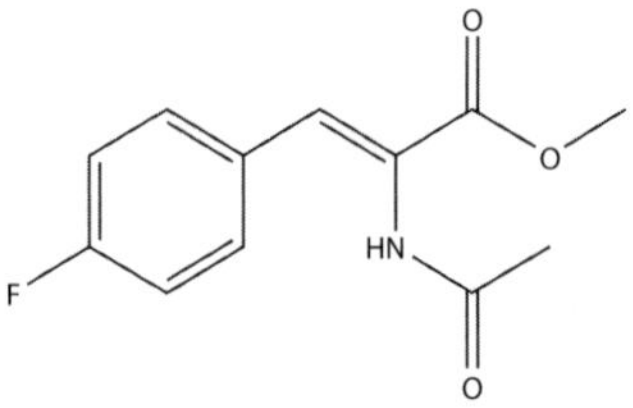	**19**[b]	1	18	100	75 (*S*)
	20[b]	1	18	85	50 (*S*)
	20[b]	1	22	100	50 (*S*)
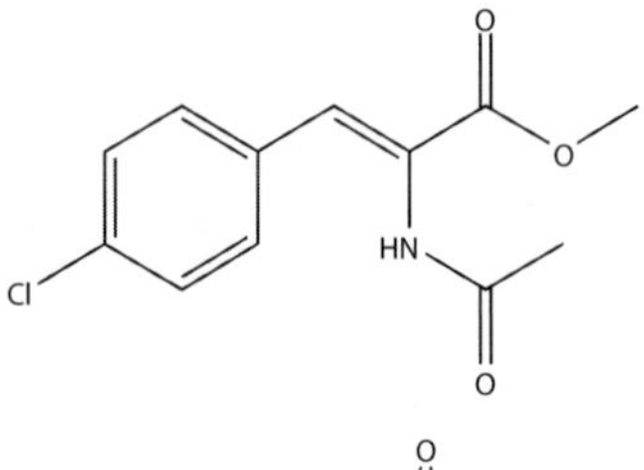	**14**	1	18	100	60 (*R*)
	15	1	18	100	47 (*S*)
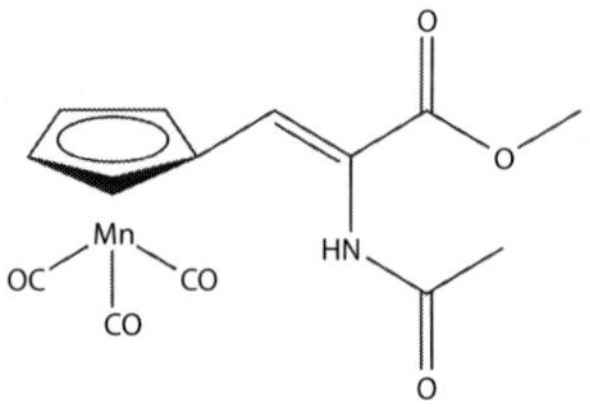	**14**	1	18	100	86 (*R*)
	15	1	18	100	90 (*R*)

[a] Reaction conditions: Ratio substrate/[Rh(cod)$_2$]BF_4/ligand = 1/0.01/0.02; solvent CH_2Cl_2; room temperature.
[b] Reaction conditions: Ratio substrate/[Rh(cod)$_2$]BF_4/ligand = 1/0.01/0.02; solvent CH_2Cl_2; room temperature.

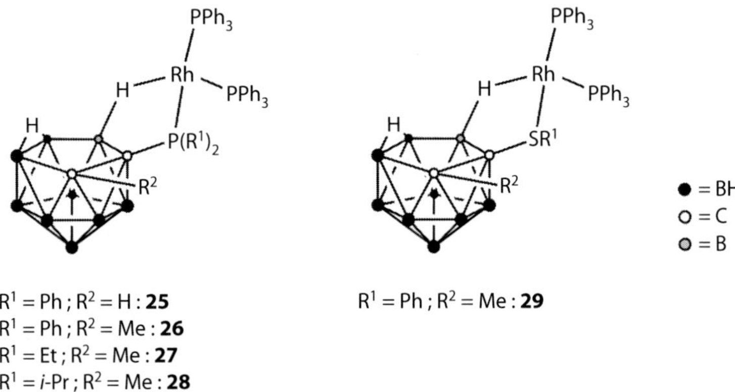

FIGURE 22.12 Rhodium complexes from *nido*-carborane(12)s used in hydrogenation.

of the 2-hexenes formed during the reaction proceeded much more slowly than that of 1-hexene, though the authors suggest that 2-hexene may be hydrogenated if high hydrogen pressure is applied. Nuclear magnetic resonance (NMR) spectroscopic studies showed that the catalytic systems seem to be recoverable after completion of the reaction. Addition of triphenylphosphine to the reaction mixture inhibited the reaction. However, the (mono-phosphinocarborane)rhodium complexes were more robust towards the presence of donor atoms belonging to the substrate than the mono-thio derivative.

The complex [*closo*-3,3-(η^2:η^3-$C_7H_7CH_2$)-3,1,2-$RhC_2B_9H_{11}$] was used in the hydrogenation of the antibiotic precursor methacycline with a high diastereomeric excess of approximately 95%, which is greater than or equal to those of reported systems (Scheme 22.2) [42]. Based on this, the (mono-phosphinocarborane)rhodium complexes **25** and **26** were tested in this reaction, and showed high conversions of 85% and 99.7%, respectively. Furthermore, the pharmacologically important doxycycline was obtained in very high diastereoselectivity (>99%), whilst the other possible diastereomer was not observed in any reaction.

Brunner et al. [43] investigated the hydrogenation of (*Z*)-α-*N*-acetamidocinnamic acid and ketopantolactone with complexes **30** and **31** (Figure 22.13, Scheme 22.3). In **30** the planar chirality of

TABLE 22.8
Hydrogenation of 1-Hexene

Complex	Temperature (°C)	Hexane (%)	2-Hexene (%)
25	25	12	0.7
26	25	6	0.3
27	25	0.2	0.1
28	25	2.4	0.1
29	25	98	2.0
25	66	90	9.3
26	66	61	20
27	66	82	11
28	66	83	14
29	66	85	11

Reaction conditions: 5.21×10^{-3} mmol catalyst; 5 mL 1-hexene; solvent: 5 mL THF; $p = 4.5$ MPa; $t = 1$ h; analysis by capillary GC.

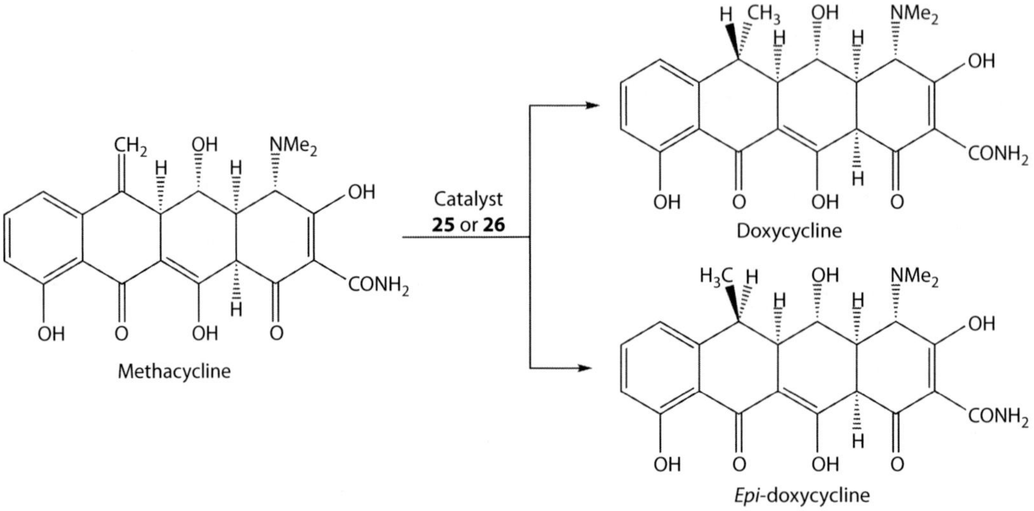

SCHEME 22.2 Hydrogenation of methacycline to doxycycline and *epi*-doxycycline.

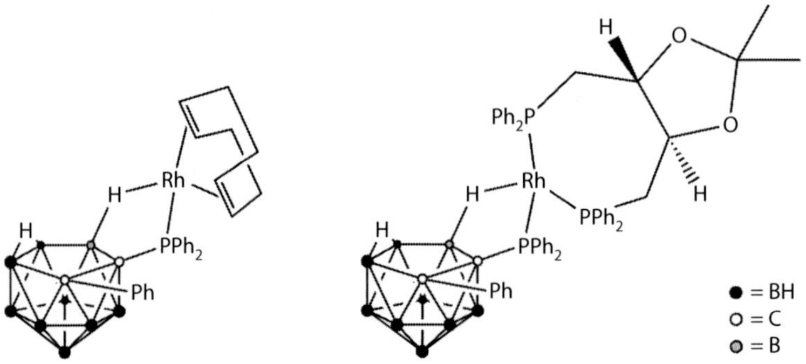

FIGURE 22.13 Rhodium complexes investigated in the hydrogenation of acetamidocinnamic acid and keto-pantolactone.

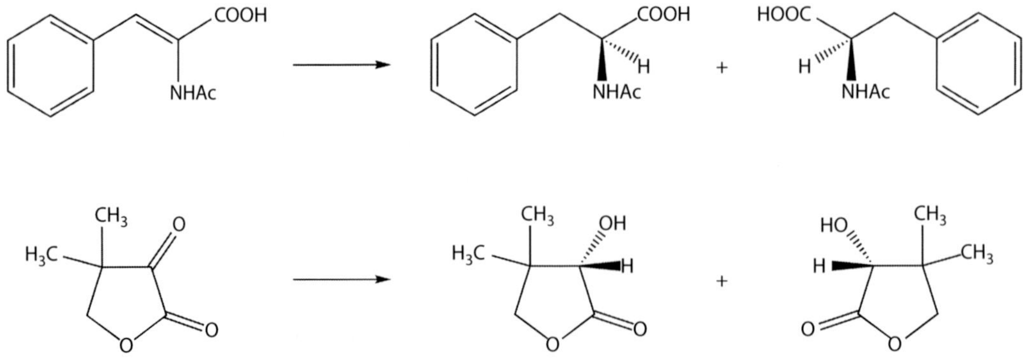

SCHEME 22.3 Hydrogenation of acetamidocinnamic acid and keto-pantolactone.

the carborane moiety is the only source of chirality and is responsible for chiral induction in the enantioselective hydrogenation reaction. With the (*R*,*R*)-DIOP ligand as second ligand at the rhodium atom, **31** has two more stereocenters that could give chiral induction.

In the asymmetric hydrogenation of acetamidocinnamic acid enantiomerically pure (*R*)-**30** gave the L isomer with 60–62% ee, whereas (*S*)-**30** led to the D isomer with a slightly reduced ee of 55–62%, resulting from the enantiomerically less pure (*S*)-**30** complex. The complex (*R*,*R*,*R*)-**31** with the *R*-configured carborane moiety in addition to the (*R*,*R*)-DIOP ligand gave the D isomer. By using the other enantiomer of the carborane cluster the D-configured product was also obtained. This result could be explained either through the dominance of the (*R*,*R*)-DIOP ligand over the carborane moiety or by the dissociation of the carborane moiety from the (*R*,*R*)-DIOP complex. Dissociation might be expected, as binding sites for the substrate and hydrogen are needed during catalysis. However, complete dissociation should give similar chiral inductions and therefore identical enantiomeric excesses for the reaction.

The hydrogenation of (*Z*)-α-*N*-acetamidocinnamic acid in ethanol–toluene with complexes **30** and **31** led to complete conversion. It was observed that the ratio of the two solvents plays an important role in the obtained enantiomeric excess of the reaction (Table 22.9). For **30**, an increase in the ee was observed with decreasing ethanol content, whereas for **31** the reverse behavior was noted. For (*R*,*R*,*R*)-**31** complex an ee of ca. 80% was obtained at an ethanol–toluene ratio of 2/1. This value is comparable with literature values for the rhodium–DIOP catalyst, and together with the solvent-dependent behavior of **31**, it is in agreement with the assumed dissociation mentioned above [44].

It was necessary to work in a mixture of solvents, as the complex is sparingly soluble in ethanol, and the acid insoluble in toluene. No conversion was observed when methanol was used as an alternative. Whilst an orange solution remained after carrying out the hydrogenation reaction with complex **31**, the black solution formed on performing the same reaction with both enantiomers of **30** indicated decomposition of the complexes.

The asymmetric hydrogenation of keto-pantolactone catalyzed by the rhodium complexes **30** and **31** resulted in complete conversion to (*R*)- and (*S*)-pantolactone (Scheme 22.3, Table 22.10). (*R*)-**30** gave (*R*)-pantolactone with 23% ee, whereas the less enantiomerically pure (*S*)-**30** afforded 16.3% ee for (*S*)-pantolactone. The (*R*,*R*-*R*)-**31** enantiomer containing the (*R*,*R*)-DIOP and (*R*)-carborane ligands and corresponding (*S*,*S*-*S*)-configured enantiomer were also tested in this reaction. These systems gave enantioselectivities between 59% and 66% for the *R* and *S* enantiomer, respectively. [{Rh(μ-Cl)(*R*,*R*)-DIOP}] was used as the standard complex and yielded 54.4% (*R*)-pantolactone, which supports (partial) dissociation of the carborane ligand (see Figure 22.17).

TABLE 22.9
Performance of the Complexes 30 and 31 in the Hydrogenation of (*Z*)-α-*N*-Acetamidocinnamic Acid

	Complex[a]		
Solvent Mixture	**10 mL Ethanol/5 mL Toluene**	**5 mL Ethanol/5 mL Toluene**	**5 mL Ethanol/10 mL Toluene**
(*R*)-**30**	12.0–12.7 (L)	48.1–50.0 (L)	60.3–62.3 (L)
(*S*)-**30**	19.5 (D)	—	54.8–62.3 (D)
(*R*,*R*-*R*)-**31**	76.1–79.6 (D)	80.5 (D)	60.2–63.0 (D)
(*R*,*R*-*S*)-**31**	57.1–62.6 (D)	—	50.0–56.6 (D)

Reaction conditions: Ratio catalyst/substrate = 1/100; substrate quantity 200 mg (0.975 mmol); $p = 5$ MPa H_2; $T = 65°C$, $t = 48$ h. Quantitative hydrogenation according to ^1H NMR spectra. ee determined by GC on a Chirasil-L-Val column.

[a] Chirality is given in the following order: chirality of bisphosphine ligand followed by planar chirality of carborane cluster.

TABLE 22.10
Performance of Complexes 30 and 31 in the Hydrogenation of Keto-Pantolactone

Complex	ee (%)
(*R*)-**30**	22.5, 23.0 (*R*)
(*S*)-**30**	16.3 (*S*)
(*R*,*R*-*R*)-**31**	63.6, 58.7, 63.4 (*R*)
(*S*,*S*-*S*)-**31**	58.9, 61.1, 66.6 (*S*)

Reaction conditions: Ratio catalyst/substrate = 1/200; keto-pantolactone quantity 258 mg (2.01 mmol); $p = 5$ MPa H_2; $T = 50°C$; $t = 48$ h. Quantitative reduction and enantiomeric excess determined by GC on a Chirasil-DEX-CB column.

Folic acid

[{Rh(μ-Cl)(cod)}$_2$]
L
50 bar H_2

5, 6, 7, 8-Tetrahydrofolic acid

DMSO/pyridine = 5/1
$MeCO_2H$, HCO_2H

5-Formyltetrahydrofolic acid

SCHEME 22.4 Hydrogenation of folic acid and further formylation to 5-formyltetrahydrofolic acid.

TABLE 22.11
Conversion and Diastereomeric Excesses of the Hydrogenation of Folic Acid to 5,6,7,8-Tetrahydrofolic Acid

Ligand	Yield (%)	de (%)[a]
(*R*)-**30**	93, 99	23.5, 26.4 (*S*)
(*S*)-**30**	99, 100	4.7, 5.4 (*R*)
(−)-DIOP	92–99	14.1–15.5 (*S*)
(+)-DIOP	95, 97	10.2, 12.3 (*R*)
(−)-DUPHOS	95, 96	9.6, 11.4 (*S*)
(+)-DUPHOS	94, 97	0.7, 1.2 (*R*)

Reaction conditions: 0.014 mmol **30** or 0.014 mmol [{Rh(μ-Cl)(cod)}$_2$] and 0.033 mmol bidentate ligand in CH_2Cl_2 and 700 mg silica gel, 0.566 mmol folic acid, 25 mL phosphate buffer (c = 0.067 M, pH = 7.0), p = 5 MPa, T = 80°C, t = 24 h.

[a] The mentioned chiral center is the 6-position of the molecule; the other stereocenter remained unchanged.

The same research group used complex **30** in the heterogeneous asymmetric hydrogenation of folic acid [45]. For immobilization, the free ligand was treated with [{Rh(μ-Cl)(cod)}$_2$] and silica gel in dichloromethane. Completion of the immobilization was confirmed by decolourization of the yellow solution. The immobilized complex is insoluble in the aqueous solution in which hydrogenation is performed. As the formed 5,6,7,8-tetrahydrofolic acid is light- and air sensitive, it was further formylated with methyl formate to obtain 5-formyltetrahydrofolic acid (Scheme 22.4). (*R*)-**30**, which can be heterogenized on silica gel, gave conversions between 93% and 99% and a diastereomeric excess of 24–26% for the *R* diastereomer. The *S* enantiomer of the complex fully converted the substrate, but only yielded 5% de (Table 22.11). From these observations it was concluded that the substrate has a matched combination when complex (*R*)-**30** is used as the catalyst, whilst use of the complex (*S*)-**30** led to a substrate–catalyst mismatched situation. The complexes are in the same range as DIOP and DUPHOS in terms of activity and selectivity. The matched–mismatched situation was also observed with these ligands, where the (−)-enantiomer yielded a higher de (matched situation) than the (+) enantiomer of the ligand (Table 22.11).

22.3.2 Hydroformylation

Hydroformylation reactions play a major role in chemical processes to obtain aldehydes from olefins [46]. Aldehyde groups are among the most versatile functional groups in organic synthesis and can be transformed into alcohols, amines, imines, and acids [47]. High temperatures are required for commercial hydroformylation reactions, as these processes proceed at a slower rate than hydrogenation reactions [48].

Hart and Owen [33] first investigated the hydroformylation of simple olefins with **32**, which was formed by the reaction of 1,2-bis(diphenylphosphino)-*closo*-carborane(12) as ligand and [RhCl(PPh$_3$)$_3$] as metal precursor (Figure 22.14). The best results were obtained in a 12 h reaction at 55°C and 4.5 MPa syngas pressure (Table 22.12). At higher temperatures, the yields were much lower (30%) and formation of condensation products was observed. Unfortunately, a lack of selectivity in the carbonylation was observed. The same effect was also observed with 1,2-bis(diphenylphosphino)-ethane (dppe) as chelating ligand.

Hey-Hawkins and coworkers [49] investigated the hydroformylation of allyl cyanide and styrene using the ferrocenyl-based phosphinitocarborane(12)s **33** and **34** (Figure 22.14, Scheme 22.5). Compound **33** is a bidentate P,N ligand bearing one phosphorus atom at the carborane moiety, whereas **34** could act as a bi- (P,P) or tridentate (P,P,N) chelating ligand. The catalytically active metal complexes were formed *in situ* by using [Rh(acac)(CO)$_2$] as the late transition metal precursor. It was shown that both ligands are active in the hydroformylation and no hydrogenation by-products were observed (Table 22.13). The regioselectivity in the hydroformylation of allyl cyanide with **33** is moderate, whereas with **34** it is very good compared to known ligands. Unfortunately, no stereoselectivity was observed and the obtained products were almost racemic. Similar results were obtained for the hydroformylation of styrene. As the branched product racemizes easily, it was converted into the corresponding imine by treatment with (*R*)-1-phenylethylamine. The results for both hydroformylation reactions are comparable with those in the literature for activity and regioselectivity [50]. In comparison to the hydroformylation of allyl cyanide, ligand **33** gave better results, showing that asymmetric hydroformylation is highly dependent on the substrate.

A second class of substances that was investigated by Hey-Hawkins and coworkers [35] were the 1,2-bis(phosphonito)-*closo*-carborane(12)s (Figure 22.15). The ligands used were sterically demanding, and ligands **7** and **8** exhibited chirality in the carbon backbone and/or at the phosphorus center. The performance of the ligands was investigated with dimethyl itaconate. The catalytically active species was formed *in situ* by treating the ligand with [Rh(acac)(CO)$_2$] as catalyst precursor. Ligand **35** was the only one that gave a reasonable activity with good regioselectivity, with a branched–linear ratio of 1/10. Ligand **7** and both enantiomers of **8** yielded no conversion to the aldehyde. Similar results were obtained with other olefin substrates (Table 22.14). Formation of 21–29% hydrogenation by-products lowered the yield. An interaction between the aromatic system in the ligand and additional π systems in the substrate and its influence on the regioselectivity was discussed [35].

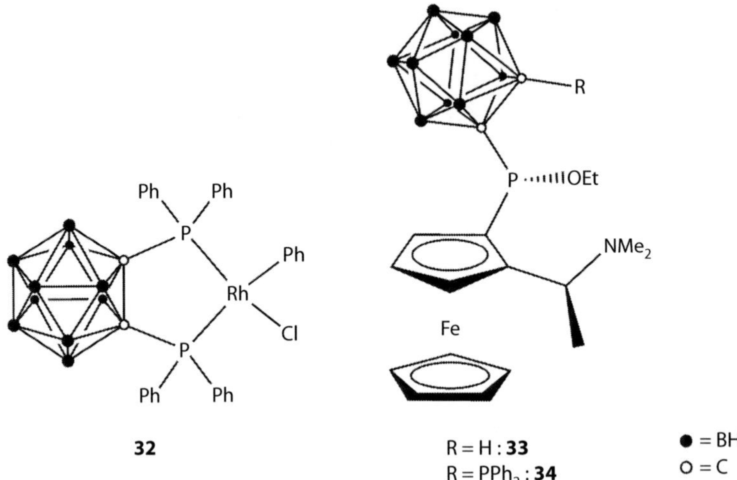

FIGURE 22.14 Substances **32–34** used in hydroformylation.

TABLE 22.12
Hydroformylation of Olefins Using 32

Substrate	Product, Yield (%)	
Hex-1-ene	2-Methylhexanal, 46%	*n*-Heptanal, 53%
Pent-1-ene	2-Methylpentanal, 43%	*n*-Hexanal, 57%
trans-Pent-2-ene	2-Methylpentanal, 44%	2-Ethylbutanal, 44%

Reaction conditions: 30 mg **32**; 8 mmol substrate; 20 mL benzene; $p = 4.5$ MPa H_2 and 4.5 MPa CO; $T = 55°C$; $t = 12$ h.

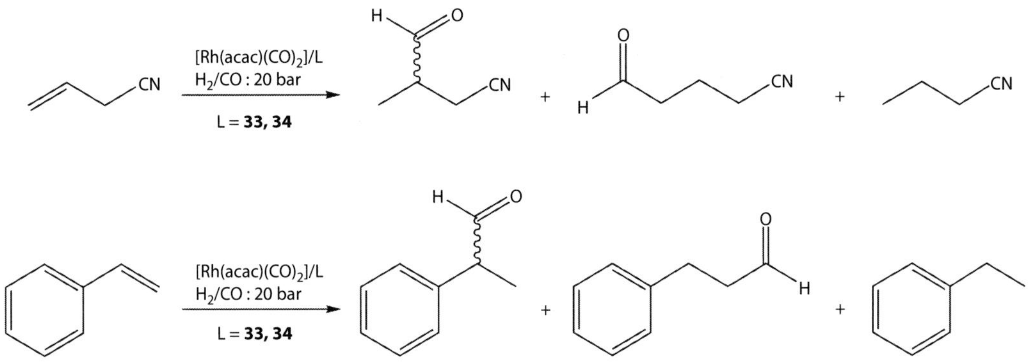

SCHEME 22.5 Hydroformylation of allyl cyanide and styrene.

TABLE 22.13
Hydroformylation of Allyl Cyanide and Styrene Using 33 and 34

Substrate	Ligand	Conversion (%)	Branched/Linear Ratio	ee (%)
Allyl cyanide	**33**	>99	3.9/1	0
	34	>99	17/1	8
Styrene	**33**	>99	16/1	17
	34	>99	12.5/1	8

Reaction conditions: Ratio [Rh(acac)$(CO)_2$]/ligand/substrate = 0.5/0.6/100; syngas CO/H_2 = 1/1; solvent: toluene; p = 2 MPa; T = 50°C; t = 12 h.

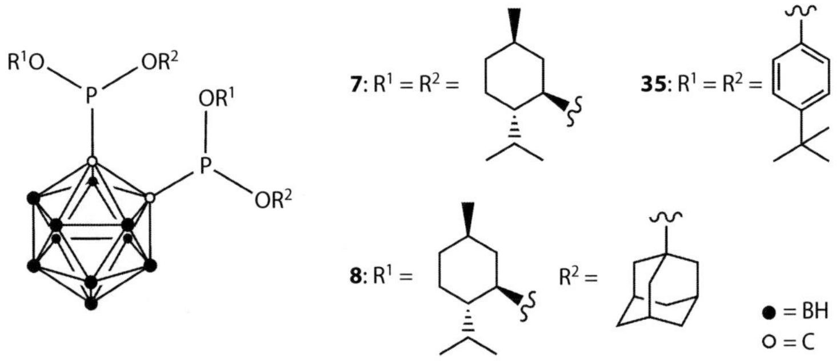

FIGURE 22.15 Bis(phosphonito)-*closo*-carboranes investigated in hydroformylation.

22.3.3 Hydrosilylation

Hydrosilanes undergo addition to carbon–carbon multiple bonds under catalysis by transition metal complexes. Nickel, rhodium, palladium, and platinum were used as catalytically active metals. By incorporating chiral ligands into the metal catalyst, the hydrosilylation can be performed analogously to other addition reactions with double bonds, for example, asymmetric hydrogenation to obtain optically active alkylsilanes.

Tamao and co-workers [51] performed a series of hydrosilylation reactions with several phosphine–nickel(II) complexes, in which the reactivity of 1,2-bis(dimethylphosphino)-1,2-dicarba-*closo*-dodecaborane as ligand was investigated (Figure 22.16). The hydrosilylation of simple sub-

TABLE 22.14
Hydroformylation of Olefins Using Rhodium Complexes of Ligands 35

Substrate	Conversion (Hydrogenation) (%)	Branched/linear Ratio (%)
MeO_2C, CO_2Me	100 (29)	1/10
Ph, CO_2Me	100 (29)	1/5
n-Bu, CO_2Me	98 (22)	3/1
N, CO_2Me	92 (26)	1/100
Ph, CO_2Me	100 (28)	1/27
Ph, CO_2Me	100 (21)	1/25

Reaction conditions: 0.2% [Rh(acac)$(CO)_2$]; 0.25% ligand **35**; solvent: toluene; syngas $H_2/CO = 1/1$; $p = 5$ MPa; $T = 100°C$; $t = 24h$.

strates with methyldichlorosilane led to yields of up to 98% depending on the substrate. The internal–terminal product ratio was virtually constant for all substrates at approximately 30/60 (Scheme 22.6, Table 22.15). Using other silicon hydrides such as trichlorosilane or dichlorophenylsilane gave a larger amount of the terminal product (Table 22.16). Remarkably, internal olefins lead also to terminal alkylsilanes, which shows that isomerization occurs during the reaction.

It was observed that Si–H/Si–Cl exchange sometimes occurred in the hydrosilylation of simple terminal olefins using tertiary phosphine–nickel complexes, depending on the electronic nature of the substrate. Nickel complexes containing electron-accepting ligands suppressed this reaction and therefore only traces of interchange products were observed. On the other hand, a tendency for

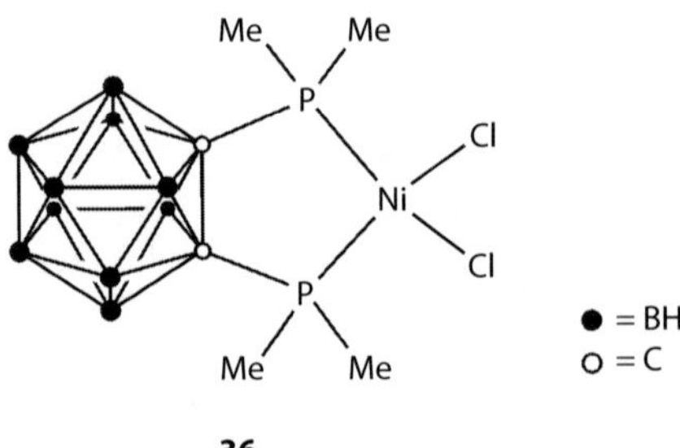

FIGURE 22.16 Nickel(II) complex of 1,2-bis(dimethylphosphino)-1,2-dicarba-*closo*-dodecaborane (**36**).

SCHEME 22.6 Hydrosilylation of olefins.

nickel–phosphine complexes was observed to give internal adduct, albeit only in traces. It seems that the carborane ligand enhances this tendency, presumably due to the strongly electron-accepting character.

A proposed mechanism for hydrosilylation catalyzed by nickel–phosphine complexes, based on the Chalk–Harrod mechanism for platinum catalysis, incorporates an equilibrium between a σ and a π complex. It is possible that four intermediates are formed: a terminal or inner σ or π complex (Scheme 22.7). It was observed that the internal–terminal ratio was consistent over a reaction period of 45 h, which suggests a similar stability of the formed inner and terminal σ complexes (Table 22.17). In these σ complexes the involved organic group can be regarded as carbanionic in

TABLE 22.15
Hydrosilylation of Various Olefins Using 36

Olefin	Yield (%)	Internal/Terminal Ratio
1-Octene[a]	59	38/62
1-Octene[b]	97	36/64
1-Hexene[a]	98	36/64
1-Pentene[a]	84	33/67
Propylene[a]	40	41/59
2-Hexene[a]	38	80/20
2-Pentene[a]	47	51/49
Isobutene[c]	20	0/100[d]
Styrene[c]	54	94/6[d]
Vinyltrimethylsilane[c]	100	0/100[d]
Vinyldichloromethylsilane[c]	19	13/87[d]

[a] Reaction conditions: $HSiMeCl_2$/olefin = 1; **36**/olefin ≈ 10^{-3}; $T = 120°C$; $t = 20$ h.
[b] Reaction conditions: $HSiMeCl_2$/olefin = 2; **36**/olefin ≈ 10^{-3}; $T = 120°C$; $t = 20$ h.
[c] Reaction conditions: $HSiMeCl_2$/olefin = 1; **36**/olefin ≈ 10^{-3}; $T = 120°C$; $t = 40$ h.
[d] α/β ratio.
Yield and ratio determined by GLC.

TABLE 22.16
Hydrosilylation of 1-Octene Using Different Silyl Hydrides

Silylhydride	Yield (%)	Internal/Terminal Ratio
$HSiMeCl_2$	59	38/62
$HSiPhCl_2$	100	35/65
$HSiCl_3$	100	15/85

Reaction conditions: Silane/1-octene = 1; **36**/olefin ≈ 10^{-3}; $T = 120°C$; $t = 20$ h; yield and ratio determined by GLC.

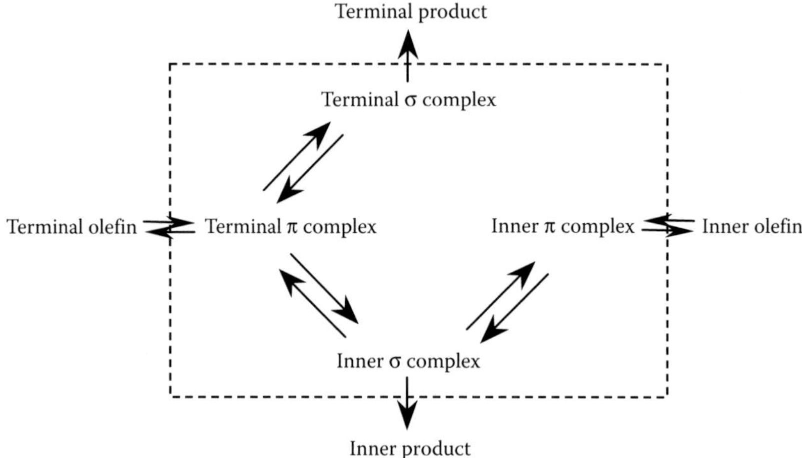

SCHEME 22.7 Proposed equilibrium step between the four intermediates.

nature. The primary and secondary carbanion should have a different stability, wherein the equalization is brought about by the strong electron-accepting nature of the carborane backbone. This can only be applied to simple olefins, as substituent effects overcome this leveling effect in more complex olefins. The hydrosilylation of isobutene led to no internal product, whereas in vinylsilane derivatives the ratio increases proportionally to the inductive effect of the silyl groups (Table 22.15).

It is noted that using 1,2-bis(diphenylphosphino)-*ortho*-carborane(12) (**1**) as the ligand in the reaction of 1-octene with methylchlorosilane led solely to the terminal product in a 3–5% yield. This is a very low catalytic activity in comparison with the use of triphenylphosphine as the ligand.

Other than the previously shown hydrogenation, Brunner et al. [43] investigated the hydrosilylation of acetophenone with diphenylsilane and enantiomerically pure, planar-chiral, anionic 7-diphenylphosphanyl-8-phenyl-7,8-dicarba-*nido*-undecaborate rhodium complexes **30** and **31** (Figure 22.17, Scheme 22.8). Resolution of anionic ligand **37** was achieved via a diastereomeric palladium complex that led to the enantiomerically pure form of the ligand. The precatalyst **30** was obtained by reaction of **37** with $[\{Rh(\mu\text{-}Cl)(cod)\}_2]$, whereas both enantiomers of **31** were obtained by the reaction of (*R*,*R*)- or (*S*,*S*)-DIOP, $[\{Rh(\mu\text{-}Cl)(cod)\}_2]$ and **37**.

TABLE 22.17
Isomerization of Olefins during Hydrosilylation

Starting Olefin	Time (h)	Residual Olefin 1-Pentene/2-Pentene	Yield (%)	Internal/Terminal Ratio
1-Pentene	7	33/67	78	34/66
	12	12/88	84	34/66
	17	14/86	87	33/67
	40	12/88	84	34/66
2-Pentene	20	6/94	36	50/50
	40	5/95	47	49/51
	45	11/89	54	49/51

Reaction conditions: $HSiMeCl_2$/olefin = 1; **36**/olefin ≈ 10^{-3}; $T = 120°C$; yield and ratio determined by GLC.

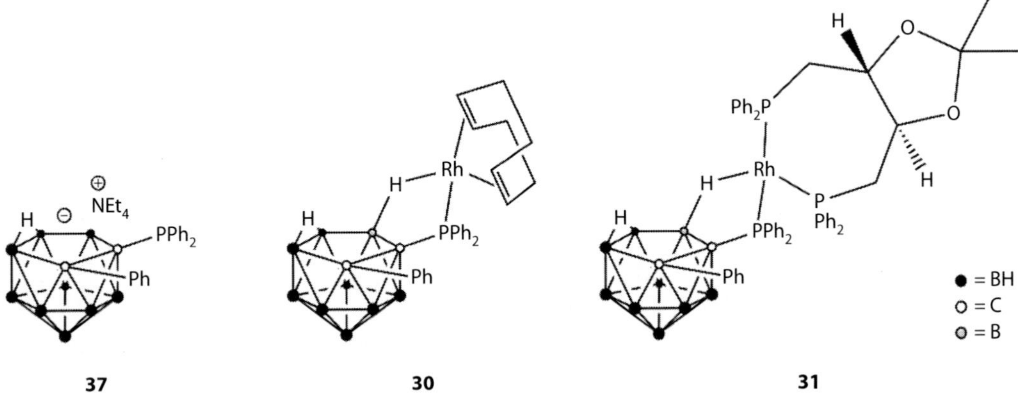

FIGURE 22.17 Ligand and rhodium complexes used in hydrosilylation reactions.

SCHEME 22.8 Hydrosilylation of acetophenone.

The hydrosilylation of acetophenone with diphenylsilane led to good conversions of up to 92% (Table 22.18). Using (*R*)-**30** only poor enantioselectivities of 7–9% were achieved, whereas the diastereomers (*R*,*R*-*S*)-**31** or (*S*,*S*-*S*)-**31** led to an enantiomeric excess of 26–28%. Unfortunately, the reaction of [{Rh(μ-Cl)(*R*,*R*)-DIOP}$_2$] under the same reaction conditions led to the same ee. This suggests dissociation of the carborane ligand from the rhodium atom and that an Rh-(*R*,*R*)-DIOP species is the enantioselective catalyst (see further p. 529).

22.3.4 Carbonylation

Kang and co-workers [52] synthesized P,S-chelating ligands that were used to form dimeric complexes using rhodium and iridium. The obtained sulphur-bridged complexes were further treated

TABLE 22.18
Hydrosilylation of Acetophenone with Diphenylsilane

			1-Phenylethanol	
Complex	**Conversion (%)**	**Enol Ether (%)**	**Yield (%)**	**ee (%)**
(*R*)-**30**	75; 83	0; 5	75; 79	8.9; 6.7 (*R*)
(*R*,*R*-*S*)-**31**	92; 92; 37	16; 9; 0	76; 83; 37	26.0; 26.3; 27.7 (*R*)
(*S*,*S*-*S*)-**31**	72; 80; 89	17; 45; 48	62; 59; 52	(20.7); 25.7; 27.5 (*S*)

Reaction conditions: Ratio catalyst/acetophenone/diphenylsilane = 1/400/400, room temperature, t = 2 d; conversion content of enol ether and silyl ether measured by ^{1}H NMR spectroscopy; enantiomeric excess determined by GC on a Chirasil-DEX-CB column.

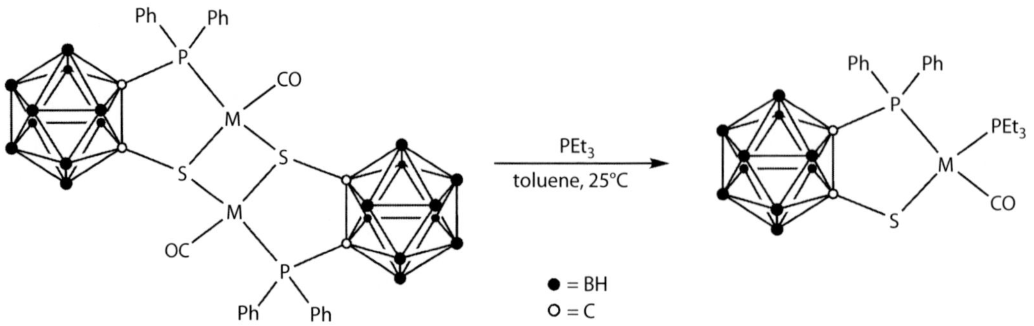

SCHEME 22.9 Metal complexes used in the carbonylation of methanol.

with triethylphosphine to obtain the monomeric complex (Scheme 22.9). These metal complexes were then tested for catalytic activity in the carbonylation of methanol [52].

This group demonstrated that both dinuclear complexes **38** and **40** exhibit a remarkably better activity than the mononuclear complexes **39** and **41** at equivalent metal–catalyst concentrations (Table 22.19). The reaction was evaluated/tested against the Monsanto catalyst $[RhI_2(CO)_2]^-$ used in industrial processes, which is formed *in situ* from $[\{RhCl(CO)_2\}_2]$. In the carbonylation of methanol, **38** outperformed the Monsanto catalyst, whereas **40** demonstrated a similar turnover number. The exceptionally good reactivity could be explained by the formation of thermally stable metal chelate complexes, due to the *ortho*-carborane(12) backbone. ^{31}P NMR spectra of the residue at the end of the reaction showed a high content of phosphine bound to rhodium with a retained bimetallic structure, indicated by the chemical shifts and coupling constants.

Cheong and Ziegler [53] investigated the oxidative addition step in the carbonylation with density functional theory (DFT) methods. This is the rate-determining step in the catalytic cycle, and reduction of the energy barrier will lead to higher catalytic activity. In their calculations they incorporated rhodium or iridium as the catalytically active metal in $[MI_2(CO_2)]^-$ in the first step. In the second step one carbonyl ligand and one iodo ligand were substituted by two (*trans*) PEt_3 ligands, dppe, dppms, and the carborane-containing dppcs (see Figure 22.18). With these ligands, it was possible to compare the calculated thermodynamic results with the kinetic experimental ones. It was shown that solvent effects could be used to cleave the iodine–methyl bond, which allowed a more practical correlation between the calculated model and the real system.

It could be shown that the oxidative addition reaction is exergonic for iridium and endergonic for rhodium. The calculated free energy of activation ($\Delta G^{\ddagger}$) increases from iridium to rhodium (19.3 and 26.9 kcal mol^{-1}, respectively). These values are in good agreement with experimental data

TABLE 22.19
Carbonylation of Methanol

Catalyst	Max. Turnover (h^{-1})	Catalyst	Max. Turnover (h^{-1})
38	1280	**39**	20
40	790	**41**	90
Monsanto $[RhI_2(CO)_2]^-$	590		

Reaction conditions: 9.0 g CH_3OH; 20.0 g CH_3COOH; 2.0 g CH_3I; catalyst load: 0.025 mmol for **38** and **40** or 0.05 mmol for **39** and **41**; $p = 4.8$ MPa; $T = 180°C$.

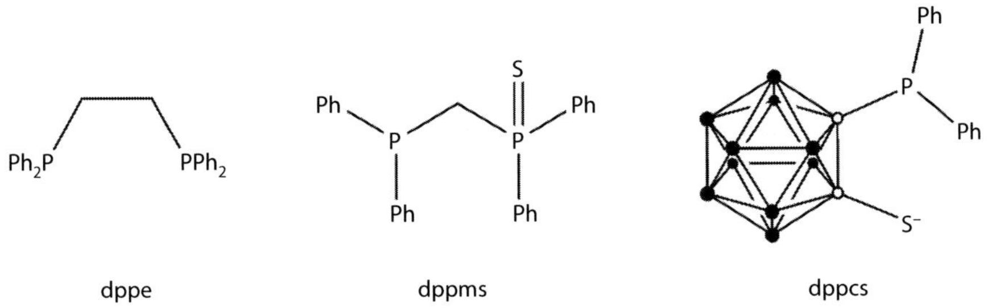

FIGURE 22.18 Chosen ligands to compare calculated thermodynamic results with experimental results.

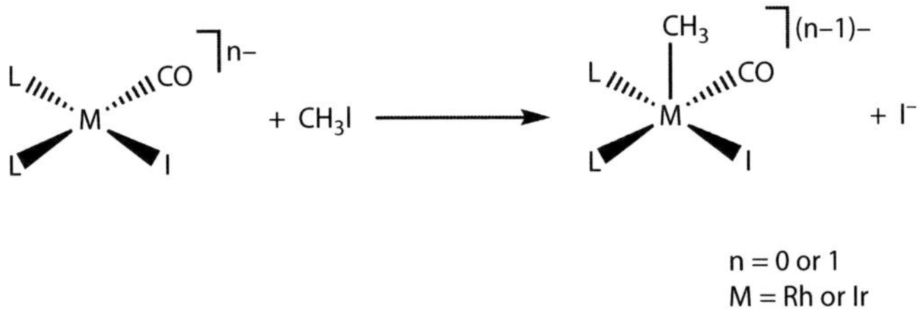

SCHEME 22.10 Nucleophilic attack of methyl iodide on $[MI(CO)L_2]$ in the oxidative addition.

(Table 22.20). This trend reflects the relativistic effects that make metal–CH_3 bonds stronger for heavier elements. The agreement between experiment and theory for the individual contributions $\Delta H^‡$ and $\Delta S^‡$ to $\Delta G^‡ = \Delta H^‡ – T\Delta S^‡$ is not as good as for $\Delta G^‡$ itself. The difference can be interpreted in terms of the strong dependence of both values on the solvent. Experimental data showed that changing the solvent had an influence on $\Delta H^‡$ and $\Delta S^‡$ but no significant effect on $\Delta G^‡$.

In the oxidative addition step of the reaction, the rate-determining step is nucleophilic attack on CH_3I by $[MI(CO)L_2]$ in an S_N2 reaction (Scheme 22.10), in which the most direct influence on the energy barrier comes from the ligands, with their ability to increase the nucleophilicity and electron density of the metal complex. This also facilitates attack on methyl iodide. For $L_2 = (CO)I^-$,

TABLE 22.20
Calculated and Experimental Reaction Activation Parameters ($\Delta S^‡$, $\Delta H^‡$, $\Delta G^‡$) for the Oxidative Addition of Methyl Iodide to $[MI(CO)L_2]^{n-}$ and Computed Reaction Thermodynamics (ΔH, ΔG) for the Formation of $[MI(CH_3)(CO)L_2]^{(n-1)-}$ from $[MI(CO)L_2]^{n-}$

Metal	Ligand L_2	$\Delta S^‡$ (calcd)	$\Delta H^‡$ (calcd)	$\Delta G^‡$ (calcd)	$\Delta S^‡$ (exptl)	$\Delta H^‡$ (exptl)	$\Delta G^‡$ (exptl)	ΔH (calcd)	ΔG (calcd)
Rh	$I(CO)^-$	–43.9	13.8	26.9	–28.7 (m)	14.3 (m)	22.9 (m)	7.5	11.8
	trans-$(PEt_3)_2$							1.8	2.5
	Dppe				–39.91 (d)	9.56 (d)	21.46 (d)	4.3	8.7
	Dppms				–34.42 (d)	11.23 (d)	21.49 (d)	1.3	5.7
	Dppcs							–2.1	1.3
Ir	$I(CO)^-$	–44.6	6.0	19.3	–27.0 (d)	12.9 (d)	20.9 (d)	2.0	5.6
	$(CO)_2$							19.8	24.0
	(CO)(solvent)							9.5	13.5

Note: All values in kcal mol^{-1} and obtained in solution; solvent in theoretical calculations methanol and in experiments methanol (m) or dichloromethane (d).

the difference of ΔG between rodium and iridium is 6.2 kcal mol^{-1} and is comparable with the calculated difference of $\Delta G^{\#}$ between the two metals in the same process. [MI(CO)$_3$] itself is known to be unreactive ($\Delta G = 24$ kcal mol^{-1} for M = Ir), because the three carbonyl ligands make the metal atom electron-poor and therefore unable to exhibit nucleophilic behavior. [MI(CO)$_3$] is often found as the major species in reactions with solvents containing traces of water and iodide. This could be the reason for the observed low activity of the complexes. On the other hand, the solvent-stabilized molecule [MI(CO)$_2$(solvent)], formed by dissociation of the starting compound under low CO pressure, gave the second highest ΔG value for the nucleophilic attack of 13.5 kcal mol^{-1} for M = Ir. This molecule is, therefore, capable of rather slow oxidative addition reactions. All investigated ligands in [RhI(CO)L$_2$] result in lower reaction enthalpies (ΔH) and lower Gibbs free energies (ΔG) with respect to the nucleophilic attack of CH_3I in solution compared to [RhI$_2$(CO)$_2$]$^-$. This can be explained by the greater donor ability of the phosphine ligands and is in agreement with experimental data. Interestingly, the carborane-containing ligand dppcs (Figure 22.18) coordinated to rhodium has the lowest reaction enthalpy (ΔH) and free energy of reaction (ΔG) in solution.

22.3.5 Amination

To perform asymmetric amination reactions the transformation of (*E*)-1,3-diphenylallyl acetate with pyrrolidine or di-*n*-propylamine was investigated by Lyubimov et al. [54] This reaction may serve as a model reaction for metal-catalyzed allylation reactions, which involve the attack of nucleophiles at an allylic metal intermediate (Scheme 22.11). C-, H-, N-, O-, or S-centerded substrates can be employed as nucleophiles, which leads to a high flexibility in the type of bonds that can be created. Thus, this reaction leads to a high degree of asymmetric induction and is tolerant to a wide range of functional groups [55].

The chiral carboranylphosphite ligands **42–47** were tested, where **42** and **44** carry (*S*)-BINOL, **43** and **45** 1,4:3,6-dianhydro-D-mannitol and **46** and **47** a diamidophosphite as backbone (Figure 22.19) [54]. As a benchmark reaction the allylic amination of 1,3-diphenylallyl acetate with pyrrolidine was performed at room temperature with a [{Pd(allyl)(μ-Cl)}$_2$] precatalyst in THF or dichloromethane, with a Pd–ligand ratio of 1/1 or 1/2 (Table 22.21). Ligand **42** showed excellent conversion by producing a nearly racemic product. **43** led to moderate conversions with a maximum enantiomeric excess of 73% of the *R*-configured product. An inverse behavior was observed using **44** and **45**. Here the (*S*)-BINOL carrying ligand **44** led to moderate yields with an ee of 83%, whereas ligand **45** yielded high conversions to the racemic product. Overall, the best results were achieved by using dichloromethane as solvent and a Pd–ligand ratio of 1/2. In the case of the carboranyl diamidophosphite ligands **46** and **47** the 9-carboranyl substituted ligand gave full conversion and up to 77% ee for the *R* enantiomer. For ligand **47** a dramatic drop in conversion is observed, whereas the enantiomeric excess remains nearly unchanged. To compare these results, the allylic amination of 1,3-diphenylallyl acetate with di-*n*-propylamine was investigated also using ligands **42–47**. This reaction followed the same trend under the optimized conditions given above. Ligand **46** led to full conversion and the highest ee values. Interestingly, the diamidophosphite ligands gave the opposite product to the ligands with the diol backbone.

SCHEME 22.11 Palladium-catalyzed allylic amination of 1,3-diphenylallyl acetate.

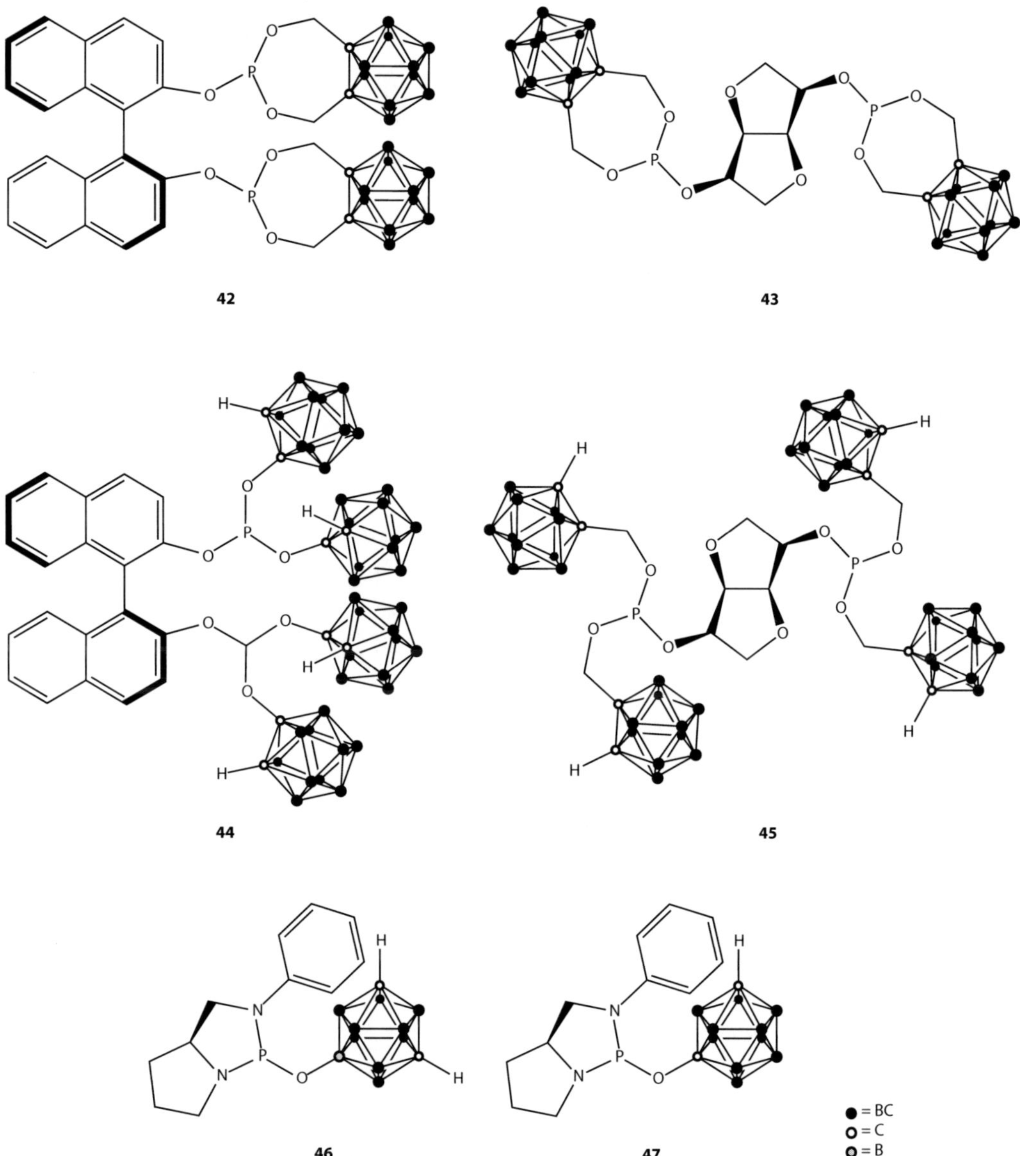

FIGURE 22.19 Ligands investigated for allylic amination.

Copper(I)-catalyzed amination of iodobenzene with aniline in the presence of KO-*t*-Bu as base was described by Dou and co-workers [56] (Scheme 22.12). They reported the use of the $[Cu_2(\mu\text{-}I)_2(L)_2]$ (L = 1,2-bis(diphenylphosphino)-*closo*-carborane(12) (**1**)) catalyst, which showed a conversion of the amine of 58%, compared to no significant conversion without ligand **1**. The conversion is in the range of the reaction of copper(I) iodide in the presence of 2,2′-dithiobis(5-nitropyridine) but lower than that of the 2,2′-bipyridine ligand [57]. It was postulated that the reaction proceeds via an insertion of the copper(I) species into the carbon–halogen bond as the initial step.

22.3.6 Alkylation and Sulfonylation

As the previous shown amination reaction, Lyubimov et al. [54b,58] investigated the palladium-catalyzed allylic alkylation with dimethyl malonate and allylic sulfonylation of (*E*)-1,3-diphenylallyl

TABLE 22.21
Allylic Amination of 1,3-Diphenylallyl Acetate with Pyrrolidine or Di-*n*-Propylamine

Substrate	Ligand	Pd/L Ratio	Solvent	Conversion (%)	ee (%)
Pyrrolidine	**42**	1/1	THF	100	3 (*R*)
	42	1/2	THF	100	5 (*R*)
	42	1/1	CH_2Cl_2	100	5 (*R*)
	42	1/2	CH_2Cl_2	100	6 (*R*)
	43	1/1	THF	30	43 (*R*)
	43	1/2	THF	20	53 (*R*)
	43	1/1	CH_2Cl_2	45	20 (*R*)
	43	1/2	CH_2Cl_2	90	73 (*R*)
	44	1/1	THF	30	63 (*R*)
	44	1/2	THF	45	82 (*R*)
	44	1/1	CH_2Cl_2	50	35 (*R*)
	44	1/2	CH_2Cl_2	54	83 (*R*)
	45	1/1	THF	100	4 (*R*)
	45	1/2	THF	100	5 (*R*)
	45	1/1	CH_2Cl_2	100	6 (*R*)
	45	1/2	CH_2Cl_2	100	7 (*R*)
	46	1/1	THF	100	75 (*R*)
	46	1/2	THF	100	77 (*R*)
	47	1/1	THF	21	74 (*R*)
	47	1/2	THF	15	60 (*R*)
Di-*n*-propylamine	**42**	1/2	CH_2Cl_2	30	20 (–)
	43	1/2	CH_2Cl_2	60	41 (–)
	44	1/2	CH_2Cl_2	90	60 (–)
	45	1/2	CH_2Cl_2	64	41 (–)
	46	1/1	THF	95	90 (+)
	46	1/2	THF	100	83 (+)
	47	1/1	THF	5	44 (+)
	47	1/2	THF	8	55 (+)

acetate (Scheme 22.13). Monodentate carboranyl diamidophosphites **46–48** were used as ligands in this reaction (Figure 22.20), where ligand **48** is an *ortho*-carborane and **46** and **47** are *meta*-carboranyl species. The catalytically active palladium complex was either prepared *in situ*, which led to a neutral species, or in the case of **48** also previously synthesized by treatment of the ligand with $[\{Pd(allyl)(\mu\text{-}Cl)\}_2]$ in the presence of $AgBF_4$ to form a cationic complex (Scheme 22.14).

SCHEME 22.12 Amination of iodobenzene with aniline.

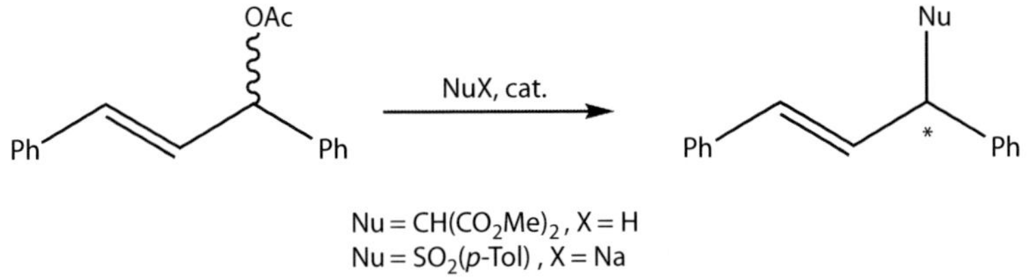

Nu = $CH(CO_2Me)_2$, X = H
Nu = SO_2(*p*-Tol), X = Na

SCHEME 22.13 Asymmetric allylic alkylation and allylic sulfonylation of (*E*)-1,3-diphenylallyl acetate.

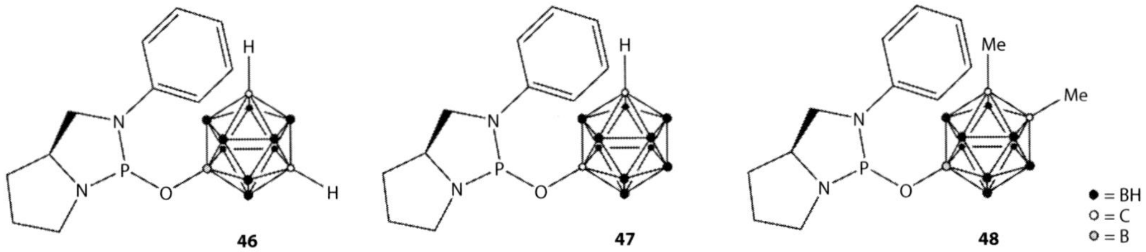

FIGURE 22.20 Ligands used in allylic alkylation and allylic sulfonylation.

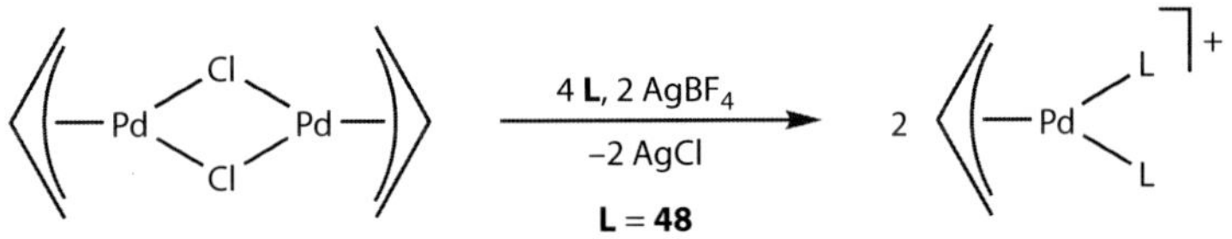

SCHEME 22.14 Synthesis of the palladium complex.

For allylic alkylation with dimethyl malonate, phase-transfer conditions and supercritical CO_2 ($scCO_2$) as solvent were applied besides standard reaction conditions with solvents such as CH_2Cl_2 or THF and BSA (*N*,*O*-bis(trimethylsilyl)acetamide) as base. Under standard reaction conditions, complete conversion and enantiomeric excesses between 90% and 98% were obtained for the electron-donating ligands (Table 22.22). These results were independent of the ligand–palladium ratio, and no difference was noted in the performance of complex **49** when it was formed *in situ* or preformed. The electron-withdrawing ligand **47** gave significantly lower conversions, which indicated the strong influence of the donor–acceptor properties of the ligand. Under phase-transfer conditions, only the preformed complex was investigated and the reaction time was halved. It was shown that the use of K_2CO_3 as the base and CH_2Cl_2 as the solvent gave comparable results in half the reaction time. Using THF as the solvent led to the same ee but only 56% conversion. The use of supercritical CO_2 under phase-transfer conditions with K_2CO_3 and 18-crown-6 led to moderate ee values and yields. The enantioselectivity could be increased by the use of Cs_2CO_3 without the crown ether, but it was not possible to reach the values of the standard reaction conditions.

Allylic sulfonylation with sodium *para*-toluenesulfinate was investigated with ligands **46** and **47**, where the catalytically active species was prepared *in situ* (Table 22.23). The same effect as in the previous allylic substitution reaction was observed. The electron-withdrawing ligand **47** led to a decrease in conversion and enantiomeric excess, whereas with the electron-donating ligand **46** full conversion and an ee of up to 92% were observed.

22.3.7 Kharasch Reaction

The addition of carbon tetrachloride to olefin double bonds catalyzed by peroxides as radical initiator or by transition metal complexes is known as Kharasch reaction [59]. The reaction proceeds in a

TABLE 22.22
Palladium-Catalyzed Allylic Alkylation of (*E*)-1,3-Diphenylallyl Acetate with Dimethyl Malonate

Catalyst	L/Pd Ratio	Solvent	Base	*t* (h)	Conversion (%)	ee (%)
46/[{Pd(allyl)(μ-Cl)}$_2$]	1/1	CH_2Cl_2	BSA, KOAc	48	98	94 (*S*)
46/[{Pd(allyl)(μ-Cl)}$_2$]	2/1	CH_2Cl_2	BSA, KOAc	48	97	98 (*S*)
47/[{Pd(allyl)(μ-Cl)}$_2$]	1/1	CH_2Cl_2	BSA, KOAc	48	38	70 (*S*)
47/[{Pd(allyl)(μ-Cl)}$_2$]	2/1	CH_2Cl_2	BSA, KOAc	48	49	92 (*S*)
48/[{Pd(allyl)(μ-Cl)}$_2$]	1/1	CH_2Cl_2	BSA	36	100	93 (*S*)
48/[{Pd(allyl)(μ-Cl)}$_2$]	1/1	THF	BSA	36	100	91 (*S*)
48/[{Pd(allyl)(μ-Cl)}$_2$]	2/1	CH_2Cl_2	BSA	36	100	95 (*S*)
48/[{Pd(allyl)(μ-Cl)}$_2$]	2/1	THF	BSA	36	100	92 (*S*)
49	2/1	CH_2Cl_2	BSA	36	100	92 (*S*)
49	2/1	THF	BSA	36	100	90 (*S*)
49	2/1	CH_2Cl_2	K_2CO_3[a]	18	100	91 (*S*)
49	2/1	THF	K_2CO_3[a]	18	56	90 (*S*)
49	2/1	$scCO_2$	K_2CO_3[b]	18	75	64 (*S*)
49	2/1	$scCO_2$	Cs_2CO_3	18	60	81 (*S*)

Reaction conditions: Room temperature; for supercritical carbon dioxide ($scCO_2$), $p = 17$ MPa, $T = 70°C$.

[a] Phase-transfer conditions: Bu_4NBr.

[b] Phase-transfer conditions: 18-crown-6.

TABLE 22.23
Palladium-Catalyzed Allylic Sulfonylation of (*E*)-1,3-diphenylallyl Acetate with Sodium *para*-Toluenesulfinate

Catalyst	L/Pd Ratio	Conversion (%)	ee (%)
46/[{Pd(allyl)(μ-Cl)}$_2$]	1/1	97	87 (*S*)
46/[{Pd(allyl)(μ-Cl)}$_2$]	2/1	98	92 (*S*)
47/[{Pd(allyl)(μ-Cl)}$_2$]	1/1	30	30 (*S*)
47/[{Pd(allyl)(μ-Cl)}$_2$]	2/1	26	46 (*S*)

Reaction conditions: Room temperature for 48 h, 2 mol% [{Pd(allyl)(μ-Cl)}$_2$].

R^2 R^1 + $CXCl_3$ → Cl R^2 R^1 $CXCl_2$

$R^1 = C_8H_{17}$; $R^2 = H$
$R^1 = Ph$; $R^2 = H$
$R^1 = CO_2n$-Bu; $R^2 = H$
$R^1 = CO_2Me$; $R^2 = CH_3$

X = H, Cl

SCHEME 22.15 Kharasch reaction.

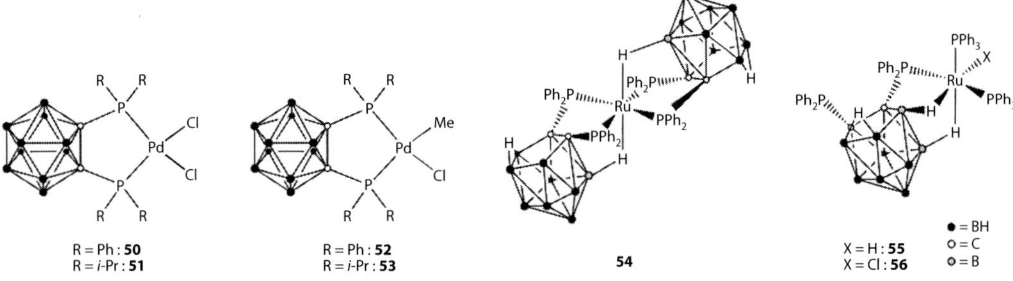

FIGURE 22.21 Metal complexes for the Kharasch reaction and radical polymerization of methyl methacrylate (Section 22.3.8).

classical anti-Markovnikov manner, whereby a free-radical mechanism is postulated (Scheme 22.15). This reaction can also be extended to olefin polymerization, which is the competing reaction to the addition reaction. Sawamoto and co-workers and Matyjaszewski et al. [60] found independently that in the presence of a halogenated compound an olefin yields polymers with highly controlled molecular weight and molecular weight distribution.

Ruthenium plays a major role in Kharasch chemistry by displaying the highest efficiency and versatility for halocarbon activation and addition to olefins. Formation of the 14-electron ruthenium species through the dissociation of the phosphine ligands takes place prior to halocarbon activation. Ligand dissociation is dependent on several factors, such as temperature and stability of the complex itself. Bulky phosphines were therefore used to protect the vacant coordination site by steric effects. In this case, agostic interactions between C–H groups and the metal atom prevent dimerization or solvent occupation of the vacant coordination site. Similar features have been observed in carborane-containing ruthenium complexes, with B–H → Ru interactions. Due to this observation, Demonceau and co-workers [61] investigated the Kharasch addition of carbon tetrachloride to olefins with the 14-electron ruthenium(II) complexes **54–56** (Figure 22.21, Scheme 22.15). It was shown that the catalytic results are highly dependent on the catalyst and substrate. For both acrylates a substrate conversion of almost 100% could be obtained, whereby complex **54** gave higher conversion to the Kharasch addition product. An inverse behavior was observed for the reactions of 1-decene and styrene. Complexes **55** and **56** gave substrate conversions between 52% and 90% with 42–78% Kharasch addition product, whereas **54** showed conversions of less than 10% (Table 22.24). Ruthenium hydride complex **55** is more efficient than chloro analogue **56**. In comparison to the most often used catalyst $[RuCl_2(PPh_3)_3]$, **54** is comparable in reactions of both *n*-butyl acrylate and methyl

TABLE 22.24
Kharasch Reaction Using Various Olefins

Substrate	Alkyl Halide	Temperature (°C)	Substrate Conversion (Kharasch Addition) (%)		
			54	55	56
1-Decene	CCl_4	60	8 (7)	64 (44)	52 (42)
Styrene	CCl_4	60	8 (3)	90 (78)	56 (46)
n-Butyl acrylate	CCl_4	85	99 (40)	99 (23)	99 (17)
Methyl methacrylate	CCl_4	85	96 (69)	91 (51)	97 (40)
Styrene	$CHCl_3$	60	0 (0)	8 (0)	18 (0)

Reaction conditions: 0.03 mmol catalyst in toluene; 9 mmol alkene, 13 mmol CCl_4 or $CHCl_3$; internal standard: 0.25 mL dodecane in 3 mL toluene; $t = 24$ h. Conversions and yields obtained by GLC using dodecane as internal standard.

methacrylate with carbon tetrachloride, whereas with the other substrates the catalyst underperforms. However, complexes **55** and **56** perform equally with styrene and 1-decene in comparison with $[RuCl_2(PPh_3)_3]$. On the other hand, modified ruthenium(II) complexes such as $[RuCl(C_5Me_5)(PPh_3)_2]$ led to the same or better conversions in shorter times or at lower temperatures. In the Kharasch addition of styrene with chloroform all of the complexes were inefficient. As by-products low-molecular-weight polymers were obtained, independent of the used halogenated substrate, which limits the yield of the Kharasch addition.

Complexes **50–53** were also tested in the Kharasch addition of carbon tetrachloride to methyl methacrylate and styrene. Instead of the expected addition reaction polymerization products were obtained (see Chapter 22.3.8) [62].

22.3.8 Polymerization

Group 4 *ansa*-metallocenes have been widely employed as catalyst precursors in olefin polymerization. In these compounds, the bridging atom and the molecular structure have a great influence on the polymerization behavior of the catalyst.

Xie and co-workers [63] presented the reaction of **57** with group 4 metal dialkylamides, which led to deprotonation of the acidic protons and the elimination of two amine groups at the metal (Scheme 22.16). The thus-obtained constrained-geometry group IV metal amides **58–61** were then tested in the polymerization of ethylene and ε-caprolactone. In the polymerization of ethylene with modified methylalumoxane as co-catalyst (Al/M = 1500) at 25°C it was shown that only **59** exhibits an activity of 3.7×10^3 g PE mol^{-1} atm^{-1} h^{-1}, whereas **58**, **60**, and **61** showed a very low activity. The molar mass and polydispersity of the polymers obtained with **59** were 2.16×10^5 g mol^{-1} and 1.87, respectively. Thus, the metallocenes were less active than the Me_2C-, Me_2Si-, or *i*-PrNB-bridged analogues, although their geometries are similar [64]. It was concluded that the electronic properties of the bridging atoms led to these differences in their catalytic activity. The electron-rich phosphorus atom could decrease the Lewis acidity of the central atom, whereas the influence of the electron-deficient carborane moiety should increase the Lewis acidity of the metal atom. The results of the polymerization of ethylene indicate that the PR moiety serves as an electron-donating group in the metallocene catalyst.

Organozirconium cations and lanthanide amide complexes initiate the ring-opening polymerization of ε-caprolactone, due to the highly electropositive nature of the metal. The zirconium amide **59** was able to polymerize ε-caprolactone to the polylactone by ring-opening polymerization. THF

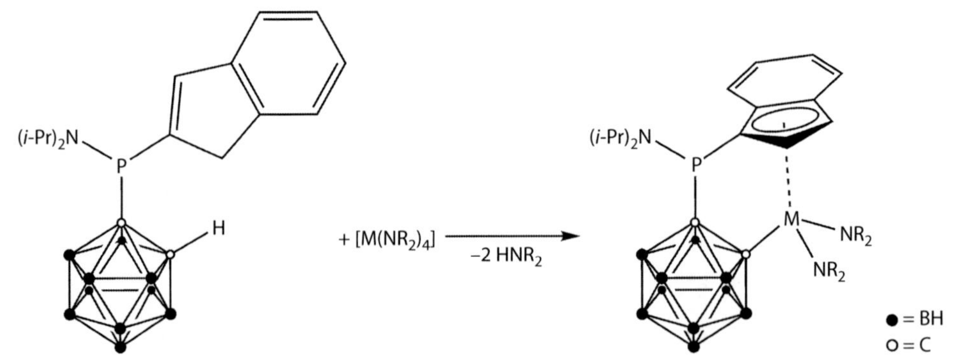

(*i*-Pr)$_2$N P H **57** + $[M(NR_2)_4]$ → −2 HNR_2 (*i*-Pr)$_2$N P M NR_2 NR_2

● = BH
○ = C

M = Ti ; R = Me : **58**
M = Zr ; R = Me : **59**
M = Hf ; R = Me : **60**
M = Hf ; R = Et : **61**

SCHEME 22.16 Synthesis of various *ansa*-metallocenes.

TABLE 22.25
Polymerization of ε-Caprolactone with 59

Solvent	[Monomer]/59	Temperature (°C)	Conversion (%)	10^{-4} M_n (g mol^{-1})	10^{-4} M_w (g mol^{-1})	M_w/M_n
THF	250	25	58	2.34	3.00	1.29
THF	500	25	55	4.23	5.34	1.26
CH_2Cl_2	500	25	4.7	10.52	14.17	1.35
toluene	500	25	33	3.27	4.56	1.39
THF	500	45	92	5.32	6.88	1.29
THF	500	60	95	5.67	7.54	1.33

Reaction conditions: **59**: 20 mg, 3.6×10^{-3} mmol, $t = 2$ h; M_n and M_w determined by GPC relative to polystyrene standards.

was the best solvent and, by raising the temperature to 45°C or 60°C, the polymerization became rapid in combination with a slight increase in the molar mass and polydispersity of the polymer (Table 22.25). In most cases, high-molecular-weight polymers ($M_w > 40{,}000$) and narrow polydispersity ($M_w/M_n < 1.4$) were obtained. By using NMR spectroscopy, it was also observed that the two NMe_2 groups of the metallocene disappear immediately after the addition of two equivalents of ε-caprolactone. These generate a $-CONMe_2$ group, which is later found in the resulting polylactone. It is likely that attack of one of the nucleophilic amido nitrogen atoms at the carbonyl carbon atom of the lactone occurs as the initial step of the polymerization and is followed by acyl bond cleavage (Scheme 22.17). These results led to the conclusion that the carborane moiety in the ligand can enhance the electrophilicity of the group IV metal.

Demonceau et al. [61,62] investigated the radical polymerization of methyl methacrylate with palladium(II) complexes **50–53** (Figure 22.21). As mentioned above, ruthenium complexes **54–56** are active in the Kharasch reaction, in which poly(methyl methacrylate) was obtained as the by-product.

Complexes **50–53** were tested under standard Kharasch reaction conditions, that is, treating carbon tetrachloride with methyl methacrylate at 85°C in the presence of the palladium catalyst. Polymers with a polydispersity of 1.8–2.0 were obtained in moderate yield (Table 22.26). Interestingly, the yield of Kharasch addition product was less than 1%. It was observed that the polymer yields are

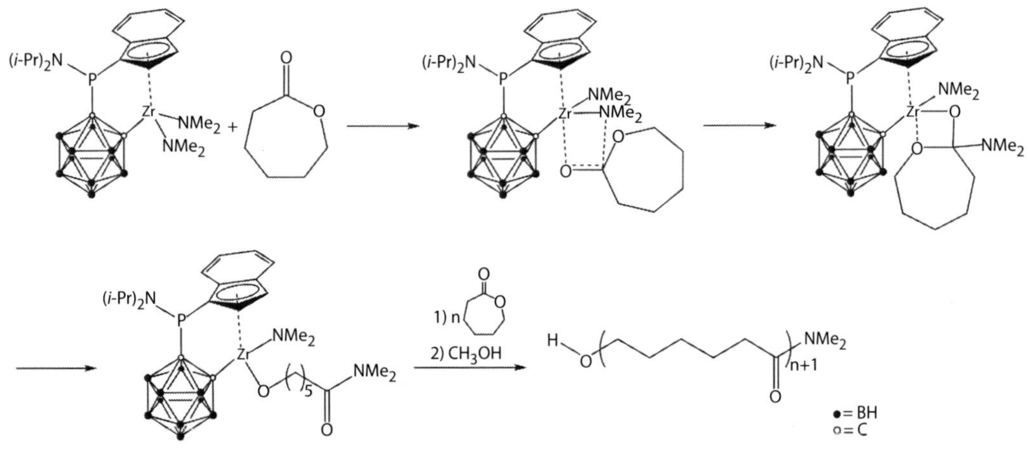

SCHEME 22.17 Proposed mechanism of the first step of ring-opening polymerization of ε-caprolactone with **59**.

TABLE 22.26
Radical Polymerization of Methyl Methacrylate

Catalyst	Polymer Yield (%)	10^{-4} M_n(g · mol^{-1})	M_w/M_n
50	47	5.5	1.8
51	39	8.7	1.75
52	76	3.25	1.8
53	43	6.0	2.0
[$PdCl_2$(cod)]	24	3.3	1.35
[$PdCl_2(PPh_3)_3$]	24	5.6	1.8
[$PdCl_2$(P(*i*-Pr)$_3$)$_3$]	5	—	—
[$PdCl_2$(dppe)]	33	8.0	1.95
[$PdCl_2$(dppf)]	19	4.6	1.7
[PdCl(CH_3)(cod)]	67	2.3	1.6

Reaction conditions: 0.03 mmol palladium complex, 4 mL toluene, 0.25 mL dodecane, 13 mmol carbon tetrachloride, 9 mmol methyl methacrylate, $T = 85°C$; conversion was determined by GLC; M_n and M_w/M_n was determined by SEC in THF at 40°C.

higher with [PdCl(Me)L_2] than with [$PdCl_2L_2$] and also higher with PPh_n than with P(*i*-Pr)$_n$ ($n = 2, 3$) groups. Palladium complex **50**, with the carborane moiety as C_2 backbone, is more reactive than the related complexes containing dppe or dppf ligands. The investigation of the dependence of the molecular weight (M_n) and molecular weight distribution (M_w/M_n) on the monomer conversion revealed a decrease in the average molecular weight and an increase in M_w/M_n. The absence of carbon tetrachloride gave poly(methyl methacrylate) with very high M_n but low yield. In the presence of radical scavengers, such as galvinoxyl or 2,2-di(4-*tert*-octylphenyl)-1-picrylhydrazyl, complete inhibition of the reaction was observed. These two results, coupled with the absence of Kharasch addition products, indicate that the polymerization does not occur via an atom transfer radical polymerization mechanism. More likely, the reaction is a redox-initiated free-radical process in which carbon tetrachloride acts as a chain-transfer agent. Polymerization of ethylene with MAO or $AlClEt_2$ in combination with or without CPh_3^+ BPh_4^- led to no polymer formation, and only minute amounts of hexenes and butenes were observed.

22.3.9 Ring-Opening Metathesis Polymerization

Instead of a conventional polymerization reaction, Demonceau et al. [65] observed ring-opening metathesis polymerization of norbornene, which was initialized by ethyl diazoacetate or trimethylsilyldiazomethane (Scheme 22.18). The used ruthenium complexes **62** and **63** gave moderate yields of the polymer (Figure 22.22, Table 22.27).

This result confirmed the presence of two vacancies in *cis* position at the metal atom, which are necessary for coordinating both the carbene and the olefin. In these complexes the agostic B–H → Ru bonds are believed to be stable. It was shown that the PPh_3 groups are labile, depending on the

n ROMP n

SCHEME 22.18 Ring-opening metathesis polymerization reaction of norbornene.

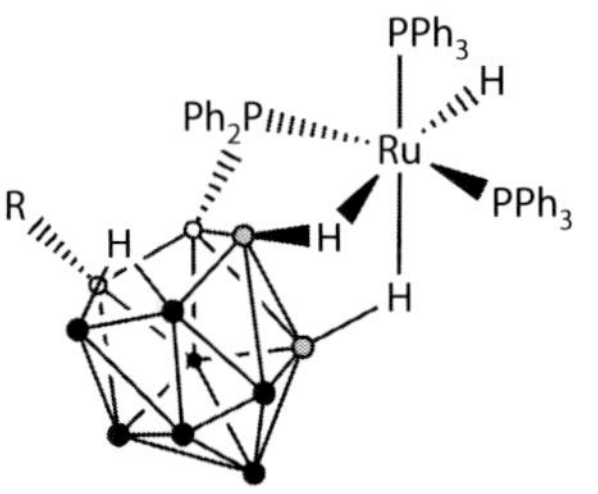

FIGURE 22.22 Ligands investigated in ROMP.

incoming ligand. It could not be clarified which of the phosphine ligands is more labile. It was postulated that at least two catalytically active species are present during ROMP, which was confirmed by the polydispersity index (M_w/M_n) and gel permeation chromatography (GPC).

22.3.10 Cyclopropanation

Cyclopropanation of olefins is currently performed by direct transition metal-catalyzed carbene transfer from a diazo compound to the olefin. Dirhodium(II) carboxylates and carboxamidates have proved to be the catalysts of choice. Other rhodium compounds, such as $Rh_6(CO)_{16}$, $Rh_2(BF_4)_4$, and rhodium(III) porphyrins, have been also investigated, but did not show better reactivity, while rhodium(I) compounds have never been successful [66]. Other complexes containing copper or ruthenium have been tested in cyclopropanation reactions, but have never shown better reactivity or selectivity than rhodium(II) compounds [67].

Demonceau and Viñas et al. [68] demonstrated cyclopropanation of several olefins with ruthenium and rhodium complexes **62–66** of *nido*-carboranes (Scheme 22.19, Figure 22.23) [**65**,68]. They observed a conversion between 90% and 98% for both of the ruthenium complexes, where activated olefins, such as styrene and derivatives thereof, were used as substrates (Table 22.28). The yields were significantly lower with cyclic olefins or terminal linear mono-olefins (51–65%). Similar results

TABLE 22.27
ROMP of Norbornene Initiated by Ethyl Diazoacetate and Trimethylsilyldiazomethane

	Polymer			
	62		63	
Catalyst	N_2CHCO_2Et	$N_2CHSiMe_3$	N_2CHCO_2Et	$N_2CHSiMe_3$
Yield	30	42	20	31
cis/trans Ratio[a]	0.75	1.35	0.90	1.65
$r_C \times r_t$[b]	1.15	2.55	1.55	2.4
M_n[c]	38,000	31,000	31,000	28,000
M_w/M_n	2.95	8.7	2.95	7.3

Reaction conditions: 0.0075 mmol catalyst; 0.5 g norbornene in 30 mL chlorobenzene; $T = 60°C$; $t = 20$ min then 0.1 mmol diazo compound diluted in 1 mL chlorobenzene added over 30 min; $T = 60°C$; $t = 5$ h.

[a] Determined by ^1H and ^{13}C NMR.

[b] Blockiness parameter determined by ^{13}C NMR.

[c] Determined by GPC using polystyrene standards.

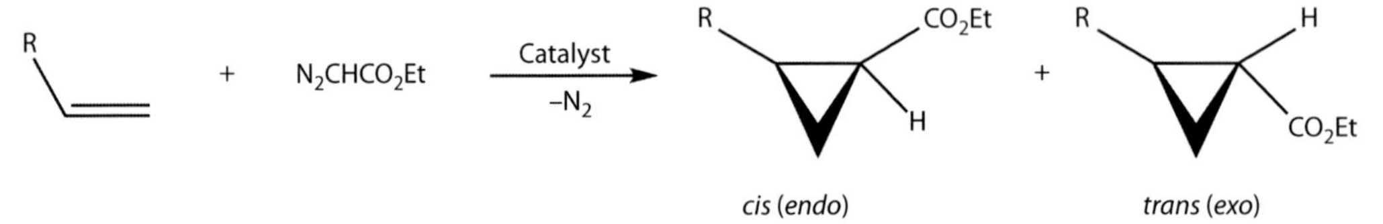

SCHEME 22.19 Cyclopropanation of olefins.

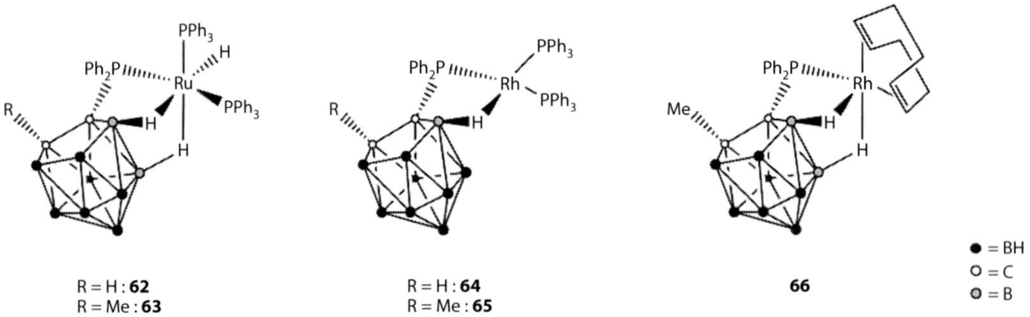

FIGURE 22.23 Catalysts applied in the catalytic cyclopropanation.

to those observed with the ruthenium complexes were obtained by using the rhodium complexes **64–66** in the cyclopropanation of activated olefins. For the reaction with inactivated olefins, the rhodium complexes performed significantly better.

The obtained stereoselectivities are in the same range as those of other ruthenium- or rhodium-based catalysts. As the carbene addition occurs with *trans* (*exo*) diastereoselectivity, the more thermodynamically stable *trans* (*exo*) isomer is preferentially formed. It was observed that two steric effects have influence on the stereoselectivity: the steric bulk of the 4-position of the styrene derivatives and the steric bulk of the diazoacetate used. In both cases the *cis*/*trans* ratio decreases with the

TABLE 22.28
Cyclopropanation of Olefins with Ethyl Diazoacetate

	Substrate				
	Cyclopropanation Yield (%), (*cis/trans* or *endo/exo* Ratio)				
Catalyst	**62**	**63**	**64**	**65**	**66**
Styrene	97 (0.63)	96 (0.64)	91 (0.88)	92 (0.91)	91 (0.86)
4-Methylstyrene	96 (0.52)	96 (0.54)	90 (0.68)	93 (0.71)	94 (0.70)
4-*tert*-Butylstyrene	93 (0.50)	91 (0.48)	93 (0.54)		
4-Methoxystyrene	90 (0.61)	89 (0.56)	87 (0.60)		
4-Chlorostyrene	94 (0.50)	93 (0.48)	89 (0.51)	92 (0.60)	87 (0.52)
α-Methylstyrene	98 (0.95)	97 (1.02)	94 (0.89 *Z*/*E*)		
Cyclooctene	51 (0.86)	65 (1.08)	86 (0.69)	85 (0.67)	88 (0.72)
1-Octene	61 (0.71)	58 (0.62)	71 (0.56)		
1-Dodecene	59 (0.73)	61 (0.73)	73 (0.66)		

Reaction conditions: 0.0075 mmol catalyst; 20 mmol olefin (also as solvent); 1 mmol ethyl diazoacetate; addition time: 4 h; $T = 100°C$; yield based on ethyl diazoacetate and determined by GLC analysis.

TABLE 22.29
Relative Reactivity of Olefins Compared to Styrene

	Catalyst		
Olefin	**62**	**63**	**64**
4-Methylstyrene	1.04 (0.61; 0.48)	1.9 (0.56; 0.49)	1.06 (0.77; 0.66)
4-Chlorostyrene	0.85 (0.54; 0.43)	0.60 (0.56; 0.46)	0.92 (0.64; 0.47)
Cyclooctene	0.12 (0.59; 0.61)	0.09 (0.60; 0.88)	0.10 (0.65; 0.64)
1-Octene	0.12 (0.62; 0.60)	0.08 (0.62; 0.55)	0.09 (0.66; 0.57)

Reaction conditions same as in Table 22.28, except the temperature $T = 90°C$. In parentheses: *cis*/*trans* ratio for styrene; *cis*/*trans* or *endo*/*exo* ratio for the co-olefin.

use of more sterically challenging groups. The influence of the steric difference in the carborane backbone is negligible. The catalysts become active above 60°C and no significant increase in stereoselectivity compared to known rhodium(II) catalysts can be obtained. Nevertheless, it can be concluded that the anionic carboranyl phosphine ligand has an influence on the rhodium(I) center, and thus the first active rhodium(I) complex has been demonstrated.

The relative reactivity of certain substrates compared to styrene (which was set as reference = 1) was also investigated (Table 22.29). 4-Methylstyrene was the only substrate that showed faster relative reactivity of 1.04 (**62**) and 1.06 (**64**). Interestingly, complex **63**, with its methyl group at the 8-position of the cluster, gave a value of 1.9, which suggests that the methyl group has an influence on the reactivity. In the competitive experiment between styrene and cyclooctene or 1-octene with complex **62**, styrene was 10 times more reactive. In the reaction of **64** and ethyl diazoacetate with *n*-butyl vinyl ether—another activated olefin—a yield of 80–90% was observed. In the competitive experiment, the *n*-butyl vinyl ether was six times less reactive than styrene.

The anionic carboranyl phosphine ligand has a multi-coordinating ability to the metal atom. Apart from the Rh–PPh_2 bond, B–H–Rh agostic bonds can be formed. In the catalytic cycle, di- and tricoordinating species can be operative. The exact nature of the catalytically active species is not known, but an equilibrium was proposed (Scheme 22.20). It was suggested that the dramatic changes in the activity of the catalyst when large amounts of ethyl diazoacetate were employed could be indicative of a gradual modification of the catalytic center. Furthermore, a participation of cyclopropanes or by-products could disturb the above-mentioned equilibrium.

22.3.11 Cross-Coupling with Grignard Reagents

Nickel bisphosphine complexes are used as catalysts in cross-coupling of aromatic and olefinic halides with Grignard reagents. Kumada and co-workers [69] reported a cross-coupling reaction

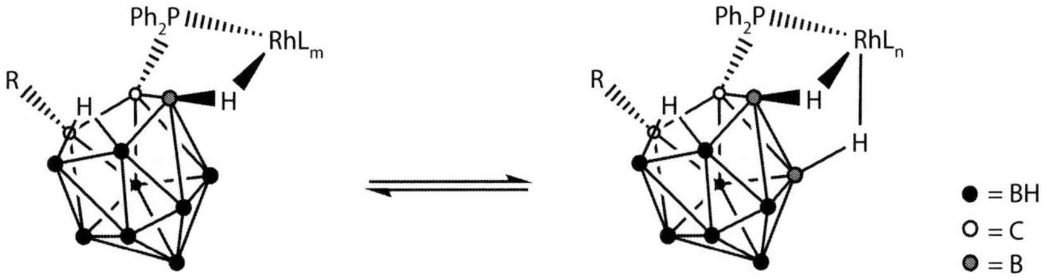

SCHEME 22.20 Proposed equilibrium.

SCHEME 22.21 Alkyl group isomerization in the cross-coupling reaction of isopropyl magnesium chloride with chlorobenzene.

between isopropylmagnesium chloride, a secondary alkyl Grignard reagent, and chlorobenzene using various nickel(II) bisphosphine complexes (Scheme 22.21). The carborane-containing ligands 1,2-bis(diphenylphosphino)-*ortho*-carborane(12) (**1**) and 1,2-bis(dimethylphosphino)-*ortho*-carborane(12) (**67**) were amongst those employed (Figure 22.24). In all cases isomerization from the secondary to the primary product was observed, but strong dependence on the electronic nature of the bisphosphine ligand was reported [69].

The nickel complex with **1** showed preferential formation of isopropylbenzene, whereas that with **67** induced alkyl group isomerization to form *n*-propylbenzene. Unfortunately, the total yields were poor, at just 18% and 7%, respectively. Nickel complexes with dppe, dmpe, dppp, and dmpf gave total conversions of up to 89% (Table 22.30). Nickel complexes with the monodentate ligands PEt_3, PBu_3, and PPh_3 led to the formation of benzene instead of the desired coupling products, with a poor overall yield. In all cases, the phenyl-substituted ligands led to isopropylbenzene as the preferred product, whereas the methyl-substituted ligands gave *n*-propylbenzene preferentially. It was suggested that the more electron-donating the ligand, the more alkyl isomerization from secondary to primary takes place.

The postulated mechanism involves σ-alkyl nickel intermediates and a hydrido–olefin nickel intermediate. An increase in electron density at the nickel atom, due to good electron-donating ligands, may facilitate the σ–π conversion and formation of a hydrido olefin intermediate. This may tend to form a primary alkyl complex, and then undergo the reverse reaction to the secondary intermediate, due to its higher stability. The postulated mechanism could be proved by the formation of benzene through decomposition of the hydrido–olefin intermediate, which can only be detected when *n*-propylbenzene is formed preferentially. The reaction of a Grignard reagent prepared from optically pure (+)-(*S*)-2-methylbutylchloride ($[\alpha^{D}_{25}] = +1.68°$) with chlorobenzene and $[NiCl_2(dppp)]$ as catalyst gave the optically active (+)-(*S*)-2-methylbutylbenzene, with a little loss of optical activity ($[\alpha^{D}_{25}] = +10.54°$ neat). In the same reaction with $[NiCl_2(dmpe)]$, the optical activity of the product was slightly lower ($[\alpha^{D}_{25}] = +9.56°$ neat). This proves the facilitating effect of the ligand on the σ–π conversion.

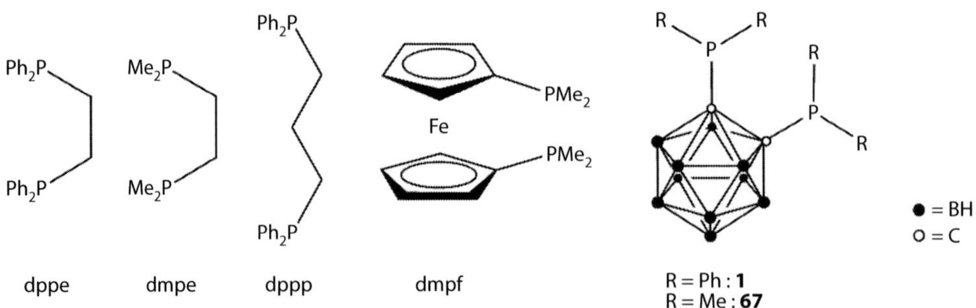

FIGURE 22.24 Ligands investigated in the cross-coupling with Grignard reagents.

TABLE 22.30
Cross-Coupling Reaction of Isopropyl Magnesium Chloride with Chlorobenzene

L in Catalyst $[NiCl_2(L)]$	Total Yield (%)	Product Distribution (%)		
1	18	78	1	21
67	7	12	88	0
dppe	74	96	4	0
dmpe	84	9	84	7
dppp	89	96	4	0
dmpf	48	8	74	18
$Ph_2PCH=CHPPh_2$	8	92	8	0
2 PEt_3	9	1	11	88
2 PBu_3	8	2	16	82
2 PPh_3	44	16	30	54

Reaction conditions: 5 mmol chlorobenzene; 0.05 mmol $[NiCl_2(L)]$; 5 mL diethyl ether, 6.9 mmol isopropyl Grignard solution, 20 h reflux.

22.3.12 Sonogashira Coupling with Hydride Transfer

Nakamura et al. [70] demonstrated the first example of transition metal-catalyzed coupling reactions using the 1,2-bis-phosphinocarborane(12)s **1** and **68**. In this reaction propargyldicyclohexylamine was transformed into an allene anion, which reacts *in situ* with various aryl halides and heterocyclic halides (Figure 22.25, Scheme 22.22). During this reaction it is transformed into various allenes in a single step through a Sonogashira coupling reaction, followed by a hydride-transfer step. A two-step reaction with a Sonogashira reaction followed by hydride transfer is necessary to synthesize allenes from organic halides. Palladium-catalyzed reactions of allenylstannanes, -indium, and -zinc with organic halides have also been reported, where it is necessary to use a stoichiometric amount of metal.

It was shown that palladium complexes with 1,2-bis(diphenylphosphino)-*ortho*-carborane(12) (**1**) and 1,2-bis(diethoxyphosphino)-*ortho*-carborane(12) (**68**) are suitable catalysts in the one-pot reaction of 4-iodoacetanilide and propargyldicyclohexylamine (Scheme 22.22). The observed yields

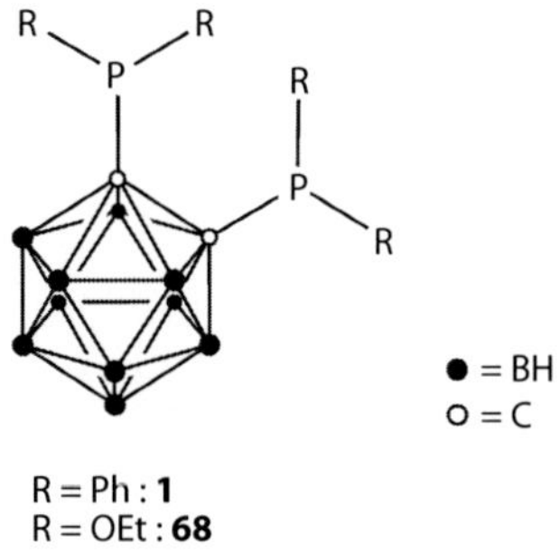

FIGURE 22.25 Ligands used in the reaction.

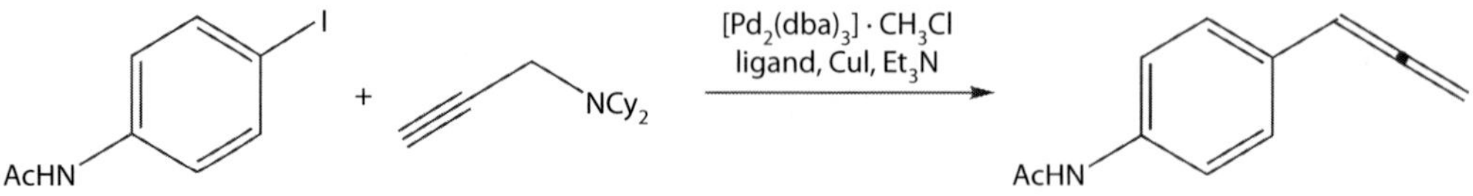

AcHN I + NCy$_2$ $[Pd_2(dba)_3] \cdot CH_3Cl$ ligand, CuI, Et_3N → AcHN

SCHEME 22.22 Sonogashira cross-coupling and hydride transfer of allenes from aryl iodides.

were 87% and 50%, respectively. Various other bidentate phosphines, such as dppe, dppm, and dppf, failed in this reaction. Monodentate ligands such as PPh_3 and $P(OPh)_3$ yielded 23% and 40% of the desired allene and 70% and 53% only of the Sonogashira coupling product, respectively. $P(C_6F_5)_3$ failed in the Sonogashira reaction but was found to be an effective catalyst in the hydride-transfer reaction. Only $(C_6F_5)_2PCH_2CH_2P(C_6F_5)_2$ was effective, with an 87% yield, similar to **1**. This indicated that electron-deficient bidentate ligands are suitable for these reactions.

Several other substrates were tested using 1,2-bis(diphenylphosphino)-*ortho*-carborane(12) as ligand. Yields between 65% and full conversion were obtained (Table 22.31). It is remarkable that with bromoiodo-substituted compounds the reaction takes place only at the iodo-substituted position. The use of $(C_6F_5)_2PCH_2CH_2P(C_6F_5)_2$ gave lower yields in some reactions, but may be suitable for other reactions.

The reaction was monitored by HPLC to clarify the reaction rate and mechanism. The reaction undergoes a transformation from the aryl iodide to the Sonogashira coupling product as an

TABLE 22.31
One-Pot Synthesis of Various Allenes from Aryl Iodides with Propargyldicyclohexylamine

Substrate	Time (h)	Yield (%)[a]	Substrate	Time (h)	Yield (%)[a]
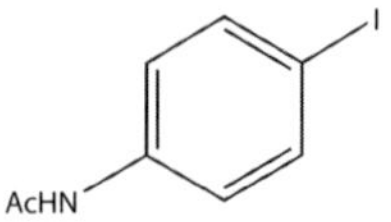 AcHN, I	23	87	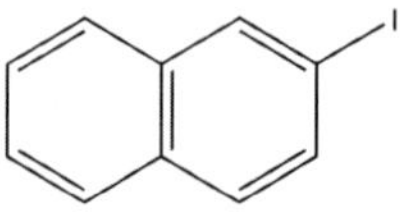 I	48	60
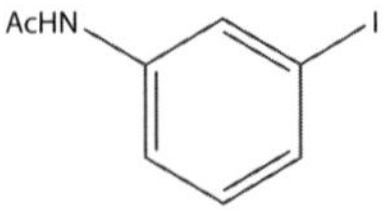 AcHN, I	24	>99	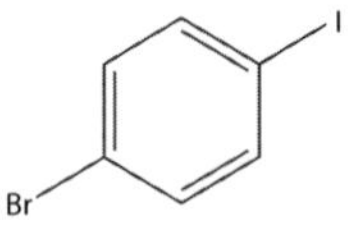 Br, I	48	68
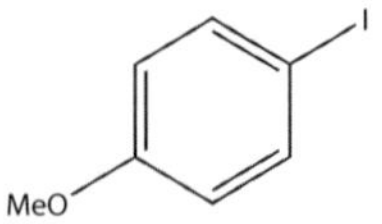 MeO, I	36	96	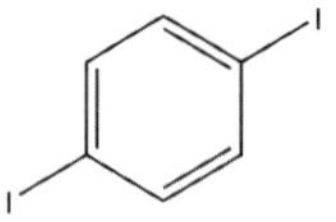 I, I	48	65[b]
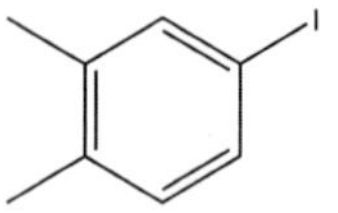 I	36	86	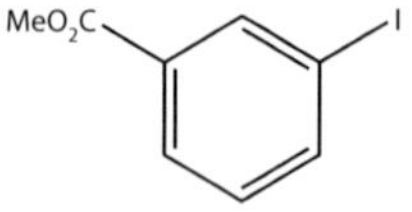 MeO_2C, I	23	99
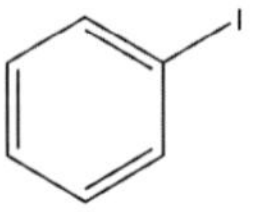 I	43	>99	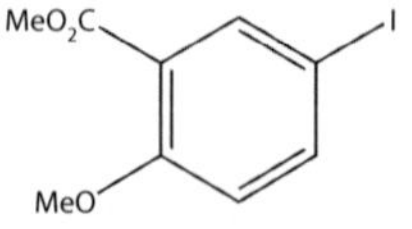 MeO_2C, MeO, I	22	90

Reaction conditions: 2.5 mol% $[Pd_2(dba)_3] \cdot CH_3Cl$; 10 mol% **1**; 15 mol% CuI; 150 mol% Et_3N; $T = 80°C$ for 3 h, then heating to 100°C.

[a] Yield of isolated product based on substrate.

[b] Diallene product was obtained exclusively.

intermediate, which can be monitored after 20 min with 25% yield, and the hydride-transfer product with 3% yield. After 3 h the reaction reached a plateau without any reason. It was thus necessary to raise the reaction temperature to 100°C to drive the reaction to high yields. It was indicated that starting the reaction at 100°C leads to an incomplete Sonogashira coupling reaction, due to decomposition of the catalyst.

The anionic palladium hydride species generated by hydride transfer from the cyclohexyl carbon atom of the coupling product is a key intermediate in the proposed mechanism. It is therefore necessary that the ligand is able to stabilise the palladium hydride intermediate and facilitate certain reactivity in the oxidative addition step at the same time. With this in mind, carboranes and $(C_6F_5)_2PCH_2CH_2P(C_6F_5)_2$ are suitable ligands for both the Sonogashira coupling and the hydride-transfer catalytic steps.

22.4 CONCLUSIONS AND FUTURE CHALLENGES

The literature presented in this chapter has shown that phosphorus-substituted carboranes have a multitude of possible applications in catalysis, ranging from hydrogenation and hydroformylation to polymerization and cross-coupling reactions. As carboranes offer a wide range of structural features, through the *closo* and *nido* structures and different substitution patterns thereof, the presented applications are also wide and varied. The reported tests have opened a new class of substances which look promising in homogeneous catalysis, and while only a small number of these substances have been tested for their catalytic activity, this leaves room for further research and improvement.

Nevertheless, research into phosphorus-substituted carboranes should not be limited to catalysis. There are numerous possible applications for these substances, not only those presented here, but also with minor modifications. Carboranes have always been quite expensive substances, costing almost half of the price of gold. Thus, use of these substances will be limited to applications where no other substances come close to outperforming them. The implementation of carboranes must be further developed, so that processes will only utilize minute amounts or permit almost complete recycling of the catalyst. BNCT, opto-electronic applications and catalysis all conform to these demands and are as such promising areas of research.

ACKNOWLEDGMENT

We are grateful to the Deutsche Forschungsgemeinschaft within the Graduate School of Excellence BuildMoNa for financial support.

ANNEX

ABBREVIATIONS

(*R*)-(+)-BINAP	(*R*)-(+)-2,2′-bis(diphenylphosphino)-1,1′-binaphthyl
(*R*)-(+)-BINOL	(*R*)-(+)-1,1′-binaphthyl-2,2′-diol
(*S*,*S*)-CHIRAPHOS	(*S*,*S*)-2,3-bis(diphenylphosphino)butane
(*S*,*S*)-(+)-DIOP	(*S*,*S*)-(+)-2,3-*O*-isopropylidene-2,3-dihydroxy-1,4-bis(diphenylphosphino) butane
(*R*,*R*)-DIPAMP	(*R*,*R*)-1,2-bis[(2-methoxyphenyl)phenylphosphino]ethane
(*R*,*R*)-Et-DuPHOS	(−)-1,2-bis((2*R*,5*R*)-2,5-diethylphospholano)benzene
acac	acetylacetonate
BSA	*N*,*O*-bis(trimethylsilyl)acetamide
cod	1,5-cyclooctadiene
DFT	density functional theory

dmpe	1,2-bis(dimethylphosphino)ethane
dmpf	1,1'-bis(dimethylphosphino)ferrocene
dppe	1,2-bis(diphenylphosphino)ethane
dppf	1,2-bis(diphenylphosphino)ferrocene
dppm	1,2-bis(diphenylphosphino)methane
dppp	1,2-bis(diphenylphosphino)propane
GPC	gel permeation chromatography
nbd	2,5-norbornadiene, bicyclo[2.2.1]hepta-2,5-diene
$Pd_2(dba)_3$	tris(dibenzylideneacetone)dipalladium(0)
PPFA	*N*,*N*-dimethyl-1-[(*S*)-2-(diphenylphosphanyl)ferrocenyl]ethylamine
ROMP	ring-opening metathesis polymerization
$scCO_2$	supercritical carbon dioxide
SEC	size exclusion chromatography

REFERENCES

1. Frost & Sullivan, *Developments in Global Chiral Technology Markets*, **2003**; available from http://www.frost.com/prod/servlet/report-brochure.pag?id=B105-01-00-00-00.
2. G. T. Whiteker, J. R. Briggs, J. E. Babin, B. A. Barner, in D. G. Morrell, ed., *Catalysis of Organic Reactions*. CRC Press, Boca Roton, FL, **2002**, p. 359.
3. R. Noyori, Asymmetrische Katalyse: Kenntnisstand und Perspektiven (Nobel-Vortrag), *Angew. Chem.* **2002**, *114*, 2108; Asymmetric catalysis: Science and opportunities (Nobel Lecture), *Angew. Chem. Int. Ed.* **2002**, *41*, 2008.
4. P. Sabatier, How I have been led to the direct hydrogenation method by metallic catalysts, *J. Ind. Eng. Chem.* **1926**, *18*, 1005.
5. J. Halpern, J. F. Harrod, B. R. James, Homogeneous catalytic hydrogenation of olefinic compounds, *J. Am. Chem. Soc.* **1961**, *83*, 753.
6. J. F. Young, J. A. Osborn, F. H. Jardine, G. Wilkinson, Hydride intermediates in homogeneous hydrogenation reactions of olefins and acetylenes using rhodium catalysts, *J. Chem. Soc., Chem. Commun.* **1965**, 131.
7. (a) W. S. Knowles, M. L. Sabacky, Catalytic asymmetric hydrogenation employing a soluble optically active rhodium complex, *J. Chem. Soc. Chem. Commun.* **1968**, 1445; (b) L. Horner, H. Siegel, H. Büthe, Asymmetrische katalytische Hydrierung mit einem homogen gelösten optisch aktiven Phosphin-Rhodium-Komplex, *Angew. Chem.* **1968**, *80*, 1034; Asymmetric catalytic hydrogenation with an optically active phosphinerhodium complex in homogeneous solution, *Angew. Chem. Int. Ed.* **1968**, *7*, 942.
8. A. Miyashita, A. Yasuda, H. Takaya, K. Toriumi, T. Ito, T. Souchi, R. Noyori, Synthesis of 2,2′-Bis(diphenylphosphino)-1,1′-binaphthyl (BINAP), an atropisomeric chiral bis(triaryl)phosphine, and its use in the rhodium(i)-catalyzed asymmetric hydrogenation of α-(acylamino)acrylic acids, *J. Am. Chem. Soc.* **1980**, *102*, 7932.
9. C. A. Tolman, Steric effects of phosphorus ligands in organometallic chemistry and homogeneous catalysis, *Chem. Rev.* **1977**, *77*, 313.
10. W. S. Knowles, Asymmetric hydrogenations (Nobel Lecture 2001), *Adv. Synth. Catal.* **2003**, *345*, 3.
11. (a) D. E. C. Corbridge, *The Structural Chemistry of Phosphorus*, Elsevier Scientific Publishing Comp., Amsterdam, **1974**; (b) R. J. Morris, G. S. Girolami, On the π-donor ability of early transition metals: Evidence that trialkylphosphines can engage in π-back-bonding and x-ray structure of the titanium(II) phenoxide $Ti(OPh)_2(dmpe)_2$, *Inorg. Chem.* **1990**, *29*, 4167.
12. (a) M. P. Magee, W. Luo, W. H. Hersh, Electron-withdrawing phosphine compounds in hydroformylation reactions. 1. Syntheses and reactions using mono- and bis(*p*-toluenesulfonylamino) phosphines, *Organometallics* **2002**, *21*, 362; (b) D. Selent, D. Hess, K.-D. Wiese, D. Röttger, C. Kunze, A. Börner, Rhodiumkatalysierte Isomerisierung/Hydroformylierung interner Octene mit neuartigen Phosphorliganden, *Angew. Chem.* **2001**, *113*, 1739; New phosphorus ligands for the rhodium-catalyzed isomerization/hydroformylation of internal octenes, *Angew. Chem. Int. Ed.* **2001**, *40*, 1696.
13. (a) J. Plešek, Extraction of alkali-metals into nitrobenzene in presence of univalent polyhedral borate anions, *Chem. Rev.* **1992**, *92*, 269; (b) R. N. Grimes, Boron clusters come of age, *J. Chem. Edu.* **2004**, *81*, 657.

14. P. Kaszynski, A. G. Douglass, Organic derivatives of *closo*-boranes: A new class of liquid crystal materials, *J. Organomet. Chem.* **1999**, *581*, 28.
15. (a) D. G. Allis, J. T. Spencer, Polyhedral-based nonlinear optical materials. 2. theoretical investigation of some new high nonlinear optical response compounds involving polyhedral bridges with charged aromatic donors and acceptors, *Inorg. Chem.* **2001**, *40*, 3373; (b) D. G. Allis, J. T. Spencer, Polyhedral-based nonlinear optical materials. Part 1. Theoretical investigation of some new high nonlinear optical response compounds involving carboranes and charged aromatic donors and accepters, *J. Organomet. Chem.* **2000**, *614–615*, 309; (c) C. D. Entwistle, T. B. Marder, Die Borchemie leuchtet: optische Eigenschaften von Molekülen und Polymeren, *Angew. Chem.* **2002**, *114*, 3051; Boron chemistry lights the way: Optical properties of molecular and polymeric systems, *Angew. Chem. Int. Ed.* **2002**, *41*, 2927; (d) K. Base, M. T. Tierney, A. Fort, J. Muller, M. W. Grinstaff, On the second-order nonlinear optical structure – property relationships of metal chromophores, *Inorg. Chem.* **1999**, *38*, 287; (e) J. Abe, N. Nemoto, Yu Nagase, Y. Shirai, T. Iyoda, A new class of carborane compounds for second-order nonlinear optics: Ab initio molecular orbital study of hyperpolarizabilities for 1-(1′,X′-dicarba-*closo*-dodecaborane-1′-yl)-*closo*-dodecaborate dianion (X = 2, 7, 12), *Inorg. Chem.* **1998**, *37*, 172; (f) R. Hamasaki, M. Ito, M. Lamrani, M. Mitsuishi, T. Myashita, Y. Yamamoto, Nonlinear optical studies of fullerene-arylethyne hybrids, *J. Mater. Chem.* **2003**, *13*, 21.
16. (a) S. Peper, Y. Qin, P. Almond, M. McKee, M. Telting-Diaz, T. Albrecht-Schmitt, E. Bakker, Ion-pairing ability, chemical stability, and selectivity behavior of halogenated dodecacarborane cation exchangers in neutral carrier-based ion-selective electrodes, *Anal. Chem.* **2003**, *75*, 2131; (b) A. Malon, A. Radu, W. Qin, Y. Qin, A. Ceresa, M. Maj-Zurawska, E. Bakker, E. Pretsch, Improving the detection limit of anion-selective electrodes: An Iodide-Selective membrane with a nanomolar detection limit", *Anal. Chem.* **2003**, *75*, 3865; (c) C. Masalles, F. Teixidor, S. Borros, C. Viñas, Cobaltabisdicarbollide anion $[Co(C_2B_9H_{11})_2]^-$ as doping agent on intelligent membranes for ion capture, *J. Organomet. Chem.* **2002**, *657*, 239.
17. (a) G. L. Jackson, J. Winter, K. H. Burell, J. C. deBoo, C. M. Greenfield, R. J. Groebner, T. Hodapp et al., Boronization in DIII-D, *J. Nuc. Mater.* **1992**, *196–198*, 236; (b) O. I. Buzhinsky, V. G. Ostroshchenko, D. G. Whyte, M. Baldwin, R. W. Conn, R. P. Doerner, R. Seraydarian, S. Luckhardt, H. Kugel, W. P. West, Plasma deposition of boron films with high growth rate and efficiency using carborane, *J. Nucl. Mater.* **2003**, *313–316*, 214; (c) D. Tafalla, F. L. Tabares, First boronization of the TJ-II Stellarator, *Vacuum* **2002**, *67*, 393.
18. (a) Thiokol Corp. U.S. Pat. 4,358,998, Igniter for a pyrotechnic gas bag inflator, Nov. 16, 1982; (b) Thiokol Corp. U.S. Pat. 5,401,340, Borohydride fuels in gas generant compositions, March 28, 1995.
19. J. de Zeeuv, J. Luong, Developments in stationary phase technology for gas chromatography, *TRAC—Trends Anal. Chem.* **2002**, *21*, 594.
20. (a) J. Rais, P. Selucky, M. Kyrs, Extraction of alkali metals into nitrobenzene in the presence of univalent polyhedral borate anions, *J. Inorg. Nucl. Chem.* **1976**, *38*, 1376; (b) P. K. Hurlburt, R. L. Miller, K. D. Abney, T. M. Foreman, R. J. Butcher, S. A. Kinkead, New synthetic routes to b-halogenated derivatives of cobalt dicarbollide, *Inorg. Chem.* **1995**, *34*, 5215; (c) C. Viñas, S. Gomez, J. Bertran, J. Barron, F. Teixidor, J.-F. Dozol, H. Rouquette, R. Kivekäs, R. Sillanpää, C-substituted bis(dicarbollide) metal compounds as sensors and extractants of radionuclides from nuclear wastes, *J. Organomet. Chem.* **1999**, *581*, 188; (d) C. Viñas, S. Gomez, J. Bertran, F. Teixidor, J.-F. Dozol, H. Rouquette, New polyether-substituted metallacarboranes as extractants for ^{137}Cs and ^{90}Sr from nuclear wastes, *Inorg. Chem.* **1998**, *37*, 3640.
21. A. J. Lunato, J. Wang, J. E. Woollard, A. K. M. Anisuzzaman, W. Ji, F.-G. Rong, S. Ikeda et al., Synthesis of 5-(carboranylalkylmercapto)-2'-deoxyuridines and 3-(carboranylalkyl)thymidines and their evaluation as substrates for human thymidine kinases 1 and 2, *J. Med. Chem.* **1999**, *42*, 3378 (therein ref. 16: M. Hofmann, University of Georgia, Athens, GA, unpublished results: According to *ab initio* computations at the MP2(fc)/ 6-31G* level, the diameter of benzene is 4.968 Å [H(C1)-H(C4)] and those of the *o*-carborane clusters are 5.48 Å [H(C1)-H(B12)], 5.75 Å [H(B3)-H(B12)], and 5.279 Å [H(B4)-H(B11)]), respectively.
22. (a) L. I. Zakharkin, V. N. Kalinin, A. P. Snyakin, B. A. Kvasov, Effect of solvents on electronic properties of 1-*o*-carboranyl, 3-*o*-carboranyl and 1-*m*-carboranyl groups, *J. Organomet. Chem.* **1969**, *18*, 19; (b) F. Teixidor, G. Barberà, A. Vaca, R. Kivekäs, R. Sillanpää, J. Oliva, C. Viñas, Are methyl groups electron-donating or electron-withdrawing in boron clusters? Permethylation of *o*-carborane, *J. Am. Chem. Soc.* **2005**, *127*, 10158.
23. F. Teixidor, C. Viñas, R. Benakki, A route to *exo*-heterodisubstituted and monosubstituted *o*-carborane derivatives, *Inorg. Chem.* **1997**, *36*, 1719.

24. L. I. Zakharkin, V. V. Guseva, V. A. Ol'shevskaya, Synthesis of dimethyl (*m*-carboran-9-yl)phosphonate and dimethyl (*p*-carboran-2-yl)phosphonate by palladium-catalyzed cross coupling of 9-I-*m*- and 2-I-*p*-carboranes with dimethyl hydrogen phosphite, *Russ. J. Gen. Chem.* **2001**, *71*, 903.
25. K. Ohta, T. Goto, H. Yamazaki, F. Pichierri, Y. Endo, facile and efficient synthesis of *C*-hydroxycarboranes and *C,C'*-dihydroxycarboranes, *Inorg. Chem.* **2007**, *46*, 3966.
26. S. E. Lyubimov, V. N. Kalinin, A. A. Tyutyunov, V. A. Olshevskaya, Y. V. Dutikova, C. S. Cheong, P. V. Petrovskii, A. S. Safronov, V. A. Davankov, Chiral phosphites derived from carboranes: Electronic effect in catalytic asymmetric hydrogenation, *Chirality* **2009**, *21*, 2.
27. (a) R. A. Wiesboeck, M. F. Harthorne, Dicarbaundecaborane(13) and derivatives, *J. Am. Chem. Soc.* **1964**, *86*, 1642; (b) L. I. Zakharkin, V. N. Kalinin, On the reaction of amines with barenes, *Tetrahedron Lett.* **1965**, *6*, 407; (c) M. A. Fox, W. R. Gill, P. L. Herbertson, J. A. H. MacBride, K. Wade, Deboronation of C-substituted *ortho-* and *meta-closo*-carboranes using "wet" fluoride ion solutions, *Polyhedron* **1996**, *15*, 565; (d) Y. Taoda, T. Sawabe, Y. Endo, K. Yamaguchi, S. Fujii, H. Kagechika, Identification of an intermediate in the deboronation of *ortho*-carborane: An adduct of *ortho*-carborane with two nucleophiles on one boron atom, *Chem. Commun.* **2008**, 2049.
28. K. Schlögl, Stereochemisty of metallocenes, *Top. Sterochem.* **1967**, *1*, 39.
29. (a) R. Núñez, C. Viñas, F. Teixidor, R. Sillanpää, R. Kivakäs, Contribution of the *o*-carboranyl fragment to the chemical stability and the ^{31}P-NMR chemical shift in *closo*-carboranylphosphines. crystal structure of bis(1-yl-2-methyl-1,2-dicarba-*closo*-dodecaborane)phenylphosphine, *J. Organomet. Chem.* **1999**, *592*, 22; (b) M. F. Hawthorne, D. C. Young, T. D. Andrews, D. V. Howe, R. L. Pilling, A. D. Pitts, M. Reintjes, L. F. Warren Jr., P. A. Wegner, π-Dicarbollyl derivatives of the transition metals. metallocene analogs, *J. Am. Chem. Soc.* **1968**, *90*, 879; (c) I. Rojo, F. Teixidor, C. Viñas, R. Kivekäs, R. Sillanpää, Relevance of the electronegativity of boron in η^5-coordinating ligands: Regioselective monoalkylation and monoarylation in cobaltabisdicarbollide $[3,3'\text{-Co}(1,2\text{-}C_2B_9H_{11})_2]^-$ Clusters, *Chem. Eur. J.* **2003**, *9*, 4311.
30. M. F. Hawthorne, J. F. Liebman, A. Greenberg, R. E. Williams, *Advances in Boron and the Boranes*, VCH Publishers Inc., New York, **1988**.
31. M. Yoshida, D. J. Crowther, R. F. Jordan, Synthesis, structure, and reactivity of a novel hafnium carboranyl hydride complex, *Organometallics* **1997**, *16*, 1349.
32. (a) G. G. Hlatky, H. W. Turner, R. R. Eckman, Ionic, base-free zirconocene catalysts for ethylene polymerization, *J. Am. Chem. Soc.* **1989**, *111*, 2728; (b) D.-H. Kim, J. H. Won, S.-J. Kim, J. Ko, S. H. Kim, S. Cho, S. O. Kang, Dicarbollide analogues of the constrained-geometry polymerization catalyst, *Organometallics* **2001**, *20*, 4298; (c) O. Tutusaus, S. Delfosse, S. Simal, A. Demonceau, A. F. Noels, R. Núñez, C. Viñas, F. Teixidor, Half-sandwich ruthenium complexes for the controlled radical polymerization of vinyl monomers, *Inorg. Chem. Commun.* **2002**, *5*, 941; (d) C. De Rosa, P. Corradini, A. Buono, F. Auriemma, A. Grassi, P. Altamura, Crystalline ethylene-norbornene copolymers: plastic crystals from macromolecules, *Macromolecules*, **2003**, *36*, 3789.
33. F. A. Hart, D. W. Owen, Chlorocarbonyl-1,2-bisdiphenylphosphino-1,2-dicarbadodecaboranerhodium(I) andchloro-1,2-bisdiphenylphosphino-1,2-dicarbadodecaboranetriphenylphosphinerhodium(I)—Preparation and catalytic properties, *Inorg. Chim. Acta* **1985**, *103*, L1.
34. I. Maulana, Synthesis, coordination chemistry and application in catalysis of chiral bidentate phosphino-*ortho*-carboranes(12), Dissertation, Universität Leipzig **2005**.
35. (a) S. Bauer, Synthese und Charakterisierung *P*-chiraler Phosphonitocarboran(12)e und ihre Anwendung, Dissertation, Universität Leipzig **2009**; (b) S. Bauer, S. Tschirschwitz, P. Lönnecke, R. Frank, B. Kirchner, M. L. Clarke, E. Hey-Hawkins, Enantiomerically pure bis(phosphanyl)carborane(12) compounds, *Eur. J. Inorg. Chem.* **2009**, 2776.
36. J.-D. Lee, T. T. Co, T.-J. Kim, S. O. Kang, New types of *o*-carborane-based chiral phosphinooxazoline (Cab-PHOX) ligand systems: synthesis and characterization of chiral Cab-PHOX ligands and their application to asymmetric hydrogenation, *Synlett* **2009**, *5*, 771.
37. H.-S. Lee, J.-Y. Bae, J. Ko, Y. S. Kang, H. S. Kim, S.-J. Kim, J.-H. Chung, S. O. Kang, New Group 9 metal complexes containing *N,P*-chelate ligand system. Synthesis, characterization and application to catalytic hydrogenation, *J. Organomet. Chem.* **2000**, *614–615*, 83.
38. (a) T. Hayashi, K. Yamamoto, M. Kumada, Asymmetric catalytic hydrosilylation of ketones preparation of chiral ferrocenylphosphines as chiral ligands, *Tetrahedron Lett.* **1974**, 4405; (b) T. Hayashi, M. Tajika, K. Tamao, M. Kumada, High stereoselectivity in asymmetric grignard cross-coupling catalyzed by nickel-complexes of chiral (aminoalkylferrocenyl)phosphines, *J. Am. Chem. Soc.* **1976**, *98*, 3718.
39. S. P. Smidt, F. Menges, A. Pfaltz, SimplePHOX, a readily available chiral ligand system for iridium-catalyzed asymmetric hydrogenation, *Org. Lett.* **2004**, *6*, 2023.

40. (a) S. E. Lyubimov, A. A. Tyutyunov, V. N. Kalinin, E. E. Said-Galiev, A. R. Khokhlov, P. V. Petrovskii, V. A. Davankov, Carboranylphosphites – New effective ligands for rhodiumcatalyzed asymmetric hydrogenation of dimethyl itaconate, *Tetrahedron Lett.* **2007**, *48*, 8217; (b) S. E. Lyubimov, V. A. Davankov, P. V. Petrovskii, E. Hey-Hawkins, A. A. Tyutyunov, E. G. Rys, V. N. Kalinin, Chiral carborane-derived thiophosphites: A new generation of ligands for rh-catalyzed asymmetric hydrogenation, *J. Organomet. Chem.* **2008**, *693*, 3689; (c) S. E. Lyubimov, I. V. Kuchurov, A. A. Tyutyunov, P. V. Petrovskii, V. N. Kalinin, S. G. Zlotin, V. A. Davankov, E. Hey-Hawkins, The use of new carboranylphosphite ligands in the asymmetric Rh-catalyzed hydrogenation, *Catal. Comm.* **2010**, *11*, 419.
41. C. Viñas, M. A. Flores, R. Núñez, F. Teixidor, R. Kivekäs, S. Sillanpää, "*exo-nido*-Monothio- and *exo-nido*-monophosphinorhodacarboranes: Synthesis, reactivity, and catalytic properties in alkene hydrogenation, *Organometallics* **1998**, *17*, 2278.
42. B. Pirotte, A. Felekidis, M. Fontaine, A. Demonceau, A. F. Noels, J. Delarge, I. T. Chizhevsky, T. V. Zinevich, I. V. Pisareva, V. I. Bregadze, Stereoselective hydrogenation of methacycline to doxycycline catalyzed by rhodium-carborane complexes, *Tetrahedron Lett.* **1993**, *34*, 1471.
43. H. Brunner, A. Apfelbacher, M. Zabel, "Palladium and rhodium complexes with planar-chiral carborane ligands", *Eur. J. Inorg. Chem.* **2001**, 917.
44. H. Brunner, D. Mijolovic, Enantioselective catalysis Part 129. A new rhodium(I) complex with a μ_2-H bridged Cp_2WH_2 ligand, *J. Organomet. Chem.* **1999**, *577*, 346.
45. H. Brunner, S. Rosenboem, Enantioselective catalyses CXXXV. Stereoselective hydrogenation of folic acid and 2-methylquinoxaline with optically active rhodium(I)-phosphane complexes, *Chem. Monthly* **2000**, *131*, 1371.
46. C. Botteghi, S. Paganelli, A. Schionato, M. Marchetti, The asymmetric hydroformylation in the synthesis of pharmaceuticals, *Chirality* **1991**, *3*, 355.
47. (a) H.-W. Bohnen, B. Cornils, Hydroformylation of alkenes: An industrial view of the status and importance, *Adv. Catal.* **2002**, *47*, 1; (b) C. J. Cobley, K. Gardner, J. Klosin, C. Praquin, C. Hill, G. T. Whiteker, A. Zanotti-Gerosa, Synthesis and application of a new bisphosphite ligand collection for asymmetric hydroformylation of allyl cyanide, *J. Org. Chem.* **2004**, *69*, 4031.
48. T. P. Clark, C. R. Landis, S. L. Freed, J. Klosin, K. A. Abboud, highly active, regioselective, and enantioselective hydroformylation with Rh catalysts ligated by bis-3,4-diazaphospholanes, *J. Am. Chem. Soc.* **2005**, *127*, 5040.
49. S. Tschirschwitz, Synthese, Charakterisierung und Anwendung neuer *P*-chiraler Aminoalkylferrocenyl phosphane, Dissertation, Universität Leipzig **2007**.
50. (a) M. M. H. Lambers-Verstappen, J. G. de Vries, Rhodium-catalyzed asymmetric hydroformylation of unsaturated nitriles, *Adv. Synth. Catal.* **2003**, *345*, 478; (b) C. J. Cobley, K. Gardner, J. Klosin, C. Praquin, C. Hill, G. T. Whiteker, A. Zanotti-Gerosa, J. L. Petersen, K. A. Abboud, Synthesis and application of a new bisphosphite ligand collection for asymmetric hydroformylation of allyl cyanide, *J. Org. Chem.* **2004**, *69*, 4031; (c) I. del Río, W. G. L. de Lange, P. W. N. M. van Leeuwen, C. Claver, *In situ* study of diphosphine rhodium systems in asymmetric hydroformylation of styrene, *J. Chem. Soc., Dalton Trans.* **2001**, 1293; (d) G. J. H. Buisman, E. J. Vos, P. C. J. Kamer, P. W. N. M. van Leeuwen, Hydridorhodium diphosphite catalysts in the asymmetric hydroformylation of styrene, *J. Chem. Soc., Dalton Trans.* **1995**, 409; (e) S. Cserépi-Szűcs, G. Huttner, L. Zsolnai, J. Bakos, Asymmetric hydroformylation of styrene using rhodium and platinum complexes of diphosphites containing atropisomeric backbones and chiral 1,3,2-dioxaphosphorinane moieties, *J. Organomet. Chem.* **1999**, *586*, 70; (f) O. Pàmies, G. Net, A. Ruiz, C. Claver, Asymmetric hydroformylation of styrene catalyzed by furanoside phosphine–phosphite–Rh(I) complexes, *Tetrahedron: Asymmetry* **2002**, *12*, 3441.
51. (a) Y. Kiso, M. Kumada, K. Tamao, M. Umeno, Silicon hydrides and nickel-complexes. 1. phosphine-nickel(ii) complexes as hydrosilylation catalysts, *J. Organomet. Chem.* **1973**, *73*, 297; (b) Y. Kiso, M. Kumada, K. Maeda, K. Sumitani, K. Tamao, Silicon hydrides and nickel-complexes. 2. Mechanism of hydrosilylation catalyzed by nickel-phosphine complexes, *J. Organomet. Chem.* **1973**, *73*, 311; (c) M. Kumada, K. Sumitani, Y. Kiso, K. Tamao, Silicon hydrides and nickel-complexes. 3. Hydrosilylation of olefins with dichloro[1,2-bis-(dimethylphosphino)-1,2-dicarba-*closo*-dodecaborane]-nickel(ii) as catalyst, *J. Organomet. Chem.* **1973**, *73*, 319.
52. (a) H.-S. Lee, J.-Y. Bae, J. Ko, Y. S. Kang, H. S. Kim, S. O. Kang, Novel bimetallic group 9 metal catalysts containing *P,S*-chelating *o*-carboranyl ligand system for the carbonylation of methanol, *Chem. Lett.* **2000**, 602; (b) H.-S. Lee, J.-Y. Bae, D.-H. Kim, H. S. Kim, S.-J. Kim, S. Cho, J. Ko, S. O. Kang, Rhodium and iridium phosphinothiolato complexes synthesis and crystal structures of mononuclear [M(cod)(*S,P*-$SC_2B_{10}H_{10}PPh_2$)] and dinuclear [$M_2(CO)_2$(S,P-μ-$SC_2B_{10}H_{10}PPh_2$)] (M = Rh, Ir) and their performance in catalytic carbonylation, *Organometallics* **2002**, *21*, 210.

53. M. Cheong, T. Ziegler, Density functional study of the oxidative addition step in the carbonylation of methanol catalyzed by $[M(CO)_2I_2]^-$ (M = Rh, Ir), *Organometallics* **2005**, *24*, 3053.
54. (a) S. E. Lyubimov, A. A. Tyutyunov, P. A. Vologzhanin, A. S. Safronov, P. V. Petrovskii, V. N. Kalinin, K. N. Gavrilov, V. A. Davankov, Carborane-derived diphosphites: new ligands for Pd-catalyzed allylic amination, *J. Organomet. Chem.* **2008**, *693*, 3321; (b) S. E. Lyubimov, V. A. Davankov, K. N. Gavrilov, T. B. Grishina, E. A. Rastorguev, A. A. Tyutyunov, T. A. Verbitskaya, V. N. Kalinin, E. Hey-Hawkins, Diamidophosphites with isomeric carborane fragments: A comparison of catalytic activity in asymmetric pd-catalyzed allylic substitution reactions, *Tetrahedron Lett.* **2010**, *51*, 1682.
55. Z. Lu, S. M. Ma, Metallkatalysierte enantioselektive Allylierungen in der asymmetrischen Synthese, *Angew. Chem.* **2008**, *120*, 264; Metal-catalyzed enantioselective allylation in asymmetric synthesis, *Angew. Chem. Int. Ed.* **2008**, *47*, 258.
56. D. Zhang, J. Dou, S. Gong, D. Li, D. Wang, Synthesis and characterization of three dinuclear copper(I) complexes of 1,2-bis(diphenylphosphino)-1,2-dicarba-*closo*-dodecaborane, *Appl. Organometal. Chem.* **2006**, *20*, 632.
57. N. M. Patil, A. A. Kelker, R. V. Chaudhari, Synthesis of triarylamines by copper-catalyzed amination of aryl halides, *J. Mol. Catal. A – Chem.* **2004**, *223*, 45.
58. S. E. Lyubimov, I. V. Kuchurov, A. A. Vasil'ev, A. A. Tyutyunov, V. N. Kalinin, V. A. Davankov, S. G. Zlotin, The use of a new carboranylamidophosphite ligand in the asymmetric Pd-catalyzed allylic alkylation in organic solvents and supercritical carbon dioxide, *J. Organomet. Chem.* **2009**, *694*, 3047.
59. (a) M. S. Kharasch, E. V. Jensen, W. H. Urry, Addition of Carbon Tetrachloride and Chloroform to Olefins, *Science* **1945**, *102*, 128; (b) M. S. Kharasch, E. V. Jensen, W. H. Urry, Addition of derivatives of chlorinated acetic acids to olefins, *J. Am. Chem. Soc.* **1945**, *67*, 1626.
60. (a) M. Kato, M. Kamigaito, M. Sawamoto, T. Higashimura, Polymerization of methyl methacrylate with the carbon tetrachloride/dichlorotris-(triphenylphosphine)ruthenium(II)/ methylaluminum bis(2,6-di-*tert*-butylphenoxide) initiating system: Possibility of living radical polymerization, *Macromolecules* **1995**, *28*, 1721; (b) J.-S. Wang, K. Matyjaszewski, Controlled living radical polymerization—atom-transfer radical polymerization in the presence of transition-metal complexes, *J. Am. Chem. Soc.* **1995**, *117*, 5614.
61. F. Simal, S. Sebille, A. Demonceau, A. F. Noels, R. Nuñez, M. Abad, F. Teixidor, C. Viñas, Highly Effcient Kharasch Addition Catalyzed by $RuCl(Cp^*)(PPh_3)_2$, *Tetrahedron Lett.* **2000**, *41*, 5347.
62. (a) S. Paavola, *Synthesis and Characterisation of 1,2-Diphosphino-o-Carborane Metal Complexes*, Academic dissertation, University of Helsinki **2002**; (b) A. Richel, S. Delfosse, A. Demonceau, A. F. Noels, S. Paavola, R. Kivekäs, C. Viñas, F. Teixidor, Radical polymerization of methyl methacrylate catalyzed by palladium(II) complexes containing chelating *o*-carboranyldiphosphine ligands, *Polym. Repr.* **2002**, *43*, 177.
63. H. Wang, H.-S. Chan, J. Okuda, Z. Xie, Synthesis, structural characterization, and catalytic properties of group 4 metal complexes incorporating a phosphorus-bridged indenyl-carboranyl constrained-geometry ligand, *Organometallics* **2005**, *24*, 3118.
64. (a) H. Wang, Y. Wang, H.-W. Li, Z. Xie, Synthesis, structural characterization, and olefin polymerization behavior of group 4 metal complexes with constrained-geometry carborane ligands, *Organometallics* **2001**, *20*, 5110; (b) G. Zi, H.-W. Li, Z. Xie, Synthesis, structural characterization, and catalytic property of group 4 metal carborane compounds with a $^i Pr_2NB$-bridged constrained-geometry ligand, *Organometallics* **2002**, *21*, 3850.
65. A. Demonceau, F. Simal, A. F. Noels, C. Viñas, R. Núñez, F. Teixidor, Cyclopropanation reactions catalyzed by ruthenium complexes with new anionic phosphine ligands, *Tetrahedron Lett.* **1997**, *38*, 4079.
66. (a) M. P. Doyle, W. H. Tamblyn, W. E. Buhro, R. L Dorow, Exceptionally effective catalysis of cyclopropanation reactions by the hexarhodium carbonyl cluster, *Tetrahedron Lett.* **1981**, *22*, 1783; (b) M. P. Doyle, D. van Leusen, W. H. Tamblyn, Efficient alternative catalysts and methods for the synthesis of cyclopropanes from olefins and diazo compounds, *Synthesis* **1981**, 787; (c) M. P. Doyle, R. L. Dorow, W. E. Buhro, J. H. Griffin, W. H. Tamblyn, M. L. Trudell, Stereoselectivity of catalytic cyclopropanation reactions—catalyst dependence in reactions of ethyl diazoacetate with alkenes, *Organometallics* **1984**, *3*, 44; (d) D. Holland, D. J. J. Milner, Stereoselective catalytic formation of cyclopropanecarboxylates from ethyl diazoacetate and 1,1-dihalogeno-4-methylpenta-1,3-dienes, *J. Chem. Res. (S)* **1979**, 317; *J. Chem. Res. (M)* **1979**, 3734; (e) H. J. Callot, C. Pieehocki, Cyclopropanation using rhodium(III)porphyrins—large *cis* vs *trans* selectivity, *Tetrahedron Lett.* **1980**, *21*, 3489; (f) S. O'Malley, T. Kodadek, Asymmetric cyclopropanation of alkenes catalyzed by a chiral wall porphyrin, *Tetrahedron Lett.* **1991**, *32*, 2445; (g) J. L. Maxwell, K. C. Brown, D. W. Bartley, T. Kodadek, Mechanism of the rhodium porphyrin catalyzed cyclopropanation of alkenes, *Science* **1992**, *256*, 1544; (h) D. W. Bartley, T. Kodadek,

Identification of the active catalyst in the rhodium porphyrin-mediated cyclopropanation of alkenes, *J. Am. Chem. Soc.* **1993**, *115*, 1656.

67. (a) W. H. Tamblyn, S. R. Hoffmann, M. P. J. Doyle, Correlation between catalytic cyclopropanation and ylide generation, *J. Organomet. Chem.* **1981**, *216*, C64; (b) A. J. Ancianx, A. J. Hubert, A. F. Noels, N. Petiniot, P. J. Teyssié, Transition-metal-catalyzed reactions of diazo-compounds. 1. Cyclopropanation of double-bonds, *J. Org. Chem.* **1980**, *45*, 695.
68. A. Demonceau, F. Simal, A. F. Noels, C. Viñas, R. Núñez, F. Teixidor, Cyclopropanation reactions catalyzed by rhodium(I) complexes with new anionic carborane phosphine ligands, *Tetrahedron Lett.* **1997**, *38*, 7879.
69. K. Tamao, Y. Kiso, K. Sumitani, M. Kumada, Alkyl group isomerization in cross-coupling reaction of secondary alkyl grignard-reagents with organic halides in presence of nickel-phosphine complexes as catalysts, *J. Am. Chem. Soc.* **1972**, *94*, 9268.
70. H. Nakamura, T. Kamakura, S. Onagi, 1,2-Bis(diphenylphosphino)carborane as a dual mode ligand for both the sonogashira coupling and hydride-transfer steps in palladium-catalyzed one-pot synthesis of allenes from aryl iodides, *Org. Lett.* **2006**, *8*, 2095.

23 Organic Synthesis Using Boron and Organoboron Halides

Min-Liang Yao and George W. Kabalka

CONTENTS

23.1 INTRODUCTION

Due to the empty *p*-orbital on boron, boron halides (BX_3:X = F, Cl, Br, I) and organoboron halides (RBX_2, R_2BX) are Lewis acids. The observed Lewis acidity of boron trihalides is in the order $BI_3 > BBr_3 > BCl_3 > BF_3$ which is exactly the reverse of that expected on the basis of the relative σ-donor strengths of the halide anions but is easily rationalized by considering back bonding through the halogen *p*-orbitals.[1] One major application of BX_3 in organic syntheses is C–O bond cleavage in ethers.[2] The notable features of boron halide ether cleavage are mild reaction conditions and high regio- and chemoselectivities. As haloboration reagents, boron trihalides are used to synthesize (Z)-2-halo-1-alkenylboron halides,[3] highly functionalized intermediates that play an important role in the syntheses of numerous vinyl halides, alkadienes, alkenynes, and olefinic products. In addition, boron trihalides are used as halogenation reagents in conversions of alcohols into the corresponding halides.[4] Halogen exchange between alkyl halides and boron trihalides has also been reported.[5] The reaction of boron trihalides with *N,O*-based chiral ligands is also an active research area because the chiral boron Lewis acids thus obtained possess the capacity to induce chirality in other molecules. The concept has been successfully used for enantioselective Diels–Alder reactions.[6]

The most important application of organoboron halides in organic syntheses is the generation of enol borinates that are highly versatile intermediates for stereocontrolled aldol reactions.[7] The organoboron halide-induced aldol reaction has been used extensively in the preparation of natural products.[8] Conversion of organoboron halides to the corresponding organoboranes via controlled hydridation also plays an important role in the preparation of mixed dialkylboron halides and trialkylboranes.[9] This stepwise hydroboration strategy successfully solves the dihydroboration problem encountered using borane (BH_3) complexes for the hydroboration of alkynes and is used widely in the stereocontrolled syntheses of natural products containing (*Z*)- and (*E*)-alkene moieties. In addition, the chiral organoboron halide (–)-diisopinocampheylchloroborane [(–)-DIP-Cl] is known for its ability to enantioselectively reduce prochiral carbonyl compounds.[10] The remarkable feature of a DIP-Cl reduction is asymmetric amplification. As Lewis acids, chiral organoboron halides are also used in enantioselective Diels–Alder[11] and Pictet–Spengler reactions.[12]

In recent years, novel reactions have been developed that utilize boron and organoboron halides.[13] These include the haloalkylation of aryl aldehydes, a transition metal-free formal Suzuki-coupling of alkoxides with vinylboron dichlorides, the boron trichloride-mediated coupling of alkoxides with allylsilane, and the boron trihalide-mediated haloallylation of aryl aldehydes using allylmetal compounds. Interestingly, in all these reactions the boron halides or organoboron halides are found to act, simultaneously, as Lewis acids and reactants. In this chapter, an overview of these reactions, along with recent advances in organic syntheses using boron halides and organoboron halides, is presented. Boron trifluoride chemistry is not included because BF_3, although an effective Lewis acid, is not particularly effective in reactions involving carbon–carbon bond formation.

23.2 BORON TRIHALIDES IN ORGANIC SYNTHESES

23.2.1 BX_3-Promoted Carbon–Oxygen Bond Cleavage in Ethers, Acetals, and Esters

Although all BX_3 reagents will cleave ethers, BBr_3 is most widely used because these cleavage reactions proceed under very mild reaction conditions. Boron trichloride has been found to be less reactive but it does provide a means for adjusting the selectivity and rate of ether cleavage. For example, the cleavage of a methylenedioxy group in the presence of an aromatic methoxy group using BCl_3 at room temperature can be achieved.[14] Interestingly, the simultaneous cleavage of a methylenedioxy group and an aromatic methoxy group using excess BCl_3 at elevated temperature has also been reported.[15] Generally, stoichiometric quantities of BX_3 reagents are required for complete cleavage. These reactions demonstrate high chemoselectivity, regioselectivity, and a compatibility with a variety of functional groups. In addition, it was discovered that the reactivity of BX_3 can be greatly enhanced when combined with NaI[16] or Bu_4NI[17]. For example, the BCl_3/*n*-Bu_4NI reagent's reactivity is between that of BBr_3 and BI_3. The resulting combination reagent system is mild and it can provide results superior to those of BBr_3 for the cleavage of numerous substrates.[17a] Furthermore, the combined reagent provides for selective dealkylation in certain cases. Since several reviews and book chapters are available on this subject,[2a,18] the current discussion is limited to a brief summary that emphasizes the important features of the cleavage reactions.

The removal of methoxy and benzoxy protecting groups in aryl alkyl ethers is the most widely used BX_3 ether cleavage reaction. Overall, the reaction produces a phenol and an alkyl halide. The chemoselective deprotection of triphenylmethyl (trityl) ethers in the presence of differentially protected diols has been reported (Scheme 23.1).[19] At –30°C, trityloxy ethers can be selectively unmasked by BCl_3 while the C–O bond in silicon-based ethers as well as benzyl ethers is unaffected.

Cleavage of aryl allyl ethers using boron trihalides is normally unsuccessful because allyl ethers are prone to Claisen rearrangements. Sterically hindered organoboron halides such as 9-bromo-9-borabicyclo[3.3.1]nonane (B-Br-9-BBN) are required to achieve aryl allyl ether cleavage.[17a,20] Interestingly, it was discovered that aryl propargyl ethers and esters can be selectively cleaved in the presence of aryl methyl ethers and esters using boron tribromide in dichloromethane solvent.[21] Originally, a reaction mechanism involving the delivery of bromine to the propargyl terminus, to

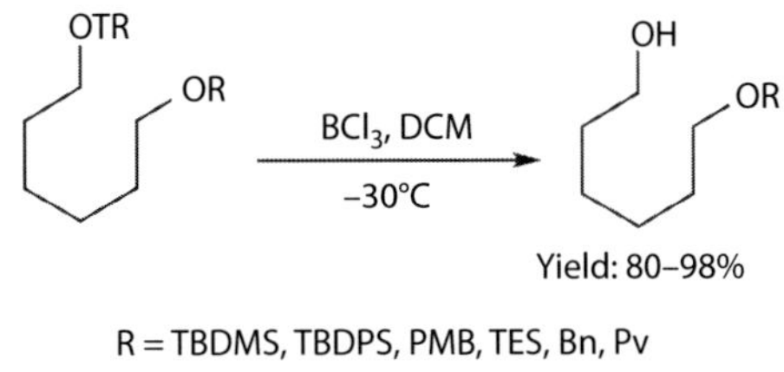

OTR OR $\xrightarrow[-30°C]{BCl_3, DCM}$ OH OR

Yield: 80–98%

R = TBDMS, TBDPS, PMB, TES, Bn, Pv

SCHEME 23.1 Chemoselective cleavage of trityloxy ethers using boron trichloride.

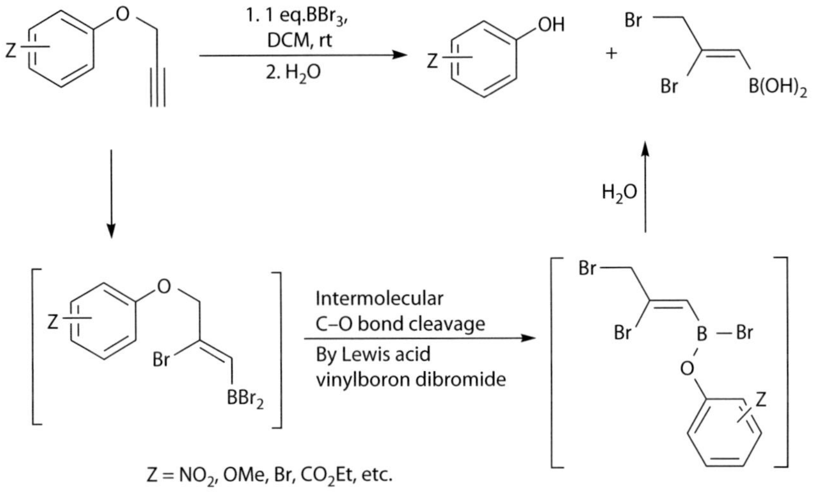

1. 1 eq. BBr_3, DCM, rt; 2. H_2O

Intermolecular C–O bond cleavage By Lewis acid vinylboron dibromide

$Z = NO_2$, OMe, Br, CO_2Et, etc.

SCHEME 23.2 Boron tribromide-mediated aryl propargyl ether cleavage.

generate a bromoallene, was proposed. Later an alternative mechanism[22] (Scheme 23.2) was suggested based on the facts that the haloboration reaction of terminal alkynes is a facile reaction at –78°C and that organoboron dibromides are powerful ether cleavage reagents.[23] This alternative mechanism was supported by the successful isolation of a key boron-containing by-product. Based on the newly proposed mechanism, it was discovered that dibromoborane also cleaves aryl propargyl ethers.

The introduction of a non-Lewis-basic cation scavenger C_6HMe_5 induces selective debenzylation in the presence of the labile thiolester group (Scheme 23.3).[24] The reaction was successfully applied in the total synthesis of yatakemycin.[25] Further investigation revealed that this selective debenzylation

R=Bn → R=H: BCl_3, C_6HMe_5, DCM, –78°C; Yield: 83%

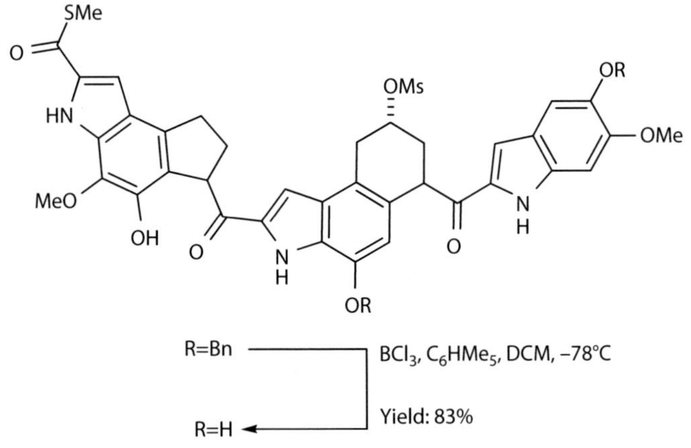

SCHEME 23.3 Selective debenzylation using non-Lewis-basic cation scavenger.

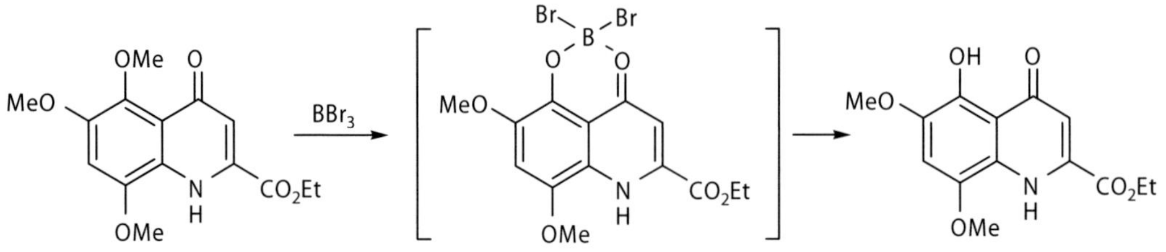

SCHEME 23.4 Regioselective deprotection of a methoxy group in a polymethoxyaryl.

is particularly useful for substrates bearing electron-rich aromatic rings such as trialkoxybenzenes or indol derivatives. Acid-sensitive functional groups such as Boc and TBS ether also survive under these novel reaction conditions.

Regioselective deprotection of a methoxy group adjacent to a carbonyl substitutent in polymethoxyaryl compounds is also possible (Scheme 23.4).[26] The selective cleavage is believed to be promoted by the formation of a six-membered boronate intermediate.

In addition to the notable regio- and chemoselectivity observed in boron halide-mediated ether cleavage reactions, the compatibility of these reagents with a wide variety of functional groups is quite remarkable. For example, boron trichloride efficiently cleaves benzyl ethers and ketals without affecting the imine *N*-oxide functionality in nitrones (Scheme 23.5).[27] The method was used in the efficient syntheses of polyhydroxylated *N*-hydroxypyrrolidines and *N*-hydroxyiminosugars.

Recently, it was discovered that a diazonium group also survives under the ether cleavage conditions (Scheme 23.6).[28] The reaction provides a practical route to arenediazonium salts bearing a catechol moiety.

While boron trihalide-mediated ether cleavage has been successfully applied in a number of organic syntheses, it was discovered that the BBr_3-mediated demethylation of 2-methoxybiaryl systems could be problematic due to an intramolecular electrophilic aromatic cyclization (Scheme 23.7).[29] In these cases, 10-hydroxy-10,9-boroxarophenanthrenes are obtained as major products.

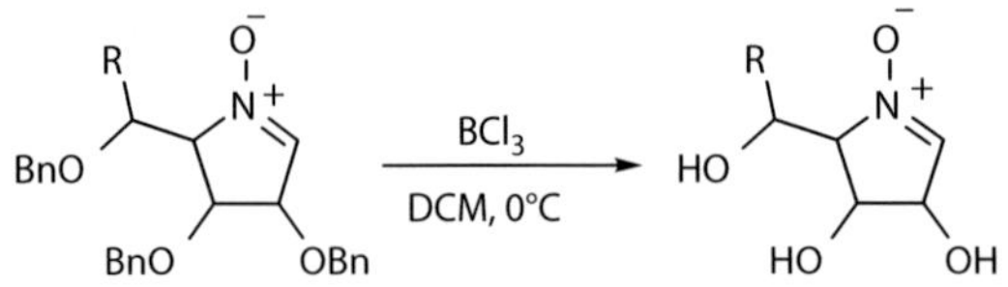

SCHEME 23.5 Debenzylation in the presence of an imine *N*-oxide.

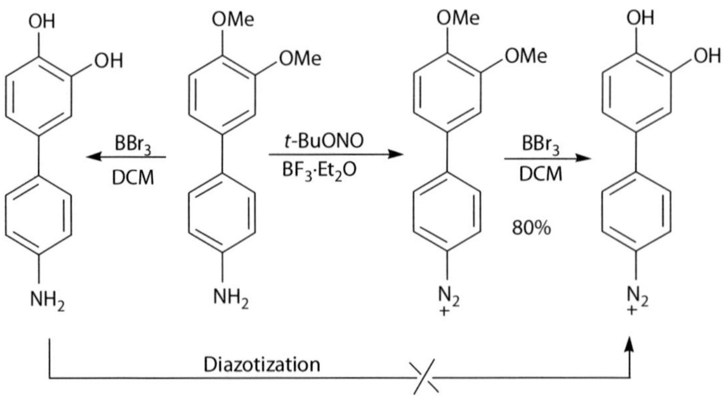

SCHEME 23.6 Aryl ether cleavage in the presence of a diazonium group.

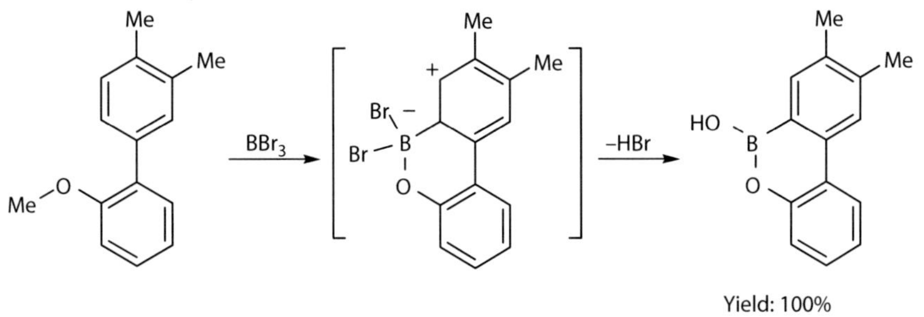

Me Me BBr3 Br Br B O Me Me -HBr HO B O Me Me Yield: 100%

SCHEME 23.7 Side reaction in BBr_3-mediated demethylation of 2-methoxybiaryl.

Compared to the well-known BBr_3-mediated cleavage of aryl alkyl ethers, reactions of cyclic ethers have rarely been investigated. One known example is the cleavage of cyclic ethers to generate ω-bromoalkanols after treatment with BBr_3/MeOH.[30] Recently, this methodology has been successfully used in the synthesis of 3,4-disubstituted isoxazole[31] and haloconduritol[32] derivatives (Scheme 23.8).

An efficient synthesis of carbonyl compounds containing remote halide functionality has also been developed. The reaction involves the ring-opening of 2-alkylidenetetrahydrofurans (a vinyl ether) with 4 equivalent of BBr_3.[33] One example is the chemo- and regioselective synthesis of 6-bromo-3-oxoalkanoates (Scheme 23.9). The reaction presumably proceeds by activation of the conjugated ester through coordination, ring-cleavage, and subsequent protonation of the enolate.

In addition, it was discovered that BCl_3-mediated ring-opening of tetrahydrofuran followed by electrophilic attack of an acid chloride was feasible (Scheme 23.10).[34] The method provides a one-pot route to ω-chloroesters.

The cleavage of noncyclic, mixed dialkyl ethers normally occurs at the most substituted carbon of the carbon–oxygen bond. Methyl ethers of secondary or tertiary alcohols give methanol and secondary or tertiary alkyl bromides after reaction with BBr_3.[35] However, the addition of NaI and a crown ether can reverse this selectivity (Scheme 23.11).[16] In contrast, methyl ethers of primary alcohols are generally cleaved at the Me–O bond.[36]

Generally speaking, ester groups, except for lactones,[37] survive the BBr_3-mediated ether cleavage reaction conditions. However, $BI_3 \cdot NEt_2Ph$, obtained by refluxing the amine-boron complex with I_2 in benzene, can be used to cleave the C–O bond in ethers, esters, and geminal diacetates.[38] Esters are cleaved to a reactive acyl intermediate RCOI that can then be used to prepare acids, esters, and amides (Scheme 23.12). It has also been discovered that $BI_3 \cdot NEt_2Ph$ cleaves the sulfur–oxygen bond in sulfinyl and sulfonyl compounds.[39]

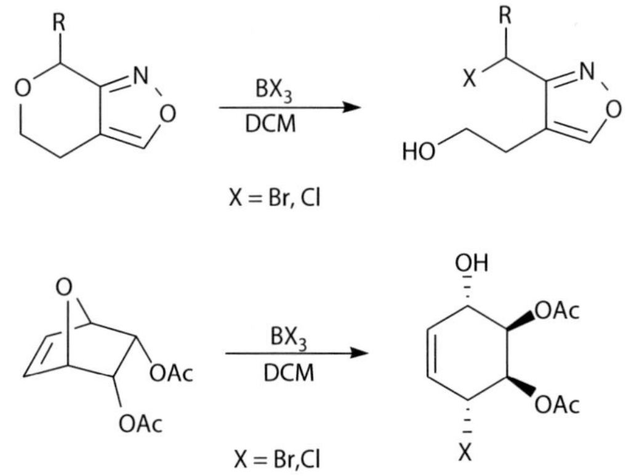

R O N O BX3 DCM R X N O HO X = Br, Cl

O OAc OAc BX3 DCM OH OAc OAc X X = Br,Cl

SCHEME 23.8 Cyclic ether cleavage to ω-haloalkanols.

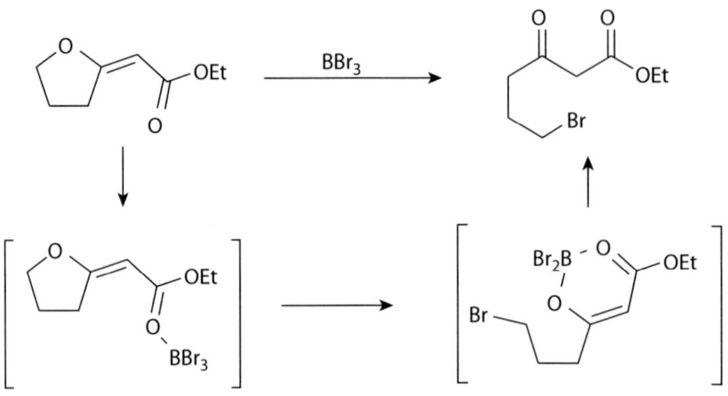

SCHEME 23.9 Boron tribromide-mediated 2-alkylidenetetrahydrofuran cleavage.

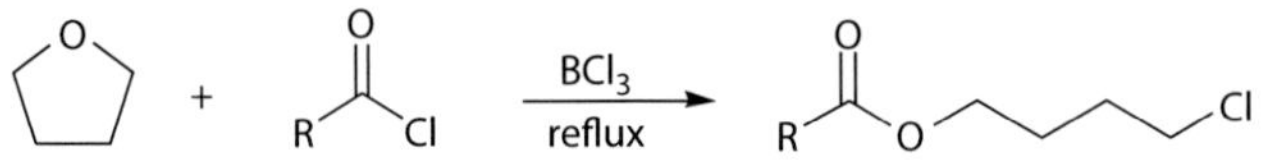

SCHEME 23.10 BCl_3-mediated ring-opening of tetrahydrofuran.

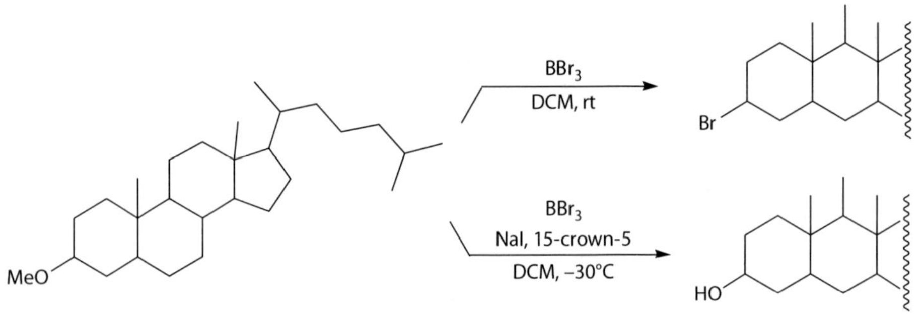

SCHEME 23.11 Reverse selectivity in cleaving noncyclic mixed dialkyl ethers.

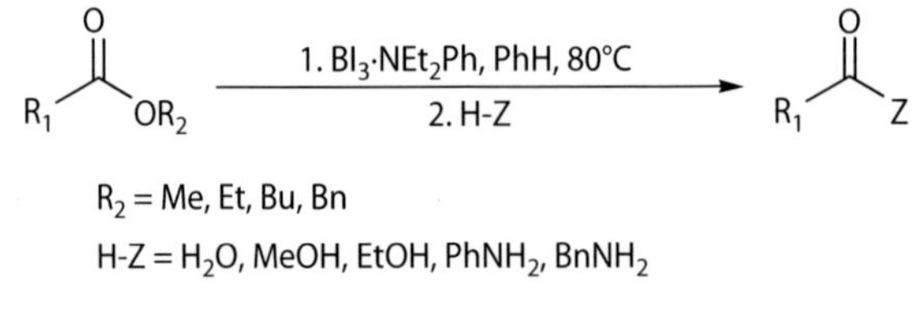

SCHEME 23.12 BI_3•NEt_2Ph-mediated ester cleavage.

Carbamate esters also react with BCl_3 in the presence of Et_3N to afford isocyanates under mild reaction conditions (Scheme 23.13).[40] In most cases, quantitative or near-quantitative conversion to isocyanates is achieved in aryl, alkyl, alicyclic, and tosyl carbamate esters.

23.2.2 Application in the Preparation of Organoboron Halides

Reaction of boron trihalides with trialkylboranes in the presence of catalytic amounts of sodium borohydride or diborane induces a redistribution to generate dialkylboron halides in good yields.[41] Generally, the reaction requires high temperature. The importance of this redistribution reaction

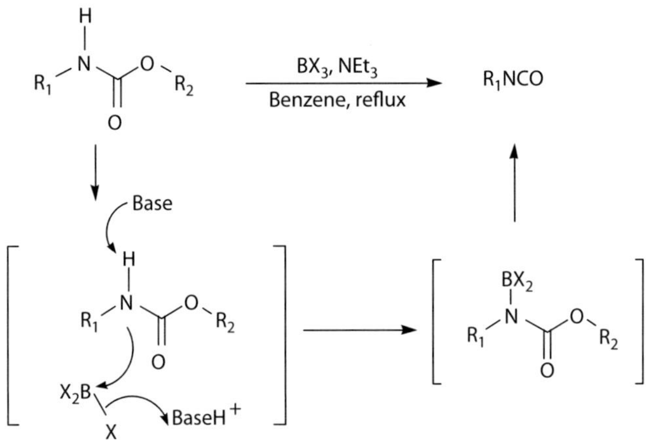

SCHEME 23.13 BCl_3-mediated carbamate ester cleavage in the presence of Et_3N.

has diminished since the discovery of an alternative route involving the hydroboration of alkenes or alkynes using haloboranes.[42] However, the redistribution method has been used to synthesize allylboron dichloride as well as trimethallylboron dichloride.[43] In addition, it was reported that reaction of anhydrides of boronic or borinic acids with boron trihalides in hydrocarbon solvents leads to the clean exchange of the oxygen and halide to form the organodihaloboranes and diorganohaloboranes in good to excellent yields.[44]

Transmetallation reactions between boron trihalides and organometallic reagents containing Si,[45] Hg,[46] Sn,[47] Zr,[48] and Cu[49] are practical and valuable routes to arylboron halides and allylboron halides since hydroboration reactions using haloboranes are not possible for these moieties.

Haloboration of terminal alkynes using boron trihalides is an important method for preparing halovinylboron halides.[3] The reactions take place in a highly stereo- and regioselective manner. Halovinylboron halides are very useful synthetic intermediates because of the presence of three useful functionalities: vinylboron, vinyl halide, and alkene. For terminal alkynes, haloboration reactions occur instantaneously at –78°C using 1 equiv. of alkyne to generate (*Z*)-halovinylboron dihalide in high purity (>98%). If a second equivalent of alkyne is available, further haloboration takes place to afford a (*Z*,*Z*)-di(halovinyl)boron halide (Scheme 23.14). During an investigation of boron trihalide-mediated alkyne–aldehyde coupling reactions, it was discovered that, at low temperatures, the reaction of BCl_3 with alkynes stops at the monovinylboron dichloride stage.[50] The reaction of a halovinylboron dichloride with a second molecule of alkyne is sluggish and elevated temperatures (refluxing DCM) are required for completion. Only boron tribromide successfully haloborates internal alkynes.

Notably, it has been observed that bromoboration of acetylene with boron tribromide proceeds rapidly to generate the unexpected (*E*)-2-bromoethenylboron dibromide (Scheme 23.15).[51] The generally accepted reaction mechanism is that the bromoboration of acetylene proceeds in a *syn* fashion and then the intermediate spontaneously rearranges to the *trans* isomer.

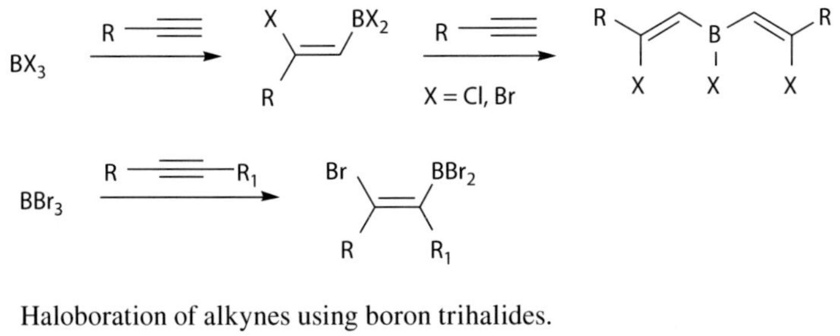

SCHEME 23.14 Haloboration of alkynes using boron trihalides.

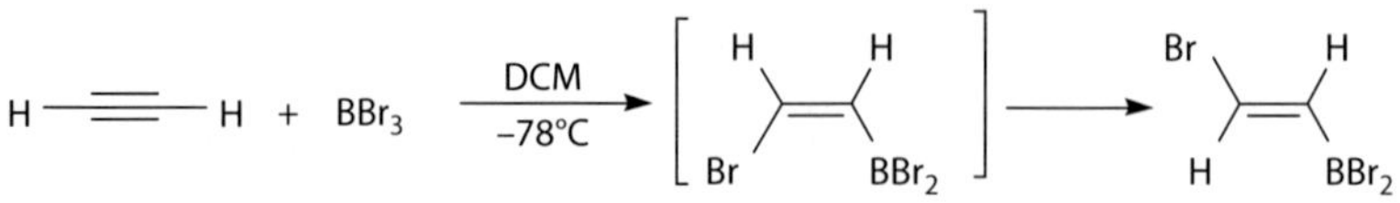

$H-C{\equiv}C-H + BBr_3 \xrightarrow[-78°C]{DCM}$ [(Z)-BrHC=CH-BBr_2] → (E)-BrHC=CH-BBr_2

SCHEME 23.15 Bromoboration of acetylene.

For the application of these (*Z*)-halovinylboron dihalides and (Z,Z)-di(halovinyl)boron halide in organic syntheses, see Section 23.3.1.

23.2.3 Reaction of Aldehydes with Boron Trichloride

Boron halides are strong Lewis acids and readily form crystalline complexes with carbonyl compounds.[52] BF_3 complexes are quite stable and have been used as reaction accelerators,[53] while BCl_3 complexes readily decompose. Boron trichloride complexes of aliphatic aldehydes, such as propanal, have been found to undergo chlorination to form bis(α-chloroethyl) ethers;[54] while aromatic aldehydes slowly form aryldichloromethanes.[55] At elevated reaction temperatures, BCl_3 readily reacts with aromatic aldehydes to form aryldichloromethanes[56] (Scheme 23.16). Aromatic aldehydes bearing a variety of functional groups are successfully converted into the corresponding geminal dichlorides in good to excellent yields. Aromatic aldehydes with strongly electron-withdrawing groups such as the nitro group proceed at a slower rate. The reaction of *p*-anisaldehyde is complicated by BCl_3 cleavage of the methoxy moiety.

The dihalogenation reaction is believed to proceed through coordination of the carbonyl group to the boron trihalide followed by a migration of the second chlorine to the carbon center. The key intermediate, benzyloxyboron dichloride, has been characterized using nuclear magnetic resonance (NMR).[56]

23.2.4 Boron Halide-Mediated Coupling of Aromatic Aldehydes with Styrene Derivatives

Lewis acid-catalyzed addition of carbonyl compounds to alkenes is an important method for forming new carbon–carbon bonds. Lewis acid-promoted ene[57] and Baylis–Hillman[58] reactions have been well documented. Notably, boron trihalide-mediated aryl aldehyde-styrene coupling reactions proceed in a different way (Scheme 23.17).[59] Boron trihalide acts simultaneously as a Lewis acid and as a reactant to regioselectively produce 1,3-dihalo-1,3-diarylpropane (*syn*/*anti* = 50/50).

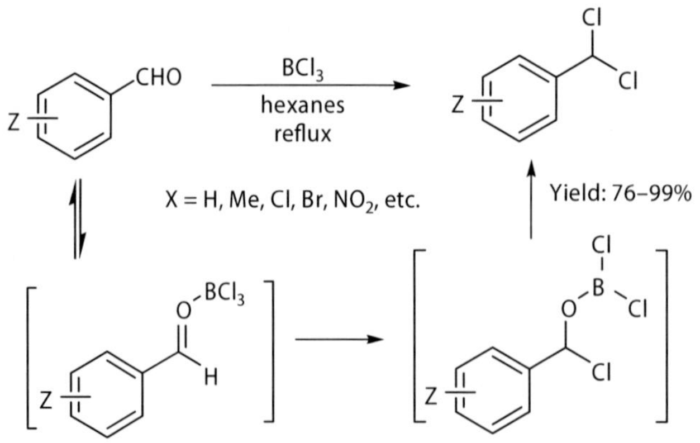

Z-C_6H_4-CHO $\xrightarrow[\text{hexanes, reflux}]{BCl_3}$ Z-C_6H_4-$CHCl_2$

X = H, Me, Cl, Br, NO_2, etc.

Yield: 76–99%

[Z-C_6H_4-CH=O→BCl_3] → [Z-C_6H_4-CHCl-O-BCl_2]

SCHEME 23.16 Reaction of aryl aldehydes with boron trichloride.

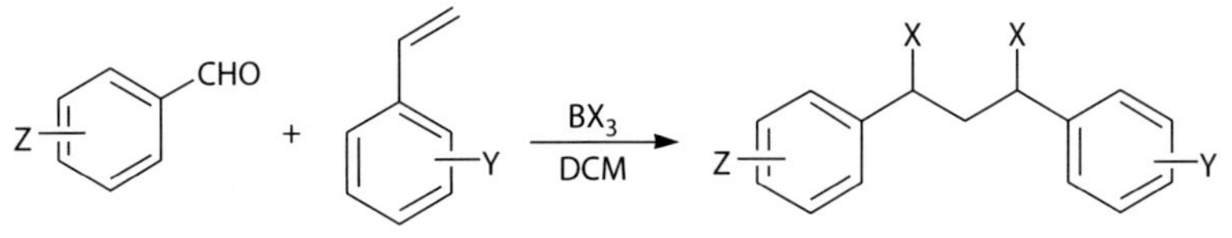

SCHEME 23.17 Boron halide-mediated coupling of aryl aldehydes with styrene.

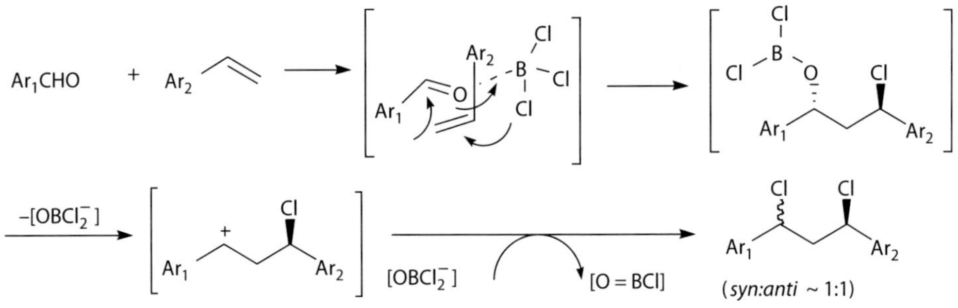

SCHEME 23.18 Plausible mechanism for BX_3-mediated aryl aldehyde-styrene coupling.

Interestingly, commercially available styrenes containing 4-*tert*-butylcatechol as a stabilizer produce the 1,3-dihalo-1,3-diarylpropane in excellent yields while freshly distilled styrenes produce only polymerization products.[60] In order to inhibit dihalogenation reaction of aryl aldehydes, reactions with BCl_3 must be carried out at 0°C and reactions with BBr_3 must be carried out at –40°C. Aldehydes containing electron-withdrawing groups such as Cl, F, CN, and NO_2 react at a slower rate.

Reactions of aryl aldehydes with styrenes presumably proceed through coordination of the carbonyl group to the boron trihalide followed by addition of the carbonyl carbon to the alkene to form a carbocation, which then adds halide to generate the final product (Scheme 23.18). A statistical distribution of the diastereoisomers (*syn*/*anti* ~ 50:50) indicates that the reaction involves a cation intermediate. A similar intermediate has been observed in the $AlCl_3$-catalyzed *ene* reactions of aldehydes with aliphatic alkenes[61] and in the fragmentation reaction of β-aryl-β-hydeoxyketones by boron trifluoride[62]. Another piece of supporting evidence for this mechanism is the isolation of *anti*-3-halo-1,3-diarylpropanols when the reactions are quenched with water prior to completion.

Under the same reaction conditions, the nonenolizable aliphatic aldehyde, tribromoacetaldehyde, also readily reacts with substituted styrenes.[59c] However, the reactions provide 3-halopropanols instead of the dihalide compounds (Scheme 23.19). The reaction is stereoselective, with the *anti* isomer being the major product (*anti*/*syn* > 10:1).

Reactions of aryl aldehydes with β-substituted styrenes such as *trans*-β-methylstyrene and stilbene also proceed well. The desired 1,3-dihalo-2-substituted propanes are isolated in good yields (Scheme 23.20). However, reactions using α-methylstyrene or 1,1-diphenylstyrene do not afford the anticipated 1,3-dihalide products. Trisubstituted and tetrasubstituted styrenes are unreactive under the reaction conditions.

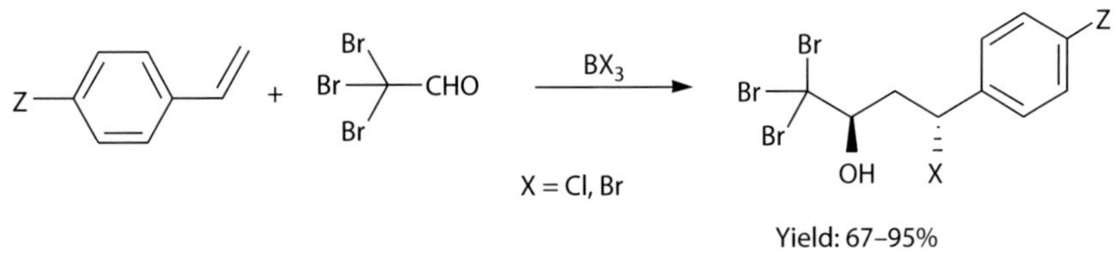

X = Cl, Br

Yield: 67–95%

SCHEME 23.19 Boron halide-mediated coupling of nonenolizable aliphatic aldehyde with styrene.

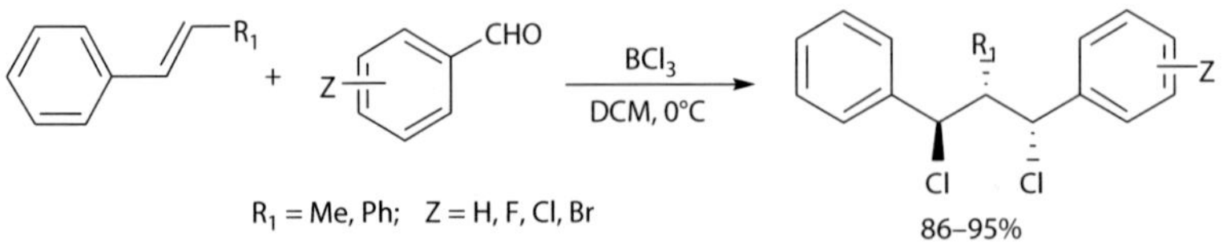

SCHEME 23.20 Reactions of aryl aldehydes with β-substituted styrenes.

Recent studies demonstrate that the 1,3-dihalides products are very useful intermediates in the syntheses of nitrogen-containing heterocycles, such as azetidines, pyrrolidines, piperidines, azepanes, *N*-substituted 2,3-dihydro-1*H*-isoindoles, 4,5-dihydropyrazoles, pyrazolidines, and 1,2-dihydrophthalazines.[63]

23.2.5 Boron Halide-Mediated Coupling of Aromatic Aldehydes with Alkynes

Boron trihalide-mediated coupling reactions of aryl aldehydes and alkynes have been reported (Scheme 23.21).[50,64] Interestingly, the stereochemistry of the diene product depends on the boron halide used. If boron trichloride is used, the reactions afford (*Z,E*)-1,4-pentadienes as major products, while (*Z,Z*)-1,4-dienes are obtained using boron tribromide. In order to avoid dihalogenation reactions of the aryl aldehydes, reactions must be carried out at low temperatures.

In addition, the stereochemistry of the major diene product is dependent on the reaction conditions employed. A slight change in either the sequence of addition of the reagents or the reaction temperature results in a dramatic change in the ratio of (*Z,E*)- and (*Z,Z*)-1,5-dichloro-1,4-pentadiene products.[50] A detailed mechanistic investigation using a model system (boron trichloride-mediated coupling of phenylacetylene with *p*-bromobenzaldehyde) has been reported.[65] The study reveals that there are three potential pathways for the boron trichloride-mediated aldehyde–alkyne coupling reaction (Scheme 23.22). Under controlled reaction conditions, one pathway predominates and thus accounts for the observed product stereochemistry. Pathway A represents the coupling of the aldehyde with *in situ* formed vinylboron dichloride which leads to the formation of the (*Z,Z*)-diene. (For simplicity, the carbocation is shown reacting directly with the second vinylboron dichloride intermediate, it is also possible that the vinyl reagent could be a more electron-rich complex formed by coordination of the vinylboron dichloride with the oxyboron dihalide fragment generated during cation formation.) Pathway A is the predominant one if the aldehyde is added to a premixed solution of phenylacetylene and boron trichloride since this sequence favors the formation of the vinylboron dichloride. Pathway B involves the coupling of the terminal alkyne with the alkoxoboron dichloride intermediate derived from reaction of the aldehyde with vinylboron dichloride. This pathway plays the predominant role when the reactions are carried out by adding boron trichloride to a mixture of aldehyde and alkyne. A mixture of (*Z,E*)-, (*Z,Z*)-1,5-dichloro-1,4-pentadiene is then generally obtained with the (*Z,E*)-isomer being favored due to its thermodynamic stability; the ratio of (*Z,E*)- to (*Z,Z*)- isomer increases as the reaction temperature decreases. Pathway C will occur only when the coupling reaction is carried out below –60°C. (*E,E*)-1,5-Dichloro-1,4-diene is isolated as a minor product since the pathway B still competes.

For the BBr_3-mediated aldehyde–alkyne coupling reaction, another pathway (pathway D) plays the dominate role due to the instant generation of the di(bromovinyl)boron bromide intermediate when the alkyne and boron tribromide are mixed (Scheme 23.23).

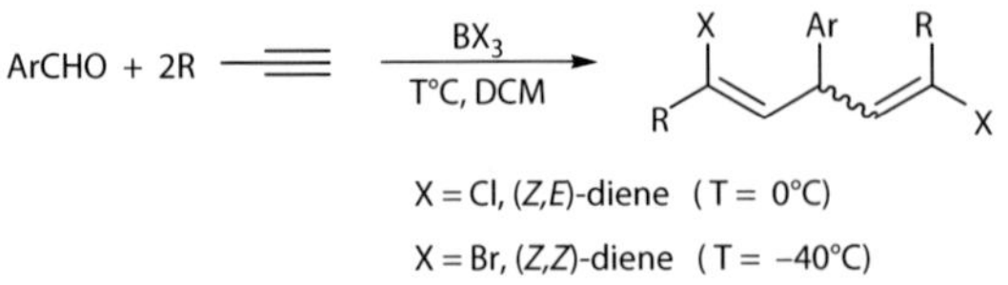

SCHEME 23.21 Boron halide-mediated aryl aldehyde–alkyne coupling.

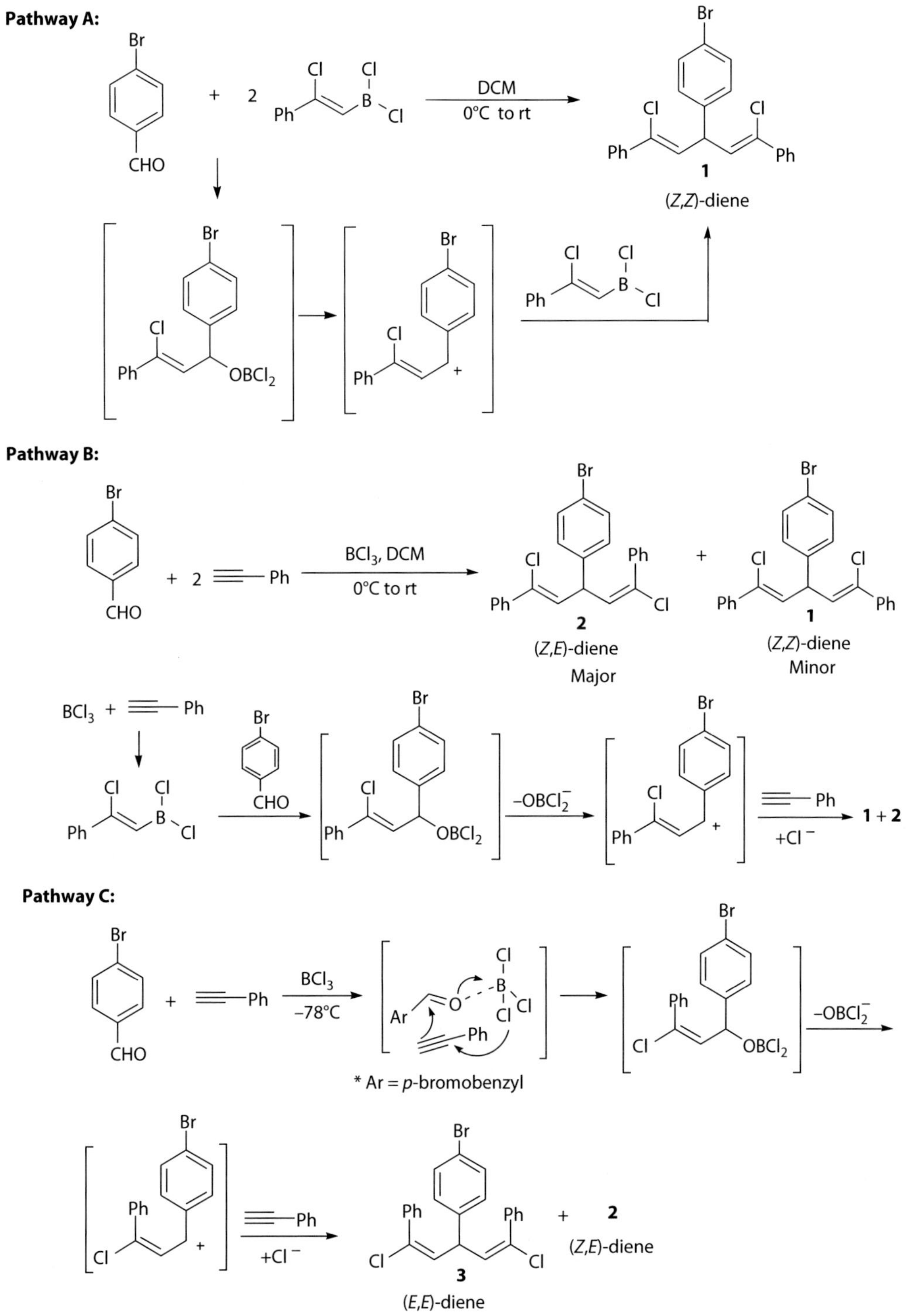

Pathway A:

DCM
0°C to rt

1
(*Z,Z*)-diene

Pathway B:

BCl_3, DCM
0°C to rt

2
(*Z,E*)-diene
Major

1
(*Z,Z*)-diene
Minor

$-OBCl_2^-$

+Cl⁻

1 + 2

Pathway C:

BCl_3
−78°C

* Ar = *p*-bromobenzyl

$-OBCl_2^-$

+Cl⁻

3
(*E,E*)-diene

+ **2**
(*Z,E*)-diene

SCHEME 23.22 Three different pathways for BCl_3-mediated aldehyde–alkyne coupling under controlled reaction conditions.

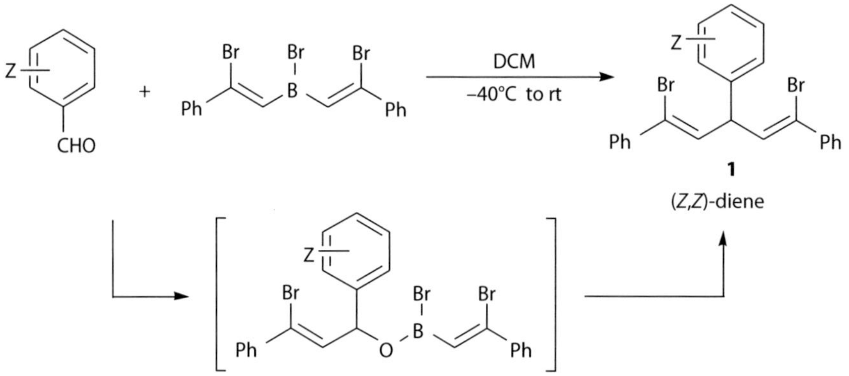

SCHEME 23.23 Additional pathway for BBr_3-mediated aldehyde–alkyne coupling.

For evidence supporting the generation of carbocation from the alkoxoboron dihalides and the stereochemistry illustrated in the pathway A–D, please see Section 23.3.3. In view of the potential value of these dihalodiene products in organic synthesis, the study has led investigators to investigate the use of other Lewis acids, including TiX_4,[66] $FeCl_3$,[67] and GaX_3.[68]

23.2.6 Boron Halide-Mediated Coupling of Aromatic Aldehydes with Allylmetals

The Lewis acid-mediated allylation of carbonyl compounds leading to homoallylic alcohols constitutes one of the most important synthetic reactions.[69] Numerous allylic organometallic reagents and Lewis acids have been investigated. The low-temperature boron halide-mediated allylation of cyclic ketones using allylstannane to give homoallylic alcohol has also been reported.[70]

The reaction of aryl aldehydes with allylsilane in the presence of excess BCl_3 does not give homoallylic alcohols, chloroallylation products are obtained instead.[71]This chloroallylation reaction was originally thought to be a consequence of the presence of excess BCl_3, which leads to halogenation of the intermediate homoallylic alcohols. Recent investigations reveal that the reaction affords chloroallylation products in good-to-excellent yields even if only a slight excess of BCl_3 is used.[72] Similar reactions using other allylmetals (Sn and Ge) have also been examined (Quinn, M. P., Yao, M.-L., Yong, L., Kabalka, G. W., unpublished results). In all cases, haloallylation products are isolated in good-to-excellent yields (Scheme 23.24). Generally, reactions with aldehydes bearing electron-donating groups take place smoothly at 0°C while reactions of aldehydes bearing electron-withdrawing groups require elevated reaction temperatures. Reactions with BBr_3 are faster than those using BCl_3. Bromoallylated products form exclusively when a mixture of BBr_3 and BCl_3 is added to *p*-chlorobenzaldehyde, demonstrating a high chemoselectivity.

After a series of control experiments, a two-step reaction mechanism was proposed: (i) conversion of the aldehyde to a homoallyloxy borate in the presence of a stoichiometric amount of boron trihalide and (ii) the boron trihalide-catalyzed conversion of homoallyloxy borates to the halogenated product in the presence of Me_3SiX (X = Br, Cl) (Scheme 23.25).

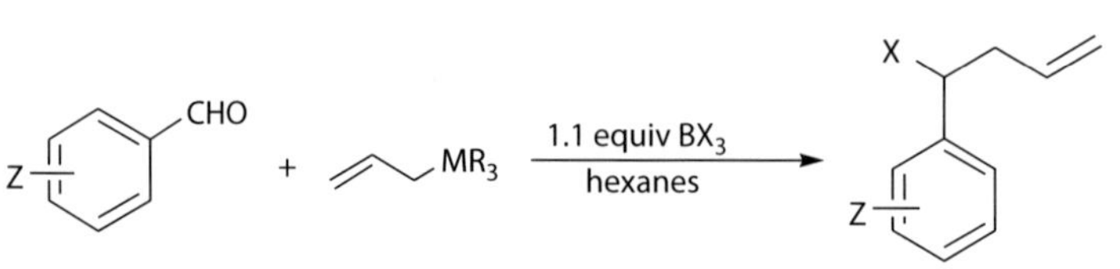

Z =, Me, Cl, CF_3, Br, CN, NO_2, etc.
MR_3 = $SiMe_3$, $GeEt_3$, $SnPh_3$
X = Cl, Br

SCHEME 23.24 Boron halide-mediated haloallylation of aryl aldehydes using allylmetals.

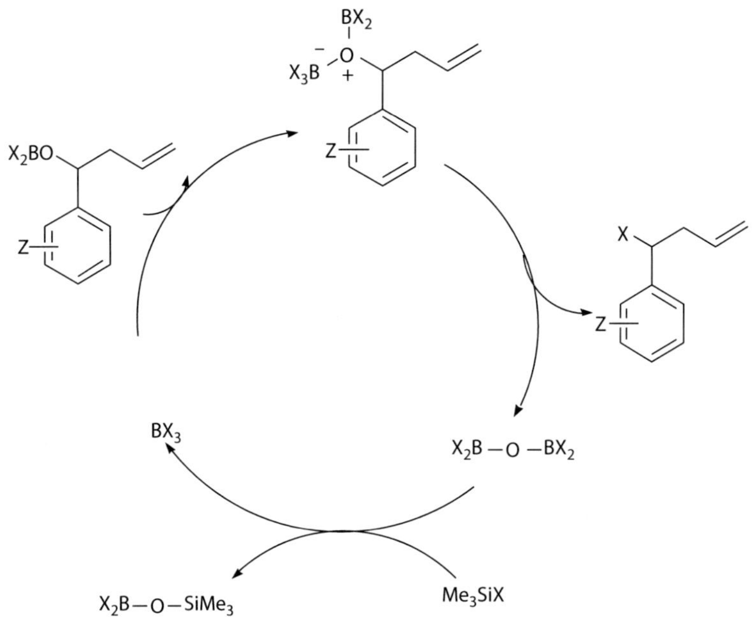

SCHEME 23.25 BX_3-catalyzed conversion of homoallyloxide borates to halogenated products.

23.2.7 Miscellaneous Reactions Mediated by Boron Trihalides

The boron trihalide-mediated halogen exchange of alkyl fluorides was first discovered in 1995.[5b] However, the reaction saw limited use in synthesis until it was realized that it could be used in the synthesis of bridged tetrahydrofluorenones (Scheme 23.26).[73] Several important advantages are attained by using fluoride in place of an alkoxide as a latent group for further elaboration: (1) alkyl fluorides are relatively stable to a variety of reaction conditions; (2) they are often more easily introduced than their alkoxy counterparts; and (3) they can be directly activated to a better leaving group without the need for a deprotection step.

It was reported that tertiary allylic amino amides, activated by BBr_3, followed by deprotonation using strong base, leads to a [2,3]-sigmatropic rearrangement (Scheme 23.27).[74] (*E*)-Crotyl and (*E*)-cinnamyl α-amino amides exhibit excellent *syn*-diastereoselectivity on rearrangement. Later it was found that BF_3•Et_2O also mediates [2,3]-sigmatropic rearrangements. Based on DFT calculations, a plausible reaction mechanism involving an *endo* transition state has been proposed.[75]

Reaction of BCl_3 with enones provides an efficient route to β-halogenated ketones (Scheme 23.28). A four-coordinated boron intermediate is proposed to explain the conjugate addition of chloride to the enone.[76]

The combination of boron halide and reducing agents (such as DIBAL and BH_3) is also known to reduce β-hydroxyketones. Due to the formation of boron-based chelating intermediates, *syn*-1,3-diols can be obtained in high diastereoselectivity (up to 95%).[77]

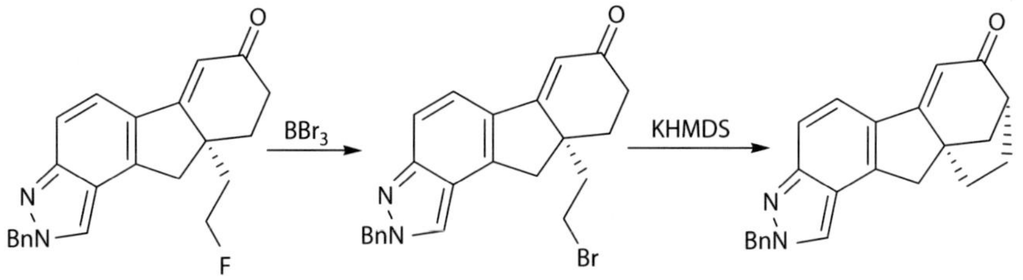

SCHEME 23.26 Bridged tetrahydrofluorenone synthesis using halogen exchange strategy.

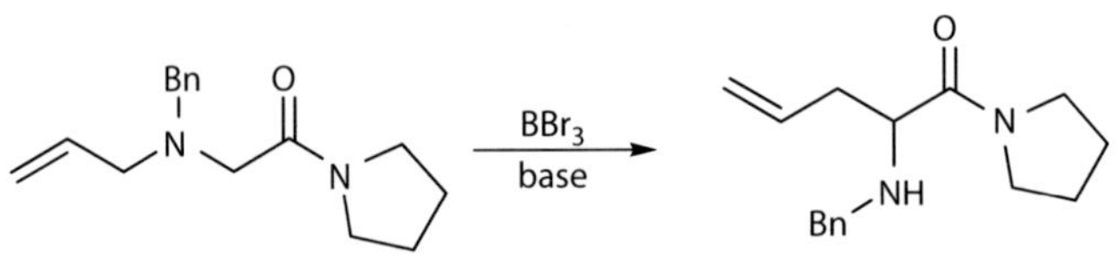

SCHEME 23.27 BBr_3-mediated [2,3]-sigmatropic rearrangement in tertiary allylic amino amides.

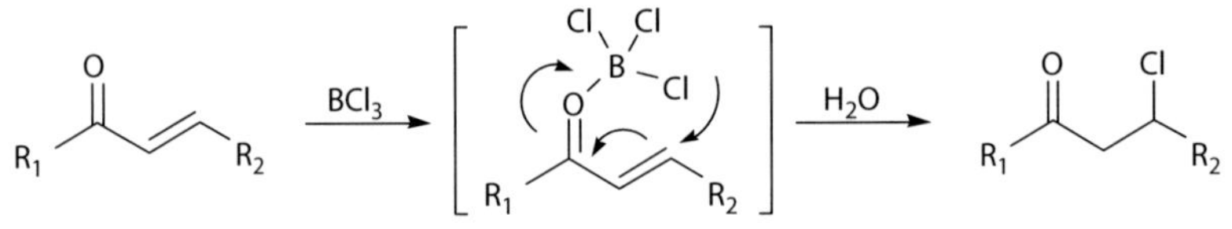

SCHEME 23.28 Reaction of BCl_3 with enones.

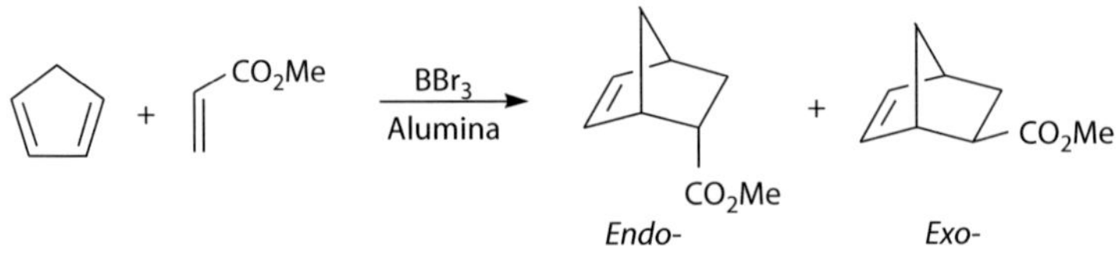

SCHEME 23.29 Diels–Alder reactions on highly acidic, boronated aluminas.

The characterization and application of a series of highly acidic boronated aluminas, prepared by the reaction of BX_3 with unactivated alumina, has been reported.[78] The modified aluminas, BX_n/Al_2O_3, were found to catalyze the Diels–Alder reactions of isoprene with methyl acrylate, and methyl acrylate with cyclopentadiene (Scheme 23.29). Both reactions proceed in high yield and with high *endo* to *exo* selectivity (12–24:1). Recently, it was discovered that when microwave-activated alumina is employed, the *endo* to *exo* selectivity can reach up to 99 to 1.[79]

23.3 ORGANOBORON HALIDES IN ORGANIC SYNTHESES

One important application of organoboron halides in organic syntheses is to generate enol borinates, which are highly versatile intermediates in stereocontrolled aldol reactions.[7] The Aldol stereoselectivity of enol borinates has been established; *Z* enol borinates give *syn* aldols and *E* enol borinates give *anti* aldols stereoselectively. Using $(c\text{-Hex})_2BCl$ in the presence of triethylamine, most ketones will selectively generate *E* enol borinates.

In addition, organoboron halides have been found to be more reactive than the corresponding trialkylboranes in reactions with organic azides[80] and ethyl diazoacetate.[81] Reactions with organic azides occur smoothly to give secondary amines and the stereochemistry of the original carbon–boron bond is retained during the reaction. Reactions using dialkylboron chlorides and ethyl diazoacetate produce nearly quantitative yields of the corresponding ethyl alkylacetates. The reaction readily accommodates even bulky alkyl groups which generally give very poor results in reactions employing the corresponding trialkylboranes. Organoboron halides are also useful intermediates in the synthesis of other boron derivatives. One important transformation is the synthesis of chiral allylborane derivatives via treatment of chiral organoboron halides with allylmagnesium bromide or allyllithium.[82]

Organoboron halides also can be converted to organoboranes via controlled hydridation reactions using lithium aluminum hydride.[9a] A stepwise hydroboration to prepare mixed dialkylboron halides or trialkylboranes has been developed. The process successfully solves the dihydroboration problem noted in early hydroboration reactions that employed borane–methyl sulfide. The

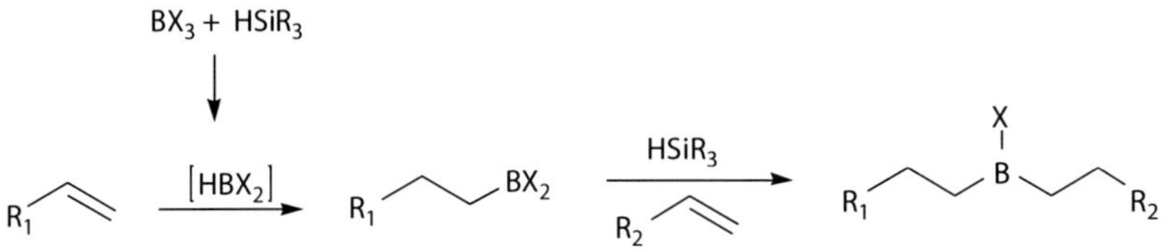

$$BX_3 + HSiR_3 \rightarrow [HBX_2];\quad R_1\text{CH=CH}_2 \xrightarrow{[HBX_2]} R_1CH_2CH_2BX_2 \xrightarrow[R_2CH=CH_2]{HSiR_3} (R_1CH_2CH_2)(R_2CH_2CH_2)BX$$

SCHEME 23.30 Syntheses of mixed dialkylboron halides via stepwise hydroboration.

methodology has been applied widely in the stereospecific syntheses of insect pheromones containing (*Z*)- and (*E*)-alkene moieties.[83]

A modified stepwise hydroboration avoiding the use of lithium aluminum hydride has also been developed (Scheme 23.30). It was discovered that trialkylsilane reacts rapidly with boron trihalide to form unsolvated dihaloborane.[9b] Subsequent reaction with alkenes in the presence of sufficient boron trihalide occurs rapidly at –78°C to afford alkylboron dihalides without detectable contamination by dialkylboron halides. If additional equivalents of both trialkylsilane and alkene are added, dialkylboron halides are obtained. The methodology has been successfully applied to the synthesis of 1,1,1-tris(dichloroboryl)alkanes, important precursors for dicarbapentaboranes.[84]

The alkylboron halide, 9-halo-9-borabicyclo[3.3.1]nonane, readily undergo haloboration reactions in a high stereo-, regio-, and chemoselective manner (Scheme 23.31).[85] Terminal carbon–carbon triple bonds are selectively haloborated in the presence of internal carbon–carbon triple bonds and terminal double bonds. Functional groups such as aldehydes, ketone, esters, and halogens are not affected by B-X-9-BBN.

The general reactivity order for haloboration reagents is shown in Scheme 23.32; B-Cl-9-BBN is very inert as a haloborating agent.

Allylboron dichlorides, generated *in situ* by the treatment of allylsilanes with BCl_3, can undergo carbometallation of allylsilane at room temperature (Scheme 23.33).[86] The allylborations are regiospecific affording, after oxidation, alcohols derived from addition of the boryl group distally to the silyl group.

The allylborations of vinyl and silyl enol ethers affords 1,4-dienes (Scheme 23.34). The dienes apparently arise from an initial allylboration followed by elimination of the β-alkoxy- or β-(silyloxy) borane.

Allylboration of terminal alkynes generates vinylboron dihalides. Recently, this chemistry has been successfully applied to the synthesis of bicyclic and cage boron compounds (Scheme 23.35).[87] The carbometallation of allenes is also known.[43]

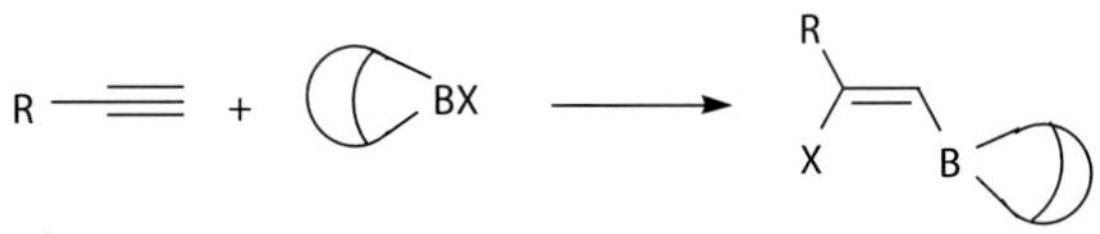

R—≡ + B-X-9-BBN → R(X)C=CH–B(9-BBN)

SCHEME 23.31 Haloboration of terminal alkynes using B-X-9-BBN.

B-I-9-BBN, BBr_3 > BCl_3 > B-Br-9-BBN >> B-Cl-9-BBN

SCHEME 23.32 Reactivity order for haloboration reagents.

$$\text{CH}_2\text{=CHCH}_2SiMe_3 \xrightarrow{BCl_3} Cl_2B\text{–CH}_2CH(CH_2CH\text{=CH}_2)CH_2SiMe_3 \xrightarrow[NaOH]{H_2O_2} HO\text{–CH}_2CH(CH_2CH\text{=CH}_2)CH_2SiMe_3$$

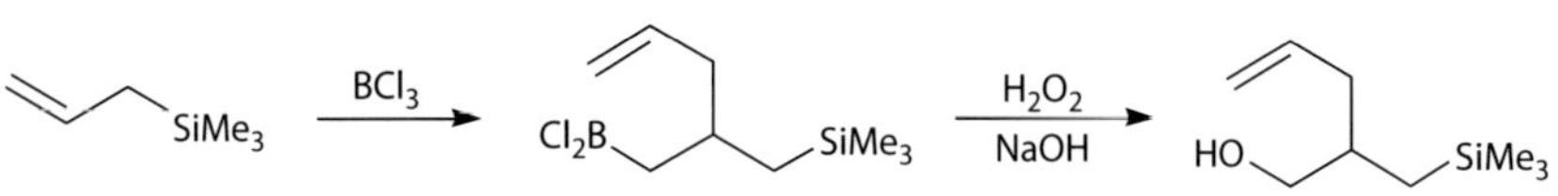

SCHEME 23.33 Allylboration of allylsilane using allylboron dichloride.

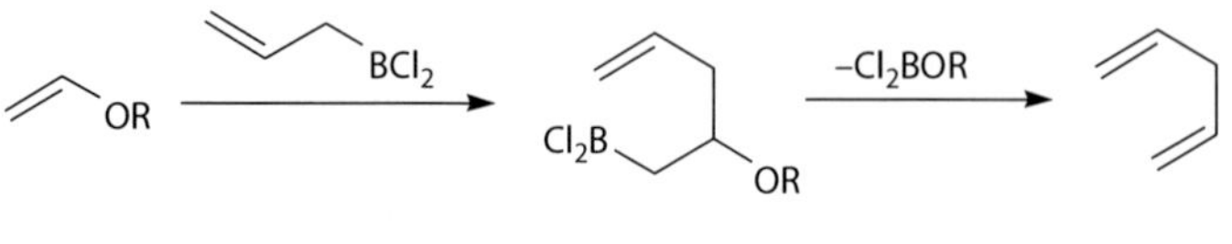

SCHEME 23.34 Allylborations of vinyl and silyl enol ethers.

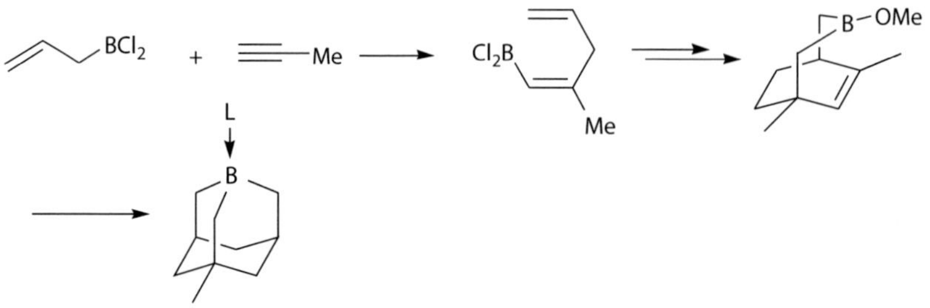

SCHEME 23.35 Synthesis of bicyclic and cage boron compounds.

However, attempts to allylborate simple alkenes using allylboron dichloride have been unsuccessful. Recent investigations reveal that allylboron dibromide can allylborate simple alkenes (Scheme 23.36).[88] The reaction provides a novel route to alkylboron dibromides that otherwise are unattainable by hydroboration. The reaction proceeds in high regio- and stereospecificity at 0°C and provides a strategy to synthesize alkylboron dibromides bearing a terminal alkene moiety.

In addition, allylboron dichloride, generated *in situ* from allylstannanes and boron trichloride, smoothly allylates indol. An "ate"-like boron intermediate has been proposed to explain the excellent regioselectivity observed. The chemistry has been applied in the total synthesis of Tryprostatin B (Scheme 23.37).[89]

It was reported that chiral oxazaborolidine, prepared from the reaction of n-BuBCl$_2$ with (S)-1,1-diphenylpyrrolidinemethanol, can be obtained in high purity (Scheme 23.38).[90] A catalyst prepared in this way was used in the reduction of the Torgov diketone with catecholborane and yielded the hydroxy ketone in 87% yield and 92% ee. Based on this observation, the role of additive *N,N*-diethylaniline was clarified.[91]

A number of advances have been made in preparative methods leading to organoboron halides. One involves the discovery of a highly reactive hydroborating reagent, dioxane-monochloroborane.[92] This adduct can be prepared in 98% purity by the reaction of dioxane and dioxane-BCl_3 with diborane or $NaBH_4$; it is stable indefinitely at room temperature. The introduction of this new reagent successfully overcomes purity concerns related to the traditional monochloroborane adducts. For example, dimethyl sulfide-monochloroborane is generally contaminated by an equilibrium between dimethylsulfide-BH_3 (12.5%) and dimethylsulfide-$BHCl_2$ (12.5%); the diethyl ether-monochloroborane adduct can only be obtained in approximately 90% purity. Dioxane-monochlorborane hydroborates unhindered olefins to form the corresponding dialkylboron halides within 30 min; moderately hindered olefins require up to 4 h at room temperature. Hindered tetrasubstituted olefins hydroborate only to the monoalkylchloroborane stage.

An efficient route to alkylboron dihalides involving the reaction of tetrachlorosilane with air- and water-stable organotrifluoroborates RBF_3K has been reported.[93] Reactions occur smoothly in THF

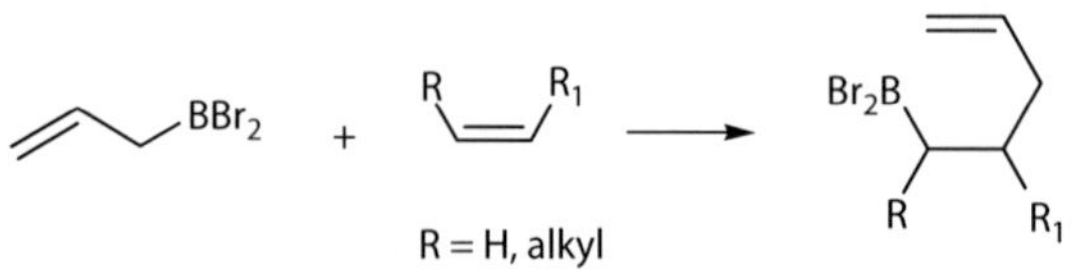

R = H, alkyl

SCHEME 23.36 Allylboration of simple alkenes using allylboron dibromide.

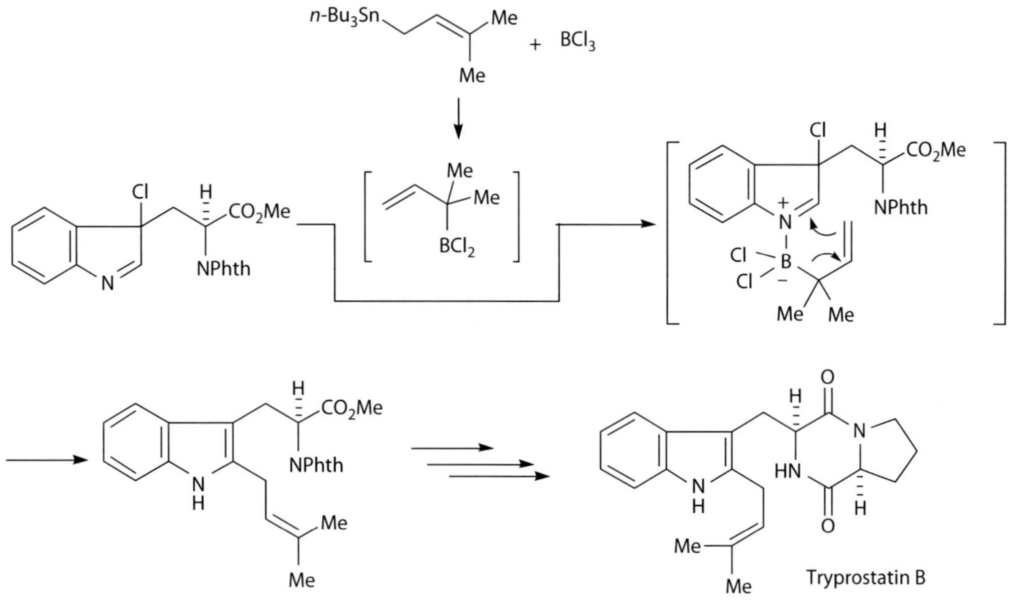

SCHEME 23.37 Allylation of indol using allylboron dichloride.

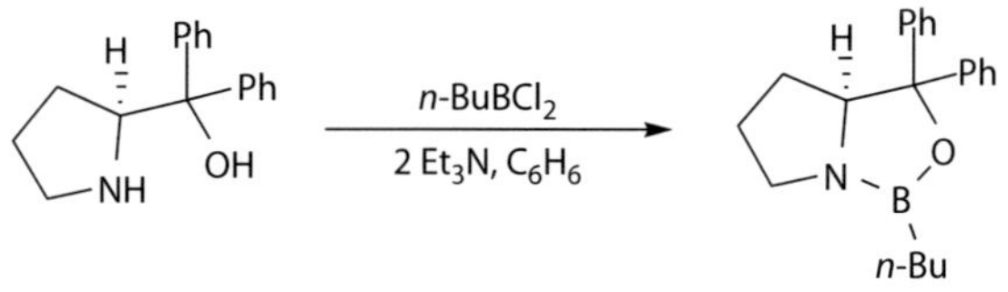

SCHEME 23.38 Chiral oxazaborolidine synthesis via reaction of n-BuBCl$_2$ with (S)-1,1-diphenyl-pyrrolidinemethanol.

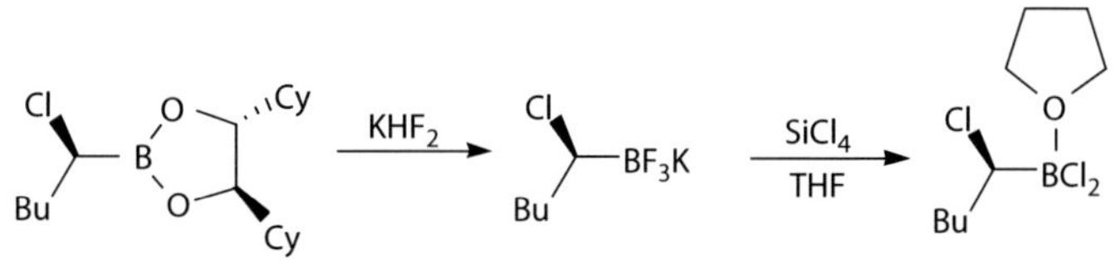

SCHEME 23.39 Conversion of organotrifluoroborates to alkylboron dihalides using tetrachlorosilane.

or acetonitrile at room temperature. Using this methodology, asymmetric (α-chloroalkyl)boronic esters, through their corresponding BF_3K salts, can be converted to α-chloroalkyl)dichloroboranes without measurable loss in stereopurity (Scheme 23.39).

23.3.1 Applications in the Preparation of Stereo-Defined Alkene Derivatives

Although haloborations of terminal alkynes using boron trihalides or B-X-9-BBN are known to occur in a highly stereo- and regioselective manner and the addition products are highly functionalized, no application of these boronated products in organic syntheses appeared prior to 1983. In that year, a systematic investigation of their application in organic syntheses was initiated.[85] The major purpose of the investigation was to develop efficient synthetic routes to stereo-defined alkenyl halides due to their importance in transition-metal-catalyzed cross-coupling reactions. The use of these halogenated borane derivatives provides convenient routes to conjugated dienes, alkenynes,

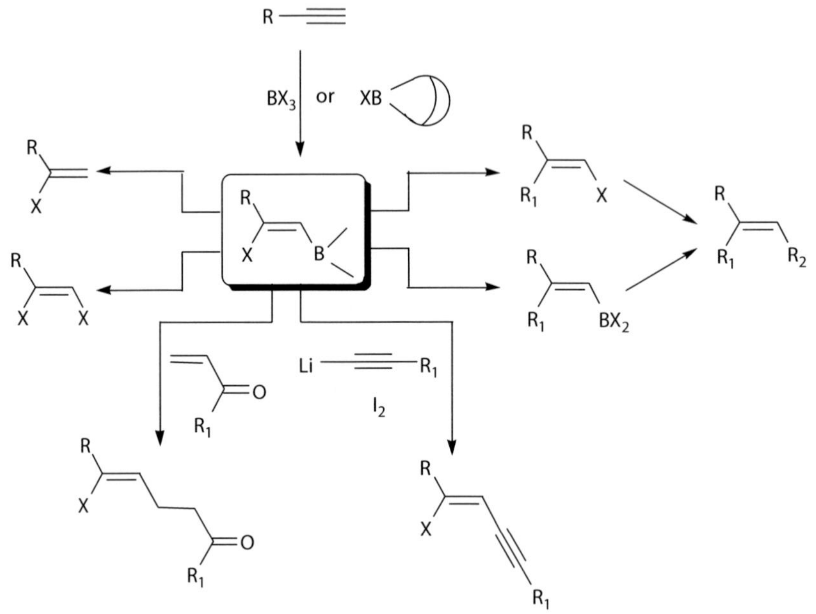

SCHEME 23.40 Preparation of stereodefined alkene derivatives based on 2-halovinylboron halides.

and other olefinic compounds (Scheme 23.40).[3b,94] Examples include the synthesis of stereo-defined trisubstituted alkenes based on a tandem Negishi–Suzuki coupling sequence[95] or double Negishi coupling process.[96] It is worth mentioning that the methylcopper-mediated, homocoupling of dialkenylboron chlorides to produce symmetrical (*E*,*E*)-1,3-dienes was already known at that time.[97]

The preparation of (*Z*,*Z*)-1-bromo-1,3-dienes by the reaction of (1-alkenyl)(2-bromoalkenyl)boron bromides with iodine in the presence of potassium acetate has also been reported (Scheme 23.41).[98] High chemospecificity and stereospecificity are achieved since the iodine reacts selectively with the 1-alkenyl group that is not deactivated by the bromine substituent. Deactivation of the alkenyl moiety by the bromine explains why di(2-bromoalkenyl)boron bromide derivatives do not undergo this reaction.

Even though these transformations are extremely useful for the synthesis of natural products,[99] the chemistry is generally not suitable for the synthesis of polyenes. In addition, the halovinylboron halides are air and water sensitive and are somewhat difficult to handle. It has been reported that the (*E*)-bromovinylboronate ester prepared via complexation of (*E*)-bromovinylboron dibromide with *N*-methyliminodiacetic acid (MIDA) is sufficiently stable to be handled on a benchtop under

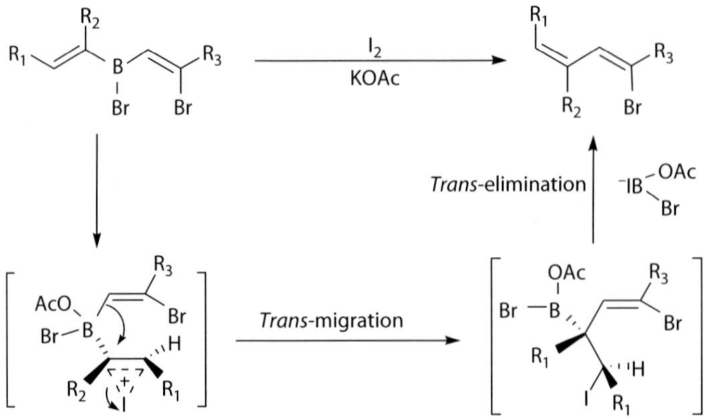

SCHEME 23.41 Synthesis of (*Z*,*Z*)-1-bromo-1,3-dienes.

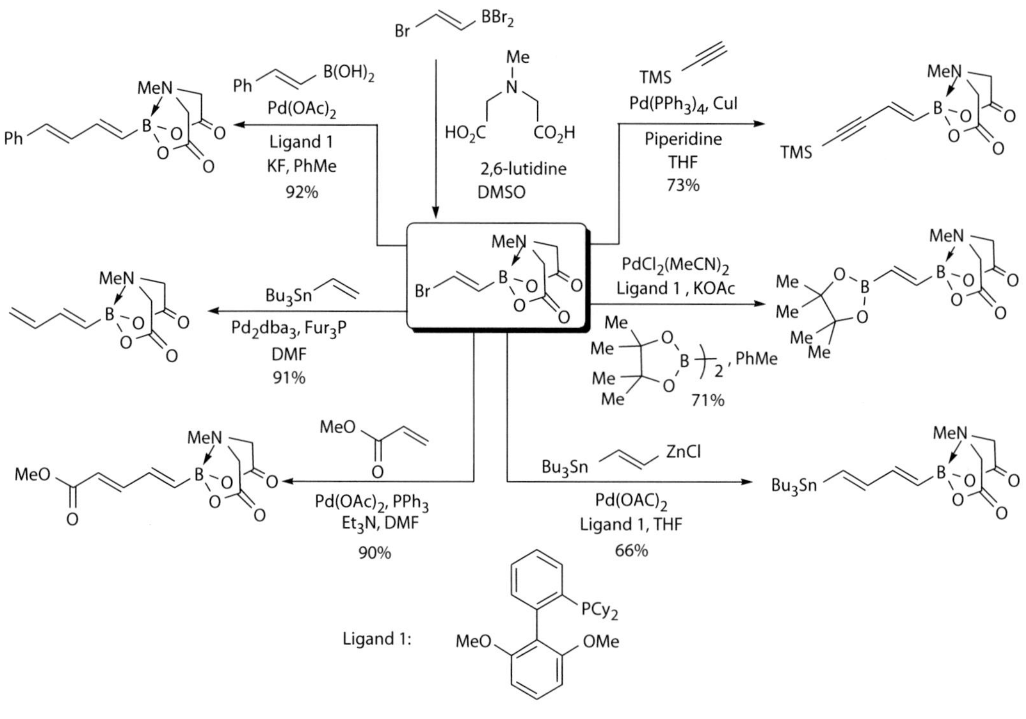

SCHEME 23.42 Syntheses of versatile building blocks using (*E*)-bromovinylboronate ester complex.

atmospheric conditions.[100] Notably, starting from this (*E*)-bromovinylboronate ester complex, versatile building blocks containing diene, enyne, and triene moieties can now be readily prepared via palladium-catalyzed coupling reactions (Scheme 23.42). A modular synthesis of polyenes, such as retinal and β-parinaric acid, via iterative cross-coupling with this (*E*)-bromovinylboronate ester building block has been reported.

The transformation of vinylboron halides, obtained from the hydroboration of terminal alkynes using haloborane complexes, to air- and water-stable trifluoroborates is known in the literature.[101] Recently, the conversion of halovinylboron dihalides into the corresponding bromovinylboronate esters and trifluoroborate derivatives has also been reported.[22] Good yields are obtained in both cases (Scheme 23.43).

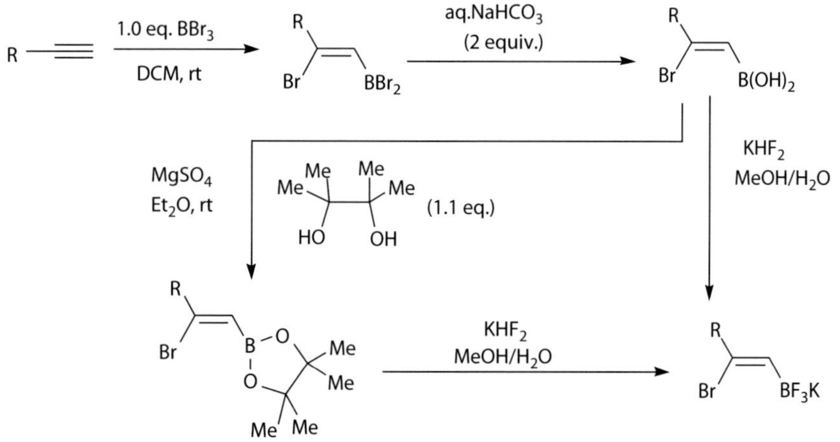

SCHEME 23.43 Conversion of halovinylboron dihalides to boronate esters and trifluoroborates.

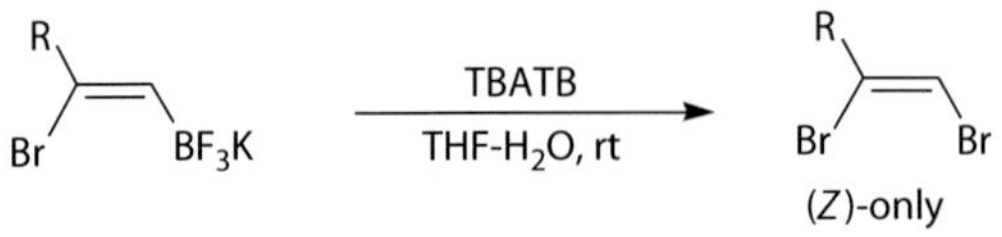

SCHEME 23.44 Improved method to prepare (Z)-dibromoalkenes using TBATB.

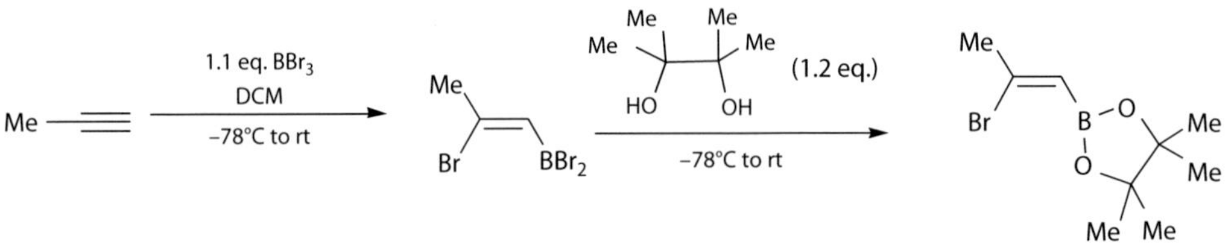

SCHEME 23.45 Direct convertion of halovinylboron dibromides into boronate esters.

Treatment of these halovinylboron trifluoroborate derivatives with tetrabutylammonium tribromide (TBATB) at room temperature affords (Z)-dibromoalkenes within 20 min (Scheme 23.44).[102] Remarkably, the base- and oxidant-free reaction conditions produce (Z)-dibromoalkenes in higher purity than can be obtained using previously reported methods.[103]

A straightforward route for converting halovinylboron dibromides into boronate esters was recently reported.[104] Utilizing this transformation, the uncertainty relating to the stereoselectivity[105] of the bromoboration of propyne using BBr_3 was resolved. The reaction proceeds with >98% *syn*-selectivity but the (Z)-2-bromo-1-propenylboron dibromide produced is prone to stereo-isomerization under a variety of conditions (Scheme 23.45). Starting from (Z)-2-bromo-1-propenylboron dibromide, an unprecedented stereoselective route to natural products containing methyl-branched (Z)-trisubstituted alkenes (such as mycolactone A and (+)-calyculin A) is now possible.

23.3.2 Reaction of Organoboron Halide with Aromatic Aldehydes

Grignard-like reactions involving organoborane reagents would possess a number of synthetic advantages over traditional Grignard reactions; these include mild reaction conditions, stereochemical control, and compatibility with a wide range of functional groups.[106] Indeed, the 1,2-addition of alkenylborane[82,107] and allylborane[35b,108] derivatives to aldehydes has been reported. Alkynylboranes also add to nonconjugated aldehydes and ketones (although more slowly) to give the corresponding propargyl alcohols.[109]

However, saturated organoborane derivatives appeared to be unreactive toward carbonyl compounds. Early studies involving alkylborane additions to aldehydes resulted in the reduction of the aldehyde via β-hydrogen elimination.[110] Grignard-like reactions were successful only after modification of the carbonyl group,[111] the trialkylborane,[112] or by adding transition metal catalysts.[113]

Recently, it was discovered that organoboron halides react smoothly with aromatic aldehyde under very mild reaction conditions. The reaction mechanism differs from that of other organometallic reagents (RMgX, RZnX, and 9-BBN derivatives[107,114]). Instead of the anticipated alcohol products, reactions of organoboron halides with aldehydes generally afford difunctionalized products. These difunctionalized products are quite useful in the pharmaceutical and agricultural industries. Compared to recent achievements in the difunctionalization of aryl aldehydes using dual-reagent catalyst systems (e.g., the diarylation using $[Ir(COD)Cl]_2$-$SnCl_4$,[115] dialkylation using $Ni(acac)_2$-Me_3SiCl,[116] and dialkynylation using $[ReBr(CO)_3(THF)]_2$-AuCl systems[117]), difunctionalization using organoboron halides is atom efficient and environmentally benign.

23.3.2.1 Reaction of Aromatic Aldehydes with Organoboron Dihalides

Chiral organoboron dihalide-induced enantioselective reactions have attracted much attention in recent years. An asymmetric Diels–Alder reaction using isopinocampheylboron dibromide ($IpcBBr_2$) was reported a number of years ago. Low enantiomeric excess (28.5% ee) was obtained for cycloaddition reactions involving cyclopentadiene and 2-methylpropenal (Scheme 23.46).[118] A modified $IpcBBr_2$ was synthesized and subjected to the asymmetric Diels–Alder reaction, but reactions still generated moderate ee.[119]

The great advance in this area is the discovery that chiral organoboron dichlorides bearing a naphthalene moiety catalyze the enantioselective Diels–Alder reactions of α,β-unsaturated esters and cyclopentadiene to produce up to 99.5% ee (Scheme 23.47).[120] A molecular complex between methyl crotononate and the catalyst was isolated. Using the same catalyst, Diels–Alder reactions between simple acyclic α,β-unsaturated ketones and α,β-unsaturated acid chlorides produce 83% and 92% ee, respectively.[121]

An enantioselective Diels–Alder reaction induced by an axial chiral arylboron dichloride bearing the binaphthyl skeleton has also been reported.[122] Although the reaction of cyclopentadiene and methyl acrylate in the presence of 10% of this catalyst gives the *endo* adduct with >99% diastereoselectivity, only moderate enantiomeric excess is obtained.

Compared to chiral organoboron dihalides, studies involving achiral organoboron dihalides are rare[80,81,123] except where they are used as intermediates in conversions of organoboron halides and trialkylboranes.

In 2000, it was noted that alkylboron dibromides such as $IpcBBr_2$ and cyclohexylboron dibromide readily convert aryl aldehydes to the corresponding benzyl bromides in excellent yields at room temperature (Scheme 23.48).[124]

The reaction is tolerant of a variety of functional groups. The new synthesis provides a direct conversion of aromatic aldehydes to the corresponding benzyl bromides. α,β-Unsaturated aldehydes also can be smoothly converted into the corresponding bromides.

Although the reductive bromination of aromatic aldehydes using cyclohexylboron dibromide proceeds at room temperature, the reaction is quite slow. A mechanistic study revealed that the

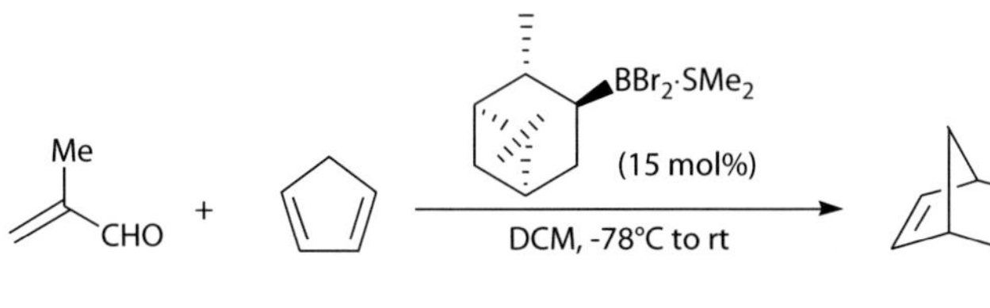

SCHEME 23.46 Asymmetric Diels–Alder reaction catalyzed by $IpcBBr_2$.

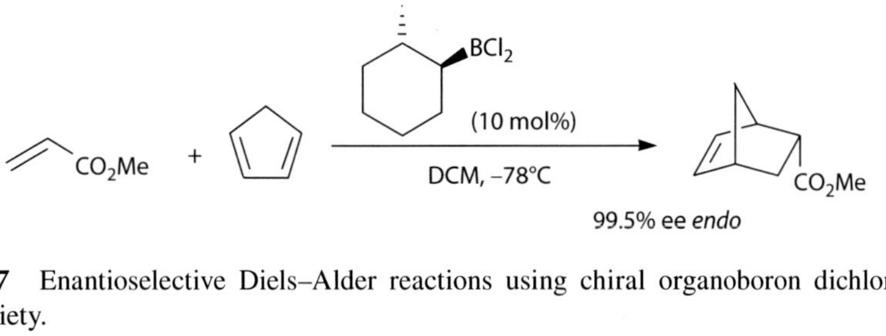

SCHEME 23.47 Enantioselective Diels–Alder reactions using chiral organoboron dichloride bearing a naphthalene moiety.

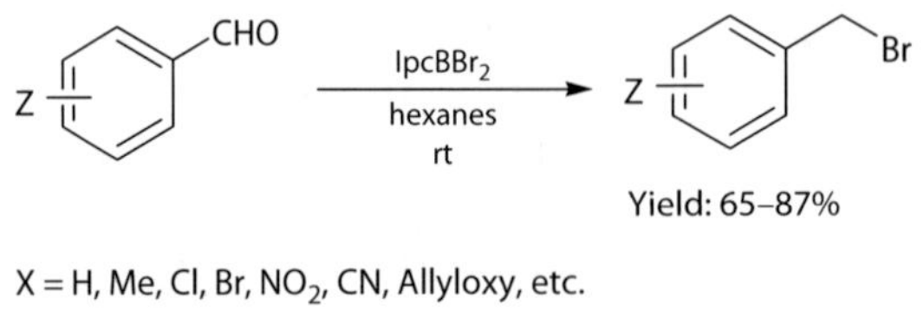

SCHEME 23.48 Reductive bromination of aryl aldehydes.

reductive bromination proceeds through pathway "a" (Scheme 23.49), whereas $IpcBBr_2$ reactions proceed via pathway "b" because the isopinocampheyl group is one of the most effective groups for inducing β-hydrogen transfer reactions.

Interestingly, the reaction of alkylboron dichlorides with aryl aldehydes in the presence of oxygen produces chloroalkylation products exclusively at room temperature (Scheme 23.50).[125]

It is known that organoboranes readily undergo autoxidation in the presence of oxygen and this reaction has been used to prepare alcohols and alkyl hydroperoxides as well as to induce free radical reactions.[126] Organoboranes can also be used to alkylate α,β-unsaturated carbonyl compounds through a free radical 1,4-addition reaction in the presence of oxygen.[127] However, organoboranes do not normally react with saturated carbonyl compounds except for the reaction of formaldehyde with trialkylboranes in the presence of air.[128]

Based on the isolation of *n*-butyl peroxide on hydrolysis of the reaction of *n*-butylboron dichloride and benzaldehyde, the oxygen-induced reaction most likely proceeds via the pathway shown in Scheme 23.51. On initial coordination to the alkylboron dichloride, a reaction occurs to generate a peroxide intermediate. Then an alkyl group transfer occurs through an intramolecular six-membered ring transition state, generating a borinate intermediate. Subsequent migration of the chlorine atom affords the observed product.

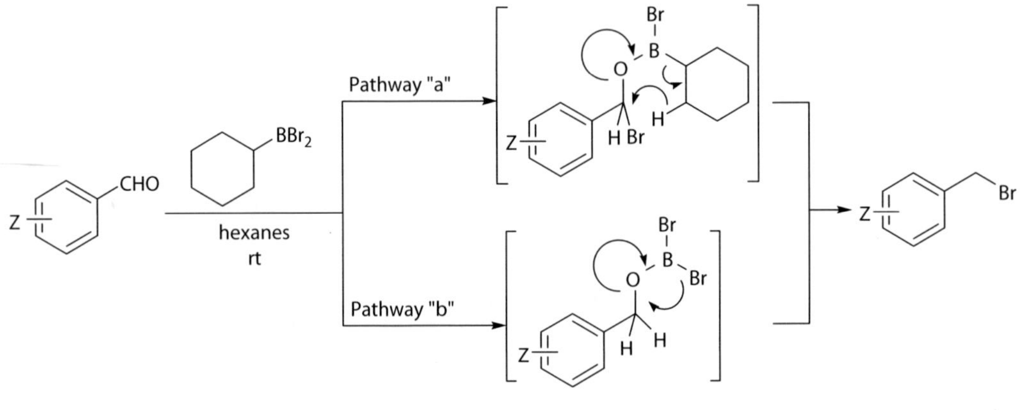

SCHEME 23.49 Pathways involved in reductive bromination reaction.

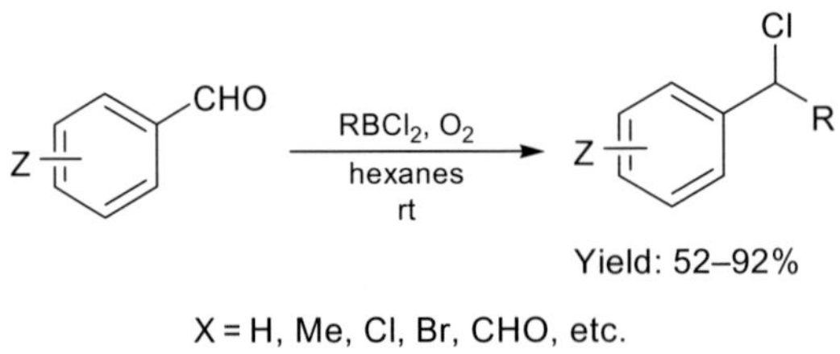

SCHEME 23.50 Alkylboron dichlorides react with aryl aldehydes.

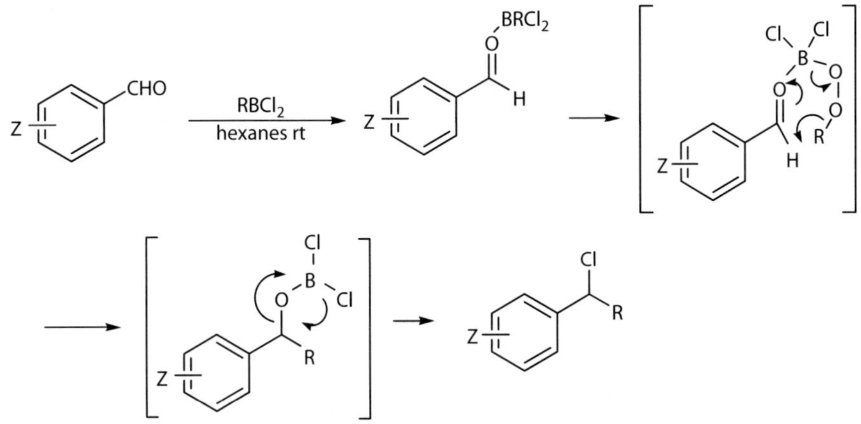

SCHEME 23.51 Plausible mechanism for reaction of alkylboron dichlorides with aryl aldehydes.

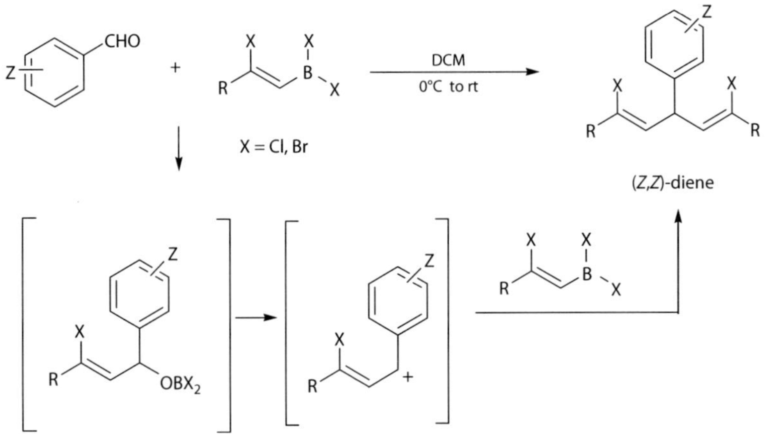

SCHEME 23.52 Reaction of vinylboron dihalides with aryl aldehyde.

In a continuation of these studies, the feasibility of utilizing vinylboron dichloride reagents for aryl aldehyde addition reactions was examined. Surprisingly, these reactions afford (*Z,Z*)-1,5-dihalo-1,4-pentadienes as major products if preformed (Z)-vinylphenylboron dihalide reagents are used (Scheme 23.52). The Grignard-like addition of the (Z)-vinylboron dichloride to the aldehyde apparently is followed by carbon–oxygen bond cleavage to afford a carbocation intermediate. The cation then adds to a second equivalent of the (Z)-vinylboron dichloride to generate the observed (*Z,Z*)-diene product.

23.3.2.2 Reaction of Aromatic Aldehydes with Diorganoboron Halides

Among the dialkylboron halides, diisopinocamphenylboron chloride (Ipc_2BCl, DIP-Cl) has been utilized the most extensively since it readily reduces ketones and aldehydes in nearly quantitative yield.[10] Furthermore, Ipc_2BCl is easily obtained in both enantiomeric forms. Using this reagent, asymmetric reduction of aryl ketones or α-quaternary alkyl ketones generally produces chiral alcohols with very high levels of enantiomeric excess (Scheme 23.53). Significantly, the reactions exhibit a strong positive nonlinear effect, (+)-NLE, (termed "asymmetric amplification").[129] When prepared from (+)-α-pinene reagents that are only 70% enantiomerically pure, DIP-Cl reduces acetophenone to the corresponding chiral alcohol in >90% enantiomeric excess. Theoretical models and practical applications of this phenomenon have been the subject of recent publications and reviews.[130] This

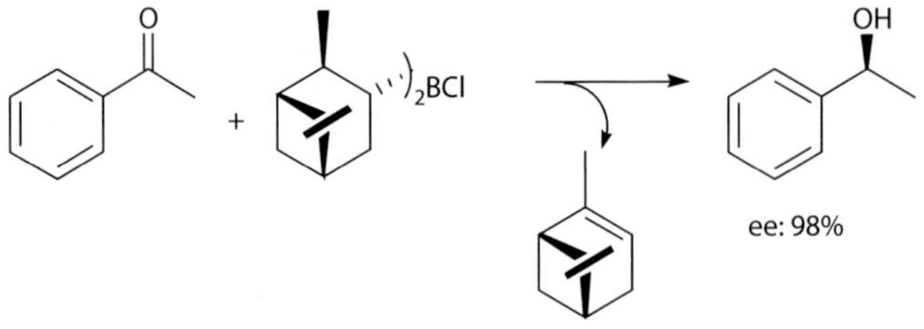

SCHEME 23.53 Asymmetric reduction of ketones using (–)-DIP-Cl.

reaction has also been applied in key steps in the preparation of several important pharmaceutical compounds such as the antidepressant fluoxetine hydrochloride,[131] a selective D1 agonist,[132] PAF-antagonists L-659,989[133] and MK-287,[134] dolastatin 10,[135] a potential antipsychotic and bronchodilator,[136] (–)-lobeline,[137] and an LTD4 antagonist.[138]

It should be noted that (–)-DIP-Cl also works efficiently for intramolecular asymmetric reductions of *o*-acylphenols, *o*-anilines, and *o*-benzoic acids (Scheme 23.54).[139] In all of these intramolecular reductions, the product alcohols are obtained in the opposite stereochemistry as that obtained in the intermolecular reductions.

The intermolecular asymmetric reduction of various keto esters with (–)-DIP-Cl has also been investigated.[140] Interestingly, the opposite stereochemistry is observed for the products from the reduction of aromatic and aliphatic α-keto esters (Scheme 23.55). Due to facile enolization, reductions of β-keto esters are unsuccessful.

The reductions of γ-keto esters and δ-keto esters are also facile using (–)-DIP-Cl. The reductions provide high % ee for the reduction of aryl ketones, while low % ee are observed in the case of aliphatic ketones (Scheme 23.56). The products obtained, γ-hydroxy esters and δ-hydroxy esters, are valuable intermediates; they can be readily converted to the corresponding lactones by treatment with a catalytic amount of trifluoroacetic acid.

The highly enantioselective reduction of aryl ketones and α-quaternary alkyl ketones can be accounted for by a transition state in which the bulky aryl or α-quaternary alkyl group interacts sterically with the methyl group at the 2-position of the isopinocampheyl moiety. When the stearic interaction between the reagent and ketone is diminished, as in the reduction of 3-methyl-2-butanone (32% ee) and 2-butanone (4% ee), the chiral induction is poor. In those cases, the use of more stearically hindered diorganoboron halides, such as Eap_2BCl,[141] is required.

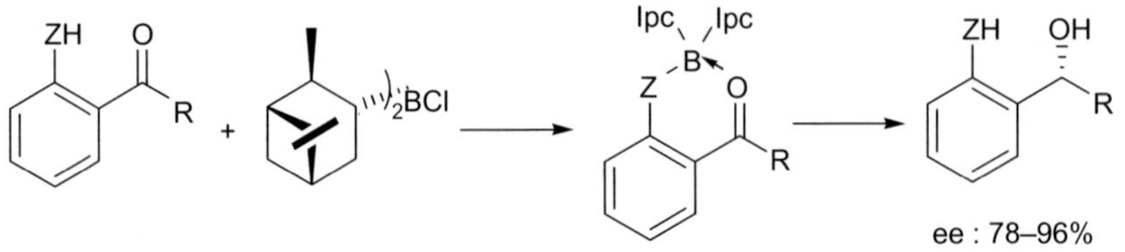

SCHEME 23.54 Intramolecular asymmetric reductions using (–)-DIP-Cl.

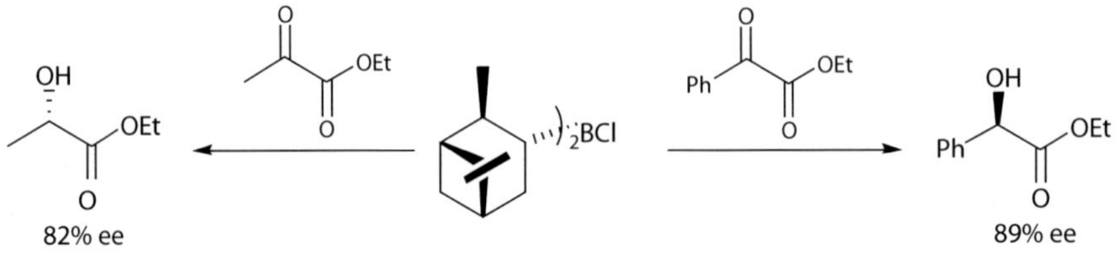

SCHEME 23.55 Reduction of aromatic and aliphatic α-keto esters using (–)-DIP-Cl.

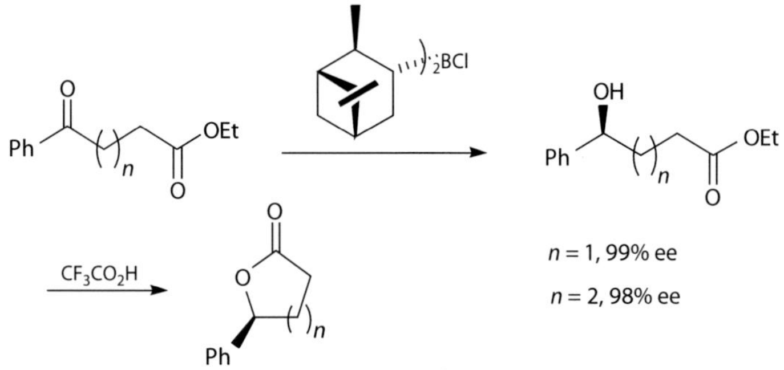

SCHEME 23.56 Reductions of γ-keto esters and δ-keto esters using (–)-DIP-Cl.

(–)-DIP-Cl is also known to cleave *meso*-epoxides in an enantioselective manner to generate synthetically important vicinal halohydrins.[142] A recent investigation[143] shows that the enantiomeric excess achieved with *B*-chlorobis(2-isocaranyl)borane ($DIcr_2B$-Cl, 78% ee) is significantly higher than that attained with (–)-DIP-Cl (41%), especially for *meso*-cyclohexene oxide.

Prior to 2000, alkylboron halide reagents had not been used to alkylate carbonyl compounds. In 2000, it was discovered that alkylation could be achieved by simply adding a base to a mixture of aldehyde and diorganoboron halide (Scheme 23.57).[144] The reaction occurs under mild reaction conditions and tolerates a variety of functional groups. This new alkylation reaction provides a useful alternative to traditional Grignard and organolithium reactions. The survival of base-sensitive functional groups under these reaction conditions is attractive.

Bases such as triethylamine, quinuclidine, pyridine, DBU, *n*-butyllithium, *t*-butoxyllithium, and 2,6-lutidine were evaluated. Both bulky and strong bases produce the highest yields. Moderate yields of alkylation products, arylalkylmethanols, are obtained in all cases. Product yields are not dependent on the electronic nature of the substituents on the aldehyde, but boranes containing secondary alkyl groups give relatively high yields compared to boranes containing primary alkyl groups. However, if more hindered organoboranes, such as diisopinocampheylboron chloride, dinorbornylboron chloride, and di-(3-methyl-2-butyl)boron chloride are utilized, benzyl alcohols are formed because reduction predominates.

When 4-bromobenzaldehyde is allowed to react with *n*-butyl(cyclohexyl)boron chloride, the major product is cyclohexyl(4-bromophenyl)methanol (Scheme 23.58).

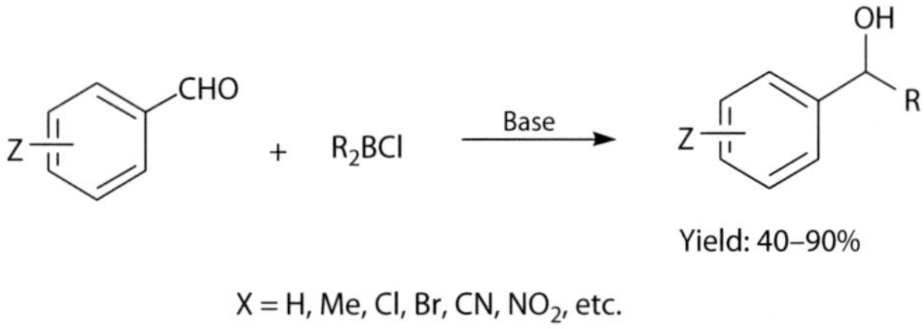

SCHEME 23.57 Alkylation of aryl aldehydes using dialkylboron chlorides in the presence of base.

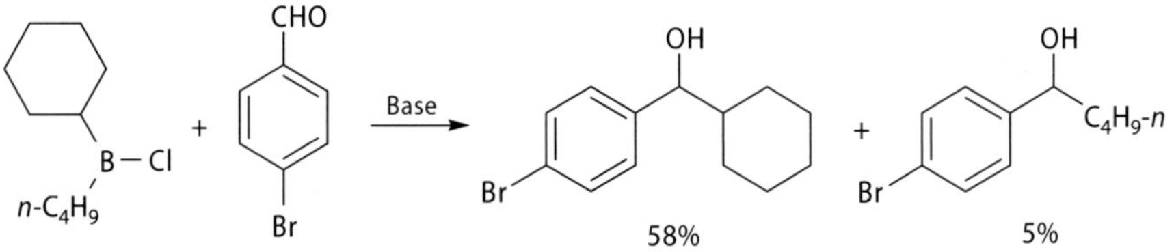

SCHEME 23.58 Reaction of mixed dialkylboron chloride with aryl aldehyde.

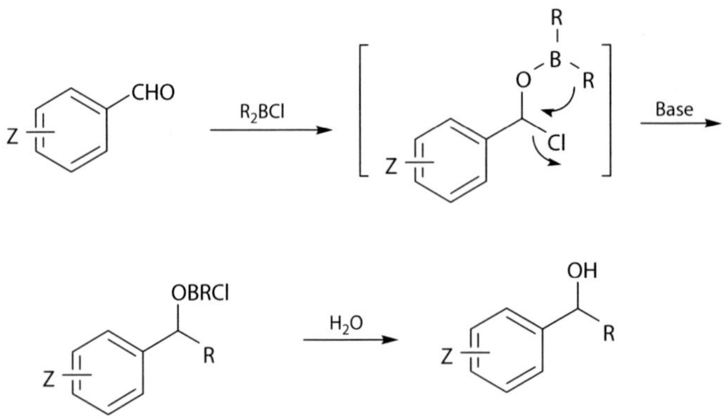

SCHEME 23.59 Plausible mechanism for alkylation of aryl aldehydes using dialkylboron chlorides in the presence of base.

In a mechanistic study, the reaction of 4-bromobenzaldehyde with dicyclohexylboron chloride was monitored by NMR spectroscopy. The NMR data suggest that the reaction proceeds via formation of a borinate ester followed by migration of the alkyl group (Scheme 23.59).

Interestingly, oxygen-induced reactions proceed more readily than the corresponding alkylations carried out in the presence of base.[145] This reaction proceeds most efficiently at 0°C, partial reduction occurs at room temperature if boranes containing secondary alkyl groups are used due to β-hydrogen transfer. In contrast to alkylation reactions carried out in the presence of base, organoboranes containing primary, secondary, and hindered alkyl groups all produce excellent yields of alkylation products in the presence of oxygen (Scheme 23.60).

The alkylation reaction is rapid but no alkylation occurs in the presence of a radical scavenger such as galvinoxyl. In addition, when both primary and secondary alkyl groups are present, the secondary group reacts preferentially. These observations support the postulation that the reaction is occurring via a radical pathway such as the one outlined in Scheme 23.61.

In 2007, the alkylation of aldimine with dicyclohexylboron halide in the presence of hydrogen peroxides was reported (Scheme 23.62).[146] The sequential addition of hydrogen peroxide and dicyclohexylboron chloride to the imine solution was found to be essential for generation of the expected cyclohexylated derivatives. Reverse addition of the reagents failed to give the desired alkylation products. This alkylation reaction also proceeds smoothly with other oxidants such as Oxone, UHP, or MCPBA. Later, a three-component reaction involving an arenecarbaldehyde, an arylamine, and a dialkylchloroborane reagent was developed. Reactions of lithium aldimines with dialkylboron chlorides in the presence of hydrogen peroxide and base have been previously reported to produce unsymmetrical ketones[147] at room temperature and partially mixed trialkylcarbinols[148] at an elevated temperature.

Reactions of aryl aldehydes with di(halovinyl)boron halides have also been evaluated. Notably, the reaction proceeds smoothly to provide dialkenylation products. The reaction is believed to proceed through a Grignard-like addition of the di(halovinyl)boron halide to the aldehyde to form an

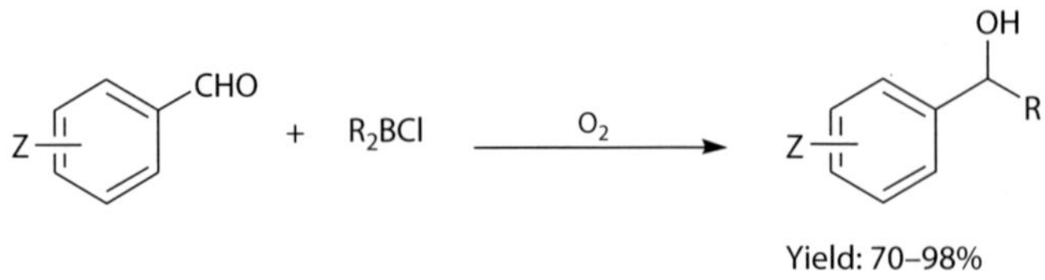

SCHEME 23.60 Alkylation of aryl aldehydes using dialkylboron chlorides in the presence of oxygen.

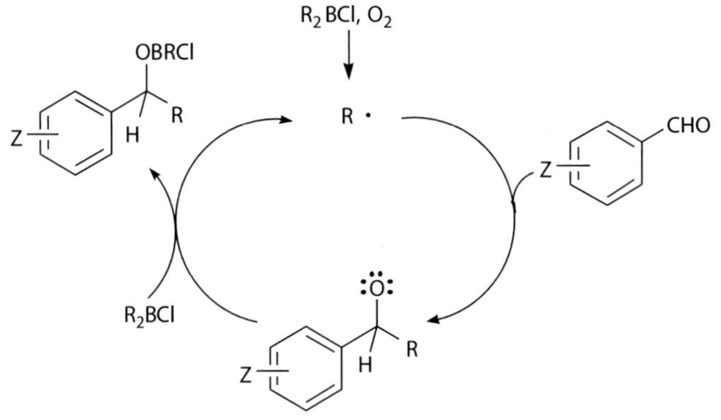

SCHEME 23.61 Plausible mechanism for alkylation of aryl aldehydes using dialkylboron chlorides in the presence of oxygen.

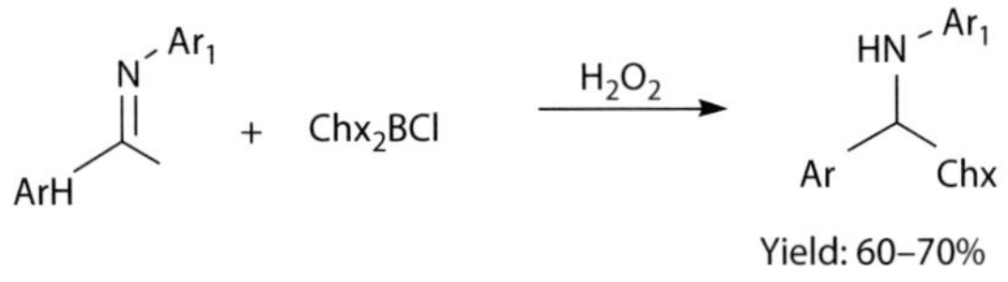

SCHEME 23.62 Alkylation of aryl aldehydes using dialkylboron chlorides in the presence of H_2O_2.

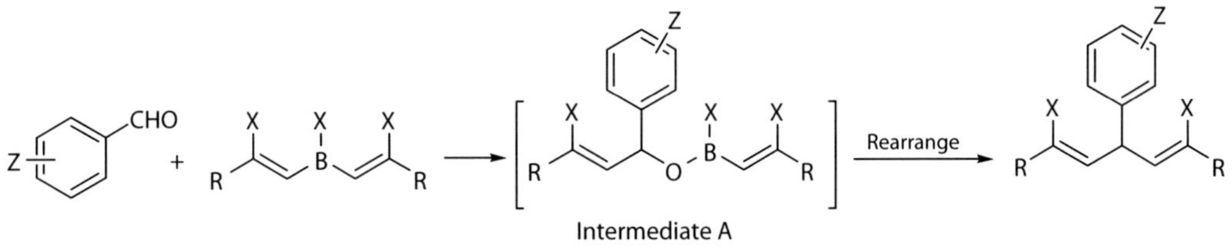

SCHEME 23.63 Reaction of aryl aldehydes with di(halovinyl)boron halides.

alkoxyboron monohalide intermediate, **A**. Then, the migration of the second halovinyl group from boron to carbon affords the final diene product (Scheme 23.63). The isolation of (*Z*)-3-halo-1,3-diarylprop-2-en-1-ol, the expected product from intermediate **A**, when the reaction was quenched with water prior to completion, supports the proposed mechanism. (For evidence supporting the proposed migration of the second halovinyl group from boron to carbon, see Section 23.3.3.)

It was then discovered that the preformed divinylboron halide reagents react smoothly with aliphatic aldehydes to generate dialkenylation products (Scheme 23.64).[50] Significantly, enolization of the aliphatic aldehyde does not occur.

Later, the reaction was extended to dialkynylation reactions (Scheme 23.65) (Yao, M.-L., Wu, Z., Kabalka, G. W., Quinn, M. P., unpublished results). 1,4-Diynes that are not attainable using earlier reported methods are now easily synthesized. The transition-metal-free feature makes this reaction far more attractive when compared with the reported dialkynylation reactions that employ rhenium or gold cocatalysts.[117]

23.3.2.3 Organoboron Halide-Mediated Coupling of Aromatic Aldehydes with Styrene Derivatives

Encouraged by the isolation of *anti*-3-halo-1,3-diarylpropanols in low yields from the BCl_3-mediated aldehyde-styrene coupling reaction, an investigation was initiated that focused on the feasibility of developing an efficient route to chloropropanol products via a replacement of one or two chlorides

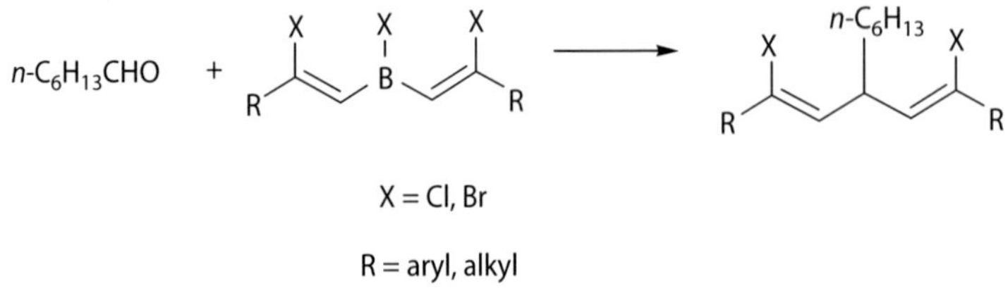

SCHEME 23.64 Reaction of preformed divinylboron halides with aliphatic aldehydes.

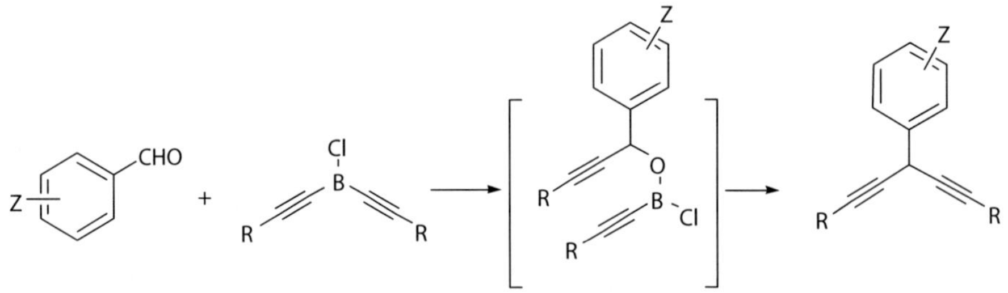

SCHEME 23.65 Reaction of aryl aldehydes with dialkynylboron chlorides.

in BCl_3 with an organic group (R). After a careful examination of $RBCl_2$ and R_2BCl derivatives, it was discovered that phenylboron dichloride promotes the reaction at –10°C (Scheme 23.66).[149] The low migratory aptitude of phenyl group was also observed recently in the reactions of mixed alkylarylboranes with benzylic sulfur ylides.[150] The desired products, 3-halo-1,3-diarylpropanols, are isolated in good to excellent yields. The reactions predominantly produce the *anti*-diastereoisomers (*anti*/*syn* > 10:1), indicating that a cation intermediate is not involved.

Heteroaromatic aldehydes such as 3-pyridinecarboxaldehyde react with styrenes to generate 3-chloro-1-(3-pyridyl)-3-phenylpropanols in excellent yields (Scheme 23.67).[59c] However, the reactions are slow compared to the reaction of aryl aldehydes. Aliphatic aldehydes and alkenes do not undergo the reaction, aliphatic aldehydes simply enolize. The reaction of aliphatic alkenes with aryl aldehydes produces mixtures of 1,3-dichloro compounds and ene-carbonyl adducts in very low yield.

Under similar conditions, the reaction of *trans*-β-methylstyrene with aryl aldehydes produces mainly *R,R,R*/*S,S,S* isomers. For aryl aldehydes with electron-withdrawing groups, the reactions give excellent yields of products (Scheme 23.68).

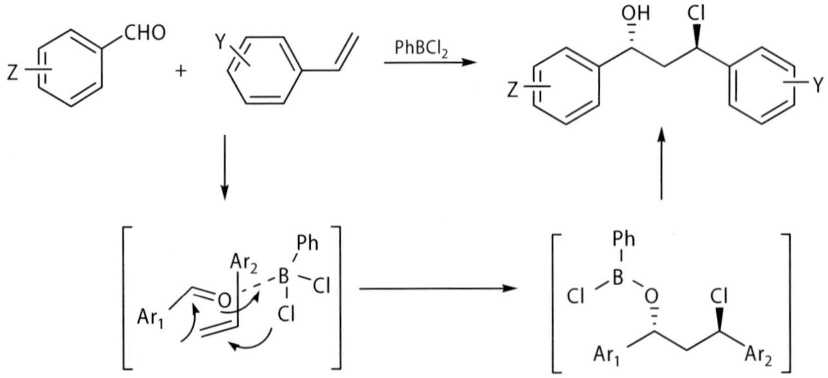

SCHEME 23.66 $PhBCl_2$-mediated coupling of aryl aldehydes with styrene derivatives.

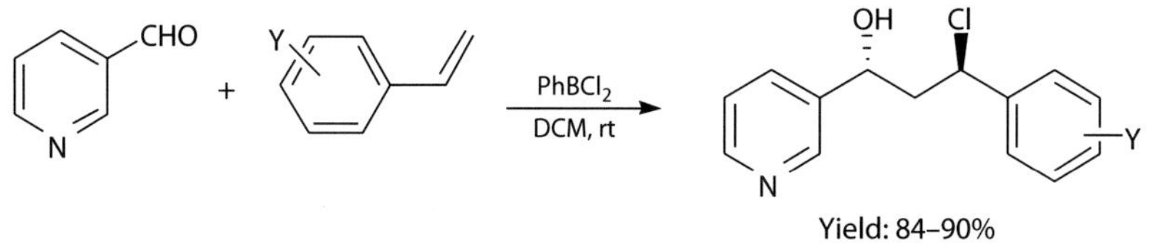

SCHEME 23.67 $PhBCl_2$-mediated coupling of 3-pyridinecarboxaldehyde with styrenes.

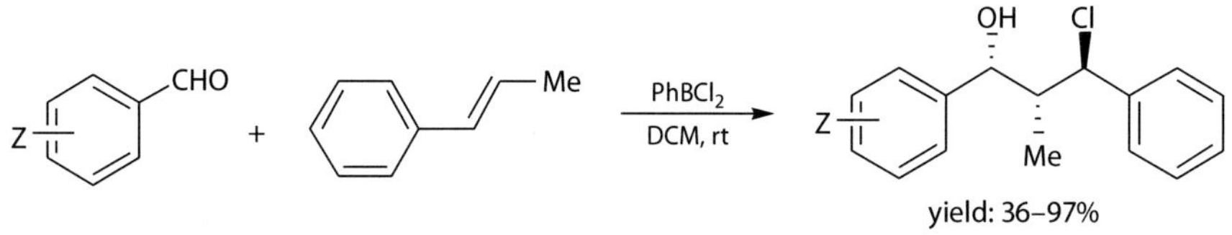

SCHEME 23.68 $PhBCl_2$-mediated coupling of aryl aldehydes with *trans*-β-methylstyrene.

23.3.3 Coupling Reaction of Alkoxides with Organoboron Halides

Substitution of the hydroxyl group in alcohols by nucleophiles is important in organic synthesis. However, because of the poor leaving ability of the hydroxyl group, these substitution reactions generally require preactivation of the hydroxyl groups by converting them to more reactive leaving groups (X, OAc, OTf, OTs, etc.). This preactivation process inevitably produces a stoichiometric quantity of by-product. The subsequent substitution of the halides and related compounds again produces a stoichiometric quantity of a by-product (Scheme 23.69). In terms of atom efficiency and environmental concerns, the direct substitution of hydroxyl groups by nucleophiles is highly desirable because only water is generated as the by-product.

In this context, the Lewis acid-catalyzed substitution of hydroxyl by an allyl group using allyltrimethylsilane[151] and the transition metal-catalyzed cross-coupling of allylic alcohols with aryl-, alkenylboronic acids[152] have been achieved in recent years. It was recently discovered that the hydroxyl group in allylic, benzylic, and propargylic alcohols, after conversion to the corresponding alkoxide, can be readily substituted by a vinyl group using boron chemistry. The advantage of using alkoxides as reactants in place of alcohols is that most alcohols are prepared from the reaction of aldehydes with organometallic reagents that results in an alkoxides product initially. The coupling reactions of alkoxides and organoboron halides became apparent during a study originally designed to probe the reaction mechanism of the boron trihalide-mediated alkyne–aldehyde coupling reactions. In order to explain the stereochemistry of the observed diene product, an unprecedented migration of a halovinyl group from boron to carbon was proposed (Scheme 23.63). In order to verify the proposed alkenyl migration, an alternative route to intermediate **B**, which is structurally similar to intermediate **A**, was designed (Scheme 23.70); the alternate route produced the expected diene product. This observation

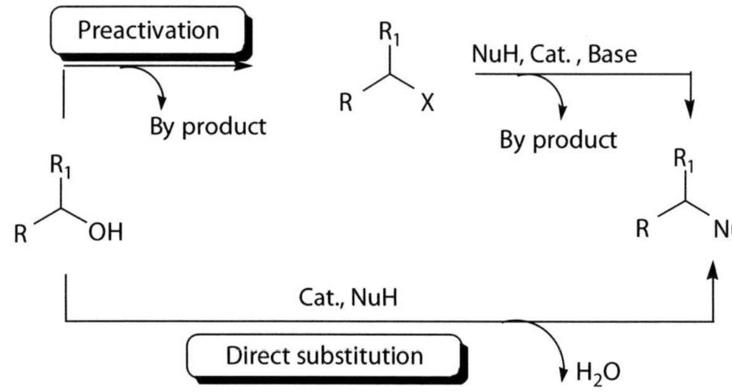

SCHEME 23.69 Substitution of the hydroxyl group in alcohols by nucleophiles.

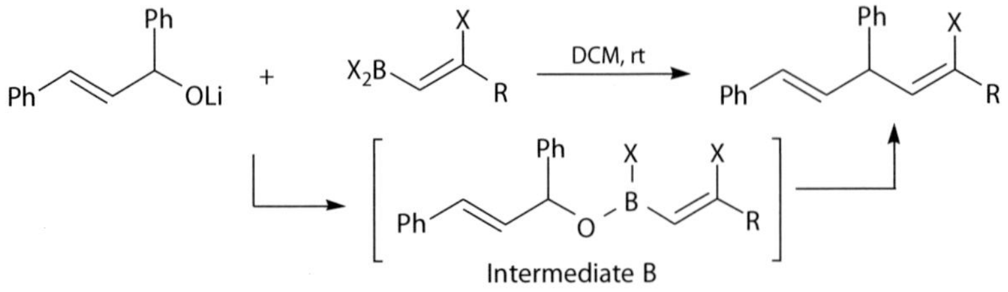

SCHEME 23.70 Reaction designed to verify the vinyl group migration in BX_3-mediated alkyne–aldehyde coupling.

not only supported the proposed alkenyl migration, but also provided a new methodology for nucleophilic substitution of hydroxyl groups by alkenyl moieties.[153] Since it involves alkenylboron reagents and the products are identical to those obtained in Suzuki coupling reactions, the new reaction can be viewed as a formal transition-metal-free Suzuki reaction.[154]

Through a designed competition reaction sequence involving the reaction of an alkenylboron dichloride with an allyloxide in the presence of allyltrimethylsilane, the migration of the alkenyl moiety was shown to proceed via a cationic pathway (Scheme 23.71).[50]

The generation of cations from alkoxides from alkoxyboron halide intermediates is significant because it is generally assumed that alkoxides (RO^-) react with Lewis acids (MCl_x) to yield the corresponding complexes, $M(OR)_nCl_{x-n}$. The C–O bond cleavage in alkoxides most probably is induced by the combination of two forces: the high electronegativity of chlorine and the strengthening of the B–O bond through p–π conjugation. The reported reaction of alkylvinylboron bromide with sodium methoxide to generate a stable boron ester [84,155] (Scheme 23.72) supports the postulated mechanism.

Reactions of vinylboron dihalides with benzyloxides[156] and propargyloxides[157] have also been examined and they proceed quite well. Isolation of racemic products from reactions employing chiral alcohols further supports a cation pathway for the alkenyl group migration. The tandem reactions of *in situ* generated alkoxides with vinylboron dihalides has also been accomplished (Scheme 23.73).

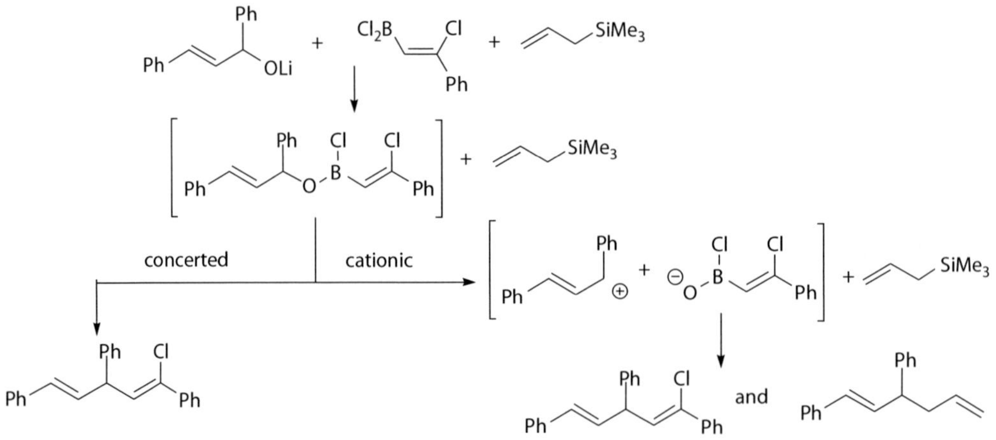

SCHEME 23.71 Designed competitive reaction to confirm the cationic pathway.

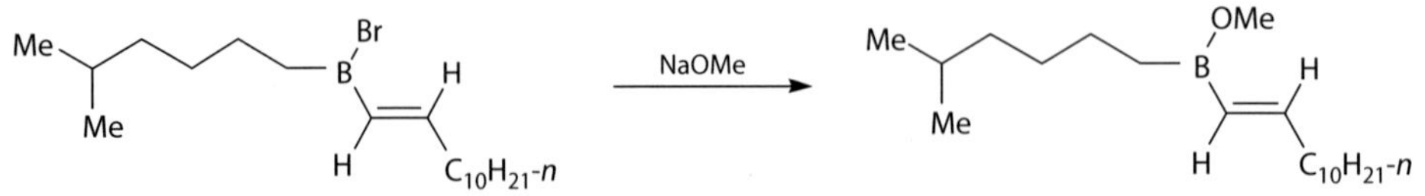

SCHEME 23.72 Reaction of alkylvinylboron bromide with sodium methoxide to form a stable boron ester.

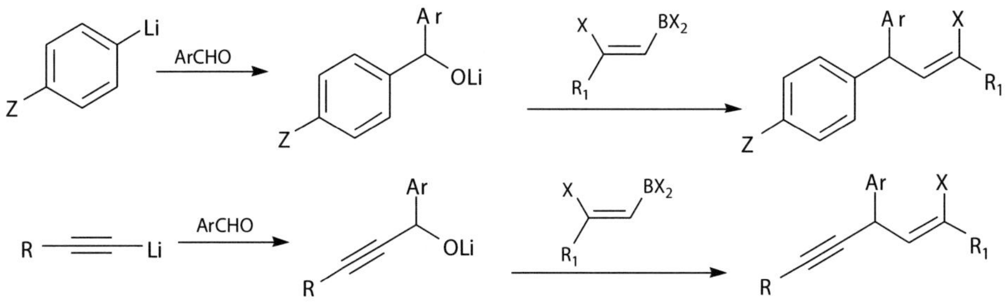

SCHEME 23.73 Tandem reactions of *in situ* generated alkoxides with vinylboron dihalides.

The reaction has been extended to alkynylboron dihalides (Scheme 23.74).[158] This new chemistry provides a simple route to internal alkynes that are otherwise not readily prepared.

It was then discovered that cations can also be generated from lithium alkoxides in the presence of boron trihalides. Thus, the reaction of a lithium alkoxide with phenylacetylene in the presence of boron trichloride generates (*E,E*)-1-chloro-1,3,5-triphenyl-1,4-pentadiene in 78% yield, only 4% of (*Z,E*)-1-chloro-1,3,5-triphenyl-1,4-pentadiene is formed (Scheme 23.75). Notably, the stereochemistry of the major product obtained using this process is opposite that of the product from the reaction shown in Scheme 23.70.

This discovery led to the development of a regioselective allylation of propargylic alcohols via their lithium salts when allyltrimethylsilane is used to capture the cation (Scheme 23.76).[159]

Encouraged by these results, the reaction of benzylic alkoxides with dichloroborane was examined.[160] These reactions provide an alternative route to diarylmethane derivatives (Scheme 23.77). The deoxygenation of alkoxides using dichloroborane provides useful evidence for understanding the importance of Lewis acidity in previously reported deoxygenation methods operating through alkoxide intermediates.[161]

The stability of the cyclopropyl group during the deoxygenation reaction (Scheme 23.78) strongly indicates that the reaction proceeds through a concerted reaction mechanism rather than a cation mechanism. In addition, vinyl groups are unaffected during the deoxygenation.

R, R_1 = aryl, alkenyl, propargyl, methyl
R_2 = aryl, alkyl

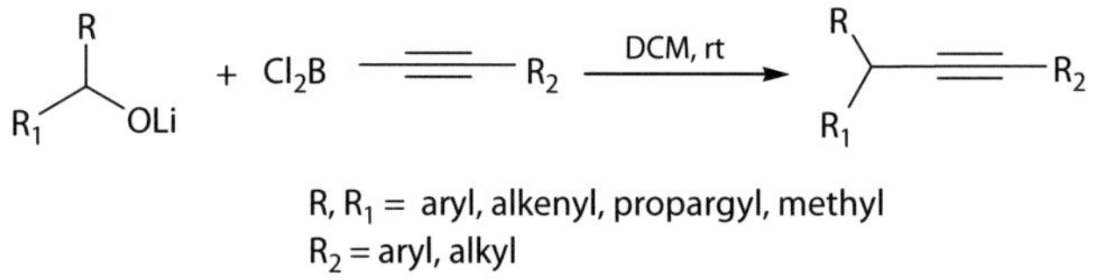

SCHEME 23.74 Coupling of alkoxides with alkynylboron dichlorides.

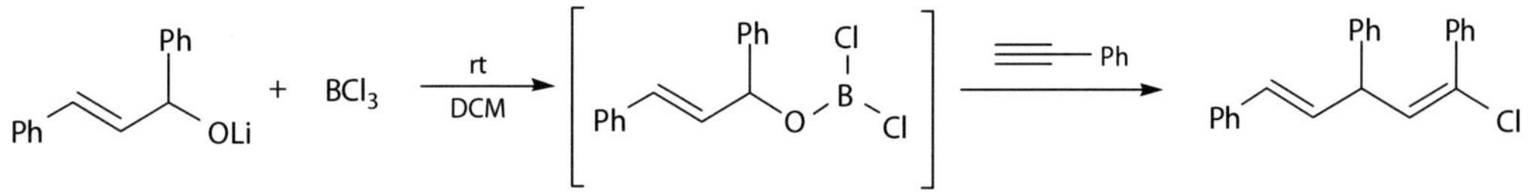

SCHEME 23.75 BCl_3-mediated coupling of lithium alkoxide with phenylacetylene.

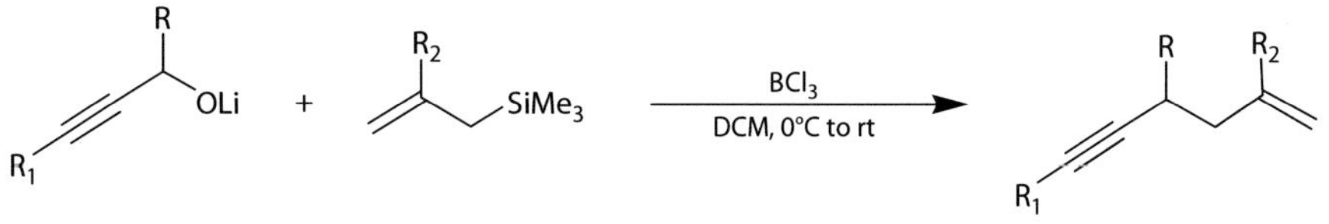

SCHEME 23.76 BCl_3-mediated coupling of lithium proparglalkoxides with allylsilanes.

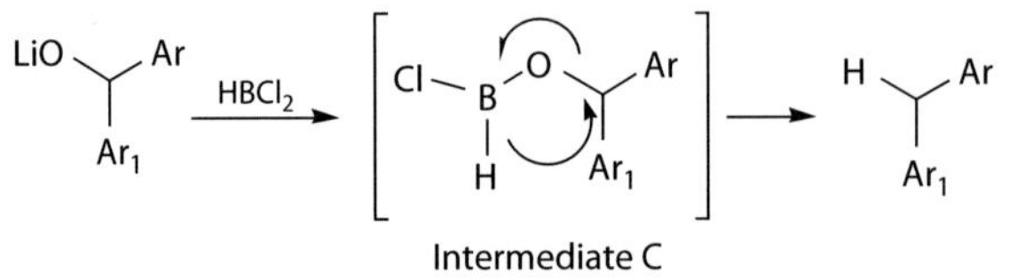

LiO Ar Ar$_1$ $HBCl_2$ Cl B O H Ar Ar$_1$ H Ar Ar$_1$

Intermediate C

SCHEME 23.77 Reaction of benzylic alkoxides with dichloroborane.

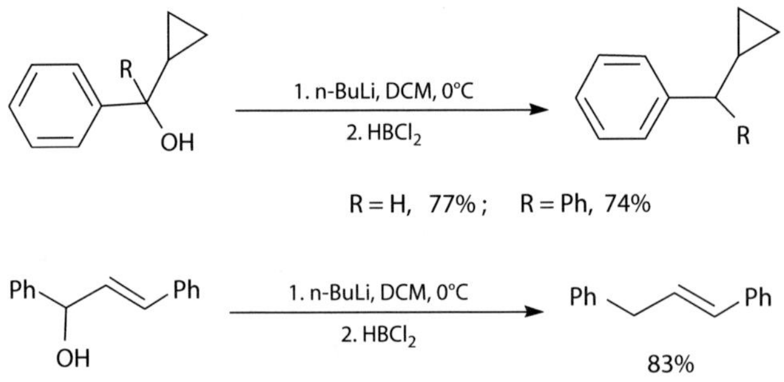

R OH 1. n-BuLi, DCM, 0°C 2. $HBCl_2$ R

R = H, 77%; R = Ph, 74%

Ph Ph OH 1. n-BuLi, DCM, 0°C 2. $HBCl_2$ Ph Ph 83%

SCHEME 23.78 Deoxygenation of alkoxides using dichloroborane.

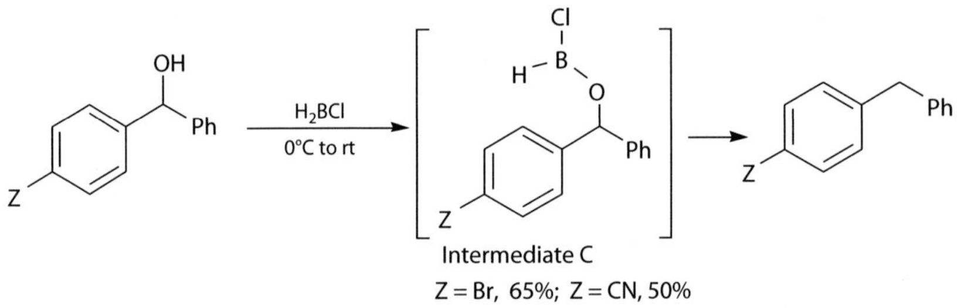

OH Ph Z H_2BCl 0°C to rt Cl H B O Ph Z Ph Z

Intermediate C

Z = Br, 65%; Z = CN, 50%

SCHEME 23.79 Deoxygenation of alcohols using monochloroborane.

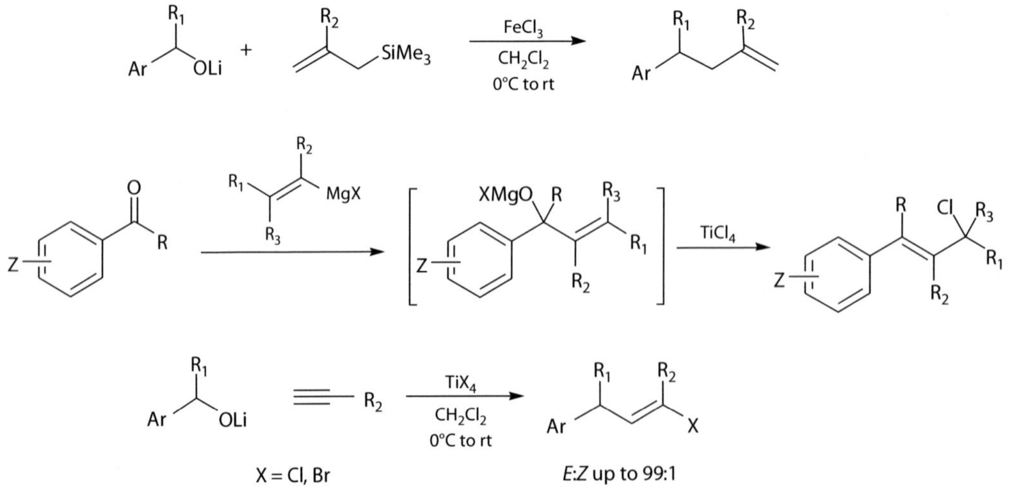

R$_1$ Ar OLi + R$_2$ $SiMe_3$ $FeCl_3$ CH_2Cl_2 0°C to rt R$_1$ R$_2$ Ar

O R Z R$_2$ R$_1$ MgX R$_3$ XMgO R R$_3$ R$_1$ R$_2$ Z $TiCl_4$ R Cl R$_3$ R$_1$ R$_2$ Z

R$_1$ Ar OLi R$_2$ TiX_4 CH_2Cl_2 0°C to rt R$_1$ R$_2$ Ar X

X = Cl, Br *E:Z* up to 99:1

SCHEME 23.80 Metal halide-mediated coupling with alkoxides.

SCHEME 23.81 C–S bond cleavage in lithium thiolate.

The possibility of generating intermediate **C** by mixing benzylic alcohols with monochloroborane was also tested for the purpose of obviating the use of butyllithium (Scheme 23.79). This modification could be beneficial when certain functional groups (–Br and –CN) are present. The direct reaction produces the desired products in moderate yields.

Although the study using alkoxides as synthetic intermediates was initiated only in recent years, it has attracted attention. Metal halides, such as FeX_3,[162] TiX_4[163], and NbX_5,[164] mediated coupling reactions involving alkoxides have been reported (Scheme 23.80).

Related work also documented a successful C–S bond cleavage (Scheme 23.81).[165] The reaction provides a practical route to tertiary amines.

23.4 CONCLUSION

Although synthetic routes to organoboron halides have been known for many years, their application in organic syntheses was limited to the asymmetric reduction of ketones using DIP-Cl as well as Aldol reactions using $(c\text{-Hex})_2BCl$. The recent achievements highlighted in this chapter demonstrate that the boron halide reagents are extremely valuable synthetic intermediates. Significantly, reactions in which organoboron halides are employed generally tolerate a wide variety of functional groups including those incompatible with more traditional organometallic reagents.

REFERENCES

1. Hirao, H., Omoto, K., Fujimoto, H. 1999. Lewis acidity of boron trihalides. *J. Phys. Chem.* A 103: 5807–5811.
2. (a) Bhatt, M. V., Kalkarni, S. U. 1983. Cleavage of ethers. *Synthesis* 1983: 249–282; (b) McOmie, J. F. W., Watts, M. L., West, D. E. 1968. Demethylation of aryl methyl ethers by boron tribromide. *Tetrahedron* 24: 2289–2292.
3. (a) Lappert, M. F., Prokai, B. 1964. Chloroboration and allied reactions of unsaturated compounds. II. Haloboration and phenylboration of acetylenes; preparation of some alkynylboranes. *J. Organomet. Chem.* 1: 384–400; (b) Suzuki, A. 1997. Haloboration of 1-alkynes and its synthetic application. *Rev. Heteroatom Chem.* 17: 271–314.
4. (a) Amrollah-Madjdababi, A., Pham, T. N., Ashby, E. C. 1989. A simple method for the conversion of adamantyl, benzyl and benzyhydryl alcohols to their corresponding bromides and chlorides and the transhalogenation of adamantyl, benzyl, benzhdryl and tertiary alkyl bromides and chlorides. *Synthesis* 1989: 614–616; (b) Pelletier, J. D., Poirier, D. 1994. Bromination of alcohols by boron tribromide. *Tetrahedron Lett.* 35: 1051–1054.
5. (a) Goldstein, M., Haines, L. I. B., Hemmings, J. A. G. 1972. Kinetics of the halogen-exchange reaction between alkyl halides and boron trihalides. formation of adducts. *J. Chem. Soc., Dalton Trans.* 2260–2263; (b) Namavari, M., Satyamurthy, N., Barrio, J. R. 1995. Halogen-exchange reactions between alkyl fluorides and boron trihalides or titanium tetrahalides. A convenient synthesis of alkyl halides from alkyl fluorides. *J. Fluorine Chem.* 72: 89–93.
6. Sprott, K. T., Corey, E. J. 2003. A new cationic, chiral catalyst for highly enantioselective Diels–Alder reactions. *Org. Lett.* 5: 2465–2467.
7. (a) Brown, H. C., Dhar, R. K., Ganesan, K., Singaram, B. 1992. Enolboration. 1. Dicyclohexylchloroborane/triethylamine as a convenient reagent for enolboration of ketones and other carbonyl derivatives. *J. Org. Chem.* 57: 499–504; (b) Brown, H. C., Ganesan, K., Dhar, R. K. 1993. Enolboration. 4. An examination of the effect of the leaving group (X) on the stereoselective enolboration of ketones with various R_2BX/Et_3N. New reagents for the selective generation of either *Z* or *E* enol borinates from representative ketones. *J. Org.*

Chem. 58: 147–153; (c) Ganesan, K., Brown, H. C. 1993. Enolboration. 5. An examination of the effects of amine, solvent, and other reaction parameters on the stereoselective enolboration of ketones with various Chx_2BX reagents. An optimized procedure to achieve the stereoselective synthesis of *E* enol borinates from representative ketones using Chx_2BCl/Et_3N. *J. Org. Chem.* 58: 7162–7169; (d) Righi, G., Spirito F., Bonini, C. 2002. Stereoselective aldol condensation of boron enolates to *trans* α,β-epoxy aldehydes. *Tetrahedron Lett.* 43: 4737–4740; (e) Brown, H. C., Dhar, R. K., Bakshi, R. K., Pandiarjan, P. K., Singaram, B. 1989. Major effect of the leaving group in dialkylboron chlorides and triflates in controlling the stereospecific conversion of ketones into either [*E*]- or [*Z*]-enol borinates. *J. Am. Chem. Soc.* 111: 3441–3442; (f) Ward, D. E., Lu, W. L. 1998. Enantioselective enolborination. *J. Am. Chem. Soc.* 120: 1098–1099.

8. (a) Patterson, I. 1992. New methods and strategies for the stereocontrolled synthesis of polypropionate-derived natural products. *Pure Appl. Chem.* 64: 1821–1830; (b) B. M. Trost, I. Fleming. *Comprehensive Organic Synthesis*, Vol. 2, Pergamon Press, Oxford, 1991; (c) Brown, H. C., Zou, M. F., Ramachandran, P. V. 1999. Efficient diastereoselective synthesis of *anti*-α-bromo-β-hydroxyketones. *Tetrahedron Lett.* 40: 7875–7877; (d) Dunetz, J. R., Roush, W. R. 2008. Concerning the synthesis of the Tedanolide C(13)-C(23) fragment via anti-aldol reaction. *Org. Lett.* 10: 2059–2062; (e) Evans, D. A., Ng, H. P., Clark, J. S., Rieger, D. L. 1992. Diastereoselective anti aldol reactions of chiral ethyl ketones. Enantioselective processes for the synthesis of polypropionate natural products. *Tetrahedron* 48: 2127–2142; (f) Majewki, M., Nowak, P. 2000. Aldol addition of lithium and boron enolates of 1,3-dioxan-5-ones to aldehydes. A new entry into monosaccharide derivatives. *J. Org. Chem.* 65: 5152–5160; (g) Galobards, M., Gascon, M., Mena, M., Romea, P., Urpi, F., Vilarrasa, J. 2000. Enolization of chiral α-silyloxy ketones with dicyclohexylchloroborane. application to stereoselective aldol reactions. *Org. Lett.* 2: 2599–2602; (h) Carda, M., Murga, J., Falomir, E., Gonzalez, F., Marco, J. A. 2000. Aldol reactions with erythrulose derivatives: Stereoselective synthesis of differentially protected *syn*-α,β-dihydroxy esters. *Tetrahedron* 56: 677–683; (i) Paterson, I., Osborne, S. 1990. Stereoselective aldol reactions of β-chlorovinyl ketones using dienol borinates: A new synthesis of dihydropyrones. *Tetrahedron Lett.* 31: 2213–2216; (j) Marco, J. A., Carda, M., Folomir, E., Palomo, C., Oiardide, M., Ortiz, J. A., Linden, A. 1999. Erythrulose as a multifunctional chiron: Highly stereoselective boron aldol additions. *Tetrahedron Lett.* 40: 1065–1068.
9. (a) Brown, H. C., Kulkarni, S. U. 1981. Organoboranes. XXV. Hydridation of dialkylhaloboranes. New practical syntheses of dialkylboranes under mild conditions. *J. Organomet. Chem.* 218: 299–307; (b) Soundararajan, R., Matteson, D. S. 1990. Hydroboration with boron halides and trialkylsilanes. *J. Org. Chem.* 55: 2274–2275.
10. (a) Brown, H. C., Ramachandran, P. V. 1991. The boron approach to asymmetric synthesis. *Pure Appl. Chem.* 63: 307–316; (b) Brown, H. C., Ramachandran, P. V. 1992. Asymmetric reduction with chiral organoboranes based on α-pinene. *Acc. Chem. Res.* 25: 16–24; (c) Singh, V. K. 1992. Practical and useful methods for the enantioselective reduction of unsymmetrical ketones. *Synthesis* 1992: 605–617; (d) Farina, V., Reeves, J. T., Senanayake, C. H., Song, J. J. 2006. Asymmetric synthesis of active pharmaceutical ingredients. *Chem. Rev.* 106: 2734–2793; (e) Zhang, P. 2007. (–)-Diisopinocampheyl chloroborane [(–)-DIP-Chloride™]: A versatile reagent in asymmetric synthesis. *Synlett* 2762–2763.
11. Ishihara, K. *Lewis Acids in Organic Synthesis*; Yamamoto, H., Ed.; Wiley-V. C. H.: Weinheim, Germany, 2000; Vol. 1. pp. 91–96.
12. Yamada, H., Kawate, T., Matsumizu, M., Nishida, A., Yamaguchi, K., Nakagawa, M. 1998. Chiral Lewis acid-mediated enantioselective Pictet–Spengler reaction of N_b-hydroxytryptamine with aldehydes. *J. Org. Chem.* 63: 6348–6354.
13. (a) Kabalka, G. W., Wu, Z., Ju, Y. 2003. Use of organoboron halides in organic synthesis. *Pure Appl. Chem.* 75: 1231–1237; (b) Kabalka, G. W., Wu, Z., Ju, Y. 2003. The use of organoboron chlorides and bromides in organic synthesis. *J. Organomet. Chem.* 680: 12–22.
14. Teitel, S., O'Brien, J., Brossi, A. 1976. Selective removal of an aromatic methylenedioxy group. *J. Org. Chem.* 41: 1657–1658.
15. (a) Williard, P. G., Fryhle, C. B. 1980. Boron trihalide-methyl sulfide complexes as convenient reagents for dealkylation of aryl ethers. *Tetrahedron Lett.* 21: 3731–3734; (b) Bonner, T. G., Lewis, D., Rutter, K. 1981. Opening of cyclic acetals by trichloro-, dichloro-, and tribromoborane. *J. Chem. Soc., Perkin Trans 1.* 1807–1810.
16. Niwa, H., Hida, T., Yamada, K. 1981. A new method for cleavage of aliphatic methyl ethers. *Tetrahedron Lett.* 22: 4239–4240.
17. (a) Brooks, P. R., Wirtz, M. C., Vetelino, M. G., Rescek, D. M., Woodworth, G. F., Morgan, B. P., Coe, J. W. 1999. Boron trichloride/tetra-*n*-butylammonium iodide: A mild, selective combination reagent for the cleavage of primary alkyl aryl ethers. *J. Org. Chem.* 64: 9719–9721; (b) Paruch, K., Vyklicky, L., Wang,

D. Z., Katz, T. J., Incarvito, C., Zakharov, L., Rheingold, A. L. 2003. Functionalizations of [6]- and [7] helicenes at their most sterically hindered positions. *J. Org. Chem.* 68: 8539–8544.

18. (a) Pons, J. M., Santelli, M., Eds.; *Lewis Acids and Selectivity in Organic Synthesis*; CRC Press: Boca Raton, FL, 1996; (b) Leach, M. R. *Lewis Acid/Base Reaction Chemistry*; Metasynthesis.com: Brighton, UK, 1999.
19. Jones, G. B., Hynd, G., Wright, J. M, Sharma, A. 2000. On the selective deprotection of trityl ethers. *J. Org. Chem.* 65: 263–265.
20. Bhatt, M. V. 1978. *B*-bromo-9-borabicyclo[3.3.1]nonane. A convenient and selective reagent for ether cleavage. *J. Organomet. Chem.* 156: 221–226.
21. Punna, S., Meunier, S., Finn, M. G. 2004. A hierarchy of aryloxide deprotection by boron tribromide. *Org. Lett.* 6: 2777–2779.
22. Yao, M.-L., Reddy, M. S., Zeng, W.-B., Hall, K., Walfish, I., Kabalka G. W. 2009. Identification of a boron-containing intermediate in the boron tribromide mediated aryl propargyl ether cleavage reaction. *J. Org. Chem.* 74: 1385–1387.
23. (a) Guindon, Y., Morton, H. E., Yoakim, C. 1983. Dimethylboron bromide and diphenylboron bromide. Acetal and ketal cleavage. Cleavage of MEM, MOM, and MTM ethers. *Tetrahedron Lett.* 24: 3969–3972; (b) Guindon, Y., Yoakim, C., Morton, H. E. 1984. Dimethylboron bromide and diphenylboron bromide: Cleavage of acetals and ketals. *J. Org. Chem.* 49: 3912–3920; (c) Gauthier, J. Y., Guindon, Y. 1987. Exceptionally mild and stereospecific ring fragmentations promoted by dimethylboron bromide. *Tetrahedron Lett.* 28: 5985–5988.
24. Okano, K., Okuyama, K.-I., Fukuyama, T., Tokuyama, H. 2008. Mild debenzylation of aryl benzyl ether with BCl_3 in the presence of pentamethylbenzene as a non-Lewis-Basic cation scavenger. *Synlett* 1977–1980.
25. Okano, K., Tokuyama, H., Fukuyama, T. 2006. Total synthesis of (+)-Yatakemycin. *J. Am. Chem. Soc.* 128: 7136–7137.
26. Schäfer, W., Franck, B. 1966. Selective ether cleavage of 4-hydroxymethoxyquinolinecarboxylic acid esters. *Chem. Ber.* 99: 160–164.
27. Desvergnes, S., Vallée, Y., Py, S. 2008. Novel polyhydroxylated cyclic nitrones and *N*-hydroxypyrrolidines through BCl_3-mediated deprotection. *Org. Lett.* 10: 2967–2970.
28. Nguyen, N. H., Cougnon, C., Gohier, F. 2009. Deprotection of arenediazonium tetrafluoroborate ethers with BBr_3. *J. Org. Chem.* 74: 3955–3957.
29. Zhou, Q. J., Worm, K., Dolle, R. E. 2004. 10-Hydroxy-10,9-boroxarophenanthrenes: Versatile synthetic intermediates to 3,4-benzocoumarins and triaryls. *J. Org. Chem.* 69: 5147–5149.
30. (a) Kulkarni, S. U., Patil, V. D. 1982. Cleavage of cyclic ethers with boron bromide. A convenient route to the bromosubstituted alcohols, aldehydes and ketones. *Heterocycles* 18: 163–167; (b) Roy, C. D. 2006. Regiocontrolled opening of 2-methyltetrahydrofuran with various boron reagents. *Aust. J. Chem.* 59: 657–659.
31. Kim H. J., Lee, Y. J. 1998. A facile synthesis of 3,4-disubstituted isoxazole derivatives by regioselective cleavage of pyrano[3,4-C]isoxazoles with boron trihalide. *Synth. Commun.* 28: 3527–3537.
32. Baran, A., Kazaz, C., Seçen, H., Sütbeyaz, Y. 2003. Synthesis of haloconduritols from an endo-cycloadduct of furan and vinylene carbonate. *Tetrahedron* 59: 3643–3648.
33. (a) Bellur, E., Langer, P. 2004. Reaction of 2-alkylidenetetrahydrofurans with boron tribromide: Chemo- and regioselective synthesis of 6-bromo-3-oxoalkanoates by application of a "cyclization-ring-opening" strategy. *Synlett* 2172–2174; (b) Bellur, E., Langer, P. 2005. Convenient synthesis of ε-halo-β-ketoesters and γ,γ'-dibromoalkanones by regio- and chemoselective reaction of 2-alkylidenetetrahydrofurans with boron trihalides: A "ring-closure/ring-cleavage" strategy. *J. Org. Chem.* 70: 3819–3825; (c) Bellur, E., Langer, P. 2005. Synthesis of benzofurans with remote bromide functionality by domino "ring-cleavage-deprotection-cyclization" reactions of 2-alkylidenetetrahydrofurans with boron tribromide. *J. Org. Chem.* 70: 7686–7693.
34. Malladi, R. R., Kabalka, G. W. 2002. One-pot synthesis of ω-chloroesters via the reaction of acid chlorides with tetrahydrofuran in the presence of trichloroborane. *Synth. Commun.* 32: 1997–2001.
35. (a) Youssefyeh, R. D., Mazur, Y. 1963. Cleavage of steroidal ethers with boron trichloride and boron tribromide. *Chem. Ind. (London)* 609–610; (b) Nordvik, T., Brinker, U. H. 2003. A novel route to geminal dibromocyclobutanes: Syntheses of 2-substituted cyclobutanone acetals and their reaction with boron tribromide. *J. Org. Chem.* 68: 9394–9399.
36. (a) Corey, E. J., Weinhensker, N. M., Schaaf, T. K., Huber, W. 1969. Stereo-controlled synthesis of prostaglandins $F_{2\alpha}$ and E_2 (*dl*). *J. Am. Chem. Soc.* 91: 5675–5677; (b) Su, Q., Panek, J. S. 2004. Total synthesis of (-)-Apicularen A. *J. Am. Chem. Soc.* 126: 2425–2430.
37. Olah, G. A., Karpeles, R., Narang, S. C. 1982. Synthetic methods and reactions; 107. Preparation of ω-haloalkylcarboxylic acids and esters or related compounds from lactones and boron trihalides. *Synthesis* 1982: 963–965.

38. (a) Narayana, C., Padmanabhan, S., Kabalka, G. W. 1990. Cleavage of ethers and geminal diacetates using the boron triiodide-*N,N*-diethylaniline complex. *Tetrahedron Lett.* 31: 6977–6978; (b) Kabalka, G. W., Narayana, C., Reddy, N. 1992. Boron triiodide-*N-N*-diethylaniline complex: A new reagent for cleaving esters. *Synth. Commun.* 22: 1793–1798.
39. Narayana, C., Padmanabhan, S., Kabalka, G. W. 1991. Reductive dimerization of sulfonyl derivatives to disulfides and deoxygenation of sulfoxides to sulfides using the boron triiodide-*N,N*-diethylaniline complex. *Synlett* 125–126.
40. Butler, D. C. D., Alper, H. 1998. Synthesis of isocyanates from carbamate esters employing boron trichloride. *Chem. Commun.* 2575–2576.
41. (a) McCusker, P. A., Hennion, G. F., Ashby, E. C. 1957. Organoboron compounds. VII. Dialkylchloroboranes from the reaction of boron chloride with trialkylboranes. Disproportionation of dialkylchloroboranes. *J. Am. Chem. Soc.* 79: 5192–5194; (b) Brown, H. C., Levy A, B. 1972. Organoboranes. XV. Simple convenient procedure for the preparation of alkyldichloroboranes via hydroboration-redistribution. *J. Organomet. Chem.* 44: 233–236; (c) Brown, H. C., Basavaiah, D., Bhat, N. G. 1983. Organoboranes. 29. A convenient synthesis of alkyldibromoboranes and dialkylbromoboranes via hydroboration-redistribution. *Organometallics* 2: 1309–1311.
42. (a) Brown, H. C., Ravindran, N., Kulkarni, S. U. 1979. Hydroboration. 52. Monohaloborane-methyl sulfide adducts as new reagents for the hydroboration of alkenes. A convenient synthesis of dialkylhaloboranes and their derivatives for organic synthesis. *J. Org. Chem.* 44: 2417–2422; (b) Brown, H. C., Ravindran, N., Kulkarni, S. U. 1980. Hydroboration. 54. New general synthesis of alkyldihaloboranes via hydroboration of alkenes with dihaloborane-dimethyl sulfide complexes. Unusual trends in the reactivities and directive effects. *J. Org. Chem.* 45: 384–389; (c) Brown, H. C., Campbell, J. B., Jr. 1980. Hydroboration. 55. Hydroboration of alkynes with dibromoborane-dimethyl sulfide. convenient preparation of alkenyldibromoboranes. *J. Org. Chem.* 45: 389–395.
43. Erdyakov, S. Yu., Ignatenko, A. V., Gurskii, M. E., Bubnov, Yu. N. 2004. Regioselectivity of 1,1-dimethylallene allylboration: Synthesis of isomeric 6- and 7-methylene-3-borabicyclo[3.3.1]nonanes. *Mendeleev Commun.* 242–244.
44. Cole, T. E., Quintanilla, R., Smith, B. M., Hurst, D. 1992. Simple conversion of anhydrides of boronic and borinic acids to the corresponding organodihaloboranes and diorganohaloboranes. *Tetrahedton Lett.* 33: 2761–2764.
45. (a) Kaufmann, D. 1987. Borylation of arylsilanes. I. A general, easy, and selective access to phenyldihaloboranes. *Chem. Ber.* 120: 853–854; (b) Kaufmann, D. 1987. Borylation of arylsilanes. II. Synthesis and reactions of silylated dihalophenylboranes. *Chem. Ber.* 120: 901–905; (c) Gross, U., Kaufmann, D. 1987. Borylation of arylsilanes. III. Reaction of silylated biphenyls and 9H-9-silafluorenes with tribromoborane. *Chem. Ber.* 120: 991–994; (d) Deck, P. A., Fisher, T. S., Downey J. S. 1997. Boron-silicon exchange reactions of boron trihalides with trimethylsilyl-substituted metallocenes. *Organometallics* 16: 1193–1196.
46. Gerrard, W., Howarth, M., Monney, E. F., Pratt, D. E. 1963. Advances in the preparation of arylboron dihalides, cyclic 2-arylboroles, and B-triarylborazoles. *J. Chem. Soc.* 1582–1584.
47. Gerwarth, U. W., Weber, W. 1983. Dichlorophenylborane. *Inorganic Syntheses* 22: 207–208.
48. (a) Cole, T. E., Quintanilla, R., Rodewald, S. 1991. Migration of 1-alkenyl groups from zirconium to boron compounds. *Organometallics* 10: 3777–3781; (b) Cole, T. E., Watson, C., Rodewald, S. 1997. Transmetalation of organic groups from zirconacycles to haloboranes: A new route to borolane compounds. *Tetrahedron Lett.* 38: 8487–8490.
49. (a) Lipshutz, B. H., Boskovic, Z. V., Aue, D. H. 2008. Synthesis of activated alkenylboronates from acetylenic esters by CuH-catalyzed 1,2-addition/transmetalation. *Angew. Chem. Int. Ed.* 47: 10183–10186; (b) Sundararaman, A., Jäkle F. 2003. A comparative study of base-free arylcopper reagents for the transfer of aryl groups to boron halides. *J. Organomet. Chem.* 681: 134–142.
50. Kabalka, G. W., Yao, M.-L., Borella, S., Wu, Z., Ju, Y. 2008. Boron trihalide mediated alkyne–aldehyde coupling reaction: A mechanistic investigation. *J. Org. Chem.* 73: 2668–2673.
51. Hyuga, S., Chiba, Y., Yamashina, N., Hara, S., Suzuki, A. 1987. Organic synthesis using haloboration reactions. 13. (*E*)-(2-Bromoethenyl)dibromoborane. A new precursor for (*E*)-1,2-disubstituted ethenes. *Chem. Lett.* 16: 1757–1760.
52. Rabinovitz, M., Grinvald, A. 1972. Boron trifluoride complexes of aromatic aldehydes. V. CHO:BF_3 pseudosubstituent. *J. Am. Chem. Soc.* 94: 2724–2729.
53. Nakamura, E., Yamanaka, M., Mori, S. 2000. Complexation of Lewis acid with trialkylcopper (III): On the origin of BF3-Acceleration of cuprate conjugate addition. *J. Am. Chem. Soc.* 122: 1826–1827.
54. Frazer, M. J., Gerrard, W., Lappert, M. F. 1957. Interaction of aldehydes with boron trichloride. *J. Chem. Soc.* 739–744.

55. Lansinger, J. M., Ronald, R. C. 1979. Reactions of aromatic aldehydes with boron halides. *Synth. Commun.* 9: 341–349.
56. Kabalka, G. W., Wu, Z. 2000. Conversion of aromatic aldehydes to gem-dichlorides using boron trichloride. A new highly efficient method for preparing dichloroarylmethanes. *Tetrahedron Lett.* 41: 579–581.
57. (a) Mikami, K., Shimizu, M. 1992. Asymmetric ene reactions in organic synthesis. *Chem. Rev.* 92: 1021–1050; (b) Snider, B. B. 1980. Lewis- acid catalyzed ene reactions. *Acc. Chem. Res.* 13: 426–432.
58. (a) Iwamura, T., Fujita, M., Kawakita, T., Kinoshita, S., Watanabe, S.-I., Kataoka, T. 2001. Dimethyl sulfide-boron trihalide-mediated reactions of α, β-unsaturated ketones with aldehydes: One-pot synthesis of Baylis–Hillman adducts and α–halomethyl enones. *Tetrahedron* 57: 8455–8462; (b) Shi, M., Jiang, J. K., Cui, S. C., Feng, Y. S. 2001. Titanium(IV) chloride, zirconium(IV) chloride or boron trichloride and phosphine-promoted Baylis–Hillman reaction of aldehydes with α, β-unsaturated ketone. *J. Chem. Soc. Perkin Trans.* 1 :390–393.
59. (a) Kabalka, G. W., Wu, Z., Ju, Y. 2001. Boron trihalide-promoted addition of aryl aldehydes to styrenes. A new convenient and highly efficient synthesis of 1,3-dihalo-1,3-diarylpropanes. *Tetrahedron Lett.* 42: 5793–5796; (b) Kabalka, G. W., Wu, Z., Ju, Y. 2004. A new convenient, efficient, and regioselective synthesis of 1,3-diaryl-1,3-dihalopropanes. *Synthesis* 2004: 2927–2929; (c) Kabalka, G. W., Wu, Z., Ju, Y., Yao, M.-L. 2005. Lewis acid mediated reactions of aldehydes with styrene derivatives: Synthesis of 1,3-dihalo-1,3-diarylpropanes and 3-chloro-1,3-diarylpropanols. *J. Org. Chem.* 70: 10285–10291.
60. Ludek, T., Jiri, S., Petr, H., Rudolf, L. 1996. Effect of PVC on low-polar isobutylene polymerization in the presence of BCl_3. Part 2. Vinyl chloride-2-chloropropene copolymer as an initiator. *Polymer Bull. (Berlin)* 36:557–562.
61. (a) Snider, B. B., Rodini, D. J., Kirk, T. C., Cordova, R. 1982. Dimethylaluminum chloride catalyzed ene reactions of aldehydes. *J. Am. Chem. Soc.* 104: 555–563; (b) Gill, G. B., Parrot, S. J., Wallace, B. 1978. Formation of abnormal products in the aluminium chloride-catalysed ene additions of chloral and bromal to mono- and 1,2-di-alkyl ethylenes. *J. Chem. Soc., Chem. Commun.* 655–656.
62. (a) Kabalka, G. W., Tejedor, D., Li, N.-S., Reddy, M., Trotman, S. 1998. Boron trifluoride induced fragmentation of β-aryl-β-hydroxy ketones. *Tetrahedron Lett.* 39: 8071–8072; (b) Kabalka, G. W., Tejedor, D., Li, N.-S., Malladi, R. R., Trotman, S. 1998. A tandem aldol-Grob reaction of ketones with aromatic aldehydes. *Tetrahedron* 54: 15525–15532; (c) Kabalka, G. W., Tejedor, D., Li, N.-S., Malladi, R. R.; Trotman, S. 1998. An unprecedented, tandem Aldol-Grob reaction sequence. *J. Org. Chem.* 63: 6438–6439; (d) Kabalka, G. W., Li, N.-S., Tejedor, D., Malladi, R. R., Trotman, S. 1999. Synthesis of (*E*)-1-aryl-1-alkenes *via* a novel BF_3. OEt_2-catalyzed aldol-Grob reaction sequence. *J. Org. Chem.* 64: 3157–3161.
63. Ju, Y., Varma, R. S. 2006. Aqueous *N*-heterocyclization of primary amines and hydrazines with dihalides: microwave-assisted syntheses of *N*-azacycloalkanes, isoindole, pyrazole, pyrazolidine, and phthalazine derivatives. *J. Org. Chem.* 71: 135–141.
64. Kabalka, G. W., Wu, Z., Ju, Y. 2002. A new reaction of aryl aldehydes with aryl acetylenes in the presence of boron trihalides. *Org. Lett.* 4: 1491–1493.
65. Yao, M.-L., Quinn, M. P., Kabalka, G. W. 2010. Boron trichloride mediated alkyne-aldehyde coupling reactions. *Heterocycles* 80: 779–785.
66. Kabalka, G. W., Wu, Z., Ju, Y. 2002. A novel titanium tetrahalide-mediated carbon–carbon bond-forming reaction: Regioselective synthesis of substituted (*E,Z*)-1,5-dihalo-1,4-dienes. *Org. Lett.* 4: 3415–3417.
67. Miranda, P. O., Diaz, D. D., Padron, J. I., Ramirez, M. A., Martin, V. S. 2005. Fe(III) halides as effective catalysts in carbon–carbon bond formation: Synthesis of 1,5-dihalo-1,4-dienes, α, β-unsaturated ketones, and cyclic ethers. *J. Org. Chem.* 70: 57–62.
68. Yadav, J. S., Reddy, B. V. S., Eeshwaraiah, B., Gupta, M. K., Biswas, S. K. 2005. Gallium(III) halide promoted synthesis of 1,3,5-triaryl-1,5-dihalo-1,4-pentadienes. *Tetrahedron Lett.* 46: 1161–1163.
69. Denmark, S. E., Fu, J. 2003. Catalytic enantioselective addition of allylic organometallic reagents to aldehydes and ketones. *Chem. Rev.* 103: 2763–2794.
70. Schkeryantz, J. M., Woo, J. C. G., Siliphaivanh, P., Depew, K. M., Danishefsky S. J. 1999. Total synthesis of Gypsetin, Deoxybrevianamide E, Brevianamide E, and Tryprostatin B: Novel constructions of 2,3-disubstituted indoles. *J. Am. Chem. Soc.* 121: 11964–11975.
71. Mayr, H., Gorath, G. 1995. Kinetics of the reactions of carboxonium ions and aldehyde boron trihalide complexes with alkenes and allylsilanes. *J. Am. Chem. Soc.* 117: 7862–7868.
72. Yao, M.-L., Borella, S., Quick, T., Kabalka, G. W. 2008. Boron trihalides mediated haloallylation of aryl aldehydes: Reaction and mechanistic insight. *J. Chem. Soc. Dalton.* 776–778.
73. Parker, D. L., Fried, A. K., Meng, D., Greenlee, M. L. 2008. Use of fluoroalkyl as a latent group for internal alkylation: Application to the synthesis of bridged tetrahydrofluorenones. *Org. Lett.* 10: 2983–2985.

74. Blid, J., Somfai, P. 2003. Lewis acid mediated [2,3]-sigmatropic rearrangement of allylic ammonium ylides. *Tetrahedron Lett.* 44: 3159–3162.
75. Blid, J., Brandt, P., Somfai, P. 2004. Lewis acid mediated [2,3]-Sigmatropic rearrangement of allylic α-amino amides. *J. Org. Chem.* 69: 3043–3049.
76. Cardillo, G., Simone, A. D., Gentilucci, L., Tomasini, C. 1994. Conjugate addition of ahloride to α, β-unsaturated chiral imides promoted by $BC1_3$-derivatives. a synthesis of 3-chlorobutanoic acid. *J. Chem. Soc., Chem. Commun.* 735–736.
77. Sarko, C. R., Collibee, S. E., Knorr, A. L., DiMare, M. 1996. BCl_3- and $TiCl_4$-Mediated Reductions of β-Hydroxy Ketones. *J. Org. Chem.* 61: 868–873.
78. McGinnis, M. B., Vagle, K., Green, J. F., Tan, L. C., Palmer, R., Siler, J., Pagni, R. M., Kabalka, G. W. 1996. Selectivities of Diels-Alder reactions catalyzed by highly acidic boronated aluminas. *J. Org. Chem.* 61: 3496–3500.
79. Dadush, E., Green, J. F., Sease, A., Naravane, A., Pagni, R. M., Kabalka, G. W. 2009. Microwave activation of alumina and its use as a catalyst in synthetic reactions. *J. Chem. Res.* 120–123.
80. (a) Brown, H. C., Midland, M. M., Levy, A. B. 1972. Facile reaction of dialkylchloroboranes with organic azides. Remarkable enhancement of reactivity relative to trialkylboranes. *J. Am. Chem. Soc.* 94: 2114–2115; (b) Brown, H. C., Midland, M. M., Levy, A. B., Suzuki, A., Sono, S., Itoh, M. 1987. Organoboranes for synthesis. 8. Reaction of organoboranes with representative organic azides. A general stereospecific synthesis of secondary amines and N-substituted aziridines. *Tetrahedron* 43: 4079–4088; (c) Carboni, B., Vaultier, M., Carrie, R. 1987. Chemoselectivity of the reaction of dichloroboranes with azides: An efficient synthesis of secondary amines. *Tetrahedron* 43: 1799–1810.
81. Brown, H. C., Midland, M. M., Levy, A. B. 1972. Reaction of dialkylchloroboranes with ethyl diazoacetate at low temperatures. Facile two-carbon homologation under exceptionally mild conditions. *J. Am. Chem. Soc.* 94: 3662–3664.
82. (a) Jadhav, P. K., Bhat, K. S., Perumal, P. T., Brown, H. C. 1986. Chiral synthesis via organoboranes. 6. Asymmetric allylboration via chiral allyldialkylboranes. Synthesis of homoallylic alcohols with exceptionally high enantiomeric excess. *J. Org. Chem.* 51: 432–439; (b) Brown, H. C., Jadhav, P. K. 1983. Asymmetric carbon-carbon bond formation via B -allyldiisopinocampheylborane. Simple synthesis of secondary homoallylic alcohols with excellent enantiomeric purities. *J. Am. Chem. Soc.* 105: 2092–2093.
83. (a) Brown, H. C., Basavaiah, D. 1982. A general and stereospecific synthesis of cis alkenes via stepwise hydroboration: a simple synthesis of Muscalure, the sex pheromone of the housefly *(Musca domestha)*. *J.* Org. *Chem.* 47: 3806–3808; (b) Brown, H. C., Basavaiah, D. 1983. Pheromone synthesis via organoboranes: A convenient stereospecific synthesis of racemic disparlure, the sex pheromone of the Gypsy Moth (*Porthetria dispar L.*). *Synthesis* 1983: 283–284.
84. Bayer,M.J.,Pritzkow,H.,Siebert,W.2002.1,1,1-Trisborylalkanesasprecursorsfordicarbapentaboranes(5)-synthesis, reactivity, and structures of closo-1,5-bis(neopentyl)-2,3,4-trichloro-1,5-dicarbapentaborane and its derivatives. *Eur. J. Inorg. Chem.* 2002: 1293–1300.
85. Hara, S., Dojo, H., Takinami, S., Suzuki, A. 1983. Organic synthesis using haloboration reaction. I. A simple and selective synthesis of 2-bromo- and 2-iodo-1-alkenes. *Tetrahedron Lett.* 24: 731–734.
86. Singleton, D. A., Waller, S. C., Zhang, Z., Frantz, D. E., Leung, S.-W. 1996. Allylboration of alkenes with allyldihaloboranes. *J. Am. Chem. Soc.* 118: 9986–9987.
87. Erdyakov, S. Yu., Ignatenko, A. V., Potapova, T., Lyssenko, K. A., Gurskii, M. E, Bubnov, Y. N. 2009. Design of bicyclic and cage boron compounds based on allylboration of acetylenes with allyldichloroboranes. *Org. Lett.* 11: 2872–2875.
88. Frantz, D. E., Singleton, D. A. 1999. Carbometalation of simple alkenes with allyldibromoborane. *Org. Lett.* 1: 485–486.
89. Depew, K. M., Danishefsky, S. J., Rosen, N., Sepp-Lorenzino, L. 1996. Total synthesis of Tryprostatin B: Generation of a nucleophilic prenylating species from a prenylstannane. *J. Am. Chem. Soc.* 118: 12463–12464.
90. Chein, R.-J., Yeung, Y.-Y., Corey, E. J. 2009.2 Highly enantioselective oxazaborolidine-catalyzed reduction of 1,3-dicarbonyl compounds: Role of the additive diethylaniline. *Org. Lett.* 11: 1611–1614.
91. Yeung, Y.-Y., Chein, R.-J., Corey, E. J. 2007. Conversion of Torgov's synthesis of estrone into a highly enantioselective and efficient process. *J. Am. Chem. Soc.* 129: 10346–10347.
92. Kanth, J. V. B., Brown, H. C. 1999. Dioxane-monochloroborane: A new and highly reactive hydroborating reagent with exceptional properities. *Org. Lett.* 1: 315–317.
93. Kim, B. J., Matteson, D. S. 2004. Conversion of alkyltrifluoroborates into alkyldichloroboranes with tetrachlorosilane in coordinating solvents. *Angew. Chem. Int. Ed.* 43: 3056–3058.

94. (a) Suzuki, A. 1986. New application of organoboron compounds in organic synthesis. *Pure Appl. Chem.* 58: 629–638; (b) Hara, S., Satoh, Y., Ishiguro, H., Suzuki, A. 1983. Organic synthesis using haloboration reaction. II. A stereo- and regioselective synthesis of [Z]-1-alkynyl-2-halo-1-alkenes. *Tetrahedron Lett.* 24: 735–738; (c) Hara, S., Kato, T., Shimizu, H., Suzuki, A. 1985. Organic synthesis using haloboration reaction. VIII. Stereo- and regioselective synthesis of (*Z*)-1,2-dihalo-1-alkenes. *Tetrahedron Lett.* 26: 1065–1068; (d) Satoh, Y., Serizawa, H., Hara, S., Suzuki, A. 1985. Organic synthesis using haloboration reaction. Part 7. A stereospecific synthesis of (Z)-δ-halo-γ, δ -unsaturated ketones via haloboration reaction of terminal alkynes. *J. Am. Chem. Soc.* 107: 5225–5228.
95. Satoh, Y., Serizawa, H., Miyaura, N., Hara, S., Suzuki, A. 1988. Organic synthesis using haloboration reactions. 11. A formal carboboration reaction of 1-alkynes and its application to di- and trisubstituted alkene synthesis. *Tetrahedron Lett.* 29: 1811–1814.
96. (a) Wang, K. K., Wang, Z. 1994. A convenient procedure for the synthesis of enyne-allenes. *Tetrahedon Lett.* 35: 1829–1832; (b) Wang, K. K., Wang, Z., Tarli, A., Gannett, P. 1996. Cascade radical cyclizations via biradicals generated from (*Z*)-1,2,4-heptatrien-6-ynes. *J. Am. Chem. Soc.* 118: 10783–10791; (c) Tarli, A., Wang, K. K. 1997. Synthesis and thermolysis of enediynyl ethyl ethers as precursors of enyne-ketenes. *J. Org. Chem.* 62: 8841–8847.
97. Yamamoto, Y., Yatagai, H., Maruyama, K., Sonoda, A., Murahashi, S. 1977. Reaction of alkenylboranes with methylcopper. A convenient new procedure for the synthesis of symmetrical (*E,E*)-1,3-dienes. *J. Am. Chem. Soc.* 99: 5652–5656.
98. Hyuga, S., Takinami, S., Hara, S., Suzuki, A. 1986. Organic synthesis using haloboration reaction. Part 9. A direct and selective synthesis of (*Z,Z*)-1-bromo-1,3-dienes and (*E,Z*)-1,3-dienes by the hydroboration-bromoboration sequence of two alkynes. *Tetrahedron Lett.* 27: 977–980.
99. (a) Hyuga, S., Hara, S., Suzuki, A. 1992. Organic synthesis using haloboration reaction. XXI. A synthesis of prostaglandin B1 methyl ester by the stepwise cross-coupling reaction using (*E*)-(2-bromoethenyl) diisopropoxyborane. *Bull. Chem. Soc. Jpn.* 65: 2303–2305; (b) Chappell, M. D., Stachel, S. J., Lee, C. B., Danishefsky, S. J. 2000. En route to a plant scale synthesis of the promising antitumor agent 12,13-desoxyepothilone B. *Org. Lett.* 2: 1633–1636.
100. (a) Lee, S. J., Gray, K. C., Paek, J. S., Burke, M. D. 2008. Simple, efficient, and modular syntheses of polyene natural products via iterative cross-coupling. *J. Am. Chem. Soc.* 130: 466–468; (b) Uno, B. E., Gillis, E. P., Burke, M. D. 2009. Vinyl MIDA boronate: A readily accessible and highly versatile building block for small molecule synthesis. *Tetrahedron* 65: 3130–3138.
101. Molander, G. A., Bernardi, C. R. 2002. Suzuki–Miyaura cross-coupling reactions of potassium alkenyltrifluoroborates. *J. Org. Chem.* 67: 8424–8429.
102. Yao, M.-L., Reddy, M. S., Yong, L., Walfish, I., Blevins, D. W., Kabalka, G. W. 2010. Chemoselective bromodeboronation of organotrifluoroborates using tetrabutylammonium tribromide: Application in (*Z*)-dibromoalkene syntheses. *Org. Lett.* 12: 700–703.
103. (a) Hara, S., Kato, T., Suzuki, A. 1983. Organic synthesis using haloboration reactions. III. Bromoboration reactions of 1-halo-1-alkynes: Synthesis of (*E*)- and (*Z*)-1,2-dihalo-1-alkenes. *Synthesis* 1005–1006; (b) Mataka, S., Liu, G.-B., Tashiro, M. 1995. Palladium(0) complex-catalyzed debrominative coupling of (tribromomethyl)- and (dibromomethyl)benzenes to diarylacetylenes and 1,2-diarylethenes. *Synthesis* 133–135. (c) Ye, C., Shreeve, J. M. 2004. Structure-dependent oxidative bromination of unsaturated C–C bonds mediated by selectfluor. *J. Org. Chem.* 69: 8561–8563.
104. Wang, C., Tobrman, T., Xu, Z., Negishi, E.-i. 2009. Highly regio- and stereoselective synthesis of (Z)-trisubstituted alkenes via propyne bromoboration and tandem Pd-catalyzed cross-coupling. *Org. Lett.* 11: 4092–4095.
105. (a) Satoh, Y., Serizawa, H., Hara, S., Suzuki, A. 1984. Organic synthesis using haloboration reaction. V. A new synthesis of *N*-phenyl-β-bromo-α,β-unsaturated amides via bromoboration reaction of terminal alkynes, followed by treatment with phenyl isocyanate. *Synth. Commun.* 14: 313–319; (b) Sato, M., Yamamoto, Y., Hara, S., Suzuki, A. 1993. A stereoselective synthesis of 3,3-disubstituted allylborane derivatives using haloboration reaction and their application to the diastereospecific synthesis of homoallylic alcohols having quaternary carbon. *Tetrahedron Lett.* 34: 7071–7014; (c) Trost, B. M., Toste, F. D. 2002. Mechanistic dichotomy in $CpRu(CH_3CN)_3PF_6$ catalyzed enyne cycloisomerizations. *J. Am. Chem. Soc.* 124: 5025–5036.
106. (a) Brown, H. C., Kramer, G. W., Ley, A. B., Midland, M. M. *Organic Syntheses via Boranes*; Wiley-Interscience: New York, 1975; (b) Suzuki, A., Brown, H. C. *Organic Syntheses via Boranes, volume 3, Suzuki coupling*; Aldrich Chemical Company: Milwaukee, Wisconsin, 2003.
107. (a) Jacob, P., Brown, H. C. 1977. A Grignard-like addition of B-alkenyl-9-borabicyclo[3.3.1]nonanes to aldehydes, a Novel synthesis of allylic alcohols with defined stereochemistry. *J. Org. Chem.* 42: 579–580; (b) Brown, H. C., Randad, R. S., Bhat, K. S., Zaidlewicz, M., Racherla, U. S. 1990. Chiral synthesis via

organoboranes. 24. B-allylbis(2-isocaranyl)borane as a superior reagent for the asymmetric allylboration of aldehydes. *J. Am. Chem. Soc.* 112: 2389–2392; (c) Rane, A. M., Vaquer, J., Colberg, J. C., Soderquist, J. A. 1995. *Trans*-3-silyl allylic alcohols via the Brown vinylation. *Tetrahedron Lett.* 36: 987–990; (d) Burgos, C. H., Canales, E., Matos, K., Soderquist, J. A. 2005. Asymmetric allyl- and crotylboration with the robust, versatile, and recyclable 10-TMS-9-borabicyclo[3.3.2]-decanes. *J. Am. Chem. Soc.* 127: 8044–8049.

108. (a) Ramachandran, P. V., Liu, H., Reddy, M. V. R., Brown, H. C. 2003. Synthesis of homoallylic chiral tertiary alcohols via chelation-controlled diastereoselective nucleophilic addition on α-alkoxyketones: Application for the synthesis of the C1-C11 subunit of 8-epi-Fostriecin. *Org. Lett.* 5: 3755–3757; (b) Pragani, R., Roush, W. R. 2008. Studies on the synthesis of Durhamycin A: Stereoselective synthesis of a model aglycone. *Org. Lett.* 10: 4613–4616.
109. (a) Brown, H. C., Molander, G. A., Singh, S. M., Racherla, U. S. 1985. Organoboranes. 38. A facile and highly efficient addition of B-1-alkynyl-9-borabicyclo[3.3.1]nonanes to aldehydes and ketones: An exceptionally chemoselective synthesis of propargylic alcohols. *J. Org. Chem.* 50: 1577–1582; (b) Evans, J. C., Goralski, C. T., Hasha, D. L. 1992. B-[2-(Trimethylsilyl)ethynyl]-9-borabicyclo[3.3.1]nonane. A new organoboron reagent for the preparation of propargylic alcohols. *J. Org. Chem.* 57: 2941–2943; (c) Corey, E. J., Cimprich, K. A. 1994. Highly enantioselective alkynylation of aldehydes promoted by chiral oxazaborolidines. *J. Am. Chem. Soc.* 116: 3151–3152.
110. (a) Midland, M. M., Tramontano, A., Zderic, S. A. 1978. The reaction of B-alkyl-9-borabicyclo[3.3.1] nonanes with aldehydes and ketones. A facile elimination of the alkyl group by aldehydes. *J. Oganomet. Chern.* 156: 203–211; (b) Brown, H. C., Ford, T. M. 1981. Synthesis of methylenecycloalkanes from cycloalkenes via borane chemistry. *J. Org. Chem.* 46: 647–648.
111. (a) Kabalka, G. W., Maddox, J. T., Bogas, E., Kelly, S. W. 1997. Alkylation of aldehyde (arenesulfonyl) hydrazones with trialkylboranes. *J. Org. Chem.* 62: 3688–3695; (b) Okada, K., Hosoda, Y., Oda, M. 1986. Facile alkylation of *o*-hydroxyarylaldehydes and α-formylketones with trialkylboranes. *Tetrahedron Lett.* 27: 6213–6216.
112. Mikhailov, B. M., Baryshnikova, T. K., Shashkov, A. S. 1981. Organoboron compounds. CCCXCIV. Reaction of 1-boraadamantane with carbonyl compounds. *J. Organomet. Chem.* 219: 301–308.
113. Hirano, K., Yorimitsu, H., Oshima, K. 2005. Nickel-catalyzed alkylation of aldehydes with trialkylboranes. *Org. Lett.* 7: 4689–4691.
114. (a) Brown, H. C., Molander, G. A., Singh, S. M., Racherla, U. S. 1985. Organoboranes. 38. A facile and highly efficient addition of B-1-alkynyl-9-borabicyclo[3.3.1]nonanes to aldehydes and ketones: An exceptionally chemoselective synthesis of propargylic alcohols. *J. Org. Chem.* 50: 1577–1582; (b) Ramachandran, P. V., Chen, G. M., Brown, H. C. 1997. Efficient synthesis of enantiomerically pure C_2-symmetric diols via the allylboration of appropriate dialdehydes. *Tetrahedron Lett.* 38: 2417–2420.
115. Podder, S., Choudhury, J., Roy, U. K., Roy, S. 2007. Dual-reagent catalysis within Ir–Sn domain: Highly selective alkylation of arenes and heteroarenes with aromatic aldehydes. *J. Org. Chem.* 72: 3100–3103.
116. Fu, Y., Wang, J.-X., Wang, K., Hu, Y. 2008. Nickel(II)-catalyzed carbon–carbon bond formation reaction of functionalized organozinc reagents with aromatic aldehydes. *Tetrahedron* 64: 11124–11128.
117. Kuninobu, Y., Ishii, E., Takai, K. 2007. Rhenium- and gold-catalyzed coupling of aromatic aldehydes with trimethyl(phenylethynyl)silane: Synthesis of diethynylmethanes. *Angew. Chem. Int. Ed.* 46: 3296–3299.
118. Kaufmann, D. E., Bir, G. 1987. Synthese von isopinocampheylhalogenboranen und ihre verwendung als chirale katalysatoren für asymmetrische diels-alder-reaktionen. *Tetrahedron Lett.* 28: 777–780.
119. Bir, G., Kaufmann, D. E. 1990. Modified isopinocampheyldibromoboranes; selective catalysts for asymmetric Diels-Alder reactions. *J. Organomet. Chem.* 390: 1–6.
120. (a) Hawkins, J. M., Loren, S. 1991. Two-point-binding asymmetric Diels-Alder catalysts: Aromatic alkyldichloroboranes. *J. Am. Chem. Soc.* 113: 7794–7795; (b) Hawkins, J. M., Loren, S., Nambu, M. 1994. Asymmetric Lewis acid-dienophile complexation: Secondary attraction versus catalyst polarizability. *J. Am. Chem. Soc.* 116: 1657–1660.
121. Hawkins, J. M., Nambu, M., Loren, S. 2003. Asymmetric Lewis acid-catalyzed Diels–Alder reactions of α, β-unsaturated ketones and α, β-unsaturated acid chlorides. *Org. Lett.* 5: 4293–4295.
122. Ishihara, K., Inanaga, K., Kondo, S., Funahashi, M., Yamamoto, H. 1998. Rational design of a new chiral Lewis acid catalyst for enantioselective Diels–Alder reaction: Optically active 2-dichloroboryl-1,1′-binaphthyl. *Synlett* 1053–1056.
123. Mortier, J., Gridnev, I. D., Guénot P. 2000. Reactions of phosphonates with organohaloboranes: New route to molecular borophosphonates. *Organometallics* 19: 4266–4275.

124. Kabalka, G. W., Wu, Z., Ju, Y. 2000. Reductive bromination of aromatic aldehydes using alkylboron dibromides. *Tetrahedron Lett.* 41: 5161–5164.
125. Kabalka, G. W., Wu, Z., Ju, Y. 2001. Chloroalkylation of aryl aldehydes using alkylboron dichlorides in the presence of oxygen. *Tetrahedron Lett.* 42: 6239–6241.
126. (a) Brown, H. C., Midland, M. M. 1987. Organoboranes for synthesis. 6.: A convenient, general synthesis of alkylhydroperoxides *via* autoxidation of organoboranes. *Tetrahedron* 43: 4059–4070; (b) Brown, H. C., Midland, M. M., Kabalka, G. W. 1986. Organoboranes for synthesis. 5: Stoichiometrically controlled reaction of organoboranes with oxygen under mild conditions to achieve quantitative conversion to alcohols. *Tetrahedron* 42: 5523–5530; (c) Devin, P., Fensterbank, L., Malacria, M. 1999. Regioselective alkylation of substituted quinones by trialkylboranes. *Tetrahedron Lett.* 40: 4473–4476; (d) Miyabe, H., Ueda, M., Yoshioka, N., Naito, T. 1999. Carbon radical addition to glyoxylic oxime ether *via* iodine atom-transfer process. *Synlett*, 465–467.
127. (a) Ollivier, C., Renaud, P. 1999. B-Alkylcatecholboranes as a source of radicals for efficient conjugate additions to unsaturated ketones and aldehydes. *Chem. Eur. J.* 5: 1468–1473; (b) Brown, H. C., Rogic, M. M., Rathke, M. W., Kabalka, G. W. 1967. Facile reaction of organoboranes with acrolein. Convenient new aldehyde synthesis via hydroboration. *J. Am. Chem. Soc.* 89: 5709–5710; (c) Brown, H. C., Kabalka, G. W. 1970. Oxygen-induced reactions of organoboranes with the inert α, β-unsaturated carbonyl derivatives. Convenient new aldehyde and ketone synthesis *via* hydroboration. *J. Am. Chem. Soc.* 92: 714–716; (d) Kabalka, G. W. 1973. 1,4-Addition reactions of organoboranes. *Intra-Sci. Chem. Rep.* 7, 57–64; (e) Beraud, V., Gnanou, Y., Walton, J. C., Maillard, B. 2000. New insight into the mechanism of the reaction between α, β-unsaturated carbonyl compounds and triethylborane (Brown's reaction). *Tetrahedron Lett.* 41: 1195–1198.
128. Miyaura, N., Itoh, M., Suzuki, A., Brown, H. C., Midland, M. M., Jacob, III, P. 1972. Two novel reactions of monomeric formaldehyde with trialkylboranes. Remarkably rapid elimination diverted by oxygen to a free-radical chain addition. *J. Am. Chem. Soc.* 94: 6549–6550.
129. Girard, C., Kagan, H. B. 1995. Nonlinear effects in the reduction of acetophenone by diisopinocampheyl chloroborane: Influence of the reagent preparation. *Tetrahedron: Asymmetry* 6: 1881–1884.
130. (a) Kagan, H. B., Girard, C. 1998. Nonlinear effects in asymmetric synthesis and stereoselective reactions: Ten years of investigation. *Angew. Chem., Int. Ed.* 37: 2922–2959; (b) Kagan, H. B., Girard, C., Guillaneux, D., Rainford, D., Samuel, O., Zhang, S. Y., Zhao, S. H. 1996. Nonlinear effects in asymmetric catalysis: Some recent aspects. *Acta Chem. Scand.* 50: 345–352; (c) Kagan, H. B. 2001. Nonlinear effects in asymmetric catalysis: A personal account. *Synlett* 888–899.
131. Srebnik, M., Ramachandran, P. V., Brown, H. C. 1988. Chiral synthesis via organoboranes. 18. Selective reductions. 43. Diisopinocampheylchloroborane as an excellent chiral reducing reagent for the synthesis of halo alcohols of high enantiomeric purity. A highly enantioselective synthesis of both optical isomers of Tomoxetine, Fluoxetine, and Nisoxetine. *J. Org. Chem.* 53: 2916–2920.
132. Deninno, M. P., Schoenleber, R., Asin, K. E., MacKenzie, R., Kebabian, J. W. 1990. (1R,3S)-1-(aminomethyl)-3,4-dihydro-5,6-dihydroxy-3-phenyl-1H-2-benzopyran: A potent and selective D1 agonist. *J. Med. Chem.* 33: 2948–2950.
133. Thompson, A. S., Tschaen, D. M., Simpson, P., McSwine, D. J., Russ, W., Little, E. D., Verhoeven, T. R., Shinkai, I. 1990. Conversion of a silylated hemiacetal into an α-bromoether using trimethylsilyl bromide. Synthesis of platelet activating factor antagonist L-659,989. *Tetrahedron Lett.* 31: 6953–6956.
134. Thompson, A. S., Tschaen, D. M., Simpson, P., McSwine, D. J., Reamer, R. A., Verhoeven, T. R., Shinkai, I. 1992. Synthesis of PAF antagonist MK-287. *J. Org. Chem.* 57: 7044–7052.
135. King, A. O., Corley, E. G., Anderson, R. K., Larsen, R. D., Verhoeven, T. R., Reider, P. J., Xiang, Y. B. et al., 1993. An efficient synthesis of LTD4 antagonist L-699,392. *J. Org. Chem.* 58: 3731–3735.
136. Ramachandran, P. V., Gong, B., Brown, H. C. 1995. Chiral synthesis via organoboranes. 41. The utility of B-chlorodiisopinocampheylborane for a general synthesis of enantiomerically pure drugs. *Chirality* 7: 103–110.
137. Felpin, F.-X., Lebreton, J. 2002. A Highly Stereoselective Asymmetric Synthesis of (−)-Lobeline and (−)-Sedamine. *J. Org. Chem.* 67: 9192–9199.
138. Shinkai, I., King, A. O., Larsen, R. D. 1994. A practical asymmetric synthesis of LTD4 antagonist. *Pure Appl. Chem.* 66: 1551–1556.
139. (a) Ramachandran, P. V., Gong, B., Brown, H. C. 1994. A remarkable inversion in configuration of the product alcohols from the asymmetric reduction of *ortho*-hydroxyacetophenones with *B*-chlorodiisopinocampheylborane. *Tetrahedron Lett.* 35: 2141–2144; (b) Ramachandran. P. V., Malhotra, S. V., Brown, H. C. 1997. Rate enhancing effect of hydrogen chloride and methanesulfonic acid on the

intramolecular asymmetric reduction of *o*-aminoaceto- and -benzophenones with diisopinocampheylborane. *Tetrahedron Lett.* 38: 957–960.

140. Ramachandran, P. V., Pitre, S., Brown, H. C. 2002. Selective reductions. 59. Effective intramolecular asymmetric reductions of α-, β-, and γ-keto acids with diisopinocampheylborane and intermolecular asymmetric reductions of the corresponding esters with *B*-chlorodiisopinocampheylborane. *J. Org. Chem.* 67: 5315–5319.
141. Brown, H. C., Ramachandran, P. V., Teodorovic, A V., Swaminathan, S. 1991. B-Chlorodiiso-2-ethylapopinocampheylborane: An extremely efficient chiral reducing agent for the reduction of prochiral ketones of intermediate steric requirements. *Tetrahedron Lett.* 32: 6691–6694.
142. (a) Joshi, N. N., Srebnik, M., Brown, H. C. 1988. Enantioselective ring cleavage of meso-epoxides with B-halodiisopinocampheylboranes. *J. Am. Chem. Soc.* 110: 6246–6248; (b) Srebnik, M., Joshi, N. N., Brown, H. C. 1989. Chiral synthesis via organoboranes. 23. Enantioselective ring opening of meso-epoxides with B-halodiisopinocampheylboranes. The first general synthesis of optically active 1,2-halohydrins. *Israel J. Chem.* 29: 229–237.
143. Roy C. D., Brown, H. C. 2006. Asymmetric ring opening of meso-epoxides with B-halobis(2-isocaranyl) boranes 2-dIcr$_2$BX. *Tetrahedron: Asymmetry* 17: 1931–1936.
144. (a) Kabalka, G. W., Wu, Z., Trotman, S. E., Gao, X. 2000. Alkylation of aromatic aldehydes with boron halide derivatives. *Org. Lett.* 2: 255–256; (b) Kabalka, G. W., Wu, Z., Ju, Y. 2001. Alkylation of aromatic aldehydes with alkylboron chloride derivatives. *Tetrahedron* 57: 1663–1670.
145. Kabalka, G. W., Wu, Z., Ju, Y. 2002. A new method for alkylation of aromatic aldehydes using alkylboron chloride derivatives in the presence of oxygen. *Tetrahedron* 58: 3243–3248.
146. Valpuesta, M., Muñoz, C., Díaz, A., Suau, R., Torres, G. 2007. Organoborane reagents in the *C*-alkylation of aromatic aldimines. *Eur. J. Org. Chem.* 4467–4470.
147. Yamamoto, Y., Kondo, K., Moritani, I. 1974. Reaction of dialkylchloroboranes with lithium aldimines. New method for synthesis of unsymmetrical ketones via organoboranes. *Tetrahedron Lett.* 15: 793–796.
148. Yamamoto, Y., Kondo, K., Moritani, I. 1975. New approach for the synthesis of partially mixed trialkylcarbinols by the reaction of dialkylchloroboranes with lithium aldimines. *Tetrahedron Lett.* 16: 2689–2692.
149. Kabalka, G. W., Wu, Z., Ju, Y. 2003. Stereo- and regioselective synthesis of 1,3-diaryl-3-chloro-1-propanols via the reaction of aryl aldehydes with styrene and (*E*)-β-methylstyrene. *Tetrahedron Lett.* 44: 1187–1189.
150. Robiette, R., Fang, G. Y., Harvey, J. N., Aggarwal, V. K. 2006. Is phenyl a good migrating group in the rearrangement of organoborates generated from sulfur ylides? *Chem. Commun.* 741–743.
151. (a) Yasuda, M., Saito, T., Ueba, M., Baba, A. 2004. Direct substitution of the hydroxy group in alcohols with silyl nucleophiles catalyzed by indium trichloride. *Angew. Chem., Int. Ed.* 43: 1414–1416; (b) Rubin, M., Gevorgyan, V. 2001. $B(C_6F_5)_3$-catalyzed allylation of secondary benzyl acetates with allylsilanes. *Org. Lett.* 3: 2705–2707.
152. (a) Kabalka, G. W., Dong, G., Venkataiah, B. 2003. Rhodium-catalyzed cross-coupling of allyl alcohols with aryl- and vinylboronic acids in ionic liquids. *Org. Lett.* 5: 893–895; (b) Yoshida, M., Gotou, T., Ihara, M. 2004. Palladium-catalyzed coupling reaction of allenic alcohols with aryl- and alkenylboronic acids. *Chem. Commun.* 1124–1125; (c) Tsukamoto, H., Sato, M., Kondo, Y. 2004. Palladium(0)-catalyzed direct cross-coupling reaction of allyl alcohols with aryl- and vinylboronic acids. *Chem. Commun.* 1200–120; (d) Navarre, L., Darses, S., Genet, J.-P. 2004. Baylis–Hillman adducts in rhodium-catalyzed 1,4-additions: unusual reactivity. *Chem. Commun.* 1108–1109.
153. Kabalka, G. W., Yao, M.-L., Borella, S., Wu, Z.-Z. 2005. Allylic alkenylation of allylic alcohols using alkenylboron dihalides: A formal transition-metal free suzuki reaction. *Chem. Commun.* 2492–2495.
154. (a) Leadbeater, N. E., Marco, M. 2003. Transition-metal-free Suzuki-type coupling reactions. *Angew. Chem., Int. Ed.* 42: 1407–1409; (b) Arvela, R. K., Leadbeater, N. E., Sangi, M. S., Williams, V. A., Granados, P., Singer, R. D. 2005. A reassessment of the transition-metal free Suzuki-type coupling methodology. *J. Org. Chem.* 70: 161–168.
155. (a) Brown, H. C., Basavaiah, D. 1982. A general synthesis of B-(*cis*-1-bromo-1-alkenyl)dialkylboranes. valuable intermediates for the synthesis of ketones, trans alkenes and trisubstituted alkenes. *J. Org. Chem.* 47,: 754–756; (b) Brown, H. C., Basavaiah, D., Kulkarni, S. U. 1982. A general and stereospecific synthesis of trans alkenes and regiospecific synthesis of ketones via stepwise hydroboration. *J. Org. Chem.* 47: 3808–3810.
156. Kabalka, G. W., Yao, M.-L., Borella, S., Wu, Z.-Z. 2005. Substitution of the benzylic hydroxyl groups with vinyl moieties using vinylboron dihalides. *Org. Lett.* 7: 2865–2867.

157. Kabalka, G. W., Wu, Z., Ju, Y. 2004. A new halopropargylation of alkynes promoted by boron trihalides. highly stereo- and regioselective syntheses of substituted (*Z*)-1-halo-1,4-enyne derivatives. *Org. Lett.* 6: 3929–3931.
158. Kabalka, G. W., Yao, M.-L., Borella, S. 2006. Substitution of the hydroxyl groups with alkynyl moieties using alkynylboron dihalides: Efficient approach to secondary alkylacetylene derivatives. *Org. Lett.* 8: 879–881.
159. Kabalka, G. W., Yao, M.-L., Borella, S. 2006. Generation of cations from alkoxides: Allylation of propargyl alcohols. *J. Am. Chem. Soc.* 128: 11320–11321.
160. Yao, M.-L., Pippin, A. B., Kabalka, G. W. 2010. Deoxygenation of benzylic alcohols using chloroboranes. *Tetrahedron Lett.* 51: 853–855.
161. (a) Eisch, J. J., Liu, Z.-R., Boleslawski, M. P. 1992. Organometallic compounds of Group III. 49. Reductive deoxygenation of ketones and secondary alcohols by organoaluminum Lewis acids. *J. Org. Chem.* 57: 2143–2147; (b) Yasuda, M., Onishi, Y., Ueba, M., Miyai, T., Baba, A. 2001. Direct reduction of alcohols: Highly chemoselective reducing system for secondary or tertiary alcohols using chlorodiphenylsilane with a catalytic amount of indium trichloride. *J. Org. Chem.* 66: 7741–7744.
162. Kabalka, G. W., Yao, M.-L., Borella, S., Goins, L. K. 2007. Iron trichloride mediated allylation of lithium alkoxides through an unusual carbon-oxygen bond cleavage. *Organometallics* 26: 4112–4114.
163. (a) Fuchter, M. J., Levy, J.-N. 2008. One-pot formation of allylic chlorides from carbonyl derivatives. *Org. Lett.* 10: 4919–4922; (b) Yao, M.-L., Quick, T., Wu, Z.-Z., Quinn, M. P., Kabalka, G. W. 2009. Titanium(IV) halide mediated coupling of alkoxide and alkyne: An efficient and stereoselective route to trisubstituted (*E*)-alkenyl halide. *Org. Lett.* 11: 2647–2649.
164. Fleming, F. F., Ravikumar, P. C., Yao, L. 2009. Direct conversion of aldehydes and ketones to allylic halides by a NbX_5-[3,3] rearrangement. *Synlett.* 1077–1080.
165. (a) Murai, T., Asai, F. 2008. Diastereoselective synthesis of *N*-secondary alkyl 2-alkoxymethylpyrrolidines via sequential addition reactions of organolithium and -magnesium reagents to *N*-thioformyl 2-alkoxymethylpyrrolidines. *J. Org. Chem.* 73: 9518–9521; (b) Murai T., Asai, F. 2007. Three-component coupling reactions of thioformamides with organolithium and Grignard reagents leading to formation of tertiary amines and a thiolating agent. *J. Am. Chem. Soc.* 129: 780–781.

24 Cyclic Oxonium Derivatives as an Efficient Synthetic Tool for the Modification of Polyhedral Boron Hydrides

Igor B. Sivaev and Vladimir I. Bregadze

CONTENTS

24.1 INTRODUCTION

Synthesis of polyhedral boron hydrides approximately 50 years ago appeared as an answer to the request for new high-energy compounds as components of powerful rocket fuels. By the end of the 1950s, new generations of jet engines and new fuels involving liquid hydrogen and hydrazine made boron fuels obsolete [1]. Nevertheless, the child was born and it was not lacking vital capacity monster: in contrast to the previously known boron hydrides, the polyhedral boron hydrides were shown to be exceptionally stable. The high stability of cluster boron hydrides open ways to their practical use in many fields starting with treatment of nuclear wastes and finishing at treatment of cancer [2,3].

It should be noted that most of these applications require modification of boron hydride clusters mainly through a substitution of various atoms or functional groups for hydrogen atoms. For the icosahedral carboranes $C_2B_{10}H_{12}$ this problem can be easily solved due to availability of their carbon atoms for "normal" organic chemistry and synthesis of a wide range of carborane-based molecules have been described [4–6]. Another situation arose with anionic boron hydrides, where controlled substitution at boron atoms was a challenge for boron chemists during a few decades. Halogenation of anionic boron hydrides is well studied [7–16] and some halogenated boron hydrides found application as weakly coordinating anions for stabilization of highly reactive complexes and reaction intermediates [14], as components of electrolytes in lithium-ion batteries [17,18] and ion-selective electrodes [19,20], in extraction of radionuclides from nuclear waste [21,22].

However, many fields of application of polyhedral boron hydrides, for example, medicinal chemistry, require their deeper modification. At present, there are two general approaches to the modification of polyhedral boron hydrides at boron atoms. The first one includes the introduction of a primary substituent (–I, –OH, –SH, $-NH_2$) followed by a modification of this substituent by using organic chemistry methodology [23–28]. The second approach is based on the synthesis of cyclic oxonium derivatives of anionic polyhedral boron hydrides followed by ring-opening with various nucleophilic reagents. An extensive application of this approach was started 10 years ago and was found to be very efficient for the synthesis of derivatives with pendant functional groups connected to the boron cage through flexible spacers of 5–6 atoms [29,30].

In this chapter we will try to overview the results of use of cyclic oxonium derivatives of polyhedral boron hydrides in the synthesis of various boron-containing molecules.

24.2 SYNTHESIS AND FEATURES OF CYCLIC OXONIUM DERIVATIVES

The general approach to the synthesis of cyclic oxonium derivatives of polyhedral boron hydrides consists in abstraction of a hydride by treatment with Lewis or Brönsted acids resulting in the formation of a carbocation-like center on the boron atom, which is then subjected to the attack of an ether solvent as nucleophile, resulting in the corresponding cyclic oxonium derivatives (Figure 24.1) [29].

At present, effective methods of synthesis of cyclic oxonium derivatives of the main types of polyhedral boron hydrides, such as *closo*-dodecaborate $[B_{12}H_{12}]^{2-}$ [31–33], *closo*-decaborate $[B_{10}H_{10}]^{2-}$ [34,35], 7,8-dicarba-*nido*-undecaborate $[7,8\text{-}C_2B_9H_{12}]^-$ [36–38], cobalt bis(dicarbollide) $[3,3'\text{-}Co(1,2\text{-}C_2B_9H_{11})_2]^-$ [39–41], and iron bis(dicarbollide) $[3,3'\text{-}Fe(1,2\text{-}C_2B_9H_{11})_2]^-$ [42] have been elaborated.

Trialkyloxonium salts are widely used in organic synthesis as powerful alkylating agents. Despite the higher stability of oxonium derivatives of boron hydrides, it was reasonable to suppose that they also could act as alkylating agents. This is especially attractive for the cyclic oxonium derivatives, where breaking one carbon–oxygen bond should result in moieties having a boron cluster separated from the functional fragment by a spacer of 5–6 atoms. In such a way, molecules with a reasonably longer space between the boron cage and the property-determining part of the molecule could be prepared. Moreover, the hydrophilic/lipophilic nature of the spacer can be affected by proper choice of the initial substituent. Thus, the ring-opening of the tetrahydrofuran- and tetrahydropyrane-based derivatives produce compounds with lipophilic spacers between the boron cage and bioactive part of molecule, whereas the 1,4-dioxane ring-opening *O*-nucleophiles gives compounds with hydrophilic—$O(CH_2CH_2O)_2$- spacer.

24.3 MODIFICATION OF POLYHEDRAL BORON HYDRIDES USING CYCLIC OXONIUM DERIVATIVES

24.3.1 Synthesis of Boron-Containing Building Blocks

There are two general strategies for the preparation of boronated biomolecules using cyclic oxonium derivatives of polyhedral boron hydrides. The first one is based on a two-step synthesis. At the first

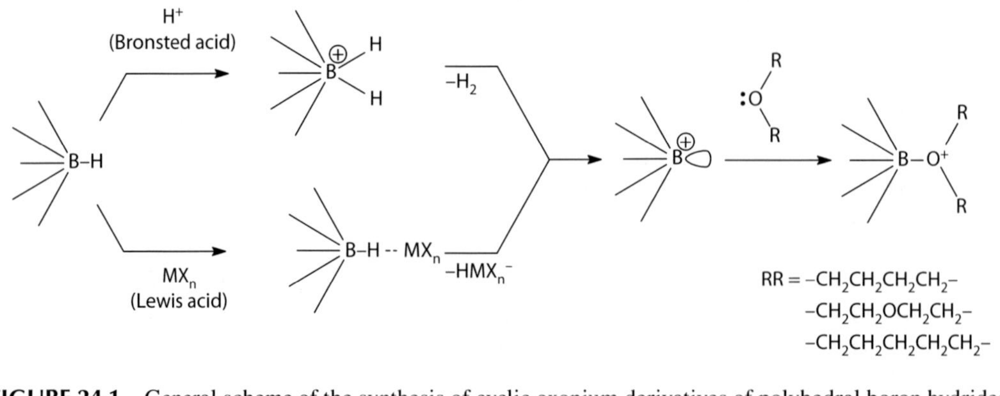

FIGURE 24.1 General scheme of the synthesis of cyclic oxonium derivatives of polyhedral boron hydrides.

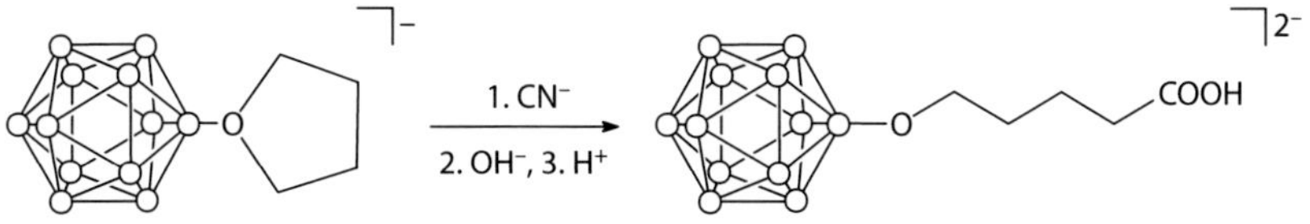

FIGURE 24.2 Synthesis of *closo*-dodecaborate acid through the ring-opening of the tetramethylene oxonium derivative.

step polyhedral boron hydride derivatives with various terminal functional groups (–OH, –SH, $-NH_2$, –COOH, $-CH(NH_2)COOH$, $-N_3$, –C≡CH) are synthesized through the oxoniun cycle disclosure, whereas the second step includes their conjugation with biomolecules using standard methods of organic/bioorganic synthesis. In the frameworks of this strategy a wide spectrum of polyhedral boron hydride building blocks containing carboxylic, amino, amino acid, azide, and acetylenic functional groups were prepared.

Carboxylic acids are of the most popular building blocks for the synthesis of complex organic and bioorganic molecules. The first carboxylic acid produced through the oxonium ring-opening was prepared by the reaction of the tetramethylene oxonium derivative of the *closo*-dodecaborate anion with cyanide ion as synthetic equivalent of carboxylic group followed by alkaline hydrolysis of the nitrile formed (Figure 24.2) [31]. Somewhat later, this approach was used for the synthesis of similar acids derived from the *closo*-decaborate anion [43].

Attempts to apply the same approach for preparation of carboxylic acids starting from the 1,4-dioxane-based derivatives failed due to the elimination of acrylonitrile under hydrolysis of the nitrile derivatives. In order to avoid this problem, the 1,4-dioxane derivative of the *closo*-dodecaborate anion was opened with diethylmalonate in the presence of potassium carbonate and the subsequent acidic hydrolysis and decarboxylation gave the carboxylic acid containing one extra methylene group (Figure 24.3) [44].

A series of benzoic acids derived from the 7,8-dicarba-*nido*-undecaborate anion was prepared by reaction of its cyclic oxonium derivatives with hydroxybenzoic acids in the presence of potassium carbonate (Figure 24.4) [37].

A similar approach was applied for the synthesis of benzoic acids derived from the *closo*-dodecaborate [45], *closo*-decaborate [43], and cobalt bis(dicarbollide) [46,47] anions.

All the boron hydrides discussed here are mono- or dianions. It should be noted that the net charge of the molecule has a strong influence on its biophysical properties, including the interactions with biological membranes. The net charge can be reduced by the introduction of an amino group forming an inner salt on protonation. In application to carboxylic acids this goal can be accomplished by oxonium ring-opening with the amino group of aminoacid ethyl esters (Figure 24.5) [48].

Some of the boron-containing acids described above were used in the synthesis of boron-containing carbohydrates (Figure 24.6) [49,50] and porphyrins (Figure 24.7) [44,46].

Polyhedral boron hydride derivatives with primary amino group were prepared by the reaction of cyclic oxonium derivatives with potassium phthalimide followed by removal of the phthalimide protection group with hydrazine hydrate (Figure 24.8) [31,51].

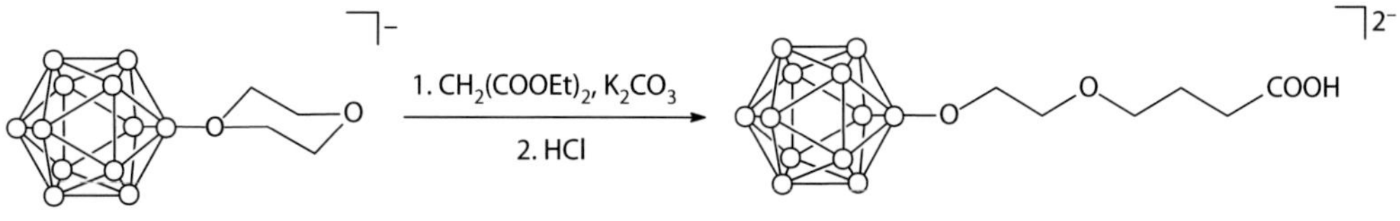

FIGURE 24.3 Synthesis of *closo*-dodecaborate acid through the ring-opening of the 1,4-dioxane derivative.

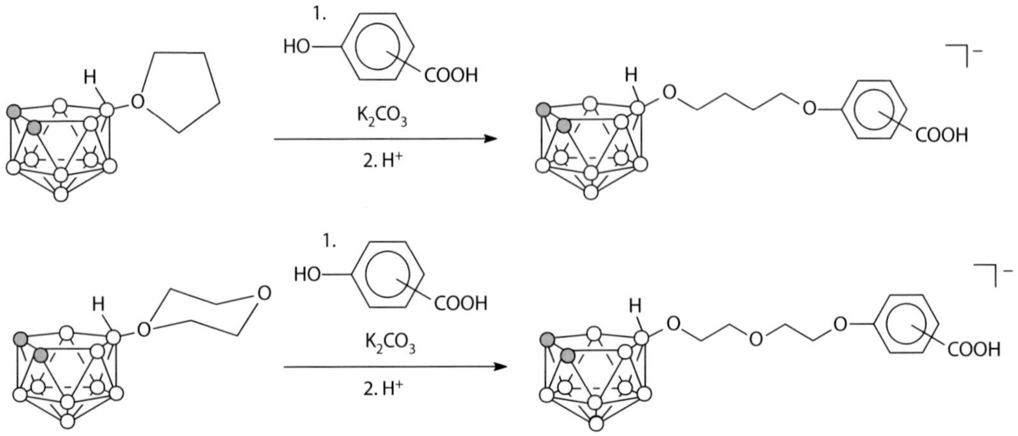

FIGURE 24.4 Synthesis of *nido*-carborane-based carboxylic acids.

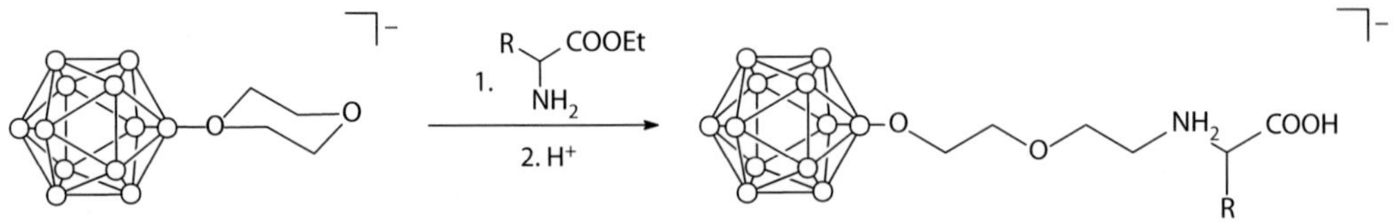

FIGURE 24.5 Synthesis of charge-compensated *closo*-dodecaborate acids.

Somewhat later, the direct synthesis of amino derivative of the *closo*-dodecaborate anion by the treatment of the 1,4-dioxane derivative with aqueous ammonia was reported (Figure 24.9) [48].

The reaction of cyclic oxonium derivatives of polyhedral boron hydrides with acetamido diethylmalonate (glycine anion equivalent for amino acid synthesis), a variation of the classical Sörensen synthesis of amino acids, followed by acidic hydrolysis and decarboxylation was proposed as a route to boron hydride-based aminoacids (Figure 24.10) [31,51].

The Cu(I)-catalyzed 1,3-dipolar cycloaddition of alkyne and azide to form a triazole, termed "click chemistry," has been recently established as a promising tool for chemical modification of

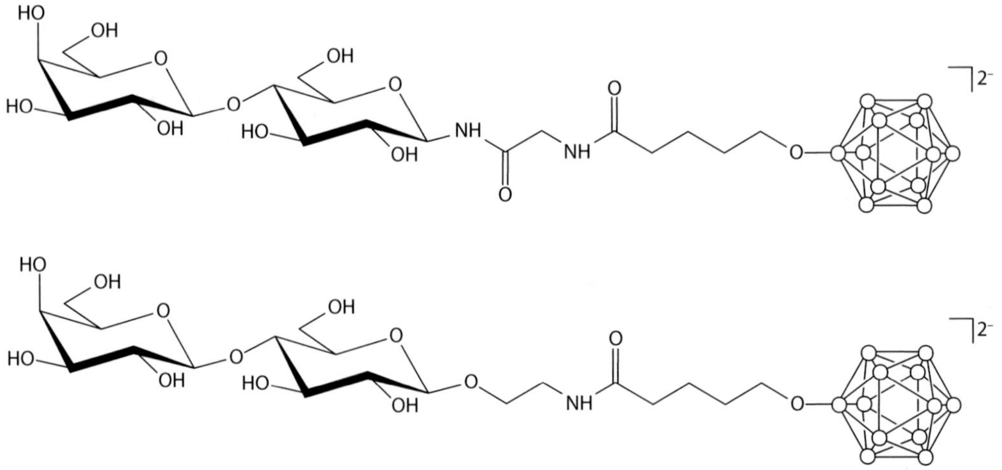

FIGURE 24.6 Conjugates of *closo*-dodecaborate anion with lactose derivatives.

FIGURE 24.7 Conjugates of *closo*-dodecaborate and 3,3-cobalt bis(1,2-dicarbollide) anions with bacteriochlorin *p* cycloimide.

biomolecules. The 1,2,3-triazole is a rigid linking unit that mimics the geometry and electronic properties of a peptide bond and is more stable to hydrolytic cleavage. The reactants, alkyne and azide, are convenient to introduce, independently stable and do not react with common organic reagents or functional groups in biomolecules [52,53]. A series of polyhedral boron hydrides with pendant azide group were prepared by the reaction of their cyclic oxonium derivatives with azide anion (Figure 24.11) [37,54,55].

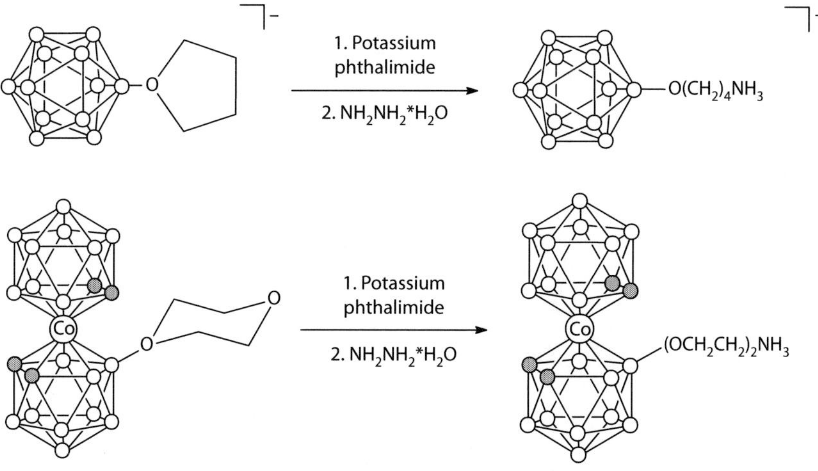

FIGURE 24.8 Synthesis of amino derivatives of *closo*-dodecaborate and 3,3′-cobalt bis(1,2-dicarbollide) anions.

aq. NH_3

$-(OCH_2CH_2)_2NH_3$

FIGURE 24.9 Synthesis of amino derivative of *closo*-dodecaborate anion.

1. $HC(COOEt)_2NHCOMe$
K_2CO_3
2. HCl

COOH
NH_3

1. $HC(COOEt)_2NHCOMe$
K_2CO_3
2. HCl

Co

COOH
NH_3

FIGURE 24.10 Synthesis of amino acid derivatives of *closo*-dodecaborate and 3,3′-cobalt- bis(1,2-dicarbollide) anions.

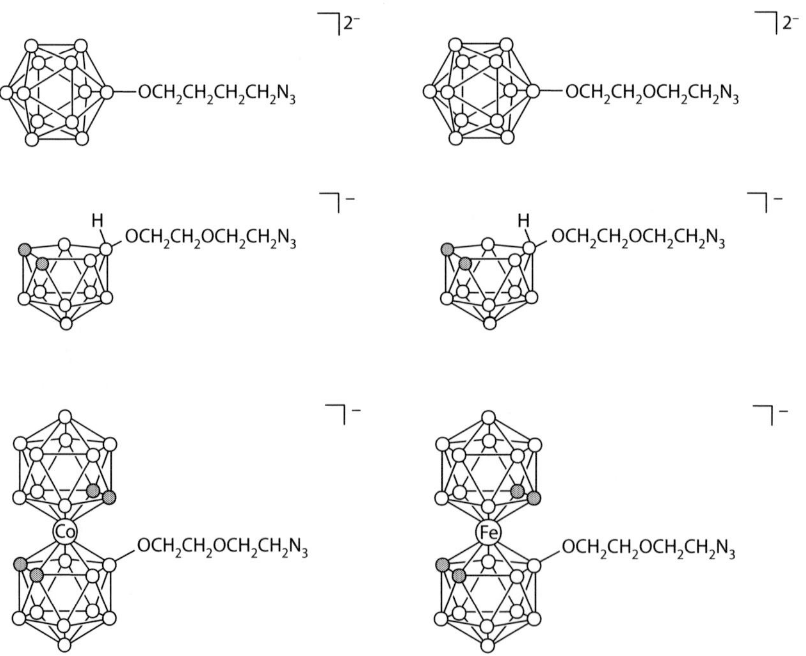

FIGURE 24.11 Azide derivatives of *closo*-dodecaborate, 7,8-dicarba-*nido*-undecaborate, and 3,3′-cobalt bis(1,2-dicarbollide) anions.

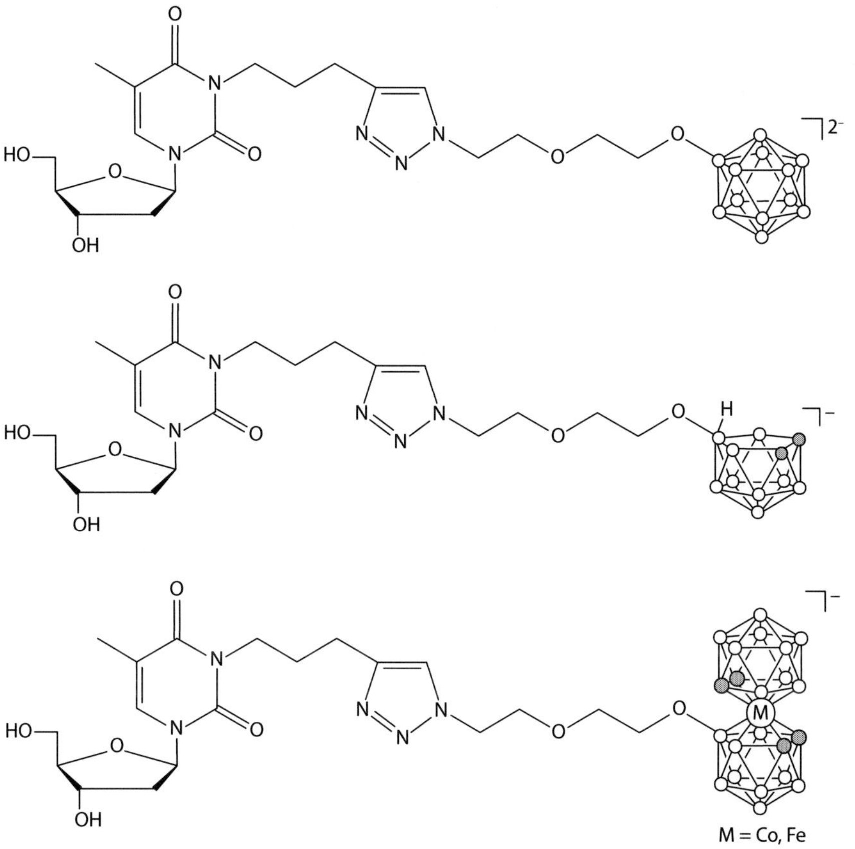

FIGURE 24.12 Conjugates of *closo*-dodecaborate, 7,8-dicarba-*nido*-undecaborate, 3,3′-cobalta-, and 3,3′-ferra bis(1,2-dicarbollide) anions with thymidine.

The azide derivatives described were successfully applied for the synthesis of boron-containing nucleosides (Figure 24.12) [55–57] and porphyrins (Figure 24.13) [58] using "click chemistry" methods.

Polyhedral boron hydride derivatives with terminal alkyne group were prepared by oxonium ring disclosure with alkynyl alcoholates or amines (Figure 24.14) [32,56,59].

The alkyne derivatives were used for the synthesis of boron-containing nucleosides [56] and porphyrins [60] using "click reaction" and Sonogashira reaction, respectively.

The above examples of "building blocks" do not exhaust the diversity of polyhedral boron hydrides prepared by disclosure of their cyclic oxonium derivatives. More examples of such compounds can be found in a comprehensive review [29] and some more recent papers [38,47,61–66].

24.3.2 Direct Boronation of Organic/Bioorganic Molecules

The second strategy of the synthesis of boronated biomolecules using cyclic oxonium derivatives of polyhedral boron hydrides is based on direct reaction of their cyclic oxonium derivatives with nucleophilic sites of premodified biomolecules or their analogs. This approach was successfully used for the preparation of boronated porphyrins and nucleosides.

In the synthesis of boron-containing synthetic porphyrins, attention was paid mainly to the modification of *meso*-tetraphenylporphyrin analogues bearing pyridine, phenol, or aniline groups as

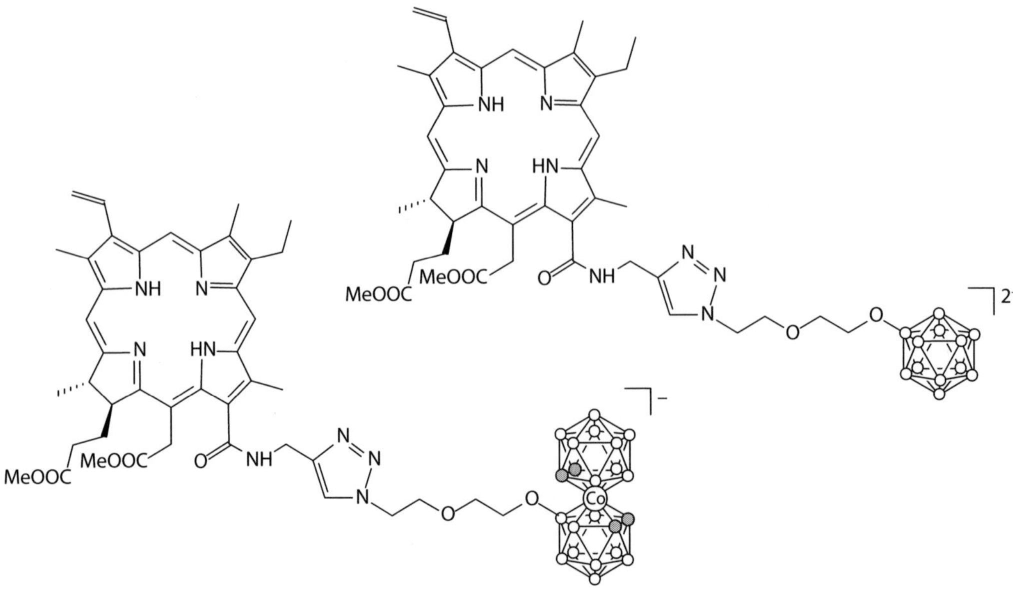

FIGURE 24.13 Conjugates of *closo*-dodecaborate and 3,3′-cobalt bis(1,2-dicarbollide) anions with chlorin e_6.

nucleophilic centers (Figure 24.15) [67–71]. Some of the prepared boron-containing porphyrins were found to be highly specific and potent inhibitors of HIV protease [69].

In the absence of peripheral nucleophilic centers (5,10,15,20-tetraphenylporphyrin, 2,3,7,8,12,13, 17,18-octaethylporphyrin), the disclosure of the oxonium ring results in mono- and disubstituted *N*-cobaltacarboranyl porphyrins [71].

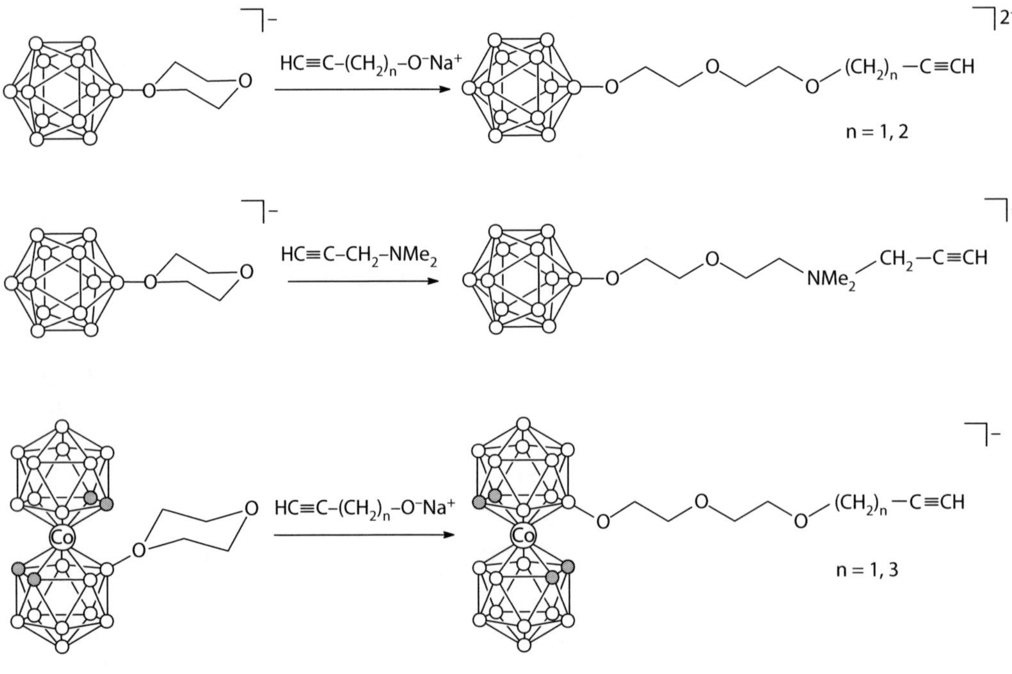

FIGURE 24.14 Synthesis of alkyne derivatives of *closo*-dodecaborate and 3,3′-cobalt- bis(1,2-dicarbollide) anions.

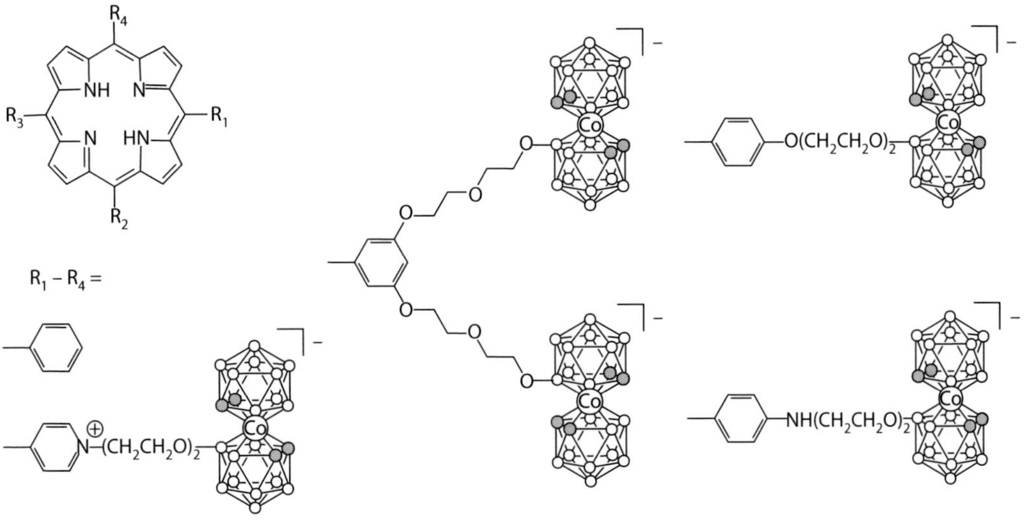

FIGURE 24.15 3,3′-cobalta bis(1,2-dicarbollide) derivatives of *meso*-tetraphenylporphyrin.

More recently, the ring-opening of the dioxane derivative of cobalt bis(dicarbollide) was applied to the modification of natural porphyrins (Figure 24.16) [72].

A series of boron-containing nucleosides was prepared by the reactions of the 1,4-dioxane derivative of cobalt bis(dicarbollide) with the canonical nucleosides—thymidine, 2′-*O*-deoxycytidine, 2′-*O*-deoxyadenosine, and 2′-*O*-deoxyguanosine (Figure 24.17). In the case of thymidine and 2′-*O*-deoxyguanosine the reactions result in mixtures of the *N*- and *O*-alkylated products, which can be separated by chromatographic methods [73,74]. The 2′-*O*-deoxyadenosine derivatives of iron and cromium bis(dicarbollides) were prepared in a similar way [75,76].

Selective formation of an *N*-alkylated product was found in the reaction of the 1,4-dioxane derivative of *closo*-dodecaborate with thymidine (Figure 24.18) [57].

Synthesis of agents for lanthanide and actinide extraction from high-level activity nuclear waste is the only one not related to the medical field where the 1,4-dioxane derivative of cobalt bis-(dicarbollide) was successfully applied. Moreover, the first described reaction of oxonium ring-opening in polyhedral boron hydrides was performed specifically in this field [77]. During the last 10 years, numerous cobalt bis(dicarbollide)-based extraction agents were synthesized, including

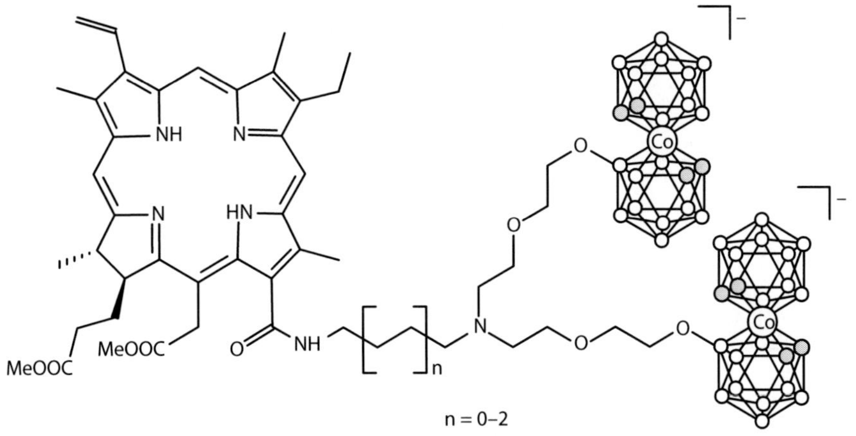

FIGURE 24.16 Conjugates of 3,3′-cobalta bis(1,2-dicarbollide) with chlorin e_6.

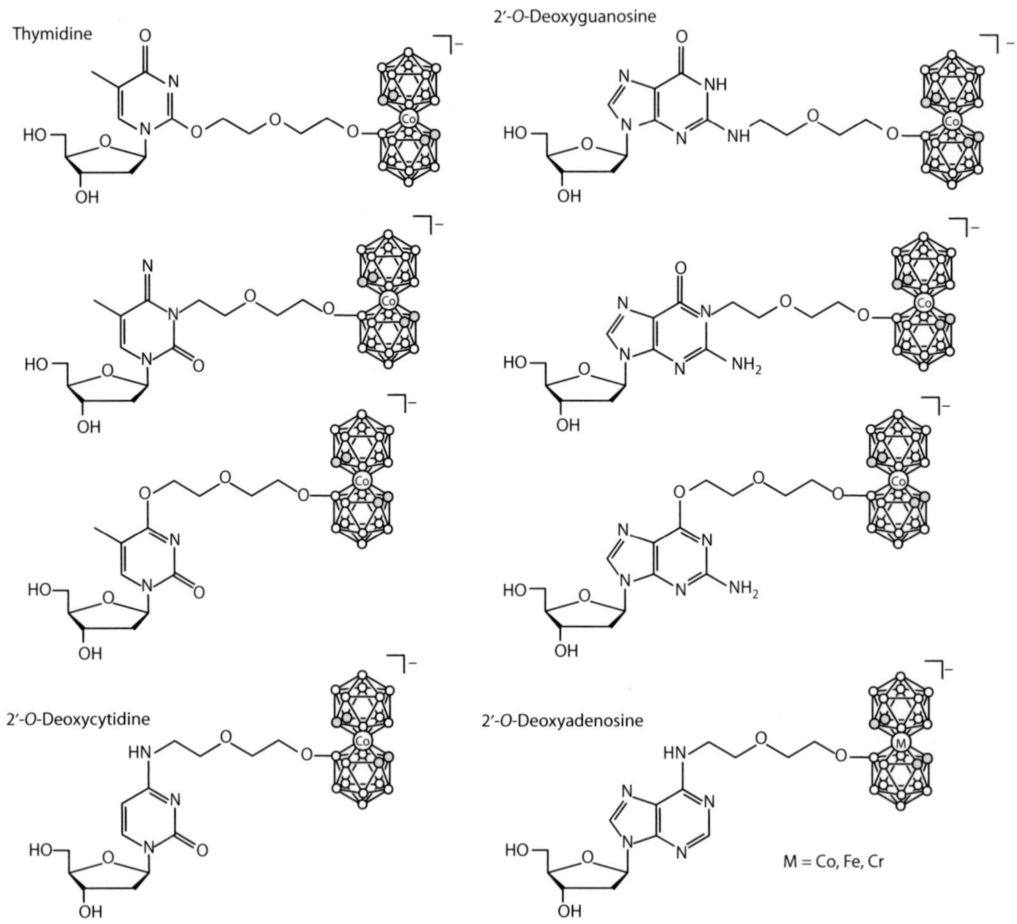

FIGURE 24.17 Nucleoside conjugates with 3,3′-cobalta-, 3,3′-ferra-, and 3,3′-croma bis(1,2-dicarbollide) anions.

simple oligo(ethylene oxide) [40,78] and crown ether [79] derivatives, alkyl diglycoldiamides [80] (*N,N*-dialkyl-carbamoylmethyl)-dialkyl phosphine oxides (CMPO) [81–83], resorc[4]arenes [84], calix[4]arenes [84], and CMPO-calix[4]arenes [85,86] (Figure 24.19). A more detailed discussion on cobalt bis(dicarbollide)-based extractants is given in Chapter 21.

Another class of compounds that should be mentioned here are cobalt bis(dicarbollide)-based HIV protease inhibitors synthesized using the oxonium ring-opening methodology (Figure 24.20)

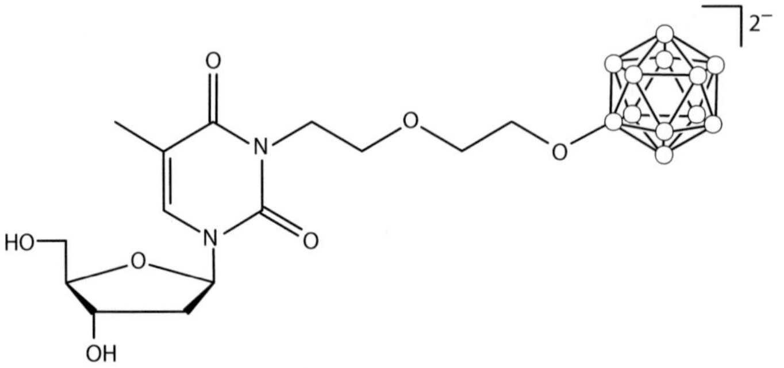

FIGURE 24.18 Conjugate of *closo*-dodecaborate anion and thymidine.

$R = H, Et, (CH_2)_nNHC(O)CH_2P(O)Ph_2$ ($n = 2–7$)

FIGURE 24.19 3,3′-cobalta bis(1,2-dicarbollide) derivatives of tert-butyl-calix[4]arenes.

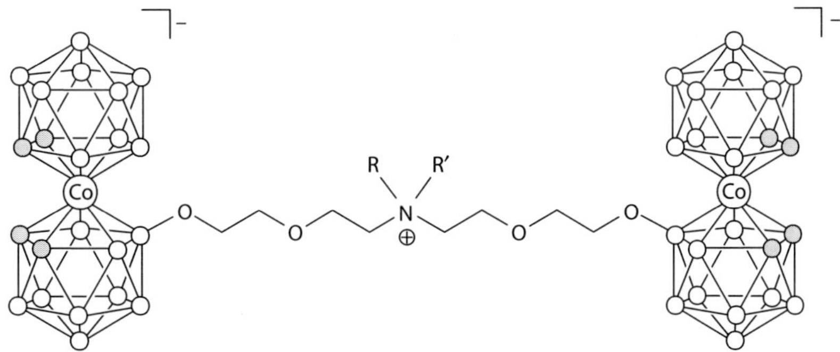

FIGURE 24.20 3,3′-cobalta bis(1,2-dicarbollide)-based HIV protease inhibitors.

[38,87,88]. A more detailed discussion on HIV protease inhibitors based on cobalt bis(dicarbollide) is available in Chapter 4.

In conclusion, the oxonium derivatives of polyhedral boron hydrides are very promising starting materials for the synthesis of various boronated organic and bioorganic materials.

ACKNOWLEDGMENTS

The authors thank the Russian Foundation for Basic Research (08-03-00463) for financial support.

REFERENCES

1. Schubert, D. 2001. From missiles to medicine: The development of boron hydrides. Available at http://www.borax.com/pioneer54.html (accessed January 07, 2010).
2. Plešek, J. 1992. Potential applications of the boron cluster compounds. *Chem. Rev.* 92:269–278.
3. Sivaev, I. B. and Bregadze, V. I. 2007. Polyhedral boron hydrides in use: Current status and perspectives. In R. P. Irwin, ed., *Organometallic Chemistry Research Perspectives*, pp. 1–59. New York: Nova Science Publishers.
4. Grimes, R. N. 1970. *Carboranes*, New York: Academic Press.
5. Bregadze, V. I. 1992. Dicarba-*closo*-dodecaboranes $C_2B_{10}H_{12}$ and their derivatives. *Chem. Rev.* 92:209–223.

6. Valliant, J. F., Guenther, K. J., King, A. S. et al. 2002. The medicinal chemistry of carboranes. *Coord. Chem. Rev.* 232:173–230.
7. Knoth, W. H., Miller, H. C., Sauer, J. C., Balthis, J. H., Chia, Y. T., and Muetterties, E. L. 1964. Chemistry of boranes. IX. Halogenation of $B_{10}H_{10}^{2-}$ and $B_{12}H_{12}^{2-}$. *Inorg. Chem.* 3:159–167.
8. Preetz, W., Srebny, H.-G., and Marsmann, H. C. 1984. Darstellung, ^{11}B-NMR- und schwingungsspektren isomerenreiner halogenohydrodecaborate $X_nB_{10}H_{10-n}^{2-}$; X = Cl, Br, I; n = 1, 2. *Z. Naturforsch.* 39B:6–13.
9. Ivanov, S. V., Ivanova, S. M., Miller, S. M., Anderson, O. P., Solntsev, K. A., and Strauss, S. H. 1996. Fluorination of $B_{10}H_{10}^{2-}$ with an *N*-fluoro reagent. A new way to transform B–H bonds into B–F bonds. *Inorg. Chem.* 35:6914–6915.
10. Einholz, W., Vaas, K., Wieloch, C. et al. 2002. Chemical and cyclovoltammetric investigation of the redox reactions of the decahalodecaborates *closo*-$[B_{10}X_{10}]^{2-}$ and *hypercloso*-$[B_{10}X_{10}]^{\bullet-}$ (X= Cl, Br). Crystal structure analysis of $Cs_2[B_{10}Br_{10}]\cdot 2H_2O$. *Z. Anorg. Allg. Chem.* 628:258–268.
11. Sivaev, I. B., Bregadze, V. I., and Sjöberg, S. 2002. Chemistry of the *closo*-dodecaborate anion $[B_{12}H_{12}]^{2-}$: A review. *Collect. Czech. Chem. Commun.* 67:679–727.
12. Lepšik, M., Srnec, M., Hnyk, D. et al. 2009. *exo*-Substituents effects in halogenated icosahedral ($B_{12}H_{12}^{2-}$) and octahedral ($B_6H_6^{2-}$) *closo*-borane skeletons: Chemical reactivity studied by experimental and quantum chemical methods. *Collect. Czech. Chem. Commun.* 74:1–27.
13. Peryshkov, D. V., Popov, A. A., and Strauss, S. H. 2009. Direct perfluorination of $K_2B_{12}H_{12}$ in acetonitrile occurs at the gas bubble-solution interface and is inhibited by HF. Experimental and DFT study of inhibition by protic acids and soft, polarizable anions. *J. Am. Chem. Soc.* 131:18393–18403.
14. Körbe, S., Schreiber, P. J., and Michl, J. 2006. Chemistry of the carba-*closo*-dodecaborate (1-) anion, $CB_{11}H_{12}^{-}$. *Chem. Rev.* 106:5208–5249.
15. Sivaev, I. B. and Bregadze, V. I. 1999. Chemistry of cobalt bis(dicarbollides). A review. *Collect. Czech. Chem. Commun.* 64:783–805.
16. Gashti, A. N., Huffman, J. C., Edwards, A. et al. 2000. Fluorination studies of the [*commo*-3,3′-Co(3,1,2-$CoC_2B_9H_{11})_2]^{-}$ ion. *J. Organomet. Chem.* 614–615:120–124.
17. Johnson, J. W. and Brody, J. F. 1982. Lithium closoborane electrolytes. III. Preparation and characterization. *J. Electrochem. Soc.* 129:2213–2219.
18. Ivanov, S. V., Casteel, Jr., W. J., Pez, G. P., and Ulman, M. 2008. Polyfluorinated boron cluster anions for lithium electrolytes. *US Patent* 7348103.
19. Peper, S., Qin, Y., Almond, P. et al. 2003. Ion-pairing ability, chemical stability, and selectivity behavior of halogenated dodecacarborane cation exchangers in neutral carrier-based ion-selective electrodes. *Anal. Chem.* 75:2131–2139.
20. Bobacka, J., Väänänen, V., Lewenstam, A. and Ivaska, A. 2004. Influence of anionic additive on Hg^{2+} interference on Ag^{+}-ISEs based on [2.2.2]*p*,*p*,*p*-cyclophane as neutral carrier. *Talanta* 63:135–138.
21. Rais, J. and Grüner, B. 2004. Extraction with metal bis(dicarbollide) anions: Metal bis(dicarbollide) extractants and their applications in separation chemistry. In Y. Marcus and A. K. Sengupta, ed., *Ion Exchange and Solvent Extraction*, Vol. 17, pp. 243–334. New York: Marcel Dekker Inc.
22. Luther, T. A., Herbst, R. S., Peterman, D. R. et al. 2006. Some aspects of fundamental chemistry of the Universal Extraction (UNEX) process for the simultaneous separation of major radionuclides (cesium, strontium, actinides, and lanthanides) from radioactive wastes. *J. Radioanal. Nucl. Chem.* 267:603–613.
23. Peymann, T., Knobler, C. B., and Hawthorne, M. F. 1998. Synthesis of alkyl and aryl derivatives of *closo*-$B_{12}H_{12}^{2-}$ by the palladium-catalyzed coupling of *closo*-$B_{12}H_{11}I^{2-}$ with Grignard reagents. *Inorg. Chem.* 37:1544–1548.
24. Gabel, D., Moller, D., Harfst, S., Rösler, J., and Ketz, H. 1993. Synthesis of *S*-alkyl and *S*-acyl derivatives of mercaptoundecahydrododecaborate, a possible boron carrier for neutron capture therapy. *Inorg. Chem.* 32:2276–2278.
25. Peymann, T., Lork, E., and Gabel, D. 1996. Hydroxoundecahydro-*closo*-dodecaborate(2-) as a nucleophile. Preparation and structural characterization of *O*-alkyl and *O*-acyl derivatives of hydroxoundecahydro-*closo*-dodecaborate(2-). *Inorg. Chem.* 35:1355–1360.
26. Sivaev, I. B., Sjöberg, S., Bregadze V. I., and Gabel, D. 1999. Synthesis of alkoxy derivatives of dodecahydro-*closo*-dodecaborate anion $[B_{12}H_{12}]^{2-}$. *Tetrahedron Lett.* 40:3451–3454.
27. Peymann, T., Lork, E., Schmidt, M., Nöth, H., and Gabel, D. 1997. *N*-Alkylation of ammine-undecahydro-*closo*-dodecaborate(1-). *Chem. Ber.* 130:795–799.
28. Sivaev, I. B., Bruskin, A. B., Nesterov, V. V., Antipin, M. Yu., Bregadze, V. I., and Sjöberg, S. 1999. Synthesis of Schiff bases derived from the ammoniaundecahydro-*closo*-dodecaborate(1-) anion, $[B_{12}H_{11}NH{=}CHR]^{-}$, and their reduction into monosubstituted amines $[B_{12}H_{11}NH_2CH_2R]^{-}$: A new route to water soluble agents for BNCT. *Inorg. Chem.* 38:5887–5893.

29. Semioshkin, A. A., Sivaev, I. B., and Bregadze, V. I. 2008. Cyclic oxonium derivatives of polyhedral boron hydrides and their synthetic applications. *Dalton Trans.* 977–992.
30. Sivaev, I. B., Semioshkin, A. A., and Bregadze, V. I. 2009. New approach to incorporation of boron in tumor-seeking molecules. *Appl. Radiat. Isotop.* 67:S91–S93.
31. Sivaev, I. B., Semioshkin, A. A., Brellochs, B., Sjöberg, S., and Bregadze, V. I. 2000. Synthesis of oxonium derivatives of the dodecahydro-*closo*-dodecaborate anion $[B_{12}H_{12}]^{2-}$. Tetramethylene oxonium derivative of $[B_{12}H_{12}]^{2-}$ as a convenient precursor for the synthesis of functional compounds for boron neutron capture therapy. *Polyhedron* 19:627–632.
32. Sivaev, I. B., Kulikova, N. Yu., Nizhnik, E. A. et al. 2008. Practical synthesis of 1,4-dioxane derivative of the *closo*-dodecaborate anion and its ring opening with acetylenic alkoxides. *J. Organomet. Chem.* 693:519–525.
33. Schaffran, T., Burghardt, A., Barnert, S. et al. 2009. Pyridinium lipids with the dodecaborate cluster as polar headgroup: Synthesis, characterization of the physical–chemical behavior, and toxicity in cell culture. *Bioconjugate Chem.* 20:2190–2198.
34. Bernard, R., Cornu, D., Perrin, M., Schaff, J.-P., and Miele, P. 2004. Synthesis and x-ray structural characterization of the tetramethylene oxonium derivative of the hydrodecaborate anion. A versatile route for derivative chemistry of $[B_{10}H_{10}]^{2-}$. *J. Organomet. Chem.* 689:2581–2585.
35. Zhizhin, K. Yu., Mustyatsa, V. N., Malinina, E. A. et al. 2004. Interaction of *closo*-decaborate anion $B_{10}H_{10}{}^{2-}$ with cyclic ethers. *Russ J. Inorg. Chem.* 49:180–189.
36. Zakharkin, L. I., Kalinin, V. N., and Zhigareva, G. G. 1979. Oxidation of dicarbadodecahydro-*nido*-undecaborate anions by mercuric chloride in tetrahydrofuran and pyridine. *Russ. Chem. Bull.* 28:2198–2199.
37. Stogniy, M. Yu., Abramova, E. N., Lobanova I. A. et al. 2007. Synthesis of functional derivatives of 7,8-dicarba-*nido*-undecaborate anion by ring-opening of its cyclic oxonium derivatives. *Collect. Czech. Chem. Commun.* 72:1676–1688.
38. Řezačova, P., Pokorna, J., Brynda, J. et al. 2009. Design of HIV protease inhibitors based on inorganic polyhedral metallacarboranes. *J. Med. Chem.* 52:7132–7141.
39. Plešek, J., Heřmanek, S., Franken, A., Cisarova, I., and Nachtigal, C. 1997. Dimethyl sulfate induced nucleophilic substitution of the [bis(1,2-dicarbollido)-3-cobalt(1-)]ate ion. Synthesis, properties and structures of its 8,8′-μ-sulfato, 8-phenyl and 8-dioxane derivatives. *Collect. Czech. Chem. Commun.* 62:47–56.
40. Teixidor, F., Pedrajas, J., Rojo, I. et al. 2003. Chameleonic capacity of $[3,3'\text{-Co}(1,2\text{-}C_2B_9H_{11})_2]^-$ in coordination. Generation of the highly uncommon *S*(thioether)–Na bond. *Organometallics* 22:3414–3423.
41. Llop, J., Masalles, C., Viñas, C., Teixidor, F., Sillanpää, R., and Kivekäs, R. 2003. The $[3,3'\text{-Co}(1,2\text{-}C_2B_9H_{11})_2]^-$ anion as a platform for new materials: Synthesis of its functionalized monosubstituted derivatives incorporating synthons for conducting organic polymers. *Dalton Trans.* 556–561.
42. Plešek, J., Grüner, B., Machaček, J., Cisařova, I., and Časlavsky, J. 2007. 8-Dioxane ferra(III) bis(dicarbollide): A paramagnetic functional molecule as versatile building block for introduction of a Fe(III) centre into organic molecules. *J. Organomet. Chem.* 692:4801–4804.
43. Prikaznov, A., Sivaev, I., and Bregadze, V. 2008. Synthesis of boron-containing acids based on decahydro-*closo*-decaborate anion. In *Abstracts of XIII International Conference on Boron Chemistry*, P095. Bellaterra: Institut de Ciencia de Materials de Barcelona.
44. Grin, M. A., Semioshkin, A. A., Titeev, R. A. et al. 2007. Synthesis of a cycloimide bacteriochlorin *p* conjugate with the *closo*-dodecaborate anion. *Mendeleev Commun.* 17:14–15.
45. Semioshkin, A. A., Las'kova, Yu. N., Zhidkova, O. B., and Bregadze, V. I. 2008. Synthesis of new building blocks based on the *closo*-dodecaborate anion. *Russ. Chem. Bull.* 57:1996–1998.
46. Grin, M. A., Titeev, R. A., Bakieva, O. M. et al. 2008. New boron-containing bacteriochlorin *p* cycloimide conjugate. *Russ. Chem. Bull.* 57:2230–2232.
47. Rak, J., Kaplanek, R., and Kral, V. 2010. Solubilization and deaggregation of cobalt bis(dicarbollide) derivatives in water by biocompatible excipients. *Bioorg. Med. Chem. Lett.* 20:1045–1048.
48. Semioshkin, A., Nizhnik, E., Godovikov, I., Starikova, Z., and Bregadze, V. 2007. Reactions of oxonium derivatives of $[B_{12}H_{12}]^{2-}$ with amines: Synthesis and structure of novel B_{12}-based ammonium salts and amino acids. *J. Organomet. Chem.* 692:4020–4028.
49. Kondakov, N. N., Orlova, A. V., Zinin, A. I. et al. 2005. Conjugates of polyhedral boron compounds with carbohydrates. 3. The first synthesis of a conjugate of the dodecaborate anion with a disaccharide lactose as a potential agent for boron neutron capture therapy of cancer. *Russ. Chem. Bull.* 54:1311–1312.
50. Orlova, A. V., Kondakov, N. N., Zinin, A. I. et al. 2006. A uniform approach to the synthesis of carbohydrate conjugates of polyhedral boron compounds as potential agents for boron neutron capture therapy. *Russ. J. Bioorg. Chem.* 32:P.568–577.

51. Sivaev, I. B., Starikova, Z. A., Sjöberg, S., and Bregadze, V. I. 2002. Synthesis of functional derivatives of [3,3′-Co(1,2-$C_2B_9H_{11})_2]^-$ anion. *J. Organomet. Chem.* 649:1–8.
52. Kolb, H. C. and Sharpless, K. B. 2003. The growing impact of click chemistry on drug discovery. *Drug Discov. Today* 8:1128–1137.
53. Moses, J. E., and Moorhouse, A. D. 2007. The growing applications of click chemistry. *Chem. Soc. Rev.* 36:1249–1262.
54. Orlova, A. V., Kondakov, N. N., Kimel, B. G. et al. 2007. Synthesis of novel derivatives of *closo*-dodecaborate anion with azido group at the terminal position of the spacer. *J. Appl. Organomet. Chem.* 21:98–100.
55. Olejniszak, A., Wojtczak, B., and Lesnikowski, Z. J. 2007. 2′-Deoxyadenosine bearing hydrophobic carborane pharmacophore. *Nucleosides, Nucleotides, Nucl. Acids* 26:1611–1613.
56. Wojtczak, B. A., Andrysiak, A., Grüner, B., and Lesnikowski, Z. J. 2008. Chemical ligation: A versatile method for nucleoside modification with boron clusters. *Chem. Eur. J.* 14:10675–10682.
57. Semioshkin, A., Laskova, J., Wojtczak, B. et al. 2009. Synthesis of *closo*-dodecaborate based nucleoside conjugates. *J. Organomet. Chem.* 694:1375–1379.
58. Bregadze, V. I., Semioshkin, A. A., Las'kova, J. N. et al. 2009. Novel types of boronated chlorin e_6 conjugates via "click chemistry." *Appl. Organomet. Chem.* 23:370–374.
59. Semioshkin, A. A., Osipov, S. N., Grebenyuk, J. N. et al. 2007. An effective approach to 1,2,3-triazole containing *closo*-dodecaborates. *Collect. Czech. Chem. Commun.* 72:1717–1724.
60. Grin, M. A., Titeev, R. A., Brittal, D. I. et al. 2010. Synthesis of bis(dicarbollide)cobalt conjugates with natural chlorines by the Sonogashira reaction. *Russ. Chem. Bull.* 59:219–224.
61. Farras, P., Teixidor, F., Kivekäs, R. et al. 2008. Metallacarborane as building blocks for polyanionic polyarmed aryl-ether materials. *Inorg. Chem.* 47:9497–9508.
62. Farras, P., Teixidor, F., Sillanpää, R., and Viñas, C. 2010. A convenient synthetic route to useful monobranched polyethoxylated halogen terminated [3,3-Co(1,2-$C_2B_9H_{11})_2]^-$ synthons. *Dalton Trans.* 39:1716–1718.
63. Schaffran, T., Lissel, F., Samatanga, B. et al. 2009. Dodecaborate cluster lipids with variable headgroups for boron neutron capture therapy: Synthesis, physical–chemical properties and toxicity. *J. Organomet. Chem.* 694:1708–1712.
64. Semioshkin, A., Laskova, J., Zhidkova, O. et al. 2010. Synthesis and structure of novel *closo*-dodecaborate-based glycerols. *J. Organomet. Chem.* 695:370–374.
65. Li, H., Fronczek, F. R., and Vicente, M. G. H. 2008. Synthesis and properties of cobaltacarborane-functionalized Zn(II)-phthalocyanines. *Tetrahedron Lett.* 49:4828–4830.
66. Li, H., Fronczek, F. R., and Vicente, M. G. H. 2009. Cobaltacarborane-phthalocyanine conjugates: Syntheses and photophysical properties. *J. Organomet. Chem.* 694:1607–1611.
67. Hao, E., Jensen, T. J., Courtney, B. H., and Vicente, M. G. H. 2005. Synthesis and cellular studies of porphyrin-cobaltacarborane conjugates. *Biooconjugate Chem.* 16:1495–1502.
68. Sibrian-Vazquez, M., Hao, E., Jensen, T. J., and Vicente, M. G. H. 2006. Enhanced cellular uptake with a cobaltacarborane-porphyrin-HIV-1 Tat 48–60 conjugate. *Biooconjugate Chem.* 17:928–934.
69. Hao, E., Sibrian-Vazquez, M., Serem, W., Garno, J. C., Fronczek, F. R., and Vicente, M. G. H. 2007. Synthesis aggregation and cellular investigations of porphyrin-cobaltacarborane conjugates. *Chem. Eur. J.* 13:9035–9047.
70. Kubat, P., Lang, K., Cigler, P. et al. 2007. Tetraphenylporphyrin-cobalt(III) bis(1,2-dicarbollide) conjugates: From the solution characteristics to inhibition of HIV protease. *J. Phys. Chem. B* 111:4539–4546.
71. Hao, E., Zhang, M., E, W. et al. 2008. Synthesis and spectroelectrochemistry of *N*-cobaltacarborane porphyrin conjugates. *Bioconjugate Chem.* 19:2171–2181.
72. Bregadze, V. I., Sivaev, I. B., Lobanova, I. A. et al. 2009. Conjugates of boron clusters with derivatives of natural chlorin and bacteriochlorin. *Appl. Radiat. Isotop.* 67:S101–S104.
73. Olejniczak, A. B., Plešek, J., Křiz, O., and Lesnikowski, Z. J. 2003. A nucleoside conjugate containing a metallacarborane group and its incorporation into a DNA oligonucleotide. *Angew. Chem. Int. Ed.* 42:5740–5743.
74. Olejniczak, A. B., Plešek, J., and Lesnikowski, Z. J. 2007. Nucleoside-metallacarborane conjugates for base-specific metal labeling of DNA. *Chem. Eur. J.* 13:311–318.
75. Olejniczak, A. B., Mucha, P., Grüner, B., and Lesnikowski, Z. J. 2007. DNA-Dinucleotides bearing a 3,3′-cobalt- or 3,3′-iron-1,2,1′,2-dicarbollide complex. *Organometallics* 26:3272–3274.
76. Olejniczak, A. B., Grüner, B., Šicha, V., Broniarek, S., and Lesnikowski, Z. J. 2009. Metallacarboranes as labels for multipotential electrochemical coding of DNA. [3-Chromium bis(dicarbollide)](1-)ate and its nucleoside conjugates. *Electroanalysis* 21:501–506.

77. Selucky, P., Plešek, J., Rais, J., Kyrs, M., and Kadlecova, L. 1991. Extraction of fission products into nitrobenzene with dicobalt tris-dicarbollide and ethyleneoxy-substituted cobalt bis-dicarbollide. *J. Radioanal. Nucl. Chem.* 149:131–140.
78. Plešek, J., Grüner, B., Heřmanek, S. et al. 2002. Synthesis of functionalized cobaltacarboranes based on the *closo*-$[(1,2\text{-}C_2B_9H_{11})_2\text{-}3,3'\text{-Co}]^-$ ion bearing polydentate ligands for separation of M^{3+} cations from nuclear waste solutions. Electrochemical and liquid–liquid extraction study of selective transfer of M^{3+} metal cations to an organic phase. Molecular structure of the *closo*-$[(8\text{-}(2\text{-}CH_3O\text{-}C_6H_4\text{-}O)\text{-}(CH_2CH_2O)_2\text{-}1,2\text{-}C_2B_9H_{10})\text{-}(1',2'\text{-}C_2B_9H_{11})\text{-}3,3'\text{-Co}]Na$ determined by x-ray diffraction analysis. *Polyhedron* 21:975–986.
79. Grüner, B., Plešek, J., Bača, J. et al. 2002. Crown ether substituted cobalta bis(dicarbollide) ions as selective extraction agents for removal of Cs^+ and Sr^{2+} from nuclear waste. *New J. Chem.* 26:867–875.
80. Grüner, B., Kvičalova, M., Selucky, P., and Lučanikova, M. 2010. Anionic alkyl diglycoldiamides with covalently bonded cobalt bis(dicarbollide)(1-) ions for lanthanide and actinide extractions. *J. Organomet. Chem.* 695:1261–1264.
81. Grüner, B., Plešek, J., Bača, J. et al. 2002. Cobalt bis(dicarbollide) ions with covalently bonded CMPO groups as selective extraction agents for lanthanide and actinide cations from highly acidic nuclear waste solutions. *New J. Chem.* 26:1519–1527.
82. Selucky, P., Rais, J., Lučanikova, M. et al. 2008. Lanthanide and actinide extractions with anionic ligands based on cobalt bis(dicarbollide) ions with covalently bonded CMPO functions. *Radiochim. Acta* 96:273–284.
83. Reinoso-Garcia, M. M., Janczewski, D., Reinhoudt, D. N. et al. 2006. CMP(O) tripodants: Synthesis, potentiometric studies and extractions. *New J. Chem.* 30:1480–1492.
84. Grüner, B., Mikulašek, L., Bača, J. et al. 2005. Cobalt bis(dicarbollides)(1-) covalently attached to the calix[4]arene platform : The first combination of organic bowl-shaped matrices and inorganic metallaborane cluster anions. *Eur. J. Org. Chem.* 2022–2039.
85. Mikulašek, L., Grüner, B., Danila, C., Bőhmer, V., Časlavsky, J., and Selucky, P. 2006. Synergistic effect of ligating and ionic functions, prearranged on a calix[4]arene. *Chem. Commun.* 4001–4003.
86. Mikulašek, L., Grüner, B., Dordea, C. et al. 2007. *tert*-Butyl-calix[4]arenes substituted at the narrow rim with cobalt bis(dicarbollide)(1-) and CMPO groups—New and efficient extractants for lanthanides and actinides. *Eur. J. Org. Chem.* 4772–4783.
87. Cigler, P., Kožišek, M., Řezačova, P. et al. 2005. From nonpeptide toward noncarbon protease inhibitors: Metallacarboranes as specific and potent inhibitors of HIV protease. *Proc. Natl. Acad. Sci. USA* 102:15394–15399.
88. Kožišek, M., Cigler, P., Lepšik, M. et al. 2008. Inorganic polyhedral metallacarborane inhibitors of HIV protease: A new approach to overcoming antiviral resistance. *J. Med. Chem.* 51:4839–4843.

25 Asymmetric Allylation of Carbonyl Compounds via Organoboranes*

Subash C. Jonnalagadda, J. Sravan Kumar, Anthony Cirri, and Venkatram R. Mereddy

CONTENTS

* Dedicated to the memory of Professor Herbert C. Brown.

25.1 INTRODUCTION

Allylboration is an important carbon–carbon bond-forming reaction that has been extensively used in synthetic organic chemistry.[1] Reaction of "allyl" boranes **1** with carbonyl compounds **2** is generally advantageous over other "allyl" metalations because of superior levels of stereoselectivity observed due to a rigid six-membered boracyclic transition state **3**. Further, the allylboration reaction is highly versatile because of five different points of substitution (a–e) theoretically possible on the "allyl" borane that could be stereoselectively transferred on to carbonyl compounds thus obtaining a wide variety of complex products (Figure 25.1).

Although several functionalized allylboranes are known in the literature, this chapter primarily focuses on the asymmetric allylation of carbonyl compounds using chiral allylboranes. Stereoselective allylboration of aldehydes/imines has been explored with various chiral auxiliaries substituted on boron furnishing the homoallylic alcohols/amines in high ee and de.[2] Hoffmann, Brown, Yamamoto, Roush, Soderquist, and several other researchers have made significant contributions in the area of asymmetric allylboration. Hoffmann pioneered a camphor-derived chiral allylborane for the first stereoselective allylboration of aldehydes.[3] This protocol found extensive utility for single- and double-diastereoselective asymmetric synthesis.[4] Tartarate esters were introduced as chiral auxiliaries by Yamamoto and coworkers for propargyl- and allenylboration of aldehydes to produce allenyl and homopropargylic alcohols, respectively.[5] The steric bulk of the alkyl group in the tartarate ester was found to play an important role in determining the enantioselectivity of the product homoallylic alcohols. Relatively less bulky ethyl and isopropyl tartarate esters provided low ee's, while sterically hindered cyclododecyl and 2,4-dimethylpentyl tartarates led to higher ee's.[6] The utility of isopropyl tartarate-derived chiral allylboronate was later extended for allyl and crotylborations by Roush et. al.[7] Other prominent asymmetric allylboranes that have also been developed for the asymmetric allylboration of aldehydes/imines include those derived from chiral 1,2-diols **5–7**,[8] oxazaborolane **8a**,[9] and diazaborolane **8b**,[10] and so on (Figure 25.2). Due to the vast and elaborate literature available on allylboration, this chapter only deals with chiral allyl dialkylboranes and does not include allylboronates. The latter topic has been extensively reviewed elsewhere.[1,2a,2b,2e]

Of all the chiral auxiliaries that have been studied so far, α-pinene-based allylboranes have proven to be one of the most effective, inexpensive, and versatile reagents based on high optical purity observed for the product homoallylic alcohols.[2,11] α-Pinene is highly inexpensive and readily obtained from pine trees. Both enantiomers of α-pinene are available in nature: (−)-α-pinene is more common in Europe, and the (+)-α-isomer is found in North American pine trees. Moreover, α-pinene-based allylboranes were found to exhibit excellent levels of reagent-controlled stereoselectivity in their reaction with a wide variety of chiral and achiral aldehydes. These factors greatly simplify the synthetic design of complex natural products as the requisite chiral centers can be

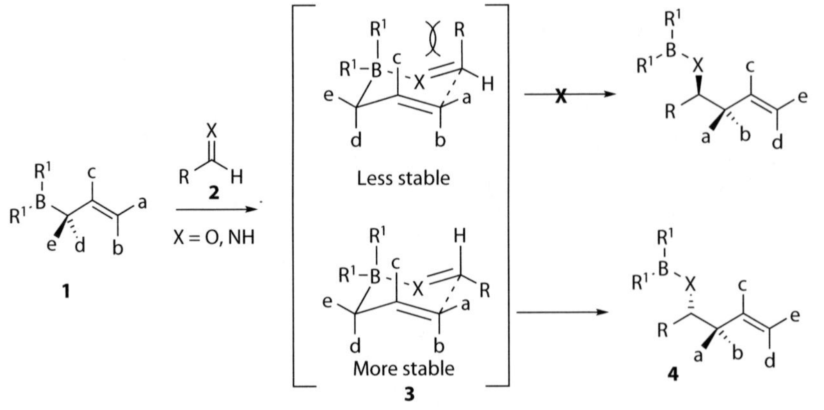

FIGURE 25.1 The scope of "allyl" boration.

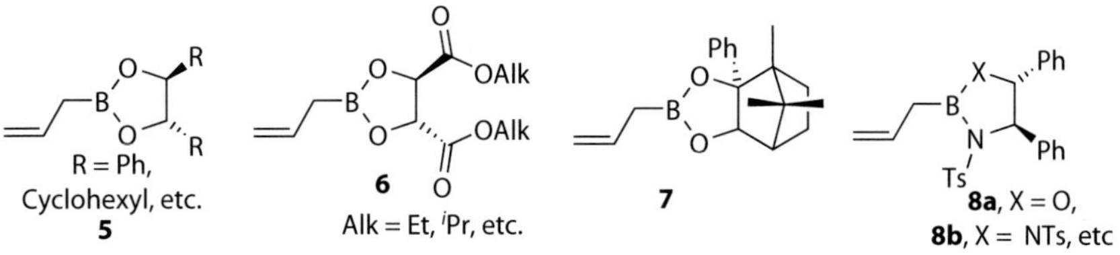

FIGURE 25.2 Various asymmetric "allyl" boranes.

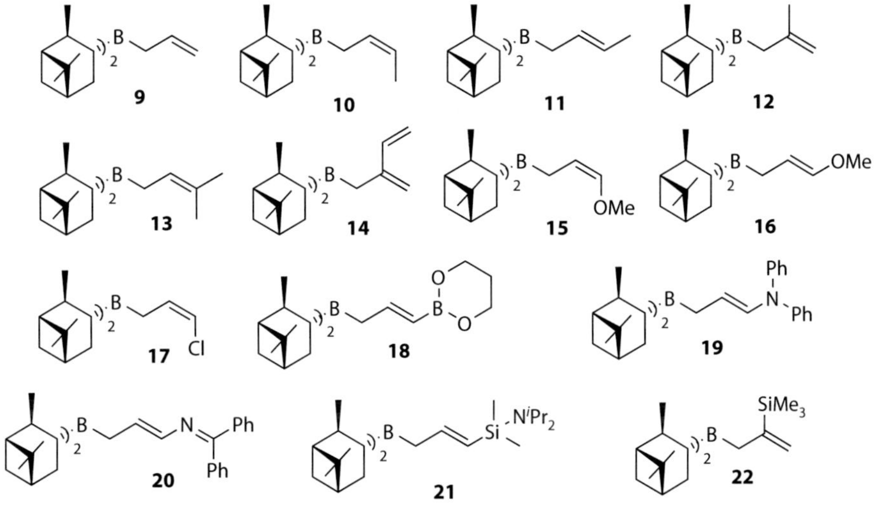

FIGURE 25.3 α-Pinene-based "allyl" boranes.

easily established solely based on the antipode of α-pinene used regardless of the inherent chirality of the starting materials. Accordingly, several higher-order "allyl" boranes (e.g., **9–22**) derived from α-pinene have been readily prepared either *in situ* or as isolable reagents (Figure 25.3).[2c–d] All these reagents provide very good yields of the corresponding homoallylic alcohols/amines on reaction with aldehydes/imines in excellent enantio- and diastereoselectivities. The detailed description of several of these reagents and their applications are provided in the following sections.

25.2 (*B*)ALLYLDIISOPINOCAMPHEYLBORANE

(*B*) Allyldiisopinocampheylborane ($Ipc_2BAllyl$, **9**) is prepared in three steps starting from α-pinene. Hydroboration of α-pinene **23** with $BH_3 \cdot Me_2S$ provides diisopinocamheylborane **24** (Ipc_2BH) that on reaction with HCl or methanol generates *B*-chlorodiisopinocampheylborane (Ipc_2BCl, **25**) or *B*-methoxydiisopinocampheylborane (Ipc_2BOMe, **26**) respectively. The reaction of either Ipc_2BCl or Ipc_2BOMe with allyl Grignard provides the requisite allylborane **9** (Scheme 25.1). This allylborane is very stable and can be stored under nitrogen atmosphere for extended periods of time.*[12]

The allylborane **9** on reaction with aldehydes provides the borinate **27** that on oxidation with NaOH and H_2O_2 provides the homoallylic alcohol **28** in high yield and ee (Scheme 25.2).

Typical experimental procedure for the preparation of *B*-Allyldiisopinocampheylborane **9** and its subsequent allylboration with aldehydes: Allylmagnesium bromide (1.0 mol) is added drop-wise to a stirred solution of *B*-chlorodiisopinocampheylborane (Ipc_2BCl, **25**) or *B*-methoxydiisopinocampheylborane (Ipc_2BOMe, **26**) (1.05 mol in 1 L ether) at 0 C. After the completion of the reaction as monitored by ^{11}B NMR (δ 79), the reaction mixture is filtered under nitrogen and concentrated under

* These reagents are commercially available from Sigma-Aldrich.

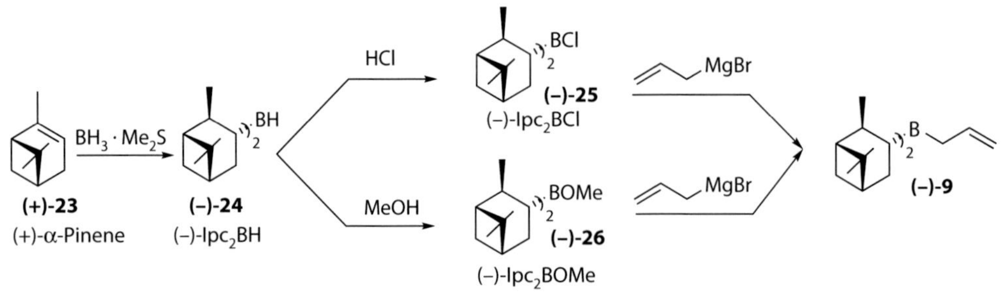

SCHEME 25.1 Preparation of *B*-Allyldiisopinocampheylborane.

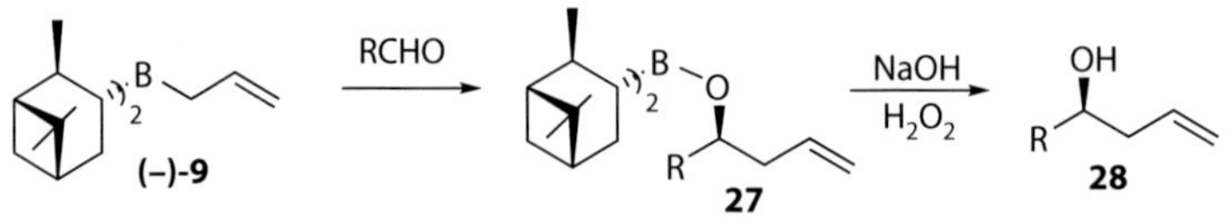

SCHEME 25.2 Preparation of *B*-Allyldiisopinocampheylborane.

vacuum. Pentane is added to the solution using a canula, stirred for 5 min and allowed to settle down. The supernatant liquid is then transferred via a canula into another round-bottom flask under nitrogen and the solvent is evaporated off under vacuum. After repeated washing with pentane, the concentrate was dissolved in an appropriate amount of pentane so as to make a 1 M stock solution. Aldehyde (1 mmol) dissolved in 1.2 mL ether is added to a stirred solution of (*B*)-allyldiisopinocampheylborane **9**, (1.2 mL, 1.0 M solution) at –100°C and maintained at that temperature until complete consumption

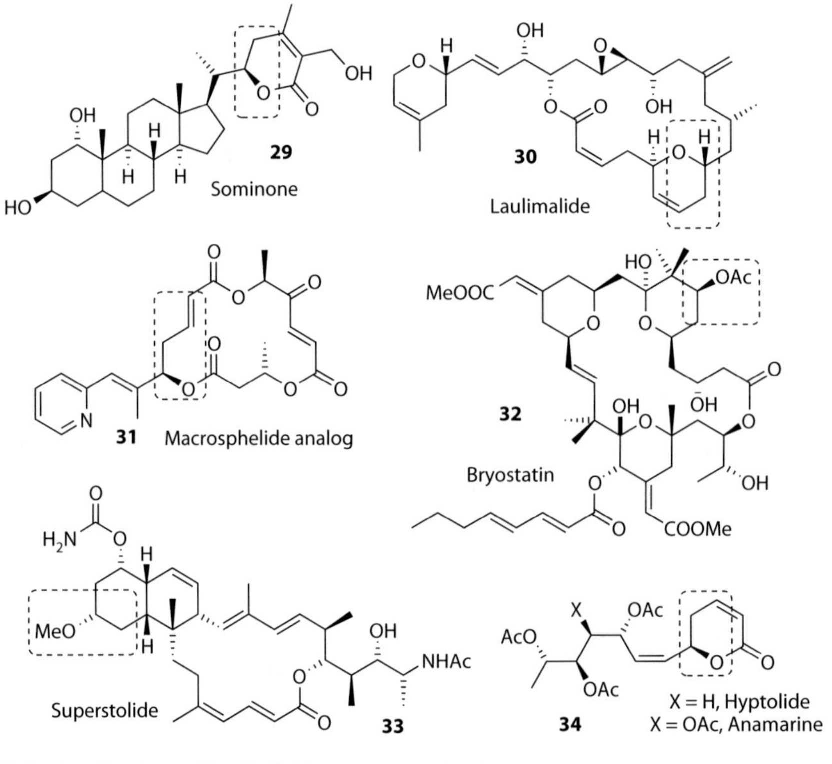

FIGURE 25.4 Applications of Ipc$_2$Ballyl in organic synthesis.

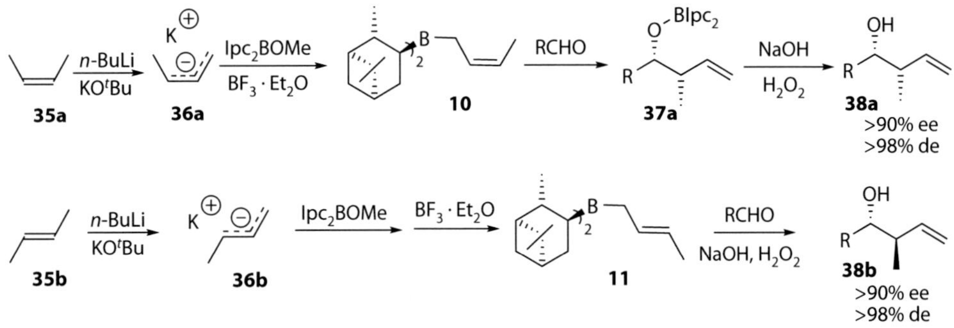

SCHEME 25.3 Preparation of (Z) and (E)-B-crotyldiisopinocampheylboranes.

of the reagent as determined by ^{11}B NMR spectroscopy (δ 56). On completion, the mixture is oxidized with 0.5 mL of 3.0 M sodium hydroxide and 0.5 mL of 30% hydrogen peroxide, stirred overnight at room temperature, and extracted with Et_2O. The product is typically purified by column chromatography or distillation to obtain the pure homoallylic alcohol.

Owing to the reliability and versatility of this reagent, it has been extensively utilized in the synthesis of complex natural products and some of the very recent examples include sominone **29**, laulimalide **30**, macrosphelide **31**, bryostatin **32**, superstolide **33**, hyptolide **34**, and so on (Figure 25.4).[12]

25.3 (B)-(Z)-γ-METHYLALLYLDIISOPINOCAMPHEYLBORANE

The reaction of *n*-BuLi and KO^{t}Bu (Schlosser's base) with *cis*-2-butene **35a** yields (Z)-crotyl potassium that further reacts with Ipc_2BOMe followed by $BF_3 \cdot Et_2O$ to generate (Z)-(B)-γ-methylallyldiisopinoca mpheylborane ((Z)-Ipc_2BCrotyl, **10**).[13] This reagent is typically prepared *in situ* and utilized for the reaction with aldehydes yielding the β-methylhomoallylic alcohols **38a** on oxidation. Similarly, (B)-(E)-(γ)-methylallyldiisopinocampheylborane ((E)-Ipc_2BCrotyl, **11**) can be prepared under similar conditions by replacing *cis*-2-butene with *trans*-2-butene (Scheme 25.3). It has been reported that temperature plays a critical role in determining the diastereoselectivity of both these reactions due to the rapid equilibration of reagents **10** and **11** especially at higher temperatures. Hence, both of these reagents are generated *in situ* at low temperatures and utilized directly for crotylboration with aldehydes.

Typical experimental procedure for the crotylboration of aldehydes with Ipc_2BCrotyl: Potassium *tert*-butoxide (100 mmol) is dissolved in 100 mL THF at −78°C and *cis* or *trans*-2-butene (300 mmol) is added to it. *n*-Butyl lithium (100 mmol) is added drop-wise to the reaction mixture and stirred for 20 min at −45°C. Warming the reaction to higher temperatures leads to reduced diastereoselectivity. The reaction mixture is cooled to −78°C and *B*-methoxydiisopino-campheylborane (Ipc_2BOMe, 110 mmol) dissolved in 50.0 mL THF is added to it and stirred for 1 h. A solution of aldehyde (90 mmol) dissolved in 20.0 mL of THF pre-cooled to −78°C is transferred to the reaction mixture via a canula and stirred for 3 h. The reaction mixture is then oxidized with 50 mL 3M NaOH and 50 mL 30% H_2O_2 and stirred overnight. The reaction mixture is worked up with ether and water and the combined organic layer dried over $MgSO_4$, concentrated *in vacuo* and purified by column chromatography to obtain the homoallylic alcohol.

The utility of these reagents for natural product synthesis is very well documented as shown in few recent representative examples such as cochleamycin **39**, amphidinolide **40**, azithromycin **41**, brevisamide **42** (Figure 25.5).[14]

25.4 (B)-β-METHYLALLYLDIISOPINOCAMPHEYLBORANE

This reagent (**12**) is obtained via the reaction of 2-methylpropene **43** with butyl lithium/TMEDA followed by the addition of Ipc_2BOMe **26**.[15] As expected, this reagent also exhibits high levels of

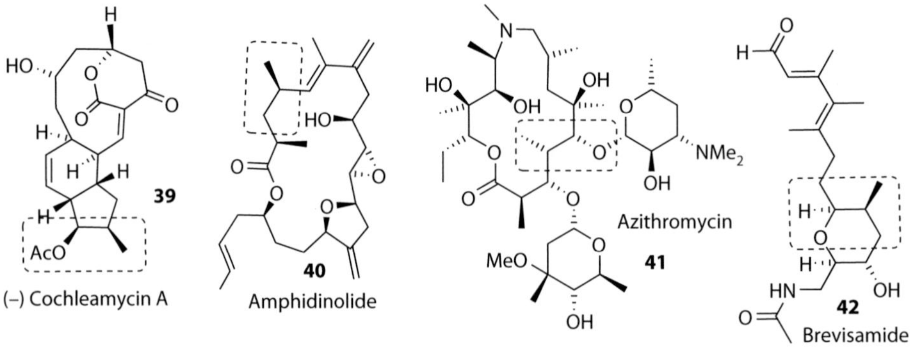

FIGURE 25.5 Utilization of crotylboration in organic synthesis.

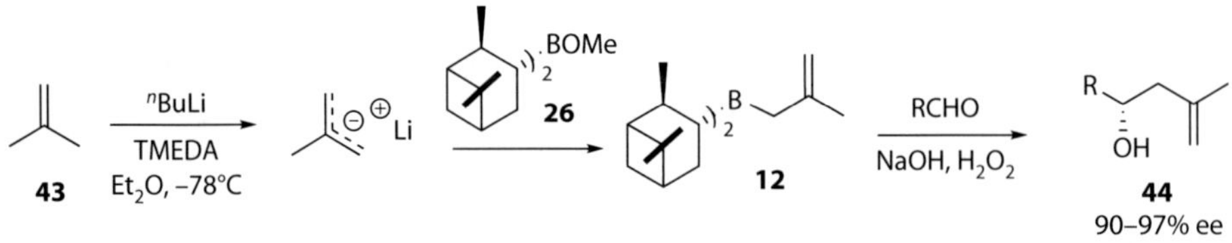

SCHEME 25.4 Preparation of (*B*)-β-Methylallyldiisopinocampheylborane.

enantiocontrol in the reaction with aldehydes to yield the product homoallylic alcohols **44** (Scheme 25.4).

Some of the representative applications of methallylboration in natural product synthesis include laulimalide **45,** neopeltolide **46**, clavosolide **47**, reidispongiolide **48** (Figure 25.6).[16]

25.5 (*B*)-γ,γ-DIMETHYLALLYLDIISOPINOCAMPHEYLBORANE

The dimethylallylborane reagent **13** can be obtained either via hydroboration of dimethylallene **49** with Ipc_2BH or via the reaction of dimethylallyl Grignard **50** with Ipc_2BOMe.[15b,17] The latter procedure is more practical considering the high cost of 1,1-dimethylallene. Subsequent allylboration with this reagent provides dimethyl homoallylic alcohols **51** in high ee (Scheme 25.5).

We were able to utilize this allylborane in the synthesis of C_1–C_6 subunit **54** of epothilone A (**55**), a potent microtubule stabilizing macrolactone (Scheme 25.6).[17a,b]

25.6 (*B*)-ISOPRENYLDIISOPINOCAMPHEYLBORANE

Brown and Randad prepared the isoprenyldiisopinocampheylborane **14** by the reaction of potassium tetramethylpiperidide **57** with isoprene **58** followed by the reaction with Ipc_2BOMe **26.**[18] Further reaction of this reagent with isobutyraldehyde **61** and 3-methylbutenal **62** provided ipsenol **63** and ipsdienol **64** respectively in high optical purity (Scheme 25.7). Apart from these molecules, the other major application of this reagent was in the synthesis of spongistatin 2 by Nakata and coworkers.[19]

25.7 (*B*)-(*Z*)-γ-[METHOXYALLYL]DIISOPINOCAMPHEYLBORANE

B-γ-methoxyallyldiisopinocampheylborane **15** is typically prepared and used *in situ*. Deprotonation of allyl methyl ether **65** with *sec*-BuLi provides allylic anion with exclusive (Z)-stereochemistry due to a five-membered ring intermediate **66** resulting from the coordination of lithium and oxygen.[20]

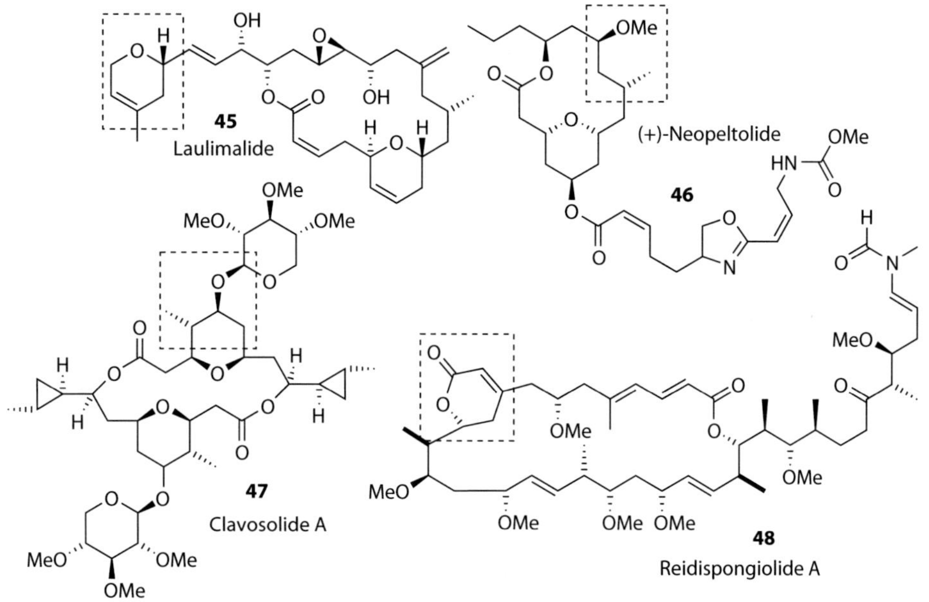

FIGURE 25.6 Utilization of methallylboration in organic synthesis.

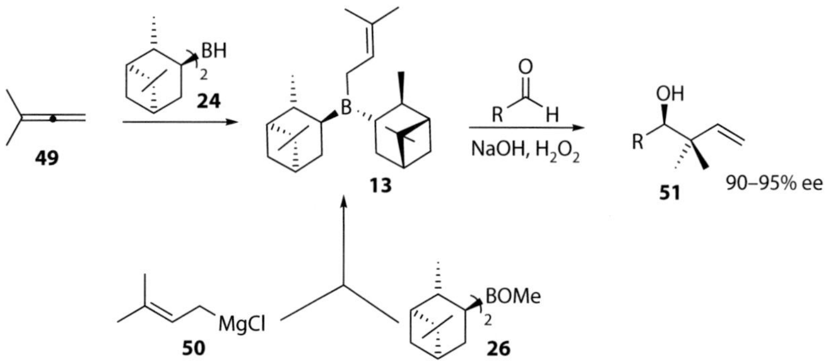

SCHEME 25.5 Preparation of (*B*)-γ,γ-Dimethylallyldiisopinocampheylborane.

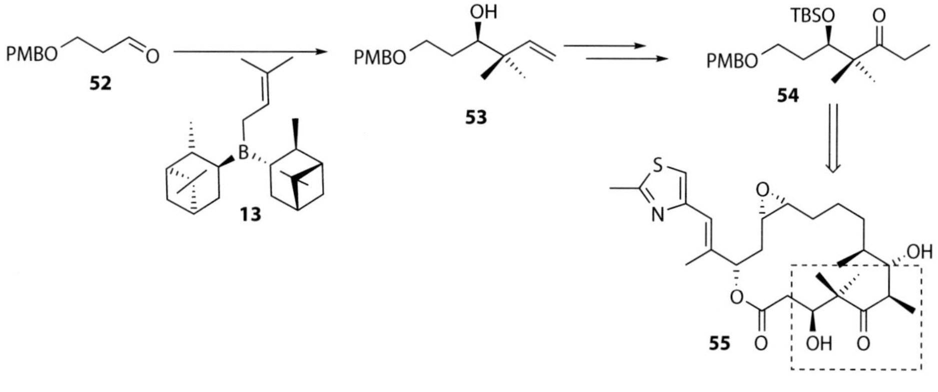

SCHEME 25.6 Utilization of Dimethylallylborane for Synthesis of Epothilone.

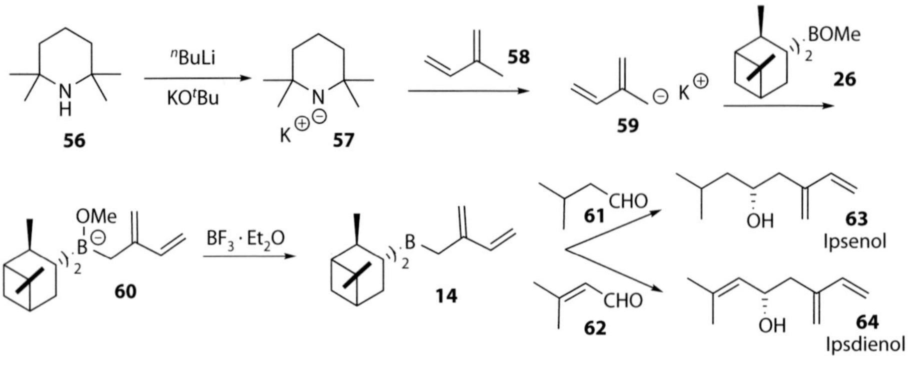

SCHEME 25.7 Preparation of (*B*)-Isoprenyldiisopinocampheylborane.

Reaction of **66** with Ipc_2BOMe followed by $BF_3 \cdot Et_2O$ provides the allylborane **15**. Allylation of aldehydes with **15** and subsequent alkaline hydroperoxide oxidation yields *syn* β-methoxy homoallylic alcohol **69** in very high de and ee (Scheme 25.8).

This reaction proceeds via a high level of reagent control and the antipode of Ipc_2BOMe employed in the reaction determines the enantiomer of the homoallylic alcohol obtained. Typically, the alkoxyallylborane obtained from (+)-Ipc_2BOMe on reaction with aldehydes yields homoallylic alcohols with α-configuration while that derived from (–)-Ipc_2BOMe gives alcohols with β-configuration as the major enantiomerm for example, reaction of allylmethyl ether **65** with *sec*-butyl lithium followed by (+)-Ipc_2BOMe, $BF_3 \cdot Et_2O$, and standard aldehydes such as acetaldehyde, propionaldehyde, isobutyraldehyde, benzaldehyde, or acrolein, provide the corresponding (*S*,*S*)-**69** while (*R*,*R*)-**69** would be obtained on reaction with the corresponding alkoxyallylborane derived from (–)-Ipc_2BOMe (Scheme 25.9).[21]

Typical experimental procedure for the methoxyallylboration of aldehydes with (*B*)-(*Z*)-γ-[methoxyallyl]diisopinocampheylborane: *sec*-Butyl lithium (100 mmol) is added drop-wise to a well-stirred solution of allyl methyl ether (110 mmol) in 200 mL THF at –78°C and stirred for 0.5 h. Ipc_2BOMe (120 mol, 1.0 M solution in THF) is added to the reaction mixture and stirred at -78°C for 1 h. $BF_3 \cdot Et_2O$ (150 mmol) is then added and the reaction mixture cooled to –100°C. A precooled solution of the appropriate aldehyde (90 mmol) in 50 mL THF is added drop-wise to the reaction mixture at –100°C and stirred until complete consumption of the reactants (analyzed based on ^{11}B NMR). The reaction mixture is oxidized with 50 mL of 3.0 M sodium hydroxide and 50 mL of 30% hydrogen peroxide and stirred overnight at room temperature. The product is extracted with Et_2O, washed with water, and dried over $MgSO_4$, concentrated *in vacuo* and purified via column chromatography to obtain the homoallylic alcohol.

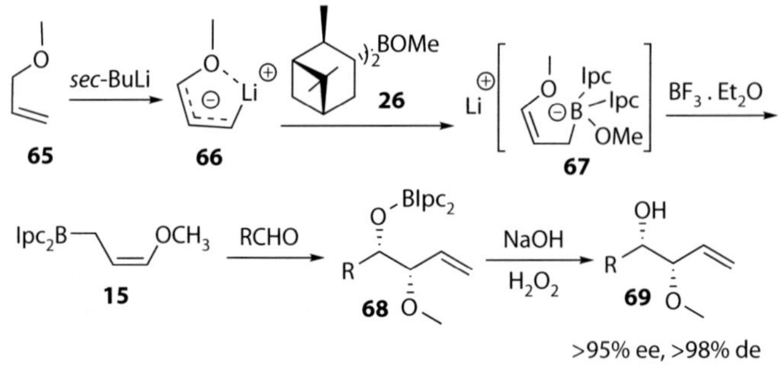

SCHEME 25.8 Preparation and Utilization of Methoxyallylboranes.

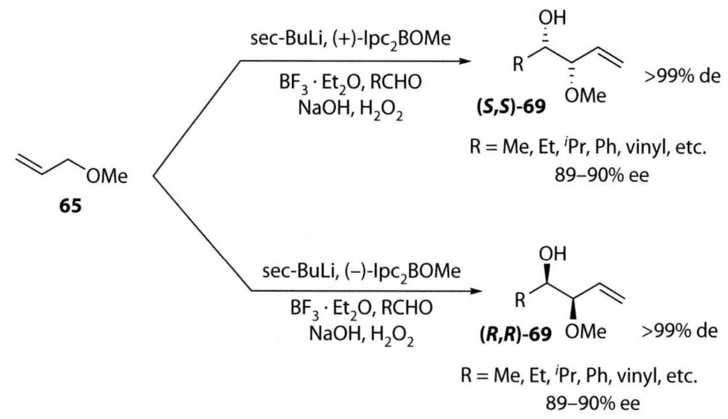

SCHEME 25.9 Reagent-controlled methoxyallylboration.

25.7.1 Applications of (*B*)-(*Z*)-γ-[Methoxyallyl]diisopinocampheylborane

Wuts and Bigelow [22] achieved a formal synthesis of carbomycin **71**, a macrolide antibiotic utilizing alkoxyallylboration as a key step (Figure 25.7).

Tatsuta and coworkers,[23a] Panek,[23b] and Cossy[23c] utilized alkoxyallylboration in the synthesis of herbimycin A **72** a potent antibiotic belonging to the class benzoquinoid ansamycin (Figure 25.8).

Kirschning and coworkers described the synthesis of proansamitocin **76**, a key biosynthetic intermediate of a highly potent antitumor agent ansamitocin, utilizing the methoxyallylboration with allylborane **15** as a key step (Figure 25.9).[24]

Ganesh and Nicholas demonstrated that the cobalt carbonyl complexes of the acetylenic aldehydes **77** undergo facile reaction with alkoxyallyborane to provide the homoallylic alcohols **78**. Oxidative deprotection of the cobalt carbonyl complexes using ceric ammonium nitrate readily furnishes the free homoallylic alcohols **79** (Scheme 25.10).[25]

Starting from the homoallylic alcohol **80** obtained via alkoxyallylbroation of *p*-anisaldehyde, Jung and coworkers developed a novel regio- and diastereoselective amination of allylic ethers using chlorosulfonyl isocyanate CSI.[26] Further application of this methodology was shown in the synthesis of novel cytokine modulators cytoxazone and epi-cytoxazone **82** (Figure 25.10).[27]

The reaction of *N*-aluminoimines **83** with alkoxyallylboranes readily furnish the homoallylic amines **84** that were further utilized for the synthesis of tetrahydropyridines **85** utilizing ring-closing metathesis as a key step (Scheme 25.11).[28]

25.8 (*B*)-(*Z*)-γ-[ALKOXYALLYL]DIISOPINOCAMPHEYLBORANE

Alkoxyallylboration using allyl methyl ether provides β-methoxyhomoallylic alcohols. The deprotection of methoxy group typically requires harsh reaction conditions that could prove detrimental

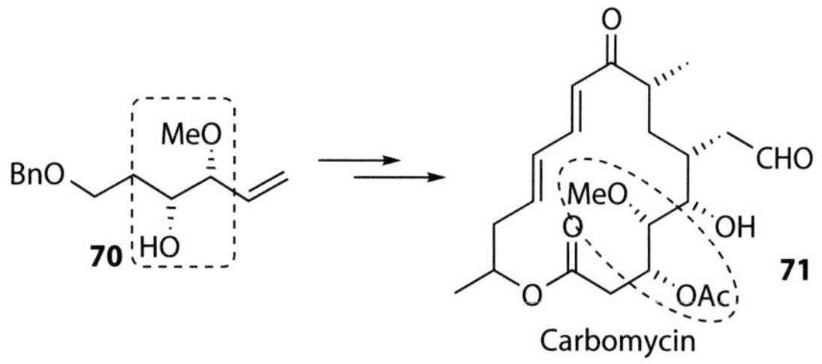

FIGURE 25.7 Synthesis of a subunit of carbomycin.

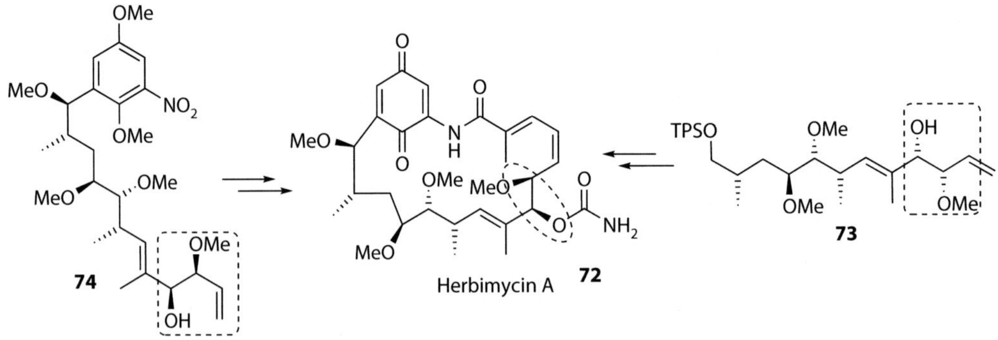

FIGURE 25.8 Synthesis of herbimycin.

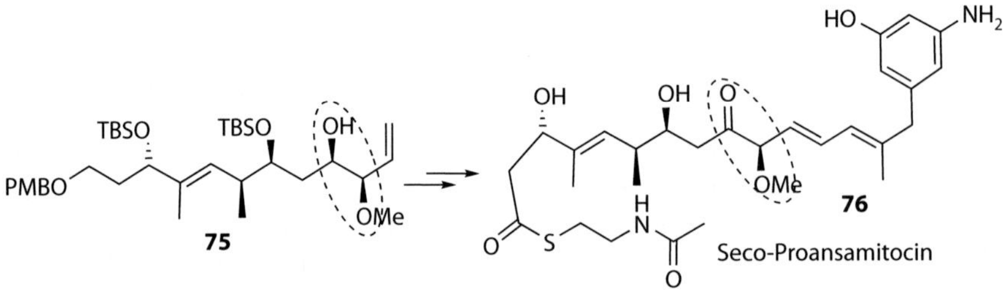

FIGURE 25.9 Synthesis of *seco*-proansamitocin.

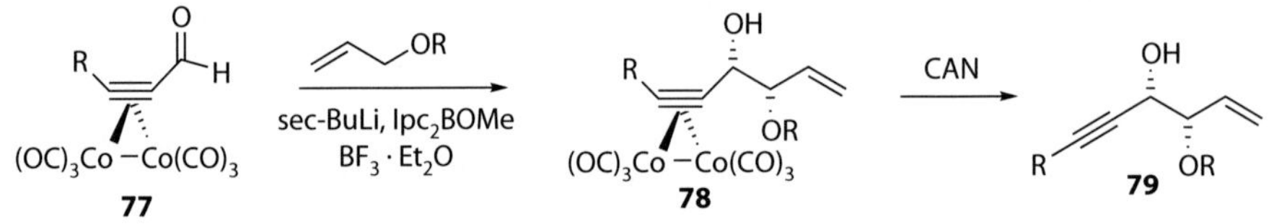

SCHEME 25.10 Alkoxyallylboration of $CO_2(CO)_6$ Complexes of acetylenic aldehydes.

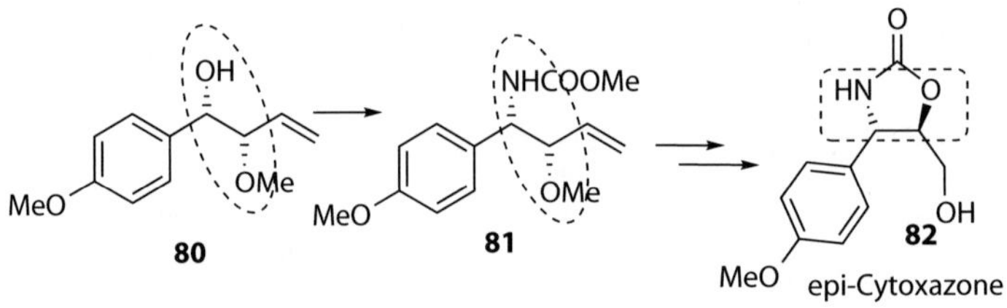

FIGURE 25.10 Synthesis of epi-cytoxazone.

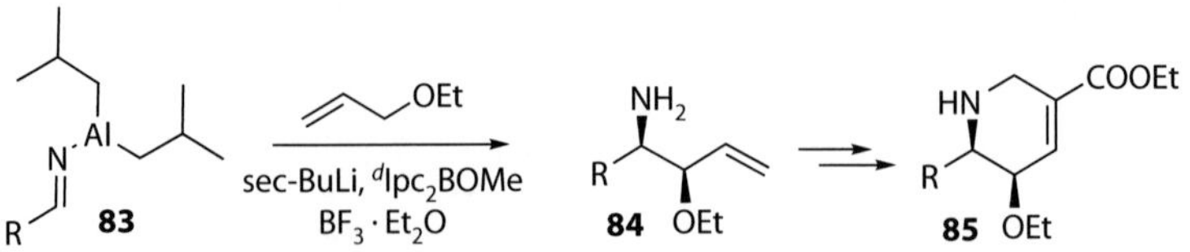

SCHEME 25.11 Alkoxyallylboration of Imines.

FIGURE 25.11 Various (*B*)-γ-alkoxyallyldiisopinocampheylboranes.

to the sensitive functional/protecting groups already present in the molecules, thereby limiting the application of this reagent in the natural product synthesis. Accordingly, extensive research has been carried out toward the modification of protecting groups on the reagent to obtain a variety of alkoxyallylborane reagents **15a–f** (Figure 25.11). Protecting groups such as methoxymethyl (MOM), 2-methoxyethoxymethyl (MEM), tetrahydropyranyl (THP), *p*-methoxyphenyl (PMP), 2-Trimethyl silylethoxymethyl (SEM), and so on have been introduced on the allylborane reagent so as to enhance the removal of the protecting group under relatively milder conditions. The following section describes the applications of these reagents in organic synthesis as well as the development of novel reaction methodologies.

25.8.1 (*B*)-(*Z*)-γ-[(Methoxymethoxy)allyl]diisopinocampheylborane

Burgess et al. developed a novel approach for the synthesis of polyhydroxylated compounds such as castanospermine **87**[29] and swainsonine **89**[30] utilizing alkoxyallylboration of as one of the key steps (Figure 25.12). Similarly, Jadhav and Woerner were able to convert the octose derivative **90** into the polyhydroxyindolizidine **91** in seven further transformations (Figure 25.12).[31] Several of these indolizidines exhibit potent *anti*-cancer, *anti*-viral, and *anti*-AIDS activities.[32]

A diverse set of *C*-trisaccharides **93** obtained from alkoxyallylboration of a disaccharide aldehyde **92** were developed as potential inhibitors for the cell surface proteins of the Helicobactor *pylori* (Scheme 25.12).[33]

An efficient method for the formation of 1,2-oxasilines **95** was developed starting from the homoallylic alcohols **94** via protection of alcohols as vinylsilyl ethers, followed by ring-closing metathesis. Fluoride-assisted deprotection of oxasilines provides the δ-substituted homoallylic

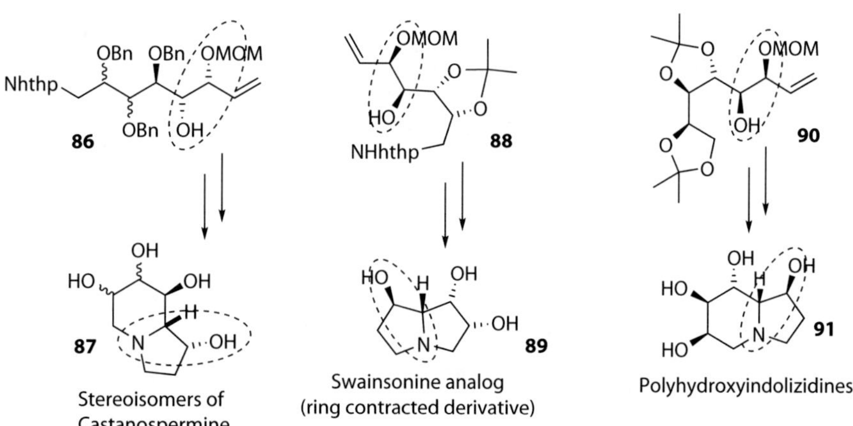

FIGURE 25.12 Synthesis of castanospermine stereoisomers. (Adapted from Burgess, K.; et al. 1992. *J. Org. Chem.* 57: 1103–1109; Burgess, K.; Chaplin, D. A. 1992. *Tetrahedron Lett.* 33: 6077–6080; Jadhav, P. K.; Woerner, F. J. 1994. *Tetrahedron Lett.* 35: 8973–8976.)

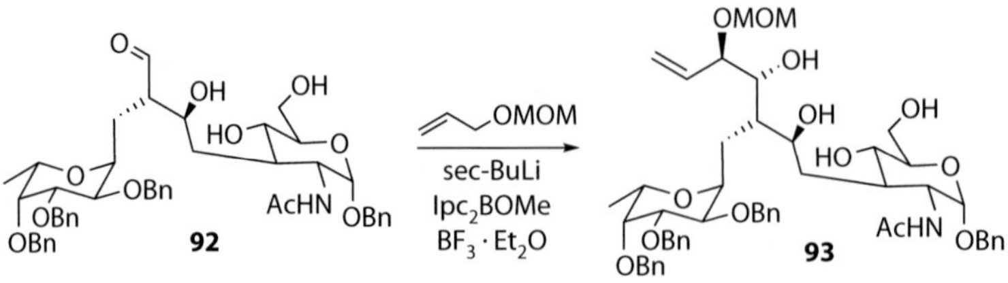

SCHEME 25.12 Synthesis of *C*-trisaccharides.

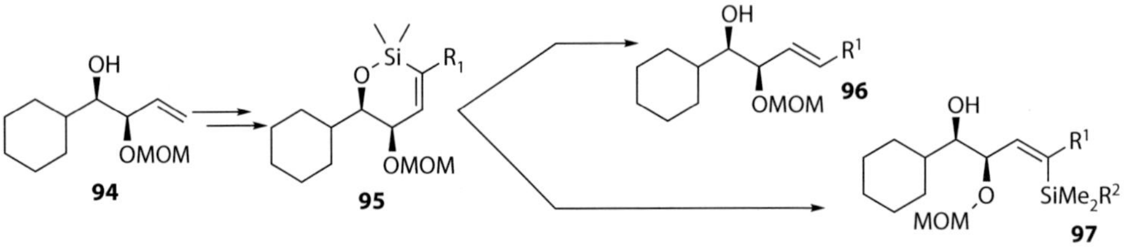

SCHEME 25.13 Synthesis and applications of 1,2-oxasilines.

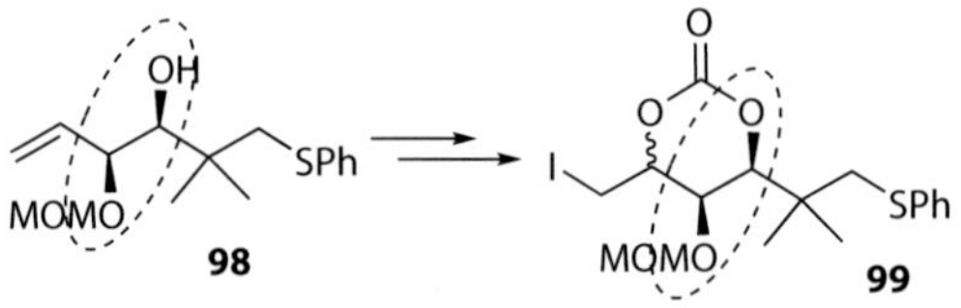

FIGURE 25.13 Iodolactonization of homoallylic alcohols.

alcohols **96**. Similarly, alkyl metals react with oxasilines **95** yielding the alkyl–silyl disubstituted homoallylic alcohols **97** (Scheme 25.13).[34]

Duan and Smith developed a diastereoselective electrophilic cyclization of carbonates derived from homoallylic alcohols **98** via reaction with iodine monobromide to afford α-iodocarbonates **99** (Figure 25.13).[35] These can further be utilized as intermediates for the synthesis of epoxy alcohols, iodohydrins, diols, triols, cyclic carbonates, and so on[36]

In addition, several research groups have employed alkoxyallylboration using reagent **15b** *en route* to their stereoselective syntheses of several biologically important molecules. Some of the prominent molecules made in this route include oxamides **100** (Jadhav and Man),[37] azinomycin **101** (Coleman and Kong)[38], restrictinol **102**[39], and glycosphingolipids **103**[40] (Barrett), calyculin A **104** (Smith et al.),[41] oximidines **105** (Porco),[42] eleuthrobin **106** (Gennari and coworkers),[43] peloruside **107** (DeBrabander and coworkers[44] and Liu and Zhou[45]), and mycolactone **109** (Negishi and coworkers)[46] (Figures 25.14–25.15).

25.8.2 (*B*)-(*Z*)-γ-[(2-Methoxyethoxymethoxy)allyl]diisopinocampheylborane

Nicolaou described the total synthesis of calicheamicinone **111** involving the alkoxyallylboration of a γ-lactol **110** with *B*-γ-methoxyethoxymethoxyallyldiisopinocampheyl-borane **15c** as a key step.[47] We reported the stereoselective synthesis of C_7–C_{21} subunit of potent anticancer agent epothilone A, **113** starting from the homoallylic alcohol **112** derived from the alkoxyallylboration of acetaldehyde.[17a–b,48] Similarly, the extension of alkoxyallylboration to perfluoroaldehydes provided fluoro-substituted homoallylic alcohols **114** that were eventually converted to the fluorinated γ-lactones and δ-lactones[49] **115** as well as a trifluoromethyl analog of blastmycinolactol **116** (Figure 25.16).[50]

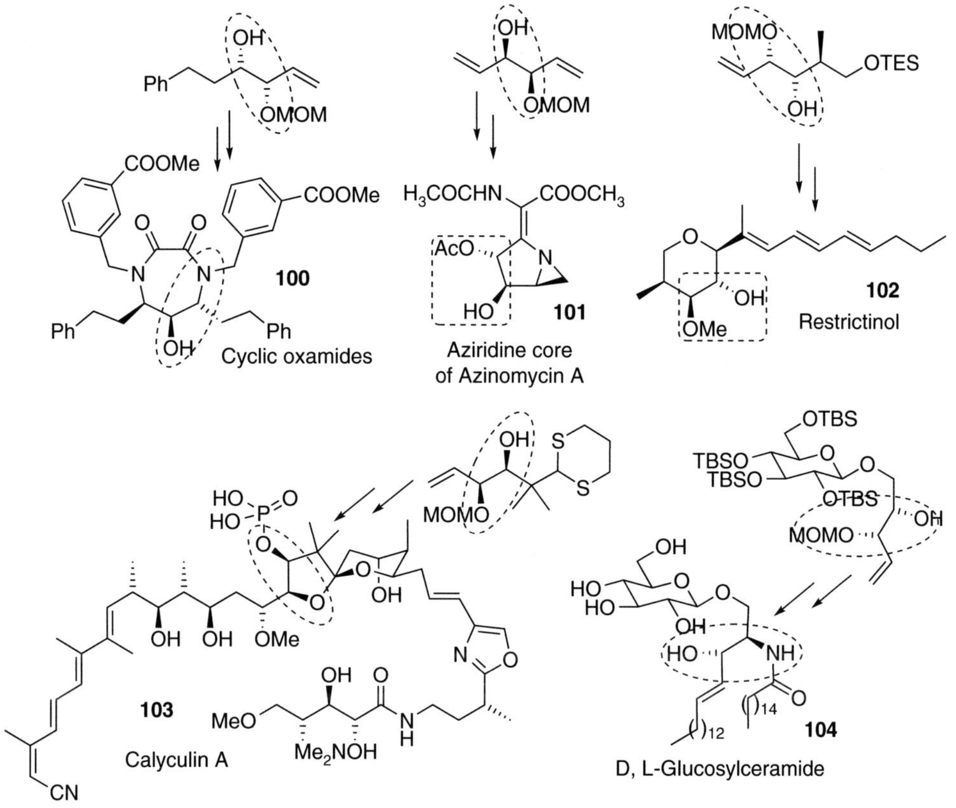

FIGURE 25.14 Applications of MOM-protected alkoxyallylborane 15b.

The utility of alkoxyallylboration was further demonstrated in the synthesis of several antitumor styryllactones such as goniodiol **120**, epigoniodiol **119**, and deoxygoniopypyrone **121**.[51] The key steps in the synthesis of these molecules included allylboration and ring-closing metathesis reactions with homoallylic alcohol ***syn*-118** or its diastereomer ***anti*-118** obtained via Mitsunobu inversion (Scheme 25.14).

The β-alkoxyhomoallylic alcohols were further utilized for the synthesis of β-hydroxy-δ-lactones.[52] The alcohols **122** were converted to α-pyrones **123** via reaction with acryloyl chloride and ring-closing metathesis. Dihydroxylation of alkene in **123** provided the 1,2-*cis* diol as a single diastereomer **124** that underwent regioselective deoxygenation under Barton–McCombie conditions to furnish β-hydroxy-δ-lactone **125** (Scheme 25.15). Stereoselective synthesis of two subunits (C_1–C_8 and C_{15}–C_{21}) of a potent anticancer agent discodermolide were later accomplished utilizing this protocol.[52]

The C_1–C_{11} subunit of epi-fostreicin **128** was synthesized via diastereoselective chelation-controlled nucleophilic addition of nucleophiles on α-alkoxyketones **127** obtained via the oxidation of the α-alkoxyhomoallylic alcohols **126** (Figure 25.17).[53]

Unsaturated acetamides **130** were synthesized via a [3,3]-sigmatropic Overman rearrangement[54] of the homoallylic alcohols **129** obtained from the alkoxyallylboration of unsaturated aldehydes (Figure 25.17).[55]

Curran and coworkers prepared the two *syn* enantiomers of α-alkoxyhomoallylic alcohols via alkoxyallylboration of *n*-tridecanal with both antipodes of *B*-γ-alkoxyallyldiisopino-campheylborane and the corresponding diastereomeric *anti*-homoallylic alcohols **131** were obtained via Mitsunobu inversion.[56] These four alcohols were later tagged with a fluorous PMB-bromide[57] and further transformed in several steps into fifteen diastereomers of murisolin **132** (Figure 25.17).

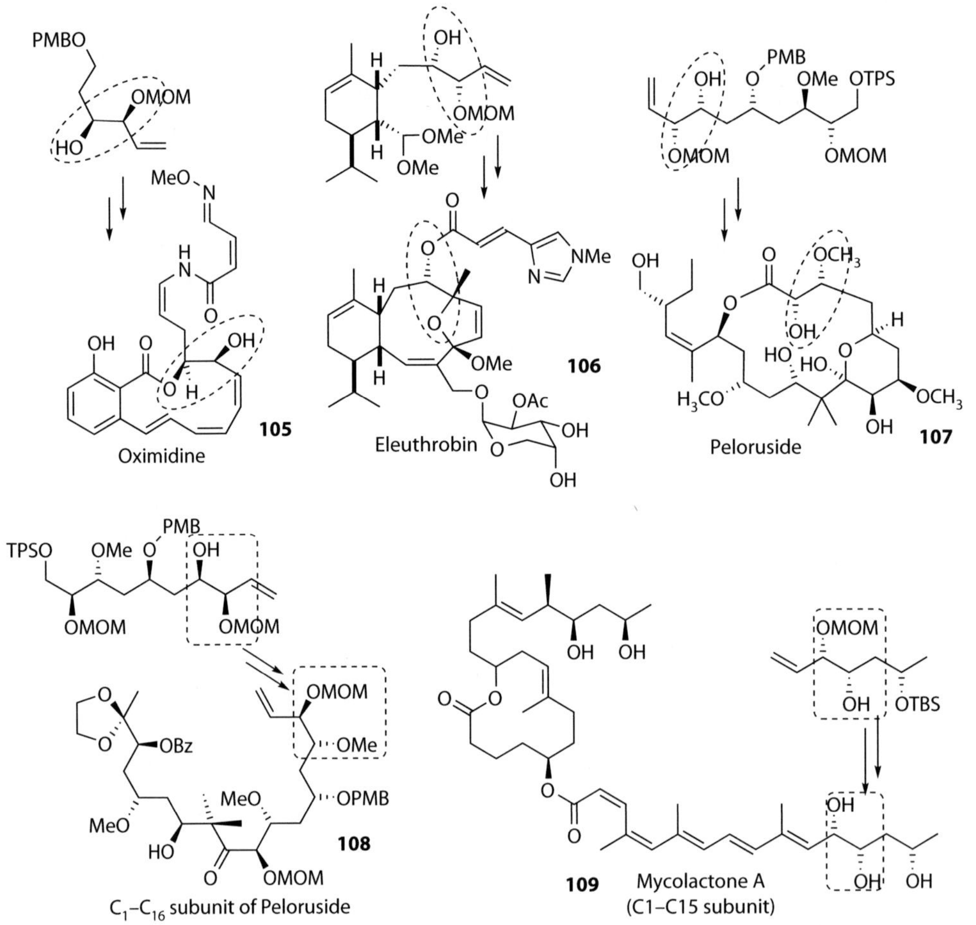

FIGURE 25.15 Applications of MOM-protected alkoxyallylborane 15b.

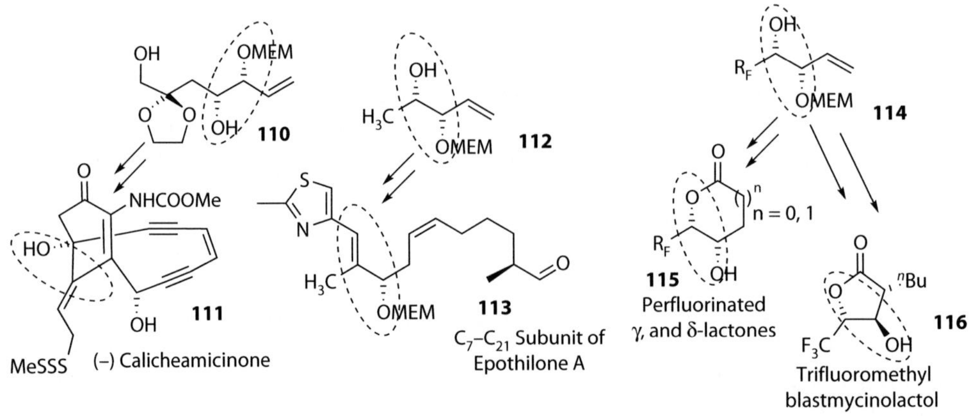

FIGURE 25.16 Applications of MEM-protected alkoxyallylborane 15c.

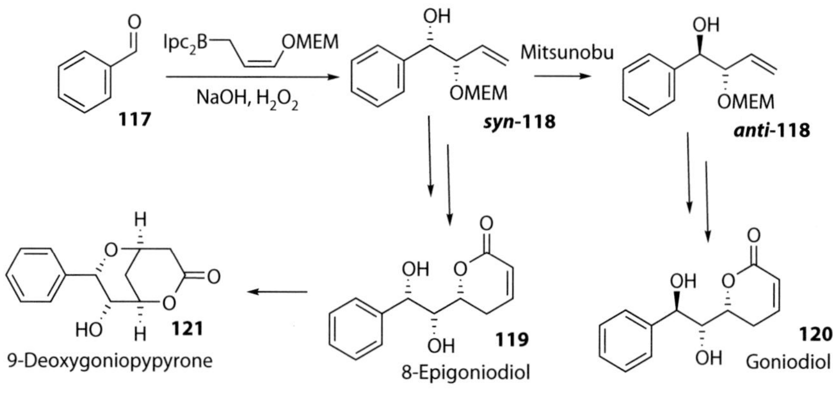

SCHEME 25.14 Stereoselective synthesis of styryllactones.

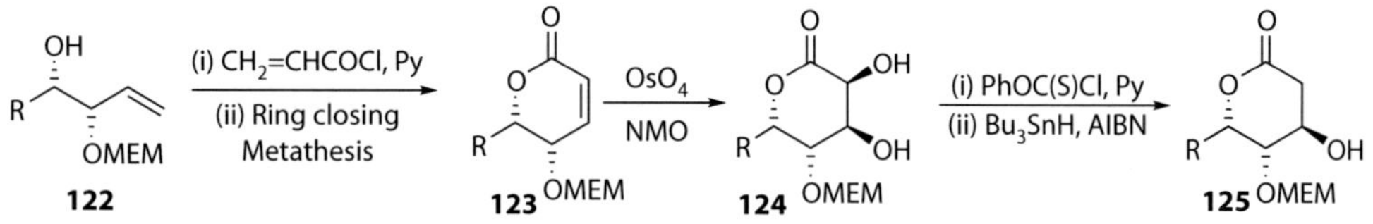

SCHEME 25.15 Diastereoselective dihydroxylation and regioselective deoxygenation.

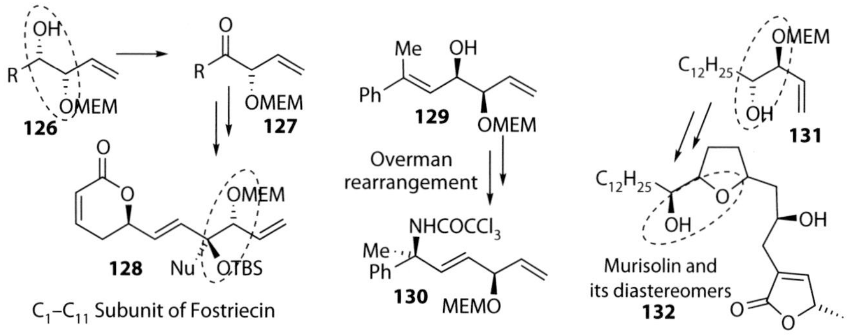

FIGURE 25.17 Applications of MEM-protected alkoxyallylborane 15c.

25.8.3 (*B*)-(*Z*)-γ-[(2-Trimethylsilylethoxymethoxy)allyl]diisopinocampheylborane

Barrett et al. realized that the formation of SEM ether substituted allylborane **15d** via the reaction of corresponding allyl SEM ether with Ipc_2BOMe. As the SEM group is sensitive to the Lewis acid ($BF_3 \cdot Et_2O$), allylboration of aldehydes was carried out without the use of $BF_3.Et_2O$ to provide the homoallylic alcohol. This methodology has been utilized by Barrett (synthesis of calyculin A **134**),[58] Roush (hydroxylamino sugar of calicheamycin **136**),[59] and Overman and coworkers (laurencin **138**)[60] as the main steps in their respective syntheses (Figure 25.18).

25.8.4 Applications of (*B*)-(*Z*)-γ-[(4-Methoxyphenoxy)allyl]diisopinocampheylborane

During our synthesis of epigoniodiol (1,2-*syn* diol) **119** and goniodiol (1,2-*anti*-diol) **120**, the introduction of PMP protecting group on the alkoxyallylborane **15e** was envisioned so as to further carry out a Mitsunobu inversion of the resulting homoallylic alcohol **139** with the *p*-methoxyphenol as nucleophile yielding the diPMP ether **140**.[51] Similarly, Gennari and coworkers utilized the PMP

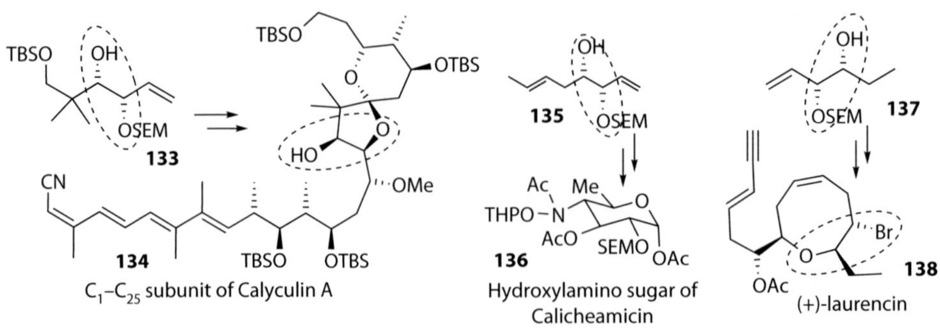

FIGURE 25.18 Applications of SEM-protected alkoxyallylborane 15d.

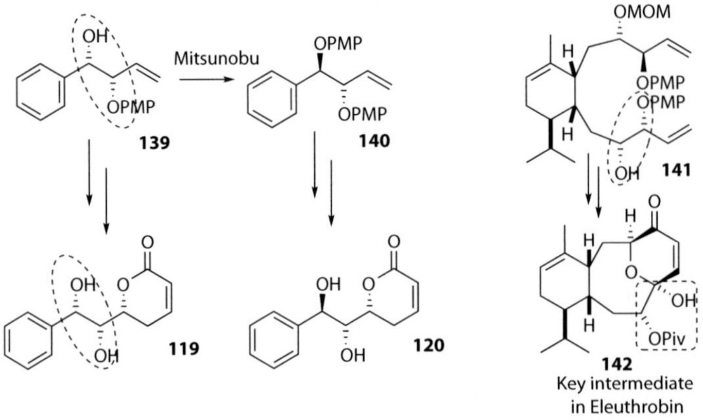

FIGURE 25.19 Applications of PMP-protected alkoxyallylborane 15e.

substituted alkoxyallylborane reagent **15e** *en route* to the synthesis of a key intermediate of eleuthrobin **142** (Figure 25.19).[43c,61]

25.8.5 (*B*)-(*Z*)-γ-[(2-Tetrahydropyranyloxy)allyl]diisopinocampheylborane

Armstrong and Sutherlin demonstrated the use of tetrahydropyranyloxyallyldiisopino campheylborane **15f** with a sugar-based aldehyde to obtain the alcohol **143** that was further utilized in the preparation of fused dihydropyranose oligosachcharides **144** (Figure 25.20).[62]

25.9 *(B)-(E)*-γ-METHOXYALLYLDIISOPINOCAMPHEYLBORANE

(B)-(E)-Methoxyallyldiisopinocampheylborane **16** was synthesized via the reduction of methoxy-3-(phenylthio)propene **145** with potassium naphthalenide under Hoffmann et al. conditions,[63] followed by the treatment of the resulting anion with Ipc_2BOMe.[64] Ganesh and Nicholas demonstrated the use of this reagent for the synthesis of *anti*-β-alkoxyhomoallylic alcohols ***anti*-78** via the alkoxyallylboration of cobalt-complexed acetylenic aldehydes **77** (Scheme 25.16).

25.10 *(B)*-(γ-TRIMETHYLSILYL)PROPARGYLDIISOPINOCAMPHEYLBORANE

Brown et al. synthesized γ-trimethylsilylpropargyldiisopinocampheylborane **148** via the reaction of 1-trimethylsilyl-1-propyne **146** with ^{t}BuLi and treatment with Ipc_2BOMe and $BF_3{\cdot}Et_2O$.[65] On reaction

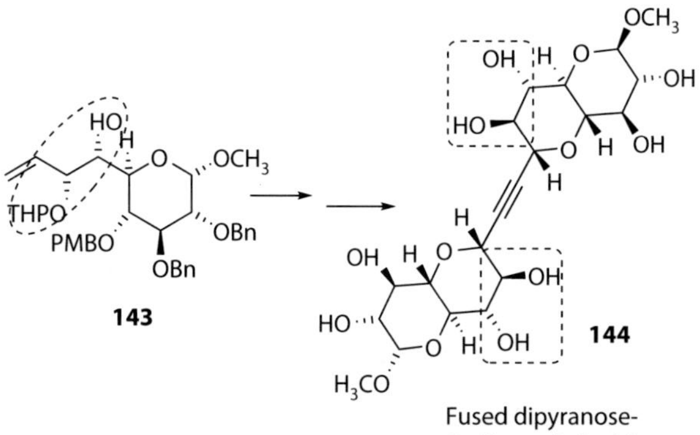

FIGURE 25.20 Synthesis of *C*-oligosachcharides.

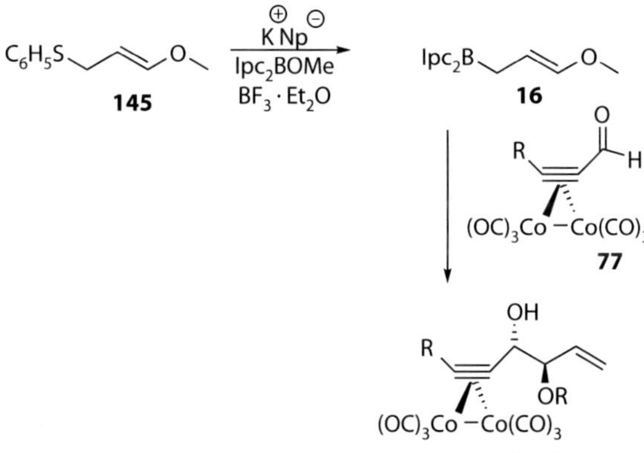

SCHEME 25.16 Synthesis of (*E*)-γ-alkoxyallyldiisopinocampheylborane.

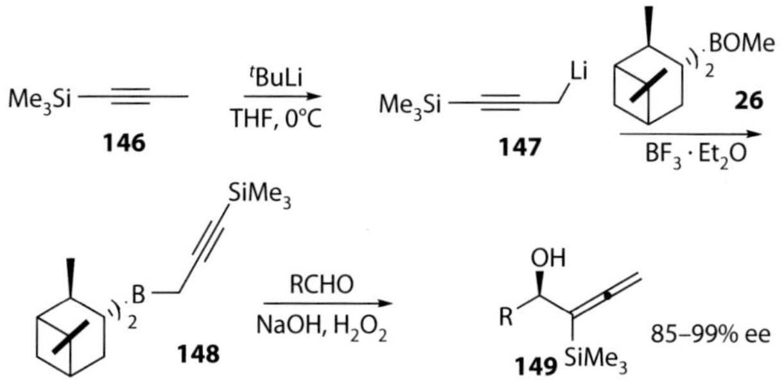

SCHEME 25.17 (*B*)-(γ-trimethylsilyl)propargyldiisopinocampheylborane.

with aldehyde, and subsequent oxidative work up, these boranes provide α-allenic alcohols **149** in good yield and high optical purity (Scheme 25.17).

25.11 (*B*)-(*Z*)γ-CHLOROALLYLDIISOPINOCAMPHEYLBORANE

Boland and Hertweck synthesized the chloroallylborane **17** via the deprotonation of allyl chloride **150** with lithium dicyclohexyl amide **151** and subsequent reaction of the resulting (*Z*)-chloroallylanion **152** with Ipc_2BOMe and $BF_3 \cdot Et_2O$ (Scheme 25.18).[66] Allylboration of aldehydes with **17** provide

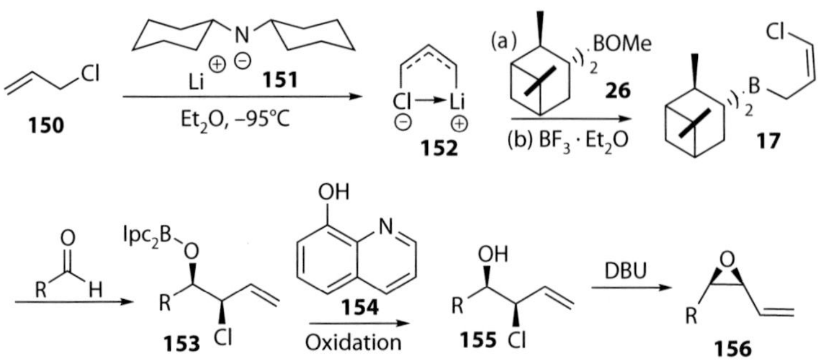

SCHEME 25.18 (*B*)-(*Z*) γ-chloroallyldiisopinocampheylborane.

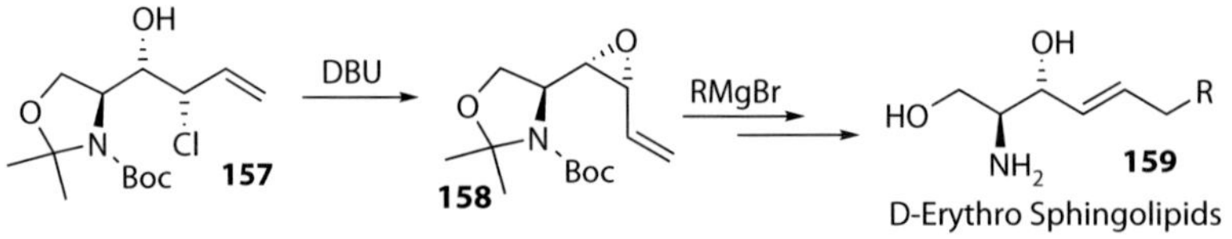

SCHEME 25.19 Preparation of D-erythro sphingolipids.

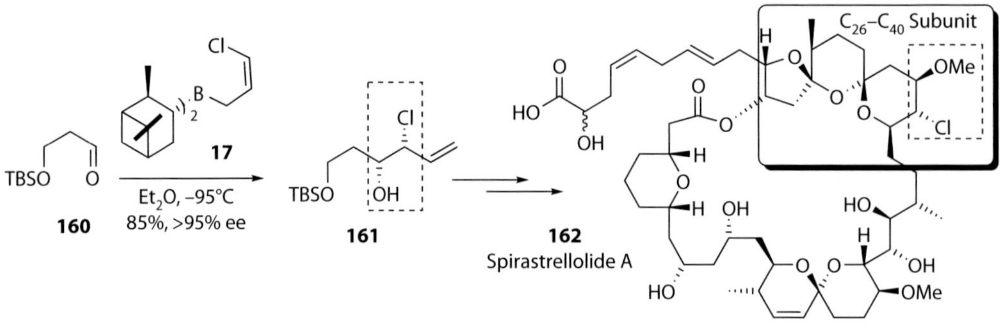

SCHEME 25.20 Preparation of spirastrellolide.

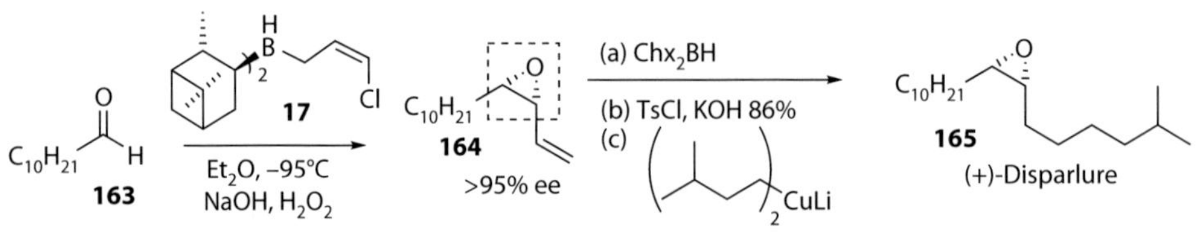

SCHEME 25.21 Preparation of disparlure.

syn-chlorohydrins **155** stereoselectively. As the products halohydrins are base sensitive, alternate nonoxidative workup conditions (e.g., 8-hydroxyquinoline **154**) are required for the oxidation of the intermediate borinates **153**. The product halohydrins could be conveniently converted to the epoxides **156** via treatment with DBU.

Several applications of this reagent are known in the literature toward the synthesis of several natural products/templates such as erthyrosphingolipids **159**[67a] (Hertweck and Boland Scheme 25.19), C_{26}–C_{40} subunit of spirastrellolide **162**[67b–c] (Paterson et al. Scheme 25.20), disparlure **165**[67d] (Oehlschlager, Scheme 25.21), exo-brevicomin **168**[67e] (Jayaraman, Scheme 25.22).

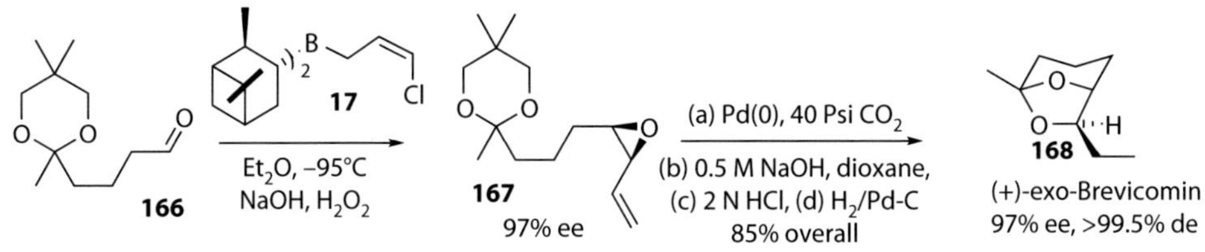

SCHEME 25.22 Preparation of brevicomin.

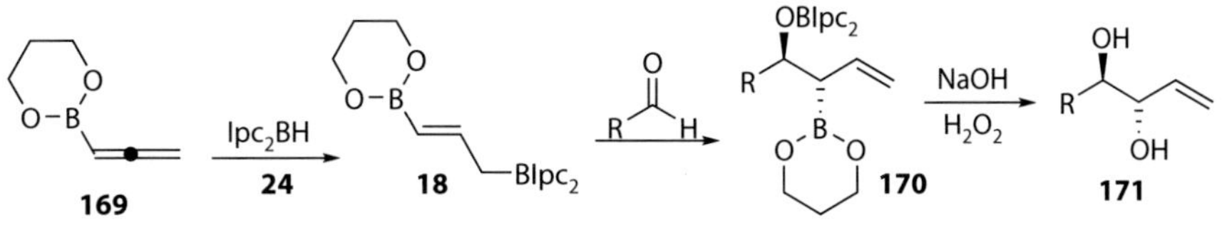

SCHEME 25.23 (*B*)-(*E*)-γ-(1,3,2-dioxaborinanyl)allyldiisopinocampheylborane.

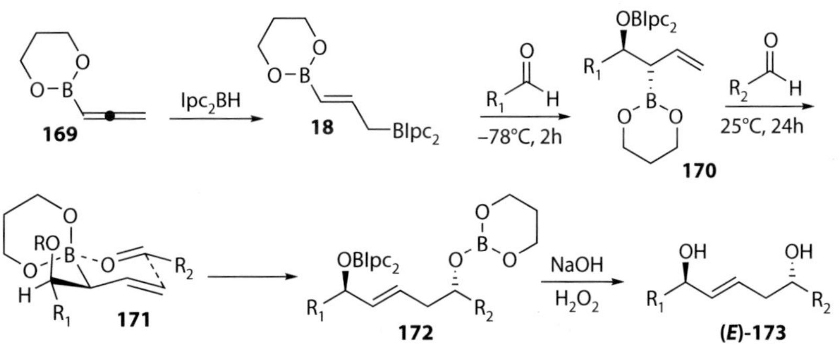

SCHEME 25.24 Preparation of (*E*)-pent-2-ene-1,5-diols.

25.12 (*B*)-(*E*)-γ-(1,3,2-DIOXABORINANYL)ALLYLDIISOPINOCAMPHEYLBORANE

γ-Borolanylallylborane **18** was first introduced by Brown and Narla for the synthesis of *anti*-diols. This reagent serves as a complimentary reagent for (Z)-alkoxyallylboranes **15** that typically generate *syn* diols stereoselectively.[68] This allylborane is prepared by the hydroboration of dioxaborinanylallene **169** with Ipc_2BH **24**. Allylboration of aldehydes with **18** followed by oxidative workup leads to the simultaneous oxidation of borinate and boronate groups in **170** to furnish the *anti*-diols **171** diastereoselectively (Scheme 25.23).

Roush and coworkers expanded the scope of this reagent **18** for the double allylation of carbonyl compounds. As expected, the first allylation (**170**) showed reagent-controlled *anti*-diastereoselection and the second allylation (**172**) proceeded via substrate-controlled stereoselection to produce optically active (*E*)-pent-2-ene-1,5-diols **(*E*)-173** in high yields (Scheme 25.24). They further noted that the increase of bulk on boronate ring led to the formation of the corresponding **(Z)-173** *Z*-diols as single isomers upon oxidative workup (Scheme 25.25).

25.13 (*B*)-(*E*)-γ-(*N*,*N*-DIPHENYLAMINO)ALLYLDIISOPINOCAMPHEYLBORANE

Barrett and Seefeld prepared the aminoallylborane **19** via the reaction of lithiated allyl diphenylamine **180** with Ipc_2BOMe **26**.[69] Allylboration of aldehydes with **19** and oxidative workup provided homoallylic alcohols **181** in *anti*-stereochemistry (Scheme 25.26).

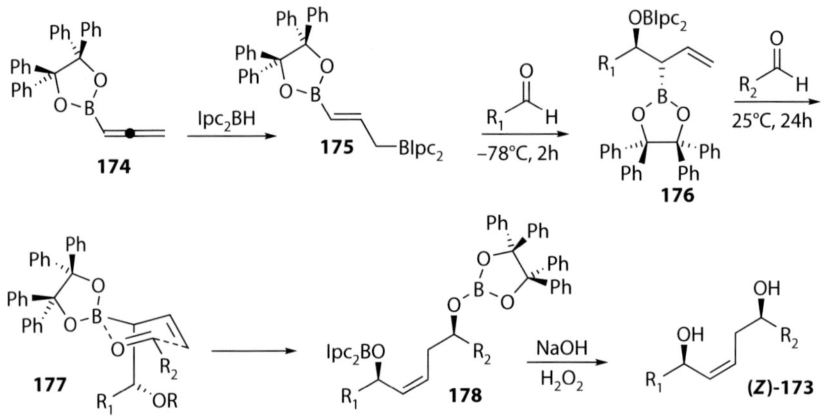

SCHEME 25.25 Preparation of (*Z*)-pent-2-ene-1,5-diols.

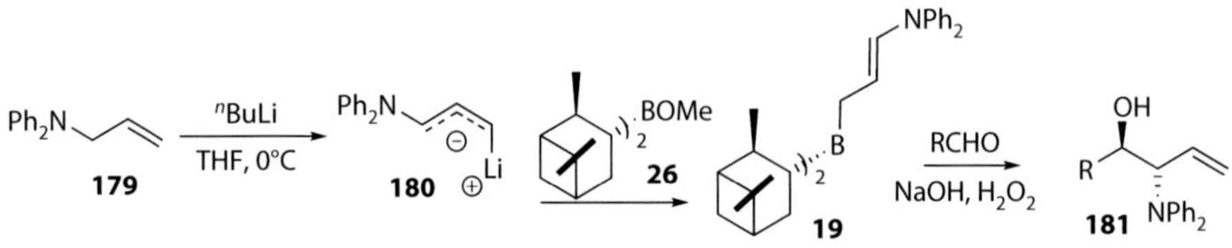

SCHEME 25.26 (*B*)-(*E*)-γ-(*N*,*N*-diphenylamino)allyldiisopinocampheylborane.

25.14 (*B*)-(*E*)-γ-(DIPHENYLMETHYLENEAMINO) ALLYLDIISOPINOCAMPHEYLBORANE

Barrett et. al. introduced the methyleneaminoallylborane **20** for the stereoselective formation of *anti*-amino alcohols **186**.[70] The reagent was obtained via the reaction of lithiated *N*-allylimine **183** with Ipc_2BCl **25**. This reagent on reaction with aldehyde followed by oxidative workup and acid–base manipulations produces the requisite *anti*-amino alcohols **186** in high ee and de (Scheme 25.27).

Further treatment of α-amino homoallylic alcohol **185** with triflic anhydride formed the intermediate aziridine **187** that undergoes regio and stereoselective ring opening to yield the α-hydroxy homoallylic amines **189** (Scheme 25.28).[70] These protocols are immensely useful owing to the versatility in obtaining complimentary homoallylic alcohols and amines.

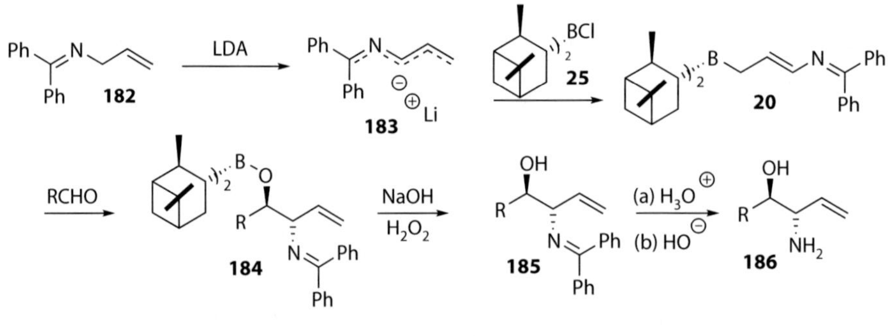

SCHEME 25.27 (*B*)-(*E*)-γ-(diphenylmethyleneamino)allyldiisopinocampheylborane.

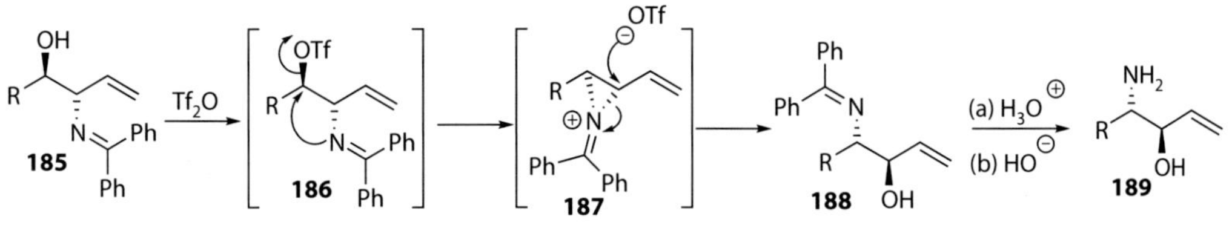

SCHEME 25.28 Conversion of homoallylic alcohols to homoallylic amines.

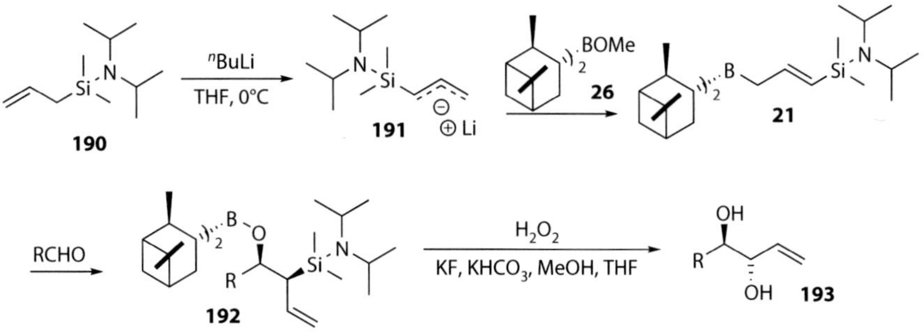

SCHEME 25.29 (*B*)-(*E*)-γ-(*N*,*N*-diisopropylamino)dimethylsilyl)allyldiisopinocampheylborane.

25.15 (*B*)-(*E*)-γ-(*N*,*N*-DIISOPROPYLAMINO)DIMETHYLSILYL) ALLYLDIISOPINOCAMPHEYLBORANE

Allylsilanes **190** were used as the precursors for the synthesis of γ-silylallylborane **21** that on further reaction with aldehydes yields *anti*-α-silyl homoallylic alcohols.[71] Tamao oxidation of the corresponding silyl group provides *anti*-diols **193** stereoselectively. Hence, this procedure is complimentary to alkoxyallylboration that typically provides *syn* diols (Scheme 25.29).

25.16 (*B*)-β-(ALKYLDIMETHYLSILYL)ALLYLDIISOPINOCAMPHEYLBORANE

Lithiated α-methylene-β-stannylsilane **195** on reaction with Ipc_2BCl **25** provides allylborane **22** that upon treatment with aldehydes and oxidative workup to yield α-methylene-γ-hydroxysilanes **196** (Scheme 25.30).[72]

25.17 γ-ALKYL-γ-SILYL DISUBSTITUTED ALLYLBORANES

In addition to the allylboranes reported above, several higher-order allylboranes have been prepared either utilizing hydroboration of allenes or reaction of the corresponding allyl Grginard/lithium reagents with the borinates. Roush utilized the γ-γ-disubstituted allylborane **198** for the synthesis of α-silylcyclopentanol **200a** and α-silylcyclohexanol **200b** via a sequential allylboration and ring-closing metathesis protocol (Scheme 25.31).[73]

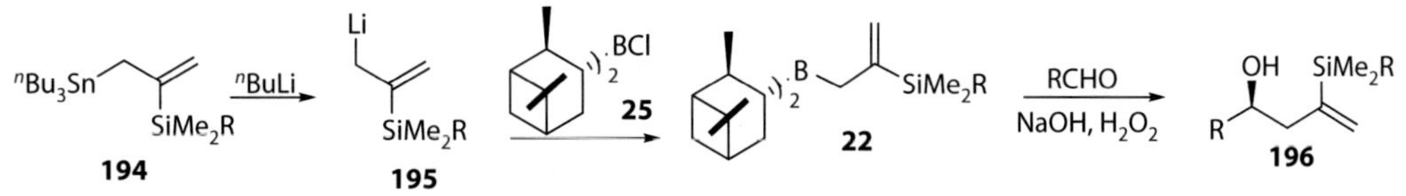

SCHEME 25.30 (*B*)-β-(Alkyldimethylsilyl)allyldiisopinocampheylborane.

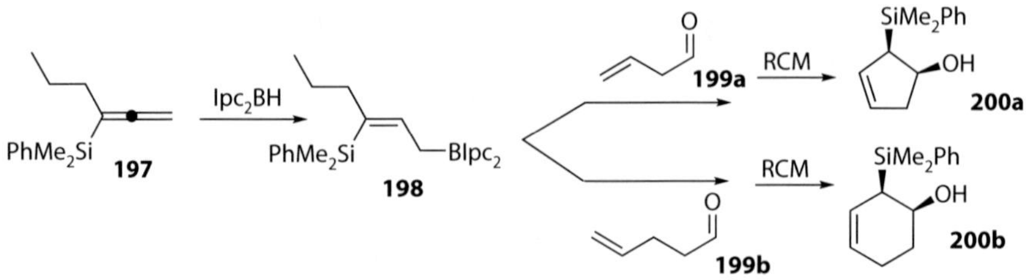

SCHEME 25.31 Preparation of α-silylcycloalkanols.

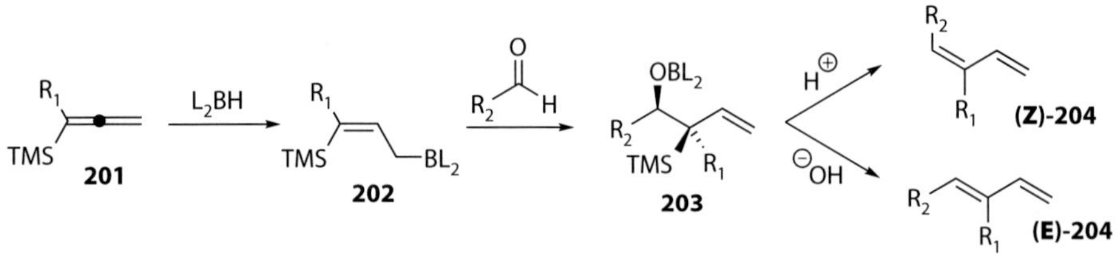

SCHEME 25.32 Stereoselective preparation of (Z) and (E)-1,3-dienes.

Wang and coworkers extended this protocol for the synthesis of (*Z*)- and (*E*) 1,3-dienes **204**. The reaction of 1-alkyl-1-trimethylsilylallene **201** with dialkylborane such as dicyclohexyl-borane or 9-borabicyclo[3.3.1]nonane (9-BBN) yields the γ–γ-disubstituted allylborane **202**. Allylboration of aldehyde with **202** yields the borinate **203** that reacts under acidic or basic conditions to furnish 1,3-dienes **(*Z*)-204** and **(*E*)-204**, respectively (Scheme 25.32).[74]

25.18 OTHER ASYMMETRIC ALLYLBORANES

Several other chiral auxiliaries have been developed for enantioselective allylboration of amines. Apart from α-pinene, the allylboranes **205b–210b** (Figure 25.21) derived from β-pinene **205a**,[75] ethylapopinene **206a**, 2-carene **207a**,[76] 3-carene **208a**,[77] limonene **209a**, and longifolene **210a** have

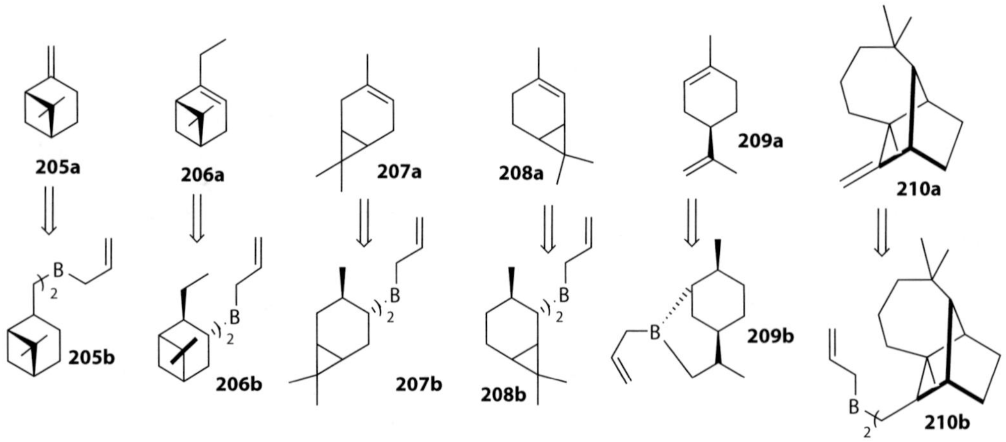

FIGURE 25.21 Asymmetric allylboranes.

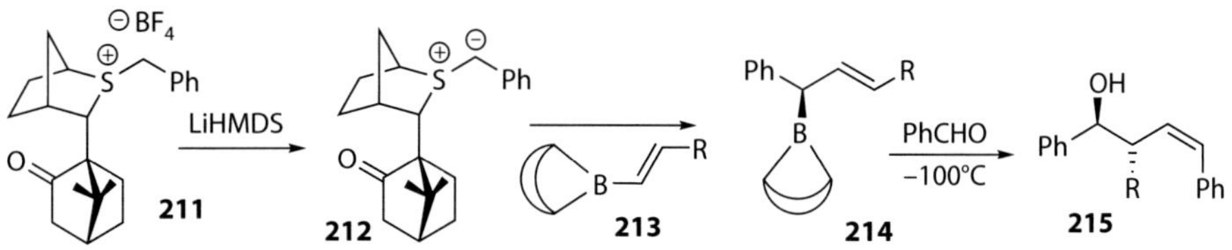

SCHEME 25.33 Asymmetric synthesis of α-substituted allyl boranes.

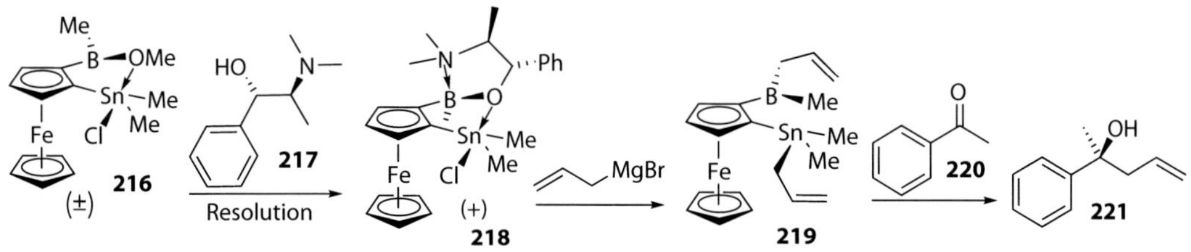

SCHEME 25.34 Chiral ferrocene-derived allylboranes.

been tested with mixed results.[15b] The reagents derived from β-pinene, limonene, and longifolene (**205b**, **209b**, & **210b**) provide low-to-moderate ee's, and the ethylapopinene-based reagent (**206b**) is tedious to prepare as compared with α-pinene. The results obtained from 2-carene- and 3-careen-derived *B*-allyldiisocaranylboranes (**207b** & **208b**) are very closely comparable to those obtained from α-pinene both in terms of stereoselection as well as the reagent control. However, α-pinene is still the auxiliary of choice because of high natural abundance of both antipodes from pine trees.

Aggarwal and Fang utilized the chiral sulfur ylide **212** to impart chirality on the allylborane **214** that underwent smooth allylboration with aldehydes to provide the corresponding homoallylic alcohol **215** (Scheme 25.33).[78]

Recently, Jakle and coworkers reported a chiral ferrocene-based allylborane **219** that has been utilized for the allylboration of ketones to afford the corresponding homoallylic alcohols **221** in moderate optical purity (~40–80%ee, Scheme 25.34).[79]

25.19 MASAMUNE'S ALLYLBORANE[80]

Other prominent allylboranes that provide significant levels of enantio- and diastereoselectivity toward the allylboration of aldehydes include those by Masamune and Soderquist. Masamune's reagent is synthesized in four steps via the hydroboration of trimethylsilylbutadiene **222** followed by methanolysis to obtain the racemic *B*-methoxyborolane **224**. Kinetic resolution with *N*-Methylpseudoephedrine **225** yields the remarkably air-stable crystalline complex **226** that can be separated from the solution at –20°C in almost quantitative yield in >95% de. The solution contains the enantiomerically pure borolane **227**. Either of these borolanes could be reacted with allyl magnesium bromide to provide the corresponding allylboranes **228** (Scheme 25.35).

Allylboration of aldehydes with these reagents provides the corresponding enantiomeric homoallylic alcohols **229** in high ee (Scheme 25.35). Although several other substituents on this borolane (**230a-d**) were tested for allylboration, the 2-trimethylsilyl group proved to be the most effective reagent in providing high ee's for the product homoallylic alcohols (Figure 25.22).

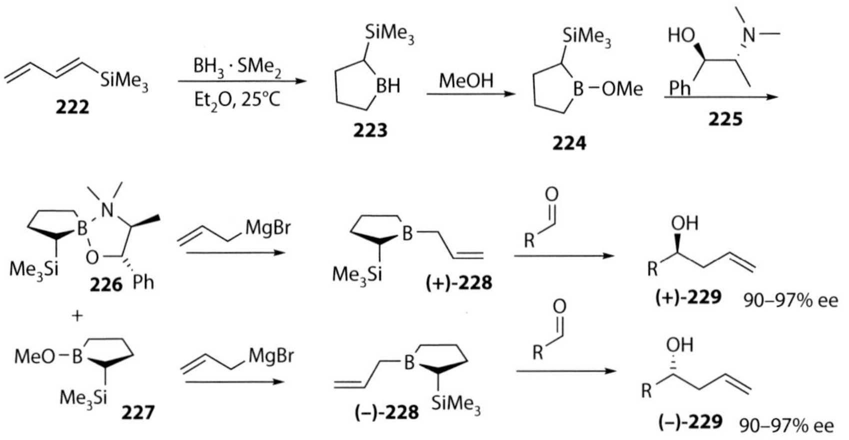

SCHEME 25.35 Masamune's *B*-allylborolanes.

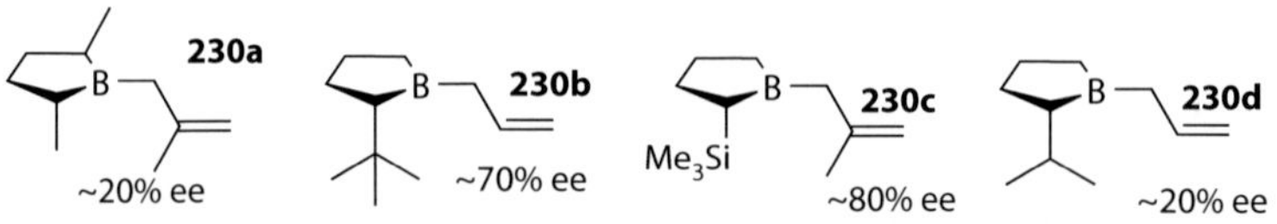

FIGURE 25.22 Masamune's *B*-allylborolanes.

25.20 SODERQUIST'S ALLYLBORANE

The major development of this decade in the area of allylboranes came from Soderquist. The borabicyclodecane (BBD)-based chiral auxiliary was introduced as an effective reagent for the stereo-selective allylboration of aldehydes and ketones.[81] This reagent is prepared very readily and provides excellent levels of stereocontrol for a wide variety of substrates. Although pinene-based reagents provide excellent stereocontrol, the byproduct isopinocampheol poses significant challenges while purifying the product homoallylic alcohols. Soderquist's reagents alleviate these issues because of ease of separation via recrystallization or standard acid base manipulations. The typical procedure for the preparation of this reagent involves the reaction of B-OMe-9-BBN **231** with trimethylsilyldiazomethane to yield the *racemic* 10-trimethylsilylborabicyclo[3.3.2]decane **233**. Thermodynamic resolution with pseudoephedrine **225** provides the crystalline air-stable borane complex **234** that can be readily separated from the solution. The supernatant liquid can further be resolved with the enantiomeric pseudoephedrine to obtain the other antipode of borane–ephedrine complex **234**. Both these reagents [(+)- and **(–)-234]** can be converted to the corresponding allylboranes **235** via treatment with allyl magnesium bromide. Allylboration of aldehydes with **235** provides enantiomeric homoallylic alcohols **236** in high ee (Scheme 25.36).

Allylborane **235** exhibits excellent reagent control and provides high ee's for a wide range of carbonyl compounds regardless of the substrate stereochemistry. For example, in the reaction of α-chiral aldehydes ***S*-237** and ***R*-237** with allylboranes (+)**-235 &** (–)**-235**, the product stereochemistry entirely depends on the antipode of the reagent used and not on the substrate stereochemistry (Scheme 25.37).

SCHEME 25.36 Soderquist allylboranes.

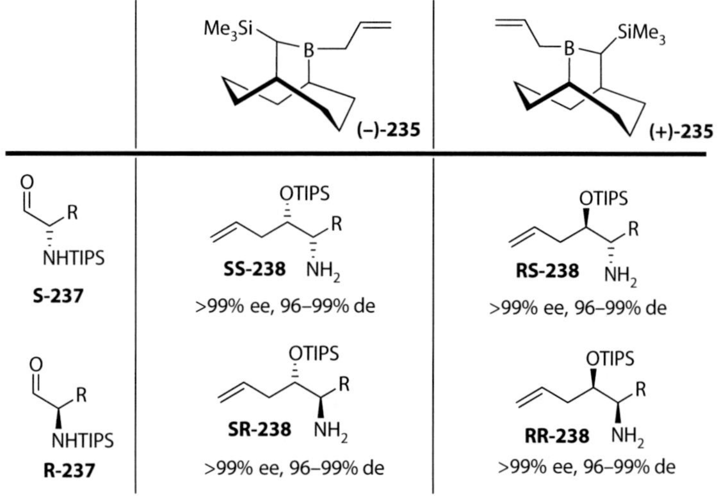

SCHEME 25.37 Allylboration of α-chiral aldehydes.

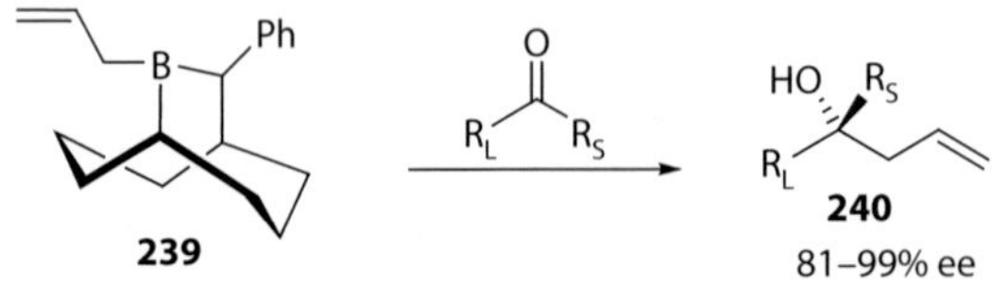

SCHEME 25.38 Allylboration of ketones.

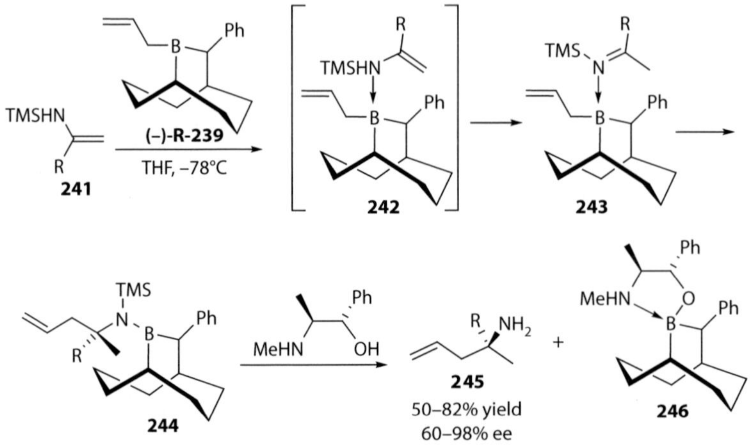

SCHEME 25.39 Allylation of ketimines.

Unlike the pinene-based reagents that fail to provide high ee's for ketones, Soderquist and coworkers succeeded in obtaining high selectivity in allylboration of ketones by replacing the TMS group with a phenyl group on the boracycle **239** (Scheme 25.38). Similarly, the ketimines also provide high levels of enantioselectivity for the corresponding homoallylic amines **246** with B-allyl-10-Ph-BBD **239** (Scheme 25.39). This is undoubtedly a remarkable discovery considering the paucity of reagents available that can carry out enantioselective allylation of ketones.

Typical experimental procedure: $TMSCHN_2$ in hexanes (110 mmol, 2 M) is added to a 1 M solution of *B*-MeO-9-BBN (100 mmol), at room temperature and refluxed for 10 h. The solvent is removed *in vacuo* and the residue distilled to obtain racemic *B*-methoxy-10-trimethylsilyl-9-borabicyclo[3.3.2]decane. 100 mmol of *B*-OMe-10-TMS-BBD is added drop-wise to a solution of pseudoephedrine (50 mmol) in acetonitrile (110 mL) and refluxed for 6 h. Slow cooling to room temperature results in large crystals after which, the supernatant liquid is decanted using a cannula. The crystals are washed with hexanes and dried *in vacuo* to give **234**. Allylmagnesium bromide (10 mmol) is added drop-wise to a solution of **234** (10 mmol) in ether (100 mL) at –78°C and stirred for 1 h at room temperature. The reaction mixture is concentrated *in vacuo* under inert atmosphere and washed with pentane. The washings are filtered through a celite pad and concentrated to obtain the allylborane reagent **235**. Aldehyde (4.1 mmol) is added drop-wise to a solution of (–)-**235** (4 mmol) in ethyl ether (40 mL) at –78°C and after 3 h, the solvent is concentrated *in vacuo* to quantitatively yield the borinate. Pseudoephedrine (4.1 mmol) and acetonitrile (8 mL) are added to the mixture and refluxed for 2 h. The crystals are separated and washed with pentane to yield the chiral homoallylic alcohol after concentration.

25.20.1 Higher-Order Allylborations with Soderquist "Allyl" borane

Further exploration in this area has led to significant advances in the area of higher-order allylboration. The entire range of pinane-based allylborane reagents such as crotylborane, alkoxyallylborane, methallylborane, and so on have now been extended to this robust BBD-based system. The schemes below portray the excellent stereocontrol achieved with the new system. For example, (*E*) and (Z)-B-crotyl-10-TMS-BBD **(*E*)-249 or (*Z*)-249** have been utilized for crotylboration (Scheme 25.40), *B*-γ-alkoxyallyl-10-TMS-BBD reagent **250** provides *syn* α-alkoxyhomoallylic alcohols **252** (Scheme 25.41). Similarly, methallylborane **258** (Scheme 25.42), allenylborane **260** (Scheme 25.43), and

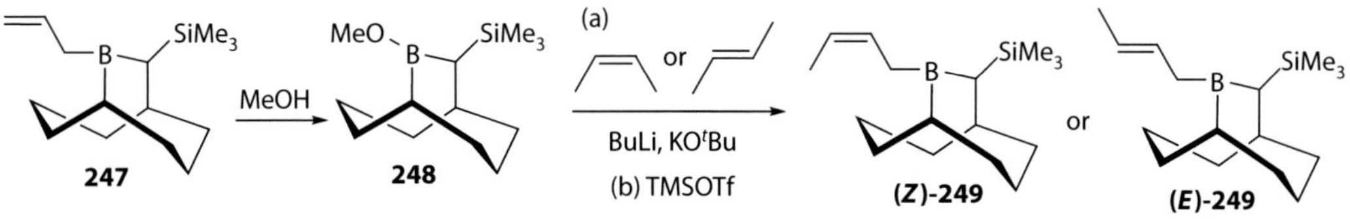

SCHEME 25.40 Crotylboration of aldehydes.

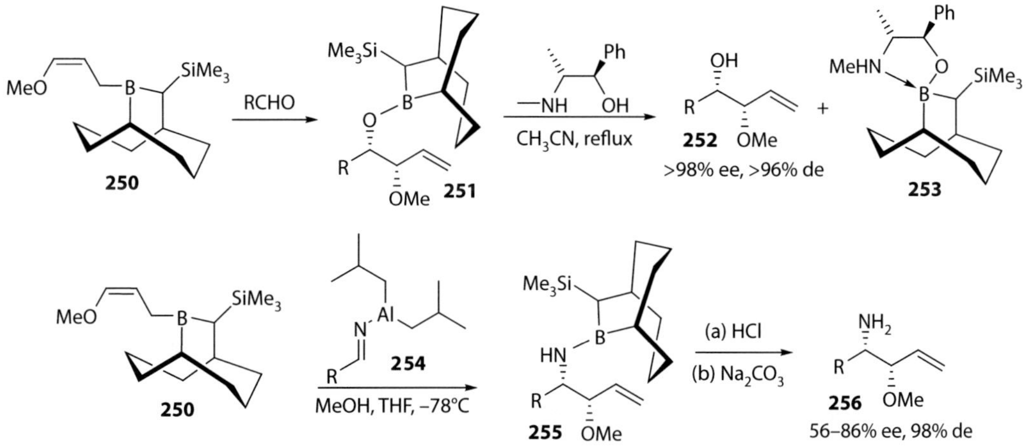

SCHEME 25.41 Alkoxyallylboration of aldehydes and aldimines.

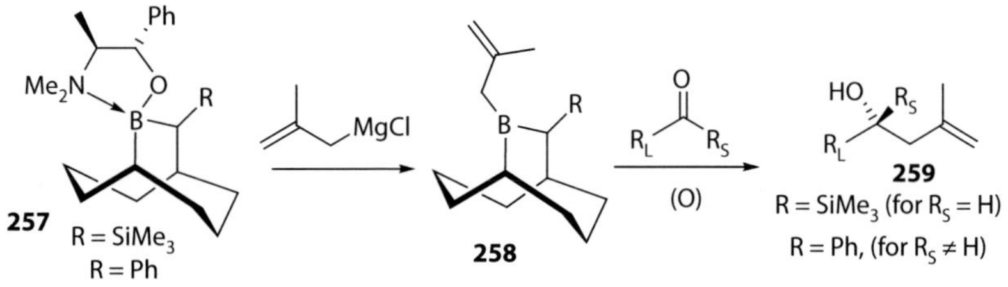

SCHEME 25.42 Methallylboration.

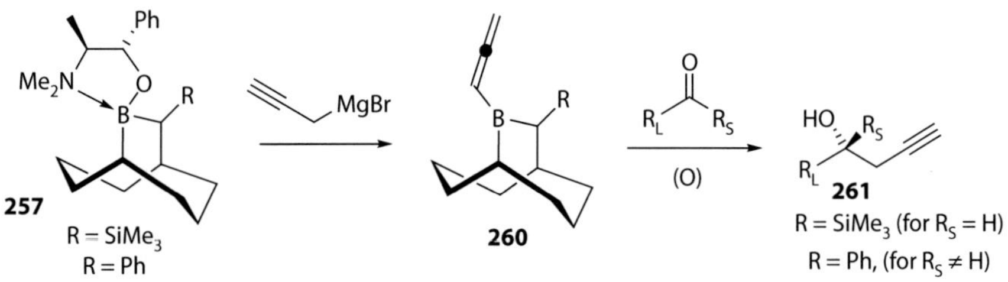

SCHEME 25.43 Allenylboration.

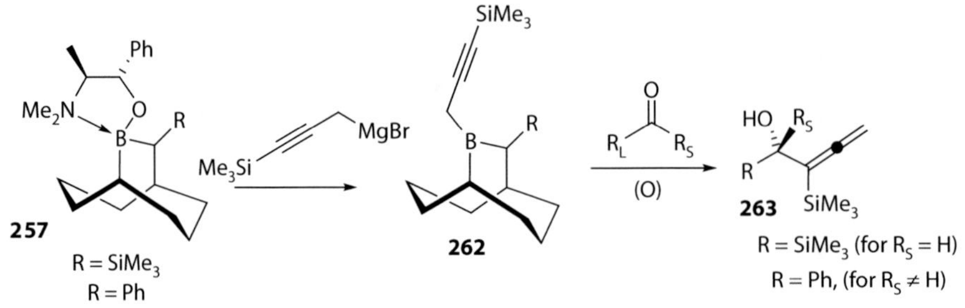

SCHEME 25.44 Propargylboration.

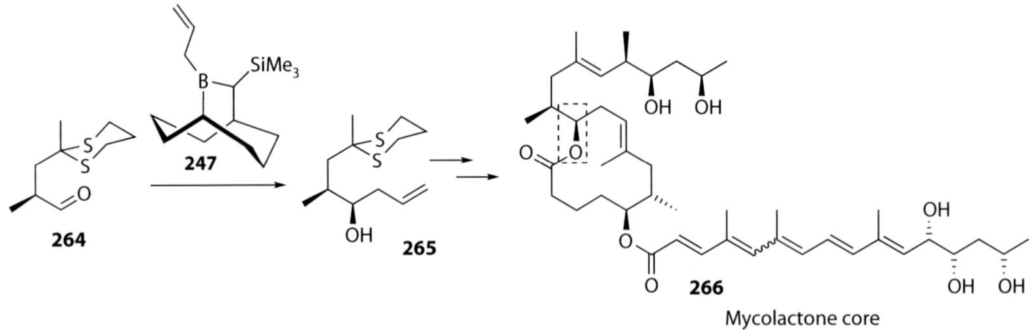

SCHEME 25.45 Application of B-allyl-10-TMS-BBD for the synthesis of mycolactone core.

propargylborane **262** (Scheme 25.44), have also been developed utilizing the borabicyclodecane-based chiral auxiliary.[81]

These new reagents are extremely robust and with their versatility, they are beginning to attract the attention of the synthetic chemists as is seen in the application of this reagent in the synthesis of mycolactone core **266** (Scheme 25.45).

25.21 CONCLUSION

In conclusion, this review describes asymmetric allylation of carbonyl compounds via dialkylallylboranes and their applications in the development of novel methodologies and stereoselective total synthesis of complex natural products. Major emphasis has been laid on α-pinene-based allylboranes as this chiral auxiliary has proven to be very versatile in obtaining very high diastereo- and enantioselectivities for allylboration with wide variety of aldehydes/imines. Some of the other important chiral auxiliaries discussed in this chapter include allylboranes derived from borabicyclo[3.3.2] decane (Soderquist), as well as borolanes (Masamune). We hope that this chapter would prove useful to the synthetic organic and medicinal chemistry community with the wide array of reagents and methodologies that are available based on these reagents.

REFERENCES

1. For reviews on allylation, see: (a) Bubnov, Y. N.; Gurskii, M. E.; Erdyakov, S. Y.; Kizas, O. A.; Kolomnikova, G. D.; Kuznetsov, N. Y.; Potapova, T. V.; Varzatskii, O. A.; Voloshin, Y. Z. 2009. Allylic boranes are chemist's best friends: Reactivity, applications, new opportunities. *J. Organomet. Chem.* 694: 1754–1763. (b) Hoffmann, R. W. 1988. α-Chiral allylboronates: Reagents for asymmetric synthesis. *Pure Appl. Chem.* 60: 123–130. (c) Yamamoto, Y.; Asao, N. 1993. Selective reactions using allylic metals. *Chem. Rev.* 93: 2207–2293. (d) Roush, W. R. In *Methods of Organic Chemistry (Houben-Weyl)*; Georg Thieme: Stuttgart, 1995; Vol. *E 21*, p 1410.

2. For reviews on boron allylations, see: (a) Hall, D. G. 2008. Preparation of allylic boronates and applications in stereoselective catalytic allylborations, *Pure Appl. Chem.* 80: 913–927. (b) Kennedy, J. W. J.; Hall, D. G. 2003. Recent advances in the activation of boron and silicon reagents for stereocontrolled allylation reactions. *Angew. Chem. Int. Ed.* 42: 4732–4739. (c) Brown, H. C.; Ramahcandran, P. V. 1995. Versatile α-pinene-based borane reagents for asymmetric syntheses. *J. Organomet. Chem.* 500: 1–19. (d) Brown, H. C.; Ramahcandran, P. V. 1994. Recent advances in the boron route to asymmetric-synthesis. *Pure Appl. Chem.* 66: 201–212. (e) Roush W. R. Allyl organometallics. In: *Comprehensive Organic Synthesis*. Trost, B. M.; Fleming, I.; Heathcock, C. H. Ed., Vol. 2. Pergamon; Oxford: 1991. pp. 1–53.
3. (a) Herold, T.; Schrott, U.; Hoffmann, R. W. 1981. Stereoselective synthesis of alcohols VI. Asymmetric synthesis of 4-penten-2-ol via allylboronates of chiral glycols. *Chem. Ber.* 114: 359–374. (b) Hoffmann, R. W.; Herold, T. 1981. stereoselective synthesis of alcohols VII. optically active homoallyl alcohols via addition of the chiral boronates to aldehydes. *Chem. Ber.* 114: 375–383.
4. (a) Hoffmann, R. W. 1982. Diastereogenic addition of crotylmetal compounds to aldehydes. *Angew. Chem. Int. Ed.* 21: 555–566. (b) Hoffmann, R. W. 1987. Stereoselective syntheses of building blocks with three consecutive stereogenic Centers: Important precursors of polyketide natural products new synthetic methods. *Angew. Chem. Int. Ed.* 26: 489–503. (c) Hoffmann, R. W.; Neil, G.; Schlapbach, A. 1990. Stereocontrol in allylboration reactions. *Pure Appl. Chem.* 62: 1993–1998.
5. Haruta, R.; Ishiguro, M.; Ikeda, N.; Yamamoto, H. 1982. Chiral allenylboronic esters: A practical reagent for enantioselective carbon-carbon bond formation. *J. Am. Chem. Soc.* 104: 7667–7669.
6. Ikeda, N.; Arai, I.; Yamamoto, H. 1986. Chiral allenylboronic esters as practical reagents for enantioselective carbon-carbon bond formation. Facile synthesis of (−)-ipsenol. *J. Am. Chem. Soc.*108: 483–486.
7. (a) Roush, W. R.; Walts, A. E.; Hoong, L. K. 1985. Diastereo- and enantioselective aldehyde addition reactions of 2-allyl-1,3,2-dioxaborolane-4,5-dicarboxylic esters, a useful class of tartrate ester modified allylboronates. *J. Am. Chem. Soc.* 107: 8186–8190. (b) Roush, W. R.; Halterman, R. L. 1986. Diisopropyl tartrate modified (E)-crotylboronates: Highly enantioselective propionate (E)-enolate equivalents. *J. Am. Chem. Soc.* 108: 294–296.
8. Ditrich, K.; Bube, T.; Strumer, R.; Hoffmann, R. W. 1986. Total synthesis of mycinolide V, the aglycone of a macrolide antibiotic of the mycinamycin series. *Angew. Chem. Int. Ed.* 25: 1028–1030.
9. Reetz, M. T.; Zierke, T. 1988. Highly enantioselective additions of a chirally modified allylboron reagent to aldehydes. *Chem. Ind.* 663–664.
10. Corey, E. J.; Yu, C.-M.; Kim, S. S. 1989. A practical and efficient method for enantioselective allylation of aldehydes. *J. Am. Chem. Soc.* 111: 5495–5496.
11. Brown, H. C.; Jadhav, P. K. 1983. Asymmetric carbon-carbon bond formation via *B*-allyldiisopinocampheylborane. Simple synthesis of secondary homoallylic alcohols with excellent enantiomeric purities. *J. Am. Chem. Soc.* 105: 2092–2093.
12. Owing to the space limitations as well as the extensive utility of this reagent in the literature, only the references from the past three years have been cited here. (a) Liu, J.; De Brabander, J. K. 2009. A concise total synthesis of saliniketal B. *J. Am. Chem. Soc.* 131: 12562–12563. (b) Matsuya, Y.; Yamakawa, Y.-I.; Tohda, C.; Teshigawara, K.; Yamada, M.; Nemoto, H. 2009. Synthesis of sominone and its derivatives based on an RCM strategy: Discovery of a novel anti-alzheimer's disease medicine candidate "Denosomin" *Org. Lett.* 11: 3970–3973. (c) Gollner, A.; Altmann, K.-H.; Gertsch, J.; Mulzer, J. 2009. The laulimalide family: Total synthesis and biological evaluation of neolaulimalide, isolaulimalide, laulimalide and a nonnatural analogue. *Chem. Eur. J.* 15: 5979–5997. (d) Matsuya, Y.; Kobayashi, Y.; Kawaguchi, T.; Hori, A.; Watanabe, Y.; Ishihara, K.; Ahmed, K.; Wei, Z-L.; Yu, D.-Y.; Zhao, Q.-L., Kondo, T.; Nemoto, H. 2009. Design, synthesis, and biological evaluation of artificial macrosphelides in the search for new apoptosis-inducing agents. *Chem. Eur. J.* 15: 5799–5813. (e) Taguchi, A.; Nishiguchi, S.; Regnier, T.; Ozeki, M.; Node, M.; Kiso, Y.; Hayashi, Y. 2009. Efficient total synthesis of (+)-negamycin and its derivatives *Peptide Science.* 45: 375–376. (f) Gopalakrishna, J.; Jonnalagadda, S. C.; Mereddy, V. R. 2008, A concise synthesis of γ-carboxy-γ-lactones. *Org. Chem. Ind. J.* 4: 513–517. (g) Trost, B. M.; Dong, G. 2008. Total synthesis of bryostatin 16 using atom-economical and chemoselective approaches. *Nature.* 456: 485–488. (h) Mereddy, V. R.; Josyula, K. V. B. 2008. Stable borane reagents and methods for their use. U.S. Pat. Appl. Publ. US 2008287708 A1 20081120. (i) Tortosa, M.; Yakelis, N. A.; Roush, W. R. 2008. Total synthesis of (+)-superstolide A. *J. Org. Chem.* 73: 9657–9667. (j) Zhang, X.-S.; Da, S.-J.; Jiao, Y.; Li, H.-Z.; Xie, Z.-X.; Li, Y. 2008. Total synthesis of (-)-(3S,6R)-3,6-dihydroxy-10-methylundecanoic acid and its trimer. *Chin. J. Chem.* 26: 1315–1322. (k) Chakraborty, T. K.; Purkait, S. 2008. Total synthesis of hyptolide. *Tetrahedron Lett.* 49: 5502–5504. (l) Zhan, W.; Jiang, Y.; Brodie, P. J.; Kingston, D. G. I.; Liotta, D. C.; Snyder, J. P. 2008. Design and synthesis of C6–C8 bridged epothilone A. *Org. Lett.* 10: 1565–1568. (m) Perkins, M. V. 2008. Product class 6: Homoallylic alcohols. *Science of*

Synthesis. 36: 667–755. (n) Hanessian, S.; Auzzas, L. 2008. Alternative and expedient asymmetric syntheses of L-(+)-noviose. *Org. Lett.* 10: 261–264. (o) Brimble, M. A.; Bachu, P.; Sperry, J. 2007. Enantioselective synthesis of an analogue of nanaomycin A. *Synthesis*. 2887–2893. (p) Gunasekera, D. S.; Gerold, D. J.; Aalderks, N. S.; Chandra, J. S.; Maanu, C. A.; Kiprof, P.; Zhdankin, V. V.; Reddy, M. V. R. 2007. Practical synthesis and applications of benzoboroxoles. *Tetrahedron*. 63: 9401–9405. (q) Kawai, N.; Mahadeo H. S.; Uenishi, J. 2007. Stereoselective synthesis of (−)-diospongins A and B and their stereoisomers at C–5. *Tetrahedron* 63: 9049–9056. (r) Mamane, V.; Garcia, A. B.; Umarye, J. D.; Lessmann, T.; Sommer, S.; Waldmann, H. 2007. Stereoselective allylation of aldehydes on solid support and its application in biology-oriented synthesis (BIOS) *Tetrahedron*. 63: 5754–5767. (s) Chen, X. S.; Da, S. J.; Yang, L. H.; Xu, B. Y.; Xie, Z. X.; Li, Y. 2007. A new convenient asymmetric approach to herbarumin III. *Chin. Chem. Lett.* 18: 255–257. (t) Umarye, J. D.; Lessmann, T.; Garcia, A. B.; Mamane, V.; Sommer S.; Waldmann, H. 2007. Biology-oriented synthesis of stereochemically diverse natural-product-derived compound collections by iterative allylations on a solid support. *Chem. Eur. J.* 13: 3305–3319.

13. (a) Brown, H. C.; Bhat, K. S. 1986. Enantiomeric *Z*- and *E*-crotyldiisopinocampheyl-boranes. Synthesis in high optical purity of all four possible stereoisomers of .beta.-methylhomoallyl alcohols. *J. Am. Chem. Soc.* 108: 293–294. (b) Brown, H. C.; Bhat, K. S. 1986. Chiral synthesis via organoboranes. 7. Diastereoselective and enantioselective synthesis of erythro- and threo-.beta.-methylhomoallyl alcohols via enantiomeric (*Z*)- and (*E*)-crotylboranes. *J. Am. Chem. Soc.* 108: 5919–5923.
14. Owing to the space limitations as well as the extensive utility of this reagent in the literature, only the references from the past four years have been cited here: (a) Mukherjee, S.; Lee, D. 2009. Application of tandem ring-closing enyne metathesis: Formal total synthesis of (−)-cochleamycin A *Org. Lett.* 11: 2916–2919. (b) Ko, H. Min; Lee, C. W.; Kwon, H. K.; Chung, H. S.; Choi, S. Y.; Chung, Y. K.; Lee, E. 2009. Total synthesis of (−)-amphidinolide K. *Angew. Chem. Int. Ed.* 48: 2364–2366. (c) Kim, H. C.; Kang, S. H. 2009. Total synthesis of azithromycin. *Angew. Chem. Int. Ed.* 48: 1827–1829. (d) Kuranaga, T.; Shirai, T.; Baden, D. G.; Wright, J. L. C.; Satake, M.; Tachibana, K. 2009. Total synthesis and structural confirmation of brevisamide, a new marine cyclic ether alkaloid from the dinoflagellate karenia brevis. *Org. Lett.* 11: 217–220. (e) Li, S.; Liang, S.; Xu, Z.; Ye, T. 2008. Total synthesis of the proposed structure of LL15G256γ. *Synlett* 569–574. (f) Jung, W.-H.; Harrison, C.; Shin, Y.; Fournier, J.-H.; Balachandran, R.; Raccor, B. S.; Sikorski, R. P.; Vogt, A.; Curran, D. P.; Day, B. W. 2007. Total synthesis and biological evaluation of C16 analogs of (−)-dictyostatin. *J. Med. Chem.* 50: 2951–2966. (g) Morita, A.; Kuwahara, S. 2007. Cross-metathesis approach to a (2*E*,4*E*)-dienoic acid intermediate for the synthesis of elaiolide. *Tetrahedron Lett.* 48: 3163–3166. (h) Lawhorn, B. G.; Boga, S. B.; Wolkenberg, S. E.; Boger, D. L. 2006. Total synthesis of cytostatin. *Heterocycles*. 70: 65–70. (i) Paterson, I.; Anderson, E. A.; Dalby, S. M.; Lim, J. Ho; Maltas, P.; Moessner, C. 2006. Synthesis of the DEF-bis-spiroacetal of spirastrellolide A exploiting a double asymmetric dihydroxylation/ spiroacetalization strategy. *Chem. Commun.* 4186–4188. (j) Fuerstner, A.; Aiessa, C.; Chevrier, C.; Teply, F.; Nevado, C.; Tremblay, M. 2006. Studies on iejimalide B: Preparation of the seco acid and identification of the molecule's "Achilles heel". *Angew. Chem. Int. Ed.* 45: 5832–5837. (k) Fuerstner, A.; Fenster, M. D. B.; Fasching, B.; Godbout, C.; Radkowski, K. 2006. Toward the total synthesis of spirastrellolide A. Part 1: Strategic considerations and preparation of the southern domain. *Angew. Chem. Int. Ed.* 45: 5506–5510. (l) Nyavanandi, V. K.; Nanduri, S.; Dev, R. V.; Naidu, A.; Iqbal, J. 2006. Studies toward the total synthesis of tedanolide: Stereoselective synthesis of the C(8) – C(17) segment. *Tetrahedron Lett.* 47: 6667–6672. (m) Mandal, A. K.; Schneekloth, J. S., Jr.; Kuramochi, K.; Crews, C. M. 2006. Synthetic studies on amphidinolide B1. *Org. Lett.* 8: 427–430.
15. (a) Nelson, Scott G.; Cheung, Wing S.; Kassick, Andrew J.; Hilfiker, Mark A. 2002. A de Novo enantioselective total synthesis of (−)-laulimalide *J. Am. Chem. Soc.* 124: 13654–13655. (b) Jadhav, P. K.; Bhat, K. S.; Perumal, P. T.; Brown, H. C. 1986. Chiral synthesis via organoboranes. 6. Asymmetric allylboration via chiral allyldialkylboranes. Synthesis of homoallylic alcohols with exceptionally high enantiomeric excess *J. Org. Chem.* 51: 432–439. (c) Brown, H. C.; Jadhav, P. K.; Perumal, P. T. 1984. Asymmetric methallylboration of prochiral aldehydes with methallyldiisopinocampheyl-borane. Synthesis of 2-methyl-1-alken-4-ols in ≥ 90% enantiomeric purities *Tetrahedron Lett.* 25: 5111–5114.
16. (a) Nelson, Scott G.; Cheung, Wing S.; Kassick, Andrew J.; Hilfiker, Mark A. 2002. A de Novo enantioselective total synthesis of (-)-laulimalide *J. Am. Chem. Soc.* 124: 13654–13655. (b) Paterson, I.; Miller, N. A. 2008. Total synthesis of the marine macrolide (+)-neopeltolide. *Chem. Commun.* 4708–4710. (c) Son, J. B.; Kim, S. N.; Kim, N. Y.; Lee, D. H. 2006. Total synthesis, structural revision, and absolute configuration of (−)-clavosolide A, *Org. Lett.* 8: 661–664. (d) Paterson, I.; Britton, R.; Ashton, K.; Knust,

H.; Stafford, J. 2004. Synthesis of antimicrofilament marine macrolides: Synthesis and configurational assignment of a C_5-C_{16} degradation fragment of reidispongiolide A. *Proc. Natl. Acad. Sci.* 101: 11986–11991.

17. (a) Ramahcandran, P. V.; Prabhudas, B.; Chandra, J. S.; Reddy, M. V. R.; Brown, H. C. 2004. Preparative-scale synthesis of both antipodes of B-γ,γ-dimethylallyldiisopinocampheylborane: application for the synthesis of C1-C6 subunit of epothilone *Tetrahedron Lett.* 45: 1011–1013. (b) Ramahcandran, P. V.; Chandra, J. S.; Prabhudas, B.; Pratihar, D.; Reddy, M. V. R. 2005. Studies towards the synthesis of epothilone A via organoboranes *Org. Biomol. Chem.* 3: 3812–3824. (c) Schinzer, D.; Limberg, A.; Boehm, O. M. 1997. Intermediate products within the total synthesis of Epothilones A and B. Ger., DE 19636343 C119971023. (d) Brown, H. C.; Jadhav, P. K. 1984. (3,3-Dimethylallyl)diisopinocampheylborane: a novel reagent for chiral isoprenylation of aldehydes. Synthesis of (+)- and (−)-artemisia alcohol in exceptionally high enantiomeric purity *Tetrahedron Lett.* 25: 1215–1218.
18. (a) Brown, H. C.; Randad, R. S. 1990. Chiral synthesis via organoboranes. 26. An efficient synthesis of isoprenyl derivatives of borane. Valuable reagents for the isoprenylboration of aldehyde. A convenient route to both enantiomers of ipsenol and ipsdienol in high optical purity. *Tetrahedron* 46: 4463–4472. (b) Brown, H. C.; Randad, R. S. 1990. B-2′-Isoprenyldiisopinocampheylborane: An efficient reagent for the chiral isoprenylation of aldehydes. A convenient route to both enantiomers of ipsenol and ipsdienol *Tetrahedron Lett.* 31: 455–458.
19. Terauchi, T.; Tanaka, T.; Terauchi, T.; Morita, M.; Kimijima, K.; Sato, I.; Shoji, W.; Nakamura, Y.; Tsukada, T.; Tsunoda, T. 2003. Formal total synthesis of altohyrtin C (spongistatin 2). Part 2: Construction of fully elaborated ABCD and EF fragments. *Tetrahedron Lett.* 44: 7747–7751.
20. (a) Evans, D. A.; Andrews, G. C.; Buckwalter, B. 1974. Metalated allylic ethers as homoenolate anion equivalents. *J. Am. Chem. Soc.* 96: 5560–5561; (b) Still, W. C.; Macdonald, T. L. 1976. Allyloxycarbanions. A synthesis of 3,4-dihydroxy-1-olefins from carbonyl compounds. *J. Org. Chem.* 41: 3620–3622.
21. Brown, H. C.; Jadhav, P. K.; Bhat, K. S. 1988. Chiral synthesis via organoboranes. 13. A highly diastereoselective and enantioselective addition of [(*Z*)-γ-alkoxyallyl]diisopino-campheylboranes to aldehydes. *J. Am. Chem. Soc.* 110: 1535–1538.
22. (a) Wuts, P. G. M.; Bigelow, S. S. 1988. Application of allylboronates to the synthesis of carbomycin B. *J. Org. Chem.* 53: 5023–5034. (b) Nicolaou, K. C.; Seitz, S. P.; Pavia, M. R. 1981. Synthesis of 16-membered-ring macrolide antibiotics. 3. Carbomycin B and leucomycin A3: retrosynthetic studies. *J. Am. Chem. Soc.* 103: 1222–1224.
23. (a) Nakata, M.; Osumi, T.; Ueno, A.; Kimura, T.; Tamai, T.; Tatsuta, K. 1991. Total synthesis of herbimycin A. *Tetrahedron Lett.* 32: 6015–6018. (b) Carter, K. D.; Panek, J. S. 2004. Total synthesis of herbimycin A. *Org. Lett.* 6: 55–57. (c) Canova, S.; Bellosta, V.; Bigot, A.; Mailliet, P.; Mignani, S.; Cossy, J. 2007. Total synthesis of herbimycin A. *Org. Lett.* 9: 145–148.
24. Frenzel,T.; Brunjes, M.; Quitschalle,M.; Kirschning, A. 2006. Synthesis of the N-acetylcysteamine thioester of seco-proansamitocin. *Org. Lett.* 8: 135–138.
25. Ganesh, P.; Nichols, K. M. 1993. Reactions of cobalt-complexed acetylenic aldehydes with chiral (γ-alkoxyally1)boranes: Enantioselective synthesis of 3,4-dioxy 1,5-enynes. *J. Org. Chem.* 58: 5587–5588.
26. Kim, J. D.; Zee, O. P.; Jung, Y. H. 2003. Regioselective and diastereoselective allylic amination using chlorosulfonyl isocyanate. A novel asymmetric synthesis of unsaturated aromatic 1,2-amino alcohols. *J. Org. Chem.* 68: 3721–3724.
27. (a) Kim, J. D.; Kim, I. S.; Hua, J. C.; Zee, O. P.; Jung, Y. H. 2005. Diastereoselective synthesis of unsaturated 1,2-amino alcohols from α-hydroxy allyl ethers using chlorosulfonyl isocyanate. *Tetrahedron Lett.* 46: 1079–1082. (b) Kim, I. S.; Kim, J. D.; Ryu, C. B.; Zee, O. P.; Jung, Y. H. 2006. A concise synthesis of (L)-cytoxazone and (L)-4-epi-cytoxazone using chlorosulfonyl isocyanate. *Tetrahedron* 62: 9349–9358.
28. Ramahcandran, P. V.; Burghardt, T. V.; Berry, L. B. 2005. Chiral synthesis of functionalized tetrahydropyridines: γ-Aminobutyric acid uptake inhibitor analogues. *J. Org. Chem.* 70: 7911–7918.
29. (a) Burgess, K.; Chaplin, D. A.; Henderson, I.; Pan, Y. T.; Elbein, A. D. 1992. A Route to several stereoisomers of castanospermine. *J. Org. Chem.* 57: 1103–1109. (b) Burgess, K.; Chaplin, D. A. 1992. An asymmetric synthesis of D-1,6-diepicastanospermine. *Tetrahedron Lett.* 33: 6077–6080.
30. Burgess, K. Henderson, I. 1990. A new approach to swainsonine and castanospermine analogues. *Tetrahedron Lett.* 31: 6949–6952.
31. Jadhav, P. K.; Woerner, F. J. 1994. Potentially general synthesis of polyhydroxy-indolizidines. *Tetrahedron Lett.* 35: 8973–8976.

32. (a) Gruters, R. A.; Neefjes, J. J.; Tersmette, M.; de Coede, R. E. Y.; Tulp. A.; Huisman, H. G.; Miedema, F.; Ploegh, H. L. 1987. Interference with HIV-induced syncytium formation and viral infectivity by inhibitors of trimming glucosidase. *Nature* 330, 74–77. (g) Walker, B. D.; Kawalski, M.; Goh, W. C.; Kozarsky, K.; Krieger, M.; Rosen, C.; Rohrschneider, L.; Haseltine, W. A.; Sodroski, J. 1987. Inhibition of human immunodeficiency virus syncytium formation and virus replication by castanospermine. *Proc. Nat. Acad. Sci. U.S.A.* 84: 8120–8124.
33. Sutherlin, D. P.; Armstrong, R. W. 1997. Synthesis of 12 stereochemically and structurally diverse C-trisaccharides. *J. Org. Chem.* 62: 5267–5283.
34. Ahmed, M.; Barrett, A. G. M.; Beall, J. C.; Braddock, D. C.; Flack, K.; Gibson, V. C.; Procopiou, P. A.; Salter, M. M. 1999. A tripartite asymmetric allylboration – Silicon tethered alkene ring closing metathesis—*in situ* ring opening protocol for the regiospecific generation of functionalized (E)-disubstituted homoallylic alcohols. *Tetrahedron* 55: 3219–3232.
35. Duan, J. J. W.; Smith, A. B. 1993. Iodine monobromide (IBr) at low temperature: Enhanced diastereoselectivity in electrophilic cyclizations of homoallylic carbonates. *J. Org. Chem.* 58: 3703–3711.
36. (a) Bartlett, P. A.; Meadow, J. D.; Brown, E. G.; Morimoto, A.; Jernstedt, K. K. 1982. Carbonate extension. A versatile procedure for functionalization of acyclic homoallylic alcohols with moderate stereocontrol. *J. Org. Chem.* 47: 4013–4018. (b) Cardillo, G.; Orena, M.; Porei, G.; Sandri, S. 1981. A new regio- and stereo-selective functionalization of allylic and homoallylic alcohols. *J. Chem. Soc. Chem. Commun.* 465–466. (c) Bongini, A.; Cardillo, G.; Orena, M.; Porzi, G.; Sandri, S. 1982. Regio- and stereocontrolled synthesis of epoxy alcohols and triols from allylic and homoallylic alcohols via iodocarbonates. *J. Org. Chem.* 47: 4626–4633.
37. Jadhav, P. K.; Man, H. W. 1996. Synthesis of 7-membered cyclic oxamides: Novel HIV-1 Protease Inhibitors. *Tetrahedron Lett.* 37: 1153–1156.
38. (a) Coleman, R. S.; Kong, J. S. 1998. Stereocontrolled synthesis of the fully elaborated aziridine core of the azinomycins. *J. Am. Chem. Soc.* 120: 3538–3539. (b) Coleman, R. S.; Kong, J. S.; Richardson, T. E. 1999. Synthesis of naturally occurring antitumor agents: Stereocontrolled synthesis of the azabicyclic ring system of the azinomycins. *J. Am. Chem. Soc.* 121: 9088–9095. (c) Coleman, R. S.; Li, J.; Navarro, A. 2001. Total synthesis of azinomycin . *Angew. Chem. Int. Ed.* 40: 1736–1739.
39. Barrett, A. G. M.; Bennett, A. J.; Menzer, S.; Smith, M. L.; White, A. J. P.; Williams, D. J. 1999. Applications of crotonyldiisopinocampheylboranes in synthesis: Total synthesis of restrictinol. *J. Org. Chem.* 64: 162–171.
40. Barrett, A. G. M.; Beall, J. C.; Braddock, D. C.; Flack, K.; Gibson, V. C.; Salter, M. M. 2000. Asymmetric allylboration and ring closing alkene metathesis: A novel strategy for the synthesis of glycosphingolipids. *J. Org. Chem.* 65: 6508–6514.
41. Smith, A. B.; Friestad, G. K.; Barbosa, J.; Bertounesque, E.; Hull, K. G.; Iwashima, M.; Qiu, Y.; Salvatore, B. A.; Spoors, P. G.; Duan, J. J. W. 1999. Total synthesis of (+)-calyculin A and (−)-calyculin B: Asymmetric synthesis of the C(9-25) spiroketal dipropionate subunit. *J. Am. Chem. Soc.* 121: 10468–10477.
42. Wang, X.; Porco, J. A. 2003. Total synthesis of the salicylate enamide macrolide oximidine II. *J. Am. Chem. Soc.* 125: 6040–6041.
43. (a) Beumer, R.; Bayon, P.; Bugada, P.; Ducki, S.; Mongelli,N.; Sirtori, F. R.; Telser, J.; Gennari, C. 2003. Synthesis of novel simplified eleutheside analogues with potent microtubule-stabilizing activity, using ring-closing metathesis as the key-step. *Tetrahedron Lett.* 44: 681–684. (b) Beumer, R.; Bayon, P.; Bugada, P.; Ducki, S.; Mongelli,N.; Sirtori, F. R.; Telser, J.; Gennari, C. 2003. Synthesis of novel simplified sarcodictyin/eleutherobin analogs with potent microtubule-stabilizing activity, using ring closing metathesis as the key-step. *Tetrahedron.* 59: 8803–8820. (c) Castoldi, D.; Caggiano, L.; Bayon, P.; Costa, A. M.; Cappella, P.; Sharon, O.; Gennari, C. 2005. Synthesis of novel, simplified, C-7 substituted eleutheside analogues with potent microtubule-stabilizing activity. *Tetrahedron. 61*, 2123–2139. (d) Castoldi, D.; Caggiano, L.; Panigada, L.; Sharon, O.; Costa, A. M.; Gennari, C. 2006. A formal total synthesis of eleutherobin using the ring-closing metathesis (RCM) reaction of a densely functionalized diene as the key step: Investigation of the unusual kinetically controlled RCM stereochemistry. *Chem. Eur. J.* 12: 51–62.
44. Liao, X.; Wu, Y.; deBrabander, J. K. 2003. Total synthesis and absolute configuration of the novel microtubule-stabilizing agent peloruside A. *Angew. Chem. Int. Ed.* 42: 1648–1652.
45. Liu, B.; Zhou, W. S. 2004. Toward the total synthesis of natural peloruside A: Stereoselective synthesis of the backbone of the core. *Org. Lett. 6*: 71–74.
46. Yin, N.; Wang, G.; Qian, M.; Negishi, E. I. 2006. Stereoselective synthesis of the side chains of mycolactones A and B featuring stepwise double substitutions of 1,1-dibromo-1-alkenes. *Angew. Chem. Int. Ed.* 45: 2916–2920.

47. (a) Smith, A. L.; Hwang, C. K.; Pitsinos, E. N.; Scarlato, G. R.; Nicolaou, K. C. 1992. Enantioselective total synthesis of (–)-calicheamicinone. *J. Am. Chem. Soc.* 114: 3134–3136. (b) Smith, A. L.; Pitsinos, E. N.; Hwang, C. K.; Mizuno, Y.; Saimoto, H.; Scarlato, G. R.; Suzuki, T.; Nicolaou, K. C. 1993. Total synthesis of calicheamicin γ_1^{I}. 2. development of an enantioselective route to (–)-calicheamicinone. *J. Am. Chem. Soc.* 115: 7612–7614.
48. Ramahcandran, P. V.; Prabhudas, B.; Pratihar, D.; Chandra, J. S.; Reddy, M. V. R. 2003. Stereoselective synthesis of the C7–C21 segment of epothilone A via asymmetric alkoxyallyl- and crotylboration. *Tetrahedron Lett.* 44: 3745–3748.
49. (a) Ramahcandran, P. V.; Padiya, K. J.; Rauniyar, V.; Reddy, M. V. R.; Brown, H. C. 2004. Asymmetric synthesis of 6-(2′,3′,4′,5′,6′-pentafluorophenyl)-δ-lactones via ''allyl''boranes: application for the synthesis of fluorinated analog of key pharmacophore of statin drugs. *J. Fluor. Chem.* 125: 615–620. (b) Ramahcandran, P. V.; Padiya, K. J.; Rauniyar, V.; Reddy, M. V. R.; Brown, H. C. 2004. Asymmetric synthesis of γ-perfluoroalkyl(aryl) butyrolactones via organoboranes. *Tetrahedron Lett.* 45: 1015–1017.
50. Ramahcandran, P. V.; Padiya, K. J.; Reddy, M. V. R.; Brown, H. C. 2004. An efficient enantioselective total synthesis of a trifluoromethyl analog of blastmycinolactol. *J. Fluor. Chem.* 125: 579–583.
51. Ramahcandran, P. V.; Chandra, J. S.; Reddy, M. V. R. 2002. Stereoselective syntheses of (+)-goniodiol, (–)-8-epigoniodiol, and (+)-9-deoxygoniopypyrone via alkoxyallylboration and ring-closing metathesis. *J. Org. Chem.* 67: 7547–7550.
52. Ramahcandran, P. V.; Prabhudas, B.; Chandra, J. S.; Reddy, M. V. R. 2004. Diastereoselective dihydroxylation and regioselective deoxygenation of dihydropyranones: A novel protocol for the stereoselective synthesis of C1 – C8 and C15 – C21 subunits of (+)-discodermolide. *J. Org. Chem.* 69: 6294–6304.
53. Ramahcandran, P. V.; Liu, H.; Reddy, M. V. R.; Brown, H. C. 2003. Synthesis of homoallylic chiral tertiary alcohols via chelation-controlled diastereoselective nucleophilic addition on α-alkoxyketones: Application for the synthesis of the C1-C11 subunit of 8-epi-fostriecin. *Org. Lett.* 5: 3755–3757.
54. (a) Overman, L. E. 1974. Thermal and mercuric ion catalyzed [3,3]-sigmatropic rearrangement of allylic trichloroacetimidates. 1,3 Transposition of alcohol and amine functions. *J. Am. Chem. Soc.* 96: 597–599. (b) Review: Overman, L. E. 1980. Allylic and propargylic imidic esters in organic synthesis. *Acc. Chem. Res.* 13: 218–224.
55. Ramahcandran, P. V.; Burghardt, T. E.; Berry, L. B. 2005, Asymmetric allylboration of α-β-enals as a surrogate for the enantioselective synthesis of allylic amines and γ-amino acids. *J. Org. Chem.* 70: 2329–2331.
56. (a) Zhang, Q.; Lu, H.; Richard, C.; Curran, D. P. 2004. Fluorous mixture synthesis of stereoisomer libraries: Total syntheses of (+)-murisolin and fifteen diastereoisomers. *J. Am. Chem. Soc.* 126: 36–37. (b) Curran, D. P.; Zhang, Q.; Richard, C.; Lu, H.; Gudipati, V.; Wilcox, C. 2006. Total synthesis of a 28-member stereoisomer library of murisolins. *J. Am. Chem. Soc.* 128: 9561–9573.
57. Curran, D. P.; Furukawa, T. 2002. Simultaneous preparation of four truncated analogues of discodermolide by fluorous mixture synthesis. *Org. Lett.* 4: 2233–2235.
58. (a) Barrett, A. G. M.; Edmunds, J. J.; Horita, K.; Parkinson, C. J.; 1992. Stereocontrolled synthesis of calyculin A: Construction of the C(15)–C(25) spiroketal unit. *J. Chem. Soc. Chem.Commun.* 1236–1238. (b) Anderson, O. P.; Barrett, A. G. M.; Edmunds, J. J.; Hachiya, S. I.; Hendrix, J. A.; Horita, K.; Malecha, J. W.; Parkinson, C. J.; VanSickle, A. 2001. Applications of crotyldiisopinocampheylboranes in synthesis: A formal total synthesis of (+)-calyculin A. *Can. J. Chem.* 79: 1562–1592.
59. Roush, W. R.; Follows, B. C. 1994. Asymmetric synthesis of the hydraxylamino sugar of calicheamicin. *Tetrahedron Lett.* 35: 4935–4938.
60. Bratz, M.; Bullock, W. H.; Overman, L. E.; Takemoto, T. 1995. Total synthesis of (+)-laurencin. Use of acetal-vinyl sulfide cyclizations for forming highly functionalized eight-membered cyclic ethers. *J. Am. Chem. Soc.* 117: 5958–5966.
61. (a) Castoldi, D.; Caggiano, L.; Panigada, L.; Sharon, O.; Costa, A. M.; Gennari, C. 2005. A formal total synthesis of eleutherobin through an unprecedented kinetically controlled ring-closing-metathesis reaction of a densely functionalized diene. *Angew. Chem. Int. Ed.* 44: 588–591.
62. Sutherlin, D. P.; Armstrong, R. W. 1993. Stereoselective synthesis of dipyranyl C-disaccharides. *Tetrahedron Lett.* 34: 4897–4900.
63. Hoffmann, R. W.; Kemper, B.; Metternich, R.; Lehmeier,T. 1985. Stereoselective synthesis of alcohols. XX. Diastereoselective addition of γ-alkoxyallylboronates to aldehydes. *Liebigs. Ann. Chem.* 2246–2260.
64. Ganesh, P.; Nicholas, K. M. 1997. Reactions of cobalt-complexed acetylenic aldehydes with chiral (γ-alkoxyallyl)boranes: Enantioselective synthesis of 3,4-dioxy 1,5-enynes and stereoselective entry to polyfunctional building blocks. *J. Org. Chem.* 62: 1737–1747.

65. Brown, H. C.; Khire, U. R.; Narla, G. 1995. B-(γ-(trimethylsily l)propargyl)diisopino-campheylborane: A new, highly efficient reagent for the enantioselective propargylboration of aldehydes. Synthesis of trimethylsilyl-substituted and parent α-allenic alcohols in high optical purity. *J. Org. Chem.* 60: 8130–8131.
66. (a) Hertweck, C.; Boland, W. 2000. Tandem reduction-chloroallylboration of esters: Asymmetric synthesis of lamoxirene, the spermatozoid releasing and attracting pheromone of the laminariales (Phaeophyceae). *J. Org. Chem.* 65: 2458–2463. (b) Hertweck, C.; Boland, W.; Goerls, H. 1998. Serinal-derived vinyl oxiranes as novel and versatile building blocks for the stereoselective synthesis of D- and L-erythro-sphingosines. *Chem. Commun.* 1955–1956.
67. (a) Hertweck, C.; Boland, W. 1999. Asymmetric α-chloroallylboration of amino aldehydes: A novel and highly versatile route to d- and l-erythro-sphingoid bases. *J. Org. Chem.* 64: 4426–4430. (b) Furstner, A.; Fenster, M. D. B.; Fasching, B.; Godbout, C.; Radkowski, K. 2006. Toward the total synthesis of spirastrellolide A. Part 2: Conquest of the northern hemisphere. *Angew. Chem. Int. Ed.* 45: 5510–5515. (c) Paterson, I.; Anderson, E. A.; Dalby, S. M.; Loiseleur, O. 2005. Toward the Synthesis of Spirastrellolide A: Construction of a tetracyclic C26-C40 subunit containing the DEF-bis-spiroacetal *Org. Lett.* 7: 4121–4124. (d) Hu, S.; Jayaraman, S.; Oehlschlager, A. C. 1999. An efficient enantioselective synthesis of (+)-Disparlure. *J. Org. Chem.* 64: 3719–3721. (e) Hu, S.; Jayaraman, S.; Oehlschlager, A. C. 1999. An efficient synthesis of (+)-exo-brevicomin via chloroallylboration. *J. Org. Chem.* 64: 2524–2526.
68. (a) Chen, M.; Handa, M.; Roush, W. R. 2009. Enantioselective synthesis of 2-methyl-1,2-syn- and 2-methyl-1,2-anti-3-butenediols via allene hydroboration-aldehyde allylboration reaction sequences. *J. Am. Chem. Soc.* 131: 14602–14603. (b) Li, F.; Roush, W. R. 2009. Stereoselective synthesis of *syn,syn*- and *syn,anti*-1,3,5-triols via intramolecular hydrosilylation of substituted pent-3-en-1,5-diols. *Org. Lett.* 11: 2932–2935. (c) Brown, H. C.; Narla, G. 1995. [(E)-γ-(1,3,2-Dioxaborinanyl)allyl]diisopinocampheyl-borane, an exceptional reagent for the stereo- and enantioselective synthesis of anti- 1-alkene-3,4-diols via a masked α-hydroxyallylboration *J. Org. Chem.* 60: 4686–4687.
69. (a) Barrett, A. G. M.; Seefeld, M. A. 1993. *B*-[*E*-3-(diphenylamino)allyl]diisopino-campheylborane: An excellent reagent for the stereoselective synthesis of anti-diphenylamino alcohols. *J. Chem. Soc. Chem. Commun.* 339–341.
70. (a) Barrett, A. G. M.; Seefeld, M. A.; Williams, D. J. 1994. A convenient asymmetric synthesis of anti-β-amino alcohols: An x-ray crystallographic study of (4R)-2,2-dimethyl-4-[(2S)-(diphenylmethyleneamino)-(1S)-hydroxy-3-buten-1-yl]-1,3-dioxolane. *J. Chem. Soc. Chem. Commun.* 1053–1054. (b) Barrett, A. G. M.; Seefeld, M. A.; White, A. J. P.; and Williams. D. J. 1996. Convenient asymmetric syntheses of anti-β-amino alcohols. *J. Org. Chem.* 61: 2677–2685.
71. (a) Barrett, A. G. M.; Malecha, J. W. 1994. Synthetic studies on calyculin A: A convenient asymmetric synthesis of anti-vicinal diols. *J. Chem. Soc., Perkin Trans. 1.* 1901–1905. (b) Niel, G.; Roux, F.; Maisonnasse, Y.; Maugras, I.; Poncet, J.; Jouin, P. 1994. Substrate-controlled crotylboration from N-(tert-butoxycarbonyl)amino aldehydes *J. Chem. Soc., Perkin Trans. 1.* 1275–1280. (c) Barrett, A. G. M.; Edmunds, J. J.; Hendrix, J. A.; Malecha, J. W.; Parkinson, C. J. 1992. Stereocontrolled synthesis of calyculin A: construction of the C(26)–C(37) amide-oxazole unit. *J. Chem. Soc. Chem. Commun.* 1240–1242. (d) Barrett, A. G. M.; Malecha, J. W. 1991. B-[3-((Diisopropylamino)dimethylsilyl)allyl]diisopino-campheylborane: An excellent reagent for the stereoselective synthesis of anti vicinal diols *J. Org. Chem.* 56: 5243–5245.
72. Barrett, A. G. M.; Wan, P. W. H. 1996. A convenient synthesis of γ-hydroxy-α-methylene silanes *J. Org. Chem.* 61: 8667–8670.
73. Heo, J. N.; Micalizio, G. C; Roush, W. R. 2003. Enantio- and diastereoselective synthesis of cyclic β-Hydroxy allylsilanes via sequential aldehyde γ-silylallylboration and ring-closing metathesis reactions, *Org. Lett.* 5: 1693–1696.
74. (a) Pearson, W. H.; Lin, K. C.; Poon, Y. F. 1989. A stereoselective route to 2-(phenylthio)-1,3-butadienes. *J. Org. Chem.* 54: 5814–5819. (b) Liu, C.; Wang, K. K. 1986. Stereoselective synthesis of 2-(trimethylsilyl)methyl-1,3-butadienes. *J. Org. Chem.* 51: 4733–4734. (c) Wang, K. K.; Gu, Y. G.; Liu, C. 1990. Stereoselective synthesis of all four geometric isomers of internal 1,3-butadienes by the condensation reaction of aldehydes with the γ-trimethylsilyl-substituted allylboranes *J. Am. Chem. Soc.* 112: 4424–4431. (d) Gu, Y. G.; Wang, K. K. 1991. Stereoselective synthesis of both geometric isomers of γ-(trimethylsilyl)allylboranes and the diastereoselective condensations with aldehydes Gu, Y. G.; Wang, K. K. *Tetrahedron Lett.* 32: 3029.
75. Mulzer, J.; Mantoulidis, A. 1996. Synthesis of the C(1)-C(9) segment of the cytotoxic macrolides epothilon A and B. *Tetrahedron Lett.* 37: 9179–9182.

76. (a) Brown, H. C.; Jadhav, P. K. 1984. *B*-Allyldiisocaranylborane: a new, remarkable enantioselective allylborating agent for prochiral aldehydes. Synthesis of homoallylic alcohols approaching 100% enantiomeric purities. *J. Org. Chem.* 49: 4089–4091. (b) Ramahcandran, P. V.; Biswas, D. 2007. Convenient synthesis of stable aldimine-borane complexes, chiral δ-amino alcohols, and γ-substituted GABA analogues from nitriles. *Org. Lett.* 9: 3025–3027. (c) Mamane, V.; Garcia, A. B.; Umarye, J. D.; Lessmann, T.; Sommer, S.; Waldmann, H. 2007. Stereoselective allylation of aldehydes on solid support and its application in biology-oriented synthesis (BIOS). *Tetrahedron* 63: 5754–5767. (d) Paterson, I.; Anderson, E. A.; Dalby, S. M.; Loiseleur, O. 2005. Toward the synthesis of spirastrellolide A: Construction of two C1-C25 diastereomers containing the BC-spiroacetal. *Org. Lett.* 7: 4125–4128. (e) Jacobs, M. F.; Glenn, M. P.; McGrath, M. J.; Zhang, H.; Brereton, I.; Kitching, W. 2001. Stereoselective synthesis and stereochemistry of seven isomeric spiroacetal structures based on the C17–C28 fragment (CD rings) of spongistatin 1 *Arkivoc.* 2: 114–137. (f) Paterson, I.; Collett, L. A. 2001. Remote 1,5-stereoinduction in boron aldol reactions of methyl ketones: application to the convergent assembly of the 1,3-polyol sequence of (+)-roxaticin. *Tetrahedron Lett.* 42: 1187–1191. (g) Kumar, D. J. S.; Madhavan, S.; Ramahcandran, P. V.; Brown, H. C. 2000. Highly enantioselective synthesis of (+)- and (−)-perfluoroalkyl(aryl) homoallyl alcohols. *Tet. Asymm.* 11: 4629–4632. (h) Reddy, M. V. R.; Rearick, J. P.; Hoch, N.; Ramahcandran, P. V. 2001. Asymmetric Synthesis of Umuravumbolide *Org. Lett.* 3: 19–20. (i) Paterson, I.; Oballa, R. M.; Norcross, R. D. 1996. Studies in marine macrolide synthesis: Stereocontrolled synthesis of the AB-spiracetal subunit of spongistatin 1 (Altohyrtin A) *Tetrahedron Lett.* 37: 8581–8584. (j) Jirousek, M. R.; Cheung, A. W. H.; Babine, R. E.; Sass, P. M.; Schow, S. R.; Wick, M. M. 1993. A synthesis of indolizidines related to castanospermine and swainsonine *Tetrahedron Lett.* 34: 3671–3674. (k) Racherla, U. S.; Liao, Y.; Brown, H. C. 1992. Chiral synthesis via organoboranes. 36. Exceptionally enantioselective allylborations of representative heterocyclic aldehydes at −100°C under salt-free conditions *J. Org. Chem.* 57: 6614–6617. (l) Racherla, U.S.; Brown, H. C. 1991. Chiral synthesis via organoboranes. 27. Remarkably rapid and exceptionally enantioselective (approaching 100% ee) allylboration of representative aldehydes at -100° under new, salt-free conditions *J. Org. Chem.* 56: 401–404. (m) Brown, H. C.; Randad, R. S.; Bhat, K. S.; Zaidlewicz, M.; Racherla, U.S. 1990. Chiral synthesis via organoboranes. 24. B-allylbis(2-isocaranyl)borane as a superior reagent for the asymmetric allylboration of aldehydes. *J. Am. Chem. Soc.* 112: 2389–2392.
77. (a) Wang, Kan; Bungard, Christopher J.; Nelson, Scott G. 2007. Stereoselective olefin isomerization leading to asymmetric quaternary carbon construction. *Org. Lett.* 9: 2325–2328. (b) Mitton-Fry, M. J.; Cullen, A. J.; Sammakia, T. 2007. The total synthesis of the oxopolyene macrolide RK-397. *Angew. Chem. Int. Ed.* 46: 1066–1070. (c) Dhokte, U. P.; Khau, V. V.; Martinelli, M. J. 2000. Crotylboration process for preparation of cryptophycin antineoplastic and antifungal agents *PCT Int. Appl.* WO2000034253 A2 20000615. (d) Itsuno, S.; Yokoi, A.; Kuroda, S. 1999. Asymmetric synthesis of homoallylamines by nucleophilic addition of chirally modified allylboron reagent to N-borylimines. *Synlett.* 1987–1989. (e) Watanabe, K.; Kuroda, S.; Yokoi, A.; Ito, K.; Itsuno, S. 1999. Enantioselective synthesis of optically active homoallylamines by allylboration of N-diisobutylaluminum imines. *J. Organomet. Chem.* 581: 103–107.(f) Bubnov, Y. N.; Lavrinovich, L. I.; Zykov, A. Yu.; Ignatenko, A. V. 1992. Synthesis of (R)- and (S)-allyloxiranes via enantioselective allylboration of bromoacetaldehyde. Transformation of (R)-allyloxirane into (−)-(R)-γ-amino-β-hydroxybutyric acid. *Mendeleev Commun.* 86–87.
78. Fang, G. Y.; Aggarwal, V. K. 2007. Asymmetric synthesis of α-substituted allyl boranes and their application in the synthesis of iso-agatharesinol. *Angew. Chem. Int. Ed.* 46: 359–362.
79. Boshra, R.; Doshi, A.; Jaekle, F. 2008. Allylation of ketones with a ferrocene-based planar chiral Lewis acid. *Angew. Chem. Int. Ed.* 47: 1134–1137.
80. (a) Short, R. P.; Masamune, S. 1989. Asymmetric allylboration with B-allyl-2-(trimethylsilyl)borolane. *J. Am. Chem. Soc.* 111: 1892–1894. (b) Masamune, S. 1987. Preparation and application of chiral 2,5-dimethylborolanes. US Patent US4644075 A 19870217.
81. (a) Munoz-Hernandez, L.; Soderquist, J. A. 2009. Asymmetric γ-Methoxyallylation with the Robust 10-TMS-9-Borabicyclo[3.3.2]decanes. *Org. Lett.* 11: 2571–2574. (b) Soto-Cairoli, B.; Soderquist, J. A. 2009. Strict reagent control in the asymmetric allylboration of N-TIPS-α-amino aldehydes with the B-allyl-10-TMS-9-borabicyclo[3.3.2]decanes. *Org. Lett.* 11: 401–404. (c) Soto-Cairoli, B.; Justo de Pomar, J.; Soderquist, J. A. 2008. Enantiomerically pure α-amino aldehydes from silylated α-amino acids. *Org. Lett.* 10: 333–336. (d) Gonzalez, Ana Z.; Soderquist, John A. 2007. β-Silylated homopropargylic amines via the asymmetric allenylboration of aldimines. *Org. Lett.* 9: 1081–1084. (d) Hernandez, E.; Burgos, C. H.; Alicea, E.; Soderquist, J. A. 2006. B-Allenyl- and B-(γ-Trimethylsilylpropargyl)-10-phenyl-9-borabicyclo[3.3.2]decanes: Asymmetric synthesis of propargyl and α-Allenyl 3°-carbinols

from ketones. *Org. Lett.* 8: 4089–4091. (e) Hernandez, E.; Soderquist, J. A. 2005. Nonracemic α-allenyl carbinols from asymmetric propargylation with the 10-Trimethylsilyl-9-borabicyclo[3.3.2]decanes. *Org. Lett.* 7: 5397–5400. (f) Lai, C.; Soderquist, J. A. 2005. Nonracemic homopropargylic alcohols via asymmetric allenylboration with the robust and versatile 10-TMS-9-borabicyclo[3.3.2]decanes. *Org. Lett.* 7: 799–802. (g) Gonzalez, J. R.; Gonzalez, A. Z.; Soderquist, J. A. 2009. (E)-2-Boryl-1,3-butadiene Derivatives of the 10-TMS-9-BBDs: Highly selective reagents for the asymmetric synthesis of anti-1,2-disubstituted 3,4-pentadien-1-ols. *J. Am. Chem. Soc.* 131: 9924–9925. (h) Gonzalez, A. Z.; Roman, J. G.; Alicea, E.; Canales, E.; Soderquist, J. A. 2009. Borabicyclo[3.3.2]decanes and the stereoselective asymmetric synthesis of 1,3-Diol stereotriads from 1,3-diborylpropenes. *J. Am. Chem. Soc.* 131: 1269–1273. (i) Canales, E.; Hernandez, E.; Soderquist, J. A. 2006. Nonracemic 3°-carbamines from the asymmetric allylboration of N-trimethylsilyl ketimines with B-allyl-10-phenyl-9-borabicyclo[3.3.2]decanes *J. Am. Chem. Soc.* 128: 8712–8713. (j) Canales, E.; Prasad, K. G.; Soderquist, J. A. 2005. B-Allyl-10-Ph-9-borabicyclo[3.3.2]decanes: Strategically designed for the asymmetric allylboration of ketones. *J. Am. Chem. Soc.* 127: 11572–11573. (k) Burgos, C. H.; Canales, E.; Matos, K.; Soderquist, J. A. 2005. asymmetric allyl- and crotylboration with the robust, versatile, and recyclable 10-TMS-9-borabicyclo[3.3.2] decanes. *J. Am. Chem. Soc.* 127: 8044–8049. (l) Roman, J. G.; Soderquist, J. A. 2007. Asymmetric Synthesis of 2°- and 3°-Carbinols via B-Methallyl-10-(TMS and Ph)-9-borabicyclo[3.3.2]decanes. *J. Org. Chem.* 72: 9772–9775. (m) Kister, J.; DeBaillie, A. C.; Lira, R.; Roush, W. R. 2009. Stereoselective synthesis of γ-substituted (*Z*)-allylic boranes via kinetically controlled hydroboration of allenes with 10-TMS-9-borabicyclo[3.3.2]decane. *J. Am. Chem. Soc.* 131: 14174–14175. (n) Alexander, M. D.; Fontaine, S. D.; La Clair, J. J.; DiPasquale, A. G.; Rheingold, A. L.; Burkart, M. D. 2006. Synthesis of the mycolactone core by ring-closing metathesis. *Chem. Commun.* 4602–4604.

26 Carborane Clusters

Versatile Synthetic Building Blocks for Dendritic, Nanostructured, and Polymeric Materials

Barada Prasanna Dash, Rashmirekha Satapathy, John A. Maguire, and Narayan S. Hosmane

CONTENTS

26.1 INTRODUCTION

Carboranes are a class of heteroboranes that contain both boron and carbon atoms in electron-delocalized clusters (Grimes, 1970). One of the most commonly encountered carborane clusters are the icosahedral dicarboranes ($C_2B_{10}H_{12}$), as shown in Figure 26.1. There are three isomers, the 1,2-$C_2B_{10}H_{12}$ (*ortho*-), 1,7-$C_2B_{10}H_{12}$ (*meta*-), and the 1,12-$C_2B_{10}H_{12}$ (*para*-), all of which are commercially available. All three icosahedral carboranes are chemically and thermally very stable. In general, the *ortho*-carborane derivatives can be prepared from the reaction of decaborane with various acetylenes; the corresponding *meta* and *para* isomers can be obtained via thermal isomerization of the respective *ortho*-carboranes. The hydrogens attached to the cage carbons of the carboranes are acidic and can be removed by reaction with strong bases, which allows further derivatization of the clusters. Similarly, these icosahedral carboranes having electron-rich boron atoms can also be functionalized by Lewis acid-catalyzed Friedel–Craft-type reactions. All these factors make the icosahedral carboranes attractive as synthetic building blocks in materials science, medicine and catalysis (Valliant et al., 2002). In addition, there are a number of smaller cage dicarboranes that are in use (see Figure 26.1). (Russell et al., 2002). In this chapter we will highlight some recent works where carborane clusters have been incorporated into dendritic, nanostructured, and polymeric compounds.

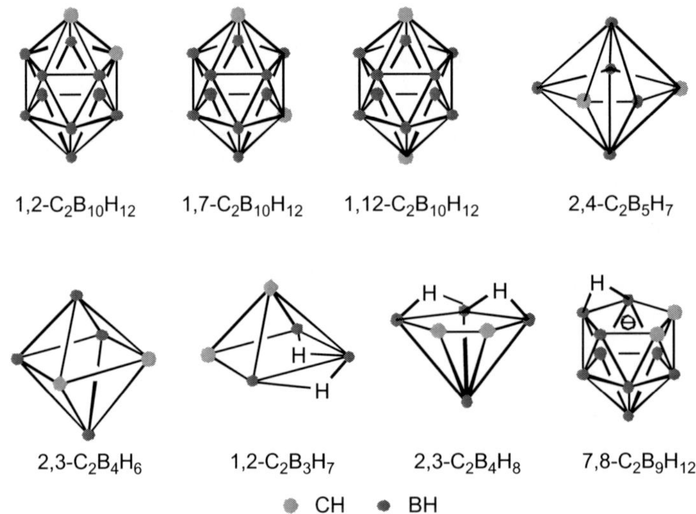

FIGURE 26.1 Structures of selected carboranes.

26.2 DENDRITIC MACROMOLECULES

Dendrimers are globular macromolecules that have distinct structural features such as a symmetrical core, layers of branching units, and surface groups. These dendritic macromolecules are conventionally synthesized by either a convergent approach in which the growth of the dendrimer starts from the exterior and progresses inward to a core or a divergent one that starts from the core and builds outward (Grayson and Frechet, 2001). Dendritic macromolecules find applications in drug delivery (Nishimaya et al., 2005), light harvesting (Lo and Burn, 2007; Saragi et al., 2007), and catalysis (Astruc et al., 2008). Carboranes are also used in medicine as sources of boron in boron neutron capture therapy (BNCT) for the treatment of cancer (Hawthorne and Maderna, 1999; Sivaev and Bregadze, 2009), boron neutron capture synvectomy (BNCS) for the treatment of rheumatoid arthritis (Watson-Clark et al., 1998) and a number of other medicinal applications (Valliant et al., 2002). Boron neutron capture therapy (BNCT) is a form of radiation therapy in which a boron-10 nucleus captures a low-energy (thermal) neutron and undergoes a rapid fission reaction, producing two high-energy charged particles, an alpha particle and a recoil lithium ion. Preferential localization of boron-10 atoms in a target cell followed by thermal neutron irradiation could lead to cell death, but not affect neighboring cells that are free of boron. It has been estimated that a concentration of 10^9 boron-10 atoms per target cell or 30 μg of boron-10 per gram tissue is necessary for effective BNCT; this is a relatively large concentration. One of the ways in which this can be accomplished is to incorporate multiple carborane clusters into macromolecular dendritic compounds. It has been reported that dendritic and macromolecular drug-delivery agents are also superior in terms of tumor accumulation and retention (Parrott et al., 2005; Yinghuai et al., 2005). Icosahedral carborane clusters are also useful in other applications, such as the synthesis of thermally stable materials and use as metal carriers and as synthetic building blocks for liquid crystalline and nonlinear optical materials (Plesek, 1992; Grimes, 2004; Zhu et al., 2007).

While the incorporation of carborane cages into macromolecules is desirable, steric and electronic considerations make the synthesis of dendritic and symmetrical star-shaped molecules containing closely placed multiple carborane clusters a difficult task (Craciun et al., 2004). There have been several approaches to the incorporation of carboranes into star-shaped molecules.

Ortho-carborane clusters can be functionalized at both the boron and carbon atoms of the cages. Selective functionalization of *ortho*-carborane clusters can be achieved at the B(9) boron atom via electrophilic alkylation with alkyl halides and benzyl halides that contain electron-withdrawing substituents such as NO_2, COOH, COOMe, and COPh (Zakharkin and Olshevskaya, 1995). We have synthesized the benzyl derivative of the *ortho*-carboranes (with benzylation at the B(9) position) containing an acetyl group on the benzene ring, as shown in Scheme 26.1. Electrophilic benzylation of the *ortho*-carborane and its derivatives were conducted by refluxing a mixture of 1-(4-(bromomethyl)phenyl)ethanone (**1**) and *ortho*-carborane in dichloromethane in the presence of aluminum chloride that led to the formation of B(9)-substituted ketone **2** (Scheme 26.1), which underwent facile trimerization with silicon tetrachloride and ethanol to produce B_{cage}-appended trimer **3** (Scheme 26.1). Mechanistically, the silicon tetrachloride/ethanol trimerizations proceed via the initial formation of a carbocation on the acetyl group, which then reacts with another acetyl group in its enol form, and ultimately the trimer forms by combining with a third acetyl carbocation (Gupta et al., 2002). Functionalization of the acidic hydrogens of trimer **3** with 1-iodoheptane or trivinylchlorosilane via its lithium salts was also possible, generating dendritic-structured macromolecules (Dash et al., 2008).

A series of C_3-symmetric C_{cage}-appended n-conjugated compounds containing three (**4–6**) to six (**7** and **8**) *ortho*-carborane clusters has also been synthesized by employing palladium-catalyzed Suzuki coupling reactions, palladium-catalyzed acetylation reactions followed by silicon tetrachloride-mediated trimerization reactions (Figure 26.2). Carborane-appended extended trimers (**5**, **6**, and **8**) were found to be blue light emitting. A 22–70% enhancement of relative fluorescence quantum yields of carborane-substituted extended n-conjugated compounds (**5** and **8**) was observed compared to the unsubstituted extended n-conjugated core. Decapitation of *ortho*-carborane clusters in methanolic sodium hydroxide solution led to the formation of monoionic *nido* carborane-appended trimers (**9–13**), which were found to be water soluble. The water-soluble extended trimers (**10**, **11**, and **13**) were also found to be fluorescent in water but with a reduced fluorescence intensity. These n-conjugated compounds containing multiple carborane clusters were found to be extremely thermal stable. It was also found that the addition of more *ortho*-carborane clusters to the n-conjugated systems increased their thermal stability. Thermal analysis showed that the C_{cage}-appended trimers were more stable than the corresponding of B_{cage}-appended trimer (**3**) (Dash et al., 2010).

Cobalt-catalyzed cycloaddition reactions of symmetrical alkynes usually generate C_6-symmetric hexaphenylbenzenes (Pisula et al., 2004; Chebny et al., 2006). Synthesis of the star-shaped compound **14**, which contains six *ortho*-carborane clusters surrounding a hexaphenylbenzene core (Figure 26.3) was synthesized in a like manner. Heximer **14** was then decapitated in a methanolic sodium hydroxide solution to convert the neutral *closo*-carboranes into their corresponding [*nido*-$C_2B_9H_{10}]^-$ cages, as shown for **15**, imparting water solubility (Dash et al., 2009).

Star-shaped hexaphenylbenzenes and their derivatives, such as hexabenzocoronenes (HBCs), show remarkable properties such as self-assembly and aggregation in solutions, high charge carrier motilities, and high thermal stability (Muller et al., 1998; Ito et al., 2000). These conjugated systems are also useful in materials science in that they can form the core structures for discotic liquid

1) $AlCl_3$, CH_2C_{l2}, 50°C
2) Br ... (1)
$SiCl_4$, EtOH

SCHEME 26.1 Synthesis of the B_{cage}-appended symmetrical core.

n = 1 **4;** n = 2 **5**; n = 3 **6**

n = 1 **7** ; n = 2 **8**

n = 1 **9** ; n = 2 **10** ; n = 3 **11**

n = 1 **12** ; n = 2 **13**

FIGURE 26.2 C_3-symmetric C_{cage}-appended n-conjugated compounds.

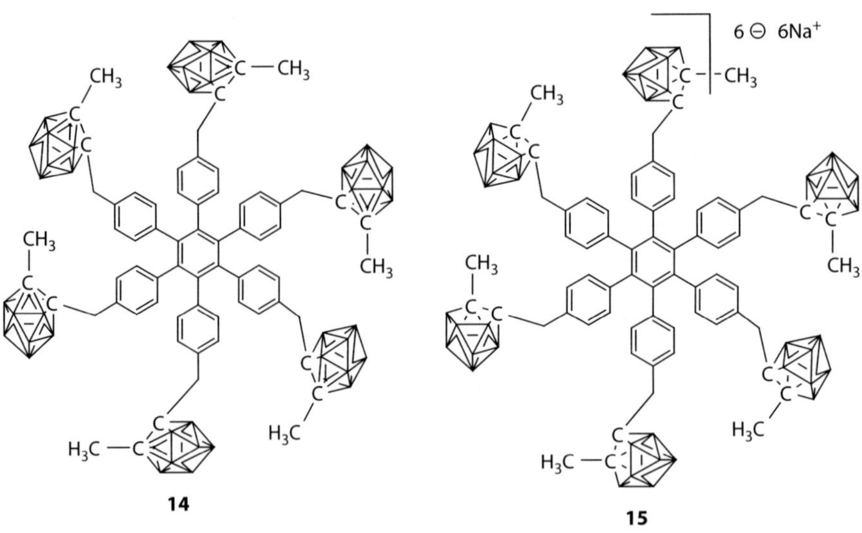

FIGURE 26.3 C_6-symmetric C_{cage}-appended π-conjugated compound.

crystals (Kumar, 2006), components in organic field effect transistors, and organic light-emitting diodes (Wu et al., 2003); heximers **14** and **15** may serve as bases of materials.

In addition, icosahedral carborane clusters are versatile building blocks in materials science. Recent reports reveal that incorporation of icosahedral carborane clusters into different systems have enhanced their thermal stability and electrical conductivity and also helped in stabilizing different liquid crystalline phases (Grimes, 2004). Therefore, carborane-containing hexaphenylbenzenes and other such derivatives such as HBCs could form the basis for a new class of materials. Due to the presence of hexaphenylbenzene core, **14** shows moderate absorption and emission properties. The electronic absorption spectra show two peaks for **14**, one at 230 nm and the other at 254 nm. The emission spectra of **14** showed a major peak at 344 nm. Since carborane clusters do not absorb in the UV–vis region, any photochemistry due to the presence of *ortho*-carboranes is ruled out (Xamena et al., 2008); therefore, the spectroscopic behavior of **14** arises from the hexaphenylbenzene core (Wang et al., 2004). DSC analysis of **14** showed its melting point to be 468°C. In addition, **14** shows very little mass loss (<10%) up to 491°C under an inert atmosphere. This exceptional thermal stability is consistent with its melting point and is in accordance with the fact that incorporation of *ortho*-carborane clusters enhances the thermal stability of materials (Dash et al., 2009).

Synthesis of carboranyl–carbosilane dendrimers in which *ortho*-carborane derivatives are bonded to the silicon atom through a propyl spacer has been reported (Figure 26.4). In this case, dendrimers **16** and **17** containing four and eight peripheral *ortho*-carborane clusters, respectively, were synthesized from their respective dendritic cores using a sequence of alkenylation, hydrosilylation and reduction (Gonzalez-Campo et al., 2007). Boron-rich carbosilane core-based polyanionic macromolecules peripherally containing four and eight cobaltabisdicarbollides have also been synthesized (Figure 26.5). In this work, the monoanionic C_c-substituted cobaltabisdicarbollides functionalized with the Si–H group, Cs[1,1′-η-SiMeH-3,3′-Co(1, 2-$C_2B_9H_{10})_2$], was used as the starting metallacarborane. The dendritic metallacarboranes were then synthesized using successive alkenylation and hydrosilylation steps in a divergent methodology (Juarez-Perez et al., 2009) (Figure 26.6).

A water-soluble polysulfate carborane dendrimer containing twelve *ortho*-carborane clusters (**20**) was synthesized from its second-generation polyalkyne dendrimer precursor by reaction with decaborane ($B_{10}H_{14}$). The water solubility imparting sulfate moieties were introduced by the reaction of the precursor carborane polyol compound with chloro-sulfonic acid (Newkome et al., 1994).

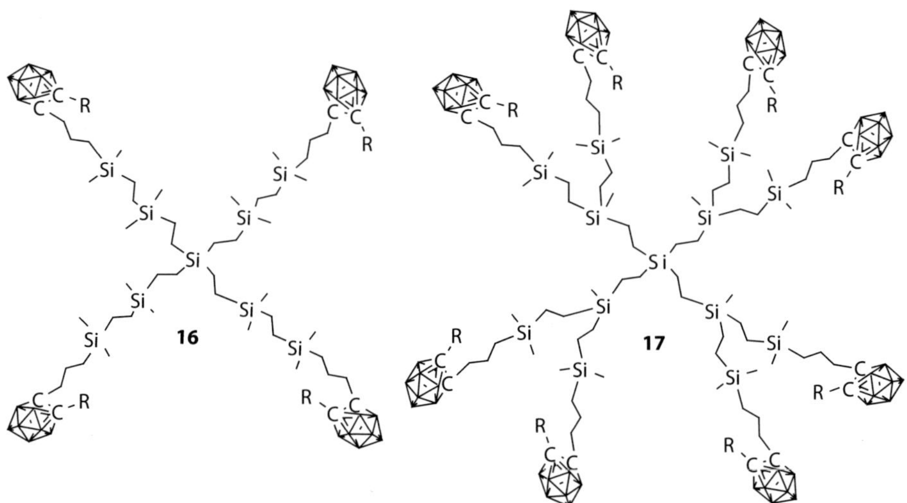

FIGURE 26.4 Carbosilane dendrimers containing *o*-carboranes.

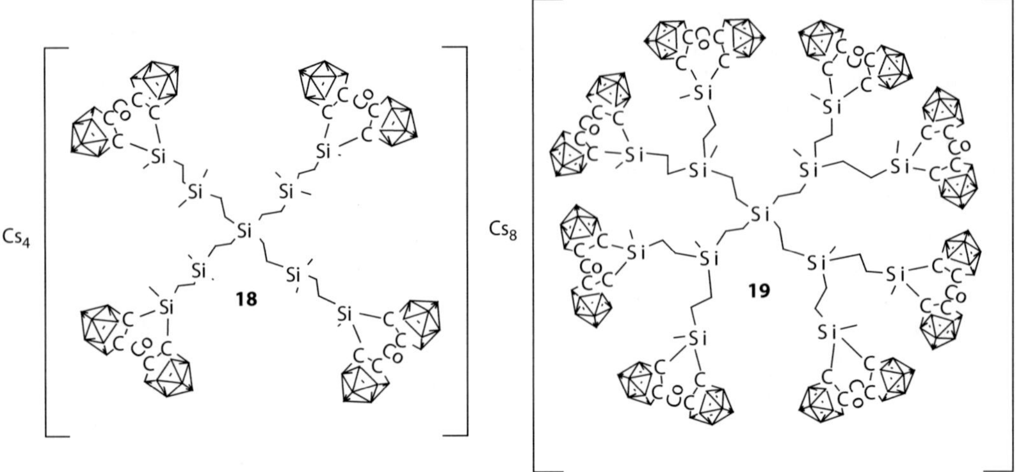

FIGURE 26.5 G1 and G2 carbosilane dendrimers containing cobaltabisdicarbollides.

Symmetrical structures involving small cage carboranes (C_2B_4 clusters) have been reported by Grimes' group. For example, the triangular metallacarborane **21** (Figure 26.7) contains three benzene-anchored, seven vertex ferracarborane clusters linked to the central phenyl ring via – C≡C – units. Attachment of the –C≡C – units to the apical boron atom of the C_2B_4 clusters was achieved by using a catalyzed Negishi-type coupling reaction. Finally, catalyzed Sonogashira-type reaction of triiodobenzene with the apically substituted terminal alkynes led to the formation of **21** (Yao et al., 2002).

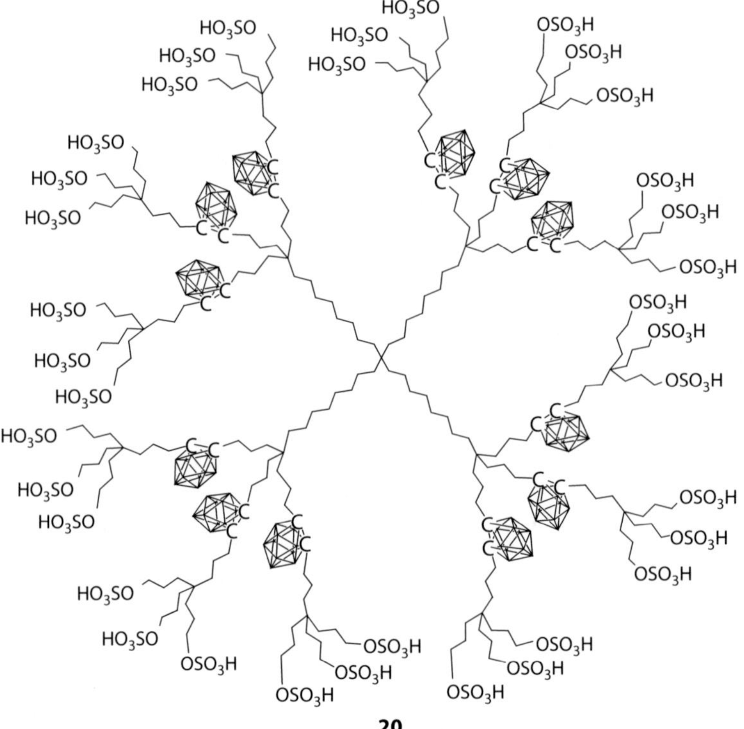

FIGURE 26.6 Water-soluble polysulfate carborane dendrimer.

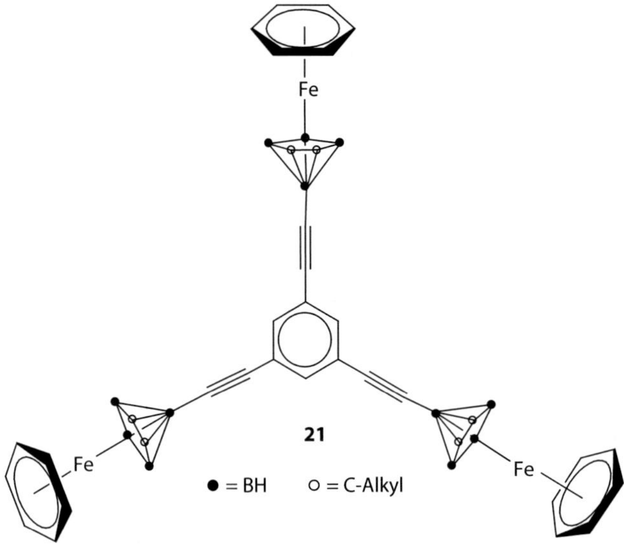

FIGURE 26.7 C_3-symmetrical structures containing C_2B_4 clusters.

The use of small cage polyhedral metal–boron clusters for the synthesis of highly branched metallodendrimers has so far received little attention. One attempt to use the decapitated open faced cobaltacarborane *nido*-[η^5-$C_5H_4C(O)Cl$]Co(2,3-$Et_2C_2B_3H_5$) for the synthesis of a 32-cobalt metallacarborane dendrimer **22** (Figure 26.8) was reported by Grime's group. This was accomplished by treating the fourth-generation dendrimer diaminobutane-dend $(NH_2)_{32}$ (DAB-32) with the decapitated cobaltacarborane in the presence of trimethyl amine (Yao et al., 2003a).

26.3 NANOSTRUCTURED MATERIALS

There are two common approaches for miniaturization and nanoscale synthesis, the so-called "top-down" and "bottom-up." The more advanced, "top-down" approach is nearing its limits in scaling. So in recent years molecular assemblies are being constructed using the bottom-up approach in which shape-persistent nanostructured molecules are assembled to build the desired nanomaterials. The synthesis of molecular building blocks for their possible use in the bottom-up synthetic applications is therefore receiving considerable attention (Shirai et al., 2006). In this section we will focus on the use of carborane clusters for the syntheses of molecular architectures such as rods, boxes, and nanomachines that could be eventually useful in nanoscale syntheses.

26.3.1 NANOMACHINES

The synthesis of nanomachines such as motors, rotors, cars, and trucks that could perform tasks at the molecular level similar to what real macroscopic machines can perform has gained considerable momentum. A nanocar that contains *para*-carborane wheels has been designed. *Para*-carborane was chosen as wheels because of their robust three-dimensional and near-spherical structure. Moreover, *para*-carborane does not absorb light at 365 nm, which is the motor's operational wavelength (Figure 26.9). Thus, the use of *para*-carboranes did not have any effect on quenching the motor's photochemical rotary process. In nanocar **23** alkynes were used as axes because they have a very-low-energy barrier to rotation. Thioxanthene and naphtha (2,1-*b*)-thiopyran units were used to build the motor. The molecular motor having hydrogen atoms on the thioxanthene in place of the axle groups showed a photostationary state (PSS) ratio of 8:92 after irradiation at 365 nm for 3 h at

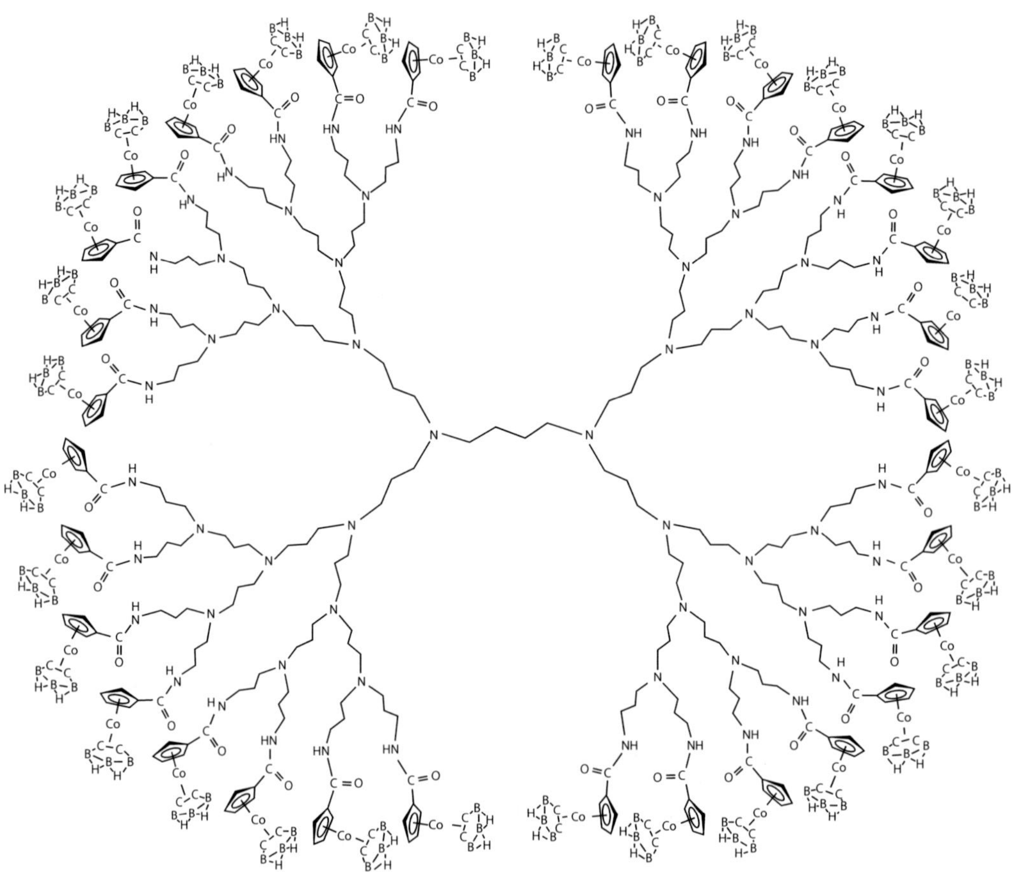

FIGURE 26.8 Metalladendrimer **22** involving small cage C_2B_4 clusters.

room temperature (Koumura et al., 2002), indicating that molecule **23** could form the basis of a good mono-directional nanocar. Kinetic studies in solution indicated the rotation of the motor on irradiation with 365 nm light (Morin et al., 2006) (Figure 26.10).

A number of nanovehicles, containing three to six *para*-carboranes, such as nanocars, **25** and **27**, nanocaterpillar, **26**, and trimer molecule, **28**, were synthesized. The placement of the molecular axles would determine the types of motion: nanocar **25** and nanocaterpillar **26** were expected to translate in a one-directional fashion; nanocar **27** was designed to make small circular motions on

FIGURE 26.9 Nanocar **23** and nanoinchworm **24**.

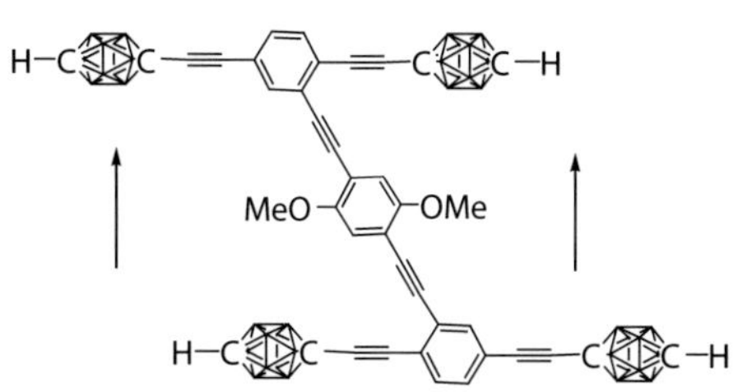

FIGURE 26.10 Structures of several nanovehicles.

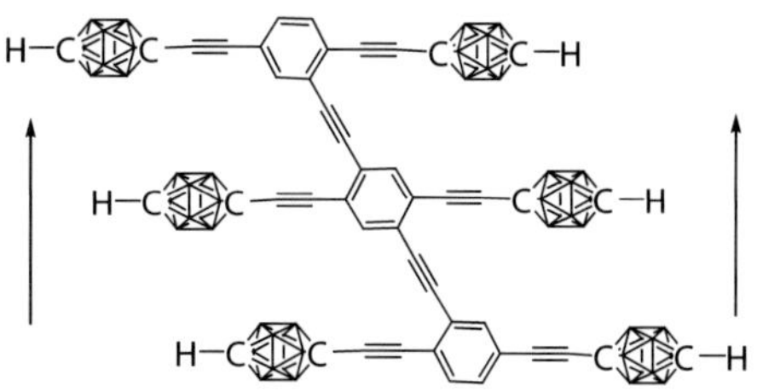

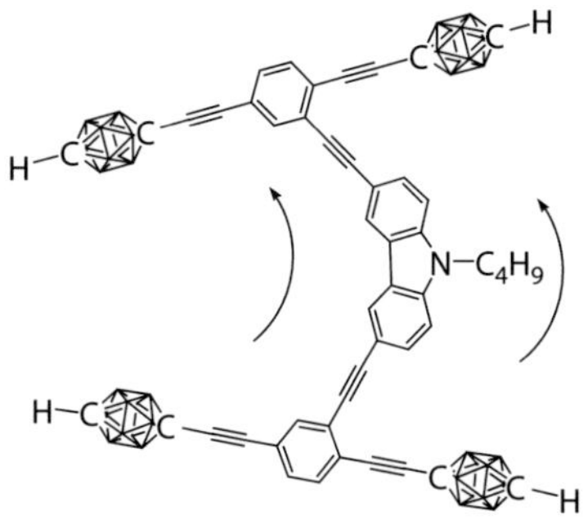

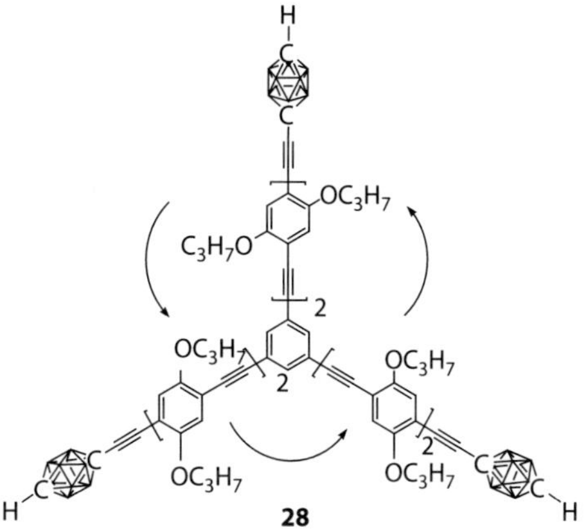

a surface; trimer **28** was designed to pivot on the surface with no translational movement (Morin et al., 2007). Molecule **24** is an example of a nanovehicle that is propelled by an inchworm motion brought about by the *cis–trans* photoisomerization of azobenzene chromophores (Sasaki and Tour, 2008). In all of the nanovehicles the *para*-carboranes were used as the "tires"; there are duplicates in which fullerenes replace the carboranes. Since the carboranes do not absorb in the spectral regions of interest, they are used in place of fullerenes in vehicles propelled by photochemical processes. Vives and Tour have recently reviewed the synthesis of such nanocars (Vives and Tour, 2009).

26.3.2 Rods, Chains, Boxes, and Macrocycles

Rigid-rod-shaped molecules with defined lengths are useful molecular building blocks (MBBS) in the synthesis of nanostructured and supramolecular constructs. The potential applications of such assemblies include the construction of nanomachines and the creation of molecular electronic devices using a "bottom-up" approach (Darling, 1995; Lehn, 1995; Regis, 1995; Tour, 1996). Due to their excellent chemical, thermal, and photochemical stabilities and their dimensional structures, *para*- and *meta*-carborane clusters are ideal choices as synthetic building blocks of such rigid-rod-shaped molecules (Schwab et al., 1999).

Michl and coworkers have synthesized a number of rod-shaped molecules (**29**–**33**) containing multiple *para*-carborane units as shown in Figure 26.11 (Müller et al., 1992). Compounds **29** and **31** were formed by the oxidative coupling of the lithium salt of the parent *para*-carborane with $CuCl_2$, in yields of 56% and 21%, respectively. Tetramer **33** was obtained by a similar oxidative coupling reaction of **29** in only a 9% yield. The dimeric and trimeric carborarods could be further functionalized to give **30** and **32** (Figure 26.12).

The B-methylated carboranes, such as the permethylated *para*-carborane $[(CH_3)_{10}C_2B_{10}H_2]$, and the *octa*-methylated *meta*-carborane $[(CH_3)_8C_2B_{10}H_4]$ possess a number of unique properties,

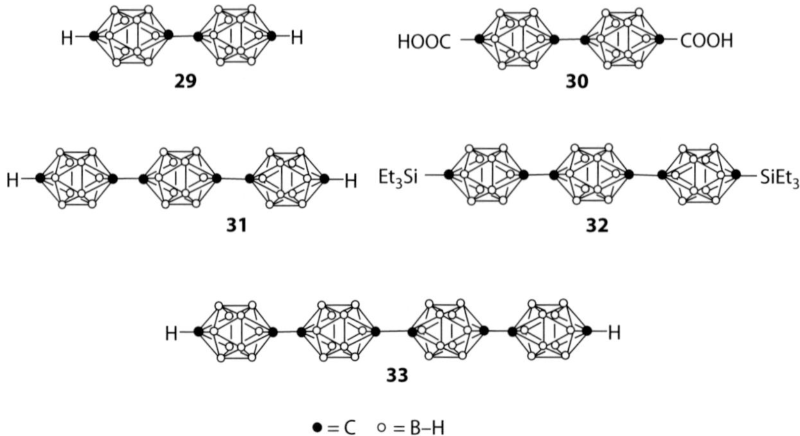

FIGURE 26.11 *Para*-carborane rods.

such as high solubility in organic solvents as well as increased stabilities toward bases and protic media compared to their nonmethylated *para*- and *meta*-carboranes. These properties are considered suitable for the synthesis of longer carborarods and chains. Transition metal fragments are widely used as end caps or junctions in *s*-acetylide complexes, as seen in **34**, linkage of the selected alkyne-substituted dimeric *para*-carborarods was carried out from the alkyne monomers by using *trans* [$(nBu_3P)_2PtCl_2$] in the presence of cuprous iodide and diethylamine. The tetrameric chain **35** containing four *octa*-methylated *meta*-carboranes was synthesized via an oxidative coupling from its precursor dimer species. This tetrameric chain **35** is found to be sparingly soluble in hexane but dissolved well in hot aromatic solvents (Herzog et al., 2005).

The selective and modular incorporation of acetylenic moieties into oligomers is useful for the synthesis of structurally well-defined molecular nanostructures. In addition to the C-substituted carborarods, a number of B-substituted rod-shaped *para*-carboranes joined by acetylenic moieties were synthesized by Hawthorne's group, as shown in Figure 26.13 (Jiang et al., 1996b). The selective and modular incorporation of acetylenic moieties into the oligomers is useful for the synthesis of structurally well-defined molecular nanostructures. Compounds **36**, **37**, and **38** were synthesized by using 2-iodo-*p*-carborane as the starting materials (Figure 26.13). 2-Iodo-*p*-carborane underwent a palladium-catalyzed Sonogashira-type coupling reaction with trimethylsilyl acety-

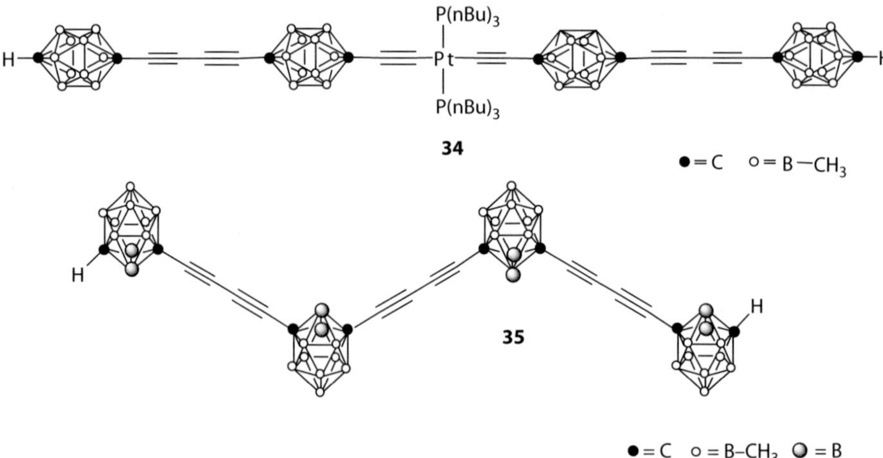

FIGURE 26.12 Carborane chains from B-permethylated *para*- and *meta*-carboranes.

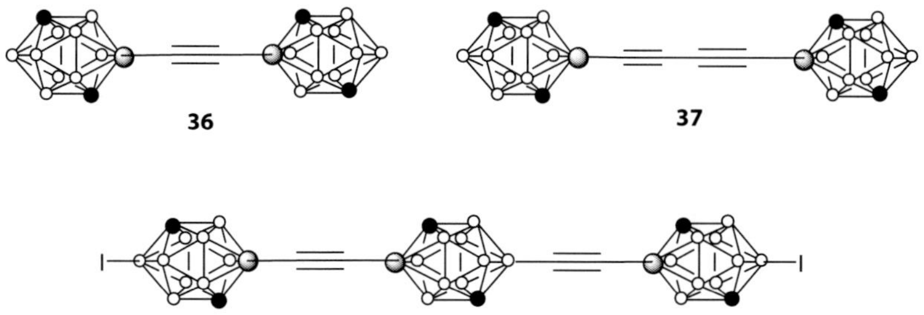

FIGURE 26.13 B-substituted rod-shaped *para*-carboranes.

lene and the subsequent deprotection of trimethylsilyl group generated 2-ethynyl-*p*-carborane, which was then coupled with 2-iodo-*p*-carborane using a palladium-catalyzed reaction to generate ethynyl-bridged **36**. Similarly, two units of 2-ethynyl-*p*-carborane underwent oxidative coupling to generate ethynyl-bridged rod-shaped **37** and oxidative coupling of two equivalents of 2,9-diiodo-*p*-carborane with 2,9-diethynyl-*p*-carborane generated the ethynyl-bridged diiodinated product **38**.

Self-assembly and self-organization of simple building units, called tectons, into finite nanoscopic two- and three-dimensional supramolecular structures of predetermined shape, size, and symmetry, such as molecular rectangles, triangles, squares, hexagons, and higher-order polygons, are of importance in the construction of nanomaterials. One such self-assembly process, coordination-driven self-assembly, has been used for the synthesis of carborane-containing nanoarchitectures (Leininger et al., 2000; Swiegers and Malefetse, 2002; Das et al., 2003). The rectangle-shaped **39** and square-shaped **40** were prepared via the self-assembly of several donor and acceptor carborane tectons (Figure 26.14). Since these platinum-containing macrocycles utilized Pt–N bonding, the resulting complexes were ionic in nature and thus the final box-shaped structures were formed as respective nitrate and triflate salts (Jude et al., 2005). Carboxylate-based neutral platinum macrocycles **41** and **42** have also been synthesized (see Figure 26.15). In this case, dicarboxylic acids of *para*- and *meta*-carboranes were used. The crystal structure analysis showed the length and width of macrocycle **41** to be 18.2 and 7.48 Å, respectively. Molecule **41** is not planar; the twist angle between the two anthracene units is 51° (Das et al., 2005).

The dimensions of the rhomboid macrocycle **42**, which consists of two $Pt(PEt_3)_2$ units connected by two *meta*-carborane dicarboxylate dianions, were found to be 9.19 and 7.01 Å, respectively (Das et al., 2005).

The use of crown ethers in the field of host–guest chemistry and their ability to trap alkali metal cations are well known. On the other hand, anticrowns were rare until Hawthorne's group reported the synthesis of a new class of carborane-containing macrocycles of the form $Hg_n(C_2B_{10}H_{10})_n$ (Figure 26.16). These mercuracarboranes are the charge-reversed analogues of crown ethers possessing Lewis acidic mercury atoms that can trap anions. The synthesis of box-shaped [12]-mercuracarborand-4 (**43**) and [9]-mercuracarborand-3 (**44**) and their use as anticrown agents in host–guest chemistry were reported (Hawthorne and Zheng, 1997; Wedge and Hawthorne, 2003). These two host molecules were synthesized by the reaction of *closo*-1,2-Li_2-1,2-$C_2B_{10}H_{10}$ with mercuric halide and mercuric acetate, respectively. The electrophilic mercury atoms of compound **43** bind to chloride anions, as shown in **45**, and thus act as an anticrown. Tetramer **43** also binds to two iodide ions and trimer **44** is known to combine with water and benzene to form n-sandwich compounds (Zinn et al., 1996, 1999; Lee et al., 2000).

The syntheses of carboracycles containing four 9,10-disubstituted *ortho*-carboranes (**46**) were described by Hawthorne and coworkers (Figure 26.17). The starting 9,10-disubstituted *o*-carboranes

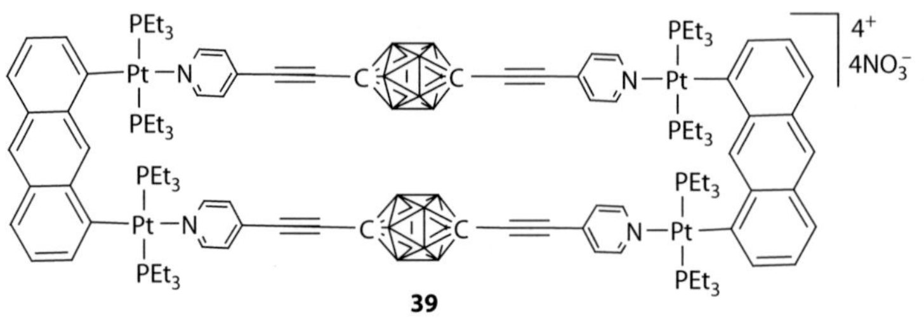

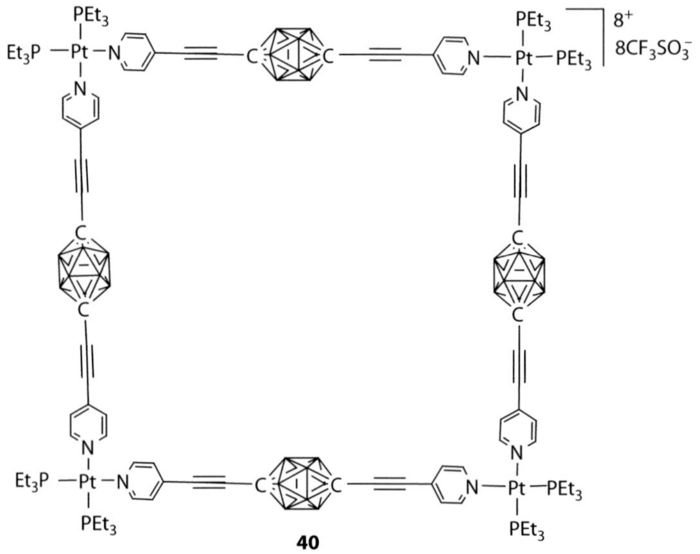

FIGURE 26.14 *Para*-carborane-containing ionic rectangles and squares.

required for these macrocycles were synthesized via two steps; first a Lewis acid-catalyzed iodination at the boron atoms followed by a Kumuda-type palladium-catalyzed coupling reactions with the respective Grignard's reagents. The synthesis of macrocycles required four additional steps, involving protection with the bulky *t*-butyl dimethyl silyl (TBDMS) group, lithiation of the cage carbons and then treatment with appropriate alkyl dibromides, such as 1,3-dibromopropane. Conversion of the methoxy groups of **46** into hydroxyl ones led to the formation of a water-soluble octasulfonate derivative (Bayer et al., 2003). A *meta*-carborane containing an arylcarborane cyclic trimer of the form $[1,7(C_2H_{10}H_{10})]_3(C_6H_4)_3$ (**47**) was reported to be formed in about 5% yield from the reaction between dicopper(I) derivative of *meta*-carborane with *meta*-diiodobenzene in dimethoxyethane (Clegg et al., 1993).

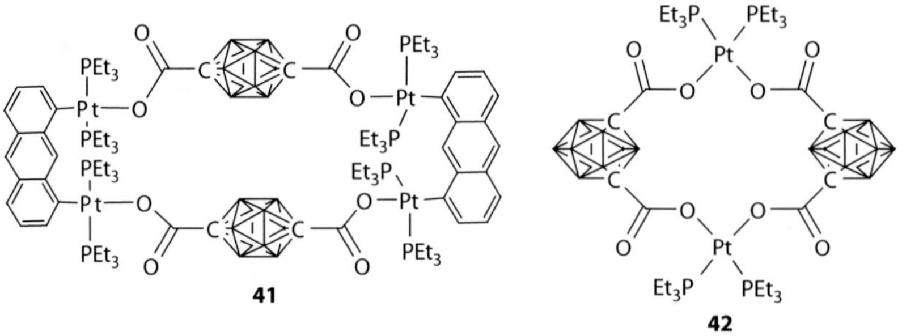

FIGURE 26.15 *Para*- and *meta*-carborane-containing neutral macrocycles.

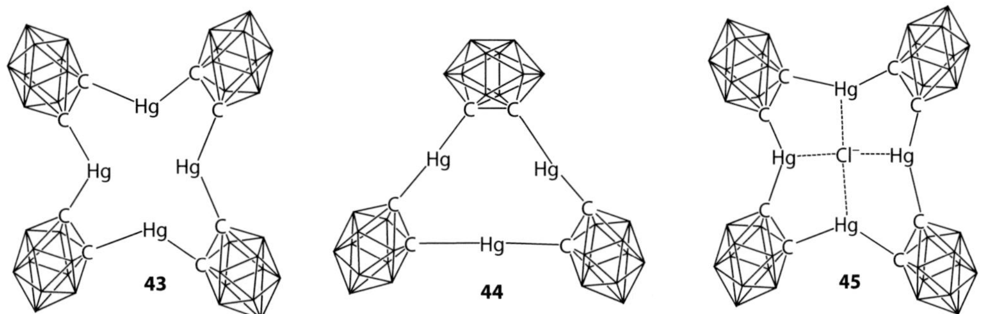

FIGURE 26.16 Mercuracarborands.

The formation of a new category of carboracycles **48** and **49** was also reported in which *ortho*-carborane subunits are joined by alkylene and aromatic linkers using a methodology similar to that described in the synthesis of the previous macrocycle **46** (Figure 26.18). For **48**, 1,3-dibromopropane was used as a linker, whereas for **49**, a,a′-dibromo-1,3-xylene was used as a linker (Jiang et al., 1996a,b).

Pentagonal-pyramidal *nido*-RR′$C_2B_4H_4^{2-}$ and *nido*-RR′$C_2B_4H_5^-$ anions (where *R*, *R′*=H, silyl or organic substituents) are versatile metal-binding ligands because of the ability of these small carborane ligands to form strong covalent bonds with a large number of the main group, transition metals and rare earth elements (Grimes, 1992; Wasczcak et al., 1997; Hosmane and Maguire, 2005). Such characteristics of these small cage carboranes have been used to build metal–ligand molecular architectures such as multidecker rods, boxes, and so on (Russell et al., 2002). Small cage metallacarboranes such as *closo*-ferra- and cobalta-carboranes underwent Negishi-type cross coupling at the B–I and B–Br bonds effecting controlled substitution at the boron, which allows the formation of polymetallated carborane systems. Some examples of rod- and box-shaped polymetallated compounds (**50–53**) are shown in Figure 26.19. The ferra-carborane (η^6-C_6H_6)Fe(2,3-$Et_2C_2B_4H_5$), when

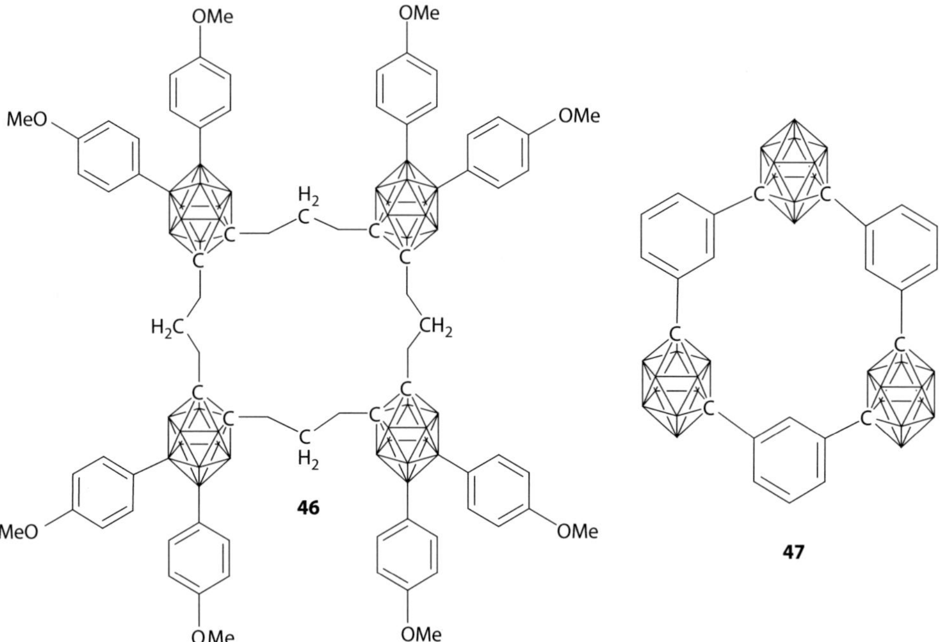

FIGURE 26.17 Macrocycles containing *ortho*- and *meta*-carboranes.

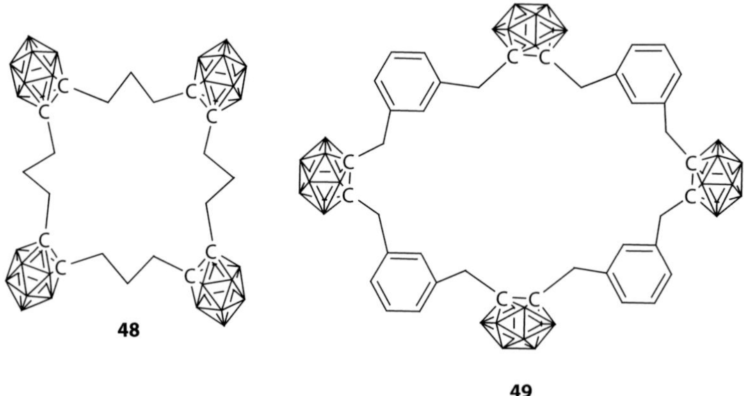

FIGURE 26.18 *Ortho*-carborane-containing macrocycles.

treated with *N*-iodosuccinimide, gave the B(5)-iodo species (η^6-C_6H_6)Fe(2,3-$Et_2C_2B_4H_3$-5-I). This B(5)-iodo species was subjected to palladium-catalyzed Negishi coupling reactions with ClZnC≡CSiMe$_3$ [((trimethylsilyl)ethynyl)zinc chloride (TMSZ)], to generate η^6-(C_6H_6)Fe(2,3-$Et_2C_2B_4H_3$5-I-C≡CSiMe$_3$). Tetrabutyl ammonium fluoride (TBAF)-mediated deprotection of the TMS group afforded the B(5)-ethynyl species. Similarly, the apically substituted B(7)-ethynyl species was also generated from the η^6-(C_6H_6)Fe(2,3-$Et_2C_2B_4H_3$)-7-X (X = Br or I) via a Negishi-type coupling with TMSZ, followed by deprotection of the TMS group. The B(5)-ethynyl species underwent a Sonogashira-type palladium-catalyzed reaction to generate the rod-shaped dimer [η^6-(C_6H_6) Fe(2,3-$Et_2C_2B_4H_3$-5-C≡C)]$_2$, **51** in moderate yield. Similarly, the apically substituted B(7)-connected dimer [η^6-(C_6H_6)Fe(2,3-$Et_2C_2B_4H_3$-5-C≡C)]$_2$, **50**, was obtained via a similar palladium-catalyzed reaction of the B(7)-ethynyl species in the presence of chloroacetone and triethyl amine. In an analogous procedure, the Negishi-type reaction of the *p-bis*(chlorozinc)triphenyl reagent,

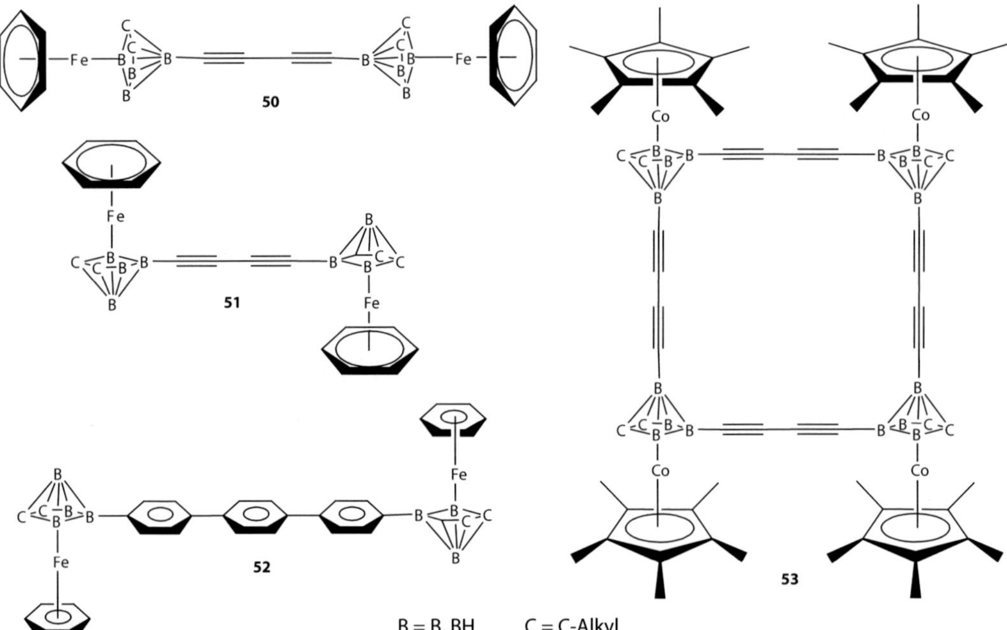

FIGURE 26.19 Rod- and box-shaped small cage (C_2B_4) carboranes.

$ClZn(C_6H_4)_3ZnCl$, with two equivalents of B(5)-iodo species led to the formation of the phenylene-linked *bis*ferra-carborane $[\eta^6\text{-}(C_6H_6)Fe(2,3\text{-}Et_2C_2B_4H_3)\text{-}5\text{-}]_2(C_6H_4)_3$, **52** (Russell et al., 2002). The synthesis of a small cage cobaltacarborane square-shaped nanostructure was also reported by Grimes' group. The compound was synthesized from $[Cp^*Co(2,3\text{-}Et_2C_2B_4H_3)\text{-}7\text{-}I]$ by repetitively using Negishi- and Sonogashira-type couplings followed by copper-catalyzed dimerization reaction. The B(5,7)-disubstituted ethynyl derivatives were synthesized from the B(7) functionalized compound by treating it with *N*-iodosuccinimide and subsequent Negishi-type coupling with $ClZn\text{-}C{\equiv}C\text{-}SiMe_3$ (TMSZ). Finally, dimerization of the rod-shaped B(5,7) ethynyl derivatives with copper led to the formation of metallacarborane molecular box **53** (Yao et al., 2003b).

26.4 POLYMERIC MATERIALS

Some unique properties of carboranes and weakly coordinating carborane anions have been exploited for the synthesis of a variety of polymeric materials. Carborane-containing polymeric materials are thermally very stable. In this section we will discuss the use of carborane clusters for the synthesis of thermally stable polymers, light-emitting luminescent polymers, conducting organic polymers (COPs), and coordination polymers.

26.4.1 Thermally Stable Polymers

The use of icosahedral carborane clusters ($C_2B_{10}H_{12}$) in the synthesis of thermally stable polymeric materials is well known. Over the last few decades, several groups have reported the incorporation of carboranes into polymeric backbones, using various well-established polymer linking processes (Stewart et al., 1973; Ichitani et al., 1999; Papetti et al., 1996). Incorporation of carborane units into polymers generally leads to an enhanced thermal stability. In addition, a few carborane–siloxane elastomers, bearing the commercial names DEXSIL and UCARSIL, have found specialized applications (Grimes, 1970). A silyl-carborane hybrid diethynylbenzene–silylene polymer, **55**, that contains *meta*-carborane units in the side chains was found to be highly thermally stable, exceeding that of other polymers in heat resistance (Figure 26.20). After it was cured at 350°C, the thermal gravimetric analysis of **55** in air showed less than 5% mass loss at temperatures exceeding 1000°C (Hideaki et al., 2003).

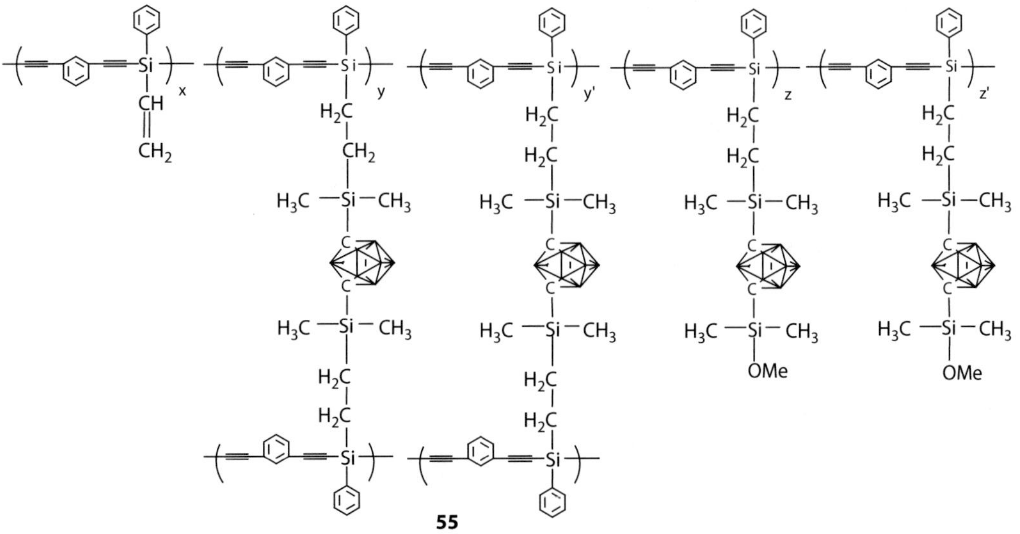

55

FIGURE 26.20 A silyl-carborane hybrid diethynylbenzene–silylene polymer.

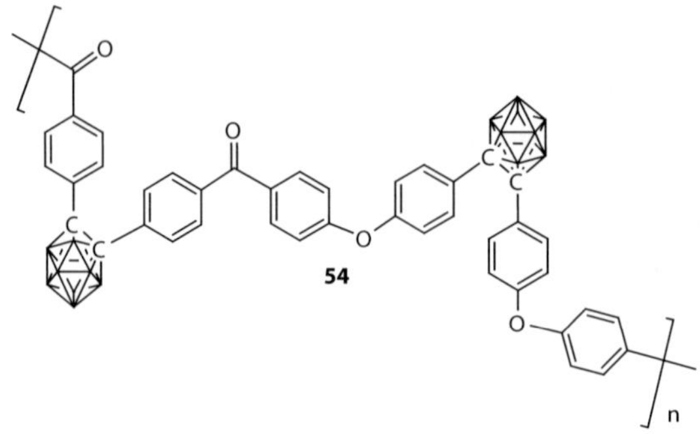

FIGURE 26.21 Phenylene ether carboranylene ketone (PECK) polymer.

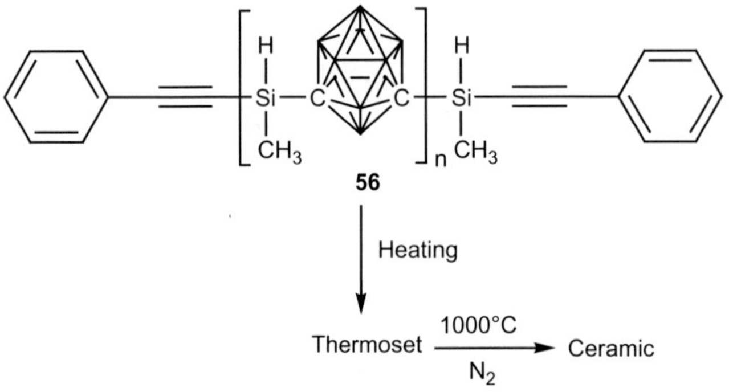

FIGURE 26.22 Phenyl acetylene-terminated poly(carborane-silane) (PACS) polymer.

In another report, polyether ketones containing carboranes in the backbone of aromatic units were synthesized by the electrophilic condensation between the ether, 1,2-(4-PhOC_6H_4)$_2$-1,2-$C_2B_{10}H_{10}$, and the dicarboxylic acid, 1,2-(4-$CO_2HC_6H_4$)$_2$-1,2-$C_2B_{10}H_{10}$, with trifluoro-methanesulfonic acid (Figure 26.21). This phenylene ether carboranylene ketone (PECK) polymer (**54**) was found to be extremely thermally stable and showed a much smaller mass loss, up to 1000°C, than was observed for conventional aromatic ether-ketones (Mark and Kenneth, 2002).

The synthesis and thermal properties of a phenyl acetylene terminated poly(carborane-silane) (PACS) polymers, containing *meta*-carborane clusters, were recently reported (Quan et al., 2007). The polymer backbone (**56** in Figure 26.22) consisted of silicon units bonded to *meta*-carboranes, with phenyl acetylene units as end groups. This polymer, found to be a viscous liquid at room temperature, was obtained from the reaction of methyldichlorosilane and lithium phenylacetylide with the dilithium salt of the *m*-carborane. Heating of polymer **56** led to the formation of both thermosetting and ceramic formation. The thermoset, which was obtained after heating the PACS polymer, showed no mass loss up to 510°C, and only 5% mass loss at 800°C under nitrogen. Similarly, the ceramics obtained from the PACS polymer were also found to be highly thermally stable (Quan et al., 2007).

26.4.2 Light-Emitting Polymers

Light-emitting n-conjugated polymers find numerous electronic applications as organic light-emitting diodes, organic lasers, organic field effect transistors, and other organic devices (McQuade et al., 2000; Thomas et al., 2007). Incorporation of boron atoms into conjugated polymers leads to new materials with improved optoelectronic properties (Entwistle and Marder, 2002; Sundararaman et al., 2005; Zhao et al., 2006; Nagata and Chujo, 2008). Recent reports showed that boron clusters, such as *meta*- and *para*-carboranes, are good candidates for the synthesis of light-emitting polymeric

FIGURE 26.23 *Meta*-carborane–based luminescent n-conjugated polymers.

materials. The incorporation of these clusters was also found to improve the optoelectronic properties of polymeric materials.

A series of *meta*-carborane-based n-conjugated polymers (**57**–**60** in Figure 26.23), possessing alternate *meta*-carborane-*p*-phenylene–ethynylene sequences, were found to exhibit intense blue light emission in solution. The UV–vis absorption spectra of these polymers indicated the extension of n-conjugation due to the presence of *meta*-carborane moieties in the polymer backbone (Kokado et al., 2009b).

The synthesis and photophysical properties of *ortho*- and *meta*-carborane-containing n-conjugated chiral polybinaphthyls were recently reported (Figure 26.24) (Kokado et al., 2009a,b). The polymers containing *meta*-carborane (**62**) exhibited intense blue luminescence, while the *ortho*-carborane-containing polymer (**61**) showed almost no emission in solution, with a quantum yield $\surd F\phi_F < 0.0001$ (Kokado et al., 2009a,b); however, a characteristic aggregation-induced emission (AIE) was observed for *ortho*-carborane-containing polymers on aggregate or film formation.

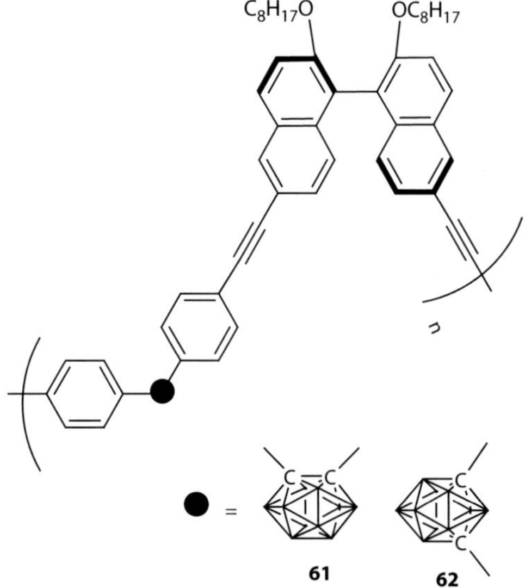

FIGURE 26.24 Luminescent chiral n-conjugated polymers with carboranes.

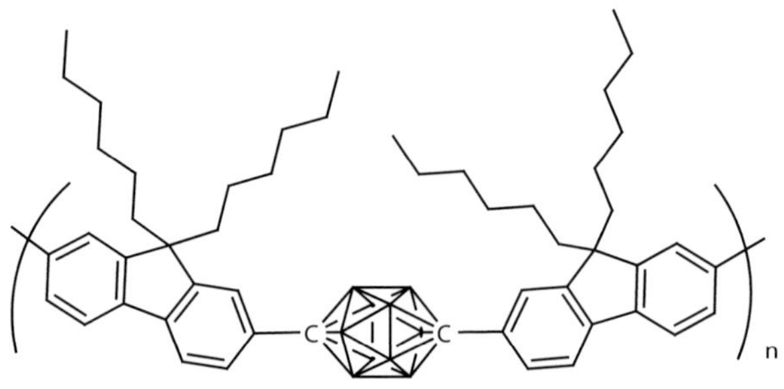

FIGURE 26.25 Polyfluorene with *p*-carborane backbone.

Intramolecular charge transfer from the electron-donating *para*-phenylene-ethylene units to the antibonding orbital of C–C bond in *ortho*-carborane cluster was thought to result in a nonradiative quenching process. The presence of the icosahedral 3D framework effectively prevented the approximation and stacking of the n-conjugated systems (Kokado et al., 2009a,b).

The synthesis of a polyfluorene with *para*-carborane in the backbone was reported (compound **63** in Figure 26.25). This polymer was found to be a blue light-emitting material. The monomer 1,12-*bis*(7-bromo-9,9-dihexyl-9H-fluoren-2-yl)-*closo*-1,12-dicarbadodecaborane was first synthesized and then polymerized via a Ni(0)-catalyzed dehalogenative polymerization. Shifts in UV absorption and fluorescence emission indicate some involvement of the p-carborane clusters in extending the conjugation (Peterson et al., 2009).

26.4.3 Conducting Polymers

Conducting organic polymers (COPs) can be a substitute for metal and inorganic semiconductors and, hence, find applications in numerous electronic and mechanical devices. The commercialization of COP-based devices faces multiple problems due to their poor stability. These polymers must be thermally stable and must have the potential to withstand rapid oxidation–reduction processes. Therefore, the synthesis of COPs with improved chemical and thermal stability is a challenging task (Gardner and Bartlett, 1995; Horowitz, 1998; Cumpston et al., 1997; Savvateev et al., 1999; Smela, 1999). In this section we will discuss the syntheses of carborane-containing conjugated polypyrroles (Ppy) and polythiophenes as COPs.

The cobalta-*bis*-dicarbollide anion $[Co(C_2B_9H_{11})_2]^-$ is a thermally and chemically stable, noncoordinating, lipophilic, rigid molecule that has been used as a doping agent for the activation of polypyrrole-based conducting polymers. This anion was uniformally distributed on the surface of the resulting Ppy polymer. The $[Co(C_2B_9H_{11})_2]^-$-containing polymer showed enhanced thermal stability and a dramatic enhancement of its overoxidation threshold (Masalles et al., 2000). Modified Ppy-$[Co(C_2B_9H_{11})_2]^-$ whose surface was covered by a layer of alkylammonium or alkylphosphonium $[Co(C_2B_9H_{11})_2]^-$ salts, resulting in highly stable layers, was also synthesized by the same group (Masalles et al., 2002a). In another report the polymerization of pyrole, bonded through a diether to a cobalt*bis*dicarbollide anion, produced a self-doped conducting COP (Figure 26.26). Derivatization was at the *N*-atom of the pyrrole, so as to leave the a-carbon atoms unsubstituted and ready for subsequent polymerization. The direct polymerization of the monomer **64**, as well as copolymerization with pyrrole units, led to the formation of self-doped COPs. This polymer showed the highest overoxidation resistance of 1.5 V ever found for any COP, thereby making it the best COP in terms of overoxidation resistance (Masalles et al., 2002b).

The synthesis, electropolymerization, and electrical conductivity properties of (2-thiophenyl) carboranes **65**–**67** containing *ortho*-, *meta*-, and *para*-carborane clusters have been studied (Hao et al., 2007) (Figure 26.27). The resulting carborane-containing polythiophenes showed high

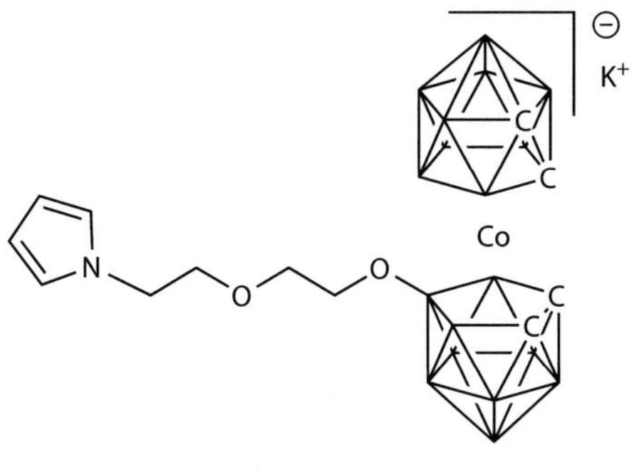

FIGURE 26.26 Cobaltbisdicarbollide anion-attached pyrrole monomer.

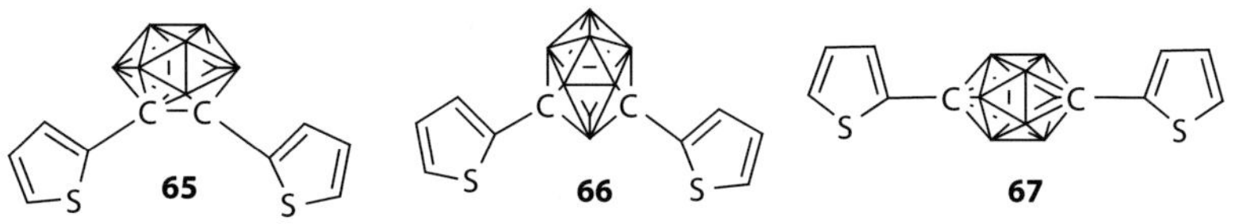

FIGURE 26.27 Carborane-containing thiophenes.

thermal and electrochemical stabilities in comparison with polythiophenes without carboranes. Among these polythiophenes the *ortho*-carborane-containing polythiophene showed highest conductivity (Hao et al., 2007).

26.4.4 Coordination Polymers

Coordination polymers and metal organic frameworks (MOFs) are becoming increasingly important in supramolecular chemistry. Coordination polymers are highly ordered coordination complexes with infinite 1D, 2D, or 3D network structures. These polymers possess properties such as high internal surface areas and permanent microporosity. The internal cavities and channels present in MOFs can hold guest molecules and counter-ions; thus, these compounds find zeolitic, magnetic, optical, and catalytic applications. The C–H bonds of carboranes are acidic in nature and hence are capable of forming hydrogen bonds. Some of the anionic carboranes and metallacarboranes, shown in Figure 26.28, have been used in the synthesis of coordination polymers (Hardie, 2007).

These compounds tend to crystallize nicely and can also be easily functionalized. Properties such as their bulky 3D structure and steric factors are also favorable for network structures (Hardie, 2007). Although carboranes are weakly coordinating in nature, they are capable of forming bonding interactions such as 3c–2e bonds (three-centered two-electron bonds). Many coordination networks with carborane anions employ additional bridging ligands as connecting units. Some examples of such bidentate (dinitriles, pyz, and bpy) and tridentate *tris*(pyamino)CTG ligands are shown in Figure 26.29 (Hardie, 2007). A series of Ag(I) coordination networks with alkane-dinitriles NC-$(CH_2)_n$-CN ($n = 1$–4) and carborane counteranions, such as $(CB_{11}H_{12})^-$ and $Co(C_2B_9H_{10})_2^-$, were reported by Hardie's group. The carborane counteranions interact with the Ag(I) centers through B-H—Ag interactions, forming either discrete, chain or 3D lattice structures. [Ag(pyz)$(CB_{11}H_{12})^-$] formed a 3D coordination network structure, whereas for [Ag(bpy)$(CH_3CN)Co(C_2B_9H_{10})_2$] two polymorphs were observed. These coordination chains pack in a manner that forms box-like channels occupying $Co(C_2B_9H_{10})_2^-$ anions (Westcott et al., 2004; Cunha-Silva et al., 2006a,b).

The rigid bowl-shaped cyclotriveratrylene (CTV in Figure 26.29) with its molecular cavity is capable of complexing guest molecules. Unsubstituted CTV forms supramolecular structures

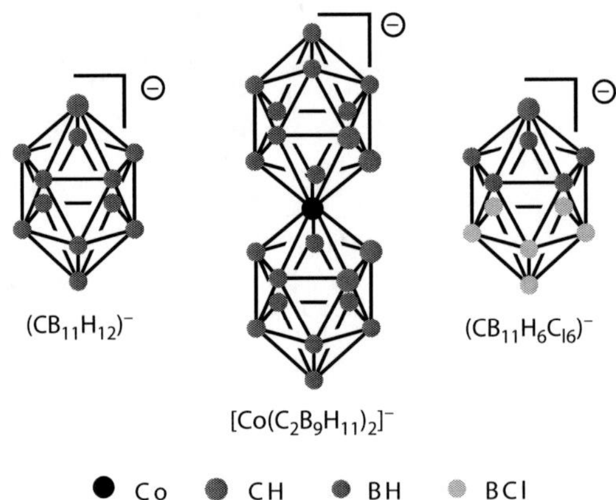

FIGURE 26.28 Anionic carboranes and metallacarboranes for coordination polymerization.

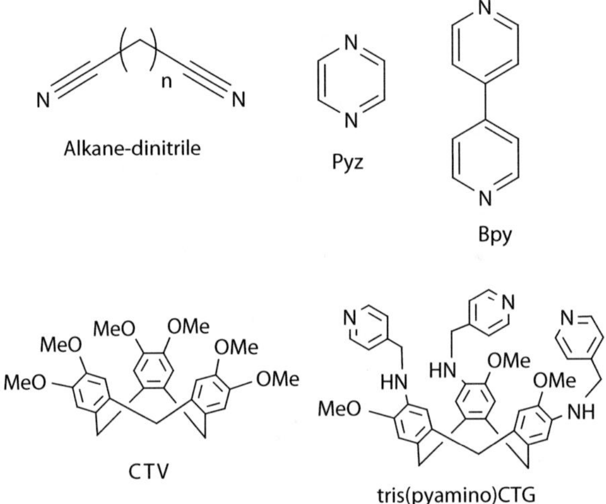

FIGURE 26.29 Useful connecting ligands for coordination polymerization.

with *o*-carboranes, anionic $(CB_{11}H_{12})^-$ and alkali metals. Similarly, CTV also forms crystalline complexes with group I metals and the halogenated monocarborane $(CB_{11}H_6Cl_6)^-$ ions. Amine functionalized CTV, such as [*tris*(pyamino)CTG], shown in Figure 26.29, was also found to be a useful ligand for making MOFs with Ag(I) having $Co(C_2B_9H_{10})_2^-$ and $(CB_{11}H_{12})^-$ as counteranions (Blanch et al., 1997; Hardie et al., 2001; Ahmad et al., 2004; Sumby et al., 2006). MOFs possessing extensive hydrogen bonds were also formed from a mixture of CTV, acetonitrile, $Sr[Co(C_2B_9H_{10})_2]_2$, and water (Hardie et al., 2001). Unsubstituted CTV also forms a series of compounds involving complex and host–guest formation with $Ag[Co(C_2B_9H_{10})_2]$ and acetonitrile (Ahmad and Hardie, 2003). Rare earth elements such as La, Ce, Gd, Tb, and Ho formed aquo-lanthanide complexes with hydrogen bonded network structures involving 4,4′-bipyridine (bpy) and $Co(C_2B_9H_{10})_2^-$ anion. The use of large $Co(C_2B_9H_{10})_2^-$ anion resulted in complexes with considerably larger rectangular channels (Cunha-Silva et al., 2006a,b). Europium-containing crystalline network structures were also synthesized from CTV, $Eu[Co(C_2B_9H_{10})_2]_3$ in a mixture of water and acetonitrile (Ahmad et al., 2003).

A *para*-carborane-containing MOF material showed hydrogen uptake properties. Dicarboxylic acid dianions of *p*-carborane (*p*-CDC) were used (Figure 26.30) and were coordinated with Zn(II) in the presence of diethyl formamide (DEF) to generate an MOF of the form $[Zn_3(OH)$

FIGURE 26.30 *Para*-carborane dicarboxylic acid dianion.

$(p\text{-CDC})_{2.5}(\text{DEF})_4]_n$. The compound lost its DEF molecules when heated under vacuum at 300°C. Although the solvent-free version was crystalline, no structure was available. This solvent-free version was found to display a surprisingly high H_2 uptake (25 mg of H_2/cm^3 at 77 K and 1 atm H_2) (Omar et al., 2007).

26.5 CONCLUSIONS

Carborane clusters are useful synthetic building blocks for synthesis of new materials. Numerous unique properties of carboranes, such as thermal stability, robust 3D inorganic cage structures, ability to bind metals firmly, and inertness toward chemicals make these compounds ideal choices for use in synthetic chemistry. Carborane cages are easy to functionalize at either the boron or carbon atoms. They have been utilized in the syntheses of polymers of high thermal and conducting properties, organic light-emitting polymeric materials, dendritic macromolecules, and nanostructured building blocks such as rods, boxes, chains, and nanomachines. More recently, boron clusters have been utilized in supramolecular constructs such as metal organic frameworks (MOFS) for specialized applications. We can look forward to an increasing use of carboranes to enhance desirable, specialized properties of materials.

ACKNOWLEDGMENT

Supports by grants from the National Science Foundation (CHE-0906179), the Robert A. Welch Foundation (N-1322), and the awards from Alexander von Humboldt Foundation and NIU Inaugural Board of Trustees Professorship are hereby acknowledged.

REFERENCES

Ahmad, R.; Hardie, M. J. 2003. Variable Ag(I) coordination modes in silver cobalt(III) bis(dicarbollide) supramolecular assemblies with cyclotriveratrylene host molecules. *Cryst. Growth Des.*, 3: 493–9.

Ahmad, R.; Dix, I.; Hardie, M. J. 2003. Hydrogen-bonded superstructures of a small host molecule and lanthanide aquo ions. *Inorg. Chem.*, 42: 2182–4.

Ahmad, R.; Franken, A.; Kennedy, J. D.; Hardie, M. J. 2004. Group 1 coordination chains and hexagonal networks of host cyclotriveratrylene with halogenated monocarborane anions. *Chem. Eur. J.*, 10: 2190–8.

Astruc, D.; Orneals, C.; Ruiz, J. 2008. Metallocenyl dendrimers and their applications in molecular electronics, sensing, and catalysis. *Acc. Chem. Res.*, 41: 841–56.

Bayer, M. J.; Herzog, A.; Diaz, M.; Harakas, G. A.; Lee, H.; Knobler, C. B.; Hawthorne, M. F. 2003. The Synthesis of carboracycles derived from B,B`-Bis(aryl) derivatives of icosahedral *ortho*-carborane. *Chem. Eur., J.* 9: 2732–44.

Blanch, R. J.; Williams, M.; Fallon, G. D.; Gardiner, M. G.; Kaddour, R.; Raston, C. L. 1997. Supramolecular complexation of 1,2-dicarbadodecaborane. *Angew. Chem., Int. Ed. Engl.*, 36: 504–6.

Chebny, V. C.; Dhar, D.; Lindeman, S. V.; Rathore, R. 2006. Simultaneous ejection of six electrons at a constant potential by hexakis(4-ferrocenylphenyl)benzene. *Org. Lett.*, 8: 5041–4.

Clegg, W.; Gill, W. R.; MacBride, J. A. H.; Wade, K. 1993. $(1,7\text{-}C_2B_{10}H_{10}\text{-}1',3'\text{-}C_6H_4)_3$, a cyclic trimer from *meta*-carboranediyl and *meta*-phenylene units: A new category of macrocycle. *Angew. Chem. Int. Ed. Engl.*, 32: 1328–9.

Craciun, L.; Ho, D. M.; Jones, M. Jr., Pascal, R. A. Jr. 2004. A Carboranylpentaphenylbenzene. *Tetrahedron Lett.*, 45: 4985–7.

Cumpston, B. H.; Parker, I. D.; Jensen, K. F. 1997. *In situ* characterization of the oxidative degradation of a polymeric light emitting device. *J. Appl. Phys.*, 81: 3716.

Cunha-Silva, L.; Ahmad, R.; Hardie, M. J. 2006a. Coordination networks with carborane anions: Ag(I) and nitrogen bridging ligands. *Aust. J. Chem.*, 59: 40–8.

Cunha-Silva, L.; Westcott, A.; Whitford, N.; Hardie, M. J. 2006b. Hydrogen-bonded 3-D network structures of lanthanide aquo ions and 4,4′-bipyridine with carborane anions. *Cryst. Growth Des.*, 3: 726–35.

Darling, D. J. 1995. *Micromachines and Nanotechnology: The Amazing New World of the Ultrasmall.* Parsippany: Dillon Press.

Das, N.; Mukherjee, P. S.; Arif, A. M.; Stang, P. J. 2003. Facile self-assembly of predesigned neutral 2D Pt-macrocycles via a new class of rigid oxygen donor linkers. *J. Am. Chem. Soc.*, 125: 13950–1.

Das, N.; Stang, P. J.; Arif, A. M.; Campana, C. F. 2005. Synthesis and structural characterization of carborane-containing neutral, self-assembled Pt-metallacycles. *J. Org. Chem.*, 70: 10440–6.

Dash, B. P.; Satapathy, R.; Gaillard, E. R.; Maguire, J. A.; Hosmane, N. S. 2010. *J. Am. Chem. Soc.,* 132, 6578–87.

Dash, B. P.; Satapathy, R.; Maguire, J. A.; Hosmane, N. S. 2008. Synthesis of a new class of carborane-containing star-shaped molecules via silicon tetrachloride promoted cyclotrimerization reactions. *Org. Lett.*, 10: 2247–50 and references therein.

Dash, B. P.; Satapathy, R.; Maguire, J. A.; Hosmane, N. S. 2009. Boron enriched star shaped molecules via cycloaddition reaction. *Chem. Commun.*, 3267–9.

Entwistle, C. D.; Marder, T. B. 2002. Boron chemistry lights the way: Optical properties of molecular and polymeric systems. *Angew. Chem., Int. Ed.*, 41: 2927–31.

Gardner, J. W.; Bartlett, P. N. 1995. Application of conducting polymer technology in microsystems. *Sens. Actuators A.*, 51: 57–66.

Gonzalez-Campo, A.; Vinas, C.; Teixidor, F.; Nunez, R.; Sillanpaa, R.; Kivekas, R. 2007. Modular construction of neutral and anionic carboranyl-containing carbosilane-based dendrimers. *Macromolecules*, 40: 5644–52.

Grayson, S. M.; Fréchet, J. M. J. 2001. Convergent dendrons and dendrimers: From synthesis to applications. *Chem. Rev.*, 101: 3819–67.

Grimes, R. N. 1970. *Carboranes*. New York: Academic Press.

Grimes, R. N. 1992. Boron–carbon ring ligands in organometallic synthesis. *Chem. Rev.*, 92: 251–68.

Grimes, R. N. 2004. Boron clusters come of age. *J. Chem. Educ.*, 81: 657–672.

Gupta, H. K.; Reginato, N.; Ogini, F. O.; Brydges, S.; McGlinchey, M. J. 2002. $SiCl_4$–ethanol as a trimerization agent for organometallics: Convenient syntheses of the symmetrically substituted arenes 1,3,5-$C_6H_3R_3$ where R=$(C_5H_4)Mn(CO)_3$ and $(C_5H_4)Fe(C_5H_5)$. *Can. J. Chem.*, 80: 1546–54.

Hao, E.; Fabre, B.; Fronczek, F. R.; Vicente, M. G. H. 2007. Poly[di(2-thiophenyl)carborane]s: Conducting polymers with high electrochemical and thermal resistance. *Chem. Commun.*, 4387–9.

Hardie, M. J. 2007. The use of carborane anions in coordination polymers and extended solids. *J. Chem. Crystallogr.,* 37: 69–80.

Hardie, M. J.; Raston, C. L. 2001. Alkali-metal – cyclotriveratrylene coordination polymers: Inclusion of neutral $C_2B_{10}H_{12}$ or anionic $[CB_{11}H_{12}]^-$ and DMF. *Cryst. Growth Des.*, 1: 53–8.

Hardie, M. J.; Raston, C. L.; Salinas, A. 2001. A 3,12-connected vertice sharing adamantoid hydrogen bonded network featuring tetrameric clusters of cyclotriveratrylene. *Chem. Commun.*, 1850–1.

Hawthorne, M. F.; Maderna, A. 1999. Applications of radiolabeled boron clusters to the diagnosis and treatment of cancer. *Chem. Rev.,* 99: 3421–34.

Hawthorne, M. F.; Zheng, Z. 1997. Recognition of electron-donating guests by carborane-supported multidentate macrocyclic Lewis acid host. *Accounts Chem. Res.*, 30: 267–76.

Herzog, A.; Jalisatgi, S. S.; Knobler, C. B.; Wedge, T. J.; Hawthorne, M. F. 2005. Camouflaged carborarods derived from B-permethyl-1,12-diethynyl-para-and B-octamethyl-1,7-diethynyl-meta-carborane modules. *Chem. Eur. J.*, 11: 7155–74.

Hideaki, K.; Koichi, O.: Motokuni, I.; Toshiya, S.; Shigeki, K.; Isao, A. 2003. Structural study of silyl–carborane hybrid diethynylbenzene–silylene polymers by high-resolution solid-state ^{11}B, ^{13}C, and ^{29}Si NMR spectroscopy. *Chem. Mater.*, 15: 355–62.

Horowitz, G. 1998. Organic field-effect transistors. *Adv. Mater.*, 10: 365–77.

Hosmane, N. S; Maguire, J. A. 2005. Evolution of C_2B_4 chemistry: From early years to the present. 24: 1356–89 and references there in.

Ichitani, M.; Yonezawa, K.; Okada, K.; Sugimoto, T. 1999. Silyl-carborane hybridized diethynylbenzene–silylene polymers. *Polym. J.*, 31: 908–12.

Ito, S.; Wehmeier, M.; Brand, J. D. et al. 2000. Synthesis and self-assembly of functionalized hexa-*peri*-hexabenzocoronenes. *Chem. Eur. J.*, 6: 4327–42.

Jiang, W.; Chizhevsky, I. T.; Mortimer, M. D. et.al. 1996a. Carboracycles: Macrocyclic compounds composed of carborane iscosahedra linked by organic bridging groups. *Inorg. Chem.*, 35: 5417–26.

Jiang, W.; Harwell, D. E.; Mortimer, M. D.; Knobler, C. B.; Hawthorne, M. F. 1996b. Palladium-catalyzed coupling of ethynylated *p*-carborane derivatives: Synthesis and structural characterization of modular ethynylated *p*-carborane molecules. *Inorg. Chem.*, 35: 4355–89.

Juarez-Perez, E. J.; Vinas, C.; Teixidor, F.; Nunez, R. 2009. Polyanionic carbosilane and carbosiloxane metallodendrimers based on cobaltabisdicarbollide derivatives. *Organometallics* 28: 5550–9.

Jude, H.; Disteldorf, H.; Fischer, S. et al. 2005. Coordination-driven self-assemblies with a carborane backbone. *J. Am. Chem. Soc.*, 127: 12131–9.

Kokado, K.; Tokoro, Y.; Chujo, Y. 2009a. Luminescent and axially chiral n-conjugated polymers linked by carboranes in the main chain. *Macromolecules,* 42: 9238–42.

Kokado, K.; Tokoro, Y.; Chujo, Y. 2009b. Luminescent *m*-carborane-based n-conjugated polymer. *Macromolecules,* 42: 2925–30.

Koumura, N.; Geertsema, E. M.; van Gelder, M. B.; Meetsa, A.; Feringa, B. L. 2002. Second generation light-driven molecular motors. Unidirectional rotation controlled by a single stereogenic center with near-perfect photoequilibria and acceleration of the speed of rotation by structural modifications. *J. Am. Chem. Soc.,* 124: 5037–5051.

Kumar, S. 2006. Self-organization of disc-like molecules: Chemical aspects. *Chem. Soc. Rev.*, 35: 83–109.

Lee, H.; Diaz, M.; Knobler, C. B.; Hawthorne, M. F. 2000. Octahedral coordination of iodide in an electrophilic sandwich. *Angew. Chem. Int. Ed.*, 39: 776–8.

Lehn, J. M. 1995. *Supramolecular Chemistry—Concepts and Perspectives.* Weinheim: Wiley-VCH.

Leininger, S.; Olenyuk, B.; Stang, P. J. 2000. Self-assembly of discrete cyclic nanostructures mediated by transition metals. *Chem. Rev.*, 100: 853–908.

Lo, S.-C.; Burn, P. L. 2007. Development of dendrimers: Macromolecules for use in organic light-emitting diodes and solar cells. *Chem. Rev.*, 107: 1097–1116.

Mark, A. F.; Kenneth, W. 2002. Model compounds and monomers for phenylene ether carboranylene ketone (PECK) polymer synthesis: Preparation and characterization of boron-arylated *ortho*-carboranes bearing carboxyphenyl, phenoxyphenyl or benzoylphenyl substituents. *J. Mater. Chem.*, 12: 1301–6.

Masalles, C.; Borros, S.; Vinas, C.; Teixidor, F. 2000. Are low-coordinating anions of interest as doping agents in organic conducting polymers? *Adv. Mater.*, 12: 1199–1202.

Masalles, C.; Borros, S.; Vinas, C.; Teixidor, F. 2002a. Surface layer formation on polypyrrole films. *Adv. Mater.*, 14: 449–52.

Masalles, C.; Llop, J.; Vinas, C.; Teixidor, F. 2002b. Overoxidation resistance increase in self-doped polypyrroles by using non-conventional low charge-density anions. *Adv. Mater.*, 14: 826–9.

McQuade, D. T.; Pullen, A. E.; Swager, T. M. 2000. Conjugated polymer-based chemical sensors. *Chem. Rev.*, 100: 2537–74.

Morin, J-F.; Shirai, Y.; Tour, J. M. 2006. En route to a motorized nanocar. *Org. Lett.*, 8: 1713–6.

Morin, J-F.; Sasaki, T.; Shirai, Y.; Guerrero, J. M.; Tour, J. M. 2007. Synthetic routes toward carborane-wheeled nanocars. *J. Org. Chem.*, 72: 9481–90.

Müller, J.; Base, K.; Magnera, T. F.; Michl, J. 1992. Rigid-rod oligo-p-carboranes for molecular tinkertoys. An inorganic Langmuir–Blodgett film with a functionalized outer surface. *J. Am. Chem. Soc.*, 114: 9721–22.

Muller, M.; Kubel, C.; Müllen, K. 1998. Giant polycyclic aromatich hydrocarbons. *Chem. Eur. J.*, 4: 2099–2109.

Nagata, Y.; Chujo, Y. 2008. Synthesis of methyl-substituted main-chain-type organoboron quinolate polymers and their emission color tuning. *Macromolecules* 41: 2809–13.

Newkome, G. R.; Moorefield, C. N.; Keith, J. M.; Baker, G. R.; Escamilla, G. S. 1994. Chemistry within a unimolecular micelle precursor: Boron superclusters by site-and depth-specific transformations of dendrimers. *Angew. Chem. Int. Ed. Engl.*, 33: 666–8.

Nishimaya, N.; Iriyama, A.; Jang, W-D. et al. 2005. Light-induced gene transfer from packaged DNA enveloped in a dendrimeric photosensitizer. *Nat. Mater.*, 4: 934–41.

Omar, K. F. Alexander, M. S. Karen, L. M. Hawthorne, M. F. Chad, A. M.; Joseph, T. H. 2007. Synthesis and hydrogen sorption properties of carborane based metal-organic framework materials. *J. Am. Chem. Soc.*, 129: 12680–1.

Papetti, S.; Schaeffer, B. B.; Gray, A. P.; Heying, T. L. 1966. A new series of organoboranes. VII. The preparation of poly-*m*-carboranylenesiloxanes. *J. Polym. Sci. Part A*-1., 4: 1623–36.

Parrott, M. C.; Marchington, E. B.; Valliant, J. F.; Adronov, A. 2005. Synthesis and properties of carborane-functionalized aliphatic polyester dendrimers. *J. Am. Chem. Soc.*, 127: 12081–9.

Peterson, J. J.; Simon, Y. C.; Coughlin, E. B.; Carter, K. R. 2009. Polyfluorene with *p*-carborane in the backbone. *Chem. Commun.*, 4950–2.

Pisula, W.; Kastler, M.; Wasserfallen, D.; Pakula, T.; Müllen, K. 2004. Exceptionally long-range self-assembly of hexa-*peri*-hexabenzocoronene with dove-tailed alkyl substituents. *J. Am. Chem. Soc.*, 126: 8074–5.

Plesek, J. 1992. Potential application of the boron cluster compounds. *Chem. Rev.*, 92: 269–78.

Quan, Z.; Zuju, M.; Lizhong, N.; Jianding, C. 2007. Novel phenyl acetylene terminated poly(carborane-silane): Synthesis, characterization, and thermal property. *J. Appl. Polym. Sci.*, 104: 2498–2503.

Regis, E. 1995. *Nano: The Emerging Science of Nanotechnology*. Boston: Back Bay Books.

Russell, J. M.; Sabat, H.; Grimes, R. N. 2002. Organotransition-metal metallacarboranes. 59. Synthesis and linkage of boron-functionalized ferracarborane clusters. *Organometallics*, 21: 4113–28.

Saragi, T. P. I.; Spehr, T.; Siebert, A. et al. 2007. Spiro compound for organic optoelectronics. *Chem. Rev.*, 107: 1011–65.

Sasaki, T.; Tour, J. M. 2008. Synthesis of a new photoactive nanovehicle: A Nanoworm. *Org. Lett.*, 10: 897–900.

Savvateev, V.; Yakimov, A.; Davidov, D. 1999. Transient electroluminescence from poly(phenylenevinylene)-based devices. *Adv. Mater.*, 11: 519–31.

Schwab, P. F. H.; Levin, M. D.; Michl, J. 1999. Molecular rods. 1. Simple axial rods. *Chem. Rev.*, 99: 1863–1934.

Shirai, Y.; Morin, J-F.; Sasaki, T.; Guerrero, J. M.; Tour, J. M. 2006. Recent progress on nanovehicles. *Chem. Soc. Rev.*, 35: 1043–55.

Sivaev, I. B.; Bregadze, V. V. 2009 Polyhedral boranes for medicinal applications: Current status and perspectives. *Eur. J. Inorg. Chem.*, 1433–1450.

Smela, E. 1999. A microfabricated movable electrochromic "pixel" based on polypyrrole. *Adv. Mater.*, 11:1343–5.

Stewart, D. D.; Peters, E. N.; Beard, C. D.; Dunks, G. B.; Hadaya, E.; Kwiatkowski, G. T.; Moffitt, R. B.; Bohan, J. J. 1973. Viscosity and normal stresses of linear and star branched polystyrene solutions. II. Shear-dependent properties. *Macromolecules*, 12: 373–7.

Sumby, C. J.; Fisher, J.; Prior, T. J.; Hardie, M. J. 2006. Tris(pyridylmethylamino)cyclotriguaiacylene cavitands: An investigation of the solution and solid-state behaviour of metallo-supramolecular cages and cavitand-based coordination polymers. *Chem. Eur. J.*, 12: 2945–59.

Sundararaman, A.; Victor, M.; Varughese, R.; Jakle, F. 2005. A family of main-chain polymeric Lewis acids: Synthesis and fluorescent sensing properties of boron-modified polythiophenes. *J. Am. Chem. Soc.*, 127: 13748–9.

Swiegers, G. F.; Malefetse, T. J. 2002. Classification of coordination polygons and polyhedra according to their mode of self-assembly. 2. Review of the literature. *Coord. Chem. Rev.* 225: 91–121.

Thomas, S. W. III; Joly, G. D.; Swager, T. M. 2007. Chemical sensors based on amplifying fluorescent conjugated polymers. *Chem. Rev.*, 107: 1339–86.

Tour, J. M. 1996. Conjugated macromolecules of precise length and constitution. Organic synthesis for the construction of nano-architectures. *Chem. Rev.*, 96: 537–53.

Valliant, J. F.; Guenther, K. J.; Arienne, S. et al. 2002. The medicinal chemistry of carboranes. *Coord. Chem. Rev.*, 232: 173–230 and references therein.

Vives, G.; Tours, J. M. 2009. Synthesis of single-molecule nanocars. *Accts. Chem. Res.*, 43: 473–87.

Wang, Z.; Watson, M. D.; Wu. J.; Müllen, K. 2004. Partially stripped insulated nanowires: A lightly substituted hexa-*peri*-hexabenzocoronene-based columnar liquid crystal. *Chem. Commun.*, 2004, 336–7.

Wasczcak, M. D.; Lee, C. C.; Hall, I. H.; Carroll, P. J.; Sneddon, L. G. 1997. *Angew. Chem., Int. Ed. Engl.*, 36: 2228–30.

Watson-Clark, R. A.; Banquerigo, M. L.; Shelly, K.; Hawthorne, M. F.; Branh, E. 1998. Model studies directed toward the application of boron neutron capture therapy to rheumatoid arthritis: Boron delivery by liposomes to rat collagen-induced arthritis. *Proc. Natl. Acad. Sci. USA*, 95: 2531–4.

Wedge, T. J.; Hawthorne, M. F. 2003. Multidentate carborane-containing Lewis acids and their chemistry: Mercuracarborands. *Coord. Chem. Rev.*, 240: 111–28, and references therein.

Westcott, A.; Whitford, N.; Michaele J.; Hardie, M. J. 2004. Coordination polymers with carborane anions: Silver dinitrile complexes. *Inorg. Chem.*, 43: 3663–72.

Wu, J.; Watson, M. D.; Müllen, K. 2003. The versatile synthesis and self-assembly of star-type hexabenzocoronenes. *Angew Chem. Int. Ed.*, 42: 5329–33.

Xamena, F. X. L.; Teruel, L.; Galletero, M. S.; Corma, A.; Garcia, H. 2008. Unexpected photochemistry and charge-transfer complexes of $[CB_{11}H_{12}]^-$ carborane. *Chem. Commun.*, 499–501.

Yao, H.; Sabat, M.; Grimes, R. N. 2002. Wired multidecker sandwich assemblies. stepwise construction of a hexanuclear benzene-centered Tris(alkynyl triple-decker) Complex. *Organometallics,* 21: 2833–5.

Yao, H.; Grimes, R. N.; Corsini, M.; Zanello, P. 2003a. Polynuclear metallacarborane–hydrocarbon assemblies: Metallacarborane dendrimers. *Organometallics*, 22: 4381–3.

Yao, H.; Sabat, M.; Grimes, R. N.; de Biani, F. F.; Zanello, P. 2003b. Metallacarborane-based nanostructures: A carbon-wired planar octagon. *Angew. Chem. Int. Ed.*, 42: 1002–5.

Yinghuai, Z.; Peng, A. T.; Carpenter, K.; Maguire, J. A.; Hosmane, N. S.; Takagaki, M. 2005. Substituted carborane-appended water-soluble single-wall carbon nanotubes: New approach to boron neutron capture therapy drug delivery. *J. Am. Chem. Soc.*, 127: 9875–80.

Zakharkin, L. I.; Olshevskaya, V. A. 1995. Synthesis of 9-benzyl-*o*- and 9-benzyl-*m*-carboranes containing functional substituents in the benzene ring by electrophilic alkylation of *o*- and *m*-carboranes by the corresponding *R*-benzyl halides. *Russ. Chem. Bull.*, 44: 1099–1101.

Zhao, C.; Wakamiya, A.; Inukai, Y.; Yamaguchi, S. 2006. Highly emissive organic solids containing 2,5-diboryl-1,4-phenylene unit. *J. Am. Chem. Soc.*, 128: 15934–5.

Zhu, Y.; Cheng, K. Y.; Maguire, J. A.; Hosmane, N. S. 2007. Recent developments in the boron neutron capture therapy (BNCT) driven by nanotechnology. *Curr. Chem. Biol.*, 1: 141–9.

Zinn, A. A; Knobler, C. B; Harwell, D. E; Hawthorne, M. F. 1999. Molecular aggregates of nitrate ion with the tetravalent Lewis acid host 12-mercuracarborand-4: Novel trihapto coordination of NO_3^-. *Inorg. Chem.*, 38: 2227–30.

Zinn, A. A.; Zheng, Z.; Knobler, C. B.; Hawthorne, M. F. 1996. A Hexamethyl derivative of [9] mercuracarborand-3: Synthesis, characterization, and host–guest chemistry. *J. Am. Chem. Soc.*, 118: 70–4.

27 Large Molecules Containing Icosahedral Boron Clusters Designed for Potential Applications

Clara Viñas, Rosario Núñez, and Francesc Teixidor

CONTENTS

27.1 INTRODUCTION

Boron and carbon are the elements that have the property to build molecules of unlimited size by covalent self-bonding. In this way, boron produces polyhedral clusters that have been developed for over 50 years. Boron clusters (Figure 27.1) display many particular characteristics that do not find parallel in their organic counterparts.

The polyhedral *closo* $[B_{12}H_{12}]^{2-}$ borane is a robust icosahedral cluster stabilized by three-dimensional (3D) delocalization of 13 bonding electron pairs.[1] This charge-delocalized ion $[B_{12}H_{12}]^{2-}$ may be considered the parent aromatic species that serves borane chemistry as benzene serves organic chemistry.[2] Isoelectronic substitution of one or two B–H vertices in *closo* $[B_{12}H_{12}]^{2-}$ by C–H provides the aromatic derivatives *closo* $[1\text{-}CB_{11}H_{12}]^-$ and *closo* $C_2B_{10}H_{12}$.[3] The high stability of the ion $[B_{12}H_{12}]^{2-}$ to strong bases, strong acids, and oxidizing agents is unique for boron hydride structures. However, the anions react smoothly with certain reagents, particularly electrophilic species, to give stable derivatives in which hydrogen atoms are replaced by the attacking group.[4]

Largely, the 12-vertex *closo* $C_2B_{10}H_{12}$ icosahedral carboranes have been the most widely studied. The three known isomers, in which the carbon atoms occupy *ortho*-(1,2), *meta*-(1,7), or *para*-(1,12) vertices are white solids that rank among the most stable molecular compounds known. The unique stabilities and geometrical properties of the isomeric *closo* $C_2B_{10}H_{12}$ carboranes suggested these species as building blocks for stereo-precise structural platforms of novel reaction centers having properties that cannot be achieved with organic hydrocarbon compounds. Each one of the three isomeric *closo* $C_2B_{10}H_{12}$ carboranes has the potential for the incorporation of a large number of substituents. These species have two moderately acidic CH vertices that are deprotonated with strong bases and subsequently can be functionalized using electrophilic reagents. The carbon substitution reaction allows the isomeric *closo* $C_2B_{10}H_{12}$ carborane compounds to be part of macromolecules such as macrocycles or dendrimers and also to be bonded at the periphery of scaffolds such as star molecules, porphyrins, a set of "Fréchet-type" dendrons, among others.

The chemistry of boron-substituted carboranes is less developed than that of the carbon-substituted analogues because of the higher difficulty of introducing functional groups at the boron atoms of the carborane cage. The 10 BH vertices present in each of the icosahedral carborane isomers have electrophilic substitution chemistry in many ways reminiscent of arenes.[2] Thus, most reactions that occur at the boron vertices do not affect the carbon vertices, and vice versa. The rigid 3D icosahedra hold substituents in well-defined spatial relationships and most transformations maintain the integrity of the underlying geometry.

The cobaltabisdicarbollide $[3,3'\text{-}Co(1,2\text{-}C_2B_9H_{11})_2]$ was synthesized in 1965.[5] This anion has a great chemical stability, high molecular volume, low nucleophilic character, and low charge density because the negative charge is distributed between 45 atoms.[6] The derivative chemistry of the

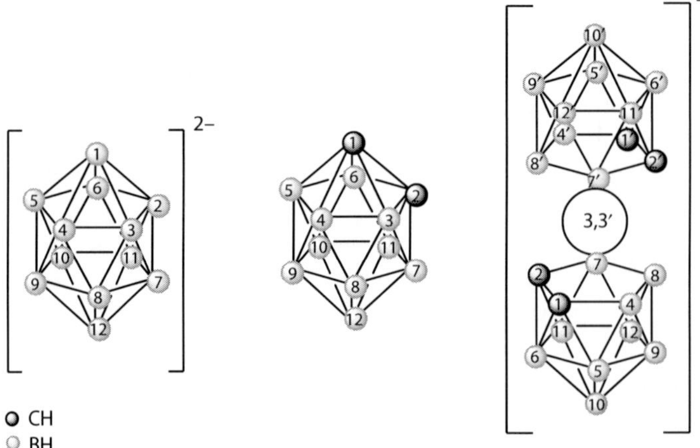

FIGURE 27.1 Icosahedral heteroboranes with their vertex numbering: dianionic *closo* $[B_{12}H_{12}]^{2-}$ borane, neutral *closo* $o\text{-}C_2B_{10}H_{12}$ carborane, and metallacarboranes $[3,3'\text{-}M(1,2\text{-}C_2B_9H_{11})_2]^-$, M=Co(III) and Fe(III).

cobaltabisdicarbollide remains very much unexplored.[7] The fundamental reason is the lack of a comprehensive synthetic strategy leading to these derivatives. As for the $C_2B_{10}H_{12}$ carborane compounds, substitutions may occur on $[3,3'\text{-Co}(1,2\text{-}C_2B_9H_{11})_2]^-$, either on carbon or on boron atoms.

With few exceptions,[8] substitutions on carbon have been achieved only at an early stage of the synthetic process, that is, on the starting *o*-carborane,[9] but not by direct reaction at the $[3,3'\text{-Co}(1,2\text{-}C_2B_9H_{11})_2]^-$ cage. Recently, the isolation of the synthon C_c-silyl-substituted cobaltabisdicarbollide $Cs[1,1'\text{-}\mu\text{-SiMeH-}3,3'\text{-Co}(1,2\text{-}C_2B_9H_{10})_2]$ that allows the preparation of boron-rich polyanionic macromolecules via hydrosilylation reactions[10] of vinyl- and allyl-terminated dendrimers has been reported.[8c]

Substitution at boron has been achieved under Friedel–Crafts conditions[11] or with strong alkylating agents.[12] Consequently, regioselective substitutions were not possible, and specific derivatives could be obtained only after careful separations of complex mixtures. The zwitterionic compound $[3,3'\text{-Co}(8\text{-}(C_2H_4O)_2\text{-}1,2\text{-}C_2B_9H_{10})(1',2'\text{-}C_2B_9H_{11})]$ was reported for the first time by the reaction of the parent $Cs[Co(C_2B_9H_{11})_2]$ with H_2SO_4–Me_2SO_4 in 1,4-dioxane in 1996.[13] Compound $[3,3'\text{-Co}(8\text{-}(C_2H_4O)_2\text{-}1,2\text{-}C_2B_9H_{10})(1',2'\text{-}C_2B_9H_{11})]$ has been shown to be susceptible to nucleophilic attack on the positively charged oxygen atom resulting in one anionic species formed by the opening of the dioxane ring. Later, the same compound was obtained in a higher yield (94% compared to 45%) by using $BF_3 \cdot OEt_2$ as a Lewis acid that permitted an easier work-up procedure.[14] The related tetrahydropyrane-based derivative $[3,3'\text{-Co}(8\text{-}(C_5H_{10}O)_2\text{-}1,2\text{-}C_2B_9H_{10})(1',2'\text{-}C_2B_9H_{11})]$ was also prepared.[15] A great impulse on the synthesis of polyanionic macromolecules incorporating cobaltabisdicarbollide was achieved after the most efficient protocol for the synthesis of $[3,3'\text{-Co}(8\text{-}(C_2H_4O)_2\text{-}1,2\text{-}C_2B_9H_{10})(1',2'\text{-}C_2B_9H_{11})]$ was reported.[14] The latter was covalently bonded to the periphery of scaffolds such as organic aromatic molecules, porphyrins, calixarenes, and resorcarenes, among others. A recent review[16] updates the synthesis of different oxonium derivatives of polyhedral boron hydrides.

In this chapter, we report herein on large high-boron-content molecules that contain several icosahedral borane, carborane, and metallacarborane clusters either as part of the scaffolds or as decorating the periphery of platforms. On purpose, we will not deal with in this chapter on polymers containing boranes, carboranes, and metallacarboranes.

27.2 ICOSAHEDRAL CARBORANES AS A PART OF MACROCYCLES OR DENDRIMER PLATFORMS

27.2.1 Novel Icosahedral Carborane-Based Macrocycles

The supramolecular chemistry of carborane clusters received significant impulses in the 1990s with the development of two types of compounds: the arene-coupled macrocyclic systems incorporating *closo*-carboranes, and the mercuracarborands.

27.2.1.1 Arene-Coupled Macrocyclic Systems Incorporating *closo*-Carboranes

Wade and coworkers[17] developed reactions whereby a variety of aromatic rings can be attached directly to the three isomeric *closo* $C_2B_{10}H_{12}$ carborane clusters by coupling reactions between aryl halides and carboranyl-copper derivatives, themselves prepared via lithio-derivatives, or using copper(I) *tert*-butoxide (Scheme 27.1). The preparation of a cyclic trimer, in which three icosahedral *meta*-carboranediyl and three planar *meta*-phenylene building blocks (C_2B_{10} carboranes and benzene rings, respectively) alternate in the novel macrocycle (Figure 27.2a), was achieved by treating *meta*-diiodobenzene with 1,7-Cu_2-1,7-*closo*-$C_2B_{10}H_{10}$.[18] The macrocycle contains a cavity that is surrounded by hydrogen atoms, of which six are bound to boron and three to carbon; this cavity thus offered potential guests an unusual environment. Such "cluster-cycles" offer some scope for use as frameworks on which to hold bi- and trinuclear arrays of metal atoms. By treatment of 2,6-$Br_2H_3C_5N$ with 1,7-Cu_2-1,7-*closo*-$C_2B_{10}H_{10}$ the cyclic six-membered macrocycle

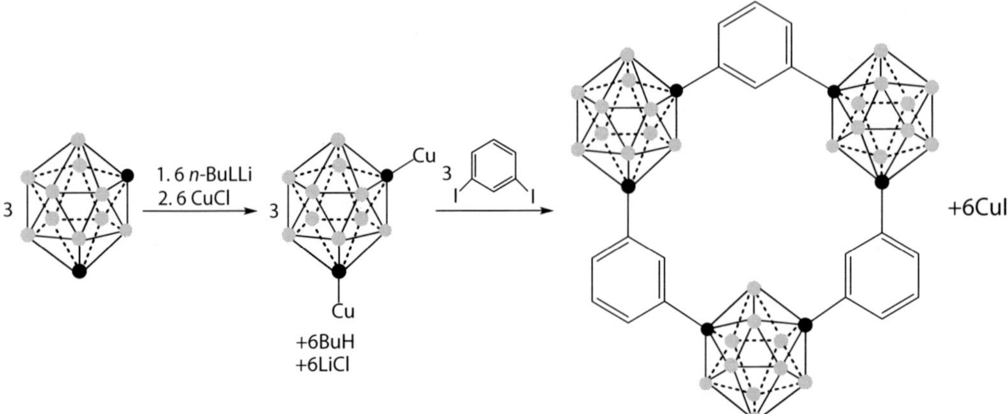

SCHEME 27.1 Preparation of a cyclic trimer by $C_{cluster}$–C(arene) coupling reaction.

system alternating *meta*-carborane cages and pyridine rings was also obtained in 10% yield.[19] In addition, the eight-membered macrocycle "big MACs" in which *ortho*- and *meta*-carborane icosahedra alternates with *para*-phenylene rings were also obtained (Figure 27.2b).[20]

Macrocycles incorporating the 1,2-*closo*-$C_2B_{10}H_{12}$ cage and linked by trimethylene or 1,3-xylyl groups were prepared in Hawthorne's laboratory, and these include both trimers and tetramers.[21] The structure of a xylyl-linked tetramer that features a remarkable 28-membered ring (Figure 27.2c) was determined by x-ray crystallography.

Morever, members of a new family of macrocyclic compounds incorporating 2, 3, and 4 icosahedral *m*-carborane moieties, linked via their carbon vertices, to *N*,*N*-dimethyldiphenylurea groups were synthesized by Endo and coworkers[22] in 2001. These carborane-containing functionalized molecules were expected to be useful for the construction of cyclic, layered, or helical molecules displaying both hydrophobic and hydrogen-bonding characters. Figure 27.3 shows the x-ray molecular structure of the macrocyclic carborane-containing urea compounds.

27.2.1.2 Mercuracarborands

Host–guest chemistry, based on the ability of Lewis base functionalities to coordinate positively charged ions, and the crown ethers ability to trap alkaline metals developed rapidly since its

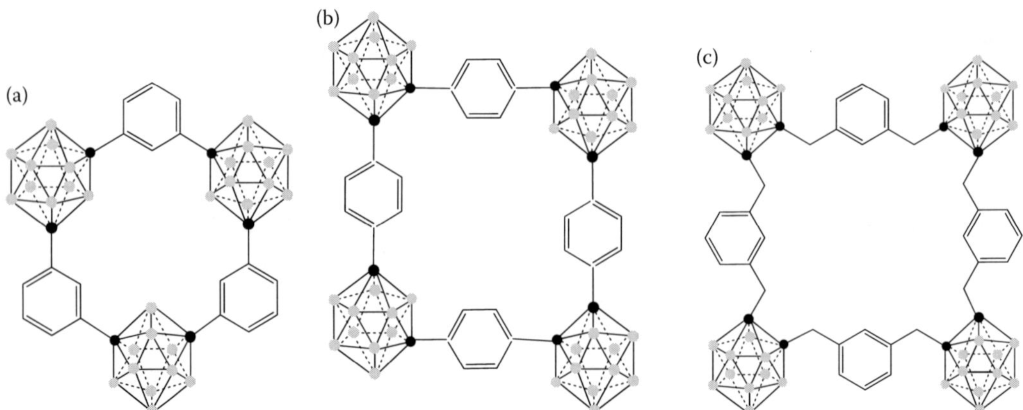

FIGURE 27.2 (a) Arene-coupled macrocyclic systems incorporating *meta*-carboranes, (b) Cyclooctaphane assembling *ortho*- and *meta*-carborane units, and (c) xylyl-linked tetramer.

FIGURE 27.3 Crystal structures of the macrocyclic carborane-containing urea compounds.

discovery by Pederson[23] in 1967. Charge reversed macrocycles named "mercuracarborands" were reported by Hawthorne and coworkers[24] in a series of papers. These macrocycles $(HgC_2B_{10}H_{10})_n$ ($n = 3, 4$), which can be considered as an anticrown, are multidentate Lewis acids that bind tightly to halide anions, Cl^-, Br^-, and I^- as well as uncharged nucleophilic species, or produce molecular aggregates with anionic *closo* $[B_{10}H_{10}]^{2-}$. The three or four *closo* C_2B_{10} carborane units in these mercuracarborands are linked to an equal number of Hg atoms by the carbon cluster atoms. It was revealed that the anions associated with the Hg(II) source profoundly affected the ring size of the resulting mercuracarborand species.[25]

The trimer mercuracarborand $(HgC_2B_{10}H_{10})_3$ was obtained by the reaction of Li_2[1,2-*closo*-$C_2B_{10}H_{10}$] with mercuric acetate in diethylether at room temperature and was isolated in 60% yield.[25] The $(HgC_2B_{10}H_{10})_3$ binds efficiently nucleophilic ions (Cl, Br, and I) and molecules (acetonitrile).

The tetramer mercuracarborand $Li[(HgC_2B_{10}H_{10})_4 \cdot X]$ (X = Cl, Br), which binds to Cl^-or Br^-ions, was produced in high yield by the reaction of Li_2[1,2-*closo*-$C_2B_{10}H_{10}$] with the corresponding HgX_2 (X = Cl, Br) salt.[24a] The reaction of HgI_2 with 1 molar equiv. of Li_2[1,2-*closo*-$C_2B_{10}H_{10}$] in dry ethylether at room temperature results in the formation of the cyclic tetramer $Li_2[(HgC_2B_{10}H_{10})_4 \cdot I_2]$ in 80% yield and of $Li[(HgC_2B_{10}H_{10})_4 \cdot I]$ in 70% yield after the 24 h reaction.[24b] It was concluded that the efficient assembly of halide ion complexes of the tetrameric $Li_n[(HgC_2B_{10}H_{10})_4 \cdot X_n]$ (X= Cl, Br, $n = 1$; X = I, $n = 2$) had to be due to a template effect with halide anion.

The halide anion can be extracted from the complex by Ag^+ ions without causing the decomposition of the host. The synthesis of the plain tetramer mercuracarborand, $(HgC_2B_{10}H_{10})_4$ (Figure 27.4a) can only be achieved by the extraction of the halide ions with AgOAc from $Li_2[(HgC_2B_{10}H_{10})_4 \cdot I_2]$ or $Li[(HgC_2B_{10}H_{10})_4 \cdot X]$ (X= Cl, Br, I).[24c]

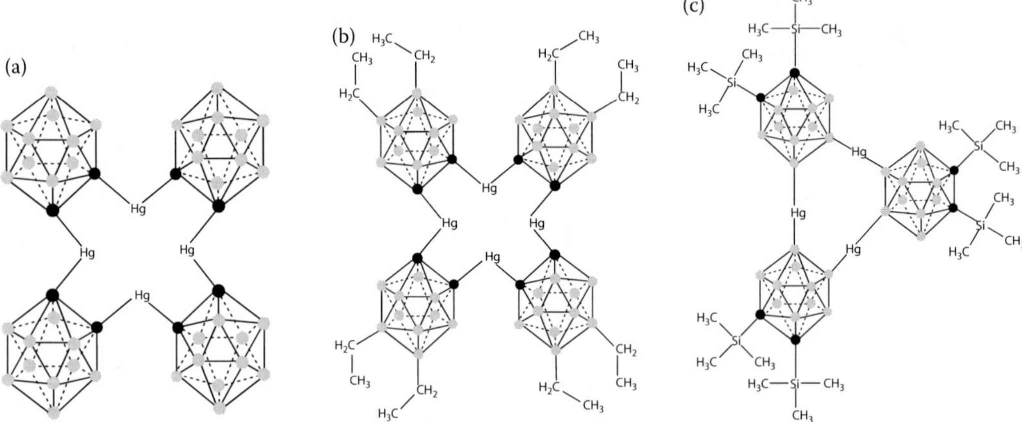

FIGURE 27.4 (a,b) Tetrameric mercuracarborands with C–Hg–C bonds and (c) trimeric mercuracarborand with B–Hg–B bonds.

These mercuracarborands are reasonably soluble in a variety of organic electron-donor solvents but insoluble in noncoordinating ones such as hydrocarbons. In order to increase the hydrocarbon solubility of mercuracarborands, hydrogen atoms of the B–H vertices were modified by lipophilic methyl, ethyl, and phenyl groups or iodine atoms.[24d,26] The reaction of Li_2[9,12-R_2-1,2-*closo*-$C_2B_{10}H_8$] (R = CH_3, C_2H_5) with mercuric acetate resulted in the formation of a variety of hydrocarbon-soluble mercuracarborand hosts such as the hexaalkyl derivatives of $(HgC_2B_{10}H_{10})_3$, [Hg(9,12-R_2-1,2-*closo*-$C_2B_{10}H_8$)]$_3$ (R = CH_3, C_2H_5) (Figure 27.4b). In general, it appears that the attachment of alkyl substituents to those boron vertices of the carborane cage, which are distant from the mercury atoms, has little effect on the coordinating properties of the mercuracarborands, while having significant effects on the solubilities of the various substituted hosts and their complexes. However, a definite, if small, trend can be discerned in the ^{199}Hg nuclear magnetic resonance (NMR) data indicating that the degree of coordination of guests to the mercury atoms decreases with increasing alkyl substitution. The borane *closo* $[B_{10}H_{10}]^{2-}$ aggregates, which are stabilized through weak 3c-2e$^-$ B–H–Hg bonds within alkyl substituted mercuracarborands, are easily dissociated in solution by the addition of halides.

These host mercuracarborand molecules were applied in Ion Selective Electrodes (ISE) and membrane formulations, as selective optical chloride sensors, as sensitive liquid/polymeric membrane electrodes for anions and as catalysts.[27]

By using a procedure which is synthetically different, but results in a structurally analogous product, it was reported the synthesis of the trimeric mercuracarborands that incorporate B–Hg instead of C–Hg bonds (Figure 27.4c).[28] The change of the mode of mercury-cage bonding from Hg–C to Hg–B clearly reverses the electronic demand upon the mercury centers and reduces the coordinating properties of the host. Presumably, the strong electron-donating effect of the 9- and 10-B vertices deactivates the mercury centers toward complexation with Lewis bases and complexation of halides guest species with such compounds were unsuccessful.

27.2.2 Icosahedral Borane and *o*-Carborane as a Core of Dendrimers through the Boron or the Carbon Vertices

The shape of the stable icosahedral clusters, $[B_{12}H_{12}]^{2-}$ and $C_2B_{10}H_{12}$, brings the possibility to build dendrimers having the cluster as the core, allowing for a maximum of 12 primary branches in a unique spatial disposition. Because the utility of the units is dependent on their functionalization,[3,29] the introduction of functional groups is a necessary target. Substitution of the carbon-bound hydrogen in carborane clusters is easy because the C–H vertices may be deprotonated with strong bases. Conversely, the chemistry of boron-substituted carboranes is more difficult. As each boron vertex in icosahedral $C_2B_{10}H_{12}$ possesses different electron density, each vertex has different reactivity.

27.2.2.1 Dianionic *closo*-$[B_{12}H_{12}]^{2-}$ Boranes

The *closo*-dodecahydrododecaborate $[NEt_3H]_2[B_{12}H_{12}]$ has been prepared on a lab scale by an improved synthesis from cheap and readily available starting materials ($Na[BH_4]$ and I_2) in diglyme.[30] The *closo* $[B_{12}H_{12}]^{2-}$ derivatives in which every available B–H vertex has been substituted by halogen are of special interest. The perhalogenated derivatives $Cs_2[B_{12}X_{12}]$ (X= Cl, Br, I) were synthesized by reaction of $Cs_2[B_{12}H_{12}]$ with the respective elemental halogens (Cl_2,[30,31] Br_2[31], and I_2[31]). Recently, the direct perfluorination of $K_2[B_{12}H_{12}]$ in acetonitrile that occurs at the gas bubble-solution interface has been reported.[32] By simple metathesis reaction, a variety of useful salts $[\text{cation}]_2[B_{12}Cl_{12}]$ (cation = NH_4^+, $[NEt_3H]^+$, $[NBu_4]^+$, Li^+, Na^+, K^+, Cs^+, $[C_n\text{mim}]^+$ (1-alkyl-3-methylimidazolium with $n = 2, 4, 6, 8, 10, 12, 14, 16$, or 18), and tetraalkylphosphonium $[PR_4]^+$) are available.[30,33]

These salts are useful starting materials, which have the potential to open up the chemistry of *closo* $[B_{12}Cl_{12}]^{2-}$ as a weakly coordinating dianion.[30,33] Highly pure and nonhygroscopic ionic liquids can be obtained by combining $[B_{12}Cl_{12}]^{2-}$ with either imidazolium or phosphonium cations in straightforward metathetic reactions (Figure 27.5).[33] One interesting property provided by the inert and weakly coordinating nature of *closo* $[B_{12}Cl_{12}]^{2-}$ is the noticeable thermal resistance to decomposition of their imidazolium salts.[33]

The perhydroxylated *closo* $[B_{12}(OH)_{12}]^{2-}$ in which the icosahedral cage is retained was achieved by treating alkali metal salts of $[B_{12}H_{12}]^{2-}$ with 30% hydrogen peroxide at reflux temperature.[2] The hydrogen-bonded arrays observed in the new clusters point to the potential that *closo* $[B_{12}(OH)_{12}]^{2-}$ has in the development of species as central cores for the divergent growth of dendrimers. The 12-fold functional icosahedral dodecaborate core is rigid and the directions of the arms within the first generation should be sterically favorable.

The total esterification of the *closo* $[B_{12}(OH)_{12}]^{2-}$ with acetic anhydride or benzoyl chloride produced organoderivatized molecules of the icosahedral cluster. The resulting scaffold, in which

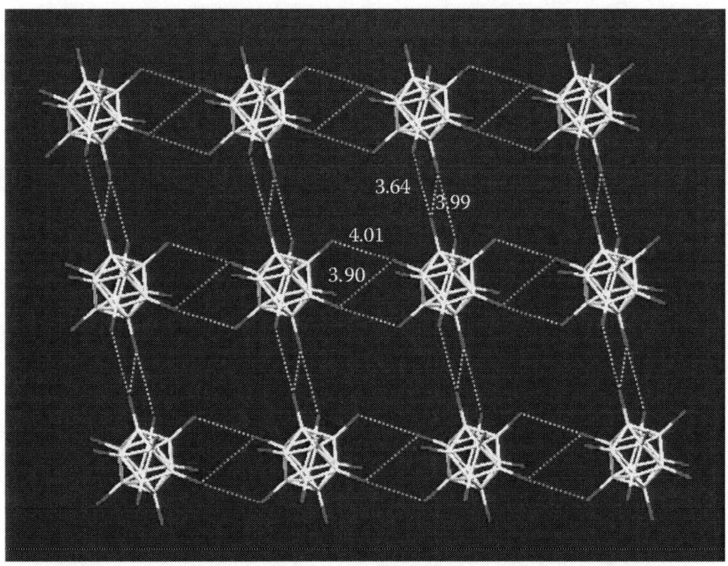

FIGURE 27.5 Packing of anions into layers in $[C_2\text{mim}]_2[B_{12}Cl_{12}]$ showing the Cl ··· Cl contacts (distances in Å).

each vertex is substituted with an organic ester moiety, represents the first example of such structural motifs known in chemistry.[2] Figure 27.6 displays the result of an x-ray analysis performed with the dodecabenzoate ester derivative. The syntheses of the dodeca(carboranyl) closomers, *closo* $[B_{12}$(1′-yl-2′-CH_3-1′,2′-*closo*-C_2B_{10}-7-$(CH_2)_6$COO-$)_{12}]^{2-}$ with an icosahedral $[B_{12}]^{2-}$ core and 12 branches ending with neutral *closo* C_2B_{10} units (Figure 27.7) or 12 anionic *nido*-$[C_2B_9]^-$ units respectively, were performed by reaction of $[NBu_4]_2[B_{12}(OH)_{12}]$ closomer core with the acid chloride 1-COCl$(CH_2)_6$-2-CH_3-1,2-*closo*-$C_2B_{10}H_{10}$.[35] Deboronation of the 12 pendant *closo*-carborane units led to the 14 anionic macromolecule $[B_{12}$((7′-yl-8′-CH_3-7′,8′-*nido*-C_2B_9)-7-$(CH_2)_6$COO-$)_{12}]^{14-}$. The high negative charge imparts hydrophilicity to the system, which attains solubility in hot water as the Cs^+ salt, while maintaining solubility in polar organics. These boron-rich macromolecular, unimolecular nanospheres possess great potential for future use as drug-delivery platforms for boron neutron capture therapy (BNCT). This is a bimodal cancer treatment that makes use of thermal neutrons of low kinetic energy and ^{10}B-containing molecules. The resulting nuclear reactions with the ^{10}B nucleus in malignant cells ultimately lead to tumour cell destruction in a very near small volume due to the high kinetic energy (approximately 2.4 MeV) of the primary fission products (^{7}Li and ^{4}He). In recent years, the search for new ^{10}B carriers for BNCT has led to the development of platforms capable of binding several boron clusters.

27.2.2.2 Neutral *closo*-$C_2B_{10}H_{12}$ Carboranes as Core through the Carbon Vertices

The versatility of the carborane cluster allows it to act as a scaffold to extend the dendron from the carbon atoms to the periphery or to be modified as a focal point. Yamamoto and coworkers[36] reported on the synthesis of polyols of a cascade type as water-solubilizing groups of carborane derivatives for BNCT. Small carbosilane dendrons in which a *closo*-carborane is located at the focal point have been prepared by a sequence of steps involving hydrosilylation and reduction reactions.[37] These compounds are used as scaffolds for peripheral functionalization with styrene, chlorovinylstyrene, or suitable carboranes, while keeping the $C_{cluster}$–Si (C_c–Si) bond (Scheme 27.2). Besides, the reaction involving nucleophilic attack by fluoride or other nucleophiles, and the subsequent cleavage of C_c–Si bonds, is a well-documented process in carboranylsilane chemistry.[38] Thus, the possibility to

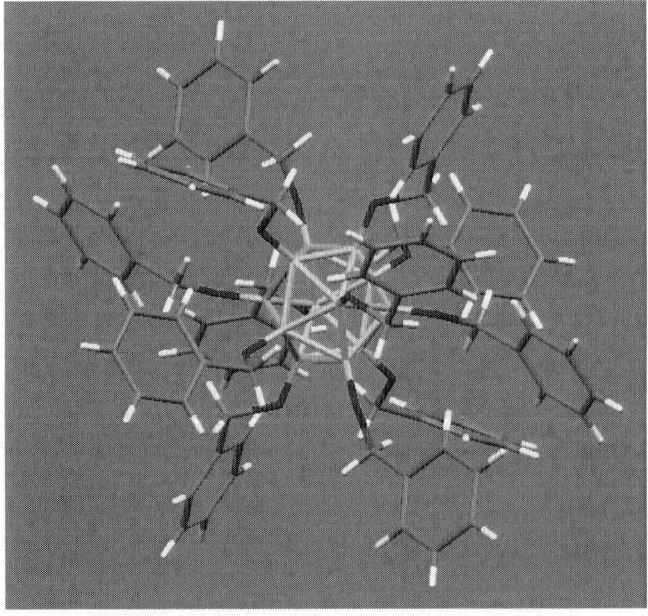

FIGURE 27.6 X-ray analysis of the dodecabenzoate ester of $[B_{12}(OH)_{12}]^{2-}$.

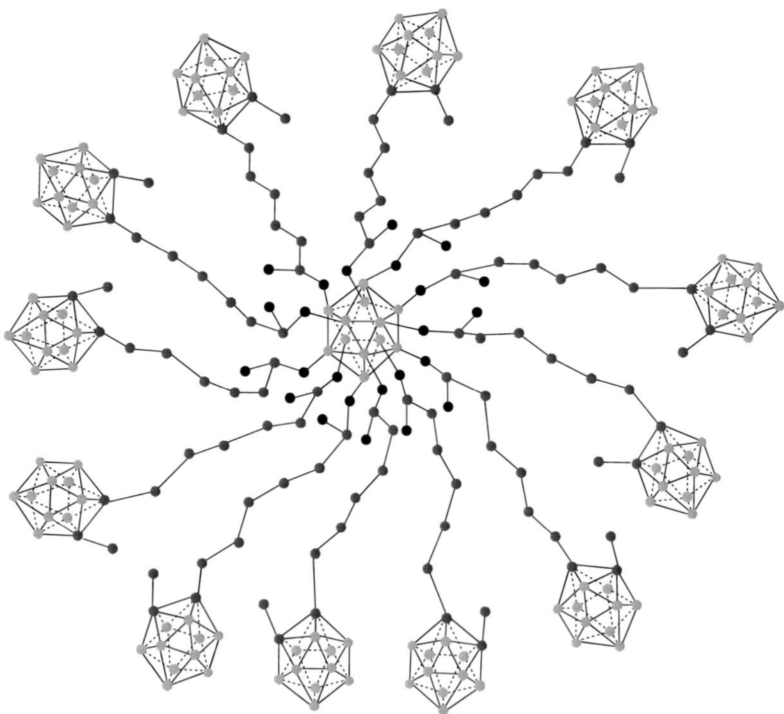

FIGURE 27.7 The icosahedral core $[B_{12}]^{2-}$ with the 12 branches ending with neutral *closo* C_2B_{10} units.

use carboranes as a platform for dendron functionalization and cluster elimination by nucleophilic cleavage of the C_c–Si bond offers a new strategy to introduce other functions. The modification of the carborane by reduction with Mg/$BrCH_2CH_2Br$ was also achieved.[37] The possibility of the cluster modification by reducing agents, while keeping the C_c–Si bonds, represents another alternative to prepare new carboranylsilane dendrons.

In addition, following our studies on cobaltabisdicarbollides' direct substitution[39] to obtain novel high-boron-content polyanionic species with enhanced water solubility, we have recently explored the possibility of using lithiated boron clusters as nucleophiles to produce a new family of high-boron-content polyanionic large molecules.[39,40] The zwitterionic compound [3,3′-Co(8-$(C_4H_8O)_2$-1,2-$C_2B_9H_{10}$)(1′,2′-$C_2B_9H_{11}$)] has been shown to be susceptible to nucleophilic attack on the positively charged oxygen atom, for example, by pyrrolyl,[15] imide, cyanide or amines,[41] phenolate, dialkyl or diarylphosphite,[42] *N*-alkylcarbamoyldiphenylphosphine oxides,[43] alkoxides,[14,44] and nucleosides[45] resulting in one anionic species formed by the opening of the dioxane ring.

We studied the nucleophilic behavior of the mono- and dilithiated salts of 1,2-*closo*-$C_2B_{10}H_{12}$, 1,7-*closo*-$C_2B_{10}H_{12}$ and 1,12-*closo*-$C_2B_{10}H_{12}$ isomers toward [3,3′-Co(8-$(C_2H_4O)_2$-1,2-$C_2B_9H_{10}$)(1′,2′-$C_2B_9H_{11}$)]. It was concluded that mono- and dilithiated salts of 1,2-*closo*-, 1,7-*closo*-, and 1,12-*closo*-$C_2B_{10}H_{12}$ isomers act as nucleophiles (Scheme 27.3) in the ring-opening reaction with [3,3′-Co(8-$(CH_2CH_2O)_2$-1,2-$C_2B_9H_{10}$)(1′,2′-$C_2B_9H_{11}$)] leading to (1) the monoanionic-type species [1-X-2-*R*-1,2-*closo*-$C_2B_{10}H_{10}]^-$, (*R* = H, CH_3, C_6H_5); [1-X-1,7-*closo*-$C_2B_{10}H_{11}]^-$ and [1-X-12-*closo*-$C_2B_{10}H_{11}]^-$ and (2) the dianionic compounds [1,2-X_2-1,2-*closo*-$C_2B_{10}H_{10}]^{2-}$, [1,7-X_2-1,7-*closo*-$C_2B_{10}H_{10}]^{2-}$, and [1,12-X_2-1,12-$C_2B_{10}H_{10}]^{2-}$, (where *X* = [3,3′-Co(8-$(CH_2CH_2O)_2$-1,2-$C_2B_9H_{10}$)(1′,2′-$C_2B_9H_{11}$)]$^-$). These two families of compounds contain both structural motifs, cobaltabisdicarbollide and *closo*-carborane, within the same molecule.

Even though the *closo*-carborane clusters are structures showing high stability with respect to strong acids, they react with Lewis bases yielding more opened structures, known as *nido*, by a partial deboronation process that implies the loss of a cluster's vertex. Several nucleophiles such as

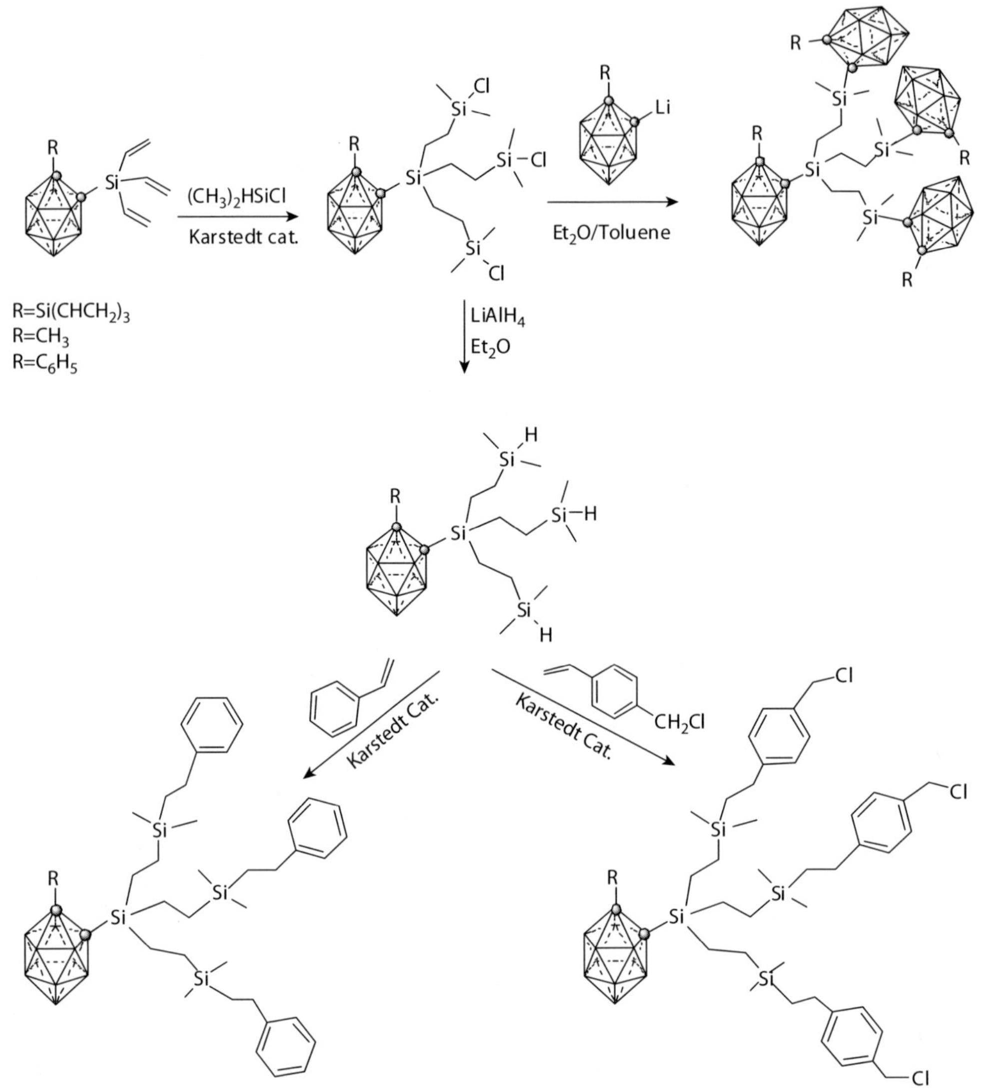

SCHEME 27.2 Preparation of peripheral carboranyl-functionalized dendrons.

alkoxides,[46] amines,[47] fluorides,[48] phosphanes[49] have been used in the deboronation reaction. The nucleophilic attack takes place at one of the boron atoms directly bonded to both carbon atoms, the B(3) or its equivalent B(6), since they both present electronic deficiency. Scheme 27.4 shows that regioselective removal of one BH vertex from the 1,2-*closo*-$C_2B_{10}H_{12}$ cluster of the double-cluster monoanion of type [1-X-2-R-1,2-*closo*-$C_2B_{10}H_{10}]^-$ species (where X = [3,3′-Co(8-$(C_2H_4O)_2$-1,2-C_2-B_9H_{10})(1′,2′-$C_2B_9H_{11}$)$]^-$ and R = H, CH_3, C_6H_5) takes place via heating with ethanolic KOH to give a series of dianions [7-X-8-R-7,8-*nido*-$C_2B_9H_{10}]^{2-}$. As anticipated, the same degradation procedure applied to the dianionic triple-cluster compound [1,2-X_2-1,2-*closo*-$C_2B_{10}H_{10}]^{2-}$ gave rise to the trianionic species [7,8-X_2-7,8-*nido*-$C_2B_9H_{10}]^{3-}$, which is substituted at both carbon atoms of the 11-vertex dicarborane cluster. The fact that these high-boron-content large molecules are polyanionic, makes them more soluble in water, and increases their potential for biological uses. The presence of *closo* or *nido*-carborane subclusters permits a ready modification of these molecules by attaching variable substituents onto the carbon and boron vertices. This may add new properties to the boron clusters

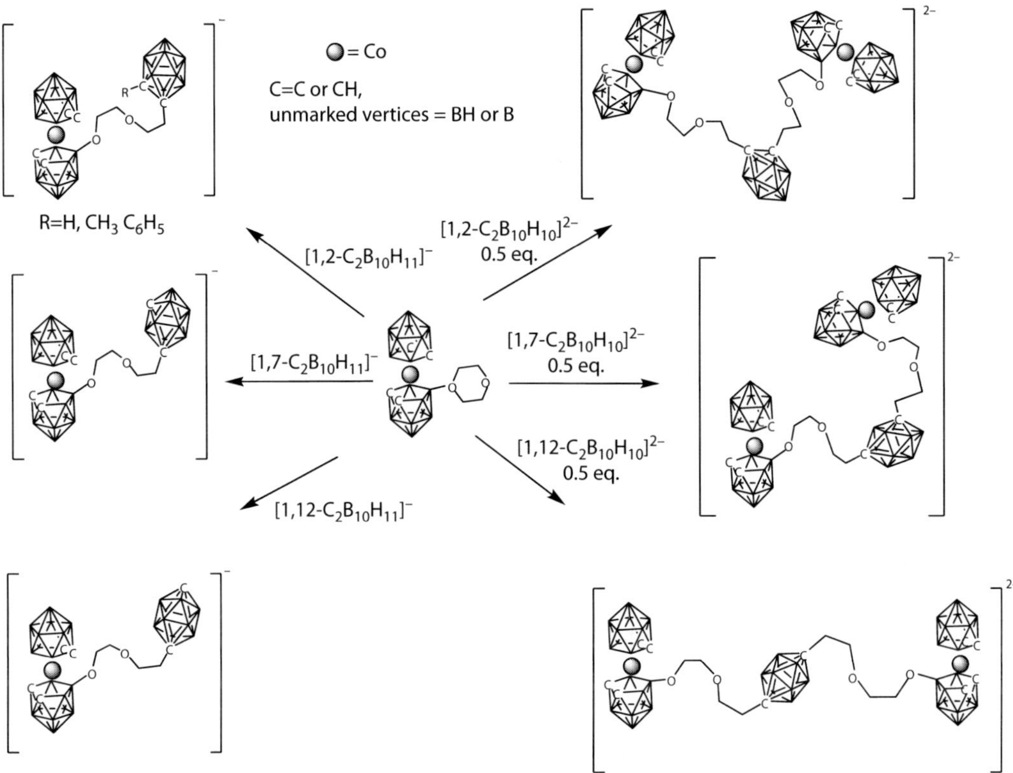

SCHEME 27.3 Simplified structures and formation of the monoanionic and dianionic 12-vertex dicarboranes modified by the 8-substituted $[3,3'\text{-Co}(1,2\text{-C}_2\text{B}_9\text{H}_{11})_2]^-$ function via a 1,4-dioxahexane interconnection chain.

and, in fact, give rise to new species that can be used potentially in medicine as good candidates for BNCT techniques' efficiency.

27.2.3 Icosahedral Carboranes Inside Dendrimeric Scaffolds

As mentioned before, carboranes are versatile building blocks for the synthesis of a wide range of compounds with tailor-made properties. The research on the chemistry of carboranyl-containing dendrimers was initiated by Newkome et al.[50] in 1994 and was focused on the integration of *o*-carborane moieties within the interior of dendrimeric structures. The first dendrimers containing carboranyl clusters inside the dendrimeric branches were prepared by treatment of polyalkyne precursors with decaborane (Scheme 27.5). In order to achieve water-soluble dendrimers, carboranyl-containing dendrimers react with $PdO \cdot xH_2O$ and EtOH to give the corresponding polyols that after further reaction with $ClSO_3H$ at 0°C for 1 h afforded water-soluble polysulfates, respectively. Aqueous solubility over a wide pH range was provided by the presence of peripheral sulfate groups, affording a unimolecular micelle-type structure.

Aliphatic polyester dendrimers that incorporate carboranes within the interior of the dendrimeric systems have been accomplished following a divergent approach.[51] In a first step, the polyester dendrimer was prepared by the reaction of excess of 2,2-bis(hydroxymethyl)propionic acid (bisMPA) anhydride with the 1,1,1-tris(hydroxyphenyl)ethane core using a catalytic amount of 4-(dimethylamino)pyridine (DMAP) in a 3:2 mixture of CH_2Cl_2 and pyridine. After purification and further hydrogenolysis of the dendrimer containing benzylidene-protecting groups, a series of hydroxyl-terminated dendrimer generations, from G-1 to G-4, were obtained. In a second step, preparation of

−

KOH/EtoH
reflux

2−

R=H, CH_3, C_6H_5

R=H, CH_3, C_6H_5

2−

KOH/EtoH
reflux

3−

SCHEME 27.4 Partial deboronation reaction of *closo* compounds yielding the corresponding di- and trianionic species.

the appropriate bifunctional 1,12-*closo*-$C_2B_{10}H_{12}$ synthon, which has a carboxylic acid and a protected alcohol functionality with a *tert*-butyldiphenylsilyl (TBDPS) group, was required to incorporate the carborane cages within the dendrimer.[51a] Finally, this synthon was coupled to the peripheral alcohols of different generations of polyester dendrimers. All generations were fully characterized by NMR and MALDI-TOF MS indicating complete functionalization of all peripheral alcohols on the starting dendrimers. Nevertheless, attempts to add dendrimer generations at the periphery were unsuccessful causing degradation of the dendrimers.

In order to avoid the degradation of the dendrimer, a benzyl ether group was used as the protecting group to obtain the carborane synthon shown in Scheme 27.6. In addition, the core of the

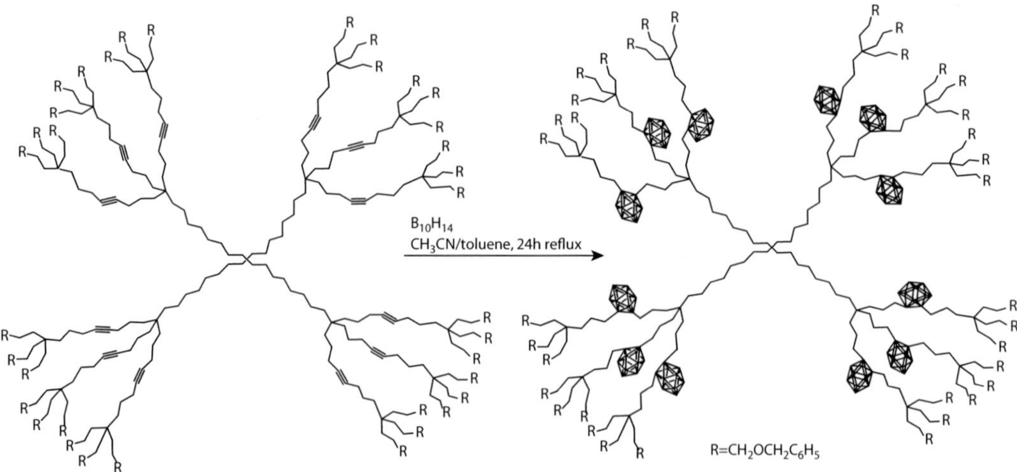

SCHEME 27.5 Preparation of dendrimers with carboranyl units inside the structure.

dendrimer was changed to lowercase for pentaerythritol that permits to increase the number of carborane cages to four, eight, and 16 within the dendrimer. Coupling reactions between the carborane synthon and protected dendrimers having pentaerythritol as core (G-1 bisMPA and G2 bisMPa dendrimers), in the presence of 1-(3-dimethylamino-propyl)-3-ethylcarbodiimide hydrochloride (EDC) and a catalytic amount of 4-dimethylaminopyridine/p-toluene sulfonic acid salt (DPTS), afforded carboranyl functionalized dendrimers in very good yields (Scheme 27.6).

Deprotection of peripheral benzyl ether groups was achieved by catalyzed hydrogenolysis to obtain the corresponding dendrimers with peripheral alcohols. Subsequently, these peripheral functionalities were reacted with the bisMPA anhydride to further dendronize the periphery of each molecule, allowing substantial increase in the number of hydroxyl groups that provide higher water solubility to the final dendrimers (Figure 27.8). In fact, it was observed that a minimum of eight hydroxyl groups per carborane unit was required to present water solubility. Characterization of end-functionalized dendrimers was accomplished by NMR spectroscopy, MALDI-TOF MS, and elemental analyses. In addition, irradiation of some dendrimers with thermal neutrons showed emission of gamma radiation that is indicative of boron neutron capture events.

Dendrimers shown in Figure 27.8 were found to exhibit reversible precipitation (or a cloud point) in aqueous solution at elevated temperatures. Due to the extreme hyrophobicity of the *p*-carborane

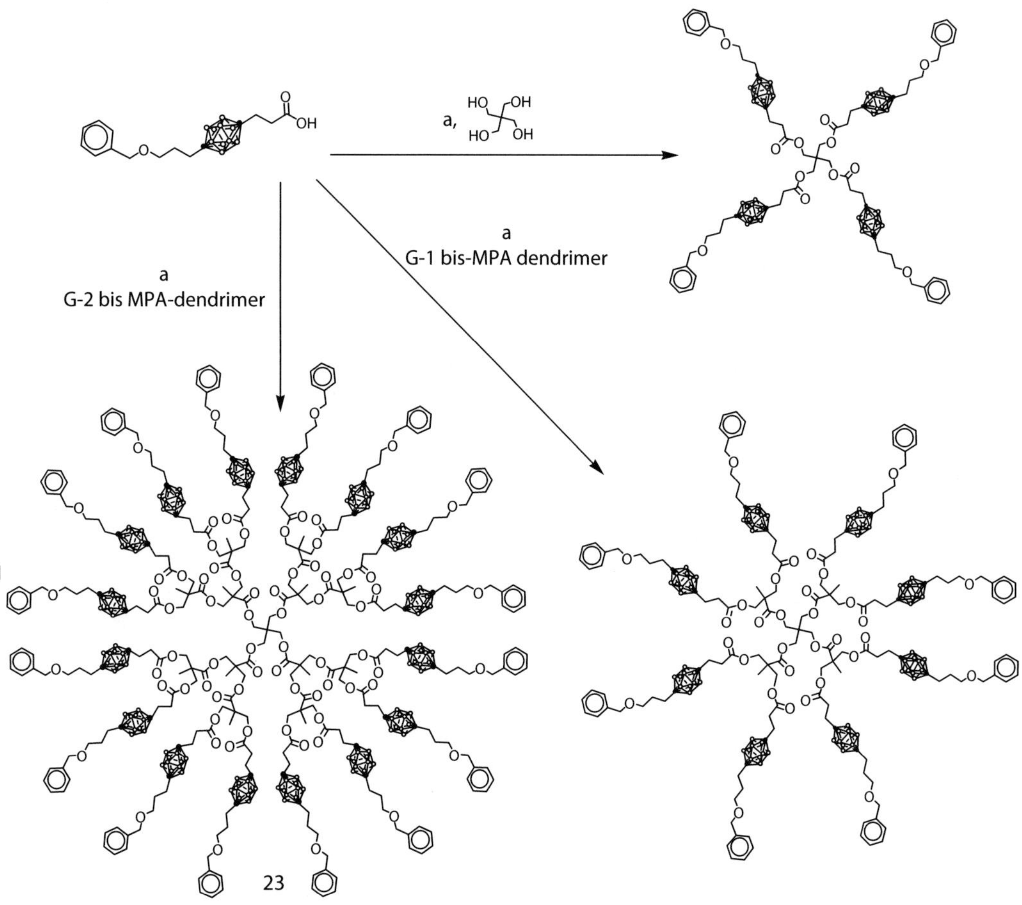

SCHEME 27.6 Preparation of aliphatic polyester dendrimes that incorporate carboranes inside the structure. Scheme transferred by John Valliant and modified by us.

moieties introduced within these hydrophilic dendrimers, it seems that the 8:1 ratio between peripheral hydroxyl groups and internal 1,12-*closo*-$C_2B_{10}H_{10}$ provides a correct balance for a thermal transition to occur. Structures with a lower ratio exhibited no aqueous solubility, and those with a higher ratio remained soluble at all temperatures. The dendrimers were found to degrade under basic conditions.[51c]

27.3 ICOSAHEDRAL CARBORANES AND METALLACARBORANES DECORATING THE PERIPHERY OF MACROMOLECULES

27.3.1 Star-Shaped Molecules Decorated with Boron Clusters

27.3.1.1 Carbosilanes as Core Molecules

Carbosilane dendrimers are kinetically and thermodynamically very stable and their structures can be easily modified for the required application.[52] In view of the emerging importance of dendrimers[53] as a new class of materials and the versatility of carboranes in applications, we were interested in functionalizing the periphery of dendrimers with carborane clusters. Our aim was to generate new star-shaped macromolecules using carbosilanes as an inert scaffold in which carborane derivatives were attached on the periphery.[54]

Two different approaches were used for their preparation: the first consists of the nucleophilic substitution of peripheral Si–Cl functions in *star-shaped* carbosilane compounds $1G_V$-Cl[55] and $1G_A$-Cl[56] with the monolithium salt of carborane derivatives, 1-R-1,2-*closo*-$C_2B_{10}H_{11}$ (R = H, CH_3, C_6H_5) (Scheme 27.7), in Et_2O/toluene or dimethoxyethane/toluene (1:2). This procedure affords compounds that incorporate four carborane moieties covalently bonded to the peripheral Si atoms as crystalline air-stable solids. In the second method, the hydrosilylation reactions of tetravinylsilane with 1-$(CH_3)_2$HSi-2-R-1,2-*closo*-$C_2B_{10}H_{10}$ (R = CH_3, C_6H_5) was performed (Scheme 27.7). The latter reagents had been previously synthesized by the reaction of the lithium salt of carboranes with $(CH_3)_2$ClHSi, catalyzed by Karsted's catalyst, followed by reduction of the SiCl group with $LiAlH_4$.[54] This was a highly efficient method to obtain the expected compounds in large yield. The structures of all these compounds were established on the basis of elemental analysis, IR, ^1H, ^{13}C, ^{11}B, ^{29}Si NMR spectroscopy, and mass spectrometry (electron spectroscopic imaging (ESI) or CI). In addition, it is important to emphasize that carboranyl-containing star-shaped carbosilanes were

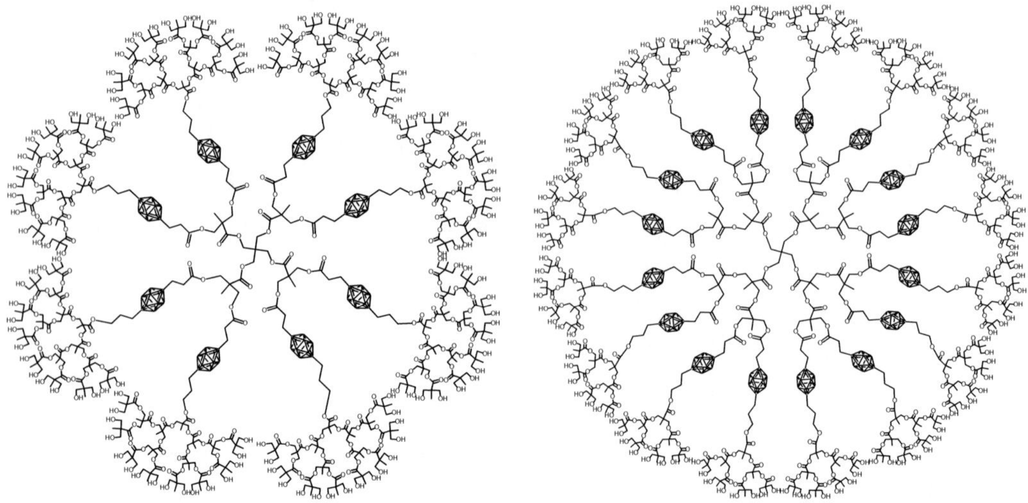

FIGURE 27.8 Schematic representation of aliphatic polyester dendrimers bearing eight and 16 1,12-$C_2B_{10}H_{10}$ cages. (Figure transferred by John Valliant.)

SCHEME 27.7 Two different strategies to synthesize carboranyl-containing star-shaped carbosilanes.

successfully isolated in crystalline form suitable for x-ray diffraction analyses. The *o*-carboranyl-containing star-shaped carbosilane molecule is represented in Figure 27.9.

27.3.1.2 Benzene as Core Molecules

A route to the synthesis of star-shaped assemblies in which three *o*-carborane icosaedra are linked to benzene (Figure 27.10a) was achieved by using two different synthetic strategies. Endo and coworkers[57] reported the preparation of the aromatic compounds composed of benzene nuclei linked through *o*-carborane by the reaction of 1,3,5-tris(2-phenylethynyl)benzene with $B_{10}H_{14}$ in the presence of SEt_2 as a Lewis base. Wade and coworkers[20] reported the same compounds by the coupling reaction between aryl trihalides and carboranyl-copper derivatives. Carboranyl-copper(I) derivatives may be generated from C-unsubstituted or C-monosubstituted *ortho-*, *meta-*, or *para*-$C_2B_{10}H_{12}$ using copper(I) *tert*-butoxide. It was concluded that aryl iodides are clearly more reactive than bromides under these coupling reaction conditions since the targeted compounds were obtained in higher yield. Then, the

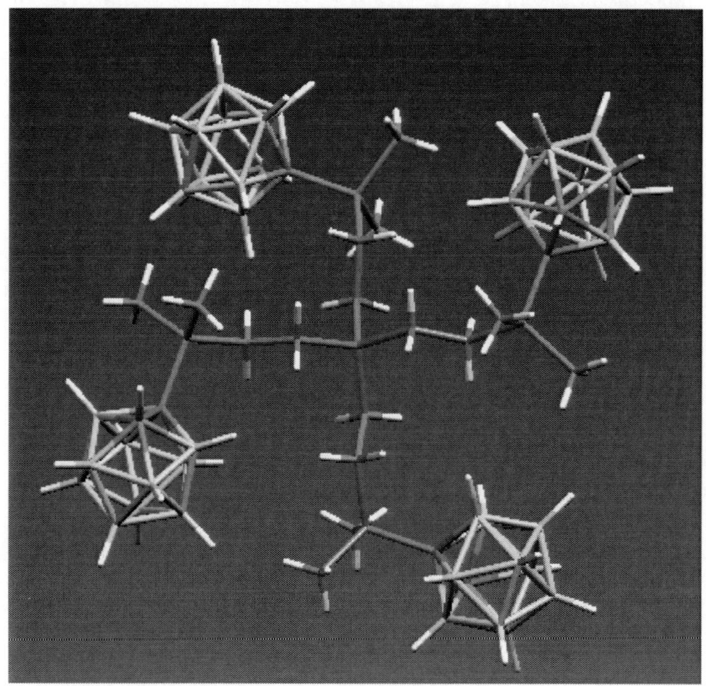

FIGURE 27.9 Molecular structure of a star-shaped carbosilane bearing four 1,2-$C_2B_{10}H_{11}$ cages.

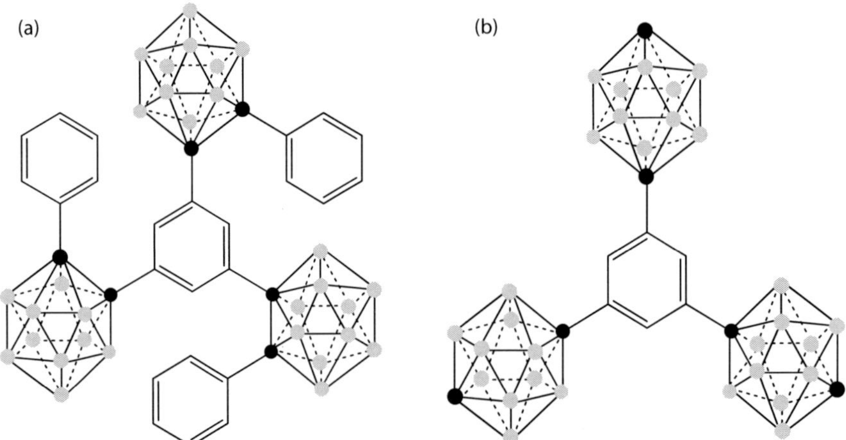

FIGURE 27.10 Star-shaped molecules of (a) *ortho*- and (b) *para*-carborane assemblies.

arylene-coupled star-shaped systems 1,3,5-(1-yl-2-C_6H_5-1,2-*closo*-$C_2B_{10}H_{10}$)$_3$-C_6H_3 or 1,3,5-(1-yl-1,12-*closo*-$C_2B_{10}H_{11}$)$_3$-C_6H_3 were achieved by the reaction between 1,3,5-triiodobenzene and Cu[1-C_6H_5-1,2-*closo*-$C_2B_{10}H_{10}$] or Cu[1,12-*closo*-$C_2B_{10}H_{10}$], respectively (Scheme 27.8).

It was in 1997, when Michl and coworkers[58] proposed the compound 1,3,5-(1-yl-1,12-*closo*-$C_2B_{10}H_{11}$)$_3$-C_6H_3 (Figure 27.10b) as a new type of trigonal star connector toward the synthesis of 2D grid polymers. The simplest route used through the synthesis of this star connector 1,3,5-(1-yl-1,12-*closo*-$C_2B_{10}H_{11}$)$_3$-C_6H_3 was offered by the palladium-catalyzed cross-coupling reaction[59] of three 1,12-*closo*-$C_2B_{10}H_{12}$ units to a 1,3,5-trihalobenzene (Scheme 27.8). The Pd-catalyzed reaction with 1,3,5-tribromobenzene yielded a mixture of (1,12-*closo*-$C_2B_{10}H_{11}$)-C_6H_5, 1,3-(1,12-*closo*-$C_2B_{10}H_{11}$)$_2$-C_6H_4, and only a small amount of the desired 1,3,5-(1,12-*closo*-$C_2B_{10}H_{11}$)$_3$-C_6H_3. However, a reaction with 1,3,5-triiodobenzene afforded 1,3,5-(1,12-*closo*-$C_2B_{10}H_{11}$)$_3$-C_6H_3 in 56% yield. Each arm of the trigonal connector 1,3,5-(1,12-*closo*-$C_2B_{10}H_{11}$)$_3$-C_6H_3 was provided with one or two sticky tentacles by conversion to 1,3,5-(12-(Si(CH_3)$_2$($C_3H_6SC_2H_5$)-1,12-*closo*-$C_2B_{10}H_{11}$)$_3$-C_6H_3 and 1,3,5-(12-(Si(CH_3) ($C_3H_6SC_2H_5$)$_2$-1,12-*closo*-$C_2B_{10}H_{11}$)$_3$-C_6H_3 (Figure 27.11).

In 2007, we reported the high-yield synthesis of *closo*- and *nido*-carborane containing "Fréchet-type" aryl ether cores, which exhibited photoluminescent (PL) properties.[60] The reaction of α, α′-bis-(3,5-bis(bromomethyl)phenoxy-p-xylene with 4 equiv. of the monolithium salt of 1-R-1,2-*closo*-$C_2B_{10}H_{12}$ (R = CH_3, C_6H_5) in THF, after 24 h. at reflux, gave the corresponding neutral carboranyl-functionalized aryl ether derivatives, in 82% and 85% yield, respectively (Scheme 27.9). Subsequently, *closo* clusters

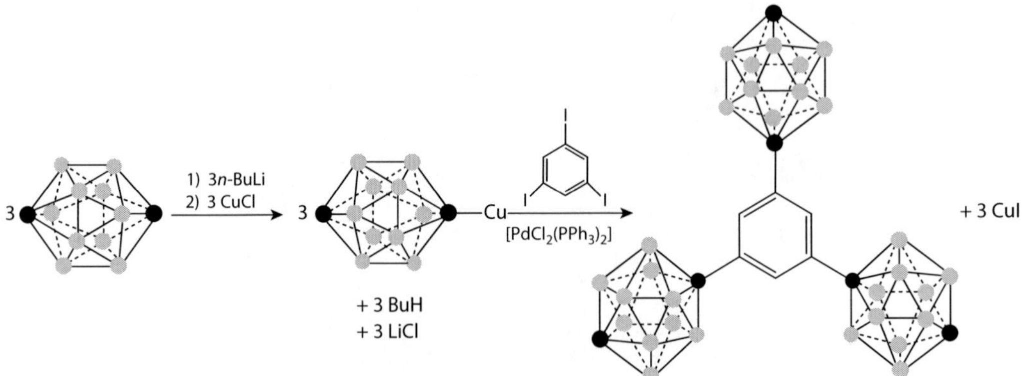

SCHEME 27.8 Synthesis of the star-shaped 1,3,5-(1-yl-1,12-*closo*-$C_2B_{10}H_{11}$)$_3$-C_6H_3 by $C_{cluster}$–C(arene) coupling reaction.

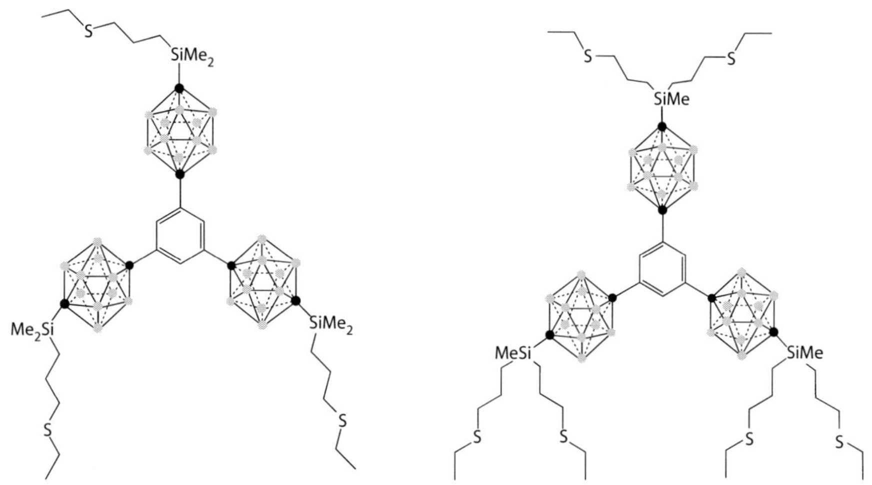

FIGURE 27.11 One or two sticky tentacles bonded to each arm of the $C_{cluster}$ star molecules.

partial degradation was successfully performed to afford the respective *nido*-carboranes, which were easily isolated as K^+ or $[N(CH_3)_4]^+$ salts (see Scheme 27.9). The potassium salts displayed good solubility in water and polar solvents.

The electronic data in different solvents indicated a solvatochromic shift for all compounds and a red shift of the absorption maxima for the *nido*-species with regard to the *closo* derivatives. These neutral and anionic carboranyl-functionalized aryl ether derivatives represent a new family of high-boron-content luminescent compounds that show strong blue emission under ultraviolet radiation in

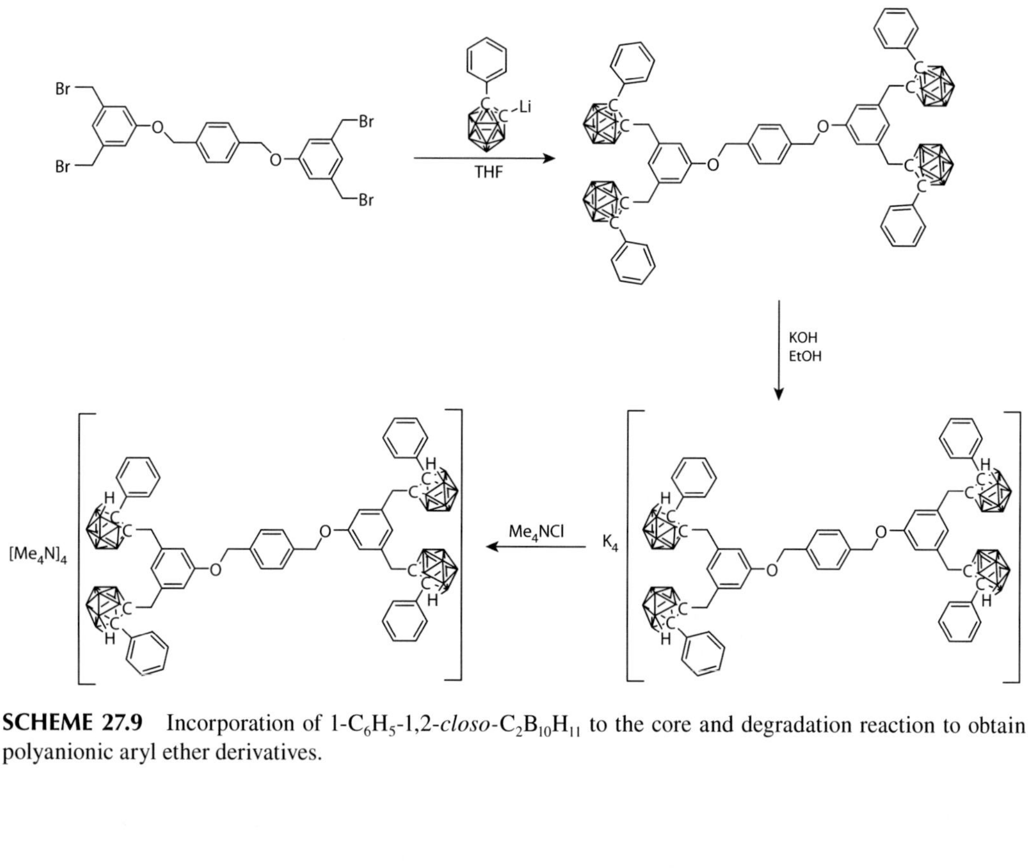

SCHEME 27.9 Incorporation of 1-C_6H_5-1,2-*closo*-$C_2B_{10}H_{11}$ to the core and degradation reaction to obtain polyanionic aryl ether derivatives.

different solvents at room temperature (Figure 27.12). This phenomenon is very interesting because none of the precursors have such property. Their properties and high content of boron make these macromolecules interesting to be used in biomedical applications.

More recently, Hosmane and coworkers[61,62] have reported the synthesis of symmetrical star-shaped molecules with carborane clusters on the periphery via $SiCl_4$-mediated cyclotrimerization[61] in EtOH of 9-benzyl derivatives of carboranes, which contain an acetyl group on the benzene ring, to obtain a 1,3,5-tris(1,2-dicarba-*closo*-dodecaborane-9-yl)benzene and 1,3,5-tris(1-CH_3-2-dicarba-*closo*-dodecaborane-9-yl)benzene.[62] The corresponding lithium salts of these symmetrical trimers were further functionalized with 1-iodoheptane and trivinylchlorosilane in THF to produce the respective carborane-functionalized compounds (Figure 27.13a), which can be used as precursors for liquid crystals[63] and for the growth of higher-order carbosilane dendrons, and dendrimers,[37,54,64] to make a new class of metallacarboranes via decapitation, or potential agents for BNCT. Due to the presence of the 1,3,5-triphenylbenzene core, these compounds could exhibit PL properties.[65]

In an attempt to further enlarge the number of carborane clusters surrounding a symmetrical core, the facile synthesis of a star-shaped molecule containing six carborane clusters on the periphery of a hexaphenylbenzene core via a cobalt-catalyzed [2 + 2 + 2] cycloaddition reactions has been reported.[66] The Sonogashira coupling reaction of 1-(4′-I-C_6H_4)-2-CH_3-1,2-*closo*-$C_2B_{10}H_{10}$ with ethynyltrimethylsilane produces a compound that after removal of TMS gives a terminal alkyne. This was reacted with 1-(4′-I-C_6H_4)-2-CH_3-1,2-*closo*-$C_2B_{10}H_{10}$ via another Sonogashira coupling-type reaction under copper-free conditions. The resulting compound then underwent facile cobalt-catalyzed cycloaddition reaction to produce the star-shaped compound shown in Figure 27.13b in good yield. Due to the presence of hexaphenylbenzene core,[67] this compound displays moderate absorbance and emission efficiency, and the presence of *o*-carborane clusters makes it highly thermally stable.[38d,68] These star-shaped carborane-appended hexaphenylbenzene macromolecular compounds possess a great potential as a boron drug delivery platform, but they can also be useful in materials science, as they can form the core structures for discotic liquid crystals, components in organic field-effect transistors, and organic light-emitting diodes.[69]

We have recently reported the synthesis of high-boron-content polyanionic multicluster macromolecules (Figure 27.14) by using Grignard reagents, carboxylic acid, carboranes, and thiocarboranes as nucleophiles to attack the positively oxygen atom of the zwitterionic [3,3′-Co(8-$(C_2H_4O)_2$-1,2-$C_2B_9H_{10}$)(1′,2′-$C_2B_9H_{11}$)].[39,40,70] The potassium, sodium, or lithium salts of the resulting high-boron-content polyanionic macromolecules containing multiple metallacarborane clusters at their periphery show high water solubility. These compounds may prove useful as new classes of BNCT compounds with enhanced water solubility and as a core to make a new class of dendrimers.[71]

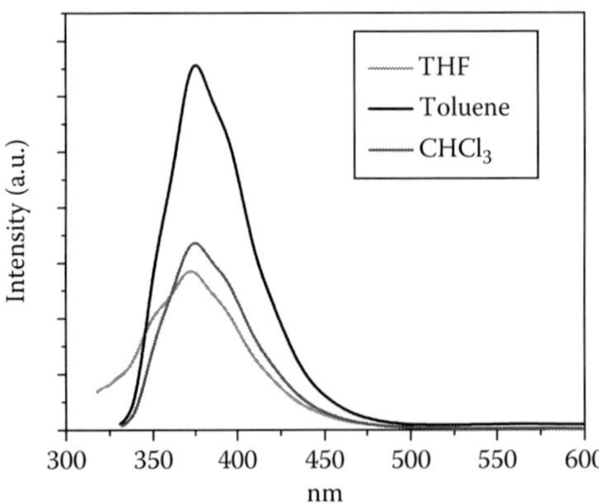

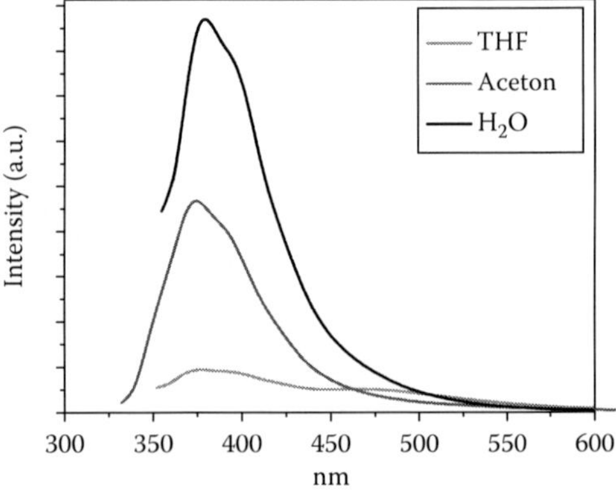

FIGURE 27.12 Fluorescence emission spectrum of a *closo*-dendron (left) and a *nido*-dendron (right) in different solvents.

FIGURE 27.13 Carboranyl containing star–shape molecules.

The crystal structure of $[Na_3(H_2O)(C_2H_5OH)][1'',3'',5''\text{-}\{3,3'\text{-}Co(8\text{-}O(C_2H_4O)_2\text{-}1,2\text{-}C_2B_9H_{10})(1',2'\text{-}C_2B_9H_{11})\}_3\text{-}C_6H_3]$ (Figure 27.15) shows that the chain contributed two or three oxygen atoms for coordination to Na^+ and, interestingly, the $[3,3'\text{-}Co(1,2\text{-}C_2B_9H_{11})_2]^-$ moiety provides extra B–H coordination sites. These B–H · Na interactions in the solid state have also been confirmed by dynamic NMR studies in solution.

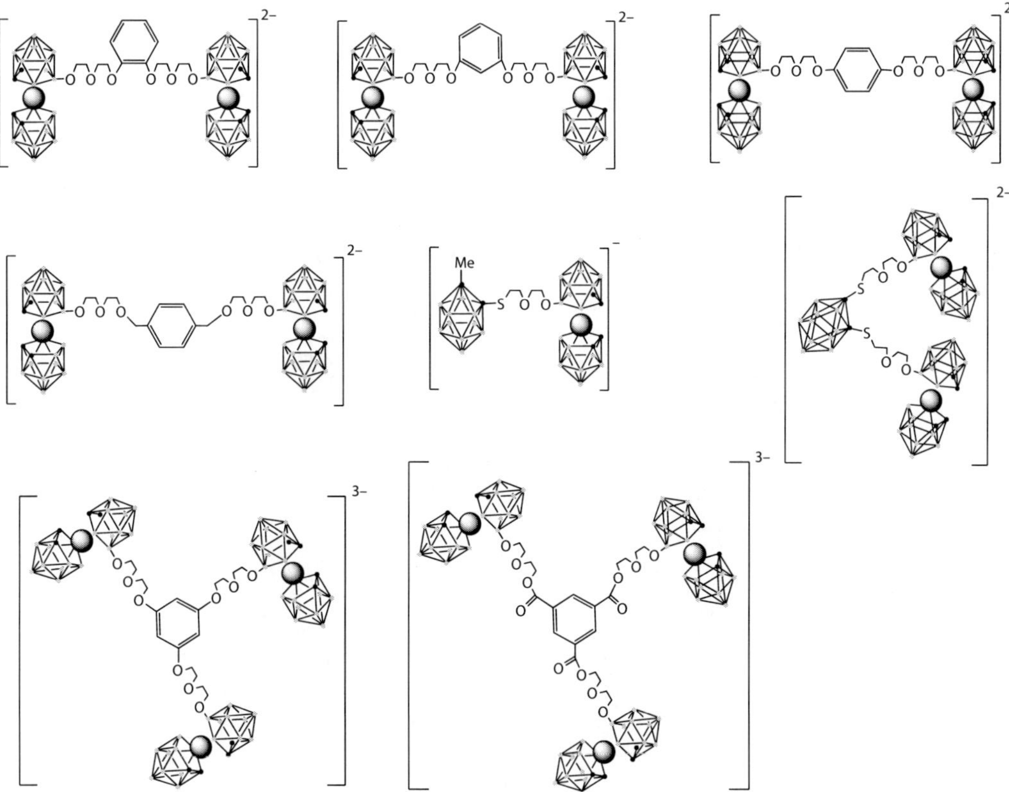

FIGURE 27.14 High-boron-content polyanionic multicluster macromolecules.

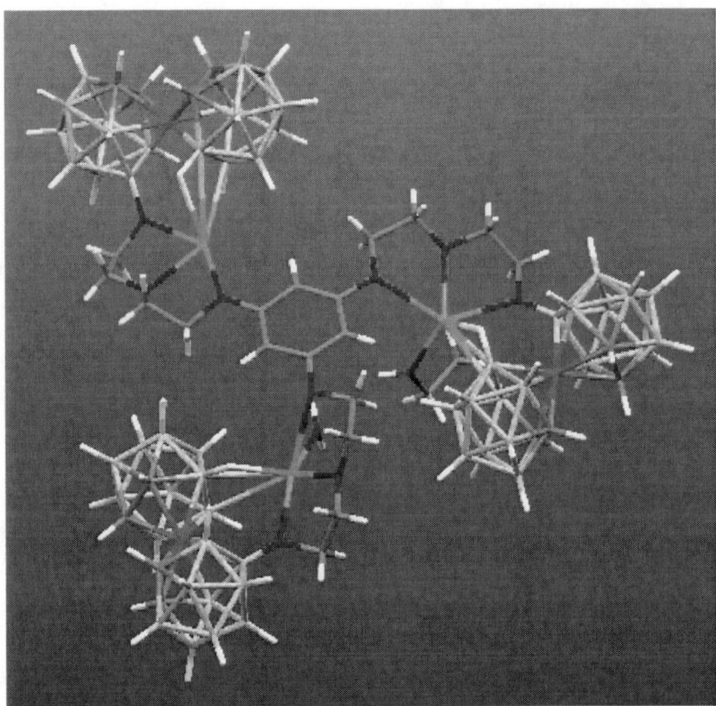

FIGURE 27.15 Crystal structure of the trianion sodium salt.

27.3.1.3 Other Cores

Other boron-rich building blocks with potential applications in BNCT or energy-filtering transmission electron microscopy have also been reported recently.[72] For that, tetracarboranylketone was prepared from the reaction of a tetraalkynylated ketone and decaborane. The crystal structure is represented in Figure 27.16. Reduction of the ketone yielded the corresponding tetracarboranylalcohol. Afterward, the tetracarboranylalcohol was functionalized with small spacer molecules, such as a propylalcohol or

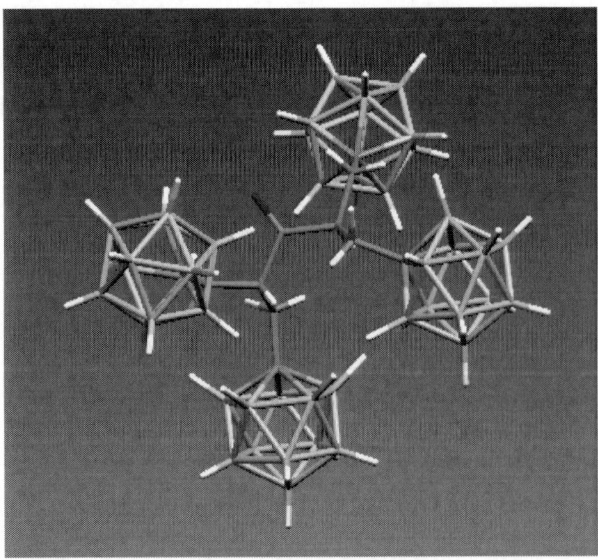

FIGURE 27.16 Molecular structure of the tetracarboranylketone.

4-bromobut-2-ynol, on the hydroxy position to be anchored to different biomolecules, such as nucleosides[73] or saccharides.[74] In a similar synthesis to the one described before,[72] the introduction of the bromobutynyl spacer into the tetracarboranylalcohol and further reaction with di-*tert*-butylmalonate in toluene leads to the formation of a compound that contains eight carborane clusters.

Synthesis of functional carborane-containing dendrons based on a polyester backbone 2,2-bis(hydroxymethyl)propanoic acid (bisMPA) scaffold,[75] in which *o*-carborane moieties are attached to the exterior of the dendron while a 10-carbon linker carrying a carboxylic group is fit at the focal point, has also been carried out.[76] To attach the *o*-carborane to the dendron, *o*-carboranyl acid was prepared by deprotection of 4-pentynoic acid with trichloroester, and further reaction of this with $B_{10}H_{14}$ by using a refluxing biphasic system of toluene and the anionic liquid 1-butyl-3-methylimidazolium chloride, (bmim)Cl, was carried out. Using a divergent approach, a second generation of a carborane-containing dendron carrying 40 boron atoms was also synthesized by coupling *ortho*-carboranyl acid to the acetonide-protected generation-2 dendritic acid. In a first step, the 10-carbon alkyl linker was attached to the focal point in CH_2Cl_2 in the presence of 1-(3-dimethylamino-propyl)-3-ethylcarbodiimide hydrochloride (EDC) and 4-dimethylaminopyridine (DMAP), followed by the attachment of the carboranyl acid in the same conditions to yield the expected dendron (Scheme 27.10). Higher generations were not synthetically available due to the sterical hindrance of the surface, confirmed by molecular dynamic simulations.

In a similar manner, tricarboranyl pentaerythritol-based building blocks have also been reported by Mollard and Zharov.[77] For that purpose, synthesis of a pentaerythritol derivative bearing three alkyne moieties and retaining a protected hydroxyl group necessary for further coupling to other molecules was accomplished. Boron clusters were introduced by reacting the terminal alkynes with slight excess of decaborane in a biphasic mixture of toluene and the ionic liquid (bmim)Cl at reflux for 3 h. After that, installation of a linker was carried out by the opening of succinic anhydride with the hydroxyl group of tricarboranyl alcohol in the presence of NaH in THF. Under these conditions, the desired tricarboranyl acid was isolated in 96% yield (Scheme 27.11) and no deboronation was observed. In the final step, the reaction of tricarboranyl acid with the 2,2,2-trichloroethyl ester of bisMPA in CH_2Cl_2 in the presence of *N*,*N′*-dicyclohexylcarbodiimide (DCC) and 4-(dimethylamino) pyridinium-4-toluenesulfonate (DPTS) led to the formation of the desired dendron with six o-carboranyl moieties (Scheme 27.11).

27.3.2 Dendrimers Functionalized with Peripheral Boranes, Carboranes, and Metallacarboranes

27.3.2.1 PAMAM, Poly-L-Lysine and Pentaerythritol-Based Dendrimers

The first dendrimeric structures peripherally functionalized with a polyhedral borane were reported by Barth et al.[78] in 1994. For that purpose, zero, first, second, and third generation of amino-terminated PAMAM startbust dendrimers were boronated by their reaction with $Na(CH_3)_3NB_{10}H_8NCO$. After that, the boronated dendrimers reacted with the monoclonal antibody MoAb IB16-6, previously derivatized with *N*-succinimidyl 3-(2-pyridyldithio)propionate, to obtain stable immunoconjugates.

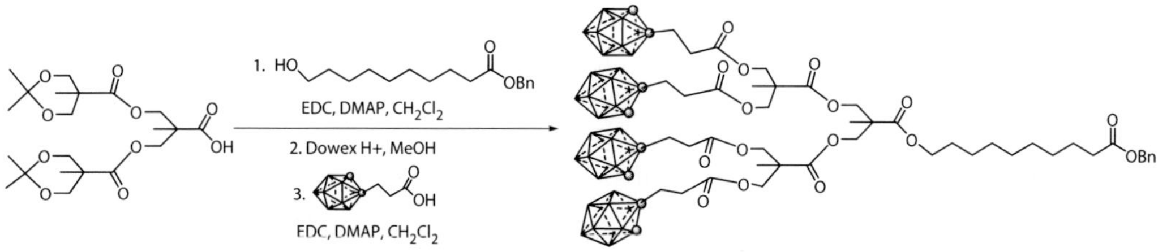

1. HO~~~~~~~~~(C=O)OBn
EDC, DMAP, CH_2Cl_2
2. Dowex H+, MeOH
3. carboranyl acid (OH)
EDC, DMAP, CH_2Cl_2

SCHEME 27.10 Synthesis of functional carborane-containing dendrons based on a polyester backbone 2,2-bis(hydroxymethyl)propanoic acid (bisMPA) scaffold.

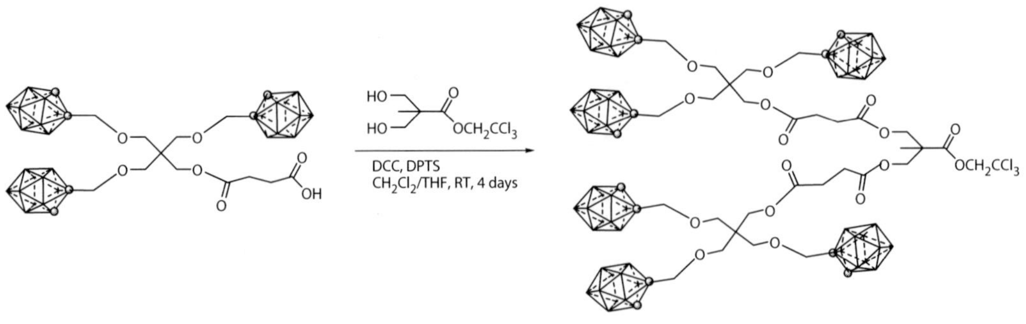

SCHEME 27.11 Preparation of hexacarboranyl pentaerythritol-based building blocks.

Later, folic acid conjugates of boronated poly(ethylene glycol) containing third-generation polyamidoamine dendrimers, as well as the synthesis of VEGF-driven boronated dendrimer (where VEGF is vascular endothelial growth factor) with a fifth generation of PAMAM dendrimer linked to 105-110 decaborate molecules, to obtain ^{10}B concentrations necessary for BNCT have been reported.[79]

Although this chapter focuses on the preparation of large molecules bearing icosahedral boron clusters, we should mention that the first example of polynuclear metallacarborane-containing dendrimers were prepared by Grimes and coworkers.[80] In his work, the preparation of PAMAM dendrimers decorated with small cobaltacarborane clusters, $[C_pCo(2,3\text{-}Et_2C_2B_3H_5)]$, as well as the study of their redox properties were carried out. Metallodendrimers that contain 16 and 32 cobaltacarboranes at the periphery were well characterized indicating a full substitution of the NH_2-terminal groups from the starting dendrimers. It is to remark that in these dendrimers, cyclic voltammetry showed a single reduction process exhibiting features of chemical reversibility. The process involves simultaneous one-electron reduction of all the peripheral cobalt centers; furthermore, electronic interaction was not observed among the cobaltacarborane units.

Boron-rich lysine dendrimers carrying the carborane clusters at the external surface were synthesized by Qualman et al.[81] in 1996, to be used as protein markers in electron microscopy. Two poly-(α,ε-L-Lys) dendrimers were prepared following standard procedures of solid-phase peptide synthesis by using the 9-fluorenylmethoxycarboranyl/tBu (Fmoc/tBu) strategy, which consists of coupling Fmoc-Lys(Dnasyl)-OH to TentaGel S-PHB or TentaGel PAP in DMF, and pegylated polystyrene resins. In the final acylation step of the lysine, (*S*)-5-(2-CH_3)-1,2-dicarba-*closo*-dodecaborane(12)-1-yl)-2-amino pentanoic acid,[82] was used to be coupled to the dendrimer. Finally, cleavage and deprotection of the resin-bound protected peptide with the appropriate agents led to the expected dendrimers with eight carborane derivatives that were well characterized by mass spectrometry (EI-MS or MALDI-MS[83]). These polylysine dendrimers showed good solubilities in polar organic solvents and in aqueous organic mixtures. They were tested as markers in electron energy loss spectroscopy experiments using a Zeiss 902 electron microscope and its suitability for immunolabeling in ESI was studied showing high-resolution properties and confirming the validity of the method in terms of exceeding the detection limits with the 80 boron atoms label.[84]

In parallel, it was reported by Housecroft and coworkers[85] a completely different metallodendritic system based on a pentaerythritol, as a core molecule, and four carboranyl-functionalized complexes of 2,2′:6′,2″-terpyridine ruthenium. The reaction of pentaerythrytol with 4′-Cl-2,2′:6′,2″'-terpyridine in dimethyl sulfoxide (DMSO) in the presence of KOH gave a molecule that contains four tpy metal-binding sites, the crucial point for the formation of the metallodendrimer. The reaction of this with the carboranyl-functionalized Ru complex [Ru{4′-[2-(*tert*-butyldimethylsilyl)-1,2-carboranyl]-2,2′:6′,2″-ter-pyridine}Cl_3], [Ru(sicarbtpy)Cl_3],[86] in ethane-1,2-diol at 120°C led to the formation of the first generation of a cationic tetranuclear metallodendrimer (Scheme 27.12), confirmed by NMR and MALDI-TOF.

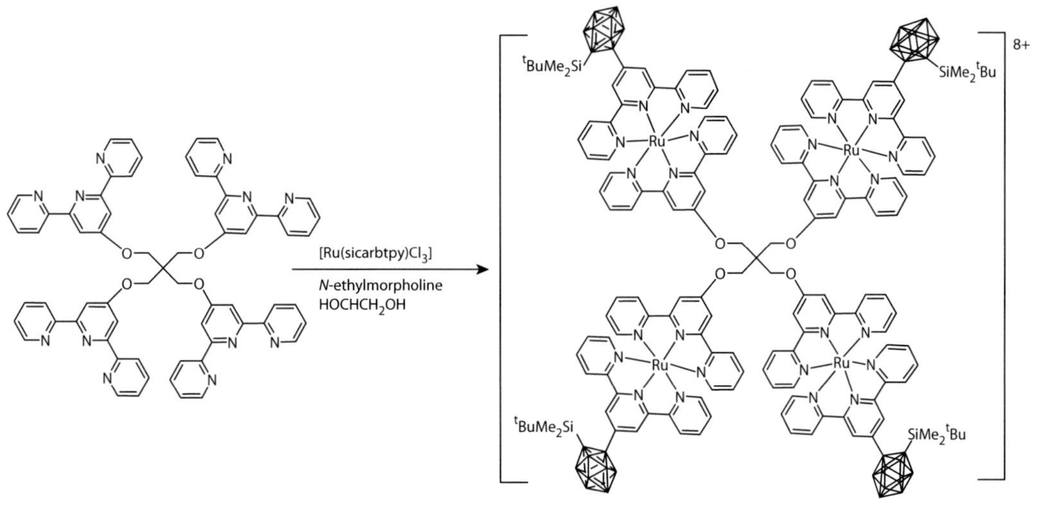

SCHEME 27.12 Preparation of metallodendrimers.

27.3.2.2 Carbosilane, Carbosiloxane, and Silsesquioxane-Based Dendrimers

More recently, following our interest in the preparation of high-boron-content molecules, we reported the modular construction of new carbosilane dendrimers that contain four and eight peripheral carborane derivatives and propyl linkages between the $C_{cluster}$ and the Si atoms.[64] The first generations of neutral carboranyl-containing dendrimers were constructed by two approaches: (a) a divergent approach via regiospecific hydrosilylation of 1-($CH_2CH{=}CH_2$)-2-R-1,2-*closo*-$C_2B_{10}H_{10}$ (R = C_6H_5, CH_3) with the first generation of a carbosilane dendrimer containing peripheral Si–H functions, Si[$(CH_2)_2(CH_3)_2SiH]_4$, 1G-H_4; and (b) a convergent strategy, the growth of which was initiated with the groups that will become the periphery of the dendrimer. In the latter, the hydrosilylation of the allyl group with $(CH_3)_2HSiCl$ followed by reduction with $LiAlH_4$ gave compounds 1-$(CH_2)_3SiH$-2-R-1,2-*closo*-$C_2B_{10}H_{10}$ (R = C_6H_5, CH_3) which were treated with tetravinylsilane to yield the first-generation dendrimers with four peripheral carborane clusters. Next-generation dendrimers were built up in a similar manner. In addition, another set of a second-generation dendrimers containing eight carborane clusters were prepared by hydrosilylation of the vinyl groups in the dendrimer Si{$(CH_2)_2(CH_3)Si[(CHCH_2)_2]\}_4$, 1G-$Vi_8$, with carboranylsilane dendrons that contain Si–H functions (Scheme 27.13). Degradation reaction of the peripheral *closo*-carboranes by using KOH/EtOH led to the formation of the corresponding polyanionic carbosilane dendrimers containing peripheral *nido*-carborane clusters.

Recently, we have also reported the synthesis of carbosilane and carbosiloxane metallodendrimers that contain one, four, and eight peripheral cobaltabisdicarbollide derivatives (Scheme 27.14), by using regiospecific hydrosilylation of vinyl-terminated dendrimers with Cs[1,1′-μ-SiMeH-3,3′-Co(1,2-$C_2B_9H_{10})_2$].[87] Furthermore, the synthesis of a trifunctional molecule having one cobaltabisdicarbollide and three vinylsilane moieties has been successful. The first and second generation of anionic metallacarborane-containing metallodendrimers were constructed via hydrosilylation of the respective carbosilane and cyclocarboxiloxane dendrimers functionalized with four or eight vinyl functions,[55,56,88] by using stoichiometric amounts of Cs[1,1′-μ-SiMeH-3,3′-Co(1,2-$C_2B_9H_{10})_2$] as the hydrosilylating agent in the presence of Karstedt catalyst (Scheme 27.14). In addition, the hydrosilylation reaction of 1 equiv. of Cs[1,1′-μ-Si(CH_3)H-3,3′-Co(1,2-$C_2B_9H_{10})_2$] with 1 equiv. of tetravinylsilane, under similar conditions yielded the earlier mentioned trifunctional molecule in 77%.

Carbosilane metallodendrimers were purified, isolated as cesium salts, and fully characterized by FTIR, ^{1}H, ^{11}B, ^{13}C, and ^{29}Si NMR and UV–vis spectroscopy. The silyl-containing cobaltabisdicarbollides show absorption bands at 310 and 462 nm in the UV–vis spectra that follow the

R = Ph, Me
Toluene
Karstedt cat.
R = Ph, Me

SCHEME 27.13 Synthesis of carbosilane dendrimers peripherally decorated with eight carborane cages.

Lambert–Beer law. Thus, the UV–vis absorptions have been a good tool for estimating the experimental number of cobaltabisdicarbollide moieties attached to the periphery. This method consists of the UV–vis absorption's measurement of the solution containing the functionalized dendrimer to study the trend of its molar absorptivity (ε), as this is proportional to the number of metallacarboranes attached to the periphery.[87] The number of cobaltabisdicarbollide fragments for each dendrimer can be estimated by comparing the different absortivities of carbosilane and carbosiloxane dendrimers that contain four and eight cobaltabisdicarbollide units with the absortivity obtained for a monomer which is represented by the low-intensity band in Figure 27.17.

Polyhedral oligomeric silsesquioxanes (POSS) $(RSiO_{1.5})_n$ ($n = 8$), are nanosize building blocks for organic/inorganic hybrid materials; their high potential for applications is based on the possibility to control and balance the inorganic and organic moieties in their architecture.[89] Therefore, they can be tuned for very different applications in accordance with the nature of the organic functionality,[90] such as octa-arms dendrimers-core.[91] Carboranyl-containing disiloxane,

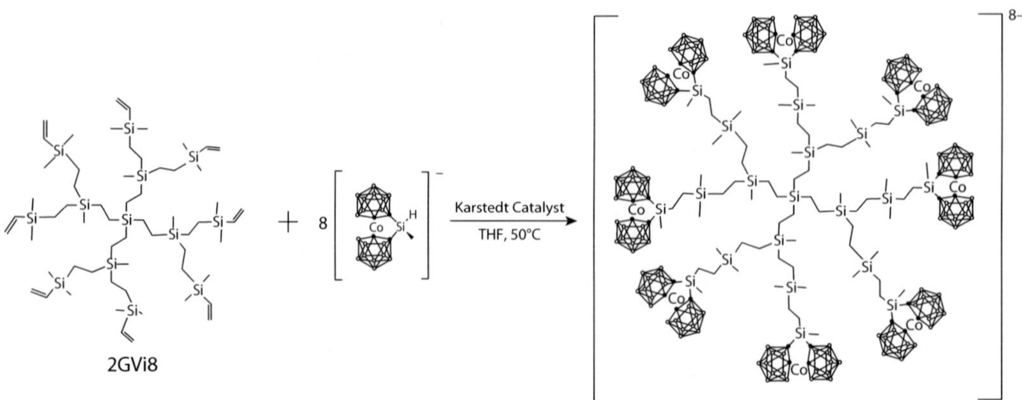

SCHEME 27.14 Preparation of a carbosilane metallodendrimer decorated with eight cobaltabisdicarbollide units.

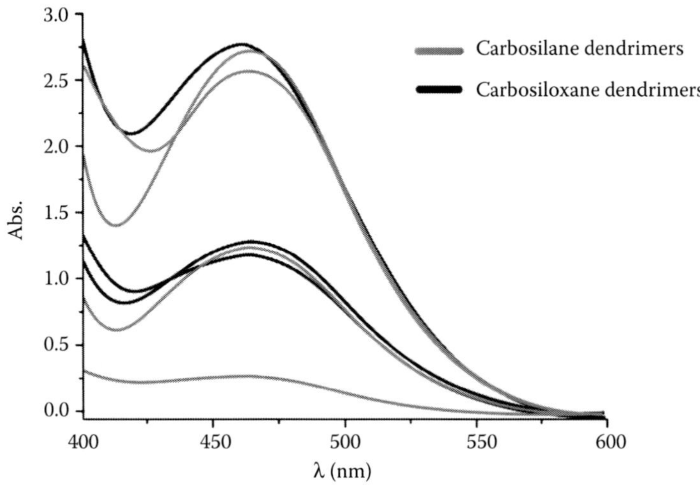

FIGURE 27.17 UV–vis spectra of carbosilane and carbosiloxane dendrimers in DMSO.

cyclic-siloxane, and cage-like octasilsesquioxane (POSS) structures were prepared in high yields by our group.[92] Two routes have been compared for their preparation, a classical hydrolytic process based on hydrolysis and condensation of the freshly prepared carboranylalkylchlorosilane or ethoxysilane precursors and a nonhydrolytic route based on the specific reactivity of carboranylalkylchorosilane toward DMSO as oxygen source.[93] The required precursors were prepared by hydrosilylation of 1-CH_3-2-$CH_2CH{=}CH_2$-1,2-*closo*-$C_2B_{10}H_{10}$ and 1-C_6H_5-2-$CH_2CH{=}CH_2$-1,2-*closo*-$C_2B_{10}H_{10}$ with the respective $HSi(CH_3)_2Cl$, $HSiCH_3Cl_2$, $HSiCl_3$, and $HSi(OCH_2CH_3)_3$, in the presence of Karstedt's catalyst, without solvent and with excess of silane. It is important to remark that the α adduct was formed in all cases. Preparation of POSS by the nonhydrolytic method from trichlorosilanes was carried out in DMSO:$CHCl_3$ mixture (2:1) or (3:1) at room temperature leading to octasilsesquioxane (T_8) cages (Scheme 27.15).[92] POSS were also obtained by the hydrolysis of carboranyltriethoxysilanes respectively, using different stoichiometric amounts of water in THF or $CHCl_3$, and tetrabutylammonium fluoride as catalyst, at room temperature for a long reaction time (133–180 days) (Scheme 27.15).[92]

Based on the typical reactivity of the carboranyl group toward nucleophiles,[94] dianionic disiloxanes and octaanionic silsesquioxanes were obtained by classical decapitation of the *closo* clusters via the reaction with KOH in ethanol at reflux (Scheme 27.15). Products were fully characterized by FT-IR, ^1H, ^{11}B, ^{13}C, and ^{29}Si NMR spectroscopy, MALDI-TOF mass spectrometry and the carboranyldisiloxane was also confirmed by x-ray diffraction analysis (Figure 27.18).

27.3.2.3 Poly(aryl ether) Dendrimers with 1,3,5-Triphenylbenzene as the Core

Fluorescent Fréchet-type poly(aryl ether) dendrimers, which incorporate the 1,3,5-triphenylbenzene as the core molecule, and three, six, nine, or 12 terminal allyl ether groups were prepared following the Fréchet convergent approach with the aim to decorate the periphery with metallacarboranes.[95] Regiospecific hydroslylation reactions on the allyl ether functions with the cobaltabisdicarbollide derivative Cs[1,1′-μ-$Si(CH_3)H$-3,3′-Co(1,2-$C_2B_9H_{10})_2$] using the ratios 3:1, 6:1, and 12:1, with the Karstedt catalyst, led to different generations of Féchet-type polyanionic metallodendrimers decorated with three, six, and 12 cobaltabisdicarbollide units (Figure 27.19).

Starting poly(aryl ether) dendrimers exhibited photoluminescence properties at room temperature under UV irradiation; nevertheless, after functionalization with cobaltabisdicarbollide derivatives, the fluorescence properties were quenched.

DMSO

$CHCl_3$

H_2O, Cat.

THF

R = Me, Ph,

40 KOH

EtOH

SCHEME 27.15 Different strategies to prepare carboranyl-containing octasilsesquioxanes (T_8), and further degradation.

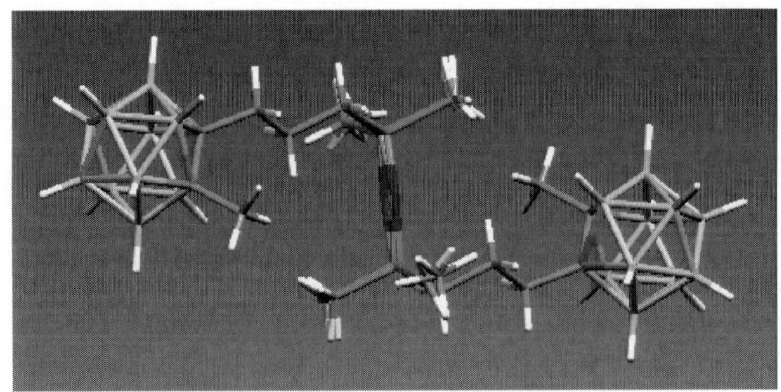

FIGURE 27.18 Molecular structure of carboranyldisiloxane $[1\text{-}CH_3\text{-}2\text{-}CH_2CH_2CH_2(CH_3)_2Si\text{-}1,2\text{-}closo\text{-}C_2B_{10}H_{10}]_2O$.

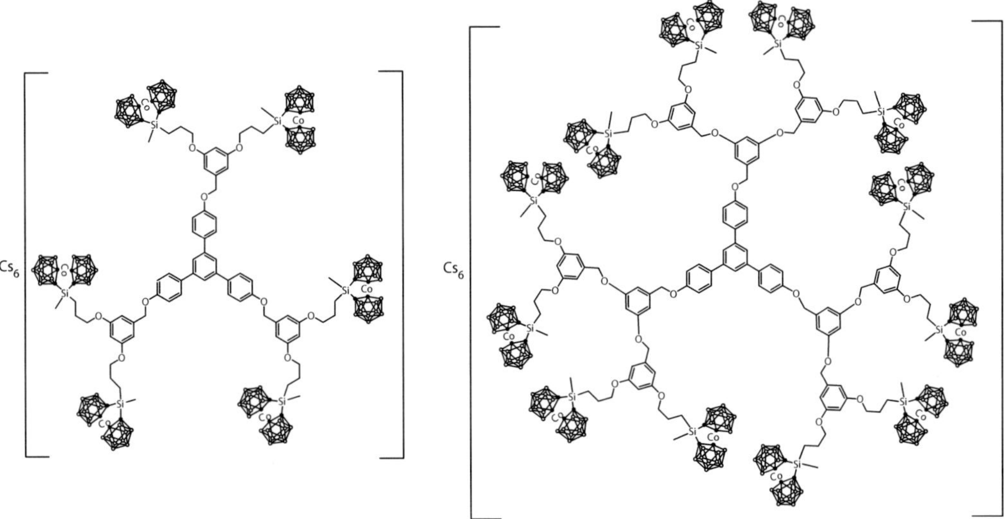

FIGURE 27.19 Schematic representation of two generations of poly(aryl ether) metallodendrimers decorated with six and 12 cobaltabisdicarbollide units.

27.3.3 Macrocycles Functionalized with Boron Clusters

27.3.3.1 Porphyrins and Phthalocyanines Functionalized with Boron Clusters

Porphyrins are promising tumor-selective compounds because they have demonstrated the tendency to accumulate in neoplastic tissue.[96] In 1978, Rudolph and coworkers[97] reported the synthesis of the first *meso*-tetracarboranylporphyrins in which the carborane cage was bonded to the macrocycle, thorough a $-CH_2-$ spacer. Their objective was to use them as catalysts in the reversible multielectron reduction of small molecules, such as O_2 and N_2. In a first attempt, the Rothemund condensation of pyrrole with 1-formylcarborane 1-CHO-1,2-*closo*-$C_2B_{10}H_{11}$ failed according to the authors due to the steric hindrance. The use of functions separated carboranylaldehides 1-CH_2CHO-1,2-*closo*-$C_2B_{10}H_{11}$ and 1-CH_2CHO-2-CH_3-1,2-*closo*-$C_2B_{10}H_{10}$ successfully yielded the expected carboranylporphyrins, although in low yield (Figure 27.20). Partial degradation of the *closo* cluster by using pyridine and piperidine produced the tetra*nido*carboranyl-porphyrins.

In 1990, Gabel and coworkers[98] reported the synthesis of carboranylporphyrins in which the carborane cluster is linked to the porphyrin via carbon–carbon bonds through a benzyl group. Later, Vicente et al.[99] used a similar strategy to prepare carboranylporphyrins. In both cases, tetraarylporphyrins (*para*- and *meta*-substituted) containing four or eight 1-CH_3-1,2-*closo*-$C_2B_{10}H_{10}$, not directly

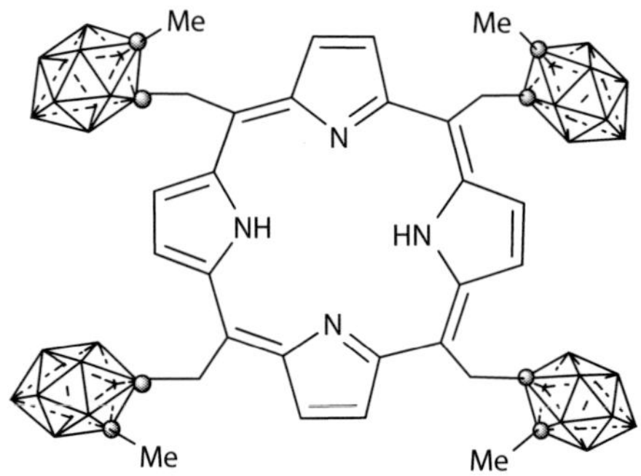

FIGURE 27.20 Schematic representation of the first *meso*-tetracarboranylporphyrin.

bonded to the porphyrin ring, were prepared by condensation of previously synthesized carboranyl-benzaldehyde or dicarboranylbenzaldehyde with pyrrole or dipyrromethane under Lindsey-type conditions,[100] respectively. In a similar manner, homologous porphyrins containing four 1,2-*closo*-$C_2B_{10}H_{12}$ clusters on the *para*- and *meta*-positions and directly linked to the *meso*-phenyl groups were also prepared. In order to achieve water solubility, the tetra(*nido*-carboranyl) derivatives of the former porphyrins were prepared by basic degradation of the *closo*-carboranes using pyridine and piperidine in a 3:1 ratio. Insertion of zinc and nickel using normal procedures afforded the corresponding metallocarboranylporphyrins. The x-ray crystal molecular structure of a metalloporphyrin covalently bonded to carborane clusters is shown in Figure 27.21. Other porphyrin derivatives, such as carboranylchlorins with dual applications in BNCT and photodynamic therapy (PDT) therapies of tumors,[101] and *closo* and *nido*-carboranyl-tetrabenzoporphyrins,[102] have also been reported. In addition, a new high-yield route to tetra- and octa-carboranylated porphyrins has also been reported that involves the Suzuki coupling of boronic acid–pyrroles and bromoporphyrins with the adequate carboranyl boronic acid derivatives.[103]

Kahl et al.[104] synthesized tetraphenylporphyrin bearing four *nido*-$[C_2B_9H_{11}]^-$ cages linked to the *o*-phenyl ring positions by anilide bonds, from tetra-(*o*-aminophenyl)porphyrin and carborane carbonyl chloride followed by base-assisted cage opening and ion exchange to give the highly water-soluble potassium salt. A binary conjugate of a boronated porphyrin with a peptide, bearing four 1,7-$C_2B_{10}H_{11}$ carboranes has been recently reported.[105] In addition, porphyrins bearing anionic boron clusters such as 5-(benzamidododecahydro-*closo*-dodecaborate)-10,15,20-triphenylporphyrin and *meso*-tetrakis-benzamidododecahydro-*closo*-dodecaborate)porphyrin, as well as porphyrins bearing $[B_{12}H_{12}X]^-$ (X = NH_2, O, S) have recently been prepared (Figure 27.22).[106]

More recently, a set of porphyrin–cobaltacarborane conjugates have also been prepared in high yields via a nucleophilic ring-opening reaction consisting of a single-step reaction between a 3,5-dihydroxyporphyrin or meso-pyridyl-containing porphyrin, and zwitterionic cobaltacarborane [3,3′-Co(8-$C_4H_8O_2$-1,2-$C_2B_9H_{10}$)(1′,2′-$C_2B_9H_{11}$)].[107] These compounds have one to four cobaltabisdicarbollide anions conjugated to the porphyrin macrocycle via $(CH_2CH_2O)_2$ chains, as shown in the crystal structure of Figure 27.23. Cell studies of these boron-rich and fluorescent compounds showed that some of them are promising agents for BNCT.

Other carboranylporphyrins and boronated chlorine e_6 for PDT, in which porphyrin is linked to the carboranyl moiety via the boron or the carbon atoms have also been reported by Ol'shevskaya et al.[108] In a recent work, boronated amide derivatives starting from 5,10,15,20-tetra(p-aminophenyl) porphyrin and 9-*o*- and 9-*m*-carborane carboxylic acid chlorides were prepared. Also, the reaction of 2-formyl-5,10,15,20-tetraphenylporphyrin with the lithium salt of 1,2-*closo*-$C_2B_{10}H_{12}$, and

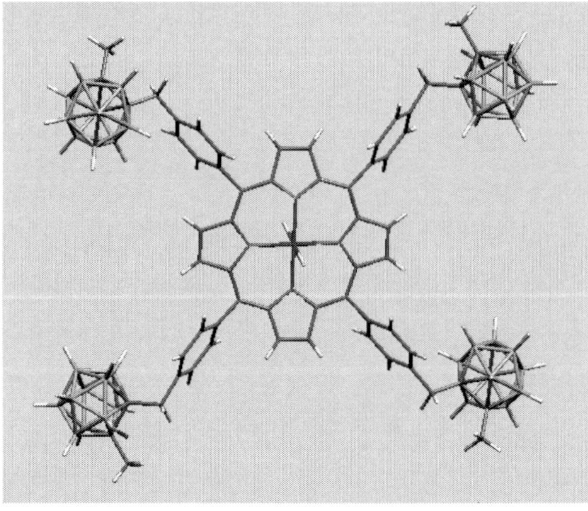

FIGURE 27.21 Crystal structure of a tetraphenylporphyrin bearing four 1-CH_3-1,2-*closo*-$C_2B_{10}H_{10}$ cages.

X = NH_2 n = 4
X = S n = 8
X = O n = 8

FIGURE 27.22 Representation of borane anion porphyrins.

$[1\text{-}closo\text{-}CB_{10}H_{12}]^-$ produces neutral and anionic boronated hydroxy derivatives of 5,10,15,20-tetraphenylporphyrin.

Biological properties,[108a,109] such as dark toxicity, the ability of porphyrin to induce DNA photodamage, interaction of porphyrins with DNA,[109a] biodistribution in tumor-bearing mice,[109c] of hydrophobic *o*-carboranylporphyrins and above all amphiphilic *nido*-carboranylporphyrins containing four carborane clusters have been studied. In addition, carboranylporphyrins bearing eight *nido*-carborane cages can still accumulate intracellularly and have low dark toxicity toward cells in culture, and therefore might have promise for application in BNCT.[110] The pharmacokinetic results have shown that some porphyrin derivatives bearing up to four carborane cages can be systemically injected *in vivo* with no detectable toxic effects.[111]

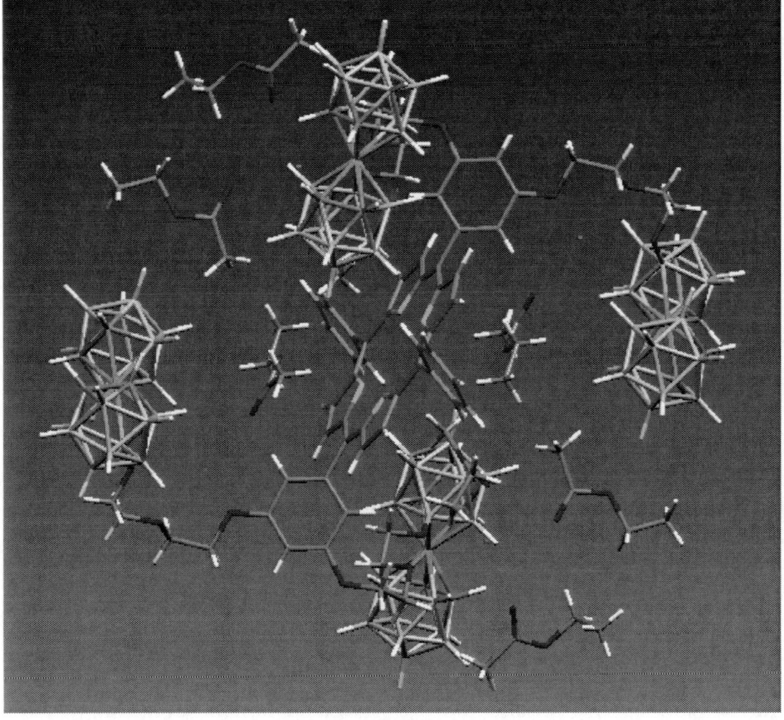

FIGURE 27.23 Molecular structure of a cobaltacarborane–porphyrin conjugate.

27.3.3.2 Calixarenes Functionalized with Boron Clusters

The name calixarenes was introduced by Gutsche[112] for the cyclic oligomers that are obtained from the condensation of formaldehyde with *p*-alkylphenols under alkaline conditions. The use of "calix," which means "beaker" in Latin and Greek, was suggested by the shape of the tetramer, which uses to adopt a bowl or beaker like conformation that indicates the possibility of the inclusion of "guest" molecules. Resorcinarenes are cyclic oligomers based on the condensation of resorcinol and aldehydes. The importance of the "cyclic oligonuclear phenolic compounds" in the development of host–guest chemistry, inclusion compounds, or even supramolecular chemistry was stimulated after Pedersen's discovery of the crown ethers.[23,113] Calixarenes and resorcinarenes[114] are platforms on which it is possible to assemble various functional groups in a well-defined arrangement in space. Numerous chemical modifications are available[115] at the upper (wide) rim or the lower (narrow) rim to fine tune their properties and their molecular shape.[112,114a] Numerous cation extractants derived from calixarenes have been developed, in which the preorganization of ligating functions adds favorably to their specificity.[116] The use of calixarenes and cavitands with organic functional groups in the partitioning of radionuclide cations from strongly nitric acid high-level activity nuclear waste is accompanied by the unwanted cotransport of nitrate ions into the organic phase. Covalent bonding of anions to these platforms can compensate intramolecularly for the cationic charge of the radionuclide and thus reduce or eliminate the cotransport of nitrate ions.

Grüner et al. reported the covalent bonding of cobaltabisdicarbollide units to the lower or upper rim rings of platforms derived from *tert*-butylcalix[4]arenes and resorcinarenes. These novel anionic compounds combine the properties of basket-like molecules with those of hydrophobic, weakly coordinating anions of strong nonoxidizing inorganic acids. These polyanionic platforms were obtained by the ring-opening reaction of the dioxane ring of the zwitterionic $[3,3'\text{-Co}(8\text{-}(C_2H_4O)_2\text{-}1,2\text{-}C_2B_9H_{10})(1',2'\text{-}C_2B_9H_{11})]$ by the phenolate groups of the calixarene/resorcinarene platforms. The mono-, di-, tri-, and tetrasubstitution of the *tert*-Butylcalix[4]arene platforms provided polyanionic large molecules with tailor-made cavities and compensating anions (Figure 27.24). Di- and tetrasubstitution was also achieved with cavitands derived from resorcin[4]arenes.[44a]

Carbamoylmethylphosphine oxides (CMPOs), and especially (*N,N*-diisobutylcarbamoylmethyl)-phenyloctylphosphine oxide, have been designed as extracting ligands for actinides and lanthanides on a technical scale (TRUEX-process).[117] When four CMPO functions have been bonded to the wide rim of calix[4]arene, tetraether extractants that are about 100–1000 times more effective than the single CMPO have been produced.[118] The covalent attachment of two CMPO functions and two anionic $[3,3'\text{-Co}(1,2\text{-}C_2B_9H_{11})_2]^-$ clusters to the narrow rim of *tert*-butylcalix[4]-arene leads to dianionic calix[4]arene tetraethers.

The study of these polyanionic platforms in liquid–liquid extraction of various trivalent lanthanide and actinide cations from high-level radioactive acidic waste solutions was tested.[119] It is interesting to emphasize that these novel ionic ligands have shown a dramatic enhanced extraction ability for trivalent actinides and lanthanides, particularly for Am^{3+} a thorough extraction (>99%) has been achieved in a single extraction step.[44b,119]

27.4 SUMMARY

The view of boranes, carboranes, and metallacarboranes as rare, strange, or esoteric compounds, in addition to having high price and an assumed, unrealistically, low stability, tends to preclude synthetic chemists in other areas to think on these clusters as real building blocks. In this chapter, we have reported on large high-boron-content molecules that contain several icosahedral borane, carborane, and metallacarborane clusters either as part of the scaffolds or as decorating the periphery of platforms. The largest part of these compounds has been proposed as highly promising boron delivery agents for the BNCT of tumors, particularly for malignant brain tumors. However, many other applications such as ISE's, ionic liquids, selective optical sensors, sensitive liquid/polymeric

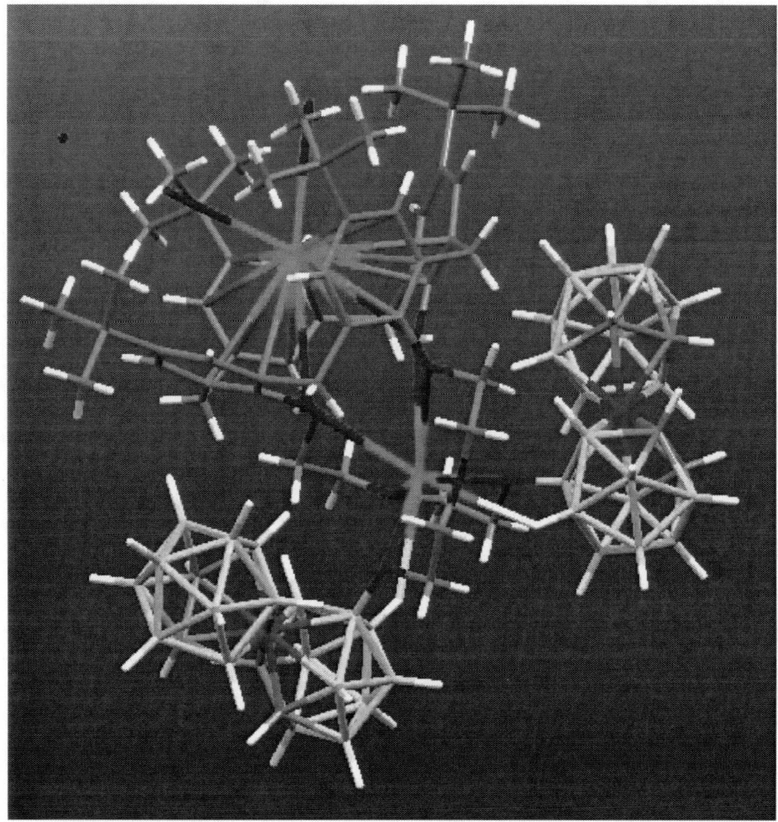

FIGURE 27.24 X-ray crystal structure of the dianionic platform derived of *tert*-Butylcalix[4]arenes with two arms of anionic $[3,3'\text{-Co}(1,2\text{-C}_2\text{B}_9\text{H}_{11})_2]^-$ clusters. A Cs^+ is located into the calixarene cavity and the second one is surrounded by the oxygen atoms of the polyether chains.

membrane electrodes, catalysts, optoelectronic devices, and extracting ligands for actinides and lanthanides, among others have been indicated in each particular section.

ACKNOWLEDGMENTS

Our own contributions to this area of research would not have been possible without the enthusiastic work of our coworkers, our postdoctoral associates, and graduate project students whose names appear in the references. The writing of this chapter has been financially supported by Generalitat de Catalunya (Project 2009 SGR 279).

The structural pictures for this review have been redrawn, mostly using atomic coordinates taken from the Cambridge Crystallographic Data Base, implemented through Mercury 1.4.2[120]—Crystal Structure Visualisation and Exploration Made Easy- 1.4.2 software, by The Cambridge Crystallographic Data Centre 12 Union Road, Cambridge, CB2 1EZ, UK, +44 1223 336408.

REFERENCES

1. (a) Pitochelli, A. R.; Hawthorne, M. F. The isolation of the icosahedral $B_{12}H_{12}^{2-}$ Ion, *J. Am. Chem. Soc.* **1960**, *82*, 3228. (b) Miller, H. C.; Miller, N. E.; Muetterties, E. L. Synthesis of polyhedral boranes, *J. Am. Chem. Soc.* **1963**, *85*, 3885.
2. (a) Peymann, T.; Herzog, A.; Knobler, C. B.; Hawthorne, M. F. Aromatic polyhedral hydroxyborates: Bridging boron oxides and boron hydrides, *Angew. Chem. Int. Ed.* **1999**, *38*, 1061. (b) Hawthorne, M. F.,

carborane chemistry at work and at play, proceedings of the ninth international meeting on boron chemistry. In *Advances in Boron Chemistry*, W. Siebert Ed. (Special Publication No. 201, Royal Society of Chemistry, London, 1997), pp. 261.
3. Grimes, R. N. *Carboranes*, Academic Press, New York, 1970.
4. Knoth, W. H.; Miller, H. C.; England, D. C.; Parshall, G. W.; Muetterties, E. L. Derivative chemistry of $B_{10}H_{10}^{2-}$ and $B_{12}H_{12}^{2-}$, *J. Am. Chem. Soc.* **1962**, *84*, 1056.
5. Hawthorne, M. F.; Young, D. C.; Wegner, P. A. Carbametallic boron hydride derivatives. i. apparent analogs of ferrocene and ferricinium ion, *J. Am. Chem. Soc.* **1965**, *87*, 1818.
6. Masalles, C.; Llop, J.; Viñas, C.; Teixidor, F. Extraodinary overoxidation resistance increase in self doped polypyrroles by using non-conventional low charge-density anions, *Adv. Mater.* **2002**, *14*, 826.
7. (a) Sivaev, I. B.; Bregadze, V. I. Chemistry of cobalt bis(dicarbollides). A review, *Collect. Czech. Chem. Commun.* **1999**, *64*, 783. (b) Juárez-Pérez, E. J.; Núñez, R.; Viñas, C.; Sillanpää, R.; Teixidor, F. The role of C-H···H-B interactions in establishing rotamer configurations in metallabis(dicarbollide) systems, *Eur. J. Inorg. Chem.* **2010**, 2385.
8. (a) Chamberlin, R. M.; Scott, B. L.; Melo, M. M.; Abney, K. D. Butyllithium deprotonation vs. alkali metal reduction of cobalt dicarbollide: A new synthetic route to c-substituted derivatives *Inorg. Chem.* **1997**, *36*, 809. (b) Rojo, I.; Teixidor, F.; Viñas, C.; Kivekäs, R.; Sillanpää, R. Synthesis and coordinating ability of an anionic cobaltabisdicarbollide ligand geometrically analogous to BINAP *Chem. Eur. J.* **2004**, *10*, 5376. (c) Juárez-Pérez, E. J.; Viñas, C.; González-Campo, A.; Teixidor, F.; Kivekäs, R.; Sillanpää R.; Núñez, R. Controlled direct synthesis of C-mono- and C-disubstituted derivatives of $[3,3'\text{-Co}(1,2\text{-}C_2B_9H_{11})_2]^-$ with organosilane groups: Theoretical calculations compared with experimental results, *Chem. Eur. J.* **2008**, *14*, 4924. (d) Juárez-Pérez, E. J.; Viñas, C.; Teixidor, F.; Núñez, R. First example of the formation of a Si–C bond from an intramolecular Si–H···H–C diyhydrogen interaction in a metallacarborane: A theoretical study, *J. Organomet. Chem.* **2009**, *694*, 1764.
9. (a) Viñas, C.; Pedrajas, J.; Bertrán, J.; Teixidor, F.; Kivekäs, R.; Sillanpää, R. Synthesis of cobaltabis(dicarbollyl) complexes incorporating exocluster SR substituents and the improved synthesis of $[3,3'\text{-Co}(1\text{-R-}2\text{-R'-}1,2\text{-}C_2B_9H_9)_2]^-$ Derivatives, *Inorg. Chem.* **1997**, *36*, 2482. (b) Viñas, C.; Gómez, S.; Bertrán, J.; Teixidor, F.; Dozol, J. F.; Rouquette, H. Cobaltabis(dicarbollide) derivatives as extractants for europium from nuclear wastes, *Chem. Commun.* **1998**, 191. (c) Viñas, C.; Gómez, S.; Bertrán, J.; Teixidor, F.; Dozol, J. F.; Rouquette, H. New polyether-substituted metallacarboranes as extractants for ^{137}Cs and ^{90}Sr from nuclear wastes, *Inorg. Chem.* **1998**, *37*, 3640. (d) Viñas, C.; Bertrán, J.; Gómez, S; Teixidor, F.; Dozol, J. F.; Rouquette, H.; Kivekäs, R.; Sillanpää, R. Aromatic substituted metallacarboranes as extractants of ^{137}Cs and ^{90}Sr from nuclear wastes, *J. Chem. Soc., Dalton Trans.* **1998**, 2849. (e) Viñas, C.; Pedrajas, J.; Teixidor, F.; Kivekäs R.; Sillanpää, R.; Welch, A. J. First Example of a bis(dicarbollide) metallacarborane containing a B,C'-heteronuclear bridge, *Inorg. Chem.* **1997,** *36*, 2988.
10. Marciniec, B. *Hydrosilylation: A Comprehensive Review on Recent Advances* In Advances in Silicon Science,. Matisons, J. Ed; Springer Netherlands, 2009. Vol. 1.
11. Francis, J. N.; Hawthorne, M. F. Synthesis and reactions of novel bridged dicarbollide complexes having electron-deficient carbon atoms, *Inorg. Chem.* **1971**, *10*, 594.
12. a) Plešek, J.; Hermánek, S.; Base, K.; Todd, L. J.; Wright, W. F. Zwitterionic compounds of the $8,8'\text{-X}(C_2B_8H_{10})_2\text{Co}$ series with nonoatomic O, S, Se, Te, and N bridges between carborane ligands, *Collect. Czech. Chem. Commun.* **1976**, *41*, 3509. b) Janousek, Z.; Plešek, J.; Hermánek, S.; Base, K.; Todd, L. J.; Wright, W. F. Preparation and characteristics of sulphur interlingad bridge-derivatives and further s-substituted compounds in the $(C_2B_9H_{11})_2\text{Co}^-$ series. conformations of $M(C_2B_9H_{11})_2^{X-}$ metallacarobrnes, *Collect. Czech. Chem. Commun.* **1981**, *46*, 2818. c) Rojo, I.; Teixidor, F.; Viñas, C.; Kivekäs R.; Sillanpää, R. Relevance of the electronegativity of boron in η5-coordinating ligands: Regioselective monoalkylation and monoarylation in cobaltabisdicarbollide $[3,3'\text{-Co}(1,2\text{-}C_2B_9H_{11})_2]^-$ clusters, *Chem. Eur. J.* **2003**, *9*, 4311. d) González-Cardoso, P., Stoica, A.I., Farràs, P., Pepiol, A., Viñas, C., Teixidor, F. Additive tuning of redox potential in metallacarboranes by sequential halogen substitution, *Chem. Eur. J.* **2010**, 16, 6660.
13. a) Plešek, J.; Hermánek, S.; Franken, A.; Cisarova I.; Nachtigal, C. Dimethyl sulfate induced nucleophilic substitution of the [Bis(1,2-dicarbollido)-3-cobalt(1-)]ate ion. Syntheses, properties and structures of its 8,8′-μ-sulfato, 8-phenyland 8-dioxane derivatives, *Collect. Czech. Chem. Commun.* **1997**, *62*, 47. b) Selucky, P.; Plešek, J.; Rais, J.; Kyrs M.; Kadlecova, L. Extraction of fission products into nitrobenzene with dicobalt bis-dicarbollide and ethyleneoxy-substituted cobalt bis-dicarbollide, *J. Radioanal. Nucl. Chem.* **1991**, *149*, 131.

14. Teixidor, F.; Pedrajas, J.; Rojo, I.; Viñas, C.; Kivekäs, R.; Sillanpää, R.; Sivaev, I.; Bregadze, V.; Sjöberg, S. Chameleonic capacity of $[3,3'\text{-Co}(1,2\text{-C}_2\text{B}_9\text{H}_{11})_2]^-$ in coordination. generation of the highly uncommon S(thioether)-Na bond, *Organometallics*, **2003**, *22*, 3414.
15. Llop, J.; Masalles C.; Viñas C.; Teixidor F.; Sillanpää R.; Kivekäs R. The $[3,3'\text{-Co}(1,2\text{-C}_2\text{B}_9\text{H}_{11})_2]^-$ anion as a platform for new materials: synthesis of its functionalized monosubstituted derivatives incorporating synthons for conducting organic polymers, *Dalton Trans.*, **2003**, 556.
16. Semioshkin, A. A.; Sivaev, I. B.; Bregadze, V. I. Cyclic oxonium derivatives of polyhedral boron hydrides and their synthetic applications, *Dalton Trans.* **2008**, 977.
17. Boyd, L. A.; Colquhoun, H. M.; Davidson, M. G.; Fox, M. A.; Gill, W. R.; Herbertson, P. L.; Hibbert, T. G. et al. Carborane rings, chains and related systems, in *Advances in Boron Chemistry*, edited by W. Siebert, The Royal Society of Chemistry, Cambridge, 1997, 289.
18. (a) Clegg, W.; Gill, W. R.; MacBride, J. A. H.; Wade, K. $(1,7\text{-C}_2\text{B}_{10}\text{H}_{10}\text{-}1',3'\text{-C}_6\text{H}_4)_3$, a cyclic trimer from meta-carboranediyl and meta-phenylene units: a new category of macrocycle, *Angew. Chem. Int. Ed. Engl.* **1993**, *32*, 1328. (b) Coult, R.; Fox, M. A.; Gill, W. R.; Herbertson, P. L.; MacBride, J. A. H.; Wade, K. C-arylation and C-heteroarylation of icosahedral carboranes via their copper(I) derivatives, *J. Organomet. Chem.* **1993**, *462*, 19.
19. Gill, W. R.; Herbertson, P. L.; MacBride, J. A. H.; Wade, K. Preparation of C-2-pyridyl derivatives of icosahedral carboranes via copper(I) intermediates *J. Organomet. Chem.* **1996**, *507*, 249.
20. Fox, M. A.; Howard, J. A. K.; MacBride J.A.H.; Mackinnon A.; Wade K. Big macrocyclic assemblies of carboranes (big MACs): synthesis and crystal structure of a macrocyclic assembly of four carboranes containing alternate ortho- and meta-carborane icosahedra linked by para-phenylene units, *J. Organomet. Chem.* **2003**, *680*, 155.
21. (a) Chizhevsky, I. T.; Johnson, S. E.; Knobler, C. B.; Gomez, F. A.; Hawthorne, M. F. Carboracycles: a family of novel macrocyclic carborane derivatives, *J. Am. Chem. Soc.* ***1993***, *115*, 6981. (b) Jiang, W.; Chizhevsky, I. T.; Mortimer, M. D.; Chen, W. L.; Knobler, C. B.; Johnson, S. E.; Gomez, F. A.; Hawthorne, M. F. Carboracycles: macrocyclic compounds composed of carborane icosahedra linked by organic bridging groups, *Inorg. Chem.* **1996**, *35*, 5417.
22. Songkram, C.; Yamasaki, R.; Tanatani, A.; Takaishi, K.; Yamaguchi, K.; Kagechika, H.; Endo, Y. Molecular construction based on icosahedral carboranes and aromatic urea groups. A new type of carboracycle, *Tetrahedron Lett.* **2001**, *42*, 5913.
23. Pederson, C. J, Cyclic polyethers and their complexes with metal salts, *J. Am. Chem. Soc.* **1967,** *89*, 7017.
24. (a) Yang, X.; Knobler, Hawthorne, M. F. "[12]Mercuracarborand-4", the first representative of a new class of rigid macrocyclic electrophiles: the chloride ion complex of a charge-reversed analogue of [12] crown-4, *Angew. Chem., Int. Ed. Engl.* **1991,** *30*, 1507. (b) Yang, X.; Knobler, C. B.; Hawthorne, M. F. Macrocyclic lewis acid host-halide ion guest species. Complexes of iodide ion, *J. Am. Chem. Soc.* **1992,** *114*, 380. (c) Yang, X.; Johnson, S. E.; Khan, S. I.; Hawthorne, M. F. Multidentate macrocyclic lewis acids: release of "[12] mercuracarborand-4" from its iodide complex and the structure of its tetra(tetrahydrofuran) dihydrate complex, *Angew. Chem., Int. Ed. Engl.* **1992,** *31*, 893. (d) Zheng, Z.; Yang, X.; Knobler, C. B.; Hawthorne, M. F. An iodide ion complex of a hydrophobic tetraphenyl [12] mercuracarborand-4 having a sterically encumbered cavity, *J. Am. Chem. Soc.* **1993,** *115*, 5320. (e) Yang, X.; Zheng, Z.; Knobler, C. B.; Hawthorne, M. F. "Anti-crown" chemistry: synthesis of [9]mercuracarborand-3 and the crystal structure of its acetonitrile complexes, *J. Am. Chem. Soc.* **1993,** *115*, 193.
25. Yang, X. G.; Knobler, C. B.; Zheng, Z. P.; Hawthorne, M. F. Host-guest chemistry of a new class of macrocyclic multidentate lewis acids comprised of carborane-supported electrophilic mercury centers, *J. Am. Chem. Soc.* **1994**, *116*, 7142.
26. (a) Zinn, A. A.; Zheng, Z.; Knobler, C. B.; Hawthorne, M. F. A hexamethyl derivative of [9]mercuracarborand-3: synthesis, characterization, and host–guest chemistry, *J. Am. Chem. Soc.* **1996**, *118*, 70. (b) Lee, H.; Knobler, C. B.; Hawthorne, M. F. Octahedral coordination of halide ions (I^-, Br^-, Cl^-) sandwich bonded with tridentate mercuracarborand-3 receptors, *J. Am. Chem. Soc.* **2001,** *123*, 8543. (c) Yang, X.; Knobler, C. B.; Hawthorne, M. F. Supramolecular chemistry: molecular aggregates of *closo*-$B_{10}H_{10}^{2-}$ with [12]mercuracarborand-4, *J. Am. Chem. Soc.* **1993,** *15*, 4904. (d) Bayer, M. J.; Herzog, A.; Diaz, M.; Harakas, G. A.; Lee, H.; Knobler, C. B.; Hawthorne, M. F. The synthesis of carboracycles derived from B,B′-bis(aryl) derivatives of icosahedral ortho-carborane, *Chem. Eur. J.* **2003**, *9*, 2732.
27. (a) Badr, I. H. A.; Johnson, R. D.; Diaz, M.; Hawthorne, M. F.; Bachas, L. G. A Selective optical sensor based on [9]mercuracarborand-3, a new type of ionophore with a chloride complexing cavity, *Anal. Chem.* **2000,** *72*, 4249. (b) Lee, H.; Diaz, M.; Hawthorne, M. F. Mercuracarborand-catalyzed diels-alder reactions of a thionoester with cyclopentadiene, *Tetrahedron Lett.* **1999,** *40*, 7651. (c) Badr, I. H. A.;

Diaz, M.; Hawthorne, M. F.; Bachas, L. G. Mercuracarborand "anti-crown ether"-based chloride-sensitive liquid/polymeric membrane electrodes, *Anal. Chem.* **1999,** *71*, 1371.

28. (a) Zheng, Z.; Diaz, M.; Knobler, C. B.; Hawthorne, M. F. A Mercuracarborand characterized by B-Hg-B bonds: synthesis and structure of cyclo-[(t-BuMe$_2$Si)$_2$C$_2$B$_{10}$H$_8$Hg]$_3$, *J. Am. Chem. Soc.* **1995**, *117*, 12338. (b) Zheng, Z.; Knobler, C. B.; Hawthorne, M. F. Stereoselective anion template effects: syntheses and molecular structures of tetraphenyl [12]mercuracarborand-4 complexes of halide ions, *J. Am. Chem. Soc.* **1995**, *117*, 5105.
29. (a)Mehta, S. C.; Lu, D. R. Targeted drug delivery for boron neutron capture therapy, *Pharm. Res.* **1996**, *13*, 344. (b) Barth, R. F.; Soloway, A. H.; Brugger, R. M. Boron neutron capture therapy of brain tumors: past history, current status, and future potential, *Cancer Invest.* **1996**, *14*, 534. (c) Grimes, R. N. Boron clusters come of age, *J. Chem. Educ.* **2004**, *81*, 658; (d) Plešek, J. Potential applications of the boron cluster compounds, *Chem. Rev.* **1992**, *92*, 269.
30. Geis, V.; Guttsche, K.; Knapp, C.; Scherer, H.; Uzun, R. Synthesis and characterization of synthetically useful salts of the weakly-coordinating dianion [B$_{12}$Cl$_{12}$]$^{2-}$, *Dalton Trans.*, **2009**, 2687.
31. (a) Knoth, W. H.; Miller, H. C.; Sauer, J. C.; Balthis, J. H.; Chia, Y. T.; Muetterties, E. L. Chemistry of boranes. IX. halogenation of $B_{10}H_{10}^{-2}$ and $B_{12}H_{12}^{-2}$, *Inorg. Chem.* **1964**, *3*, 159. (b) Tiritiris, I.; Schleid, T. Die kristallstrukturen der dicaesium-dodekahalogeno-*closo*-dodekaborate Cs$_2$[B$_{12}$X$_{12}$] (X = Cl, Br, I) und ihrer hydrate, *Z. Anorg. Allg. Chem.*, **2004**, *630*, 1555.
32. Peryshkov, D. V.; Popov, A. A.; Strauss, S. H. Direct perfluorination of K$_2$B$_{12}$H$_{12}$ in acetonitrile occurs at the gas bubble–solution interface and is inhibited by HF. experimental and DFT study of inhibition by protic acids and soft, polarizable anions, *J. Am. Chem. Soc.* **2009**, *131*, 18393.
33. Nieuwenhuyzen, M.; Seddon, K. R.; Teixidor, F.; Puga, A. V.; Viñas, C. Ionic liquids containing boron cluster anions, *Inorg. Chem.* **2009**, *48*, 889.
34. (a) Maderna, A.; Knobler, C. B.; Hawthorne, M. F. Twelvefold functionalization of an icosahedral surface by total esterification of [B$_{12}$(OH)$_{12}$]$^{2-}$: 12(12)-closomers, *Angew. Chem. Int. Ed.* **2001**, *40*, 1661. (b) Peymann, T.; Knobler, C. B.; Khan, S. I.; Hawthorne, M. F. *Angew. Chem. Int. Ed.* **2001**, *40*, 1664.
35. Thomas, J.; Hawthorne, M. F. Dodeca(carboranyl)-substituted closomers: toward unimolecular nanoparticles as delivery vehicles for BNCT, *Chem. Commun.*, **2001**, 1884.
36. (a) Nemoto, H.; Wilson, J. G.; Nakamura, H.; Yamamoto, Y. Polyols of a cascade type as a water-solubilizing element of carborane derivatives for boron neutron capture therapy, *J. Org. Chem.* **1992**, *57*, 435. (b) Nemoto, H.; Cai, J.; Yamamoto, Y. Synthesis of a water-soluble *o*-carborane bearing a uracil moiety via a palladium-catalysed reaction under essentially neutral conditions *J. Chem. Soc., Chem. Commun.* **1994**, 577.
37. Núñez, R.; González-Campo, A.; Laromaine, A.; Teixidor, F.; Sillanpää, R.; Kivekäs, R.; Viñas, C. Synthesis of small carboranylsilane dendrons as scaffolds for multiple functionalizations, *Org. Letters*, **2006**, 8, 4549.
38. (a) Gómez, F. A.; Hawthorne, M. F. A simple route to C-monosubstituted carborane derivatives, *J. Org. Chem.* **1992**, *57*, 1384. (b) Wang, S.; Yang, Q.; Mak, T. C. W.; Xie, Z. Carbon versus silicon bridges. synthesis of a new versatile ligand and its applications in organolanthanide chemistry, *Organometallics* **2000**, *19*, 334. (c) Xie, Z. Cyclopentadienyl–carboranyl hybrid compounds: a new class of versatile ligands for organometallic chemistry, *Acc. Chem. Res*, **2003**, *36*, 1. (d) González-Campo, A.; Boury, B.; Teixidor, F.; Núñez, R. Carboranyl units bringing unusual thermal and structural properties to hybrid materials prepared by sol–gel process, *Chem. Mater.* **2006**, *18*, 4344.
39. Farràs, P.; Teixidor, F.; Kivekäs, R.; Sillanpää, R.; Viñas, C; Grüner, B.; Cisarova, I. Metallacarboranes as building blocks for polyanionic polyarmed aryl-ether materials, *Inorg. Chem.*, **2008**, 47, 9497.
40. Šícha, V.; Farràs, P.; Štíbr, B.; Teixidor, F.; Grüner, B.; Viñas, C. Syntheses of C-substituted icosahedral dicarboranes bearing the 8-dioxane-cobalt bisdicarbollide moiety, *J. Organomet. Chem.*, **2009**, *694*, 1599.
41. (a) Sivaev, I. B.; Starikova, Z. A.; Sjöberg S.; Bregadze, V. I. Synthesis of functional derivatives of the [3,3′-Co(1,2-C$_2$B$_9$H$_{11}$)$_2$]$^-$ anion, *J. Organomet. Chem.*, **2002**, *649*, 1. (b) Sivaev, I. B.; Sjöberg S.; Bregadze, V. I., Materials of the next century, nizhny novgorod, Russia, May–June 29, **2000**.
42. Plešek, J.; Grüner, B.; Heřmánek, S.; Báča, J.; Mareček, V.; Jänchenová, J.; Lhotský, A. et al. Synthesis of functionalized cobaltacarboranes based on the *closo*-[(1,2-C$_2$B$_9$H$_{11}$)2-3,3′-Co]$^-$ ion bearing polydentate ligands for separation of M3$^+$ cations from nuclear waste solutions. Electrochemical and liquid–liquid extraction study of selective transfer of M3$^+$ metal cations to an organic phase. Molecular structure of the *closo*-[(8-(2-CH$_3$O···C$_6$H$_4$···O)-(CH$_2$CH$_2$O)$_2$-1,2-C$_2$B$_9$H$_{10}$)-(1′,2′-C$_2$B$_9$H$_{11}$)-3,3′-Co]Na determined by X-ray diffraction analysis *Polyhedron* **2002**, *21*, 975.

43. Grüner, B.; Plešek, J.; Báća, J.; Císařová, I.; Dozol, J.-F.; Rouquette, H.; Viňas, C.; Selucký, P.; Rais, J. Cobalt bis(dicarbollide) ions with covalently bonded CMPO groups as selective extraction agents for lanthanide and actinide cations from highly acidic nuclear waste solutions, *New J. Chem.* **2002,** *26*, 1519.
44. (a) Grüner, B.; Mikulašek, L.; Bača, J.; Cisařova, I.; Böhmer, V.; Danila, C.; Reinoso-Garcia, M. M. et al. Cobalt bis(dicarbollides)(1–) covalently attached to the calix[4]arene platform: the first combination of organic bowl-shaped matrices and inorganic metallaborane cluster anions, *Eur. J. Org. Chem.* **2005**, 2022. (b) Mikulašek, L.; Grüner, B.; Danila, C.; Bőhmer, V.; Ćaslavsky, J.; Selucky, P. Synergistic effect of ligating and ionic functions, prearranged on a calix[4]arene, *Chem. Commun.* **2006**, 4001.
45. (a) Olejniczak, A. B.; Plešek, J.; Křiž, O.; Lesnikowski, Z. A nucleoside conjugate containing a metallacarborane group and its incorporation into a DNA oligonucleotide, J. *Angew. Chem., Int. Ed.* **2003**, *42*, 5740. (b) Lesnikowski, Z. J.; Paradowska, E.; Olejniczak, A. B.; Studzinska, M.; Seekamp, P.; Schüßler, U.; Gabel, D.; Schinazi, R. F.; Plešek, J. Towards new boron carriers for boron neutron capture therapy: metallacarboranes and their nucleoside conjugates, *Bioorg. Med. Chem.* **2005**, *13*, 4168. (c) Olejniczak, A. B.; Plešek, J.; Lesnikowski, Z. J. Nucleoside–metallacarborane conjugates for base-specific metal labeling of DNA, *Chem.–Eur. J.* **2007**, *13*, 311.
46. (a) Wiesboeck, R. A.; Hawthorne, M. F. Dicarbaundecaborane(13) and derivatives, *J. Am. Chem. Soc.*, **1964**, *86*, 1642. (b) Garret, P. M.; Tebbe, F. N.; Hawthorne, M. F. The thermal isomerization of C-phenyldicarbaundecaborate(12), *J. Am. Chem. Soc.*, **1964**, *86*, 5016. (c) Hawthorne, M. F.; Young, D. C.; Garret, P. M.; Owen, D. A.; Schwerin, S. G.; Tebbe, F. N.; Wegner, P. M. Preparation and characterization of the (3)-1,2- and (3)-1,7-dicarbadodecahydroundecaborate(-1) ions, *J. Am. Chem. Soc.*, **1968**, *90*, 862.
47. (a) Zakharkin, L. I.; Kalinin, U. N. On the reaction of amines with borenes, *Tetrahedron Lett.*, **1965**, 407. (b) Zakharkin, L. I.; Kirillova, V. S. Cleavage of *o*-carborane to (3)-1,2-dicarbaundecaborates in the presence of amines. *Izv. Akad. Nauk SSSR, Ser. Khim.*, **1975**, *11*, 2596. (c) Taoda, Y.; Sawabe, T.; Endo, Y.; Yamaguchi, K.; Fujii, S.; Kagechika, H. Identification of an intermediate in the deboronation of ortho-carborane: an adduct of ortho-carborane with two nucleophiles on one boron atom, *Chem. Commun.*, **2008**, 2049.
48. (a) Fox, M. A.; Gill, W. R.; Herbertson, P. L.; MacBride, J. A. H.; Wade, K. Deboronation of C-substituted ortho- and meta-*closo*-carboranes using "wet" fluoride ion solutions, *Polyhedron*, **1996**, *15*, 565. (b) Fox, M. A.; MacBride, J. A. H.; Wade, K. Fluoride-ion deboronation of *p*-fluorophenyl-ortho- and -meta-carboranes. NMR evidence for the new fluoroborate, $HOBHF_2^-$, *Polyhedron*, **1997**, *16*, 2499. (c) Fox, M.A.; Wade, K. Cage-fluorination during deboronation of meta-carboranes, *Polyhedron*, **1997**, *16*, 2517. (d) Yoo, J.; Hwang, J. W.; Do, Y. Facile and mild deboronation of *o*-carboranes using cesium fluoride, *Inorg. Chem.*, **2001**, *40*, 568.
49. Davidson, M. G.; Fox, M. A.; Hibbert, T.G.; Howard, J.A.K.; Mackinnon, A.; Neretin, I.S.; Wade, K. Deboronation of ortho-carborane by an iminophosphorane: crystal structures of the novel carborane adduct *nido*-$C_2B_{10}H_{12}{\cdot}HNP(NMe_2)_3$ and the borenium salt $[(Me_2N)_3PNHBNP(NMe_2)_3]_2O^{2+}(C_2B_9H_{12}^-)_2$, *Chem. Commun.*, **1999**, 1649.
50. Newkome, G. R.; Moorefield, C. N.; Keith, J. M.; Baker, G. R.; Escamilla, G. H. Chemistry within a unimolecular micelle precursor: boron superclusters by site- and depth-specific transformations of dendrimers, *Angew. Chem., Int. Ed. Engl.* **1994**, *33*, 666.
51. (a) Parrott, M. C.; Marchington, E. B.; Valliant, J. F.; Adronov, A. preparation of synthons for carborane containing macromolecules, *Macromol. Symp.* **2003**, *196*, 201. (b) Parrott, M. C.; Marchington, E. B.; Valliant, J. F.; Adronov, A. J. Synthesis and properties of carborane-functionalized aliphatic polyester dendrimers, *J. Am. Chem. Soc.* **2005**, *127*, 12081–12089. (c) Parrott, M. C.; Valliant, J. F.; Adronov, A. Thermally induced phase transition of carborane-functionalized aliphatic polyester dendrimers in aqueous media, *Langmuir*, **2006**, *22*, 5251.
52. (a) Gudat, D. Inorganic cauliflower: functional main group element dendrimers constructed from phosphorous- and silicon-based building blocks, *Angew. Chem., Int. Ed. Engl.* **1997**, *36*, 1951. (b) Frey, H.; Schlenk, C. Silicon-based dendrimers, *Top. Curr. Chem.* **2000**, *210*, 69. (c) Kreiter, R.; Kleij, A. W.; Gebbink, R. J. M. K.; van Koten G. Dendritic catalysts, *Top. Curr. Chem.* **2001**, *217*, 163.
53. (a) Fréchet, J. M., Tomalia, D. A., Eds. *Dendrimers and other dendritic polymers.* Wiley Series in Polymer Science, Wiley, West Sussex 2001. (b) Newkome, G. R.; Moorefield, C. N.; Vötgle, F. *Dendrimers and Dendrons: Concepts, Synthesis, Applications*; Wiley: New York, 2002. (c) Beletskaya, I. P.; Chuchurjukin, A. V. Synthesis and properties of functionalised dendrimers, *Russ. Chem. Rev.* **2000**, *69*, 639. (d) Astruc, D.; Chardac, F. Dendritic catalysts and dendrimers in catalysis, *Chem. Rev.* **2001**, *101*, 2991.
54. (a) Núñez, R.; González, A.; Viñas, C.; Teixidor, F.; Sillanpää, R.; Kivekäs, R. Approaches to the preparation of carborane-containing carbosilane compounds, *Org. Letters* **2005**, *7*, 231. (b) Núñez, R.; González-Campo, A.; Viñas, C.; Teixidor, F.; Sillanpää, R.; Kivekäs, R. Boron-functionalized carbosilanes: insertion

of carborane cluster into peripheral si atoms of carbosilane compounds, *Organometallics*, **2005**, *24*, 6351.

55. Seyferth, D.; Son, D. Y.; Rheingold, A. L.; Ostrader, R. L. Synthesis of an organosilicon dendrimer containing 324 Si-H bonds, *Organometallics* **1994**, *13*, 2682.
56. (a) van der Made, A. W.; van Leeuwen, P. W. N. M. Silane dendrimers, *J. Chem. Soc., Chem. Commun.* **1992**, 1400. (b) Alonso, B.; Cuadrado, I.; Morán, M.; Losada, J. Organometallic silicon dendrimers, *J. Chem. Soc., Chem. Commun.* **1994**, 2575.
57. Songkram, C.; Takaishi, K.; Yamaguchi, K.; Kagechika, H.; Endo, Y., Structures of bis- and tris(2-phenyl-*o*-carboran-1-yl)benzenes. Construction of three-dimensional structures converted from planar arylacetylenic arrays, *Tetrahedron Lett.*, **2001**, *42*, 6365.
58. Schöberl, U.; Magnera, T. F.; Harrison, R. M.; Fleischer, F.; Pflug, J. L.; Schwab, P. F. H.; Meng, X. et al., Toward a hexagonal grid polymer: synthesis, coupling, and chemically reversible surface-pinning of the star connectors, 1,3,5-$C_6H_3(CB_{10}H_{10}CX)_3$, *J. Am. Chem. Soc.* **1997**, *119*, 3907.
59. (a) Kalinin, V. N., Carbon-carbon bond formation in heterocycles using Ni- and Pd-Catalyzed reactions, *Synthesis* **1992**, 413. (b) Li, J.; Logan, C. M.; Jones, N., Jr., Simple syntheses and alkylation reactions of 3-iodo-*o*-carborane and 9,12-diiodo-*o*-carborane, *Inorg. Chem.*, **1991**, *30*, 4866.
60. Lerouge, F.; Viñas, C. Teixidor, F.; Núñez, R.; Abreu, A.; Xochitiotzi, E.; Santillán, R.; Farfán, N. High-boron content carboranyl-functionalized aryl ether derivatives displaying photoluminiscent properties, *Dalton Trans.*, **2007**, 1898.
61. (a) Bao, C.; Lu, R.; Jin, M.; Xue, P.; Tan, C.; Xu, T.; Liu, G.; Zhao, Y. Helical stacking tuned by alkoxy side chains in π-conjugated triphenylbenzene discotic derivatives, *Chem.—Eur. J.* **2006**, *12*, 3287. (b) Yang, J.-X.; Tao, X.-T.; Yuan, C. X.; Yan, X. Y.; Wang, L.; Liu, Z.; Ren, Y.; Jiang, M. H. A facile synthesis and properties of multicarbazole molecules containing multiple vinylene bridges, *J. Am. Chem. Soc.* **2005**, *127*, 3278.
62. (a) Dash, B.P.; Satapathy, R.; Maguire, J.A.; Hosmane, N.S. Synthesis of new class of carborane-containing star-shaped molecules via silicon tetrachloride promoted cyclotrimerization reactions, *Org. Letters* **2008**, *10*, 2247. (b) Hosmane, N.S.; Yinghuai, Z.; Maguire, J.A.; Kaim, W.; Takagaki, M. Nano and dendritic structured carboranes and metallacarboranes: from materials to cancer therapy, *J. Organomet. Chem.* **2009**, *694*, 1690.
63. Kaszynski, P.; Pakhomov, S.; Tesh, K. F.; Young, V. G., Jr. Carborane-containing liquid crystals: synthesis and structural, conformational, thermal, and spectroscopic characterization of diheptyl and diheptynyl derivatives of *p*-carboranes, *Inorg. Chem.* **2001**, *40*, 6622.
64. González-Campo, A.; Viñas, C.; Teixidor, F.; Núñez, R.; Kivekäs, R.; Sillanpää, R. Modular construction of neutral and anionic carboranyl-containing carbosilane dendrimers, *Macromolecules* **2007**, *40*, 5644.
65. (a) Lo, S.-C.; Burn, P. L. Development of dendrimers: macromolecules for use in organic light-emitting diodes and solar cells, *Chem. Rev.* **2007**, *107*, 1097. (b) Cao, X.-Y.; Liu, X.-H.; Zhou, X.-H.; Zhang, Y.; Jiang, Y.; Cao, Y.; Cui, Y.-X.; Pei, J. Giant extended π-conjugated dendrimers containing the 10,15-dihydro-5h-diindeno[1,2-a;1′,2′-c]fluorene chromophore: synthesis, NMR behaviors, optical properties, and electroluminescence, *J. Org. Chem.* **2004**, *69*, 6050, and references therein. (c) Kotha, S.; Kasinath, D.; Lahiri, K.; Sunoj, R. B. Synthesis of C3-symmetric nano-sized polyaromatic compounds by trimerization and suzuki–miyaura cross-coupling reactions, *Eur. J. Org. Chem.* **2004**, 4003, and references therein. (d) Kaafarani, B. R.; Wex, B.; Wang, F.; Catanescu, O.; Chien, L. C.; Neckers, D. C. Synthesis of highly fluorescent Y-enyne dendrimers with four and six arms, *J. Org. Chem.* **2003**, *68*, 5378.
66. Dash, B. P.; Satapathy, R.; Maguire, J. A.; Hosmane, N. S. Boron-enriched star-shaped molecule via cycloaddition reaction, *Chem. Commun.* **2009**, 3267.
67. Wang, Z., Watson, M. D., Wu, J. and Mullen, K. Partially stripped insulated nanowires: a lightly substituted hexa-peri-hexabenzocoronene-based columnar liquid crystal, *Chem. Commun.* **2004**, 336.
68. (a) Kimura, H.; Okita, K.; Ichitani, M.; Sugimoto, T.; Kuroki S.; Ando, I. Structural study of silylcarborane hybrid diethynylbenzene-silylene polymers by high-resolution solid-state ^{11}B, ^{13}C, and ^{29}Si NMR spectroscopy, *Chem. Mater.* **2003**, *15*, 355. (b) Hao, E.; Fabre, B.; Fronczek F. R.; Vicente, M. G. H. Synteheses and electropolymerization of carboranyl-functionalized pyrroles and thiophenes, *Chem. Mater.* **2007**, *19*, 6195. (c) Farha, O. K.; Spokoyny, A. M.; Mulfort, K. L.; Hawthorne, M. F.; Mirkin C. A.; Hupp, J. T. Synthesis and hydrogen sorption properties of carborane based metal-organic framework materials, *J. Am. Chem. Soc.* **2007**, *129*, 12680.
69. (a) Wu, J.; Watson M. D.; Mullen, K. The versatile synthesis and self-assembly of star-type hexabenzocoronenes, *Angew. Chem. Int. Ed.*, **2003**, *42*, 5329. (b) Wu, J.; Watson, M. D.; Zhang, L.; Wang Z.; Mullen, K. Hexakis(4-iodophenyl)-peri-hexabenzocoronene A versatile building block for highly ordered discotic liquid crystalline materials, *J. Am. Chem. Soc.* **2004**, *126*, 177. (c) Pisula, W.; Kastler, M.;

Wasserfallen, D.; Pakula, T.; Mullen, K. Exceptionally long-range self-assembly of hexa-peri-hexabenzocoronene with dove-tailed alkyl substituents, *J. Am. Chem. Soc.*, **2004**, *126*, 8074. (d) Kumar, S. Self-organization of disc-like molecules: chemical aspects, *Chem. Soc. Rev.* **2006**, *35*, 83.

70. Farràs, P.; Cioran, A.; Šícha, V.; Teixidor, F.; Štíbr, B.; Grüner B.; Viñas C. Toward the synthesis of high boron content polyanionic multicluster macromolecules, *Inorg. Chem.*, **2009**, *48*, 8210.
71. Fanning, J. C. The solubilities of the alkali metal salts and the precipitation of Cs^+ from aqueous solution, *Coord. Chem. Rev.* **1995**, *140*, 27.
72. Raddatz, S.; Marcello, M.; Kliem, H.-C.; Troester, H.; Trendelenburg, M. F.; Oeser, T.; Granzow, C.; Wiessler, M. Synthesis of new boron-rich building blocks for boron neutron capture therapy or energy-filtering transmission electron microscopy, *ChemBioChem* **2004**, *5*, 474.
73. Kane, R. R.; Beno, C. L.; Romano, S.; Mendez, G.; Hawthorne, M. F.; Kim, Y. S. Synthesis of new building blocks for boron-rich oligomers in boron neutron capture therapy (BNCT). II. Monomers derived from 2,2-disubstituted-1,3-diols, *Tetrahedron Lett.* **1995**, *36*, 5147.
74. S. M. Kerwin, Synthesis of a DNA-cleaving bis(propargylic) sulfone crown ether, *Tetrahedron Lett.* **1994**, *35*, 1023.
75. (a) Ihre, H.; Hult, A. Double-stage convergent approach for the synthesis of functionalized dendritic aliphatic polyesters based on 2,2-bis(hydroxymethyl)propionic acid, *Macromolecules* **1998**, *31*, 4061. (b) Malkoch, M.; Malmström, E.; Hult, A. Rapid and efficient synthesis of aliphatic ester dendrons and dendrimers, *Macromolecules* **2002**, *35*, 8307.
76. Galie, K. M.; Mollard, A.; Zharov. I. Polyester-based carborane-containing dendrons, *Inorg. Chem.* **2006**, *45*, 7815.
77. Mollard, A.; Zharov, I. Tricarboranyl pentaerythritol-based building block, *Inorg. Chem.* **2006**, *45*, 10172.
78. Barth, R. F.; Adams, D. M.; Soloway, A. H.; Alam, F.; Darby, M. V. Boronated starburst dendrimer-monoclonal antibody immunoconjugates: evaluation as a potential delivery system for neutron capture therapy, *Bioconjugate Chem.* **1994**, *5*, 58.
79. (a) Shukla, S.; Wu, G.; Chatterjee, M.; Yang, W.; Sekido, M.; Diop, L. A.; Müller, R. et al., Synthesis and biological evaluation of folate receptor-targeted boronated PAMAM dendrimers as potential agents for neutron capture therapy, *Bioconjugate Chem.* **2003**, *14*, 158. (b) Backer, M. V.; Gaynutdinov, T. I., Patel, V.; Bandyopadhyaya, A. K.; Thirumamagal, B. T. S.; Tjarks, W.; Barth, R. F.; Claffey, K.; Backer, J. M. Vascular endothelial growth factor selectively targets boronated dendrimers to tumor vasculature, *Mol. Cancer Ther.*, **2005**, *4*, 1423.
80. Yao, H. J.; Grimes, R. N.; Corsini, M.; Zanello, P. Polynuclear metallacarborane-hydrocarbon assemblies: metallacarborane dendrimers, *Organometallics* **2003**, *22*, 4381.
81. Qualman, B.; Kessels, M. M.; Musiol, H.-J.; Sierralta, W. D.; Jungblut, P. W.; Moroder, L. Synthesis of boron-rich lysine dendrimers as protein labels in electron microscopy, *Angew. Chem., Int. Ed. Engl.* **1996**, *35*, 909.
82. Kessels, M. M.; Qualmann, B. Facile enantioselective synthesis of (S)-5-(2-methyl-1,2-dicarba-*closo*-dodecaborane(12)-1-yl)-2-aminopentanoic acid (L-MeCBA) using the bislactim ether method, *J. Prakt. Chem.* **1996**, 89.
83. Felix, A. M.; Lu, Y.-A.; Campbell, R. M. Pegylated peptides IV. Enhanced biological activity of site-directed pegylated GRF analogs, *Int. J. Peptide Protein Res.* **1995**, *46*, 253.
84. Qualmann, B.; Kessels, M. M.; Jungblut, P. W.; Sierralta, Electron spectroscopic imaging of antigens by reaction with boronated antibodies, *J. Microsc.* **1996**, *183*, 69.
85. (a) Armspach, D.; Cattalini, M.; Constable, E. C.; Housecroft, C. E.; Phillips, D. Boron-rich metallodendrimers-mix-and-match assembly of multifunctional metallosupramolecules. *Chem. Commun.* **1996**, 1823. (b) Housecroft, C. E. Icosahedral building blocks: towards dendrimers with twelve primary branches?. *Angew. Chem. Int. Ed.* **1999**, *38*, 2717.
86. Armspach, D.; Constable, E. C.; Housecroft, C. E.; Neuburger, M.; Zehnder, M. Cluster functionalized ligands: metal-cluster interaction and nuclearity changes in carboranyl-2,2′:6′2″-terpyridines, *New J. Chem.* **1996**, *20*, 331.
87. Juárez-Pérez, E. J.; Viñas, C.; Teixidor, F.; Núñez, R. Polyanionic carbosilane and carbosiloxane metallodendrimers based on cobaltabisdicarbollide derivatives, *Organometallics* **2009**, *28*, 5550.
88. (a) Zhan, L. -L.; Roovers, J. Synthesis of novel carbosilane dendritic macromolecules, *Macromolecules* **1993**, *26*, 963. (b) Krsda, S. W.; Seyferth, D. Synthesis of water-soluble carbosilane dendrimers, *J. Am. Chem. Soc.* **1998**, *120*, 3604. (c) Casado, M. A.; Hack, V.; Camerano, J. A.; Ciriano, M. A.; Tejel, C.; Oro, L. A. Unprecedented hybrid scorpionate/phosphine ligand able to be anchored to carbosilane dendrimers, *Inorg. Chem.* **2005**, *44*, 9122. (d) Donnio, B.; Buathong, S.; Bury, I.; Guillon, D. Liquid crystalline

dendrimers, *Chem. Soc. Rev.* **2007**, *36*, 1495. (e) Zamora, M.; Alonso, B.; Pastor, C.; Cuadrado, I. Multiredox heterometallic carbosilane dendrimers, *Organometallics*, **2007**, *26*, 5153. (f) Kim, C.; An, K. Preparation and termination of carbosilane dendrimers based on a siloxane tetramer as a core molecule: silane arborols, part VIII, *J. Organomet. Chem.* **1997**, *547*, 55.

89. (a) Feher, F. J.; Wyndham K. D. Amine and ester-substituted silsesquioxanes: synthesis, characterization and use as a core for starburst dendrimers, *Chem. Commun.* **1998**, 323. (b) Zhang, C.; Bunning, T. J.; Laine, R. M. Synthesis and characterization of liquid crystalline silsesquioxanes, *Chem. Mater.* **2001**, *13*, 3653. (c) Neumann, D.; Fisher, M.; Tran, M.; Matisons, J. G. Synthesis and characterization of an isocyanate functionalized polyhedral oligosilsesquioxane and the subsequent formation of an organic–inorganic hybrid polyurethane, *J. Am. Chem. Soc.* **2002**, *124*, 13998. (d) Constable, G. S.; Lesser, A. J.; Coughlin, B. Morphological and mechanical evaluation of hybrid organic–inorganic thermoset copolymers of dicyclopentadiene and mono- or tris(norbornenyl)-substituted polyhedral oligomeric silsesquioxanes, *Macromolecules* **2004**, *37*, 1276. (e) Baker, E. S.; Gidden, J.; Anderson, S. E.; Haddad, T. S.; Bowers, M. T. omeric structural characterization of polyhedral oligomeric silsesquioxanes (POSS) with styryl and epoxy phenyl capping agents, *Nano Lett.* **2004**, *4*, 779. (f) Blanc, F.; Copéret, C.; Thivolle-Cazat, J.; Basset, J. –M. Lesage, A.; Emsley, L.; Sinha, A.; Schrock, R. R. Surface versus molecular siloxy ligands in well-defined olefin metathesis catalysts: [{(RO)$_3$SiO}Mo(NAr)(CHtBu)(CH$_2$tBu), *Angew. Chem. Int. Ed.* **2006**, *45*, 1216.

90. (a) Kannan, R. Y.; Salacinski, H. J.; Butler, P. E. Seifalian, A. M. Polyhedral oligomeric silsesquioxane nanocomposites: the next generation material for biomedical applications, *Acc. Chem. Res.* **2005**, *38*, 879. (b) Frankamp, B. L.; Fischer, N. O.; Hong, R.; Srivastava, S.; Rotello, V. M. Surface modification using cubic silsesquioxane ligands. facile synthesis of water-soluble metal oxide nanoparticles, *Chem. Mater.* **2006**, *18*, 956.

91. (a) Lang, H.; Lühmann, B. Carbosiloxane based dendrimers: synthesis, reaction chemistry, and potential applications, *Adv. Mater.* **2001**, *13*, 1523. (b) Dvornic, P. R.; Hartmann-Thompson, C. H.; Keinath S. E.; Hill, E. J. Organic–inorganic polyamidoamine (PAMAM) dendrimer–polyhedral oligosilsesquioxane (POSS) nanohybrids, M*acromolecules* **2004**, *37*, 7818. (c) Laine, R. M. Nanobuilding blocks based on the [OSiO1.5]x (x = 6, 8, 10) octasilsesquioxanes, *J. Mater. Chem.* **2005**, *15*, 3725.

92. González-Campo, A.; Juárez-Pérez, E. J.; Viñas, C.; Boury, B.; Kivekäs, R.; Sillanpää, R.; Núñez, R. Carboranyl substituted linear and cyclic siloxanes and octasilsesquioxanes: synthesis, characterization and reactivity, *Macromolecules* **2008**, *41*, 8458.

93. (a) Bassindale, A. R.; Mackinnon, I. A.; Maesano, M. G.; Taylor. P. G. The preparation of hexasilsesquioxane (T6) cages by "non aqueous" hydrolysis of trichlorosilanes, *Chem. Commun.* **2003**, 1382. (b) Brook M. A. Silicones. In *Silicon in Organic, Organometallic, and Polymer Chemistry*, John Wiley & Sons, Inc.: New York, 2000; p.256. (c) Le Roux, C.; Yang, H.; Wenzel, S.; Brook, M. A. Using "Anhydrous" hydrolysis to favor formation of hexamethylcyclotrisiloxane from dimethyldichlorosilane, *Organometallics* **1998**, *17*, 556. (d) Lu, P.; Paulasaari, J. K.; Weber, W. P. Reaction of dimethyldichlorosilane, phenylmethyldichlorosilane, or diphenyldichlorosilane with dimethyl sulfoxide. Evidence for silanone and cyclodisiloxane intermediates, *Organometallics* **1996**, *15*, 4649. (e) Arkhireeva, A.; Hay, J. N.; Manzano, M. Preparation of silsesquioxane particles via a nonhydrolytic sol–gel route, *Chem. Mater.* **2005**, *17*, 875.

94. Valliant, J. F.; Guenther, K. J.; King, A. S.; Morel, P.; Schaffer, P.; Sogbein, O. O.; Stephenson, K. A. The medicinal chemistry of carboranes, *Coord. Chem. Rev.* **2002**, *232*, 173.

95. Juárez-Pérez, E. J.; Viñas, C.; Teixidor, F.; Santillan, R.; Farfán, N.; Abreu, A.; Yépez, R. Núñez, R. Polyanionic aryl-ether metallodendrimers based on cobaltabisdicarbollide derivatives. Photoluminescent properties, *Macromolecules*, **2010**, *43*, 150.

96. (a) Renner, M. W.; Miura, M. W.; Easson, M. W.; Vicente, M. G. H. Recent progress in the syntheses and biological evaluation of boronated porphyrins for boron neutron capture therapy, *Anti-Cancer Agents Med. Chem.* **2006**, *6*, 145. (b) Ratajski, M.; Osterloh, J.; Gabel, D. Boron-containing chlorins and tetraazaporphyrins: synthesis and cell uptake of boronated pyropheophorbide a derivatives, *Anti-Cancer Agents Med. Chem.*, **2006**, *6*, 159.

97. (a) Haushalter, R. C.; Rudolph, R. W. meso-tetracarboranylporphyrins, *J. Am. Chem. Soc.* **1978**, *103*, 4628. (b) Haushalter, R. C.; Butler, W. M.; Rudolph, R. W. The preparation and characterization of several meso-tetracarboranylporphyrins, *J. Am. Chem. Soc.*, **1981**, 103, 2620.

98. (a) Miura, M.; Gabel, D.; Oenbrink, G.; Fairchild, R. G. Preparation of carboranyl porphyrins for boron neutron capture therapy, *Tetrahedron Lett.* **1990**, *31*, 2247. (b) Bregadze, V. I.; Sivaev, I. B.; Gabel, D.; Wöhrle, D. *J.* Polyhedral boron derivatives of porphyrins and phthalocyanines, *Porphyrins Phthalocyanines* **2001**, *5*, 767 and references therein.

99. (a) Vicente, M. G. H.; Shetty, J. S.; Wickramasinghe, A.; Smith K. M. Syntheses of carbon-carbon linked carboranylated porphyrins for boron neutron capture therapy of cancer, *Tetrahedron Lett.* **2000**, *41*, 7623. (b) Vicente, M. G. H.; Nurco, D. J.; Shetty, S. J.; Medforth, C. J.; Smith, K. M. First structural characterization of a covalently bonded porphyrin–carborane system, *Chem. Commun.* **2001**, 483.
100. Lindsey, J. S.; Schreiman, I. C.; Hsu, H. C.; Kearney, P. C.; Marguerettaz, A. M. Rothemund and Adler-Longo reactions revisited: synthesis of tetraphenylporphyrins under equilibrium conditions, *J. Org. Chem.* **1987**, *52*, 827.
101. Hao, E.; Friso, E.; Miotto, G.; Jori, G.; Soncin, M.; Fabris, C.; Sibrian-Vazquez, M.; Vicente, M. G. H. Synthesis and biological investigations of tetrakis(*p*-carboranylthio-tetrafluorophenyl)chlorin (TPFC), *Org. Biomol. Chem*,. **2008**, *6*, 3732.
102. Gottumukkala, V.; Ongayi, O.; Baker, David G.; Lomax, L. G.; Vicente, M. G. H. Synthesis, cellular uptake and animal toxicity of a tetra(carboranylphenyl)-tetrabenzoporphyrin, *Bioorg. Med. Chem.*, **2006**, *14*, 1871.
103. Hao, E.; Fronczek, F. R.; Vicente, M. G. H. Carborane functionalized pyrroles and porphyrins via the suzuki crosscoupling reaction, *Chem. Commun.* **2006**, 4900.
104. (a) Kahl, S. B.; Joel, D. D.; Nawrocky, M. M.; Micca, P. L.; Tran, K. P.; Finkel, G. C.; Slatkin, D. N. Uptake of a *nido*-carboranylporphyrin by human glioma xenografts in athymic nude mice and by syngeneic ovarian carcinomas in immunocompetent mice, *Proc. Natl. Acad. Sci. U.S.A.* **1990**, *87*, 7265. (b) Rosenthal, M. A.; Kavar, B.; Hill, J. S.; Morgan, D. J.; Nation, R. L.; Stylli, S. S.; Basser, R. et al. Phase I and pharmacokinetic study of photodynamic therapy for high-grade gliomas using a novel boronated porphyrin, *J. Clin. Oncol.* **2001**, *19*, 519, and references therein.
105. Dozzo, P.; Koo, M.-S.; Berger, S.; Forte, T. M.; Kahl, S. B. Synthesis, characterization, and plasma lipoprotein association of a nucleus-targeted boronated porphyrin, *J. Med. Chem.* **2005**, *48*, 357.
106. (a) Genady, A. R.; El-Zaria, M. E.; Gabel, D. Non-covalent assemblies of negatively charged boronated porphyrins with different cationic moieties, *J. Organomet. Chem.* **2004**, *689*, 3242. (b) Koo, M.–S.; Ozawa, T.; Santos, R. A.; Lamborn, K. R.; Bollen, A. W.; Deen, D. F.; Kahl, S. B. Synthesis and comparative toxicology of a series of polyhedral borane anion-substituted tetraphenyl porphyrins, *J. Med. Chem.* **2007**, *50*, 820.
107. Hao, E.; Fronczek, F. R.; Sibrian-Vazquez, M.; Serem, W.; Garno, J. C.; Vicente, M. G. H. Synthesis, aggregation and cellular investigations of porphyrin–cobaltacarborane conjugates, *Chem. Eur. J.* **2007**, *13*, 9035.
108. (a) Ol'shevskaya, V. A.; Zaitsev, A. V.; Luzgina, V. N.; Kondratieva, T. T.; Ivanov, O. G.; Kononova, E. G.; Petrovskii, P. P. et al. Novel boronated derivatives of 5,10,15,20-tetraphenylporphyrin: synthesis and toxicity for drug-resistant tumor cells. *Bioorg. Med. Chem.*, **2006**, *14*, 109, and references therein. (b) Ol'shevskaya, V. A.; Savchenko, A. N.; Zaitsev, A. V.; Kononova, E. G.; Petrovskii, P. V.; Ramonova, A. A.; Tatarskiy Jr., V. V. et al. Novel metal complexes of boronated chlorin e_6 for photodynamic therapy, *J. Organomet. Chem.* **2009**, *694*, 1632.
109. (a) Miura, M.; Micca, P. L.; Heinrichs, J. C.; Gabel, D.; Fairchild, R. G.; Slatkin, D. N. Biodistribution and toxicity of 2,4-divinyl-*nido-o*-carboranyldeuteroporphyrin IX in mice. *Biochem. Pharmacol.*, **1992**, *43*, 467. (b) Lauceri, R.; Purrello, R.; Shetty, S. J.; Vicente, M. G. H. Interactions of anionic carboranylated porphyrins with DNA, *J. Am. Chem. Soc.* **2001**, *123*, 5835. (c) Vicente, M. G. H.; Wickramasinghe, A.; Nurco, D. J.; Wang, H. J. H.; Nawrocky, M. M.; Makar, M. S.; Miura, M. Synthesis, toxicity and biodistribution of two 5,15-Di[3,5-(*nido*-carboranylmethyl) phenyl]porphyrins in EMT-6 tumor bearing mice, *Bioorg. Med. Chem.* **2003**, *11*, 3101.
110. Gottumukkala, V.; Luguya, R.; Fronczek F. R.; Vicente, M. G. H. Synthesis and cellular studies of an octa-anionic 5,10,15,20-tetra[3,5-(*nido*-carboranylmethyl)phenyl]porphyrin (H_2OCP) for application in BNCT, *Bioorg. Med. Chem.*, **2005**, *13*, 1633.
111. (a) Miura, M.; Morris, G. M.; Micca, P. L.; Lombardo, D. T.; Youngs, K. M.; Kalef-Ezra, J. A.; Hoch, D. A.; Slatkin, D. N.; Ma, R.; Coderre, J. A. Boron neutron capture therapy of murine mammary carcinoma using a lipophylic carboranyltetraphenylporphyrin, *Radiat. Res.* **2001**, *155*, 603. (b) Jori, G.; Soncin, M.; Friso, E. Vicente, M. G. H.; Hao, E., Miotto, G.; Colautti, P. et al. A novel boronated-porphyrin as a radio-sensitizing agent for boron neutron capture therapy of tumours: *In vitro* and *in vivo* studies, *Appl. Radiat. Isot.* **2009**, *67*, S321.
112. (a) Gutsche, C. D. *Calixarenes*, The Royal Society of Chemistry. Cambridge. England. 1989. (b) *Calixarenes. A Versatile Class of Macrocyclic Compounds*, (Eds.: J. Vicens, V. Bohmer), Kluwer, Dordrecht, **1991.** (c) Kappe. T. The early history of calixarene chemistry, *J. Incl. Phenom. Macrocycl. Chem.* **1994**, *19*, 3.

113. Pedersen, C.J.; *"The discovery of crown ethers"*, Nobel lecture, December 8, **1987**. http://nobelprize.org/nobel_prizes/chemistry/laureates/1987/pedersen-lecture.pdf
114. (a) Verboom, W. *Cavitands. In Calixarenes 2001* (Eds.:Z. Asfari, V. Böhmer, J. Harrowfield, J. Vicens), Kluwer Academic Publishers, Dordrecht, Boston, London, **2001**, pp. 181. (b) Timmerman, P.; Verboom, W.; Reinhoudt, D. N. Resorcinarenes, *Tetrahedron* **1996**, *52*, 2663.
115. Thorndorf, A.; Shivanyuk, A.; Böhmer, V. *Chemical Modification of Calix[4]arenes and Resorcarenes.* In *Calixarenes 2001* (Eds.: Z. Asfari, V. Böhmer, J. Harrowfield, J. Vicens), Kluwer Academic Publishers, Dordrecht, Boston, London, **2001**, pp. 26.
116. McKervey, M. A.; Schwing-Weill M. J.; Arnaud-Neu, F. in Comprehensive Supramolecular Chemistry, ed. G. W. Gokel, Pergamon, Oxford, 1996, vol. 1, p. 537.
117. Horwitz, E. P.; Kalina, D. G.; Diamond, H.; Vandegrift G. F.; Schulz, W. W. The Truex process: a process for the extraction of the transuranic elements from nitric acid in wastes utilizing modified purex solvent, *Solvent Extr. Ion Exch.*, **1985**, *3*, 75.
118. (a) Arnaud-Neu, F.; Böhmer, V.; Dozol, J.-F.; Grüttner, C.; Jakobi, R. A.; Kraft, D.; Mauprivez, O. et al. Calixarenes with diphenylphosphoryl acetamide functions at the upper rim. A new class of highly efficient extractants for lanthanides and actinides, *J. Chem. Soc., Perkin Trans. 2*, **1996**, 1175. (b) Matthews, S. E.; Saadioui, M.; Böhmer, V.; Barboso, S.; Arnaud-Neu, F.; Schwing-Weill, M.-J.; Garcia Carrera A.; Dozol, J.-F. Conformationally mobile wide rim carbamoylmethylphosphine oxide (CMPO)-calixarenes, *J. Prakt. Chem.*, **1999**, *341*, 264. (c) Delmau, L. H.; Simon, N.; Schwing-Weill, M.-J.; Arnaud-Neu, F.; Dozol, J.-F.; Eymard, S.; Tournois, B. et al. Extraction of trivalent lanthanides and actinides by "CMPO-Like" calixarenes, *Sep. Sci. Technol.*, **1999**, 34, 863. (d) Delmau, L. H.; Simon, N.; Schwing-Weill, M.-J.; Arnaud-Neu, F.; Dozol, J.-F.; Eymard, S.; Tournois, B. et al. CMPO-substituted calix[4]arenes, extractants with selectivity among trivalent lanthanides and between trivalent actinides and lanthanides, *Chem. Commun.*, **1998**, 1627.
119. Mikulasek, L.; Grüner, B.; Dordea, C.; Rudzevich, V.; Böhmer,V.; Haddaoui, J.; Hubscher-Bruder, V.; Arnaud-Neu, F.; Caslavsky, J.; Selucky, P. tert-Butyl-calix[4]arenes substituted at the narrow rim with cobalt bis(dicarbollide)(1–) and CMPO groups – new and efficient extractants for lanthanides and actinides, *Eur. J. Org. Chem.* **2007**, 4772.
120. Bruno, I. J.; Cole, J. C.; Edgington, P. R.; Kessler, M. K.; Macrae, C. F.; McCabe, P.; Pearson, J.; Taylor, R. New software for searching the cambridge structural database and visualizing crystal structures, *Acta Crystallogr., Sect. B*, **2002**, *58*, 389.

28 Synthetic Applications of Suzuki–Miyaura Cross-Coupling Reaction*

Subash C. Jonnalagadda, Michael A. Corsello, Brandon R. Hetzell, and Venkatram R. Mereddy

CONTENTS

* Dedicated to Professor Akira Suzuki on his 80th birthday.

28.1 INTRODUCTION

Carbon–carbon bond formation is central to synthetic organic chemistry and the transition metal-catalyzed cross-coupling of aryl/vinyl halides with organometallic reagents has proven to be a highly effective strategy for the construction of C–C bonds.[1] Initially, Kochi reported the utility of Fe(III) catalyst[2] and Li_2CuCl_4[3] for the cross-coupling of Grignard reagents with vinyl halides and alkyl halides, respectively. Later, Kumada and coworkers[4] and Corriu and Masse[5] observed the reaction of organomagnesium reagents with vinyl/aryl halides catalyzed by Ni(II) complex. The first Pd-catalyzed coupling of Grignard reagents was reported by Murahashi and coworkers.[6] The utility of palladium catalysis was further expanded for the coupling of other organometallic reagents by several researchers including Negishi (aluminum,[7] zinc,[8] zirconium[9]), Suzuki and coworkers[10] (boron), Murahashi et al.[11] (lithium), Milstein and Stille[12] and Migita and coworkers[13] (tin), Alexakis and Normant[14] (copper), and Hatanaka and Hiyama[15] (silicon). Although most of these methods have been utilized in a wide variety of applications, the boron-based Suzuki–Miyaura cross-coupling reaction has proven to be the most popular among all these.* This reaction offers distinct advantages over others in terms of optimal reactivity of organoboron compounds as opposed to the highly reactive Grignards or the relatively unreactive copper/silicon reagents). Further benefits of this protocol include the use of air- and water-stable, relatively nontoxic, and environmentally benign starting materials and by-products (borates and boric acid as opposed to heavy metals such as tin, etc.).

The present chapter exclusively deals with palladium-catalyzed cross-coupling of organoboron reagents (Suzuki–Miyaura reactions)[16] and their synthetic applications. The wide appeal of the Suzuki–Miyaura cross-coupling can be attributed to the extreme versatility of boron reagents (e.g., alkyl/alkenyl/alkynyl/arylboronates/boranes/trifluoroborates), as well as halide counterpart (vinyl/aryl/alkyl chloride/bromide/iodide/tosylates/triflates) that can be readily employed for this reaction. Organoboron compounds can be readily obtained via hydroboration, metalation, borylation, and other protocols under relatively mild conditions. In addition, several hundred boronic acids/boronates/boranes are commercially available. In fact, several small companies produce boronic acids as the sole products.† The enormous diversity of this coupling reaction has few parallels in the annals of synthetic organic chemistry. Hence, the nature of this chapter cannot be exhaustive, and we have attempted to provide brief general outlines of the research carried out in this area. Several reviews, book chapters, books, and so on provide a more detailed account of various aspects of Suzuki–Miyaura cross-coupling reaction.[17]

28.2 SYNTHESIS OF ORGANOBORANES

Typically, boronic acids/boronates, dialkylboranes, or trifluoroborates can all be used for Suzuki–Miyaura cross-coupling reaction. The following section details several protocols reported in the literature for the synthesis of these boron congeners.

* A general search on Scifinder Scholar for the keywords Suzuki cross-coupling revealed ~3000 references, and Suzuki–Miyaura cross-coupling revealed ~1800 references in the past ten years.

† Some of the companies selling boronic acids include Boron Molecular, Synthonix, Oakwood Products Inc., Combi-Blocks, Optima Chemical, and so on.

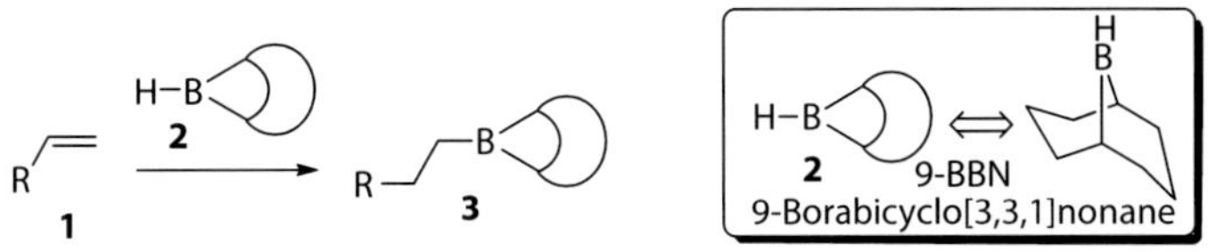

SCHEME 28.1 Hydroboration of alkenes with 9-BBN.

28.2.1 Hydroboration

28.2.1.1 Hydroboration with 9-Borabicyclo[3,3,1]nonane

Hydroboration of alkenes **1** with commercially available 9-borabicyclo[3,3,1]nonane (9-BBN) **2** progresses very smoothly to provide alkyl-9-BBN derivatives **3** (Scheme 28.1). 9-BBN provides superior levels of regioselectivity as compared with $BH_3.SMe_2$ and the boron atom is added to less-substituted carbon in the double bond.[18]

The regioselectivity for hydroboration of alkenes with 9-BBN is controlled by steric factors. Selective hydroboration of terminal alkenes in dienes **4**, **6**, and **8** can be readily achieved in the presence of internal olefins (Scheme 28.2).[19]

Although 9-BBN, **2** routinely provides high regiocontrol for several categories of alkenes, diterminal alkenes such as 2-methylhexa-1,5-diene only provide modest selectivity (Scheme 28.3). In such cases, bulkier boranes such as disiamylborane (Sia_2BH), dicyclohexylborane (Chx_2BH), and diisopinocampheylborane (Ipc_2BH) can be used to furnish the trialkylboranes regioselectively, based on the sterics of the alkene. For example, monosubstituted alkenes (**10**) can be selectively hydroborated in the presence of 1,1-disubstituted alkenes using any of these bulky dialkylboranes (Scheme 28.3).[19]

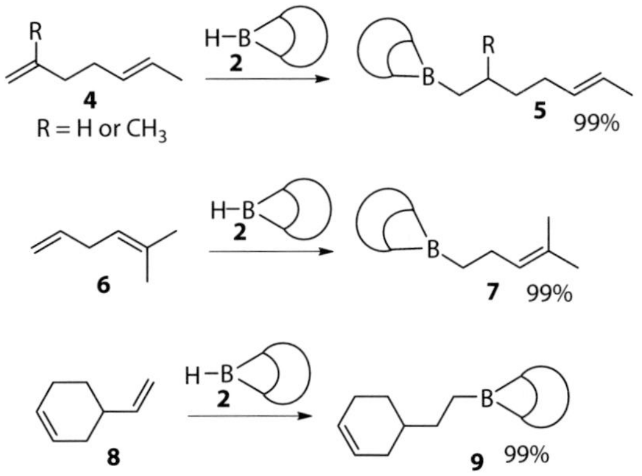

SCHEME 28.2 Regioselectivity in hydroboration of alkenes with 9-BBN.

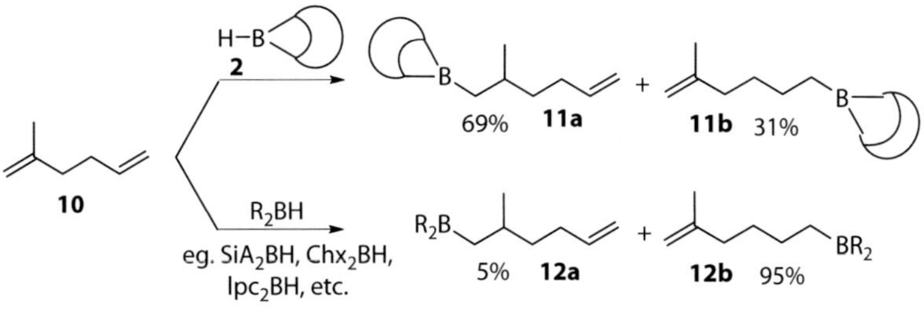

SCHEME 28.3 Reversal in regioselectivity in hydroboration with 9-BBN and Sia2BH.

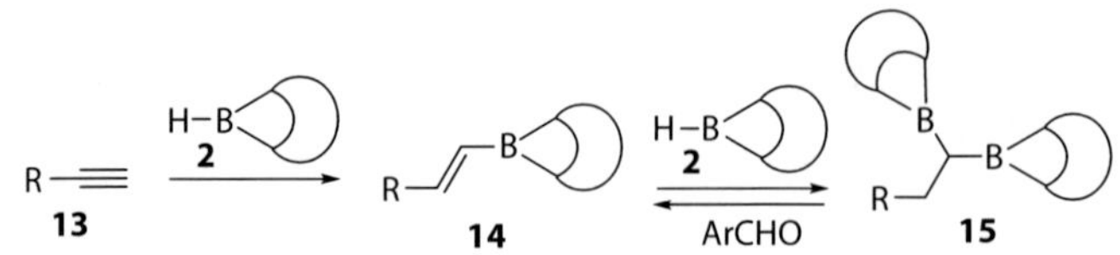

SCHEME 28.4 Hydroboration of alkynes with 9-BBN.

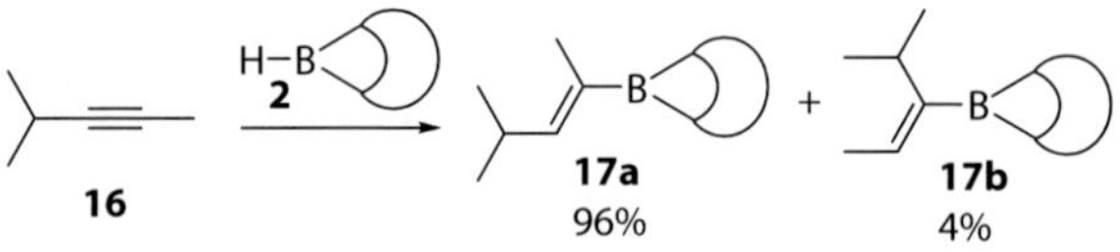

SCHEME 28.5 Regioselectivity in hydroboration of alkynes with 9-BBN.

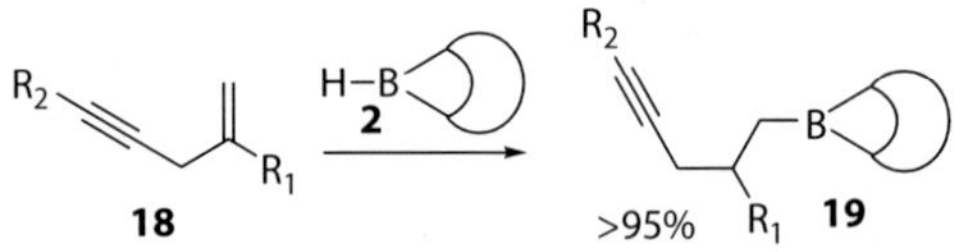

SCHEME 28.6 Chemoselectivity in hydroboration of alkynes with 9-BBN.

Hydroboration of terminal alkynes **13** with 9-BBN typically leads to diborylated product **15**. However, the vinylboranes **14** can be easily regenerated by reacting **15** with an aldehyde (Scheme 28.4).[20] Internal alkynes **16** undergo clean hydroboration with 9-BBN and furnish the vinylboranes **17a** from the less hindered side of the alkyne (Scheme 28.5).[21]

Alkenes undergo rapid hydroboration with 9-BBN as compared with internal alkynes and hence chemoselective hydroboration of enynes **18** can be achieved under these circumstances to obtain the borane **19** using 9-BBN as the hydroborating agent (Scheme 28.6).[22]

28.2.1.2 Hydroboration with Diisopinocampheylborane and Dibromoborane

Hydroboration of alkenes and alkynes (**20**) with Ipc_2BH proceeds regioselectively to provide the trialkylborane **21** that readily undergoes β-hydride transfer when treated with an aldehyde to release two equivalents of α-pinene **23** and the corresponding boronate **22** (Scheme 28.7).[23]

Alternatively, the boronates **25** can be prepared via hydroboration of alkenes/alkynes **20** with dibromoborane followed by alcoholysis of the resulting alkyldibromoborane **24** (Scheme 28.8). As

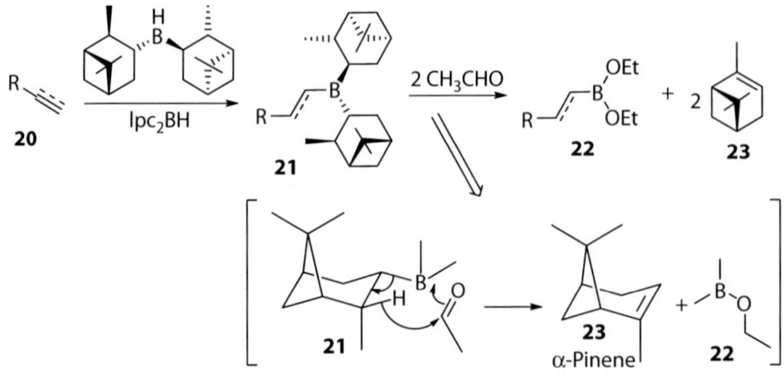

SCHEME 28.7 Hydroboration of alkenes and alkynes with diisopinocampheylborane.

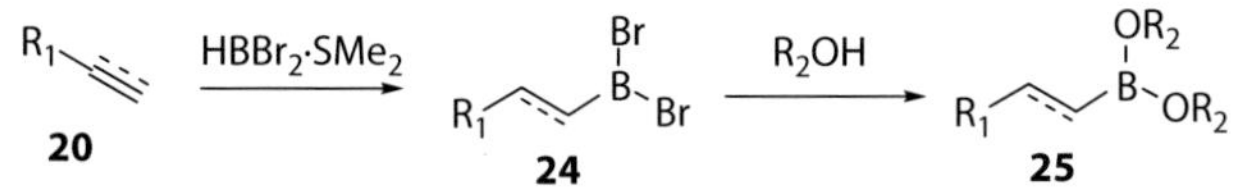

SCHEME 28.8 Preparation of *E*-boronates.

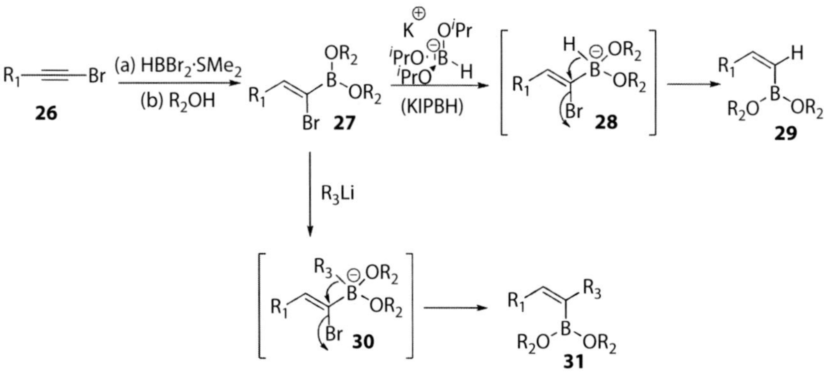

SCHEME 28.9 Preparation of *Z*-boronates.

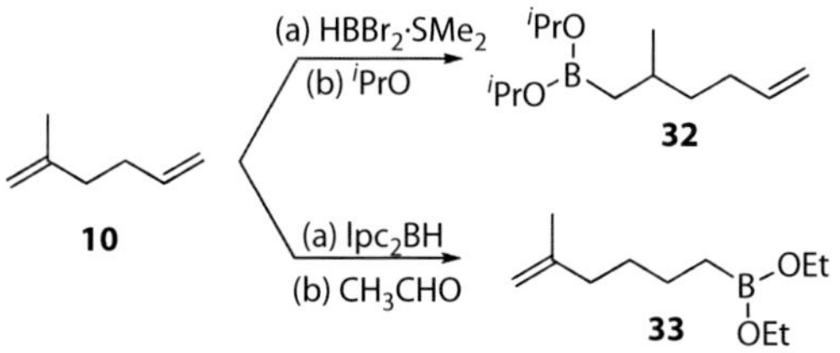

SCHEME 28.10 Reversal in regioselectivity with Ipc_2BH and $HBBr_2$.

expected, hydroboration of alkynes with dibromoborane provides (*E*)-vinylboronates, exclusively. The corresponding (*Z*)-vinylboronates **29–31** can be obtained by reaction of dibromoborane with bromoacetylenes **26** followed by reaction of the resulting (*E*)-boronate **27** with potassium triisopropoxyborohydride (KIPBH). Similarly, treatment of **27** with alkyl-lithiums furnish alkyl substituted (*Z*)-boronates **31** (Scheme 28.9).[24]

In the case of diterminal dienes such as 2-methylhexa-1,5-diene **10**, electrophilic boranes (e.g., $HBBr_2$) hydroborate at the electron rich 1,1-disubstituted alkene to afford the boronate **32** while the sterically bulky boranes (e.g., Ipc_2BH) prefer the less hindered monosubstituted C=C to furnish the alkylboronate **33** (Scheme 28.10).

28.2.1.3 Asymmetric Hydroboration with Mono and Diisopinocampheylboranes

Diisopinocampheylborane reacts with sterically less hindered (*Z*)-alkenes **34** at a considerably faster rate even at low temperatures (–25°C) to provide chiral alkyldiisopinocampheylboranes **36**. Reaction of **36** with an aldehyde leads to the elimination of α-pinene and furnishes the chiral alkylboronates **37** in moderate-to-high ee (Scheme 28.11).[25] The enantioselectivity can be further enhanced via recrystallization of the chelates **42** of alkylborinates **40** with chiral amino alcohols such as prolinol (Scheme 28.11).[23]

While diisopinocampheylborane (Ipc_2BH) provides excellent enantioselectivity for the hydroboration of less hindered cis-alkenes **44–48** (Scheme 28.12), monoisopinocampheylborane ($IpcBH_2$) is

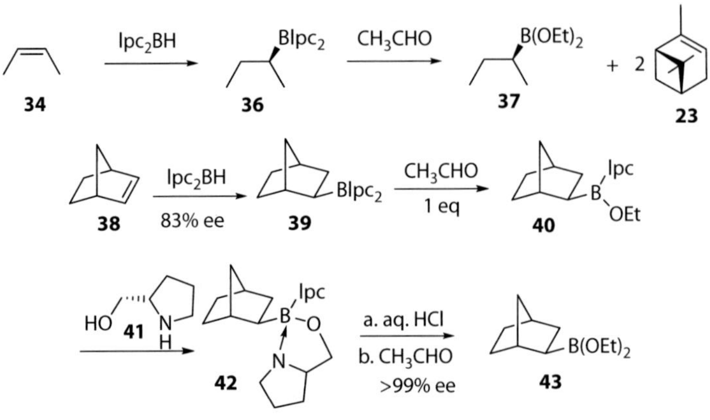

SCHEME 28.11 Asymmetric hydroboration with Ip_2BH and optical upgradation.

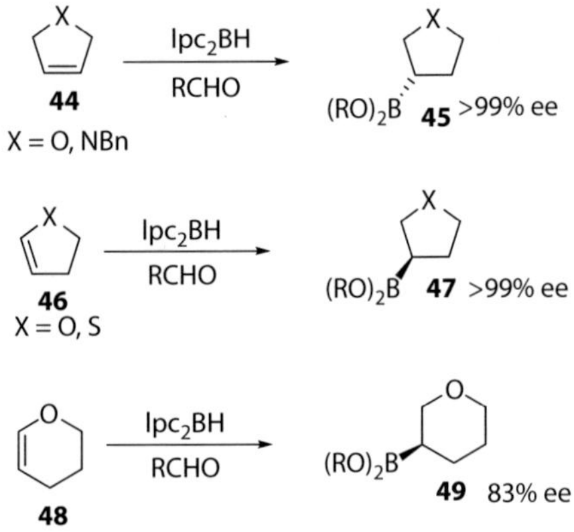

SCHEME 28.12 Asymmetric hydroboration of *cis*-disubstituted alkenes with Ipc_2BH.

SCHEME 28.13 Hydroboration of *trans*-disubstituted and trisubstituted alkenes with $IpcBH_2$.

the complimentary reagent that furnishes high enantioselectivites for *trans*-alkenes **50** as well as trisubstituted alkenes **52** (Scheme 28.13).[26]

28.2.1.4 Catalytic Hydroboration with Pinacol/Catecholboranes

Catalytic hydroboration of alkenes and alkynes can be achieved by the reaction of dialkoxyboranes such as pinacolborane **55** and catecholborane **56** in the presence of transition metal catalysts such as Cp_2ZrHCl, $Rh(PPh_3)_3Cl$, or $Rh(PPh_3)_2COCl$, (Scheme 28.14).[27] This reaction is of particular significance because it readily provides alkyl/vinylboronates directly in one step for further utility in Suzuki–Miyaura cross-coupling reaction.

Reversal in regioselectivity was observed for the reaction of electron deficient alkenes such as perfluoroalkenes **59** on catalytic hydroboration with neutral (e.g., $Rh(PPh_3)_3Cl$) or cationic ($[Rh(COD)(dppb)]^+[BF_4]^-$) transition metal catalysts (Scheme 28.15).[28]

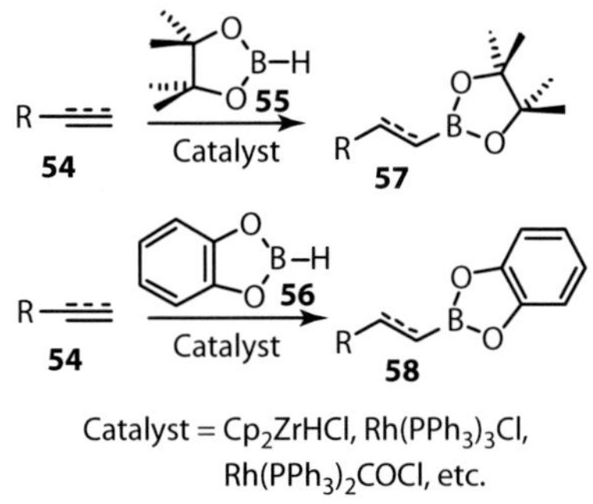

Catalyst = Cp_2ZrHCl, $Rh(PPh_3)_3Cl$, $Rh(PPh_3)_2COCl$, etc.

SCHEME 28.14 Catalytic hydroboration of alkenes and alkynes.

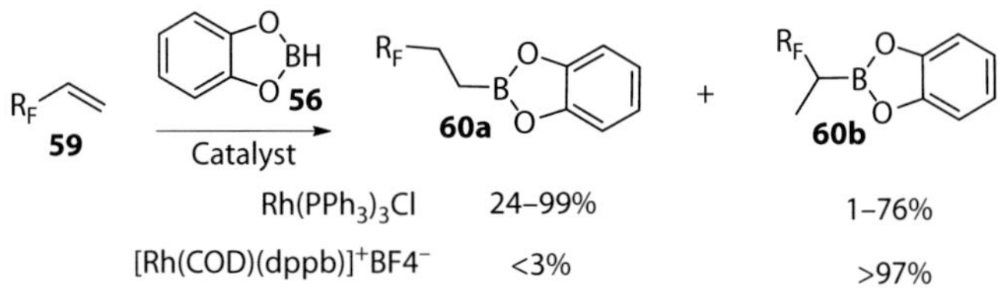

Catalyst	60a	60b
$Rh(PPh_3)_3Cl$	24–99%	1–76%
$[Rh(COD)(dppb)]^+BF4^-$	<3%	>97%

SCHEME 28.15 Reversal in regioselectivity in hydroboration of perflorinated alkenes.

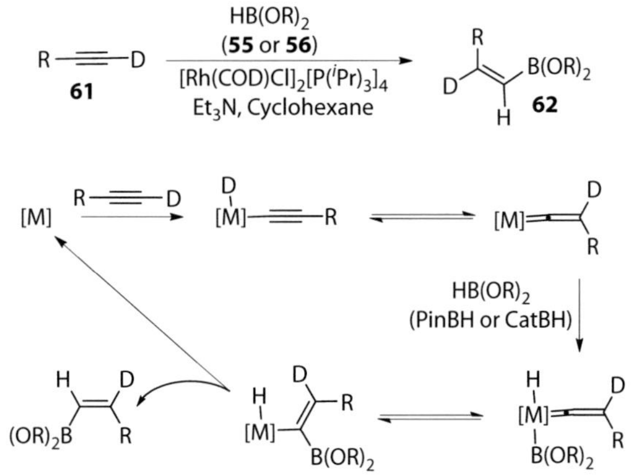

SCHEME 28.16 *trans*-Hydroboration of alkynes.

During the reaction of pinacol/catecholborane with alkynes in the presence of catalysts such as $[Rh(COD)Cl]_2[P(^iPr)_3]_4$ and $[Ir(COD)Cl]_2[P(^iPr)_3]_4$, Miyauraand coworkers noticed the formation of (Z)-boronates as opposed to the expected (E)-isomer as shown in Scheme 28.16.[29]

28.2.1.5 Catalytic Asymmetric Hydroboration with Pinacol/Catecholboranes

Hayashi and coworkers reported catalytic asymmetric hydroboration of aryl alkenes **63** with catecholborane **56** using chiral phosphine ligands in the presence of transition metal catalysts to obtain good ee's in the product boronates **64** (Scheme 28.17).[30]

28.2.2 Reaction of Organolithiums and Grignard Reagents

Apart from hydroboration, another frequently utilized procedure for the preparation of organoboranes involves the treatment of organolithium or Grignard reagents with alkoxyboranes or

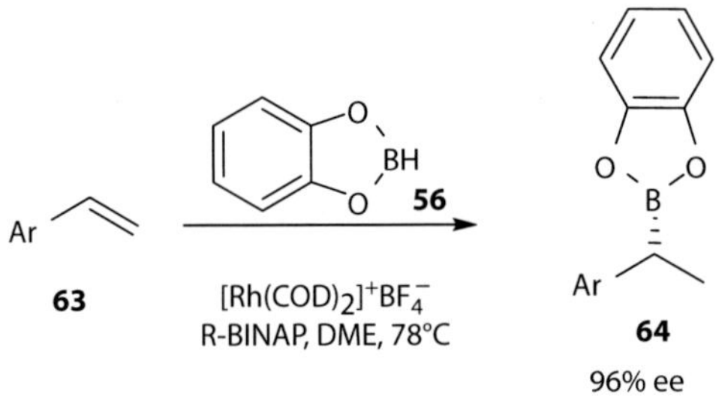

SCHEME 28.17 Catalytic asymmetric hydroboration of alkenes.

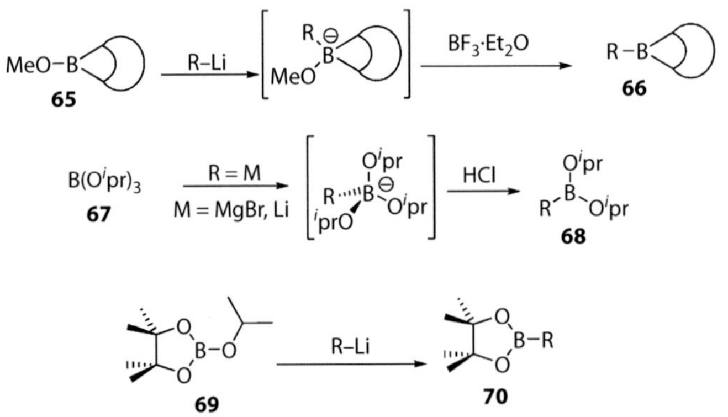

SCHEME 28.18 Reaction of organolithium and Grignard reagents with alkoxyboranes.

trialkylborates. For example, the reaction of alkyl-lithium with B-methoxy-9-BBN **65** provides the "ate" complex that on addition of a Lewis acid ($BF_3{\cdot}Et_2O$) releases the nascent trialkylborane **66**.[31] Alternatively, treatment of organolithiums or Grignard reagents with triisopropylborate **67** followed by acidification provides alkylboronates **66**.[32] *B*-isopropoxypinacolborane **69** is also commonly used under these conditions to provide B-alkylpinacolboronates **70** in high yields (Scheme 28.18).[31–32]

28.2.3 Haloboration of Alkynes

Reaction of terminal alkynes **71** with boron tribromide or *B*-bromo-9-BBN **74** leads to the addition of *B-Br* across the triple bond. This reaction furnishes β-bromoalkenylboronates **72–73** stereoselectively as the major diastereomer (Scheme 28.19).[33]

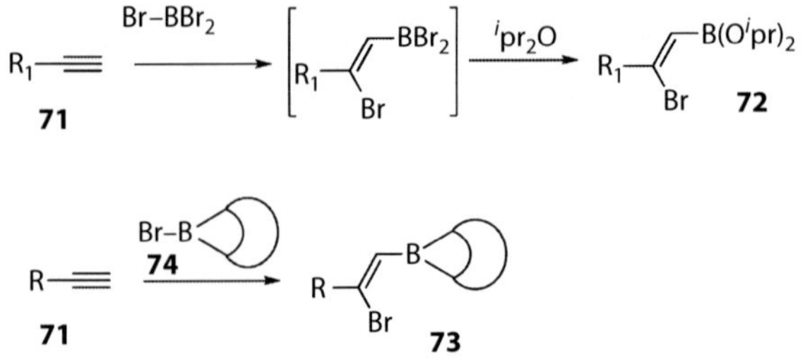

SCHEME 28.19 Bromoboration of alkynes.

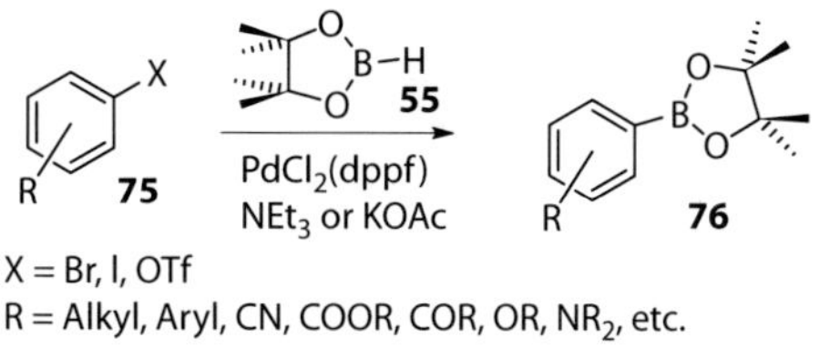

SCHEME 28.20 Borylation of aryl halides.

28.2.4 Borylation

Pinacolborane is a mild organoboron reagent and has excellent compatibility with a multitude of functional groups including esters, ketones, ethers, tertiary amines, and nitriles. The *B*-organopinacolboronates are relatively stable to air and moisture and, in many cases, it is possible to purify them via column chromatography. Aryl halides **75** react readily with pinacolborane in the presence of a mild base (e.g., Et_3N, or KOAc) and $PdCl_2$(dppf) catalyst to undergo borylation and provide the corresponding arylboronates **76** (Scheme 28.20).[34]

Organohalides readily react with pinacolborane to provide the corresponding boronates in high yields. For example, vinyl iodides/triflates (**77**),[35] aryl chlorides,[36] benzyl halides (**79**),[37] and allyl halides (**81**)[38] undergo efficient borylation with pinacolborane to afford the corresponding boronates **78**, **80**, and **82**, respectively (Scheme 28.21). It has also been shown that arylalkenes (**83**)[39] and alkoxyalkenes (**85**)[40] undergo dehydrogenative borylation to afford the corresponding vinylboronates **84** and **86** (Scheme 28.21). Similar studies on arenes (**87**)[41], and alkylarenes (**89**)[41d] yield the arylboronates **88** and **90** (Scheme 28.21).

28.2.5 Synthesis of Alkyl/Aryl Trifluoroborates Using KHF_2

Apart from the boronic acids/boronates, and trialkylboranes, another important class of boron compounds that have been utilized for Suzuki–Miyaura cross-coupling reaction include the alkyl/

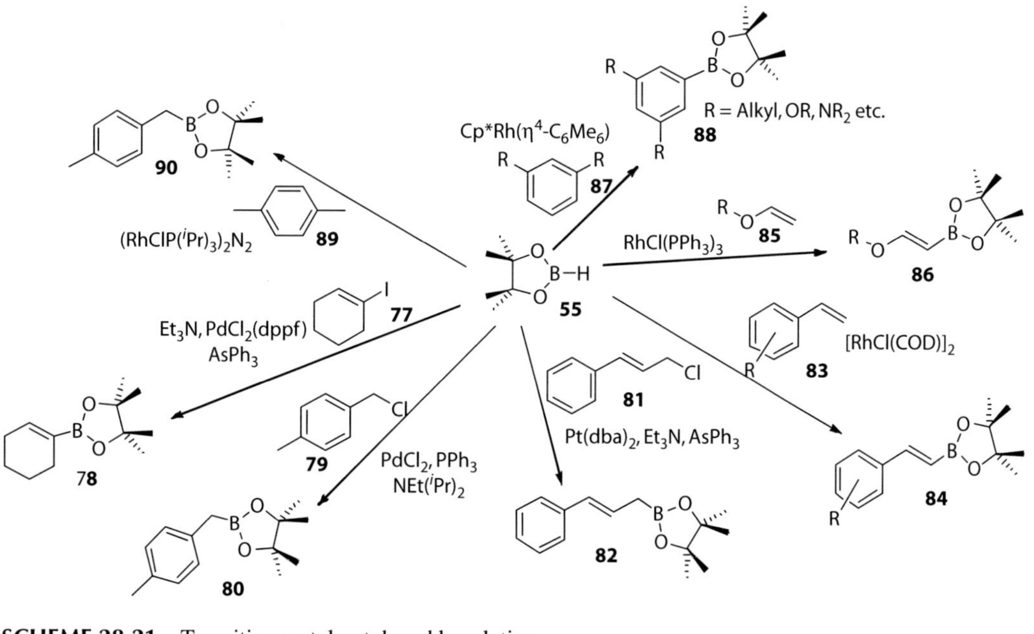

SCHEME 28.21 Transition metal-catalyzed borylation.

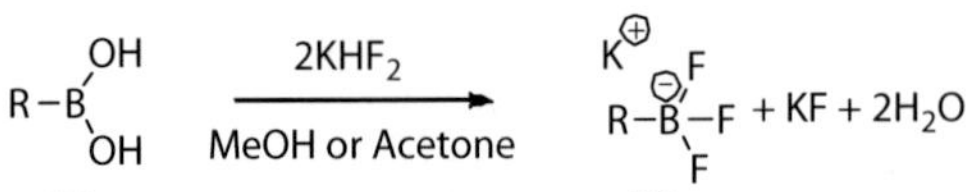

SCHEME 28.22 Preparation of alkyl/aryl trifluoroborates.

aryltrifluoroborates. These boron reagents **92** can be conveniently prepared via the reaction of the corresponding boronic acids **91** with potassium dihydrogen fluoride in methanol or acetone (Scheme 28.22).[42]

28.3 MECHANISM OF SUZUKI–MIYAURA CROSS-COUPLING REACTION

The general mechanism of Suzuki–Miyaura cross-coupling involves the oxidative addition of organohalide to the palladium catalyst followed by transmetallation with organoboron derivatives. Since sp^3 organoboron compounds are not nucleophilic, they do not undergo transmetallation readily in the absence of a base. It has been found that the addition of the base results in the formation of the more nucleophilic boron "ate" complexes for transmetallation with the palladium catalyst.[43] Finally, reductive elimination leads to the formation of the cross-coupled product followed by the regeneration of the catalyst (Scheme 28.23). The first step (oxidative addition) is typically the rate-determining step and the relative leaving group abilities of the halides determines their reactivity towards cross-coupling. Hence, iodides react faster than triflates followed by bromides and chlorides. Further, the stereochemistry is retained and the structural integrity of the vinyl halides is retained during the oxidative addition. However, allylic and benzylic halides typically undergo cross-coupling with inversion.

28.4 CATALYSTS AND LIGANDS IN THE SUZUKI–MIYAURA CROSS COUPLING REACTION

The most commonly used catalytic system is the inexpensive tetrakistriphenylphosphine palladium $Pd(PPh_3)_4$. However, other palladium sources such as $Pd_2(dba)_3/PPh_3$, $Pd(OAc)_2/PPh_3$, and $PdCl_2(dppf)$ (for sp^3-sp^2 couplings) have also been employed for this reaction. These latter catalyst systems generate Pd^0 *in situ* from the respective Pd^{II} salts. In addition to triphenyl phosphine, other electron-rich bulky phosphines such as **93–95**[44–45] and *N*-heterocyclic carbene (NHC)[46] ligands have also been commonly employed for the cross-coupling purposes, especially for less reactive substrates such as aryl chlorides (Figure 28.1).[47]

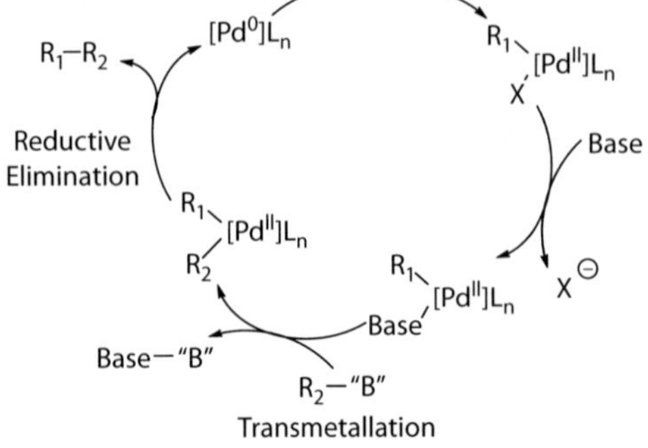

SCHEME 28.23 Mechansim of Suzuki cross-coupling.

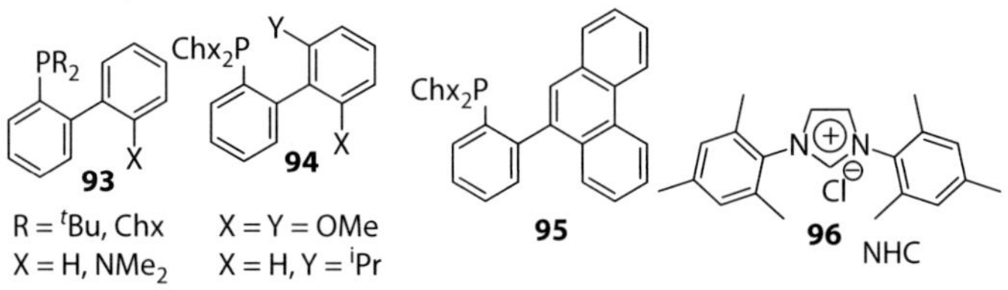

FIGURE 28.1 Common ligands used in Suzuki cross-coupling.

28.5 CROSS-COUPLING REACTION OF ORGANOBORANES

Due to the sheer volume and variety of Suzuki–Miyaura cross-coupling, it is unrealistic to summarize all the developments in one book chapter. Hence, in this section, we have categorized the cross-coupling reaction based on the nature of organo "boron" and organo "halide." In the subsequent section, we have summarized several applications of these reactions in the total synthesis of natural products. Even in the synthesis section, many of the important applications could not be included due to the space constraints.

28.5.1 Coupling with Vinyl Halides

This section deals with the cross-coupling of vinyl halides with various types of organoboranes such as vinyl/aryl/alkylboranes.

28.5.1.1 Coupling of Vinyl Halides and Vinyl Boranes

The coupling between vinyl halides and vinylboronates was first reported by Suzuki and Miyaura in 1979.[10,16] The general conditions employed for the coupling of vinylbromides use catalytic $Pd(PPh_3)_4$ along with sodium ethoxide (base) while $PdCl_2(PPh_3)_2$ and aqueous NaOH are generally recommended for coupling vinylboronates with vinyl iodides (Scheme 28.24).[48]

For simple conjugated dienes such as **100a-b**,[49] **100c**,[50] and **100d–f**,[51] (Figure 28.2) strong bases such as NaOH or NaOEt provide good yields of the cross-coupled products. However, mild bases

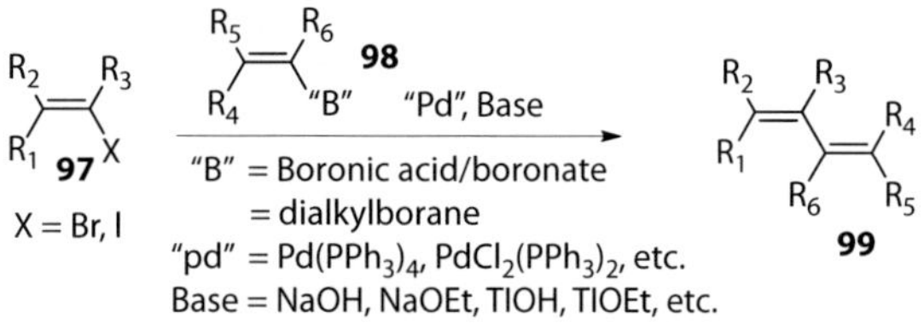

SCHEME 28.24 Coupling of vinyl halides and vinyl "Boron."

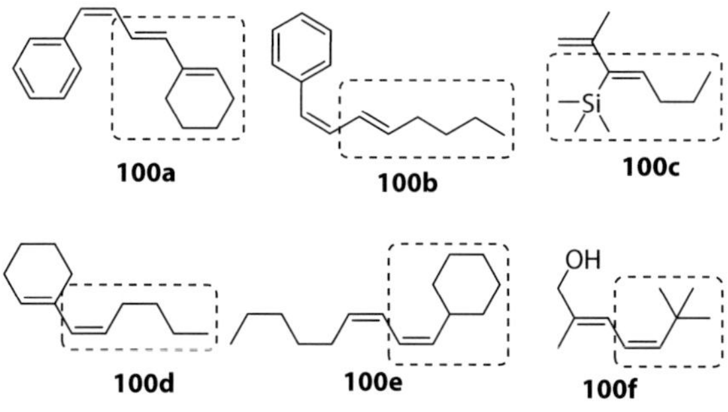

FIGURE 28.2 Polyenes obtained from Suzuki–Miyaura cross-coupling.

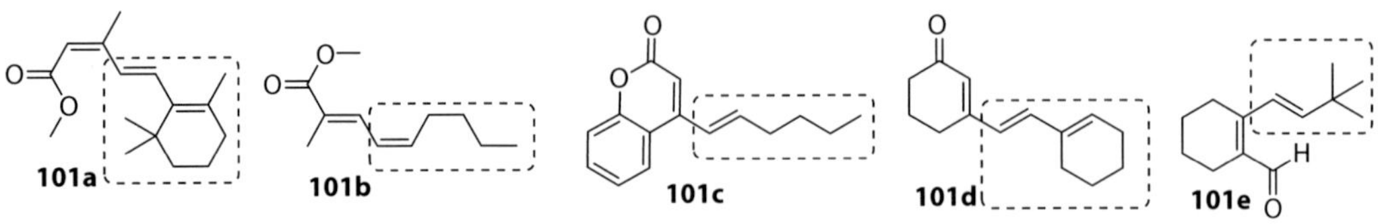

FIGURE 28.3 Base-sensitive polyenes obtained from Suzuki–Miyaura cross-coupling.

such as sodium acetate, triethylamine, or sodium carbonate are utilized for base-sensitive substrates **101a–e** (Figure 28.3).[52]

A dramatic rate enhancement was realized by Kishi using TlOH as a base.[53] The coupling can be achieved even at 0°C and a wide variety of sensitive functional groups can be readily incorporated into the synthetic protocols because of these mild conditions. Kishi utilized these conditions in the synthesis of Palytoxin via the coupling of vinylboronic acid **102** with the vinyl iodide **103** (Scheme 28.25). However, TlOH is air and moisture sensitive and has a limited shelf life. In this regard, Roush found TlOEt to be superior to TlOH due to its ready availability and greater stability.[54]

28.5.1.2 Coupling of Vinyl Halides and Aryl Boranes

Similar to vinylboronates, arylboronate species **106** also undergo very facile cross-coupling with vinyl halides **105** to provide vinylarenes **107** (Scheme 28.26). The stereo-integrity is maintained during this process similar to the prior cross-coupling reactions.

Some representative examples in this category include (1) synthesis of arylglucals **109** via the coupling of arylboronic acids with vinyl iodide **108** (Scheme 28.27),[55] (2) synthesis of 1,1-diarylalkenes **111a–i** via reaction of arylboronic acids with vinylidene dibromides **110** (Scheme 28.28),[56]

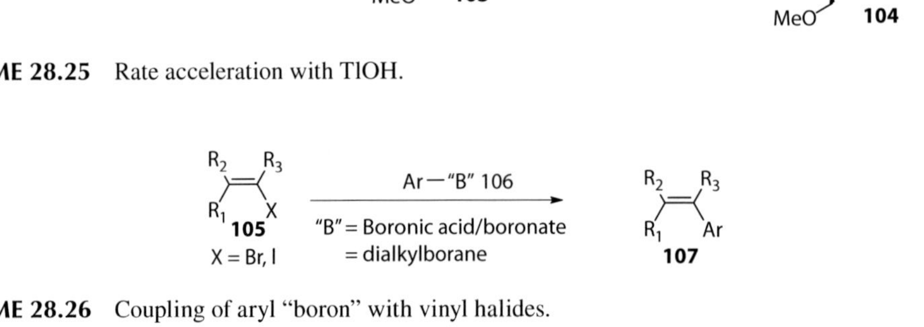

SCHEME 28.25 Rate acceleration with TlOH.

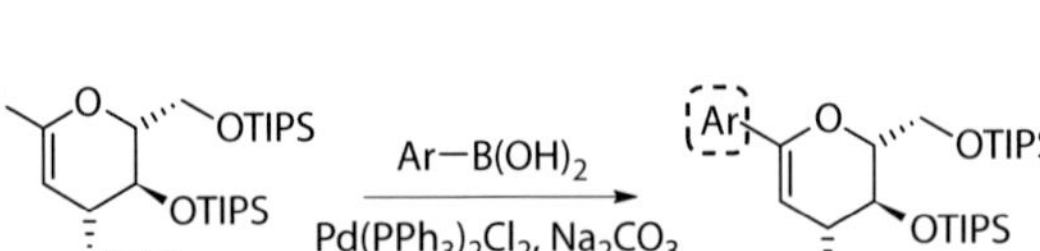

SCHEME 28.26 Coupling of aryl "boron" with vinyl halides.

SCHEME 28.27 Synthesis of arylglucals.

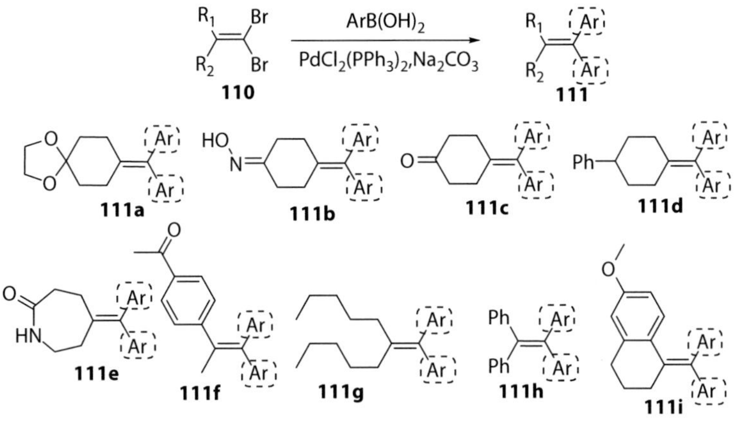

Ar = Phenyl, (*o*/*m*/*p*)-tolyl, (*o*/*m*/*p*)-Anisyl, (*o*/*m*/*p*)-fluorophenyl, (*o*/*m*/*p*)-trifluoromethylphenyl, (*o*/*m*/*p*)-thiomethylphenyl, 4-Pyridyl, 3-Thiophenyl, etc.

SCHEME 28.28 Synthesis of 1,1-diarylalkenes.

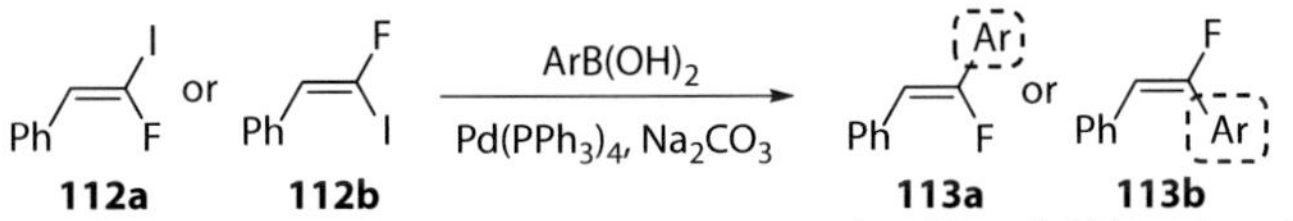

Ar = Phenyl, Chlorophenyl, Anisyl, naphthyl, etc.

SCHEME 28.29 Synthesis of 1-aryl-1-fluoroalkenes.

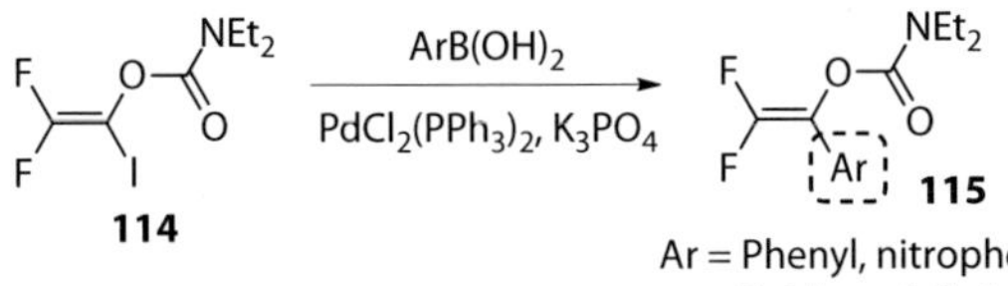

Ar = Phenyl, nitrophenyl, Tolyl, naphthyl, etc.

SCHEME 28.30 Synthesis of α-aryldifluoroenolcarbamates.

(3) synthesis of 1-aryl-1-fluoroalkenes such as **113a–b** via the coupling of 1-fluoro-1-iodoalkenes **112a–b** with arylboronic acids (Scheme 28.29),[57] and (4) the synthesis of α-aryldifluoroenolcarbamates **115** via the reaction of **114** with arylboronic acids (Scheme 28.30).[58]

28.5.1.3 Coupling of Vinyl Halides and Alkyl Boranes

Vinyl halides **116** also undergo facile *B*-Alkyl Suzuki (sp^2-sp^3) coupling with B-alkylboranes (Scheme 28.31).[59] Although several palladium catalysts can be utilized for this coupling, the preferred catalyst for *B*-Alkyl Suzuki reaction is diphenylphosphinoferrocenyl palladium dichloride ($PdCl_2$(dppf)) **118** (Figure 28.4) due to its favorable bite angle and reduced possibility of β-elimination.

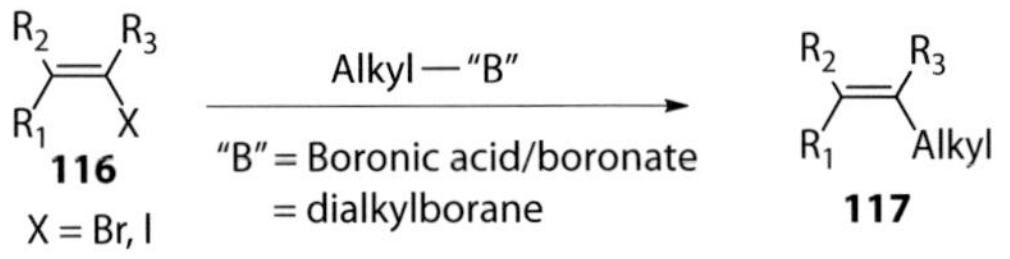

SCHEME 28.31 Coupling of vinyl halides and alkyl "Boron."

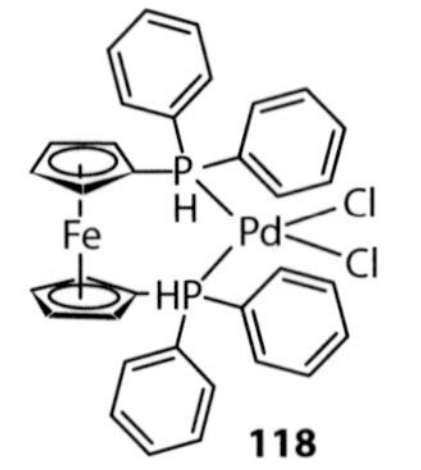
P
H
Pd
Cl
Cl
Fe
HP
118

FIGURE 28.4 Diphenylphosphinoferrocenyl palladium dichloride $PdCl_2(dppf)$.

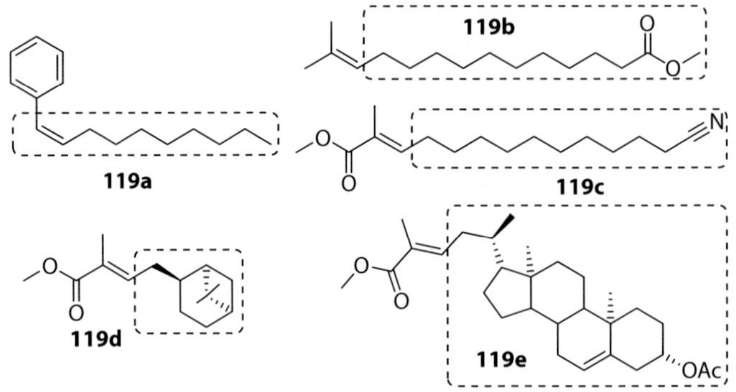
119a
119b
119c
119d
119e
OAc

FIGURE 28.5 B-alkyl Suzuki cross-coupling.

Suzuki and Miyaura first demonstrated the utility of B-alkyl-9-BBN derivatives for coupling with vinyl halides (Figure 28.4).[60] The reaction conditions are mild enough to tolerate many functional groups such as esters, nitriles, amides, amines, and sulfides and a wide variety of compounds (e.g., **119a–e)** have been synthesized using this protocol (Figure 28.5).

Suzuki and coworkersfurther elaborated this protocol for the coupling of B-alkyl-9-BBN derivatives **121** and **124** with functionalized vinyl bromides such as β-acetoxy (**120**)[61] and β-mercapto (**123**)[62] bromoalkenes to afford the diketone **122** and vinyl sulfides **125a–c**, respectively (Schemes 28.32, and 28.33).

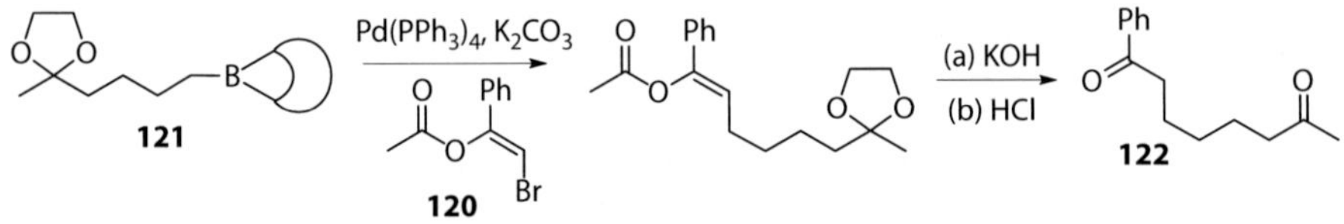
Pd(PPh3)4, K2CO3
Ph
Br
121
120
(a) KOH
(b) HCl
122

SCHEME 28.32 Coupling of β-acetoxyvinyl bromides with B-alkyl-9-BBN.

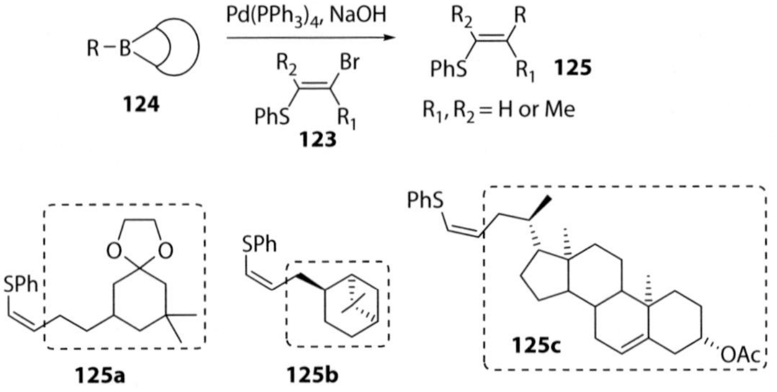
R−B
124
Pd(PPh3)4, NaOH
R2
Br
PhS
R1
123
R
125
R1, R2 = H or Me
SPh
PhS
125a
125b
125c
OAc

SCHEME 28.33 Coupling of β-mercaptovinyl bromides with B-alkyl-9-BBN.

Ar–Ar **128** ← Ar–B(OH)$_2$ / "Pd", Base — Ar–X, X = Br, I **126** — R–CH=CH–B(OH)$_2$ / "Pd", Base → Ar–CH=CH–R **127**
R = ^{n}Bu, Chx, ^{t}Bu, etc.
Ar = Ph, Mesityl, Naphthyl, etc.

SCHEME 28.34 Coupling of aryl halides and vinyl/aryl "Boron."

28.5.2 Coupling with Aryl Halides

28.5.2.1 Coupling of Aryl Halides and Vinyl/Aryl Boranes

Suzuki and Miyaura demonstrated the coupling of aryl halides with vinyl/arylboronates for the preparation of biaryls/vinyl arenes (Scheme 28.34)[50,63] and this particular topic has been extensively reviewed in the literature.[17c, 64]

Traditionally, aryl bromides/iodides/triflates are employed for coupling with boronates. For example, α-trifluoromethyl vinylboronic acids **129** were utilized for coupling with wide variety of aryl halide substrates such as phenyl, naphthyl, and pyrazyl halides under standard conditions to provide trifluoromethyl-α-arylalkenes **130** (Scheme 28.35).[65] Similarly, *B*-vinylboronates **131** were also coupled with aryl halides to provide arylpiperidines **132** (Scheme 28.36).[66]

Aryl chlorides are typically less expensive and more readily available than the corresponding bromides and iodides. However, they are comparatively less reactive in cross-coupling reactions. However, recent developments in catalyst and ligand design have led to the successful utilization of aryl chlorides as well in coupling reactions (Scheme 28.37). The majority of the research in this area has concentrated on the use of bulky electron-rich organophosphine and N-heterocyclic carbene ligands.[46,67] Use of aryl chlorides for palladium-catalyzed cross-coupling has been reviewed elsewhere.[68–69]

28.5.2.2 Coupling of Aryl Halides and Alkyl Boranes

Aryl halides react readily with a variety of alkylboranes such as 9-BBN **137a**, disiamylborane **137b**, dicyclohexylborane **137c** in the presence of palladium catalysts (Scheme 28.38).[60] The *B*-alkyl-9-BBNs provides the best yields under these conditions, and the ease with which a variety of 9-BBN

CH_2=C(CF$_3$)B(OH)$_2$ **129** — Ar–X / Pd(PPh$_3$)$_4$, Na$_2$CO$_3$ → CH_2=C(CF$_3$)Ar **130**
Ar = Phenyl, nitrotolyl, Chloronitrophenyl, trifloromethylphenyl, Naphthyl, Pyrazyl, etc.

SCHEME 28.35 Coupling of α-trifluoromethylvinylboronic acid with aryl halides.

131 (R = Cbz, Boc) — Ar–X / PdCl$_2$(dppf), K$_2$CO$_3$ → **132**
Ar = Phenyl, tolyl, anisyl, Pyridyl, etc.

SCHEME 28.36 Coupling of *B*-vinylpinacolboronates with aryl halides.

133 + (HO)$_2$B–Ph **134** — Pd$_2$(dba)$_3$, Cs$_2$CO$_3$ / NHC, **96** → **135**

SCHEME 28.37 Coupling of aryl chloride with aryl boronic acids.

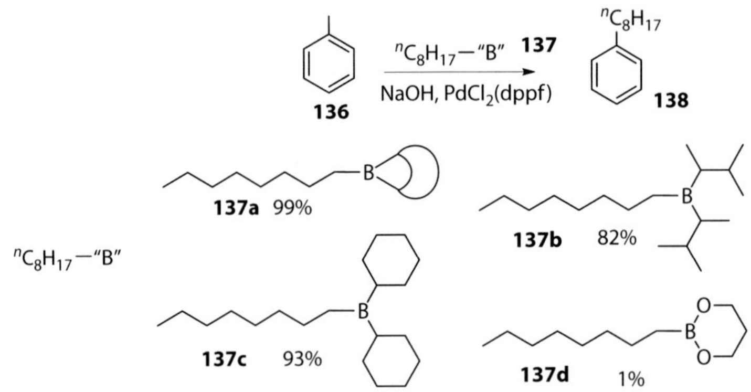

SCHEME 28.38 Coupling of aryl halides with alkylboranes.

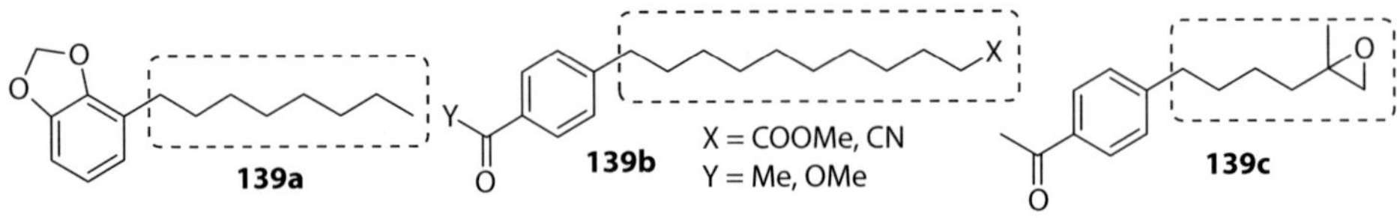

FIGURE 28.6 Alkylarenes obtained from Suzuki–Miyaura cross-coupling.

derivatives can be obtained makes them highly preferred choice of reagents for *B*-alkyl Suzuki coupling reactions. The milder reaction conditions further enhance the applicability of this protocol and a plethora of functional groups readily survive these conditions (Figure 28.6).[60]

Allylboranes are extremely labile and undergo allyl transfer to carbonyl groups even at very low temperatures. In this regard, it is interesting to note that Fürstner and Seidel were able to carry out chemoselective cross-coupling of B-Allyl-9-BBN **141** with aryl bromides **140, 143** (Scheme 28.39).[70]

Gray et al. reported the coupling of aryl halides with trimethylboroxine **145** in the presence of $Pd(PPh_3)_4$ and Na_2CO_3 to cause methyl transfer thereby avoiding the use of air-sensitive *B*-methyl-9-BBN (Scheme 28.40).[71]

B-Alkyl Suzuki coupling offers tremendous opportunities for one pot cross-coupling between any alkene and aryl halides, as the intermediate alkylboranes are typically not isolated and can be used *in situ* for further cross-coupling. Some of the representative examples of this coupling include the synthesis of pyridyl alcohols **148** (Scheme 28.41),[72] synthesis of alkyl arenes **149** (Scheme 28.42),[73] and coupling of cyclopropyl boronic acid with aryl halides to obtain cyclopropyl arenes **153** (Scheme 28.43).[74]

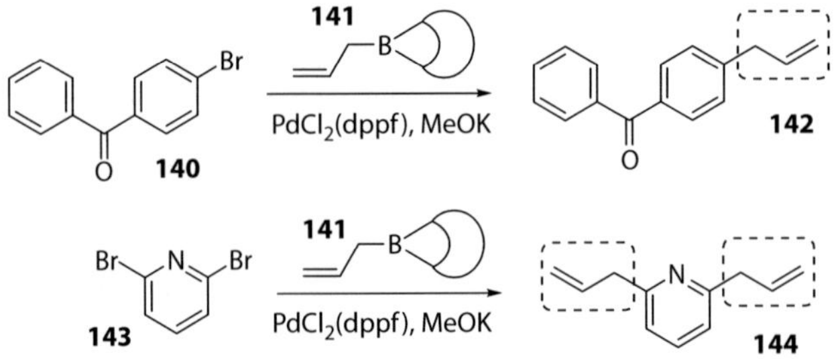

SCHEME 28.39 Cross-coupling with B-allyl-9-BBN.

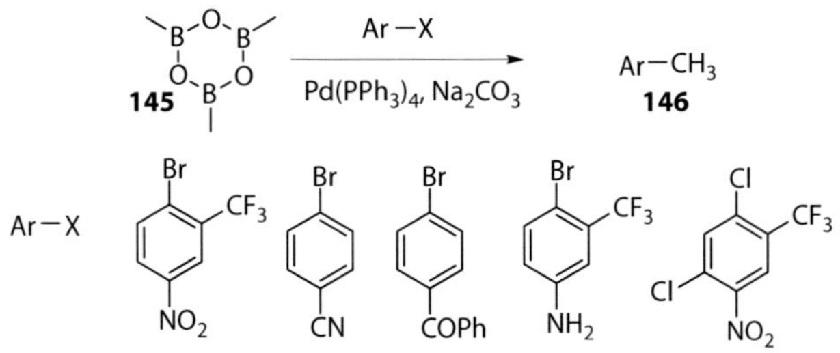

SCHEME 28.40 Coupling of trimethylboroxine with aryl halides.

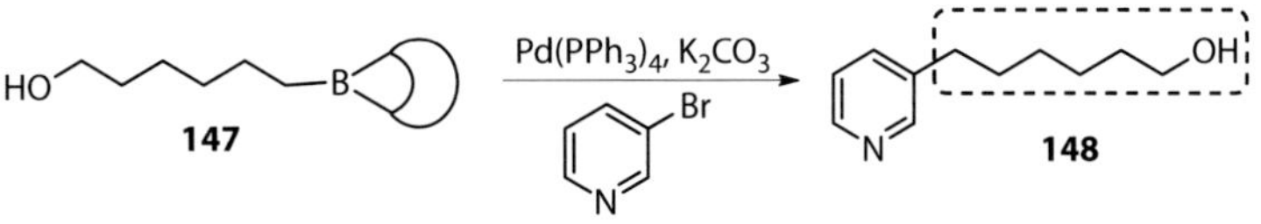

SCHEME 28.41 Preparation of pyridyl alcohols.

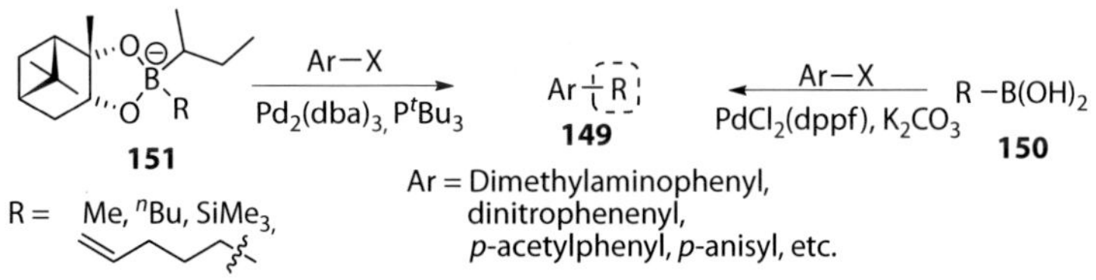

Ar = Dimethylaminophenyl, dinitrophenenyl, *p*-acetylphenyl, *p*-anisyl, etc.

R = Me, ^{n}Bu, $SiMe_3$,

SCHEME 28.42 Preparation of alkyl arenes by coupling alkyl boronates with aryl halides.

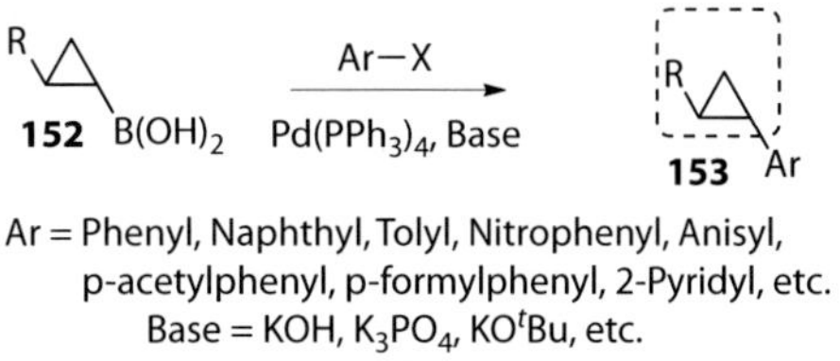

Ar = Phenyl, Naphthyl, Tolyl, Nitrophenyl, Anisyl, p-acetylphenyl, p-formylphenyl, 2-Pyridyl, etc.
Base = KOH, K_3PO_4, KO^{t}Bu, etc.

SCHEME 28.43 Coupling of aryl halides with cyclopropyl boronic acids.

28.5.3 Coupling with Alkyl Halides

Alkyl halides typically undergo β-elimination under the cross-coupling conditions, and are used less frequently than the vinyl/aryl halides. Suzuki and Miyaura were successful in coupling the alkyl iodides with boranes such as *B*-phenyl-9-BBN **155** and *B*-octyl-9-BBN **157** in the presence of $Pd(PPh_3)_4$ to obtain the C(sp^3-sp^3)-coupled products **156** and **158** (Scheme 28.44).[75]

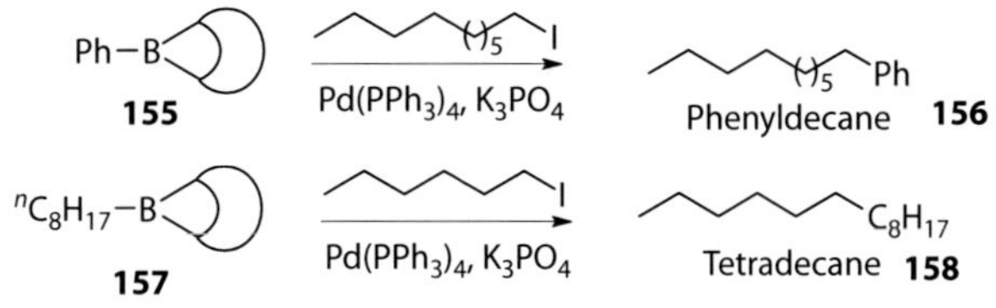

SCHEME 28.44 Coupling between alkyl iodide and B-alkyl-9-BBN.

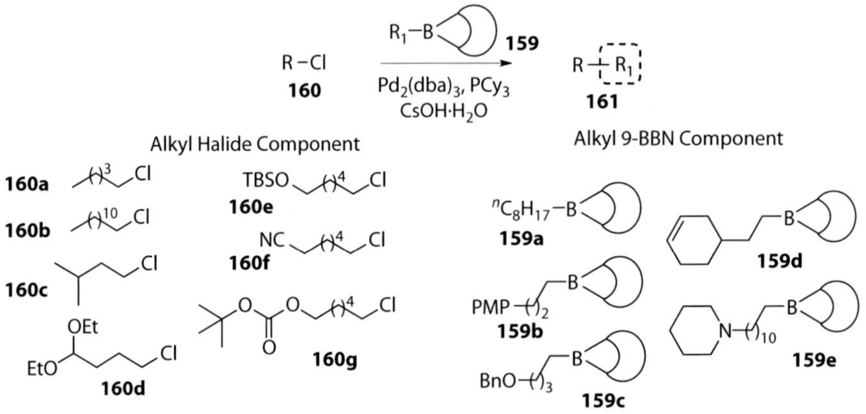

SCHEME 28.45 Coupling of alkyl chlorides and B-alkyl-9-BBN.

Fu and coworkers reported a facile coupling of alkyl chlorides with *B*-alkyl-9-BBN employing sterically bulky tricyclohexylphosphine as a ligand in the presence of $Pd_2(dba)_3$ as a catalyst. A number of alkyl chlorides (**160a–g**) were coupled with several 9-BBN derivatives (**159a–e**) to obtain the corresponding coupled products **161** under these conditions (Scheme 28.45).[76] Arylboronic acids were also utilized for coupling with alkyl chlorides and bromides by Fu and Gonzalez-Bobes in the presence of nickel catalysts for the preparation of alkylarenes.[77]

28.5.4 Intramolecular Cross-Coupling

Intramolecular cross-coupling offers an important extension for the preparation of various ring-size molecules. For example, several five and six membered cyclic compounds have been synthesized via *B*-alkyl Suzuki cross-coupling of vinyl bromides shown in Scheme 28.46.[60] More complex ring systems can also be generated with utmost ease utilizing this protocol as demonstrated in the total synthesis of complex natural products such as rutamycin (Scheme 28.47)[78] and RP66453 (Scheme 28.48).[79]

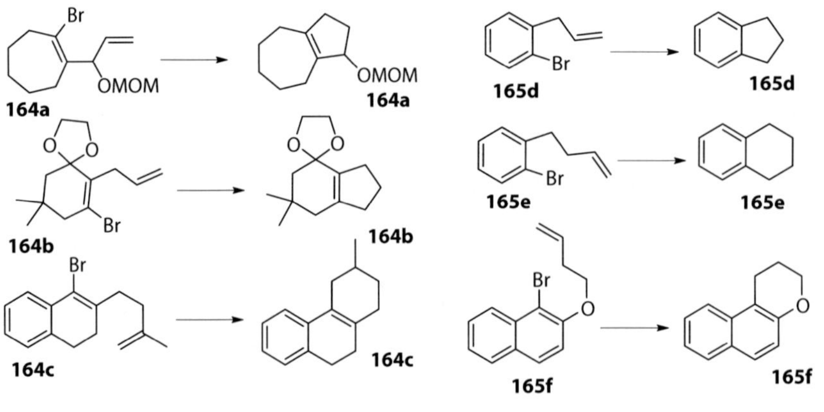

SCHEME 28.46 Intramolecular B-alkyl Suzuki cross-coupling.

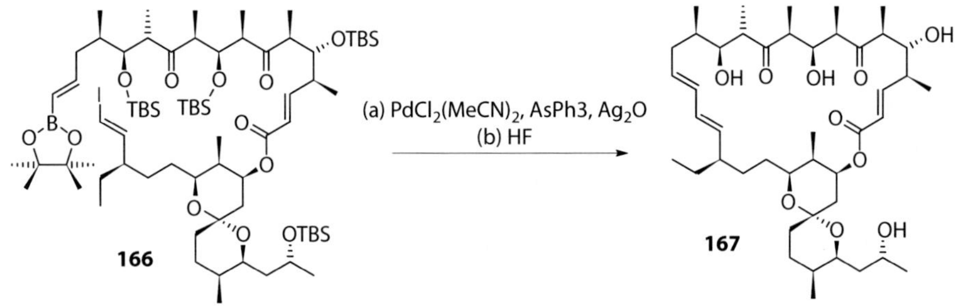

(a) $PdCl_2(MeCN)_2$, AsPh3, Ag_2O
(b) HF

166 → 167

SCHEME 28.47 Synthesis of rutamycin.

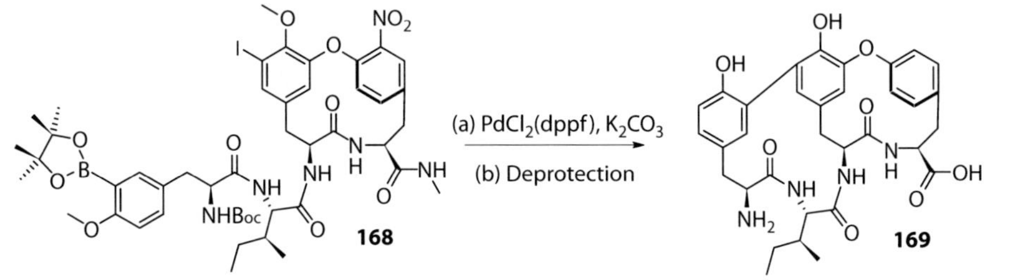

(a) $PdCl_2$(dppf), K_2CO_3
(b) Deprotection

168 → 169

SCHEME 28.48 Synthesis of RP66453.

28.5.5 Cross-Coupling Using Organotrifluoroborates

Organotrifluoroborates are a robust class of organoboranes because of their stability as compared with the corresponding boronic acids.[80] Several aryltrifluoroborates (**170a–e**) have been prepared and utilized for Suzuki–Miyaura cross-coupling reaction with a variety of electrophiles such as aryl halides,[81] benzyl halides,[82] heteroaryl halides,[83] aryl triflate,[84] aryl diazonium salts,[85] aryl tosylates,[86] and mesylates.[87] In addition, several alkyl[88] and cycloalkyl (e.g., cyclopropyl/cyclobutyl, **170g**)[89] trifluoroborates have also been introduced for the efficient cross-coupling with aryl halides. Molander and Petrillo have contributed extensively to this field and they have been able to introduce further functional groups into the alkyltrifluoroborates and carried out cross-coupling with all these borates as well. Some of the representative examples in this regard include trifluoroboratohomoenolates **170f**,[90] β- and α-aminoalkylfluoroborates **170h–i**,[91] and (α-alkoxyalkyl)-fluoroborates **170j**[92] (Scheme 28.49).

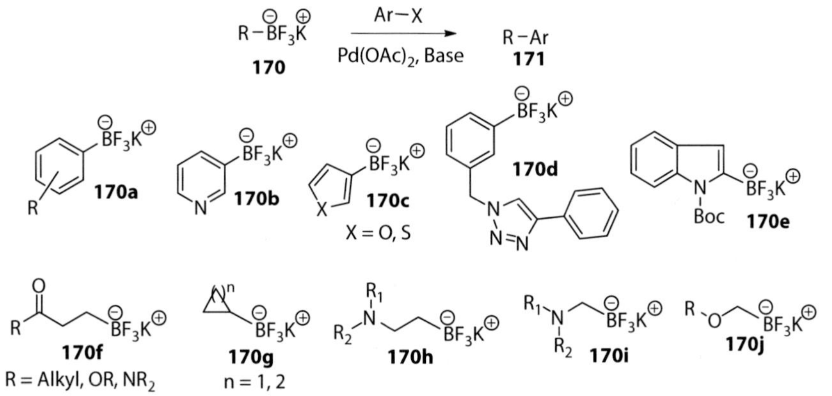

$R{-}BF_3^{\ominus}K^{\oplus}$ (170) → [Ar–X, $Pd(OAc)_2$, Base] → R–Ar (171)

170a; 170b; 170c (X = O, S); 170d; 170e; 170f (R = Alkyl, OR, NR_2); 170g (n = 1, 2); 170h; 170i; 170j

SCHEME 28.49 Organotrifluoroborates for Suzuki–Miyaura cross-coupling.

28.6 APPLICATIONS OF CROSS-COUPLING REACTION FOR ORGANIC SYNTHESIS

Over the past three decades, there has been an extensive development of cross-coupling reaction under Suzuki–Miyaura conditions and thousands of papers have been published in this area. This reaction is one of the most heavily used protocols in organic chemistry with a plethora of applications in several different fields. We have limited ourselves to the following few examples that demonstrate the wide applicability of this protocol in organic synthesis. The examples shown here are representative and several others could not be included due to the space limitations. This section has been categorized based on the type of coupling agents used. The reaction conditions including the catalyst and the base along with the two coupling partners (organoboron and organohalide) have been clearly identified for all the examples in the subsections.

28.6.1 Applications of Cross-Coupling between Vinyl Halides and Vinyl Boranes

Schemes 28.50 through 28.75 detail the application of Suzuki–Miyaura cross-coupling reaction between vinyl halides and vinylboron derivatives.

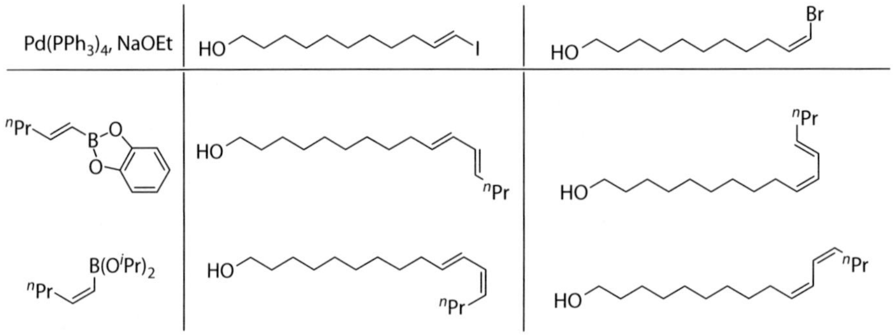

$Pd(PPh_3)_4$, NaOEt

HO I

Br HO

^nPr B O O

HO ^nPr

^nPr HO

$B(O^iPr)_2$ ^nPr

HO ^nPr

^nPr HO

SCHEME 28.50 Preparation of diastereomers of bombykol.[93]

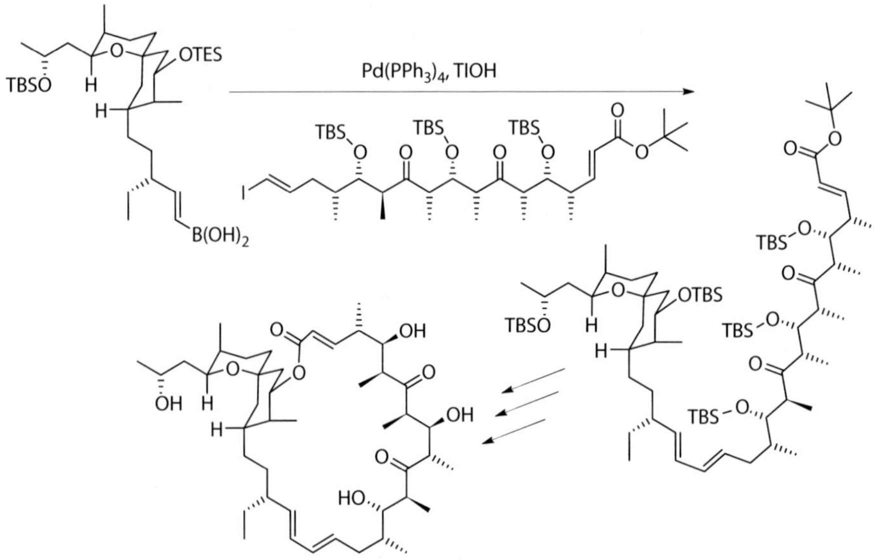

OTES TBSO H H B(OH)$_2$

$Pd(PPh_3)_4$, TlOH

TBS TBS TBS O O O O O O I O O

O O TBS O TBS O O TBS O O TBS O OTBS TBSO H H

O O OH O OH O HO OH H H

SCHEME 28.51 Preparation of rutamycin.[94]

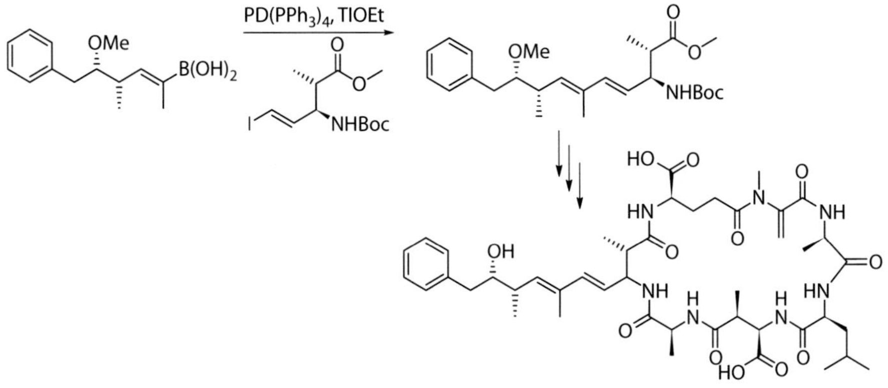

SCHEME 28.52 Preparation of microcystin.[95]

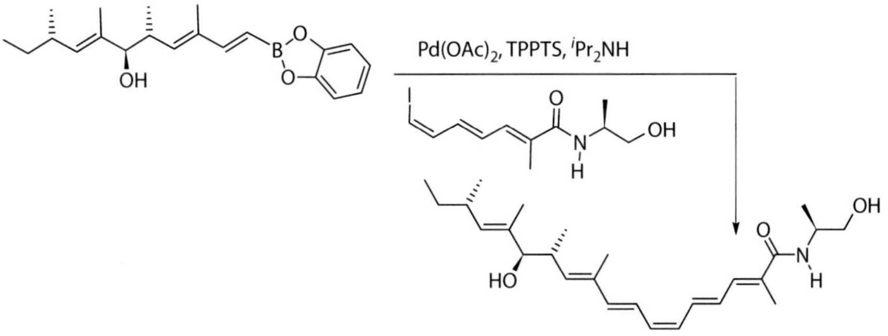

SCHEME 28.53 Preparation of myxalamide A.[96]

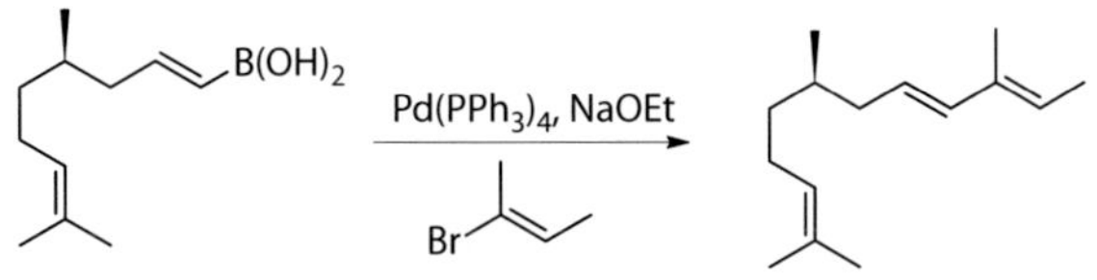

SCHEME 28.54 Preparation of caparratriene.[97]

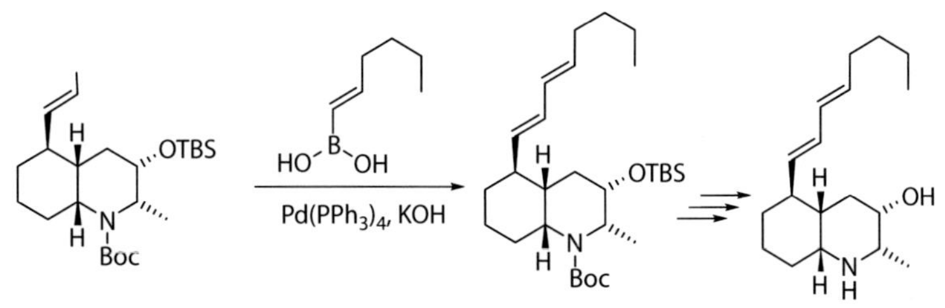

SCHEME 28.55 Preparation of lepadin B.[98]

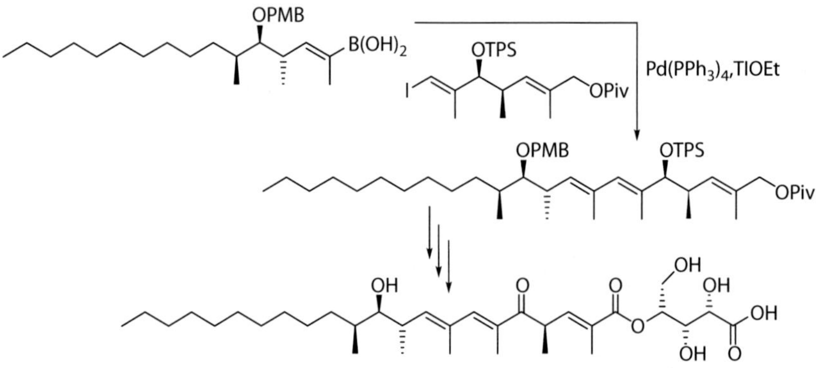

SCHEME 28.56 Preparation of khafrefungin.[99]

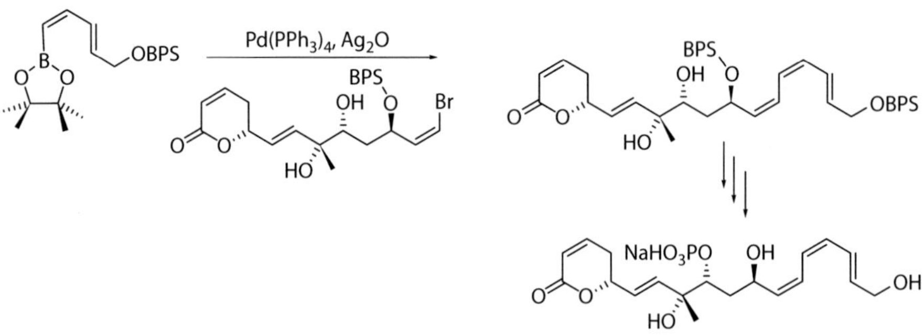

SCHEME 28.57 Preparation of fostriecin.[100]

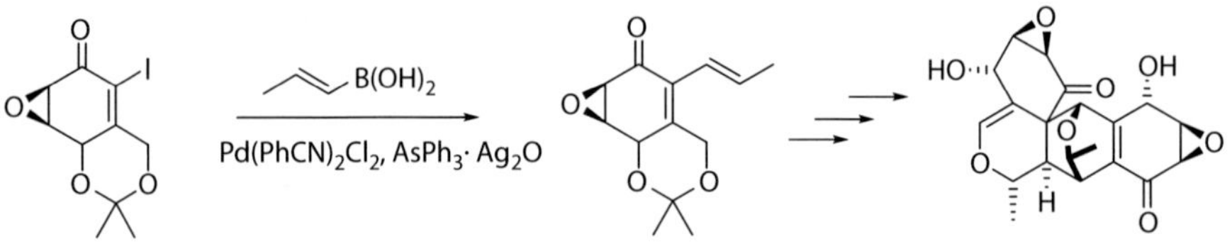

SCHEME 28.58 Preparation of epoxyquinols.[101]

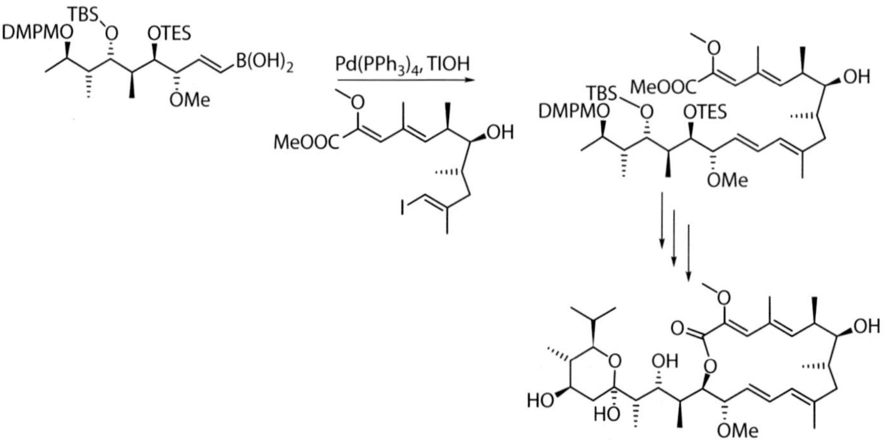

SCHEME 28.59 Preparation of bafilomycin A.[102]

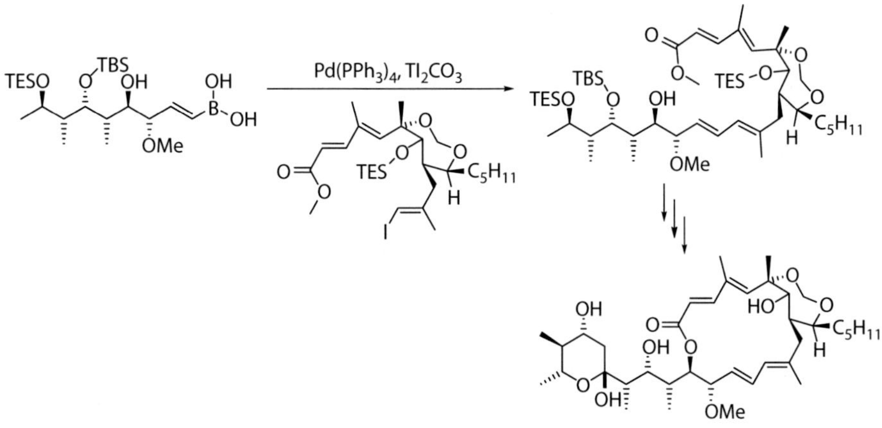

SCHEME 28.60 Preparation of formamicinone.[103]

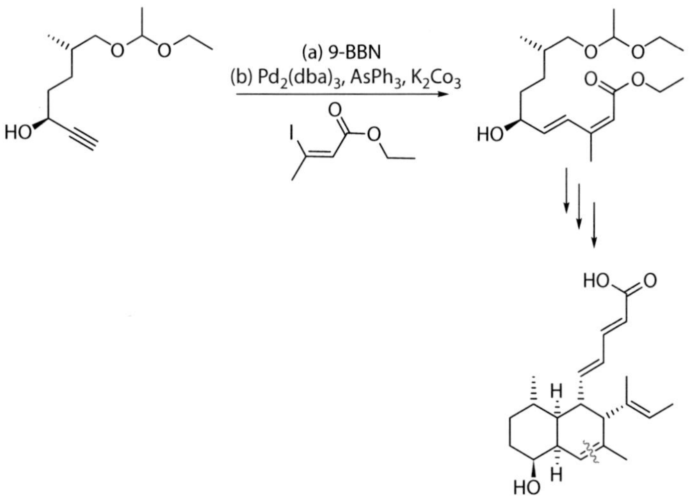

SCHEME 28.61 Preparation of phomopsidin.[104]

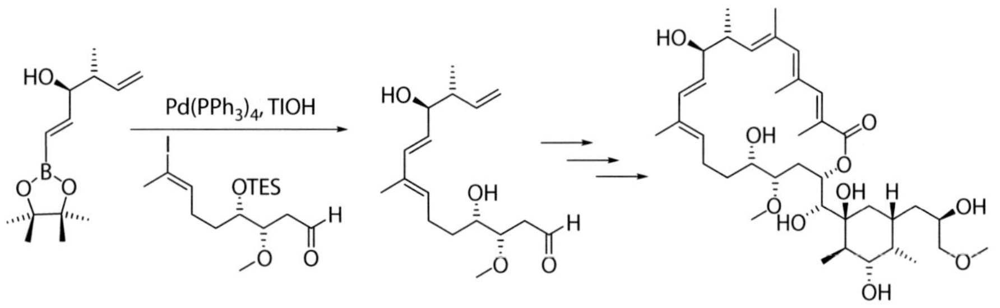

SCHEME 28.62 Preparation of apoptolidine.[105]

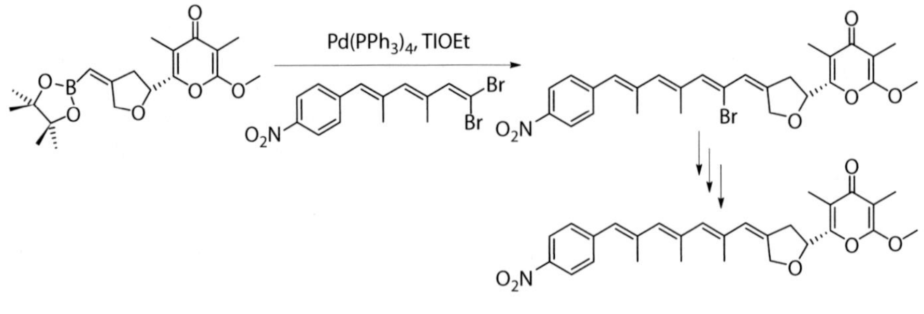

Pd(PPh$_3$)$_4$, TlOEt

SCHEME 28.63 Preparation of spectinabilin.[106]

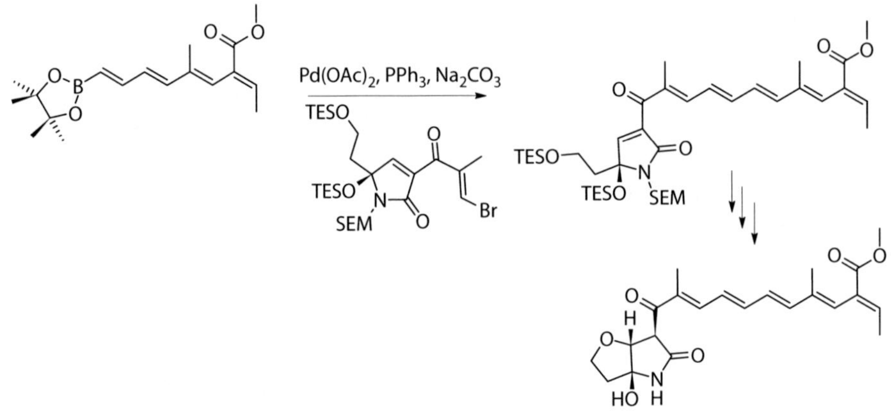

Pd(OAc)$_2$, PPh$_3$, Na$_2$CO$_3$

SCHEME 28.64 Preparation of lucilactaene.[107]

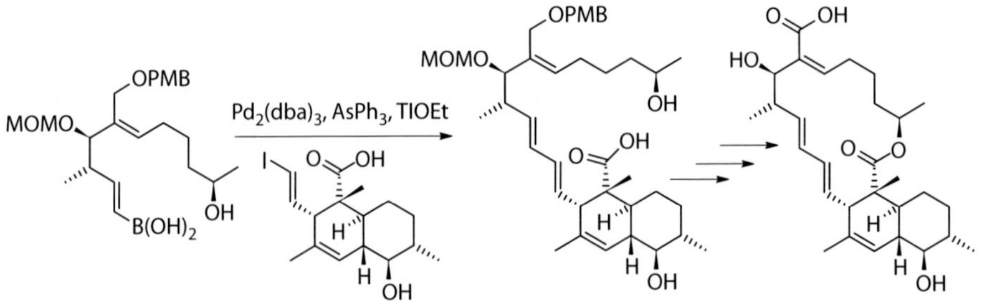

Pd$_2$(dba)$_3$, AsPh$_3$, TlOEt

SCHEME 28.65 Preparation of tubelactomycin.[108]

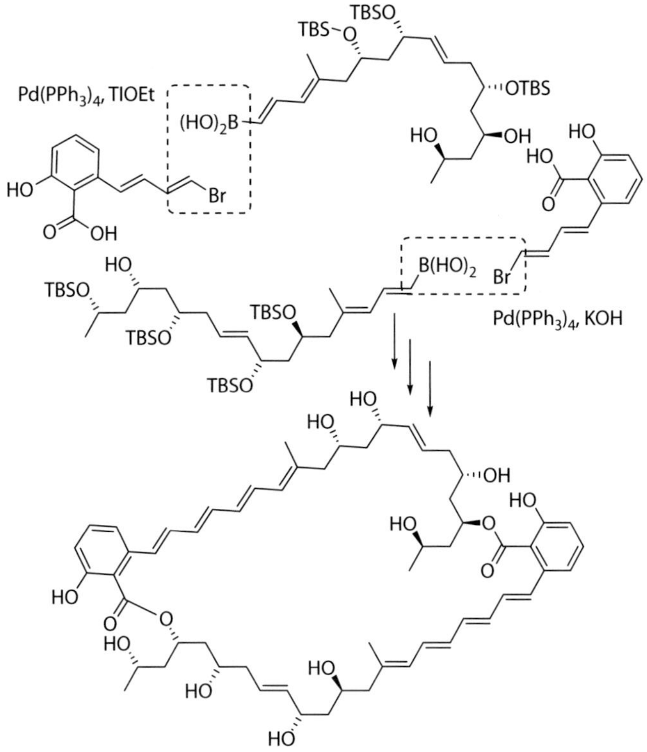

SCHEME 28.66 Preparation of marinomycin A.[109]

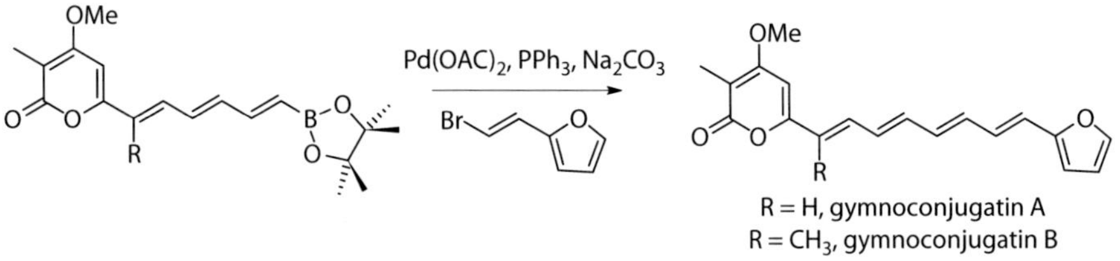

SCHEME 28.67 Preparation of gymnoconjugatins.[110]

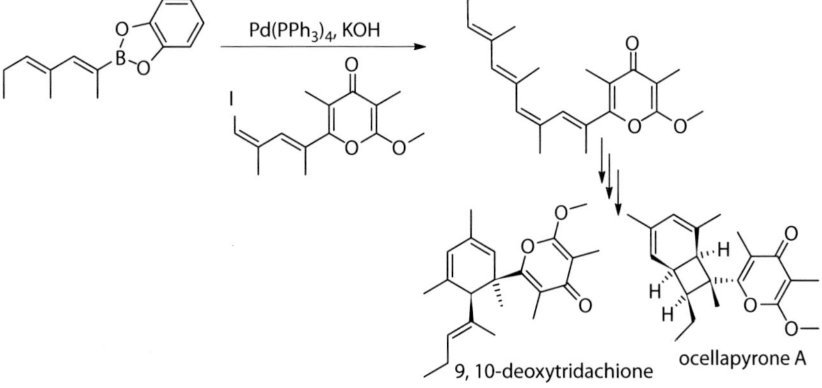

SCHEME 28.68 Preparation of (±)-9,10-deoxytridachione and (±)-ocellapyrone A.[111]

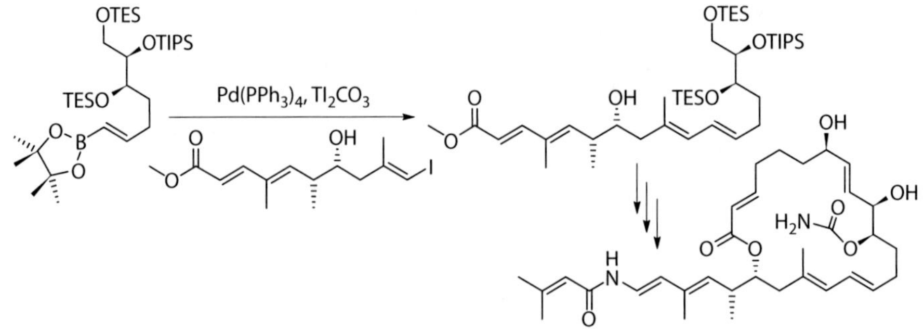

SCHEME 28.69 Preparation of palmerolide.[112]

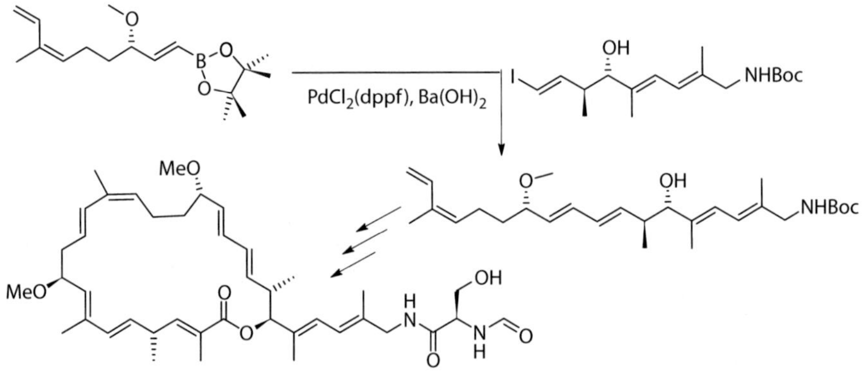

SCHEME 28.70 Preparation of iejimalide.[113]

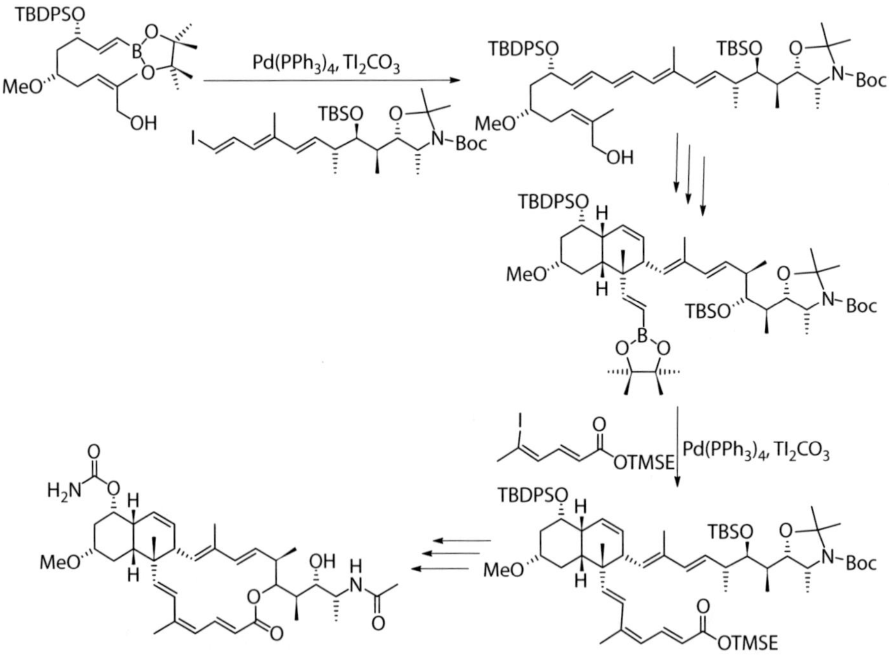

SCHEME 28.71 Preparation of superstolide.[114]

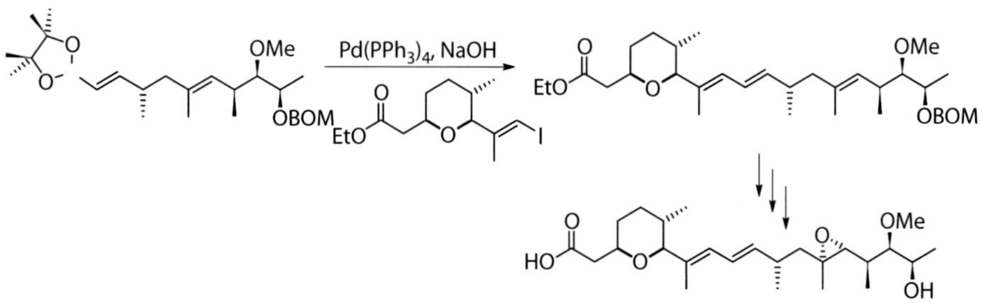

SCHEME 28.72 Preparation of GEX1A.[115]

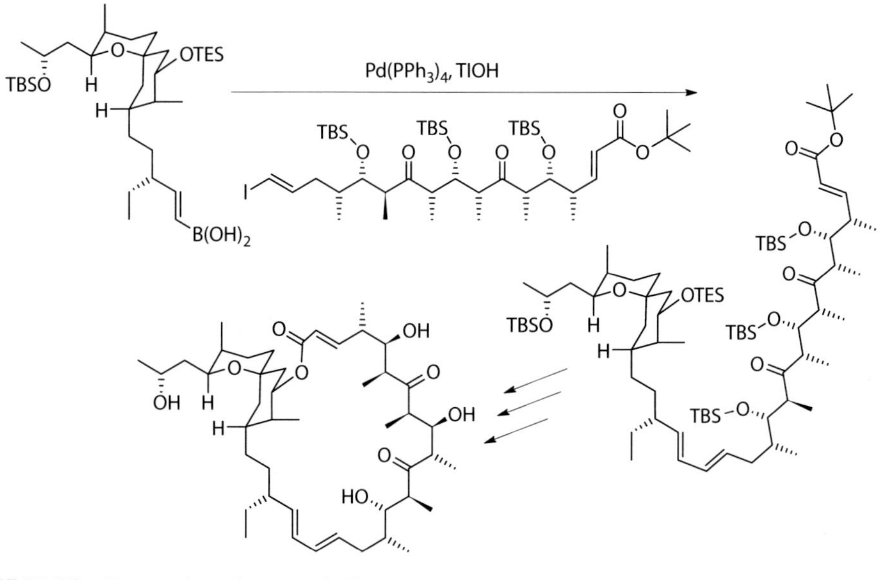

SCHEME 28.73 Preparation of rutamycin B.

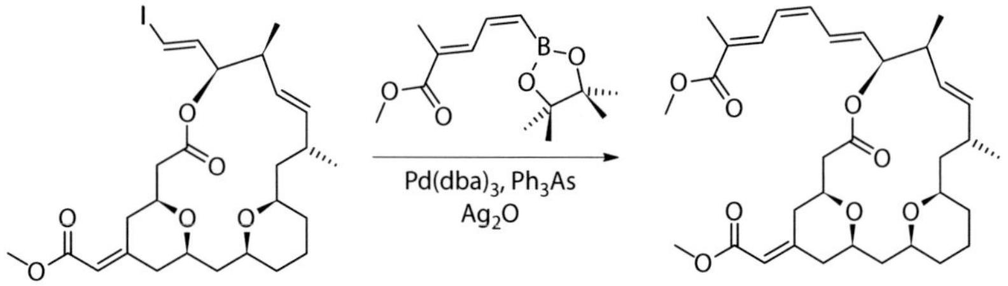

SCHEME 28.74 Preparation of exiguolide.[116]

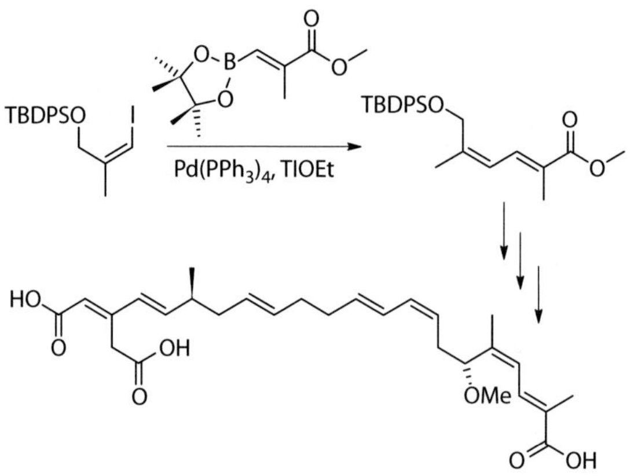

SCHEME 28.75 Preparation of bongkrekic acid.[117]

28.6.2 Applications of Cross-Coupling between Vinyl Halides and Aryl Boranes

Schemes 28.76 through 28.84 detail the cross-coupling between vinyl halides and aryl boron and their application in the synthesis of complex natural products.

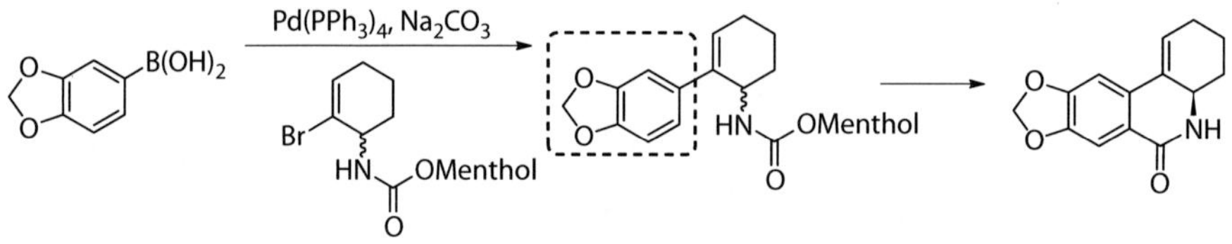

SCHEME 28.76 Preparation of lycoricidine and narciclasine.[118]

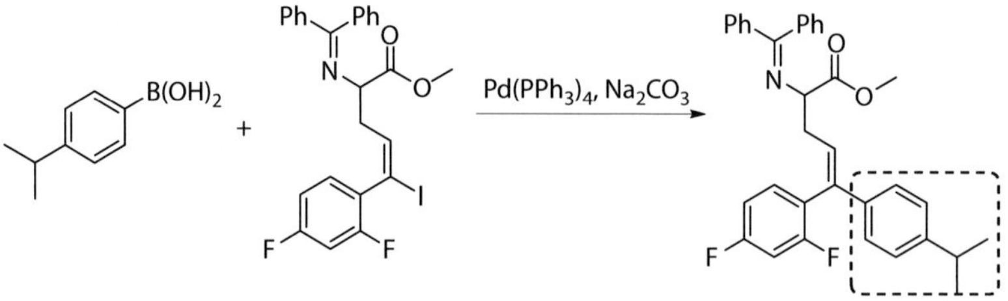

SCHEME 28.77 Preparation of unnatural amino acids.[119]

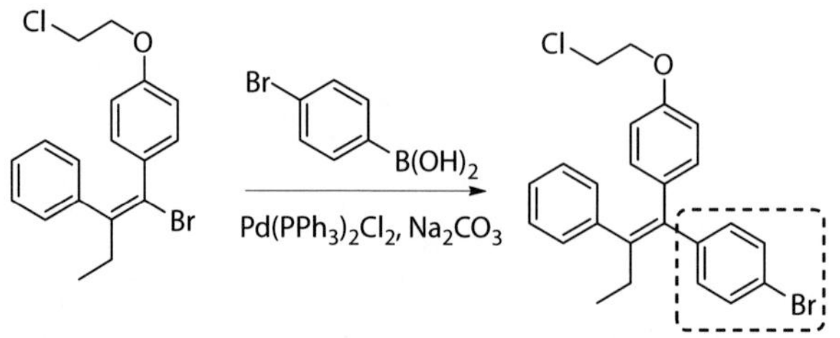

SCHEME 28.78 Preparation of tamoxifen analogs.[120]

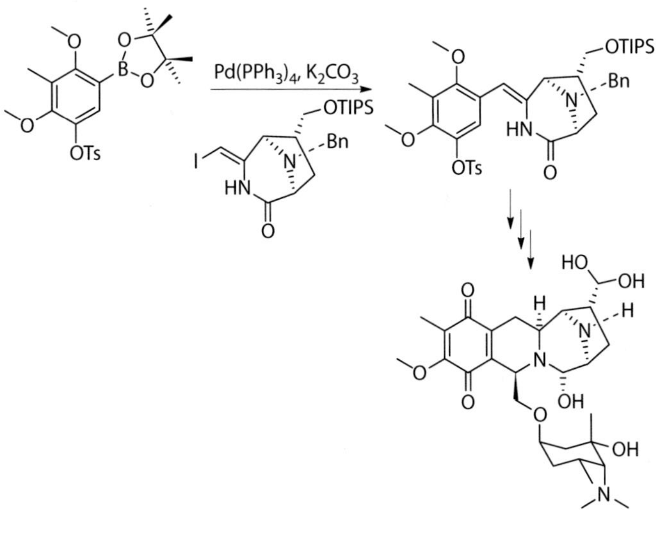
Pd(PPh3)4, K2CO3
OTIPS
Bn
HN
O
I
OTs
HO
OH
H
N
OH
O
N

SCHEME 28.79 Preparation of lemonomycin.[121]

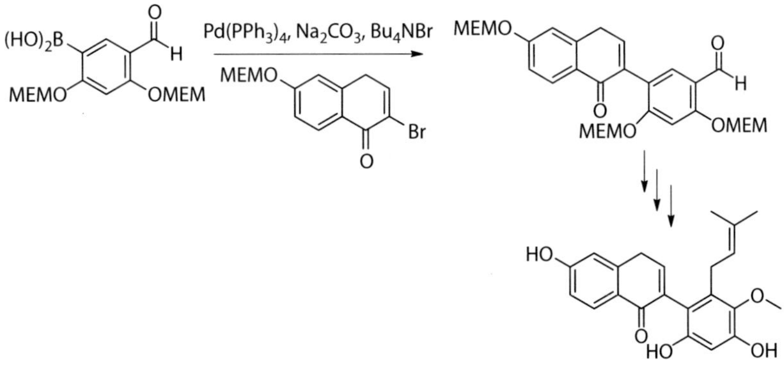
(HO)2B
H
MEMO
OMEM
Pd(PPh3)4, Na2CO3, Bu4NBr
Br
O
HO
OH

SCHEME 28.80 Preparation of kwakhurin.[122]

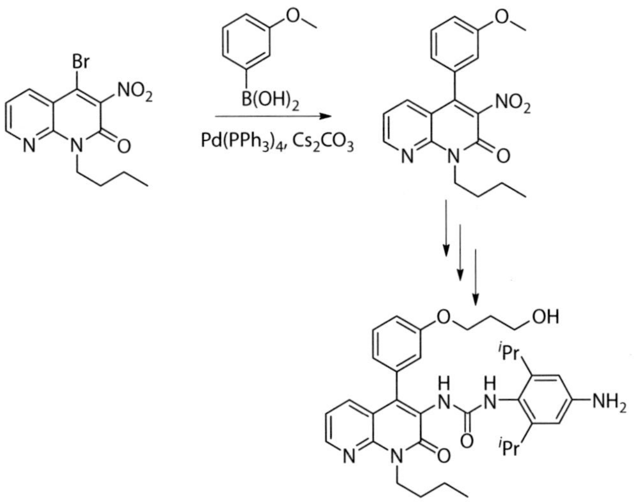
Br
NO2
N
O
B(OH)2
Pd(PPh3)4, Cs2CO3
OH
iPr
H
NH2

SCHEME 28.81 Preparation of SMP-797.[123]

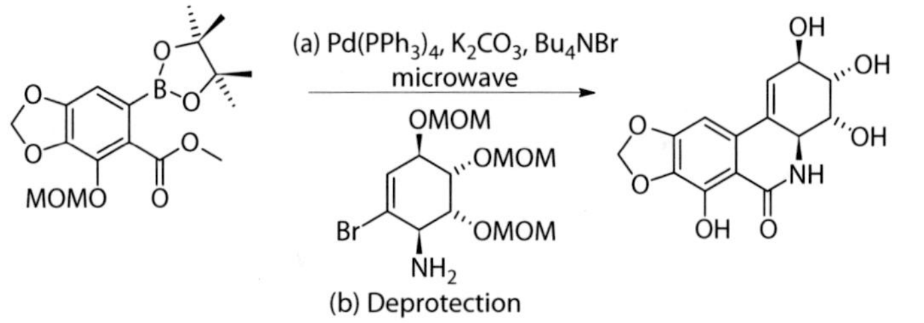

SCHEME 28.82 Preparation of narciclasine.[124]

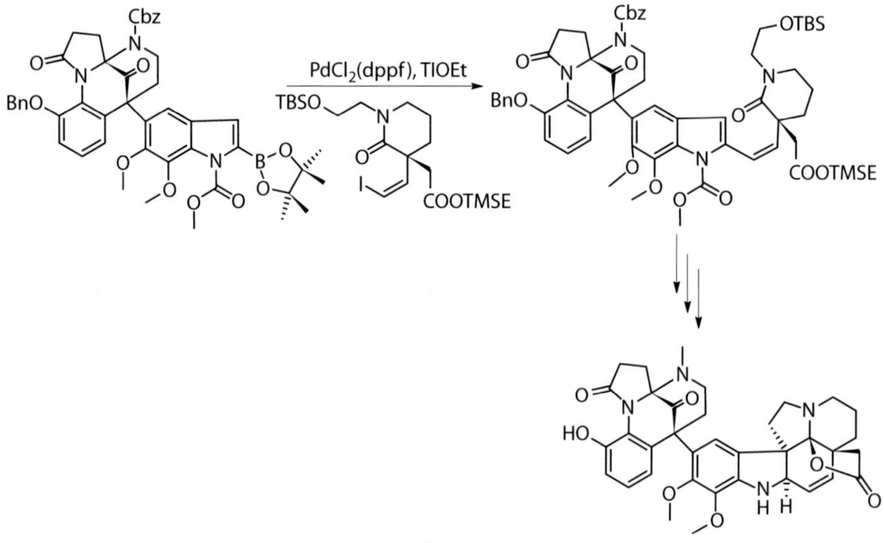

SCHEME 28.83 Preparation of halophytine and aspidophytine.[125]

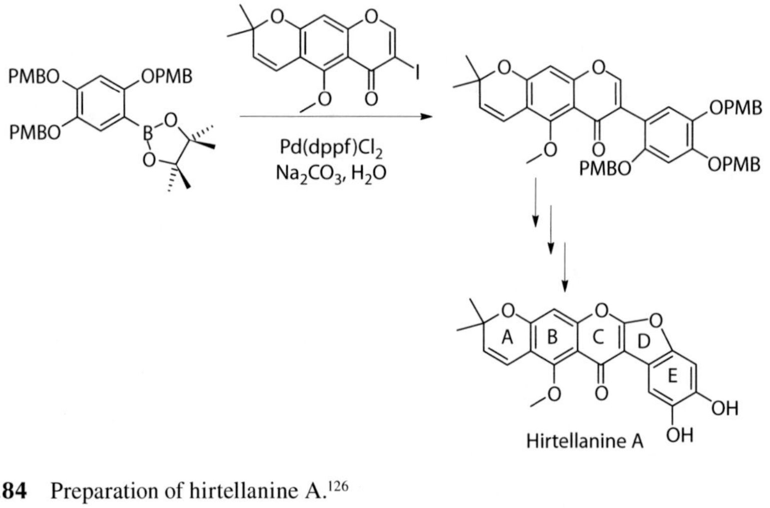

SCHEME 28.84 Preparation of hirtellanine A.[126]

28.6.3 Applications of Cross-Coupling between Vinyl Halides and Alkyl Boranes

Schemes 28.85 through 28.109 include some of the representative examples of cross-coupling between vinyl halides and alkyl boron and their applications in organic synthesis.

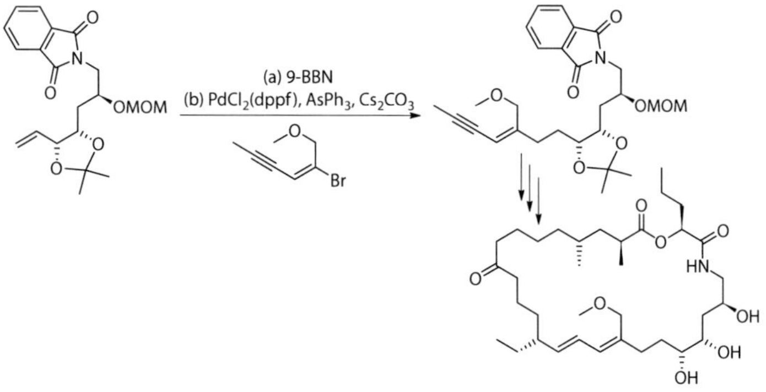

(a) 9-BBN
(b) $PdCl_2(dppf)$, $AsPh_3$, Cs_2CO_3

SCHEME 28.85 Preparation of myxovirescin.[127]

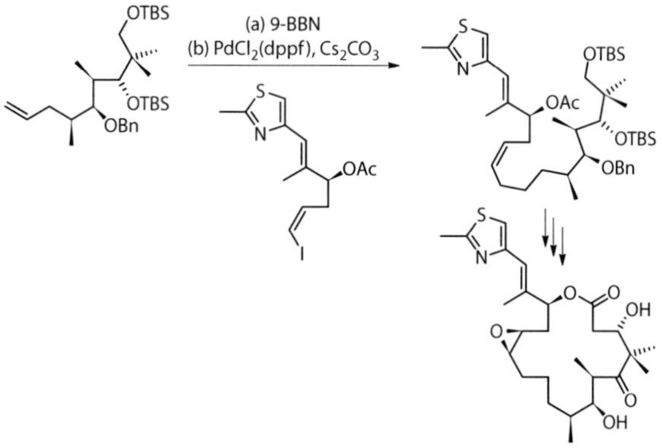

(a) 9-BBN
(b) $PdCl_2(dppf)$, Cs_2CO_3

SCHEME 28.86 Preparation of epothilone A.[128]

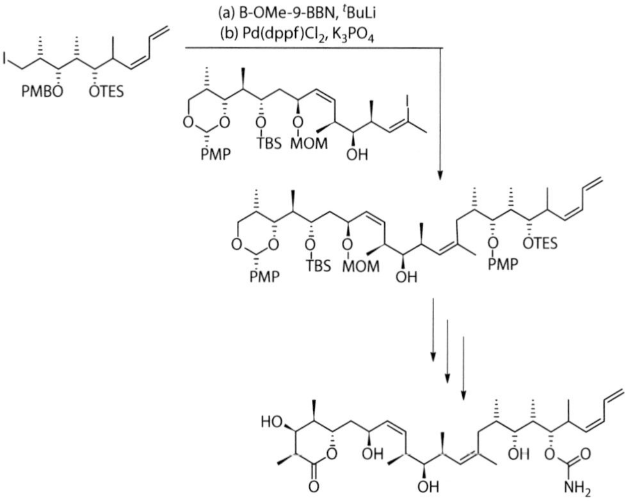

(a) B-OMe-9-BBN, ^tBuLi
(b) $Pd(dppf)Cl_2$, K_3PO_4

SCHEME 28.87 Preparation of discodermolide.[33a]

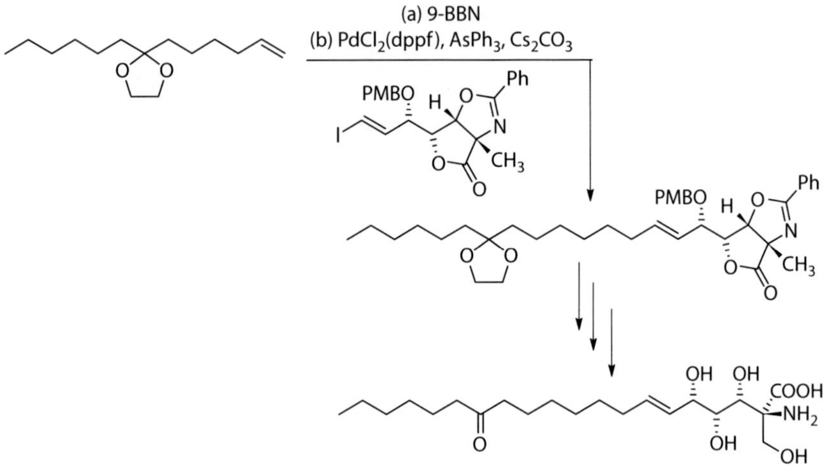

SCHEME 28.88 Preparation of sphingofungin.[129]

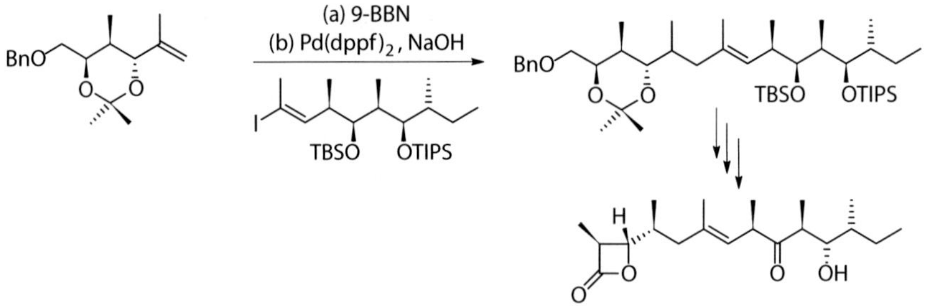

SCHEME 28.89 Preparation of ebelactone A.[130]

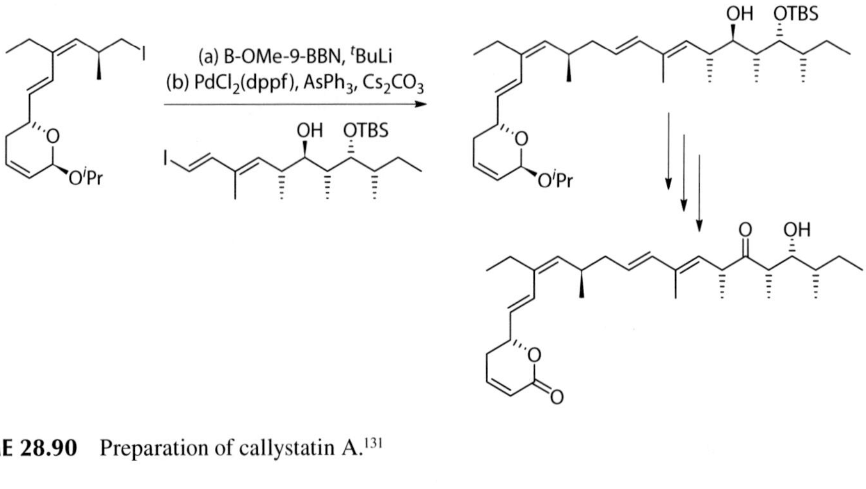

SCHEME 28.90 Preparation of callystatin A.[131]

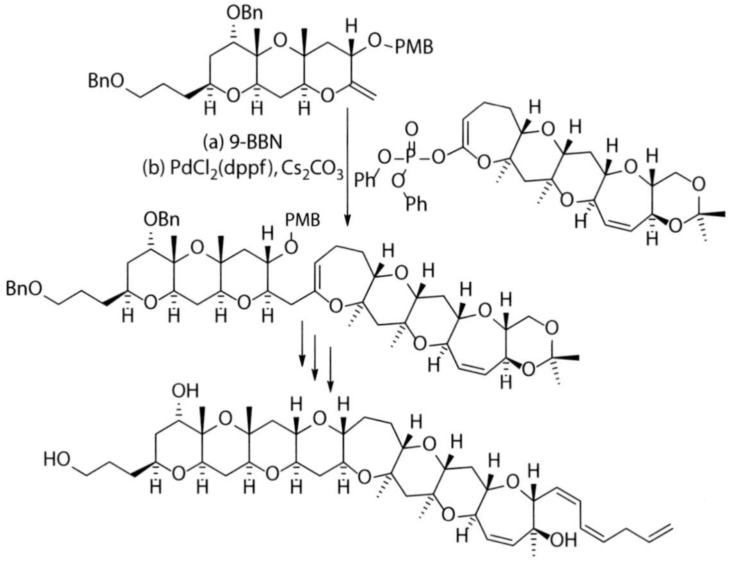

SCHEME 28.91 Preparation of gambierol.[132]

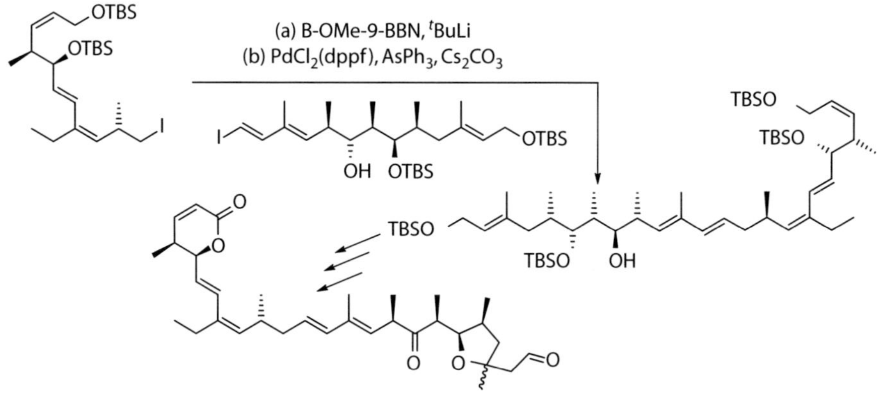

SCHEME 28.92 Preparation of leptofuranin D.[133]

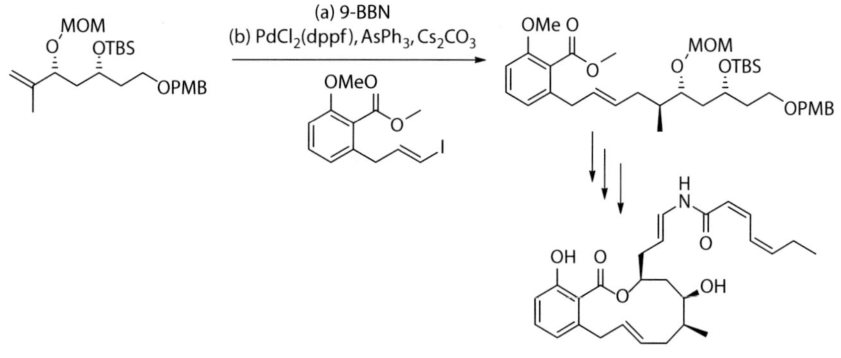

SCHEME 28.93 Formal total synthesis of salicylihalamides.[134]

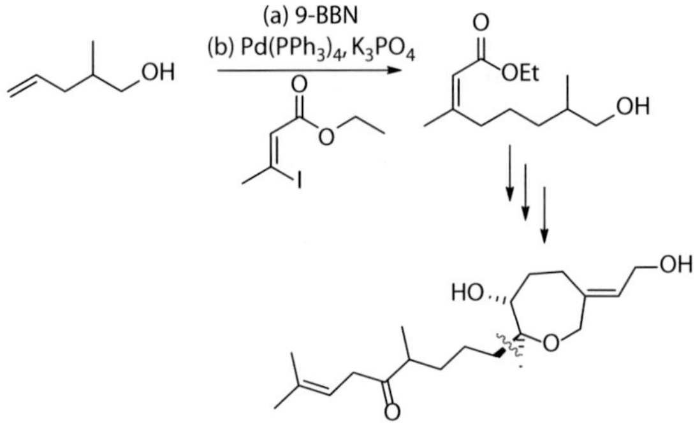

(a) 9-BBN
(b) $Pd(PPh_3)_4$, K_3PO_4

SCHEME 28.94 Preparation of zoapatanol.[135]

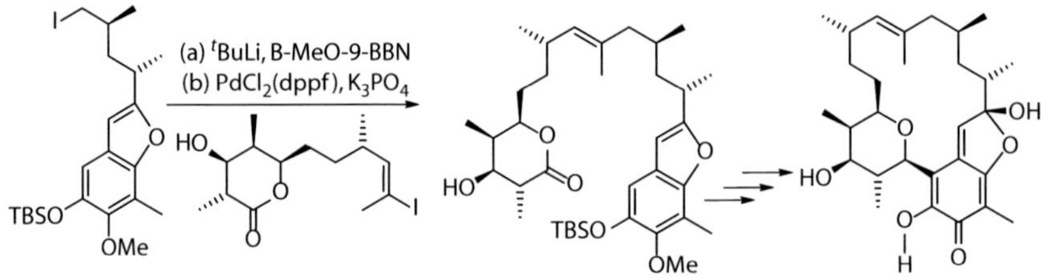

(a) ^tBuLi, B-MeO-9-BBN
(b) $PdCl_2(dppf)$, K_3PO_4

SCHEME 28.95 Preparation of kendomycin.[136]

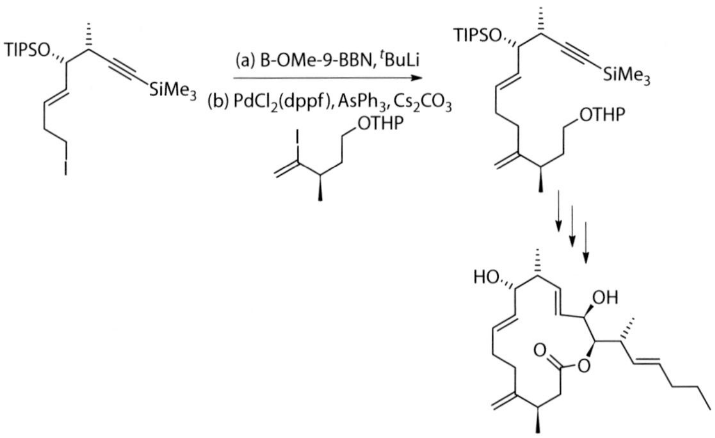

(a) B-OMe-9-BBN, ^tBuLi
(b) $PdCl_2(dppf)$, $AsPh_3$, Cs_2CO_3

SCHEME 28.96 Preparation of amphidinolide.[137]

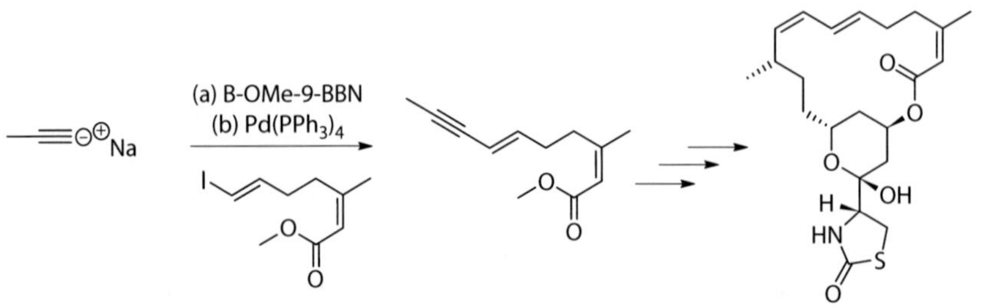

(a) B-OMe-9-BBN
(b) $Pd(PPh_3)_4$

SCHEME 28.97 Preparation of latrunculins.[138]

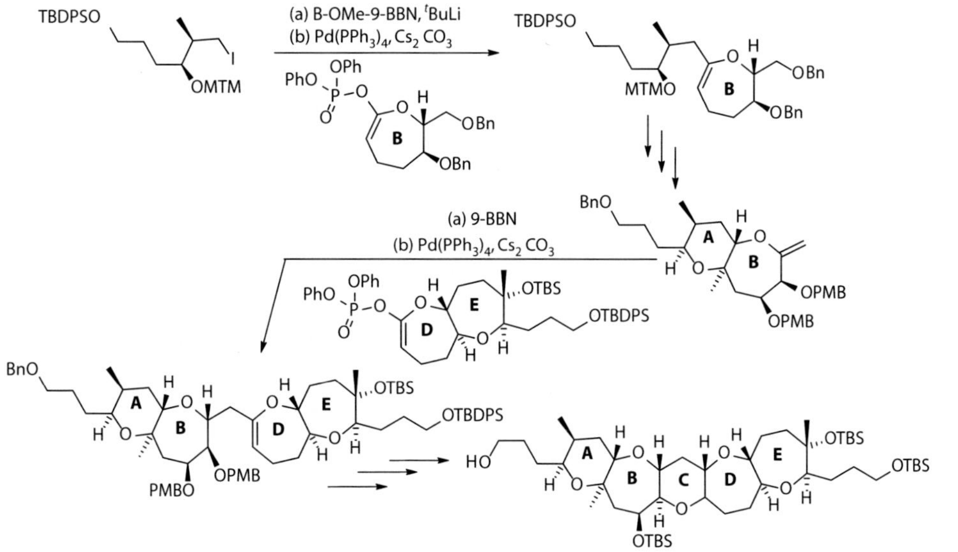

SCHEME 28.98 Preparation of brevenal.[139]

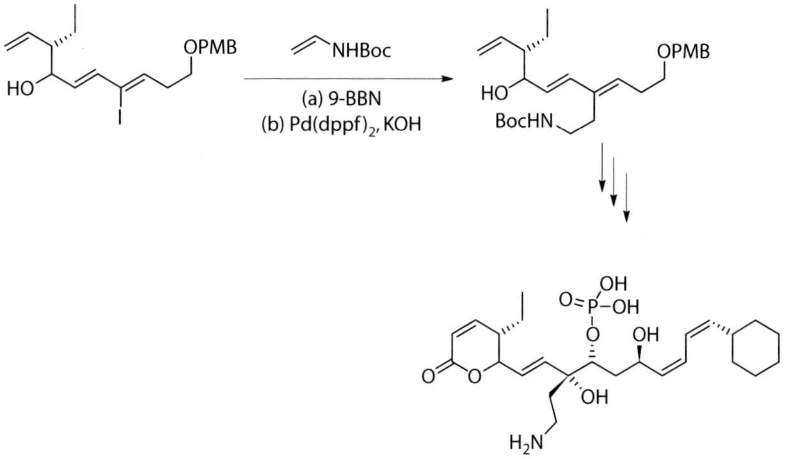

SCHEME 28.99 Preparation of phoslactomycin.[140]

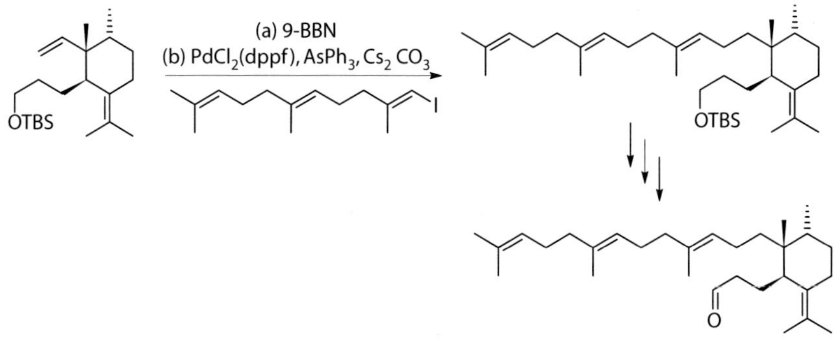

SCHEME 28.100 Preparation of marneral.[141]

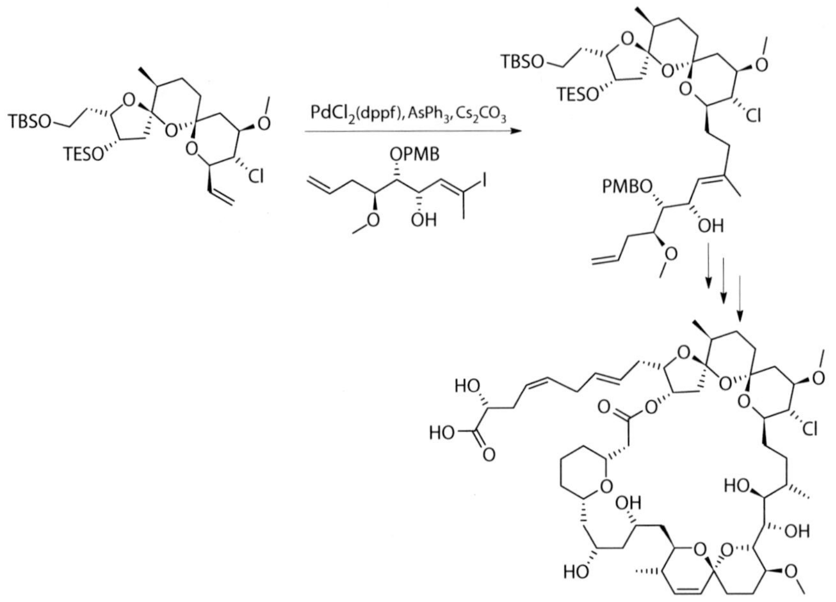

SCHEME 28.101 Preparation of spirastrellolide.[142]

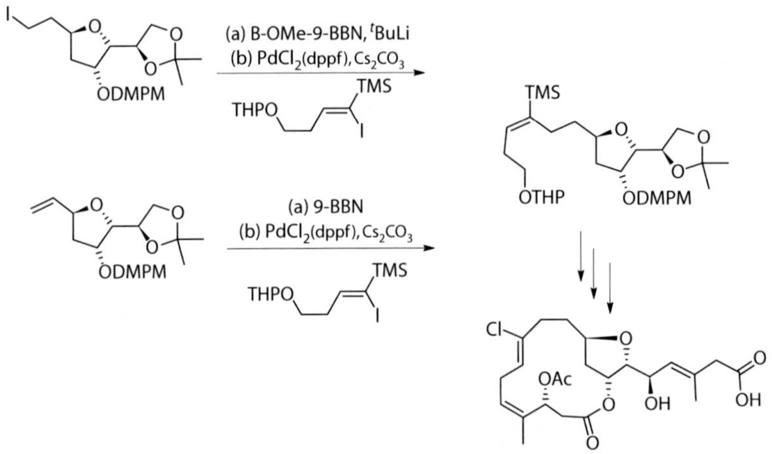

SCHEME 28.102 Preparation of haterumalide NA/Oocydin.[143]

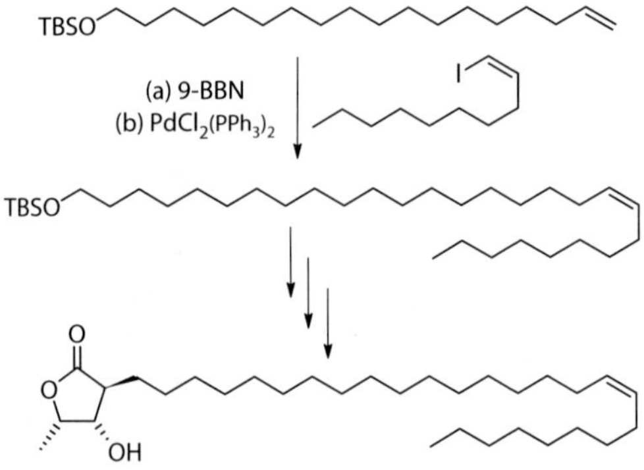

SCHEME 28.103 Preparation of megislactone.[144]

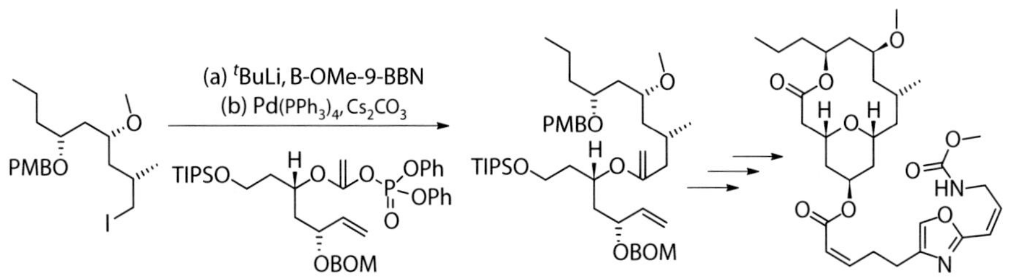

SCHEME 28.104 Preparation of neopeltolide.[145]

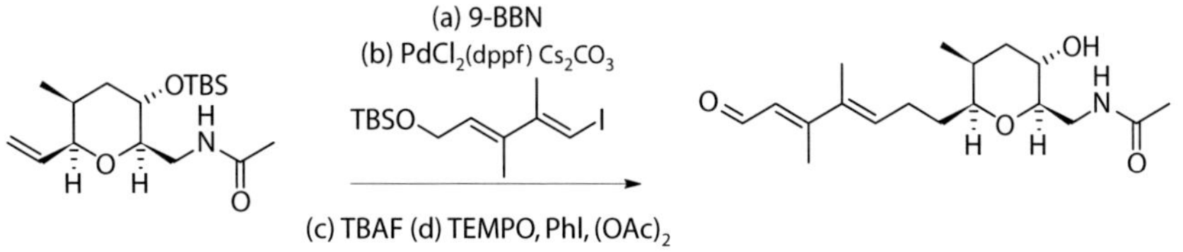

SCHEME 28.105 Preparation of brevisamide.[146]

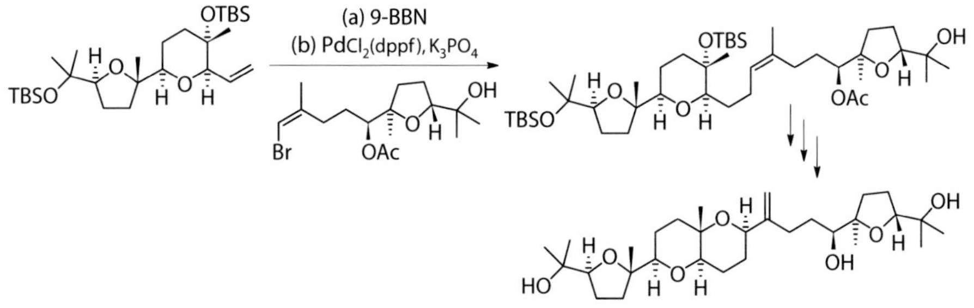

SCHEME 28.106 Preparation of pseudodehydrothyrsiferol.[147]

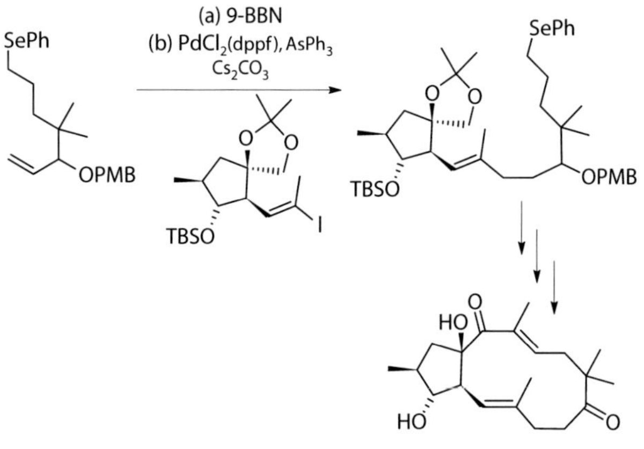

SCHEME 28.107 Preparation of jatrophane diterpenes (characiol).[148]

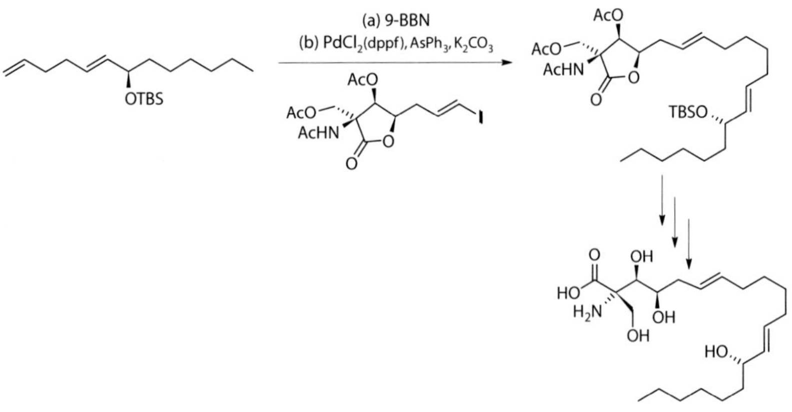
(a) 9-BBN
(b) PdCl2(dppf), AsPh3, K2CO3
AcO
AcO
AcHN
O
O
I
OTBS
TBSO
HO
H2N
OH
OH
OH
HO

SCHEME 28.108 Preparation of mycestericin.[149]

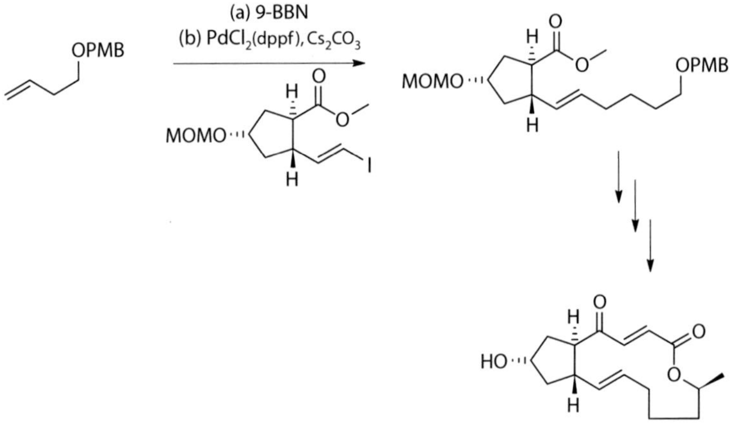
(a) 9-BBN
(b) PdCl2(dppf), Cs2CO3
OPMB
MOMO
H
O
I
HO

SCHEME 28.109 Preparation of brefeldin.[150]

28.6.4 Applications of Cross-Coupling between Aryl Halides and Vinyl Boranes

Schemes 28.110 through 28.112 detail the cross-coupling of vinyl boron and aryl halides and their application in the synthesis of complex natural products.

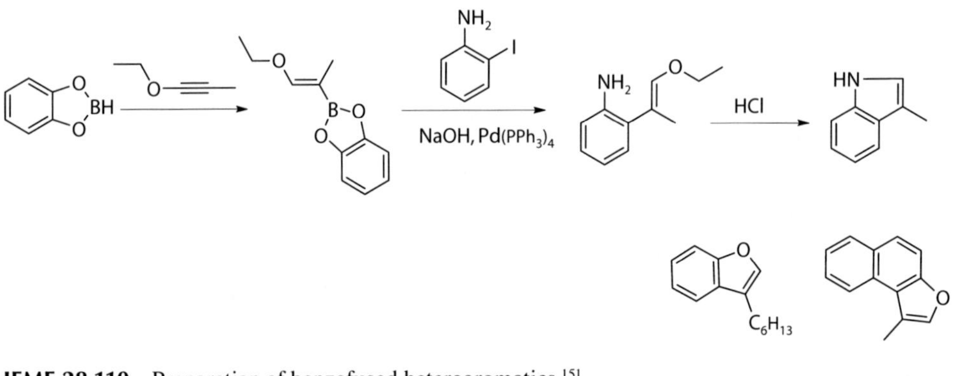
NH2
I
BH
B–O
NaOH, Pd(PPh3)4
HCl
HN
C6H13
O

SCHEME 28.110 Preparation of benzofused heteroaromatics.[151]

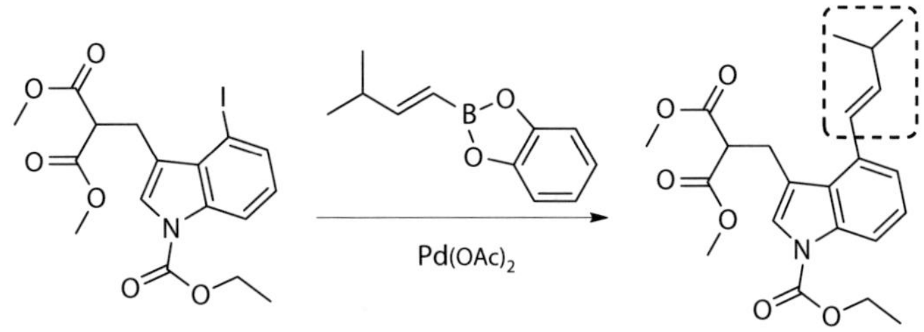

SCHEME 28.111 Preparation of substituted indoles.[152]

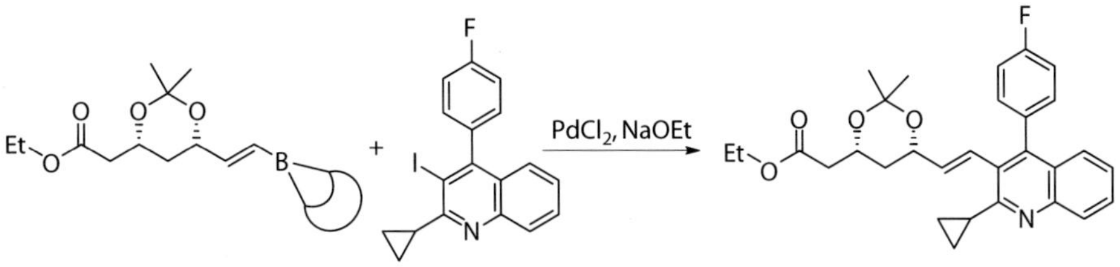

SCHEME 28.112 Preparation of NK-104.[153]

28.6.5 Applications of Cross-Coupling between Aryl Halides and Aryl Boranes

Schemes 28.113 through 28.143 detail the coupling between aryl halides and aryl boron derivatives and their applications in organic synthesis.

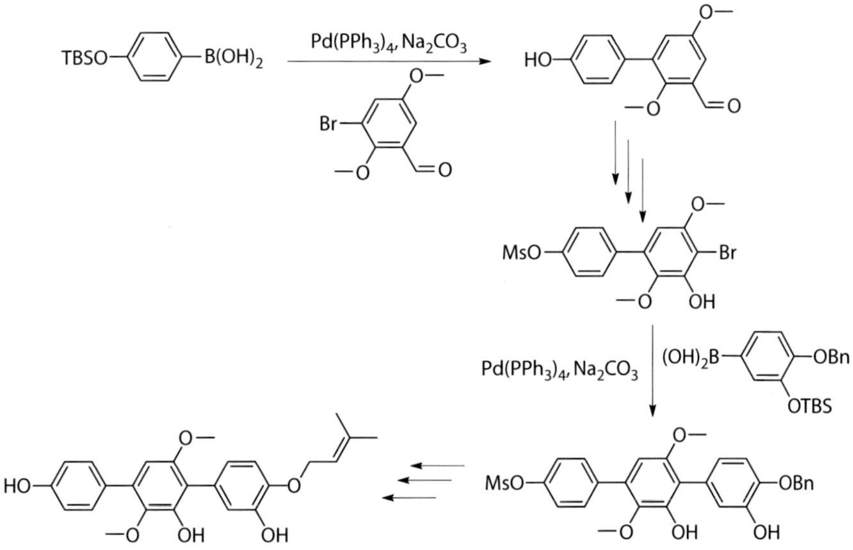

SCHEME 28.113 Synthesis of terprenin.[154]

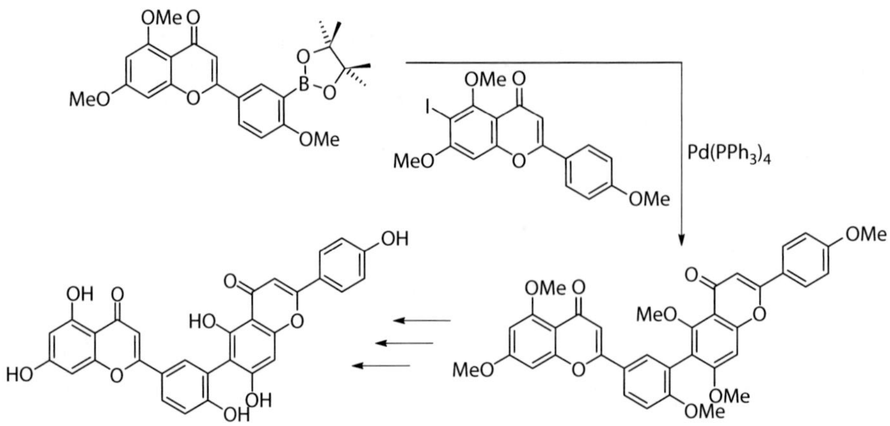

SCHEME 28.114 Synthesis of robustaflavone.[155]

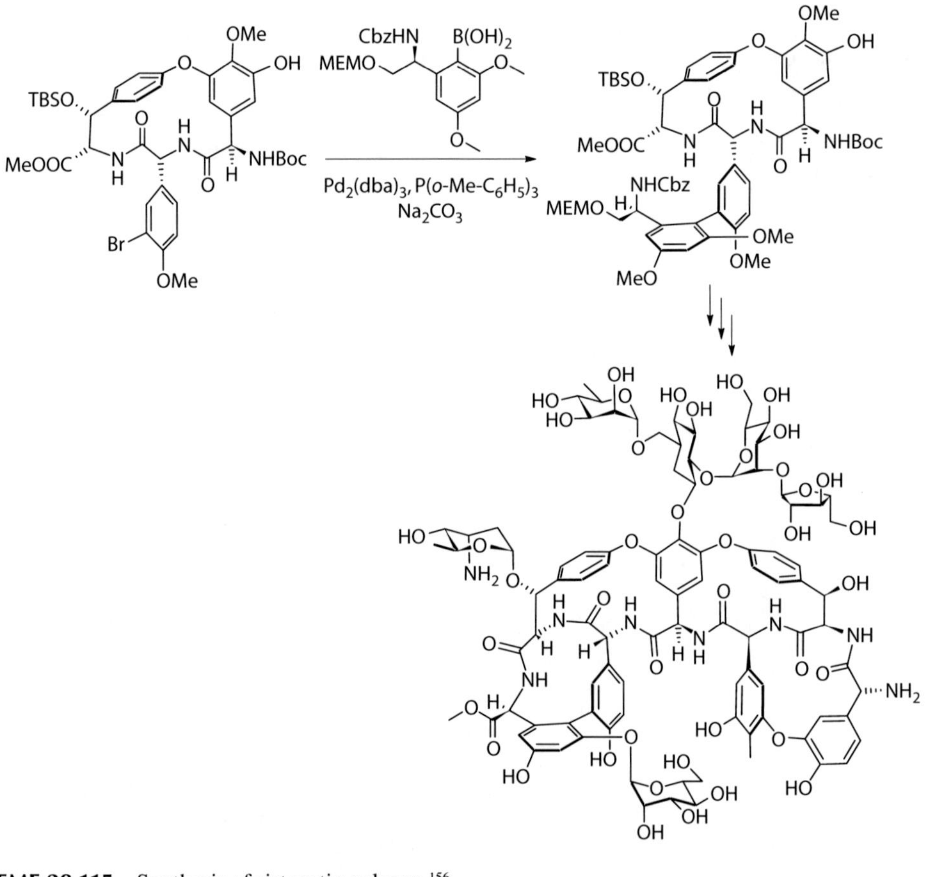

SCHEME 28.115 Synthesis of ristocetin aglycon.[156]

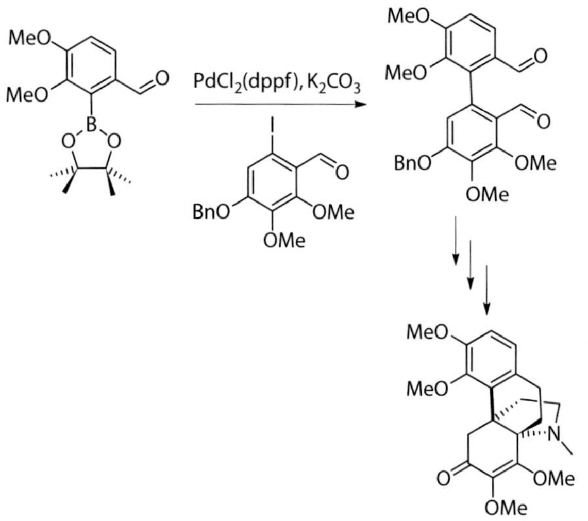

SCHEME 28.116 Synthesis of hasubanonine.[157]

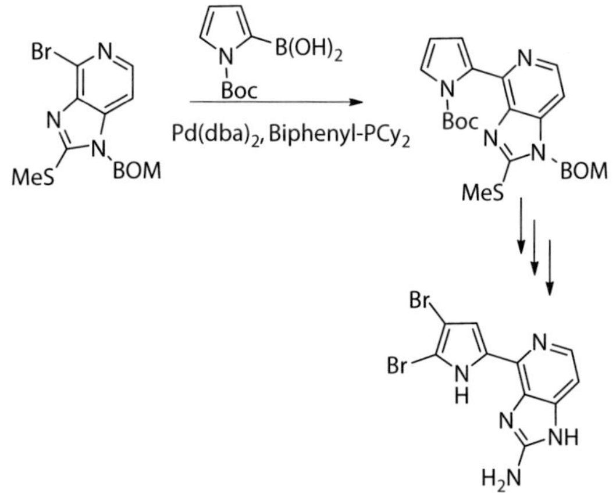

SCHEME 28.117 Synthesis of ageladine A.[158]

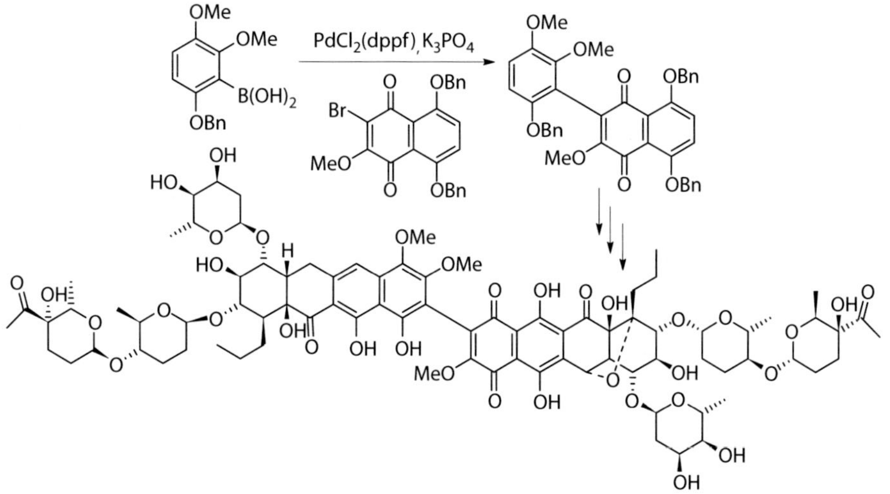

SCHEME 28.118 Synthesis of angelmicin.[159]

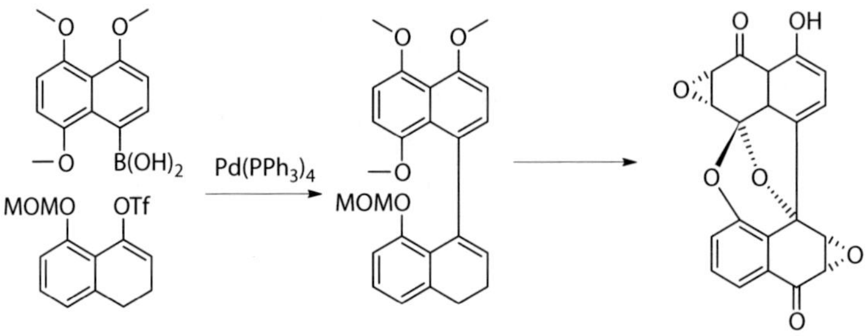

SCHEME 28.119 Synthesis of spiroxin C.[160]

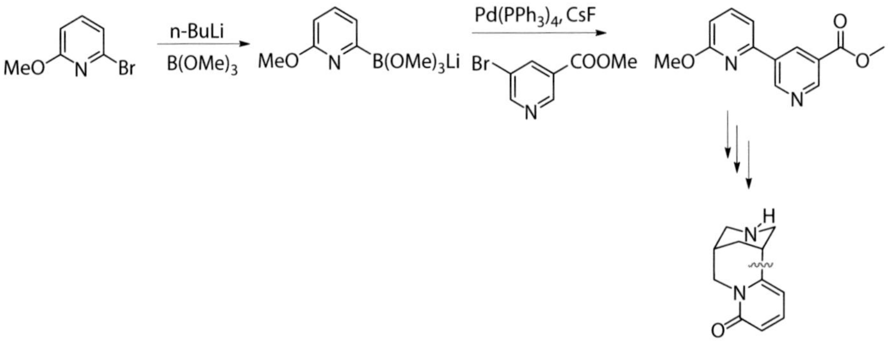

SCHEME 28.120 Synthesis of cytisine.[161]

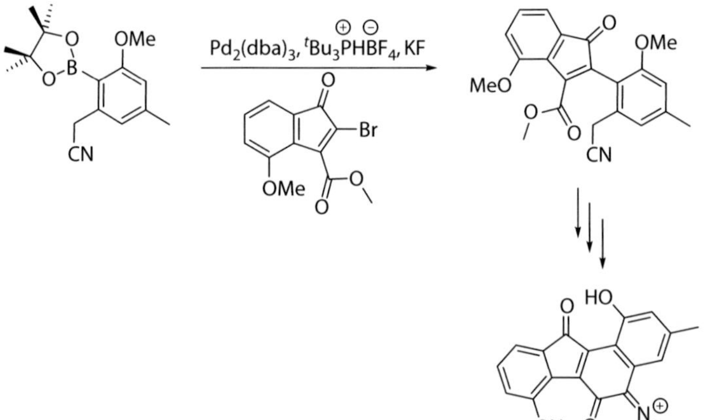

SCHEME 28.121 Synthesis of isoprekinamycin.[162]

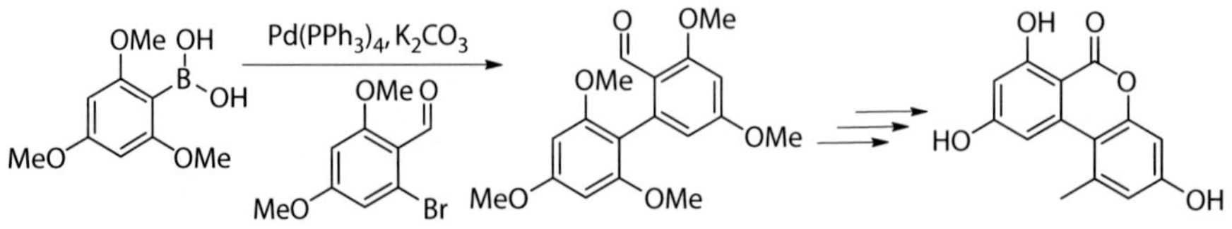

SCHEME 28.122 Synthesis of alternariol.[163]

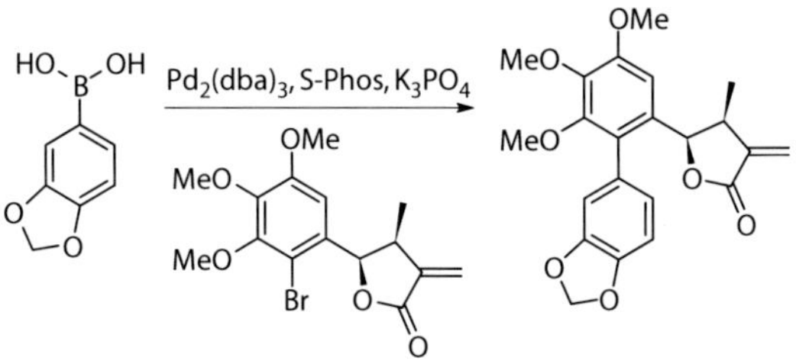

SCHEME 28.123 Synthesis of eupomatilones.[164]

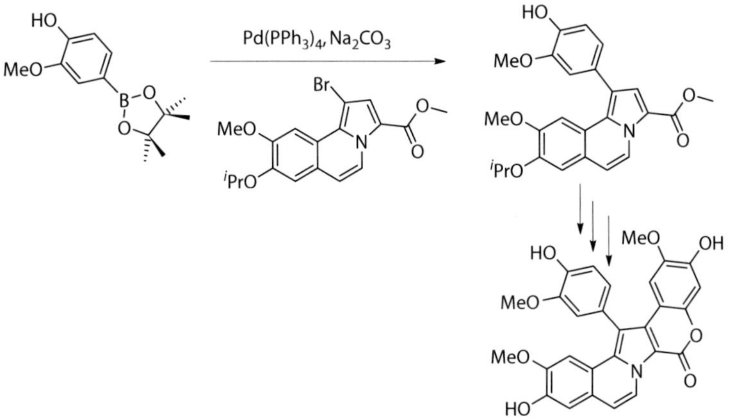

SCHEME 28.124 Synthesis of lamellarin D.[165]

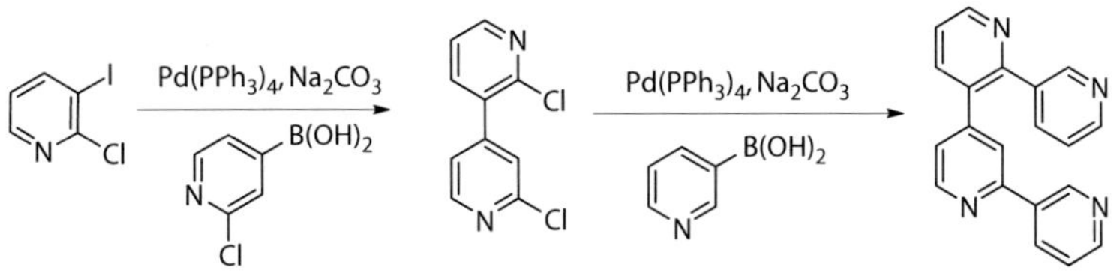

SCHEME 28.125 Synthesis of nemertelline.[166]

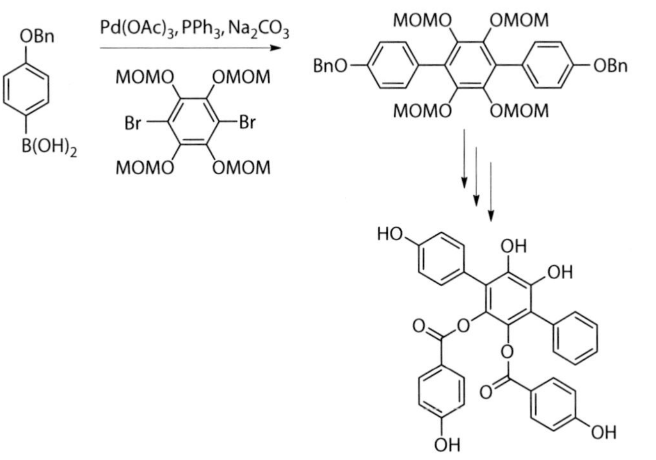

SCHEME 28.126 Synthesis of thelephantin G.[167]

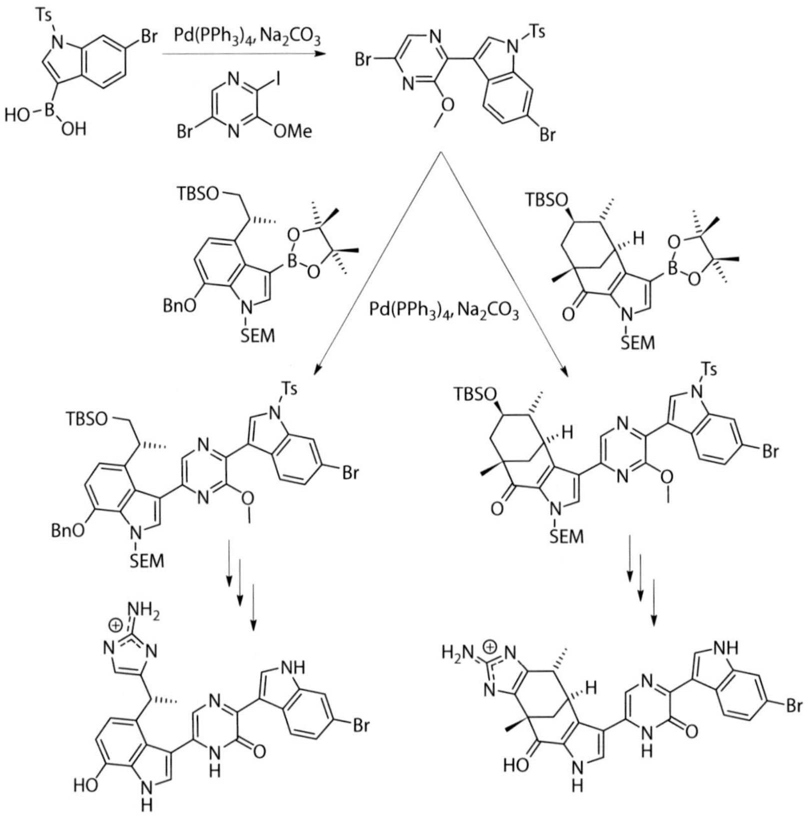

Ts
Br
Pd(PPh3)4, Na2CO3
HO-B-OH
N I
Br N OMe
Br N Ts
N O Br
TBSO
TBSO H
BnO N SEM
Pd(PPh3)4, Na2CO3
O N SEM
Ts
TBSO
TBSO H
N Br
N Br
BnO N SEM O
O N SEM
NH2
N N
NH
H2N N
N H
NH
Br
Br
HO N H
N H O
HO N H
N H O

SCHEME 28.127 Synthesis of dragmacidin D–F.[168]

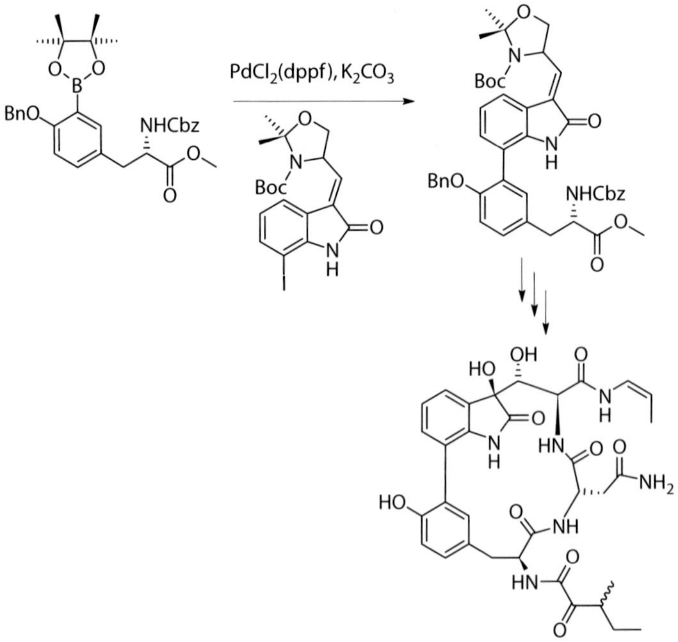

O B O
BnO NHCbz
O O
PdCl2(dppf), K2CO3
O
N Boc
O
N H
I
O
N Boc
N H O
BnO NHCbz
O O
HO OH O
N H
O
N H
HN O O
NH2
HO
O NH
HN O
O

SCHEME 28.128 Synthesis of TMC-95 A-B.[169]

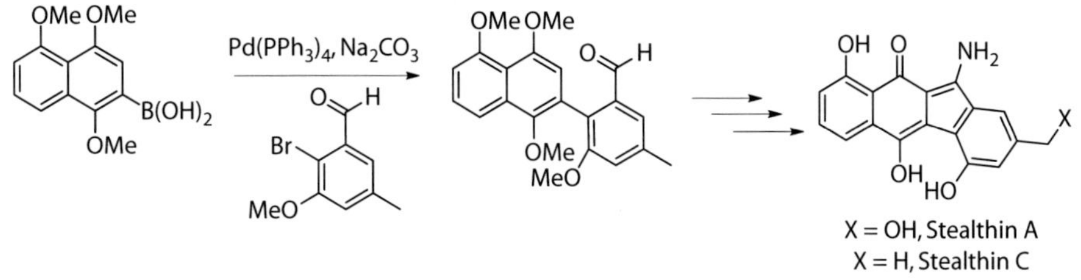

OMe OMe B(OH)$_2$ OMe Pd(PPh$_3$)$_4$, Na$_2$CO$_3$ O H Br MeO OMe OMe O H OMe MeO OH O NH$_2$ X OH HO

X = OH, Stealthin A
X = H, Stealthin C

SCHEME 28.129 Synthesis of stealthins.[170]

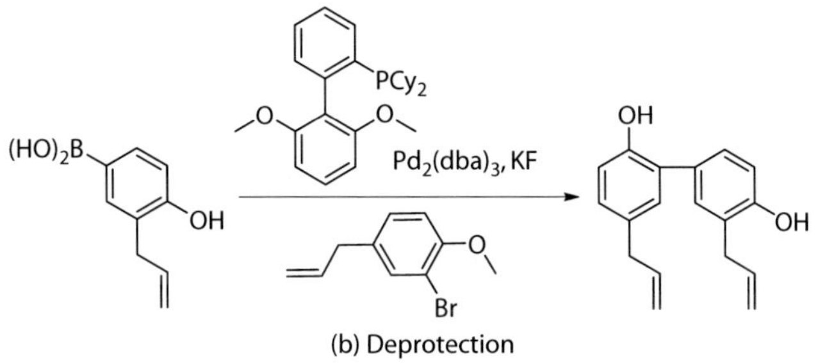

PCy$_2$ O O (HO)$_2$B OH Pd$_2$(dba)$_3$, KF O Br (b) Deprotection OH OH

SCHEME 28.130 Synthesis of honokiol.[171]

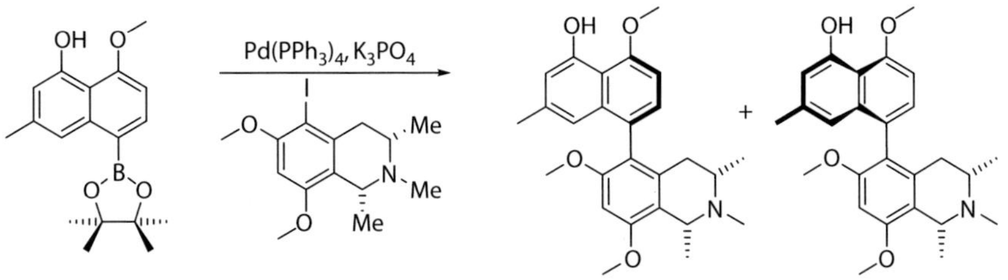

OH O O B O Pd(PPh$_3$)$_4$, K$_3$PO$_4$ I O Me N Me O Me OH O O N O + OH O O N O

SCHEME 28.131 Synthesis of *O*-demethylancistrobertsonine C.[172]

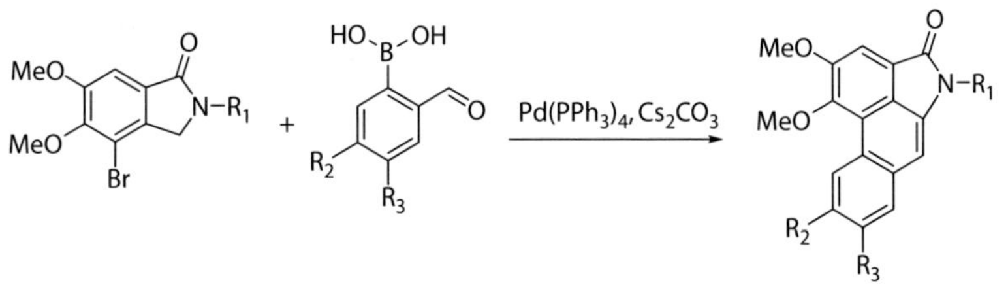

MeO MeO Br O N–R$_1$ + HO B OH O R$_2$ R$_3$ Pd(PPh$_3$)$_4$, Cs$_2$CO$_3$ MeO MeO O N–R$_1$ R$_2$ R$_3$

SCHEME 28.132 Synthesis of aristolactams.[173]

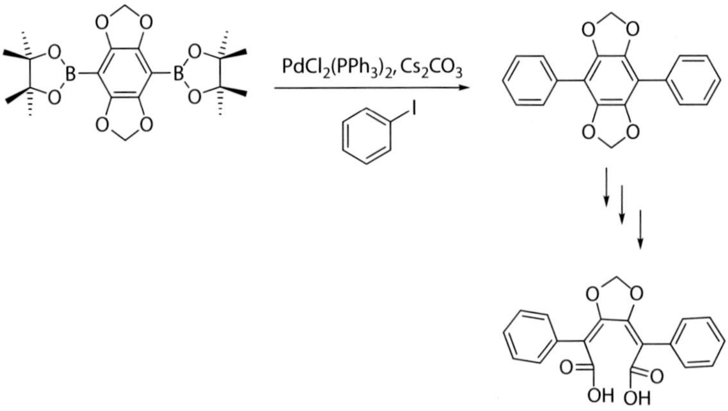

SCHEME 28.133 Synthesis of ustalic acid.[174]

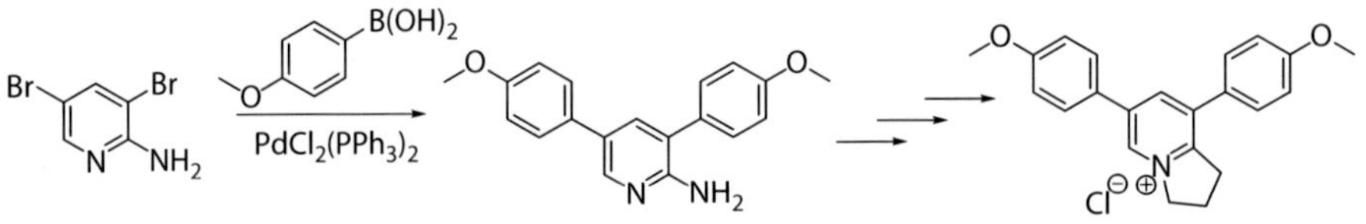

SCHEME 28.134 Synthesis of ficuseptine.[175]

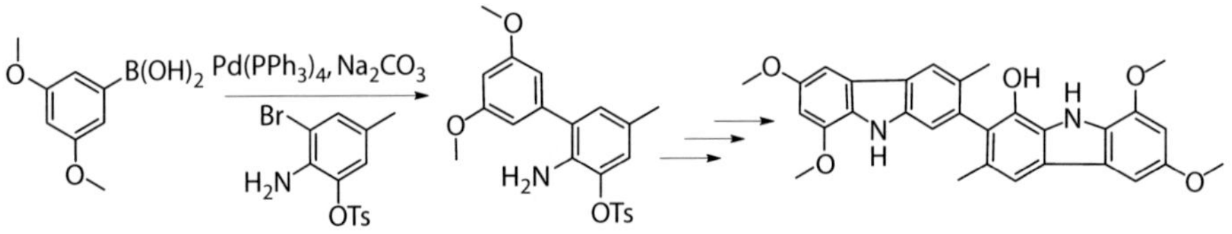

SCHEME 28.135 Synthesis of clausenamine.[176]

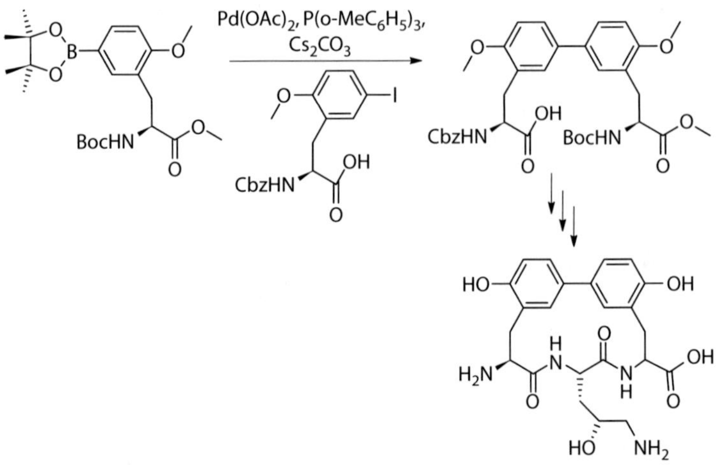

SCHEME 28.136 Synthesis of biphenomycin B.[177]

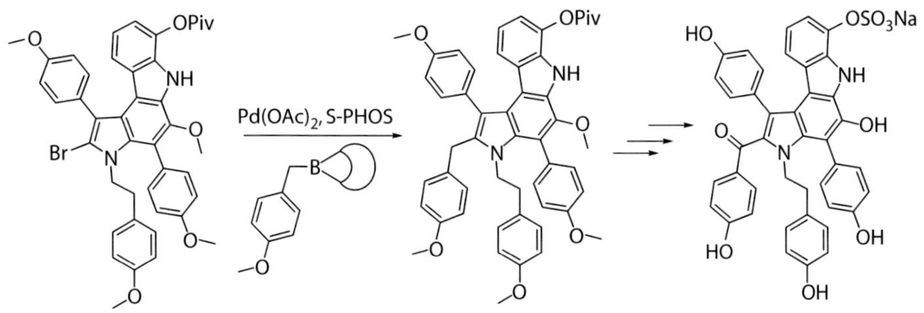

SCHEME 28.137 Synthesis of dictyodendrins.[178]

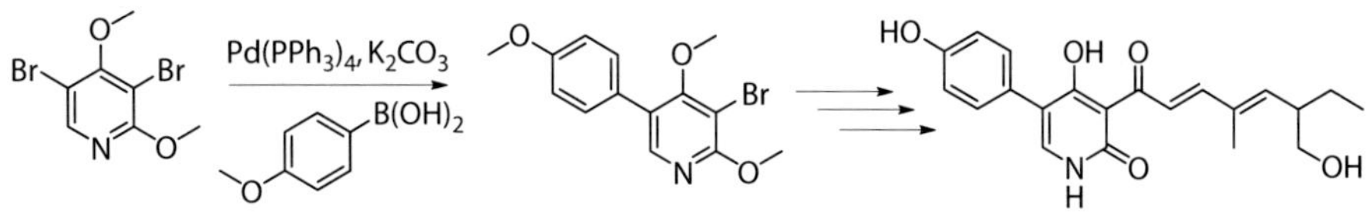

SCHEME 28.138 Synthesis of pyridovericin.[179]

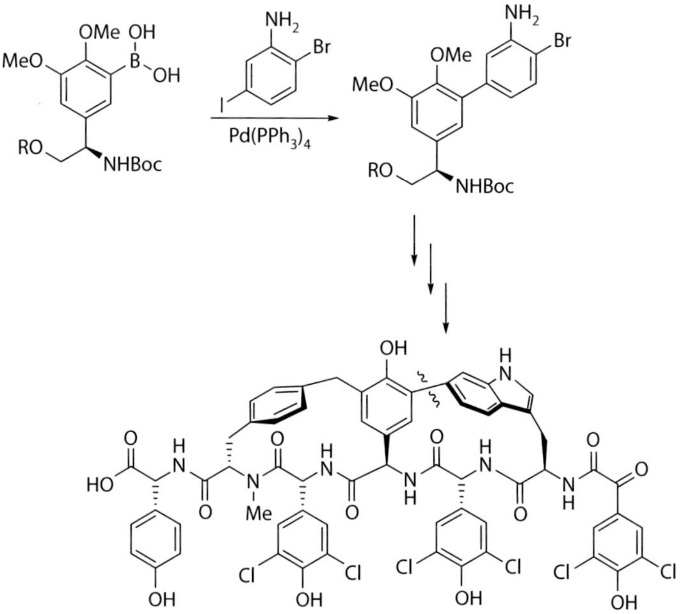

SCHEME 28.139 Synthesis of chloropeptin/complestatin.[180]

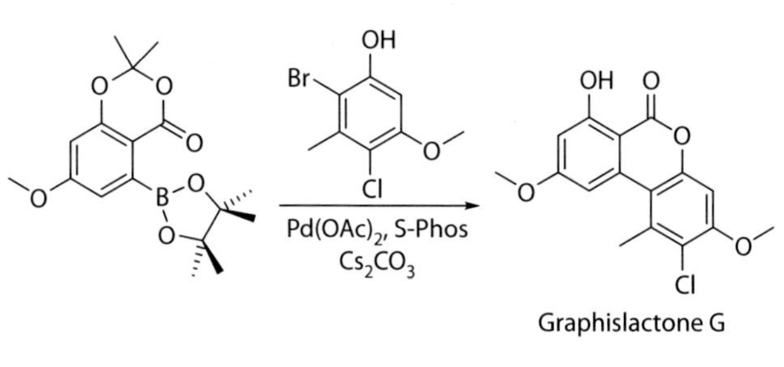

SCHEME 28.140 Synthesis of graphislactone G.[181]

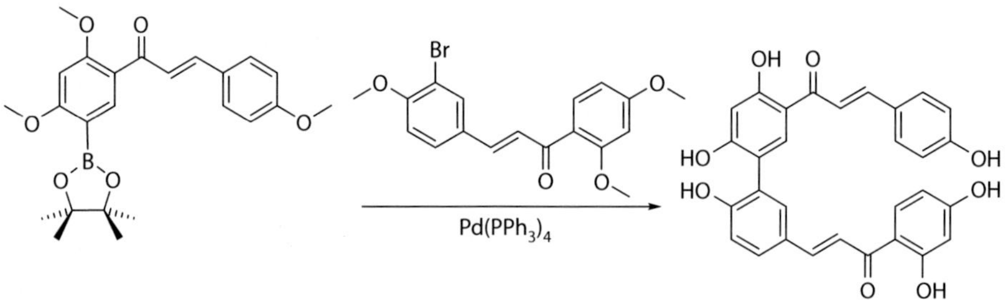

SCHEME 28.141 Synthesis of rhuschalcone VI.[182]

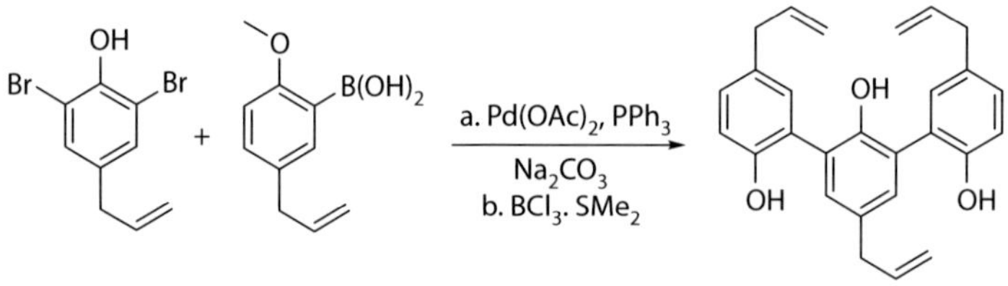

SCHEME 28.142 Synthesis of dunnianol.[183]

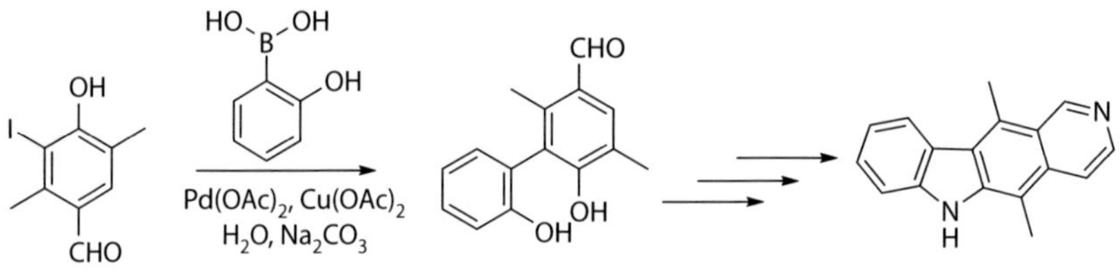

SCHEME 28.143 Synthesis of ellipticin.[184]

28.6.6 Applications of Cross-Coupling between Aryl Halides and Alkyl Boranes

Schemes 28.144 through 28.145 detail the applications of coupling of aryl halides with alkylboron species.

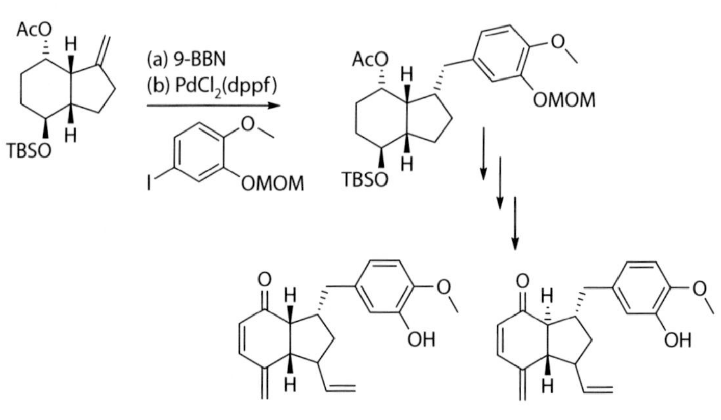

SCHEME 28.144 Synthesis of ottelione A-B.[185]

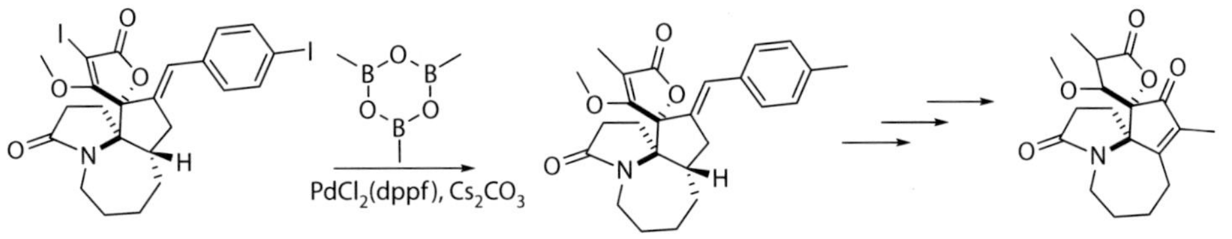

PdCl$_2$(dppf), Cs$_2$CO$_3$

SCHEME 28.145 Synthesis of stemonamide.[186]

28.6.7 Applications of Cross-Coupling with Aryl Triflates

The following section details the applications of the reaction of aryl triflates with organoboron compounds (Schemes 28.146 through 28.159).

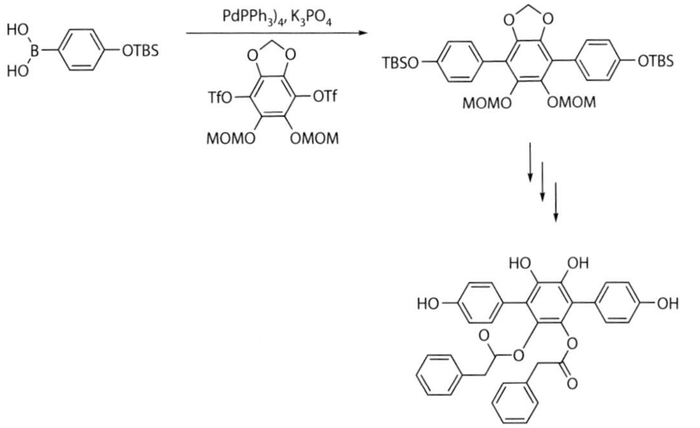

PdPPh$_3$)$_4$, K$_3$PO$_4$

SCHEME 28.146 Synthesis of vialinin A.[187]

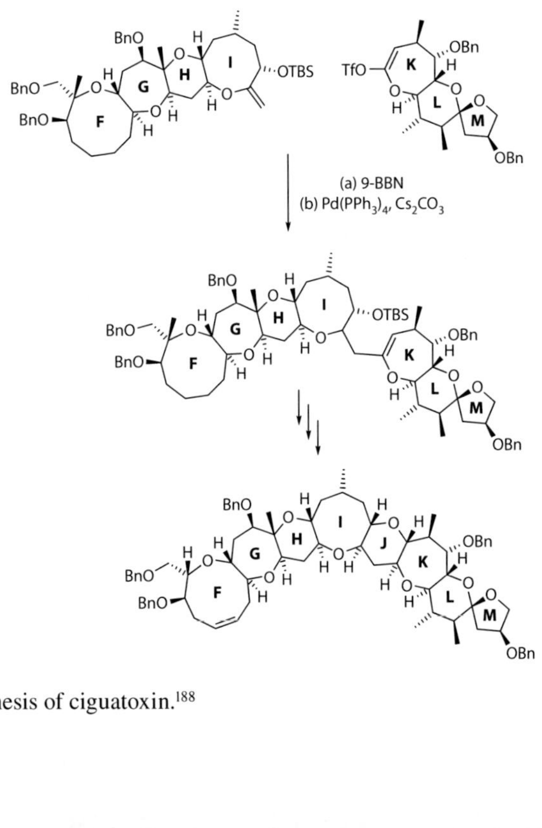

(a) 9-BBN
(b) Pd(PPh$_3$)$_4$, Cs$_2$CO$_3$

SCHEME 28.147 Synthesis of ciguatoxin.[188]

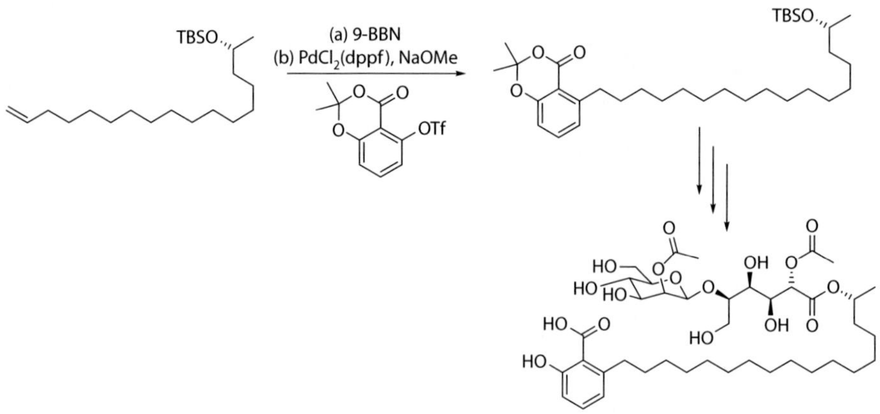

TBSO (a) 9-BBN (b) $PdCl_2$(dppf), NaOMe O O OTf O TBSO O O O HO HO HO O O O OH O O O O HO O HO OH O HO HO

SCHEME 28.148 Synthesis of caloporoside.[189]

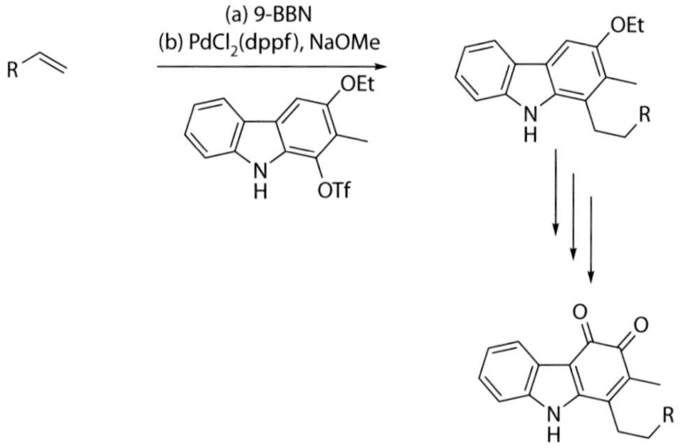

R (a) 9-BBN (b) $PdCl_2$(dppf), NaOMe OEt N H OTf OEt N H R O O N H R

SCHEME 28.149 Synthesis of carbazoquinocins.[190]

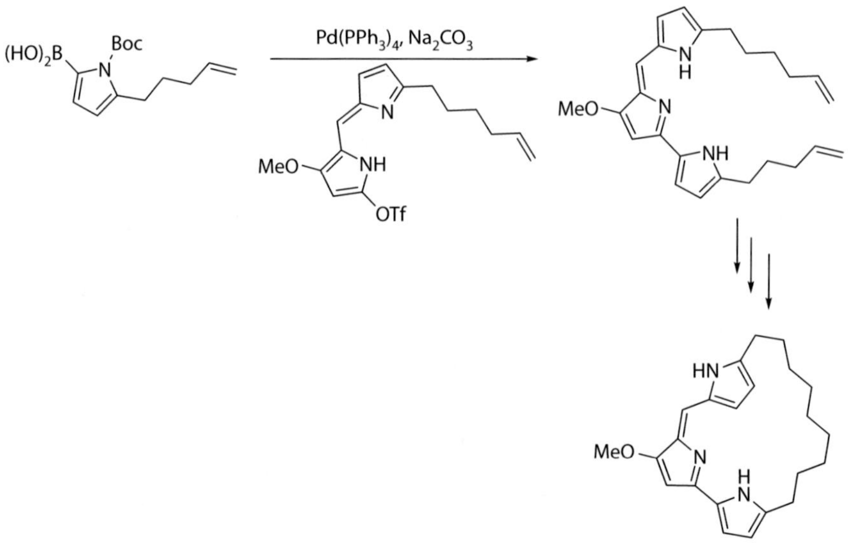

$(HO)_2B$ Boc N $Pd(PPh_3)_4$, Na_2CO_3 N MeO NH OTf N H MeO N NH HN MeO N H N

SCHEME 28.150 Synthesis of nonylprodigiosin.[191]

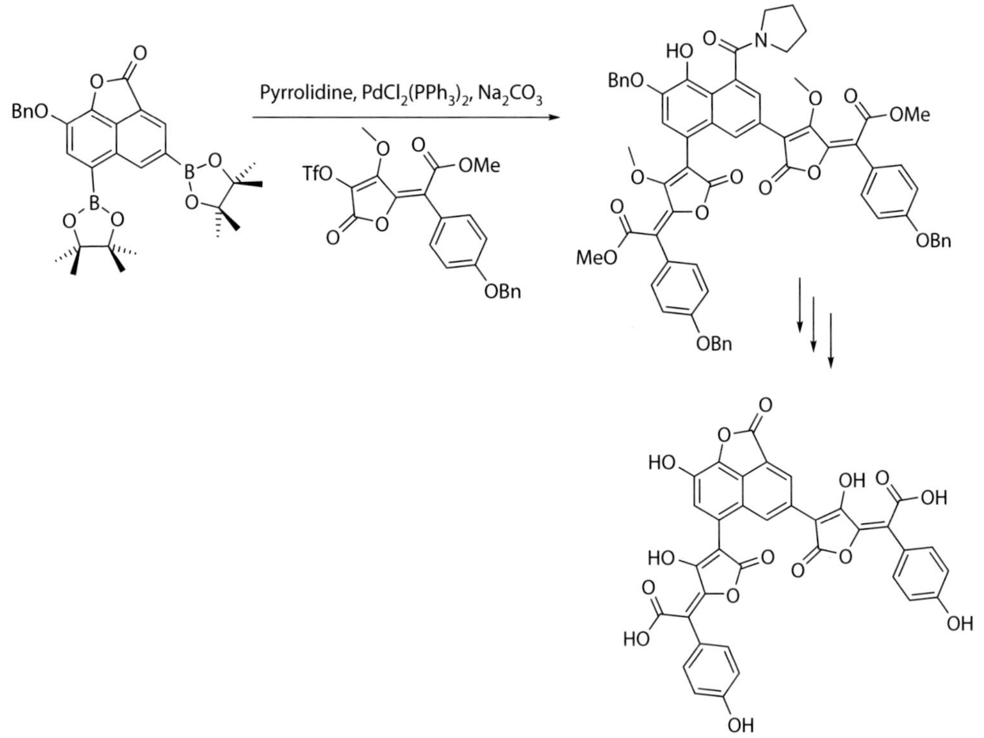

Pyrrolidine, $PdCl_2(PPh_3)_2$, Na_2CO_3

SCHEME 28.151 Synthesis of norbadione A.[192]

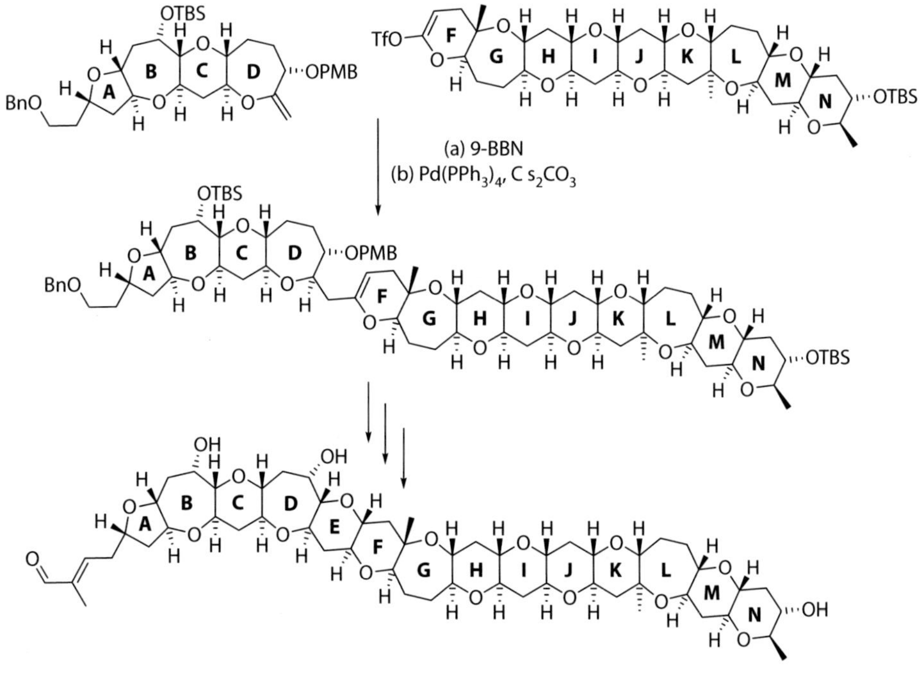

(a) 9-BBN
(b) $Pd(PPh_3)_4$, Cs_2CO_3

SCHEME 28.152 Synthesis of gymnocin.[193]

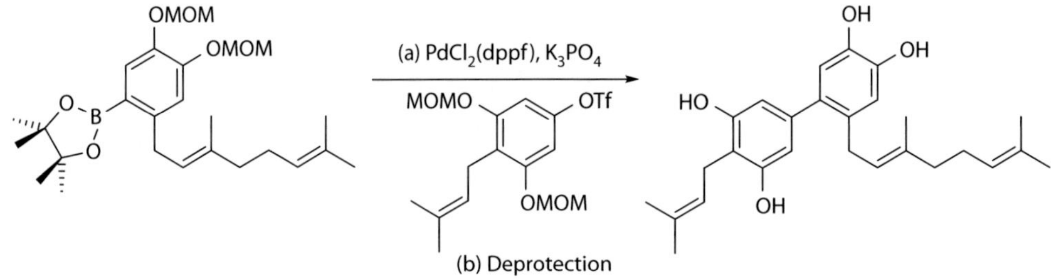

SCHEME 28.153 Synthesis of clusiparalicoline A.[194]

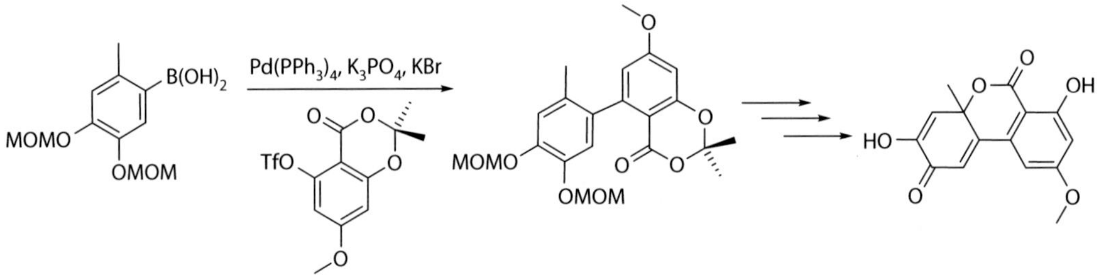

SCHEME 28.154 Synthesis of dehydroaltenusin.[195]

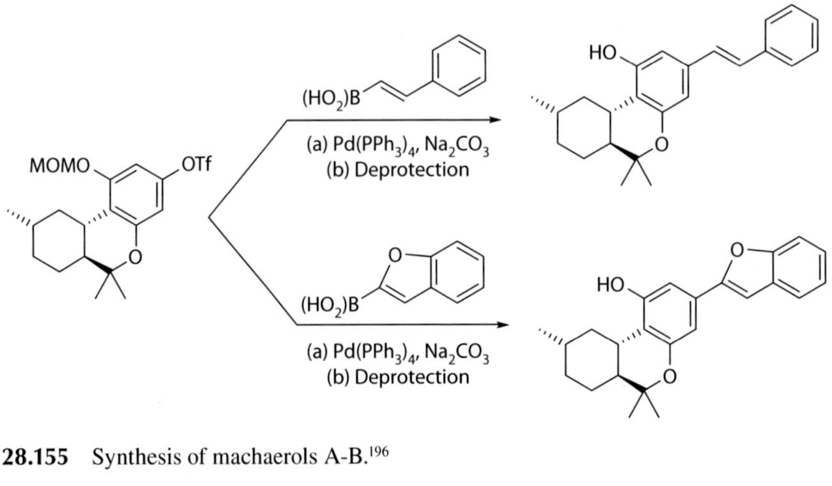

SCHEME 28.155 Synthesis of machaerols A-B.[196]

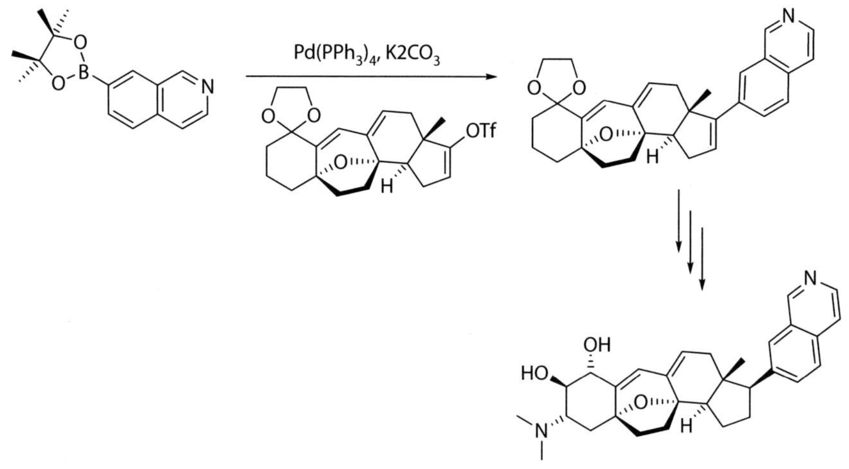

Pd(PPh$_3$)$_4$, K2CO$_3$

SCHEME 28.156 Synthesis of coristatin.[197]

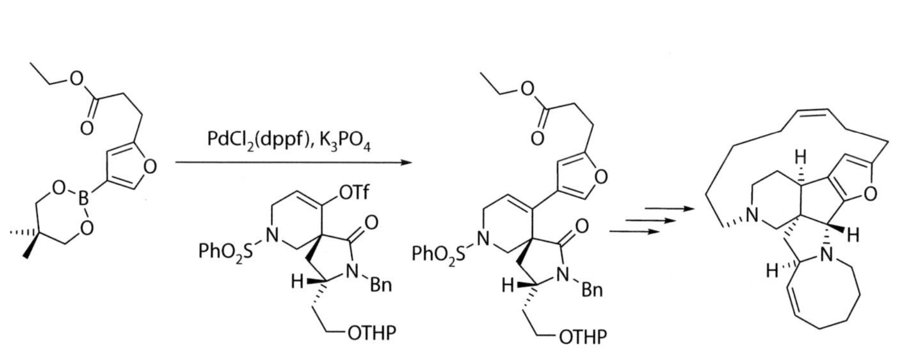

PdCl$_2$(dppf), K$_3$PO$_4$

SCHEME 28.157 Synthesis of nakadomarin.[198]

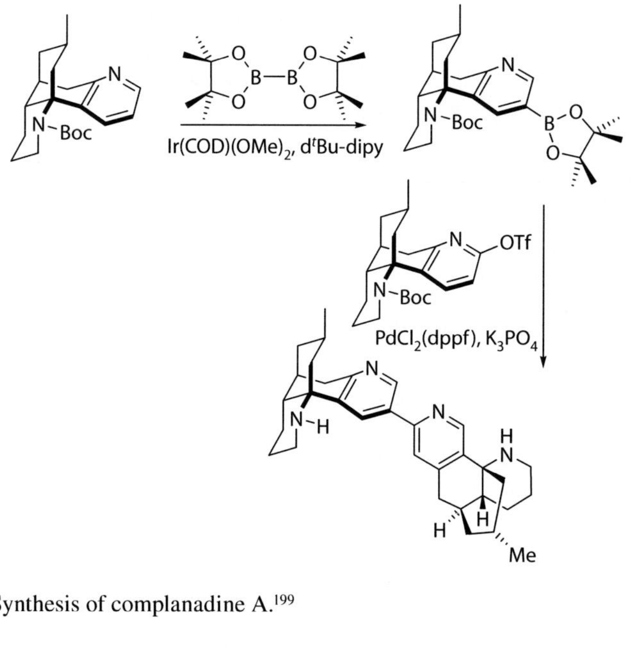

Ir(COD)(OMe)$_2$, d^{t}Bu-dipy

PdCl$_2$(dppf), K$_3$PO$_4$

SCHEME 28.158 Synthesis of complanadine A.[199]

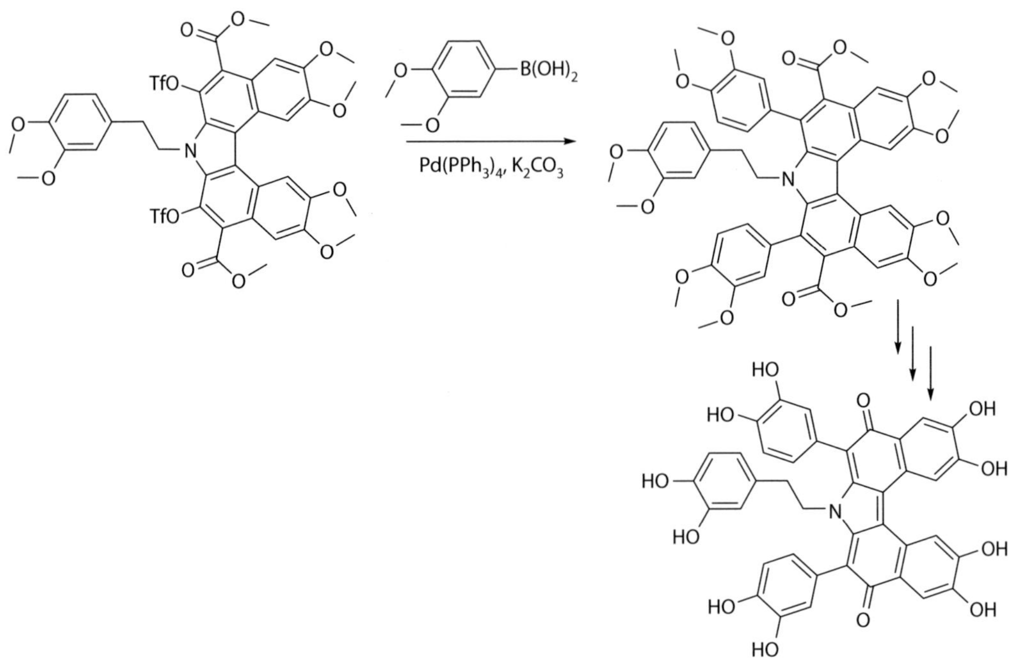

SCHEME 28.159 Synthesis of ningalin.[200]

28.7 CONCLUSIONS

This chapter focuses on the transition metal-catalyzed cross-coupling reaction between organic halides/pseudohalides and organoboron compounds. This reaction is probably one of the most studied and applied reaction in the last three decades with several thousands of research publications focusing on specific aspects of this reaction. While it is impractical to cover all the developments, this review provides general descriptions of the Suzuki–Miyaura cross-coupling reaction and documents its applications to organic synthesis. The review mainly addresses the use of various types of carbon electrophiles (e.g.. vinyl/aryl/alkyl halides or triflates) and boron partners (vinyl/aryl/alkyl boronates/boranes) these combinations provide many efficient and more direct routes to a wide variety of important synthetic targets. This review chronicles many of these applications.

ACKNOWLEDGMENTS

Supports by grants from the National Institutes of Health (CA129993) (VRM), and Non-Salary Financial Support Grants (NSFSG) (SCJ) from Rowan University are hereby acknowledged.

REFERENCES

1. (a) Metal-Catalyzed Cross-Coupling Reactions, 2nd ed. (Eds.: A. de Meijere, F. Diederich), Wiley-VCH, Weinheim, 2004. (b) Tsuji, J. Palladium Reagents and Catalysts, 2nd ed., Wiley, Chichester, 2004. (c) Hassan, J.; Sevignon, M.; Gozzi, C.; Schulz, E.; Lemaire, M. Chem. Rev. 2002, 102, 1359. (d) Kochi, J. K. *Organometallic Mechanisms and Catalysis;* Academic: New York, 1978. (e) Heck, R. F. *Palladium Reagents in Organic Syntheses;* Academic: New York, 1985. (f) Hartley, F. R.; Patai, S. *The Chemistry of Metal-Carbon Bond;* Wiley: New York, 1985; Vol. 3. (g) McQuillin, F. J.; Parker, D. G.; Stephenson, G. R. *Transition Metal Organometallics for Organic Synthesis;* Cambridge University Press: Cambridge, 1991. (h) Tamao, K. *Comprehensive Organic Synthesis;* Trost, B. M.; Fleming, I.; Pattenden, G.; Eds.; Pergaman:

New York, 1991; Vol. 3, p 435. (i) Hegedus, L. S. *Organometallics in Organic Synthesis;* Schlosser, M., Ed.; Wiley: New York, 1994; p 383.

2. (a) Tamura, M.; Kochi, J. K. 1971. Vinylation of grignard reagents. Catalysis by iron. *J. Am. Chem. Soc.* 93: 1487–1489. (b) Kochi, J. K. 1974. Electron-transfer mechanisms for organometallic intermediates in catalytic reactions. *Acc. Chem. Res.* 7: 351–360.
3. Tamura, M.; Kochi, J. K. 1971. Alkylcopper(I) in the coupling of Grignard reagents with alkyl halides. *J. Am. Chem. Soc.* 93: 1485–1487.
4. (a) Tamao, K.; Sumitani, K.; Kumada, M. 1972. Selective carbon-carbon bond formation by cross-coupling of Grignard reagents with organic halides. Catalysis by nickel-phosphine complexes. *J. Am. Chem. Soc.* 94: 4374–4376. (b) Kumada, M. 1980. Nickel and palladium complex catalyzed cross-coupling reactions of organometallic reagents with organic halides. *Pure Appl. Chem.* 52: 669–680.
5. Corriu, R. J. P.; Masse, J. P. 1972. Activation of grignard reagents by transition-metal complexes. A new and simple synthesis of trans-stilbenes and polyphenyls. *J. Chem. Soc., Chem. Commun.* 144a.
6. Yamamura, M.; Moritani, I.; Murahashi, S. 1975. The reaction of σ-vinylpalladium complexes with alkyllithiums. Stereospecific syntheses of olefins from vinyl halides and alkyllithiums. *J. Orgamomet. Chem.* 91: C39–C42.
7. Baba, S.; Negishi, E. 1976. A novel stereospecific alkenyl-alkenyl cross-coupling by a palladium- or nickel-catalyzed reaction of alkenylalanes with alkenyl halides. *J. Am. Chem. Soc.* 98: 6729–6731.
8. (a) Negishi, E.; King, A. O.; Okukado, N. 1977. Selective carbon-carbon bond formation via transition metal catalysis. 3. A highly selective synthesis of unsymmetrical biaryls and diarylmethanes by the nickel- or palladium-catalyzed reaction of aryl- and benzylzinc derivatives with aryl halides. *J. Org. Chem.* 42, 1821. (b) Negishi, E. 1982. Palladium- or nickel-catalyzed cross-coupling. A new selective method for carbon-carbon bond formation. *Acc. Chem. Res.* 15: 340–348. (c) Erdik, E. 1992. Transition metal catalyzed reactions of organozinc reagents. *Tetrahedron* 48: 9577–9643.
9. Negishi, E.; Van Horn, D. E. 1977. Selective carbon-carbon bond formation via transition metal catalysis. 4. A novel approach to cross-coupling exemplified by the nickel-catalyzed reaction of alkenylzirconium derivatives with aryl halides. *J. Am. Chem. Soc.* 99: 3168–3170.
10. Miyaura, N.; Yamada, K.; Suzuki, A. 1979. A new stereospecific cross-coupling by the palladium-catalyzed reaction of 1-alkenylboranes with 1-alkenyl or 1-alkynyl halides. *Tetrahedron Lett.* 20: 3437–3440.
11. Murahashi, S.; Yamamura, M.; Yanagisawa, K.; Mita, N.; Kondo, K. 1979. Stereoselective synthesis of alkenes and alkenyl sulfides from alkenyl halides using palladium and ruthenium catalysts. *J. Org. Chem.* 44: 2408–2417.
12. (a) Milstein, D.; Stille, J. K. 1979. Palladium-catalyzed coupling of tetraorganotin compounds with aryl and benzyl halides. Synthetic utility and mechanism. *J. Am. Chem. Soc.* 101: 4992–4998. (b) Scott, W. J.; Crisp, G. T.; Stille, J. K. 1984. Palladium-catalyzed coupling of vinyl triflates with organostannanes. A short synthesis of pleraplysillin-1. *J. Am. Chem. Soc.* 106: 4630–4632. (c) Scott, W. J.; Stille, J. K. 1986. Palladium-catalyzed coupling of vinyl triflates with organostannanes. Synthetic and mechanistic studies. *J. Am. Chem. Soc.* 108, 3033–3040. (d) Echavarren, A. M.; Stille, J. K. 1987. Palladium-catalyzed coupling of aryl triflates with organostannanes. *J. Am. Chem. Soc.* 109: 5478–5486. (e) Stille, B. J. 1986. The palladium-catalyzed cross-coupling reactions of organotin reagents with organic electrophiles. *Angew. Chem., Int. Ed.* 25: 508–524.
13. (a) Kosugi, M.; Simizu, Y.; Migita, T. 1977. Alkylation arylation, and vinylation of acyl chlorides by means of organotin compounds in the presence of catalytic amouts ot tetrakis(triuphenylphosphine)palladium. *Chem. Lett.* 6: 1423–1424. (b) Kosugi, M.; Hagiwara, I.; Migita, T. 1983. 1-Alkenylation on α-position of ketone: Palladium catalyzed reaction of tin enolates and 1-bromo-1-alkenes. *Chem. Lett.* 12:839–840.
14. Alexakis, N. J. A.; Normant, J. F. 1981. Vinyl-copper derivatives XIII: Synthesis of conjugated dienes of very high stereoisomeric purity. *Tetrahedron Lett.* 22: 959–962.
15. (a) Hatanaka, Y.; Hiyama, T. 1988. Cross-coupling of organosilanes with organic halides mediated by a palladium catalyst and tris(diethylamino)sulfonium difluorotrimethylsilicate. *J. Org. Chem.* 53: 918–920. (b) Hatanaka, Y.; Hiyama, T. 1989. Alkenylfluorosilanes as widely applicable substrates for the palladium-catalyzed coupling of alkenylsilane/fluoride reagents with alkenyl iodides. *J. Org. Chem.* 54: 268–270. (c) Hatanaka, Y.; Hiyama, T. 1990. Stereochemistry of the cross-coupling reaction of chiral alkylsilanes with aryl triflates: A novel approach to optically active compounds. *J. Am. Chem. Soc.* 112: 7793–7794. (d) Hatanaka, Y.; Hiyama, T. 1991. Highly selective cross-coupling reactions of organosilicon compounds mediated by fluoride ion and a palladium catalyst. *Synlett* 845–853.
16. (a) Miyaura, N.; Suzuki, A. 1995. Palladium-catalyzed cross-coupling reactions of organoboron compounds. *Chem. Rev.* 95, 2457–2483. (b) Suzuki, A., 1998. Metal-catalyzed cross-couplings reactions, diederich, F.; Stang, P. J. (Eds.), Wiley-VCH, Weinheim, pp. 49–97.

17. (a) Dembitsky, V. M.; Abu-Ali, H.; Srebnik, M. 2006. Chapter 3 Applied Suzuki cross-coupling reaction for syntheses of biologically active compounds in *Studies in Inorganic Chemistry: Contemporary Aspects of Boron: Chemistry and Biological Applications*. Abu-Ali, H.; Dembitsky, V. M.; Srebnik, M. (Eds). Elsevier, pp. 119–297. (b) Suzuki, A.; Brown, H. C. 2003. Organic Syntheses via Boranes: Volume 3 Suzuki Coupling. Aldrich Chemical Company; Milwaukee, WI, USA (Product# Z514306). (c) Kotha, S.; Lahiri, K.; Kashinath, D. 2002. Recent applications of the Suzuki–Miyaura cross-coupling reaction in organic synthesis. *Tetrahedron*. 58: 9633–9695.
18. Soderquist, J. A.; Roush, W. R.; Heo, J. N. 2004. 9-Borabicyclo[3.3.1]nonane Dimer, *Encyclopedia of Reagents for Organic Synthesis* John Wiley & Sons, Ltd. DOI: 10.1002/047084289X.rb235.pub2 Article Online Posting Date: October 15, 2004
19. Liotta, R.; Brown, H. C. 1977. Hydroboration. 48. Effect of structure on selective monohydroboration of representative nonconjugated dienes by 9-borabicyclo[3.3.l]nonane. *J. Org. Chem.* 42, 2836–2839.
20. Colberg, J. C.; Rane, A.; Vaquer, J.; Soderquist, J. A. 1993. Trans-vinylboranes from 9-borabicyclo[3.3.l] nonane through dehydroborylation. *J. Am. Chem. Soc.* 115: 6065–6071.
21. Brown, H. C.; Scouten, C. G.; Liotta, R. 1979. Hydroboration. 50. Hydroboration of representative alkynes with 9-borabicyclo[3.3. llnonane-a simple synthesis of versatile vinyl bora and gem-dibora intermediates *J. Am. Chem. Soc.* 101: 96–99.
22. Brown, C. A.; Coleman, R. A. 1979. Selective hydroboration of double bonds in the presence of triple bonds by 9-borabicyclo[3.3.1]nonane. New route to acetylenic organoboranes and alcohols *J. Org. Chem.* 44: 2328–2329.
23. Brown, H. C.; Prasad, J. V. R. V. 1986. Chiral synthesis via organoboranes. 9. crystalline "Chelates" from borinic and boronic esters. A simple procedure for upgrading borinates and boronates to materials approaching 100% optical purity. *J. Org. Chem.* 51: 4526–4530.
24. Brown, H. C.; Imai, T. 1984. Organoboranes. 37. Synthesis and properties of (Z)1– alkenylboronic esters. *Organometallics* 3: 1392–1395.
25. Brown, H. C.; Desai, M. C.; Jadhav, P. K. 1982. Hydroboration. 61. Diisopinocampheylborane of high optical purity. improved preparation and asymmetric hydroboration of representative cis-disubstituted alkenes. *J. Org. Chem.* 47: 5065–5069.
26. Brown, H. C.; Jadhav, P. K.; Mandal, A. K. 1982. Hydroboration. 62. Monoisopinocampheylborane, an excellent chiral hydroborating agent for trans-disubstituted and trisubstituted alkenes. Evidence for a strong steric dependence in such asymmetric hydroborations. *J. Org. Chem.* 47: 5074–5083.
27. (a) Pereira, S.; Srebnik, M. 1995. Hydroboration of alkynes with pinacolborane catalyzed by HZrCp2Cl. *Organometallics* 14: 3127–3128. (b) Pereira, S.; Srebnik, M. 1996. Transition metal-catalyzed hydroboration of and CCl4 addition to alkenes. *J. Am. Chem. Soc.* 118: 909–910. (c) Pereira, S.; Srebnik, M., 1996. A study of hydroboration of alkenes and alkynes with pinacolborane catalyzed by transition metals. *Tetrahedron Lett.* 37: 3283–3286.
28. Ramachandran, P. V.; Jennings, M. P.; Brown, H. C., 1999. Critical role of catalysts and boranes for controlling the regioselectivity in the rhodium-catalyzed hydroboration of fluoroolefins. *Org. Lett.* 1: 1399–1402.
29. Ohmura, T.; Yamamoto, Y.; Miyaura, N. 2000. Rhodium- or iridium-catalyzed trans-hydroboration of terminal alkynes, giving (Z)-1-alkenylboron compounds. *J. Am. Chem. Soc.* 122: 4990–4991.
30. Hayashi, T.; Matsumoto, Y.; Ito, Y. 1991. Asymmetric hydroboration of styrenes catalyzed by cationic chiral phosphine-rhodium(I) complexes. *Tetrahedron Asymm.* 2: 601–612.
31. (a) Marshall, J. A.; Johns, B. A. 1998. Total synthesis of (+)-discodermolide. *J. Org. Chem.* 63: 7885–7892. (b) Soderquist, J. A.; Justo de Pomar, J. C. 2000. A versatile synthesis of 9-BBN derivatives from organometallic reagents and 9-(triisopropylsilyl)thio-9-borabicyclo[3.3.1]nonane. *Tetrahedron Lett.* 41: 3537–3539. (c) Matteson, D. S. 1989. Boronic esters in stereodirected synthesis. *Tetrahedron* 45: 1859–1885.
32. (a) Brown, H. C.; Cole, T. E. 1983. Organoboranes. 31. A Simple preparation of boronic esters from organolithium reagents and selected trialkoxyboranes. *Organometallics* 2: 1316. (b) Brown, H. C.; Bhat, N. G.; Srebnik, M. 1988. A simple, general synthesis of 1-alkynyldiisopropoxyboranes. *Tetrahedron Lett.* 29: 2631–2634. (c) Brown, H. C.; Rangaishenvi, M. V. 1990. Successful application of α-haloallyllithium for a simple, convenient preparation of α-haloallylboronate ester. *Tetrahedron Lett.* 49: 7113–7114. (d) Brown, H. C.; Rangaishenvi, M. V. 1990. A simple procedure for the synthesis of three-carbon homologated boronate esters and terminal alkenes via nucleopbilic displacement in α-haloallylboronate ester. *Tetrahedron Lett.* 49: 7115–7118.
33. (a) Yamashina, N.; Hyuga, S.; Hara, S.; Suzuki, A. 1989. Organic synthesis using haloboration reaction XVIII. A stereoselective synthesis of β-mono- and β,β-disubstituted α,β-unsaturated esters. *Tetrahedron Lett.* 30: 6555–6558. (b) S. Hara, S.; Hyuga, M.; Aoyama, M.; Sato, A. Suzuki. 1990. BF3 etherate medi-

ated 1,4-addition of 1-alkenyldialkoxyboranes to α,β-unsaturated ketones. A stereoselective synthesis of γ, δ-Unsaturated ketones. *Tetrahedron Lett.* 31: 247–250.

34. (a) Murata, M.; Oyama, T.; Watanabe, S.; Masuda, Y. 2000. Palladium-catalyzed borylation of aryl halides or triflates with dialkoxyborane: A novel and facile synthetic route to arylboronates. *J. Org. Chem.* 65: 164–168. (b) Murata, M.; Watanabe, S.; Masuda, Y. 1997. Novel palladium(0)-catalyzed coupling reaction of dialkoxyborane with aryl halides: Convenient synthetic route to arylboronates. *J. Org. Chem.* 62: 6458–6459.
35. Murata, M.; Oyama, T.; Watanabe, S.; Masuda, Y. 2000. Synthesis of alkenylboronates via palladium-catalyzed borylation of alkenyl triflates (or iodides) with pinacolborane. *Synthesis* 6: 778–780.
36. Ishiyama, T.; Ishida, K.; Miyaura, N. 2001. Synthesis of pinacol arylboronates via cross-coupling reaction of bis(pinacolato)diboron with chloroarenes catalyzed by palladium(0)–tricyclohexylphosphine complexes. *Tetrahedron* 57: 9813–9816.
37. Murata, M.; Oyama, T.; Watanabe, S.; Masuda, Y. 2001. Synthesis of benzylboronates via palladium catalyzed borylation of benzyl halides with pinacolborane. *Synthetic Commun.* 32: 2513–2517.
38. Murata, M.; Watanabe, S.; Masuda, Y. 2000. Regio- and stereoselective synthesis of allylboranes via platinum(0)-catalyzed borylation of allyl halides with pinacolborane. *Tetrahedron Lett.* 41: 5877–5880.
39. (a) Murata, M.; Kawakita, K.; Asana, T.; Watanabe, S.; Masuda, Y. 2002. Rhodium- and ruthenium-catalyzed dehydrogenative borylation of vinylarenes with pinacolborane: Stereoselective synthesis of vinylboronates. *Bull. Chem. Soc. Jpn.* 75: 825–829. (b) Murata, M.; Watanabe, S.; Masuda, Y. 1999. Rhodium-catalyzed dehydrogenative coupling reaction of vinylarenes with pinacolborane to vinylboronates. *Tetrahedron Lett.* 40: 2585–2588.
40. Vogels, C. M.; Hayes, P. G.; Shaver, M. P.; Westcott, S. A. 2000. 'Metal-catalysed addition of B-H And N-H bonds to aminopropyl vinyl ethers. *Chem. Commun.* 51–52.
41. (a) Tse, M. K.; Cho, J.-Y.; Smith, M. R. 2001. Regioselective aromatic borylation in an inert solvent. *Org. Lett.* 3: 2831–2833. (b) Cho, J.-Y.; Iverson, C. N.; Smith, M. R. 2000. Steric and chelate directing effects in aromatic borylation. *J. Am. Chem. Soc.* 122: 12868–12869. (c) Iverson, C. N.; Smith, M. R., 1999. Stoichiometric and Catalytic B–C Bond Formation from Unactivated Hydrocarbons and Boranes. *J. Am. Chem. Soc.* 121: 7696–7697. (d) Shimada, S.; Batsanov, A. S.; Howard, J. A. K.; Marder, T. B. 2001. Formation of aryl- and benzylboronate esters by rhodium-catalyzed C-H bond functionalization with pinacolborane. *Angew. Chem. Int. Ed.* 40: 2168–2171.
42. (a) Chambers, R. D.; Clark, H. C.; Willis, C. J. 1960. Some salts of trifluoromethylfluoroboric acid. *J. Am. Chem. Soc.* 5298–5301. (b) Vedejs, E.; Chapman, R. W.; Fields, S. C.; Lin, S.; Schrimpf, M. R. 1995. Conversion of arylboronic acids into potassium aryltrifluoroborates: Convenient precursors of arylboron difluoride Lewis acids. *J. Org. Chem.* 60: 3020–3027.
43. Matos, K.; Soderquist, J. A. 1998. Alkylboranes in the Suzuki-Miyaura coupling: Stereochemical and mechanistic studies. *J. Org. Chem.* 63: 461–470.
44. Wolfe, J. P.; Singer, R. A.; Yang, B. H.; Buchwald, S. L. 1999. Highly active palladium catalysts for suzuki coupling reactions. *J. Am. Chem. Soc.* 121: 9550–9561.
45. Walker, S. D.; Barder, T. E.; Martinelli, J. R.; Buchwald, S. L. 2004. A rationally designed universal catalyst for suzuki–miyaura coupling processes. *Angew. Chem. Int. Ed.* 116: 1907–1912.
46. Zhang, C.; Huang, J.; Trudell, M. L.; Nolan, S. P. 1999. Palladium-imidazol-2-ylidene complexes as catalysts for facile and efficient suzuki cross-coupling reactions of aryl chlorides with arylboronic acids. *J. Org. Chem.* 64: 3804–3805.
47. (a) Martin, R.; Buchwald, S. L. 2008. Palladium-catalyzed Suzuki-Miyaura cross-coupling reactions employing dialkylbiaryl phosphine ligands. *Acc. Chem. Res.* 41: 1461–1473. (b) Hillier, A. C.; Grasa, G. A.; Viciu, M. S.; Lee, H. M.; Yang, C.; Nolan, S. P. 2002. Catalytic cross-coupling reactions mediated by palladium/nucleophilic carbene systems. *J. Organomet. Chem.* 653: 69–82.
48. Miyaura, N.; Suzuki, A. 1981. The palladium-catalyzed "head-to-tail" cross-coupling reaction of 1-alkenylboranes with phenyl or 1-alkenyl iodides. A novel synthesis of 2-phenyl-1-alkenes or 2-alkyl-1,3-alkadienes via organoboranes. *J. Organomet. Chem.* 213: C53–C56.
49. (a) Miyaura, N.; Suzuki, A. 1990. Palladium-catalyzed reaction of 1-alkenylboronates with vinylic halides: (1*Z*, 3*E*)-1-phenyl-1,3-octadiene *Organic Syntheses* 68: 130–137. (b) Suzuki, A. 1985. Organoboron compounds in new synthetic reactions. *Pure & Applied Chem.* 57: 1749–1758.
50. Soderquist, J. A.; Leon-Colon, G. 1991. *Z*-Trimethylsilylstyrenes and 1,3-dienes via the Suzuki reaction. *Tetrahedron Lett.* 32: 43–44.
51. Miyaura, N.; Satoh, M.; Suzuki, A. 1986. Stereo- and regiospecific synthesis to provide conjugated (E, Z)- and (Z, Z)-alkadienes and arylated (Z)-alkenes in excellent yields via the palladium-catalyzed

cross-coupling reactions or (Z)-1-alkenylboronates with 1-bromoalkenes and aryl iodides. *Tetrahedron Lett.* 27: 3745–3748.

52. (a) Satoh, N.; Ishiyama, T.; Miyaura, N.; Suzuki, A. 1987. Stereoselective synthesis of conjugated dienones via the palladium-catalyzed cross-coupling reaction of 1-alkenylboronates with 3-halo-2-alken-1-ones. *Bull. Chem. Soc. Jpn.* 60: 3471–3473. (b) Yanagi, T.; Oh-e, T.; Miyaura, N.; Suzuki, A. 1989. Stereoselective synthesis of conjugated 2,4-alkadienoates via the palladium-catalyzed cross-coupling of 1-alkenylboronates with 3-bromo-2-alkenoates. *Bull. Chem. Soc. Jpn.* 62: 3892–3895.
53. Uenishi, J.-I.; Beau, J.-M.; Armstrong, R. W.; Kishi, Y. 1987. Dramatic rate enhancement of Suzuki Diene synthesis: Its application to palytoxin synthesis. *J. Am. Chem. Soc.* 109: 4756–4758.
54. Frank, S. A.; Chen, H.; Kunz, R. K.; Schnaderbeck, M. J.; Roush, W. R. 2000. Use of thallium(I) ethoxide in Suzuki cross-coupling reactions. *Org. Lett.* 2: 2691–2694.
55. Friesen, R. W.; Loo, R. W. 1991. Preparation of c-aryl glucals via the palladium-catalyzed coupling of metalated aromatics with l-Iodo-3,4,6-tri-(triisopropylsily1)-D-glucal. *J. Org. Chem.* 56: 4821–4823.
56. Bauer, A.; Miller, M. W.; Vice, S. F.; McCombie, S. W. 2001. Suzuki arylation of 1,1-dibromo-1-alkenes: Synthesis of tetra-substituted alkenes. *Synlett* 254–256.
57. Chen, C.; Wilcoxen, K.; Strack, N.; McCarthy, J. R. 1999. Synthesis of fluorinated olefins via the palladium catalyzed cross-coupling reaction of 1-fluorovinyl halides with organoboranes. *Tetrahedron Lett.* 40: 827–830.
58. DeBoos, G. A.; Fullbrook, J. J.; Owton, W. M.; Percy, J. M.; Thomas, A. C. 2000. Palladium-catalyzed couplings of difluoroenol carbamate derivatives. *Synlett* 963–966.
59. Chemler, S. R.; Trauner, D. R.; Danishefsky, S. J. 2001. The b-alkyl Suzuki-Miyaura cross-coupling reaction: Development, mechanistic study, and applications in natural product synthesis. *Angew. Chem. Int. Ed.* 40: 4544–4568.
60. (a) Miyaura, N.; Ishiyama, T.; Ishikawa, M.; Suzuki, A. 1986. Palladium catalyzed cross-coupling reactions of b-alkyl-9-BBN or trialkylboranes with aryl and 1-alkenyl halides. *Tetrahedron Lett.* 27: 6369–6372. (b) Miyaura, N.; Ishiyama, T.; Sasaki, H.; Ishikawa, M.; Satoh, M.; Suzuki, A. 1989. Palladium-catalyzed inter- and intramolecular cross-coupling reactions of b-alkyl-9-borabicyclo[3.3.1] nonane derivatives with 1 -Halo- 1 -alkenes or haloarenes. Syntheses of functionalized alkenes, arenes, and cycloalkenes via a hydroboration-coupling sequence. *J. Am. Chem. Soc.* 111, 314–321.
61. Abe, S.; Miyaura, N.; Suzuki, A. 1992. The palladium–catalyzed cross-coupling reaction of enol acetates of α-bromo ketones with 1-alkenyl-, aryl-, or alkylboron compounds; A facile synthesis of ketones and their enol acetates. *J. Am. Chem. Soc.* 65: 2863–2865.
62. Hoshino, Y.; Ishiyama, T.; Miyaura, N.; Suzuki, A. 1988. A stereoselective route to alkenyl sulfides through the palladium-catalyzed cross-coupling reaction of 9-Alkyl-9-BBN with 1-Bromo-1-phenylthioethene or (E)- and (Z)-2-Bromo-1-phenylthioalkenes. *Tetrahedron Lett.* 29: 3983–3986.
63. Miyaura, N.; Suzuki, A. 1979. Stereoselective synthesis of arylated (E) -alkenes by the reaction of alk-1-enylboranes with aryl halides in the presence of palladium catalyst. *J. Chem. Soc. Chem. Commun.* 866–867.
64. (a) Hassan, J.; Sevignon, M.; Gozzi, C.; Schulz, E.; Lemaire, M. 2002. Aryl-aryl bond formation one century after the discovery of the ullmann reaction. *Chem. Rev.* 102: 1359–1469. (b) Anctil, E. J. G.; Snieckus, V. 2002. The directed ortho metalation—cross-coupling symbiosis. Regioselective methodologies for biaryls and heterobiaryls. Deployment in aromatic and heteroaromatic natural product synthesis. *J. Organomet. Chem.* 653: 150–160.
65. Jiang, B.; Wang, Q-F.; Yang, C-G.; Xu, M. 2001. α-(Trifluoromethyl)ethenyl boronic acid as a useful trifluoromethyl containing building block. Preparation and palladium-catalysed coupling with aryl halides. *Tetrahedron Lett.* 42: 4083–4085.
66. Eastwood, P. R. 2000. A versatile synthesis of 4-aryl tetrahydropyridines via palladium mediated Suzuki cross-coupling with cyclic vinyl boronates. *Tetrahedron Lett.* 41: 3705–3708.
67. (a) Littke, A. F.; Fu, G. C. 1998. A convenient and general method for Pd-catalyzed Suzuki cross-couplings of aryl chlorides and arylboronic acids. *Angew. Chem. Int. Ed.* 37: 3387–3388. (b) Littke, A. F.; Dai, C.; Fu, G. C. 2000. Versatile catalysts for the Suzuki cross-coupling of arylboronic acids with aryl and vinyl halides and triflates under mild conditions. *J. Am. Chem. Soc.* 122: 4020–4028. (c) Clarke, M. L.; Cole-Hamilton, D. J.; Woollins, J. D. 2001. Synthesis of bulky, electron rich hemilabile phosphines and their application in the Suzuki coupling reaction of aryl chlorides. *J. Chem. Soc. Dalton Trans.* 2721–2723. (d) Navarro, Kelly, R. A.; Nolan, S. P. 2003. A General method for the Suzuki-Miyaura cross-coupling of sterically hindered aryl chlorides: Synthesis of di- and tri-ortho-substituted biaryls in 2-propanol at room temperature. *J. Am. Chem. Soc.* 125: 16194–16195. (e) Colacot, T. J.; Shea, H. A. 2004. Cp2Fe(PR2)2PdCl2 (R = *i*-Pr, *t*-Bu) complexes as air-stable catalysts for challenging Suzuki coupling reactions. *Org. Lett.* 6:

3731–3734. (f) Song, C.; Ma, Y.; Chai, Q.; Ma, C.; Jiang, W.; Andrus, M. B. 2005. Palladium catalyzed Suzuki–Miyaura coupling with aryl chlorides using a bulky phenanthryl N-heterocyclic carbene ligand. *Tetrahedron* 61: 7438–7446. (g) Iwasawa, T.; Komano, T.; Tajima, A.; Tokunaga, M.; Obora, Y.; Fujihara, T.; Tsuji, Y. 2006. Phosphines having a 2,3,4,5-tetraphenylphenyl moiety: Effective ligands in palladium-catalyzed transformations of aryl chlorides. *Organometallics* 25: 4665–4669. (h) Diebolt, O.; Braunstein, P.; Nolan, S. P.; Cazin, C. S. J. 2008. Room-temperature activation of aryl chlorides in Suzuki–Miyaura coupling using a [Pd(l-Cl)Cl(NHC)]2 complex (NHC = N-heterocyclic carbene). *Chem. Commun.* 3190–3912.

68. Christmann, U.; Vilar, R. 2005. Monoligated palladium species as catalysts in cross-coupling reactions. *Angew. Chem., Int. Ed.* 44: 366–374.
69. Littke, A. F.; Fu, G. C. 2002. Palladium-catalyzed coupling reactions of aryl chlorides. *Angew. Chem. Int. Ed.* 41: 4176–4211.
70. Fürstner, A.; Seidel, G. 1998. Suzuki reactions with B-allyl-9-borabicyclo[3.3.1]nonane (B-allyl-9-BBN). *Synlett* 161–162.
71. (a) Gray, M.; Andrews, I. P.; Hook, D. F.; Kitteringham, J.; Voyle, M. 2000. Practical methylation of aryl halides by Suzuki-Miyaura coupling. *Tetrahedron Lett.* 41: 6237–6240. (b) Soderquist, J. A.; Santiago, B. 1990. Methylation via the suzuki reaction. *Tetrahedron Lett.* 31: 5541–5542.
72. Iglesias, B.; Alvarez, R.; de Lera, A. R. 2001. A general synthesis of alkylpyridines. *Tetrahedron* 57: 3125–3130.
73. (a) Zou, G.; Falck, J. R. 2001. Suzuki–Miyaura cross-coupling of lithium n-alkylborates. *Tetrahedron Lett.* 42: 5817–5819. (b) Zou, G.; Reddy, Y. K.; Falck, J. R. 2001. Ag(I)-promoted Suzuki–Miyaura cross-couplings of n-alkylboronic acids. *Tetrahedron Lett.* 42: 7213–7215.
74. (a) Hildebrand, J. P.; Marsden, S. P. 1996. A novel, stereocontrolled synthesis of 1,2-trans-cyclopropanes. Cyclopropyl boronate esters as partners in Suzuki couplings with aryl halides. *Synlett* 893–894. (b) Luithle, J. E. A.; Pietruszka, J. 1999. Synthesis of enantiomerically pure cyclopropanes from cyclopropylboronic acids. *J. Org. Chem.* 64: 8287–8297. (c) Zhou, S.-M.; Deng, M.-Z.; Xia, L.-J.; Tang. M.-H. 1998. Efficient Suzuki-type cross-coupling of enantiomerically pure cyclopropylboronic acids. *Angew. Chem. Int. Ed.* 37: 2845–2847. (d) Soderquist, J. A.; Huertas, R.; Leon-Colon, G. 2000. Aryl and vinyl cyclopropanes through the in situ generation of B-cyclopropyl-9-BBN and its Suzuki-Miyaura coupling. *Tetrahedron Lett.* 41: 4251–4255.
75. Ishiyama, T.; Abe, S.; Miyaura, N.; Suzuki, A. 1992. Palladium-catalyzed alkyl-alkyl cross-coupling reaction of 9-alkyl-9-BBN derivatives with iodoalkanes possessing β-Hydrogens. *Chem. Lett.* 21: 691–694.
76. Kirchhoff, J. H.; Dai, C. Y.; Fu, G. C. 2002. A Method for palladium-catalyzed cross-couplings of simple alkyl chlorides: Suzuki reactions catalyzed by [Pd2(dba)3]/PCy3. *Angew. Chem. Int. Ed.* 41: 1945–1947.
77. Gonzalez-Bobes, F.; Fu, G. C. 2006. Amino alcohols as ligands for nickel-catalyzed Suzuki reactions of unactivated alkyl halides, including secondary alkyl chlorides, with arylboronic acids. *J. Am. Chem. Soc.* 128: 5360–5361.
78. White, J. D.; Hanselmann, R.; Jackson, R. W.; Porter, W. J.; Ohba, Y.; Tiller, T.; Wang, S. 2001. Total synthesis of rutamycin B, a macrolide antibiotic from streptomyces aureofaciens. *J. Org. Chem.* 66: 5217–5231.
79. Bois-Choussy, M.; Cristau, P.; Zhu, J. 2003. Total synthesis of an atropdiastereomer of RP-66453 and determination of its absolute configuration. *Angew. Chem. Int. Ed.* 42: 4238–4241.
80. Molander, G. A.; Ellis, N. 2007. Organotrifluoroborates: Protected boronic acids that expand the versatility of the Suzuki coupling reaction. *Acc. Chem. Res.* 40: 275–286.
81. Molander, G. A.; Biolatto, B. 2003. Palladium-catalyzed Suzuki-Miyaura cross-coupling reactions of potassium aryl- and heteroaryltrifluoroborates. *J. Org. Chem.* 68: 4302–4314.
82. Molander, G. A.; Ellia, M. D. 2006. Suzuki-Miyaura cross-coupling reactions of benzyl halides with potassium aryltrifluoroborates. *J. Org. Chem.* 71: 9198–9202.
83. Mizuta, M.; Seio, K.; Miyata, K.; Sekine, M. 2007. Fluorescent pyrimidopyrimidoindole nucleosides: Control of photophysical characterizations by substituent effects. *J. Org. Chem.* 72: 5046–5055.
84. Molander, G. A.; Biolatto, B. 2002. Efficient ligandless palladium-catalyzed Suzuki reactions of potassium aryltrifluoroborates. *Org. Lett.* 4: 1867–1870.
85. (a) Darses, S.; Genet, J.-P.; Brayer, J.-L.; Demoute, J.-P. 1997. Cross-coupling reactions of arenediazonium tetrafluoroborates with potassium aryl- or alkenyltrifluoroborates catalyzed by palladium. *Tetrahedron Lett.* 38: 4393–4396. (b) Darses, S.; Michaud, G.; Genet, J.-P. 1999. Potassium Organotrifluoroborates: New partners in palladium-catalysed cross-coupling reactions. *Eur. J. Org. Chem.* 1875–1883. (c) Kabalka, G. W.; Zhou, L.-L.; Zhou, Naravane, A. 2007. Microwave accelerated cross-coupling reaction of arenediazonium tetrafluoroborates with potassium aryltrifluoroborates. *Lett. Org. Chem.* 4: 325–328.
86. (a) Nguyen, H. N.; Huang, X.; Buchwald, S. L. 2003. The first general palladium catalyst for the Suzuki-Miyaura and carbonyl enolate coupling of aryl arenesulfonates. *J. Am. Chem. Soc.* 125: 11818–11819.

(b) Zhang, L.; Meng. T.; Wu, J. 2007. Palladium-catalyzed Suzuki-Miyaura cross-couplings of aryl tosylates with potassium aryltrifluoroborates. *J. Org. Chem.* 72: 9346–9349. (c) So, C. M.; Lau, C. P.; Chan, A. S. C.; Kwong, F. Y. 2008. Suzuki-Miyaura coupling of aryl tosylates catalyzed by an array of indolyl phosphine-palladium catalysts. *J. Org. Chem.* 73: 7731–7734.

87. So, C. M.; Lau, C. P.; Kwong, F. Y. 2008. A general palladium-catalyzed Suzuki–Miyaura coupling of aryl mesylates. *Angew. Chem. Int. Ed.* 120: 8179–8183.
88. Drehrer, S. D.; Lin, S. E.; Sandrock, D. L.; Molander, G. A. 2009. Suzuki-Miyaura cross-coupling reactions of primary alkyltrifluoroborates with aryl chlorides. *J. Org. Chem.* 74: 3626–3631.
89. (a) Fang, G.-H.; Yan, Z.-J.; Deng, D.-Z. 2004. Palladium-catalyzed cross-coupling of stereospecific potassium cyclopropyl trifluoroborates with aryl bromides. *Org. Lett.* 6: 357–360. (b) Molander, G. A.; Gormisky, P. E. 2008. Cross-coupling of cyclopropyl- and cyclobutyltrifluoroborates with aryl and heteroaryl chlorides. *J. Org. Chem.* 73: 7481–7485.
90. (a) Molander, G. A.; Petrillo, D. E. 2008. Suzuki-Miyaura cross-coupling of potassium trifluoroboratohomoenolates. *Org. Lett.* 10: 1795–1798. (b) Molander, G. A.; Jean-Gerard, L. 2009. Scope of the Suzuki-Miyaura cross-couplingrReaction of potassium trifluoroboratoketohomoenolates. *J. Org. Chem.* 74: 1297–1303. (c) Molander, G. A.; Jean-Gerard, L. 2009. Use of potassium β-Trifluoroborato amides in Suzuki-Miyaura cross-coupling reactions. *J. Org. Chem.* 74: 5446–5450.
91. (a) Molander, G. A.; Vargas, F. 2007. β-Aminoethyltrifluoroborates: Efficient aminoethylations via Suzuki-Miyaura cross-coupling. *Org. Lett.* 9: 203–206. (b) Molander, G. A.; Jean-Gerard, L. 2007. Scope of the Suzuki-Miyaura aminoethylation reaction using organotrifluoroborates. *J. Org. Chem.* 72: 8422–8426. (c) Molander, G. A.; Sandrock, D. L. 2007. Aminomethylations via cross-coupling of potassium organotrifluoroborates with aryl bromides. *Org. Lett.* 9: 1597–1600. (d) Molander, G. A.; Gormisky, P. E.; Sandrock, D. L. 2008. Scope of aminomethylations via Suzuki-Miyaura cross-coupling of organotrifluoroborates. *J. Org. Chem.* 73: 2052–2057.
92. Molander, G. A.; Canturk, B. 2008. Preparation of potassium alkoxymethyltrifluoroborates and their cross-coupling with aryl chlorides. *Org. Lett.* 10: 2135–2138.
93. (a) Miyaura, N.; Suginome, H.; Suzuki, A. 1983. New stereospecific synthesis of pheromone bombykol and its three geometric isomers. *Tetrahedron Lett.* 24:1527–1530. (b) Miyaura, N.; Suginome, H.; Suzuki, A. 1983. New stereospecific synthesis of pheromone bombykol and its three geometric isomers. *Tetrahedron* 39: 3271–3277.
94. Evans, D. A.; Ng, H. P.; Rieger, D. L. 1993. Total synthesis of the macrolide antibiotic rutamycin B. *J. Am. Chem. Soc.* 115: 11446–11459.
95. Humphrey, J. M.; Aggen, J. B.; Chamberlin, A. R. 1996. Total synthesis of the serine-threonine phosphatase inhibitor microcystin-LA. *J. Am. Chem. Soc.* 118: 11759–11770.
96. Mapp, A. K.; Heathcock, C. H. 1999. Total synthesis of myxalamide A. *J. Org. Chem.* 64: 23–27.
97. Vyvyan, J. R.; Peterson, E. A.; Stephan, M. L. 1999. An expedient total synthesis of (±)-caparratriene. *Tetrahedron Lett.* 40: 4947–4949.
98. (a) Ozawa, T.; Aoyagi, S.; Kibayashi, C. 2000. Total synthesis of the marine alkaloid (-)-lepadin B. *Org. Lett.* 2: 2955–2958. (b) Ozawa, T.; Aoyagi, S.; Kibayashi, C. 2001. Total synthesis of the marine alkaloids (-)-lepadins a, b, and c based on stereocontrolled intramolecular acylnitroso-Diels-Alder reaction. *J. Org. Chem.* 66: 3338–3347.
99. Kobayashi, S.; Mori, K.; Wakabayashi, T.; Yasuda, S.; Hanada, K. 2001. Convergent total synthesis of khafrefungin and its inhibitory activity of fungal sphingolipid syntheses. *J. Org. Chem.* 66: 5580–5584.
100. Reddy, Y. K.; Falck, J. R. 2002. Asymmetric total synthesis of (+)-fostriecin. *Org. Lett.* 4: 969–971.
101. Shoji, M.; Yamaguchi, J.; Kakeya, H.; Osada, H.; Hayashi, Y. 2002. Total synthesis of (+)-epoxyquinols A and B. *Angew. Chem. Int. Ed.* 41: 3192–3194.
102. Scheidt, K. A.; Bannister, T. D.; Tasaka, A.; Wendt, M. D.; Savall, B. M.; Fegley, G. J.; Roush, W. R. 2002. Total synthesis of (-)-bafilomycin A1. *J. Am. Chem. Soc.* 124: 6981–6990.
103. (a) Savall B. M; Blanchard N.; Roush W. R. 2003. Total synthesis of the formamicin aglycon, formamicinone. *Org. Lett.* 5: 377–379. (b) Durham T. B.; Blanchard N.; Savall B. M.; Powell N. A.; Roush W. R. 2004. Total synthesis of formamicin. *J. Am. Chem. Soc.* 126: 9307–9317.
104. Suzuki, T.; Usui, K.; Miyake, Y.; Namikoshi, M.; Nakada, M. 2004. First total synthesis of antimitotic compound, (+)-phomopsidin. *Org. Lett.* 6: 553–556.
105. Wu, B.; Liu, Q.; Sulikowski, G. A. 2004. Total synthesis of apoptolidinone. *Angew. Chem. Int. Ed.* 43: 6673–6675.
106. Jacobsen, M. F.; Moses, J. E.; Adlington, R. M.; Baldwin, J. E. 2005. The total synthesis of spectinabilin and its biomimetic conversion to SNF4435C and SNF4435D. *Org. Lett.* 7: 2473–2476.

107. Coleman, R. S.; Walczak, M. C.; Campbell, E. L. 2005. Total synthesis of lucilactaene, A cell cycle inhibitor active in p53-inactive cells. *J. Am. Chem. Soc.* 127: 16038–16039.
108. Hosokawa, S.; Seki, M.; Fukuda, H.; Tatsuta, K. 2006. Total synthesis of an antitubercular lactone antibiotic, (+)-tubelactomicin A. *Tetrahedron Lett.* 47, 2439–2442.
109. (a) Nicolaou, K. C.; Nold, A. L.; Milburn, R. R.; Schindler, C. S. 2006. Total synthesis of marinomycins A-C. *Angew. Chem. Int. Ed.* 45: 6527–6532. (b) Nicolaou, K. C.; Nold, A. L.; Milburn, R. R.; Schindler, C. S.; Cole, K. P.; Yamaguchi, J. 2007. Total synthesis of marinomycins A-C and of their monomeric counterparts monomarinomycin A and iso-monomarinomycin A. *J. Am. Chem. Soc.* 129: 1760–1768.
110. Coleman, R. S.; Walczak, M. C. 2006. Total Synthesis of Gymnoconjugatins A and B. *J. Org. Chem.* 71: 9841–9844.
111. Rodriguez, R.; Adlington, R. M.; Eade, S. J.; Walter, M. W.; Baldwin, J. E.; Moses, J. E. 2007. Total synthesis of cyercene A and the biomimetic synthesis of (±)-9,10-deoxytridachione and (±)-ocellapyrone A. *Tetrahedron* 63: 4500–4509.
112. Jiang, X.; Liu, B.; Lebreton, S.; De Brabander, J. K. 2007. Total synthesis and structure revision of the marine metabolite palmerolide A. *J. Am. Chem. Soc.* 129: 6386–6387.
113. Fürstner, A.; Nevado, C.; Waser, M.; Tremblay, M.; Chevrier, C.; Teply, F.; Aiessa, C.; Moulin, E.; Mueller, O. 2007. Total synthesis of iejimalide A-D and assessment of the remarkable actin-depolymerizing capacity of these polyene macrolides. *J. Am. Chem. Soc.* 129: 9150–9161.
114. (a) Tortosa, M.; Yakelis, N. A.; Roush, W. R. 2008. Total synthesis of (+)-superstolide A. *J. Am. Chem. Soc.* 130: 2722–2723. (b) Tortosa, M.; Yakelis, N. A.; Roush, W. R. 2008. Total synthesis of (+)-superstolide A. *J. Org. Chem.* 73: 9657–9667.
115. Murray, T. J.; Forsyth, C. J. 2008. Total synthesis of GEX1A. *Org. Lett.* 10: 3429–3431.
116. (a) Fuwa, H.; Sasaki, M. 2010. Total synthesis of (-)-exiguolide. *Org. Lett.* 12: 584–587. (b) Cook, C.; Guinchard, X.; Liron, F.; Roulland, E. 2010. Total Synthesis of (-)-exiguolide. *Org. Lett.* 12: 744–747.
117. Francais, A.; Leyva, A.; Etxebarria-Jardi, G.; Ley, S. V. 2010. Total synthesis of the anti-apoptotic agents iso- and bongkrekic acids. *Org. Lett.* 12: 340–343.
118. Banwell, M. G.; Cowden, C. J.; Mackay, M. F. 1994. Concise synthetic route to both enantiomeric forms of 2,3,4,4a-tetrahydro[1,3]dioxolo[4,5-j]phenanthridin-6(5H)-one. Tetracyclic skeleton associated with the narcissus alkaloids lycoricidine and narciclasine. *J. Chem. Soc. Chem. Commun.* 61–62.
119. Isaac, M.; Slassi, A.; DaSilva, K.; Xin, T. 2001. Synthesis of chiral and geometrically defined 5,5-diaryl-2-amino-4-pentenoates: Novel amino acid derivatives. *Tetrahedron Lett.* 42: 2957–2960.
120. Potter, G. A.; McCague, R. 1990. Highly stereoselective access to an (E)-vinyl bromide from an aryl ketone leads to short syntheses of (2)-tamoxifen and important substituted derivatives. *J. Org. Chem.* 55: 6184–6187.
121. Ashley, E. R.; Cruz, E. G.; Stoltz, B. M. 2003. The total synthesis of (-)-lemonomycin. *J. Am. Chem. Soc.* 125: 15000–15001.
122. Ito, F.; Iwasaki, M.; Watanabe, T.; Ishikawa, T.; Higuchi, Y. 2005. The first total synthesis of kwakhurin, a characteristic component of a rejuvenating plant, "kwao keur": Toward an efficient synthetic route to phytoestrogenic isoflavones. *Org. Biomol. Chem.* 3: 674–681.
123. Ban, H.; Muraoka, M.; Ohashi, N. 2005. Synthesis of SMP-797: A new potent ACAT inhibitor. *Tetrahedron* 61: 10081–10092.
124. Matveenko, M.; Banwell, M. G.; Willis, A. C. 2008. A chemoenzymatic total synthesis of ent-narciclasine. *Tetrahedron* 64: 4817–4826.
125. (a) Nicolaou, K. C.; Dalby, S. M.; Majumder, U. 2008. A concise asymmetric total synthesis of aspidophytine. *J. Am. Chem. Soc.* 130: 14942–14943. (b) Nicolaou, K. C.; Dalby, S. M.; Li, S.; Suzuki, T.; Chen, D. Y.-K. 2009. Total synthesis of (+)-haplophytine. *Angew. Chem. Int. Ed.* 48: 7616–7620.
126. Zheng, S-Y.; Shen, Z-W. 2010. Total synthesis of Hirtellanine A. *Tetrahedron Lett.* 51: 2883–2887.
127. (a) Seebach, D.; Maestro, M. A.; Sefkow, M.; Adam, G.; Hintermann, S.; Neidlein, A. 1994. Total synthesis of Myxovirescins. 1. Strategy and construction of the "southeastern" part [O(1)-C(14)]. *Liebigs. Annal. Chemie.* 701–717. (b) Fürstner, A.; Bonnekessel, M.; Blank, J. T; Radkowski, K.; Seidel, G.; Lacombe, F.; Gabor, B.; Mynott, R. 2007. Total synthesis of myxovirescin A1. *Chem. Eur. J.* 13: 8762–8783.
128. (a) Su, D-S.; Meng, D.; Bertinato, P.; Balog, A.; Sorensen, E. J.; Danishefsky, S. J.; Zheng, Y-H.; Chou, T-C.; He, L.; Horwitz, S. B. 1997. Total synthesis of (-)-epothilone B: An extension of the Suzuki coupling method and insights into structure-activity relationships of the epothilones. *Angew. Chem. Int. Ed.* 36: 757–759. (b) Balog, A.; Meng, D.; Kamenecka, T.; Bertinato, P.; Su, D-S.; Sorensen, E. J.; Danishefsky, S. J. 1997. Total synthesis of (-)-epothilone A. *Angew. Chem. Int. Ed.* 35(23/24), 2801–2803. (c) Meng, D.; Bertinato, P.; Balog, A.; Su, D-S.; Kamenecka, T.; Sorensen, E.; Danishefsky, S. J. 1997. Total syntheses

of epothilones A and B. *J. Am. Chem. Soc.* 119: 10073–10092. (d) Zhu, B.; Panek, J. S. 2000. Total synthesis of epothilone A. *Org. Lett.* 2: 2575–2578. (e) Lee, C. B.; Chou, T-C.; Zhang, X-G.; Wang, Z-G.; Kuduk, S. D.; Chappell, M. D.; Stachel, S. J.; Danishefsky, S. J. 2000. Total synthesis and antitumor activity of 12,13-desoxyepothilone F: An unexpected solvolysis problem at C15, mediated by remote substitution at C21. *J. Org. Chem.* 65: 6525–6533. (f) Altmann, K.-H.; Bold, G.; Caravatti, G.; Denni, D.; Florsheimer, A.; Schmidt, A.; Rihs, G.; Wartmann, M. 2002. The total synthesis and biological assessment of trans-epothilone A. *Helv. Chim. Acta.* 85: 4086–4110. (g) End, N.; Furet, P.; van Campenhout, N.; Wartmann, M.; Altmann, K-H. 2004. Total synthesis and biological evaluation of a C(10)/C(12)-phenylene-bridged analog of epothilone D. *Chem. Biodiversity* 1: 1771–1784. (h) Broadrup, R. L.; Sundar, H. M.; Swindell, C. S. 2005. Total synthesis of 12,13-desoxyepothilone B (Epothilone D). *Bioorg. Chem.* 33: 116–133. (i) Hutt, O. E.; Reddy, B. S.; Nair, S. K.; Reiff, E. A.; Henri, J. T.; Greiner, J. F.; Chiu, T.-L. et al. 2008. Total synthesis and evaluation of C25-benzyloxyepothilone C for tubulin assembly and cytotoxicity against MCF-7 breast cancer cells. *Bioorg. Med. Chem. Lett.* 18: 4904–4906.

129. (a) Trost, B. M.; Lee, C. 2001. Gem-diacetates as carbonyl surrogates for asymmetric synthesis. total syntheses of sphingofungins E and F. *J. Am. Chem. Soc.* 123: 12191–12201. (b) Nakamura, T.; Shiozaki, M. 2002. Total synthesis of sphingofungin E from D-glucose derivative. *Tetrahedron* 58: 8779–8791.
130. Mandal, A. K. 2002. Stereocontrolled total synthesis of (-)-ebelactone A. *Org. Lett.* 4: 2043–2045.
131. (a) Marshall, J. A.; Bourbeau, M. P. 2002. Total synthesis of (-)-callystatin A. *J. Org. Chem.* 67: 2751–2754. (b) Dias, L. C.; Meira, P. R. R. 2005. Total synthesis of the potent antitumor polyketide (-)-callystatin A. *J. Org. Chem.* 70: 4762–4773.
132. Fuwa, H.; Kainuma, N.; Tachibana, K.; Sasaki, M. 2002. Total synthesis of (-)-gambierol. *J. Am. Chem. Soc.* 124: 14983–14992.
133. Marshall, J. A.; Schaaf, G. M. 2003. Total synthesis and structure confirmation of leptofuranin D. *J. Org. Chem.* 68: 7428–7432.
134. (a) Herb, C.; Maier M. E. 2003. A formal total synthesis of the salicylihalamides. *J. Org. Chem.* 68: 8129–8135. (b) Herb, C.; Bayer, A.; Maier, M. E. 2004. Total synthesis of salicylihalamides A and B. *Chem. Eur. J.* 10: 5649–5660.
135. Taillier, C.; Bellosta, V.; Cossy, J. 2004. Total synthesis of natural (+)-(2'S,3′R)-Zoapatanol. *Org. Lett.* 6: 2149–2151.
136. Yuan, Y.; Men, H.; Lee, C. 2004. Total synthesis of kendomycin: A macro-C-glycosidation approach. *J. Am. Chem. Soc.* 126: 14720–14721.
137. (a) Lepage, O.; Kattnig, E.; Fürstner, A. 2004. Total synthesis of amphidinolide X. *J. Am. Chem. Soc.* 126: 15970–15971. (b) Fürstner, A.; Kattnig, E.; Lepage, O. 2006. Total syntheses of amphidinolide X and Y. *J. Am. Chem. Soc.* 128: 9194–9204. (c) Barbazanges, M.; Meyer, C.; Cossy, J. 2008. Total synthesis of amphidinolide. *J. Org. Lett.* 10: 4489–4492.
138. Fürstner, A.; De Souza, D.; Turet, L.; Fenster, M. D. B.; Parra-Rapado, L.; Wirtz, C.; Myott, R.; Lehmann, C. W. 2006. Total syntheses of the actin-binding macrolides latrunculin A, B, C, M, S and 16-epi-latrunculin B. *Chem. Eur. J.* 13: 115–134.
139. (a) Fuwa, H.; Ebine, M.; Sasaki, M. 2006. Total synthesis of the proposed structure of brevenal. *J. Am. Chem. Soc.* 128: 9648–9650. (b) Fuwa, H.; Ebine, M.; Bourdelais, A. J.; Baden, D. G.; Sasaki, M. 2006. Total synthesis, structure revision, and absolute configuration of (-)-brevenal. *J. Am. Chem. Soc.* 128: 16989–16999. (c) Ebine, M.; Fuwa, H.; Sasaki, M. 2008. Total synthesis of (-)-brevenal: A concise synthetic entry to the pentacyclic polyether core. *Org. Lett.* 10: 2275–2278.
140. Shibahara S.; Fujino M.; Tashiro Y.; Takahashi K.; Ishihara J.; Hatakeyama S. 2008. Asymmetric total synthesis of (+)-phoslactomycin B. *Org. Lett.* 10: 2139–2142.
141. Corbu, A.; Perez, M.; Aquino, M.; Retailleau, P.; Arseniyadis, S. 2008. Total synthesis and structural confirmation of ent-galbanic acid and marneral. *Org. Lett.* 10: 2853–2856.
142. (a) Paterson, I.; Anderson, E. A.; Dalby, S. M.; Lim, J. H.; Genovino, J.; Maltas, P.; Moessner, C. 2008. Total synthesis of spirastrellolide A methyl ester-part 1: Synthesis of an advanced C17-C40 bis-spiroacetal subunit. *Angew. Chem. Int. Ed.* 47: 3016–3020. (b) O'Neil, G. W.; Ceccon, J.; Benson, S.; Collin, M.-P.; Fasching, B.; Fürstner, A. 2009. Total synthesis of spirastrellolide F methyl ester. Part 1: Strategic considerations and revised approach to the southern hemisphere. *Angew. Chem. Int. Ed.* 48: 9940–9945.
143. (a) Roulland, E. 2008. Total synthesis of (+)-oocydin A: Application of the Suzuki-Miyaura cross-coupling of 1,1-dichloro-1-alkenes with 9-BBN. *Angew. Chem. Int. Ed.* 47: 3762–3765. (b) Ueda, M.; Yamaura, M.; Ikeda, Y.; Suzuki, Y.; Yoshizato, K.; Hayakawa, I.; Kigoshi, H. 2009. Total synthesis and cytotoxicity of haterumalides NA and B and their artificial analogues. *J. Org. Chem.* 74: 3370–3377.

144. Ren, G.-B.; Wu, Y. 2008. Enantioselective total synthesis and correction of the absolute configuration of megislactone. *Tetrahedron* 64: 4408–4415.
145. (a) Fuwa, H.; Naito, S.; Goto, T.; Sasaki, M. 2008. Total synthesis of (+)-neopeltolide. *Angew. Chem. Int. Ed.* 47: 4737–4739. (b) Fuwa, H.; Saito, A.; Naito, S.; Konoki, K.; Yotsu-Yamashita, M.; Sasaki, M. 2009. Total synthesis and biological evaluation of (+)-neopeltolide and its analogues. *Chem. Eur. J.* 15: 12807–12818.
146. (a) Kuranaga T.; Shirai T.; Baden D. G; Wright J. L. C.; Satake M.; Tachibana K. 2009. Total synthesis and structural confirmation of brevisamide, a new marine cyclic ether alkaloid from the dinoflagellate Karenia brevis. *Org. Lett.* 11: 217–220. (b) Ghosh A. K; Li J. 2009. An asymmetric total synthesis of brevisamide. *Org. Lett.* 11: 4164–4167.
147. Hioki, H.; Motosue, M.; Mizutani, Y.; Noda, A.; Shimoda, T.; Kubo, M.; Harada, K.; Fukuyama, Y.; Kodama, M. 2009. Total synthesis of pseudodehydrothyrsiferol. *Org. Lett.* 11: 579–582.
148. Schnabel, C.; Hiersemann, M. 2009. Total synthesis of jatrophane diterpenes from euphorbia characias. *Org. Lett.* 11: 2555–2558.
149. Yamanaka, H.; Sato, K.; Sato, H.; Iida, M.; Oishi, T.; Chida, N. 2009. Total synthesis of mycestericin A and its 14-epimer. *Tetrahedron* 65: 9188–9201.
150. Archambaud, S.; Legrand, F.; Aphecetche-Julienne, K.; Collet, S.; Guingant, A.; Evain, M. 2010. Total synthesis of (+)-brefeldin C, (+)-nor-me brefeldin A and (+)-4-epi-nor-Me brefeldin A. *Eur. J. Org. Chem.* 1364–1380.
151. Satoh, M.; Miyaura, N.; Suzuki, A. 1987. Palladium catalyzed cross-coupling reaction of (1-ethoxy-1-alken-2-yl)boranes with ortho-functionalized iodoarenes. A novel and convenient synthesis of benzo fused heteroaromatic compounds. *Synthesis* 373–377.
152. Tidwell, J. H.; Peat, A. J.; Buchwald, S. L. 1994. Synthesis and reactions of 3- (bromomethyl)-1-carbethoxy-4-iodoindole: The preparation of 3,4-differentially substituted indoles. *J. Org. Chem.* 59: 7164–7168.
153. Miyachi, N.; Yanagawa, Y.; Iwasaki, H.; Ohara, Y.; Hiyama, T. 1993. A novel synthetic method of HMG-CoA reductase inhibitor NK-104 via a hydroboration-cross-coupling sequence. *Tetrahedron Lett.* 34: 8267–8270.
154. Yonezawa, S.; Komurasaki, T.; Kawada, K.; Tsuri, T.; Fuji, M.; Kugimiya, A.; Haga, N. et al. 1998. Total synthesis of terprenin, a novel immunosuppressive p-terphenyl derivative. *J. Org. Chem.* 63: 5831–5837.
155. (a) Zembower, D. E.; Zhang, H. 1998. Total synthesis of robustaflavone, a potential anti-hepatitis B agent. *J. Org. Chem.* 63: 9300–9305. (b) Zembower, D. E.; Zhang, H.; Flavin, M. T.; Lin, Y-M. 1999. Total synthesis of robustaflavone and its analogs. *PCT Int. Appl.* WO 9910292 A2 19990304.
156. (a) Nicolaou, K. C.; Li, H.; Boddy, C. N. C.; Ramanjulu, J. M.; Yue, T.-Y.; Natarajan, S.; Chu, X.-J.; Braese, S.; Rubsam, F. 1999. Total synthesis of vancomycin-part 1: Design and development of methodology. *Chem. Eur. J.* 5: 2584–2601. (b) Nicolaou, K. C.; Koumbis, A. E.; Takayanagi, M.; Natarajan, S.; Jain, N. F.; Bando, T.; Li, H.; Hughes, R. 1999. Total synthesis of vancomycin-part 3: Synthesis of the aglycon. *Chem. Eur. J.* 5: 2622–2647. (c) Crowley, B. M.; Mori, Y.; McComas, C. C.; Tang, D.; Boger, D. L. 2004. Total synthesis of the ristocetin aglycon. *J. Am. Chem. Soc.* 126: 4310–4317. (d) Crowley, B. M; Boger D. L. 2006. Total synthesis and evaluation of [Psi[CH2NH]Tpg4]vancomycin aglycon: Reengineering vancomycin for dual D-Ala-D-Ala and D-Ala-D-Lac binding. *J. Am. Chem. Soc.* 128: 2885–2892.
157. Jones, S. B.; He, L.; Castle, S. L. 2006. Total synthesis of (±)-hasubanonine. *Org. Lett.* 8: 3757–3760.
158. Meketa, M. L.; Weinreb, S. M. 2006. Total synthesis of ageladine A, an angiogenesis inhibitor from the marine sponge agelas nakamurai. *Org. Lett.* 8: 1443–1446.
159. Narayan, S.; Roush, W. R. 2004. Studies toward the total synthesis of angelmicin B (Hibarimicin B): Synthesis of a model CD-D' arylnaphthoquinone. *Org. Lett.* 6: 3789–3792.
160. Miyashita, K.; Sakai, T.; Imanishi, T. 2003. Total synthesis of (±)-spiroxin C. *Org. Lett.* 5: 2683–2686.
161. O'Neill, B. T.; Yohannes, D.; Bundesmann, M. W.; Arnold, E. P. 2000. Total synthesis of (±)-cytisine. *Org. Lett.* 2: 4201–4204.
162. Liu W.; Buck M.; Chen N.; Shang M.; Taylor N. J; Asoud J.; Wu X.; Hasinoff B. B.; Dmitrienko G. I. 2007. Total synthesis of isoprekinamycin: structural evidence for enhanced diazonium ion character and growth inhibitory activity toward cancer cells. *Org. Lett.* 9: 2915–2918.
163. Koch, K.; Podlech, J.; Pfeiffer, E.; Metzler, M. 2005. Total synthesis of alternariol. *J. Org. Chem.* 70: 3275–3276.
164. (a) Mitra, S.; Gurrala, S. R.; Coleman, R. S. 2007. Total synthesis of the eupomatilones. *J. Org. Chem.* 72: 8724–8736. (b) Kabalka, G. W.; Venkataiah, B. 2005. The total synthesis of eupomatilones 2 and 5. *Tetrahedron Lett.* 46: 7325–7328.

165. (a) Pla, D.; Marchal, A.; Olsen, C. A.; Albericio, F.; Alvarez, M. 2005. Modular total synthesis of lamellarin D. *J. Org. Chem.* 70: 8231–8234. (b) Yamaguchi, T.; Fukuda, T.; Ishibashi, F.; Iwao, M. 2006. The first total synthesis of lamellarin α 20-sulfate, a selective inhibitor of HIV-1 integrase. *Tetrahedron Lett.* 47: 3755–3757. (c) Fujikawa, N.; Ohta, T.; Yamaguchi, T.; Fukuda, T.; Ishibashi, F.; Iwao, M. 2006. Total synthesis of lamellarins D, L, and N. *Tetrahedron* 62: 594–604.
166. Bouillon, A.; Voisin, A. S.; Robic, A.; Lancelot, J.-C.; Collot, V.; Rault, S. 2003. An efficient two-step total synthesis of the quaterpyridine nemertelline. *J. Org. Chem.* 68: 10178–10180.
167. Ye, Y. Q.; Koshino, H.; Onose, J.-I.; Negishi, C.; Yoshikawa, K.; Abe, N.; Takahashi, S. 2009. Structural revision of thelephantin G by total synthesis and the inhibitory activity against TNF-α production. *J. Org. Chem.* 74: 4642–4645.
168. (a) Garg, N. K.; Sarpong, R.; Stoltz, B. M. 2002. The first total synthesis of dragmacidin d. *J. Am. Chem. Soc.* 124: 13179–13184. (b) Garg, N. K.; Caspi, D. D.; Stoltz, B. M. 2004. The total synthesis of (+)-Dragmacidin F. *J. Am. Chem. Soc.* 126: 9552–9553.
169. (a) Lin, S.; Yang, Z.-Q.; Kwok, B. H. B.; Koldobskiy, M.; Crews, C. M.; Danishefsky, S. J. 2004. Total synthesis of TMC-95A and -B via a new reaction leading to Z-enamides. Some preliminary findings as to SAR. *J. Am. Chem. Soc.* 126: 6347–6355. (b) Inoue, M.; Sakazaki, H.; Furuyama, H.; Hirama, M. 2003. Total synthesis of TMC-95A. *Angew. Chem. Int. Ed.* 42: 2654–2657. (c) Lin, S.; Danishefsky, S. J. 2002. The total synthesis of proteasome inhibitors TMC-95A and TMC-95B: Discovery of a new method to generate cis-propenyl amides. *Angew. Chem. Int. Ed.* 41: 512–515.
170. Koyama, H.; Kamikawa, T. 1997. Total syntheses of O4,9-dimethyl stealthins A and C. *Tetrahedron Lett.* 38: 3973–3976.
171. Chen, C.-M.; Liu, Y.-C. 2009. A concise synthesis of honokiol. *Tetrahedron Lett.* 50: 1151–1152.
172. Bringmann, G.; Ruedenauer, S.; Bruhn, T.; Benson, L.; Brun, R. 2008. Total synthesis of the antimalarial naphthylisoquinoline alkaloid 5-epi-4'-O-demethylancistrobertsonine C by asymmetric Suzuki cross-coupling. *Tetrahedron* 64: 5563–5568.
173. Kim, J. K.; Kim, Y. H.; Nam, H. T.; Kim, B. T.; Heo, J. N. 2008. Total synthesis of aristolactams via a one-pot Suzuki-Miyaura coupling/aldol condensation cascade reaction. *Org. Lett.* 10: 3543–3546.
174. Hayakawa, I.; Watanabe, H.; Kigoshi, H. 2008. Synthesis of ustalic acid, an inhibitor of Na+ , K+ -ATPase. *Tetrahedron* 64: 5873–5877.
175. (a) Fürstner, A.; Kennedy, J. W. J. 2006. Total syntheses of the tylophora alkaloids cryptopleurine, (-)-antofine, (-)-tylophorine, and (-)-ficuseptine C. *Chem. Eur. J.* 12: 7398–7410. (b) Bracher, F.; Daab, J. 2002. Total synthesis of the indolizidinium alkaloid ficuseptine. *Eur. J. Org. Chem.* 2288–2291.
176. Zhang, A.; Lin, G. 2000. The first synthesis of clausenamine-A and cytotoxic activities of three biscarbazole analogues against cancer cells. *Bioorg. Med. Chem. Lett.* 10: 1021–1023.
177. Waldmann, H.; He, Y.-P.; Tan, H.; Arve, L.; Arndt, H.-D. 2008. Flexible total synthesis of biphenomycin B. *Chem. Commun.* 5562–5564.
178. Fürstner, A.; Domostoj, M. M.; Scheiper, B. 2006. Total syntheses of the telomerase inhibitors dictyodendrin B, C, and E. *J. Am. Chem. Soc.* 128: 8087–8094.
179. Irlapati, N. R.; Adlington, R. M.; Conte, A.; Pritchard, G. J.; Marquez, R.; Baldwin, J. E. 2004. Total synthesis of pyridovericin. *Tetrahedron* 60: 9307–9317.
180. (a) Garfunkle, J.; Kimball, F. S.; Trzupek, J. D.; Takizawa, S.; Shimamura, H.; Tomishima, M.; Boger, D. L. 2009. Total synthesis of chloropeptin II (complestatin) and chloropeptin I. *J. Am. Chem. Soc.* 131: 16036–16038. (b) Shimamura, H.; Breazzano, S. P.; Garfunkle, J.; Kimball, F. S.; Trzupek, J. D.; Boger, D. L. 2010. Total synthesis of complestatin: Development of a Pd(0)-mediated indole annulation for macrocyclization. *J. Am. Chem. Soc.* 132: 7776–7783. (c) Wang, Z.; Bois-Choussy, M.; Jia, Y.; Zhu, J. 2010. Total synthesis of complestatin (Chloropeptin II). *Angew. Chem. Int. Ed.* 49: 2018–2022.
181. (a) Cudaj, J.; Podlech, J. 2010. Total synthesis of graphislactone G. *Tetrahedron Lett.* 51: 3092–3094. (b) Altemoeller, M.; Gehring, T.; Cudaj, J.; Podlech, J.; Goesmann, H.; Feldmann, C.; Rothenberger, A. 2009. Total synthesis of graphislactones A, C, D, and H, of ulocladol, and of the originally proposed and revised structures of graphislactones E and F. *Eur. J. Org. Chem.* 2130–2140.
182. Mihigo, S. O.; Mammo, W.; Bezabih, M.; Andrae-Marobela, K.; Abegaz, B. M. 2010. Total synthesis, antiprotozoal and cytotoxicity activities of rhuschalcone VI and analogs. *Bioorg. Med. Chem.* 18: 2464–2473.
183. Denton, R. M.; Scragg, J. T. 2010. A concise synthesis of dunnianol. *Synlett* 633–635.
184. Konakahara, T.; Kiran, Y. B.; Okuno, Y.; Ikeda, R.; Sakai, N. 2010. An expedient synthesis of ellipticine via Suzuki-Miyaura coupling. *Tetrahedron Lett.* 51: 2335–2338.

185. Chen, C.-H.; Chen, Y.-K.; Sha, C.-K. 2010. Enantioselective total synthesis of otteliones A and B. *Org. Lett.* 12: 1377–1379.
186. Taniguchi, T.; Ishibashi, H. 2008. Total synthesis of (±)-stemonamide, (±)-isostemonamide, (±)-stemonamine, and (±)-isostemonamine using a radical cascade. *Tetrahedron* 64: 8773–8779.
187. (a) Ye, Y. Q.; Koshino, H.; Onose, J.-I.; Yoshikawa, K.; Abe, N.; Takahashi, S. 2007. First total synthesis of vialinin A, a novel and extremely potent inhibitor of TNF-α production. *Org. Lett.* 9: 4131–4134. (b) Ye, Y. Q.; Koshino, H.; Onose, J.-I.; Yoshikawa, K.; Abe, N.; Takahashi, S. 2007. Expeditious synthesis of vialinin B, an extremely potent inhibitor of TNF-α production. *Org. Lett.* 9: 5074–5077.
188. Takakura, H.; Sasaki, M.; Honda, S.; Tachibana, K. 2002. Progress toward the Total synthesis of ciguatoxins: A convergent synthesis of the FGHIJKLM ring fragment. *Org. Lett.* 4: 2771–2774.
189. Fürstner, A.; Konetzki, I. 1998. Total synthesis of caloporoside. *J. Org. Chem.* 63: 3072–3080.
190. Choshi, T.; Sada, T.; Fujimoto, H.; Nagayama, C.; Sugino, E.; Hibino, S. 1997. Total syntheses of carazostatin, hyellazole, and carbazoquinocins B.-F. *J. Org. Chem.* 62: 2535–2543.
191. (a) Fürstner, A.; Grabowski, J.; Lehmann, C. W. 1999. Total synthesis and structural refinement of the cyclic tripyrrole pigment nonylprodigiosin. *J. Org. Chem.* 64: 8275–8280. (b) Fürstner, A.; Radkowski, K.; Peters, H.; Seidel, G.; Wirtz, C.; Mynott, R.; Lehmann, C. W. 2007. Total synthesis, molecular editing and evaluation of a tripyrrolic natural product: the case of "butylcycloheptylprodigiosin". *Chem. Eur. J.* 13: 1929–1945.
192. Bourdreux, Y.; Nowaczyk, S.; Billaud, C.; Mallinger, A.; Willis, C.; Murr, M. D.-E.; Toupet, L.; Lion, C.; Gall, T. L.; Mioskowski, C. 2008. Total synthesis of norbadione A. *J. Org. Chem.* 73: 22–26.
193. (a) Tsukano, C.; Sasaki, M. 2003. Total synthesis of gymnocin-A. *J. Am. Chem. Soc.* 125: 14294–14295. (b) Tsukano, C.; Ebine, M.; Sasaki, M. 2005. Convergent total synthesis of gymnocin-A and evaluation of synthetic analogues. *J. Am. Chem. Soc.* 127: 4326–4335.
194. Takaoka, S.; Nakade, K.; Fukuyama, Y. 2002. The first total synthesis and neurotrophic activity of clusiparalicoline A, a prenylated and geranylated biaryl from Clusia paralicola. *Tetrahedron Lett.* 43: 6919–6923.
195. (a) Kamisuki, S.; Takahashi, S.; Mizushina, Y.; Hanashima, S.; Kuramochi, K.; Kobayashi, S.; Sakaguchi, K.; Nakata, T.; Sugawara, F. 2004. Total synthesis of dehydroaltenusin. *Tetrahedron* 60: 5695–5700. (b) Takahashi, S.; Kamisuki, S.; Mizushina, Y.; Sakaguchi, K.; Sugawara, F.; Nakata, T. 2003. Total synthesis of dehydroaltenusin. *Tetrahedron Lett.* 44: 1875–1877. (c) Altemoeller, M.; Podlech, J. 2009. Total synthesis of neoaltenuene. *Eur. J. Org. Chem.* 2275–2282. (d) Altemoeller, M.; Podlech, J.; Fenske, D. 2006. Total synthesis of altenuene and isoaltenuene. *Eur. J. Org. Chem.* 1678–1684.
196. Chittiboyina, A. G.; Reddy, C. R.; Watkins, E. B.; Avery, M. A. 2004. First synthesis of antimalarial Machaeriols A and B. *Tetrahedron Lett.* 45: 1689–1691.
197. (a) Nicolaou, K. C.; Sun, Y.-P.; Peng, X.-S.; Polet, D.; Chen, D. Y.-K. 2008. Total synthesis of (+)-cortistatin A. *Angew. Chem. Int. Ed.* 47: 7310–7313. (b) Nicolaou, K. C.; Peng, X.-S.; Sun, Y.-P.; Polet, D.; Zou, B.; Lim, C. S.; Chen, D. Y.-K. 2009. Total synthesis and biological evaluation of cortistatins A and J and analogues thereof. *J. Am. Chem. Soc.* 131: 10587–10597.
198. Nagata, T.; Nakagawa, M.; Nishida, A. 2003. The first total synthesis of nakadomarin A. *J. Am. Chem. Soc.* 125: 7484–7485.
199. Fischer, D. F.; Sarpong, R. 2010. Total synthesis of (+)-complanadine a ising an iridium-catalyzed pyridine C-H functionalization. *J. Am. Chem. Soc.* 132: 5926–5927.
200. (a) Hamasaki A.; Zimpleman J. M; Hwang I.; Boger D. L 2005. Total synthesis of ningalin D. *J. Am. Chem. Soc.* 127: 10767–10770. (b) Hasse, K.; Willis, A. C.; Banwell, M. G. A. 2009. Total synthesis of the marine alkaloid ningalin B from (S)-proline. *Australian J. Chem.* 62: 683–691.

29 Boron in Weakly Coordinating Anions and Ionic Liquids

Andrea Vöge and Detlef Gabel

CONTENTS

29.1 INTRODUCTION

Ionic liquids (ILs) represent a novel class of solvents: they are molten salts with melting points below 100°C, composed only of ions, in which at least one ion is weakly coordinating. In some ILs, only the cation is weakly coordinating, in some only the anion; rarely, both ions are weakly coordinating. Many ILs are even liquid at room temperature, the so-called room temperature ionic liquids (RTILs).

In 1914, Paul Walden made a seminal development to pave the way for ILs: he discovered this novel class of liquids. He found that the salt ethylammonium nitrate, $[EtNH_3][NO_3]$, has a low melting point of 13°C. The interest in further investigations and possible applications for these low melting salts had not yet awakened at that time. It needed several decades for further publications in this research area, to reach a maximum in this age, i.a. in the context of "Green Chemistry," which has as goal to reduce and prevent the generation and use of hazardous substances.

One advantage of ILs is their negligible vapor pressure; hence, they do not evaporate like liquids normally do. Over the last years the interest in ILs as environment-friendly "green" alternative to traditional highly volatile organic compounds (VOCs) has increased immensely. VOCs contribute to a great deal to atmospheric pollution by industrial processes and hence contribute to climate changes and affect human health.

Most ILs show high thermal stability and some decompose first at temperatures above 400°C. Further positive attributes are their chemical stability, nonflammability, a wide electrochemical window, and high electrical conductivity. The latter typically ranges, for example, in the case of 1-alkyl-3-methylimidazolium salts, between 1.3 and 8.5 mS cm^{-1} (Ignatév et al., 2005). The conductivity is determined by the mobility of the ions, which in turn is i.a. dependent on ion size and association, and viscosity. Because of their ability to conduct current, they can replace commonly used, solvent-based, volatile, and flammable electrolytes, which tend to be corrosive. Examples of electrochemical applications of ILs as electrolytes are in fuel cells (Doyle et al., 2000; Gang et al., 1993), solar cells (Matsumoto and Matsuda, 2002a; Papageorgiou et al., 1996), batteries (Carlin et al., 1994; Fuller et al., 1997a,b; Ito and Nohira, 2000; MacFarlane et al., 1999; Schmidt et al., 2001; Webber and Blomgren, 2002), and capacitors (McEwen et al., 1997; Nanjundiah et al., 1997; Ue et al., 1994).

Commonly used solvents, such as water, organic liquids, and alcohols have a narrow region between freezing and boiling points, the "liquidus region," thus chemical processing is only possible in small temperature ranges. In contrast, many ILs have wide operation temperature ranges, some as great as 400°C (Bonhôte et al., 1996), where they are liquids (Brennecke and Maginn, 2001).

ILs can find use in reactions and in electrochemical and separation processes. Their negligible vapor pressure permits high-vacuum applications without any loss. ILs show remarkable solvation behavior; they dissolve a wide range of organic, inorganic, and polymeric materials, and hence offer the possibility to bring very different reagents into one phase so that the compounds are able to react with each other. In several cases ILs are able to increase product yields, simplify product retrieval, improve reaction rates and selectivities, and oftentimes recycling of the ILs is possible, leading to reduced production costs.

Several reactions could be carried out in ILs solvents; examples are transition-metal-catalyzed hydrogenation, hydroformylation, isomerization, dimerization, and coupling reactions (Holbrey and Seddon, 1999b). Carborane derivatives could also be synthesized successfully in ILs such as 1-*n*-butyl-3-methylimidazolium chloride (Li et al., 2008); the same IL was used for the preparation of functional carboranyl-containing dendrons (Galie et al., 2006; Mollard and Zharov, 2006).

In general, ILs are more viscous than water, which in some cases can be disadvantageous for applications, for example, in organic reactions. Molar mass, van der Waals interactions, and hydrogen bonds, among other factors, have influence on viscosity (Chiappe and Pieraccini, 2005).

ILs can be classified into two main groups: the first group contains simple salts with simple cation and anion, for example, ethylammonium nitrate, and the second group consists of binary systems, that is, mixtures of different ionic species, for instance, aluminum(III) chloride and 1,3-dialkylimidazolium chloride, where the melting point is dependent on the salt composition (Earle and Seddon, 2000).

In the past chloroaluminate(III) ILs underwent very extensive investigations. Because of the Lewis acidity of the anion these ILs promote reactions in which a Lewis acidic catalyst is needed, for example, Friedel–Crafts reactions (Earle and Seddon, 2000). A wide variety of reactions can be carried out in these ILs, often at low temperatures, in good yields, and with high selectivity, and in which the IL acts as a catalyst. This kind of IL displays, however, several disadvantages, such as moisture sensitivity—the contact with water causes exothermic evolution of hydrogen chloride—and difficulties in recycling of the ILs. Hence the design of water- and air-stable ILs is of increasing interest (Earle and Seddon, 2000).

In recent years, various investigations of three-dimensional *closo*-carboranes and polyhedral boron cluster anions have elucidated that some of them are, despite of their excellent stability, extremely weakly coordinating (Reed, 1998, 2005, 2009; Reed et al., 1999, 2000, 2003; Shelly and Reed, 1986; Stasko and Reed, 2002), hence carboranes and further boranes are of interest as anions in ILs (Dymon et al., 2008; Justus et al., 2008a,b; Larsen et al., 2000; Matsumi et al., 2009; Nieuwenhuyzen et al., 2009; Ronig et al., 2002; Zhu et al,. 2003).

29.1.1 The Choice of Cations and Anions for Ionic Liquids (ILs)

Chemical and physical properties of ILs depend on the combination of cations and anions, and their nature and possible substituents. In ILs, forces between the ions are predominantly Coulomb forces. Examples for bulky organic and often nonsymmetrical cations are imidazolium, pyridinium, pyrrolidinium, piperidinium, ammonium, and phosphonium, which all can carry different alkyl chains (Figure 29.1). The modification of the cation by *N*-alkylation plays an important role in the design of ILs as well. In general, an increasing chain length and hence an increasing size of the cation

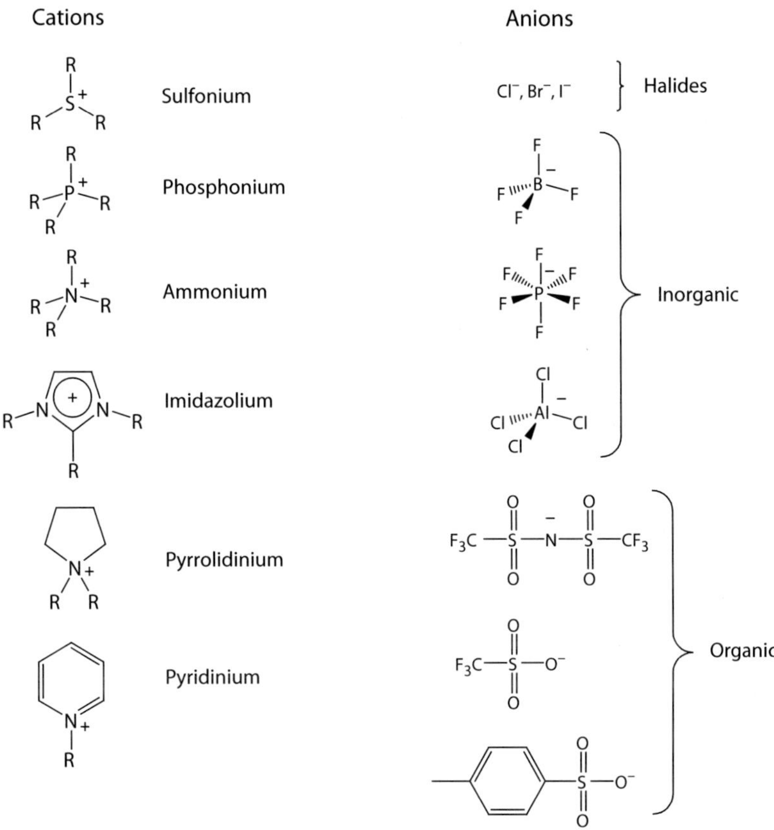

FIGURE 29.1 Examples for cations and anions of ILs.

results in a decrease of melting points, but in higher viscosity (Wasserscheid and Keim, 2000). With branched alkyl chains, higher melting points are found (Carda-Broch et al., 2003; Dzyuba and Bartsch, 2002; Huddleston et al., 2001).

The reason for the fact that ILs melt at low temperatures is not completely understood, but the asymmetry of the cation seems to play an important role (Seddon, 1996a,b); asymmetry causes packing inefficiency and hence prevents crystallization. Thus, for example, salts of the asymmetrical 1-butyl-3-methylimidazolium cation with only C_1 symmetry exhibit lower melting points than analogous salts with cations of higher symmetry, such as 1-butylpyridinium with C_{2v} symmetry (Holbrey et al., 2003). The same is true for symmetrically 1,3-dialkylsubstituted imidazolium, and imidazolium with two different *N,N′*-alkyl residues (Plechkova and Seddon, 2008; Seddon, 1997).

The absence of strong hydrogen bonds contributes to a decrease in the melting points as well (Suarez et al., 1998; Wilkes and Zaworotko, 1992). An example of the occurrence of hydrogen bonds is 1-butyl-3-methylimidazolium which forms hydrogen bonds with halide anions ($CH \cdots X^-$), with the hydrogen in position 2; in the case of *N*-methylated imidazolium, one hydrogen of the methyl group interacts with the halide as well (Abdul-Sada et al., 1986). As a consequence, halides as anions lead to higher melting points. In 2004, Kölle and Dronskowski found that hydrogen bonds are more important than Coulombic and dispersive forces for controlling ion packing in a crystal lattice.

The choice of anion has a remarkable influence on the balance of hydrophobicity and hydrophilicity, respectively, and hence the water miscibility of an IL. But the melting point is also affected by the anion (Wasserscheid and Keim, 2000). Besides the halides, frequently used inorganic anions are tetrafluoroborate and hexafluorophosphate. Fluorination of the anion leads to enhanced hydrophobicity and weaker hydrogen bonding and thus to lowered melting points; at the same time, it increases the thermal and electrochemical stability (Koch et al., 1995; Matsumoto et al., 2002b; McEwen et al., 1999; Pringle et al., 2003; Visser et al., 2002).

Large organic anions, for example, triflate (trifluoromethanesulfonate), tosylate (*p*-toluenesulfonate), and bistriflimide (bis(trifluoromethylsulfonyl)imide, can also be components of ILs. Figure 29.1 shows some of the commonly used cations and anions for ILs.

There are around one billion (10^{12}) possibilities for binary combinations of cation and anion, and one trillion (10^{18}) possibilities for ternary systems (Holbrey and Seddon, 1999b). This offers the perspective for the design of tailor-made ILs. For ILs, the term "designer solvents" was coined (Freemantle, 1998): not only the melting point of an IL can be varied by changes in structure, but also properties such as hydrophobicity, water miscibility, conductivity, viscosity, and density can be optimized according to the actual demands. Roughly estimated, the cation is accountable for the physical features of an IL, for instance density, viscosity, and melting point, the anion for the chemical properties and reactivity (Plechkova and Seddon, 2008).

The most popular cations in ILs are the *N,N′*-dialkylated imidazolium ions, because of their suitable physical properties and their ease of preparation. Several studies could confirm the adequacy of imidazolium as cations, as they afford high conductivities and low viscosities together with low melting points. In general, the melting point decreases with increasing alkyl chain length in the region between three and five methylene units. By increasing the chain length above five carbon atoms, some salts have higher melting points because of their liquid crystalline behavior (Forsyth et al., 2004). For 1,3-dialkylated imidazolium, increasing alkyl chain length decreases the density in the corresponding salts (Wasserscheid and Keim, 2000).

In comparison with the oftentimes preferred imidazolium cations, the choice of a quarternary ammonium as cation is advisable for some applications; the use of ILs with 1,3-dialkylimidazolium cation in high-energy electrochemical devices, for example, 4 V lithium batteries, is not possible because of the electrochemical instability of the cation. ILs containing quaternary ammonium ions are more resistant against reduction and oxidation, and the ILs with electrochemically stable anions display much wider electrochemical windows (Zhou et al., 2005).

The melting points of many ILs are often difficult to determine, because of supercooling phenomena; the temperature for phase transition during heating is higher than during cooling, that is, an IL below its freezing point can still be a liquid (Chiappe and Pieraccini, 2005).

29.1.1.1 Investigations of ILs Containing Imidazolium or Ammonium Cations

During the last decades the number of publications on ILs has dramatically increased because of the growing interest in this novel class of liquids with their exceptional properties. Therefore, in the following, only a few examples of the many investigations reported can be mentioned.

Ngo et al. (2000) investigated thermal properties of several imidazolium ILs with varying anions (ordered by anion stability: $PF_6^- > N(SO_2C_2F_5)_2^- > N(SO_2CF_3)_2^- \approx BF_4^- > C(SO_2CF_3)_3^- \approx AsF_6^- >> I^-$, Br^-, Cl^-). With inorganic anions the ILs decompose endothermically, organic anions lead to exothermic thermal decomposition. The thermal stability is much reduced for halides (<300°C). Many of these imidazolium salts show supercooling and are liquid at room temperature. The melting points range from −15°C for 1-ethyl-3-methylimidazolium, $[EMIM]^+$, with $N(SO_2CF_3)_2^-$ as anion to 213°C for the permethylated imidazolium with iodide as anion. More symmetric cations lead to higher melting points and *vice versa*. A longer alkyl chain (propyl as compared to ethyl) lowers the melting points, a branching of the chain (*i*-propyl) increases the melting point, for example, the melting point of 1-propyl-3-methylimidazolium with PF_6^- is 40°C, that of 1-ethyl-3-methylimidazolium is 62°C, and the corresponding *i*-propyl derivative shows a melting point of 102°C.

Suarez et al. (1998) gave an account of the physical properties of air- and water-stable ILs bearing 1-butyl-3-methylimidazolium cation, $[BMIM]^+$, with tetrafluoroborate BF_4^- and hexafluorophosphate PF_6^-, respectively, as anions. These viscous liquids show a wide liquidus range (down to −81°C) and a wide electrochemical window (up to 7.0 V). Infrared spectroscopy shows a characteristic stretching of an aromatic CH···F⁻ hydrogen bond. The hydrogen bond to the PF_6^- anion is stronger than to BF_4^-; as a consequence, the salt with hexafluorophosphate as anion displays higher viscosity and density. The stronger cation/anion interactions of these salts are also reflected in their conductivities above 5.9°C: the compound with BF_4^- as anion has higher conductivity than the PF_6^- containing salt. The glass transition temperatures for [BMIM][BF_4] and [BMIM][PF_6] are −81 and −61°C, respectively.

Holbrey and Seddon (1999a) studied 1-alkyl-3-methylimidazolium salts with tetrafluoroborate as counterion ([C_nMIM][BF_4] with $n = 0$–18), which are air and water stable. With alkyl chains of $n = 2$–10, they are isotropic RTILs and show a wide liquid range. Prolongation of the alkyl chain results in low-melting mesomorphic crystalline solids which exhibit an enantiotropic smetic A mesophase. By increasing the chain length, the thermal range of the mesophase increases. The non-1-alkylated ($n = 0$) and 1-methylated ($n = 1$) imidazolium salts are crystalline solids with low melting points. The ILs with $n = 2$–9 display a tendency to supercool and form glasses on cooling. Above $n = 9$ the melting points rapidly increase.

29.2 BORON IN ILs

29.2.1 Tetrahedral Boron in ILs

Most of the tetrahedral boron compounds bear a negative charge on the boron atom, the so-called borates. They are anions, except when the negative charge is compensated by other, positively charged groups such as ammonium. Therefore, most ILs with a tetrahedral boron contain the boron in the anion of the salt. Some of the compounds described in the literature are shown in Figure 29.2.

ILs with BF_4^- as anion (**A** in Figure 29.2) are widely commercially available, combined with the usual cations such as $EMIM^+$. These ILs, when reacted with the trimethylsilyl ethers of acids, can react to mixed borates (**B** and **C**) (Schreiner et al., 2009).

Perfluorinated side chains have been introduced in ILs to reduce viscosity (**F** in Figure 29.2). This is motivated by the general observation that perfluorinated molecules show greatly reduced van der Waals interactions. As an analog of the commercially available tetrafluoroborate anion, substitution of one of the fluorine atoms by a perfluorinated alkyl side chain has been reported (Zhou et al., 2004, 2005). With imidazolium as cation, they have lower melting points than the BF_4^- salts, and low viscosity also for longer alkyl chains of the cation.

Wasserscheid et al. (2002) have described a highly Brønstedt-acidic IL consisting of tetrakis(hydrogensulfato)borate as anion (**G** in Figure 29.2) and $[BMIM]^+$ or $[OMIM]^+$ as cation. In a two-phase system of sulfuric acid and benzene, the ILs promoted the electrophilic alkylation of benzene with 1-decene, by increasing the contents of acidic Friedel–Crafts catalyst in the organic phase.

A single class of ILs with boron in the cation part of an IL has been described by Fox et al. (2005) (**I** in Figure 29.2). In these imidazolium-based ILs, the ethylene moiety of a 2,2-dimethylpropyl side

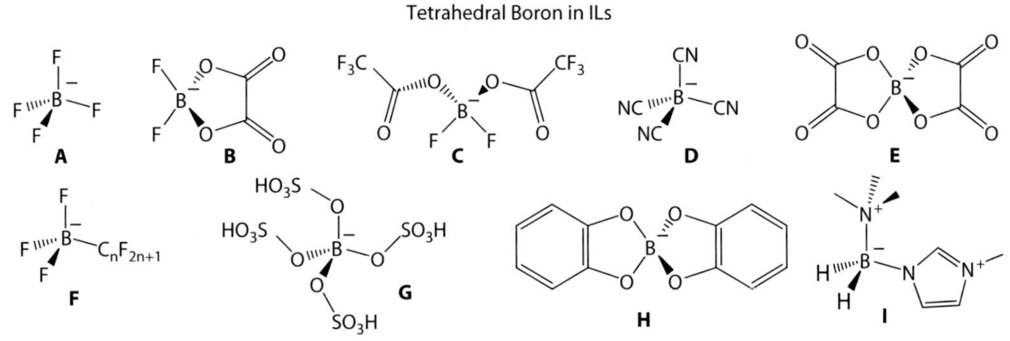

Tetrahedral Boron in ILs

A B C D E F G H I

FIGURE 29.2 Examples of tetrahedral boron in ILs.

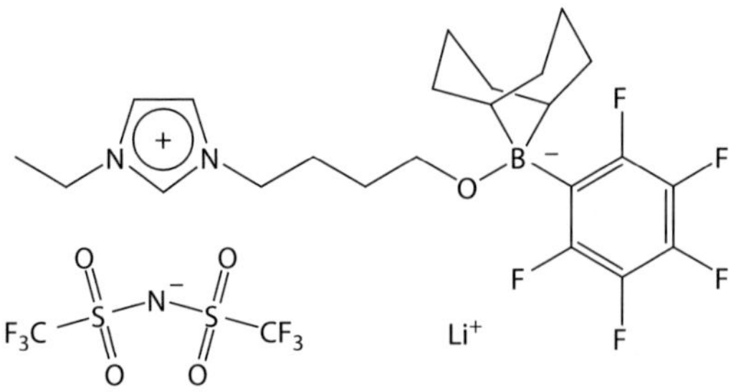

FIGURE 29.3 Blend of zwitterionic organoborate and lithium bis(trifluoromethylsulfonyl) imide. (Adapted from A. Narita et al., *Chem. Commun.* 2006, 1926).

chain of the imidazolium is replaced by an isosteric and isoelectronic B–N unit. In these cations, the overall positive charge of the cation is more delocalized than in the carbon analogue.

A zwitterionic tetracoordinate organoborate compound with lithium bis(trifluoromethylsulfonyl) imide as lithium ion conductive IL has been described by Narita et al. (2006) (Figure 29.3).

The idea was to increase the lithium transfer number, which is low in most of ILs, IL-salt mixtures, or IL-acid mixtures. Zwitterions with imidazolium cations covalently bound to anions, such as sulfonate, carbonate, imide, or borate, are solids at room temperature, despite their IL-like structure; mixing with bis(trifluoromethylsulfonyl) imide, however, leads to blends that are liquid at room temperature. The organoborate-containing zwitterion-lithium salt mixture (Figure 29.3) displays a high lithium transfer number of 0.69, a high-ionic conductivity of 3.0×10^{-5} S cm^{-1} at 50°C, and a low glass transition temperature (–35°C).

A different tetracoordinate boron species, where the boron bears a positive charge is called boronium cation (Figure 29.4, left side). In 2010, Rüther et al. described boronium-cation-based RTILs as novel electrolytes for rechargeable Lithium batteries (Figure 29.4, right side).

These boronium-containing RTILs exhibit sufficient conductivities and electrochemical windows (4.3–5.8 V), they are stable up to 238 and 335°C, respectively, and reversible charge–discharge cycling of batteries with good capacity retention was possible. Thus, these compounds are of interest as electrolytes for various electrochemical applications.

29.2.2 Weakly Coordinating Boron Clusters as Anions in ILs

29.2.2.1 Weakly Coordinating Boron Clusters

Rosenthal published in 1973 his studies to disprove the myth of "noncoordinating" anions. He could show that anions such as ClO_4^-, NO_3^-, and BF_4^- are noncoordinating only in aqueous solution; hence, it is more appropriate to speak about weakly coordinating anions.

There are several desired properties for this kind of anion: a small negative charge and a high charge delocalization over the entire anion; the latter implies that anions with large size show better

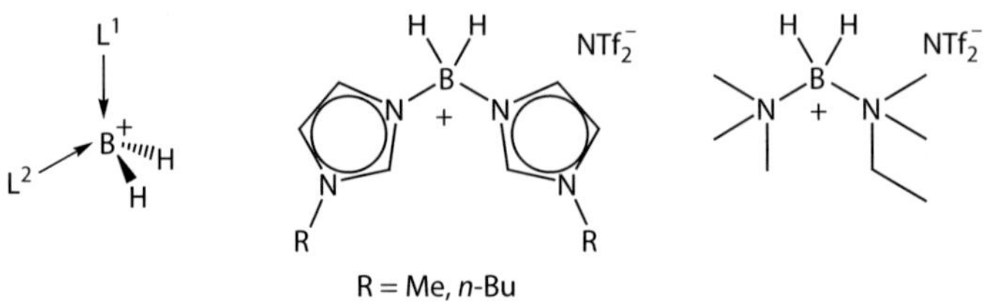

FIGURE 29.4 Boronium cation (L = Lewis base, e.g., NR_3, *N*-Alkyl-imidazole) and boronium-cation-based RTILs. (Adapted from T. Rüther, et al., *Chem. Mater.* 2010, 22, 1038.)

delocalization. Nonnucleophilicity and a minimally basic surface of the anion lead to weak coordination as well. Kinetic and thermodynamic stability is also a very important point; anions should not dissociate into smaller fragments (as an example, fluorinated anions, for instance BF_4^-, PF_6^-, and SbF_6^-, lose a fluoride when attacked by a nucleophile (Seppelt, 1993)). Anions should show resistance against oxidation as well, because the corresponding electrophilic cation often acts as oxidizing agent; as an example, BPh_4^-, can undergo oxidation by metal ions, such as Ce(IV), Fe(III), or Ir(IV) (Strauss, 1993).

An alternative to the Lewis acid/base pairs anions with a Lewis acidic central atom is opened by derivatives of the polyhedral 1-carba-*closo*-dodecaborate(1-) anions, such as $[CB_{11}H_{12}]^-$, due to their specific properties: despite of their excellent stability, they are some of the most inert and least nucleophilic anions presently known (Larsen et al., 2000). The resultant extremely weak coordination is founded in a charge distribution over the entire 12-vertex carborane anion. The 10-vertex anion $[CB_9H_{10}]^-$ and its halogenated and methylated derivatives coordinate more strongly than 12-vertex carboranes (Tsang et al., 2000). (In this chapter, monocarbon carboranes will be referred to as carboranes.)

Highly reactive cations, for example, R_3Si^+ and R_2Al^+ (R = Me, Et, *i*-Pr), generally decompose weakly coordinating anions; in contrast, with halogenated carboranes they form coordination compounds in the solid state (Si- and Al-distances to the anion show molecular character and not ionic), but in solution they behave "ion-like" (Krossing and Raabe, 2004).

Coordination of cations to carboranes is worth noting. Electrostatic potential calculations (Zharov et al., 2004) show that the coordination of the cation to the carborane anion, $[CB_{11}H_{12}]^-$, takes place in the region of boron atoms 7 to 12, despite the fact that most of the negative charge is located at the most electronegative atom, the carbon in position 1. Not only the cation coordination, but also electrophilic substitutions occur in this region. An explanation, given by Zharov et al. (2004), for this unexpected behavior is shown in Figure 29.5; nearly half of the negative charge is located on vertex 1, but the decisive factor for the coordination of the cation in the area of boron 7 to 12 is that there exists a local dipole caused by the hydrogen atoms with the positive center in the periphery of the carborane as well. Thus, the location of lowest energy for a cation is in the region of boron atoms 7 to 12 (Körbe et al., 2006).

The first reports about carboranes $[CB_{11}H_{12}]^-$ as weakly coordinating anions were published in 1986 by Shelly and Reed. A very weakly coordinating anion should offer the possibility to stabilize the most reactive, coordinatively unsaturated cations. Carboranes are promising candidates as anions because of their large size, spherical shape, remarkable chemical stability, and because their B–H groups are only weakly coordinating. Shelly and Reed had chosen $[Fe(TPP)]^+$ (TPP = tetraphenylporphyrinate)

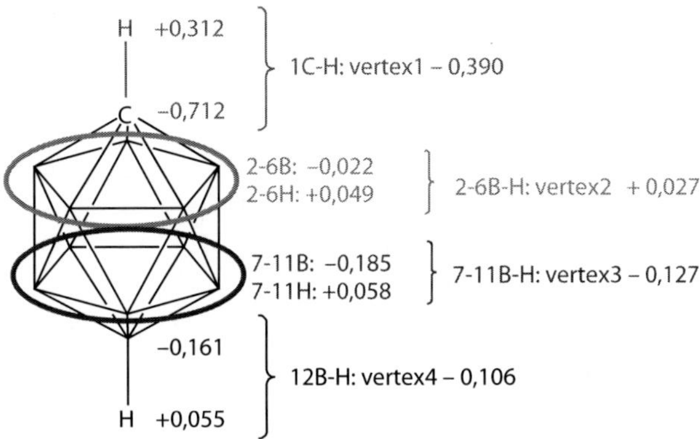

FIGURE 29.5 Natural atomic charges for $[CB_{11}H_{12}]^-$ anion. In this and subsequent figures, an unsubstituted corner of the polyhedron without symbol or letter represents a BH unit, a substituted corner without symbol or letter, a B atom. (Adapted from I. Zharov et al., *J. Am. Chem. Soc.* 2004, 126, 12033.)

as cation and could demonstrate that the carborane anion in combination with this cation was the least coordinating anion known at that time.

Reed (1998) published further work about carboranes as a new class of weakly coordinating anions. Carboranes have a great potential because of their favorable properties: they show inertness despite easy possible functionalizations, they can be seen as three-dimensional analogs of benzene, they show high symmetry and dipolarity and so on. Because carboranes have no lone electron pairs and no π-electrons, they exhibit very low nucleophilicity. Reactions at the carbon atom of the cluster are possible; after deprotonation the nucleophilic C atom can be alkylated. Further functionalization by electrophilic substitution, for example, hexahalogenation, is possible as well, resulting in $[CB_{11}H_6X_6]^-$ (X = Cl, Br, I) with the halogen atoms on the boron vertex opposite the carbon atom and in the lower belt. Hexahalogenated carborane anions are more stable toward oxidation and are less coordinating than nonhalogenated carboranes. They exhibit greater acid stability than, for example, BAr_4^{F-}: the carborane-acids are stable at room temperature, whereas fluorinated tetraarylborates decompose.

Several investigations have been carried out to find the least coordinating anion; hexabromocarborane $[CB_{11}H_6Br_6]^-$ and hexachlorocarborane $[CB_{11}H_6Cl_6]^-$, respectively—dependent on the experiment in question—seemed to be the least coordinating ones in 1998 (Reed, 1998). An even less coordinating anion is the silver–carborane complex ion $[Ag(CB_{11}H_6Br_6)_2]^-$. The η^3-coordination of the two carboranes to silver takes place via three bromine atoms, and thus the silver has an octahedral coordination. As the negative charge is distributed over an even bigger volume, this anion is even more weakly coordinating than an icosahedral carborane anion (Xie et al., 1994).

The weakest coordination was found in 2002 for the fluorinated $[i\text{-}Pr_3Si][1\text{-}MeCB_{11}F_{11}]$ (Stasko and Reed, 2002). Tsang et al. (2000) observed a similar behavior for the 10-vertex carboranes: with increasing halogenation level the coordination decreases, with the fluorinated anion showing the least, the iodinated anion the strongest coordination. The synthesis of a permethylated carborane has been described (King et al., 1996). They can stabilize trimethylgermanium, trimethyltin, and trimethyllead cations (Zharov et al., 2004).

Ammonioundecafluoro-*closo*-dodecaborate(1-) and some *N*-trialkylated derivatives have been described by Ivanov et al. (2003a) as weakly coordinating anions. Equally, the dodecafluorododecaborate has been described as weakly coordinating, despite it being a dianion (Ivanov et al., 2003b). Both salts crystallize with the triphenylmethyl cation (Ph_3C^+), and interaction of a cluster fluorine atom with the carbocation is found in the crystal.

With hexachlorinated carborane as a counterion, benzene could be protonated and isolated as exceptionally stable benzenium ion salt: $[C_6H_7][CB_{11}H_6Cl_6]$. This superacid is nonoxidizing and has a very weakly coordinating anion (Reed et al., 1999). Various further arene salts could be isolated (Reed et al., 2003): protonated toluene, *m*-xylene, mesitylene, and hexamethylbenzene with different halogenated and partly *B*-methylated carboranes as anions. Figure 29.6 depicts the possible substitution variations, such as halogenation and alkylation, on a carborane anion. The methylation

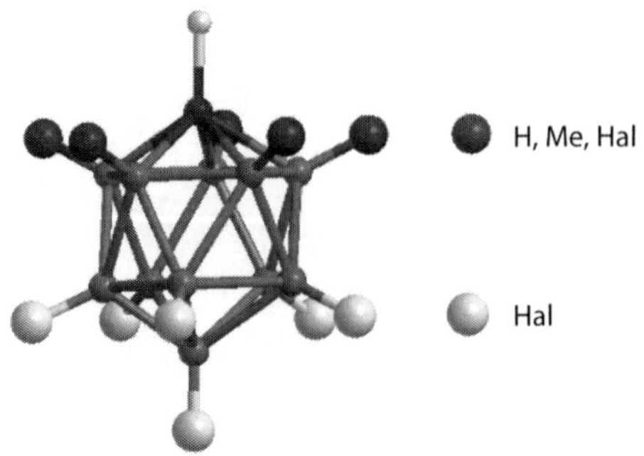

FIGURE 29.6 Weakly coordinating carborane anions $[CHB_{11}R_5X_6]^-$ (R = H, Me or Hal; X = Hal). (Adapted from C. A. Reed, *Acc. Chem. Res.* 2009, DOI: 10.1021/ar900159e.)

in position 2 to 6 of hexahalogenated carboranes, resulting in better solubility in organic solvents, was described by Stasko and Reed (2002).

Superacids have an acidity greater than 100% sulfuric acid and they are used for, for example, stabilizing carbocations. In 1998, Reed described the first hydronium salts, such as $[H_9O_4][CB_{11}H_6Br_6]$, where the hydronium ion is trihydrated. Further results about carborane acids were reported by Reed et al. in 2000 and 2005. Because of their very weakly coordinating anions, acids of halogenated carboranes, for example, $H[CB_{11}H_6X_6]$ (X = Cl, Br) and $H[CHB_{11}Cl_{11}]$ are the strongest and most robust Brønsted acids known so far. These superacids are able to protonate neutral molecules.

Reed et al. (2000) described the reaction of superacids $H[CB_{11}H_6X_6]$ (X = Cl or Br) with fullerene; the fullerene carbocations HC_{60}^+ (addition of an electrophile) and $C_{60}^{\cdot+}$ (removal of an electron) could be obtained, which had previously not been accessible because of the decomposition of fullerene by superacids (Bausch et al., 1991).

In addition to the carborane-stabilized cations mentioned above, the following listing shows further cations that could be stabilized by carborane anions:

- Carbocations (e.g., Me^+, Et^+, *i*-Pr^+, $EtMe_2C^+$, $MeC_5H_8^+$ with $[CHB_{11}Me_5X_6]^-$ (X = Br, Cl), Kato and Reed, 2004a; Kato et al., 2004b),
- Vinyl cations, $R_2C = C^+R$, with $[CB_{11}H_6Br_6]^-$ (Müller et al., 2004),
- Silylium cations (R_3Si^+, e.g., R = *i*-Pr, Et with $[CB_{11}H_6Cl_6]^-$ and $[CHB_{11}Cl_{11}]^-$, Hoffmann et al., 2006; Kim et al., 2002a),
- Phosphazene cations (e.g., $[H(NPCl_2)_3]^+$ and $[MeN_3P_3Cl_6]^+$ with $[CHB_{11}R_5X_6]^-$ (R = H, Me; X = Br, Cl), Zhang et al., 2006),
- Hydronium ions ($H_5O_2^+$, $H_7O_3^+$ $H_9O_4^+$ with $[CB_{11}H_6X_6]^-$, Reed, 1998),
- fluorinated benzyl-type carbocations (e.g., $[(p\text{-}FC_6H_4)(CH_3)CF]^+$, $[(p\text{-}FC_6H_4)_2CF]^+$ with $[CHB_{11}I_{11}]^-$, Douvris et al., 2007),
- Aluminum cations $[Et_2Al]^+$ with $[CB_{11}H_6X_6]^-$ (X = Br, Cl) (Kim et al., 2002b), and
- Me_3M^+ (M = Sn, Ge, Pb) with permethylated carboranes $[CB_{11}Me_{12}]^-$ (Zharov et al., 2004).

Kato and Reed (2004a) described carborane compounds as strong R^+ alkylating agents, where the weakly coordinating carborane anion acts as a leaving group. Reaction of $CH_3[CHB_{11}Me_5X_6]$ (X = Cl, Br) with alkanes produces various carbenium ions by hydride abstraction. In addition, the alkylation of benzene is possible without Friedel–Crafts catalyst, and phosphorus compounds can be alkylated (Kato et al., 2004b).

Similar results were reported by Reed in 2009: previously, triflic acid, methyl triflate, and trialkylsilyl triflate have been used as electrophilic sources of H^+ for protonation, of CH_3^+ for alkylation, and of R_3Si^+ for silylation, respectively. Replacement of these anions by $[CHB_{11}R_5X_6]^-$ (R = H, Me; X = Cl, Br) increases the electrophilicity of the cations and avoids the nucleophilic chemistry of the anion; this allows various reactions that are not possible with commonly used reagents.

29.2.2.2 Boron Clusters as Anions in ILs

The three-dimensional *closo*-carboranes and polyhedral borane clusters are of special interest as materials and building blocks because of their high chemical and thermal stability. In recent years, triggered by a growing interest in ILs, various investigations of numerous boron cluster anions have been described. Studies of divalent anions, such as perchlorinated *closo*-dodecaborate and *closo*-decaborate (Nieuwenhuyzen et al., 2009), have been carried out as well as studies of monovalent anions, for instance, alkylated stannadodecaborates (Ronig et al., 2002), alkylated ammoniododecaborates (Justus et al., 2008a,b), a cobaltabis(dicarbollide) (Nieuwenhuyzen et al., 2009), (*C*-alkylated) monocarbon-carboranes (Dymon et al., 2008; Larsen et al., 2000; Zhu et al., 2003;), a *nido*-carborate (Dymon et al., 2008; Nieuwenhuyzen et al., 2009), and an *ortho*-carborate

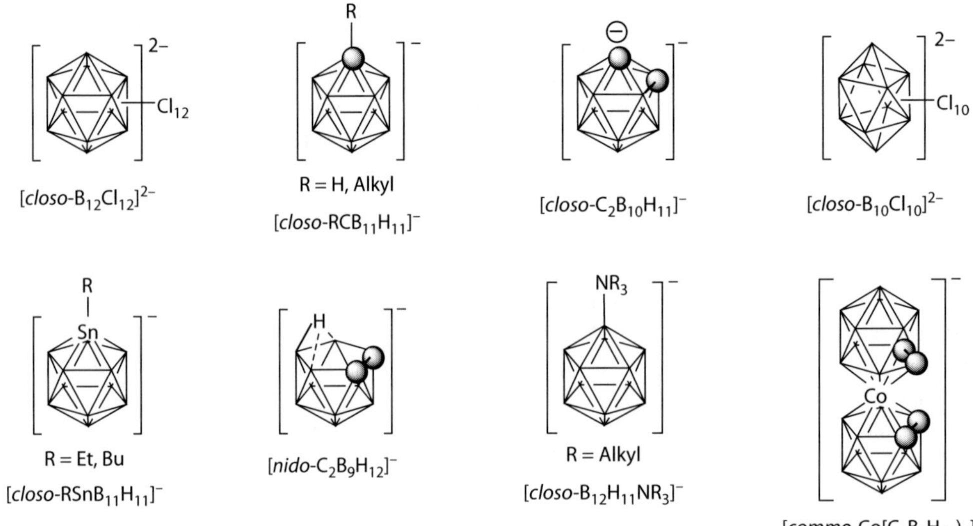

FIGURE 29.7 Boron cluster anions for ILs. In this chapter, the polyhedral boron clusters are represented such that every corner carries one B–H unit, unless a substituent is drawn explicitly.

(Matsumi et al., 2009) (Figure 29.7). (In this chapter, the CB_{11} unit will be named carborane, not reflecting its charge in the name.)

A a circle represents a C–H unit or a C in the cases where a charge is specified.

On the basis of the knowledge that carboranes are the least coordinating anions and hence their salts have low melting points, Larsen et al. (2000) introduced for the first time carboranes as anions into the repertoire of ILs. In combination with unsymmetrically substituted *N,N′*-dialkylimidazolium cations, carboranes show low melting points in the range of 45–156°C, mainly dependent on the alkylation pattern of the imidazolium cation (Figure 29.8). Carborane salts with cations other than imidazolium usually melt or decompose at temperatures above 300°C.

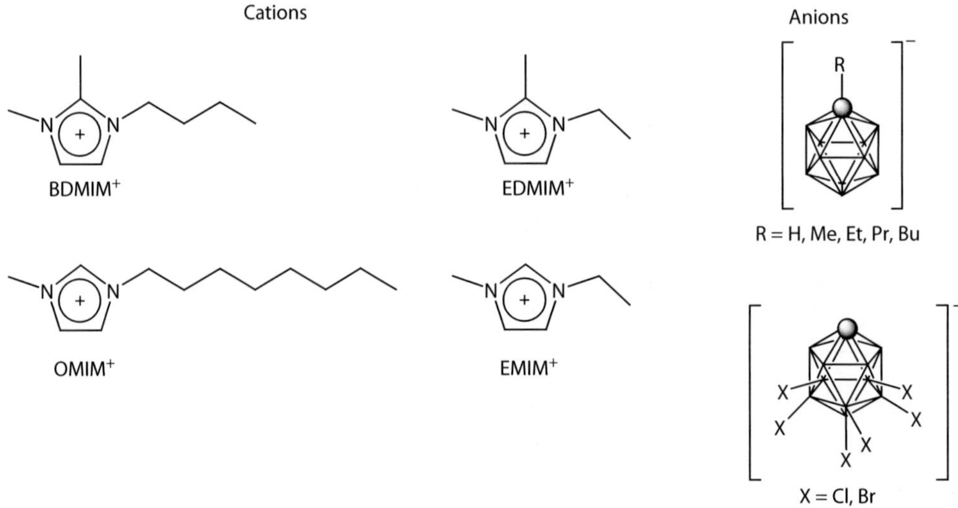

FIGURE 29.8 *N,N′*-Dialkylimidazolium cations and carborane anions. (Adapted from A. S. Larsen, *J. Am. Chem. Soc.* 2000, 122, 7264.)

The different alkylation pattern of the imidazolium ion used (1-butyl-2,3-dimethyl- $[BDMIM]^+$, 1-ethyl-2,3-dimethyl- $[EDMIM]^+$, 1-octyl-3-methyl- $[OMIM]^+$, and 1-ethyl-3-methylimidazolium, $[EMIM]^+$), combined with (*C*-alkylated) $[CB_{11}H_{12}]^-$, and $[CB_{11}H_6X_6]^-$ (X = Cl, Br), respectively, as anions, were used to identify the factors which contribute to low melting points. Carboranes offer many possibilities to be functionalized.

C-Alkylated carboranes coordinated with $[EMIM]^+$ cation exhibit quite low melting points in the range of 45–64°C, in contrast to the related nonalkylated carborane salts, which have higher melting points, between 114 and 139°C. Chlorination of the cluster slightly lower the melting points, for example, the salt with nonhalogenated carborane, $[EMIM][CB_{11}H_{12}]$, exhibits a higher melting point (122°C) than the hexachloro substituted $[EMIM][CB_{11}H_6Cl_6]$ (114°C).

For imidazolium cations with longer alkyl chains, the salts have lower melting points, for example, $[OMIM][CB_{11}H_{12}]$ has a melting point of 70°C. *C*-Methylation of the cation in position 2 unexpectedly leads to higher melting points, for example, *C*-alkylated $[EDMIM][CB_{11}H_6Cl_6]$ exhibits a higher melting point (137°C) than $[EMIM][CB_{11}H_6Cl_6]$ (melting point 114°C). This increase of the melting point by *C*-alkylation is unexpected because the C(2)-H of the imidazolium cation is supposed to form hydrogen bonds to anions. One explanation for this phenomenon is that the *C*-methylation masks the asymmetric *N*-alkylation in position 1 and 3 and hence increases the packing efficiency.

X-ray structural investigations arrive at the conclusion that packing inefficiency is the main cause for the low melting points of imidazolium–carborane salts (Larsen et al., 2000).

In 2003, Zhu et al. reported on a 1-carba-*closo*-dodecaborate(1-) with *N*-pentylpyridinium as the cation which is liquid at room-temperature (Figure 29.9). This RTIL exhibits a quite low melting point of 19°C, lower than the melting points of corresponding imidazolium salts (70°C and higher) described by Larsen et al. (2000). They carried out several palladium-catalyzed dehalogenation reactions in this new solvent. In comparison with conventional organic solvents, the reaction times are shorter. In the case of 1,2,4-trichlorobenzene, dechlorination shows very high product selectivity, resulting in 1,2-dichlorobenzene.

Küppers et al. (2007) reported on fluorinated *closo*-carboranes with silyl cations, $Me_3Si[RCB_{11}F_{11}]$ with R = H or Et. These compounds are salts and not molecular adducts, as the conductivity of $Me_3Si[EtCB_{11}F_{11}]$ in the melt (90°C) is 2.5 mS cm^{-1}, a conductivity typical for ILs. Because of its conductivity and a melting point below 100°C, the ethylated carborane salt can be called IL; its use as IL is, however, limited because of its high reactivity.

Dymon et al. (2008) designed ILs with carboranes as anions and alkylimidazolium and pyridinium, respectively, as cations. In addition to the already known $[closo\text{-}CB_{11}H_{12}]^-$, they investigated the corresponding *C*-alkylated carborane $[closo\text{-}RCB_{11}H_{11}]^-$ (R = Me, Bu) and *nido*-carborane $[nido\text{-}C_2B_9H_{12}]^-$ salts (Figure 29.10).

nido-salts have the advantage that they exhibit melting points below 100°C (in the range of 47–97°C) and they melt at lower temperatures than the corresponding *closo*-carborane salts (melting points range from 70 to 134°C). A possible explanation could be the larger asymmetry of the *nido*- compared to the *closo*-carborane, similar to the influence of the cation asymmetry on melting

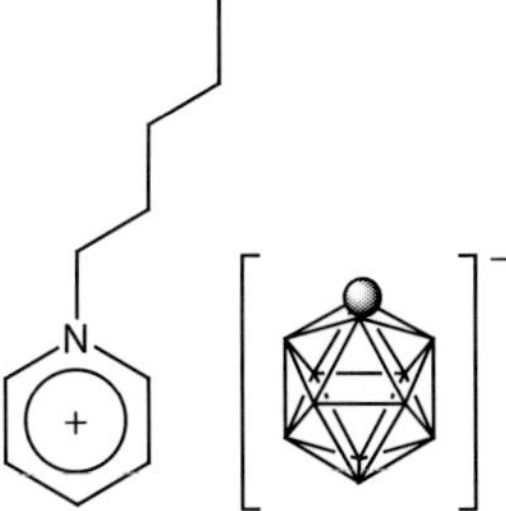

FIGURE 29.9 RTIL with 1-carba-*closo*-dodecaborate(1-) and *N*-pentylpyridinium. (Adapted from Y. Zhu, et al., *Appl. Organometal. Chem.* 2003, 17, 346.)

Cations

Anions

$[EMIM]^+$: $R = CH_3$, $R' = H$, $R'' = C_2H_5$
$[OMIM]^+$: $R = CH_3$, $R' = H$, $R'' = C_8H_{17}$
$[DMEIM]^+$: $R = CH_3$, $R' = CH_3$, $R'' = C_2H_5$

$R = C_4H_9, C_6H_{13}, C_8H_{17}$

R = H, Me, Bu

FIGURE 29.10 Alkylated imidazolium and pyridinium cations with (*C*-alkylated) carborane and *nido*-carborane anions. (Adapted from J. Dymon et al., *Dalton Trans.* 2008, 2999.)

points. The same can be said about the *C*-alkylated *closo*-carboranes; they have lower symmetry hence the melting points are lower than for nonalkylated [*closo*-$CB_{11}H_{12}]^-$. Unexpectedly, in the case of alkylpyridinium, the melting points increase with increasing chain length in contrast to the alkylimidazolium cations, where in general an increase in chain length causes a decrease in melting points (Holbrey and Seddon, 1999a).

In 2002, low-melting Sn-alkyl-stannadodecaborate derivatives were reported by Ronig et al., where one boron–hydrogen group in the icosahedral boron cluster is substituted by a tin atom, $[SnB_{11}H_{11}]^{2-}$. The tin atom can be alkylated; with ethyl and butyl groups, respectively, and combined with $[EMIM]^+$ and $[BMIM]^+$ as cations, the salts have low melting points (55–106°C) (Figure 29.11).

As expected, the lowest melting point was obtained with the longest alkyl chain (butyl) on both cation and anion: [BMIM][1-Bu$SnB_{11}H_{11}$]. The salts are stable toward air and water, and nonhygroscopic.

In 2009, Nieuwenhuyzen et al. published new nonhygroscopic ILs containing different boron clusters as counterions: [*commo*-3,3′-Co(1,2-$C_2B_9H_{11})_2]^-$, [*nido*-$C_2B_9H_{12}]^-$, [*closo*-$B_{10}Cl_{10}]^{2-}$, and [*closo*-$B_{12}Cl_{12}]^{2-}$ combined with the commonly used imidazolium ($[C_nMIM]^+$ = 1-alkyl-3-methylimidazolium, $n = 2, 4, 6, 8, 10, 12, 14, 16, 18$) and phosphonium ($[P(C_nH_{2n+1})_3C_mH_{2m+1}]^+$) cations (Figure 29.12).

The melting points of ILs with different carborane anions increase in the following order: $[Co(C_2B_9H_{11})_2]^- < [C_2B_9H_{12}]^- < [B_{10}Cl_{10}]^{2-} < [B_{12}Cl_{12}]^{2-}$.

The imidazolium/boron cluster salts can be separated into two groups: the first one includes the monovalent anions. These compounds exhibit low melting points, around or below 100°C and no thermotropic metaphases. The imidazolium salts of cobaltabis(dicarbollide) with 4, 6, 8, 10, 12, and 14 carbon atoms in the side chain of the imidazolium cation are liquid at room temperature.

The second group contains the divalent anions, which melt at higher temperatures, as expected from the Kapustinskii equation (Kapustinskii, 1956) which predicts higher melting points and higher melting enthalpies for doubly negatively charged ions because of higher Coulomb interactions.

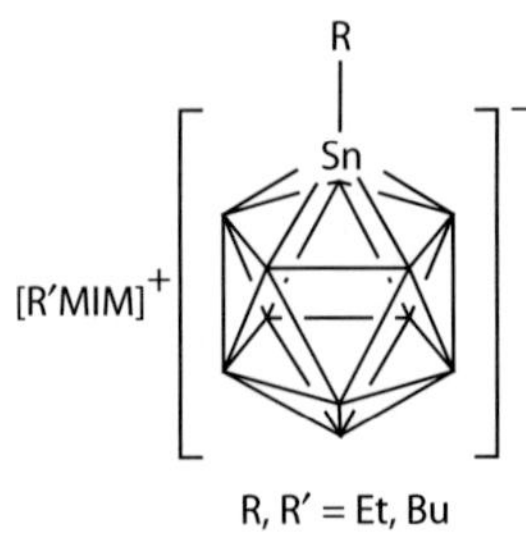

FIGURE 29.11 ILs with alkylated stanna-*closo*-dodecaborate(1-) and *N*-alkylated 1-methylimidazolium. (Adapted from B. Ronig, I. Pantenburg, L. Wesemann, *Eur. J. Inorg. Chem.* 2002, 319.)

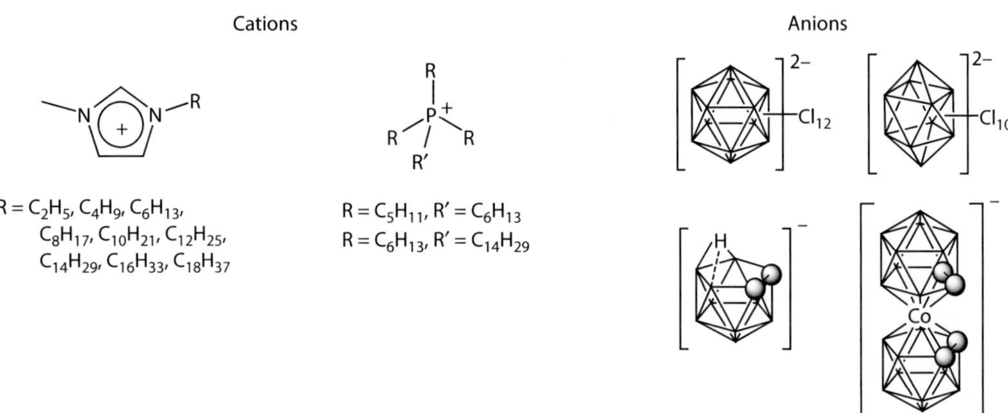

FIGURE 29.12 ILs containing 1-alkyl-3-methylimidazolium and tetraalkylphosphonium, respectively, and perchlorinated *closo*-dodecaborate and decaborate, *nido*-carborane and cobaltabis(dicarbollide). (Adapted from M. Nieuwenhuyzen et al., *Inorg. Chem.* 2009, 48, 889.)

$$U_L = 1202.5 \cdot \left(\frac{\nu \cdot z_C \cdot z_A}{r_C + r_A} \right) \cdot \left(1 - \frac{0.345}{r_C + r_A} \right)$$

The lattice energy U_L in ionic crystals can be calculated by this equation. The lattice energy is dependent on the number of ions ν, the charge of the cation z_C (>0) and the anion z_A (<0) and the radii of the ions, r_C and r_A.

When the imidazolium bears longer alkyl chains, thermotropic interphases were observed, because of microphase segregation.

Comparison of the imidazolium/cobaltabis(dicarbollide) salts shows that packing inefficiency seems to be the main cause for low melting points, as described by Larsen et al. (2000) (see above). The imidazolium/cobaltabis(dicarbollide) salts with $n = 4, 6, 8, 10, 12, 14$, exhibit the lowest melting points, in the range of −18–34°C; they are viscous liquids. The crystalline salt with $n = 2$ shows the highest melting point (113°C). With imidazolium cations bearing hexadecyl ($n = 16$) and octadecyl ($n = 18$) chains, they are crystalline at room temperature as well, with melting points of 58 and 66°C, respectively. The imidazolium/*nido*-carborane salts have melting points ranging between 43 and 98°C, much higher than those of the cobaltabis(dicarbollides), due to a more dispersed charge in the cobalt cluster and its less spherical geometry.

$[C_nMIM][Co(C_2B_9H_{11})_2]$ with $n = 2$, 4, 6, 8, 12, and 14 are viscous at room temperature. With increasing temperature, density, and viscosity decrease. The viscosity depends on the molecular weights of cation and anion, electrostatic forces, van der Waals interactions, hydrogen bonds, or geometry of the anions (Bonhôte et al., 1996; Noda et al., 2001; Wasserscheid and Keim, 2000; Wasserscheid and Welton, 2007).

The melting points of doubly negatively charged boron anions [*closo*-$B_{10}Cl_{10}]^{2-}$ and [*closo*-$B_{12}Cl_{12}]^{2-}$ with imidazolium cations are higher than those of monovalent anions, as expected (Nieuwenhuyzen et al., 2009). The imidazolium salt with $n = 16$ of the perchlorinated decaborate (71°C) exhibits the lowest melting point in this group; the highest is for the salt with $n = 2$ of the perchlorinated dodecaborate (265°C).

Thermogravimetric analyses of the imidazolium salts were carried out, and it could be shown that the thermal stability is dependent on the anion (decomposition temperatures: $[C_4MIM][Co(C_2B_9H_{11})_2]$ = 340°C; $[C_4MIM][C_2B_9H_{12}]$ = 280°C; $[C_2MIM]_2[B_{10}Cl_{10}]$ = 440°C; $[C_{18}MIM]_2[B_{10}Cl_{10}]$ = 380°C; $[C_2MIM]_2[B_{12}Cl_{12}]$ = 480°C). The chlorinated dianions have a higher stability compared to the monoanionic carborane derivatives (Nieuwenhuyzen et al., 2009).

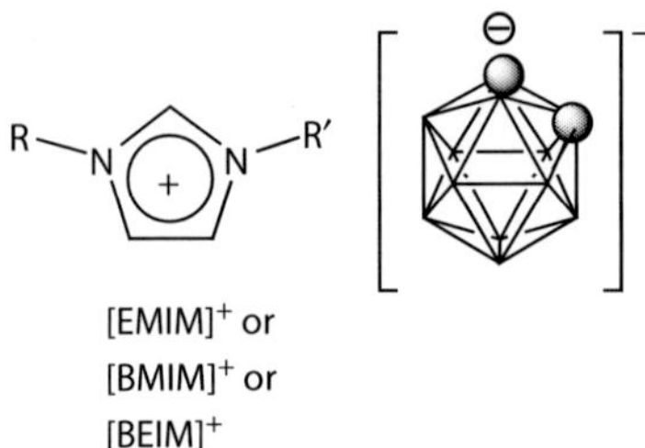

FIGURE 29.13 Imidazolium salts of *ortho*-carborate(1-). (Adapted from N. Matsumi, et al., *J. Organomet. Chem.* 2009, 694, 1612.)

The decomposition temperatures can be compared with data described in the literature for imidazolium salts: $Cl^- < I^- < [C_2B_9H_{12}]^- < [PF_6]^- < [Co(C_2B_9H_{11})_2]^- < [BF_4]^- < [(CF_3SO_2)_2N]^- \sim [B_{10}Cl_{10}]^{2-} < [B_{12}Cl_{12}]^{2-}$ (Bonhôte et al., 1996; Holbrey and Seddon, 1999a; Huddleston et al., 2001; Ngo et al., 2000). Less nucleophilic anions are thermally more stable, because the decomposition of the imidazolium cation occurs by nucleophilic attack of the anion on the cation via a S_N2 reaction.

Two different tetraalkylphosphonium salts, $[P(C_5H_{11})_3C_6H_{13}]^+$ and $[P(C_6H_{13})_3C_{14}H_{29}]^+$, of the divalent perchlorinated decaborate, [*closo*-$B_{10}Cl_{10}]^{2-}$, were investigated as well, to elucidate the influence of the cation (Nieuwenhuyzen et al., 2009). As expected, an increase in length of alkyl chain of the cation results in lower melting points because of the larger distance between the ions, and because of more conformational freedom: the compound with a smaller cation, $[P(C_5H_{11})_3C_6H_{13}]_2$[*closo*-$B_{10}Cl_{10}$], has a melting point of 239°C and that with a larger cation, $[P(C_6H_{13})_3C_{14}H_{29}]_2$[*closo*-$B_{10}Cl_{10}$], a greatly reduced melting point of 53°C. It could be shown that a variation of the cation can also allow ILs with divalent cluster anions; the choice of the right cation is of importance.

Matsumi et al. (2009) published novel ILs bearing an *ortho*-carborane anion with different 1,3-dialkylated imidazolium cations (1-ethyl-3-methyl, 1-butyl-3-methyl and 1-butyl-3-ethyl residues) (Figure 29.13). They are prepared easily from the carbene obtained from imidazolium halides. The ionic conductivity of the $[EMIM]^+$ salt, 2.9×10^{-5} S cm^{-1} at 51°C, is quite low compared to other ILs. The melting point of this *ortho*-carborane salt (30°C) is lower than the melting points for $[EMIM]^+$ salts of the carboranes $[1\text{-}C_3H_7CB_{11}H_{11}]^-$ (45°C) and $[1\text{-}C_4H_9CB_{11}H_{11}]^-$ (40°C) described by Larsen et al. (2000).

Justus et al. (2008a) published *N,N,N*-trialkylammonioundecahydro-*closo*-dodecaborates(1-) as anions for ILs, which were obtained by alkylation of deprotonated $[B_{12}H_{11}NH_3]^-$, with cations such as lithium, potassium, and protons (Figure 29.14). This anion structure allows to vary to a large extent the properties of the anion, which is not readily achieved with the other anionic cluster compounds.

As expected, the viscosity of the 1-ethyl-3-methylimidazolium, $[EMIM]^+$, and the methyltrioctylammonium, $[NMeOc_3]^+$, salt of the trihexylammoniododecaborate, $[B_{12}H_{11}N(Hex)_3]^-$, decrease with rising temperatures. DSC measurements provided melting points, for example, for the potassium salt 79°C, for the tetrabutylammonium salt 87°C and for the $[EMIM]^+$ salt 60°C.

NR_3

C^+

R = H, CH_3, C_2H_5, C_3H_7, C_4H_9, *i*-C_5H_{11}, C_6H_{13}, $C_{12}H_{25}$, $C_3H_6CH=CH_2$

$C^+ = K^+, H^+, Li^+, N(CH_3)_4^+, N(C_4H_9)_4^+, N(CH_3)(C_8H_{17})_3^+, N(CH_3)(C_2H_5)_2(C_3H_7)^+$, *N*-hexylpyridinium, 1-ethyl-3-methylimidazolium $[EMIM]^+$, 1-butyl-3-methylimidazolium$[BMIM]^+$

FIGURE 29.14 Alkylated ammonioundecahydro-*closo*-dodecaborates(1-) as anions for ILs. (Adapted from E. Justus, et al. *Chem. Eur. J.* 2008a, 14, 1918.)

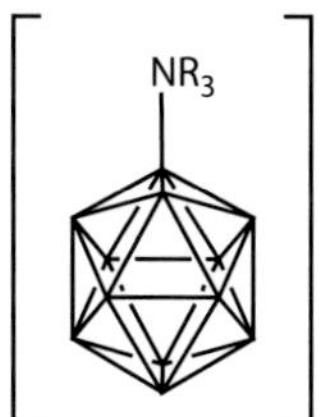

bis-alkylation
R = H, $(C_{12}H_{25})_2$
H, $[(CH_2)_3CCH]_2$
H, $[(CH_2)_2CH(CH_3)_2]_2$

tris-alkylation,
two different residues
R = CH_3, $(C_2H_5)_2$
CH_3, $(C_{12}H_{25})_2$
$CH_2C_6H_5$, $(C_2H_5)_2$

FIGURE 29.15 Bis- and tris-alkylated derivatives of ammonioundecahydro-*closo*-dodecaborate. (Adapted from E. Justus, A. Vöge, D. Gabel, *Eur. J. Inorg. Chem.* 2008b, 5245.)

Thermogravimetric analyses show that the salts of $[B_{12}H_{11}N(Hex)_3]^-$ all are stable at temperatures up to 200°C and higher. The conductivities follow Arrhenius behavior; they are quite low, probably because of high viscosity of the ILs.

Melting points of various cation/anion combinations have been measured. Many of the salts prepared exhibit melting points below 100°C and can be considered as ionic liquids. The melting points are highest with tetramethylammonium as cation. Even liquid potassium and lithium salts can be obtained; thus, for example, in combination with trihexyl substituted ammoniododecaborate as anion, both salts display melting points of 65°C. The proton salt can also be obtained, and its trihexylammonio derivative has a melting point of 60°C. In contrast to the proton salts of carboranes obtained by Reed et al. (2000, 2005), it is not able to protonate benzene.

As it could be expected that asymmetric substitution yields salts with lower viscosity and melting points than those with the same number of carbon atoms in a symmetric arrangement, Justus et al. (2008b) have also described the synthesis of asymmetrically substituted ammoniododecaborates (Figure 29.15).

29.3 SUMMARY

Boron anions in the form of tetravalent boron compounds can serve as anions in ILs, combined with the conventionally used cations such as *N,N′*-dialkylimidazolium, *N*-alkylpyridinium, tetraalkylammonium, and tetraalkylphosphonium. The properties of the salts are very similar to those of salts with other anions. Boronium-cation-based RTILs are also promising candidates as electrolytes in various electrochemical applications.

Anionic cluster compounds are different. They generally show a very weak coordination of cations, enabling them to stabilize cations which otherwise would be very reactive. Several of the cations used regularly for ILs form salts with ionic boron clusters with low melting points. Even with small, usually strongly coordinating cations, for instance, proton and lithium, they are liquid below 100°C. A drawback of the presently available ILs is their high viscosity. With the limited number of compounds investigated so far, there appears to be room for improvement. With this, new applications will open for these ILs.

REFERENCES

A. K. Abdul-Sada, A. M. Greenway, P. B. Hitchcock, T. J. Mohammed, K. R. Seddon, J. A. Zora, Upon the structure of room temperature halogenoaluminate ionic liquids. *J. Chem. Soc., Chem. Commun.* 1986, 1753.

J. W. Bausch, G. K. S. Prakash, G. A. Olah, Diamagnetic polyanions of the C60 and C70 fullerenes: preparation, 13C and 7LI NMR spectroscopic. *J. Am. Chem. Soc.* 1991, 113, 3205.

P. Bonhôte, A.-P. Dias, N. Papageorgiou, K. Kalyanasundaram, M. Grätzel, Hydrophobic, highly conductive ambient-temperature molten salts. *Inorg. Chem.* 1996, 35, 1168.

J. F. Brennecke, E. J. Maginn, Ionic liquids: innovative fluids for chemical processing. *AIChE J.* 2001, 47(11), 2384.

S. Carda-Broch, A. Berthold, D. W. Armstrong, Solvent properties of the 1-butyl-3-methylimidazolium hexafluorophosphate ionic liquid. *Anal. Bioanal. Chem.* 2003, 375, 191.

R. T. Carlin, H. C. De Long, J. Fuller, P. C. Trulove, Dual intercalating molten electrolyte batteries. *J. Electrochem. Soc.* 1994, 141(7), L73.

C. Chiappe, D. Pieraccini, Ionic liquids: solvent properties and organic reactivity. *J. Phys. Org. Chem.* 2005, 18, 275.

C. Douvris, E. S. Stoyanov, F. S. Tham, C. A. Reed, Isolating fluorinated carbocations *Chem. Commun.* 2007, 1145.

M. Doyle, S. K. Choi, G. Proulx, High-temperature proton conducting membranes based on perfluorinated ionomer membrane-ionic liquid composites. *J. Electrochem. Soc.* 2000, 147(1), 34.

S. Dzyuba, R. A. Bartsch, Influence of structural variations in 1-alkyl(aralkyl)-3-methylimidazolium hexafluorophosphates and bis(trifluoromethylsulfonyl) imides on physical properties of the ionic liquids. *Chem. Phys. Chem.* 2002, 3, 161.

J. Dymon, R. Wibby, J. Kleingardner, J. M. Tanski, I. A. Guzei, J. D. Holbrey, A. S. Larsen, Designing ionic liquids with boron cluster anions: alkylpyridinium and imidazolium [*nido*-$C_2B_9H_{11}$] and [*closo*-$CB_{11}H_{12}$] carborane salts. *Dalton Trans.* 2008, 2999.

M. J. Earle, K. R. Seddon, Ionic liquids. Green solvents for the future. *Pure Appl. Chem.* 2000, 72(7), 1391.

S. A. Forsyth, J. M. Pringle, D. R. MacFarlane, Ionic liquids. Green solvents for the future. *Aust. J. Chem.* 2004, 57, 113.

P. A. Fox, S. T. Griffin, W. M. Reichert, E. A. Salter, A. B. Smith, M. D. Tickell, B. F. Wicker, E. A. Cioffi, J. H. Davis, R. D. Rogers, A. Wierzbicki, Exploiting isolobal relationships to create new ionic liquids: novel room-temperature ionic liquids based upon (N-alkylimidazole)(amine)BH_2^+ "boronium" ions. *Chem. Commun.* 2005, 3679.

M. Freemantle, Ionic liquids show promise for clean separation technology. *Chem. Eng. News* 1998, 76, 32.

J. Fuller, A. C. Breda, R. T. Carlin, Ionic liquid-polymer gel electrolytes. *J. Electrochem. Soc.* 1997a, 144(4), L67.

J. Fuller, R. T. Carlin, R. A. Osteryoung, The room temperature ionic liquid 1 -ethyl-3-methylimidazolium tetrafluoroborate: electrochemical couples and physical properties. *J. Electrochem. Soc.* 1997b, 144(11), 3881.

K. M. Galie, A. Mollard, I. Zharov, Polyester-based carborane-containing dendrons. *Inorg. Chem.* 2006, 45, 7815.

X. Gang, H. A. Hjuler, C. Olsen, R. W. Berg, N. J. Bjerrum, Electrolyte additives for phosphoric acid fuel cells. *J. Electrochem. Soc.* 1993, 140(4), 896.

S. P. Hoffmann, T. Kato, F. S. Tham, C. A. Reed, novel weak coordination to silylium ions: formation of nearly linear Si–H–Si bonds. *Chem. Commun.* 2006, 767.

J. D. Holbrey, W. M. Reichert, M. Nieuwenhuyzen, S. Johnston, K. R. Seddon, R. D. Rogers, Crystal polymorphism in 1-butyl-3-methylimidazolium halides: supporting ionic liquid formation by inhibition of crystallization. *Chem. Commun.* 2003, 1636.

J. D. Holbrey, K. R. Seddon, The phase behaviour of 1-alkyl-3-methylimidazolium tetrafluoroborates; ionic liquids and ionic liquid crystals. *J. Chem. Soc., Dalton Trans.* 1999a, 2133.

J. D. Holbrey, K. R. Seddon, Review. Ionic liquids. *Clean Technol. Environ. Pol.* 1999b, 223.

J. G. Huddleston, A. E. Visser, W. M. Reichert, H. D. Willauer, G. A. Broker, R. D. Rogers, Characterization and comparison of hydrophilic and hydrophobic room temperature ionic liquids incorporating the imidazolium cation. *Green Chem.* 2001, 3, 156.

N. V. Ignatév, U. Welz-Biermann, A. Kucheryna, G. Bissky, H. Willner, New ionic liquids with tris(perfluoroalkyl) trifluorophosphate (FAP) anions. *J. Fluor. Chem.* 2005, 126, 1150.

J. Ito, T. Nohira, Non-conventional electrolytes for electrochemical applications. *Electrochim. Acta* 2000, 45, 2611.

S. V. Ivanov, J. A. Davis, S. M. Miller, O. P. Anderson, S. H. Strauss, Synthesis and stability of reactive salts of dodecafluoro-closo-dodecaborate(2-). *Inorg. Chem.* 2003a, 42, 4489.

S. V. Ivanov, S. M. Miller, O. P. Anderson, K. A. Solntsev, S. H. Strauss, Synthesis and stability of reactive salts of dodecafluoro-closo-dodecaborate(2-). *J. Am. Chem Soc.* 2003b, 125, 4694.

E. Justus, K. Rischka, J. F. Wishart, K. Werner, D. Gabel, Trialkylammoniododecaborates: anions for ionic liquids with potassium, lithium and protons as cations. *Chem. Eur. J.* 2008a, 14, 1918.

E. Justus, A. Vöge, D. Gabel, N-alkylation of ammonioundecahydro-closo-dodecaborate(1–) for the preparation of anions for ionic liquids. *Eur. J. Inorg. Chem.* 2008b, 5245.

A. F. Kapustinskii, Lattice energy of ionic crystals. *Quart. Rev. Chem. Soc.* 1956, 10, 283.

T. Kato, C. A. Reed, Carbocations. Putting tert-butyl cation in a bottle, *Angew. Chem. Int. Ed.* 2004a, 43, 2908.

T. Kato, E. Stoyanov, J. Geier, H. Grützmacher, C. A. Reed, Alkylating agents stronger than alkyl triflates. *J. Am. Chem. Soc.* 2004b, 126, 12451.

K.-C. Kim, C. A. Reed, D. W. Elliott, L. J. Mueller, F. Tham, L. Lin, J. B. Lambert, Crystallographic evidence for a free silylium ion. *Science* 2002a, 297, 825.

K.-C. Kim, C. A. Reed, G. S. Long, A. Sen, Et2Al+ Alumenium ion-like chemistry. Synthesis and reactivity toward alkenes and alkene oxides. *J. Am. Chem. Soc.* 2002b, 124, 7662.

B. T. King, Z. Janoušek, B. Grüner, M. Trammell, B. C. Noll, J. Michl, Dodecamethylcarba-closo-dodecaborate(-) Anion, $CB_{11}Me_{12}^-$. *J. Am. Chem. Soc.* 1996, 118, 3313.

V. R. Koch, C. Nanjundiah, G. B. Appetecchi, B. Scrosati, The interfacial stability of li with two new solvent-free ionic liquids: 1,2-dimethyl-3-propylimidazolium imide and methide. *J. Electrochem. Soc.* 1995, 142, L116.

P. Kölle, R. Dronskowski, Hydrogen bonding in the crystal structures of the ionic liquid compounds butyldimethylimidazolium hydrogen sulfate, chloride, and chloroferrate(II,III). *Inorg. Chem.* 2004, 43, 2803.

S. Körbe, P. J. Schreiber, J. Michl, Chemistry of the Carba-closo-dodecaborate(-) Anion, $CB_{11}H_{12}^-$.*Chem. Rev.* 2006, 106, 5208.

I. Krossing, I. Raabe, Noncoordinating anions - fact or fiction? A survey of likely candidates. *Angew. Chem. Int. Ed.* 2004, 43, 2066.

T. Küppers, E. Bernhardt, R. Eujen, H. Willner, C. W. Lehmann, Silyl cations. [Me3Si][R-$CB_{11}F_{11}$] - synthesis and properties. *Angew. Chem. Int. Ed.* 2007, 46, 6346.

A. S. Larsen, J. D. Holbrey, F. S. Tham, C. A. Reed, Designing ionic liquids: imidazolium melts with inert carborane anions. *J. Am. Chem. Soc.* 2000, 122, 7264.

Y. Li, P. J. Carroll, L. G. Sneddon, Ionic-liquid-promoted decaborane dehydrogenative alkyne-insertion reactions: a new route to o-carboranes. *Inorg. Chem.* 2008, 47, 9193.

D. R. MacFarlane, J. Huang, M. Forsyth, Lithium-doped plastic crystal electrolytes exhibiting fast ion conduction for secondary batteries. *Nature* 1999, 402, 792.

N. Matsumi, M. Miyamoto, K. Aoi, Preparation of ionic liquids bearing o-carborane anion via N,N'-dialkylimidazol-2-ylidene carbene. *J. Organomet. Chem.* 2009, 694, 1612.

H. Matsumoto, H. Kageyama, Y. Miyazaki, Room temperature ionic liquids based on small aliphatic ammonium cations and asymmetric amide anions. *Chem. Commun.* 2002b, 1726.

H. Matsumoto, T. Matsuda, (Ionic liquids as electrolytes in wet solar cells) (Japanese). *Electrochem.* (Tokyo, Japan) 2002a, 70, 190.

A. B. McEwen, S. F. McDevitt, V. R. Koch, Nonaqueous electrolytes for electrochemical capacitors: imidazolium cations and inorganic fluorides with organic carbonates. *J. Electrochem. Soc.* 1997, 144(4), L84.

A. B. McEwen, J. L. Goldman, D. Wasel, L. Hargens in Molten Salts XII, Eds. H. C. De Long, S. Deki, G. R. Stafford, P. C. Trulove, *Proc. Electrochem. Soc.* 1999, 99, 222.

A. Mollard, I. Zharov, Tricarboranyl pentaerythritol-based building block. *Inorg. Chem.* 2006, 45, 10172.

T. Müller, M. Juhasz, C. A. Reed, Vinyl cations. The X-ray structure of a vinyl cation. *Angew. Chem. Int. Ed. Engl.* 2004, 43, 1543.

C. Nanjundiah, S. F. McDevitt, V. R. Koch, Differential capacitance measurements in solvent-free ionic liquids at hg and c interfaces. *J. Electrochem. Soc.* 1997, 144(10), 3392.

A. Narita, W. Shibayame, K. Sakamoto, T. Mizumo, N. Matsumi, H. Ohno, Lithium ion conduction in an organoborate zwitterion–LiTFSI mixture. *Chem. Commun.* 2006, 1926.

H. L. Ngo, K. LeCompte, L. Hargens, A. B. McEwan, Thermal properties of imidazolium ionic liquids. *Thermochim. Acta* 2000, 357–358, 97.

M. Nieuwenhuyzen, K. R. Seddon, F. Teixidor, A. V. Puga, C. Viñas, Ionic Liquids Containing Boron Cluster Anions. *Inorg. Chem.* 2009, 48, 889.

A. Noda, K. Hayamizu, M. Watanabe, Pulsed-Gradient Spin-Echo 1H and 19F NMR ionic diffusion coefficient, viscosity, and ionic conductivity of non-chloroaluminate room-temperature ionic liquids. *J. Phys. Chem.* B 2001, 105, 4603.

N. Papageorgiou, Y. Athanassov, M. Armand, P. Bonhôte, H. Pettersson, A. Azam, M. Grätzel, The performance and stability of ambient temperature molten salts for solar cell applications. *J. Electrochem. Soc.* 1996, 143, 3099.

N. Plechkova, K. R. Seddon, Applications of ionic liquids in the chemical industry. *Chem. Soc. Rev*. 2008, 37, 123.

J. M. Pringle, J. Golding, K. Baranyai, C. M. Forsyth, G. B. Deacon, J. L. Scott, D. R. MacFarlane, The effect of anion fluorination in ionic liquids—physical properties of a range of bis(methanesulfonyl)amide salts. *New J. Chem*. 2003, 27, 1504.

C. A. Reed, Carboranes: A new class of weakly coordinating anions for strong electrophiles, oxidants, and superacids. *Acc. Chem. Res*. 1998, 31, 133.

C. A. Reed, Carborane acids. New "strong yet gentle" acids for organic and inorganic chemistry. *Chem. Commun*. 2005, 1669.

C. A. Reed, *Acc. Chem. Res*. 2009, DOI: 10.1021/ar900159e.

C. A. Reed, N. L. P. Fackler, K. C. Kim, D. Stasko, D. R. Evans, P. D. W. Boyd, C. E. F. Rickard, Isolation of Protonated Arenes (Wheland Intermediates) with BArF and Carborane Anions. A Novel Crystalline Superacid. *J. Am. Chem. Soc*. 1999, 121, 6314.

C. A. Reed, K. C. Kim, R. D. Bolskar, L. J. Mueller, Taming superacids: Stabilization of the fullerene cations HC60+ and C60+. *Science* 2000, 289, 101.

C. A. Reed, K. C. Kim, E. S. Stoyanov, D. Stasko, F. S. Tham, L. J. Mueller, P. D. W. Boyd, Isolating benzenium ion salts. *J. Am. Chem. Soc*. 2003, 125, 1796.

B. Ronig, I. Pantenburg, L. Wesemann, meltable stannaborate salts. *Eur. J. Inorg. Chem*. 2002, 319.

M. R. Rosenthal, The myth of the non-coordinating anion. *J. Chem. Educ*. 1973, 50, 531.

T. Rüther, T. D. Huynh, J. Huang, A. F. Hollenkamp, E. A. Salter, A. Wierzbicki, K. Mattson, A. Lewis, J. H. Davis, Stable Cycling of Lithium Batteries Using Novel Boronium-Cation-Based Ionic Liquid Electrolytes. *Chem. Mater*. 2010, 22, 1038.

M. Schmidt, U. Heider, A. Kuehner, R. Oesten, M. Jungnitz, N. Ignatév, P. Sartori, *J. Power Sources* 2001, 97, 557.

C. Schreiner, M. Amereller, H. J. Gores, *Chem. Eur. J*. 2009, 15, 2270.

K. R. Seddon, Room-temperature ionic liquids: Neoteric solvents for clean catalysis. *Kinet. Catal*. 1996a, 37, 743.

K. R. Seddon, Room-temperature ionic liquids: Neoteric solvents for clean Ccatalysis. *Kinet. Catal*. 1996b, 37, 693.

K. R. Seddon, Review. Ionic liquids for clean technology. *J. Chem. Technol. Biotechnol*. 1997, 68, 351.

K. Seppelt, "Noncoordinating" anions, II. *Angew. Chem. Int. Ed*. 1993, 32(7), 1025.

K. Shelly, C. A. Reed, The least coordinating anion. *J. Am. Chem. Soc*. 1986, 108, 3117.

D. Stasko, C. A. Reed, Optimizing the least nucleophilic anion. A new, strong methyl+ reagent. *J. Am. Chem. Soc*. 2002, 124, 1148.

P. A. Z. Suarez, S. Einloft, J. E. L. Dullius, R. F. de Souza, J. Dupont, Synthesis and physical-chemical properties of ionic liquids based on l-n-butyl-3-methylimidazolium cation. *J. Chim. Phys. Phys.-Chim. Biol*. 1998, 95, 1626.

S. H. Strauss, The search for larger and more weakly coordinating anions. *Chem. Rev*. 1993, 93, 927.

C.-W. Tsang, Q. Yang, E. T.-P. Sze, T. C. W. Mak, D. T. W. Chan, Z. Xie, Synthesis and structural characterization of highly chlorinated, brominated, iodinated, and methylated carborane anions, 1-H-CB9X9-, 1-NH2-CB9X9- (X) Cl, Br, I), and 1-H-CB9(CH3)9-. *Inorg. Chem*. 2000, 39, 3582.

M. Ue, K. Ida, S. Mori, Electrochemical properties of organic liquid electrolytes based on quaternary onium salts for electrical double-layer capacitors. *J. Electrochem. Soc*. 1994, 141, 2989.

A. E. Visser, J. G. Huddleston, J. D. Holbrey, R. D. Rogers, Hydrophobic n-alkyl-isoquinolinium ionic liquids: Characterization, solvent properties, and use in separations. in Proceedings 224th *Am. Chem. Soc*. National Meeting 2002, ACS: Washington, DC.

P. Walden, Molecular weights and electrical conductivity of several fused salts. *Bull. Acad. Impér. Sci. St. Pétersbourg* 1914, 8, 405.

P. Wasserscheid, W. Keim, Ionic Liquids–New "Solutions" for Transition Metal Catalysis. *Angew. Chem. Int. Ed*. 2000, 39, 3773.

P. Wasserscheid, M. Sesing, W. Korth, Hydrogensulfate and tetrakis(hydrogensulfato)borate ionic liquids: synthesis and catalytic application in highly brønsted-acidic systems for friedel–crafts alkylation. *Green Chem*. 2002, 4, 134.

P. Wasserscheid, T. Welton, Eds., Ionic Liquids in Synthesis, 2nd ed., Wiley-VCH: Weinheim Germany, 2007.

A. Webber, G. E. Blomgren, in Advances in Lithium-Ion Batteries 2002, Eds. W. A. Schalkwijk, B. Scrosati, Kluwer Academics/Plenum Publishers, New York, 185.

J. S. Wilkes, M. J. Zaworotko, Dialkylimidazolium chloroaluminate melts: A new class of room-temperature ionic liquids for electrochemistry, spectroscopy, and synthesis. *J. Chem. Soc., Chem. Comm*. 1992, 965.

Z. Xie, R. Bau, C. A. Reed, "Free" $[Fe(tpp)]^+$ cation: A new concept in the search for the least coordinating anion. *Angew. Chem. Int. Ed*. 1994, 33, 2433.

Y. Zhang, F. S. Tham, C. A. Reed, Phosphazene cations. *Inorg. Chem*. 2006, 43, 10446.

I. Zharov, T.-C. Weng, A. M. Orendt, D. H. Barich, J. Penner-Hahn, D. M. Grant, Z. Havlas, J. Michl, Metal cation-methyl interactions in $CB_{11}Me_{12}$–Salts of Me3Ge+, Me3Sn+, and Me3Pb+. *J. Am. Chem. Soc.* 2004, 126, 12033.

Z. B. Zhou, H. Matsumoto, K. Tatsumi, Low-melting, low-viscous, hydrophobic ionic liquids: 1-alkyl(alkyl ether)-3-methylimidazolium perfluoroalkyltrifluoroborate. *Chem. Eur. J.* 2004, 10, 6581.

Z. B. Zhou, H. Matsumoto, K. Tatsumi, Low-melting, low-viscous, hydrophobic ionic liquids: aliphatic quaternary ammonium salts with perfluoroalkyltrifluoroborates. *Chem. Eur. J.* 2005, 11, 752.

Y. Zhu, C. Ching, K. Carpenter, R. Xu, S. Selvaratnam, N. S. Hosmane, J. A. Maguire, Synthesis of the novel ionic liquid [N-pentylpyridinium]+ $[closo\text{-}CB_{11}H_{12}]^-$ and its usage as a reaction medium in catalytic dehalogenation of aromatic halides. *Appl. Organometal. Chem.* 2003, 17, 346.

Index

Note: n = Footnote

B

C

D

E

F

I

J

K

N

O

P

S

T

U

V

W

X

Z